A Brief Guide to **Getting the Most** from this Book

1 Read the Book

Feature	Description	Benefit
Applications Using Real-World Data	From the chapter and section openers through the examples and exercises, interesting applications from nearly every discipline, supported by up-to-date real-world data, are included in every section.	Ever wondered how you'll use algebra? This feature will show you how algebra can solve real problems.
Detailed Worked-Out Examples	Examples are clearly written and provide step-by-step solutions. No steps are omitted, and key steps are thoroughly explained to the right of the mathematics.	The blue annotations will help you to understand the solutions by providing the reason why the algebraic steps are true.
Explanatory Voice Balloons	Voice balloons help to demystify algebra. They translate algebraic language into plain English, clarify problem-solving procedures, and present alternative ways of understanding.	Does math ever look foreign to you? This feature translates math into everyday English.
Great Question!	Answers to students' questions offer suggestions for problem solving, point out common errors to avoid, and provide informal hints and suggestions.	This feature should help you not to feel anxious or threatened when asking questions in class.
Achieving Success	The book's Achieving Success boxes offer strategies for success in learning algebra.	Follow these suggestions to help achieve your full academic potential in mathematics.

2 Work the Problems

Feature	Description	Benefit
Check Point Examples	Each example is followed by a similar problem, called a Check Point, that offers you the opportunity to work a similar exercise. Answers to all Check Points are provided in the answer section.	You learn best by doing. You'll solidify your understanding of worked examples if you try a similar problem right away to be sure you understand what you've just read.
Concept and Vocabulary Checks	These short-answer questions, mainly fill-in-the blank and true/false items, assess your understanding of the definitions and concepts presented in each section.	It is difficult to learn algebra without knowing its special language. These exercises test your understanding of the vocabulary and concepts.
Extensive and Varied Exercise Sets	An abundant collection of exercises is included in an Exercise Set at the end of each section. Exercises are organized within several categories. Practice Exercises follow the same order as the section's worked examples. Practice PLUS Exercises contain more challenging problems that often require you to combine several skills or concepts.	The parallel order of the Practice Exercises lets you refer to the worked examples and use them as models for solving these problems. Practice PLUS provides you with ample opportunity to dig in and develop your problem-solving skills.

3 Review for Quizzes and Tests

Feature	Description	Benefit
Mid-Chapter Check Points	Near the midway point in the chapter, an integrated set of review exercises allows you to review the skills and concepts you learned separately over several sections.	Combining exercises from the first half of the chapter gives you a comprehensive review before you continue on.
Chapter Review Charts	Each chapter contains a review chart that summarizes the definitions and concepts in every section of the chapter, complete with examples.	Review this chart and you'll know the most important material in the chapter.
Chapter Tests	Each chapter contains a practice test with problems that cover the important concepts in the chapter. Take the test, check your answers, and then watch the Chapter Test Prep Videos.	You can use the Chapter Test to determine whether you have mastered the material covered in the chapter.
Chapter Test Prep Videos	These videos contain worked-out solutions to every exercise in each chapter test.	These videos let you review any exercises you miss on the chapter test.
Lecture Series on DVD	These interactive lecture videos highlight key examples from every section of the textbook. A new interface allows easy navigation to sections, objectives, and examples.	These videos let you review each objective from the textbook that you need extra help on.

Prepare for Exams with Blitzer's
New Interactive Video Lecture Series
with Chapter Test Prep Videos.

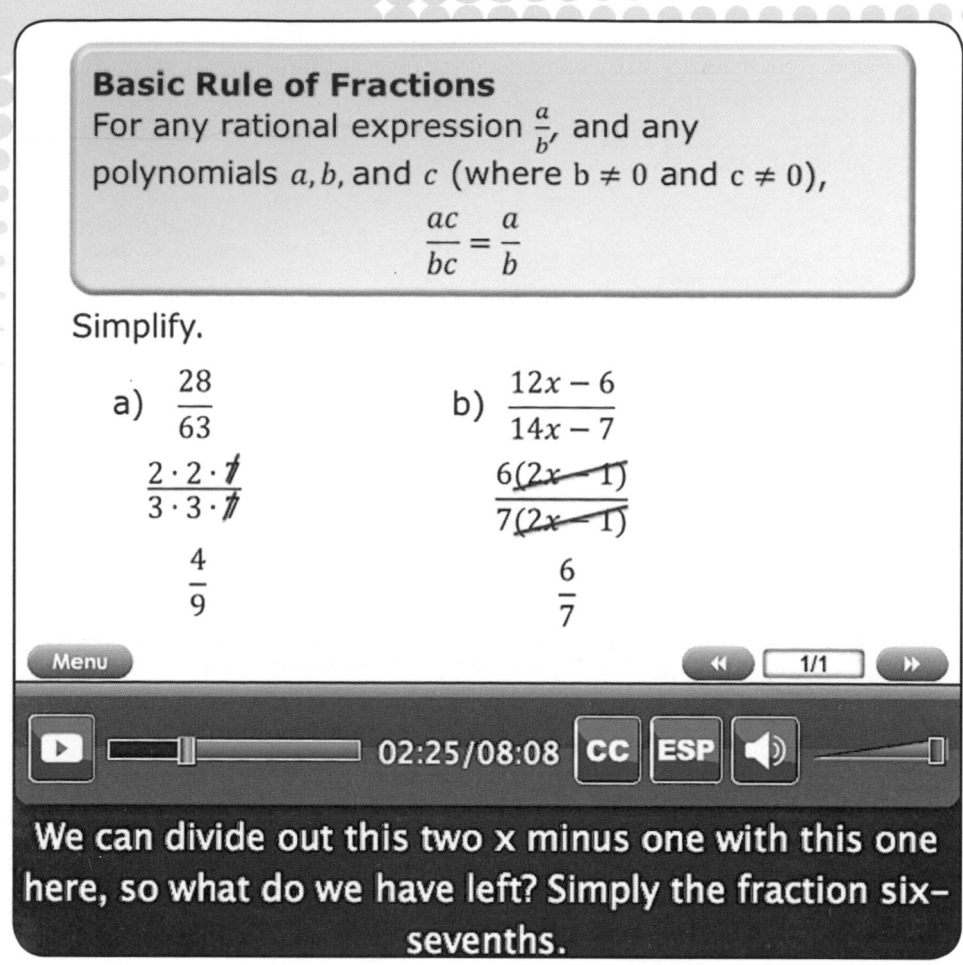

Basic Rule of Fractions
For any rational expression $\frac{a}{b}$, and any polynomials $a, b,$ and c (where $b \neq 0$ and $c \neq 0$),

$$\frac{ac}{bc} = \frac{a}{b}$$

Simplify.

a) $\frac{28}{63}$

$\frac{2 \cdot 2 \cdot 7}{3 \cdot 3 \cdot 7}$

$\frac{4}{9}$

b) $\frac{12x - 6}{14x - 7}$

$\frac{6(2x - 1)}{7(2x - 1)}$

$\frac{6}{7}$

Menu ◀◀ 1/1 ▶▶

▶ 02:25/08:08 CC ESP ◀))

We can divide out this two x minus one with this one here, so what do we have left? Simply the fraction six–sevenths.

Students can make the most of their study time by preparing for exams with the new Interactive Lecture Series with Chapter Test Prep Videos:

▸ **Interactive Lectures** highlight key examples and exercises from every section of the textbook. A new interface allows easy navigation to sections, objectives, and examples. These lectures are available in MyMathLab and on the Lecture Series on DVD.

▸ **Chapter Test Prep Videos** provide step-by-step video solutions for every problem from the Chapter Tests in the textbook. The Chapter Test Prep videos are available in MyMathLab, on the Lecture Series DVD, and on You Tube

Robert Blitzer

Introductory & Intermediate Algebra for College Students

Custom Edition for Cuesta College

Taken from:
Introductory & Intermediate Algebra for College Students, Fourth Edition
by Robert Blitzer

Cover Art: Courtesy of Photodisc/Getty Images.

Taken from:

Introductory & Intermediate Algebra for College Students, Fourth Edition
by Robert Blitzer
Copyright © 2009, 2006, 2002 by Pearson Education, Inc.
Published by Prentice Hall
Upper Saddle River, New Jersey 07458

This special edition published in cooperation with Pearson Learning Solutions.

Pearson Learning Solutions, 501 Boylston Street, Suite 900, Boston, MA 02116
A Pearson Education Company
www.pearsoned.com

Printed in the United States of America

1 12 13 14 V0UD 19 18 17 16 15

0200010271296104

PEARSON ISBN 10: 1-256-75306-8
ISBN 13: 978-1-256-75306-3

Table of Contents

Preface

Introductory and Intermediate Algebra for College Students, Fourth Edition, provides comprehensive, in-depth coverage of the topics required in a course combining the study of introductory and intermediate algebra. The book is written for college students who have no previous experience in algebra and for those who need a review of basic algebraic concepts before moving on to intermediate algebra. I wrote the book to help diverse students, with different backgrounds and career plans, to succeed in a combined introductory and intermediate algebra course. *Introductory and Intermediate Algebra for College Students,* Fourth Edition, has two primary goals:

1. To help students acquire a solid foundation in the skills and concepts of introductory and intermediate algebra, without the repetition of topics in two separate texts.

2. To show students how algebra can model and solve authentic real-world problems.

One major obstacle in the way of achieving these goals is the fact that very few students actually read their textbook. This has been a regular source of frustration for me and for my colleagues in the classroom. Anecdotal evidence gathered over years highlights two basic reasons why students do not take advantage of their textbook:

*   "I'll never use this information."
*   "I can't follow the explanations."

I've written every page of the Fourth Edition with the intent of eliminating these two objections. The ideas and tools I've used to do so are described in the features that follow. These features and their benefits are highlighted for the student in "A Brief Guide to Getting the Most from This Book," which appears inside the front cover.

What's New in the Fourth Edition?

*   **New Applications and Real-World Data.** I'm on a constant search for data that can be used to illustrate unique algebraic applications. I researched hundreds of books, magazines, newspapers, almanacs, and online sites to prepare the Fourth Edition. Among the worked-out examples and exercises based on new data sets, you'll find applications involving stretching one's life span, the emotional health of college freshmen, changing attitudes toward marriage, *American Idol*'s viewership, and the year humans become immortal.

*   **Concept and Vocabulary Checks.** The Fourth Edition contains more than 1500 new short-answer exercises, mainly fill-in-the-blank and true/false items, that assess students' understanding of the definitions and concepts presented in each section. The Concept and Vocabulary Checks appear as separate features preceding the Exercise Sets.

- **Great Question!** This feature takes the content of each Study Tip in the Third Edition and presents it in the context of a student question. Answers to questions offer suggestions for problem solving, point out common errors to avoid, and provide informal hints and suggestions. "Great Question!" should draw students' attention and curiosity more than the "Study Tips." As a secondary benefit, this new feature should help students not to feel anxious or threatened when asking questions in class.

- **Achieving Success.** The book's Achieving Success boxes offer strategies for success in math courses, as well as suggestions for developmental students on achieving one's full academic potential.

- **New Chapter-Opening and Section-Opening Scenarios.** Every chapter and every section open with a scenario based on an application, many of which are unique to the Fourth Edition. These scenarios are revisited in the course of the chapter or section in one of the book's new examples, exercises, or discussions. The often-humorous tone of these openers is intended to help fearful and reluctant students overcome their negative perceptions about math.

- **New The Lecture Series on DVD** has been completely revised to provide students with extra help for each section of the textbook. The Lecture Series DVD includes:
  - **Interactive Lectures** that highlight key examples and exercises for every section of the textbook. A new interface allows easy navigation to sections, objectives, and examples.
  - **Chapter Test Prep Videos** provide step-by-step solutions to every problem in each Chapter Test in the textbook. The Chapter Test Prep Videos are now also available on You-Tube™.

What Content Changes Have Been Made to the Fourth Edition?

- Section 2.3 (Solving Linear Equations) contains a new discussion on solving linear equations with decimals.

- Section 2.4 (Formulas and Percents) speeds up the discussion on basics of percent, a prealgebra topic, with the focus on algebraic applications.

- Section 2.7 (Solving Linear Inequalities) includes an additional application involving an example about staying within a budget.

- Section 6.1 (The Greatest Common Factor and Factoring by Grouping) includes a new objective on factoring out the negative of a polynomial's greatest common factor.

- Section 6.5 (A General Factoring Strategy) slows down the pace of the Practice Exercises in the Exercise Set. The Exercise Set now opens with problems that require students to use only one factoring technique before moving on to multiple-step factorizations.

- Section 7.6 (Solving Rational Equations) expands the coverage of solving a formula with a rational expression for a variable with a new worked example and Check Point.

- Section 9.3 (Equations and Inequalities Involving Absolute Value) solves absolute value inequalities using equivalent compound inequalities, rather than boundary points. With the section's focus on this one method, the pace has been both slowed down and expanded.

- Section 11.1 (The Square Root Property and Completing the Square) contains a new example on solving a quadratic equation with imaginary solutions by completing the square.

- Section 11.2 (The Quadratic Formula) has a new example on writing a quadratic equation whose solution set contains irrational numbers.

- Section 11.5 (Polynomial and Rational Inequalities) includes a new example on solving a polynomial inequality with irrational boundary points.

What Familiar Features Have Been Retained in the Fourth Edition?

- **Detailed Worked-Out Examples.** Each worked example is titled, making clear the purpose of the example. Examples are clearly written and provide students with detailed step-by-step solutions. No steps are omitted and key steps are thoroughly explained to the right of the mathematics.

- **Explanatory Voice Balloons.** Voice balloons are used in a variety of ways to demystify mathematics. They translate algebraic ideas into everyday English, help clarify problem-solving procedures, present alternative ways of understanding concepts, and connect problem solving to concepts students have already learned.

- **Check Point Examples.** Each example is followed by a similar matched problem, called a Check Point, offering students the opportunity to test their understanding of the example by working a similar exercise. The answers to the Check Points are provided in the answer section.

- **Extensive and Varied Exercise Sets.** An abundant collection of exercises is included in an Exercise Set at the end of each section. Exercises are organized within eight category types: Practice Exercises, Practice Plus Exercises, Application Exercises, Writing in Mathematics, Critical Thinking Exercises, Technology Exercises, Review Exercises, and Preview Exercises. This format makes it easy to create well-rounded homework assignments. The order of the Practice Exercises is exactly the same as the order of the section's worked examples. This parallel order enables students to refer to the titled examples and their detailed explanations to achieve success working the Practice Exercises.

- **Practice Plus Problems.** This category of exercises contains more challenging practice problems that often require students to combine several skills or concepts. With an average of ten Practice Plus problems per Exercise Set, instructors are provided with the option of creating assignments that take Practice Exercises to a more challenging level.

- **Mid-Chapter Check Points.** At approximately the midway point in each chapter, an integrated set of Review Exercises allows students to review and assimilate the skills and concepts they learned separately over several sections.

- **Early Graphing.** Chapter 1 connects formulas and mathematical models to data displayed by bar and line graphs. The rectangular coordinate system is introduced in Chapter 3. Graphs appear in nearly every section and Exercise Set. Examples and exercises use graphs to explore relationships between data and to provide ways of visualizing a problem's solution.

- **Geometric Problem Solving.** Chapter 2 (Linear Equations and Inequalities in One Variable) contains a section that teaches geometric concepts that are important to a student's understanding of algebra. There is frequent emphasis on problem solving in geometric situations, as well as on geometric models that allow students to visualize algebraic formulas.

- **Section Objectives.** Learning objectives are clearly stated at the beginning of each section. These objectives help students recognize and focus on the section's most important ideas. The objectives are restated in the margin at their point of use.

- **Integration of Technology Using Graphic and Numerical Approaches to Problems.** Side-by-side features in the technology boxes connect algebraic solutions to graphic and numerical approaches to problems. Although the use of graphing utilities is optional, students can use the explanatory voice balloons to understand different approaches to problems even if they are not using a graphing utility in the course.

- **Chapter Review Grids.** Each chapter contains a review chart that summarizes the definitions and concepts in every section of the chapter. Examples that illustrate these key concepts are also included in the chart.

- **End-of-Chapter Materials.** A comprehensive collection of Review Exercises for each of the chapter's sections follows the review grid. This is followed by a Chapter Test that enables students to test their understanding of the material covered in the chapter. Beginning with Chapter 2, each chapter concludes with a comprehensive collection of mixed Cumulative Review Exercises.

- **Blitzer Bonuses.** These enrichment essays provide historical, interdisciplinary, and otherwise interesting connections to the algebra under study, showing students that math is an interesting and dynamic discipline.

- **Discovery.** Discover for Yourself boxes, found throughout the text, encourage students to further explore algebraic concepts. These explorations are optional and their omission does not interfere with the continuity of the topic under consideration.

- **Chapter Projects.** At the end of each chapter is a collaborative activity that gives students the opportunity to work cooperatively as they think and talk about mathematics. Additional group projects can be found in the *Instructor's Resource Manual.* Many of these exercises should result in interesting group discussions.

I hope that my passion for teaching, as well as my respect for the diversity of students I have taught and learned from over the years, is apparent throughout this new edition. By connecting algebra to the whole spectrum of learning, it is my intent to show students that their world is profoundly mathematical, and indeed, π is in the sky.

Robert Blitzer

Resources for the Fourth Edition

For Students

Student Solutions Manual. Fully-worked solutions to the odd-numbered section exercises plus all Check Points, Review/Preview Exercises, Mid-Chapter Check Points, Chapter Reviews, Chapter Tests, and Cumulative Reviews.

NEW Learning Guide, organized by the textbook's learning objectives, helps students learn how to make the most of their textbook and all of the learning tools, including MyMathLab, while also providing additional practice for each section and guidance for test preparation. Published in an unbound, binder-ready format, the Learning Guide can serve as the foundation for a course notebook for students.

NEW Lecture Series on DVD has been completely revised to provide students with extra help for each section of the textbook. The Lecture Series on DVD includes Interactive Lectures and Chapter Test Prep videos.

For Instructors

Annotated Instructor's Edition. Answers to exercises printed on the same text page with graphing answers in a special Graphing Answer Section in the back of the text.

Instructor's Solutions Manuals. This manual contains fully-worked solutions to the even-numbered section exercises plus all Check Points, Review/Preview Exercises, Mid-Chapter Check Points, Chapter Reviews, Chapter Tests, and Cumulative Reviews. Available in MyMathLab® and on the Instructor's Resource Center.

Instructor's Resources Manual with Tests. Includes a Mini-Lecture, Skill Builder, and Additional Exercises for every section of the text; two short group Activities per chapter, several chapter test forms, both free-response and multiple-choice, as well as cumulative tests and final exams. Answers to all items also included. Available in MyMath Lab® and on the Instructor's Resource Center.

MyMathLab® Online Course (access code required). MyMathLab® delivers proven results in helping individual students succeed. It provides engaging experiences that personalize, stimulate, and measure learning for each student. And it comes from a trusted partner with educational expertise and an eye on the future.

MyMathLab® Ready to Go Course (access code required). These new Ready to Go courses provide students with all the same great MyMathLab® features that you're used to, but make it easier for instructors to get started. Each course includes preassigned homework and quizzes to make creating your course even simpler. Ask your Pearson representative about the details for this particular course or to see a copy of this course.

To learn more about how MyMathLab® combines proven learning applications with powerful assessment, visit www.mymathlab.com or contact your Pearson representative.

MathXL® Online Courses (access code required) is the homework and assessment engine that runs MyMathLab. (MyMathLab is MathXL plus a learning management system.)

MathXL is available to qualified adopters. For more information, visit our Web site at www.mathxl.com, or contact your Pearson representative.

TestGen® (www.pearsoned.com/testgen) enables instructors to build, edit, print, and administer tests using a computerized bank of questions developed to cover all the objectives of the text. TestGen is algorithmically based, allowing instructors to create multiple but equivalent versions of the same question or test with the click of a button. Instructors can also modify test bank questions or add new questions. The software and testbank are available for download from Pearson Education's online catalog.

PowerPoint Lecture Slides (download only). Available through www.pearsonhighered.com or inside your MyMathLab® course, these fully editable lecture slides include definitions, key concepts, and examples for use in a lecture setting and are available for each section of the text.

Acknowledgments

An enormous benefit of authoring a successful series is the broad-based feedback I receive from the students, dedicated users, and reviewers. Every change to this edition is the result of their thoughtful comments and suggestions. I would like to express my appreciation to all the reviewers, whose collective insights form the backbone of this revision. In particular, I would like to thank the following people for reviewing *Introductory and Intermediate Algebra for College Students.*

Gwen P. Aldridge, *Northwest Mississippi Community College*

Howard Anderson, *Skagit Valley College*

John Anderson, *Illinois Valley Community College*

Michael H. Andreoli, *Miami Dade College—North Campus*

Jan Archibald, *Ventura College*

Jana Barnard, *Angelo State University*

Donna Beatty, *Ventura College*

Michael S. Bowen, *Ventura College*

Gale Brewer, *Amarillo College*

Carmen Buhler, *Minneapolis Community & Technical College*

Hien Bui, *Hillsborough Community College*

Warren J. Burch, *Brevard Community College*

Alice Burstein, *Middlesex Community College*

Edie Carter, *Amarillo College*

Thomas B. Clark, *Trident Technical College*

Sandra Pryor Clarkson, *Hunter College*

Sally Copeland, *Johnson County Community College*

Carol Curtis, *Fresno City College*

Bettyann Daley, *University of Delaware*

Robert A. Davies, *Cuyahoga Community College*

Paige Davis, *Lurleen B. Wallace Community College*

Jill DeWitt, *Baker College*

Ben Divers, Jr., *Ferrum College*

Irene Doo, *Austin Community College*

Charles C. Edgar, *Onondaga Community College*

Karen Edwards, *Diablo Valley College*

Rhoderick Fleming, *Wake Technical Community College*

Susan Forman, *Bronx Community College*

Wendy Fresh, *Portland Community College*

Donna Gerken, *Miami Dade College*

Marion K. Glasby, *Anne Arundel Community College*

Sue Glascoe, *Mesa Community College*

Gary Glaze, *Eastern Washington University*

Jay Graening, *University of Arkansas*

Christina M. Gundlach, *Eastern Connecticut State University*

Robert B. Hafer, *Brevard Community College*

Mary Lou Hammond, *Spokane Community College*

Andrea Hendricks, *Georgia Perimeter College*

Donald Herrick, *Northern Illinois University*

Beth Hooper, *Golden West College*

Sandee House, *Georgia Perimeter College*

Tracy Hoy, *College of Lake County*

Laura Hoye, *Trident Community College*

Margaret Huddleston, *Schreiner University*

Marcella Jones, *Minneapolis Community and Technical College*

Shelbra B, Jones, *Wake Technical Community College*

Judy Kasabian, *Lansing Community College*

Sharon Keenee, *Georgia Perimeter College*

Gary Kersting, *North Central Michigan College*

Dennis Kimzey, *Rogue Community College*

Kandace Kling, *Portland Community College*

Gary Knippenberg, *Lansing Community College*

Mary A. Koehler, *Cuyahoga Community College*

Kristi Laird, *Jackson Community College*

Scot Leavitt, *Portland Community College*

Robert Leibman, *Austin Community College*

Jennifer Lempke, *North Central Michigan College*

Sandy Lofstock, *St. Petersburg College*

Ann M. Loving, *J. Sargent Reynolds Community College*

Kent MacDougall, *Temple College*

Hank Martel, *Broward Community College*

Diana Martelly, *Miami Dade College*

Kim Martin, *Southeastern Illinois College*

John Robert Martin, *Tarrant County College*

Mikal McDowell, *Cedar Valley College*

Lisa McMillen, *Baker College*

Irwin Metviner, *State University of New York at Old Westbury*

Jean P. Millen, *Georgia Perimeter College*

Lawrence Morales, *Seattle Central Community College*

Terri Moser, *Austin Community College*

Robert Musselman, *California State University, Fresno*

Kamilia Nemri, *Spokane Community College*

Allen R. Newhart, *Parkersburg Community College*

Lois Jean Nieme, *Minneapolis Community and Technical College*

Steve O'Donnell, *Rogue Community College*

Peg Pankowski, *Community College of Allegheny County—South Campus*

Jeff Parent, *Oakland Community College*

Robert Patenade, *College of the Canyons*

Jill Rafael, *Sierra College*

James Razavi, *Sierra College*

Christopher Reisch, *The State University of New York at Buffalo*

Tian Ren, *Queensborough Community College*

Nancy Ressler, *Oakton Community College*

Katalin Rozsa, *Mesa Community College*

Haazim Sabree, *Georgia Perimeter College*

Scott W. Satake, *Eastern Washington University*

Mike Schramm, *Indian River Community College*

Chris Schultz, *Iowa State University*

Shannon Schumann, *University of Phoenix*

Terri Seiver, *San Jacinto College*

Kathy Shepard, *Monroe County Community College*

Gayle Smith, *Lane Community College*

Linda Smoke, *Central Michigan University*

Dick Spangler, *Tacoma Community College*

Janette Summers, *University of Arkansas*

Kory Swart, *Kirkwood Community College*

Robert Thornton, *Loyola University*

Lucy C. Thrower, *Francis Marion College*

Richard Townsend, *North Carolina Central University*

Andrew Walker, *North Seattle Community College*

Kim Ward, *Eastern Connecticut State University*

Kathryn Wetzel, *Amarillo College*

Margaret Williamson, *Milwaukee Area Technical College*

Peggy Wright, *Cuesta College*

Roberta Yellott, *McNeese State University*

Marilyn Zopp, *McHenry County College*

Additional acknowledgments are extended to Dan Miller and Kelly Barber, for preparing the solutions manuals and the new Learning Guide, Brad Davis, for preparing the answer section and serving as accuracy checker, the Preparé, Inc. formatting team, for the book's brilliant paging, Brian Morris at Scientific Illustrators, for superbly illustrating the book, Sheila Norman, photo researcher, for obtaining the book's new photographs, and Rebecca Dunn, project manager, and Kathleen Manley, production editor, whose collective talents kept every aspect of this complex project moving through its many stages.

I would like to thank my editors at Pearson, Paul Murphy and Dawn Giovanniello, who with the assistance of Chelsea Pingree guided and coordinated the book from manuscript through production. Thanks to Beth Paquin and Nancy Goulet for the cover and Ellen Pettengell for the interior design. Finally, thanks to the entire Pearson sales force for your confidence and enthusiasm about the book.

Robert Blitzer

The bar graph shows some of the qualities that students say make a great teacher.

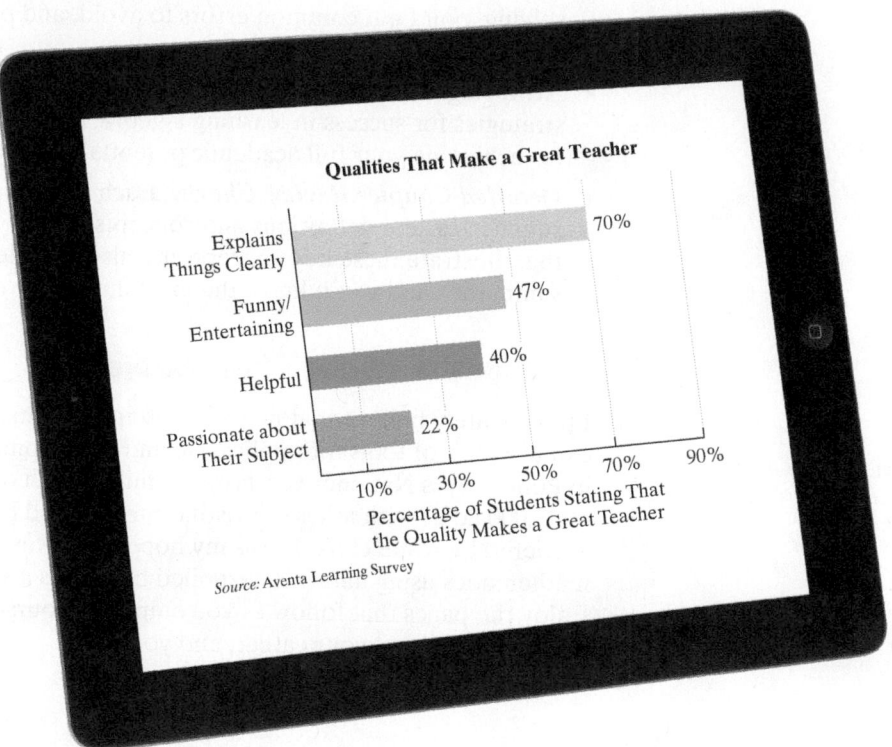

Qualities That Make a Great Teacher

Source: Aventa Learning Survey

It was my goal to incorporate each of the qualities that make a great teacher throughout the pages of this book.

Explains Things Clearly

I understand that your primary purpose in reading *Introductory and Intermediate Algebra for College Students* is to acquire a solid understanding of the required topics in your algebra course. In order to achieve this goal, I've carefully explained each topic. Important definitions and procedures are set off in boxes, and worked-out examples that present solutions in a step-by-step manner appear in every section. Each example is followed by a similar matched problem, called a Check Point, for you to try so that you can actively participate in the learning process as you read the book. (Answers to all Check Points appear in the back of the book.)

Funny/Entertaining

Who says that an algebra textbook can't be entertaining? From our quirky cover to the photos in the chapter and section openers, prepare to expect the unexpected. I hope some of the book's enrichment essays, called Blitzer Bonuses, will put a smile on your face from time to time.

Helpful

I designed the book's features to help you acquire knowledge of introductory and intermediate algebra, as well as to show you how algebra can solve authentic problems that apply to your life. These helpful features include:

- *Explanatory Voice Balloons:* Voice balloons are used in a variety of ways to make math less intimidating. They translate algebraic language into everyday English, help clarify problem-solving procedures, present alternative ways of understanding concepts, and connect new concepts to concepts you have already learned.

- *Great Question!:* The book's Great Question! boxes are based on questions students ask in class. The answers to these questions give suggestions for problem solving, point out common errors to avoid, and provide informal hints and suggestions.

- *Achieving Success:* The book's Achieving Success boxes give you helpful strategies for success in learning algebra, as well as suggestions that can be applied for achieving your full academic potential in future college coursework.

- *Detailed Chapter Review Charts:* Each chapter contains a review chart that summarizes the definitions and concepts in every section of the chapter. Examples that illustrate these key concepts are also included in the chart. Review these summaries and you'll know the most important material in the chapter!

Passionate about Their Subject

I passionately believe that no other discipline comes close to math in offering a more extensive set of tools for application and development of your mind. I wrote the book in Point Reyes National Seashore, 40 miles north of San Francisco. The park consists of 75,000 acres with miles of pristine surf-washed beaches, forested ridges, and bays bordered by white cliffs. It was my hope to convey the beauty and excitement of mathematics using nature's unspoiled beauty as a source of inspiration and creativity. Enjoy the pages that follow as you empower yourself with the algebra needed to succeed in college, your career, and your life.

Regards
Bob

Robert Blitzer

About the Author

Bob Blitzer is a native of Manhattan and received a Bachelor of Arts degree with dual majors in mathematics and psychology (minor: English literature) from the City College of New York. His unusual combination of academic interests led him toward a Master of Arts in mathematics from the University of Miami and a doctorate in behavioral sciences from Nova University. Bob's love for teaching mathematics was nourished for nearly 30 years at Miami Dade College, where he received numerous teaching awards, including Innovator of the Year from the League for Innovations in the Community College and an endowed chair based on excellence in the classroom. In addition to *Introductory and Intermediate Algebra for College Students*, Bob has written textbooks covering introductory algebra, intermediate algebra, college algebra, algebra and trigonometry, precalculus, and liberal arts mathematics, all published by Pearson. When not secluded in his Northern California writer's cabin, Bob can be found hiking the beaches and trails of Point Reyes National Seashore, and tending to the chores required by his beloved entourage of horses, chickens, and irritable roosters.

Bob and his horse Jerid

Variables, Real Numbers, and Mathematical Models

What can algebra possibly tell me about

- the rising cost of movie ticket prices over the years?
- how I can stretch or shrink my life span?
- the widening imbalance between numbers of women and men on college campuses?
- the number of calories I need to maintain energy balance?
- how much I can expect to earn at my first job after college?

In this chapter, you will learn how the special language of algebra describes your world.

Here's where you will find these applications:

- Movie ticket prices: Exercise Set 1.1, Exercises 83–84
- Stretching or shrinking my life span: Exercise Set 1.6, Exercises 101–110
- College gender imbalance: Section 1.7, Example 9
- Caloric needs: Section 1.8, Example 12; Exercise Set 1.8, Exercises 97–98
- Anticipated earnings at my first job after college:
  Exercise Set 1.8, Exercises 99–100.

1958	1967	1980	1990	2000	2010
Cat on a Hot Tin Roof TICKET PRICE $0.68	Cool Hand Luke TICKET PRICE $1.22	Ordinary People TICKET PRICE $2.69	Dances with Wolves TICKET PRICE $4.23	Erin Brockovich TICKET PRICE $5.39	Alice in Wonderland TICKET PRICE $7.85

Sources: Motion Picture Association of America, National Association of Theater Owners (NATO), and Bureau of Labor Statistics (BLS)

1.1

Introduction to Algebra: Variables and Mathematical Models

Objectives

1 Evaluate algebraic expressions.

2 Translate English phrases into algebraic expressions.

3 Determine whether a number is a solution of an equation.

4 Translate English sentences into algebraic equations.

5 Evaluate formulas.

You are thinking about buying a high-definition television. How much distance should you allow between you and the TV for pixels to be undetectable and the image to appear smooth?

Algebraic Expressions

Let's see what the distance between you and your TV has to do with algebra. The biggest difference between arithmetic and algebra is the use of *variables* in algebra. A **variable** is a letter that represents a variety of different numbers. For example, we can let x represent the diagonal length, in inches, of a high-definition television. The industry rule for most of the current HDTVs on the market is to multiply this diagonal length by 2.5 to get the distance, in inches, at which a person with perfect vision can see a smooth image. This can be written $2.5 \cdot x$, but it is usually expressed as $2.5x$. Placing a number and a letter next to one another indicates multiplication.

Notice that $2.5x$ combines the number 2.5 and the variable x using the operation of multiplication. A combination of variables and numbers using the operations of addition, subtraction, multiplication, or division, as well as powers or roots, is called an **algebraic expression**. Here are some examples of algebraic expressions:

$$x + 2.5 \qquad x - 2.5 \qquad 2.5x \qquad \frac{x}{2.5} \qquad 3x + 5 \qquad \sqrt{x} + 7.$$

| The variable x increased by **2.5** | The variable x decreased by **2.5** | **2.5** times the variable x | The variable x divided by **2.5** | **5** more than **3** times the variable x | **7** more than the square root of the variable x |

Great Question!

Are variables always represented by *x*?

No. As you progress through algebra, you will often see x, y, and z used, but any letter can be used to represent a variable. For example, if we use l to represent a TV's diagonal length, your ideal distance from the screen is described by the algebraic expression $2.5l$.

1 Evaluate algebraic expressions.

Evaluating Algebraic Expressions

We can replace a variable that appears in an algebraic expression by a number. We are **substituting** the number for the variable. The process is called **evaluating the expression**. For example, we can evaluate $2.5x$ (the ideal distance between you and your x-inch TV) for $x = 50$. We substitute 50 for x. We obtain $2.5 \cdot 50$, or 125. This means that if the diagonal length of your TV is 50 inches, your distance from the screen should be 125 inches. Because 12 inches = 1 foot, this distance is $\frac{125}{12}$ feet, or approximately 10.4 feet.

Many algebraic expressions involve more than one operation. The order in which we add, subtract, multiply, and divide is important. In Section 1.8, we will discuss the rules for the order in which operations should be done. For now, follow this order:

A First Look at Order of Operations

1. Perform all operations within grouping symbols, such as parentheses.
2. Do all multiplications in the order in which they occur from left to right.
3. Do all additions and subtractions in the order in which they occur from left to right.

EXAMPLE 1 Evaluating Expressions

Evaluate each algebraic expression for $x = 5$:

a. $3 + 4x$ **b.** $4(x + 3)$.

Solution

a. We begin by substituting 5 for x in $3 + 4x$. Then we follow the order of operations: Multiply first, and then add.

$$3 + 4x$$

Replace x with 5.

$$= 3 + 4 \cdot 5$$
$$= 3 + 20 \qquad \text{Perform the multiplication: } 4 \cdot 5 = 20.$$
$$= 23 \qquad \text{Perform the addition.}$$

b. We begin by substituting 5 for x in $4(x + 3)$. Then we follow the order of operations: Perform the addition in parentheses first, and then multiply.

$$4(x + 3)$$

Replace x with 5.

$$= 4(5 + 3)$$
$$= 4(8) \qquad \text{Perform the addition inside the parentheses: } 5 + 3 = 8.$$
$$ \qquad 4(8) \text{ can also be written as } 4 \cdot 8.$$
$$= 32 \qquad \text{Multiply.} \quad \blacksquare$$

✓ **CHECK POINT 1** Evaluate each expression for $x = 10$:

a. $6 + 2x$ **b.** $2(x + 6)$.

Example 2 illustrates that algebraic expressions can contain more than one variable.

EXAMPLE 2 Evaluating Expressions

Evaluate each algebraic expression for $x = 6$ and $y = 4$:

a. $5x - 3y$ **b.** $\dfrac{3x + 5y + 2}{2x - y}$.

Solution

a. $5x - 3y$ This is the given algebraic expression.

| Replace x with 6. | Replace y with 4. |

$= 5 \cdot 6 - 3 \cdot 4$ We are evaluating the expression for $x = 6$ and $y = 4$.

$= 30 - 12$ Multiply: $5 \cdot 6 = 30$ and $3 \cdot 4 = 12$.

$= 18$ Subtract.

b. $\dfrac{3x + 5y + 2}{2x - y}$ This is the given algebraic expression.

| Replace x with 6. | Replace y with 4. |

$= \dfrac{3 \cdot 6 + 5 \cdot 4 + 2}{2 \cdot 6 - 4}$

We are evaluating the expression for $x = 6$ and $y = 4$.

$= \dfrac{18 + 20 + 2}{12 - 4}$ Multiply: $3 \cdot 6 = 18, 5 \cdot 4 = 20$, and $2 \cdot 6 = 12$.

$= \dfrac{40}{8}$ Add in the numerator.
                                                 Subtract in the denominator.

$= 5$ Simplify by dividing 40 by 8.   ■

✓ **CHECK POINT 2**   Evaluate each algebraic expression for $x = 3$ and $y = 8$:

a. $7x + 2y$ **b.** $\dfrac{6x - y}{2y - x - 8}$.

---

2 Translate English phrases into algebraic expressions.

Translating to Algebraic Expressions

Problem solving in algebra often requires the ability to translate word phrases into algebraic expressions. **Table 1.1** lists some key words associated with the operations of addition, subtraction, multiplication, and division.

Table 1.1   Key Words for Addition, Subtraction, Multiplication, and Division

Operation	Addition (+)	Subtraction (−)	Multiplication (·)	Division (÷)
	plus	minus	times	divide
	sum	difference	product	quotient
Key words	more than	less than	twice	ratio
	increased by	decreased by	multiplied by	divided by

EXAMPLE 3   Translating English Phrases into Algebraic Expressions

Write each English phrase as an algebraic expression. Let the variable x represent the number.

  a. the sum of a number and 7

  b. ten less than a number

  c. twice a number, decreased by 6

  d. the product of 8 and a number

  e. three more than the quotient of a number and 11

Solution

a.

The sum of	a number and 7

$$x + 7$$

The algebraic expression for "the sum of a number and 7" is $x + 7$.

b.

Ten less than	a number

$$x - 10$$

The algebraic expression for "ten less than a number" is $x - 10$.

c.

Twice a number,	decreased by 6

$$2x - 6$$

The algebraic expression for "twice a number, decreased by 6" is $2x - 6$.

d.

The product of	8 and a number

$$8 \cdot x$$

The algebraic expression for "the product of 8 and a number" is $8 \cdot x$, or $8x$.

e.

Three more than	the quotient of a number and 11

$$\frac{x}{11} + 3$$

The algebraic expression for "three more than the quotient of a number and 11" is $\frac{x}{11} + 3$. ∎

Great Question!

Are there any helpful hints I can use when writing algebraic expressions for English phrases involving subtraction?

Yes. Pay close attention to order when translating phrases involving subtraction.

Phrase	Translation
A number decreased by 7	$x - 7$
A number subtracted from 7	$7 - x$
Seven less than a number	$x - 7$
Seven less a number	$7 - x$

Think carefully about what is expressed in English before you translate into the language of algebra.

✓ **CHECK POINT 3** Write each English phrase as an algebraic expression. Let the variable x represent the number.

   a. the product of 6 and a number

   b. a number added to 4

   c. three times a number, increased by 5

   d. twice a number subtracted from 12

   e. the quotient of 15 and a number

3 Determine whether a number is a solution of an equation.

Equations

An **equation** is a statement that two algebraic expressions are equal. **An equation always contains the equality symbol =.** Here are some examples of equations:

$$6x + 16 = 46 \qquad 4y + 2 = 2y + 6 \qquad 2(z + 1) = 5(z - 2).$$

Solutions of an equation are values of the variable that make the equation a true statement. To determine whether a number is a solution, substitute that number for the variable and evaluate each side of the equation. If the values on both sides of the equation are the same, the number is a solution.

EXAMPLE 4 Determining Whether Numbers Are Solutions of Equations

Determine whether the given number is a solution of the equation.

a. $6x + 16 = 46; 5$ **b.** $2(z + 1) = 5(z - 2); 7$

Solution

a.

$$6x + 16 = 46$$ This is the given equation.

Is 5 a solution?

$$6 \cdot 5 + 16 \stackrel{?}{=} 46$$ To determine whether 5 is a solution, substitute 5 for x. The question mark over the equal sign indicates that we do not yet know if the statement is true.

$$30 + 16 \stackrel{?}{=} 46$$ Multiply: $6 \cdot 5 = 30$.

$$46 = 46$$ Add: $30 + 16 = 46$.

This statement is true.

Because the values on both sides of the equation are the same, the number 5 is a solution of the equation.

b.

$$2(z + 1) = 5(z - 2)$$ This is the given equation.

Is 7 a solution?

$$2(7 + 1) \stackrel{?}{=} 5(7 - 2)$$ To determine whether 7 is a solution, substitute 7 for z.

$$2(8) \stackrel{?}{=} 5(5)$$ Perform operations in parentheses: $7 + 1 = 8$ and $7 - 2 = 5$.

$$16 = 25$$ Multiply: $2 \cdot 8 = 16$ and $5 \cdot 5 = 25$.

This statement is false.

Because the values on both sides of the equation are not the same, the number 7 is not a solution of the equation. ■

✓ **CHECK POINT 4** Determine whether the given number is a solution of the equation.

a. $9x - 3 = 42; 6$ **b.** $2(y + 3) = 5y - 3; 3$

4 Translate English sentences into algebraic equations.

Translating to Equations

Earlier in the section, we translated English phrases into algebraic expressions. Now we will translate English sentences into equations. You'll find that there are a number of different words and phrases for an equation's equality symbol.

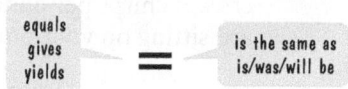

EXAMPLE 5   Translating English Sentences into Algebraic Equations

Write each sentence as an equation. Let the variable x represent the number.

a. The product of 6 and a number is 30.

b. Seven less than 3 times a number gives 17.

Solution

a.

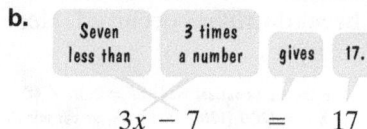

$$6x = 30$$

The equation for "the product of 6 and a number is 30" is $6x = 30$.

b.

| Seven less than | 3 times a number | gives | 17. |

$$3x - 7 = 17$$

The equation for "seven less than 3 times a number gives 17" is $3x - 7 = 17$. ∎

✓ **CHECK POINT 5**   Write each sentence as an equation. Let the variable x represent the number.

a. The quotient of a number and 6 is 5.

b. Seven decreased by twice a number yields 1.

Great Question!

I don't know how to find the solutions of the equations in Example 5. Should I be worried?

Do not be concerned. We'll discuss how to solve these equations in Chapter 2.

Great Question!

When translating English sentences containing commas into algebraic equations, should I pay attention to the commas?

Commas make a difference. In English, sentences and phrases can take on different meanings depending on the way words are grouped with commas. Some examples:

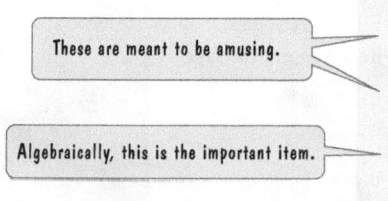

- Woman, without her man, is nothing.
  Woman, without her, man is nothing.
- Do not break your bread or roll in your soup.
  Do not break your bread, or roll in your soup.
- The product of 6 and a number increased by 5 is 30: $6(x + 5) = 30$.
  The product of 6 and a number, increased by 5, is 30: $6x + 5 = 30$.

5 Evaluate formulas.

Formulas and Mathematical Models

One aim of algebra is to provide a compact, symbolic description of the world. These descriptions involve the use of *formulas*. A **formula** is an equation that expresses a relationship between two or more variables. For example, one variety of crickets chirps faster as the temperature rises. You can calculate the temperature by counting the number of times a cricket chirps per minute and applying the following formula:

$$T = 0.3n + 40.$$

In the formula, T is the temperature, in degrees Fahrenheit, and n is the number of cricket chirps per minute. We can use this formula to determine the temperature if you are sitting on your porch and count 80 chirps per minute. Here is how to do so:

$T = 0.3n + 40$	This is the given formula.
$T = 0.3(80) + 40$	Substitute 80 for n.
$T = 24 + 40$	Multiply: $0.3(80) = 24$.
$T = 64.$	Add.

When there are 80 cricket chirps per minute, the temperature is 64 degrees.

The process of finding formulas to describe real-world phenomena is called **mathematical modeling**. Such formulas, together with the meaning assigned to the variables, are called **mathematical models**. We often say that these formulas model, or describe, the relationship among the variables.

In creating mathematical models, we strive for both accuracy and simplicity. For example, the formula $T = 0.3n + 40$ is relatively simple to use. However, you should not get upset if you count 80 cricket chirps and the actual temperature is 62 degrees, rather than 64 degrees, as predicted by the formula. Many mathematical models give an approximate, rather than an exact, description of the relationship between variables.

Sometimes a mathematical model gives an estimate that is not a good approximation or is extended to include values of the variable that do not make sense. In these cases, we say that **model breakdown** has occurred. Here is an example:

Use the mathematical model $T = 0.3n + 40$ with $n = 1200$ (1200 cricket chirps per minute).

$$T = 0.3(1200) + 40 = 360 + 40 = 400$$

At 400° F, forget about 1200 chirps per minute! At this temperature, the cricket would "cook" and, alas, all chirping would cease.

EXAMPLE 6   Age at Marriage and the Probability of Divorce

Divorce rates are considerably higher for couples who marry in their teens. The line graphs in **Figure 1.1** show the percentages of marriages ending in divorce based on the wife's age at marriage.

Figure 1.1

Source: B. E. Pruitt et al., *Human Sexuality*, Prentice Hall, 2007

Here are two mathematical models that approximate the data displayed by the line graphs:

Wife is under 18 at time of marriage.	Wife is over 25 at time of marriage.
$d = 4n + 5$	$d = 2.3n + 1.5.$

In each model, the variable n is the number of years after marriage and the variable d is the percentage of marriages ending in divorce.

a. Use the appropriate formula to determine the percentage of marriages ending in divorce after 10 years when the wife is over 25 at the time of marriage.

b. Use the appropriate line graph in **Figure 1.1** to determine the percentage of marriages ending in divorce after 10 years when the wife is over 25 at the time of marriage.

c. Does the value given by the mathematical model underestimate or overestimate the actual percentage of marriages ending in divorce after 10 years as shown by the graph? By how much?

Solution

a. Because the wife is over 25 at the time of marriage, we use the formula on the right, $d = 2.3n + 1.5$. To find the percentage of marriages ending in divorce after 10 years, we substitute 10 for n and evaluate the formula.

$$d = 2.3n + 1.5$$ This is one of the two given mathematical models.

$$d = 2.3(10) + 1.5$$ Replace n with 10.

$$d = 23 + 1.5$$ Multiply: $2.3(10) = 23$.

$$d = 24.5$$ Add.

The model indicates that 24.5% of marriages end in divorce after 10 years when the wife is over 25 at the time of marriage.

b. Now let's use the line graph that shows the percentage of marriages ending in divorce when the wife is over 25 at the time of marriage. The graph is shown again in **Figure 1.2**. To find the percentage of marriages ending in divorce after 10 years:

- Locate 10 on the horizontal axis and locate the point above 10.
- Read across to the corresponding percent on the vertical axis.

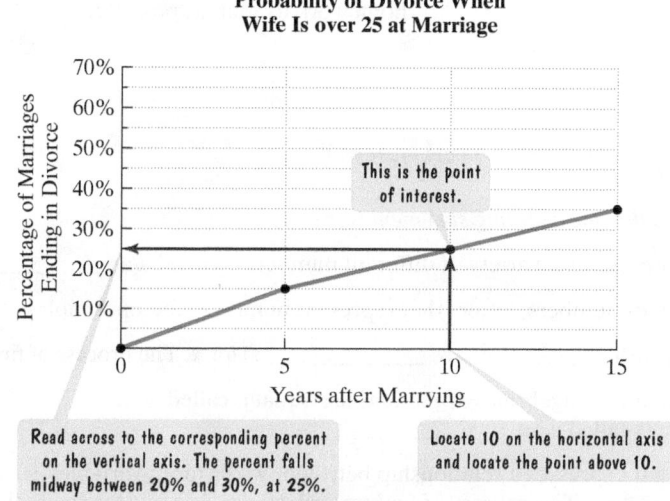

Probability of Divorce When Wife Is over 25 at Marriage

This is the point of interest.

Read across to the corresponding percent on the vertical axis. The percent falls midway between 20% and 30%, at 25%.

Locate 10 on the horizontal axis and locate the point above 10.

Figure 1.2

The actual data displayed by the graph indicate that 25% of these marriages end in divorce after 10 years.

c. Here's a summary of what we found in parts (a) and (b).

Percentage of Marriages Ending in Divorce after 10 Years (Wife over 25 at Marriage)
Mathematical model: 24.5%
Actual data displayed by graph: 25.0%

The value obtained by evaluating the mathematical model, 24.5%, is close to, but slightly less than, the actual percentage of divorces, 25.0%. The difference between these percents is 25.0% − 24.5%, or 0.5%. The value given by the mathematical model underestimates the actual percent by only 0.5%, providing a fairly accurate description of the data. ∎

✓ **CHECK POINT 6**

a. Use the appropriate formula from Example 6 on page 9 to determine the percentage of marriages ending in divorce after 15 years when the wife is under 18 at the time of marriage.

b. Use the appropriate line graph in **Figure 1.1** on page 8 to determine the percentage of marriages ending in divorce after 15 years when the wife is under 18 at the time of marriage.

c. Does the value given by the mathematical model underestimate or overestimate the actual percentage of marriages ending in divorce after 15 years as shown by the graph? By how much?

Achieving Success

Practice! Practice! Practice!

The way to learn algebra is by seeing solutions to examples and **doing exercises**. This means working the Check Points and the assigned exercises in the Exercise Sets. There are no alternatives. It's easy to read a solution, or watch your professor solve an example, and believe you know what to do. However, learning algebra requires that you actually **perform solutions by yourself**. Get in the habit of working exercises every day. The more time you spend solving exercises, the easier the process becomes. It's okay to take a short break after class, but start reviewing and working the assigned homework as soon as possible.

CONCEPT AND VOCABULARY CHECK

Fill in each blank so that the resulting statement is true.

1. A letter that represents a variety of different numbers is called a/an _____.

2. A combination of numbers, letters that represent numbers, and operation symbols is called an algebraic _____.

3. By replacing x with 60 in $2.5x$, we are _____ 60 for x. The process of finding $2.5 \cdot 60$ is called _____ $2.5x$ for $x = 60$.

4. A statement that two algebraic expressions are equal is called a/an _____. A value of the variable that makes such a statement true is called a/an _____.

5. A statement that expresses a relationship between two or more variables, such as $T = 0.3x + 40$, is called a/an _____. The process of finding such statements to describe real-world phenomena is called mathematical _____. Such statements, together with the meaning assigned to the variables, are called mathematical _____.

1.1 EXERCISE SET MyMathLab®

Watch the videos
in MyMathLab

Download the
MyDashBoard App

Practice Exercises

In Exercises 1–14, evaluate each expression for x = 4.

1. $x + 8$
2. $x + 10$
3. $12 - x$
4. $16 - x$
5. $5x$
6. $6x$
7. $\dfrac{28}{x}$
8. $\dfrac{36}{x}$
9. $5 + 3x$
10. $3 + 5x$
11. $2(x + 5)$
12. $5(x + 3)$
13. $\dfrac{12x - 8}{2x}$
14. $\dfrac{5x + 52}{3x}$

In Exercises 15–24, evaluate each expression for x = 7 and y = 5.

15. $2x + y$
16. $3x + y$
17. $2(x + y)$
18. $3(x + y)$
19. $4x - 3y$
20. $5x - 4y$
21. $\dfrac{21}{x} + \dfrac{35}{y}$
22. $\dfrac{50}{y} - \dfrac{14}{x}$
23. $\dfrac{2x - y + 6}{2y - x}$
24. $\dfrac{2y - x + 24}{2x - y}$

In Exercises 25–42, write each English phrase as an algebraic expression. Let the variable x represent the number.

25. four more than a number
26. six more than a number
27. four less than a number
28. six less than a number
29. the sum of a number and 4
30. the sum of a number and 6
31. nine subtracted from a number
32. three subtracted from a number
33. nine decreased by a number
34. three decreased by a number
35. three times a number, decreased by 5
36. five times a number, decreased by 3
37. one less than the product of 12 and a number
38. three less than the product of 13 and a number
39. the sum of 10 divided by a number and that number divided by 10
40. the sum of 20 divided by a number and that number divided by 20
41. six more than the quotient of a number and 30
42. four more than the quotient of 30 and a number

In Exercises 43–58, determine whether the given number is a solution of the equation.

43. $x + 14 = 20; 6$
44. $x + 17 = 22; 5$
45. $30 - y = 10; 20$
46. $50 - y = 20; 30$
47. $4z = 20; 10$
48. $5z = 30; 8$
49. $\dfrac{r}{6} = 8; 48$
50. $\dfrac{r}{9} = 7; 63$
51. $4m + 3 = 23; 6$
52. $3m + 4 = 19; 6$
53. $5a - 4 = 2a + 5; 3$
54. $5a - 3 = 2a + 6; 3$
55. $6(p - 4) = 3p; 8$
56. $4(p + 3) = 6p; 6$
57. $2(w + 1) = 3(w - 1); 7$
58. $3(w + 2) = 4(w - 3); 10$

In Exercises 59–74, write each sentence as an equation. Let the variable x represent the number.

59. Four times a number is 28.
60. Five times a number is 35.
61. The quotient of 14 and a number is $\frac{1}{2}$.
62. The quotient of a number and 8 is $\frac{1}{4}$.
63. The difference between 20 and a number is 5.
64. The difference between 40 and a number is 10.
65. The sum of twice a number and 6 is 16.
66. The sum of twice a number and 9 is 29.
67. Five less than 3 times a number gives 7.
68. Three less than 4 times a number gives 29.
69. The product of 4 and a number, increased by 5, is 33.
70. The product of 6 and a number, increased by 3, is 33.
71. The product of 4 and a number increased by 5 is 33.
72. The product of 6 and a number increased by 3 is 33.
73. Five times a number is equal to 24 decreased by the number.
74. Four times a number is equal to 25 decreased by the number.

Practice PLUS

75. Evaluate $\dfrac{x - y}{4}$ when x is 2 more than 7 times y and $y = 5$.
76. Evaluate $\dfrac{x - y}{3}$ when x is 2 more than 5 times y and $y = 4$.

77. Evaluate $4x + 3(y + 5)$ when x is 1 less than the quotient of y and 4 and $y = 12$.

78. Evaluate $3x + 4(y + 6)$ when x is 1 less than the quotient of y and 3, and $y = 15$.

79. a. Evaluate $2(x + 3y)$ for $x = 4$ and $y = 1$.

    b. Is the number you obtained in part (a) a solution of $5z - 30 = 40$?

80. a. Evaluate $3(2x + y)$ for $x = 1$ and $y = 5$.

    b. Is the number you obtained in part (a) a solution of $4z - 30 = 54$?

81. a. Evaluate $6x - 2y$ for $x = 3$ and $y = 6$.

    b. Is the number you obtained in part (a) a solution of $7w = 45 - 2w$?

82. a. Evaluate $5x - 14y$ for $x = 3$ and $y = \frac{1}{2}$.

    b. Is the number you obtained in part (a) a solution of $4w = 54 - 5w$?

Application Exercises

We opened the chapter with what it would have cost to see six classic movies on the big screen. The bar graph shows the average price of a movie ticket for selected years from 1980 through 2010.

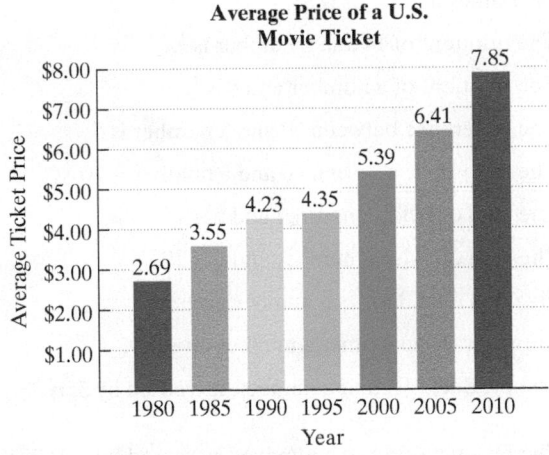

Average Price of a U.S. Movie Ticket

Sources: Motion Picture Association of America, National Association of Theater Owners (NATO), and Bureau of Labor Statistics (BLS)

Here is a mathematical model that approximates the data displayed by the bar graph:

$$T = 0.15n + 2.72.$$

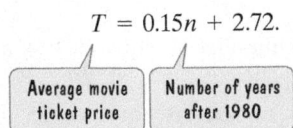

Use this formula to solve Exercises 83–84.

83. a. Use the formula to find the average ticket price 10 years after 1980, or in 1990. Does the mathematical model underestimate or overestimate the average ticket price shown by the bar graph for 1990? By how much?

b. Does the mathematical model underestimate or overestimate the average ticket price shown by the bar graph for 2010? By how much?

84. a. Use the formula to find the average ticket price 5 years after 1980, or in 1985. Does the mathematical model underestimate or overestimate the average ticket price shown by the bar graph in the previous column for 1985? By how much?

b. Does the mathematical model underestimate or overestimate the average ticket price shown by the bar graph for 2005? By how much?

We're just not that into marriage. *Among the 2691 American adults surveyed by the Pew Research Center in 2010, 39% said marriage is optional and becoming obsolete, up from 28% who responded to the same survey in 1978. The bar graph shows the percentage of Americans for four selected ages who say that marriage is obsolete.*

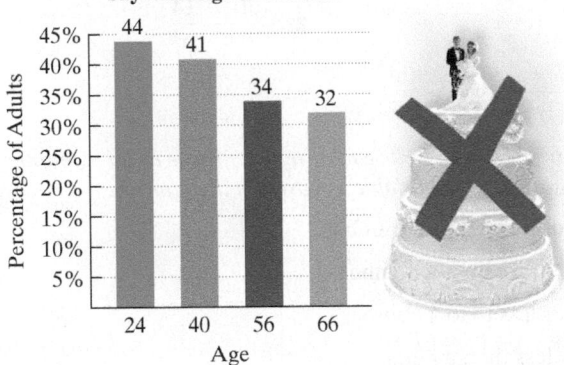

Percentage of Americans Who Say Marriage Is Obsolete

Source: Pew Research Center, "For Millennials, Parenthood Trumps Marriage," March 9, 2011 (www.pewresearch.org)

Here is a mathematical model that approximates the data displayed by the bar graph:

$$p = 52 - 0.3a.$$

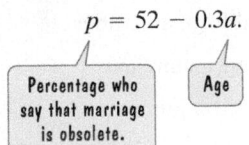

Use this formula to solve Exercises 85–86.

85. Does the mathematical model underestimate or overestimate the percentage of 24-year-olds who say that marriage is obsolete? By how much?

86. Does the mathematical model underestimate or overestimate the percentage of 66-year-olds who say that marriage is obsolete? By how much?

A bowler's handicap, H, is often found using the following formula:

$$H = 0.8(200 - A).$$

Bowler's handicap   Bowler's average score

A bowler's final score for a game is the score for that game increased by the handicap.

Use this information and the formula at the bottom of the previous page to solve Exercises 87–88.

87. a. If your average bowling score is 145, what is your handicap?

b. What would your final score be if you bowled 120 in a game?

88. a. If your average bowling score is 165, what is your handicap?

b. What would your final score be if you bowled 140 in a game?

Writing in Mathematics

Writing about mathematics will help you to learn mathematics. For all writing exercises in this book, use complete sentences to respond to the questions. Some writing exercises can be answered in a sentence; others require a paragraph or two. You can decide how much you need to write as long as your writing clearly and directly answers the question in the exercise. Standard references such as a dictionary and a thesaurus should be helpful.

89. What is a variable?

90. What is an algebraic expression?

91. Explain how to evaluate $2 + 5x$ for $x = 3$.

92. If x represents the number, explain the difference between translating the following phrases:
a number decreased by 5
a number subtracted from 5.

93. What is an equation?

94. How do you tell the difference between an algebraic expression and an equation?

95. How do you determine whether a given number is a solution of an equation?

96. What is a mathematical model?

97. The bar graph for Exercises 85–86 shows a decline with increasing age in the percentage of Americans who say that marriage is obsolete. What explanations can you offer for this trend?

98. In Exercises 87–88, we used the formula $H = 0.8(200 - A)$ to find a bowler's handicap, H, where the variable A represents the bowler's average score. Describe what happens to the handicap when the average score is 200.

Critical Thinking Exercises

Make Sense? *In Exercises 99–102, determine whether each statement "makes sense" or "does not make sense" and explain your reasoning.*

99. As I read this book, I write questions in the margins that I might ask in class.

100. I'm solving a problem that requires me to determine if 5 is a solution of $4x + 7$.

101. The model $T = 0.15n + 2.72$ describes the average movie ticket price, T, n years after 1980, so I can use it to estimate the average movie ticket price in 1980.

102. Because there are four quarters in a dollar, I can use the formula $q = 4d$ to determine the number of quarters, q, in d dollars.

In Exercises 103–106, determine whether each statement is true or false. If the statement is false, make the necessary change(s) to produce a true statement.

103. The algebraic expression for "3 less than a number" is the same as the algebraic expression for "a number decreased by 3."

104. Some algebraic expressions contain the equality symbol, $=$.

105. The algebraic expressions $3 + 2x$ and $(3 + 2)x$ do not mean the same thing.

106. The algebraic expression for "the quotient of a number and 6" is the same as the algebraic expression for "the quotient of 6 and a number."

In Exercises 107–108, define variables and write a formula that describes the relationship in each table.

107.

Number of Hours Worked	Salary
3	$60
4	$80
5	$100
6	$120

108.

Number of Workers	Number of Televisions Assembled
3	30
4	40
5	50
6	60

Preview Exercises

Exercises 109–111 will help you prepare for the material covered in the next section. In each exercise, use the given formula to perform the indicated operation with the two fractions.

109. $\dfrac{a}{b} \cdot \dfrac{c}{d} = \dfrac{a \cdot c}{b \cdot d};\ \dfrac{3}{7} \cdot \dfrac{2}{5}$

110. $\dfrac{a}{b} \div \dfrac{c}{d} = \dfrac{a}{b} \cdot \dfrac{d}{c} = \dfrac{a \cdot d}{b \cdot c};\ \dfrac{2}{3} \div \dfrac{7}{5}$

111. $\dfrac{a}{b} - \dfrac{c}{b} = \dfrac{a - c}{b};\ \dfrac{9}{17} - \dfrac{5}{17}$

SECTION

1.2

Fractions in Algebra

Objectives

1 Convert between mixed numbers and improper fractions.

2 Write the prime factorization of a composite number.

3 Reduce or simplify fractions.

4 Multiply fractions.

5 Divide fractions.

6 Add and subtract fractions with identical denominators.

7 Add and subtract fractions with unlike denominators.

8 Solve problems involving fractions in algebra.

Had a good workout lately? If so, could you tell from your heart rate if you were overdoing it or not pushing yourself hard enough?

Couch-Potato Exercise

$$H = \frac{2}{5}(220 - a)$$

Heart rate, in beats per minute   Age

Working It

$$H = \frac{9}{10}(220 - a)$$

Heart rate, in beats per minute   Age

The fractions $\frac{2}{5}$ and $\frac{9}{10}$ provide the difference between these formulas. Recall that in a fraction, the number that is written above the fraction bar is called the **numerator**. The number below the fraction bar is called the **denominator**.

Numerator $\frac{2}{5}$ Fraction bar $\frac{9}{10}$ Numerator

Denominator   Denominator

The numerators and denominators of these fractions, 2, 5, 9, and 10, are examples of *natural numbers*. The **natural numbers** are the numbers that we use for counting.

Natural numbers   1, 2, 3, 4, 5, . . .

The three dots after the 5 indicate that the list continues in the same manner without ending.

Fractions appear throughout algebra. The first part of this section provides a review of the arithmetic of fractions. Later in the section, we focus on fractions in algebra.

Mixed Numbers and Improper Fractions

1 Convert between mixed numbers and improper fractions.

A **mixed number** consists of the sum of a natural number and a fraction, expressed without the use of an addition sign. Here is an example of a mixed number:

$3\frac{4}{5}$.   The natural number is 3 and the fraction is $\frac{4}{5}$. $3\frac{4}{5}$ means $3 + \frac{4}{5}$.

An **improper fraction** is a fraction whose numerator is greater than its denominator. An example of an improper fraction is $\frac{19}{5}$.

The mixed number $3\frac{4}{5}$ can be converted to the improper fraction $\frac{19}{5}$ using the following procedure:

Converting a Mixed Number to an Improper Fraction

1. Multiply the denominator of the fraction by the natural number and add the numerator to this product.
2. Place the result from step 1 over the denominator of the original mixed number.

EXAMPLE 1   Converting from Mixed Number to Improper Fraction

Convert $3\frac{4}{5}$ to an improper fraction.

Solution

$$3\frac{4}{5} = \frac{5 \cdot 3 + 4}{5}$$

Multiply the denominator by the natural number and add the numerator.

Place the result over the mixed number's denominator.

$$= \frac{15 + 4}{5} = \frac{19}{5}$$

Great Question!

I'm a visual learner. How can I "see" that $3\frac{4}{5}$ and $\frac{19}{5}$ represent the same number?

Figure 1.3 illustrates that shading $3\frac{4}{5}$ circles is the same as shading $\frac{19}{5}$ of the circles.

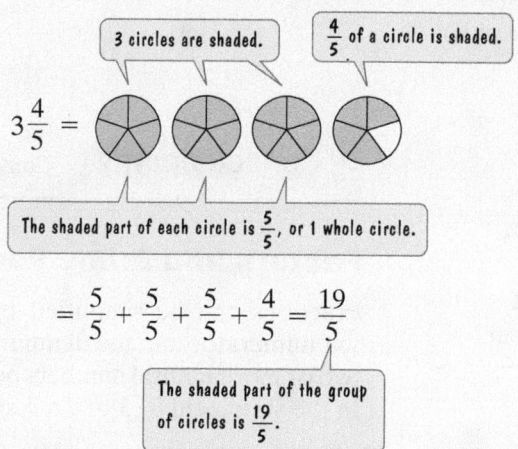

3 circles are shaded.

$\frac{4}{5}$ of a circle is shaded.

$$3\frac{4}{5} =$$

The shaded part of each circle is $\frac{5}{5}$, or 1 whole circle.

$$= \frac{5}{5} + \frac{5}{5} + \frac{5}{5} + \frac{4}{5} = \frac{19}{5}$$

The shaded part of the group of circles is $\frac{19}{5}$.

Figure 1.3 Supporting the conclusion that $3\frac{4}{5} = \frac{19}{5}$

✓   **CHECK POINT 1**   Convert $2\frac{5}{8}$ to an improper fraction.

An **improper fraction** can be converted to a mixed number using the following procedure:

Converting an Improper Fraction to a Mixed Number

1. Divide the denominator into the numerator. Record the quotient (the result of the division) and the remainder.
2. Write the mixed number using the following form:

$$\text{quotient } \frac{\text{remainder}}{\text{original denominator}}.$$

natural number part   fraction part

EXAMPLE 2 Converting from Improper Fraction to Mixed Number

Convert $\frac{42}{5}$ to a mixed number.

Solution We use two steps to convert $\frac{42}{5}$ to a mixed number.

Step 1. Divide the denominator into the numerator.

$$\begin{array}{r} 8 \\ 5\overline{)42} \\ 40 \\ \hline 2 \end{array}$$

quotient

remainder

Step 2. Write the mixed number using quotient $\dfrac{\text{remainder}}{\text{original denominator}}$. Thus,

$$\frac{42}{5} = 8\frac{2}{5}.$$

remainder

original denominator

quotient

✓ **CHECK POINT 2** Convert $\frac{5}{3}$ to a mixed number.

Great Question!

Should I express a fraction whose numerator is greater than its denominator as a mixed number or as an improper fraction?

In applied problems, answers are usually expressed as mixed numbers, which many people find more meaningful than improper fractions. Improper fractions are often easier to work with when performing operations with fractions.

2 Write the prime factorization of a composite number.

Factors and Prime Factorizations

Fractions can be simplified by first factoring the natural numbers that make up the numerator and the denominator. To **factor** a natural number means to write it as two or more natural numbers being multiplied. For example, 21 can be factored as $7 \cdot 3$. In the statement $7 \cdot 3 = 21$, 7 and 3 are called the **factors** and 21 is the **product**.

7 is a factor of 21. $7 \cdot 3 = 21$ The product of 7 and 3 is 21.

3 is a factor of 21.

Are 7 and 3 the only factors of 21? The answer is no because 21 can also be factored as $1 \cdot 21$. Thus, 1 and 21 are also factors of 21. The factors of 21 are 1, 3, 7, and 21.

Unlike the number 21, some natural numbers have only two factors: the number itself and 1. For example, the number 7 has only two factors: 7 (the number itself) and 1. The only way to factor 7 is $1 \cdot 7$ or, equivalently, $7 \cdot 1$. For this reason, 7 is called a *prime number*.

Prime Numbers

A **prime number** is a natural number greater than 1 that has only itself and 1 as factors.

The first ten prime numbers are

$$2, 3, 5, 7, 11, 13, 17, 19, 23, \text{ and } 29.$$

Can you see why the natural number 15 is not in this list? In addition to having 15 and 1 as factors ($15 = 1 \cdot 15$), it also has factors of 3 and 5 ($15 = 3 \cdot 5$). The number 15 is an example of a *composite number*.

Composite Numbers

A **composite number** is a natural number greater than 1 that is not a prime number.

Every composite number can be expressed as the product of prime numbers. For example, the composite number 45 can be expressed as

$$45 = 3 \cdot 3 \cdot 5.$$

> This product contains only prime numbers: 3 and 5.

Expressing a composite number as the product of prime numbers is called the **prime factorization** of that composite number. The prime factorization of 45 is $3 \cdot 3 \cdot 5$. The order in which we write these factors does not matter. This means that

$$45 = 3 \cdot 3 \cdot 5 \quad \text{or} \quad 45 = 5 \cdot 3 \cdot 3 \quad \text{or} \quad 45 = 3 \cdot 5 \cdot 3.$$

To find the prime factorization of a composite number, begin by selecting any two numbers, excluding 1 and the number itself, whose product is the number to be factored. If one or both of the factors are not prime numbers, continue by factoring each composite number. Stop when all numbers in the factorization are prime.

EXAMPLE 3   Prime Factorization of a Composite Number

Find the prime factorization of 100.

Solution   Begin by selecting any two numbers, excluding 1 and 100, whose product is 100. Here is one possibility:

$$100 = 4 \cdot 25.$$

Because the factors 4 and 25 are not prime, we factor each of these composite numbers.

$$100 = 4 \cdot 25 \qquad \text{This is our first factorization.}$$
$$= 2 \cdot 2 \cdot 5 \cdot 5 \qquad \text{Factor 4 and 25.}$$

Notice that 2 and 5 are both prime. The prime factorization of 100 is $2 \cdot 2 \cdot 5 \cdot 5$. ∎

☑ **CHECK POINT 3**   Find the prime factorization of 36.

Great Question!

Is 1 a prime number or a composite number?

It's neither. The number 1 is the only natural number that is neither a prime number nor a composite number.

③ Reduce or simplify
fractions.

Figure 1.4

Reducing Fractions

Two fractions are **equivalent** if they represent the same value. Writing a fraction as an equivalent fraction with a smaller denominator is called **reducing a fraction**. A fraction is **reduced to its lowest terms** when the numerator and denominator have no common factors other than 1.

Look at the rectangle in **Figure 1.4**. Can you see that it is divided into 6 equal parts? Of these 6 parts, 4 of the parts are red. Thus, $\frac{4}{6}$ of the rectangle is red.

The rectangle in **Figure 1.4** is also divided into 3 equal stacks and 2 of the stacks are red. Thus, $\frac{2}{3}$ of the rectangle is red. Because both $\frac{4}{6}$ and $\frac{2}{3}$ of the rectangle are red, we can conclude that $\frac{4}{6}$ and $\frac{2}{3}$ are equivalent fractions.

How can we show that $\frac{4}{6} = \frac{2}{3}$ without using **Figure 1.4**? Prime factorizations of 4 and 6 play an important role in the process. So does the **Fundamental Principle of Fractions**.

Fundamental Principle of Fractions

In words: The value of a fraction does not change if both the numerator and the denominator are divided (or multiplied) by the same nonzero number.

In algebraic language: If $\frac{a}{b}$ is a fraction and c is a nonzero number, then

$$\frac{a \cdot c}{b \cdot c} = \frac{a}{b}.$$

We use prime factorizations and the Fundamental Principle to reduce $\frac{4}{6}$ to its lowest terms as follows:

$$\frac{4}{6} \quad = \quad \frac{2 \cdot 2}{3 \cdot 2} \quad = \quad \frac{2}{3}.$$

Write prime factorizations of 4 and 6.

Divide the numerator and the denominator by the common prime factor, 2.

Here is a procedure for writing a fraction in lowest terms:

Reducing a Fraction to Its Lowest Terms

1. Write the prime factorizations of the numerator and the denominator.
2. Divide the numerator and the denominator by the greatest common factor, the product of all factors common to both.

Division lines can be used to show dividing out common factors from a fraction's numerator and denominator:

$$\frac{4}{6} = \frac{2 \cdot \cancel{2}}{3 \cdot \cancel{2}} = \frac{2}{3}.$$

Great Question!

When can I divide out numbers that are common to a fraction's numerator and denominator?

When reducing a fraction to its lowest terms, only *factors* that are common to the numerator and the denominator can be divided out. **If you have not factored** and expressed the numerator and denominator in terms of multiplication, **do not divide out**.

Correct:
$$\frac{2 \cdot \cancel{2}}{3 \cdot \cancel{2}} = \frac{2}{3}$$

Incorrect:
$$\frac{2 + \cancel{2}}{3 + \cancel{2}} = \frac{2}{3}$$

Note that $\frac{2+2}{3+2}$ $= \frac{4}{5}$, not $\frac{2}{3}$.

EXAMPLE 4 Reducing Fractions

Reduce each fraction to its lowest terms:

a. $\dfrac{6}{14}$ **b.** $\dfrac{15}{75}$ **c.** $\dfrac{25}{11}$ **d.** $\dfrac{11}{33}$.

Solution For each fraction, begin with the prime factorization of the numerator and the denominator.

a. $\dfrac{6}{14} = \dfrac{3 \cdot 2}{7 \cdot 2} = \dfrac{3}{7}$ 2 is the greatest common factor of 6 and 14. Divide the numerator and the denominator by 2.

> Including 1 as a factor is helpful when all other factors can be divided out.

b. $\dfrac{15}{75} = \dfrac{3 \cdot 5}{3 \cdot 25} = \dfrac{1 \cdot 3 \cdot 5}{3 \cdot 5 \cdot 5} = \dfrac{1}{5}$ $3 \cdot 5$, or 15, is the greatest common factor of 15 and 75. Divide the numerator and the denominator by $3 \cdot 5$.

c. $\dfrac{25}{11} = \dfrac{5 \cdot 5}{1 \cdot 11}$

Because 25 and 11 share no common factor (other than 1), $\dfrac{25}{11}$ is already reduced to its lowest terms.

d. $\dfrac{11}{33} = \dfrac{1 \cdot 11}{3 \cdot 11} = \dfrac{1}{3}$ 11 is the greatest common factor of 11 and 33. Divide the numerator and denominator by 11. ■

When reducing fractions, it may not always be necessary to write prime factorizations. In some cases, you can use inspection to find the greatest common factor of the numerator and the denominator. For example, when reducing $\dfrac{15}{75}$, you can use 15 rather than $3 \cdot 5$:

$$\dfrac{15}{75} = \dfrac{1 \cdot 15}{5 \cdot 15} = \dfrac{1}{5}.$$

✓ **CHECK POINT 4** Reduce each fraction to its lowest terms:

a. $\dfrac{10}{15}$ **b.** $\dfrac{42}{24}$ **c.** $\dfrac{13}{15}$ **d.** $\dfrac{9}{45}$.

4 Multiply fractions.

Multiplying Fractions

The result of multiplying two fractions is called their **product**.

Multiplying Fractions

In words: The product of two or more fractions is the product of their numerators divided by the product of their denominators.

In algebraic language: If $\dfrac{a}{b}$ and $\dfrac{c}{d}$ are fractions, then

$$\dfrac{a}{b} \cdot \dfrac{c}{d} = \dfrac{a \cdot c}{b \cdot d}.$$

Here is an example that illustrates the rule in the previous box:

$$\dfrac{3}{8} \cdot \dfrac{5}{11} = \dfrac{3 \cdot 5}{8 \cdot 11} = \dfrac{15}{88}.$$

> The product of $\dfrac{3}{8}$ and $\dfrac{5}{11}$ is $\dfrac{15}{88}$.

> Multiply numerators and multiply denominators.

EXAMPLE 5 Multiplying Fractions

Multiply. If possible, reduce the product to its lowest terms:

a. $\dfrac{3}{7} \cdot \dfrac{2}{5}$ **b.** $5 \cdot \dfrac{7}{12}$ **c.** $\dfrac{2}{3} \cdot \dfrac{9}{4}$ **d.** $\left(3\dfrac{2}{3}\right)\left(1\dfrac{1}{4}\right)$.

Solution

a. $\dfrac{3}{7} \cdot \dfrac{2}{5} = \dfrac{3 \cdot 2}{7 \cdot 5} = \dfrac{6}{35}$ Multiply numerators and multiply denominators.

b. $5 \cdot \dfrac{7}{12} = \dfrac{5}{1} \cdot \dfrac{7}{12} = \dfrac{5 \cdot 7}{1 \cdot 12} = \dfrac{35}{12}$ or $2\dfrac{11}{12}$ Write 5 as $\frac{5}{1}$. Then multiply numerators and multiply denominators.

c. $\dfrac{2}{3} \cdot \dfrac{9}{4} = \dfrac{2 \cdot 9}{3 \cdot 4} = \dfrac{18}{12} = \dfrac{3 \cdot \cancel{6}}{2 \cdot \cancel{6}} = \dfrac{3}{2}$ or $1\dfrac{1}{2}$

> Simplify $\frac{18}{12}$; 6 is the greatest common factor of 18 and 12.

d. $\left(3\dfrac{2}{3}\right)\left(1\dfrac{1}{4}\right) = \dfrac{11}{3} \cdot \dfrac{5}{4} = \dfrac{11 \cdot 5}{3 \cdot 4} = \dfrac{55}{12}$ or $4\dfrac{7}{12}$ ∎

✓ **CHECK POINT 5** Multiply. If possible, reduce the product to its lowest terms:

a. $\dfrac{4}{11} \cdot \dfrac{2}{3}$ **b.** $6 \cdot \dfrac{3}{5}$ **c.** $\dfrac{3}{7} \cdot \dfrac{2}{3}$ **d.** $\left(3\dfrac{2}{5}\right)\left(1\dfrac{1}{2}\right)$.

Great Question!

Can I divide numerators and denominators by common factors before I multiply fractions?

Yes, you can divide numerators and denominators by common factors before performing multiplication. Then multiply the remaining factors in the numerators and multiply the remaining factors in the denominators. For example,

$$\dfrac{7}{15} \cdot \dfrac{20}{21} = \dfrac{\cancel{7} \cdot 1}{5 \cdot 3} \cdot \dfrac{5 \cdot 4}{\cancel{7} \cdot 3} = \dfrac{1 \cdot 4}{3 \cdot 3} = \dfrac{4}{9}.$$

> 7 is the greatest common factor of 7 and 21.

> 5 is the greatest common factor of 15 and 20.

The divisions involving the common factors, 7 and 5, are often shown as follows:

$$\dfrac{7}{15} \cdot \dfrac{20}{21} = \dfrac{\overset{1}{\cancel{7}}}{\underset{3}{\cancel{15}}} \cdot \dfrac{\overset{4}{\cancel{20}}}{\underset{3}{\cancel{21}}} = \dfrac{1 \cdot 4}{3 \cdot 3} = \dfrac{4}{9}.$$

> Divide by 7.

> Divide by 5.

---

5 Divide fractions.

Dividing Fractions

The result of dividing two fractions is called their **quotient**. A geometric figure is useful for developing a process for determining the quotient of two fractions.

Consider the division

$$\dfrac{4}{5} \div \dfrac{1}{10}.$$

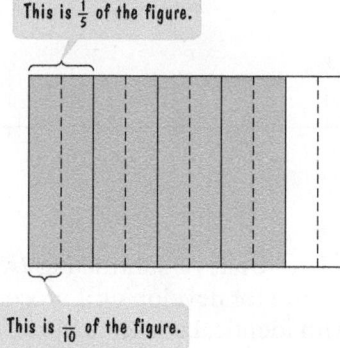

This is $\frac{1}{5}$ of the figure.

This is $\frac{1}{10}$ of the figure.

Figure 1.5

We want to know how many $\frac{1}{10}$'s are in $\frac{4}{5}$. We can use **Figure 1.5** to find this quotient. The rectangle is divided into fifths. The dashed lines further divide the rectangle into tenths.

Figure 1.5 shows that $\frac{4}{5}$ of the rectangle is red. How many $\frac{1}{10}$'s of the rectangle does this include? Can you see that this includes eight of the $\frac{1}{10}$ pieces? Thus, there are eight $\frac{1}{10}$'s in $\frac{4}{5}$:

$$\frac{4}{5} \div \frac{1}{10} = 8.$$

We can obtain the quotient 8 in the following way:

$$\frac{4}{5} \div \frac{1}{10} = \frac{4}{5} \cdot \frac{10}{1} = \frac{4 \cdot 10}{5 \cdot 1} = \frac{40}{5} = 8.$$

Change the division to multiplication.

Invert the divisor, $\frac{1}{10}$.

By inverting the divisor, $\frac{1}{10}$, and obtaining $\frac{10}{1}$, we are writing the divisor's *reciprocal*. Two fractions are **reciprocals** of each other if their product is 1. Thus, $\frac{1}{10}$ and $\frac{10}{1}$ are reciprocals because $\frac{1}{10} \cdot \frac{10}{1} = 1$.

Generalizing from the result above and using the word *reciprocal*, we obtain the following rule for dividing fractions:

Dividing Fractions

In words: The quotient of two fractions is the first fraction multiplied by the reciprocal of the second fraction.

In algebraic language: If $\frac{a}{b}$ and $\frac{c}{d}$ are fractions and $\frac{c}{d}$ is not 0, then

$$\frac{a}{b} \div \frac{c}{d} = \frac{a}{b} \cdot \frac{d}{c}.$$

Change division to multiplication.

Invert $\frac{c}{d}$ and write its reciprocal.

EXAMPLE 6   Dividing Fractions

a. $\frac{2}{3} \div \frac{7}{15}$ **b.** $\frac{3}{4} \div 5$ **c.** $4\frac{3}{4} \div 1\frac{1}{2}$.

Solution

a. $\frac{2}{3} \div \frac{7}{15} = \frac{2}{3} \cdot \frac{15}{7} = \frac{2 \cdot 15}{3 \cdot 7} = \frac{30}{21} = \frac{10 \cdot 3}{7 \cdot 3} = \frac{10}{7}$ or $1\frac{3}{7}$

Change division to multiplication.   Invert $\frac{7}{15}$ and write its reciprocal.   Simplify: 3 is the greatest common factor of 30 and 21.

b. $\frac{3}{4} \div 5 = \frac{3}{4} \div \frac{5}{1} = \frac{3}{4} \cdot \frac{1}{5} = \frac{3 \cdot 1}{4 \cdot 5} = \frac{3}{20}$

Change division to multiplication.   Invert $\frac{5}{1}$ and write its reciprocal.

c. $4\frac{3}{4} \div 1\frac{1}{2} = \frac{19}{4} \div \frac{3}{2} = \frac{19}{4} \cdot \frac{2}{3} = \frac{19 \cdot 2}{4 \cdot 3} = \frac{38}{12} = \frac{19 \cdot 2}{6 \cdot 2} = \frac{19}{6}$ or $3\frac{1}{6}$ ∎

✓ **CHECK POINT 6**   Divide:

a. $\dfrac{5}{4} \div \dfrac{3}{8}$ **b.** $\dfrac{2}{3} \div 3$ **c.** $3\dfrac{3}{8} \div 2\dfrac{1}{4}$

6 Add and subtract fractions with identical denominators.

Adding and Subtracting Fractions with Identical Denominators

The result of adding two fractions is called their **sum**. The result of subtracting two fractions is called their **difference**. A geometric figure is useful for developing a process for determining the sum or difference of two fractions with identical denominators.

Consider the addition

$$\frac{3}{7} + \frac{2}{7}.$$

We can use **Figure 1.6** to find this sum. The rectangle is divided into sevenths. On the left, $\frac{3}{7}$ of the rectangle is red. On the right, $\frac{2}{7}$ of the rectangle is red. Including both the left and the right, a total of $\frac{5}{7}$ of the rectangle is red. Thus,

$$\frac{3}{7} + \frac{2}{7} = \frac{5}{7}.$$

We can obtain the sum $\frac{5}{7}$ in the following way:

$$\frac{3}{7} + \frac{2}{7} = \frac{3+2}{7} = \frac{5}{7}.$$

Add numerators and put this result over the common denominator.

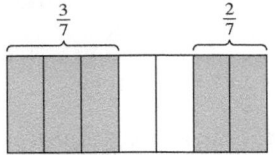

Figure 1.6

Generalizing from this result gives us the following rule:

Adding and Subtracting Fractions with Identical Denominators

In words: The sum or difference of two fractions with identical denominators is the sum or difference of their numerators over the common denominator.

In algebraic language: If $\frac{a}{b}$ and $\frac{c}{b}$ are fractions, then

$$\frac{a}{b} + \frac{c}{b} = \frac{a+c}{b} \qquad \text{and} \qquad \frac{a}{b} - \frac{c}{b} = \frac{a-c}{b}.$$

EXAMPLE 7   Adding and Subtracting Fractions with Identical Denominators

Perform the indicated operations:

a. $\dfrac{3}{11} + \dfrac{4}{11}$ **b.** $\dfrac{11}{12} - \dfrac{5}{12}$ **c.** $5\dfrac{1}{4} - 2\dfrac{3}{4}.$

Solution

a. $\dfrac{3}{11} + \dfrac{4}{11} = \dfrac{3+4}{11} = \dfrac{7}{11}$

b. $\dfrac{11}{12} - \dfrac{5}{12} = \dfrac{11-5}{12} = \dfrac{6}{12} = \dfrac{1 \cdot \cancel{6}}{2 \cdot \cancel{6}} = \dfrac{1}{2}$

c. $5\dfrac{1}{4} - 2\dfrac{3}{4} = \dfrac{21}{4} - \dfrac{11}{4} = \dfrac{21-11}{4} = \dfrac{10}{4} = \dfrac{2 \cdot 5}{2 \cdot 2} = \dfrac{5}{2}$ or $2\dfrac{1}{2}$ ∎

✔ **CHECK POINT 7**   Perform the indicated operations:

a. $\dfrac{2}{11} + \dfrac{3}{11}$ **b.** $\dfrac{5}{6} - \dfrac{1}{6}$ **c.** $3\dfrac{3}{8} - 1\dfrac{1}{8}$.

7 Add and subtract fractions with unlike denominators.

Adding and Subtracting Fractions with Unlike Denominators

How do we add or subtract fractions with different denominators? We must first rewrite them as equivalent fractions with the same denominator. We do this by using the Fundamental Principle of Fractions: The value of a fraction does not change if the numerator and the denominator are multiplied by the same nonzero number. Thus, if $\frac{a}{b}$ is a fraction and c is a nonzero number, then

$$\frac{a}{b} = \frac{a \cdot c}{b \cdot c}.$$

EXAMPLE 8   Writing an Equivalent Fraction

Write $\frac{3}{4}$ as an equivalent fraction with a denominator of 16.

Solution   To obtain a denominator of 16, we must multiply the denominator of the given fraction, $\frac{3}{4}$, by 4. So that we do not change the value of the fraction, we also multiply the numerator by 4.

$$\frac{3}{4} = \frac{3 \cdot 4}{4 \cdot 4} = \frac{12}{16} \quad \blacksquare$$

✔ **CHECK POINT 8**   Write $\frac{2}{3}$ as an equivalent fraction with a denominator of 21.

Equivalent fractions can be used to add fractions with different denominators, such as $\frac{1}{2}$ and $\frac{1}{3}$. **Figure 1.7** indicates that the sum of half the whole figure and one-third of the whole figure results in 5 parts out of 6, or $\frac{5}{6}$, of the figure. Thus,

$$\frac{1}{2} + \frac{1}{3} = \frac{5}{6}.$$

We can obtain the sum $\frac{5}{6}$ if we rewrite each fraction as an equivalent fraction with a denominator of 6.

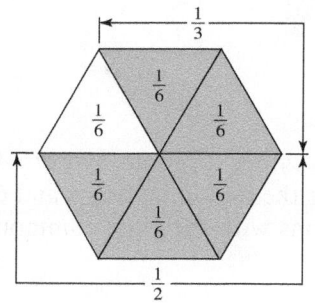

Figure 1.7 $\frac{1}{2} + \frac{1}{3} = \frac{5}{6}$

$$\frac{1}{2} + \frac{1}{3} = \frac{1 \cdot 3}{2 \cdot 3} + \frac{1 \cdot 2}{3 \cdot 2} \qquad \text{Rewrite each fraction as an equivalent fraction with a denominator of 6.}$$

$$= \frac{3}{6} + \frac{2}{6} \qquad \text{Perform the multiplications. We now have a common denominator.}$$

$$= \frac{3 + 2}{6} \qquad \text{Add the numerators and place this sum over the common denominator.}$$

$$= \frac{5}{6} \qquad \text{Perform the addition.}$$

When adding $\frac{1}{2}$ and $\frac{1}{3}$, there are many common denominators that we can use, such as 6, 12, 18, and so on. The given denominators, 2 and 3, divide into all of these numbers. However, the denominator 6 is the smallest number that 2 and 3 divide into. For this reason, 6 is called the *least common denominator*, abbreviated LCD.

Adding and Subtracting Fractions with Unlike Denominators

1. Rewrite the fractions as equivalent fractions with the least common denominator.
2. Add or subtract the numerators, putting this result over the common denominator.

EXAMPLE 9 Adding and Subtracting Fractions with Unlike Denominators

Perform the indicated operation:

a. $\dfrac{1}{5} + \dfrac{3}{4}$ **b.** $\dfrac{3}{4} - \dfrac{1}{6}$ **c.** $2\dfrac{7}{15} - 1\dfrac{4}{5}.$

Solution

a. Just by looking at $\frac{1}{5} + \frac{3}{4}$, can you tell that the smallest number divisible by both 5 and 4 is 20? Thus, the least common denominator for the denominators 5 and 4 is 20. We rewrite both fractions as equivalent fractions with the least common denominator, 20.

$$\frac{1}{5} + \frac{3}{4} = \frac{1 \cdot 4}{5 \cdot 4} + \frac{3 \cdot 5}{4 \cdot 5}$$

To obtain denominators of 20, multiply the numerator and denominator of the first fraction by 4 and the second fraction by 5.

$$= \frac{4}{20} + \frac{15}{20}$$

Perform the multiplications.

$$= \frac{4 + 15}{20}$$

Add the numerators and put this sum over the least common denominator.

$$= \frac{19}{20}$$

Perform the addition.

Discover for Yourself

Try Example 9(a), $\frac{1}{5} + \frac{3}{4}$, using a common denominator of 40. Because both 5 and 4 divide into 40, 40 is a common denominator, although not the least common denominator. Describe what happens. What is the advantage of using the least common denominator?

b. By looking at $\frac{3}{4} - \frac{1}{6}$, can you tell that the smallest number divisible by both 4 and 6 is 12? Thus, the least common denominator for the denominators 4 and 6 is 12. We rewrite both fractions as equivalent fractions with the least common denominator, 12.

$$\frac{3}{4} - \frac{1}{6} = \frac{3 \cdot 3}{4 \cdot 3} - \frac{1 \cdot 2}{6 \cdot 2}$$

To obtain denominators of 12, multiply the numerator and denominator of the first fraction by 3 and the second fraction by 2.

$$= \frac{9}{12} - \frac{2}{12}$$

Perform the multiplications.

$$= \frac{9 - 2}{12}$$

Subtract the numerators and put this difference over the least common denominator.

$$= \frac{7}{12}$$

Perform the subtraction.

c. $2\frac{7}{15} - 1\frac{4}{5} = \frac{37}{15} - \frac{9}{5}$ Convert each mixed number to an improper fraction.

> The smallest number divisible by both 15 and 5 is 15. Thus, the least common denominator for the denominators 15 and 5 is 15. Because the first fraction already has a denominator of 15, we only have to rewrite the second fraction.

$= \frac{37}{15} - \frac{9 \cdot 3}{5 \cdot 3}$ To obtain a denominator of 15, multiply the numerator and denominator of the second fraction by 3.

$= \frac{37}{15} - \frac{27}{15}$ Perform the multiplications.

$= \frac{37 - 27}{15}$ Subtract the numerators and put this difference over the common denominator.

$= \frac{10}{15}$ Perform the subtraction.

$= \frac{2 \cdot 5}{3 \cdot 5} = \frac{2}{3}$ Reduce to lowest terms. ■

✓ **CHECK POINT 9**   Perform the indicated operation:

a. $\frac{1}{2} + \frac{3}{5}$ **b.** $\frac{4}{3} - \frac{3}{4}$ **c.** $3\frac{1}{6} - 1\frac{11}{12}$.

EXAMPLE 10   Using Prime Factorizations to Find the LCD

Perform the indicated operation: $\frac{1}{15} + \frac{7}{24}$.

Solution   We need to find the least common denominator first. Using inspection, it is difficult to determine the smallest number divisible by both 15 and 24. We will use their prime factorizations to find the least common denominator:

$$15 = 5 \cdot 3 \quad \text{and} \quad 24 = 8 \cdot 3 = 2 \cdot 2 \cdot 2 \cdot 3.$$

The different prime factors are 5, 3, and 2. The least common denominator is obtained by using the greatest number of times each factor appears in any prime factorization. Because 5 and 3 appear as prime factors and 2 is a factor of 24 three times, the least common denominator is

$$5 \cdot 3 \cdot 2 \cdot 2 \cdot 2 = 5 \cdot 3 \cdot 8 = 120.$$

Now we can rewrite both fractions as equivalent fractions with the least common denominator, 120.

$\frac{1}{15} + \frac{7}{24} = \frac{1 \cdot 8}{15 \cdot 8} + \frac{7 \cdot 5}{24 \cdot 5}$ To obtain denominators of 120, multiply the numerator and denominator of the first fraction by 8 and the second fraction by 5.

$= \frac{8}{120} + \frac{35}{120}$ Perform the multiplications.

$= \frac{8 + 35}{120}$ Add the numerators and put this sum over the least common denominator.

$= \frac{43}{120}$ Perform the addition. ■

✓ **CHECK POINT 10**   Perform the indicated operation: $\frac{3}{10} + \frac{7}{12}$.

Great Question!

I see that least common denominators are used to add and subtract fractions with unlike denominators. Is it ever necessary to use least common denominators when multiplying or dividing fractions?

No. You do not need least common denominators to multiply or divide fractions.

8 Solve problems involving fractions in algebra.

Fractions in Algebra

Fractions appear throughout algebra. Operations with fractions can be used to determine whether a particular fraction is a solution of an equation.

> **EXAMPLE 11**   Determining Whether Fractions Are Solutions of Equations

Determine whether the given number is a solution of the equation.

a. $x + \dfrac{1}{4}x = 7; 6\dfrac{2}{5}$ **b.** $\dfrac{1}{7} - w = \dfrac{1}{2}w; \dfrac{2}{21}$

Solution

a. To determine whether $6\frac{2}{5}$ is a solution of $x + \frac{1}{4}x = 7$, we begin by converting $6\frac{2}{5}$ from a mixed number to an improper fraction.

$$6\frac{2}{5} = \frac{5 \cdot 6 + 2}{5} = \frac{30 + 2}{5} = \frac{32}{5}$$

Now we substitute $\frac{32}{5}$ for x.

$$x + \frac{1}{4}x = 7 \qquad \text{This is the given equation.}$$

Is $\frac{32}{5}$ a solution?

$$\frac{32}{5} + \frac{1}{4} \cdot \frac{32}{5} \stackrel{?}{=} 7 \qquad \text{Substitute } \frac{32}{5} \text{ for x.}$$

$$\frac{32}{5} + \frac{8}{5} \stackrel{?}{=} 7 \qquad \text{Multiply: } \frac{1}{\underset{1}{4}} \cdot \frac{\overset{8}{32}}{5} = \frac{1 \cdot 8}{5} = \frac{8}{5}.$$

$$\frac{40}{5} \stackrel{?}{=} 7 \qquad \text{Add: } \frac{32}{5} + \frac{8}{5} = \frac{32 + 8}{5} = \frac{40}{5}.$$

This statement is false. $8 = 7 \qquad \text{Simplify: } \frac{40}{5} = 8.$

Because the values on both sides of the equation are not the same, the fraction $\frac{32}{5}$, or equivalently $6\frac{2}{5}$, is not a solution of the equation.

b.

$$\frac{1}{7} - w = \frac{1}{2}w \qquad \text{This is the given equation.}$$

Is $\frac{2}{21}$ a solution?

$$\frac{1}{7} - \frac{2}{21} \stackrel{?}{=} \frac{1}{2} \cdot \frac{2}{21} \qquad \text{Substitute } \frac{2}{21} \text{ for w.}$$

$$\frac{1}{7} - \frac{2}{21} \stackrel{?}{=} \frac{1}{21} \qquad \text{Multiply: } \frac{1}{\underset{1}{2}} \cdot \frac{\overset{1}{2}}{21} = \frac{1 \cdot 1}{1 \cdot 21} = \frac{1}{21}.$$

$$\frac{1 \cdot 3}{7 \cdot 3} - \frac{2}{21} \stackrel{?}{=} \frac{1}{21} \qquad \text{Turn to the subtraction. The LCD is 21, so multiply the numerator and denominator of the first fraction by 3.}$$

$$\frac{3}{21} - \frac{2}{21} \stackrel{?}{=} \frac{1}{21} \qquad \text{Perform the multiplications.}$$

This statement is true. $\dfrac{1}{21} = \dfrac{1}{21} \qquad \text{Subtract: } \dfrac{3}{21} - \dfrac{2}{21} = \dfrac{3 - 2}{21} = \dfrac{1}{21}.$

Because the values on both sides of the equation are the same, the fraction $\frac{2}{21}$ is a solution of the equation. ∎

✓ **CHECK POINT 11**   Determine whether the given number is a solution of the equation.

a. $x - \frac{2}{9}x = 1; 1\frac{2}{7}$ **b.** $\frac{1}{5} - w = \frac{1}{3}w; \frac{3}{20}$

In Section 1.1, we translated phrases into algebraic expressions and sentences into equations. When these phrases and sentences involve fractions, the word *of* frequently appears. **When used with fractions, the word *of* represents multiplication.** For example, the phrase "$\frac{2}{5}$ of a number" can be represented by the algebraic expression $\frac{2}{5} \cdot x$, or $\frac{2}{5}x$.

EXAMPLE 12   Algebraic Representations of Phrases and Sentences with Fractions

Translate from English to an algebraic expression or equation, whichever is appropriate. Let the variable x represent the number.

a. $\frac{1}{3}$ of a number increased by 5 is half of that number.

b. $\frac{1}{4}$ of a number, decreased by 7

Solution   Part (a) is a sentence, so we will translate into an equation. Part (b) is a phrase, so we will translate into an algebraic expression.

a.

| $\frac{1}{3}$ | of | a number increased by 5 | is | half | of | that number. |

$$\frac{1}{3} \quad \cdot \quad (x + 5) \quad = \quad \frac{1}{2} \quad \cdot \quad x$$

The equation for "$\frac{1}{3}$ of a number increased by 5 is half of that number" is $\frac{1}{3}(x + 5) = \frac{1}{2}x$.

b.

| $\frac{1}{4}$ | of | a number | , | decreased by 7 |

$$\frac{1}{4} \quad \cdot \quad x \quad - \quad 7$$

The algebraic expression for "$\frac{1}{4}$ of a number, decreased by 7" is $\frac{1}{4}x - 7$. ∎

✓ **CHECK POINT 12**   Translate from English to an algebraic expression or equation, whichever is appropriate. Let the variable x represent the number.

a. $\frac{2}{3}$ of a number decreased by 6

b. $\frac{3}{4}$ of a number, decreased by 2, is $\frac{1}{5}$ of that number.

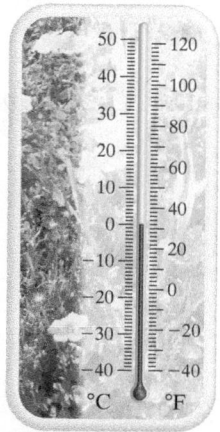

Figure 1.8 The Celsius scale is on the left and the Fahrenheit scale is on the right.

Many formulas and mathematical models contain fractions. For example, consider temperatures on the Celsius scale and on the Fahrenheit scale, as shown in **Figure 1.8**. The formula $C = \frac{5}{9}(F - 32)$ expresses the relationship between Fahrenheit temperature, F, and Celsius temperature, C.

The Formula	What the Formula Tells Us
$C = \dfrac{5}{9}(F - 32)$	If 32 is subtracted from the Fahrenheit temperature, $F - 32$, and this difference is multiplied by $\frac{5}{9}$, the resulting product, $\frac{5}{9}(F - 32)$, gives the Celsius temperature.

EXAMPLE 13 Evaluating a Formula Containing a Fraction

The temperature on a warm summer day is 86°F. Use the formula $C = \frac{5}{9}(F - 32)$ to find the equivalent temperature on the Celsius scale.

Solution Because the temperature is 86°F, we substitute 86 for F in the given formula. Then we determine the value of C.

$$C = \frac{5}{9}(F - 32) \qquad \text{This is the given formula.}$$

$$C = \frac{5}{9}(86 - 32) \qquad \text{Replace } F \text{ with 86.}$$

$$C = \frac{5}{9}(54) \qquad \text{Work inside parentheses first: } 86 - 32 = 54.$$

$$C = 30 \qquad \text{Multiply: } \frac{5}{9}(54) = \frac{5}{\cancel{9}} \cdot \frac{\overset{6}{\cancel{54}}}{1} = \frac{5 \cdot 6}{1 \cdot 1} = \frac{30}{1} = 30.$$

Thus, 86°F is equivalent to 30°C. ∎

✓ **CHECK POINT 13** The temperature on a warm spring day is 77°F. Use the formula $C = \frac{5}{9}(F - 32)$ to find the equivalent temperature on the Celsius scale.

CONCEPT AND VOCABULARY CHECK

Fill in each blank so that the resulting statement is true.

1. In the fraction $\frac{2}{5}$, the number 2 is called the _____ and the number 5 is called the _____.

2. The number $3\frac{2}{5}$ is called a/an _____ number and the number $\frac{17}{5}$ is called a/an _____ fraction.

3. The number $3\frac{2}{5}$ can be converted to $\frac{17}{5}$ by multiplying _____ and _____, adding _____, and placing the result over _____.

4. The numbers that we use for counting $(1, 2, 3, 4, 5, \ldots)$ are called _____ numbers.

5. Among the numbers $1, 2, 3, 4, 5, \ldots$, a number greater than 1 that has only itself and 1 as factors is called a/an _____ number.

6. In $7 \cdot 5 = 35$, the numbers 7 and 5 are called _____ of 35 and the number 35 is called the _____ of 7 and 5.

7. If $\dfrac{a}{b}$ is a fraction and c is a nonzero number, then $\dfrac{a \cdot c}{b \cdot c} = $ _____.

8. If $\dfrac{a}{b}$ and $\dfrac{c}{d}$ are fractions, then $\dfrac{a}{b} \cdot \dfrac{c}{d} = $ _____.

9. Two fractions whose product is 1, such as $\frac{2}{3}$ and $\frac{3}{2}$, are called _____ of each other.

10. If $\dfrac{a}{b}$ and $\dfrac{c}{d}$ are fractions and $\dfrac{c}{d}$ is not 0, then $\dfrac{a}{b} \div \dfrac{c}{d} = \dfrac{a}{b}$ _____.

11. If $\dfrac{a}{b}$ and $\dfrac{c}{b}$ are fractions, then $\dfrac{a}{b} + \dfrac{c}{b} = $ _____.

12. In order to add $\frac{1}{5}$ and $\frac{3}{4}$, we use 20 as the _____.

1.2 EXERCISE SET MyMathLab®

Watch the videos
in MyMathLab

Download the
MyDashBoard App

Practice Exercises

In Exercises 1–6, convert each mixed number to an improper fraction.

1. $2\dfrac{3}{8}$ **2.** $2\dfrac{7}{9}$ **3.** $7\dfrac{3}{5}$

4. $6\dfrac{2}{5}$ **5.** $8\dfrac{7}{16}$ **6.** $9\dfrac{5}{16}$

In Exercises 7–12, convert each improper fraction to a mixed number.

7. $\dfrac{23}{5}$ **8.** $\dfrac{47}{8}$ **9.** $\dfrac{76}{9}$

10. $\dfrac{59}{9}$ **11.** $\dfrac{711}{20}$ **12.** $\dfrac{788}{25}$

In Exercises 13–28, identify each natural number as prime or composite. If the number is composite, find its prime factorization.

13. 22   **14.** 15   **15.** 20

16. 75   **17.** 37   **18.** 23

19. 36   **20.** 100   **21.** 140

22. 110   **23.** 79   **24.** 83

25. 81   **26.** 64

27. 240   **28.** 360

In Exercises 29–40, simplify each fraction by reducing it to its lowest terms.

29. $\dfrac{10}{16}$ **30.** $\dfrac{8}{14}$ **31.** $\dfrac{15}{18}$ **32.** $\dfrac{18}{45}$

33. $\dfrac{35}{50}$ **34.** $\dfrac{45}{50}$ **35.** $\dfrac{32}{80}$ **36.** $\dfrac{75}{80}$

37. $\dfrac{44}{50}$ **38.** $\dfrac{38}{50}$ **39.** $\dfrac{120}{86}$ **40.** $\dfrac{116}{86}$

In Exercises 41–90, perform the indicated operation. Where possible, reduce the answer to its lowest terms.

41. $\dfrac{2}{5} \cdot \dfrac{1}{3}$ **42.** $\dfrac{3}{7} \cdot \dfrac{1}{4}$ **43.** $\dfrac{3}{8} \cdot \dfrac{7}{11}$

44. $\dfrac{5}{8} \cdot \dfrac{3}{11}$ **45.** $9 \cdot \dfrac{4}{7}$ **46.** $8 \cdot \dfrac{3}{7}$

47. $\dfrac{1}{10} \cdot \dfrac{5}{6}$ **48.** $\dfrac{1}{8} \cdot \dfrac{2}{3}$ **49.** $\dfrac{5}{4} \cdot \dfrac{6}{7}$

50. $\dfrac{7}{4} \cdot \dfrac{6}{11}$ **51.** $\left(3\dfrac{3}{4}\right)\left(1\dfrac{3}{5}\right)$

52. $\left(2\dfrac{4}{5}\right)\left(1\dfrac{1}{4}\right)$ **53.** $\dfrac{5}{4} \div \dfrac{4}{3}$

54. $\dfrac{7}{8} \div \dfrac{2}{3}$ **55.** $\dfrac{18}{5} \div 2$ **56.** $\dfrac{12}{7} \div 3$

57. $2 \div \dfrac{18}{5}$ **58.** $3 \div \dfrac{12}{7}$ **59.** $\dfrac{3}{4} \div \dfrac{1}{4}$

60. $\dfrac{3}{7} \div \dfrac{1}{7}$ **61.** $\dfrac{7}{6} \div \dfrac{5}{3}$ **62.** $\dfrac{7}{4} \div \dfrac{3}{8}$

63. $\dfrac{1}{14} \div \dfrac{1}{7}$ **64.** $\dfrac{1}{8} \div \dfrac{1}{4}$ **65.** $6\dfrac{3}{5} \div 1\dfrac{1}{10}$

66. $1\dfrac{3}{4} \div 2\dfrac{5}{8}$ **67.** $\dfrac{2}{11} + \dfrac{4}{11}$ **68.** $\dfrac{5}{13} + \dfrac{2}{13}$

69. $\dfrac{7}{12} + \dfrac{1}{12}$ **70.** $\dfrac{5}{16} + \dfrac{1}{16}$ **71.** $\dfrac{5}{8} + \dfrac{5}{8}$

72. $\dfrac{3}{8} + \dfrac{3}{8}$ **73.** $\dfrac{7}{12} - \dfrac{5}{12}$ **74.** $\dfrac{13}{18} - \dfrac{5}{18}$

75. $\dfrac{16}{7} - \dfrac{2}{7}$ **76.** $\dfrac{17}{5} - \dfrac{2}{5}$ **77.** $\dfrac{1}{2} + \dfrac{1}{5}$

78. $\dfrac{1}{3} + \dfrac{1}{5}$ **79.** $\dfrac{3}{4} + \dfrac{3}{20}$ **80.** $\dfrac{2}{5} + \dfrac{2}{15}$

81. $\dfrac{3}{8} + \dfrac{5}{12}$ **82.** $\dfrac{3}{10} + \dfrac{2}{15}$ **83.** $\dfrac{11}{18} - \dfrac{2}{9}$

84. $\dfrac{17}{18} - \dfrac{4}{9}$ **85.** $\dfrac{4}{3} - \dfrac{3}{4}$ **86.** $\dfrac{3}{2} - \dfrac{2}{3}$

87. $\dfrac{7}{10} - \dfrac{3}{16}$ **88.** $\dfrac{7}{30} - \dfrac{5}{24}$

89. $3\dfrac{3}{4} - 2\dfrac{1}{3}$ **90.** $3\dfrac{2}{3} - 2\dfrac{1}{2}$

In Exercises 91–102, determine whether the given number is a solution of the equation.

91. $\dfrac{7}{2}x = 28; 8$ **92.** $\dfrac{5}{3}x = 30; 18$

93. $w - \dfrac{2}{3} = \dfrac{3}{4}; 1\dfrac{5}{12}$

94. $w - \dfrac{3}{4} = \dfrac{7}{4}; 2\dfrac{1}{2}$

95. $20 - \dfrac{1}{3}z = \dfrac{1}{2}z; 12$

96. $12 - \dfrac{1}{4}z = \dfrac{1}{2}z; 20$

97. $\dfrac{2}{9}y + \dfrac{1}{3}y = \dfrac{3}{7}; \dfrac{27}{35}$

98. $\dfrac{2}{3}y + \dfrac{5}{6}y = 2; 1\dfrac{1}{3}$

99. $\dfrac{1}{3}(x - 2) = \dfrac{1}{5}(x + 4) + 3; 26$

100. $\dfrac{1}{2}(x - 2) + 3 = \dfrac{3}{8}(3x - 4); 4$

101. $(y \div 6) + \dfrac{2}{3} = (y \div 2) - \dfrac{7}{9}; 4\dfrac{1}{3}$

102. $(y \div 6) + \dfrac{1}{3} = (y \div 2) - \dfrac{5}{9}; 2\dfrac{2}{3}$

In Exercises 103–114, translate from English to an algebraic expression or equation, whichever is appropriate. Let the variable x represent the number.

103. $\frac{1}{5}$ of a number   **104.** $\frac{1}{6}$ of a number

105. A number decreased by $\frac{1}{4}$ of itself

106. A number decreased by $\frac{1}{3}$ of itself

107. A number decreased by $\frac{1}{4}$ is half of that number.

108. A number decreased by $\frac{1}{3}$ is half of that number.

109. The sum of $\frac{1}{7}$ of a number and $\frac{1}{8}$ of that number gives 12.

110. The sum of $\frac{1}{9}$ of a number and $\frac{1}{10}$ of that number gives 15.

111. The product of $\frac{2}{3}$ and a number increased by 6

112. The product of $\frac{3}{4}$ and a number increased by 9

113. The product of $\frac{2}{3}$ and a number, increased by 6, is 3 less than the number.

114. The product of $\frac{3}{4}$ and a number, increased by 9, is 2 less than the number.

Practice PLUS

In Exercises 115–118, perform the indicated operation. Write the answer as an algebraic expression.

115. $\dfrac{3}{4} \cdot \dfrac{a}{5}$ **116.** $\dfrac{2}{3} \div \dfrac{a}{7}$

117. $\dfrac{11}{x} + \dfrac{9}{x}$ **118.** $\dfrac{10}{y} - \dfrac{6}{y}$

In Exercises 119–120, perform the indicated operations. Begin by performing operations in parentheses.

119. $\left(\dfrac{1}{2} - \dfrac{1}{3}\right) \div \dfrac{5}{8}$

120. $\left(\dfrac{1}{2} + \dfrac{1}{4}\right) \div \left(\dfrac{1}{2} + \dfrac{1}{3}\right)$

In Exercises 121–122, determine whether the given number is a solution of the equation.

121. $\dfrac{1}{5}(x + 2) = \dfrac{1}{2}\left(x - \dfrac{1}{5}\right); \dfrac{5}{8}$

122. $12 - 3(x - 2) = 4x - (x + 3); 3\dfrac{1}{2}$

Application Exercises

The formula

$$C = \dfrac{5}{9}(F - 32)$$

expresses the relationship between Fahrenheit temperature, F, and Celsius temperature, C. In Exercises 123–124, use the formula to convert the given Fahrenheit temperature to its equivalent temperature on the Celsius scale.

123. 68°F   **124.** 41°F

The maximum heart rate, in beats per minute, that you should achieve during exercise is 220 minus your age:

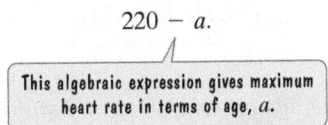

$$220 - a.$$

This algebraic expression gives maximum heart rate in terms of age, a.

The bar graph shows the target heart rate ranges for four types of exercise goals. The lower and upper limits of these ranges are fractions of the maximum heart rate, 220 – a. Exercises 125–128 are based on the information in the graph.

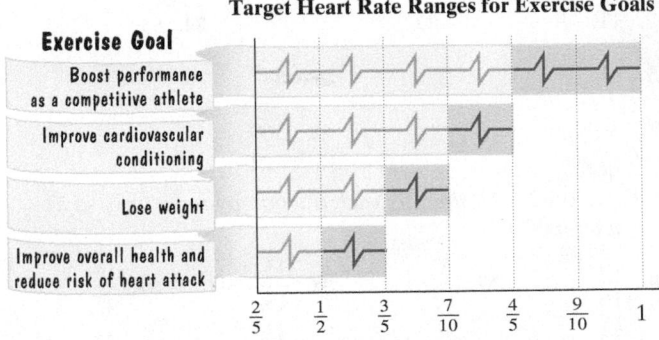

Target Heart Rate Ranges for Exercise Goals

Fraction of Maximum Heart Rate, 220 – a

125. If your exercise goal is to improve cardiovascular conditioning, the graph shows the following range for target heart rate, *H*, in beats per minute:

Lower limit of range $H = \dfrac{7}{10}(220 - a)$

Upper limit of range $H = \dfrac{4}{5}(220 - a).$

a. What is the lower limit of the heart range, in beats per minute, for a 20-year-old with this exercise goal?

b. What is the upper limit of the heart range, in beats per minute, for a 20-year-old with this exercise goal?

126. If your exercise goal is to improve overall health, the graph on the previous page shows the following range for target heart rate, H, in beats per minute:

Lower limit of range — $H = \dfrac{1}{2}(220 - a)$

Upper limit of range — $H = \dfrac{3}{5}(220 - a)$.

a. What is the lower limit of the heart range, in beats per minute, for a 30-year-old with this exercise goal?

b. What is the upper limit of the heart range, in beats per minute, for a 30-year-old with this exercise goal?

127. a. Write a formula that models the heart rate, H, in beats per minute, for a person who is a years old and would like to achieve $\frac{9}{10}$ of maximum heart rate during exercise.

b. Use your formula from part (a) to find the heart rate during exercise for a 40-year-old with this goal.

128. a. Write a formula that models the heart rate, H, in beats per minute, for a person who is a years old and would like to achieve $\frac{7}{8}$ of maximum heart rate during exercise.

b. Use your formula from part (a) to find the heart rate during exercise for a 20-year-old with this goal.

Making a list and trimming it twice. *The graph shows the average number of holiday presents bought by U.S. shoppers from 2007 through 2010.*

Average Number of Holiday Presents Bought by U.S. Shoppers

Source: Deloitte's 25th Annual Holiday Survey, © 2010 Deloitte Development LLC

Here is a mathematical model that approximates the data displayed by the bar graph:

$$P = 23\frac{1}{5} - 2\frac{1}{5}n.$$

Average number of holiday presents   Number of years after 2007

Use the formula at the bottom of the previous column to solve Exercises 129–130.

129. Use the formula to find the average number of holiday presents bought by U.S. shoppers in 2008. Does the mathematical model underestimate or overestimate the actual number shown in the bar graph for 2008? By how much?

130. Use the formula to find the average number of holiday presents bought by U.S. shoppers in 2010. Does the mathematical model underestimate or overestimate the actual number shown in the bar graph for 2010? By how much?

Writing in Mathematics

131. Explain how to convert a mixed number to an improper fraction and give an example.

132. Explain how to convert an improper fraction to a mixed number and give an example.

133. Describe the difference between a prime number and a composite number.

134. What is meant by the prime factorization of a composite number?

135. What is the Fundamental Principle of Fractions?

136. Explain how to reduce a fraction to its lowest terms. Give an example with your explanation.

137. Explain how to multiply fractions and give an example.

138. Explain how to divide fractions and give an example.

139. Describe how to add or subtract fractions with identical denominators. Provide an example with your description.

140. Explain how to add fractions with different denominators. Use $\frac{5}{6} + \frac{1}{2}$ as an example.

Critical Thinking Exercises

Make Sense? *In Exercises 141–144, determine whether each statement "makes sense" or "does not make sense" and explain your reasoning.*

141. I find it easier to multiply $\frac{1}{5}$ and $\frac{3}{4}$ than to add them.

142. Fractions frustrated me in arithmetic, so I'm glad I won't have to use them in algebra.

143. I need to be able to perform operations with fractions to determine whether $\frac{3}{2}$ is a solution of $8x = 12\left(x - \frac{1}{2}\right)$.

144. I saved money by buying a computer for $\frac{3}{2}$ of its original price.

In Exercises 145–148, determine whether each statement is true or false. If the statement is false, make the necessary change(s) to produce a true statement.

145. $\dfrac{1}{2} + \dfrac{1}{5} = \dfrac{2}{7}$

146. $\dfrac{1}{2} \div 4 = 2$

147. Every fraction has infinitely many equivalent fractions.

148. $\dfrac{3 + 7}{30} = \dfrac{\overset{1}{3 + 7}}{\underset{10}{30}} = \dfrac{8}{10} = \dfrac{4}{5}$

149. Shown below is a short excerpt from "The Star-Spangled Banner." The time is $\frac{3}{4}$, which means that each measure must contain notes that add up to $\frac{3}{4}$. The values of the different notes tell musicians how long to hold each note.

$$\circ = 1 \qquad \textbf{\textit{d}} = \frac{1}{2} \qquad \textbf{\textit{♩}} = \frac{1}{4} \qquad \textbf{\textit{♪}} = \frac{1}{8}$$

Use vertical lines to divide this line of "The Star-Spangled Banner" into measures.

say does that Star-span-gled Ban-ner yet wave O'er the

Preview Exercises

Exercises 150–152 will help you prepare for the material covered in the next section. Consider the following "infinite ruler" that shows numbers that lie to the left and to the right of zero.

$$\begin{array}{c} \quad\ (c) \qquad\qquad\qquad\qquad (b)\quad\ (a) \\ \overset{\longleftarrow}{\underset{-5\ \ -4\ \ -3\ \ -2\ \ -1\ \ \ 0\ \ \ 1\ \ \ 2\ \ \ 3\ \ \ 4\ \ \ 5}{\rule{9cm}{0.4pt}}}\!\!\!\!\longrightarrow \end{array}$$

150. What number is represented by point (a)?

151. What number is represented by point (b)? Express the number as an improper fraction.

152. What number is represented by point (c)?

The Real Numbers

Objectives

1. Define the sets that make up the real numbers.
2. Graph numbers on a number line.
3. Express rational numbers as decimals.
4. Classify numbers as belonging to one or more sets of the real numbers.
5. Understand and use inequality symbols.
6. Find the absolute value of a real number.

The United Nations Building in New York was designed to represent its mission of promoting world harmony. Viewed from the front, the building looks like three rectangles stacked upon each other. In each rectangle, the ratio of the width to height is $\sqrt{5} + 1$ to 2, approximately 1.618 to 1. The ancient Greeks believed that such a rectangle, called a **golden rectangle**, was the most visually pleasing of all rectangles.

The ratio 1.618 to 1 is approximate because $\sqrt{5}$ is an irrational number, a special kind of real number. Irrational? Real? Let's make sense of all this by describing the kinds of numbers you will encounter in this course.

The U.N. building is designed with three golden rectangles.

1 Define the sets that make up the real numbers.

Natural Numbers and Whole Numbers

Before we describe the set of real numbers, let's be sure you are familiar with some basic ideas about sets. A **set** is a collection of objects whose contents can be clearly determined. The objects in a set are called the **elements** of the set. For example, the set of numbers used for counting can be represented by

$$\{1, 2, 3, 4, 5, \dots \}.$$

The braces, $\{\ \}$, indicate that we are representing a set. This form of representing a set uses commas to separate the elements of the set. Remember that the three dots after the 5 indicate that there is no final element and that the listing goes on forever.

We have seen that the set of numbers used for counting is called the set of **natural numbers**. When we combine the number 0 with the natural numbers, we obtain the set of **whole numbers**.

Natural Numbers and Whole Numbers

The set of **natural numbers** is $\{1, 2, 3, 4, 5, \dots \}$.

The set of **whole numbers** is $\{0, 1, 2, 3, 4, 5, \dots \}$.

Integers and the Number Line

The whole numbers do not allow us to describe certain everyday situations. For example, if the balance in your checking account is $30 and you write a check for $35, your checking account is overdrawn by $5. We can write this as -5, read *negative* 5. The set consisting of the natural numbers, 0, and the negatives of the natural numbers is called the set of **integers**.

Integers

The set of **integers** is

$$\{\dots, \underbrace{-4, -3, -2, -1,}_{\text{Negative integers}} 0, \underbrace{1, 2, 3, 4. \dots}_{\text{Positive integers}} \}.$$

Notice that the term **positive integers** is another name for the natural numbers. The positive integers can be written in two ways:

1. Use a "+" sign. For example, +4 is "positive four."
2. Do not write any sign. For example, 4 is assumed to be "positive four."

EXAMPLE 1   Practical Examples of Negative Integers

Write a negative integer that describes each of the following situations:

a. A debt of $10
b. The shore surrounding the Dead Sea is 1312 feet below sea level.

Solution

a. A debt of $10 can be expressed by the negative integer -10 (negative ten).
b. The shore surrounding the Dead Sea is 1312 feet below sea level, expressed as -1312. ∎

✓ **CHECK POINT 1**   Write a negative integer that describes each of the following situations:

a. A debt of $500

b. Death Valley, the lowest point in North America, is 282 feet below sea level.

---

2 Graph numbers on a number line.

The **number line** is a graph we use to visualize the set of integers, as well as other sets of numbers. The number line is shown in **Figure 1.9**.

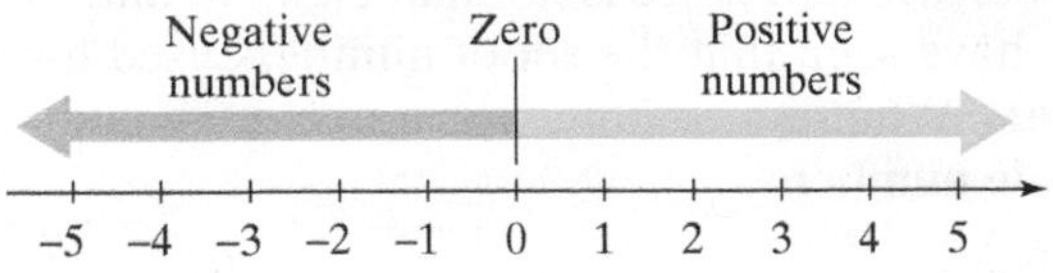

Negative numbers   Zero   Positive numbers

$$-5 \quad -4 \quad -3 \quad -2 \quad -1 \quad 0 \quad 1 \quad 2 \quad 3 \quad 4 \quad 5$$

Figure 1.9   The number line

The number line extends indefinitely in both directions. Zero separates the positive numbers from the negative numbers on the number line. The positive integers are located to the right of 0, and the negative integers are located to the left of 0. Zero is neither positive nor negative. For every positive integer on a number line, there is a corresponding negative integer on the opposite side of 0.

Integers are graphed on a number line by placing a dot at the correct location for each number.

EXAMPLE 2   Graphing Integers on a Number Line

Graph:

a. -3 **b.** 4   **c.** 0.

Solution   Place a dot at the correct location for each integer.

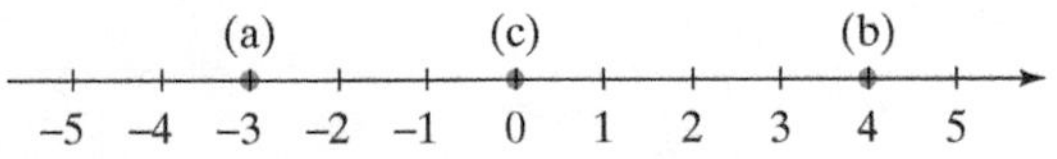

(a)   (c)   (b)

$$-5 \quad -4 \quad -3 \quad -2 \quad -1 \quad 0 \quad 1 \quad 2 \quad 3 \quad 4 \quad 5$$

✓ **CHECK POINT 2**   Graph: **a.** -4 **b.** 0   **c.** 3.

Rational Numbers

If two integers are added, subtracted, or multiplied, the result is always another integer. This, however, is not always the case with division. For example, 10 divided by 5 is the integer 2. By contrast, 5 divided by 10 is $\frac{1}{2}$, and $\frac{1}{2}$ is not an integer. To permit divisions such as $\frac{5}{10}$, we enlarge the set of integers, calling the new collection the *rational numbers*. The set of **rational numbers** consists of all the numbers that can be expressed as a quotient of two integers, with the denominator not 0.

The Rational Numbers

The set of **rational numbers** is the set of all numbers that can be expressed in the form $\frac{a}{b}$, where a and b are integers and b is not equal to 0, written $b \neq 0$. The integer a is called the **numerator** and the integer b is called the **denominator**.

Here are two examples of rational numbers:

- $\dfrac{1}{2}$ $a = 1$ $b = 2$

- $\dfrac{-3}{4}$ $a = -3$ $b = 4$

Great Question!

Is there another way to express the rational number $\frac{-3}{4}$?

In Section 1.7, you will learn that a negative number divided by a positive number gives a negative result. Thus, $\frac{-3}{4}$ can also be written as $-\frac{3}{4}$.

Is the integer 5 another example of a rational number? Yes. The integer 5 can be written with a denominator of 1.

$$5 = \frac{5}{1}$$

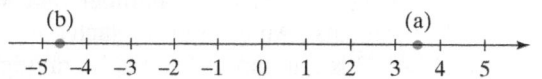

a = 5

b = 1

All integers are also rational numbers because they can be written with a denominator of 1.

How can we express a negative mixed number, such as $-2\frac{3}{4}$, in the form $\frac{a}{b}$? Copy the negative sign and then follow the procedure discussed in the previous section:

$$-2\frac{3}{4} = -\frac{4 \cdot 2 + 3}{4} = -\frac{8 + 3}{4} = -\frac{11}{4} = \frac{-11}{4}.$$

a = -11

b = 4

Copy the negative sign from step to step and convert $2\frac{3}{4}$ to an improper fraction.

Rational numbers are graphed on a number line by placing a dot at the correct location for each number.

EXAMPLE 3   Graphing Rational Numbers on a Number Line

Graph: **a.** $\frac{7}{2}$ **b.** -4.6.

Solution   Place a dot at the correct location for each rational number.

a. Because $\frac{7}{2} = 3\frac{1}{2}$, its graph is midway between 3 and 4.

b. Because $-4.6 = -4\frac{6}{10}$, its graph is $\frac{6}{10}$, or $\frac{3}{5}$, of a unit to the left of -4.

(b)   (a)

-5 -4 -3 -2 -1   0   1   2   3   4   5

☑ **CHECK POINT 3**   Graph: **a.** $\frac{9}{2}$ **b.** -1.2.

---

3 Express rational numbers as decimals.

Every rational number can be expressed as a fraction and as a decimal. To express the fraction $\frac{a}{b}$ as a decimal, divide the denominator, b, into the numerator, a.

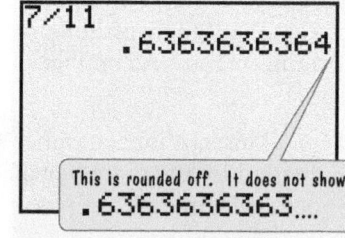
EXAMPLE 4   Expressing Rational Numbers as Decimals

Express each rational number as a decimal:

a. $\frac{5}{8}$ **b.** $\frac{7}{11}$.

Solution   In each case, divide the denominator into the numerator.

a.
```
    0.625
8)5.000
  48
  ──
   20
   16
   ──
    40
    40
    ──
     0
```

b.
```
    0.6363 ...
11)7.0000 ...
   66
   ──
    40
    33
    ──
     70
     66
     ──
      40
      33
      ──
       70
       ⋮
```

In Example 4, the decimal for $\frac{5}{8}$, namely 0.625, stops and is called a **terminating decimal**. Other examples of terminating decimals are

$$\frac{1}{4} = 0.25, \qquad \frac{2}{5} = 0.4, \qquad \text{and} \qquad \frac{7}{8} = 0.875.$$

By contrast, the division process for $\frac{7}{11}$ results in 0.6363..., with the digits 63 repeating over and over indefinitely. To indicate this, write a bar over the digits that repeat. Thus,

$$\frac{7}{11} = 0.\overline{63}.$$

The decimal for $\frac{7}{11}$, $0.\overline{63}$, is called a **repeating decimal**. Other examples of repeating decimals are

$$\frac{1}{3} = 0.333\ldots = 0.\overline{3} \qquad \text{and} \qquad \frac{2}{3} = 0.666\ldots = 0.\overline{6}.$$

Rational Numbers and Decimals

Any rational number can be expressed as a decimal. The resulting decimal will either terminate (stop), or it will have a digit that repeats or a block of digits that repeat.

✓ **CHECK POINT 4**   Express each rational number as a decimal:

a. $\dfrac{3}{8}$ **b.** $\dfrac{5}{11}$.

Irrational Numbers

Can you think of a number that, when written in decimal form, neither terminates nor repeats? An example of such a number is $\sqrt{2}$ (read: "the square root of 2"). The number $\sqrt{2}$ is a number that can be multiplied by itself to obtain 2. No terminating or repeating decimal can be multiplied by itself to get 2. However, some approximations come close to 2.

- 1.4 is an approximation of $\sqrt{2}$:

$$1.4 \times 1.4 = 1.96.$$

- 1.41 is an approximation of $\sqrt{2}$:

$$1.41 \times 1.41 = 1.9881.$$

- 1.4142 is an approximation of $\sqrt{2}$:

$$1.4142 \times 1.4142 = 1.99996164.$$

Can you see how each approximation in the list is getting better? This is because the products are getting closer and closer to 2.

The number $\sqrt{2}$, whose decimal representation does not come to an end and does not have a block of repeating digits, is an example of an **irrational number**.

The Irrational Numbers

Any number that can be represented on the number line that is not a rational number is called an **irrational number**. Thus, the set of irrational numbers is the set of numbers whose decimal representations are neither terminating nor repeating.

Perhaps the best known of all the irrational numbers is π (pi). This irrational number represents the distance around a circle (its circumference) divided by the diameter of the circle. In the *Star Trek* episode "Wolf in the Fold," Spock foils an evil computer by telling it to "compute the last digit in the value of π." Because π is an irrational number, there is no last digit in its decimal representation:

$$\pi = 3.14159265358979323846264338332795\ldots.$$

Because irrational numbers cannot be represented by decimals that come to an end, mathematicians use symbols such as $\sqrt{2}$, $\sqrt{3}$, and π to represent these numbers. However, **not all square roots are irrational**. For example, $\sqrt{25} = 5$ because 5 multiplied by itself is 25. Thus, $\sqrt{25}$ is a natural number, a whole number, an integer, and a rational number $\left(\sqrt{25} = \frac{5}{1}\right)$.

Using Technology

You can obtain decimal approximations for irrational numbers using a calculator. For example, to approximate $\sqrt{2}$, use the following keystrokes:

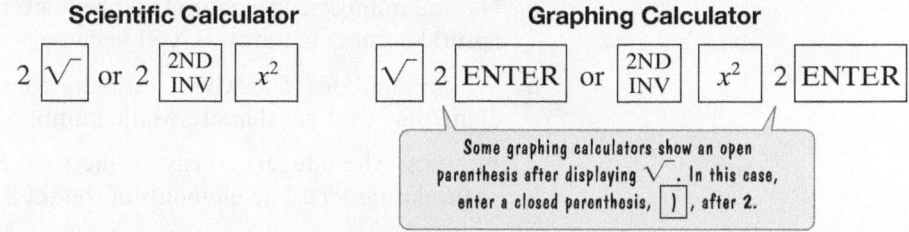

Scientific Calculator   **Graphing Calculator**

Some graphing calculators show an open parenthesis after displaying $\sqrt{}$. In this case, enter a closed parenthesis, $)$, after **2.**

The display may read 1.41421356237, although your calculator may show more or fewer digits. Between which two integers would you graph $\sqrt{2}$ on a number line?

4 Classify numbers as belonging to one or more sets of the real numbers.

The Set of Real Numbers

All numbers that can be represented by points on the number line are called **real numbers**. Thus, the set of real numbers is formed by combining the rational numbers and the irrational numbers. Every real number is either rational or irrational.

The sets that make up the real numbers are summarized in **Table 1.2**. Notice the use of the symbol $\approx$ in the examples of irrational numbers. The symbol $\approx$ means "is approximately equal to."

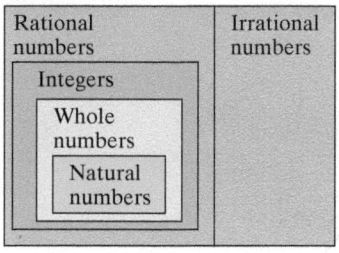

Real numbers

Rational numbers	Irrational numbers
Integers	
Whole numbers	
Natural numbers	

This diagram shows that every real number is rational or irrational.

Table 1.2	The Sets that Make Up the Real Numbers	
Name	**Description**	**Examples**
Natural numbers	$\{1, 2, 3, 4, 5, \ldots\}$ These numbers are used for counting.	$2, 3, 5, 17$
Whole numbers	$\{0, 1, 2, 3, 4, 5, \ldots\}$ The set of whole numbers is formed by adding 0 to the set of natural numbers.	$0, 2, 3, 5, 17$
Integers	$\{\ldots, -5, -4, -3, -2, -1, 0, 1, 2, 3, 4, 5, \ldots\}$ The set of integers is formed by adding negatives of the natural numbers to the set of whole numbers.	$-17, -5, -3, -2, 0,$ $2, 3, 5, 17$
Rational numbers	The set of rational numbers is the set of all numbers that can be expressed in the form $\frac{a}{b}$, where a and b are integers and b is not equal to 0, written $b \neq 0$. Rational numbers can be expressed as terminating or repeating decimals.	$-17 = \frac{-17}{1}, -5 = \frac{-5}{1}, -3, -2,$ $0, 2, 3, 5, 17,$ $\frac{2}{5} = 0.4,$ $\frac{-2}{3} = -0.6666\cdots = -0.\overline{6}$
Irrational numbers	The set of irrational numbers is the set of all numbers whose decimal representations are neither terminating nor repeating. Irrational numbers cannot be expressed as a quotient of integers.	$\sqrt{2} \approx 1.414214$ $-\sqrt{3} \approx -1.73205$ $\pi \approx 3.142$ $-\frac{\pi}{2} \approx -1.571$

EXAMPLE 5 Classifying Real Numbers

Consider the following set of numbers:

$$\left\{-7, -\frac{3}{4}, 0, 0.\overline{6}, \sqrt{5}, \pi, 7.3, \sqrt{81}\right\}.$$

List the numbers in the set that are

a. natural numbers.   b. whole numbers.   c. integers.

d. rational numbers.   e. irrational numbers. f. real numbers.

Solution

a. Natural numbers: The natural numbers are the numbers used for counting. The only natural number in the set is $\sqrt{81}$ because $\sqrt{81} = 9$. (9 multiplied by itself is 81.)

b. Whole numbers: The whole numbers consist of the natural numbers and 0. The elements of the set that are whole numbers are 0 and $\sqrt{81}$.

c. Integers: The integers consist of the natural numbers, 0, and the negatives of the natural numbers. The elements of the set that are integers are $\sqrt{81}$, 0, and -7.

d. Rational numbers: All numbers in the set that can be expressed as the quotient of integers are rational numbers. These include $-7 \left(-7 = \frac{-7}{1}\right), -\frac{3}{4}, 0 \left(0 = \frac{0}{1}\right),$ and $\sqrt{81} \left(\sqrt{81} = \frac{9}{1}\right)$. Furthermore, all numbers in the set that are terminating or repeating decimals are also rational numbers. These include $0.\overline{6}$ and 7.3.

e. Irrational numbers: The irrational numbers in the set are $\sqrt{5} \left(\sqrt{5} \approx 2.236\right)$ and $\pi (\pi \approx 3.14)$. Both $\sqrt{5}$ and π are only approximately equal to 2.236 and 3.14, respectively. In decimal form, $\sqrt{5}$ and π neither terminate nor have blocks of repeating digits.

f. Real numbers: All the numbers in the given set are real numbers. ■

✓ **CHECK POINT 5** Consider the following set of numbers:

$$\left\{-9, -1.3, 0, 0.\overline{3}, \frac{\pi}{2}, \sqrt{9}, \sqrt{10}\right\}.$$

List the numbers in the set that are

a. natural numbers.   b. whole numbers.

c. integers.   d. rational numbers.

e. irrational numbers.   f. real numbers.

---

5 Understand and use inequality symbols.

Ordering the Real Numbers

On the real number line, the real numbers increase from left to right. The lesser of two real numbers is the one farther to the left on a number line. The greater of two real numbers is the one farther to the right on a number line.

Look at the number line in **Figure 1.10**. The integers 2 and 5 are graphed. Observe that 2 is to the left of 5 on the number line. This means that 2 is less than 5.

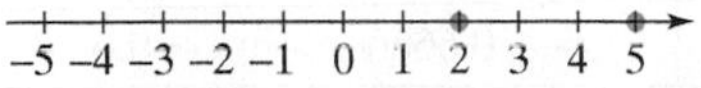

$$-5\ -4\ -3\ -2\ -1\ \ 0\ \ 1\ \ 2\ \ 3\ \ 4\ \ 5$$

Figure 1.10

$2 < 5$:   2 is less than 5 because 2 is to the *left* of 5 on the number line.

In **Figure 1.10**, we can also observe that 5 is to the *right* of 2 on the number line. This means that 5 is greater than 2.

5 > 2:   5 is greater than 2 because 5 is to the *right* of 2 on the number line.

The symbols < and > are called **inequality symbols**. These symbols always point to the lesser of the two real numbers when the inequality is true.

> **2 is less than 5.**
>
> 2 < 5   The symbol points to 2, the lesser number.
>
> 5 > 2   The symbol points to 2, the lesser number.
>
> **5 is greater than 2.**

EXAMPLE 6   Using Inequality Symbols

Insert either < or > in the shaded area between each pair of numbers to make a true statement:

a. 3 ▮ 17   **b.** −4.5 ▮ 1.2   **c.** −5 ▮ −83   **d.** $\dfrac{4}{5}$ ▮ $\dfrac{2}{3}$.

Solution   In each case, mentally compare the graph of the first number to the graph of the second number. If the first number is to the left of the second number, insert the symbol < for "is less than." If the first number is to the right of the second number, insert the symbol > for "is greater than."

a. Compare the graphs of 3 and 17 on the number line.

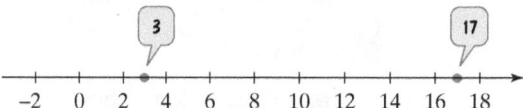

Because 3 is to the left of 17, this means that 3 is less than 17: 3 < 17.

b. Compare the graphs of −4.5 and 1.2.

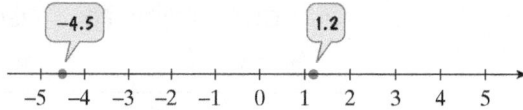

Because −4.5 is to the left of 1.2, this means that −4.5 is less than 1.2: −4.5 < 1.2.

c. Compare the graphs of −5 and −83.

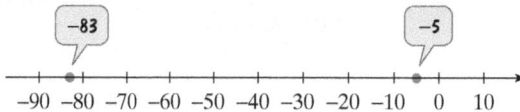

Because −5 is to the right of −83, this means that −5 is greater than −83: −5 > −83.

d. Compare the graphs of $\frac{4}{5}$ and $\frac{2}{3}$. To do so, convert to decimal notation or use a common denominator. Using decimal notation, $\frac{4}{5} = 0.8$ and $\frac{2}{3} = 0.\overline{6}$.

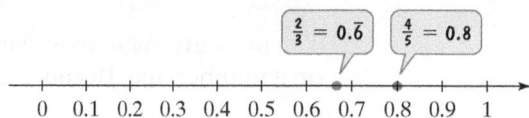

Because 0.8 is to the right of $0.\overline{6}$, this means that $\frac{4}{5}$ is greater than $\frac{2}{3}$: $\frac{4}{5} > \frac{2}{3}$. ∎

☑ **CHECK POINT 6**   Insert either $<$ or $>$ in the shaded area between each pair of numbers to make a true statement:

a. 14 ▬ 5   **b.** −5.4 ▬ 2.3   **c.** −19 ▬ −6   **d.** $\dfrac{1}{4}$ ▬ $\dfrac{1}{2}$.

The symbols $<$ and $>$ may be combined with an equal sign, as shown in the table.

	Symbols	Meaning	Examples	Explanation
This inequality is true if either the $<$ part or the $=$ part is true.	$a \leq b$	a is less than or equal to b.	$3 \leq 7$ $7 \leq 7$	Because $3 < 7$ Because $7 = 7$
This inequality is true if either the $>$ part or the $=$ part is true.	$b \geq a$	b is greater than or equal to a.	$7 \geq 3$ $-5 \geq -5$	Because $7 > 3$ Because $-5 = -5$

When using the symbol $\leq$ (is less than or equal to), the inequality is a true statement if either the $<$ part or the $=$ part is true. When using the symbol $\geq$ (is greater than or equal to), the inequality is a true statement if either the $>$ part or the $=$ part is true.

EXAMPLE 7   Using Inequality Symbols

Determine whether each inequality is true or false:

a. $-7 \leq 4$ **b.** $-7 \leq -7$ **c.** $-9 \geq 6$.

Solution

a. $-7 \leq 4$ is true because $-7 < 4$ is true.

b. $-7 \leq -7$ is true because $-7 = -7$ is true.

c. $-9 \geq 6$ is false because neither $-9 > 6$ nor $-9 = 6$ is true.   ■

☑ **CHECK POINT 7**   Determine whether each inequality is true or false:

a. $-2 \leq 3$ **b.** $-2 \geq -2$ **c.** $-4 \geq 1$.

6 Find the absolute value of real number.

Absolute Value

Absolute value describes distance from 0 on a number line. If a represents a real number, the symbol $|a|$ represents its absolute value, read "the absolute value of a." For example,

$$|-5| = 5.$$

The absolute value of −5 is 5 because −5 is 5 units from 0 on a number line.

Absolute Value

The **absolute value** of a real number a, denoted by $|a|$, is the distance from 0 to a on a number line. Because absolute value describes a distance, it is never negative.

EXAMPLE 8   Finding Absolute Value

Find the absolute value:

a. $|-3|$ **b.** $|5|$ **c.** $|0|$.

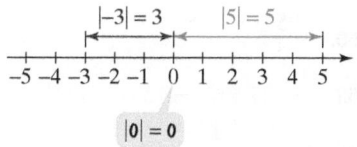

Figure 1.11 Absolute value describes distance from 0 on a number line.

Solution The solution is illustrated in **Figure 1.11**.

a. $|-3| = 3$ The absolute value of −3 is 3 because −3 is 3 units from 0.

b. $|5| = 5$ 5 is 5 units from 0.

c. $|0| = 0$ 0 is 0 units from itself. ∎

Example 8 illustrates that the absolute value of a positive real number or 0 is the number itself. The absolute value of a negative real number, such as −3, is the number without the negative sign. Zero is the only real number whose absolute value is 0: $|0| = 0$. **The absolute value of any real number other than 0 is always positive.**

✓ **CHECK POINT 8** Find the absolute value:

a. $|-4|$ **b.** $|6|$ **c.** $|-\sqrt{2}|$.

Achieving Success

Check! Check! Check!

After completing each Check Point or odd-numbered exercise, compare your answer with the one given in the answer section at the back of the book. To make this process more convenient, place a Post-it® or some other marker at the appropriate page of the answer section. If your answer is different from the one given in the answer section, try to figure out your mistake. Then correct the error. If you cannot determine what went wrong, show your work to your professor. **By recording each step neatly and using as much paper as you need**, your professor will find it easier to determine where you had trouble.

CONCEPT AND VOCABULARY CHECK

Fill in each blank so that the resulting statement is true.

1. The set $\{1, 2, 3, 4, 5, \ldots\}$ is called the set of _____ numbers.

2. The set $\{0, 1, 2, 3, 4, 5, \ldots\}$ is called the set of _____ numbers.

3. The set $\{\ldots, -4, -3, -2, -1, 0, 1, 2, 3, 4, \ldots\}$ is called the set of _____.

4. The set of numbers in the form $\frac{a}{b}$, where a and b belong to the set in statement 3 above and $b \neq 0$, is called the set of _____ numbers.

5. The set of numbers whose decimal representations are neither terminating nor repeating is called the set of _____ numbers.

6. Every real number is either a /an _____ number or a/an _____ number.

7. The notation $2 < 5$ means that 2 is to the _____ of 5 on a number line.

8. The distance from 0 to a on a number line is called the _____ of a, denoted by _____.

1.3 EXERCISE SET

MyMathLab®

Watch the videos
in MyMathLab

Download the
MyDashBoard App

Practice Exercises

In Exercises 1–8, write a positive or negative integer that describes each situation.

1. Meteorology: 20° below zero

2. Navigation: 65 feet above sea level

3. Health: A gain of 8 pounds

4. Economics: A loss of $12,500.00

5. Banking: A withdrawal of $3000.00

6. Physics: An automobile slowing down at a rate of 3 meters per second each second

7. Economics: A budget deficit of 4 billion dollars

8. Football: A 14-yard loss

In Exercises 9–20, start by drawing a number line that shows integers from −5 to 5. Then graph each real number on your number line.

9. 2   **10.** 5   **11.** −5   **12.** −2   **13.** $3\frac{1}{2}$ **14.** $2\frac{1}{4}$

15. $\frac{11}{3}$ **16.** $\frac{7}{3}$ **17.** −1.8   **18.** −3.4   **19.** $-\frac{16}{5}$ **20.** $-\frac{11}{5}$

In Exercises 21–32, express each rational number as a decimal.

21. $\frac{3}{4}$ **22.** $\frac{3}{5}$ **23.** $\frac{7}{20}$

24. $\frac{3}{20}$ **25.** $\frac{7}{8}$ **26.** $\frac{5}{16}$

27. $\frac{9}{11}$ **28.** $\frac{3}{11}$ **29.** $-\frac{1}{2}$

30. $-\frac{1}{4}$ **31.** $-\frac{5}{6}$ **32.** $-\frac{7}{6}$

In Exercises 33–36, list all numbers from the given set that are: **a.** *natural numbers,* **b.** *whole numbers,* **c.** *integers,* **d.** *rational numbers,* **e.** *irrational numbers,* **f.** *real numbers.*

33. $\left\{-9, -\frac{4}{5}, 0, 0.25, \sqrt{3}, 9.2, \sqrt{100}\right\}$

34. $\left\{-7, -0.\overline{6}, 0, \sqrt{49}, \sqrt{50}\right\}$

35. $\left\{-11, -\frac{5}{6}, 0, 0.75, \sqrt{5}, \pi, \sqrt{64}\right\}$

36. $\left\{-5, -0.\overline{3}, 0, \sqrt{2}, \sqrt{4}\right\}$

37. Give an example of a whole number that is not a natural number.

38. Give an example of an integer that is not a whole number.

39. Give an example of a rational number that is not an integer.

40. Give an example of a rational number that is not a natural number.

41. Give an example of a number that is an integer, a whole number, and a natural number.

42. Give an example of a number that is a rational number, an integer, and a real number.

43. Give an example of a number that is an irrational number and a real number.

44. Give an example of a number that is a real number, but not an irrational number.

In Exercises 45–62, insert either < *or* > *in the shaded area between each pair of numbers to make a true statement.*

45. $\frac{1}{2}$ ▓ 2   **46.** 4 ▓ −3

47. 3 ▓ $-\frac{5}{2}$ **48.** 3 ▓ $\frac{3}{2}$

49. −4 ▓ −6   **50.** $-\frac{5}{2}$ ▓ $-\frac{5}{3}$

51. −2.5 ▓ 1.5   **52.** −1.25 ▓ −0.5

53. $-\frac{3}{4}$ ▓ $-\frac{5}{4}$ **54.** 0 ▓ $-\frac{1}{2}$

55. −4.5 ▓ 3   **56.** −5.5 ▓ 2.5

57. $\sqrt{2}$ ▓ 1.5   **58.** $\sqrt{3}$ ▓ 2

59. $0.\overline{3}$ ▓ 0.3   **60.** 0.6 ▓ $0.\overline{6}$

61. −π ▓ −3.5   **62.** $-\frac{\pi}{2}$ ▓ −2.3

In Exercises 63–70, determine whether each inequality is true or false.

63. $-5 \geq -13$ **64.** $-5 \leq -8$

65. $-9 \geq -9$ **66.** $-14 \leq -14$

67. $0 \geq -6$ **68.** $0 \geq -13$

69. $-17 \geq 6$ **70.** $-14 \geq 8$

In Exercises 71–78, find each absolute value.

71. $|6|$ **72.** $|3|$ **73.** $|-7|$

74. $|-9|$ **75.** $\left|\frac{5}{6}\right|$ **76.** $\left|\frac{4}{5}\right|$

77. $|-\sqrt{11}|$ **78.** $|-\sqrt{29}|$

Practice PLUS

In Exercises 79–86, insert either <, >, *or* = *in the shaded area to make a true statement.*

79. $|-6|$ ▓ $|-3|$ **80.** $|-20|$ ▓ $|-50|$

81. $\left|\frac{3}{5}\right|$ ▓ $|-0.6|$ **82.** $\left|\frac{5}{2}\right|$ ▓ $|-2.5|$

83. $\frac{30}{40} - \frac{3}{4}$ ▓ $\frac{14}{15} \cdot \frac{15}{14}$ **84.** $\frac{17}{18} \cdot \frac{18}{17}$ ▓ $\frac{50}{60} - \frac{5}{6}$

85. $\frac{8}{13} \div \frac{8}{13}$ ▓ $|-1|$ **86.** $|-2|$ ▓ $\frac{4}{17} \div \frac{4}{17}$

Application Exercises

In Exercises 87–94, determine whether natural numbers, whole numbers, integers, rational numbers, or all real numbers are appropriate for each situation.

87. Shoe sizes of students on campus

88. Recorded heights of students on campus

89. Temperatures in weather reports

90. Class sizes of algebra courses

91. Values of d given by the formula $d = \sqrt{1.5h}$, where d is the distance, in miles, that you can see to the horizon from a height of h feet

92. Values of C given by the formula $C = 2\pi r$, where C is the circumference of a circle with radius r

93. The number of pets a person has

94. The number of siblings a person has

95. The table shows the record low temperatures for five U.S. states.

State	Record Low (°F)	Date
Florida	−2	Feb. 13, 1899
Georgia	−17	Jan. 27, 1940
Hawaii	12	May 17, 1979
Louisiana	−16	Feb. 13, 1899
Rhode Island	−25	Feb. 5, 1996

Source: National Climatic Data Center

   a. Graph the five record low temperatures on a number line.

   b. Write the names of the states in order from the coldest record low to the warmest record low.

96. The table shows the record low temperatures for five U.S. states.

State	Record Low (°F)	Date
Virginia	−30	Jan. 22, 1985
Washington	−48	Dec. 30, 1968
West Virginia	−37	Dec. 30, 1917
Wisconsin	−55	Feb. 4, 1996
Wyoming	−66	Feb. 9, 1933

Source: National Climatic Data Center

   a. Graph the five record low temperatures on a number line.

   b. Write the names of the states in order from the coldest record low to the warmest record low.

Writing in Mathematics

97. What is a set?

98. What are the natural numbers?

99. What are the whole numbers?

100. What are the integers?

101. How does the set of integers differ from the set of whole numbers?

102. Describe how to graph a number on the number line.

103. What is a rational number?

104. Explain how to express $\frac{3}{8}$ as a decimal.

105. Describe the difference between a rational number and an irrational number.

106. If you are given two different real numbers, explain how to determine which one is the lesser.

107. Describe what is meant by the absolute value of a number. Give an example with your explanation.

Critical Thinking Exercises

Make Sense? *In Exercises 108–111, determine whether each statement "makes sense" or "does not make sense" and explain your reasoning.*

108. The humor in this joke is based on the fact that the football will never be hiked.

Foxtrot copyright © 2003, 2009 by Bill Amend/Distributed by Universal Uclick

109. *Titanic* came to rest 12,500 feet below sea level and *Bismarck* came to rest 15,617 feet below sea level, so *Bismarck's* resting place is higher than *Titanic's*.

110. I expressed a rational number as a decimal and the decimal neither terminated nor repeated.

111. I evaluated the formula $d = \sqrt{1.5h}$ for a value of h that resulted in a rational number for d.

In Exercises 112–117, determine whether each statement is true or false. If the statement is false, make the necessary change(s) to produce a true statement.

112. Every rational number is an integer.

113. Some whole numbers are not integers.

114. Some rational numbers are not positive.

115. Irrational numbers cannot be negative.

116. Some real numbers are not rational numbers.

117. Some integers are not rational numbers.

In Exercises 118–119, write each phrase as an algebraic expression.

118. a loss of $\frac{1}{3}$ of an investment of d dollars

119. a loss of half of an investment of d dollars

Technology Exercises

In Exercises 120–123, use a calculator to find a decimal approximation for each irrational number, correct to three decimal places. Between which two integers should you graph each of these numbers on the number line?

120. $\sqrt{3}$

121. $-\sqrt{12}$

122. $1 - \sqrt{2}$

123. $2 - \sqrt{5}$

Preview Exercises

Exercises 124–126 will help you prepare for the material covered in the next section. In each exercise, evaluate both expressions for $x = 4$. What do you observe?

124. $3(x + 5); 3x + 15$

125. $3x + 5x; 8x$

126. $9x - 2x; 7x$

Objectives

1. Understand and use the vocabulary of algebraic expressions.

2. Use commutative properties.

3. Use associative property.

4. Use the distributive property.

5. Combine like terms.

6. Simplify algebraic expressions.

Basic Rules of Algebra

Starting as a link among U.S. research scientists, more than 1.8 billion people worldwide now use the Internet. Some random Internet factoids:

- Projected number of Facebook users by 2012: one billion
- Peak time for sex-related searches: 11 P.M.
- Fraction of people who use the word "password" as their password: $\frac{1}{8}$
- Vanity searchers: Fraction of people who typed their own name into a search engine: $\frac{1}{4}$
- Social interactions on Facebook every minute: 231,605 messages sent; 135,849 photos added; 98,604 friendships approved

(*Sources*: Facebook; Paul Grobman, *Vital Statistics*, Plume, 2005)

In this section, we move from these quirky tidbits to mathematical models that describe the remarkable growth of the Internet in the United States and worldwide. To use these models efficiently (you'll work with them in the Exercise Set), they should be simplified using basic rules of algebra. Before turning to these rules, we open the section with a closer look at algebraic expressions.

1. Understand and use the vocabulary of algebraic expressions.

The Vocabulary of Algebraic Expressions

We have seen that an algebraic expression combines numbers and variables. Here is an example of an algebraic expression:

$$7x + 3.$$

The **terms** of an algebraic expression are those parts that are separated by addition. For example, the algebraic expression $7x + 3$ contains two terms, namely $7x$ and 3. Notice that a term is a number, a variable, or a number multiplied by one or more variables.

The numerical part of a term is called its **coefficient**. In the term $7x$, the 7 is the coefficient. If a term containing one or more variables is written without a coefficient, the coefficient is understood to be 1. Thus, x means $1x$ and ab means $1ab$.

A term that consists of just a number is called a **constant term**. The constant term of $7x + 3$ is 3.

The parts of each term that are multiplied are called the **factors of the term**. The factors of the term $7x$ are 7 and x.

Like terms are terms that have exactly the same variable factors. Here are two examples of like terms:

$7x$ and $3x$ These terms have the same variable factor, x.

$4y$ and $9y$ These terms have the same variable factor, y.

By contrast, here are some examples of terms that are not like terms. These terms do not have the same variable factor.

$7x$ and 3 The variable factor of the first term is x.
The second term has no variable factor.

$7x$ and $3y$ The variable factor of the first term is x.
The variable factor of the second term is y.

Constant terms are like terms. Thus, the constant terms 7 and -12 are like terms.

EXAMPLE 1   Using the Vocabulary of Algebraic Expressions

Use the algebraic expression

$$4x + 7 + 5x$$

to answer the following questions:

a. How many terms are there in the algebraic expression?

b. What is the coefficient of the first term?

c. What is the constant term?

d. What are the like terms in the algebraic expression?

Solution

a. Because terms are separated by addition, the algebraic expression $4x + 7 + 5x$ contains three terms.

$$4x + 7 + 5x$$

First term   Second term   Third term

b. The coefficient of the first term, $4x$, is 4.

c. The constant term in $4x + 7 + 5x$ is 7.

d. The like terms in $4x + 7 + 5x$ are $4x$ and $5x$. These terms have the same variable factor, x. ■

✓ **CHECK POINT 1**   Use the algebraic expression $6x + 2x + 11$ to answer each of the four questions in Example 1.

Equivalent Algebraic Expressions

In Example 1, we considered the algebraic expression

$$4x + 7 + 5x.$$

Let's compare this expression with a second algebraic expression

$$9x + 7.$$

Evaluate each expression for some choice of x. We will select $x = 2$.

$4x + 7 + 5x$ $9x + 7$

Replace x with 2.   Replace x with 2.

$= 4 \cdot 2 + 7 + 5 \cdot 2$ $= 9 \cdot 2 + 7$

$= 8 + 7 + 10$ $= 18 + 7$

$= 25$ $= 25$

Both algebraic expressions have the same value when $x = 2$. Regardless of what number you select for x, the algebraic expressions $4x + 7 + 5x$ and $9x + 7$ will have the same value. These expressions are called *equivalent algebraic expressions*. Two algebraic expressions that have the same value for all replacements are called **equivalent algebraic expressions**. Because $4x + 7 + 5x$ and $9x + 7$ are equivalent algebraic expressions, we write

$$4x + 7 + 5x = 9x + 7.$$

Properties of Real Numbers and Algebraic Expressions

We now turn to basic properties, or rules, that you know from past experiences in working with whole numbers and fractions. These properties will be extended to include all real numbers and algebraic expressions. We will give each property a name so that we can refer to it throughout the study of algebra.

2 Use commutative properties.

The Commutative Properties

The addition or multiplication of two real numbers can be done in any order. For example, $3 + 5 = 5 + 3$ and $3 \cdot 5 = 5 \cdot 3$. Changing the order does not change the answer of a sum or a product. These facts are called **commutative properties**.

Great Question!

Are there commutative properties for subtraction and division?

No. The commutative property does not hold for subtraction or division.

$$6 - 1 \neq 1 - 6$$
$$8 \div 4 \neq 4 \div 8$$

The Commutative Properties

Let a and b represent real numbers, variables, or algebraic expressions.

Commutative Property of Addition

$$a + b = b + a$$

Changing order when adding does not affect the sum.

Commutative Property of Multiplication

$$ab = ba$$

Changing order when multiplying does not affect the product.

EXAMPLE 2 Using the Commutative Properties

Use the commutative properties to write an algebraic expression equivalent to each of the following:

a. $y + 6$ **b.** $5x$.

Solution

a. By the commutative property of addition, an algebraic expression equivalent to $y + 6$ is $6 + y$. Thus,

$$y + 6 = 6 + y.$$

b. By the commutative property of multiplication, an algebraic expression equivalent to $5x$ is $x5$. Thus,

$$5x = x5. \quad \blacksquare$$

✓ **CHECK POINT 2** Use the commutative properties to write an algebraic expression equivalent to each of the following:

a. $x + 14$ **b.** $7y$.

EXAMPLE 3 Using the Commutative Properties

Write an algebraic expression equivalent to $13x + 8$ using

a. the commutative property of addition.

b. the commutative property of multiplication.

Solution

a. By the commutative property of addition, we change the order of the terms being added. This means that an algebraic expression equivalent to $13x + 8$ is $8 + 13x$:

$$13x + 8 = 8 + 13x.$$

b. By the commutative property of multiplication, we change the order of the factors being multiplied. This means that an algebraic expression equivalent to $13x + 8$ is $x13 + 8$:

$$13x + 8 = x13 + 8. \quad \blacksquare$$

✓ **CHECK POINT 3** Write an algebraic expression equivalent to $5x + 17$ using

a. the commutative property of addition.

b. the commutative property of multiplication.

Blitzer Bonus

Commutative Words and Sentences

The commutative property states that a change in order produces no change in the answer. The words and sentences listed here suggest a characteristic of the commutative property; they read the same from left to right and from right to left!

- dad
- repaper
- never odd or even
- Six is a six is a six is a six is a six is . . . .
- Go deliver a dare, vile dog!
- May a moody baby doom a yam?
- Madam, in Eden I'm Adam.
- Ma is a nun, as I am.
- A man, a plan, a canal: Panama
- Was it a rat I saw?

- Deb sat in Anita's bed.
  Ned sat in Anita's den.
  But Anita sat in a tub.
- Ed, is Nik inside?
  Ed, is Deb bedside?
  Ed is busy—subside!

---

3 Use associative properties.

The Associative Properties

Parentheses indicate groupings. As we have seen, we perform operations within the parentheses first. For example,

$$(2 + 5) + 10 = 7 + 10 = 17$$

and

$$2 + (5 + 10) = 2 + 15 = 17.$$

In general, the way in which three numbers are grouped does not change their sum. It also does not change their product. These facts are called the **associative properties**.

Great Question!

Are there associative properties for subtraction and division?

No. The associative property does not hold for subtraction or division.

$$(6 - 3) - 1 \neq 6 - (3 - 1)$$
$$(8 \div 4) \div 2 \neq 8 \div (4 \div 2)$$

The Associative Properties

Let a, b, and c represent real numbers, variables, or algebraic expressions.

Associative Property of Addition

$$(a + b) + c = a + (b + c)$$

Changing grouping when adding does not affect the sum.

Associative Property of Multiplication

$$(ab)c = a(bc)$$

Changing grouping when multiplying does not affect the product.

The associative properties can be used to simplify some algebraic expressions by removing the parentheses.

EXAMPLE 4 Simplifying Using the Associative Properties

Simplify:

a. $3 + (8 + x)$ **b.** $8(4x)$.

Solution

a. $3 + (8 + x)$ This is the given algebraic expression.

     $= (3 + 8) + x$ Use the associative property of addition to group the first two numbers.

     $= 11 + x$ Add within parentheses.

Using the commutative property of addition, this simplified algebraic expression can also be written as $x + 11$.

b. $8(4x)$ This is the given algebraic expression.

     $= (8 \cdot 4)x$ Use the associative property of multiplication to group the first two numbers.

     $= 32x$ Multiply within parentheses.

We can use the commutative property of multiplication to write this simplified algebraic expression as $x32$ or $x \cdot 32$. However, it is customary to express a term with its coefficient on the left. Thus, we use $32x$ as the simplified form of the algebraic expression. ■

✓ **CHECK POINT 4** Simplify:

a. $8 + (12 + x)$ **b.** $6(5x)$.

The next example involves the use of both basic properties to simplify an algebraic expression.

EXAMPLE 5 Using the Commutative and Associative Properties

Simplify: $7 + (x + 2)$.

Solution

         $7 + (x + 2)$ This is the given algebraic expression.

      $= 7 + (2 + x)$ Use the commutative property to change the order of the addition.

      $= (7 + 2) + x$ Use the associative property to group the first two numbers.

      $= 9 + x$ Add within parentheses.

Using the commutative property of addition, an equivalent algebraic expression is $x + 9$. ■

✓ **CHECK POINT 5** Simplify: $8 + (x + 4)$.

Great Question!

Is there an easy way to distinguish between the commutative and associative properties?

Commutative: changes *order*

Associative: changes *grouping*

④ Use the distributive property.

The Distributive Property

The **distributive property** involves both multiplication and addition. The property shows how to multiply the sum of two numbers by a third number. Consider, for example, $4(7 + 3)$, which can be calculated in two ways. One way is to perform the addition within the grouping symbols and then multiply:

$$4(7 + 3) = 4(10) = 40.$$

The other way to find $4(7 + 3)$ is to *distribute* the multiplication by 4 over the addition: First multiply each number within the parentheses by 4 and then add:

$$4 \cdot 7 + 4 \cdot 3 = 28 + 12 = 40.$$

The result in both cases is 40. Thus,

$$\overset{\frown}{4(7 + 3)} = 4 \cdot 7 + 4 \cdot 3. \qquad \text{Multiplication distributes over addition.}$$

The distributive property allows us to rewrite the product of a number and a sum as the sum of two products.

> ### The Distributive Property
>
> Let a, b, and c represent real numbers, variables, or algebraic expressions.
>
> $$\overset{\frown}{a(b + c)} = ab + ac$$
>
> Multiplication distributes over addition.

Great Question!

Can you give examples so that I don't confuse the distributive property with the associative property of multiplication?

Distributive:

$$4(5 + x) = 4 \cdot 5 + 4x$$
$$= 20 + 4x$$

Associative:

$$4(5 \cdot x) = (4 \cdot 5)x$$
$$= 20x$$

EXAMPLE 6 Using the Distributive Property

Multiply: $6(x + 4)$.

Solution Multiply *each term* inside the parentheses, x and 4, by the multiplier outside, 6.

$$6(x + 4) = 6x + 6 \cdot 4 \qquad \text{Use the distributive property to remove parentheses.}$$
$$= 6x + 24 \qquad \text{Multiply: } 6 \cdot 4 = 24. \quad \blacksquare$$

✓ **CHECK POINT 6** Multiply: $5(x + 3)$.

EXAMPLE 7 Using the Distributive Property

Multiply: $5(3y + 7)$.

Solution Multiply *each term* inside the parentheses, $3y$ and 7, by the multiplier outside, 5.

$$5(3y + 7) = 5 \cdot 3y + 5 \cdot 7 \qquad \text{Use the distributive property to remove parentheses.}$$
$$= 15y + 35 \qquad \text{Multiply. Use the associative property of multiplication to find } 5 \cdot 3y\text{: } 5 \cdot 3y = (5 \cdot 3)y = 15y. \quad \blacksquare$$

✓ **CHECK POINT 7** Multiply: $6(4y + 7)$.

Great Question!

What's the bottom line when using the distributive property?

When using the distributive property to remove parentheses, be sure to multiply *each term* inside the parentheses by the multiplier outside.

Incorrect!

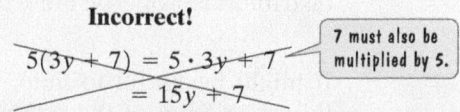

7 must also be multiplied by 5.

Table 1.3 shows a number of other forms of the distributive property.

Table 1.3	Other Forms of the Distributive Property	
Property	**Meaning**	**Example**
$a(b - c) = ab - ac$	Multiplication distributes over subtraction.	$5(4x - 3) = 5 \cdot 4x - 5 \cdot 3$ $= 20x - 15$
$a(b + c + d) = ab + ac + ad$	Multiplication distributes over three or more terms in parentheses.	$4(x + 10 + 3y)$ $= 4x + 4 \cdot 10 + 4 \cdot 3y$ $= 4x + 40 + 12y$
$(b + c)a = ba + ca$	Multiplication on the right distributes over addition (or subtraction).	$(x + 7)9 = x \cdot 9 + 7 \cdot 9$ $= 9x + 63$

5 Combine like terms.

Combining Like Terms

The distributive property

$$a(b + c) = ab + ac$$

lets us add and subtract like terms. To do this, we will usually apply the property in the form

$$ax + bx = (a + b)x$$

and then combine a and b. For example,

$$3x + 7x = (3 + 7)x = 10x.$$

This process is called **combining like terms**.

EXAMPLE 8 Combining Like Terms

Combine like terms:

a. $4x + 15x$ **b.** $7a - 2a.$

Solution

a. $4x + 15x$ These are like terms because 4x and 15x have identical variable factors.

     $= (4 + 15)x$ Apply the distributive property.

     $= 19x$ Add within parentheses.

b. $7a - 2a$ These are like terms because 7a and 2a have identical variable factors.

     $= (7 - 2)a$ Apply the distributive property.

     $= 5a$ Subtract within parentheses. ■

✓ **CHECK POINT 8** Combine like terms:

   a. $7x + 3x$ **b.** $9a - 4a.$

When combining like terms, you may find yourself leaving out the details of the distributive property. For example, you may simply write

$$7x + 3x = 10x.$$

It might be useful to think along these lines: Seven things plus three of the (same) things give ten of those things. To add like terms, add the coefficients and copy the common variable.

Combining Like Terms Mentally

1. Add or subtract the coefficients of the terms.

2. Use the result of step 1 as the coefficient of the term's variable factor.

When an expression contains three or more terms, use the commutative and associative properties to group like terms. Then combine the like terms.

EXAMPLE 9 Grouping and Combining Like Terms

Simplify:

a. $7x + 5 + 3x + 8$ **b.** $4x + 7y + 2x + 3y$.

Solution

a. $7x + 5 + 3x + 8$

$= (7x + 3x) + (5 + 8)$ Rearrange terms and group the like terms using the commutative and associative properties. This step is often done mentally.

$= 10x + 13$ Combine like terms: $7x + 3x = 10x$. Combine constant terms: $5 + 8 = 13$.

b. $4x + 7y + 2x + 3y$

$= (4x + 2x) + (7y + 3y)$ Group like terms.

$= 6x + 10y$ Combine like terms by adding coefficients and keeping the variable factor. ∎

✓ **CHECK POINT 9** Simplify:

a. $8x + 7 + 10x + 3$ **b.** $9x + 6y + 5x + 2y$.

Great Question!

What do like objects, such as apples and apples, have to do with like terms?

Combining like terms should remind you of adding and subtracting numbers of like objects.

7 apples + 3 apples = 10 apples $7a + 3a = 10a$

9 feet − 5 feet = 4 feet $9f - 5f = 4f$

6 apples + 10 feet = ? $6a$ and $10f$ are not like terms and cannot be added.

6 Simplify algebraic expressions.

Simplifying Algebraic Expressions

An algebraic expression is **simplified** when parentheses have been removed and like terms have been combined.

Simplifying Algebraic Expressions

1. Use the distributive property to remove parentheses.

2. Rearrange terms and group like terms using the commutative and associative properties. This step may be done mentally.

3. Combine like terms by combining the coefficients of the terms and keeping the same variable factor.

EXAMPLE 10 Simplifying an Algebraic Expression

Simplify: $5(3x + 7) + 6x$.

Solution

$$5(3x + 7) + 6x$$
$$= 5 \cdot 3x + 5 \cdot 7 + 6x \qquad \text{Use the distributive property to remove the parentheses.}$$
$$= 15x + 35 + 6x \qquad \text{Multiply.}$$
$$= (15x + 6x) + 35 \qquad \text{Group like terms.}$$
$$= 21x + 35 \qquad \text{Combine like terms.} \quad \blacksquare$$

✓ **CHECK POINT 10** Simplify: $7(2x + 3) + 11x$.

EXAMPLE 11 Simplifying an Algebraic Expression

Simplify: $6(2x + 4y) + 10(4x + 3y)$.

Solution

$$6(2x + 4y) + 10(4x + 3y)$$
$$= 6 \cdot 2x + 6 \cdot 4y + 10 \cdot 4x + 10 \cdot 3y \qquad \text{Use the distributive property to remove the parentheses.}$$
$$= 12x + 24y + 40x + 30y \qquad \text{Multiply.}$$
$$= (12x + 40x) + (24y + 30y) \qquad \text{Group like terms.}$$
$$= 52x + 54y \qquad \text{Combine like terms.} \quad \blacksquare$$

✓ **CHECK POINT 11** Simplify: $7(4x + 3y) + 2(5x + y)$.

Great Question!

Do I need to use the distributive property to simplify an algebraic expression such as 4(3x + 5x) that contains like terms within the parentheses?

Use the distributive property to remove parentheses when the terms inside parentheses are not like terms:

$$4(3x + 5y) = 4 \cdot 3x + 4 \cdot 5y = 12x + 20y.$$

$3x$ and $5y$ are not like terms.

It is not necessary to use the distributive property to remove parentheses when the terms inside parentheses are like terms:

$$4(3x + 5x) = 4(8x) = 32x.$$

$3x$ and $5x$ are like terms and can be combined: $3x + 5x = 8x$.

We mentally applied the associative property: $4(8x) = (4 \cdot 8)x = 32x$.

CONCEPT AND VOCABULARY CHECK

Fill in each blank so that the resulting statement is true.

1. Terms that have the same variable factors, such as $7x$ and $5x$, are called _____ terms.
2. If a and b are real numbers, the commutative property of addition states that $a + b =$ _____.
3. If a and b are real numbers, the commutative property of multiplication states that _____ $= ba$.
4. If a, b, and c are real numbers, the associative property of addition states that $(a + b) + c =$ _____.
5. If a, b, and c are real numbers, the associative property of multiplication states that _____ $= a(bc)$.
6. If a, b, and c are real numbers, the distributive property states that $a(b + c) =$ _____.
7. An algebraic expression is _____ when parentheses have been removed and like terms have been combined.

1.4 EXERCISE SET   MyMathLab®   Watch the videos in MyMathLab   Download the MyDashBoard App

Practice Exercises

In Exercises 1–6, an algebraic expression is given. Use each expression to answer the following questions.
  a. *How many terms are there in the algebraic expression?*
  b. *What is the numerical coefficient of the first term?*
  c. *What is the constant term?*
  d. *Does the algebraic expression contain like terms? If so, what are the like terms?*

1. $3x + 5$
2. $9x + 4$
3. $x + 2 + 5x$
4. $x + 6 + 7x$
5. $4y + 1 + 3x$
6. $8y + 1 + 10x$

In Exercises 7–14, use the commutative property of addition to write an equivalent algebraic expression.

7. $y + 4$ 8. $x + 7$
9. $5 + 3x$ 10. $4 + 9x$
11. $4x + 5y$ 12. $10x + 9y$
13. $5(x + 3)$ 14. $6(x + 4)$

In Exercises 15–22, use the commutative property of multiplication to write an equivalent algebraic expression.

15. $9x$ 16. $8x$
17. $x + y6$ 18. $x + y7$
19. $7x + 23$ 20. $13x + 11$
21. $5(x + 3)$ 22. $6(x + 4)$

In Exercises 23–26, use an associative property to rewrite each algebraic expression. Once the grouping has been changed, simplify the resulting algebraic expression.

23. $7 + (5 + x)$
24. $9 + (3 + x)$
25. $7(4x)$
26. $8(5x)$

In Exercises 27–46, use a form of the distributive property to rewrite each algebraic expression without parentheses.

27. $3(x + 5)$ 28. $4(x + 6)$
29. $8(2x + 3)$ 30. $9(2x + 5)$
31. $\frac{1}{3}(12 + 6r)$ 32. $\frac{1}{4}(12 + 8r)$
33. $5(x + y)$ 34. $7(x + y)$
35. $3(x - 2)$ 36. $4(x - 5)$
37. $2(4x - 5)$ 38. $6(3x - 2)$
39. $\frac{1}{2}(5x - 12)$ 40. $\frac{1}{3}(7x - 21)$
41. $(2x + 7)4$ 42. $(5x + 3)6$
43. $6(x + 3 + 2y)$
44. $7(2x + 4 + y)$
45. $5(3x - 2 + 4y)$
46. $4(5x - 3 + 7y)$

In Exercises 47–64, simplify each algebraic expression.

47. $7x + 10x$ 48. $5x + 13x$
49. $11a - 3a$ 50. $14b - 5b$
51. $3 + (x + 11)$ 52. $7 + (x + 10)$
53. $5y + 3 + 6y$ 54. $8y + 7 + 10y$
55. $2x + 5 + 7x - 4$ 56. $7x + 8 + 2x - 3$
57. $11a + 12 + 3a + 2$
58. $13a + 15 + 2a + 11$
59. $5(3x + 2) - 4$ 60. $2(5x + 4) - 3$
61. $12 + 5(3x - 2)$ 62. $14 + 2(5x - 1)$
63. $7(3a + 2b) + 5(4a + 2b)$
64. $11(6a + 3b) + 4(12a + 5b)$

Practice PLUS

In Exercises 65–66, name the property used to go from step to step each time that "(why?)" occurs.

65. $7 + 2(x + 9)$
$= 7 + (2x + 18)$ (why?)
$= 7 + (18 + 2x)$ (why?)
$= (7 + 18) + 2x$ (why?)
$= 25 + 2x$
$= 2x + 25$ (why?)

66. $5(x + 4) + 3x$
$= (5x + 20) + 3x$ (why?)
$= (20 + 5x) + 3x$ (why?)
$= 20 + (5x + 3x)$ (why?)
$= 20 + (5 + 3)x$ (why?)
$= 20 + 8x$
$= 8x + 20$ (why?)

In Exercises 67–76, write each English phrase as an algebraic expression. Then simplify the expression. Let x represent the number.

67. the sum of 7 times a number and twice the number

68. the sum of 8 times a number and twice the number

69. the product of 3 and a number, which is then subtracted from the product of 12 and a number

70. the product of 5 and a number, which is then subtracted from the product of 11 and a number

71. six times the product of 4 and a number

72. nine times the product of 3 and a number

73. six times the sum of 4 and a number

74. nine times the sum of 3 and a number

75. eight increased by the product of 5 and one less than a number

76. nine increased by the product of 3 and 2 less than a number

Application Exercises

The graph shows the number of Internet users in the United States and worldwide for four selected years. Exercises 77–78 involve mathematical models for the data.

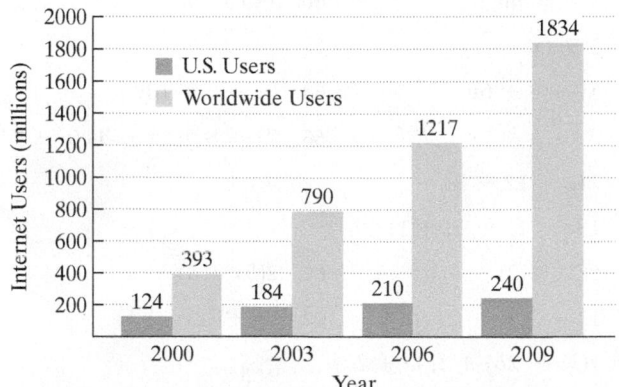

Number of Internet Users in the United States and Worldwide

Source: International Telecommunication Union

77. The number of Internet users in the United States, U, in millions, n years after 2000 can be modeled by the formula
$$U = 4(2n + 32) + 5(n + 1).$$
  a. Simplify the formula.
  b. Use the simplified form of the mathematical model to find the number of Internet users in the United States in 2006. Does the model underestimate or overestimate the actual number shown by the bar graph for 2006? By how much?

78. The number of worldwide Internet users, W, in millions, n years after 2000 can be modeled by the formula
$$W = 2(19n + 38) + 3(40n + 90).$$
  a. Simplify the formula.
  b. Use the simplified form of the mathematical model to find the number of worldwide Internet users in 2009. Does the model underestimate or overestimate the actual number shown by the bar graph for 2009? By how much?

Writing in Mathematics

79. What is a term? Provide an example with your description.

80. What are like terms? Provide an example with your description.

81. What are equivalent algebraic expressions?

82. State a commutative property and give an example.

83. State an associative property and give an example.

84. State a form of the distributive property and give an example.

85. Explain how to add like terms. Give an example.

86. What does it mean to simplify an algebraic expression?

87. An algebra student incorrectly used the distributive property and wrote $3(5x + 7) = 15x + 7$. If you were that student's teacher, what would you say to help the student avoid this kind of error?

88. You can transpose the letters in the word "conversation" to form the phrase "voices rant on." From "total abstainers" we can form "sit not at ale bars." What two algebraic properties do each of these transpositions (called *anagrams*) remind you of? Explain your answer.

Critical Thinking Exercises

Make Sense?   In Exercises 89–92, determine whether each statement "makes sense" or "does not make sense" and explain your reasoning.

89. I applied the commutative property and rewrote $x - 4$ as $4 - x$.

90. Just as the commutative properties change groupings, the associative properties change order.

91. I did not use the distributive property to simplify $3(2x + 5x)$.

92. The commutative, associative, and distributive properties remind me of the rules of a game.

In Exercises 93–96, determine whether each statement is true or false. If the statement is false, make the necessary change(s) to produce a true statement.

93. $(24 \div 6) \div 2 = 24 \div (6 \div 2)$

94. $2x + 5 = 5x + 2$

95. $a + (bc) = (a + b)(a + c)$; in words, addition can be distributed over multiplication.

96. Like terms contain the same coefficients.

Preview Exercises

Exercises 97–99 will help you prepare for the material covered in the next section. In each exercise, write an integer that is the result of the given situation.

97. You earn $150, but then you misplace $90.

98. You lose $50 and then you misplace $10.

99. The temperature is 30 degrees, and then it drops by 35 degrees.

MID-CHAPTER CHECK POINT | Section 1.1–Section 1.4

 What You Know: Algebra uses variables, or letters, that represent a variety of different numbers. These variables appear in algebraic expressions, equations, and formulas. Mathematical models use variables to describe real-world phenomena. We reviewed operations with fractions and saw how fractions are used in algebra. We defined the real numbers and represented them as points on a number line. Finally, we introduced some basic rules of algebra and used the commutative, associative, and distributive properties to simplify algebraic expressions.

1. Evaluate for $x = 6$: $2 + 10x$

2. Evaluate for $x = \frac{2}{5}$: $10x - 4$.

3. Evaluate for $x = 3$ and $y = 10$:
$$\frac{xy}{2} + 4(y - x).$$

In Exercises 4–5, translate from English to an algebraic expression or equation, whichever is appropriate. Let the variable x represent the number.

4. Two less than $\frac{1}{4}$ of a number

5. Five more than the quotient of a number and 6 gives 19.

In Exercises 6–7, determine whether the given number is a solution of the equation.

6. $3(x + 2) = 4x - 1$; 6

7. $8y = 12\left(y - \frac{1}{2}\right); \frac{3}{4}$

8. The average number of miles that a passenger car is driven per year is declining. The bar graph shows the average number of miles per car in the United States for four selected years.

Average Number of Miles per U.S. Passenger Car

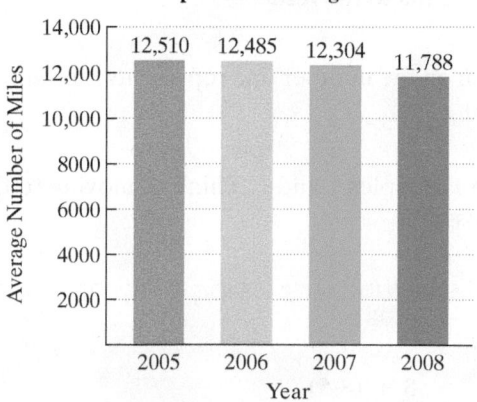

Source: Energy Information Administration

Here is a mathematical model that approximates the data displayed by the bar graph:

$$M = 12{,}624 - 235n.$$

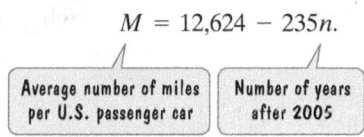

a. Use the formula to find the average number of miles per passenger car in 2008. Does the mathematical model underestimate or overestimate the actual number shown by the bar graph for 2008? By how much?

b. If trends from 2005 through 2008 continue, use the formula to project the average number of miles per U.S passenger car in 2015.

In Exercises 9–15, perform the indicated operation. Where possible, reduce answers to lowest terms.

9. $\frac{7}{10} - \frac{8}{15}$

10. $\frac{2}{3} \cdot \frac{3}{4}$

11. $\frac{5}{22} + \frac{5}{33}$

12. $\frac{3}{5} \div \frac{9}{10}$

13. $\frac{23}{105} - \frac{2}{105}$

14. $2\frac{7}{9} \div 3$

15. $5\frac{2}{9} - 3\frac{1}{6}$

16. The formula $C = \frac{5}{9}(F - 32)$ expresses the relationship between Fahrenheit temperature, F, and Celsius temperature, C. If the temperature is 50°F, use the formula to find the equivalent temperature on the Celsius scale.

17. Insert either $<$ or $>$ in the shaded area to make a true statement:
$$-8000 \quad \blacksquare \quad -8\frac{1}{4}.$$

18. Express $\frac{1}{11}$ as a decimal.

19. Find the absolute value: $|-19.3|$.

20. List all the rational numbers in this set:
$$\left\{-11, -\frac{3}{7}, 0, 0.45, \sqrt{23}, \sqrt{25}\right\}.$$

In Exercises 21–23, rewrite $5(x + 3)$ as an equivalent expression using the given property.

21. the commutative property of multiplication

22. the commutative property of addition

23. the distributive property

In Exercises 24–25, simplify each algebraic expression.

24. $7(9x + 3) + \frac{1}{3}(6x)$

25. $2(3x + 5y) + 4(x + 6y)$

SECTION

1.5

Objectives

1. Add numbers with a number line.

2. Find sums using identity and inverse properties.

3. Add numbers without a number line.

4. Use addition rules to simplify algebraic expressions.

5. Solve applied problems using a series of additions.

Addition of Real Numbers

It has not been a good day! First, you lost your wallet with $50 in it. Then, to get through the day, you borrowed $10, which you somehow misplaced. Your loss of $50 followed by a loss of $10 is an overall loss of $60. This can be written

$$-50 + (-10) = -60.$$

The result of adding two or more numbers is called the **sum** of the numbers. The sum of -50 and -10 is -60. You can think of gains and losses of money to find sums. For example, to find $17 + (-13)$, think of a gain of $17 followed by a loss of $13. There is an overall gain of $4. Thus, $17 + (-13) = 4$. In the same way, to find $-17 + 13$, think of a loss of $17 followed by a gain of $13. There is an overall loss of $4, so $-17 + 13 = -4$.

Adding with a Number Line

We use the number line to help picture the addition of real numbers. Here is the procedure for finding $a + b$, the sum of a and b, using the number line:

1. Add numbers with a number line.

Using the Number Line to Find a Sum

Let a and b represent real numbers. To find $a + b$ using a number line,

1. Start at a.
2. a. If b is **positive**, move b units to the **right**.
   b. If b is **negative**, move $|b|$ units to the **left**.
   c. If b is **0, stay** at a.
3. The number where we finish on the number line represents the sum of a and b.

This procedure is illustrated in Examples 1 and 2. Think of moving to the right as a gain and moving to the left as a loss.

EXAMPLE 1 Adding Real Numbers Using a Number Line

Find the sum using a number line:

$$3 + (-5).$$

Solution   We find $3 + (-5)$ using the number line in **Figure 1.12**.

Step 1. We consider 3 to be the first number, represented by a in the box on the previous page. We start at a, or 3.

Step 2. We consider -5 to be the second number, represented by b. Because this number is negative, we move 5 units to the left.

Step 3. We finish at -2 on the number line. The number where we finish represents the sum of 3 and -5. Thus,

$$3 + (-5) = -2.$$

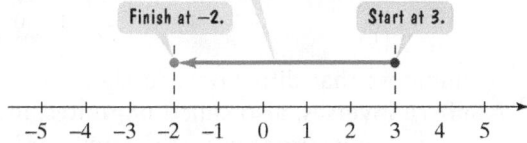

Figure 1.12 $3 + (-5) = -2$

Observe that if there is a gain of $3 followed by a loss of $5, there is an overall loss of $2. ∎

✓ **CHECK POINT 1**   Find the sum using a number line:

$$4 + (-7).$$

EXAMPLE 2   Adding Real Numbers Using a Number Line

Find each sum using a number line:

a. $-3 + (-4)$ **b.** $-6 + 2$.

Solution

a. To find $-3 + (-4)$, start at -3. Move 4 units to the left. We finish at -7. Thus,

$$-3 + (-4) = -7.$$

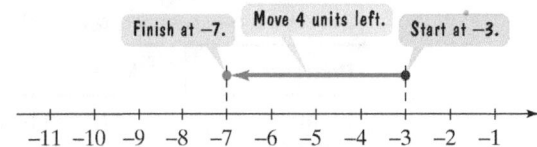

Observe that if there is a loss of $3 followed by a loss of $4, there is an overall loss of $7.

b. To find $-6 + 2$, start at -6. Move 2 units to the right because 2 is positive. We finish at -4. Thus,

$$-6 + 2 = -4.$$

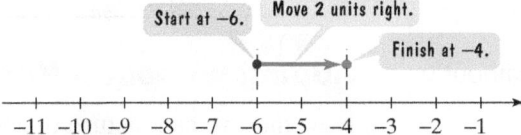

Observe that if there is a loss of $6 followed by a gain of $2, there is an overall loss of $4. ∎

✓ **CHECK POINT 2**   Find each sum using a number line:
a. $-1 + (-3)$ **b.** $-5 + 3$.

2 Find sums using identity and inverse properties.

The Number Line and Properties of Addition

The number line can be used to picture some useful properties of addition. For example, let's see what happens if we add two numbers with different signs but the same absolute value. Two such numbers are 3 and −3. To find $3 + (-3)$ on a number line, we start at 3 and move 3 units to the left. We finish at 0. Thus,

$$3 + (-3) = 0.$$

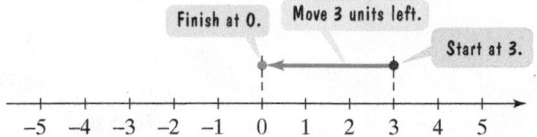

Numbers that differ only in sign, such as 3 and −3, are called *additive inverses*. **Additive inverses**, also called **opposites**, are pairs of real numbers that are the same number of units from zero on the number line, but are on opposite sides of zero. Thus, −3 is the additive inverse, or opposite, of 3, and 5 is the additive inverse, or opposite, of −5. The additive inverse of 0 is 0. Other additive inverses come in pairs.

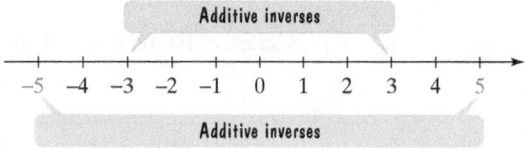

In general, the sum of any real number, denoted by a, and its additive inverse, denoted by $-a$, is zero:

$$a + (-a) = 0.$$

This property is called the **inverse property of addition**.

Table 1.4 shows the identity and inverse properties of addition.

Table 1.4	Identity and Inverse Properties of Addition

Let a be a real number, a variable, or an algebraic expression.

Property	Meaning	Examples
Identity Property of Addition	Zero can be deleted from a sum. $a + 0 = a$ $0 + a = a$	• $4 + 0 = 4$ • $-3x + 0 = -3x$ • $0 + (5a + b) = 5a + b$
Inverse Property of Addition	The sum of a real number and its additive inverse, or opposite, gives 0, the additive identity. $a + (-a) = 0$ $(-a) + a = 0$	• $6 + (-6) = 0$ • $3x + (-3x) = 0$ • $[-(2y + 1)] + (2y + 1) = 0$

3 Add numbers without a number line.

Adding without a Number Line

Now that we can picture the addition of real numbers, we look at two rules for using absolute value to add signed numbers.

Adding Two Numbers with the Same Sign

1. Add the absolute values.

2. Use the common sign as the sign of the sum.

EXAMPLE 3 Adding Real Numbers

Add without using a number line:

a. $-11 + (-15)$ **b.** $-0.2 + (-0.8)$ **c.** $-\dfrac{3}{4} + \left(-\dfrac{1}{2}\right)$.

Solution In each part of this example, we are adding numbers with the same sign.

a. $-11 + (-15) = -26$ — Add absolute values: $11 + 15 = 26$.

Use the common sign.

b. $-0.2 + (-0.8) = -1$ — Add absolute values: $0.2 + 0.8 = 1.0$ or 1.

Use the common sign.

c. $-\dfrac{3}{4} + \left(-\dfrac{1}{2}\right) = -\dfrac{5}{4}$ — Add absolute values: $\frac{3}{4} + \frac{1}{2} = \frac{3}{4} + \frac{2}{4} = \frac{5}{4}$.

Use the common sign.

◼

✓ **CHECK POINT 3** Add without using a number line:

a. $-10 + (-25)$ **b.** $-0.3 + (-1.2)$ **c.** $-\dfrac{2}{3} + \left(-\dfrac{1}{6}\right)$.

We also use absolute value to add two real numbers with different signs.

Adding Two Numbers with Different Signs

1. Subtract the smaller absolute value from the greater absolute value.
2. Use the sign of the number with the greater absolute value as the sign of the sum.

Using Technology

You can use a calculator to add signed numbers. Here are the keystrokes for finding $-11 + (-15)$:

Scientific Calculator

11 [+/−] [+] 15 [+/−] [=]

Graphing Calculator

[(−)] 11 [+] [(−)] 15 [ENTER]

Here are the keystrokes for finding $-13 + 4$:

Scientific Calculator

13 [+/−] [+] 4 [=]

Graphing Calculator

[(−)] 13 [+] 4 [ENTER]

EXAMPLE 4 Adding Real Numbers

Add without using a number line:

a. $-13 + 4$ **b.** $-0.2 + 0.8$ **c.** $-\dfrac{3}{4} + \dfrac{1}{2}$.

Solution In each part of this example, we are adding numbers with different signs.

a. $-13 + 4 = -9$ — Subtract absolute values: $13 - 4 = 9$.

Use the sign of the number with the greater absolute value.

b. $-0.2 + 0.8 = 0.6$ — Subtract absolute values: $0.8 - 0.2 = 0.6$.

Use the sign of the number with the greater absolute value. The sign is assumed to be positive.

c. $-\dfrac{3}{4} + \dfrac{1}{2} = -\dfrac{1}{4}$ — Subtract absolute values: $\frac{3}{4} - \frac{1}{2} = \frac{3}{4} - \frac{2}{4} = \frac{1}{4}$.

Use the sign of the number with the greater absolute value.

◼

☑ **CHECK POINT 4**   Add without using a number line:

a. $-15 + 2$ **b.** $-0.4 + 1.6$ **c.** $-\dfrac{2}{3} + \dfrac{1}{6}$.

Great Question!

What's the bottom line for determining the sign of a sum?

- The sum of two positive numbers is always positive.
  Example: $0.8 + 0.3$ is a positive number.
- The sum of two negative numbers is always negative.
  Example: $-0.8 + (-0.3)$ is a negative number.
- The sum of two numbers with different signs may be positive or negative. The sign of the sum is the sign of the number with the greater absolute value.
  Examples: $-0.8 + 0.3$ is a negative number.
  $0.8 + (-0.3)$ is a positive number.
- The sum of a number and its additive inverse, or opposite, is always zero.
  Example: $0.8 + (-0.8) = 0$.

4 Use addition rules to simplify algebraic expressions.

Algebraic Expressions

The rules for adding real numbers can be used to simplify certain algebraic expressions.

EXAMPLE 5   Simplifying Algebraic Expressions

Simplify:

a. $-11x + 7x$ **b.** $7y + (-12z) + (-9y) + 15z$ **c.** $3(7x + 8) + (-25x)$.

Solution

a. $-11x + 7x$ The given algebraic expression has two like terms: $-11x$ and $7x$ have identical variable factors.

$= (-11 + 7)x$ Apply the distributive property.

$= -4x$ Add within parentheses: $-11 + 7 = -4$.

b. $7y + (-12z) + (-9y) + 15z$ The colors indicate that there are two pairs of like terms.

$= 7y + (-9y) + (-12z) + 15z$ Arrange like terms so that they are next to one another.

$= [7 + (-9)]y + [(-12) + 15]z$ Apply the distributive property.

$= -2y + 3z$ Add within the grouping symbols:
$7 + (-9) = -2$ and $-12 + 15 = 3$.

c. $3(\overbrace{7x + 8}) + (-25x)$

$= 3 \cdot 7x + 3 \cdot 8 + (-25x)$ Use the distributive property to remove the parentheses.

$= 21x + 24 + (-25x)$ Multiply.

$= 21x + (-25x) + 24$ Arrange like terms so that they are next to one another.

$= [21 + (-25)]x + 24$ Apply the distributive property.

$= -4x + 24$ Add within the grouping symbols: $21 + (-25) = -4$. ∎

☑ **CHECK POINT 5**   Simplify:

a. $-20x + 3x$

b. $3y + (-10z) + (-10y) + 16z$

c. $5(2x + 3) + (-30x)$

5 Solve applied problems using a series of additions.

Applications

Positive and negative numbers are used in everyday life to represent such things as gains and losses in the stock market, rising and falling temperatures, deposits and withdrawals on bank statements, and ascending and descending motion. Positive and negative numbers are used to solve applied problems involving a series of additions.

One way to add a series of positive and negative numbers is to use the commutative and associative properties.

- Add all the positive numbers.
- Add all the negative numbers.
- Add the sums obtained in the first two steps.

The next example illustrates this idea.

EXAMPLE 6 An Application of Adding Signed Numbers

A glider was towed 1000 meters into the air and then let go. It descended 70 meters into a thermal (rising bubble of warm air), which took it up 2100 meters. At this point it dropped 230 meters into a second thermal. Then it rose 1200 meters. What was its altitude at that point?

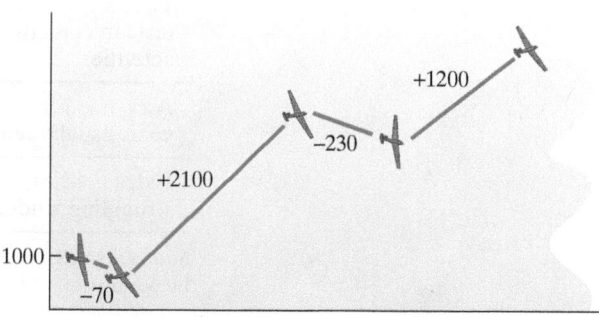

Solution We use the problem's conditions to write a sum. The altitude of the glider is expressed by the following sum:

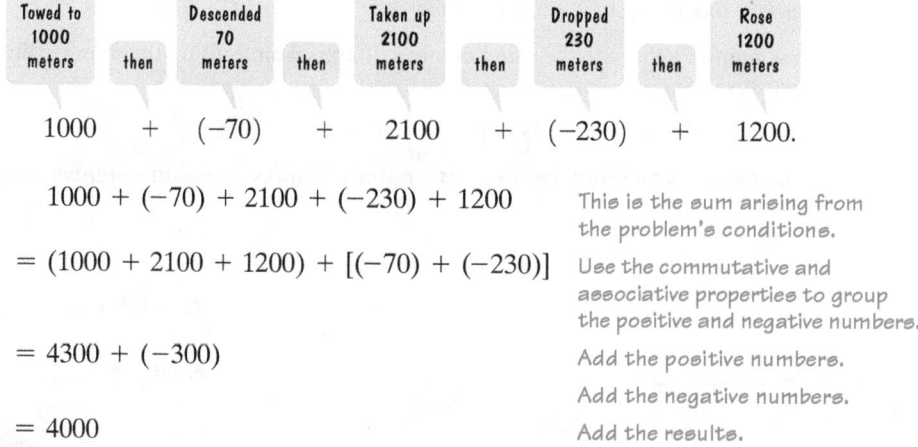

$$1000 + (-70) + 2100 + (-230) + 1200$$

This is the sum arising from the problem's conditions.

$$= (1000 + 2100 + 1200) + [(-70) + (-230)]$$

Use the commutative and associative properties to group the positive and negative numbers.

$$= 4300 + (-300)$$

Add the positive numbers.

Add the negative numbers.

$$= 4000$$

Add the results.

Discover for Yourself

Try working Example 6 by adding from left to right. You should still obtain 4000 for the sum. Which method do you find easier?

The altitude of the glider is 4000 meters. ∎

✓ **CHECK POINT 6** The water level of a reservoir is measured over a five-month period. During this time, the level rose 2 feet, then fell 4 feet, then rose 1 foot, then fell 5 feet, and then rose 3 feet. What was the change in the water level at the end of the five months?

Achieving Success

Some of the difficulties facing students who take a college math class for the first time are a result of the differences between a high school math class and a college math course. It is helpful to understand these differences throughout your college math courses.

High School Math Class	College Math Course
Attendance is required.	Attendance may be optional.
Teachers monitor progress and performance closely.	Students receive grades, but may not be informed by the professor if they are in trouble.
There are frequent tests, as well as make-up tests if grades are poor.	There are usually no more than three or four tests per semester. Make-up tests are rarely allowed.
Grades are often based on participation and effort.	Grades are usually based exclusively on test grades.
Students have contact with their instructor every day.	Students usually meet with their instructor two or three times per week.
Teachers cover all material for tests in class through lectures and activities.	Students are responsible for information whether or not it is covered in class.
A course is covered over the school year, usually ten months.	A course is covered in a semester, usually four months.
Extra credit is often available for struggling students.	Extra credit is almost never offered.

Source: BASS, ALAN, *MATH STUDY SKILLS*, 1st, © 2008. Printed and Electronically reproduced by permission of Pearson Education, Inc., Upper Saddle River, New Jersey.

CONCEPT AND VOCABULARY CHECK

Fill in each blank so that the resulting statement is true.

1. Pairs of real numbers that are the same number of units from zero on the number line, but are on opposite sides of zero, are called opposites or _____.

2. The sum of any real number and its opposite is always _____.

In the remaining items, state whether each sum is a positive number, a negative number, or 0. Do not actually find the sum.

3. $-20 + (-18)$ _____

4. $-20 + 28$ _____

5. $-0.8 + 0.8$ _____

6. $-\dfrac{4}{5} + \dfrac{1}{2}$ _____

7. $-1.3 + 2.7$ _____

8. $-\dfrac{2}{5} + \dfrac{2}{5}$ _____

1.5 EXERCISE SET   MyMathLab®

 Watch the videos in MyMathLab

 Download the MyDashBoard App

Practice Exercises

In Exercises 1–8, find each sum using a number line.

1. $7 + (-3)$
2. $7 + (-2)$
3. $-2 + (-5)$
4. $-1 + (-5)$
5. $-6 + 2$
6. $-8 + 3$
7. $3 + (-3)$
8. $5 + (-5)$

In Exercises 9–46, find each sum without the use of a number line.

9. $-7 + 0$
10. $-5 + 0$

11. $30 + (-30)$
12. $15 + (-15)$
13. $-30 + (-30)$
14. $-15 + (-15)$
15. $-8 + (-10)$
16. $-4 + (-6)$
17. $-0.4 + (-0.9)$
18. $-1.5 + (-5.3)$
19. $-\frac{7}{10} + \left(-\frac{3}{10}\right)$
20. $-\frac{7}{8} + \left(-\frac{1}{8}\right)$
21. $-9 + 4$
22. $-7 + 3$
23. $12 + (-8)$
24. $13 + (-5)$
25. $6 + (-9)$
26. $3 + (-11)$
27. $-3.6 + 2.1$
28. $-6.3 + 5.2$
29. $-3.6 + (-2.1)$
30. $-6.3 + (-5.2)$
31. $\frac{9}{10} + \left(-\frac{3}{5}\right)$
32. $\frac{7}{10} + \left(-\frac{2}{5}\right)$
33. $-\frac{5}{8} + \frac{3}{4}$
34. $-\frac{5}{6} + \frac{1}{3}$
35. $-\frac{3}{7} + \left(-\frac{4}{5}\right)$
36. $-\frac{3}{8} + \left(-\frac{2}{3}\right)$
37. $4 + (-7) + (-5)$
38. $10 + (-3) + (-8)$
39. $85 + (-15) + (-20) + 12$
40. $60 + (-50) + (-30) + 25$
41. $17 + (-4) + 2 + 3 + (-10)$
42. $19 + (-5) + 1 + 8 + (-13)$
43. $-45 + \left(-\frac{3}{7}\right) + 25 + \left(-\frac{4}{7}\right)$
44. $-50 + \left(-\frac{7}{9}\right) + 35 + \left(-\frac{11}{9}\right)$
45. $3.5 + (-45) + (-8.4) + 72$
46. $6.4 + (-35) + (-2.6) + 14$

In Exercises 47–60, simplify each algebraic expression.

47. $-10x + 2x$
48. $-19x + 10x$
49. $25y + (-12y)$
50. $26y + (-14y)$
51. $-8a + (-15a)$
52. $-9a + (-13a)$
53. $4y + (-13z) + (-10y) + 17z$
54. $5y + (-11z) + (-15y) + 20z$
55. $-7b + 10 + (-b) + (-6)$
56. $-10b + 13 + (-b) + (-4)$
57. $7x + (-5y) + (-9x) + 19y$
58. $13x + (-9y) + (-17x) + 20y$
59. $8(4y + 3) + (-35y)$
60. $7(3y + 5) + (-25y)$

Practice PLUS

In Exercises 61–64, find each sum.

61. $|-3 + (-5)| + |2 + (-6)|$
62. $|4 + (-11)| + |-3 + (-4)|$
63. $-20 + [-|15 + (-25)|]$
64. $-25 + [-|18 + (-26)|]$

In Exercises 65–66, insert either <, >, or = in the shaded area to make a true statement.

65. $6 + [2 + (-13)]$ ▒ $-3 + [4 + (-8)]$
66. $[(-8) + (-6)] + 10$ ▒ $-8 + [9 + (-2)]$

In Exercises 67–70, write each English phrase as an algebraic expression. Then simplify the expression. Let x represent the number.

67. The product of −6 and a number, which is then increased by the product of −13 and the number
68. The product of −9 and a number, which is then increased by the product of −11 and the number
69. The quotient of −20 and a number, increased by the quotient of 3 and the number
70. The quotient of −15 and a number, increased by the quotient of 4 and the number

Application Exercises

Solve Exercises 71–78 by writing a sum of signed numbers and adding.

71. The greatest temperature variation recorded in a day is 100 degrees in Browning, Montana, on January 23, 1916. The low temperature was −56°F. What was the high temperature?
72. In Spearfish, South Dakota, on January 22, 1943, the temperature rose 49 degrees in two minutes. If the initial temperature was −4°F, what was the temperature two minutes later?
73. The Dead Sea is the lowest elevation on Earth, 1312 feet below sea level. What is the elevation of a person standing 712 feet above the Dead Sea?
74. Lake Assal in Africa is 512 feet below sea level. What is the elevation of a person standing 642 feet above Lake Assal?
75. The temperature at 8:00 A.M. was −7°F. By noon it had risen 15°F, but by 4:00 P.M. it had fallen 5°F. What was the temperature at 4:00 P.M.?
76. On three successive plays, a football team lost 15 yards, gained 13 yards, and then lost 4 yards. What was the team's total gain or loss for the three plays?
77. A football team started with the football at the 27-yard line, advancing toward the center of the field (the 50-yard line). Four successive plays resulted in a 4-yard gain, a 2-yard loss, an 8-yard gain, and a 12-yard loss. What was the location of the football at the end of the fourth play?
78. The water level of a reservoir is measured over a five-month period. At the beginning, the level is 20 feet. During this time, the level rose 3 feet, then fell 2 feet, then fell 1 foot, then fell 4 feet, and then rose 2 feet. What is the reservoir's water level at the end of the five months?

The bar graph shows that in 2000 and 2001, the U.S. government collected more in taxes than it spent, so there was a budget surplus for each of these years. By contrast, in 2002 through 2009, the government spent more than it collected, resulting in budget deficits. Exercises 79–80 involve these deficits.

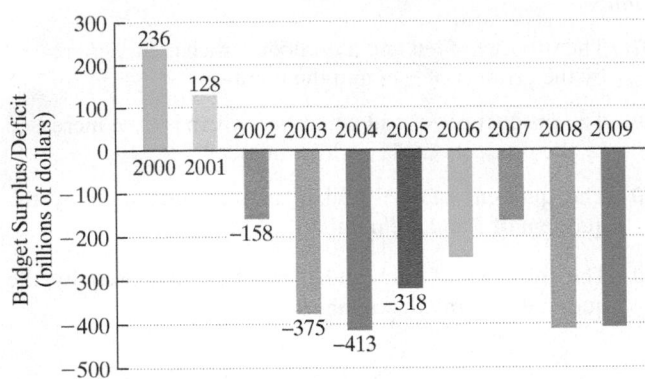

U.S. Government Budget Surplus/Deficit

Source: Budget of the U.S. Government

79. a. In 2008, the government collected $2521 billion and spent $2931 billion. Find $2521 + (-2931)$ and determine the deficit, in billions of dollars, for 2008.

   b. In 2009, the government collected $2700 billion and spent $3107 billion. Find the deficit, in billions of dollars, for 2009.

   c. Use your answers from parts (a) and (b) to determine the combined deficit, in billions of dollars, for 2008 and 2009.

80. a. In 2006, the government collected $2407 billion and spent $2655 billion. Find $2407 + (-2655)$ and determine the deficit, in billions of dollars, for 2006.

   b. In 2007, the government collected $2568 billion and spent $2730 billion. Find the deficit, in billions of dollars, for 2007.

   c. Use your answers from part (a) and (b) to determine the combined deficit, in billions of dollars, for 2006 and 2007.

Writing in Mathematics

81. Explain how to add two numbers with a number line. Provide an example with your explanation.

82. What are additive inverses?

83. Describe how the inverse property of addition

$$a + (-a) = 0$$

can be shown on a number line.

84. Without using a number line, describe how to add two numbers with the same sign. Give an example.

85. Without using a number line, describe how to add two numbers with different signs. Give an example.

86. Write a problem that can be solved by finding the sum of at least three numbers, some positive and some negative. Then explain how to solve the problem.

87. Without a calculator, you can add numbers using a number line, using absolute value, or using gains and losses. Which method do you find most helpful? Why is this so?

Critical Thinking Exercises

Make Sense? *In Exercises 88–91, determine whether each statement "makes sense" or "does not make sense" and explain your reasoning.*

88. It takes me too much time to add real numbers with a number line.

89. I found the sum of -13 and 4 by thinking of temperatures above and below zero: If it's 13 below zero and the temperature rises 4 degrees, the new temperature will be 9 below zero, so $-13 + 4 = -9$.

90. I added two negative numbers and obtained a positive sum.

91. Without adding numbers, I can see that the sum of -227 and 319 is greater than the sum of 227 and -319.

In Exercises 92–95, determine whether each statement is true or false. If the statement is false, make the necessary change(s) to produce a true statement.

92. $\dfrac{3}{4} + \left(-\dfrac{3}{5}\right) = -\dfrac{3}{20}$

93. The sum of zero and a negative number is always a negative number.

94. If one number is positive and the other negative, then the absolute value of their sum equals the sum of their absolute values.

95. The sum of a positive number and a negative number is always a positive number.

In Exercises 96–99, use the number line to determine whether each expression is positive or negative.

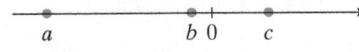

96. $a + b$ **97.** $a + c$

98. $b + c$ **99.** $|a + c|$

Technology Exercises

100. Use a calculator to verify any five of the sums that you found in Exercises 9–46.

101. Use a calculator to verify any three of the answers that you obtained in Application Exercises 71–78.

Review Exercises

From here on, each exercise set will contain three review exercises. It is essential to review previously covered topics to improve your understanding of the topics and to help maintain your mastery of the material. If you are not certain how to solve a review exercise, turn to the section and the worked example given in parentheses at the end of each exercise.

102. Consider the set
$$\{-6, -\pi, 0, 0.\overline{7}, \sqrt{3}, \sqrt{4}\}.$$
List all numbers from the set that are **a.** natural numbers, **b.** whole numbers, **c.** integers, **d.** rational numbers, **e.** irrational numbers, **f.** real numbers. (Section 1.3, Example 5)

103. Determine whether this inequality is true or false: $19 \geq -18$. (Section 1.3, Example 7)

104. Determine whether 18 is a solution of $16 = 2(x - 1) - x$. (Section 1.1, Example 4)

Preview Exercises

Exercises 105–107 will help you prepare for the material covered in the next section. In each exercise, a subtraction is expressed as addition of an opposite. Find this sum, indicated by a question mark.

105. $7 - 10 = 7 + (-10) = ?$

106. $-8 - 13 = -8 + (-13) = ?$

107. $-8 - (-13) = -8 + 13 = ?$

SECTION

1.6

Subtraction of Real Numbers

Objectives

1. Subtract real numbers.
2. Simplify a series of additions and subtractions.
3. Use the definition of subtraction to identify terms.
4. Use the subtraction definition to simplify algebraic expressions.
5. Solve problems involving subtraction.

Everybody complains about the weather, but where is it really the worst? **Table 1.5** shows that in the United States, Alaska and Montana are the record-holders for the coldest temperatures.

Table 1.5	Lowest Recorded U.S. Temperatures	
State	**Year**	**Temperature (°F)**
Alaska	1971	−80
Montana	1954	−70

Source: National Climatic Data Center

The table shows that Montana's record low temperature, −70, decreased by 10 degrees gives Alaska's record low temperature, −80. This can be expressed in two ways:

$$-70 - 10 = -80 \quad \text{or} \quad -70 + (-10) = -80.$$

This means that

$$-70 - 10 = -70 + (-10).$$

To subtract 10 from −70, we add −70 and the opposite, or additive inverse, of 10. Generalizing from this situation, we define subtraction as addition of an opposite.

Definition of Subtraction

For all real numbers a and b,

$$a - b = a + (-b).$$

In words: To subtract b from a, add the opposite, or additive inverse, of b to a. The result of subtraction is called the **difference**.

1 Subtract real numbers.

A Procedure for Subtracting Real Numbers

The definition of subtraction gives us a procedure for subtracting real numbers.

Subtracting Real Numbers

1. Change the subtraction operation to addition.
2. Change the sign of the number being subtracted.
3. Add, using one of the rules for adding numbers with the same sign or different signs.

Using Technology

You can use a calculator to subtract signed numbers. Here are the keystrokes for finding $5 - (-6)$:

Scientific Calculator

$5 \boxed{-} 6 \boxed{+/_-} \boxed{=}$

Graphing Calculator

$5 \boxed{-} \boxed{(-)} 6 \boxed{\text{ENTER}}$

Here are the keystrokes for finding $-9 - (-3)$:

Scientific Calculator

$9 \boxed{+/_-} \boxed{-} 3 \boxed{+/_-} \boxed{=}$

Graphing Calculator

$\boxed{(-)} 9 \boxed{-} \boxed{(-)} 3 \boxed{\text{ENTER}}$

Don't confuse the subtraction key on a graphing calculator, $\boxed{-}$, with the sign change or additive inverse key, $\boxed{(-)}$. What happens if you do?

EXAMPLE 1 Using the Definition of Subtraction

Subtract:

a. $7 - 10$ **b.** $5 - (-6)$ **c.** $-9 - (-3)$.

Solution

a. $7 - 10 = 7 + (-10) = -3$

Change the subtraction to addition. Replace 10 with its opposite.

b. $5 - (-6) = 5 + 6 = 11$

Change the subtraction to addition. Replace −6 with its opposite.

c. $-9 - (-3) = -9 + 3 = -6$

Change the subtraction to addition. Replace −3 with its opposite.

✓ **CHECK POINT 1** Subtract:

a. $3 - 11$ **b.** $4 - (-5)$ **c.** $-7 - (-2)$.

The definition of subtraction can be applied to real numbers that are not integers.

EXAMPLE 2 Using the Definition of Subtraction

Subtract:

a. $-5.2 - (-11.4)$ **b.** $-\dfrac{3}{4} - \dfrac{2}{3}$ **c.** $4\pi - (-9\pi)$.

Solution

a. $-5.2 - (-11.4) = -5.2 + 11.4 = 6.2$

> Change the subtraction to addition.

> Replace -11.4 with its opposite.

b. $-\dfrac{3}{4} - \dfrac{2}{3} = -\dfrac{3}{4} + \left(-\dfrac{2}{3}\right) = -\dfrac{9}{12} + \left(-\dfrac{8}{12}\right) = -\dfrac{17}{12}$

> Change the subtraction to addition.

> Replace $\frac{2}{3}$ with its opposite.

c. $4\pi - (-9\pi) = 4\pi + 9\pi = (4 + 9)\pi = 13\pi$

> Change the subtraction to addition.

> Replace -9π with its opposite.

Reading the symbol "$-$" can be a bit tricky. The way you read it depends on where it appears. For example,

$$-5.2 - (-11.4)$$

is read "negative five point two minus negative eleven point four." Read parts (b) and (c) of Example 2 aloud. When is "$-$" read "negative" and when is it read "minus"?

✓ **CHECK POINT 2**   Subtract:

a. $-3.4 - (-12.6)$ **b.** $-\dfrac{3}{5} - \dfrac{1}{3}$ **c.** $5\pi - (-2\pi)$.

---

2 Simplify a series of additions and subtractions.

Problems Containing a Series of Additions and Subtractions

In some problems, several additions and subtractions occur together. We begin by converting all subtractions to additions of opposites, or additive inverses.

> ### Simplifying a Series of Additions and Subtractions
>
> 1. Change all subtractions to additions of opposites.
> 2. Group and then add all the positive numbers.
> 3. Group and then add all the negative numbers.
> 4. Add the results of steps 2 and 3.

EXAMPLE 3   Simplifying a Series of Additions and Subtractions

Simplify: $7 - (-5) - 11 - (-6) - 19$.

Solution

$$7 - (-5) - 11 - (-6) - 19$$
$$= 7 + 5 + (-11) + 6 + (-19) \qquad \text{Write subtractions as additions of opposites.}$$
$$= (7 + 5 + 6) + [(-11) + (-19)] \qquad \text{Group the positive numbers. Group the negative numbers.}$$
$$= 18 + (-30) \qquad \text{Add the positive numbers. Add the negative numbers.}$$
$$= -12 \qquad \text{Add the results.} \quad ■$$

✓ **CHECK POINT 3**   Simplify: $10 - (-12) - 4 - (-3) - 6$.

3 Use the definition of subtraction to identify terms.

Subtraction and Algebraic Expressions

We know that the terms of an algebraic expression are separated by addition signs. Let's use this idea to identify the terms of the following algebraic expression:

$$9x - 4y - 5.$$

Because terms are separated by addition, we rewrite the subtractions in the algebraic expression as additions of opposites. Thus,

$$9x - 4y - 5 = 9x + (-4y) + (-5).$$

The three terms of the algebraic expression are $9x$, $-4y$, and -5.

> **EXAMPLE 4** Using the Definition of Subtraction to Identify Terms

Identify the terms of the algebraic expression:

$$2xy - 13y - 6.$$

Solution Rewrite the subtractions in the algebraic expression as additions of opposites.

$$2xy - 13y - 6 = 2xy + (-13y) + (-6)$$

First term   Second term   Third term

Because terms are separated by addition, the terms are $2xy$, $-13y$, and -6. ∎

> ✓ **CHECK POINT 4** Identify the terms of the algebraic expression:
>
> $$-6 + 4a - 7ab.$$

---

4 Use the subtraction definition to simplify algebraic expressions.

The procedure for subtracting real numbers can be used to simplify certain algebraic expressions that involve subtraction.

> **EXAMPLE 5** Simplifying Algebraic Expressions

Simplify:

a. $2 + 3x - 8x$ **b.** $-4x - 9y - 2x + 12y.$

Solution

a. $2 + 3x - 8x$ This is the given algebraic expression.

$\quad = 2 + 3x + (-8x)$ Write the subtraction as the addition of an opposite.

$\quad = 2 + [3 + (-8)]x$ Apply the distributive property.

$\quad = 2 + (-5x)$ Add within the grouping symbols.

$\quad = 2 - 5x$ Be concise and express as subtraction.

b. $-4x - 9y - 2x + 12y$ This is the given algebraic expression.

$\quad = -4x + (-9y) + (-2x) + 12y$ Write the subtractions as the additions of opposites.

$\quad = -4x + (-2x) + (-9y) + 12y$ Arrange like terms so that they are next to one another.

$\quad = [-4 + (-2)]x + (-9 + 12)y$ Apply the distributive property.

$\quad = -6x + 3y$ Add within the grouping symbols. ∎

Great Question!

How can I speed up the process of subtracting like terms?

You can think of gains and losses of money to work some of the steps in Example 5 mentally:

- $3x - 8x = -5x$ *A gain of 3 dollars followed by a loss of 8 dollars is a net loss of 5 dollars.*

- $-9y + 12y = 3y$ *A loss of 9 dollars followed by a gain of 12 dollars is a net gain of 3 dollars.*

✓ **CHECK POINT 5**   Simplify:

   a. $4 + 2x - 9x$ **b.** $-3x - 10y - 6x + 14y$.

---

5 Solve problems involving subtraction.

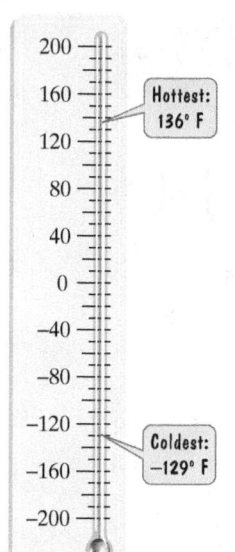

Figure 1.13

Applications

Subtraction is used to solve problems in which the word *difference* appears. The difference between real numbers a and b is expressed as $a - b$.

EXAMPLE 6   An Application of Subtraction Using the Word *Difference*

Figure 1.13 shows the hottest and coldest temperatures ever recorded on Earth. The record high was 136°F, recorded in Libya in 1922. The record low was −129°F, at the Vostok Scientific Station in Antarctica, in 1983. What is the difference between these highest and lowest temperatures? (*Source: The World Almanac Book of Records*, 2006)

Solution

The difference | is | the record high | minus | the record low.

$$= \quad 136 \quad - \quad (-129)$$
$$= \quad 136 \quad + \quad 129 \; = \; 265$$

The difference between the two temperatures is 265°F. We can also say that the variation in temperature between these two world extremes is 265°F. ■

✓ **CHECK POINT 6**   The peak of Mount Everest is 8848 meters above sea level. The Marianas Trench, on the floor of the Pacific Ocean, is 10,915 meters below sea level. What is the difference in elevation between the peak of Mount Everest and the Marianas Trench?

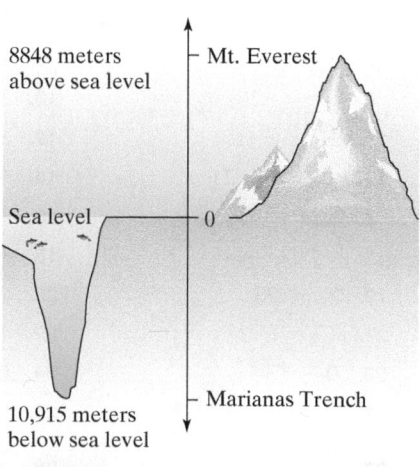

CONCEPT AND VOCABULARY CHECK

Fill in each blank so that the resulting statement is true.

1. $8 - 14 = 8 +$ _____

2. $8 - (-14) = 8 +$ _____

3. $-8 - (-14) = -8 +$ _____

4. $-8 - 14 =$ _____ $+$ _____

5. $7 - (-3) - 12 - 23 = 7 +$ _____ $+$ _____ $+$ _____

6. $9x - 4y - (-6) = 9x +$ _____ $+$ _____

7. The algebraic expression in Concept Check 6 contains _____ terms because terms are separated by _____.

1.6 EXERCISE SET   MyMathLab®

Watch the videos in MyMathLab

Download the MyDashBoard App

Practice Exercises

1. Consider the subtraction $5 - 12$.

   a. Find the opposite, or additive inverse, of 12.

   b. Rewrite the subtraction as the addition of the opposite of 12.

2. Consider the subtraction $4 - 10$.

   a. Find the opposite, or additive inverse, of 10.

   b. Rewrite the subtraction as the addition of the opposite of 10.

3. Consider the subtraction $5 - (-7)$.

   a. Find the opposite, or additive inverse, of -7.

   b. Rewrite the subtraction as the addition of the opposite of -7.

4. Consider the subtraction $2 - (-8)$.

   a. Find the opposite, or additive inverse, of -8.

   b. Rewrite the subtraction as the addition of the opposite of -8.

In Exercises 5–50, perform the indicated subtraction.

5. $14 - 8$ **6.** $15 - 2$

7. $8 - 14$ **8.** $2 - 15$

9. $3 - (-20)$ **10.** $5 - (-17)$

11. $-7 - (-18)$ **12.** $-5 - (-19)$

13. $-13 - (-2)$ **14.** $-21 - (-3)$

15. $-21 - 17$ **16.** $-29 - 21$

17. $-45 - (-45)$ **18.** $-65 - (-65)$

19. $23 - 23$ **20.** $26 - 26$

21. $13 - (-13)$ **22.** $15 - (-15)$

23. $0 - 13$ **24.** $0 - 15$

25. $0 - (-13)$ **26.** $0 - (-15)$

27. $\dfrac{3}{7} - \dfrac{5}{7}$ **28.** $\dfrac{4}{9} - \dfrac{7}{9}$

29. $\dfrac{1}{5} - \left(-\dfrac{3}{5}\right)$ **30.** $\dfrac{1}{7} - \left(-\dfrac{3}{7}\right)$

31. $-\dfrac{4}{5} - \dfrac{1}{5}$ **32.** $-\dfrac{4}{9} - \dfrac{1}{9}$

33. $-\dfrac{4}{5} - \left(-\dfrac{1}{5}\right)$ **34.** $-\dfrac{4}{9} - \left(-\dfrac{1}{9}\right)$

35. $\dfrac{1}{2} - \left(-\dfrac{1}{4}\right)$ **36.** $\dfrac{2}{5} - \left(-\dfrac{1}{10}\right)$

37. $\dfrac{1}{2} - \dfrac{1}{4}$ **38.** $\dfrac{2}{5} - \dfrac{1}{10}$

39. $9.8 - 2.2$ **40.** $5.7 - 3.3$

41. $-3.1 - (-1.1)$ **42.** $-4.6 - (-1.1)$

43. $1.3 - (-1.3)$ **44.** $1.4 - (-1.4)$

45. $-2.06 - (-2.06)$ **46.** $-3.47 - (-3.47)$

47. $5\pi - 2\pi$ **48.** $9\pi - 7\pi$

49. $3\pi - (-10\pi)$ **50.** $4\pi - (-12\pi)$

In Exercises 51–68, simplify each series of additions and subtractions.

51. $13 - 2 - (-8)$ **52.** $14 - 3 - (-7)$

53. $9 - 8 + 3 - 7$ **54.** $8 - 2 + 5 - 13$

55. $-6 - 2 + 3 - 10$ **56.** $-9 - 5 + 4 - 17$

57. $-10 - (-5) + 7 - 2$ **58.** $-6 - (-3) + 8 - 11$

59. $-23 - 11 - (-7) + (-25)$

60. $-19 - 8 - (-6) + (-21)$

61. $-823 - 146 - 50 - (-832)$

62. $-726 - 422 - 921 - (-816)$

63. $1 - \dfrac{2}{3} - \left(-\dfrac{5}{6}\right)$ **64.** $2 - \dfrac{3}{4} - \left(-\dfrac{7}{8}\right)$

65. $-0.16 - 5.2 - (-0.87)$

66. $-1.9 - 3 - (-0.26)$

67. $-\dfrac{3}{4} - \dfrac{1}{4} - \left(-\dfrac{5}{8}\right)$ **68.** $-\dfrac{1}{2} - \dfrac{2}{3} - \left(-\dfrac{1}{3}\right)$

In Exercises 69–72, identify the terms in each algebraic expression.

69. $-3x - 8y$

70. $-9a - 4b$

71. $12x - 5xy - 4$

72. $8a - 7ab - 13$

In Exercises 73–84, simplify each algebraic expression.

73. $3x - 9x$

74. $2x - 10x$

75. $4 + 7y - 17y$

76. $5 + 9y - 29y$

77. $2a + 5 - 9a$

78. $3a + 7 - 11a$

79. $4 - 6b - 8 - 3b$

80. $5 - 7b - 13 - 4b$

81. $13 - (-7x) + 4x - (-11)$

82. $15 - (-3x) + 8x - (-10)$

83. $-5x - 10y - 3x + 13y$

84. $-6x - 9y - 4x + 15y$

Practice PLUS

In Exercises 85–90, find the value of each expression.

85. $-|-9 - (-6)| - (-12)$

86. $-|-8 - (-2)| - (-6)$

87. $\dfrac{5}{8} - \left(\dfrac{1}{2} - \dfrac{3}{4}\right)$

88. $\dfrac{9}{10} - \left(\dfrac{1}{4} - \dfrac{7}{10}\right)$

89. $|-9 - (-3 + 7)| - |-17 - (-2)|$

90. $|24 - (-16)| - |-51 - (-31 + 2)|$

In Exercises 91–94, write each English phrase as an algebraic expression. Then simplify the expression. Let x represent the number.

91. The difference between 6 times a number and -5 times the number

92. The difference between 9 times a number and -4 times the number

93. The quotient of -2 and a number, subtracted from the quotient of -5 and the number

94. The quotient of -7 and a number, subtracted from the quotient of -12 and the number

Application Exercises

95. The peak of Mount Kilimanjaro, the highest point in Africa, is 19,321 feet above sea level. Qattara Depression, Egypt, one of the lowest points in Africa, is 436 feet below sea level. What is the difference in elevation between the peak of Mount Kilimanjaro and the Qattara Depression?

96. The peak of Mount Whitney is 14,494 feet above sea level. Mount Whitney can be seen directly above Death Valley, which is 282 feet below sea level. What is the difference in elevation between these geographic locations?

The bar graph shows the average daily low temperature for each month in Fairbanks, Alaska. Use the graph to solve Exercises 97–100.

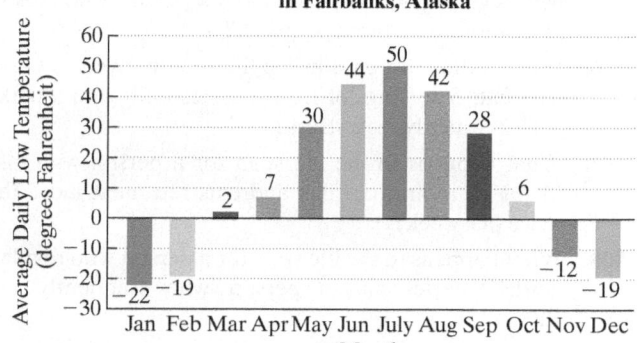

Source: The Weather Channel Enterprises, Inc.

97. What is the difference between the average daily low temperatures for March and February?

98. What is the difference between the average daily low temperatures for October and November?

99. How many degrees warmer is February's average low temperature than January's average low temperature?

100. How many degrees warmer is November's average low temperature than December's average low temperature?

Life expectancy for the average American man is 75.2 years; for a woman, it's 80.4 years. The number line, with points representing eight integers, indicates factors, many within our control, that can stretch or shrink one's probable life span. Use this information to solve Exercises 101–110.

Stretching or Shrinking One's Life Span

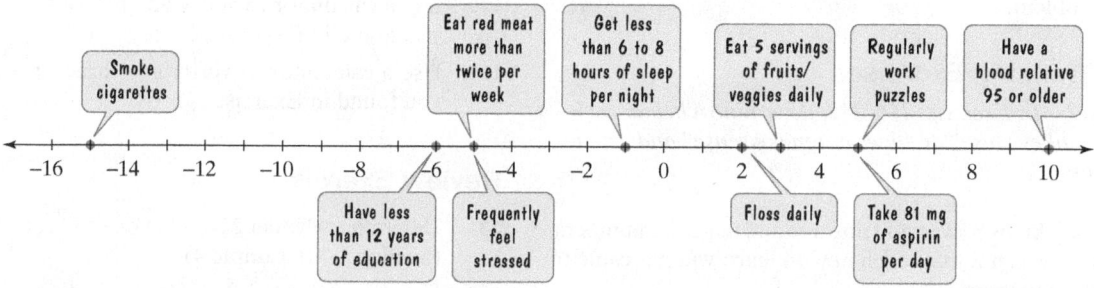

Years of Life Gained or Lost

Source: Newsweek

(In Exercises 101–110, be sure to refer to the number line at the bottom of the previous page. Apologies from your author for this awkward, but unavoidable, page split.)

101. If you have a blood relative 95 or older and you smoke cigarettes, do you stretch or shrink your life span? By how many years?

102. If you floss daily and eat red meat more than twice per week, do you stretch or shrink your life span? By how many years?

103. If you frequently feel stressed and have less than 12 years of education, do you stretch or shrink your life span? By how many years?

104. If you get less than 6 to 8 hours of sleep per night and smoke cigarettes, do you stretch or shrink your life span? By how many years?

105. What is the difference in the life span between a person who regularly works puzzles and a person who eats red meat more than twice per week?

106. What is the difference in the life span between a person who eats 5 servings of fruits/veggies daily and a person who frequently feels stressed?

107. What happens to the life span for a person who takes 81 mg of aspirin per day and eats red meat more than twice per week?

108. What happens to the life span for a person who regularly works puzzles and a person who frequently feels stressed?

109. What is the difference in the life span between a person with less than 12 years of education and a person who smokes cigarettes?

110. What is the difference in the life span between a person who gets less than 6 to 8 hours of sleep per night and a person who smokes cigarettes?

Writing in Mathematics

111. Explain how to subtract real numbers.

112. How is $4 - (-2)$ read?

113. Explain how to simplify a series of additions and subtractions. Provide an example with your explanation.

114. Explain how to find the terms of the algebraic expression $5x - 2y - 7$.

115. Write a problem that can be solved by finding the difference between two numbers. At least one of the numbers should be negative. Then explain how to solve the problem.

Critical Thinking Exercises

Make Sense? *In Exercises 116–119, determine whether each statement "makes sense" or "does not make sense" and explain your reasoning.*

116. I already knew how to add positive and negative numbers, so there was not that much new to learn when it came to subtracting them.

117. I can find the closing price of stock PQR on Wednesday by subtracting the change in price, −1.23, from the closing price on Thursday, 47.19.

		C3
STOCK PRICES		
Thursday		
Stock	**Close**	**Change**
ABC	31.54	0.47
XYZ	16.23	−0.87
PQR	47.19	−1.23
DEF	21.54	0.21

118. I found the variation in elevation between two heights by taking the difference between the high point and the low point.

119. I found the variation in U.S. temperature by subtracting the record low temperature, a negative number, from the record high temperature, a positive number.

In Exercises 120–123, determine whether each statement is true or false. If the statement is false, make the necessary change(s) to produce a true statement.

120. If a and b are negative numbers, then $a - b$ is sometimes a negative number.

121. $7 - (-2) = 5$

122. The difference between 0 and a negative number is always a positive number.

123. $|a - b| = |b - a|$

In Exercises 124–127, use the number line to determine whether each difference is positive or negative.

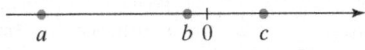

124. $c - a$ 125. $a - b$

126. $b - c$ 127. $0 - b$

128. Order the expressions $|x - y|$, $|x| - |y|$, and $|x + y|$ from least to greatest for $x = -6$ and $y = -8$.

Technology Exercises

129. Use a calculator to verify any five of the differences that you found in Exercises 5–46.

130. Use a calculator to verify any three of the answers that you found in Exercises 51–68.

Review Exercises

131. Determine whether 2 is a solution of $13x + 3 = 3(5x - 1)$. (Section 1.1, Example 4)

132. Simplify: $5(3x + 2y) + 6(5y)$. (Section 1.4, Example 11)

133. Give an example of an integer that is not a natural number. (Section 1.3; Example 5)

Preview Exercises

Exercises 134–136 will help you prepare for the material covered in the next section.

In Exercises 134–135, a multiplication is expressed as a repeated addition. Find this sum, indicated by a question mark.

134. $4(-3) = (-3) + (-3) + (-3) + (-3) = ?$

135. $3(-3) = (-3) + (-3) + (-3) = ?$

136. The list shows a pattern for various products.

$2(-3) = (-3) + (-3) = -6$

$$2(-3) = -6$$
$$1(-3) = -3$$
$$0(-3) = 0$$
$$-1(-3) = 3$$
$$-2(-3) = 6$$
$$-3(-3) = 9$$
$$-4(-3) = ?$$

Reading down the list, products keep increasing by 3.

Use this pattern to find $-4(-3)$.

Multiplication and Division of Real Numbers

Objectives

1. Multiply real numbers.
2. Multiply more than two real numbers.
3. Find multiplicative inverses.
4. Use the definition of division.
5. Divide real numbers.
6. Simplify algebraic expressions involving multiplication.
7. Determine whether a number is a solution of an equation.
8. Use mathematical models involving multiplication and division.

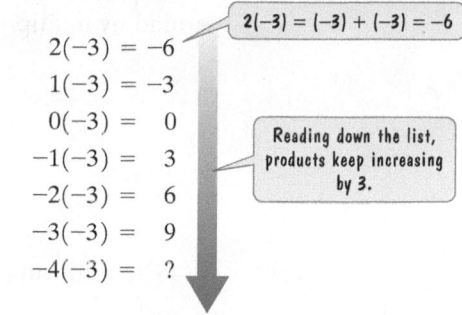

Where the Boys Are, released in 1960, was the first college spring-break movie. Today, visitors to college campuses can't help but ask: Where are the boys?

In 2007, 135 women received bachelor's degrees for every 100 men. In this section, we use data from a report by the U.S. Department of Education to develop mathematical models showing that this gender imbalance will widen in the coming years. Working with these models requires that we know how to multiply and divide real numbers.

Multiplying Real Numbers

Multiplication is repeated addition. For example, $5(-3)$ means that -3 is added five times:

$$5(-3) = (-3) + (-3) + (-3) + (-3) + (-3) = -15.$$

The result of the multiplication, -15, is called the **product** of 5 and -3. The numbers being multiplied, 5 and -3, are called the **factors** of the product.

Rules for multiplying real numbers are described in terms of absolute value. For example, $5(-3) = -15$ illustrates that the product of numbers with different signs is found by multiplying their absolute values. The product is negative.

$$5(-3) = -15$$

Multiply absolute values:
$|5| \cdot |-3| = 5 \cdot 3 = 15.$

Factors have different signs and the product is negative.

The following rules are used to determine the sign of the product of two numbers:

1 Multiply real numbers.

The Product of Two Real Numbers

- The product of two real numbers with **different signs** is found by multiplying their absolute values. The product is **negative**.
- The product of two real numbers with the **same sign** is found by multiplying their absolute values. The product is **positive**.
- The product of 0 and any real number is 0. Thus, for any real number a,
$$a \cdot 0 = 0 \qquad \text{and} \qquad 0 \cdot a = 0.$$

EXAMPLE 1 Multiplying Real Numbers

Multiply:

a. $6(-3)$ **b.** $-\dfrac{1}{5} \cdot \dfrac{2}{3}$ **c.** $(-9)(-10)$ **d.** $(-1.4)(-2)$ **e.** $(-372)(0)$.

Solution

Multiply absolute values: $6 \cdot 3 = 18$.

a. $6(-3) = -18$

Different signs: negative product

b. $-\dfrac{1}{5} \cdot \dfrac{2}{3} = -\dfrac{2}{15}$ Multiply absolute values: $\frac{1}{5} \cdot \frac{2}{3} = \frac{1 \cdot 2}{5 \cdot 3} = \frac{2}{15}$.

Different signs: negative product

Multiply absolute values: $9 \cdot 10 = 90$.

c. $(-9)(-10) = 90$

Same sign: positive product

Multiply absolute values: $(1.4)(2) = 2.8$.

d. $(-1.4)(-2) = 2.8$

Same sign: positive product

The product of 0 and any real number is 0: $a \cdot 0 = 0$.

e. $(-372)(0) = 0$ ∎

✓ CHECK POINT 1 Multiply:

a. $8(-5)$ **b.** $-\dfrac{1}{3} \cdot \dfrac{4}{7}$ **c.** $(-12)(-3)$

d. $(-1.1)(-5)$ **e.** $(-543)(0)$.

2 Multiply more than two real numbers.

Multiplying More Than Two Numbers

How do we perform more than one multiplication, such as

$$-4(-3)(-2)?$$

Because of the associative and commutative properties, we can order and group the numbers in any manner. Each pair of negative numbers will produce a positive product. Thus, the product of an even number of negative numbers is always positive. By contrast, the product of an odd number of negative numbers is always negative.

$$-4(-3)(-2) = -24$$

Multiply absolute values:
$4 \cdot 3 \cdot 2 = 24.$

Odd number of negative numbers (three): negative product

Multiplying More Than Two Numbers

1. Assuming that no factor is zero,
   - The product of an **even** number of **negative numbers** is **positive**.
   - The product of an **odd** number of **negative numbers** is **negative**.

   The multiplication is performed by multiplying the absolute values of the given numbers.
2. If any factor is 0, the product is 0.

EXAMPLE 2   Multiplying More Than Two Numbers

Multiply:

a. $(-3)(-1)(2)(-2)$ **b.** $(-1)(-2)(-2)(3)(-4).$

Solution

a. $(-3)(-1)(2)(-2) = -12$

Multiply absolute values:
$3 \cdot 1 \cdot 2 \cdot 2 = 12.$

Odd number of negative numbers (three): negative product

b. $(-1)(-2)(-2)(3)(-4) = 48$

Multiply absolute values:
$1 \cdot 2 \cdot 2 \cdot 3 \cdot 4 = 48.$

Even number of negative numbers (four): positive product

✓ **CHECK POINT 2**   Multiply:

a. $(-2)(3)(-1)(4)$ **b.** $(-1)(-3)(2)(-1)(5).$

Is it always necessary to count the number of negative factors when multiplying more than two numbers? No. If any factor is 0, you can immediately write 0 for the product. For example,

$$(-37)(423)(0)(-55)(-3.7) = 0.$$

If any factor is 0, the product is 0.

3 Find multiplicative inverses.

The Meaning of Division

The result of dividing the real number a by the nonzero real number b is called the **quotient** of a and b. We can write this quotient as $a \div b$ or $\frac{a}{b}$.

We know that subtraction is defined in terms of addition of an additive inverse, or opposite:

$$a - b = a + (-b).$$

In a similar way, we can define division in terms of multiplication. For example, the quotient of 8 and 2 can be written as multiplication:

$$8 \div 2 = 8 \cdot \frac{1}{2}.$$

We call $\frac{1}{2}$ the *multiplicative inverse*, or *reciprocal*, of 2. Two numbers whose product is 1 are called **multiplicative inverses** or **reciprocals** of each other. Thus, the multiplicative inverse of 2 is $\frac{1}{2}$ and the multiplicative inverse of $\frac{1}{2}$ is 2 because $2 \cdot \frac{1}{2} = 1$.

> **EXAMPLE 3** Finding Multiplicative Inverses
>
> Find the multiplicative inverse of each number:
>
> **a.** 5   **b.** $\frac{1}{3}$ **c.** -4 **d.** $-\frac{4}{5}$.

Solution

a. The multiplicative inverse of 5 is $\frac{1}{5}$ because $5 \cdot \frac{1}{5} = 1$.

b. The multiplicative inverse of $\frac{1}{3}$ is 3 because $\frac{1}{3} \cdot 3 = 1$.

c. The multiplicative inverse of -4 is $-\frac{1}{4}$ because $(-4)\left(-\frac{1}{4}\right) = 1$.

d. The multiplicative inverse of $-\frac{4}{5}$ is $-\frac{5}{4}$ because $\left(-\frac{4}{5}\right)\left(-\frac{5}{4}\right) = 1$. ∎

> ✓ **CHECK POINT 3** Find the multiplicative inverse of each number:
>
> **a.** 7   **b.** $\frac{1}{8}$ **c.** -6 **d.** $-\frac{7}{13}$.

Can you think of a real number that has no multiplicative inverse? The number **0 has no multiplicative inverse** because 0 multiplied by any number is never 1, but always 0. We now define division in terms of multiplication by a multiplicative inverse.

4 Use the definition of division.

Definition of Division

If a and b are real numbers and b is not 0, then the quotient of a and b is defined as

$$a \div b = a \cdot \frac{1}{b}.$$

In words: The quotient of two real numbers is the product of the first number and the multiplicative inverse of the second number.

> **EXAMPLE 4** Using the Definition of Division
>
> Use the definition of division to find each quotient:
>
> **a.** $-15 \div 3$ **b.** $\frac{-20}{-4}$.

Solution

a. $-15 \div 3 = -15 \cdot \dfrac{1}{3} = -5$

> Change the division to multiplication.

> Replace 3 with its multiplicative inverse.

b. $\dfrac{-20}{-4} = -20 \cdot \left(-\dfrac{1}{4}\right) = 5$

> Change the division to multiplication.

> Replace −4 with its multiplicative inverse.

✓ **CHECK POINT 4** Use the definition of division to find each quotient:

a. $-28 \div 7$

b. $\dfrac{-16}{-2}$.

A Procedure for Dividing Real Numbers

Because the quotient $a \div b$ is defined as the product $a \cdot \frac{1}{b}$, the sign rules for dividing numbers are the same as the sign rules for multiplying them.

The Quotient of Two Real Numbers

- The quotient of two real numbers with **different signs** is found by dividing their absolute values. The quotient is **negative**.
- The quotient of two real numbers with the **same sign** is found by dividing their absolute values. The quotient is **positive**.
- Division of any number by zero is undefined.
- Any nonzero number divided into 0 is 0.

5 Divide real numbers.

EXAMPLE 5 Dividing Real Numbers

Divide:

a. $\dfrac{8}{-2}$ **b.** $-\dfrac{3}{4} \div \left(-\dfrac{5}{9}\right)$ **c.** $\dfrac{-20.8}{4}$ **d.** $\dfrac{0}{-7}$.

Solution

a. $\dfrac{8}{-2} = -4$ > Divide absolute values: $\frac{8}{2} = 4$.

> Different signs: negative quotient

b. $-\dfrac{3}{4} \div \left(-\dfrac{5}{9}\right) = \dfrac{27}{20}$ > Divide absolute values: $\frac{3}{4} \div \frac{5}{9} = \frac{3}{4} \cdot \frac{9}{5} = \frac{27}{20}$.

> Same sign: positive quotient

c. $\dfrac{-20.8}{4} = -5.2$ > Divide absolute values: $4\overline{)20.8}^{\,5.2}$.

> Different signs: negative quotient

d. $\dfrac{0}{-7} = 0$ > Any nonzero number divided into 0 is 0.

Great Question!

Why is zero under the line undefined?

Here's another reason why division by zero is undefined. We know that

$$\frac{12}{4} = 3 \text{ because } 3 \cdot 4 = 12.$$

Now think about $\frac{-7}{0}$. If

$$\frac{-7}{0} = ? \quad \text{then} \quad ? \cdot 0 = -7.$$

However, any real number multiplied by 0 is 0 and not -7. No matter what number we use to replace the question mark in $? \cdot 0 = -7$, we can never obtain -7.

Can you see why $\frac{0}{-7}$ must be 0? The definition of division tells us that

$$\frac{0}{-7} = 0 \cdot \left(-\frac{1}{7}\right)$$

and the product of 0 and any real number is 0. By contrast, the definition of division does not allow for division by 0 because 0 does not have a multiplicative inverse. It is incorrect to write

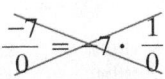

$$\frac{-7}{0} = -7 \cdot \frac{1}{0}.$$

0 does not have a multiplicative inverse.

Division by zero is not allowed, or not defined. Thus, $\frac{-7}{0}$ does not represent a real number. A real number can never have a denominator of 0.

✓ **CHECK POINT 5** Divide:

a. $\dfrac{-32}{-4}$ **b.** $-\dfrac{2}{3} \div \dfrac{5}{4}$ **c.** $\dfrac{21.9}{-3}$ **d.** $\dfrac{0}{-5}$.

6 Simplify algebraic expressions involving multiplication.

Multiplication and Algebraic Expressions

In Section 1.4, we discussed the commutative and associative properties of multiplication. We also know that multiplication distributes over addition and subtraction. We now add some additional properties to our previous list (**Table 1.6**). These properties are frequently helpful in simplifying algebraic expressions.

Table 1.6   Additional Properties of Multiplication

Let *a* be a real number, a variable, or an algebraic expression.

Property	Meaning	Examples
Identity Property of Multiplication	1 can be deleted from a product. $a \cdot 1 = a$ $1 \cdot a = a$	• $\sqrt{3} \cdot 1 = \sqrt{3}$ • $1x = x$ • $1(2x + 3) = 2x + 3$
Inverse Property of Multiplication	If a is not 0: $a \cdot \dfrac{1}{a} = 1$ $\dfrac{1}{a} \cdot a = 1$ The product of a nonzero number and its multiplicative inverse, or reciprocal, gives 1, the multiplicative identity.	• $6 \cdot \dfrac{1}{6} = 1$ • $3x \cdot \dfrac{1}{3x} = 1$ (x is not 0.) • $\dfrac{1}{(y-2)} \cdot (y-2) = 1$ (y is not 2.)
Multiplication Property of -1	Negative 1 times a is the opposite, or additive inverse, of a. $-1 \cdot a = -a$ $a(-1) = -a$	• $-1 \cdot \sqrt{3} = -\sqrt{3}$ • $-1\left(-\dfrac{3}{4}\right) = \dfrac{3}{4}$ • $-1x = -x$ • $-(x + 4) = -1(x + 4)$ $= -x - 4$
Double Negative Property	The opposite of $-a$ is a. $-(-a) = a$	• $-(-4) = 4$ • $-(-6y) = 6y$

In the table on the previous page, we used two steps to remove the parentheses from $-(x + 4)$. First, we used the multiplication property of -1.

$$-(x + 4) = -1(x + 4)$$

Then we used the distributive property, distributing -1 to each term in parentheses.

$$-1(x + 4) = (-1)x + (-1)4 = -x + (-4) = -x - 4$$

There is a fast way to obtain $-(x + 4) = -x - 4$ in just one step.

Negative Signs and Parentheses

If a negative sign precedes parentheses, remove the parentheses and change the sign of every term within the parentheses.

Here are some examples that illustrate this method.

$$-(11x + 5) = -11x - 5$$
$$-(11x - 5) = -11x + 5$$
$$-(-11x + 5) = 11x - 5$$
$$-(-11x - 5) = 11x + 5$$

EXAMPLE 6 Simplifying Algebraic Expressions

Simplify:

a. $-2(3x)$ **b.** $6x + x$ **c.** $8a - 9a$ **d.** $-3(2x - 5)$ **e.** $-(3y - 8)$.

Discover for Yourself

Verify each simplification in Example 6 by substituting 5 for the variable. The value of the given expression should be the same as the value of the simplified expression. Which expression is easier to evaluate?

Solution We will show all steps in the solution process. However, you probably are working many of these steps mentally.

a. $-2(3x)$ — This is the given algebraic expression.

$= (-2 \cdot 3)x$ — Use the associative property and group the first two numbers.

$= -6x$ — Numbers with opposite signs have a negative product.

b. $6x + x$ — This is the given algebraic expression.

$= 6x + 1x$ — Use the multiplication property of 1.

$= (6 + 1)x$ — Apply the distributive property.

$= 7x$ — Add within parentheses.

c. $8a - 9a$ — This is the given algebraic expression.

$= (8 - 9)a$ — Apply the distributive property.

$= -1a$ — Subtract within parentheses: $8 - 9 = 8 + (-9) = -1$.

$= -a$ — Apply the multiplication property of -1.

d. $-3(2x - 5)$ — This is the given algebraic expression.

$= -3(2x) - (-3) \cdot (5)$ — Apply the distributive property.

$= -6x - (-15)$ — Multiply.

$= -6x + 15$ — Subtraction is the addition of an opposite.

e. $-(3y - 8)$ — This is the given algebraic expression.

$= -3y + 8$ — Remove parentheses by changing the sign of every term inside the parentheses. ∎

☑ **CHECK POINT 6**   Simplify:

a. $-4(5x)$ **b.** $9x + x$ **c.** $13b - 14b$
d. $-7(3x - 4)$ **e.** $-(7y - 6)$.

EXAMPLE 7   Simplifying an Algebraic Expression

Simplify: $5(2y - 9) - (9y - 8)$.

Solution

$$5(2y - 9) - (9y - 8)$$ This is the given algebraic expression.

$$= 5 \cdot 2y - 5 \cdot 9 - (9y - 8)$$ Apply the distributive property over the first parentheses.

$$= 10y - 45 - (9y - 8)$$ Multiply.

$$= 10y - 45 - 9y + 8$$ Remove the remaining parentheses by changing the sign of each term within parentheses.

$$= (10y - 9y) + (-45 + 8)$$ Group like terms.

$$= 1y + (-37)$$ Combine like terms. For the variable terms, $10y - 9y = 10y + (-9y) = [10 + (-9)]y = 1y$.

$$= y + (-37)$$ Use the multiplication property of 1: $1y = y$.

$$= y - 37$$ Express addition of an opposite as subtraction. ∎

☑ **CHECK POINT 7**   Simplify: $4(3y - 7) - (13y - 2)$.

A Summary of Operations with Real Numbers

Operations with real numbers are summarized in **Table 1.7**.

Table 1.7   Summary of Operations with Real Numbers

Signs of Numbers	Addition	Subtraction	Multiplication	Division
Both numbers are positive. **Examples** 8 and 2 2 and 8	Sum is always positive. $8 + 2 = 10$ $2 + 8 = 10$	Difference may be either positive or negative. $8 - 2 = 6$ $2 - 8 = -6$	Product is always positive. $8 \cdot 2 = 16$ $2 \cdot 8 = 16$	Quotient is always positive. $8 \div 2 = 4$ $2 \div 8 = \frac{1}{4}$
One number is positive and the other number is negative. **Examples** 8 and −2 −8 and 2	Sum may be either positive or negative. $8 + (-2) = 6$ $-8 + 2 = -6$	Difference may be either positive or negative. $8 - (-2) = 10$ $-8 - 2 = -10$	Product is always negative. $8(-2) = -16$ $-8(2) = -16$	Quotient is always negative. $8 \div (-2) = -4$ $-8 \div 2 = -4$
Both numbers are negative. **Examples** −8 and −2 −2 and −8	Sum is always negative. $-8 + (-2) = -10$ $-2 + (-8) = -10$	Difference may be either positive or negative. $-8 - (-2) = -6$ $-2 - (-8) = 6$	Product is always positive. $-8(-2) = 16$ $-2(-8) = 16$	Quotient is always positive. $-8 \div (-2) = 4$ $-2 \div (-8) = \frac{1}{4}$

7 Determine whether a number is a solution of an equation.

To determine if a number is a solution of an equation, it is often necessary to perform more than one operation with real numbers.

EXAMPLE 8   Determining Whether a Number Is a Solution of an Equation

Determine whether -3 is a solution of

$$6x + 14 = -7 - x.$$

Solution

$$6x + 14 = -7 - x \qquad \text{This is the given equation.}$$

Is −3 a solution?

$$6(-3) + 14 \overset{?}{=} -7 - (-3) \qquad \text{Substitute } -3 \text{ for } x.$$

$$-18 + 14 \overset{?}{=} -7 - (-3) \qquad \text{Multiply: } 6(-3) = -18.$$

$$-18 + 14 \overset{?}{=} -7 + 3 \qquad \text{Rewrite subtraction as addition of an opposite.}$$

This statement is true $\qquad -4 = -4 \qquad \text{Add: } -18 + 14 = -4 \text{ and } -7 + 3 = -4.$

Because the values on both sides of the equation are the same, −3 is a solution of the equation. ■

✓ **CHECK POINT 8** Determine whether −3 is a solution of $2x - 5 = 8x + 7$.

---

8 Use mathematical models involving multiplication and division.

Applications

Mathematical models frequently involve multiplication and division.

EXAMPLE 9 Big (Lack of) Men on Campus

In 2007, 135 women received bachelor's degrees for every 100 men. According to the U.S. Department of Education, that gender imbalance will widen in the coming years, as shown by the bar graphs in **Figure 1.14**.

Percentage of College Degrees Awarded to United States Men and Women

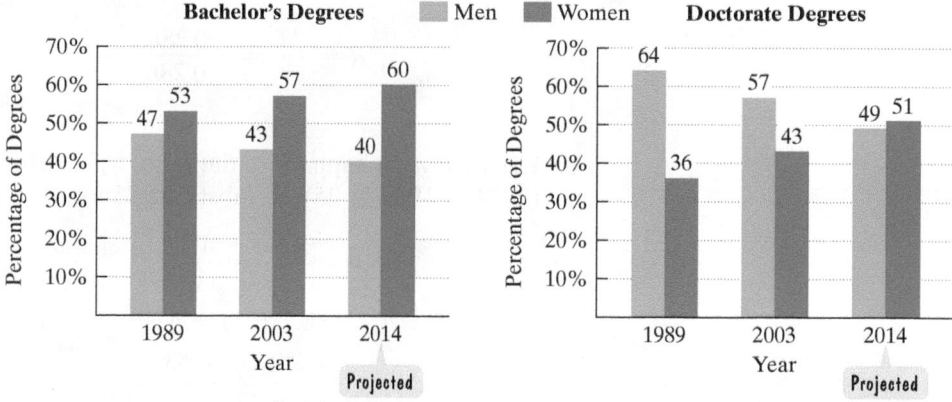

Figure 1.14

Source: U.S. Department of Education

The data for bachelor's degrees can be described by the following mathematical models:

Percentage of bachelor's degrees awarded to men $\qquad M = -0.28n + 47$

Number of years after 1989

Percentage of bachelor's degrees awarded to women $\qquad W = 0.28n + 53.$

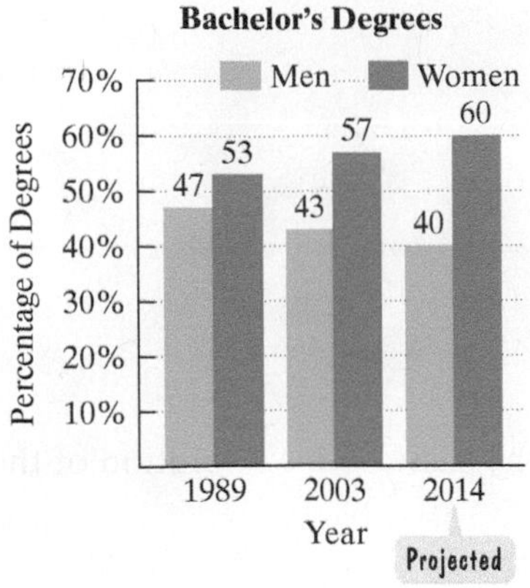

Percentage of College Degrees Awarded to United States Men and Women

Bachelor's Degrees

Figure 1.14 (repeated)

a. According to the first formula, $M = -0.28n + 47$, what percentage of bachelor's degrees were awarded to men in 2003? Does this underestimate or overestimate the actual percent shown in **Figure 1.14** for 2003? By how much?

b. A **ratio** is a comparison by division. Write a formula that describes the ratio of the percentage of bachelor's degrees received by men, $M = -0.28n + 47$, to the percentage of bachelor's degrees received by women, $W = 0.28n + 53$. Name this new mathematical model R, for ratio.

c. Use the formula for R to find the projected ratio of bachelor's degrees received by men to degrees received by women in 2014.

Solution

a. We are interested in the percentage of bachelor's degrees awarded to men in 2003. The year 2003 is 14 years after 1989: $2003 - 1989 = 14$. We substitute 14 for n in the formula for men and evaluate the formula.

$$M = -0.28n + 47$$ This is the model for the percentage of degrees awarded to men.

$$M = -0.28(14) + 47$$ Replace n with 14.

$$M = -3.92 + 47$$ Multiply: $-0.28(14) = -3.92$.

$$M = 43.08$$ Add. Use the sign of 47 and subtract absolute values:

$$\begin{array}{r} {\scriptstyle 6\ 9\ 10} \\ 47.0\cancel{0} \\ -\ \ 3.92 \\ \hline 43.08 \end{array}$$

The model indicates that 43.08% of bachelor's degrees were awarded to men in 2003. Because the number displayed in **Figure 1.14** is 43%, the formula overestimates the actual percent by 0.08%.

b. Now we use division to write a model comparing the ratio, R, of bachelor's degrees received by men to degrees received by women.

$$R = \frac{M}{W} = \frac{-0.28n + 47}{0.28n + 53}$$

Formula for percentage of degrees awarded to men

Formula for percentage of degrees awarded to women

c. Let's see what happens to the ratio, R, in 2014. The year 2014 is 25 years after 1989 ($2014 - 1989 = 25$), so substitute 25 for n.

$$R = \frac{-0.28n + 47}{0.28n + 53}$$ This is the model for the ratio of the percentage of degrees awarded to men and to women.

$$R = \frac{-0.28(25) + 47}{0.28(25) + 53}$$ Replace each occurrence of n with 25.

$$R = \frac{-7 + 47}{7 + 53}$$ Multiply: $-0.28(25) = -7$ and $0.28(25) = 7$.

The models for M and W give the projected percents for 2014 in **Figure 1.14**.

$$R = \frac{40}{60}$$ Add: $-7 + 47 = 40$ and $7 + 53 = 60$.

$$R = \frac{2}{3}$$ Reduce: $\frac{40}{60} = \frac{2 \cdot 20}{3 \cdot 20} = \frac{2}{3}$.

Our model for R, consistent with the actual data, shows that in 2014, the ratio of bachelor's degrees received by men to degrees received by women is projected to be 2 to 3. If model breakdown does not occur, three women will receive bachelor's degrees for every two men. ∎

✓ **CHECK POINT 9**   The data for doctorate degrees shown in **Figure 1.14** on page 81 can be described by $M = -0.6n + 64.4$, where M is the percentage of doctorate degrees awarded to men n years after 1989. According to this mathematical model, what percentage of doctorate degrees are projected to be received by men in 2014? Does this underestimate or overestimate the projection shown in **Figure 1.14**? By how much?

Achieving Success

Ask! Ask! Ask!
Do not be afraid to ask questions in class. Your professor may not realize that a concept is unclear until you raise your hand and ask a question. Other students who have problems asking questions in class will be appreciative that you have spoken up. Be polite and professional, but ask as many questions as required.

CONCEPT AND VOCABULARY CHECK

Use the choices below to fill in each blank so that the resulting statement is true:

positive   negative   0   undefined.

1. The product of two negative numbers is a/an _____ number.

2. The product of a negative number and a positive number is a/an _____ number.

3. The product of three negative numbers is a/an _____ number.

4. The product of an even number of negative numbers is a/an _____ number.

5. The product of two negative numbers and 0 is _____.

6. The multiplicative inverse, or reciprocal, of a negative number is a/an _____ number.

7. The quotient of a positive number and a negative number is a/an _____ number.

8. The quotient of two negative numbers is a/an _____ number.

9. The quotient of 0 and a negative number is _____.

10. The quotient of a negative number and 0 is _____.

11. The opposite of a negative number is a/an _____ number.

1.7 EXERCISE SET

Watch the videos in MyMathLab   Download the MyDashBoard App

Practice Exercises

In Exercises 1–34, perform the indicated multiplication.

1. $5(-9)$
2. $10(-7)$
3. $(-8)(-3)$
4. $(-9)(-5)$
5. $(-3)(7)$
6. $(-4)(8)$
7. $(-19)(-1)$
8. $(-11)(-1)$
9. $0(-19)$
10. $0(-11)$
11. $\frac{1}{2}(-24)$
12. $\frac{1}{3}(-21)$
13. $\left(-\frac{3}{4}\right)(-12)$
14. $\left(-\frac{4}{5}\right)(-30)$
15. $-\frac{3}{5} \cdot \left(-\frac{4}{7}\right)$
16. $-\frac{5}{7} \cdot \left(-\frac{3}{8}\right)$

17. $-\frac{7}{9} \cdot \frac{2}{3}$
18. $-\frac{5}{11} \cdot \frac{2}{7}$
19. $3(-1.2)$
20. $4(-1.2)$
21. $-0.2(-0.6)$
22. $-0.3(-0.7)$
23. $(-5)(-2)(3)$
24. $(-6)(-3)(10)$
25. $(-4)(-3)(-1)(6)$ -72
26. $(-2)(-7)(-1)(3)$
27. $-2(-3)(-4)(-1)$
28. $-3(-2)(-5)(-1)$
29. $(-3)(-3)(-3)$
30. $(-4)(-4)(-4)$
31. $5(-3)(-1)(2)(3)$
32. $2(-5)(-2)(3)(1)$
33. $(-8)(-4)(0)(-17)(-6)$
34. $(-9)(-12)(-18)(0)(-3)$

In Exercises 35–42, find the multiplicative inverse of each number.

35. 4   **36.** 3   **37.** $\dfrac{1}{5}$

38. $\dfrac{1}{7}$ **39.** -10 **40.** -12

41. $-\dfrac{2}{5}$ **42.** $-\dfrac{4}{9}$

In Exercises 43–46,

    a. *Rewrite the division as multiplication involving a multiplicative inverse.*

    b. *Use the multiplication from part (a) to find the given quotient.*

43. $-32 \div 4$

44. $-18 \div 6$

45. $\dfrac{-60}{-5}$

46. $\dfrac{-30}{-5}$

In Exercises 47–76, perform the indicated division or state that the expression is undefined.

47. $\dfrac{12}{-4}$ **48.** $\dfrac{40}{-5}$ **49.** $\dfrac{-21}{3}$

50. $\dfrac{-60}{6}$ **51.** $\dfrac{-90}{-3}$ **52.** $\dfrac{-66}{-6}$

53. $\dfrac{0}{-7}$ **54.** $\dfrac{0}{-8}$ **55.** $\dfrac{7}{0}$

56. $\dfrac{-8}{0}$ **57.** $-15 \div 3$ **58.** $-80 \div 8$

59. $120 \div (-10)$ **60.** $130 \div (-10)$

61. $(-180) \div (-30)$ **62.** $(-150) \div (-25)$

63. $0 \div (-4)$ **64.** $0 \div (-10)$

65. $-4 \div 0$ **66.** $-10 \div 0$

67. $\dfrac{-12.9}{3}$ **68.** $\dfrac{-21.6}{3}$

69. $-\dfrac{1}{2} \div \left(-\dfrac{3}{5}\right)$ **70.** $-\dfrac{1}{2} \div \left(-\dfrac{7}{9}\right)$

71. $-\dfrac{14}{9} \div \dfrac{7}{8}$ **72.** $-\dfrac{5}{16} \div \dfrac{25}{8}$

73. $\dfrac{1}{3} \div \left(-\dfrac{1}{3}\right)$ **74.** $\dfrac{1}{5} \div \left(-\dfrac{1}{5}\right)$

75. $6 \div \left(-\dfrac{2}{5}\right)$ **76.** $8 \div \left(-\dfrac{2}{9}\right)$

In Exercises 77–96, simplify each algebraic expression.

77. $-5(2x)$ **78.** $-9(3x)$

79. $-4\left(-\dfrac{3}{4}y\right)$ **80.** $-5\left(-\dfrac{3}{5}y\right)$

81. $8x + x$ **82.** $12x + x$

83. $-5x + x$ **84.** $-6x + x$

85. $6b - 7b$ **86.** $12b - 13b$

87. $-y + 4y$ **88.** $-y + 9y$

89. $-4(2x - 3)$ **90.** $-3(4x - 5)$

91. $-3(-2x + 4)$ **92.** $-4(-3x + 2)$

93. $-(2y - 5)$ **94.** $-(3y - 1)$

95. $4(2y - 3) - (7y + 2)$

96. $5(3y - 1) - (14y - 2)$

In Exercises 97–108, determine whether the given number is a solution of the equation.

97. $4x = 2x - 10; -5$

98. $5x = 3x - 6; -3$

99. $-7y + 18 = -10y + 6; -4$

100. $-4y + 21 = -7y + 15; -2$

101. $5(w + 3) = 2w - 21; -10$

102. $6(w + 2) = 4w - 10; -9$

103. $4(6 - z) + 7z = 0; -8$

104. $5(7 - z) + 12z = 0; -5$

105. $14 - 2x = -4x + 7; -2\dfrac{1}{2}$

106. $16 - 4x = -2x + 21; -3\dfrac{1}{2}$

107. $\dfrac{5m - 1}{6} = \dfrac{3m - 2}{4}; -4$

108. $\dfrac{6m - 5}{11} = \dfrac{3m - 2}{5}; -1$

Practice PLUS

In Exercises 109–116, write a numerical expression for each phrase. Then simplify the numerical expression by performing the given operations.

109. 8 added to the product of 4 and -10

110. 14 added to the product of 3 and -15

111. The product of -9 and -3, decreased by -2

112. The product of -6 and -4, decreased by -5

113. The quotient of -18 and the sum of -15 and 12

114. The quotient of -25 and the sum of -21 and 16

115. The difference between -6 and the quotient of 12 and -4

116. The difference between -11 and the quotient of 20 and -5

Application Exercises

In Exercises 117–118, use the formula $C = \dfrac{5}{9}(F - 32)$ to express each Fahrenheit temperature, F, as its equivalent Celsius temperature, C.

117. $-22°F$

118. $-31°F$

In the years after warning labels were put on cigarettes packs, the number of smokers dropped from approximately two in five adults to one in five. The bar graph shows the percentage of American adults who smoked cigarettes for selected years from 1965 through 2009.

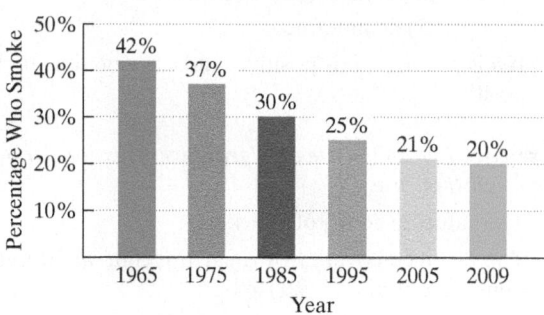

Percentage of American Adults Who Smoke Cigarettes

Source: Centers for Disease Control and Prevention

Here is a mathematical model that approximates the data displayed by the bar graph:

$$C = -0.5x + 41.$$

Percentage of Americans who smoked cigarettes | Number of years after 1965

Use this formula to solve Exercises 119–120.

119. a. Does the mathematical model underestimate or overestimate the percentage of American adults who smoked cigarettes in 2009? By how much?

b. Use the mathematical model to project the percentage of American adults who will smoke cigarettes in 2015.

120. a. Use the mathematical model to determine the percentage of American adults who smoked cigarettes in 2005. How does this compare with the actual percentage displayed by the bar graph?

b. Use the mathematical model to project the percentage of American adults who will smoke cigarettes in 2019.

The graph shows the average number of hours per week women and men in the United States devoted to household chores and child care for five selected years from 1965 through 2005.

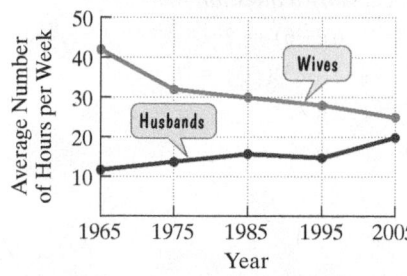

Time Devoted to Household Chores and Child Care

Source: Davis and Palladino, *Psychology*, Ninth Edition, Pearson, 2010.

The data can be described by the following mathematical models:

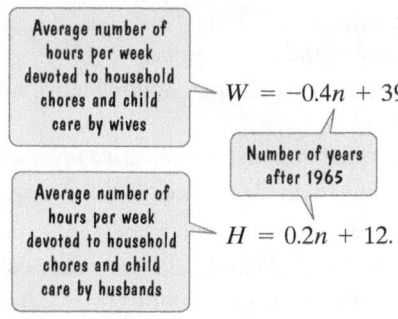

Average number of hours per week devoted to household chores and child care by wives — $W = -0.4n + 39$

Number of years after 1965

Average number of hours per week devoted to household chores and child care by husbands — $H = 0.2n + 12.$

Use this information to solve Exercises 121–122.

121. a. Use the appropriate graph to estimate the average number of hours per week devoted to household chores and child care by wives in 2005.

b. Use the appropriate formula to determine the average number of hours per week devoted to household chores and child care by wives in 2005. How does this compare with your estimate in part (a)?

c. Write a formula that describes the difference between the average number of hours per week devoted to household chores and child care between wives and husbands n years after 1965. Name this new mathematical model D, for difference. Then simplify the algebraic expression in the model.

d. Use the simplified form of the mathematical model from part (c) to determine the difference in the number of hours per week women and men spent on household chores and child care in 2005. Does this overestimate or underestimate the difference displayed by the graphs?

122. a. Use the appropriate graph to estimate the average number of hours per week devoted to household chores and child care by wives in 1995.

b. Use the appropriate formula to determine the average number of hours per week devoted to household chores and child care by wives in 1995. How does this compare with your estimate in part (a)?

c. Write a formula that describes the ratio of the average number of hours per week devoted to household chores and child care by husbands to the average number of hours per week devoted by wives. Name this new mathematical model R, for ratio.

d. Use the mathematical model from part (c) to determine the fraction of time men spent on household chores and child care in 1995 compared with women. Express the fraction reduced to its lowest terms.

Writing in Mathematics

123. Explain how to multiply two real numbers. Provide examples with your explanation.

124. Explain how to determine the sign of a product that involves more than two numbers.

125. Explain how to find the multiplicative inverse of a number.

126. Why is it that 0 has no multiplicative inverse?

127. Explain how to divide real numbers.

128. Why is division by zero undefined?

129. Explain how to simplify an algebraic expression in which a negative sign precedes parentheses.

130. Do you believe that the trend in the graphs showing a decline in male college attendance (**Figure 1.14** on page 81) should be reversed by providing admissions preferences for men? Explain your position on this issue.

Critical Thinking Exercises

Make Sense? *In Exercises 131–134, determine whether each statement "makes sense" or "does not make sense" and explain your reasoning.*

131. I've noticed that the sign rules for dividing real numbers are slightly different than the sign rules for multiplying real numbers.

132. Just as two negative factors give a positive product, I've seen the same thing occur with double negatives in English:

$$\boxed{\text{That rumor is not unknown}} \text{ means } \boxed{\text{That rumor is known.}}$$

133. This pattern suggests that multiplying two negative numbers results in a positive answer:

Decreasing by 1 from left to right

$$2(-3) = -6; \; 1(-3) = -3; \; 0(-3) = 0; \; -1(-3) = 3; \; -2(-3) = 6.$$

Increasing by 3 from left to right

134. When I used

$$R = \frac{M}{W} = \frac{-0.28n + 47}{0.28n + 53}$$

Number of years after 1989

to project the ratio of bachelor's degrees received by men to degrees received by women in 2020, I had to find the quotient of two negative numbers.

In Exercises 135–138, determine whether each statement is true or false. If the statement is false, make the necessary change(s) to produce a true statement.

135. Both the addition and the multiplication of two negative numbers result in a positive number.

136. Multiplying a negative number by a nonnegative number will always give a negative number.

137. $0 \div (-\sqrt{2})$ is undefined.

138. If a is negative, b is positive, and c is negative, then $\dfrac{a}{bc}$ is positive.

In Exercises 139–142, write an algebraic expression for the given English phrase.

139. The value, in cents, of x nickels

140. The distance covered by a car traveling at 50 miles per hour for x hours

141. The monthly salary, in dollars, for a person earning x dollars per year

142. The fraction of people in a room who are women if there are 40 women and x men in the room

Technology Exercises

143. Use a calculator to verify any five of the products that you found in Exercises 1–34.

144. Use a calculator to verify any five of the quotients that you found in Exercises 47–76.

145. Simplify using a calculator:
$$0.3(4.7x - 5.9) - 0.07(3.8x - 61).$$

146. Use your calculator to attempt to find the quotient of -3 and 0. Describe what happens. Does the same thing occur when finding the quotient of 0 and -3? Explain the difference. Finally, what happens when you enter the quotient of 0 and itself?

Review Exercises

In Exercises 147–149, perform the indicated operation.

147. $-6 + (-3)$ (Section 1.5, Example 3)

148. $-6 - (-3)$ (Section 1.6, Example 1)

149. $-6 \div (-3)$ (Section 1.7, Example 4)

Preview Exercises

Exercises 150–152 will help you prepare for the material covered in the next section. In each exercise, an expression with an exponent is written as a repeated multiplication. Find this product, indicated by a question mark.

150. $(-6)^2 = (-6)(-6) = \;?$

151. $(-5)^3 = (-5)(-5)(-5) = \;?$

152. $(-2)^4 = (-2)(-2)(-2)(-2) = \;?$

Exponents and Order of Operations

Eat, drink, and be merry—for tomorrow we diet. But is there a way to avoid the extremes of merriment and dieting? In this section, we continue to see how mathematical models describe your world, including formulas that provide daily caloric needs for maintaining energy balance based on age and lifestyle.

Objectives

1 Evaluate exponential expressions.

2 Simplify algebraic expressions with exponents.

3 Use the order of operations agreement.

4 Evaluate mathematical models.

1 Evaluate exponential expressions.

Natural Number Exponents

Although people do a great deal of talking, the total output since the beginning of gabble to the present day, including all baby talk, love songs, and congressional debates, only amounts to about 10 million billion words. This can be expressed as 16 factors of 10, or 10^{16} words.

Exponents such as 2, 3, 4, and so on are used to indicate repeated multiplication. For example,

$$10^2 = 10 \cdot 10 = 100,$$
$$10^3 = 10 \cdot 10 \cdot 10 = 1000, \quad 10^4 = 10 \cdot 10 \cdot 10 \cdot 10 = 10,000.$$

The 10 that is repeated when multiplying is called the **base**. The small numbers above and to the right of the base are called **exponents** or **powers**. The exponent tells the number of times the base is to be used when multiplying. In 10^3, the base is 10 and the exponent is 3.

Any number with an exponent of 1 is the number itself. Thus, $10^1 = 10$.

Multiplications that are expressed in exponential notation are read as follows:

10^1:   "ten to the first power"

10^2:   "ten to the second power" or "ten squared"

10^3:   "ten to the third power" or "ten cubed"

10^4:   "ten to the fourth power"

10^5:   "ten to the fifth power"

etc.

Any real number can be used as the base. Thus,

$$7^2 = 7 \cdot 7 = 49 \quad \text{and} \quad (-3)^4 = (-3)(-3)(-3)(-3) = 81.$$

The bases are 7 and -3, respectively. Do not confuse $(-3)^4$ and -3^4.

$$-3^4 = -(3 \cdot 3 \cdot 3 \cdot 3) = -81$$

The negative is not taken to the power because it is not inside parentheses.

An exponent applies only to a base. A negative sign is not part of a base unless it appears in parentheses.

> **EXAMPLE 1**   Evaluating Exponential Expressions

Evaluate:

a. 4^2 **b.** $(-5)^3$ **c.** $(-2)^4$ **d.** -2^4.

Solution

> Exponent is 2.

a. $4^2 = 4 \cdot 4 = 16$ The exponent indicates that the base is used as a factor two times.

> Base is 4.

We read $4^2 = 16$ as "4 to the second power is 16" or "4 squared is 16."

> Exponent is 3.

b. $(-5)^3 = (-5)(-5)(-5)$ The exponent indicates that the base is used as a factor three times.

> Base is −5.

$\quad\quad = -125$ An odd number of negative factors yields a negative product.

We read $(-5)^3 = -125$ as "the number negative 5 to the third power is negative 125" or "negative 5 cubed is negative 125."

> Exponent is 4.

c. $(-2)^4 = (-2)(-2)(-2)(-2)$ The exponent indicates that the base is used as a factor four times.

> Base is −2.

$\quad\quad = 16$ An even number of negative factors yields a positive product.

We read $(-2)^4 = 16$ as "the number negative 2 to the fourth power is 16."

> Exponent is 4.

d. $-2^4 = -(2 \cdot 2 \cdot 2 \cdot 2)$ The negative is not inside parentheses and is not taken to the fourth power.

> Base is 2.

$\quad\quad = -16$ Multiply the twos and copy the negative.

We read $-2^4 = -16$ as "the negative of 2 raised to the fourth power is negative 16" or "the opposite, or additive inverse, of 2 raised to the fourth power is negative 16." ∎

✓ **CHECK POINT 1**   Evaluate:

a. 6^2 **b.** $(-4)^3$ **c.** $(-1)^4$ **d.** -1^4.

The formal algebraic definition of a natural number exponent summarizes our discussion:

Definition of a Natural Number Exponent

If b is a real number and n is a natural number,

$$b^n = \underbrace{b \cdot b \cdot b \cdots b.}_{b \text{ appears as a factor } n \text{ times.}}$$

b^n is read "the nth power of b" or "b to the nth power." Thus, the nth power of b is defined as the product of n factors of b. The expression b^n is called an **exponential expression**.

Furthermore, $b^1 = b$.

Blitzer Bonus

Integers, Karma, and Exponents

On Friday the 13th, are you a bit more careful crossing the street even if you don't consider yourself superstitious? Numerology, the belief that certain integers have greater significance and can be lucky or unlucky, is widespread in many cultures.

Integer	Connotation	Culture	Origin	Example
4	Negative	Chinese	The word for the number 4 sounds like the word for death.	Many buildings in China have floor-numbering systems that skip 40–49.
7	Positive	United States	In dice games, this prime number is the most frequently rolled number with two dice.	There was a spike in the number of couples getting married on 7/7/07.
8	Positive	Chinese	It's considered a sign of prosperity.	The Beijing Olympics began at 8 P.M. on 8/8/08.
13	Negative	Various	Various reasons, including the number of people at the Last Supper	Many buildings around the world do not label any floor "13."
18	Positive	Jewish	The Hebrew letters spelling *chai*, or living, are the 8th and 10th in the alphabet, adding up to 18	Monetary gifts for celebrations are often given in multiples of 18.
666	Negative	Christian	The New Testament's Book of Revelation identifies 666 as the "number of the beast," which some say refers to Satan.	In 2008, Reeves, Louisiana, eliminated 666 as the prefix of its phone numbers.

Although your author is not a numerologist, he is intrigued by curious exponential representations for 666:

$$666 = 6 + 6 + 6 + 6^3 + 6^3 + 6^3$$
$$666 = 1^3 + 2^3 + 3^3 + 4^3 + 5^3 + 6^3 + 5^3 + 4^3 + 3^3 + 2^3 + 1^3$$
$$666 = 2^2 + 3^2 + 5^2 + 7^2 + 11^2 + 13^2 + 17^2$$

Sum of the squares of the first seven prime numbers

$$666 = 1^6 - 2^6 + 3^6.$$

2 Simplify algebraic expressions with exponents.

Exponents and Algebraic Expressions

The distributive property can be used to simplify certain algebraic expressions that contain exponents. For example, we can use the distributive property to combine like terms in the algebraic expression $4x^2 + 6x^2$:

$$4x^2 + 6x^2 = (4 + 6)x^2 = 10x^2.$$

First term with variable factor x^2 Second term with variable factor x^2 The common variable factor is x^2.

Great Question!

When I add like terms, do I add exponents?

When adding algebraic expressions, if you have like terms you add only the numerical coefficients—not the exponents. **Exponents are never added when the operation is addition.** Avoid these common errors.

Incorrect

- $7x^3 + 2x^3 = 9x^6$
- $5x^2 + x^2 = 6x^4$
- $3x^2 + 4x^3 = 7x^5$

EXAMPLE 2   Simplifying Algebraic Expressions

Simplify, if possible:

a. $7x^3 + 2x^3$ **b.** $5x^2 + x^2$ **c.** $3x^2 + 4x^3$.

Solution

a. $7x^3 + 2x^3$ There are two like terms with the same variable factor, namely x^3.
$= (7 + 2)x^3$ Apply the distributive property.
$= 9x^3$ Add within parentheses.

b. $5x^2 + x^2$ There are two like terms with the same variable factor, namely x^2.
$= 5x^2 + 1x^2$ Use the multiplication property of 1.
$= (5 + 1)x^2$ Apply the distributive property.
$= 6x^2$ Add within parentheses.

c. $3x^2 + 4x^3$ cannot be simplified. The terms $3x^2$ and $4x^3$ are not like terms because they have different variable factors, namely x^2 and x^3. ∎

✓ **CHECK POINT 2**   Simplify, if possible:

a. $16x^2 + 5x^2$ **b.** $7x^3 + x^3$ **c.** $10x^2 + 8x^3$.

3 Use the order of operations agreement.

Order of Operations

Suppose that you want to find the value of $3 + 7 \cdot 5$. Which procedure shown is correct?

$$3 + 7 \cdot 5 = 3 + 35 = 38 \quad \text{or} \quad 3 + 7 \cdot 5 = 10 \cdot 5 = 50$$

You know the answer because we introduced certain rules, called the **order of operations**, at the beginning of the chapter. One of these rules stated that if a problem contains no parentheses or other grouping symbols, perform multiplication before addition. Thus, the procedure on the left is correct because the multiplication of 7 and 5 is done first. Then the addition is performed. The correct answer is 38.

Some problems contain grouping symbols, such as parentheses, ( ); brackets, [ ]; braces, { }; absolute value symbols, | |; or fraction bars. These grouping symbols tell us what to do first. Here are two examples:

- $(3 + 7) \cdot 5 = 10 \cdot 5 = 50$

First, perform operations in grouping symbols.

- $8|6 - 16| = 8|-10| = 8 \cdot 10 = 80.$

Here are the rules for determining the order in which operations should be performed:

Order of Operations

1. Perform all operations within grouping symbols.
2. Evaluate all exponential expressions.
3. Do all multiplications and divisions in the order in which they occur, working from left to right.
4. Finally, do all additions and subtractions using one of the following procedures:

   a. Work from left to right and do additions and subtractions in the order in which they occur.

   or

   b. Rewrite subtractions as additions of opposites. Combine positive and negative numbers separately, and then add these results.

The last step in the order of operations indicates that you have a choice when working with additions and subtractions, although in this section we will perform these operations from left to right. However, when working with multiplications and divisions, you must perform these operations *as they occur* from left to right. For example,

$$8 \div 4 \cdot 2 = 2 \cdot 2 = 4$$

Do the division first because it occurs first.

$$8 \cdot 4 \div 2 = 32 \div 2 = 16.$$

Do the multiplication first because it occurs first.

EXAMPLE 3 Using the Order of Operations

Simplify: $18 + 2 \cdot 3 - 10$.

Solution There are no grouping symbols or exponential expressions. In cases like this, we multiply and divide before adding and subtracting.

$$18 + 2 \cdot 3 - 10 = 18 + 6 - 10 \qquad \text{Multiply: } 2 \cdot 3 = 6.$$
$$= 24 - 10 \qquad \text{Add and subtract from left to right: } 18 + 6 = 24.$$
$$= 14 \qquad \text{Subtract: } 24 - 10 = 14. \blacksquare$$

✓ **CHECK POINT 3** Simplify: $20 + 4 \cdot 3 - 17$.

EXAMPLE 4 Using the Order of Operations

Simplify: $6^2 - 24 \div 2^2 \cdot 3 - 1$.

Solution There are no grouping symbols. Thus, we begin by evaluating exponential expressions. Then we multiply or divide. Finally, we add or subtract.

$$6^2 - 24 \div 2^2 \cdot 3 - 1$$
$$= 36 - 24 \div 4 \cdot 3 - 1 \qquad \text{Evaluate exponential expressions: } 6^2 = 6 \cdot 6 = 36 \text{ and } 2^2 = 2 \cdot 2 = 4.$$
$$= 36 - 6 \cdot 3 - 1 \qquad \text{Perform the multiplications and divisions from left to right. Start with } 24 \div 4 = 6.$$
$$= 36 - 18 - 1 \qquad \text{Now do the multiplication: } 6 \cdot 3 = 18.$$
$$= 18 - 1 \qquad \text{Finally, perform the subtraction from left to right: } 36 - 18 = 18.$$
$$= 17 \qquad \text{Complete the subtraction: } 18 - 1 = 17. \blacksquare$$

✓ **CHECK POINT 4** Simplify: $7^2 - 48 \div 4^2 \cdot 5 - 2$.

EXAMPLE 5 Using the Order of Operations

Simplify:

a. $(2 \cdot 5)^2$ **b.** $2 \cdot 5^2$.

Solution

a. Because $(2 \cdot 5)^2$ contains grouping symbols, namely parentheses, we perform the operation within parentheses first.

$$(2 \cdot 5)^2 = 10^2 \qquad \text{Multiply within parentheses: } 2 \cdot 5 = 10.$$
$$= 100 \qquad \text{Evaluate the exponential expression: } 10^2 = 10 \cdot 10 = 100.$$

b. Because $2 \cdot 5^2$ does not contain grouping symbols, we begin by evaluating the exponential expression.

$$2 \cdot 5^2 = 2 \cdot 25 \qquad \text{Evaluate the exponential expression: } 5^2 = 5 \cdot 5 = 25.$$
$$= 50 \qquad \text{Now do the multiplication: } 2 \cdot 25 = 50. \quad \blacksquare$$

✓ **CHECK POINT 5** Simplify:

a. $(3 \cdot 2)^2$ **b.** $3 \cdot 2^2$.

EXAMPLE 6 Using the Order of Operations

Simplify: $\left(\dfrac{1}{2}\right)^3 - \left(\dfrac{1}{2} - \dfrac{3}{4}\right)^2 (-4)$.

Solution Because grouping symbols appear, we perform the operation within parentheses first.

$$\left(\frac{1}{2}\right)^3 - \left(\frac{1}{2} - \frac{3}{4}\right)^2 (-4)$$

$$= \left(\frac{1}{2}\right)^3 - \left(-\frac{1}{4}\right)^2 (-4) \qquad \text{Work inside parentheses first:}$$
$$\frac{1}{2} - \frac{3}{4} = \frac{2}{4} - \frac{3}{4} = \frac{2}{4} + \left(-\frac{3}{4}\right) = -\frac{1}{4}.$$

$$= \frac{1}{8} - \frac{1}{16}(-4) \qquad \text{Evaluate exponential expressions:}$$
$$\left(\frac{1}{2}\right)^3 = \frac{1}{2} \cdot \frac{1}{2} \cdot \frac{1}{2} = \frac{1}{8} \text{ and } \left(-\frac{1}{4}\right)^2 = \left(-\frac{1}{4}\right)\left(-\frac{1}{4}\right) = \frac{1}{16}.$$

$$= \frac{1}{8} - \left(-\frac{1}{4}\right) \qquad \text{Multiply: } \frac{1}{16} \cdot \left(\frac{-4}{1}\right) = -\frac{4}{16} = -\frac{1}{4}.$$

$$= \frac{3}{8} \qquad \text{Subtract: } \frac{1}{8} - \left(-\frac{1}{4}\right) = \frac{1}{8} + \frac{1}{4} = \frac{1}{8} + \frac{2}{8} = \frac{3}{8}. \quad \blacksquare$$

✓ **CHECK POINT 6** Simplify: $\left(-\dfrac{1}{2}\right)^2 - \left(\dfrac{7}{10} - \dfrac{8}{15}\right)^2 (-18)$.

Some expressions contain many grouping symbols. An example of such an expression is $2[5(4 - 7) + 9]$. The grouping symbols are the parentheses and the brackets.

The parentheses, the innermost grouping symbols, group $4 - 7$.

$$2[5(4 - 7) + 9]$$

The brackets, the outermost grouping symbols, group $5(4 - 7) + 9$.

When combinations of grouping symbols appear, **perform operations within the innermost grouping symbols first**. Then work to the outside, performing operations within the outermost grouping symbols.

EXAMPLE 7   Using the Order of Operations

Simplify: $2[5(4 - 7) + 9]$.

Solution

$2[5(4 - 7) + 9]$

$= 2[5(-3) + 9]$ Work inside parentheses first:
$\quad\quad\quad\quad\quad\quad$ $4 - 7 = 4 + (-7) = -3$.

$= 2[-15 + 9]$ Work inside brackets and multiply: $5(-3) = -15$.

$= 2[-6]$ Add inside brackets: $-15 + 9 = -6$. The resulting
$\quad\quad\quad\quad$ problem can also be expressed as $2(-6)$.

$= -12$ Multiply: $2[-6] = -12$. ∎

Parentheses can be used for both innermost and outermost grouping symbols. For example, the expression $2[5(4 - 7) + 9]$ can also be written $2(5(4 - 7) + 9)$. However, too many parentheses can be confusing. The use of both parentheses and brackets makes it easier to identify inner and outer groupings.

✓ **CHECK POINT 7**   Simplify: $4[3(6 - 11) + 5]$.

EXAMPLE 8   Using the Order of Operations

Simplify: $18 \div 6 + 4[5 + 2(8 - 10)^3]$.

Solution

$18 \div 6 + 4[5 + 2(8 - 10)^3]$

$= 18 \div 6 + 4[5 + 2(-2)^3]$ Work inside parentheses first:
$\quad\quad\quad\quad\quad\quad\quad\quad\quad\quad$ $8 - 10 = 8 + (-10) = -2$.

$= 18 \div 6 + 4[5 + 2(-8)]$ Work inside brackets and evaluate
$\quad\quad\quad\quad\quad\quad\quad\quad\quad\quad$ the exponential expression:
$\quad\quad\quad\quad\quad\quad\quad\quad\quad\quad$ $(-2)^3 = (-2)(-2)(-2) = -8$.

$= 18 \div 6 + 4[5 + (-16)]$ Work inside brackets and multiply:
$\quad\quad\quad\quad\quad\quad\quad\quad\quad\quad$ $2(-8) = -16$.

$= 18 \div 6 + 4[-11]$ Work inside brackets and add:
$\quad\quad\quad\quad\quad\quad\quad\quad\quad\quad$ $5 + (-16) = -11$.

$= 3 + 4[-11]$ Perform the multiplications and divisions
$\quad\quad\quad\quad\quad\quad\quad\quad\quad\quad$ from left to right. Start with $18 \div 6 = 3$.

$= 3 + (-44)$ Now do the multiplication: $4(-11) = -44$.

$= -41$ Finally, perform the addition:
$\quad\quad\quad\quad\quad\quad\quad\quad\quad\quad$ $3 + (-44) = -41$. ∎

✓ **CHECK POINT 8**   Simplify: $25 \div 5 + 3[4 + 2(7 - 9)^3]$.

Fraction bars are grouping symbols that separate expressions into two parts, the numerator and the denominator. Consider, for example,

The numerator is one
part of the expression.

The fraction bar is the
grouping symbol. $\dfrac{2(3 - 12) + 6 \cdot 4}{2^4 + 1}$.

The denominator is the other
part of the expression.

We can use brackets instead of the fraction bar. An equivalent expression for $\dfrac{2(3 - 12) + 6 \cdot 4}{2^4 + 1}$ is

$$[2(3 - 12) + 6 \cdot 4] \div [2^4 + 1].$$

The grouping suggests a method for simplifying expressions with fraction bars as grouping symbols:

- Simplify the numerator.
- Simplify the denominator.
- If possible, simplify the fraction.

EXAMPLE 9   Using the Order of Operations

Simplify: $\dfrac{2(3 - 12) + 6 \cdot 4}{2^4 + 1}$.

Solution

$$\dfrac{2(3 - 12) + 6 \cdot 4}{2^4 + 1}$$

$$= \dfrac{2(-9) + 6 \cdot 4}{16 + 1}$$ Work inside parentheses in the numerator: $3 - 12 = 3 + (-12) = -9$. Evaluate the exponential expression in the denominator: $2^4 = 2 \cdot 2 \cdot 2 \cdot 2 = 16$.

$$= \dfrac{-18 + 24}{16 + 1}$$ Multiply in the numerator: $2(-9) = -18$ and $6 \cdot 4 = 24$.

$$= \dfrac{6}{17}$$ Perform the addition in the numerator and in the denominator. ∎

✓ **CHECK POINT 9**   Simplify: $\dfrac{5(4 - 9) + 10 \cdot 3}{2^3 - 1}$.

EXAMPLE 10   Using the Order of Operations

Evaluate: $-x^2 - 7x$ for $x = -2$.

Solution   We begin by substituting -2 for each occurrence of x in the algebraic expression. Then we use the order of operations to evaluate the expression.

$$-x^2 - 7x$$

Replace x with -2. Place parentheses around -2.

$$= -(-2)^2 - 7(-2)$$

$$= -4 - 7(-2)$$ Evaluate the exponential expression: $(-2)^2 = (-2)(-2) = 4$.

$$= -4 - (-14)$$ Multiply: $7(-2) = -14$.

$$= 10$$ Subtract: $-4 - (-14) = -4 + 14 = 10$. ∎

✓ **CHECK POINT 10**   Evaluate: $-x^2 - 4x$ for $x = -5$.

Some algebraic expressions contain two sets of grouping symbols. Using the order of operations, grouping symbols are removed from innermost (parentheses) to outermost (brackets).

EXAMPLE 11   Simplifying an Algebraic Expression

Simplify: $18x^2 + 4 - [6(x^2 - 2) + 5]$.

Solution

$18x^2 + 4 - [6(x^2 - 2) + 5]$

$= 18x^2 + 4 - [6x^2 - 12 + 5]$ Use the distributive property to remove parentheses:

$6(x^2 - 2) = 6x^2 - 6 \cdot 2 = 6x^2 - 12.$

$= 18x^2 + 4 - [6x^2 - 7]$ Add inside brackets: $-12 + 5 = -7$.

$= 18x^2 + 4 - 6x^2 + 7$ Remove brackets by changing the sign of each term within brackets.

$= (18x^2 - 6x^2) + 4 + 7$ Group like terms.

$= 12x^2 + 11$ Combine like terms.   ■

✓ **CHECK POINT 11**   Simplify: $14x^2 + 5 - [7(x^2 - 2) + 4]$.

4 Evaluate mathematical models.

Applications

In Examples 12 and 13, we use the order of operations to evaluate mathematical models.

EXAMPLE 12   Modeling Caloric Needs

The bar graph in **Figure 1.15** shows the estimated number of calories per day needed to maintain energy balance for various gender and age groups for moderately active lifestyles. (Moderately active means a lifestyle that includes physical activity equivalent to walking 1.5 to 3 miles per day at 3 to 4 miles per hour, in addition to the light physical activity associated with typical day-to-day life.)

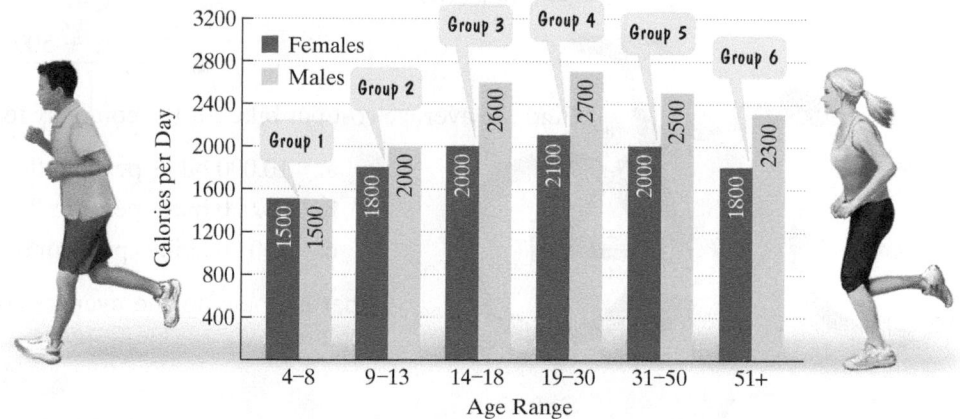

Calories Needed to Maintain Energy Balance for Moderately Active Lifestyles

Figure 1.15

Source: U.S.D.A.

The mathematical model

$$F = -66x^2 + 526x + 1030$$

describes the number of calories needed per day, F, by females in age group x with moderately active lifestyles. According to the model, how many calories per day are needed by females between the ages of 19 and 30, inclusive, with this lifestyle? Does this underestimate or overestimate the number shown by the graph in **Figure 1.15**? By how much?

Solution Because the 19–30 age range is designated as group 4, we substitute 4 for x in the given model. Then we use the order of operations to find F, the number of calories needed per day by females between the ages of 19 and 30.

$F = -66x^2 + 526x + 1030$	This is the given mathematical model.
$F = -66 \cdot 4^2 + 526 \cdot 4 + 1030$	Replace each occurrence of x with 4.
$F = -66 \cdot 16 + 526 \cdot 4 + 1030$	Evaluate the exponential expression: $4^2 = 4 \cdot 4 = 16$.
$F = -1056 + 2104 + 1030$	Multiply from left to right: $-66 \cdot 16 = -1056$ and $526 \cdot 4 = 2104$.
$F = 2078$	Add: $-1056 + 2104 = 1048$ and $1048 + 1030 = 2078$.

The formula indicates that females in the 19–30 age range with moderately active lifestyles need 2078 calories per day. **Figure 1.15** on the previous page indicates that 2100 calories are needed. Thus, the mathematical model underestimates caloric needs by $2100 - 2078$ calories, or by 22 calories per day. ∎

Great Question!

In the solution to Example 12, $-1056 + 2104 + 1030$ was simplified by performing the addition from left to right. Do I have to do it that way?

No. Here's another way to simplify the expression. First add the positive numbers

$$2104 + 1030 = 3134$$

and then add the negative number to that result:

$$3134 + (-1056) = 2078.$$

✓ **CHECK POINT 12** The mathematical model

$$M = -120x^2 + 998x + 590$$

describes the number of calories needed per day, M, by males in age group x with moderately active lifestyles. According to the model, how many calories per day are needed by males between the ages of 19 and 30, inclusive, with this lifestyle? Does this underestimate or overestimate the number shown by the graph in **Figure 1.15** on the previous page? By how much?

EXAMPLE 13 Shrinky-Dink Size without Rinky-Dink Style

A company has decided to jump into the pedal game with foldable, ultraportable bikes. Its fixed monthly costs are $500,000 and it will cost $400 to manufacture each bike. The average cost per bike, $\overline{C}$, for the company to manufacture x foldable bikes per month is modeled by

$$\overline{C} = \frac{400x + 500{,}000}{x}.$$

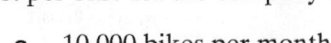

Find the average cost per bike for the company to manufacture

   a. 10,000 bikes per month.

   b. 50,000 bikes per month.

   c. 100,000 bikes per month.

What happens to the average cost per bike as the production level increases?

Solution

   a. We are interested in the average cost per bike for the company if 10,000 bikes are manufactured per month. Because x represents the number of bikes manufactured per month, we substitute 10,000 for x in the given mathematical model.

$$\overline{C} = \frac{400x + 500{,}000}{x} = \frac{400(10{,}000) + 500{,}000}{10{,}000} = \frac{4{,}000{,}000 + 500{,}000}{10{,}000}$$

$$= \frac{4{,}500{,}000}{10{,}000} = 450$$

The average cost per bike of producing 10,000 bikes per month is $450.

b. Now, 50,000 bikes are manufactured per month. We find the average cost per bike by substituting 50,000 for x in the given mathematical model.

$$\overline{C} = \frac{400x + 500,000}{x} = \frac{400(50,000) + 500,000}{50,000} = \frac{20,000,000 + 500,000}{50,000}$$

$$= \frac{20,500,000}{50,000} = 410$$

The average cost per bike of producing 50,000 bikes per month is $410.

c. Finally, the production level has increased to 100,000 bikes per month. We find the average cost per bike for the company by substituting 100,000 for x in the given mathematical model.

$$\overline{C} = \frac{400x + 500,000}{x} = \frac{400(100,000) + 500,000}{100,000} = \frac{40,000,000 + 500,000}{100,000}$$

$$= \frac{40,500,000}{100,000} = 405$$

The average cost per bike of producing 100,000 bikes per month is $405.

As the production level increases, the average cost of producing each foldable bike decreases. This illustrates the difficulty with small businesses. It is nearly impossible to have competitively low prices when production levels are low. ∎

The graph in **Figure 1.16** shows the relationship between production level and cost. The three points with the voice balloons illustrate our evaluations in Example 13. The other unlabeled points along the smooth blue curve represent the company's average costs for various production levels. The symbol ⌇ on the vertical axis indicates that there is a break in the values between 0 and 400. Thus, values for the average cost per bike begin at $400.

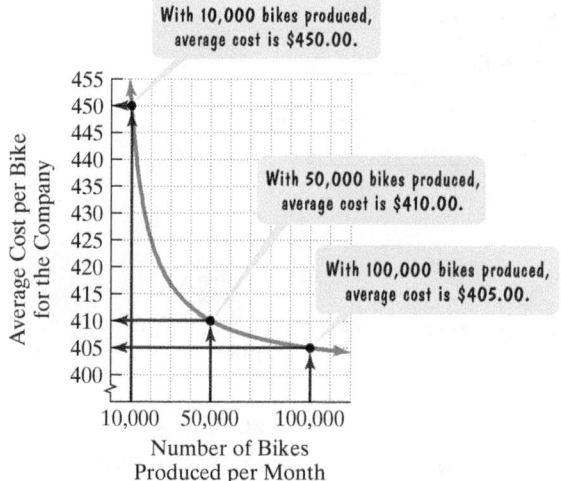

The points that lie along the blue curve in **Figure 1.16** are falling from left to right. Can you see how this shows that the company's cost per foldable bike is decreasing as the production level increases?

Figure 1.16

✓ **CHECK POINT 13**   A company that manufactures running shoes has weekly fixed costs of $300,000 and it costs $30 to manufacture each pair of running shoes. The average cost per pair of running shoes, $\overline{C}$, for the company to manufacture x pairs per week is modeled by

$$\overline{C} = \frac{30x + 300,000}{x}.$$

Find the average cost per pair of running shoes for the company to manufacture

a. 1000 pairs per week.   **b.** 10,000 pairs per week.

c. 100,000 pairs per week.

Achieving Success

Do not wait until the last minute to study for an exam. Cramming is a high-stress activity that forces your brain to make a lot of weak connections. No wonder crammers tend to forget everything they learned minutes after taking a test.

Preparing for Tests Using the Book
- Study the appropriate sections from the review chart in the Chapter Summary. The chart contains definitions, concepts, procedures, and examples. Review this chart and you'll know the most important material in each section!
- Work the assigned exercises from the Review Exercises. The Review Exercises contain the most significant problems for each of the chapter's sections.
- Find a quiet place to take the Chapter Test. Do not use notes, index cards, or any other resources. Check your answers and ask your professor to review any exercises you missed.

CONCEPT AND VOCABULARY CHECK

Fill in each blank so that the resulting statement is true.

1. In the expression b^n, b is called the _____ and n is called the _____.

2. The expression b^n is read _____.

In the remaining items, use the choices below to fill in each blank:

add   subtract   multiply   divide.

3. To simplify the expression $10 + 4 \cdot 5 - 30$, first _____.

4. To simplify the expression $(10 + 4) \cdot 5 - 30$, first _____.

5. To simplify the expression $36 - 24 \div 4 \cdot 3$, first _____.

6. To simplify the expression $4[8 + 3(2 - 6)^2]$, first _____.

7. To simplify the expression $8|5 - 4 \cdot 6|$, first _____.

1.8 EXERCISE SET

MyMathLab®

Watch the videos
in MyMathLab

Download the
MyDashBoard App

Practice Exercises

In Exercises 1–14, evaluate each exponential expression.

1. 9^2
2. 3^2
3. 4^3
4. 6^3
5. $(-4)^2$
6. $(-10)^2$
7. $(-4)^3$
8. $(-10)^3$
9. $(-5)^4$
10. $(-1)^6$
11. -5^4
12. -1^6
13. -10^2
14. -8^2

In Exercises 15–28, simplify each algebraic expression, or explain why the expression cannot be simplified.

15. $7x^2 + 12x^2$
16. $6x^2 + 18x^2$
17. $10x^3 + 5x^3$
18. $14x^3 + 8x^3$
19. $8x^4 + x^4$
20. $14x^4 + x^4$
21. $26x^2 - 27x^2$
22. $29x^2 - 30x^2$
23. $27x^3 - 26x^3$
24. $30x^3 - 29x^3$
25. $5x^2 + 5x^3$
26. $8x^2 + 8x^3$
27. $16x^2 - 16x^2$
28. $34x^2 - x^2$

In Exercises 29–72, use the order of operations to simplify each expression.

29. $7 + 6 \cdot 3$
30. $3 + 4 \cdot 5$
31. $45 \div 5 \cdot 3$
32. $40 \div 4 \cdot 2$
33. $6 \cdot 8 \div 4$
34. $8 \cdot 6 \div 2$
35. $14 - 2 \cdot 6 + 3$
36. $36 - 12 \div 4 + 2$
37. $8^2 - 16 \div 2^2 \cdot 4 - 3$
38. $10^2 - 100 \div 5^2 \cdot 2 - 1$

39. $3(-2)^2 - 4(-3)^2$ **40.** $5(-3)^2 - 2(-4)^2$

41. $(4 \cdot 5)^2 - 4 \cdot 5^2$ **42.** $(3 \cdot 5)^2 - 3 \cdot 5^2$

43. $(2 - 6)^2 - (3 - 7)^2$ **44.** $(4 - 6)^2 - (5 - 9)^2$

45. $6(3 - 5)^3 - 2(1 - 3)^3$

46. $-3(-6 + 8)^3 - 5(-3 + 5)^3$

47. $[2(6 - 2)]^2$ **48.** $[3(4 - 6)]^3$

49. $2[5 + 2(9 - 4)]$ **50.** $3[4 + 3(10 - 8)]$

51. $[7 + 3(2^3 - 1)] \div 21$ **52.** $[11 - 4(2 - 3^3)] \div 37$

53. $\dfrac{10 + 8}{5^2 - 4^2}$ **54.** $\dfrac{6^2 - 4^2}{2 - (-8)}$

55. $\dfrac{37 + 15 \div (-3)}{2^4}$ **56.** $\dfrac{22 + 20 \div (-5)}{3^2}$

57. $\dfrac{(-11)(-4) + 2(-7)}{7 - (-3)}$ **58.** $\dfrac{-5(7 - 2) - 3(4 - 7)}{-13 - (-5)}$

59. $4|10 - (8 - 20)|$ **60.** $6|7 - 4 \cdot 3|$

61. $8(-10) + |4(-5)|$ **62.** $4(-15) + |3(-10)|$

63. $-2^2 + 4[16 \div (3 - 5)]$

64. $-3^2 + 2[20 \div (7 - 11)]$

65. $24 \div \dfrac{3^2}{8 - 5} - (-6)$

66. $30 \div \dfrac{5^2}{7 - 12} - (-9)$

67. $\dfrac{\dfrac{1}{4} - \dfrac{1}{2}}{\dfrac{1}{3}}$ **68.** $\dfrac{\dfrac{3}{5} - \dfrac{7}{10}}{\dfrac{1}{2}}$

69. $-\dfrac{9}{4}\left(\dfrac{1}{2}\right) + \dfrac{3}{4} \div \dfrac{5}{6}$

70. $\left[-\dfrac{4}{7} - \left(-\dfrac{2}{5}\right)\right]\left[-\dfrac{3}{8} + \left(-\dfrac{1}{9}\right)\right]$

71. $\dfrac{\dfrac{7}{9} - 3}{\dfrac{5}{6}} \div \dfrac{3}{2} + \dfrac{3}{4}$

72. $\dfrac{\dfrac{17}{25}}{\dfrac{3}{5} - 4} \div \dfrac{1}{5} + \dfrac{1}{2}$

In Exercises 73–80, evaluate each algebraic expression for the given value of the variable.

73. $x^2 + 5x; x = 3$ **74.** $x^2 - 2x; x = 6$

75. $3x^2 - 8x; x = -2$ **76.** $4x^2 - 2x; x = -3$

77. $-x^2 - 10x; x = -1$ **78.** $-x^2 - 14x; x = -1$

79. $\dfrac{6y - 4y^2}{y^2 - 15}; y = 5$ **80.** $\dfrac{3y - 2y^2}{y(y - 2)}; y = 5$

In Exercises 81–88, simplify each algebraic expression by removing parentheses and brackets.

81. $3[5(x - 2) + 1]$

82. $4[6(x - 3) + 1]$

83. $3[6 - (y + 1)]$

84. $5[2 - (y + 3)]$

85. $7 - 4[3 - (4y - 5)]$

86. $6 - 5[8 - (2y - 4)]$

87. $2(3x^2 - 5) - [4(2x^2 - 1) + 3]$

88. $4(6x^2 - 3) - [2(5x^2 - 1) + 1]$

Practice PLUS

In Exercises 89–92, express each sentence as a single numerical expression. Then use the order of operations to simplify the expression.

89. Cube -2. Subtract this exponential expression from -10.

90. Cube -5. Subtract this exponential expression from -100.

91. Subtract 10 from 7. Multiply this difference by 2. Square this product.

92. Subtract 11 from 9. Multiply this difference by 2. Raise this product to the fourth power.

In Exercises 93–96, let x represent the number. Express each sentence as a single algebraic expression. Then simplify the expression.

93. Multiply a number by 5. Add 8 to this product. Subtract this sum from the number.

94. Multiply a number by 3. Add 9 to this product. Subtract this sum from the number.

95. Cube a number. Subtract 4 from this exponential expression. Multiply this difference by 5.

96. Cube a number. Subtract 6 from this exponential expression. Multiply this difference by 4.

Application Exercises

The bar graph shows the estimated number of calories per day needed to maintain energy balance for various gender and age groups for sedentary lifestyles. (Sedentary means a lifestyle that includes only the light physical activity associated with typical day-to-day life.)

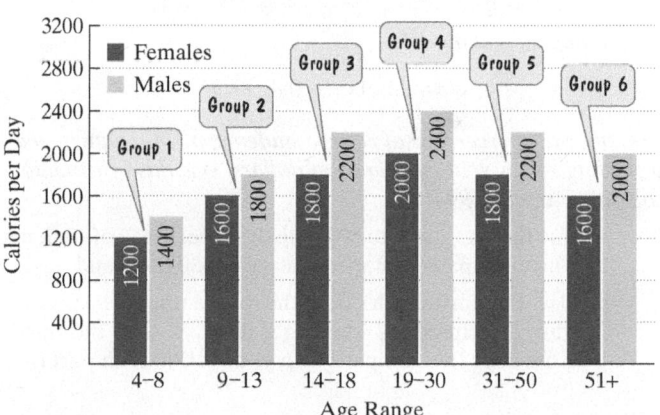

Calories Needed to Maintain Energy Balance for Sedentary Lifestyles

Source: U.S.D.A.

Use the appropriate information displayed by the graph at the bottom of the previous page to solve Exercises 97–98.

97. The mathematical model

$$F = -82x^2 + 654x + 620$$

describes the number of calories needed per day, F, by females in age group x with sedentary lifestyles. According to the model, how many calories per day are needed by females between the ages of 19 and 30, inclusive, with this lifestyle? Does this underestimate or overestimate the number shown by the graph? By how much?

98. The mathematical model

$$M = -96x^2 + 802x + 660$$

describes the number of calories needed per day, M, by males in age group x with sedentary lifestyles. According to the model, how many calories per day are needed by males between the ages of 19 and 30, inclusive, with this lifestyle? Does this underestimate or overestimate the number shown by the graph? By how much?

Salary after College. *In 2010, MonsterCollege surveyed 1250 U.S. college students expecting to graduate in the next several years. Respondents were asked the following question:*

What do you think your starting salary will be at your first job after college?

The line graph shows the percentage of college students who anticipated various starting salaries.

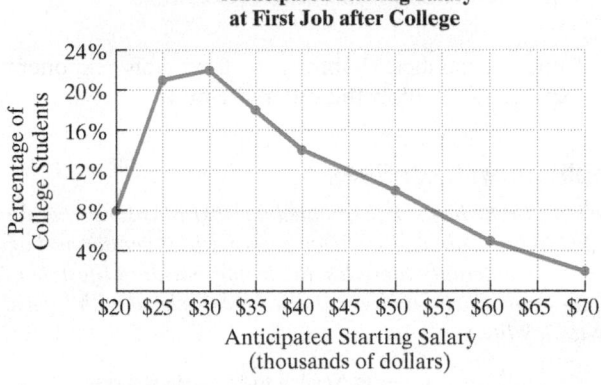

Anticipated Starting Salary at First Job after College

Source: MonsterCollege™

The mathematical model

$$p = -0.01s^2 + 0.8s + 3.7$$

describes the percentage of college students, p, who anticipated a starting salary s, in thousands of dollars. Use this information to solve Exercises 99–100.

99. a. Use the line graph to estimate the percentage of students who anticipated a starting salary of $30 thousand.

b. Use the formula to find the percentage of students who anticipated a starting salary of $30 thousand. How does this compare with your estimate in part (a)?

100. a. Use the line graph to estimate the percentage of students who anticipated a starting salary of $40 thousand.

b. Use the formula to find the percentage of students who anticipated a starting salary of $40 thousand. How does this compare with your estimate in part (a)?

In Palo Alto, California, a government agency ordered computer-related companies to contribute to a pool of money to clean up underground water supplies. (The companies had stored toxic chemicals in leaking underground containers.) The mathematical model

$$C = \frac{200x}{100 - x}$$

describes the cost, C, in tens of thousands of dollars, for removing x percent of the contaminants. Use this formula to solve Exercises 101–102.

101. a. Find the cost, in tens of thousands of dollars, for removing 50% of the contaminants.

b. Find the cost, in tens of thousands of dollars, for removing 80% of the contaminants.

c. Describe what is happening to the cost of the cleanup as the percentage of contaminant removed increases.

102. a. Find the cost, in tens of thousands of dollars, for removing 60% of the contaminants.

b. Find the cost, in tens of thousands of dollars, for removing 90% of the contaminants.

c. Describe what is happening to the cost of the cleanup as the percentage of contaminants removed increases.

Writing in Mathematics

103. Describe what it means to raise a number to a power. In your description, include a discussion of the difference between -5^2 and $(-5)^2$.

104. Explain how to simplify $4x^2 + 6x^2$. Why is the sum not equal to $10x^4$?

105. Why is the order of operations agreement needed?

Critical Thinking Exercises

Make Sense? *In Exercises 106–109, determine whether each statement "makes sense" or "does not make sense" and explain your reasoning.*

106. Without parentheses, an exponent has only the number next to it as its base.

107. I read that a certain star is 10^4 light-years from Earth, which means 100,000 light-years.

108. When I evaluated $(-1)^n$, I obtained positive numbers when n was even and negative numbers when n was odd.

109. The rules for the order of operations avoid the confusion of obtaining different results when I simplify the same expression.

In Exercises 110–113, determine whether each statement is true or false. If the statement is false, make the necessary change(s) to produce a true statement.

110. If x is -3, then the value of $-3x - 9$ is -18.

111. The algebraic expression $\dfrac{6x + 6}{x + 1}$ cannot have the same value when two different replacements are made for x such as $x = -3$ and $x = 2$.

112. The value of $\dfrac{|3 - 7| - 2^3}{(-2)(-3)}$ is the fraction that results when $\frac{1}{3}$ is subtracted from $-\frac{1}{3}$.

113. $-2(6 - 4^2)^3 = -2(6 - 16)^3$

$\qquad\qquad = -2(-10)^3 = (-20)^3 = -8000$

114. Simplify: $\dfrac{1}{4} - 6(2 + 8) \div \left(-\dfrac{1}{3}\right)\left(-\dfrac{1}{9}\right)$.

In Exercises 115–116, insert parentheses in each expression so that the resulting value is 45.

115. $2 \cdot 3 + 3 \cdot 5$

116. $2 \cdot 5 - \dfrac{1}{2} \cdot 10 \cdot 9$

Review Exercises

117. Simplify: $-8 - 2 - (-5) + 11$. (Section 1.6, Example 3)

118. Multiply: $-4(-1)(-3)(2)$. (Section 1.7, Example 2)

119. Give an example of a real number that is not an irrational number. (Section 1.3, Example 5).

Preview Exercises

Exercises 120–122 will help you prepare for the material covered in the first section of the next chapter. In each exercise, determine whether the given number is a solution of the equation.

120. $-\dfrac{1}{2} = x - \dfrac{2}{3}; \dfrac{1}{6}$

121. $5y + 3 - 4y - 8 = 15; 20$

122. $4x + 2 = 3(x - 6) + 8; -11$

GROUP PROJECT

CHAPTER 1

One measure of physical fitness is your *resting heart rate*. Generally speaking, the more fit you are, the lower your resting heart rate. The best time to take this measurement is when you first awaken in the morning, before you get out of bed. Lie on your back with no body parts crossed and take your pulse in your neck or wrist. Use your index and second fingers and count your pulse beat for one full minute to get your resting heart rate. A resting heart rate under 48 to 57 indicates high fitness, 58 to 62, above average fitness, 63 to 70, average fitness, 71 to 82, below average fitness, and 83 or more, low fitness.

Another measure of physical fitness is your percentage of body fat. You can estimate your body fat using the following formulas:

For men: Body fat $= -98.42 + 4.15w - 0.082b$

For women: Body fat $= -76.76 + 4.15w - 0.082b$

where w = waist measurement, in inches, and b = total body weight, in pounds. Then divide your body fat by your total weight to get your body fat percentage. For men, less than 15% is considered athletic, 25% about average. For women, less than 22% is considered athletic, 30% about average.

Each group member should bring his or her age, resting heart rate, and body fat percentage to the group. Using the data, the group should create three graphs.

a. Create a graph that shows age and resting heart rate for group members.

b. Create a graph that shows age and body fat percentage for group members.

c. Create a graph that shows resting heart rate and body fat percentage for group members.

For each graph, select the style (line or bar) that is most appropriate.

Chapter 1 Summary

Definitions and Concepts	**Examples**

Section 1.1 Introduction to Algebra: Variables and Mathematical Models

A letter that represents a variety of different numbers is called a variable. An algebraic expression is a combination of variables, numbers, and operation symbols. To evaluate an algebraic expression, substitute a given number for the variable and simplify:

1. Perform calculations within parentheses first.
2. Perform multiplication before addition or subtraction.

(A more detailed order of operations is given in Section 1.8.)

Evaluate $6(x - 3) + 4x$ for $x = 5$.

Replace x with 5.

$$6(5 - 3) + 4 \cdot 5$$
$$= 6(2) + 4 \cdot 5$$
$$= 12 + 20$$
$$= 32$$

Here are some key words for translating into algebraic expressions:

- Addition: plus, sum, more than, increased by
- Subtraction: minus, difference, less than, decreased by
- Multiplication: times, product, twice, multiplied by
- Division: divide, quotient, ratio, divided by

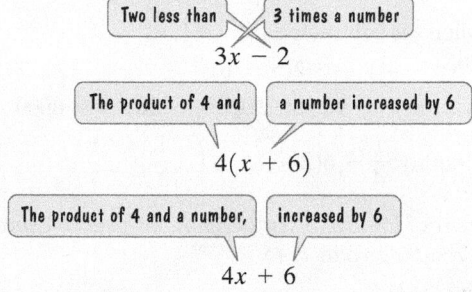

Two less than / 3 times a number
$$3x - 2$$

The product of 4 and / a number increased by 6
$$4(x + 6)$$

The product of 4 and a number, / increased by 6
$$4x + 6$$

An equation is a statement that two algebraic expressions are equal. Solutions of an equation are values of the variable that make the equation a true statement. To determine whether a number is a solution, substitute that number for the variable and evaluate each side of the equation. If the values on both sides of the equation are the same, the number is a solution.

Is 5 a solution of

$$3(3x + 5) = 10x + 10?$$

Substitute 5 for x.

$$3(3 \cdot 5 + 5) \stackrel{?}{=} 10 \cdot 5 + 10$$
$$3(15 + 5) \stackrel{?}{=} 50 + 10$$
$$3(20) \stackrel{?}{=} 60$$
$$60 = 60, \qquad \text{true}$$

The true statement indicates that 5 is a solution.

A formula is an equation that expresses a relationship between two or more variables. The process of finding formulas to describe real-world phenomena is called mathematical modeling. Such formulas, together with the meaning assigned to the variables, are called mathematical models. These formulas model, or describe, the relationship among the variables.

The formulas

$$M = 0.2n + 12$$

and

$$W = -0.4n + 39$$

model the time each week that men, M, and women, W, devoted to housework n years after 1965.

Definitions and Concepts	**Examples**

Section 1.2 Fractions in Algebra

Mixed Numbers and Improper Fractions

A mixed number consists of the addition of a natural number $(1, 2, 3, \ldots)$ and a fraction, expressed without the use of an addition sign. An improper fraction has a numerator that is greater than its denominator. To convert a mixed number to an improper fraction, multiply the denominator by the natural number and add the numerator. Then place this result over the original denominator.

To convert an improper fraction to a mixed number, divide the denominator into the numerator and write the mixed number using

$$\text{quotient} \frac{\text{remainder}}{\text{original denominator}}.$$

Convert $5\frac{3}{7}$ to an improper fraction.

$$5\frac{3}{7} = \frac{7 \cdot 5 + 3}{7} = \frac{35 + 3}{7} = \frac{38}{7}$$

Convert $\frac{14}{3}$ to a mixed number.

$$\frac{14}{3} = 4\frac{2}{3} \qquad \begin{array}{r} 4 \\ 3\overline{)14} \\ \underline{12} \\ 2 \end{array}$$

A prime number is a natural number greater than 1 that has only itself and 1 as factors. A composite number is a natural number greater than 1 that is not a prime number. The prime factorization of a composite number is the expression of the composite number as the product of prime numbers.

Find the prime factorization:

$$\begin{aligned} 60 &= 6 \cdot 10 \\ &= 2 \cdot 3 \cdot 2 \cdot 5 \end{aligned}$$

A fraction is reduced to its lowest terms when the numerator and denominator have no common factors other than 1. To reduce a fraction to its lowest terms, divide both the numerator and the denominator by their greatest common factor. The greatest common factor can be found by inspection or prime factorizations of the numerator and the denominator.

Reduce to lowest terms:

$$\frac{8}{14} = \frac{2 \cdot 4}{2 \cdot 7} = \frac{4}{7}$$

Multiplying Fractions

The product of two or more fractions is the product of their numerators divided by the product of their denominators.

Multiply:

$$\frac{2}{7} \cdot \frac{5}{9} = \frac{2 \cdot 5}{7 \cdot 9} = \frac{10}{63}$$

Dividing Fractions

The quotient of two fractions is the first multiplied by the reciprocal (or multiplicative inverse) of the second.

Divide:

$$\frac{4}{9} \div \frac{3}{7} = \frac{4}{9} \cdot \frac{7}{3} = \frac{4 \cdot 7}{9 \cdot 3} = \frac{28}{27}$$

Adding and Subtracting Fractions with Identical Denominators

Add or subtract numerators. Put this result over the common denominator.

Subtract:

$$\frac{5}{8} - \frac{3}{8} = \frac{5 - 3}{8} = \frac{2}{8} = \frac{2 \cdot 1}{2 \cdot 4} = \frac{1}{4}$$

Adding and Subtracting Fractions with Unlike Denominators

Rewrite the fractions as equivalent fractions with the least common denominator. Then add or subtract numerators, putting this result over the common denominator.

Add:

$$\frac{3}{8} + \frac{5}{12} = \frac{3 \cdot 3}{8 \cdot 3} + \frac{5 \cdot 2}{12 \cdot 2}$$

The LCD is 24.

$$= \frac{9}{24} + \frac{10}{24} = \frac{19}{24}$$

Definitions and Concepts	**Examples**

Section 1.2 Fractions in Algebra (continued)

Fractions appear throughout algebra. Many formulas and mathematical models contain fractions. Operations with fractions can be used to determine whether a particular fraction is a solution of an equation. When used with fractions, the word *of* represents multiplication.

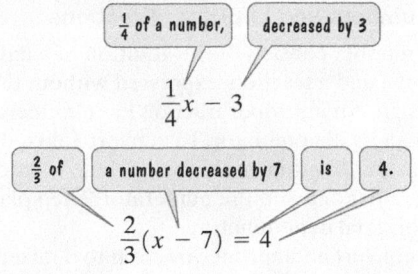

Section 1.3 The Real Numbers

A set is a collection of objects, called elements, whose contents can be clearly determined.

$$\{a, b, c\}$$

A line used to visualize numbers is called a number line.

Real Numbers: the set of all numbers that can be represented by points on the number line

The Sets That Make Up the Real Numbers

- Natural Numbers: $\{1, 2, 3, 4, \dots\}$
- Whole Numbers: $\{0, 1, 2, 3, 4, \dots\}$
- Integers: $\{\dots, -3, -2, -1, 0, 1, 2, 3, \dots\}$
- Rational Numbers: the set of numbers that can be expressed as the quotient of an integer and a nonzero integer; can be expressed as terminating or repeating decimals
- Irrational Numbers: the set of numbers that cannot be expressed as the quotient of integers; decimal representations neither terminate nor repeat.

Given the set

$$\left\{-1.4, 0, 0.\overline{7}, \frac{9}{10}, \sqrt{2}, \sqrt{4}\right\}$$

list the

- natural numbers: $\sqrt{4}$, or 2
- whole numbers: $0, \sqrt{4}$
- integers: $0, \sqrt{4}$
- rational numbers: $-1.4, 0, 0.\overline{7}, \frac{9}{10}, \sqrt{4}$
- irrational numbers: $\sqrt{2}$
- real numbers:

$$-1.4, 0, 0.\overline{7}, \frac{9}{10}, \sqrt{2}, \sqrt{4}$$

For any two real numbers, a and b, a is less than b if a is to the left of b on the number line.

Inequality Symbols

$<$: is less than $>$: is greater than

$\leq$: is less than or equal to $\geq$: is greater than or equal to

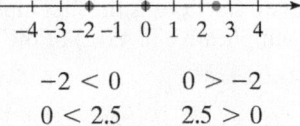

$-2 < 0$ $0 > -2$

$0 < 2.5$ $2.5 > 0$

The absolute value of a, written $|a|$, is the distance from 0 to a on the number line.

$|4| = 4$ $|0| = 0$ $|-6| = 6$

Definitions and Concepts	**Examples**

Section 1.4 Basic Rules of Algebra

Terms of an algebraic expression are separated by addition. The parts of each term that are multiplied are its factors. The numerical part of a term is its coefficient. Like terms have the same variable factors raised to the same powers.

- An expression with three terms:

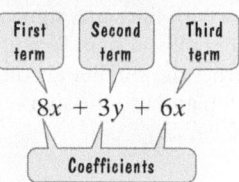

The like terms are $8x$ and $6x$.

Properties of Real Numbers and Algebraic Expressions

- Commutative Properties:

$$a + b = b + a$$
$$ab = ba$$

- Associative Properties:

$$(a + b) + c = a + (b + c)$$
$$(ab)c = a(bc)$$

- Distributive Properties:

$$a(b + c) = ab + ac$$
$$(b + c)a = ba + ca$$
$$a(b - c) = ab - ac$$
$$(b - c)a = ba - ca$$
$$a(b + c + d) = ab + ac + ad$$

Commutative of Addition:

$$5x + 4 = 4 + 5x$$

Commutative of Multiplication:

$$5x + 4 = x5 + 4$$

Associative of Addition:

$$6 + (4 + x) = (6 + 4) + x = 10 + x$$

Associative of Multiplication:

$$7(10x) = (7 \cdot 10)x = 70x$$

Distributive:

$$8(x + 5 + 4y) = 8x + 40 + 32y$$

Distributive to Combine Like Terms:

$$8x + 12x = (8 + 12)x = 20x$$

Simplifying Algebraic Expressions

Use the distributive property to remove grouping symbols. Then combine like terms.

$$4(5x + 7) + 13x$$
$$= 20x + 28 + 13x$$
$$= (20x + 13x) + 28$$
$$= 33x + 28$$

Section 1.5 Addition of Real Numbers

Sums on a Number Line

To find $a + b$, the sum of a and b, on a number line, start at a. If b is positive, move b units to the right. If b is negative, move $|b|$ units to the left. If b is 0, stay at a. The number where we finish on the number line represents $a + b$.

$$-7 + 5 = -2$$

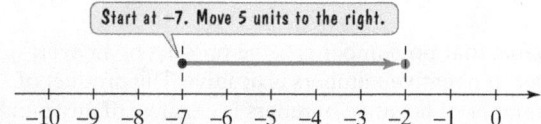

Additive inverses, or opposites, are pairs of real numbers that are the same number of units from zero on the number line, but on opposite sides of zero.

- Identity Property of Addition:

$$a + 0 = 0 \qquad 0 + a = a$$

- Inverse Property of Addition:

$$a + (-a) = 0 \qquad (-a) + a = 0$$

The additive inverse (or opposite) of 4 is -4. The additive inverse of -1.7 is 1.7.

Identity Property of Addition:

$$4x + 0 = 4x$$

Inverse Property of Addition:

$$4x + (-4x) = 0$$

Definitions and Concepts	**Examples**

Section 1.5 Addition of Real Numbers (continued)

Addition without a Number Line

To add two numbers with the same sign, add their absolute values and use their common sign. To add two numbers with different signs, subtract the smaller absolute value from the greater absolute value and use the sign of the number with the greater absolute value.

Add:

$$10 + 4 = 14$$
$$-4 + (-6) = -10$$
$$-30 + 5 = -25$$
$$12 + (-8) = 4$$

To add a series of positive and negative numbers, add all the positive numbers and add all the negative numbers. Then add the resulting positive and negative sums.

$$5 + (-3) + (-7) + 2$$
$$= (5 + 2) + [(-3) + (-7)]$$
$$= 7 + (-10)$$
$$= -3$$

Section 1.6 Subtraction of Real Numbers

To subtract b from a, add the opposite, or additive inverse, of b to a:

$$a - b = a + (-b).$$

The result is called the difference between a and b.

Subtract:

$$-7 - (-5) = -7 + 5 = -2$$
$$-\frac{3}{4} - \frac{1}{2} = -\frac{3}{4} + \left(-\frac{1}{2}\right)$$
$$= -\frac{3}{4} + \left(-\frac{2}{4}\right) = -\frac{5}{4}$$

To simplify a series of additions and subtractions, change all subtractions to additions of opposites. Then use the procedure for adding a series of positive and negative numbers.

Simplify:

$$-6 - 2 - (-3) + 10$$
$$= -6 + (-2) + 3 + 10$$
$$= -8 + 13$$
$$= 5$$

Section 1.7 Multiplication and Division of Real Numbers

The result of multiplying a and b, ab, is called the product of a and b. If the two numbers have different signs, the product is negative. If the two numbers have the same sign, the product is positive. If either number is 0, the product is 0.

Multiply:

$$-5(-10) = 50$$
$$\frac{3}{4}\left(-\frac{5}{7}\right) = -\frac{3}{4} \cdot \frac{5}{7} = -\frac{15}{28}$$

Assuming that no number is 0, the product of an even number of negative numbers is positive. The product of an odd number of negative numbers is negative. If any number is 0, the product is 0.

Multiply:

$$(-3)(-2)(-1)(-4) = 24$$
$$(-3)(2)(-1)(-4) = -24$$

The result of dividing the real number a by the nonzero real number b is called the quotient of a and b. If two numbers have different signs, their quotient is negative. If two numbers have the same sign, their quotient is positive. Division by zero is undefined.

Divide:

$$\frac{21}{-3} = -7$$
$$-\frac{1}{3} \div (-3) = \frac{1}{3} \cdot \frac{1}{3} = \frac{1}{9}$$

Definitions and Concepts	**Examples**

Section 1.7 Multiplication and Division of Real Numbers (continued)

Two numbers whose product is 1 are called multiplicative inverses, or reciprocals, of each other. The number 0 has no multiplicative inverse.

- Identity Property of Multiplication

$$a \cdot 1 = a \qquad 1 \cdot a = a$$

- Inverse Property of Multiplication
  If a is not 0:

$$a \cdot \frac{1}{a} = 1 \qquad \frac{1}{a} \cdot a = 1$$

- Multiplication Property of -1

$$-1a = -a \qquad a(-1) = -a$$

- Double Negative Property

$$-(-a) = a$$

The multiplicative inverse of 4 is $\frac{1}{4}$.
The multiplicative inverse of $-\frac{1}{3}$ is -3.
Simplify:

$$1x = x$$

$$7x \cdot \frac{1}{7x} = 1, \quad x \neq 0$$

$$4x - 5x = -1x = -x$$

$$-(-7y) = 7y$$

If a negative sign precedes parentheses, remove parentheses and change the sign of every term within parentheses.

Simplify:

$$-(7x - 3y + 2) = -7x + 3y - 2$$

Section 1.8 Exponents and Order of Operations

If b is a real number and n is a natural number, b^n, the nth power of b, is the product of n factors of b. Furthermore, $b^1 = b$.

Evaluate:

$$8^2 = 8 \cdot 8 = 64$$

$$(-5)^3 = (-5)(-5)(-5) = -125$$

Order of Operations

1. Perform operations within grouping symbols, starting with the innermost grouping symbols. Grouping symbols include parentheses, brackets, fraction bars, and absolute value symbols.

2. Evaluate exponential expressions.

3. Multiply and divide, from left to right.

4. Add and subtract. In this step, you have a choice. You can add and subtract in order from left to right. You can also rewrite subtractions as additions of opposites, combine positive and negative numbers separately, and then add these results.

Simplify:

$$5(4 - 6)^2 - 2(1 - 3)^3$$
$$= 5(-2)^2 - 2(-2)^3$$
$$= 5(4) - 2(-8)$$
$$= 20 - (-16)$$
$$= 20 + 16 = 36$$

Some algebraic expressions contain two sets of grouping symbols: parentheses, the inner grouping symbols, and brackets, the outer grouping symbols. To simplify such expressions, use the order of operations and remove grouping symbols from innermost (parentheses) to outermost (brackets).

Simplify:

$$5 - 3[2(x + 1) - 7]$$
$$= 5 - 3[2x + 2 - 7]$$
$$= 5 - 3[2x - 5]$$
$$= 5 - 6x + 15$$
$$= -6x + 20$$

CHAPTER 1 REVIEW EXERCISES

1.1 *In Exercises 1–2, evaluate each expression for x = 6.*

1. $10 + 5x$

2. $8(x - 2) + 3x$

In Exercises 3–4, evaluate each expression for x = 8 and y = 10.

3. $\dfrac{40}{x} - \dfrac{y}{5}$

4. $3(2y + x)$

In Exercises 5–8, translate from English to an algebraic expression or equation, whichever is appropriate. Let the variable x represent the number.

5. Six subtracted from the product of 7 and a number

6. The quotient of a number and 5, decreased by 2, is 18.

7. Nine less twice a number is 14.

8. The product of 3 and 7 more than a number

In Exercises 9–11, determine whether the given number is a solution of the equation.

9. $4x + 5 = 13; 3$

10. $2y + 7 = 4y - 5; 6$

11. $3(w + 1) + 11 = 2(w + 8); 2$

Exercises 12–13 involve an experiment on memory. Students in a language class are asked to memorize 40 vocabulary words in Latin, a language with which they are not familiar. After studying the words for one day, students are tested each day after to see how many words they remember. The class average is taken and the results are shown as ten points on a line graph.

**Average Number of Words
Remembered Over Time**

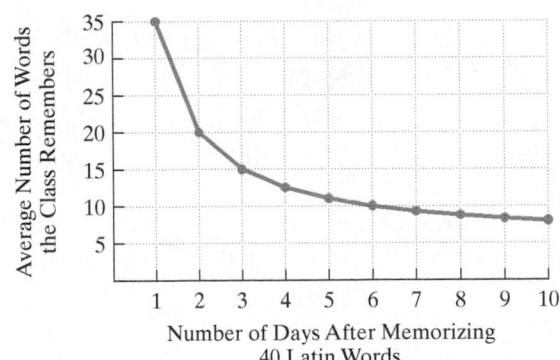

Number of Days After Memorizing
40 Latin Words

12. Use the line graph to estimate the average number of Latin words the class remembered after 5 days.

13. The mathematical model

$$L = \dfrac{5n + 30}{n}$$

describes the average number of Latin words remembered by the students, L, after n days. Use the formula to find the average number of words remembered after 5 days. How well does this compare with your estimate from Exercise 12?

14. Average annual premiums for employer-sponsored family health policies more than doubled in the past decade. The bar graph shows the average cost of a family health insurance plan in the United States for six selected years from 2000 through 2009.

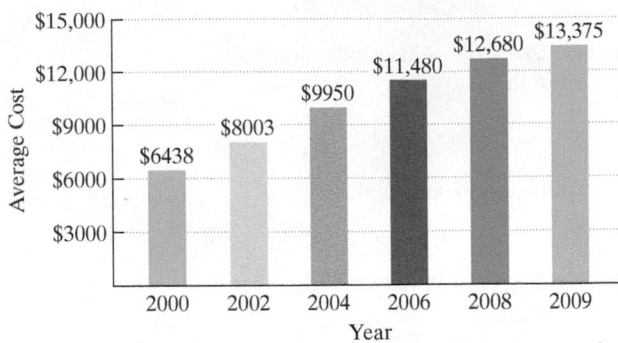

**Average Cost of a Family
Health Insurance Plan**

Source: Kaiser Family Foundation

Here is a mathematical model that approximates the data displayed by the bar graph:

$$C = 783n + 6522.$$

Average cost of a family health insurance plan

Number of years after 2000

a. Use the formula to find the average cost of a family health insurance plan in 2009. Does the mathematical model underestimate or overestimate the actual cost shown by the bar graph? By how much?

b. If trends from 2000 through 2009 continue, use the formula to project the average cost of a family health insurance plan in 2020.

1.2 *In Exercises 15–16, convert each mixed number to an improper fraction.*

15. $3\dfrac{2}{7}$

16. $5\dfrac{9}{11}$

In Exercises 17–18, convert each improper fraction to a mixed number.

17. $\dfrac{17}{9}$

18. $\dfrac{27}{5}$

In Exercises 19–21, identify each natural number as prime or composite. If the number is composite, find its prime factorization.

19. 60

20. 63

21. 67

In Exercises 22–23, simplify each fraction by reducing it to its lowest terms.

22. $\dfrac{15}{33}$

23. $\dfrac{40}{75}$

In Exercises 24–29, perform the indicated operation. Where possible, reduce the answer to its lowest terms.

24. $\dfrac{3}{5} \cdot \dfrac{7}{10}$

25. $\dfrac{4}{5} \div \dfrac{3}{10}$

26. $1\dfrac{2}{3} \div 6\dfrac{2}{3}$

27. $\dfrac{2}{9} + \dfrac{4}{9}$

28. $\dfrac{5}{6} + \dfrac{7}{9}$

29. $\dfrac{3}{4} - \dfrac{2}{15}$

In Exercises 30–31, determine whether the given number is a solution of the equation.

30. $x - \dfrac{3}{4} = \dfrac{7}{4}; 2\dfrac{1}{2}$

31. $\dfrac{2}{3}w = \dfrac{1}{15}w + \dfrac{3}{5}; 2$

In Exercises 32–33, translate from English to an algebraic expression or equation, whichever is appropriate. Let the variable x represent the number.

32. Two decreased by half of a number is $\frac{1}{4}$ of the number.

33. $\frac{3}{5}$ of a number increased by 6

34. Suppose that the target heart rate, H, in beats per minute, for your exercise goal is given by

$$H = \frac{4}{5}(220 - a),$$

where a is your age. If you are 30 years old, what is your target heart rate?

1.3 *In Exercises 35–36, graph each real number on a number line.*

35. -2.5

36. $4\dfrac{3}{4}$

In Exercises 37–38, express each rational number as a decimal.

37. $\dfrac{5}{8}$

38. $\dfrac{3}{11}$

39. Consider the set

$$\left\{-17, -\frac{9}{13}, 0, 0.75, \sqrt{2}, \pi, \sqrt{81}\right\}.$$

List all numbers from the set that are: **a.** natural numbers, **b.** whole numbers, **c.** integers, **d.** rational numbers, **e.** irrational numbers, **f.** real numbers.

40. Give an example of an integer that is not a natural number.

41. Give an example of a rational number that is not an integer.

42. Give an example of a real number that is not a rational number.

In Exercises 43–46, insert either $<$ or $>$ in the shaded area between each pair of numbers to make a true statement.

43. $-93 \quad \blacksquare \quad 17$

44. $-2 \quad \blacksquare \quad -200$

45. $0 \quad \blacksquare \quad -\dfrac{1}{3}$

46. $-\dfrac{1}{4} \quad \blacksquare \quad -\dfrac{1}{5}$

In Exercises 47–48, determine whether each inequality is true or false.

47. $-13 \geq -11$

48. $-126 \leq -126$

In Exercises 49–50, find each absolute value.

49. $|-58|$

50. $|2.75|$

1.4

51. Use the commutative property of addition to write an equivalent algebraic expression: $7 + 13y$.

52. Use the commutative property of multiplication to write an equivalent algebraic expression: $9(x + 7)$.

In Exercises 53–54, use an associative property to rewrite each algebraic expression. Then simplify the resulting algebraic expression.

53. $6 + (4 + y)$

54. $7(10x)$

55. Use the distributive property to rewrite without parentheses: $6(4x - 2 + 5y)$.

In Exercises 56–57, simplify each algebraic expression.

56. $4a + 9 + 3a - 7$

57. $6(3x + 4) + 5(2x - 1)$

1.5

58. Use a number line to find the sum: $-6 + 8$.

In Exercises 59–61, find each sum without the use of a number line.

59. $8 + (-11)$

60. $-\dfrac{3}{4} + \dfrac{1}{5}$

61. $7 + (-5) + (-13) + 4$

In Exercises 62–63, simplify each algebraic expression.

62. $8x + (-6y) + (-12x) + 11y$

63. $10(3y + 4) + (-40y)$

64. The Dead Sea is the lowest elevation on Earth, 1312 feet below sea level. If a person is standing 512 feet above the Dead Sea, what is that person's elevation?

65. The water level of a reservoir is measured over a five-month period. At the beginning, the level is 25 feet. During this time, the level fell 3 feet, then rose 2 feet, then rose 1 foot, then fell 4 feet, and then rose 2 feet. What is the reservoir's water level at the end of the five months?

1.6

66. Rewrite $9 - 13$ as the addition of an opposite.

In Exercises 67–69, perform the indicated subtraction.

67. $-9 - (-13)$

68. $-\dfrac{7}{10} - \dfrac{1}{2}$

69. $-3.6 - (-2.1)$

In Exercises 70–71, simplify each series of additions and subtractions.

70. $-7 - (-5) + 11 - 16$

71. $-25 - 4 - (-10) + 16$

72. Simplify: $3 - 6a - 8 - 2a$.

73. What is the difference in elevation between a plane flying 26,500 feet above sea level and a submarine traveling 650 feet below sea level?

1.7 *In Exercises 74–76, perform the indicated multiplication.*

74. $-7(-12)$

75. $\frac{3}{5}\left(-\frac{5}{11}\right)$

76. $5(-3)(-2)(-4)$

In Exercises 77–79, perform the indicated division or state that the expression is undefined.

77. $\frac{45}{-5}$

78. $-17 \div 0$

79. $-\frac{4}{5} \div \left(-\frac{2}{5}\right)$

In Exercises 80–81, simplify each algebraic expression.

80. $-4\left(-\frac{3}{4}x\right)$

81. $-3(2x - 1) - (4 - 5x)$

In Exercises 82–83, determine whether the given number is a solution of the equation.

82. $5x + 16 = -8 - x; -6$

83. $2(x + 3) - 18 = 5x; -4$

84. Are you expecting a tax refund this year? The bar graph shows the percentage of taxpayers expecting a tax refund for six selected ages.

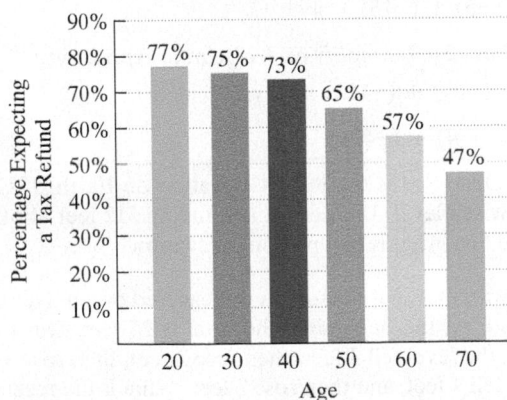

Percentage of U.S. Taxpayers Expecting a Tax Refund, by Age

Source: National Retail Federation

The percentage of U.S. taxpayers expecting a tax refund, p, can be described by the mathematical model

$$p = -0.6a + 93,$$

where a is the age of the taxpayer. Use the formula to find the percentage of 20-year-old taxpayers expecting a refund. Does the mathematical model underestimate or overestimate the actual percent shown by the bar graph? By how much?

1.8 *In Exercises 85–87, evaluate each exponential expression.*

85. $(-6)^2$

86. -6^2

87. $(-2)^5$

In Exercises 88–89, simplify each algebraic expression, or explain why the expression cannot be simplified.

88. $4x^3 + 2x^3$

89. $4x^3 + 4x^2$

In Exercises 90–98, use the order of operations to simplify each expression.

90. $-40 \div 5 \cdot 2$

91. $-6 + (-2) \cdot 5$

92. $6 - 4(-3 + 2)$

93. $28 \div (2 - 4^2)$

94. $36 - 24 \div 4 \cdot 3 - 1$

95. $-8[-4 - 5(-3)]$

96. $\dfrac{6(-10 + 3)}{2(-15) - 9(-3)}$

97. $\left(\dfrac{1}{2} + \dfrac{1}{3}\right) \div \left(\dfrac{1}{4} - \dfrac{3}{8}\right)$

98. $\dfrac{1}{2} - \dfrac{2}{3} \div \dfrac{5}{9} + \dfrac{3}{10}$

In Exercises 99–100, evaluate each algebraic expression for the given value of the variable.

99. $x^2 - 2x + 3; x = -1$

100. $-x^2 - 7x; x = -2$

In Exercises 101–102, simplify each algebraic expression.

101. $4[7(a - 1) + 2]$

102. $-6[4 - (y + 2)]$

103. The bar graph shows the percentage of people 25 years of age and older who were college graduates in the United States for seven selected years.

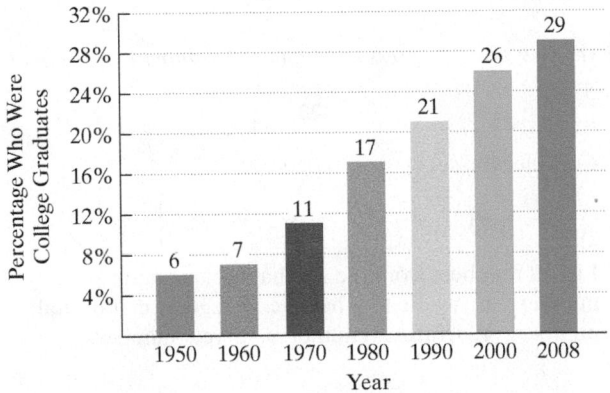

Percentage of College Graduates, among People Ages 25 and Older, in the United States

Source: U.S. Census Bureau

The percentage of people 25 years of age and older who were college graduates in the United States, p, can be described by the mathematical model

$$p = 0.002n^2 + 0.3n + 5,$$

where n is the number of years after 1950.

a. Does the mathematical model underestimate or overestimate the percentage of college graduates in 2000? By how much?

b. If the trend shown by the graph continues, use the formula to project the percentage of people 25 years of age and older who will be college graduates in 2020.

CHAPTER
Test Prep
VIDEOS

Step-by-step test solutions are found on the Chapter Test Prep Videos available in MyMathLab® or on YouTube (search "BlitzerIntroAlg" and click on "Channels").

CHAPTER 1 TEST

In Exercises 1–10, perform the indicated operation or operations.

1. $1.4 - (-2.6)$

2. $-9 + 3 + (-11) + 6$

3. $3(-17)$

4. $\left(-\dfrac{3}{7}\right) \div \left(-\dfrac{15}{7}\right)$

5. $\left(3\dfrac{1}{3}\right)\left(-1\dfrac{3}{4}\right)$

6. $-50 \div 10$

7. $-6 - (5 - 12)$

8. $(-3)(-4) \div (7 - 10)$

9. $(6 - 8)^2(5 - 7)^3$

10. $\dfrac{3(-2) - 2(2)}{-2(8 - 3)}$

In Exercises 11–13, simplify each algebraic expression.

11. $11x - (7x - 4)$

12. $5(3x - 4y) - (2x - y)$

13. $6 - 2[3(x + 1) - 5]$

14. List all the rational numbers in this set.

$$\left\{-7, -\dfrac{4}{5}, 0, 0.25, \sqrt{3}, \sqrt{4}, \dfrac{22}{7}, \pi\right\}$$

15. Insert either $<$ or $>$ in the shaded area to make a true statement: -1 ▨ -100.

16. Find the absolute value: $|-12.8|$.

In Exercises 17–18, evaluate each algebraic expression for the given value of the variable.

17. $5(x - 7); x = 4$

18. $x^2 - 5x; x = -10$

19. Use the commutative property of addition to write an equivalent algebraic expression: $2(x + 3)$.

20. Use the associative property of multiplication to rewrite $-6(4x)$. Then simplify the expression.

21. Use the distributive property to rewrite without parentheses: $7(5x - 1 + 2y)$.

22. What is the difference in elevation between a plane flying 16,200 feet above sea level and a submarine traveling 830 feet below sea level?

In Exercises 23–24, determine whether the given number is a solution of the equation.

23. $\dfrac{1}{5}(x + 2) = \dfrac{1}{10}x + \dfrac{3}{5}; 3$

24. $3(x + 2) - 15 = 4x; -9$

In Exercises 25–26, translate from English to an algebraic expression or equation, whichever is appropriate. Let the variable x represent the number.

25. $\frac{1}{4}$ of a number, decreased by 5, is 32.

26. Seven subtracted from the product of 5 and 4 more than a number

27. In 2009, Marian Robinson, President Obama's mother-in-law, moved into the White House, creating a multigenerational first family. The bar graph shows the number of multigenerational households in the United States for selected years from 1950 through 2008.

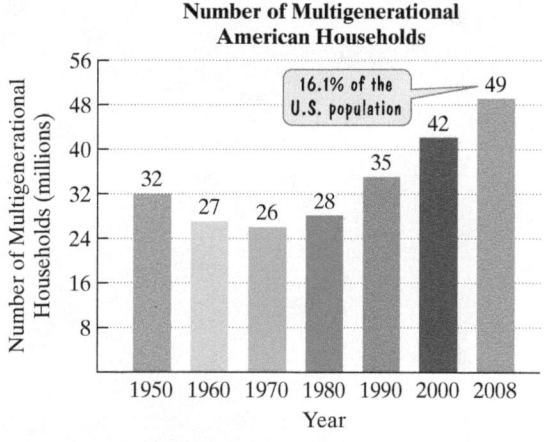

Number of Multigenerational American Households

Source: Pew Research Center

The number of multigenerational American households, H, in millions, can be described by the mathematical model

$$H = 0.01n^2 - 0.5n + 31,$$

where n is the number of years after 1950. Does the mathematical model underestimate or overestimate the actual number of multigenerational households shown by the graph in 1970? By how many million?

28. Electrocardiograms are used in exercise stress tests to determine a person's fitness for strenuous exercise. The target heart rate, in beats per minute, for such tests depends on a person's age. The line graph shows target heart rates for stress tests for people of various ages.

Target Heart Rates for Stress Tests

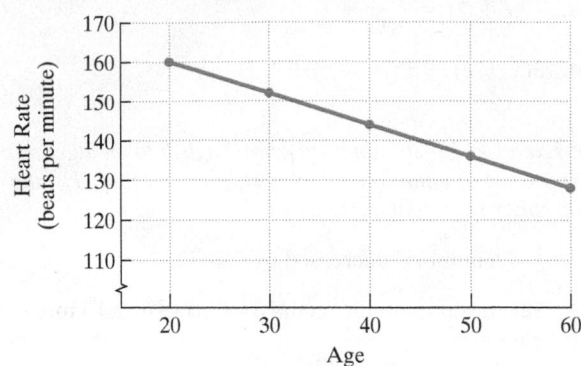

Use the graph to estimate the target heart rate for a 40-year-old taking a stress test.

29. The formula $H = \frac{4}{5}(220 - a)$ gives the target heart rate, H, in beats per minute, on a stress test for a person of age a. Use the formula to find the target heart rate for a 40-year-old. How does this compare with your estimate from Exercise 28?

Linear Equations and Inequalities in One Variable

The belief that humor and laughter can have positive effects on our lives is not new. The Bible tells us, "A merry heart doeth good like a medicine, but a broken spirit drieth the bones." **(Proverbs 17:22)**

Some random humor factoids: ■ The average adult laughs 15 times each day. (**Newhouse News Service**) ■ Forty-six percent of people who are telling a joke laugh more than the people they are telling it to. (**U.S. News and World Report**) ■ Eighty percent of adult laughter does not occur in response to jokes or funny situations. (**Independent**) ■ Algebra can be used to model the influence that humor plays in our responses to negative life events. (**Bob Blitzer, Introductory Algebra**)

That last tidbit that your author threw into the list is true. Based on our sense of humor, there is actually a formula that predicts how we will respond to difficult life events.

Formulas can be used to explain what is happening in the present and to make predictions about what might occur in the future. In this chapter, you will learn to use formulas in new ways that will help you to recognize patterns, logic, and order in a world that can appear chaotic to the untrained eye.

A mathematical model that includes sense of humor as a variable is developed in Example 8 of Section 2.3.

SECTION

2.1

Objectives

1. Identify linear equations in one variable.

2. Use the addition property of equality to solve equations.

3. Solve applied problems using formulas.

The Addition Property of Equality

Credit cards: Are they convenient tools or snakes in your wallet? Before President Obama signed legislation that prohibited issuing credit cards to college students younger than 21, the average credit-card debt for college seniors exceeded $4000. In this section's Exercise Set, you will work with a mathematical model that describes credit-card debt. Your work with this model will involve solving linear equations.

Linear Equations in One Variable

In Chapter 1, we learned that an equation is a statement that two algebraic expressions are equal. We determined whether a given number is an equation's solution by substituting that number for each occurrence of the variable. When the substitution resulted in a true statement, that number was a solution. When the substitution resulted in a false statement, that number was not a solution.

In the next three sections, we will study how to solve equations in one variable. **Solving an equation** is the process of finding the number (or numbers) that make the equation a true statement. These numbers are called the **solutions**, or **roots**, of the equation, and we say that they **satisfy** the equation.

In this chapter, you will learn to solve the simplest type of equation, called a *linear equation in one variable*.

1. Identify linear equations in one variable.

Definition of a Linear Equation in One Variable

A **linear equation in one variable** x is an equation that can be written in the form

$$ax + b = c,$$

where a, b, and c are real numbers, and $a \neq 0$ (a is not equal to 0).

Linear Equations in One Variable (x)

$$3x + 7 = 9 \qquad -15x = 45 \qquad x = 6.8$$

| $ax + b = c$, with $a = 3$, $b = 7$, and $c = 9$ | $ax + b = c$, with $a = -15$, $b = 0$, and $c = 45$ | $ax + b = c$ ($1x + 0 = 6.8$), with $a = 1$, $b = 0$, and $c = 6.8$ |

Nonlinear Equations in One Variable (x)

$$3x^2 + 7 = 9 \qquad -\frac{15}{x} = 45 \qquad |x| = 6.8$$

| Nonlinear because x is squared | Nonlinear because x is in the denominator | Nonlinear because of the absolute value bars around x |

The adjective *linear* contains the word *line*. As we shall see, linear equations are related to graphs whose points lie along a straight line.

2 Use the addition property of equality to solve equations.

Using the Addition Property of Equality to Solve Equations

Consider the equation

$$x = 11.$$

By inspection, we can see that the solution to this equation is 11. If we substitute 11 for x, we obtain the true statement $11 = 11$.

Now consider the equation

$$x - 3 = 8.$$

If we substitute 11 for x, we obtain $11 - 3 \overset{?}{=} 8$. Subtracting on the left side, we get the true statement $8 = 8$.

The equations $x - 3 = 8$ and $x = 11$ both have the same solution, namely 11, and are called *equivalent equations*. **Equivalent equations** are equations that have the same solution.

The idea in solving a linear equation is to get an equivalent equation with the variable (the letter) by itself on one side of the equal sign and a number by itself on the other side. For example, consider the equation $x - 3 = 8$. To get x by itself on the left side, add 3 to the left side, because $x - 3 + 3$ gives $x + 0$, or just x. You must then add 3 to the right side also. By doing this, we are using the **addition property of equality**.

The Addition Property of Equality

The same real number (or algebraic expression) may be added to both sides of an equation without changing the equation's solution. This can be expressed symbolically as follows:

$$\text{If } a = b, \text{ then } a + c = b + c.$$

EXAMPLE 1 Solving an Equation Using the Addition Property

Solve the equation: $x - 3 = 8$.

Solution We can isolate the variable, x, by adding 3 to both sides of the equation.

$$x - 3 = 8 \qquad \text{This is the given equation.}$$
$$x - 3 + 3 = 8 + 3 \qquad \text{Add 3 to both sides.}$$
$$x + 0 = 11 \qquad \text{This step is often done mentally and not listed.}$$
$$x = 11$$

By inspection, we can see that the solution to $x = 11$ is 11. To check this proposed solution, replace x with 11 in the original equation.

Check
$$x - 3 = 8 \qquad \text{This is the original equation.}$$
$$11 - 3 \overset{?}{=} 8 \qquad \text{Substitute 11 for } x.$$
$$8 = 8 \qquad \text{Subtract: } 11 - 3 = 8.$$

This statement is true.

Because the check results in a true statement, we conclude that the solution to the given equation is 11. ∎

Great Question!

Is there another way to show that I'm adding the same number to both sides of an equation?

Some people prefer to show the number below the equation:

$$\begin{array}{rcr} x - 3 = & & 8 \\ + 3 & & + 3 \\ \hline x & = & 11. \end{array}$$

Great Question!

In a high school algebra course, I remember my teacher talking about balancing an equation. What does the addition property of equality have to do with a balanced equation?

You can think of an equation as a balanced scale—balanced because its two sides are equal. To maintain this balance, whatever you do to one side must also be done to the other side.

The set of an equation's solutions is called its **solution set**. Thus, the solution set of the equation in Example 1, $x - 3 = 8$, is {11}. The solution can be expressed as 11 or, using set notation, {11}.

✓ **CHECK POINT 1** Solve the equation and check your proposed solution:
$$x - 5 = 12.$$

When we use the addition property of equality, we add the same number to both sides of an equation. We know that subtraction is the addition of an opposite, or additive inverse. Thus, the addition property also lets us subtract the same number from both sides of an equation without changing the equation's solution.

EXAMPLE 2 Subtracting the Same Number from Both Sides

Solve and check: $z + 1.4 = 2.06$.

Solution

$$z + 1.4 = 2.06$$ This is the given equation.

$$z + 1.4 - 1.4 = 2.06 - 1.4$$ Subtract 1.4 from both sides. This is equivalent to adding −1.4 to both sides.

$$z = 0.66$$ Subtracting 1.4 from both sides eliminates 1.4 on the left.

Can you see that the solution to $z = 0.66$ is 0.66? To check this proposed solution, replace z with 0.66 in the original equation.

Check $z + 1.4 = 2.06$ This is the original equation.

$$0.66 + 1.4 \stackrel{?}{=} 2.06$$ Substitute 0.66 for z.

$$2.06 = 2.06$$ This statement is true.

This true statement indicates that the solution is 0.66, or the solution set is {0.66}. ∎

✓ **CHECK POINT 2** Solve and check: $z + 2.8 = 5.09$.

When isolating the variable, we can isolate it on either the left side or the right side of an equation.

EXAMPLE 3   Isolating the Variable on the Right

Solve and check: $-\dfrac{1}{2} = x - \dfrac{2}{3}$.

Solution   We can isolate the variable, x, on the right side by adding $\frac{2}{3}$ to both sides of the equation.

$$-\frac{1}{2} = x - \frac{2}{3}$$ This is the given equation.

$$-\frac{1}{2} + \frac{2}{3} = x - \frac{2}{3} + \frac{2}{3}$$ Add $\frac{2}{3}$ to both sides, isolating x on the right.

$$-\frac{3}{6} + \frac{4}{6} = x$$ Rewrite each fraction as an equivalent fraction with a denominator of 6: $-\frac{1}{2} + \frac{2}{3} = -\frac{1}{2} \cdot \frac{3}{3} + \frac{2}{3} \cdot \frac{2}{2} = -\frac{3}{6} + \frac{4}{6}$.

$$\frac{1}{6} = x$$ Add on the left side: $-\frac{3}{6} + \frac{4}{6} = \frac{-3 + 4}{6} = \frac{1}{6}$.

Take a moment to check the proposed solution, $\frac{1}{6}$. Substitute $\frac{1}{6}$ for x in the original equation. You should obtain $-\frac{1}{2} = -\frac{1}{2}$. This true statement indicates that the solution is $\frac{1}{6}$, or the solution set is $\left\{\frac{1}{6}\right\}$. ■

✓ **CHECK POINT 3**   Solve and check: $-\dfrac{1}{2} = x - \dfrac{3}{4}$.

In Example 4, we combine like terms before using the addition property.

EXAMPLE 4   Combining Like Terms before Using the Addition Property

Solve and check: $5y + 3 - 4y - 8 = 6 + 9$.

Solution

$$5y + 3 - 4y - 8 = 6 + 9$$ This is the given equation.

$$y - 5 = 15$$ Combine like terms:
$5y - 4y = y$, $3 - 8 = -5$, and $6 + 9 = 15$.

$$y - 5 + 5 = 15 + 5$$ Add 5 to both sides.

$$y = 20$$ Simplify.

To check the proposed solution, 20, replace y with 20 in the original equation.

Check $5y + 3 - 4y - 8 = 6 + 9$ Be sure to use the original equation and not the simplified form from the second step above. (Why?)

$5(20) + 3 - 4(20) - 8 \overset{?}{=} 6 + 9$ Substitute 20 for y.

$100 + 3 - 80 - 8 \overset{?}{=} 6 + 9$ Multiply on the left.

$103 - 88 \overset{?}{=} 6 + 9$ Combine positive numbers and combine negative numbers on the left.

$15 = 15$ This statement is true.

This true statement verifies that the solution is 20, or the solution set is {20}. ■

✓ **CHECK POINT 4**   Solve and check: $8y + 7 - 7y - 10 = 6 + 4$.

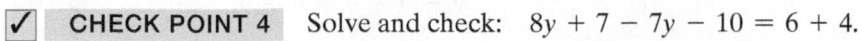

Great Question!

Am I allowed to "flip" the two sides of an equation?

The equations $a = b$ and $b = a$ have the same meaning. If you prefer, you can solve

$$-\frac{1}{2} = x - \frac{2}{3}$$

by reversing the two sides and solving

$$x - \frac{2}{3} = -\frac{1}{2}.$$

Adding and Subtracting Variable Terms on Both Sides of an Equation

In some equations, variable terms appear on both sides. Here is an example:

$$4x = 7 + 3x.$$

A variable term, $4x$, is on the left side.

A variable term, $3x$, is on the right side.

Our goal is to isolate all the variable terms on one side of the equation. We can use the addition property of equality to do this. The property allows us to add or subtract the same variable term on both sides of an equation without changing the solution. Let's see how we can use this idea to solve $4x = 7 + 3x$.

EXAMPLE 5 Using the Addition Property to Isolate Variable Terms

Solve and check: $4x = 7 + 3x$.

Solution In the given equation, variable terms appear on both sides. We can isolate them on one side by subtracting $3x$ from both sides of the equation.

$4x = 7 + 3x$	This is the given equation.
$4x - 3x = 7 + 3x - 3x$	Subtract $3x$ from both sides and isolate variable terms on the left.
$x = 7$	Subtracting $3x$ from both sides eliminates $3x$ on the right. On the left, $4x - 3x = 1x = x$.

To check the proposed solution, 7, replace x with 7 in the original equation.

Check		
	$4x = 7 + 3x$	Use the original equation.
	$4(7) \stackrel{?}{=} 7 + 3(7)$	Substitute 7 for x.
	$28 \stackrel{?}{=} 7 + 21$	Multiply: $4(7) = 28$ and $3(7) = 21$.
	$28 = 28$	This statement is true.

This true statement verifies that the solution is 7, or the solution set is {7}. ∎

✓ **CHECK POINT 5** Solve and check: $7x = 12 + 6x$.

EXAMPLE 6 Solving an Equation by Isolating the Variable

Solve and check: $3y - 9 = 2y + 6$.

Solution Our goal is to isolate variable terms on one side and constant terms on the other side. Let's begin by isolating the variable on the left.

$3y - 9 = 2y + 6$	This is the given equation.
$3y - 2y - 9 = 2y - 2y + 6$	Isolate the variable terms on the left by subtracting $2y$ from both sides.
$y - 9 = 6$	Subtracting $2y$ from both sides eliminates $2y$ on the right. On the left, $3y - 2y = 1y = y$.

Now we isolate the constant terms on the right by adding 9 to both sides.

$y - 9 + 9 = 6 + 9$	Add 9 to both sides.
$y = 15$	Simplify.

Check $3y - 9 = 2y + 6$ *Use the original equation.*

$3(15) - 9 \stackrel{?}{=} 2(15) + 6$ *Substitute 15 for y.*

$45 - 9 \stackrel{?}{=} 30 + 6$ *Multiply: 3(15) = 45 and 2(15) = 30.*

$36 = 36$ *This statement is true.*

The solution is 15, or the solution set is {15}. ∎

✓   **CHECK POINT 6**   Solve and check: $3x - 6 = 2x + 5$.

3   Solve applied problems using formulas.

Applications

Our next example shows how the addition property of equality can be used to find the value of a variable in a mathematical model.

EXAMPLE 7   An Application: Vocabulary and Age

There is a relationship between the number of words in a child's vocabulary, V, and the child's age, A, in months, for ages between 15 and 50 months, inclusive. This relationship can be modeled by the formula

$$V + 900 = 60A.$$

Use the formula to find the number of words in a child's vocabulary at the age of 30 months.

Solution   In the formula, A represents the child's age, in months. We are interested in a 30-month-old child. Thus, we substitute 30 for A. Then we use the addition property of equality to find V, the number of words in the child's vocabulary.

$V + 900 = 60A$ *This is the given formula.*

$V + 900 = 60(30)$ *Substitute 30 for A.*

$V + 900 = 1800$ *Multiply: 60(30) = 1800.*

$V + 900 - 900 = 1800 - 900$ *Subtract 900 from both sides and solve for V.*

$V = 900$

At the age of 30 months, a child has a vocabulary of 900 words. ∎

✓   **CHECK POINT 7**   Use the formula $V + 900 = 60A$ to find the number of words in a child's vocabulary at the age of 50 months.

The line graph in **Figure 2.1** allows us to "see" the formula $V + 900 = 60A$. The two points labeled with voice balloons illustrate what we learned about vocabulary and age by solving equations in Example 7 and Check Point 7.

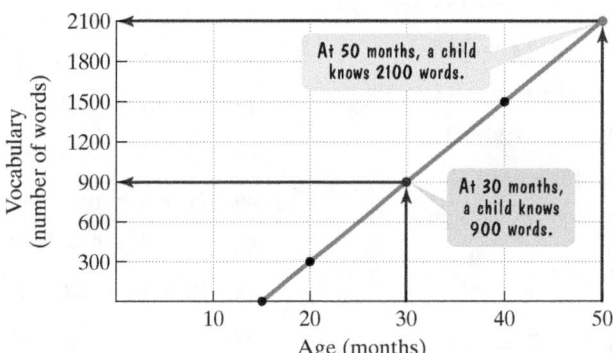

Vocabulary and Age

At 50 months, a child knows 2100 words.

At 30 months, a child knows 900 words.

Vocabulary (number of words)

Age (months)

Figure 2.1

Later in the book, you will learn to create graphs for mathematical models and formulas. Linear equations are related to models whose graphs are straight lines, like the one in **Figure 2.1**. For this reason, these equations are called *linear* equations.

Achieving Success

Algebra is cumulative. This means that the topics build on one another. Understanding each topic depends on understanding the previous material. Do not let yourself fall behind.

CONCEPT AND VOCABULARY CHECK

Fill in each blank so that the resulting statement is true.

1. The process of finding the number or numbers that make an equation a true statement is called _____ the equation.

2. An equation in the form $ax + b = c$, such as $7x + 9 = 13$, is called a/an _____ equation in one variable.

3. Equations that have the same solution are called _____ equations.

4. The addition property of equality states that if $a = b$, then $a + c =$ _____.

5. The addition property of equality lets us add or _____ the same number on both sides of an equation without changing the equation's _____.

6. The equation $x - 7 = 13$ can be solved by _____ to both sides.

7. The equation $7x = 5 + 6x$ can be solved by _____ from both sides.

2.1 EXERCISE SET

MyMathLab®   Watch the videos in MyMathLab    Download the MyDashBoard App

Practice Exercises

In Exercises 1–10, identify the linear equations in one variable.

1. $x - 9 = 13$
2. $x - 15 = 20$
3. $x^2 - 9 = 13$
4. $x^2 - 15 = 20$
5. $\dfrac{9}{x} = 13$
6. $\dfrac{15}{x} = 20$
7. $\sqrt{2}x + \pi = 0.\overline{3}$
8. $\sqrt{3}x + \pi = 0.\overline{6}$
9. $|x + 2| = 5$
10. $|x + 5| = 8$

Solve each equation in Exercises 11–54 using the addition property of equality. Be sure to check your proposed solutions.

11. $x - 4 = 19$
12. $y - 5 = -18$
13. $z + 8 = -12$
14. $z + 13 = -15$
15. $-2 = x + 14$
16. $-13 = x + 11$
17. $-17 = y - 5$
18. $-21 = y - 4$
19. $7 + z = 11$
20. $18 + z = 14$
21. $-6 + y = -17$
22. $-8 + y = -29$
23. $x + \dfrac{1}{3} = \dfrac{7}{3}$
24. $x + \dfrac{7}{8} = \dfrac{9}{8}$
25. $t + \dfrac{5}{6} = -\dfrac{7}{12}$
26. $t + \dfrac{2}{3} = -\dfrac{7}{6}$
27. $x - \dfrac{3}{4} = \dfrac{9}{2}$
28. $x - \dfrac{3}{5} = \dfrac{7}{10}$
29. $-\dfrac{1}{5} + y = -\dfrac{3}{4}$
30. $-\dfrac{1}{8} + y = -\dfrac{1}{4}$
31. $3.2 + x = 7.5$
32. $-2.7 + w = -5.3$
33. $x + \dfrac{3}{4} = -\dfrac{9}{2}$
34. $r + \dfrac{3}{5} = -\dfrac{7}{10}$
35. $5 = -13 + y$
36. $-11 = 8 + x$
37. $-\dfrac{3}{5} = -\dfrac{3}{2} + s$
38. $\dfrac{7}{3} = -\dfrac{5}{2} + z$
39. $830 + y = 520$
40. $-90 + t = -35$
41. $r + 3.7 = 8$
42. $x + 10.6 = -9$
43. $-3.7 + m = -3.7$
44. $y + \dfrac{7}{11} = \dfrac{7}{11}$
45. $6y + 3 - 5y = 14$
46. $-3x - 5 + 4x = 9$
47. $7 - 5x + 8 + 2x + 4x - 3 = 2 + 3 \cdot 5$
48. $13 - 3r + 2 + 6r - 2r - 1 = 3 + 2 \cdot 9$
49. $7y + 4 = 6y - 9$
50. $4r - 3 = 5 + 3r$
51. $12 - 6x = 18 - 7x$
52. $20 - 7s = 26 - 8s$
53. $4x + 2 = 3(x - 6) + 8$
54. $7x + 3 = 6(x - 1) + 9$

Practice PLUS

The equations in Exercises 55–58 contain small geometric figures that represent real numbers. Use the addition property of equality to isolate x on one side of the equation and the geometric figures on the other side.

55. $x - \square = \triangle$

56. $x + \square = \triangle$

57. $2x + \triangle = 3x + \square$

58. $6x - \triangle = 7x - \square$

In Exercises 59–62, use the given information to write an equation. Let x represent the number described in each exercise. Then solve the equation and find the number.

59. If 12 is subtracted from a number, the result is −2. Find the number.

60. If 23 is subtracted from a number, the result is −8. Find the number.

61. The difference between $\frac{2}{5}$ of a number and 8 is $\frac{7}{5}$ of that number. Find the number.

62. The difference between 3 and $\frac{2}{7}$ of a number is $\frac{5}{7}$ of that number. Find the number.

Application Exercises

Formulas frequently appear in the business world. For example, the cost, C, of an item (the price paid by a retailer) plus the markup, M, on that item (the retailer's profit) equals the selling price, S, of the item. The formula is

$$C + M = S.$$

Use the formula to solve Exercises 63–64.

63. The selling price of a computer is $1850. If the markup on the computer is $150, find the cost to the retailer for the computer.

64. The selling price of a television is $650. If the cost to the retailer for the television is $520, find the markup.

The bar graph shows the average credit-card debt per U.S. household for selected years from 2000 through 2008.

Average Credit-Card Debt per U.S. Household

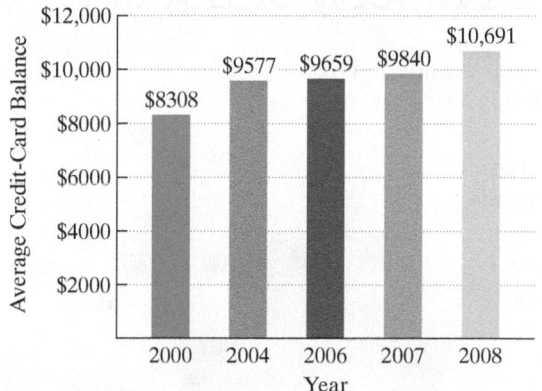

Source: CardTrak.com

The data displayed by the bar graph can be described by the mathematical model

$$d - 257x = 8328,$$

where d is the average credit-card debt per U.S. household x years after 2000. Use this information to solve Exercises 65–66.

65. According to the formula at the bottom of the previous column, what was the average credit-card debt per U.S. household for 2008? Does this underestimate or overestimate the number displayed by the bar graph? By how much?

66. According to the formula at the bottom of the previous column, what was the average credit-card debt per U.S. household for 2007? Does this underestimate or overestimate the number displayed by the bar graph? By how much?

Diversity Index. *The diversity index, from 0 (no diversity) to 100, measures the chance that two randomly selected people are a different race or ethnicity. The diversity index in the United States varies widely from region to region, from as high as 79 in Hawaii and 68 in California to as low as 10 in Maine and Vermont and 13 in West Virginia. The line graph shows the national diversity index for the United States for four years in the period from 1980 through 2009.*

Chance That Two Randomly Selected Americans Are a Different Race or Ethnicity

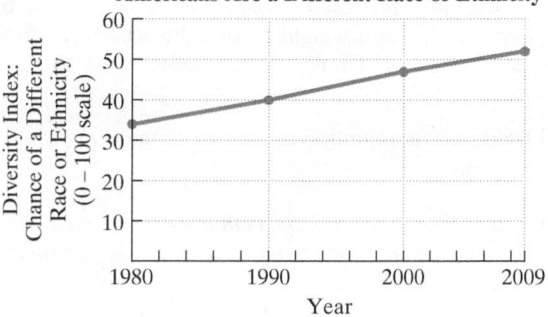

Source: USA Today

The data displayed by the line graph can be described by the mathematical model

$$I - 0.6x = 34,$$

where I is the national diversity index in the United States x years after 1980. Use this information to solve Exercises 67–68.

67. a. Use the line graph to estimate the U.S. diversity index in 2009.

b. Use the formula to determine the U.S. diversity index in 2009. How does this compare with your graphical estimate from part (a)?

68. a. Use the line graph to estimate the U.S. diversity index in 2000.

b. Use the formula to determine the U.S. diversity index in 2000. How does this compare with your graphical estimate from part (a)?

Writing in Mathematics

69. State the addition property of equality and give an example.

70. Explain why $x + 2 = 9$ and $x + 2 = -6$ are not equivalent equations.

71. What is the difference between solving an equation such as

$$5y + 3 - 4y - 8 = 6 + 9$$

and simplifying an algebraic expression such as

$$5y + 3 - 4y - 8?$$

If there is a difference, which topic should be taught first? Why?

72. Look, again, at the graph for Exercises 67–68 that shows the diversity index in the United States over time. You used a *linear* equation to solve Exercise 67 or 68. What does the adjective *linear* have to do with the relationship among the four data points in the graph?

Critical Thinking Exercises

Make Sense? *In Exercises 73–76, determine whether each statement "makes sense" or "does not make sense" and explain your reasoning.*

73. The book is teaching me totally different things than my instructor: The book adds the number to both sides beside the equation, but my instructor adds the number underneath.

74. There are times that I prefer to check an equation's solution in my head and not show the check.

75. Solving an equation reminds me of keeping a barbell balanced: If I add weight to or subtract weight from one side of the bar, I must do the same thing to the other side.

76. I used a linear equation to explore data points lying on the same line.

In Exercises 77–80, determine whether each statement is true or false. If the statement is false, make the necessary change(s) to produce a true statement.

77. If $y - a = -b$, then $y = a + b$.

78. If $y + 7 = 0$, then $y = 7$.

79. If $2x - 5 = 3x$, then $x = -5$.

80. If $3x = 18$, then $x = 18 - 3$.

81. Write an equation with a negative solution that can be solved by adding 100 to both sides.

Use a calculator to solve each equation in Exercises 82–83.

82. $x - 7.0463 = -9.2714$

83. $6.9825 = 4.2296 + y$

Review Exercises

84. Write as an algebraic expression in which x represents the number: the quotient of 9 and a number, decreased by 4 times the number. (Section 1.1, Example 3)

85. Simplify: $-16 - 8 \div 4 \cdot (-2)$. (Section 1.8, Example 4)

86. Simplify: $3[7x - 2(5x - 1)]$. (Section 1.8, Example 11)

Preview Exercises

Exercises 87–89 will help you prepare for the material covered in the next section.

87. Multiply and simplify: $5 \cdot \dfrac{x}{5}$.

88. Divide and simplify: $\dfrac{-7y}{-7}$.

89. Is 4 a solution of $3x - 14 = -2x + 6$?

SECTION

2.2

Objectives

1. Use the multiplication property of equality to solve equations.

2. Solve equations in the form $-x = c$.

3. Use the addition and multiplication properties to solve equations.

4. Solve applied problems using formulas.

The Multiplication Property of Equality

How much was that doggie in the window? This purebred Westie, selling for $2000 in 2009, was a real bargain for $50 back in 1940.

Prices and salaries have soared over the decades. If present trends continue, what will you spend and earn in the future? In this section, we introduce a new property for solving equations to address these questions.

1 Use the multiplication property of equality to solve equations.

Using the Multiplication Property of Equality to Solve Equations

Can the addition property of equality be used to solve every linear equation in one variable? No. For example, consider the equation

$$\frac{x}{5} = 9.$$

We cannot isolate the variable x by adding or subtracting 5 on both sides. To get x by itself on the left side, multiply the left side by 5:

$$5 \cdot \frac{x}{5} = \left(5 \cdot \frac{1}{5}\right)x = 1x = x.$$

5 is the multiplicative inverse of $\frac{1}{5}$.

You must then multiply the right side by 5 also. By doing this, we are using the **multiplication property of equality**.

The Multiplication Property of Equality

The same nonzero real number (or algebraic expression) may multiply both sides of an equation without changing the solution. This can be expressed symbolically as follows:

If $a = b$ and $c \neq 0$, then $ac = bc$.

EXAMPLE 1   Solving an Equation Using the Multiplication Property

Solve the equation: $\frac{x}{5} = 9.$

Solution   We can isolate the variable, x, by multiplying both sides of the equation by 5.

$$\frac{x}{5} = 9 \qquad \text{This is the given equation.}$$

$$5 \cdot \frac{x}{5} = 5 \cdot 9 \qquad \text{Multiply both sides by 5.}$$

$$1x = 45 \qquad \text{Simplify.}$$

$$x = 45 \qquad 1x = x$$

By substituting 45 for x in the original equation, we obtain the true statement $9 = 9$. This verifies that the solution is 45, or the solution set is {45}. ■

✓ **CHECK POINT 1**   Solve the equation: $\frac{x}{3} = 12.$

When we use the multiplication property of equality, we multiply both sides of an equation by the same nonzero number. We know that division is multiplication by a multiplicative inverse. Thus, the multiplication property also lets us divide both sides of an equation by a nonzero number without changing the solution.

EXAMPLE 2   Dividing Both Sides by the Same Nonzero Number

Solve: **a.** $6x = 30$ **b.** $-7y = 56$ **c.** $-18.9 = 3z$.

Solution   In each equation, the variable is multiplied by a number. We can isolate the variable by dividing both sides of the equation by that number.

a. $6x = 30$ This is the given equation.

$\dfrac{6x}{6} = \dfrac{30}{6}$ Divide both sides by 6.

$1x = 5$ Simplify.

$x = 5$ $1x = x$

By substituting 5 for x in the original equation, we obtain the true statement $30 = 30$. The solution is 5, or the solution set is {5}.

b. $-7y = 56$ This is the given equation.

$\dfrac{-7y}{-7} = \dfrac{56}{-7}$ Divide both sides by -7.

$1y = -8$ Simplify.

$y = -8$ $1y = y$

By substituting -8 for y in the original equation, we obtain the true statement $56 = 56$. The solution is -8, or the solution set is $\{-8\}$.

c. $-18.9 = 3z$ This is the given equation.

$\dfrac{-18.9}{3} = \dfrac{3z}{3}$ Divide both sides by 3.

$-6.3 = 1z$ Simplify.

$-6.3 = z$ $1z = z$

By substituting -6.3 for z in the original equation, we obtain the true statement $-18.9 = -18.9$. The solution is -6.3, or the solution set is $\{-6.3\}$.   ∎

✓ **CHECK POINT 2**   Solve:

a. $4x = 84$ **b.** $-11y = 44$

c. $-15.5 = 5z$.

Some equations have a variable term with a fractional coefficient. Here is an example:

The coefficient of the term $\frac{3}{4}y$ is $\frac{3}{4}$. $\dfrac{3}{4}y = 12$.

To isolate the variable, multiply both sides of the equation by the multiplicative inverse of the fraction. For the equation $\frac{3}{4}y = 12$, the multiplicative inverse of $\frac{3}{4}$ is $\frac{4}{3}$. Thus, we solve $\frac{3}{4}y = 12$ by multiplying both sides by $\frac{4}{3}$.

EXAMPLE 3   Using the Multiplication Property to Eliminate a Fractional Coefficient

Solve: **a.** $\dfrac{3}{4}y = 12$ **b.** $9 = -\dfrac{3}{5}x$.

Solution

a. $\dfrac{3}{4}y = 12$ This is the given equation.

$\dfrac{4}{3}\left(\dfrac{3}{4}y\right) = \dfrac{4}{3} \cdot 12$ Multiply both sides by $\frac{4}{3}$, the multiplicative inverse of $\frac{3}{4}$.

$1y = 16$ On the left, $\frac{4}{3}\left(\frac{3}{4}y\right) = \left(\frac{4}{3} \cdot \frac{3}{4}\right)y = 1y$.

                         On the right, $\frac{4}{3} \cdot \frac{12}{1} = \frac{48}{3} = 16$.

$y = 16$ $1y = y$

By substituting 16 for y in the original equation, we obtain the true statement $12 = 12$. The solution is 16, or the solution set is {16}.

b. $9 = -\dfrac{3}{5}x$ This is the given equation.

$-\dfrac{5}{3} \cdot 9 = -\dfrac{5}{3}\left(-\dfrac{3}{5}x\right)$ Multiply both sides by $-\frac{5}{3}$, the multiplicative inverse of $-\frac{3}{5}$.

$-15 = 1x$ Simplify.

$-15 = x$ $1x = x$

By substituting -15 for x in the original equation, we obtain the true statement $9 = 9$. The solution is -15, or the solution set is $\{-15\}$. ∎

✓ **CHECK POINT 3**   Solve:

a. $\dfrac{2}{3}y = 16$ **b.** $28 = -\dfrac{7}{4}x$.

---

② Solve equations in the form $-x = c$.

Equations and Coefficients of -1

How do we solve an equation in the form $-x = c$, such as $-x = 4$? Because the equation means $-1x = 4$, we have not yet obtained a solution. The solution of an equation is obtained from the form $x =$ some number. The equation $-x = 4$ is not yet in this form. We still need to isolate x. We can do this by multiplying or dividing both sides of the equation by -1. We will multiply by -1.

EXAMPLE 4   Solving Equations in the Form $-x = c$

Solve: **a.** $-x = 4$ **b.** $-x = -7$.

Solution   We multiply both sides of each equation by -1. This will isolate x on the left side.

a. $-x = 4$ This is the given equation.

$-1x = 4$ Rewrite $-x$ as $-1x$.

$(-1)(-1x) = (-1)(4)$ Multiply both sides by -1.

$1x = -4$ On the left, $(-1)(-1) = 1$. On the right, $(-1)(4) = -4$.

$x = -4$ $1x = x$

Check $-x = 4$ This is the original equation.

$-(-4) \stackrel{?}{=} 4$ Substitute -4 for x.

$4 = 4$ $-(-a) = a$, so $-(-4) = 4$.

This true statement indicates that the solution is -4, or the solution set is $\{-4\}$.

Great Question!

Is there a fast way to solve equations in the form $-x = c$?

If $-x = c$, then the equation's solution is the opposite, or additive inverse, of c. For example, the solution of $-x = -7$ is the opposite of -7, which is 7.

b.

$-x = -7$	This is the given equation.
$-1x = -7$	Rewrite $-x$ as $-1x$.
$(-1)(-1x) = (-1)(-7)$	Multiply both sides by -1.
$1x = 7$	$(-1)(-1) = 1$ and $(-1)(-7) = 7$.
$x = 7$	$1x = x$

By substituting 7 for x in the original equation, we obtain the true statement $-7 = -7$. The solution is 7, or the solution set is $\{7\}$. ■

✓ **CHECK POINT 4** Solve: **a.** $-x = 5$ **b.** $-x = -3$.

3 Use the addition and multiplication properties to solve equations.

Equations Requiring Both the Addition and Multiplication Properties

When an equation does not contain fractions, we will often use the addition property of equality before the multiplication property of equality. Our overall goal is to isolate the variable with a coefficient of 1 on either the left or right side of the equation.

Here is the procedure that we will be using to solve the equations in the next three examples:

- Use the addition property of equality to isolate the variable term.
- Use the multiplication property of equality to isolate the variable.

EXAMPLE 5 Using Both the Addition and Multiplication Properties

Solve: $3x + 1 = 7$.

Solution We first isolate the variable term, $3x$, by subtracting 1 from both sides. Then we isolate the variable, x, by dividing both sides by 3.

- **Use the addition property of equality to isolate the variable term.**

$3x + 1 = 7$	This is the given equation.
$3x + 1 - 1 = 7 - 1$	Use the addition property, subtracting 1 from both sides.
$3x = 6$	Simplify.

- **Use the multiplication property of equality to isolate the variable.**

$\dfrac{3x}{3} = \dfrac{6}{3}$	Divide both sides of $3x = 6$ by 3.
$x = 2$	Simplify.

By substituting 2 for x in the original equation, $3x + 1 = 7$, we obtain the true statement $7 = 7$. The solution is 2, or the solution set is $\{2\}$. ■

✓ **CHECK POINT 5** Solve: $4x + 3 = 27$.

EXAMPLE 6 Using Both the Addition and Multiplication Properties

Solve: $-2y - 28 = 4$.

Solution We first isolate the variable term, $-2y$, by adding 28 to both sides. Then we isolate the variable, y, by dividing both sides by -2.

- **Use the addition property of equality to isolate the variable term.**

$$-2y - 28 = 4$$ This is the given equation.

$$-2y - 28 + 28 = 4 + 28$$ Use the addition property, adding 28 to both sides.

$$-2y = 32$$ Simplify.

- **Use the multiplication property of equality to isolate the variable.**

$$\frac{-2y}{-2} = \frac{32}{-2}$$ Divide both sides by −2.

$$y = -16$$ Simplify.

Take a moment to substitute −16 for y in the given equation. Do you obtain the true statement $4 = 4$? The solution is −16, or the solution set is {−16}. ■

✓ **CHECK POINT 6**   Solve: $-4y - 15 = 25$.

EXAMPLE 7   Using Both the Addition and Multiplication Properties

Solve: $3x - 14 = -2x + 6$.

Solution   We will use the addition property to collect all terms involving x on the left and all numerical terms on the right. Then we will isolate the variable, x, by dividing both sides by its coefficient.

- **Use the addition property of equality to isolate the variable term.**

$$3x - 14 = -2x + 6$$ This is the given equation.

$$3x + 2x - 14 = -2x + 2x + 6$$ Add 2x to both sides.

$$5x - 14 = 6$$ Simplify.

$$5x - 14 + 14 = 6 + 14$$ Add 14 to both sides.

$$5x = 20$$ Simplify. The variable term, 5x, is isolated on the left. The numerical term, 20, is isolated on the right.

- **Use the multiplication property of equality to isolate the variable.**

$$\frac{5x}{5} = \frac{20}{5}$$ Divide both sides by 5.

$$x = 4$$ Simplify.

Check $3x - 14 = -2x + 6$ Use the original equation.

$$3(4) - 14 \overset{?}{=} -2(4) + 6$$ Substitute the proposed solution, 4, for x.

$$12 - 14 \overset{?}{=} -8 + 6$$ Multiply.

$$-2 = -2$$ Simplify.

The true statement $-2 = -2$ verifies that the solution is 4, or the solution set is {4}. ■

✓ **CHECK POINT 7**   Solve: $2x - 15 = -4x + 21$.

4 Solve applied problems using formulas.

Applications

Most, but not all, salaries and prices have soared over the decades. To make it easier to compare, **Figure 2.2** converts historical prices into today's dollars, with adjustments based on the consumer price index. Our next example, as well as forthcoming exercises, shows how solving linear equations can be used to explore mathematical models based on some of the data displayed by the bar graphs.

Comparing Salaries and Prices Using Today's Dollars

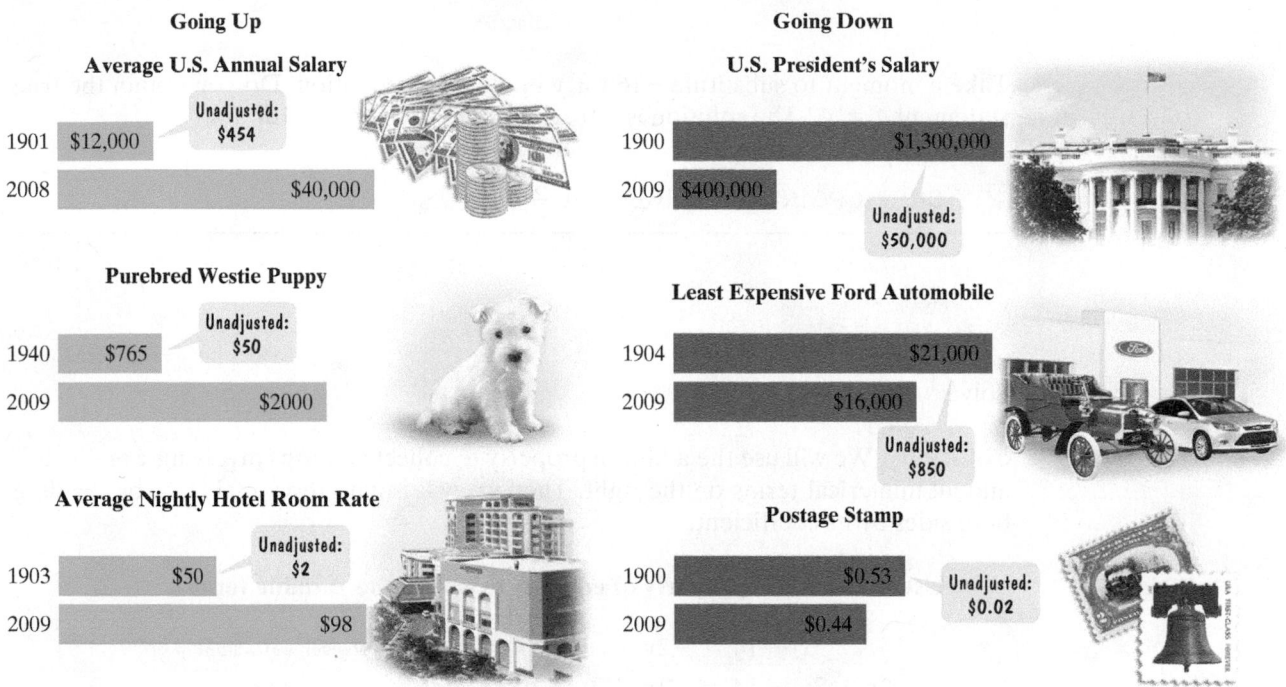

Figure 2.2

Source: Newsweek

EXAMPLE 8   Average Annual Salary

The data in **Figure 2.2** for average annual salary can be described by the mathematical model

$$S = 262n + 12{,}000,$$

where S is the average U.S. annual salary, in dollars, n years after 1901.

a. Does the formula underestimate or overestimate the average annual salary in 2008? By how much?

b. If trends shown by the formula continue, when will the average annual salary be $43,000? Round to the nearest year.

Solution

a. **Figure 2.2** indicates that the average annual salary was $40,000 in 2008. Let's see if the formula underestimates or overestimates this value. The year 2008 is 107 years after 1901 (2008 − 1901 = 107), so we substitute 107 for n.

$$S = 262n + 12{,}000 \qquad \text{This is the given formula.}$$
$$S = 262(107) + 12{,}000 \qquad \text{Replace } n \text{ with 107.}$$
$$S = 28{,}034 + 12{,}000 \qquad \text{Multiply: } 262(107) = 28{,}034.$$
$$S = 40{,}034 \qquad \text{Add: } 28{,}034 + 12{,}000 = 40{,}034.$$

The model indicates that the average annual salary was $40,034 in 2008. This overestimates the number displayed by the graph, $40,000, by $34.

b. We are interested in when the average annual salary will be $43,000. We substitute 43,000 for the average annual salary, S, in the given formula. Then we solve for n, the number of years after 1901.

$$S = 262n + 12,000$$ This is the given formula.

$$43,000 = 262n + 12,000$$ Replace S with 43,000.

Our goal is to isolate n.

$$43,000 - 12,000 = 262n + 12,000 - 12,000$$ Isolate the term containing n by subtracting 12,000 from both sides.

$$31,000 = 262n$$ Simplify.

$$\frac{31,000}{262} = \frac{262n}{262}$$ Divide both sides by 262.

$$118 \approx n$$ Simplify: $\dfrac{31,000}{262} \approx 118.3 \approx 118.$

The formula indicates that approximately 118 years after 1901, or in 2019 (1901 + 118 = 2019), the average annual salary in the United States will be $43,000. ∎

Great Question!

In Example 8, did you use the formula $S = 262n + 12,000$ in two different ways?

Good observation! In Example 8(a), we substituted a value for n and used the order of operations to determine S. In Example 8(b), we substituted a value for S and solved a linear equation to determine n.

✓ **CHECK POINT 8**   The data in **Figure 2.2** for the price of a Westie puppy can be described by the mathematical model

$$P = 18n + 765,$$

where P is the puppy's price n years after 1940.

a. Does the formula underestimate or overestimate the price of a Westie puppy in 2009? By how much?

b. If trends shown by the formula continue, when will the price of a Westie puppy be $2151?

CONCEPT AND VOCABULARY CHECK

Fill in each blank so that the resulting statement is true.

1. The multiplication property of equality states that if $a = b$ and $c \neq 0$, then $ac =$ _____.

2. The multiplication property of equality lets us multiply or _____ both sides of an equation by the same nonzero number.

3. The equation $\dfrac{x}{7} = 13$ can be solved by _____ both sides by _____.

4. The equation $-8y = 32$ can be solved by _____ both sides by _____.

5. The equation $\dfrac{3}{5}x = 20$ can be solved by _____ both sides by _____.

6. The equation $-x = 12$ can be solved by _____ both sides by _____.

7. The equation $5x + 2 = 17$ can be solved by first _____ from both sides and then _____ both sides by _____.

MyMathLab®   Watch the videos in MyMathLab   Download the MyDashBoard App

Practice Exercises

Solve each equation in Exercises 1–28 using the multiplication property of equality. Be sure to check your proposed solutions.

1. $\dfrac{x}{6} = 5$

2. $\dfrac{x}{7} = 4$

3. $\dfrac{x}{-3} = 11$

4. $\dfrac{x}{-5} = 8$

5. $5y = 35$

6. $6y = 42$

7. $-7y = 63$

8. $-4y = 32$

9. $-28 = 8z$

10. $-36 = 8z$

11. $-18 = -3z$

12. $-54 = -9z$

13. $-8x = 6$

14. $-8x = 4$

15. $17y = 0$

16. $-16y = 0$

17. $\dfrac{2}{3}y = 12$

18. $\dfrac{3}{4}y = 15$

19. $28 = -\dfrac{7}{2}x$

20. $20 = -\dfrac{5}{8}x$

21. $-x = 17$

22. $-x = 23$

23. $-47 = -y$

24. $-51 = -y$

25. $-\dfrac{x}{5} = -9$

26. $-\dfrac{x}{5} = -1$

27. $2x - 12x = 50$

28. $8x - 3x = -45$

Solve each equation in Exercises 29–54 using both the addition and multiplication properties of equality. Check proposed solutions.

29. $2x + 1 = 11$

30. $2x + 5 = 13$

31. $2x - 3 = 9$

32. $3x - 2 = 9$

33. $-2y + 5 = 7$

34. $-3y + 4 = 13$

35. $-3y - 7 = -1$

36. $-2y - 5 = 7$

37. $12 = 4z + 3$

38. $14 = 5z - 21$

39. $-x - 3 = 3$

40. $-x - 5 = 5$

41. $6y = 2y - 12$

42. $8y = 3y - 10$

43. $3z = -2z - 15$

44. $2z = -4z + 18$

45. $-5x = -2x - 12$

46. $-7x = -3x - 8$

47. $8y + 4 = 2y - 5$

48. $5y + 6 = 3y - 6$

49. $6z - 5 = z + 5$

50. $6z - 3 = z + 2$

51. $6x + 14 = 2x - 2$

52. $9x + 2 = 6x - 4$

53. $-3y - 1 = 5 - 2y$

54. $-3y - 2 = -5 - 4y$

Practice PLUS

The equations in Exercises 55–58 contain small geometric figures that represent nonzero real numbers. Use the multiplication property of equality to isolate x on one side of the equation and the geometric figures on the other side.

55. $\dfrac{x}{\square} = \triangle$

56. $\triangle = \square\, x$

57. $\triangle = -x$

58. $\dfrac{-x}{\square} = \triangle$

In Exercises 59–66, use the given information to write an equation. Let x represent the number described in each exercise. Then solve the equation and find the number.

59. If a number is multiplied by 6, the result is 10. Find the number.

60. If a number is multiplied by −6, the result is 20. Find the number.

61. If a number is divided by −9, the result is 5. Find the number.

62. If a number is divided by −7, the result is 8. Find the number.

63. Eight subtracted from the product of 4 and a number is 56.

64. Ten subtracted from the product of 3 and a number is 23.

65. Negative three times a number, increased by 15, is −6.

66. Negative five times a number, increased by 11, is −29.

Application Exercises

The formula

$$M = \dfrac{n}{5}$$

models your distance, M, in miles, from a lightning strike in a thunderstorm if it takes n seconds to hear thunder after seeing the lightning. Use this formula to solve Exercises 67–68.

67. If you are 2 miles away from the lightning flash, how long will it take the sound of thunder to reach you?

68. If you are 3 miles away from the lightning flash, how long will it take the sound of thunder to reach you?

The Mach number is a measurement of speed, named after the man who suggested it, Ernst Mach (1838–1916). The formula

$$M = \frac{A}{740}$$

indicates that the speed of an aircraft, A, in miles per hour, divided by the speed of sound, approximately 740 miles per hour, results in the Mach number, M. Use the formula to determine the speed, in miles per hour, of the aircrafts in Exercises 69–70. (Note: When an aircraft's speed increases beyond Mach 1, it is said to have broken the sound barrier.)

69. **70.**

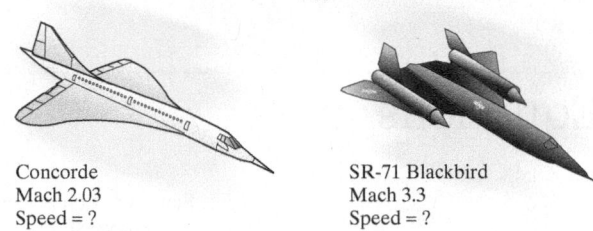

Concorde
Mach 2.03
Speed = ?

SR-71 Blackbird
Mach 3.3
Speed = ?

In Exercises 71–72, we continue with our historical comparisons using today's dollars.

71. **Least Expensive Ford Automobile**

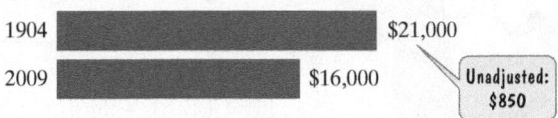

1904 $21,000

2009 $16,000 Unadjusted: $850

Using today's dollars, the data in the bar graph can be described by the mathematical model

$$F = -48n + 21{,}000,$$

where F is the price of the least expensive Ford n years after 1904.

a. Does the formula underestimate or overestimate the price of the least expensive Ford in 2009? By how much?

b. If trends shown by the formula continue, when will the least expensive Ford cost $15,000?

72. **Average Nightly Hotel Room Rate**

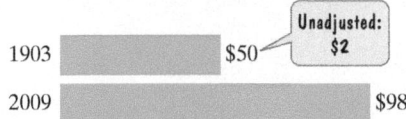

1903 $50 Unadjusted: $2

2009 $98

Using today's dollars, the data in the bar graph can be described by the mathematical model

$$H = 0.5n + 50,$$

where H is the average nightly hotel room rate n years after 1903.

a. Does the formula underestimate or overestimate the average nightly hotel room rate in 2009? By how much?

b. If trends shown by the formula continue, when will the average nightly hotel room rate cost $110?

Writing in Mathematics

73. State the multiplication property of equality and give an example.

74. Explain how to solve the equation $-x = -50$.

75. Explain how to solve the equation $2x + 8 = 5x - 3$.

Critical Thinking Exercises

Make Sense? *In Exercises 76–79, determine whether each statement "makes sense" or "does not make sense" and explain your reasoning.*

76. I used the addition and multiplication properties of equality to solve $3x = 20 + 4$.

77. I used the addition and multiplication properties of equality to solve $12 - 3x = 15$ as follows:

$$12 - 3x = 15$$
$$3x = 3$$
$$x = 1.$$

78. When I use the addition and multiplication properties to solve $2x + 5 = 17$, I undo the operations in the opposite order in which they are performed.

79. The model $P = 18n + 765$ describes the price of a Westie puppy, P, n years after 1940, so I have to solve a linear equation to determine the puppy's price in 2009.

In Exercises 80–83, determine whether each statement is true or false. If the statement is false, make the necessary change(s) to produce a true statement.

80. If $7x = 21$, then $x = 21 - 7$.

81. If $3x - 4 = 16$, then $3x = 12$.

82. If $3x + 7 = 0$, then $x = \dfrac{7}{3}$.

83. The solution of $6x = 0$ is not a natural number.

In Exercises 84–85, write an equation with the given characteristics.

84. The solution is a positive integer and the equation can be solved by dividing both sides by -60.

85. The solution is a negative integer and the equation can be solved by multiplying both sides by $\frac{4}{5}$.

Technology Exercises

Solve each equation in Exercises 86–87. Use a calculator to help with the arithmetic. Check your solution using the calculator.

86. $3.7x - 19.46 = -9.988$

87. $-72.8y - 14.6 = -455.43 - 4.98y$

Review Exercises

88. Evaluate: $(-10)^2$. (Section 1.8, Example 1)
89. Evaluate: -10^2. (Section 1.8, Example 1)
90. Evaluate $x^3 - 4x$ for $x = -1$. (Section 1.8, Example 10)

Preview Exercises

Exercises 91–93 will help you prepare for the material covered in the next section.

91. Simplify: $13 - 3(x + 2)$.
92. Is 6 a solution of $2(x - 3) - 17 = 13 - 3(x + 2)$?
93. Multiply and simplify: $10\left(\dfrac{x}{5} - \dfrac{39}{5}\right)$.

SECTION

2.3

Solving Linear Equations

Objectives

1. Solve linear equations.
2. Solve linear equations containing fractions.
3. Solve linear equations containing decimals.
4. Identify equations with no solution or infinitely many solutions.
5. Solve applied problems using formulas.

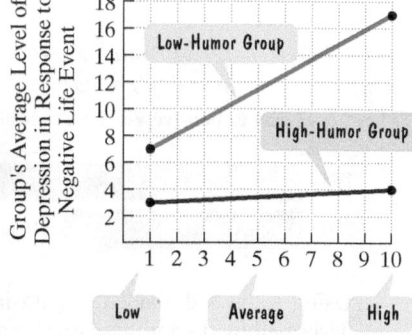

Sense of Humor and Depression

Figure 2.3

Source: Steven Davis and Joseph Palladino. *Psychology,* Fifth Edition, Prentice Hall, 2007.

The belief that humor and laughter can have positive benefits on our lives is not new. The graphs in **Figure 2.3** indicate that persons with a low sense of humor have higher levels of depression in response to negative life events than those with a high sense of humor. In this section, we will see how algebra models these relationships. To use these mathematical models, it would be helpful to have a systematic procedure for solving linear equations. We open this section with such a procedure.

1. Solve linear equations.

A Step-by-Step Procedure for Solving Linear Equations

Here is a step-by-step procedure for solving a linear equation in one variable. Not all of these steps are necessary to solve every equation.

Solving a Linear Equation

1. Simplify the algebraic expression on each side.
2. Collect all the variable terms on one side and all the constant terms on the other side.
3. Isolate the variable and solve.
4. Check the proposed solution in the original equation.

EXAMPLE 1   Solving a Linear Equation

Solve and check: $2x - 8x + 40 = 13 - 3x - 3$.

Solution

Step 1. Simplify the algebraic expression on each side.

$$2x - 8x + 40 = 13 - 3x - 3 \qquad \text{This is the given equation.}$$
$$-6x + 40 = 10 - 3x \qquad \text{Combine like terms: } 2x - 8x = -6x$$
$$\text{and } 13 - 3 = 10.$$

Step 2. Collect variable terms on one side and constant terms on the other side. The simplified equation is $-6x + 40 = 10 - 3x$. We will collect variable terms on the left by adding $3x$ to both sides. We will collect the numbers on the right by subtracting 40 from both sides.

$$-6x + 40 + 3x = 10 - 3x + 3x \qquad \text{Add } 3x \text{ to both sides.}$$
$$-3x + 40 = 10 \qquad \text{Simplify: } -6x + 3x = -3x.$$
$$-3x + 40 - 40 = 10 - 40 \qquad \text{Subtract 40 from both sides.}$$
$$-3x = -30 \qquad \text{Simplify.}$$

Step 3. Isolate the variable and solve. We isolate the variable, x, by dividing both sides by -3.

$$\frac{-3x}{-3} = \frac{-30}{-3} \qquad \text{Divide both sides by } -3.$$
$$x = 10 \qquad \text{Simplify.}$$

Step 4. Check the proposed solution in the original equation. Substitute 10 for x in the original equation.

$$2x - 8x + 40 = 13 - 3x - 3 \qquad \text{This is the original equation.}$$
$$2 \cdot 10 - 8 \cdot 10 + 40 \stackrel{?}{=} 13 - 3 \cdot 10 - 3 \qquad \text{Substitute 10 for } x.$$
$$20 - 80 + 40 \stackrel{?}{=} 13 - 30 - 3 \qquad \text{Perform the indicated multiplications.}$$
$$-60 + 40 \stackrel{?}{=} -17 - 3 \qquad \text{Subtract: } 20 - 80 = -60 \text{ and}$$
$$13 - 30 = -17.$$
$$-20 = -20 \qquad \text{Simplify.}$$

By substituting 10 for x in the original equation, we obtain the true statement $-20 = -20$. This verifies that the solution is 10, or the solution set is $\{10\}$. ∎

☑   **CHECK POINT 1**   Solve and check: $-7x + 25 + 3x = 16 - 2x - 3$.

Discover for Yourself

Solve the equation in Example 1 by collecting terms with the variable on the right and numbers on the left. What do you observe?

EXAMPLE 2   Solving a Linear Equation

Solve and check: $5x = 8(x + 3)$.

Solution

Step 1. Simplify the algebraic expression on each side. Use the distributive property to remove parentheses on the right.

$$5x = 8(x + 3) \qquad \text{This is the given equation.}$$
$$5x = 8x + 24 \qquad \text{Use the distributive property.}$$

Step 2. Collect variable terms on one side and constant terms on the other side. We will work with $5x = 8x + 24$ and collect variable terms on the left by subtracting $8x$ from both sides. The only constant term, 24, is already on the right.

$$5x - 8x = 8x + 24 - 8x \qquad \text{Subtract } 8x \text{ from both sides.}$$
$$-3x = 24 \qquad \text{Simplify: } 5x - 8x = -3x.$$

Step 3. Isolate the variable and solve. We isolate the variable, x, by dividing both sides of $-3x = 24$ by -3.

$$\frac{-3x}{-3} = \frac{24}{-3} \qquad \text{Divide both sides by } -3.$$

$$x = -8 \qquad \text{Simplify.}$$

Step 4. Check the proposed solution in the original equation. Substitute -8 for x in the original equation.

$$5x = 8(x + 3) \qquad \text{This is the original equation.}$$
$$5(-8) \stackrel{?}{=} 8(-8 + 3) \qquad \text{Substitute } -8 \text{ for } x.$$
$$5(-8) \stackrel{?}{=} 8(-5) \qquad \text{Perform the addition in parentheses: } -8 + 3 = -5.$$
$$-40 = -40 \qquad \text{Multiply.}$$

The true statement $-40 = -40$ verifies that -8 is the solution, or the solution set is $\{-8\}$. ∎

✓ **CHECK POINT 2**   Solve and check: $8x = 2(x + 6)$.

EXAMPLE 3   Solving a Linear Equation

Solve and check: $2(x - 3) - 17 = 13 - 3(x + 2)$.

Solution

Step 1. Simplify the algebraic expression on each side.

> Do not begin with $13 - 3$. Multiplication (the distributive property) is applied before subtraction.

$$2(x - 3) - 17 = 13 - 3(x + 2) \qquad \text{This is the given equation.}$$
$$2x - 6 - 17 = 13 - 3x - 6 \qquad \text{Use the distributive property.}$$
$$2x - 23 = -3x + 7 \qquad \text{Combine like terms.}$$

Step 2. Collect variable terms on one side and constant terms on the other side. We will collect variable terms on the left by adding $3x$ to both sides. We will collect the numbers on the right by adding 23 to both sides.

$$2x - 23 + 3x = -3x + 7 + 3x \qquad \text{Add } 3x \text{ to both sides.}$$
$$5x - 23 = 7 \qquad \text{Simplify: } 2x + 3x = 5x.$$
$$5x - 23 + 23 = 7 + 23 \qquad \text{Add 23 to both sides.}$$
$$5x = 30 \qquad \text{Simplify.}$$

Step 3. Isolate the variable and solve. We isolate the variable, x, by dividing both sides by 5.

$$\frac{5x}{5} = \frac{30}{5} \qquad \text{Divide both sides by 5.}$$

$$x = 6 \qquad \text{Simplify.}$$

Step 4. Check the proposed solution in the original equation. Substitute 6 for x in the original equation.

$$2(x - 3) - 17 = 13 - 3(x + 2) \qquad \text{This is the original equation.}$$
$$2(6 - 3) - 17 \stackrel{?}{=} 13 - 3(6 + 2) \qquad \text{Substitute 6 for } x.$$
$$2(3) - 17 \stackrel{?}{=} 13 - 3(8) \qquad \text{Simplify inside parentheses.}$$
$$6 - 17 \stackrel{?}{=} 13 - 24 \qquad \text{Multiply.}$$
$$-11 = -11 \qquad \text{Subtract.}$$

The true statement $-11 = -11$ verifies that 6 is the solution, or the solution set is $\{6\}$. ∎

✓   **CHECK POINT 3**   Solve and check: $4(2x + 1) - 29 = 3(2x - 5)$.

2 Solve linear equations containing fractions.

Linear Equations with Fractions

Equations are easier to solve when they do not contain fractions. How do we remove fractions from an equation? We begin by multiplying both sides of the equation by the least common denominator of any fractions in the equation. The least common denominator is the smallest number that all denominators will divide into. Multiplying every term on both sides of the equation by the least common denominator will eliminate the fractions in the equation. Example 4 shows how we "clear an equation of fractions."

EXAMPLE 4   Solving a Linear Equation Involving Fractions

Solve and check: $\dfrac{3x}{2} = \dfrac{x}{5} - \dfrac{39}{5}$.

Solution   The denominators are 2, 5, and 5. The smallest number that is divisible by 2, 5, and 5 is 10. We begin by multiplying both sides of the equation by 10, the least common denominator.

$$\frac{3x}{2} = \frac{x}{5} - \frac{39}{5}$$ This is the given equation.

$$10 \cdot \frac{3x}{2} = 10 \left(\frac{x}{5} - \frac{39}{5} \right)$$ Multiply both sides by 10.

$$10 \cdot \frac{3x}{2} = 10 \cdot \frac{x}{5} - 10 \cdot \frac{39}{5}$$ Use the distributive property. Be sure to multiply all terms by 10.

$$\overset{5}{\cancel{10}} \cdot \frac{3x}{\underset{1}{\cancel{2}}} = \overset{2}{\cancel{10}} \cdot \frac{x}{\underset{1}{\cancel{5}}} - \overset{2}{\cancel{10}} \cdot \frac{39}{\underset{1}{\cancel{5}}}$$ Divide out common factors in the multiplications.

$$15x = 2x - 78$$ Complete the multiplications. The fractions are now cleared.

At this point, we have an equation similar to those we previously have solved. Collect the variable terms on one side and the constant terms on the other side.

$$15x - 2x = 2x - 2x - 78$$ Subtract 2x to get the variable terms on the left.

$$13x = -78$$ Simplify.

Isolate x by dividing both sides by 13.

$$\frac{13x}{13} = \frac{-78}{13}$$ Divide both sides by 13.

$$x = -6$$ Simplify.

Check $\dfrac{3x}{2} = \dfrac{x}{5} - \dfrac{39}{5}$ This is the original equation.

$\dfrac{3(-6)}{2} \overset{?}{=} \dfrac{-6}{5} - \dfrac{39}{5}$ Substitute -6 for x.

$-9 \overset{?}{=} \dfrac{-6}{5} - \dfrac{39}{5}$ Simplify the left side: $\dfrac{3(-6)}{2} = \dfrac{-18}{2} = -9$.

$-9 \overset{?}{=} \dfrac{-45}{5}$ Subtract on the right side:
$\dfrac{-6}{5} - \dfrac{39}{5} = \dfrac{-6}{5} + \left(\dfrac{-39}{5}\right) = \dfrac{-45}{5}$.

$-9 = -9$ Simplify: $\dfrac{-45}{5} = -9$.

The true statement $-9 = -9$ verifies that -6 is the solution, or the solution set is $\{-6\}$.

■

✓ **CHECK POINT 4** Solve and check: $\dfrac{x}{4} = \dfrac{2x}{3} + \dfrac{5}{6}$.

---

3 Solve linear equations containing decimals.

Linear Equations with Decimals

It is not a requirement to clear decimals in an equation. However, if you are not using a calculator, eliminating decimal numbers can make calculations easier.

Multiplying a decimal number by 10^n, where n is a positive integer, has the effect of moving the decimal point n places to the right. For example,

$$0.3 \times 10 = 3 \qquad 0.37 \times 10^2 = 37 \qquad 0.408 \times 10^3 = 408.$$

| One decimal place | Two decimal places | Three decimal places |

These numerical examples suggest a procedure for clearing decimals in an equation.

Clearing an Equation of Decimals

Multiply every term on both sides of the equation by a power of 10. The exponent on 10 will equal the greatest number of decimal places in the equation.

EXAMPLE 5   Solving a Linear Equation Involving Decimals

Solve and check: $0.3(x - 6) = 0.37x - 1.1$.

Solution   We will first apply the distributive property to remove parentheses on the left. Then we will clear the equation of decimals.

$0.3(x - 6) = 0.37x - 1.1$ This is the given equation.

$0.3x - 0.3(6) = 0.37x - 1.1$ Use the distributive property.

$0.3x - 1.8 = 0.37x - 1.1$ Multiply: $0.3(6) = 1.8$.

> The number 0.37 has two decimal places, the greatest number of decimal places in the equation. Multiply both sides by 10^2, or 100, to clear the decimals.

$100(0.3x - 1.8) = 100(0.37x - 1.1)$ Multiply both sides by 100.

$100(0.3x) - 100(1.8) = 100(0.37x) - 100(1.1)$ Use the distributive property.

$30x - 180 = 37x - 110$ Simplify by moving decimal points two places to the right.

At this point, we have an equation similar to those we previously solved. We will solve $30x - 180 = 37x - 110$ by collecting the variable terms on one side and the constant terms on the other side.

$$30x - 37x - 180 = 37x - 37x - 110$$ Subtract 37x to get the variable terms on the left.

$$-7x - 180 = -110$$ Simplify.

$$-7x - 180 + 180 = -110 + 180$$ Add 180 to get the constant terms on the right.

$$-7x = 70$$ Simplify.

$$\frac{-7x}{-7} = \frac{70}{-7}$$ Isolate x by dividing both sides by −7.

$$x = -10$$ Simplify.

Check $0.3(x - 6) = 0.37x - 1.1$ This is the original equation.

$$0.3(-10 - 6) = 0.37(-10) - 1.1$$ Substitute −10 for x.

$$0.3(-16) = 0.37(-10) - 1.1$$ Simplify the left side: −10 − 6 = −16.

$$-4.8 = -3.7 - 1.1$$ Multiply: 0.3(−16) = −4.8 and 0.37(−10) = −3.7.

$$-4.8 = -4.8$$ Subtract on the right side: −3.7 − 1.1 = −3.7 + (−1.1) = −4.8.

The true statement $-4.8 = -4.8$ verifies that -10 is the solution, or the solution set is $\{-10\}$. ∎

Great Question!

In the solution to Example 5, you simplified using the distributive property. Then you multiplied by 100 and cleared the equation of decimals. Can I clear the equation of decimals before using the distributive property?

Yes. Decimals can be cleared at any step in the process of solving an equation. Here's how it works for the equation in Example 5:

$$0.3(x - 6) = 0.37x - 1.1$$ This is the given equation.

$$100[0.3(x - 6)] = 100(0.37x - 1.1)$$ Multiply both sides by 100.

> 0.3(x − 6) is one term with two factors, 0.3 and x − 6. To multiply the term by 100, we need only multiply one factor, in this case 0.3, by 100.

$$[100(0.3)](x - 6) = 100(0.37x) - 100(1.1)$$ Use the associative property on the left and the distributive property on the right.

$$30(x - 6) = 37x - 110$$ Simplify by moving decimal points two places to the right.

$$30x - 180 = 37x - 110.$$ Use the distributive property.

This last equation, $30x - 180 = 37x - 110$, is the same equation cleared of decimals that we obtained in the solution to Example 5.

✓ **CHECK POINT 5**   Solve and check: $0.48x + 3 = 0.2(x - 6)$.

4 Identify equations with no solution or infinitely many solutions.

Equations with No Solution or Infinitely Many Solutions

Thus far, each equation that we have solved has had a single solution. However, some equations are not true for even one real number. Such an equation is called an **inconsistent equation**. Here is an example of such an equation:

$$x = x + 4.$$

There is no number that is equal to itself plus 4. This equation has no solution.

You can express the fact that an equation has no solution using words or set notation.

- Use the phrase "no solution."
- Use set notation: $\varnothing$.

> This symbol stands for the empty set, a set with no elements.

An equation that is true for all real numbers is called an **identity**. An example of an identity is

$$x + 3 = x + 2 + 1.$$

Every number plus 3 is equal to that number plus 2 plus 1. Every real number is a solution to this equation.

You can express the fact that every real number is a solution of an equation using words or set notation.

- Use the phrase "all real numbers."
- Use set notation: $\{x \mid x \text{ is a real number}\}$.

> The set of   all x such that x is a real number

Recognizing Inconsistent Equations and Identities

If you attempt to solve an equation with no solution or one that is true for every real number, you will eliminate the variable.

- An inconsistent equation with no solution results in a false statement, such as $2 = 5$.
- An identity that is true for every real number results in a true statement, such as $4 = 4$.

EXAMPLE 6   Solving an Equation

Solve: $2x + 6 = 2(x + 4)$.

Solution

$2x + 6 = 2(x + 4)$	This is the given equation.
$2x + 6 = 2x + 8$	Use the distributive property.
$2x + 6 - 2x = 2x + 8 - 2x$	Subtract 2x from both sides.
$6 = 8$	Simplify.

> Keep reading. 6 = 8 is not the solution.

The original equation is equivalent to the statement $6 = 8$, which is false for every value of x. The equation is inconsistent and has no solution. You can express this by writing "no solution" or using the symbol for the empty set, $\varnothing$. ∎

✓ **CHECK POINT 6**   Solve: $3x + 7 = 3(x + 1)$.

EXAMPLE 7   Solving an Equation

Solve: $-3x + 5 + 5x = 4x - 2x + 5$.

Solution

$$-3x + 5 + 5x = 4x - 2x + 5 \qquad \text{This is the given equation.}$$

$$2x + 5 = 2x + 5 \qquad \text{Combine like terms: } -3x + 5x = 2x$$
$$\text{and } 4x - 2x = 2x.$$

$$2x + 5 - 2x = 2x + 5 - 2x \qquad \text{Subtract } 2x \text{ from both sides.}$$

> Keep reading. 5 = 5
> is not the solution.

$$5 = 5 \qquad \text{Simplify.}$$

The original equation is equivalent to the statement $5 = 5$, which is true for every value of x. The equation is an identity and all real numbers are solutions. You can express this by writing "all real numbers" or using set notation: $\{x \mid x \text{ is a real number}\}$. ∎

✓ **CHECK POINT 7** Solve: $3(x - 1) + 9 = 8x + 6 - 5x.$

---

5 Solve applied problems using formulas.

Applications

The next example shows how our procedure for solving equations with fractions can be used to find the value of a variable in a mathematical model.

EXAMPLE 8 An Application: Responding to Negative Life Events

Sense of Humor and Depression

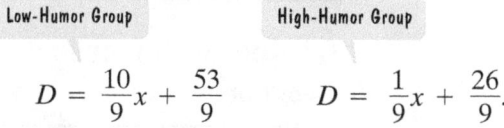

In the section opener, we introduced line graphs, repeated in **Figure 2.3**, indicating that persons with a low sense of humor have higher levels of depression in response to negative life events than those with a high sense of humor. These graphs can be modeled by the following formulas:

Low-Humor Group   High-Humor Group

$$D = \frac{10}{9}x + \frac{53}{9} \qquad D = \frac{1}{9}x + \frac{26}{9}.$$

In each formula, x represents the intensity of a negative life event (from 1, low, to 10, high) and D is the level of depression in response to that event. If the high-humor group averages a level of depression of 3.5, or $\frac{7}{2}$, in response to a negative life event, what is the intensity of that event? How is the solution shown on the red line graph in **Figure 2.3**?

Figure 2.3 (repeated)

Solution We are interested in the intensity of a negative life event with an average level of depression of $\frac{7}{2}$ for the high-humor group. We substitute $\frac{7}{2}$ for D in the high-humor model and solve for x, the intensity of the negative life event.

$$D = \frac{1}{9}x + \frac{26}{9} \qquad \text{This is the given formula for the high-humor group.}$$

$$\frac{7}{2} = \frac{1}{9}x + \frac{26}{9} \qquad \text{Replace } D \text{ with } \frac{7}{2}.$$

$$18 \cdot \frac{7}{2} = 18\left(\frac{1}{9}x + \frac{26}{9}\right) \qquad \text{Multiply both sides by 18, the least common denominator.}$$

$$18 \cdot \frac{7}{2} = 18 \cdot \frac{1}{9}x + 18 \cdot \frac{26}{9} \qquad \text{Use the distributive property.}$$

$$\overset{9}{\cancel{18}} \cdot \frac{7}{\underset{1}{\cancel{2}}} = \overset{2}{\cancel{18}} \cdot \frac{1}{\underset{1}{\cancel{9}}}x + \overset{2}{\cancel{18}} \cdot \frac{26}{\underset{1}{\cancel{9}}} \qquad \text{Divide out common factors in the multiplications.}$$

$$63 = 2x + 52$$ Complete the multiplications. The fractions are now cleared.

$$63 - 52 = 2x + 52 - 52$$ Subtract 52 from both sides to get constants on the left.

$$11 = 2x$$ Simplify.

$$\frac{11}{2} = \frac{2x}{2}$$ Divide both sides by 2.

$$\frac{11}{2} = x$$ Simplify.

The formula indicates that if the high-humor group averages a level of depression of 3.5 in response to a negative life event, the intensity of that event is $\frac{11}{2}$, or 5.5. This is illustrated on the line graph for the high-humor group in **Figure 2.4**.

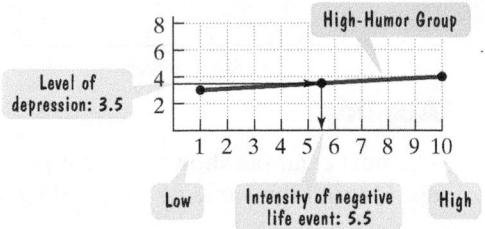

Figure 2.4

✓ **CHECK POINT 8**   Use the model for the low-humor group given in Example 8 to solve this problem. If the low-humor group averages a level of depression of 10 in response to a negative life event, what is the intensity of that event? How is the solution shown on the blue line graph in **Figure 2.3**?

Achieving Success

FoxTrot

Foxtrot copyright © 2003, 2009 by Bill Amend/Distributed by Universal Uclick

Because concepts in mathematics build on each other, **it is extremely important that you complete all homework assignments**. This requires more than attempting a few of the assigned exercises. When it comes to assigned homework, you need to do four things and to do these things consistently throughout any math course:

1. Attempt to work *every* assigned problem.
2. Check your answers.
3. Correct your errors.
4. Ask for help with the problems you have attempted but do not understand.

Having said this, **don't panic at the length of the Exercise Sets**. You are not expected to work all, or even most, of the problems. Your professor will provide guidance on which exercises to work by assigning those problems that are consistent with the goals and objectives of your course.

CONCEPT AND VOCABULARY CHECK

Fill in each blank so that the resulting statement is true.

1. The first step in solving $3x - 9x + 30 = 15 - 2x - 4$ is to _____.

2. The equation $\frac{x}{5} - \frac{1}{2} = \frac{x}{6}$ can be cleared of fractions by multiplying both sides by _____.

3. The equation $0.9x - 4.3 = 0.47$ can be cleared of decimals by multiplying both sides by _____.

4. A linear equation that is not true for even one real number, and therefore has no solution, is called a/an _____ equation.

5. A linear equation that is true for all real numbers is called a/an _____.

6. In solving an equation, if you eliminate the variable and obtain a false statement such as $2 = 5$, the equation is a/an _____ equation.

7. In solving an equation, if you eliminate the variable and obtain a true statement such as $5 = 5$, the equation is a/an _____.

2.3 EXERCISE SET

MyMathLab®

Watch the videos in MyMathLab

Download the MyDashBoard App

Practice Exercises

In Exercises 1–30, solve each equation. Be sure to check your proposed solution by substituting it for the variable in the original equation.

1. $5x + 3x - 4x = 10 + 2$

2. $4x + 8x - 2x = 20 - 15$

3. $4x - 9x + 22 = 3x + 30$

4. $3x + 2x + 64 = 40 - 7x$

5. $3x + 6 - x = 8 + 3x - 6$

6. $3x + 2 - x = 6 + 3x - 8$

7. $4(x + 1) = 20$

8. $3(x - 2) = -6$

9. $7(2x - 1) = 42$

10. $4(2x - 3) = 32$

11. $38 = 30 - 2(x - 1)$

12. $20 = 44 - 8(2 - x)$

13. $2(4z + 3) - 8 = 46$

14. $3(3z + 5) - 7 = 89$

15. $6x - (3x + 10) = 14$

16. $5x - (2x + 14) = 10$

17. $5(2x + 1) = 12x - 3$

18. $3(x + 2) = x + 30$

19. $3(5 - x) = 4(2x + 1)$

20. $3(3x - 1) = 4(3 + 3x)$

21. $8(y + 2) = 2(3y + 4)$

22. $8(y + 3) = 3(2y + 12)$

23. $3(x + 1) = 7(x - 2) - 3$

24. $5x - 4(x + 9) = 2x - 3$

25. $5(2x - 8) - 2 = 5(x - 3) + 3$

26. $7(3x - 2) + 5 = 6(2x - 1) + 24$

27. $6 = -4(1 - x) + 3(x + 1)$

28. $100 = -(x - 1) + 4(x - 6)$

29. $10(z + 4) - 4(z - 2) = 3(z - 1) + 2(z - 3)$

30. $-2(z - 4) - (3z - 2) = -2 - (6z - 2)$

Solve each equation and check your proposed solution in Exercises 31–46. Begin your work by rewriting each equation without fractions.

31. $\frac{x}{5} - 4 = -6$

32. $\frac{x}{2} + 13 = -22$

33. $\frac{2x}{3} - 5 = 7$

34. $\frac{3x}{4} - 9 = -6$

35. $\frac{2y}{3} - \frac{3}{4} = \frac{5}{12}$

36. $\frac{3y}{4} - \frac{2}{3} = \frac{7}{12}$

37. $\frac{x}{3} + \frac{x}{2} = \frac{5}{6}$

38. $\frac{x}{4} - \frac{x}{5} = 1$

39. $20 - \frac{z}{3} = \frac{z}{2}$

40. $\frac{z}{5} - \frac{1}{2} = \frac{z}{6}$

41. $\frac{y}{3} + \frac{2}{5} = \frac{y}{5} - \frac{2}{5}$

42. $\frac{y}{12} + \frac{1}{6} = \frac{y}{2} - \frac{1}{4}$

43. $\frac{3x}{4} - 3 = \frac{x}{2} + 2$

44. $\frac{3x}{5} - \frac{2}{5} = \frac{x}{3} + \frac{2}{5}$

45. $\frac{x - 3}{5} - 1 = \frac{x - 5}{4}$

46. $\frac{x - 2}{3} - 4 = \frac{x + 1}{4}$

Solve each equation and check your proposed solution in Exercises 47–58.

47. $3.6x = 2.9x + 6.3$

48. $1.2x - 3.6 = 2.4 - 0.3x$

49. $0.92y + 2 = y - 0.4$

50. $0.15y - 0.1 = 2.5y - 1.04$

51. $0.3x - 4 = 0.1(x + 10)$

52. $0.1(x + 80) = 14 - 0.2x$

53. $0.4(2z + 6) + 0.1 = 0.5(2z - 3)$

54. $1.4(z - 5) - 0.2 = 0.5(6z - 8)$

55. $0.01(x + 4) - 0.04 = 0.01(5x + 4)$

56. $0.02(x - 2) = 0.06 - 0.01(x + 1)$

57. $0.6(x + 300) = 0.65x - 205$

58. $0.05(7x + 36) = 0.4x + 1.2$

In Exercises 59–78, solve each equation. Use words or set notation to identify equations that have no solution, or equations that are true for all real numbers.

59. $3x - 7 = 3(x + 1)$

60. $2(x - 5) = 2x + 10$

61. $2(x + 4) = 4x + 5 - 2x + 3$

62. $3(x - 1) = 8x + 6 - 5x - 9$

63. $7 + 2(3x - 5) = 8 - 3(2x + 1)$

64. $2 + 3(2x - 7) = 9 - 4(3x + 1)$

65. $4x + 1 - 5x = 5 - (x + 4)$

66. $5x - 5 = 3x - 7 + 2(x + 1)$

67. $4(x + 2) + 1 = 7x - 3(x - 2)$

68. $5x - 3(x + 1) = 2(x + 3) - 5$

69. $3 - x = 2x + 3$

70. $5 - x = 4x + 5$

71. $\dfrac{x}{3} + 2 = \dfrac{x}{3}$

72. $\dfrac{x}{4} + 3 = \dfrac{x}{4}$

73. $\dfrac{x}{2} - \dfrac{x}{4} + 4 = x + 4$

74. $\dfrac{x}{2} + \dfrac{2x}{3} + 3 = x + 3$

75. $\dfrac{2}{3}x = 2 - \dfrac{5}{6}x$

76. $\dfrac{2}{3}x = \dfrac{1}{4}x - 8$

77. $0.06(x + 5) = 0.03(2x + 7) + 0.09$

78. $0.04(x - 2) = 0.02(6x - 3) - 0.02$

Practice PLUS

The equations in Exercises 79–80 contain small figures ($\square$, $\triangle$, and $\$$) that represent nonzero real numbers. Use properties of equality to isolate x on one side of the equation and the small figures on the other side.

79. $\dfrac{x}{\square} + \triangle = \$$

80. $\dfrac{x}{\square} - \triangle = -\$$

81. If $\dfrac{x}{5} - 2 = \dfrac{x}{3}$, evaluate $x^2 - x$.

82. If $\dfrac{3x}{2} + \dfrac{3x}{4} = \dfrac{x}{4} - 4$, evaluate $x^2 - x$.

In Exercises 83–86, use the given information to write an equation. Let x represent the number described in each exercise. Then solve the equation and find the number.

83. When one-third of a number is added to one-fifth of the number, the sum is 16. What is the number?

84. When two-fifths of a number is added to one-fourth of the number, the sum is 13. What is the number?

85. When 3 is subtracted from three-fourths of a number, the result is equal to one-half of the number. What is the number?

86. When 30 is subtracted from seven-eighths of a number, the result is equal to one-half of the number. What is the number?

Application Exercises

In Massachusetts, speeding fines are determined by the formula

$$F = 10(x - 65) + 50,$$

where F is the cost, in dollars, of the fine if a person is caught driving x miles per hour. Use this formula to solve Exercises 87–88.

87. If a fine comes to $250, how fast was that person driving?

88. If a fine comes to $400, how fast was that person driving?

The latest guidelines, which apply to both men and women, give healthy weight ranges, rather than specific weights, for your height. The further you are above the upper limit of your range, the greater are the risks of developing weight-related health problems. The bar graph shows these ranges for various heights for people between the ages of 19 and 34, inclusive.

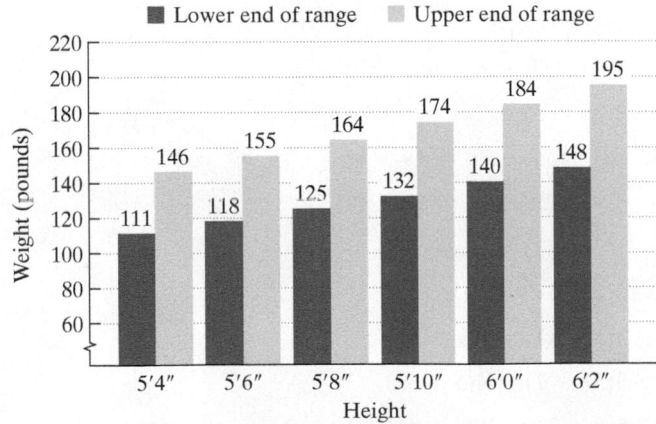

Healthy Weight Ranges for Men and Women, Ages 19 to 34

Source: U.S. Department of Health and Human Services

The mathematical model

$$\frac{W}{2} - 3H = 53$$

describes a weight, W, in pounds, that lies within the healthy weight range for a person whose height is H inches over 5 feet. Use this information to solve Exercises 89–90.

89. Use the formula to find a healthy weight for a person whose height is 5′6″. (*Hint:* $H = 6$ because this person's height is 6 inches over 5 feet.) How many pounds is this healthy weight below the upper end of the range shown by the bar graph?

90. Use the formula to find a healthy weight for a person whose height is 6′0″. (*Hint:* $H = 12$ because this person's height is 12 inches over 5 feet.) How many pounds is this healthy weight below the upper end of the range shown by the bar graph?

The formula

$$p = 15 + \frac{5d}{11}$$

describes the pressure of sea water, p, in pounds per square foot, at a depth of d feet below the surface. Use the formula to solve Exercises 91–92.

91. The record depth for breath-held diving, by Francisco Ferreras (Cuba) off Grand Bahama Island, on November 14, 1993, involved pressure of 201 pounds per square foot. To what depth did Ferreras descend on this ill-advised venture? (He was underwater for 2 minutes and 9 seconds!)

92. At what depth is the pressure 20 pounds per square foot?

Writing in Mathematics

93. In your own words, describe how to solve a linear equation.

94. Explain how to solve a linear equation containing fractions.

95. Suppose that you solve $\frac{x}{5} - \frac{x}{2} = 1$ by multiplying both sides by 20, rather than the least common denominator, 10. Describe what happens. If you get the correct solution, why do you think we clear the equation of fractions by multiplying by the *least* common denominator?

96. Explain how to clear decimals in a linear equation.

97. Suppose you are an algebra teacher grading the following solution on an examination:

Solve: $\qquad -3(x - 6) = 2 - x$

Solution: $\quad -3x - 18 = 2 - x$

$\qquad\qquad\qquad -2x - 18 = 2$

$\qquad\qquad\qquad\qquad -2x = -16$

$\qquad\qquad\qquad\qquad\quad x = 8.$

You should note that 8 checks, and the solution is 8. The student who worked the problem therefore wants full credit. Can you find any errors in the solution? If full credit is 10 points, how many points should you give the student? Justify your position.

Critical Thinking Exercises

Make Sense? *In Exercises 98–101, determine whether each statement "makes sense" or "does not make sense" and explain your reasoning.*

98. Although I can solve $3x + \frac{1}{5} = \frac{1}{4}$ by first subtracting $\frac{1}{5}$ from both sides, I find it easier to begin by multiplying both sides by 20, the least common denominator.

99. I can use any common denominator to clear an equation of fractions, but using the least common denominator makes the arithmetic easier.

100. When I substituted 5 for x in the equation

$$4x + 6 = 6(x + 1) - 2x$$

I obtained a true statement, so the equation's solution is 5.

101. I cleared the equation $0.5x + 8.3 = 12.4$ of decimals by multiplying both sides by 100.

In Exercises 102–105, determine whether each statement is true or false. If the statement is false, make the necessary change(s) to produce a true statement.

102. The equation $3(x + 4) = 3(4 + x)$ has precisely one solution.

103. The equation $2y + 5 = 0$ is equivalent to $2y = 5$.

104. If $2 - 3y = 11$ and the solution to the equation is substituted into $y^2 + 2y - 3$, a number results that is neither positive nor negative.

105. The equation $x + \frac{1}{3} = \frac{1}{2}$ is equivalent to $x + 2 = 3$.

106. A woman's height, h, is related to the length of her femur, f (the bone from the knee to the hip socket), by the formula $f = 0.432h - 10.44$. Both h and f are measured in inches. A partial skeleton is found of a woman in which the femur is 16 inches long. Police find the skeleton in an area where a woman slightly over 5 feet tall has been missing for over a year. Can the partial skeleton be that of the missing woman? Explain.

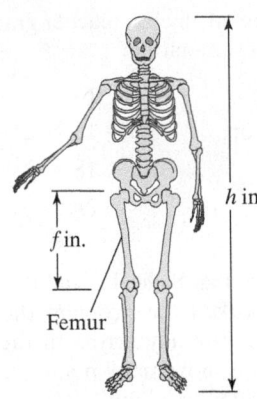

h in.

f in.

Femur

Solve each equation in Exercises 107–108.

107. $\dfrac{2x - 3}{9} + \dfrac{x - 3}{2} = \dfrac{x + 5}{6} - 1$

108. $2(3x + 4) = 3x + 2[3(x - 1) + 2]$

Review Exercises

In Exercises 109–110, insert either $<$ or $>$ in the shaded area between each pair of numbers to make a true statement.

109. -24 ▨ -20 (Section 1.3, Example 6)

110. $-\dfrac{1}{3}$ ▨ $-\dfrac{1}{5}$ (Section 1.3, Example 6)

111. Simplify: $-9 - 11 + 7 - (-3)$. (Section 1.6, Example 3)

Preview Exercises

Exercises 112–114 will help you prepare for the material covered in the next section.

112. Consider the formula

$$T = D + pm.$$

  a. Subtract D from both sides and write the resulting formula.

  b. Divide both sides of your formula from part (a) by p and write the resulting formula.

113. Solve: $4 = 0.25B$.

114. Solve: $1.3 = P \cdot 26$.

SECTION

2.4

Formulas and Percents

Objectives

1 Solve a formula for a variable.

2 Use the percent formula.

3 Solve applied problems involving percent change.

"And, if elected, it is my solemn pledge to cut your taxes by 10% for each of my first three years in office, for a total cut of 30%."

Did you know that one of the most common ways that you are given numerical information is with percents? In this section, you will learn to use a formula that will help you to understand percents from an algebraic perspective, enabling you to make sense of the politician's promise.

1 Solve a formula for a variable.

Solving a Formula for One of Its Variables

We know that solving an equation is the process of finding the number (or numbers) that make the equation a true statement. All of the equations we have solved contained only one letter, such as x or y.

By contrast, the formulas we have seen contain two or more letters, representing two or more variables. Here is an example:

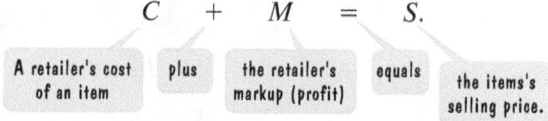

$$C \quad + \quad M \quad = \quad S.$$

| A retailer's cost of an item | plus | the retailer's markup (profit) | equals | the items's selling price. |

We say that this formula is solved for the variable S because S is alone on one side of the equation and the other side does not contain an S.

Solving a formula for a variable means rewriting the formula so that the variable is isolated on one side of the equation. It does not mean obtaining a numerical value for that variable.

The addition and multiplication properties of equality are used to solve a formula for one of its variables. Consider the retailer's formula, $C + M = S$. How do we solve this formula for C? Use the addition property to isolate C by subtracting M from both sides:

> We need to isolate C.

$$C + M = S \qquad \text{This is the given formula.}$$
$$C + M - M = S - M \qquad \text{Subtract } M \text{ from both sides.}$$
$$C = S - M \qquad \text{Simplify.}$$

Solved for C, the formula $C = S - M$ tells us that the cost of an item for a retailer is the item's selling price minus its markup.

To solve a formula for one of its variables, treat that variable as if it were the only variable in the equation. Think of the other variables as if they were numbers. Use the addition property of equality to isolate all terms with the specified variable on one side of the equation and all terms without the specified variable on the other side. Then use the multiplication property of equality to get the specified variable alone.

Our first example involves the formula for the area of a rectangle. The **area of a two-dimensional figure** is the number of square units it takes to fill the interior of the figure. A **square unit** is a square, each of whose sides is one unit in length, as illustrated in **Figure 2.5**. The figure shows that there are 12 square units contained within the rectangle. The area of the rectangle is 12 square units. Notice that the area can be determined in the following manner:

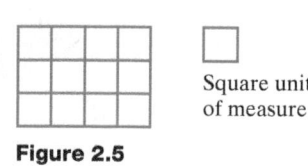

Square unit of measure

Figure 2.5

| Across | Down |

$$4 \text{ units} \cdot 3 \text{ units} = 4 \cdot 3 \text{ units} \cdot \text{units} = 12 \text{ square units.}$$

The area of a rectangle is the product of the distance across, its length, and the distance down, its width.

Area of a Rectangle

The area, A, of a rectangle with length l and width w is given by the formula

$$A = lw.$$

EXAMPLE 1 Solving a Formula for a Variable

Solve the formula $A = lw$ for w.

Solution Our goal is to get w by itself on one side of the formula. There is only one term with w, lw, and it is already isolated on the right side. We isolate w on the right by using the multiplication property of equality and dividing both sides by l.

We need to isolate w.

$$A = lw \qquad \text{This is the given formula.}$$

$$\frac{A}{l} = \frac{lw}{l} \qquad \text{Isolate } w \text{ by dividing both sides by } l.$$

$$\frac{A}{l} = w \qquad \text{Simplify: } \frac{lw}{l} = 1w = w.$$

The formula solved for w is $\dfrac{A}{l} = w$ or $w = \dfrac{A}{l}$. Thus, the area of a rectangle divided by its length is equal to its width. ■

✓ **CHECK POINT 1** Solve the formula $A = lw$ for l.

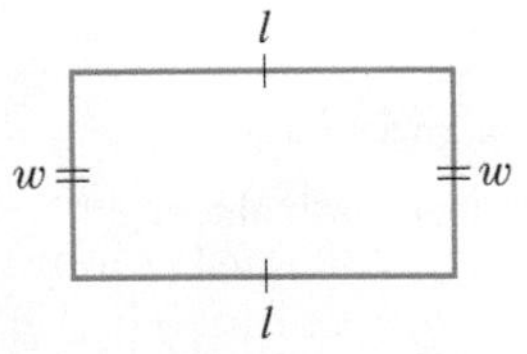

Figure 2.6 A rectangle with length *l* and width *w*

The **perimeter of a two-dimensional figure** is the sum of the lengths of its sides. Perimeter is measured in linear units, such as inches, feet, yards, meters, or kilometers.

Example 2 involves the perimeter, P, of a rectangle. Because perimeter is the sum of the lengths of the sides, the perimeter of the rectangle shown in **Figure 2.6** is $l + w + l + w$. This can be expressed as

$$P = 2l + 2w.$$

Perimeter of a Rectangle

The perimeter, P, of a rectangle with length l and width w is given by the formula

$$P = 2l + 2w.$$

The perimeter of a rectangle is the sum of twice the length and twice the width.

EXAMPLE 2 Solving a Formula for a Variable

Solve the formula $2l + 2w = P$ for w.

Solution First, isolate $2w$ on the left by subtracting $2l$ from both sides. Then solve for w by dividing both sides by 2.

We need to isolate w.

$$2l + 2w = P \qquad \text{This is the given formula.}$$

$$2l - 2l + 2w = P - 2l \qquad \text{Isolate } 2w \text{ by subtracting } 2l \text{ from both sides.}$$

$$2w = P - 2l \qquad \text{Simplify.}$$

$$\frac{2w}{2} = \frac{P - 2l}{2} \qquad \text{Isolate } w \text{ by dividing both sides by 2.}$$

$$w = \frac{P - 2l}{2} \qquad \text{Simplify.} ■$$

✓ **CHECK POINT 2** Solve the formula $2l + 2w = P$ for l.

EXAMPLE 3 Solving a Formula for a Variable

The total price of an article purchased on a monthly deferred payment plan is described by the following formula:

$$T = D + pm.$$

In this formula, T is the total price, D is the down payment, p is the amount of the monthly payment, and m is the number of payments. Solve the formula for p.

Solution First, isolate pm on the right by subtracting D from both sides. Then isolate p from pm by dividing both sides of the formula by m.

> We need to isolate p.

$T = D + pm$	This is the given formula. We want p alone.
$T - D = D - D + pm$	Isolate pm by subtracting D from both sides.
$T - D = pm$	Simplify.
$\dfrac{T - D}{m} = \dfrac{pm}{m}$	Now isolate p by dividing both sides by m.
$\dfrac{T - D}{m} = p$	Simplify: $\dfrac{pm}{m} = \dfrac{p\overset{1}{m}}{\underset{1}{m}} = p \cdot 1 = p.$ ∎

✓ **CHECK POINT 3** Solve the formula $T = D + pm$ for m.

The next example has a formula that contains a fraction. To solve for a variable in a formula involving fractions, we begin by multiplying both sides by the least common denominator of all fractions in the formula. This will eliminate the fractions. Then we solve for the specified variable.

EXAMPLE 4 Solving a Formula Containing a Fraction for a Variable

Solve the formula $\dfrac{W}{2} - 3H = 53$ for W.

Solution Do you remember seeing this formula in the last section's Exercise Set? It models a person's healthy weight, W, where H represents that person's height, in inches, in excess of 5 feet. We begin by multiplying both sides of the formula by 2 to eliminate the fraction. Then we isolate the variable W.

$\dfrac{W}{2} - 3H = 53$	This is the given formula.
$2\left(\dfrac{W}{2} - 3H\right) = 2 \cdot 53$	Multiply both sides by 2.
$2 \cdot \dfrac{W}{2} - 2 \cdot 3H = 2 \cdot 53$	Use the distributive property.

> We need to isolate W.

$W - 6H = 106$	Simplify.
$W - 6H + 6H = 106 + 6H$	Isolate W by adding $6H$ to both sides.
$W = 106 + 6H$	Simplify.

This form of the formula makes it easy to find a person's healthy weight, W, if we know that person's height, H, in inches, in excess of 5 feet. ∎

☑ **CHECK POINT 4**   Solve for x: $\dfrac{x}{3} - 4y = 5.$

---

2 Use the percent formula.

A Formula Involving Percent

Great Question!

Before presenting this formula involving percent, can you briefly review what I should already know about the basics of percent?

- **Percents** are the result of expressing numbers as part of 100. The word *percent* means *per hundred*. For example, the bar graph in **Figure 2.7** shows that 47% of Americans find "whatever" most annoying in conversation. Thus, 47 out of every 100 Americans are most annoyed by "whatever" in conversation: $47\% = \frac{47}{100}.$

Which Is Most Annoying In Conversation?

"Whatever" 47%
"You know" 25%
"It is what it is" 11%
"Anyway" 7%
"At the end of the day" 2%

Figure 2.7

Source: Marist Poll Survey of 938 Americans ages 16 and older, August 3–6, 2010.

- To convert a number from percent form to decimal form, move the decimal point two places to the left and drop the percent sign. Example:

$$47\% = 47.\% = 0.47\,\%$$

Thus, $47\% = 0.47.$

- To convert a number from decimal form to a percent, move the decimal point two places to the right and attach a percent sign.

$$0.86 = 086.\%$$

Thus, $0.86 = 86\%.$

- Dictionaries indicate that the word *percentage* has the same meaning as the word *percent*. Use the word that sounds better in a given circumstance.

Percents are useful in comparing two numbers. To compare the number A to the number B using a percent P, the following formula is used:

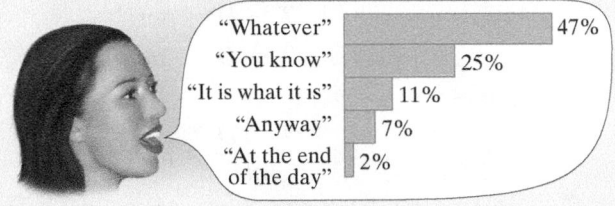

$$A \quad \text{is} \quad P \text{ percent} \quad \text{of} \quad B.$$
$$A \quad = \quad P \quad \cdot \quad B.$$

In the formula

$$A = PB,$$

$B =$ the base number, $P =$ the percent (in decimal form), and $A =$ the number being compared to B.

There are three basic types of percent problems that can be solved using the percent formula

$$A = PB. \quad \text{A is P percent of B.}$$

Question	Given	Percent Formula
What is P percent of B?	P and B	Solve for A.
A is P percent of what?	A and P	Solve for B.
A is what percent of B?	A and B	Solve for P.

Let's look at an example of each type of problem.

EXAMPLE 5 Using the Percent Formula: What Is P Percent of B?

What is 8% of 20?

Solution We use the formula $A = PB$: A is P percent of B. We are interested in finding the quantity A in this formula.

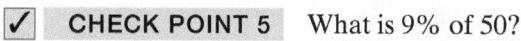

What | is | 8% | of | 20?

$$A = 0.08 \cdot 20 \quad \text{Express 8\% as 0.08.}$$
$$A = 1.6 \quad \text{Multiply:} \quad \begin{array}{r} 0.08 \\ \times\ 20 \\ \hline 1.60 \end{array}$$

Thus, 1.6 is 8% of 20. The answer is 1.6. ∎

✓ **CHECK POINT 5** What is 9% of 50?

EXAMPLE 6 Using the Percent Formula: A Is P Percent of What?

4 is 25% of what?

Solution We use the formula $A = PB$: A is P percent of B. We are interested in finding the quantity B in this formula.

4 | is | 25% | of | what?

$$4 = 0.25 \cdot B \quad \text{Express 25\% as 0.25.}$$
$$\frac{4}{0.25} = \frac{0.25B}{0.25} \quad \text{Divide both sides by 0.25.}$$
$$16 = B \quad \text{Simplify:} \quad 0.25\overline{)4.00}^{\,16.}$$

Thus, 4 is 25% of 16. The answer is 16. ∎

✓ **CHECK POINT 6** 9 is 60% of what?

EXAMPLE 7 Using the Percent Formula:
A Is What Percent of *B*?

1.3 is what percent of 26?

Solution We use the formula $A = PB$: A is P percent of B. We are interested in finding the quantity P in this formula.

1.3   is   what percent   of   26?

$$1.3 \;=\; P \;\cdot\; 26$$

$$\frac{1.3}{26} = \frac{P \cdot 26}{26}$$ Divide both sides by 26.

$$0.05 = P$$ Simplify: $\dfrac{0.05}{26\overline{)1.30}}$

We change 0.05 to a percent by moving the decimal point two places to the right and adding a percent sign: $0.05 = 5\%$. Thus, 1.3 is 5% of 26. The answer is 5%. ▪

✓ **CHECK POINT 7**   18 is what percent of 50?

 Solve applied problems involving percent change.

Applications

Percents are used for comparing changes, such as increases or decreases in sales, population, prices, and production. If a quantity changes, its percent increase or percent decrease can be determined by asking the following question:

The change is what percent of the original amount?

The question is answered using the percent formula as follows:

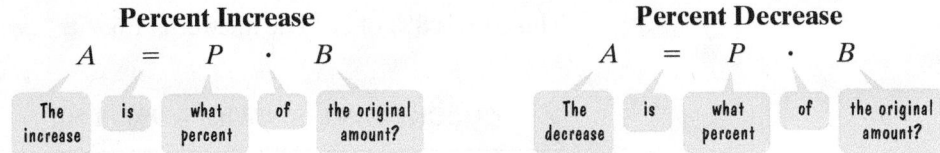

EXAMPLE 8 Finding Percent Decrease

A jacket regularly sells for $135.00. The sale price is $60.75. Find the percent decrease in the jacket's price.

Solution The percent decrease in price can be determined by asking the following question:

The price decrease is what percent of the original price ($135.00)?

The price decrease is the difference between the original price and the sale price ($60.75):

$$\$135.00 - \$60.75 = \$74.25.$$

Now we use the percent formula to find the percent decrease.

The price decrease   is   what percent   of   the original price?

$$74.25 \;=\; P \;\cdot\; 135$$

$$\frac{74.25}{135} = \frac{P \cdot 135}{135}$$ Divide both sides by 135.

$$0.55 = P$$ Simplify: $\dfrac{0.55}{135\overline{)74.25}}$

We change 0.55 to a percent by moving the decimal point two places to the right and adding a percent sign: $0.55 = 55\%$. Thus, the percent decrease in the jacket's price is 55%. ▪

✓ **CHECK POINT 8** A television regularly sells for $940. The sale price is $611. Find the percent decrease in the television's price.

In our next example, we look at one of the many ways that percent can be used incorrectly.

EXAMPLE 9 Promises of a Politician

A politician states, "If you elect me to office, I promise to cut your taxes for each of my first three years in office by 10% each year, for a total reduction of 30%." Evaluate the accuracy of the politician's statement.

Solution To make things simple, let's assume that a taxpayer paid $100 in taxes in the year previous to the politician's election. A 10% reduction during year 1 is 10% of $100.

$$10\% \text{ of previous year's tax} = 10\% \text{ of } \$100 = 0.10 \cdot \$100 = \$10$$

With a 10% reduction the first year, the taxpayer will pay only $100 − $10, or $90, in taxes during the politician's first year in office.

The table below shows how we calculate the new, reduced tax for each of the first three years in office.

Year	Tax Paid the Year Before	10% Reduction	Taxes Paid This Year
1	$100	$0.10 \cdot \$100 = \$10$	$\$100 - \$10 = \$90$
2	$ 90	$0.10 \cdot \$90 = \$9$	$\$90 - \$9 = \$81$
3	$ 81	$0.10 \cdot \$81 = \$8.10$	$\$81 - \$8.10 = \$72.90$

The tax reduction is the amount originally paid, $100.00, minus the amount paid during the politician's third year in office, $72.90:

$$\$100.00 - \$72.90 = \$27.10.$$

Now we use the percent formula to determine the percent decrease in taxes over the three years.

The tax decrease | is | what percent | of | the original tax?

$$27.1 = P \cdot 100$$

$$\frac{27.1}{100} = \frac{P \cdot 100}{100} \qquad \text{Divide both sides by 100.}$$

$$0.271 = P \qquad \text{Simplify.}$$

Change 0.271 to a percent: 0.271 = 27.1%. The percent decrease is 27.1%. The taxes decline by 27.1%, not by 30%. The politician is ill-informed in saying that three consecutive 10% cuts add up to a total tax cut of 30%. In our calculation, which serves as a counterexample to the promise, the total tax cut is only 27.1%. ∎

✓ **CHECK POINT 9** Suppose you paid $1200 in taxes. During year 1, taxes decrease by 20%. During year 2, taxes increase by 20%.

  a. What do you pay in taxes for year 2?

  b. How do your taxes for year 2 compare with what you originally paid, namely $1200? If the taxes are not the same, find the percent increase or decrease.

CONCEPT AND VOCABULARY CHECK

Fill in each blank so that the resulting statement is true.

1. Solving a formula for a variable means rewriting the formula so that the variable is _____.
2. The area, A, of a rectangle with length l and width w is given by the formula _____.
3. The perimeter, P, of a rectangle with length l and width w is given by the formula _____.
4. The sentence "A is P percent of B" is expressed by the formula _____.
5. In order to solve $y = mx + b$ for x, we first _____ and then _____.

2.4 EXERCISE SET   MyMathLab®

Watch the videos in MyMathLab

Download the MyDashBoard App

Practice Exercises

In Exercises 1–26, solve each formula for the specified variable. Do you recognize the formula? If so, what does it describe?

1. $d = rt$ for r
2. $d = rt$ for t
3. $I = Prt$ for P
4. $I = Prt$ for r
5. $C = 2\pi r$ for r
6. $C = \pi d$ for d
7. $E = mc^2$ for m
8. $V = \pi r^2 h$ for h
9. $y = mx + b$ for m
10. $y = mx + b$ for x
11. $T = D + pm$ for D
12. $P = C + MC$ for M
13. $A = \frac{1}{2}bh$ for b
14. $A = \frac{1}{2}bh$ for h
15. $M = \frac{n}{5}$ for n
16. $M = \frac{A}{740}$ for A
17. $\frac{c}{2} + 80 = 2F$ for c
18. $p = 15 + \frac{5d}{11}$ for d
19. $A = \frac{1}{2}(a + b)$ for a
20. $A = \frac{1}{2}(a + b)$ for b
21. $S = P + Prt$ for r
22. $S = P + Prt$ for t
23. $A = \frac{1}{2}h(a + b)$ for b
24. $A = \frac{1}{2}h(a + b)$ for a
25. $Ax + By = C$ for x
26. $Ax + By = C$ for y

Use the percent formula, $A = PB$: A is P percent of B, to solve Exercises 27–42.

27. What is 3% of 200?
28. What is 8% of 300?
29. What is 18% of 40?
30. What is 16% of 90?
31. 3 is 60% of what?
32. 8 is 40% of what?
33. 24% of what number is 40.8?
34. 32% of what number is 51.2?
35. 3 is what percent of 15?
36. 18 is what percent of 90?
37. What percent of 2.5 is 0.3?
38. What percent of 7.5 is 0.6?
39. If 5 is increased to 8, the increase is what percent of the original number?
40. If 5 is increased to 9, the increase is what percent of the original number?
41. If 4 is decreased to 1, the decrease is what percent of the original number?
42. If 8 is decreased to 6, the decrease is what percent of the original number?

Practice PLUS

In Exercises 43–50, solve each equation for x.

43. $y = (a + b)x$

44. $y = (a - b)x$

45. $y = (a - b)x + 5$

46. $y = (a + b)x - 8$

47. $y = cx + dx$

48. $y = cx - dx$

49. $y = Ax - Bx - C$

50. $y = Ax + Bx + C$

Application Exercises

51. The average, or mean, A, of three exam grades, x, y, and z, is given by the formula
$$A = \frac{x + y + z}{3}.$$
  a. Solve the formula for z.
  b. Use the formula in part (a) to solve this problem. On your first two exams, your grades are 86% and 88%: $x = 86$ and $y = 88$. What must you get on the third exam to have an average of 90%?

52. The average, or mean, A, of four exam grades, x, y, z, and w, is given by the formula
$$A = \frac{x + y + z + w}{4}.$$
  a. Solve the formula for w.
  b. Use the formula in part (a) to solve this problem. On your first three exams, your grades are 76%, 78%, and 79%: $x = 76$, $y = 78$, and $z = 79$. What must you get on the fourth exam to have an average of 80%?

53. If you are traveling in your car at an average rate of r miles per hour for t hours, then the distance, d, in miles, that you travel is described by the formula $d = rt$: distance equals rate times time.
  a. Solve the formula for t.
  b. Use the formula in part (a) to find the time that you travel if you cover a distance of 100 miles at an average rate of 40 miles per hour.

54. The formula $F = \frac{9}{5}C + 32$ expresses the relationship between Celsius temperature, C, and Fahrenheit temperature, F.
  a. Solve the formula for C.
  b. Use the formula from part (a) to find the equivalent Celsius temperature for a Fahrenheit temperature of 59°.

An Adecco survey of 1800 workers asked participants about taboo topics to discuss at work. The circle graph shows the results. Use this information to solve Exercises 55–56.

What Is the Most Taboo Topic to Discuss at Work?

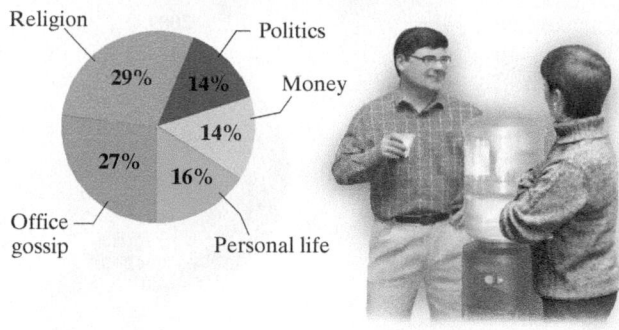

Source: Adecco

55. Among the 1800 workers who participated in the poll, how many stated that religion is the most taboo topic to discuss at work?

56. Among the 1800 workers who participated in the poll, how many stated that politics is the most taboo topic to discuss at work?

The graph shows the composition of a typical American community's trash. Use this information to solve Exercises 57–58.

Types of Trash in an American Community by Percentage of Total Weight

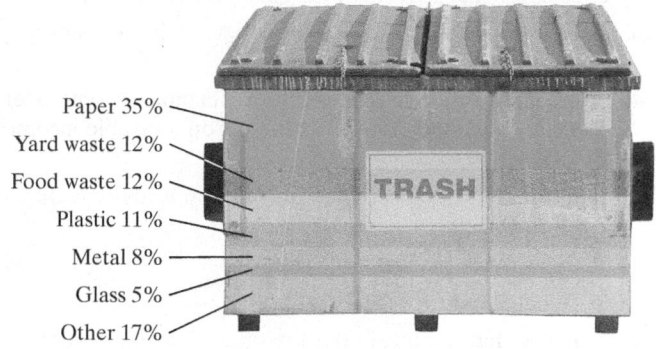

Source: U.S. Environmental Protection Agency

57. Across the United States, people generate approximately 175 billion pounds of trash in the form of paper each year. (That's approximately 580 pounds per person per year.) How many billions of pounds of trash do we throw away each year?

58. Across the United States, people generate approximately 55 billion pounds of trash in the form of plastic each year. (That's approximately 180 pounds per person per year.) How many billions of pounds of trash do we throw away each year?

The circle graphs show the number of free, partly free, and not free countries in 1974 and 2009. Use this information to solve Exercises 59–60. Round answers to the nearest percent.

Global Trends in Freedom

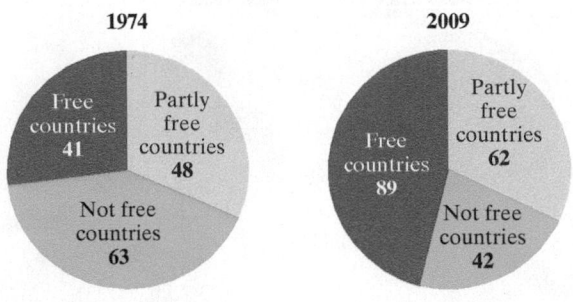

59. **a.** What percentage of the world's countries were free in 1974?

    b. What percentage of the world's countries were free in 2009?

    c. Find the percent increase in the number of free countries from 1974 to 2009.

60. **a.** What percentage of the world's countries were not free in 1974?

    b. What percentage of the world's countries were not free in 2009?

    c. Find the percent decrease in the number of not free countries from 1974 to 2009.

61. A charity has raised $7500, with a goal of raising $60,000. What percent of the goal has been raised?

62. A charity has raised $225,000, with a goal of raising $500,000. What percent of the goal has been raised?

63. A restaurant bill came to $60. If 15% of this amount was left as a tip, how much was the tip?

64. If income tax is $3502 plus 28% of taxable income over $23,000, how much is the income tax on a taxable income of $35,000?

65. Suppose that the local sales tax rate is 6% and you buy a car for $16,800.

    a. How much tax is due?

    b. What is the car's total cost?

66. Suppose that the local sales tax rate is 7% and you buy a graphing calculator for $96.

    a. How much tax is due?

    b. What is the calculator's total cost?

67. An exercise machine with an original price of $860 is on sale at 12% off.

    a. What is the discount amount?

    b. What is the exercise machine's sale price?

68. A dictionary that normally sells for $16.50 is on sale at 40% off.

    a. What is the discount amount?

    b. What is the dictionary's sale price?

69. A sofa regularly sells for $840. The sale price is $714. Find the percent decrease in the sofa's price.

70. A fax machine regularly sells for $380. The sale price is $266. Find the percent decrease in the machine's price.

71. Suppose that you put $10,000 in a rather risky investment recommended by your financial advisor. During the first year, your investment decreases by 30% of its original value. During the second year, your investment increases by 40% of its first-year value. Your advisor tells you that there must have been a 10% overall increase of your original $10,000 investment. Is your financial advisor using percentages properly? If not, what is the actual percent gain or loss on your original $10,000 investment?

72. The price of a color printer is reduced by 30% of its original price. When it still does not sell, its price is reduced by 20% of the reduced price. The salesperson informs you that there has been a total reduction of 50%. Is the salesperson using percentages properly? If not, what is the actual percent reduction from the original price?

Writing in Mathematics

73. Explain what it means to solve a formula for a variable.

74. What does the percent formula, $A = PB$, describe? Give an example of how the formula is used.

Critical Thinking Exercises

Make Sense? *In Exercises 75–78, determine whether each statement "makes sense" or "does not make sense" and explain your reasoning.*

75. To help me get started, I circle the variable that I need to solve for in a formula.

76. By solving a formula for one of its variables, I find a numerical value for that variable.

77. I have $100 and my restaurant bill comes to $80, which is not enough to leave a 20% tip.

78. I found the percent decrease in a jacket's price to be 120%.

In Exercises 79–82, determine whether each statement is true or false. If the statement is false, make the necessary change(s) to produce a true statement.

79. If $ax + b = 0$, then $x = \dfrac{b}{a}$.

80. If $A = lw$, then $w = \dfrac{l}{A}$.

81. If $A = \dfrac{1}{2}bh$, then $b = \dfrac{A}{2h}$.

82. Solving $x - y = -7$ for y gives $y = x + 7$.

83. In psychology, an intelligence quotient, Q, also called IQ, is measured by the formula

$$Q = \frac{100M}{C},$$

where M = mental age and C = chronological age. Solve the formula for C.

Review Exercises

84. Solve and check: $5x + 20 = 8x - 16$.
(Section 2.2, Example 7)

85. Solve and check: $5(2y - 3) - 1 = 4(6 + 2y)$.
(Section 2.3, Example 3)

86. Simplify: $x - 0.3x$.
(Section 1.4, Example 8)

Preview Exercises

Exercises 87–89 will help you prepare for the material covered in the next section. In each exercise, let x represent the number and write the phrase as an algebraic expression.

87. The quotient of 13 and a number, decreased by 7 times the number

88. Eight times the sum of a number and 14

89. Nine times the difference of a number and 5

MID-CHAPTER CHECK POINT   Section 2.1–Section 2.4

 What You Know: We learned a step-by-step procedure for solving linear equations, including equations with fractions and decimals. We saw that some equations have no solution, whereas others have all real numbers as solutions. We used the addition and multiplication properties of equality to solve formulas for variables. Finally, we worked with the percent formula $A = PB$: A is P percent of B.

1. Solve: $\dfrac{x}{2} = 12 - \dfrac{x}{4}$.

2. Solve: $5x - 42 = -57$.

3. Solve for C: $H = \dfrac{EC}{825}$.

4. What is 6% of 140?

5. Solve: $\dfrac{-x}{10} = -3$.

6. Solve: $1 - 3(y - 5) = 4(2 - 3y)$.

7. Solve for r: $S = 2\pi rh$.

8. 12 is 30% of what?

9. Solve: $\dfrac{3y}{5} + \dfrac{y}{2} = \dfrac{5y}{4} - 3$.

10. Solve: $2.4x + 6 = 1.4x + 0.5(6x - 9)$.

11. Solve: $5z + 7 = 6(z - 2) - 4(2z - 3)$.

12. Solve for x: $Ax - By = C$.

13. Solve: $6y + 7 + 3y = 3(3y - 1)$.

14. Solve: $10\left(\dfrac{1}{2}x + 3\right) = 10\left(\dfrac{3}{5}x - 1\right)$.

15. 50 is what percent of 400?

16. Solve: $\dfrac{3(m + 2)}{4} = 2m + 3$.

17. If 40 is increased to 50, the increase is what percent of the original number?

18. Solve: $12x - 4 + 8x - 4 = 4(5x - 2)$.

19. The bar graph indicates that reading books for fun loses value as kids age.

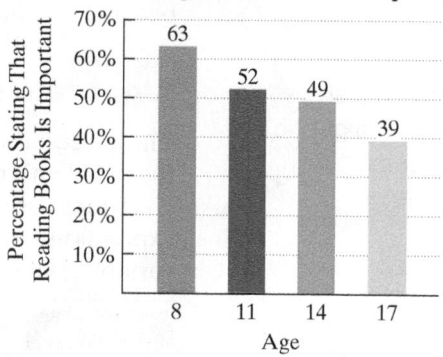

Percentage of U.S. Kids Who Believe Reading Books Not Required for School Is Important

Source of data: Scholastic Kids and Family Reading Report 2010, Scholastic Inc.

The data can be described by the mathematical model

$$B = -\dfrac{5}{2}a + 82,$$

where B is the percentage at age a who believe that reading books is important.

a. Does the formula underestimate or overestimate the percentage of 14-year-olds who believe that reading books is important? By how much?

b. If trends shown by the formula continue, at which age will 22% believe that reading books is important?

An Introduction to Problem Solving

Objectives

1 Translate English phrases into algebraic expressions.

2 Solve algebraic word problems using linear equations.

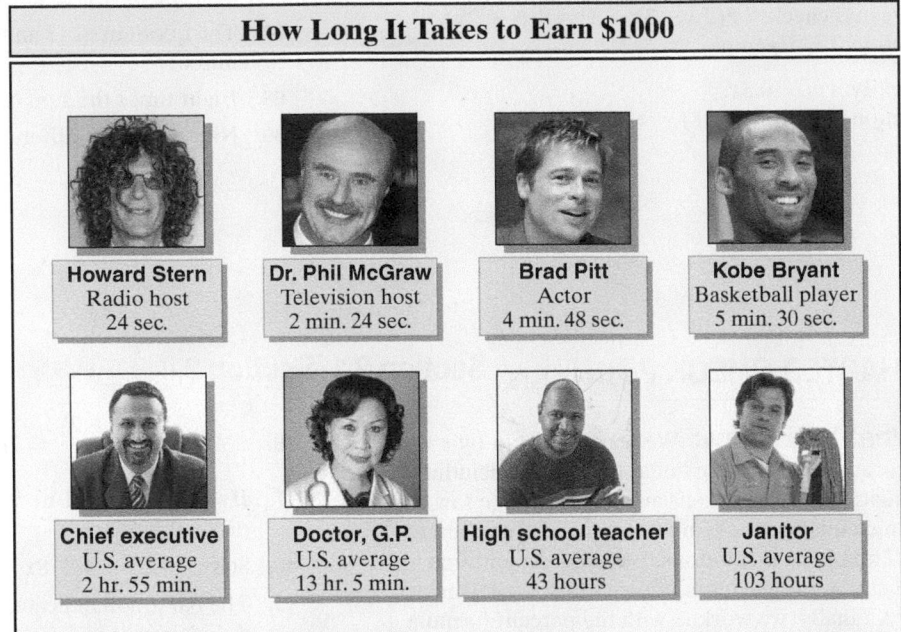

How Long It Takes to Earn $1000

Howard Stern
Radio host
24 sec.

Dr. Phil McGraw
Television host
2 min. 24 sec.

Brad Pitt
Actor
4 min. 48 sec.

Kobe Bryant
Basketball player
5 min. 30 sec.

Chief executive
U.S. average
2 hr. 55 min.

Doctor, G.P.
U.S. average
13 hr. 5 min.

High school teacher
U.S. average
43 hours

Janitor
U.S. average
103 hours

Source of data: TIME Magazine, October 30, 2006

In this section, you'll see examples and exercises focused on how much money Americans earn. These situations illustrate a step-by-step strategy for solving problems. As you become familiar with this strategy, you will learn to solve a wide variety of problems.

Algebraic Expressions for English Phrases

1 Translate English phrases into algebric expressions.

The hardest thing about word problems is writing an equation that translates, or models, the problem's conditions. Throughout Chapter 1, you wrote algebraic expressions and equations for conditions about numbers. **Table 2.1** summarizes many of the algebraic expressions that you wrote for these conditions. We are using x to represent the variable, but we can use any letter.

Great Question!

Table 2.1 looks long and intimidating. What's the best way to get through the table?

Cover the right column with a sheet of paper and attempt to formulate the algebraic expression for the English phrase in the left column on your own. Then slide the paper down and check your answer. Work through the entire table in this manner.

Table 2.1 Algebraic Translations of English Phrases

English Phrase	Algebraic Expression
Addition	
The sum of a number and 7	$x + 7$
Five more than a number; a number plus 5	$x + 5$
A number increased by 6; 6 added to a number	$x + 6$
Subtraction	
A number minus 4	$x - 4$
A number decreased by 5	$x - 5$
A number subtracted from 8	$8 - x$
The difference between a number and 6	$x - 6$
The difference between 6 and a number	$6 - x$
Seven less than a number	$x - 7$
Seven minus a number	$7 - x$
Nine fewer than a number	$x - 9$

Table 2.1 continued

English Phrase	Algebraic Expression
Multiplication	
Five times a number	$5x$
The product of 3 and a number	$3x$
Two-thirds of a number (used with fractions)	$\frac{2}{3}x$
Seventy-five percent of a number (used with decimals)	$0.75x$
Thirteen multiplied by a number	$13x$
A number multiplied by 13	$13x$
Twice a number	$2x$
Division	
A number divided by 3	$\dfrac{x}{3}$
The quotient of 7 and a number	$\dfrac{7}{x}$
The quotient of a number and 7	$\dfrac{x}{7}$
The reciprocal of a number	$\dfrac{1}{x}$
More than one operation	
The sum of twice a number and 7	$2x + 7$
Twice the sum of a number and 7	$2(x + 7)$
Three times the sum of 1 and twice a number	$3(1 + 2x)$
Nine subtracted from 8 times a number	$8x - 9$
Twenty-five percent of the sum of 3 times a number and 14	$0.25(3x + 14)$
Seven times a number, increased by 24	$7x + 24$
Seven times the sum of a number and 24	$7(x + 24)$

A Strategy for Solving Word Problems Using Equations

Here are some general steps we will follow in solving word problems:

Strategy for Solving Word Problems

Step 1. Read the problem carefully several times until you can state in your own words what is given and what the problem is looking for. Let x (or any variable) represent one of the unknown quantities in the problem.

Step 2. If necessary, write expressions for any other unknown quantities in the problem in terms of x.

Step 3. Write an equation in x that translates, or models, the conditions of the problem.

Step 4. Solve the equation and answer the problem's question.

Step 5. Check the solution *in the original wording* of the problem, not in the equation obtained from the words.

Great Question!

Why are word problems important?

There is great value in reasoning through the steps for solving a word problem. This value comes from the problem-solving skills that you will attain and is often more important than the specific problem or its solution.

2 Solve algebraic word problems using linear equations.

Applying the Strategy for Solving Word Problems

Now that you've read why word problems are important, let's apply our five-step strategy for solving these problems.

EXAMPLE 1 Solving a Word Problem

Nine subtracted from eight times a number is 39. Find the number.

Solution

Step 1. Let x represent one of the quantities. Because we are asked to find a number, let

$$x = \text{the number.}$$

Step 2. Represent other unknown quantities in terms of x. There are no other unknown quantities to find, so we can skip this step.

Step 3. Write an equation in x that models the conditions.

Nine subtracted from	eight times a number	is	39.
$8x$	$- 9$	$=$	39

Step 4. Solve the equation and answer the question.

$$8x - 9 = 39 \qquad \text{This is the equation that models the problem's conditions.}$$
$$8x - 9 + 9 = 39 + 9 \qquad \text{Add 9 to both sides.}$$
$$8x = 48 \qquad \text{Simplify.}$$
$$\frac{8x}{8} = \frac{48}{8} \qquad \text{Divide both sides by 8.}$$
$$x = 6 \qquad \text{Simplify.}$$

The number is 6.

Step 5. Check the proposed solution in the original wording of the problem. "Nine subtracted from eight times a number is 39." The proposed number is 6. Eight times 6 is $8 \cdot 6$, or 48. Nine subtracted from 48 is $48 - 9$, or 39. The proposed solution checks in the problem's wording, verifying that the number is 6. ■

✓ **CHECK POINT 1** Four subtracted from six times a number is 68. Find the number.

EXAMPLE 2 Starting Salaries for College Graduates with Undergraduate Degrees

The bar graph in **Figure 2.8** shows the ten most popular college majors with median, or middlemost, starting salaries for recent college graduates.

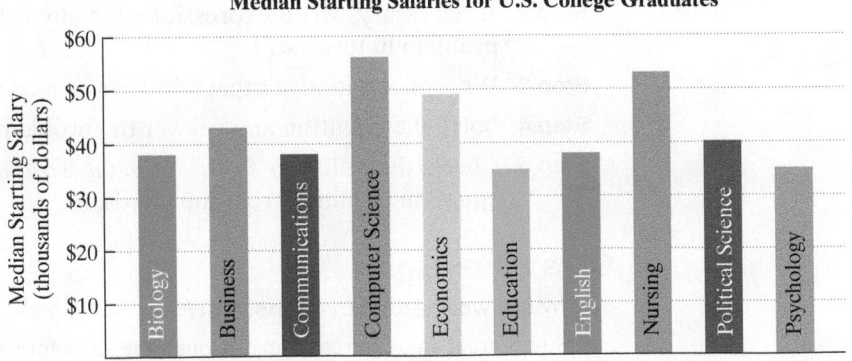

Median Starting Salaries for U.S. College Graduates

Figure 2.8

Source: PayScale (2010 data)

The median starting salary of a nursing major exceeds that of a business major by $10 thousand. Combined, their median starting salaries are $96 thousand. Determine the median starting salaries of business and nursing majors with bachelor's degrees.

Solution

Step 1.   Let x represent one of the quantities. We know something about the median starting salary of a nursing major: It exceeds that of a business major by $10 thousand. This means that nursing majors with bachelor's degrees earn $10 thousand more than business majors. We will let

x = the median starting salary, in thousands of dollars, of business majors.

Step 2.   Represent other unknown quantities in terms of x. The other unknown quantity is the median starting salary of nursing majors. Because it is $10 thousand more than that of business majors, let

$x + 10$ = the median starting salary, in thousands of dollars, of nursing majors.

Step 3.   Write an equation in x that models the conditions. Combined, the median starting salaries for business and nursing majors are $96 thousand.

The median starting salary for business majors	plus	the median starting salary for nursing majors	is	$96 thousand.
x	$+$	$(x + 10)$	$=$	96

Step 4.   Solve the equation and answer the question.

$$x + (x + 10) = 96 \qquad \text{This is the equation that models the problem's conditions.}$$
$$2x + 10 = 96 \qquad \text{Regroup and combine like terms on the left side.}$$
$$2x + 10 - 10 = 96 - 10 \qquad \text{Subtract 10 from both sides.}$$
$$2x = 86 \qquad \text{Simplify.}$$
$$\frac{2x}{2} = \frac{86}{2} \qquad \text{Divide both sides by 2.}$$
$$x = 43 \qquad \text{Simplify.}$$

Because x represents the median starting salary, in thousands of dollars, of business majors, we see that business majors with bachelor's degrees have a median starting salary of $43 thousand, or $43,000. Because $x + 10$ represents the median starting salary, in thousands of dollars, of nursing majors, their median starting salary is $43 + 10$, or $53 thousand ($53,000).

Step 5.   Check the proposed solution in the original wording of the problem. The problem states that combined, the median starting salaries are $96 thousand. By adding $43 thousand, the median starting salary of business majors, and $53 thousand, the median starting salary of nursing majors, we do, indeed, obtain a sum of $96 thousand.   ■

Great Question!

Example 2 involves using the word *exceeds* to represent one of the unknown quantities. Can you help me to write algebraic expressions for quantities described using *exceeds*?

Modeling with the word *exceeds* can be a bit tricky. It's helpful to identify the smaller quantity. Then add to this quantity to represent the larger quantity. For example, suppose that Tim's height exceeds Tom's height by a inches. Tom is the shorter person. If Tom's height is represented by x, then Tim's height is represented by $x + a$.

✓   **CHECK POINT 2**   Two of the bars in **Figure 2.8** on page 158 represent starting salaries of computer science and English majors. The median starting salary of a computer science major exceeds that of an English major by $18 thousand. Combined, their median starting salaries are $94 thousand. Determine the median starting salaries of English and computer science majors with bachelor's degrees.

EXAMPLE 3   Consecutive Integers and the Super Bowl

Only once, in 1991, were the winning and losing scores in the Super Bowl consecutive integers. The New York Giants beat the Buffalo Bills in a nearly error-free game. If the sum of the scores was 39, determine the points scored by the losing team, the Bills, and the winning team, the Giants.

Solution

Step 1. Let x represent one of the quantities. We will let

$$x = \text{points scored by the losing team, the Bills.}$$

Step 2. Represent other unknown quantities in terms of x. The other unknown quantity involves points scored by the winning team, the Giants. Because the scores were consecutive integers, the winning team scored one point more than the losing team. Thus,

$$x + 1 = \text{points scored by the winning team, the Giants.}$$

Step 3. Write an equation in x that models the conditions. We are told that the sum of the scores was 39.

Points scored by the losing team	plus	points scored by the winning team	result in	39 points.
x	$+$	$(x + 1)$	$=$	39

Step 4. Solve the equation and answer the question.

$$x + (x + 1) = 39 \qquad \text{This is the equation that models the problem's conditions.}$$
$$2x + 1 = 39 \qquad \text{Regroup and combine like terms.}$$
$$2x + 1 - 1 = 39 - 1 \qquad \text{Subtract 1 from both sides.}$$
$$2x = 38 \qquad \text{Simplify.}$$
$$\frac{2x}{2} = \frac{38}{2} \qquad \text{Divide both sides by 2.}$$
$$x = 19 \qquad \text{Simplify.}$$

Thus,

$$\text{points scored by the losing team, the Bills} = x = 19$$

and

$$\text{points scored by the winning team, the Giants} = x + 1 = 20.$$

With the closest final score in Super Bowl history, the Giants scored 20 points and the Bills scored 19 points.

Step 5. Check the proposed solution in the original wording of the problem. The problem states that the sum of the scores was 39. With a final score of 20 to 19, we see that the sum of these numbers is, indeed, 39. ■

✓   **CHECK POINT 3**   Page numbers on facing pages of a book are consecutive integers. Two pages that face each other have 145 as the sum of their page numbers. What are the page numbers?

Example 3 and Check Point 3 involved consecutive integers. By contrast, some word problems involve consecutive odd integers, such as 5, 7, and 9. Other word problems involve consecutive even integers, such as 6, 8, and 10. When working with consecutive even or consecutive odd integers, we must continually add 2 to move from one integer to the next successive integer in the list.

Table 2.2 should be helpful in solving consecutive integer problems.

Table 2.2	Consecutive Integers	
English Phrase	**Algebraic Expressions**	**Example**
Two consecutive integers	$x, x + 1$	13, 14
Three consecutive integers	$x, x + 1, x + 2$	$-8, -7, -6$
Two consecutive even integers	$x, x + 2$	40, 42
Two consecutive odd integers	$x, x + 2$	$-37, -35$
Three consecutive even integers	$x, x + 2, x + 4$	30, 32, 34
Three consecutive odd integers	$x, x + 2, x + 4$	9, 11, 13

EXAMPLE 4   Renting a Car

Rent-a-Heap Agency charges \$125 per week plus \$0.20 per mile to rent a small car. How many miles can you travel for \$335?

Solution

Step 1. Let x represent one of the quantities. Because we are asked to find the number of miles we can travel for \$335, let

$$x = \text{the number of miles.}$$

Step 2. Represent other unknown quantities in terms of x. There are no other unknown quantities to find, so we can skip this step.

Step 3. Write an equation in x that models the conditions. Before writing the equation, let us consider a few specific values for the number of miles traveled. The rental charge is \$125 plus \$0.20 for each mile.

3 miles: The rental charge is \$125 + \$0.20(3).

30 miles: The rental charge is \$125 + \$0.20(30).

100 miles: The rental charge is \$125 + \$0.20(100).

x miles: The rental charge is \$125 + \$0.20x.

The weekly charge of \$125	plus	the charge of \$0.20 per mile for x miles	equals	the total \$335 rental charge.
125	+	0.20x	=	335

Step 4. Solve the equation and answer the question.

$$125 + 0.20x = 335 \qquad \text{This is the equation that models the conditions of the problem.}$$

$$125 + 0.20x - 125 = 335 - 125 \qquad \text{Subtract 125 from both sides.}$$

$$0.20x = 210 \qquad \text{Simplify.}$$

$$\frac{0.20x}{0.20} = \frac{210}{0.20} \qquad \text{Divide both sides by 0.20.}$$

$$x = 1050 \qquad \text{Simplify.}$$

You can travel 1050 miles for \$335.

Step 5. Check the proposed solution in the original wording of the problem. Traveling 1050 miles should result in a total rental charge of $335. The mileage charge of $0.20 per mile is

$$\$0.20(1050) = \$210.$$

Adding this to the $125 weekly charge gives a total rental charge of

$$\$125 + \$210 = \$335.$$

Because this results in the given rental charge of $335, this verifies that you can travel 1050 miles. ∎

✓ **CHECK POINT 4**   A taxi charges $2.00 to turn on the meter plus $0.25 for each eighth of a mile. If you have $10.00, how many eighths of a mile can you go? How many miles is that?

We will be using the formula for the perimeter of a rectangle, $P = 2l + 2w$, in our next example. Twice the rectangle's length plus twice the rectangle's width is its perimeter.

EXAMPLE 5   Finding the Dimensions of a Soccer Field

A rectangular soccer field is twice as long as it is wide. If the perimeter of a soccer field is 300 yards, what are the field's dimensions?

Solution

Step 1. Let x represent one of the quantities. We know something about the length; the field is twice as long as it is wide. We will let

$$x = \text{the width.}$$

Step 2. Represent other unknown quantities in terms of x. Because the field is twice as long as it is wide, let

$$2x = \text{the length.}$$

Figure 2.9 illustrates the soccer field and its dimensions.

Great Question!

Should I draw pictures like Figure 2.9 when solving geometry problems?

When solving word problems, particularly problems involving geometric figures, drawing a picture of the situation is often helpful. Label x on your drawing and, where appropriate, label other parts of the drawing in terms of x.

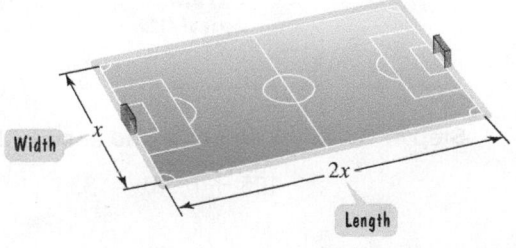

Width x

$2x$

Length

Figure 2.9

Step 3. Write an equation in x that models the conditions. Because the perimeter of a soccer field is 300 yards,

Twice the length	plus	twice the width	is	the perimeter.
$2 \cdot 2x$	$+$	$2 \cdot x$	$=$	$300.$

Step 4. Solve the equation and answer the question.

$$2 \cdot 2x + 2 \cdot x = 300 \quad \text{This is the equation that models the problem's conditions.}$$
$$4x + 2x = 300 \quad \text{Multiply.}$$
$$6x = 300 \quad \text{Combine like terms.}$$
$$\frac{6x}{6} = \frac{300}{6} \quad \text{Divide both sides by 6.}$$
$$x = 50 \quad \text{Simplify.}$$

Thus,
$$\text{width} = x = 50$$
$$\text{length} = 2x = 2(50) = 100.$$

The dimensions of a soccer field are 50 yards by 100 yards.

Step 5. Check the proposed solution in the original wording of the problem. The perimeter of the soccer field, using the dimensions that we found, is 2(50 yards) + 2(100 yards) = 100 yards + 200 yards, or 300 yards. Because the problem's wording tells us that the perimeter is 300 yards, our dimensions are correct. ■

✓ **CHECK POINT 5** A rectangular swimming pool is three times as long as it is wide. If the perimeter of the pool is 320 feet, what are the pool's dimensions?

EXAMPLE 6 A Price Reduction

Your local computer store is having a sale. After a 30% price reduction, you purchase a new computer for $980. What was the computer's price before the reduction?

Solution

Step 1. Let x represent one of the quantities. We will let
$$x = \text{the original price of the computer.}$$

Step 2. Represent other unknown quantities in terms of x. There are no other unknown quantities to find, so we can skip this step.

Great Question!

Is there a common error that I can avoid when solving problems about price reductions?

Yes. In Example 6, notice that the original price, x, reduced by 30% is $x - 0.3x$ and *not* $x - 0.3$.

Step 3. Write an equation in x that models the conditions. The computer's original price minus the 30% reduction is the reduced price, $980.

Original price	minus	the reduction (30% of the original price)	is	the reduced price, $980.
x	$-$	$0.3x$	$=$	980

Step 4. Solve the equation and answer the question.

$$x - 0.3x = 980 \quad \text{This is the equation that models the problem's conditions.}$$
$$0.7x = 980 \quad \text{Combine like terms: } x - 0.3x = 1x - 0.3x = 0.7x.$$
$$\frac{0.7x}{0.7} = \frac{980}{0.7} \quad \text{Divide both sides by 0.7.}$$
$$x = 1400 \quad \text{Simplify: } 0.7\overline{)980.0} \; \frac{1400}{}$$

The computer's price before the reduction was $1400.

Step 5. Check the proposed solution in the original wording of the problem. The price before the reduction, $1400, minus the reduction in price should equal the reduced price given in the original wording, $980. The reduction in price is equal to 30% of the price before the reduction, $1400. To find the reduction, we multiply the decimal equivalent of 30%, 0.30 or 0.3, by the original price, $1400:

$$30\% \text{ of } \$1400 = (0.3)(\$1400) = \$420.$$

Now we can determine whether the calculation for the price before the reduction, $1400, minus the reduction, $420, is equal to the reduced price given in the problem, $980. We subtract:

$$\$1400 - \$420 = \$980.$$

This verifies that the price of the computer before the reduction was $1400. ■

Great Question!

Can I solve the equation in Example 6, $x - 0.3x = 980$, by first clearing the decimal?

Yes. Because 0.3 has one decimal place, the greatest number of decimal places in the equation, multiply both sides by 10^1, or 10, to clear the decimal.

$x - 0.3x = 980$	This is the equation that models the conditions in Example 6.
$10(x - 0.3x) = 10(980)$	Multiply both sides by 10.
$10x - 10(0.3x) = 10(980)$	Use the distributive property.
$10x - 3x = 9800$	Simplify by moving decimal points one place to the right.
$7x = 9800$	Combine like terms.
$\dfrac{7x}{7} = \dfrac{9800}{7}$	Divide both sides by 7.
$x = 1400$	Simplify.

This is the same value for x that we obtained in Example 6 when we did not clear the decimal. Which method do you prefer?

✓ **CHECK POINT 6** After a 40% price reduction, an exercise machine sold for $564. What was the exercise machine's price before this reduction?

Achieving Success

Do not expect to solve every word problem immediately. As you read each problem, underline the important parts. It's a good idea to read the problem at least twice. Be persistent, but use the **"Ten Minutes of Frustration" Rule**. If you have exhausted every possible means for solving a problem and you are still bogged down, stop after ten minutes. Put a question mark by the exercise and move on. When you return to class, ask your professor for assistance.

CONCEPT AND VOCABULARY CHECK

Fill in each blank so that the resulting statement is true.

1. If x represents a number, six subtracted from four times the number can be represented by _____.

2. According to *Forbes* magazine, the top-earning dead celebrities in 2010 were Michael Jackson and Elvis Presley. Jackson's earnings exceeded Presley's earnings by $215 million. If x represents Presley's earnings, in millions of dollars, Jackson's earnings can be represented by _____.

3. If the number of any left-hand page in this book is represented by x, the number on the facing page can be represented by _____.

4. I can rent a car for $125 per week plus $0.15 for each mile driven. If I drive x miles in a week, the cost for the rental can be represented by _____.

5. If x represents a rectangle's width and $4x$ represents its length, the perimeter of the rectangle can be represented by _____.

6. I purchased a computer after a 35% price reduction. If x represents the computer's original price, the reduced price can be represented by _____.

2.5 EXERCISE SET   MyMathLab®

Watch the videos in MyMathLab

Download the MyDashBoard App

Practice Exercises

In Exercises 1–20, let x represent the number. Use the given conditions to write an equation. Solve the equation and find the number.

1. A number increased by 60 is equal to 410. Find the number.

2. The sum of a number and 43 is 107. Find the number.

3. A number decreased by 23 is equal to 214. Find the number.

4. The difference between a number and 17 is 96. Find the number.

5. The product of 7 and a number is 126. Find the number.

6. The product of 8 and a number is 272. Find the number.

7. The quotient of a number and 19 is 5. Find the number.

8. The quotient of a number and 14 is 8. Find the number.

9. The sum of four and twice a number is 56. Find the number.

10. The sum of five and three times a number is 59. Find the number.

11. Seven subtracted from five times a number is 178. Find the number.

12. Eight subtracted from six times a number is 298. Find the number.

13. A number increased by 5 is two times the number. Find the number.

14. A number increased by 12 is four times the number. Find the number.

15. Twice the sum of four and a number is 36. Find the number.

16. Three times the sum of five and a number is 48. Find the number.

17. Nine times a number is 30 more than three times that number. Find the number.

18. Five more than four times a number is that number increased by 35. Find the number.

19. If the quotient of three times a number and five is increased by four, the result is 34. Find the number.

20. If the quotient of three times a number and four is decreased by three, the result is nine. Find the number.

Application Exercises

In Exercises 21–46, use the five-step strategy to solve each problem.

How will you spend your average life expectancy of 78 years? The bar graph shows the average number of years you will devote to each of your most time-consuming activities. Exercises 21–22 are based on the data displayed by the graph.

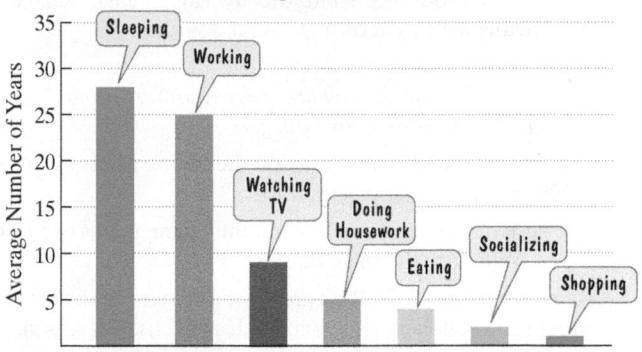

Source: U.S. Bureau of Labor Statistics

21. According to the American Bureau of Labor Statistics, you will devote 37 years to sleeping and watching TV. The number of years sleeping will exceed the number of years watching TV by 19. Over your lifetime, how many years will you spend on each of these activities?

22. According to the American Bureau of Labor Statistics, you will devote 32 years to sleeping and eating. The number of years sleeping will exceed the number of years eating by 24. Over your lifetime, how many years will you spend on each of these activities?

The bar graph shows average yearly earnings in the United States for people with a college education, by final degree earned. Exercises 23–24 are based on the data displayed by the graph.

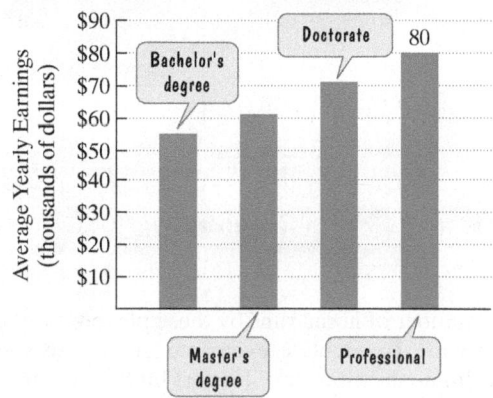

Source: U.S. Census Bureau

23. The average yearly salary of an American whose final degree is a master's is $49 thousand less than twice that of an American whose final degree is a bachelor's. Combined, two people with each of these educational attainments earn $116 thousand. Find the average yearly salary of Americans with each of these final degrees.

24. The average yearly salary of an American whose final degree is a doctorate is $39 thousand less than twice that of an American whose final degree is a bachelor's. Combined, two people with each of these educational attainments earn $126 thousand. Find the average yearly salary of Americans with each of these final degrees.

In Exercises 25–26, use the fact that page numbers on facing pages of a book are consecutive integers.

25. The sum of the page numbers on the facing pages of a book is 629. What are the page numbers?

26. The sum of the page numbers on the facing pages of a book is 525. What are the page numbers?

27. Roger Maris and Babe Ruth are among the ten baseball players with the most home runs in a major league baseball season.

The number of home runs by these players for the seasons shown are consecutive odd integers whose sum is 120. Determine the number of homers hit by Ruth and by Maris.

28. Babe Ruth and Hank Greenberg are among the ten baseball players with the most home runs in a major league baseball season.

The number of home runs by these players for the seasons shown are consecutive even integers whose sum is 118. Determine the number of homers hit by Greenberg and by Ruth.

29. A car rental agency charges $200 per week plus $0.15 per mile to rent a car. How many miles can you travel in one week for $320?

30. A car rental agency charges $180 per week plus $0.25 per mile to rent a car. How many miles can you travel in one week for $395?

The bar graph shows that average rent and mortgage payments in the United States have increased since 1975, even after taking inflation into account. Exercises 31–32 are based on the information displayed by the graph.

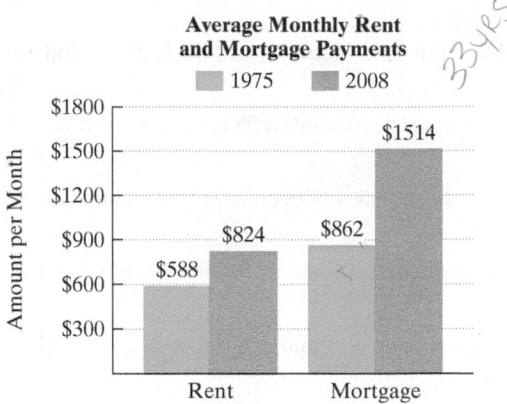

Source: U.S. Department of Housing and Urban Development

31. In 2008, mortgage payments averaged $1514 per month. For the period shown, monthly mortgage payments increased by approximately $20 per year. If this trend continues, how many years after 2008 will mortgage payments average $1714? In which year will this occur?

32. In 2008, rent payments averaged $824 per month. For the period shown, monthly rent payments increased by approximately $7 per year. If this trend continues, how many years after 2008 will rent payments average $929? In which year will this occur?

33. A rectangular field is four times as long as it is wide. If the perimeter of the field is 500 yards, what are the field's dimensions?

34. A rectangular field is five times as long as it is wide. If the perimeter of the field is 288 yards, what are the field's dimensions?

35. An American football field is a rectangle with a perimeter of 1040 feet. The length is 200 feet more than the width. Find the width and length of the rectangular field.

36. A basketball court is a rectangle with a perimeter of 86 meters. The length is 13 meters more than the width. Find the width and length of the basketball court.

37. A bookcase is to be constructed as shown in the figure. The length is to be 3 times the height. If 60 feet of lumber is available for the entire unit, find the length and height of the bookcase.

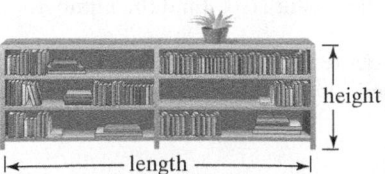

38. The height of the bookcase in the figure is 3 feet longer than the length of a shelf. If 18 feet of lumber is available for the entire unit, find the length and height of the unit.

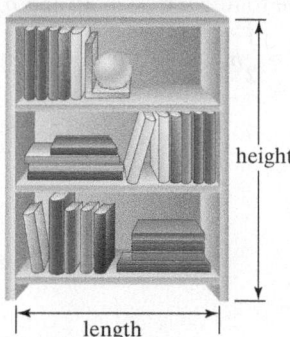

39. After a 20% reduction, you purchase a television for $320. What was the television's price before the reduction?

40. After a 30% reduction, you purchase a DVD player for $98. What was the price before the reduction?

41. This year's salary, $50,220, is an 8% increase over last year's salary. What was last year's salary?

42. This year's salary, $42,074, is a 9% increase over last year's salary. What was last year's salary?

43. Including 6% sales tax, a car sold for $23,850. Find the price of the car before the tax was added.

44. Including 8% sales tax, a bed-and-breakfast inn charges $172.80 per night. Find the inn's nightly cost before the tax is added.

45. An automobile repair shop charged a customer $448, listing $63 for parts and the remainder for labor. If the cost of labor is $35 per hour, how many hours of labor did it take to repair the car?

46. A repair bill on a sailboat came to $1603, including $532 for parts and the remainder for labor. If the cost of labor is $63 per hour, how many hours of labor did it take to repair the sailboat?

Writing in Mathematics

47. In your own words, describe a step-by-step approach for solving algebraic word problems.

48. Many students find solving linear equations much easier than solving algebraic word problems. Discuss some of the reasons why this is the case.

49. Did you have some difficulties solving some of the problems that were assigned in this Exercise Set? Discuss what you did if this happened to you. Did your course of action enhance your ability to solve algebraic word problems?

50. Write an original word problem that can be solved using a linear equation. Then solve the problem.

Critical Thinking Exercises

Make Sense?   In Exercises 51–54, determine whether each statement "makes sense" or "does not make sense" and explain your reasoning.

51. Rather than struggling with the assigned word problems, I'll ask my instructor to solve them all in class and then study the solutions.

52. By reasoning through word problems, I can increase my problem-solving skills in general.

53. I find the hardest part in solving a word problem is writing the equation that models the verbal conditions.

54. I made a mistake when I used x and $x + 2$ to represent two consecutive odd integers, because 2 is even.

In Exercises 55–58, determine whether each statement is true or false. If the statement is false, make the necessary change(s) to produce a true statement.

55. Ten pounds less than Bill's weight, x, equals 160 pounds is modeled by $10 - x = 160$.

56. After a 35% reduction, a computer's price is $780, so its original price, x, can be found by solving $x - 0.35 = 780$.

57. If the length of a rectangle is 6 inches more than its width, and its perimeter is 24 inches, the distributive property must be used to solve the equation that determines the length.

58. On a number line, consecutive integers do not have any other integers between them.

59. An HMO pamphlet contains the following recommended weight for women: "Give yourself 100 pounds for the first 5 feet plus 5 pounds for every inch over 5 feet tall." Using this description, which height corresponds to an ideal weight of 135 pounds?

60. The rate for a particular international telephone call is $0.55 for the first minute and $0.40 for each additional minute. Determine the length of a call that costs $6.95.

61. In a film, the actor Charles Coburn played an elderly "uncle" character criticized for marrying a woman when he is 3 times her age. He wittily replies, "Ah, but in 20 years time I shall only be twice her age." How old is the "uncle" and the woman?

62. Answer the question in the following *Peanuts* cartoon strip. (*Note*: You may not use the answer given in the cartoon!)

Peanuts copyright © 1979, 2011 by United Features Syndicate, Inc.

Review Exercises

63. Solve and check: $\frac{4}{5}x = -16$.
(Section 2.2, Example 3)

64. Solve and check: $6(y - 1) + 7 = 9y - y + 1$.
(Section 2.3, Example 3)

65. Solve for w: $V = \frac{1}{3}lwh$. (Section 2.4, Example 4)

Preview Exercises

Exercises 66–68 will help you prepare for the material covered in the next section.

66. Use $A = \frac{1}{2}bh$ to find h if $A = 30$ and $b = 12$.

67. Evaluate $A = \frac{1}{2}h(a + b)$ for $a = 10$, $b = 16$, and $h = 7$.

68. Solve: $x = 4(90 - x) - 40$.

2.6

Problem Solving in Geometry

Objectives

1 Solve problems using formulas for perimeter and area.

2 Solve problems using formulas for a circle's area and circumference.

3 Solve problems using formulas for volume.

4 Solve problems involving the angles of a triangle.

5 Solve problems involving complementary and supplementary angles.

Geometry is about the space you live in and the shapes that surround you. You're even made of it. The human lung consists of nearly 300 spherical air sacs, geometrically designed to provide the greatest surface area within the limited volume of our bodies. Viewed in this way, geometry becomes an intimate experience.

For thousands of years, people have studied geometry in some form to obtain a better understanding of the world in which they live. A study of the shape of your world will provide you with many practical applications that will help to increase your problem-solving skills.

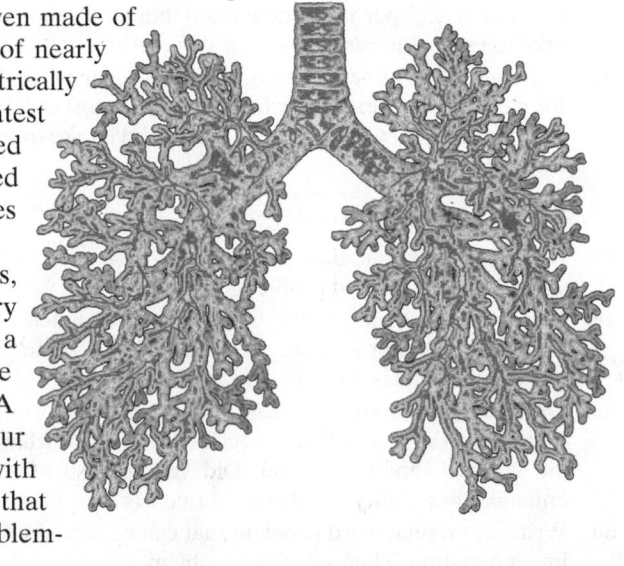

Geometric Formulas for Perimeter and Area

1 Solve problems using formulas for perimeter and area.

Solving geometry problems often requires using basic geometric formulas. Formulas for perimeter and area are summarized in **Table 2.3**. Remember that perimeter is measured in linear units, such as feet or meters, and area is measured in square units, such as square feet, ft^2, or square meters, m^2.

| Table 2.3 | Common Formulas for Perimeter and Area |

Square	Rectangle	Triangle	Trapezoid
$A = s^2$ $P = 4s$	$A = lw$ $P = 2l + 2w$	$A = \frac{1}{2}bh$	$A = \frac{1}{2}h(a + b)$

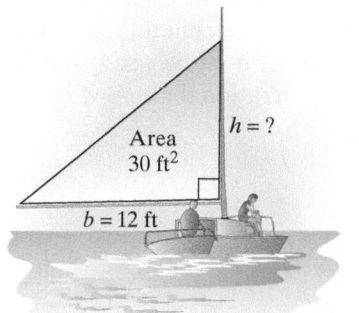

Figure 2.10 Finding the height of a triangular sail

EXAMPLE 1   Using the Formula for the Area of a Triangle

A sailboat has a triangular sail with an area of 30 square feet and a base that is 12 feet long. (See **Figure 2.10**.) Find the height of the sail.

Solution   We begin with the formula for the area of a triangle given in **Table 2.3**.

$$A = \frac{1}{2}bh \qquad \text{The area of a triangle is } \tfrac{1}{2} \text{ the product of its base and height.}$$

$$30 = \frac{1}{2}(12)h \qquad \text{Substitute 30 for } A \text{ and 12 for } b.$$

$$30 = 6h \qquad \text{Simplify.}$$

$$\frac{30}{6} = \frac{6h}{6} \qquad \text{Divide both sides by 6.}$$

$$5 = h \qquad \text{Simplify.}$$

The height of the sail is 5 feet.

Check
The area is $A = \frac{1}{2}bh = \frac{1}{2}(12 \text{ feet})(5 \text{ feet}) = 30$ square feet. ∎

✓ **CHECK POINT 1**   A sailboat has a triangular sail with an area of 24 square feet and a base that is 4 feet long. Find the height of the sail.

---

2 Solve problems using formulas for a circle's area and circumference.

Geometric Formulas for Circumference and Area of a Circle

It's a good idea to know your way around a circle. Clocks, angles, maps, and compasses are based on circles. Circles occur everywhere in nature: in ripples on water, patterns on a butterfly's wings, and cross sections of trees. Some consider the circle to be the most pleasing of all shapes.

The point at which a pebble hits a flat surface of water becomes the center of a number of circular ripples.

A **circle** is a set of points in the plane equally distant from a given point, its center. **Figure 2.11** shows two circles. A **radius** (plural: radii), *r*, is a line segment from the center to any point on the circle. For a given circle, all radii have the same length. A **diameter**, *d*, is a line segment through the center whose endpoints both lie on the circle. For a given circle, all diameters have the same length. In any circle, **the length of a diameter is twice the length of a radius**.

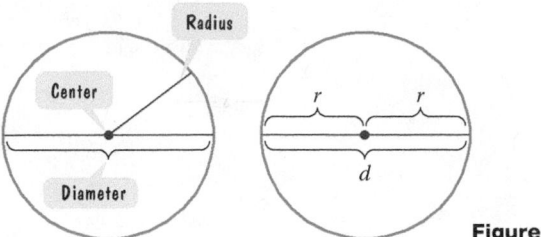

Figure 2.11

The words *radius* and *diameter* refer to both the line segments in **Figure 2.11** as well as their linear measures. The distance around a circle (its perimeter) is called its **circumference**. Formulas for the area and circumference of a circle are given in terms of π and appear in **Table 2.4**. We have seen that π is an irrational number and is approximately equal to 3.14.

Table 2.4 Formulas for Circles

Circle	Area	Circumference
	$A = \pi r^2$	$C = 2\pi r$

When computing a circle's area or circumference by hand, round π to 3.14. When using a calculator, use the $\boxed{\pi}$ key, which gives the value of π rounded to approximately 11 decimal places. In either case, calculations involving π give approximate answers. These answers can vary slightly depending on how π is rounded. The symbol ≈ (is approximately equal to) will be written in these calculations.

EXAMPLE 2 Finding the Area and Circumference of a Circle

Find the area and circumference of a circle with a diameter measuring 20 inches.

Solution The radius is half the diameter, so $r = \frac{20}{2} = 10$ inches.

$$A = \pi r^2 \qquad C = 2\pi r \qquad \text{Use the formulas for area and circumference of a circle.}$$
$$A = \pi(10)^2 \qquad C = 2\pi(10) \qquad \text{Substitute 10 for } r.$$
$$A = 100\pi \qquad C = 20\pi$$

The area of the circle is 100π square inches and the circumference is 20π inches. Using the fact that π ≈ 3.14, the area is approximately 100(3.14), or 314 square inches. The circumference is approximately 20(3.14), or 62.8 inches. ■

✓ **CHECK POINT 2** The diameter of a circular landing pad for helicopters is 40 feet. Find the area and circumference of the landing pad. Express answers in terms of π. Then round answers to the nearest square foot and foot, respectively.

| EXAMPLE 3 | Problem Solving Using the Formula for a Circle's Area |

Which one of the following is the better buy: a large pizza with a 16-inch diameter for $15.00 or a medium pizza with an 8-inch diameter for $7.50?

Solution The better buy is the pizza with the lower price per square inch. The radius of the large pizza is $\frac{1}{2} \cdot 16$ inches, or 8 inches, and the radius of the medium pizza is $\frac{1}{2} \cdot 8$ inches, or 4 inches. The area of the surface of each circular pizza is determined using the formula for the area of a circle.

$$\text{Large pizza:} \quad A = \pi r^2 = \pi(8 \text{ in.})^2 = 64\pi \text{ in.}^2 \approx 201 \text{ in.}^2$$
$$\text{Medium pizza:} \quad A = \pi r^2 = \pi(4 \text{ in.})^2 = 16\pi \text{ in.}^2 \approx 50 \text{ in.}^2$$

For each pizza, the price per square inch is found by dividing the price by the area:

$$\text{Price per square inch for large pizza} = \frac{\$15.00}{64\pi \text{ in.}^2} \approx \frac{\$15.00}{201 \text{ in.}^2} \approx \frac{\$0.07}{\text{in.}^2}$$

$$\text{Price per square inch for medium pizza} = \frac{\$7.50}{16\pi \text{ in.}^2} \approx \frac{\$7.50}{50 \text{ in.}^2} = \frac{\$0.15}{\text{in.}^2}.$$

The large pizza costs approximately $0.07 per square inch and the medium pizza costs approximately $0.15 per square inch. Thus, the large pizza is the better buy. ■

In Example 3, did you at first think that the price per square inch would be the same for the large and the medium pizzas? After all, the radius of the large pizza is twice that of the medium pizza, and the cost of the large is twice that of the medium. However, the large pizza's area, 64π square inches, is *four times the area* of the medium pizza, 16π square inches. Doubling the radius of a circle increases its area by four times the original amount.

✓ **CHECK POINT 3** Which one of the following is the better buy: a large pizza with an 18-inch diameter for $20.00 or a medium pizza with a 14-inch diameter for $14.00?

Using Technology

You can use your calculator to obtain the price per square inch for each pizza in Example 3. The price per square inch for the large pizza, $\frac{15}{64\pi}$, is approximated by one of the following keystrokes:

Many Scientific Calculators

$15 \div (64 \times \pi) =$

Many Graphing Calculators

$15 \div (64 \pi) $ ENTER

3 Solve problems using formulas for volume.

Geometric Formulas for Volume

A shoe box and a basketball are examples of three-dimensional figures. **Volume** refers to the amount of space occupied by such a figure. To measure this space, we begin by selecting a cubic unit. One such cubic unit, 1 cubic centimeter (cm^3), is shown in **Figure 2.12**.

The edges of a cube all have the same length. Other cubic units used to measure volume include 1 cubic inch (in.^3) and 1 cubic foot (ft^3). The volume of a solid is the number of cubic units that can be contained in the solid.

Formulas for volumes of three-dimensional figures are given in **Table 2.5**.

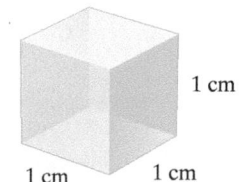

Figure 2.12

| Table 2.5 | Common Formulas for Volume |

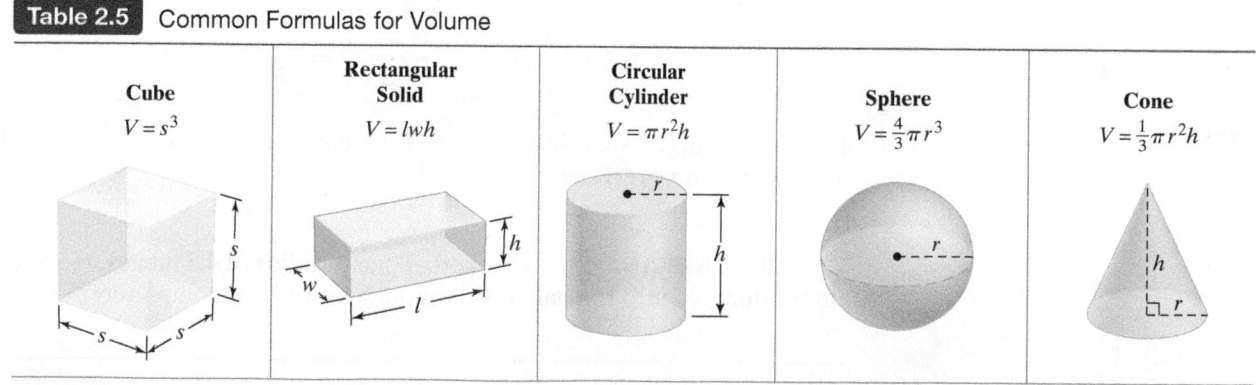

Cube	Rectangular Solid	Circular Cylinder	Sphere	Cone
$V = s^3$	$V = lwh$	$V = \pi r^2 h$	$V = \frac{4}{3}\pi r^3$	$V = \frac{1}{3}\pi r^2 h$

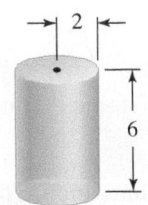

Radius: 2 inches
Height: 6 inches

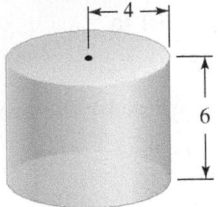

Radius: 4 inches
Height: 6 inches

Figure 2.13 Doubling a cylinder's radius

EXAMPLE 4 Using the Formula for the Volume of a Cylinder

A cylinder with a radius of 2 inches and a height of 6 inches has its radius doubled. (See **Figure 2.13**.) How many times greater is the volume of the larger cylinder than the volume of the smaller cylinder?

Solution We begin with the formula for the volume of a cylinder, $V = \pi r^2 h$, given on the previous page in **Table 2.5**. Find the volume of the smaller cylinder and the volume of the larger cylinder. To compare the volumes, divide the volume of the larger cylinder by the volume of the smaller cylinder.

$$V = \pi r^2 h \qquad \text{Use the formula for the volume of a cylinder.}$$

Radius is doubled.

$$V_{\text{Smaller}} = \pi(2)^2(6) \quad V_{\text{Larger}} = \pi(4)^2(6) \qquad \text{Substitute the given values.}$$

$$V_{\text{Smaller}} = \pi(4)(6) \quad V_{\text{Larger}} = \pi(16)(6)$$

$$V_{\text{Smaller}} = 24\pi \qquad V_{\text{Larger}} = 96\pi$$

The volume of the smaller cylinder is 24π cubic inches. The volume of the larger cylinder is 96π cubic inches. We use division to compare the volumes:

$$\frac{V_{\text{Larger}}}{V_{\text{Smaller}}} = \frac{96\pi}{24\pi} = \frac{4}{1}.$$

Thus, the volume of the larger cylinder is 4 times the volume of the smaller cylinder. ■

✓ **CHECK POINT 4** A cylinder with a radius of 3 inches and a height of 5 inches has its height doubled. How many times greater is the volume of the larger cylinder than the volume of the smaller cylinder?

EXAMPLE 5 Applying Volume Formulas

An ice cream cone is 5 inches deep and has a radius of 1 inch. A spherical scoop of ice cream also has a radius of 1 inch. (See **Figure 2.14**.) If the ice cream melts into the cone, will it overflow?

Solution The ice cream will overflow if the volume of the ice cream, a sphere, is greater than the volume of the cone. Find the volume of each.

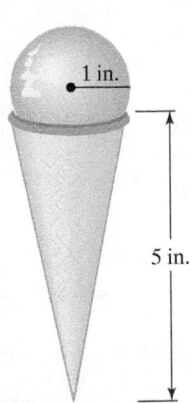

1 in.

5 in.

Figure 2.14

$$V_{\text{cone}} = \frac{1}{3}\pi r^2 h = \frac{1}{3}\pi(1 \text{ in.})^2 \cdot 5 \text{ in.} = \frac{5\pi}{3}\text{in.}^3 \approx 5 \text{ in.}^3$$

$$V_{\text{sphere}} = \frac{4}{3}\pi r^3 = \frac{4}{3}\pi(1 \text{ in.})^3 = \frac{4\pi}{3}\text{in.}^3 \approx 4 \text{ in.}^3$$

The volume of the spherical scoop of ice cream is less than the volume of the cone, so there will be no overflow. ■

✓ **CHECK POINT 5** A basketball has a radius of 4.5 inches. If 350 cubic inches of air are pumped into the ball, is this enough air to fill it completely?

Blitzer Bonus

Deceptions in Visual Displays of Data

Graphs can be used to distort data, making it difficult for the viewer to learn the truth. One potential source of misunderstanding involves geometric figures whose lengths are in the correct ratios for the displayed data, but whose areas or volumes are then varied to create a misimpression about how the data are changing over time. Here are two examples of misleading visual displays.

Graphic Display	Presentation Problems
Purchasing Power of the Diminishing Dollar $1.00 63¢ 54¢ 48¢ 43¢ 1980 1990 1995 2000 2005 *Source:* Bureau of Labor Statistics	Although the length of each dollar bill is proportional to its spending power, the visual display varies both the length *and* width of the bills to show the diminishing power of the dollar over time. Because our eyes focus on the *areas* of the dollar-shaped bars, this creates the impression that the purchasing power of the dollar diminished even more than it really did. If the area of the dollar were drawn to reflect its purchasing power, the 2005 dollar would be approximately twice as large as the one shown in the graphic display.
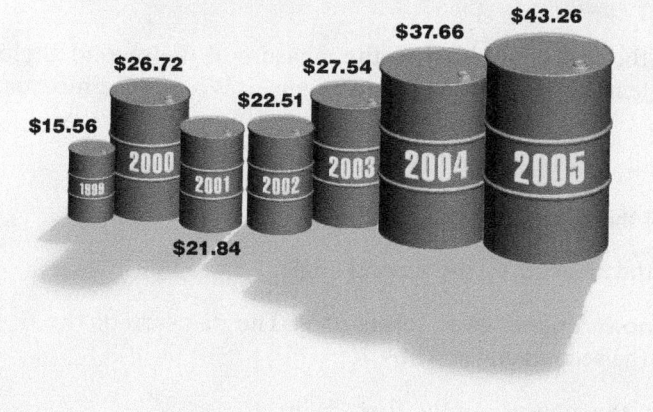 **Average Daily Price per Barrel of Oil** $15.56 $26.72 $22.51 $27.54 $37.66 $43.26 $21.84 1999 2000 2001 2002 2003 2004 2005 *Source:* U.S. Department of Energy	The height of each barrel is proportional to its price. However, the graph varies both the height *and radius* of the barrels to show the increase in the price of oil over time. Our eyes focus on the *volumes* of the cylinders: $V = \pi r^2 h$. By varying the radii, the volumes are not proportional to the barrel prices, creating the impression that oil prices increased even more than they really did. Cosmetic effects, including dates that are printed in different sizes and shadows under the barrels, further exaggerate how the data are changing over time.

---

4 Solve problems involving the angles of a triangle.

The Angles of a Triangle

The hour hand of a clock moves from 12 to 2. The hour hand suggests a **ray**, a part of a line that has only one endpoint and extends forever in the opposite direction. An *angle* is formed as the ray in **Figure 2.15** rotates from 12 to 2.

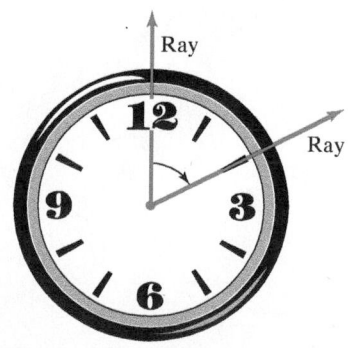

Figure 2.15 Clock with a ray rotating to form an angle

An **angle**, symbolized ∡, is made up of two rays that have a common endpoint. **Figure 2.16** shows an angle. The common endpoint, B in the figure, is called the **vertex** of the angle. The two rays that form the angle are called its **sides**. The four ways of naming the angle are shown to the right of **Figure 2.16**.

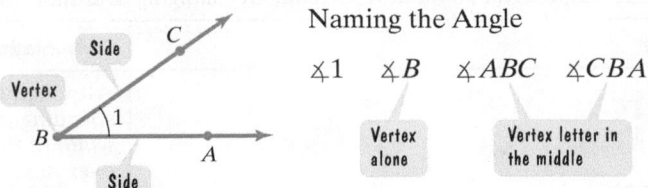

Figure 2.16 An angle: two rays with a common endpoint

One way to measure angles is in **degrees**, symbolized by a small, raised circle °. Think of the hour hand of a clock. From 12 noon to 12 midnight, the hour hand moves around in a complete circle. By definition, the ray has rotated through 360 degrees, or 360°. Using 360° as the amount of rotation of a ray back onto itself, **a degree, 1°, is $\frac{1}{360}$ of a complete rotation**.

Our next problem is based on the relationship among the three angles of any triangle.

The Angles of a Triangle

The sum of the measures of the three angles of any triangle is 180°.

EXAMPLE 6   Angles of a Triangle

In a triangle, the measure of the first angle is twice the measure of the second angle. The measure of the third angle is 20° less than the second angle. What is the measure of each angle?

Solution

Step 1. Let x represent one of the quantities. Let

$$x = \text{the measure of the second angle.}$$

Step 2. Represent other unknown quantities in terms of x. The measure of the first angle is twice the measure of the second angle. Thus, let

$$2x = \text{the measure of the first angle.}$$

The measure of the third angle is 20° less than the second angle. Thus, let

$$x - 20 = \text{the measure of the third angle.}$$

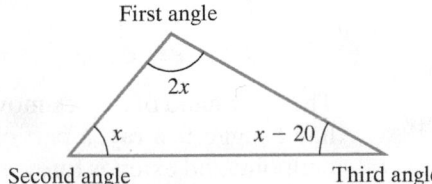

Step 3. Write an equation in x that models the conditions. Because we are working with a triangle, the sum of the measures of its three angles is 180°.

Measure of first angle	plus	measure of second angle	plus	measure of third angle	equals	180°.
$2x$	$+$	x	$+$	$(x - 20)$	$=$	180

Step 4. Solve the equation and answer the question.

$$2x + x + (x - 20) = 180 \qquad \text{This is the equation that models the sum of the measures of the angles.}$$

$$4x - 20 = 180 \qquad \text{Regroup and combine like terms.}$$

$$4x - 20 + 20 = 180 + 20 \qquad \text{Add 20 to both sides.}$$

$$4x = 200 \qquad \text{Simplify.}$$

$$\frac{4x}{4} = \frac{200}{4} \qquad \text{Divide both sides by 4.}$$

$$x = 50 \qquad \text{Simplify.}$$

$$\text{Measure of first angle} = 2x = 2 \cdot 50 = 100$$

$$\text{Measure of second angle} = x = 50$$

$$\text{Measure of third angle} = x - 20 = 50 - 20 = 30$$

The angles measure 100°, 50°, and 30°.

Step 5. Check the proposed solution in the original wording of the problem. The problem tells us that we are working with a triangle's angles. Thus, the sum of the measures should be 180°. Adding the three measures, we obtain 100° + 50° + 30°, giving the required sum of 180°. ∎

✓ **CHECK POINT 6** In a triangle, the measure of the first angle is three times the measure of the second angle. The measure of the third angle is 20° less than the second angle. What is the measure of each angle?

5 Solve problems involving complementary and supplementary angles.

Complementary and Supplementary Angles

Two angles with measures having a sum of 90° are called **complementary angles**. For example, angles measuring 70° and 20° are complementary angles because 70° + 20° = 90°. For angles such as those measuring 70° and 20°, each angle is a **complement** of the other: The 70° angle is the complement of the 20° angle and the 20° angle is the complement of the 70° angle. The measure of the complement can be found by subtracting the angle's measure from 90°. For example, we can find the complement of a 25° angle by subtracting 25° from 90°: 90° − 25° = 65°. Thus, an angle measuring 65° is the complement of one measuring 25°.

Two angles with measures having a sum of 180° are called **supplementary angles**. For example, angles measuring 110° and 70° are supplementary angles because 110° + 70° = 180°. For angles such as those measuring 110° and 70°, each angle is a **supplement** of the other: The 110° angle is the supplement of the 70° angle and the 70° angle is the supplement of the 110° angle. The measure of the supplement can be found by subtracting the angle's measure from 180°. For example, we can find the supplement of a 25° angle by subtracting 25° from 180°: 180° − 25° = 155°. Thus, an angle measuring 155° is the supplement of one measuring 25°.

Algebraic Expressions for Complements and Supplements

Measure of an angle: x

Measure of the angle's complement: $90 - x$

Measure of the angle's supplement: $180 - x$

EXAMPLE 7 Angle Measures and Complements

The measure of an angle is 40° less than four times the measure of its complement. What is the angle's measure?

Solution

Step 1. Let x represent one of the quantities. Let

$$x = \text{the measure of the angle.}$$

Step 2. Represent other unknown quantities in terms of x. Because this problem involves an angle and its complement, let

$$90 - x = \text{the measure of the complement.}$$

Step 3. Write an equation in x that models the conditions.

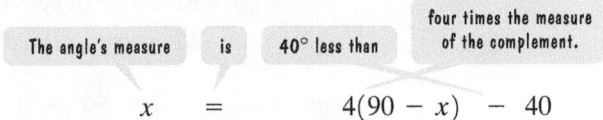

$$x \quad = \quad 4(90 - x) \quad - \quad 40$$

Step 4. Solve the equation and answer the question.

$x = 4(90 - x) - 40$	This is the equation that models the problem's conditions.
$x = 360 - 4x - 40$	Use the distributive property.
$x = 320 - 4x$	Simplify: $360 - 40 = 320$.
$x + 4x = 320 - 4x + 4x$	Add 4x to both sides.
$5x = 320$	Simplify.
$\dfrac{5x}{5} = \dfrac{320}{5}$	Divide both sides by 5.
$x = 64$	Simplify.

The angle measures 64°.

Step 5. Check the proposed solution in the original wording of the problem. The measure of the complement is $90° - 64° = 26°$. Four times the measure of the complement is $4 \cdot 26°$, or $104°$. The angle's measure, $64°$, is 40° less than $104°$: $104° - 40° = 64°$. As specified by the problem's wording, the angle's measure is 40° less than four times the measure of its complement. ∎

✓ **CHECK POINT 7** The measure of an angle is twice the measure of its complement. What is the angle's measure?

CONCEPT AND VOCABULARY CHECK

Fill in each blank so that the resulting statement is true.

1. The area, A, of a triangle with base b and height h is given by the formula _____.

2. The area, A, of a circle with radius r is given by the formula _____.

3. The circumference, C, of a circle with radius r is given by the formula _____.

4. In any circle, twice the length of the _____ is the length of the _____.

5. The volume, V, of a rectangular solid with length l, width w, and height h is given by the formula _____.

6. The volume, V, of a circular cylinder with radius r and height h is given by the formula _____.

7. The sum of the measures of the three angles of any triangle is _____.

8. Two angles with measures having a sum of 90° are called _____ angles.

9. Two angles with measures having a sum of 180° are called _____ angles.

10. If the measure of an angle is represented by x, the measure of its complement is represented by _____ and the measure of its supplement is represented by _____.

2.6 EXERCISE SET MyMathLab®

Practice Exercises

*Use the formulas for perimeter and area in **Table 2.3** on page 169 to solve Exercises 1–12.*

In Exercises 1–2, find the perimeter and area of each rectangle.

1.
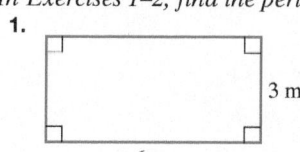
3 m
6 m

2.
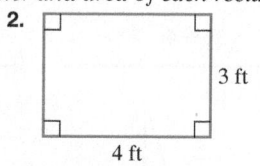
3 ft
4 ft

In Exercises 3–4, find the area of each triangle.

3.
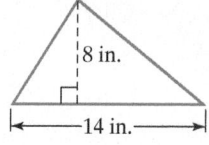
8 in.
14 in.

4.
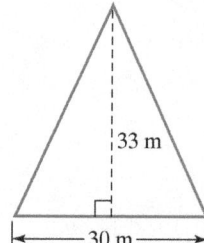
33 m
30 m

In Exercises 5–6, find the area of each trapezoid.

5.
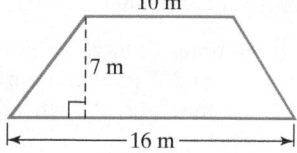
10 m
7 m
16 m

6.
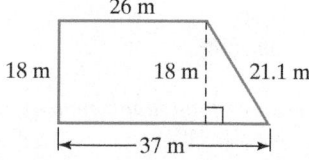
26 m
18 m   18 m   21.1 m
37 m

7. A rectangular swimming pool has a width of 25 feet and an area of 1250 square feet. What is the pool's length?

8. A rectangular swimming pool has a width of 35 feet and an area of 2450 square feet. What is the pool's length?

9. A triangle has a base of 5 feet and an area of 20 square feet. Find the triangle's height.

10. A triangle has a base of 6 feet and an area of 30 square feet. Find the triangle's height.

11. A rectangle has a width of 44 centimeters and a perimeter of 188 centimeters. What is the rectangle's length?

12. A rectangle has a width of 46 centimeters and a perimeter of 208 centimeters. What is the rectangle's length?

*Use the formulas for the area and circumference of a circle in **Table 2.4** on page 170 to solve Exercises 13–18.*

In Exercises 13–16, find the area and circumference of each circle. Express answers in terms of π. Then round to the nearest whole number.

13.
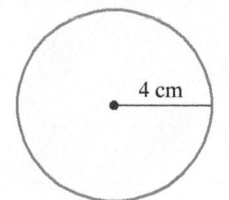
4 cm

14.
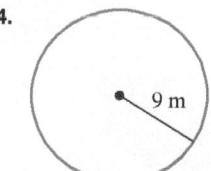
9 m

15.
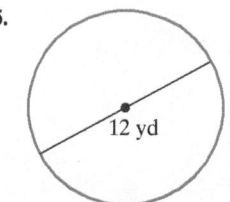
12 yd

16.
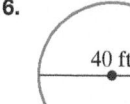
40 ft

17. The circumference of a circle is 14π inches. Find the circle's radius and diameter.

18. The circumference of a circle is 16π inches. Find the circle's radius and diameter.

*Use the formulas for volume in **Table 2.5** on page 171 to solve Exercises 19–30.*

In Exercises 19–26, find the volume of each figure. Where applicable, express answers in terms of π. Then round to the nearest whole number.

19.
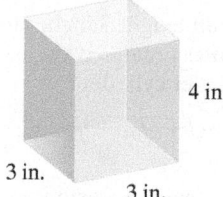
4 in.
3 in.
3 in.

20.
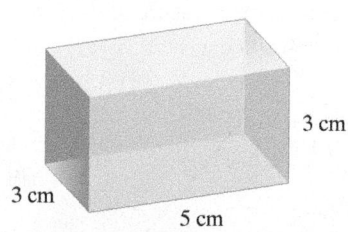
3 cm
3 cm
5 cm

21.
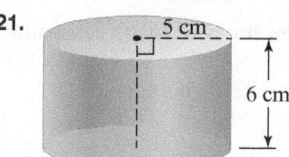
5 cm
6 cm

22.

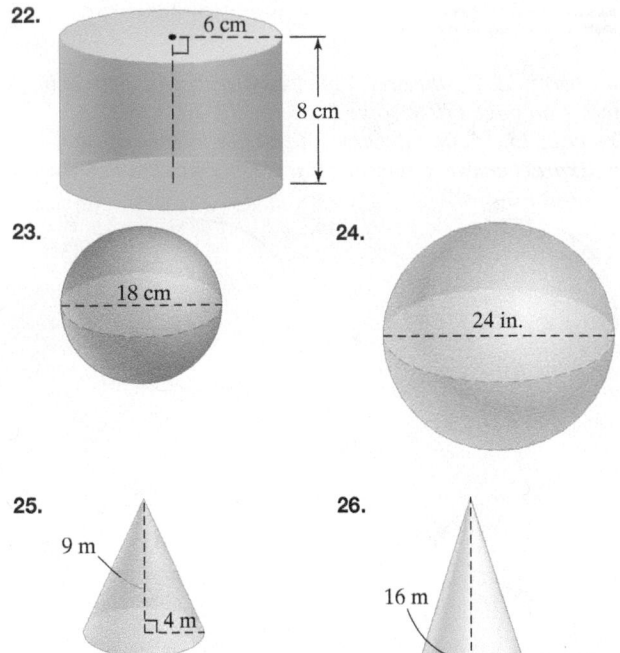

23. **24.**

25. **26.**

27. Solve the formula for the volume of a circular cylinder for *h*.

28. Solve the formula for the volume of a cone for *h*.

29. A cylinder with radius 3 inches and height 4 inches has its radius tripled. How many times greater is the volume of the larger cylinder than the smaller cylinder?

30. A cylinder with radius 2 inches and height 3 inches has its radius quadrupled. How many times greater is the volume of the larger cylinder than the smaller cylinder?

Use the relationship among the three angles of any triangle to solve Exercises 31–36.

31. Two angles of a triangle have the same measure and the third angle is 30° greater than the measure of the other two. Find the measure of each angle.

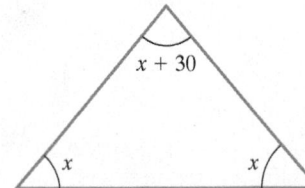

32. One angle of a triangle is three times as large as another. The measure of the third angle is 40° more than that of the smallest angle. Find the measure of each angle.

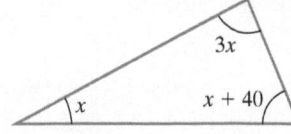

Find the measure of each angle whose degree measure is represented in terms of x in the triangles in Exercises 33–34.

33.

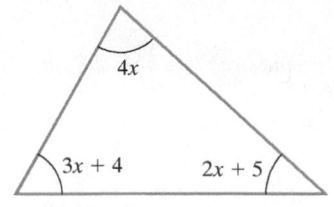

34.

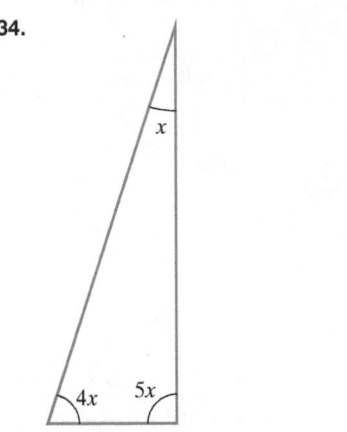

35. One angle of a triangle is twice as large as another. The measure of the third angle is 20° more than that of the smallest angle. Find the measure of each angle.

36. One angle of a triangle is three times as large as another. The measure of the third angle is 30° greater than that of the smallest angle. Find the measure of each angle.

In Exercises 37–40, find the measure of the complement of each angle.

37. 58°   **38.** 41°   **39.** 88°   **40.** 2°

In Exercises 41–44, find the measure of the supplement of each angle.

41. 132°   **42.** 93°

43. 90°   **44.** 179.5°

In Exercises 45–50, use the five-step problem-solving strategy to find the measure of the angle described.

45. The angle's measure is 60° more than that of its complement.

46. The angle's measure is 78° less than that of its complement.

47. The angle's measure is three times that of its supplement.

48. The angle's measure is 16° more than triple that of its supplement.

49. The measure of the angle's supplement is 10° more than three times that of its complement.

50. The measure of the angle's supplement is 52° more than twice that of its complement.

Practice PLUS

In Exercises 51–53, find the area of each figure.

51.

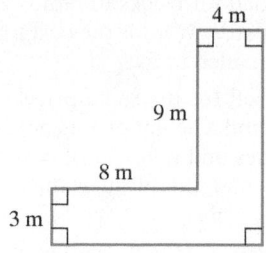

52.

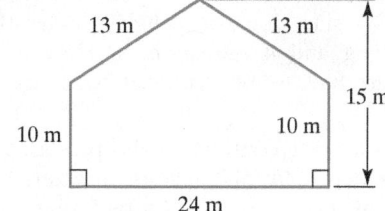

53.

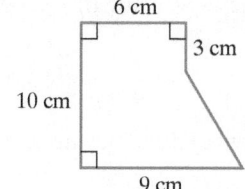

54. Find the area of the shaded region in terms of π.

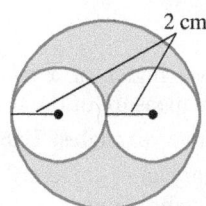

55. Find the volume of the cement block in the figure shown.

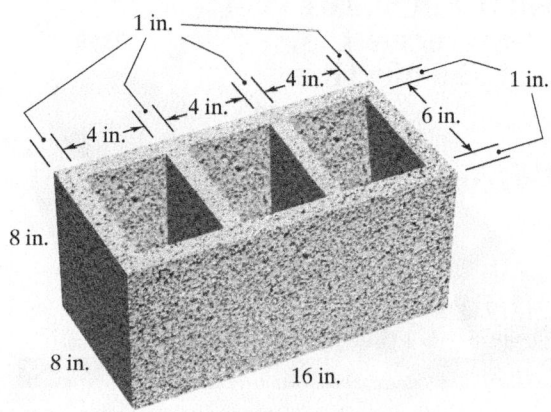

56. Find the volume of the darkly shaded region. Express the answer in terms of π.

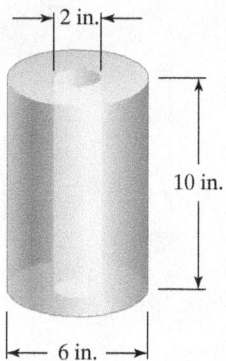

Application Exercises

*Use the formulas for perimeter and area in **Table 2.3** on page 169 to solve Exercises 57–58.*

57. Taxpayers with an office in their home may deduct a percentage of their home-related expenses. This percentage is based on the ratio of the office's area to the area of the home. A taxpayer with a 2200-square-foot home maintains a 20-foot by 16-foot office. If the yearly electricity bills for the home come to $4800, how much of this is deductible?

58. The lot in the figure shown, except for the house, shed, and driveway, is lawn. One bag of lawn fertilizer costs $25.00 and covers 4000 square feet.

a. Determine the minimum number of bags of fertilizer needed for the lawn.

b. Find the total cost of the fertilizer.

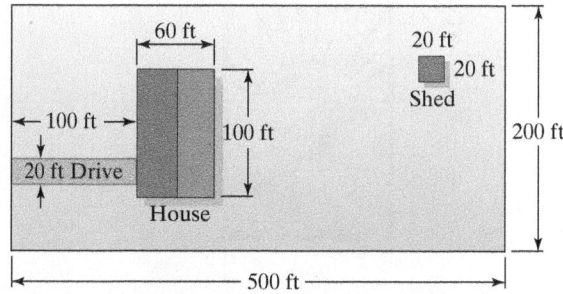

*Use the formulas for the area and the circumference of a circle in **Table 2.4** on page 170 to solve Exercises 59–64. Unless otherwise indicated, round all circumference and area calculations to the nearest whole number.*

59. Which one of the following is a better buy: a large pizza with a 14-inch diameter for $12.00 or a medium pizza with a 7-inch diameter for $5.00?

60. Which one of the following is a better buy: a large pizza with a 16-inch diameter for $12.00 or two small pizzas, each with a 10-inch diameter, for $12.00?

61. If asphalt pavement costs $0.80 per square foot, find the cost to pave the circular road in the figure shown. Round to the nearest dollar.

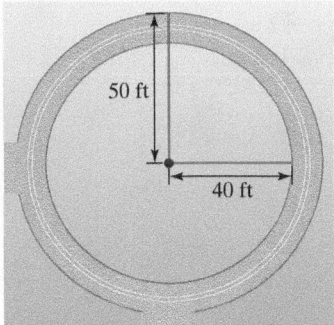

50 ft

40 ft

62. Hardwood flooring costs $10.00 per square foot. How much will it cost (to the nearest dollar) to cover the dance floor shown in the figure with hardwood flooring?

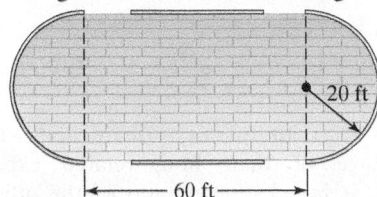

20 ft

60 ft

63. A glass window is to be placed in a house. The window consists of a rectangle, 6 feet high by 3 feet wide, with a semicircle at the top. Approximately how many feet of stripping, to the nearest tenth of a foot, will be needed to frame the window?

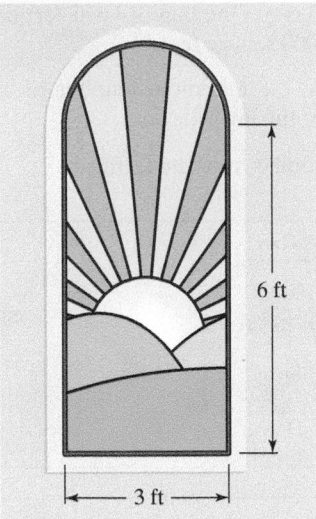

6 ft

3 ft

64. How many plants spaced every 6 inches are needed to surround a circular garden with a 30-foot radius?

*Use the formulas for volume in **Table 2.5** on page 171 to solve Exercises 65–69. When necessary, round all volume calculations to the nearest whole number.*

65. A water reservoir is shaped like a rectangular solid with a base that is 50 yards by 30 yards, and a vertical height of 20 yards. At the start of a three-month period of no rain, the reservoir was completely full. At the end of this period, the height of the water was down to 6 yards. How much water was used in the three-month period?

66. A building contractor is to dig a foundation 4 yards long, 3 yards wide, and 2 yards deep for a toll booth's foundation. The contractor pays $10 per load for trucks to remove the dirt. Each truck holds 6 cubic yards. What is the cost to the contractor to have all the dirt hauled away?

67. Two cylindrical cans of soup sell for the same price. One can has a diameter of 6 inches and a height of 5 inches. The other has a diameter of 5 inches and a height of 6 inches. Which can contains more soup and, therefore, is the better buy?

68. The tunnel under the English Channel that connects England and France is one of the world's longest tunnels. The Chunnel, as it is known, consists of three separate tunnels built side by side. Each is a half-cylinder that is 50,000 meters long and 4 meters high. How many cubic meters of dirt had to be removed to build the Chunnel?

69. You are about to sue your contractor who promised to install a water tank that holds 500 gallons of water. You know that 500 gallons is the capacity of a tank that holds 67 cubic feet. The cylindrical tank has a radius of 3 feet and a height of 2 feet 4 inches. Does the evidence indicate you can win the case against the contractor if it goes to court?

Writing in Mathematics

70. Using words only, describe how to find the area of a triangle.

71. Describe the difference between the following problems: How much fencing is needed to enclose a garden? How much fertilizer is needed for the garden?

72. Describe how volume is measured. Explain why linear or square units cannot be used.

73. What is an angle?

74. If the measures of two angles of a triangle are known, explain how to find the measure of the third angle.

75. Can a triangle contain two 90° angles? Explain your answer.

76. What are complementary angles? Describe how to find the measure of an angle's complement.

77. What are supplementary angles? Describe how to find the measure of an angle's supplement.

78. Describe what is misleading in this visual display of data.

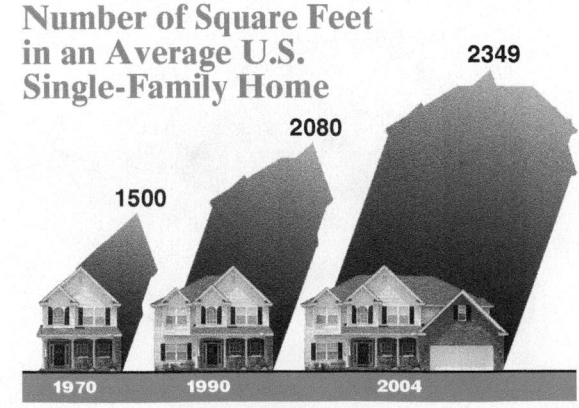

Number of Square Feet in an Average U.S. Single-Family Home

2349

2080

1500

1970   1990   2004

Source: National Association of Home Builders

Critical Thinking Exercises

Make Sense? *In Exercises 79–82, determine whether each statement "makes sense" or "does not make sense" and explain your reasoning.*

79. There is nothing that is misleading in this visual display of data.

Book Title Output in the United States

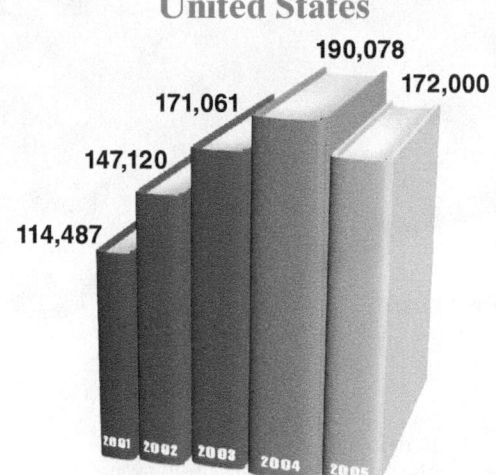

190,078

172,000

171,061

147,120

114,487

2001 2002 2003 2004 2005

Source: R. R. Bowker

80. I solved a word problem and determined that a triangle had angles measuring 37°, 58°, and 86°.

81. I paid $10 for a pizza, so I would expect to pay approximately $20 for the same kind of pizza with twice the radius.

82. I find that my answers involving π can vary slightly depending on whether I round π mid-calculation or use the π key on my calculator and then round at the very end.

In Exercises 83–86, determine whether each statement is true or false. If the statement is false, make the necessary change(s) to produce a true statement.

83. It is possible to have a circle whose circumference is numerically equal to its area.

84. When the measure of a given angle is added to three times the measure of its complement, the sum equals the sum of the measures of the complement and supplement of the angle.

85. The complement of an angle that measures less than 90° is an angle that measures more than 90°.

86. Two complementary angles can be equal in measure.

87. Suppose you know the cost for building a rectangular deck measuring 8 feet by 10 feet. If you decide to increase the dimensions to 12 feet by 15 feet, by how many times will the cost increase?

88. A rectangular swimming pool measures 14 feet by 30 feet. The pool is surrounded on all four sides by a path that is 3 feet wide. If the cost to resurface the path is $2 per square foot, what is the total cost of resurfacing the path?

89. What happens to the volume of a sphere if its radius is doubled?

90. A scale model of a car is constructed so that its length, width, and height are each $\frac{1}{10}$ the length, width, and height of the actual car. By how many times does the volume of the car exceed its scale model?

91. Find the measure of the angle of inclination, denoted by x in the figure, for the road leading to the bridge.

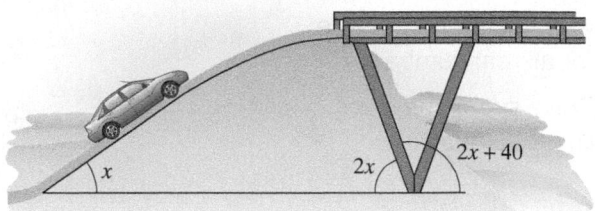

x $2x$ $2x + 40$

Review Exercises

92. Solve for s: $P = 2s + b$.
(Section 2.4, Example 3)

93. Solve for x: $\dfrac{x}{2} + 7 = 13 - \dfrac{x}{4}$.
(Section 2.3, Example 4)

94. Simplify: $\left[3\left(12 \div 2^2 - 3\right)^2\right]^2$.
(Section 1.8, Example 8)

Preview Exercises

Exercises 95–97 will help you prepare for the material covered in the next section.

95. Is 2 a solution of $x + 3 < 8$?

96. Is 6 a solution of $4y - 7 \geq 5$?

97. Solve: $2(x - 3) + 5x = 8(x - 1)$.

2.7

Solving Linear Inequalities

Objectives

1. Graph the solutions of an inequality on a number line.

2. Use interval notation.

3. Understand properties used to solve linear inequalities.

4. Solve linear inequalities.

5. Identify inequalities with no solution or true for all real numbers.

6. Solve problems using linear inequalities.

Do you remember Rent-a-Heap, the car rental company that charged $125 per week plus $0.20 per mile to rent a small car? In Example 4 on page 161, we asked the question: How many miles can you travel for $335? We let x represent the number of miles and set up a linear equation as follows:

The weekly charge of $125	plus	the charge of $0.20 per mile for x miles	equals	the total $335 rental charge.
125	+	0.20x	=	335.

Because we are limited by how much money we can spend on everything from buying clothing to renting a car, it is also possible to ask: How many miles can you travel if you can spend *at most* $335? We again let x represent the number of miles. Spending *at most* $335 means that the amount spent on the weekly rental must be *less than or equal to* $335:

The weekly charge of $125	plus	the charge of $0.20 per mile for x miles	must be less than or equal to	$335.
125	+	0.20x	≤	335.

Using the commutative property of addition, we can express this inequality as

$$0.20x + 125 \leq 335.$$

The form of this inequality is $ax + b \leq c$, with $a = 0.20$, $b = 125$, and $c = 335$. Any inequality in this form is called a **linear inequality in one variable**. The symbol between $ax + b$ and c can be ≤ (is less than or equal to), < (is less than), ≥ (is greater than or equal to), or > (is greater than). The greatest exponent on the variable in such an inequality is 1.

In this section, we will study how to solve linear inequalities such as $0.20x + 125 \leq 335$. **Solving an inequality** is the process of finding the set of numbers that will make the inequality a true statement. These numbers are called the **solutions** of the inequality, and we say that they **satisfy** the inequality. The set of all solutions is called the **solution set** of the inequality. We begin by discussing how to graph and how to represent these solution sets.

1 Graph the solutions of an inequality on a number line.

Graphs of Inequalities

There are infinitely many solutions to the inequality $x < 3$, namely, all real numbers that are less than 3. Although we cannot list all the solutions, we can make a drawing on a number line that represents these solutions. Such a drawing is called the **graph of the inequality**.

Graphs of solutions to linear inequalities are shown on a number line by shading all points representing numbers that are solutions. **Square brackets, [ ], indicate endpoints that are solutions. Parentheses, ( ), indicate endpoints that are not solutions.**

Great Question!

Is there another way that I can write $x < 3$?

Because an inequality symbol points to the smaller number, $x < 3$ (x is less than 3) may be expressed as $3 > x$ (3 is greater than x).

EXAMPLE 1 Graphing Inequalities

Graph the solutions of each inequality:

a. $x < 3$ **b.** $x \geq -1$ **c.** $-1 < x \leq 3$.

Solution

a. The solutions of $x < 3$ are all real numbers that are less than 3. They are graphed on a number line by shading all points to the left of 3. The parenthesis at 3 indicates that 3 is not a solution, but numbers such as 2.9999 and 2.6 are. The arrow shows that the graph extends indefinitely to the left.

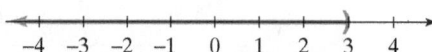

b. The solutions of $x \geq -1$ are all real numbers that are greater than or equal to -1. We shade all points to the right of -1 and the point for -1 itself. The square bracket at -1 shows that -1 is a solution of the given inequality. The arrow shows that the graph extends indefinitely to the right.

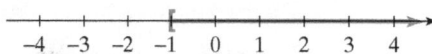

c. The inequality $-1 < x \leq 3$ is read "-1 is less than x *and* x is less than or equal to 3," or "x is greater than -1 *and* less than or equal to 3." The solutions of $-1 < x \leq 3$ are all real numbers between -1 and 3, not including -1 but including 3. The parenthesis at -1 indicates that -1 is not a solution. The square bracket at 3 shows that 3 is a solution. Shading indicates the other solutions.

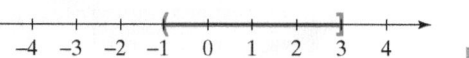

✓ **CHECK POINT 1** Graph the solutions of each inequality:
a. $x < 4$ **b.** $x \geq -2$ **c.** $-4 \leq x < 1$.

---

2 Use interval notation.

Interval Notation

The solutions of $x < 3$ are all real numbers that are less than 3:

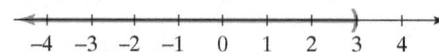

These numbers form an *interval* on the number line. The solution set of $x < 3$ can be expressed in **interval notation** as

$$(-\infty, 3).$$

> The negative infinity symbol indicates that the interval extends indefinitely to the left.

> The parenthesis indicates that 3 is excluded from the interval.

The solution set of $x < 3$ can also be expressed in set-builder notation as

$$\{ \ x \ | \ x \ < \ 3 \ \}.$$

The set of   all x such that x is less than 3.

Table 2.6 shows four inequalities, their solution sets using interval and set-builder notations, and graphs of the solution sets.

| Table 2.6 | Solution Sets of Inequalities |

Let a and b be real numbers.

Inequality	Interval Notation	Set-Builder Notation	Graph
$x > a$	(a, ∞)	$\{x \mid x > a\}$	(graph, open at a, extends right)
$x \geq a$	$[a, \infty)$	$\{x \mid x \geq a\}$	(graph, closed at a, extends right)
$x < b$	$(-\infty, b)$	$\{x \mid x < b\}$	(graph, open at b, extends left)
$x \leq b$	$(-\infty, b]$	$\{x \mid x \leq b\}$	(graph, closed at b, extends left)

Parentheses and Brackets in Interval Notation

Parentheses indicate endpoints that are not included in an interval. Square brackets indicate endpoints that are included in an interval. Parentheses are always used with ∞ or $-\infty$.

EXAMPLE 2   Using Interval Notation

Express the solution set of each inequality in interval notation and graph the interval:

a. $x \leq -1$ **b.** $x > 2$.

Solution

Inequality	Interval Notation	Graph
a. $x \leq -1$	$(-\infty, -1]$	(number line from -4 to 4, bracket at -1 extends left)

x is less than or equal to -1.   The interval extends indefinitely to the left.   The square bracket on the graph and in interval notation shows -1 is included in the interval.

| **b.** $x > 2$ | $(2, \infty)$ | (number line from -4 to 4, parenthesis at 2 extends right) |

x is greater than 2.   The interval extends indefinitely to the right.   The parenthesis on the graph and in interval notation shows 2 is excluded from the interval.

✓ **CHECK POINT 2**   Express the solution set of each inequality in interval notation and graph the interval:

   a. $x \geq 0$ **b.** $x < 5$.

---

3 Understand properties used to solve linear inequalities.

Properties Used to Solve Linear Inequalities

Back to our question that opened this section: How many miles can you drive your Rent-a-Heap car if you can spend at most \$335 per week? We answer the question by solving

$$0.20x + 125 \leq 335$$

for x. The solution procedure is nearly identical to that for solving

$$0.20x + 125 = 335.$$

Our goal is to get x by itself on the left side. We do this by first subtracting 125 from both sides to isolate $0.20x$:

$$0.20x + 125 \leq 335 \qquad \text{This is the given inequality.}$$
$$0.20x + 125 - 125 \leq 335 - 125 \qquad \text{Subtract 125 from both sides.}$$
$$0.20x \leq 210. \qquad \text{Simplify.}$$

Finally, we isolate x from $0.20x$ by dividing both sides of the inequality by 0.20:

$$\frac{0.20x}{0.20} \leq \frac{210}{0.20} \qquad \text{Divide both sides by 0.20.}$$

$$x \leq 1050. \qquad \text{Simplify.}$$

Great Question!

What are some common English phrases and sentences that I can model with linear inequalities?

English phrases such as "at least" and "at most" can be represented by inequalities.

English Sentence	Inequality
x is at least 5.	$x \geq 5$
x is at most 5.	$x \leq 5$
x is no more than 5.	$x \leq 5$
x is no less than 5.	$x \geq 5$

With at most \$335 per week to spend, you can travel at most 1050 miles.

We started with the inequality $0.20x + 125 \leq 335$ and obtained the inequality $x \leq 1050$ in the final step. Both of these inequalities have the same solution set, namely $\{x \mid x \leq 1050\}$. Inequalities such as these, with the same solution set, are said to be **equivalent**.

We isolated x from $0.20x$ by dividing both sides of $0.20x \leq 210$ by 0.20, a positive number. Let's see what happens if we divide both sides of an inequality by a negative number. Consider the inequality $10 < 14$. Divide both 10 and 14 by -2:

$$\frac{10}{-2} = -5 \quad \text{and} \quad \frac{14}{-2} = -7.$$

Because -5 lies to the right of -7 on the number line, -5 is greater than -7:

$$-5 > -7.$$

Notice that the direction of the inequality symbol is reversed:

$$10 < 14$$
$$\updownarrow$$
$$-5 > -7.$$

Dividing by -2 changes the direction of the inequality symbol.

In general, **when we multiply or divide both sides of an inequality by a negative number, the direction of the inequality symbol is reversed**. When we reverse the direction of the inequality symbol, we say that we change the *sense* of the inequality.

We can isolate a variable in a linear inequality the same way we can isolate a variable in a linear equation. The following properties are used to create equivalent inequalities:

Properties of Inequalities

Property	The Property in Words	Example
The Addition Property of Inequality If $a < b$, then $a + c < b + c$. If $a < b$, then $a - c < b - c$.	If the same quantity is added to or subtracted from both sides of an inequality, the resulting inequality is equivalent to the original one.	$2x + 3 < 7$ Subtract 3: $2x + 3 - 3 < 7 - 3$. Simplify: $2x < 4$.
The Positive Multiplication Property of Inequality If $a < b$ and c is positive, then $ac < bc$. If $a < b$ and c is positive, then $\dfrac{a}{c} < \dfrac{b}{c}$.	If we multiply or divide both sides of an inequality by the same positive quantity, the resulting inequality is equivalent to the original one.	$2x < 4$ Divide by 2: $\dfrac{2x}{2} < \dfrac{4}{2}$. Simplify: $x < 2$.
The Negative Multiplication Property of Inequality If $a < b$ and c is negative, then $ac > bc$. If $a < b$ and c is negative, then $\dfrac{a}{c} > \dfrac{b}{c}$.	If we multiply or divide both sides of an inequality by the same negative quantity and reverse the direction of the inequality symbol, the resulting inequality is equivalent to the original one.	$-4x < 20$ Divide by -4 and reverse the direction of the inequality symbol: $\dfrac{-4x}{-4} > \dfrac{20}{-4}$. Simplify: $x > -5$.

4 Solve linear inequalities.

Solving Linear Inequalities Involving Only One Property of Inequality

If you can solve a linear equation, it is likely that you can solve a linear inequality. Why? The procedure for solving linear inequalities is nearly the same as the procedure for solving linear equations, with one important exception: **When multiplying or dividing by a negative number, reverse the direction of the inequality symbol, changing the sense of the inequality**.

EXAMPLE 3 Solving a Linear Inequality

Solve and graph the solution set on a number line:

$$x + 3 < 8.$$

Solution Our goal is to isolate x. We can do this by using the addition property, subtracting 3 from both sides.

$x + 3 < 8$	This is the given inequality.
$x + 3 - 3 < 8 - 3$	Subtract 3 from both sides.
$x < 5$	Simplify.

The solution set consists of all real numbers that are less than 5. We express this in interval notation as $(-\infty, 5)$, or in set-builder notation as $\{x \mid x < 5\}$. The graph of the solution set is shown as follows:

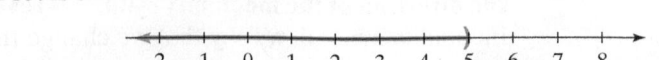

Discover for Yourself

Can you check all solutions to Example 3 in the given inequality? Is a partial check possible? Select a real number that is less than 5 and show that it satisfies $x + 3 < 8$.

✓ **CHECK POINT 3**   Solve and graph the solution set on a number line:
$$x + 6 < 9.$$

EXAMPLE 4   Solving a Linear Inequality

Solve and graph the solution set on a number line:
$$4x - 1 \geq 3x - 6.$$

Solution   Our goal is to isolate all terms involving x on one side and all numerical terms on the other side, exactly as we did when solving equations. Let's begin by using the addition property to isolate variable terms on the left.

$4x - 1 \geq 3x - 6$	This is the given inequality.
$4x - 3x - 1 \geq 3x - 3x - 6$	Subtract 3x from both sides.
$x - 1 \geq -6$	Simplify.

Now we isolate the numerical terms on the right. Use the addition property and add 1 to both sides.

$x - 1 + 1 \geq -6 + 1$	Add 1 to both sides.
$x \geq -5$	Simplify.

The solution set consists of all real numbers that are greater than or equal to -5. We express this in interval notation as $[-5, \infty)$, or in set-builder notation as $\{x \mid x \geq -5\}$.
   The graph of the solution set is shown as follows:

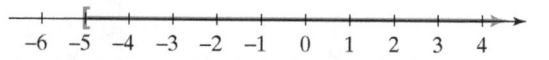

✓ **CHECK POINT 4**   Solve and graph the solution set on a number line:
$$8x - 2 \geq 7x - 4.$$

   We solved the inequalities in Examples 3 and 4 using the addition property of inequality. Now let's practice using the multiplication property of inequality. Do not forget to reverse the direction of the inequality symbol when multiplying or dividing both sides by a negative number.

EXAMPLE 5   Solving Linear Inequalities

Solve and graph the solution set on a number line:

a. $\dfrac{1}{3}x < 5$ **b.** $-3x < 21.$

Solution   In each case, our goal is to isolate x. In the first inequality, this is accomplished by multiplying both sides by 3. In the second inequality, we can do this by dividing both sides by -3.

a. $\dfrac{1}{3}x < 5$ This is the given inequality.

$3 \cdot \dfrac{1}{3}x < 3 \cdot 5$ Isolate x by multiplying by 3 on both sides.

   The symbol $<$ stays the same because we are multiplying both sides by a positive number.

$x < 15$ Simplify.

The solution set consists of all real numbers that are less than 15. We express this in interval notation as $(-\infty, 15)$, or in set-builder notation as $\{x \mid x < 15\}$. The graph of the solution set is shown as follows:

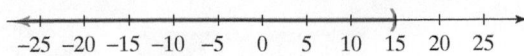

b. $-3x < 21$ This is the given inequality.

$$\frac{-3x}{-3} > \frac{21}{-3}$$ Isolate x by dividing by -3 on both sides.
The symbol $<$ must be reversed because we are dividing both sides by a negative number.

$\quad x > -7$ Simplify.

The solution set consists of all real numbers that are greater than -7. We express this in interval notation as $(-7, \infty)$, or in set-builder notation as $\{x \mid x > -7\}$. The graph of the solution set is shown as follows:

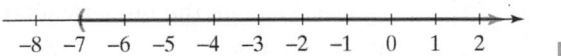

✓ **CHECK POINT 5**   Solve and graph the solution set on a number line:

a. $\frac{1}{4}x < 2$ 　　　　　　　　**b.** $-6x < 18.$

Inequalities Requiring Both the Addition and Multiplication Properties

If a linear inequality does not contain fractions, it can be solved using the following procedure. Notice, again, how similar this procedure is to the procedure for solving a linear equation.

> ### Solving a Linear Inequality
>
> 1. Simplify the algebraic expression on each side.
> 2. Use the addition property of inequality to collect all the variable terms on one side and all the constant terms on the other side.
> 3. Use the multiplication property of inequality to isolate the variable and solve. Change the sense of the inequality when multiplying or dividing both sides by a negative number.
> 4. Express the solution set in interval or set-builder notation, and graph the solution set on a number line.

EXAMPLE 6　Solving a Linear Inequality

Solve and graph the solution set on a number line:

$$4y - 7 \geq 5.$$

Solution

Step 1. Simplify each side. Because each side is already simplified, we can skip this step.

Step 2. Collect variable terms on one side and constant terms on the other side. The only variable term, $4y$, is already on the left side of $4y - 7 \geq 5$. We will collect constant terms on the right by adding 7 to both sides.

$$4y - 7 \geq 5 \qquad \text{This is the given inequality.}$$
$$4y - 7 + 7 \geq 5 + 7 \qquad \text{Add 7 to both sides.}$$
$$4y \geq 12 \qquad \text{Simplify.}$$

Step 3. Isolate the variable and solve. We isolate the variable, y, by dividing both sides by 4. Because we are dividing by a positive number, we do not reverse the inequality symbol.

$$\frac{4y}{4} \geq \frac{12}{4} \qquad \text{Divide both sides by 4.}$$
$$y \geq 3 \qquad \text{Simplify.}$$

Step 4. Express the solution set in interval or set-builder notation, and graph the set on a number line. The solution set consists of all real numbers that are greater than or equal to 3, expressed in interval notation as $[3, \infty)$, or in set-builder notation as $\{y \mid y \geq 3\}$. The graph of the solution set is shown as follows:

$$\begin{array}{c} \xleftarrow{\quad} \begin{matrix} + & + & + & + & + & + & + & + & [& + & + \\ -5 & -4 & -3 & -2 & -1 & 0 & 1 & 2 & 3 & 4 & 5 \end{matrix} \xrightarrow{\quad} \end{array} \quad \blacksquare$$

> ✓ **CHECK POINT 6** Solve and graph the solution set on a number line:
>
> $$5y - 3 \geq 17.$$

EXAMPLE 7 Solving a Linear Inequality

Solve and graph the solution set on a number line:

$$7x + 15 \geq 13x + 51.$$

Solution

Step 1. Simplify each side. Because each side is already simplified, we can skip this step.

Step 2. Collect variable terms on one side and constant terms on the other side. We will collect variable terms on the left and constant terms on the right.

$$7x + 15 \geq 13x + 51 \qquad \text{This is the given inequality.}$$
$$7x + 15 - 13x \geq 13x + 51 - 13x \qquad \text{Subtract 13x from both sides.}$$
$$-6x + 15 \geq 51 \qquad \text{Simplify.}$$
$$-6x + 15 - 15 \geq 51 - 15 \qquad \text{Subtract 15 from both sides.}$$
$$-6x \geq 36 \qquad \text{Simplify.}$$

Step 3. Isolate the variable and solve. We isolate the variable, x, by dividing both sides by -6. Because we are dividing by a negative number, we must reverse the inequality symbol.

$$\frac{-6x}{-6} \leq \frac{36}{-6} \qquad \begin{array}{l} \text{Divide both sides by } -6 \text{ and change the} \\ \text{sense of the inequality.} \end{array}$$

$$x \leq -6 \qquad \text{Simplify.}$$

Step 4. Express the solution set in interval or set-builder notation, and graph the set on a number line. The solution set consists of all real numbers that are less than or equal to -6, expressed in interval notation as $(-\infty, -6]$, or in set-builder notation as $\{x \mid x \leq -6\}$. The graph of the solution set is shown as follows:

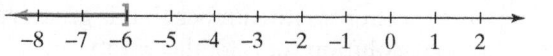

$$\begin{array}{c} \xleftarrow{\quad} \begin{matrix} + &] & + & + & + & + & + & + & + & + & + \\ -8 & -7 & -6 & -5 & -4 & -3 & -2 & -1 & 0 & 1 & 2 \end{matrix} \xrightarrow{\quad} \end{array} \quad \blacksquare$$

Great Question!

Is there a way to check that I obtained the correct solution set for a linear inequality?

It is possible to perform a partial check for an inequality. Select one number from the solution set. Substitute that number into the original inequality and perform the resulting computations. You should obtain a true statement.

Great Question!

Do I have to solve the inequality in Example 7 by isolating the variable on the left?

No. You can solve

$$7x + 15 \geq 13x + 51$$

by isolating x on the right side. Subtract $7x$ from both sides:

$$7x + 15 - 7x$$
$$\geq 13x + 51 - 7x$$
$$15 \geq 6x + 51.$$

Now subtract 51 from both sides:

$$15 - 51 \geq 6x + 51 - 51$$
$$-36 \geq 6x.$$

Finally, divide both sides by 6:

$$\frac{-36}{6} \geq \frac{6x}{6}$$
$$-6 \geq x.$$

This last inequality means the same thing as

$$x \leq -6.$$

✓ **CHECK POINT 7** Solve and graph the solution set: $6 - 3x \leq 5x - 2$.

EXAMPLE 8 Solving a Linear Inequality

Solve and graph the solution set on a number line:

$$2(x - 3) + 5x \leq 8(x - 1).$$

Solution

Step 1. Simplify each side. We use the distributive property to remove parentheses. Then we combine like terms.

$$2(x - 3) + 5x \leq 8(x - 1) \qquad \text{This is the given inequality.}$$

$$2x - 6 + 5x \leq 8x - 8 \qquad \text{Use the distributive property.}$$

$$7x - 6 \leq 8x - 8 \qquad \text{Add like terms on the left.}$$

Step 2. Collect variable terms on one side and constant terms on the other side. We will collect variable terms on the left and constant terms on the right.

$$7x - 8x - 6 \leq 8x - 8x - 8 \qquad \text{Subtract 8x from both sides.}$$

$$-x - 6 \leq -8 \qquad \text{Simplify.}$$

$$-x - 6 + 6 \leq -8 + 6 \qquad \text{Add 6 to both sides.}$$

$$-x \leq -2 \qquad \text{Simplify.}$$

Step 3. Isolate the variable and solve. To isolate x in $-x \leq -2$, we must eliminate the negative sign in front of the x. Because $-x$ means $-1x$, we can do this by multiplying (or dividing) both sides of the inequality by -1. We are multiplying by a negative number. Thus, we must reverse the inequality symbol.

$$(-1)(-x) \geq (-1)(-2) \qquad \text{Multiply both sides of } -x \leq -2 \text{ by } -1 \text{ and change the sense of the inequality.}$$

$$x \geq 2 \qquad \text{Simplify.}$$

Step 4. Express the solution set in interval or set-builder notation, and graph the set on a number line. The solution set consists of all real numbers that are greater than or equal to 2, expressed in interval notation as $[2, \infty)$, or in set-builder notation as $\{x \mid x \geq 2\}$. The graph of the solution set is shown as follows:

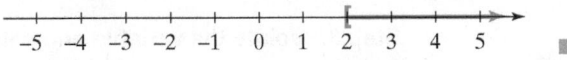

✓ **CHECK POINT 8** Solve and graph the solution set on a number line:

$$2(x - 3) - 1 \leq 3(x + 2) - 14.$$

---

5 Identify inequalities with no solution or true for all real numbers.

Inequalities with Unusual Solution Sets

We have seen that some equations have no solution. This is also true for some inequalities. An example of such an inequality is

$$x > x + 1.$$

There is no number that is greater than itself plus 1. This inequality has no solution. Its solution set is $\varnothing$, the empty set.

By contrast, some inequalities are true for all real numbers. An example of such an inequality is

$$x < x + 1.$$

Every real number is less than itself plus 1. The solution set is expressed in interval notation as $(-\infty, \infty)$, or in set-builder notation as $\{x \mid x \text{ is a real number}\}$.

Recognizing Inequalities with No Solution or True for All Real Numbers

If you attempt to solve an inequality with no solution or one that is true for every real number, you will eliminate the variable.

- An inequality with no solution results in a false statement, such as $0 > 1$. The solution set is $\varnothing$, the empty set.
- An inequality that is true for every real number results in a true statement, such as $0 < 1$. The solution set is $(-\infty, \infty)$ or $\{x \mid x \text{ is a real number}\}$.

EXAMPLE 9   Solving a Linear Inequality

Solve: $3(x + 1) > 3x + 5$.

Solution

$$3(x + 1) > 3x + 5 \qquad \text{This is the given inequality.}$$
$$3x + 3 > 3x + 5 \qquad \text{Apply the distributive property.}$$
$$3x + 3 - 3x > 3x + 5 - 3x \qquad \text{Subtract 3x from both sides.}$$

Keep reading. $3 > 5$ is not the solution. $3 > 5$ Simplify.

The original inequality is equivalent to the statement $3 > 5$, which is false for every value of x. The inequality has no solution. The solution set is $\varnothing$, the empty set.   ■

✓ **CHECK POINT 9**   Solve: $4(x + 2) > 4x + 15$.

EXAMPLE 10   Solving a Linear Inequality

Solve: $2(x + 5) \le 5x - 3x + 14$.

Solution

$$2(x + 5) \le 5x - 3x + 14 \qquad \text{This is the given inequality.}$$
$$2x + 10 \le 5x - 3x + 14 \qquad \text{Apply the distributive property.}$$
$$2x + 10 \le 2x + 14 \qquad \text{Combine like terms.}$$
$$2x + 10 - 2x \le 2x + 14 - 2x \qquad \text{Subtract 2x from both sides.}$$

Keep reading. $10 \le 14$ is not the solution. $10 \le 14$ Simplify.

The original inequality is equivalent to the statement $10 \le 14$, which is true for every value of x. The solution is the set of all real numbers, expressed in interval notation as $(-\infty, \infty)$, or in set-builder notation as $\{x \mid x \text{ is a real number}\}$.   ■

✓ **CHECK POINT 10**   Solve: $3(x + 1) \ge 2x + 1 + x$.

6 Solve problems using linear inequalities.

Applications

As you know, different professors may use different grading systems to determine your final course grade. Some professors require a final examination; others do not. In our next example, a final exam is required *and* it also counts as two grades.

EXAMPLE 11 An Application: Final Course Grade

To earn an A in a course, you must have a final average of at least 90%. On the first four examinations, you have grades of 86%, 88%, 92%, and 84%. If the final examination counts as two grades, what must you get on the final to earn an A in the course?

Solution We will use our five-step strategy for solving algebraic word problems.

Steps 1 and 2. Represent unknown quantities in terms of *x*. Let

$$x = \text{your grade on the final examination.}$$

Step 3. Write an inequality in *x* that models the conditions. The average of the six grades is found by adding the grades and dividing the sum by 6.

$$\text{Average} = \frac{86 + 88 + 92 + 84 + x + x}{6}$$

Because the final counts as two grades, the *x* (your grade on the final examination) is added twice. This is also why the sum is divided by 6.

To get an A, your average must be at least 90. This means that your average must be greater than or equal to 90.

Your average | must be greater than or equal to | 90.

$$\frac{86 + 88 + 92 + 84 + x + x}{6} \geq 90$$

Step 4. Solve the inequality and answer the problem's question.

$\dfrac{86 + 88 + 92 + 84 + x + x}{6} \geq 90$	This is the inequality that models the given conditions.
$\dfrac{350 + 2x}{6} \geq 90$	Combine like terms in the numerator.
$6\left(\dfrac{350 + 2x}{6}\right) \geq 6(90)$	Multiply both sides by 6, clearing the fraction.
$350 + 2x \geq 540$	Multiply.
$350 + 2x - 350 \geq 540 - 350$	Subtract 350 from both sides.
$2x \geq 190$	Simplify.
$\dfrac{2x}{2} \geq \dfrac{190}{2}$	Divide both sides by 2.
$x \geq 95$	Simplify.

You must get at least 95% on the final examination to earn an A in the course.

Step 5. Check. We can perform a partial check by computing the average with any grade that is at least 95. We will use 96. If you get 96% on the final examination, your average is

$$\frac{86 + 88 + 92 + 84 + 96 + 96}{6} = \frac{542}{6} = 90\frac{1}{3}.$$

Because $90\frac{1}{3} > 90$, you earn an A in the course. ■

✓ **CHECK POINT 11** To earn a B in a course, you must have a final average of at least 80%. On the first three examinations, you have grades of 82%, 74%, and 78%. If the final examination counts as two grades, what must you get on the final to earn a B in the course?

EXAMPLE 12 An Application: Staying within a Budget

You can spend at most $1000 to have a picnic catered. The caterer charges a $150 setup fee and $25 per person. How many people can you invite while staying within your budget?

Solution

Steps 1 and 2. Represent unknown quantities in terms of x. Let

$$x = \text{the number of people you invite to the picnic.}$$

Step 3. Write an inequality in x that models the conditions. You can spend at most $1000. This means that the caterer's setup fee plus the cost of the meals must be less than or equal to $1000.

The setup fee: $150	plus	the cost of the meals: $25 per person for x people	must be less than or equal to	$1000.
150	+	25x	≤	1000

Step 4. Solve the inequality and answer the problem's question.

$$150 + 25x \le 1000 \qquad \text{This is the inequality that models the given conditions.}$$

$$150 + 25x - 150 \le 1000 - 150 \qquad \text{Subtract 150 from both sides.}$$

$$25x \le 850 \qquad \text{Simplify.}$$

$$\frac{25x}{25} \le \frac{850}{25} \qquad \text{Divide both sides by 25.}$$

$$x \le 34 \qquad \text{Simplify.}$$

You can invite at most 34 people to the picnic and still stay within your budget.

Step 5. Check. We can perform a partial check. Because $x \le 34$, let's see if you stay within your $1000 budget by inviting 33 guests.

The setup fee: $150	plus	the cost of the meals: $25 per person for 33 people
$150	+	$25(33) = $150 + $825 = $975

Inviting 33 people results in catering costs of $975, which is within your $1000 budget. ■

✓ **CHECK POINT 12** You can spend at most $1600 to have a picnic catered. The caterer charges a $95 setup fee and $35 per person. How many people can you invite while staying within your budget?

Achieving Success

Assuming that you have done very well preparing for an exam, **there are certain things you can do that will make you a better test taker**.

- Get a good sleep the night before the exam.
- Have a good breakfast that balances protein, carbohydrates, and fruit.
- Just before the exam, briefly review the relevant material in the chapter summary.
- Bring everything you need to the exam, including two pencils, an eraser, scratch paper (if permitted), a calculator (if you're allowed to use one), water, and a watch.
- Survey the entire exam quickly to get an idea of its length.
- Read the directions to each problem carefully. Make sure that you have answered the specific question asked.
- Work the easy problems first. Then return to the hard problems you are not sure of. Doing the easy problems first will build your confidence. If you get bogged down on any one problem, you may not be able to complete the exam and receive credit for the questions you can easily answer.
- Attempt every problem. There may be partial credit even if you do not obtain the correct answer.
- Work carefully. Show your step-by-step solutions neatly. Check your work and answers.
- Watch the time. Pace yourself and be aware of when half the time is up. Determine how much of the exam you have completed. This will indicate if you're moving at a good pace or need to speed up. Prepare to spend more time on problems worth more points.
- Never turn in a test early. Use every available minute you are given for the test. If you have extra time, double check your arithmetic and look over your solutions.

CONCEPT AND VOCABULARY CHECK

Fill in each blank so that the resulting statement is true.

1. The solution set of $x < 5$ can be expressed in interval notation as _____.

2. The solution set of $x \geq 2$ can be expressed in interval notation as _____.

3. The addition property of inequality states that if $a < b$, then $a + c$ _____.

4. The positive multiplication property of inequality states that if $a < b$ and c is positive, then ac _____.

5. The negative multiplication property of inequality states that if $a < b$ and c is negative, then ac _____.

6. The linear inequality $-3x + 4 > 13$ can be solved by first _____ from both sides and then _____ both sides by _____, which changes the _____ of the inequality symbol from _____ to _____.

7. In solving an inequality, if you eliminate the variable and obtain a false statement such as $0 > 1$, the solution set is _____.

8. In solving an inequality, if you eliminate the variable and obtain a true statement such as $0 < 1$, the solution set in interval notation is _____.

2.7 EXERCISE SET MyMathLab®

Watch the videos
in MyMathLab

Download the
MyDashBoard App

Practice Exercises

In Exercises 1–12, graph the solutions of each inequality on a number line.

1. $x > 5$
2. $x > -3$
3. $x < -2$
4. $x < 0$
5. $x \geq -4$
6. $x \geq -6$
7. $x \leq 4.5$
8. $x \leq 7.5$
9. $-2 < x \leq 6$
10. $-3 \leq x < 6$
11. $-1 < x < 3$
12. $-2 \leq x \leq 0$

In Exercises 13–20, express the solution set of each inequality in interval notation and graph the interval.

13. $x \leq 3$
14. $x \leq 5$
15. $x > \dfrac{5}{2}$
16. $x > \dfrac{7}{2}$
17. $x \leq 0$
18. $x \leq 1$
19. $x < 4$
20. $x < 5$

Use the addition property of inequality to solve each inequality in Exercises 21–38 and graph the solution set on a number line.

21. $x - 3 > 4$
22. $x + 1 < 6$
23. $x + 4 \leq 10$
24. $x - 5 \geq 2$
25. $y - 2 < 0$
26. $y + 3 \geq 0$
27. $3x + 4 \leq 2x + 7$
28. $2x + 9 \leq x + 2$
29. $5x - 9 < 4x + 7$
30. $3x - 8 < 2x + 11$
31. $7x - 7 > 6x - 3$
32. $8x - 9 > 7x - 3$
33. $x - \dfrac{2}{3} > \dfrac{1}{2}$
34. $x - \dfrac{1}{3} \geq \dfrac{5}{6}$
35. $y + \dfrac{7}{8} \leq \dfrac{1}{2}$
36. $y + \dfrac{1}{3} \leq \dfrac{3}{4}$
37. $-15y + 13 > 13 - 16y$
38. $-12y + 17 > 20 - 13y$

Use the multiplication property of inequality to solve each inequality in Exercises 39–56 and graph the solution set on a number line.

39. $\dfrac{1}{2}x < 4$
40. $\dfrac{1}{2}x > 3$
41. $\dfrac{x}{3} > -2$
42. $\dfrac{x}{4} < -1$
43. $4x < 20$
44. $6x < 18$
45. $3x \geq -21$
46. $7x \geq -56$
47. $-3x < 15$
48. $-7x > 21$
49. $-3x \geq 15$
50. $-7x \leq 21$
51. $-16x > -48$
52. $-20x > -140$
53. $-4y \leq \dfrac{1}{2}$
54. $-2y \leq \dfrac{1}{2}$
55. $-x < 4$
56. $-x > -3$

Use both the addition and multiplication properties of inequality to solve each inequality in Exercises 57–80 and graph the solution set on a number line.

57. $2x - 3 > 7$
58. $3x + 2 \leq 14$
59. $3x + 3 < 18$
60. $8x - 4 > 12$
61. $3 - 7x \leq 17$
62. $5 - 3x \geq 20$
63. $-2x - 3 < 3$
64. $-3x + 14 < 5$
65. $5 - x \leq 1$
66. $3 - x \geq -3$
67. $2x - 5 > -x + 6$
68. $6x - 2 \geq 4x + 6$
69. $2y - 5 < 5y - 11$
70. $4y - 7 > 9y - 2$
71. $3(2y - 1) < 9$
72. $4(2y - 1) > 12$
73. $3(x + 1) - 5 < 2x + 1$
74. $4(x + 1) + 2 \geq 3x + 6$
75. $8x + 3 > 3(2x + 1) - x + 5$
76. $7 - 2(x - 4) < 5(1 - 2x)$
77. $\dfrac{x}{3} - 2 \geq 1$
78. $\dfrac{x}{4} - 3 \geq 1$
79. $1 - \dfrac{x}{2} > 4$
80. $1 - \dfrac{x}{2} < 5$

In Exercises 81–90, solve each inequality.

81. $4x - 4 < 4(x - 5)$
82. $3x - 5 < 3(x - 2)$

83. $x + 3 < x + 7$

84. $x + 4 < x + 10$

85. $7x \leq 7(x - 2)$

86. $3x + 1 \leq 3(x - 2)$

87. $2(x + 3) > 2x + 1$

88. $5(x + 4) > 5x + 10$

89. $5x - 4 \leq 4(x - 1)$

90. $6x - 3 \leq 3(x - 1)$

Practice PLUS

In Exercises 91–94, use properties of inequality to rewrite each inequality so that x is isolated on one side.

91. $3x + a > b$

92. $-2x - a \leq b$

93. $y \leq mx + b$ and $m < 0$

94. $y > mx + b$ and $m > 0$

We know that $|x|$ represents the distance from 0 to x on a number line. In Exercises 95–98, use each sentence to describe all possible locations of x on a number line. Then rewrite the given sentence as an inequality involving $|x|$.

95. The distance from 0 to x on a number line is less than 2.

96. The distance from 0 to x on a number line is less than 3.

97. The distance from 0 to x on a number line is greater than 2.

98. The distance from 0 to x on a number line is greater than 3.

Application Exercises

An online test of English spelling looked at how well people spelled difficult words. The bar graph shows the percentage of people who spelled each word correctly. Let x represent the percentage who spelled a word correctly. In Exercises 99–104, write the word or words described by the given inequality. (Yes, spelling counts!)

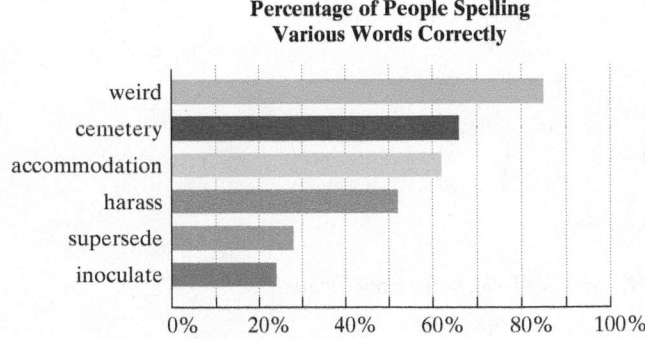

Percentage of People Spelling Various Words Correctly

Source: Vivian Cook. *Accomodating Brocolli in the Cemetary or Why Can't Anybody Spell?.* Simon and Schuster, 2004.

99. $x > 55\%$

100. $x \geq 70\%$

101. $x \leq 30\%$

102. $x \leq 50\%$

103. $40\% \leq x < 60\%$

104. $50\% < x \leq 70\%$

The graph shows the decline in the number of stamped letters mailed in the United States from 2000 to 2010. The data shown by the graph can be modeled by

$$S = 55 - 2.5x,$$

where S is the number of stamped letters mailed, in billions, x years after 2000. Use this formula to solve Exercises 105–106.

Number of Stamped Letters Mailed in the United States

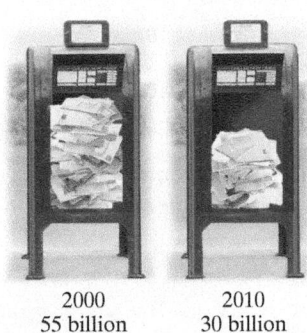

2000
55 billion

2010
30 billion

Source: U.S. Postal Service

105. Describe how many years after 2000 there will be no more than 15 billion stamped letters mailed in the United States. Which years are included in your description?

106. Describe how many years after 2000 there will be no more than 10 billion stamped letters mailed in the United States. Which years are included in your description?

107. On two examinations, you have grades of 86 and 88. There is an optional final examination, which counts as one grade. You decide to take the final in order to get a course grade of A, meaning a final average of at least 90.

    a. What must you get on the final to earn an A in the course?

    b. By taking the final, if you do poorly, you might risk the B that you have in the course based on the first two exam grades. If your final average is less than 80, you will lose your B in the course. Describe the grades on the final that will cause this to happen.

108. On three examinations, you have grades of 88, 78, and 86. There is still a final examination, which counts as one grade.

    a. In order to get an A, your average must be at least 90. If you get 100 on the final, compute your average and determine if an A in the course is possible.

    b. To earn a B in the course, you must have a final average of at least 80. What must you get on the final to earn a B in the course?

109. A car can be rented from Continental Rental for $80 per week plus 25 cents for each mile driven. How many miles can you travel if you can spend at most $400 for the week?

110. A car can be rented from Basic Rental for $60 per week plus 50 cents for each mile driven. How many miles can you travel if you can spend at most $600 for the week?

111. An elevator at a construction site has a maximum capacity of 3000 pounds. If the elevator operator weighs 245 pounds and each cement bag weighs 95 pounds, how many bags of cement can be safely lifted on the elevator in one trip?

112. An elevator at a construction site has a maximum capacity of 2800 pounds. If the elevator operator weighs 265 pounds and each cement bag weighs 65 pounds, how many bags of cement can be safely lifted on the elevator in one trip?

Writing in Mathematics

113. When graphing the solutions of an inequality, what is the difference between a parenthesis and a bracket?

114. When solving an inequality, when is it necessary to change the direction of the inequality symbol? Give an example.

115. Describe ways in which solving a linear inequality is similar to solving a linear equation.

116. Describe ways in which solving a linear inequality is different from solving a linear equation.

Critical Thinking Exercises

Make Sense? *In Exercises 117–120, determine whether each statement "makes sense" or "does not make sense" and explain your reasoning.*

117. I prefer interval notation over set-builder notation because it takes less space to write solution sets.

118. I can check inequalities by substituting 0 for the variable: When 0 belongs to the solution set, I should obtain a true statement, and when 0 does not belong to the solution set, I should obtain a false statement.

119. In an inequality such as $5x + 4 < 8x - 5$, I can avoid division by a negative number depending on which side I collect the variable terms and on which side I collect the constant terms.

120. I solved $-2x + 5 \geq 13$ and concluded that -4 is the greatest integer in the solution set.

In Exercises 121–124, determine whether each statement is true or false. If the statement is false, make the necessary change(s) to produce a true statement.

121. The inequality $x - 3 > 0$ is equivalent to $x < 3$.

122. The statement "x is at most 5" is written $x < 5$.

123. The inequality $-4x < -20$ is equivalent to $x > -5$.

124. The statement "the sum of x and 6% of x is at least 80" is modeled by $x + 0.06x \geq 80$.

125. A car can be rented from Basic Rental for $260 per week with no extra charge for mileage. Continental charges $80 per week plus 25 cents for each mile driven to rent the same car. How many miles should be driven in a week to make the rental cost for Basic Rental a better deal than Continental's?

126. Membership in a fitness club costs $500 yearly plus $1 per hour spent working out. A competing club charges $440 yearly plus $1.75 per hour for use of their equipment. How many hours must a person work out yearly to make membership in the first club cheaper than membership in the second club?

Technology Exercises

Solve each inequality in Exercises 127–128. Use a calculator to help with the arithmetic.

127. $1.45 - 7.23x > -1.442$

128. $126.8 - 9.4y \leq 4.8y + 34.5$

Review Exercises

129. 8 is 40% of what number? (Section 2.4, Example 6)

130. The length of a rectangle exceeds the width by 5 inches. The perimeter is 34 inches. What are the rectangle's dimensions? (Section 2.5, Example 5)

131. Solve and check: $5x + 16 = 3(x + 8)$. (Section 2.3, Example 2)

Preview Exercises

Exercises 132–134 will help you prepare for the material covered in the first section of the next chapter.

132. Is $x - 4y = 14$ a true statement for $x = 2$ and $y = -3$?

133. Is $x - 4y = 14$ a true statement for $x = 12$ and $y = 1$?

134. If $y = \frac{2}{3}x + 1$, find the value of y for $x = -6$.

GROUP PROJECT

CHAPTER 2

One of the best ways to learn how to *solve* a word problem in algebra is to *design* word problems of your own. Creating a word problem makes you very aware of precisely how much information is needed to solve the problem. You must also focus on the best way to present information to a reader and on how much information to give. As you write your problem, you gain skills that will help you solve problems created by others.

The group should design five different word problems that can be solved using an algebraic equation. All of the problems should be on different topics. For example, the group should not have more than one problem on finding a number. The group should turn in both the problems and their algebraic solutions.

Chapter 2 Summary

Definitions and Concepts	**Examples**

Section 2.1 The Addition Property of Equality

A linear equation in one variable can be written in the form $ax + b = c$, where a is not zero.

$3x + 7 = 9$ is a linear equation.

Equivalent equations have the same solution.

$2x - 4 = 6$, $2x = 10$, and $x = 5$ are equivalent equations.

The Addition Property of Equality

Adding the same number (or algebraic expression) to both sides of an equation or subtracting the same number (or algebraic expression) from both sides of an equation does not change its solution.

- $$x - 3 = 8$$
$$x - 3 + 3 = 8 + 3$$
$$x = 11$$

- $$x + 4 = 10$$
$$x + 4 - 4 = 10 - 4$$
$$x = 6$$

Section 2.2 The Multiplication Property of Equality

The Multiplication Property of Equality

Multiplying both sides of an equation or dividing both sides of an equation by the same nonzero real number (or algebraic expression) does not change the solution.

- $$\frac{x}{-5} = 6$$
$$-5\left(\frac{x}{-5}\right) = -5(6)$$
$$x = -30$$

- $$-50 = -5y$$
$$\frac{-50}{-5} = \frac{-5y}{-5}$$
$$10 = y$$

Equations and Coefficients of -1

If $-x = c$, multiply both sides by -1 to solve for x. The solution is the opposite, or additive inverse, of c.

$$-x = -12$$
$$(-1)(-x) = (-1)(-12)$$
$$x = 12$$

Definitions and Concepts	**Examples**

Section 2.2 The Multiplication Property of Equality (continued)

Using the Addition and Multiplication Properties

If an equation does not contain fractions,

- Use the addition property to isolate the variable term.
- Use the multiplication property to isolate the variable.

$$-2x - 5 = 11$$
$$-2x - 5 + 5 = 11 + 5$$
$$-2x = 16$$
$$\frac{-2x}{-2} = \frac{16}{-2}$$
$$x = -8$$

Section 2.3 Solving Linear Equations

Solving a Linear Equation

1. Simplify each side.

2. Collect all the variable terms on one side and all the constant terms on the other side.

3. Isolate the variable and solve. (If the variable is eliminated and a false statement results, the inconsistent equation has no solution. If a true statement results, all real numbers are solutions of the identity.)

4. Check the proposed solution in the original equation.

Solve: $\quad 7 - 4(x - 1) = x + 1.$

$$7 - 4x + 4 = x + 1$$
$$-4x + 11 = x + 1$$

$$-4x - x + 11 = x - x + 1$$
$$-5x + 11 = 1$$
$$-5x + 11 - 11 = 1 - 11$$
$$-5x = -10$$

$$\frac{-5x}{-5} = \frac{-10}{-5}$$
$$x = 2$$

$$7 - 4(x - 1) = x + 1$$
$$7 - 4(2 - 1) \stackrel{?}{=} 2 + 1$$
$$7 - 4(1) \stackrel{?}{=} 2 + 1$$
$$7 - 4 \stackrel{?}{=} 2 + 1$$
$$3 = 3, \text{ true}$$

The solution is 2, or the solution set is $\{2\}$.

Equations Containing Fractions

Multiply both sides (all terms) by the least common denominator. This clears the equation of fractions.

Solve: $\quad \dfrac{x}{5} + \dfrac{1}{2} = \dfrac{x}{2} - 1.$

$$10\left(\frac{x}{5} + \frac{1}{2}\right) = 10\left(\frac{x}{2} - 1\right)$$
$$10 \cdot \frac{x}{5} + 10 \cdot \frac{1}{2} = 10 \cdot \frac{x}{2} - 10 \cdot 1$$
$$2x + 5 = 5x - 10$$
$$-3x = -15$$
$$x = 5$$

The solution is 5, or the solution set is $\{5\}$.

Equations Containing Decimals

An equation may be cleared of decimals by multiplying every term on both sides by a power of 10. The exponent on 10 will equal the greatest number of decimal places in the equation.

Solve: $\quad 1.4(x - 5) = 3x - 3.8$

$$1.4x - 7 = 3x - 3.8$$
$$10(1.4x) - 10(7) = 10(3x) - 10(3.8)$$
$$14x - 70 = 30x - 38$$
$$-16x = 32$$
$$x = -2$$

The solution is -2, or the solution set is $\{-2\}$.

Definitions and Concepts	**Examples**

Section 2.3 Solving Linear Equations (continued)

Inconsistent equations with no solution (solution set: $\varnothing$) result in false statements in the solution process. Identities, true for every real number (solution set: $\{x \mid x$ is a real number$\}$) result in true statements in the solution process.	Solve: $\qquad\qquad 3x + 2 = 3(x + 5).$ $$3x + 2 = 3x + 15$$ $$3x + 2 - 3x = 3x + 15 - 3x$$ $$2 = 15 \quad \text{(false)}$$ No solution: $\varnothing$ Solve: $\qquad\qquad 2(x + 4) = x + x + 8.$ $$2x + 8 = 2x + 8$$ $$2x + 8 - 2x = 2x + 8 - 2x$$ $$8 = 8 \quad \text{(true)}$$ $\{x \mid x$ is a real number$\}$

Section 2.4 Formulas and Percents

To solve a formula for one of its variables, use the steps for solving a linear equation and isolate the specified variable on one side of the equation.	Solve for l: $\qquad\qquad w = \dfrac{P - 2l}{2}.$ $$2w = 2\left(\frac{P - 2l}{2}\right)$$ $$2w = P - 2l$$ $$2w - P = P - P - 2l$$ $$2w - P = -2l$$ $$\frac{2w - P}{-2} = \frac{-2l}{-2}$$ $$\frac{2w - P}{-2} = l$$

A Formula Involving Percent

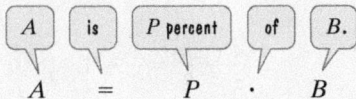

$$A = P \cdot B$$

In the formula $A = PB$, P is expressed as a decimal.

What is 5% of 20?

$$A = 0.05 \cdot 20$$
$$A = 1$$

Thus, 1 is 5% of 20.

6 is 30% of what?

$$6 = 0.3 \cdot B$$
$$\frac{6}{0.3} = B$$
$$20 = B$$

Thus, 6 is 30% of 20.

33 is what percent of 75?

$$33 = P \cdot 75$$
$$\frac{33}{75} = P$$
$$P = 0.44 = 44\%$$

Thus, 33 is 44% of 75.

Definitions and Concepts	**Examples**

Section 2.5 An Introduction to Problem Solving

Strategy for Solving Word Problems

Step 1. Let x represent one of the quantities.

Step 2. Represent other unknown quantities in terms of x.

Step 3. Write an equation that models the conditions.

Step 4. Solve the equation and answer the question.

Step 5. Check the proposed solution in the original wording of the problem.

The length of a rectangle exceeds the width by 3 inches. The perimeter is 26 inches. What are the rectangle's dimensions?

$$\text{Let } x = \text{ the width.}$$

$$x + 3 = \text{ the length}$$

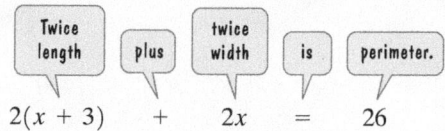

$$2(x + 3) \quad + \quad 2x \quad = \quad 26$$

$$2(x + 3) + 2x = 26$$

$$2x + 6 + 2x = 26$$

$$4x + 6 = 26$$

$$4x = 20$$

$$x = 5$$

The width (x) is 5 inches and the length ($x + 3$) is 5 + 3, or 8 inches.

$$\text{Perimeter} = 2(5 \text{ in.}) + 2(8 \text{ in.})$$

$$= 10 \text{ in.} + 16 \text{ in.} = 26 \text{ in.}$$

This checks with the given perimeter.

Section 2.6 Problem Solving in Geometry

Solving geometry problems often requires using basic geometric formulas. Formulas for perimeter, area, circumference, and volume are given in **Table 2.3** (page 169), **Table 2.4** (page 170), and **Table 2.5** (page 171) in Section 2.6.

A sailboat's triangular sail has an area of 24 ft^2 and a base of 8 ft. Find its height.

$$A = \frac{1}{2}bh$$

$$24 = \frac{1}{2}(8)h$$

$$24 = 4h$$

$$6 = h$$

The sail's height is 6 ft.

Definitions and Concepts	**Examples**

Section 2.6 Problem Solving in Geometry (continued)

The sum of the measures of the three angles of any triangle is 180°.

In a triangle, the first angle measures 3 times the second and the third measures 40° less than the second. Find each angle's measure.

$$\text{Second angle} = x$$

$$\text{First angle} = 3x$$

$$\text{Third angle} = x - 40$$

Sum of measures is 180°.

$$x + 3x + (x - 40) = 180$$

$$5x - 40 = 180$$

$$5x = 220$$

$$x = 44$$

The angles measure $x = 44$, $3x = 3 \cdot 44 = 132$, and $x - 40 = 44 - 40 = 4$.

The angles measure 44°, 132°, and 4°.

Two complementary angles have measures whose sum is 90°. Two supplementary angles have measures whose sum is 180°. If an angle measures x, its complement measures $90 - x$, and its supplement measures $180 - x$.

An angle measures five times its complement. Find the angle's measure.

$$x = \text{angle's measure}$$

$$90 - x = \text{measure of complement}$$

$$x = 5(90 - x)$$

$$x = 450 - 5x$$

$$6x = 450$$

$$x = 75$$

The angle measures 75°.

Section 2.7 Solving Linear Inequalities

A linear inequality in one variable can be written in one of these forms:

$$ax + b < c \quad ax + b \le c.$$

$$ax + b > c \quad ax + b \ge c.$$

$3x + 6 > 12$ is a linear inequality.

Graphs of solutions to linear inequalities are shown on a number line by shading all points representing numbers that are solutions. Square brackets, [ ], indicate endpoints that are solutions. Parentheses, ( ), indicate endpoints that are not solutions.

- Graph the solutions of $x < 4$.

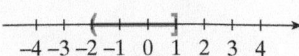

- Graph the solutions of $-2 < x \le 1$.

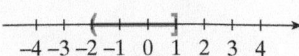

| **Definitions and Concepts** | **Examples** |

Section 2.7 Solving Linear Inequalities (continued)

Solutions of inequalities can be expressed in interval notation or set-builder notation.

Inequality	Interval Notation	Set-Builder Notation	Graph
$x > b$	(b, ∞)	$\{x \mid x > b\}$	$a \quad b$
$x \leq a$	$(-\infty, a]$	$\{x \mid x \leq a\}$	$a \quad b$

Express the solution set in interval notation and graph the interval:

- $x \leq 1$
  $(-\infty, 1]$

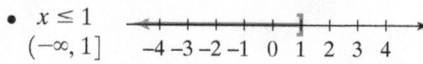

- $x > -2$
  $(-2, \infty)$

The Addition Property of Inequality

Adding the same number to both sides of an inequality or subtracting the same number from both sides of an inequality does not change the solutions.

$$x + 3 < 8$$
$$x + 3 - 3 < 8 - 3$$
$$x < 5$$

The Positive Multiplication Property of Inequality

Multiplying or dividing both sides of an inequality by the same positive number does not change the solutions.

$$\frac{x}{6} \geq 5$$
$$6 \cdot \frac{x}{6} \geq 6 \cdot 5$$
$$x \geq 30$$

The Negative Multiplication Property of Inequality

Multiplying or dividing both sides of an inequality by the same negative number and reversing the direction of the inequality sign does not change the solutions.

$$-3x \leq 12$$
$$\frac{-3x}{-3} \geq \frac{12}{-3}$$
$$x \geq -4$$

Solving Linear Inequalities

Use the procedure for solving linear equations. When multiplying or dividing by a negative number, reverse the direction of the inequality symbol. Express the solution set in interval or set-builder notation, and graph the set on a number line. If the variable is eliminated and a false statement results, the inequality has no solution. The solution set is $\varnothing$, the empty set. If a true statement results, the solution is the set of all real numbers: $(-\infty, \infty)$ or $\{x \mid x \text{ is a real number}\}$.

Solve:
$$x + 4 \geq 6x - 16$$
$$x + 4 - 6x \geq 6x - 16 - 6x$$
$$-5x + 4 \geq -16$$
$$-5x + 4 - 4 \geq -16 - 4$$
$$-5x \geq -20$$
$$\frac{-5x}{-5} \leq \frac{-20}{-5}$$
$$x \leq 4$$

Solution set: $(-\infty, 4]$ or $\{x \mid x \leq 4\}$

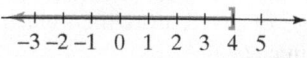

CHAPTER 2 REVIEW EXERCISES

2.1 *Solve each equation in Exercises 1–5 using the addition property of equality. Be sure to check proposed solutions.*

1. $x - 10 = 22$

2. $-14 = y + 8$

3. $7z - 3 = 6z + 9$

4. $4(x + 3) = 3x - 10$

5. $6x - 3x - 9 + 1 = -5x + 7x - 3$

2.2 *Solve each equation in Exercises 6–13 using the multiplication property of equality. Be sure to check proposed solutions.*

6. $\dfrac{x}{8} = 10$

7. $\dfrac{y}{-8} = 7$

8. $7z = 77$

9. $-36 = -9y$

10. $\dfrac{3}{5}x = -9$

11. $30 = -\dfrac{5}{2}y$

12. $-x = 25$

13. $\dfrac{-x}{10} = -1$

Solve each equation in Exercises 14–18 using both the addition and multiplication properties of equality. Check proposed solutions.

14. $4x + 9 = 33$

15. $-3y - 2 = 13$

16. $5z + 20 = 3z$

17. $5x - 3 = x + 5$

18. $3 - 2x = 9 - 8x$

19. The percentage of tax returns filed electronically in the United States exceeded 50% for the first time in 2005. The bar graph shows the percentage of electronically filed tax returns from 2005 through 2010.

Percentage of U.S. Tax Returns Filed Electronically

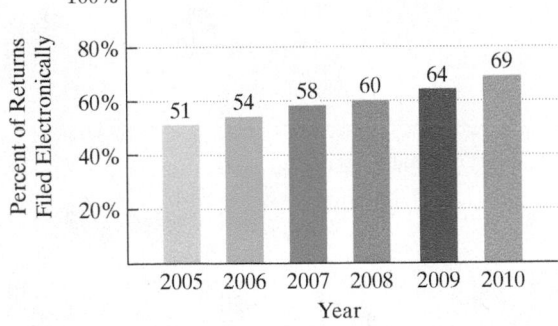

Source: IRS

The mathematical model

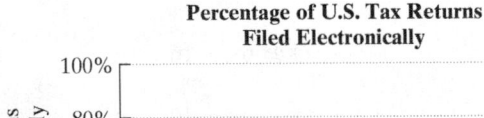
$$p = 3.5n + 51$$

describes the percentage of U.S. tax returns filed electronically, p, n years after 2005.

a. Does the formula underestimate or overestimate the percentage of tax returns filed electronically in 2009? By how much?

b. If trends shown by the formula continue, when will 93% of tax returns be filed electronically?

2.3 *Solve and check each equation in Exercises 20–30.*

20. $5x + 9 - 7x + 6 = x + 18$

21. $3(x + 4) = 5x - 12$

22. $1 - 2(6 - y) = 3y + 2$

23. $2(x - 4) + 3(x + 5) = 2x - 2$

24. $-2(y - 4) - (3y - 2) = -2 - (6y - 2)$

25. $\dfrac{2x}{3} = \dfrac{x}{6} + 1$

26. $\dfrac{x}{2} - \dfrac{1}{10} = \dfrac{x}{5} + \dfrac{1}{2}$

27. $0.5x + 8.75 = 13.25$

28. $0.1(x - 3) = 1.1 - 0.25x$

29. $3(8x - 1) = 6(5 + 4x)$

30. $4(2x - 3) + 4 = 8x - 8$

31. The formula $H = 0.7(220 - a)$ can be used to determine target heart rate during exercise, H, in beats per minute, by a person of age a. If the target heart rate is 133 beats per minute, how old is that person?

2.4 *In Exercises 32–36, solve each formula for the specified variable.*

32. $I = Pr$ for r

33. $V = \dfrac{1}{3}Bh$ for h

34. $P = 2l + 2w$ for w

35. $A = \dfrac{B + C}{2}$ for B

36. $T = D + pm$ for m

37. What is 8% of 120?

38. 90 is 45% of what?

39. 36 is what percent of 75?

40. If 6 is increased to 12, the increase is what percent of the original number?

41. If 5 is decreased to 3, the decrease is what percent of the original number?

42. A college that had 40 students for each lecture course increased the number to 45 students. What is the percent increase in the number of students in a lecture course?

43. Consider the following statement:

> My portfolio fell 10% last year, but then it rose 10% this year, so at least I recouped my losses.

Is this statement true? In particular, suppose you invested $10,000 in the stock market last year. How much money would be left in your portfolio with a 10% fall and then a 10% rise? If there is a loss, what is the percent decrease in your portfolio?

44. The radius is one of two bones that connect the elbow and the wrist. The formula $r = \dfrac{h}{7}$ models the length of a woman's radius, r, in inches, and her height, h, in inches.

a. Solve the formula for h.

b. Use the formula in part (a) to find a woman's height if her radius is 9 inches long.

45. Every day, the average U.S. household uses 91 gallons of water flushing toilets. The circle graph shows that this represents 26% of the total number of gallons of water used per day. How many gallons of water does the average U.S. household use per day?

Where United States Households Use Water

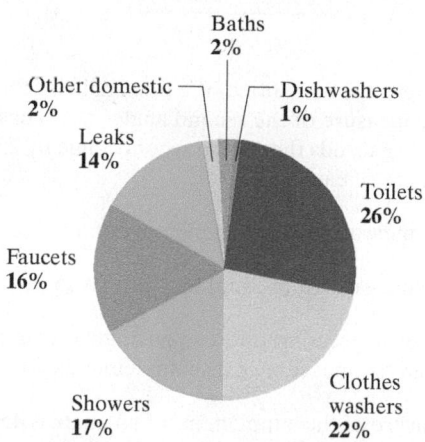

Source: American Water Works Association

2.5 *In Exercises 46–53, use the five-step strategy to solve each problem.*

46. Six times a number, decreased by 20, is four times the number. Find the number.

47. In 2010, the wealthiest Americans were Bill Gates and Warren Buffett. At that time, Gates's net worth exceeded that of Buffett's by $9 billion. Combined, the two men were worth $99 billion. Determine each man's net worth in 2010. (*Source: Forbes*)

48. Two pages that face each other in a book have 93 as the sum of their page numbers. What are the page numbers?

49. The graph shows the gender breakdown of the U.S. population at each end of the age spectrum.

Gender Breakdown of the American Population

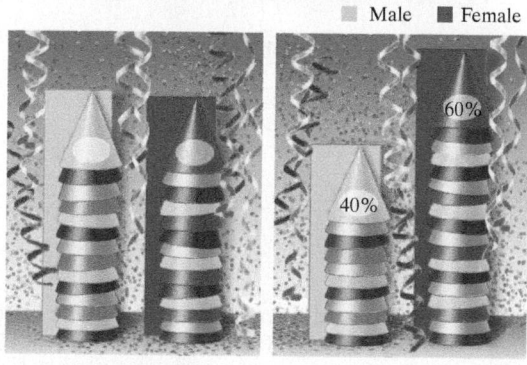

Source: U.S. Census Bureau

For Americans under 20, the percentage of males is greater than the percentage of females and these percents are consecutive odd integers. What percentage of Americans younger than 20 are females and what percentage are males?

50. In 2001, the U.S. defense budget was $316 billion, increasing by approximately $42 billion per year for the period shown by the bar graph. If this trend continues, in how many years after 2001 will the defense budget be $904 billion? In which year will that be?

U.S. Defense Budget

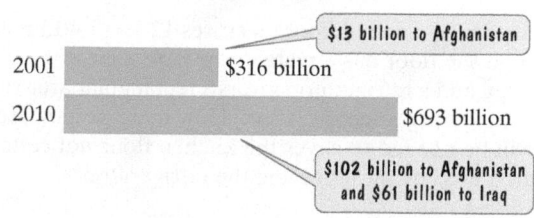

Source: Office of Management and Budget, www.whitehouse.gov <http://www.whitehouse.gov>

51. A bank's total monthly charge for a checking account is $6 plus $0.05 per check. If your total monthly charge is $6.90, how many checks did you write during that month?

52. A rectangular field is three times as long as it is wide. If the perimeter of the field is 400 yards, what are the field's dimensions?

53. After a 25% reduction, you purchase a table for $180. What was the table's price before the reduction?

2.6 *Use a formula for area to find the area of each figure in Exercises 54–57.*

54.

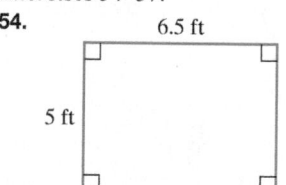

55.

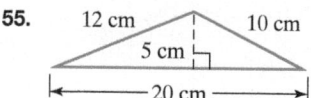

56.

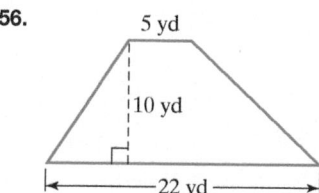

57.

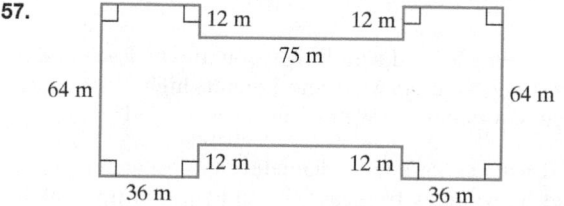

58. Find the circumference and the area of a circle with a diameter of 20 meters. Round answers to the nearest whole number.

59. A sailboat has a triangular sail with an area of 42 square feet and a base that measures 14 feet. Find the height of the sail.

60. A rectangular kitchen floor measures 12 feet by 15 feet. A stove on the floor has a rectangular base measuring 3 feet by 4 feet, and a refrigerator covers a rectangular area of the floor measuring 3 feet by 4 feet. How many square feet of tile will be needed to cover the kitchen floor not counting the area used by the stove and the refrigerator?

61. A yard that is to be covered with mats of grass is shaped like a trapezoid. The bases are 80 feet and 100 feet, and the height is 60 feet. What is the cost of putting the grass mats on the yard if the landscaper charges $0.35 per square foot?

62. Which one of the following is a better buy: a medium pizza with a 14-inch diameter for $6.00 or two small pizzas, each with an 8-inch diameter, for $6.00?

Use a formula for volume to find the volume of each figure in Exercises 63–65. Where applicable, express answers in terms of π. Then round to the nearest whole number.

63.

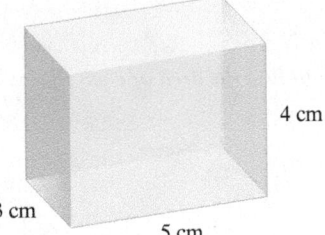

4 cm

3 cm

5 cm

64.

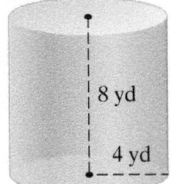

8 yd

4 yd

65.

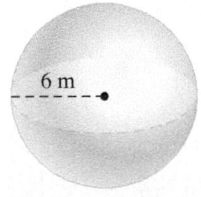

6 m

66. A train is being loaded with freight containers. Each box is 8 meters long, 4 meters wide, and 3 meters high. If there are 50 freight containers, how much space is needed?

67. A cylindrical fish tank has a diameter of 6 feet and a height of 3 feet. How many tropical fish can be put in the tank if each fish needs 5 cubic feet of water?

68. Find the measure of each angle of the triangle shown in the figure.

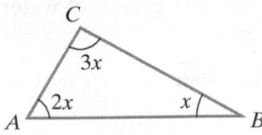

69. In a triangle, the measure of the first angle is 15° more than twice the measure of the second angle. The measure of the third angle exceeds that of the second angle by 25°. What is the measure of each angle?

70. Find the measure of the complement of a 57° angle.

71. Find the measure of the supplement of a 75° angle.

72. How many degrees are there in an angle that measures 25° more than the measure of its complement?

73. The measure of the supplement of an angle is 45° less than four times the measure of the angle. Find the measure of the angle and its supplement.

2.7 *In Exercises 74–75, graph the solution of each inequality on a number line.*

74. $x < -1$

75. $-2 < x \leq 4$

In Exercises 76–77, express the solution set of each inequality in interval notation and graph the interval.

76. $x \geq \dfrac{3}{2}$

77. $x < 0$

In Exercises 78–85, solve each inequality and graph the solution set on a number line. It is not necessary to provide graphs if the inequality has no solution or is true for all real numbers.

78. $2x - 5 < 3$

79. $\dfrac{x}{2} > -4$

80. $3 - 5x \leq 18$

81. $4x + 6 < 5x$

82. $6x - 10 \geq 2(x + 3)$

83. $4x + 3(2x - 7) \leq x - 3$

84. $2(2x + 4) > 4(x + 2) - 6$

85. $-2(x - 4) \leq 3x + 1 - 5x$

86. To pass a course, a student must have an average on three examinations of at least 60. If a student scores 42 and 74 on the first two tests, what must be earned on the third test to pass the course?

87. You can spend at most $2000 for a catered party. The caterer charges a setup fee of $350 and $55 per person. How many people can you invite while staying within your budget?

CHAPTER 2 TEST

CHAPTER
Test Prep
VIDEOS

Step-by-step test solutions are found on the Chapter Test Prep Videos available in MyMathLab® or on YouTube (search "BlitzerIntroAlg" and click on "Channels").

In Exercises 1–7, solve each equation.

1. $4x - 5 = 13$

2. $12x + 4 = 7x - 21$

3. $8 - 5(x - 2) = x + 26$

4. $3(2y - 4) = 9 - 3(y + 1)$

5. $\frac{3}{4}x = -15$

6. $\frac{x}{10} + \frac{1}{3} = \frac{x}{5} + \frac{1}{2}$

7. $9.2x - 80.1 = 21.3x - 19.6$

8. The formula $P = 2.4x + 180$ models U.S. population, P, in millions x years after 1960. How many years after 1960 is the U.S. population expected to reach 324 million? In which year is this expected to occur?

In Exercises 9–10, solve each formula for the specified variable.

9. $V = \pi r^2 h$ for h

10. $l = \dfrac{P - 2w}{2}$ for w

11. What is 6% of 140?

12. 120 is 80% of what?

13. 12 is what percent of 240?

In Exercises 14–18, solve each problem.

14. The product of 5 and a number, decreased by 9, is 306. What is the number?

15. Compared with other major countries, American employees have less vacation time. The bar graph shows the average number of vacation days per person for selected countries.

Average Vacation Time for Selected Countries

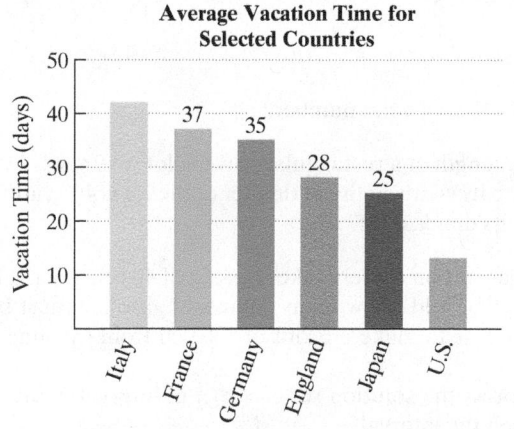

Source: World Development Indicators

The average time Italians spend on vacation exceeds the average American vacation time by 29 days. The combined average vacation time for Americans and Italians is 55 days. On average, how many days do Americans spend on vacation and how many days do Italians spend on vacation?

16. A phone plan has a monthly fee of $15.00 and a rate of $0.05 per minute. How many calling minutes can you use in a month for a total cost, including the $15.00, of $45.00?

17. A rectangular field is twice as long as it is wide. If the perimeter of the field is 450 yards, what are the field's dimensions?

18. After a 20% reduction, you purchase a new Stephen King novel for $28. What was the book's price before the reduction?

In Exercises 19–21, find the area of each figure.

19.

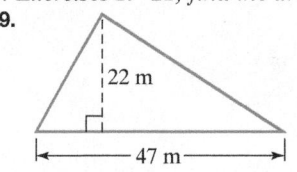

20.

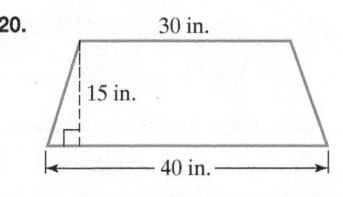

21.

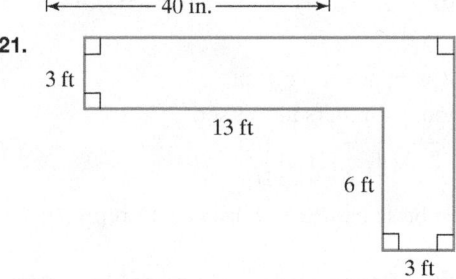

In Exercises 22–23, find the volume of each figure. Where applicable, express answers in terms of π. Then round to the nearest whole number.

22.

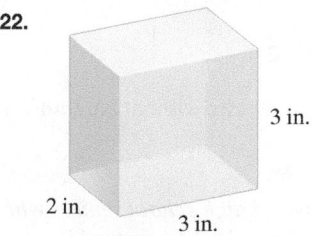

23.

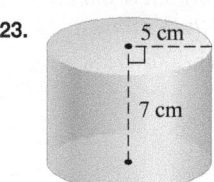

24. What will it cost to cover a rectangular floor measuring 40 feet by 50 feet with square tiles that measure 2 feet on each side if a package of 10 tiles costs $13 per package?

25. A sailboat has a triangular sail with an area of 56 square feet and a base that measures 8 feet. Find the height of the sail.

26. In a triangle, the measure of the first angle is three times that of the second angle. The measure of the third angle is 30° less than the measure of the second angle. What is the measure of each angle?

27. How many degrees are there in an angle that measures $16°$ more than the measure of its complement?

In Exercises 28–29, express the solution set of each inequality in interval notation and graph the interval.

28. $x > -2$ **29.** $x \leq 3$

Solve each inequality in Exercises 30–32 and graph the solution set on a number line.

30. $\dfrac{x}{2} < -3$

31. $6 - 9x \geq 33$

32. $4x - 2 > 2(x + 6)$

33. A student has grades on three examinations of 76, 80, and 72. What must the student earn on a fourth examination to have an average of at least 80?

34. The length of a rectangle is 20 inches. For what widths is the perimeter greater than 56 inches?

CUMULATIVE REVIEW EXERCISES (CHAPTERS 1–2)

In Exercises 1–3, perform the indicated operation or operations.

1. $-8 - (12 - 16)$ **2.** $(-3)(-2) + (-2)(4)$

3. $(8 - 10)^3(7 - 11)^2$

4. Simplify: $2 - 5[x + 3(x + 7)]$.

5. List all the rational numbers in this set:

$$\left\{-4, -\frac{1}{3}, 0, \sqrt{2}, \sqrt{4}, \frac{\pi}{2}, 1063\right\}.$$

6. Write as an algebraic expression, using x to represent the number:

The quotient of 5 and a number, decreased by 2 more than the number.

7. Insert either $<$ or $>$ in the shaded area to make a true statement:

$$-10,000 \;\;\blacksquare\;\; -2.$$

8. Use the distributive property to rewrite without parentheses:

$$6(4x - 1 - 5y).$$

The bar graph shows the percentage of high school seniors who used alcohol during the 30 days prior to being surveyed for the University of Michigan's Monitoring the Future study.

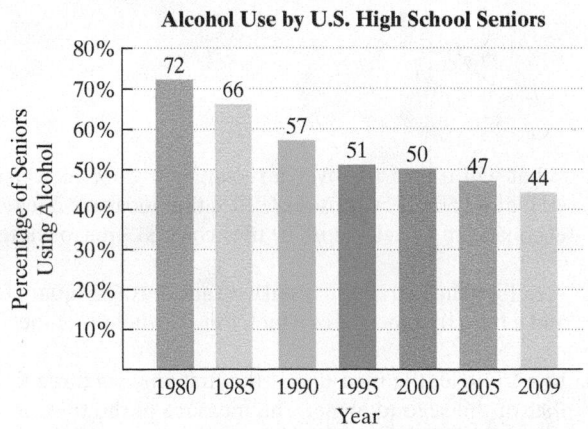

Alcohol Use by U.S. High School Seniors

Source of data: "Marijuana use is rising; ecstasy use is begining to rise; and alcohol use is declining among U.S. teens," December 14, 2010, University of Michigan

The mathematical model

$$A = -0.9n + 69$$

describes the percentage of seniors who used alcohol, A, n years after 1980. Use this information to solve Exercises 9–10.

9. Does the formula underestimate or overestimate the percentage of seniors who used alcohol in 2000? By how much?

10. If trends shown by the formula continue, when will only 33% of high school seniors use alcohol?

In Exercises 11–12, solve each equation.

11. $5 - 6(x + 2) = x - 14$

12. $\dfrac{x}{5} - 2 = \dfrac{x}{3}$

13. Solve for A: $V = \dfrac{1}{3}Ah$.

14. 48 is 30% of what number?

15. The length of a rectangular parking lot is 10 yards less than twice its width. If the perimeter of the lot is 400 yards, what are its dimensions?

16. A gas station owner makes a profit of 40 cents per gallon of gasoline sold. How many gallons of gasoline must be sold in a year to make a profit of $30,000 from gasoline sales?

17. Express the solution set of $x \leq \frac{1}{2}$ in interval notation and graph the interval.

Solve each inequality in Exercises 18–19 and graph the solution set on a number line.

18. $3 - 3x > 12$

19. $5 - 2(3 - x) \leq 2(2x + 5) + 1$

20. You take a summer job selling medical supplies. You are paid $600 per month plus 4% of the sales price of all the supplies you sell. If you want to earn more than $2500 per month, what value of medical supplies must you sell?

Linear Equations in Two Variables

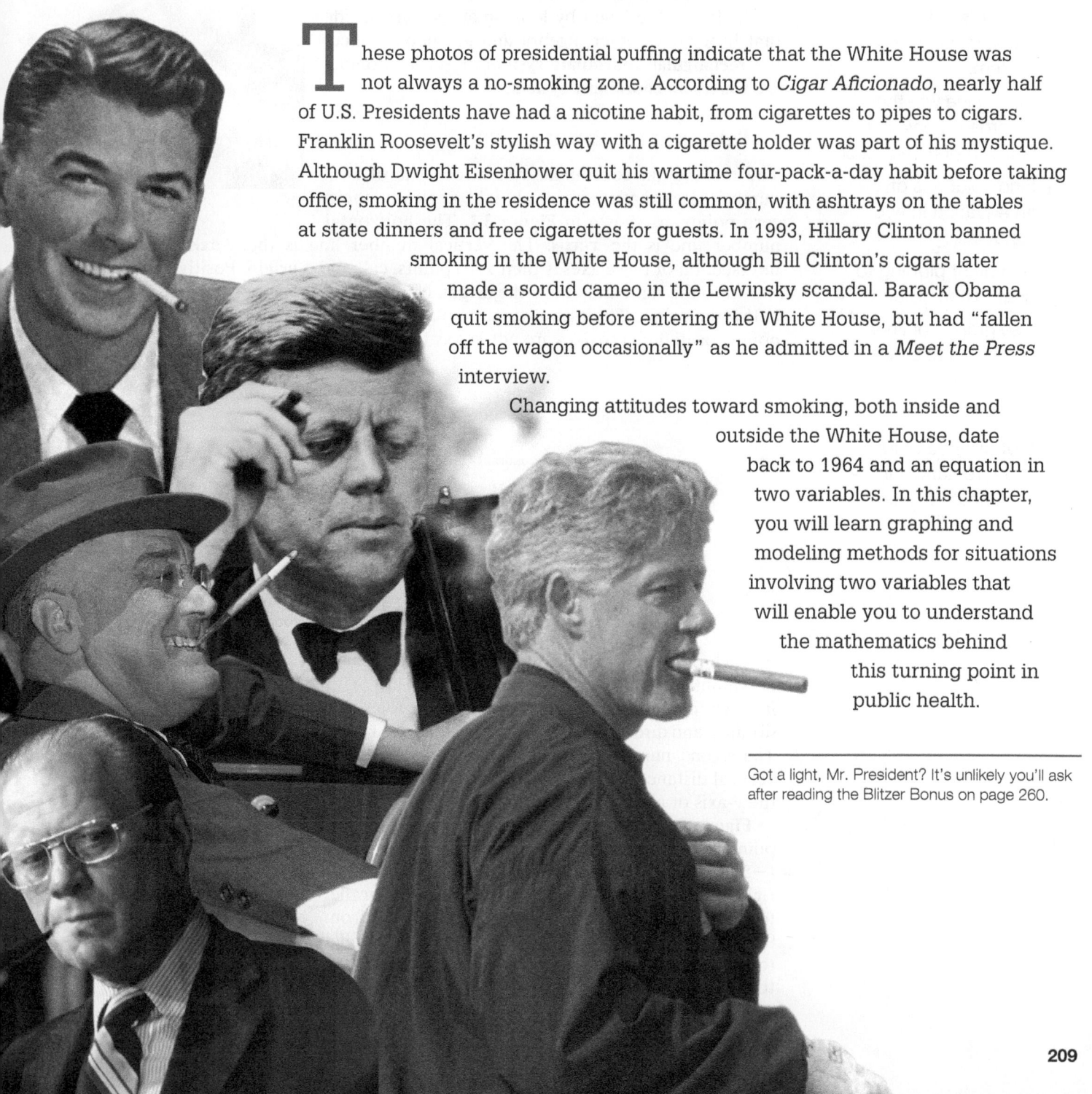

These photos of presidential puffing indicate that the White House was not always a no-smoking zone. According to *Cigar Aficionado*, nearly half of U.S. Presidents have had a nicotine habit, from cigarettes to pipes to cigars. Franklin Roosevelt's stylish way with a cigarette holder was part of his mystique. Although Dwight Eisenhower quit his wartime four-pack-a-day habit before taking office, smoking in the residence was still common, with ashtrays on the tables at state dinners and free cigarettes for guests. In 1993, Hillary Clinton banned smoking in the White House, although Bill Clinton's cigars later made a sordid cameo in the Lewinsky scandal. Barack Obama quit smoking before entering the White House, but had "fallen off the wagon occasionally" as he admitted in a *Meet the Press* interview.

Changing attitudes toward smoking, both inside and outside the White House, date back to 1964 and an equation in two variables. In this chapter, you will learn graphing and modeling methods for situations involving two variables that will enable you to understand the mathematics behind this turning point in public health.

Got a light, Mr. President? It's unlikely you'll ask after reading the Blitzer Bonus on page 260.

SECTION

3.1

Graphing Linear Equations in Two Variables

Objectives

1 Plot ordered pairs in the rectangular coordinate system.

2 Find coordinates of points in the rectangular coordinate system.

3 Determine whether an ordered pair is a solution of an equation.

4 Find solutions of an equation in two variables.

5 Use point plotting to graph linear equations.

6 Use graphs of linear equations to solve problems.

Despite high interest rates, fees, and penalties, American consumers are paying more with credit cards, and less with checks and cash. In this section, we will use geometric figures to visualize mathematical models describing our primary spending methods. The idea of visualizing equations as geometric figures was developed by the French philosopher and mathematician René Descartes (1596–1650). We begin by looking at Descartes's idea that brought together algebra and geometry, called the **rectangular coordinate system** or (in his honor) the **Cartesian coordinate system**.

Points and Ordered Pairs

The rectangular coordinate system consists of two number lines that intersect at right angles at their zero points, as shown in **Figure 3.1**. The horizontal number line is the **x-axis**. The vertical number line is the **y-axis**. The point of intersection of these axes is their zero points, called the **origin**. Positive numbers are shown to the right and above the origin. Negative numbers are shown to the left and below the origin. The axes divide the plane into four regions, called **quadrants**. The points located on the axes are not in any quadrant.

1 Plot ordered pairs in the rectangular coordinate system.

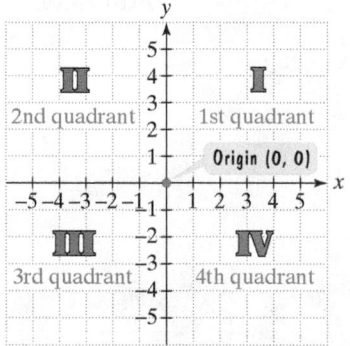

Figure 3.1 The rectangular coordinate system

Each point in the rectangular coordinate system corresponds to an **ordered pair** of real numbers, (x, y). Examples of such pairs are $(4, 2)$ and $(-5, -3)$. The first number in each pair, called the **x-coordinate**, denotes the distance and direction from the origin along the x-axis. The second number, called the **y-coordinate**, denotes vertical distance and direction along a line parallel to the y-axis or along the y-axis itself.

Figure 3.2 shows how we **plot**, or locate, the points corresponding to the ordered pairs $(4, 2)$ and $(-5, -3)$. We plot $(4, 2)$ by going 4 units from 0 to the right along the x-axis. Then we go 2 units up parallel to the y-axis. We plot $(-5, -3)$ by going 5 units from 0 to the left along the x-axis and 3 units down parallel to the y-axis. The phrase "the point corresponding to the ordered pair $(-5, -3)$" is often abbreviated as "the point $(-5, -3)$."

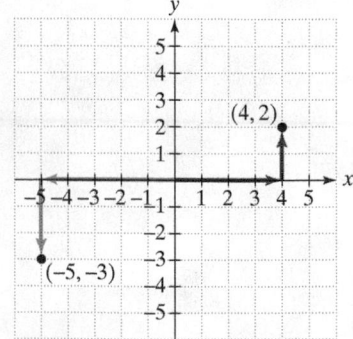

Figure 3.2 Plotting $(4, 2)$ and $(-5, -3)$

EXAMPLE 1   Plotting Points in the Rectangular Coordinate System

Plot the points: $A(-3, 5)$, $B(2, -4)$, $C(5, 0)$, $D(-5, -2)$, $E(0, 4)$, and $F(0, 0)$.

Solution   See **Figure 3.3**. We move from the origin and plot the points in the following way:

$A(-3, 5)$:	3 units left, 5 units up
$B(2, -4)$:	2 units right, 4 units down
$C(5, 0)$:	5 units right, 0 units up or down
$D(-5, -2)$:	5 units left, 2 units down
$E(0, 4)$:	0 units right or left, 4 units up
$F(0, 0)$:	0 units right or left, 0 units up or down

Notice that the origin is represented by (0, 0).

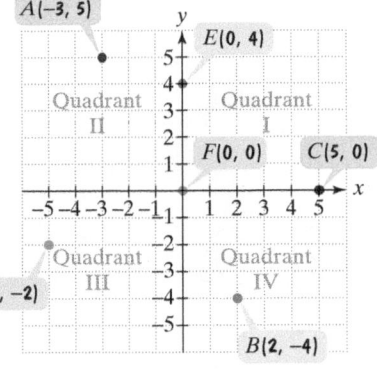

Figure 3.3 Plotting points

Great Question!

If a point is located on an axis, is one of its coordinates always zero?

Any point on the x-axis, such as $(5, 0)$, has a y-coordinate of 0.
Any point on the y-axis, such as $(0, 4)$, has an x-coordinate of 0.

The phrase *ordered pair* is used because **order is important**. For example, the points $(2, 5)$ and $(5, 2)$ are not the same. To plot $(2, 5)$, move 2 units right and 5 units up. To plot $(5, 2)$, move 5 units right and 2 units up. The points $(2, 5)$ and $(5, 2)$ are in different locations. **The order in which coordinates appear makes a difference in a point's location.**

✓ **CHECK POINT 1**   Plot the points: $A(-2, 4)$, $B(4, -2)$, $C(-3, 0)$, and $D(0, -3)$.

2 Find coordinates of points in the rectangular coordinate system.

In the rectangular coordinate system, each ordered pair corresponds to exactly one point. Example 2 illustrates that each point in the rectangular coordinate system corresponds to exactly one ordered pair.

EXAMPLE 2   Finding Coordinates of Points

Determine the coordinates of points A, B, C, and D shown in **Figure 3.4**.

Solution

Point	Position from the Origin	Coordinates
A	6 units left, 0 units up or down	$(-6, 0)$
B	0 units right or left, 2 units up	$(0, 2)$
C	2 units right, 0 units up or down	$(2, 0)$
D	4 units right, 2 units down	$(4, -2)$

✓ **CHECK POINT 2**   Determine the coordinates of points E, F, and G shown in **Figure 3.4**.

Figure 3.4 Finding coordinates of points

3 Determine whether an ordered pair is a solution of an equation.

Solutions of Equations in Two Variables

The rectangular coordinate system allows us to visualize relationships between two variables by connecting any equation in two variables with a geometric figure. Consider, for example, the following equation in two variables:

$$x + y = 10.$$

The sum of two numbers, x and y, is 10.

Many pairs of numbers fit the description in the voice balloon, such as $x = 1$ and $y = 9$, or $x = 3$ and $y = 7$. The phrase "$x = 1$ and $y = 9$" is abbreviated using the ordered pair (1, 9). Similarly, the phrase "$x = 3$ and $y = 7$" is abbreviated using the ordered pair (3, 7).

A **solution of an equation in two variables**, x and y, is an ordered pair of real numbers with the following property: When the x-coordinate is substituted for x and the y-coordinate is substituted for y in the equation, we obtain a true statement. For example, (1, 9) is a solution of the equation $x + y = 10$. When 1 is substituted for x and 9 is substituted for y, we obtain the true statement $1 + 9 = 10$, or $10 = 10$. Because there are infinitely many pairs of numbers that have a sum of 10, the equation $x + y = 10$ has infinitely many solutions. Each ordered-pair solution is said to **satisfy** the equation. Thus, (1, 9) satisfies the equation $x + y = 10$.

EXAMPLE 3 Deciding Whether an Ordered Pair Satisfies an Equation

Determine whether each ordered pair is a solution of the equation

$$x - 4y = 14:$$

a. $(2, -3)$ **b.** $(12, 1)$.

Solution

a. To determine whether $(2, -3)$ is a solution of the equation, we substitute 2 for x and -3 for y.

$x - 4y = 14$	This is the given equation.
$2 - 4(-3) \stackrel{?}{=} 14$	Substitute 2 for x and −3 for y.
$2 - (-12) \stackrel{?}{=} 14$	Multiply: 4(−3) = −12.
$14 = 14$	Subtract: 2 − (−12) = 2 + 12 = 14.

This statement is true.

Because we obtain a true statement, we conclude that $(2, -3)$ is a solution of the equation $x - 4y = 14$. Thus, $(2, -3)$ satisfies the equation.

b. To determine whether $(12, 1)$ is a solution of the equation, we substitute 12 for x and 1 for y.

$x - 4y = 14$	This is the given equation.
$12 - 4(1) \stackrel{?}{=} 14$	Substitute 12 for x and 1 for y.
$12 - 4 \stackrel{?}{=} 14$	Multiply: 4(1) = 4.
$8 = 14$	Subtract: 12 − 4 = 8.

This statement is false.

Because we obtain a false statement, we conclude that $(12, 1)$ is not a solution of $x - 4y = 14$. The ordered pair $(12, 1)$ does not satisfy the equation. ∎

✓ **CHECK POINT 3**   Determine whether each ordered pair is a solution of the equation $x - 3y = 9$:

a. $(3, -2)$ **b.** $(-2, 3)$.

In this chapter, we will use x and y to represent the variables of an equation in two variables. However, any two letters can be used. Solutions are still ordered pairs. Although it depends on the role variables play in equations, the first number in an ordered pair often replaces the variable that occurs first alphabetically. The second number in an ordered pair often replaces the variable that occurs last alphabetically.

4 Find solutions of an equation in two variables.

How do we find ordered pairs that are solutions of an equation in two variables, x and y?

- Select a value for one of the variables.
- Substitute that value into the equation and find the corresponding value of the other variable.
- Use the values of the two variables to form an ordered pair (x, y). This pair is a solution of the equation.

EXAMPLE 4   Finding Solutions of an Equation

Find five solutions of

$$y = 2x - 1.$$

Select integers for x, starting with -2 and ending with 2.

Solution   We organize the process of finding solutions in the following table of values.

Start with these values of x.

Substitute x into $y = 2x - 1$ and compute y.

Use values for x and y to form an ordered-pair solution.

Any numbers can be selected for x. There is nothing special about integers from **-2 to 2**, inclusive. We chose these values to include two negative numbers, 0, and two positive numbers. We also wanted to keep the resulting computations for y relatively simple.

x	$y = 2x - 1$	(x, y)
-2	$y = 2(-2) - 1 = -4 - 1 = -5$	$(-2, -5)$
-1	$y = 2(-1) - 1 = -2 - 1 = -3$	$(-1, -3)$
0	$y = 2 \cdot 0 - 1 = 0 - 1 = -1$	$(0, -1)$
1	$y = 2 \cdot 1 - 1 = 2 - 1 = 1$	$(1, 1)$
2	$y = 2 \cdot 2 - 1 = 4 - 1 = 3$	$(2, 3)$

Look at the ordered pairs in the last column. Five solutions of $y = 2x - 1$ are $(-2, -5), (-1, -3), (0, -1), (1, 1),$ and $(2, 3)$. ∎

✓ **CHECK POINT 4**   Find five solutions of $y = 3x + 2$. Select integers for x, starting with -2 and ending with 2.

---

5 Use point plotting to graph linear equations.

Graphing Linear Equations in the Form $y = mx + b$

In Example 4, we found five solutions of $y = 2x - 1$. We can generate as many ordered-pair solutions as desired to $y = 2x - 1$ by substituting numbers for x and then finding the corresponding values for y. The **graph of the equation** is the set of all points whose coordinates satisfy the equation.

One method for graphing an equation such as $y = 2x - 1$ is the **point-plotting method**.

The Point-Plotting Method for Graphing an Equation in Two Variables

1. Find several ordered pairs that are solutions of the equation.
2. Plot these ordered pairs as points in the rectangular coordinate system.
3. Connect the points with a smooth curve or line.

EXAMPLE 5   Graphing an Equation Using the Point-Plotting Method

Graph the equation: $y = 3x$.

Solution

Step 1. Find several ordered pairs that are solutions of the equation. Because there are infinitely many solutions, we cannot list them all. To find a few solutions of the equation, we select integers for x, starting with -2 and ending with 2.

Start with these values of x.	Substitute x into $y = 3x$ and compute y.	These are solutions of $y = 3x$.
x	$y = 3x$	(x, y)
-2	$y = 3(-2) = -6$	$(-2, -6)$
-1	$y = 3(-1) = -3$	$(-1, -3)$
0	$y = 3 \cdot 0 = 0$	$(0, 0)$
1	$y = 3 \cdot 1 = 3$	$(1, 3)$
2	$y = 3 \cdot 2 = 6$	$(2, 6)$

Step 2. Plot these ordered pairs as points in the rectangular coordinate system. The five ordered pairs in the table of values are plotted in **Figure 3.5(a)**.

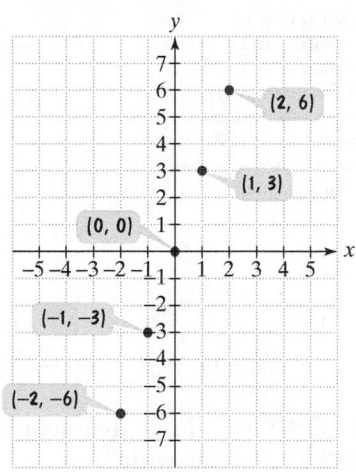

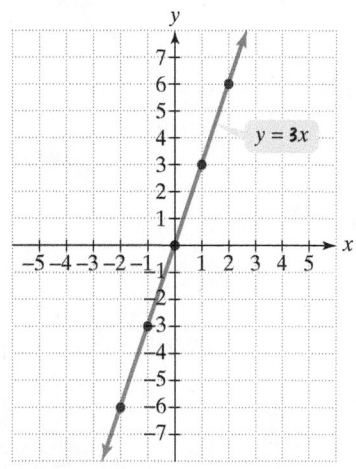

Figure 3.5(a) A few solutions of $y = 3x$ plotted as points

Figure 3.5(b) The graph of $y = 3x$

Step 3. Connect the points with a smooth curve or line. The points lie along a straight line. The graph of $y = 3x$ is shown in **Figure 3.5(b)**. The arrows on both ends of the line indicate that it extends indefinitely in both directions. ■

✓ **CHECK POINT 5**   Graph the equation: $y = 2x$.

Equations like $y = 3x$ and $y = 2x$ are called **linear equations in two variables** because the graph of each equation is a line. Any equation that can be written in the form $y = mx + b$, where m and b are constants, is a linear equation in two variables. Here are examples of linear equations in two variables:

$$y = 3x \qquad\qquad y = 3x - 2$$
$$\text{or} \quad y = 3x + 0 \qquad\qquad \text{or} \quad y = 3x + (-2).$$

> This is in the form of $y = mx + b$, with $m = 3$ and $b = 0$.

> This is in the form of $y = mx + b$, with $m = 3$ and $b = -2$.

Great Question!

What's the big deal about the constants m and b in the linear equation $y = mx + b$?

In Section 3.4, you will see how these numbers affect the equation's graph in special ways. The number m determines the line's steepness. The number b determines the point where the line crosses the y-axis. Any real numbers can be used for m and b.

Can you guess how the graph of the linear equation $y = 3x - 2$ compares with the graph of $y = 3x$? In Example 5, we graphed $y = 3x$. Now, let's graph the equation $y = 3x - 2$.

EXAMPLE 6   Graphing a Linear Equation in Two Variables

Graph the equation: $y = 3x - 2$.

Solution

Step 1. Find several ordered pairs that are solutions of the equation. To find a few solutions, we select integers for x, starting with -2 and ending with 2.

> Start with x.

> Compute y.

> Form the ordered pair (x, y).

x	$y = 3x - 2$	(x, y)
-2	$y = 3(-2) - 2 = -6 - 2 = -8$	$(-2, -8)$
-1	$y = 3(-1) - 2 = -3 - 2 = -5$	$(-1, -5)$
0	$y = 3 \cdot 0 - 2 = 0 - 2 = -2$	$(0, -2)$
1	$y = 3 \cdot 1 - 2 = 3 - 2 = 1$	$(1, 1)$
2	$y = 3 \cdot 2 - 2 = 6 - 2 = 4$	$(2, 4)$

Step 2. Plot these ordered pairs as points in the rectangular coordinate system. The five ordered pairs in the table of values are plotted in **Figure 3.6(a)**.

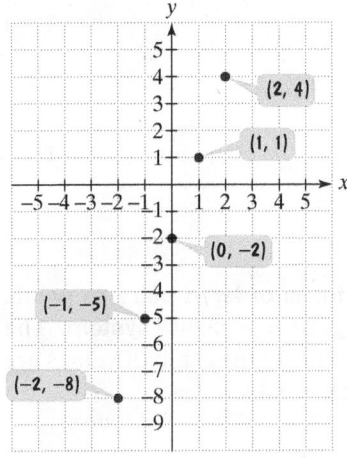

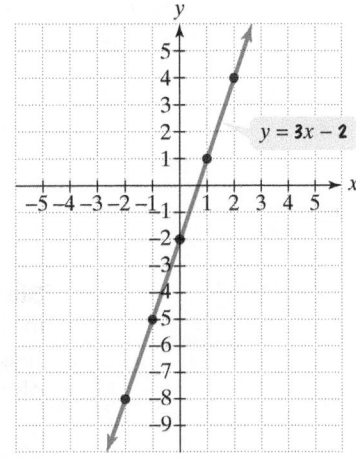

Figure 3.6(a) A few solutions of $y = 3x - 2$ plotted as points

Figure 3.6(b) The graph of $y = 3x - 2$

Step 3. Connect the points with a smooth curve or line. The points lie along a straight line. The graph of $y = 3x - 2$ is shown in **Figure 3.6(b)**. ∎

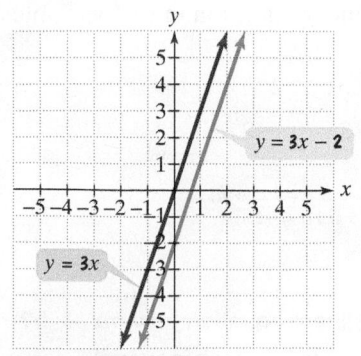

Figure 3.7

Now we are ready to compare the graphs of $y = 3x - 2$ and $y = 3x$. The graphs of both linear equations are shown in the same rectangular coordinate system in **Figure 3.7**. Can you see that the blue graph of $y = 3x - 2$ is parallel to the red graph of $y = 3x$ and 2 units lower than the graph of $y = 3x$? Instead of crossing the y-axis at $(0, 0)$, the graph of $y = 3x - 2$ crosses the y-axis at $(0, -2)$.

Comparing Graphs of Linear Equations

If the value of m does not change,

- The graph of $y = mx + b$ is the graph of $y = mx$ shifted b units up when b is a positive number.
- The graph of $y = mx + b$ is the graph of $y = mx$ shifted $|b|$ units down when b is a negative number.

✓ **CHECK POINT 6** Graph the equation: $y = 2x - 2$.

EXAMPLE 7 Graphing a Linear Equation in Two Variables

Graph the equation: $y = \dfrac{2}{3}x + 1$.

Solution

Step 1. Find several ordered pairs that are solutions of the equation. Notice that m, the coefficient of x, is $\frac{2}{3}$. When m is a fraction, we will select values of x that are multiples of the denominator. In this way, we can avoid values of y that are fractions. Because the denominator of $\frac{2}{3}$ is 3, we select multiples of 3 for x. Let's use $-6, -3, 0, 3,$ and 6.

Start with multiples of 3 for x.	Compute y.	Form the ordered pair (x, y).
x	$y = \frac{2}{3}x + 1$	(x, y)
-6	$y = \frac{2}{3}(-6) + 1 = -4 + 1 = -3$	$(-6, -3)$
-3	$y = \frac{2}{3}(-3) + 1 = -2 + 1 = -1$	$(-3, -1)$
0	$y = \frac{2}{3} \cdot 0 + 1 = 0 + 1 = 1$	$(0, 1)$
3	$y = \frac{2}{3} \cdot 3 + 1 = 2 + 1 = 3$	$(3, 3)$
6	$y = \frac{2}{3} \cdot 6 + 1 = 4 + 1 = 5$	$(6, 5)$

Step 2. Plot these ordered pairs as points in the rectangular coordinate system. The five ordered pairs in the table of values are plotted in **Figure 3.8**.

Step 3. Connect the points with a smooth curve or line. The points lie along a straight line. The graph of $y = \frac{2}{3}x + 1$ is shown in **Figure 3.8**.

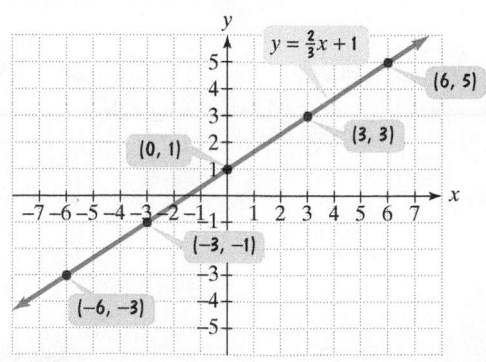

Figure 3.8 The graph of $y = \frac{2}{3}x + 1$ ■

✓ **CHECK POINT 7**   Graph the equation: $y = \frac{1}{2}x + 2$.

---

6 Use graphs of linear equations to solve problems.

Applications

Part of the beauty of the rectangular coordinate system is that it allows us to "see" mathematical models and visualize the solution to a problem. This idea is demonstrated in Example 8.

EXAMPLE 8   Graphing a Mathematical Model

The graph in **Figure 3.9** shows that over time fewer consumers are paying with checks and cash, and more are paying with plastic.

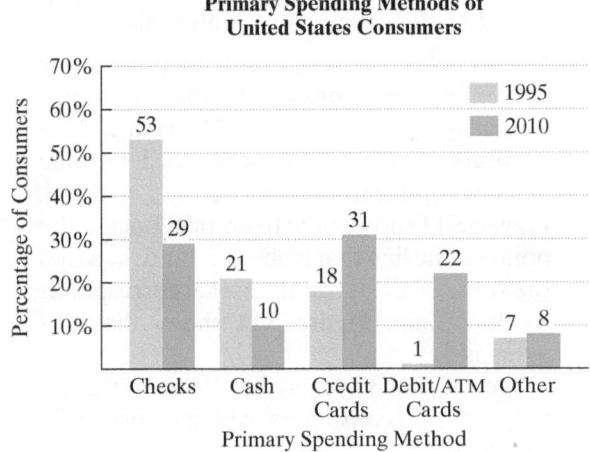

Figure 3.9

Source: Mintel/Visa USA Research Services

The mathematical model

$$C = -1.6n + 53$$

describes the percentage of consumers, C, who paid primarily with checks n years after 1995.

a. Let $n = 0, 5, 10, 15,$ and 20. Make a table of values showing five solutions of the equation.

b. Graph the formula in a rectangular coordinate system.

c. Use the graph to estimate the percentage of consumers who will pay primarily by check in 2025.

d. Use the formula to project the percentage of consumers who will pay primarily by check in 2025.

Solution

a. The table of values shows five solutions of $C = -1.6n + 53$.

Start with these values of n.	Substitute n into $C = -1.6n + 53$ and compute C.	Form the ordered pair (n, C).
n	**$C = -1.6n + 53$**	**(n, C)**
0	$C = -1.6(0) + 53 = 0 + 53 = 53$	$(0, 53)$
5	$C = -1.6(5) + 53 = -8 + 53 = 45$	$(5, 45)$
10	$C = -1.6(10) + 53 = -16 + 53 = 37$	$(10, 37)$
15	$C = -1.6(15) + 53 = -24 + 53 = 29$	$(15, 29)$
20	$C = -1.6(20) + 53 = -32 + 53 = 21$	$(20, 21)$

b. Now we are ready to graph $C = -1.6n + 53$. Because of the letters used to represent the variables, we label the horizontal axis n and the vertical axis C. We plot the five ordered pairs from the table of values and connect them with a line. The graph of $C = -1.6n + 53$ is shown in **Figure 3.10**.

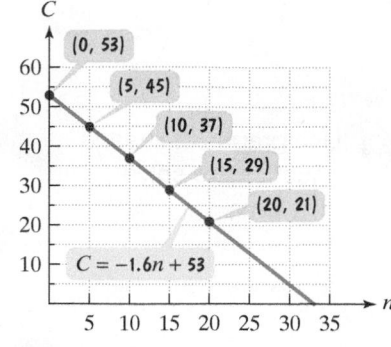

Figure 3.10

As you examine **Figure 3.10**, are you wondering why we did not extend the graph into the other quadrants of the rectangular coordinate system? Keep in mind that n represents the number of years after 1995 and C represents the percentage of consumers paying by check. Only nonnegative values of each variable are meaningful. Thus, when drawing the graph, we started on the positive portion of the vertical axis and ended on the positive horizontal axis.

c. Now we can use the graph to project the percentage of consumers who will pay primarily by check in 2025. The year 2025 is 30 years after 1995. We need to determine what second coordinate is paired with 30. **Figure 3.11** shows how to do this. Locate the point on the line that is above 30 and then find the value on the vertical axis that corresponds to that point. **Figure 3.11** shows that this value is 5.

We see that 30 years after 1995, or in 2025, only 5% of consumers will pay primarily by check.

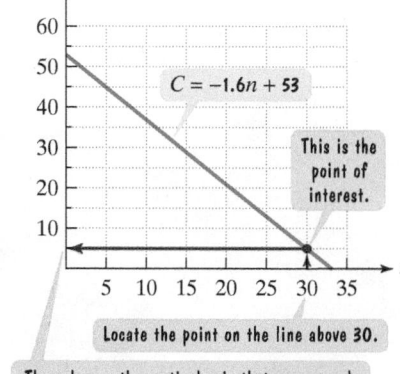

Figure 3.11

d. Now let's use the mathematical model to verify our graphical observation in part (c).

$$C = -1.6n + 53 \qquad \text{This is the given mathematical model.}$$
$$C = -1.6(30) + 53 \qquad \text{Substitute 30 for } n.$$
$$C = -48 + 53 \qquad \text{Multiply: } -1.6(30) = -48.$$
$$C = 5 \qquad \text{Add.}$$

This verifies that 30 years after 1995, in 2025, 5% of consumers will pay primarily by check. ■

✓ **CHECK POINT 8** The mathematical model

$$D = 1.4n + 1$$

describes the percentage of consumers, D, who paid primarily with debit cards n years after 1995.

a. Let $n = 0, 5, 10,$ and 15. Make a table of values showing four solutions of the equation.

b. Graph the formula in a rectangular coordinate system.

c. Use the graph to estimate the percentage of consumers who will pay primarily with debit cards in 2015.

d. Use the formula to project the percentage of consumers who will pay primarily by debit cards in 2015.

Using Technology

Graphing calculators or graphing software packages for computers are referred to as **graphing utilities** or graphers. A graphing utility is a powerful tool that quickly generates the graph of an equation in two variables. **Figure 3.12** shows two such graphs for the equations in Examples 5 and 6.

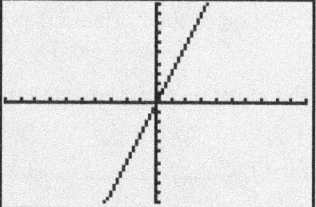

Figure 3.12(a) The graph of $y = 3x$

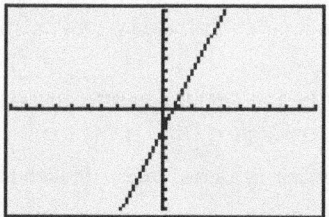

Figure 3.12(b) The graph of $y = 3x - 2$

What differences do you notice between these graphs and the graphs that we drew by hand? They do seem a bit "jittery." Arrows do not appear on both ends of the graphs. Furthermore, numbers are not given along the axes. For both graphs in **Figure 3.12**, the x-axis extends from -10 to 10 and the y-axis also extends from -10 to 10. The distance represented by each consecutive tick mark is one unit. We say that the **viewing rectangle**, or the **viewing window**, is $[-10, 10, 1]$ by $[-10, 10, 1]$.

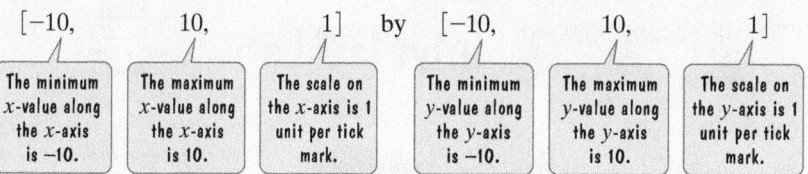

To graph an equation in x and y using a graphing utility, enter the equation and specify the size of the viewing rectangle. The size of the viewing rectangle sets minimum and maximum values for both the x- and y-axes. Enter these values, as well as the values for the distances between consecutive tick marks on the respective axes. The $[-10, 10, 1]$ by $[-10, 10, 1]$ viewing rectangle used in **Figure 3.12** is called the **standard viewing rectangle**.

On most graphing utilities, the display screen is two-thirds as high as it is wide. By using a square setting, you can equally space the x and y tick marks. (This does not occur in the standard viewing rectangle.) Graphing utilities can also *zoom in* and *zoom out*. When you zoom in, you see a smaller portion of the graph, but you do so in greater detail. When you zoom out, you see a larger portion of the graph. Thus, zooming out may help you to develop a better understanding of the overall character of the graph. With practice, you will become more comfortable with graphing equations in two variables using your graphing utility. You will also develop a better sense of the size of the viewing rectangle that will reveal needed information about a particular graph.

Achieving Success

An effective way to understand something is to explain it to someone else. You can do this by using the Writing in Mathematics exercises that ask you to respond with verbal or written explanations. Speaking about a new concept uses a different part of your brain than thinking about the concept. Explaining new ideas verbally will quickly reveal any gaps in your understanding. It will also help you to remember new concepts for longer periods of time.

CONCEPT AND VOCABULARY CHECK

Fill in each blank so that the resulting statement is true.

1. In the rectangular coordinate system, the horizontal number line is called the _____.

2. In the rectangular coordinate system, the vertical number line is called the _____.

3. In the rectangular coordinate system, the point of intersection of the horizontal axis and the vertical axis is called the _____.

4. The axes of the rectangular coordinate system divide the plane into regions, called _____. There are _____ of these regions.

5. The first number in an ordered pair such as (3, 8) is called the _____. The second number in such an ordered pair is called the _____.

6. The ordered pair (1, 3) is a/an _____ of the equation $y = 5x - 2$ because when 1 is substituted for x and 3 is substituted for y, we obtain a true statement. We also say that (1, 3) _____ the equation.

7. Each ordered pair of numbers corresponds to _____ point in the rectangular coordinate system.

8. A linear equation in two variables can be written in the form $y =$ _____, where m and b are constants.

3.1 EXERCISE SET  MyMathLab®

Watch the videos in MyMathLab

Download the MyDashBoard App

Practice Exercises

In Exercises 1–8, plot the given point in a rectangular coordinate system. Indicate in which quadrant each point lies.

1. $(3, 5)$	**2.** $(5, 3)$
3. $(-5, 1)$	**4.** $(1, -5)$
5. $(-3, -1)$	**6.** $(-1, -3)$
7. $(6, -3.5)$	**8.** $(-3.5, 6)$

In Exercises 9–24, plot the given point in a rectangular coordinate system.

9. $(-3, -3)$	**10.** $(-5, -5)$
11. $(-2, 0)$	**12.** $(-5, 0)$
13. $(0, 2)$	**14.** $(0, 5)$
15. $(0, -3)$	**16.** $(0, -5)$
17. $\left(\frac{5}{2}, \frac{7}{2}\right)$	**18.** $\left(\frac{7}{2}, \frac{5}{2}\right)$
19. $\left(-5, \frac{3}{2}\right)$	**20.** $\left(-\frac{9}{2}, -4\right)$
21. $(0, 0)$	**22.** $\left(-\frac{5}{2}, 0\right)$
23. $\left(0, -\frac{5}{2}\right)$	**24.** $\left(0, \frac{7}{2}\right)$

In Exercises 25–32, give the ordered pairs that correspond to the points labeled in the figure.

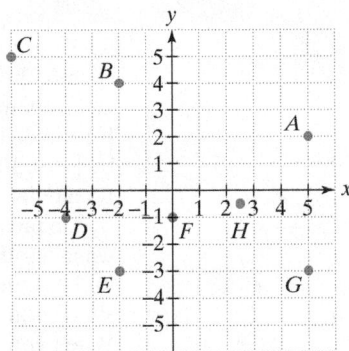

25. A	**26.** B	**27.** C
28. D	**29.** E	**30.** F
31. G	**32.** H	

33. In which quadrants are the y-coordinates positive?

34. In which quadrants are the x-coordinates negative?

35. In which quadrants do the x-coordinates and the y-coordinates have the same sign?

36. In which quadrants do the x-coordinates and the y-coordinates have opposite signs?

In Exercises 37–48, determine whether each ordered pair is a solution of the given equation.

37. $y = 3x$ $(2, 3), (3, 2), (-4, -12)$

38. $y = 4x$ $(3, 12), (12, 3), (-5, -20)$

39. $y = -4x$ $(-5, -20), (0, 0), (9, -36)$

40. $y = -3x$ $(-5, 15), (0, 0), (7, -21)$

41. $y = 2x + 6$ $(0, 6), (-3, 0), (2, -2)$

42. $y = 8 - 4x$ $(8, 0), (16, -2), (3, -4)$

43. $3x + 5y = 15$ $(-5, 6), (0, 5), (10, -3)$

44. $2x - 5y = 0$ $(-2, 0), (-10, 6), (5, 0)$

45. $x + 3y = 0$ $(0, 0), \left(1, \dfrac{1}{3}\right), \left(2, -\dfrac{2}{3}\right)$

46. $x + 5y = 0$ $(0, 0), \left(1, \dfrac{1}{5}\right), \left(2, -\dfrac{2}{5}\right)$

47. $x - 4 = 0$ $(4, 7), (3, 4), (0, -4)$

48. $y + 2 = 0$ $(0, 2), (2, 0), (0, -2)$

In Exercises 49–56, find five solutions of each equation. Select integers for x, starting with −2 and ending with 2. Organize your work in a table of values.

49. $y = 12x$

50. $y = 14x$

51. $y = -10x$

52. $y = -20x$

53. $y = 8x - 5$

54. $y = 6x - 4$

55. $y = -3x + 7$

56. $y = -5x + 9$

In Exercises 57–80, graph each linear equation in two variables. Find at least five solutions in your table of values for each equation.

57. $y = x$

58. $y = x + 1$

59. $y = x - 1$

60. $y = x - 2$

61. $y = 2x + 1$

62. $y = 2x - 1$

63. $y = -x + 2$

64. $y = -x + 3$

65. $y = -3x - 1$

66. $y = -3x - 2$

67. $y = \dfrac{1}{2}x$

68. $y = -\dfrac{1}{2}x$

69. $y = -\dfrac{1}{4}x$

70. $y = \dfrac{1}{4}x$

71. $y = \dfrac{1}{3}x + 1$

72. $y = \dfrac{1}{3}x - 1$

73. $y = -\dfrac{3}{2}x + 1$

74. $y = -\dfrac{3}{2}x + 2$

75. $y = -\dfrac{5}{2}x - 1$

76. $y = -\dfrac{5}{2}x + 1$

77. $y = x + \dfrac{1}{2}$

78. $y = x - \dfrac{1}{2}$

79. $y = 4$, or $y = 0x + 4$

80. $y = 3$, or $y = 0x + 3$

Practice PLUS

In Exercises 81–84, write each sentence as a linear equation in two variables. Then graph the equation.

81. The *y*-variable is 3 more than the *x*-variable.

82. The *y*-variable exceeds the *x*-variable by 4.

83. The *y*-variable exceeds twice the *x*-variable by 5.

84. The *y*-variable is 2 less than 3 times the *x*-variable.

85. At the beginning of a semester, a student purchased eight pens and six pads for a total cost of $14.50.

   a. If *x* represents the cost of one pen and *y* represents the cost of one pad, write an equation in two variables that reflects the given conditions.

   b. If pads cost $0.75 each, find the cost of one pen.

86. A nursery offers a package of three small orange trees and four small grapefruit trees for $22.

   a. If *x* represents the cost of one orange tree and *y* represents the cost of one grapefruit tree, write an equation in two variables that reflects the given conditions.

   b. If a grapefruit tree costs $2.50, find the cost of an orange tree.

Application Exercises

A football is thrown by a quarterback to a receiver. The points in the figure show the height of the football, in feet, above the ground in terms of its distance, in yards, from the quarterback. Use this information to solve Exercises 87–92.

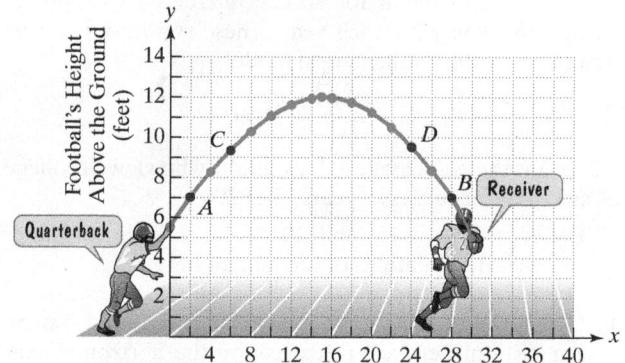

Distance of the Football from the Quarterback (yards)

87. Find the coordinates of point *A*. Then interpret the coordinates in terms of the information given.

(In Exercises 88–92, be sure to refer to the figure at the bottom of the previous page.)

88. Find the coordinates of point *B*. Then interpret the coordinates in terms of the information given.

89. Estimate the coordinates of point *C*.

90. Estimate the coordinates of point *D*.

91. What is the football's maximum height? What is its distance from the quarterback when it reaches its maximum height?

92. What is the football's height when it is caught by the receiver? What is the receiver's distance from the quarterback when he catches the football?

Even as Americans increasingly view a college education as essential for success, many believe that a college education is becoming less available to qualified students. Exercises 93–94 are based on the data displayed by the graph.

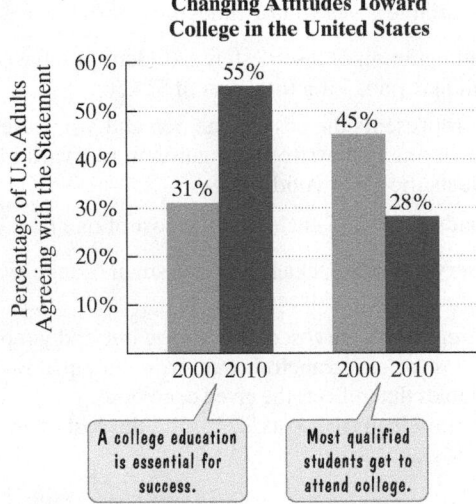

Changing Attitudes Toward College in the United States

Source: Public Agenda

93. The graph shows that in 2000, 31% of U.S. adults viewed a college education as essential for success. For the period from 2000 through 2010, the percentage viewing a college education as essential for success increased on average by approximately 2.4 each year. These conditions can be described by the mathematical model

$$S = 2.4n + 31,$$

where *S* is the percentage of U.S. adults who viewed college as essential for success *n* years after 2000.

a. Let $n = 0, 5, 10, 15,$ and 20. Make a table of values showing five solutions of the equation.

b. Graph the formula in a rectangular coordinate system. Suggestion: Let each tick mark on the horizontal axis, labeled *n*, represent 5 units. Extend the horizontal axis to include $n = 25$. Let each tick mark on the vertical axis, labeled *S*, represent 10 units and extend the axis to include $S = 100$.

c. Use your graph from part (b) to estimate the percentage of U.S. adults who will view college as essential for success in 2018.

d. Use the formula to project the percentage of U.S. adults who will view college as essential for success in 2018.

94. The graph shows that in 2000, 45% of U.S. adults believed that most qualified students get to attend college. For the period from 2000 through 2010, the percentage who believed that a college education is available to most qualified students decreased by approximately 1.7 each year. These conditions can be described by the mathematical model

$$Q = -1.7n + 45,$$

where *Q* is the percentage believing that a college education is available to most qualified students *n* years after 2000.

a. Let $n = 0, 5, 10, 15,$ and 20. Make a table of values showing five solutions of the equation.

b. Graph the formula in a rectangular coordinate system. Suggestion: Let each tick mark on the horizontal axis, labeled *n*, represent 5 units. Extend the horizontal axis to include $n = 25$. Let each tick mark on the vertical axis, labeled *Q*, represent 5 units and extend the axis to include $Q = 50$.

c. Use your graph from part (b) to estimate the percentage of U.S. adults who will believe that a college education is available to most qualified students in 2018.

d. Use the formula to project the percentage of U.S. adults who will believe that a college education is available to most qualified students in 2018.

Writing in Mathematics

95. What is the rectangular coordinate system?

96. Explain how to plot a point in the rectangular coordinate system. Give an example with your explanation.

97. Explain why $(5, -2)$ and $(-2, 5)$ do not represent the same point.

98. Explain how to find the coordinates of a point in the rectangular coordinate system.

99. How do you determine whether an ordered pair is a solution of an equation in two variables, *x* and *y*?

100. Explain how to find ordered pairs that are solutions of an equation in two variables, *x* and *y*.

101. What is the graph of an equation?

102. Explain how to graph an equation in two variables in the rectangular coordinate system.

Critical Thinking Exercises

Make Sense? *In Exercises 103–106, determine whether each statement "makes sense" or "does not make sense" and explain your reasoning.*

103. When I know that an equation's graph is a straight line, I don't need to plot more than two points, although I sometimes plot three just to check that the points line up.

104. The graph that I'm looking at is U-shaped, so its equation cannot be of the form $y = mx + b$.

105. I'm working with a linear equation in two variables and found that $(-2, 2)$, $(0, 0)$, and $(2, 2)$ are solutions.

106. When a real-world situation is modeled with a linear equation in two variables, I can use its graph to predict specific information about the situation.

In Exercises 107–110, determine whether each statement is true or false. If the statement is false, make the necessary change(s) to produce a true statement.

107. The graph of $y = 3x + 1$ is the graph of $y = 2x$ shifted up 1 unit.

108. The graph of any equation in the form $y = mx + b$ passes through the point $(0, b)$.

109. The ordered pair $(3, 4)$ satisfies the equation
$$2y - 3x = -6.$$

110. If $(2, 5)$ satisfies an equation, then $(5, 2)$ also satisfies the equation.

111. a. Graph each of the following points:
$$\left(1, \frac{1}{2}\right), (2, 1), \left(3, \frac{3}{2}\right), (4, 2).$$

Parts (b)–(d) can be answered by changing the sign of one or both coordinates of the points in part (a).

b. What must be done to the coordinates so that the resulting graph is a mirror-image reflection about the y-axis of your graph in part (a)?

c. What must be done to the coordinates so that the resulting graph is a mirror-image reflection about the x-axis of your graph in part (a)?

d. What must be done to the coordinates so that the resulting graph is a straight-line extension of your graph in part (a)?

Technology Exercises

Use a graphing utility to graph each equation in Exercises 112–115 in a standard viewing rectangle, $[-10, 10, 1]$ by $[-10, 10, 1]$. Then use the TRACE *feature to trace along the line and find the coordinates of two points.*

112. $y = 2x - 1$

113. $y = -3x + 2$

114. $y = \frac{1}{2}x$

115. $y = \frac{3}{4}x - 2$

116. Use a graphing utility to verify any five of your hand-drawn graphs in Exercises 57–80. Use an appropriate viewing rectangle and the ZOOM SQUARE feature to make the graph look like the one you drew by hand.

Review Exercises

117. Solve: $3x + 5 = 4(2x - 3) + 7$. (Section 2.3, Example 3)

118. Simplify: $3(1 - 2 \cdot 5) - (-28)$. (Section 1.8, Example 7)

119. Solve for h: $V = \frac{1}{3}Ah$. (Section 2.4, Example 4)

Preview Exercises

Exercises 120–122 will help you prepare for the material covered in the next section. Remember that a solution of an equation in two variables is an ordered pair.

120. Let $y = 0$ and find a solution of $3x - 4y = 24$.

121. Let $x = 0$ and find a solution of $3x - 4y = 24$.

122. Let $x = 0$ and find a solution of $x + 2y = 0$.

SECTION

3.2

Graphing Linear Equations Using Intercepts

Objectives

1 Use a graph to identify intercepts.

2 Graph a linear equation in two variables using intercepts.

3 Graph horizontal or vertical lines.

It's hard to believe that this gas-guzzler, with its huge fins and overstated design, was available in 1957 for approximately $1800. Sadly, its elegance quickly faded, depreciating linearly by $300 per year, often sold for scrap just six years after its glorious emergence from the dealer's showroom.

From the casual observations on the previous page, we can obtain a mathematical model and its graph. The model is

$$y = -300x + 1800.$$

| The car depreciated by $300 per year for x years. | The new car was worth $1800. |

In this model, y is the car's value after x years. **Figure 3.13** shows the equation's graph.

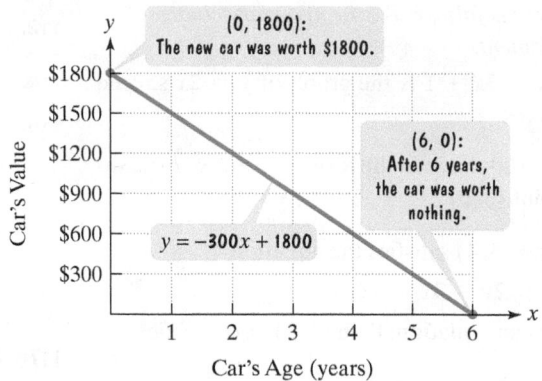

Figure 3.13

Here are some important observations about the model and its graph:

- The points in **Figure 3.13** show where the graph intersects the y-axis and where it intersects the x-axis. These important points are the topic of this section.
- We can rewrite the model, $y = -300x + 1800$, by adding $300x$ to both sides:

$$300x + y = 1800.$$

The form of this equation is $Ax + By = C$.

$$300x + y = 1800$$

| A, the coefficient of x, is 300. | B, the coefficient of y, is 1. | C, the constant on the right, is 1800. |

All equations of the form $Ax + By = C$ are straight lines when graphed as long as A and B are not both zero. To graph linear equations of this form, we will use the intercepts.

1 Use a graph to identify intercepts.

Intercepts

An **x-intercept** of a graph is the x-coordinate of a point where the graph intersects the x-axis. For example, look at the graph of $2x - 4y = 8$ in **Figure 3.14**. The graph crosses the x-axis at $(4, 0)$. Thus, the x-intercept is 4. **The y-coordinate corresponding to an x-intercept is always zero.**

A **y-intercept** of a graph is the y-coordinate of a point where the graph intersects the y-axis. The graph of $2x - 4y = 8$ in **Figure 3.14** shows that the graph crosses the y-axis at $(0, -2)$. Thus, the y-intercept is -2. **The x-coordinate corresponding to a y-intercept is always zero.**

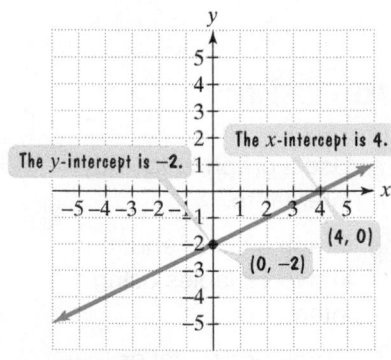

Figure 3.14
The graph of $2x - 4y = 8$

Great Question!

Are single numbers the only way to represent intercepts? Can ordered pairs also be used?

Mathematicians tend to use two ways to describe intercepts. Did you notice that we are using single numbers? If a graph's x-intercept is a, it passes through the point $(a, 0)$. If a graph's y-intercept is b, it passes through the point $(0, b)$.

Some books state that the x-intercept is the *point* $(a, 0)$ and the x-intercept is *at a* on the x-axis. Similarly, the y-intercept is the *point* $(0, b)$ and the y-intercept is *at b* on the y-axis. In these descriptions, the intercepts are the actual points where a graph intersects the axes.

Although we'll describe intercepts as single numbers, we'll immediately state the point on the x- or y-axis that the graph passes through. Here's the important thing to keep in mind:

x-intercept: The corresponding y is 0.

y-intercept: The corresponding x is 0.

EXAMPLE 1   Identifying Intercepts

Identify the x- and y-intercepts.

a.

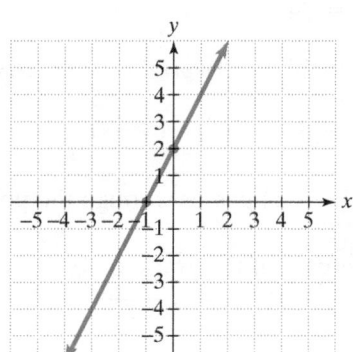

b.

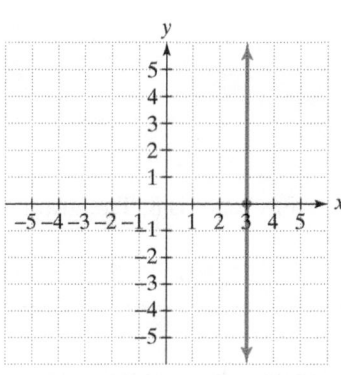

c.

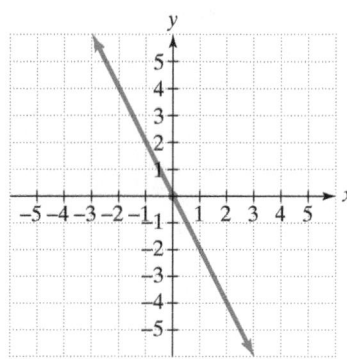

Solution

a. The graph crosses the x-axis at $(-1, 0)$. Thus, the x-intercept is -1. The graph crosses the y-axis at $(0, 2)$. Thus, the y-intercept is 2.

b. The graph crosses the x-axis at $(3, 0)$, so the x-intercept is 3. This vertical line does not cross the y-axis. Thus, there is no y-intercept.

c. This graph crosses the x- and y-axes at the same point, the origin. Because the graph crosses both axes at $(0, 0)$, the x-intercept is 0 and the y-intercept is 0.   ■

✓ **CHECK POINT 1**   Identify the x- and y-intercepts.

a.

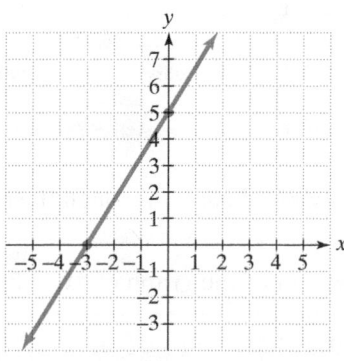

b.

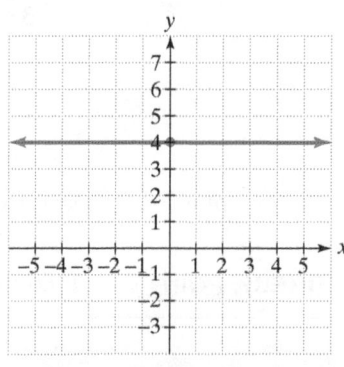

c.

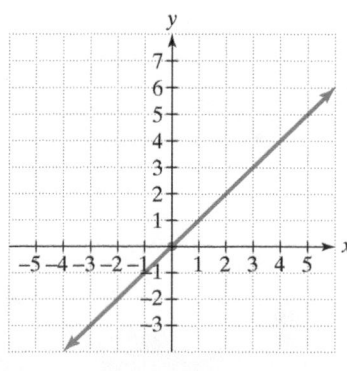

2 Graph a linear equation in two variables using intercepts.

Graphing Using Intercepts

An equation of the form $Ax + By = C$, where A, B, and C are integers, is called the **standard form** of the equation of a line. The equation can be graphed by finding the x- and y-intercepts, plotting the intercepts, and drawing a straight line through these points. How do we find the intercepts of a line, given its equation? Because the y-coordinate of the x-intercept is 0, to find the x-intercept,

* Substitute 0 for y in the equation.
* Solve for x.

Great Question!

Can x or y have exponents in the standard form of the equation of a line?

In the form $Ax + By = C$, the exponent that is understood on both x and y is 1. Neither x nor y can have exponents other than 1.

Forms of Equations That Do Not Represent Lines

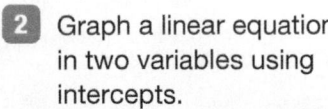

$Ax^2 + By^2 = C \quad Ax + By^2 = C$

Exponents on x and y are not both 1.

EXAMPLE 2   Finding the x-Intercept

Find the x-intercept of the graph of $3x - 4y = 24$.

Solution   To find the x-intercept, let $y = 0$ and solve for x.

$$3x - 4y = 24 \qquad \text{This is the given equation.}$$

$$3x - 4 \cdot 0 = 24 \qquad \text{Let } y = 0.$$

$$3x = 24 \qquad \text{Simplify: } 4 \cdot 0 = 0 \text{ and } 3x - 0 = 3x.$$

$$x = 8 \qquad \text{Divide both sides by 3.}$$

The x-intercept is 8. The graph of $3x - 4y = 24$ passes through the point $(8, 0)$. ∎

✓ **CHECK POINT 2**   Find the x-intercept of the graph of $4x - 3y = 12$.

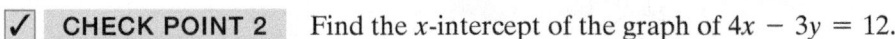

Because the x-coordinate of the y-intercept is 0, to find the y-intercept,

* Substitute 0 for x in the equation.
* Solve for y.

EXAMPLE 3   Finding the y-Intercept

Find the y-intercept of the graph of $3x - 4y = 24$.

Solution   To find the y-intercept, let $x = 0$ and solve for y.

$$3x - 4y = 24 \qquad \text{This is the given equation.}$$

$$3 \cdot 0 - 4y = 24 \qquad \text{Let } x = 0.$$

$$-4y = 24 \qquad \text{Simplify: } 3 \cdot 0 = 0 \text{ and } 0 - 4y = -4y.$$

$$y = -6 \qquad \text{Divide both sides by } -4.$$

The y-intercept is -6. The graph of $3x - 4y = 24$ passes through the point $(0, -6)$. ∎

✓ **CHECK POINT 3**   Find the y-intercept of the graph of $4x - 3y = 12$.

When graphing using intercepts, it is a good idea to use a third point, a checkpoint, before drawing the line. A checkpoint can be obtained by selecting a value for either variable, other than 0, and finding the corresponding value for the other variable. The checkpoint should lie on the same line as the x- and y-intercepts. If it does not, recheck your work and find the error.

Using Intercepts to Graph $Ax + By = C$

1. Find the x-intercept. Let $y = 0$ and solve for x.
2. Find the y-intercept. Let $x = 0$ and solve for y.
3. Find a checkpoint, a third ordered-pair solution.
4. Graph the equation by drawing a line through the three points.

EXAMPLE 4 Using Intercepts to Graph a Linear Equation

Graph: $3x + 2y = 6$.

Solution

Step 1. Find the x-intercept. Let $y = 0$ and solve for x.

$$3x + 2 \cdot 0 = 6 \quad \text{Replace y with 0 in } 3x + 2y = 6.$$

$$3x = 6 \quad \text{Simplify.}$$

$$x = 2 \quad \text{Divide both sides by 3.}$$

The x-intercept is 2, so the line passes through $(2, 0)$.

Step 2. Find the y-intercept. Let $x = 0$ and solve for y.

$$3 \cdot 0 + 2y = 6 \quad \text{Replace x with 0 in } 3x + 2y = 6.$$

$$2y = 6 \quad \text{Simplify.}$$

$$y = 3 \quad \text{Divide both sides by 2.}$$

The y-intercept is 3, so the line passes through $(0, 3)$.

Step 3. Find a checkpoint, a third ordered-pair solution. For our checkpoint, we will let $x = 1$ and find the corresponding value for y.

$$3x + 2y = 6 \quad \text{This is the given equation.}$$

$$3 \cdot 1 + 2y = 6 \quad \text{Substitute 1 for x.}$$

$$3 + 2y = 6 \quad \text{Simplify.}$$

$$2y = 3 \quad \text{Subtract 3 from both sides.}$$

$$y = \frac{3}{2} \quad \text{Divide both sides by 2.}$$

The checkpoint is the ordered pair $\left(1, \dfrac{3}{2}\right)$, or $(1, 1.5)$.

Step 4. Graph the equation by drawing a line through the three points. The three points in **Figure 3.15** lie along the same line. Drawing a line through the three points results in the graph of $3x + 2y = 6$. ∎

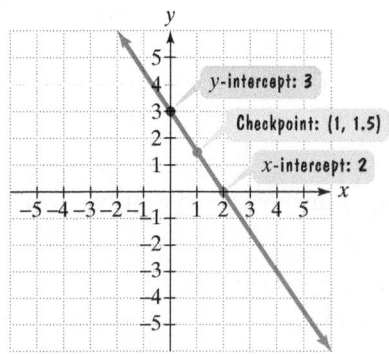

Figure 3.15
The graph of $3x + 2y = 6$

Using Technology

You can use a graphing utility to graph equations of the form $Ax + By = C$. Begin by solving the equation for y. For example, to graph $3x + 2y = 6$, solve the equation for y.

$$3x + 2y = 6 \qquad \text{This is the equation to be graphed.}$$
$$3x - 3x + 2y = -3x + 6 \qquad \text{Subtract 3x from both sides.}$$
$$2y = -3x + 6 \qquad \text{Simplify.}$$
$$\frac{2y}{2} = \frac{-3x + 6}{2} \qquad \text{Divide both sides by 2.}$$
$$y = -\frac{3}{2}x + 3 \qquad \text{Divide each term of } -3x + 6 \text{ by 2.}$$

This is the equation to enter into your graphing utility. The graph of $y = -\frac{3}{2}x + 3$ or, equivalently, $3x + 2y = 6$, is shown below in a $[-6, 6, 1]$ by $[-6, 6, 1]$ viewing rectangle.

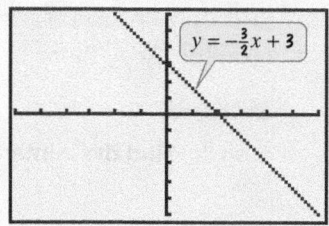

$$y = -\tfrac{3}{2}x + 3$$

✓ **CHECK POINT 4** Graph: $2x + 3y = 6$.

EXAMPLE 5 Using Intercepts to Graph a Linear Equation

Graph: $2x - y = 4$.

Solution

Step 1. Find the x-intercept. Let $y = 0$ and solve for x.

$$2x - 0 = 4 \qquad \text{Replace y with 0 in } 2x - y = 4.$$
$$2x = 4 \qquad \text{Simplify.}$$
$$x = 2 \qquad \text{Divide both sides by 2.}$$

The x-intercept is 2, so the line passes through $(2, 0)$.

Step 2. Find the y-intercept. Let $x = 0$ and solve for y.

$$2 \cdot 0 - y = 4 \qquad \text{Replace x with 0 in } 2x - y = 4.$$
$$-y = 4 \qquad \text{Simplify.}$$
$$y = -4 \qquad \text{Divide (or multiply) both sides by } -1.$$

The y-intercept is -4, so the line passes through $(0, -4)$.

Step 3. Find a checkpoint, a third ordered-pair solution. For our checkpoint, we will let $x = 1$ and find the corresponding value for y.

$$2x - y = 4 \qquad \text{This is the given equation.}$$
$$2 \cdot 1 - y = 4 \qquad \text{Substitute 1 for x.}$$
$$2 - y = 4 \qquad \text{Simplify.}$$
$$-y = 2 \qquad \text{Subtract 2 from both sides.}$$
$$y = -2 \qquad \text{Divide (or multiply) both sides by } -1.$$

The checkpoint is $(1, -2)$.

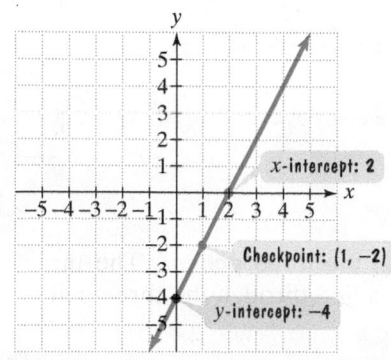

Figure 3.16
The graph of $2x - y = 4$

Step 4. Graph the equation by drawing a line through the three points. The three points in **Figure 3.16** lie along the same line. Drawing a line through the three points results in the graph of $2x - y = 4$. ∎

✓ **CHECK POINT 5**   Graph: $x - 2y = 4$.

We have seen that not all lines have two different intercepts. Some lines pass through the origin. Thus, they have an x-intercept of 0 and a y-intercept of 0. Is it possible to recognize these lines by their equations? Yes. **The graph of the linear equation** $Ax + By = 0$ **passes through the origin.** Notice that the constant on the right side of this equation is 0.

An equation of the form $Ax + By = 0$ can be graphed by using the origin as one point on the line. Find two other points by finding two other solutions of the equation. Select values for either variable, other than 0, and find the corresponding values for the other variable.

| **EXAMPLE 6** | Graphing a Linear Equation of the Form $Ax + By = 0$ |

Graph: $x + 2y = 0$.

Solution   Because the constant on the right is 0, the graph passes through the origin. The x- and y-intercepts are both 0. Remember that we are using two points and a checkpoint to graph a line. Thus, we still want to find two other points. Let $y = -1$ to find a second ordered-pair solution. Let $y = 1$ to find a third ordered-pair (checkpoint) solution.

$$x + 2y = 0 \qquad\qquad x + 2y = 0$$

Let $y = -1$. $\qquad\qquad$ Let $y = 1$.

$$x + 2(-1) = 0 \qquad x + 2 \cdot 1 = 0$$
$$x + (-2) = 0 \qquad\quad x + 2 = 0$$
$$x = 2 \qquad\qquad\quad x = -2$$

The solutions are $(2, -1)$ and $(-2, 1)$. Plot these two points, as well as the origin—that is, $(0, 0)$. The three points in **Figure 3.17** lie along the same line. Drawing a line through the three points results in the graph of $x + 2y = 0$. ∎

✓ **CHECK POINT 6**   Graph: $x + 3y = 0$.

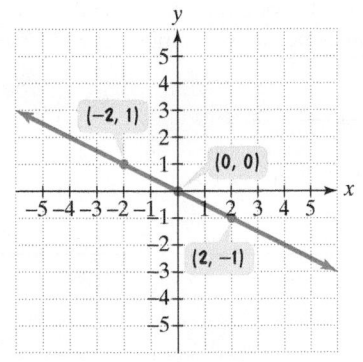

Figure 3.17
The graph of $x + 2y = 0$

3 Graph horizontal or vertical lines.

Equations of Horizontal and Vertical Lines

We know that the graph of any equation of the form $Ax + By = C$ is a line as long as A and B are not both zero. What happens if A or B, but not both, is zero? Example 7 shows that if $A = 0$, the equation $Ax + By = C$ has no x-term and the graph is a horizontal line.

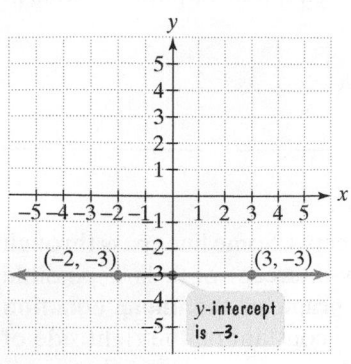

Figure 3.18
The graph of $y = -3$

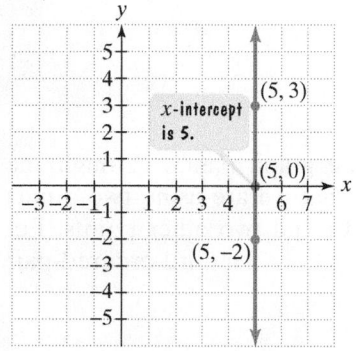

Figure 3.19
The graph of $x = 5$

EXAMPLE 7 Graphing a Horizontal Line

Graph: $y = -3$.

Solution All ordered pairs that are solutions of $y = -3$ have a value of y that is -3. Any value can be used for x. In the table on the right, we have selected three of the possible values for x: $-2, 0,$ and 3. The table shows that three ordered pairs that are solutions of $y = -3$ are $(-2, -3), (0, -3),$ and $(3, -3)$. Drawing a line that passes through the three points gives the horizontal line shown in **Figure 3.18**. ■

x	$y = -3$	(x, y)
-2	-3	$(-2, -3)$
0	-3	$(0, -3)$
3	-3	$(3, -3)$

For all choices of x, y is a constant -3.

✓ CHECK POINT 7 Graph: $y = 3$.

Example 8 shows that if $B = 0$, the equation $Ax + By = C$ has no y-term and the graph is a vertical line.

EXAMPLE 8 Graphing a Vertical Line

Graph: $x = 5$.

Solution All ordered pairs that are solutions of $x = 5$ have a value of x that is 5. Any value can be used for y. In the table on the right, we have selected three of the possible values for y: $-2, 0,$ and 3. The table shows that three ordered pairs that are solutions of $x = 5$ are $(5, -2), (5, 0),$ and $(5, 3)$. Drawing a line that passes through the three points gives the vertical line shown in **Figure 3.19**. ■

For all choices of y,

x is always 5.

$x = 5$	y	(x, y)
5	-2	$(5, -2)$
5	0	$(5, 0)$
5	3	$(5, 3)$

Great Question!

Why isn't the graph of $x = 5$ just a single point at 5 on a number line?

Do not confuse two-dimensional graphing and one-dimensional graphing of $x = 5$. The graph of $x = 5$ in a two-dimensional rectangular coordinate system is the vertical line in **Figure 3.19**. By contrast, the graph of $x = 5$ on a one-dimensional number line representing values of x is a single point at 5:

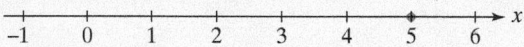

✓ CHECK POINT 8 Graph: $x = -2$.

Horizontal and Vertical Lines

The graph of $y = b$ is a horizontal line. The y-intercept is b.

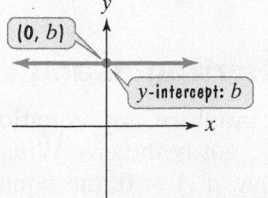

The graph of $x = a$ is a vertical line. The x-intercept is a.

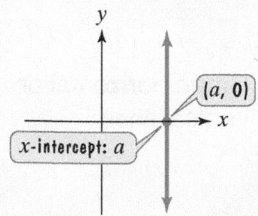

CONCEPT AND VOCABULARY CHECK

Fill in each blank so that the resulting statement is true.

1. The x-coordinate of a point where a graph crosses the x-axis is called a/an _____.

2. The y-coordinate of a point where a graph crosses the y-axis is called a/an _____.

3. The point $(4, 0)$ lies along a line, so 4 is the _____ of that line.

4. The point $(0, 3)$ lies along a line, so 3 is the _____ of that line.

5. An equation that can be written in the form $Ax + By = C$, where A and B are not both zero, is called the _____ form of the equation of a line.

6. Given the equation $Ax + By = C$, to find the x-intercept (if there is one), let _____ = 0 and solve for _____.

7. Given the equation $Ax + By = C$, to find the y-intercept (if there is one), let _____ = 0 and solve for _____.

8. The graph of the equation $y = 3$ is a/an _____ line.

9. The graph of the equation $x = -2$ is a/an _____ line.

3.2 EXERCISE SET

MyMathLab®

Watch the videos in MyMathLab   Download the MyDashBoard App

Practice Exercises

In Exercises 1–8, use the graph to identify the

a. *x-intercept, or state that there is no x-intercept;*

b. *y-intercept, or state that there is no y-intercept.*

1.
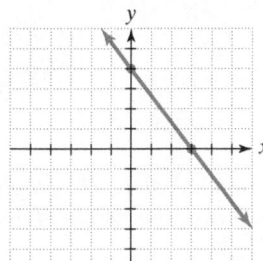

2.
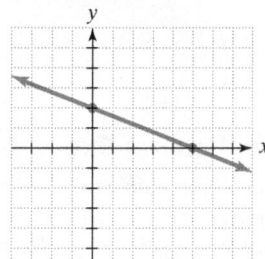

3.
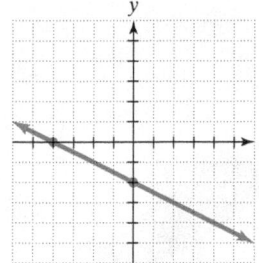

4.
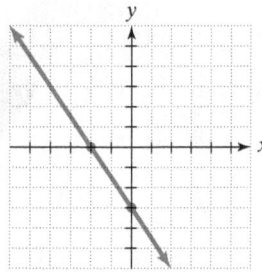

5.
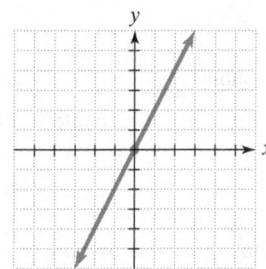

6.
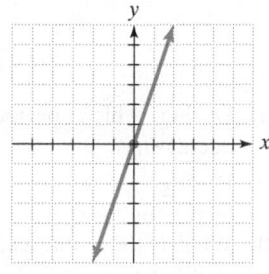

7.

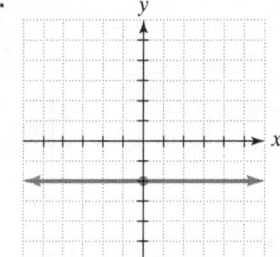

8.

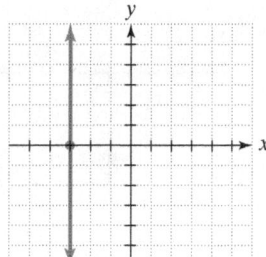

In Exercises 9–18, find the x-intercept and the y-intercept of the graph of each equation. Do not graph the equation.

9. $2x + 5y = 20$

10. $2x + 6y = 30$

11. $2x - 3y = 15$

12. $4x - 5y = 10$

13. $-x + 3y = -8$

14. $-x + 3y = -10$

15. $7x - 9y = 0$

16. $8x - 11y = 0$

17. $2x = 3y - 11$

18. $2x = 4y - 13$

In Exercises 19–40, use intercepts and a checkpoint to graph each equation.

19. $x + y = 5$
20. $x + y = 6$
21. $x + 3y = 6$
22. $2x + y = 4$
23. $6x - 9y = 18$
24. $6x - 2y = 12$
25. $-x + 4y = 6$
26. $-x + 3y = 10$
27. $2x - y = 7$
28. $2x - y = 5$

29. $3x = 5y - 15$

30. $2x = 3y + 6$

31. $25y = 100 - 50x$

32. $10y = 60 - 40x$

33. $2x - 8y = 12$

34. $3x - 6y = 15$

35. $x + 2y = 0$

36. $2x + y = 0$

37. $y - 3x = 0$

38. $y - 4x = 0$

39. $2x - 3y = -11$

40. $3x - 2y = -7$

In Exercises 41–46, write an equation for each graph.

41.

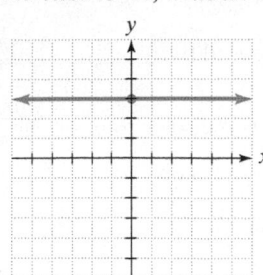

42.

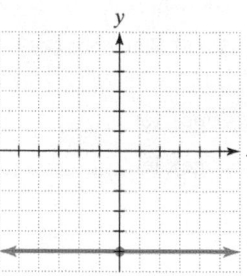

43.

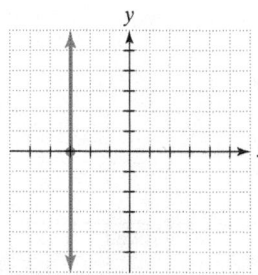

44.

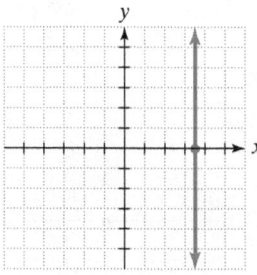

45.

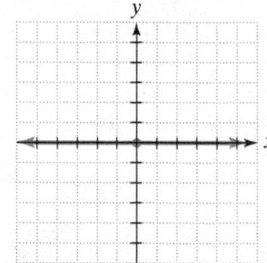

46.

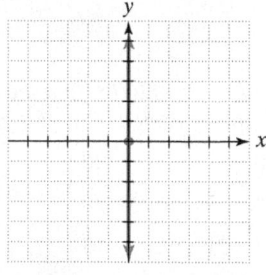

In Exercises 47–62, graph each equation.

47. $y = 4$

48. $y = 2$

49. $y = -2$

50. $y = -3$

51. $x = 2$

52. $x = 4$

53. $x + 1 = 0$

54. $x + 5 = 0$

55. $y - 3.5 = 0$

56. $y - 2.5 = 0$

57. $x = 0$

58. $y = 0$

59. $3y = 9$

60. $5y = 20$

61. $12 - 3x = 0$

62. $12 - 4x = 0$

Practice PLUS

In Exercises 63–68, match each equation with one of the graphs shown in Exercises 1–8.

63. $3x + 2y = -6$

64. $x + 2y = -4$

65. $y = -2$

66. $x = -3$

67. $4x + 3y = 12$

68. $2x + 5y = 10$

In Exercises 69–70,

a. *Write a linear equation in standard form satisfying the given condition. Assume that all measurements shown in each figure are in feet.*

b. *Graph the equation in part (a). Because x and y must be nonnegative (why?), limit your final graph to quadrant I and its boundaries.*

69. The perimeter of the larger rectangle is 58 feet.

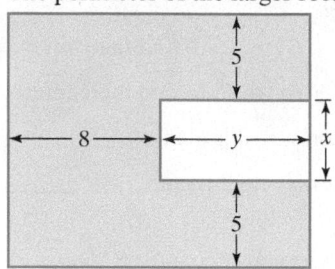

70. The perimeter of the shaded trapezoid is 84 feet.

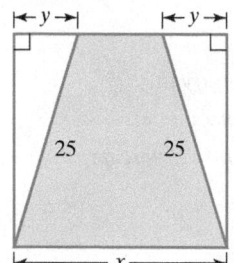

Application Exercises

The flight of an eagle is observed for 30 seconds. The graph shows the eagle's height, in meters, during this period of time. Use the graph to solve Exercises 71–75.

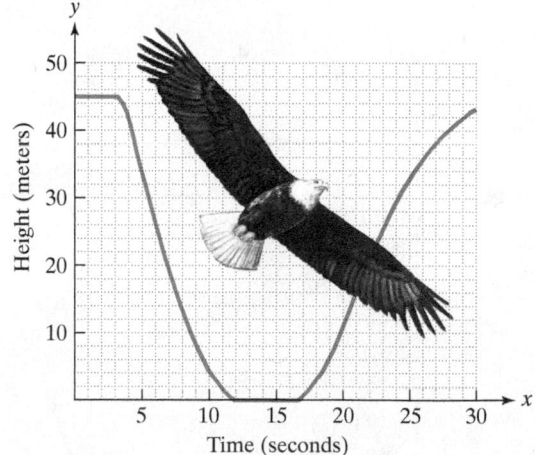

71. During which period of time is the eagle's height decreasing?

72. During which period of time is the eagle's height increasing?

73. What is the *y*-intercept? What does this mean about the eagle's height at the beginning of the observation?

(In Exercises 74–75, be sure to refer to the graph at the bottom of the previous page.)

74. During the first three seconds of observation, the eagle's flight is graphed as a horizontal line. Write the equation of the line. What does this mean about the eagle's flight pattern during this time?

75. Use integers to write five x-intercepts of the graph. What is the eagle doing during these times?

76. A new car worth $24,000 is depreciating in value by $3000 per year. The mathematical model

$$y = -3000x + 24,000$$

describes the car's value, y, in dollars, after x years.

  a. Find the x-intercept. Describe what this means in terms of the car's value.

  b. Find the y-intercept. Describe what this means in terms of the car's value.

  c. Use the intercepts to graph the linear equation. Because x and y must be nonnegative (why?), limit your graph to quadrant I and its boundaries.

  d. Use your graph to estimate the car's value after five years.

77. A new car worth $45,000 is depreciating in value by $5000 per year. The mathematical model

$$y = -5000x + 45,000$$

describes the car's value, y, in dollars, after x years.

  a. Find the x-intercept. Describe what this means in terms of the car's value.

  b. Find the y-intercept. Describe what this means in terms of the car's value.

  c. Use the intercepts to graph the linear equation. Because x and y must be nonnegative (why?), limit your graph to quadrant I and its boundaries.

  d. Use your graph to estimate the car's value after five years.

Writing in Mathematics

78. What is an x-intercept of a graph?

79. What is a y-intercept of a graph?

80. If you are given an equation of the form $Ax + By = C$, explain how to find the x-intercept.

81. If you are given an equation of the form $Ax + By = C$, explain how to find the y-intercept.

82. Explain how to graph $Ax + By = C$ if C is not equal to zero.

83. Explain how to graph a linear equation of the form $Ax + By = 0$.

84. How many points are needed to graph a line? How many should actually be used? Explain.

85. Describe the graph of $y = 200$.

86. Describe the graph of $x = -100$.

Critical Thinking Exercises

Make Sense?   *In Exercises 87–90, determine whether each statement "makes sense" or "does not make sense" and explain your reasoning.*

87. If I could be absolutely certain that I have not made an algebraic error in obtaining intercepts, I would not need to use checkpoints.

88. I like to select a point represented by one of the intercepts as my checkpoint.

89. The graphs of $2x - 3y = -18$ and $-2x + 3y = 18$ must have the same intercepts because I can see that the equations are equivalent.

90. From 1997 through 2007, the federal minimum wage remained constant at $5.15 per hour, so I modeled the situation with $y = 5.15$ and the graph of a vertical line.

In Exercises 91–92, find the coefficients that must be placed in each shaded area so that the equation's graph will be a line with the specified intercepts.

91. ▨ $x + $ ▨ $y = 10$; x-intercept $= 5$; y-intercept $= 2$

92. ▨ $x + $ ▨ $y = 12$; x-intercept $= -2$; y-intercept $= 4$

Technology Exercises

93. Use a graphing utility to verify any five of your hand-drawn graphs in Exercises 19–40. Solve the equation for y before entering it.

In Exercises 94–97, use a graphing utility to graph each equation. You will need to solve the equation for y before entering it. Use the graph displayed on the screen to identify the x-intercept and the y-intercept.

94. $2x + y = 4$

95. $3x - y = 9$

96. $2x + 3y = 30$

97. $4x - 2y = -40$

Review Exercises

98. Find the absolute value: $|-13.4|$.
(Section 1.3, Example 8)

99. Simplify: $7x - (3x - 5)$.
(Section 1.7, Example 7)

100. Solve: $8(x - 2) - 2(x - 3) \leq 8x$.
(Section 2.7, Example 8)

Preview Exercises

Exercises 101–103 will help you prepare for the material covered in the next section. In each exercise, evaluate

$$\frac{y_2 - y_1}{x_2 - x_1}$$

for the given ordered pairs (x_1, y_1) and (x_2, y_2).

101. $(x_1, y_1) = (1, 3)$; $(x_2, y_2) = (6, 13)$

102. $(x_1, y_1) = (4, -2)$; $(x_2, y_2) = (6, -4)$

103. $(x_1, y_1) = (3, 4)$; $(x_2, y_2) = (5, 4)$

SECTION

3.3

Slope

Objectives

1. Compute a line's slope.
2. Use slope to show that lines are parallel.
3. Use slope to show that lines are perpendicular.
4. Calculate rate of change in applied situations.

A best guess at the future of our nation indicates that the numbers of men and women living alone will increase each year. **Figure 3.20** shows that in 2008, 14.7 million men and 18.3 million women lived alone, an increase over the numbers displayed in the graph for 1990.

By looking at **Figure 3.20**, can you tell that the green graph representing men is steeper than the red graph representing women? This indicates a greater rate of increase in the millions of men living alone than in the millions of women living alone over the period from 1990 through 2008. In this section, we will study the idea of a line's steepness and see what this has to do with how its variables are changing.

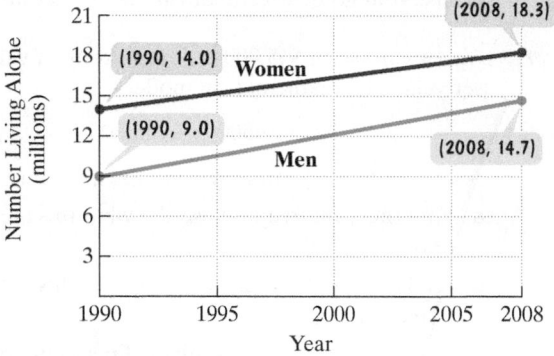

**Number of U.S. Adults
Ages 18 and Older Living Alone**

(1990, 14.0) Women (2008, 18.3)
(1990, 9.0) Men (2008, 14.7)

Figure 3.20

Source: U.S. Census Bureau

1. Compute a line's slope.

The Slope of a Line

Mathematicians have developed a useful measure of the steepness of a line, called the **slope** of the line. Slope compares the vertical change (the **rise**) to the horizontal change (the **run**) when moving from one fixed point to another along the line. To calculate the slope of a line, we use a ratio that compares the change in y (the rise) to the change in x (the run).

Definition of Slope

The **slope** of the line through the distinct points (x_1, y_1) and (x_2, y_2) is

$$\frac{\text{Change in } y}{\text{Change in } x} = \frac{\text{Rise}}{\text{Run}} \xleftarrow{\text{Vertical change}}_{\text{Horizontal change}}$$

$$= \frac{y_2 - y_1}{x_2 - x_1}$$

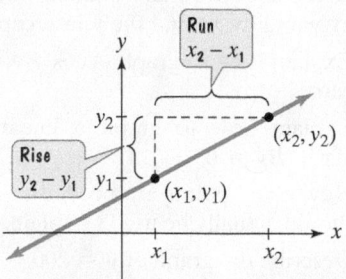

where $x_2 - x_1 \neq 0$.

It is common notation to let the letter m represent the slope of a line. The letter m is used because it is the first letter of the French verb *monter*, meaning to rise, or to ascend.

EXAMPLE 1   Using the Definition of Slope

Find the slope of the line passing through each pair of points:

a. $(-3, -1)$ and $(-2, 4)$ **b.** $(-3, 4)$ and $(2, -2)$.

Solution

a. To find the slope of the line passing through $(-3, -1)$ and $(-2, 4)$, we let $(x_1, y_1) = (-3, -1)$ and $(x_2, y_2) = (-2, 4)$. We obtain the slope as follows:

$$m = \frac{\text{Change in } y}{\text{Change in } x} = \frac{y_2 - y_1}{x_2 - x_1} = \frac{4 - (-1)}{-2 - (-3)} = \frac{5}{1} = 5.$$

The situation is illustrated in **Figure 3.21**. The slope of the line is 5, indicating that there is a vertical change, a rise, of 5 units for each horizontal change, a run, of 1 unit. The slope is positive and the line rises from left to right.

Great Question!

When using the definition of slope, how do I know which point to call (x_1, y_1) and which point to call (x_2, y_2)?

When computing slope, it makes no difference which point you call (x_1, y_1) and which point you call (x_2, y_2). If we let $(x_1, y_1) = (-2, 4)$ and $(x_2, y_2) = (-3, -1)$, the slope is still 5:

$$m = \frac{\text{Change in } y}{\text{Change in } x} = \frac{y_2 - y_1}{x_2 - x_1} = \frac{-1 - 4}{-3 - (-2)} = \frac{-5}{-1} = 5.$$

However, you should not subtract in one order in the numerator $(y_2 - y_1)$ and then in the opposite order in the denominator $(x_1 - x_2)$.

$$\frac{-1 - 4}{-2 - (-3)} = \frac{-5}{1} = -5 \quad \text{Incorrect! The slope is not } -5.$$

b. To find the slope of the line passing through $(-3, 4)$ and $(2, -2)$, we can let $(x_1, y_1) = (-3, 4)$ and $(x_2, y_2) = (2, -2)$. The slope is computed as follows:

$$m = \frac{\text{Change in } y}{\text{Change in } x} = \frac{y_2 - y_1}{x_2 - x_1} = \frac{-2 - 4}{2 - (-3)} = \frac{-6}{5} = -\frac{6}{5}.$$

The situation is illustrated in **Figure 3.22**. The slope of the line is $-\frac{6}{5}$. For every vertical change of -6 units (6 units down), there is a corresponding horizontal change of 5 units. The slope is negative and the line falls from left to right.

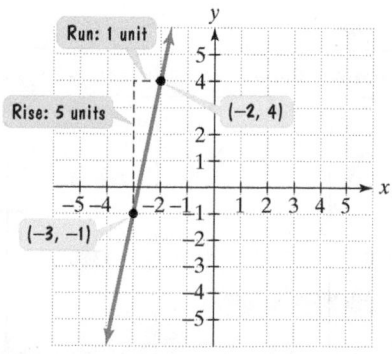

Figure 3.21 Visualizing a slope of 5

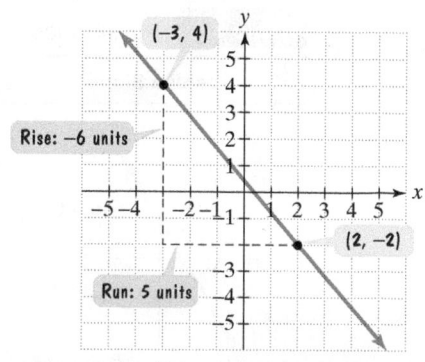

Figure 3.22 Visualizing a slope of $-\frac{6}{5}$ ∎

✓ **CHECK POINT 1** Find the slope of the line passing through each pair of points:

a. $(-3, 4)$ and $(-4, -2)$ **b.** $(4, -2)$ and $(-1, 5)$.

EXAMPLE 2 Using the Definition of Slope for Horizontal and Vertical Lines

Find the slope of the line passing through each pair of points:

a. $(5, 4)$ and $(3, 4)$ **b.** $(2, 5)$ and $(2, 1)$.

Solution

a. Let $(x_1, y_1) = (5, 4)$ and $(x_2, y_2) = (3, 4)$. We obtain the slope as follows:

$$m = \frac{\text{Change in } y}{\text{Change in } x} = \frac{y_2 - y_1}{x_2 - x_1} = \frac{4 - 4}{3 - 5} = \frac{0}{-2} = 0.$$

The situation is illustrated in **Figure 3.23**. Can you see that the line is horizontal? Because any two points on a horizontal line have the same y-coordinate, these lines neither rise nor fall from left to right. The change in y, $y_2 - y_1$, is always zero. Thus, **the slope of any horizontal line is zero.**

b. We can let $(x_1, y_1) = (2, 5)$ and $(x_2, y_2) = (2, 1)$. **Figure 3.24** shows that these points are on a vertical line. We attempt to compute the slope as follows:

$$m = \frac{\text{Change in } y}{\text{Change in } x} = \frac{1 - 5}{2 - 2} = \frac{-4}{0}. \quad \text{Division by zero is undefined.}$$

Because division by zero is undefined, the slope of the vertical line in **Figure 3.24** is undefined. In general, **the slope of any vertical line is undefined.**

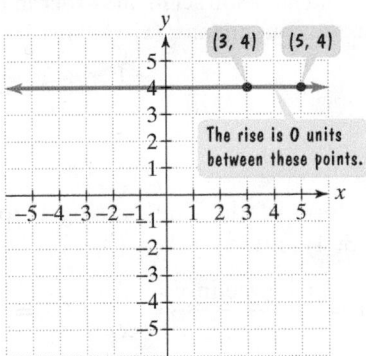

Figure 3.23 Horizontal lines have no vertical change.

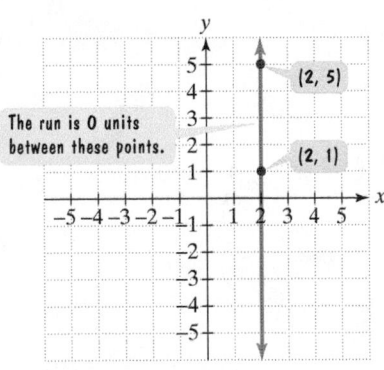

Figure 3.24 Vertical lines have no horizontal change. ■

Table 3.1 summarizes four possibilities for the slope of a line.

Table 3.1 Possibilities for a Line's Slope

Positive Slope	Negative Slope	Zero Slope	Undefined Slope
$m > 0$	$m < 0$	$m = 0$	m is undefined.
Line rises from left to right.	Line falls from left to right.	Line is horizontal.	Line is vertical.

✓ **CHECK POINT 2** Find the slope of the line passing through each pair of points or state that the slope is undefined:

a. $(6, 5)$ and $(2, 5)$ **b.** $(1, 6)$ and $(1, 4)$.

2 Use slope to show that lines are parallel.

Slope and Parallel Lines

Two nonintersecting lines that lie in the same plane are **parallel**. If two lines do not intersect, the ratio of the vertical change to the horizontal change is the same for each line. Because two parallel lines have the same "steepness," they must have the same slope.

Slope and Parallel Lines

1. If two nonvertical lines are parallel, then they have the same slope.
2. If two distinct nonvertical lines have the same slope, then they are parallel.
3. Two distinct vertical lines, each with undefined slope, are parallel.

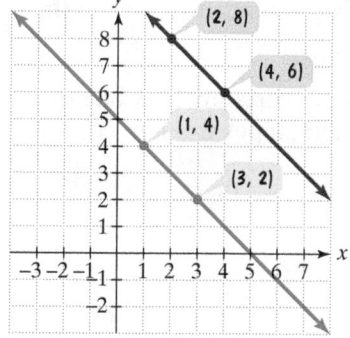

Figure 3.25 Using slope to show that lines are parallel

| EXAMPLE 3 | Using Slope to Show That Lines Are Parallel |

Show that the line passing through $(1, 4)$ and $(3, 2)$ is parallel to the line passing through $(2, 8)$ and $(4, 6)$.

Solution The situation is illustrated in **Figure 3.25**. The lines certainly look like they are parallel. Let's use equal slopes to confirm this fact. For each line, we compute the ratio of the difference in y-coordinates to the difference in x-coordinates. Be sure to subtract the coordinates in the same order.

We begin with the slope of the blue line through $(1, 4)$ and $(3, 2)$.

Change in y is $2 - 4$.

$$(1, 4) \qquad (3, 2)$$

Change in x is $3 - 1$.

$$m = \frac{\text{Change in } y}{\text{Change in } x} = \frac{2 - 4}{3 - 1} = \frac{-2}{2} = -1$$

Now we find the slope of the red line through $(2, 8)$ and $(4, 6)$.

Change in y is $6 - 8$.

$$(2, 8) \qquad (4, 6)$$

Change in x is $4 - 2$.

$$m = \frac{\text{Change in } y}{\text{Change in } x} = \frac{6 - 8}{4 - 2} = \frac{-2}{2} = -1$$

Because the slopes are equal, the lines are parallel. ∎

✓ **CHECK POINT 3** Show that the line passing through $(4, 2)$ and $(6, 6)$ is parallel to the line passing through $(0, -2)$ and $(1, 0)$.

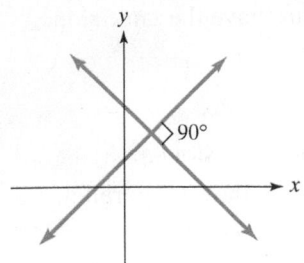

3 Use slope to show that lines are perpendicular.

Slope and Perpendicular Lines

Two lines that intersect at a right angle (90°) are said to be **perpendicular**, as shown in **Figure 3.26**. There is a relationship between the slopes of perpendicular lines.

Figure 3.26
Perpendicular lines

Slope and Perpendicular Lines

1. If two nonvertical lines are perpendicular, then the product of their slopes is -1.
2. If the product of the slopes of two lines is -1, then the lines are perpendicular.
3. A horizontal line having zero slope is perpendicular to a vertical line having undefined slope.

EXAMPLE 4 Using Slope to Show That Lines Are Perpendicular

Show that the line passing through $(-6, -9)$ and $(3, 6)$ is perpendicular to the line passing through $(10, -8)$ and $(-5, 1)$.

Solution The situation is illustrated in **Figure 3.27**. The lines certainly look like they intersect at a right angle. Let's show that the product of their slopes is -1 to confirm this fact.

We begin with the slope of the red line through $(-6, -9)$ and $(3, 6)$.

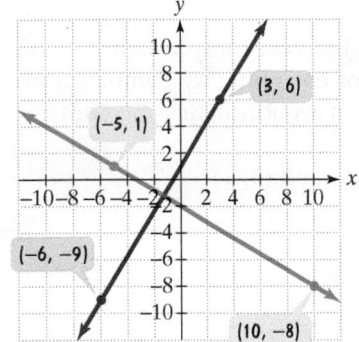

Figure 3.27
Using slope to show that lines are perpendicular

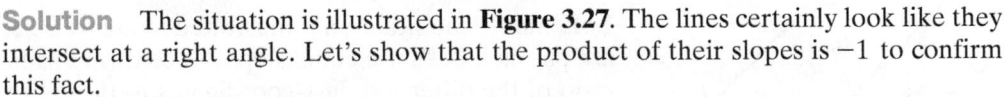

Change in y is $6 - (-9)$.

$$(-6, -9) \qquad (3, 6)$$

Change in x is $3 - (-6)$.

$$m = \frac{\text{Change in } y}{\text{Change in } x} = \frac{6 - (-9)}{3 - (-6)} = \frac{6 + 9}{3 + 6} = \frac{15}{9} = \frac{5}{3}$$

Now we find the slope of the blue line through $(10, -8)$ and $(-5, 1)$.

Change in y is $1 - (-8)$.

$$(10, -8) \qquad (-5, 1)$$

Change in x is $-5 - 10$.

$$m = \frac{\text{Change in } y}{\text{Change in } x} = \frac{1 - (-8)}{-5 - 10} = \frac{1 + 8}{-5 - 10} = \frac{9}{-15} = -\frac{3}{5}$$

We have determined that the slope of the red line in **Figure 3.27** is $\frac{5}{3}$ and the slope of the blue line is $-\frac{3}{5}$. Now we can find the product of the slopes:

$$\frac{5}{3}\left(-\frac{3}{5}\right) = -1.$$

Because the product of the slopes is -1, the lines are perpendicular. ∎

In this example, the slopes of the perpendicular lines, $\frac{5}{3}$ and $-\frac{3}{5}$, are reciprocals with opposite signs, called **negative reciprocals**. This gives us an equivalent way of stating the relationship between slope and perpendicular lines:

Two nonvertical lines are perpendicular if the slope of one is the negative reciprocal of the slope of the other.

✓ **CHECK POINT 4** Show that the line passing through $(-1, 4)$ and $(3, 2)$ is perpendicular to the line passing through $(-2, -1)$ and $(2, 7)$.

---

4 Calculate rate of change in applied situations.

Slope as Rate of Change

Slope is defined as the ratio of a change in y to a corresponding change in x. It tells how fast y is changing with respect to x. Thus, the slope of a line represents its rate of change.

Our next example shows how slope can be interpreted as a rate of change in an applied situation. When calculating slope in applied problems, keep track of the units in the numerator and the denominator.

EXAMPLE 5 Slope as a Rate of Change

The line graphs for the numbers of women and men living alone are shown again in **Figure 3.28**. Find the slope of the line segment for the women. Describe what this slope represents.

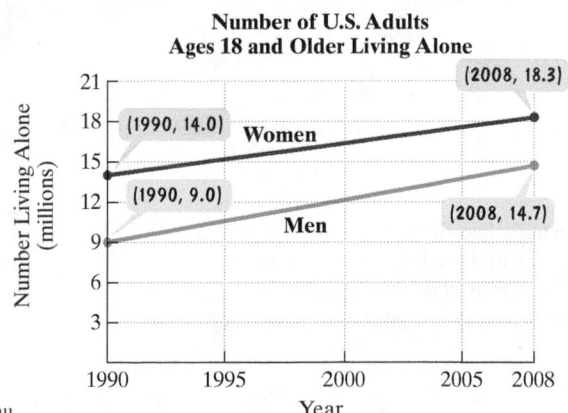

**Number of U.S. Adults
Ages 18 and Older Living Alone**

Figure 3.28

Source: U.S. Census Bureau

Solution We let x represent a year and y the number of women living alone in that year. Use the two points shown on the red line segment for women.

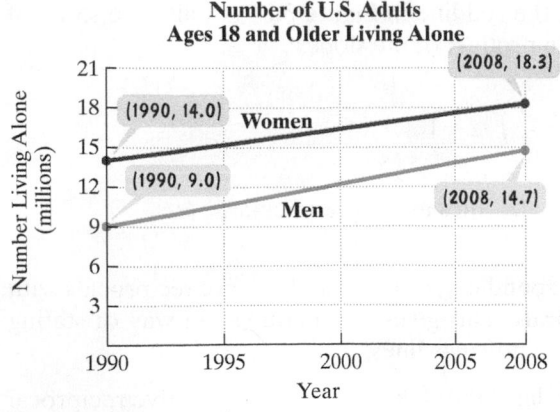

Number of U.S. Adults Ages 18 and Older Living Alone

(1990, 14.0) Women (2008, 18.3)

(1990, 9.0) Men (2008, 14.7)

Figure 3.28 (repeated)

The two points shown on the line segment for women have the following coordinates:

$$(1990, 14.0) \qquad (2008, 18.3).$$

In 1990, 14 million U.S. women lived alone.

In 2008, 18.3 million U.S. women lived alone.

Now we compute the slope:

$$m = \frac{\text{Change in } y}{\text{Change in } x} = \frac{18.3 - 14.0}{2008 - 1990}$$

The unit in the numerator is *million women*.

The unit in the denominator is *year*.

$$= \frac{4.3}{18} \approx \frac{0.24 \text{ million women}}{\text{year}}.$$

The slope indicates that the number of American women living alone increased at a rate of approximately 0.24 million each year for the period from 1990 through 2008. The rate of change is 0.24 million women per year. ■

✓ **CHECK POINT 5**   Use the ordered pairs in **Figure 3.28** to find the slope of the green line segment for the men. Express the slope correct to two decimal places and describe what it represents.

In Check Point 5, did you find that the slope of the line segment for the men is different from that of the women? The rate of change for men living alone is greater than the rate of change for women living alone. The line segment representing men in **Figure 3.28** is steeper than the line segment representing women. If you extend the line segments far enough, the resulting lines will intersect. They are not parallel.

Blitzer Bonus

Railroads and Highways

The steepest part of Mt. Washington Cog Railway in New Hampshire has a 37% grade. This is equivalent to a slope of $\frac{37}{100}$. For every horizontal change of 100 feet, the railroad ascends 37 feet vertically. Engineers denote slope by grade, expressing slope as a percentage.

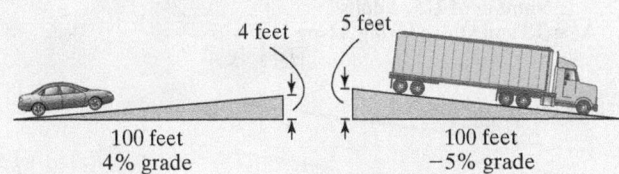

4 feet

5 feet

100 feet
4% grade

100 feet
−5% grade

Railroad grades are usually less than 2%, although in the mountains they may go as high as 4%. The grade of the Mount Washington Cog Railway is phenomenal, making it necessary for locomotives to *push* single cars up its steepest part.

A Mount Washington Cog Railway locomotive pushing a single car up the steepest part of the railroad. The locomotive is about 120 years old.

Achieving Success

According to the Ebbinghaus retention model, you forget 50% of processed information within one hour of leaving the classroom. You lose 60% within 24 hours. After 30 days, 70% is gone. Reviewing and rewriting class notes is an effective way to counteract this phenomenon. At the very least, read your lecture notes at the end of each day. The more you engage with the material, the more you retain.

CONCEPT AND VOCABULARY CHECK

Fill in each blank so that the resulting statement is true.

1. The slope, m, of the line through the distinct points (x_1, y_1) and (x_2, y_2) is given by the formula $m =$ _____ .

2. The slope of the line through the distinct points (x_1, y_1) and (x_2, y_2) can be interpreted as the rate of change in _____ with respect to _____ .

3. If a line rises from left to right, the line has _____ slope.

4. If a line falls from left to right, the line has _____ slope.

5. The slope of a horizontal line is _____ .

6. The slope of a vertical line is _____ .

7. If two distinct nonvertical lines have the same slope, then the lines are _____ .

8. If the product of the slopes of two lines is -1, then the lines are _____ .

3.3 EXERCISE SET   MyMathLab®

Practice Exercises

In Exercises 1–10, find the slope of the line passing through each pair of points or state that the slope is undefined. Then indicate whether the line through the points rises, falls, is horizontal, or is vertical.

1. $(4, 7)$ and $(8, 10)$

2. $(2, 1)$ and $(3, 4)$

3. $(-2, 1)$ and $(2, 2)$

4. $(-1, 3)$ and $(2, 4)$

5. $(4, -2)$ and $(3, -2)$

6. $(4, -1)$ and $(3, -1)$

7. $(-2, 4)$ and $(-1, -1)$

8. $(6, -4)$ and $(4, -2)$

9. $(5, 3)$ and $(5, -2)$

10. $(3, -4)$ and $(3, 5)$

In Exercises 11–22, find the slope of each line, or state that the slope is undefined.

11.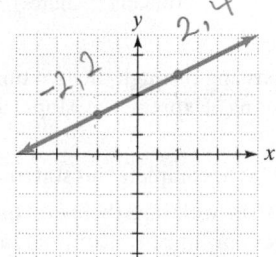

12.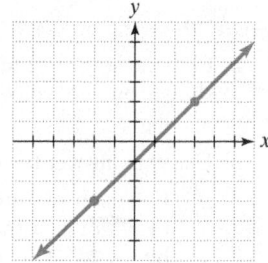

13.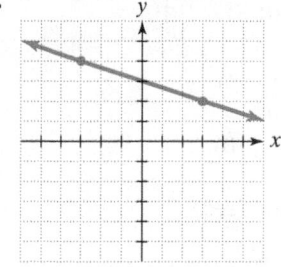

14.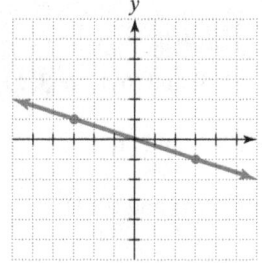

15.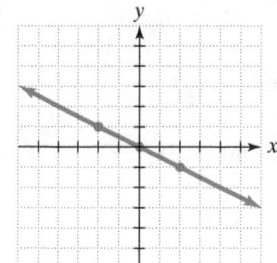

16.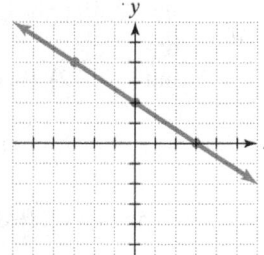

17.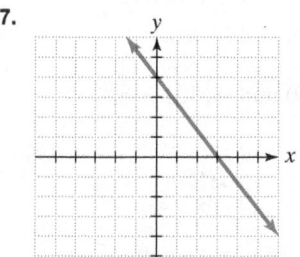

18.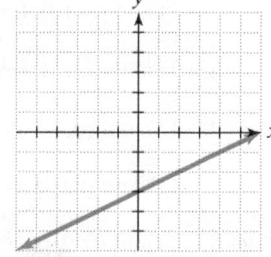

19.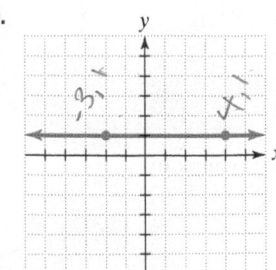

20.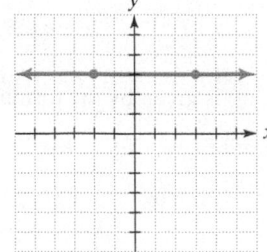

21.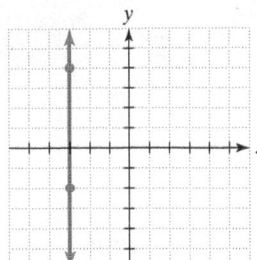

22.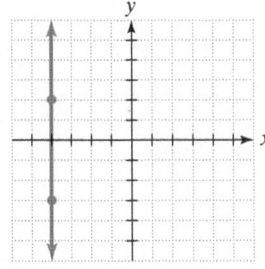

In Exercises 23–26, determine whether the distinct lines through each pair of points are parallel.

23. $(-2, 0)$ and $(0, 6)$; $(1, 8)$ and $(0, 5)$

24. $(2, 4)$ and $(6, 1)$; $(-3, 1)$ and $(1, -2)$

25. $(0, 3)$ and $(1, 5)$; $(-1, 7)$ and $(1, 10)$

26. $(-7, 6)$ and $(0, 4)$; $(-9, -3)$ and $(1, 5)$

In Exercises 27–30, determine whether the lines through each pair of points are perpendicular.

27. $(1, 5)$ and $(0, 3)$; $(-2, 8)$ and $(2, 6)$

28. $(3, 2)$ and $(-2, -2)$; $(3, -2)$ and $(-1, 3)$

29. $(-1, -6)$ and $(2, 9)$; $(-15, -1)$ and $(5, 3)$

30. $(-1, -6)$ and $(2, 6)$; $(-8, -1)$ and $(4, 2)$

In Exercises 31–36, determine whether the lines through each pair of points are parallel, perpendicular, or neither.

31. $(-2, -5)$ and $(3, 10)$; $(-1, -9)$ and $(4, 6)$

32. $(-2, -7)$ and $(3, 13)$; $(-1, -9)$ and $(5, 15)$

33. $(-4, -12)$ and $(0, -4)$; $(0, -5)$ and $(2, -4)$

34. $(-1, -11)$ and $(0, -5)$; $(0, -8)$ and $(12, -6)$

35. $(-5, -1)$ and $(0, 2)$; $(-6, 9)$ and $(3, -6)$

36. $(-2, -15)$ and $(0, -3)$; $(-12, 6)$ and $(6, 3)$

Practice PLUS

37. On the same set of axes, draw lines passing through the origin with slopes $-1, -\frac{1}{2}, 0, \frac{1}{3}$, and 2.

38. On the same set of axes, draw lines with y-intercept 4 and slopes $-1, -\frac{1}{2}, 0, \frac{1}{3}$, and 2.

Use slopes to solve Exercises 39–40.

39. Show that the points whose coordinates are $(-3, -3)$, $(2, -5)$, $(5, -1)$, and $(0, 1)$ are the vertices of a four-sided figure whose opposite sides are parallel. (Such a figure is called a *parallelogram.*)

40. Show that the points whose coordinates are $(-3, 6)$, $(2, -3)$, $(11, 2)$, and $(6, 11)$ are the vertices of a four-sided figure whose opposite sides are parallel.

41. The line passing through $(5, y)$ and $(1, 0)$ is parallel to the line joining $(2, 3)$ and $(-2, 1)$. Find y.

42. The line passing through $(1, y)$ and $(7, 12)$ is parallel to the line joining $(-3, 4)$ and $(-5, -2)$. Find y.

43. The line passing through $(-1, y)$ and $(1, 0)$ is perpendicular to the line joining $(2, 3)$ and $(-2, 1)$. Find y.

44. The line passing through $(-2, y)$ and $(-4, 4)$ is perpendicular to the line passing through $(-1, -2)$ and $(4, -1)$. Find y.

Application Exercises

45. Exercise is useful not only in preventing depression, but also as a treatment. The graphs show the percentage of patients with depression in remission when exercise (brisk walking) was used as a treatment. (The control group that engaged in no exercise had 11% of the patients in remission.)

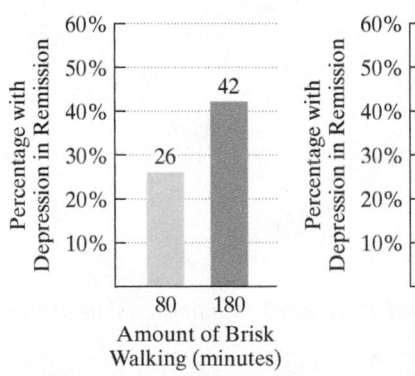

Exercise and Percentage of Patients with Depression in Remission

Source: Newsweek, March 26, 2007

a. Find the slope of the line passing through the two points shown by the voice balloons. Express the slope as a decimal.

b. Use your answer from part (a) to complete this statement:
   For each minute of brisk walking, the percentage of patients with depression in remission increased by ___%. The rate of change is ___% per _____.

46. Older, Calmer. As we age, daily stress and worry decrease and happiness increases, according to an analysis of 340,847 U.S. adults, ages 18–85, in the journal *Proceedings of the National Academy of Sciences*. The graphs show a portion of the research.

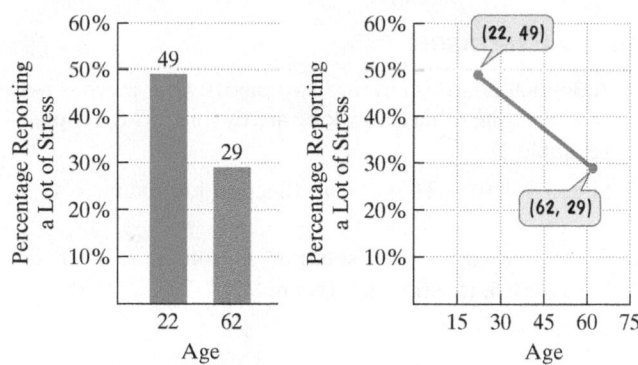

Percentage of Americans Reporting "a Lot" of Stress, by Age

Source: National Academy of Sciences

a. Find the slope of the line passing through the two points shown by the voice balloons. Express the slope as a decimal.

b. Use your answer from part (a) to complete the statement:

For each year of aging, the percentage of Americans reporting "a lot" of stress decreases by __%. The rate of change is ___% per _____.

The pitch of a roof refers to the absolute value of its slope. In Exercises 47–48, find the pitch of each roof shown.

47.

6 feet ←18 feet→

48.

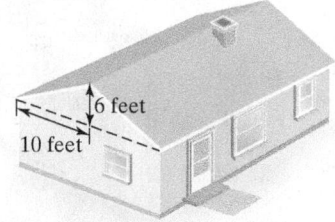

6 feet
10 feet

The grade of a road or ramp refers to its slope expressed as a percent. Use this information to solve Exercises 49–50.

49. Construction laws are very specific when it comes to access ramps for the disabled. Every vertical rise of 1 foot requires a horizontal run of 12 feet. What is the grade of such a ramp? Round to the nearest tenth of a percent.

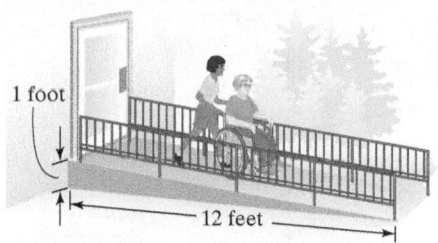

1 foot

12 feet

50. A college campus goes beyond the standards described in Exercise 49. All wheelchair ramps on campus are designed so that every vertical rise of 1 foot is accompanied by a horizontal run of 14 feet. What is the grade of such a ramp? Round to the nearest tenth of a percent.

Writing in Mathematics

51. What is the slope of a line?

52. Describe how to calculate the slope of a line passing through two points.

53. What does it mean if the slope of a line is zero?

54. What does it mean if the slope of a line is undefined?

55. If two lines are parallel, describe the relationship between their slopes.

56. If two lines are perpendicular, describe the relationship between their slopes.

Critical Thinking Exercises

Make Sense? *In Exercises 57–60, determine whether each statement "makes sense" or "does not make sense" and explain your reasoning.*

57. When finding the slope of the line passing through $(-1, 5)$ and $(2, -3)$, I must let (x_1, y_1) be $(-1, 5)$ and (x_2, y_2) be $(2, -3)$.

58. When applying the slope formula, it is important to subtract corresponding coordinates in the same order.

59. I visualize slope as walking along a line from left to right. If I'm walking uphill, the slope is positive, and if I'm walking downhill, the slope is negative.

60. I computed the slope of one line to be $-\frac{3}{5}$ and the slope of a second line to be $-\frac{5}{3}$, so the lines must be perpendicular.

In Exercises 61–64, determine whether each statement is true or false. If the statement is false, make the necessary change(s) to produce a true statement.

61. Slope is run divided by rise.

62. The line through $(2, 2)$ and the origin has slope 1.

63. A line with slope 3 can be parallel to a line with slope -3.

64. The line through $(3, 1)$ and $(3, -5)$ has zero slope.

In Exercises 65–66, use the figure shown to make the indicated list.

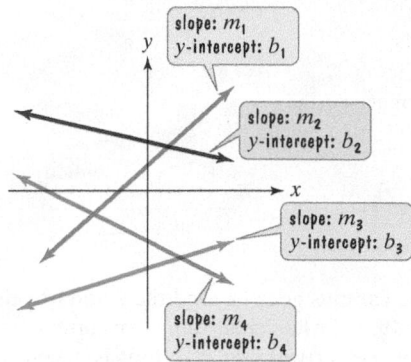

65. List the slopes $m_1, m_2, m_3,$ and m_4 in order of decreasing size.

66. List the y-intercepts $b_1, b_2, b_3,$ and b_4 in order of decreasing size.

Technology Exercises

Use a graphing utility to graph each equation in Exercises 67–70. Then use the $\boxed{\text{TRACE}}$ *feature to trace along the line and find the coordinates of two points. Use these points to compute the line's slope.*

67. $y = 2x + 4$

68. $y = -3x + 6$

69. $y = -\frac{1}{2}x - 5$

70. $y = \frac{3}{4}x - 2$

71. In Exercises 67–70, compare the slope that you found with the line's equation. What relationship do you observe between the line's slope and one of the constants in the equation?

Review Exercises

72. A 36-inch board is cut into two pieces. One piece is twice as long as the other. How long are the pieces? (Section 2.5, Example 2)

73. Simplify: $-10 + 16 \div 2(-4)$. (Section 1.8, Example 4)

74. Solve and graph the solution set on a number line: $2x - 3 \le 5$. (Section 2.7, Example 6)

Preview Exercises

Exercises 75–77 will help you prepare for the material covered in the next section.

75. From $(0, -3)$, move 4 units up and 1 unit to the right. What point do you obtain?

76. From $(0, 1)$, move 2 units down and 3 units to the right. What point do you obtain?

77. Solve for y: $2x + 5y = 0$.

SECTION

3.4

The Slope-Intercept Form of the Equation of a Line

Objectives

1 Find a line's slope and y-intercept from its equation.

2 Graph lines in slope-intercept form.

3 Use slope and y-intercept to graph $Ax + By = C$.

4 Use slope and y-intercept to model data.

Every day in the United States, 75.2 million students gather in elementary and secondary schools, as well as in 6.6 million colleges, to learn things from 5.1 million teachers. (*Source*: Department of Education) These numbers are fueled by the reality that employers use diplomas and degrees to determine who is eligible for a job. This section contains two mathematical models based on this reality. To develop these models, we turn to a new form for a line's equation.

1 Find a line's slope and *y*-intercept from its equation.

The Slope-Intercept Form of the Equation of a Line

Let's begin with an example that shows how easy it is to find a line's slope and *y*-intercept from its equation.

Figure 3.29 shows the graph of $y = 2x + 4$. Verify that the *x*-intercept is -2 by setting *y* equal to 0 and solving for *x*. Similarly, verify that the *y*-intercept is 4 by setting *x* equal to 0 and solving for *y*.

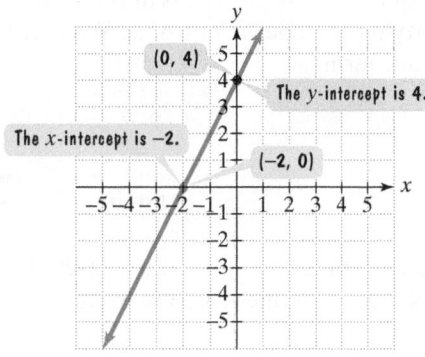

The *x*-intercept is -2.

The *y*-intercept is 4.

$(0, 4)$ $(-2, 0)$

Figure 3.29
The graph of $y = 2x + 4$

Now that we have two points on the line, $(-2, 0)$ and $(0, 4)$, we can calculate the slope of the graph of $y = 2x + 4$.

$$\text{Slope} = \frac{\text{Change in } y}{\text{Change in } x}$$

$$= \frac{4 - 0}{0 - (-2)} = \frac{4}{2} = 2$$

We see that the slope of the line is 2, the same as the coefficient of *x* in the equation $y = 2x + 4$. The *y*-intercept is 4, the same as the constant in the equation $y = 2x + 4$.

$$y = 2x + 4$$

The slope is 2.   The *y*-intercept is 4.

A linear equation like $y = 2x + 4$ that is solved for *y* is said to be in *slope-intercept form*. This is because the slope and the *y*-intercept can be immediately determined from the equation. The *x*-coefficient is the line's slope and the constant term is the *y*-intercept.

Slope-Intercept Form of the Equation of a Line

The **slope-intercept form of the equation** of a nonvertical line with slope *m* and *y*-intercept *b* is

$$y = mx + b.$$

EXAMPLE 1   Finding a Line's Slope and *y*-Intercept from Its Equation

Find the slope and the *y*-intercept of the line with the given equation:

a. $y = 2x - 4$ **b.** $y = \frac{1}{2}x + 2$ **c.** $5x + y = 4.$

Great Question!

Which are the constants and which are the variables in $y = mx + b$?

The variables in $y = mx + b$ vary in different ways. The variables for slope, m, and y-intercept, b, vary from one line's equation to another. However, they remain constant in the equation of a single line. By contrast, the variables x and y represent the infinitely many points, (x, y), on a single line. Thus, these variables vary in both the equation of a single line, as well as from one equation to another.

Solution

a. We write $y = 2x - 4$ as $y = 2x + (-4)$. The slope is the x-coefficient and the y-intercept is the constant term.

$$y = 2x + (-4)$$

The slope is 2.   The y-intercept is -4.

b. The equation $y = \frac{1}{2}x + 2$ is in the form $y = mx + b$. We can find the slope, m, by identifying the coefficient of x. We can find the y-intercept, b, by identifying the constant term.

$$y = \frac{1}{2}x + 2$$

The slope is $\frac{1}{2}$.   The y-intercept is 2.

c. The equation $5x + y = 4$ is not in the form $y = mx + b$. We can obtain this form by isolating y on one side. We isolate y on the left side by subtracting $5x$ from both sides.

$$5x + y = 4 \qquad \text{This is the given equation.}$$
$$5x - 5x + y = -5x + 4 \qquad \text{Subtract } 5x \text{ from both sides.}$$
$$y = -5x + 4 \qquad \text{Simplify.}$$

Now, the equation is in the form $y = mx + b$. The slope is the coefficient of x and the y-intercept is the constant term.

$$y = -5x + 4$$

The slope is -5.   The y-intercept is 4. ■

✓ **CHECK POINT 1**   Find the slope and the y-intercept of the line with the given equation:

a. $y = 5x - 3$ **b.** $y = \frac{2}{3}x + 4$

c. $7x + y = 6$.

---

2 Graph lines in slope-intercept form.

Graphing $y = mx + b$ by Using the Slope and y-Intercept

If a line's equation is written with y isolated on one side, we can use the y-intercept and the slope to obtain its graph.

Graphing $y = mx + b$ by Using the Slope and y-Intercept

1. Plot the point containing the y-intercept on the y-axis. This is the point $(0, b)$.
2. Obtain a second point using the slope, m. Write m as a fraction, and use rise over run, starting at the point on the y-axis, to plot this point.
3. Use a straightedge to draw a line through the two points. Draw arrowheads at the ends of the line to show that the line continues indefinitely in both directions.

EXAMPLE 2   Graphing by Using the Slope and y-Intercept

Graph the line whose equation is $y = 4x - 3$.

Solution   We write $y = 4x - 3$ in the form $y = mx + b$.

$$y = 4x + (-3)$$

The slope is 4.   The y-intercept is -3.

Now that we have identified the slope and the y-intercept, we use the three steps in the preceding box to graph the equation.

Step 1. Plot the point containing the y-intercept on the y-axis. The y-intercept is -3. We plot the point $(0, -3)$, shown in **Figure 3.30(a)**.

Step 2. Obtain a second point using the slope, m. Write m as a fraction, and use rise over run, starting at the point on the y-axis, to plot this point. The slope, 4, written as a fraction is $\frac{4}{1}$.

$$m = \frac{4}{1} = \frac{\text{Rise}}{\text{Run}}$$

We plot the second point on the line by starting at $(0, -3)$, the first point. Based on the slope, we move 4 units *up* (the rise) and 1 unit to the *right* (the run). This puts us at a second point on the line, $(1, 1)$, shown in **Figure 3.30(b)**.

Step 3. Use a straightedge to draw a line through the two points. The graph of $y = 4x - 3$ is shown in **Figure 3.30(c)**.

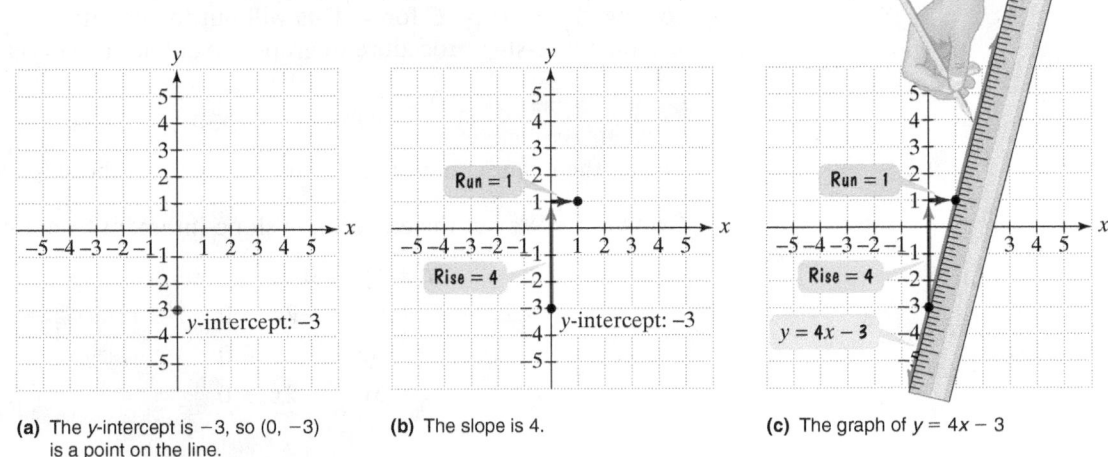

(a) The y-intercept is -3, so $(0, -3)$ is a point on the line.

(b) The slope is 4.

(c) The graph of $y = 4x - 3$

Figure 3.30 Graphing $y = 4x - 3$ using the y-intercept and slope ■

✓   **CHECK POINT 2**   Graph the line whose equation is $y = 3x - 2$.

EXAMPLE 3   Graphing by Using the Slope and y-Intercept

Graph the line whose equation is $y = \frac{2}{3}x + 2$.

Solution   The equation of the line is in the form $y = mx + b$. We can find the slope, m, by identifying the coefficient of x. We can find the y-intercept, b, by identifying the constant term.

$$y = \frac{2}{3}x + 2$$

The slope is $\frac{2}{3}$.   The y-intercept is **2.**

Now that we have identified the slope and the y-intercept, we use the three-step procedure to graph the equation.

$$y = \frac{2}{3}x + 2$$

The slope is $\frac{2}{3}$.

The y-intercept is 2.

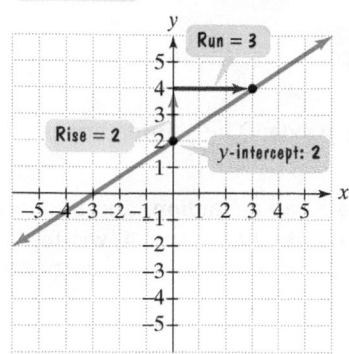

Figure 3.31
The graph of $y = \frac{2}{3}x + 2$

3 Use slope and y-intercept to graph $Ax + By = C$.

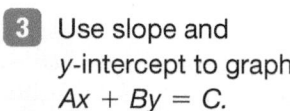

Step 1. Plot the point containing the y-intercept on the y-axis. The y-intercept is 2. We plot $(0, 2)$, shown in **Figure 3.31**.

Step 2. Obtain a second point using the slope, m. Write m as a fraction, and use rise over run, starting at the point on the y-axis, to plot this point. The slope, $\frac{2}{3}$, is already written as a fraction.

$$m = \frac{2}{3} = \frac{\text{Rise}}{\text{Run}}$$

We plot the second point on the line by starting at $(0, 2)$, the first point. Based on the slope, we move 2 units *up* (the rise) and 3 units to the *right* (the run). This puts us at a second point on the line, $(3, 4)$, shown in **Figure 3.31**.

Step 3. Use a straightedge to draw a line through the two points. The graph of $y = \frac{2}{3}x + 2$ is shown in **Figure 3.31**. ∎

✓ **CHECK POINT 3** Graph the line whose equation is $y = \frac{3}{5}x + 1$.

Graphing $Ax + By = C$ by Using the Slope and y-Intercept

Earlier in this chapter, we considered linear equations of the form $Ax + By = C$. We used x- and y-intercepts, as well as checkpoints, to graph these equations. It is also possible to obtain the graphs by using the slope and y-intercept. To do this, begin by solving $Ax + By = C$ for y. This will put the equation in slope-intercept form. Then use the three-step procedure to graph the equation. This is illustrated in Example 4.

EXAMPLE 4   Graphing by Using the Slope and y-Intercept

Graph the linear equation $2x + 5y = 0$ by using the slope and y-intercept.

Solution   We put the equation in slope-intercept form by solving for y.

$2x + 5y = 0$	This is the given equation.
$2x - 2x + 5y = -2x + 0$	Subtract 2x from both sides.
$5y = -2x + 0$	Simplify.
$\dfrac{5y}{5} = \dfrac{-2x + 0}{5}$	Divide both sides by 5.
$y = \dfrac{-2x}{5} + \dfrac{0}{5}$	Divide each term in the numerator by 5.
$y = -\dfrac{2}{5}x + 0$	Simplify.

Now that the equation is in slope-intercept form, we can use the slope and y-intercept to obtain its graph. Examine the slope-intercept form:

$$y = -\frac{2}{5}x + 0.$$

slope: $-\frac{2}{5}$ y-intercept: 0

Note that the slope is $-\frac{2}{5}$ and the y-intercept is 0. Use the y-intercept to plot $(0, 0)$ on the y-axis. Then locate a second point by using the slope.

$$m = -\frac{2}{5} = \frac{-2}{5} = \frac{\text{Rise}}{\text{Run}}$$

Because the rise is -2 and the run is 5, move *down* 2 units and to the *right* 5 units, starting at the point $(0, 0)$. This puts us at a second point on the line, $(5, -2)$. The graph of $2x + 5y = 0$ is the line drawn through these points, shown in **Figure 3.32**. ∎

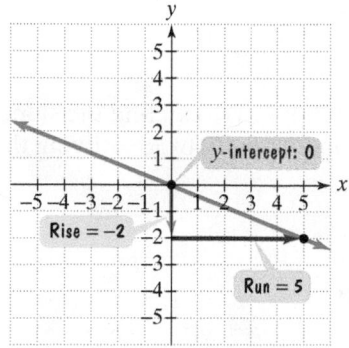

Figure 3.32 The graph of $2x + 5y = 0$, or $y = -\frac{2}{5}x + 0$

Discover for Yourself

You can obtain a second point in Example 4 by writing the slope as follows:

$$m = \frac{2}{-5} = \frac{\text{Rise}}{\text{Run}}.$$

$\frac{2}{5}$ can be expressed as $\frac{-2}{5}$ or $\frac{2}{-5}$.

Now obtain this second point in **Figure 3.32** by moving *up* 2 units and to the *left* 5 units, starting at $(0, 0)$. What do you observe once you graph the line?

✓ **CHECK POINT 4**   Graph the linear equation $3x + 4y = 0$ by using the slope and y-intercept.

---

4 Use slope and y-intercept to model data.

Modeling with the Slope-Intercept Form of the Equation of a Line

The slope-intercept form for equations of lines is useful for obtaining mathematical models for data that fall on or near a line. For example, the bar graph in **Figure 3.33(a)** shows the percentage of the U.S. population who had graduated from high school and from college in 1960 and 2010. The data are displayed as points in a rectangular coordinate system in **Figure 3.33(b)**.

Percentage of High School Graduates and College Graduates in the U.S. Population

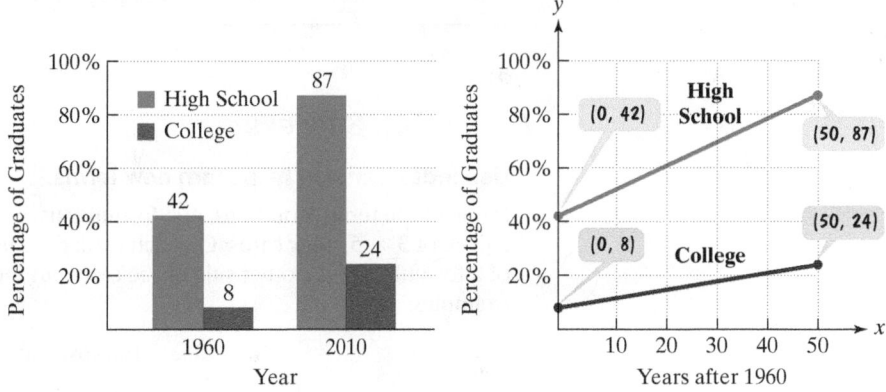

Figure 3.33(a)   **Figure 3.33(b)**

Source: James M. Henslin, *Essentials of Sociology*, Ninth Edition, Pearson, 2011.

Example 5 illustrates how we can use the equation $y = mx + b$ to obtain a model for the data and make predictions about what might occur in the future.

EXAMPLE 5   Modeling with the Slope-Intercept Form of the Equation

a. Use the two points for high school in **Figure 3.33(b)** to find an equation in the form $y = mx + b$ that models the percentage of high school graduates in the U.S. poulation, y, x years after 1960.

b. Use the model to project the percentage of high school graduates in 2020.

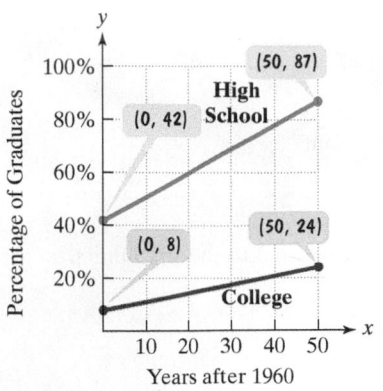

Figure 3.33(b) (repeated)

Solution

a. We will use the line segment for high school using the points (0, 42) and (50, 87) to obtain a model. We need values for m, the slope, and b, the y-intercept.

$$y = mx + b$$

$m = \dfrac{\text{Change in } y}{\text{Change in } x}$

$= \dfrac{87 - 42}{50 - 0} = 0.9$

The point (0, 42) lies on the line segment, so the y-intercept is 42: $b = 42.$

The percentage of the U.S. population who graduated from high school, y, x years after 1960 can be modeled by the linear equation

$$y = 0.9x + 42.$$

The slope, 0.9, indicates an increase in the percentage of high school graduates of 0.9% per year from 1960 through 2010.

b. Now let's use this model to project the percentage of high school graduates in 2020. Because 2020 is 60 years after 1960, substitute 60 for x in $y = 0.9x + 42$ and evaluate the formula.

$$y = 0.9(60) + 42 = 54 + 42 = 96$$

Our model projects that 96% of the U.S. population will have graduated from high school in 2020. ∎

✓ **CHECK POINT 5**

a. Use the two points for college in **Figure 3.33(b)** to find an equation in the form $y = mx + b$ that models the percentage of college graduates in the U.S. population, y, x years after 1960.

b. Use the model to project the percentage of college graduates in 2020.

Achieving Success

Use index cards to help learn new terms.

Many of the terms, notations, and formulas used in this book will be new to you. Buy a pack of 3 × 5 index cards. On each card, list a new vocabulary word, symbol, or title of a formula. On the other side of the card, put the definition or formula. Here are two examples:

Effective Index Cards

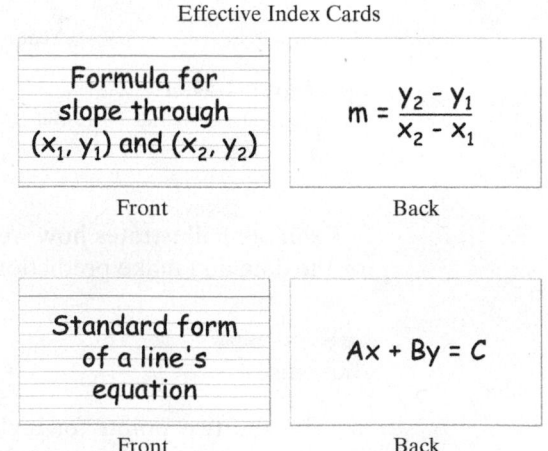

Review these cards frequently. Use the cards to quiz yourself and prepare for exams.

CONCEPT AND VOCABULARY CHECK

Fill in each blank so that the resulting statement is true.

1. The slope-intercept form of the equation of a line is _____, where m represents the _____ and b represents the _____.

2. In order to graph the line whose equation is $y = \frac{2}{5}x + 3$, begin by plotting the point _____. From this point, we move _____ units up (the rise) and _____ units to the right (the run).

3. In order to graph equations of the form $Ax + By = C$ using the slope and y-intercept, we begin by solving the equation for _____.

3.4 EXERCISE SET

MyMathLab®

Watch the videos
in MyMathLab

Download the
MyDashBoard App

Practice Exercises

In Exercises 1–12, find the slope and the y-intercept of the line with the given equation.

1. $y = 3x + 2$
2. $y = 9x + 4$

3. $y = 3x - 5$
4. $y = 4x - 2$

5. $y = -\frac{1}{2}x + 5$
6. $y = -\frac{3}{4}x + 6$

7. $y = 7x$
8. $y = 10x$

9. $y = 10$
10. $y = 7$

11. $y = 4 - x$
12. $y = 5 - x$

In Exercises 13–26, begin by solving the linear equation for y. This will put the equation in slope-intercept form. Then find the slope and the y-intercept of the line with this equation.

13. $-5x + y = 7$

14. $-9x + y = 5$

15. $x + y = 6$

16. $x + y = 8$

17. $6x + y = 0$

18. $8x + y = 0$

19. $3y = 6x$

20. $3y = -9x$

21. $2x + 7y = 0$

22. $2x + 9y = 0$

23. $3x + 2y = 3$

24. $4x + 3y = 4$

25. $3x - 4y = 12$

26. $5x - 2y = 10$

In Exercises 27–38, graph each linear equation using the slope and y-intercept

27. $y = 2x + 4$
28. $y = 3x + 1$

29. $y = -3x + 5$
30. $y = -2x + 4$

31. $y = \frac{1}{2}x + 1$
32. $y = \frac{1}{3}x + 2$

33. $y = \frac{2}{3}x - 5$
34. $y = \frac{3}{4}x - 4$

35. $y = -\frac{3}{4}x + 2$
36. $y = -\frac{2}{3}x + 4$

37. $y = -\frac{5}{3}x$
38. $y = -\frac{4}{3}x$

In Exercises 39–46,

  a. *Put the equation in slope-intercept form by solving for y.*

  b. *Identify the slope and the y-intercept.*

  c. *Use the slope and y-intercept to graph the equation.*

39. $3x + y = 0$
40. $2x + y = 0$

41. $3y = 4x$
42. $4y = 5x$

43. $2x + y = 3$
44. $3x + y = 4$

45. $7x + 2y = 14$

46. $5x + 3y = 15$

In Exercises 47–56, graph both linear equations in the same rectangular coordinate system. If the lines are parallel or perpendicular, explain why.

47. $y = 3x + 1$
    $y = 3x - 3$
48. $y = 2x + 4$
    $y = 2x - 3$

49. $y = -3x + 2$
    $y = 3x + 2$
50. $y = -2x + 1$
    $y = 2x + 1$

51. $y = x + 3$
$y = -x + 1$

52. $y = x + 2$
$y = -x - 1$

53. $x - 2y = 2$
$2x - 4y = 3$

54. $x - 3y = 9$
$3x - 9y = 18$

55. $2x - y = -1$
$x + 2y = -6$

56. $3x - y = -2$
$x + 3y = -9$

58. The y-intercept is -4 and the line is parallel to the line whose equation is $2x + y = 8$.

59. The y-intercept is 6 and the line is perpendicular to the line whose equation is $y = 5x - 1$.

60. The y-intercept is 7 and the line is perpendicular to the line whose equation is $y = 8x - 3$.

61. The line has the same y-intercept as the line whose equation is $16y = 8x + 32$ and is parallel to the line whose equation is $3x + 3y = 9$.

62. The line has the same y-intercept as the line whose equation is $2y = 6x + 8$ and is parallel to the line whose equation is $4x + 4y = 20$.

Practice PLUS

In Exercises 57–64, write an equation in the form $y = mx + b$ of the line that is described.

57. The y-intercept is 5 and the line is parallel to the line whose equation is $3x + y = 6$.

63. The line rises from left to right. It passes through the origin and a second point with equal x- and y-coordinates.

64. The line falls from left to right. It passes through the origin and a second point with opposite x- and y-coordinates.

Application Exercises

The bar graph shows the racial and ethnic composition of the United States population in 2008, with projections for 2050. Exercises 65–66 involve the graphs shown in the rectangular coordinate system for two of the racial/ethnic groups.

Racial and Ethnic Composition of the United States Population

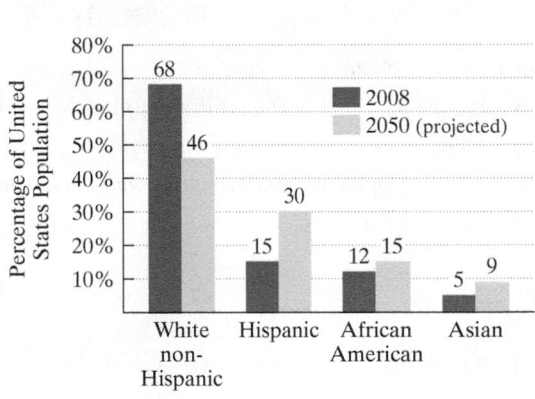

Source: Urban Institute

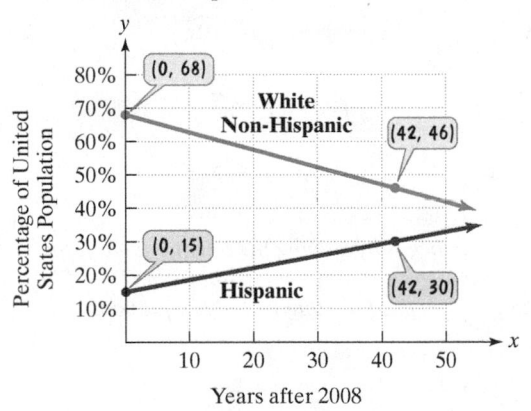

65. a. Use the two points for white non-Hispanics to find an equation in the form $y = mx + b$ that models the percentage of white non-Hispanics, y, in the United States population x years after 2008. Round m to two decimal places.

  b. Use the model from part (a) to project the percentage of white non-Hispanics in the United States in 2108.

66. a. Use the two points for Hispanics to find an equation in the form $y = mx + b$ that models the percentage of Hispanics, y, in the United States population x years after 2008. Round m to two decimal places.

  b. Use the model from part (a) to project the percentage of Hispanics in the United States in 2108.

Writing in Mathematics

67. Describe how to find the slope and the y-intercept of a line whose equation is given.

68. Describe how to graph a line using the slope and y-intercept. Provide an original example with your description.

69. A formula in the form $y = mx + b$ models the cost, y, of a four-year college x years after 2010. Would you expect m to be positive, negative, or zero? Explain your answer.

Critical Thinking Exercises

Make Sense? *In Exercises 70–73, determine whether each statement "makes sense" or "does not make sense" and explain your reasoning.*

70. The slope-intercept form of a line's equation makes it possible for me to determine immediately the slope and the *y*-intercept.

71. Because the variable *m* does not appear in $Ax + By = C$, equations in this form make it impossible to determine the line's slope.

72. If I drive *m* miles in a year, the formula $c = 0.25m + 3500$ models the annual cost, *c*, in dollars, of operating my car, so the equation shows that with no driving at all, the cost is $3500, and the rate of increase in this cost is $0.25 for each mile that I drive.

73. Some lines are impossible to graph because when obtaining a second point using the slope, this point lands outside the edge of my graph paper.

In Exercises 74–77, determine whether each statement is true or false. If the statement is false, make the necessary change(s) to produce a true statement.

74. The equation $y = mx + b$ shows that no line can have a *y*-intercept that is numerically equal to its slope.

75. Every line in the rectangular coordinate system has an equation that can be expressed in slope-intercept form.

76. The line $3x + 2y = 5$ has slope $-\frac{3}{2}$.

77. The line $2y = 3x + 7$ has a *y*-intercept of 7.

78. The relationship between Celsius temperature, *C*, and Fahrenheit temperature, *F*, can be described by a linear equation in the form $F = mC + b$. The graph of this equation contains the point (0, 32): Water freezes at 0°C or at 32°F. The line also contains the point (100, 212): Water boils at 100°C or at 212°F. Write the linear equation expressing Fahrenheit temperature in terms of Celsius temperature.

Review Exercises

79. Solve: $\dfrac{x}{2} + 7 = 13 - \dfrac{x}{4}$. (Section 2.3, Example 4)

80. Simplify: $3(12 \div 2^2 - 3)^2$. (Section 1.8, Example 6)

81. 14 is 25% of what number? (Section 2.4, Example 6)

Preview Exercises

Exercises 82–84 will help you prepare for the material covered in the next section. In each exercise, solve for y and put the equation in slope-intercept form.

82. $y - 3 = 4(x + 1)$

83. $y + 3 = -\frac{3}{2}(x - 4)$

84. $y - 30.0 = 0.265(x - 10)$

MID-CHAPTER CHECK POINT Section 3.1–Section 3.4

 What You Know: We learned to graph equations in two variables using point plotting, as well as a variety of other techniques. We used intercepts and a checkpoint to graph linear equations in the form $Ax + By = C$. We saw that $y = b$ graphs as a horizontal line and $x = a$ graphs as a vertical line. We determined a line's steepness, or rate of change, by computing its slope. We saw that lines with the same slope are parallel and lines with slopes that have a product of -1 are perpendicular. Finally, we learned to graph linear equations in slope-intercept form, $y = mx + b$, using the slope, *m*, and the *y*-intercept, *b*.

In Exercises 1–3, use each graph to determine

a. *the x-intercept, or state that there is no x-intercept.*

b. *the y-intercept, or state that there is no y-intercept.*

c. *the line's slope, or state that the slope is undefined.*

1.

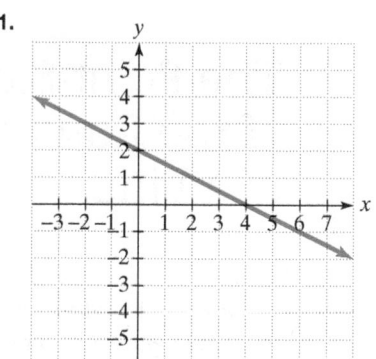

2.

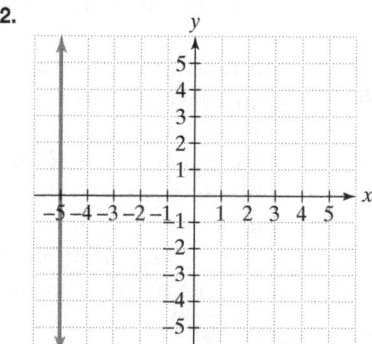

3.

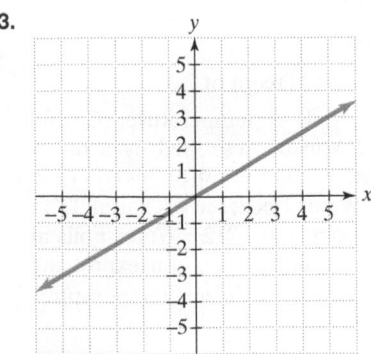

In Exercises 4–15, graph each equation in a rectangular coordinate system.

4. $y = -2x$　　　**5.** $y = -2$　　　**6.** $x + y = -2$

7. $y = \frac{1}{3}x - 2$　　**8.** $x = 3.5$　　　**9.** $4x - 2y = 8$

10. $y = 3x + 2$　　**11.** $3x + y = 0$　　**12.** $y = -x + 4$

13. $y = x - 4$　　**14.** $5y = -3x$　　**15.** $5y = 20$

16. Find the slope and the y-intercept of the line whose equation is $5x - 2y = 10$.

In Exercises 17–19, determine whether the lines through each pair of points are parallel, perpendicular, or neither.

17. $(-5, -3)$ and $(0, -4)$; $(-2, -8)$ and $(1, 7)$

18. $(-4, 1)$ and $(2, 7)$; $(-5, 13)$ and $(4, -5)$

19. $(2, -4)$ and $(7, 0)$; $(-4, 2)$ and $(1, 6)$

20. The graphs indicate that a smaller percentage of new military recruits graduated high school in 2009 than in 1999.

Percentage of New U.S. Military Recruits Who Graduated High School

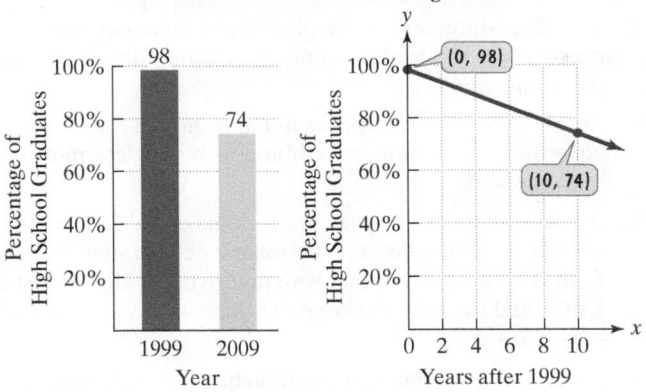

Source: Department of Defense

a. Use the two points labeled by the voice balloons to find an equation in the form $y = mx + b$ that models the percentage of new U.S. military recruits who graduated high school, y, x years after 1999.

b. If trends shown by the graphs continue, use the model in part (a) to project the percentage of new military recruits who graduated high school in 2014.

3.5

The Point-Slope Form of the Equation of a Line

Objectives

1 Use the point-slope form to write equations of a line.

2 Find slopes and equations of parallel and perpendicular lines.

3 Write linear equations that model data and make predictions.

Surprised by the number of people smoking cigarettes in movies and television shows made in the 1940s and 1950s? At that time, there was little awareness of the relationship between tobacco use and numerous diseases. Cigarette smoking was seen as a healthy way to relax and help digest a hearty meal. Then, in 1964, a linear equation changed everything. To understand the mathematics behind this turning point in public health, we explore another form of a line's equation.

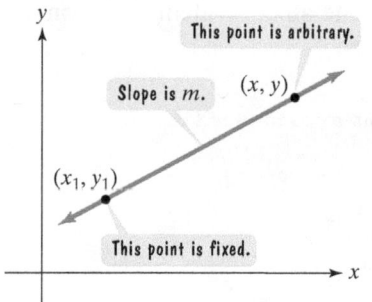

Figure 3.34 A line passing through (x_1, y_1) with slope m

Point-Slope Form

We can use the slope of a line to obtain another useful form of the line's equation. Consider a nonvertical line that has slope m and contains the point (x_1, y_1). Now let (x, y) represent any other point on the line, shown in **Figure 3.34**. Keep in mind that the point (x, y) is arbitrary and is not in one fixed position. By contrast, the point (x_1, y_1) is fixed.

Regardless of where the point (x, y) is located on the line, the steepness of the line in **Figure 3.34** remains the same. Thus, the ratio for slope stays a constant m. This means that for all points (x, y) along the line,

$$m = \frac{\text{Change in } y}{\text{Change in } x} = \frac{y - y_1}{x - x_1}.$$

We can clear the fraction by multiplying both sides by $x - x_1$, the least common denominator, where $x - x_1 \neq 0$.

$$m = \frac{y - y_1}{x - x_1} \qquad \text{This is the slope of the line in Figure 3.34.}$$

$$m(x - x_1) = \frac{y - y_1}{x - x_1} \cdot (x - x_1) \qquad \text{Multiply both sides by } x - x_1.$$

$$m(x - x_1) = y - y_1 \qquad \text{Simplify: } \frac{y - y_1}{x - x_1} \cdot (x - x_1) = y - y_1.$$

Now, if we reverse the two sides, we obtain the **point-slope form** of the equation of a line.

Great Question!

When using $y - y_1 = m(x - x_1)$, for which variables do I substitute numbers?

When writing the point-slope form of a line's equation, you will never substitute numbers for x and y. You will substitute values for x_1, y_1, and m.

Point-Slope Form of the Equation of a Line

The **point-slope form of the equation** of a nonvertical line with slope m that passes through the point (x_1, y_1) is

$$y - y_1 = m(x - x_1).$$

For example, the point-slope form of the equation of the line passing through $(1, 4)$ with slope $2\,(m = 2)$ is

$$y - 4 = 2(x - 1).$$

1 Use the point-slope form to write equations of a line.

Using the Point-Slope Form to Write a Line's Equation

If we know the slope of a line and a point not containing the y-intercept through which the line passes, the point-slope form is the equation that we should use. Once we have obtained this equation, it is customary to solve for y and write the equation in slope-intercept form. Examples 1 and 2 illustrate these ideas.

EXAMPLE 1   Writing the Point-Slope Form and the Slope-Intercept Form

Write the point-slope form and the slope-intercept form of the equation of the line with slope 4 that passes through the point $(-1, 3)$.

Solution   We begin with the point-slope form of the equation of a line with $m = 4$, $x_1 = -1$, and $y_1 = 3$.

$$y - y_1 = m(x - x_1) \qquad \text{This is the point-slope form of the equation.}$$

$$y - 3 = 4[x - (-1)] \qquad \text{Substitute: } (x_1, y_1) = (-1, 3) \text{ and } m = 4.$$

$$y - 3 = 4(x + 1) \qquad \text{We now have the point-slope form of the equation of the given line.}$$

Now we solve the equation $y - 3 = 4(x + 1)$ for y and write an equivalent equation in slope-intercept form ($y = mx + b$).

We need to isolate y. $y - 3 = 4(x + 1)$ This is the point-slope form of the equation.

$$y - 3 = 4x + 4 \quad \text{Use the distributive property.}$$
$$y = 4x + 7 \quad \text{Add 3 to both sides.}$$

The slope-intercept form of the line's equation is $y = 4x + 7$. ∎

✓ **CHECK POINT 1**   Write the point-slope form and the slope-intercept form of the equation of the line with slope 6 that passes through the point $(2, -5)$.

EXAMPLE 2   Writing the Point-Slope Form and the Slope-Intercept Form

A line passes through the points $(4, -3)$ and $(-2, 6)$. (See **Figure 3.35**.) Find the equation of the line

a. in point-slope form.   **b.** in slope-intercept form.

Solution

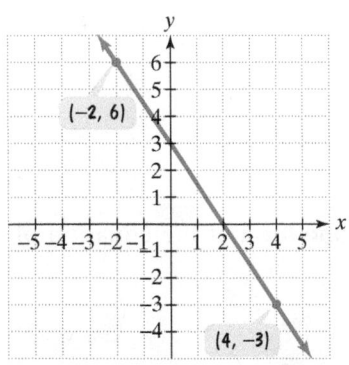

Figure 3.35

a. To use the point-slope form, we need to find the slope. The slope is the change in the y-coordinates divided by the corresponding change in the x-coordinates.

$$m = \frac{6 - (-3)}{-2 - 4} = \frac{9}{-6} = -\frac{3}{2} \quad \begin{array}{l}\text{This is the definition of slope} \\ \text{using } (4, -3) \text{ and } (-2, 6).\end{array}$$

We can take either point on the line to be (x_1, y_1). Let's use $(x_1, y_1) = (4, -3)$. Now, we are ready to write the point-slope form of the equation.

$$y - y_1 = m(x - x_1) \quad \text{This is the point-slope form of the equation.}$$

$$y - (-3) = -\frac{3}{2}(x - 4) \quad \text{Substitute: } (x_1, y_1) = (4, -3) \text{ and } m = -\frac{3}{2}.$$

$$y + 3 = -\frac{3}{2}(x - 4) \quad \text{Simplify.}$$

This equation is one point-slope form of the equation of the line shown in **Figure 3.35**.

Discover for Yourself

If you are given two points on a line, you can use either point for (x_1, y_1) when you write its point-slope equation. Rework Example 2 using $(-2, 6)$ for (x_1, y_1). Once you solve for y, you should obtain the same slope-intercept equation as the one shown in the last line of the solution to Example 2.

b. Now, we solve this equation for y and write an equivalent equation in slope-intercept form ($y = mx + b$).

We need to isolate y. $y + 3 = -\frac{3}{2}(x - 4)$ This is the point-slope form of the equation.

$$y + 3 = -\frac{3}{2}x + 6 \quad \text{Use the distributive property.}$$

$$y = -\frac{3}{2}x + 3 \quad \text{Subtract 3 from both sides.}$$

This equation is the slope-intercept form of the equation of the line shown in **Figure 3.35**. ∎

✓ **CHECK POINT 2**   A line passes through the points $(-2, -1)$ and $(-1, -6)$. Find the equation of the line

a. in point-slope form.

b. in slope-intercept form.

The many forms of a line's equation can be a bit overwhelming. **Table 3.2** summarizes the various forms and contains the most important things you should remember about each form.

Table 3.2 Equations of Lines	
Form	**What You Should Know**
Standard Form $Ax + By = C$	• Graph equations in this form using intercepts (x-intercept: set $y = 0$; y-intercept: set $x = 0$) and a checkpoint.
$y = b$	• Graph equations in this form as horizontal lines with b as the y-intercept.
$x = a$	• Graph equations in this form as vertical lines with a as the x-intercept.
Slope-Intercept Form $y = mx + b$	• Graph equations in this form using the y-intercept, b, and the slope, m. • Start with this form when writing a linear equation if you know a line's slope and y-intercept.

Point-Slope Form $y - y_1 = m(x - x_1)$	• Start with this form when writing a linear equation if you know
	     the slope of the line and a point on the line not containing the y-intercept
	     or
	     two points on the line, neither of which contains the y-intercept. Calculate the slope using $$m = \frac{\text{Change in } y}{\text{Change in } x} = \frac{y_2 - y_1}{x_2 - x_1}.$$ Although you begin with point-slope form, you usually solve for y and convert to slope-intercept form.

2 Find slopes and equations of parallel and perpendicular lines.

Parallel and Perpendicular Lines

The next example uses the fact that parallel lines have the same slope.

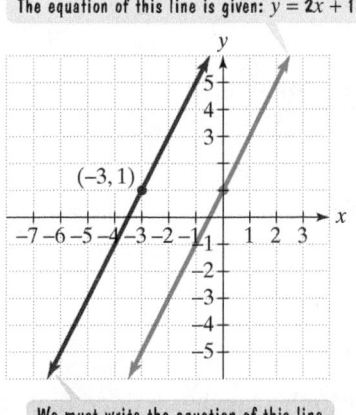

The equation of this line is given: $y = 2x + 1$.

$(-3, 1)$

We must write the equation of this line.

Figure 3.36

EXAMPLE 3 Writing Equations of a Line Parallel to a Given Line

Write an equation of the line passing through $(-3, 1)$ and parallel to the line whose equation is $y = 2x + 1$. Express the equation in point-slope form and slope-intercept form.

Solution The situation is illustrated in **Figure 3.36**. We are looking for the equation of the red line shown on the left. How do we obtain this equation? Notice that the line passes through the point $(-3, 1)$. Using the point-slope form of the line's equation, we have $x_1 = -3$ and $y_1 = 1$.

$$y - y_1 = m(x - x_1)$$

$y_1 = 1 \qquad x_1 = -3$

Now the only thing missing from the equation of the red line is m, the slope. Do we know anything about the slope of either line in **Figure 3.36**? The answer is yes; we know the slope of the blue line on the right, whose equation is given.

$$y = 2x + 1$$

The slope of the blue line on the right in Figure 3.36 is 2.

The equation of this line is given: $y = 2x + 1$.

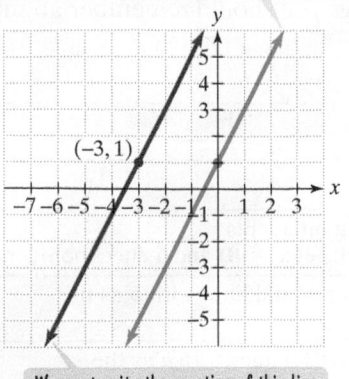

$(-3, 1)$

We must write the equation of this line.

Figure 3.36   (repeated)

Parallel lines have the same slope. Because the slope of the blue line in **Figure 3.36** is 2, the slope of the red line, the line whose equation we must write, is also 2: $m = 2$. We now have values for x_1, y_1, and m for the red line.

$$y - y_1 = m(x - x_1)$$

$$\boxed{y_1 = 1} \quad \boxed{m = 2} \quad \boxed{x_1 = -3}$$

The point-slope form of the red line's equation is

$$y - 1 = 2[x - (-3)] \text{ or }$$
$$y - 1 = 2(x + 3).$$

Solving for y, we obtain the slope-intercept form of the equation.

$$y - 1 = 2x + 6 \qquad \text{Apply the distributive property.}$$
$$y = 2x + 7 \qquad \text{Add 1 to both sides. This is the}$$
$$\text{slope-intercept form, } y = mx + b, \text{ of the}$$
$$\text{equation.} \blacksquare$$

☑ **CHECK POINT 3**   Write an equation of the line passing through $(-2, 5)$ and parallel to the line whose equation is $y = 3x + 1$. Express the equation in point-slope form and slope-intercept form.

We have seen that if two nonvertical lines are perpendicular, then the product of their slopes is -1. The next example uses the fact that perpendicular lines have slopes that are negative reciprocals.

EXAMPLE 4   Writing Equations of a Line Perpendicular to a Given Line

a. Find the slope of any line that is perpendicular to the line whose equation is $x + 4y = 8$.

b. Write the equation of the line passing through $(3, -5)$ and perpendicular to the line whose equation is $x + 4y = 8$. Express the equation in point-slope form and slope-intercept form.

Solution

a. We begin by writing the equation of the given line, $x + 4y = 8$, in slope-intercept form. Solve for y.

$$x + 4y = 8 \qquad \text{This is the given equation.}$$
$$4y = -x + 8 \qquad \text{To isolate the } y\text{-term, subtract } x \text{ from both sides.}$$
$$y = -\frac{1}{4}x + 2 \qquad \text{Divide both sides by 4.}$$

$$\boxed{\text{Slope is } -\tfrac{1}{4}.}$$

The given line has slope $-\frac{1}{4}$. Any line perpendicular to this line has a slope that is the negative reciprocal of $-\frac{1}{4}$. Thus, the slope of any perpendicular line is 4.

b. Let's begin by writing the point-slope form of the perpendicular line's equation. Because the line passes through the point $(3, -5)$, we have $x_1 = 3$ and $y_1 = -5$. In part (a), we determined that the slope of any line perpendicular to $x + 4y = 8$ is 4, so the slope of this particular perpendicular line must be 4: $m = 4$.

$$y - y_1 = m(x - x_1)$$

$$\boxed{y_1 = -5} \quad \boxed{m = 4} \quad \boxed{x_1 = 3}$$

The point-slope form of the perpendicular line's equation is

$$y - (-5) = 4(x - 3) \text{ or}$$
$$y + 5 = 4(x - 3).$$

How can we express this equation in slope-intercept form, $y = mx + b$? We need to solve for y.

$y + 5 = 4(x - 3)$	This is the point-slope form of the line's equation.
$y + 5 = 4x - 12$	Apply the distributive property.
$y = 4x - 17$	Subtract 5 from both sides of the equation and solve for y.

The slope-intercept form of the perpendicular line's equation is $y = 4x - 17$. ∎

✓ **CHECK POINT 4**

a. Find the slope of any line that is perpendicular to the line whose equation is $x + 3y = 12$.

b. Write the equation of the line passing through $(-2, -6)$ and perpendicular to the line whose equation is $x + 3y = 12$. Express the equation in point-slope form and slope-intercept form.

---

3 Write linear equations that model data and make predictions.

Applications

Linear equations are useful for modeling data that fall on or near a line. For example, the bar graph in **Figure 3.37(a)** gives the median age of the U.S. population in the indicated year. (The median age is the age in the middle when all the ages of the U.S. population are arranged from youngest to oldest.) The data are displayed as a set of five points in a rectangular coordinate system in **Figure 3.37(b)**.

The Graying of America: Median Age of the United States Population

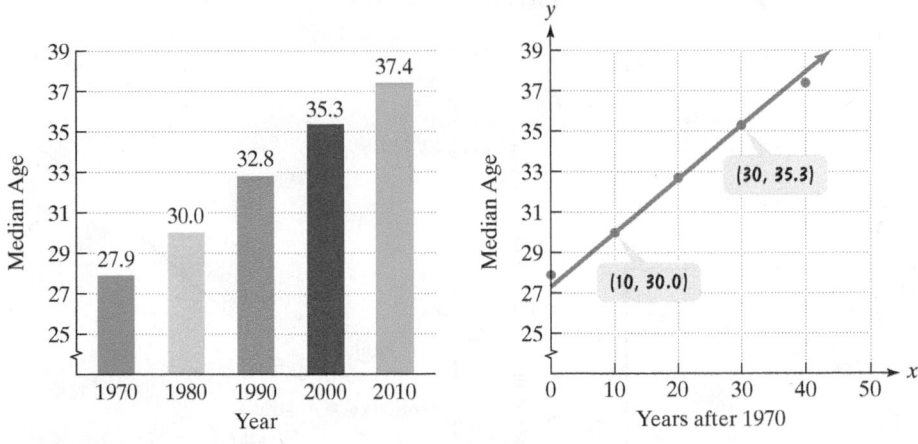

Figure 3.37(a)

Source: U.S. Census Bureau

Figure 3.37(b)

A set of points representing data is called a **scatter plot**. Also shown on the scatter plot in **Figure 3.37(b)** is a line that passes through or near the five points. By writing the equation of this line, we can obtain a model of the data and make predictions about the median age of the U.S. population in the future.

Median Age of the United States Population

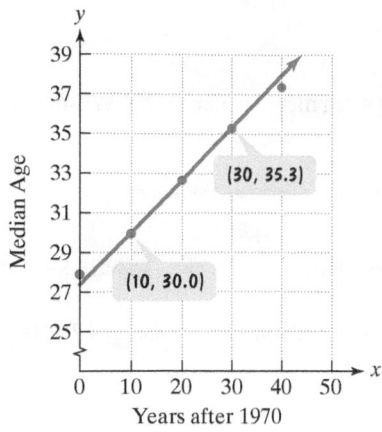

Figure 3.37(b) (repeated)

EXAMPLE 5 Modeling the Graying of America

Write the slope-intercept form of the equation of the line shown in **Figure 3.37(b)**. Use the equation to predict the median age of the U.S. population in 2020.

Solution   The line in **Figure 3.37(b)** passes through (10, 30.0) and (30, 35.3). We start by finding its slope.

$$m = \frac{\text{Change in } y}{\text{Change in } x} = \frac{35.3 - 30.0}{30 - 10} = \frac{5.3}{20} = 0.265$$

The slope indicates that each year the median age of the U.S. population is increasing by 0.265 year.

Now, we write the slope-intercept form of the equation for the line.

$y - y_1 = m(x - x_1)$	Begin with the point-slope form.
$y - 30.0 = 0.265(x - 10)$	Either ordered pair can be (x_1, y_1). Let $(x_1, y_1) = (10, 30.0)$. From above, $m = 0.265$.
$y - 30.0 = 0.265x - 2.65$	Apply the distributive property.
$y = 0.265x + 27.35$	Add 30 to both sides and solve for y.

A linear equation that models the median age of the U.S. population, y, x years after 1970 is

$$y = 0.265x + 27.35.$$

Now, let's use this equation to predict the median age in 2020. Because 2020 is 50 years after 1970, substitute 50 for x and compute y.

$$y = 0.265(50) + 27.35 = 40.6$$

Our model predicts that the median age of the U.S. population in 2020 will be 40.6.   ∎

✓ **CHECK POINT 5**   Use the data points (10, 30.0) and (20, 32.8) from **Figure 3.37(b)** to write the slope-intercept form of an equation that models the median age of the U.S. population x years after 1970. Use this model to predict the median age in 2020.

Using Technology

You can use a graphing utility to obtain a model for a scatter plot in which the data points fall on or near a straight line. The line that best fits the data is called the **regression line**. After entering the data in **Figure 3.37(b)**, a graphing utility displays a scatter plot of the data and the regression line.

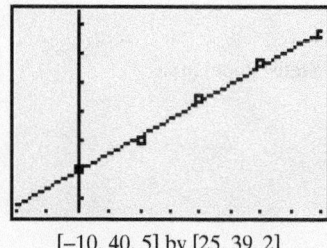

[−10, 40, 5] by [25, 39, 2]

Also displayed is the regression line's equation.

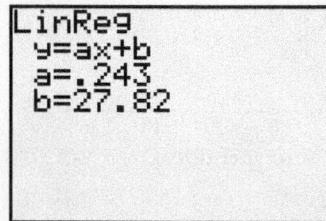

Blitzer Bonus

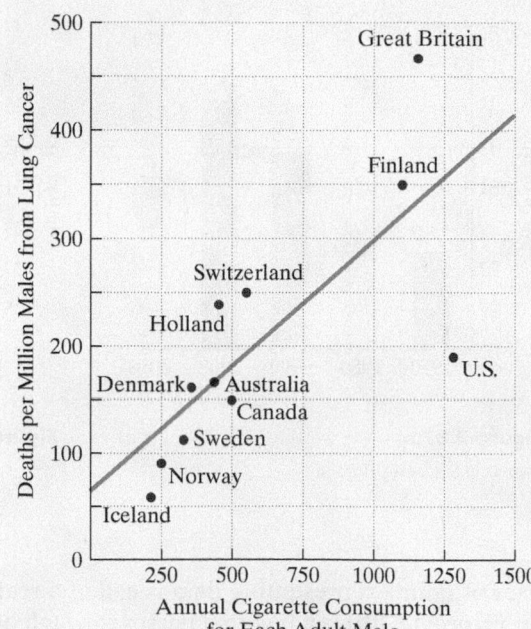

Source: *Smoking and Health*, Washington, D.C., 1964

Cigarettes and Lung Cancer

This scatter plot shows a relationship between cigarette consumption among males and deaths due to lung cancer per million males. The data are from 11 countries and date back to a 1964 report by the U.S. Surgeon General. The scatter plot can be modeled by a line whose slope indicates an increasing death rate from lung cancer with increased cigarette consumption. At that time, the tobacco industry argued that in spite of this regression line, tobacco use is not the cause of cancer. Recent data do, indeed, show a causal effect between tobacco use and numerous diseases.

Achieving Success

The Secret of Math Success

What's the secret of math success? The bar graph in **Figure 3.38** shows that Japanese teachers and students are more likely than their American counterparts to believe that the key to doing well in math is working hard. Americans tend to think that either you have mathematical intelligence or you don't. Alan Bass, author of *Math Study Skills* (Pearson Education, 2008), strongly disagrees with this American perspective:

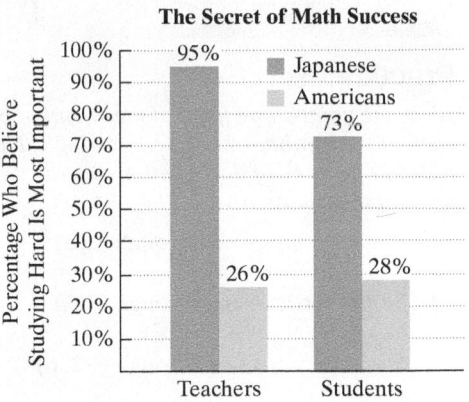

Figure 3.38

Source: Wade and Tavris, *Psychology*, Ninth Edition, Pearson, 2008.

"Human beings are easily intelligent enough to understand the basic principles of math. I cannot repeat this enough, but I'll try ... **Poor performance in math is not due to a lack of intelligence!** The fact is that the key to success in math is in taking an intelligent approach. Students come up to me and say, 'I'm just not good at math.' Then I ask to see their class notebooks and they show me a chaotic mess of papers jammed into a folder. In math, that's a lot like taking apart your car's engine, dumping the heap of disconnected parts back under the hood, and then going to a mechanic to ask why it won't run. Students come to me and say, 'I'm just not good at math.' Then I ask them about their study habits and they say, 'I have to do all my studying on the weekend.' In math, that's a lot like trying to do all your eating or sleeping on the weekends and wondering why you're so tired and hungry all the time. **How you approach math is much more important than how smart you are.**"

—Alan Bass

CONCEPT AND VOCABULARY CHECK

Fill in each blank so that the resulting statement is true.

1. The point-slope form of the equation of a nonvertical line with slope m that passes through the point (x_1, y_1) is _____.

2. The equation $5x + 3y = 10$ is written in _____ form.

3. The equation $y = 3x + 7$ is written in _____ form.

4. The equation $y - 6 = 4(x + 1)$ is written in _____ form.

5. The graph of $y = 3$ is a/an _____ line.

6. The graph of $x = -1$ is a/an _____ line.

7. The slope of the line whose equation is $y = -4x + 3$ is _____. The slope of any line parallel to $y = -4x + 3$ is _____.

8. The slope of the line whose equation is $y = \frac{1}{2}x - 5$ is _____. The slope of any line perpendicular to $y = \frac{1}{2}x - 5$ is _____.

3.5 EXERCISE SET MyMathLab®

Watch the videos
in MyMathLab

Download the
MyDashBoard App

Practice Exercises

Write the point-slope form of the equation of the line satisfying each of the conditions in Exercises 1–28. Then use the point-slope form of the equation to write the slope-intercept form of the equation.

1. Slope = 3, passing through $(2, 5)$

2. Slope = 6, passing through $(3, 1)$

3. Slope = 5, passing through $(-2, 6)$

4. Slope = 7, passing through $(-4, 9)$

5. Slope = -8, passing through $(-3, -2)$

6. Slope = -4, passing through $(-5, -2)$

7. Slope = -12, passing through $(-8, 0)$

8. Slope = -11, passing through $(0, -3)$

9. Slope = -1, passing through $\left(-\frac{1}{2}, -2\right)$

10. Slope = -1, passing through $\left(-4, -\frac{1}{4}\right)$

11. Slope = $\frac{1}{2}$, passing through the origin

12. Slope = $\frac{1}{3}$, passing through the origin

13. Slope = $-\frac{2}{3}$, passing through $(6, -2)$

14. Slope = $-\frac{3}{5}$, passing through $(10, -4)$

15. Passing through $(1, 2)$ and $(5, 10)$

16. Passing through $(3, 5)$ and $(8, 15)$

17. Passing through $(-3, 0)$ and $(0, 3)$

18. Passing through $(-2, 0)$ and $(0, 2)$

19. Passing through $(-3, -1)$ and $(2, 4)$

20. Passing through $(-2, -4)$ and $(1, -1)$

21. Passing through $(-4, -1)$ and $(3, 4)$

22. Passing through $(-6, 1)$ and $(2, -5)$

23. Passing through $(-3, -1)$ and $(4, -1)$

24. Passing through $(-2, -5)$ and $(6, -5)$

25. Passing through $(2, 4)$ with x-intercept $= -2$

26. Passing through $(1, -3)$ with x-intercept $= -1$

27. x-intercept $= -\frac{1}{2}$ and y-intercept $= 4$

28. x-intercept $= 4$ and y-intercept $= -2$

*In Exercises 29–44, the equation of a line is given. Find the slope of a line that is **a.** parallel to the line with the given equation; and **b.** perpendicular to the line with the given equation.*

29. $y = 5x$

30. $y = 3x$

31. $y = -7x$

32. $y = -9x$

33. $y = \frac{1}{2}x + 3$

34. $y = \frac{1}{4}x - 5$

35. $y = -\frac{2}{5}x - 1$

36. $y = -\frac{3}{7}x - 2$

37. $4x + y = 7$

38. $8x + y = 11$

39. $2x + 4y = 8$

40. $3x + 2y = 6$

41. $2x - 3y = 5$

42. $3x - 4y = -7$

43. $x = 6$

44. $y = 9$

In Exercises 45–48, write an equation for line L in point-slope form and slope-intercept form.

45.

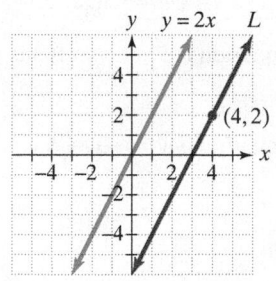

L is parallel to y = 2x.

46.

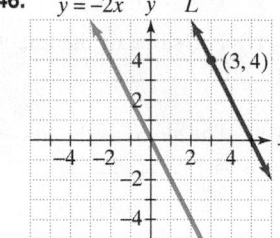

L is parallel to y = -2x.

47.

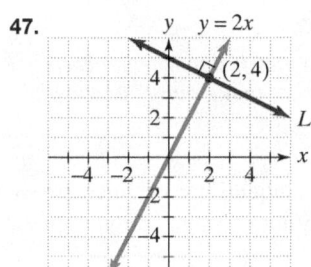

L is perpendicular to y = 2x.

48.

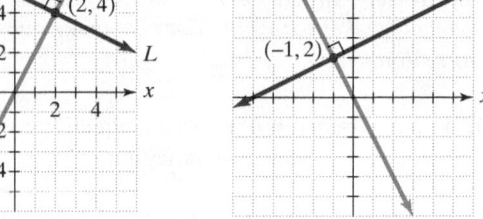

L is perpendicular to y = –2x.

In Exercises 49–56, use the given conditions to write an equation for each line in point-slope form and slope-intercept form.

49. Passing through $(-8, -10)$ and parallel to the line whose equation is $y = -4x + 3$

50. Passing through $(-2, -7)$ and parallel to the line whose equation is $y = -5x + 4$

51. Passing through $(2, -3)$ and perpendicular to the line whose equation is $y = \frac{1}{5}x + 6$

52. Passing through $(-4, 2)$ and perpendicular to the line whose equation is $y = \frac{1}{3}x + 7$

53. Passing through $(-2, 2)$ and parallel to the line whose equation is $2x - 3y = 7$

54. Passing through $(-1, 3)$ and parallel to the line whose equation is $3x - 2y = 5$

55. Passing through $(4, -7)$ and perpendicular to the line whose equation is $x - 2y = 3$

56. Passing through $(5, -9)$ and perpendicular to the line whose equation is $x + 7y = 12$

Practice PLUS

In Exercises 57–66, write an equation in slope-intercept form of the line satisfying the given conditions.

57. The line passes through $(2, 4)$ and has the same y-intercept as the line whose equation is $x - 4y = 8$.

58. The line passes through $(2, 6)$ and has the same y-intercept as the line whose equation is $x - 3y = 18$.

59. The line has an x-intercept at -4 and is parallel to the line containing $(3, 1)$ and $(2, 6)$.

60. The line has an x-intercept at -6 and is parallel to the line containing $(4, -3)$ and $(2, 2)$.

61. The line passes through $(-1, 5)$ and is perpendicular to the line whose equation is $x = 6$.

62. The line passes through $(-2, 6)$ and is perpendicular to the line whose equation is $x = -4$.

63. The line passes through $(-6, 4)$ and is perpendicular to the line that has an x-intercept of 2 and a y-intercept of -4.

64. The line passes through $(-5, 6)$ and is perpendicular to the line that has an x-intercept of 3 and a y-intercept of -9.

65. The line is perpendicular to the line whose equation is $3x - 2y = 4$ and has the same y-intercept as this line.

66. The line is perpendicular to the line whose equation is $4x - y = 6$ and has the same y-intercept as this line.

67. What is the slope of a line that is parallel to the line whose equation is $Ax + By = C, B \neq 0$?

68. What is the slope of a line that is perpendicular to the line whose equation is $Ax + By = C, A \neq 0$ and $B \neq 0$?

Application Exercises

69. Studies show that texting while driving is as risky as driving with a 0.08 blood alcohol level, the standard for drunk driving. The bar graph shows the number of fatalities in the United States involving distracted driving from 2004 through 2008. Although the distracted category involves such activities as talking on cellphones, conversing with passengers, and eating, experts at the National Highway Traffic Safety Administration claim that texting while driving is the clearest menace because it requires looking away from the road.

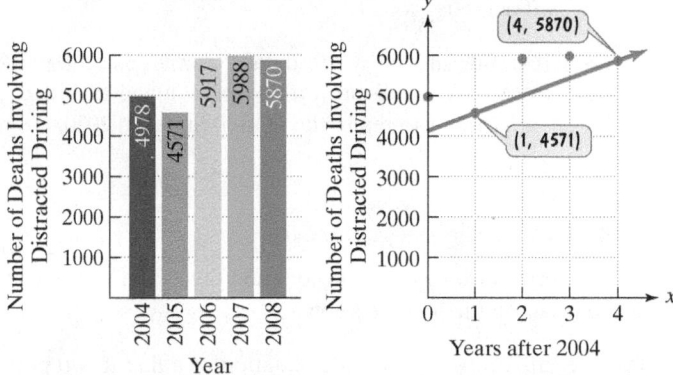

Number of Highway Fatalities in the United States Involving Distracted Driving

Source: National Highway Traffic Safety Administration

a. Shown to the right of the bar graph is a scatter plot with a line passing through two of the data points. Use the two points whose coordinates are shown by the voice balloons to write the slope-intercept form of an equation that models the number of highway fatalities involving distracted driving, y, in the United States x years after 2004.

b. In 2010, surveys showed overwhelming public support to ban texting while driving, although at that time only 19 states and Washington, D.C., outlawed the practice. Without additional laws that penalize texting drivers, use the model you obtained from part (a) on the previous page to project the number of fatalities in the United States in 2014 involving distracted driving.

70. The bar graph shows the rise in the percentage of births to U.S. unmarried women from 1960 through 2009.

Percentage of Births to Unmarried Women in the U.S.

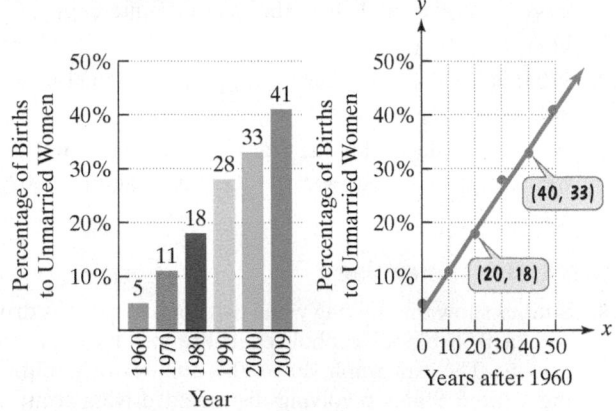

Source: National Center for Health Statistics

a. Shown to the right of the bar graph is a scatter plot with a line passing through two of the data points. Use the two points whose coordinates are shown by the voice balloons to write the slope-intercept form of an equation that models the percentage of births to unmarried women, *y*, in the United States *x* years after 1960.

b. If trends shown by the data continue, use the model from part (a) to project the percentage of births to unmarried women in the United States in 2020.

Writing in Mathematics

71. Describe how to write the equation of a line if its slope and a point on the line are known.

72. Describe how to write the equation of a line if two points on the line are known.

Critical Thinking Exercises

Make Sense? *In Exercises 73–76, determine whether each statement "makes sense" or "does not make sense" and explain your reasoning.*

73. I use $y = mx + b$ to write equations of lines passing through two points when neither contains the *y*-intercept.

74. In many examples, I use the slope-intercept form of a line's equation to obtain an equivalent equation in point-slope form.

75. I have linear models that describe changes for men and women over the same time period. The models have the same slope, so the graphs are parallel lines, indicating that the rate of change for men is the same as the rate of change for women.

76. The shape of this scatter plot showing per capita U.S. adult wine consumption suggests that I should not use a line or any of its equations to obtain a model.

Wine Consumption per United States Adult

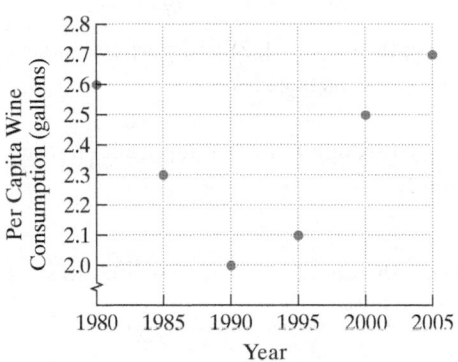

Source: Adams Business Media

In Exercises 77–80, determine whether each statement is true or false. If the statement is false, make the necessary change(s) to produce a true statement.

77. If a line has undefined slope, then it has no equation.

78. The line whose equation is $y - 3 = 7(x + 2)$ passes through $(-3, 2)$.

79. The point-slope form can be applied to obtain the equation of the line through the points $(2, -5)$ and $(2, 6)$.

80. The slope of the line whose equation is $3x + y = 7$ is 3.

81. Excited about the success of celebrity stamps, post office officials were rumored to have put forth a plan to institute two new types of thermometers. On these new scales, $°E$ represents degrees Elvis and $°M$ represents degrees Madonna. If it is known that $40°E = 25°M$, $280°E = 125°M$, and degrees Elvis is linearly related to degrees Madonna, write an equation expressing E in terms of M.

Technology Exercises

82. Use a graphing utility to graph $y = 1.75x - 2$. Select the best viewing rectangle possible by experimenting with the range settings to show that the line's slope is $\frac{7}{4}$.

83. a. Use the statistical menu of a graphing utility to enter the six data points shown in the scatter plot in Exercise 70. Refer to the bar graph to obtain coordinates of points that are not given in the scatter plot.

   b. Use the $\boxed{\text{DRAW}}$ menu and the scatter plot capability to draw a scatter plot of the data points like the one shown in Exercise 70.

   c. Select the linear regression option. Use your utility to obtain values for a and b for the equation of the regression line, $y = ax + b$. You may also be given a *correlation coefficient, r*. Values of r close to 1 indicate that the points can be described by a linear relationship and the regression line has a positive slope. Values of r close to -1 indicate that the points can be described by a linear relationship and the regression line has a negative slope. Values of r close to 0 indicate no linear relationship between the variables.

   d. Use the appropriate sequence (consult your manual) to graph the regression equation on top of the points in the scatter plot.

Review Exercises

84. How many sheets of paper, weighing 2 grams each, can be put in an envelope weighing 4 grams if the total weight must not exceed 29 grams? (Section 2.7, Example 11)

85. List all the natural numbers in this set:

$$\left\{ -2, 0, \frac{1}{2}, 1, \sqrt{3}, \sqrt{4} \right\}.$$

(Section 1.3, Example 5)

86. Use intercepts to graph $3x - 5y = 15$. (Section 3.2, Example 4)

Preview Exercises

Exercises 87–89 will help you prepare for the material covered in the first section of the next chapter.

87. Is $(4, -1)$ a solution of both $x + 2y = 2$ and $x - 2y = 6$?

88. Is $(-4, 3)$ a solution of both $x + 2y = 2$ and $x - 2y = 6$?

89. Determine the point of intersection of the graphs of $2x + 3y = 6$ and $2x + y = -2$ by graphing both equations in the same rectangular coordinate system.

GROUP PROJECT

CHAPTER 3

In Example 5 on page 260, we used the data in **Figure 3.37** on page 259 to develop a linear equation that modeled the graying of America. For this group exercise, you might find it helpful to pattern your work after **Figure 3.37** and the solution to Example 5. Group members should begin by consulting an almanac, newspaper, magazine, or the Internet to find data that lie approximately on or near a straight line. Working by hand or using a graphing utility, construct a scatter plot for the data. If working by hand, draw a line that passes through or near the data points and then write its equation. If using a graphing utility, obtain the equation of the regression line. Then use the equation of the line to make a prediction about what might happen in the future. Are there circumstances that might affect the accuracy of this prediction? List some of these circumstances.

Chapter 3 Summary

Definitions and Concepts	**Examples**

Section 3.1 Graphing Linear Equations in Two Variables

The rectangular coordinate system consists of a horizontal number line, the x-axis, and a vertical number line, the y-axis, intersecting at their zero points, the origin. Each point in the system corresponds to an ordered pair of real numbers (x, y). The first number in the pair is the x-coordinate; the second number is the y-coordinate.

Plot: $(2, 3), (-5, 4), (-4, -3),$ and $(5, -2)$.

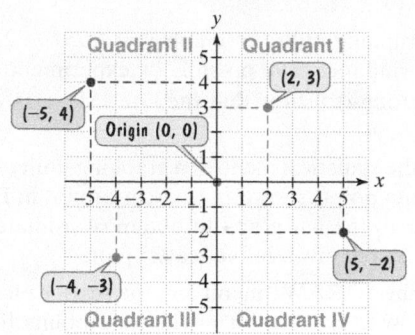

An ordered pair is a solution of an equation in two variables if replacing the variables by the coordinates of the ordered pair results in a true statement.

Is $(-1, 4)$ a solution of $2x + 5y = 18$?
$$2(-1) + 5 \cdot 4 \overset{?}{=} 18$$
$$-2 + 20 \overset{?}{=} 18$$
$$18 = 18, \text{ true}$$

Thus, $(-1, 4)$ is a solution.

One method for graphing an equation in two variables is point plotting. Find several ordered-pair solutions, plot them as points, and connect the points with a smooth curve or line.

Graph: $y = 2x + 1$.

x	$y = 2x + 1$	(x, y)
-2	$y = 2(-2) + 1 = -3$	$(-2, -3)$
-1	$y = 2(-1) + 1 = -1$	$(-1, -1)$
0	$y = 2 \cdot 0 + 1 = 1$	$(0, 1)$
1	$y = 2 \cdot 1 + 1 = 3$	$(1, 3)$
2	$y = 2 \cdot 2 + 1 = 5$	$(2, 5)$

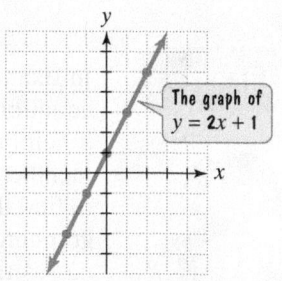

Section 3.2 Graphing Linear Equations Using Intercepts

If a graph intersects the x-axis at $(a, 0)$, then a is an x-intercept.

If a graph intersects the y-axis at $(0, b)$, then b is a y-intercept.

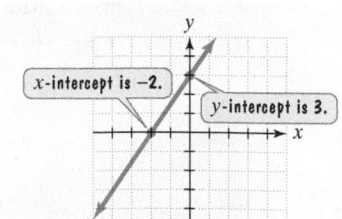

Definitions and Concepts	**Examples**

Section 3.2 Graphing Linear Equations Using Intercepts (continued)

An equation of the form $Ax + By = C$, where A, B, and C are integers, is called the standard form of the equation of a line. The graph of $Ax + By = C$ is a line that can be obtained using intercepts. To find the x-intercept, let $y = 0$ and solve for x. To find the y-intercept, let $x = 0$ and solve for y. Find a checkpoint, a third ordered-pair solution. Graph the equation by drawing a line through the three points.

Graph using intercepts: $4x + 3y = 12$.

x-intercept: $\quad\quad 4x = 12$

$\quad\quad\quad\quad\quad\quad x = 3$

y-intercept: $\quad\quad 3y = 12$

$\quad\quad\quad\quad\quad\quad y = 4$

Checkpoint: $\quad$ Let $x = 2$.

$\quad\quad\quad\quad 8 + 3y = 12$

$\quad\quad\quad\quad\quad 3y = 4$

$\quad\quad\quad\quad\quad y = \frac{4}{3}$

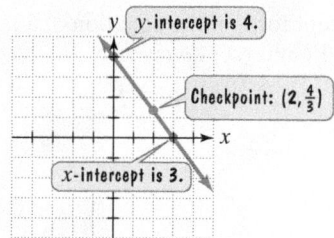

The graph of $Ax + By = 0$ is a line that passes through the origin. Find two other points by finding two other solutions of the equation. Graph the equation by drawing a line through the origin and these two points.

Graph: $x + 2y = 0$.

$x = 2$: $\quad\quad 2 + 2y = 0$

$\quad\quad\quad\quad\quad 2y = -2$

$\quad\quad\quad\quad\quad y = -1$

$y = 1$: $\quad x + 2(1) = 0$

$\quad\quad\quad\quad\quad x = -2$

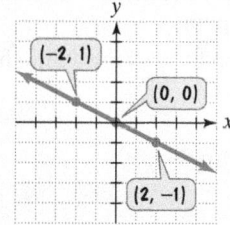

Horizontal and Vertical Lines

The graph of $y = b$ is a horizontal line. The y-intercept is b.

The graph of $x = a$ is a vertical line. The x-intercept is a.

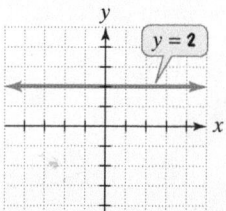
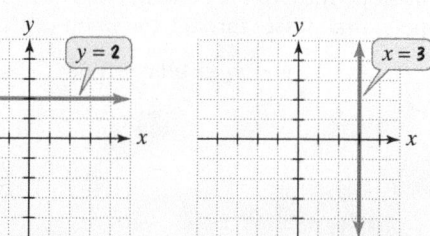

Definitions and Concepts	**Examples**

Section 3.3 Slope

The slope, m, of the line through the points (x_1, y_1) and (x_2, y_2) is

$$m = \frac{y_2 - y_1}{x_2 - x_1}, \quad x_2 - x_1 \neq 0.$$

If the slope is positive, the line rises from left to right. If the slope is negative, the line falls from left to right.
The slope of a horizontal line is 0. The slope of a vertical line is undefined.
If two distinct nonvertical lines have the same slope, then the lines are parallel.
If the product of the slopes of two lines is -1, then the lines are perpendicular. Equivalently, if the slopes are negative reciprocals, then the lines are perpendicular.
The slope of a line represents its rate of change.

Find the slope of the line passing through the points shown.

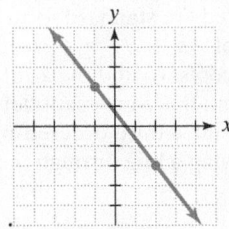

Let $(x_1, y_1) = (-1, 2)$ and $(x_2, y_2) = (2, -2)$.

$$m = \frac{y_2 - y_1}{x_2 - x_1} = \frac{-2 - 2}{2 - (-1)} = \frac{-4}{3} = -\frac{4}{3}$$

Section 3.4 The Slope-Intercept Form of the Equation of a Line

The slope-intercept form of the equation of a nonvertical line with slope m and y-intercept b is

$$y = mx + b.$$

Find the slope and the y-intercept of the line with the given equation.

- $y = -2x + 5$

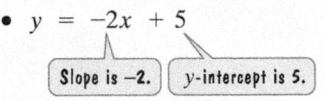

- $2x + 3y = 9$ (Solve for y.)

$$3y = -2x + 9 \quad \text{Subtract 2x.}$$

$$y = -\frac{2}{3}x + 3 \quad \text{Divide by 3.}$$

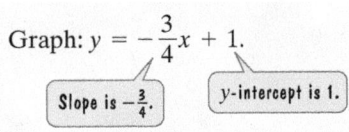

Graphing $y = mx + b$ Using the Slope and y-Intercept

1. Plot the point containing the y-intercept on the y-axis. This is the point $(0, b)$.

2. Use the slope, m, to obtain a second point. Write m as a fraction, and use rise over run, starting at the point on the y-axis, to plot this point.

3. Graph the equation by drawing a line through the two points.

Graph: $y = -\dfrac{3}{4}x + 1$.

Slope is $-\frac{3}{4}$. y-intercept is 1.

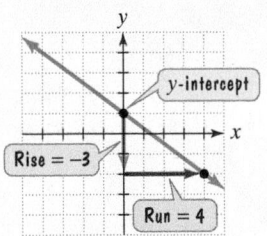

Section 3.5 The Point-Slope Form of the Equation of a Line

The point-slope form of the equation of a nonvertical line with slope m that passes through the point (x_1, y_1) is

$$y - y_1 = m(x - x_1).$$

Slope $= -3$, passing through $(-1, 5)$

$m = -3$ $x_1 = -1$ $y_1 = 5$

The point-slope form of the line's equation is

$$y - 5 = -3[x - (-1)].$$

Simplify:

$$y - 5 = -3(x + 1).$$

Definitions and Concepts	**Examples**

Section 3.5 The Point-Slope Form of the Equation of a Line (continued)

To write the point-slope form of the equation of a line passing through two points, begin by using the points to compute the slope, m. Use either given point as (x_1, y_1) and write the point-slope form of the equation:

$$y - y_1 = m(x - x_1).$$

Solving this equation for y gives the slope-intercept form of the line's equation.

Write an equation in point-slope form and slope-intercept form of the line passing through $(-1, -3)$ and $(4, 2)$.

$$m = \frac{2 - (-3)}{4 - (-1)} = \frac{2 + 3}{4 + 1} = \frac{5}{5} = 1$$

Using $(4, 2)$ as (x_1, y_1), the point-slope form of the equation is

$$y - 2 = 1(x - 4).$$

Solve for y to obtain the slope-intercept form.

$$y = x - 2 \quad \text{Add 2 to both sides.}$$

Parallel lines with the same slope and perpendicular lines with slopes that are negative reciprocals often appear in problems involving the point-slope form of the equation of a line.

Write equations in point-slope form and in slope-intercept form of the line passing through $(2, -1)$

$$\boxed{x_1} \quad \boxed{y_1}$$

and perpendicular to $y = -\dfrac{1}{5}x + 6.$

$$\boxed{\text{slope}}$$

The slope, m, of the perpendicular line is 5, the negative reciprocal of $-\frac{1}{5}$.

$$y - (-1) = 5(x - 2) \quad \boxed{\text{Point-slope form of the equation}}$$

$$y + 1 = 5(x - 2)$$

$$y + 1 = 5x - 10$$

$$y = 5x - 11 \quad \boxed{\text{Slope-intercept form of the equation}}$$

CHAPTER 3 REVIEW EXERCISES

3.1 *In Exercises 1–4, plot the given point in a rectangular coordinate system. Indicate in which quadrant each point lies.*

1. $(1, -5)$

2. $(4, -3)$

3. $\left(\dfrac{7}{2}, \dfrac{3}{2}\right)$

4. $(-5, 2)$

5. Give the ordered pairs that correspond to the points labeled in the figure.

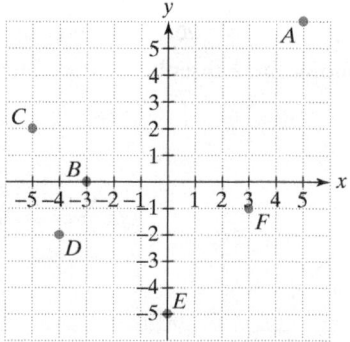

In Exercises 6–7, determine whether each ordered pair is a solution of the given equation.

6. $y = 3x + 6$ $(-3, 3)$, $(0, 6)$, $(1, 9)$

7. $3x - y = 12$ $(0, 4)$, $(4, 0)$, $(-1, 15)$

In Exercises 8–9,

a. *Find five solutions of each equation. Organize your work in a table of values.*

b. *Use the five solutions in the table to graph each equation.*

8. $y = 2x - 3$ **9.** $y = \frac{1}{2}x + 1$

3.2 *In Exercises 10–12, use the graph to identify the*

a. *x-intercept, or state that there is no x-intercept.*

b. *y-intercept, or state that there is no y-intercept.*

10.   **11.**

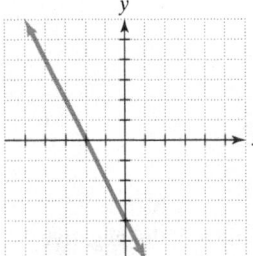 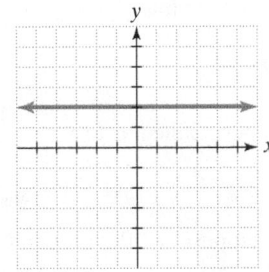

12.

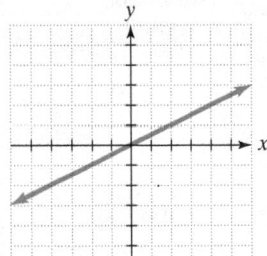

In Exercises 13–16, use intercepts to graph each equation.

13. $2x + y = 4$ **14.** $3x - 2y = 12$

15. $3x = 6 - 2y$ **16.** $3x - y = 0$

In Exercises 17–20, graph each equation.

17. $x = 3$ **18.** $y = -5$

19. $y + 3 = 5$ **20.** $2x = -8$

21. The graph shows the Fahrenheit temperature, y, x hours after noon.

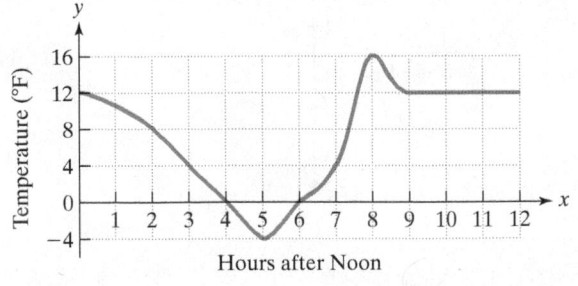

Hours after Noon

a. At what time did the minimum temperature occur? What is the minimum temperature?

b. At what time did the maximum temperature occur? What is the maximum temperature?

c. What are the x-intercepts? In terms of time and temperature, interpret the meaning of these intercepts.

d. What is the y-intercept? What does this mean in terms of time and temperature?

e. From 9 P.M. until midnight, the graph is shown as a horizontal line. What does this mean about the temperature over this period of time?

3.3 *In Exercises 22–25, calculate the slope of the line passing through the given points. If the slope is undefined, so state. Then indicate whether the line rises, falls, is horizontal, or is vertical.*

22. $(3, 2)$ and $(5, 1)$

23. $(-1, 2)$ and $(-3, -4)$

24. $(-3, 4)$ and $(6, 4)$

25. $(5, 3)$ and $(5, -3)$

In Exercises 26–29, find the slope of each line, or state that the slope is undefined.

26.   **27.**

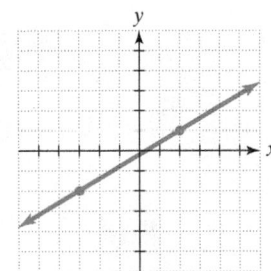 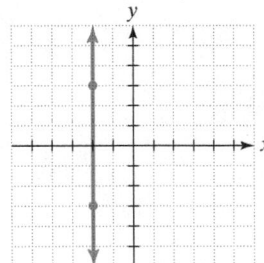

28.   **29.**

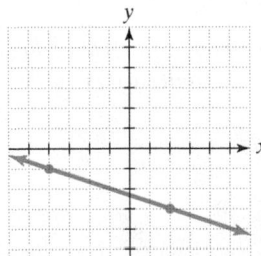 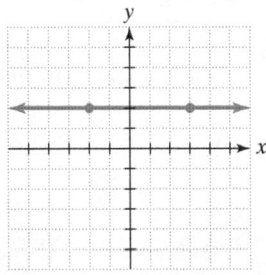

In Exercises 30–32, determine whether the lines through each pair of points are parallel, perpendicular, or neither.

30. $(-1, -3)$ and $(2, -8)$; $(8, -7)$ and $(9, 10)$

31. $(0, -4)$ and $(5, -1)$; $(-6, 8)$ and $(3, -7)$

32. $(5, 4)$ and $(9, 7)$; $(-6, 0)$ and $(-2, 3)$

33. In a 2010 survey of more than 200,000 freshmen at 279 colleges, only 52% rated their emotional health high or above average, a drop from 64% in 1985.

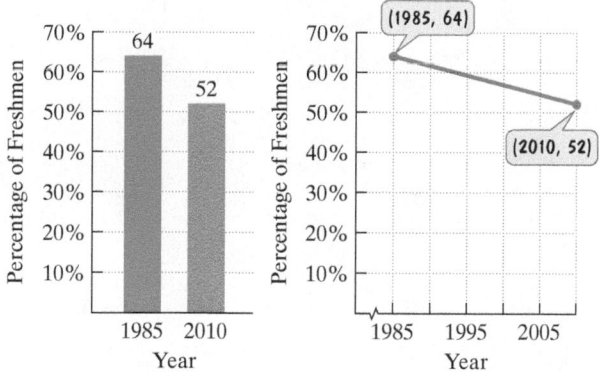

Percentage of U.S. College Freshmen Rating Their Emotional Health High or Above Average

Source: UCLA Higher Education Research Institute

a. Find the slope of the line passing through the two points shown by the voice balloons.

b. Use your answer from part (a) to complete this statement:

For each year from 1985 through 2010, the percentage of U.S. college freshmen rating their emotional health high or above average decreased by ___ . The rate of change was _____ per ___ .

3.4 *In Exercises 34–37, find the slope and the y-intercept of the line with the given equation.*

34. $y = 5x - 7$ **35.** $y = 6 - 4x$

36. $y = 3$ **37.** $2x + 3y = 6$

In Exercises 38–40, graph each linear equation using the slope and y-intercept.

38. $y = 2x - 4$ **39.** $y = \frac{1}{2}x - 1$

40. $y = -\frac{2}{3}x + 5$

In Exercises 41–42, write each equation in slope-intercept form. Then use the slope and y-intercept to graph the equation.

41. $y - 2x = 0$

42. $\frac{1}{3}x + y = 2$

43. Graph $y = -\frac{1}{2}x + 4$ and $y = -\frac{1}{2}x - 1$ in the same rectangular coordinate system. Are the lines parallel? If so, explain why.

44. No matter who hosts, the Miss America pageant keeps losing viewers. The scatter plot shows the number of viewers of the pageant, in millions, along with some of the MCs from 1989 through 2009. Also shown is a line passing through two of the data points.

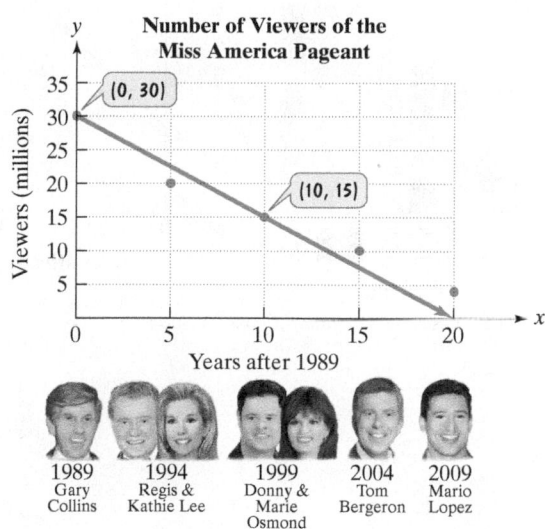

Source: Entertainment Weekly

a. Use the two points shown by the voice balloons to find an equation in the form $y = mx + b$ that models the number of viewers of the Miss America pageant, y, in millions, x years after 1989.

b. Use the model from part (a) to determine the number of viewers of the pageant in 1994. Does this underestimate or overestimate the actual number of viewers shown by the scatter plot? By how many millions of viewers?

3.5 *Write the point-slope form of the equation of the line satisfying the conditions in Exercises 45–48. Then use the point-slope form of the equation to write the slope-intercept form.*

45. Slope = 6, passing through $(-4, 7)$

46. Passing through $(3, 4)$ and $(2, 1)$

47. Passing through $(4, -7)$ and parallel to the line whose equation is $3x + y = 9$

48. Passing through $(-2, 6)$ and perpendicular to the line whose equation is $y = \frac{1}{3}x + 4$

49. The bar graph shows world population, in billions, for seven selected years from 1950 through 2010. Also shown is a scatter plot with a line passing through two of the data points.

World Population, 1950–2010

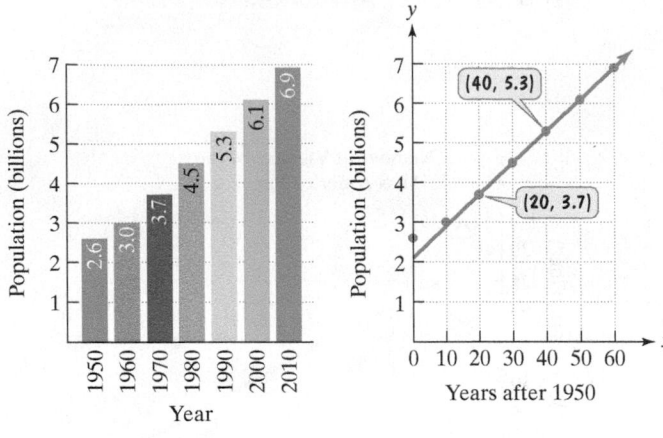

Source: U.S. Census Bureau, International Database

a. Use the two points whose coordinates are shown by the voice balloons to write the slope-intercept form of an equation that models world population, *y*, in billions, *x* years after 1950.

b. Use the model from part (a) to project world population in 2025.

CHAPTER 3 TEST

CHAPTER
Test Prep
VIDEOS

Step-by-step test solutions are found on the Chapter Test Prep Videos available in MyMathLab® or on YouTube (search "BlitzerCombinedAlg" and click on "Channels").

1. Determine whether each ordered pair is a solution of $4x - 2y = 10$:

$$(0, -5), \quad (-2, 1), \quad (4, 3).$$

2. Find five solutions of $y = 3x + 1$. Organize your work in a table of values. Then use the five solutions in the table to graph the equation.

3. Use the graph to identify the

  a. *x*-intercept, or state that there is no *x*-intercept.

  b. *y*-intercept, or state that there is no *y*-intercept.

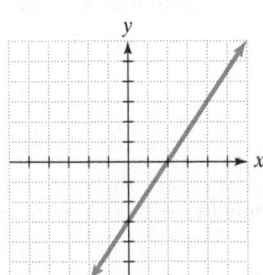

4. Use intercepts to graph $4x - 2y = -8$.

5. Graph $y = 4$ in a rectangular coordinate system.

In Exercises 6–7, calculate the slope of the line passing through the given points. If the slope is undefined, so state. Then indicate whether the line rises, falls, is horizontal, or is vertical.

6. $(-3, 4)$ and $(-5, -2)$

7. $(6, -1)$ and $(6, 3)$

8. Find the slope of the line in the figure shown or state that the slope is undefined.

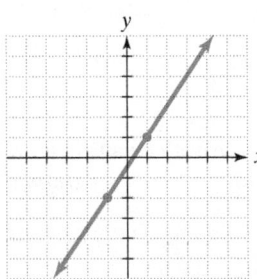

In Exercises 9–10 determine whether the lines through each pair of points are parallel, perpendicular, or neither.

9. $(-2, 10)$ and $(0, 2)$; $(-8, -7)$ and $(24, 1)$

10. $(2, 4)$ and $(6, 1)$; $(-3, 1)$ and $(1, -2)$

In Exercises 11–12, find the slope and the y-intercept of the line with the given equation.

11. $y = -x + 10$

12. $2x + y = 6$

In Exercises 13–14, graph each linear equation using the slope and y-intercept.

13. $y = \dfrac{2}{3}x - 1$

14. $y = -2x + 3$

In Exercises 15–18, use the given conditions to write an equation for each line in point-slope form and slope-intercept form.

15. Slope $= -2$, passing through $(-1, 4)$

16. Passing through $(2, 1)$ and $(-1, -8)$

17. Passing through $(-2, 3)$ and perpendicular to the line whose equation is $y = -\frac{1}{2}x - 4$

18. Passing through $(6, -4)$ and parallel to the line whose equation is $x + 2y = 5$

19. Could Cool Hand Luke bust out today? Not likely. The bar graph shows the decrease in the number of inmates who escaped from U.S. prisons for selected years from 1993 through 2008.

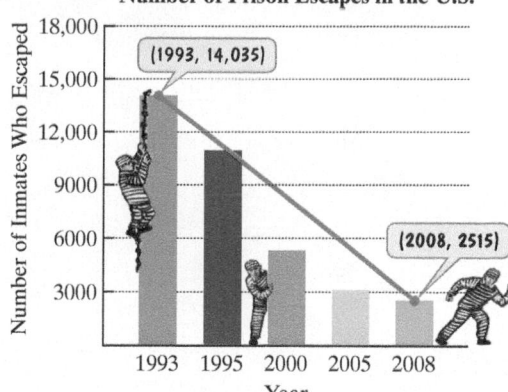

Number of Prison Escapes in the U.S.

Source: Bureau of Justice Statistics

a. Find the slope of the line passing through the two points shown by the voice balloons.

b. Use your answer from part (a) to complete this statement:

For the period shown, the number of inmates who escaped from U.S. prisons decreased each year by approximately ___. The rate of change was _____ per ___.

CUMULATIVE REVIEW EXERCISES (CHAPTERS 1–3)

1. Perform the indicated operations:
$$\frac{10 - (-6)}{3^2 - (4 - 3)}.$$

2. Simplify: $6 - 2[3(x - 1) + 4]$.

3. List all the irrational numbers in this set:
$\left\{-3, 0, 1, \sqrt{4}, \sqrt{5}, \frac{11}{2}\right\}$.

In Exercises 4–5, solve each equation.

4. $6(2x - 1) - 6 = 11x + 7$

5. $x - \dfrac{3}{4} = \dfrac{1}{2}$

6. Solve for x: $y = mx + b$.

7. 120 is 15% of what number?

8. The formula $y = 4.5x - 46.7$ models the stopping distance, y, in feet, for a car traveling x miles per hour. If the stopping distance is 133.3 feet, how fast was the car traveling?

In Exercises 9–10, solve each inequality and graph the solution set on a number line.

9. $2 - 6x \geq 2(5 - x)$

10. $6(2 - x) > 12$

11. A plumber charged a customer $228, listing $18 for parts and the remainder for labor. If the cost of the labor is $35 per hour, how many hours did the plumber work?

12. The length of a rectangular football field is 14 meters more than twice the width. If the perimeter is 346 meters, find the field's dimensions.

13. After a 10% weight loss, a person weighed 180 pounds. What was the weight before the loss?

14. In a triangle, the measure of the second angle is 20° greater than the measure of the first angle. The measure of the third angle is twice that of the first. What is the measure of each angle?

15. Evaluate $x^2 - 10x$ for $x = -3$.

16. Insert either $<$ or $>$ in the shaded area to make a true statement:

$$-2000 \;\rule{1cm}{0.4pt}\; -3$$

In Exercises 17–20, graph each equation in the rectangular coordinate system.

17. $2x - y = 4$ **18.** $x = -5$ **19.** $y = -4x + 3$

20. $y = -1$

Systems of Linear Equations

Television, movies, and magazines place great emphasis on physical beauty. Our culture emphasizes physical appearance to such an extent that it is a central factor in the perception and judgment of others. The modern emphasis on thinness as the ideal body shape has been suggested as a major cause of eating disorders among adolescent women. In this chapter, you will learn how systems of linear equations in two variables reveal the hidden patterns of your world, including a relationship between our changing cultural values of physical attractiveness and undernutrition.

The Blitzer Bonus on page 284 contains a graphical look at the shifting shape of that icon of American beauty, Miss America. Is it your imagination,or does she have that lean and hungry look?

SECTION

4.1

Solving Systems of Linear Equations by Graphing

Objectives

1. Decide whether an ordered pair is a solution of a linear system.

2. Solve systems of linear equations by graphing.

3. Use graphing to identify systems with no solution or infinitely many solutions.

4. Use graphs of linear systems to solve problems.

Visitors to the world's great bridges are frequently inspired by their beauty. This feeling is not always shared by daily commuters dealing with escalating toll costs and peak-hour congestion. For these frequent users, most bridge authorities provide the option of a fixed cost that reduces the toll. In this section, you will see how toll options in these situations can be modeled by two linear equations in two variables and their graphs.

1. Decide whether an ordered pair is a solution of a linear system.

Systems of Linear Equations and Their Solutions

We have seen that all equations in the form $Ax + By = C$ are straight lines when graphed. Two such equations are called a **system of linear equations** or a **linear system**. A **solution to a system of two linear equations in two variables** is an ordered pair that satisfies both equations in the system. For example, (3, 4) satisfies the system

$$\begin{cases} x + y = 7 & \text{(3 + 4 is, indeed, 7.)} \\ x - y = -1. & \text{(3 - 4 is, indeed, -1.)} \end{cases}$$

Thus, (3, 4) satisfies both equations and is a solution of the system. The solution can be described by saying that $x = 3$ and $y = 4$. The solution can also be described using the ordered pair (3, 4). The solution set of the system is $\{(3, 4)\}$—that is, the set consisting of the ordered pair (3, 4).

A system of linear equations can have exactly one solution, no solution, or infinitely many solutions. We begin with systems having exactly one solution.

EXAMPLE 1   Determining Whether Ordered Pairs Are Solutions of a System

Consider the system:

$$\begin{cases} x + 2y = 2 \\ x - 2y = 6 \end{cases}$$

Determine if each ordered pair is a solution of the system:

a. (4, −1)   **b.** (−4, 3).

Solution

a. We begin by determining whether $(4, -1)$ is a solution. Because 4 is the x-coordinate and -1 is the y-coordinate of $(4, -1)$, we replace x with 4 and y with -1.

$$x + 2y = 2 \qquad\qquad\qquad x - 2y = 6$$
$$4 + 2(-1) \overset{?}{=} 2 \qquad\qquad 4 - 2(-1) \overset{?}{=} 6$$
$$4 + (-2) \overset{?}{=} 2 \qquad\qquad 4 - (-2) \overset{?}{=} 6$$
$$2 = 2, \quad \text{true} \qquad\qquad 4 + 2 \overset{?}{=} 6$$
$$6 = 6, \quad \text{true}$$

The pair $(4, -1)$ satisfies both equations: It makes each equation true. Thus, the ordered pair is a solution of the system.

b. To determine whether $(-4, 3)$ is a solution, we replace x with -4 and y with 3.

$$x + 2y = 2 \qquad\qquad\qquad x - 2y = 6$$
$$-4 + 2 \cdot 3 \overset{?}{=} 2 \qquad\qquad -4 - 2 \cdot 3 \overset{?}{=} 6$$
$$-4 + 6 \overset{?}{=} 2 \qquad\qquad -4 - 6 \overset{?}{=} 6$$
$$2 = 2, \quad \text{true} \qquad\qquad -10 = 6, \quad \text{false}$$

The pair $(-4, 3)$ fails to satisfy *both* equations: It does not make both equations true. Thus, the ordered pair is not a solution of the system. ■

✓ **CHECK POINT 1**   Consider the system:

$$\begin{cases} 2x - 3y = -4 \\ 2x + y = 4 \end{cases}$$

Determine if each ordered pair is a solution of the system:

a. $(1, 2)$ **b.** $(7, 6)$.

---

2 Solve systems of linear equations by graphing.

Solving Linear Systems by Graphing

The solution of a system of two linear equations in two variables can be found by graphing both of the equations in the same rectangular coordinate system. For a system with one solution, **the coordinates of the point of intersection give the system's solution.** For example, the system in Example 1,

$$\begin{cases} x + 2y = 2 \\ x - 2y = 6 \end{cases}$$

is graphed in **Figure 4.1.** The solution of the system, $(4, -1)$, corresponds to the point of intersection of the lines.

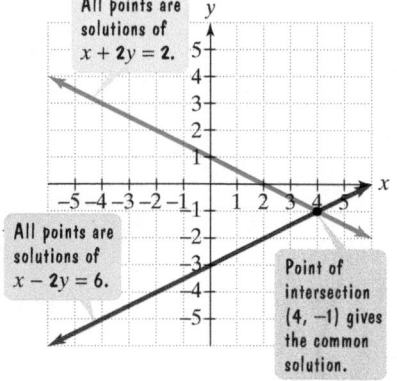

Figure 4.1
Visualizing a system's solution

Solving Systems of Two Linear Equations in Two Variables,
x and *y*, by Graphing

1. Graph the first equation.
2. Graph the second equation on the same set of axes.
3. If the lines representing the two graphs intersect at a point, determine the coordinates of this point of intersection. The ordered pair is the solution of the system.
4. Check the solution in both equations.

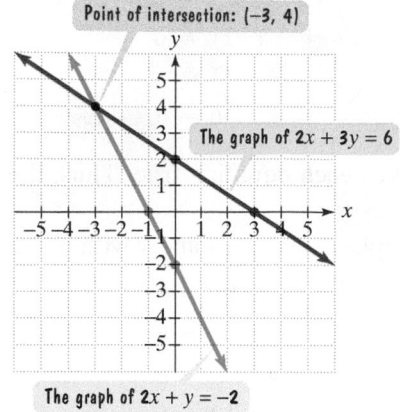

Point of intersection: (−3, 4)

The graph of 2x + 3y = 6

The graph of 2x + y = −2

Figure 4.2

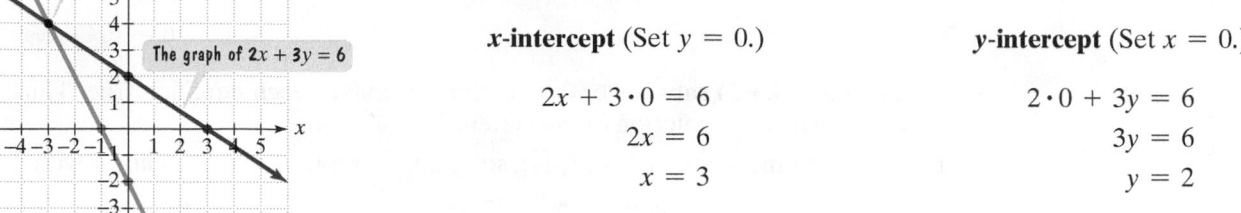

EXAMPLE 2 Solving a Linear System by Graphing

Solve by graphing:

$$\begin{cases} 2x + 3y = 6 \\ 2x + y = -2. \end{cases}$$

Solution

Step 1. Graph the first equation. We use intercepts to graph $2x + 3y = 6$.

x-intercept (Set $y = 0$.)	**y-intercept** (Set $x = 0$.)
$2x + 3 \cdot 0 = 6$	$2 \cdot 0 + 3y = 6$
$2x = 6$	$3y = 6$
$x = 3$	$y = 2$

The x-intercept is 3, so the line passes through $(3, 0)$. The y-intercept is 2, so the line passes through $(0, 2)$. The graph of $2x + 3y = 6$ is shown as the red line in **Figure 4.2**.

Step 2. Graph the second equation on the same axes. We use intercepts to graph $2x + y = -2$.

x-intercept (Set $y = 0$.)	**y-intercept** (Set $x = 0$.)
$2x + 0 = -2$	$2 \cdot 0 + y = -2$
$2x = -2$	$y = -2$
$x = -1$	

The x-intercept is -1, so the line passes through $(-1, 0)$. The y-intercept is -2, so the line passes through $(0, -2)$. The graph of $2x + y = -2$ is shown as the blue line in **Figure 4.2**.

Step 3. Determine the coordinates of the intersection point. This ordered pair is the system's solution. Using **Figure 4.2**, it appears that the lines intersect at $(-3, 4)$. The "apparent" solution of the system is $(-3, 4)$.

Step 4. Check the solution in both equations.

Check $(-3, 4)$ in	**Check $(-3, 4)$ in**
$2x + 3y = 6$:	$2x + y = -2$:
$2(-3) + 3 \cdot 4 \stackrel{?}{=} 6$	$2(-3) + 4 \stackrel{?}{=} -2$
$-6 + 12 \stackrel{?}{=} 6$	$-6 + 4 \stackrel{?}{=} -2$
$6 = 6,$ true	$-2 = -2,$ true

Because both equations are satisfied, $(-3, 4)$ is the solution of the system and $\{(-3, 4)\}$ is the solution set. ∎

✓ **CHECK POINT 2** Solve by graphing:

$$\begin{cases} 2x + y = 6 \\ 2x - y = -2. \end{cases}$$

Discover for Yourself

Must two lines intersect at exactly one point? Sketch two lines that have less than one intersection point. Now sketch two lines that have more than one intersection point. What does this say about each of these systems?

Great Question!

Can I use a rough sketch on scratch paper to solve a linear system by graphing?

No. When solving linear systems by graphing, neatly drawn graphs are essential for determining points of intersection.

- Use rectangular coordinate graph paper.
- Use a ruler or straightedge.
- Use a pencil with a sharp point.

EXAMPLE 3   Solving a Linear System by Graphing

Solve by graphing:

$$\begin{cases} y = -3x + 2 \\ y = 5x - 6. \end{cases}$$

Solution   Each equation is in the form $y = mx + b$. Thus, we use the y-intercept, b, and the slope, m, to graph each line.

Step 1. Graph the first equation.

$$y = -3x + 2$$

The slope is -3.   The y-intercept is 2.

The y-intercept is 2, so the line passes through $(0, 2)$. The slope is $-\frac{3}{1}$. Start at the y-intercept and move 3 units down (the rise) and 1 unit to the right (the run). The graph of $y = -3x + 2$ is shown as the red line in **Figure 4.3**.

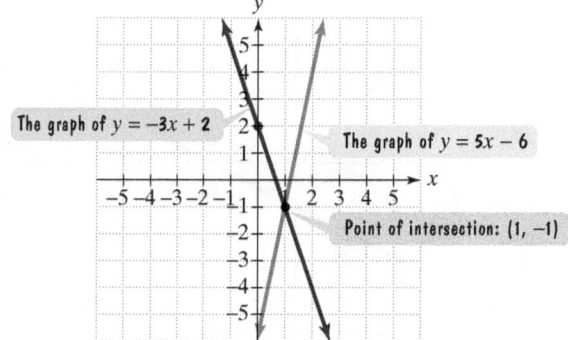

Figure 4.3

Step 2. Graph the second equation on the same axes.

$$y = 5x - 6$$

The slope is 5.   The y-intercept is -6.

The y-intercept is -6, so the line passes through $(0, -6)$. The slope is $\frac{5}{1}$. Start at the y-intercept and move 5 units up (the rise) and 1 unit to the right (the run). The graph of $y = 5x - 6$ is shown as the blue line in **Figure 4.3**.

Step 3. Determine the coordinates of the intersection point. This ordered pair is the system's solution.   Using **Figure 4.3**, it appears that the lines intersect at $(1, -1)$. The "apparent" solution of the system is $(1, -1)$.

Step 4. Check the solution in both equations.

Check $(1, -1)$ in   Check $(1, -1)$ in
$y = -3x + 2$: $y = 5x - 6$:
$-1 \overset{?}{=} -3 \cdot 1 + 2$ $-1 \overset{?}{=} 5 \cdot 1 - 6$
$-1 \overset{?}{=} -3 + 2$ $-1 \overset{?}{=} 5 - 6$
$-1 = -1,$ true $-1 = -1,$ true

Because both equations are satisfied, $(1, -1)$ is the solution and the solution set is $\{(1, -1)\}$. ∎

✓ **CHECK POINT 3**   Solve by graphing:

$$\begin{cases} y = -x + 6 \\ y = 3x - 6. \end{cases}$$

③ Use graphing to identify systems with no solution or infinitely many solutions.

Linear Systems Having No Solution or Infinitely Many Solutions

We have seen that a system of linear equations in two variables represents a pair of lines. The lines either intersect at one point, are parallel, or are identical. Thus, there are three possibilities for the number of solutions to a system of two linear equations.

The Number of Solutions to a System of Two Linear Equations

The number of solutions to a system of two linear equations in two variables is given by one of the following. (See **Figure 4.4**.)

Number of Solutions	What This Means Graphically
Exactly one ordered-pair solution	The two lines intersect at one point.
No solution	The two lines are parallel.
Infinitely many solutions	The two lines are identical.

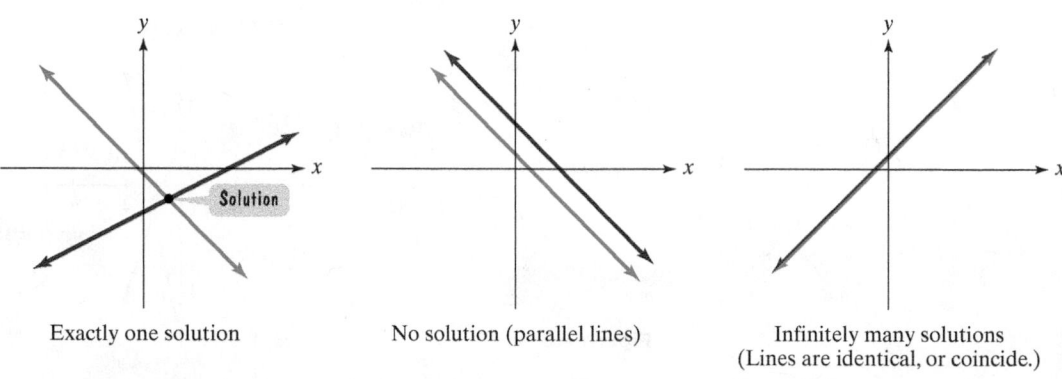

Exactly one solution   No solution (parallel lines)   Infinitely many solutions (Lines are identical, or coincide.)

Figure 4.4 Possible graphs for a system of two linear equations in two variables

A linear system with no solution is called an **inconsistent system**. If you attempt to solve such a system by graphing, you will obtain two parallel lines. The solution set is the empty set, $\varnothing$.

EXAMPLE 4   A System with No Solution

Solve by graphing:

$$\begin{cases} y = 2x - 1 \\ y = 2x + 3 \end{cases}.$$

Solution   Compare the slopes and y-intercepts in the two equations.

> The lines have the same slope, **2**.

$$y = 2x - 1 \qquad y = 2x + 3$$

> The lines have different y-intercepts, **−1** and **3**.

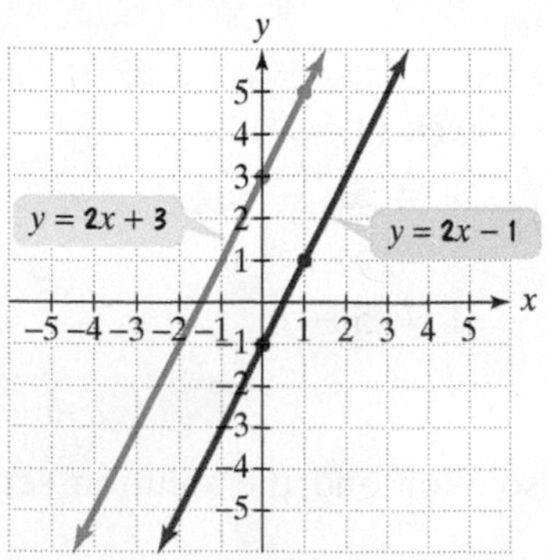

Figure 4.5 shows the graphs of the two equations. Because both equations have the same slope, 2, but different y-intercepts, the lines are parallel. The system is inconsistent and has no solution. The solution set is the empty set, $\varnothing$. ■

$y = 2x + 3$ $y = 2x - 1$

Figure 4.5 The graph of an inconsistent system

✓ CHECK POINT 4   Solve by graphing:

$$\begin{cases} y = 3x - 2 \\ y = 3x + 1 \end{cases}.$$

EXAMPLE 5 A System with Infinitely Many Solutions

Solve by graphing:

$$\begin{cases} 2x + y = 3 \\ 4x + 2y = 6 \end{cases}$$

Solution We use intercepts to graph each equation.

• $2x + y = 3$

x-intercept	**y-intercept**
$2x + 0 = 3$	$2 \cdot 0 + y = 3$
$2x = 3$	$y = 3$
$x = \dfrac{3}{2}$	

Graph $(\frac{3}{2}, 0)$ and $(0, 3)$.

• $4x + 2y = 6$

x-intercept	**y-intercept**
$4x + 2 \cdot 0 = 6$	$4 \cdot 0 + 2y = 6$
$4x = 6$	$2y = 6$
$x = \dfrac{6}{4} = \dfrac{3}{2}$	$y = 3$

Graph $(\frac{3}{2}, 0)$ and $(0, 3)$.

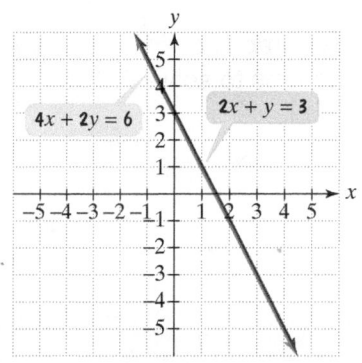

Figure 4.6 The graph of a system with infinitely many solutions

Both lines have the same x-intercept, $\frac{3}{2}$ or 1.5, and the same y-intercept, 3. Thus, the graphs of the two equations in the system are the same line, shown in **Figure 4.6**. The two equations have the same solutions. Any ordered pair that is a solution of one equation is a solution of the other, and, consequently, a solution of the system. The system has an infinite number of solutions, namely all points that are solutions of either line.

We express the solution set for the system in one of two equivalent ways:

$$\{(x, y) \mid 2x + y = 3\} \quad \text{or} \quad \{(x, y) \mid 4x + 2y = 6\}.$$

The set of all ordered pairs (x, y) such that $2x + y = 3$

The set of all ordered pairs (x, y) such that $4x + 2y = 6$

Great Question!

The system in Example 5 has infinitely many solutions. Does that mean that any ordered pair of numbers is a solution?

No. Although the system in Example 5 has infinitely many solutions, this does not mean that any ordered pair of numbers you can form will be a solution. The ordered pair (x, y) must satisfy one of the system's equations, $2x + y = 3$ or $4x + 2y = 6$, and there are infinitely many such ordered pairs. Because the graphs are coinciding lines, the ordered pairs that are solutions of one of the equations are also solutions of the other equation.

Take a second look at the two equations, $2x + y = 3$ and $4x + 2y = 6$, in Example 5. If you multiply both sides of the first equation, $2x + y = 3$, by 2, you will obtain the second equation, $4x + 2y = 6$.

$2x + y = 3$	This is the first equation in the system.
$2(2x + y) = 2 \cdot 3$	Multiply both sides by 2.
$2 \cdot 2x + 2y = 2 \cdot 3$	Use the distributive property.
$4x + 2y = 6$	Simplify.

This is the second equation in the system.

Because $2x + y = 3$ and $4x + 2y = 6$ are different forms of the same equation, these equations are called **dependent equations**. In general, the equations in a linear system with infinitely many solutions are called **dependent equations**.

✓ **CHECK POINT 5**   Solve by graphing:

$$\begin{cases} x + y = 3 \\ 2x + 2y = 6. \end{cases}$$

4 Use graphs of linear systems to solve problems.

Applications

One advantage of solving linear systems by graphing is that it allows us to "see" the equations and visualize the solution. This idea is demonstrated in Example 6.

EXAMPLE 6   Modeling Options for a Toll

The toll to a bridge costs $2.50. Commuters who use the bridge frequently have the option of purchasing a monthly discount pass for $21.00. With the discount pass, the toll is reduced to $1.00. The monthly cost, y, of using the bridge x times can be modeled by the following linear equations:

Without the discount pass:

$$y = 2.50x \qquad \text{The monthly cost, y, is \$2.50 times the number of times, x, that the bridge is used.}$$

With the discount pass:

$$y = 21 + 1 \cdot x \qquad \text{The monthly cost, y, is \$21 for the pass plus \$1 times the number of times, x, that the bridge is used.}$$

$$y = 21 + x \qquad \text{Simplify: } 1 \cdot x = x.$$

Expressing 2.50 as $\frac{5}{2}$, the options for the toll can be modeled by the linear system

$$\begin{cases} y = \dfrac{5}{2}x & \boxed{\text{Without discount pass}} \\ y = x + 21. & \boxed{\text{With discount pass}} \end{cases}$$

  a. Use graphing to solve the system.

  b. Interpret the coordinates of the solution in practical terms.

Solution

  a. Notice that x, the number of times the bridge is used in a month, and y, the monthly cost, are nonnegative. Thus, we will use only the first quadrant and its boundary to graph the system. We begin by identifying slopes and y-intercepts.

$$y = \frac{5}{2}x + 0 \qquad\qquad y = x + 21$$

The slope is $\frac{5}{2}$. The y-intercept is 0. The slope is 1. The y-intercept is 21.

Using slopes and y-intercepts, we graph the two lines, as shown in **Figure 4.7**.

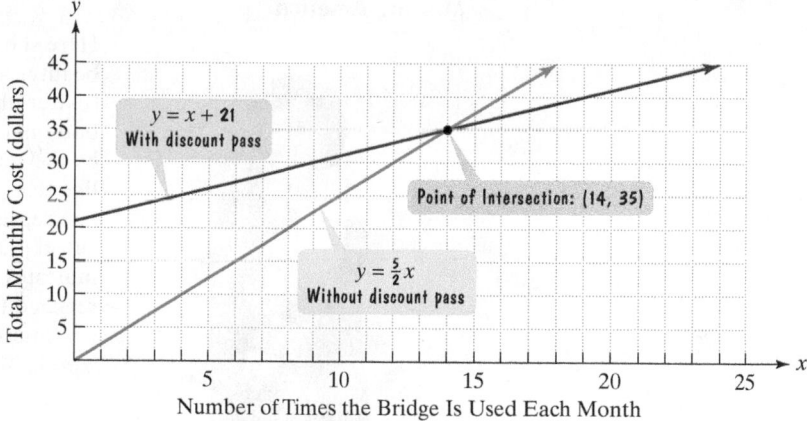

Figure 4.7 Options for a Toll

Using **Figure 4.7**, it appears that the lines intersect at (14, 35). Let's check this solution in both equations.

<table>
<tr><td align="center">

Check (14, 35) in

$$y = \frac{5}{2}x:$$

$$35 \overset{?}{=} \frac{5}{2} \cdot 14$$

$$35 \overset{?}{=} \frac{5}{2} \cdot \frac{\overset{7}{14}}{\underset{1}{1}}$$

$$35 = 35, \quad \text{true}$$

</td><td align="center">

Check (14, 35) in

$$y = x + 21:$$

$$35 \overset{?}{=} 14 + 21$$

$$35 = 35, \quad \text{true}$$

</td></tr>
</table>

Because both equations are satisfied, (14, 35) is the solution and the solution set is $\{(14, 35)\}$.

b. The graphs intersect at (14, 35). This means that if the bridge is used 14 times in a month, the total monthly cost without the discount pass is the same as the total monthly cost with the discount pass, namely $35. ∎

In **Figure 4.7**, look at the two graphs to the right of the intersection point (14, 35). The red graph of $y = x + 21$ lies below the blue graph of $y = \frac{5}{2}x$. This means that if the bridge is used more than 14 times in a month ($x > 14$), the (red) monthly cost, y, with the discount pass is cheaper than the (blue) monthly cost, y, without the discount pass.

✓ **CHECK POINT 6** The toll to a bridge costs $2.00. If you use the bridge x times in a month, the monthly cost, y, is $y = 2x$. With a $10 discount pass, the toll is reduced to $1.00. The monthly cost, y, of using the bridge x times in a month with the discount pass is $y = x + 10$.

a. Solve the system by graphing:

$$\begin{cases} y = 2x \\ y = x + 10. \end{cases}$$

Suggestion: Let the x-axis extend from 0 to 20 and let the y-axis extend from 0 to 40.

b. Interpret the coordinates of the solution in practical terms.

Blitzer Bonus

Missing America

Here she is, Miss America, the icon of American beauty. Always thin, she is becoming more so. The scatter plot in the figure shows Miss America's body-mass index, a ratio comparing weight divided by the square of height. Two lines are also shown: a line that passes near the data points and a horizontal line representing the World Health Organization's cutoff point for undernutrition. The intersection point indicates that in approximately 1978, Miss America reached this cutoff. There she goes: If the trend continues, Miss America's body-mass index could reach zero in about 320 years.

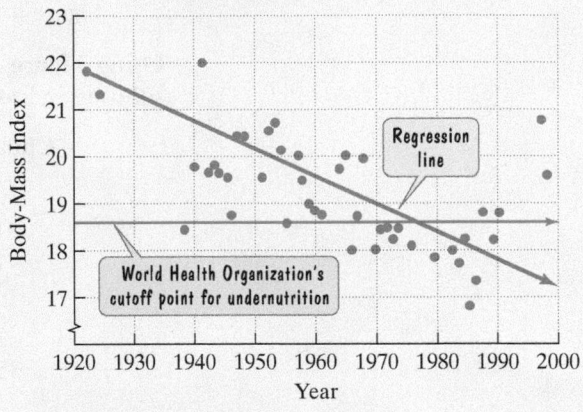

Body-Mass Index of Miss America

Source: Johns Hopkins School of Public Health

Mary Katherine Campbell,
Miss America 1922

Achieving Success

Learn from your mistakes. Being human means making mistakes. By finding and understanding your errors, you will become a better math student.

Source of Error	Remedy
Not Understanding a Concept	Review the concept by finding a similar example in your textbook or class notes. Ask your professor questions to help clarify the concept.
Skipping Steps	Show clear step-by-step solutions. Detailed solution procedures help organize your thoughts and enhance understanding. Doing too many steps mentally often results in preventable mistakes.
Carelessness	Write neatly. Not being able to read your own math writing leads to errors. Avoid writing in pen so you won't have to put huge marks through incorrect work.

"You can achieve your goal if you persistently pursue it."

—Cha Sa-Soon, a 68-year-old South Korean woman who passed her country's written driver's-license exam on her 950th try (*Source: Newsweek*)

CONCEPT AND VOCABULARY CHECK

Fill in each blank so that the resulting statement is true.

1. A solution to a system of linear equations in two variables is an ordered pair that _____.
2. When solving a system of linear equations by graphing, the system's solution is determined by using _____.
3. A system of two linear equations that has no solution is called a/an _____ system. If you attempt to solve such a system by graphing, you will obtain two lines that are _____.
4. The equations in a system of two linear equations with infinitely many solutions are called _____ equations. If you attempt to solve such a system by graphing, you will obtain two lines that _____.

4.1 EXERCISE SET

MyMathLab®

Watch the videos in MyMathLab

Download the MyDashBoard App

Practice Exercises

In Exercises 1–10, determine whether the given ordered pair is a solution of the system.

1. $(2, -3)$
$$\begin{cases} 2x + 3y = -5 \\ 7x - 3y = 23 \end{cases}$$

2. $(-2, -5)$
$$\begin{cases} 6x - 2y = -2 \\ 3x + y = -11 \end{cases}$$

3. $\left(\dfrac{2}{3}, \dfrac{1}{9}\right)$
$$\begin{cases} x + 3y = 1 \\ 4x + 3y = 3 \end{cases}$$

4. $\left(\dfrac{7}{25}, -\dfrac{1}{25}\right)$
$$\begin{cases} 4x + 3y = 1 \\ 3x - 4y = 1 \end{cases}$$

5. $(-5, 9)$
$$\begin{cases} 5x + 3y = 2 \\ x + 4y = 14 \end{cases}$$

6. $(10, 7)$
$$\begin{cases} 6x - 5y = 25 \\ 4x + 15y = 13 \end{cases}$$

7. $(1400, 450)$
$$\begin{cases} x - 2y = 500 \\ 0.03x + 0.02y = 51 \end{cases}$$

8. $(200, 700)$
$$\begin{cases} -4x + y = -100 \\ 0.05x - 0.06y = -32 \end{cases}$$

9. $(8, 5)$
$$\begin{cases} 5x - 4y = 20 \\ 3y = 2x + 1 \end{cases}$$

10. $(5, -2)$
$$\begin{cases} 4x - 3y = 26 \\ x = 15 - 5y \end{cases}$$

In Exercises 11–42, solve each system by graphing. If there is no solution or an infinite number of solutions, so state. Use set notation to express solution sets.

11. $\begin{cases} x + y = 6 \\ x - y = 2 \end{cases}$

12. $\begin{cases} x + y = 2 \\ x - y = 4 \end{cases}$

13. $\begin{cases} x + y = 1 \\ y - x = 3 \end{cases}$

14. $\begin{cases} x + y = 4 \\ y - x = 4 \end{cases}$

15. $\begin{cases} 2x - 3y = 6 \\ 4x + 3y = 12 \end{cases}$

16. $\begin{cases} x + 2y = 2 \\ x - y = 2 \end{cases}$

17. $\begin{cases} 4x + y = 4 \\ 3x - y = 3 \end{cases}$

18. $\begin{cases} 5x - y = 10 \\ 2x + y = 4 \end{cases}$

19. $\begin{cases} y = x + 5 \\ y = -x + 3 \end{cases}$

20. $\begin{cases} y = x + 1 \\ y = 3x - 1 \end{cases}$

21. $\begin{cases} y = 2x \\ y = -x + 6 \end{cases}$

22. $\begin{cases} y = 2x + 1 \\ y = -2x - 3 \end{cases}$

23. $\begin{cases} y = -2x + 3 \\ y = -x + 1 \end{cases}$

24. $\begin{cases} y = 3x - 4 \\ y = -2x + 1 \end{cases}$

25. $\begin{cases} y = 2x - 1 \\ y = 2x + 1 \end{cases}$

26. $\begin{cases} y = 3x - 1 \\ y = 3x + 2 \end{cases}$

27. $\begin{cases} x + y = 4 \\ x = -2 \end{cases}$

28. $\begin{cases} x + y = 6 \\ y = -3 \end{cases}$

29. $\begin{cases} x - 2y = 4 \\ 2x - 4y = 8 \end{cases}$

30. $\begin{cases} 2x + 3y = 6 \\ 4x + 6y = 12 \end{cases}$

31. $\begin{cases} y = 2x - 1 \\ x - 2y = -4 \end{cases}$

32. $\begin{cases} y = -2x - 4 \\ 4x - 2y = 8 \end{cases}$

33. $\begin{cases} x + y = 5 \\ 2x + 2y = 12 \end{cases}$

34. $\begin{cases} x - y = 2 \\ 3x - 3y = -6 \end{cases}$

35. $\begin{cases} x - y = 0 \\ y = x \end{cases}$

36. $\begin{cases} 2x - y = 0 \\ y = 2x \end{cases}$

37. $\begin{cases} x = 2 \\ y = 4 \end{cases}$

38. $\begin{cases} x = 3 \\ y = 5 \end{cases}$

39. $\begin{cases} x = 2 \\ x = -1 \end{cases}$

40. $\begin{cases} x = 3 \\ x = -2 \end{cases}$

41. $\begin{cases} y = 0 \\ y = 4 \end{cases}$

42. $\begin{cases} y = 0 \\ y = 5 \end{cases}$

Practice PLUS

In Exercises 43–50, find the slope and the y-intercept for the graph of each equation in the given system. Use this information (and not the equations' graphs) to determine if the system has no solution, one solution, or an infinite number of solutions.

43. $$\begin{cases} y = \dfrac{1}{2}x - 3 \\ y = \dfrac{1}{2}x - 5 \end{cases}$$

44. $\begin{cases} y = \dfrac{3}{4}x - 2 \\ y = \dfrac{3}{4}x + 1 \end{cases}$

45. $\begin{cases} y = -\dfrac{1}{2}x + 4 \\ 3x - y = -4 \end{cases}$

46. $\begin{cases} y = -\dfrac{1}{4}x + 3 \\ 4x - y = -3 \end{cases}$

47. $\begin{cases} 3x - y = 6 \\ x = \dfrac{y}{3} + 2 \end{cases}$

48. $\begin{cases} 2x - y = 4 \\ x = \dfrac{y}{2} + 2 \end{cases}$

49. $\begin{cases} 3x + y = 0 \\ y = -3x + 1 \end{cases}$

50. $\begin{cases} 2x + y = 0 \\ y = -2x + 1 \end{cases}$

Application Exercises

51. A rental company charges $40.00 a day plus $0.35 per mile to rent a moving truck. The total cost, y, for a day's rental if x miles are driven is described by $y = 0.35x + 40$. A second company charges $36.00 a day plus $0.45 per mile, so the daily cost, y, if x miles are driven is described by $y = 0.45x + 36$. The graphs of the two equations are shown in the same rectangular coordinate system.

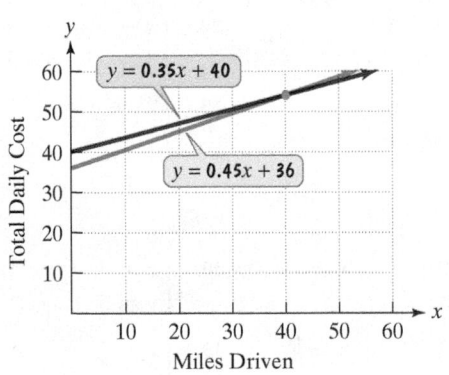

a. What is the x-coordinate of the intersection point of the graphs? Describe what this x-coordinate means in practical terms.

b. What is a reasonable estimate for the y-coordinate of the intersection point?

c. Substitute the x-coordinate of the intersection point into each of the equations and find the corresponding value for y. Describe what this value represents in practical terms. How close is this value to your estimate from part (b)?

52. A band plans to record a demo. Studio A rents for $100 plus $50 per hour. Studio B rents for $50 plus $75 per hour. The total cost, y, in dollars, of renting the studios for x hours can be modeled by the linear system

$$\begin{cases} y = 50x + 100 & \text{Studio A} \\ y = 75x + 50. & \text{Studio B} \end{cases}$$

a. Use graphing to solve the system. Extend the x-axis from 0 to 4 and let each tick mark represent 1 unit (one hour in a recording studio). Extend the y-axis from 0 to 400 and let each tick mark represent 100 units (a rental cost of $100).

b. Interpret the coordinates of the solution in practical terms.

53. You plan to start taking an aerobics class. Nonmembers pay $4 per class. Members pay a $10 monthly fee plus an additional $2 per class. The monthly cost, y, of taking x classes can be modeled by the linear system

$$\begin{cases} y = 4x & \text{Nonmembers} \\ y = 2x + 10. & \text{Members} \end{cases}$$

a. Use graphing to solve the system.

b. Interpret the coordinates of the solution in practical terms.

Writing in Mathematics

54. What is a system of linear equations? Provide an example with your description.

55. What is a solution of a system of linear equations?

56. Explain how to determine if an ordered pair is a solution of a system of linear equations.

57. Explain how to solve a system of linear equations by graphing.

58. What is an inconsistent system? What happens if you attempt to solve such a system by graphing?

59. Explain how a linear system can have infinitely many solutions.

60. What are dependent equations? Provide an example with your description.

61. The following system models the winning times, y, in seconds, in the Olympic 500-meter speed skating event x years after 1970:

$$\begin{cases} y = -0.19x + 43.7 & \text{Women} \\ y = -0.16x + 39.9. & \text{Men} \end{cases}$$

Use the slope of each model to explain why the system has a solution. What does this solution represent?

Critical Thinking Exercises

Make Sense?   *In Exercises 62–65, determine whether each statement "makes sense" or "does not make sense" and explain your reasoning.*

62. Each equation in a system of linear equations has infinitely many ordered-pair solutions.

63. Every linear system has infinitely many ordered-pair solutions.

64. In dependent systems, the two equations represent the same line.

65. When I use graphing to solve an inconsistent system, the lines should look parallel, and I can always use slope to confirm that they really are.

In Exercises 66–69, determine whether each statement is true or false. If the statement is false, make the necessary change(s) to produce a true statement.

66. If a linear system has graphs with equal slopes, the system must be inconsistent.

67. If a linear system has graphs with equal y-intercepts, the system must have infinitely many solutions.

68. If a linear system has two distinct points that are solutions, then the graphs of the system's equations have equal slopes and equal y-intercepts.

69. It is possible for a linear system with one solution to have graphs with equal slopes.

70. Write a system of linear equations whose solution is (5, 1). How many different systems are possible? Explain.

71. Write a system of equations with one solution, a system of equations with no solution, and a system of equations with infinitely many solutions. Explain how you were able to think of these systems.

Technology Exercises

72. Verify your solutions to any five exercises from Exercises 11 through 36 by using a graphing utility to graph the two equations in the system in the same viewing rectangle. After entering the two equations, one as y_1 and the other as y_2, and graphing them, use the $\boxed{\text{INTERSECTION}}$ feature to find the system's solution. (It may first be necessary to solve the equations for y before entering them.)

Read Exercise 72. Then use a graphing utility to solve the systems in Exercises 73–80.

73. $\begin{cases} y = 2x + 2 \\ y = -2x + 6 \end{cases}$

74. $\begin{cases} y = -x + 5 \\ y = x - 7 \end{cases}$

75. $\begin{cases} x + 2y = 4 \\ x - y = 4 \end{cases}$

76. $\begin{cases} 2x - 3y = 10 \\ 4x + 3y = 20 \end{cases}$

77. $\begin{cases} 3x - y = 5 \\ -5x + 2y = -10 \end{cases}$

78. $\begin{cases} 2x - 3y = 7 \\ 3x + 5y = 1 \end{cases}$

79. $\begin{cases} y = \dfrac{1}{3}x + \dfrac{2}{3} \\ y = \dfrac{5}{7}x - 2 \end{cases}$

80. $\begin{cases} y = -\dfrac{1}{2}x + 2 \\ y = \dfrac{3}{4}x + 7 \end{cases}$

Review Exercises

In Exercises 81–83, perform the indicated operation.

81. $-3 + (-9)$ (Section 1.7, **Table 1.7**)

82. $-3 - (-9)$ (Section 1.7, **Table 1.7**)

83. $-3(-9)$ (Section 1.7, **Table 1.7**)

Preview Exercises

Exercises 84–86 will help you prepare for the material covered in the next section. In each exercise, solve the given equation.

84. $4x - 3(-x - 1) = 24$

85. $5(2y - 3) - 4y = 9$

86. $(5x - 1) + 1 = 5x + 5$

4.2

Solving Systems of Linear Equations by the Substitution Method

Objectives

1 Solve linear systems by the substitution method.

2 Use the substitution method to identify systems with no solution or infinitely many solutions.

3 Solve problems using the substitution method.

1 Solve linear systems by the substitution method.

Other than outrage, what is going on at the gas pumps? Is surging demand creating the increasing oil prices? Like all things in a free market economy, the price of a commodity is based on supply and demand. In this section, we use a second method for solving linear systems, the *substitution method*, to understand this economic phenomenon.

Eliminating a Variable Using the Substitution Method

Finding the solution of a linear system by graphing equations may not be easy to do. For example, a solution of $\left(-\frac{2}{3}, \frac{157}{29}\right)$ would be difficult to "see" as an intersection point on a graph.

Let's consider a method that does not depend on finding a system's solution visually: the substitution method. This method involves converting the system to one equation in one variable by an appropriate substitution.

BEST GAS & DIESEL PRICES IN TOWN

Great Question!

In the first step of the substitution method, how do I know which variable to isolate and in which equation?

You can choose both the variable and the equation. If possible, solve for a variable whose coefficient is 1 or −1 to avoid working with fractions.

Solving Linear Systems by Substitution

1. Solve either of the equations for one variable in terms of the other. (If one of the equations is already in this form, you can skip this step.)

2. Substitute the expression found in step 1 into the *other* equation. This will result in an equation in one variable.

3. Solve the equation containing one variable.

4. Back-substitute the value found in step 3 into the equation from step 1. Simplify and find the value of the remaining variable.

5. Check the proposed solution in both of the system's given equations.

EXAMPLE 1   Solving a System by Substitution

Solve by the substitution method:

$$\begin{cases} y = -x - 1 \\ 4x - 3y = 24. \end{cases}$$

Solution

Step 1. Solve either of the equations for one variable in terms of the other. This step has already been done for us. The first equation, $y = -x - 1$, is solved for y in terms of x.

Step 2. Substitute the expression from step 1 into the other equation. We substitute the expression $-x - 1$ for y into the second equation, $4x - 3y = 24$:

$$y = \boxed{-x - 1} \qquad 4x - 3\boxed{y} = 24. \qquad \text{Substitute } -x - 1 \text{ for } y.$$

This gives us an equation in one variable, namely

$$4x - 3(-x - 1) = 24.$$

The variable y has been eliminated.

Step 3. Solve the resulting equation containing one variable.

$$4x - 3(-x - 1) = 24 \qquad \text{This is the equation containing one variable.}$$
$$4x + 3x + 3 = 24 \qquad \text{Apply the distributive property.}$$
$$7x + 3 = 24 \qquad \text{Combine like terms.}$$
$$7x = 21 \qquad \text{Subtract 3 from both sides.}$$
$$x = 3 \qquad \text{Divide both sides by 7.}$$

Step 4. Back-substitute the obtained value into the equation from step 1. We now know that the x-coordinate of the solution is 3. To find the y-coordinate, we back-substitute the x-value in the equation from step 1,

$$y = -x - 1. \qquad \text{This is the equation from step 1.}$$

Substitute **3** for x.

$$y = -3 - 1$$
$$y = -4 \qquad \text{Simplify.}$$

With $x = 3$ and $y = -4$, the proposed solution is $(3, -4)$.

Step 5. Check the proposed solution in both of the system's given equations. Replace x with 3 and y with -4.

$$y = -x - 1 \qquad\qquad\qquad 4x - 3y = 24$$
$$-4 \overset{?}{=} -3 - 1 \qquad\qquad 4(3) - 3(-4) \overset{?}{=} 24$$
$$-4 = -4, \ \ \text{true} \qquad\qquad 12 + 12 \overset{?}{=} 24$$
$$24 = 24, \ \ \text{true}$$

The pair $(3, -4)$ satisfies both equations. The system's solution is $(3, -4)$ and the solution set is $\{(3, -4)\}$. ■

✓ **CHECK POINT 1**   Solve by the substitution method:

$$\begin{cases} y = 5x - 13 \\ 2x + 3y = 12. \end{cases}$$

Using Technology

A graphing utility can be used to solve the system in Example 1. Graph each equation and use the intersection feature. The utility displays the solution $(3, -4)$ as $x = 3, y = -4$.

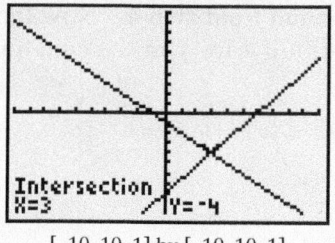

[−10, 10, 1] by [−10, 10, 1]

EXAMPLE 2   Solving a System by Substitution

Solve by the substitution method:

$$\begin{cases} 5x - 4y = 9 \\ x - 2y = -3. \end{cases}$$

Solution

Step 1. Solve either of the equations for one variable in terms of the other. We begin by isolating one of the variables in either of the equations. By solving for x in the second equation, which has a coefficient of 1, we can avoid fractions.

$$x - 2y = -3 \qquad \text{This is the second equation in the given system.}$$
$$x = 2y - 3 \qquad \text{Solve for x by adding 2y to both sides.}$$

Step 2. Substitute the expression from step 1 into the other equation. We substitute $2y - 3$ for x in the first equation, $5x - 4y = 9$.

$$x = \boxed{2y - 3} \qquad 5\boxed{x} - 4y = 9$$

This gives us an equation in one variable, namely

$$5(2y - 3) - 4y = 9.$$

The variable x has been eliminated.

Step 3. Solve the resulting equation containing one variable.

$5(2y - 3) - 4y = 9$	This is the equation containing one variable.
$10y - 15 - 4y = 9$	Apply the distributive property.
$6y - 15 = 9$	Combine like terms.
$6y = 24$	Add 15 to both sides.
$y = 4$	Divide both sides by 6.

Step 4. Back-substitute the obtained value into the equation from step 1. Now that we have the y-coordinate of the solution, we back-substitute 4 for y in the equation $x = 2y - 3$.

$x = 2y - 3$	Use the equation obtained in step 1.
$x = 2(4) - 3$	Substitute 4 for y.
$x = 8 - 3$	Multiply.
$x = 5$	Subtract.

With $x = 5$ and $y = 4$, the proposed solution is $(5, 4)$.

Step 5. Check. Take a moment to show that $(5, 4)$ satisfies both given equations. The solution is $(5, 4)$ and the solution set is $\{(5, 4)\}$. ∎

Great Question!

If my solution satisfies one of the equations in the system, do I have to check the solution in the other equation?

Yes. Get into the habit of checking ordered-pair solutions in both equations of the system.

✓ **CHECK POINT 2**   Solve by the substitution method:

$$\begin{cases} 3x + 2y = -1 \\ x - y = 3. \end{cases}$$

2 Use the substitution method to identify systems with no solution or infinitely many solutions.

The Substitution Method with Linear Systems Having No Solution or Infinitely Many Solutions

Recall that a linear system with no solution is called an **inconsistent system**. If you attempt to solve such a system by substitution, you will eliminate both variables. A false statement such as $0 = 5$ will be the result.

EXAMPLE 3   Using the Substitution Method on an Inconsistent System

Solve the system:

$$\begin{cases} y + 1 = 5(x + 1) \\ y = 5x - 1. \end{cases}$$

Using Technology

A graphing utility was used to graph the equations in Example 3. The lines are parallel and have no point of intersection. This verifies that the system is inconsistent.

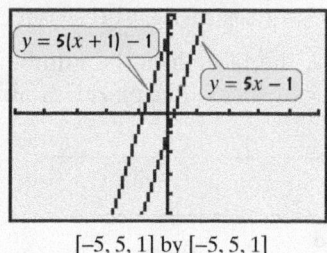

$[-5, 5, 1]$ by $[-5, 5, 1]$

Solution The variable y is isolated in the second equation, $y = 5x - 1$. We use the substitution method and substitute the expression for y in the first equation, $y + 1 = 5(x + 1)$.

$$\boxed{y} + 1 = 5(x + 1) \qquad y = \boxed{5x - 1}$$ Substitute $5x - 1$ for y.

$$(5x - 1) + 1 = 5(x + 1)$$ This substitution gives an equation in one variable.

$$5x = 5x + 5$$ Simplify on the left side. Use the distributive property on the right side.

There are no values of x and y for which $0 = 5$. $$0 = 5, \quad \text{false}$$ Subtract $5x$ from both sides.

The false statement $0 = 5$ indicates that the system is inconsistent and has no solution. The solution set is the empty set, $\varnothing$. ∎

✓ **CHECK POINT 3**   Solve the system:

$$\begin{cases} 3x + y = -5 \\ \quad\quad y = -3x + 3. \end{cases}$$

Do you remember that the equations in a linear system with infinitely many solutions are called **dependent**? If you attempt to solve such a system by substitution, you will eliminate both variables. However, a true statement such as $6 = 6$ will be the result.

EXAMPLE 4   Using the Substitution Method on a System with Infinitely Many Solutions

Solve the system:

$$\begin{cases} \quad\quad y = 3 - 2x \\ 4x + 2y = 6. \end{cases}$$

Solution The variable y is isolated in the first equation. We use the substitution method and substitute the expression for y in the second equation.

$$y = \boxed{3 - 2x} \qquad 4x + 2\boxed{y} = 6$$ Substitute $3 - 2x$ for y.

$$4x + 2(3 - 2x) = 6$$ This substitution gives an equation in one variable.

$$4x + 6 - 4x = 6$$ Apply the distributive property:

$$2(3 - 2x) = 2 \cdot 3 - 2 \cdot 2x = 6 - 4x.$$

$$6 = 6, \quad \text{true}$$ Simplify: $4x - 4x = 0$.

This true statement indicates that the system contains dependent equations and has infinitely many solutions. We express the solution set for the system in one of two equivalent ways:

$$\{(x, y) \,|\, y = 3 - 2x\}$$ The set of all ordered pairs (x, y) such that $y = 3 - 2x$

or $$\{(x, y) \,|\, 4x + 2y = 6\}.$$ The set of all ordered pairs (x, y) such that $4x + 2y = 6$ ∎

✓ **CHECK POINT 4**   Solve the system:

$$\begin{cases} \quad\quad y = 3x - 4 \\ 9x - 3y = 12. \end{cases}$$

3 Solve problems using the substitution method.

Applications

An important application of systems of equations arises in connection with supply and demand. As the price of a product increases, the demand for that product decreases. However, at higher prices suppliers are willing to produce greater quantities of the product.

EXAMPLE 5 Supply and Demand Models

Table 4.1 shows the price of a slice of pizza. For each price, the table lists the number of slices that consumers are willing to buy and the number of slices that pizzerias are willing to supply.

Table 4.1 Supply and Demand for Pizza Slices

Price of a slice of pizza	Quantity demanded per day	Quantity supplied per day	Result
$0.50	300	100	There is a shortage from excess demand.
$1.00	250	150	
$2.00	150	250	There is a surplus from excess supply.
$3.00	50	350	

Source: O'Sullivan and Sheffrin, *Economics*, Prentice Hall, 2007.

The data in **Table 4.1** can be described by the following demand and supply models:

Demand Model

$$p = -0.01x + 3.5$$

Price per slice   Number of slices demanded per day

Supply Model

$$p = 0.01x - 0.5.$$

Price per slice   Number of slices supplied per day

The price at which supply and demand are equal is called the **equilibrium price**. The quantity supplied and demanded at that price is called the **equilibrium quantity**. Find the equilibrium quantity and the equilibrium price for pizza slices.

Solution We can find the equilibrium quantity and the equilibrium price by solving the demand-supply linear system. We will use substitution, substituting $-0.01x + 3.5$ for p in the second equation.

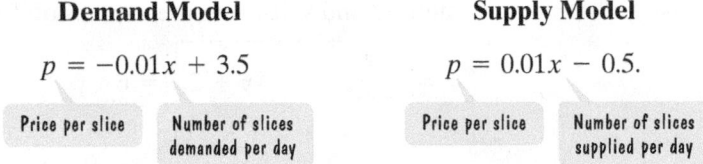

$$p = \boxed{-0.01x + 3.5} \quad \boxed{p} = 0.01x - 0.5 \qquad \text{Substitute } -0.01x + 3.5 \text{ for } p.$$

$$-0.01x + 3.5 = 0.01x - 0.5 \qquad \text{The resulting equation contains only one variable.}$$

$$-0.02x + 3.5 = -0.5 \qquad \text{Subtract 0.01x from both sides.}$$

$$-0.02x = -4.0 \qquad \text{Subtract 3.5 from both sides:} \\ -0.5 - 3.5 = -4.0.$$

$$x = 200 \qquad \text{Divide both sides by } -0.02: \dfrac{-4.0}{-0.02} = 200.$$

Because $x = 200$, the equilibrium quantity is 200 slices per day. To find the equilibrium price, we back-substitute 200 for x into either the demand or the supply model. We'll use both models to make sure we get the same number in each case.

Demand Model	**Supply Model**
$p = -0.01x + 3.5$	$p = 0.01x - 0.5$
Substitute **200** for x.	Substitute **200** for x.
$p = -0.01(200) + 3.5$	$p = 0.01(200) - 0.5$
$= -2 + 3.5 = 1.5$	$= 2 - 0.5 = 1.5$

The equilibrium price is $1.50 per slice. At that price, and only at that price, the quantity demanded and the quantity supplied are equal, at 200 slices per day. **Figure 4.8** shows the equilibrium point, (200, 1.50), as the intersection point for the graphs of the demand and supply models.

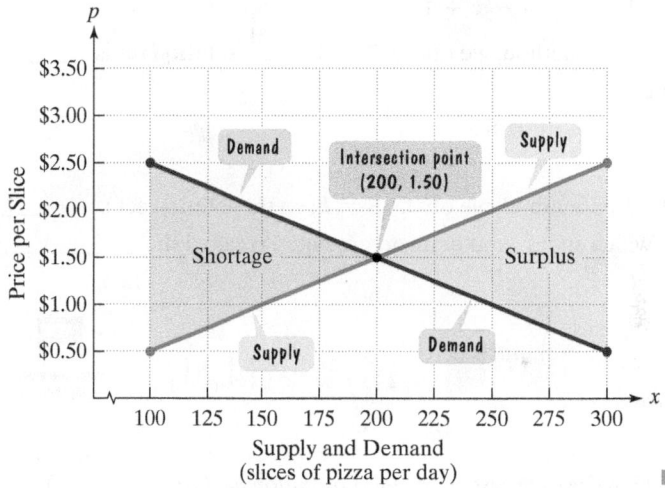

Figure 4.8

✓ **CHECK POINT 5** The following models describe demand and supply for two-bedroom rental apartments.

Demand Model	**Supply Model**
$p = -30x + 1800$	$p = 30x$
Monthly rental price — Number of apartments demanded, in thousands	Monthly rental price — Number of apartments supplied, in thousands

a. Solve the system and find the equilibrium quantity and the equilibrium price.

b. Use your answer from part (a) to complete this statement:

When rents are ___ per month, consumers will demand ____ apartments and suppliers will offer ____ apartments for rent.

Achieving Success

In college, it is recommended that students study and do homework for at least two hours for each hour of class time. For many students, finding the necessary time to study is not always easy. In order to make education a top priority, efficient time management is essential. You can continue to improve your time management by identifying your biggest time-wasters. Do you really need to spend hours on the phone, update your Facebook page throughout the night, or watch that sitcom rerun? My biggest time-wasters: _____

CONCEPT AND VOCABULARY CHECK

Fill in each blank so that the resulting statement is true.

1. When solving
$$\begin{cases} x = 5 - y \\ 2x - y = 7 \end{cases}$$
   by the substitution method, we obtain $y = 1$, so the solution set is _____.

2. When solving
$$\begin{cases} x - 7y = -22 \\ 5x + 2y = 1 \end{cases}$$
   by the substitution method, we obtain $x = -1$, so the solution set is _____.

3. When solving
$$\begin{cases} x + 2y = 4 \\ y = -\frac{1}{2}x + 1 \end{cases}$$
   by the substitution method, we obtain $2 = 4$, so the solution set is _____.

4. When solving
$$\begin{cases} 2x - 6y = 8 \\ x = 3y + 4 \end{cases}$$
   by the substitution method, we obtain $8 = 8$, so the solution set is _____.

5. The price at which supply and demand are equal is called the _____ price.

4.2 EXERCISE SET MyMathLab®

Watch the videos in MyMathLab   Download the MyDashBoard App

Practice Exercises

In Exercises 1–32, solve each system by the substitution method. If there is no solution or an infinite number of solutions, so state. Use set notation to express solution sets.

1. $\begin{cases} x + y = 4 \\ y = 3x \end{cases}$

2. $\begin{cases} x + y = 6 \\ y = 2x \end{cases}$

3. $\begin{cases} x + 3y = 8 \\ y = 2x - 9 \end{cases}$

4. $\begin{cases} 2x - 3y = -13 \\ y = 2x + 7 \end{cases}$

5. $\begin{cases} x + 3y = 5 \\ 4x + 5y = 13 \end{cases}$

6. $\begin{cases} x + 2y = 5 \\ 2x - y = -15 \end{cases}$

7. $\begin{cases} 2x - y = -5 \\ x + 5y = 14 \end{cases}$

8. $\begin{cases} 2x + 3y = 11 \\ x - 4y = 0 \end{cases}$

9. $\begin{cases} 2x - y = 3 \\ 5x - 2y = 10 \end{cases}$

10. $\begin{cases} -x + 3y = 10 \\ 2x + 8y = -6 \end{cases}$

11. $\begin{cases} -3x + y = -1 \\ x - 2y = 4 \end{cases}$

12. $\begin{cases} -4x + y = -11 \\ 2x - 3y = 5 \end{cases}$

13. $\begin{cases} x = 9 - 2y \\ x + 2y = 13 \end{cases}$

14. $\begin{cases} 6x + 2y = 7 \\ y = 2 - 3x \end{cases}$

15. $\begin{cases} y = 3x - 5 \\ 21x - 35 = 7y \end{cases}$

16. $\begin{cases} 9x - 3y = 12 \\ y = 3x - 4 \end{cases}$

17. $\begin{cases} 5x + 2y = 0 \\ x - 3y = 0 \end{cases}$

18. $\begin{cases} 4x + 3y = 0 \\ 2x - y = 0 \end{cases}$

19. $\begin{cases} 2x - y = 6 \\ 3x + 2y = 5 \end{cases}$

20. $\begin{cases} 2x - y = 4 \\ 3x - 5y = 2 \end{cases}$

21. $\begin{cases} 2(x - 1) - y = -3 \\ y = 2x + 3 \end{cases}$

22. $\begin{cases} x + y - 1 = 2(y - x) \\ y = 3x - 1 \end{cases}$

23. $\begin{cases} x = 2y + 9 \\ x = 7y + 10 \end{cases}$

24. $\begin{cases} x = 5y - 3 \\ x = 8y + 4 \end{cases}$

25. $\begin{cases} 4x - y = 100 \\ 0.05x - 0.06y = -32 \end{cases}$

26. $\begin{cases} x + 6y = 8000 \\ 0.3x - 0.6y = 0 \end{cases}$

27. $\begin{cases} y = \frac{1}{3}x + \frac{2}{3} \\ y = \frac{5}{7}x - 2 \end{cases}$

28. $\begin{cases} y = -\frac{1}{2}x + 2 \\ y = \frac{3}{4}x + 7 \end{cases}$

29. $\begin{cases} \frac{x}{6} - \frac{y}{2} = \frac{1}{3} \\ x + 2y = -3 \end{cases}$

30. $\begin{cases} \frac{x}{4} - \frac{y}{4} = -1 \\ x + 4y = -9 \end{cases}$

31. $\begin{cases} 2x - 3y = 8 - 2x \\ 3x + 4y = x + 3y + 14 \end{cases}$

32. $\begin{cases} 3x - 4y = x - y + 4 \\ 2x + 6y = 5y - 4 \end{cases}$

Practice PLUS

In Exercises 33–38, write a system of equations modeling the given conditions. Then solve the system by the substitution method and find the two numbers.

33. The sum of two numbers is 81. One number is 41 more than the other. Find the numbers.

34. The sum of two numbers is 62. One number is 12 more than the other. Find the numbers.

35. The difference between two numbers is 5. Four times the larger number is 6 times the smaller number. Find the numbers.

36. The difference between two numbers is 25. Two times the larger number is 12 times the smaller number. Find the numbers.

37. The difference between two numbers is 1. The sum of the larger number and twice the smaller number is 7. Find the numbers.

38. The difference between two numbers is 5. The sum of the larger number and twice the smaller number is 14. Find the numbers.

In Exercises 39–40, multiply each equation in the system by an appropriate number so that the coefficients are integers. Then solve the system by the substitution method.

39. $\begin{cases} 0.7x - 0.1y = 0.6 \\ 0.8x - 0.3y = -0.8 \end{cases}$

40. $\begin{cases} 1.25x - 0.01y = 4.5 \\ 0.5x - 0.02y = 1 \end{cases}$

Application Exercises

41. The following models describe wages for low-skilled labor.

Demand Model

$$p = -0.325x + 5.8$$

Price of labor (per hour)   Millions of workers employers will hire

Supply Model

$$p = 0.375x + 3$$

Price of labor (per hour)   Millions of available workers

(*Source:* O'Sullivan and Sheffrin, *Economics*, Prentice Hall, 2007.)

a. Solve the system, and find the equilibrium number of workers, in millions, and the equilibrium hourly wage.

b. Use your answer from part (a) to complete this statement:

If workers are paid ____ per hour, there will be __ million available workers and __ million workers will be hired.

c. In 2007, the federal minimum wage was set at $5.15 per hour. Substitute 5.15 for p in the demand model, $p = -0.325x + 5.8$, and determine the millions of workers employers will hire at this price.

d. At a minimum wage of $5.15 per hour, use the supply model, $p = 0.375x + 3$, to determine the millions of available workers. Round to one decimal place.

e. At a minimum wage of $5.15 per hour, use your answers from parts (c) and (d) to determine how many more people are looking for work than employers are willing to hire.

42. The music business is evolving into a digital marketplace. The bar graph shows that from 2004 through 2009, CD album sales declined, while digital track sales grew.

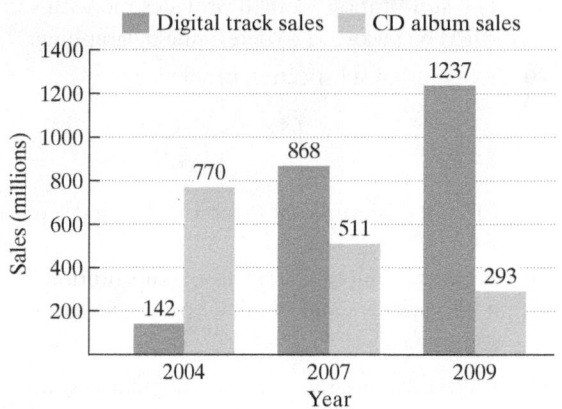

Sales of Albums and Digital Tracks in the United States

Source: RIAA

The data can be described by the following mathematical models:

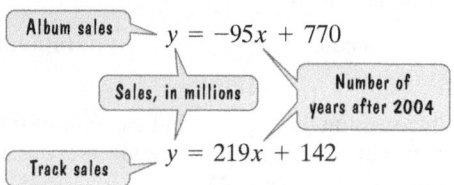

Album sales — $y = -95x + 770$

Sales, in millions   Number of years after 2004

Track sales — $y = 219x + 142$

a. Solve the system.

b. Based on your solution from part (a), when did digital track sales catch up with album sales? How many digital tracks and how many albums were sold in that year?

Writing in Mathematics

43. Describe a problem that might arise when solving a system of equations using graphing. Assume that both equations in the system have been graphed correctly and the system has exactly one solution.

44. Explain how to solve a system of equations using the substitution method. Use $y = 3 - 3x$ and $3x + 4y = 6$ to illustrate your explanation.

45. When using the substitution method, how can you tell if a system of linear equations has no solution?

46. When using the substitution method, how can you tell if a system of linear equations has infinitely many solutions?

47. The law of supply and demand states that, in a free market economy, a commodity tends to be sold at its equilibrium price. At this price, the amount that the seller will supply is the same amount that the consumer will buy. Explain how systems of equations can be used to determine the equilibrium price.

Critical Thinking Exercises

Make Sense? *In Exercises 48–51, determine whether each statement "makes sense" or "does not make sense" and explain your reasoning.*

48. The substitution method provides me with solutions that might be awkward to determine by graphing.

49. When using substitution to solve

$$\begin{cases} 5x - 4y = 9 \\ x - 2y = -3, \end{cases}$$

I find it easiest to solve for x in the first equation.

50. While solving a system using substitution, I eliminated both variables and obtained $6 = 6$, so the system has no solution.

51. I think of equilibrium as the point at which quantity demanded exceeds quantity supplied.

In Exercises 52–55, determine whether each statement is true or false. If the statement is false, make the necessary change(s) to produce a true statement.

52. Solving an inconsistent system by substitution results in a true statement.

53. The line passing through the intersection of the graphs of $x + y = 4$ and $x - y = 0$ with slope $= 3$ has an equation given by $y - 2 = 3(x - 2)$.

54. Unlike the graphing method, where solutions cannot be seen, the substitution method provides a way to visualize solutions as intersection points.

55. To solve the system

$$\begin{cases} 2x - y = 5 \\ 3x + 4y = 7 \end{cases}$$

by substitution, replace y in the second equation with $5 - 2x$.

56. If $x = 3 - y - z$, $2x + y - z = -6$, and $3x - y + z = 11$, find the values for $x, y,$ and z.

57. Find the value of m that makes

$$\begin{cases} y = mx + 3 \\ 5x - 2y = 7 \end{cases}$$

an inconsistent system.

Review Exercises

58. Graph: $4x + 6y = 12$. (Section 3.2, Example 4)

59. Solve: $4(x + 1) = 25 + 3(x - 3)$. (Section 2.3, Example 3)

60. List all the integers in this set: $\left\{ -73, -\frac{2}{3}, 0, \frac{3}{1}, \frac{3}{2}, \frac{\pi}{1} \right\}$. (Section 1.3, Example 5)

Preview Exercises

Exercises 61–63 will help you prepare for the material covered in the next section.

61. Use both equations in the system

$$\begin{cases} 3x + 2y = 48 \\ 9x - 8y = -24 \end{cases}$$

to find x for $y = 12$. What do you observe?

62. Solve: $-14y = -168$.

63. Multiply both sides of $x - 5y = 3$ by -4 and simplify.

SECTION
4.3

Solving Systems of Linear Equations by the Addition Method

Objectives

1️⃣ Solve linear systems by the addition method.

2️⃣ Use the addition method to identify systems with no solution or infinitely many solutions.

3️⃣ Determine the most efficient method for solving a linear system.

Researchers identified college students who generally were procrastinators or nonprocrastinators. The students were asked to report throughout the semester how many symptoms of physical illness they had experienced. **Figure 4.9** shows that by late in the semester, all students experienced increases in symptoms. Early in the semester, procrastinators reported fewer symptoms, but late in the semester, as work came due, they reported more symptoms than their nonprocrastinating peers.

The data in **Figure 4.9** can be analyzed using a pair of linear models in two variables. The figure shows that by week 6, both groups reported the same number of symptoms of illness, an average of approximately 3.5 symptoms per group. In this section, you will learn a third method, called *addition*, that will reinforce this graphic observation, verifying (6, 3.5) as the point of intersection.

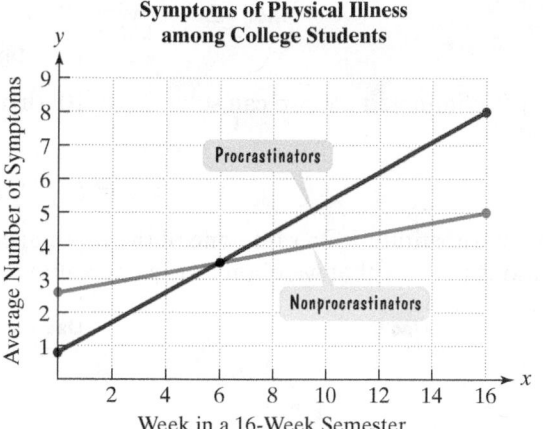

Figure 4.9
Source: Gerrig and Zimbardo, *Psychology and Life*, 18th Edition, Allyn and Bacon, 2008.

1️⃣ Solve linear systems by the addition method.

Eliminating a Variable Using the Addition Method

The substitution method is most useful if one of the given equations has an isolated variable. A third, and frequently the easiest, method for solving a linear system is the addition method. Like the substitution method, the addition method involves eliminating a variable and ultimately solving an equation containing only one variable. However, this time we eliminate a variable by adding the equations.

For example, consider the following system of linear equations:

$$\begin{cases} 3x - 4y = 11 \\ -3x + 2y = -7. \end{cases}$$

When we add these two equations, the x-terms are eliminated. This occurs because the coefficients of the x-terms, 3 and -3, are opposites (additive inverses) of each other.

$$\begin{cases} 3x - 4y = 11 \\ -3x + 2y = -7 \end{cases}$$

Add: $-2y = 4$ The sum is an equation in one variable.

$$y = -2 \quad \text{Divide both sides by } -2 \text{ and solve for } y.$$

Now we can back-substitute -2 for y into one of the original equations to find x. It does not matter which equation you use; you will obtain the same value for x in either case. If we use either equation, we can show that $x = 1$. The solution $(1, -2)$ satisfies both equations in the system.

When we use the addition method, we want to obtain two equations whose sum is an equation containing only one variable. The key step is to **obtain, for one of the variables, coefficients that differ only in sign**. To do this, we may need to multiply one or both equations by some nonzero number so that the coefficients of one of the variables, x or y, become opposites. Then when the two equations are added, this variable is eliminated.

EXAMPLE 1   Solving a System by the Addition Method

Solve by the addition method:

$$\begin{cases} x + y = 4 \\ x - y = 6. \end{cases}$$

Great Question!

Isn't the addition method also called the elimination method?

Although the addition method is also known as the elimination method, variables are eliminated when using both the substitution and addition methods. The name *addition method* specifically tells us that the elimination of a variable is accomplished by adding two equations.

Solution   The coefficients of y in the two equations, 1 and -1, differ only in sign and are opposites. Therefore, by adding the two left sides and the two right sides, we can eliminate the y-terms.

$$\begin{cases} x + y = 4 \\ \underline{x - y = 6} \end{cases}$$

Add: $2x + 0y = 10$

$2x = 10$ Simplify.

Now y is eliminated and we can solve $2x = 10$ for x.

$$2x = 10$$
$$x = 5 \qquad \text{Divide both sides by 2 and solve for } x.$$

We back-substitute 5 for x into one of the original equations to find y. We will use both equations to show that we obtain the same value for y in either case.

Use the first equation:	**Use the second equation:**
$x + y = 4$	$x - y = 6$
$5 + y = 4$	$5 - y = 6$ Replace x with 5.
$y = -1.$	$-y = 1$ Solve for y.
	$y = -1.$

Thus, $x = 5$ and $y = -1$. The proposed solution, $(5, -1)$, can be shown to satisfy both equations in the system. Consequently, the solution is $(5, -1)$ and the solution set is $\{(5, -1)\}$. ∎

✓ CHECK POINT 1   Solve by the addition method:
$$\begin{cases} x + y = 5 \\ x - y = 9. \end{cases}$$

EXAMPLE 2   Solving a System by the Addition Method

Solve by the addition method:

$$\begin{cases} 3x - y = 11 \\ 2x + 5y = 13 \end{cases}.$$

Solution   We must rewrite one or both equations in equivalent forms so that the coefficients of the same variable (either x or y) are opposites. Consider the terms in y in each equation, that is, $-1y$ and $5y$. To eliminate y, we can multiply each term of the first equation by 5 and then add the equations.

$$\begin{cases} 3x - y = 11 \\ 2x + 5y = 13 \end{cases} \xrightarrow[\text{No change}]{\text{Multiply by 5.}} \begin{cases} 15x - 5y = 55 \\ \underline{2x + 5y = 13} \end{cases}$$

$$\text{Add:} \quad 17x + 0y = 68$$

$$17x = 68 \quad \text{Simplify.}$$

$$x = 4 \quad \begin{array}{l}\text{Divide both sides by 17} \\ \text{and solve for x.}\end{array}$$

Thus, $x = 4$. To find y, we back-substitute 4 for x into either one of the given equations. We'll use the second equation.

$$2x + 5y = 13 \quad \text{This is the second equation in the given system.}$$

$$2 \cdot 4 + 5y = 13 \quad \text{Substitute 4 for x.}$$

$$8 + 5y = 13 \quad \text{Multiply: } 2 \cdot 4 = 8.$$

$$5y = 5 \quad \text{Subtract 8 from both sides.}$$

$$y = 1 \quad \text{Divide both sides by 5.}$$

The solution is (4, 1). Check to see that it satisfies both of the original equations in the system. The solution set is {(4, 1)}. ∎

✓ **CHECK POINT 2** Solve by the addition method:

$$\begin{cases} 4x - y = 22 \\ 3x + 4y = 26 \end{cases}$$

Before considering additional examples, let's summarize the steps for solving linear systems by the addition method.

Solving Linear Systems by Addition

1. If necessary, rewrite both equations in the form $Ax + By = C$.
2. If necessary, multiply either equation or both equations by appropriate nonzero numbers so that the sum of the x-coefficients or the sum of the y-coefficients is 0.
3. Add the equations in step 2. The sum is an equation in one variable.
4. Solve the equation in one variable.
5. Back-substitute the value obtained in step 4 into either of the given equations and solve for the other variable.
6. Check the solution in both of the original equations.

EXAMPLE 3 Solving a System by the Addition Method

Solve by the addition method:

$$\begin{cases} 3x + 2y = 48 \\ 9x - 8y = -24. \end{cases}$$

Solution

Step 1. Rewrite both equations in the form $Ax + By = C$. Both equations are already in this form. Variable terms appear on the left and constants appear on the right.

Step 2. If necessary, multiply either equation or both equations by appropriate numbers so that the sum of the *x*-coefficients or the sum of the *y*-coefficients is 0. We can eliminate *x* or *y*. Let's eliminate *x*. Consider the terms in *x* in each equation, that is, $3x$ and $9x$. To eliminate *x*, we can multiply each term of the first equation by -3 and then add the equations.

$$\begin{cases} 3x + 2y = 48 \\ 9x - 8y = -24 \end{cases} \xrightarrow[\text{No change}]{\text{Multiply by } -3.} \begin{cases} -9x - 6y = -144 \\ 9x - 8y = -24 \end{cases}$$

Step 3. Add the equations. Add: $-14y = -168$

Step 4. Solve the equation in one variable. We solve $-14y = -168$ by dividing both sides by -14.

$$\frac{-14y}{-14} = \frac{-168}{-14} \qquad \text{Divide both sides by } -14.$$

$$y = 12 \qquad \text{Simplify.}$$

Step 5. Back-substitute and find the value for the other variable. We can back-substitute 12 for *y* in either one of the given equations. We'll use the first one.

$$3x + 2y = 48 \qquad \text{This is the first equation in the given system.}$$
$$3x + 2(12) = 48 \qquad \text{Substitute 12 for } y.$$
$$3x + 24 = 48 \qquad \text{Multiply.}$$
$$3x = 24 \qquad \text{Subtract 24 from both sides.}$$
$$x = 8 \qquad \text{Divide both sides by 3.}$$

The solution is $(8, 12)$.

Step 6. Check. Take a minute or so to show that $(8, 12)$ satisfies both of the original equations in the system. The solution set is $\{(8, 12)\}$. ∎

✓ **CHECK POINT 3** Solve by the addition method:

$$\begin{cases} 4x + 5y = 3 \\ 2x - 3y = 7 \end{cases}.$$

Some linear systems have solutions that are not integers. If the value of one variable turns out to be a "messy" fraction, back-substitution might lead to cumbersome arithmetic. If this happens, you can return to the original system and use addition to find the value of the other variable.

EXAMPLE 4 Solving a System by the Addition Method

Solve by the addition method:

$$\begin{cases} 2x = 7y - 17 \\ 5y = 17 - 3x \end{cases}.$$

Solution

Step 1. Rewrite both equations in the form *Ax + By = C*. We first arrange the system so that variable terms appear on the left and constants appear on the right. We obtain

$$\begin{cases} 2x - 7y = -17 \qquad \text{Subtract } 7y \text{ from both sides of the first equation.} \\ 3x + 5y = 17. \qquad \text{Add } 3x \text{ to both sides of the second equation.} \end{cases}$$

Step 2. If necessary, multiply either equation or both equations by appropriate numbers so that the sum of the *x*-coefficients or the sum of the *y*-coefficients is 0. We can eliminate *x* or *y*. Let's eliminate *x* by multiplying the first equation by 3 and the second equation by −2.

$$\begin{cases} 2x - 7y = -17 \\ 3x + 5y = 17 \end{cases} \xrightarrow[\text{Multiply by } -2.]{\text{Multiply by 3.}} \begin{cases} 6x - 21y = -51 \\ -6x - 10y = -34 \end{cases}$$

Step 3. Add the equations.   Add: $-31y = -85$

Step 4. Solve the equation in one variable. We solve $-31y = -85$ by dividing both sides by −31.

$$\frac{-31y}{-31} = \frac{-85}{-31} \qquad \text{Divide both sides by } -31.$$

$$y = \frac{85}{31} \qquad \text{Simplify.}$$

Step 5. Back-substitute and find the value for the other variable. Back-substitution of $\frac{85}{31}$ for *y* into either of the given equations results in cumbersome arithmetic. Instead, let's use the addition method on the given system in the form $Ax + By = C$ to find the value for *x*. Thus, we eliminate *y* by multiplying the first equation by 5 and the second equation by 7.

$$\begin{cases} 2x - 7y = -17 \\ 3x + 5y = 17 \end{cases} \xrightarrow[\text{Multiply by 7.}]{\text{Multiply by 5.}} \begin{cases} 10x - 35y = -85 \\ 21x + 35y = \underline{119} \end{cases}$$

$$\text{Add: } 31x = 34$$

$$x = \frac{34}{31} \qquad \text{Divide both sides by 31.}$$

The solution is $\left(\dfrac{34}{31}, \dfrac{85}{31}\right)$.

Step 6. Check. For this system, a calculator is helpful in showing that $\left(\frac{34}{31}, \frac{85}{31}\right)$ satisfies both of the original equations in the system. The solution set is $\left\{ \left(\frac{34}{31}, \frac{85}{31}\right) \right\}$. ∎

✓ **CHECK POINT 4**   Solve by the addition method:

$$\begin{cases} 2x = 9 + 3y \\ 4y = 8 - 3x \end{cases}.$$

---

2 Use the addition method to identify systems with no solution or infinitely many solutions.

The Addition Method with Linear Systems Having No Solution or Infinitely Many Solutions

As with the substitution method, if the addition method results in a false statement, the linear system is inconsistent and has no solution.

EXAMPLE 5   Using the Addition Method on an Inconsistent System

Solve the system:

$$\begin{cases} 4x + 6y = 12 \\ 6x + 9y = 12. \end{cases}$$

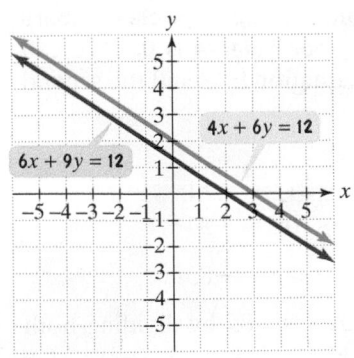

Figure 4.10 Visualizing the inconsistent system in Example 5

Solution　We can eliminate x or y. Let's eliminate x by multiplying the first equation, $4x + 6y = 12$, by 3 and the second equation, $6x + 9y = 12$, by -2.

$$\begin{cases} 4x + 6y = 12 \\ 6x + 9y = 12 \end{cases} \xrightarrow[\text{Multiply by } -2.]{\text{Multiply by 3.}} \begin{cases} 12x + 18y = 36 \\ -12x - 18y = -24 \end{cases}$$

Add:　$0 = 12$

There are no values of x and y for which $0 = 12$.

The false statement $0 = 12$ indicates that the system is inconsistent and has no solution. The solution set is the empty set, $\varnothing$. The graphs of the system's equations are shown in **Figure 4.10**. The lines are parallel and have no point of intersection. ∎

✓ **CHECK POINT 5**　Solve the system:

$$\begin{cases} x + 2y = 4 \\ 3x + 6y = 13. \end{cases}$$

If you use the addition method, how can you tell if a system has infinitely many solutions? As with the substitution method, you will eliminate both variables and obtain a true statement.

EXAMPLE 6　Using the Addition Method on a System with Infinitely Many Solutions

Solve by the addition method:

$$\begin{cases} 2x - y = 3 \\ -4x + 2y = -6. \end{cases}$$

Solution　We can eliminate y by multiplying the first equation by 2.

$$\begin{cases} 2x - y = 3 \\ -4x + 2y = -6 \end{cases} \xrightarrow[\text{No change}]{\text{Multiply by 2.}} \begin{cases} 4x - 2y = 6 \\ -4x + 2y = -6 \end{cases}$$

Add:　$0 = 0$

The true statement $0 = 0$ indicates that the system contains dependent equations and has infinitely many solutions. Any ordered pair that satisfies the first equation also satisfies the second equation. We express the solution set for the system in one of two equivalent ways:

$$\{(x, y) \mid 2x - y = 3\} \text{ or } \{(x, y) \mid -4x + 2y = -6\}. ∎$$

✓ **CHECK POINT 6**　Solve by the addition method:

$$\begin{cases} x - 5y = 7 \\ 3x - 15y = 21. \end{cases}$$

3 Determine the most efficient method for solving a linear system.

Comparing the Three Solution Methods

Table 4.2 at the top of the next page compares the graphing, substitution, and addition methods for solving systems of linear equations. With increased practice, it will become easier for you to select the best method for solving a particular linear system.

Table 4.2	Comparing Solution Methods	
Method	**Advantages**	**Disadvantages**
Graphing	You can see the solutions.	If the solutions do not involve integers or are too large to be seen on the graph, it's impossible to tell exactly what the solutions are.
Substitution	Gives exact solutions. Easy to use if a variable is on one side by itself.	Solutions cannot be seen. Can introduce extensive work with fractions when no variable has a coefficient of 1 or -1.
Addition	Gives exact solutions. Easy to use even if no variable has a coefficient of 1 or -1.	Solutions cannot be seen.

Achieving Success

We have seen that **the best way to achieve success in math is through practice.** Keeping up with your homework, preparing for tests, asking questions of your professor, reading your textbook, and attending all classes will help you learn the material and boost your confidence. Use language in a proactive way that reflects a sense of responsibility for your own successes and failures.

Reactive Language	**Proactive Language**
I'll try.	I'll do it.
That's just the way I am.	I can do better than that.
There's not a thing I can do.	I have options for improvement.
I have to.	I choose to.
I can't.	I can find a way.

CONCEPT AND VOCABULARY CHECK

Fill in each blank so that the resulting statement is true.

1. When solving
$$\begin{cases} x + 5y = 18 \\ 3x + 2y = -11 \end{cases}$$
by the addition method, we can eliminate x by multiplying the first equation by _____ and then adding the equations.

2. When solving
$$\begin{cases} 2x + 10y = 9 \\ 8x + 5y = 7 \end{cases}$$
by the addition method, we can eliminate y by multiplying the second equation by _____ and then adding the equations.

3. When solving
$$\begin{cases} 2x + 5y = 7 \\ 3x + 4y = 9 \end{cases}$$
by the addition method, we can eliminate x by multiplying the first equation by 3 and the second equation by _____, and then adding the equations.

4. When solving
$$\begin{cases} 4x - 3y = 2 \\ -5x + 8y = 1 \end{cases}$$
by the addition method, we can eliminate y by multiplying the first equation by 8 and the second equation by _____, and then adding the equations.

4.3 EXERCISE SET MyMathLab®

Watch the videos in MyMathLab

Download the MyDashBoard App

Practice Exercises

In Exercises 1–44, solve each system by the addition method. If there is no solution or an infinite number of solutions, so state. Use set notation to express solution sets.

1. $\begin{cases} x + y = -3 \\ x - y = 11 \end{cases}$

2. $\begin{cases} x + y = 6 \\ x - y = -2 \end{cases}$

3. $\begin{cases} 2x + 3y = 6 \\ 2x - 3y = 6 \end{cases}$

4. $\begin{cases} 3x + 2y = 14 \\ 3x - 2y = 10 \end{cases}$

5. $\begin{cases} x + 2y = 7 \\ -x + 3y = 18 \end{cases}$

6. $\begin{cases} 2x + y = -2 \\ -2x - 3y = -6 \end{cases}$

7. $\begin{cases} 5x - y = x\ 14 \\ -5x + 2y = -13 \end{cases}$

8. $\begin{cases} 7x - 4y = 13 \\ -7x + 6y = -11 \end{cases}$

9. $\begin{cases} 3x + y = 7 \\ 2x - 5y = -1 \end{cases}$

10. $\begin{cases} 3x - y = 11 \\ 2x + 5y = 13 \end{cases}$

11. $\begin{cases} x + 3y = 4 \\ 4x + 5y = 2 \end{cases}$

12. $\begin{cases} x + 2y = -1 \\ 4x - 5y = 22 \end{cases}$

13. $\begin{cases} -3x + 7y = 14 \\ 2x - y = -13 \end{cases}$

14. $\begin{cases} 2x - 5y = -1 \\ 3x + y = 7 \end{cases}$

15. $\begin{cases} 3x - 14y = 6 \\ 5x + 7y = 10 \end{cases}$

16. $\begin{cases} 5x - 4y = 19 \\ 3x + 2y = 7 \end{cases}$

17. $\begin{cases} 3x - 4y = 11 \\ 2x + 3y = -4 \end{cases}$

18. $\begin{cases} 2x + 3y = -16 \\ 5x - 10y = 30 \end{cases}$

19. $\begin{cases} 3x + 2y = -1 \\ -2x + 7y = 9 \end{cases}$

20. $\begin{cases} 5x + 3y = 27 \\ 7x - 2y = 13 \end{cases}$

21. $\begin{cases} 3x = 2y + 7 \\ 5x = 2y + 13 \end{cases}$

22. $\begin{cases} 9x = 25 + y \\ 2y = 4 - 9x \end{cases}$

23. $\begin{cases} 2x = 3y - 4 \\ -6x + 12y = 6 \end{cases}$

24. $\begin{cases} 5x = 4y - 8 \\ 3x + 7y = 14 \end{cases}$

25. $\begin{cases} 2x - y = 3 \\ 4x + 4y = -1 \end{cases}$

26. $\begin{cases} 3x - y = 22 \\ 4x + 5y = -21 \end{cases}$

27. $\begin{cases} 4x = 5 + 2y \\ 2x + 3y = 4 \end{cases}$

28. $\begin{cases} 3x = 4y + 1 \\ 4x + 3y = 1 \end{cases}$

29. $\begin{cases} 3x - y = 1 \\ 3x - y = 2 \end{cases}$

30. $\begin{cases} 4x - 9y = -2 \\ -4x + 9y = -2 \end{cases}$

31. $\begin{cases} x + 3y = 2 \\ 3x + 9y = 6 \end{cases}$

32. $\begin{cases} 4x - 2y = 2 \\ 2x - y = 1 \end{cases}$

33. $\begin{cases} 7x - 3y = 4 \\ -14x + 6y = -7 \end{cases}$

34. $\begin{cases} 2x + 4y = 5 \\ 3x + 6y = 6 \end{cases}$

35. $\begin{cases} 5x + y = 2 \\ 3x + y = 1 \end{cases}$

36. $\begin{cases} 2x - 5y = -1 \\ 2x - y = 1 \end{cases}$

37. $\begin{cases} x = 5 - 3y \\ 2x + 6y = 10 \end{cases}$

38. $\begin{cases} 4x = 36 + 8y \\ 3x - 6y = 27 \end{cases}$

39. $\begin{cases} 4(3x - y) = 0 \\ 3(x + 3) = 10y \end{cases}$

40. $\begin{cases} 2(2x + 3y) = 0 \\ 7x = 3(2y + 3) + 2 \end{cases}$

41. $\begin{cases} x + y = 11 \\ \dfrac{x}{5} + \dfrac{y}{7} = 1 \end{cases}$

42. $\begin{cases} x - y = -3 \\ \dfrac{x}{9} - \dfrac{y}{7} = -1 \end{cases}$

43. $\begin{cases} \dfrac{4}{5}x - y = -1 \\ \dfrac{2}{5}x + y = 1 \end{cases}$

44. $\begin{cases} \dfrac{x}{3} + y = 3 \\ \dfrac{x}{2} - \dfrac{y}{4} = 1 \end{cases}$

In Exercises 45–56, solve each system by the method of your choice. If there is no solution or an infinite number of solutions, so state. Use set notation to express solution sets. Explain why you selected one method over the other two.

45. $\begin{cases} 3x - 2y = 8 \\ x = -2y \end{cases}$

46. $\begin{cases} 2x - y = 10 \\ y = 3x \end{cases}$

47. $\begin{cases} 3x + 2y = -3 \\ 2x - 5y = 17 \end{cases}$

48. $\begin{cases} 2x - 7y = 17 \\ 4x - 5y = 25 \end{cases}$

49. $\begin{cases} 3x - 2y = 6 \\ y = 3 \end{cases}$

50. $\begin{cases} 2x + 3y = 7 \\ x = 2 \end{cases}$

51. $\begin{cases} y = 2x + 1 \\ y = 2x - 3 \end{cases}$

52. $\begin{cases} y = 2x + 4 \\ y = 2x - 1 \end{cases}$

53. $\begin{cases} 2(x + 2y) = 6 \\ 3(x + 2y - 3) = 0 \end{cases}$

54. $\begin{cases} 2(x + y) = 4x + 1 \\ 3(x - y) = x + y - 3 \end{cases}$

55. $\begin{cases} 3y = 2x \\ 2x + 9y = 24 \end{cases}$

56. $\begin{cases} 4y = -5x \\ 5x + 8y = 20 \end{cases}$

Practice PLUS

In Exercises 57–60, write a system of equations modeling the given conditions. Then solve the system by the addition method and find the two numbers.

57. Five times a first number increased by a second number is 14. The difference between four times the first number and the second number is 4. Find the numbers.

58. Three times a first number increased by twice a second number is 11. The difference between the first number and twice the second number is 9. Find the numbers.

59. If four times a first number is decreased by three times a second number, the result is 0. The sum of the numbers is −7. Find the numbers.

60. If three times a first number is decreased by six times a second number, the result is 15. The sum of the numbers is 2. Find the numbers.

In Exercises 61–68, solve each system or state that the system is inconsistent or dependent.

61. $\begin{cases} \dfrac{3x}{5} + \dfrac{4y}{5} = 1 \\ \dfrac{x}{4} - \dfrac{3y}{8} = -1 \end{cases}$

62. $\begin{cases} \dfrac{x}{3} - \dfrac{y}{2} = \dfrac{2}{3} \\ \dfrac{2x}{3} + y = \dfrac{4}{3} \end{cases}$

63. $\begin{cases} 5(x + 1) = 7(y + 1) - 7 \\ 6(x + 1) + 5 = 5(y + 1) \end{cases}$

64. $\begin{cases} 6x = 5(x + y + 3) - x \\ 3(x - y) + 4y = 5(y + 1) \end{cases}$

65. $\begin{cases} 0.4x + y = 2.2 \\ 0.5x - 1.2y = 0.3 \end{cases}$

66. $\begin{cases} 1.25x - 1.5y = 2 \\ 3.5x - 1.75y = 10.5 \end{cases}$

67. $\begin{cases} \dfrac{x}{2} = \dfrac{y + 8}{3} \\ \dfrac{x + 2}{2} = \dfrac{y + 11}{3} \end{cases}$

68. $\begin{cases} \dfrac{x}{2} = \dfrac{y + 8}{4} \\ \dfrac{x + 3}{2} = \dfrac{y + 5}{4} \end{cases}$

Application Exercises

69. We opened this section with a study showing that late in the semester, procrastinating students reported more symptoms of physical illness than their nonprocrastinating peers. The data can be modeled by the following system of equations:

$\begin{cases} -0.45x + y = 0.8 \\ -0.15x + y = 2.6. \end{cases}$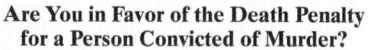

The average number of symptoms for procrastinators, y, after x weeks

The average number of symptoms for nonprocrastinators, y, after x weeks

Use the addition method to determine by which week in the semester both groups report the same number of symptoms of physical illness. For that week, how many symptoms were reported by each group? How is this shown in **Figure 4.9** on page 297?

70. The bar graph shows the percentage of Americans for and against the death penalty for a person convicted of murder.

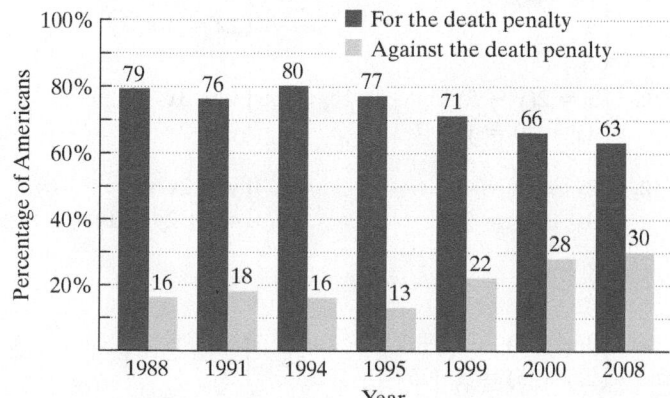

Are You in Favor of the Death Penalty for a Person Convicted of Murder?

■ For the death penalty
▨ Against the death penalty

Source: Newsweek poll

The data can be modeled by the following system of equations:

$\begin{cases} 13x + 12y = 992 \\ -x + y = 16. \end{cases}$

The percent, y, in favor of the death penalty x years after 1988

The percent, y, against the death penalty x years after 1988

Use the addition method to determine in which year the percentage of Americans in favor of the death penalty will be the same as the percentage of Americans who oppose it. For that year, what percent will be for the death penalty and what percent will be against it?

Writing in Mathematics

71. Explain how to solve a system of equations using the addition method. Use $3x + 5y = -2$ and $2x + 3y = 0$ to illustrate your explanation.

72. When using the addition method, how can you tell if a system of linear equations has no solution?

73. When using the addition method, how can you tell if a system of linear equations has infinitely many solutions?

74. Take a second look at the data about the death penalty shown in Exercise 70. Do you think that these trends will continue? Explain your answer.

75. The formula $3239x + 96y = 134{,}014$ models the number of daily evening newspapers, y, x years after 1980. The formula $-665x + 36y = 13{,}800$ models the number of daily morning newspapers, y, x years after 1980. What is the most efficient method for solving this system? Explain why. What does the solution mean in terms of the variables in the formulas? (It is not necessary to actually solve the system.)

Critical Thinking Exercises

Make Sense? *In Exercises 76–79, determine whether each statement "makes sense" or "does not make sense" and explain your reasoning.*

76. When I use the addition method, I add the equations only if the x-coefficients or the y-coefficients are opposites.

77. Unlike substitution, the addition method lets me see solutions as intersection points of graphs.

78. When I use the addition method, I sometimes need to multiply more than one equation by a nonzero number before adding the equations.

79. I find it easiest to use the addition method when one of the equations has a variable on one side by itself.

In Exercises 80–83, determine whether each statement is true or false. If the statement is false, make the necessary change(s) to produce a true statement.

80. If x can be eliminated by the addition method, y cannot be eliminated by using the original equations of the system.

81. If $Ax + 2y = 2$ and $2x + By = 10$ have graphs that intersect at $(2, -2)$, then $A = -3$ and $B = 3$.

82. The equations $y = x - 1$ and $x = y + 1$ are dependent.

83. If the two equations in a linear system are $5x - 3y = 7$ and $4x + 9y = 11$, multiplying the first equation by 4, the second by 5, and then adding equations will eliminate x.

84. Solve by expressing x and y in terms of a and b:
$$\begin{cases} x - y = a \\ y = 2x + b \end{cases}.$$

85. The point of intersection of the graphs of the equations $Ax - 3y = 16$ and $3x + By = 7$ is $(5, -2)$. Find A and B.

Review Exercises

86. For which number is 5 times the number equal to the number increased by 40? (Section 2.5, Example 1)

87. In which quadrant is $\left(-\frac{3}{2}, 15\right)$ located? (Section 3.1, Example 1)

88. Solve: $29{,}700 + 150x = 5000 + 1100x$. (Section 2.2, Example 7)

Preview Exercises

Exercises 89–91 will help you prepare for the material covered in the next section.

89. The sum of two numbers, x and y, is 28. The difference between the numbers is 6.

    a. Write a system of linear equations that models these conditions.

    b. Solve the system and find the numbers.

90. If a slice of cheese contains x calories and a glass of wine contains y calories, write an algebraic expression for the number of calories in 3 slices of cheese and 2 glasses of wine.

91. A telephone plan has a monthly fee of $20 with a charge of $0.05 per minute.

    a. What is the total monthly cost for the plan if there are 200 minutes of calls?

    b. Write a formula that describes the total monthly cost of the plan, y, for x minutes of calls.

| MID-CHAPTER CHECK POINT | Section 4.1–Section 4.3 |

 What You Know: We learned how to solve systems of linear equations by graphing, by the substitution method, and by the addition method. We saw that some systems, called inconsistent systems, have no solution, whereas other systems, called dependent systems, have infinitely many solutions.

In Exercises 1–3, solve each system by graphing.

1. $\begin{cases} 3x + 2y = 6 \\ 2x - y = 4 \end{cases}$

2. $\begin{cases} y = 2x - 1 \\ y = 3x - 2 \end{cases}$

3. $\begin{cases} y = 2x - 1 \\ 6x - 3y - 12 \end{cases}$

In Exercises 4–15, solve each system by the method of your choice.

4. $\begin{cases} 5x - 3y = 1 \\ y = 3x - 7 \end{cases}$

5. $\begin{cases} 6x + 5y = 7 \\ 3x - 7y = 13 \end{cases}$

6. $\begin{cases} x = \dfrac{y}{3} - 1 \\ 6x + y = 21 \end{cases}$

7. $\begin{cases} 3x - 4y = 6 \\ 5x - 6y = 8 \end{cases}$

8. $\begin{cases} 3x - 2y = 32 \\ \dfrac{x}{5} + 3y = -1 \end{cases}$

9. $\begin{cases} x - y = 3 \\ 2x = 4 + 2y \end{cases}$

10. $\begin{cases} x = 2(y - 5) \\ 4x + 40 = y - 7 \end{cases}$

11. $\begin{cases} y = 3x - 2 \\ y = 2x - 9 \end{cases}$

12. $\begin{cases} 2x - 3y = 4 \\ 3x + 4y = 0 \end{cases}$

13. $\begin{cases} y - 2x = 7 \\ 4x = 2y - 14 \end{cases}$

14. $\begin{cases} 4(x + 3) = 3y + 7 \\ 2(y - 5) = x + 5 \end{cases}$

15. $\begin{cases} \dfrac{x}{2} - \dfrac{y}{5} = 1 \\ y - \dfrac{x}{3} = 8 \end{cases}$

Objectives

1 Solve problems using linear systems.

2 Solve simple interest problems.

3 Solve mixture problems.

4 Solve motion problems.

1 Solve problems using linear systems.

Problem Solving Using Systems of Equations

"Bad television is three things: a bullet train to a morally bankrupt youth, a slow spiral into an intellectual void, and of course, a complete blast to watch."

—*Dennis Miller*

A blast, indeed. Americans spend more of their free time watching TV than any other activity. We open this section with a system of equations that models how we spend free time, including how many of those 1440 minutes in a day we spend glued to the tube.

A Strategy for Solving Word Problems Using Systems of Equations

When we solved problems in Chapter 2, we let *x* represent a quantity that was unknown. Problems in this section involve two unknown quantities. We will let *x* and *y* represent these quantities. We then translate from the verbal conditions of the problem to a *system* of linear equations.

EXAMPLE 1   Free Time

Figure 4.11 shows the average time per day on weekends that Americans spend on various leisure and sports activities.

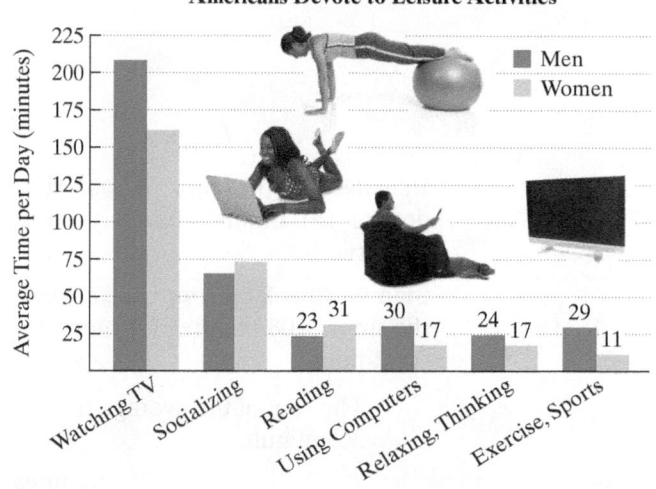

Average Time per Day on Weekends Americans Devote to Leisure Activities

Men
Women

23 31   30 17   24 17   29 11

Watching TV   Socializing   Reading   Using Computers   Relaxing, Thinking   Exercise, Sports

Figure 4.11

Source: Bureau of Labor Statistics American Time Use Survey

It appears that TV is our preferred way of goofing off. Each weekend day, the sum of the average times spent watching TV for men and women is 369 minutes. (That's more than 6 hours.) The difference between the average times spent watching TV for men and women is 47 minutes. How many minutes per day on weekends do men and women devote to watching TV?

Solution

Step 1. Use variables to represent unknown quantities.

Let x = average time per day men watch TV.

Let y = average time per day women watch TV.

Step 2. Write a system of equations that models the problem's conditions.

The sum of average times for men and women | is | 369 minutes.

$$x + y = 369$$

The difference between average times for men and women | is | 47 minutes.

$$x - y = 47$$

Step 3. Solve the system and answer the problem's question. The system

$$\begin{cases} x + y = 369 \\ x - y = 47 \end{cases}$$

can easily be solved by addition because the y-coefficients are opposites. Adding equations will eliminate y.

$$\begin{cases} x + y = 369 \\ \underline{x - y = 47} \end{cases}$$

Add: $\quad 2x \qquad = 416$

$\qquad\qquad\quad x = 208 \qquad$ Divide both sides by 2.

Because x represents average time per day men watch TV, we see that men devote 208 minutes per day on weekends (or 3 hours 28 minutes) to this leisure activity. Now we can find y, the time for women. We do so by back-substituting 208 for x in either of the system's equations.

$$x + y = 369 \qquad \text{We'll use the first equation.}$$
$$208 + y = 369 \qquad \text{Back-substitute 208 for x.}$$
$$y = 161 \qquad \text{Subtract 208 from both sides.}$$

Because y represents average time per day women watch TV, we see that women devote 161 minutes per day on weekends (or 2 hours 41 minutes) to this leisure activity.

Step 4. Check the proposed answers in the original wording of the problem. We check the times, 208 minutes for men and 161 minutes for women, in each of the problem's conditions.

The sum of the average times spent watching TV for men and women is 369 minutes:

$$208 \text{ minutes} + 161 \text{ minutes} = 369 \text{ minutes.}$$

The difference between the average times spent watching TV for men and women is 47 minutes:

$$208 \text{ minutes} - 161 \text{ minutes} = 47 \text{ minutes.}$$

This verifies that men spend 208 minutes per day on weekends watching TV and women spend 161 minutes per day. ■

✓ CHECK POINT 1 **Figure 4.11** on page 307 indicates that socializing is our second-favorite leisure activity. Each weekend day, the sum of the average times spent socializing for men and women is 138 minutes. The difference between the average times spent socializing for women and men is 8 minutes. How many minutes per day on weekends do men and women devote to socializing?

EXAMPLE 2 Cholesterol and Heart Disease

The verdict is in. After years of research, the nation's health experts agree that high cholesterol in the blood is a major contributor to heart disease. Thus, cholesterol intake should be limited to 300 milligrams or less each day. Fast foods provide a cholesterol carnival. All together, two McDonald's Quarter Pounders and three Burger King Whoppers with cheese contain 520 milligrams of cholesterol. Three Quarter Pounders and one Whopper with cheese exceed the suggested daily cholesterol intake by 53 milligrams. Determine the cholesterol content in each item.

Solution

Step 1. Use variables to represent the unknown quantities.

Let x = the cholesterol content, in milligrams, of a Quarter Pounder.

Let y = the cholesterol content, in milligrams, of a Whopper with cheese.

Step 2. Write a system of equations that models the problem's conditions.

The amount of cholesterol in 2 Quarter Pounders	plus	the amount of cholesterol in 3 Whoppers with cheese	is	520 mg.
$2x$	$+$	$3y$	$=$	520

The amount of cholesterol in 3 Quarter Pounders	plus	the amount of cholesterol in 1 Whopper with cheese	is	the suggested daily limit plus 53 mg.
$3x$	$+$	y	$=$	$300 + 53$

Step 3. Solve the system and answer the problem's question. The system

$$\begin{cases} 2x + 3y = 520 \\ 3x + y = 353 \end{cases}$$

can be solved by substitution or addition. We will use addition; if we multiply the second equation by -3, adding equations will eliminate y.

$$\begin{cases} 2x + 3y = 520 \\ 3x + y = 353 \end{cases} \xrightarrow[\text{Multiply by } -3.]{\text{No change}} \begin{cases} 2x + 3y = 520 \\ -9x - 3y = -1059 \end{cases}$$

$$\text{Add: } -7x = -539$$

$$x = \frac{-539}{-7} = 77$$

Because x represents the cholesterol content of a Quarter Pounder, we see that a Quarter Pounder contains 77 milligrams of cholesterol. Now we can find y, the cholesterol content of a Whopper with cheese. We do so by back-substituting 77 for x in either of the system's equations.

$3x + y = 353$ We'll use the second equation.

$3(77) + y = 353$ Back-substitute 77 for x.

$231 + y = 353$ Multiply.

$y = 122$ Subtract 231 from both sides.

Because $x = 77$ and $y = 122$, a Quarter Pounder contains 77 milligrams of cholesterol and a Whopper with cheese contains 122 milligrams of cholesterol.

Step 4. Check the proposed answers in the original wording of the problem. Two Quarter Pounders and three Whoppers with cheese contain

$$2(77 \text{ mg}) + 3(122 \text{ mg}) = 520 \text{ mg},$$

which checks with the given conditions. Furthermore, three Quarter Pounders and one Whopper with cheese contain

$$3(77 \text{ mg}) + 1(122 \text{ mg}) = 353 \text{ mg},$$

which does exceed the daily limit of 300 milligrams by 53 milligrams. ∎

✓ **CHECK POINT 2**   How do the Quarter Pounder and Whopper with cheese measure up in the calorie department? Actually, not too well. Two Quarter Pounders and three Whoppers with cheese provide 2607 calories. Even combining one of each provides enough calories to bring tears to Jenny Craig's eyes—9 calories in excess of what is allowed on a 1000-calorie-a-day diet. Find the caloric content of each item.

EXAMPLE 3   Fencing a Waterfront Lot

You just purchased a rectangular waterfront lot along a river's edge. The perimeter of the lot is 1000 feet. To create a sense of privacy, you decide to fence along three sides, excluding the side that fronts the river (see **Figure 4.12**). An expensive fencing along the lot's front length costs $25 per foot. An inexpensive fencing along the two side widths costs only $5 per foot. The total cost of the fencing along three sides comes to $9500.

a. What are the lot's dimensions?

b. You are considering using the expensive fencing on all three sides of the lot. However, you are limited by a $16,000 budget. Can this be done within your budget constraints?

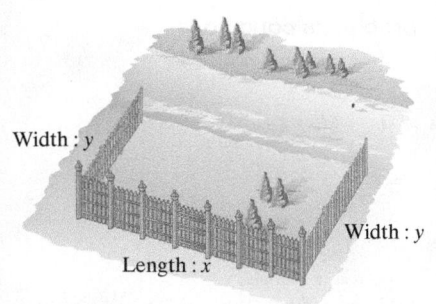

Width: y

Width: y

Length: x

Figure 4.12

Solution

a. We begin by finding the lot's dimensions.

Step 1. Use variables to represent unknown quantities.

Let x = the lot's length, in feet.
Let y = the lot's width, in feet.

(If you prefer, you can use the variable l for length and w for width.)

Step 2. Write a system of equations that models the problem's conditions.

• The lot's perimeter is 1000 feet.

Twice the length	plus	twice the width	is	the perimeter.
$2x$	$+$	$2y$	$=$	1000

• The cost of fencing three sides of the lot is $9500.

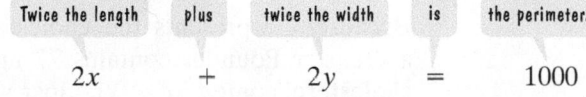

Fencing along the front length	plus	fencing along the two side widths	costs	$9500.

Cost per foot	·	number of feet	+	cost per foot	·	number of feet	is	$9500.
25	·	x	$+$	5	·	$2y$	$=$	9500

Step 3. Solve the system and answer the problem's question. The system

$$\begin{cases} 2x + 2y = 1000 \\ 25x + 10y = 9500 \end{cases}$$

can be solved most easily by addition. If we multiply the first equation by −5, adding equations will eliminate y.

$$\begin{cases} 2x + 2y = 1000 \\ 25x + 10y = 9500 \end{cases} \xrightarrow[\text{No change}]{\text{Multiply by −5.}} \begin{cases} -10x - 10y = -5000 \\ \underline{25x + 10y = 9500} \end{cases}$$

$$\text{Add: } 15x = 4500$$

$$x = \frac{4500}{15} = 300$$

Because x represents length, we see that the lot is 300 feet long. Now we can find y, the lot's width. We do so by back-substituting 300 for x in either of the system's equations.

$$2x + 2y = 1000 \qquad \text{We'll use the first equation.}$$

$$2 \cdot 300 + 2y = 1000 \qquad \text{Back-substitute 300 for x.}$$

$$600 + 2y = 1000 \qquad \text{Multiply: } 2 \cdot 300 = 600.$$

$$2y = 400 \qquad \text{Subtract 600 from both sides.}$$

$$y = 200 \qquad \text{Divide both sides by 2.}$$

Because $x = 300$ and $y = 200$, the lot is 300 feet long and 200 feet wide. Its dimensions are 300 feet by 200 feet.

Step 4. Check the proposed answers in the original wording of the problem. The perimeter should be 1000 feet:

$$2(300 \text{ ft}) + 2(200 \text{ ft}) = 600 \text{ ft} + 400 \text{ ft} = 1000 \text{ ft}.$$

The cost of fencing three sides of the lot should be $9500:

Fencing cost: front length   Fencing cost: two side widths

$$300 \cdot \$25 + 400 \cdot \$5 = \$7500 + \$2000 = \$9500.$$

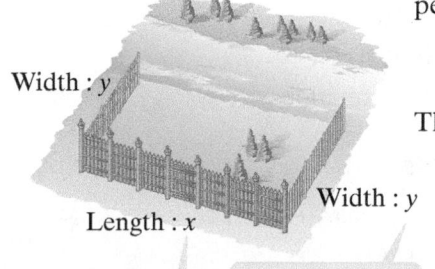

Width : y

Width : y

Length : x

proposed: $x = 300$

proposed: $y = 200$

Figure 4.12 (repeated) Showing proposed values for x and y

b. Can you use the expensive fencing along all three sides of the lot and stay within a $16,000 budget? The fencing is to be placed along the length, 300 feet, and the two side widths, $2 \cdot 200$ feet, or 400 feet. Thus, you will fence 700 feet. The expensive fencing costs $25 per foot. The cost is

$$700 \text{ feet} \cdot \frac{\$25}{\text{feet}} = 700 \cdot \$25 = \$17,500.$$

Limited by a $16,000 budget, you cannot use the expensive fencing on three sides of the lot. ∎

✓ **CHECK POINT 3** A rectangular lot whose perimeter is 360 feet is fenced along three sides. An expensive fencing along the lot's length costs $20 per foot. An inexpensive fencing along the two side widths costs only $8 per foot. The total cost of the fencing along the three sides comes to $3280. What are the lot's dimensions?

EXAMPLE 4   Solar and Electric Heating Systems

The costs for two different kinds of heating systems for a two-bedroom home are given in the following table.

System	Cost to Install	Operating Cost/Year
Solar	$29,700	$150
Electric	$5000	$1100

After how many years will total costs for solar heating and electric heating be the same? What will be the cost at that time?

Solution

Step 1.   Use variables to represent unknown quantities.

Let $x =$ the number of years the heating system is used.

Let $y =$ the total cost for the heating system.

Step 2.   Write a system of equations that models the problem's conditions.

Total cost for the solar system	equals	installation cost	plus	yearly operating cost	times	the number of years the system is used.
y	$=$	29,700	$+$	150	$\cdot$	x

Total cost for the electric system	equals	installation cost	plus	yearly operating cost	times	the number of years the system is used.
y	$=$	5000	$+$	1100	$\cdot$	x

Step 3.   Solve the system and answer the problem's question.   We want to know after how many years the total costs for the two heating systems will be the same. We must solve the system of equations

$$\begin{cases} y = 29{,}700 + 150x \\ y = 5000 + 1100x. \end{cases}$$

Substitution works well because y is isolated in each equation.

$y = \boxed{29{,}700 + 150x}$ $\boxed{y} = 5000 + 1100x$ Substitute 29,700 + 150x for y.

$29{,}700 + 150x = 5000 + 1100x$ This substitution gives an equation in one variable.

$29{,}700 = 5000 + 950x$ Subtract 150x from both sides.

$24{,}700 = 950x$ Subtract 5000 from both sides.

$26 = x$ Divide both sides by 950: $\dfrac{24{,}700}{950} = 26.$

Because x represents the number of years the heating system is used, we see that after 26 years, the total costs for the two systems will be the same. Now we can find y, the total cost. Back-substitute 26 for x in either of the system's equations. We will use the second equation, $y = 5000 + 1100x$.

$$y = 5000 + 1100 \cdot 26 = 5000 + 28{,}600 = 33{,}600$$

Because $x = 26$ and $y = 33{,}600$, after 26 years, the total costs for the two systems will be the same. The cost for each system at that time will be $33,600.

Step 4.   Check the proposed answers in the original wording of the problem.   Let's verify that after 26 years, the two systems will cost the same amount. The installation cost for the solar system is $29,700 and the yearly operating cost is $150. Thus, the total cost after 26 years is

$$\$29{,}700 + \$150(26) = \$29{,}700 + \$3900 = \$33{,}600.$$

The installation cost for the electric system is $5000 and the yearly operating cost is $1100. Thus, the total cost after 26 years is

$$\$5000 + \$1100(26) = \$5000 + \$28,600 = \$33,600.$$

This verifies that after 26 years the two systems will cost the same amount, $33,600. ∎

The graphs in **Figure 4.13** give us a way of visualizing the solution to Example 4. The total cost of solar heating over 40 years is represented by the blue line. The total cost of electric heating over 40 years is represented by the red line. The lines intersect at (26, 33,600): After 26 years, the cost for each system is the same, $33,600.

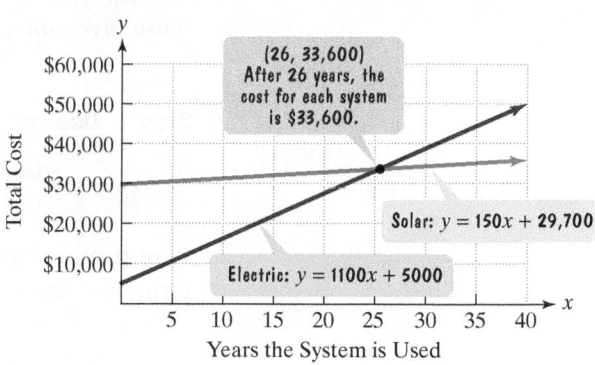

Can you see that to the right of the intersection point, (26, 33,600), the blue graph representing solar costs lies below the red graph representing electric costs? Thus, after 26 years, or when $x > 26$, the cost for solar heating is less than the cost for electric heating.

Figure 4.13 Visualizing total costs of solar and electric heating

✓ **CHECK POINT 4** Costs for two different kinds of heating systems for a home are given in the following table.

System	Cost to Install	Operating Cost/Year
Electric	$5000	$1100
Gas	$12,000	$700

After how long will total costs for electric heating and gas heating be the same? What will be the cost at that time?

Next, we will solve problems involving investments, mixtures, and motion with systems of equations. We will continue using our four-step problem-solving strategy. We will also use tables to help organize the information in the problems.

2 Solve simple interest problems.

Dual Investments with Simple Interest

Interest is the amount of money that we get paid for investing money or that we pay for borrowing money. Simple interest involves interest calculated only on the amount of money that we invest, called the **principal**.

Calculating Simple Interest for One Year

$$I = Pr$$

I is the interest earned for one year when the principal, P, is invested at an annual interest rate r. The rate, r, is expressed as a decimal when calculating simple interest.

For example, if you deposit $2000 at a rate of 6%, the interest at the end of the first year is

$$I = Pr = (2000)(0.06) = 120, \text{ or } \$120.$$

Dual investment problems involve different amounts of money in two or more investments, each paying a different rate.

EXAMPLE 5 Solving a Dual Investment Problem

Your grandmother needs your help. She has $50,000 to invest. Part of this money is to be invested in noninsured bonds paying 15% annual interest. The rest of this money is to be invested in a government-insured certificate of deposit paying 7% annual interest. She told you that she requires $6000 per year in extra income from both of these investments. How much money should be placed in each investment?

Solution

Step 1. Use variables to represent unknown quantities.

Let x = the amount invested in the 15% noninsured bonds.

Let y = the amount invested in the 7% certificate of deposit.

Step 2. Write a system of equations that models the problem's conditions. Because Grandma has $50,000 to invest,

The amount invested at 15%	plus	the amount invested at 7%	equals	$50,000.
x	$+$	y	$=$	$50{,}000$

Furthermore, Grandma requires $6000 in total interest. We can use a table to organize the information in the problem and obtain a second equation.

	Principal (amount invested)	$\times$	Interest rate	$=$	Interest earned
15% Investment	x		0.15		$0.15x$
7% Investment	y		0.07		$0.07y$

The interest for the two investments combined must be $6000.

Interest from the 15% investment	plus	interest from the 7% investment	is	$6000.
$0.15x$	$+$	$0.07y$	$=$	6000

Step 3. Solve the system and answer the problem's question. The system

$$\begin{cases} x + y = 50{,}000 \\ 0.15x + 0.07y = 6000 \end{cases}$$

can be solved by substitution or addition. Substitution works well because both variables in the first equation have coefficients of 1. Addition also works well; if we multiply the first equation by -0.15 or -0.07, adding equations will eliminate a variable. We will use addition.

$$\begin{cases} x + y = 50{,}000 \\ 0.15x + 0.07y = 6000 \end{cases} \xrightarrow[\text{No change}]{\text{Multiply by } -0.07.} \begin{cases} -0.07x - 0.07y = -3500 \\ 0.15x + 0.07y = 6000 \end{cases}$$

$$0.08x \qquad\qquad = 2500$$

$$x = \frac{2500}{0.08}$$

$$x = 31{,}250$$

Because x represents the amount that should be invested at 15%, Grandma should place $31,250 in 15% noninsured bonds. Now we can find y, the amount that she should place in the 7% certificate of deposit. We do so by back-substituting 31,250 for x in either of the system's equations.

$$x + y = 50{,}000 \qquad \text{We'll use the first equation.}$$

$$31{,}250 + y = 50{,}000 \qquad \text{Back-substitute 31,250 for } x.$$

$$y = 18{,}750 \qquad \text{Subtract 31,250 from both sides.}$$

Because $x = 31,250$ and $y = 18,750$, Grandma should invest $31,250 at 15% and $18,750 at 7%.

Step 4. Check the proposed answers in the original wording of the problem. Has Grandma invested $50,000?

$$\$31,250 + \$18,750 = \$50,000$$

Yes, all her money was placed in the dual investments. Can she count on $6000 interest? The interest earned on $31,250 at 15% is ($31,250)(0.15), or $4687.50. The interest earned on $18,750 at 7% is ($18,750)(0.07), or $1312.50. The total interest is $4687.50 + $1312.50, or $6000, exactly as it should be. You've made your grandmother happy. (Now if you would just visit her more often ...) ■

☑ CHECK POINT 5   You inherited $5000 with the stipulation that for the first year the money had to be invested in two funds paying 9% and 11% annual interest. How much did you invest at each rate if the total interest earned for the year was $487?

---

3 Solve mixture problems.

Problems Involving Mixtures

Chemists and pharmacists often have to change the concentration of solutions and other mixtures. In these situations, the amount of a particular ingredient in the solution or mixture is expressed as a percentage of the total.

EXAMPLE 6   Solving a Mixture Problem

A chemist working on a flu vaccine needs to mix a 10% sodium-iodine solution with a 60% sodium-iodine solution to obtain 50 milliliters of a 30% sodium-iodine solution. How many milliliters of the 10% solution and of the 60% solution should be mixed?

Solution

Step 1. Use variables to represent unknown quantities.

Let $x =$ the number of milliliters of the 10% solution to be used in the mixture.

Let $y =$ the number of milliliters of the 60% solution to be used in the mixture.

Step 2. Write a system of equations that models the problem's conditions. The situation is illustrated in **Figure 4.14**. The chemist needs 50 milliliters of a 30% sodium-iodine solution. We form a table that shows the amount of sodium-iodine in each of the three solutions.

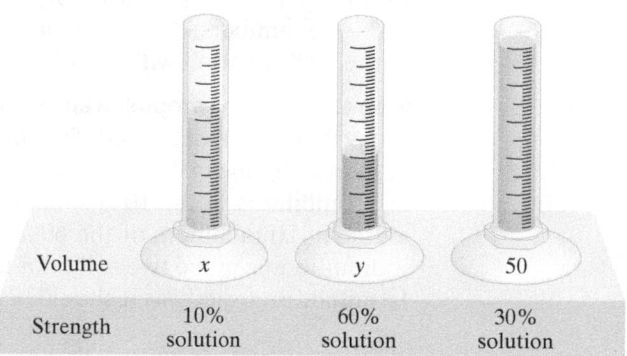

| Volume | x | y | 50 |
| Strength | 10% solution | 60% solution | 30% solution |

Figure 4.14

Solution	Number of milliliters	×	Percent of Sodium-Iodine	=	Amount of Sodium-Iodine
10% Solution	x		10% = 0.1		$0.1x$
60% Solution	y		60% = 0.6		$0.6y$
30% Mixture	50		30% = 0.3		$0.3(50) = 15$

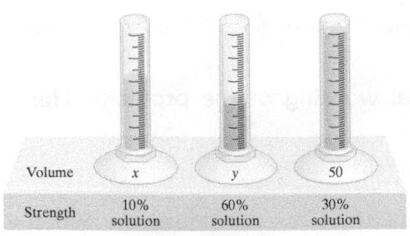

Volume x y 50

Strength 10% solution   60% solution   30% solution

Figure 4.14 (repeated)

The chemist needs to obtain a 50-milliliter mixture.

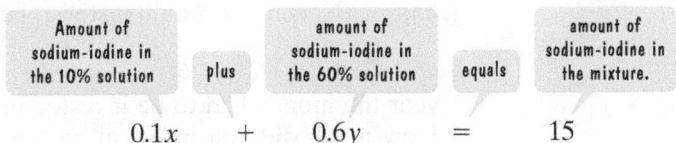

The number of milliliters used of the 10% solution	plus	the number of milliliters used of the 60% solution	must equal	50 milliliters.
x	$+$	y	$=$	50

The 50-milliliter mixture must be 30% sodium-iodine. The amount of sodium-iodine must be 30% of 50, or $(0.3)(50) = 15$ milliliters.

Amount of sodium-iodine in the 10% solution	plus	amount of sodium-iodine in the 60% solution	equals	amount of sodium-iodine in the mixture.
$0.1x$	$+$	$0.6y$	$=$	15

Step 3. Solve the system and answer the problem's question. The system

$$\begin{cases} x + y = 50 \\ 0.1x + 0.6y = 15 \end{cases}$$

can be solved by substitution or addition. Let's use substitution. The first equation can easily be solved for x or y. Solving for y, we obtain $y = 50 - x$.

$$y = \boxed{50 - x} \qquad 0.1x + 0.6\boxed{y} = 15$$

We substitute $50 - x$ for y in the second equation. This gives us an equation in one variable.

$0.1x + 0.6(50 - x) = 15$	This equation contains one variable, x.
$0.1x + 30 - 0.6x = 15$	Apply the distributive property.
$-0.5x + 30 = 15$	Combine like terms.
$-0.5x = -15$	Subtract 30 from both sides.
$x = \dfrac{-15}{-0.5} = 30$	Divide both sides by -0.5.

Back-substituting 30 for x in either of the system's equations ($x + y = 50$ is easier to use) gives $y = 20$. Because x represents the number of milliliters of the 10% solution and y the number of milliliters of the 60% solution, the chemist should mix 30 milliliters of the 10% solution with 20 milliliters of the 60% solution.

Step 4. Check the proposed answers in the original wording of the problem. The problem states that the chemist needs 50 milliliters of a 30% sodium-iodine solution. The amount of sodium-iodine in this mixture is 0.3(50), or 15 milliliters. The amount of sodium-iodine in 30 milliliters of the 10% solution is 0.1(30), or 3 milliliters. The amount of sodium-iodine in 20 milliliters of the 60% solution is 0.6(20) = 12 milliliters. The amount of sodium-iodine in the two solutions used in the mixture is 3 milliliters + 12 milliliters, or 15 milliliters, exactly as it should be. ■

✓ **CHECK POINT 6** A chemist needs to mix a 12% acid solution with a 20% acid solution to obtain 160 ounces of a 15% acid solution. How many ounces of each of the acid solutions must be used?

Great Question!

Are there similarities between dual investment problems and mixture problems?

Problems involving dual investments and problems involving mixtures are both based on the same idea: The total amount times the rate gives the amount.

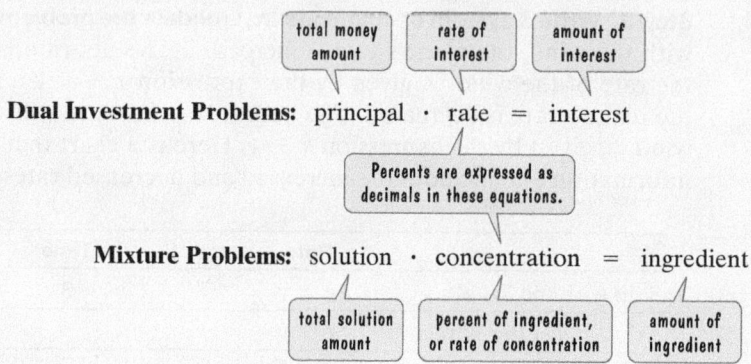

Our dual investment problem involved mixing two investments. Our mixture problem involved mixing two liquids. The equations in these problems are obtained from similar conditions:

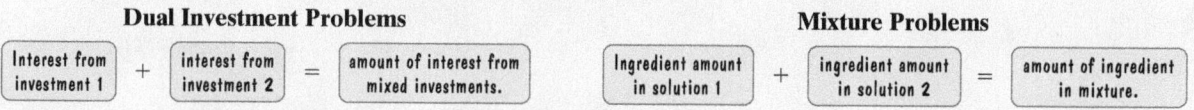

Being aware of the similarities between dual investment and mixture problems should make you a better problem solver in a variety of situations that involve mixtures.

---

4 Solve motion problems.

Problems Involving Motion

Suppose that you ride your bike at an average speed of 12 miles per hour. What distance do you cover in 2 hours? Your distance is the product of your speed and the time that you travel:

$$\frac{12 \text{ miles}}{\text{hour}} \times 2 \text{ hours} = 24 \text{ miles}$$

Your distance is 24 miles. Notice how the hour units cancel. The distance is expressed in miles.

In general, the distance covered by any moving body is the product of its average speed, or rate, and its time in motion:

A Formula for Motion

$$d = rt$$

Distance equals rate times time.

Wind and water current have the effect of increasing or decreasing a traveler's rate.

EXAMPLE 7 Solving a Motion Problem

When a small airplane flies with the wind, it can travel 450 miles in 3 hours. When the same airplane flies in the opposite direction against the wind, it takes 5 hours to fly the same distance. Find the average rate of the plane in still air and the average rate of the wind.

Great Question!

Is it always necessary to use *x* and *y* to represent a problem's variables?

No. Select letters that help you remember what the variables represent. For example, in Example 7, you may prefer using *p* and *w* rather than *x* and *y*:

p = plane's average rate in still air

w = wind's average rate.

Solution

Step 1. Use variables to represent unknown quantities.

Let x = the average rate of the plane in still air.

Let y = the average rate of the wind.

Step 2. Write a system of equations that models the problem's conditions. As it travels with the wind, the plane's rate is increased. The net rate is its rate in still air, x, plus the rate of the wind, y, given by the expression $x + y$. As it travels against the wind, the plane's rate is decreased. The net rate is its rate in still air, x, minus the rate of the wind, y, given by the expression $x - y$. Here is a chart that summarizes the problem's information and includes the increased and decreased rates.

	Rate	×	Time	=	Distance
Trip with the Wind	$x + y$		3		$3(x + y)$
Trip against the Wind	$x - y$		5		$5(x - y)$

The problem states that the distance in each direction is 450 miles. We use this information to write our system of equations.

The distance of the trip with the wind is 450 miles.

$$3(x + y) = 450$$

The distance of the trip against the wind is 450 miles.

$$5(x - y) = 450$$

Step 3. Solve the system and answer the problem's question. We can simplify the system by dividing both sides of the equations by 3 and 5, respectively.

$$\begin{cases} 3(x + y) = 450 \\ 5(x - y) = 450 \end{cases} \xrightarrow{\begin{array}{c}\text{Divide by 3.}\\\text{Divide by 5.}\end{array}} \begin{cases} x + y = 150 \\ x - y = 90 \end{cases}$$

Solve the system on the right by the addition method.

$$\begin{cases} x + y = 150 \\ x - y = 90 \end{cases}$$

Add: $2x = 240$

$x = 120$ Divide both sides by 2.

Back-substituting 120 for x in either of the system's equations gives $y = 30$. Because $x = 120$ and $y = 30$, the average rate of the plane in still air is 120 miles per hour and the average rate of the wind is 30 miles per hour.

Step 4. Check the proposed solution in the original wording of the problem. The problem states that the distance in each direction is 450 miles. The average rate of the plane with the wind is $120 + 30 = 150$ miles per hour. In 3 hours, it travels $150 \cdot 3$, or 450 miles, which checks with the stated condition. Furthermore, the average rate of the plane against the wind is $120 - 30 = 90$ miles per hour. In 5 hours, it travels $90 \cdot 5 = 450$ miles, which is the stated distance. ∎

✓ **CHECK POINT 7**   With the current, a motorboat can travel 84 miles in 2 hours. Against the current, the same trip takes 3 hours. Find the average rate of the boat in still water and the average rate of the current.

Achieving Success

Avoid asking for help on a problem that you have not thought about. This is basically asking your teacher to do the work for you. First try solving the problem on your own!

CONCEPT AND VOCABULARY CHECK

Fill in each blank so that the resulting statement is true.

1. Suppose that x represents the cost of one apple and y represents the cost of one banana. The cost of five apples and six bananas is represented by _____.

2. Suppose that x represents the length of a rectangular lot, in feet, and y represents its width, in feet. The cost of fencing the entire lot at $10 per foot for the length and $15 per foot for the width is represented by _____.

3. A solar heating system costs $25,600 to install and has operating costs of $225 per year. The total cost for the solar system after x years is represented by _____.

4. The combined yearly interest for x dollars invested at 4% and y dollars invested at 5% is represented by _____.

5. The total amount of acid in x milliliters of a 7% acid solution and y milliliters of a 15% acid solution is represented by _____.

6. If x represents the average rate of a plane in still air and y represents the average rate of the wind, the plane's rate with the wind is represented by _____ and the plane's rate against the wind is represented by _____.

7. If $x + y$ represents a motorboat's rate with the current, in miles per hour, its distance after 4 hours is represented by _____.

4.4 EXERCISE SET MyMathLab®

 Watch the videos in MyMathLab

 Download the MyDashBoard App

Practice Exercises

In Exercises 1–4, let x represent one number and let y represent the other number. Use the given conditions to write a system of equations. Solve the system and find the numbers.

1. The sum of two numbers is 17. If one number is subtracted from the other, their difference is −3. Find the numbers.

2. The sum of two numbers is 5. If one number is subtracted from the other, their difference is 13. Find the numbers.

3. Three times a first number decreased by a second number is −1. The first number increased by twice the second number is 23. Find the numbers.

4. The sum of three times a first number and twice a second number is 43. If the second number is subtracted from twice the first number, the result is −4. Find the numbers.

Application Exercises

The bar graph shows the average time per day that Americans devote to sprucing up. Exercises 5–6 are based on the graph.

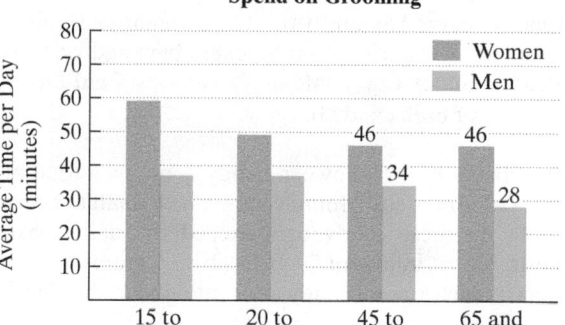

Average Time per Day Americans Spend on Grooming

Source: Bureau of Labor Statistics American Time Use Survey

5. Each day, the sum of the average times spent on grooming for 20- to 24-year-old women and men is 86 minutes. The difference between grooming times for 20- to 24-year-old women and men is 12 minutes. How many minutes per day do 20- to 24-year-old women and men spend on grooming?

6. Each day, the sum of the average times spent on grooming for 15- to 19-year-old women and men is 96 minutes. The difference between grooming times for 15- to 19-year-old women and men is 22 minutes. How many minutes per day do 15- to 19-year-old women and men spend on grooming?

Looking for Mr. Goodbar? It's probably not a good idea if you want to look like Mr. Universe or Kate Winslet. The graph shows the four candy bars with the highest fat content, representing grams of fat and calories in each bar. Exercises 7–10 are based on the graph.

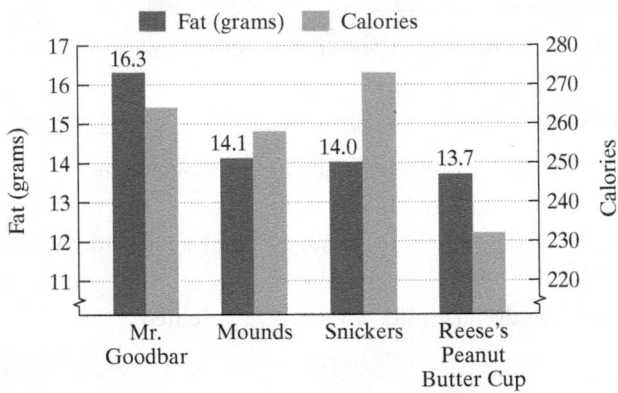

Candy Bars with the Highest Fat Content

Source: Krantz and Sveum, The World's Worsts, HarperCollins, 2005.

7. One Mr. Goodbar and two Mounds bars contain 780 calories. Two Mr. Goodbars and one Mounds bar contain 786 calories. Find the caloric content of each candy bar.

(In Exercises 8–10 refer to the bar graph to visualize that your solution is reasonable).

8. One Snickers bar and two Reese's Peanut Butter Cups contain 737 calories. Two Snickers bars and one Reese's Peanut Butter Cup contain 778 calories. Find the caloric content of each candy bar.

9. A collection of Halloween candy contains a total of five Mr. Goodbars and Mounds bars. Chew on this: The grams of fat in these candy bars exceed the daily maximum desirable fat intake of 70 grams by 7.1 grams. How many bars of each kind of candy are contained in the Halloween collection?

10. A collection of Halloween candy contains a total of 12 Snickers bars and Reese's Peanut Butter Cups. Chew on this: The grams of fat in these candy bars exceed twice the daily maximum desirable fat intake of 70 grams by 26.5 grams. How many bars of each kind of candy are contained in the Halloween collection?

11. In a discount clothing store, all sweaters are sold at one fixed price and all shirts are sold at another fixed price. If one sweater and three shirts cost $42, while three sweaters and two shirts cost $56, find the price of one sweater and the price of one shirt.

12. A restaurant purchased eight tablecloths and five napkins for $106. A week later, a tablecloth and six napkins were bought for $24. Find the cost of one tablecloth and the cost of one napkin, assuming the same prices for both purchases.

13. The perimeter of a badminton court is 128 feet. After a game of badminton, a player's coach estimates that the athlete has run a total of 444 feet, which is equivalent to six times the court's length plus nine times its width. What are the dimensions of a standard badminton court?

14. The perimeter of a tennis court is 228 feet. After a round of tennis, a player's coach estimates that the athlete has run a total of 690 feet, which is equivalent to 7 times the court's length plus four times its width. What are the dimensions of a standard tennis court?

15. A rectangular lot whose perimeter is 320 feet is fenced along three sides. An expensive fencing along the lot's length costs $16 per foot. An inexpensive fencing along the two side widths costs only $5 per foot. The total cost of the fencing along the three sides comes to $2140. What are the lot's dimensions?

16. A rectangular lot whose perimeter is 1600 feet is fenced along three sides. An expensive fencing along the lot's length costs $20 per foot. An inexpensive fencing along the two side widths costs only $5 per foot. The total cost of the fencing along the three sides comes to $13,000. What are the lot's dimensions?

17. You are choosing between two telephone plans. Plan A has a monthly fee of $20 with a charge of $0.05 per minute for all calls. Plan B has a monthly fee of $5 with a charge of $0.10 per minute for all calls.

    a. For how many minutes of calls will the costs for the two plans be the same? What will be the cost for each plan?

    b. If you make approximately 10 calls per month, each averaging 20 minutes, which plan should you select? Explain your answer.

18. You are choosing between two telephone plans. Plan A has a monthly fee of $15 with a charge of $0.08 per minute for all calls. Plan B has a monthly fee of $3 with a charge of $0.12 per minute for all calls.

    a. For how many minutes of calls will the costs for the two plans be the same? What will be the cost for each plan?

    b. If you make approximately 15 calls per month, each averaging 30 minutes, which plan should you select? Explain your answer.

19. You are choosing between two plans at a discount warehouse. Plan A offers an annual membership fee of $100 and you pay 80% of the manufacturer's recommended list price. Plan B offers an annual membership fee of $40 and you pay 90% of the manufacturer's recommended list price. How many dollars of merchandise would you have to purchase in a year to pay the same amount under both plans? What will be the cost for each plan?

20. You are choosing between two plans at a discount warehouse. Plan A offers an annual membership fee of $300 and you pay 70% of the manufacturer's recommended list price. Plan B offers an annual membership fee of $40 and you pay 90% of the manufacturer's recommended list price. How many dollars of merchandise would you have to purchase in a year to pay the same amount under both plans? What will be the cost for each plan?

21. A community center sells a total of 301 tickets for a basketball game. An adult ticket costs $3. A student ticket costs $1. The sponsors collect $487 in ticket sales. Find the number of each type of ticket sold.

22. On a special day, tickets for a minor league baseball game cost $5 for adults and $1 for students. The attendance that day was 1281 and $3425 was collected. Find the number of each type of ticket sold.

23. At the 1987 annual meeting of the National Council of Teachers of Mathematics, the system below was found on a menu card at a restaurant in California. Find the cost of each item in column A and the cost of each item in column B.

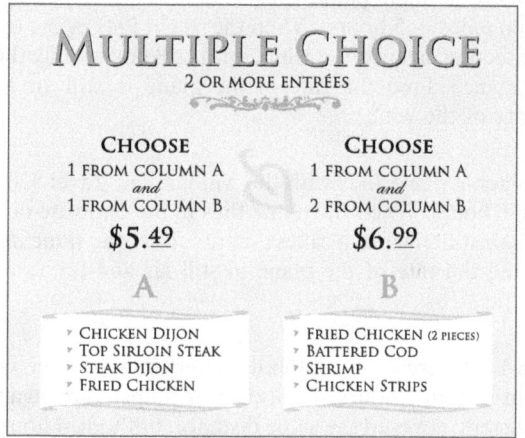

24. The perimeter of the rectangle is 34 inches. The perimeter of the triangle is 30 inches.

    a. Find the values of x and y.

    b. Find the area of the rectangle and the area of the triangle.

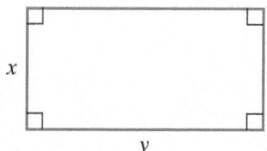 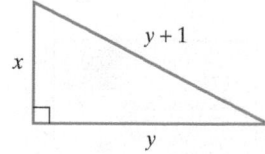

25. Nutritional information for macaroni and broccoli is given in the table at the top of the next column. How many servings of each would it take to get exactly 14 grams of protein and 48 grams of carbohydrates?

	Macaroni	Broccoli
Protein (grams/serving)	3	2
Carbohydrates (grams/serving)	16	4

26. The calorie-nutrient information for an apple and an avocado is given in the table. How many of each should be eaten to get exactly 1000 calories and 100 grams of carbohydrates?

	One Apple	One Avocado
Calories	100	350
Carbohydrates (grams)	24	14

In Exercises 27–28, an isosceles triangle in which angles B and C have the same measure is shown. Find the measure of each angle whose degree measure is represented with variables.

27.

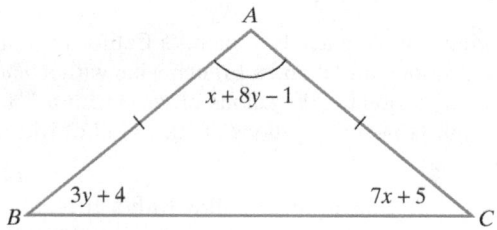

28.

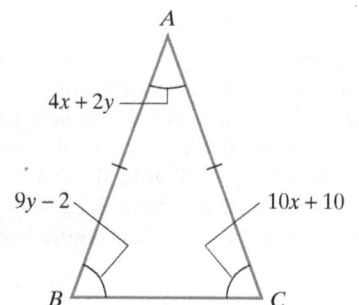

Exercises 29–34 involve dual investments.

29. You invested $7000 in two accounts paying 6% and 8% annual interest. If the total interest earned for the year was $520, how much was invested at each rate?

30. You invested $11,000 in stocks and bonds, paying 5% and 8% annual interest. If the total interest earned for the year was $730, how much was invested in stocks and how much was invested in bonds?

31. You invested money in two funds. Last year, the first fund paid a dividend of 9% and the second a dividend of 3%, and you received a total of $900. This year, the first fund paid a 10% dividend and the second only 1%, and you received a total of $860. How much money did you invest in each fund?

32. You invested money in two funds. Last year, the first fund paid a dividend of 8% and the second a dividend of 5%, and you received a total of $1330. This year, the first fund paid a 12% dividend and the second only 2%, and you received a total of $1500. How much money did you invest in each fund?

33. Things did not go quite as planned. You invested $20,000, part of it in a stock with a 12% annual return. However, the rest of the money suffered a 5% loss. If the total annual income from both investments was $1890, how much was invested at each rate?

34. Things did not go quite as planned. You invested $30,000, part of it in a stock with a 14% annual return. However, the rest of the money suffered a 6% loss. If the total annual income from both investments was $200, how much was invested at each rate?

Exercises 35–42 involve mixtures.

35. A wine company needs to blend a California wine with a 5% alcohol content and a French wine with a 9% alcohol content to obtain 200 gallons of wine with a 7% alcohol content. How many gallons of each kind of wine must be used?

36. A jeweler needs to mix an alloy with a 16% gold content and an alloy with a 28% gold content to obtain 32 ounces of a new alloy with a 25% gold content. How many ounces of each of the original alloys must be used?

37. For thousands of years, gold has been considered one of Earth's most precious metals. One hundred percent pure gold is 24-karat gold, which is too soft to be made into jewelry. In the United States, most gold jewelry is 14-karat gold, approximately 58% gold. If 18-karat gold is 75% gold and 12-karat gold is 50% gold, how much of each should be used to make a 14-karat gold bracelet weighing 300 grams?

38. In the "Peanuts" cartoon shown, solve the problem that is sending Peppermint Patty into an agitated state. How much cream and how much milk, to the nearest hundredth of a gallon, must be mixed together to obtain 50 gallons of cream that contains 12.5% butterfat?

39. The manager of a candystand at a large multiplex cinema has a popular candy that sells for $1.60 per pound. The manager notices a different candy worth $2.10 per pound that is not selling well. The manager decides to form a mixture of both types of candy to help clear the inventory of the more expensive type. How many pounds of each kind of candy should be used to create a 75-pound mixture selling for $1.90 per pound?

40. A grocer needs to mix raisins at $2.00 per pound with granola at $3.25 per pound to obtain 10 pounds of a mixture that costs $2.50 per pound. How many pounds of raisins and how many pounds of granola must be used?

41. A coin purse contains a mixture of 15 coins in nickels and dimes. The coins have a total value of $1.10. Determine the number of nickels and the number of dimes in the purse.

42. A coin purse contains a mixture of 15 coins in dimes and quarters. The coins have a total value of $3.30. Determine the number of dimes and the number of quarters in the purse.

Exercises 43–48 involve motion.

43. When a small plane flies with the wind, it can travel 800 miles in 5 hours. When the plane flies in the opposite direction, against the wind, it takes 8 hours to fly the same distance. Find the rate of the plane in still air and the rate of the wind.

44. When a plane flies with the wind, it can travel 4200 miles in 6 hours. When the plane flies in the opposite direction, against the wind, it takes 7 hours to fly the same distance. Find the rate of the plane in still air and the rate of the wind.

45. A boat's crew rowed 16 kilometers downstream, with the current, in 2 hours. The return trip upstream, against the current, covered the same distance, but took 4 hours. Find the crew's rowing rate in still water and the rate of the current.

46. A motorboat traveled 36 miles downstream, with the current, in 1.5 hours. The return trip upstream, against the current, covered the same distance, but took 2 hours. Find the boat's rate in still water and the rate of the current.

PEANUTS © United Feature Syndicate, Inc.

47. With the current, you can canoe 24 miles in 4 hours. Against the same current, you can canoe only $\frac{3}{4}$ of this distance in 6 hours. Find your rate in still water and the rate of the current.

48. With the current, you can row 24 miles in 3 hours. Against the same current, you can row only $\frac{2}{3}$ of this distance in 4 hours. Find your rowing rate in still water and the rate of the current.

Writing in Mathematics

49. Describe the conditions in a problem that enable it to be solved using a system of linear equations.

50. Use **Figure 4.11** on page 307 to write a word problem similar to Example 1 or Check Point 1 that involves the leisure activity of reading. Then solve the problem.

51. Exercises 17–20 involve using systems of linear equations to compare costs of telephone plans and plans at a discount warehouse. Describe another situation that involves choosing between two options that can be modeled and solved with a linear system.

52. Describe two similarities between interest and mixture problems.

53. Must the concentration of a mixture always be greater than the concentration of an ingredient in one of the solutions and less than the concentration of the ingredient in the other solution being mixed? Explain your answer.

Critical Thinking Exercises

Make Sense?   *In Exercises 54–57, determine whether each statement "makes sense" or "does not make sense" and explain your reasoning.*

54. The Sea Drift Hotel charges $650 for 7 nights and 10 meals, so I modeled these conditions using $7n + 10m = 650$, where n is the cost of one night and m is the cost of one meal.

55. A telephone calling plan charges $4 per month plus 7¢ per minute, so I modeled these conditions using $y = 0.07 + 4x$, where y is the monthly charge when x minutes of calls are made.

56. The perimeter of the large rectangle shown in the figure is 58 meters, so I modeled this condition using $2(x + 8) + 2(y + 10) = 58$.

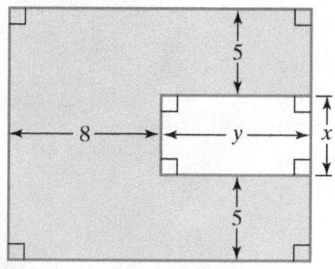

57. Angles A and B are supplementary, so I modeled this condition using

$$4x - 2y + 4 = 12x + 6y + 12 + 180.$$

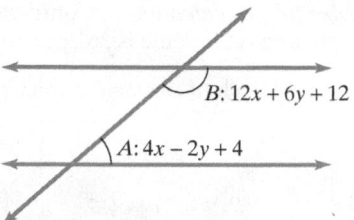

58. A set of identical twins can only be distinguished by the characteristic that one always tells the truth and the other always lies. One twin tells you of a lucky number pair: "When I multiply my first lucky number by 3 and my second lucky number by 6, the addition of the resulting numbers produces a sum of 12. When I add my first lucky number and twice my second lucky number, the sum is 5." Which twin is talking?

59. Tourist: "How many birds and lions do you have in your zoo?" Zookeeper: "There are 30 heads and 100 feet." Tourist: "I can't tell from that." Zookeeper: "Oh, yes, you can!" Can you? Find the number of each.

60. Find the measure of each angle whose degree measure is represented with a variable.

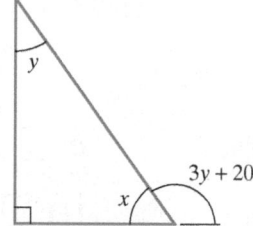

61. One apartment is directly above a second apartment. The resident living downstairs calls his neighbor living above him and states, "If one of you is willing to come downstairs, we'll have the same number of people in both apartments." The upstairs resident responds, "We're all too tired to move. Why don't one of you come up here? Then we'll have twice as many people up here as you've got down there." How many people are in each apartment?

62. In Lewis Carroll's *Through the Looking Glass*, the following dialogue takes place:

> *Tweedledum (to Tweedledee):* The sum of your weight and twice mine is 361 pounds.
>
> *Tweedledee (to Tweedledum):* Contrawise, the sum of your weight and twice mine is 362 pounds.

Find the weight of each of the two characters.

63. You have $70,000 to invest. Part of the money is to be placed in a certificate of deposit paying 8% per year. The rest is to be placed in corporate bonds paying 12% per year. If you wish to obtain an overall return of 9% per year, how much should you place in each investment?

Technology Exercise

64. Select any two problems that you solved from Exercises 5–26. Use a graphing utility to graph the system of equations that you wrote for that problem. Then use the ⎡TRACE⎤ or ⎡INTERSECTION⎤ feature to show the point on the graphs that corresponds to the problem's solution.

Review Exercises

65. Solve: $2(x + 3) = 24 - 2(x + 4)$.
(Section 2.3, Example 3)

66. Simplify: $5 + 6(x + 1)$.
(Section 1.7, Example 7)

67. Write the slope-intercept form of the equation of the line passing through $(-5, 6)$ and $(3, -10)$.
(Section 3.5, Example 2)

Preview Exercises

Exercises 68–70 will help you prepare for the material covered in the next section.

68. If $x = 3, y = 2$, and $z = -3$, does the ordered triple (x, y, z) satisfy the equation $2x - y + 4z = -8$?

69. Consider the following equations:

$$\begin{cases} 5x - 2y - 4z = 3 & \text{Equation 1} \\ 3x + 3y + 2z = -3. & \text{Equation 2} \end{cases}$$

Eliminate z by copying Equation 1, multiplying Equation 2 by 2, and then adding the equations.

70. Write an equation involving a, b, and c based on the following description:

> When the value of x in $y = ax^2 + bx + c$ is 4, the value of y is 1682.

Systems of Linear Equations in Three Variables

Objectives

1 Verify the solution of a system of linear equations in three variables.

2 Solve systems of linear equations in three variables.

3 Identify inconsistent and dependent systems.

4 Solve problems using systems in three variables.

All animals sleep, but the length of time they sleep varies widely: Cattle sleep for only a few minutes at a time. We humans seem to need more sleep than other animals, up to eight hours a day. Without enough sleep, we have difficulty concentrating, make mistakes in routine tasks, lose energy, and feel bad-tempered. There is a relationship between hours of sleep and death rate per year per 100,000 people. How many hours of sleep will put you in the group with the minimum death rate? In this section, we will answer this question by solving a system of linear equations with more than two variables.

① Verify the solution of a system of linear equations in three variables.

Systems of Linear Equations in Three Variables and Their Solutions

An equation such as $x + 2y - 3z = 9$ is called a *linear equation in three variables*. In general, any equation of the form

$$Ax + By + Cz = D,$$

where A, B, C, and D are real numbers such that A, B, and C are not all 0, is a **linear equation in three variables: x, y, and z.** The graph of this linear equation in three variables is a plane in three-dimensional space.

The process of solving a system of three linear equations in three variables is geometrically equivalent to finding the point of intersection (assuming that there is one) of three planes in space (see **Figure 4.15**). A **solution** of a system of linear equations in three variables is an ordered triple of real numbers that satisfies all equations in the system. The **solution set** of the system is the set of all its solutions.

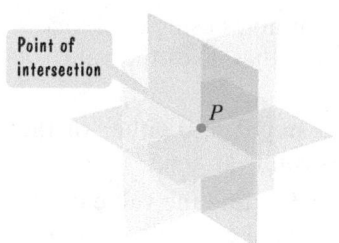

Point of intersection

P

Figure 4.15

| EXAMPLE 1 | Determining Whether an Ordered Triple Satisfies a System |

Show that the ordered triple $(-1, 2, -2)$ is a solution of the system:

$$\begin{cases} x + 2y - 3z = 9 \\ 2x - y + 2z = -8 \\ -x + 3y - 4z = 15. \end{cases}$$

Solution   Because -1 is the x-coordinate, 2 is the y-coordinate, and -2 is the z-coordinate of $(-1, 2, -2)$, we replace x with -1, y with 2, and z with -2 in each of the three equations.

$x + 2y - 3z = 9$	$2x - y + 2z = -8$	$-x + 3y - 4z = 15$
$-1 + 2(2) - 3(-2) \overset{?}{=} 9$	$2(-1) - 2 + 2(-2) \overset{?}{=} -8$	$-(-1) + 3(2) - 4(-2) \overset{?}{=} 15$
$-1 + 4 + 6 \overset{?}{=} 9$	$-2 - 2 - 4 \overset{?}{=} -8$	$1 + 6 + 8 \overset{?}{=} 15$
$9 = 9$, true	$-8 = -8$, true	$15 = 15$, true

The ordered triple $(-1, 2, -2)$ satisfies the three equations: It makes each equation true. Thus, the ordered triple is a solution of the system. ■

✓ **CHECK POINT 1**   Show that the ordered triple $(-1, -4, 5)$ is a solution of the system:

$$\begin{cases} x - 2y + 3z = 22 \\ 2x - 3y - z = 5 \\ 3x + y - 5z = -32. \end{cases}$$

② Solve systems of linear equations in three variables.

Solving Systems of Linear Equations in Three Variables by Eliminating Variables

The method for solving a system of linear equations in three variables is similar to that used on systems of linear equations in two variables. We use addition to eliminate any variable, reducing the system to two equations in two variables. Once we obtain a system of two equations in two variables, we use addition or substitution to eliminate a variable. The result is a single equation in one variable. We solve this equation to get the value of the remaining variable. Other variable values are found by back-substitution.

Great Question!

When solving a linear system in three variables, which variable should I eliminate first?

It does not matter which variable you eliminate first, as long as you eliminate the same variable in two different pairs of equations.

Solving Linear Systems in Three Variables by Eliminating Variables

1. Reduce the system to two equations in two variables. This is usually accomplished by taking two different pairs of equations and using the addition method to eliminate the same variable from both pairs.

2. Solve the resulting system of two equations in two variables using addition or substitution. The result is an equation in one variable that gives the value of that variable.

3. Back-substitute the value of the variable found in step 2 into either of the equations in two variables to find the value of the second variable.

4. Use the values of the two variables from steps 2 and 3 to find the value of the third variable by back-substituting into one of the original equations.

5. Check the proposed solution in each of the original equations.

EXAMPLE 2   Solving a System in Three Variables

Solve the system:

$$\begin{cases} 5x - 2y - 4z = 3 & \text{Equation 1} \\ 3x + 3y + 2z = -3 & \text{Equation 2} \\ -2x + 5y + 3z = 3. & \text{Equation 3} \end{cases}$$

Solution   There are many ways to proceed. Because our initial goal is to reduce the system to two equations in two variables, **the central idea is to take two different pairs of equations and eliminate the same variable from both pairs.**

Step 1. Reduce the system to two equations in two variables. We choose any two equations and use the addition method to eliminate a variable. Let's eliminate z using Equations 1 and 2. We do so by multiplying Equation 2 by 2. Then we add equations.

$$\text{(Equation 1)} \begin{cases} 5x - 2y - 4z = 3 & \xrightarrow{\text{No change}} \\ \text{(Equation 2)} \ 3x + 3y + 2z = -3 & \xrightarrow{\text{Multiply by 2.}} \end{cases} \begin{cases} 5x - 2y - 4z = 3 \\ 6x + 6y + 4z = -6 \end{cases}$$

$$\text{Add: } 11x + 4y = -3 \quad \text{Equation 4}$$

Now we must eliminate the *same* variable from another pair of equations. We can eliminate z using Equations 2 and 3. First, we multiply Equation 2 by -3. Next, we multiply Equation 3 by 2. Finally, we add equations.

$$\text{(Equation 2)} \begin{cases} 3x + 3y + 2z = -3 & \xrightarrow{\text{Multiply by } -3.} \\ \text{(Equation 3)} \ -2x + 5y + 3z = 3 & \xrightarrow{\text{Multiply by 2.}} \end{cases} \begin{cases} -9x - 9y - 6z = 9 \\ -4x + 10y + 6z = 6 \end{cases}$$

$$\text{Add: } -13x + y = 15 \quad \text{Equation 5}$$

Equations 4 and 5 give us a system of two equations in two variables.

Step 2. Solve the resulting system of two equations in two variables. We will use the addition method to solve Equations 4 and 5 for x and y. To do so, we multiply Equation 5 by -4 and add this to Equation 4.

$$\text{(Equation 4)} \begin{cases} 11x + 4y = -3 & \xrightarrow{\text{No change}} \\ \text{(Equation 5)} \ -13x + y = 15 & \xrightarrow{\text{Multiply by } -4.} \end{cases} \begin{cases} 11x + 4y = -3 \\ 52x - 4y = -60 \end{cases}$$

$$\text{Add: } 63x = -63$$

$$x = -1 \quad \text{Divide both sides by 63.}$$

Step 3. Use back-substitution in one of the equations in two variables to find the value of the second variable. We back-substitute -1 for x in either Equation 4 or 5 to find the value of y. We will use Equation 5.

$$-13x + y = 15 \quad \text{Equation 5}$$
$$-13(-1) + y = 15 \quad \text{Substitute } -1 \text{ for } x.$$
$$13 + y = 15 \quad \text{Multiply.}$$
$$y = 2 \quad \text{Subtract 13 from both sides.}$$

Step 4. Back-substitute the values found for two variables into one of the original equations to find the value of the third variable. We can now use any one of the original equations and back-substitute the values of x and y to find the value for z. We will use Equation 2.

$$3x + 3y + 2z = -3 \quad \text{Equation 2}$$
$$3(-1) + 3(2) + 2z = -3 \quad \text{Substitute } -1 \text{ for } x \text{ and 2 for } y.$$
$$3 + 2z = -3 \quad \text{Multiply and then add:}$$
$$3(-1) + 3(2) = -3 + 6 = 3.$$
$$2z = -6 \quad \text{Subtract 3 from both sides.}$$
$$z = -3 \quad \text{Divide both sides by 2.}$$

With $x = -1$, $y = 2$, and $z = -3$, the proposed solution is the ordered triple $(-1, 2, -3)$.

Step 5. Check. Check the proposed solution, $(-1, 2, -3)$, by substituting the values for $x, y,$ and z into each of the three original equations. These substitutions yield three true statements. Thus, the solution is $(-1, 2, -3)$ and the solution set is $\{(-1, 2, -3)\}$. ∎

✓ **CHECK POINT 2** Solve the system:

$$\begin{cases} x + 4y - z = 20 \\ 3x + 2y + z = 8 \\ 2x - 3y + 2z = -16. \end{cases}$$

In some examples, one of the variables is missing from a given equation. In this case, the missing variable should be eliminated from the other two equations, thereby making it possible to omit one of the elimination steps. We illustrate this idea in Example 3.

EXAMPLE 3 Solving a System of Equations with a Missing Term

Solve the system:

$$\begin{cases} x + \quad z = 8 \quad \text{Equation 1} \\ x + y + 2z = 17 \quad \text{Equation 2} \\ x + 2y + z = 16. \quad \text{Equation 3} \end{cases}$$

Solution

Step 1. Reduce the system to two equations in two variables. Because Equation 1 contains only x and z, we can omit one of the elimination steps by eliminating y using Equations 2 and 3. This will give us two equations in x and z. To eliminate y using Equations 2 and 3, we multiply Equation 2 by -2 and add Equation 3.

(Equation 2) $\begin{cases} x + y + 2z = 17 \end{cases}$ $\xrightarrow{\text{Multiply by } -2.}$ $\begin{cases} -2x - 2y - 4z = -34 \end{cases}$
(Equation 3) $\begin{cases} x + 2y + z = 16 \end{cases}$ $\xrightarrow{\text{No change}}$ $\begin{cases} x + 2y + z = 16 \end{cases}$

$$\text{Add:} \quad -x \quad - 3z = -18 \quad \text{Equation 4}$$

Equation 4 and the given Equation 1 provide us with a system of two equations in two variables:

$$\begin{cases} x + z = 8 \quad \text{Equation 1} \\ -x - 3z = -18. \quad \text{Equation 4} \end{cases}$$

Step 2. Solve the resulting system of two equations in two variables. We will solve Equations 1 and 4 for x and z.

$$\begin{cases} x + z = 8 & \text{Equation 1} \\ -x - 3z = -18 & \text{Equation 4} \end{cases}$$

$$\text{Add:} \quad -2z = -10$$
$$ \quad z = 5 \quad \text{Divide both sides by } -2.$$

Step 3. Use back-substitution in one of the equations in two variables to find the value of the second variable. To find x, we back-substitute 5 for z in either Equation 1 or 4. We will use Equation 1.

$$x + z = 8 \quad \text{Equation 1}$$
$$x + 5 = 8 \quad \text{Substitute 5 for } z.$$
$$x = 3 \quad \text{Subtract 5 from both sides.}$$

Step 4. Back-substitute the values found for two variables into one of the original equations to find the value of the third variable. To find y, we back-substitute 3 for x and 5 for z into Equation 2, $x + y + 2z = 17$, or Equation 3, $x + 2y + z = 16$. We can't use Equation 1, $x + z = 8$, because y is missing in this equation. We will use Equation 2.

$$x + y + 2z = 17 \quad \text{Equation 2}$$
$$3 + y + 2(5) = 17 \quad \text{Substitute 3 for } x \text{ and 5 for } z.$$
$$y + 13 = 17 \quad \text{Multiply and add.}$$
$$y = 4 \quad \text{Subtract 13 from both sides.}$$

We found that $z = 5$, $x = 3$, and $y = 4$. Thus, the proposed solution is the ordered triple $(3, 4, 5)$.

Step 5. Check. Substituting 3 for x, 4 for y, and 5 for z into each of the three original equations yields three true statements. Consequently, the solution is $(3, 4, 5)$ and the solution set is $\{(3, 4, 5)\}$. ∎

✓ **CHECK POINT 3** Solve the system:

$$\begin{cases} 2y - z = 7 \\ x + 2y + z = 17 \\ 2x - 3y + 2z = -1. \end{cases}$$

3 Identify inconsistent and dependent systems.

Inconsistent and Dependent Systems

A system of three linear equations in three variables represents three planes. The three planes need not intersect at one point. The planes may have no common point of intersection and represent an **inconsistent system** with no solution. **Figure 4.16** illustrates some of the geometric possibilities for inconsistent systems.

Three planes are parallel with no common intersection point.

Two planes are parallel with no common intersection point.

Planes intersect two at a time. There is no intersection point common to all three planes.

Figure 4.16 Three planes may have no common point of intersection.

If you attempt to solve an inconsistent system algebraically, at some point in the solution process you will eliminate all three variables. A false statement, such as $0 = -10$, will be the result. For example, consider the system

$$\begin{cases} 2x + 5y + z = 12 & \text{Equation 1} \\ x - 2y + 4z = -10 & \text{Equation 2} \\ -3x + 6y - 12z = 20. & \text{Equation 3} \end{cases}$$

Suppose we reduce the system to two equations in two variables by eliminating x. To eliminate x using Equations 2 and 3, we multiply Equation 2 by 3 and add Equation 3:

$$\begin{cases} x - 2y + 4z = -10 \\ -3x + 6y - 12z = 20 \end{cases} \xrightarrow[\text{No change}]{\text{Multiply by 3.}} \begin{cases} 3x - 6y + 12z = -30 \\ -3x + 6y - 12z = 20 \end{cases}$$

$$\text{Add:} \qquad\qquad 0 = -10$$

There are no values of x, y, and z for which $0 = -10$. The false statement $0 = -10$ indicates that the system is inconsistent and has no solution. The solution set is the empty set, $\varnothing$.

We have seen that a linear system that has at least one solution is called a **consistent system**. Planes that intersect at one point and planes that intersect at infinitely many points both represent consistent systems. **Figure 4.17** illustrates two different cases of three planes that intersect at infinitely many points. The equations in these linear systems with infinitely many solutions are called **dependent**.

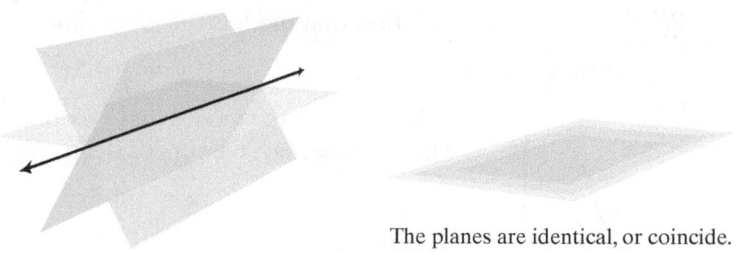

The planes intersect
along a common line.

The planes are identical, or coincide.

Figure 4.17 Three planes may intersect at infinitely many points.

If you attempt to solve a system with dependent equations algebraically, at some point in the solution process you will eliminate all three variables. A true statement, such as $0 = 0$, will be the result. If this occurs as you are solving a linear system, simply state that the equations are dependent.

4 Solve problems using systems in three variables.

Applications

Systems of equations may allow us to find models for data without using a graphing utility. Three data points that do not lie on or near a line determine the graph of an equation of the form

$$y = ax^2 + bx + c, a \neq 0.$$

Such an equation has a graph that is shaped like a bowl, making it ideal for modeling situations in which values of y are decreasing and then increasing.

The process of determining an equation whose graph contains given points is called **curve fitting**. In our next example, we fit the curve whose equation is $y = ax^2 + bx + c$ to three data points. Using a system of equations, we find values for a, b, and c.

> **EXAMPLE 4** Modeling Data Relating Sleep and Death Rate

In a study relating sleep and death rate, the following data were obtained. Use the equation $y = ax^2 + bx + c$ to model the data.

x (Average Number of Hours of Sleep)	y (Death Rate per Year per 100,000 Males)
4	1682
7	626
9	967

Solution We need to find values for a, b, and c in $y = ax^2 + bx + c$. We can do so by solving a system of three linear equations in a, b, and c. We obtain the three equations by using the values of x and y from the data as follows:

$$y = ax^2 + bx + c \qquad \text{Use the quadratic function to model the data.}$$

When $x = 4$, $y = 1682$:

When $x = 7$, $y = 626$:

When $x = 9$, $y = 967$:

$$\begin{cases} 1682 = a \cdot 4^2 + b \cdot 4 + c \\ 626 = a \cdot 7^2 + b \cdot 7 + c \\ 967 = a \cdot 9^2 + b \cdot 9 + c \end{cases} \text{ or } \begin{cases} 16a + 4b + c = 1682 \\ 49a + 7b + c = 626 \\ 81a + 9b + c = 967. \end{cases}$$

The easiest way to solve this system is to eliminate c from two pairs of equations, obtaining two equations in a and b. Solving this system gives $a = 104.5$, $b = -1501.5$, and $c = 6016$. We now substitute the values for a, b, and c into $y = ax^2 + bx + c$. The equation that models the given data is

$$y = 104.5x^2 - 1501.5x + 6016. \quad \blacksquare$$

We can use the model that we obtained in Example 4 to find the death rate of males who average, say, 6 hours of sleep. Substitute 6 for x in

$$y = 104.5x^2 - 1501.5x + 6016.$$

We obtain

$$y = 104.5(6)^2 - 1501.5(6) + 6016 = 769.$$

According to the model, the death rate for males who average 6 hours of sleep is 769 deaths per 100,000 males.

> ✓ **CHECK POINT 4** Find the equation $y = ax^2 + bx + c$ whose graph passes through the points $(1, 4)$, $(2, 1)$, and $(3, 4)$.

Problems involving three unknowns can be solved using the same strategy for solving problems with two unknown quantities. You can let x, y, and z represent the unknown quantities. We then translate from the verbal conditions of the problem to a system of three equations in three variables. Problems of this type are included in the Exercise Set that follows.

Achieving Success

Organizing and creating your own compact chapter summaries can reinforce what you know and help with the retention of this information. Imagine that your professor will permit two index cards of notes (3 by 5; front and back) on all exams. Organize and create such a two-card summary for the test on this chapter. Begin by determining what information you would find most helpful to include on the cards. Take as long as you need to create the summary. Based on how effective you find this strategy, you may decide to use the technique to help prepare for future exams.

Using Technology

The graph of

$$y = 104.5x^2 - 1501.5x + 6016$$

is displayed in a [3, 12, 1] by [500, 2000, 100] viewing rectangle. The utility shows that the lowest point on the graph is approximately (7.2, 622.5). Men who average 7.2 hours of sleep are in the group with the lowest death rate, approximately 622.5 deaths per 100,000 males.

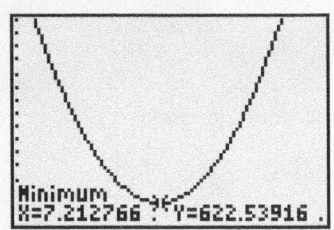

Minimum
X=7.212766   Y=622.53916

CONCEPT AND VOCABULARY CHECK

Fill in each blank so that the resulting statement is true.

1. A solution of a system of linear equations in three variables is an ordered _____ of real numbers that satisfies <u>all/some</u> of the equations in the system.

   Circle the correct choice.

2. Consider the following system:
$$\begin{cases} x + y - z = -1 & \text{Equation 1} \\ 2x - 2y - 5z = 7 & \text{Equation 2} \\ 4x + y - 2z = 7. & \text{Equation 3} \end{cases}$$
   We can eliminate x from Equations 1 and 2 by multiplying Equation 1 by _____ and adding equations. We can eliminate x from Equations 1 and 3 by multiplying Equation 1 by _____ and adding equations.

3. Consider the following system:
$$\begin{cases} x + y + z = 2 & \text{Equation 1} \\ 2x - 3y = 3 & \text{Equation 2} \\ 10y - z = 12. & \text{Equation 3} \end{cases}$$

   Equation 2 does not contain the variable _____. To obtain a second equation that does not contain this variable, we can _____.

4. The process of determining an equation whose graph contains given points is called _____

4.5 EXERCISE SET   MyMathLab®

Watch the videos in MyMathLab   Download the MyDashBoard App

Practice Exercises

In Exercises 1–4, determine if the given ordered triple is a solution of the system.

1. $(2, -1, 3)$
$$\begin{cases} x + y + z = 4 \\ x - 2y - z = 1 \\ 2x - y - z = -1 \end{cases}$$

2. $(5, -3, -2)$
$$\begin{cases} x + y + z = 0 \\ x + 2y - 3z = 5 \\ 3x + 4y + 2z = -1 \end{cases}$$

3. $(4, 1, 2)$
$$\begin{cases} x - 2y = 2 \\ 2x + 3y = 11 \\ y - 4z = -7 \end{cases}$$

4. $(-1, 3, 2)$
$$\begin{cases} x - 2z = -5 \\ y - 3z = -3 \\ 2x - z = -4 \end{cases}$$

Solve each system in Exercises 5–22. If there is no solution or if there are infinitely many solutions and a system's equations are dependent, so state.

5. $\begin{cases} x + y + 2z = 11 \\ x + y + 3z = 14 \\ x + 2y - z = 5 \end{cases}$

6. $\begin{cases} 2x + y - 2z = -1 \\ 3x - 3y - z = 5 \\ x - 2y + 3z = 6 \end{cases}$

7. $\begin{cases} 4x - y + 2z = 11 \\ x + 2y - z = -1 \\ 2x + 2y - 3z = -1 \end{cases}$

8. $\begin{cases} x - y + 3z = 8 \\ 3x + y - 2z = -2 \\ 2x + 4y + z = 0 \end{cases}$

9. $\begin{cases} 3x + 2y - 3z = -2 \\ 2x - 5y + 2z = -2 \\ 4x - 3y + 4z = 10 \end{cases}$

10. $\begin{cases} 2x + 3y + 7z = 13 \\ 3x + 2y - 5z = -22 \\ 5x + 7y - 3z = -28 \end{cases}$

11. $\begin{cases} 2x - 4y + 3z = 17 \\ x + 2y - z = 0 \\ 4x - y - z = 6 \end{cases}$

12. $\begin{cases} x + z = 3 \\ x + 2y - z = 1 \\ 2x - y + z = 3 \end{cases}$

13. $\begin{cases} 2x + y = 2 \\ x + y - z = 4 \\ 3x + 2y + z = 0 \end{cases}$

14. $\begin{cases} x + 3y + 5z = 20 \\ y - 4z = -16 \\ 3x - 2y + 9z = 36 \end{cases}$

15. $\begin{cases} x + y = -4 \\ y - z = 1 \\ 2x + y + 3z = -21 \end{cases}$

16. $\begin{cases} x + y = 4 \\ x + z = 4 \\ y + z = 4 \end{cases}$

17. $\begin{cases} 2x + y + 2z = 1 \\ 3x - y + z = 2 \\ x - 2y - z = 0 \end{cases}$

18. $\begin{cases} 3x + 4y + 5z = 8 \\ x - 2y + 3z = -6 \\ 2x - 4y + 6z = 8 \end{cases}$

19. $\begin{cases} 5x - 2y - 5z = 1 \\ 10x - 4y - 10z = 2 \\ 15x - 6y - 15z = 3 \end{cases}$

20. $\begin{cases} x + 2y + z = 4 \\ 3x - 4y + z = 4 \\ 6x - 8y + 2z = 8 \end{cases}$

21. $\begin{cases} 3(2x + y) + 5z = -1 \\ 2(x - 3y + 4z) = -9 \\ 4(1 + x) = -3(z - 3y) \end{cases}$

22. $\begin{cases} 7z - 3 = 2(x - 3y) \\ 5y + 3z - 7 = 4x \\ 4 + 5z = 3(2x - y) \end{cases}$

In Exercises 23–26, find the equation $y = ax^2 + bx + c$ whose graph passes through the given points.

23. $(-1, 6), (1, 4), (2, 9)$

24. $(-2, 7), (1, -2), (2, 3)$

25. $(-1, -4), (1, -2), (2, 5)$

26. $(1, 3), (3, -1), (4, 0)$

In Exercises 27–28, let x represent the first number, y the second number, and z the third number. Use the given conditions to write a system of equations. Solve the system and find the numbers.

27. The sum of three numbers is 16. The sum of twice the first number, 3 times the second number, and 4 times the third number is 46. The difference between 5 times the first number and the second number is 31. Find the three numbers.

28. The following is known about three numbers: Three times the first number plus the second number plus twice the third number is 5. If 3 times the second number is subtracted from the sum of the first number and 3 times the third number, the result is 2. If the third number is subtracted from the sum of 2 times the first number and 3 times the second number, the result is 1. Find the numbers.

Practice PLUS

Solve each system in Exercises 29–30.

29. $\begin{cases} \dfrac{x + 2}{6} - \dfrac{y + 4}{3} + \dfrac{z}{2} = 0 \\ \dfrac{x + 1}{2} + \dfrac{y - 1}{2} - \dfrac{z}{4} = \dfrac{9}{2} \\ \dfrac{x - 5}{4} + \dfrac{y + 1}{3} + \dfrac{z - 2}{2} = \dfrac{19}{4} \end{cases}$

30. $\begin{cases} \dfrac{x + 3}{2} - \dfrac{y - 1}{2} + \dfrac{z + 2}{4} = \dfrac{3}{2} \\ \dfrac{x - 5}{2} + \dfrac{y + 1}{3} - \dfrac{z}{4} = -\dfrac{25}{6} \\ \dfrac{x - 3}{4} - \dfrac{y + 1}{2} + \dfrac{z - 3}{2} = -\dfrac{5}{2} \end{cases}$

In Exercises 31–32, find the equation $y = ax^2 + bx + c$ whose graph is shown. Select three points whose coordinates appear to be integers.

31. 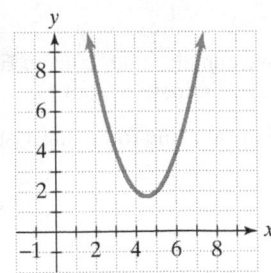   **32.**

In Exercises 33–34, solve each system for (x, y, z) in terms of the nonzero constants a, b, and c.

33. $\begin{cases} ax - by - 2cz = 21 \\ ax + by + cz = 0 \\ 2ax - by + cz = 14 \end{cases}$

34. $\begin{cases} ax - by + 2cz = -4 \\ ax + 3by - cz = 1 \\ 2ax + by + 3cz = 2 \end{cases}$

Application Exercises

35. The bar graph shows the percentage of U.S. parents willing to pay for all, some, or none of their child's college education.

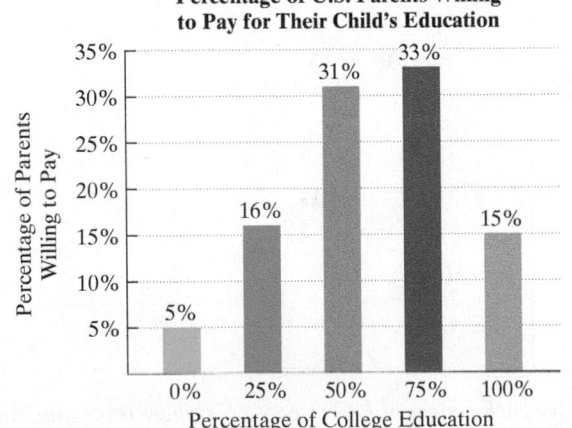

Percentage of U.S. Parents Willing to Pay for Their Child's Education

Source: Student Monitor LLC

a. Use the bars that represent none, half, and all of a college education. Represent the data for each bar as an ordered pair (x, y), where x is the percentage of college education and y is the percentage of parents willing to pay for that percent of their child's education.

b. The three data points in part (a) can be modeled by the equation $y = ax^2 + bx + c$, where $a < 0$. Substitute each ordered pair into this equation, one ordered pair at a time, and write a system of linear equations in three variables that can be used to find values for a, b, and c. It is not necessary to solve the system.

36. How much time do you spend on hygiene/grooming in the morning (including showering, washing face and hands, brushing teeth, shaving, and applying makeup)? The bar graph shows the time American adults spend on morning grooming.

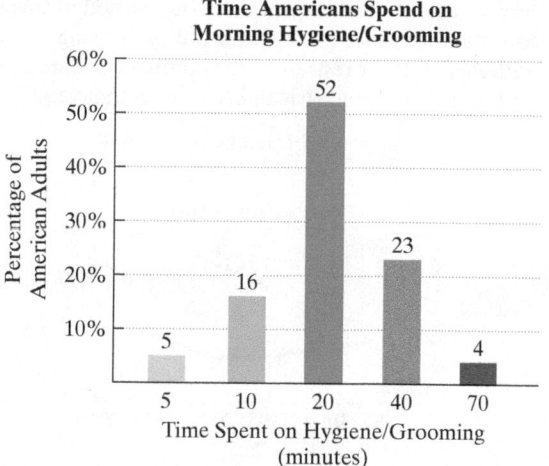

Time Americans Spend on Morning Hygiene/Grooming

Source: SCA *Hygiene Matters*

a. Write the data for 10 minutes, 20 minutes, and 40 minutes as ordered pairs (x, y), where x is the time spent on morning grooming and y is the percentage of American adults spending that much time on grooming.

b. The three data points in part (a) can be modeled by the equation $y = ax^2 + bx + c$, where $a < 0$. Substitute each ordered pair into this equation, one ordered pair at a time, and write a system of linear equations in three variables that can be used to find values for a, b, and c. It is not necessary to solve the system.

37. You throw a ball straight up from a rooftop. The ball misses the rooftop on its way down and eventually strikes the ground. A mathematical model can be used to describe the ball's height above the ground, y, after x seconds. Consider the following data.

x, seconds after the ball is thrown	y, ball's height, in feet, above the ground
1	224
3	176
4	104

a. Find the equation $y = ax^2 + bx + c$ whose graph passes through the given points.

b. Use the equation in part (a) to find the value for y when $x = 5$. Describe what this means.

38. A mathematical model can be used to describe the relationship between the number of feet a car travels once the brakes are applied, y, and the number of seconds the car is in motion after the brakes are applied, x. A research firm collects the data shown below.

x, seconds in motion after brakes are applied	y, feet car travels once the brakes are applied
1	46
2	84
3	114

a. Find the equation $y = ax^2 + bx + c$ whose graph passes through the given points.

b. Use the equation in part (a) to find the value for y when $x = 6$. Describe what this means.

In Exercises 39–46, use the four-step strategy to solve each problem. Use x, y, and z to represent unknown quantities. Then translate from the verbal conditions of the problem to a system of three equations in three variables.

The bar graph shows the average annual spending per person on selected items in 1980 and 2010. All dollar amounts are adjusted for inflation. Use this display to solve Exercises 39–40.

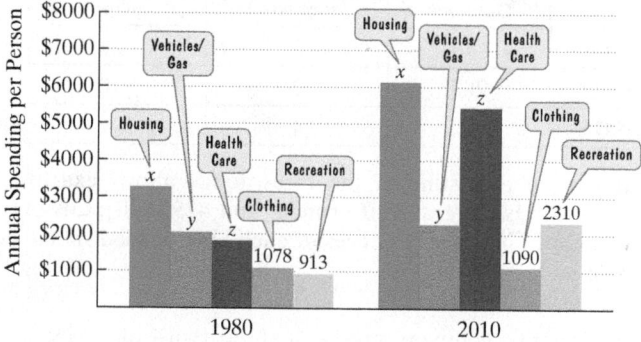

Annual Spending per Person in the United States, Adjusted for Inflation

Source: Bureau of Economic Analysis

39. In this exercise, we refer to annual spending per person in 2010. The combined spending on housing, vehicles/gas, and health care was $13,840. The difference between spending on housing and spending on vehicles/gas was $3864. The difference between spending on housing and spending on health care was $695. Find the average per-person spending on housing, vehicles/gas, and health care in 2010.

40. In this exercise, we refer to annual spending per person in 1980. The combined spending on housing, vehicles/gas, and health care was $7073. The difference between spending on housing and spending on vehicles/gas was $1247. The difference between spending on housing and spending on health care was $1466. Find the average per-person spending on housing, vehicles/gas, and health care in 1980.

41. A person invested $6700 for one year, part at 8%, part at 10%, and the remainder at 12%. The total annual income from these investments was $716. The amount of money invested at 12% was $300 more than the amounts invested at 8% and 10% combined. Find the amount invested at each rate.

42. A person invested $17,000 for one year, part at 10%, part at 12%, and the remainder at 15%. The total annual income from these investments was $2110. The amount of money invested at 12% was $1000 less than the amounts invested at 10% and 15% combined. Find the amount invested at each rate.

43. At a college production of *Streetcar Named Desire*, 400 tickets were sold. The ticket prices were $8, $10, and $12, and the total income from ticket sales was $3700. How many tickets of each type were sold if the combined number of $8 and $10 tickets sold was 7 times the number of $12 tickets sold?

44. A certain brand of razor blades comes in packages of 6, 12, and 24 blades, costing $2, $3, and $4 per package, respectively. A store sold 12 packages containing a total of 162 razor blades and took in $35. How many packages of each type were sold?

45. Three foods have the following nutritional content per ounce.

	Calories	Protein (in grams)	Vitamin C (in milligrams)
Food A	40	5	30
Food B	200	2	10
Food C	400	4	300

If a meal consisting of the three foods allows exactly 660 calories, 25 grams of protein, and 425 milligrams of vitamin C, how many ounces of each kind of food should be used?

46. A furniture company produces three types of desks: a children's model, an office model, and a deluxe model. Each desk is manufactured in three stages: cutting, construction, and finishing. The time requirements for each model and manufacturing stage are given in the following table.

	Children's Model	Office Model	Deluxe Model
Cutting	2 hr	3 hr	2 hr
Construction	2 hr	1 hr	3 hr
Finishing	1 hr	1 hr	2 hr

Each week the company has available a maximum of 100 hours for cutting, 100 hours for construction, and 65 hours for finishing. If all available time must be used, how many of each type of desk should be produced each week?

Writing in Mathematics

47. What is a system of linear equations in three variables?

48. How do you determine whether a given ordered triple is a solution of a system of linear equations in three variables?

49. Describe in general terms how to solve a system in three variables.

50. Describe what happens when using algebraic methods to solve an inconsistent system.

51. Describe what happens when using algebraic methods to solve a system with dependent equations.

52. AIDS is taking a deadly toll on southern Africa. Describe how to use the techniques that you learned in this section to obtain a model for African life span using projections with AIDS. Let x represent the number of years after 1985 and let y represent African life span in that year.

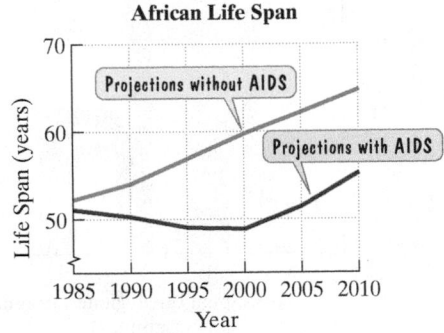

Source: United Nations

Technology Exercises

53. Does your graphing utility have a feature that allows you to solve linear systems by entering coefficients and constant terms? If so, use this feature to verify the solutions to any five exercises that you worked by hand from Exercises 5–16.

54. Verify your results in Exercises 23–26 by using a graphing utility to graph the equation. Trace along the curve and convince yourself that the three points given in the exercise lie on the equation's graph.

Critical Thinking Exercises

Make Sense? *In Exercises 55–58, determine whether each statement "makes sense" or "does not make sense" and explain your reasoning.*

55. Solving a system in three variables, I found that $x = 3$ and $y = -1$. Because z represents a third variable, z cannot equal 3 or −1.

56. A system of linear equations in three variables, x, y, and z, cannot contain an equation in the form $y = mx + b$.

57. I'm solving a three-variable system in which one of the given equations has a missing term, so it will not be necessary to use any of the original equations twice when I reduce the system to two equations in two variables.

58. Because the percentage of the U.S. population that was foreign-born decreased from 1910 through 1970 and then increased after that, an equation of the form $y = ax^2 + bx + c$, rather than a linear equation of the form $y = mx + b$, should be used to model the data.

In Exercises 59–62, determine whether each statement is true or false. If the statement is false, make the necessary change(s) to produce a true statement.

59. The ordered triple (2, 15, 14) is the only solution of the equation $x + y - z = 3$.

60. The equation $x - y - z = -6$ is satisfied by $(2, -3, 5)$.

61. If two equations in a system are $x + y - z = 5$ and $x + y - z = 6$, then the system must be inconsistent.

62. An equation with four variables, such as $x + 2y - 3z + 5w = 2$, cannot be satisfied by real numbers.

63. In the following triangle, the degree measures of the three interior angles and two of the exterior angles are represented with variables. Find the measure of each interior angle.

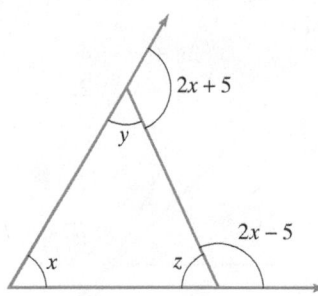

64. A modernistic painting consists of triangles, rectangles, and pentagons, all drawn so as to not overlap or share sides. Within each rectangle are drawn 2 red roses, and each pentagon contains 5 carnations. How many triangles, rectangles, and pentagons appear in the painting if the painting contains a total of 40 geometric figures, 153 sides of geometric figures, and 72 flowers?

65. Two blocks of wood having the same length and width are placed on the top and bottom of a table, as shown in (a). Length A measures 32 centimeters. The blocks are rearranged as shown in (b). Length B measures 28 centimeters. Determine the height of the table.

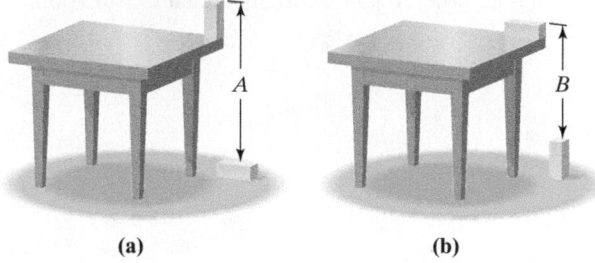

      (a)   **(b)**

Review Exercises

In Exercises 66–68, graph each equation.

66. $y = -\dfrac{3}{4}x + 3$ (Section 3.4, Example 3)

67. $-2x + y = 6$ (Section 3.2, Example 4)

68. $y = -5$ (Section 3.2, Example 7)

Preview Exercises

Exercises 69–71 will help you prepare for the material covered in the first section of the next chapter.

69. Add: $5x^3 + 12x^3$.

70. Add: $-8x^2 + 6x^2$.

71. Subtract: $-9y^4 - (-2y^4)$.

GROUP PROJECT

CHAPTER 4

Group members who have cellphone plans should describe the total monthly cost of the plan as follows:

    $_____ per month buys _____ minutes.

    Additional time costs $_____ per minute.

(For simplicity, ignore other charges.) The group should select any three plans, from "basic" to "premier." For each plan selected, write equations that describe the plan in terms of the total monthly cost, y, for x minutes of use. Compare the plans, two at a time. After how many minutes of use will the costs for the two plans be the same? Solve a linear system to obtain your answer. What will the be the cost for each plan? Graph the equations in the same rectangular coordinate system. For any given time of use, the best plan is the one whose graph is lowest at that point.

    Now compare all three plans. Is one plan always a better deal than the other two? If not, determine the number of minutes of use for which each plan is the better deal. (You can check out cellphone plans by visiting www.point.com.)

Chapter 4 Summary

Definitions and Concepts	Examples

Section 4.1 Solving Systems of Linear Equations by Graphing

A system of linear equations in two variables, x and y, consists of two equations of the form $Ax + By = C$. A solution is an ordered pair of numbers that satisfies both equations.

Determine whether $(3, -1)$ is a solution of

$$\begin{cases} 2x + 5y = 1 \\ 4x + y = 11. \end{cases}$$

Replace x with 3 and y with -1 in both equations.

$$2x + 5y = 1 \qquad\qquad 4x + y = 11$$
$$2 \cdot 3 + 5(-1) \stackrel{?}{=} 1 \qquad 4 \cdot 3 + (-1) \stackrel{?}{=} 11$$
$$6 + (-5) \stackrel{?}{=} 1 \qquad\quad 12 + (-1) \stackrel{?}{=} 11$$
$$1 = 1, \text{ true} \qquad\qquad 11 = 11, \text{ true}$$

Thus, $(3, -1)$ is a solution of the system.

To solve a linear system by graphing,

1. Graph the first equation.
2. Graph the second equation on the same set of axes.
3. If the graphs intersect, determine the coordinates of this point. The ordered pair is the solution of the system.
4. Check the solution in both equations.

If the graphs are parallel lines, the system has no solution and is called inconsistent. If the graphs are the same line, the system has infinitely many solutions. The equations are called dependent.

Solve by graphing:

$$\begin{cases} 2x + y = 4 \\ x + y = 2. \end{cases}$$

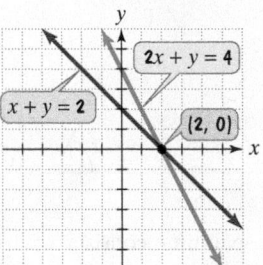

The solution is $(2, 0)$ and the solution set is $\{(2, 0)\}$.

Section 4.2 Solving Systems of Linear Equations by the Substitution Method

To solve a linear system by the substitution method,

1. Solve one equation for one variable in terms of the other.
2. Substitute the expression for that variable into the other equation. This will result in an equation in one variable.
3. Solve the equation in one variable.
4. Back-substitute the value of the variable found in step 3 in the equation from step 1. Simplify and find the value of the remaining variable.
5. Check the proposed solution in both of the system's given equations.

If both variables are eliminated and a false statement results, the system has no solution. If both variables are eliminated and a true statement results, the system has infinitely many solutions.

Solve by the substitution method:

$$\begin{cases} y = 2x + 3 \\ 7x - 5y = -18. \end{cases}$$

Substitute $2x + 3$ for y in the second equation.

$$7x - 5(2x + 3) = -18$$
$$7x - 10x - 15 = -18$$
$$-3x - 15 = -18$$
$$-3x = -3$$
$$x = 1$$

Find y. Substitute 1 for x in $y = 2x + 3$.

$$y = 2 \cdot 1 + 3 = 2 + 3 = 5$$

The solution, $(1, 5)$, checks. The solution set is $\{(1, 5)\}$.

Definitions and Concepts	**Examples**

Section 4.3 Solving Systems of Linear Equations by the Addition Method

To solve a linear system by the addition method,

1. Write the equations in $Ax + By = C$ form.

2. Multiply one or both equations by nonzero numbers so that coefficients of one of the variables are opposites.

3. Add the equations.

4. Solve the resulting equation for the remaining variable.

5. Back-substitute the value of the variable into either original equation and find the value of the other variable.

6. Check the proposed solution in both of the original equations.

If both variables are eliminated and a false statement results, the system has no solution. If both variables are eliminated and a true statement results, the system has infinitely many solutions.

Solve by the addition method:
$$\begin{cases} 3x + y = -11 \\ 6x - 2y = -2. \end{cases}$$

Eliminate y. Multiply both sides of the first equation by 2.
$$\begin{cases} 6x + 2y = -22 \\ \underline{6x - 2y = \;\;-2} \end{cases}$$
$$\text{Add:} \quad 12x \qquad = -24$$
$$x = -2$$

Find y. Back-substitute -2 for x. We'll use the first equation.
$$3(-2) + y = -11$$
$$-6 + y = -11$$
$$y = -5$$

The solution, $(-2, -5)$, checks. The solution set is $\{(-2, -5)\}$.

Section 4.4 Problem Solving Using Systems of Equations

A Problem-Solving Strategy

1. Use variables, usually x and y, to represent unknown quantities.

2. Write a system of equations that models the problem's conditions.

3. Solve the system and answer the problem's question.

4. Check proposed answers in the problem's wording.

You invested $14,000 in two funds paying 7% and 9% interest. Total year-end interest was $1180. How much was invested at each rate?

Let $x =$ amount invested at 7% and
$y =$ amount invested at 9%.

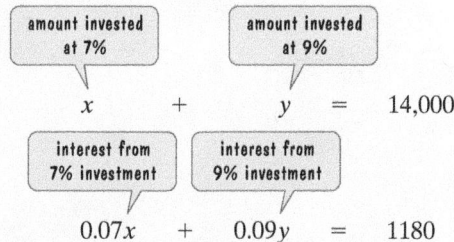

Solving by substitution or addition, $x = 4000$ and $y = 10,000$. Thus, 4000 was invested at 7% and $10,000$ at 9%.

Section 4.5 Systems of Linear Equations in Three Variables

A system of linear equations in three variables, x, y, and z, consists of three equations of the form $Ax + By + Cz = D$. The solution set is the set consisting of the ordered triple that satisfies all three equations. The solution represents the point of intersection of three planes in space.

Is $(2, -1, 3)$ a solution of
$$\begin{cases} 3x + 5y - 2z = -5 \\ 2x + 3y - \;\;z = -2 \\ 2x + 4y + 6z = 18? \end{cases}$$

Replace x with 2, y with -1, and z with 3. Using the first equation, we obtain:

$$3 \cdot 2 + 5(-1) - 2(3) \overset{?}{=} -5$$
$$6 - 5 - 6 \overset{?}{=} -5$$
$$-5 = -5, \quad \text{true}$$

The ordered triple $(2, -1, 3)$ satisfies the first equation. In a similar manner, you can verify that it satisfies the other two equations and is a solution.

Definitions and Concepts	**Examples**

Section 4.5 Systems of Linear Equations in Three Variables (continued)

To solve a linear system in three variables by eliminating variables,

1. Reduce the system to two equations in two variables.

2. Solve the resulting system of two equations in two variables.

3. Use back-substitution in one of the equations in two variables to find the value of the second variable.

4. Back-substitute the values for two variables into one of the original equations to find the value of the third variable.

5. Check.

If all variables are eliminated and a false statement results, the system is inconsistent and has no solution. If a true statement results, the system contains dependent equations and has infinitely many solutions.

Solve

$$\begin{cases} 2x + 3y - 2z = 0 & \text{Equation 1} \\ x + 2y - z = 1 & \text{Equation 2} \\ 3x - y + z = -15. & \text{Equation 3} \end{cases}$$

Add Equations 2 and 3 to eliminate z.

$$4x + y = -14 \quad \text{Equation 4}$$

Eliminate z again. Multiply Equation 3 by 2 and add to Equation 1.

$$8x + y = -30 \quad \text{Equation 5}$$

Multiply Equation 4 by -1 and add to Equation 5.

$$\begin{cases} -4x - y = 14 \\ 8x + y = -30 \end{cases}$$
$$\text{Add:} \quad 4x = -16$$
$$x = -4$$

Substitute -4 for x in Equation 4.

$$4(-4) + y = -14$$
$$y = 2$$

Substitute -4 for x and 2 for y in Equation 3.

$$3(-4) - 2 + z = -15$$
$$-14 + z = -15$$
$$z = -1$$

Checking verifies that $(-4, 2, -1)$ is the solution and $\{(-4, 2, -1)\}$ is the solution set.

Curve Fitting

Curve fitting is determining an equation whose graph contains given points. Three points that do not lie on a line determine the graph of an equation in the form

$$y = ax^2 + bx + c.$$

Use the three given points to create a system of three equations. Solve the system to find a, b, and c.

Find the equation $y = ax^2 + bx + c$ whose graph passes through the points $(-1, 2)$, $(1, 8)$, and $(2, 14)$.
Use $y = ax^2 + bx + c$.

When $x = -1, y = 2$: $2 = a(-1)^2 + b(-1) + c$

When $x = 1, y = 8$: $8 = a \cdot 1^2 + b \cdot 1 + c$

When $x = 2, y = 14$: $14 = a \cdot 2^2 + b \cdot 2 + c$

Solving

$$\begin{cases} a - b + c = 2 \\ a + b + c = 8 \\ 4a + 2b + c = 14, \end{cases}$$

$a = 1$, $b = 3$, and $c = 4$. The equation, $y = ax^2 + bx + c$, is $y = x^2 + 3x + 4$.

CHAPTER 4 REVIEW EXERCISES

4.1 *In Exercises 1–2, determine whether the given ordered pair is a solution of the system.*

1. $(1, -5)$

$$\begin{cases} 4x - y = 9 \\ 2x + 3y = -13 \end{cases}$$

2. $(-5, 2)$

$$\begin{cases} 2x + 3y = -4 \\ x - 4y = -10 \end{cases}$$

3. Does the graphing-utility screen show the solution for the following system? Explain.

$$\begin{cases} x + y = 2 \\ 2x + y = -5 \end{cases}$$

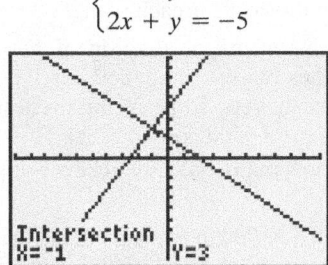

In Exercises 4–14, solve each system by graphing. If there is no solution or an infinite number of solutions, so state. Use set notation to express solution sets.

4. $\begin{cases} x + y = 2 \\ x - y = 6 \end{cases}$

5. $\begin{cases} 2x - 3y = 12 \\ -2x + y = -8 \end{cases}$

6. $\begin{cases} 3x + 2y = 6 \\ 3x - 2y = 6 \end{cases}$

7. $\begin{cases} y = \dfrac{1}{2}x \\ y = 2x - 3 \end{cases}$

8. $\begin{cases} x + 2y = 2 \\ y = x - 5 \end{cases}$

9. $\begin{cases} x + 2y = 8 \\ 3x + 6y = 12 \end{cases}$

10. $\begin{cases} 2x - 4y = 8 \\ x - 2y = 4 \end{cases}$

11. $\begin{cases} y = 3x - 1 \\ y = 3x + 2 \end{cases}$

12. $\begin{cases} x - y = 4 \\ x = -2 \end{cases}$

13. $\begin{cases} x = 2 \\ y = 5 \end{cases}$

14. $\begin{cases} x = 2 \\ x = 5 \end{cases}$

4.2 *In Exercises 15–23, solve each system by the substitution method. If there is no solution or an infinite number of solutions, so state. Use set notation to express solution sets.*

15. $\begin{cases} 2x - 3y = 7 \\ y = 3x - 7 \end{cases}$

16. $\begin{cases} 2x - 3y = 6 \\ x = 13 - 2y \end{cases}$

17. $\begin{cases} 2x - 5y = 1 \\ 3x + y = -7 \end{cases}$

18. $\begin{cases} 3x + 4y = -13 \\ 5y - x = -21 \end{cases}$

19. $\begin{cases} y = 39 - 3x \\ y = 2x - 61 \end{cases}$

20. $\begin{cases} 4x + y = 5 \\ 12x + 3y = 15 \end{cases}$

21. $\begin{cases} 4x - 2y = 10 \\ y = 2x + 3 \end{cases}$

22. $\begin{cases} x - 4 = 0 \\ 9x - 2y = 0 \end{cases}$

23. $\begin{cases} 8y = 4x \\ 7x + 2y = -8 \end{cases}$

24. The following models describe demand and supply for three-bedroom rental apartments.

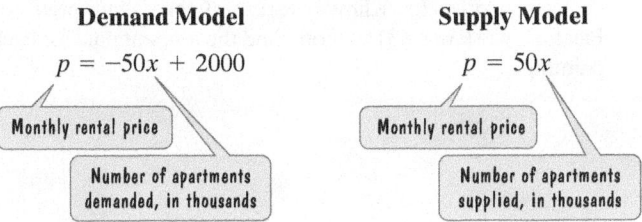

Demand Model $p = -50x + 2000$

Supply Model $p = 50x$

a. Solve the system and find the equilibrium quantity and the equilibrium price.

b. Use your answer from part (a) to complete this statement:

When rents are _____ per month, consumers will demand _____ apartments and suppliers will offer _____ apartments for rent.

4.3 *In Exercises 25–35, solve each system by the addition method. If there is no solution or an infinite number of solutions, so state. Use set notation to express solution sets.*

25. $\begin{cases} x + y = 6 \\ 2x + y = 8 \end{cases}$

26. $\begin{cases} 3x - 4y = 1 \\ 12x - y = -11 \end{cases}$

27. $\begin{cases} 3x - 7y = 13 \\ 6x + 5y = 7 \end{cases}$

28. $\begin{cases} 8x - 4y = 16 \\ 4x + 5y = 22 \end{cases}$

29. $\begin{cases} 5x - 2y = 8 \\ 3x - 5y = 1 \end{cases}$

30. $\begin{cases} 2x + 7y = 0 \\ 7x + 2y = 0 \end{cases}$

31. $\begin{cases} x + 3y = -4 \\ 3x + 2y = 3 \end{cases}$

32. $\begin{cases} 2x + y = 5 \\ 2x + y = 7 \end{cases}$

33. $\begin{cases} 3x - 4y = -1 \\ -6x + 8y = 2 \end{cases}$

34. $\begin{cases} 2x = 8y + 24 \\ 3x + 5y = 2 \end{cases}$

35. $\begin{cases} 5x - 7y = 2 \\ 3x = 4y \end{cases}$

In Exercises 36–41, solve each system by the method of your choice. If there is no solution or an infinite number of solutions, so state. Use set notation to express solution sets.

36. $\begin{cases} 3x + 4y = -8 \\ 2x + 3y = -5 \end{cases}$

37. $\begin{cases} 6x + 8y = 39 \\ y = 2x - 2 \end{cases}$

38. $\begin{cases} x + 2y = 7 \\ 2x + y = 8 \end{cases}$

39. $\begin{cases} y = 2x - 3 \\ y = -2x - 1 \end{cases}$

40. $\begin{cases} 3x - 6y = 7 \\ 3x = 6y \end{cases}$

41. $\begin{cases} y - 7 = 0 \\ 7x - 3y = 0 \end{cases}$

4.4

42. Talk about paintings by numbers: In 2007, this glittering Klimt set the record for the most ever paid for a painting, a title that had been held by Picasso's *Boy with a Pipe*. Combined, the two paintings sold for $239 million. The difference between the selling price for Klimt's work and the selling price for Picasso's work was $31 million. Find the amount paid for each painting.

Gustav Klimt (1826–1918) "Mrs. Adele Bloch-Bauer, I", 1907. Oil on canvas, 138 × 138 cm. Private Collection. Photo: Erich Lessing/Art Resource, NY.

Pablo Picasso (1881–1973) "Boy with a Pipe", 1905 (oil on canvas). © 2011 Picasso Estate, ARS.

43. Health experts agree that cholesterol intake should be limited to 300 milligrams or less each day. Three ounces of shrimp and 2 ounces of scallops contain 156 milligrams of cholesterol. Five ounces of shrimp and 3 ounces of scallops contain 45 milligrams of cholesterol less than the suggested maximum daily intake. Determine the cholesterol content in an ounce of each item.

44. The perimeter of a table tennis top is 28 feet. The difference between 4 times the length and 3 times the width is 21 feet. Find the dimensions.

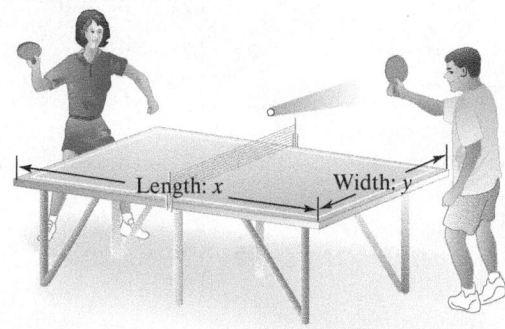

Length: x Width: y

45. A rectangular garden has a perimeter of 24 yards. Fencing across the length costs $3 per yard and along the width $2 per yard. The total cost of the fencing is $62. Find the length and width of the garden.

46. A travel agent offers two package vacation plans. The first plan costs $360 and includes 3 days at a hotel and a rental car for 2 days. The second plan costs $500 and includes 4 days at a hotel and a rental car for 3 days. The daily charge for the room is the same under each plan, as is the daily charge for the car. Find the cost per day for the room and for the car.

47. You are choosing between two telephone plans. One plan has a monthly fee of $15 with a charge of $0.05 per minute for all calls. The other plan has a monthly fee of $10 with a charge of $0.075 per minute for all calls. For how many minutes of calls will the costs for the two plans be the same? What will be the cost for each plan?

48. Tickets for a touring production of *Sweeney Todd* cost $90 for orchestra seats and $60 for balcony seats. You purchase nine tickets, some in the orchestra and some in the balcony, for a total cost of $720. How many orchestra tickets and how many balcony tickets were purchased?

49. You invested $10,000 in two accounts paying 8% and 10% annual interest. At the end of the year, the total interest from these investments was $940. How much was invested at each rate?

50. A chemist needs to mix a 75% saltwater solution with a 50% saltwater solution to obtain 10 gallons of a 60% saltwater solution. How many gallons of each of the solutions must be used?

51. When a plane flies with the wind, it can travel 2160 miles in 3 hours. When the plane flies in the opposite direction, against the wind, it takes 4 hours to fly the same distance. Find the rate of the plane in still air and the rate of the wind.

4.5

52. Is $(-3, -2, 5)$ a solution of the system

$$\begin{cases} x + y + z = 0 \\ 2x - 3y + z = 5 \\ 4x + 2y + 4z = 3? \end{cases}$$

Solve each system in Exercises 53–55 by eliminating variables using the addition method. If there is no solution or if there are infinitely many solutions and a system's equations are dependent, so state.

53. $\begin{cases} 2x - y + z = 1 \\ 3x - 3y + 4z = 5 \\ 4x - 2y + 3z = 4 \end{cases}$

54. $\begin{cases} x + 2y - z = 5 \\ 2x - y + 3z = 0 \\ 2y + z = 1 \end{cases}$

55. $\begin{cases} 3x - 4y + 4z = 7 \\ x - y - 2z = 2 \\ 2x - 3y + 6z = 5 \end{cases}$

56. Find the equation $y = ax^2 + bx + c$ whose graph passes through the points $(1, 4)$, $(3, 20)$, and $(-2, 25)$.

57. 20th Century Death The greatest cause of death in the 20th century was disease, killing 1390 million people. The bar graph shows the five leading causes of death in that century, excluding disease.

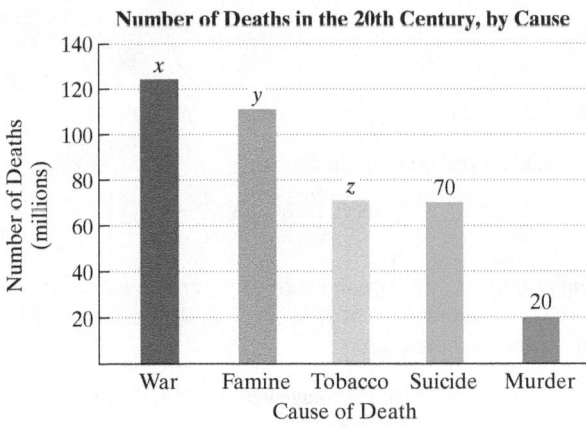

Number of Deaths in the 20th Century, by Cause

Source: Wikipedia

War, famine, and tobacco combined resulted in 306 million deaths. The difference between the number of deaths from war and famine was 13 million. The difference between the number of deaths from war and tobacco was 53 million. Find the number of 20th century deaths from war, famine, and tobacco.

CHAPTER 4 TEST

CHAPTER Test Prep VIDEOS

Step-by-step test solutions are found on the Chapter Test Prep Videos available in MyMathLab° or on YouTube (search "BlitzerCombinedAlg" and click on "Channels").

In Exercises 1–2, determine whether the given ordered pair is a solution of the system.

1. $(5, -5)$

$$\begin{cases} 2x + y = 5 \\ x + 3y = -10 \end{cases}$$

2. $(-3, 2)$

$$\begin{cases} x + 5y = 7 \\ 3x - 4y = 1 \end{cases}$$

In Exercises 3–4, solve each system by graphing. If there is no solution or an infinite number of solutions, so state. Use set notation to express solution sets.

3. $\begin{cases} x + y = 6 \\ 4x - y = 4 \end{cases}$

4. $\begin{cases} 2x + y = 8 \\ y = 3x - 2 \end{cases}$

In Exercises 5–7, solve each system by the substitution method. If there is no solution or an infinite number of solutions, so state. Use set notation to express solution sets.

5. $\begin{cases} x = y + 4 \\ 3x + 7y = -18 \end{cases}$

6. $\begin{cases} 2x - y = 7 \\ 3x + 2y = 0 \end{cases}$

7. $\begin{cases} 2x - 4y = 3 \\ x = 2y + 4 \end{cases}$

In Exercises 8–10, solve each system by the addition method. If there is no solution or an infinite number of solutions, so state. Use set notation to express solution sets.

8. $\begin{cases} 2x + y = 2 \\ 4x - y = -8 \end{cases}$

9. $\begin{cases} 2x + 3y = 1 \\ 3x + 2y = -6 \end{cases}$

10. $\begin{cases} 3x - 2y = 2 \\ -9x + 6y = -6 \end{cases}$

11. According to the U.S. Census Bureau, the two most popular female first names in the United States are Mary and Patricia. Combined, these two names account for 3.7% of all female first names. The difference between the percentage of women named Mary and the percentage of women named Patricia is 1.5%. What percentage of all female first names in the United States are Mary and what percentage are Patricia?

12. A rectangular garden has a perimeter of 34 yards. Fencing across the length costs $2 per yard and fencing along the width costs $1 per yard. The total cost of the fencing is $58. Find the length and width of the garden.

13. You are choosing between two telephone plans. One plan has a monthly fee of $15 with a charge of $0.05 per minute. The other plan has a monthly fee of $5 with a charge of $0.07 per minute. For how many minutes of calls will the costs for the two plans be the same? What will be the cost for each plan?

14. You invested $9000 in two funds paying 6% and 7% annual interest. At the end of the year, the total interest from these investments was $610. How much was invested at each rate?

15. You need to mix a 6% peroxide solution with a 9% peroxide solution to obtain 36 ounces of an 8% peroxide solution. How many ounces of each of the solutions must be used?

16. A paddleboat on the Mississippi River travels 48 miles downstream, with the current, in 3 hours. The return trip, against the current, takes the paddleboat 4 hours. Find the boat's rate in still water and the rate of the current.

17. Solve by eliminating variables using the addition method:

$$\begin{cases} x + y + z = 6 \\ 3x + 4y - 7z = 1 \\ 2x - y + 3z = 5. \end{cases}$$

CUMULATIVE REVIEW EXERCISES (CHAPTERS 1–4)

1. Perform the indicated operations:
$$-14 - [18 - (6 - 10)].$$

2. Simplify: $6(3x - 2) - (x - 1)$.

In Exercises 3–4, solve each equation.

3. $17(x + 3) = 13 + 4(x - 10)$

4. $\dfrac{x}{4} - 1 = \dfrac{x}{5}$

5. Solve for t: $A = P + Prt$.

6. Solve and graph the solution set on a number line: $2x - 5 < 5x - 11$.

In Exercises 7–9, graph each equation in the rectangular coordinate system.

7. $x - 3y = 6$

8. $y = 4$

9. $y = -\dfrac{3}{5}x + 2$

In Exercises 10–11, solve each linear system.

10. $\begin{cases} 3x - 4y = 8 \\ 4x + 5y = -10 \end{cases}$

11. $\begin{cases} 2x - 3y = 9 \\ y = 4x - 8 \end{cases}$

12. Find the slope of the line passing through $(5, -6)$ and $(6, -5)$.

13. Write the point-slope form and the slope-intercept form of the equation of the line passing through $(-1, 6)$ with slope $= -4$.

14. The area of a triangle is 80 square feet. Find the height if the base is 16 feet.

15. If 10 pens and 15 pads of paper cost $26, and 5 of the same pens and 10 of the same pads cost $16, find the cost of a pen and a pad.

16. List all the integers in this set:
$$\left\{ -93, -\dfrac{7}{3}, 0, \sqrt{3}, \dfrac{7}{1}, \sqrt{100} \right\}.$$

The graphs show the percentage of American adults in two living arrangements from 1960 through 2008. Exercises 17–20 are based on these graphs.

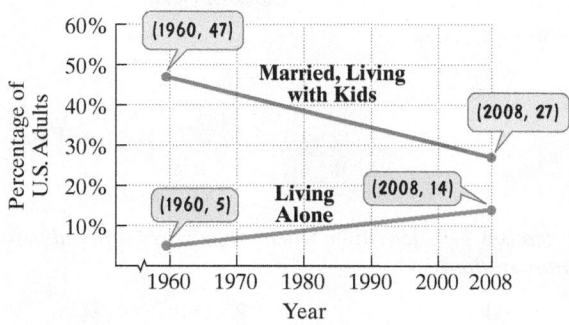

Living Arrangements of U.S. Adults

Source: U.S. Census Bureau

17. Which line has a positive slope? What does this mean in terms of the variables in this situation?

18. Which line has a negative slope? What does this mean in terms of the variables in this situation?

19. In 1960, 47% of U.S. adults were married, living with kids. If the trend shown by the graph continues and this percentage decreases by 0.4 each year, in how many years after 1960 will 19% of U.S. adults be married, living with kids? In which year will that occur?

20. The mathematical model $y = 0.2x + 5$ describes the percentage of U.S. adults living alone, y, x years after 1960. Use the formula to project in which year 19% of U.S. adults will be living alone. Based on your work in Exercise 19, what do you observe?

Exponents and Polynomials

Listening to the radio on the way to campus, you hear politicians discussing the problem of the national debt that exceeds $12 trillion. They state that it's more than the gross domestic product of China, the world's second-richest nation, and four times greater than the combined net worth of America's 691 billionaires. They make it seem like the national debt is a real problem, but later you realize that you don't really know what a number like 12 trillion means. If the national debt were evenly divided among all citizens of the country, how much would every man, woman, and child have to pay? Is economic doomsday about to arrive?

Literacy with numbers, called *numeracy*, is a prerequisite for functioning in a meaningful way personally, professionally, and as a citizen. In this chapter, you will learn to use exponents to provide a way of putting large and small numbers into perspective. The problem of the national debt appears as Example 10 in Section 5.7.

Adding and Subtracting Polynomials

Objectives

1. Understand the vocabulary used to describe polynomials.

2. Add polynomials.

3. Subtract polynomials.

4. Graph equations defined by polynomials of degree 2.

More education results in a higher income. The mathematical models

$$M = -18x^3 + 923x^2 - 9603x + 48,446$$

and $\quad W = 17x^3 - 450x^2 + 6392x - 14,764$

describe the median, or middlemost, annual income for men, M, and women, W, who have completed x years of education. We'll be working with these models and the data upon which they are based in the exercise set.

The algebraic expressions that appear on the right side of the models are examples of *polynomials*. A **polynomial** is a single term or the sum of two or more terms containing variables with whole-number exponents. These particular polynomials each contain four terms. Equations containing polynomials are used in such diverse areas as science, business, medicine, psychology, and sociology. In this section, we present basic ideas about polynomials. We then use our knowledge of combining like terms to find sums and differences of polynomials.

1. Understand the vocabulary used to describe polynomials.

Describing Polynomials

Consider the polynomial

$$7x^3 - 9x^2 + 13x - 6.$$

We can express this polynomial as

$$7x^3 + (-9x^2) + 13x + (-6).$$

The polynomial contains four terms. It is customary to write the terms in the order of descending powers of the variables. This is the **standard form** of a polynomial.

We begin this chapter by limiting our discussion to polynomials containing only one variable. Each term of such a polynomial in x is of the form ax^n. The **degree** of ax^n is n. For example, the degree of the term $7x^3$ is 3.

Great Question!

If the degree of a nonzero constant is 0, why doesn't the constant 0 also have degree 0?

We can express 0 in many ways, including $0x$, $0x^2$, and $0x^3$. It is impossible to assign a unique exponent to the variable. This is why 0 has no defined degree.

The Degree of ax^n

If $a \neq 0$ and n is a whole number, the degree of ax^n is n. The degree of a nonzero constant term is 0. The constant 0 has no defined degree.

Here is an example of a polynomial and the degree of each of its four terms:

$$6x^4 - 3x^3 - 2x - 5.$$

| degree 4 | degree 3 | degree 1 | degree of nonzero constant: 0 |

Notice that the exponent on x for the term $-2x$, meaning $-2x^1$, is understood to be 1. For this reason, the degree of $-2x$ is 1.

A polynomial is simplified when it contains no grouping symbols and no like terms. A simplified polynomial that has exactly one term is called a **monomial**. A **binomial** is a simplified polynomial that has two terms. A **trinomial** is a simplified polynomial with three terms. Simplified polynomials with four or more terms have no special names.

The **degree of a polynomial** is the greatest degree of all the terms of the polynomial. For example, $4x^2 + 3x$ is a binomial of degree 2 because the degree of the first term is 2, and the degree of the other term is less than 2. Also, $7x^5 - 2x^2 + 4$ is a trinomial of degree 5 because the degree of the first term is 5, and the degrees of the other terms are less than 5.

Up to now, we have used x to represent the variable in a polynomial. However, any letter can be used. For example,

- $7x^5 - 3x^3 + 8$ is a polynomial (in x) of degree 5. Because there are three terms, the polynomial is a trinomial.

- $6y^3 + 4y^2 - y + 3$ is a polynomial (in y) of degree 3. Because there are four terms, the polynomial has no special name.

- $z^7 + \sqrt{2}$ is a polynomial (in z) of degree 7. Because there are two terms, the polynomial is a binomial.

2 Add polynomials.

Adding Polynomials

Recall that *like terms* are terms containing exactly the same variables to the same powers. Polynomials are added by combining like terms. For example, we can add the monomials $-9x^3$ and $13x^3$ as follows:

$$-9x^3 + 13x^3 = (-9 + 13)x^3 = 4x^3.$$

These like terms both contain x to the third power.	Add coefficients and keep the same variable factor, x^3.

EXAMPLE 1   Adding Polynomials

Add: $(-9x^3 + 7x^2 - 5x + 3) + (13x^3 + 2x^2 - 8x - 6)$.

Solution   The like terms are $-9x^3$ and $13x^3$, containing the same variable to the same power (x^3), as well as $7x^2$ and $2x^2$ (both containing x^2), $-5x$ and $-8x$ (both containing x), and the constant terms 3 and -6. We begin by grouping these pairs of like terms.

$(-9x^3 + 7x^2 - 5x + 3) + (13x^3 + 2x^2 - 8x - 6)$

$= (-9x^3 + 13x^3) + (7x^2 + 2x^2) + (-5x - 8x) + (3 - 6)$ Group like terms.

$= 4x^3 + 9x^2 + (-13x) + (-3)$ Combine like terms.

$= 4x^3 + 9x^2 - 13x - 3$ Express addition of opposites as subtraction. ∎

✓ **CHECK POINT 1**   Add: $(-11x^3 + 7x^2 - 11x - 5) + (16x^3 - 3x^2 + 3x - 15)$.

Polynomials can also be added by arranging like terms in columns. Then combine like terms, column by column.

EXAMPLE 2 Adding Polynomials Vertically

Add: $(-9x^3 + 7x^2 - 5x + 3) + (13x^3 + 2x^2 - 8x - 6)$.

Solution

$$
\begin{array}{l}
-9x^3 + 7x^2 - 5x + 3 \qquad \text{Line up like terms vertically.}\\
\underline{13x^3 + 2x^2 - 8x - 6}\\
4x^3 + 9x^2 - 13x - 3 \qquad \text{Add the like terms in each column.}
\end{array}
$$

This is the same answer that we found in Example 1. ∎

✓ **CHECK POINT 2** Add the polynomials in Check Point 1 using a vertical format. Begin by arranging like terms in columns.

> **Great Question!**
>
> **What's the advantage of using a vertical format to add polynomials instead of a horizontal format?**
>
> A vertical format often makes it easier to see the like terms.

3 Subtract polynomials.

Subtracting Polynomials

We subtract real numbers by adding the opposite, or additive inverse, of the number being subtracted. For example,

$$8 - 3 = 8 + (-3) = 5.$$

Subtraction of polynomials also involves opposites. If the sum of two polynomials is 0, the polynomials are **opposites**, or **additive inverses**, of each other. Here is an example:

$$(4x^2 - 6x - 7) + (-4x^2 + 6x + 7) = 0.$$

The opposite of $4x^2 - 6x - 7$ is $-4x^2 + 6x + 7$, and vice versa.

Observe that the opposite of $4x^2 - 6x - 7$ can be obtained by changing the sign of each of its coefficients:

Polynomial	Change 4 to -4, change -6 to 6, and change -7 to 7.	**Opposite**
$4x^2 - 6x - 7$		$-4x^2 + 6x + 7.$

In general, **the opposite of a polynomial is that polynomial with the sign of every coefficient changed**. Just as we did with real numbers, we subtract one polynomial from another by adding the opposite of the polynomial being subtracted.

> **Subtracting Polynomials**
>
> To subtract two polynomials, add the first polynomial and the opposite of the polynomial being subtracted.

EXAMPLE 3 Subtracting Polynomials

Subtract: $(7x^2 + 3x - 4) - (4x^2 - 6x - 7)$.

Solution

$$(7x^2 + 3x - 4) - (4x^2 - 6x - 7)$$

Change the sign of each coefficient.

$$
\begin{array}{ll}
= (7x^2 + 3x - 4) + (-4x^2 + 6x + 7) & \text{Add the opposite of the polynomial being subtracted.}\\
= (7x^2 - 4x^2) + (3x + 6x) + (-4 + 7) & \text{Group like terms.}\\
= 3x^2 + 9x + 3 & \text{Combine like terms.} \quad ∎
\end{array}
$$

✓ **CHECK POINT 3** Subtract: $(9x^2 + 7x - 2) - (2x^2 - 4x - 6)$.

Great Question!

I'm confused by what it means to subtract one polynomial from a second polynomial. Which polynomial should I write first?

Be careful of the order in Example 4. For example, subtracting 2 from 5 means $5 - 2$. In general, subtracting B from A means $A - B$. The order of the resulting algebraic expression is not the same as the order in English. The polynomial following the word *from* is the one to write first.

EXAMPLE 4 Subtracting Polynomials

Subtract $2x^3 - 6x^2 - 3x + 9$ from $7x^3 - 8x^2 + 9x - 6$.

Solution

$$(7x^3 - 8x^2 + 9x - 6) - (2x^3 - 6x^2 - 3x + 9)$$

Change the sign of each coefficient.

$$= (7x^3 - 8x^2 + 9x - 6) + (-2x^3 + 6x^2 + 3x - 9)$$ Add the opposite of the polynomial being subtracted.

$$= (7x^3 - 2x^3) + (-8x^2 + 6x^2)$$
$$\quad + (9x + 3x) + (-6 - 9)$$ Group like terms.

$$= 5x^3 + (-2x^2) + 12x + (-15)$$ Combine like terms.

$$= 5x^3 - 2x^2 + 12x - 15$$ Express addition of opposites as subtraction. ∎

✓ **CHECK POINT 4**   Subtract $3x^3 - 8x^2 - 5x + 6$ from $10x^3 - 5x^2 + 7x - 2$.

Subtraction can also be performed in columns.

EXAMPLE 5 Subtracting Polynomials Vertically

Use the method of subtracting in columns to find

$$(12y^3 - 9y^2 - 11y - 3) - (4y^3 - 5y + 8).$$

Solution   Arrange like terms in columns.

$$
\begin{array}{l}
12y^3 - 9y^2 - 11y - 3 \\
-(4y^3 \qquad - 5y + 8)
\end{array}
$$

Add the opposite of the polynomial being subtracted.

Leave space for the missing term.

$$
\begin{array}{l}
 12y^3 - 9y^2 - 11y - 3 \\
+ -4y^3 \qquad + 5y - 8 \\
\hline
 8y^3 - 9y^2 - 6y - 11
\end{array}
$$

Change the sign of each coefficient of $4y^3 - 5y + 8$.

Combine like terms. ∎

✓ **CHECK POINT 5**   Use the method of subtracting in columns to find

$$(8y^3 - 10y^2 - 14y - 2) - (5y^3 - 3y + 6).$$

Graphing Equations Defined by Polynomials

Look at the picture of this gymnast. He has created a perfect balance in which the two halves of his body are mirror images of each other. Graphs of equations defined by polynomials of degree 2, such as $y = x^2 - 4$, have this mirrorlike quality. We can obtain their graphs, shaped like bowls or inverted bowls, using the point-plotting method for graphing an equation in two variables.

4 Graph equations defined by polynomials of degree 2.

Graphing an Equation Defined by a Polynomial of Degree 2

Graph the equation: $y = x^2 - 4$.

Solution   The given equation involves two variables, x and y. However, because the variable x is squared, it is not a linear equation in two variables.

$$y = x^2 - 4$$

This is not in the form $y = mx + b$ because x is squared.

Although the graph is not a line, it is still a picture of all the ordered-pair solutions of $y = x^2 - 4$. Thus, we can use the point-plotting method to obtain the graph.

Step 1. Find several ordered pairs that are solutions of the equation. To find some solutions of $y = x^2 - 4$, we select integers for x, starting with -3 and ending with 3.

Great Question!

When I graphed lines, I used two, or possibly three, points. Why isn't this enough to use when graphing equations that are not linear?

If the graph of an equation is not a straight line, extra points are needed to get a better general idea of the graph's shape.

Start with x.	Compute y.	Form the ordered pair (x, y).
x	$y = x^2 - 4$	(x, y)
-3	$y = (-3)^2 - 4 = 9 - 4 = 5$	$(-3, 5)$
-2	$y = (-2)^2 - 4 = 4 - 4 = 0$	$(-2, 0)$
-1	$y = (-1)^2 - 4 = 1 - 4 = -3$	$(-1, -3)$
0	$y = 0^2 - 4 = 0 - 4 = -4$	$(0, -4)$
1	$y = 1^2 - 4 = 1 - 4 = -3$	$(1, -3)$
2	$y = 2^2 - 4 = 4 - 4 = 0$	$(2, 0)$
3	$y = 3^2 - 4 = 9 - 4 = 5$	$(3, 5)$

Step 2. Plot these ordered pairs as points in the rectangular coordinate system. The seven ordered pairs in the table of values are plotted in **Figure 5.1(a)**.

Step 3. Connect the points with a smooth curve. The seven points are joined with a smooth curve in **Figure 5.1(b)**. The graph of $y = x^2 - 4$ is a curve where the part of the graph to the right of the y-axis is a reflection of the part to the left of it, and vice versa. The arrows on both ends of the curve indicate that it extends indefinitely in both directions.

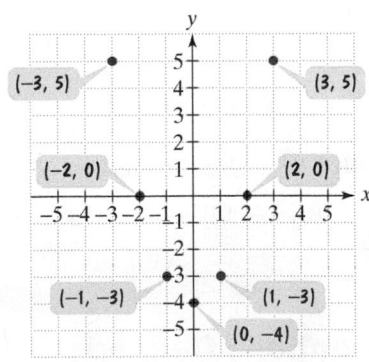

Figure 5.1(a) Some solutions of $y = x^2 - 4$ plotted as points

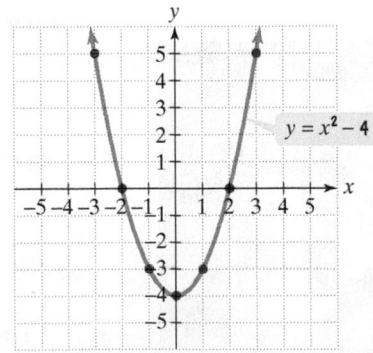

Figure 5.1(b)

The graph of $y = x^2 - 4$ ■

✓ **CHECK POINT 6**   Graph the equation: $y = x^2 - 1$. Select integers for x, starting with -3 and ending with 3.

Blitzer Bonus

Modeling *American Idol* with a Polynomial of Degree 2

The graph in **Figure 5.2** indicates that the ratings of *American Idol* from season 1 (2002) through season 9 (2010) have a mirrorlike quality. This suggests modeling the show's average number of viewers with a polynomial of degree 2.

American Idol: Each Season's Champion and Average Number of Viewers

Figure 5.2

Source: Nielsen

The equation

$$y = -0.68x^2 + 7.94x + 6.78$$

models *American Idol*'s average number of viewers, *y*, in millions, where *x* is the show's season number. The graph of the model in **Figure 5.3** is shaped like an inverted bowl. Can you see why projections based on this graph have the show's producers looking for a shake-up?

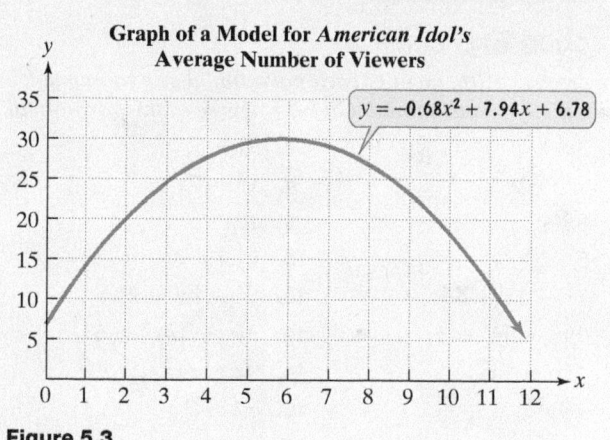

Graph of a Model for *American Idol's* Average Number of Viewers

$y = -0.68x^2 + 7.94x + 6.78$

Figure 5.3

Achieving Success

Address your stress. Stress levels can help or hinder performance. The graph of the polynomial equation of degree 2 in **Figure 5.4** serves as a model that shows people under both low stress and high stress perform worse than their moderate-stress counterparts.

Figure 5.4

Source: Herbert Benson, *Your Maximum Mind*, Random House, 1987.

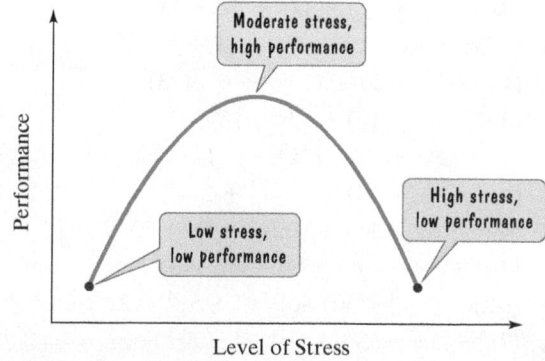

CONCEPT AND VOCABULARY CHECK

Fill in each blank so that the resulting statement is true.

1. A polynomial is a single term or the sum of two or more terms containing variables with exponents that are _____ numbers.

2. It is customary to write the terms of a polynomial in the order of descending powers of the variable. This is called the _____ form of a polynomial.

3. A simplified polynomial that has exactly one term is called a/an _____.

4. A simplified polynomial that has two terms is called a/an _____.

5. A simplified polynomial that has three terms is called a/an _____.

6. The degree of ax^n is _____, provided $a \neq 0$.

7. The degree of a polynomial is the _____ degree of all the terms of the polynomial.

8. Polynomials are added by combining _____ terms.

9. To subtract two polynomials, add the first polynomial and the _____ of the polynomial being subtracted.

5.1 EXERCISE SET

MyMathLab®

Watch the videos in MyMathLab

Download the MyDashBoard App

Practice Exercises

In Exercises 1–16, identify each polynomial as a monomial, a binomial, or a trinomial. Give the degree of the polynomial.

1. $3x + 7$
2. $5x - 2$
3. $x^3 - 2x$
4. $x^5 - 7x$
5. $8x^2$
6. $10x^2$
7. 5
8. 9
9. $x^2 - 3x + 4$
10. $x^2 - 9x + 2$
11. $7y^2 - 9y^4 + 5$
12. $3y^2 - 14y^5 + 6$
13. $15x - 7x^3$
14. $9x - 5x^3$
15. $-9y^{23}$
16. $-11y^{26}$

In Exercises 17–38, add the polynomials.

17. $(9x + 8) + (-17x + 5)$
18. $(8x - 5) + (-13x + 9)$
19. $(4x^2 + 6x - 7) + (8x^2 + 9x - 2)$
20. $(11x^2 + 7x - 4) + (27x^2 + 10x - 20)$
21. $(7x^2 - 11x) + (3x^2 - x)$
22. $(-3x^2 + x) + (4x^2 + 8x)$
23. $(4x^2 - 6x + 12) + (x^2 + 3x + 1)$
24. $(-7x^2 + 8x + 3) + (2x^2 + x + 8)$
25. $(4y^3 + 7y - 5) + (10y^2 - 6y + 3)$
26. $(2y^3 + 3y + 10) + (3y^2 + 5y - 22)$
27. $(2x^2 - 6x + 7) + (3x^3 - 3x)$
28. $(4x^3 + 5x + 13) + (-4x^2 + 22)$
29. $(4y^2 + 8y + 11) + (-2y^3 + 5y + 2)$
30. $(7y^3 + 5y - 1) + (2y^2 - 6y + 3)$
31. $(-2y^6 + 3y^4 - y^2) + (-y^6 + 5y^4 + 2y^2)$
32. $(7r^4 + 5r^2 + 2r) + (-18r^4 - 5r^2 - r)$

33. $\left(9x^3 - x^2 - x - \dfrac{1}{3}\right) + \left(x^3 + x^2 + x + \dfrac{4}{3}\right)$

34. $\left(12x^3 - x^2 - x + \dfrac{4}{3}\right) + \left(x^3 + x^2 + x - \dfrac{1}{3}\right)$

35. $\left(\dfrac{1}{5}x^4 + \dfrac{1}{3}x^3 + \dfrac{3}{8}x^2 + 6\right) +$
    $\left(-\dfrac{3}{5}x^4 + \dfrac{2}{3}x^3 - \dfrac{1}{2}x^2 - 6\right)$

36. $\left(\dfrac{2}{5}x^4 + \dfrac{2}{3}x^3 + \dfrac{5}{8}x^2 + 7\right) +$
    $\left(-\dfrac{4}{5}x^4 + \dfrac{1}{3}x^3 - \dfrac{1}{4}x^2 - 7\right)$

37. $(0.03x^5 - 0.1x^3 + x + 0.03) +$
    $(-0.02x^5 + x^4 - 0.7x + 0.3)$

38. $(0.06x^5 - 0.2x^3 + x + 0.05) +$
    $(-0.04x^5 + 2x^4 - 0.8x + 0.5)$

In Exercises 39–54, use a vertical format to add the polynomials.

39. $5y^3 - 7y^2$
    $6y^3 + 4y^2$

40. $13x^4 - x^2$
    $7x^4 + 2x^2$

41. $3x^2 - 7x + 4$
    $-5x^2 + 6x - 3$

42. $7x^2 - 5x - 6$
    $-9x^2 + 4x + 6$

43. $\frac{1}{4}x^4 - \frac{2}{3}x^3 - 5$
    $-\frac{1}{2}x^4 + \frac{1}{5}x^3 + 4.7$

44. $\frac{1}{3}x^9 - \frac{1}{5}x^5 - 2.7$
  $-\frac{3}{4}x^9 + \frac{2}{3}x^5 + 1$

45. $\quad y^3 + 5y^2 - 7y - 3$
  $-2y^3 + 3y^2 + 4y - 11$

46. $\quad y^3 + \ y^2 - 7y + 9$
  $-y^3 - 6y^2 - 8y + 11$

47. $\quad 4x^3 - 6x^2 + 5x - 7$
  $-9x^3 \qquad - 4x + 3$

48. $-4y^3 + 6y^2 - 8y + 11$
  $\quad 2y^3 \qquad + 9y - 3$

49. $7x^4 - 3x^3 + x^2$
  $\qquad x^3 - x^2 + 4x - 2$

50. $7y^5 - 3y^3 + y^2$
  $\qquad 2y^3 - y^2 - 4y - 3$

51. $\quad 7x^2 - \ 9x + 3$
  $\quad 4x^2 + 11x - 2$
  $-3x^2 + \ 5x - 6$

52. $\quad 7y^2 - 11y - 6$
  $\quad 8y^2 + \ 3y + 4$
  $-9y^2 - \ 5y + 2$

53. $1.2x^3 - \ 3x^2 + 9.1$
  $7.8x^3 - 3.1x^2 + \ 8$
  $\qquad 1.2x^2 - \ 6$

54. $7.9x^3 - 6.8x^2 + 3.3$
  $6.1x^3 - 2.2x^2 + 7$
  $\qquad 4.3x^2 - 5$

In Exercises 55–74, subtract the polynomials.

55. $(x - 8) - (3x + 2)$

56. $(x - 2) - (7x + 9)$

57. $(x^2 - 5x - 3) - (6x^2 + 4x + 9)$

58. $(3x^2 - 8x - 2) - (11x^2 + 5x + 4)$

59. $(x^2 - 5x) - (6x^2 - 4x)$

60. $(3x^2 - 2x) - (5x^2 - 6x)$

61. $(x^2 - 8x - 9) - (5x^2 - 4x - 3)$

62. $(x^2 - 5x + 3) - (x^2 - 6x - 8)$

63. $(y - 8) - (3y - 2)$

64. $(y - 2) - (7y - 9)$

65. $(6y^3 + 2y^2 - y - 11) - (y^2 - 8y + 9)$

66. $(5y^3 + y^2 - 3y - 8) - (y^2 - 8y + 11)$

67. $(7n^3 - n^7 - 8) - (6n^3 - n^7 - 10)$

68. $(2n^2 - n^7 - 6) - (2n^3 - n^7 - 8)$

69. $(y^6 - y^3) - (y^2 - y)$

70. $(y^5 - y^3) - (y^4 - y^2)$

71. $(7x^4 + 4x^2 + 5x) - (-19x^4 - 5x^2 - x)$

72. $(-3x^6 + 3x^4 - x^2) - (-x^6 + 2x^4 + 2x^2)$

73. $\left(\frac{3}{7}x^3 - \frac{1}{5}x - \frac{1}{3}\right) - \left(-\frac{2}{7}x^3 + \frac{1}{4}x - \frac{1}{3}\right)$

74. $\left(\frac{3}{8}x^2 - \frac{1}{3}x - \frac{1}{4}\right) - \left(-\frac{1}{8}x^2 + \frac{1}{2}x - \frac{1}{4}\right)$

In Exercises 75–88, use a vertical format to subtract the polynomials.

75. $\quad 7x + 1$
  $-(3x - 5)$

76. $\quad 4x + 2$
  $-(3x - 5)$

77. $\quad 7x^2 - 3$
  $-(-3x^2 + 4)$

78. $\quad 9y^2 - 6$
  $-(-5y^2 + 2)$

79. $\quad 7y^2 - 5y + 2$
  $-(11y^2 + 2y - 3)$

80. $\quad 3x^5 - 5x^3 + 6$
  $-(7x^5 + 4x^3 - 2)$

81. $\quad 7x^3 + 5x^2 - 3$
  $-(-2x^3 - 6x^2 + 5)$

82. $\quad 3y^4 - 4y^2 + \ 7$
  $-(-5y^4 - 6y^2 - 13)$

83. $\quad 5y^3 + 6y^2 - 3y + 10$
  $-(6y^3 - 2y^2 - 4y - \ 4)$

84. $\quad 4y^3 + 5y^2 + 7y + 11$
  $-(-5y^3 + 6y^2 - 9y - \ 3)$

85. $\quad 7x^4 - 3x^3 + 2x^2$
  $-(\quad - x^3 - x^2 + x - 2)$

86. $\quad 5y^6 - 3y^3 + 2y^2$
  $-(\quad - y^3 - y^2 - y - 1)$

87. $\quad 0.07x^3 - 0.01x^2 + 0.02x$
  $-(0.02x^3 - 0.03x^2 - \qquad x)$

88. $\quad 0.04x^3 - 0.03x^2 + 0.05x$
  $-(0.02x^3 - 0.06x^2 - \qquad x)$

Graph each equation in Exercises 89–94. Find seven solutions in your table of values for each equation by using integers for x, starting with −3 and ending with 3.

89. $y = x^2$ **90.** $y = x^2 - 2$

91. $y = x^2 + 1$ **92.** $y = x^2 + 2$

93. $y = 4 - x^2$ **94.** $y = 9 - x^2$

Practice PLUS

In Exercises 95–98, perform the indicated operations.

95. $\left[(4x^2 + 7x - 5) - (2x^2 - 10x + 3)\right] - (x^2 + 5x - 8)$

96. $\left[(10x^3 - 5x^2 + 4x + 3) - (-3x^3 - 4x^2 + x)\right] -$
  $(7x^3 - 5x + 4)$

97. $\left[(4y^2 - 3y + 8) - (5y^2 + 7y - 4)\right] -$
  $\left[(8y^2 + 5y - 7) + (-10y^2 + 4y + 3)\right]$

98. $\left[(7y^2 - 4y + 2) - (12y^2 + 3y - 5)\right] -$
  $\left[(5y^2 - 2y - 8) + (-7y^2 + 10y - 13)\right]$

99. Subtract $x^3 - 2x^2 + 2$ from the sum of $4x^3 + x^2$ and $-x^3 + 7x - 3$.

100. Subtract $-3x^3 - 7x + 5$ from the sum of $2x^2 + 4x - 7$ and $-5x^3 - 2x - 3$.

101. Subtract $-y^2 + 7y^3$ from the difference between $-5 + y^2 + 4y^3$ and $-8 - y + 7y^3$. Express the answer in standard form.

102. Subtract $-2y^2 + 8y^3$ from the difference between $-6 + y^2 + 5y^3$ and $-12 - y + 13y^3$. Express the answer in standard form.

Application Exercises

As you complete more years of education, you can count on a greater income. The bar graph shows the median, or middlemost, annual income for Americans, by level of education, for a recent year.

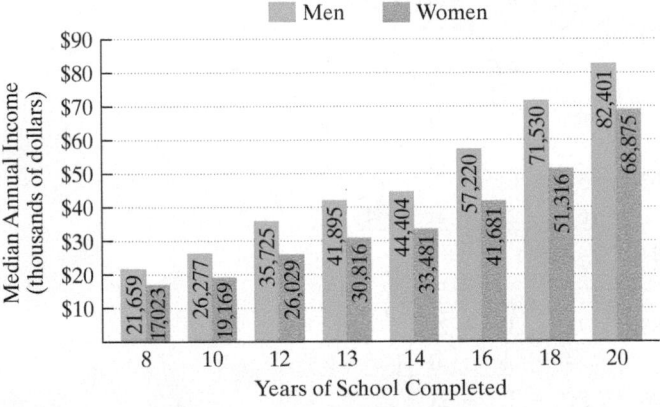

Median Annual Income, by Level of Education

■ Men ■ Women

Source: Bureau of the Census

Here are polynomial models that describe the median annual income for men, M, and for women, W, who have completed x years of education:

$$M = 177x^2 + 288x + 7075$$
$$W = 255x^2 - 2956x + 24,336$$

$$M = -18x^3 + 923x^2 - 9603x + 48,446$$
$$W = 17x^3 - 450x^2 + 6392x - 14,764.$$

Exercises 103–106 are based on these models and the data displayed by the graph.

103. a. Use the equations defined by polynomials of degree 3 to find a mathematical model for $M - W$.

b. According to the model in part (a), what is the difference in the median annual income between men and women with 14 years of education?

c. According to the data displayed by the graph, what is the actual difference in the median annual income between men and women with 14 years of education? Did the model in part (b) underestimate or overestimate this difference? By how much?

104. a. Use the equations defined by polynomials of degree 3 to find a mathematical model for $M - W$.

b. According to the model in part (a), what is the difference in the median annual income between men and women with 16 years of education?

c. According to the data displayed by the graph, what is the actual difference in the median annual income between men and women with 16 years of education? Did the model in part (b) underestimate or overestimate this difference? By how much?

105. a. Use the equation defined by a polynomial of degree 2 to find the median annual income for a man with 16 years of education. Does this underestimate or overestimate the median income shown by the bar graph in the previous column? By how much?

b. Shown in a rectangular coordinate system are the graphs of the polynomial models of degree 2 that describe median annual income, by level of education. Identify your solution from part (a) as a point on the appropriate graph.

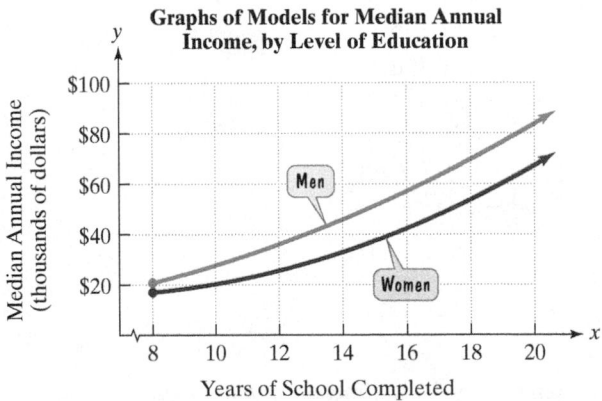

Graphs of Models for Median Annual Income, by Level of Education

c. Use the appropriate graph in part (b) to estimate, to the nearest thousand dollars, the median annual income for a woman with 16 years of education.

106. a. Use the equation defined by a polynomial of degree 2 to find the median annual income for a woman with 18 years of education. Does this underestimate or overestimate the median income shown by the bar graph in the previous column? By how much?

b. Shown in Exercise 105(b) are rectangular coordinate graphs of the polynomial models of degree 2 that describe median annual income, by level of education. Identify your solution from part (a) as a point on the appropriate graph.

c. Use the appropriate graph in Exercise 105(b) to estimate, to the nearest thousand dollars, the median annual income for a man with 18 years of education.

Writing in Mathematics

107. What is a polynomial?

108. What is a monomial? Give an example with your explanation.

109. What is a binomial? Give an example with your explanation.

110. What is a trinomial? Give an example with your explanation.

111. What is the degree of a polynomial? Provide an example with your explanation.

112. Explain how to add polynomials.

113. Explain how to subtract polynomials.

Critical Thinking Exercises

Make Sense? *In Exercises 114–117, determine whether each statement "makes sense" or "does not make sense" and explain your reasoning.*

114. I add like monomials by adding both their coefficients and the exponents that appear on their common variable factor.

115. By looking at the first term of a polynomial, I can determine its degree.

116. As long as I understand how to add and subtract polynomials, I can select the format, horizontal or vertical, that works best for me.

117. I used two points and a checkpoint to graph $y = x^2 - 4$.

In Exercises 118–121, determine whether each statement is true or false. If the statement is false, make the necessary change(s) to produce a true statement.

118. It is not possible to write a binomial with degree 0.

119. $\dfrac{1}{5x^2} + \dfrac{1}{3x}$ is a binomial.

120. $(2x^2 - 8x + 6) - (x^2 - 3x + 5) = x^2 - 5x + 1$ for any value of x.

121. In the polynomial $3x^2 - 5x + 13$, the coefficient of x is 5.

122. What polynomial must be subtracted from $5x^2 - 2x + 1$ so that the difference is $8x^2 - x + 3$?

123. The number of people who catch a cold t weeks after January 1 is $5t - 3t^2 + t^3$. The number of people who recover t weeks after January 1 is $t - t^2 + \frac{1}{3}t^3$. Write a polynomial in standard form for the number of people who are still ill with a cold t weeks after January 1.

124. Explain why it is not possible to add two polynomials of degree 3 and get a polynomial of degree 4.

Review Exercises

125. Simplify: $(-10)(-7) \div (1 - 8)$.
     (Section 1.8, Example 8)

126. Subtract: $-4.6 - (-10.2)$.
     (Section 1.6, Example 2)

127. Solve: $3(x - 2) = 9(x + 2)$.
     (Section 2.3, Example 3)

Preview Exercises

Exercises 128–130 will help you prepare for the material covered in the next section.

128. Find the missing exponent, designated by the question mark, in the final step.
     $$x^3 \cdot x^4 = (x \cdot x \cdot x) \cdot (x \cdot x \cdot x \cdot x) = x^?$$

129. Use the distributive property to multiply: $3x(x + 5)$.

130. Simplify: $x(x + 2) + 3(x + 2)$.

Multiplying Polynomials

Objectives

1. Use the product rule for exponents.

2. Use the power rule for exponents.

3. Use the products-to-powers rule.

4. Multiply monomials.

5. Multiply a monomial and a polynomial.

6. Multiply polynomials when neither is a monomial.

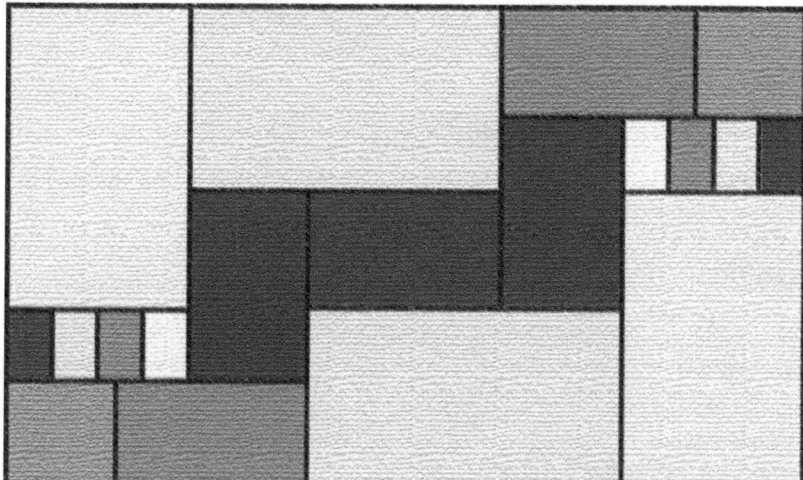

The ancient Greeks believed that the most visually pleasing rectangles have a ratio of length to width of approximately 1.618 to 1. With the exception of the squares on the lower left and the upper right, the interior of this geometric figure is filled entirely with these *golden rectangles*. Furthermore, the large rectangle is also a golden rectangle.

The total area of the large rectangle shown on the previous page can be found in many ways. This is because the area of any large rectangular region is related to the areas of the smaller rectangles that make up that region. In this section, we apply areas of rectangles as a way to picture the multiplication of polynomials. Before studying how polynomials are multiplied, we must develop some rules for working with exponents.

1 Use the product rule for exponents.

The Product Rule for Exponents

We have seen that exponents are used to indicate repeated multiplication. For example, 2^4, where 2 is the base and 4 is the exponent, indicates that 2 occurs as a factor four times:

$$2^4 = 2 \cdot 2 \cdot 2 \cdot 2.$$

Now consider the multiplication of two exponential expressions, such as $2^4 \cdot 2^3$. We are multiplying 4 factors of 2 and 3 factors of 2. We have a total of 7 factors of 2:

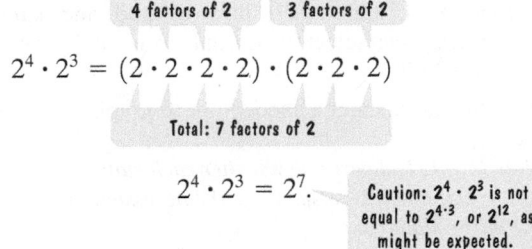

4 factors of 2   3 factors of 2

$$2^4 \cdot 2^3 = (2 \cdot 2 \cdot 2 \cdot 2) \cdot (2 \cdot 2 \cdot 2)$$

Total: 7 factors of 2

Thus,

$$2^4 \cdot 2^3 = 2^7.$$

Caution: $2^4 \cdot 2^3$ is not equal to $2^{4 \cdot 3}$, or 2^{12}, as might be expected.

We can quickly find the exponent, 7, of the product by adding 4 and 3, the original exponents:

$$2^4 \cdot 2^3 = 2^{4+3} = 2^7.$$

This suggests the following rule:

> **The Product Rule**
>
> $$b^m \cdot b^n = b^{m+n}$$
>
> When multiplying exponential expressions with the same base, add the exponents. Use this sum as the exponent of the common base.

EXAMPLE 1   Using the Product Rule

Multiply each expression using the product rule:

a. $2^2 \cdot 2^3$ **b.** $x^7 \cdot x^9$ **c.** $y \cdot y^5$ **d.** $y^3 \cdot y^2 \cdot y^5$.

Great Question!

Can I use the product rule to multiply an expression such as $x^7 \cdot y^9$?

The product rule does not apply to exponential expressions with different bases:

$x^7 \cdot y^9$, or $x^7 y^9$, cannot be simplified.

Solution

a. $2^2 \cdot 2^3 = 2^{2+3} = 2^5$ or 32

b. $x^7 \cdot x^9 = x^{7+9} = x^{16}$

c. $y \cdot y^5 = y^1 \cdot y^5 = y^{1+5} = y^6$

d. $y^3 \cdot y^2 \cdot y^5 = y^{3+2+5} = y^{10}$ ∎

✓ **CHECK POINT 1**   Multiply each expression using the product rule:

a. $2^2 \cdot 2^4$ **b.** $x^6 \cdot x^4$ **c.** $y \cdot y^7$ **d.** $y^4 \cdot y^3 \cdot y^2$.

2 Use the power rule for exponents.

The Power Rule for Exponents

The next property of exponents applies when an exponential expression is raised to a power. Here is an example:

$$(3^2)^4.$$

> The exponential expression 3^2 is raised to the fourth power.

There are 4 factors of 3^2. Thus,

$$(3^2)^4 = 3^2 \cdot 3^2 \cdot 3^2 \cdot 3^2 = 3^{2+2+2+2} = 3^8.$$

> Add exponents when multiplying with the same base.

We can obtain the answer, 3^8, by multiplying the exponents:

$$(3^2)^4 = 3^{2 \cdot 4} = 3^8.$$

By generalizing $(3^2)^4 = 3^{2 \cdot 4} = 3^8$, we obtain the following rule:

The Power Rule (Powers to Powers)

$$(b^m)^n = b^{mn}$$

When an exponential expression is raised to a power, multiply the exponents. Place the product of the exponents on the base and remove the parentheses.

EXAMPLE 2 Using the Power Rule

Simplify each expression using the power rule:

a. $(2^3)^5$ **b.** $(x^6)^4$ **c.** $[(-3)^7]^5$.

Solution

a. $(2^3)^5 = 2^{3 \cdot 5} = 2^{15}$

b. $(x^6)^4 = x^{6 \cdot 4} = x^{24}$

c. $[(-3)^7]^5 = (-3)^{7 \cdot 5} = (-3)^{35}$ ■

✓ **CHECK POINT 2** Simplify each expression using the power rule:

a. $(3^4)^5$ **b.** $(x^9)^{10}$ **c.** $[(-5)^7]^3$.

Great Question!

Can you show me examples that illustrate the difference between the product rule and the power rule?

Do not confuse the product and power rules. Note the following differences:

- $x^4 \cdot x^7 = x^{4+7} = x^{11}$
- $(x^4)^7 = x^{4 \cdot 7} = x^{28}$.

3 Use the products-to-powers rule.

The Products-to-Powers Rule for Exponents

The next property of exponents applies when we are raising a product to a power. Here is an example:

$$(2x)^4.$$

> The product $2x$ is raised to the fourth power.

There are four factors of $2x$. Thus,

$$(2x)^4 = 2x \cdot 2x \cdot 2x \cdot 2x = 2 \cdot 2 \cdot 2 \cdot 2 \cdot x \cdot x \cdot x \cdot x = 2^4 x^4.$$

We can obtain the answer, $2^4 x^4$, by raising each factor within the parentheses to the fourth power:

$$(2x)^4 = 2^4 x^4.$$

Generalizing from $(2x)^4 = 2^4 x^4$ suggests the following rule:

Products to Powers

$$(ab)^n = a^n b^n$$

When a product is raised to a power, raise each factor to the power.

EXAMPLE 3 Using the Products-to-Powers Rule

Simplify each expression using the products-to-powers rule:

a. $(5x)^3$ **b.** $(-2y^4)^5$.

Solution

a. $(5x)^3 = 5^3 x^3$ Raise each factor to the third power.

        $= 125x^3$ $5^3 = 5 \cdot 5 \cdot 5 = 125$

b. $(-2y^4)^5 = (-2)^5 (y^4)^5$ Raise each factor to the fifth power.

            $= (-2)^5 y^{4 \cdot 5}$ To raise an exponential expression to a power, multiply exponents: $(b^m)^n = b^{mn}$.

            $= -32y^{20}$ $(-2)^5 = (-2)(-2)(-2)(-2)(-2) = -32$ ∎

✓ **CHECK POINT 3** Simplify each expression using the products-to-powers rule:

a. $(2x)^4$ **b.** $(-4y^2)^3$.

Great Question!

What are some common errors to avoid when simplifying exponential expressions?

Here's a partial list. The first column shows the correct simplification. The second column illustrates a common error.

Correct	Incorrect	Description of Error
$b^3 \cdot b^4 = b^{3+4} = b^7$	$b^3 \cdot b^4 = b^{12}$	Exponents should be added, not multiplied.
$3^2 \cdot 3^4 = 3^{2+4} = 3^6$	$3^2 \cdot 3^4 = 9^{2+4} = 9^6$	The common base should be retained, not multiplied.
$(x^5)^3 = x^{5 \cdot 3} = x^{15}$	$(x^5)^3 = x^{5+3} = x^8$	Exponents should be multiplied, not added, when raising a power to a power.
$(4x)^3 = 4^3 x^3 = 64x^3$	$(4x)^3 = 4x^3$	Both factors should be cubed.

---

4 Multiply monomials.

Multiplying Monomials

Now that we have developed three properties of exponents, we are ready to turn to polynomial multiplication. We begin with the product of two monomials, such as $-8x^6$ and $5x^3$. This product is obtained by multiplying the coefficients, -8 and 5, and then multiplying the variables using the product rule for exponents.

$$(-8x^6)(5x^3) = -8 \cdot 5 \cdot x^6 \cdot x^3 = -8 \cdot 5x^{6+3} = -40x^9$$

Multiply coefficients and add exponents.

Multiplying Monomials

To multiply monomials with the same variable base, multiply the coefficients and then multiply the variables. Use the product rule for exponents to multiply the variables: Keep the variable and add the exponents.

EXAMPLE 4 Multiplying Monomials

Multiply: **a.** $(2x)(4x^2)$ **b.** $(-10x^6)(6x^{10})$.

Solution

a. $(2x)(4x^2) = (2 \cdot 4)(x \cdot x^2)$ Multiply the coefficients and multiply the variables.

$= 8x^{1+2}$ Add exponents: $b^m \cdot b^n = b^{m+n}$.

$= 8x^3$ Simplify.

b. $(-10x^6)(6x^{10}) = (-10 \cdot 6)(x^6 \cdot x^{10})$ Multiply the coefficients and multiply the variables.

$= -60x^{6+10}$ Add exponents: $b^m \cdot b^n = b^{m+n}$.

$= -60x^{16}$ Simplify. ■

✓ **CHECK POINT 4**   Multiply: **a.** $(7x^2)(10x)$ **b.** $(-5x^4)(4x^5)$.

Multiplying a Monomial and a Polynomial That Is Not a Monomial

We use the distributive property to multiply a monomial and a polynomial that is not a monomial. For example,

$$3x^2(2x^3 + 5x) = 3x^2 \cdot 2x^3 + 3x^2 \cdot 5x = 3 \cdot 2x^{2+3} + 3 \cdot 5x^{2+1} = 6x^5 + 15x^3.$$

Monomial   Binomial   Multiply coefficients and add exponents.

Multiplying a Monomial and a Polynomial That Is Not a Monomial

To multiply a monomial and a polynomial, use the distributive property to multiply each term of the polynomial by the monomial.

EXAMPLE 5 Multiplying a Monomial and a Polynomial

Multiply: **a.** $2x(x + 4)$ **b.** $3x^2(4x^3 - 5x + 2)$.

Solution

a. $2x(x + 4) = 2x \cdot x + 2x \cdot 4$ Use the distributive property.

$= 2 \cdot 1x^{1+1} + 2 \cdot 4x$ To multiply the monomials, multiply coefficients and add exponents.

$= 2x^2 + 8x$ Simplify.

b. $3x^2(4x^3 - 5x + 2)$

$= 3x^2 \cdot 4x^3 - 3x^2 \cdot 5x + 3x^2 \cdot 2$ Use the distributive property.

$= 3 \cdot 4x^{2+3} - 3 \cdot 5x^{2+1} + 3 \cdot 2x^2$ To multiply the monomials, multiply coefficients and add exponents.

$= 12x^5 - 15x^3 + 6x^2$ Simplify. ■

Great Question!

Because monomials with the same base and different exponents can be multiplied, can they also be added?

No. Don't confuse adding and multiplying monomials.

Addition:

$$5x^4 + 6x^4 = 11x^4$$

Multiplication:

$$(5x^4)(6x^4) = (5 \cdot 6)(x^4 \cdot x^4)$$
$$= 30x^{4+4}$$
$$= 30x^8$$

Only like terms can be added or subtracted, but unlike terms may be multiplied.

Addition:

$5x^4 + 3x^2$ cannot be simplified.

Multiplication:

$$(5x^4)(3x^2) = (5 \cdot 3)(x^4 \cdot x^2)$$
$$= 15x^{4+2}$$
$$= 15x^6$$

5 Multiply a monomial and a polynomial.

Rectangles often make it possible to visualize polynomial multiplication. For example, **Figure 5.5** shows a rectangle with length $2x$ and width $x + 4$. The area of the large rectangle is

$$2x(x + 4).$$

The sum of the areas of the two smaller rectangles is

$$2x^2 + 8x.$$

Conclusion:

$$2x(x + 4) = 2x^2 + 8x$$

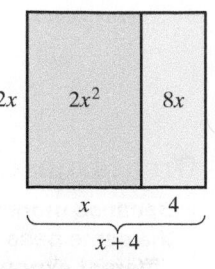

Figure 5.5

✓ **CHECK POINT 5**   Multiply:
   a. $3x(x + 5)$ **b.** $6x^2(5x^3 - 2x + 3)$.

---

6 Multiply polynomials when neither is a monomial.

Multiplying Polynomials when Neither Is a Monomial

How do we multiply two polynomials if neither is a monomial? For example, consider

$$(2x + 3)(x^2 + 4x + 5).$$

Binomial   Trinomial

One way to perform this multiplication is to distribute $2x$ throughout the trinomial

$$2x(x^2 + 4x + 5)$$

and 3 throughout the trinomial

$$3(x^2 + 4x + 5).$$

Then combine the like terms that result. In general, the product of two polynomials is the polynomial obtained by multiplying each term of one polynomial by each term of the other polynomial and then combining like terms.

Using Technology

Graphic Connections

The graphs of

$$y_1 = (x + 3)(x + 2)$$

and $y_2 = x^2 + 5x + 6$

are the same.

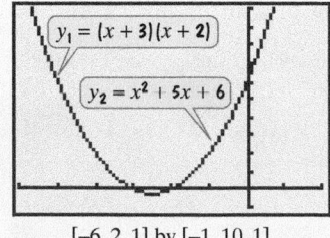

$y_1 = (x + 3)(x + 2)$

$y_2 = x^2 + 5x + 6$

[−6, 2, 1] by [−1, 10, 1]

This verifies that

$(x + 3)(x + 2) = x^2 + 5x + 6.$

Multiplying Polynomials when Neither Is a Monomial

Multiply each term of one polynomial by each term of the other polynomial. Then combine like terms.

EXAMPLE 6   Multiplying Binomials

Multiply: **a.** $(x + 3)(x + 2)$ **b.** $(3x + 7)(2x - 4)$.

Solution   We begin by multiplying each term of the second binomial by each term of the first binomial.

a. $(x + 3)(x + 2)$

$= x(x + 2) + 3(x + 2)$ Multiply the second binomial by each term of the first binomial.

$= x \cdot x + x \cdot 2 + 3 \cdot x + 3 \cdot 2$ Use the distributive property.

$= x^2 + 2x + 3x + 6$ Multiply. Note that $x \cdot x = x^1 \cdot x^1 = x^{1+1} = x^2$.

$= x^2 + 5x + 6$ Combine like terms.

b. $(3x + 7)(2x - 4)$

$= 3x(2x - 4) + 7(2x - 4)$ Multiply the second binomial by each term of the first binomial.

$= 3x \cdot 2x - 3x \cdot 4 + 7 \cdot 2x - 7 \cdot 4$ Use the distributive property.

$= 6x^2 - 12x + 14x - 28$ Multiply.

$= 6x^2 + 2x - 28$ Combine like terms. ∎

✓ **CHECK POINT 6**   Multiply:

   a. $(x + 4)(x + 5)$ **b.** $(5x + 3)(2x - 7)$.

You can visualize the polynomial multiplication in Example 6(a), $(x + 3)(x + 2) = x^2 + 5x + 6$, by analyzing the areas in **Figure 5.6**.

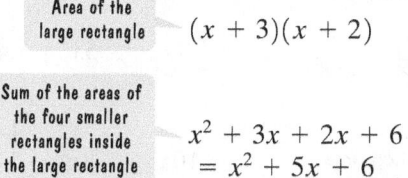

$(x + 3)(x + 2)$

Area of the large rectangle

Sum of the areas of the four smaller rectangles inside the large rectangle

$x^2 + 3x + 2x + 6$
$= x^2 + 5x + 6$

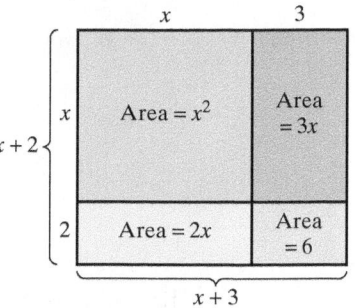

Figure 5.6

Conclusion:

$$(x + 3)(x + 2) = x^2 + 5x + 6$$

EXAMPLE 7   Multiplying a Binomial and a Trinomial

Multiply: $(2x + 3)(x^2 + 4x + 5)$.

Solution

$(2x + 3)(x^2 + 4x + 5)$

$= 2x(x^2 + 4x + 5) + 3(x^2 + 4x + 5)$ Multiply the trinomial by each term of the binomial.

$= 2x \cdot x^2 + 2x \cdot 4x + 2x \cdot 5 + 3x^2 + 3 \cdot 4x + 3 \cdot 5$ Use the distributive property.

$= 2x^3 + 8x^2 + 10x + 3x^2 + 12x + 15$ Multiply monomials: Multiply coefficients and add exponents.

$= 2x^3 + 11x^2 + 22x + 15$ Combine like terms: $8x^2 + 3x^2 = 11x^2$ and $10x + 12x = 22x$. ∎

✓ **CHECK POINT 7**   Multiply: $(5x + 2)(x^2 - 4x + 3)$.

Another method for solving Example 7 is to use a vertical format similar to that used for multiplying whole numbers.

$$
\begin{array}{r}
x^2 + 4x + 5 \\
2x + 3 \\
\hline
3x^2 + 12x + 15 \\
2x^3 + 8x^2 + 10x \\
\hline
2x^3 + 11x^2 + 22x + 15
\end{array}
$$

Write like terms in the same column.

$3(x^2 + 4x + 5)$

$2x(x^2 + 4x + 5)$

Combine like terms.

EXAMPLE 8   Multiplying Polynomials Using a Vertical Format

Multiply: $(2x^2 - 3x)(5x^3 - 4x^2 + 7x)$.

Solution To use the vertical format to find $(2x^2 - 3x)(5x^3 - 4x^2 + 7x)$, it is most convenient to write the polynomial with the greater number of terms in the top row.

$$5x^3 - 4x^2 + 7x$$
$$\underline{2x^2 - 3x}$$

We now multiply each term in the top polynomial by the last term in the bottom polynomial.

$$5x^3 - 4x^2 + 7x$$
$$\underline{\phantom{5x^3 - {}} 2x^2 - 3x}$$
$$-15x^4 + 12x^3 - 21x^2 \quad \boxed{-3x(5x^3 - 4x^2 + 7x)}$$

Then we multiply each term in the top polynomial by $2x^2$, the first term in the bottom polynomial. Like terms are placed in columns because the final step involves combining them.

Write like terms in the same column.

$$5x^3 - 4x^2 + 7x$$
$$\underline{\phantom{5x^3 - {}} 2x^2 - 3x}$$
$$-15x^4 + 12x^3 - 21x^2 \quad \boxed{-3x(5x^3 - 4x^2 + 7x)}$$
$$\underline{10x^5 - 8x^4 + 14x^3 \quad \boxed{2x^2(5x^3 - 4x^2 + 7x)}}$$
$$10x^5 - 23x^4 + 26x^3 - 21x^2 \quad \boxed{\text{Combine like terms, which are lined up in columns.}}$$

✓ **CHECK POINT 8** Multiply using a vertical format: $(3x^2 - 2x)(2x^3 - 5x^2 + 4x)$.

CONCEPT AND VOCABULARY CHECK

Fill in each blank so that the resulting statement is true.

1. The product rule for exponents states that $b^m \cdot b^n =$ _____. When multiplying exponential expressions with the same base, _____ the exponents.

2. The power rule for exponents states that $(b^m)^n =$ _____. When an exponential expression is raised to a power, _____ the exponents.

3. The products-to-powers rule for exponents states that $(ab)^n =$ _____. When a product is raised to a power, raise each _____ to the power.

4. To multiply $2x^2(x^2 + 5x + 7)$, use the _____ property to multiply each term of the polynomial _____ by the monomial _____.

5. To multiply $(4x + 7)(x^2 + 8x + 3)$, begin by multiplying each term of $x^2 + 8x + 3$ by _____. Then multiply each term of $x^2 + 8x + 3$ by _____. Then combine _____ terms.

5.2 EXERCISE SET

MyMathLab®

Watch the videos in MyMathLab

Download the MyDashBoard App

Practice Exercises

In Exercises 1–8, multiply each expression using the product rule.

1. $x^{15} \cdot x^3$
2. $x^{12} \cdot x^4$
3. $y \cdot y^{11}$
4. $y \cdot y^{19}$
5. $x^2 \cdot x^6 \cdot x^3$
6. $x^4 \cdot x^3 \cdot x^5$
7. $7^9 \cdot 7^{10}$
8. $8^7 \cdot 8^{10}$

In Exercises 9–14, simplify each expression using the power rule.

9. $(6^9)^{10}$
10. $(6^7)^{10}$
11. $(x^{15})^3$
12. $(x^{12})^4$
13. $[(-20)^3]^3$
14. $[(-50)^4]^4$

In Exercises 15–24, simplify each expression using the products-to-powers rule.

15. $(2x)^3$

16. $(4x)^3$

17. $(-5x)^2$

18. $(-6x)^2$

19. $(4x^3)^2$

20. $(6x^3)^2$

21. $(-2y^6)^4$

22. $(-2y^5)^4$

23. $(-2x^7)^5$

24. $(-2x^{11})^5$

In Exercises 25–34, multiply the monomials.

25. $(7x)(2x)$

26. $(8x)(3x)$

27. $(6x)(4x^2)$

28. $(10x)(3x^2)$

29. $(-5y^4)(3y^3)$

30. $(-6y^4)(2y^3)$

31. $\left(-\dfrac{1}{2}a^3\right)\left(-\dfrac{1}{4}a^2\right)$

32. $\left(-\dfrac{1}{3}a^4\right)\left(-\dfrac{1}{2}a^2\right)$

33. $(2x^2)(-3x)(8x^4)$

34. $(3x^3)(-2x)(5x^6)$

In Exercises 35–54, find each product of the monomial and the polynomial.

35. $4x(x + 3)$

36. $6x(x + 5)$

37. $x(x - 3)$

38. $x(x - 7)$

39. $2x(x - 6)$

40. $3x(x - 5)$

41. $-4y(3y + 5)$

42. $-5y(6y + 7)$

43. $4x^2(x + 2)$

44. $5x^2(x + 6)$

45. $2y^2(y^2 + 3y)$

46. $4y^2(y^2 + 2y)$

47. $2y^2(3y^2 - 4y + 7)$

48. $4y^2(5y^2 - 6y + 3)$

49. $(3x^3 + 4x^2)(2x)$

50. $(4x^3 + 5x^2)(2x)$

51. $(x^2 + 5x - 3)(-2x)$

52. $(x^3 - 2x + 2)(-4x)$

53. $-3x^2(-4x^2 + x - 5)$

54. $-6x^2(3x^2 - 2x - 7)$

In Exercises 55–78, find each product. In each case, neither factor is a monomial.

55. $(x + 3)(x + 5)$

56. $(x + 4)(x + 6)$

57. $(2x + 1)(x + 4)$

58. $(2x + 5)(x + 3)$

59. $(x + 3)(x - 5)$

60. $(x + 4)(x - 6)$

61. $(x - 11)(x + 9)$

62. $(x - 12)(x + 8)$

63. $(2x - 5)(x + 4)$

64. $(3x - 4)(x + 5)$

65. $\left(\dfrac{1}{4}x + 4\right)\left(\dfrac{3}{4}x - 1\right)$

66. $\left(\dfrac{1}{5}x + 5\right)\left(\dfrac{3}{5}x - 1\right)$

67. $(x + 1)(x^2 + 2x + 3)$

68. $(x + 2)(x^2 + x + 5)$

69. $(y - 3)(y^2 - 3y + 4)$

70. $(y - 2)(y^2 - 4y + 3)$

71. $(2a - 3)(a^2 - 3a + 5)$

72. $(2a - 1)(a^2 - 4a + 3)$

73. $(x + 1)(x^3 + 2x^2 + 3x + 4)$

74. $(x + 1)(x^3 + 4x^2 + 7x + 3)$

75. $\left(x - \dfrac{1}{2}\right)(4x^3 - 2x^2 + 5x - 6)$

76. $\left(x - \dfrac{1}{3}\right)(3x^3 - 6x^2 + 5x - 9)$

77. $(x^2 + 2x + 1)(x^2 - x + 2)$

78. $(x^2 + 3x + 1)(x^2 - 2x - 1)$

In Exercises 79–92, use a vertical format to find each product.

79. $\begin{array}{r} x^2 - 5x + 3 \\ x + 8 \\ \hline \end{array}$

80. $\begin{array}{r} x^2 - 7x + 9 \\ x + 4 \\ \hline \end{array}$

81. $\begin{array}{r} x^2 - 3x + 9 \\ 2x - 3 \\ \hline \end{array}$

82. $\begin{array}{r} y^2 - 5y + 3 \\ 4y - 5 \\ \hline \end{array}$

83. $\begin{array}{r} 2x^3 + x^2 + 2x + 3 \\ x + 4 \\ \hline \end{array}$

84. $\begin{array}{r} 3y^3 + 2y^2 + y + 4 \\ y + 3 \\ \hline \end{array}$

85. $\begin{array}{r} 4z^3 - 2z^2 + 5z - 4 \\ 3z - 2 \\ \hline \end{array}$

86. $\begin{array}{r} 5z^3 - 3z^2 + 4z - 3 \\ 2z - 4 \\ \hline \end{array}$

87. $\begin{array}{r} 7x^3 - 5x^2 + 6x \\ 3x^2 - 4x \\ \hline \end{array}$

88. $\begin{array}{r} 9y^3 - 7y^2 + 5y \\ 3y^2 + 5y \\ \hline \end{array}$

89. $\begin{array}{r} 2y^5 - 3y^3 + y^2 - 2y + 3 \\ 2y - 1 \\ \hline \end{array}$

90. $\begin{array}{r} n^4 - n^3 + n^2 - n + 1 \\ 2n + 3 \\ \hline \end{array}$

91. $\begin{array}{r} x^2 + 7x - 3 \\ x^2 - x - 1 \\ \hline \end{array}$

92. $\begin{array}{r} x^2 + 6x - 4 \\ x^2 - x - 2 \\ \hline \end{array}$

Practice PLUS

In Exercises 93–100, perform the indicated operations.

93. $(x + 4)(x - 5) - (x + 3)(x - 6)$

94. $(x + 5)(x - 6) - (x + 2)(x - 9)$

95. $4x^2(5x^3 + 3x - 2) - 5x^3(x^2 - 6)$

96. $3x^2(6x^3 + 2x - 3) - 4x^3(x^2 - 5)$

97. $(y + 1)(y^2 - y + 1) + (y - 1)(y^2 + y + 1)$

98. $(y + 1)(y^2 - y + 1) - (y - 1)(y^2 + y + 1)$

99. $(y + 6)^2 - (y - 2)^2$

100. $(y + 5)^2 - (y - 4)^2$

Application Exercises

101. Find a trinomial for the area of the rectangular rug shown below whose sides are $x + 5$ feet and $2x - 3$ feet.

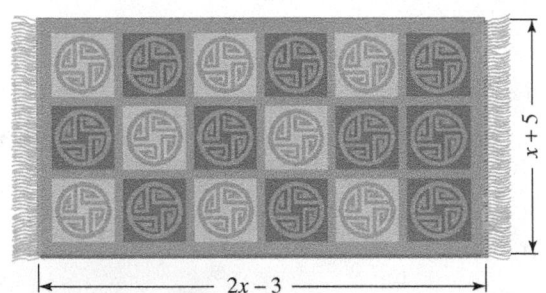

102. The base of a triangular sail is $4x$ feet and its height is $3x + 10$ feet. Write a binomial in terms of x for the area of the sail.

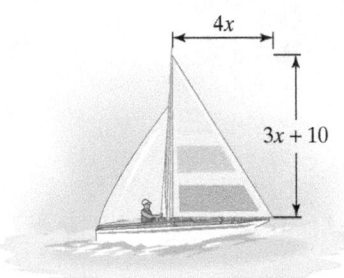

In Exercises 103–104,

a. *Express the area of the large rectangle as the product of two binomials.*

b. *Find the sum of the areas of the four smaller rectangles.*

c. *Use polynomial multiplication to show that your expressions for area in parts (a) and (b) are equal.*

103.

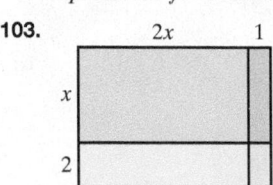

104.

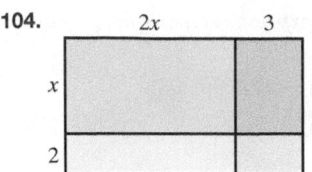

Writing in Mathematics

105. Explain the product rule for exponents. Use $2^3 \cdot 2^5$ in your explanation.

106. Explain the power rule for exponents. Use $(3^2)^4$ in your explanation.

107. Explain how to simplify an expression that involves a product raised to a power. Provide an example with your explanation.

108. Explain how to multiply monomials. Give an example.

109. Explain how to multiply a monomial and a polynomial that is not a monomial. Give an example.

110. Explain how to multiply polynomials when neither is a monomial. Give an example.

111. Explain the difference between performing these two operations:

$$2x^2 + 3x^2 \quad \text{and} \quad (2x^2)(3x^2).$$

112. Discuss situations in which a vertical format, rather than a horizontal format, is useful for multiplying polynomials.

Critical Thinking Exercises

Make Sense? *In Exercises 113–116, determine whether each statement "makes sense" or "does not make sense" and explain your reasoning.*

113. I'm working with two monomials that I cannot add, although I can multiply them.

114. I'm working with two monomials that I can add, although I cannot multiply them.

115. Other than multiplying monomials, the distributive property is used to multiply other kinds of polynomials.

116. I used the product rule for exponents to multiply x^7 and y^9.

In Exercises 117–120, determine whether each statement is true or false. If the statement is false, make the necessary change(s) to produce a true statement.

117. $4x^3 \cdot 3x^4 = 12x^{12}$

118. $5x^2 \cdot 4x^6 = 9x^8$

119. $(y - 1)(y^2 + y + 1) = y^3 - 1$

120. Some polynomial multiplications can only be performed by using a vertical format.

121. Find a polynomial in descending powers of x representing the area of the shaded region.

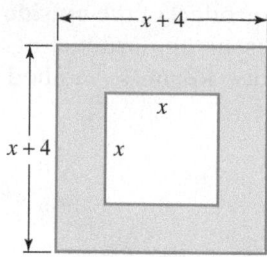

122. Find each of the products in parts (a)–(c).

a. $(x - 1)(x + 1)$

b. $(x - 1)(x^2 + x + 1)$

c. $(x - 1)(x^3 + x^2 + x + 1)$

d. Using the pattern found in parts (a)–(c), find $(x - 1)(x^4 + x^3 + x^2 + x + 1)$ without actually multiplying.

123. Find the missing factor.

$$(\underline{})\left(-\frac{1}{4}xy^3\right) = 2x^5y^3$$

Review Exercises

124. Solve: $4x - 7 > 9x - 2$. (Section 2.7, Example 7)

125. Graph $3x - 2y = 6$ using intercepts. (Section 3.2, Example 4)

126. Find the slope of the line passing through the points $(-2, 8)$ and $(1, 6)$. (Section 3.3, Example 1)

Preview Exercises

Exercises 127–129 will help you prepare for the material covered in the next section. In each exercise, find the indicated products. Then, if possible, state a fast method for finding these products. (You may already be familiar with some of these methods from a high school algebra course.)

127. a. $(x + 3)(x + 4)$

b. $(x + 5)(x + 20)$

128. a. $(x + 3)(x - 3)$

b. $(x + 5)(x - 5)$

129. a. $(x + 3)^2$

b. $(x + 5)^2$

5.3

Special Products

Objectives

1. Use FOIL in polynomial multiplication.

2. Multiply the sum and difference of two terms.

3. Find the square of a binomial sum.

4. Find the square of a binomial difference.

Let's cut to the chase. Are there fast methods for finding products of polynomials? The answer is beepingly "yes." (Or should that be (BEEP)^2 yes?) In this section, we'll cut to the chase by using the distributive property to develop patterns that will let you multiply certain binomials quite rapidly.

The Product of Two Binomials: FOIL

1. Use FOIL in polynomial multiplication.

Frequently, we need to find the product of two binomials. One way to perform this multiplication is to distribute each term in the first binomial through the second binomial. For example, we can find the product of the binomials $3x + 2$ and $4x + 5$ as follows:

$$(3x + 2)(4x + 5) = 3x(4x + 5) + 2(4x + 5)$$

Distribute $3x$ over $4x + 5$.   *Distribute 2 over $4x + 5$.*

$$= 3x(4x) + 3x(5) + 2(4x) + 2(5)$$

$$= 12x^2 + 15x + 8x + 10.$$

We'll combine these like terms later. For now, our interest is in how to obtain each of these four terms.

We can also find the product of $3x + 2$ and $4x + 5$ using a method called FOIL, which is based on our work shown on the previous page. Any two binomials can be quickly multiplied by using the FOIL method, in which **F** represents the product of the **first** terms in each binomial, **O** represents the product of the **outside** terms, **I** represents the product of the **inside** terms, and **L** represents the product of the **last**, or second, terms in each binomial. For example, we can use the FOIL method to find the product of the binomials $3x + 2$ and $4x + 5$ as follows:

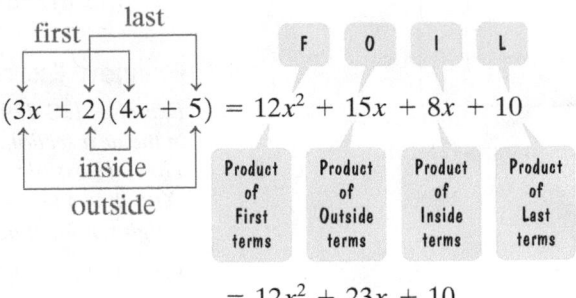

$$= 12x^2 + 23x + 10 \qquad \text{Combine like terms.}$$

In general, here's how to use the FOIL method to find the product of $ax + b$ and $cx + d$:

Using the FOIL Method to Multiply Binomials

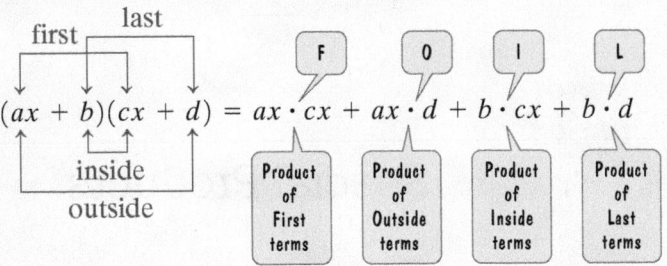

EXAMPLE 1 Using the FOIL Method

Multiply: $(x + 3)(x + 4)$.

Solution

F: First terms $= x \cdot x = x^2$ $(x + 3)(x + 4)$

O: Outside terms $= x \cdot 4 = 4x$ $(x + 3)(x + 4)$

I: Inside terms $= 3 \cdot x = 3x$ $(x + 3)(x + 4)$

L: Last terms $= 3 \cdot 4 = 12$ $(x + 3)(x + 4)$

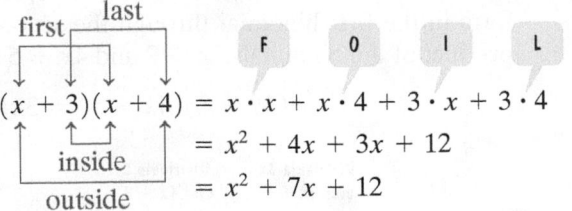

$$\begin{aligned}(x + 3)(x + 4) &= x \cdot x + x \cdot 4 + 3 \cdot x + 3 \cdot 4 \\ &= x^2 + 4x + 3x + 12 \\ &= x^2 + 7x + 12 \qquad \text{Combine like terms.} \ \blacksquare\end{aligned}$$

✓ **CHECK POINT 1** Multiply: $(x + 5)(x + 6)$.

EXAMPLE 2 Using the FOIL Method

Multiply: $(3x + 4)(5x - 3)$.

Solution

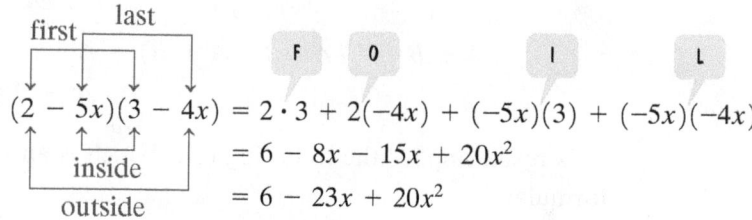

$$(3x + 4)(5x - 3) = 3x \cdot 5x + 3x(-3) + 4 \cdot 5x + 4(-3)$$
$$= 15x^2 - 9x + 20x - 12$$
$$= 15x^2 + 11x - 12 \qquad \text{Combine like terms.} \blacksquare$$

✓ **CHECK POINT 2** Multiply: $(7x + 5)(4x - 3)$.

EXAMPLE 3 Using the FOIL Method

Multiply: $(2 - 5x)(3 - 4x)$.

Solution

$$(2 - 5x)(3 - 4x) = 2 \cdot 3 + 2(-4x) + (-5x)(3) + (-5x)(-4x)$$
$$= 6 - 8x - 15x + 20x^2$$
$$= 6 - 23x + 20x^2 \qquad \text{Combine like terms.}$$

The product can also be expressed in standard form as $20x^2 - 23x + 6$. $\blacksquare$

✓ **CHECK POINT 3** Multiply: $(4 - 2x)(5 - 3x)$.

2 Multiply the sum and difference of two terms.

Multiplying the Sum and Difference of Two Terms

We can use the FOIL method to multiply $A + B$ and $A - B$ as follows:

$$(A + B)(A - B) = A^2 - AB + AB - B^2 = A^2 - B^2.$$

Notice that the outside and inside products have a sum of 0 and the terms cancel. The FOIL multiplication provides us with a quick rule for multiplying the sum and difference of two terms, referred to as a *special-product formula*.

The Product of the Sum and Difference of Two Terms

$$(A + B)(A - B) = A^2 - B^2$$

| The product of the sum and the difference of the same two terms | is | the square of the first term minus the square of the second term. |

EXAMPLE 4 Finding the Product of the Sum and Difference of Two Terms

Multiply: **a.** $(4y + 3)(4y - 3)$ **b.** $(3x - 7)(3x + 7)$ **c.** $(5a^4 + 6)(5a^4 - 6)$.

Solution Use the special-product formula shown.

$$(A + B)(A - B) = A^2 - B^2$$

First term squared − Second term squared = Product

a. $(4y + 3)(4y - 3) = (4y)^2 - 3^2 = 16y^2 - 9$

b. $(3x - 7)(3x + 7) = (3x)^2 - 7^2 = 9x^2 - 49$

c. $(5a^4 + 6)(5a^4 - 6) = (5a^4)^2 - 6^2 = 25a^8 - 36$ ∎

✓ **CHECK POINT 4** Multiply: **a.** $(7y + 8)(7y - 8)$
b. $(4x - 5)(4x + 5)$ **c.** $(2a^3 + 3)(2a^3 - 3)$.

---

3 Find the square of a binomial sum.

The Square of a Binomial

Let's now find $(A + B)^2$, the square of a binomial sum. To do so, we begin with the FOIL method and look for a general rule.

$$
\overset{\text{F}\quad\text{O}\quad\text{I}\quad\text{L}}{(A + B)^2 = (A + B)(A + B) = A \cdot A + A \cdot B + A \cdot B + B \cdot B}
$$
$$= A^2 + 2AB + B^2$$

This result implies the following rule, which is another example of a special-product formula:

The Square of a Binomial Sum

$$(A + B)^2 = A^2 + 2AB + B^2$$

The square of a binomial sum | is | first term squared | plus | 2 times the product of the terms | plus | last term squared.

Great Question!

When finding $(x + 3)^2$, why can't I just write $x^2 + 3^2$, or $x^2 + 9$?

Caution! The square of a sum is *not* the sum of the squares.

$(A + B)^2 \neq A^2 + B^2$

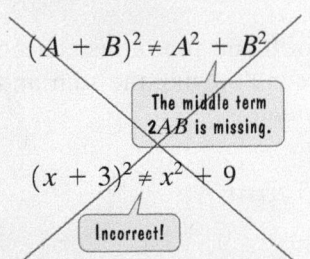

The middle term 2AB is missing.

$(x + 3)^2 \neq x^2 + 9$

Incorrect!

Show that $(x + 3)^2$ and $x^2 + 9$ are not equal by substituting 5 for x in each expression and simplifying.

EXAMPLE 5 Finding the Square of a Binomial Sum

Multiply:

a. $(x + 3)^2$ **b.** $(3x + 7)^2$.

Solution Use the special-product formula shown.

$$(A + B)^2 = A^2 + 2AB + B^2$$

	(First Term)^2	+	2 · Product of the Terms	+	(Last Term)^2	= Product
a. $(x + 3)^2 =$	x^2	+	$2 \cdot x \cdot 3$	+	3^2	$= x^2 + 6x + 9$
b. $(3x + 7)^2 =$	$(3x)^2$	+	$2(3x)(7)$	+	7^2	$= 9x^2 + 42x + 49$

∎

✓ **CHECK POINT 5** Multiply:
a. $(x + 10)^2$ **b.** $(5x + 4)^2$.

The formula for the square of a binomial sum can be interpreted geometrically by analyzing the areas in **Figure 5.7**.

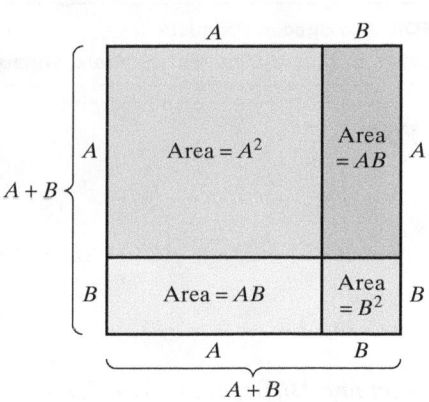

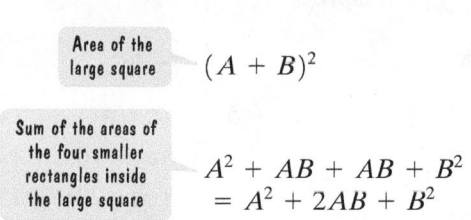

Area of the large square $(A + B)^2$

Sum of the areas of the four smaller rectangles inside the large square
$$A^2 + AB + AB + B^2$$
$$= A^2 + 2AB + B^2$$

Figure 5.7

Conclusion:

$$(A + B)^2 = A^2 + 2AB + B^2$$

4 Find the square of a binomial difference.

A similar pattern occurs for $(A - B)^2$, the square of a binomial difference. Using the FOIL method on $(A - B)^2$, we obtain the following rule:

The Square of a Binomial Difference

$$(A - B)^2 \quad = \quad A^2 \quad - \quad 2AB \quad + \quad B^2$$

The square of a binomial difference — is — first term squared — minus — 2 times the product of the terms — plus — last term squared.

EXAMPLE 6   Finding the Square of a Binomial Difference

Multiply:

a. $(x - 4)^2$ **b.** $(5y - 6)^2$.

Solution
Use the special-product formula shown.

$$(A - B)^2 = \quad A^2 \quad - \quad 2AB \quad + \quad B^2$$

	(First Term)^2	$-$	2 · Product of the Terms	$+$	(Last Term)^2	= Product
a. $(x - 4)^2 =$	x^2	$-$	$2 \cdot x \cdot 4$	$+$	4^2	$= x^2 - 8x + 16$
b. $(5y - 6)^2 =$	$(5y)^2$	$-$	$2(5y)(6)$	$+$	6^2	$= 25y^2 - 60y + 36$

☑ **CHECK POINT 6**   Multiply:

a. $(x - 9)^2$ **b.** $(7x - 3)^2$.

The table at the top of the next page summarizes the FOIL method and the three special products. The special products occur so frequently in algebra that it is convenient to memorize the form or pattern of these formulas.

FOIL and Special Products

Let A, B, C, and D be real numbers, variables, or algebraic expressions.

FOIL	*Example*
$(A + B)(C + D) = AC + AD + BC + BD$	$(2x + 3)(4x + 5) = (2x)(4x) + (2x)(5) + (3)(4x) + (3)(5)$ $= 8x^2 + 10x + 12x + 15$ $= 8x^2 + 22x + 15$
Sum and Difference of Two Terms $(A + B)(A - B) = A^2 - B^2$	*Example* $(2x + 3)(2x - 3) = (2x)^2 - 3^2$ $= 4x^2 - 9$
Square of a Binomial $(A + B)^2 = A^2 + 2AB + B^2$ $(A - B)^2 = A^2 - 2AB + B^2$	*Example* $(2x + 3)^2 = (2x)^2 + 2(2x)(3) + 3^2$ $= 4x^2 + 12x + 9$ $(2x - 3)^2 = (2x)^2 - 2(2x)(3) + 3^2$ $= 4x^2 - 12x + 9$

Achieving Success

Manage your time. Use a day planner such as the one shown below. (Go online and search "day planner" or "day scheduler" to find a schedule grid that you can print and use.)

Sample Day Planner

	Monday	Tuesday	Wednesday	Thursday	Friday	Saturday	Sunday
5:00 A.M.							
6:00 A.M.							
7:00 A.M.							
8:00 A.M.							
9:00 A.M.							
10:00 A.M.							
11:00 A.M.							
12:00 P.M.							
1:00 P.M.							
2:00 P.M.							
3:00 P.M.							
4:00 P.M.							
5:00 P.M.							
6:00 P.M.							
7:00 P.M.							
8:00 P.M.							
9:00 P.M.							
10:00 P.M.							
11:00 P.M.							
Midnight							

- On the Sunday before the week begins, fill in the time slots with fixed items such as school, extracurricular activities, work, etc.
- Because your education should be a top priority, decide what times you would like to study and do homework. Fill in these activities on the day planner.
- Plan other flexible activities such as exercise, socializing, etc. around the times already established.
- Be flexible. Things may come up that are unavoidable or some items may take longer than planned. However, be honest with yourself: Playing video games with friends is not "unavoidable."
- Stick to your schedule. At the end of the week, you will look back and be impressed at all the things you have accomplished.

CONCEPT AND VOCABULARY CHECK

Fill in each blank so that the resulting statement is true.

1. For $(x + 5)(2x + 3)$, the product of the first terms is _____, the product of the outside terms is _____, the product of the inside terms is _____, and the product of the last terms is _____.

2. $(A + B)(A - B) =$ _____. The product of the sum and the difference of the same two terms is the square of the first term _____ the square of the second term.

3. $(A + B)^2 =$ _____. The square of a binomial sum is the first term _____ plus 2 times the _____ plus the last term _____.

4. $(A - B)^2 =$ _____. The square of a binomial difference is the first term squared _____ 2 times the _____ plus the last term _____.

5. True or false: $(x + 5)(x - 5) = x^2 - 25.$ _____

6. True or false: $(x + 5)^2 = x^2 + 25.$ _____

5.3 EXERCISE SET

MyMathLab®

Watch the videos
in MyMathLab

Download the
MyDashBoard App

Practice Exercises

In Exercises 1–24, use the FOIL method to find each product. Express the product in descending powers of the variable.

1. $(x + 4)(x + 6)$
2. $(x + 8)(x + 2)$
3. $(y - 7)(y + 3)$
4. $(y - 3)(y + 4)$
5. $(2x - 3)(x + 5)$
6. $(3x - 5)(x + 7)$
7. $(4y + 3)(y - 1)$
8. $(5y + 4)(y - 2)$
9. $(2x - 3)(5x + 3)$
10. $(2x - 5)(7x + 2)$
11. $(3y - 7)(4y - 5)$
12. $(4y - 5)(7y - 4)$
13. $(7 + 3x)(1 - 5x)$
14. $(2 + 5x)(1 - 4x)$
15. $(5 - 3y)(6 - 2y)$
16. $(7 - 2y)(10 - 3y)$
17. $(5x^2 - 4)(3x^2 - 7)$
18. $(7x^2 - 2)(3x^2 - 5)$
19. $(6x - 5)(2 - x)$
20. $(4x - 3)(2 - x)$
21. $(x + 5)(x^2 + 3)$
22. $(x + 4)(x^2 + 5)$
23. $(8x^3 + 3)(x^2 + 5)$
24. $(7x^3 + 5)(x^2 + 2)$

In Exercises 25–44, multiply using the rule for finding the product of the sum and difference of two terms.

25. $(x + 3)(x - 3)$
26. $(y + 5)(y - 5)$
27. $(3x + 2)(3x - 2)$
28. $(2x + 5)(2x - 5)$
29. $(3r - 4)(3r + 4)$
30. $(5z - 2)(5z + 2)$
31. $(3 + r)(3 - r)$
32. $(4 + s)(4 - s)$
33. $(5 - 7x)(5 + 7x)$
34. $(4 - 3y)(4 + 3y)$
35. $\left(2x + \dfrac{1}{2}\right)\left(2x - \dfrac{1}{2}\right)$
36. $\left(3y + \dfrac{1}{3}\right)\left(3y - \dfrac{1}{3}\right)$
37. $(y^2 + 1)(y^2 - 1)$
38. $(y^2 + 2)(y^2 - 2)$
39. $(r^3 + 2)(r^3 - 2)$
40. $(m^3 + 4)(m^3 - 4)$
41. $(1 - y^4)(1 + y^4)$
42. $(2 - s^5)(2 + s^5)$
43. $(x^{10} + 5)(x^{10} - 5)$
44. $(x^{12} + 3)(x^{12} - 3)$

In Exercises 45–62, multiply using the rules for the square of a binomial.

45. $(x + 2)^2$
46. $(x + 5)^2$
47. $(2x + 5)^2$
48. $(5x + 2)^2$

49. $(x - 3)^2$

50. $(x - 6)^2$

51. $(3y - 4)^2$

52. $(4y - 3)^2$

53. $(4x^2 - 1)^2$

54. $(5x^2 - 3)^2$

55. $(7 - 2x)^2$

56. $(9 - 5x)^2$

57. $\left(2x + \dfrac{1}{2}\right)^2$

58. $\left(3x + \dfrac{1}{3}\right)^2$

59. $\left(4y - \dfrac{1}{4}\right)^2$

60. $\left(2y - \dfrac{1}{2}\right)^2$

61. $(x^8 + 3)^2$

62. $(x^8 + 5)^2$

In Exercises 63–82, multiply using the method of your choice.

63. $(x - 1)(x^2 + x + 1)$

64. $(x + 1)(x^2 - x + 1)$

65. $(x - 1)^2$

66. $(x + 1)^2$

67. $(3y + 7)(3y - 7)$

68. $(4y + 9)(4y - 9)$

69. $3x^2(4x^2 + x + 9)$

70. $5x^2(7x^2 + x + 6)$

71. $(7y + 3)(10y - 4)$

72. $(8y + 3)(10y - 5)$

73. $(x^2 + 1)^2$

74. $(x^2 + 2)^2$

75. $(x^2 + 1)(x^2 + 2)$

76. $(x^2 + 2)(x^2 + 3)$

77. $(x^2 + 4)(x^2 - 4)$

78. $(x^2 + 5)(x^2 - 5)$

79. $(2 - 3x^5)^2$

80. $(2 - 3x^6)^2$

81. $\left(\dfrac{1}{4}x^2 + 12\right)\left(\dfrac{3}{4}x^2 - 8\right)$

82. $\left(\dfrac{1}{4}x^2 + 16\right)\left(\dfrac{3}{4}x^2 - 4\right)$

In Exercises 83–88, find the area of each shaded region. Write the answer as a polynomial in descending powers of x.

83.

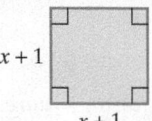

$x + 1$

$x + 1$

84.

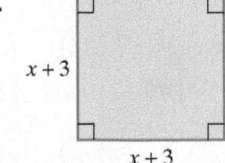

$x + 3$

$x + 3$

85.

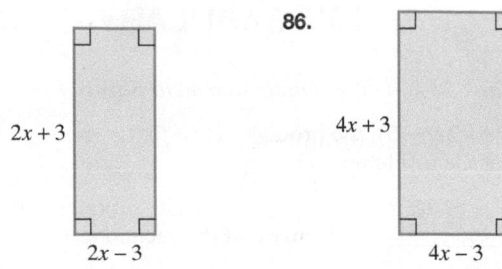

$2x + 3$

$2x - 3$

86.

$4x + 3$

$4x - 3$

87.

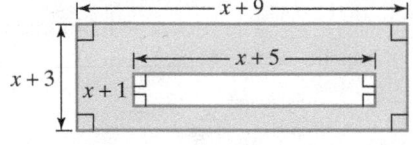

$x + 9$

$x + 3$

$x + 5$

$x + 1$

88.

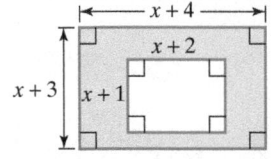

$x + 4$

$x + 2$

$x + 3$

$x + 1$

Practice PLUS

In Exercises 89–96, multiply by the method of your choice.

89. $\left[(2x + 3)(2x - 3)\right]^2$

90. $\left[(3x + 2)(3x - 2)\right]^2$

91. $(4x^2 + 1)\left[(2x + 1)(2x - 1)\right]$

92. $(9x^2 + 1)[(3x + 1)(3x - 1)]$

93. $(x + 2)^3$

94. $(x + 4)^3$

95. $\left[(x + 3) - y\right]\left[(x + 3) + y\right]$

96. $\left[(x + 5) - y\right]\left[(x + 5) + y\right]$

Application Exercises

The square garden shown in the figure measures x yards on each side. The garden is to be expanded so that one side is increased by 2 yards and an adjacent side is increased by 1 yard.

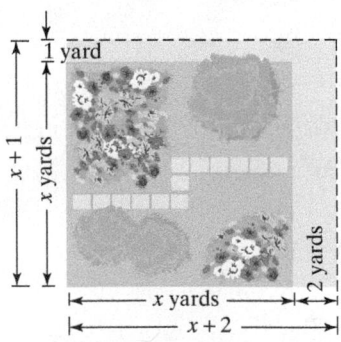

1 yard

$x + 1$

x yards

x yards

2 yards

$x + 2$

The graph shows the area of the expanded garden, y, in terms of the length of one of its original sides, x. Use this information to solve Exercises 97–100.

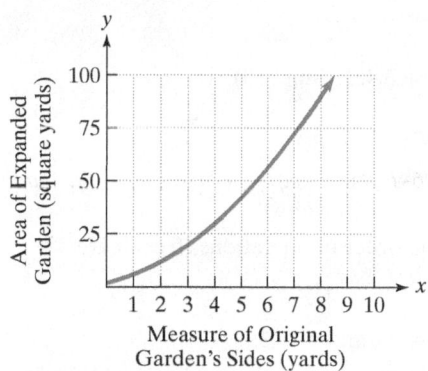

97. Use the gardens pictured at the bottom of the previous page to write a product of two binomials that expresses the area of the larger garden.
98. Use the gardens pictured at the bottom of the previous page to write a polynomial in descending powers of x that expresses the area of the larger garden.

99. If the original garden measures 6 yards on a side, use your expression from Exercise 97 to find the area of the larger garden. Then identify your solution as a point on the graph shown above.
100. If the original garden measures 8 yards on a side, use your polynomial from Exercise 98 to find the area of the larger garden. Then identify your solution as a point on the graph shown above.

The square painting in the figure measures x inches on each side. The painting is uniformly surrounded by a frame that measures 1 inch wide. Use this information to solve Exercises 101–102.

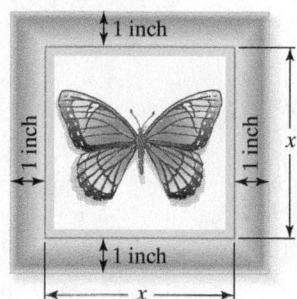

101. Write a polynomial in descending powers of x that expresses the area of the square that includes the painting and the frame.
102. Write an algebraic expression that describes the area of the frame. (*Hint:* The area of the frame is the area of the square that includes the painting and the frame minus the area of the painting.)

Writing in Mathematics

103. Explain how to multiply two binomials using the FOIL method. Give an example with your explanation.

104. Explain how to find the product of the sum and difference of two terms. Give an example with your explanation.
105. Explain how to square a binomial sum. Give an example with your explanation.
106. Explain how to square a binomial difference. Give an example with your explanation.
107. Explain why the graph for Exercises 97–100 is shown only in quadrant I.

Critical Thinking Exercises

Make Sense? *In Exercises 108–111, determine whether each statement "makes sense" or "does not make sense" and explain your reasoning.*

108. Squaring a binomial sum is as simple as squaring each of the two terms and then writing their sum.

109. I can distribute the exponent 2 on each factor of $(5x)^2$, but I cannot do the same thing on each term of $(x + 5)^2$.

110. Instead of using the formula for the square of a binomial sum, I prefer to write the binomial sum twice and then apply the FOIL method.

111. Special-product formulas for $(A + B)(A - B)$, $(A + B)^2$, and $(A - B)^2$ have patterns that make their multiplications quicker than using the FOIL method.

In Exercises 112–115, determine whether each statement is true or false. If the statement is false, make the necessary change(s) to produce a true statement.

112. $(3 + 4)^2 = 3^2 + 4^2$
113. $(2y + 7)^2 = 4y^2 + 28y + 49$
114. $(3x^2 + 2)(3x^2 - 2) = 9x^2 - 4$
115. $(x - 5)^2 = x^2 - 5x + 25$
116. What two binomials must be multiplied using the FOIL method to give a product of $x^2 - 8x - 20$?

117. Express the volume of the box as a polynomial in standard form.

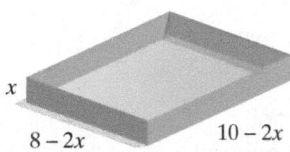

118. Express the area of the plane figure shown as a polynomial in standard form.

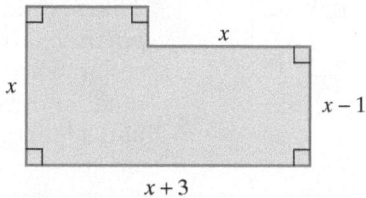

Technology Exercises

In Exercises 119–122, use a graphing utility to graph each side of the equation in the same viewing rectangle. (Call the left side y_1 and the right side y_2.) If the graphs coincide, verify that the multiplication has been performed correctly. If the graphs do not appear to coincide, this indicates that the multiplication is incorrect. In these exercises, correct the right side of the equation. Then graph the left side and the corrected right side to verify that the graphs coincide.

119. $(x + 1)^2 = x^2 + 1$; Use a $[-5, 5, 1]$ by $[0, 20, 1]$ viewing rectangle.

120. $(x + 2)^2 = x^2 + 2x + 4$; Use a $[-6, 5, 1]$ by $[0, 20, 1]$ viewing rectangle.

121. $(x + 1)(x - 1) = x^2 - 1$; Use a $[-6, 5, 1]$ by $[-2, 18, 1]$ viewing rectangle.

122. $(x - 2)(x + 2) + 4 = x^2$; Use a $[-6, 5, 1]$ by $[-2, 18, 1]$ viewing rectangle.

Review Exercises

In Exercises 123–124, solve each system by the method of your choice.

123. $\begin{cases} 2x + 3y = 1 \\ y = 3x - 7 \end{cases}$
(Section 4.2, Example 1)

124. $\begin{cases} 3x + 4y = 7 \\ 2x + 7y = 9 \end{cases}$
(Section 4.3, Example 3)

125. Graph: $y = \dfrac{1}{3}x$.

(Section 3.4, Example 3)

Preview Exercises

Exercises 126–128 will help you prepare for the material covered in the next section.

126. Use the order of operations to evaluate

$$x^3y + 2xy^2 + 5x - 2$$

for $x = -2$ and $y = 3$.

127. Use the second step to combine the like terms.

$$5xy + 6xy = (5 + 6)xy = \,?$$

128. Multiply using FOIL: $(x + 2y)(3x + 5y)$.

Polynomials in Several Variables

Objectives

1 Evaluate polynomials in several variables.

2 Understand the vocabulary of polynomials in two variables.

3 Add and subtract polynomials in several variables.

4 Multiply polynomials in several variables.

The next time you visit a lumberyard and go rummaging through piles of wood, think *polynomials*, although polynomials a bit different from those we have encountered so far. The construction industry uses a polynomial in two variables to determine the number of board feet that can be manufactured from a tree with a diameter of x inches and a length of y feet. This polynomial is

$$\frac{1}{4}x^2y - 2xy + 4y.$$

We call a polynomial containing two or more variables a **polynomial in several variables**. These polynomials can be evaluated, added, subtracted, and multiplied just like polynomials that contain only one variable.

1 Evaluate polynomials in several variables.

Evaluating a Polynomial in Several Variables

Two steps can be used to evaluate a polynomial in several variables.

Evaluating a Polynomial in Several Variables

1. Substitute the given value for each variable.
2. Perform the resulting computation using the order of operations.

EXAMPLE 1   Evaluating a Polynomial in Two Variables

Evaluate $2x^3y + xy^2 + 7x - 3$ for $x = -2$ and $y = 3$.

Solution   We begin by substituting -2 for x and 3 for y in the polynomial.

$$2x^3y + xy^2 + 7x - 3$$ This is the given polynomial.

$$= 2(-2)^3 \cdot 3 + (-2) \cdot 3^2 + 7(-2) - 3$$ Replace x with -2 and y with 3.

$$= 2(-8) \cdot 3 + (-2) \cdot 9 + 7(-2) - 3$$ Evaluate exponential expressions: $(-2)^3 = (-2)(-2)(-2) = -8$ and $3^2 = 3 \cdot 3 = 9$.

$$= -48 + (-18) + (-14) - 3$$ Perform the indicated multiplications.

$$= -83$$ Add from left to right.   ■

✓ **CHECK POINT 1**   Evaluate $3x^3y + xy^2 + 5y + 6$ for $x = -1$ and $y = 5$.

---

2 Understand the vocabulary of polynomials in two variables.

Describing Polynomials in Two Variables

In this section, we will limit our discussion of polynomials in several variables to two variables.

In general, a **polynomial in two variables**, x and y, contains the sum of one or more monomials in the form ax^ny^m. The constant, a, is the **coefficient**. The exponents, n and m, represent whole numbers. The **degree** of the monomial ax^ny^m is $n + m$. We'll use the polynomial from the construction industry to illustrate these ideas.

The coefficients are $\frac{1}{4}$, -2, and 4.

$$\frac{1}{4}x^2y \quad - 2xy \quad + 4y$$

| Degree of monomial: $2 + 1 = 3$ | Degree of monomial: $1 + 1 = 2$ | Degree of monomial: 1 |

The **degree of a polynomial in two variables** is the highest degree of all its terms. For the preceding polynomial, the degree is 3.

EXAMPLE 2   Using the Vocabulary of Polynomials

Determine the coefficient of each term, the degree of each term, and the degree of the polynomial:

$$7x^2y^3 - 17x^4y^2 + xy - 6y^2 + 9.$$

Solution

Term	Coefficient	Degree (Sum of Exponents on the Variables)
$7x^2y^3$	7	$2 + 3 = 5$
$-17x^4y^2$	-17	$4 + 2 = 6$
xy	1	$1 + 1 = 2$
$-6y^2$	-6	2
9	9	0

Think of xy as $1x^1y^1$.

The degree of the polynomial is the highest degree of all its terms, which is 6. ■

✓ **CHECK POINT 2**   Determine the coefficient of each term, the degree of each term, and the degree of the polynomial:

$$8x^4y^5 - 7x^3y^2 - x^2y - 5x + 11.$$

3 Add and subtract polynomials in several variables.

Adding and Subtracting Polynomials in Several Variables

Polynomials in several variables are added by combining like terms. For example, we can add the monomials $-7xy^2$ and $13xy^2$ as follows:

$$-7xy^2 + 13xy^2 = (-7 + 13)xy^2 = 6xy^2.$$

These like terms both contain the variable factors x and y^2.

Add coefficients and keep the same variable factors, xy^2.

EXAMPLE 3   Adding Polynomials in Two Variables

Add: $(6xy^2 - 5xy + 7) + (9xy^2 + 2xy - 6)$.

Solution

$$(6xy^2 - 5xy + 7) + (9xy^2 + 2xy - 6)$$
$$= (6xy^2 + 9xy^2) + (-5xy + 2xy) + (7 - 6) \quad \text{Group like terms.}$$
$$= 15xy^2 - 3xy + 1 \quad \text{Combine like terms by adding coefficients and keeping the same variable factors.} ■$$

✓ **CHECK POINT 3**   Add: $(-8x^2y - 3xy + 6) + (10x^2y + 5xy - 10)$.

We subtract polynomials in two variables just as we did when subtracting polynomials in one variable. Add the first polynomial and the opposite of the polynomial being subtracted.

EXAMPLE 4   Subtracting Polynomials in Two Variables

Subtract:

$$(5x^3 - 9x^2y + 3xy^2 - 4) - (3x^3 - 6x^2y - 2xy^2 + 3)$$

Solution

$$(5x^3 - 9x^2y + 3xy^2 - 4) - (3x^3 - 6x^2y - 2xy^2 + 3)$$

Change the sign of each coefficient.

$$= (5x^3 - 9x^2y + 3xy^2 - 4) + (-3x^3 + 6x^2y + 2xy^2 - 3)$$
Add the opposite of the polynomial being subtracted.

$$= (5x^3 - 3x^3) + (-9x^2y + 6x^2y) + (3xy^2 + 2xy^2) + (-4 - 3)$$
Group like terms.

$$= 2x^3 - 3x^2y + 5xy^2 - 7$$
Combine like terms by adding coefficients and keeping the same variable factors. ■

✓ **CHECK POINT 4**   Subtract:
$$(7x^3 - 10x^2y + 2xy^2 - 5) - (4x^3 - 12x^2y - 3xy^2 + 5)$$

---

4 Multiply polynomials in several variables.

Multiplying Polynomials in Several Variables

The product of monomials forms the basis of polynomial multiplication. As with monomials in one variable, multiplication can be done mentally by multiplying coefficients and adding exponents on variables with the same base.

EXAMPLE 5   Multiplying Monomials

Multiply: $(7x^2y)(5x^3y^2)$.

Solution

$$(7x^2y)(5x^3y^2)$$
$$= (7 \cdot 5)(x^2 \cdot x^3)(y \cdot y^2)$$
This regrouping can be worked mentally.

$$= 35x^{2+3}y^{1+2}$$
Multiply coefficients and add exponents on variables with the same base.

$$= 35x^5y^3$$
Simplify. ■

✓ **CHECK POINT 5**   Multiply: $(6xy^3)(10x^4y^2)$.

How do we multiply a monomial and a polynomial that is not a monomial? As we did with polynomials in one variable, multiply each term of the polynomial by the monomial.

EXAMPLE 6   Multiplying a Monomial and a Polynomial

Multiply: $3x^2y(4x^3y^2 - 6x^2y + 2)$.

Solution

$$3x^2y(4x^3y^2 - 6x^2y + 2)$$
$$= 3x^2y \cdot 4x^3y^2 - 3x^2y \cdot 6x^2y + 3x^2y \cdot 2$$
Use the distributive property.

$$= 12x^{2+3}y^{1+2} - 18x^{2+2}y^{1+1} + 6x^2y$$
Multiply coefficients and add exponents on variables with the same base.

$$= 12x^5y^3 - 18x^4y^2 + 6x^2y$$
Simplify. ■

✓ **CHECK POINT 6** Multiply: $6xy^2(10x^4y^5 - 2x^2y + 3)$.

FOIL and the special-products formulas can be used to multiply polynomials in several variables.

EXAMPLE 7 Multiplying Polynomials in Two Variables

Multiply: **a.** $(x + 4y)(3x - 5y)$ **b.** $(5x + 3y)^2$.

Solution We will perform the multiplication in part (a) using the FOIL method. We will multiply in part (b) using the formula for the square of a binomial, $(A + B)^2$.

a. $(x + 4y)(3x - 5y)$ — Multiply these binomials using the FOIL method.

$$= (x)(3x) + (x)(-5y) + (4y)(3x) + (4y)(-5y)$$
$$= 3x^2 - 5xy + 12xy - 20y^2$$
$$= 3x^2 + 7xy - 20y^2 \qquad \text{Combine like terms.}$$

$$(A + B)^2 = A^2 + 2 \cdot A \cdot B + B^2$$

b. $(5x + 3y)^2 = (5x)^2 + 2(5x)(3y) + (3y)^2$
$$= 25x^2 + 30xy + 9y^2 \quad \blacksquare$$

✓ **CHECK POINT 7** Multiply:
a. $(7x - 6y)(3x - y)$ **b.** $(2x + 4y)^2$.

EXAMPLE 8 Multiplying Polynomials in Two Variables

Multiply: **a.** $(4x^2y + 3y)(4x^2y - 3y)$ **b.** $(x + y)(x^2 - xy + y^2)$.

Solution We perform the multiplication in part (a) using the formula for the product of the sum and difference of two terms. We perform the multiplication in part (b) by multiplying each term of the trinomial, $x^2 - xy + y^2$, by x and y, respectively, and then adding like terms.

$$(A + B) \cdot (A - B) = A^2 - B^2$$

a. $(4x^2y + 3y)(4x^2y - 3y) = (4x^2y)^2 - (3y)^2$
$$= 16x^4y^2 - 9y^2$$

b. $(x + y)(x^2 - xy + y^2)$
$$= x(x^2 - xy + y^2) + y(x^2 - xy + y^2) \qquad \text{Multiply the trinomial by each term of the binomial.}$$
$$= x \cdot x^2 - x \cdot xy + x \cdot y^2 + y \cdot x^2 - y \cdot xy + y \cdot y^2 \qquad \text{Use the distributive property.}$$
$$= x^3 - x^2y + xy^2 + x^2y - xy^2 + y^3 \qquad \text{Add exponents on variables with the same base.}$$
$$= x^3 + y^3 \qquad \text{Combine like terms:}\ -x^2y + x^2y = 0\ \text{and}\ xy^2 - xy^2 = 0. \quad \blacksquare$$

✓ **CHECK POINT 8** Multiply:
a. $(6xy^2 + 5x)(6xy^2 - 5x)$ **b.** $(x - y)(x^2 + xy + y^2)$.

CONCEPT AND VOCABULARY CHECK

Fill in each blank so that the resulting statement is true.

1. The coefficient of the monomial $-18x^4y^2$ is _____.

2. The degree of the monomial $-18x^4y^2$ is _____.

3. The coefficient of the monomial ax^ny^m is _____ and the degree is _____.

4. The degree of x^3y^2 is _____ and the degree of x^2y^7 is _____, so the degree of $x^3y^2 - 8x^2y^7$ is _____.

5. True or false: The monomials $7xy^2$ and $2x^2y$ can be added. _____

6. True or false: The monomials $7xy^2$ and $2x^2y$ can be multiplied. _____

5.4 EXERCISE SET   MyMathLab®

Watch the videos in MyMathLab   Download the MyDashBoard App

Practice Exercises

In Exercises 1–6, evaluate each polynomial for $x = 2$ and $y = -3$.

1. $x^2 + 2xy + y^2$
2. $x^2 + 3xy + y^2$
3. $xy^3 - xy + 1$
4. $x^3y - xy + 2$
5. $2x^2y - 5y + 3$
6. $3x^2y - 4y + 5$

In Exercises 7–8, determine the coefficient of each term, the degree of each term, and the degree of the polynomial.

7. $x^3y^2 - 5x^2y^7 + 6y^2 - 3$
8. $12x^4y - 5x^3y^7 - x^2 + 4$

In Exercises 9–20, add or subtract as indicated.

9. $(5x^2y - 3xy) + (2x^2y - xy)$
10. $(-2x^2y + xy) + (4x^2y + 7xy)$
11. $(4x^2y + 8xy + 11) + (-2x^2y + 5xy + 2)$
12. $(7x^2y + 5xy + 13) + (-3x^2y + 6xy + 4)$
13. $(7x^4y^2 - 5x^2y^2 + 3xy) + (-18x^4y^2 - 6x^2y^2 - xy)$
14. $(6x^4y^2 - 10x^2y^2 + 7xy) + (-12x^4y^2 - 3x^2y^2 - xy)$
15. $(x^3 + 7xy - 5y^2) - (6x^3 - xy + 4y^2)$
16. $(x^4 - 7xy - 5y^3) - (6x^4 - 3xy + 4y^3)$
17. $(3x^4y^2 + 5x^3y - 3y) - (2x^4y^2 - 3x^3y - 4y + 6x)$
18. $(5x^4y^2 + 6x^3y - 7y) - (3x^4y^2 - 5x^3y - 6y + 8x)$
19. $(x^3 - y^3) - (-4x^3 - x^2y + xy^2 + 3y^3)$
20. $(x^3 - y^3) - (-6x^3 + x^2y - xy^2 + 2y^3)$

21. Add: $5x^2y^2 - 4xy^2 + 6y^2$
     $\underline{-8x^2y^2 + 5xy^2 - y^2}$

22. Add: $7a^2b^2 - 5ab^2 + 6b^2$
     $\underline{-10a^2b^2 + 6ab^2 + 6b^2}$

23. Subtract: $3a^2b^4 - 5ab^2 + 7ab$
     $\underline{-(-5a^2b^4 - 8ab^2 - ab)}$

24. Subtract: $13x^2y^4 - 17xy^2 + xy$
     $\underline{-(-7x^2y^4 - 8xy^2 - xy)}$

25. Subtract $11x - 5y$ from the sum of $7x + 13y$ and $-26x + 19y$.

26. Subtract $23x - 5y$ from the sum of $6x + 15y$ and $x - 19y$.

In Exercises 27–76, find each product.

27. $(5x^2y)(8xy)$
28. $(10x^2y)(5xy)$
29. $(-8x^3y^4)(3x^2y^5)$
30. $(7x^4y^5)(-10x^7y^{11})$
31. $9xy(5x + 2y)$
32. $7xy(8x + 3y)$
33. $5xy^2(10x^2 - 3y)$
34. $6x^2y(5x^2 - 9y)$
35. $4ab^2(7a^2b^3 + 2ab)$
36. $2ab^2(20a^2b^3 + 11ab)$
37. $-b(a^2 - ab + b^2)$
38. $-b(a^3 - ab + b^3)$
39. $(x + 5y)(7x + 3y)$
40. $(x + 9y)(6x + 7y)$
41. $(x - 3y)(2x + 7y)$
42. $(3x - y)(2x + 5y)$
43. $(3xy - 1)(5xy + 2)$
44. $(7xy + 1)(2xy - 3)$
45. $(2x + 3y)^2$
46. $(2x + 5y)^2$
47. $(xy - 3)^2$

48. $(xy - 5)^2$
49. $(x^2 + y^2)^2$
50. $(2x^2 + y^2)^2$
51. $(x^2 - 2y^2)^2$
52. $(x^2 - y^2)^2$
53. $(3x + y)(3x - y)$
54. $(x + 5y)(x - 5y)$
55. $(ab + 1)(ab - 1)$
56. $(ab + 2)(ab - 2)$
57. $(x + y^2)(x - y^2)$
58. $(x^2 + y)(x^2 - y)$
59. $(3a^2b + a)(3a^2b - a)$
60. $(5a^2b + a)(5a^2b - a)$
61. $(3xy^2 - 4y)(3xy^2 + 4y)$
62. $(7xy^2 - 10y)(7xy^2 + 10y)$
63. $(a + b)(a^2 - b^2)$
64. $(a - b)(a^2 + b^2)$
65. $(x + y)(x^2 + 3xy + y^2)$
66. $(x + y)(x^2 + 5xy + y^2)$
67. $(x - y)(x^2 - 3xy + y^2)$
68. $(x - y)(x^2 - 4xy + y^2)$
69. $(xy + ab)(xy - ab)$
70. $(xy + ab^2)(xy - ab^2)$
71. $(x^2 + 1)(x^4y + x^2 + 1)$
72. $(x^2 + 1)(xy^4 + y^2 + 1)$
73. $(x^2y^2 - 3)^2$
74. $(x^2y^2 - 5)^2$
75. $(x + y + 1)(x + y - 1)$
76. $(x + y + 1)(x - y + 1)$

In Exercises 77–80, write a polynomial in two variables that describes the total area of each region shaded in blue. Express each polynomial as the sum or difference of terms.

77.
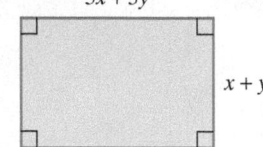
3x + 5y
x + y

78.
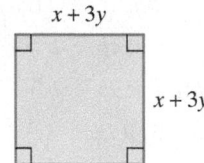
x + 3y
x + 3y

79.
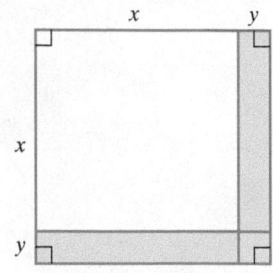
x   y
x
y

80.
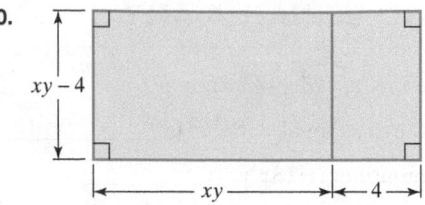
xy − 4
xy   4

Practice PLUS

In Exercises 81–86, find each product. As we said in the Section 5.3 opener, cut to the chase in each part of the polynomial multiplication: Use only the special-product formula for the sum and difference of two terms or the formulas for the square of a binomial.

81. $[(x^3y^3 + 1)(x^3y^3 - 1)]^2$
82. $[(1 - a^3b^3)(1 + a^3b^3)]^2$
83. $(xy - 3)^2(xy + 3)^2$ (Do not begin by squaring a binomial.)
84. $(ab - 4)^2(ab + 4)^2$ (Do not begin by squaring a binomial.)
85. $[x + y + z][x - (y + z)]$
86. $(a - b - c)(a + b + c)$

Application Exercises

87. The number of board feet, N, that can be manufactured from a tree with a diameter of x inches and a length of y feet is modeled by the formula

$$N = \frac{1}{4}x^2y - 2xy + 4y.$$

A building contractor estimates that 3000 board feet of lumber is needed for a job. The lumber company has just milled a fresh load of timber from 20 trees that averaged 10 inches in diameter and 16 feet in length. Is this enough to complete the job? If not, how many additional board feet of lumber is needed?

88. The storage shed shown in the figure has a volume given by the polynomial

$$2x^2y + \frac{1}{2}\pi x^2y.$$

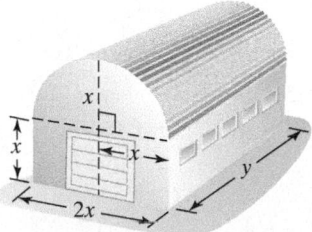

a. A small business is considering having a shed installed like the one shown in the figure. The shed's height, $2x$, is 26 feet and its length, y, is 27 feet. Using $x = 13$ and $y = 27$, find the volume of the storage shed.

b. The business requires at least 18,000 cubic feet of storage space. Should they construct the storage shed described in part (a)?

An object that is falling or vertically projected into the air has its height, in feet, above the ground given by

$$s = -16t^2 + v_0t + s_0,$$

where s is the height, in feet, v_0 is the original velocity of the object, in feet per second, t is the time the object is in motion, in seconds, and s_0 is the height, in feet, from which the object is dropped or projected. The figure shows that a ball is thrown straight up from a rooftop at an original velocity of 80 feet per second from a height of 96 feet. The ball misses the rooftop on its way down and eventually strikes the ground. Use the formula and this information to solve Exercises 89–91.

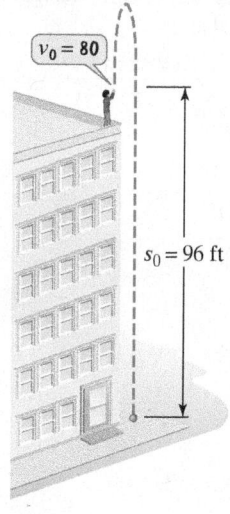

89. How high above the ground will the ball be 2 seconds after being thrown?

90. How high above the ground will the ball be 4 seconds after being thrown?

91. How high above the ground will the ball be 6 seconds after being thrown? Describe what this means in practical terms.

The graph visually displays the information about the thrown ball described in Exercises 89–91. The horizontal axis represents the ball's time in motion, in seconds. The vertical axis represents the ball's height above the ground, in feet. Use the graph to solve Exercises 92–97.

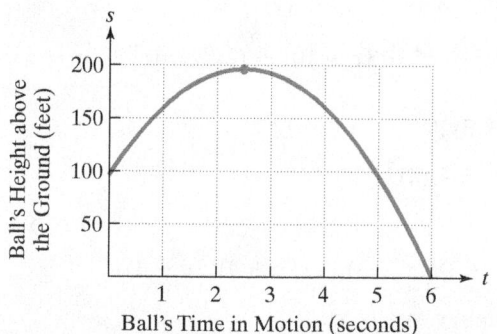

92. During which time period is the ball rising?

93. During which time period is the ball falling?

94. Identify your answer from Exercise 90 as a point on the graph.

95. Identify your answer from Exercise 89 as a point on the graph.

96. After how many seconds does the ball strike the ground?

97. After how many seconds does the ball reach its maximum height above the ground? What is a reasonable estimate of this maximum height?

Writing in Mathematics

98. What is a polynomial in two variables? Provide an example with your description.

99. Explain how to find the degree of a polynomial in two variables.

100. Suppose that you take up sky diving. Explain how to use the formula for Exercises 89–91 to determine your height above the ground at every instant of your fall.

Critical Thinking Exercises

Make Sense? *In Exercises 101–104, determine whether each statement "makes sense" or "does not make sense" and explain your reasoning.*

101. I use the same procedures for operations with polynomials in two variables as I did when performing these operations with polynomials in one variable.

102. Adding polynomials in several variables is the same as adding like terms.

103. I used FOIL to find the product of $x + y$ and $x^2 - xy + y^2$.

104. I used FOIL to multiply $5xy$ and $3xy + 4$.

In Exercises 105–108, determine whether each statement is true or false. If the statement is false, make the necessary change(s) to produce a true statement.

105. The degree of $5x^{24} - 3x^{16}y^9 - 7xy^2 + 6$ is 24.

106. In the polynomial $4x^2y + x^3y^2 + 3x^2y^3 + 7y$, the term x^3y^2 has degree 5 and no numerical coefficient.

107. $(2x + 3 - 5y)(2x + 3 + 5y) = 4x^2 + 12x + 9 - 25y^2$

108. $(6x^2y - 7xy - 4) - (6x^2y + 7xy - 4) = 0$

In Exercises 109–110, find a polynomial in two variables that describes the area of the region of each figure shaded in blue. Write the polynomial as the sum or difference of terms.

109.

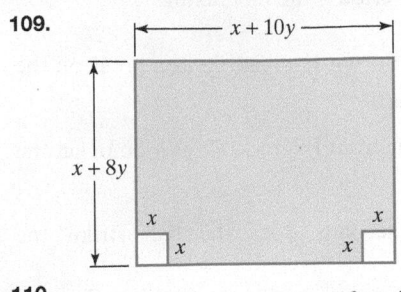

110.

111. Use the formulas for the volume of a rectangular solid and a cylinder to derive the polynomial in Exercise 88 that describes the volume of the storage building.

Review Exercises

112. Solve for W: $R = \dfrac{L + 3W}{2}$. (Section 2.4, Example 4)

113. Subtract: $-6.4 - (-10.2)$. (Section 1.6, Example 2)

114. Solve: $0.02(x - 5) = 0.03 - 0.03(x + 7)$. (Section 2.3, Example 5)

Preview Exercises

Exercises 115–117 will help you prepare for the material covered in the next section.

115. Find the missing exponent, designated by the question mark, in the final step.
$$\frac{x^7}{x^3} = \frac{\cancel{x} \cdot \cancel{x} \cdot \cancel{x} \cdot x \cdot x \cdot x \cdot x}{\cancel{x} \cdot \cancel{x} \cdot \cancel{x}} = x^?$$

116. Simplify: $\dfrac{(x^2)^3}{5^3}$.

117. Simplify: $\dfrac{(2a^3)^5}{(b^4)^5}$.

MID-CHAPTER CHECK POINT   Section 5.1–Section 5.4

 What You Know: We learned to add, subtract, and multiply polynomials. We used a number of fast methods for finding products of polynomials, including the FOIL method for multiplying binomials, a special-product formula for the product of the sum and difference of two terms $[(A + B)(A - B) = A^2 - B^2]$, and special-product formulas for squaring binomials $[(A + B)^2 = A^2 + 2AB + B^2; (A - B)^2 = A^2 - 2AB + B^2]$. Finally, we applied all of these operations to polynomials in several variables.

In Exercises 1–21, perform the indicated operations.

1. $(11x^2y^3)(-5x^2y^3)$

2. $11x^2y^3 - 5x^2y^3$

3. $(3x + 5)(4x - 7)$

4. $(3x + 5) - (4x - 7)$

5. $(2x - 5)(x^2 - 3x + 1)$

6. $(2x - 5) + (x^2 - 3x + 1)$

7. $(8x - 3)^2$

8. $(-10x^4)(-7x^5)$

9. $(x^2 + 2)(x^2 - 2)$

10. $(x^2 + 2)^2$

11. $(9a - 10b)(2a + b)$

12. $7x^2(10x^3 - 2x + 3)$

13. $(3a^2b^3 - ab + 4b^2) - (-2a^2b^3 - 3ab + 5b^2)$

14. $2(3y - 5)(3y + 5)$

15. $(-9x^3 + 5x^2 - 2x + 7) + (11x^3 - 6x^2 + 3x - 7)$

16. $10x^2 - 8xy - 3(y^2 - xy)$

17. $(-2x^5 + x^4 - 3x + 10) - (2x^5 - 6x^4 + 7x - 13)$

18. $(x + 3y)(x^2 - 3xy + 9y^2)$

19. $(5x^4 + 4)(2x^3 - 1)$

20. $(y - 6z)^2$

21. $(2x + 3)(2x - 3) - (5x + 4)(5x - 4)$

22. Graph: $y = 1 - x^2$.

5.5

Objectives

1 Use the quotient rule for exponents.

2 Use the zero-exponent rule.

3 Use the quotients-to-powers rule.

4 Divide monomials.

5 Check polynomial division.

6 Divide a polynomial by a monomial.

1 Use the quotient rule for exponents.

Dividing Polynomials

In the dramatic arts, ours is the era of the movies. As individuals and as a nation, we've grown up with them. Our images of love, war, family, country—even of things that terrify us—owe much to what we've seen on screen. In this section's exercise set, we'll model our love for movies with polynomials and polynomial division. Before discussing polynomial division, we must develop some additional rules for working with exponents.

The Quotient Rule for Exponents

Consider the quotient of two exponential expressions, such as the quotient of 2^7 and 2^3. We are dividing 7 factors of 2 by 3 factors of 2. We are left with 4 factors of 2:

$$\frac{2^7}{2^3} = \frac{2 \cdot 2 \cdot 2 \cdot 2 \cdot 2 \cdot 2 \cdot 2}{2 \cdot 2 \cdot 2} = \frac{\cancel{2} \cdot \cancel{2} \cdot \cancel{2} \cdot 2 \cdot 2 \cdot 2 \cdot 2}{\cancel{2} \cdot \cancel{2} \cdot \cancel{2}} = 2 \cdot 2 \cdot 2 \cdot 2$$

7 factors of 2

3 factors of 2

Divide out pairs of factors: $\frac{2}{2} = 1$.

4 factors of 2

Thus,

$$\frac{2^7}{2^3} = 2^4.$$

We can quickly find the exponent, 4, on the quotient by subtracting the original exponents:

$$\frac{2^7}{2^3} = 2^{7-3}.$$

This suggests the following rule:

The Quotient Rule

$$\frac{b^m}{b^n} = b^{m-n}, \quad b \neq 0$$

When dividing exponential expressions with the same nonzero base, subtract the exponent in the denominator from the exponent in the numerator. Use this difference as the exponent of the common base.

EXAMPLE 1 Using the Quotient Rule

Divide each expression using the quotient rule:

a. $\dfrac{2^8}{2^4}$ **b.** $\dfrac{x^{13}}{x^3}$ **c.** $\dfrac{y^{15}}{y}$.

Solution

a. $\dfrac{2^8}{2^4} = 2^{8-4} = 2^4$ or 16

b. $\dfrac{x^{13}}{x^3} = x^{13-3} = x^{10}$

c. $\dfrac{y^{15}}{y} = \dfrac{y^{15}}{y^1} = y^{15-1} = y^{14}$ ∎

✓ **CHECK POINT 1** Divide each expression using the quotient rule:

a. $\dfrac{5^{12}}{5^4}$ **b.** $\dfrac{x^9}{x^2}$ **c.** $\dfrac{y^{20}}{y}$.

---

2 Use the zero-exponent rule.

Zero as an Exponent

A nonzero base can be raised to the 0 power. The quotient rule can be used to help determine what zero as an exponent should mean. Consider the quotient of b^4 and b^4, where b is not zero. We can determine this quotient in two ways.

$$\frac{b^4}{b^4} = 1 \qquad \frac{b^4}{b^4} = b^{4-4} = b^0$$

Any nonzero expression divided by itself is 1.

Use the quotient rule and subtract exponents.

This means that b^0 must equal 1.

The Zero-Exponent Rule

If b is any real number other than 0,

$$b^0 = 1.$$

EXAMPLE 2 Using the Zero-Exponent Rule

Use the zero-exponent rule to simplify each expression:

a. 7^0 **b.** $(-5)^0$ **c.** -5^0 **d.** $10x^0$ **e.** $(10x)^0$.

Solution

a. $7^0 = 1$ Any nonzero number raised to the 0 power is 1.

b. $(-5)^0 = 1$ Any nonzero number raised to the 0 power is 1.

c. $-5^0 = -1$ $\qquad$ $-5^0 = -(5^0) = -1$

> Only 5 is raised to the 0 power.

$\qquad$ **d.** $10x^0 = 10 \cdot 1 = 10$

> Only x is raised to the 0 power.

$\qquad$ **e.** $(10x)^0 = 1$

> The entire expression, $10x$, is raised to the 0 power. ∎

✓ **CHECK POINT 2** Use the zero-exponent rule to simplify each expression:
a. 14^0 $\qquad$ **b.** $(-10)^0$ $\qquad$ **c.** -10^0 $\qquad$ **d.** $20x^0$ $\qquad$ **e.** $(20x)^0$.

---

3 Use the quotients-to-powers rule.

The Quotients-to-Powers Rule for Exponents

We have seen that when a product is raised to a power, we raise every factor in the product to the power:

$$(ab)^n = a^n b^n.$$

There is a similar property for raising a quotient to a power.

> **Quotients to Powers**
>
> If a and b are real numbers and b is nonzero, then
>
> $$\left(\frac{a}{b}\right)^n = \frac{a^n}{b^n}.$$

When a quotient is raised to a power, raise the numerator to the power and divide by the denominator raised to the power.

EXAMPLE 3 Using the Quotients-to-Powers Rule

Simplify each expression using the quotients-to-powers rule:

a. $\left(\dfrac{x}{4}\right)^2$ $\qquad$ **b.** $\left(\dfrac{x^2}{5}\right)^3$ $\qquad$ **c.** $\left(\dfrac{2a^3}{b^4}\right)^5$.

Solution

a. $\left(\dfrac{x}{4}\right)^2 = \dfrac{x^2}{4^2} = \dfrac{x^2}{16}$ $\qquad$ Square the numerator and the denominator.

b. $\left(\dfrac{x^2}{5}\right)^3 = \dfrac{(x^2)^3}{5^3} = \dfrac{x^{2 \cdot 3}}{5 \cdot 5 \cdot 5} = \dfrac{x^6}{125}$ $\qquad$ Cube the numerator and the denominator.

c. $\left(\dfrac{2a^3}{b^4}\right)^5 = \dfrac{(2a^3)^5}{(b^4)^5}$ $\qquad$ Raise the numerator and the denominator to the fifth power.

$\qquad = \dfrac{2^5(a^3)^5}{(b^4)^5}$ $\qquad$ Raise each factor in the numerator to the fifth power.

$\qquad = \dfrac{2^5 a^{3 \cdot 5}}{b^{4 \cdot 5}}$ $\qquad$ To raise exponential expressions to powers, multiply exponents: $(b^m)^n = b^{mn}$.

$\qquad = \dfrac{32a^{15}}{b^{20}}$ $\qquad$ Simplify. ∎

✓ **CHECK POINT 3**   Simplify each expression using the quotients-to-powers rule:

a. $\left(\dfrac{x}{5}\right)^2$ **b.** $\left(\dfrac{x^4}{2}\right)^3$ **c.** $\left(\dfrac{2a^{10}}{b^3}\right)^4$.

Great Question!

What are some common errors to avoid when using the quotient rule or the zero-exponent rule?

Here's a partial list. The first column shows the correct simplification. The second column illustrates a common error.

Correct	Incorrect	Description of Error
$\dfrac{2^{20}}{2^4} = 2^{20-4} = 2^{16}$	$\dfrac{2^{20}}{2^4} = 2^5$	Exponents should be subtracted, not divided.
$-8^0 = -1$	$-8^0 = 1$	Only 8 is raised to the 0 power.
$\left(\dfrac{x}{5}\right)^2 = \dfrac{x^2}{5^2} = \dfrac{x^2}{25}$	$\left(\dfrac{x}{5}\right)^2 = \dfrac{x^2}{5}$	The numerator and denominator must both be squared.

4 Divide monomials.

Dividing Monomials

Now that we have developed three additional properties of exponents, we are ready to turn to polynomial division. We begin with the quotient of two monomials, such as $16x^{14}$ and $8x^2$. This quotient is obtained by dividing the coefficients, 16 and 8, and then dividing the variables using the quotient rule for exponents.

$$\frac{16x^{14}}{8x^2} = \frac{16}{8}x^{14-2} = 2x^{12}$$

> Divide coefficients and subtract exponents.

Dividing Monomials

To divide monomials, divide the coefficients and then divide the variables. Use the quotient rule for exponents to divide the variables: Keep the variable and subtract the exponents.

Great Question!

I notice that the exponents on x are the same in Example 4(b). Do I have to subtract exponents since I know the quotient of x^3 and x^3 is 1?

No. Rather than subtracting exponents for division that results in a 0 exponent, you might prefer to divide out x^3.

$$\frac{2x^3}{8x^3} = \frac{2}{8} = \frac{1}{4}$$

EXAMPLE 4   Dividing Monomials

Divide: **a.** $\dfrac{-12x^8}{4x^2}$ **b.** $\dfrac{2x^3}{8x^3}$ **c.** $\dfrac{15x^5y^4}{3x^2y}$.

Solution

a. $\dfrac{-12x^8}{4x^2} = \dfrac{-12}{4}x^{8-2} = -3x^6$

b. $\dfrac{2x^3}{8x^3} = \dfrac{2}{8}x^{3-3} = \dfrac{1}{4}x^0 = \dfrac{1}{4} \cdot 1 = \dfrac{1}{4}$

c. $\dfrac{15x^5y^4}{3x^2y} = \dfrac{15}{3}x^{5-2}y^{4-1} = 5x^3y^3$ ∎

✓ **CHECK POINT 4**   Divide:

a. $\dfrac{-20x^{12}}{10x^4}$ **b.** $\dfrac{3x^4}{15x^4}$ **c.** $\dfrac{9x^6y^5}{3xy^2}$.

5 Check polynomial division.

Checking Division of Polynomial Problems

The answer to a division problem can be checked. For example, consider the following problem:

Dividend: the polynomial you are dividing into

$$\frac{15x^5y^4}{3x^2y} = 5x^3y^3.$$

Quotient: the answer to your division problem

Divisor: the polynomial you are dividing by

The quotient is correct if the product of the divisor and the quotient is the dividend. Is the quotient shown in the preceding equation correct?

$$(3x^2y)(5x^3y^3) = 3 \cdot 5x^{2+3}y^{1+3} = 15x^5y^4$$

Divisor   Quotient   This is the dividend.

Because the product of the divisor and the quotient is the dividend, the answer to the division problem is correct.

Checking Division of Polynomials

To check a quotient in a division problem, multiply the divisor and the quotient. If this product is the dividend, the quotient is correct.

6 Divide a polynomial by a monomial.

Great Question!

Can I cancel identical terms in the dividend and the divisor?

No. Try to avoid this common error:

Incorrect:

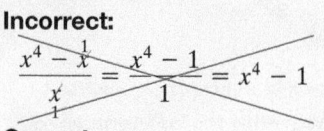

Correct:

$$\frac{x^4 - x}{x} = \frac{x^4}{x} - \frac{x}{x}$$

$$= x^{4-1} - x^{1-1}$$

$$= x^3 - x^0$$

$$= x^3 - 1$$

Don't leave out the 1.

Dividing a Polynomial That Is Not a Monomial by a Monomial

To divide a polynomial by a monomial, we divide each term of the polynomial by the monomial. For example,

polynomial dividend

$$\frac{10x^8 + 15x^6}{5x^3} = \frac{10x^8}{5x^3} + \frac{15x^6}{5x^3} = \frac{10}{5}x^{8-3} + \frac{15}{5}x^{6-3} = 2x^5 + 3x^3.$$

monomial divisor   Divide the first term by $5x^3$.   Divide the second term by $5x^3$.

Is the quotient correct? Multiply the divisor and the quotient.

$$5x^3(2x^5 + 3x^3) = 5x^3 \cdot 2x^5 + 5x^3 \cdot 3x^3$$

$$= 5 \cdot 2x^{3+5} + 5 \cdot 3x^{3+3} = 10x^8 + 15x^6$$

Because this product gives the dividend, the quotient is correct.

Dividing a Polynomial That Is Not a Monomial by a Monomial

To divide a polynomial by a monomial, divide each term of the polynomial by the monomial.

EXAMPLE 5 Dividing a Polynomial by a Monomial

Find the quotient: $(-12x^8 + 4x^6 - 8x^3) \div 4x^2$.

Solution

$$\frac{-12x^8 + 4x^6 - 8x^3}{4x^2}$$ Rewrite the division in a vertical format.

$$= \frac{-12x^8}{4x^2} + \frac{4x^6}{4x^2} - \frac{8x^3}{4x^2}$$ Divide each term of the polynomial by the monomial.

$$= \frac{-12}{4}x^{8-2} + \frac{4}{4}x^{6-2} - \frac{8}{4}x^{3-2}$$ Divide coefficients and subtract exponents.

$$= -3x^6 + x^4 - 2x$$ Simplify.

To check the answer, multiply the divisor and the quotient.

$$4x^2(-3x^6 + x^4 - 2x) = 4x^2(-3x^6) + 4x^2 \cdot x^4 - 4x^2(2x)$$

$$= 4(-3)x^{2+6} + 4x^{2+4} - 4 \cdot 2x^{2+1}$$

$$= -12x^8 + 4x^6 - 8x^3$$

Divisor   Quotient   This is the dividend.

Because the product of the divisor and the quotient is the dividend, the answer—that is, the quotient—is correct. ∎

✓ **CHECK POINT 5** Find the quotient: $(-15x^9 + 6x^5 - 9x^3) \div 3x^2$.

EXAMPLE 6 Dividing a Polynomial by a Monomial

Divide: $\dfrac{16x^5 - 9x^4 + 8x^3}{2x^3}$.

Solution

$$\frac{16x^5 - 9x^4 + 8x^3}{2x^3}$$ This is the given polynomial division.

$$= \frac{16x^5}{2x^3} - \frac{9x^4}{2x^3} + \frac{8x^3}{2x^3}$$ Divide each term by $2x^3$.

$$= \frac{16}{2}x^{5-3} - \frac{9}{2}x^{4-3} + \frac{8}{2}x^{3-3}$$ Divide coefficients and subtract exponents. Did you immediately write the last term as 4?

$$= 8x^2 - \frac{9}{2}x + 4x^0$$ Simplify.

$$= 8x^2 - \frac{9}{2}x + 4$$ $x^0 = 1$, so $4x^0 = 4 \cdot 1 = 4$.

Check the answer by showing that the product of the divisor and the quotient is the dividend. ∎

✓ **CHECK POINT 6** Divide: $\dfrac{25x^9 - 7x^4 + 10x^3}{5x^3}$.

EXAMPLE 7 Dividing Polynomials in Two Variables

Divide: $(15x^5y^4 - 3x^3y^2 + 9x^2y) \div 3x^2y$.

Solution

$$\frac{15x^5y^4 - 3x^3y^2 + 9x^2y}{3x^2y}$$

Rewrite the division in a vertical format.

$$= \frac{15x^5y^4}{3x^2y} - \frac{3x^3y^2}{3x^2y} + \frac{9x^2y}{3x^2y}$$

Divide each term of the polynomial by the monomial.

$$= \frac{15}{3}x^{5-2}y^{4-1} - \frac{3}{3}x^{3-2}y^{2-1} + \frac{9}{3}x^{2-2}y^{1-1}$$

Divide coefficients and subtract exponents.

$$= 5x^3y^3 - xy + 3$$

Simplify.

Check the answer by showing that the product of the divisor and the quotient is the dividend. ■

✓ **CHECK POINT 7**   Divide: $(18x^7y^6 - 6x^2y^3 + 60xy^2) \div 6xy^2$

CONCEPT AND VOCABULARY CHECK

Fill in each blank so that the resulting statement is true.

1. The quotient rule for exponents states that $\dfrac{b^m}{b^n} =$ _____, $b \neq 0$. When dividing exponential expressions with the same nonzero base, _____ the exponents.

2. If $b \neq 0$, $b^0 =$ _____.

3. The quotients-to-powers rule for exponents states that $\left(\dfrac{a}{b}\right)^n =$ _____, $b \neq 0$. When a quotient is raised to a power, raise both the _____ and the _____ to the power.

4. To divide monomials, _____ the coefficients and _____ the exponents.

5. Consider the following division problem:

$$\frac{20x^6y^4}{10x^2y} = 2x^4y^3.$$

The polynomial in the numerator, $20x^6y^4$, is called the _____. The polynomial in the denominator, $10x^2y$, is called the _____. The answer to the problem, $2x^4y^3$, is called the _____.

6. To check the answer to a division problem, multiply the _____ and the _____. If this product is the _____, the answer is correct.

7. To perform the division $\dfrac{20x^8 - 10x^4 + 6x^3}{2x^3}$, divide each term of _____ by _____.

5.5 EXERCISE SET   MyMathLab®    Watch the videos in MyMathLab   Download the MyDashBoard App

Practice Exercises

In Exercises 1–10, divide each expression using the quotient rule. Express any numerical answers in exponential form.

1. $\dfrac{3^{20}}{3^5}$

2. $\dfrac{3^{30}}{3^{10}}$

3. $\dfrac{x^6}{x^2}$

4. $\dfrac{x^8}{x^4}$

5. $\dfrac{y^{13}}{y^5}$

6. $\dfrac{y^{19}}{y^6}$

7. $\dfrac{5^6 \cdot 2^8}{5^3 \cdot 2^4}$

8. $\dfrac{3^6 \cdot 2^8}{3^3 \cdot 2^4}$

9. $\dfrac{x^{100}y^{50}}{x^{25}y^{10}}$

10. $\dfrac{x^{200}y^{40}}{x^{25}y^{10}}$

In Exercises 11–24, use the zero-exponent rule to simplify each expression.

11. 2^0

12. 4^0

13. $(-2)^0$

14. $(-4)^0$

15. -2^0

16. -4^0

17. $100y^0$

18. $200y^0$

19. $(100y)^0$

20. $(200y)^0$

21. $-5^0 + (-5)^0$

22. $-6^0 + (-6)^0$

23. $-\pi^0 - (-\pi)^0$

24. $-\sqrt{3}^0 - (-\sqrt{3})^0$

In Exercises 25–36, simplify each expression using the quotients-to-powers rule. If possible, evaluate exponential expressions.

25. $\left(\dfrac{x}{3}\right)^2$

26. $\left(\dfrac{x}{5}\right)^2$

27. $\left(\dfrac{x^2}{4}\right)^3$

28. $\left(\dfrac{x^2}{3}\right)^3$

29. $\left(\dfrac{2x^3}{5}\right)^2$

30. $\left(\dfrac{3x^4}{7}\right)^2$

31. $\left(\dfrac{-4}{3a^3}\right)^3$

32. $\left(\dfrac{-5}{2a^3}\right)^3$

33. $\left(\dfrac{-2a^7}{b^4}\right)^5$

34. $\left(\dfrac{-2a^8}{b^3}\right)^5$

35. $\left(\dfrac{x^2y^3}{2z}\right)^4$

36. $\left(\dfrac{x^3y^2}{2z}\right)^4$

In Exercises 37–52, divide the monomials. Check each answer by showing that the product of the divisor and the quotient is the dividend.

37. $\dfrac{30x^{10}}{10x^5}$

38. $\dfrac{45x^{12}}{15x^4}$

39. $\dfrac{-8x^{22}}{4x^2}$

40. $\dfrac{-15x^{40}}{3x^4}$

41. $\dfrac{-9y^8}{18y^5}$

42. $\dfrac{-15y^{13}}{45y^9}$

43. $\dfrac{7y^{17}}{5y^5}$

44. $\dfrac{9y^{19}}{7y^{11}}$

45. $\dfrac{30x^7y^5}{5x^2y}$

46. $\dfrac{40x^9y^5}{2x^2y}$

47. $\dfrac{-18x^{14}y^2}{36x^2y^2}$

48. $\dfrac{-15x^{16}y^2}{45x^2y^2}$

49. $\dfrac{9x^{20}y^{20}}{7x^{20}y^{20}}$

50. $\dfrac{7x^{30}y^{30}}{15x^{30}y^{30}}$

51. $\dfrac{-5x^{10}y^{12}z^6}{50x^2y^3z^2}$

52. $\dfrac{-8x^{12}y^{10}z^4}{40x^2y^3z^2}$

In Exercises 53–78, divide the polynomial by the monomial. Check each answer by showing that the product of the divisor and the quotient is the dividend.

53. $\dfrac{10x^4 + 2x^3}{2}$

54. $\dfrac{20x^4 + 5x^3}{5}$

55. $\dfrac{14x^4 - 7x^3}{7x}$

56. $\dfrac{24x^4 - 8x^3}{8x}$

57. $\dfrac{y^7 - 9y^2 + y}{y}$

58. $\dfrac{y^8 - 11y^3 + y}{y}$

59. $\dfrac{24x^3 - 15x^2}{-3x}$

60. $\dfrac{10x^3 - 20x^2}{-5x}$

61. $\dfrac{18x^5 + 6x^4 + 9x^3}{3x^2}$

62. $\dfrac{18x^5 + 24x^4 + 12x^3}{6x^2}$

63. $\dfrac{12x^4 - 8x^3 + 40x^2}{4x}$

64. $\dfrac{49x^4 - 14x^3 + 70x^2}{-7x}$

65. $(4x^2 - 6x) \div x$

66. $(16y^2 - 8y) \div y$

67. $\dfrac{30z^3 + 10z^2}{-5z}$

68. $\dfrac{12y^4 - 42y^2}{-4y}$

69. $\dfrac{8x^3 + 6x^2 - 2x}{2x}$

70. $\dfrac{9x^3 + 12x^2 - 3x}{3x}$

71. $\dfrac{25x^7 - 15x^5 - 5x^4}{5x^3}$

72. $\dfrac{49x^7 - 28x^5 - 7x^4}{7x^3}$

73. $\dfrac{18x^7 - 9x^6 + 20x^5 - 10x^4}{-2x^4}$

74. $\dfrac{25x^8 - 50x^7 + 3x^6 - 40x^5}{-5x^5}$

75. $\dfrac{12x^2y^2 + 6x^2y - 15xy^2}{3xy}$

76. $\dfrac{18a^3b^2 - 9a^2b - 27ab^2}{9ab}$

77. $\dfrac{20x^7y^4 - 15x^3y^2 - 10x^2y}{-5x^2y}$

78. $\dfrac{8x^6y^3 - 12x^8y^2 - 4x^{14}y^6}{-4x^6y^2}$

Practice PLUS

In Exercises 79–82, simplify each expression.

79. $\dfrac{2x^3(4x + 2) - 3x^2(2x - 4)}{2x^2}$

80. $\dfrac{6x^3(3x - 1) + 5x^2(6x - 3)}{3x^2}$

81. $\left(\dfrac{18x^2y^4}{9xy^2}\right) - \left(\dfrac{15x^5y^6}{5x^4y^4}\right)$

82. $\left(\dfrac{9x^3 + 6x^2}{3x}\right) - \left(\dfrac{12x^2y^2 - 4xy^2}{2xy^2}\right)$

83. Divide the sum of $(y + 5)^2$ and $(y + 5)(y - 5)$ by $2y$.

84. Divide the sum of $(y + 4)^2$ and $(y + 4)(y - 4)$ by $2y$.

In Exercises 85–86, the variable n in each exponent represents a natural number. Divide the polynomial by the monomial. Then use polynomial multiplication to check the quotient.

85. $\dfrac{12x^{15n} - 24x^{12n} + 8x^{3n}}{4x^{3n}}$

86. $\dfrac{35x^{10n} - 15x^{8n} + 25x^{2n}}{5x^{2n}}$

Application Exercises

The bar graphs show U.S. film box-office receipts, in millions of dollars, and box-office admissions, in millions of tickets sold, for five selected years.

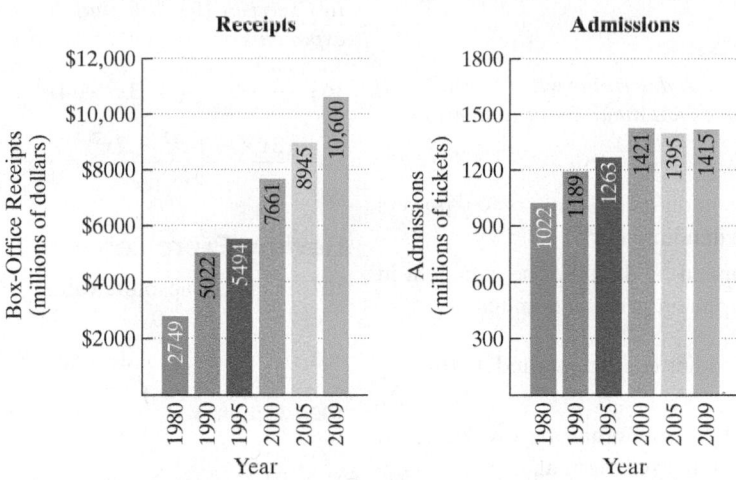

United States Film Box-Office Receipts and Admissions

Sources: U.S. Department of Commerce, Motion Picture Association of America, National Association of Theatre Owners

The following polynomial models of degree 2 can be used to describe the data in the bar graphs:

Receipts, in millions of dollars

Admissions, in millions of tickets sold

$$R = 3.6x^2 + 158x + 2790 \qquad A = -0.2x^2 + 21x + 1015.$$

The variable x represents the number of years after 1980.

Use this information to solve Exercises 87–88.

87. a. Use the data displayed by the bar graphs to find the average admission charge for a film ticket in 2000. Round to two decimal places, or to the nearest cent.

b. Use the models to write an algebraic expression that describes the average admission charge for a film ticket x years after 1980.

c. Use the model from part (b) to find the average admission charge for a film ticket in 2000. Round to the nearest cent. Does the model underestimate or overestimate the actual average charge that you found in part (a)? By how much?

d. Can the polynomial division for the model in part (b) be performed using the methods that you learned in this section? Explain your answer.

88. a. Use the data displayed by the bar graphs to find the average admission charge for a film ticket in 2005. Round to two decimal places, or to the nearest cent.

b. Use the models to write an algebraic expression that describes the average admission charge for a film ticket x years after 1980.

c. Use the model from part (b) to find the average admission charge for a film ticket in 2005. Round to the nearest cent. Does the model underestimate or overestimate the actual average charge that you found in part (a)? By how much?

d. Can the polynomial division for the model in part (b) be performed using the methods that you learned in this section? Explain your answer.

Writing in Mathematics

89. Explain the quotient rule for exponents. Use $\dfrac{3^6}{3^2}$ in your explanation.

90. Explain how to find any nonzero number to the 0 power.

91. Explain the difference between $(-7)^0$ and -7^0.

92. Explain how to simplify an expression that involves a quotient raised to a power. Provide an example with your explanation.

93. Explain how to divide monomials. Give an example.

94. Explain how to divide a polynomial that is not a monomial by a monomial. Give an example.

95. Are the expressions

$$\frac{12x^2 + 6x}{3x} \quad \text{and} \quad 4x + 2$$

equal for every value of x? Explain.

Critical Thinking Exercises

Make Sense? *In Exercises 96–99, determine whether each statement "makes sense" or "does not make sense" and explain your reasoning.*

96. Because division by 0 is undefined, numbers to 0 powers should not be written in denominators.

97. The quotient rule is applied by dividing the exponent in the numerator by the exponent in the denominator.

98. I divide monomials by dividing coefficients and subtracting exponents.

99. I divide a polynomial by a monomial by dividing each term of the monomial by the polynomial.

In Exercises 100–103, determine whether each statement is true or false. If the statement is false, make the necessary change(s) to produce a true statement.

100. $x^{10} \div x^2 = x^5$ for all nonzero real numbers x.

101. $\dfrac{12x^3 - 6x}{2x} = 6x^2 - 6x$

102. $\dfrac{x^2 + x}{x} = x$

103. If a polynomial in x of degree 6 is divided by a monomial in x of degree 2, the degree of the quotient is 4.

104. What polynomial, when divided by $3x^2$, yields the trinomial $6x^6 - 9x^4 + 12x^2$ as a quotient?

In Exercises 105–106, find the missing coefficients and exponents designated by question marks.

105. $\dfrac{?x^8 - ?x^6}{3x^?} = 3x^5 - 4x^3$

106. $\dfrac{3x^{14} - 6x^{12} - ?x^7}{?x^?} = -x^7 + 2x^5 + 3$

Review Exercises

107. Find the absolute value: $|-20.3|$. (Section 1.3, Example 8)

108. Express $\frac{7}{8}$ as a decimal. (Section 1.3, Example 4)

109. Graph: $y = \frac{1}{3}x + 2$. (Section 3.4, Example 3)

Preview Exercises

Exercises 110–112 will help you prepare for the material covered in the next section. In each exercise, perform the long division without using a calculator, and then state the quotient and the remainder.

110. $19\overline{)494}$

111. $24\overline{)2958}$

112. $98\overline{)25{,}187}$

SECTION 5.6

Long Division of Polynomials; Synthetic Division

Objectives

1. Use long division to divide by a polynomial containing more than one term.
2. Divide polynomials using synthetic division.

The Family (1963), Marisol Escobar. © 2011 Marisol Escobar/VAGA

"He was an arithmetician rather than a mathematician. None of the humor, the music, or the mysticism of higher mathematics ever entered his head."

—*John Steinbeck*

"Arithmetic has a very great and elevating effect. He who can properly divide is to be considered a god."

—*Plato*

"You cannot ask us to take sides against arithmetic."

—*Winston Churchill*

So, what's the deal? Will performing the repetitive procedure of long division (don't reach for that calculator!) have an elevating effect? Or will it confine you to a computational box that allows neither humor nor music to enter? Forget the box: Mathematician Wilhelm Leibniz believed that music is nothing but unconscious arithmetic. But do think elevation, if not to the level of an ancient Greek god, then to new, algebraic highs. The bottom line: Understanding long division of whole numbers lays the foundation for performing the division of a polynomial by a binomial, such as

$$x + 3 \overline{)x^2 + 10x + 21}.$$

Divisor has two terms and is a binomial.   The polynomial dividend has three terms and is a trinomial.

In this section, you will learn how to perform such divisions.

The Steps in Dividing a Polynomial by a Binomial

Dividing a polynomial by a binomial may remind you of long division. Let's review long division of whole numbers by dividing 3983 by 26.

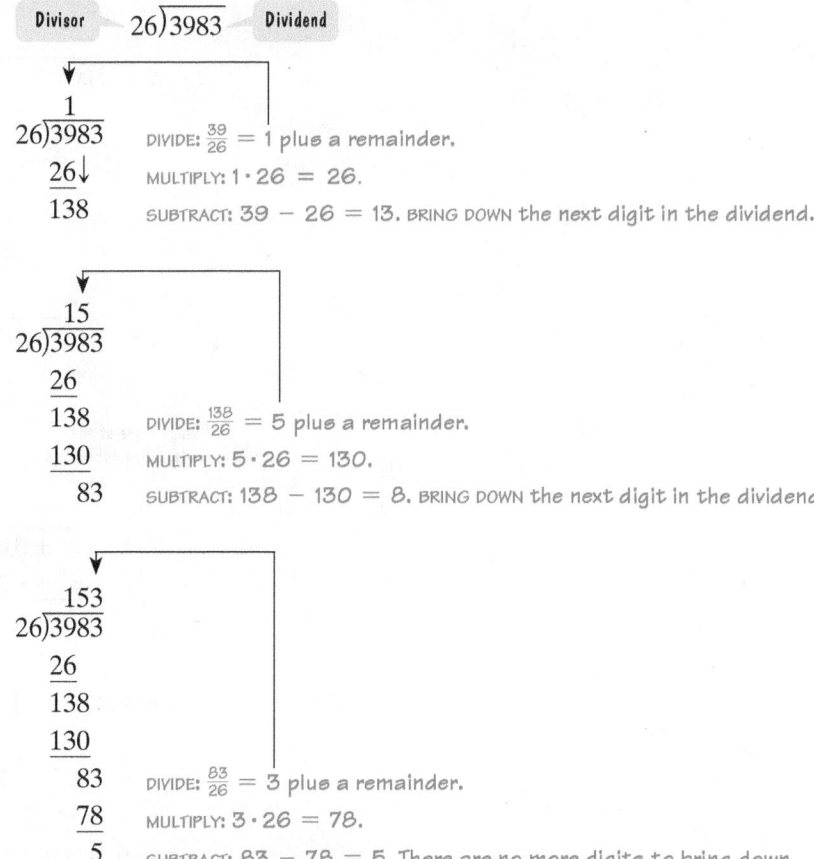

The quotient is 153 and the remainder is 5. This can be written as

Quotient $153\dfrac{5}{26}.$ Remainder   Divisor

We see that $26\overline{)3983}$ $153\frac{5}{26}$.

This answer can be checked. Multiply the divisor and the quotient. Then add the remainder. If the result is the dividend, the answer is correct. In this case, we have

$$26(153) \;+\; 5 \;=\; 3978 + 5 \;=\; 3983.$$

Divisor | Quotient | Remainder | This is the dividend

1 Use long division to divide by a polynomial containing more than one term.

Because we obtained the dividend, the answer to the division problem, $153\frac{5}{26}$, is correct.

When a divisor is a binomial, the four steps used to divide whole numbers—**divide, multiply, subtract, bring down the next term**—form the repetitive procedure for dividing a polynomial by a binomial.

EXAMPLE 1 Dividing a Polynomial by a Binomial

Divide $x^2 + 10x + 21$ by $x + 3$.

Solution The following steps illustrate how polynomial division is very similar to numerical division.

$$x + 3\overline{)x^2 + 10x + 21}$$

Arrange the terms of the dividend ($x^2 + 10x + 21$) and the divisor ($x + 3$) in descending powers of x.

$$\begin{array}{r} x \\ x + 3\overline{)x^2 + 10x + 21} \end{array}$$

DIVIDE x^2 (the first term in the dividend) by x (the first term in the divisor): $\dfrac{x^2}{x} = x$. Align like terms.

$$x(x+3) = x^2 + 3x \qquad \begin{array}{r} x \\ x + 3\overline{)x^2 + 10x + 21} \\ x^2 + 3x \end{array}$$

MULTIPLY each term in the divisor ($x + 3$) by x, aligning terms of the product under like terms in the dividend.

$$\begin{array}{r} x \\ x + 3\overline{)x^2 + 10x + 21} \\ \ominus x^2 \ominus 3x \\ \hline 7x \end{array}$$

SUBTRACT $x^2 + 3x$ from $x^2 + 10x$ by changing the sign of each term in the lower expression and adding.

Change signs of the polynomial being subtracted.

$$\begin{array}{r} x \\ x + 3\overline{)x^2 + 10x + 21} \\ x^2 + 3x \downarrow \\ \hline 7x + 21 \end{array}$$

BRING DOWN 21 from the original dividend and add algebraically to form a new dividend.

$$\begin{array}{r} x + 7 \\ x + 3\overline{)x^2 + 10x + 21} \\ x^2 + 3x \\ \hline 7x + 21 \end{array}$$

FIND the second term of the quotient. Divide the first term of $7x + 21$ by x, the first term of the divisor: $\dfrac{7x}{x} = 7$.

$$7(x + 3) = 7x + 21 \qquad \begin{array}{r} x + 7 \\ x + 3\overline{)x^2 + 10x + 21} \\ x^2 + 3x \\ \hline 7x + 21 \\ \ominus 7x \ominus 21 \\ \hline 0 \end{array}$$

MULTIPLY the divisor ($x + 3$) by 7, aligning under like terms in the new dividend. Then subtract to obtain the remainder of 0.

Remainder

The quotient is $x + 7$ and the remainder is 0. We will not list a remainder of 0 in the answer. Thus,

$$\frac{x^2 + 10x + 21}{x + 3} = x + 7. \quad \blacksquare$$

When dividing polynomials by binomials, the answer can be checked. Find the product of the divisor and the quotient and add the remainder. If the result is the dividend, the answer to the division problem is correct. For example, let's check our work in Example 1.

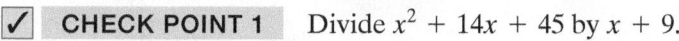

Dividend

$$\frac{x^2 + 10x + 21}{x + 3} = x + 7$$

Divisor

Quotient to be checked

Multiply the divisor and the quotient and add the remainder, 0:

$$(x + 3)(x + 7) + 0 = x^2 + 7x + 3x + 21 + 0 = x^2 + 10x + 21.$$

Divisor   Quotient   Remainder

This is the dividend.

Because we obtained the dividend, the quotient is correct.

☑ **CHECK POINT 1**   Divide $x^2 + 14x + 45$ by $x + 9$.

Before considering additional examples, let's summarize the general procedure for dividing a polynomial by a binomial.

Dividing a Polynomial by a Binomial

1. **Arrange the terms** of both the dividend and the divisor in descending powers of the variable.
2. **Divide** the first term in the dividend by the first term in the divisor. The result is the first term of the quotient.
3. **Multiply** every term in the divisor by the first term in the quotient. Write the resulting product beneath the dividend with like terms lined up.
4. **Subtract** the product from the dividend.
5. **Bring down** the next term in the original dividend and write it next to the remainder to form a new dividend.
6. Use this new expression as the dividend and repeat this process until the remainder can no longer be divided. This will occur when the degree of the remainder (the highest exponent on a variable in the remainder) is less than the degree of the divisor.

In our next division, we will obtain a nonzero remainder.

EXAMPLE 2   Dividing a Polynomial by a Binomial

Divide: $\dfrac{7x - 9 - 4x^2 + 4x^3}{2x - 1}$.

Solution We begin by writing the dividend, $7x - 9 - 4x^2 + 4x^3$, in descending powers of x.

$$7x - 9 - 4x^2 + 4x^3 = 4x^3 - 4x^2 + 7x - 9$$

Think of 9 as $9x^0$. The powers descend from 3 to 0.

$$2x - 1 \overline{)4x^3 - 4x^2 + 7x - 9}$$

This is the problem with the dividend in descending powers of x.

$$\begin{array}{r} 2x^2 \\ 2x - 1 \overline{)4x^3 - 4x^2 + 7x - 9} \end{array}$$

DIVIDE: $\dfrac{4x^3}{2x} = 2x^2$.

$$\begin{array}{r} 2x^2 \\ 2x - 1 \overline{)4x^3 - 4x^2 + 7x - 9} \\ 4x^3 - 2x^2 \end{array}$$

$2x^2(2x - 1) = 4x^3 - 2x^2$

MULTIPLY: $2x^2(2x - 1) = 4x^3 - 2x^2$.

$$\begin{array}{r} 2x^2 \\ 2x - 1 \overline{)4x^3 - 4x^2 + 7x - 9} \\ \ominus 4x^3 \oplus 2x^2 \\ -2x^2 \end{array}$$

SUBTRACT: $4x^3 - 4x^2 - (4x^3 - 2x^2)$
$ = 4x^3 - 4x^2 - 4x^3 + 2x^2$
$ = -2x^2$.

Change signs of the polynomial being subtracted.

$$\begin{array}{r} 2x^2 \\ 2x - 1 \overline{)4x^3 - 4x^2 + 7x - 9} \\ 4x^3 - 2x^2 \downarrow \\ -2x^2 + 7x \end{array}$$

BRING DOWN 7x. The new dividend is $-2x^2 + 7x$.

$$\begin{array}{r} 2x^2 - x \\ 2x - 1 \overline{)4x^3 - 4x^2 + 7x - 9} \\ 4x^3 - 2x^2 \\ -2x^2 + 7x \end{array}$$

DIVIDE: $\dfrac{-2x^2}{2x} = -x$.

$-x(2x - 1) = -2x^2 + x$

$$\begin{array}{r} 2x^2 - x \\ 2x - 1 \overline{)4x^3 - 4x^2 + 7x - 9} \\ 4x^3 - 2x^2 \\ -2x^2 + 7x \\ -2x^2 + x \end{array}$$

MULTIPLY: $-x(2x - 1) = -2x^2 + x$.

$$\begin{array}{r} 2x^2 - x \\ 2x - 1 \overline{)4x^3 - 4x^2 + 7x - 9} \\ 4x^3 - 2x^2 \\ -2x^2 + 7x \\ \oplus \ominus \\ -2x^2 + x \\ 6x \end{array}$$

SUBTRACT: $-2x^2 + 7x - (-2x^2 + x)$
$ = -2x^2 + 7x + 2x^2 - x$
$ = 6x$.

$$\begin{array}{r} 2x^2 - x \\ 2x - 1 \overline{)4x^3 - 4x^2 + 7x - 9} \\ 4x^3 - 2x^2 \\ -2x^2 + 7x \\ -2x^2 + x \\ 6x - 9 \end{array}$$

BRING DOWN -9. The new dividend is $6x - 9$.

$$
\begin{array}{r}
2x^2 - x + 3 \\
2x - 1 \overline{\smash{)}\, 4x^3 - 4x^2 + 7x - 9} \\
\underline{4x^3 - 2x^2} \\
-2x^2 + 7x \\
\underline{-2x^2 + x} \\
6x - 9
\end{array}
$$

DIVIDE: $\dfrac{6x}{2x} = 3.$

$3(2x - 1) = 6x - 3$

$$
\begin{array}{r}
2x^2 - x + 3 \\
2x - 1 \overline{\smash{)}\, 4x^3 - 4x^2 + 7x - 9} \\
\underline{4x^3 - 2x^2} \\
-2x^2 + 7x \\
\underline{-2x^2 + x} \\
6x - 9 \\
6x - 3
\end{array}
$$

MULTIPLY: $3(2x - 1) = 6x - 3.$

$$
\begin{array}{r}
2x^2 - x + 3 \\
2x - 1 \overline{\smash{)}\, 4x^3 - 4x^2 + 7x - 9} \\
\underline{4x^3 - 2x^2} \\
-2x^2 + 7x \\
\underline{-2x^2 + x} \\
6x - 9 \\
\ominus 6x \oplus 3 \\
\underline{} \\
-6
\end{array}
$$

SUBTRACT: $6x - 9 - (6x - 3)$
$\qquad = 6x - 9 - 6x + 3$
$\qquad = -6.$

Remainder

The quotient is $2x^2 - x + 3$ and the remainder is -6. When there is a nonzero remainder, as in this example, list the quotient, plus the remainder above the divisor. Thus,

$$
\frac{7x - 9 - 4x^2 + 4x^3}{2x - 1} = 2x^2 - x + 3 + \frac{-6}{2x - 1}
$$

Remainder above divisor

Quotient

or

$$
\frac{7x - 9 - 4x^2 + 4x^3}{2x - 1} = 2x^2 - x + 3 - \frac{6}{2x - 1}.
$$

Check this result by showing that the product of the divisor and the quotient,

$$
(2x - 1)(2x^2 - x + 3),
$$

plus the remainder, -6, is the dividend, $7x - 9 - 4x^2 + 4x^3$. ∎

✓ **CHECK POINT 2** Divide: $\dfrac{6x + 8x^2 - 12}{2x + 3}.$

If a power of the variable is missing in a dividend, add that power of the variable with a coefficient of 0 and then divide. In this way, like terms will be aligned as you carry out the division.

EXAMPLE 3 Dividing a Polynomial with Missing Terms

Divide: $\dfrac{8x^3 - 1}{2x - 1}$.

Solution We write the dividend, $8x^3 - 1$, as

$$8x^3 + 0x^2 + 0x - 1.$$

> Use a coefficient of 0 with missing terms.

By doing this, we will keep all like terms aligned.

$4x^2(2x-1)=8x^3-4x^2$

$$
\begin{array}{r}
4x^2 \\
2x - 1 \overline{)\,8x^3 + 0x^2 + 0x - 1} \\
8x^3 - 4x^2 \\
\hline
4x^2 + 0x
\end{array}
$$

Divide $\left(\dfrac{8x^3}{2x} = 4x^2\right)$, multiply $[4x^2(2x-1)=8x^3-4x^2]$, subtract, and bring down the next term.
The new dividend is $4x^2 + 0x$.

$2x(2x-1)=4x^2-2x$

$$
\begin{array}{r}
4x^2 + 2x \\
2x - 1 \overline{)\,8x^3 + 0x^2 + 0x - 1} \\
8x^3 - 4x^2 \\
\hline
4x^2 + 0x \\
4x^2 - 2x \\
\hline
2x - 1
\end{array}
$$

Divide $\left(\dfrac{4x^2}{2x} = 2x\right)$, multiply $[2x(2x-1)=4x^2-2x]$, subtract, and bring down the next term.
The new dividend is $2x - 1$.

$1(2x-1)=2x-1$

$$
\begin{array}{r}
4x^2 + 2x + 1 \\
2x - 1 \overline{)\,8x^3 + 0x^2 + 0x - 1} \\
8x^3 - 4x^2 \\
\hline
4x^2 + 0x \\
4x^2 - 2x \\
\hline
2x - 1 \\
2x - 1 \\
\hline
0
\end{array}
$$

Divide $\left(\dfrac{2x}{2x} = 1\right)$, multiply $[1(2x-1)=2x-1]$, and subtract.
The remainder is 0.

Thus,

$$\frac{8x^3 - 1}{2x - 1} = 4x^2 + 2x + 1.$$

Check this result by showing that the product of the divisor and the quotient

$$(2x - 1)(4x^2 + 2x + 1)$$

plus the remainder, 0, is the dividend, $8x^3 - 1$. ∎

CHECK POINT 3 Divide: $\dfrac{x^3 - 1}{x - 1}$.

Using Technology

Graphic Connections

The graphs of $y_1 = \dfrac{8x^3 - 1}{2x - 1}$ and $y_2 = 4x^2 + 2x + 1$ are shown below.

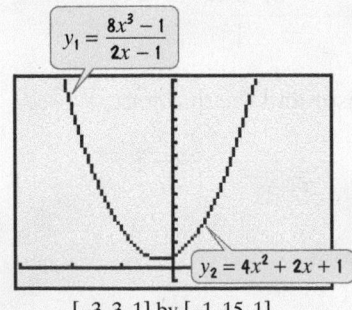

$y_1 = \dfrac{8x^3 - 1}{2x - 1}$

$y_2 = 4x^2 + 2x + 1$

$[-3, 3, 1]$ by $[-1, 15, 1]$

The graphs coincide. Thus,

$$\frac{8x^3 - 1}{2x - 1} = 4x^2 + 2x + 1.$$

EXAMPLE 4 Long Division of Polynomials

Divide $6x^4 + 5x^3 + 3x - 5$ by $3x^2 - 2x$.

Solution We write the dividend, $6x^4 + 5x^3 + 3x - 5$, as $6x^4 + 5x^3 + 0x^2 + 3x - 5$ to keep all like terms aligned.

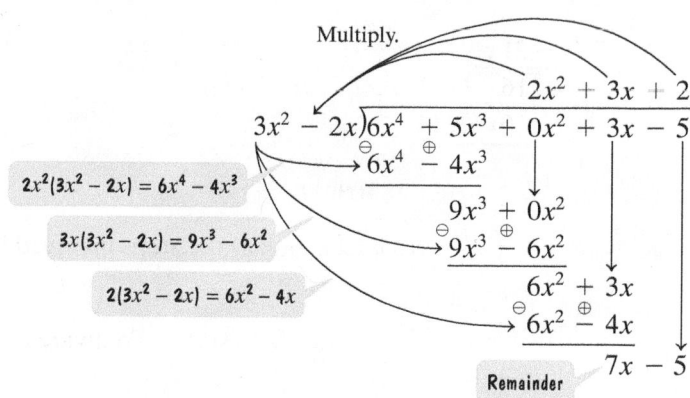

The division process is finished because the degree of $7x - 5$, which is 1, is less than the degree of the divisor $3x^2 - 2x$, which is 2. The answer is

$$\frac{6x^4 + 5x^3 + 3x - 5}{3x^2 - 2x} = 2x^2 + 3x + 2 + \frac{7x - 5}{3x^2 - 2x}. \blacksquare$$

✓ **CHECK POINT 4** Divide $2x^4 + 3x^3 - 7x - 10$ by $x^2 - 2x$.

---

2 Divide polynomials using synthetic division.

Dividing Polynomials Using Synthetic Division

We can use **synthetic division** to divide polynomials if the divisor is of the form $x - c$. This method provides a quotient more quickly than long division. Let's compare the two methods showing $x^3 + 4x^2 - 5x + 5$ divided by $x - 3$.

Long Division

Quotient

$$x - 3 \overline{)\begin{array}{r} x^2 + 7x + 16 \\ x^3 + 4x^2 - 5x + 5 \end{array}}$$

Divisor
$x - c$;
$c = 3$

$$\underset{\ominus}{x^3} \underset{\oplus}{-} 3x^2$$ Dividend

$$\underset{\ominus}{7x^2} \underset{\oplus}{-} 5x$$
$$7x^2 - 21x$$

$$16x + 5$$
$$\underset{\ominus}{16x} \underset{\oplus}{-} 48$$

$$53$$ Remainder

Synthetic Division

$$\begin{array}{r|rrrr} 3 & 1 & 4 & -5 & 5 \\ & & 3 & 21 & 48 \\ \hline & 1 & 7 & 16 & 53 \end{array}$$

On the next page, we observe the relationship between the polynomials in the long division process and the numbers that appear in synthetic division.

Long Division

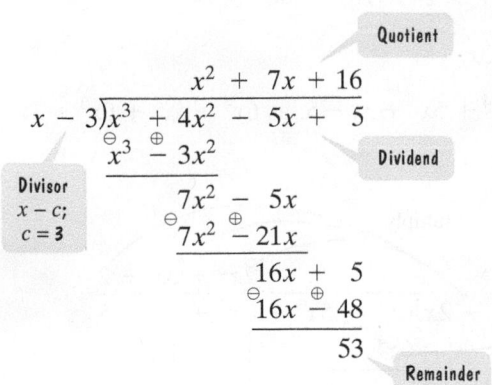

Synthetic Division

These are the coefficients of the dividend $x^3 + 4x^2 - 5x + 5$.

The divisor is $x - 3$. This is 3, or c, in $x - c$.

$$3 \,\rvert\; \begin{array}{rrrr} 1 & 4 & -5 & 5 \\ & 3 & 21 & 48 \\ \hline 1 & 7 & 16 & 53 \end{array}$$

These are the coefficients of the quotient $x^2 + 7x + 16$.

This is the remainder.

Now let's look at the steps involved in synthetic division.

Synthetic Division To divide a polynomial by $x - c$:

Example

1. Arrange polynomials in descending powers, with a 0 coefficient for any missing term.

$$x - 3\overline{)x^3 + 4x^2 - 5x + 5}$$

2. Write c for the divisor, $x - c$. To the right, write the coefficients of the dividend.

$$3 \,\rvert\; \begin{array}{rrrr} 1 & 4 & -5 & 5 \end{array}$$

3. Write the leading coefficient of the dividend on the bottom row.

$$3 \,\rvert\; \begin{array}{rrrr} 1 & 4 & -5 & 5 \end{array}$$

Bring down 1.

1

4. Multiply c (in this case, 3) times the value just written on the bottom row. Write the product in the next column in the second row.

$$3 \,\rvert\; \begin{array}{rrrr} 1 & 4 & -5 & 5 \\ & 3 \\ \hline 1 \end{array}$$

Multiply by 3: $3 \cdot 1 = 3$.

5. Add the values in this new column, writing the sum in the bottom row.

$$3 \,\rvert\; \begin{array}{rrrr} 1 & 4 & -5 & 5 \\ & 3 \\ \hline 1 & 7 \end{array}$$

Add.

6. Repeat this series of multiplications and additions until all columns are filled in.

$$3 \,\rvert\; \begin{array}{rrrr} 1 & 4 & -5 & 5 \\ & 3 & 21 \\ \hline 1 & 7 & 16 \end{array}$$

Add.

Multiply by 3: $3 \cdot 7 = 21$.

$$3 \,\rvert\; \begin{array}{rrrr} 1 & 4 & -5 & 5 \\ & 3 & 21 & 48 \\ \hline 1 & 7 & 16 & 53 \end{array}$$

Add.

Multiply by 3: $3 \cdot 16 = 48$.

7. Use the numbers in the last row to write the quotient, plus the remainder above the divisor. **The degree of the first term of the quotient is one less than the degree of the first term of the dividend.** The final value in this row is the remainder.

Written from
1 7 16 53
the last row of the synthetic division

$$1x^2 + 7x + 16 + \dfrac{53}{x - 3}$$

$$x - 3\overline{)x^3 + 4x^2 - 5x + 5}$$

EXAMPLE 5 Using Synthetic Division

Use synthetic division to divide $5x^3 + 6x + 8$ by $x + 2$.

Solution The divisor must be in the form $x - c$. Thus, we write $x + 2$ as $x - (-2)$. This means that $c = -2$. Writing a 0 coefficient for the missing x^2-term in the dividend, we can express the division as follows:

$$x - (-2) \overline{)5x^3 + 0x^2 + 6x + 8.}$$

Now we are ready to set up the problem so that we can use synthetic division.

Use the coefficients of the dividend
$5x^3 + 0x^2 + 6x + 8$ in descending powers of x.

This is c in
$x - (-2)$. $-2 \mid$ 5 0 6 8

We begin the synthetic division process by bringing down 5. This is followed by a series of multiplications and additions.

1. Bring down 5.

$-2 \mid$ 5 0 6 8

5

2. Multiply: $-2(5) = -10$.

$-2 \mid$ 5   0 6 8
           -10

5

Multiply 5 by -2.

3. Add: $0 + (-10) = -10$.

$-2 \mid$ 5   0 6 8
           -10 | Add.

5 -10

4. Multiply: $-2(-10) = 20$.

$-2 \mid$ 5   0   6 8
          -10 20

5 -10

Multiply -10 by -2.

5. Add: $6 + 20 = 26$.

$-2 \mid$ 5   0   6 8
          -10 20 | Add.

5 -10 26

6. Multiply: $-2(26) = -52$.

$-2 \mid$ 5   0   6   8
          -10 20 -52

5 -10 26

Multiply 26 by -2.

7. Add: $8 + (-52) = -44$.

$-2 \mid$ 5   0   6   8
          -10 20 -52 | Add.

5 -10 26 -44

The numbers in the last row represent the coefficients of the quotient and the remainder. The degree of the first term of the quotient is one less than that of the dividend. Because the degree of the dividend, $5x^3 + 6x + 8$, is 3, the degree of the quotient is 2. This means that the 5 in the last row represents $5x^2$.

$-2 \mid$ 5   0   6   8
          -10 20 -52

5 -10 26 -44

The quotient is
$5x^2 - 10x + 26$.

The remainder
is -44.

Thus,

$$x + 2 \overline{)5x^3 + 6x + 8}} \quad 5x^2 - 10x + 26 - \frac{44}{x + 2}$$

✓ **CHECK POINT 5** Use synthetic division to divide $x^3 - 7x - 6$ by $x + 2$.

CONCEPT AND VOCABULARY CHECK

Fill in each blank so that the resulting statement is true.

1. Consider the following long division problem:

$$2x + 3 \overline{\smash{)}4x^2 + 9 + 10x^3}.$$

We begin the division process by rewriting the dividend as _____.

2. Consider the following long division problem:

$$2x + 1 \overline{\smash{)}8x^2 + 10x - 1}.$$

We begin the division process by dividing _____ by _____. We obtain _____. We write this result above _____ in the dividend.

3. In the following long division problem, the first step has been completed:

$$\begin{array}{r} 5x \\ 3x - 2 \overline{\smash{)}15x^2 - 22x + 19}. \end{array}$$

The next step is to multiply _____ and _____. We obtain _____. We write this result below _____.

4. In the following long division problem, the first steps have been completed:

$$\begin{array}{r} 2x \\ 3x - 5 \overline{\smash{)}6x^2 + 8x - 4}. \\ \underline{6x^2 - 10x} \end{array}$$

The next step is to subtract _____ from _____. We obtain _____. Then we bring down _____ and form the new dividend _____.

5. In the following long division problem, most of the steps have been completed:

$$\begin{array}{r} x - 12 \\ x - 5 \overline{\smash{)}x^2 - 17x + 74}. \\ \underline{x^2 - 5x} \\ -12x + 74 \\ \underline{-12x + 60} \\ ? \end{array}$$

Completing the step designated by the question mark, we obtain _____. Thus, the quotient is _____ and the remainder is _____. The answer to this long division problem is _____.

6. To divide $x^3 + 5x^2 - 7x + 1$ by $x - 4$ using synthetic division, the first step is to write

$$\underline{}\rfloor \ \underline{} \ \underline{} \ \underline{} \ \underline{}.$$

7. To divide $4x^3 - 8x - 2$ by $x + 5$ using synthetic division, the first step is to write

$$\underline{}\rfloor \ \underline{} \ \underline{} \ \underline{} \ \underline{}.$$

8. True or false:

$$\begin{array}{r} -1\rfloor \ \ 3 \ \ -4 \ \ \ 2 \ \ -1 \\ -3 \ \ \ \ 7 \ \ -9 \\ \hline 3 \ \ -7 \ \ \ 9 \ \ -10 \end{array} \quad \text{means} \quad \frac{3x^3 - 4x^2 + 2x - 1}{x + 1} = 3x^2 - 7x + 9 - \frac{10}{x + 1}. \ _____$$

5.6 EXERCISE SET

MyMathLab®

Watch the videos
in MyMathLab

Download the
MyDashBoard App

Practice Exercises

In Exercises 1–40, divide as indicated. Check each answer by showing that the product of the divisor and the quotient, plus the remainder, is the dividend.

1. $\dfrac{x^2 + 6x + 8}{x + 2}$

2. $\dfrac{x^2 + 7x + 10}{x + 5}$

3. $\dfrac{2x^2 + x - 10}{x - 2}$

4. $\dfrac{2x^2 + 13x + 15}{x + 5}$

5. $\dfrac{x^2 - 5x + 6}{x - 3}$

6. $\dfrac{x^2 - 2x - 24}{x + 4}$

7. $\dfrac{2y^2 + 5y + 2}{y + 2}$

8. $\dfrac{2y^2 - 13y + 21}{y - 3}$

9. $\dfrac{x^2 - 3x + 4}{x + 2}$

10. $\dfrac{x^2 - 7x + 5}{x + 3}$

11. $\dfrac{5y + 10 + y^2}{y + 2}$

12. $\dfrac{-8y + y^2 - 9}{y - 3}$

13. $\dfrac{x^3 - 6x^2 + 7x - 2}{x - 1}$

14. $\dfrac{x^3 + 3x^2 + 5x + 3}{x + 1}$

15. $\dfrac{12y^2 - 20y + 3}{2y - 3}$

16. $\dfrac{4y^2 - 8y - 5}{2y + 1}$

17. $\dfrac{4a^2 + 4a - 3}{2a - 1}$

18. $\dfrac{2b^2 - 9b - 5}{2b + 1}$

19. $\dfrac{3y - y^2 + 2y^3 + 2}{2y + 1}$

20. $\dfrac{9y + 18 - 11y^2 + 12y^3}{4y + 3}$

21. $\dfrac{6x^2 - 5x - 30}{2x - 5}$

22. $\dfrac{4y^2 + 8y + 3}{2y - 1}$

23. $\dfrac{x^3 + 4x - 3}{x - 2}$

24. $\dfrac{x^3 + 2x^2 - 3}{x - 2}$

25. $\dfrac{4y^3 + 8y^2 + 5y + 9}{2y + 3}$

26. $\dfrac{2y^3 - y^2 + 3y + 2}{2y + 1}$

27. $\dfrac{6y^3 - 5y^2 + 5}{3y + 2}$

28. $\dfrac{4y^3 + 3y + 5}{2y - 3}$

29. $\dfrac{27x^3 - 1}{3x - 1}$

30. $\dfrac{8x^3 + 27}{2x + 3}$

31. $\dfrac{81 - 12y^3 + 54y^2 + y^4 - 108y}{y - 3}$

32. $\dfrac{8y^3 + y^4 + 16 + 32y + 24y^2}{y + 2}$

33. $\dfrac{4y^2 + 6y}{2y - 1}$

34. $\dfrac{10x^2 - 3x}{x + 3}$

35. $\dfrac{y^4 - 2y^2 + 5}{y - 1}$

36. $\dfrac{y^4 - 6y^2 + 3}{y - 1}$

37. $(4x^4 + 3x^3 + 4x^2 + 9x - 6) \div (x^2 + 3)$

38. $(3x^5 - x^3 + 4x^2 - 12x - 8) \div (x^2 - 2)$

39. $(15x^4 + 3x^3 + 4x^2 + 4) \div (3x^2 - 1)$

40. $(18x^4 + 9x^3 + 3x^2) \div (3x^2 + 1)$

In Exercises 41–58, divide using synthetic division. In the first two exercises, begin the process as shown.

41. $(2x^2 + x - 10) \div (x - 2)$ $\underline{2}|$ 2 1 −10

42. $(x^2 + x - 2) \div (x - 1)$ $\underline{1}|$ 1 1 −2

43. $(3x^2 + 7x - 20) \div (x + 5)$

44. $(5x^2 - 12x - 8) \div (x + 3)$

45. $(4x^3 - 3x^2 + 3x - 1) \div (x - 1)$

46. $(5x^3 - 6x^2 + 3x + 11) \div (x - 2)$

47. $(6x^5 - 2x^3 + 4x^2 - 3x + 1) \div (x - 2)$

48. $(x^5 + 4x^4 - 3x^2 + 2x + 3) \div (x - 3)$

49. $(x^2 - 5x - 5x^3 + x^4) \div (5 + x)$

50. $(x^2 - 6x - 6x^3 + x^4) \div (6 + x)$

51. $(3x^3 + 2x^2 - 4x + 1) \div \left(x - \dfrac{1}{3}\right)$

52. $(2x^4 - x^3 + 2x^2 - 3x + 1) \div \left(x - \dfrac{1}{2}\right)$

53. $\dfrac{x^5 + x^3 - 2}{x - 1}$

54. $\dfrac{x^7 + x^5 - 10x^3 + 12}{x + 2}$

55. $\dfrac{x^4 - 256}{x - 4}$

56. $\dfrac{x^7 - 128}{x - 2}$

57. $\dfrac{2x^5 - 3x^4 + x^3 - x^2 + 2x - 1}{x + 2}$

58. $\dfrac{x^5 - 2x^4 - x^3 + 3x^2 - x + 1}{x - 2}$

Practice PLUS

In Exercises 59–68, divide as indicated.

59. $\dfrac{x^4 + y^4}{x + y}$

60. $\dfrac{x^5 + y^5}{x + y}$

61. $\dfrac{3x^4 + 5x^3 + 7x^2 + 3x - 2}{x^2 + x + 2}$

62. $\dfrac{x^4 - x^3 - 7x^2 - 7x - 2}{x^2 - 3x - 2}$

63. $\dfrac{4x^3 - 3x^2 + x + 1}{x^2 + x + 1}$

64. $\dfrac{x^4 - x^2 + 1}{x^2 + x + 1}$

65. $\dfrac{x^5 - 1}{x^2 - x + 2}$

66. $\dfrac{5x^5 - 7x^4 + 3x^3 - 20x^2 + 28x - 12}{x^3 - 4}$

67. $\dfrac{4x^3 - 7x^2y - 16xy^2 + 3y^3}{x - 3y}$

68. $\dfrac{12x^3 - 19x^2y + 13xy^2 - 10y^3}{4x - 5y}$

69. Divide the difference between $4x^3 + x^2 - 2x + 7$ and $3x^3 - 2x^2 - 7x + 4$ by $x + 1$.

70. Divide the difference between $4x^3 + 2x^2 - x - 1$ and $2x^3 - x^2 + 2x - 5$ by $x + 2$.

Application Exercises

71. Write a simplified polynomial that represents the length of the rectangle.

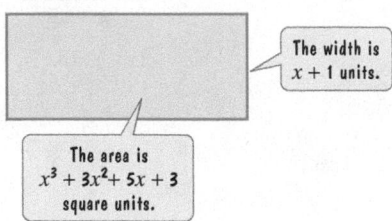

The width is $x + 1$ units.

The area is $x^3 + 3x^2 + 5x + 3$ square units.

72. Write a simplified polynomial that represents the measure of the base of the parallelogram.

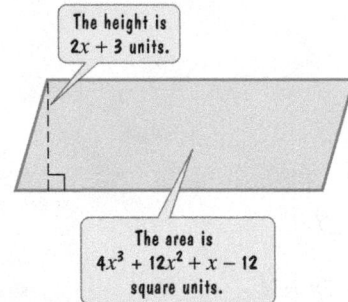

The height is $2x + 3$ units.

The area is $4x^3 + 12x^2 + x - 12$ square units.

You just signed a contract for a new job. The salary for the first year is $30,000 and there is to be a percent increase in your salary each year. The algebraic expression

$$\frac{30{,}000x^n - 30{,}000}{x - 1}$$

describes your total salary over n years, where x is the sum of 1 and the yearly percent increase, expressed as a decimal. Use this information to solve Exercises 73–74.

73. a. Use the given expression and write a quotient of polynomials that describes your total salary over three years.

   b. Simplify the expression in part (a) by performing the division.

   c. Suppose you are to receive an increase of 5% per year. Thus, x is the sum of 1 and 0.05, or 1.05. Substitute 1.05 for x in the expression in part (a) as well as in the simplified form of the expression in part (b). Evaluate each expression. What is your total salary over the three-year period?

74. a. Use the expression given above and write a quotient of polynomials that describes your total salary over four years.

   b. Simplify the expression in part (a) by performing the division.

c. Suppose you are to receive an increase of 8% per year. Thus, x is the sum of 1 and 0.08, or 1.08. Substitute 1.08 for x in the expression in part (a) as well as in the simplified form of the expression in part (b). Evaluate each expression. What is your total salary over the four-year period?

Writing in Mathematics

75. In your own words, explain how to divide a polynomial by a binomial. Use $\dfrac{x^2 + 4}{x + 2}$ in your explanation.

76. When dividing a polynomial by a binomial, explain when to stop dividing.

77. After dividing a polynomial by a binomial, explain how to check the answer.

78. When dividing a binomial into a polynomial with missing terms, explain the advantage of writing the missing terms with zero coefficients.

79. Explain how to perform synthetic division. Use the division problem

$$(2x^3 - 3x^2 - 11x + 7) \div (x - 3)$$

to support your explanation.

Critical Thinking Exercises

Make Sense? *In Exercises 80–83, determine whether each statement "makes sense" or "does not make sense" and explain your reasoning. Each statement applies to the division problem*

$$\frac{x^3 + 1}{x + 1}.$$

80. The purpose of writing $x^3 + 1$ as $x^3 + 0x^2 + 0x + 1$ is to keep all like terms aligned.

81. Rewriting $x^3 + 1$ as $x^3 + 0x^2 + 0x + 1$ can change the value of the variable expression for certain values of x.

82. There's no need to apply the long-division process to this problem because I can work the problem in my head and see that the quotient must be $x^2 + 1$.

83. The degree of the quotient must be $3 - 1$.

In Exercises 84–87, determine whether each statement is true or false. If the statement is false, make the necessary change(s) to produce a true statement.

84. If $4x^2 + 25x - 3$ is divided by $4x + 1$, the remainder is 9.

85. If polynomial division results in a remainder of zero, then the product of the divisor and the quotient is the dividend.

86. A nonzero remainder indicates that the answer to a polynomial long-division problem is not a polynomial.

87. When a polynomial is divided by a binomial, the division process stops when the last term of the dividend is brought down.

88. When a certain polynomial is divided by $2x + 4$, the quotient is

$$x - 3 + \frac{17}{2x + 4}.$$

What is the polynomial?

89. Find the number k such that when $16x^2 - 2x + k$ is divided by $2x - 1$, the remainder is 0.

90. Describe the pattern that you observe in the following quotients and remainders.

$$\frac{x^3 - 1}{x + 1} = x^2 - x + 1 - \frac{2}{x + 1}$$

$$\frac{x^5 - 1}{x + 1} = x^4 - x^3 + x^2 - x + 1 - \frac{2}{x + 1}$$

Use this pattern to find $\dfrac{x^7 - 1}{x + 1}$. Verify your result by dividing.

Technology Exercises

In Exercises 91–95, use a graphing utility to determine whether the divisions have been performed correctly. Graph each side of the given equation in the same viewing rectangle. The graphs should coincide. If they do not, correct the expression on the right side by using polynomial division. Then use your graphing utility to show that the division has been performed correctly.

91. $\dfrac{x^2 - 4}{x - 2} = x + 2$

92. $\dfrac{x^2 - 25}{x - 5} = x - 5$

93. $\dfrac{2x^2 + 13x + 15}{x - 5} = 2x + 3$

94. $\dfrac{6x^2 + 16x + 8}{3x + 2} = 2x - 4$

95. $\dfrac{x^3 + 3x^2 + 5x + 3}{x + 1} = x^2 - 2x + 3$

Review Exercises

96. Solve the system:

$$\begin{cases} 7x - 6y = 17 \\ 3x + y = 18. \end{cases}$$

(Section 4.3, Example 2)

97. What is 6% of 20? (Section 2.4, Example 5)

98. Solve: $\dfrac{x}{3} + \dfrac{2}{5} = \dfrac{x}{5} - \dfrac{2}{5}$. (Section 2.3, Example 4)

Preview Exercises

Exercises 99–101 will help you prepare for the material covered in the next section.

99. **a.** Find the missing exponent, designated by the question mark, in each final step.

$$\frac{7^3}{7^5} = \frac{7 \cdot 7 \cdot 7}{7 \cdot 7 \cdot 7 \cdot 7 \cdot 7} = \frac{1}{7^?}$$

$$\frac{7^3}{7^5} = 7^{3-5} = 7^?$$

b. Based on your two results for $\frac{7^3}{7^5}$, what can you conclude?

100. Simplify: $\dfrac{(2x^3)^4}{x^{10}}$.

101. Simplify: $\left(\dfrac{x^5}{x^2}\right)^3$.

SECTION

5.7

Objectives

1. Use the negative exponent rule.
2. Simplify exponential expressions.
3. Convert from scientific notation to decimal notation.
4. Convert from decimal notation to scientific notation.
5. Compute with scientific notation.
6. Solve applied problems using scientific notation.

Negative Exponents and Scientific Notation

Bigger than the biggest thing ever and then some. Much bigger than that in fact, really amazingly immense, a totally stunning size, real 'wow, that's big', time … Gigantic multiplied by colossal multiplied by staggeringly huge is the sort of concept we're trying to get across here.

Douglas Adams, *The Restaurant at the End of the Universe*

Although Adams's description may not quite apply to this $12.4 trillion national debt, exponents can be used to explore the meaning of this "staggeringly huge" number. In this section, you will learn to use exponents to provide a way of putting large and small numbers in perspective.

Negative Integers as Exponents

A nonzero base can be raised to a negative power. The quotient rule can be used to help determine what a negative integer as an exponent should mean. Consider the quotient of b^3 and b^5, where b is not zero. We can determine this quotient in two ways.

$$\frac{b^3}{b^5} = \frac{1 \cdot \cancel{b} \cdot \cancel{b} \cdot \cancel{b}}{\cancel{b} \cdot \cancel{b} \cdot \cancel{b} \cdot b \cdot b} = \frac{1}{b^2}$$

After dividing out pairs of factors, we have two factors of b in the denominator.

$$\frac{b^3}{b^5} = b^{3-5} = b^{-2}$$

Use the quotient rule and subtract exponents.

Notice that $\dfrac{b^3}{b^5}$ equals both b^{-2} and $\dfrac{1}{b^2}$. This means that b^{-2} must equal $\dfrac{1}{b^2}$. This example is a special case of the **negative exponent rule**.

1 Use the negative exponent rule.

The Negative Exponent Rule

If b is any real number other than 0 and n is a natural number, then

$$b^{-n} = \frac{1}{b^n}.$$

Great Question!

Does a negative exponent make the value of an expression negative?

No. For example,

$$7^{-2} = \frac{1}{7^2} = \frac{1}{49}$$

is positive. Avoid these common errors:

Incorrect!

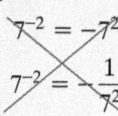

EXAMPLE 1   Using the Negative Exponent Rule

Use the negative exponent rule to write each expression with a positive exponent. Then simplify the expression.

a. 7^{-2} **b.** 4^{-3} **c.** $(-2)^{-4}$ **d.** -2^{-4} **e.** 5^{-1}

Solution

a. $7^{-2} = \dfrac{1}{7^2} = \dfrac{1}{7 \cdot 7} = \dfrac{1}{49}$

b. $4^{-3} = \dfrac{1}{4^3} = \dfrac{1}{4 \cdot 4 \cdot 4} = \dfrac{1}{64}$

c. $(-2)^{-4} = \dfrac{1}{(-2)^4} = \dfrac{1}{(-2)(-2)(-2)(-2)} = \dfrac{1}{16}$

d. $-2^{-4} = -\dfrac{1}{2^4} = -\dfrac{1}{2 \cdot 2 \cdot 2 \cdot 2} = -\dfrac{1}{16}$

> The negative is not inside parentheses and is not taken to the −4 power.

e. $5^{-1} = \dfrac{1}{5^1} = \dfrac{1}{5}$ ∎

✓ **CHECK POINT 1**   Use the negative exponent rule to write each expression with a positive exponent. Then simplify the expression.

a. 6^{-2} **b.** 5^{-3} **c.** $(-3)^{-4}$

d. -3^{-4} **e.** 8^{-1}

Negative exponents can also appear in denominators. For example,

$$\frac{1}{2^{-10}} = \frac{1}{\dfrac{1}{2^{10}}} = 1 \div \frac{1}{2^{10}} = 1 \cdot \frac{2^{10}}{1} = 2^{10}.$$

In general, if a negative exponent appears in a denominator, an expression can be written with a positive exponent using

$$\frac{1}{b^{-n}} = b^n.$$

For example,

$$\frac{1}{2^{-3}} = 2^3 = 8 \qquad \text{and} \qquad \frac{1}{(-6)^{-2}} = (-6)^2 = 36.$$

> Change only the sign of the exponent and not the sign of the base, −6.

Negative Exponents in Numerators and Denominators

If b is any real number other than 0 and n is a natural number, then

$$b^{-n} = \frac{1}{b^n} \quad \text{and} \quad \frac{1}{b^{-n}} = b^n.$$

When a negative number appears as an exponent, switch the position of the base (from numerator to denominator or from denominator to numerator) and make the exponent positive. The sign of the base does not change.

EXAMPLE 2   Using Negative Exponents

Write each expression with positive exponents only. Then simplify, if possible.

a. $\dfrac{4^{-3}}{5^{-2}}$ b. $\left(\dfrac{3}{4}\right)^{-2}$ c. $\dfrac{1}{4x^{-3}}$ d. $\dfrac{x^{-5}}{y^{-1}}$

Solution

a. $\dfrac{4^{-3}}{5^{-2}} = \dfrac{5^2}{4^3} = \dfrac{5 \cdot 5}{4 \cdot 4 \cdot 4} = \dfrac{25}{64}$

Switch the position of the bases
and make the exponents positive.

b. $\left(\dfrac{3}{4}\right)^{-2} = \dfrac{3^{-2}}{4^{-2}} = \dfrac{4^2}{3^2} = \dfrac{4 \cdot 4}{3 \cdot 3} = \dfrac{16}{9}$

Switch the position of the bases
and make the exponents positive.

c. $\dfrac{1}{4x^{-3}} = \dfrac{x^3}{4}$

Switch the position of the base
and make the exponent positive.
Note that only x is raised to the
-3 power.

d. $\dfrac{x^{-5}}{y^{-1}} = \dfrac{y^1}{x^5} = \dfrac{y}{x^5}$ ∎

☑ **CHECK POINT 2**   Write each expression with positive exponents only. Then simplify, if possible.

a. $\dfrac{2^{-3}}{7^{-2}}$ b. $\left(\dfrac{4}{5}\right)^{-2}$ c. $\dfrac{1}{7y^{-2}}$ d. $\dfrac{x^{-1}}{y^{-8}}$

2 Simplify exponential expressions.

Simplifying Exponential Expressions

Properties of exponents are used to simplify exponential expressions. An exponential expression is **simplified** when

- Each base occurs only once.
- No parentheses appear.
- No powers are raised to powers.
- No negative or zero exponents appear.

Simplifying Exponential Expressions

	Example
1. If necessary, be sure that each base appears only once, using $$b^m \cdot b^n = b^{m+n} \quad \text{or} \quad \frac{b^m}{b^n} = b^{m-n}.$$	$x^4 \cdot x^3 = x^{4+3} = x^7$
2. If necessary, remove parentheses using $$(ab)^n = a^n b^n \quad \text{or} \quad \left(\frac{a}{b}\right)^n = \frac{a^n}{b^n}.$$	$(xy)^3 = x^3 y^3$
3. If necessary, simplify powers to powers using $$(b^m)^n = b^{mn}.$$	$(x^4)^3 = x^{4\cdot3} = x^{12}$
4. If necessary, rewrite exponential expressions with zero powers as 1 ($b^0 = 1$). Furthermore, write the answer with positive exponents using $$b^{-n} = \frac{1}{b^n} \quad \text{or} \quad \frac{1}{b^{-n}} = b^n.$$	$\frac{x^5}{x^8} = x^{5-8} = x^{-3} = \frac{1}{x^3}$

The following examples show how to simplify exponential expressions. In each example, assume that any variable in a denominator is not equal to zero.

Great Question!

Is the procedure in Example 3 the only way to simplify $x^{-9} \cdot x^4$?

There is often more than one way to simplify an exponential expression. For example, you may prefer to simplify Example 3 as follows:

$$x^{-9} \cdot x^4 = \frac{x^4}{x^9} = x^{4-9} = x^{-5} = \frac{1}{x^5}.$$

EXAMPLE 3 Simplifying an Exponential Expression

Simplify: $x^{-9} \cdot x^4$.

Solution

$$x^{-9} \cdot x^4 = x^{-9+4} \quad b^m \cdot b^n = b^{m+n}$$
$$= x^{-5} \quad \text{The base, x, now appears only once.}$$
$$= \frac{1}{x^5} \quad b^{-n} = \frac{1}{b^n} \quad \blacksquare$$

✓ **CHECK POINT 3** Simplify: $x^{-12} \cdot x^2$.

EXAMPLE 4 Simplifying Exponential Expressions

Simplify:

a. $\frac{x^4}{x^{20}}$ **b.** $\frac{25x^6}{5x^8}$ **c.** $\frac{10y^7}{-2y^{10}}.$

Solution

a. $\frac{x^4}{x^{20}} = x^{4-20} = x^{-16} = \frac{1}{x^{16}}$

b. $\frac{25x^6}{5x^8} = \frac{25}{5} \cdot \frac{x^6}{x^8} = 5x^{6-8} = 5x^{-2} = \frac{5}{x^2}$

c. $\frac{10y^7}{-2y^{10}} = \frac{10}{-2} \cdot \frac{y^7}{y^{10}} = -5y^{7-10} = -5y^{-3} = -\frac{5}{y^3}$ $\blacksquare$

✓ **CHECK POINT 4** Simplify:

a. $\frac{x^2}{x^{10}}$ **b.** $\frac{75x^3}{5x^9}$ **c.** $\frac{50y^8}{-25y^{14}}.$

EXAMPLE 5 Simplifying an Exponential Expression

Simplify: $\dfrac{(5x^3)^2}{x^{10}}$.

Solution

$$\dfrac{(5x^3)^2}{x^{10}} = \dfrac{5^2(x^3)^2}{x^{10}}$$ Raise each factor in the product to the second power using $(ab)^n = a^n b^n$.

$$= \dfrac{5^2 x^{3 \cdot 2}}{x^{10}}$$ Raise powers to powers using $(b^m)^n = b^{mn}$.

$$= \dfrac{25 x^6}{x^{10}}$$ Simplify.

$$= 25 x^{6-10}$$ When dividing with the same base, subtract exponents: $\dfrac{b^m}{b^n} = b^{m-n}$.

$$= 25 x^{-4}$$ Simplify. The base, x, now appears only once.

$$= \dfrac{25}{x^4}$$ Rewrite with a positive exponent using $b^{-n} = \dfrac{1}{b^n}$. ∎

✓ **CHECK POINT 5** Simplify: $\dfrac{(6x^4)^2}{x^{11}}$.

EXAMPLE 6 Simplifying an Exponential Expression

Simplify: $\left(\dfrac{x^5}{x^2}\right)^{-3}$.

Solution

Method 1. First perform the division within the parentheses.

$$\left(\dfrac{x^5}{x^2}\right)^{-3} = (x^{5-2})^{-3}$$ Within parentheses, divide by subtracting exponents: $\dfrac{b^m}{b^n} = b^{m-n}$.

$$= (x^3)^{-3}$$ Simplify. The base, x, now appears only once.

$$= x^{3(-3)}$$ Raise powers to powers: $(b^m)^n = b^{mn}$.

$$= x^{-9}$$ Simplify.

$$= \dfrac{1}{x^9}$$ Rewrite with a positive exponent using $b^{-n} = \dfrac{1}{b^n}$.

Method 2. Remove parentheses first by raising the numerator and the denominator to the -3 power.

$$\left(\dfrac{x^5}{x^2}\right)^{-3} = \dfrac{(x^5)^{-3}}{(x^2)^{-3}}$$ Use $\left(\dfrac{a}{b}\right)^n = \dfrac{a^n}{b^n}$ and raise the numerator and denominator to the -3 power.

$$= \dfrac{x^{5(-3)}}{x^{2(-3)}}$$ Raise powers to powers using $(b^m)^n = b^{mn}$.

$$= \dfrac{x^{-15}}{x^{-6}}$$ Simplify.

$$= x^{-15-(-6)}$$ When dividing with the same base, subtract the exponent in the denominator from the exponent in the numerator: $\dfrac{b^m}{b^n} = b^{m-n}$.

$$= x^{-9}$$ Subtract: $-15 - (-6) = -15 + 6 = -9$. The base, x, now appears only once.

$$= \dfrac{1}{x^9}$$ Rewrite with a positive exponent using $b^{-n} = \dfrac{1}{b^n}$.

Which method do you prefer? ∎

✓ **CHECK POINT 6** Simplify: $\left(\dfrac{x^8}{x^4}\right)^{-5}$.

Scientific Notation

Table 5.1	Names of Large Numbers
10^2	hundred
10^3	thousand
10^6	million
10^9	billion
10^{12}	trillion
10^{15}	quadrillion
10^{18}	quintillion
10^{21}	sextillion
10^{24}	septillion
10^{27}	octillion
10^{30}	nonillion
10^{100}	googol
10^{googol}	googolplex

In 2009, the United States government spent more than it had collected in taxes, resulting in a budget deficit of $1.35 trillion. Put into perspective, the $1.35 trillion could pay for 40,000 players like Alex Rodriguez, whose $33 million salary in 2009 made him baseball's richest man. Because a trillion is 10^{12} (see **Table 5.1**), the 2009 budget deficit can be expressed as

$$\$1.35 \times 10^{12}.$$

The number 1.35×10^{12} is written in a form called *scientific notation*.

Scientific Notation

A positive number is written in **scientific notation** when it is expressed in the form

$$a \times 10^n,$$

where a is a number greater than or equal to 1 and less than 10 ($1 \le a < 10$) and n is an integer.

It is customary to use the multiplication symbol, $\times$, rather than a dot, when writing a number in scientific notation.

Here are two examples of numbers in scientific notation:

- In 2010, humankind generated 1.2×10^{21} bytes, or 1.2 zettabytes, of digital information. (Put into perspective, if all 6.8 billion (6.8×10^9) people on Earth joined Twitter and continuously tweeted for a century, they would crank out one zettabyte of data.) (*Source*: LiveScience.com)

- The length of the AIDS virus is 1.1×10^{-4} millimeter.

3 Convert from scientific notation to decimal notation.

We can use n, the exponent on the 10 in $a \times 10^n$, to change a number in scientific notation to decimal notation. If n is **positive**, move the decimal point in a to the **right** n places. If n is **negative**, move the decimal point in a to the **left** $|n|$ places.

EXAMPLE 7 Converting from Scientific to Decimal Notation

Write each number in decimal notation:

a. 2.6×10^7 **b.** 1.1×10^{-4}.

Solution In each case, we use the exponent on the 10 to move the decimal point. In part (a), the exponent is positive, so we move the decimal point to the right. In part (b), the exponent is negative, so we move the decimal point to the left.

a. $2.6 \times 10^7 = 26,000,000$

$n = 7$

Move the decimal point 7 places to the right.

b. $1.1 \times 10^{-4} = 0.00011$

$n = -4$

Move the decimal point $|-4|$ places, or 4 places, to the left.

✓ **CHECK POINT 7** Write each number in decimal notation:
a. 7.4×10^9 **b.** 3.017×10^{-6}.

---

4 Convert from decimal notation to scientific notation.

To convert a positive number from decimal notation to scientific notation, we reverse the procedure of Example 7.

Converting from Decimal to Scientific Notation

Write the number in the form $a \times 10^n$.

- Determine a, the numerical factor. Move the decimal point in the given number to obtain a number greater than or equal to 1 and less than 10.
- Determine n, the exponent on 10^n. The absolute value of n is the number of places the decimal point was moved. The exponent n is positive if the given number is greater than 10 and negative if the given number is between 0 and 1.

EXAMPLE 8 Converting from Decimal Notation to Scientific Notation

Write each number in scientific notation:

a. 4,600,000   **b.** 0.000023.

Solution

a. $4{,}600{,}000 = 4.6 \times 10^6$

| This number is greater than 10, so n is positive in $a \times 10^n$. | Move the decimal point in 4,600,000 to get $1 \le a < 10$. | The decimal point moved 6 places from 4,600,000 to 4.6. |

b. $0.000023 = 2.3 \times 10^{-5}$

| This number is less than 1, so n is negative in $a \times 10^n$. | Move the decimal point in 0.000023 to get $1 \le a < 10$. | The decimal point moved 5 places from 0.000023 to 2.3. |

Using Technology

You can use your calculator's EE (enter exponent) or EXP key to convert from decimal to scientific notation. Here is how it's done for 0.000023:

Many Scientific Calculators

Keystrokes	Display
.000023 EE =	2.3 − 05

Many Graphing Calculators

Use the mode setting for scientific notation.

Keystrokes	Display
.000023 ENTER	2.3E-5

✓ **CHECK POINT 8** Write each number in scientific notation:
a. 7,410,000,000   **b.** 0.000000092.

---

5 Compute with scientific notation.

Computations with Scientific Notation

Properties of exponents are used to perform computations with numbers that are expressed in scientific notation.

Computations with Numbers in Scientific Notation

Multiplication

$$(a \times 10^n) \times (b \times 10^m) = (a \times b) \times 10^{n+m}$$

> Add the exponents on 10 and multiply the other parts of the numbers separately.

Division

$$\frac{a \times 10^n}{b \times 10^m} = \left(\frac{a}{b}\right) \times 10^{n-m}$$

> Subtract the exponents on 10 and divide the other parts of the numbers separately.

Exponentiation

$$(a \times 10^n)^m = a^n \times 10^{nm}$$

> Multiply exponents on 10 and raise the other part of the number to the power.

After the computation is completed, the answer may require an adjustment before it is expressed in scientific notation.

EXAMPLE 9   Computations with Scientific Notation

Perform the indicated computations, writing the answers in scientific notation:

a. $(4 \times 10^5)(2 \times 10^9)$ **b.** $\dfrac{1.2 \times 10^6}{4.8 \times 10^{-3}}$ **c.** $(5 \times 10^{-4})^3$.

Solution

a. $(4 \times 10^5)(2 \times 10^9) = (4 \times 2) \times (10^5 \times 10^9)$ Regroup.

$= 8 \times 10^{5+9}$ Add the exponents on 10 and multiply the other parts.

$= 8 \times 10^{14}$ Simplify.

b. $\dfrac{1.2 \times 10^6}{4.8 \times 10^{-3}} = \left(\dfrac{1.2}{4.8}\right) \times \left(\dfrac{10^6}{10^{-3}}\right)$ Regroup.

$= 0.25 \times 10^{6-(-3)}$ Subtract the exponents on 10 and divide the other parts.

$= 0.25 \times 10^9$ Simplify. Because 0.25 is not between 1 and 10, it must be written in scientific notation.

$= 2.5 \times 10^{-1} \times 10^9$ $0.25 = 2.5 \times 10^{-1}$

$= 2.5 \times 10^{-1+9}$ Add the exponents on 10.

$= 2.5 \times 10^8$ Simplify.

c. $(5 \times 10^{-4})^3 = 5^3 \times (10^{-4})^3$ $(ab)^n = a^n b^n$. Cube each factor in parentheses.

$= 125 \times 10^{-4\cdot3}$ Multiply the exponents and cube the other part of the number.

$= 125 \times 10^{-12}$ Simplify. 125 must be written in scientific notation.

$= 1.25 \times 10^2 \times 10^{-12}$ $125 = 1.25 \times 10^2$

$= 1.25 \times 10^{2+(-12)}$ Add the exponents on 10.

$= 1.25 \times 10^{-10}$ Simplify. ■

Using Technology

$(4 \times 10^5)(2 \times 10^9)$ On a Calculator:

Many Scientific Calculators

4 EE 5 × 2 EE 9 =

Display: 8. 14

Many Graphing Calculators

4 EE 5 × 2 EE 9 ENTER

Display: 8E14

✓ **CHECK POINT 9** Perform the indicated computations, writing the answers in scientific notation:

a. $(3 \times 10^8)(2 \times 10^2)$

b. $\dfrac{8.4 \times 10^7}{4 \times 10^{-4}}$

c. $(4 \times 10^{-2})^3$.

---

6 Solve applied problems using scientific notation.

Applications: Putting Numbers in Perspective

Due to tax cuts and spending increases, the United States began accumulating large deficits in the 1980s. To finance the deficit, the government had borrowed $12.3 trillion as of December 2009. The graph in **Figure 5.8** shows the national debt increasing over time.

The National Debt

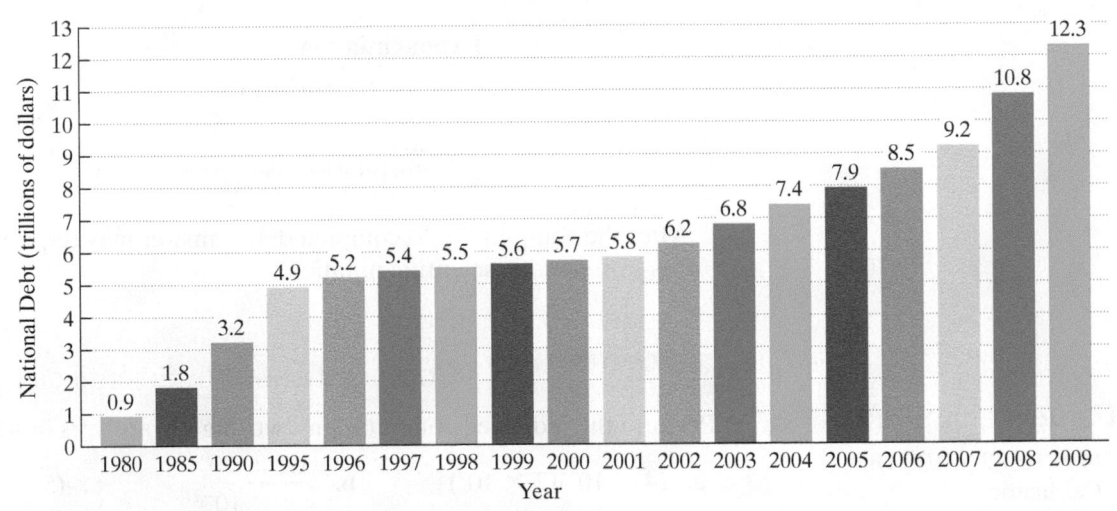

Figure 5.8

Source: Office of Management and Budget

Example 10 shows how we can use scientific notation to comprehend the meaning of a number such as 12.3 trillion.

EXAMPLE 10 The National Debt

As of December 2009, the national debt was $12.3 trillion, or 12.3×10^{12} dollars. At that time, the U.S. population was approximately 307,000,000 (307 million), or 3.07×10^8. If the national debt was evenly divided among every individual in the United States, how much would each citizen have to pay?

Solution Although it is not necessary to do so, let's express the total debt, 12.3×10^{12} dollars, in scientific notation as 1.23×10^{13}. The amount each citizen must pay is the total debt, 1.23×10^{13} dollars, divided by the number of citizens, 3.07×10^8.

$$\frac{1.23 \times 10^{13}}{3.07 \times 10^8} = \left(\frac{1.23}{3.07}\right) \times \left(\frac{10^{13}}{10^8}\right)$$
$$\approx 0.401 \times 10^{13-8}$$
$$= 0.401 \times 10^5$$
$$= (4.01 \times 10^{-1}) \times 10^5$$
$$= 4.01 \times 10^4$$
$$= 40,100$$

Every U.S. citizen would have to pay approximately $40,100 to the federal government to pay off the national debt. ∎

If a number is written in scientific notation, $a \times 10^n$, the digits in a are called **significant digits**.

National Debt: 1.23×10^{13} U.S. Population: 3.07×10^8

Three significant digits   Three significant digits

Because these were the given numbers in Example 10, we rounded the answer, 4.01×10^4, to three significant digits. When multiplying or dividing in scientific notation where rounding is necessary and rounding instructions are not given, **round the scientific notation answer to the least number of significant digits found in any of the given numbers.**

✓ **CHECK POINT 10**   The cost of President Obama's 2009 economic stimulus package was $787 billion, or 7.87×10^{11} dollars. If this cost were evenly divided among every individual in the United States (approximately 3.07×10^8 people), how much would each citizen have to pay?

Blitzer Bonus

Seven Ways to Spend $1 Trillion

Confronting a national debt of $12.3 trillion starts with grasping just how colossal $1 trillion ($1 \times 10^{12}$) actually is. To help you wrap your head around this mind-boggling number, and to put the national debt in further perspective, consider what $1 trillion will buy:

- 40,816,326 new cars based on an average sticker price of $24,500 each

- 5,574,136 homes based on the national median price of $179,400 for existing single-family homes

- one year's salary for 14.7 million teachers based on the average teacher salary of $68,000 in California

- the annual salaries of all 535 members of Congress for the next 10,742 years based on current salaries of $174,000 per year

- the salary of basketball superstar LeBron James for 50,000 years based on an annual salary of $20 million

- annual base pay for 59.5 million U.S. privates (that's 100 times the total number of active-duty soldiers in the Army) based on basic pay of $16,794 per year

- salaries to hire all 2.8 million residents of the state of Kansas in full-time minimum-wage jobs for the next 23 years based on the federal minimum wage of $7.25 per hour

Source: Kiplinger.com

Image © photobank.kiev.ua, 2009

Achieving Success

Form a study group with other students in your class. Working in small groups often serves as an excellent way to learn and reinforce new material. Set up helpful procedures and guidelines for the group. "Talk" math by discussing and explaining the concepts and exercises to one another.

CONCEPT AND VOCABULARY CHECK

Fill in each blank so that the resulting statement is true.

1. The negative exponent rule states that $b^{-n} =$ _____, $b \neq 0$.

2. True or false: $4^{-2} = -4^2$ _____

3. True or false: $4^{-2} = \dfrac{1}{4^2}$ _____

4. Negative exponents in denominators can be evaluated using $\dfrac{1}{b^{-n}} =$ _____, $b \neq 0$.

5. True or false: $\dfrac{1}{5^{-2}} = 5^2$ _____

6. True or false: $\dfrac{1}{5^{-2}} = -5^2$ _____

7. A positive number is written in scientific notation when it is expressed in the form $a \times 10^n$, where a is _____ and n is a/an _____.

8. True or false: 4×10^3 is written in scientific notation. _____

9. True or false: 40×10^2 is written in scientific notation. _____

5.7 EXERCISE SET   MyMathLab®

Watch the videos in MyMathLab   Download the MyDashBoard App

Practice Exercises

In Exercises 1–28, write each expression with positive exponents only. Then simplify, if possible.

1. 8^{-2}
2. 9^{-2}

3. 5^{-3}
4. 4^{-3}

5. $(-6)^{-2}$
6. $(-7)^{-2}$

7. -6^{-2}
8. -7^{-2}

9. 4^{-1}
10. 6^{-1}

11. $2^{-1} + 3^{-1}$

12. $3^{-1} - 6^{-1}$

13. $\dfrac{1}{3^{-2}}$
14. $\dfrac{1}{4^{-3}}$

15. $\dfrac{1}{(-3)^{-2}}$
16. $\dfrac{1}{(-2)^{-2}}$

17. $\dfrac{2^{-3}}{8^{-2}}$
18. $\dfrac{4^{-3}}{2^{-2}}$

19. $\left(\dfrac{1}{4}\right)^{-2}$
20. $\left(\dfrac{1}{5}\right)^{-2}$

21. $\left(\dfrac{3}{5}\right)^{-3}$
22. $\left(\dfrac{3}{4}\right)^{-3}$

23. $\dfrac{1}{6x^{-5}}$
24. $\dfrac{1}{8x^{-6}}$

25. $\dfrac{x^{-8}}{y^{-1}}$
26. $\dfrac{x^{-12}}{y^{-1}}$

27. $\dfrac{3}{(-5)^{-3}}$

28. $\dfrac{4}{(-3)^{-3}}$

In Exercises 29–78, simplify each exponential expression. Assume that variables represent nonzero real numbers.

29. $x^{-8} \cdot x^3$
30. $x^{-11} \cdot x^5$

31. $(4x^{-5})(2x^2)$
32. $(5x^{-7})(3x^3)$

33. $\dfrac{x^3}{x^9}$
34. $\dfrac{x^5}{x^{12}}$

35. $\dfrac{y}{y^{100}}$
36. $\dfrac{y}{y^{50}}$

37. $\dfrac{30z^5}{10z^{10}}$
38. $\dfrac{45z^4}{15z^{12}}$

39. $\dfrac{-8x^3}{2x^7}$
40. $\dfrac{-15x^4}{3x^9}$

41. $\dfrac{-9a^5}{27a^8}$
42. $\dfrac{-15a^8}{45a^{13}}$

43. $\dfrac{7w^5}{5w^{13}}$
44. $\dfrac{7w^8}{9w^{14}}$

45. $\dfrac{x^3}{(x^4)^2}$

46. $\dfrac{x^5}{(x^3)^2}$

47. $\dfrac{y^{-3}}{(y^4)^2}$

48. $\dfrac{y^{-5}}{(y^3)^2}$

49. $\dfrac{(4x^3)^2}{x^8}$

50. $\dfrac{(5x^3)^2}{x^7}$

51. $\dfrac{(6y^4)^3}{y^{-5}}$

52. $\dfrac{(4y^5)^3}{y^{-4}}$

53. $\left(\dfrac{x^4}{x^2}\right)^{-3}$

54. $\left(\dfrac{x^6}{x^2}\right)^{-3}$

55. $\left(\dfrac{4x^5}{2x^2}\right)^{-4}$

56. $\left(\dfrac{6x^7}{2x^2}\right)^{-4}$

57. $(3x^{-1})^{-2}$

58. $(4x^{-1})^{-2}$

59. $(-2y^{-1})^{-3}$

60. $(-3y^{-1})^{-3}$

61. $\dfrac{2x^5 \cdot 3x^7}{15x^6}$

62. $\dfrac{3x^3 \cdot 5x^{14}}{20x^{14}}$

63. $(x^3)^5 \cdot x^{-7}$

64. $(x^4)^3 \cdot x^{-5}$

65. $(2y^3)^4 y^{-6}$

66. $(3y^4)^3 y^{-7}$

67. $\dfrac{(y^3)^4}{(y^2)^7}$

68. $\dfrac{(y^2)^5}{(y^3)^4}$

69. $(y^{10})^{-5}$

70. $(y^{20})^{-5}$

71. $(a^4 b^5)^{-3}$

72. $(a^5 b^3)^{-4}$

73. $(a^{-2} b^6)^{-4}$

74. $(a^{-7} b^2)^{-5}$

75. $\left(\dfrac{x^2}{2}\right)^{-2}$

76. $\left(\dfrac{x^2}{2}\right)^{-3}$

77. $\left(\dfrac{x^2}{y^3}\right)^{-3}$

78. $\left(\dfrac{x^3}{y^2}\right)^{-4}$

In Exercises 79–90, write each number in decimal notation without the use of exponents.

79. 8.7×10^2

80. 2.75×10^3

81. 9.23×10^5

82. 7.24×10^4

83. 3.4×10^0

84. 9.115×10^0

85. 7.9×10^{-1}

86. 8.6×10^{-1}

87. 2.15×10^{-2}

88. 3.14×10^{-2}

89. 7.86×10^{-4}

90. 4.63×10^{-5}

In Exercises 91–106, write each number in scientific notation.

91. 32,400

92. 327,000

93. 220,000,000

94. 370,000,000,000

95. 713

96. 623

97. 6751

98. 9832

99. 0.0027

100. 0.00083

101. 0.0000202

102. 0.00000103

103. 0.005

104. 0.006

105. 3.14159

106. 2.71828

In Exercises 107–126, perform the indicated computations. Write the answers in scientific notation.

107. $(2 \times 10^3)(3 \times 10^2)$

108. $(3 \times 10^4)(3 \times 10^2)$

109. $(2 \times 10^5)(8 \times 10^3)$

110. $(4 \times 10^3)(5 \times 10^4)$

111. $\dfrac{12 \times 10^6}{4 \times 10^2}$

112. $\dfrac{20 \times 10^{20}}{10 \times 10^{10}}$

113. $\dfrac{15 \times 10^4}{5 \times 10^{-2}}$

114. $\dfrac{18 \times 10^2}{9 \times 10^{-3}}$

115. $\dfrac{15 \times 10^{-4}}{5 \times 10^2}$

116. $\dfrac{18 \times 10^{-2}}{9 \times 10^3}$

117. $\dfrac{180 \times 10^6}{2 \times 10^3}$

118. $\dfrac{180 \times 10^8}{2 \times 10^4}$

119. $\dfrac{3 \times 10^4}{12 \times 10^{-3}}$

120. $\dfrac{5 \times 10^2}{20 \times 10^{-3}}$

121. $(5 \times 10^2)^3$

122. $(4 \times 10^3)^2$

123. $(3 \times 10^{-2})^4$

124. $(2 \times 10^{-3})^5$

125. $(4 \times 10^6)^{-1}$

126. $(5 \times 10^4)^{-1}$

Practice PLUS

In Exercises 127–134, simplify each exponential expression. Assume that variables represent nonzero real numbers.

127. $\dfrac{(x^{-2}y)^{-3}}{(x^2 y^{-1})^3}$

128. $\dfrac{(xy^{-2})^{-2}}{(x^{-2}y)^{-3}}$

129. $(2x^{-3}yz^{-6})(2x)^{-5}$

130. $(3x^{-4}yz^{-7})(3x)^{-3}$

131. $\left(\dfrac{x^3 y^4 z^5}{x^{-3} y^{-4} z^{-5}}\right)^{-2}$

132. $\left(\dfrac{x^4 y^5 z^6}{x^{-4} y^{-5} z^{-6}}\right)^{-4}$

133. $\dfrac{(2^{-1}x^{-2}y^{-1})^{-2}(2x^{-4}y^3)^{-2}(16x^{-3}y^3)^0}{(2x^{-3}y^{-5})^2}$

134. $\dfrac{(2^{-1}x^{-3}y^{-1})^{-2}(2x^{-6}y^4)^{-2}(9x^3y^{-3})^0}{(2x^{-4}y^{-6})^2}$

In Exercises 135–138, perform the indicated computations. Express answers in scientific notation.

135. $(5 \times 10^3)(1.2 \times 10^{-4}) \div (2.4 \times 10^2)$

136. $(2 \times 10^2)(2.6 \times 10^{-3}) \div (4 \times 10^3)$

137. $\dfrac{(1.6 \times 10^4)(7.2 \times 10^{-3})}{(3.6 \times 10^8)(4 \times 10^{-3})}$

138. $\dfrac{(1.2 \times 10^6)(8.7 \times 10^{-2})}{(2.9 \times 10^6)(3 \times 10^{-3})}$

Application Exercises

We have seen that in 2009, the United States government spent more than it had collected in taxes, resulting in a budget deficit of $1.35 trillion. In Exercises 139–142, you will use scientific notation to put a number like 1.35 trillion in perspective.

139. **a.** Express 1.35 trillion in scientific notation.

  b. Express the 2009 U.S. population, 307 million, in scientific notation.

  c. Use your scientific notation answers from parts (a) and (b) to answer this question: If the 2009 budget deficit was evenly divided among every individual in the United States, how much would each citizen have to pay? Express the answer in scientific and decimal notations.

140. **a.** Express 1.35 trillion in scientific notation.

  b. A trip around the world at the Equator is approximately 25,000 miles. Express this number in scientific notation.

  c. Use your scientific notation answers from parts (a) and (b) to answer this question: How many times can you circle the world at the Equator by traveling 1.35 trillion miles?

141. If there are approximately 3.2×10^7 seconds in a year, approximately how many years is 1.35 trillion seconds? (*Note:* 1.35 trillion seconds would take us back in time to a period when Neanderthals were using stones to make tools.)

142. The Washington Monument, overlooking the U.S. Capitol, stands about 555 feet tall. Stacked end to end, how many monuments would it take to reach 1.35 trillion feet? (*Note:* That's more than twice the distance from Earth to the sun.)

Use the motion formula $d = rt$, distance equals rate times time, and the fact that light travels at the rate of 1.86×10^5 miles per second, to solve Exercises 143–144.

143. If the moon is approximately 2.325×10^5 miles from Earth, how many seconds does it take moonlight to reach Earth?

144. If the sun is approximately 9.14×10^7 miles from Earth, how many seconds, to the nearest tenth of a second, does it take sunlight to reach Earth?

145. Refer to the Blitzer Bonus on page 413. Use scientific notation to verify any three of the bulleted items on ways to spend $1 trillion.

Writing in Mathematics

146. Explain the negative exponent rule and give an example.

147. How do you know if an exponential expression is simplified?

148. How do you know if a number is written in scientific notation?

149. Explain how to convert from scientific to decimal notation and give an example.

150. Explain how to convert from decimal to scientific notation and give an example.

151. Describe one advantage of expressing a number in scientific notation over decimal notation.

Critical Thinking Exercises

Make Sense? *In Exercises 152–155, determine whether each statement "makes sense" or "does not make sense" and explain your reasoning.*

152. There are many exponential expressions that are equal to $36x^{12}$, such as $(6x^6)^2$, $(6x^3)(6x^9)$, $36(x^3)^9$, and $6^2(x^2)^6$.

153. If 5^{-2} is raised to the third power, the result is a number between 0 and 1.

154. The population of Colorado is approximately 4.6×10^{12}.

155. I wrote a number where there is no advantage to using scientific notation instead of decimal notation.

In Exercises 156–163, determine whether each statement is true or false. If the statement is false, make the necessary change(s) to produce a true statement.

156. $4^{-2} < 4^{-3}$

157. $5^{-2} > 2^{-5}$

158. $(-2)^4 = 2^{-4}$

159. $5^2 \cdot 5^{-2} > 2^5 \cdot 2^{-5}$

160. $534.7 = 5.347 \times 10^3$

161. $\dfrac{8 \times 10^{30}}{4 \times 10^{-5}} = 2 \times 10^{25}$

162. $(7 \times 10^5) + (2 \times 10^{-3}) = 9 \times 10^2$

163. $(4 \times 10^3) + (3 \times 10^2) = 4.3 \times 10^3$

164. The mad Dr. Frankenstein has gathered enough bits and pieces (so to speak) for $2^{-1} + 2^{-2}$ of his creature-to-be. Write a fraction that represents the amount of his creature that must still be obtained.

Technology Exercises

165. Use a calculator in a fraction mode to check any five of your answers in Exercises 1–22.

166. Use a calculator to check any three of your answers in Exercises 79–90.

167. Use a calculator to check any three of your answers in Exercises 91–106.

168. Use a calculator with an EE or EXP key to check any four of your computations in Exercises 107–126. Display the result of the computation in scientific notation.

Review Exercises

169. Solve: $8 - 6x > 4x - 12$. (Section 2.7, Example 7)

170. Simplify: $24 \div 8 \cdot 3 + 28 \div (-7)$. (Section 1.8, Example 8)

171. List the whole numbers in this set:

$$\left\{ -4, -\frac{1}{5}, 0, \pi, \sqrt{16}, \sqrt{17} \right\}.$$

(Section 1.3, Example 5)

Preview Exercises

Exercises 172–174 will help you prepare for the material covered in the first section of the next chapter. In each exercise, find the product.

172. $4x^3(4x^2 - 3x + 1)$

173. $9xy(3xy^2 - y + 9)$

174. $(x + 3)(x^2 + 5)$

GROUP PROJECT

CHAPTER 5

A large number can be put into perspective by comparing it with another number. For example, we put the $12.3 trillion national debt (Example 10) into perspective by comparing this number to the number of U.S. citizens. In Exercises 139–142, we put the $1.35 trillion budget deficit into perspective by comparing 1.35 trillion to the number of U.S. citizens, the distance around the world, the number of seconds in a year, and the height of the Washington Monument.

For this project, each group member should consult an almanac, a newspaper, or the Internet to find a number greater than one million. Explain to other members of the group the context in which the large number is used. Express the number in scientific notation. Then put the number into perspective by comparing it with another number.

Chapter 5 Summary

Definitions and Concepts	Examples

Section 5.1 Adding and Subtracting Polynomials

A polynomial is a single term or the sum of two or more terms containing variables with whole number exponents. A monomial is a polynomial with exactly one term; a binomial has exactly two terms; a trinomial has exactly three terms. The degree of a polynomial is the highest power of all the terms. The standard form of a polynomial is written in descending powers of the variable.

Polynomials

Monomial: $2x^5$

> Degree is 5.

Binomial: $6x^3 + 5x$

> Degree is 3.

Trinomial: $7x + 4x^2 - 5$

> Degree is 2.

To add polynomials, add like terms.

$(6x^3 + 5x^2 - 7x) + (-9x^3 + x^2 + 6x)$
$= (6x^3 - 9x^3) + (5x^2 + x^2) + (-7x + 6x)$
$= -3x^3 + 6x^2 - x$

The opposite, or additive inverse, of a polynomial is that polynomial with the sign of every coefficient changed. To subtract two polynomials, add the first polynomial and the opposite of the polynomial being subtracted.

$(5y^3 - 9y^2 - 4) - (3y^3 - 12y^2 - 5)$
$= (5y^3 - 9y^2 - 4) + (-3y^3 + 12y^2 + 5)$
$= (5y^3 - 3y^3) + (-9y^2 + 12y^2) + (-4 + 5)$
$= 2y^3 + 3y^2 + 1$

Definitions and Concepts	**Examples**

Section 5.1 Adding and Subtracting Polynomials (continued)

The graphs of equations defined by polynomials of degree 2, shaped like bowls or inverted bowls, can be obtained using the point-plotting method.

Graph: $y = x^2 - 1$.

x	$y = x^2 - 1$
-2	$(-2)^2 - 1 = 3$
-1	$(-1)^2 - 1 = 0$
0	$0^2 - 1 = -1$
1	$1^2 - 1 = 0$
2	$2^2 - 1 = 3$

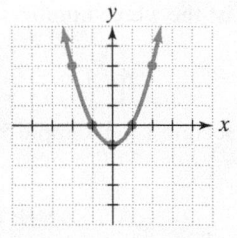

Section 5.2 Multiplying Polynomials

Properties of Exponents

Product Rule: $b^m \cdot b^n = b^{m+n}$

Power Rule: $(b^m)^n = b^{mn}$

Products to Powers: $(ab)^n = a^n b^n$

$x^3 \cdot x^8 = x^{3+8} = x^{11}$

$(x^3)^8 = x^{3 \cdot 8} = x^{24}$

$(-5x^2)^3 = (-5)^3(x^2)^3 = -125x^6$

To multiply monomials, multiply coefficients and add exponents.

$(-6x^4)(3x^{10}) = -6 \cdot 3x^{4+10} = -18x^{14}$

To multiply a monomial and a polynomial, multiply each term of the polynomial by the monomial.

$2x^4(3x^2 - 6x + 5)$

$= 2x^4 \cdot 3x^2 - 2x^4 \cdot 6x + 2x^4 \cdot 5$

$= 6x^6 - 12x^5 + 10x^4$

To multiply polynomials when neither is a monomial, multiply each term of one polynomial by each term of the other polynomial. Then combine like terms.

$(2x + 3)(5x^2 - 4x + 2)$

$= 2x(5x^2 - 4x + 2) + 3(5x^2 - 4x + 2)$

$= 10x^3 - 8x^2 + 4x + 15x^2 - 12x + 6$

$= 10x^3 + 7x^2 - 8x + 6$

Section 5.3 Special Products

The FOIL method may be used when multiplying two binomials: First terms multiplied. Outside terms multiplied. Inside terms multiplied. Last terms multiplied.

F   O   I   L

$(3x + 7)(2x - 5) = 3x \cdot 2x + 3x(-5) + 7 \cdot 2x + 7(-5)$

$= 6x^2 - 15x + 14x - 35$

$= 6x^2 - x - 35$

The Product of the Sum and Difference of Two Terms

$(A + B)(A - B) = A^2 - B^2$

$(4x + 7)(4x - 7) = (4x)^2 - 7^2$

$= 16x^2 - 49$

Definitions and Concepts	**Examples**

Section 5.3 Special Products (continued)

The Square of a Binomial Sum

$$(A + B)^2 = A^2 + 2AB + B^2$$

$$(x^2 + 6)^2 = (x^2)^2 + 2 \cdot x^2 \cdot 6 + 6^2$$
$$= x^4 + 12x^2 + 36$$

The Square of a Binomial Difference

$$(A - B)^2 = A^2 - 2AB + B^2$$

$$(9x - 3)^2 = (9x)^2 - 2 \cdot 9x \cdot 3 + 3^2$$
$$= 81x^2 - 54x + 9$$

Section 5.4 Polynomials in Several Variables

To evaluate a polynomial in several variables, substitute the given value for each variable and perform the resulting computation.

Evaluate $4x^2y + 3xy - 2x$ for $x = -1$ and $y = -3$.

$$4x^2y + 3xy - 2x$$
$$= 4(-1)^2(-3) + 3(-1)(-3) - 2(-1)$$
$$= 4(1)(-3) + 3(-1)(-3) - 2(-1)$$
$$= -12 + 9 + 2 = -1$$

For a polynomial in two variables, the degree of a term is the sum of the exponents on its variables. The degree of the polynomial is the highest degree of all its terms.

$$7x^2y + 12x^4y^3 - 17x^5 + 6$$

degree:	degree:	degree:	degree:
$2 + 1 = 3$	$4 + 3 = 7$	5	0

Degree of polynomial $= 7$

Polynomials in several variables are added, subtracted, and multiplied using the same rules for polynomials in one variable.

$$(5x^2y^3 - xy + 4y^2) - (8x^2y^3 - 6xy - 2y^2)$$
$$= (5x^2y^3 - xy + 4y^2) + (-8x^2y^3 + 6xy + 2y^2)$$
$$= (5x^2y^3 - 8x^2y^3) + (-xy + 6xy) + (4y^2 + 2y^2)$$
$$= -3x^2y^3 + 5xy + 6y^2$$

F   O   I   L

$$(3x - 2y)(x - y) = 3x \cdot x + 3x(-y) + (-2y)x + (-2y)(-y)$$
$$= 3x^2 - 3xy - 2xy + 2y^2$$
$$= 3x^2 - 5xy + 2y^2$$

Section 5.5 Dividing Polynomials

Additional Properties of Exponents

Quotient Rule: $\dfrac{b^m}{b^n} = b^{m-n}, \quad b \neq 0$

Zero-Exponent Rule: $b^0 = 1, \quad b \neq 0$

Quotients to Powers: $\left(\dfrac{a}{b}\right)^n = \dfrac{a^n}{b^n}, \quad b \neq 0$

$$\frac{x^{12}}{x^4} = x^{12-4} = x^8$$

$$(-3)^0 = 1 \qquad -3^0 = -(3^0) = -1$$

$$\left(\frac{y^2}{4}\right)^3 = \frac{(y^2)^3}{4^3} = \frac{y^{2 \cdot 3}}{4 \cdot 4 \cdot 4} = \frac{y^6}{64}$$

To divide monomials, divide coefficients and subtract exponents.

$$\frac{-40x^{40}}{20x^{20}} = \frac{-40}{20}x^{40-20} = -2x^{20}$$

Definitions and Concepts	**Examples**

Section 5.5 Dividing Polynomials (continued)

To divide a polynomial by a monomial, divide each term of the polynomial by the monomial.

$$\frac{8x^6 - 4x^3 + 10x}{2x}$$

$$= \frac{8x^6}{2x} - \frac{4x^3}{2x} + \frac{10x}{2x}$$

$$= 4x^{6-1} - 2x^{3-1} + 5x^{1-1} = 4x^5 - 2x^2 + 5$$

Section 5.6 Long Division of Polynomials; Synthetic Division

To divide a polynomial by another polynomial, begin by arranging the dividend in descending powers of the variable. If a power of a variable is missing, add that power with a coefficient of 0. Repeat the four steps—divide, multiply, subtract, bring down the next term—until the degree of the remainder is less than the degree of the divisor.

Divide: $\dfrac{10x^2 + 13x + 8}{2x + 3}$.

$$
\begin{array}{r}
5x - 1 + \dfrac{11}{2x+3} \\[2pt]
2x + 3 \overline{)\,10x^2 + 13x + 8\,} \\
\underline{10x^2 + 15x} \\
-2x + 8 \\
\underline{-2x - 3} \\
11
\end{array}
$$

A shortcut to long division, called synthetic division, can be used to divide a polynomial by a binomial of the form $x - c$.

Divide: $(2x^3 - x^2 - 7) \div (x - 2)$.

Coefficients of the dividend,
$2x^3 - x^2 + 0x - 7$

This is c in $x - c$.
For $x - 2$, c is 2.

$$
\begin{array}{c|rrrr}
2 & 2 & -1 & 0 & -7 \\
  &   & 4 & 6 & 12 \\
\hline
  & 2 & 3 & 6 & 5
\end{array}
$$

Coefficients of quotient   Remainder

The answer is $2x^2 + 3x + 6 + \dfrac{5}{x - 2}$.

Section 5.7 Negative Exponents and Scientific Notation

Negative Exponents in Numerators and Denominators

If $b \neq 0$, $b^{-n} = \dfrac{1}{b^n}$ and $\dfrac{1}{b^{-n}} = b^n$.

$$6^{-2} = \frac{1}{6^2} = \frac{1}{36}$$

$$\frac{1}{(-2)^{-4}} = (-2)^4 = 16$$

$$\left(\frac{2}{3}\right)^{-3} = \frac{2^{-3}}{3^{-3}} = \frac{3^3}{2^3} = \frac{27}{8}$$

An exponential expression is simplified when

- Each base occurs only once.
- No parentheses appear.
- No powers are raised to powers.
- No negative or zero exponents appear.

Simplify: $\dfrac{(2x^4)^3}{x^{18}}$.

$$\frac{(2x^4)^3}{x^{18}} = \frac{2^3(x^4)^3}{x^{18}} = \frac{8x^{4\cdot3}}{x^{18}} = \frac{8x^{12}}{x^{18}} = 8x^{12-18} = 8x^{-6} = \frac{8}{x^6}$$

Definitions and Concepts	**Examples**

Section 5.7 Negative Exponents and Scientific Notation (continued)

A positive number in scientific notation is expressed as $a \times 10^n$, where $1 \le a < 10$ and n is an integer.

Write 2.9×10^{-3} in decimal notation.

$$2.9 \times 10^{-3} = \underset{\underrightarrow{\quad}}{.0029} = 0.0029$$

Write 16,000 in scientific notation.

$$16,000 = 1.6 \times 10^4$$

Use properties of exponents with base 10

$$10^m \cdot 10^n = 10^{m+n}, \quad \frac{10^m}{10^n} = 10^{m-n}, \quad \text{and} \quad (10^m)^n = 10^{mn}$$

to perform computations with scientific notation.

$(5 \times 10^3)(4 \times 10^{-8})$

$= 5 \cdot 4 \times 10^{3-8}$

$= 20 \times 10^{-5}$

$= 2 \times 10^1 \times 10^{-5} = 2 \times 10^{-4}$

CHAPTER 5 REVIEW EXERCISES

5.1 *In Exercises 1–3, identify each polynomial as a monomial, binomial, or trinomial. Give the degree of the polynomial.*

1. $7x^4 + 9x$

2. $3x + 5x^2 - 2$

3. $16x$

In Exercises 4–8, add or subtract as indicated.

4. $(-6x^3 + 7x^2 - 9x + 3) + (14x^3 + 3x^2 - 11x - 7)$

5. $(9y^3 - 7y^2 + 5) + (4y^3 - y^2 + 7y - 10)$

6. $(5y^2 - y - 8) - (-6y^2 + 3y - 4)$

7. $(13x^4 - 8x^3 + 2x^2) - (5x^4 - 3x^3 + 2x^2 - 6)$

8. Subtract $x^4 + 7x^2 - 11x$ from $-13x^4 - 6x^2 + 5x$.

In Exercises 9–11, add or subtract as indicated.

9. Add. $7y^4 - 6y^3 + 4y^2 - 4y$
   $\underline{\qquad\quad y^3 - \; y^2 + 3y - 4}$

10. Subtract. $7x^2 - 9x + 2$
    $\underline{-(4x^2 - 2x - 7)}$

11. Subtract. $5x^3 - 6x^2 - 9x + 14$
    $\underline{-(-5x^3 + 3x^2 - 11x + 3)}$

In Exercises 12–13, graph each equation.

12. $y = x^2 + 3$

13. $y = 1 - x^2$

5.2 *In Exercises 14–18, simplify each expression.*

14. $x^{20} \cdot x^3$

15. $y \cdot y^5 \cdot y^8$

16. $(x^{20})^5$

17. $(10y)^2$

18. $(-4x^{10})^3$

In Exercises 19–27, find each product.

19. $(5x)(10x^3)$

20. $(-12y^7)(3y^4)$

21. $(-2x^5)(-3x^4)(5x^3)$

22. $7x(3x^2 + 9)$

23. $5x^3(4x^2 - 11x)$

24. $3y^2(-7y^2 + 3y - 6)$

25. $2y^5(8y^3 - 10y^2 + 1)$

26. $(x + 3)(x^2 - 5x + 2)$

27. $(3y - 2)(4y^2 + 3y - 5)$

In Exercises 28–29, use a vertical format to find each product.

28. $y^2 - 4y + 7$
    $\underline{\qquad\quad 3y - 5}$

29. $4x^3 - 2x^2 - 6x - 1$
    $\underline{\qquad\qquad\quad 2x + 3}$

5.3 *In Exercises 30–42, find each product.*

30. $(x + 6)(x + 2)$

31. $(3y - 5)(2y + 1)$

32. $(4x^2 - 2)(x^2 - 3)$

33. $(5x + 4)(5x - 4)$

34. $(7 - 2y)(7 + 2y)$

35. $(y^2 + 1)(y^2 - 1)$

36. $(x + 3)^2$

37. $(3y + 4)^2$

38. $(y - 1)^2$

39. $(5y - 2)^2$

40. $(x^2 + 4)^2$

41. $(x^2 + 4)(x^2 - 4)$

42. $(x^2 + 4)(x^2 - 5)$

43. Write a polynomial in descending powers of x that represents the area of the shaded region.

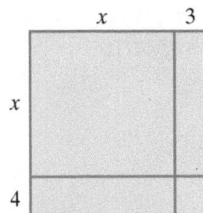

44. The parking garage shown in the figure measures 30 yards by 20 yards. The length and the width are each increased by a fixed amount, x yards. Write a trinomial that describes the area of the expanded garage.

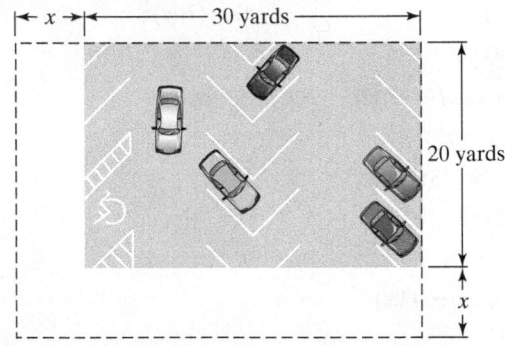

5.4

45. Evaluate $2x^3y - 4xy^2 + 5y + 6$ for $x = -1$ and $y = 2$.

46. Determine the coefficient of each term, the degree of each term, and the degree of the polynomial:

$$4x^2y + 9x^3y^2 - 17x^4 - 12.$$

In Exercises 47–56, perform the indicated operations.

47. $(7x^2 - 8xy + y^2) + (-8x^2 - 9xy + 4y^2)$

48. $(13x^3y^2 - 5x^2y - 9x^2) - (11x^3y^2 - 6x^2y - 3x^2 + 4)$

49. $(-7x^2y^3)(5x^4y^6)$

50. $5ab^2(3a^2b^3 - 4ab)$

51. $(x + 7y)(3x - 5y)$

52. $(4xy - 3)(9xy - 1)$

53. $(3x + 5y)^2$

54. $(xy - 7)^2$

55. $(7x + 4y)(7x - 4y)$

56. $(a - b)(a^2 + ab + b^2)$

5.5   *In Exercises 57–63, simplify each expression.*

57. $\dfrac{6^{40}}{6^{10}}$

58. $\dfrac{x^{18}}{x^3}$

59. $(-10)^0$

60. -10^0

61. $400x^0$

62. $\left(\dfrac{x^4}{2}\right)^3$

63. $\left(\dfrac{-3}{2y^6}\right)^4$

In Exercises 64–68, divide and check each answer.

64. $\dfrac{-15y^8}{3y^2}$

65. $\dfrac{40x^8y^6}{5xy^3}$

66. $\dfrac{18x^4 - 12x^2 + 36x}{6x}$

67. $\dfrac{30x^8 - 25x^7 - 40x^5}{-5x^3}$

68. $\dfrac{27x^3y^2 - 9x^2y - 18xy^2}{3xy}$

5.6   *In Exercises 69–73, divide and check each answer.*

69. $\dfrac{2x^2 + 3x - 14}{x - 2}$

70. $\dfrac{2x^3 - 5x^2 + 7x + 5}{2x + 1}$

71. $\dfrac{x^3 - 2x^2 - 33x - 7}{x - 7}$

72. $\dfrac{y^3 - 27}{y - 3}$

73. $(4x^4 + 6x^3 + 3x - 1) \div (2x^2 + 1)$

In Exercises 74–76, divide using synthetic division.

74. $(4x^3 - 3x^2 - 2x + 1) \div (x + 1)$

75. $(3x^4 - 2x^2 - 10x - 20) \div (x - 2)$

76. $(x^4 + 16) \div (x + 4)$

5.7   *In Exercises 77–81, write each expression with positive exponents only and then simplify.*

77. 7^{-2}

78. $(-4)^{-3}$

79. $2^{-1} + 4^{-1}$

80. $\dfrac{1}{5^{-2}}$

81. $\left(\dfrac{2}{5}\right)^{-3}$

In Exercises 82–90, simplify each exponential expression. Assume that variables in denominators do not equal zero.

82. $\dfrac{x^3}{x^9}$

83. $\dfrac{30y^6}{5y^8}$

84. $(5x^{-7})(6x^2)$

85. $\dfrac{x^4 \cdot x^{-2}}{x^{-6}}$

86. $\dfrac{(3y^3)^4}{y^{10}}$

87. $\dfrac{y^{-7}}{(y^4)^3}$

88. $(2x^{-1})^{-3}$

89. $\left(\dfrac{x^7}{x^4}\right)^{-2}$

90. $\dfrac{(y^3)^4}{(y^{-2})^4}$

In Exercises 91–93, write each number in decimal notation without the use of exponents.

91. 2.3×10^4

92. 1.76×10^{-3}

93. 9×10^{-1}

In Exercises 94–97, write each number in scientific notation.

94. 73,900,000

95. 0.00062

96. 0.38

97. 3.8

In Exercises 98–100, perform the indicated computation. Write the answers in scientific notation.

98. $(6 \times 10^{-3})(1.5 \times 10^6)$

99. $\dfrac{2 \times 10^2}{4 \times 10^{-3}}$

100. $(4 \times 10^{-2})^2$

In Exercises 101–103, use 10^6 for one million and 10^9 for one billion to rewrite the number in each statement in scientific notation.

101. The 2009 economic stimulus package allocated $53.6 billion for grants to states for education.

102. The population of the United States at the time the economic stimulus package was voted into law was approximately 307 million.

103. Use your scientific notation answers from Exercises 101 and 102 to answer this question:

If the cost for grants to states for education was evenly divided among every individual in the United States, how much would each citizen have to pay?

CHAPTER 5 TEST

CHAPTER
Test Prep
VIDEOS

Step-by-step test solutions are found on the Chapter Test Prep Videos available in MyMathLab® or on YouTube (search "BlitzerCombinedAlg" and click on "Channels").

1. Identify $9x + 6x^2 - 4$ as a monomial, binomial, or trinomial. Give the degree of the polynomial.

In Exercises 2–3, add or subtract as indicated.

2. $(7x^3 + 3x^2 - 5x - 11) + (6x^3 - 2x^2 + 4x - 13)$

3. $(9x^3 - 6x^2 - 11x - 4) - (4x^3 - 8x^2 - 13x + 5)$

4. Graph the equation: $y = x^2 - 3$. Select integers for x, starting with -3 and ending with 3.

In Exercises 5–11, find each product.

5. $(-7x^3)(5x^8)$

6. $6x^2(8x^3 - 5x - 2)$

7. $(3x + 2)(x^2 - 4x - 3)$

8. $(3y + 7)(2y - 9)$

9. $(7x + 5)(7x - 5)$

10. $(x^2 + 3)^2$

11. $(5x - 3)^2$

12. Evaluate $4x^2y + 5xy - 6x$ for $x = -2$ and $y = 3$.

In Exercises 13–15, perform the indicated operations.

13. $(8x^2y^3 - xy + 2y^2) - (6x^2y^3 - 4xy - 10y^2)$

14. $(3a - 7b)(4a + 5b)$

15. $(2x + 3y)^2$

In Exercises 16–18, divide and check each answer.

16. $\dfrac{-25x^{16}}{5x^4}$

17. $\dfrac{15x^4 - 10x^3 + 25x^2}{5x}$

18. $\dfrac{2x^3 - 3x^2 + 4x + 4}{2x + 1}$

19. Divide using synthetic division:

$$(3x^4 + 11x^3 - 20x^2 + 7x + 35) \div (x + 5).$$

In Exercises 20–21, write each expression with positive exponents only and then simplify.

20. 10^{-2}

21. $\dfrac{1}{4^{-3}}$

In Exercises 22–27, simplify each expression.

22. $(-3x^2)^3$

23. $\dfrac{20x^3}{5x^8}$

24. $(-7x^{-8})(3x^2)$

25. $\dfrac{(2y^3)^4}{y^8}$

26. $(5x^{-4})^{-2}$

27. $\left(\dfrac{x^{10}}{x^5}\right)^{-3}$

28. Write 3.7×10^{-4} in decimal notation.

29. Write 7,600,000 in scientific notation.

In Exercises 30–31, perform the indicated computation. Write the answers in scientific notation.

30. $(4.1 \times 10^2)(3 \times 10^{-5})$

31. $\dfrac{8.4 \times 10^6}{4 \times 10^{-2}}$

32. Write a polynomial in descending powers of x that represents the area of the figure.

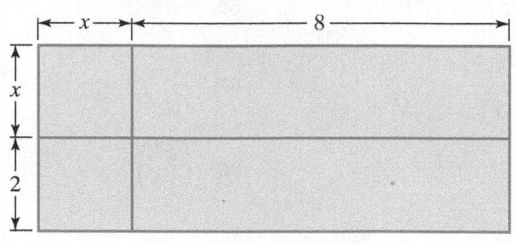

CUMULATIVE REVIEW EXERCISES (CHAPTERS 1–5)

In Exercises 1–2, perform the indicated operation or operations.

1. $(-7)(-5) \div (12 - 3)$

2. $(3 - 7)^2(9 - 11)^3$

3. What is the difference in elevation between a plane flying 14,300 feet above sea level and a submarine traveling 750 feet below sea level?

In Exercises 4–5, solve each equation.

4. $2(x + 3) + 2x = x + 4$

5. $\dfrac{x}{5} - \dfrac{1}{3} = \dfrac{x}{10} - \dfrac{1}{2}$

6. The length of a rectangular sign is 2 feet less than three times its width. If the perimeter of the sign is 28 feet, what are its dimensions?

7. Solve $7 - 8x \le -6x - 5$ and graph the solution set on a number line.

8. You invested $6000 in two accounts paying 12% and 14% annual interest. At the end of the year, the total interest from these investments was $772. How much invested at each rate?

9. You need to mix a solution that is 70% antifreeze with one that is 30% antifreeze to obtain 20 liters of a mixture that is 60% antifreeze. How many liters of each of the solutions must be used?

10. Graph $y = -\frac{2}{5}x + 2$ using the slope and y-intercept.

11. Graph $x - 2y = 4$ using intercepts.

12. Find the slope of the line passing through the points $(-3, 2)$ and $(2, -4)$. Is the line rising, falling, horizontal, or vertical?

13. The slope of a line is -2 and the line passes through the point $(3, -1)$. Write the line's equation in point-slope form and slope-intercept form.

In Exercises 14–15, solve each system by the method of your choice.

14. $\begin{cases} 3x + 2y = 10 \\ 4x - 3y = -15 \end{cases}$

15. $\begin{cases} 2x + 3y = -6 \\ y = 3x - 13 \end{cases}$

16. You are choosing between two telephone plans. One has a monthly fee of $15 with a charge of $0.05 per minute for all calls. The other plan has a monthly fee of $5 with a charge of $0.07 per minute for all calls. For how many minutes of calls will the costs for the two plans be the same? What will be the cost for each plan?

17. Write in scientific notation: 0.0024.

18. Subtract: $(9x^5 - 3x^3 + 2x - 7) - (6x^5 + 3x^3 - 7x - 9)$.

19. Divide: $\dfrac{x^3 + 3x^2 + 5x + 3}{x + 1}$.

20. Simplify: $\dfrac{(3x^2)^4}{x^{10}}$.

Factoring Polynomials

Motion and change are the very essence of life. Moving air brushes against our faces; rain falls on our heads; birds fly past us; plants spring from the earth, grow, and then die; and rocks thrown upward reach a maximum height before falling to the ground. In this chapter, we analyze the where and when of moving objects by writing polynomials as products and using equations in which the highest exponent on the variable is 2.

Motion is modeled in Example 7 of Section 6.6, as well as in the visual discussion that follows the example.

The Greatest Common Factor and Factoring by Grouping

Objectives

1 Find the greatest common factor.

2 Factor out the greatest common factor of a polynomial.

3 Factor out the negative of the greatest common factor of a polynomial.

4 Factor by grouping.

A two-year-old boy is asked, "Do you have a brother?" He answers, "Yes." "What is your brother's name?" "Tom." Asked if Tom has a brother, the two-year-old replies, "No." The child can go in the direction from self to brother, but he cannot reverse this direction and move from brother back to self.

As our intellects develop, we learn to reverse the direction of our thinking. Reversibility of thought is found throughout algebra. For example, we can multiply polynomials and show that

$$5x(2x + 3) = 10x^2 + 15x.$$

We can also reverse this process and express the resulting polynomial as

$$10x^2 + 15x = 5x(2x + 3).$$

Factoring a polynomial containing the sum of monomials means finding an equivalent expression that is a product.

Factoring $10x^2 + 15x$

Sum of monomials   Equivalent expression that is a product

$$10x^2 + 15x = 5x(2x + 3)$$

The factors of $10x^2 + 15x$ are $5x$ and $2x + 3$.

In this chapter, we will be factoring over the set of integers, meaning that the coefficients in the factors are integers. Polynomials that cannot be factored using integer coefficients are called **prime polynomials** over the set of integers.

Factoring Out the Greatest Common Factor

We use the distributive property to multiply a monomial and a polynomial of two or more terms. When we factor, we reverse this process, expressing the polynomial as a product.

Multiplication	Factoring
$a(b + c) = ab + ac$	$ab + ac = a(b + c)$

Here is a specific example:

Multiplication	Factoring
$5x(2x + 3)$	$10x^2 + 15x$
$= 5x \cdot 2x + 5x \cdot 3$	$= 5x \cdot 2x + 5x \cdot 3$
$= 10x^2 + 15x$	$= 5x(2x + 3).$

In the process of finding an equivalent expression for $10x^2 + 15x$ that is a product, we used the fact that $5x$ is a factor of the monomials $10x^2$ and $15x$. The factoring on the right shows that $5x$ is a *common factor* for both terms of the binomial $10x^2 + 15x$.

1 Find the greatest common factor.

In any factoring problem, the first step is to look for the *greatest common factor*. The **greatest common factor**, abbreviated GCF, is an expression of the highest degree that divides each term of the polynomial. Can you see that $5x$ is the greatest common factor of $10x^2 + 15x$? 5 is the greatest integer that divides both 10 and 15. Furthermore, x is the greatest expression that divides both x^2 and x.

The variable part of the greatest common factor always contains the smallest power of a variable that appears in all terms of the polynomial. For example, consider the polynomial

$$10x^2 + 15x.$$

> x^1, or x, is the variable raised to the smaller exponent.

We see that x is the variable part of the greatest common factor, $5x$.

EXAMPLE 1 Finding the Greatest Common Factor

Find the greatest common factor of each list of monomials:

a. $6x^3$ and $10x^2$ **b.** $15y^5, -9y^4,$ and $27y^3$ **c.** $x^5y^3, x^4y^4,$ and $x^3y^2.$

Solution Use numerical coefficients to determine the coefficient of the GCF. Use variable factors to determine the variable factor of the GCF.

> 2 is the greatest integer that divides 6 and 10.

a. $6x^3$ and $10x^2$

> x^2 is the variable raised to the smaller exponent.

We see that 2 is the coefficient of the GCF and x^2 is the variable factor of the GCF. Thus, the GCF of $6x^3$ and $10x^2$ is $2x^2$.

> 3 is the greatest integer that divides 15, −9, and 27.

b. $15y^5, \quad -9y^4, \quad$ and $\quad 27y^3$

> y^3 is the variable raised to the smallest exponent.

We see that 3 is the coefficient of the GCF and y^3 is the variable factor of the GCF. Thus, the GCF of $15y^5, -9y^4,$ and $27y^3$ is $3y^3$.

> x^3 is the variable, x, raised to the smallest exponent.

c. $x^5y^3, \quad x^4y^4, \quad$ and $\quad x^3y^2$

> y^2 is the variable, y, raised to the smallest exponent.

Because all terms have coefficients of 1, 1 is the greatest integer that divides these coefficients. Thus, 1 is the coefficient of the GCF. The voice balloons show that x^3 and y^2 are the variable factors of the GCF. Thus, the GCF of $x^5y^3, x^4y^4,$ and x^3y^2 is x^3y^2. ∎

✓ **CHECK POINT 1** Find the greatest common factor of each list of monomials:

a. $18x^3$ and $15x^2$ **b.** $-20x^2, 12x^4,$ and $40x^3$

c. $x^4y, x^3y^2,$ and $x^2y.$

2 Factor out the greatest common factor of a polynomial.

When we factor a monomial from a polynomial, we determine the greatest common factor of all terms in the polynomial. Sometimes there may not be a GCF other than 1. When a GCF other than 1 exists, we use the following procedure:

Factoring a Monomial From a Polynomial

1. Determine the greatest common factor of all terms in the polynomial.
2. Express each term as the product of the GCF and its other factor.
3. Use the distributive property to factor out the GCF.

Great Question!

Is $5 \cdot x^2 + 5 \cdot 6$ a factorization of $5x^2 + 30$?

No. When we express $5x^2 + 30$ as $5 \cdot x^2 + 5 \cdot 6$, we have factored the *terms* of the binomial, but not the binomial itself. The factorization of the binomial is not complete until we write

$$5(x^2 + 6).$$

Now we have expressed the binomial *as a product*.

EXAMPLE 2 Factoring Out the Greatest Common Factor

Factor: $5x^2 + 30$.

Solution The GCF of $5x^2$ and 30 is 5.

$$5x^2 + 30$$
$$= 5 \cdot x^2 + 5 \cdot 6 \qquad \text{Express each term as the product of the GCF and its other factor.}$$
$$= 5(x^2 + 6) \qquad \text{Factor out the GCF.}$$

Because factoring reverses the process of multiplication, all factorizations can be checked by multiplying.

$$5(x^2 + 6) = 5 \cdot x^2 + 5 \cdot 6 = 5x^2 + 30$$

The factorization is correct because multiplication gives us the original polynomial. ■

☑ **CHECK POINT 2** Factor: $6x^2 + 18$.

EXAMPLE 3 Factoring Out the Greatest Common Factor

Factor: $18x^3 + 27x^2$.

Solution We begin by determining the greatest common factor.

9 is the greatest integer that divides 18 and 27.

$$18x^3 \quad \text{and} \quad 27x^2$$

x^2 is the variable raised to the smaller exponent.

Discover for Yourself

What happens if you factor out $3x^2$ rather than $9x^2$ from $18x^3 + 27x^2$? Although $3x^2$ is a common factor of the two terms, it is not the *greatest* common factor. Factor out $3x^2$ from $18x^3 + 27x^2$ and describe what happens with the second factor. Now factor again. Make the final result look like the factorization in Example 3. What is the advantage of factoring out the greatest common factor rather than just a common factor?

The GCF of the two terms in the polynomial is $9x^2$.

$$18x^3 + 27x^2$$
$$= 9x^2 \cdot 2x + 9x^2 \cdot 3 \qquad \text{Express each term as the product of the GCF and its other factor.}$$
$$= 9x^2(2x + 3) \qquad \text{Factor out the GCF.}$$

We can check this factorization by multiplying $9x^2$ and $2x + 3$, obtaining the original polynomial as the answer. ■

☑ **CHECK POINT 3** Factor: $25x^2 + 35x^3$.

EXAMPLE 4 Factoring Out the Greatest Common Factor

Factor: $16x^5 - 12x^4 + 4x^3$.

Solution First, determine the greatest common factor.

4 is the greatest integer that divides 16, −12, and 4.

$16x^5, \quad -12x^4, \quad \text{and} \quad 4x^3$

x^3 is the variable raised to the smallest exponent.

The GCF of the three terms of the polynomial is $4x^3$.

$$16x^5 - 12x^4 + 4x^3$$

$$= 4x^3 \cdot 4x^2 - 4x^3 \cdot 3x + 4x^3 \cdot 1 \qquad \text{Express each term as the product of the GCF and its other factor.}$$

You can obtain the factors shown in black by dividing each term of the given polynomial by $4x^3$, the GCF:

$$\frac{16x^5}{4x^3} = 4x^2 \qquad \frac{12x^4}{4x^3} = 3x \qquad \frac{4x^3}{4x^3} = 1.$$

$$= 4x^3(4x^2 - 3x + 1) \qquad \text{Factor out the GCF.} \quad \blacksquare$$

Don't leave out the 1.

✓ **CHECK POINT 4** Factor: $15x^5 + 12x^4 - 27x^3$.

EXAMPLE 5 Factoring Out the Greatest Common Factor

Factor: $27x^2y^3 - 9xy^2 + 81xy$.

Solution First, determine the greatest common factor.

9 is the greatest integer that divides 27, −9, and 81.

$27x^2y^3, \quad -9xy^2, \quad \text{and} \quad 81xy$

The variables raised to the smallest exponents are x and y.

The GCF of the three terms of the polynomial is $9xy$.

$$27x^2y^3 - 9xy^2 + 81xy$$

$$= 9xy \cdot 3xy^2 - 9xy \cdot y + 9xy \cdot 9 \qquad \text{Express each term as the product of the GCF and its other factor.}$$

You can obtain the factors shown in black by dividing each term of the given polynomial by $9xy$, the GCF:

$$\frac{27x^2y^3}{9xy} = 3xy^2 \qquad \frac{9xy^2}{9xy} = y \qquad \frac{81xy}{9xy} = 9.$$

$$= 9xy(3xy^2 - y + 9) \qquad \text{Factor out the GCF.} \quad \blacksquare$$

✓ **CHECK POINT 5** Factor: $8x^3y^2 - 14x^2y + 2xy$.

3 Factor out the negative of the greatest common factor of a polynomial.

Factoring with a Negative Coefficient in the First Term

Suppose we are interested in factoring

$$-5x^2 + 30.$$

> The first term has a negative coefficient.

When factoring polynomials, it is preferable to have a first term with a positive coefficient inside parentheses. We can do this with $-5x^2 + 30$ by factoring out -5, the negative of the GCF.

$$-5x^2 + 30$$
$$= -5 \cdot x^2 - 5(-6) \qquad \text{Express each term as the product of the negative of the GCF and its other factor.}$$
$$= -5(x^2 - 6) \qquad \text{Factor out the negative of the GCF.}$$

Factoring with a Negative Coefficient in the First Term

Express each term as the product of the negative of the GCF and its other factor. Then use the distributive property to factor out the negative of the GCF.

EXAMPLE 6   Factoring Out the Negative of the GCF

Factor: $-18a^4b^3 + 6a^2b^2 - 15a^3b$.

Solution   First, determine the greatest common factor.

> 3 is the greatest integer that divides −18, 6, and −15.

$$-18a^4b^3, \qquad 6a^2b^2, \qquad \text{and} \qquad -15a^3b$$

> The variables raised to the smallest exponents are a^2 and b.

The GCF of the three terms of the polynomial is $3a^2b$. Because the polynomial has a negative coefficient in the first term, -18, we will factor out the negative of the GCF. Thus, we will factor out $-3a^2b$.

$$-18a^4b^3 + 6a^2b^2 - 15a^3b$$

$$= -3a^2b \cdot 6a^2b^2 - 3a^2b(-2b) - 3a^2b \cdot 5a \qquad \text{Express each term as the product of the negative of the GCF and its other factor.}$$

$$= -3a^2b(6a^2b^2 - 2b + 5a) \qquad \text{Factor out the negative of the GCF.} \quad ∎$$

✓ **CHECK POINT 6**   Factor: $-16a^4b^5 + 24a^3b^4 - 20ab^2$.

4 Factor by grouping.

Factoring by Grouping

Up to now, we have factored a monomial from a polynomial. By contrast, in our next example, the greatest common factor of the polynomial is a binomial.

EXAMPLE 7   Factoring Out the Greatest Common Binomial Factor

Factor:
  a. $x^2(x + 3) + 5(x + 3)$ **b.** $x(y + 1) - 2(y + 1)$.

Solution   Let's identify the common binomial factor in each part of the problem.

$$x^2(x + 3) \quad \text{and} \quad 5(x + 3) \qquad\qquad x(y + 1) \quad \text{and} \quad -2(y + 1)$$

The GCF, a binomial, is $x + 3$.   The GCF, a binomial, is $y + 1$.

We factor out these common binomial factors as follows.

a. $x^2(x + 3) + 5(x + 3)$

$= (x + 3)x^2 + (x + 3)5$ Express each term as the product of the GCF and its other factor, in that order. Hereafter, we omit this step.

$= (x + 3)(x^2 + 5)$ Factor out the GCF, $x + 3$.

b. $x(y + 1) - 2(y + 1)$ The GCF is $y + 1$.

$= (y + 1)(x - 2)$ Factor out the GCF. ∎

✓ **CHECK POINT 7**   Factor:

a. $x^2(x + 1) + 7(x + 1)$ **b.** $x(y + 4) - 7(y + 4)$.

Some polynomials have only a greatest common factor of 1. However, by a suitable grouping of the terms, it still may be possible to factor. This process, called **factoring by grouping**, is illustrated in Example 8.

EXAMPLE 8   Factoring by Grouping

Factor: $x^3 + 4x^2 + 3x + 12$.

Solution   There is no factor other than 1 common to all four terms. However, we can group terms that have a common factor:

$$\boxed{x^3 + 4x^2} \;+\; \boxed{3x + 12}.$$

Common factor is x^2.   Common factor is 3.

Discover for Yourself

In Example 8, group the terms as follows:

$$(x^3 + 3x) + (4x^2 + 12).$$

Factor out the greatest common factor from each group and complete the factoring process. Describe what happens. What can you conclude?

We now factor the given polynomial as follows:

$x^3 + 4x^2 + 3x + 12$

$= (x^3 + 4x^2) + (3x + 12)$ Group terms with common factors.

$= x^2(x + 4) + 3(x + 4)$ Factor out the greatest common factor from the grouped terms. The remaining two terms have $x + 4$ as a common binomial factor.

$= (x + 4)(x^2 + 3)$.   Factor out the GCF, $x + 4$.

Thus, $x^3 + 4x^2 + 3x + 12 = (x + 4)(x^2 + 3)$. Check the factorization by multiplying the right side of the equation using the FOIL method. Because the factorization is correct, you should obtain the original polynomial. ∎

✓ **CHECK POINT 8**   Factor: $x^3 + 5x^2 + 2x + 10$.

Factoring by Grouping

1. Group terms that have a common monomial factor. There will usually be two groups. Sometimes the terms must be rearranged.
2. Factor out the common monomial factor from each group.
3. Factor out the remaining common binomial factor (if one exists).

EXAMPLE 9 Factoring by Grouping

Factor: $xy + 5x - 4y - 20$.

Solution There is no factor other than 1 common to all four terms. However, we can group terms that have a common factor:

$$\boxed{xy + 5x} \quad + \quad \boxed{-4y - 20}.$$

Common factor is x:
$xy + 5x = x(y + 5)$.

Use -4, rather than 4, as the common factor:
$-4y - 20 = -4(y + 5)$. In this way, the common binomial factor, $y + 5$, appears.

The voice balloons illustrate that it is sometimes necessary to factor out a negative number from a grouping to obtain a common binomial factor for the two groupings. We now factor the given polynomial as follows:

$$xy + 5x - 4y - 20$$
$$= x(y + 5) - 4(y + 5) \qquad \text{Factor } x \text{ and } -4, \text{ respectively, from each grouping.}$$
$$= (y + 5)(x - 4). \qquad \text{Factor out the GCF, } y + 5.$$

Thus, $xy + 5x - 4y - 20 = (y + 5)(x - 4)$. Using the commutative property of multiplication, the factorization can also be expressed as $(x - 4)(y + 5)$. Multiply these factors using the FOIL method to verify that, regardless of the order, these are the correct factors. ∎

☑ **CHECK POINT 9** Factor: $xy + 3x - 5y - 15$.

Achieving Success

When using your professor's office hours, show up prepared. If you are having difficulty with a concept or problem, bring your work so that your instructor can determine where you are having trouble. If you miss a lecture, read the appropriate section in the textbook, borrow class notes, and attempt the assigned homework before your office visit. Because this text has an accompanying video lesson for every section, you might find it helpful to view the CD-ROM covering the material you missed. It is not realistic to expect your professor to rehash all or part of a class lecture during office hours.

CONCEPT AND VOCABULARY CHECK

Fill in each blank so that the resulting statement is true.

1. The process of writing a polynomial containing the sum of monomials as a product is called _____ .

2. An expression of the highest degree that divides each term of a polynomial is called the _____ . The variable part of this expression contains the _____ power of a variable that appears in all terms of the polynomial.

3. True or false: The factorization of $15x + 20$ is $5 \cdot 3x + 5 \cdot 4$. _____

4. True or false: The factorization of $x^2 + 3x + 5x + 15$ is $x(x + 3) + 5(x + 3)$. _____

6.1 EXERCISE SET MyMathLab®

Practice Exercises

In Exercises 1–12, find the greatest common factor of each list of monomials.

1. 4 and $8x$
2. 5 and $15x$
3. $12x^2$ and $8x$
4. $20x^2$ and $15x$
5. $-2x^4$ and $6x^3$
6. $-3x^4$ and $6x^3$
7. $9y^5$, $18y^2$, and $-3y$
8. $10y^5$, $20y^2$, and $-5y$
9. xy, xy^2, and xy^3
10. x^2y, $3x^3y$, and $6x^2$
11. $16x^5y^4$, $8x^6y^3$, and $20x^4y^5$
12. $18x^5y^4$, $6x^6y^3$, and $12x^4y^5$

In Exercises 13–48, factor each polynomial using the greatest common factor. If there is no common factor other than 1 and the polynomial cannot be factored, so state.

13. $8x + 8$
14. $9x + 9$
15. $4y - 4$
16. $5y - 5$
17. $5x + 30$
18. $10x + 30$
19. $30x - 12$
20. $32x - 24$
21. $x^2 + 5x$
22. $x^2 + 6x$
23. $18y^2 + 12$
24. $20y^2 + 15$
25. $14x^3 + 21x^2$
26. $6x^3 + 15x^2$
27. $13y^2 - 25y$
28. $11y^2 - 30y$
29. $9y^4 + 27y^6$
30. $10y^4 + 15y^6$
31. $8x^2 - 4x^4$
32. $12x^2 - 4x^4$
33. $12y^2 + 16y - 8$
34. $15y^2 - 3y + 9$
35. $9x^4 + 18x^3 + 6x^2$
36. $32x^4 + 2x^3 + 8x^2$
37. $100y^5 - 50y^3 + 100y^2$
38. $26y^5 - 13y^3 + 39y^2$
39. $10x - 20x^2 + 5x^3$
40. $6x - 4x^2 + 2x^3$
41. $11x^2 - 23$
42. $12x^2 - 25$
43. $6x^3y^2 + 9xy$
44. $4x^2y^3 + 6xy$
45. $30x^2y^3 - 10xy^2 + 20xy$
46. $27x^2y^3 - 18xy^2 + 45x^2y$
47. $32x^3y^2 - 24x^3y - 16x^2y$
48. $18x^3y^2 - 12x^3y - 24x^2y$

In Exercises 49–56, factor each polynomial using the negative of the greatest common factor.

49. $-12x^2 + 18$
50. $-15x^2 + 20$
51. $-8x^4 + 32x^3 + 16x^2$
52. $-18x^4 + 9x^3 + 6x^2$
53. $-4a^3b^2 + 6ab$
54. $-9a^2b^3 + 12ab$
55. $-12x^3y^2 - 18x^3y + 24x^2y$
56. $-24x^3y^2 - 32x^3y + 16x^2y$

In Exercises 57–68, factor each polynomial using the greatest common binomial factor.

57. $x(x + 5) + 3(x + 5)$
58. $x(x + 7) + 10(x + 7)$
59. $x(x + 2) - 4(x + 2)$
60. $x(x + 3) - 8(x + 3)$
61. $x(y + 6) - 7(y + 6)$
62. $x(y + 9) - 11(y + 9)$
63. $3x(x + y) - (x + y)$
64. $7x(x + y) - (x + y)$
65. $4x(3x + 1) + 3x + 1$
66. $5x(2x + 1) + 2x + 1$
67. $7x^2(5x + 4) + 5x + 4$
68. $9x^2(7x + 2) + 7x + 2$

In Exercises 69–86, factor by grouping.

69. $x^2 + 2x + 4x + 8$
70. $x^2 + 3x + 5x + 15$
71. $x^2 + 3x - 5x - 15$
72. $x^2 + 7x - 4x - 28$
73. $x^3 - 2x^2 + 5x - 10$
74. $x^3 - 3x^2 + 4x - 12$
75. $x^3 - x^2 + 2x - 2$
76. $x^3 + 6x^2 - 2x - 12$
77. $xy + 5x + 9y + 45$
78. $xy + 6x + 2y + 12$
79. $xy - x + 5y - 5$
80. $xy - x + 7y - 7$
81. $3x^2 - 6xy + 5xy - 10y^2$
82. $10x^2 - 12xy + 35xy - 42y^2$
83. $3x^3 - 2x^2 - 6x + 4$
84. $4x^3 - x^2 - 12x + 3$
85. $x^2 - ax - bx + ab$
86. $x^2 + ax + bx + ab$

Practice PLUS

In Exercises 87–94, factor each polynomial.

87. $24x^3y^3z^3 + 30x^2y^2z + 18x^2yz^2$
88. $16x^2y^2z^2 + 32x^2yz^2 + 24x^2yz$
89. $x^3 - 4 + 3x^3y - 12y$
90. $x^3 - 5 + 2x^3y - 10y$
91. $4x^5(x + 1) - 6x^3(x + 1) - 8x^2(x + 1)$

92. $8x^5(x + 2) - 10x^3(x + 2) - 2x^2(x + 2)$

93. $3x^5 - 3x^4 + x^3 - x^2 + 5x - 5$

94. $7x^5 - 7x^4 + x^3 - x^2 + 3x - 3$

The figures for Exercises 95–96 show one or more circles drawn inside a square. Write a polynomial that represents the shaded blue area in each figure. Then factor the polynomial.

95.
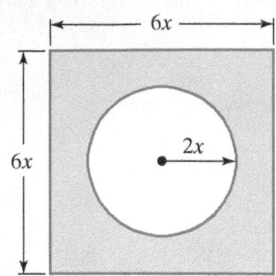

96.
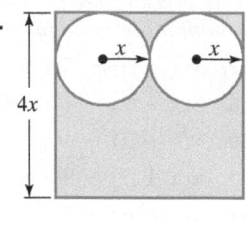

Application Exercises

97. An explosion causes debris to rise vertically with an initial velocity of 64 feet per second. The polynomial $64x - 16x^2$ describes the height of the debris above the ground, in feet, after x seconds.

   a. Find the height of the debris after 3 seconds.

   b. Factor the polynomial.

   c. Use the factored form of the polynomial in part (b) to find the height of the debris after 3 seconds. Do you get the same answer as you did in part (a)? If so, does this prove that your factorization is correct? Explain.

98. An explosion causes debris to rise vertically with an initial velocity of 72 feet per second. The polynomial $72x - 16x^2$ describes the height of the debris above the ground, in feet, after x seconds.

   a. Find the height of the debris after 4 seconds.

   b. Factor the polynomial.

   c. Use the factored form of the polynomial in part (b) to find the height of the debris after 4 seconds. Do you get the same answer as you did in part (a)? If so, does this prove that your factorization is correct? Explain

In Exercises 99–100, write a polynomial for the length of each rectangle.

99.
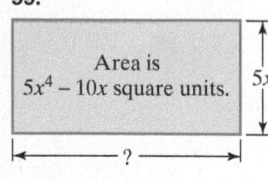

100.
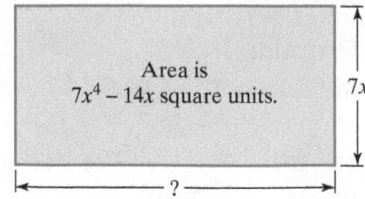

Writing in Mathematics

101. What is factoring?

102. What is a prime polynomial?

103. Explain how to find the greatest common factor of a list of terms. Give an example with your explanation.

104. Use an example and explain how to factor out the greatest common factor of a polynomial.

105. Suppose that a polynomial contains four terms and can be factored by grouping. Explain how to obtain the factorization.

106. Write a sentence that uses the word "factor" as a noun. Then write a sentence that uses the word "factor" as a verb.

Critical Thinking Exercises

Make Sense? *In Exercises 107–110, determine whether each statement "makes sense" or "does not make sense" and explain your reasoning.*

107. After factoring $20x^3 + 8x^2$ and $20x^3 + 10x$, I noticed that I factored the monomial $20x^3$ in two different ways.

108. After I've factored a polynomial, my answer cannot always be checked by multiplication.

109. The word *greatest* in greatest common factor is helpful because it tells me to look for the greatest power of a variable appearing in all terms.

110. You grouped the polynomial's terms using different groupings than I did, yet we both obtained the same factorization.

In Exercises 111–114, determine whether each statement is true or false. If the statement is false, make the necessary change(s) to produce a true statement.

111. Since the GCF of $9x^3 + 6x^2 + 3x$ is $3x$, it is not necessary to write the 1 when $3x$ is factored from the last term.

112. $a(x - 7) + b(7 - x) = a(x - 7) + b(-1)(x - 7)$
$$= a(x - 7) - b(x - 7)$$
$$= (x - 7)(a - b)$$

113. $a^2 + b^2 = a^2 + ab - ab + b^2$
$$= a(a + b) - b(a + b)$$
$$= (a + b)(a - b)$$

114. $-4x^2 + 12x$ can be factored as $-4x(x - 3)$ or $4x(-x + 3)$.

115. Suppose you receive x dollars in January. Each month thereafter, you receive $100 more than you received the month before. Write a factored polynomial that describes the total dollar amount you receive from January through April.

In Exercises 116–117, write a polynomial that fits the given description. Do not use a polynomial that appears in this section or in the Exercise Set.

116. The polynomial has four terms and can be factored using a greatest common factor that has both a coefficient and a variable.

117. The polynomial has four terms and can be factored by grouping.

Technology Exercises

In Exercises 118–120, use a graphing utility to graph each side of the equation in the same viewing rectangle. Do the graphs coincide? If so, this means that the polynomial on the left side has been factored correctly. If not, factor the polynomial correctly and then use your graphing utility to verify the factorization.

118. $-3x - 6 = -3(x - 2)$

119. $x^2 - 2x + 5x - 10 = (x - 2)(x - 5)$

120. $x^2 + 2x + x + 2 = x(x + 2) + 1$

Review Exercises

121. Multiply: $(x + 7)(x + 10)$. (Section 5.3, Example 1)

122. Solve the system by graphing:
$$\begin{cases} 2x - y = -4 \\ x - 3y = 3. \end{cases}$$
(Section 4.1, Example 2)

123. Write the point-slope form of the equation of the line passing through $(-7, 2)$ and $(-4, 5)$. Then use the point-slope form of the equation to write the slope-intercept form of the equation. (Section 3.5, Example 2)

Preview Exercises

Exercises 124–126 will help you prepare for the material covered in the next section.

124. Find two factors of 8 whose sum is 6.

125. Find two factors of 6 whose sum is -5.

126. Find two factors of -35 whose sum is 2.

SECTION

6.2

Factoring Trinomials Whose Leading Coefficient Is 1

Objective

1️⃣ Factor trinomials of the form $x^2 + bx + c$.

Not afraid of heights and cutting-edge excitement? How about sky diving? Behind your exhilarating experience is the world of algebra. After you jump from the airplane, your height above the ground at every instant of your fall can be described by a formula involving a variable that is squared. At a height of approximately 2000 feet, you'll need to open your parachute. How can you determine when you must do so?

The answer to this critical question involves using the factoring technique presented in this section. In Section 6.6, in which applications are discussed, this technique is applied to models involving the height of any free-falling object—in this case, you.

1 Factor trinomials of the form $x^2 + bx + c$.

A Strategy for Factoring $x^2 + bx + c$

In Section 5.3, we used the FOIL method to multiply two binomials. The product was often a trinomial. The following are some examples:

Factored Form	F	O	I	L	Trinomial Form

$$(x + 3)(x + 4) = x^2 + 4x + 3x + 12 = x^2 + 7x + 12$$
$$(x - 3)(x - 4) = x^2 - 4x - 3x + 12 = x^2 - 7x + 12$$
$$(x + 3)(x - 5) = x^2 - 5x + 3x - 15 = x^2 - 2x - 15.$$

Observe that each trinomial is of the form $x^2 + bx + c$, where the coefficient of the squared term, called the **leading coefficient**, is 1. Our goal in this section is to start with the trinomial form and, assuming that it is factorable, return to the factored form.

The first FOIL multiplication shown above indicates that $(x + 3)(x + 4) = x^2 + 7x + 12$. Let's reverse the sides of this equation:

$$x^2 + 7x + 12 = (x + 3)(x + 4).$$

We can make several important observations about the factors on the right side.

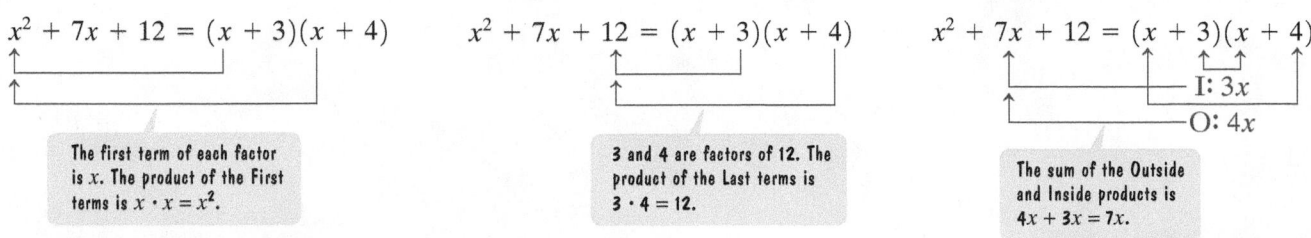

The first term of each factor is x. The product of the First terms is $x \cdot x = x^2$.

3 and 4 are factors of 12. The product of the Last terms is $3 \cdot 4 = 12$.

The sum of the Outside and Inside products is $4x + 3x = 7x$.

These observations provide us with a procedure for factoring $x^2 + bx + c$.

A Strategy for Factoring $x^2 + bx + c$

1. Enter x as the first term of each factor.

$$(x \quad)(x \quad) = x^2 + bx + c$$

2. List pairs of factors of the constant c.

3. Try various combinations of these factors as the second term in each set of parentheses. Select the combination in which the sum of the Outside and Inside products is equal to bx.

$$(x + \Box)(x + \Box) = x^2 + bx + c$$

  I
  O
Sum of O + I

4. Check your work by multiplying the factors using the FOIL method. You should obtain the original trinomial.

If none of the possible combinations yield an Outside product and an Inside product whose sum is equal to bx, the trinomial cannot be factored using integers and is called **prime** over the set of integers.

EXAMPLE 1 Factoring a Trinomial in $x^2 + bx + c$ Form

Factor: $x^2 + 6x + 8$.

Solution

Step 1. Enter x as the first term of each factor.

$$x^2 + 6x + 8 = (x \quad)(x \quad)$$

Step 2. List pairs of factors of the constant, 8.

Factors of 8	8, 1	4, 2	$-8, -1$	$-4, -2$

Step 3. Try various combinations of these factors. The correct factorization of $x^2 + 6x + 8$ is the one in which the sum of the Outside and Inside products is equal to $6x$. Here is a list of the possible factorizations:

Possible Factorizations of $x^2 + 6x + 8$	Sum of Outside and Inside Products (Should Equal 6x)	
$(x + 8)(x + 1)$	$x + 8x = 9x$	This is the required middle term.
$(x + 4)(x + 2)$	$2x + 4x = 6x$	
$(x - 8)(x - 1)$	$-x - 8x = -9x$	
$(x - 4)(x - 2)$	$-2x - 4x = -6x$	

Thus, $x^2 + 6x + 8 = (x + 4)(x + 2)$.

Step 4. Check this result by multiplying the right side using the FOIL method. You should obtain the original trinomial. Because of the commutative property, the factorization can also be expressed as

$$x^2 + 6x + 8 = (x + 2)(x + 4). \quad \blacksquare$$

Using Technology

Graphic and Numeric Connections

If a polynomial contains one variable, a graphing utility can be used to check its factorization. For example, the factorization in Example 1,

$$x^2 + 6x + 8 = (x + 4)(x + 2),$$

can be checked graphically or numerically.

Graphic Check

Use the $\boxed{\text{GRAPH}}$ feature. Graph $y_1 = x^2 + 6x + 8$ and $y_2 = (x + 4)(x + 2)$ on the same screen. Because the graphs are identical, the factorization appears to be correct.

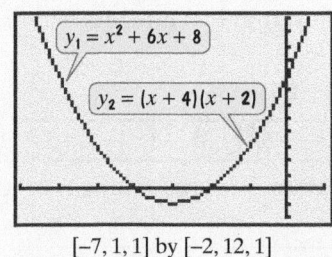

$y_1 = x^2 + 6x + 8$

$y_2 = (x + 4)(x + 2)$

$[-7, 1, 1]$ by $[-2, 12, 1]$

Numeric Check

Use the $\boxed{\text{TABLE}}$ feature. Enter $y_1 = x^2 + 6x + 8$ and $y_2 = (x + 4)(x + 2)$ and press $\boxed{\text{TABLE}}$. Two columns of values are shown, one for y_1 and one for y_2. Because the corresponding values are equal regardless of how far up or down we scroll, the factorization is correct.

X	Y1	Y2
-3	-1	-1
-2	0	0
-1	3	3
0	8	8
1	15	15
2	24	24
3	35	35

X= -3

✓ **CHECK POINT 1** Factor: $x^2 + 5x + 6$.

EXAMPLE 2 Factoring a Trinomial in $x^2 + bx + c$ Form

Factor: $x^2 - 5x + 6$.

Solution

Step 1. Enter x as the first term of each factor.
$$x^2 - 5x + 6 = (x \quad)(x \quad)$$

Step 2. List pairs of factors of the constant, 6.

Factors of 6	6, 1	3, 2	−6, −1	−3, −2

Step 3. Try various combinations of these factors. The correct factorization of $x^2 - 5x + 6$ is the one in which the sum of the Outside and Inside products is equal to $-5x$. Here is a list of the possible factorizations:

Possible Factorizations of $x^2 - 5x + 6$	Sum of Outside and Inside Products (Should Equal −5x)	
$(x + 6)(x + 1)$	$x + 6x = 7x$	
$(x + 3)(x + 2)$	$2x + 3x = 5x$	
$(x - 6)(x - 1)$	$-x - 6x = -7x$	
$(x - 3)(x - 2)$	$-2x - 3x = -5x$	This is the required middle term.

Thus, $x^2 - 5x + 6 = (x - 3)(x - 2)$. Verify this result using the FOIL method. ∎

In factoring a trinomial of the form $x^2 + bx + c$, you can speed things up by listing the factors of c and then finding their sums. We are interested in a sum of b. For example, in factoring $x^2 - 5x + 6$, we are interested in the factors of 6 whose sum is -5.

Factors of 6	6, 1	3, 2	−6, −1	−3, −2
Sum of Factors	7	5	−7	−5

This is the desired sum.

Thus, $x^2 - 5x + 6 = (x - 3)(x - 2)$. ∎

✓ **CHECK POINT 2** Factor: $x^2 - 6x + 8$.

EXAMPLE 3 Factoring a Trinomial in $x^2 + bx + c$ Form

Factor: $x^2 + 2x - 35$.

Solution

Step 1. Enter x as the first term of each factor.
$$x^2 + 2x - 35 = (x \quad)(x \quad)$$

To find the second term of each factor, we must find two integers whose product is -35 and whose sum is 2.

Step 2. List pairs of factors of the constant, −35.

Factors of −35	−35, 1	−7, 5	35, −1	7, −5

Step 3. Try various combinations of these factors. We are looking for the pair of factors whose sum is 2.

Factors of −35	−35, 1	−7, 5	35, −1	7, −5
Sum of Factors	−34	−2	34	2

This is the desired sum.

Thus, $x^2 + 2x - 35 = (x + 7)(x - 5)$.

Great Question!

Is there a way to eliminate some of the combinations of factors for $x^2 + bx + c$ when c is positive?

Yes. To factor $x^2 + bx + c$ when c is positive, find two numbers with the same sign as the middle term.

$$x^2 + 6x + 8 = (x + 2)(x + 4)$$

Same signs

$$x^2 - 5x + 6 = (x - 3)(x - 2)$$

Same signs

Step 4. Verify the factorization using the FOIL method.

$$(x + 7)(x - 5) = x^2 - 5x + 7x - 35 = x^2 + 2x - 35$$

Because the product of the factors is the original polynomial, the factorization is correct. ∎

✓ **CHECK POINT 3**   Factor: $x^2 + 3x - 10$.

EXAMPLE 4   Factoring a Trinomial Whose Leading Coefficient Is 1

Factor: $y^2 - 2y - 99$.

Solution

Step 1. Enter y as the first term of each factor.

$$y^2 - 2y - 99 = (y \quad)(y \quad)$$

To find the second term of each factor, we must find two integers whose product is -99 and whose sum is -2.

Step 2. List pairs of factors of the constant, -99.

Factors of -99	$-99, 1$	$-11, 9$	$-33, 3$	$99, -1$	$11, -9$	$33, -3$

Step 3. Try various combinations of these factors. In order to factor $y^2 - 2y - 99$, we are interested in the pair of factors of -99 whose sum is -2.

Factors of -99	$-99, 1$	$-11, 9$	$-33, 3$	$99, -1$	$11, -9$	$33, -3$
Sum of Factors	-98	-2	-30	98	2	30

This is the desired sum.

Thus, $y^2 - 2y - 99 = (y - 11)(y + 9)$. Verify this result using the FOIL method. ∎

✓ **CHECK POINT 4**   Factor: $y^2 - 6y - 27$.

EXAMPLE 5   Trying to Factor a Trinomial in $x^2 + bx + c$ Form

Factor: $x^2 + x - 5$.

Solution

Step 1. Enter x as the first term of each factor.

$$x^2 + x - 5 = (x \quad)(x \quad)$$

To find the second term of each factor, we must find two integers whose product is -5 and whose sum is 1.

Steps 2 and 3. List pairs of factors of the constant, -5, and try various combinations of these factors. We are interested in a pair of factors whose sum is 1.

Factors of -5	$-5, 1$	$5, -1$
Sum of Factors	-4	4

No pair gives the desired sum, 1.

Because neither pair has a sum of 1, $x^2 + x - 5$ cannot be factored using integers. This trinomial is prime. ∎

✓ **CHECK POINT 5**   Factor: $x^2 + x - 7$.

Great Question!

Is there a way to eliminate some of the combinations of factors for $x^2 + bx + c$ when c is negative?

Yes. To factor $x^2 + bx + c$ when c is negative, find two numbers with opposite signs whose sum is the coefficient of the middle term.

$$x^2 + 2x - 35 = (x + 7)(x - 5)$$

Negative   Opposite signs

$$y^2 - 2y - 99 = (y - 11)(y + 9)$$

Negative   Opposite signs

EXAMPLE 6 Factoring a Trinomial in Two Variables

Factor: $x^2 - 5xy + 6y^2$.

Solution

Step 1. Enter x as the first term of each factor. Because the last term of the trinomial contains y^2, the second term of each factor must contain y.

$$x^2 - 5xy + 6y^2 = (x \quad ?y)(x \quad ?y)$$

The question marks indicate that we are looking for the coefficients of y in each factor. To find these coefficients, we must find two integers whose product is 6 and whose sum is -5.

Steps 2 and 3. List pairs of factors of the coefficient of the last term, 6, and try various combinations of these factors. We are interested in the pair of factors whose sum is -5.

Factors of 6	6, 1	3, 2	−6, −1	−3, −2
Sum of Factors	7	5	−7	−5

This is the desired sum.

Thus, $x^2 - 5xy + 6y^2 = (x - 3y)(x - 2y)$.

Step 4. Verify the factorization using the FOIL method.

$$(x - 3y)(x - 2y) = x^2 - 2xy - 3xy + 6y^2 = x^2 - 5xy + 6y^2$$

Because the product of the factors is the original polynomial, the factorization is correct. ∎

✓ **CHECK POINT 6** Factor: $x^2 - 4xy + 3y^2$.

Some polynomials can be factored using more than one technique. **Always begin by looking for a greatest common factor** and, if there is one, factor it out. A polynomial is **factored completely** when it is written as the product of prime polynomials.

EXAMPLE 7 Factoring Completely

Factor: $3x^3 - 15x^2 - 42x$.

Solution The GCF of the three terms of the polynomial is $3x$. We begin by factoring out $3x$. Then we factor the remaining trinomial by the methods of this section.

$$3x^3 - 15x^2 - 42x$$
$$= 3x(x^2 - 5x - 14) \qquad \text{Factor out the GCF.}$$
$$= 3x(x \quad)(x \quad) \qquad \text{Begin factoring } x^2 - 5x - 14. \text{ Find two integers whose product is } -14 \text{ and whose sum is } -5.$$
$$= 3x(x - 7)(x + 2) \qquad \text{The integers are } -7 \text{ and } 2.$$

Thus,

$$3x^3 - 15x^2 - 42x = 3x(x - 7)(x + 2).$$

Be sure to include the GCF in the factorization.

How can we check this factorization? We will multiply the binomials using the FOIL method. Then use the distributive property and multiply each term of this product by $3x$. If the factorization is correct, we should obtain the original polynomial.

$$3x(x - 7)(x + 2) = 3x(x^2 + 2x - 7x - 14) = 3x(x^2 - 5x - 14) - 3x^3 - 15x^2 - 42x$$

Use the FOIL method on $(x - 7)(x + 2)$.

This is the original polynomial.

The factorization is correct. ■

✓ **CHECK POINT 7**   Factor: $2x^3 + 6x^2 - 56x$.

EXAMPLE 8   Factoring Completely by Factoring Out the Negative of the GCF

Factor: $-16t^2 + 16t + 96$.

Solution   The GCF of the three terms of the polynomial is 16. Because the polynomial has a negative coefficient in the first term, we will factor out the negative of the GCF. Thus, we will factor out -16.

$$-16t^2 + 16t + 96$$
$$= -16(t^2 - t - 6)$$ Factor out the negative of the GCF.
$$= -16(t\ \ )(t\ \ )$$ Begin factoring $t^2 - t - 6$. Find two integers whose product is -6 and whose sum is -1.
$$= -16(t - 3)(t + 2)$$ The integers are -3 and 2.

Thus,

$$-16t^2 + 16t + 96 = -16(t - 3)(t + 2).$$

Verify this factorization using the FOIL method to multiply $t - 3$ and $t + 2$. Then use the distributive property and multiply each term of this product by -16. You should obtain the original polynomial. ■

✓ **CHECK POINT 8**   Factor: $-2y^2 - 10y + 28$.

CONCEPT AND VOCABULARY CHECK

Fill in each blank so that the resulting statement is true.

1. To factor $x^2 - 12x + 20$, we must find two integers whose product is _____ and whose sum is _____.

2. A polynomial is factored _____ when it is written as a product of prime polynomials.

3. $x^2 + 13x + 30 = (x + 3)(x\ \_\_\_)$

4. $x^2 - 9x + 18 = (x - 3)(x\ \_\_\_)$

5. $x^2 - x - 30 = (x - 6)(x\ \_\_\_)$

6. $x^2 - 5x - 14 = (x + 2)(x\ \_\_\_)$

7. $x^2 - 10xy + 16y^2 = (x - 8y)(x\ \_\_\_)$

6.2 EXERCISE SET

MyMathLab®

Watch the videos
in MyMathLab

Download the
MyDashBoard App

Practice Exercises

In Exercises 1–42, factor each trinomial, or state that the trinomial is prime. Check each factorization using FOIL multiplication.

1. $x^2 + 7x + 6$
2. $x^2 + 9x + 8$
3. $x^2 + 7x + 10$
4. $x^2 + 9x + 14$
5. $x^2 + 11x + 10$
6. $x^2 + 13x + 12$
7. $x^2 - 7x + 12$
8. $x^2 - 13x + 40$
9. $x^2 - 12x + 36$
10. $x^2 - 8x + 16$
11. $y^2 - 8y + 15$
12. $y^2 - 8y + 7$
13. $x^2 + 3x - 10$
14. $x^2 + 3x - 28$
15. $y^2 + 10y - 39$
16. $y^2 + 5y - 24$
17. $x^2 - 2x - 15$
18. $x^2 - 4x - 5$
19. $x^2 - 2x - 8$
20. $x^2 - 5x - 6$
21. $x^2 + 4x + 12$
22. $x^2 + 4x + 5$
23. $y^2 - 16y + 48$
24. $y^2 - 10y + 21$
25. $x^2 - 3x + 6$
26. $x^2 + 4x - 10$
27. $w^2 - 30w - 64$
28. $w^2 + 12w - 64$
29. $y^2 - 18y + 65$
30. $y^2 - 22y + 72$
31. $r^2 + 12r + 27$
32. $r^2 - 15r - 16$
33. $y^2 - 7y + 5$
34. $y^2 - 15y + 5$
35. $x^2 + 7xy + 6y^2$
36. $x^2 + 6xy + 8y^2$
37. $x^2 - 8xy + 15y^2$
38. $x^2 - 9xy + 14y^2$
39. $x^2 - 3xy - 18y^2$
40. $x^2 - xy - 30y^2$
41. $a^2 - 18ab + 45b^2$
42. $a^2 - 18ab + 80b^2$

In Exercises 43–66, factor completely.

43. $3x^2 + 15x + 18$
44. $3x^2 + 21x + 36$
45. $4y^2 - 4y - 8$
46. $3y^2 + 3y - 18$
47. $10x^2 - 40x - 600$
48. $2x^2 + 10x - 48$
49. $3x^2 - 33x + 54$
50. $2x^2 - 14x + 24$
51. $2r^3 + 6r^2 + 4r$
52. $2r^3 + 8r^2 + 6r$
53. $4x^3 + 12x^2 - 72x$
54. $3x^3 - 15x^2 + 18x$
55. $2r^3 + 8r^2 - 64r$
56. $3r^3 - 9r^2 - 54r$
57. $y^4 + 2y^3 - 80y^2$
58. $y^4 - 12y^3 + 35y^2$
59. $x^4 - 3x^3 - 10x^2$
60. $x^4 - 22x^3 + 120x^2$
61. $2w^4 - 26w^3 - 96w^2$
62. $3w^4 + 54w^3 + 135w^2$
63. $15xy^2 + 45xy - 60x$
64. $20x^2y - 100xy + 120y$
65. $x^5 + 3x^4y - 4x^3y^2$
66. $x^3y - 2x^2y^2 - 3xy^3$

In Exercises 67–74, use the negative of the greatest common factor to factor completely.

67. $-16t^2 + 64t + 80$
68. $-16t^2 + 80t + 96$
69. $-5x^2 + 50x - 45$
70. $-3x^2 + 36x - 33$
71. $-x^2 - 3x + 40$
72. $-x^2 - 4x + 45$
73. $-2x^3 - 6x^2 + 8x$
74. $-3x^3 + 6x^2 + 24x$

Practice PLUS

In Exercises 75–82, factor completely.

75. $2x^2y^2 - 32x^2yz + 30x^2z^2$
76. $2x^2y^2 - 30x^2yz + 28x^2z^2$
77. $(a + b)x^2 + (a + b)x - 20(a + b)$
78. $(a + b)x^2 - 13(a + b)x + 36(a + b)$

(Hint on Exercises 79–82: Factors contain rational numbers.)

79. $x^2 + 0.5x + 0.06$
80. $x^2 - 0.5x - 0.06$
81. $x^2 - \dfrac{2}{5}x + \dfrac{1}{25}$
82. $x^2 + \dfrac{2}{3}x + \dfrac{1}{9}$

Application Exercises

83. You dive directly upward from a board that is 32 feet high. After t seconds, your height above the water is described by the polynomial

$$-16t^2 + 16t + 32.$$

a. Factor the polynomial completely.

b. Evaluate both the original polynomial and its factored form for $t = 2$. Do you get the same answer for each evaluation? Describe what this answer means.

84. You dive directly upward from a board that is 48 feet high. After t seconds, your height above the water is described by the polynomial

$$-16t^2 + 32t + 48.$$

a. Factor the polynomial completely.

b. Evaluate both the original polynomial and its factored form for $t = 3$. Do you get the same answer for each evaluation? Describe what this answer means.

Writing in Mathematics

85. Explain how to factor $x^2 + 8x + 15$.

86. Give two helpful suggestions for factoring $x^2 - 5x + 6$.

87. In factoring $x^2 + bx + c$, describe how the last terms in each factor are related to b and c.

88. Without actually factoring and without multiplying the given factors, explain why the following factorization is not correct:

$$x^2 + 46x + 513 = (x - 27)(x - 19).$$

Critical Thinking Exercises

Make Sense? *In Exercises 89–92, determine whether each statement "makes sense" or "does not make sense" and explain your reasoning.*

89. If I have a correct factorization, switching the order of the factors can give me an incorrect factorization.

90. I began factoring $x^2 - 17x + 72$ by finding all number pairs with a sum of -17.

91. It's easy to factor $x^2 + x + 1$ because of the relatively small numbers for the constant term and the coefficient of x.

92. I factor $x^2 + bx + c$ by finding two numbers that have a product of c and a sum of b.

In Exercises 93–96, determine whether each statement is true or false. If the statement is false, make the necessary change(s) to produce a true statement.

93. One factor of $x^2 + x + 20$ is $x + 5$.

94. A trinomial can never have two identical factors.

95. One factor of $y^2 + 5y - 24$ is $y - 3$.

96. $x^2 + 4 = (x + 2)(x + 2)$

In Exercises 97–98, find all positive integers b so that the trinomial can be factored.

97. $x^2 + bx + 15$

98. $x^2 + 4x + b$

99. Write a trinomial of the form $x^2 + bx + c$ that is prime.

100. Factor: $x^{2n} + 20x^n + 99$.

101. Factor $x^3 + 3x^2 + 2x$. If x represents an integer, use the factorization to describe what the trinomial represents.

102. A box with no top is to be made from an 8-inch by 6-inch piece of metal by cutting identical squares from each corner and turning up the sides (see the figure). The volume of the box is modeled by the polynomial $4x^3 - 28x^2 + 48x$. Factor the polynomial completely. Then use the dimensions given on the box and show that its volume is equivalent to the factorization that you obtain.

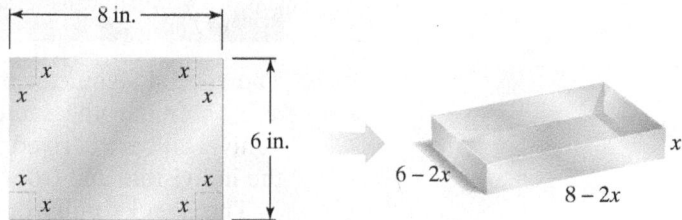

Technology Exercises

In Exercises 103–106, use the $\boxed{\text{GRAPH}}$ or $\boxed{\text{TABLE}}$ feature of a graphing utility to determine if the polynomial on the left side of each equation has been correctly factored. If the graphs of y_1 and y_2 coincide, or if their corresponding table values are equal, this means that the polynomial on the left side has been correctly factored. If not, factor the trinomial correctly and then use your graphing utility to verify the factorization.

103. $x^2 - 5x + 6 = (x - 2)(x - 3)$

104. $2x^2 + 2x - 12 = 2(x - 3)(x + 2)$

105. $x^2 - 2x + 1 = (x + 1)(x - 1)$

106. $2x^2 + 8x + 6 = (x + 3)(x + 1)$

Review Exercises

107. Solve: $4(x - 2) = 3x + 5$. (Section 2.3, Example 2)

108. Graph: $6x - 5y = 30$. (Section 3.2, Example 4)

109. Graph: $y = -\frac{1}{2}x + 2$. (Section 3.4, Example 3)

Preview Exercises

Exercises 110–112 will help you prepare for the material covered in the next section.

110. Multiply: $(2x + 3)(x - 2)$.

111. Multiply: $(3x + 4)(3x + 1)$.

112. Factor by grouping: $8x^2 - 2x - 20x + 5$.

Factoring Trinomials Whose Leading Coefficient Is Not 1

Objectives

1 Factor trinomials by trial and error.

2 Factor trinomials by grouping.

The False Mirror (Le Faux Miroir) (1928), René Magritte. © 2011 MoMA/Herscovici/ARS.

The special significance of the number 1 is reflected in our language. "One," "an," and "a" mean the same thing. The words "unit," "unity," "union," "unique," and "universal" are derived from the Latin word for "one." For the ancient Greeks, 1 was the indivisible unit from which all other numbers arose.

The Greeks' philosophy of 1 applies to our work in this section. Factoring trinomials whose leading coefficient is 1 is the basic technique from which other methods of factoring $ax^2 + bx + c$, where the leading coefficient a is not equal to 1, follow.

1 Factor trinomials by trial and error.

Factoring by the Trial-and-Error Method

How do we factor a trinomial such as $3x^2 - 20x + 28$? Notice that the leading coefficient is 3. We must find two binomials whose product is $3x^2 - 20x + 28$. The product of the First terms must be $3x^2$:

$$(3x \quad)(x \quad).$$

From this point on, the factoring strategy is exactly the same as the one we use to factor trinomials whose leading coefficient is 1.

A Strategy for Factoring $ax^2 + bx + c$

Assume, for the moment, that there is no greatest common factor other than 1.

1. Find two First terms whose product is ax^2:

$$(\Box x + \quad)(\Box x + \quad) = ax^2 + bx + c.$$

2. Find two Last terms whose product is c:

$$(\Box x + \Box)(\Box x + \Box) = ax^2 + bx + c.$$

3. By trial and error, perform steps 1 and 2 until the sum of the Outside product and the Inside product is bx:

$$(\Box x + \Box)(\Box x + \Box) = ax^2 + bx + c.$$
$$\underset{\text{Sum of O + I}}{\underbrace{\overset{\text{I}}{\overbrace{\qquad}}\quad\overset{\text{O}}{\overbrace{\qquad\qquad}}}}$$

If no such combinations exist, the polynomial is prime.

EXAMPLE 1   Factoring a Trinomial Whose Leading Coefficient Is Not 1

Factor: $3x^2 - 20x + 28$.

Solution

Step 1. Find two First terms whose product is $3x^2$.

$$3x^2 - 20x + 28 = (3x \quad)(x \quad)$$

Step 2. Find two Last terms whose product is 28. The number 28 has pairs of factors that are either both positive or both negative. Because the middle term, $-20x$, is negative, both factors must be negative. The negative factorizations of 28 are $-1(-28)$, $-2(-14)$, and $-4(-7)$.

Step 3. Try various combinations of these factors. The correct factorization of $3x^2 - 20x + 28$ is the one in which the sum of the Outside and Inside products is equal to $-20x$. Here is a list of the possible factorizations:

Possible Factorizations of $3x^2 - 20x + 28$	Sum of Outside and Inside Products (Should Equal $-20x$)	
$(3x - 1)(x - 28)$	$-84x - x = -85x$	
$(3x - 28)(x - 1)$	$-3x - 28x = -31x$	
$(3x - 2)(x - 14)$	$-42x - 2x = -44x$	
$(3x - 14)(x - 2)$	$-6x - 14x = -20x$	This is the required middle term.
$(3x - 4)(x - 7)$	$-21x - 4x = -25x$	
$(3x - 7)(x - 4)$	$-12x - 7x = -19x$	

Thus,

$$3x^2 - 20x + 28 = (3x - 14)(x - 2) \quad \text{or} \quad (x - 2)(3x - 14).$$

Show that this factorization is correct by multiplying the factors using the FOIL method. You should obtain the original trinomial.   ∎

✓ CHECK POINT 1   Factor: $5x^2 - 14x + 8$.

Great Question!

When factoring trinomials, must I list every possible factorization before getting the correct one?

With practice, you will find that it is not necessary to list every possible factorization of the trinomial. As you practice factoring, you will be able to narrow down the list of possible factors to just a few. When it comes to factoring, practice makes perfect.

EXAMPLE 2   Factoring a Trinomial Whose Leading Coefficient Is Not 1

Factor: $8x^2 - 10x - 3$.

Solution

Step 1. Find two First terms whose product is $8x^2$.

$$8x^2 - 10x - 3 \stackrel{?}{=} (8x \quad)(x \quad)$$

$$8x^2 - 10x - 3 \stackrel{?}{=} (4x \quad)(2x \quad)$$

Step 2. Find two Last terms whose product is -3. The possible factorizations are $1(-3)$ and $-1(3)$.

Step 3. Try various combinations of these factors. The correct factorization of $8x^2 - 10x - 3$ is the one in which the sum of the Outside and Inside products is equal to $-10x$. Here is a list of the possible factorizations:

These four factorizations use $(8x\quad)(x\quad)$ with $1(-3)$ and $-1(3)$ as factorizations of -3.

These four factorizations use $(4x\quad)(2x\quad)$ with $1(-3)$ and $-1(3)$ as factorizations of -3.

Possible Factorizations of $8x^2 - 10x - 3$	Sum of Outside and Inside Products (Should Equal $-10x$)
$(8x + 1)(x - 3)$	$-24x + x = -23x$
$(8x - 3)(x + 1)$	$8x - 3x = 5x$
$(8x - 1)(x + 3)$	$24x - x = 23x$
$(8x + 3)(x - 1)$	$-8x + 3x = -5x$
$(4x + 1)(2x - 3)$	$-12x + 2x = -10x$
$(4x - 3)(2x + 1)$	$4x - 6x = -2x$
$(4x - 1)(2x + 3)$	$12x - 2x = 10x$
$(4x + 3)(2x - 1)$	$-4x + 6x = 2x$

This is the required middle term.

Thus,

$$8x^2 - 10x - 3 = (4x + 1)(2x - 3) \quad \text{or} \quad (2x - 3)(4x + 1).$$

Use FOIL multiplication to check either of these factorizations. ■

✓ **CHECK POINT 2** Factor: $6x^2 + 19x - 7$.

Great Question!

I zone out reading your long lists of possible factorizations. Are there any rules for shortening these lists?

Here are some suggestions for reducing the list of possible factorizations for $ax^2 + bx + c$:

1. If b is relatively small, avoid the larger factors of a.

2. If c is positive, the signs in both binomial factors must match the sign of b.

3. If the trinomial has no common factor, no binomial factor can have a common factor.

4. Reversing the signs in the binomial factors changes the sign of bx, the middle term.

EXAMPLE 3 Factoring a Trinomial in Two Variables

Factor: $2x^2 - 7xy + 3y^2$.

Solution

Step 1. Find two First terms whose product is $2x^2$.

$$2x^2 - 7xy + 3y^2 = (2x\quad)(x\quad)$$

Step 2. Find two Last terms whose product is $3y^2$. The possible factorizations are $(y)(3y)$ and $(-y)(-3y)$.

Step 3. Try various combinations of these factors. The correct factorization of $2x^2 - 7xy + 3y^2$ is the one in which the sum of the Outside and Inside products is equal to $-7xy$. Here is a list of possible factorizations:

Possible Factorizations of $2x^2 - 7xy + 3y^2$	Sum of Outside and Inside Products (Should Equal $-7xy$)
$(2x + 3y)(x + y)$	$2xy + 3xy = 5xy$
$(2x + y)(x + 3y)$	$6xy + xy = 7xy$
$(2x - 3y)(x - y)$	$-2xy - 3xy = -5xy$
$(2x - y)(x - 3y)$	$-6xy - xy = -7xy$

This is the required middle term.

Thus,

$$2x^2 - 7xy + 3y^2 = (2x - y)(x - 3y) \quad \text{or} \quad (x - 3y)(2x - y).$$

Use FOIL multiplication to check either of these factorizations. ■

✓ **CHECK POINT 3** Factor: $3x^2 - 13xy + 4y^2$.

2 Factor trinomials by grouping.

Factoring by the Grouping Method

A second method for factoring $ax^2 + bx + c, a \neq 0$, is called the **grouping method**. The method involves both trial and error, as well as grouping. The trial and error in factoring $ax^2 + bx + c$ depends on finding two numbers, p and q, for which $p + q = b$. Then we factor $ax^2 + px + qx + c$ using grouping.

Let's see how this works by looking at our factorization in Example 2:

$$8x^2 - 10x - 3 = (2x - 3)(4x + 1).$$

If we multiply using FOIL on the right, we obtain:

$$(2x - 3)(4x + 1) = 8x^2 + 2x - 12x - 3.$$

In this case, the desired numbers, p and q, are $p = 2$ and $q = -12$. Compare these numbers to ac and b in the given polynomial:

$$\boxed{a=8} \quad \boxed{b=-10} \quad \boxed{c=-3}$$

$$8x^2 - 10x - 3.$$

$$\boxed{ac = 8(-3) = -24}$$

Can you see that p and q, 2 and -12, are factors of ac, or -24? Furthermore, p and q have a sum of b, namely -10. By expressing the middle term, $-10x$, in terms of p and q, we can factor by grouping as follows:

$$
\begin{aligned}
8x^2 - 10x - 3 & \\
= 8x^2 + (2x - 12x) - 3 & \qquad \text{Rewrite } -10x \text{ as } 2x - 12x. \\
= (8x^2 + 2x) + (-12x - 3) & \qquad \text{Group terms.} \\
= 2x(4x + 1) - 3(4x + 1) & \qquad \text{Factor from each group.} \\
= (4x + 1)(2x - 3) & \qquad \text{Factor out the common binomial factor.}
\end{aligned}
$$

As we obtained in Example 2,

$$8x^2 - 10x - 3 = (4x + 1)(2x - 3).$$

Generalizing from this example, here's how to factor a trinomial by grouping:

Factoring $ax^2 + bx + c$ Using Grouping ($a \neq 1$)

1. Multiply the leading coefficient, a, and the constant, c.
2. Find the factors of ac whose sum is b.
3. Rewrite the middle term, bx, as a sum or difference using the factors from step 2.
4. Factor by grouping.

EXAMPLE 4   Factoring by Grouping

Factor by grouping: $2x^2 - x - 6$.

Solution   The trinomial is of the form $ax^2 + bx + c$.

$$2x^2 - x - 6$$

$$\boxed{a=2} \quad \boxed{b=-1} \quad \boxed{c=-6}$$

Step 1. Multiply the leading coefficient, a, and the constant, c. Using $a = 2$ and $c = -6$,

$$ac = 2(-6) = -12.$$

Step 2. Find the factors of ac whose sum is b. We want the factors of -12 whose sum is b, or -1. The factors of -12 whose sum is -1 are -4 and 3.

Discover for Yourself

In step 2, we discovered that the desired numbers were -4 and 3, and in step 3 we wrote $-x$ as $-4x + 3x$. What happens if we write $-x$ as $3x - 4x$? Use factoring by grouping on

$$2x^2 - x - 6$$
$$= 2x^2 + 3x - 4x - 6.$$

Describe what happens.

Step 3. Rewrite the middle term, $-x$, as a sum or difference using the factors from step 2, -4 and 3.

$$2x^2 - x - 6 = 2x^2 - 4x + 3x - 6$$

Step 4. Factor by grouping.

$$= (2x^2 - 4x) + (3x - 6) \qquad \text{Group terms.}$$
$$= 2x(x - 2) + 3(x - 2) \qquad \text{Factor from each group.}$$
$$= (x - 2)(2x + 3) \qquad \text{Factor out the common binomial factor.}$$

Thus,

$$2x^2 - x - 6 = (x - 2)(2x + 3) \quad \text{or} \quad (2x + 3)(x - 2). \quad \blacksquare$$

✓ **CHECK POINT 4** Factor by grouping: $3x^2 - x - 10$.

EXAMPLE 5 Factoring by Grouping

Factor by grouping: $8x^2 - 22x + 5$.

Solution The trinomial is of the form $ax^2 + bx + c$.

$$8x^2 - 22x + 5$$

$$\boxed{a = 8} \quad \boxed{b = -22} \quad \boxed{c = 5}$$

Step 1. Multiply the leading coefficient, a, and the constant, c. Using $a = 8$ and $c = 5$, $ac = 8 \cdot 5 = 40$.

Step 2. Find the factors of ac whose sum is b. We want the factors of 40 whose sum is b, or -22. The factors of 40 whose sum is -22 are -2 and -20.

Step 3. Rewrite the middle term, $-22x$, as a sum or difference using the factors from step 2, -2 and -20.

$$8x^2 - 22x + 5 = 8x^2 - 2x - 20x + 5$$

Step 4. Factor by grouping.

$$= (8x^2 - 2x) + (-20x + 5) \qquad \text{Group terms.}$$
$$= 2x(4x - 1) - 5(4x - 1) \qquad \text{Factor from each group.}$$
$$= (4x - 1)(2x - 5) \qquad \text{Factor out the common binomial factor.}$$

Thus,

$$8x^2 - 22x + 5 = (4x - 1)(2x - 5) \quad \text{or} \quad (2x - 5)(4x - 1). \quad \blacksquare$$

✓ **CHECK POINT 5** Factor by grouping: $8x^2 - 10x + 3$.

Factoring Completely

Always begin the process of factoring a polynomial by looking for a greatest common factor other than 1. If there is one, **factor out the GCF first**. After doing this, you should attempt to factor the remaining polynomial, using one of the methods presented in this section, if appropriate.

EXAMPLE 6 Factoring Completely

Factor completely: $15y^4 + 26y^3 + 7y^2$.

Solution We will first factor out a common monomial factor from the polynomial and then factor the resulting trinomial by the methods of this section. The GCF of the three terms is y^2.

$$15y^4 + 26y^3 + 7y^2 = y^2(15y^2 + 26y + 7) \qquad \text{Factor out the GCF.}$$
$$= y^2(5y + 7)(3y + 1) \qquad \text{Factor } 15y^2 + 26y + 7 \text{ using trial and error or grouping.}$$

Thus,

$$15y^4 + 26y^3 + 7y^2 = y^2(5y + 7)(3y + 1) \quad \text{or} \quad y^2(3y + 1)(5y + 7).$$

Be sure to include the GCF, y^2, in the factorization.

✓ **CHECK POINT 6**   Factor completely: $5y^4 + 13y^3 + 6y^2.$

Great Question!

As I approach the next Exercise Set (it looks long!), can you give me any other hints for shortening those oppressive lists of possible factorizations?

Here is an observation that sometimes helps narrow down the list of possible factorizations. If a polynomial does not have a GCF other than 1 or if you have factored out the GCF, there will be no common factor within any of its binomial factors. Here is an example:

$$6x^2 - 17x + 12. \quad \boxed{\text{There is no GCF other than 1.}}$$

$$(2x - 4)(3x - 3) \quad \boxed{\text{This is not a possible factorization.}}$$

This binomial has a common factor of **2**.   This binomial has a common factor of **3**.

CONCEPT AND VOCABULARY CHECK

Fill in each blank so that the resulting statement is true.

1. We begin the process of factoring a polynomial by first factoring out the _____, assuming that there is one other than 1.

2. $8x^2 - 10x - 3 = (4x + 1)(2x \underline{\quad})$

3. $12x^2 - x - 20 = (4x + 5)(3x \underline{\quad})$

4. $2x^2 - 5x + 3 = (x - 1)(\underline{\quad})$

5. $6x^2 + 17x + 12 = (2x + 3)(\underline{\quad})$

6. $5x^2 - 8xy - 4y^2 = (5x + 2y)(\underline{\quad})$

6.3 EXERCISE SET   MyMathLab®

Watch the videos in MyMathLab   Download the MyDashBoard App

Practice Exercises

In Exercises 1–58, use the method of your choice to factor each trinomial, or state that the trinomial is prime. Check each factorization using FOIL multiplication.

1. $2x^2 + 5x + 3$

2. $3x^2 + 5x + 2$

3. $3x^2 + 13x + 4$

4. $2x^2 + 7x + 3$

5. $2x^2 + 11x + 12$

6. $2x^2 + 19x + 35$

7. $5y^2 - 16y + 3$

8. $5y^2 - 17y + 6$

9. $3y^2 + y - 4$

10. $3y^2 - y - 4$

11. $3x^2 + 13x - 10$

12. $3x^2 + 14x - 5$

13. $3x^2 - 22x + 7$

14. $3x^2 - 10x + 7$

15. $5y^2 - 16y + 3$

16. $5y^2 - 8y + 3$

17. $3x^2 - 17x + 10$

18. $3x^2 - 25x - 28$

19. $6w^2 - 11w + 4$

20. $6w^2 - 17w + 12$

21. $8x^2 + 33x + 4$

22. $7x^2 + 43x + 6$

23. $5x^2 + 33x - 14$

24. $3x^2 + 22x - 16$

25. $14y^2 + 15y - 9$

26. $6y^2 + 7y - 24$

27. $6x^2 - 7x + 3$

28. $9x^2 + 3x + 2$

29. $25z^2 - 30z + 9$

30. $9z^2 + 12z + 4$

31. $15y^2 - y - 2$

32. $15y^2 + 13y - 2$

33. $5x^2 + 2x + 9$

34. $3x^2 - 5x + 1$

35. $10y^2 + 43y - 9$

36. $16y^2 - 46y + 15$

37. $8x^2 - 2x - 1$

38. $8x^2 - 22x + 5$

39. $9y^2 - 9y + 2$

40. $9y^2 + 5y - 4$

41. $20x^2 + 27x - 8$

42. $15x^2 - 19x + 6$

43. $2x^2 + 3xy + y^2$

44. $3x^2 + 4xy + y^2$

45. $3x^2 + 5xy + 2y^2$

46. $3x^2 + 11xy + 6y^2$

47. $2x^2 - 9xy + 9y^2$

48. $3x^2 + 5xy - 2y^2$

49. $6x^2 - 5xy - 6y^2$

50. $6x^2 - 7xy - 5y^2$

51. $15x^2 + 11xy - 14y^2$

52. $15x^2 - 31xy + 10y^2$

53. $2a^2 + 7ab + 5b^2$

54. $2a^2 + 5ab + 2b^2$

55. $15a^2 - ab - 6b^2$

56. $3a^2 - ab - 14b^2$

57. $12x^2 - 25xy + 12y^2$

58. $12x^2 + 7xy - 12y^2$

In Exercises 59–88, factor completely.

59. $4x^2 + 26x + 30$

60. $4x^2 - 18x - 10$

61. $9x^2 - 6x - 24$

62. $12x^2 - 33x + 21$

63. $4y^2 + 2y - 30$

64. $36y^2 + 6y - 12$

65. $9y^2 + 33y - 60$

66. $16y^2 - 16y - 12$

67. $3x^3 + 4x^2 + x$

68. $3x^3 + 14x^2 + 8x$

69. $2x^3 - 3x^2 - 5x$

70. $6x^3 + 4x^2 - 10x$

71. $9y^3 - 39y^2 + 12y$

72. $10y^3 + 12y^2 + 2y$

73. $60z^3 + 40z^2 + 5z$

74. $80z^3 + 80z^2 - 60z$

75. $15x^4 - 39x^3 + 18x^2$

76. $24x^4 + 10x^3 - 4x^2$

77. $10x^5 - 17x^4 + 3x^3$

78. $15x^5 - 2x^4 - x^3$

79. $6x^2 - 3xy - 18y^2$

80. $4x^2 + 14xy + 10y^2$

81. $12x^2 + 10xy - 8y^2$

82. $24x^2 + 3xy - 27y^2$

83. $8x^2y + 34xy - 84y$

84. $6x^2y - 2xy - 60y$

85. $12a^2b - 46ab^2 + 14b^3$

86. $12a^2b - 34ab^2 + 14b^3$

87. $-32x^2y^4 + 20xy^4 + 12y^4$

88. $-10x^2y^4 + 14xy^4 + 12y^4$

Practice PLUS

In Exercises 89–90, factor completely.

89. $30(y + 1)x^2 + 10(y + 1)x - 20(y + 1)$

90. $6(y + 1)x^2 + 33(y + 1)x + 15(y + 1)$

91. a. Factor $2x^2 - 5x - 3$.

    b. Use the factorization in part (a) to factor
$$2(y + 1)^2 - 5(y + 1) - 3.$$
Then simplify each factor.

92. a. Factor $3x^2 + 5x - 2$.

    b. Use the factorization in part (a) to factor
$$3(y + 1)^2 + 5(y + 1) - 2.$$
Then simplify each factor.

93. Divide $3x^3 - 11x^2 + 12x - 4$ by $x - 2$. Use the quotient to factor $3x^3 - 11x^2 + 12x - 4$ completely.

94. Divide $2x^3 + x^2 - 13x + 6$ by $x - 2$. Use the quotient to factor $2x^3 + x^2 - 13x + 6$ completely.

Application Exercises

It is possible to construct geometric models for factorizations so that you can see the factoring. This idea is developed in Exercises 95–96.

95. Consider the following figure.

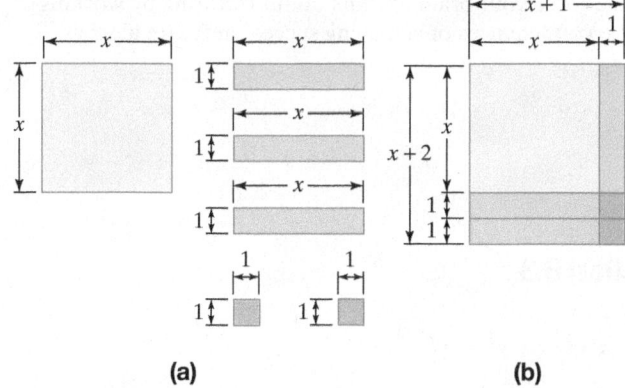

(a)   **(b)**

    a. Write a trinomial that expresses the sum of the areas of the six rectangular pieces shown in figure (a).

    b. Express the area of the large rectangle in figure (b) as the product of two binomials.

    c. Are the pieces in figures (a) and (b) the same? Set the expressions that you wrote in parts (a) and (b) equal to each other. What factorization is illustrated?

96. Copy the figure and cut out the six pieces. Use the pieces to create a geometric model for the factorization

$$2x^2 + 3x + 1 = (2x + 1)(x + 1)$$

by forming a large rectangle using all the pieces.

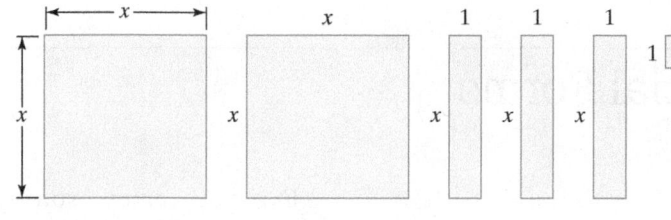

Writing in Mathematics

97. Explain how to factor $2x^2 - x - 1$.

98. Why is it a good idea to factor out the GCF first and then use other methods of factoring? Use $3x^2 - 18x + 15$ as an example. Discuss what happens if one first uses trial and error to factor as two binomials rather than first factoring out the GCF.

99. In factoring $3x^2 - 10x - 8$, a student lists $(3x - 2)(x + 4)$ as a possible factorization. Use FOIL multiplication to determine if this factorization is correct. If it is not correct, describe how the correct factorization can quickly be obtained using these factors.

100. Explain why $2x - 10$ cannot be one of the factors in the correct factorization of $6x^2 - 19x + 10$.

Critical Thinking Exercises

Make Sense? *In Exercises 101–104, determine whether each statement "makes sense" or "does not make sense" and explain your reasoning.*

101. I'm often able to use an incorrect factorization to lead me to the correct factorization.

102. First factoring out the GCF makes it easier for me to determine how to factor the remaining factor, assuming it is not prime.

103. I'm working with a polynomial that has a GCF other than 1, but then it doesn't factor further, so the polynomial that I'm working with is prime.

104. My graphing calculator showed the same graphs for $y_1 = 4x^2 - 20x + 24$ and $y_2 = 4(x^2 - 5x + 6)$, so I can conclude that the complete factorization of $4x^2 - 20x + 24$ is $4(x^2 - 5x + 6)$.

In Exercises 105–108, determine whether each statement is true or false. If the statement is false, make the necessary change(s) to produce a true statement.

105. Once the GCF is factored from $18y^2 - 6y + 6$, the remaining trinomial factor is prime.

106. One factor of $12x^2 - 13x + 3$ is $4x + 3$.

107. One factor of $4y^2 - 11y - 3$ is $y + 3$.

108. The trinomial $3x^2 + 2x + 1$ has relatively small coefficients and therefore can be factored.

In Exercises 109–110, find all integers b so that the trinomial can be factored.

109. $3x^2 + bx + 2$

110. $2x^2 + bx + 3$

111. Factor: $3x^{10} - 4x^5 - 15$.

112. Factor: $2x^{2n} - 7x^n - 4$.

Review Exercises

113. Solve the system:
$$\begin{cases} 4x - y = 105 \\ x + 7y = -10. \end{cases}$$ (Section 4.3, Example 3)

114. Write 0.00086 in scientific notation. (Section 5.7, Example 8)

115. Solve: $8x - \dfrac{x}{6} = \dfrac{1}{6} - 8$. (Section 2.3, Example 4)

Preview Exercises

Exercises 116–118 will help you prepare for the material covered in the next section. In each exercise, perform the indicated operation.

116. $(9x + 10)(9x - 10)$

117. $(4x + 5y)^2$

118. $(x + 2)(x^2 - 2x + 4)$

Achieving Success

In the next chapter, you will see how mathematical models involving quotients of polynomials describe environmental issues. Factoring is an essential skill for working with such models. **Success in mathematics cannot be achieved without a complete understanding of factoring.** Be sure to work all the exercises in the Mid-Chapter Check Point so that you can apply each of the factoring techniques discussed in the first half of the chapter. The more deeply you force your brain to think about factoring by working many exercises, the better will be your chances of achieving success in future algebra courses.

MID-CHAPTER CHECK POINT　Section 6.1–Section 6.3

 What You Know: We learned to factor out a polynomial's greatest common factor and to use grouping to factor polynomials with four terms. We factored polynomials with three terms, beginning with trinomials with leading coefficient 1 and moving on to $ax^2 + bx + c$, with $a \neq 1$. We saw that the factoring process should begin by looking for a GCF other than 1 and, if there is one, factoring it out first.

In Exercises 1–13, factor completely, or state that the polynomial is prime.

1. $x^5 + x^4$
2. $x^2 + 7x - 18$

3. $x^2y^3 - x^2y^2 + x^2y$
4. $x^2 - 2x + 4$
5. $7x^2 - 22x + 3$
6. $x^3 + 5x^2 + 3x + 15$
7. $2x^3 - 11x^2 + 5x$
8. $xy - 7x - 4y + 28$
9. $x^2 - 17xy + 30y^2$
10. $25x^2 - 25x - 14$
11. $16x^2 - 70x + 24$
12. $3x^2 + 10xy + 7y^2$
13. $-6x^3 + 8x^2 + 30x$

SECTION

6.4

Factoring Special Forms

Objectives

1　Factor the difference of two squares.

2　Factor perfect square trinomials.

3　Factor the sum or difference of two cubes.

Do you enjoy solving puzzles? The process is a natural way to develop problem-solving skills that are important to every area of our lives. Engaging in problem solving for sheer pleasure releases chemicals in the brain that enhance our feeling of well-being. Perhaps this is why puzzles date back 12,000 years.

In this section, we develop factoring techniques by reversing the formulas for special products discussed in Chapter 5. These factorizations can be visualized by fitting pieces of a puzzle together to form rectangles.

1 Factor the difference of two squares.

Factoring the Difference of Two Squares

A method for factoring the difference of two squares is obtained by reversing the special product for the sum and difference of two terms.

> ### The Difference of Two Squares
>
> If A and B are real numbers, variables, or algebraic expressions, then
>
> $$A^2 - B^2 = (A + B)(A - B).$$
>
> In words: The difference of the squares of two terms factors as the product of a sum and a difference of those terms.

EXAMPLE 1 Factoring the Difference of Two Squares

Factor:

a. $x^2 - 4$　　　　　**b.** $81x^2 - 49$.

Solution We must express each term as the square of some monomial. Then we use the formula for factoring $A^2 - B^2$.

a.　　$x^2 - 4 = x^2 - 2^2 = (x + 2)(x - 2)$

$$A^2 - B^2 = (A + B)(A - B)$$

b. $81x^2 - 49 = (9x)^2 - 7^2 = (9x + 7)(9x - 7)$ ∎

In order to apply the factoring formula for $A^2 - B^2$, each term must be the square of an integer or a polynomial.

- A number that is the square of an integer is called a **perfect square**. For example, 100 is a perfect square because $100 = 10^2$.

- Any monomial involving a perfect-square coefficient and variables to even powers is a perfect square. For example, $16x^{10}$ is a perfect square because $16x^{10} = (4x^5)^2$.

Great Question!

You mentioned that because $100 = 10^2$, 100 is a perfect square. What are some other perfect squares that I should recognize?

It's helpful to identify perfect squares. Here are 16 perfect squares, printed in boldface.

$\mathbf{1} = 1^2$	$\mathbf{25} = 5^2$	$\mathbf{81} = 9^2$	$\mathbf{169} = 13^2$
$\mathbf{4} = 2^2$	$\mathbf{36} = 6^2$	$\mathbf{100} = 10^2$	$\mathbf{196} = 14^2$
$\mathbf{9} = 3^2$	$\mathbf{49} = 7^2$	$\mathbf{121} = 11^2$	$\mathbf{225} = 15^2$
$\mathbf{16} = 4^2$	$\mathbf{64} = 8^2$	$\mathbf{144} = 12^2$	$\mathbf{256} = 16^2$

✓ **CHECK POINT 1** Factor:

a. $x^2 - 81$　　　　　**b.** $36x^2 - 25$.

Be careful when determining whether or not to apply the factoring formula for the difference of two squares.

Prime Over the Integers   **Factorable**

- $x^2 - 5$ • $x^7 - 25$ • $9 - 16x^{10}$

| 5 is not a perfect square. | 7 is an odd power. x^7 is not the square of any integer power of x. | Perfect square: $9 = 3^2$ | Perfect square: $16x^{10} = (4x^5)^2$ |

EXAMPLE 2 Factoring the Difference of Two Squares

Factor:

a. $9 - 16x^{10}$ **b.** $25x^2 - 4y^2$.

Solution Begin by expressing each term as the square of some monomial. Then use the formula for factoring $A^2 - B^2$.

a. $9 - 16x^{10} = 3^2 - (4x^5)^2 = (3 + 4x^5)(3 - 4x^5)$

$$A^2 - B^2 = (A + B)(A - B)$$

b. $25x^2 - 4y^2 = (5x)^2 - (2y)^2 = (5x + 2y)(5x - 2y)$ ∎

✓ **CHECK POINT 2** Factor:

     a. $25 - 4x^{10}$ **b.** $100x^2 - 9y^2$.

When factoring, always check first for common factors. If there are common factors other than 1, factor out the GCF and then factor the resulting polynomial.

EXAMPLE 3 Factoring Out the GCF and Then Factoring the Difference of Two Squares

Factor:

a. $12x^3 - 3x$ **b.** $80 - 125x^2$.

Solution

a. $12x^3 - 3x = 3x(4x^2 - 1) = 3x[(2x)^2 - 1^2] = 3x(2x + 1)(2x - 1)$

Factor out the GCF. $A^2 - B^2 = (A + B)(A - B)$

b. $80 - 125x^2 = 5(16 - 25x^2) = 5[4^2 - (5x)^2] = 5(4 + 5x)(4 - 5x)$ ∎

Great Question!

If I rewrite 80 − 125x^2 as −125x^2 + 80, will I still get the same factorization?

Yes. Because the coefficient of $-125x^2$ is negative, factor out -5, the negative of the GCF.

$$-125x^2 + 80 = -5(25x^2 - 16) = -5[(5x)^2 - 4^2] = -5(5x + 4)(5x - 4)$$

Factor out the negative of the GCF.

To show that this is the same factorization as the one in the solution to part (b), factor -1 from $(5x - 4)$.

✓ **CHECK POINT 3** Factor:

a. $18x^3 - 2x$ **b.** $72 - 18x^2$.

We have seen that a polynomial is factored completely when it is written as the product of prime polynomials. To be sure that you have factored completely, check to see whether any factors with more than one term in the factored polynomial can be factored further. If so, continue factoring.

EXAMPLE 4 A Repeated Factorization

Factor completely: $x^4 - 81$.

Solution

$$
\begin{aligned}
x^4 - 81 &= (x^2)^2 - 9^2 && \text{Express as the difference of two squares.} \\
&= (x^2 + 9)(x^2 - 9) && \text{The factors are the sum and the difference of} \\
& && \text{the expressions being squared.} \\
&= (x^2 + 9)(x^2 - 3^2) && \text{The factor } x^2 - 9 \text{ is the difference of two} \\
& && \text{squares and can be factored.} \\
&= (x^2 + 9)(x + 3)(x - 3) && \text{The factors of } x^2 - 9 \text{ are the sum and the} \\
& && \text{difference of the expressions being squared.} \blacksquare
\end{aligned}
$$

Are you tempted to factor $x^2 + 9$ further, the sum of two squares, in Example 4? Resist the temptation! **The sum of two squares, $A^2 + B^2$, with no common factor other than 1 is a prime polynomial over the integers.**

✓ **CHECK POINT 4** Factor completely: $81x^4 - 16$.

In Examples 1–4, we used the formula for factoring the difference of two squares. Although we obtained the formula by reversing the special product for the sum and difference of two terms, it can also be obtained geometrically.

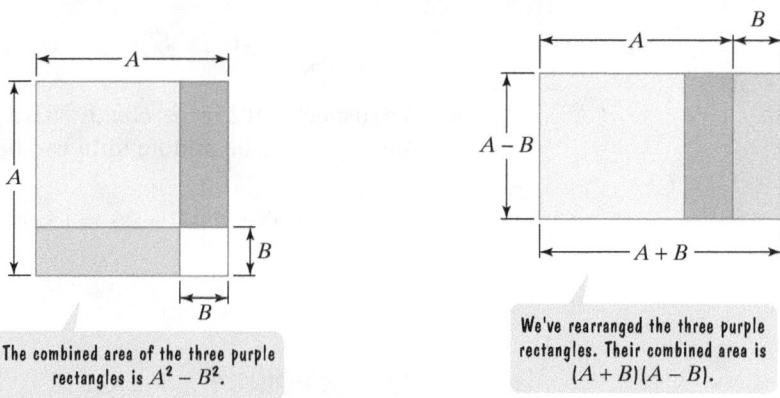

The combined area of the three purple rectangles is $A^2 - B^2$.

We've rearranged the three purple rectangles. Their combined area is $(A + B)(A - B)$.

Because the three purple rectangles make up the same combined area in both figures,

$$A^2 - B^2 = (A + B)(A - B).$$

Great Question!

Why isn't factoring $x^4 - 81$ as $(x^2 + 9)(x^2 - 9)$ a complete factorization?

The second factor, $x^2 - 9$, is itself a difference of two squares and can be factored.

2 Factor perfect square trinomials.

Factoring Perfect Square Trinomials

Our next factoring technique is obtained by reversing the special products for squaring binomials. The trinomials that are factored using this technique are called **perfect square trinomials**.

Factoring Perfect Square Trinomials

Let A and B be real numbers, variables, or algebraic expressions.

1. $A^2 + 2AB + B^2 = (A + B)^2$

Same sign

2. $A^2 - 2AB + B^2 = (A - B)^2$

Same sign

The two formulas in the box show that perfect square trinomials, $A^2 + 2AB + B^2$ and $A^2 - 2AB + B^2$, come in two forms: one in which the coefficient of the middle term is positive and one in which the coefficient of the middle term is negative. Here's how to recognize a perfect square trinomial:

1. The first and last terms are squares of monomials or integers.

2. The middle term is twice the product of the expressions being squared in the first and last terms.

EXAMPLE 5   Factoring Perfect Square Trinomials

Factor:

a. $x^2 + 6x + 9$ **b.** $x^2 - 16x + 64$ **c.** $25x^2 - 60x + 36$.

Solution

a. $x^2 + 6x + 9 = x^2 + 2 \cdot x \cdot 3 + 3^2 = (x + 3)^2$

The middle term has a positive sign.

$A^2 + 2AB + B^2 = (A + B)^2$

b. $x^2 - 16x + 64 = x^2 - 2 \cdot x \cdot 8 + 8^2 = (x - 8)^2$

The middle term has a negative sign.

$A^2 - 2AB + B^2 = (A - B)^2$

c. We suspect that $25x^2 - 60x + 36$ is a perfect square trinomial because $25x^2 = (5x)^2$ and $36 = 6^2$. The middle term can be expressed as twice the product of $5x$ and 6.

$$25x^2 - 60x + 36 = (5x)^2 - 2 \cdot 5x \cdot 6 + 6^2 = (5x - 6)^2$$

$A^2 - 2AB + B^2 = (A - B)^2$ ∎

✓ CHECK POINT 5   Factor:

a. $x^2 + 14x + 49$ **b.** $x^2 - 6x + 9$

c. $16x^2 - 56x + 49$.

EXAMPLE 6   Factoring a Perfect Square Trinomial in Two Variables

Factor: $16x^2 + 40xy + 25y^2$.

Solution Observe that $16x^2 = (4x)^2$, $25y^2 = (5y)^2$, and $40xy$ is twice the product of $4x$ and $5y$. Thus, we have a perfect square trinomial.

$$16x^2 + 40xy + 25y^2 = (4x)^2 + 2 \cdot 4x \cdot 5y + (5y)^2 = (4x + 5y)^2$$

$$A^2 + 2AB + B^2 = (A + B)^2$$

■

✓ **CHECK POINT 6** Factor: $4x^2 + 12xy + 9y^2$.

In Examples 5 and 6, we factored perfect square trinomials using $A^2 + 2AB + B^2 = (A + B)^2$. Although we obtained the formula by reversing the special product for $(A + B)^2$, it can also be obtained geometrically.

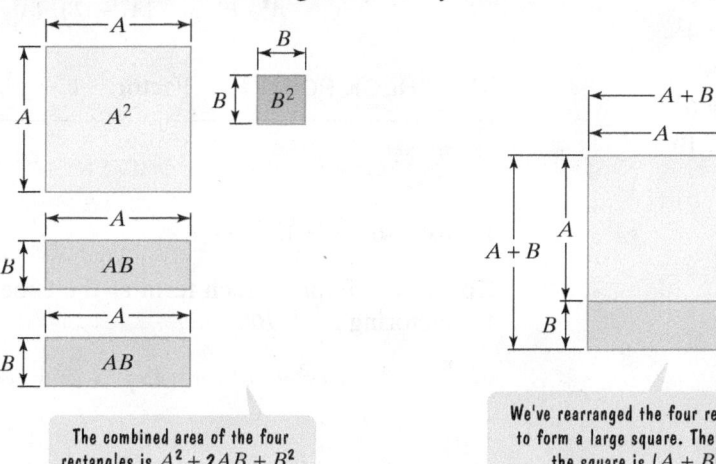

The combined area of the four rectangles is $A^2 + 2AB + B^2$.

We've rearranged the four rectangles to form a large square. The area of the square is $(A + B)^2$.

Because the four rectangles make up the same combined area in both figures,

$$A^2 + 2AB + B^2 = (A + B)^2.$$

3 Factor the sum or difference of two cubes.

Factoring the Sum or Difference of Two Cubes

We can use the following formulas to factor the sum or the difference of two cubes:

Factoring the Sum or Difference of Two Cubes

1. Factoring the Sum of Two Cubes

$$A^3 + B^3 = (A + B)(A^2 - AB + B^2)$$

Same signs Opposite signs

2. Factoring the Difference of Two Cubes

$$A^3 - B^3 = (A - B)(A^2 + AB + B^2)$$

Same signs Opposite signs

EXAMPLE 7 Factoring the Sum of Two Cubes

Factor: $x^3 + 8$.

Solution We must express each term as the cube of some monomial. Then we use the formula for factoring $A^3 + B^3$.

$$x^3 + 8 = x^3 + 2^3 = (x + 2)(x^2 - x \cdot 2 + 2^2) = (x + 2)(x^2 - 2x + 4)$$

$$A^3 + B^3 = (A + B)(A^2 - AB + B^2)$$

■

Great Question!

What are some cubes that I should be able to identify?

When factoring the sum or difference of cubes, it is helpful to recognize the following cubes:

$$1 = 1^3$$
$$8 = 2^3$$
$$27 = 3^3$$
$$64 = 4^3$$
$$125 = 5^3$$
$$216 = 6^3$$
$$1000 = 10^3.$$

✓ **CHECK POINT 7** Factor: $x^3 + 27$.

EXAMPLE 8 Factoring the Difference of Two Cubes

Factor: $27 - y^3$.

Solution Express each term as the cube of some monomial. Then use the formula for factoring $A^3 - B^3$.

$$27 - y^3 = 3^3 - y^3 = (3 - y)(3^2 + 3y + y^2) = (3 - y)(9 + 3y + y^2)$$

$$A^3 - B^3 = (A - B)(A^2 + AB + B^2)$$

■

✓ **CHECK POINT 8** Factor: $1 - y^3$.

EXAMPLE 9 Factoring the Sum of Two Cubes

Factor: $64x^3 + 125$.

Solution Express each term as the cube of some monomial. Then use the formula for factoring $A^3 + B^3$.

$$64x^3 + 125 = (4x)^3 + 5^3 = (4x + 5)[(4x)^2 - (4x)(5) + 5^2]$$

$$A^3 + B^3 = (A + B)(A^2 - AB + B^2)$$

$$= (4x + 5)(16x^2 - 20x + 25)$$

■

✓ **CHECK POINT 9** Factor: $125x^3 + 8$.

Great Question!

A Cube of SOAP

The formulas for factoring $A^3 + B^3$ and $A^3 - B^3$ are difficult to remember and easy to confuse. Can you help me out?

When factoring sums or differences of cubes, observe the sign patterns.

Same signs

$$A^3 + B^3 = (A + B)(A^2 - AB + B^2)$$

Opposite signs   Always positive

Same signs

$$A^3 - B^3 = (A - B)(A^2 + AB + B^2)$$

Opposite signs   Always positive

The word *SOAP* is a way to remember these patterns:

S   O   A   P.

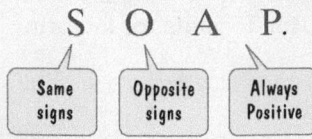

| Same signs | Opposite signs | Always Positive |

CONCEPT AND VOCABULARY CHECK

Fill in each blank so that the resulting statement is true.

1. The formula for factoring the difference of two squares is $A^2 - B^2 = $ _____.

2. A formula for factoring a perfect square trinomial is $A^2 + 2AB + B^2 = $ _____.

3. A formula for factoring a perfect square trinomial is $A^2 - 2AB + B^2 = $ _____.

4. The formula for factoring the sum of two cubes is $A^3 + B^3 = $ _____.

5. The formula for factoring the difference of two cubes is $A^3 - B^3 = $ _____.

6. $36x^2 - 49 = ($ _____ $+ 7)($ _____ $- 7)$

7. $x^2 - 12x + 36 = (x$ _____ $)^2$

8. $16x^2 - 24x + 9 = ($ _____ $- 3)^2$

9. $x^3 + 8 = (x$ _____ $)(x^2 - 2x + 4)$

10. $x^3 - 27 = (x$ _____ $)(x^2 + 3x$ _____ $)$

11. True or false: $x^2 - 8$ is the difference of two perfect squares. _____

12. True or false: $x^2 + 8x + 16$ is a perfect square trinomial. _____

13. True or false: $x^2 - 5x + 25$ is a perfect square trinomial. _____

14. True or false: $x^3 + 1000$ is the sum of two cubes. _____

15. True or false: $x^3 - 100$ is the difference of two cubes. _____

6.4 EXERCISE SET   MyMathLab®

 Watch the videos in MyMathLab

 Download the MyDashBoard App

Practice Exercises

In Exercises 1–26, factor each difference of two squares.

1. $x^2 - 25$
2. $x^2 - 16$
3. $y^2 - 1$
4. $y^2 - 9$
5. $4x^2 - 9$
6. $9x^2 - 25$
7. $25 - x^2$
8. $16 - x^2$
9. $1 - 49x^2$
10. $1 - 64x^2$
11. $9 - 25y^2$
12. $16 - 49y^2$
13. $x^4 - 9$
14. $x^4 - 25$
15. $49y^4 - 16$
16. $49y^4 - 25$
17. $x^{10} - 9$
18. $x^{10} - 1$
19. $25x^2 - 16y^2$
20. $9x^2 - 25y^2$
21. $x^4 - y^{10}$
22. $x^{14} - y^4$
23. $x^4 - 16$
24. $x^4 - 1$
25. $16x^4 - 81$
26. $81x^4 - 1$

In Exercises 27–44, factor completely, or state that the polynomial is prime.

27. $2x^2 - 18$
28. $5x^2 - 45$
29. $2x^3 - 72x$
30. $2x^3 - 8x$
31. $x^2 + 36$
32. $x^2 + 4$
33. $3x^3 + 27x$
34. $3x^3 + 15x$
35. $18 - 2y^2$
36. $32 - 2y^2$
37. $3y^3 - 48y$
38. $3y^3 - 75y$
39. $18x^3 - 2x$
40. $20x^3 - 5x$
41. $-3x^2 + 75$
42. $-4x^2 + 4$
43. $-5y^3 + 20y$
44. $-54y^3 + 6y$

In Exercises 45–66, factor any perfect square trinomials, or state that the polynomial is prime.

45. $x^2 + 2x + 1$
46. $x^2 + 4x + 4$
47. $x^2 - 14x + 49$
48. $x^2 - 10x + 25$
49. $x^2 - 2x + 1$
50. $x^2 - 4x + 4$

51. $x^2 + 22x + 121$

52. $x^2 + 24x + 144$

53. $4x^2 + 4x + 1$

54. $9x^2 + 6x + 1$

55. $25y^2 - 10y + 1$

56. $64y^2 - 16y + 1$

57. $x^2 - 10x + 100$ **58.** $x^2 - 7x + 49$

59. $x^2 + 14xy + 49y^2$

60. $x^2 + 16xy + 64y^2$

61. $x^2 - 12xy + 36y^2$

62. $x^2 - 18xy + 81y^2$

63. $x^2 - 8xy + 64y^2$ **64.** $x^2 + 9xy + 16y^2$

65. $16x^2 - 40xy + 25y^2$

66. $9x^2 + 48xy + 64y^2$

In Exercises 67–78, factor completely.

67. $12x^2 - 12x + 3$

68. $18x^2 + 24x + 8$

69. $9x^3 + 6x^2 + x$

70. $25x^3 - 10x^2 + x$

71. $2y^2 - 4y + 2$

72. $2y^2 - 40y + 200$

73. $2y^3 + 28y^2 + 98y$

74. $50y^3 + 20y^2 + 2y$

75. $-6x^2 + 24x - 24$

76. $-5x^2 + 30x - 45$

77. $-16y^3 - 16y^2 - 4y$

78. $-45y^3 - 30y^2 - 5y$

In Exercises 79–96, factor using the formula for the sum or difference of two cubes.

79. $x^3 + 1$

80. $x^3 + 64$

81. $x^3 - 27$

82. $x^3 - 64$

83. $8y^3 - 1$

84. $27y^3 - 1$

85. $27x^3 + 8$

86. $125x^3 + 8$

87. $x^3y^3 - 64$

88. $x^3y^3 - 27$

89. $27y^4 + 8y$

90. $64y - y^4$

91. $54 - 16y^3$

92. $128 - 250y^3$

93. $64x^3 + 27y^3$

94. $8x^3 + 27y^3$

95. $125x^3 - 64y^3$

96. $125x^3 - y^3$

Practice PLUS

In Exercises 97–104, factor completely. (Hint on Exercises 97–102: Factors contain rational numbers.)

97. $25x^2 - \dfrac{4}{49}$

98. $16x^2 - \dfrac{9}{25}$

99. $y^4 - \dfrac{y}{1000}$

100. $y^4 - \dfrac{y}{8}$

101. $0.25x - x^3$

102. $0.64x - x^3$

103. $(x + 1)^2 - 25$

104. $(x + 2)^2 - 49$

105. Divide $x^3 - x^2 - 5x - 3$ by $x - 3$. Use the quotient to factor $x^3 - x^2 - 5x - 3$ completely.

106. Divide $x^3 + 4x^2 - 3x - 18$ by $x - 2$. Use the quotient to factor $x^3 + 4x^2 - 3x - 18$ completely.

Application Exercises

In Exercises 107–110, find the formula for the area of the shaded blue region and express it in factored form.

107.   **108.**

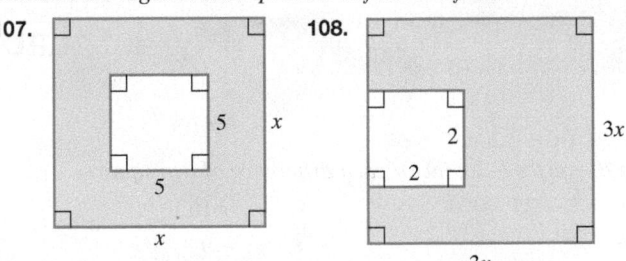

109.   **110.**

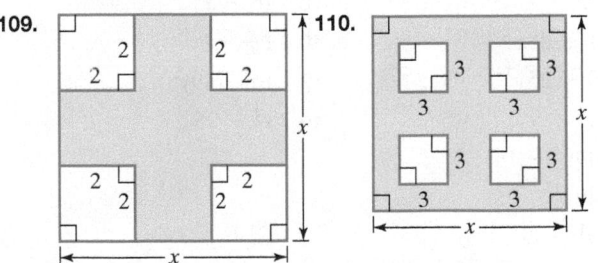

Writing in Mathematics

111. Explain how to factor the difference of two squares. Provide an example with your explanation.

112. What is a perfect square trinomial and how is it factored?

113. Explain why $x^2 - 1$ is factorable, but $x^2 + 1$ is not.

114. Explain how to factor $x^3 + 1$.

Critical Thinking Exercises

Make Sense? *In Exercises 115–118, determine whether each statement "makes sense" or "does not make sense" and explain your reasoning.*

115. I factored $9x^2 - 36$ completely and obtained

$$(3x + 6)(3x - 6).$$

116. Although I can factor the difference of squares and perfect square trinomials using trial-and-error, recognizing these special forms shortens the process.

117. I factored $9 - 25x^2$ as $(3 + 5x)(3 - 5x)$ and then applied the commutative property to rewrite the factorization as $(5x + 3)(5x - 3)$.

118. I compared the factorization for the sum of cubes with the factorization for the difference of cubes and noticed that the only difference between them is the positive and negative signs.

In Exercises 119–122, determine whether each statement is true or false. If the statement is false, make the necessary change(s) to produce a true statement.

119. Because $x^2 - 25 = (x + 5)(x - 5)$, then $x^2 + 25 = (x - 5)(x + 5)$.

120. All perfect square trinomials are squares of binomials.

121. Any polynomial that is the sum of two squares is prime.

122. The polynomial $16x^2 + 20x + 25$ is a perfect square trinomial.

123. Where is the error in this "proof" that $2 = 0$?

$a = b$	Suppose that a and b are any equal real numbers.
$a^2 = b^2$	Square both sides of the equation.
$a^2 - b^2 = 0$	Subtract b^2 from both sides.
$2(a^2 - b^2) = 2 \cdot 0$	Multiply both sides by 2.
$2(a^2 - b^2) = 0$	On the right side, $2 \cdot 0 = 0$.
$2(a + b)(a - b) = 0$	Factor $a^2 - b^2$.
$2(a + b) = 0$	Divide both sides by $a - b$.
$2 = 0$	Divide both sides by $a + b$.

In Exercises 124–127, factor each polynomial.

124. $x^2 - y^2 + 3x + 3y$

125. $x^{2n} - 25y^{2n}$

126. $4x^{2n} + 12x^n + 9$

127. $(x + 3)^2 - 2(x + 3) + 1$

In Exercises 128–129, find all integers k so that the trinomial is a perfect square trinomial.

128. $9x^2 + kx + 1$

129. $64x^2 - 16x + k$

Technology Exercises

In Exercises 130–133, use the GRAPH *or* TABLE *feature of a graphing utility to determine if the polynomial on the left side of each equation has been correctly factored. If the graphs of y_1 and y_2 coincide, or if their corresponding table values are equal, this means that the polynomial on the left side has been correctly factored. If not, factor the polynomial correctly and then use your graphing utility to verify the factorization.*

130. $4x^2 - 9 = (4x + 3)(4x - 3)$

131. $x^2 - 6x + 9 = (x - 3)^2$

132. $4x^2 - 4x + 1 = (4x - 1)^2$

133. $x^3 - 1 = (x - 1)(x^2 - x + 1)$

Review Exercises

134. Simplify: $(2x^2y^3)^4(5xy^2)$. (Section 5.7, Example 5)

135. Subtract: $(10x^2 - 5x + 2) - (14x^2 - 5x - 1)$. (Section 5.1, Example 3)

136. Divide: $\dfrac{6x^2 + 11x - 10}{3x - 2}$. (Section 5.6, Example 1)

Preview Exercises

Exercises 137–139 will help you prepare for the material covered in the next section. In each exercise, factor completely.

137. $3x^3 - 75x$

138. $2x^2 - 20x + 50$

139. $x^3 - 2x^2 - x + 2$

A General Factoring Strategy

Objectives

1. Recognize the appropriate method for factoring a polynomial.

2. Use a general strategy for factoring polynomials.

Yogi Berra, catcher and renowned hitter for the New York Yankees (1946–1963), said it best: "If you don't know where you're going, you'll probably end up someplace else." When it comes to factoring, it's easy to know where you're going. Why? In this section, you will learn a step-by-step strategy that provides a plan and direction for solving factoring problems.

Man with a Bowler Hat (1964), René Magritte. © 2011 Herscovici/ARS

A Strategy for Factoring Polynomials

It is important to practice factoring a wide variety of polynomials so that you can quickly select the appropriate technique. The polynomial is factored completely when all its polynomial factors, except possibly the monomial factor, are prime. Because of the commutative property, the order of the factors does not matter.

Here is a general strategy for factoring polynomials:

1. Recognize the appropriate method for factoring a polynomial.

A Strategy for Factoring a Polynomial

1. If there is a common factor other than 1, factor out the GCF.

2. Determine the number of terms in the polynomial and try factoring as follows:

   a. If there are two terms, can the binomial be factored by one of the following special forms?

   Difference of two squares: $A^2 - B^2 = (A + B)(A - B)$
   Sum of two cubes: $A^3 + B^3 = (A + B)(A^2 - AB + B^2)$
   Difference of two cubes: $A^3 - B^3 = (A - B)(A^2 + AB + B^2)$

   b. If there are three terms, is the trinomial a perfect square trinomial? If so, factor by one of the following special forms:

   $$A^2 + 2AB + B^2 = (A + B)^2$$
   $$A^2 - 2AB + B^2 = (A - B)^2.$$

   If the trinomial is not a perfect square trinomial, try factoring by trial and error or grouping.

   c. If there are four or more terms, try factoring by grouping.

3. Check to see if any factors with more than one term in the factored polynomial can be factored further. If so, factor completely.

4. Check by multiplying.

2 Use a general strategy for factoring polynomials.

The following examples and those in the exercise set are similar to the previous factoring problems. One difference is that although these polynomials may be factored using the techniques we have studied in this chapter, most must be factored using at least two techniques. Also different is that these factorizations are not all of the same type. They are intentionally mixed to promote the development of a general factoring strategy.

EXAMPLE 1 Factoring a Polynomial

Factor: $4x^4 - 16x^2$.

Solution

Step 1. If there is a common factor, factor out the GCF. Because $4x^2$ is common to both terms, we factor it out.

$$4x^4 - 16x^2 = 4x^2(x^2 - 4) \qquad \text{Factor out the GCF.}$$

Step 2. Determine the number of terms and factor accordingly. The factor $x^2 - 4$ has two terms. It is the difference of two squares: $x^2 - 2^2$. We factor using the special form for the difference of two squares and rewrite the GCF.

$$4x^4 - 16x^2 = 4x^2(x + 2)(x - 2) \qquad \text{Use } A^2 - B^2 = (A + B)(A - B) \text{ on}$$
$$x^2 - 4: A = x \text{ and } B = 2.$$

Step 3. Check to see if any factors with more than one term can be factored further. No factor with more than one term can be factored further, so we have factored completely.

Step 4. Check by multiplying.

$$4x^2(x + 2)(x - 2) = 4x^2(x^2 - 4) = 4x^4 - 16x^2$$

This is the original polynomial, so the factorization is correct.

✓ **CHECK POINT 1** Factor: $5x^4 - 45x^2$.

EXAMPLE 2 Factoring a Polynomial

Factor: $3x^2 - 6x - 45$.

Solution

Step 1. If there is a common factor, factor out the GCF. Because 3 is common to all terms, we factor it out.

$$3x^2 - 6x - 45 = 3(x^2 - 2x - 15) \qquad \text{Factor out the GCF.}$$

Step 2. Determine the number of terms and factor accordingly. The factor $x^2 - 2x - 15$ has three terms, but it is not a perfect square trinomial. We factor it using trial and error.

$$3x^2 - 6x - 45 = 3(x^2 - 2x - 15) = 3(x - 5)(x + 3)$$

Step 3. Check to see if factors can be factored further. In this case, they cannot, so we have factored completely.

Step 4. Check by multiplying.

$$3(x - 5)(x + 3) = 3(x^2 - 2x - 15) = 3x^2 - 6x - 45$$

FOIL

This is the original polynomial, so the factorization is correct.

☑ **CHECK POINT 2** Factor: $4x^2 - 16x - 48$.

EXAMPLE 3 Factoring a Polynomial

Factor: $7x^5 - 7x$.

Solution

Step 1. If there is a common factor, factor out the GCF. Because $7x$ is common to both terms, we factor it out.

$$7x^5 - 7x = 7x(x^4 - 1) \qquad \text{Factor out the GCF.}$$

Step 2. Determine the number of terms and factor accordingly. The factor $x^4 - 1$ has two terms. This binomial can be expressed as $(x^2)^2 - 1^2$, so it can be factored as the difference of two squares.

$$7x^5 - 7x = 7x(x^4 - 1) = 7x(x^2 + 1)(x^2 - 1) \qquad \text{Use } A^2 - B^2 = (A + B)(A - B)$$
$$\text{on } x^4 - 1\text{: } A = x^2 \text{ and } B = 1.$$

Step 3. Check to see if factors can be factored further. We note that $(x^2 - 1)$ is also the difference of two squares, $x^2 - 1^2$, so we continue factoring.

$$7x^5 - 7x = 7x(x^2 + 1)(x + 1)(x - 1) \qquad \text{Factor } x^2 - 1 \text{ as the difference}$$
$$\text{of two squares.}$$

Step 4. Check by multiplying.

$$7x(x^2 + 1)(x + 1)(x - 1) = 7x(x^2 + 1)(x^2 - 1) = 7x(x^4 - 1) = 7x^5 - 7x$$

We obtain the original polynomial, so the factorization is correct. ∎

☑ **CHECK POINT 3** Factor: $4x^5 - 64x$.

EXAMPLE 4 Factoring a Polynomial

Factor: $x^3 - 5x^2 - 4x + 20$.

Solution

Step 1. If there is a common factor, factor out the GCF. Other than 1, there is no common factor.

Step 2. Determine the number of terms and factor accordingly. There are four terms. We try factoring by grouping.

$$x^3 - 5x^2 - 4x + 20$$
$$= (x^3 - 5x^2) + (-4x + 20) \qquad \text{Group terms with common factors.}$$
$$= x^2(x - 5) - 4(x - 5) \qquad \text{Factor from each group.}$$
$$= (x - 5)(x^2 - 4) \qquad \text{Factor out the common binomial factor, } x - 5.$$

Step 3. Check to see if factors can be factored further. We note that $(x^2 - 4)$ is the difference of two squares, $x^2 - 2^2$, so we continue factoring.

$$x^3 - 5x^2 - 4x + 20 = (x - 5)(x + 2)(x - 2) \qquad \text{Factor } x^2 - 4 \text{ as the difference}$$
$$\text{of two squares.}$$

We have factored completely because no factor with more than one term can be factored further.

Step 4. Check by multiplying.

F   O   I   L

$$(x - 5)(x + 2)(x - 2) = (x - 5)(x^2 - 4) = x^3 - 4x - 5x^2 + 20$$

$$= x^3 - 5x^2 - 4x + 20$$

We obtain the original polynomial, so the factorization is correct.   ■

✓ **CHECK POINT 4**   Factor: $x^3 - 4x^2 - 9x + 36$.

EXAMPLE 5   Factoring a Polynomial

Factor: $2x^3 - 24x^2 + 72x$.

Solution

Step 1. If there is a common factor, factor out the GCF. Because $2x$ is common to all terms, we factor it out.

$$2x^3 - 24x^2 + 72x = 2x(x^2 - 12x + 36) \qquad \text{Factor out the GCF.}$$

Step 2. Determine the number of terms and factor accordingly. The factor $x^2 - 12x + 36$ has three terms. Is it a perfect square trinomial? Yes. The first term, x^2, is the square of a monomial. The last term, 36 or 6^2, is the square of an integer. The middle term involves twice the product of x and 6. We factor using $A^2 - 2AB + B^2 = (A - B)^2$.

$$2x^3 - 24x^2 + 72x = 2x(x^2 - 12x + 36)$$

$$= 2x(x^2 - 2 \cdot x \cdot 6 + 6^2) \qquad \text{The second factor is a perfect square trinomial.}$$

$$\underbrace{\quad\quad}_{A^2 - 2\ A\ B\ +\ B^2}$$

$$= 2x(x - 6)^2 \qquad A^2 - 2AB + B^2 = (A - B)^2$$

Step 3. Check to see if factors can be factored further. In this problem, they cannot, so we have factored completely.

Step 4. Check by multiplying. Let's verify that $2x^3 - 24x^2 + 72x = 2x(x - 6)^2$.

$$2x(x - 6)^2 = 2x(x^2 - 12x + 36) = 2x^3 - 24x^2 + 72x \quad ■$$

We obtain the original polynomial, so the factorization is correct.

✓ **CHECK POINT 5**   Factor: $3x^3 - 30x^2 + 75x$.

EXAMPLE 6   Factoring a Polynomial

Factor: $3x^5 + 24x^2$.

Solution

Step 1. If there is a common factor, factor out the GCF. Because $3x^2$ is common to both terms, we factor it out.

$$3x^5 + 24x^2 = 3x^2(x^3 + 8) \qquad \text{Factor out the GCF.}$$

Step 2. Determine the number of terms and factor accordingly. Our factorization up to this point, $3x^5 + 24x^2 = 3x^2(x^3 + 8)$, is not complete. The factor $x^3 + 8$ has two terms. This binomial can be expressed as $x^3 + 2^3$, so it can be factored as the sum of two cubes.

$$3x^5 + 24x^2 = 3x^2(\underbrace{x^3 + 2^3}_{A^3 + B^3})$$

Express $x^3 + 8$ as the sum of two cubes.

$$= 3x^2\underbrace{(x + 2)(x^2 - 2x + 4)}_{(A + B)\,(A^2 - AB + B^2)}$$

Factor the sum of two cubes.

Step 3. Check to see if factors can be factored further. In this problem, they cannot, so we have factored completely.

Step 4. Check by multiplying.

$$3x^2(x + 2)(x^2 - 2x + 4) = 3x^2[x(x^2 - 2x + 4) + 2(x^2 - 2x + 4)]$$
$$= 3x^2(x^3 - 2x^2 + 4x + 2x^2 - 4x + 8)$$
$$= 3x^2(x^3 + 8) = 3x^5 + 24x^2$$

We obtain the original polynomial, so the factorization is correct. ■

✓ **CHECK POINT 6** Factor: $2x^5 + 54x^2$.

Discover for Yourself

In Examples 1–6, substitute 1 for the variable in both the given polynomial and in its factored form. Evaluate each expression. What do you observe? Do this for a second value of the variable. Is this a complete check or only a partial check of the factorization? Explain.

EXAMPLE 7 Factoring a Polynomial in Two Variables

Factor: $32x^4y - 2y^5$.

Solution

Step 1. If there is a common factor, factor out the GCF. Because $2y$ is common to both terms, we factor it out.

$$32x^4y - 2y^5 = 2y(16x^4 - y^4) \qquad \text{Factor out the GCF.}$$

Step 2. Determine the number of terms and factor accordingly. The factor $16x^4 - y^4$ has two terms. It is the difference of two squares: $(4x^2)^2 - (y^2)^2$. We factor using the special form for the difference of two squares.

$$32x^4y - 2y^5 = 2y[\underbrace{(4x^2)^2 - (y^2)^2}_{A^2 - B^2}]$$

Express $16x^4 - y^4$ as the difference of two squares.

$$= 2y\underbrace{(4x^2 + y^2)}_{(A + B)}\underbrace{(4x^2 - y^2)}_{(A - B)} \qquad A^2 - B^2 = (A + B)(A - B)$$

Step 3. Check to see if factors can be factored further. We note that the last factor, $4x^2 - y^2$, is also the difference of two squares, $(2x)^2 - y^2$, so we continue factoring.

$$32x^4y - 2y^5 = 2y(4x^2 + y^2)(2x + y)(2x - y)$$

Step 4. Check by multiplying. Multiply the factors in the factorization and verify that you obtain the original polynomial. ■

✓ **CHECK POINT 7** Factor: $3x^4y - 48y^5$.

EXAMPLE 8 Factoring a Polynomial in Two Variables

Factor: $18x^3 + 48x^2y + 32xy^2$.

Solution

Step 1. If there is a common factor, factor out the GCF. Because $2x$ is common to all terms, we factor it out.

$$18x^3 + 48x^2y + 32xy^2 = 2x(9x^2 + 24xy + 16y^2)$$

Step 2. Determine the number of terms and factor accordingly. The factor $9x^2 + 24xy + 16y^2$ has three terms. Is it a perfect square trinomial? Yes. The first term, $9x^2$ or $(3x)^2$, and the last term, $16y^2$ or $(4y)^2$, are squares of monomials. The middle term, $24xy$, is twice the product of $3x$ and $4y$. We factor using $A^2 + 2AB + B^2 = (A + B)^2$.

$$18x^3 + 48x^2y + 32xy^2 = 2x(9x^2 + 24xy + 16y^2)$$
$$= 2x[(3x)^2 + 2 \cdot 3x \cdot 4y + (4y)^2] \quad \text{The second factor is a perfect square trinomial.}$$

$$\underbrace{A^2 + 2 \cdot A \cdot B + B^2}$$

$$= 2x(3x + 4y)^2 \quad A^2 + 2AB + B^2 = (A + B)^2$$

Step 3. Check to see if factors can be factored further. In this problem, they cannot, so we have factored completely.

Step 4. Check by multiplication. Multiply the factors in the factorization and verify that you obtain the original polynomial. ∎

✓ **CHECK POINT 8** Factor: $12x^3 + 36x^2y + 27xy^2$.

CONCEPT AND VOCABULARY CHECK

Here is a list of the factoring techniques that we have discussed.

  a. Factoring out the GCF

  b. Factoring out the negative of the GCF

  c. Factoring by grouping

  d. Factoring trinomials by trial and error or grouping

  e. Factoring the difference of two squares
   $$A^2 - B^2 = (A + B)(A - B)$$

  f. Factoring perfect square trinomials
   $$A^2 + 2AB + B^2 = (A + B)^2$$
   $$A^2 - 2AB + B^2 = (A - B)^2$$

  g. Factoring the sum of two cubes
   $$A^3 + B^3 = (A + B)(A^2 - AB + B^2)$$

  h. Factoring the difference of two cubes
   $$A^3 - B^3 = (A - B)(A^2 + AB + B^2)$$

Fill in each blank by writing the letter of the technique (a through h) for factoring the polynomial.

  1. $-3x^2 + 21x$ _____

  2. $16x^2 - 25$ _____

  3. $27x^3 - 1$ _____

  4. $x^2 + 7x + xy + 7y$ _____

  5. $4x^2 + 8x + 3$ _____

  6. $9x^2 + 24x + 16$ _____

  7. $5x^2 + 10x$ _____

  8. $x^3 + 1000$ _____

6.5 EXERCISE SET   MyMathLab®

Watch the videos
in MyMathLab

Download the
MyDashBoard App

Practice Exercises

Before getting to multiple-step factorizations, let's be sure that you are comfortable with exercises requiring only one of the factoring techniques. In Exercises 1–16, factor each polynomial.

1. $-7x^2 + 35x$
2. $-6x^2 + 24x$
3. $25x^2 - 49$
4. $100x^2 - 81$
5. $27x^3 - 1$
6. $64x^3 - 1$
7. $5x + 5y + x^2 + xy$
8. $7x + 7y + x^2 + xy$
9. $14x^2 - 9x + 1$
10. $3x^2 + 2x - 5$
11. $x^2 - 2x + 1$
12. $x^2 - 4x + 4$
13. $27x^3y^3 + 8$
14. $216x^3y^3 + 125$
15. $6x^2 + x - 15$
16. $4x^2 - x - 5$

Now let's move on to factorizations that may require two or more techniques. In Exercises 17–80, factor completely, or state that the polynomial is prime. Check factorizations using multiplication or a graphing utility.

17. $5x^3 - 20x$
18. $4x^3 - 100x$
19. $7x^3 + 7x$
20. $6x^3 + 24x$
21. $5x^2 - 5x - 30$
22. $5x^2 - 15x - 50$
23. $2x^4 - 162$
24. $7x^4 - 7$
25. $x^3 + 2x^2 - 9x - 18$
26. $x^3 + 3x^2 - 25x - 75$
27. $3x^3 - 24x^2 + 48x$
28. $5x^3 - 20x^2 + 20x$
29. $2x^5 + 2x^2$
30. $2x^5 + 128x^2$
31. $6x^2 + 8x$
32. $21x^2 - 35x$
33. $-2y^2 + 2y + 112$
34. $-6x^2 + 6x + 12$
35. $7y^4 + 14y^3 + 7y^2$
36. $2y^4 + 28y^3 + 98y^2$
37. $y^2 + 8y - 16$
38. $y^2 - 18y - 81$
39. $16y^2 - 4y - 2$
40. $32y^2 + 4y - 6$
41. $r^2 - 25r$
42. $3r^2 - 27r$
43. $4w^2 + 8w - 5$
44. $35w^2 - 2w - 1$
45. $x^3 - 4x$
46. $9x^3 - 9x$

47. $x^2 + 64$
48. $y^2 + 36$
49. $9y^2 + 13y + 4$
50. $20y^2 + 12y + 1$
51. $y^3 + 2y^2 - 4y - 8$
52. $y^3 + 2y^2 - y - 2$
53. $16y^2 + 24y + 9$
54. $25y^2 + 20y + 4$
55. $-4y^3 + 28y^2 - 40y$
56. $-7y^3 + 21y^2 - 14y$
57. $y^5 - 81y$
58. $y^5 - 16y$
59. $20a^4 - 45a^2$
60. $48a^4 - 3a^2$
61. $9x^4 + 18x^3 + 6x^2$
62. $10x^4 + 20x^3 + 15x^2$
63. $12y^2 - 11y + 2$
64. $21x^2 - 25x - 4$
65. $9y^2 - 64$
66. $100y^2 - 49$
67. $9y^2 + 64$
68. $100y^2 + 49$
69. $2y^3 + 3y^2 - 50y - 75$
70. $12y^3 + 16y^2 - 3y - 4$
71. $2r^3 + 30r^2 - 68r$
72. $3r^3 - 27r^2 - 210r$
73. $8x^5 - 2x^3$
74. $y^9 - y^5$
75. $3x^2 + 243$
76. $27x^2 + 75$
77. $x^4 + 8x$
78. $x^4 + 27x$
79. $2y^5 - 2y^2$
80. $2y^5 - 128y^2$

Exercises 81–112 contain polynomials in several variables. Factor each polynomial completely and check using multiplication.

81. $6x^2 + 8xy$
82. $21x^2 - 35xy$
83. $xy - 7x + 3y - 21$
84. $xy - 5x + 2y - 10$
85. $x^2 - 3xy - 4y^2$
86. $x^2 - 4xy - 12y^2$
87. $72a^3b^2 + 12a^2 - 24a^4b^2$
88. $24a^4b + 60a^3b^2 + 150a^2b^3$
89. $3a^2 + 27ab + 54b^2$
90. $3a^2 + 15ab + 18b^2$
91. $48x^4y - 3x^2y$
92. $16a^3b^2 - 4ab^2$
93. $6a^2b + ab - 2b$

94. $16a^2 - 32ab + 12b^2$

95. $7x^5y - 7xy^5$

96. $3x^4y^2 - 3x^2y^2$

97. $10x^3y - 14x^2y^2 + 4xy^3$

98. $18x^3y + 57x^2y^2 + 30xy^3$

99. $2bx^2 + 44bx + 242b$

100. $3xz^2 - 72xz + 432x$

101. $15a^2 + 11ab - 14b^2$

102. $25a^2 + 25ab + 6b^2$

103. $-36x^3y + 62x^2y^2 - 12xy^3$

104. $-10a^4b^2 + 15a^3b^3 + 25a^2b^4$

105. $a^2y - b^2y - a^2x + b^2x$

106. $bx^2 - 4b + ax^2 - 4a$

107. $9ax^3 + 15ax^2 - 14ax$

108. $4ay^3 - 12ay^2 + 9ay$

109. $2x^4 + 6x^3y + 2x^2y^2$

110. $3x^4 - 9x^3y + 3x^2y^2$

111. $81x^4y - y^5$

112. $16x^4y - y^5$

Practice PLUS

In Exercises 113–122, factor completely.

113. $10x^2(x + 1) - 7x(x + 1) - 6(x + 1)$

114. $12x^2(x - 1) - 4x(x - 1) - 5(x - 1)$

115. $6x^4 + 35x^2 - 6$

116. $7x^4 + 34x^2 - 5$

117. $(x - 7)^2 - 4a^2$

118. $(x - 6)^2 - 9a^2$

119. $x^2 + 8x + 16 - 25a^2$

120. $x^2 + 14x + 49 - 16a^2$

121. $y^7 + y$

122. $(y + 1)^3 + 1$

Application Exercises

123. A rock is dropped from the top of a 256-foot cliff. The height, in feet, of the rock above the water after t seconds is modeled by the polynomial $256 - 16t^2$. Factor this expression completely.

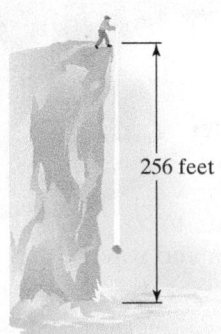

256 feet

124. The building shown in the figure has a height represented by x feet. The building's base is a square and the building's volume is $x^3 - 60x^2 + 900x$ cubic feet. Express the building's dimensions in terms of x.

125. Express the area of the blue shaded ring shown in the figure in terms of π. Then factor this expression completely.

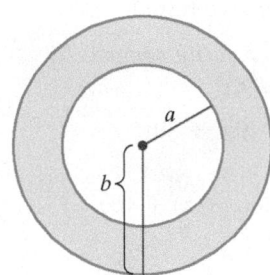

Writing in Mathematics

126. Describe a strategy that can be used to factor polynomials.

127. Describe some of the difficulties in factoring polynomials. What suggestions can you offer to overcome these difficulties?

128. You are about to take a great picture of fog rolling into San Francisco from the middle of the Golden Gate Bridge, 400 feet above the water. Whoops! You accidently lean too far over the safety rail and drop your camera. The height, in feet, of the camera after t seconds is modeled by the polynomial $400 - 16t^2$. The factored form of the polynomial is $16(5 + t)(5 - t)$. Describe something about your falling camera that is easier to see from the factored form, $16(5 + t)(5 - t)$, than from the form $400 - 16t^2$.

Critical Thinking Exercises

Make Sense? *In Exercises 129–132, determine whether each statement "makes sense" or "does not make sense" and explain your reasoning.*

129. It takes a great deal of practice to get good at factoring a wide variety of polynomials.

130. Multiplying polynomials is relatively mechanical, but factoring often requires a great deal of thought.

131. The factorable trinomial $4x^2 + 8x + 3$ and the prime trinomial $4x^2 + 8x + 1$ are in the form $ax^2 + bx + c$, but $b^2 - 4ac$ is a perfect square only in the case of the factorable trinomial.

132. When a factorization requires two factoring techniques, I'm less likely to make errors if I show one technique at a time rather than combining the two factorizations into one step.

In Exercises 133–136, determine whether each statement is true or false. If the statement is false, make the necessary change(s) to produce a true statement.

133. $x^2 - 9 = (x - 3)^2$ for any real number x.

134. The polynomial $4x^2 + 100$ is the sum of two squares and therefore cannot be factored.

135. If the general factoring strategy is used to factor a polynomial, at least two factorizations are necessary before the given polynomial is factored completely.

136. Once a common monomial factor is removed from $3xy^3 + 9xy^2 + 21xy$, the remaining trinomial factor cannot be factored further.

In Exercises 137–141, factor completely.

137. $3x^5 - 21x^3 - 54x$

138. $5y^5 - 5y^4 - 20y^3 + 20y^2$

139. $4x^4 - 9x^2 + 5$

140. $(x + 5)^2 - 20(x + 5) + 100$

141. $3x^{2n} - 27y^{2n}$

Technology Exercises

In Exercises 142–146, use the $\boxed{\text{GRAPH}}$ *or* $\boxed{\text{TABLE}}$ *feature of a graphing utility to determine if the polynomial on the left side of each equation has been correctly factored. If not, factor*

the polynomial correctly and then use your graphing utility to verify the factorization.

142. $4x^2 - 12x + 9 = (4x - 3)^2$; $[-5, 5, 1]$ by $[0, 20, 1]$

143. $3x^3 - 12x^2 - 15x = 3x(x + 5)(x - 1)$; $[-5, 7, 1]$ by $[-80, 80, 10]$

144. $6x^2 + 10x - 4 = 2(3x - 1)(x + 2)$; $[-5, 5, 1]$ by $[-20, 20, 2]$

145. $x^4 - 16 = (x^2 + 4)(x + 2)(x - 2)$; $[-5, 5, 1]$ by $[-20, 20, 2]$

146. $2x^3 + 10x^2 - 2x - 10 = 2(x + 5)(x^2 + 1)$; $[-8, 4, 1]$ by $[-100, 100, 10]$

Review Exercises

147. Factor: $9x^2 - 16$. (Section 6.4, Example 1)

148. Graph using intercepts: $5x - 2y = 10$. (Section 3.2, Example 4)

149. The second angle of a triangle measures three times that of the first angle's measure. The third angle measures 80° more than the first. Find the measure of each angle. (Section 2.6, Example 6)

Preview Exercises

Exercises 150–152 will help you prepare for the material covered in the next section.

150. Evaluate $(3x - 1)(x + 2)$ for $x = \dfrac{1}{3}$.

151. Evaluate $2x^2 + 7x - 4$ for $x = \dfrac{1}{2}$.

152. Factor: $(x - 2)(x + 3) - 6$.

Solving Quadratic Equations by Factoring

Objectives

1 Use the zero-product principle.

2 Solve quadratic equations by factoring.

3 Solve problems using quadratic equations.

The alligator, at one time an endangered species, was the subject of a protection program at Florida's Everglades National Park. Park rangers used the formula

$$P = -10x^2 + 475x + 3500$$

to estimate the alligator population, P, after x years of the protection program. Their goal was to bring the population up to 7250. To find out how long the program had to be continued for this to happen, we substitute 7250 for P in the formula and solve for x:

$$7250 = -10x^2 + 475x + 3500.$$

Do you see how this equation differs from a linear equation? The highest exponent on x is 2. Solving such an equation involves finding the numbers that will make the equation a true statement. In this section, we use factoring to solve equations in the form $ax^2 + bx + c = 0$. We also look at applications of these equations.

The Standard Form of a Quadratic Equation

We begin by defining a quadratic equation.

Definition of a Quadratic Equation

A **quadratic equation** in x is an equation that can be written in the **standard form**

$$ax^2 + bx + c = 0,$$

where a, b, and c are real numbers, with $a \neq 0$. A quadratic equation in x is also called a **second-degree polynomial equation** in x.

Here is an example of a quadratic equation in standard form:

$$x^2 - 7x + 10 = 0.$$

$\boxed{a = 1}$ $\boxed{b = -7}$ $\boxed{c = 10}$

1 Use the zero-product principle.

Solving Quadratic Equations by Factoring

We can factor the left side of the quadratic equation $x^2 - 7x + 10 = 0$. We obtain $(x - 5)(x - 2) = 0$. If a quadratic equation has zero on one side and a factored expression on the other side, it can be solved using the **zero-product principle**.

The Zero-Product Principle

If the product of two algebraic expressions is zero, then at least one of the factors is equal to zero.

If $AB = 0$, then $A = 0$ or $B = 0$.

For example, consider the equation $(x - 5)(x - 2) = 0$. According to the zero-product principle, this product can be zero only if at least one of the factors is zero. We set each individual factor equal to zero and solve each resulting equation for x.

$$(x - 5)(x - 2) = 0$$
$$x - 5 = 0 \quad \text{or} \quad x - 2 = 0$$
$$x = 5 \qquad\qquad x = 2$$

We can check each of the proposed solutions, 5 and 2, in the original quadratic equation, $x^2 - 7x + 10 = 0$. Substitute each one separately for x in the equation.

Check 5:	**Check 2:**
$x^2 - 7x + 10 = 0$	$x^2 - 7x + 10 = 0$
$5^2 - 7 \cdot 5 + 10 \overset{?}{=} 0$	$2^2 - 7 \cdot 2 + 10 \overset{?}{=} 0$
$25 - 35 + 10 \overset{?}{=} 0$	$4 - 14 + 10 \overset{?}{=} 0$
$0 = 0, \text{ true}$	$0 = 0, \text{ true}$

The resulting true statements indicate that the solutions are 5 and 2. Note that with a quadratic equation, we can have two solutions, compared to the linear equation that usually had one.

EXAMPLE 1 Using the Zero-Product Principle

Solve the equation: $(3x - 1)(x + 2) = 0$.

Solution The product $(3x - 1)(x + 2)$ is equal to zero. By the zero-product principle, the only way that this product can be zero is if at least one of the factors is zero. Thus,

$$3x - 1 = 0 \quad \text{or} \quad x + 2 = 0.$$

$$3x = 1 \qquad\qquad x = -2 \quad \text{Solve each equation for } x.$$

$$x = \frac{1}{3}$$

Because each linear equation has a solution, the original equation, $(3x - 1)(x + 2) = 0$, has two solutions, $\frac{1}{3}$ and -2. Check these solutions by substituting each one separately into the given equation. The equation's solution set is $\left\{-2, \frac{1}{3}\right\}$. ∎

✓ **CHECK POINT 1** Solve the equation: $(2x + 1)(x - 4) = 0$.

---

2 Solve quadratic equations by factoring.

In Example 1 and Check Point 1, the given equations were in factored form. Here is a procedure for solving a quadratic equation when we must first do the factoring.

Solving a Quadratic Equation by Factoring

1. If necessary, rewrite the equation in the standard form $ax^2 + bx + c = 0$, moving all terms to one side, thereby obtaining zero on the other side.
2. Factor.
3. Apply the zero-product principle, setting each factor equal to zero.
4. Solve the equations formed in step 3.
5. Check the solutions in the original equation.

EXAMPLE 2 Solving a Quadratic Equation by Factoring

Solve: $2x^2 + 7x - 4 = 0$.

Solution

Step 1. Move all terms to one side and obtain zero on the other side. All terms are already on the left and zero is on the other side, so we can skip this step.

Step 2. Factor.

$$2x^2 + 7x - 4 = 0$$

$$(2x - 1)(x + 4) = 0$$

Steps 3 and 4. Set each factor equal to zero and solve each resulting equation.

$$2x - 1 = 0 \quad \text{or} \quad x + 4 = 0$$

$$2x = 1 \qquad\qquad x = -4$$

$$x = \frac{1}{2}$$

Step 5. Check the solutions in the original equation.

Check $\frac{1}{2}$:

$2x^2 + 7x - 4 = 0$

$2\left(\frac{1}{2}\right)^2 + 7\left(\frac{1}{2}\right) - 4 \stackrel{?}{=} 0$

$2\left(\frac{1}{4}\right) + 7\left(\frac{1}{2}\right) - 4 \stackrel{?}{=} 0$

$\frac{1}{2} + \frac{7}{2} - 4 \stackrel{?}{=} 0$

$4 - 4 \stackrel{?}{=} 0$

$0 = 0, \quad \text{true}$

Check -4:

$2x^2 + 7x - 4 = 0$

$2(-4)^2 + 7(-4) - 4 \stackrel{?}{=} 0$

$2(16) + 7(-4) - 4 \stackrel{?}{=} 0$

$32 + (-28) - 4 \stackrel{?}{=} 0$

$4 - 4 \stackrel{?}{=} 0$

$0 = 0, \quad \text{true}$

The solutions are -4 and $\frac{1}{2}$, and the solution set is $\left\{-4, \frac{1}{2}\right\}$. ∎

✓ **CHECK POINT 2**   Solve: $x^2 - 6x + 5 = 0$.

Great Question!

After factoring a polynomial, should I set each factor equal to zero?

No. Do not confuse factoring a polynomial with solving a quadratic equation by factoring.

Factoring a Polynomial

Factor: $2x^2 + 7x - 4$.

> This is not an equation.
> There is no equal sign.

Solution $(2x - 1)(x + 4)$

> Stop! Avoid the common
> error of setting each
> factor equal to zero.

Solving a Quadratic Equation

Solve: $2x^2 + 7x - 4 = 0$.

> This is an equation.
> There is an equal sign.

Solution $(2x - 1)(x + 4) = 0$

$2x - 1 = 0 \quad \text{or} \quad x + 4 = 0$

$x = \frac{1}{2} \qquad\qquad x = -4$

The solution set is $\left\{-4, \frac{1}{2}\right\}$.

Using Technology

Graphic Connections

You can use a graphing utility to check the real number solutions of a quadratic equation. **The solutions of $ax^2 + bx + c = 0$ correspond to the x-intercepts for the graph of $y = ax^2 + bx + c$.** For example, to check the solutions of $2x^2 + 7x - 4 = 0$, graph $y = 2x^2 + 7x - 4$. The U-shaped, bowl-like, graph is shown below. The x-intercepts are -4 and $\frac{1}{2}$, verifying -4 and $\frac{1}{2}$ as the solutions.

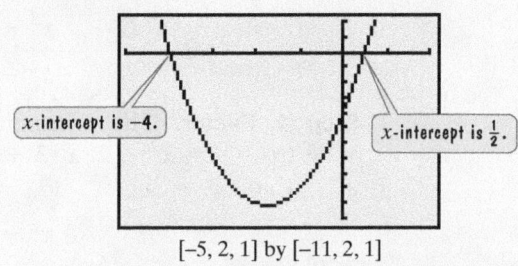

> x-intercept is -4.

> x-intercept is $\frac{1}{2}$.

$[-5, 2, 1]$ by $[-11, 2, 1]$

EXAMPLE 3 Solving a Quadratic Equation by Factoring

Solve: $3x^2 = 2x$.

Solution

Step 1. Move all terms to one side and obtain zero on the other side. Subtract $2x$ from both sides and write the equation in standard form.

$$3x^2 - 2x = 2x - 2x$$

$$3x^2 - 2x = 0$$

Step 2. Factor. We factor out x from the two terms on the left side.

$$3x^2 - 2x = 0$$

$$x(3x - 2) = 0$$

Steps 3 and 4. Set each factor equal to zero and solve the resulting equations.

$$x = 0 \quad \text{or} \quad 3x - 2 = 0$$

$$3x = 2$$

$$x = \frac{2}{3}$$

Great Question!

Can I simplify $3x^2 = 2x$ by dividing both sides by x?

No. If you divide both sides of $3x^2 = 2x$ by x, you will obtain $3x = 2$ and, consequently, $x = \frac{2}{3}$. The other solution, 0, is lost. We can divide both sides of an equation by any *nonzero* real number. If x is zero, we lose the second solution.

Step 5. Check the solutions in the original equation.

Check 0:

$$3x^2 = 2x$$

$$3 \cdot 0^2 \stackrel{?}{=} 2 \cdot 0$$

$$0 = 0, \quad \text{true}$$

Check $\frac{2}{3}$:

$$3x^2 = 2x$$

$$3\left(\frac{2}{3}\right)^2 \stackrel{?}{=} 2\left(\frac{2}{3}\right)$$

$$3\left(\frac{4}{9}\right) \stackrel{?}{=} 2\left(\frac{2}{3}\right)$$

$$\frac{4}{3} = \frac{4}{3}, \quad \text{true}$$

The solutions are 0 and $\frac{2}{3}$, and the solution set is $\left\{0, \frac{2}{3}\right\}$. ∎

✓ **CHECK POINT 3** Solve: $4x^2 = 2x$.

EXAMPLE 4 Solving a Quadratic Equation by Factoring

Solve: $x^2 = 6x - 9$.

Solution

Step 1. Move all terms to one side and obtain zero on the other side. To obtain zero on the right, we subtract $6x$ and add 9 on both sides.

$$x^2 - 6x + 9 = 6x - 6x - 9 + 9$$

$$x^2 - 6x + 9 = 0$$

Step 2. Factor. The trinomial on the left side is a perfect square trinomial: $x^2 - 6x + 9 = x^2 - 2 \cdot x \cdot 3 + 3^2$. We factor using $A^2 - 2AB + B^2 = (A - B)^2$: $A = x$ and $B = 3$.

$$x^2 - 6x + 9 = 0$$

$$(x - 3)^2 = 0$$

Using Technology

Graphic Connections

The graph of
$y = x^2 - 6x + 9$ is shown
below. Notice that there is
only one x-intercept, namely
3, verifying that the solution
of

$$x^2 - 6x + 9 = 0$$

is 3.

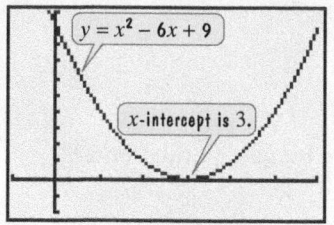

$y = x^2 - 6x + 9$

x-intercept is 3.

$[-1, 6, 1]$ by $[-2, 10, 1]$

Steps 3 and 4. Set each factor equal to zero and solve the resulting equations. Because both factors are the same, it is only necessary to set one of them equal to zero.

$$x - 3 = 0$$
$$x = 3$$

Step 5. Check the solution in the original equation.

Check 3:

$$x^2 = 6x - 9$$
$$3^2 \overset{?}{=} 6 \cdot 3 - 9$$
$$9 \overset{?}{=} 18 - 9$$
$$9 = 9, \quad \text{true}$$

The solution is 3 and the solution set is {3}. ■

✓ **CHECK POINT 4** Solve: $x^2 = 10x - 25$.

EXAMPLE 5 Solving a Quadratic Equation by Factoring

Solve: $9x^2 = 16$.

Solution

Step 1. Move all terms to one side and obtain zero on the other side. Subtract 16 from both sides and write the equation in standard form.

$$9x^2 - 16 = 16 - 16$$
$$9x^2 - 16 = 0$$

Step 2. Factor. The binomial on the left side is the difference of two squares: $9x^2 - 16 = (3x)^2 - 4^2$. We factor using $A^2 - B^2 = (A + B)(A - B)$: $A = 3x$ and $B = 4$.

$$9x^2 - 16 = 0$$
$$(3x + 4)(3x - 4) = 0$$

Steps 3 and 4. Set each factor equal to zero and solve the resulting equations. We use the zero-product principle to solve $(3x + 4)(3x - 4) = 0$.

$$3x + 4 = 0 \quad \text{or} \quad 3x - 4 = 0$$
$$3x = -4 \qquad\qquad 3x = 4$$
$$x = -\frac{4}{3} \qquad\qquad x = \frac{4}{3}$$

Step 5. Check the solutions in the original equation. Do this now and verify that the solutions of $9x^2 = 16$ are $-\frac{4}{3}$ and $\frac{4}{3}$. The equation's solution set is $\left\{-\frac{4}{3}, \frac{4}{3}\right\}$. ■

✓ **CHECK POINT 5** Solve: $16x^2 = 25$.

EXAMPLE 6 Solving a Quadratic Equation by Factoring

Solve: $(x - 2)(x + 3) = 6$.

Solution

Step 1. Move all terms to one side and obtain zero on the other side. We write $(x - 2)(x + 3) = 6$ in standard form by multiplying out the product on the left side and then subtracting 6 from both sides.

$$(x - 2)(x + 3) = 6 \qquad \text{This is the given equation.}$$
$$x^2 + 3x - 2x - 6 = 6 \qquad \text{Use the FOIL method.}$$
$$x^2 + x - 6 = 6 \qquad \text{Simplify.}$$
$$x^2 + x - 6 - 6 = 6 - 6 \qquad \text{Subtract 6 from both sides.}$$
$$x^2 + x - 12 = 0 \qquad \text{Simplify.}$$

Step 2. Factor.

$$x^2 + x - 12 = 0$$
$$(x + 4)(x - 3) = 0$$

Steps 3 and 4. Set each factor equal to zero and solve the resulting equations.

$$x + 4 = 0 \quad \text{or} \quad x - 3 = 0$$
$$x = -4 \qquad\qquad x = 3$$

Step 5. Check the solutions in the original equation. Do this now and verify that the solutions are -4 and 3. The equation's solution set is $\{-4, 3\}$. ∎

✓ **CHECK POINT 6** Solve: $(x - 5)(x - 2) = 28$.

---

3 Solve problems using quadratic equations.

Applications of Quadratic Equations

Solving quadratic equations by factoring can be used to answer questions about variables contained in mathematical models.

EXAMPLE 7 Modeling Motion

You throw a ball straight up from a rooftop 160 feet high with an initial speed of 48 feet per second. The formula

$$h = -16t^2 + 48t + 160$$

describes the ball's height above the ground, h, in feet, t seconds after you throw it. The ball misses the rooftop on its way down and eventually strikes the ground. The situation is illustrated in **Figure 6.1**. How long will it take for the ball to hit the ground?

Solution The ball hits the ground when h, its height above the ground, is 0 feet. Thus, we substitute 0 for h in the given formula and solve for t.

Figure 6.1

$$h = -16t^2 + 48t + 160 \qquad \text{This is the formula that models the ball's height.}$$
$$0 = -16t^2 + 48t + 160 \qquad \text{Substitute 0 for } h.$$
$$0 = -16(t^2 - 3t - 10) \qquad \text{Factor out the negative of the GCF.}$$
$$0 = -16(t - 5)(t + 2) \qquad \text{Factor the trinomial.}$$

Do not set the constant, -16, equal to zero: $-16 \neq 0$.

$$t - 5 = 0 \text{ or } t + 2 = 0 \qquad \text{Set each variable factor equal to 0.}$$
$$t = 5 \qquad\quad t = -2 \qquad \text{Solve for } t.$$

Because we begin describing the ball's height at $t = 0$, we discard the solution $t = -2$. The ball hits the ground after 5 seconds. ∎

Great Question!

You began solving $0 = -16t^2 + 48t + 160$ by factoring out -16, the negative of the GCF. What happens if I begin by multiplying both sides of the equation by -1?

You will still get the same solutions, 5 and -2. Here's how it works:

$-1 \cdot 0 = -1(-16t^2 + 48t + 160)$	Multiply both sides of $0 = -16t^2 + 48t + 160$ by -1.
$0 = 16t^2 - 48t - 160$	Multiply each term on the right side by -1.
$0 = 16(t^2 - 3t - 10)$	Factor out the GCF, 16.
$0 = 16(t - 5)(t + 2)$	Factor the trinomial.
$t - 5 = 0 \text{ or } t + 2 = 0$	Set each variable factor equal to O.
$t = 5 \qquad t = -2.$	Solve for t.

Figure 6.2 shows the graph of the formula $h = -16t^2 + 48t + 160$. The horizontal axis is labeled t, for the ball's time in motion. The vertical axis is labeled h, for the ball's height above the ground. Because time and height are both positive, the model is graphed in quadrant I and its boundaries only.

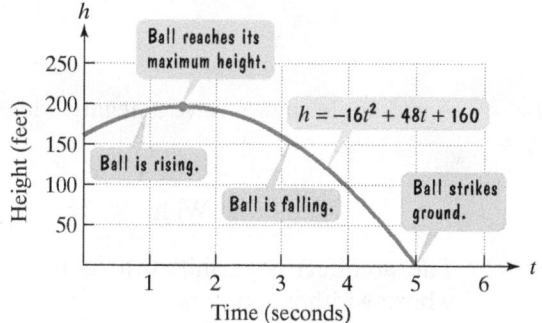

Figure 6.2

The graph visually shows what we discovered algebraically: The ball hits the ground after 5 seconds. The graph also reveals that the ball reaches its maximum height, nearly 200 feet, after 1.5 seconds. Then the ball begins to fall.

✓ **CHECK POINT 7** Use the formula $h = -16t^2 + 48t + 160$ to determine when the ball's height is 192 feet. Identify your solutions as points on the graph in **Figure 6.2**.

In our next example, we use our five-step strategy for solving word problems.

EXAMPLE 8 Solving a Problem About a Rectangle's Area

An architect is allowed no more than 15 square meters to add a small bedroom to a house. Because of the room's design in relationship to the existing structure, the width of its rectangular floor must be 7 meters less than two times the length. Find the precise length and width of the rectangular floor of maximum area that the architect is permitted.

Solution

Step 1. Let x represent one of the quantities. We know something about the width: It must be 7 meters less than two times the length. We will let

$$x = \text{the length of the floor.}$$

Step 2. Represent other unknown quantities in terms of x. Because the width must be 7 meters less than two times the length, let

$$2x - 7 = \text{the width of the floor.}$$

The problem is illustrated in **Figure 6.3**.

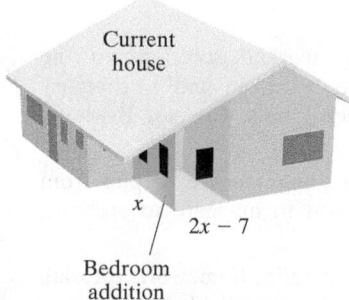

Figure 6.3

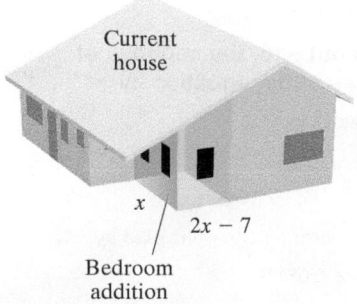

Current house

x

$2x - 7$

Bedroom addition

Figure 6.3 (repeated)

Step 3. Write an equation that models the conditions. Because the architect is allowed no more than 15 square meters, an area of 15 square meters is the maximum area permitted. The area of a rectangle is the product of its length and its width.

Length of the floor	times	Width of the floor	is	the area.
x	$\cdot$	$(2x - 7)$	$=$	15

Step 4. Solve the equation and answer the question.

$$x(2x - 7) = 15 \quad \text{This is the equation that models the problem's conditions.}$$
$$2x^2 - 7x = 15 \quad \text{Use the distributive property.}$$
$$2x^2 - 7x - 15 = 0 \quad \text{Subtract 15 from both sides.}$$
$$(2x + 3)(x - 5) = 0 \quad \text{Factor.}$$
$$2x + 3 = 0 \quad \text{or} \quad x - 5 = 0 \quad \text{Set each factor equal to zero.}$$
$$2x = -3 \qquad\qquad x = 5 \quad \text{Solve the resulting equations.}$$
$$x = -\frac{3}{2}$$

A rectangle cannot have a negative length. Thus,

$$\text{Length} = x = 5$$
$$\text{Width} = 2x - 7 = 2 \cdot 5 - 7 = 10 - 7 = 3.$$

The architect is permitted a room of maximum area whose length is 5 meters and whose width is 3 meters.

Step 5. Check the proposed solution in the original wording of the problem. The area of the floor using the dimensions that we found is

$$A = lw = (5 \text{ meters})(3 \text{ meters}) = 15 \text{ square meters.}$$

Because the problem's wording tells us that the maximum area permitted is 15 square meters, our dimensions are correct. ■

✓ **CHECK POINT 8** The length of a rectangular sign is 3 feet longer than the width. If the sign's area is 54 square feet, find its length and width.

Achieving Success

Be sure to use the Chapter Test Prep on YouTube for each chapter test. The Chapter Test Prep videos provide step-by-step solutions to every exercise in the test and let you review any exercises you miss.

Are you using any of the other textbook supplements for help and additional study? These include:

- The Student Solutions Manual. This contains fully worked solutions to the odd-numbered section exercises plus all Check Points, Concept and Vocabulary Checks, Review/Preview Exercises, Mid-Chapter Check Points, Chapter Reviews, Chapter Tests, and Cumulative Reviews.

- Lecture Videos on DVD. These interactive lectures highlight key examples from every section of the textbook. The interface allows you to navigate to sections, objectives, and examples.

- MyMathLab is a text-specific online course. Math XL is an online homework, tutorial, and assessment system. Ask your instructor whether these are available to you.

CONCEPT AND VOCABULARY CHECK

Fill in each blank so that the resulting statement is true.

1. An equation that can be written in the standard form $ax^2 + bx + c = 0, a \neq 0$, is called a/an _____.

2. The zero-product principle states that if $AB = 0$, then _____.

3. The solutions of $ax^2 + bx + c = 0$ correspond to the _____ for the graph of $y = ax^2 + bx + c$.

4. The equation $3x^2 = 5x$ can be written in standard form by _____ on both sides.

5. The equation $9x^2 = 30x - 25$ can be written in standard form by _____ and _____ on both sides.

6.6 EXERCISE SET

MyMathLab®

Watch the videos in MyMathLab

Download the MyDashBoard App

Practice Exercises

In Exercises 1–8, solve each equation using the zero-product principle.

1. $x(x + 7) = 0$
2. $x(x - 3) = 0$
3. $(x - 6)(x + 4) = 0$
4. $(x - 3)(x + 8) = 0$
5. $(x - 9)(5x + 4) = 0$
6. $(x + 7)(3x - 2) = 0$
7. $10(x - 4)(2x + 9) = 0$
8. $8(x - 5)(3x + 11) = 0$

In Exercises 9–56, use factoring to solve each quadratic equation. Check by substitution or by using a graphing utility and identifying x-intercepts.

9. $x^2 + 8x + 15 = 0$
10. $x^2 + 5x + 6 = 0$
11. $x^2 - 2x - 15 = 0$
12. $x^2 + x - 42 = 0$
13. $x^2 - 4x = 21$
14. $x^2 + 7x = 18$
15. $x^2 + 9x = -8$
16. $x^2 - 11x = -10$
17. $x^2 + 4x = 0$
18. $x^2 - 6x = 0$
19. $x^2 - 5x = 0$
20. $x^2 + 3x = 0$
21. $x^2 = 4x$
22. $x^2 = 8x$
23. $2x^2 = 5x$
24. $3x^2 = 5x$
25. $3x^2 = -5x$
26. $2x^2 = -3x$
27. $x^2 + 4x + 4 = 0$
28. $x^2 + 6x + 9 = 0$
29. $x^2 = 12x - 36$
30. $x^2 = 14x - 49$
31. $4x^2 = 12x - 9$
32. $9x^2 = 30x - 25$
33. $2x^2 = 7x + 4$
34. $3x^2 = x + 4$
35. $5x^2 = 18 - x$
36. $3x^2 = 15 + 4x$
37. $x^2 - 49 = 0$
38. $x^2 - 25 = 0$
39. $4x^2 - 25 = 0$
40. $9x^2 - 100 = 0$
41. $81x^2 = 25$
42. $25x^2 = 49$
43. $x(x - 4) = 21$

44. $x(x - 3) = 18$
45. $4x(x + 1) = 15$
46. $x(3x + 8) = -5$
47. $(x - 1)(x + 4) = 14$
48. $(x - 3)(x + 8) = -30$
49. $(x + 1)(2x + 5) = -1$
50. $(x + 3)(3x + 5) = 7$
51. $y(y + 8) = 16(y - 1)$
52. $y(y + 9) = 4(2y + 5)$
53. $4y^2 + 20y + 25 = 0$
54. $4y^2 + 44y + 121 = 0$
55. $64w^2 = 48w - 9$
56. $25w^2 = 80w - 64$

Practice PLUS

In Exercises 57–66, solve each equation and check your solutions.

57. $(x - 4)(x^2 + 5x + 6) = 0$
58. $(x - 5)(x^2 - 3x + 2) = 0$
59. $x^3 - 36x = 0$
60. $x^3 - 4x = 0$
61. $y^3 + 3y^2 + 2y = 0$
62. $y^3 + 2y^2 - 3y = 0$
63. $2(x - 4)^2 + x^2 = x(x + 50) - 46x$
64. $(x - 4)(x - 5) + (2x + 3)(x - 1) = x(2x - 25) - 13$
65. $(x - 2)^2 - 5(x - 2) + 6 = 0$
66. $(x - 3)^2 + 2(x - 3) - 8 = 0$

Application Exercises

A ball is thrown straight up from a rooftop 300 feet high. The formula

$$h = -16t^2 + 20t + 300$$

describes the ball's height above the ground, h, in feet, t seconds after it was thrown. The ball misses the rooftop on its way down and eventually strikes the ground. The graph of the formula is shown, with tick marks omitted along the horizontal axis. Use the formula to solve Exercises 67–69.

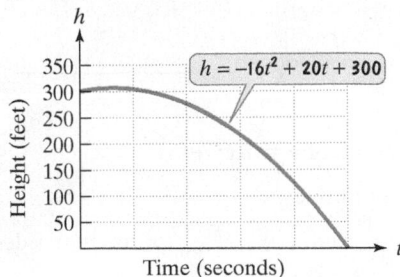

$$h = -16t^2 + 20t + 300$$

67. How long will it take for the ball to hit the ground? Use this information to provide tick marks with appropriate numbers along the horizontal axis in the figure shown.

68. When will the ball's height be 304 feet? Identify the solution as a point on the graph.

69. When will the ball's height be 276 feet? Identify the solution as a point on the graph.

An explosion causes debris to rise vertically with an initial speed of 72 feet per second. The formula

$$h = -16t^2 + 72t$$

describes the height of the debris above the ground, h, in feet, t seconds after the explosion. Use this information to solve Exercises 70–71.

70. How long will it take for the debris to hit the ground?

71. When will the debris be 32 feet above the ground?

The formula

$$S = 2x^2 - 12x + 82$$

models spending by international travelers to the United States, S, in billions of dollars, x years after 2000. Use this formula to solve Exercises 72–73.

72. In which years did international travelers spend $72 billion?

73. In which years did international travelers spend $66 billion?

The graph of the formula modeling spending by international travelers is shown below. Use the graph to solve Exercises 74–75.

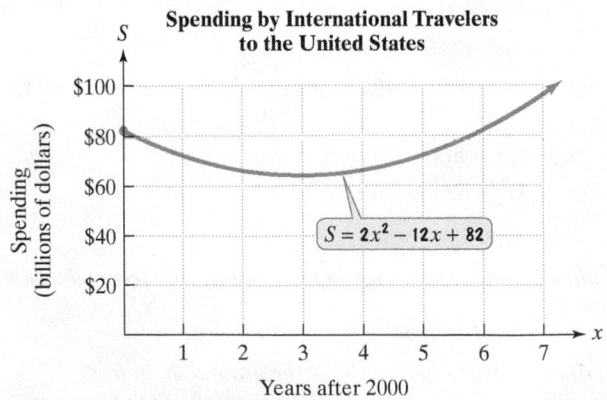

Spending by International Travelers to the United States

$$S = 2x^2 - 12x + 82$$

Years after 2000

Source: Travel Industry Association of America

74. Identify your solutions from Exercise 72 as points on the graph.

75. Identify your solutions from Exercise 73 as points on the graph.

The alligator, at one time an endangered species, is the subject of a protection program. The formula

$$P = -10x^2 + 475x + 3500$$

models the alligator population, P, after x years of the protection program, where $0 \le x \le 12$. Use the formula to solve Exercises 76–77.

76. After how long is the population up to 5990?

77. After how long is the population up to 7250?

The graph of the alligator population is shown over time. Use the graph to solve Exercises 78–79.

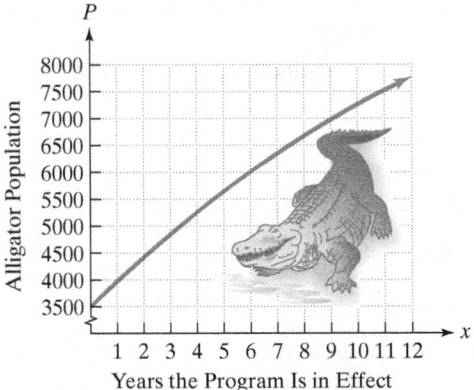

Years the Program Is in Effect

78. Identify your solution from Exercise 76 as a point on the graph.

79. Identify your solution from Exercise 77 as a point on the graph.

The formula

$$N = \frac{t^2 - t}{2}$$

describes the number of football games, N, that must be played in a league with t teams if each team is to play every other team once. Use this information to solve Exercises 80–81.

80. If a league has 36 games scheduled, how many teams belong to the league, assuming that each team plays every other team once?

81. If a league has 45 games scheduled, how many teams belong to the league, assuming that each team plays every other team once?

82. The length of a rectangular garden is 5 feet greater than the width. The area of the rectangle is 300 square feet. Find the length and the width.

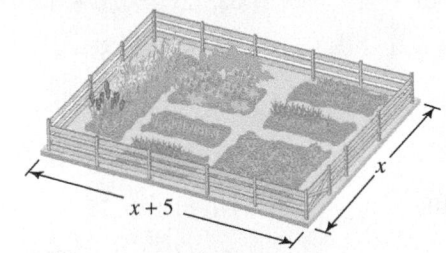

83. A rectangular parking lot has a length that is 3 yards greater than the width. The area of the parking lot is 180 square yards. Find the length and the width.

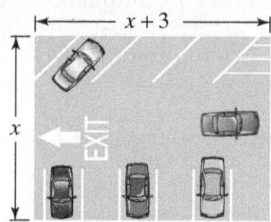

84. Each end of a glass prism is a triangle with a height that is 1 inch shorter than twice the base. If the area of the triangle is 60 square inches, how long are the base and height?

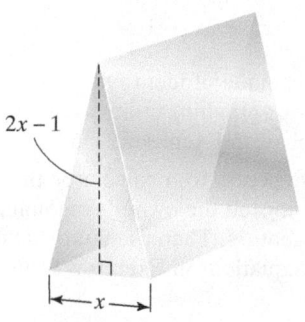

85. Great white sharks have triangular teeth with a height that is 1 centimeter longer than the base. If the area of one tooth is 15 square centimeters, find its base and height.

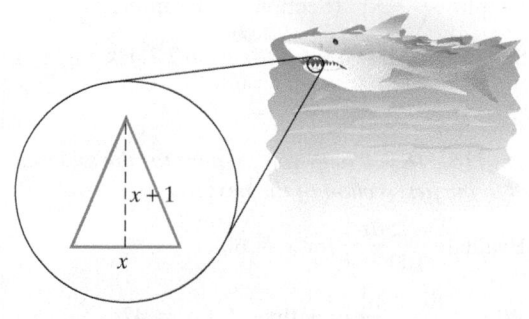

86. A vacant rectangular lot is being turned into a community vegetable garden measuring 15 meters by 12 meters. A path of uniform width is to surround the garden. If the area of the lot is 378 square meters, find the width of the path surrounding the garden.

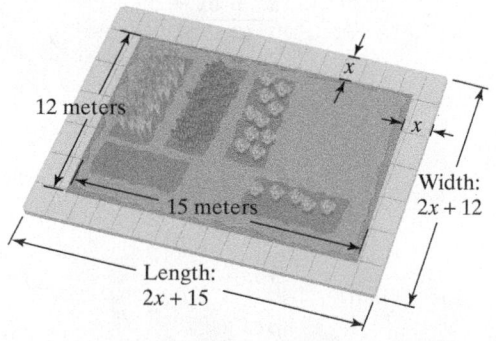

87. As part of a landscaping project, you put in a flower bed measuring 10 feet by 12 feet. You plan to surround the bed with a uniform border of low-growing plants.

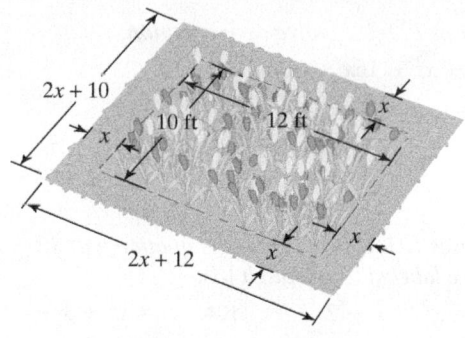

a. Write a polynomial that describes the area of the uniform border that surrounds your flower bed. (*Hint*: The area of the border is the area of the large rectangle shown in the figure minus the area of the flower bed.)

b. The low-growing plants surrounding the flower bed require 1 square foot each when mature. If you have 168 of these plants, how wide a strip around the flower bed should you prepare for the border?

Writing in Mathematics

88. What is a quadratic equation?

89. Explain how to solve $x^2 + 6x + 8 = 0$ using factoring and the zero-product principle.

90. If $(x + 2)(x - 4) = 0$ indicates that $x + 2 = 0$ or $x - 4 = 0$, explain why $(x + 2)(x - 4) = 6$ does not mean $x + 2 = 6$ or $x - 4 = 6$. Could we solve the equation using $x + 2 = 3$ and $x - 4 = 2$ because $3 \cdot 2 = 6$?

Critical Thinking Exercises

Make Sense? *In Exercises 91–94, determine whether each statement "makes sense" or "does not make sense" and explain your reasoning.*

91. When solving $4(x - 3)(x + 2) = 0$ and $4x(x + 2) = 0$, I can ignore the monomial factors.

92. I set the quadratic equation $2x^2 - 5x = 12$ equal to zero and obtained $2x^2 - 5x = 0$.

93. Because some trinomials are prime, some quadratic equations cannot be solved by factoring.

94. I'm looking at a graph with one x-intercept, so it must be the graph of a linear equation.

In Exercises 95–98, determine whether each statement is true or false. If the statement is false, make the necessary change(s) to produce a true statement.

95. If $(x + 3)(x - 4) = 2$, then $x + 3 = 0$ or $x - 4 = 0$.

96. The solutions of the equation $4(x - 5)(x + 3) = 0$ are 4, 5, and -3.

97. Equations solved by factoring always have two different solutions.

98. Both 0 and $-\pi$ are solutions of the equation $x(x + \pi) = 0$.

99. Write a quadratic equation in standard form whose solutions are -3 and 5.

In Exercises 100–102, solve each equation.

100. $x^3 - x^2 - 16x + 16 = 0$

101. $3^{x^2 - 9x + 20} = 1$

102. $(x^2 - 5x + 5)^3 = 1$

In Exercises 103–106, match each equation with its graph. The graphs are labeled (a) through (d).

103. $y = x^2 - x - 2$ **104.** $y = x^2 + x - 2$

105. $y = x^2 - 4$ **106.** $y = x^2 - 4x$

a.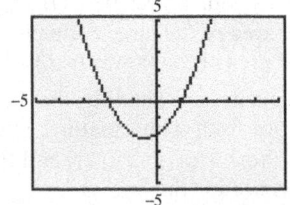

b.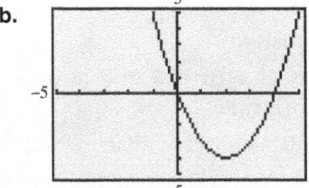

c.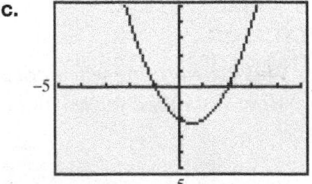

d.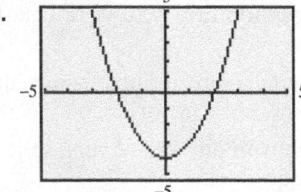

Technology Exercises

In Exercises 107–110, use the x-intercepts for the graph in a $[-10, 10, 1]$ by $[-13, 10, 1]$ viewing rectangle to solve the quadratic equation. Check by substitution.

107. Use the graph of $y = x^2 + 3x - 4$ to solve
$$x^2 + 3x - 4 = 0.$$

108. Use the graph of $y = x^2 + x - 6$ to solve
$$x^2 + x - 6 = 0.$$

109. Use the graph of $y = (x - 2)(x + 3) - 6$ to solve
$$(x - 2)(x + 3) - 6 = 0.$$

110. Use the graph of $y = x^2 - 2x + 1$ to solve
$$x^2 - 2x + 1 = 0.$$

111. Use the technique of identifying x-intercepts on a graph generated by a graphing utility to check any five equations that you solved in Exercises 9–56.

112. If you have access to a calculator that solves quadratic equations, consult the owner's manual to determine how to use this feature. Then use your calculator to solve any five of the equations in Exercises 9–56.

Review Exercises

113. Graph: $y = -\dfrac{2}{3}x + 1$. (Section 3.4, Example 3)

114. Simplify: $\left(\dfrac{8x^4}{4x^7}\right)^2$. (Section 5.7, Example 6)

115. Solve: $5x + 28 = 6 - 6x$. (Section 2.2, Example 7)

Preview Exercises

Exercises 116–118 will help you prepare for the material covered in the first section of the next chapter.

116. Evaluate $\dfrac{250x}{100 - x}$ for $x = 60$.

117. Why is $\dfrac{6x + 12}{7x - 28}$ undefined for $x = 4$?

118. Factor the numerator and the denominator. Then simplify by dividing out the common factor in the numerator and the denominator.
$$\frac{x^2 + 6x + 5}{x^2 - 25}$$

GROUP PROJECT

CHAPTER 6

Group members are on the board of a condominium association. The condominium has just installed a 35-foot-by-30-foot pool. Your job is to choose a material to surround the pool to create a border of uniform width.

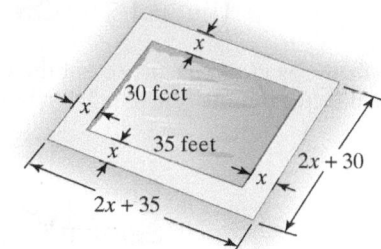

a. Begin by writing an algebraic expression for the area, in square feet, of the border around the pool. (*Hint*: The border's area is the combined area of the pool and border minus the area of the pool.)

b. You must select one of the following options for the border.

Options for the Border	Price
Cement	$6 per square foot
Outdoor carpeting	$5 per square foot plus $10 per foot to install edging around the rectangular border
Brick	$8 per square foot plus a $60 charge for delivering the bricks

Write an algebraic expression for the cost of installing the border for each of these options.

c. You would like the border to be 5 feet wide. Use the algebraic expressions in part (b) to find the cost of the border for each of the three options.

d. You would prefer not to use cement. However, the condominium association is limited by a $5000 budget. Given this limitation, approximately how wide can the border be using outdoor carpeting or brick? Which option should you select and why?

Chapter 6 Summary

Definitions and Concepts	Examples

Section 6.1 The Greatest Common Factor and Factoring by Grouping

Factoring a polynomial containing the sum of monomials means finding an equivalent expression that is a product. The greatest common factor, GCF, is an expression that divides every term of the polynomial. The variable part of the GCF contains the smallest power of a variable that appears in all terms of the polynomial.

Find the GCF of $16x^2y$, $20x^3y^2$, and $8x^2y^3$.

The GCF of 16, 20, and 8 is 4.

The GCF of x^2, x^3, and x^2 is x^2.

The GCF of y, y^2, and y^3 is y.

$$\text{GCF} = 4 \cdot x^2 \cdot y = 4x^2y$$

To factor a monomial from a polynomial, express each term as the product of the GCF and its other factor. Then use the distributive property to factor out the GCF.

$$16x^2y + 20x^3y^2 + 8x^2y^3$$
$$= 4x^2y \cdot 4 + 4x^2y \cdot 5xy + 4x^2y \cdot 2y^2$$
$$= 4x^2y(4 + 5xy + 2y^2)$$

To factor a monomial from a polynomial with a negative coefficient in the first term, express each term as the product of the negative of the GCF and its other factor. Then use the distributive property to factor out the negative of the GCF.

$$-20x^4y^3 + 10x^2y^2 - 15x^3y$$
$$= -5x^2y \cdot 4x^2y^2 - 5x^2y(-2y) - 5x^2y \cdot 3x$$

The negative of the GCF is $-5x^2y$.

$$= -5x^2y(4x^2y^2 - 2y + 3x)$$

Definitions and Concepts	**Examples**

Section 6.1 The Greatest Common Factor and Factoring by Grouping (continued)

To factor by grouping, factor out the GCF from each group. Then factor out the remaining common factor.

$$xy + 5x - 3y - 15$$
$$= x(y + 5) - 3(y + 5)$$
$$= (y + 5)(x - 3)$$

Section 6.2 Factoring Trinomials Whose Leading Coefficient Is 1

To factor a trinomial of the form $x^2 + bx + c$, find two numbers whose product is c and whose sum is b. The factorization is

$$(x + \text{one number})(x + \text{other number}).$$

Factor: $x^2 + 9x + 20$.

Find two numbers whose product is 20 and whose sum is 9. The numbers are 4 and 5.

$$x^2 + 9x + 20 = (x + 4)(x + 5)$$

Section 6.3 Factoring Trinomials Whose Leading Coefficient Is Not 1

To factor $ax^2 + bx + c$ by trial and error, try various combinations of factors of ax^2 and c until a middle term of bx is obtained for the sum of outside and inside products.

Factor: $3x^2 + 7x - 6$.

Factors of $3x^2$: $3x, x$

Factors of -6: 1 and -6, -1 and 6, 2 and -3, -2 and 3

A possible combination of these factors is

$$(3x - 2)(x + 3).$$

Sum of outside and inside products should equal $7x$.

$$9x - 2x = 7x$$

Thus, $3x^2 + 7x - 6 = (3x - 2)(x + 3)$.

To factor $ax^2 + bx + c$ by grouping, find the factors of ac whose sum is b. Write bx using these factors. Then factor by grouping.

Factor: $3x^2 + 7x - 6$.

Find the factors of $3(-6)$, or -18, whose sum is 7. They are 9 and -2.

$$3x^2 + 7x - 6$$
$$= 3x^2 + 9x - 2x - 6$$
$$= 3x(x + 3) - 2(x + 3) = (x + 3)(3x - 2)$$

Section 6.4 Factoring Special Forms

The Difference of Two Squares
$$A^2 - B^2 = (A + B)(A - B)$$

$$9x^2 - 25y^2$$
$$= (3x)^2 - (5y)^2 = (3x + 5y)(3x - 5y)$$

Perfect Square Trinomials
$$A^2 + 2AB + B^2 = (A + B)^2$$
$$A^2 - 2AB + B^2 = (A - B)^2$$

$$x^2 + 16x + 64 = x^2 + 2 \cdot x \cdot 8 + 8^2 = (x + 8)^2$$
$$25x^2 - 30x + 9 = (5x)^2 - 2 \cdot 5x \cdot 3 + 3^2 = (5x - 3)^2$$

Sum or Difference of Cubes
$$A^3 + B^3 = (A + B)(A^2 - AB + B^2)$$
$$A^3 - B^3 = (A - B)(A^2 + AB + B^2)$$

$$8x^3 - 125 = (2x)^3 - 5^3$$
$$= (2x - 5)[(2x)^2 + 2x \cdot 5 + 5^2]$$
$$= (2x - 5)(4x^2 + 10x + 25)$$

| **Definitions and Concepts** | **Examples** |

Section 6.5 A General Factoring Strategy

A Factoring Strategy

1. Factor out the GCF.

2. a. If two terms, try

$$A^2 - B^2 = (A + B)(A - B)$$
$$A^3 + B^3 = (A + B)(A^2 - AB + B^2)$$
$$A^3 - B^3 = (A - B)(A^2 + AB + B^2).$$

b. If three terms, try

$$A^2 + 2AB + B^2 = (A + B)^2$$
$$A^2 - 2AB + B^2 = (A - B)^2.$$

If not a perfect square trinomial, try trial and error or grouping.

c. If four terms, try factoring by grouping.

3. See if any factors can be factored further.

4. Check by multiplying.

Factor: $2x^4 + 10x^3 - 8x^2 - 40x$.
The GCF is $2x$.

$$2x^4 + 10x^3 - 8x^2 - 40x$$
$$= 2x(x^3 + 5x^2 - 4x - 20)$$

> Four terms: Try grouping.

$$= 2x[x^2(x + 5) - 4(x + 5)]$$
$$= 2x(x + 5)(x^2 - 4)$$

> This can be factored further.

$$= 2x(x + 5)(x + 2)(x - 2)$$

Section 6.6 Solving Quadratic Equations by Factoring

The Zero-Product Principle

If $AB = 0$, then $A = 0$ or $B = 0$.

Solve: $(x - 6)(x + 10) = 0$.
$$x - 6 = 0 \quad \text{or} \quad x + 10 = 0$$
$$x = 6 \qquad\qquad x = -10$$
The solutions are -10 and 6, and the solution set is $\{-10, 6\}$.

A quadratic equation in x is an equation that can be written in the standard form

$$ax^2 + bx + c = 0, \quad a \neq 0.$$

To solve by factoring, write the equation in standard form, factor, set each factor equal to zero, and solve each resulting equation. Check proposed solutions in the original equation.

Solve: $4x^2 + 9x = 9$.
$$4x^2 + 9x - 9 = 0$$
$$(4x - 3)(x + 3) = 0$$
$$4x - 3 = 0 \quad \text{or} \quad x + 3 = 0$$
$$x = \frac{3}{4} \qquad\qquad x = -3$$
The solutions are -3 and $\frac{3}{4}$, and the solution set is $\left\{-3, \frac{3}{4}\right\}$.

CHAPTER 6 REVIEW EXERCISES

6.1 *In Exercises 1–5, factor each polynomial using the greatest common factor. If there is no common factor other than 1 and the polynomial cannot be factored, so state.*

1. $30x - 45$

2. $-12x^3 - 16x^2 + 400x$

3. $30x^4y + 15x^3y + 5x^2y$

4. $7(x + 3) - 2(x + 3)$

5. $7x^2(x + y) - (x + y)$

In Exercises 6–9, factor by grouping.

6. $x^3 + 3x^2 + 2x + 6$

7. $xy + y + 4x + 4$

8. $x^3 + 5x + x^2 + 5$

9. $xy + 4x - 2y - 8$

6.2 *In Exercises 10–17, factor completely, or state that the trinomial is prime.*

10. $x^2 - 3x + 2$

11. $x^2 - x - 20$

12. $x^2 + 19x + 48$

13. $x^2 - 6xy + 8y^2$

14. $x^2 + 5x - 9$

15. $x^2 + 16xy - 17y^2$

16. $3x^2 + 6x - 24$

17. $3x^3 - 36x^2 + 33x$

6.3 *In Exercises 18–26, factor completely, or state that the trinomial is prime.*

18. $3x^2 + 17x + 10$

19. $5y^2 - 17y + 6$

20. $4x^2 + 4x - 15$

21. $5y^2 + 11y + 4$

22. $-8x^2 - 8x + 6$

23. $2x^3 + 7x^2 - 72x$

24. $12y^3 + 28y^2 + 8y$

25. $2x^2 - 7xy + 3y^2$

26. $5x^2 - 6xy - 8y^2$

6.4 *In Exercises 27–30, factor each difference of two squares completely.*

27. $4x^2 - 1$

28. $81 - 100y^2$

29. $25a^2 - 49b^2$

30. $z^4 - 16$

In Exercises 31–34, factor completely, or state that the polynomial is prime.

31. $2x^2 - 18$ 32. $x^2 + 1$

33. $9x^3 - x$

34. $18xy^2 - 8x$

In Exercises 35–41, factor any perfect square trinomials, or state that the polynomial is prime.

35. $x^2 + 22x + 121$

36. $x^2 - 16x + 64$ 37. $9y^2 + 48y + 64$

38. $16x^2 - 40x + 25$

39. $25x^2 + 15x + 9$

40. $36x^2 + 60xy + 25y^2$

41. $25x^2 - 40xy + 16y^2$

In Exercises 42–45, factor using the formula for the sum or difference of two cubes.

42. $x^3 - 27$

43. $64x^3 + 1$

44. $54x^3 - 16y^3$

45. $27x^3y + 8y$

In Exercises 46–47, find the formula for the area of the blue shaded region and express it in factored form.

46.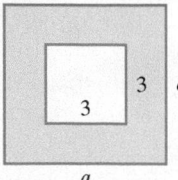

47.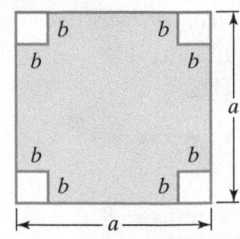

48. The figure shows a geometric interpretation of a factorization. Use the sum of the areas of the four pieces on the left and the area of the square on the right to write the factorization that is illustrated.

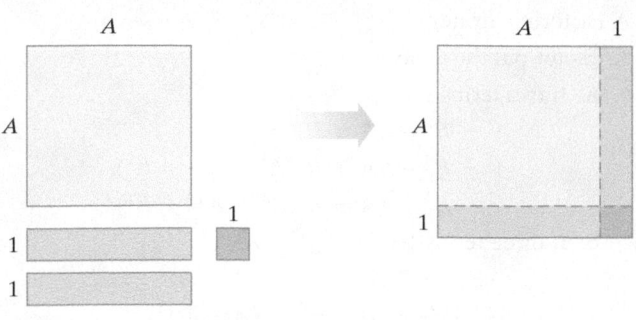

6.5 *In Exercises 49–81, factor completely, or state that the polynomial is prime.*

49. $x^3 - 8x^2 + 7x$

50. $10y^2 + 9y + 2$

51. $128 - 2y^2$

52. $9x^2 + 6x + 1$

53. $-20x^7 + 36x^3$

54. $x^3 - 3x^2 - 9x + 27$

55. $y^2 + 16$

56. $2x^3 + 19x^2 + 35x$

57. $3x^3 - 30x^2 + 75x$

58. $3x^5 - 24x^2$

59. $4y^4 - 36y^2$

60. $5x^2 + 20x - 105$

61. $9x^2 + 8x - 3$

62. $-10x^5 + 44x^4 - 16x^3$

63. $100y^2 - 49$

64. $9x^5 - 18x^4$

65. $x^4 - 1$

66. $2y^3 - 16$

67. $x^3 + 64$

68. $6x^2 + 11x - 10$

69. $3x^4 - 12x^2$

70. $x^2 - x - 90$

71. $25x^2 + 25xy + 6y^2$

72. $x^4 + 125x$

73. $32y^3 + 32y^2 + 6y$

74. $-2y^2 + 16y - 32$

75. $x^2 - 2xy - 35y^2$

76. $x^2 + 7x + xy + 7y$

77. $9x^2 + 24xy + 16y^2$

78. $2x^4y - 2x^2y$

79. $100y^2 - 49z^2$

80. $x^2 + xy + y^2$

81. $3x^4y^2 - 12x^2y^4$

6.6 *In Exercises 82–83, solve each equation using the zero-product principle.*

82. $x(x - 12) = 0$

83. $3(x - 7)(4x + 9) = 0$

In Exercises 84–92, use factoring to solve each quadratic equation.

84. $x^2 + 5x - 14 = 0$

85. $5x^2 + 20x = 0$

86. $2x^2 + 15x = 8$

87. $x(x - 4) = 32$

88. $(x + 3)(x - 2) = 50$

89. $x^2 = 14x - 49$

90. $9x^2 = 100$

91. $3x^2 + 21x + 30 = 0$

92. $3x^2 = 22x - 7$

93. You dive from a board that is 32 feet above the water. The formula

$$h = -16t^2 + 16t + 32$$

describes your height above the water, h, in feet, t seconds after you dive. How long will it take you to hit the water?

94. The length of a rectangular sign is 3 feet longer than the width. If the sign has space for 40 square feet of advertising, find its length and its width.

95. The square lot shown here is being turned into a garden with a 3-meter path at one end. If the area of the garden is 88 square meters, find the dimensions of the square lot.

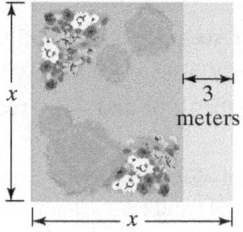

CHAPTER 6 TEST

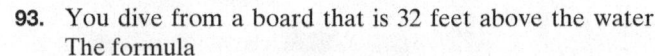

Step-by-step test solutions are found on the Chapter Test Prep Videos available in MyMathLab® or on YouTube (search "BlitzerCombinedAlg" and click on "Channels").

In Exercises 1–21, factor completely, or state that the polynomial is prime.

1. $x^2 - 9x + 18$

2. $x^2 - 14x + 49$

3. $15y^4 - 35y^3 + 10y^2$

4. $x^3 + 2x^2 + 3x + 6$

5. $x^2 - 9x$

6. $x^3 + 6x^2 - 7x$

7. $14x^2 + 64x - 30$

8. $25x^2 - 9$ **9.** $x^3 + 8$

10. $x^2 - 4x - 21$

11. $x^2 + 4$

12. $6y^3 + 9y^2 + 3y$

13. $4y^2 - 36$

14. $16x^2 + 48x + 36$

15. $2x^4 - 32$

16. $36x^2 - 84x + 49$

17. $7x^2 - 50x + 7$

18. $x^3 + 2x^2 - 5x - 10$

19. $-12y^3 + 12y^2 + 45y$

20. $y^3 - 125$

21. $5x^2 - 5xy - 30y^2$

In Exercises 22–27, solve each quadratic equation.

22. $x^2 + 2x - 24 = 0$

23. $3x^2 - 5x = 2$

24. $x(x - 6) = 16$

25. $6x^2 = 21x$

26. $16x^2 = 81$

27. $(5x + 4)(x - 1) = 2$

28. Find a formula for the area of the shaded blue region and express it in factored form.

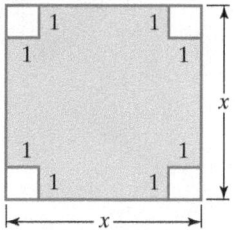

29. A model rocket is launched from a height of 96 feet. The formula

$$h = -16t^2 + 80t + 96$$

describes the rocket's height, h, in feet, t seconds after it was launched. How long will it take the rocket to reach the ground?

30. The length of a rectangular garden is 6 feet longer than its width. If the area of the garden is 55 square feet, find its length and its width.

CUMULATIVE REVIEW EXERCISES (CHAPTERS 1–6)

1. Simplify: $6[5 + 2(3 - 8) - 3]$.
2. Solve: $4(x - 2) = 2(x - 4) + 3x$.

3. Solve: $\dfrac{x}{2} - 1 = \dfrac{x}{3} + 1$.

4. Solve and graph the solution set on a number line.

$$5 - 5x > 2(5 - x) + 1$$

5. Find the measures of the angles of a triangle whose two base angles have equal measure and whose third angle is 10° less than three times the measure of a base angle.

6. A dinner for six people cost $159, including a 6% tax. What was the dinner's cost before tax?

7. Graph using the slope and y-intercept: $y = -\frac{3}{5}x + 3$.

8. Write the point-slope form of the equation of the line passing through $(2, -4)$ and $(3, 1)$. Then use the point-slope form of the equation to write the slope-intercept form of the equation.

9. Solve the system:

$$\begin{cases} x - 2y + 2z = 4 \\ 3x - y + 4z = 4 \\ 2x + y - 3z = 5. \end{cases}$$

10. Solve the system:

$$\begin{cases} 5x + 2y = 13 \\ y = 2x - 7. \end{cases}$$

11. Solve the system:

$$\begin{cases} 2x + 3y = 5 \\ 3x - 2y = -4. \end{cases}$$

12. Subtract: $\dfrac{4}{5} - \dfrac{9}{8}$.

In Exercises 13–15, perform the indicated operations.

13. $\dfrac{6x^5 - 3x^4 + 9x^2 + 27x}{3x}$

14. $(3x - 5y)(2x + 9y)$

15. $\dfrac{6x^3 + 5x^2 - 34x + 13}{3x - 5}$

16. Write 0.0071 in scientific notation.

In Exercises 17–19, factor completely.

17. $3x^2 + 11x + 6$

18. $y^5 - 16y$

19. $4x^2 + 12x + 9$

20. The length of a rectangle is 2 feet greater than its width. If the rectangle's area is 24 square feet, find its dimensions.

CHAPTER

7

Rational Expressions

We are plagued by questions about the environment. Will we run out of oil? How hot will it get? Will there be neighborhoods where the air is pristine? Can we make garbage disappear? Will there be any wilderness left? Which wild animals will become extinct? How much will it cost to clean up toxic wastes from our rivers so that they can safely provide food, recreation, and enjoyment of wildlife for the millions who live along and visit their banks?

When making decisions on public policies dealing with the environment, algebraic fractions play an important role in modeling the costs. By learning to work with these fractional expressions, you will gain new insights into phenomena as diverse as the dosage of drugs prescribed for children, inventory costs for a business, the cost of environmental cleanup, and even the shape of our heads.

Here's where you'll find these applications:

- Children and drug dosage: Exercise Set 7.4, Exercises 93–100
- Inventory costs for a business: Exercise Set 7.6, Exercises 75–76
- Costs of environmental cleanup: Section 7.1 opener; Section 7.1, page 496; Exercise Set 7.1, Exercises 85–86; Section 7.6, Example 6; Exercise Set 7.6, Exercises 71–72
- Shape of our heads: Section 7.3 opener; Exercise Set 7.3, Exercise 73.

489

SECTION

7.1

Rational Expressions and Their Simplification

Objectives

1 Find numbers for which a rational expression is undefined.

2 Simplify rational expressions.

3 Solve applied problems involving rational expressions.

How do we describe the costs of reducing environmental pollution? We often use algebraic expressions involving quotients of polynomials. For example, the algebraic expression

$$\frac{250x}{100 - x}$$

Discover for Yourself

What happens if you try substituting 100 for x in

$$\frac{250x}{100 - x}?$$

What does this tell you about the cost of cleaning up all of the river's pollutants?

describes the cost, in millions of dollars, to remove x percent of the pollutants that are discharged into a river. Removing a modest percentage of pollutants, say 40%, is far less costly than removing a substantially greater percentage, such as 95%. We see this by evaluating the algebraic expression for $x = 40$ and $x = 95$.

$$\text{Evaluating } \frac{250x}{100 - x} \text{ for}$$

$x = 40$:

Cost is $\frac{250(40)}{100 - 40} \approx 167.$

$x = 95$:

Cost is $\frac{250(95)}{100 - 95} = 4750.$

The cost increases from approximately \$167 million to a possibly prohibitive \$4750 million, or \$4.75 billion. Costs spiral upward as the percentage of removed pollutants increases.

Many algebraic expressions that describe costs of environmental projects are examples of *rational expressions*. In this section, we introduce rational expressions and their simplification.

1 Find numbers for which a rational expression is undefined.

Excluding Numbers from Rational Expressions

A **rational expression** is the quotient of two polynomials. Some examples are

$$\frac{x - 2}{4}, \quad \frac{4}{x - 2}, \quad \frac{x}{x^2 - 1}, \quad \text{and} \quad \frac{x^2 + 1}{x^2 + 2x - 3}.$$

Rational expressions indicate division and division by zero is undefined. This means that **we must exclude any value or values of the variable that make a denominator zero**. For example, consider the rational expression

$$\frac{4}{x - 2}.$$

When x is replaced with 2, the denominator is 0 and the expression is undefined.

If $x = 2$: $\frac{4}{x - 2} = \frac{4}{2 - 2} = \frac{4}{0}.$ **Division by zero is undefined.**

Notice that if x is replaced by a number other than 2, such as 1, the expression is defined because the denominator is nonzero.

$$\text{If } x = 1: \quad \frac{4}{x-2} = \frac{4}{1-2} = \frac{4}{-1} = -4.$$

Thus, only 2 must be excluded as a replacement for x in the rational expression $\dfrac{4}{x-2}$.

Using Technology

We can use the $\boxed{\text{TABLE}}$ feature of a graphing utility to verify our work with $\dfrac{4}{x-2}$.
Enter

$$y_1 = 4 \;\boxed{\div}\; \boxed{(}\; \boxed{x}\; \boxed{-}\; 2\; \boxed{)}$$

and press $\boxed{\text{TABLE}}$.

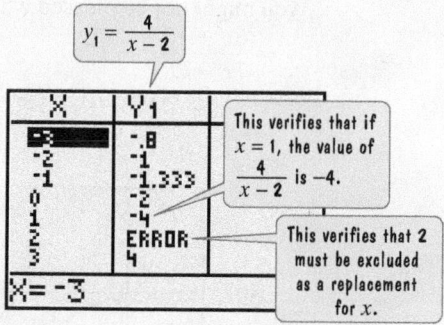

$y_1 = \dfrac{4}{x-2}$

This verifies that if $x = 1$, the value of $\dfrac{4}{x-2}$ is -4.

This verifies that 2 must be excluded as a replacement for x.

Excluding Values from Rational Expressions

If a variable in a rational expression is replaced by a number that causes the denominator to be 0, that number must be excluded as a replacement for the variable. The rational expression is undefined at any value that produces a denominator of 0.

How do we determine the value or values of the variable for which a rational expression is undefined? Set the denominator equal to 0 and then solve the resulting equation for the variable.

EXAMPLE 1   Determining Numbers for Which Rational Expressions Are Undefined

Find all the numbers for which the rational expression is undefined:

a. $\dfrac{6x + 12}{7x - 28}$ **b.** $\dfrac{2x + 6}{x^2 + 3x - 10}$.

Solution   In each case, we set the denominator equal to 0 and solve.

$$\dfrac{6x + 12}{7x - 28} \qquad \text{Exclude values of } x \text{ that make these denominators 0.} \qquad \dfrac{2x + 6}{x^2 + 3x - 10}$$

a. $7x - 28 = 0$ Set the denominator of $\dfrac{6x + 12}{7x - 28}$ equal to 0.

   $\qquad 7x = 28$ Add 28 to both sides.

   $\qquad\; x = 4$ Divide both sides by 7.

Thus, $\dfrac{6x + 12}{7x - 28}$ is undefined for $x = 4$.

b. $x^2 + 3x - 10 = 0$ Set the denominator of $\dfrac{2x + 6}{x^2 + 3x - 10}$ equal to 0.

$(x + 5)(x - 2) = 0$ Factor.

$x + 5 = 0$ or $x - 2 = 0$ Set each factor equal to 0.

$x = -5$ $x = 2$ Solve the resulting equations.

Thus, $\dfrac{2x + 6}{x^2 + 3x - 10}$ is undefined for $x = -5$ and $x = 2$.   ■

Using Technology

Graphic Connections

When using a graphing utility to graph an equation containing a rational expression, you might not be pleased with the quality of the display. Compare these two graphs of

$$y = \frac{6x + 12}{7x - 28}.$$

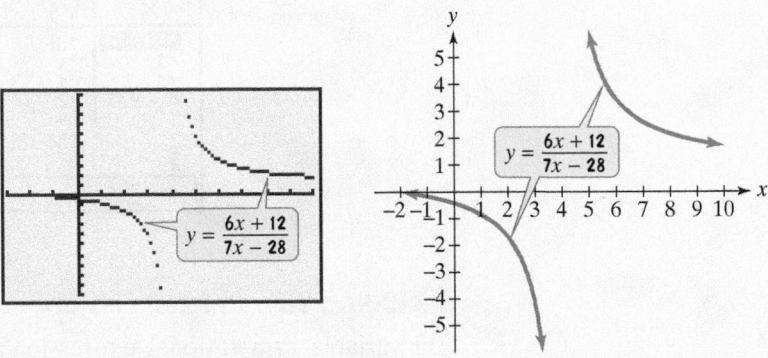

The graph on the left was obtained using the ⃞ DOT ⃞ mode in a $[-3, 10, 1]$ by $[-10, 10, 1]$ viewing rectangle. Examine the behavior of the graph near $x = 4$, the number for which the rational expression is undefined. The values of the rational expression are decreasing as the values of x get closer to 4 on the left and increasing as the values of x get closer to 4 on the right. However, there is no point on the graph corresponding to $x = 4$. Would you agree that this behavior is better illustrated in the hand-drawn graph on the right?

✓ **CHECK POINT 1**   Find all the numbers for which the rational expression is undefined:

a. $\dfrac{7x - 28}{8x - 40}$

b. $\dfrac{8x - 40}{x^2 + 3x - 28}.$

Is every rational expression undefined for at least one number? No. Consider

$$\frac{x - 2}{x^2 + 1}.$$

Because the denominator, $x^2 + 1$, is not zero for any real-number replacement for x, the rational expression is defined for all real numbers. Thus, it is not necessary to exclude any values for x.

2 Simplify rational expressions.

Simplifying Rational Expressions

A rational expression is **simplified** if its numerator and denominator have no common factors other than 1 or -1. The following principle is used to simplify a rational expression:

Fundamental Principle of Rational Expressions

If P, Q, and R are polynomials, and Q and R are not 0,

$$\frac{PR}{QR} = \frac{P}{Q}.$$

As you read the Fundamental Principle, can you see why $\dfrac{PR}{QR}$ is not simplified? The numerator and denominator have a common factor, the polynomial R. By dividing the numerator and the denominator by the common factor, R, we obtain the simplified form $\dfrac{P}{Q}$. This is often shown as follows:

$$\frac{P\overset{1}{\cancel{R}}}{Q\underset{1}{\cancel{R}}} = \frac{P}{Q}.$$

Observe that
$$\frac{PR}{QR} = \frac{P}{Q} \cdot \frac{R}{R} = \frac{P}{Q} \cdot 1 = \frac{P}{Q}.$$

The following procedure can be used to simplify rational expressions:

Simplifying Rational Expressions

1. Factor the numerator and the denominator completely.
2. Divide both the numerator and the denominator by any common factors.

EXAMPLE 2   Simplifying a Rational Expression

Simplify: $\dfrac{5x + 35}{20x}$.

Solution

$$\frac{5x + 35}{20x} = \frac{5(x + 7)}{5 \cdot 4x} \qquad \text{Factor the numerator and denominator.}$$
$$\text{Because the denominator is 20x, } x \neq 0.$$

$$= \frac{\overset{1}{\cancel{5}}(x + 7)}{\underset{1}{\cancel{5}} \cdot 4x} \qquad \text{Divide out the common factor of 5.}$$

$$= \frac{x + 7}{4x} \qquad ■$$

✓ **CHECK POINT 2**   Simplify: $\dfrac{7x + 28}{21x}$.

EXAMPLE 3   Simplifying a Rational Expression

Simplify: $\dfrac{x^3 + x^2}{x + 1}$.

Solution

$$\frac{x^3 + x^2}{x + 1} = \frac{x^2(x + 1)}{x + 1} \qquad \text{Factor the numerator. Because the}$$
$$\text{denominator is } x + 1, x \neq -1.$$

$$= \frac{x^2\overset{1}{\cancel{(x + 1)}}}{\underset{1}{\cancel{x + 1}}} \qquad \text{Divide out the common factor of } x + 1.$$

$$= x^2 \qquad ■$$

Simplifying a rational expression can change the numbers that make it undefined. For example, we just showed that

$$\frac{x^3 + x^2}{x + 1} = x^2.$$

This is undefined for $x = -1$.

This simplified form is defined for all real numbers.

Thus, to equate the two expressions, we must restrict the values for x in the simplified expression to exclude -1. We can write

$$\frac{x^3 + x^2}{x + 1} = x^2, \quad x \neq -1.$$

Hereafter, we will assume that the simplified rational expression is equal to the original rational expression for all real numbers except those for which either denominator is 0.

Using Technology

Graphic and Numeric Connections

A graphing utility can be used to verify that

$$\frac{x^3 + x^2}{x + 1} = x^2, \quad x \neq -1.$$

Enter $y_1 = \dfrac{x^3 + x^2}{x + 1}$ and $y_2 = x^2$.

Graphic Check

The graphs of y_1 and y_2 appear to be identical. You can use the TRACE feature to trace y_1 and show that it is undefined for $x = -1$.

Numeric Check

No matter how far up or down we scroll, if $x \neq -1$, $y_1 = y_2$. If $x = -1$, y_1 is undefined, although the value of y_2 is 1.

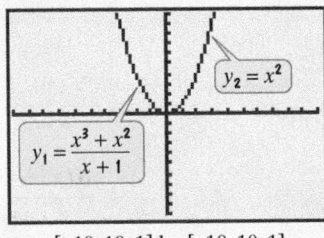

[−10, 10, 1] by [−10, 10, 1]

✓ **CHECK POINT 3**   Simplify: $\dfrac{x^3 - x^2}{7x - 7}$.

EXAMPLE 4   Simplifying a Rational Expression

Simplify: $\dfrac{x^2 + 6x + 5}{x^2 - 25}$.

Solution

$$\frac{x^2 + 6x + 5}{x^2 - 25} = \frac{(x + 5)(x + 1)}{(x + 5)(x - 5)}$$

Factor the numerator and denominator. Because the denominator is $(x + 5)(x - 5)$, $x \neq -5$ and $x \neq 5$.

$$= \frac{\overset{1}{\cancel{(x + 5)}}(x + 1)}{\underset{1}{\cancel{(x + 5)}}(x - 5)}$$

Divide out the common factor of $x + 5$.

$$= \frac{x + 1}{x - 5} \quad \blacksquare$$

Great Question!

Can I simplify $\dfrac{x+1}{x-5}$ by dividing the numerator and the denominator by x?

No. When simplifying rational expressions, only *factors* that are common to the *entire numerator* and the *entire denominator* can be divided out. **It is incorrect to divide out common terms from the numerator and denominator.**

Incorrect!

$$\frac{\overset{1}{\cancel{x}}+1}{\underset{1}{\cancel{x}}-5}=\frac{1}{-5}=-\frac{1}{5} \qquad \frac{x^2-\overset{1}{\cancel{4}}}{\underset{1}{\cancel{4}}}=x^2-1 \qquad \frac{\overset{3}{\cancel{x^2}}-\overset{3}{\cancel{9}}}{\underset{1}{\cancel{x}}-\underset{1}{\cancel{3}}}=x-3$$

The first two expressions, $\dfrac{x+1}{x-5}$ and $\dfrac{x^2-4}{4}$, have no common factors in their numerators and denominators. Thus, these rational expressions are in simplified form. The rational expression $\dfrac{x^2-9}{x-3}$ can be simplified as follows:

Correct

$$\frac{x^2-9}{x-3}=\frac{(x+3)(\overset{1}{\cancel{x-3}})}{\underset{1}{\cancel{x-3}}}=x+3.$$

> Divide out the common factor, $x-3$.

✓ **CHECK POINT 4**   Simplify: $\dfrac{x^2-1}{x^2+2x+1}$.

Factors That Are Opposites

How do we simplify rational expressions that contain factors in the numerator and denominator that are opposites, or additive inverses? Here is an example of such an expression:

$$\frac{x-3}{3-x}. \qquad \boxed{\text{The numerator and denominator are opposites. They differ only in their signs.}}$$

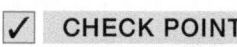

Factor out -1 from either the numerator or the denominator. Then divide out the common factor.

$$\frac{x-3}{3-x}=\frac{-1(-x+3)}{3-x} \qquad \text{Factor } -1 \text{ from the numerator. Notice how the sign of each term in the polynomial } x-3 \text{ changes.}$$

$$=\frac{-1(3-x)}{3-x} \qquad \text{In the numerator, use the commutative property to rewrite } -x+3 \text{ as } 3-x.$$

$$=\frac{-1(\overset{1}{\cancel{3-x}})}{\underset{1}{\cancel{3-x}}} \qquad \text{Divide out the common factor of } 3-x.$$

$$=-1$$

Our result, -1, suggests the following useful property:

Simplifying Rational Expressions with Opposite Factors in the Numerator and Denominator

The quotient of two polynomials that have opposite signs and are additive inverses is -1.

EXAMPLE 5   Simplifying a Rational Expression

Simplify: $\dfrac{4x^2 - 25}{15 - 6x}$.

Solution

$$\frac{4x^2 - 25}{15 - 6x} = \frac{(2x + 5)(2x - 5)}{3(5 - 2x)} \qquad \text{Factor the numerator and denominator.}$$

$$= \frac{(2x + 5)(2x \overset{-1}{\cancel{-5}})}{3(\cancel{5 - 2x})} \qquad \text{The quotient of polynomials with opposite signs is } -1.$$

$$= \frac{-(2x + 5)}{3} \quad \text{or} \quad -\frac{2x + 5}{3} \quad \text{or} \quad \frac{-2x - 5}{3}$$

> Each of these forms is an acceptable answer.

☑ **CHECK POINT 5**   Simplify: $\dfrac{9x^2 - 49}{28 - 12x}$.

3 Solve applied problems involving rational expressions.

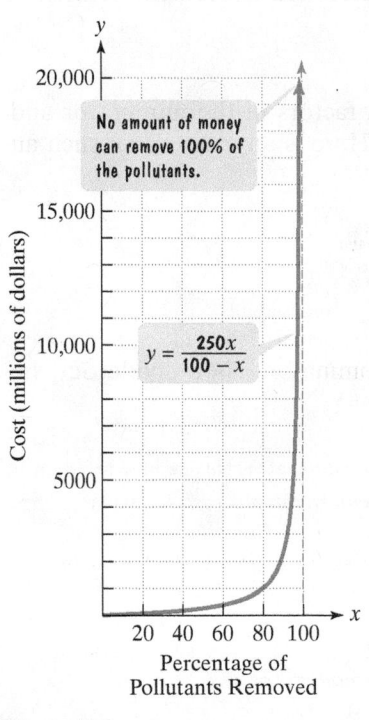

Figure 7.1

Applications

The equation

$$y = \frac{250x}{100 - x}$$

models the cost, in millions of dollars, to remove x percent of the pollutants that are discharged into a river. This equation contains the rational expression that we looked at in the opening to this section. Do you remember how costs were spiraling upward as the percentage of removed pollutants increased?

Is it possible to clean up the river completely? To do this, we must remove 100% of the pollutants. The problem is that the rational expression is undefined for $x = 100$.

$$y = \frac{250x}{100 - x} \quad \boxed{\text{If } x = 100, \text{ the value of the denominator is 0.}}$$

Notice how the graph of $y = \dfrac{250x}{100 - x}$, shown in **Figure 7.1**, approaches but never touches the dashed vertical line $x = 100$, where 100 is the value of x for which $\dfrac{250\,x}{100 - x}$ is undefined. The graph continues to rise more and more steeply, visually showing the escalating costs. By never touching the dashed vertical line, the graph illustrates that no amount of money will be enough to remove all pollutants from the river.

Achieving Success

Two Ways to Stay Sharp

- **Concentrate on one task at a time.** Do not multitask. Doing several things at once can cause confusion and can take longer to complete the tasks than tackling them sequentially.
- **Get enough sleep.** Fatigue impedes the ability to learn and do complex tasks.

CONCEPT AND VOCABULARY CHECK

Fill in each blank so that the resulting statement is true.

1. A rational expression is the quotient of two _____.

2. A rational expression is undefined for values that make the denominator _____.

3. We simplify a rational expression by _____ the numerator and the denominator completely. Then we divide the numerator and the denominator by any _____.

4. The rational expression $\dfrac{x-7}{x-7}$ simplifies to _____.

5. The rational expression $\dfrac{x-7}{7-x}$ simplifies to _____.

6. True or false: $\dfrac{x+5}{x+10}$ can be simplified. _____

7. True or false: $\dfrac{5x}{x+10}$ can be simplified. _____

8. True or false: $\dfrac{-x-10}{x+10}$ can be simplified. _____

9. True or false: The rational expression $\dfrac{x-2}{7x}$ is undefined for $x=2$. _____

7.1 EXERCISE SET   MyMathLab®

Watch the videos in MyMathLab   Download the MyDashBoard App

Practice Exercises

In Exercises 1–20, find all numbers for which each rational expression is undefined. If the rational expression is defined for all real numbers, so state.

1. $\dfrac{5}{2x}$

2. $\dfrac{11}{3x}$

3. $\dfrac{x}{x-8}$

4. $\dfrac{x}{x-6}$

5. $\dfrac{13}{5x-20}$

6. $\dfrac{17}{6x-30}$

7. $\dfrac{x+3}{(x+9)(x-2)}$

8. $\dfrac{x+5}{(x+7)(x-9)}$

9. $\dfrac{4x}{(3x-17)(x+3)}$

10. $\dfrac{8x}{(4x-19)(x+2)}$

11. $\dfrac{x+5}{x^2+x-12}$

12. $\dfrac{7x-14}{x^2-9x+20}$

13. $\dfrac{x+5}{5}$

14. $\dfrac{x+7}{7}$

15. $\dfrac{y+3}{4y^2+y-3}$

16. $\dfrac{y+8}{6y^2-y-2}$

17. $\dfrac{y+5}{y^2-25}$

18. $\dfrac{y+7}{y^2-49}$

19. $\dfrac{5}{x^2+1}$

20. $\dfrac{8}{x^2+4}$

In Exercises 21–76, simplify each rational expression. If the rational expression cannot be simplified, so state.

21. $\dfrac{14x^2}{7x}$

22. $\dfrac{9x^2}{6x}$

23. $\dfrac{5x-15}{25}$

24. $\dfrac{7x+21}{49}$

25. $\dfrac{2x-8}{4x}$

26. $\dfrac{3x-9}{6x}$

27. $\dfrac{3}{3x-9}$

28. $\dfrac{12}{6x-18}$

29. $\dfrac{-15}{3x-9}$

30. $\dfrac{-21}{7x-14}$

31. $\dfrac{3x+9}{x+3}$

32. $\dfrac{5x-10}{x-2}$

33. $\dfrac{x + 5}{x^2 - 25}$

34. $\dfrac{x + 4}{x^2 - 16}$

35. $\dfrac{2y - 10}{3y - 15}$

36. $\dfrac{6y + 18}{11y + 33}$

37. $\dfrac{x + 1}{x^2 - 2x - 3}$

38. $\dfrac{x + 2}{x^2 - x - 6}$

39. $\dfrac{4x - 8}{x^2 - 4x + 4}$

40. $\dfrac{x^2 - 12x + 36}{4x - 24}$

41. $\dfrac{y^2 - 3y + 2}{y^2 + 7y - 18}$

42. $\dfrac{y^2 + 5y + 4}{y^2 - 4y - 5}$

43. $\dfrac{2y^2 - 7y + 3}{2y^2 - 5y + 2}$

44. $\dfrac{3y^2 + 4y - 4}{6y^2 - y - 2}$

45. $\dfrac{2x + 3}{2x + 5}$

46. $\dfrac{3x + 7}{3x + 10}$

47. $\dfrac{x^2 + 12x + 36}{x^2 - 36}$

48. $\dfrac{x^2 - 14x + 49}{x^2 - 49}$

49. $\dfrac{x^3 - 2x^2 + x - 2}{x - 2}$

50. $\dfrac{x^3 + 4x^2 - 3x - 12}{x + 4}$

51. $\dfrac{x^3 - 8}{x - 2}$

52. $\dfrac{x^3 - 125}{x^2 - 25}$

53. $\dfrac{(x - 4)^2}{x^2 - 16}$

54. $\dfrac{(x + 5)^2}{x^2 - 25}$

55. $\dfrac{x}{x + 1}$

56. $\dfrac{x}{x + 7}$

57. $\dfrac{x + 4}{x^2 + 16}$

58. $\dfrac{x + 5}{x^2 + 25}$

59. $\dfrac{x - 5}{5 - x}$

60. $\dfrac{x - 7}{7 - x}$

61. $\dfrac{2x - 3}{3 - 2x}$

62. $\dfrac{5x - 4}{4 - 5x}$

63. $\dfrac{x - 5}{x + 5}$

64. $\dfrac{x - 7}{x + 7}$

65. $\dfrac{4x - 6}{3 - 2x}$

66. $\dfrac{9x - 15}{5 - 3x}$

67. $\dfrac{4 - 6x}{3x^2 - 2x}$

68. $\dfrac{9 - 15x}{5x^2 - 3x}$

69. $\dfrac{x^2 - 1}{1 - x}$

70. $\dfrac{x^2 - 4}{2 - x}$

71. $\dfrac{y^2 - y - 12}{4 - y}$

72. $\dfrac{y^2 - 7y + 12}{3 - y}$

73. $\dfrac{x^2y - x^2}{x^3 - x^3y}$

74. $\dfrac{xy - 2x}{3y - 6}$

75. $\dfrac{x^2 + 2xy - 3y^2}{2x^2 + 5xy - 3y^2}$

76. $\dfrac{x^2 + 3xy - 10y^2}{3x^2 - 7xy + 2y^2}$

Practice PLUS

In Exercises 77–84, simplify each rational expression.

77. $\dfrac{x^2 - 9x + 18}{x^3 - 27}$

78. $\dfrac{x^3 - 8}{x^2 + 2x - 8}$

79. $\dfrac{9 - y^2}{y^2 - 3(2y - 3)}$

80. $\dfrac{16 - y^2}{y(y - 8) + 16}$

81. $\dfrac{xy + 2y + 3x + 6}{x^2 + 5x + 6}$

82. $\dfrac{xy + 4y - 7x - 28}{x^2 + 11x + 28}$

83. $\dfrac{8x^2 + 4x + 2}{1 - 8x^3}$

84. $\dfrac{x^3 - 3x^2 + 9x}{x^3 + 27}$

Application Exercises

85. The rational expression

$$\frac{130x}{100 - x}$$

describes the cost, in millions of dollars, to inoculate x percent of the population against a particular strain of flu.

a. Evaluate the expression for $x = 40$, $x = 80$, and $x = 90$. Describe the meaning of each evaluation in terms of percentage inoculated and cost.

b. For what value of x is the expression undefined?

c. What happens to the cost as x approaches 100%? How can you interpret this observation?

86. The rational expression

$$\frac{60,000x}{100 - x}$$

describes the cost, in dollars, to remove x percent of the air pollutants in the smokestack emissions of a utility company that burns coal to generate electricity.

a. Evaluate the expression for $x = 20$, $x = 50$, and $x = 80$. Describe the meaning of each evaluation in terms of percentage of pollutants removed and cost.

b. For what value of x is the expression undefined?

c. What happens to the cost as x approaches 100%? How can you interpret this observation?

Doctors use the rational expression

$$\frac{DA}{A + 12}$$

to determine the dosage of a drug prescribed for children. In this expression, A = the child's age and D = the adult dosage. Use the expression to solve Exercises 87–88.

87. If the normal adult dosage of medication is 1000 milligrams, what dosage should an 8-year-old child receive?

88. If the normal adult dosage of medication is 1000 milligrams, what dosage should a 4-year-old child receive?

89. A company that manufactures bicycles has costs given by the equation

$$C = \frac{100x + 100{,}000}{x},$$

in which x is the number of bicycles manufactured and C is the cost to manufacture each bicycle.

a. Find the cost per bicycle when manufacturing 500 bicycles.

b. Find the cost per bicycle when manufacturing 4000 bicycles.

c. Does the cost per bicycle increase or decrease as more bicycles are manufactured? Explain why this happens.

90. A company that manufactures small canoes has costs given by the equation

$$C = \frac{20x + 20{,}000}{x},$$

in which x is the number of canoes manufactured and C is the cost to manufacture each canoe.

a. Find the cost per canoe when manufacturing 100 canoes.

b. Find the cost per canoe when manufacturing 10,000 canoes.

c. Does the cost per canoe increase or decrease as more canoes are manufactured? Explain why this happens.

A drug is injected into a patient and the concentration of the drug in the bloodstream is monitored. The drug's concentration, y, in milligrams per liter, after x hours is modeled by

$$y = \frac{5x}{x^2 + 1}.$$

The graph of this equation, obtained with a graphing utility, is shown in the figure in a [0, 10, 1] *by* [0, 3, 1] *viewing rectangle. Use this information to solve Exercises 91–92.*

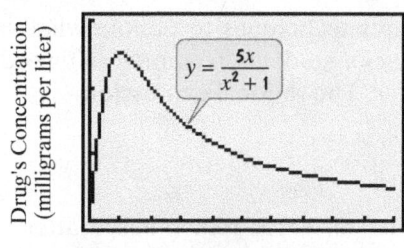

Hours after Injection
[0, 10, 1] by [0, 3, 1]

91. Use the equation to find the drug's concentration after 3 hours. Then identify the point on the equation's graph that conveys this information.

92. Use the graph of the equation to find after how many hours the drug reaches its maximum concentration. Then use the equation to find the drug's concentration at this time.

Writing in Mathematics

93. What is a rational expression? Give an example with your explanation.

94. Explain how to find the number or numbers, if any, for which a rational expression is undefined.

95. Explain how to simplify a rational expression.

96. Explain how to simplify a rational expression with opposite factors in the numerator and denominator.

97. Use the graph shown for Exercises 91–92 to write a description of the drug's concentration over time. In your description, try to convey as much information as possible that is displayed visually by the graph.

Critical Thinking Exercises

Make Sense? In Exercises 98–101, determine whether each statement "makes sense" or "does not make sense" and explain your reasoning.

98. Simplifying rational expressions is similar to reducing fractions.

99. I cannot simplify rational expressions without knowing how to factor polynomials.

100. The rational expressions

$$\frac{7}{14x} \quad \text{and} \quad \frac{7}{14 + x}$$

can both be simplified by dividing each numerator and each denominator by 7.

101. I evaluated $\dfrac{3x - 3}{4x(x - 1)}$ for $x = 1$ and obtained 0.

In Exercises 102–105, determine whether each statement is true or false. If the statement is false, make the necessary change(s) to produce a true statement.

102. $\dfrac{3x + 1}{3} = x + 1$

103. $\dfrac{x^2 + 3}{3} = x^2 + 1$

104. The expression $\dfrac{-3y - 6}{y + 2}$ simplifies to the consecutive integer that follows -4.

105. $\dfrac{3x + 7}{3x + 10} = \dfrac{8}{11}$

106. Write a rational expression that cannot be simplified.

107. Write a rational expression that is undefined for $x = -4$.

108. Write a rational expression with $x^2 - x - 6$ in the numerator that can be simplified to $x - 3$.

Technology Exercises

In Exercises 109–111, use the GRAPH *or* TABLE *feature of a graphing utility to determine if the rational expression has been correctly simplified. If the simplification is wrong, correct it and then verify your answer using the graphing utility.*

109. $\dfrac{3x + 15}{x + 5} = 3, \quad x \neq -5$

110. $\dfrac{2x^2 - x - 1}{x - 1} = 2x^2 - 1, x \neq 1$

111. $\dfrac{x^2 - x}{x} = x^2 - 1, \quad x \neq 0$

112. Use a graphing utility to verify the graph in **Figure 7.1** on page 496. TRACE along the graph as x approaches 100. What do you observe?

Review Exercises

113. Multiply: $\dfrac{5}{6} \cdot \dfrac{9}{25}$. (Section 1.2, Example 5)

114. Divide: $\dfrac{2}{3} \div 4$. (Section 1.2, Example 6)

115. Solve by the addition method:
$$\begin{cases} 2x - 5y = -2 \\ 3x + 4y = 20. \end{cases} \text{(Section 4.3, Example 3)}$$

Preview Exercises

Exercises 116–118 will help you prepare for the material covered in the next section. In each exercise, perform the indicated operation.

116. $\dfrac{2}{5} \cdot \dfrac{3}{7}$ **117.** $\dfrac{3}{4} \div \dfrac{1}{2}$ **118.** $\dfrac{5}{4} \div \dfrac{15}{8}$

Multiplying and Dividing Rational Expressions

Objectives

1 Multiply rational expressions.

2 Divide rational expressions.

Source: Wizard of Id copyright © 2011 by Johnny L. Hart FLP/Distributed by Creators Syndicate

Your psychology class is learning various techniques to double what we remember over time. At the beginning of the course, students memorize 40 words in Latin, a language with which they are not familiar. The rational expression

$$\frac{5t + 30}{t}$$

models the class average for the number of words remembered after t days, where $t \geq 1$. If the techniques are successful, what will be the new memory model?

The new model can be found by multiplying the given rational expression by 2. In this section, you will see that we multiply rational expressions in the same way that we multiply rational numbers. Thus, we multiply numerators and multiply denominators. The rational expression for doubling what the class remembers over time is

$$\frac{2}{1} \cdot \frac{5t + 30}{t} = \frac{2(5t + 30)}{1 \cdot t} = \frac{2 \cdot 5t + 2 \cdot 30}{t} = \frac{10t + 60}{t}.$$

1 Multiply rational
expressions.

Multiplying Rational Expressions

The product of two rational expressions is the product of their numerators divided by the product of their denominators.

Multiplying Rational Expressions

If P, Q, R, and S are polynomials, where $Q \neq 0$ and $S \neq 0$, then

$$\frac{P}{Q} \cdot \frac{R}{S} = \frac{PR}{QS}.$$

EXAMPLE 1   Multiplying Rational Expressions

Multiply: $\dfrac{7}{x+3} \cdot \dfrac{x-2}{5}$.

Solution

$$\frac{7}{x+3} \cdot \frac{x-2}{5} = \frac{7(x-2)}{(x+3)5} \qquad \text{Multiply numerators. Multiply denominators. } (x \neq -3)$$

$$= \frac{7x-14}{5x+15} \qquad \blacksquare$$

✓ CHECK POINT 1   Multiply: $\dfrac{9}{x+4} \cdot \dfrac{x-5}{2}$.

Here is a step-by-step procedure for multiplying rational expressions. Before multiplying, divide out any factors common to both a numerator and a denominator.

Multiplying Rational Expressions

1. Factor all numerators and denominators completely.
2. Divide numerators and denominators by common factors.
3. Multiply the remaining factors in the numerators and multiply the remaining factors in the denominators.

EXAMPLE 2   Multiplying Rational Expressions

Multiply: $\dfrac{x-3}{x+5} \cdot \dfrac{10x+50}{7x-21}$.

Solution

$$\frac{x-3}{x+5} \cdot \frac{10x+50}{7x-21}$$

$$= \frac{x-3}{x+5} \cdot \frac{10(x+5)}{7(x-3)} \qquad \text{Factor as many numerators and denominators as possible.}$$

$$= \frac{\overset{1}{\cancel{x-3}}}{\underset{1}{\cancel{x+5}}} \cdot \frac{10\overset{1}{\cancel{(x+5)}}}{7\underset{1}{\cancel{(x-3)}}} \qquad \text{Divide numerators and denominators by common factors.}$$

$$= \frac{10}{7} \qquad \text{Multiply the remaining factors in the numerators and in the denominators. } \blacksquare$$

✓ **CHECK POINT 2**   Multiply: $\dfrac{x+4}{x-7} \cdot \dfrac{3x-21}{8x+32}$.

EXAMPLE 3   Multiplying Rational Expressions

Multiply: $\dfrac{x-7}{x-1} \cdot \dfrac{x^2-1}{3x-21}$.

Solution

$$\dfrac{x-7}{x-1} \cdot \dfrac{x^2-1}{3x-21}$$

$$= \dfrac{x-7}{x-1} \cdot \dfrac{(x+1)(x-1)}{3(x-7)} \qquad \text{Factor as many numerators and denominators as possible.}$$

$$= \dfrac{\overset{1}{\cancel{x-7}}}{\underset{1}{\cancel{x-1}}} \cdot \dfrac{(x+1)\overset{1}{\cancel{(x-1)}}}{3\underset{1}{\cancel{(x-7)}}} \qquad \text{Divide numerators and denominators by common factors.}$$

$$= \dfrac{x+1}{3} \qquad \text{Multiply the remaining factors in the numerators and in the denominators.} \quad \blacksquare$$

✓ **CHECK POINT 3**   Multiply: $\dfrac{x-5}{x-2} \cdot \dfrac{x^2-4}{9x-45}$.

EXAMPLE 4   Multiplying Rational Expressions

Multiply: $\dfrac{4x+8}{6x-3x^2} \cdot \dfrac{3x^2-4x-4}{9x^2-4}$.

Solution

$$\dfrac{4x+8}{6x-3x^2} \cdot \dfrac{3x^2-4x-4}{9x^2-4}$$

$$= \dfrac{4(x+2)}{3x(2-x)} \cdot \dfrac{(3x+2)(x-2)}{(3x+2)(3x-2)} \qquad \text{Factor as many numerators and denominators as possible.}$$

$$= \dfrac{4(x+2)}{3x\underset{1}{\cancel{(2-x)}}} \cdot \dfrac{\overset{1}{\cancel{(3x+2)}}\overset{-1}{\cancel{(x-2)}}}{\underset{1}{\cancel{(3x+2)}}(3x-2)} \qquad \text{Divide numerators and denominators by common factors. Because } 2-x \text{ and } x-2 \text{ have opposite signs, their quotient is } -1.$$

$$= \dfrac{-4(x+2)}{3x(3x-2)} \quad \text{or} \quad -\dfrac{4(x+2)}{3x(3x-2)} \qquad \text{Multiply the remaining factors in the numerators and in the denominators.} \quad \blacksquare$$

It is not necessary to carry out these multiplications.

✓ **CHECK POINT 4**   Multiply: $\dfrac{5x+5}{7x-7x^2} \cdot \dfrac{2x^2+x-3}{4x^2-9}$.

---

2   Divide rational expressions.

Dividing Rational Expressions

The quotient of two rational expressions is the product of the first expression and the multiplicative inverse, or reciprocal, of the second. The reciprocal is found by interchanging the numerator and the denominator.

Dividing Rational Expressions

If P, Q, R, and S are polynomials, where $Q \neq 0$, $R \neq 0$, and $S \neq 0$, then

$$\frac{P}{Q} \div \frac{R}{S} = \frac{P}{Q} \cdot \frac{S}{R} = \frac{PS}{QR}.$$

Change division to multiplication.

Replace $\frac{R}{S}$ with its reciprocal by interchanging numerator and denominator.

Thus, **we find the quotient of two rational expressions by inverting the divisor and multiplying**. For example,

$$\frac{x}{7} \div \frac{6}{y} = \frac{x}{7} \cdot \frac{y}{6} = \frac{xy}{42}.$$

Change the division to multiplication.

Replace $\frac{6}{y}$ with its reciprocal by interchanging numerator and denominator.

Great Question!

If I'm multiplying or dividing rational expressions and an expression appears without a denominator, what should I do?

When performing operations with rational expressions, if a rational expression is written without a denominator, it is helpful to write the expression with a denominator of 1. In Example 5, we wrote $x + 5$ as

$$\frac{x+5}{1}.$$

EXAMPLE 5 Dividing Rational Expressions

Divide: $(x + 5) \div \dfrac{x - 2}{x + 9}$.

Solution

$$(x + 5) \div \frac{x - 2}{x + 9} = \frac{x + 5}{1} \cdot \frac{x + 9}{x - 2} \qquad \text{Invert the divisor and multiply.}$$

$$= \frac{(x + 5)(x + 9)}{x - 2} \qquad \begin{array}{l}\text{Multiply the factors in the numerators and}\\\text{in the denominators. We need not carry out}\\\text{the multiplication in the numerator.} \blacksquare\end{array}$$

✓ **CHECK POINT 5** Divide: $(x + 3) \div \dfrac{x - 4}{x + 7}$.

EXAMPLE 6 Dividing Rational Expressions

Divide: $\dfrac{x^2 - 2x - 8}{x^2 - 9} \div \dfrac{x - 4}{x + 3}$.

Solution

$$\frac{x^2 - 2x - 8}{x^2 - 9} \div \frac{x - 4}{x + 3}$$

$$= \frac{x^2 - 2x - 8}{x^2 - 9} \cdot \frac{x + 3}{x - 4} \qquad \text{Invert the divisor and multiply.}$$

$$= \frac{(x - 4)(x + 2)}{(x + 3)(x - 3)} \cdot \frac{x + 3}{x - 4} \qquad \begin{array}{l}\text{Factor as many numerators and}\\\text{denominators as possible.}\end{array}$$

$$= \frac{\overset{1}{\cancel{(x - 4)}}(x + 2)}{\cancel{(x + 3)}(x - 3)} \cdot \frac{\overset{1}{\cancel{(x + 3)}}}{\cancel{(x - 4)}} \qquad \begin{array}{l}\text{Divide numerators and denominators}\\\text{by common factors.}\end{array}$$

$$= \frac{x + 2}{x - 3} \qquad \begin{array}{l}\text{Multiply the remaining factors in the}\\\text{numerators and in the denominators.} \blacksquare\end{array}$$

✓ **CHECK POINT 6** Divide: $\dfrac{x^2 + 5x + 6}{x^2 - 25} \div \dfrac{x + 2}{x + 5}$.

EXAMPLE 7 Dividing Rational Expressions

Divide: $\dfrac{y^2 + 7y + 12}{y^2 + 9} \div (7y^2 + 21y)$.

Solution

$$\dfrac{y^2 + 7y + 12}{y^2 + 9} \div \dfrac{7y^2 + 21y}{1}$$ It is helpful to write the divisor with a denominator of 1.

$$= \dfrac{y^2 + 7y + 12}{y^2 + 9} \cdot \dfrac{1}{7y^2 + 21y}$$ Invert the divisor and multiply.

$$= \dfrac{(y + 4)(y + 3)}{y^2 + 9} \cdot \dfrac{1}{7y(y + 3)}$$ Factor as many numerators and denominators as possible.

$$= \dfrac{(y + 4)\overset{1}{\cancel{(y + 3)}}}{y^2 + 9} \cdot \dfrac{1}{7y\underset{1}{\cancel{(y + 3)}}}$$ Divide numerators and denominators by common factors.

$$= \dfrac{y + 4}{7y(y^2 + 9)}$$ Multiply the remaining factors in the numerators and in the denominators. ∎

✓ **CHECK POINT 7** Divide: $\dfrac{y^2 + 3y + 2}{y^2 + 1} \div (5y^2 + 10y)$.

CONCEPT AND VOCABULARY CHECK

Fill in each blank so that the resulting statement is true.

1. The product of two rational expressions is the product of their _____ divided by the product of their _____:
$\dfrac{P}{Q} \cdot \dfrac{R}{S} = $ _____, $Q \neq 0, S \neq 0$.

2. The quotient of two rational expressions is the product of the first expression and the _____ of the second:
$\dfrac{P}{Q} \div \dfrac{R}{S} = \dfrac{P}{Q} \cdot$ _____ $= $ _____, $Q \neq 0, R \neq 0, S \neq 0$.

3. $\dfrac{x}{5} \cdot \dfrac{x}{3} = $ _____

4. $\dfrac{x}{5} \div \dfrac{x}{3} = $ _____, $x \neq 0$

7.2 EXERCISE SET MyMathLab®  Watch the videos in MyMathLab   Download the MyDashBoard App

Practice Exercises

In Exercises 1–32, multiply as indicated.

1. $\dfrac{4}{x + 3} \cdot \dfrac{x - 5}{9}$

2. $\dfrac{8}{x - 2} \cdot \dfrac{x + 5}{3}$

3. $\dfrac{x}{3} \cdot \dfrac{12}{x + 5}$

4. $\dfrac{x}{5} \cdot \dfrac{30}{x - 4}$

5. $\dfrac{3}{x} \cdot \dfrac{4x}{15}$

6. $\dfrac{7}{x} \cdot \dfrac{5x}{35}$

7. $\dfrac{x - 3}{x + 5} \cdot \dfrac{4x + 20}{9x - 27}$

8. $\dfrac{x - 2}{x + 9} \cdot \dfrac{5x + 45}{2x - 4}$

9. $\dfrac{x^2 + 9x + 14}{x + 7} \cdot \dfrac{1}{x + 2}$

10. $\dfrac{x^2 + 9x + 18}{x + 6} \cdot \dfrac{1}{x + 3}$

11. $\dfrac{x^2 - 25}{x^2 - 3x - 10} \cdot \dfrac{x + 2}{x}$

12. $\dfrac{x^2 - 49}{x^2 - 4x - 21} \cdot \dfrac{x + 3}{x}$

13. $\dfrac{4y + 30}{y^2 - 3y} \cdot \dfrac{y - 3}{2y + 15}$

14. $\dfrac{9y + 21}{y^2 - 2y} \cdot \dfrac{y - 2}{3y + 7}$

15. $\dfrac{y^2 - 7y - 30}{y^2 - 6y - 40} \cdot \dfrac{2y^2 + 5y + 2}{2y^2 + 7y + 3}$

16. $\dfrac{3y^2 + 17y + 10}{3y^2 - 22y - 16} \cdot \dfrac{y^2 - 4y - 32}{y^2 - 8y - 48}$

17. $(y^2 - 9) \cdot \dfrac{4}{y - 3}$

18. $(y^2 - 16) \cdot \dfrac{3}{y - 4}$

19. $\dfrac{x^2 - 5x + 6}{x^2 - 2x - 3} \cdot \dfrac{x^2 - 1}{x^2 - 4}$

20. $\dfrac{x^2 + 5x + 6}{x^2 + x - 6} \cdot \dfrac{x^2 - 9}{x^2 - x - 6}$

21. $\dfrac{x^3 - 8}{x^2 - 4} \cdot \dfrac{x + 2}{3x}$

22. $\dfrac{x^2 + 6x + 9}{x^3 + 27} \cdot \dfrac{1}{x + 3}$

23. $\dfrac{(x - 2)^3}{(x - 1)^3} \cdot \dfrac{x^2 - 2x + 1}{x^2 - 4x + 4}$

24. $\dfrac{(x + 4)^3}{(x + 2)^3} \cdot \dfrac{x^2 + 4x + 4}{x^2 + 8x + 16}$

25. $\dfrac{6x + 2}{x^2 - 1} \cdot \dfrac{1 - x}{3x^2 + x}$

26. $\dfrac{8x + 2}{x^2 - 9} \cdot \dfrac{3 - x}{4x^2 + x}$

27. $\dfrac{25 - y^2}{y^2 - 2y - 35} \cdot \dfrac{y^2 - 8y - 20}{y^2 - 3y - 10}$

28. $\dfrac{2y}{3y - y^2} \cdot \dfrac{2y^2 - 9y + 9}{8y - 12}$

29. $\dfrac{x^2 - y^2}{x} \cdot \dfrac{x^2 + xy}{x + y}$

30. $\dfrac{4x - 4y}{x} \cdot \dfrac{x^2 + xy}{x^2 - y^2}$

31. $\dfrac{x^2 + 2xy + y^2}{x^2 - 2xy + y^2} \cdot \dfrac{4x - 4y}{3x + 3y}$

32. $\dfrac{x^2 - y^2}{x + y} \cdot \dfrac{x + 2y}{2x^2 - xy - y^2}$

In Exercises 33–64, divide as indicated.

33. $\dfrac{x}{7} \div \dfrac{5}{3}$

34. $\dfrac{x}{3} \div \dfrac{3}{8}$

35. $\dfrac{3}{x} \div \dfrac{12}{x}$

36. $\dfrac{x}{5} \div \dfrac{20}{x}$

37. $\dfrac{15}{x} \div \dfrac{3}{2x}$

38. $\dfrac{9}{x} \div \dfrac{3}{4x}$

39. $\dfrac{x + 1}{3} \div \dfrac{3x + 3}{7}$

40. $\dfrac{x + 5}{7} \div \dfrac{4x + 20}{9}$

41. $\dfrac{7}{x - 5} \div \dfrac{28}{3x - 15}$

42. $\dfrac{4}{x - 6} \div \dfrac{40}{7x - 42}$

43. $\dfrac{x^2 - 4}{x} \div \dfrac{x + 2}{x - 2}$

44. $\dfrac{x^2 - 4}{x - 2} \div \dfrac{x + 2}{4x - 8}$

45. $(y^2 - 16) \div \dfrac{y^2 + 3y - 4}{y^2 + 4}$

46. $(y^2 + 4y - 5) \div \dfrac{y^2 - 25}{y + 7}$

47. $\dfrac{y^2 - y}{15} \div \dfrac{y - 1}{5}$

48. $\dfrac{y^2 - 2y}{15} \div \dfrac{y - 2}{5}$

49. $\dfrac{4x^2 + 10}{x - 3} \div \dfrac{6x^2 + 15}{x^2 - 9}$

50. $\dfrac{x^2 + x}{x^2 - 4} \div \dfrac{x^2 - 1}{x^2 + 5x + 6}$

51. $\dfrac{x^2 - 25}{2x - 2} \div \dfrac{x^2 + 10x + 25}{x^2 + 4x - 5}$

52. $\dfrac{x^2 - 4}{x^2 + 3x - 10} \div \dfrac{x^2 + 5x + 6}{x^2 + 8x + 15}$

53. $\dfrac{y^3 + y}{y^2 - y} \div \dfrac{y^3 - y^2}{y^2 - 2y + 1}$

54. $\dfrac{3y^2 - 12}{y^2 + 4y + 4} \div \dfrac{y^3 - 2y^2}{y^2 + 2y}$

55. $\dfrac{y^2 + 5y + 4}{y^2 + 12y + 32} \div \dfrac{y^2 - 12y + 35}{y^2 + 3y - 40}$

56. $\dfrac{y^2 + 4y - 21}{y^2 + 3y - 28} \div \dfrac{y^2 + 14y + 48}{y^2 + 4y - 32}$

57. $\dfrac{2y^2 - 128}{y^2 + 16y + 64} \div \dfrac{y^2 - 6y - 16}{3y^2 + 30y + 48}$

58. $\dfrac{3y + 12}{y^2 + 3y} \div \dfrac{y^2 + y - 12}{9y - y^3}$

59. $\dfrac{2x + 2y}{3} \div \dfrac{x^2 - y^2}{x - y}$ **60.** $\dfrac{5x + 5y}{7} \div \dfrac{x^2 - y^2}{x - y}$

61. $\dfrac{x^2 - y^2}{8x^2 - 16xy + 8y^2} \div \dfrac{4x - 4y}{x + y}$

62. $\dfrac{4x^2 - y^2}{x^2 + 4xy + 4y^2} \div \dfrac{4x - 2y}{3x + 6y}$

63. $\dfrac{xy - y^2}{x^2 + 2x + 1} \div \dfrac{2x^2 + xy - 3y^2}{2x^2 + 5xy + 3y^2}$

64. $\dfrac{x^2 - 4y^2}{x^2 + 3xy + 2y^2} \div \dfrac{x^2 - 4xy + 4y^2}{x + y}$

Practice PLUS

In Exercises 65–72, perform the indicated operation or operations.

65. $\left(\dfrac{y - 2}{y^2 - 9y + 18} \cdot \dfrac{y^2 - 4y - 12}{y + 2}\right) \div \dfrac{y^2 - 4}{y^2 + 5y + 6}$

66. $\left(\dfrac{6y^2 + 31y + 18}{3y^2 - 20y + 12} \cdot \dfrac{2y^2 - 15y + 18}{6y^2 + 35y + 36}\right) \div \dfrac{2y^2 - 13y + 15}{9y^2 + 15y + 4}$

67. $\dfrac{3x^2 + 3x - 60}{2x - 8} \div \left(\dfrac{30x^2}{x^2 - 7x + 10} \cdot \dfrac{x^3 + 3x^2 - 10x}{25x^3}\right)$

68. $\dfrac{5x^2 - x}{3x + 2} \div \left(\dfrac{6x^2 + x - 2}{10x^2 + 3x - 1} \cdot \dfrac{2x^2 - x - 1}{2x^2 - x}\right)$

69. $\dfrac{x^2 + xz + xy + yz}{x - y} \div \dfrac{x + z}{x + y}$

70. $\dfrac{x^2 - xz + xy - yz}{x - y} \div \dfrac{x - z}{y - x}$

71. $\dfrac{3xy + ay + 3xb + ab}{9x^2 - a^2} \div \dfrac{y^3 + b^3}{6x - 2a}$

72. $\dfrac{5xy - ay - 5xb + ab}{25x^2 - a^2} \div \dfrac{y^3 - b^3}{15x + 3a}$

Application Exercises

In Exercises 73–76, write a simplified rational expression for the area of each figure. Assume that all measures are in inches.

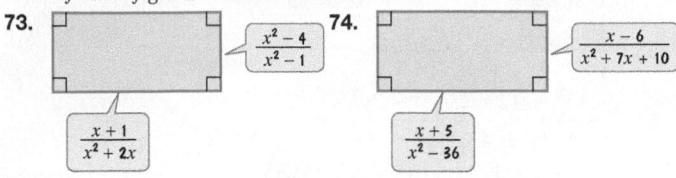

73. $\dfrac{x^2 - 4}{x^2 - 1}$ $\dfrac{x + 1}{x^2 + 2x}$

74. $\dfrac{x - 6}{x^2 + 7x + 10}$ $\dfrac{x + 5}{x^2 - 36}$

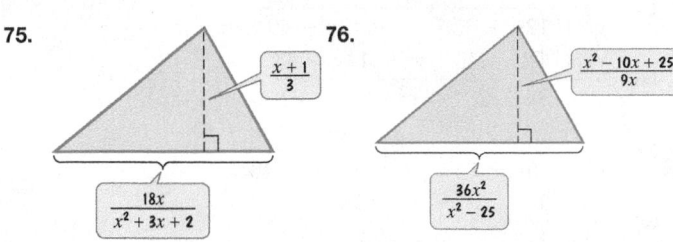

75. $\dfrac{x + 1}{3}$ $\dfrac{18x}{x^2 + 3x + 2}$

76. $\dfrac{x^2 - 10x + 25}{9x}$ $\dfrac{36x^2}{x^2 - 25}$

Writing in Mathematics

77. Explain how to multiply rational expressions.

78. Explain how to divide rational expressions.

79. In dividing polynomials

$$\frac{P}{Q} \div \frac{R}{S},$$

why is it necessary to state that polynomial R is not equal to 0?

Critical Thinking Exercises

Make Sense? *In Exercises 80–83, determine whether each statement "makes sense" or "does not make sense" and explain your reasoning.*

80. When dividing rational expressions, I multiply by the reciprocal of the divisor, just as I did when dividing rational numbers.

81. When opposite factors appear in the numerator and the denominator of a multiplication problem, I can put their quotient, −1, in either the numerator or the denominator.

82. When performing the division

$$\frac{7x}{x + 3} \div \frac{(x + 3)^2}{x - 5},$$

I began by dividing the numerator and the denominator by the common factor, $x + 3$.

83. The quotient

$$\frac{x + 2}{x - 5} \div \frac{x - 4}{x + 3}$$

is undefined for $x = 5$, $x = -3$, and $x = 4$.

In Exercises 84–87, determine whether each statement is true or false. If the statement is false, make the necessary change(s) to produce a true statement.

84. $5 \div x = \dfrac{1}{5} \cdot x$ for any nonzero number x.

85. $\dfrac{4}{x} \div \dfrac{x - 2}{x} = \dfrac{4}{x - 2}$ if $x \neq 0$ and $x \neq 2$.

86. $\dfrac{x - 5}{6} \cdot \dfrac{3}{5 - x} = \dfrac{1}{2}$ for any value of x except 5.

87. The quotient of two rational expressions can be found by inverting each expression and multiplying.

88. Find the missing polynomials: $\dfrac{\blacksquare\blacksquare}{\blacksquare\blacksquare} \cdot \dfrac{3x - 12}{2x} = \dfrac{3}{2}.$

89. Find the missing polynomials: $-\dfrac{1}{2x - 3} \div \dfrac{\blacksquare\blacksquare}{\blacksquare\blacksquare} = \dfrac{1}{3}.$

90. Divide:

$$\frac{9x^2 - y^2 + 15x - 5y}{3x^2 + xy + 5x} \div \frac{3x + y}{9x^3 + 6x^2y + xy^2}.$$

Technology Exercises

In Exercises 91–94, use the $\boxed{\text{GRAPH}}$ *or* $\boxed{\text{TABLE}}$ *feature of a graphing utility to determine if the multiplication or division has been performed correctly. If the answer is wrong, correct it and then verify your correction using the graphing utility.*

91. $\dfrac{x^2 + x}{3x} \cdot \dfrac{6x}{x + 1} = 2x$

92. $\dfrac{x^3 - 25x}{x^2 - 3x - 10} \cdot \dfrac{x + 2}{x} = x + 5$

93. $\dfrac{x^2 - 9}{x + 4} \div \dfrac{x - 3}{x + 4} = x - 3$

94. $(x - 5) \div \dfrac{2x^2 - 11x + 5}{4x^2 - 1} = 2x - 1$

Review Exercises

95. Solve: $2x + 3 < 3(x - 5)$. (Section 2.7, Example 8)

96. Factor completely: $3x^2 - 15x - 42$. (Section 6.5, Example 2)

97. Solve: $x(2x + 9) = 5$. (Section 6.6, Example 6)

Preview Exercises

Exercises 98–100 will help you prepare for the material covered in the next section.

98. Subtract: $\dfrac{7}{9} - \dfrac{1}{9}$.

99. Add: $\dfrac{2x}{3} + \dfrac{x}{3}$.

100. Simplify: $\dfrac{x^2 - 6x + 9}{x^2 - 9}$.

SECTION

7.3

Objectives

1 Add rational expressions with the same denominator.

2 Subtract rational expressions with the same denominator.

3 Add and subtract rational expressions with opposite denominators.

1 Add rational expressions with the same denominator.

Adding and Subtracting Rational Expressions with the Same Denominator

Are you long, medium, or round? Your skull, that is? The varying shapes of the human skull create physical diversity in the human species. By learning to add and subtract rational expressions with the same denominator, you will obtain an expression that models this diversity.

Addition when Denominators Are the Same

To add rational numbers having the same denominators, such as $\frac{2}{9}$ and $\frac{5}{9}$, we add the numerators and place the sum over the common denominator:

$$\frac{2}{9} + \frac{5}{9} = \frac{2+5}{9} = \frac{7}{9}.$$

We add rational expressions with the same denominator in an identical manner.

Adding Rational Expressions with Common Denominators

If $\dfrac{P}{R}$ and $\dfrac{Q}{R}$ are rational expressions, then

$$\frac{P}{R} + \frac{Q}{R} = \frac{P+Q}{R}.$$

To add rational expressions with the same denominator, add numerators and place the sum over the common denominator. If possible, factor and simplify the result.

EXAMPLE 1 Adding Rational Expressions when Denominators Are the Same

Add: $\dfrac{2x-1}{3} + \dfrac{x+4}{3}$.

Solution

$$\frac{2x-1}{3} + \frac{x+4}{3} = \frac{2x-1+x+4}{3} \qquad \text{Add numerators. Place this sum over the common denominator.}$$

$$= \frac{3x+3}{3} \qquad \text{Combine like terms.}$$

$$= \frac{\overset{1}{\cancel{3}}(x+1)}{\underset{1}{\cancel{3}}} \qquad \text{Factor and simplify.}$$

$$= x+1 \quad \blacksquare$$

✓ **CHECK POINT 1** Add: $\dfrac{3x-2}{5} + \dfrac{2x+12}{5}$.

EXAMPLE 2 Adding Rational Expressions when Denominators Are the Same

Add: $\dfrac{x^2}{x^2-9} + \dfrac{9-6x}{x^2-9}$.

Solution

$$\dfrac{x^2}{x^2-9} + \dfrac{9-6x}{x^2-9} = \dfrac{x^2+9-6x}{x^2-9}$$

Add numerators. Place this sum over the common denominator.

$$= \dfrac{x^2-6x+9}{x^2-9}$$

Write the numerator in descending powers of x.

$$= \dfrac{(x-3)\overset{1}{\cancel{(x-3)}}}{(x+3)\underset{1}{\cancel{(x-3)}}}$$

Factor and simplify. What values of x are not permitted?

$$= \dfrac{x-3}{x+3} \quad \blacksquare$$

✓ **CHECK POINT 2** Add: $\dfrac{x^2}{x^2-25} + \dfrac{25-10x}{x^2-25}$.

2 Subtract rational expressions with the same denominator.

Subtraction when Denominators Are the Same

The following box shows how to subtract rational expressions with the same denominator:

Subtracting Rational Expressions with Common Denominators

If $\dfrac{P}{R}$ and $\dfrac{Q}{R}$ are rational expressions, then

$$\dfrac{P}{R} - \dfrac{Q}{R} = \dfrac{P-Q}{R}.$$

To subtract rational expressions with the same denominator, subtract numerators and place the difference over the common denominator. If possible, factor and simplify the result.

EXAMPLE 3 Subtracting Rational Expressions when Denominators Are the Same

Subtract:

a. $\dfrac{2x+3}{x+1} - \dfrac{x}{x+1}$ **b.** $\dfrac{5x+1}{x^2-9} - \dfrac{4x-2}{x^2-9}$.

Solution

a. $\dfrac{2x+3}{x+1} - \dfrac{x}{x+1} = \dfrac{2x+3-x}{x+1}$

Subtract numerators. Place this difference over the common denominator.

$= \dfrac{x+3}{x+1}$

Combine like terms.

b. $\dfrac{5x+1}{x^2-9} - \dfrac{4x-2}{x^2-9} = \dfrac{5x+1-(4x-2)}{x^2-9}$

Subtract numerators and include parentheses to indicate that both terms are subtracted. Place this difference over the common denominator.

$= \dfrac{5x+1-4x+2}{x^2-9}$

Remove parentheses and then change the sign of each term.

$= \dfrac{x+3}{x^2-9}$

Combine like terms.

$= \dfrac{\overset{1}{\cancel{x+3}}}{\underset{1}{\cancel{(x+3)}}(x-3)}$

Factor and simplify ($x \neq -3$, and $x \neq 3$).

$= \dfrac{1}{x-3}$ ∎

✓ **CHECK POINT 3**   Subtract:

a. $\dfrac{4x+5}{x+7} - \dfrac{x}{x+7}$ **b.** $\dfrac{3x^2+4x}{x-1} - \dfrac{11x-4}{x-1}.$

Great Question!

When subtracting a numerator containing more than one term, do I really need to insert parentheses like you did in Example 3(b)?

Yes. When a numerator is being subtracted, we need to be sure that we **subtract every term in that expression**. Parentheses indicate that every term inside is being subtracted.

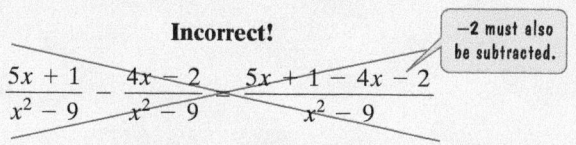

The entire numerator of the second rational expression must be subtracted. Avoid the common error of subtracting only the first term.

Incorrect!

−2 must also be subtracted.

$\dfrac{5x+1}{x^2-9} - \dfrac{4x-2}{x^2-9} \; \cancel{\dfrac{5x+1-4x-2}{x^2-9}}$

EXAMPLE 4 Subtracting Rational Expressions when Denominators Are the Same

Subtract: $\dfrac{20y^2 + 5y + 1}{6y^2 + y - 2} - \dfrac{8y^2 - 12y - 5}{6y^2 + y - 2}$.

Solution

$$\dfrac{20y^2 + 5y + 1}{6y^2 + y - 2} - \dfrac{8y^2 - 12y - 5}{6y^2 + y - 2}$$

Don't forget the parentheses.

$$= \dfrac{20y^2 + 5y + 1 - (8y^2 - 12y - 5)}{6y^2 + y - 2}$$

Subtract numerators. Place this difference over the common denominator.

$$= \dfrac{20y^2 + 5y + 1 - 8y^2 + 12y + 5}{6y^2 + y - 2}$$

Remove parentheses and then change the sign of each term.

$$= \dfrac{(20y^2 - 8y^2) + (5y + 12y) + (1 + 5)}{6y^2 + y - 2}$$

Group like terms. This step is usually performed mentally.

$$= \dfrac{12y^2 + 17y + 6}{6y^2 + y - 2}$$

Combine like terms.

$$= \dfrac{(3y + 2)^1(4y + 3)}{(3y + 2)_1(2y - 1)}$$

Factor and simplify.

$$= \dfrac{4y + 3}{2y - 1}$$ ∎

✓ **CHECK POINT 4** Subtract: $\dfrac{y^2 + 3y - 6}{y^2 - 5y + 4} - \dfrac{4y - 4 - 2y^2}{y^2 - 5y + 4}$.

3 Add and subtract rational expressions with opposite denominators.

Addition and Subtraction when Denominators Are Opposites

How do we add or subtract rational expressions when denominators are opposites, or additive inverses? Here is an example of this type of addition problem:

$$\dfrac{x^2}{x - 5} + \dfrac{4x + 5}{5 - x}.$$

These denominators are opposites. They differ only in their signs.

Multiply the numerator and the denominator of either of the rational expressions by −1. Then they will both have the same denominator.

EXAMPLE 5   Adding Rational Expressions when Denominators Are Opposites

Add: $\dfrac{x^2}{x-5} + \dfrac{4x+5}{5-x}$.

Solution

$$\dfrac{x^2}{x-5} + \dfrac{4x+5}{5-x}$$

$$= \dfrac{x^2}{x-5} + \dfrac{(-1)}{(-1)} \cdot \dfrac{4x+5}{5-x}$$ Multiply the numerator and denominator of the second rational expression by −1.

$$= \dfrac{x^2}{x-5} + \dfrac{-4x-5}{-5+x}$$ Perform the multiplications by −1 by changing every term's sign.

$$= \dfrac{x^2}{x-5} + \dfrac{-4x-5}{x-5}$$ Rewrite −5 + x as x − 5. Both rational expressions have the same denominator.

$$= \dfrac{x^2 + (-4x-5)}{x-5}$$ Add numerators. Place this sum over the common denominator.

$$= \dfrac{x^2 - 4x - 5}{x-5}$$ Remove parentheses.

$$= \dfrac{\overset{1}{\cancel{(x-5)}}(x+1)}{\underset{1}{\cancel{x-5}}}$$ Factor and simplify.

$$= x + 1 \quad\blacksquare$$

☑ **CHECK POINT 5**   Add: $\dfrac{x^2}{x-7} + \dfrac{4x+21}{7-x}$.

Adding and Subtracting Rational Expressions with Opposite Denominators

When one denominator is the opposite, or additive inverse, of the other, first multiply either rational expression by $\frac{-1}{-1}$ to obtain a common denominator.

EXAMPLE 6   Subtracting Rational Expressions when Denominators Are Opposites

Subtract: $\dfrac{5x - x^2}{x^2 - 4x - 3} - \dfrac{3x - x^2}{3 + 4x - x^2}$.

Solution   We note that $x^2 - 4x - 3$ and $3 + 4x - x^2$ are opposites. We multiply the second rational expression by $\frac{-1}{-1}$.

$$\dfrac{(-1)}{(-1)} \cdot \dfrac{3x - x^2}{3 + 4x - x^2} = \dfrac{-3x + x^2}{-3 - 4x + x^2}$$ Multiply the numerator and denominator by −1 by changing every term's sign.

$$= \dfrac{x^2 - 3x}{x^2 - 4x - 3}$$ Write the numerator and the denominator in descending powers of x.

We now return to the original subtraction problem.

$$\frac{5x - x^2}{x^2 - 4x - 3} - \frac{3x - x^2}{3 + 4x - x^2}$$

This is the given problem.

$$= \frac{5x - x^2}{x^2 - 4x - 3} - \frac{x^2 - 3x}{x^2 - 4x - 3}$$

Replace the second rational expression by the form obtained through multiplication by $\frac{-1}{-1}$.

$$= \frac{5x - x^2 - (x^2 - 3x)}{x^2 - 4x - 3}$$

Subtract numerators. Place this difference over the common denominator. Don't forget parentheses!

$$= \frac{5x - x^2 - x^2 + 3x}{x^2 - 4x - 3}$$

Remove parentheses and then change the sign of each term.

$$= \frac{-2x^2 + 8x}{x^2 - 4x - 3}$$

Combine like terms in the numerator. Although the numerator can be factored, further simplification is not possible. ■

☑ **CHECK POINT 6** Subtract: $\dfrac{7x - x^2}{x^2 - 2x - 9} - \dfrac{5x - 3x^2}{9 + 2x - x^2}$.

CONCEPT AND VOCABULARY CHECK

Fill in each blank so that the resulting statement is true.

1. $\dfrac{P}{R} + \dfrac{Q}{R} =$ _____ : To add rational expressions with the same denominator, add _____ and place the sum over the _____ .

2. $\dfrac{P}{R} - \dfrac{Q}{R} =$ _____ : To subtract rational expressions with the same denominator, subtract _____ and place the difference over the _____ .

3. When adding and subtracting rational expressions with opposite denominators, multiply either expression by _____ to obtain a common denominator.

4. $\dfrac{x}{3} + \dfrac{5}{3} =$ _____

5. $\dfrac{x}{3} - \dfrac{5}{3} =$ _____

6. $\dfrac{x}{3} - \dfrac{5 - y}{3} =$ _____

7.3 EXERCISE SET   MyMathLab®

Watch the videos in MyMathLab   Download the MyDashBoard App

Practice Exercises

In Exercises 1–38, add or subtract as indicated. Simplify the result, if possible.

1. $\dfrac{7x}{13} + \dfrac{2x}{13}$

2. $\dfrac{3x}{17} + \dfrac{8x}{17}$

3. $\dfrac{8x}{15} + \dfrac{x}{15}$

4. $\dfrac{9x}{24} + \dfrac{x}{24}$

5. $\dfrac{x - 3}{12} + \dfrac{5x + 21}{12}$

6. $\dfrac{x + 4}{9} + \dfrac{2x - 25}{9}$

7. $\dfrac{4}{x} + \dfrac{2}{x}$

8. $\dfrac{5}{x} + \dfrac{13}{x}$

9. $\dfrac{8}{9x} + \dfrac{13}{9x}$

10. $\dfrac{4}{9x} + \dfrac{11}{9x}$

11. $\dfrac{5}{x + 3} + \dfrac{4}{x + 3}$

12. $\dfrac{8}{x + 6} + \dfrac{10}{x + 6}$

13. $\dfrac{x}{x - 3} + \dfrac{4x + 5}{x - 3}$

14. $\dfrac{x}{x - 4} + \dfrac{9x + 7}{x - 4}$

15. $\dfrac{4x + 1}{6x + 5} + \dfrac{8x + 9}{6x + 5}$

16. $\dfrac{3x + 2}{3x + 4} + \dfrac{3x + 6}{3x + 4}$

17. $\dfrac{y^2 + 7y}{y^2 - 5y} + \dfrac{y^2 - 4y}{y^2 - 5y}$

18. $\dfrac{y^2 - 2y}{y^2 + 3y} + \dfrac{y^2 + y}{y^2 + 3y}$

19. $\dfrac{4y - 1}{5y^2} + \dfrac{3y + 1}{5y^2}$

20. $\dfrac{y + 2}{6y^3} + \dfrac{3y - 2}{6y^3}$

21. $\dfrac{x^2 - 2}{x^2 + x - 2} + \dfrac{2x - x^2}{x^2 + x - 2}$

22. $\dfrac{x^2 + 9x}{4x^2 - 11x - 3} + \dfrac{3x - 5x^2}{4x^2 - 11x - 3}$

23. $\dfrac{x^2 - 4x}{x^2 - x - 6} + \dfrac{4x - 4}{x^2 - x - 6}$

24. $\dfrac{x}{2x + 7} - \dfrac{2}{2x + 7}$

25. $\dfrac{3x}{5x - 4} - \dfrac{4}{5x - 4}$

26. $\dfrac{x}{x - 1} - \dfrac{1}{x - 1}$

27. $\dfrac{4x}{4x - 3} - \dfrac{3}{4x - 3}$

28. $\dfrac{2y + 1}{3y - 7} - \dfrac{y + 8}{3y - 7}$

29. $\dfrac{14y}{7y + 2} - \dfrac{7y - 2}{7y + 2}$

30. $\dfrac{2x + 3}{3x - 6} - \dfrac{3 - x}{3x - 6}$

31. $\dfrac{3x + 1}{4x - 2} - \dfrac{x + 1}{4x - 2}$

32. $\dfrac{x^3 - 3}{2x^4} - \dfrac{7x^3 - 3}{2x^4}$

33. $\dfrac{3y^2 - 1}{3y^3} - \dfrac{6y^2 - 1}{3y^3}$

34. $\dfrac{y^2 + 3y}{y^2 + y - 12} - \dfrac{y^2 - 12}{y^2 + y - 12}$

35. $\dfrac{4y^2 + 5}{9y^2 - 64} - \dfrac{y^2 - y + 29}{9y^2 - 64}$

36. $\dfrac{2y^2 + 6y + 8}{y^2 - 16} - \dfrac{y^2 - 3y - 12}{y^2 - 16}$

37. $\dfrac{6y^2 + y}{2y^2 - 9y + 9} - \dfrac{2y + 9}{2y^2 - 9y + 9} - \dfrac{4y - 3}{2y^2 - 9y + 9}$

38. $\dfrac{3y^2 - 2}{3y^2 + 10y - 8} - \dfrac{y + 10}{3y^2 + 10y - 8} - \dfrac{y^2 - 6y}{3y^2 + 10y - 8}$

In Exercises 39–64, denominators are opposites, or additive inverses. Add or subtract as indicated. Simplify the result, if possible.

39. $\dfrac{4}{x - 3} + \dfrac{2}{3 - x}$

40. $\dfrac{6}{x - 5} + \dfrac{2}{5 - x}$

41. $\dfrac{6x + 7}{x - 6} + \dfrac{3x}{6 - x}$

42. $\dfrac{6x + 5}{x - 2} + \dfrac{4x}{2 - x}$

43. $\dfrac{5x - 2}{3x - 4} + \dfrac{2x - 3}{4 - 3x}$

44. $\dfrac{9x - 1}{7x - 3} + \dfrac{6x - 2}{3 - 7x}$

45. $\dfrac{x^2}{x - 2} + \dfrac{4}{2 - x}$

46. $\dfrac{x^2}{x - 3} + \dfrac{9}{3 - x}$

47. $\dfrac{y - 3}{y^2 - 25} + \dfrac{y - 3}{25 - y^2}$

48. $\dfrac{y - 7}{y^2 - 16} + \dfrac{7 - y}{16 - y^2}$

49. $\dfrac{6}{x - 1} - \dfrac{5}{1 - x}$

50. $\dfrac{10}{x - 2} - \dfrac{6}{2 - x}$

51. $\dfrac{10}{x + 3} - \dfrac{2}{-x - 3}$

52. $\dfrac{11}{x + 7} - \dfrac{5}{-x - 7}$

53. $\dfrac{y}{y - 1} - \dfrac{1}{1 - y}$

54. $\dfrac{y}{y - 4} - \dfrac{4}{4 - y}$

55. $\dfrac{3 - x}{x - 7} - \dfrac{2x - 5}{7 - x}$

56. $\dfrac{4 - x}{x - 9} - \dfrac{3x - 8}{9 - x}$

57. $\dfrac{x - 2}{x^2 - 25} - \dfrac{x - 2}{25 - x^2}$

58. $\dfrac{x - 8}{x^2 - 16} - \dfrac{x - 8}{16 - x^2}$

59. $\dfrac{x}{x - y} + \dfrac{y}{y - x}$

60. $\dfrac{2x - y}{x - y} + \dfrac{x - 2y}{y - x}$

61. $\dfrac{2x}{x^2 - y^2} + \dfrac{2y}{y^2 - x^2}$

62. $\dfrac{2y}{x^2 - y^2} + \dfrac{2x}{y^2 - x^2}$

63. $\dfrac{x^2 - 2}{x^2 + 6x - 7} + \dfrac{19 - 4x}{7 - 6x - x^2}$

64. $\dfrac{2x + 3}{x^2 - x - 30} + \dfrac{x - 2}{30 + x - x^2}$

Practice PLUS

In Exercises 65–72, perform the indicated operation or operations. Simplify the result, if possible.

65. $\dfrac{6b^2 - 10b}{16b^2 - 48b + 27} + \dfrac{7b^2 - 20b}{16b^2 - 48b + 27} - \dfrac{6b - 3b^2}{16b^2 - 48b + 27}$

66. $\dfrac{22b + 15}{12b^2 + 52b - 9} + \dfrac{30b - 20}{12b^2 + 52b - 9} - \dfrac{4 - 2b}{12b^2 + 52b - 9}$

67. $\dfrac{2y}{y - 5} - \left(\dfrac{2}{y - 5} + \dfrac{y - 2}{y - 5}\right)$

68. $\dfrac{3x}{(x + 1)^2} - \left[\dfrac{5x + 1}{(x + 1)^2} - \dfrac{3x + 2}{(x + 1)^2}\right]$

69. $\dfrac{b}{ac + ad - bc - bd} - \dfrac{a}{ac + ad - bc - bd}$

70. $\dfrac{y}{ax + bx - ay - by} - \dfrac{x}{ax + bx - ay - by}$

71. $\dfrac{(y - 3)(y + 2)}{(y + 1)(y - 4)} - \dfrac{(y + 2)(y + 3)}{(y + 1)(4 - y)} - \dfrac{(y + 5)(y - 1)}{(y + 1)(4 - y)}$

72. $\dfrac{(y + 1)(2y - 1)}{(y - 2)(y - 3)} + \dfrac{(y + 2)(y - 1)}{(y - 2)(y - 3)} - \dfrac{(y + 5)(2y + 1)}{(3 - y)(2 - y)}$

Application Exercises

73. Anthropologists and forensic scientists classify skulls using

$$\frac{L + 60W}{L} - \frac{L - 40W}{L},$$

where L is the skull's length and W is its width.

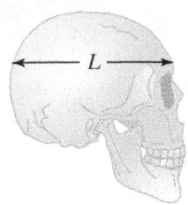

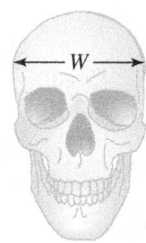

a. Express the classification as a single rational expression.

b. If the value of the rational expression in part (a) is less than 75, a skull is classified as long. A medium skull has a value between 75 and 80, and a round skull has a value over 80. Use your rational expression from part (a) to classify a skull that is 5 inches wide and 6 inches long.

74. The temperature, in degrees Fahrenheit, of a dessert placed in a freezer for t hours is modeled by

$$\frac{t + 30}{t^2 + 4t + 1} - \frac{t - 50}{t^2 + 4t + 1}.$$

a. Express the temperature as a single rational expression.

b. Use your rational expression from part (a) to find the temperature of the dessert, to the nearest hundredth of a degree, after 1 hour and after 2 hours.

In Exercises 75–76, find the perimeter of each rectangle.

75.

$\frac{5}{x+3}$ meters

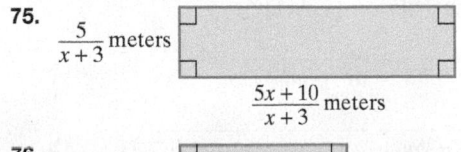

$\frac{5x+10}{x+3}$ meters

76.

$\frac{7}{x+4}$ inches

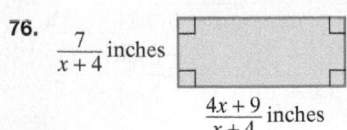

$\frac{4x+9}{x+4}$ inches

Writing in Mathematics

77. Explain how to add rational expressions when denominators are the same. Give an example with your explanation.

78. Explain how to subtract rational expressions when denominators are the same. Give an example with your explanation.

79. Describe two similarities between the following problems:

$$\frac{3}{8} + \frac{1}{8} \quad \text{and} \quad \frac{x}{x^2 - 1} + \frac{1}{x^2 - 1}.$$

80. Explain how to add rational expressions when denominators are opposites. Use an example to support your explanation.

Critical Thinking Exercises

Make Sense? *In Exercises 81–84, determine whether each statement "makes sense" or "does not make sense" and explain your reasoning.*

81. After adding $\frac{3x + 1}{4}$ and $\frac{x + 2}{4}$, I simplified the sum by dividing the numerator and the denominator by 4.

82. I use similar procedures to find each of the following sums:

$$\frac{3}{8} + \frac{1}{8} \quad \text{and} \quad \frac{x}{x^2 - 1} + \frac{1}{x^2 - 1}.$$

83. I subtracted $\frac{3x - 5}{x - 1}$ from $\frac{x - 3}{x - 1}$ and obtained a constant.

84. I added $\frac{5}{x - 7}$ and $\frac{3}{7 - x}$ by first multiplying the second rational expression by -1.

In Exercises 85–88, determine whether each statement is true or false. If the statement is false, make the necessary change(s) to produce a true statement.

85. The sum of two rational expressions with the same denominator can be found by adding numerators, adding denominators, and then simplifying.

86. $\frac{4}{b} - \frac{2}{-b} = -\frac{2}{b}$

87. The difference between two rational expressions with the same denominator can always be simplified.

88. $\frac{2x + 1}{x - 7} + \frac{3x + 1}{x - 7} - \frac{5x + 2}{x - 7} = 0$

In Exercises 89–90, perform the indicated operations. Simplify the result, if possible.

89. $\left(\frac{3x - 1}{x^2 + 5x - 6} - \frac{2x - 7}{x^2 + 5x - 6} \right) \div \frac{x + 2}{x^2 - 1}$

90. $\left(\frac{3x^2 - 4x + 4}{3x^2 + 7x + 2} - \frac{10x + 9}{3x^2 + 7x + 2} \right) \div \frac{x - 5}{x^2 - 4}$

In Exercises 91–95, find the missing expression.

91. $\frac{2x}{x + 3} + \frac{\blacksquare}{x + 3} = \frac{4x + 1}{x + 3}$

92. $\frac{3x}{x + 2} - \frac{\blacksquare}{x + 2} = \frac{6 - 17x}{x + 2}$

93. $\frac{6}{x - 2} + \frac{\blacksquare}{2 - x} = \frac{13}{x - 2}$

94. $\dfrac{a^2}{a-4} - \dfrac{\blacksquare\blacksquare}{a-4} = a+3$

95. $\dfrac{3x}{x-5} + \dfrac{\blacksquare\blacksquare}{5-x} = \dfrac{7x+1}{x-5}$

Technology Exercises

In Exercises 96–98, use the $\boxed{\text{GRAPH}}$ *or* $\boxed{\text{TABLE}}$ *feature of a graphing utility to determine if the subtraction has been performed correctly. If the answer is wrong, correct it and then verify your correction using the graphing utility.*

96. $\dfrac{3x+6}{2} - \dfrac{x}{2} = x+3$

97. $\dfrac{x^2+4x+3}{x+2} - \dfrac{5x+9}{x+2} = x-2, x \neq -2$

98. $\dfrac{x^2-13}{x+4} - \dfrac{3}{x+4} = x+4, x \neq -4$

Review Exercises

99. Subtract: $\dfrac{13}{15} - \dfrac{8}{45}$. (Section 1.2, Example 9)

100. Factor completely: $81x^4 - 1$. (Section 6.4, Example 4)

101. Divide: $\dfrac{3x^3 + 2x^2 - 26x - 15}{x+3}$. (Section 5.6, Example 2)

Preview Exercises

Exercises 102–104 will help you prepare for the material covered in the next section.
In Exercises 102–103, perform the indicated operation. Where possible, reduce the answer to its lowest terms.

102. $\dfrac{1}{2} + \dfrac{2}{3}$

103. $\dfrac{1}{8} - \dfrac{5}{6}$

104. Simplify: $\dfrac{(y+2)y - 2 \cdot 4}{4y(y+4)}$.

Adding and Subtracting Rational Expressions with Different Denominators

Objectives

1 Find the least common denominator.

2 Add and subtract rational expressions with different denominators.

When my aunt asked how I liked my five-year-old nephew, I replied "medium rare." Unfortunately, my little joke did not get me out of baby sitting for the Dennis the Menace of our family. Now the little squirt doesn't want to go to bed because his head hurts. Does my aunt have any Tylenol®? What is the proper dosage for a child his age?

In this section's Exercise Set, you will use two formulas that model drug dosage for children. Before working with these models, we continue drawing on your experience from arithmetic to add and subtract rational expressions that have different denominators.

1 Find the least common denominator.

Finding the Least Common Denominator

We can gain insight into adding rational expressions with different denominators by looking closely at what we do when adding fractions with different denominators. For example, suppose that we want to add $\frac{1}{2}$ and $\frac{2}{3}$. We must first write the fractions with the same denominator. We look for the smallest number that contains both 2 and 3 as factors. This number, 6, is then used as the *least common denominator*, or LCD.

The **least common denominator** of several rational expressions is a polynomial consisting of the product of all prime factors in the denominators, with each factor raised to the greatest power of its occurrence in any denominator.

Finding the Least Common Denominator

1. Factor each denominator completely.
2. List the factors of the first denominator.
3. Add to the list in step 2 any factors of the second denominator that do not appear in the list.
4. Form the product of the factors from the list in step 3. This product is the least common denominator.

EXAMPLE 1 Finding the Least Common Denominator

Find the LCD of $\dfrac{7}{6x^2}$ and $\dfrac{2}{9x}$.

Solution

Step 1. Factor each denominator completely.

$$6x^2 = 3 \cdot 2 \cdot x \cdot x$$
$$9x = 3 \cdot 3 \cdot x$$

$$\dfrac{7}{6x^2} \qquad\qquad \dfrac{2}{9x}$$

Factors are 3, 2, x, and x.   Factors are 3, 3, and x.

Step 2. List the factors of the first denominator.

$$3, 2, x, x$$

Step 3. Add any unlisted factors from the second denominator. Two factors from $9x$ are already in our list. These factors include one factor of 3 and x. We add the other factor of 3 to our list. We have

$$3, 3, 2, x, x.$$

Step 4. The least common denominator is the product of all factors in the final list. Thus,

$$3 \cdot 3 \cdot 2 \cdot x \cdot x,$$

or $18x^2$ is the least common denominator. ∎

✓ **CHECK POINT 1**   Find the LCD of $\dfrac{3}{10x^2}$ and $\dfrac{7}{15x}$.

EXAMPLE 2 Finding the Least Common Denominator

Find the LCD of $\dfrac{3}{x+1}$ and $\dfrac{5}{x-1}$.

Solution

Step 1. Factor each denominator completely.

$$x + 1 = 1(x + 1)$$
$$x - 1 = 1(x - 1)$$

$$\dfrac{3}{x+1} \qquad\qquad \dfrac{5}{x-1}$$

Factors are 1 and $x + 1$.   Factors are 1 and $x - 1$.

Step 2. List the factors of the first denominator.

$$1, x + 1$$

Step 3. Add any unlisted factors from the second denominator. One factor of $x - 1$ is already in our list. That factor is 1. However, the other factor, $x - 1$, is not listed in step 2. We add $x - 1$ to the list. We have

$$1, x + 1, x - 1.$$

Step 4. The least common denominator is the product of all factors in the final list. Thus

$$1(x + 1)(x - 1)$$

or $(x + 1)(x - 1)$ is the least common denominator of $\dfrac{3}{x + 1}$ and $\dfrac{5}{x - 1}$. ■

✓ **CHECK POINT 2**   Find the LCD of $\dfrac{2}{x + 3}$ and $\dfrac{4}{x - 3}$.

EXAMPLE 3   Finding the Least Common Denominator

Find the LCD of

$$\frac{7}{5x^2 + 15x} \quad \text{and} \quad \frac{9}{x^2 + 6x + 9}.$$

Solution

Step 1. Factor each denominator completely.

$$5x^2 + 15x = 5x(x + 3)$$
$$x^2 + 6x + 9 = (x + 3)^2 \quad \text{or} \quad (x + 3)(x + 3)$$

$$\frac{7}{5x^2 + 15x} \qquad \frac{9}{x^2 + 6x + 9}$$

Factors are **5**, x, and $x + $ **3**.   Factors are $x + $ **3** and $x + $ **3**.

Step 2. List the factors of the first denominator.

$$5, x, x + 3$$

Step 3. Add any unlisted factors from the second denominator. One factor of $x^2 + 6x + 9$ is already in our list. That factor is $x + 3$. However, the other factor of $x + 3$ is not listed in step 2. We add $x + 3$ to the list. We have

$$5, x, x + 3, x + 3.$$

Step 4. The least common denominator is the product of all factors in the final list. Thus,

$$5x(x + 3)(x + 3) \quad \text{or} \quad 5x(x + 3)^2$$

is the least common denominator. ■

✓ **CHECK POINT 3**   Find the LCD of $\dfrac{9}{7x^2 + 28x}$ and $\dfrac{11}{x^2 + 8x + 16}$.

---

2 Add and subtract rational expressions with different denominators.

Adding and Subtracting Rational Expressions with Different Denominators

Finding the least common denominator for two (or more) rational expressions is the first step needed to add or subtract the expressions. For example, to add $\frac{1}{2}$ and $\frac{2}{3}$, we first determine that the LCD is 6. Then we write each fraction in terms of the LCD.

$$\frac{1}{2} + \frac{2}{3} = \frac{1}{2} \cdot \frac{3}{3} + \frac{2}{3} \cdot \frac{2}{2}$$

Multiply the numerator and denominator of each fraction by whatever extra factors are required to form 6, the LCD.

$\frac{3}{3} = 1$ and $\frac{2}{2} = 1$. Multiplying by 1 does not change a fraction's value.

$$= \frac{3}{6} + \frac{4}{6}$$

$$= \frac{3 + 4}{6} = \frac{7}{6}$$

Add numerators. Place this sum over the LCD.

We follow the same steps in adding or subtracting rational expressions with different denominators.

Adding and Subtracting Rational Expressions That Have Different Denominators

1. Find the LCD of the rational expressions.
2. Rewrite each rational expression as an equivalent expression whose denominator is the LCD. To do so, multiply the numerator and the denominator of each rational expression by any factor(s) needed to convert the denominator into the LCD.
3. Add or subtract numerators, placing the resulting expression over the LCD.
4. If possible, simplify the resulting rational expression.

EXAMPLE 4 Adding Rational Expressions with Different Denominators

Add: $\dfrac{7}{6x^2} + \dfrac{2}{9x}$.

Solution

Step 1. Find the least common denominator. In Example 1, we found that the LCD for these rational expressions is $18x^2$.

Step 2. Write equivalent expressions with the LCD as denominators. We must rewrite each rational expression with a denominator of $18x^2$.

$$\frac{7}{6x^2} \cdot \frac{3}{3} = \frac{21}{18x^2} \qquad \frac{2}{9x} \cdot \frac{2x}{2x} = \frac{4x}{18x^2}$$

Multiply the numerator and denominator by 3 to get $18x^2$, the LCD.

Multiply the numerator and denominator by $2x$ to get $18x^2$, the LCD.

Because $\dfrac{3}{3} = 1$ and $\dfrac{2x}{2x} = 1$, we are not changing the value of either rational expression, only its appearance.

Now we are ready to perform the indicated addition.

$$\frac{7}{6x^2} + \frac{2}{9x}$$

This is the given problem. The LCD is $18x^2$.

$$= \frac{7}{6x^2} \cdot \frac{3}{3} + \frac{2}{9x} \cdot \frac{2x}{2x}$$

Write equivalent expressions with the LCD.

$$= \frac{21}{18x^2} + \frac{4x}{18x^2}$$

Steps 3 and 4. Add numerators, putting this sum over the LCD. Simplify, if possible.

$$= \frac{21 + 4x}{18x^2} \qquad \text{or} \qquad \frac{4x + 21}{18x^2}$$

The numerator is prime and further simplification is not possible. ∎

Great Question!

Can't I just simplify things and add rational expressions by adding numerators and adding denominators?

No. It is incorrect to add rational expressions by adding numerators and adding denominators. Avoid this common error.

Incorrect!

$$\frac{7}{6x^2} + \frac{2}{9x}$$

$$= \frac{7 + 2}{6x^2 + 9x}$$

$$\neq \frac{9}{6x^2 + 9x}$$

✓ **CHECK POINT 4** Add: $\dfrac{3}{10x^2} + \dfrac{7}{15x}$.

EXAMPLE 5 Adding Rational Expressions with Different Denominators

Add: $\dfrac{3}{x+1} + \dfrac{5}{x-1}$.

Solution

Step 1. Find the least common denominator. The factors of the denominators are 1, $x + 1$, and $x - 1$. In Example 2, we found that the LCD is $(x + 1)(x - 1)$.

Step 2. Write equivalent expressions with the LCD as denominators.

$$\dfrac{3}{x+1} + \dfrac{5}{x-1}$$

$$= \dfrac{3(x-1)}{(x+1)(x-1)} + \dfrac{5(x+1)}{(x+1)(x-1)}$$

Multiply each numerator and denominator by the extra factor required to form $(x + 1)(x - 1)$, the LCD.

Steps 3 and 4. Add numerators, putting this sum over the LCD. Simplify, if possible.

$$= \dfrac{3(x-1) + 5(x+1)}{(x+1)(x-1)}$$

$$= \dfrac{3x - 3 + 5x + 5}{(x+1)(x-1)}$$

Use the distributive property to multiply and remove grouping symbols.

$$= \dfrac{8x + 2}{(x+1)(x-1)}$$

Combine like terms: $3x + 5x = 8x$ and $-3 + 5 = 2$. ∎

We can factor 2 from the numerator of the answer in Example 5 to obtain

$$\dfrac{2(4x+1)}{(x+1)(x-1)}.$$

Because the numerator and denominator do not have any common factors, further simplification is not possible. In this section, unless there is a common factor in the numerator and denominator, we will leave an answer's numerator in unfactored form and the denominator in factored form.

✓ **CHECK POINT 5** Add: $\dfrac{2}{x+3} + \dfrac{4}{x-3}$.

EXAMPLE 6 Subtracting Rational Expressions with Different Denominators

Subtract: $\dfrac{x}{x+3} - 1$.

Solution

Step 1. Find the least common denominator. We know that 1 means $\frac{1}{1}$. The factors of the first denominator are 1 and $x + 3$:

$$1, x + 3.$$

The factor of the second denominator, 1, is already in this list. Thus, the LCD is $1(x + 3)$, or $x + 3$.

Step 2. Write equivalent expressions with the LCD as denominators.

$$\frac{x}{x + 3} - 1$$

$$= \frac{x}{x + 3} - \frac{1}{1} \qquad \text{Write 1 as } \tfrac{1}{1}.$$

$$= \frac{x}{x + 3} - \frac{1(x + 3)}{1(x + 3)} \qquad \text{Multiply the numerator and denominator of } \tfrac{1}{1} \text{ by the extra factor required to form } x + 3, \text{ the LCD.}$$

Steps 3 and 4. Subtract numerators, putting this difference over the LCD. Simplify, if possible.

$$= \frac{x - (x + 3)}{x + 3}$$

$$= \frac{x - x - 3}{x + 3} \qquad \text{Remove parentheses and then change the sign of each term.}$$

$$= \frac{-3}{x + 3} \quad \text{or} \quad -\frac{3}{x + 3} \qquad \text{Simplify.} \ \blacksquare$$

✓ **CHECK POINT 6** Subtract: $\dfrac{x}{x + 5} - 1.$

EXAMPLE 7 Subtracting Rational Expressions with Different Denominators

Subtract: $\dfrac{y + 2}{4y + 16} - \dfrac{2}{y^2 + 4y}.$

Solution

Step 1. Find the least common denominator. Start by factoring the denominators.

$$4y + 16 = 4(y + 4)$$
$$y^2 + 4y = y(y + 4)$$

The factors of the first denominator are 4 and $y + 4$. The only factor from the second denominator that is unlisted is y. Thus, the least common denominator is $4y(y + 4)$.

Step 2. Write equivalent expressions with the LCD as denominators.

$$\frac{y + 2}{4y + 16} - \frac{2}{y^2 + 4y}$$

$$= \frac{y + 2}{4(y + 4)} - \frac{2}{y(y + 4)} \qquad \text{Factor denominators. The LCD is } 4y(y + 4).$$

$$= \frac{(y + 2)y}{4y(y + 4)} - \frac{2 \cdot 4}{4y(y + 4)} \qquad \text{Multiply each numerator and denominator by the extra factor required to form } 4y(y + 4), \text{ the LCD.}$$

Steps 3 and 4. Subtract numerators, putting this difference over the LCD. Simplify, if possible.

$$= \frac{(y + 2)y - 2 \cdot 4}{4y(y + 4)}$$

$$= \frac{y^2 + 2y - 8}{4y(y + 4)} \qquad \text{Use the distributive property.}$$

$$= \frac{\overset{1}{\cancel{(y + 4)}}(y - 2)}{4y\underset{1}{\cancel{(y + 4)}}} \qquad \text{Factor and simplify.}$$

$$= \frac{y - 2}{4y} \ \blacksquare$$

✓ **CHECK POINT 7** Subtract: $\dfrac{5}{y^2 - 5y} - \dfrac{y}{5y - 25}$.

In some situations, after factoring denominators, a factor in one denominator is the opposite of a factor in the other denominator. When this happens, we can use the following procedure:

Adding and Subtracting Rational Expressions when Denominators Contain Opposite Factors

When one denominator contains the opposite factor of the other, first multiply either rational expression by $\frac{-1}{-1}$. Then apply the procedure for adding or subtracting rational expressions that have different denominators to the rewritten problem.

EXAMPLE 8 Adding Rational Expressions with Opposite Factors in the Denominators

Add: $\dfrac{x^2 - 2}{2x^2 - x - 3} + \dfrac{x - 2}{3 - 2x}$.

Solution

Step 1. Find the least common denominator. Start by factoring the denominators.

$$2x^2 - x - 3 = (2x - 3)(x + 1)$$

$$3 - 2x = 1(3 - 2x)$$

Do you see that $2x - 3$ and $3 - 2x$ are opposite factors? Thus, we multiply either rational expression by $\frac{-1}{-1}$. We will use the second rational expression, resulting in $2x - 3$ in the denominator.

$$\frac{x^2 - 2}{2x^2 - x - 3} + \frac{x - 2}{3 - 2x}$$

$$= \frac{x^2 - 2}{(2x - 3)(x + 1)} + \frac{(-1)}{(-1)} \cdot \frac{x - 2}{3 - 2x} \qquad \text{Factor the first denominator. Multiply the second rational expression by } \tfrac{-1}{-1}.$$

$$= \frac{x^2 - 2}{(2x - 3)(x + 1)} + \frac{-x + 2}{-3 + 2x} \qquad \text{Perform the multiplications by } -1 \text{ by changing every term's sign.}$$

$$= \frac{x^2 - 2}{(2x - 3)(x + 1)} + \frac{2 - x}{2x - 3} \qquad \text{Rewrite } -3 + 2x \text{ as } 2x - 3.$$

The LCD of our rewritten addition problem is $(2x - 3)(x + 1)$.

Step 2. Write equivalent expressions with the LCD as denominators.

$$= \frac{x^2 - 2}{(2x - 3)(x + 1)} + \frac{(2 - x)(x + 1)}{(2x - 3)(x + 1)} \qquad \text{Multiply the numerator and denominator of the second rational expression by the extra factor required to form } (2x - 3)(x + 1), \text{ the LCD.}$$

Here is the original addition problem and our equivalent expressions with the LCD:

$$\frac{x^2 - 2}{2x^2 - x - 3} + \frac{x - 2}{3 - 2x}$$

$$= \frac{x^2 - 2}{(2x - 3)(x + 1)} + \frac{(2 - x)(x + 1)}{(2x - 3)(x + 1)}.$$

Discover for Yourself

In Example 8, the denominators can be factored as follows:

$$2x^2 - x - 3 = (2x - 3)(x + 1)$$

$$3 - 2x = -1(2x - 3).$$

Using these factorizations, what is the LCD? Solve Example 8 by obtaining this LCD in each rational expression. Then combine the expressions. How does your solution compare with the one shown on the right?

Steps 3 and 4. Add numerators, putting this sum over the LCD. Simplify, if possible.

$$= \frac{x^2 - 2 + (2 - x)(x + 1)}{(2x - 3)(x + 1)}$$

$$= \frac{x^2 - 2 + 2x + 2 - x^2 - x}{(2x - 3)(x + 1)} \qquad \text{Use the FOIL method to multiply } (2 - x)(x + 1).$$

$$= \frac{(x^2 - x^2) + (2x - x) + (-2 + 2)}{(2x - 3)(x + 1)} \qquad \text{Group like terms.}$$

$$= \frac{x}{(2x - 3)(x + 1)} \qquad \text{Combine like terms.} \qquad ■$$

✓ **CHECK POINT 8**   Add: $\dfrac{4x}{x^2 - 25} + \dfrac{3}{5 - x}$.

Achieving Success

Take some time to consider how you are doing in this course. Check your performance by answering the following questions:

- Are you attending all lectures?
- For each hour of class time, are you spending at least two hours outside of class completing all homework assignments, checking answers, correcting errors, and using all resources to get the help that you need?
- Are you reviewing for quizzes and tests?
- Are you reading the textbook? In all college courses, you are responsible for the information in the text, whether or not it is covered in class.
- Are you keeping an organized notebook? Does each page have the appropriate section number from the text on top? Do the pages contain examples your instructor works during lecture and other relevant class notes? Have you included your worked-out homework exercises? Do you keep a special section for graded exams?
- Are you analyzing your mistakes and learning from your errors?
- Are there ways you can improve how you are doing in the course?

CONCEPT AND VOCABULARY CHECK

Fill in each blank so that the resulting statement is true.

1. The first step in finding the least common denominator of $\dfrac{7}{x^2 - 1}$ and $\dfrac{x}{x^2 - 2x + 1}$ is to _____.

2. Consider the following addition problem:

$$\frac{7}{x(x + 3)} + \frac{10}{(x + 3)(x + 5)}.$$

The factors of the first denominator are _____. The factors of the second denominator are _____. The LCD is _____.

3. An equivalent expression for $\dfrac{7}{8x}$ with a denominator of $24x$ can be obtained by multiplying the numerator and denominator by _____.

4. An equivalent expression for $\dfrac{x}{x + 5}$ with a denominator of $x^2 - 25$ can be obtained by multiplying the numerator and denominator by _____.

5. An equivalent expression for $\dfrac{7}{9 - 5x}$ with a denominator of $5x - 9$ can be obtained by multiplying the numerator and denominator by _____.

7.4 EXERCISE SET MyMathLab®

Practice Exercises

In Exercises 1–16, find the least common denominator of the rational expressions.

1. $\dfrac{7}{15x^2}$ and $\dfrac{13}{24x}$

2. $\dfrac{11}{25x^2}$ and $\dfrac{17}{35x}$

3. $\dfrac{8}{15x^2}$ and $\dfrac{5}{6x^5}$

4. $\dfrac{7}{15x^2}$ and $\dfrac{11}{24x^5}$

5. $\dfrac{4}{x-3}$ and $\dfrac{7}{x+1}$

6. $\dfrac{2}{x-5}$ and $\dfrac{3}{x+7}$

7. $\dfrac{5}{7(y+2)}$ and $\dfrac{10}{y}$

8. $\dfrac{8}{11(y+5)}$ and $\dfrac{12}{y}$

9. $\dfrac{17}{x+4}$ and $\dfrac{18}{x^2-16}$

10. $\dfrac{3}{x-6}$ and $\dfrac{4}{x^2-36}$

11. $\dfrac{8}{y^2-9}$ and $\dfrac{14}{y(y+3)}$

12. $\dfrac{14}{y^2-49}$ and $\dfrac{12}{y(y-7)}$

13. $\dfrac{7}{y^2-1}$ and $\dfrac{y}{y^2-2y+1}$

14. $\dfrac{9}{y^2-25}$ and $\dfrac{y}{y^2-10y+25}$

15. $\dfrac{3}{x^2-x-20}$ and $\dfrac{x}{2x^2+7x-4}$

16. $\dfrac{7}{x^2-5x-6}$ and $\dfrac{x}{x^2-4x-5}$

In Exercises 17–82, add or subtract as indicated. Simplify the result, if possible.

17. $\dfrac{3}{x}+\dfrac{5}{x^2}$

18. $\dfrac{4}{x}+\dfrac{8}{x^2}$

19. $\dfrac{2}{9x}+\dfrac{11}{6x}$

20. $\dfrac{5}{6x}+\dfrac{7}{8x}$

21. $\dfrac{4}{x}+\dfrac{7}{2x^2}$

22. $\dfrac{10}{x}+\dfrac{3}{5x^2}$

23. $6+\dfrac{1}{x}$

24. $3+\dfrac{1}{x}$

25. $\dfrac{2}{x}+9$

26. $\dfrac{7}{x}+4$

27. $\dfrac{x-1}{6}+\dfrac{x+2}{3}$

28. $\dfrac{x+3}{2}+\dfrac{x+5}{4}$

29. $\dfrac{4}{x}+\dfrac{3}{x-5}$

30. $\dfrac{3}{x}+\dfrac{4}{x-6}$

31. $\dfrac{2}{x-1}+\dfrac{3}{x+2}$

32. $\dfrac{3}{x-2}+\dfrac{4}{x+3}$

33. $\dfrac{2}{y+5}+\dfrac{3}{4y}$

34. $\dfrac{3}{y+1}+\dfrac{2}{3y}$

35. $\dfrac{x}{x+7}-1$

36. $\dfrac{x}{x+6}-1$

37. $\dfrac{7}{x+5}-\dfrac{4}{x-5}$

38. $\dfrac{8}{x+6}-\dfrac{2}{x-6}$

39. $\dfrac{2x}{x^2-16}+\dfrac{x}{x-4}$

40. $\dfrac{4x}{x^2-25}+\dfrac{x}{x+5}$

41. $\dfrac{5y}{y^2-9}-\dfrac{4}{y+3}$

42. $\dfrac{8y}{y^2-16}-\dfrac{5}{y+4}$

43. $\dfrac{7}{x-1}-\dfrac{3}{(x-1)^2}$

44. $\dfrac{5}{x+3}-\dfrac{2}{(x+3)^2}$

45. $\dfrac{3y}{4y-20}+\dfrac{9y}{6y-30}$

46. $\dfrac{4y}{5y-10}+\dfrac{3y}{10y-20}$

47. $\dfrac{y+4}{y}-\dfrac{y}{y+4}$

48. $\dfrac{y}{y-5}-\dfrac{y-5}{y}$

49. $\dfrac{2x+9}{x^2-7x+12}-\dfrac{2}{x-3}$

50. $\dfrac{3x+7}{x^2-5x+6}-\dfrac{3}{x-3}$

51. $\dfrac{3}{x^2-1}+\dfrac{4}{(x+1)^2}$

52. $\dfrac{6}{x^2-4}+\dfrac{2}{(x+2)^2}$

53. $\dfrac{3x}{x^2+3x-10}-\dfrac{2x}{x^2+x-6}$

54. $\dfrac{x}{x^2-2x-24}-\dfrac{x}{x^2-7x+6}$

55. $\dfrac{y}{y^2+2y+1}+\dfrac{4}{y^2+5y+4}$

56. $\dfrac{y}{y^2 + 5y + 6} + \dfrac{4}{y^2 - y - 6}$

57. $\dfrac{x - 5}{x + 3} + \dfrac{x + 3}{x - 5}$

58. $\dfrac{x - 7}{x + 4} + \dfrac{x + 4}{x - 7}$

59. $\dfrac{5}{2y^2 - 2y} - \dfrac{3}{2y - 2}$

60. $\dfrac{7}{5y^2 - 5y} - \dfrac{2}{5y - 5}$

61. $\dfrac{4x + 3}{x^2 - 9} - \dfrac{x + 1}{x - 3}$

62. $\dfrac{2x - 1}{x + 6} - \dfrac{6 - 5x}{x^2 - 36}$

63. $\dfrac{y^2 - 39}{y^2 + 3y - 10} - \dfrac{y - 7}{y - 2}$

64. $\dfrac{y^2 - 6}{y^2 + 9y + 18} - \dfrac{y - 4}{y + 6}$

65. $4 + \dfrac{1}{x - 3}$ **66.** $7 + \dfrac{1}{x - 5}$

67. $3 - \dfrac{3y}{y + 1}$ **68.** $7 - \dfrac{4y}{y + 5}$

69. $\dfrac{9x + 3}{x^2 - x - 6} + \dfrac{x}{3 - x}$

70. $\dfrac{x^2 + 9x}{x^2 - 2x - 3} + \dfrac{5}{3 - x}$

71. $\dfrac{x + 3}{x^2 + x - 2} - \dfrac{2}{x^2 - 1}$

72. $\dfrac{x}{x^2 - 10x + 25} - \dfrac{x - 4}{2x - 10}$

73. $\dfrac{y + 3}{5y^2} - \dfrac{y - 5}{15y}$

74. $\dfrac{y - 7}{3y^2} - \dfrac{y - 2}{12y}$

75. $\dfrac{x + 3}{3x + 6} + \dfrac{x}{4 - x^2}$

76. $\dfrac{x + 7}{4x + 12} + \dfrac{x}{9 - x^2}$

77. $\dfrac{y}{y^2 - 1} + \dfrac{2y}{y - y^2}$

78. $\dfrac{y}{y^2 - 1} + \dfrac{5y}{y - y^2}$

79. $\dfrac{x - 1}{x} + \dfrac{y + 1}{y}$

80. $\dfrac{x + 2}{y} + \dfrac{y - 2}{x}$

81. $\dfrac{3x}{x^2 - y^2} - \dfrac{2}{y - x}$

82. $\dfrac{7x}{x^2 - y^2} - \dfrac{3}{y - x}$

Practice PLUS

In Exercises 83–92, perform the indicated operation or operations. Simplify the result, if possible.

83. $\dfrac{x + 6}{x^2 - 4} - \dfrac{x + 3}{x + 2} + \dfrac{x - 3}{x - 2}$

84. $\dfrac{x + 8}{x^2 - 9} - \dfrac{x + 2}{x + 3} + \dfrac{x - 2}{x - 3}$

85. $\dfrac{5}{x^2 - 25} + \dfrac{4}{x^2 - 11x + 30} - \dfrac{3}{x^2 - x - 30}$

86. $\dfrac{3}{x^2 - 49} + \dfrac{2}{x^2 - 15x + 56} - \dfrac{5}{x^2 - x - 56}$

87. $\dfrac{x + 6}{x^3 - 27} - \dfrac{x}{x^3 + 3x^2 + 9x}$

88. $\dfrac{x + 8}{x^3 - 8} - \dfrac{x}{x^3 + 2x^2 + 4x}$

89. $\dfrac{9y + 3}{y^2 - y - 6} + \dfrac{y}{3 - y} + \dfrac{y - 1}{y + 2}$

90. $\dfrac{7y - 2}{y^2 - y - 12} + \dfrac{2y}{4 - y} + \dfrac{y + 1}{y + 3}$

91. $\dfrac{3}{x^2 + 4xy + 3y^2} - \dfrac{5}{x^2 - 2xy - 3y^2} + \dfrac{2}{x^2 - 9y^2}$

92. $\dfrac{5}{x^2 + 3xy + 2y^2} - \dfrac{7}{x^2 - xy - 2y^2} + \dfrac{4}{x^2 - 4y^2}$

Application Exercises

Two formulas that approximate the dosage of a drug prescribed for children are

$$\text{Young's rule: } C = \dfrac{DA}{A + 12}$$

$$\text{and Cowling's rule: } C = \dfrac{D(A + 1)}{24}.$$

In each formula, A = the child's age, in years, D = an adult dosage, and C = the proper child's dosage. The formulas apply for ages 2 through 13, inclusive. Use the formulas to solve Exercises 93–96.

93. Use Young's rule to find the difference in a child's dosage for an 8-year-old child and a 3-year-old child. Express the answer as a single rational expression in terms of D. Then describe what your answer means in terms of the variables in the model.

94. Use Young's rule to find the difference in a child's dosage for a 10-year-old child and a 3-year-old child. Express the answer as a single rational expression in terms of D. Then describe what your answer means in terms of the variables in the model.

95. For a 12-year-old child, what is the difference in the dosage given by Cowling's rule and Young's rule? Express the answer as a single rational expression in terms of D. Then describe what your answer means in terms of the variables in the models.

96. Use Cowling's rule to find the difference in a child's dosage for a 12-year-old child and a 10-year-old child. Express the answer as a single rational expression in terms of D. Then describe what your answer means in terms of the variables in the model.

The graphs illustrate Young's rule and Cowling's rule when the dosage of a drug prescribed for an adult is 1000 milligrams. Use the graphs to solve Exercises 97–100.

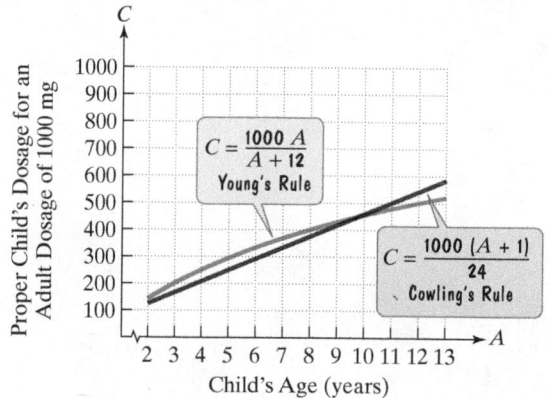

97. Does either formula consistently give a smaller dosage than the other? If so, which one?

98. Is there an age at which the dosage given by one formula becomes greater than the dosage given by the other? If so, what is a reasonable estimate of that age?

99. For what age under 11 is the difference in dosage given by the two formulas the greatest?

100. For what age over 11 is the difference in dosage given by the two formulas the greatest?

In Exercises 101–102, express the perimeter of each rectangle as a single rational expression.

101.

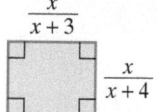

102.

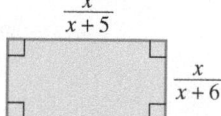

Writing in Mathematics

103. Explain how to find the least common denominator for denominators of $x^2 - 100$ and $x^2 - 20x + 100$.

104. Explain how to add rational expressions that have different denominators. Use $\dfrac{3}{x+5} + \dfrac{7}{x+2}$ in your explanation.

Explain the error in Exercises 105–106. Then rewrite the right side of the equation to correct the error that now exists.

105. $\dfrac{1}{x} + \dfrac{2}{5} = \dfrac{3}{x+5}$

106. $\dfrac{1}{x} + 7 = \dfrac{1}{x+7}$

107. The formulas in Exercises 93–96 relate the dosage of a drug prescribed for children to the child's age. Describe another factor that might be used when determining a child's dosage. Is this factor more or less important than age? Explain why.

Critical Thinking Exercises

Make Sense? *In Exercises 108–111, determine whether each statement "makes sense" or "does not make sense" and explain your reasoning.*

108. It takes me more steps to find $\dfrac{2}{x+5} + \dfrac{7}{x-3}$ than it does to find $\dfrac{2x^3}{x+5} + \dfrac{7x^3}{x+5}$.

109. The reason I can rewrite rational expressions with a common denominator is that 1 is the multiplicative identity.

110. The fastest way for me to find $\dfrac{4}{x-5} + \dfrac{9}{5-x}$ is by using $(x-5)(5-x)$ as the LCD.

111. After adding rational expressions with different denominators, I factored the numerator and found no common factors in the numerator and denominator, so my final answer is incorrect if I leave the numerator in factored form.

In Exercises 112–115, determine whether each statement is true or false. If the statement is false, make the necessary change(s) to produce a true statement.

112. $x - \dfrac{1}{5} = \dfrac{4}{5}x$

113. The LCD of $\dfrac{1}{x}$ and $\dfrac{2x}{x-1}$ is $x^2 - 1$.

114. $\dfrac{1}{x} + \dfrac{x}{1} = \dfrac{1}{\cancel{x}} + \dfrac{\cancel{x}}{1} = 1 + 1 = 2$

115. $\dfrac{2}{x} + 1 = \dfrac{2+x}{x}, x \neq 0$

In Exercises 116–117, perform the indicated operations. Simplify the result, if possible.

116. $\dfrac{y^2 + 5y + 4}{y^2 + 2y - 3} \cdot \dfrac{y^2 + y - 6}{y^2 + 2y - 3} - \dfrac{2}{y-1}$

117. $\left(\dfrac{1}{x+h} - \dfrac{1}{x}\right) \div h$

In Exercises 118–119, find the missing rational expression.

118. $\dfrac{2}{x-1} + \dfrac{\blacksquare}{\blacksquare} = \dfrac{2x^2 + 3x - 1}{x^2(x-1)}$

119. $\dfrac{4}{x-2} - \dfrac{\blacksquare}{\blacksquare} = \dfrac{2x+8}{(x-2)(x+1)}$

Review Exercises

120. Multiply: $(3x + 5)(2x - 7)$. (Section 5.3, Example 2)

121. Graph: $3x - y = 3$. (Section 3.2, Example 5)

122. Write the slope-intercept form of the equation of the line passing through $(-3, -4)$ and $(1, 0)$. (Section 3.5, Example 2)

Preview Exercises

Exercises 123–125 will help you prepare for the material covered in the next section.

123 a. Add: $\dfrac{1}{3} + \dfrac{2}{5}$.

     b. Subtract: $\dfrac{2}{5} - \dfrac{1}{3}$.

c. Use your answers from parts (a) and (b) to find
$$\left(\frac{1}{3} + \frac{2}{5}\right) \div \left(\frac{2}{5} - \frac{1}{3}\right).$$

124. a. Add: $\dfrac{1}{x} + \dfrac{1}{y}$.

     b. Use your answer from part (a) to find
$$\frac{1}{xy} \div \left(\frac{1}{x} + \frac{1}{y}\right).$$

125. Multiply and simplify: $xy\left(\dfrac{1}{x} + \dfrac{1}{y}\right)$.

MID-CHAPTER CHECK POINT   Section 7.1–Section 7.4

What You Know: We learned that it is necessary to exclude any value or values of a variable that make the denominator of a rational expression zero. We learned to simplify rational expressions by dividing the numerator and the denominator by common factors. We performed a variety of operations with rational expressions, including multiplication, division, addition, and subtraction.

1. Find all numbers for which $\dfrac{x^2 - 4}{x^2 - 2x - 8}$ is undefined.

In Exercises 2–4, simplify each rational expression.

2. $\dfrac{3x^2 - 7x + 2}{6x^2 + x - 1}$

3. $\dfrac{9 - 3y}{y^2 - 5y + 6}$

4. $\dfrac{16w^3 - 24w^2}{8w^4 - 12w^3}$

In Exercises 5–20, perform the indicated operations. Simplify the result, if possible.

5. $\dfrac{7x - 3}{x^2 + 3x - 4} - \dfrac{3x + 1}{x^2 + 3x - 4}$

6. $\dfrac{x + 2}{2x - 4} \cdot \dfrac{8}{x^2 - 4}$

7. $1 + \dfrac{7}{x - 2}$

8. $\dfrac{2x^2 + x - 1}{2x^2 - 7x + 3} \div \dfrac{x^2 - 3x - 4}{x^2 - x - 6}$

9. $\dfrac{1}{x^2 + 2x - 3} + \dfrac{1}{x^2 + 5x + 6}$

10. $\dfrac{17}{x - 5} + \dfrac{x + 8}{5 - x}$

11. $\dfrac{4y^2 - 1}{9y - 3y^2} \cdot \dfrac{y^2 - 7y + 12}{2y^2 - 7y - 4}$

12. $\dfrac{y}{y + 1} - \dfrac{2y}{y + 2}$

13. $\dfrac{w^2 + 6w + 5}{7w^2 - 63} \div \dfrac{w^2 + 10w + 25}{7w + 21}$

14. $\dfrac{2z}{z^2 - 9} - \dfrac{5}{z^2 + 4z + 3}$

15. $\dfrac{z + 2}{3z - 1} + \dfrac{5}{(3z - 1)^2}$

16. $\dfrac{8}{x^2 + 4x - 21} + \dfrac{3}{x + 7}$

17. $\dfrac{x^4 - 27x}{x^2 - 9} \cdot \dfrac{x + 3}{x^2 + 3x + 9}$

18. $\dfrac{x - 1}{x^2 - x - 2} - \dfrac{x + 2}{x^2 + 4x + 3}$

19. $\dfrac{x^2 - 2xy + y^2}{x + y} \div \dfrac{x^2 - xy}{5x + 5y}$

20. $\dfrac{5}{x + 5} + \dfrac{x}{x - 4} - \dfrac{11x - 8}{x^2 + x - 20}$

<table>
<tr><td>

SECTION

7.5

Objectives

1 Simplify complex rational expressions by dividing.

2 Simplify complex rational expressions by multiplying by the LCD.

</td><td></td></tr>
</table>

Complex Rational Expressions

Do you drive to and from campus each day? If the one-way distance of your round-trip commute is d, then your average rate, or speed, is given by the expression

$$\frac{2d}{\dfrac{d}{r_1} + \dfrac{d}{r_2}},$$

in which r_1 and r_2 are your average rates on the outgoing and return trips, respectively. Do you notice anything unusual about this expression? It has two separate rational expressions in its denominator.

Numerator

Main fraction bar

$$\frac{2d}{\dfrac{d}{r_1} + \dfrac{d}{r_2}}$$

Denominator

Separate rational expressions occur in the denominator.

Complex rational expressions, also called **complex fractions**, have numerators or denominators containing one or more rational expressions. Here is another example of such an expression:

Numerator

Main fraction bar

$$\frac{1 + \dfrac{1}{x}}{1 - \dfrac{1}{x}}.$$

Denominator

Separate rational expressions occur in the numerator and denominator.

In this section, we study two methods for *simplifying* complex rational expressions. A complex rational expression is **simplified** when it is in the form $\frac{P}{Q}$, where P and Q are polynomials that have no common factors.

1 Simplify complex rational expressions by dividing.

Simplifying by Rewriting Complex Rational Expressions as a Quotient of Two Rational Expressions

One method for simplifying a complex rational expression is to combine its numerator into a single expression and combine its denominator into a single expression. Then perform the division by inverting the denominator and multiplying.

Simplifying a Complex Rational Expression by Dividing

1. If necessary, add or subtract to get a single rational expression in the numerator.
2. If necessary, add or subtract to get a single rational expression in the denominator.
3. Perform the division indicated by the main fraction bar: Invert the denominator of the complex rational expression and multiply.
4. If possible, simplify.

The following examples illustrate the use of this first method.

EXAMPLE 1 Simplifying a Complex Rational Expression

Simplify:

$$\dfrac{\dfrac{1}{3} + \dfrac{2}{5}}{\dfrac{2}{5} - \dfrac{1}{3}}.$$

Solution Let's first identify the parts of this complex rational expression.

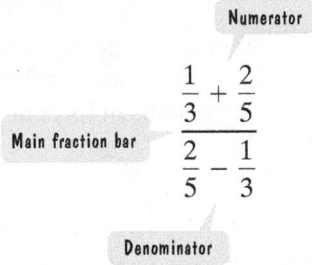

Numerator

Main fraction bar

Denominator

Step 1. Add to get a single rational expression in the numerator.

$$\frac{1}{3} + \frac{2}{5} = \frac{1 \cdot 5}{3 \cdot 5} + \frac{2 \cdot 3}{5 \cdot 3} = \frac{5}{15} + \frac{6}{15} = \frac{11}{15}$$

The LCD is 3 · 5, or 15.

Step 2. Subtract to get a single rational expression in the denominator.

$$\frac{2}{5} - \frac{1}{3} = \frac{2 \cdot 3}{5 \cdot 3} - \frac{1 \cdot 5}{3 \cdot 5} = \frac{6}{15} - \frac{5}{15} = \frac{1}{15}$$

The LCD is 15.

Steps 3 and 4. Perform the division indicated by the main fraction bar: Invert and multiply. If possible, simplify.

$$\dfrac{\dfrac{1}{3} + \dfrac{2}{5}}{\dfrac{2}{5} - \dfrac{1}{3}} = \dfrac{\dfrac{11}{15}}{\dfrac{1}{15}} = \frac{11}{15} \cdot \frac{15}{1} = \frac{11}{\overset{1}{\cancel{15}}} \cdot \frac{\overset{1}{\cancel{15}}}{1} = 11$$

Invert and multiply. ∎

✓ **CHECK POINT 1** Simplify: $\dfrac{\dfrac{1}{4} + \dfrac{2}{3}}{\dfrac{2}{3} - \dfrac{1}{4}}.$

EXAMPLE 2 Simplifying a Complex Rational Expression

Simplify:

$$\frac{1 + \dfrac{1}{x}}{1 - \dfrac{1}{x}}.$$

Solution

Step 1. Add to get a single rational expression in the numerator.

$$1 + \frac{1}{x} = \frac{1}{1} + \frac{1}{x} = \frac{1 \cdot x}{1 \cdot x} + \frac{1}{x} = \frac{x}{x} + \frac{1}{x} = \frac{x+1}{x}$$

The LCD is $1 \cdot x$, or x.

Step 2. Subtract to get a single rational expression in the denominator.

$$1 - \frac{1}{x} = \frac{1}{1} - \frac{1}{x} = \frac{1 \cdot x}{1 \cdot x} - \frac{1}{x} = \frac{x}{x} - \frac{1}{x} = \frac{x-1}{x}$$

The LCD is $1 \cdot x$, or x.

Steps 3 and 4. Perform the division indicated by the main fraction bar: Invert and multiply. If possible, simplify.

$$\frac{1 + \dfrac{1}{x}}{1 - \dfrac{1}{x}} = \frac{\dfrac{x+1}{x}}{\dfrac{x-1}{x}} = \frac{x+1}{x} \cdot \frac{x}{x-1} = \frac{x+1}{\overset{1}{\cancel{x}}} \cdot \frac{\overset{1}{\cancel{x}}}{x-1} = \frac{x+1}{x-1}$$

Invert and multiply. ■

✓ **CHECK POINT 2** Simplify: $\dfrac{2 - \dfrac{1}{x}}{2 + \dfrac{1}{x}}.$

EXAMPLE 3 Simplifying a Complex Rational Expression

Simplify:

$$\frac{\dfrac{1}{xy}}{\dfrac{1}{x} + \dfrac{1}{y}}.$$

Solution

Step 1. Get a single rational expression in the numerator. The numerator, $\dfrac{1}{xy}$, already contains a single rational expression, so we can skip this step.

Step 2. Add to get a single rational expression in the denominator.

$$\frac{1}{x} + \frac{1}{y} = \frac{1 \cdot y}{x \cdot y} + \frac{1 \cdot x}{y \cdot x} = \frac{y}{xy} + \frac{x}{xy} = \frac{y+x}{xy}$$

The LCD is xy.

Steps 3 and 4. Perform the division indicated by the main fraction bar: Invert and multiply. If possible, simplify.

$$\frac{\dfrac{1}{xy}}{\dfrac{1}{x}+\dfrac{1}{y}} = \frac{\dfrac{1}{xy}}{\dfrac{y+x}{xy}} = \frac{1}{xy} \cdot \frac{xy}{y+x} = \frac{1}{\cancel{xy}} \cdot \frac{\overset{1}{\cancel{xy}}}{y+x} = \frac{1}{y+x}$$

Invert and multiply. ∎

☑ **CHECK POINT 3**   Simplify: $\dfrac{\dfrac{1}{x}-\dfrac{1}{y}}{\dfrac{1}{xy}}$.

---

2 Simplify complex rational expressions by multiplying by the LCD.

Simplifying Complex Rational Expressions by Multiplying by the LCD

A second method for simplifying a complex rational expression is to find the least common denominator of all the rational expressions in its numerator and denominator. Then multiply each term in its numerator and denominator by this least common denominator. Because we are multiplying by a form of 1, we will obtain an equivalent expression that does not contain fractions in its numerator or denominator.

> **Simplifying a Complex Rational Expression by Multiplying by the LCD**
>
> 1. Find the LCD of all rational expressions within the complex rational expression.
> 2. Multiply both the numerator and the denominator of the complex rational expression by this LCD.
> 3. Use the distributive property and multiply each term in the numerator and denominator by this LCD. Simplify each term. No fractional expressions should remain.
> 4. If possible, factor and simplify.

We now rework Examples 1, 2, and 3 using the method of multiplying by the LCD. Compare the two simplification methods to see if there is one method that you prefer.

EXAMPLE 4   Simplifying a Complex Rational Expression by the LCD Method

Simplify:

$$\frac{\dfrac{1}{3}+\dfrac{2}{5}}{\dfrac{2}{5}-\dfrac{1}{3}}.$$

Solution The denominators in the complex rational expression are 3, 5, 5, and 3. The LCD is $3 \cdot 5$, or 15. Multiply both the numerator and denominator of the complex rational expression by 15.

$$\frac{\dfrac{1}{3} + \dfrac{2}{5}}{\dfrac{2}{5} - \dfrac{1}{3}} = \frac{15}{15} \cdot \frac{\left(\dfrac{1}{3} + \dfrac{2}{5}\right)}{\left(\dfrac{2}{5} - \dfrac{1}{3}\right)} = \frac{15 \cdot \dfrac{1}{3} + 15 \cdot \dfrac{2}{5}}{15 \cdot \dfrac{2}{5} - 15 \cdot \dfrac{1}{3}} = \frac{5 + 6}{6 - 5} = \frac{11}{1} = 11$$

$\frac{15}{15} = 1$, so we are not changing the complex fraction's value. ∎

✓ **CHECK POINT 4** Simplify by the LCD method: $\dfrac{\dfrac{1}{4} + \dfrac{2}{3}}{\dfrac{2}{3} - \dfrac{1}{4}}$.

EXAMPLE 5 Simplifying a Complex Rational Expression by the LCD Method

Simplify:

$$\frac{1 + \dfrac{1}{x}}{1 - \dfrac{1}{x}}.$$

Solution The denominators in the complex rational expression are 1, x, 1, and x.

$$\frac{1 + \dfrac{1}{x}}{1 - \dfrac{1}{x}} = \frac{\dfrac{1}{1} + \dfrac{1}{x}}{\dfrac{1}{1} - \dfrac{1}{x}}. \quad \text{Denominators}$$

Denominators

The LCD is $1 \cdot x$, or x. Multiply both the numerator and denominator of the complex rational expression by x.

$$\frac{1 + \dfrac{1}{x}}{1 - \dfrac{1}{x}} = \frac{x}{x} \cdot \frac{\left(1 + \dfrac{1}{x}\right)}{\left(1 - \dfrac{1}{x}\right)} = \frac{x \cdot 1 + x \cdot \dfrac{1}{x}}{x \cdot 1 - x \cdot \dfrac{1}{x}} = \frac{x + 1}{x - 1}$$

∎

✓ **CHECK POINT 5** Simplify by the LCD method: $\dfrac{2 - \dfrac{1}{x}}{2 + \dfrac{1}{x}}$.

EXAMPLE 6 Simplifying a Complex Rational Expression by the LCD Method

Simplify:

$$\frac{\dfrac{1}{xy}}{\dfrac{1}{x} + \dfrac{1}{y}}.$$

Solution The denominators in the complex rational expression are xy, x, and y. The LCD is xy. Multiply both the numerator and denominator of the complex rational expression by xy.

$$\frac{\frac{1}{xy}}{\frac{1}{x}+\frac{1}{y}} = \frac{xy}{xy}\cdot\frac{\left(\frac{1}{xy}\right)}{\left(\frac{1}{x}+\frac{1}{y}\right)} = \frac{xy\cdot\frac{1}{xy}}{xy\cdot\frac{1}{x}+xy\cdot\frac{1}{y}} = \frac{1}{y+x}.$$

■

☑ **CHECK POINT 6** Simplify by the LCD method: $\dfrac{\frac{1}{x}-\frac{1}{y}}{\frac{1}{xy}}$.

CONCEPT AND VOCABULARY CHECK

Fill in each blank so that the resulting statement is true.

1. A rational expression whose numerator or denominator or both contains rational expressions is called a/an _____ rational expression or a/an _____ fraction.

2. $\dfrac{\frac{3}{4}+\frac{1}{2}}{\frac{3}{8}-\frac{1}{6}} = \dfrac{24}{24}\cdot\dfrac{\left(\frac{3}{4}+\frac{1}{2}\right)}{\left(\frac{3}{8}-\frac{1}{6}\right)} = \dfrac{24\cdot\frac{3}{4}+24\cdot\frac{1}{2}}{24\cdot\frac{3}{8}-24\cdot\frac{1}{6}} = \dfrac{\underline{}+\underline{}}{\underline{}-\underline{}} = \dfrac{}{}=\dfrac{}{}=\underline{}$

3. $\dfrac{\frac{5}{x}+\frac{x}{3}}{\frac{4}{x}-\frac{x}{6}} = \dfrac{6x}{6x}\cdot\dfrac{\left(\frac{5}{x}+\frac{x}{3}\right)}{\left(\frac{4}{x}-\frac{x}{6}\right)} = \dfrac{6x\cdot\frac{5}{x}+6x\cdot\frac{x}{3}}{6x\cdot\frac{4}{x}-6x\cdot\frac{x}{6}} = \dfrac{\underline{}+\underline{}}{\underline{}-\underline{}}$

4. $\dfrac{\frac{2}{x}+\frac{3}{x^2}}{\frac{5}{x}+1} = \dfrac{x^2}{x^2}\cdot\dfrac{\left(\frac{2}{x}+\frac{3}{x^2}\right)}{\left(\frac{5}{x}+1\right)} = \dfrac{x^2\cdot\frac{2}{x}+x^2\cdot\frac{3}{x^2}}{x^2\cdot\frac{5}{x}+x^2\cdot1} = \dfrac{\underline{}+\underline{}}{\underline{}+\underline{}}$

7.5 EXERCISE SET   MyMathLab®

 Watch the videos in MyMathLab

 Download the MyDashBoard App

Practice Exercises

In Exercises 1–40, simplify each complex rational expression by the method of your choice.

1. $\dfrac{\frac{1}{2}+\frac{1}{4}}{\frac{1}{2}+\frac{1}{3}}$

2. $\dfrac{\frac{1}{3}+\frac{1}{4}}{\frac{1}{3}+\frac{1}{6}}$

3. $\dfrac{5+\frac{2}{5}}{7-\frac{1}{10}}$

4. $\dfrac{1+\frac{3}{5}}{2-\frac{1}{4}}$

5. $\dfrac{\frac{2}{5}-\frac{1}{3}}{\frac{2}{3}-\frac{3}{4}}$

6. $\dfrac{\frac{1}{2}-\frac{1}{4}}{\frac{3}{8}+\frac{1}{16}}$

7. $\dfrac{\frac{3}{4}-x}{\frac{3}{4}+x}$

8. $\dfrac{\frac{2}{3}-x}{\frac{2}{3}+x}$

9. $\dfrac{7-\frac{2}{x}}{5+\frac{1}{x}}$

10. $\dfrac{8+\frac{3}{x}}{1-\frac{7}{x}}$

11. $\dfrac{2+\frac{3}{y}}{1-\frac{7}{y}}$

12. $\dfrac{4-\frac{7}{y}}{3-\frac{2}{y}}$

13. $\dfrac{\frac{1}{y}-\frac{3}{2}}{\frac{1}{y}+\frac{3}{4}}$

14. $\dfrac{\frac{1}{y}-\frac{3}{4}}{\frac{1}{y}+\frac{2}{3}}$

15. $\dfrac{\dfrac{x}{5} - \dfrac{5}{x}}{\dfrac{1}{5} + \dfrac{1}{x}}$

16. $\dfrac{\dfrac{3}{x} + \dfrac{x}{3}}{\dfrac{x}{3} - \dfrac{3}{x}}$

17. $\dfrac{1 + \dfrac{1}{x}}{1 - \dfrac{1}{x^2}}$

18. $\dfrac{1 + \dfrac{2}{x}}{1 - \dfrac{4}{x^2}}$

19. $\dfrac{\dfrac{1}{7} - \dfrac{1}{y}}{\dfrac{7 - y}{7}}$

20. $\dfrac{\dfrac{1}{9} - \dfrac{1}{y}}{\dfrac{9 - y}{9}}$

21. $\dfrac{x + \dfrac{2}{y}}{\dfrac{x}{y}}$

22. $\dfrac{x - \dfrac{2}{y}}{\dfrac{x}{y}}$

23. $\dfrac{\dfrac{1}{x} + \dfrac{1}{y}}{xy}$

24. $\dfrac{\dfrac{1}{x} + \dfrac{1}{y}}{x + y}$

25. $\dfrac{\dfrac{x}{y} + \dfrac{1}{x}}{\dfrac{y}{x} + \dfrac{1}{x}}$

26. $\dfrac{\dfrac{1}{x} + \dfrac{1}{y}}{\dfrac{1}{x} - \dfrac{1}{y}}$

27. $\dfrac{\dfrac{1}{y} + \dfrac{2}{y^2}}{\dfrac{2}{y} + 1}$

28. $\dfrac{\dfrac{1}{y} + \dfrac{3}{y^2}}{\dfrac{3}{y} + 1}$

29. $\dfrac{\dfrac{12}{x^2} - \dfrac{3}{x}}{\dfrac{15}{x} - \dfrac{9}{x^2}}$

30. $\dfrac{\dfrac{8}{x^2} - \dfrac{2}{x}}{\dfrac{10}{x} - \dfrac{6}{x^2}}$

31. $\dfrac{2 + \dfrac{6}{y}}{1 - \dfrac{9}{y^2}}$

32. $\dfrac{3 + \dfrac{12}{y}}{1 - \dfrac{16}{y^2}}$

33. $\dfrac{\dfrac{1}{x + 2}}{1 + \dfrac{1}{x + 2}}$

34. $\dfrac{\dfrac{1}{x - 2}}{1 - \dfrac{1}{x - 2}}$

35. $\dfrac{x - 5 + \dfrac{3}{x}}{x - 7 + \dfrac{2}{x}}$

36. $\dfrac{x + 9 - \dfrac{7}{x}}{x - 6 + \dfrac{4}{x}}$

37. $\dfrac{\dfrac{3}{xy^2} + \dfrac{2}{x^2y}}{\dfrac{1}{x^2y} + \dfrac{2}{xy^3}}$

38. $\dfrac{\dfrac{2}{x^3y} + \dfrac{5}{xy^4}}{\dfrac{5}{x^3y} - \dfrac{3}{xy}}$

39. $\dfrac{\dfrac{3}{x + 1} - \dfrac{3}{x - 1}}{\dfrac{5}{x^2 - 1}}$

40. $\dfrac{\dfrac{3}{x + 2} - \dfrac{3}{x - 2}}{\dfrac{5}{x^2 - 4}}$

Practice PLUS

In Exercises 41–48, simplify each complex rational expression.

41. $\dfrac{\dfrac{6}{x^2 + 2x - 15} - \dfrac{1}{x - 3}}{\dfrac{1}{x + 5} + 1}$

42. $\dfrac{\dfrac{1}{x - 2} - \dfrac{6}{x^2 + 3x - 10}}{1 + \dfrac{1}{x - 2}}$

43. $\dfrac{y^{-1} - (y + 5)^{-1}}{5}$

44. $\dfrac{y^{-1} - (y + 2)^{-1}}{2}$

45. $\dfrac{\dfrac{1}{1 - \dfrac{1}{x}} - 1}{}$

46. $\dfrac{\dfrac{1}{1 - \dfrac{1}{x + 1}} - 1}{}$

47. $\dfrac{1}{1 + \dfrac{1}{1 + \dfrac{1}{x}}}$

48. $\dfrac{1}{1 + \dfrac{1}{1 + \dfrac{1}{2}}}$

Application Exercises

49. The average rate on a round-trip commute having a one-way distance d is given by the complex rational expression

$$\dfrac{2d}{\dfrac{d}{r_1} + \dfrac{d}{r_2}},$$

in which r_1 and r_2 are the average rates on the outgoing and return trips, respectively. Simplify the expression. Then find your average rate if you drive to campus averaging 40 miles per hour and return home on the same route averaging 30 miles per hour.

50. If two electrical resistors with resistances R_1 and R_2 are connected in parallel (see the figure), then the total resistance in the circuit is given by the complex rational expression

$$\dfrac{1}{\dfrac{1}{R_1} + \dfrac{1}{R_2}}.$$

Simplify the expression. Then find the total resistance if $R_1 = 10$ ohms and $R_2 = 20$ ohms.

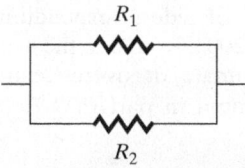

The bar graph shows the amount, in billions of dollars, that the United States government spent on human resources and total budget outlays for six selected years. (Human resources include education, health, Medicare, Social Security, and veterans benefits and services.)

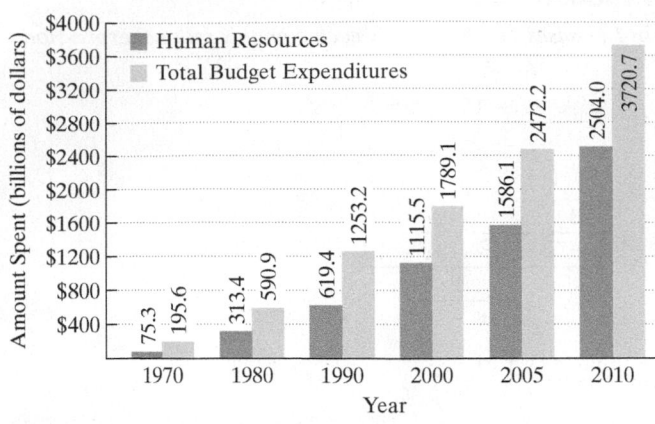

Federal Budget Expenditures on Human Resources

Source: Office of Management and Budget

The fraction of total budget outlays spent on human resources can be modeled by the complex rational expression

$$\dfrac{175 - \dfrac{1585}{x} + \dfrac{15{,}993}{x^2}}{206 - \dfrac{349}{x} + \dfrac{25{,}984}{x^2}},$$

where x is the number of years after 1970. Use this information to solve Exercises 51–52.

51. a. Simplify the complex rational expression for the fraction of federal expenditures spent on human resources.

b. Use the data displayed by the bar graph to find the percentage of federal expenditures spent on human resources in 2010. Round to the nearest percent.

c. Use the simplified rational expression from part (a) to find the percentage of federal expenditures spent on human resources in 2010. Round to the nearest percent. Does this underestimate or overestimate the actual percent that you found in part (b)? By how much?

52. a. Simplify the complex rational expression for the fraction of federal expenditures spent on human resources.

b. Use the data displayed by the bar graph to find the percentage of federal expenditures spent on human resources in 2005. Round to the nearest percent.

c. Use the simplified rational expression from part (a) to find the percentage of federal expenditures spent on human resources in 2005. Round to the nearest percent. Does this underestimate or overestimate the actual percent that you found in part (b)? By how much?

Writing in Mathematics

53. What is a complex rational expression? Give an example with your explanation.

54. Describe two ways to simplify $\dfrac{\dfrac{3}{x} + \dfrac{2}{x^2}}{\dfrac{1}{x^2} + \dfrac{2}{x}}$.

55. Which method do you prefer for simplifying complex rational expressions? Why?

Critical Thinking Exercises

Make Sense? *In Exercises 56–59, determine whether each statement "makes sense" or "does not make sense" and explain your reasoning.*

56. I simplified $\dfrac{\dfrac{1}{2} + \dfrac{x}{3}}{4}$ by multiplying the numerator by 6.

57. I added 1 to $\dfrac{1}{1 + \dfrac{1}{2}}$ and obtained $\dfrac{5}{3}$.

58. I used the LCD method to simplify

$$\dfrac{3 - \dfrac{6}{x}}{1 + \dfrac{7}{y}}$$

and obtained $\dfrac{3 - 6y}{1 + 7x}$.

59. Because x^{-1} means $\dfrac{1}{x}$ and y^{-1} means $\dfrac{1}{y}$, I simplified $\dfrac{x^{-1} + y^{-1}}{x^{-1} - y^{-1}}$ and obtained $\dfrac{x - y}{x + y}$.

In Exercises 60–63, determine whether each statement is true or false. If the statement is false, make the necessary change(s) to produce a true statement.

60. The fraction $\dfrac{31{,}729{,}546}{72{,}578{,}112}$ is a complex rational expression.

61. $\dfrac{y - \dfrac{1}{2}}{y + \dfrac{3}{4}} = \dfrac{4y - 2}{4y + 3}$ for any value of y except $-\dfrac{3}{4}$.

62. $\dfrac{\dfrac{1}{4} - \dfrac{1}{3}}{\dfrac{1}{3} + \dfrac{1}{6}} = \dfrac{1}{12} \div \dfrac{3}{6} = \dfrac{1}{6}$

63. Some complex rational expressions cannot be simplified by both methods discussed in this section.

64. In one short sentence, five words or less, explain what

$$\dfrac{\dfrac{1}{x} + \dfrac{1}{x^2} + \dfrac{1}{x^3}}{\dfrac{1}{x^4} + \dfrac{1}{x^5} + \dfrac{1}{x^6}}$$

does to each number x.

In Exercises 65–66, simplify completely.

65. $\dfrac{\dfrac{2y}{2 + \dfrac{2}{y}} + \dfrac{y}{1 + \dfrac{1}{y}}}{}$

66. $\dfrac{1 + \dfrac{1}{y} - \dfrac{6}{y^2}}{1 - \dfrac{5}{y} + \dfrac{6}{y^2}} - \dfrac{1 - \dfrac{1}{y}}{1 - \dfrac{2}{y} - \dfrac{3}{y^2}}$

Technology Exercises

In Exercises 67–69, use the $\boxed{\text{GRAPH}}$ or $\boxed{\text{TABLE}}$ feature of a graphing utility to determine if the simplification is correct. If the answer is wrong, correct it and then verify your corrected simplification using the graphing utility.

67. $\dfrac{x - \dfrac{1}{2x + 1}}{1 - \dfrac{x}{2x + 1}} = 2x - 1$

68. $\dfrac{\dfrac{1}{x} + 1}{\dfrac{1}{x}} = 2$

69. $\dfrac{\dfrac{1}{x} + \dfrac{1}{3}}{\dfrac{1}{3x}} = x + \dfrac{1}{3}$

Review Exercises

70. Factor completely: $2x^3 - 20x^2 + 50x$. (Section 6.5, Example 2)

71. Solve: $2 - 3(x - 2) = 5(x + 5) - 1$. (Section 2.3, Example 3)

72. Multiply: $(x + y)(x^2 - xy + y^2)$. (Section 5.2, Example 7)

Preview Exercises

Exercises 73–75 will help you prepare for the material covered in the next section.

73. Solve: $\dfrac{x}{3} + \dfrac{x}{2} = \dfrac{5}{6}$.

74. Solve: $\dfrac{2x}{3} = \dfrac{14}{3} - \dfrac{x}{2}$.

75. Solve: $2x^2 + 2 = 5x$.

SECTION

7.6

Objectives

1 Solve rational equations.

2 Solve problems involving formulas with rational expressions.

3 Solve a formula with a rational expression for a variable.

Solving Rational Equations

The time has come to clean up the river. Suppose that the government has committed $375 million for this project. We know that

$$y = \dfrac{250x}{100 - x}$$

models the cost, y, in millions of dollars, to remove x percent of the river's pollutants. What percentage of pollutants can be removed for $375 million?

In order to determine the percentage, we use the given model. The government has committed $375 million, so substitute 375 for y:

$$375 = \dfrac{250x}{100 - x} \quad \text{or} \quad \dfrac{250x}{100 - x} = 375.$$

The equation contains a rational expression.

Now we need to solve the equation and find the value for x. This variable represents the percentage of pollutants that can be removed for $375 million.

A **rational equation**, also called a **fractional equation**, is an equation containing one or more rational expressions. The preceding equation, $\frac{250x}{100 - x} = 375$, is an example of a rational equation. Do you see that there is a variable in a denominator? This is a characteristic of many rational equations. In this section, you will learn a procedure for solving such equations.

① Solve rational equations.

Solving Rational Equations

We have seen that the LCD is used to add and subtract rational expressions. By contrast, when solving rational equations, **the LCD is used as a multiplier that clears an equation of fractions.**

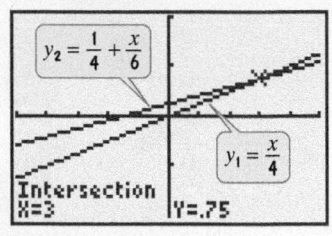
> **EXAMPLE 1** Solving a Rational Equation

Solve: $\dfrac{x}{4} = \dfrac{1}{4} + \dfrac{x}{6}$.

Solution   The denominators are 4, 4, and 6. The least common denominator is 12. To clear the equation of fractions, we multiply both sides by 12.

$$\frac{x}{4} = \frac{1}{4} + \frac{x}{6} \qquad \text{This is the given equation.}$$

$$12\left(\frac{x}{4}\right) = 12\left(\frac{1}{4} + \frac{x}{6}\right) \qquad \begin{array}{l}\text{Multiply both sides by 12, the LCD of all the}\\\text{fractions in the equation.}\end{array}$$

$$12 \cdot \frac{x}{4} = 12 \cdot \frac{1}{4} + 12 \cdot \frac{x}{6} \qquad \text{Use the distributive property on the right side.}$$

$$3x = 3 + 2x \qquad \text{Simplify: } \frac{\overset{3}{\cancel{12}}}{1} \cdot \frac{x}{\cancel{4}} = 3x;\ \frac{\overset{3}{\cancel{12}}}{1} \cdot \frac{1}{\cancel{4}} = 3;\ \frac{\overset{2}{\cancel{12}}}{1} \cdot \frac{x}{\cancel{6}} = 2x.$$

$$x = 3 \qquad \text{Subtract 2x from both sides.}$$

Substitute 3 for x in the original equation. You should obtain the true statement $\dfrac{3}{4} = \dfrac{3}{4}$. This verifies that the solution is 3 and the solution set is {3}. ∎

☑ **CHECK POINT 1**   Solve: $\dfrac{x}{6} = \dfrac{1}{6} + \dfrac{x}{8}$.

In Example 1, we solved a rational equation with constants in the denominators. Now, let's consider an equation such as

$$\frac{1}{x} = \frac{1}{5} + \frac{3}{2x}.$$

Can you see how this equation differs from the rational equation that we solved earlier? The variable, x, appears in two of the denominators. The procedure for solving this equation still involves multiplying each side by the least common denominator. However, we must avoid any values of the variable that make a denominator zero. For example, examine the denominators in the equation:

$$\frac{1}{x} = \frac{1}{5} + \frac{3}{2x}.$$

This denominator would equal zero if $x = 0$.   This denominator would equal zero if $x = 0$.

We see that x cannot equal zero. With this in mind, let's solve the equation.

EXAMPLE 2   Solving a Rational Equation

Solve: $\dfrac{1}{x} = \dfrac{1}{5} + \dfrac{3}{2x}$.

Solution   The denominators are x, 5, and $2x$. The least common denominator is $10x$. We begin by multiplying both sides of the equation by $10x$. We will also write the restriction that x cannot equal zero to the right of the equation.

$$\frac{1}{x} = \frac{1}{5} + \frac{3}{2x}, \quad x \neq 0 \qquad \text{This is the given equation.}$$

$$10x \cdot \frac{1}{x} = 10x\left(\frac{1}{5} + \frac{3}{2x}\right) \qquad \text{Multiply both sides by 10x.}$$

$$10x \cdot \frac{1}{x} = 10x \cdot \frac{1}{5} + 10x \cdot \frac{3}{2x} \qquad \begin{array}{l}\text{Use the distributive property. Be sure to}\\ \text{multiply all terms by 10x.}\end{array}$$

$$10\overset{}{x} \cdot \frac{1}{\overset{}{x}} = \overset{2}{10}x \cdot \frac{1}{\underset{1}{5}} + \overset{5}{10}x \cdot \frac{3}{\underset{1}{2x}} \qquad \begin{array}{l}\text{Divide out common factors in the}\\ \text{multiplications.}\end{array}$$

$$10 = 2x + 15 \qquad \text{Simplify.}$$

Observe that the resulting equation,

$$10 = 2x + 15$$

is now cleared of fractions. With the variable term, $2x$, already on the right, we will collect constant terms on the left by subtracting 15 from both sides.

$$-5 = 2x \qquad \text{Subtract 15 from both sides.}$$

$$-\frac{5}{2} = x \qquad \text{Divide both sides by 2.}$$

We check our solution by substituting $-\frac{5}{2}$ into the original equation or by using a calculator. With a calculator, evaluate each side of the equation for $x = -\frac{5}{2}$, or for $x = -2.5$. Note that the original restriction that $x \neq 0$ is met. The solution is $-\frac{5}{2}$ and the solution set is $\left\{-\frac{5}{2}\right\}$. ∎

✓ CHECK POINT 2   Solve: $\dfrac{5}{2x} = \dfrac{17}{18} - \dfrac{1}{3x}$.

The following steps may be used to solve a rational equation:

Solving Rational Equations

1. List restrictions on the variable. Avoid any values of the variable that make a denominator zero.
2. Clear the equation of fractions by multiplying both sides by the LCD of all rational expressions in the equation.
3. Solve the resulting equation.
4. Reject any proposed solution that is in the list of restrictions on the variable. Check other proposed solutions in the original equation.

EXAMPLE 3 Solving a Rational Equation

Solve: $x + \dfrac{1}{x} = \dfrac{5}{2}$.

Solution

Step 1. List restrictions on the variable.

$$x + \dfrac{1}{x} = \dfrac{5}{2}$$

This denominator would
equal 0 if $x = 0$.

The restriction is $x \neq 0$.

Step 2. Multiply both sides by the LCD. The denominators are x and 2. Thus, the LCD is $2x$. We multiply both sides by $2x$.

$$x + \dfrac{1}{x} = \dfrac{5}{2}, \quad x \neq 0 \qquad \text{This is the given equation.}$$

$$2x\left(x + \dfrac{1}{x}\right) = 2x\left(\dfrac{5}{2}\right) \qquad \text{Multiply both sides by the LCD.}$$

$$2x \cdot x + 2x \cdot \dfrac{1}{x} = 2x \cdot \dfrac{5}{2} \qquad \text{Use the distributive property on the left side.}$$

$$2x^2 + 2 = 5x \qquad \text{Simplify.}$$

Step 3. Solve the resulting equation. Can you see that we have a quadratic equation? Write the equation in standard form and solve for x.

$$2x^2 - 5x + 2 = 0 \qquad \text{Subtract 5x from both sides.}$$

$$(2x - 1)(x - 2) = 0 \qquad \text{Factor.}$$

$$2x - 1 = 0 \quad \text{or} \quad x - 2 = 0 \quad \text{Set each factor equal to 0.}$$

$$2x = 1 \qquad\qquad x = 2 \quad \text{Solve the resulting equations.}$$

$$x = \dfrac{1}{2}$$

Step 4. Check proposed solutions in the original equation. The proposed solutions, $\frac{1}{2}$ and 2, are not part of the restriction that $x \neq 0$. Neither makes a denominator in the original equation equal to zero.

Check $\dfrac{1}{2}$:

$$x + \dfrac{1}{x} = \dfrac{5}{2}$$

$$\dfrac{1}{2} + \dfrac{1}{\frac{1}{2}} \overset{?}{=} \dfrac{5}{2}$$

$$\dfrac{1}{2} + 2 \overset{?}{=} \dfrac{5}{2}$$

$$\dfrac{1}{2} + \dfrac{4}{2} \overset{?}{=} \dfrac{5}{2}$$

$$\dfrac{5}{2} = \dfrac{5}{2}, \text{ true}$$

Check 2:

$$x + \dfrac{1}{x} = \dfrac{5}{2}$$

$$2 + \dfrac{1}{2} \overset{?}{=} \dfrac{5}{2}$$

$$\dfrac{4}{2} + \dfrac{1}{2} \overset{?}{=} \dfrac{5}{2}$$

$$\dfrac{5}{2} = \dfrac{5}{2}, \text{ true}$$

The solutions are $\dfrac{1}{2}$ and 2, and the solution set is $\left\{\dfrac{1}{2}, 2\right\}$. ■

✓ **CHECK POINT 3** Solve: $x + \dfrac{6}{x} = -5$.

EXAMPLE 4 Solving a Rational Equation

Solve: $\dfrac{3x}{x^2 - 9} + \dfrac{1}{x - 3} = \dfrac{3}{x + 3}$.

Solution

Step 1. List restrictions on the variable. By factoring denominators, it makes it easier to see values that make denominators zero.

$$\dfrac{3x}{(x + 3)(x - 3)} + \dfrac{1}{x - 3} = \dfrac{3}{x + 3}$$

| This denominator is zero if $x = -3$ or $x = 3$. | This denominator is zero if $x = 3$. | This denominator is zero if $x = -3$. |

The restrictions are $x \neq -3$ and $x \neq 3$.

Step 2. Multiply both sides by the LCD. The LCD is $(x + 3)(x - 3)$.

$$\dfrac{3x}{(x + 3)(x - 3)} + \dfrac{1}{x - 3} = \dfrac{3}{x + 3}, \quad x \neq -3, x \neq 3 \qquad \text{This is the given equation with a denominator factored.}$$

$$(x + 3)(x - 3)\left[\dfrac{3x}{(x + 3)(x - 3)} + \dfrac{1}{x - 3}\right] = (x + 3)(x - 3) \cdot \dfrac{3}{x + 3} \qquad \text{Multiply both sides by the LCD.}$$

$$(x + 3)(x - 3) \cdot \dfrac{3x}{(x + 3)(x - 3)} + (x + 3)(x - 3) \cdot \dfrac{1}{x - 3}$$

$$= (x + 3)(x - 3) \cdot \dfrac{3}{x + 3} \qquad \text{Use the distributive property on the left side.}$$

$$3x + (x + 3) = 3(x - 3) \qquad \text{Simplify.}$$

Step 3. Solve the resulting equation.

$$3x + (x + 3) = 3(x - 3) \qquad \text{This is the equation cleared of fractions.}$$

$$4x + 3 = 3x - 9 \qquad \begin{array}{l}\text{Combine like terms on the left side.} \\ \text{Use the distributive property on the right side.}\end{array}$$

$$x + 3 = -9 \qquad \text{Subtract 3x from both sides.}$$

$$x = -12 \qquad \text{Subtract 3 from both sides.}$$

Step 4. Check proposed solutions in the original equation. The proposed solution, -12, is not part of the restriction that $x \neq -3$ and $x \neq 3$. Substitute -12 for x in the given equation and show that -12 is the solution. The equation's solution set is $\{-12\}$. ■

✓ **CHECK POINT 4** Solve: $\dfrac{11}{x^2 - 25} + \dfrac{4}{x + 5} = \dfrac{3}{x - 5}$.

EXAMPLE 5 Solving a Rational Equation

Solve: $\dfrac{8x}{x + 1} = 4 - \dfrac{8}{x + 1}$.

Solution

Step 1. List restrictions on the variable.

| This denominator is zero if $x = -1$. | $\dfrac{8x}{x + 1} = 4 - \dfrac{8}{x + 1}$ | This denominator is zero if $x = -1$. |

The restriction is $x \neq -1$.

Step 2. Multiply both sides by the LCD. The LCD is $x + 1$.

$$\frac{8x}{x + 1} = 4 - \frac{8}{x + 1}, \quad x \neq -1 \qquad \text{This is the given equation.}$$

$$(x + 1) \cdot \frac{8x}{x + 1} = (x + 1)\left[4 - \frac{8}{x + 1}\right] \qquad \text{Multiply both sides by the LCD.}$$

$$(\cancel{x + 1}) \cdot \frac{8x}{\cancel{x + 1}} = (x + 1) \cdot 4 - (\cancel{x + 1}) \cdot \frac{8}{\cancel{x + 1}} \qquad \begin{array}{l}\text{Use the distributive property}\\ \text{on the right side.}\end{array}$$

$$8x = 4(x + 1) - 8 \qquad \text{Simplify.}$$

Step 3. Solve the resulting equation.

$$8x = 4(x + 1) - 8 \qquad \text{This is the equation cleared of fractions.}$$

$$8x = 4x + 4 - 8 \qquad \text{Use the distributive property on the right side.}$$

$$8x = 4x - 4 \qquad \text{Simplify.}$$

$$4x = -4 \qquad \text{Subtract 4x from both sides.}$$

$$x = -1 \qquad \text{Divide both sides by 4.}$$

Step 4. Check proposed solutions. The proposed solution, -1, is *not* a solution because of the restriction that $x \neq -1$. Notice that -1 makes both of the denominators zero in the original equation. There is *no solution to this equation*. The solution set is $\varnothing$, the empty set. ■

Great Question!

Give me the bottom line: When do I get rid of proposed solutions in rational equations?

Reject any proposed solution that causes any denominator in a rational equation to equal 0.

✓ **CHECK POINT 5**   Solve: $\dfrac{x}{x - 3} = \dfrac{3}{x - 3} + 9.$

Great Question!

What's the difference between adding or subtracting rational expressions and solving equations containing the addition or subtraction of rational expressions?

We simplify rational expressions. We solve rational equations. Notice the differences between the procedures.

Adding Rational Expressions

Simplify: $\dfrac{5}{3x} + \dfrac{3}{x}$.

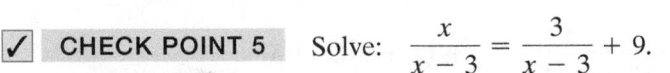

This is not an equation.
There is no equal sign.

Solution

The LCD is $3x$. Rewrite each expression with this LCD and retain the LCD.

$$\frac{5}{3x} + \frac{3}{x}$$

$$= \frac{5}{3x} + \frac{3}{x} \cdot \frac{3}{3}$$

$$= \frac{5}{3x} + \frac{9}{3x}$$

$$= \frac{5 + 9}{3x}$$

$$= \frac{14}{3x}$$

Solving a Rational Equation

Solve: $\dfrac{5}{3x} + \dfrac{3}{x} = 1$.

This is an equation.
There is an equal sign.

Solution

The LCD is $3x$. Multiply both sides by this LCD and clear the fractions.

$$3x\left(\frac{5}{3x} + \frac{3}{x}\right) = 3x \cdot 1$$

$$3x \cdot \frac{5}{3x} + 3x \cdot \frac{3}{x} = 3x$$

$$5 + 9 = 3x$$

$$14 = 3x$$

$$\frac{14}{3} = x$$

The solution set is $\left\{\dfrac{14}{3}\right\}$.

You only eliminate the denominators when solving a rational equation with an equal sign. You should never begin by eliminating the denominators when simplifying a rational expression involving addition or subtraction with no equal sign.

2 Solve problems involving formulas with rational expressions.

Applications of Rational Equations

Rational equations can be solved to answer questions about variables contained in mathematical models.

EXAMPLE 6   A Government-Funded Cleanup

The formula

$$y = \frac{250x}{100 - x}$$

models the cost, y, in millions of dollars, to remove x percent of a river's pollutants. If the government commits $375 million for this project, what percentage of pollutants can be removed?

Solution   Substitute 375 for y and solve the resulting rational equation for x.

$$375 = \frac{250x}{100 - x}$$ The LCD is $100 - x$.

$$(100 - x)375 = (100 - x) \cdot \frac{250x}{100 - x}$$ Multiply both sides by the LCD.

$$375(100 - x) = 250x$$ Simplify.

$$37{,}500 - 375x = 250x$$ Use the distributive property on the left side.

$$37{,}500 = 625x$$ Add 375x to both sides.

$$\frac{37{,}500}{625} = \frac{625x}{625}$$ Divide both sides by 625.

$$60 = x$$ Simplify.

If the government spends $375 million, 60% of the river's pollutants can be removed.   ∎

✓ **CHECK POINT 6**   Use the model in Example 6 to answer this question: If government funding is increased to $750 million, what percentage of pollutants can be removed?

3 Solve a formula with a rational expression for a variable.

Solving a Formula for a Variable

Formulas and mathematical models frequently contain rational expressions. We solve for a specified variable using the procedure for solving rational equations. The goal is to get the specified variable alone on one side of the equation. To do so, collect all terms with this variable on one side and all other terms on the other side. It is sometimes necessary to factor out the variable you are solving for.

EXAMPLE 7   Solving for a Variable in a Formula

The formula

$$S = \frac{C}{1 - r}$$

describes a product's selling price, S, in terms of its cost to the seller, C, and its markup rate, r, in decimal form.

a. Solve the formula for r.

b. What is the markup rate on a textbook costing the campus bookstore $140 and selling for $200?

Solution

a. Our goal is to isolate the variable r.

$$S = \frac{C}{1-r}$$

> We need r by itself on one side of the formula.

We begin by multiplying both sides by the least common denominator, $1 - r$, to clear the equation of fractions.

$$S = \frac{C}{1-r}$$ This is the given formula.

$$(1-r)S = (1-r)\left(\frac{C}{1-r}\right)$$ Multiply both sides by $1 - r$, the LCD.

$$S - Sr = C$$ Use the distributive property on the left side. On the right side, $(1-r)\left(\frac{C}{1-r}\right) = C$.

Observe that the formula is now cleared of fractions. The term with r, the specified variable, is already on the left side of the equation. To isolate this term, subtract S from both sides.

$$S - Sr = C$$ This is the equation cleared of fractions.

$$-Sr = C - S$$ Subtract S from both sides.

$$\frac{-Sr}{-S} = \frac{C - S}{-S}$$ Divide both sides by $-S$ and solve for r.

$$r = \frac{C - S}{-S}$$ Simplify.

Great Question!

Now that we've solved the formula for r, it looks weird with a negative sign in the denominator. Is there another way to write this formula?

Yes. Multiply the numerator and the denominator by -1.

$$r = \frac{C - S}{-S} = \frac{(-1)}{(-1)} \cdot \frac{C - S}{-S} = \frac{-C + S}{S} = \frac{S - C}{S}$$

b. Now we find the markup rate on a textbook costing the campus bookstore \$140 and selling for \$200.

$C = 140$

$$r = \frac{C - S}{-S}$$ $S = 200$ This is the formula solved for r from part (a).

$$r = \frac{140 - 200}{-200} = \frac{-60}{-200} = \frac{3}{10} = 0.3 = 30\%$$

The bookstore's markup rate is 30%. ■

✓ **CHECK POINT 7** Solve for x: $a = \dfrac{b}{x + 2}$.

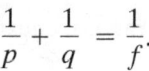

EXAMPLE 8   Solving for a Variable in a Formula

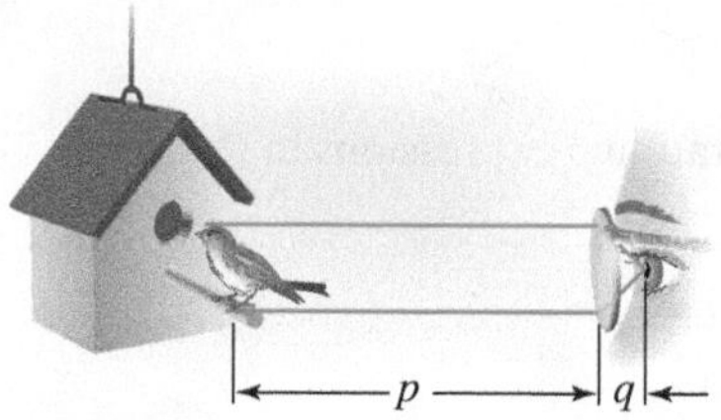

If you wear glasses, did you know that each lens has a measurement called its focal length, f? When an object is in focus, its distance from the lens, p, and the distance from the lens to your retina, q, satisfy the formula

$$\frac{1}{p} + \frac{1}{q} = \frac{1}{f}.$$

(See **Figure 7.2**.) Solve this formula for p.

Figure 7.2

Solution   Our goal is to isolate the variable p. We begin by multiplying both sides by the least common denominator, pqf, to clear the equation of fractions.

$$\frac{1}{p} + \frac{1}{q} = \frac{1}{f} \qquad \text{This is the given formula.}$$

$$pqf\left(\frac{1}{p} + \frac{1}{q}\right) = pqf\left(\frac{1}{f}\right) \qquad \text{Multiply both sides by } pqf, \text{ the LCD.}$$

$$pqf\left(\frac{1}{p}\right) + pqf\left(\frac{1}{q}\right) = pqf\left(\frac{1}{f}\right) \qquad \text{Use the distributive property on the left side.}$$

$$qf + pf = pq \qquad \text{Simplify.}$$

Observe that the formula is now cleared of fractions. Collect terms with p, the specified variable, on one side of the equation. To do so, subtract pf from both sides.

$$qf + pf = pq \qquad \text{This is the equation cleared of fractions.}$$

$$qf = pq - pf \qquad \text{Subtract } pf \text{ from both sides.}$$

$$qf = p(q - f) \qquad \text{Factor out } p, \text{ the specified variable.}$$

$$\frac{qf}{q - f} = \frac{p(q - f)}{q - f} \qquad \text{Divide both sides by } q - f \text{ and solve for } p.$$

$$\frac{qf}{q - f} = p \qquad \text{Simplify.} \ \blacksquare$$

✓ CHECK POINT 8   Solve $\dfrac{1}{x} + \dfrac{1}{y} = \dfrac{1}{z}$ for x.

Achieving Success

Analyze the errors you make on quizzes and tests.

For each error, write out the correct solution along with a description of the concept needed to solve the problem correctly. Do your mistakes indicate gaps in understanding concepts or do you at times believe that you are just not a good test taker? Are you repeatedly making the same kinds of mistakes on tests? Keeping track of errors should increase your understanding of the material, resulting in improved test scores.

CONCEPT AND VOCABULARY CHECK

Fill in each blank so that the resulting statement is true.

1. We clear a rational equation of fractions by multiplying both sides by the _____ of all rational expressions in the equation.

2. We reject any proposed solution of a rational equation that causes a denominator to equal _____.

3. The first step in solving

$$\frac{8}{x} + \frac{1}{4x} = \frac{17}{8}, x \neq 0,$$

is to multiply both sides by _____.

4. The first step in solving

$$\frac{x+1}{3(x+3)} + \frac{x}{2(x+3)} = \frac{5}{4(x+3)}, x \neq -3,$$

is to multiply both sides by _____.

5.

$$\frac{1}{3x} - \frac{1}{3} = x$$

$$3x\left(\frac{1}{3x} - \frac{1}{3}\right) = 3x \cdot x$$

The resulting equation cleared of fractions is _____.

6.

$$\frac{6}{(x+1)(x-1)} = \frac{5}{x-1} + \frac{3}{x+1}$$

$$(x+1)(x-1) \cdot \frac{6}{(x+1)(x-1)} = (x+1)(x-1)\left[\frac{5}{x-1} + \frac{3}{x+1}\right]$$

The resulting equation cleared of fractions is _____.

7. The restriction on the variable in the rational equation

$$\frac{2}{x+2} + 2 = \frac{7}{x+2}$$

is _____.

8. The restrictions on the variable in the rational equation

$$\frac{4}{x-3} - \frac{3}{x-2} = \frac{4x+9}{(x-3)(x-2)}$$

are _____ and _____.

9. True or false: A rational equation can have no solution. _____

10. True or false: There are no restrictions on the variable in the rational equation $\frac{5}{6} + \frac{3}{x} = \frac{1}{4}$. _____

7.6 EXERCISE SET MyMathLab®

 Watch the videos in MyMathLab

 Download the MyDashBoard App

Practice Exercises

In Exercises 1–46, solve each rational equation.

1. $\frac{x}{3} = \frac{x}{2} - 2$

2. $\frac{x}{5} = \frac{x}{6} + 1$

3. $\frac{4x}{3} = \frac{x}{18} - \frac{x}{6}$

4. $\frac{5x}{4} = \frac{x}{12} - \frac{x}{2}$

5. $2 - \frac{8}{x} = 6$

6. $1 - \frac{9}{x} = 4$

7. $\frac{2}{x} + \frac{1}{3} = \frac{4}{x}$

8. $\frac{5}{x} + \frac{1}{3} = \frac{6}{x}$

9. $\frac{2}{x} + 3 = \frac{5}{2x} + \frac{13}{4}$

10. $\frac{7}{2x} = \frac{5}{3x} + \frac{22}{3}$

11. $\frac{2}{3x} + \frac{1}{4} = \frac{11}{6x} - \frac{1}{3}$

12. $\frac{5}{2x} - \frac{8}{9} = \frac{1}{18} - \frac{1}{3x}$

13. $\frac{6}{x+3} = \frac{4}{x-3}$

14. $\frac{7}{x+1} = \frac{5}{x-3}$

15. $\dfrac{x-2}{2x} + 1 = \dfrac{x+1}{x}$

16. $\dfrac{7x-4}{5x} = \dfrac{9}{5} - \dfrac{4}{x}$

17. $x + \dfrac{6}{x} = -7$

18. $x + \dfrac{7}{x} = -8$

19. $\dfrac{x}{5} - \dfrac{5}{x} = 0$

20. $\dfrac{x}{4} - \dfrac{4}{x} = 0$

21. $x + \dfrac{3}{x} = \dfrac{12}{x}$

22. $x + \dfrac{3}{x} = \dfrac{19}{x}$

23. $\dfrac{4}{y} - \dfrac{y}{2} = \dfrac{7}{2}$

24. $\dfrac{4}{3y} - \dfrac{1}{3} = y$

25. $\dfrac{x-4}{x} = \dfrac{15}{x+4}$

26. $\dfrac{x-1}{2x+3} = \dfrac{6}{x-2}$

27. $\dfrac{2}{x^2-1} = \dfrac{4}{x+1}$

28. $\dfrac{3}{x+1} = \dfrac{1}{x^2-1}$

29. $\dfrac{1}{x-1} + 5 = \dfrac{11}{x-1}$

30. $\dfrac{3}{x+4} - 7 = \dfrac{-4}{x+4}$

31. $\dfrac{8y}{y+1} = 4 - \dfrac{8}{y+1}$

32. $\dfrac{2}{y-2} = \dfrac{y}{y-2} - 2$

33. $\dfrac{3}{x-1} + \dfrac{8}{x} = 3$

34. $\dfrac{2}{x-2} + \dfrac{4}{x} = 2$

35. $\dfrac{3y}{y-4} - 5 = \dfrac{12}{y-4}$

36. $\dfrac{10}{y+2} = 3 - \dfrac{5y}{y+2}$

37. $\dfrac{1}{x} + \dfrac{1}{x-3} = \dfrac{x-2}{x-3}$

38. $\dfrac{1}{x-1} + \dfrac{2}{x} = \dfrac{x}{x-1}$

39. $\dfrac{x+1}{3x+9} + \dfrac{x}{2x+6} = \dfrac{2}{4x+12}$

40. $\dfrac{3}{2y-2} + \dfrac{1}{2} = \dfrac{2}{y-1}$

41. $\dfrac{4y}{y^2-25} + \dfrac{2}{y-5} = \dfrac{1}{y+5}$

42. $\dfrac{1}{x+4} + \dfrac{1}{x-4} = \dfrac{22}{x^2-16}$

43. $\dfrac{1}{x-4} - \dfrac{5}{x+2} = \dfrac{6}{x^2-2x-8}$

44. $\dfrac{6}{x+3} - \dfrac{5}{x-2} = \dfrac{-20}{x^2+x-6}$

45. $\dfrac{2}{x+3} - \dfrac{2x+3}{x-1} = \dfrac{6x-5}{x^2+2x-3}$

46. $\dfrac{x-3}{x-2} + \dfrac{x+1}{x+3} = \dfrac{2x^2-15}{x^2+x-6}$

In Exercises 47–60, solve each formula for the specified variable.

47. $\dfrac{V_1}{V_2} = \dfrac{P_2}{P_1}$ for P_1 (chemistry)

48. $\dfrac{V_1}{V_2} = \dfrac{P_2}{P_1}$ for V_2 (chemistry)

49. $\dfrac{1}{p} + \dfrac{1}{q} = \dfrac{1}{f}$ for f (optics)

50. $\dfrac{1}{p} + \dfrac{1}{q} = \dfrac{1}{f}$ for q (optics)

51. $P = \dfrac{A}{1+r}$ for r (investment)

52. $S = \dfrac{a}{1-r}$ for r (mathematics)

53. $F = \dfrac{Gm_1m_2}{d^2}$ for m_1 (physics)

54. $F = \dfrac{Gm_1m_2}{d^2}$ for m_2 (physics)

55. $z = \dfrac{x-\bar{x}}{s}$ for x (statistics)

56. $z = \dfrac{x-\bar{x}}{s}$ for s (statistics)

57. $I = \dfrac{E}{R+r}$ for R (electronics)

58. $I = \dfrac{E}{R+r}$ for r (electronics)

59. $f = \dfrac{f_1f_2}{f_1+f_2}$ for f_1 (optics)

60. $f = \dfrac{f_1f_2}{f_1+f_2}$ for f_2 (optics)

Practice PLUS

In Exercises 61–68, solve or simplify, whichever is appropriate.

61. $\dfrac{x^2-10}{x^2-x-20} = 1 + \dfrac{7}{x-5}$

62. $\dfrac{x^2+4x-2}{x^2-2x-8} = 1 + \dfrac{4}{x-4}$

63. $\dfrac{x^2-10}{x^2-x-20} - 1 - \dfrac{7}{x-5}$

64. $\dfrac{x^2+4x-2}{x^2-2x-8} - 1 - \dfrac{4}{x-4}$

65. $5y^{-2} + 1 = 6y^{-1}$

66. $3y^{-2} + 1 = 4y^{-1}$

67. $\dfrac{3}{y+1} - \dfrac{1}{1-y} = \dfrac{10}{y^2-1}$

68. $\dfrac{4}{y-2} - \dfrac{1}{2-y} = \dfrac{25}{y+6}$

Application Exercises

A company that manufactures wheelchairs has fixed costs of $500,000. The average cost per wheelchair, C, for the company to manufacture x wheelchairs per month is modeled by the formula

$$C = \dfrac{400x + 500{,}000}{x}.$$

Use this mathematical model to solve Exercises 69–70.

69. How many wheelchairs per month can be produced at an average cost of $450 per wheelchair?

70. How many wheelchairs per month can be produced at an average cost of $405 per wheelchair?

In Palo Alto, California, a government agency ordered computer-related companies to contribute to a pool of money to clean up underground water supplies. (The companies had stored toxic chemicals in leaking underground containers.) The formula

$$C = \frac{2x}{100 - x}$$

models the cost, C, in millions of dollars, for removing x percent of the contaminants. Use this mathematical model to solve Exercises 71–72.

71. What percentage of the contaminants can be removed for $2 million?

72. What percentage of the contaminants can be removed for $8 million?

We have seen that Young's rule

$$C = \frac{DA}{A + 12}$$

can be used to approximate the dosage of a drug prescribed for children. In this formula, A = the child's age, in years, D = an adult dosage, and C = the proper child's dosage. Use this formula to solve Exercises 73–74.

73. When the adult dosage is 1000 milligrams, a child is given 300 milligrams. What is that child's age? Round to the nearest year.

74. When the adult dosage is 1000 milligrams, a child is given 500 milligrams. What is that child's age?

A grocery store sells 4000 cases of canned soup per year. By averaging costs to purchase soup and to pay storage costs, the owner has determined that if x cases are ordered at a time, the yearly inventory cost, C, can be modeled by

$$C = \frac{10,000}{x} + 3x.$$

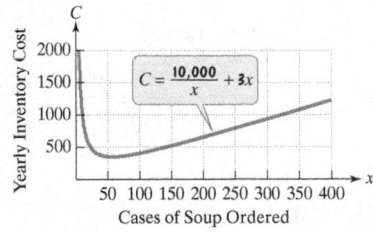

The graph of this model is shown on the right. Use this information to solve Exercises 75–76.

75. How many cases should be ordered at a time for yearly inventory costs to be $350? Identify your solutions as points on the graph.

76. How many cases should be ordered at a time for yearly inventory costs to be $790? Identify your solutions as points on the graph.

Writing in Mathematics

77. What is a rational equation?

78. Explain how to solve a rational equation.

79. Explain how to find restrictions on the variable in a rational equation.

80. Why should restrictions on the variable in a rational equation be listed before you begin solving the equation?

81. Describe similarities and differences between the procedures needed to solve the following problems:

$$\text{Add: } \frac{2}{x} + \frac{3}{4}. \qquad \text{Solve: } \frac{2}{x} + \frac{3}{4} = 1.$$

82. Without showing the details, explain how to solve the formula

$$\frac{1}{R} = \frac{1}{R_1} + \frac{1}{R_2}$$

for R_1. (The formula is used in electronics.)

Critical Thinking Exercises

Make Sense? *In Exercises 83–86, determine whether each statement "makes sense" or "does not make sense" and explain your reasoning.*

83. I added two rational expressions and found the solution set.

84. I can solve the equation $\frac{6}{x + 3} = \frac{4}{x - 3}$ by multiplying both sides by the LCD.

85. I'm solving a rational equation that became a quadratic equation, so my rational equation must have two solutions.

86. I must have made an error if a rational equation produces no solution.

In Exercises 87–90, determine whether each statement is true or false. If the statement is false, make the necessary change(s) to produce a true statement.

87. $\frac{1}{x} + \frac{1}{6} = 6x\left(\frac{1}{x} + \frac{1}{6}\right) = 6 + x$

88. If a is any real number, the equation $\frac{a}{x} + 1 = \frac{a}{x}$ has no solution.

89. All real numbers satisfy the equation $\frac{3}{x} - \frac{1}{x} = \frac{2}{x}$.

90. To solve $\frac{5}{3x} + \frac{3}{x} = 1$, we must first add the rational expressions on the left side.

In Exercises 91–92, solve each rational equation.

91. $\frac{x + 1}{2x^2 - 11x + 5} = \frac{x - 7}{2x^2 + 9x - 5} - \frac{2x - 6}{x^2 - 25}$

92. $\left(\frac{x + 1}{x + 7}\right)^2 \div \left(\frac{x + 1}{x + 7}\right)^4 = 0.$

93. Find b so that the solution of

$$\frac{7x + 4}{b} + 13 = x$$

is -6.

Technology Exercises

In Exercises 94–96, use a graphing utility to solve each rational equation. Graph each side of the equation in the given viewing rectangle. The first coordinate of each point of intersection is a solution. Check by direct substitution.

94. $\dfrac{x}{2} + \dfrac{x}{4} = 6$

[−5, 10, 1] by [−5, 10, 1]

95. $\dfrac{50}{x} = 2x$

[−10, 10, 1] by [−20, 20, 2]

96. $x + \dfrac{6}{x} = -5$

[−10, 10, 1] by [−10, 10, 1]

Review Exercises

97. Factor completely: $x^4 + 2x^3 - 3x - 6$. (Section 6.1, Example 8)

98. Simplify: $(3x^2)(-4x^{-10})$. (Section 5.7, Example 3)

99. Simplify: $-5[4(x - 2) - 3]$. (Section 1.8, Example 11)

Preview Exercises

Exercises 100–102 will help you prepare for the material covered in the next section.

100. Solve: $\dfrac{15}{8 + x} = \dfrac{9}{8 - x}$.

101. If you can complete a job in 5 hours, what fractional part of the job can you complete in 1 hour? in 3 hours? in x hours?

102. Write as an equation, where x represents the number: The quotient of 63 and a number is equal to the quotient of 7 and 5.

Applications Using Rational Equations and Proportions

Objectives

1 Solve problems involving motion.

2 Solve problems involving work.

3 Solve problems involving proportions.

4 Solve problems involving similar triangles.

The possibility of seeing a blue whale, the largest mammal ever to grace the earth, increases the excitement of gazing out over the ocean's swell of waves. Blue whales were hunted to near extinction in the last half of the nineteenth and the first half of the twentieth centuries. Using a method for estimating wildlife populations that we discuss in this section, by the mid-1960s it was determined that the world population of blue whales was less than 1000. This led the International Whaling Commission to ban the killing of blue whales to prevent their extinction. A dramatic increase in blue whale sightings indicates an ongoing increase in their population and the success of the killing ban.

1 Solve problems involving motion.

Problems Involving Motion

We have seen that the distance, d, covered by any moving body is the product of its average rate, r, and its time in motion, t: $d = rt$.

Rational expressions appear in motion problems when the conditions of the problem involve the time traveled. We can obtain an expression for t, the time traveled, by dividing both sides of $d = rt$ by r.

$$d = rt \qquad \text{Distance equals rate times time.}$$

$$\frac{d}{r} = \frac{rt}{r} \qquad \text{Divide both sides by } r.$$

$$\frac{d}{r} = t \qquad \text{Simplify.}$$

Time in Motion

$$t = \frac{d}{r}$$

$$\text{Time traveled} = \frac{\text{Distance traveled}}{\text{Rate of travel}}$$

Downstream (with the current)

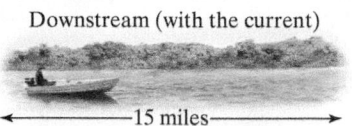

←——— 15 miles ———→

Upstream (against the current)

←——— 9 miles ———→

Traveling 15 miles downstream takes the same time as traveling 9 miles upstream.

EXAMPLE 1   A Motion Problem Involving Time

In still water, your small boat averages 8 miles per hour. It takes you the same amount of time to travel 15 miles downstream, with the current, as 9 miles upstream, against the current. What is the rate of the water's current?

Solution

Step 1. Let x represent one of the quantities. Let

$$x = \text{the rate of the current.}$$

Step 2. Represent other unknown quantities in terms of x. We still need expressions for the rate of your boat with the current and the rate against the current. Traveling with the current, the boat's rate in still water, 8 miles per hour, is increased by the current's rate, x miles per hour. Thus,

$$8 + x = \text{the boat's rate with the current.}$$

Traveling against the current, the boat's rate in still water, 8 miles per hour, is decreased by the current's rate, x miles per hour. Thus,

$$8 - x = \text{the boat's rate against the current.}$$

Step 3. Write an equation that models the conditions. By reading the problem again, we discover that the crucial idea is that the time spent going 15 miles with the current equals the time spent going 9 miles against the current. This information is summarized in the following table.

	Distance	Rate	Time = $\dfrac{\text{Distance}}{\text{Rate}}$
With the current	15	$8 + x$	$\dfrac{15}{8 + x}$
Against the current	9	$8 - x$	$\dfrac{9}{8 - x}$

These two times are equal.

We are now ready to write an equation that models the problem's conditions.

The time spent going 15 miles with the current $\dfrac{15}{8 + x}$ equals $\dfrac{9}{8 - x}$ the time spent going 9 miles against the curre

Step 4. Solve the equation and answer the question.

$$\frac{15}{8+x} = \frac{9}{8-x}$$

This is the equation that models the problem's conditions.

$$(8+x)(8-x) \cdot \frac{15}{8+x} = (8+x)(8-x) \cdot \frac{9}{8-x}$$

Multiply both sides by the LCD, $(8+x)(8-x)$.

$$15(8-x) = 9(8+x)$$

Simplify.

$$120 - 15x = 72 + 9x$$

Use the distributive property.

$$120 = 72 + 24x$$

Add 15x to both sides.

$$48 = 24x$$

Subtract 72 from both sides.

$$2 = x$$

Divide both sides by 24.

The rate of the water's current is 2 miles per hour.

Step 5. Check the proposed solution in the original wording of the problem. Does it take you the same amount of time to travel 15 miles downstream as 9 miles upstream if the current is 2 miles per hour? Keep in mind that your rate in still water is 8 miles per hour.

$$\text{Time required to travel 15 miles with the current} = \frac{\text{Distance}}{\text{Rate}} = \frac{15}{8+2} = \frac{15}{10} = 1\frac{1}{2} \text{ hours}$$

$$\text{Time required to travel 9 miles against the current} = \frac{\text{Distance}}{\text{Rate}} = \frac{9}{8-2} = \frac{9}{6} = 1\frac{1}{2} \text{ hours}$$

These times are the same, which checks with the original conditions of the problem. ■

✓ **CHECK POINT 1**   Forget the small boat! This time we have you canoeing on the Colorado River. In still water, your average canoeing rate is 3 miles per hour. It takes you the same amount of time to travel 10 miles downstream, with the current, as 2 miles upstream, against the current. What is the rate of the water's current?

2 Solve problems involving work.

Problems Involving Work

You are thinking of designing your own Web site. You estimate that it will take 30 hours to do the job. In 1 hour, $\frac{1}{30}$ of the job is completed. In 2 hours, $\frac{2}{30}$, or $\frac{1}{15}$, of the job is completed. In 3 hours, the fractional part of the job done is $\frac{3}{30}$, or $\frac{1}{10}$. In x hours, the fractional part of the job that you can complete is $\frac{x}{30}$.

Your friend, who has experience developing Web sites, took 20 hours working on his own to design an impressive site. You wonder about the possibility of working together. How long would it take both of you to design your Web site?

Problems involving work usually have two people working together to complete a job. The amount of time it takes each person to do the job working alone is frequently known. The question deals with how long it will take both people working together to complete the job.

In work problems, **the number 1 represents one whole job completed.** For example, the completion of your Web site is represented by 1. Equations in work problems are based on the following condition:

| Fractional part of the job done by the first person | + | fractional part of the job done by the second person | = | 1 (one whole job completed). |

EXAMPLE 2 Solving a Problem Involving Work

You can design a Web site in 30 hours. Your friend can design the same site in 20 hours. How long will it take to design the Web site if you both work together?

Solution

Step 1. Let x represent one of the quantities. Let $x =$ the time, in hours, for you and your friend to design the Web site together.

Step 2. Represent other unknown quantities in terms of x. Because there are no other unknown quantities, we can skip this step.

Step 3. Write an equation that models the conditions. We construct a table to help find the fractional part of the task completed by you and your friend in x hours.

		Fractional part of job completed in 1 hour	Time working together	Fractional part of job completed in x hours
You can design the site in 30 hours.	You	$\dfrac{1}{30}$	x	$\dfrac{x}{30}$
Your friend can design the site in 20 hours.	Your friend	$\dfrac{1}{20}$	x	$\dfrac{x}{20}$

Fractional part of the job done by you	+	fractional part of the job done by your friend	=	one whole job.
$\dfrac{x}{30}$	+	$\dfrac{x}{20}$	=	1

Step 4. Solve the equation and answer the question.

$$\frac{x}{30} + \frac{x}{20} = 1 \qquad \text{This is the equation that models the problem's conditions.}$$

$$60\left(\frac{x}{30} + \frac{x}{20}\right) = 60 \cdot 1 \qquad \text{Multiply both sides by 60, the LCD.}$$

$$60 \cdot \frac{x}{30} + 60 \cdot \frac{x}{20} = 60 \qquad \text{Use the distributive property on the left side.}$$

$$2x + 3x = 60 \qquad \text{Simplify: } \frac{\overset{2}{60}}{1} \cdot \frac{x}{\underset{1}{30}} = 2x \text{ and } \frac{\overset{3}{60}}{1} \cdot \frac{x}{\underset{1}{20}} = 3x.$$

$$5x = 60 \qquad \text{Combine like terms.}$$

$$x = 12 \qquad \text{Divide both sides by 5.}$$

If you both work together, you can design your Web site in 12 hours.

Step 5. Check the proposed solution in the original wording of the problem. Will you both complete the job in 12 hours? Because you can design the site in 30 hours, in 12 hours, you can complete $\frac{12}{30}$, or $\frac{2}{5}$, of the job. Because your friend can design the site in 20 hours, in 12 hours, he can complete $\frac{12}{20}$, or $\frac{3}{5}$, of the job. Notice that $\frac{2}{5} + \frac{3}{5} = 1$, which represents the completion of the entire job, or one whole job. ∎

Great Question!

Is there an equation that models all the work problems in this section?

Yes. Let

$$a = \text{the time it takes person A to do a job working alone}$$
$$b = \text{the time it takes person B to do the same job working alone.}$$

If x represents the time it takes for A and B to complete the entire job working together, then the situation can be modeled by the rational equation

$$\frac{x}{a} + \frac{x}{b} = 1.$$

✓ **CHECK POINT 2** One person can paint the outside of a house in 8 hours. A second person can do it in 4 hours. How long will it take them to do the job if they work together?

---

3 Solve problems involving proportions.

Problems Involving Proportions

A **ratio** compares quantities by division. For example, this year's entering class at a medical school contains 60 women and 30 men. The ratio of women to men is $\frac{60}{30}$. We can express this ratio as a fraction reduced to lowest terms:

$$\frac{60}{30} = \frac{30 \cdot 2}{30 \cdot 1} = \frac{2}{1}.$$

This ratio can be expressed as 2:1, or 2 to 1.

A **proportion** is a statement that says that two ratios are equal. If the ratios are $\frac{a}{b}$ and $\frac{c}{d}$, then the proportion is

$$\frac{a}{b} = \frac{c}{d}.$$

We can clear this equation of fractions by multiplying both sides by bd, the least common denominator:

$$\frac{a}{b} = \frac{c}{d} \qquad \text{This is the given proportion.}$$

$$bd \cdot \frac{a}{b} = bd \cdot \frac{c}{d} \qquad \text{Multiply both sides by } bd \ (b \neq 0 \text{ and } d \neq 0). \text{ Then simplify.}$$

$$ad = bc \qquad \text{On the left, } \frac{bd}{1} \cdot \frac{a}{b} = da = ad. \text{ On the right, } \frac{bd}{1} \cdot \frac{c}{d} = bc.$$

We see that the following principle is true for any proportion:

The Cross-Products Principle for Proportions

If $\frac{a}{b} = \frac{c}{d}$, then $ad = bc$. ($b \neq 0$ and $d \neq 0$)

The cross products ad and bc are equal.

For example, since $\frac{2}{3} = \frac{6}{9}$, we see that $2 \cdot 9 = 3 \cdot 6$, or $18 = 18$. We can also use $\frac{2}{3} = \frac{6}{9}$ and conclude that $3 \cdot 6 = 2 \cdot 9$. When using the cross-products principle, it does not matter on which side of the equation each product is placed.

Here is a procedure for solving problems involving proportions:

Solving Applied Problems Using Proportions

1. Read the problem and represent the unknown quantity by x (or any letter).
2. Set up a proportion by listing the given ratio on one side and the ratio with the unknown quantity on the other side. Each respective quantity should occupy the same corresponding position on each side of the proportion.
3. Drop units and apply the cross-products principle.
4. Solve for x and answer the question.

$$\frac{a}{b} = \frac{c}{d}$$

The cross-products principle:
$ad = bc$

EXAMPLE 3 Applying Proportions: Calculating Taxes

The property tax on a house with an assessed value of $480,000 is $5760. Determine the property tax on a house with an assessed value of $600,000, assuming the same tax rate.

Great Question!

Are there other proportions that I can use in step 2 to model the problem's conditions?

Yes. Here are three other correct proportions you can use:

- $\dfrac{\$480{,}000 \text{ value}}{\$5760 \text{ tax}} = \dfrac{\$600{,}000 \text{ value}}{\$x \text{ tax}}$

- $\dfrac{\$480{,}000 \text{ value}}{\$600{,}000 \text{ value}} = \dfrac{\$5760 \text{ tax}}{\$x \text{ tax}}$

- $\dfrac{\$600{,}000 \text{ value}}{\$480{,}000 \text{ value}} = \dfrac{\$x \text{ tax}}{\$5760 \text{ tax}}$.

Each proportion gives the same cross product obtained in step 3.

Solution

Step 1. Represent the unknown by x. Let x = the tax on the $600,000 house.

Step 2. Set up a proportion. We will set up a proportion comparing taxes to assessed value.

$$\underbrace{\frac{\text{Tax on \$480,000 house}}{\text{Assessed value (\$480,000)}}}_{} \quad \text{equals} \quad \underbrace{\frac{\text{Tax on \$600,000 house}}{\text{Assessed value (\$600,000)}}}_{}$$

$$\text{Given ratio} \left\{ \frac{\$5760}{\$480{,}000} \right. = \frac{\$x \leftarrow \text{Unknown}}{\$600{,}000 \leftarrow \text{Given quantity}}$$

Step 3. Drop the units and apply the cross-products principle. We drop the dollar signs and begin to solve for x.

$$\frac{5760}{480{,}000} = \frac{x}{600{,}000} \qquad \text{This is the proportion that models the problem's conditions.}$$

$$480{,}000x = (5760)(600{,}000) \qquad \text{Apply the cross-products principle.}$$

$$480{,}000x = 3{,}456{,}000{,}000 \qquad \text{Multiply.}$$

Step 4. Solve for x and answer the question.

$$\frac{480{,}000x}{480{,}000} = \frac{3{,}456{,}000{,}000}{480{,}000} \qquad \text{Divide both sides by 480,000.}$$

$$x = 7200 \qquad \text{Simplify.}$$

The property tax on the $600,000 house is $7200. ■

☑ **CHECK POINT 3** The property tax on a house with an assessed value of $250,000 is $3500. Determine the property tax on a house with an assessed value of $420,000, assuming the same tax rate.

Sampling in Nature

The method that was used to estimate the blue whale population described in the section opener is called the **capture-recapture method.** Because it is impossible to count each individual animal within a population, wildlife biologists randomly catch and tag a given number of animals. Sometime later they recapture a second sample of animals and count the number of recaptured tagged animals. The total size of the wildlife population is then estimated using the following proportion:

$$\underset{\substack{\text{Initially} \\ \text{unknown} \\ (x) \longrightarrow}}{} \left. \frac{\substack{\text{Original number of} \\ \text{tagged animals}}}{\substack{\text{Total number} \\ \text{of animals in the} \\ \text{population}}} = \frac{\substack{\text{Number of recaptured} \\ \text{tagged animals}}}{\substack{\text{Number of animals} \\ \text{in second sample}}} \right\} \substack{\text{Known} \\ \text{ratio}}$$

Although this is called the capture-recapture method, it is not necessary to recapture animals to observe whether or not they are tagged. This could be done from a distance, with binoculars for instance.

EXAMPLE 4 Applying Proportions: Estimating Wildlife Population

Wildlife biologists catch, tag, and then release 135 deer back into a wildlife refuge. Two weeks later they observe a sample of 140 deer, 30 of which are tagged. Assuming the ratio of tagged deer in the sample holds for all deer in the refuge, approximately how many deer are in the refuge?

Solution

Step 1. Represent the unknown by x. Let x = the total number of deer in the refuge.

Step 2. Set up a proportion. Recall that the biologists tagged 135 deer. They later observed 30 tagged deer in a sample of 140 animals.

Unknown ⟶ $\dfrac{\text{Original number of tagged deer → Total number of deer}}{x}$ equals $\dfrac{\text{Number of tagged deer in the observed sample → Total number of deer in the observed sample}}{}$ } Known ratio

$$\frac{135}{x} = \frac{30}{140}$$

Steps 3 and 4. Apply the cross-products principle, solve, and answer the question.

$$\frac{135}{x} = \frac{30}{140} \qquad \text{This is the proportion that models the problem's conditions.}$$

$$(135)(140) = 30x \qquad \text{Apply the cross-products principle.}$$

$$18{,}900 = 30x \qquad \text{Multiply.}$$

$$\frac{18{,}900}{30} = \frac{30x}{30} \qquad \text{Divide both sides by 30.}$$

$$630 = x \qquad \text{Simplify.}$$

There are approximately 630 deer in the refuge.   ∎

✓ **CHECK POINT 4**   Wildlife biologists catch, tag, and then release 120 deer back into a wildlife refuge. Two weeks later they observe a sample of 150 deer, 25 of which are tagged. Assuming the ratio of tagged deer in the sample holds for all deer in the refuge, approximately how many deer are in the refuge?

---

4 Solve problems involving similar triangles.

Pedestrian Crossing

Similar Triangles

Shown in the margin is an international road sign. This sign is shaped just like the actual sign, although its size is smaller. Figures that have the same shape, but not the same size, are used in **scale drawings**. A scale drawing always pictures the exact shape of the object that the drawing represents. Architects, engineers, landscape gardeners, and interior decorators use scale drawings in planning their work.

Figures that have the same shape, but not necessarily the same size, are called similar figures. In **Figure 7.3**, triangles *ABC* and *DEF* are similar. Angles *A* and *D* measure the same number of degrees and are called **corresponding angles**. Angles *C* and *F* are corresponding angles, as are angles *B* and *E*. Angles with the same number of tick marks in **Figure 7.3** are the corresponding angles.

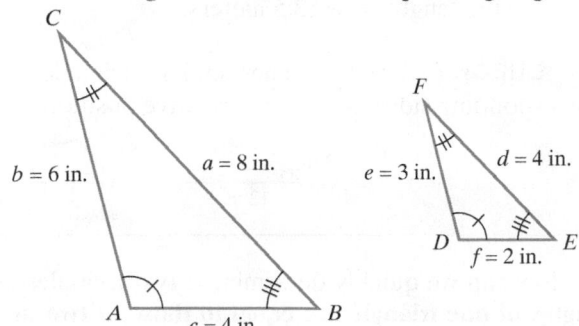

Figure 7.3

The sides opposite the corresponding angles are called **corresponding sides**. Although the measures of corresponding angles are equal, corresponding sides may or may not be the same length. For the triangles in **Figure 7.3**, each side in the smaller triangle is half the length of the corresponding side in the larger triangle.

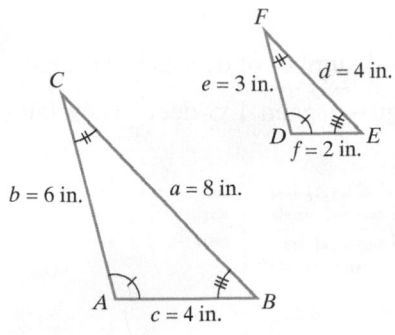

Figure 7.3 (repeated)

The triangles in **Figure 7.3** illustrate what it means to be **similar triangles.** **Corresponding angles have the same measure and the ratios of the lengths of the corresponding sides are equal.** For the triangles in **Figure 7.3**, each of these ratios is equal to 2:

$$\frac{a}{d} = \frac{8}{4} = 2 \qquad \frac{b}{e} = \frac{6}{3} = 2 \qquad \frac{c}{f} = \frac{4}{2} = 2.$$

In similar triangles, the lengths of the corresponding sides are proportional. Thus,

$$\frac{a}{d} = \frac{b}{e} = \frac{c}{f}.$$

If we know that two triangles are similar, we can set up a proportion to solve for the length of an unknown side.

> **EXAMPLE 5** Using Similar Triangles

The triangles in **Figure 7.4** are similar. Find the missing length, x.

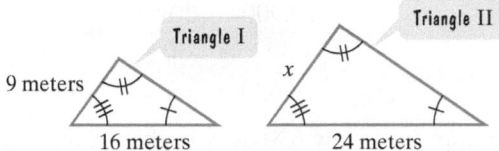

Figure 7.4

Solution Because the triangles in **Figure 7.4** are similar, their corresponding sides are proportional.

$$\underset{\substack{\text{Left side}\\ \text{of } \Delta \text{ I} \\ \\ \text{Corresponding side}\\ \text{on left of } \Delta \text{ II}}}{} \quad \frac{9}{x} = \frac{16}{24} \quad \underset{\substack{\text{Bottom side}\\ \text{of } \Delta \text{ I} \\ \\ \text{Corresponding side}\\ \text{on bottom of } \Delta \text{ II}}}{}$$

We solve this rational equation by multiplying both sides by the LCD, $24x$. (You can also apply the cross-products principle for solving proportions.)

$$24x \cdot \frac{9}{x} = 24x \cdot \frac{16}{24} \qquad \text{Multiply both sides by the LCD, } 24x.$$
$$24 \cdot 9 = 16x \qquad \text{Simplify.}$$
$$216 = 16x \qquad \text{Multiply: } 24 \cdot 9 = 216.$$
$$13.5 = x \qquad \text{Divide both sides by 16.}$$

The missing length, x, is 13.5 meters. ■

✓ **CHECK POINT 5** The similar triangles in the figure are shown with corresponding sides in the same relative position. Find the missing length, x.

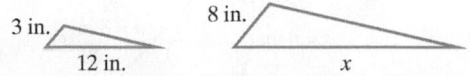

How can we quickly determine if two triangles are similar? **If the measures of two angles of one triangle are equal to those of two angles of a second triangle, then the two triangles are similar.** If the triangles are similar, then their corresponding sides are proportional.

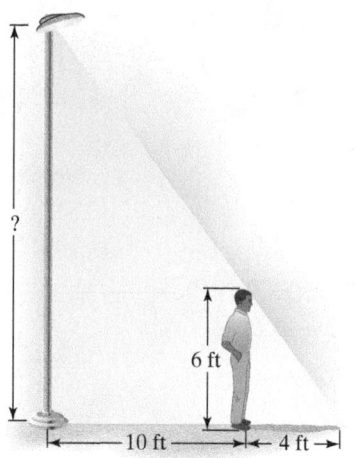

Figure 7.5

EXAMPLE 6 Problem Solving Using Similar Triangles

A man who is 6 feet tall is standing 10 feet from the base of a lamppost (see **Figure 7.5**). The man's shadow has a length of 4 feet. How tall is the lamppost?

Solution The drawing in **Figure 7.6** makes the similarity of the triangles easier to see. The large triangle with the lamppost on the left and the small triangle with the man on the left both contain 90° angles. They also share an angle. Thus, two angles of the large triangle are equal in measure to two angles of the small triangle. This means that the triangles are similar and their corresponding sides are proportional. We begin by letting x represent the height of the lamppost, in feet. Because corresponding sides of similar triangles are proportional,

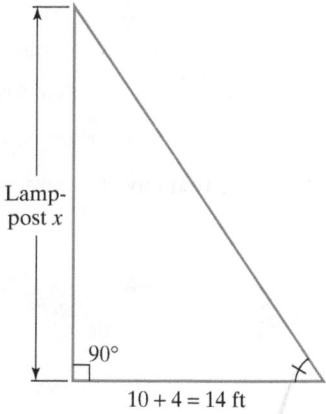

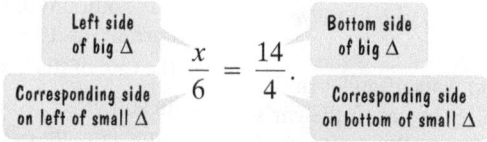

We solve for x by multiplying both sides by the LCD, 12.

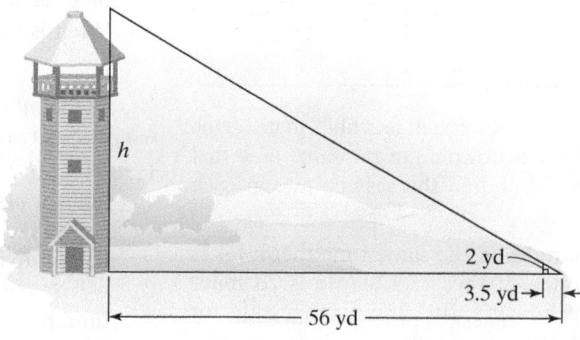

Figure 7.6

$$12 \cdot \frac{x}{6} = 12 \cdot \frac{14}{4} \qquad \text{Multiply both sides by the LCD, 12.}$$

$$2x = 42 \qquad \text{Simplify: } \frac{\overset{2}{\cancel{12}}}{1} \cdot \frac{x}{\underset{1}{\cancel{6}}} = 2x \text{ and } \frac{\overset{3}{\cancel{12}}}{1} \cdot \frac{14}{\underset{1}{\cancel{4}}} = 42.$$

$$x = 21 \qquad \text{Divide both sides by 2.}$$

The lamppost is 21 feet tall. ∎

✓ **CHECK POINT 6** Find the height of the lookout tower shown in **Figure 7.7** using the figure that lines up the top of the tower with the top of a stick that is 2 yards long.

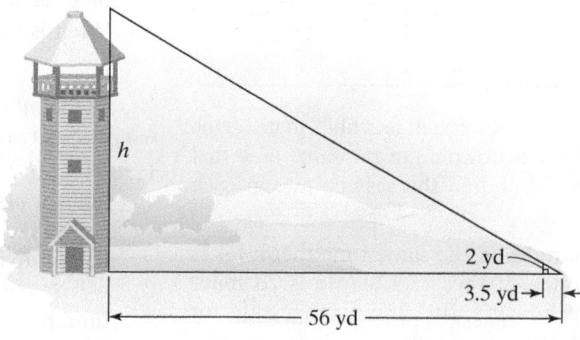

Figure 7.7

CONCEPT AND VOCABULARY CHECK

Fill in each blank so that the resulting statement is true.

1. The formula $t = \dfrac{d}{r}$ states that time traveled is _____ divided by _____.

2. In work problems, the number _____ represents the whole job completed.

3. If you can complete a job in 5 hours, the fractional part of the job that you can complete in x hours is represented by _____.

4. The cross-products principle for proportions states that if $\dfrac{a}{b} = \dfrac{c}{d}$, then _____, $b \neq 0$ and $d \neq 0$.

5. Triangles that have the same shape, but not necessarily the same size, are called _____ triangles.

7.7 EXERCISE SET   MyMathLab®

 Watch the videos in MyMathLab

 Download the MyDashBoard App

Practice and Application Exercises

Use rational equations to solve Exercises 1–10. Each exercise is a problem involving motion.

1. How bad is the heavy traffic? You can walk 10 miles in the same time that it takes to travel 15 miles by car. If the car's rate is 3 miles per hour faster than your walking rate, find the average rate of each.

	Distance	Rate	Time = $\dfrac{\text{Distance}}{\text{Rate}}$
Walking	10	x	$\dfrac{10}{x}$
Car in Heavy Traffic	15	$x + 3$	$\dfrac{15}{x+3}$

2. You can travel 40 miles on motorcycle in the same time that it takes to travel 15 miles on bicycle. If your motorcycle's rate is 20 miles per hour faster than your bicycle's, find the average rate for each.

	Distance	Rate	Time = $\dfrac{\text{Distance}}{\text{Rate}}$
Motorcycle	40	$x + 20$	$\dfrac{40}{x+20}$
Bicycle	15	x	$\dfrac{15}{x}$

3. A jogger runs 4 miles per hour faster downhill than uphill. If the jogger can run 5 miles downhill in the same time that it takes to run 3 miles uphill, find the jogging rate in each direction.

4. A truck can travel 120 miles in the same time that it takes a car to travel 180 miles. If the truck's rate is 20 miles per hour slower than the car's, find the average rate for each.

5. In still water, a boat averages 15 miles per hour. It takes the same amount of time to travel 20 miles downstream, with the current, as 10 miles upstream, against the current. What is the rate of the water's current?

6. In still water, a boat averages 18 miles per hour. It takes the same amount of time to travel 33 miles downstream, with the current, as 21 miles upstream, against the current. What is the rate of the water's current?

7. As part of an exercise regimen, you walk 2 miles on an indoor track. Then you jog at twice your walking speed for another 2 miles. If the total time spent walking and jogging is 1 hour, find the walking and jogging rates.

8. The joys of the Pacific Coast! You drive 90 miles along the Pacific Coast Highway and then take a 5-mile run along a hiking trail in Point Reyes National Seashore. Your driving rate is nine times that of your running rate. If the total time for driving and running is 3 hours, find the average rate driving and the average rate running.

9. The water's current is 2 miles per hour. A boat can travel 6 miles downstream, with the current, in the same amount of time it travels 4 miles upstream, against the current. What is the boat's average rate in still water?

10. The water's current is 2 miles per hour. A canoe can travel 6 miles downstream, with the current, in the same amount of time it travels 2 miles upstream, against the current. What is the canoe's average rate in still water?

Use a rational equation to solve Exercises 11–16. Each exercise is a problem involving work.

11. You must leave for campus in 10 minutes or you will be late for class. Unfortunately, you are snowed in. You can shovel the driveway in 20 minutes and your brother claims he can do it in 15 minutes. If you shovel together, how long will it take to clear the driveway? Will this give you enough time before you have to leave?

12. You promised your parents that you would wash the family car. You have not started the job and they are due home in 16 minutes. You can wash the car in 40 minutes and your sister claims she can do it in 30 minutes. If you work together, how long will it take to do the job? Will this give you enough time before your parents return?

13. The MTV crew will arrive in one week and begin filming the city for *The Real World Kalamazoo*. The mayor is desperate to clean the city streets before filming begins. Two teams are available, one that requires 400 hours and one that requires 300 hours. If the teams work together, how long will it take to clean all of Kalamazoo's streets? Is this enough time before the cameras begin rolling?

14. A hurricane strikes and a rural area is without food or water. Three crews arrive. One can dispense needed supplies in 10 hours, a second in 15 hours, and a third in 20 hours. How long will it take all three crews working together to dispense food and water?

15. A pool can be filled by one pipe in 4 hours and by a second pipe in 6 hours. How long will it take using both pipes to fill the pool?

16. A pool can be filled by one pipe in 3 hours and by a second pipe in 6 hours. How long will it take using both pipes to fill the pool?

Use a proportion to solve each problem in Exercises 17–24.

17. The tax on a property with an assessed value of $65,000 is $720. Find the tax on a property with an assessed value of $162,500.

18. The maintenance bill for a shopping center containing 180,000 square feet is $45,000. What is the bill for a store in the center that is 4800 square feet?

19. St. Paul Island in Alaska has 12 fur seal rookeries (breeding places). In 1961, to estimate the fur seal pup population in the Gorbath rookery, 4963 fur seal pups were tagged in early August. In late August, a sample of 900 pups was observed and 218 of these were found to have been previously tagged. Estimate the total number of fur seal pups in this rookery.

20. To estimate the number of bass in a lake, wildlife biologists tagged 50 bass and released them in the lake. Later they netted 108 bass and found that 27 of them were tagged. Approximately how many bass are in the lake?

21. According to the authors of *Number Freaking*, in a global village of 200 people, 28 suffer from malnutrition. How many people of the world's 6.9 billion people (2010 population) suffer from malnutrition? Round to the nearest hundredth of a billion.

22. According to the authors of *Number Freaking*, in a global village of 200 people, 9 get drunk every day. How many of the world's 6.9 billion people (2010 population) get drunk everyday? Round to the nearest hundredth of a billion.

23. Height is proportional to foot length. A person whose foot length is 10 inches is 67 inches tall. In 1951, photos of large footprints were published. Some believed that these footprints were made by the "Abominable Snowman." Each footprint was 23 inches long. If indeed they belonged to the Abominable Snowman, how tall is the critter?

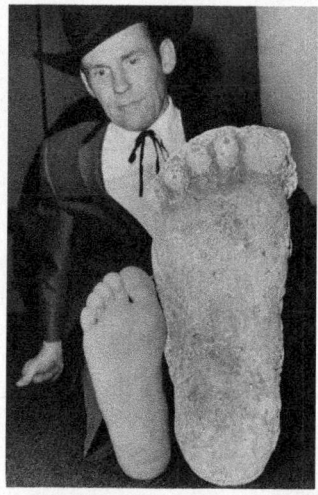

Roger Patterson comparing his foot with a plaster cast of a footprint of the purported "Bigfoot" that Mr. Patterson said he sighted in a California forest in 1967.

24. A person's hair length is proportional to the number of years it has been growing. After 2 years, a person's hair grows 8 inches. The longest moustache on record was grown by Kalyan Sain of India. Sain grew his moustache for 17 years. How long was each side of the moustache?

In Exercises 25–30, use similar triangles and the fact that corresponding sides are proportional to find the length of the side marked with an x.

25.
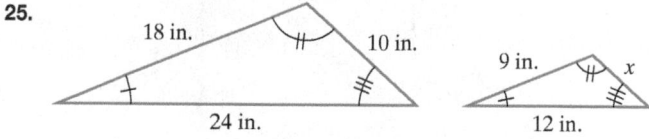

26.
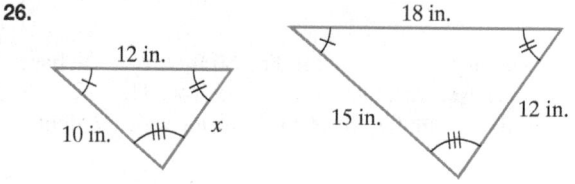

27.
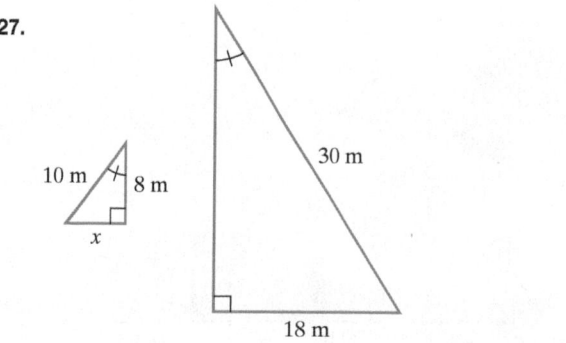

28.

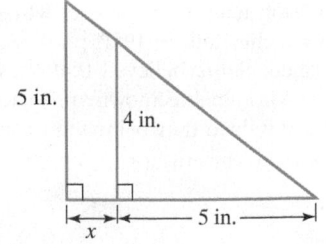

29.

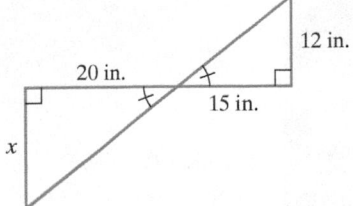

30.

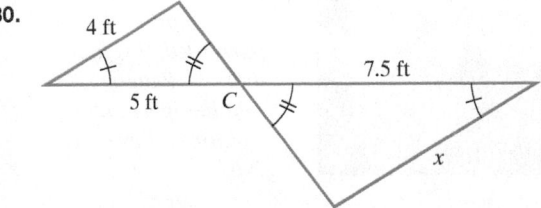

Use similar triangles to solve Exercises 31–32.

31. A tree casts a shadow 12 feet long. At the same time, a vertical rod 8 feet high casts a shadow 6 feet long. How tall is the tree?

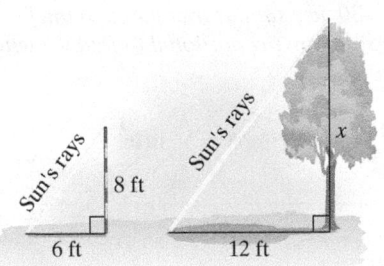

32. A person who is 5 feet tall is standing 80 feet from the base of a tree. The tree casts an 86-foot shadow. The person's shadow is 6 feet in length. What is the tree's height?

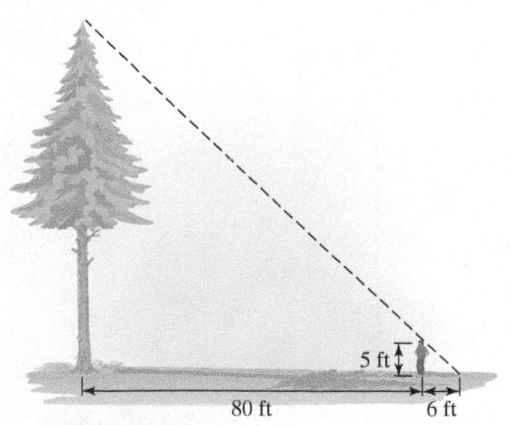

Writing in Mathematics

33. What is the relationship among time traveled, distance traveled, and rate of travel?

34. If you know how many hours it takes for you to do a job, explain how to find the fractional part of the job you can complete in x hours.

35. If you can do a job in 6 hours and your friend can do the same job in 3 hours, explain how to find how long it takes to complete the job working together. It is not necessary to solve the problem.

36. When two people work together to complete a job, describe one factor that can result in more or less time than the time given by the rational equations we have been using.

37. What is a proportion? Give an example with your description.

38. What are similar triangles?

39. If the ratio of the corresponding sides of two similar triangles is 1 to 1 $\left(\frac{1}{1}\right)$, what must be true about the triangles?

40. What are corresponding angles in similar triangles?

41. Describe how to identify the corresponding sides in similar triangles.

Critical Thinking Exercises

Make Sense? *In Exercises 42–45, determine whether each statement "makes sense" or "does not make sense" and explain your reasoning.*

42. I can solve $\frac{x}{9} = \frac{4}{6}$ by using the cross-products principle or by multiplying both sides by 18, the least common denominator.

43. It took me the same amount of time to travel 10 miles with the current as it did to travel 15 miles against the current.

44. I used $\frac{a}{d} = \frac{b}{e}$ to show that corresponding sides of similar triangles are proportional, but I could also use $\frac{a}{b} = \frac{d}{e}$ or $\frac{d}{a} = \frac{e}{b}$.

45. I can clean my house in 3 hours and my sloppy friend can completely mess it up in 6 hours, so if we both "work" together, the time, x, it takes to clean the house can be modeled by $\frac{x}{3} - \frac{x}{6} = 1$.

46. Two skiers begin skiing along a trail at the same time. The faster skier averages 9 miles per hour and the slower skier averages 6 miles per hour. The faster skier completes the trail $\frac{1}{4}$ hour before the slower skier. How long is the trail?

47. A snowstorm causes a bus driver to decrease the usual average rate along a 60-mile route by 15 miles per hour. As a result, the bus takes two hours longer than usual to complete the route. At what average rate does the bus usually cover the 60-mile route?

48. One pipe can fill a swimming pool in 2 hours, a second can fill the pool in 3 hours, and a third pipe can fill the pool in 4 hours. How many minutes, to the nearest minute, would it take to fill the pool with all three pipes operating?

49. Ben can prepare a company report in 3 hours. Shane can prepare a report in 4.2 hours. How long will it take them, working together, to prepare *four* company reports?

50. An experienced carpenter can panel a room 3 times faster than an apprentice can. Working together, they can panel the room in 6 hours. How long would it take each person working alone to do the job?

51. It normally takes 2 hours to fill a swimming pool. The pool has developed a slow leak. If the pool were full, it would take 10 hours for all the water to leak out. If the pool is empty, how long will it take to fill it?

52. Two investments have interest rates that differ by 1%. An investment for 1 year at the lower rate earns $175. The same principal invested for a year at the higher rate earns $200. What are the two interest rates?

Review Exercises

53. Factor: $25x^2 - 81$. (Section 6.4, Example 1)

54. Solve: $x^2 - 12x + 36 = 0$. (Section 6.6, Example 4)

55. Graph: $y = -\dfrac{2}{3}x + 4$. (Section 3.4, Example 3)

Preview Exercises

Exercises 56–58 will help you prepare for the material covered in the next section.

56. a. If $y = kx^2$, find the value of k using $x = 2$ and $y = 64$.

    b. Substitute the value for k into $y = kx^2$ and write the resulting equation.

    c. Use the equation from part (b) to find y when $x = 5$.

57. a. If $y = \dfrac{k}{x}$, find the value of k using $x = 8$ and $y = 12$.

    b. Substitute the value for k into $y = \dfrac{k}{x}$ and write the resulting equation.

    c. Use the equation from part (b) to find y when $x = 3$.

58. If $S = \dfrac{kA}{P}$, find the value of k using $A = 60{,}000$, $P = 40$, and $S = 12{,}000$.

SECTION

7.8

Modeling Using Variation

Objectives

1 Solve direct variation problems.

2 Solve inverse variation problems.

3 Solve combined variation problems.

4 Solve problems involving joint variation.

Have you ever wondered how telecommunication companies estimate the number of phone calls expected per day between two cities? The formula

$$C = \frac{0.02P_1P_2}{d^2}$$

shows that the daily number of phone calls, C, increases as the populations of the cities, P_1 and P_2, in thousands, increase, and decreases as the distance, d, between the cities increases.

Certain formulas occur so frequently in applied situations that they are given special names. Variation formulas show how one quantity changes in relation to other quantities. Quantities can vary *directly, inversely*, or *jointly*. In this section, we look at situations that can be modeled by each of these kinds of variation.

1 Solve direct variation problems.

Direct Variation

When you swim underwater, the pressure in your ears depends on the depth at which you are swimming. The formula

$$p = 0.43d$$

describes the water pressure, p, in pounds per square inch, at a depth of d feet. We can use this linear function to determine the pressure in your ears at various depths.

In each case, use $p = 0.43d$:

If $d = 20$, $p = 0.43(20) = 8.6$.   At a depth of 20 feet, water pressure is 8.6 pounds per square inch.

Doubling the depth doubles the pressure.

If $d = 40$, $p = 0.43(40) = 17.2$.   At a depth of 40 feet, water pressure is 17.2 pounds per square inch.

Doubling the depth doubles the pressure.

If $d = 80$, $p = 0.43(80) = 34.4$.   At a depth of 80 feet, water pressure is 34.4 pounds per square inch.

The formula $p = 0.43d$ illustrates that water pressure is a constant multiple of your underwater depth. If your depth is doubled, the pressure is doubled; if your depth is tripled, the pressure is tripled; and so on. Because of this, the pressure in your ears is said to **vary directly** as your underwater depth. The **equation of variation** is

$$p = 0.43d.$$

Generalizing, we obtain the following statement:

Direct Variation

If a situation is described by an equation in the form

$$y = kx,$$

where k is a nonzero constant, we say that **y varies directly as x** or **y is directly proportional to x**. The number k is called the **constant of variation** or the **constant of proportionality**.

Can you see that **the direct variation equation, $y = kx$, is a special case of the linear equation $y = mx + b$?** When $m = k$ and $b = 0$, $y = mx + b$ becomes $y = kx$. Thus, the slope of a direct variation equation is k, the constant of variation. Because b, the y-intercept, is 0, the graph of a direct variation equation is a line passing through the origin. This is illustrated in **Figure 7.8,** which shows the graph of $p = 0.43d$: Water pressure varies directly as depth.

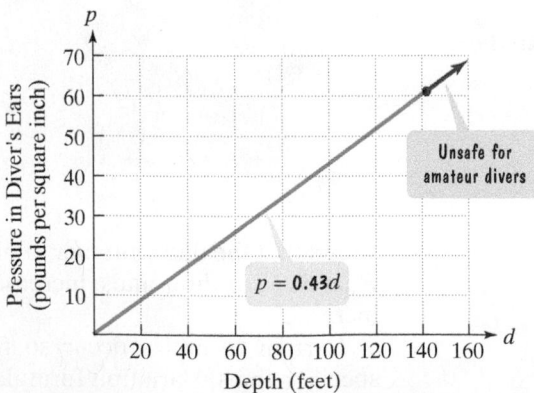

Figure 7.8 Water pressure at various depths

Problems involving direct variation can be solved using the following procedure. This procedure applies to direct variation problems, as well as to the other kinds of variation problems that we will discuss.

Solving Variation Problems

1. Write an equation that models the given English statement.
2. Substitute the given pair of values into the equation in step 1 and find the value of *k*, the constant of variation.
3. Substitute the value of *k* into the equation in step 1.
4. Use the equation from step 3 to answer the problem's question.

> **EXAMPLE 1**   Solving a Direct Variation Problem

Many areas of Northern California depend on the snowpack of the Sierra Nevada mountain range for their water supply. The volume of water produced from melting snow varies directly as the volume of snow. Meteorologists have determined that 250 cubic centimeters of snow will melt to 28 cubic centimeters of water. How much water does 1200 cubic centimeters of melting snow produce?

Solution

Step 1. Write an equation. We know that *y varies directly as x* is expressed as

$$y = kx.$$

By changing letters, we can write an equation that models the following English statement: Volume of water, *W*, varies directly as volume of snow, *S*.

$$W = kS$$

Step 2. Use the given values to find *k*. We are told that 250 cubic centimeters of snow will melt to 28 cubic centimeters of water. Substitute 28 for *W* and 250 for *S* in the direct variation equation. Then solve for *k*.

$$W = kS$$ Volume of water varies directly as volume of melting snow.

$$28 = k(250)$$ 250 cubic centimeters of snow melt to 28 cubic centimeters of water.

$$\frac{28}{250} = \frac{k(250)}{250}$$ Divide both sides by 250.

$$0.112 = k$$ Simplify.

Step 3. Substitute the value of *k* into the equation.

$$W = kS$$ This is the equation from step 1.

$$W = 0.112S$$ Replace k, the constant of variation, with 0.112.

Step 4. Answer the problem's question. How much water does 1200 cubic centimeters of melting snow produce? Substitute 1200 for *S* in $W = 0.112S$ and solve for *W*.

$$W = 0.112S$$ Use the equation from step 3.

$$W = 0.112(1200)$$ Substitute 1200 for S.

$$W = 134.4$$ Multiply.

A snowpack measuring 1200 cubic centimeters will produce 134.4 cubic centimeters of water. ∎

✓ **CHECK POINT 1** The number of gallons of water, W, used when taking a shower varies directly as the time, t, in minutes, in the shower. A shower lasting 5 minutes uses 30 gallons of water. How much water is used in a shower lasting 11 minutes?

The direct variation equation $y = kx$ is a linear equation. If $k > 0$, then the slope of the line is positive. Consequently, as x increases, y also increases.

A direct variation situation can involve variables to higher powers. For example, y can vary directly as x^2 ($y = kx^2$) or as x^3 ($y = kx^3$).

Direct Variation with Powers

y **varies directly as the nth power of x** if there exists some nonzero constant k such that

$$y = kx^n.$$

We also say that y **is directly proportional to the nth power of x.**

EXAMPLE 2 Solving a Direct Variation Problem

The distance, s, that a body falls from rest varies directly as the square of the time, t, of the fall. If skydivers fall 64 feet in 2 seconds, how far will they fall in 4.5 seconds?

Solution

Step 1. Write an equation. We know that y *varies directly as the square of* x is expressed as

$$y = kx^2.$$

By changing letters, we can write an equation that models the following English statement: Distance, s, varies directly as the square of time, t, of the fall.

$$s = kt^2$$

Step 2. Use the given values to find k. Skydivers fall 64 feet in 2 seconds. Substitute 64 for s and 2 for t in the direct variation equation. Then solve for k.

$$s = kt^2 \qquad \text{Distance varies directly as the square of time.}$$

$$64 = k \cdot 2^2 \qquad \text{Skydivers fall 64 feet in 2 seconds.}$$

$$64 = 4k \qquad \text{Simplify: } 2^2 = 4.$$

$$\frac{64}{4} = \frac{4k}{4} \qquad \text{Divide both sides by 4.}$$

$$16 = k \qquad \text{Simplify.}$$

Step 3. Substitute the value of k into the equation.

$$s = kt^2 \qquad \text{Use the equation from step 1.}$$

$$s = 16t^2 \qquad \text{Replace } k, \text{ the constant of variation, with 16.}$$

Step 4. Answer the problem's question. How far will the skydivers fall in 4.5 seconds? Substitute 4.5 for t in $s = 16t^2$ and solve for s.

$$s = 16(4.5)^2 = 16(20.25) = 324$$

Thus, in 4.5 seconds, the skydivers will fall 324 feet. ■

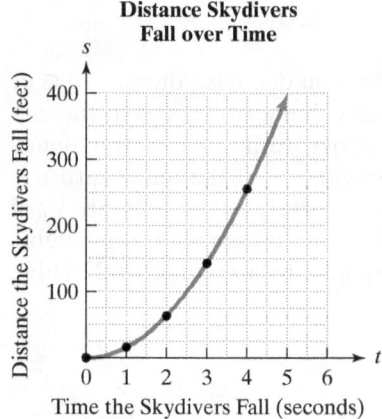

Figure 7.9 The graph of $s = 16t^2$

2 Solve inverse variation problems.

The distance that a body falls from rest depends on the time, t, of the fall. The graph of this quadratic equation is shown in **Figure 7.9**. The graph increases rapidly from left to right, showing the effects of the acceleration of gravity.

✓ **CHECK POINT 2** The distance required to stop a car varies directly as the square of its speed. If it requires 200 feet to stop a car traveling 60 miles per hour, how many feet are required to stop a car traveling 100 miles per hour?

Inverse Variation

The distance from San Francisco to Los Angeles is 420 miles. The time that it takes to drive from San Francisco to Los Angeles depends on the average rate at which one drives and is given by

$$\text{Time} = \frac{420}{\text{Rate}}.$$

For example, if you average 30 miles per hour, the time for the drive is

$$\text{Time} = \frac{420}{30} = 14,$$

or 14 hours. If you average 50 miles per hour, the time for the drive is

$$\text{Time} = \frac{420}{50} = 8.4,$$

or 8.4 hours. As your rate (or speed) increases, the time for the trip decreases and vice versa. This is illustrated by the graph in **Figure 7.10**.

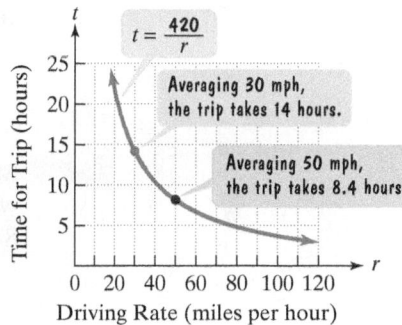

Figure 7.10

We can express the time for the San Francisco–Los Angeles trip using t for time and r for rate:

$$t = \frac{420}{r}.$$

This equation is an example of an **inverse variation** equation. Time, t, **varies inversely** as rate, r. When two quantities vary inversely, one quantity increases as the other decreases, and vice versa.

Generalizing, we obtain the following statement:

Inverse Variation

If a situation is described by an equation in the form

$$y = \frac{k}{x},$$

where k is a nonzero constant, we say that y **varies inversely as x** or y **is inversely proportional to x**. The number k is called the **constant of variation**.

Notice that the inverse variation equation

$$y = \frac{k}{x}$$

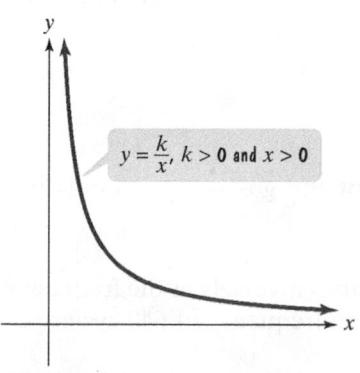

Figure 7.11 The graph of the inverse variation equation

involves a rational expression $\frac{k}{x}$. For $k > 0$ and $x > 0$, the graph of the equation takes on the shape shown in **Figure 7.11**. Under these conditions, as x increases, y decreases.

We use the same procedure to solve inverse variation problems as we did to solve direct variation problems. Example 3 illustrates this procedure.

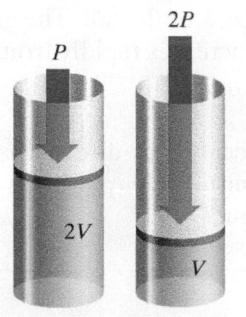

Doubling the pressure halves the volume.

EXAMPLE 3 Solving an Inverse Variation Problem

When you use a spray can and press the valve at the top, you decrease the pressure of the gas in the can. This decrease of pressure causes the volume of the gas in the can to increase. Because the gas needs more room than is provided in the can, it expands in spray form through the small hole near the valve. In general, if the temperature is constant, the pressure, P, of a gas in a container varies inversely as the volume, V, of the container. The pressure of a gas sample in a container whose volume is 8 cubic inches is 12 pounds per square inch. If the sample expands to a volume of 22 cubic inches, what is the new pressure of the gas?

Solution

Step 1. Write an equation. We know that y *varies inversely as* x is expressed as

$$y = \frac{k}{x}.$$

By changing letters, we can write an equation that models the following English statement: The pressure, P, of a gas in a container varies inversely as the volume, V.

$$P = \frac{k}{V}$$

Step 2. Use the given values to find k. The pressure of a gas sample in a container whose volume is 8 cubic inches is 12 pounds per square inch. Substitute 12 for P and 8 for V in the inverse variation equation. Then solve for k.

$$P = \frac{k}{V} \qquad \text{Pressure varies inversely as volume.}$$

$$12 = \frac{k}{8} \qquad \begin{array}{l}\text{The pressure in an 8-cubic-inch}\\\text{container is 12 pounds per square inch.}\end{array}$$

$$12 \cdot 8 = \frac{k}{8} \cdot 8 \qquad \text{Multiply both sides by 8.}$$

$$96 = k \qquad \text{Simplify.}$$

Step 3. Substitute the value of k into the equation.

$$P = \frac{k}{V} \qquad \text{Use the equation from step 1.}$$

$$P = \frac{96}{V} \qquad \begin{array}{l}\text{Replace } k, \text{ the constant of variation,}\\\text{with 96.}\end{array}$$

Step 4. Answer the problem's question. We need to find the pressure when the volume expands to 22 cubic inches. Substitute 22 for V and solve for P.

$$P = \frac{96}{V} = \frac{96}{22} = 4\frac{4}{11}$$

When the volume is 22 cubic inches, the pressure of the gas is $4\frac{4}{11}$ pounds per square inch. ∎

✓ **CHECK POINT 3**   The length of a violin string varies inversely as the frequency of its vibrations. A violin string 8 inches long vibrates at a frequency of 640 cycles per second. What is the frequency of a 10-inch string?

3 Solve combined variation problems.

Combined Variation

In **combined variation**, direct and inverse variation occur at the same time. For example, as the advertising budget, A, of a company increases, its monthly sales, S, also increase. Monthly sales vary directly as the advertising budget:

$$S = kA.$$

By contrast, as the price of the company's product, P, increases, its monthly sales, S, decrease. Monthly sales vary inversely as the price of the product:

$$S = \frac{k}{P}.$$

We can combine these two variation equations into one equation:

$$S = \frac{kA}{P}.$$

> Monthly sales , S, vary directly as the advertising budget, A, and inversely as the price of the product, P.

The following example illustrates an application of combined variation.

EXAMPLE 4   Solving a Combined Variation Problem

The owners of Rollerblades Now determined that the monthly sales, S, of its skates vary directly as its advertising budget, A, and inversely as the price of the skates, P. When \$60,000 is spent on advertising and the price of the skates is \$40, the monthly sales are 12,000 pairs of rollerblades.

a. Write an equation of variation that models this situation.

b. Determine monthly sales if the amount of the advertising budget is increased to \$70,000.

Solution

a. Write an equation.

$$S = \frac{kA}{P}.$$

> Translate "sales vary directly as the advertising budget and inversely as the skates' price."

Use the given values to find k.

$$12,000 = \frac{k(60,000)}{40}$$

When \$60,000 is spent on advertising ($A = 60,000$) and the price is \$40 ($P = 40$), monthly sales are 12,000 units ($S = 12,000$).

$$12,000 = k \cdot 1500$$

Divide 60,000 by 40.

$$\frac{12,000}{1500} = \frac{k \cdot 1500}{1500}$$

Divide both sides of the equation by 1500.

$$8 = k$$

Simplify.

Therefore, the equation of variation that models monthly sales is

$$S = \frac{8A}{P}.$$

Substitute 8 for k in $S = \frac{kA}{P}$.

b. The advertising budget is increased to $70,000, so $A = 70,000$. The skates' price is still $40, so $P = 40$.

$$S = \frac{8A}{P}$$ This is the equation from part (a).

$$S = \frac{8(70,000)}{40}$$ Substitute 70,000 for A and 40 for P.

$$S = 14,000$$ Simplify.

With a $70,000 advertising budget and $40 price, the company can expect to sell 14,000 pairs of rollerblades in a month (up from 12,000). ∎

✓ **CHECK POINT 4** The number of minutes needed to solve an Exercise Set of variation problems varies directly as the number of problems and inversely as the number of people working to solve the problems. It takes 4 people 32 minutes to solve 16 problems. How many minutes will it take 8 people to solve 24 problems?

---

4 Solve problems involving joint variation.

Joint Variation

Joint variation is a variation in which a variable varies directly as the product of two or more other variables. Thus, the equation $y = kxz$ is read "y varies jointly as x and z."

Joint variation plays a critical role in Isaac Newton's formula for gravitation:

$$F = G\frac{m_1 m_2}{d^2}.$$

The formula states that the force of gravitation, F, between two bodies varies jointly as the product of their masses, m_1 and m_2, and inversely as the square of the distance between them, d. (G is the gravitational constant.) The formula indicates that gravitational force exists between any two objects in the universe, increasing as the distance between the bodies decreases. One practical result is that the pull of the moon on the oceans is greater on the side of Earth closer to the moon. This gravitational imbalance is what produces tides.

EXAMPLE 5 Modeling Centrifugal Force

The centrifugal force, C, of a body moving in a circle varies jointly with the radius of the circular path, r, and the body's mass, m, and inversely with the square of the time, t, it takes to move about one full circle. A 6-gram body moving in a circle with radius 100 centimeters at a rate of 1 revolution in 2 seconds has a centrifugal force of 6000 dynes. Find the centrifugal force of an 18-gram body moving in a circle with radius 100 centimeters at a rate of 1 revolution in 3 seconds.

Solution

$$C = \frac{krm}{t^2}$$ Translate "Centrifugal force, C, varies jointly with radius, r, and mass, m, and inversely with the square of time, t."

$$6000 = \frac{k(100)(6)}{2^2}$$ A 6-gram body (m = 6) moving in a circle with radius 100 centimeters (r = 100) at 1 revolution in 2 seconds (t = 2) has a centrifugal force of 6000 dynes (C = 6000).

$$6000 = 150k$$

Simplify: $\dfrac{100(6)}{2^2} = \dfrac{600}{4} = 150.$

$$40 = k$$

Divide both sides by 150 and solve for k.

$$C = \dfrac{40rm}{t^2}$$

Substitute 40 for k in the model for centrifugal force.

$$C = \dfrac{40(100)(18)}{3^2}$$

Find centrifugal force, C, of an 18-gram body ($m = 18$) moving in a circle with radius 100 centimeters ($r = 100$) at 1 revolution in 3 seconds ($t = 3$).

$$= 8000$$

Simplify.

The centrifugal force is 8000 dynes. ∎

✓ **CHECK POINT 5** The volume of a cone, V, varies jointly as its height, h, and the square of its radius, r. A cone with a radius measuring 6 feet and a height measuring 10 feet has a volume of 120π cubic feet. Find the volume of a cone having a radius of 12 feet and a height of 2 feet.

CONCEPT AND VOCABULARY CHECK

Fill in each blank so that the resulting statement is true.

1. y varies directly as x can be modeled by the equation _____, where k is called the _____.

2. y varies directly as the nth power of x can be modeled by the equation _____.

3. y varies inversely as x can be modeled by the equation _____.

4. y varies directly as x and inversely as z can be modeled by the equation _____.

5. y varies jointly as x and z can be modeled by the equation _____.

6. In the equation $S = \dfrac{8A}{P}$, S varies _____ as A and _____ as P.

7. In the equation $C = \dfrac{0.02P_1P_2}{d^2}$, C varies _____ as P_1 and P_2 and _____ as the square of d.

7.8 EXERCISE SET MyMathLab® Watch the videos in MyMathLab  Download the MyDashBoard App

Practice Exercises

Use the four-step procedure for solving variation problems given on page 561 to solve Exercises 1–10.

1. y varies directly as x. $y = 65$ when $x = 5$. Find y when $x = 12$.

2. y varies directly as x. $y = 45$ when $x = 5$. Find y when $x = 13$.

3. y varies inversely as x. $y = 12$ when $x = 5$. Find y when $x = 2$.

4. y varies inversely as x. $y = 6$ when $x = 3$. Find y when $x = 9$.

5. y varies directly as x and inversely as the square of z. $y = 20$ when $x = 50$ and $z = 5$. Find y when $x = 3$ and $z = 6$.

6. a varies directly as b and inversely as the square of c. $a = 7$ when $b = 9$ and $c = 6$. Find a when $b = 4$ and $c = 8$.

7. y varies jointly as x and z. $y = 25$ when $x = 2$ and $z = 5$. Find y when $x = 8$ and $z = 12$.

8. C varies jointly as A and T. $C = 175$ when $A = 2100$ and $T = 4$. Find C when $A = 2400$ and $T = 6$.

9. y varies jointly as a and b, and inversely as the square root of c. $y = 12$ when $a = 3$, $b = 2$, and $c = 25$. Find y when $a = 5$, $b = 3$, and $c = 9$.

10. y varies jointly as m and the square of n, and inversely as p. $y = 15$ when $m = 2$, $n = 1$, and $p = 6$. Find y when $m = 3$, $n = 4$, and $p = 10$.

Practice PLUS

In Exercises 11–20, write an equation that expresses each relationship. Then solve the equation for y.

11. x varies jointly as y and z.

12. x varies jointly as y and the square of z.

13. x varies directly as the cube of z and inversely as y.

14. x varies directly as the cube root of z and inversely as y.

15. x varies jointly as y and z and inversely as the square root of w.

16. x varies jointly as y and z and inversely as the square of w.

17. x varies jointly as z and the sum of y and w.

18. x varies jointly as z and the difference between y and w.

19. x varies directly as z and inversely as the difference between y and w.

20. x varies directly as z and inversely as the sum of y and w.

Application Exercises

Use the four-step procedure for solving variation problems given on page 561 to solve Exercises 21–28.

21. An alligator's tail length, T, varies directly as its body length, B. An alligator with a body length of 4 feet has a tail length of 3.6 feet. What is the tail length of an alligator whose body length is 6 feet?

|←———— Body length, B ————→|←———— Tail length, T ————→|

22. An object's weight on the moon, M, varies directly as its weight on Earth, E. Neil Armstrong, the first person to step on the moon on July 20, 1969, weighed 360 pounds on Earth (with all of his equipment on) and 60 pounds on the moon. What is the moon weight of a person who weighs 186 pounds on Earth?

23. The height that a ball bounces varies directly as the height from which it was dropped. A tennis ball dropped from 12 inches bounces 8.4 inches. From what height was the tennis ball dropped if it bounces 56 inches?

24. The distance that a spring will stretch varies directly as the force applied to the spring. A force of 12 pounds is needed to stretch a spring 9 inches. What force is required to stretch the spring 15 inches?

25. If all men had identical body types, their weight would vary directly as the cube of their height. Shown below is Robert Wadlow, who reached a record height of 8 feet 11 inches (107 inches) before his death at age 22. If a man who is 5 feet 10 inches tall (70 inches) with the same body type as Mr. Wadlow weighs 170 pounds, what was Robert Wadlow's weight shortly before his death?

26. On a dry asphalt road, a car's stopping distance varies directly as the square of its speed. A car traveling at 45 miles per hour can stop in 67.5 feet. What is the stopping distance for a car traveling at 60 miles per hour?

27. The figure shows that a bicyclist tips the cycle when making a turn. The angle B, formed by the vertical direction and the bicycle, is called the banking angle. The banking angle varies inversely as the cycle's turning radius. When the turning radius is 4 feet, the banking angle is 28°. What is the banking angle when the turning radius is 3.5 feet?

28. The water temperature of the Pacific Ocean varies inversely as the water's depth. At a depth of 1000 meters, the water temperature is 4.4° Celsius. What is the water temperature at a depth of 5000 meters?

Heart rates and life spans of most mammals can be modeled using inverse variation. The bar graph shows the average heart rate and the average life span of five mammals. You will use the data to solve Exercises 29–30.

Heart Rate and Life Span

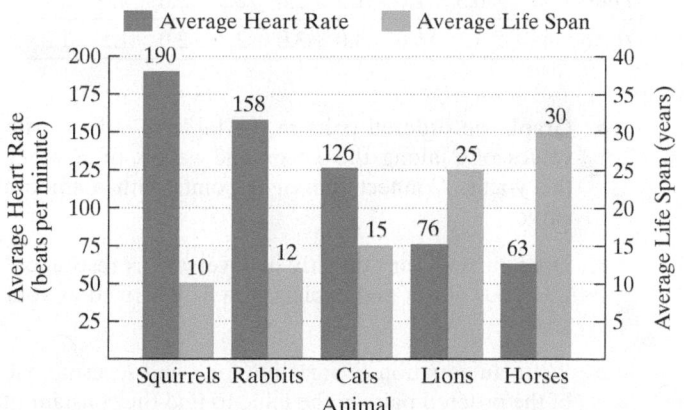

Source: The Carnegie Library of Pittsburg, THE HANDY SCIENCE ANSWER BOOK, Centennial Edition. © 2003

29. a. A mammal's average life span, L, in years, varies inversely as its average heart rate, R, in beats per minute. Use the data shown for horses to write the equation that models this relationship.

b. Is the inverse variation equation in part (a) an exact model or an approximate model for the data shown for lions?

c. Elephants have an average heart rate of 27 beats per minute. Determine their average life span.

30. a. A mammal's average life span, L, in years, varies inversely as its average heart rate, R, in beats per minute. Use the data shown for cats at the bottom of the previous page to write the equation that models this relationship.

b. Is the inverse variation equation in part (a) an exact model or an approximate model for the data shown for squirrels?

c. Mice have an average heart rate of 634 beats per minute. Determine their average life span, rounded to the nearest year.

The figure shows the graph of the inverse variation model that you wrote in Exercise 29(a) or Exercise 30(a). Use the graph of this rational function to solve Exercises 31–32.

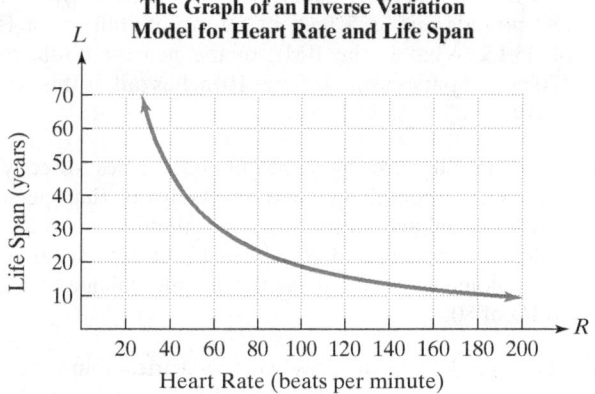

31. a. If a mammal has a life span of 20 years, use the graph to estimate its heart rate, rounded to the nearest 10 beats per minute.

b. Use the inverse variation equation that you wrote in Exercise 29(a) or Exercise 30(a) to determine the heart rate, rounded to the nearest beat per minute, for a mammal with a life span of 20 years.

c. The bar graph in the previous column uses two bars to display the data for horses. How is this information shown on the graph of the inverse variation model?

32. a. If a mammal has a life span of 50 years, use the graph to estimate its heart rate, rounded to the nearest 10 beats per minute.

b. Use the inverse variation equation that you wrote in Exercise 29(a) or Exercise 30(a) to determine the heart rate, rounded to the nearest beat per minute, for a mammal with a life span of 50 years.

c. The bar graph in the previous column uses two bars to display the data for lions. How is this information shown on the graph of the inverse variation model?

Continue to use the four-step procedure for solving variation problems given on page 561 to solve Exercises 33–40.

33. Radiation machines, used to treat tumors, produce an intensity of radiation that varies inversely as the square of the distance from the machine. At 3 meters, the radiation intensity is 62.5 milliroentgens per hour. What is the intensity at a distance of 2.5 meters?

34. The illumination provided by a car's headlight varies inversely as the square of the distance from the headlight. A car's headlight produces an illumination of 3.75 footcandles at a distance of 40 feet. What is the illumination when the distance is 50 feet?

35. Body-mass index, or BMI, takes both weight and height into account when assessing whether an individual is underweight or overweight. BMI varies directly as one's weight, in pounds, and inversely as the square of one's height, in inches. In adults, normal values for the BMI are between 20 and 25, inclusive. Values below 20 indicate that an individual is underweight and values above 30 indicate that an individual is obese. A person who weighs 180 pounds and is 5 feet, or 60 inches, tall has a BMI of 35.15. What is the BMI, to the nearest tenth, for a 170 pound person who is 5 feet 10 inches tall. Is this person overweight?

36. One's intelligence quotient, or IQ, varies directly as a person's mental age and inversely as that person's chronological age. A person with a mental age of 25 and a chronological age of 20 has an IQ of 125. What is the chronological age of a person with a mental age of 40 and an IQ of 80?

37. The heat loss of a glass window varies jointly as the window's area and the difference between the outside and inside temperatures. A window 3 feet wide by 6 feet long loses 1200 Btu per hour when the temperature outside is 20° colder than the temperature inside. Find the heat loss through a glass window that is 6 feet wide by 9 feet long when the temperature outside is 10° colder than the temperature inside.

38. Kinetic energy varies jointly as the mass and the square of the velocity. A mass of 8 grams and velocity of 3 centimeters per second has a kinetic energy of 36 ergs. Find the kinetic energy for a mass of 4 grams and velocity of 6 centimeters per second.

39. Sound intensity varies inversely as the square of the distance from the sound source. If you are in a movie theater and you change your seat to one that is twice as far from the speakers, how does the new sound intensity compare to that of your original seat?

40. Many people claim that as they get older, time seems to pass more quickly. Suppose that the perceived length of a period of time is inversely proportional to your age. How long will a year seem to be when you are three times as old as you are now?

41. The average number of daily phone calls, C, between two cities varies jointly as the product of their populations, P_1 and P_2, and inversely as the square of the distance, d, between them.

    a. Write an equation that expresses this relationship.

    b. The distance between San Francisco (population: 777,000) and Los Angeles (population: 3,695,000) is 420 miles. If the average number of daily phone calls between the cities is 326,000, find the value of k to two decimal places and write the equation of variation.

    c. Memphis (population: 650,000) is 400 miles from New Orleans (population: 490,000). Find the average number of daily phone calls, to the nearest whole number, between these cities.

42. The force of wind blowing on a window positioned at a right angle to the direction of the wind varies jointly as the area of the window and the square of the wind's speed. It is known that a wind of 30 miles per hour blowing on a window measuring 4 feet by 5 feet exerts a force of 150 pounds. During a storm with winds of 60 miles per hour, should hurricane shutters be placed on a window that measures 3 feet by 4 feet and is capable of withstanding 300 pounds of force?

43. The table shows the values for the current, I, in an electric circuit and the resistance, R, of the circuit.

I (amperes)	0.5	1.0	1.5	2.0	2.5	3.0	4.0	5.0
R (ohms)	12	6.0	4.0	3.0	2.4	2.0	1.5	1.2

    a. Graph the ordered pairs in the table of values, with values of I along the x-axis and values of R along the y-axis. Connect the eight points with a smooth curve.

    b. Does current vary directly or inversely as resistance? Use your graph and explain how you arrived at your answer.

    c. Write an equation of variation for I and R, using one of the ordered pairs in the table to find the constant of variation. Then use your variation equation to verify the other seven ordered pairs in the table.

Writing in Mathematics

44. What does it mean if two quantities vary directly?

45. In your own words, explain how to solve a variation problem.

46. What does it mean if two quantities vary inversely?

47. Explain what is meant by combined variation. Give an example with your explanation.

48. Explain what is meant by joint variation. Give an example with your explanation.

In Exercises 49–50, describe in words the variation shown by the given equation.

49. $z = \dfrac{k\sqrt{x}}{y^2}$

50. $z = kx^2\sqrt{y}$

51. We have seen that the daily number of phone calls between two cities varies jointly as their populations and inversely as the square of the distance between them. This model, used by telecommunication companies to estimate the line capacities needed among various cities, is called the *gravity model*. Compare the model to Newton's formula for gravitation on page 566 and describe why the name *gravity model* is appropriate.

Technology Exercise

52. Use a graphing utility to graph any three of the variation equations in Exercises 21–28. Then $\boxed{\text{TRACE}}$ along each curve and identify the point that corresponds to the problem's solution.

Critical Thinking Exercises

Make Sense? *In Exercises 53–56, determine whether each statement "makes sense" or "does not make sense" and explain your reasoning.*

53. I'm using an inverse variation equation and I need to determine the value of the dependent variable when the independent variable is zero.

54. The graph of this direct variation equation has a positive constant of variation and shows one variable increasing as the other variable decreases.

55. When all is said and done, it seems to me that direct variation equations are special kinds of linear equations and inverse variation equations are special kinds of rational equations.

56. Using the language of variation, I can now state the formula for the area of a trapezoid, $A = \frac{1}{2}h(b_1 + b_2)$, as, "A trapezoid's area varies jointly with its height and the sum of its bases."

57. In a hurricane, the wind pressure varies directly as the square of the wind velocity. If wind pressure is a measure of a hurricane's destructive capacity, what happens to this destructive power when the wind speed doubles?

58. The heat generated by a stove element varies directly as the square of the voltage and inversely as the resistance. If the voltage remains constant, what needs to be done to triple the amount of heat generated?

59. Galileo's telescope brought about revolutionary changes in astronomy. A comparable leap in our ability to observe the universe took place as a result of the Hubble Space Telescope. The space telescope can see stars and galaxies whose brightness is $\frac{1}{50}$ of the faintest objects now observable using ground-based telescopes. Use the fact that the brightness of a point source, such as a star, varies inversely as the square of its distance from an observer to show that the space telescope can see about seven times farther than a ground-based telescope.

Review Exercises

60. Solve:

$$8(2 - x) = -5x.$$

(Section 2.3, Example 2)

61. Divide:

$$\frac{27x^3 - 8}{3x + 2}.$$

(Section 5.6, Example 3)

62. Factor:

$$6x^3 - 6x^2 - 120x.$$

(Section 6.5, Example 2)

Preview Exercises

Exercises 63–65 will help you prepare for the material covered in the first section of the next chapter.

63. Here are two sets of ordered pairs:

set 1: $\{(1, 5), (2, 5)\}$

set 2: $\{(5, 1), (5, 2)\}$

In which set is each x-coordinate paired with only one y-coordinate?

64. Evaluate $r^3 - 2r^2 + 5$ for $r = -5$.

65. Evaluate $5x + 7$ for $x = a + h$.

A cost-benefit analysis compares the estimated costs of a project with the benefits that will be achieved. Costs and benefits are given monetary values and compared using a benefit-cost ratio. As shown in the figure, a favorable ratio for a project means that the benefits outweigh the costs and the project is cost-effective. As a group, select an environmental project that was implemented in your area of the country. Research the cost and benefit graphs that resulted in the implementation of this project. How were the benefits converted into monetary terms? Is there an equation for either the cost model or the benefit model? Group members may need to interview members of environmental groups and businesses that were part of this project. You may wish to consult an environmental science textbook to find out more about cost-benefit analyses. After doing your research, the group should write or present a report explaining why the cost-benefit analysis resulted in the project's implementation.

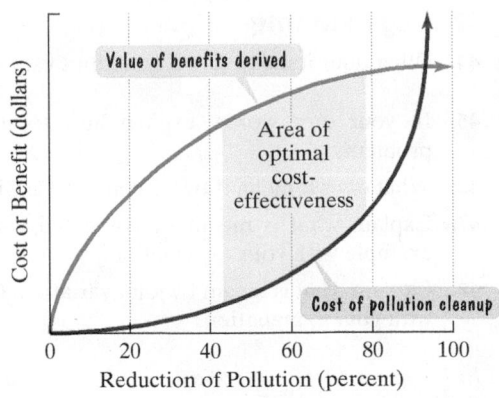

Chapter 7 Summary

Definitions and Concepts	Examples

Section 7.1 Rational Expressions and Their Simplification

A rational expression is the quotient of two polynomials. To find values for which a rational expression is undefined, set the denominator equal to 0 and solve.

Find all numbers for which

$$\frac{7x}{x^2 - 3x - 4}$$

is undefined.

$$x^2 - 3x - 4 = 0$$

$$(x - 4)(x + 1) = 0$$

$$x - 4 = 0 \quad \text{or} \quad x + 1 = 0$$

$$x = 4 \qquad\qquad x = -1$$

The expression is undefined for 4 and −1.

To simplify a rational expression:

1. Factor the numerator and the denominator completely.
2. Divide the numerator and the denominator by any common factors.

If factors in the numerator and denominator are opposites, their quotient is −1.

Simplify: $\dfrac{3x + 18}{x^2 - 36}$.

$$\frac{3x + 18}{x^2 - 36} = \frac{3\overset{1}{\cancel{(x + 6)}}}{\underset{1}{\cancel{(x + 6)}}(x - 6)} = \frac{3}{x - 6}$$

Definitions and Concepts	**Examples**

Section 7.2 Multiplying and Dividing Rational Expressions

Multiplying Rational Expressions

1. Factor completely.
2. Divide numerators and denominators by common factors.
3. Multiply remaining factors in the numerators and multiply the remaining factors in the denominators.

$$\frac{x^2 + 3x - 10}{x^2 - 2x} \cdot \frac{x^2}{x^2 - 25}$$

$$= \frac{(x + 5)(x - 2)}{x(x - 2)} \cdot \frac{x \cdot x}{(x + 5)(x - 5)}$$

$$= \frac{x}{x - 5}$$

Dividing Rational Expressions

Invert the divisor and multiply.

$$\frac{3y + 3}{(y + 2)^2} \div \frac{y^2 - 1}{y + 2}$$

$$= \frac{3y + 3}{(y + 2)^2} \cdot \frac{y + 2}{y^2 - 1}$$

$$= \frac{3(y + 1)}{(y + 2)(y + 2)} \cdot \frac{(y + 2)}{(y + 1)(y - 1)}$$

$$= \frac{3}{(y + 2)(y - 1)}$$

Section 7.3 Adding and Subtracting Rational Expressions with the Same Denominator

To add or subtract rational expressions with the same denominator, add or subtract the numerators and place the result over the common denominator. If possible, factor and simplify the resulting expression.

$$\frac{y^2 - 3y + 4}{y^2 + 8y + 15} - \frac{y^2 - 5y - 2}{y^2 + 8y + 15}$$

$$= \frac{y^2 - 3y + 4 - (y^2 - 5y - 2)}{y^2 + 8y + 15}$$

$$= \frac{y^2 - 3y + 4 - y^2 + 5y + 2}{y^2 + 8y + 15}$$

$$= \frac{2y + 6}{(y + 5)(y + 3)}$$

$$= \frac{2(y + 3)}{(y + 5)(y + 3)} = \frac{2}{y + 5}$$

To add or subtract rational expressions with opposite denominators, multiply either rational expression by $\frac{-1}{-1}$ to obtain a common denominator.

$$\frac{7}{x - 6} + \frac{x + 4}{6 - x}$$

$$= \frac{7}{x - 6} + \frac{(-1)}{(-1)} \cdot \frac{x + 4}{6 - x}$$

$$= \frac{7}{x + 6} + \frac{-x - 4}{x - 6}$$

$$= \frac{7 - x - 4}{x - 6} = \frac{3 - x}{x - 6}$$

Definitions and Concepts	**Examples**

Section 7.4 Adding and Subtracting Rational Expressions with Different Denominators

Finding the Least Common Denominator (LCD)

1. Factor denominators completely.
2. List factors of the first denominator.
3. Add to the list any factors of the second denominator that are not already in the list.
4. The LCD is the product of factors in step 3.

Find the LCD of

$$\frac{x+1}{2x-2} \quad \text{and} \quad \frac{2x}{x^2+2x-3}.$$
$$2x-2 = 2(x-1)$$
$$x^2+2x-3 = (x-1)(x+3)$$

Factors of first denominator: $2, x-1$

Factors of second denominator not in the list: $x+3$

LCD: $2(x-1)(x+3)$

Adding and Subtracting Rational Expressions with Different Denominators

1. Find the LCD.
2. Rewrite each rational expression as an equivalent expression with the LCD.
3. Add or subtract numerators, placing the resulting expression over the LCD.
4. If possible, simplify.

$$\frac{x+1}{2x-2} - \frac{2x}{x^2+2x-3}$$
$$= \frac{x+1}{2(x-1)} - \frac{2x}{(x-1)(x+3)}$$

LCD is $2(x-1)(x+3)$.

$$= \frac{(x+1)(x+3)}{2(x-1)(x+3)} - \frac{2x \cdot 2}{2(x-1)(x+3)}$$
$$= \frac{x^2+4x+3-4x}{2(x-1)(x+3)}$$
$$= \frac{x^2+3}{2(x-1)(x+3)}$$

Section 7.5 Complex Rational Expressions

Complex rational expressions have numerators or denominators containing one or more rational expressions. Complex rational expressions can be simplified by obtaining single expressions in the numerator and denominator and then dividing. They can also be simplified by multiplying the numerator and denominator by the LCD of all rational expressions within the complex rational expression.

Simplify by dividing: $\dfrac{\dfrac{1}{x}+5}{\dfrac{1}{x}-\dfrac{1}{3}}.$

$$= \frac{\dfrac{1}{x}+\dfrac{5x}{x}}{\dfrac{3}{3x}-\dfrac{x}{3x}} = \frac{\dfrac{1+5x}{x}}{\dfrac{3-x}{3x}} = \frac{1+5x}{\overset{1}{x}} \cdot \frac{\overset{1}{3x}}{3-x}$$
$$= \frac{3(1+5x)}{3-x} \quad \text{or} \quad \frac{3+15x}{3-x}$$

Simplify by the LCD method: $\dfrac{\dfrac{1}{x}+5}{\dfrac{1}{x}-\dfrac{1}{3}}.$

LCD is $3x$.

$$\frac{3x}{3x} \cdot \frac{\left(\dfrac{1}{x}+5\right)}{\left(\dfrac{1}{x}-\dfrac{1}{3}\right)} = \frac{3x\cdot\dfrac{1}{x}+3x\cdot5}{3x\cdot\dfrac{1}{x}-3x\cdot\dfrac{1}{3}}$$

$$= \frac{3+15x}{3-x}$$

Definitions and Concepts	**Examples**

Section 7.6 Solving Rational Equations

A rational equation is an equation containing one or more rational expressions.

Solving Rational Equations

1. List restrictions on the variable.
2. Clear fractions by multiplying both sides by the LCD.
3. Solve the resulting equation.
4. Reject any proposed solution in the list of restrictions. Check other proposed solutions in the original equation.

Solve: $\dfrac{7x}{x^2 - 4} + \dfrac{5}{x - 2} = \dfrac{2x}{x^2 - 4}$.

$$\dfrac{7x}{(x + 2)(x - 2)} + \dfrac{5}{x - 2} = \dfrac{2x}{(x + 2)(x - 2)}$$

> Denominators would equal 0 if $x = -2$ or $x = 2$.
> Restrictions: $x \neq -2$ and $x \neq 2$.

LCD is $(x + 2)(x - 2)$.

$$(x + 2)(x - 2)\left[\dfrac{7x}{(x + 2)(x - 2)} + \dfrac{5}{x - 2}\right]$$

$$= (x + 2)(x - 2) \cdot \dfrac{2x}{(x + 2)(x - 2)}$$

$$7x + 5(x + 2) = 2x$$

$$7x + 5x + 10 = 2x$$

$$12x + 10 = 2x$$

$$10 = -10x$$

$$-1 = x$$

The proposed solution, -1, is not part of the restriction $x \neq -2$ and $x \neq 2$. It checks. The solution is -1 and the solution set is $\{-1\}$.

To solve a formula for a variable, get the specified variable alone on one side of the formula. When working with formulas containing rational expressions, it is sometimes necessary to factor out the variable you are solving for.

Solve: $\dfrac{e}{E} = \dfrac{r}{r + R}$ for r.

$$E(r + R) \cdot \dfrac{e}{E} = E(r + R) \cdot \dfrac{r}{r + R} \qquad \text{LCD is } E(r + R).$$

$$e(r + R) = Er$$

$$er + eR = Er$$

$$eR = Er - er$$

$$eR = (E - e)r$$

$$\dfrac{eR}{E - e} = r$$

Definitions and Concepts	**Examples**

Section 7.7 Applications Using Rational Equations and Proportions

Motion problems involving time are solved using

$$t = \frac{d}{r}.$$

$$\text{Time traveled} = \frac{\text{Distance traveled}}{\text{Rate of travel}}$$

It takes a cyclist who averages 16 miles per hour in still air the same time to travel 48 miles with the wind as 16 miles against the wind. What is the wind's rate?

$$x = \text{wind's rate}$$

$$16 + x = \text{cyclist's rate with wind}$$

$$16 - x = \text{cyclist's rate against wind}$$

	Distance	Rate	Time = $\dfrac{\text{Distance}}{\text{Rate}}$
With wind	48	$16 + x$	$\dfrac{48}{16 + x}$
Against wind	16	$16 - x$	$\dfrac{16}{16 - x}$

Two times are equal.

$$\frac{48}{16 + x} = \frac{16}{16 - x}$$

$$(16 + x)(16 - x) \cdot \frac{48}{16 + x} = \frac{16}{16 - x} \cdot (16 + x)(16 - x)$$

$$48(16 - x) = 16(16 + x)$$

Solving this equation, $x = 8$.

The wind's rate is 8 miles per hour.

Work problems are solved using the following condition:

$$\boxed{\text{Fraction of job done by the first}} + \boxed{\text{fraction of job done by the second}} = \boxed{1.}$$

One pipe fills a pool in 20 hours and a second pipe in 15 hours. How long will it take to fill the pool using both pipes?

$$x = \text{time using both pipes}$$

$$\boxed{\text{Fraction of pool filled by pipe 1 in } x \text{ hours}} + \boxed{\text{fraction of pool filled by pipe 2 in } x \text{ hours}} = \boxed{1.}$$

$$\frac{x}{20} \quad + \quad \frac{x}{15} \quad = \quad 1$$

$$60\left(\frac{x}{20} + \frac{x}{15}\right) = 60 \cdot 1$$

$$3x + 4x = 60$$

$$7x = 60$$

$$x = \frac{60}{7} = 8\frac{4}{7} \text{ hours}$$

It will take $8\frac{4}{7}$ hours for both pipes to fill the pool.

Definitions and Concepts	**Examples**

Section 7.7 Applications Using Rational Equations and Proportions (continued)

A proportion is a statement in the form $\dfrac{a}{b} = \dfrac{c}{d}$. The cross-products principle states that if $\dfrac{a}{b} = \dfrac{c}{d}$, then $ad = bc$ ($b \neq 0$ and $d \neq 0$).

Solving Applied Problems Using Proportions

1. Read the problem and represent the unknown quantity by x (or any letter).

2. Set up a proportion by listing the given ratio on one side and the ratio with the unknown quantity on the other side.

3. Drop units and apply the cross-products principle.

4. Solve for x and anwer the question.

30 elk are tagged and released. Sometime later, a sample of 80 elk is observed and 10 are tagged. How many elk are there?

$$x = \text{number of elk}$$

$$\text{Tagged} \rightarrow \frac{30}{x} = \frac{10}{80} \leftarrow \text{Total}$$

$$10x = 30 \cdot 80$$

$$10x = 2400$$

$$x = 240$$

There are 240 elk.

Similar triangles have the same shape, but not necessarily the same size. Corresponding angles have the same measure, and corresponding sides are proportional. If the measures of two angles of one triangle are equal to those of two angles of a second triangle, then the two triangles are similar.

Find x for these similar triangles.

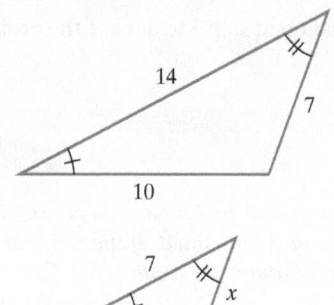

Corresponding sides are proportional:

$$\frac{7}{x} = \frac{10}{5}. \quad \left(\text{or } \frac{7}{x} = \frac{14}{7} \right)$$

$$5x \cdot \frac{7}{x} = \frac{10}{5} \cdot 5x$$

$$35 = 10x$$

$$x = \frac{35}{10} = 3.5$$

Definitions and Concepts	**Examples**

Section 7.8 Modeling Using Variation

English Statement	**Equation**
y varies directly as x. y is directly proportional to x.	$y = kx$
y varies directly as x^n. y is directly proportional to x^n.	$y = kx^n$
y varies inversely as x. y is inversely proportional to x.	$y = \dfrac{k}{x}$
y varies inversely as x^n. y is inversely proportional to x^n.	$y = \dfrac{k}{x^n}$
y varies directly as x and inversely as z.	$y = \dfrac{kx}{z}$
y varies jointly as x and z.	$y = kxz$

Solving Variation Problems

1. Write an equation that models the given English statement.
2. Substitute the pair of values into the equation in step 1 and find k.
3. Substitute k into the equation in step 1.
4. Use the equation in step 3 to answer the problem's question.

The time that it takes you to drive a certain distance varies inversely as your driving rate. Averaging 40 miles per hour, it takes you 10 hours to drive the distance. How long would the trip take averaging 50 miles per hour?

1.
$$t = \frac{k}{r} \quad \boxed{\text{Time, } t, \text{ varies inversely as rate, } r.}$$

2. It takes 10 hours at 40 miles per hour.
$$10 = \frac{k}{40}$$
$$k = 10(40) = 400$$

3. $t = \dfrac{400}{r}$

4. How long at 50 miles per hour? Substitute 50 for r.
$$t = \frac{400}{50} = 8$$

It takes 8 hours at 50 miles per hour.

CHAPTER 7 REVIEW EXERCISES

7.1 *In Exercises 1–4, find all numbers for which each rational expression is undefined. If the rational expression is defined for all real numbers, so state.*

1. $\dfrac{5x}{6x - 24}$

2. $\dfrac{x + 3}{(x - 2)(x + 5)}$

3. $\dfrac{x^2 + 3}{x^2 - 3x + 2}$

4. $\dfrac{7}{x^2 + 81}$

In Exercises 5–12, simplify each rational expression. If the rational expression cannot be simplified, so state.

5. $\dfrac{16x^2}{12x}$

6. $\dfrac{x^2 - 4}{x - 2}$

7. $\dfrac{x^3 + 2x^2}{x + 2}$

8. $\dfrac{x^2 + 3x - 18}{x^2 - 36}$

9. $\dfrac{x^2 - 4x - 5}{x^2 + 8x + 7}$

10. $\dfrac{y^2 + 2y}{y^2 + 4y + 4}$

11. $\dfrac{x^2}{x^2 + 4}$

12. $\dfrac{2x^2 - 18y^2}{3y - x}$

7.2 *In Exercises 13–17, multiply as indicated.*

13. $\dfrac{x^2 - 4}{12x} \cdot \dfrac{3x}{x + 2}$

14. $\dfrac{5x + 5}{6} \cdot \dfrac{3x}{x^2 + x}$

15. $\dfrac{x^2 + 6x + 9}{x^2 - 4} \cdot \dfrac{x - 2}{x + 3}$

16. $\dfrac{y^2 - 2y + 1}{y^2 - 1} \cdot \dfrac{2y^2 + y - 1}{5y - 5}$

17. $\dfrac{2y^2 + y - 3}{4y^2 - 9} \cdot \dfrac{3y + 3}{5y - 5y^2}$

In Exercises 18–22, divide as indicated.

18. $\dfrac{x^2 + x - 2}{10} \div \dfrac{2x + 4}{5}$

19. $\dfrac{6x + 2}{x^2 - 1} \div \dfrac{3x^2 + x}{x - 1}$

20. $\dfrac{1}{y^2 + 8y + 15} \div \dfrac{7}{y + 5}$

21. $\dfrac{y^2 + y - 42}{y - 3} \div \dfrac{y + 7}{(y - 3)^2}$

22. $\dfrac{8x + 8y}{x^2} \div \dfrac{x^2 - y^2}{x^2}$

7.3 *In Exercises 23–28, add or subtract as indicated. Simplify the result, if possible.*

23. $\dfrac{4x}{x + 5} + \dfrac{20}{x + 5}$

24. $\dfrac{8x - 5}{3x - 1} + \dfrac{4x + 1}{3x - 1}$

25. $\dfrac{3x^2 + 2x}{x - 1} - \dfrac{10x - 5}{x - 1}$

26. $\dfrac{6y^2 - 4y}{2y - 3} - \dfrac{12 - 3y}{2y - 3}$

27. $\dfrac{x}{x - 2} + \dfrac{x - 4}{2 - x}$

28. $\dfrac{x + 5}{x - 3} - \dfrac{x}{3 - x}$

7.4   *In Exercises 29–31, find the least common denominator of the rational expressions.*

29. $\dfrac{7}{9x^3}$ and $\dfrac{5}{12x}$

30. $\dfrac{3}{x^2(x-1)}$ and $\dfrac{11}{x(x-1)^2}$

31. $\dfrac{x}{x^2+4x+3}$ and $\dfrac{17}{x^2+10x+21}$

In Exercises 32–42, add or subtract as indicated. Simplify the result, if possible.

32. $\dfrac{7}{3x}+\dfrac{5}{2x^2}$

33. $\dfrac{5}{x+1}+\dfrac{2}{x}$

34. $\dfrac{7}{x+3}+\dfrac{4}{(x+3)^2}$

35. $\dfrac{6y}{y^2-4}-\dfrac{3}{y+2}$

36. $\dfrac{y-1}{y^2-2y+1}-\dfrac{y+1}{y-1}$

37. $\dfrac{x+y}{y}-\dfrac{x-y}{x}$

38. $\dfrac{2x}{x^2+2x+1}+\dfrac{x}{x^2-1}$

39. $\dfrac{5x}{x+1}-\dfrac{2x}{1-x^2}$

40. $\dfrac{4}{x^2-x-6}-\dfrac{4}{x^2-4}$

41. $\dfrac{7}{x+3}+2$

42. $\dfrac{2y-5}{6y+9}-\dfrac{4}{2y^2+3y}$

7.5   *In Exercises 43–47, simplify each complex rational expression.*

43. $\dfrac{\dfrac{1}{2}+\dfrac{3}{8}}{\dfrac{3}{4}-\dfrac{1}{2}}$

44. $\dfrac{\dfrac{1}{x}}{1-\dfrac{1}{x}}$

45. $\dfrac{\dfrac{1}{x}+\dfrac{1}{y}}{\dfrac{1}{xy}}$

46. $\dfrac{\dfrac{1}{x}-\dfrac{1}{2}}{\dfrac{1}{3}-\dfrac{x}{6}}$

47. $\dfrac{3+\dfrac{12}{x}}{1-\dfrac{16}{x^2}}$

7.6   *In Exercises 48–55, solve each rational equation.*

48. $\dfrac{3}{x}-\dfrac{1}{6}=\dfrac{1}{x}$

49. $\dfrac{3}{4x}=\dfrac{1}{x}+\dfrac{1}{4}$

50. $x+5=\dfrac{6}{x}$

51. $4-\dfrac{x}{x+5}=\dfrac{5}{x+5}$

52. $\dfrac{2}{x-3}=\dfrac{4}{x+3}+\dfrac{8}{x^2-9}$

53. $\dfrac{2}{x}=\dfrac{2}{3}+\dfrac{x}{6}$

54. $\dfrac{13}{y-1}-3=\dfrac{1}{y-1}$

55. $\dfrac{1}{x+3}-\dfrac{1}{x-1}=\dfrac{x+1}{x^2+2x-3}$

56. Park rangers introduce 50 elk into a wildlife preserve. The formula

$$P=\dfrac{250(3t+5)}{t+25}$$

models the elk population, P, after t years. How many years will it take for the population to increase to 125 elk?

57. The formula

$$S=\dfrac{C}{1-r}$$

describes the selling price, S, of a product in terms of its cost to the retailer, C, and its markup, r, usually expressed as a percent. A small television cost a retailer \$140 and was sold for \$200. Find the markup. Express the answer as a percent.

In Exercises 58–62, solve each formula for the specified variable.

58. $P=\dfrac{R-C}{n}$ for C

59. $\dfrac{P_1V_1}{T_1}=\dfrac{P_2V_2}{T_2}$ for T_1

60. $T=\dfrac{A-P}{Pr}$ for P

61. $\dfrac{1}{R}=\dfrac{1}{R_1}+\dfrac{1}{R_2}$ for R

62. $I=\dfrac{nE}{R+nr}$ for n

7.7

63. In still water, a paddle boat averages 20 miles per hour. It takes the boat the same amount of time to travel 72 miles downstream, with the current, as 48 miles upstream, against the current. What is the rate of the water's current?

64. A car travels 60 miles in the same time that a car traveling 10 miles per hour faster travels 90 miles. What is the rate of each car?

65. A painter can paint a fence around a house in 6 hours. Working alone, the painter's apprentice can paint the same fence in 12 hours. How many hours would it take them to do the job if they worked together?

66. If a school board determines that there should be 3 teachers for every 50 students, how many teachers are needed for an enrollment of 5400 students?

67. To determine the number of trout in a lake, a conservationist catches 112 trout, tags them, and returns them to the lake. Later, 82 trout are caught, and 32 of them are found to be tagged. How many trout are in the lake?

68. The triangles shown in the figure are similar. Find the length of the side marked with an x.

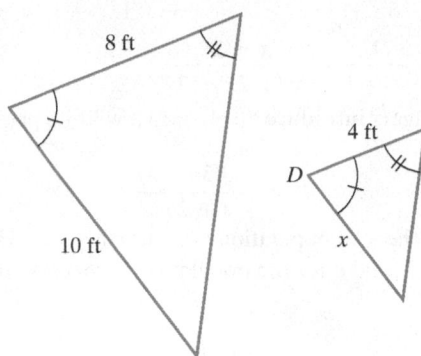

69. Find the height of the lamppost in the figure.

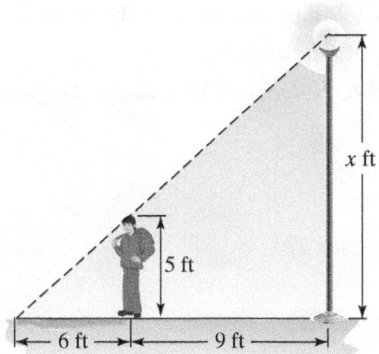

7.8 *Solve the variation problems in Exercises 70–75.*

70. A company's profit varies directly as the number of products it sells. The company makes a profit of $1175 on the sale of 25 products. What is the company's profit when it sells 105 products?

71. The distance that a body falls from rest varies directly as the square of the time of the fall. If skydivers fall 144 feet in 3 seconds, how far will they fall in 10 seconds?

72. The pitch of a musical tone varies inversely as its wavelength. A tone has a pitch of 660 vibrations per second and a wavelength of 1.6 feet. What is the pitch of a tone that has a wavelength of 2.4 feet?

73. The loudness of a stereo speaker, measured in decibels, varies inversely as the square of your distance from the speaker. When you are 8 feet from the speaker, the loudness is 28 decibels. What is the loudness when you are 4 feet from the speaker?

74. The time required to assemble computers varies directly as the number of computers assembled and inversely as the number of workers. If 30 computers can be assembled by 6 workers in 10 hours, how long would it take 5 workers to assemble 40 computers?

75. The volume of a pyramid varies jointly as its height and the area of its base. A pyramid with a height of 15 feet and a base with an area of 35 square feet has a volume of 175 cubic feet. Find the volume of a pyramid with a height of 20 feet and a base with an area of 120 square feet.

CHAPTER
Test Prep
VIDEOS

CHAPTER 7 TEST

Step-by-step test solutions are found on the Chapter Test Prep Videos available in MyMathLab® or on YouTube (search "BlitzerCombinedAlg" and click on "Channels").

1. Find all numbers for which

$$\frac{x + 7}{x^2 + 5x - 36}$$

is undefined.

In Exercises 2–3, simplify each rational expression.

2. $\dfrac{x^2 + 2x - 3}{x^2 - 3x + 2}$

3. $\dfrac{4x^2 - 20x}{x^2 - 4x - 5}$

In Exercises 4–16, perform the indicated operations. Simplify the result, if possible.

4. $\dfrac{x^2 - 16}{10} \cdot \dfrac{5}{x + 4}$

5. $\dfrac{x^2 - 7x + 12}{x^2 - 4x} \cdot \dfrac{x^2}{x^2 - 9}$

6. $\dfrac{2x + 8}{x - 3} \div \dfrac{x^2 + 5x + 4}{x^2 - 9}$

7. $\dfrac{5y + 5}{(y - 3)^2} \div \dfrac{y^2 - 1}{y - 3}$

8. $\dfrac{2y^2 + 5}{y + 3} + \dfrac{6y - 5}{y + 3}$

9. $\dfrac{y^2 - 2y + 3}{y^2 + 7y + 12} - \dfrac{y^2 - 4y - 5}{y^2 + 7y + 12}$

10. $\dfrac{x}{x + 3} + \dfrac{5}{x - 3}$

11. $\dfrac{2}{x^2 - 4x + 3} + \dfrac{6}{x^2 + x - 2}$

12. $\dfrac{4}{x - 3} + \dfrac{x + 5}{3 - x}$

13. $1 + \dfrac{3}{x - 1}$

14. $\dfrac{2x + 3}{x^2 - 7x + 12} - \dfrac{2}{x - 3}$

15. $\dfrac{8y}{y^2 - 16} - \dfrac{4}{y - 4}$

16. $\dfrac{(x - y)^2}{x + y} \div \dfrac{x^2 - xy}{3x + 3y}$

In Exercises 17–18, simplify each complex rational expression.

17. $\dfrac{5 + \dfrac{5}{x}}{2 + \dfrac{1}{x}}$

18. $\dfrac{\dfrac{1}{x} - \dfrac{1}{y}}{\dfrac{1}{x}}$

In Exercises 19–21, solve each rational equation.

19. $\dfrac{5}{x} + \dfrac{2}{3} = 2 - \dfrac{2}{x} - \dfrac{1}{6}$

20. $\dfrac{3}{y + 5} - 1 = \dfrac{4 - y}{2y + 10}$

21. $\dfrac{2}{x - 1} = \dfrac{3}{x^2 - 1} + 1$

22. Solve for a: $R = \dfrac{as}{a + s}$.

23. In still water, a boat averages 30 miles per hour. It takes the boat the same amount of time to travel 16 miles downstream, with the current, as 14 miles upstream, against the current. What is the rate of the water's current?

24. One pipe can fill a hot tub in 20 minutes and a second pipe can fill it in 30 minutes. If the hot tub is empty, how long will it take both pipes to fill it?

25. Park rangers catch, tag, and release 200 tule elk back into a wildlife refuge. Two weeks later they observe a sample of 150 elk, of which 5 are tagged. Assuming that the ratio of tagged elk in the sample holds for all elk in the refuge, how many elk are there in the park?

26. The triangles in the figure are similar. Find the length of the side marked with an x.

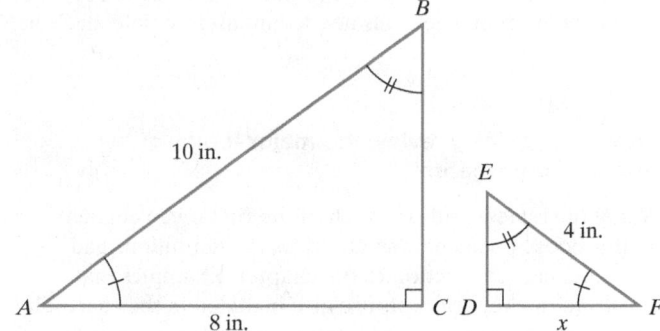

27. The amount of current flowing in an electrical circuit varies inversely as the resistance in the circuit. When the resistance in a particular circuit is 5 ohms, the current is 42 amperes. What is the current when the resistance is 4 ohms?

CUMULATIVE REVIEW EXERCISES (CHAPTERS 1–7)

In Exercises 1–6, solve each equation, inequality, or system of equations.

1. $2(x - 3) + 5x = 8(x - 1)$

2. $-3(2x - 4) > 2(6x - 12)$

3. $x^2 + 3x = 18$

4. $\dfrac{2x}{x^2 - 4} + \dfrac{1}{x - 2} = \dfrac{2}{x + 2}$

5. $\begin{cases} y = 2x - 3 \\ x + 2y = 9 \end{cases}$

6. $\begin{cases} 3x + 2y = -2 \\ -4x + 5y = 18 \end{cases}$

In Exercises 7–9, graph each equation in a rectangular coordinate system.

7. $3x - 2y = 6$

8. $y = -2x + 3$

9. $y = -3$

In Exercises 10–12, simplify each expression.

10. $-21 - 16 - 3(2 - 8)$

11. $\left(\dfrac{4x^5}{2x^2} \right)^3$

12. $\dfrac{\dfrac{1}{x} - 2}{4 - \dfrac{1}{x}}$

In Exercises 13–15, factor completely.

13. $4x^2 - 13x + 3$

14. $4x^2 - 20x + 25$

15. $3x^2 - 75$

In Exercises 16–18, perform the indicated operations.

16. $(4x^2 - 3x + 2) - (5x^2 - 7x - 6)$

17. $\dfrac{-8x^6 + 12x^4 - 4x^2}{4x^2}$

18. $\dfrac{x + 6}{x - 2} + \dfrac{2x + 1}{x + 3}$

19. You invested $4000, part at 5% and the remainder at 9% annual interest. At the end of the year, the total interest from these investments was $311. How much was invested at each rate?

20. A 68-inch board is to be cut into two pieces. If one piece must be three times as long as the other, find the length of each piece.

| MID-TEXTBOOK CHECK POINT | Are You Prepared for Intermediate Algebra? |

Algebra is cumulative. This means that **your performance in intermediate algebra depends heavily on the skills you acquired in introductory algebra.** Do you need a quick review of introductory algebra topics before starting the intermediate algebra portion of this book? This mid-textbook Check Point provides a fast way to review and practice the prerequisite skills needed in intermediate algebra.

Great Question!

How can I quickly review the major topics of introductory algebra?

Study the review grids for each of the first seven chapters in this book. Each chart summarizes the definitions and concepts in every section of the chapter. Examples that illustrate the key concepts are also included in the chart. The review charts, with worked-out examples, for each of the book's first seven chapters begin on pages 102, 198, 267, 336, 417, 483, and 572.

A Diagnostic Test for Your Introductory Algebra Skills.
You can use the 36 exercises in this mid-textbook Check Point to test your understanding of introductory algebra topics. These exercises cover the fundamental algebra skills upon which the intermediate algebra portion of this book is based. Here are some suggestions for using these exercises as a diagnostic test:

1. Work through all 36 items at your own pace.

2. Use the answer section in the back of the book to check your work.

3. If your answer differs from that in the answer section or if you are not certain how to proceed with a particular item, turn to the section and the worked-out example given in parentheses at the end of each exercise. Study the step-by-step solution of the example that parallels the exercise and then try working the exercise again. If you feel that you need more assistance, study the entire section in which the example appears and work on a selected group of exercises in the exercise set for that section.

In Exercises 1–7, solve each equation, inequality, or system of equations.

1. $2 - 4(x + 2) = 5 - 3(2x + 1)$
   (Section 2.3, Example 3)

2. $\dfrac{x}{2} - 3 = \dfrac{x}{5}$
   (Section 2.3, Example 4)

3. $3x + 9 \geq 5(x - 1)$ (Section 2.7, Example 7)

4. $\begin{cases} 2x + 3y = 6 \\ x + 2y = 5 \end{cases}$ (Section 4.3, Example 2)

5. $\begin{cases} 3x - 2y = 1 \\ y = 10 - 2x \end{cases}$ (Section 4.2, Example 1)

6. $\dfrac{3}{x + 5} - 1 = \dfrac{4 - x}{2x + 10}$ (Section 7.6, Example 4)

7. $x + \dfrac{6}{x} = -5$ (Section 7.6, Example 3)

In Exercises 8–16, perform the indicated operations. If possible, simplify the answer.

8. $\dfrac{12x^3}{3x^{12}}$ (Section 5.7, Example 4)

9. $4 \cdot 6 \div 2 \cdot 3 + (-5)$ (Section 1.8, Example 4)

10. $(6x^2 - 8x + 3) - (-4x^2 + x - 1)$
    (Section 5.1, Example 3)

11. $(7x + 4)(3x - 5)$ (Section 5.3, Example 2)

12. $(5x - 2)^2$ (Section 5.3, Example 6)

13. $(x + y)(x^2 + xy + y^2)$
    (Section 5.4, Example 8)

14. $\dfrac{x^2 + 6x + 8}{x^2} \div (3x^2 + 6x)$ (Section 7.2, Example 7)

15. $\dfrac{x}{x^2 + 2x - 3} - \dfrac{x}{x^2 - 5x + 4}$

    (Section 7.4, Example 7)

16. $\dfrac{x - \dfrac{1}{5}}{5 - \dfrac{1}{x}}$ (Section 7.5, Examples 2 and 5)

In Exercises 17–22, factor completely.

17. $4x^2 - 49$ (Section 6.4, Example 1)

18. $x^3 + 3x^2 - x - 3$
    (Section 6.5, Example 4)

19. $2x^2 + 8x - 42$ (Section 6.5, Example 2)

20. $x^5 - 16x$ (Section 6.4, Example 4)

21. $x^3 - 10x^2 + 25x$ (Section 6.4, Example 5)

22. $x^3 - 8$ (Section 6.4, Example 8)

In Exercises 23–25, graph each equation in a rectangular coordinate system.

23. $y = \dfrac{1}{3}x - 1$ (Section 3.4, Example 3)

24. $3x + 2y = -6$ (Section 3.2, Example 4)

25. $y = -2$ (Section 3.2, Example 7)

26. Find the slope of the line passing through the points $(-1, 3)$ and $(2, -3)$. (Section 3.3, Example 1)

27. Write the point-slope form of the equation of the line passing through the points $(1, 2)$ and $(3, 6)$. Then use the point-slope form of the equation to write the slope-intercept form of the line's equation. (Section 3.5, Example 2)

In Exercises 28–36, use an equation or a system of equations to solve each problem.

28. Seven subtracted from five times a number is 208. Find the number. (Section 2.5, Example 1)

29. After a 20% reduction, a digital camera sold for $256. What was the price before the reduction? (Section 2.5, Example 6)

30. A rectangular field is three times as long as it is wide. If the perimeter of the field is 400 yards, what are the field's dimensions? (Section 2.5, Example 5)

31. You invested $20,000 in two accounts paying 7% and 9% annual interest. If the total interest earned for the year was $1550, how much was invested at each rate? (Section 4.4, Example 5)

32. A chemist needs to mix a 40% acid solution with a 70% acid solution to obtain 12 liters of a 50% acid solution. How many liters of each solution should be used? (Section 4.4, Example 6)

33. A sailboat has a triangular sail with an area of 120 square feet and a base that is 15 feet long. Find the height of the sail. (Section 2.6, Example 1)

34. In a triangle, the measure of the first angle is 10° more than the measure of the second angle. The measure of the third angle is 20° more than four times that of the second angle. What is the measure of each angle? (Section 2.6, Example 6)

35. A salesperson works in the TV and stereo department of an electronics store. One day she sold 3 TVs and 4 stereos for $2530. The next day, she sold 4 of the same TVs and 3 of the same stereos for $2510. Find the price of a TV and a stereo. (Section 4.4, Example 2)

36. The length of a rectangle is 6 meters more than the width. The area is 55 square meters. Find the rectangle's dimensions. (Section 6.6, Example 8)

"Total honesty with yourself regarding math skills

is essential to good performance and understanding.

Bluffing yourself into thinking you remember

or know more than you do

only wastes time in the long run.

The more honestly you can admit what you do know and do not know,

the lower your level of anxiety will be."

—Cheryl Ooten, *Managing the Mean Math Blues,*
Pearson Education, Inc., 2003

Basics of Functions

A vast expanse of open water at the top of our world was once covered with ice. The melting of the Arctic ice caps has forced polar bears to swim as far as 40 miles, causing them to drown in significant numbers. Such deaths were rare in the past.

There is strong scientific consensus that human activities are changing the Earth's climate. Scientists now believe that there is a striking correlation between atmospheric carbon dioxide concentration and global temperature. As both of these variables increase at significant rates, there are warnings of a planetary emergency that threatens to condemn coming generations to a catastrophically diminished future.*

In this chapter, you'll learn to approach our climate crisis mathematically by using formulas, called functions, that model data for average global temperature and carbon dioxide concentration over time. Understanding the concept of a function will give you a new perspective on many situations, ranging from global warming to the movies that have won the most Oscars.

*Sources: Al Gore, An Inconvenient Truth, Rodale, 2006; Time, April.3, 2006

Mathematical models involving global warming appear in Exercises 47–50 in Exercise Set 8.2. Oscar-winning films open our discussion in Section 8.4.

Introduction to Functions

Objectives

1 Find the domain and range of a relation.

2 Determine whether a relation is a function.

3 Evaluate a function.

Top U.S. Last Names

Name	% of All Names
Smith	1.006%
Johnson	0.810%
Williams	0.699%
Brown	0.621%
Jones	0.621%

Source: Russell Ash, *The Top 10 of Everything*

Venus and Serena Williams

The top five U.S. last names shown above account for nearly 4% of the entire population. The table indicates a correspondence between a last name and the percentage of Americans who share that name. We can write this correspondence using a set of ordered pairs:

{(Smith, 1.006%), (Johnson, 0.810%), (Williams, 0.699%), (Brown, 0.621%), (Jones, 0.621%)}.

> These braces indicate we are representing a set.

The mathematical term for a set of ordered pairs is a *relation*.

Definition of a Relation

A **relation** is any set of ordered pairs. The set of all first components of the ordered pairs is called the **domain** of the relation and the set of all second components is called the **range** of the relation.

1 Find the domain and range of a relation.

EXAMPLE 1   Finding the Domain and Range of a Relation

Find the domain and range of the relation:

{(Smith, 1.006%), (Johnson, 0.810%), (Williams, 0.699%), (Brown, 0.621%), (Jones, 0.621%)}.

Solution   The domain is the set of all first components. Thus, the domain is

{Smith, Johnson, Williams, Brown, Jones}.

The range is the set of all second components. Thus, the range is

{1.006%, 0.810%, 0.699%, 0.621%}.

> Although Brown and Jones are both shared by 0.621% of the U.S. population, it is not necessary to list 0.621% twice.

✓ **CHECK POINT 1**   Find the domain and the range of the relation:

{(0, 9.1), (10, 6.7), (20, 10.7), (30, 13.2), (38, 19.6)}.

As you worked Check Point 1, did you wonder if there was a rule that assigned the "inputs" in the domain to the "outputs" in the range? For example, for the ordered pair (30, 13.2), how does the output 13.2 depend on the input 30? The ordered pair is based on the data in **Figure 8.1(a)**, which shows the percentage of first-year U.S. college students claiming no religious affiliation.

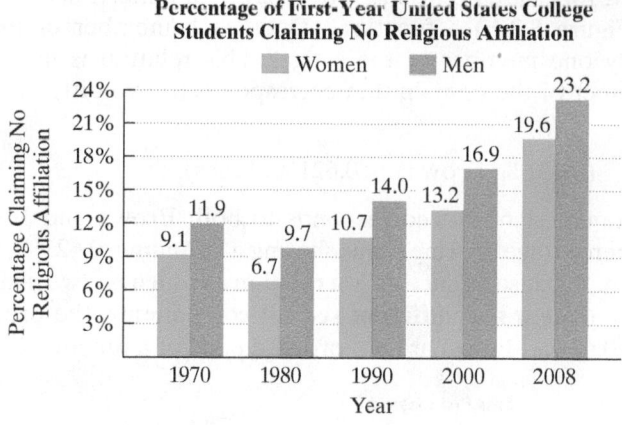

Figure 8.1(a) Data for women and men

Source: John Macionis, *Sociology*, 13th Edition, Prentice Hall, 2010.

Figure 8.1(b) Visually representing the relation for women's data

In **Figure 8.1(b)**, we used the data for college women to create the following ordered pairs:

$$\left(\text{years after 1970,} \quad \begin{array}{l} \text{percentage of first-year college} \\ \text{women claiming no religious} \\ \text{affiliation} \end{array} \right).$$

Consider, for example, the ordered pair (30, 13.2).

(30, 13.2)

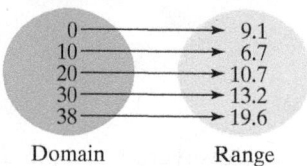

30 years after 1970, or in 2000,

13.2% of first-year college women claimed no religious affiliation.

The five points in **Figure 8.1(b)** visually represent the relation formed from the women's data. Another way to visually represent the relation is as follows:

Domain	Range
0	9.1
10	6.7
20	10.7
30	13.2
38	19.6

2 Determine whether a relation is a function.

Functions

Shown, again, in the margin are the top five U.S. last names and the percentage of Americans who share those names. We've used this information to define two relations. **Figure 8.2(a)** shows a correspondence between last names and percents sharing those names. **Figure 8.2(b)** shows a correspondence between percents sharing last names and those last names.

Top U.S. Last Names

Name	% of All Names
Smith	1.006%
Johnson	0.810%
Williams	0.699%
Brown	0.621%
Jones	0.621%

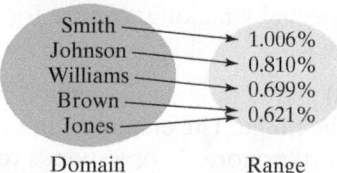

Smith, Johnson, Williams, Brown, Jones → 1.006%, 0.810%, 0.699%, 0.621%

Domain   Range

Figure 8.2 (a) Names correspond to percents.

1.006%, 0.810%, 0.699%, 0.621% → Smith, Johnson, Williams, Brown, Jones

Domain   Range

Figure 8.2 (b) Percents correspond to names.

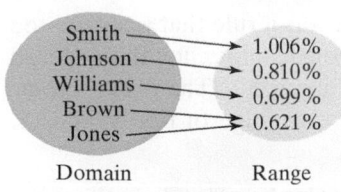

Domain   Range

Figure 8.2(a) (repeated)

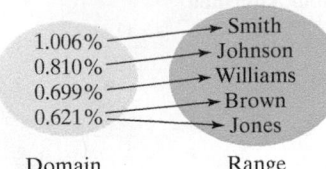

Domain   Range

Figure 8.2(b) (repeated)

A relation in which each member of the domain corresponds to exactly one member of the range is a **function**. Can you see that the relation in **Figure 8.2(a)** is a function? Each last name in the domain corresponds to exactly one percent in the range. If we know the last name, we can be sure of the percentage of Americans sharing that name. Notice that more than one element in the domain can correspond to the same element in the range: Brown and Jones are both shared by 0.621% of Americans.

Is the relation in **Figure 8.2(b)** a function? Does each member of the domain correspond to precisely one member of the range? This relation is not a function because there is a member of the domain that corresponds to two different members of the range:

$$(0.621\%, \text{Brown}) \quad (0.621\%, \text{Jones}).$$

The member of the domain, 0.621%, corresponds to both Brown and Jones in the range. If we know the percentage of Americans sharing a last name, 0.621%, we cannot be sure of that last name. Because **a function is a relation in which no two ordered pairs have the same first component and different second components**, the ordered pairs (0.621%, Brown) and (0.621%, Jones) are not ordered pairs of a function.

Same first component

$$(0.621\%, \text{Brown}) \quad (0.621\%, \text{Jones})$$

Different second components

Definition of a Function

A **function** is a correspondence from a first set, called the **domain**, to a second set, called the **range**, such that each element in the domain corresponds to *exactly one* element in the range.

In Check Point 1, we considered a relation that gave a correspondence between years after 1970 and the percentage of first-year college women claiming no religious affiliation. Can you see that this relation is a function?

Each element in the domain

$$\{(0, 9.1), (10, 6.7), (20, 10.7), (30, 13.2), (38, 19.6)\}$$

corresponds to exactly one element in the range.

However, Example 2 illustrates that not every correspondence between sets is a function.

EXAMPLE 2   Determining Whether a Relation Is a Function

Determine whether each relation is a function:

a. $\{(1, 5), (2, 5), (3, 7), (4, 8)\}$ **b.** $\{(5, 1), (5, 2), (7, 3), (8, 4)\}$.

Solution   We begin by making a figure for each relation that shows the domain and the range (**Figure 8.3**).

a. **Figure 8.3(a)** shows that every element in the domain corresponds to exactly one element in the range. The element 1 in the domain corresponds to the element 5 in the range. Furthermore, 2 corresponds to 5, 3 corresponds to 7, and 4 corresponds to 8. No two ordered pairs in the given relation have the same first component and different second components. Thus, the relation is a function.

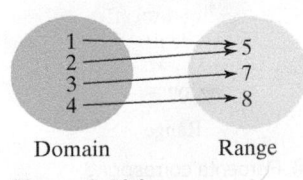

Domain   Range

Figure 8.3(a)

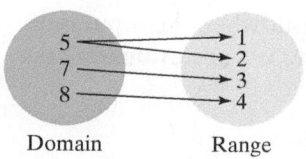

Domain   Range

Figure 8.3(b)

b. Figure 8.3(b) shows that 5 corresponds to both 1 and 2. If any element in the domain corresponds to more than one element in the range, the relation is not a function. This relation is not a function because two ordered pairs have the same first component and different second components.

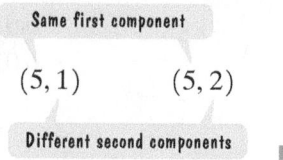

Same first component

$$(5, 1) \qquad (5, 2)$$

Different second components ∎

Look at **Figure 8.3(a)** again. The fact that 1 and 2 in the domain correspond to the same number, 5, in the range does not violate the definition of a function. **A function can have two different first components with the same second component.** By contrast, a relation is not a function when two different ordered pairs have the same first component and different second components. Thus, the relation in Example 2(b) is not a function.

Great Question!

If I reverse a function's components, will this new relation be a function?

If a relation is a function, reversing the components in each of its ordered pairs may result in a relation that is not a function.

✓ **CHECK POINT 2** Determine whether each relation is a function:

a. $\{(1, 2),(3, 4),(5, 6),(5, 7)\}$

b. $\{(1, 2),(3, 4),(6, 5),(7, 5)\}$.

3 Evaluate a function.

Functions as Equations and Function Notation

Functions are usually given in terms of equations rather than as sets of ordered pairs. For example, here is an equation that models the percentage of first-year college women claiming no religious affiliation as a function of time:

$$y = 0.014x^2 - 0.24x + 8.8.$$

The variable x represents the number of years after 1970. The variable y represents the percentage of first-year college women claiming no religious affiliation. The variable y is a function of the variable x. For each value of x, there is one and only one value of y. The variable x is called the **independent variable** because it can be assigned any value from the domain. Thus, x can be assigned any nonnegative integer representing the number of years after 1970. The variable y is called the **dependent variable** because its value depends on x. The percentage claiming no religious affiliation depends on the number of years after 1970. The value of the dependent variable, y, is calculated after selecting a value for the independent variable, x.

If an equation in x and y gives one and only one value of y for each value of x, then the variable y is a function of the variable x. When an equation represents a function, the function is often named by a letter such as f, g, h, F, G, or H. Any letter can be used to name a function. Suppose that f names a function. Think of the domain as the set of the function's inputs and the range as the set of the function's outputs. As shown in **Figure 8.4**, the input is represented by x and the output by $f(x)$. The special notation $f(x)$, read "f of x" or "f at x," represents the **value of the function at the number x.**

Let's make this clearer by considering a specific example. We know that the equation

$$y = 0.014x^2 - 0.24x + 8.8$$

defines y as a function of x. We'll name the function f. Now, we can apply our new function notation.

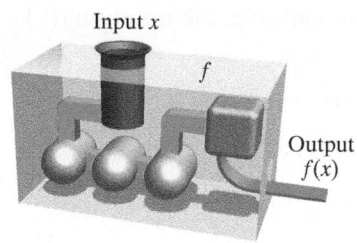

Input x

f

Output $f(x)$

Figure 8.4 A "function machine" with inputs and outputs

We read this equation as "f of x equals $0.014x^2 - 0.24x + 8.8$."

Input	Output	Equation
x	$f(x)$	$f(x) = 0.014x^2 - 0.24x + 8.8$

Great Question!

Doesn't $f(x)$ indicate that I need to multiply f and x?

The notation $f(x)$ does *not* mean "f times x." The notation describes the value of the function at x.

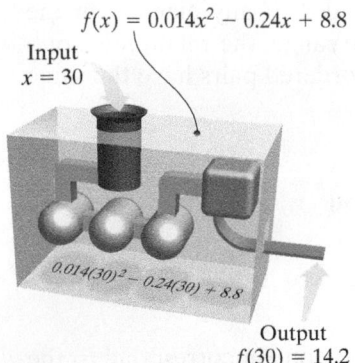

$f(x) = 0.014x^2 - 0.24x + 8.8$

Input
$x = 30$

$0.014(30)^2 - 0.24(30) + 8.8$

Output
$f(30) = 14.2$

Figure 8.5 A function machine at work

Suppose we are interested in finding $f(30)$, the function's output when the input is 30. To find the value of the function at 30, we substitute 30 for x. We are **evaluating the function** at 30.

$f(x) = 0.014x^2 - 0.24x + 8.8$	This is the given function.
$f(30) = 0.014(30)^2 - 0.24(30) + 8.8$	Replace each occurrence of x with 30.
$\qquad = 0.014(900) - 0.24(30) + 8.8$	Evaluate the exponential expression: $30^2 = 30 \cdot 30 = 900$.
$\qquad = 12.6 - 7.2 + 8.8$	Perform the multiplications.
$f(30) = 14.2$	Subtract and add from left to right.

The statement $f(30) = 14.2$, read "f of 30 equals 14.2," tells us that the value of the function at 30 is 14.2. When the function's input is 30, its output is 14.2. **Figure 8.5** illustrates the input and output in terms of a function machine.

$$f(30) = 14.2$$

30 years after 1970, or in 2000,	14.2% of first-year college women claimed no religious affiliation.

We have seen that in 2000, 13.2% actually claimed nonaffiliation, so our function that models the data overestimates the data value for 2000 by 1%.

Using Technology

Graphing utilities can be used to evaluate functions. The screens on the right show the evaluation of

$$f(x) = 0.014x^2 - 0.24x + 8.8$$

at 30 on a TI-84 Plus graphing calculator. The function f is named Y_1.

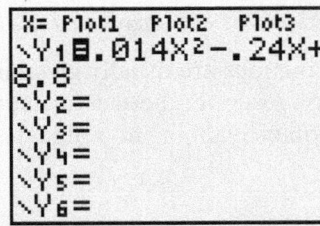

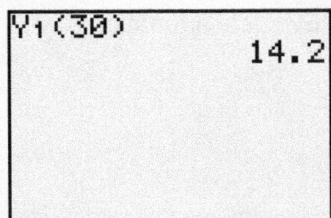

We used $f(x) = 0.014x^2 - 0.24x + 8.8$ to find $f(30)$. To find other function values, such as $f(40)$ or $f(55)$, substitute the specified input value, 40 or 55, for x in the function's equation.

If a function is named f and x represents the independent variable, the notation $f(x)$ corresponds to the y-value for a given x. Thus,

$$f(x) = 0.014x^2 - 0.24x + 8.8 \quad \text{and} \quad y = 0.014x^2 - 0.24x + 8.8$$

define the same function. This function may be written as

$$y = f(x) = 0.014x^2 - 0.24x + 8.8.$$

EXAMPLE 3 Using Function Notation

Find the indicated function value:

a. $f(4)$ for $f(x) = 2x + 3$
b. $g(-2)$ for $g(x) = 2x^2 - 1$
c. $h(-5)$ for $h(r) = r^3 - 2r^2 + 5$
d. $F(a + h)$ for $F(x) = 5x + 7$.

Solution

a. $f(x) = 2x + 3$	This is the given function.
$f(4) = 2 \cdot 4 + 3$	To find f of 4, replace x with 4.
$\qquad = 8 + 3$	Multiply: $2 \cdot 4 = 8$.
$f(4) = 11$ f of 4 is 11.	Add.

b. $g(x) = 2x^2 - 1$ This is the given function.

$g(-2) = 2(-2)^2 - 1$ To find g of -2, replace x with -2.

$= 2(4) - 1$ Evaluate the exponential expression: $(-2)^2 = 4$.

$= 8 - 1$ Multiply: $2(4) = 8$.

$g(-2) = 7$ _g of −2 is 7._   Subtract.

c. $h(r) = r^3 - 2r^2 + 5$ The function's name is h and r represents the independent variable.

$h(-5) = (-5)^3 - 2(-5)^2 + 5$ To find h of -5, replace each occurrence of r with -5.

$= -125 - 2(25) + 5$ Evaluate exponential expressions.

$= -125 - 50 + 5$ Multiply.

$h(-5) = -170$ _h of −5 is −170._ $-125 - 50 = -175$ and $-175 + 5 = -170$.

d. $F(x) = 5x + 7$ This is the given function.

$F(a + h) = 5(a + h) + 7$ Replace x with $a + h$.

$F(a + h) = 5a + 5h + 7$ Apply the distributive property.   ■

_F of a + h is 5a + 5h + 7._

✓ **CHECK POINT 3**   Find the indicated function value:

a. $f(6)$ for $f(x) = 4x + 5$

b. $g(-5)$ for $g(x) = 3x^2 - 10$

c. $h(-4)$ for $h(r) = r^2 - 7r + 2$

d. $F(a + h)$ for $F(x) = 6x + 9$.

The functions in Check Point 3(a) and 3(d), $f(x) = 4x + 5$ and $F(x) = 6x + 9$, are examples of *linear functions*. **Linear functions** have equations of the form $f(x) = mx + b$. By contrast, the functions in Check Point 3(b) and 3(c), $g(x) = 3x^2 - 10$ and $h(r) = r^2 - 7r + 2$, are examples of *quadratic functions*. **Quadratic functions** have equations of the form $f(x) = ax^2 + bx + c, a \neq 0$.

Functions Represented by Tables and Function Notation

Function notation can be applied to functions that are represented by tables.

EXAMPLE 4   Using Function Notation

Function f is defined by the following table:

x	f(x)
−2	5
−1	0
0	3
1	1
2	4

a. Explain why the table defines a function.

b. Find the domain and the range of the function.

Find the indicated function value:

c. $f(-1)$

d. $f(0)$

e. Find x such that $f(x) = 4$.

x	f(x)
−2	5
−1	0
0	3
1	1
2	4

The table defining f (repeated)

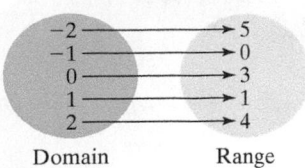

Domain   Range

Figure 8.6

Solution

a. Values in the first column of the table make up the domain, or input values. Values in the second column of the table make up the range, or output values. We see that every element in the domain corresponds to exactly one element in the range, shown in **Figure 8.6**. Therefore, the relation given by the table is a function.

The voice balloons pointing to appropriate parts of the table illustrate the solution to parts (b)–(e).

x	f(x)
−2	5
−1	0
0	3
1	1
2	4

c. $f(-1) = 0$: When the input is −1, the output is 0.

d. $f(0) = 3$: When the input is 0, the output is 3.

e. $f(x) = 4$ when $x = 2$: The output, $f(x)$, is 4 when the input, x, is 2.

b. The domain is the set of inputs: {−2, −1, 0, 1, 2}.

b. The range is the set of outputs: {5, 0, 3, 1, 4}.

✓ **CHECK POINT 4** Function g is defined by the following table:

x	g(x)
0	3
1	0
2	1
3	2
4	3

a. Explain why the table defines a function.

b. Find the domain and the range of the function.

Find the indicated function value:

c. $g(1)$

d. $g(3)$

e. Find x such that $g(x) = 3$.

Achieving Success

Check out Professor Dan Miller's *Learning Guide* that accompanies this textbook.

Benefits of using the *Learning Guide* include:

- It will help you become better organized. This includes organizing your class notes, assigned homework, quizzes, and tests.
- It will enable you to use your textbook more efficiently.
- It will bring together the learning tools for this course, including the textbook, the Video Lecture Series, and the PowerPoint Presentation.
- It will help increase your study skills.
- It will help you prepare for the chapter tests.

Ask your professor about the availability of this textbook supplement.

CONCEPT AND VOCABULARY CHECK

Fill in each blank so that the resulting statement is true.

1. Any set of ordered pairs is called a/an _____. The set of all first components of the ordered pairs is called the _____. The set of all second components of the ordered pairs is called the _____.

2. A set of ordered pairs in which each member of the set of first components corresponds to exactly one member of the set of second components is called a/an _____.

3. The notation $f(x)$ describes the value of _____ at _____.

4. If $h(r) = -r^2 + 4r - 7$, we can find $h(-2)$ by replacing each occurrence of _____ by _____.

8.1 EXERCISE SET

MyMathLab®

Practice Exercises

In Exercises 1–8, determine whether each relation is a function.
Give the domain and range for each relation.

1. $\{(1, 2), (3, 4), (5, 5)\}$
2. $\{(4, 5), (6, 7), (8, 8)\}$
3. $\{(3, 4), (3, 5), (4, 4), (4, 5)\}$

4. $\{(5, 6), (5, 7), (6, 6), (6, 7)\}$

5. $\{(-3, -3), (-2, -2), (-1, -1), (0, 0)\}$

6. $\{(-7, -7), (-5, -5), (-3, -3), (0, 0)\}$

7. $\{(1, 4), (1, 5), (1, 6)\}$
8. $\{(4, 1), (5, 1), (6, 1)\}$

In Exercises 9–24, find the indicated function values.

9. $f(x) = x + 1$
    a. $f(0)$ b. $f(5)$ c. $f(-8)$
    d. $f(2a)$ e. $f(a + 2)$

10. $f(x) = x + 3$
    a. $f(0)$ b. $f(5)$ c. $f(-8)$
    d. $f(2a)$ e. $f(a + 2)$

11. $g(x) = 3x - 2$
    a. $g(0)$ b. $g(-5)$ c. $g\left(\dfrac{2}{3}\right)$
    d. $g(4b)$ e. $g(b + 4)$

12. $g(x) = 4x - 3$
    a. $g(0)$ b. $g(-5)$ c. $g\left(\dfrac{3}{4}\right)$
    d. $g(5b)$ e. $g(b + 5)$

13. $h(x) = 3x^2 + 5$
    a. $h(0)$ b. $h(-1)$ c. $h(4)$
    d. $h(-3)$ e. $h(4b)$

14. $h(x) = 2x^2 - 4$
    a. $h(0)$ b. $h(-1)$ c. $h(5)$
    d. $h(-3)$ e. $h(5b)$

15. $f(x) = 2x^2 + 3x - 1$
    a. $f(0)$ b. $f(3)$ c. $f(-4)$
    d. $f(b)$
    e. $f(5a)$

16. $f(x) = 3x^2 + 4x - 2$
    a. $f(0)$ b. $f(3)$ c. $f(-5)$
    d. $f(b)$
    e. $f(5a)$

17. $f(x) = (-x)^3 - x^2 - x + 7$
    a. $f(0)$ b. $f(2)$
    c. $f(-2)$ d. $f(1) + f(-1)$

18. $f(x) = (-x)^3 - x^2 - x + 10$
    a. $f(0)$ b. $f(2)$
    c. $f(-2)$ d. $f(1) + f(-1)$

19. $f(x) = \dfrac{2x - 3}{x - 4}$
    a. $f(0)$ b. $f(3)$ c. $f(-4)$
    d. $f(-5)$ e. $f(a + h)$
    f. Why must 4 be excluded from the domain of f?

20. $f(x) = \dfrac{3x - 1}{x - 5}$
    a. $f(0)$ b. $f(3)$ c. $f(-3)$
    d. $f(10)$ e. $f(a + h)$
    f. Why must 5 be excluded from the domain of f?

21.

x	f(x)
−4	3
−2	6
0	9
2	12
4	15

a. $f(-2)$
b. $f(2)$
c. For what value of x is $f(x) = 9$?

22.

x	f(x)
−5	4
−3	8
0	12
3	16
5	20

a. $f(-3)$
b. $f(3)$
c. For what value of x is $f(x) = 12$?

23.

x	h(x)
−2	2
−1	1
0	0
1	1
2	2

a. $h(-2)$
b. $h(1)$
c. For what values of x is $h(x) = 1$?

24.

x	h(x)
−2	−2
−1	−1
0	0
1	−1
2	−2

a. $h(-2)$
b. $h(1)$
c. For what values of x is $h(x) = -1$?

Practice PLUS

In Exercises 25–26, let $f(x) = x^2 - x + 4$ and $g(x) = 3x - 5$.

25. Find $g(1)$ and $f(g(1))$.
26. Find $g(-1)$ and $f(g(-1))$.

In Exercises 27–28, let f and g be defined by the following table:

x	f(x)	g(x)
−2	6	0
−1	3	4
0	−1	1
1	−4	−3
2	0	−6

27. Find $\sqrt{f(-1) - f(0)} - [g(2)]^2 + f(-2) \div g(2) \cdot g(-1)$.

28. Find $|f(1) - f(0)| - [g(1)]^2 + g(1) \div f(-1) \cdot g(2)$.

In Exercises 29–30, find $f(-x) - f(x)$ for the given function f. Then simplify the expression.

29. $f(x) = x^3 + x - 5$

30. $f(x) = x^2 - 3x + 7$

In Exercises 31–32, each function is defined by two equations. The equation in the first row gives the output for negative numbers in the domain. The equation in the second row gives the output for nonnegative numbers in the domain. Find the indicated function values.

31. $f(x) = \begin{cases} 3x + 5 & \text{if } x < 0 \\ 4x + 7 & \text{if } x \geq 0 \end{cases}$

    a. $f(-2)$ **b.** $f(0)$

    c. $f(3)$ **d.** $f(-100) + f(100)$

32. $f(x) = \begin{cases} 6x - 1 & \text{if } x < 0 \\ 7x + 3 & \text{if } x \geq 0 \end{cases}$

    a. $f(-3)$ **b.** $f(0)$

    c. $f(4)$ **d.** $f(-100) + f(100)$

Application Exercises

The Corruption Perceptions Index uses perceptions of the general public, business people, and risk analysts to rate countries by how likely they are to accept bribes. The ratings are on a scale from 0 to 10, where higher scores represent less corruption. The graph shows the corruption ratings for the world's least corrupt and most corrupt countries. (The rating for the United States is 7.6.) Use the graph to solve Exercises 33–34.

Top Four Least Corrupt and Most Corrupt Countries

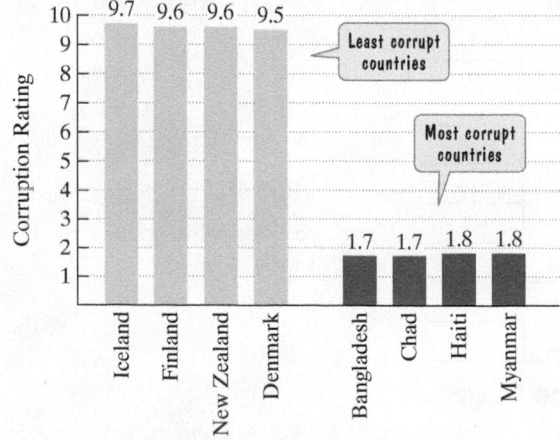

Source: Transparency International, *Corruption Perceptions Index*

33. a. Write a set of four ordered pairs in which each of the least corrupt countries corresponds to a corruption rating. Each ordered pair should be in the form

    (country, corruption rating).

b. Is the relation in part (a) a function? Explain your answer.

c. Write a set of four ordered pairs in which corruption ratings for the least corrupt countries correspond to countries. Each ordered pair should be in the form

    (corruption rating, country).

d. Is the relation in part (c) a function? Explain your answer.

34. a. Write a set of four ordered pairs in which each of the most corrupt countries corresponds to a corruption rating. Each ordered pair should be in the form

    (country, corruption rating).

b. Is the relation in part (a) a function? Explain your answer.

c. Write a set of four ordered pairs in which corruption ratings for the least corrupt countries correspond to countries. Each ordered pair should be in the form

    (corruption rating, country).

d. Is the relation in part (c) a function? Explain your answer.

Writing in Mathematics

35. What is a relation? Describe what is meant by its domain and its range.

36. Explain how to determine whether a relation is a function. What is a function?

37. Does $f(x)$ mean f times x when referring to function f? If not, what does $f(x)$ mean? Provide an example with your explanation.

38. For people filing a single return, federal income tax is a function of adjusted gross income because for each value of adjusted gross income there is a specific tax to be paid. By contrast, the price of a house is not a function of the lot size on which the house sits because houses on same-sized lots can sell for many different prices.

    a. Describe an everyday situation between variables that is a function.

    b. Describe an everyday situation between variables that is not a function.

Critical Thinking Exercises

Make Sense? *In Exercises 39–42, determine whether each statement "makes sense" or "does not make sense" and explain your reasoning.*

39. Today's temperature is a function of the time of day.

40. My height is a function of my age.

41. Although I presented my function as a set of ordered pairs, I could have shown the correspondences using a table or using points plotted in a rectangular coordinate system.

42. My function models how the chance of divorce depends on the number of years of marriage, so the range is $\{x \mid x \text{ is the number of years of marriage}\}$.

In Exercises 43–48, determine whether each statement is true or false. If the statement is false, make the necessary change(s) to produce a true statement.

43. All relations are functions.

44. No two ordered pairs of a function can have the same second components and different first components.

Using the tables that define f and g, determine whether each statement in Exercises 45–48 is true or false.

x	$f(x)$		x	$g(x)$
-4	-1		-1	-4
-3	-2		-2	-3
-2	-3		-3	-2
-1	-4		-4	-1

45. The domain of f = the range of f

46. The range of f = the domain of g

47. $f(-4) - f(-2) = 2$

48. $g(-4) + f(-4) = 0$

49. If $f(x) = 3x + 7$, find $\dfrac{f(a + h) - f(a)}{h}$.

50. Give an example of a relation with the following characteristics: The relation is a function containing two ordered pairs. Reversing the components in each ordered pair results in a relation that is not a function.

51. If $f(x + y) = f(x) + f(y)$ and $f(1) = 3$, find $f(2)$, $f(3)$, and $f(4)$. Is $f(x + y) = f(x) + f(y)$ for all functions?

Review Exercises

52. Simplify: $24 \div 4[2 - (5 - 2)]^2 - 6$. (Section 1.8, Example 8)

53. Simplify: $\left(\dfrac{3x^2y^{-2}}{y^3}\right)^{-2}$. (Section 5.7, Example 6)

54. Solve: $\dfrac{x}{3} = \dfrac{3x}{5} + 4$. (Section 2.3, Example 4)

Preview Exercises

Exercises 55–57 will help you prepare for the material covered in the next section. Use the following graph to solve each exercise.

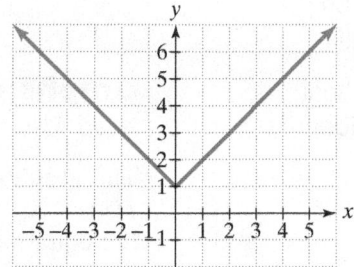

55. What are the x-coordinates when the y-coordinate is 4?

56. Use interval notation to describe the x-coordinates of all points on the graph.

57. Use interval notation to describe the y-coordinates of all points on the graph.

SECTION

8.2

Graphs of Functions

Objectives

1. Use the vertical line test to identify functions.

2. Obtain information about a function from its graph.

3. Review interval notation.

4. Identify the domain and range of a function from its graph.

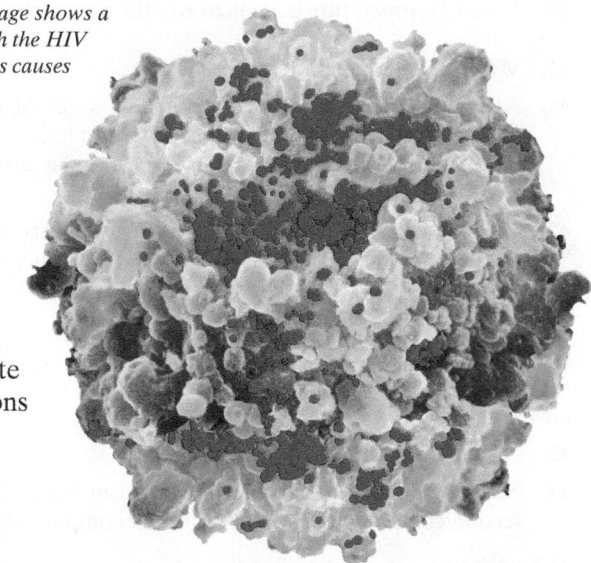

Magnified 6000 times, this color-scanned image shows a T-lymphocyte blood cell (green) infected with the HIV virus (red). Depletion of the number of T cells causes destruction of the immune system.

The number of T cells in a person with HIV is a function of time after infection. In this section, we'll analyze the graph of this function, using the rectangular coordinate system to visualize what functions look like.

1. Use the vertical line test to identify functions.

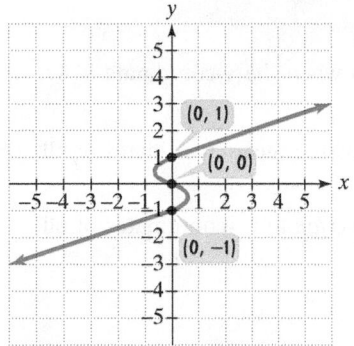

Figure 8.7 *y* is not a function of *x* because 0 is paired with three values of *y*, namely, 1, 0, and −1.

Graphs of Functions and the Vertical Line Test

The **graph of a function** is the graph of its ordered pairs. For example, the graph of $f(x) = 3x + 1$ is the set of points (x, y) in the rectangular coordinate system satisfying the equation $y = 3x + 1$. Thus, the graph of f is a line with slope 3 and y-intercept 1. Similarly, the graph of $f(x) = x^2 + 3x + 5$ is the set of points (x, y) in the rectangular coordinate system satisfying the equation $y = x^2 + 3x + 5$.

Not every graph in the rectangular coordinate system is the graph of a function. The definition of a function specifies that no value of x can be paired with two or more different values of y. Consequently, if a graph contains two or more different points with the same first coordinate, the graph cannot represent a function. This is illustrated in **Figure 8.7**. Observe that points sharing a common first coordinate are vertically above or below each other.

This observation is the basis of a useful test for determining whether a graph defines y as a function of x. The test is called the **vertical line test**.

The Vertical Line Test for Functions

If any vertical line intersects a graph in more than one point, the graph does not define y as a function of x.

EXAMPLE 1 Using the Vertical Line Test

Use the vertical line test to identify graphs in which y is a function of x.

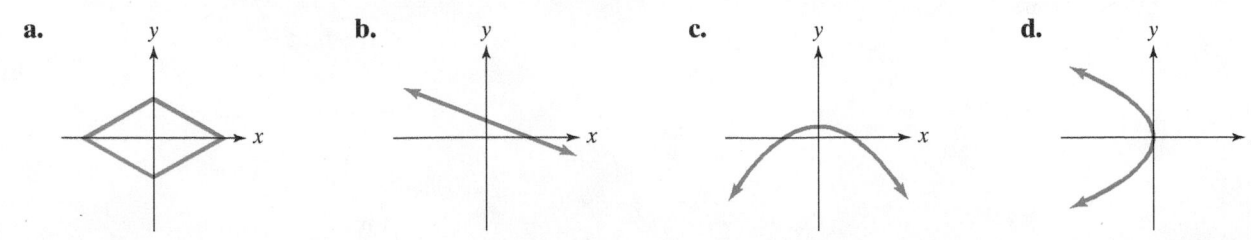

Solution y is a function of x for the graphs in (b) and (c).

a.

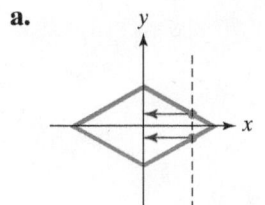

y **is not a function** of x.
Two values of y correspond to one x-value.

b.

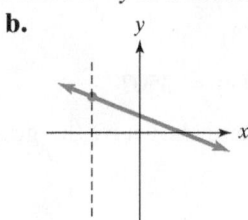

y **is a function** of x.

c.

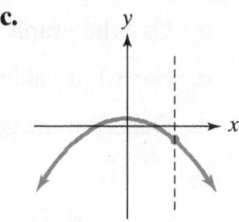

y **is a function** of x.

d.

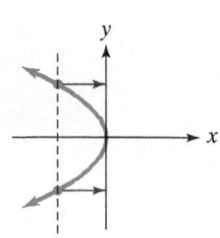

y **is not a function** of x.
Two values of y correspond to one x-value.   ■

✓ **CHECK POINT 1**   Use the vertical line test to identify graphs in which y is a function of x.

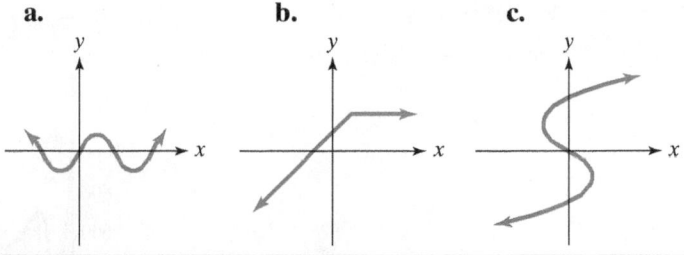

a.

b.

c.

---

2 Obtain information about a function from its graph.

Obtaining Information from Graphs

You can obtain information about a function from its graph. At the right or left of a graph, you will often find closed dots, open dots, or arrows.

- A closed dot indicates that the graph does not extend beyond this point and the point belongs to the graph.

- An open dot indicates that the graph does not extend beyond this point and the point does not belong to the graph.

- An arrow indicates that the graph extends indefinitely in the direction in which the arrow points.

EXAMPLE 2   Analyzing the Graph of a Function

The human immunodeficiency virus, or HIV, infects and kills helper T cells. Because T cells stimulate the immune system to produce antibodies, their destruction disables the body's defenses against other pathogens. By counting the number of T cells that remain active in the body, the progression of HIV can be monitored. The fewer helper T cells, the more advanced the disease. **Figure 8.8** shows a graph that is used to monitor the average progression of the disease. The number of T cells, $f(x)$, is a function of time after infection, x.

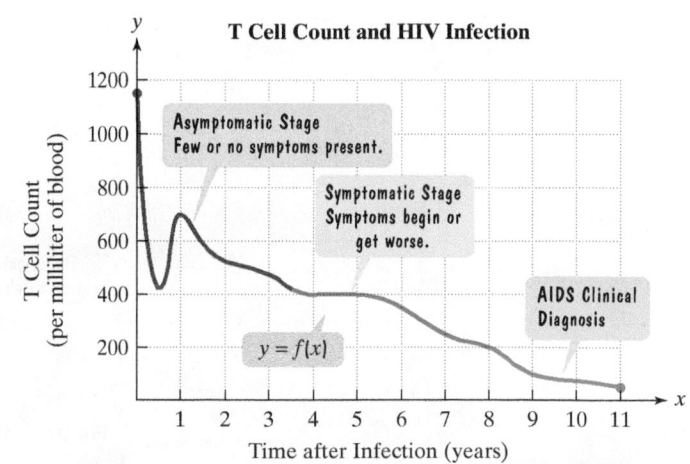

Figure 8.8

Source: HUMAN SEXUALITY by B. Pruitt, et al. © 1997 Prentice Hall

  a. Explain why f represents the graph of a function.

  b. Use the graph to find $f(8)$.

  c. For what value of x is $f(x) = 350$?

  d. Describe the general trend shown by the graph.

Solution

  a. No vertical line can be drawn that intersects the graph of f in **Figure 8.8** more than once. By the vertical line test, f represents the graph of a function.

  b. To find $f(8)$, or f of 8, we locate 8 on the x-axis. **Figure 8.9** shows the point on the graph of f for which 8 is the first coordinate. From this point, we look to the y-axis to find the corresponding y-coordinate. We see that the y-coordinate is 200. Thus,

$$f(8) = 200.$$

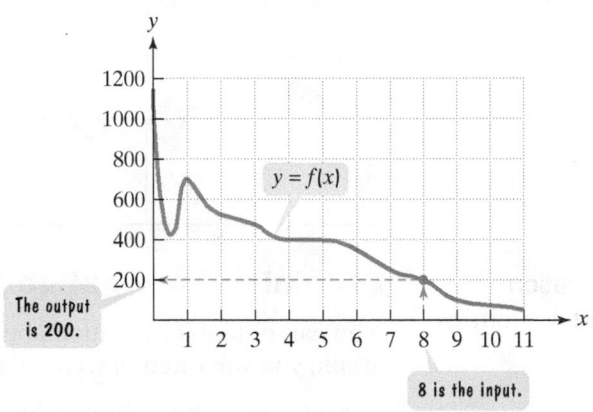

Figure 8.9 Finding $f(8)$

When the time after infection is 8 years, the T cell count is 200 cells per milliliter of blood. (AIDS clinical diagnosis is given at a T cell count of 200 or below.)

  c. To find the value of x for which $f(x) = 350$, we approximately locate 350 on the y-axis. **Figure 8.10** shows that there is one point on the graph of f for which 350 is the second coordinate. From this point, we look to the x-axis to find the corresponding x-coordinate. We see that the x-coordinate is 6. Thus,

$$f(x) = 350 \text{ for } x = 6.$$

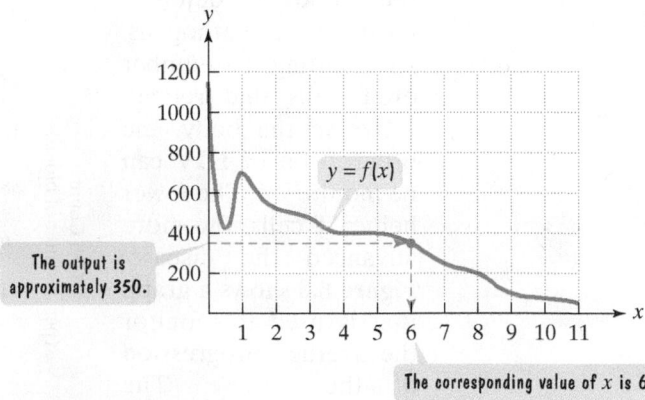

Figure 8.10 Finding x for which $f(x) = 350$

A T cell count of 350 occurs 6 years after infection.

d. Figure 8.11 uses voice balloons to describe the general trend shown by the graph.

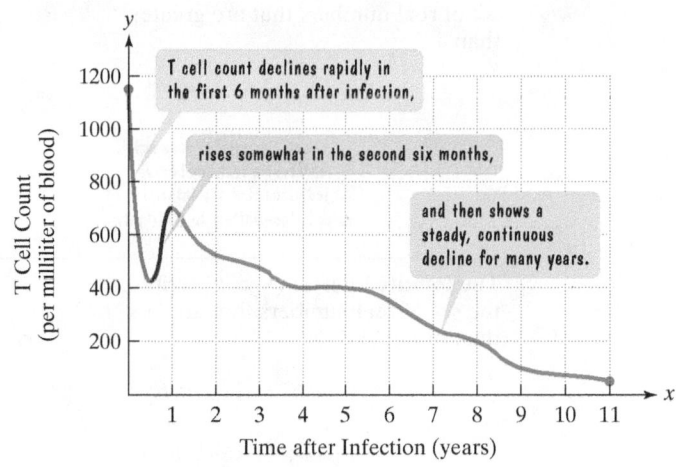

Figure 8.11 Describing changing T cell count over time in a person infected with HIV ▪

✓   **CHECK POINT 2**

a. Use the graph of f in **Figure 8.8** on page 597 to find $f(5)$.

b. For what value of x is $f(x) = 100$?

c. Estimate the minimum T cell count during the asymptomatic stage.

---

3 Review interval notation.

Reviewing Interval Notation

In Chapter 2, Section 2.7, we used interval notation to express solution sets of inequalities. Interval notation can also be used to identify the set of real numbers in a function's domain and the set of real numbers in its range.

Suppose that a and b are two real numbers such that $a < b$.

Interval Notation	Graph
The **open interval** (a, b) represents the set of real numbers between, but not including, a and b. $$(a, b) = \{x \mid a < x < b\}$$ x is greater than a $(a < x)$ and x is less than b $(x < b)$.	The parentheses in the graph and in interval notation indicate that a and b, the endpoints, are excluded from the interval.
The **closed interval** $[a, b]$ represents the set of real numbers between, and including, a and b. $$[a, b] = \{x \mid a \leq x \leq b\}$$ x is greater than or equal to a $(a \leq x)$ and x is less than or equal to b $(x \leq b)$.	The square brackets in the graph and in interval notation indicate that a and b, the endpoints, are included in the interval.

Interval Notation	Graph
The **infinite interval** (a, ∞) represents the set of real numbers that are greater than a. $$(a, \infty) = \{x \mid x > a\}$$ The infinity symbol does not represent a real number. It indicates that the interval extends indefinitely to the right.	The parenthesis indicates that a is excluded from the interval.
The **infinite interval** $(-\infty, b]$ represents the set of real numbers that are less than or equal to b. $$(-\infty, b] = \{x \mid x \le b\}$$ The negative infinity symbol indicates that the interval extends indefinitely to the left.	The square bracket indicates that b is included in the interval.

Parentheses and Brackets in Interval Notation

Parentheses, ( ), indicate endpoints that are not included in an interval. Square brackets, [ ], indicate endpoints that are included in an interval. Parentheses are always used with ∞ or $-\infty$.

Table 8.1 lists nine possible types of intervals used to describe sets of real numbers. When identifying the domain and range of a function from its graph, we will be interested in locating these intervals on the x- and y-axes.

Table 8.1 Intervals on the Real Number Line

Let a and b be real numbers such that $a < b$.

Interval Notation	Set-Builder Notation	Graph
(a, b)	$\{x \mid a < x < b\}$	
$[a, b]$	$\{x \mid a \le x \le b\}$	
$[a, b)$	$\{x \mid a \le x < b\}$	
$(a, b]$	$\{x \mid a < x \le b\}$	
(a, ∞)	$\{x \mid x > a\}$	
$[a, \infty)$	$\{x \mid x \ge a\}$	
$(-\infty, b)$	$\{x \mid x < a\}$	
$(-\infty, b]$	$\{x \mid x \le a\}$	
$(-\infty, \infty)$	$\{x \mid x$ is a real number$\}$ or $\mathbb{R}$ (set of real numbers)	

EXAMPLE 3 Interpreting Interval Notation

Express each interval in set-builder notation and graph:

a. $(-1, 4]$ **b.** $[2.5, 4]$ **c.** $(-4, \infty)$.

Solution

a. $(-1, 4] = \{x \mid -1 < x \le 4\}$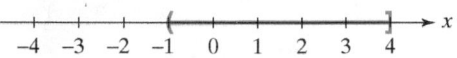

b. $[2.5, 4] = \{x \mid 2.5 \le x \le 4\}$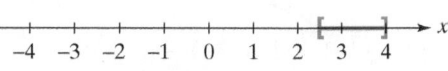

c. $(-4, \infty) = \{x \mid x > -4\}$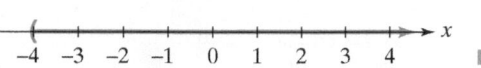

✓ **CHECK POINT 3** Express each interval in set-builder notation and graph:

a. $[-2, 5)$ **b.** $[1, 3.5]$ **c.** $(-\infty, -1)$.

---

4 Identify the domain and range of a function from its graph.

Identifying Domain and Range from a Function's Graph

Figure 8.12 illustrates how the graph of a function is used to determine the function's domain and its range.

Domain: set of inputs

> Found on the *x*-axis

Range: set of outputs

> Found on the *y*-axis

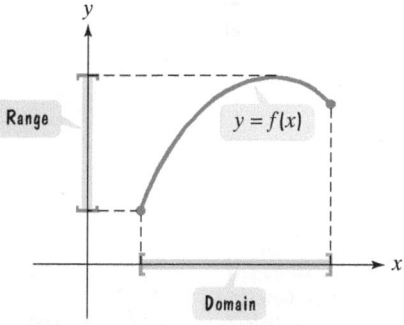

Figure 8.12 Domain and range of *f*

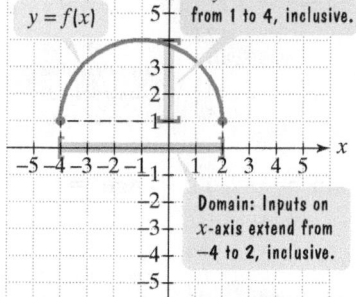

Figure 8.13 Domain and range of *f*

Let's apply these ideas to the graph of the function shown in **Figure 8.13**. To find the domain, look for all the inputs on the *x*-axis that correspond to points on the graph. Can you see that they extend from −4 to 2, inclusive? Using interval notation, the function's domain can be represented as follows:

$$[-4, 2].$$

> The square brackets indicate −4 and 2 are included. Note the square brackets on the *x*-axis in **Figure 8.13**.

To find the range, look for all the outputs on the *y*-axis that correspond to points on the graph. They extend from 1 to 4, inclusive. Using interval notation, the function's range can be represented as follows:

$$[1, 4].$$

> The square brackets indicate 1 and 4 are included. Note the square brackets on the *y*-axis in **Figure 8.13**.

EXAMPLE 4   Identifying the Domain and Range of a Function from Its Graph

Use the graph of each function to identify its domain and its range.

a.
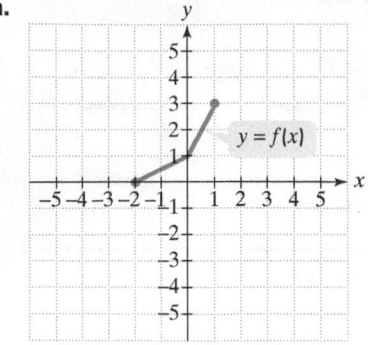

b.
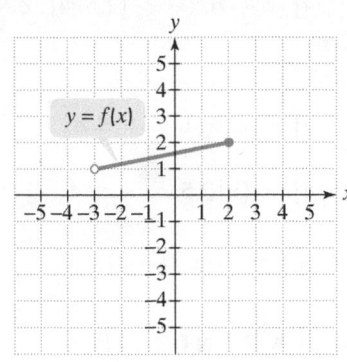

c.
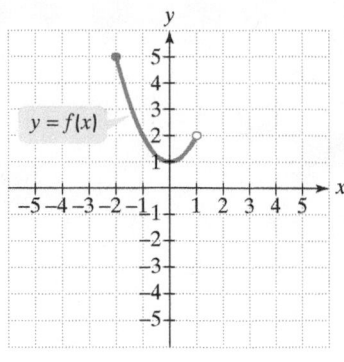

d.
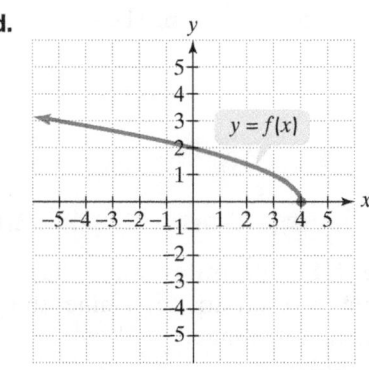

e.
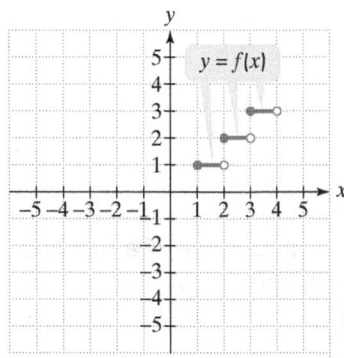

Solution   For the graph of each function, the domain is highlighted in purple on the x-axis and the range is highlighted in green on the y-axis.

a.
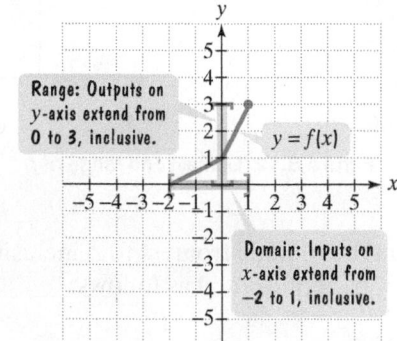

Domain = $[-2, 1]$; Range = $[0, 3]$

b.
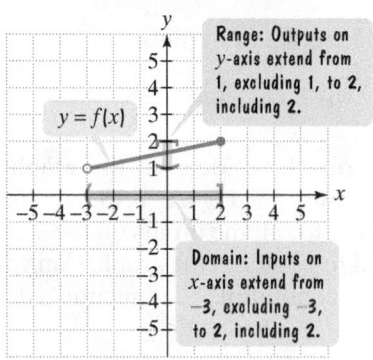

Domain = $(-3, 2]$; Range = $(1, 2]$

c.
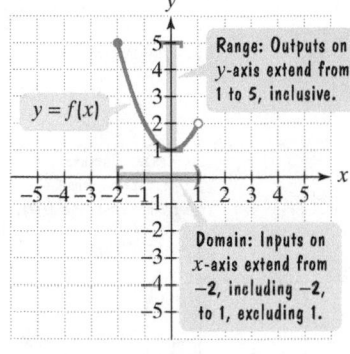

Domain = $[-2, 1)$; Range = $[1, 5]$

d.
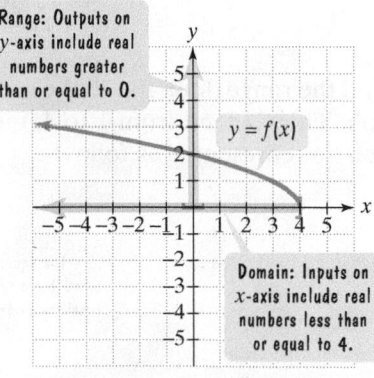

Domain = $(-\infty, 4]$; Range = $[0, \infty)$

e.
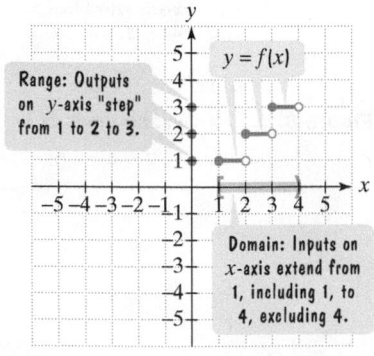

Domain = $[1, 4)$; Range = $\{1, 2, 3\}$ ∎

Great Question!

The range in Example 4(e) was identified as {1, 2, 3}. Why didn't you use interval notation like you did in the other parts of Example 4?

Interval notation is not appropriate for describing a set of distinct numbers such as {1, 2, 3}. Interval notation, [1, 3], would mean that numbers such as 1.5 and 2.99 are in the range, but they are not. That's why we used the roster method.

☑ **CHECK POINT 4** Use the graph of each function to identify its domain and its range.

a.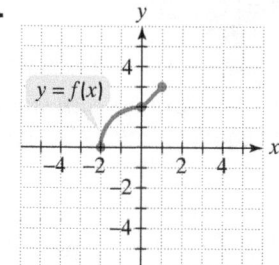

b.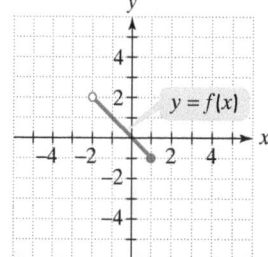

c.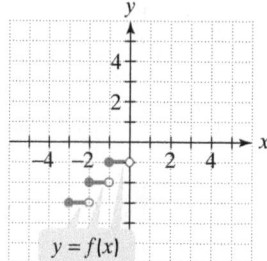

<hr>

CONCEPT AND VOCABULARY CHECK

Fill in each blank so that the resulting statement is true.

1. The graph of a function is the graph of its _____.

2. If any vertical line intersects a graph _____, the graph does not define *y* as a/an _____ of *x*.

3. The shaded set of numbers shown on the *x*-axis can be expressed in interval notation as _____. This set represents the function's _____.

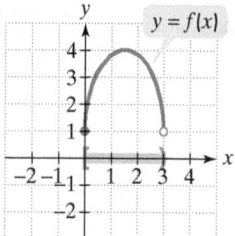

4. The shaded set of numbers shown on the *y*-axis can be expressed in interval notation as _____. This set represents the function's _____.

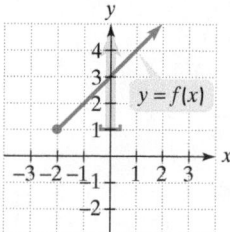

Practice Exercises

In Exercises 1–8, use the vertical line test to identify graphs in which y is a function of x.

1.
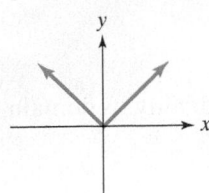

2.
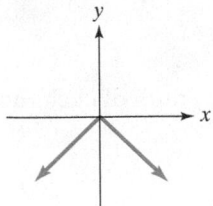

3.
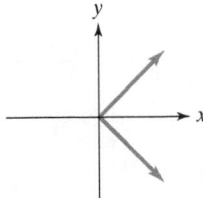

4.
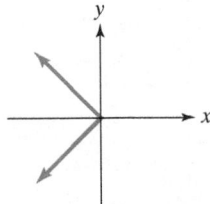

5.
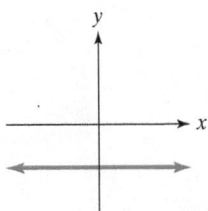

6.
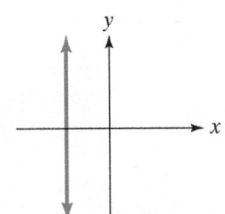

7.
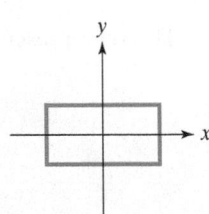

8.
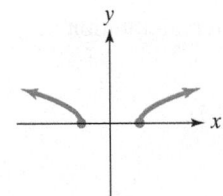

In Exercises 9–14, use the graph of f to find each indicated function value.

9. $f(-2)$

10. $f(2)$

11. $f(4)$

12. $f(-4)$

13. $f(-3)$

14. $f(-1)$

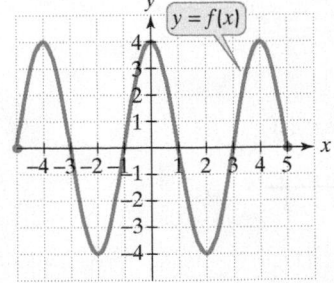

Use the graph of g to solve Exercises 15–20.

15. Find $g(-4)$.

16. Find $g(2)$.

17. Find $g(-10)$.

18. Find $g(10)$.

19. For what value of x is $g(x) = 1$?

20. For what value of x is $g(x) = -1$?

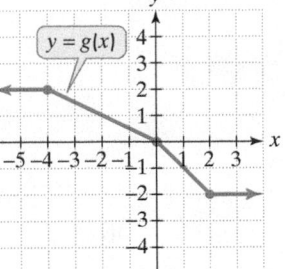

In Exercises 21–34, express each interval in set-builder notation and graph the interval on a number line.

21. $(1, 6]$

22. $(-2, 4]$

23. $[-5, 2)$

24. $[-4, 3)$

25. $[-3, 1]$

26. $[-2, 5]$

27. $(2, \infty)$

28. $(3, \infty)$

29. $[-3, \infty)$

30. $[-5, \infty)$

31. $(-\infty, 3)$

32. $(-\infty, 2)$

33. $(-\infty, 5.5)$

34. $(-\infty, 3.5]$

In Exercises 35–44, use the graph of each function to identify its domain and its range.

35.
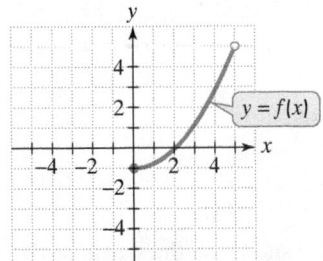

36.
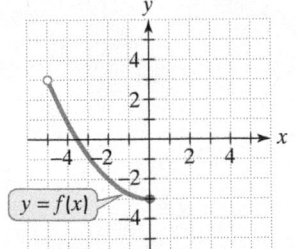

37.

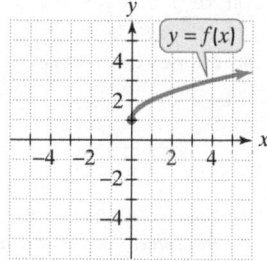

38.

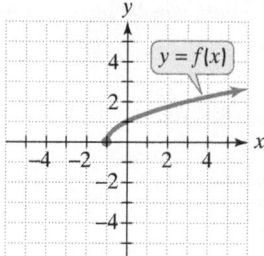

39.

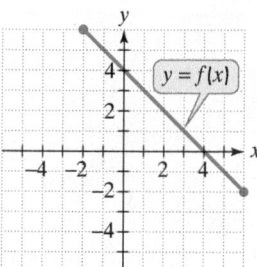

40.

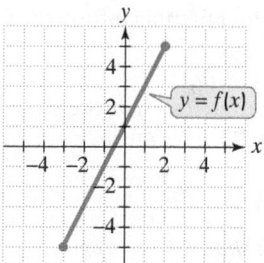

41.

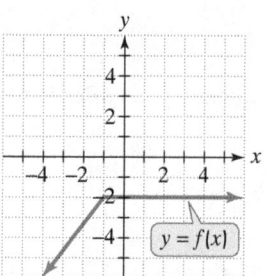

42.

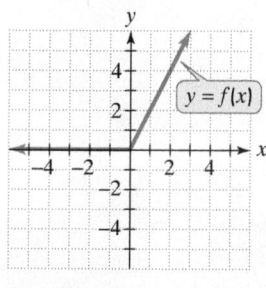

43.

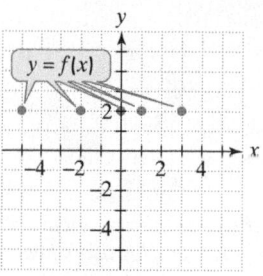

44.

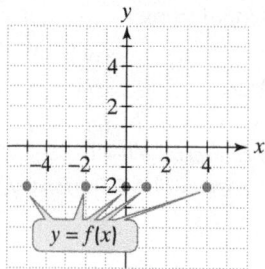

Practice PLUS

45. Use the graph of f to determine each of the following. Where applicable, use interval notation.

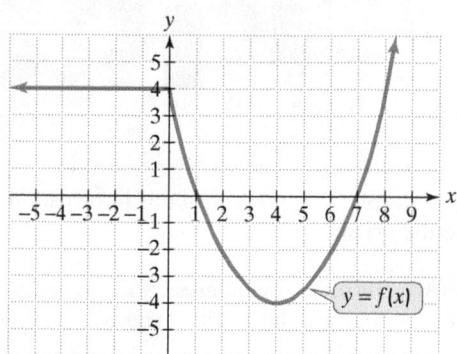

   a. the domain of f

   b. the range of f

   c. $f(-3)$

   d. the values of x for which $f(x) = -2$

   e. the points where the graph of f crosses the x-axis

   f. the point where the graph of f crosses the y-axis

   g. values of x for which $f(x) < 0$

   h. Is $f(-8)$ positive or negative?

46. Use the graph of f to determine each of the following. Where applicable, use interval notation.

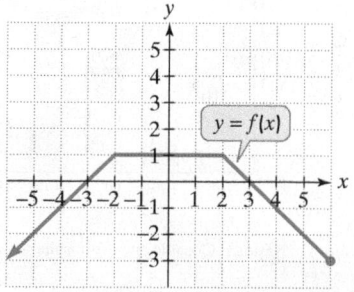

a. the domain of f

b. the range of f

c. $f(-4)$

d. the values of x for which $f(x) = -3$

e. the points where the graph of f crosses the x-axis

f. the point where the graph of f crosses the y-axis

g. values of x for which $f(x) > 0$

h. Is $f(-2)$ positive or negative?

Application Exercises

The amount of carbon dioxide in the atmosphere, measured in parts per million, has been increasing as a result of the burning of oil and coal. The buildup of gases and particles traps heat and raises the planet's temperature, a phenomenon called the greenhouse effect. Exercises 47–50 involve linear and quadratic functions related to global warming.

The bar graph shows the average global temperature, in degrees Fahrenheit, for seven selected years. The data can be modeled by the linear function

$$f(x) = 0.013x + 56.46,$$

where $f(x)$ is the average global temperature, in degrees Fahrenheit, x years after 1900. The graph of f is shown to the right of the data. Use this information to solve Exercises 47–48.

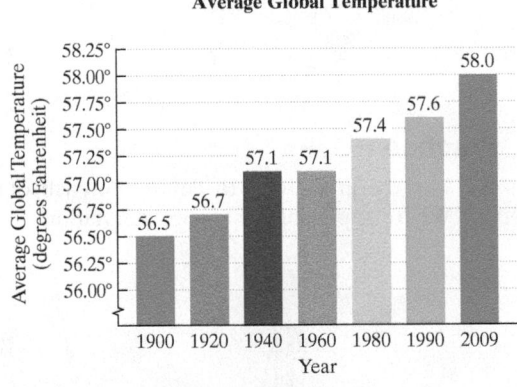

Source: National Oceanic and Atmospheric Administration

47. **a.** Find and interpret $f(80)$. Identify this information as a point on the graph.

    b. Does $f(80)$ overestimate or underestimate the actual data shown by the bar graph? By how much?

48. **a.** Find and interpret $f(90)$. Identify this information as a point on the graph.

    b. Does $f(90)$ overestimate or underestimate the actual data shown by the bar graph? By how much?

The pre-industrial concentration of atmospheric carbon dioxide was 280 parts per million. The bar graph shows the average atmospheric concentration of carbon dioxide, in parts per million, for seven selected years. The data can be modeled by the functions

$$f(x) = 1.27x + 305$$
$$and \ g(x) = 0.01x^2 + 0.65x + 310,$$

where $f(x)$ and $g(x)$ represent the average atmospheric concentration of carbon dioxide, in parts per million, x years after 1950. The graphs of f and g are shown to the right of the data. Use this information to solve Exercises 49–50.

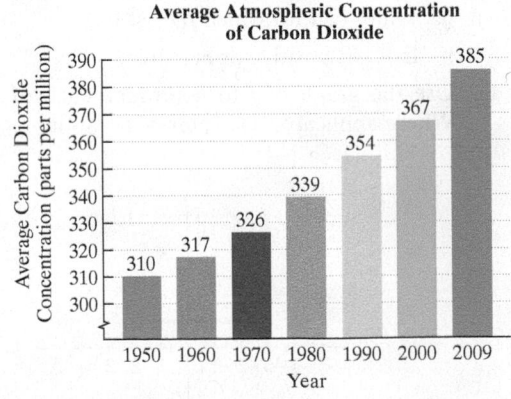

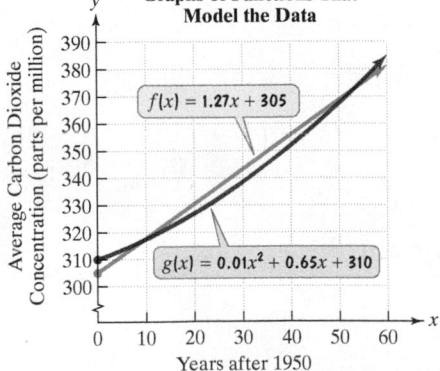

Source: National Oceanic and Atmospheric Administration

49. a. Find and interpret $f(50)$. Identify this information as a point on the appropriate graph.

b. Find and interpret $g(50)$. Identify this information as a point on the appropriate graph.

c. Which function, the linear or the quadratic, serves as a better description for the actual data shown by the bar graph?

50. a. Find and interpret $f(40)$. Identify this information as a point on the appropriate graph.

b. Find and interpret $g(40)$. Identify this information as a point on the appropriate graph.

c. Which function, the linear or the quadratic, serves as a better description for the actual data shown by the bar graph?

The function $f(x) = 0.4x^2 - 36x + 1000$ models the number of accidents, $f(x)$, per 50 million miles driven as a function of a driver's age, x, in years, for drivers from ages 16 through 74, inclusive. The graph of f is shown. Use the equation for f to solve Exercises 51–54.

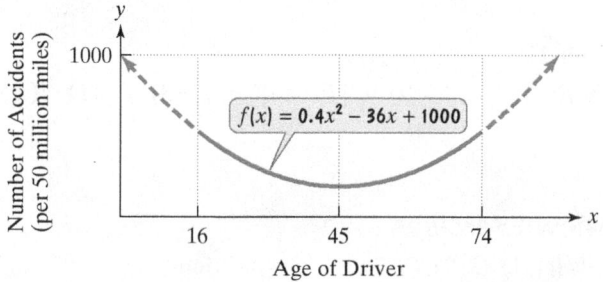

51. Find and interpret $f(20)$. Identify this information as a point on the graph of f.

52. Find and interpret $f(50)$. Identify this information as a point on the graph of f.

53. For what value of x does the graph reach its lowest point? Use the equation for f to find the minimum value of y. Describe the practical significance of this minimum value.

54. Use the graph to identify two different ages for which drivers have the same number of accidents. Use the equation for f to find the number of accidents for drivers at each of these ages.

The figure shows the cost of mailing a first-class letter, $f(x)$, as a function of its weight, x, in ounces, for weights not exceeding 3.5 ounces. Use the graph to solve Exercises 55–58.

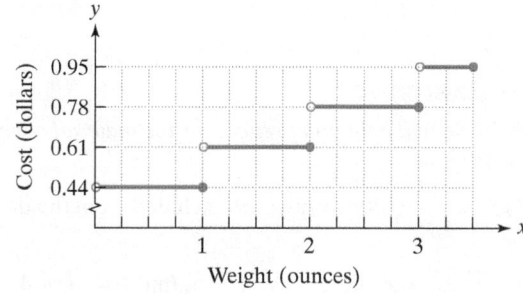

55. Find $f(3)$. What does this mean in terms of the variables in this situation?

56. Find $f(3.5)$. What does this mean in terms of the variables in this situation?

57. What is the cost of mailing a letter that weighs 1.5 ounces?

58. What is the cost of mailing a letter that weighs 1.8 ounces?

Writing in Mathematics

59. What is the graph of a function?

60. Explain how the vertical line test is used to determine whether a graph represents a function.

61. Explain how to identify the domain and range of a function from its graph.

Technology Exercise

62. The function

$$f(x) = -0.00002x^3 + 0.008x^2 - 0.3x + 6.95$$

models the number of annual physician visits, $f(x)$, by a person of age x. Graph the function in a [0, 100, 5] by [0, 40, 2] viewing rectangle. What does the shape of the graph indicate about the relationship between one's age and the number of annual physician visits? Use the $\boxed{\text{TRACE}}$ or minimum function capability to find the coordinates of the minimum point on the graph of the function. What does this mean?

Critical Thinking Exercises

Make Sense?   *In Exercises 63–66, determine whether each statement "makes sense" or "does not make sense" and explain your reasoning.*

63. I knew how to use point plotting to graph the equation $y = x^2 - 1$, so there was really nothing new to learn when I used the same technique to graph the function $f(x) = x^2 - 1$.

64. The graph of my function revealed aspects of its behavior that were not obvious by just looking at its equation.

65. I graphed a function showing how paid vacation days depend on the number of years a person works for a company. The domain was the number of paid vacation days.

66. I graphed a function showing how the number of annual physician visits depends on a person's age. The domain was the number of annual physician visits.

In Exercises 67–72, determine whether each statement is true or false. If the statement is false, make the necessary change(s) to produce a true statement.

67. The graph of every line is a function.

68. A horizontal line can intersect the graph of a function in more than one point.

Use the graph of f to determine whether each statement in Exercises 69–72 is true or false.

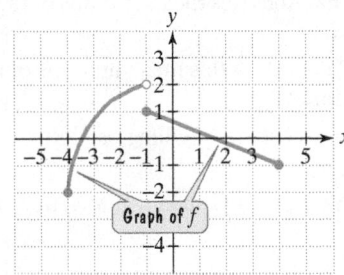

69. The domain of f is $[-4, 4]$.
70. The range of f is $[-2, 2]$.
71. $f(-1) - f(4) = 2$
72. $f(0) = 2.1$

In Exercises 73–74, let f be defined by the following graph:

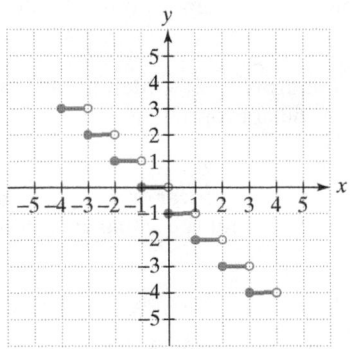

73. Find
$$\sqrt{f(-1.5) + f(-0.9)} - [f(\pi)]^2 + f(-3) \div f(1) \cdot f(-\pi).$$

74. Find
$$\sqrt{f(-2.5) - f(1.9)} - [f(-\pi)]^2 + f(-3) \div f(1) \cdot f(\pi).$$

Review Exercises

75. Is $\{(1, 1), (2, 2), (3, 3), (4, 4)\}$ a function? (Section 8.1, Example 2)

76. Solve: $12 - 2(3x + 1) = 4x - 5$.
(Section 2.3, Example 3)

77. The length of a rectangle exceeds 3 times the width by 8 yards. If the perimeter of the rectangle is 624 yards, what are its dimensions? (Section 2.5, Example 5)

Preview Exercises

Exercises 78–80 will help you prepare for the material covered in the next section.

78. If $f(x) = \dfrac{4}{x - 3}$, why must 3 be excluded from the domain of f?

79. If $f(x) = x^2 + x$ and $g(x) = x - 5$, find $f(4) + g(4)$.

80. Simplify: $-2.6x^2 + 49x + 3994 - (-0.6x^2 + 7x + 2412)$.

SECTION

8.3

The Algebra of Functions

Objectives

1 Find the domain of a function.

2 Use the algebra of functions to combine functions and determine domains.

We're born. We die. **Figure 8.14** quantifies these statements by showing the number of births and deaths in the United States from 2000 through 2009.

Number of Births and Deaths in the United States

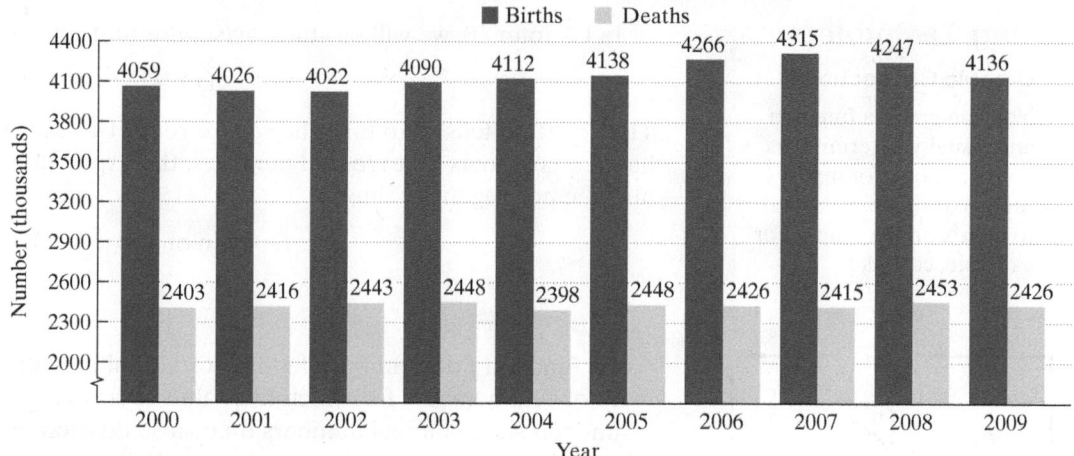

Figure 8.14
Source: U.S. Department of Health and Human Services

In this section, we look at these data from the perspective of functions. By considering the yearly change in the U.S. population, you will see that functions can be subtracted using procedures that will remind you of combining algebraic expressions.

1 Find the domain of a function.

The Domain of a Function

We begin with two functions that model the data in **Figure 8.14**.

$$B(x) = -2.6x^2 + 49x + 3994 \qquad D(x) = -0.6x^2 + 7x + 2412$$

Number of births, $B(x)$, in thousands, x years after **2000**

Number of deaths, $D(x)$, in thousands, x years after **2000**

The years in **Figure 8.14** extend from 2000 through 2009. Because x represents the number of years after 2000,

$$\text{Domain of } B = \{0, 1, 2, 3, \ldots, 9\}$$

and

$$\text{Domain of } D = \{0, 1, 2, 3, \ldots, 9\}.$$

Functions that model data often have their domains explicitly given with the function's equation. However, for most functions, only an equation is given and the domain is not specified. In cases like this, the domain of a function f is the largest set of real numbers for which the value of $f(x)$ is a real number. For example, consider the function

$$f(x) = \frac{1}{x - 3}.$$

Because division by 0 is undefined (and not a real number), the denominator, $x - 3$, cannot be 0. Thus, x cannot equal 3. The domain of the function consists of all real numbers other than 3, represented by

$$\text{Domain of } f = (-\infty, 3) \text{ or } (3, \infty).$$

Great Question!

Can you explain how $(-\infty, 3)$ or $(3, \infty)$ represents all real numbers other than 3?

The following voice balloons should help:

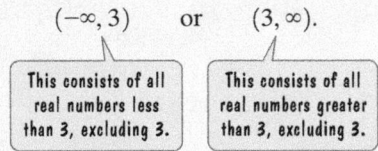

$$(-\infty, 3) \quad \text{or} \quad (3, \infty).$$

This consists of all real numbers less than 3, excluding 3.

This consists of all real numbers greater than 3, excluding 3.

Using Technology

Graphic Connections

You can graph a function and visually determine its domain. Look for inputs on the x-axis that correspond to points on the graph. For example, consider

$$g(x) = \sqrt{x}, \text{ or } y = \sqrt{x}.$$

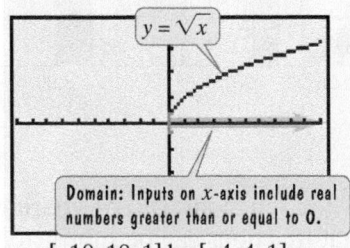

$y = \sqrt{x}$

Domain: Inputs on x-axis include real numbers greater than or equal to 0.

$[-10, 10, 1]$ by $[-4, 4, 1]$

This verifies that $[0, \infty)$ is the domain.

In Chapter 10, we will be studying square root functions such as

$$g(x) = \sqrt{x}.$$

The equation tells us to take the square root of x. Because only nonnegative numbers have square roots that are real numbers, the expression under the square root sign, x, must be nonnegative. Thus,

$$\text{Domain of } g = [0, \infty).$$

Finding a Function's Domain

If a function f does not model data or verbal conditions, its domain is the largest set of real numbers for which the value of $f(x)$ is a real number. Exclude from a function's domain real numbers that cause division by zero and real numbers that result in a square root of a negative number.

EXAMPLE 1   Finding the Domain of a Function

Find the domain of each function:

a. $f(x) = 3x + 2$ **b.** $g(x) = \dfrac{3x + 2}{x + 1}$.

Solution

a. The function $f(x) = 3x + 2$ contains neither division nor a square root. For every real number, x, the algebraic expression $3x + 2$ is a real number. Thus, the domain of f is the set of all real numbers.

$$\text{Domain of } f = (-\infty, \infty).$$

b. The function $g(x) = \dfrac{3x + 2}{x + 1}$ contains division. Because division by 0 is undefined, we must exclude from the domain the value of x that causes $x + 1$ to be 0. Thus, x cannot equal -1.

$$\text{Domain of } g = (-\infty, -1) \text{ or } (-1, \infty). \quad \blacksquare$$

☑ **CHECK POINT 1**   Find the domain of each function:

a. $f(x) = \dfrac{1}{2}x + 3$

b. $g(x) = \dfrac{7x + 4}{x + 5}$.

---

2 Use the algebra of functions to combine functions and determine domains.

The Algebra of Functions

We can combine functions using addition, subtraction, multiplication, and division by performing operations with the algebraic expressions that appear on the right side of the equations. For example, the functions $f(x) = 2x$ and $g(x) = x - 1$ can be combined to form the sum, difference, product, and quotient of f and g. Here's how it's done:

For each function, $f(x) = 2x$ and $g(x) = x - 1$.

Sum: $f + g$
$$(f + g)(x) = f(x) + g(x)$$
$$= 2x + (x - 1) = 3x - 1$$

Difference: $f - g$
$$(f - g)(x) = f(x) - g(x)$$
$$= 2x - (x - 1) = 2x - x + 1 = x + 1$$

Product: fg
$$(fg)(x) = f(x) \cdot g(x)$$
$$= 2x(x - 1) = 2x^2 - 2x$$

Quotient: $\dfrac{f}{g}$
$$\left(\dfrac{f}{g}\right)(x) = \dfrac{f(x)}{g(x)} = \dfrac{2x}{x - 1}, x \neq 1.$$

The domain for each of these functions consists of all real numbers that are common to the domains of f and g. In the case of the quotient function $\dfrac{f(x)}{g(x)}$, we must remember not to divide by 0, so we add the further restriction that $g(x) \neq 0$.

> **The Algebra of Functions: Sum, Difference, Product, and Quotient of Functions**
>
> Let f and g be two functions. The **sum** $f + g$, the **difference** $f - g$, the **product** fg, and the **quotient** $\dfrac{f}{g}$ are functions whose domains are the set of all real numbers common to the domains of f and g, defined as follows:
>
> **1.** Sum: $(f + g)(x) = f(x) + g(x)$
> **2.** Difference: $(f - g)(x) = f(x) - g(x)$
> **3.** Product: $(fg)(x) = f(x) \cdot g(x)$
> **4.** Quotient: $\left(\dfrac{f}{g}\right)(x) = \dfrac{f(x)}{g(x)}$, provided $g(x) \neq 0$.

EXAMPLE 2   Using the Algebra of Functions

Let $f(x) = x^2 - 3$ and $g(x) = 4x + 5$. Find each of the following:

a. $(f + g)(x)$ **b.** $(f + g)(3)$.

Solution

a. $(f + g)(x) = f(x) + g(x) = (x^2 - 3) + (4x + 5) = x^2 + 4x + 2$

Thus,

$$(f + g)(x) = x^2 + 4x + 2.$$

b. We find $(f + g)(3)$ by substituting 3 for x in the equation for $f + g$.

$$(f + g)(x) = x^2 + 4x + 2 \quad \text{This is the equation for } f + g.$$

Substitute 3 for x.

$$(f + g)(3) = 3^2 + 4 \cdot 3 + 2 = 9 + 12 + 2 = 23 \quad \blacksquare$$

✓ **CHECK POINT 2** Let $f(x) = 3x^2 + 4x - 1$ and $g(x) = 2x + 7$. Find each of the following:
   a. $(f + g)(x)$ **b.** $(f + g)(4)$.

EXAMPLE 3   Using the Algebra of Functions

Let $f(x) = \dfrac{4}{x}$ and $g(x) = \dfrac{3}{x + 2}$. Find each of the following:

a. $(f - g)(x)$

b. the domain of $f - g$.

Solution

a. $(f - g)(x) = f(x) - g(x) = \dfrac{4}{x} - \dfrac{3}{x + 2}$

(In Chapter 7, we discussed how to perform the subtraction with these algebraic fractions. In our study of the algebra of functions, we will leave these fractions in the form shown.)

b. The domain of $f - g$ is the set of all real numbers that are common to the domain of f and the domain of g. Thus, we must find the domains of f and g. We will do so for f first.

Note that $f(x) = \dfrac{4}{x}$ is a function involving division. Because division by 0 is undefined, x cannot equal 0.

The function $g(x) = \dfrac{3}{x + 2}$ is also a function involving division. Because division by 0 is undefined, x cannot equal -2.

To be in the domain of $f - g$, x must be in both the domain of f and the domain of g. This means that $x \neq 0$ and $x \neq -2$. **Figure 8.15** shows these excluded values on a number line.

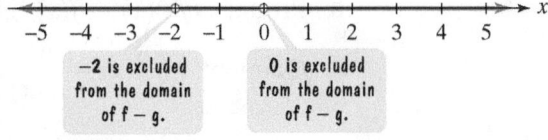

Figure 8.15

The number line in **Figure 8.15** is helpful in writing each of the three intervals that describe the domain of $f - g$.

Domain of $f - g = (-\infty, -2)$ or $(-2, 0)$ or $(0, \infty)$

All real numbers less than -2, excluding -2

All real numbers between -2 and 0, excluding -2 and excluding 0

All real numbers greater than 0, excluding 0

$\blacksquare$

Great Question!

I find the interval notation in Example 3(b), where you excluded −2 and 0 from the domain of f − g, somewhat confusing. Is there another way to describe such a domain?

You could use set-builder notation:

Domain of $f - g = \{x | x$ is a real number and $x \neq -2$ and $x \neq 0\}$.

Ask your professor if set-builder notation is an acceptable alternative for describing domains that exclude two or more numbers. The downside: We'll be using interval notation throughout the book to indicate both domains and ranges.

☑ **CHECK POINT 3**   Let $f(x) = \dfrac{5}{x}$ and $g(x) = \dfrac{7}{x - 8}$. Find each of the following:

a. $(f - g)(x)$

b. the domain of $f - g$.

EXAMPLE 4   Using the Algebra of Functions

Let $f(x) = x^2 + x$ and $g(x) = x - 5$. Find each of the following:

a. $(f + g)(4)$

b. $(f - g)(x)$ and $(f - g)(-3)$

c. $(fg)(x)$ and $(fg)(-2)$.

d. $\left(\dfrac{f}{g}\right)(x)$ and $\left(\dfrac{f}{g}\right)(7)$

Solution

a. We can find $(f + g)(4)$ using $f(4)$ and $g(4)$.

$$f(x) = x^2 + x \qquad\qquad g(x) = x - 5$$
$$f(4) = 4^2 + 4 = 20 \qquad g(4) = 4 - 5 = -1$$

Thus,

$$(f + g)(4) = f(4) + g(4) = 20 + (-1) = 19.$$

We can also find $(f + g)(4)$ by first finding $(f + g)(x)$ and then substituting 4 for x:

$$
\begin{aligned}
(f + g)(x) &= f(x) + g(x) & \text{\small This is the definition of the sum } f + g.\\
&= (x^2 + x) + (x - 5) & \text{\small Substitute the given functions.}\\
&= x^2 + 2x - 5. & \text{\small Simplify.}
\end{aligned}
$$

Using $(f + g)(x) = x^2 + 2x - 5$, we have

$$(f + g)(4) = 4^2 + 2 \cdot 4 - 5 = 16 + 8 - 5 = 19.$$

b.
$$
\begin{aligned}
(f - g)(x) &= f(x) - g(x) & \text{\small This is the definition of the difference } f - g.\\
&= (x^2 + x) - (x - 5) & \text{\small Substitute the given functions.}\\
&= x^2 + x - x + 5 & \text{\small Remove parentheses and change the sign of each}\\
& & \text{\small term in the second set of parentheses.}\\
&= x^2 + 5 & \text{\small Simplify.}
\end{aligned}
$$

Using $(f - g)(x) = x^2 + 5$, we have

$$(f - g)(-3) = (-3)^2 + 5 = 9 + 5 = 14.$$

Great Question!

How did you multiply $f(x) = x^2 + x$ and $g(x) = x - 5$?

Here are the details of the FOIL method we used to multiply the binomials:

$$(x^2 + x)(x - 5)$$

$$\boxed{F} \quad \boxed{0} \quad \boxed{I} \quad \boxed{L}$$

$$= x^2 \cdot x + x^2(-5) + x \cdot x + x(-5)$$
$$= x^3 - 5x^2 + x^2 - 5x$$

Special products of polynomials are reviewed in the Section 5.3 summary on pages 418–419.

c.
$$
\begin{aligned}
(fg)(x) &= f(x) \cdot g(x) & \text{\small This is the definition of the product } fg.\\
&= (x^2 + x)(x - 5) & \text{\small Substitute the given functions.}\\
&= x^3 - 5x^2 + x^2 - 5x & \text{\small Multiply using the FOIL method.}\\
&= x^3 - 4x^2 - 5x & \text{\small Combine like terms: } -5x^2 + x^2 = -4x^2.
\end{aligned}
$$

Using $(fg)(x) = x^3 - 4x^2 - 5x$, we have

$$(fg)(-2) = (-2)^3 - 4(-2)^2 - 5(-2)$$
$$= -8 - 4(4) - 5(-2) \qquad \text{\small Evaluate exponential expressions.}$$
$$= -8 - 16 + 10 = -14.$$

We can also find $(fg)(-2)$ using the fact that
$$(fg)(-2) = f(-2) \cdot g(-2).$$

$$f(x) = x^2 + x \qquad\qquad\qquad g(x) = x - 5$$
$$f(-2) = (-2)^2 + (-2) = 4 - 2 = 2 \qquad g(-2) = -2 - 5 = -7$$

Thus,
$$(fg)(-2) = f(-2) \cdot g(-2) = 2(-7) = -14.$$

d. $\left(\dfrac{f}{g}\right)(x) = \dfrac{f(x)}{g(x)}$ *This is the definition of the quotient* $\dfrac{f}{g}$.

$$= \dfrac{x^2 + x}{x - 5}$$ *Substitute the given functions.*

Using $\left(\dfrac{f}{g}\right)(x) = \dfrac{x^2 + x}{x - 5}$, we have

$$\left(\dfrac{f}{g}\right)(7) = \dfrac{7^2 + 7}{7 - 5} = \dfrac{56}{2} = 28. \quad \blacksquare$$

✓ **CHECK POINT 4**   Let $f(x) = x^2 - 2x$ and $g(x) = x + 3$. Find each of the following:

a. $(f + g)(5)$

b. $(f - g)(x)$ and $(f - g)(-1)$

c. $(fg)(x)$ and $(fg)(-4)$

d. $\left(\dfrac{f}{g}\right)(x)$ and $\left(\dfrac{f}{g}\right)(7)$.

EXAMPLE 5   Applying the Algebra of Functions

We opened the section with functions that model the number of births and deaths in the United States from 2000 through 2009:

$$B(x) = -2.6x^2 + 49x + 3994 \qquad D(x) = -0.6x^2 + 7x + 2412.$$

Number of births, $B(x)$, in thousands, x years after 2000

Number of deaths, $D(x)$, in thousands, x years after 2000

a. Write a function that models the change in U.S. population for the years from 2000 through 2009.

b. Use the function from part (a) to find the change in U.S. population in 2008.

c. Does the result in part (b) overestimate or underestimate the actual population change in 2008 obtained from the data in **Figure 8.14** on page 609? By how much?

Solution

a. The change in population is the number of births minus the number of deaths. Thus, we will find the difference function, $B - D$.

$(B - D)(x)$

$= B(x) - D(x)$

$= (-2.6x^2 + 49x + 3994) - (-0.6x^2 + 7x + 2412)$ *Substitute the given functions.*

$= -2.6x^2 + 49x + 3994 + 0.6x^2 - 7x - 2412$ *Remove parentheses and change the sign of each term in the second set of parentheses.*

$= (-2.6x^2 + 0.6x^2) + (49x - 7x) + (3994 - 2412)$ *Group like terms.*

$= -2x^2 + 42x + 1582$ *Combine like terms.*

The function

$$(B - D)(x) = -2x^2 + 42x + 1582$$

models the change in U.S. population, in thousands, x years after 2000.

b. Because 2008 is 8 years after 2000, we substitute 8 for x in the difference function $(B - D)(x)$.

$$(B - D)(x) = -2x^2 + 42x + 1582 \qquad \text{Use the difference function } B - D.$$
$$(B - D)(8) = -2(8)^2 + 42(8) + 1582 \qquad \text{Substitute 8 for x.}$$
$$= -2(64) + 42(8) + 1582 \qquad \text{Evaluate the exponential expression: } 8^2 = 64.$$
$$= -128 + 336 + 1582 \qquad \text{Perform the multiplications.}$$
$$= 1790 \qquad \text{Add from left to right.}$$

We see that $(B - D)(8) = 1790$. The model indicates that there was a population increase of 1790 thousand, or approximately 1,790,000 people, in 2008.

c. The data for 2008 in **Figure 8.14** on page 609 show 4247 thousand births and 2453 thousand deaths.

$$\text{population change} = \text{births} - \text{deaths}$$
$$= 4247 - 2453 = 1794$$

The actual population increase was 1794 thousand, or 1,794,000. Our model gave us an increase of 1790 thousand. Thus, the model underestimates the actual increase by $1794 - 1790$, or 4 thousand people. ∎

✓ **CHECK POINT 5**   Use the birth and death models from Example 5.

a. Write a function that models the total number of births and deaths in the United States for the years from 2000 through 2009.

b. Use the function from part (a) to find the total number of births and deaths in the United States in 2003.

c. Does the result in part (b) overestimate or underestimate the actual number of total births and deaths in 2003 obtained from the data in **Figure 8.14** on page 609? By how much?

Achieving Success

Do you get to choose your seat during an exam? If so, select a desk that minimizes distractions and puts you in the right frame of mind. Many students can focus better if they do not sit near friends. If possible, sit near a window, next to a wall, or in the front row. In these locations, you can glance up from time to time without looking at another student's work.

CONCEPT AND VOCABULARY CHECK

Fill in each blank so that the resulting statement is true.

1. We exclude from a function's domain real numbers that cause division by _____.

2. We exclude from a function's domain real numbers that result in a square root of a/an _____ number.

3. $(f + g)(x) = $ _____

4. $(f - g)(x) = $ _____

5. $(fg)(x) = $ _____

6. $\dfrac{f}{g}(x) = $ _____, provided _____ $\neq 0$

7. The domain of $f(x) = 5x + 7$ consists of all real numbers, represented in interval notation as _____.

8. The domain of $g(x) = \dfrac{3}{x - 2}$ consists of all real numbers except 2, represented in interval notation as $(-\infty, 2)$ or _____.

9. The domain of $h(x) = \dfrac{1}{x} + \dfrac{7}{x - 3}$ consists of all real numbers except 0 and 3, represented in interval notation as $(-\infty, 0)$ or _____ or _____.

Practice Exercises

In Exercises 1–10, find the domain of each function.

1. $f(x) = 3x + 5$

2. $f(x) = 4x + 7$

3. $g(x) = \dfrac{1}{x + 4}$

4. $g(x) = \dfrac{1}{x + 5}$

5. $f(x) = \dfrac{2x}{x - 3}$

6. $f(x) = \dfrac{4x}{x - 2}$

7. $g(x) = x + \dfrac{3}{5 - x}$

8. $g(x) = x + \dfrac{7}{6 - x}$

9. $f(x) = \dfrac{1}{x + 7} + \dfrac{3}{x - 9}$

10. $f(x) = \dfrac{1}{x + 8} + \dfrac{3}{x - 10}$

In Exercises 11–16, find $(f + g)(x)$ and $(f + g)(5)$.

11. $f(x) = 3x + 1, g(x) = 2x - 6$

12. $f(x) = 4x + 2, g(x) = 2x - 9$

13. $f(x) = x - 5, g(x) = 3x^2$

14. $f(x) = x - 6, g(x) = 2x^2$

15. $f(x) = 2x^2 - x - 3, g(x) = x + 1$

16. $f(x) = 4x^2 - x - 3, g(x) = x + 1$

17. Let $f(x) = 5x$ and $g(x) = -2x - 3$. Find $(f + g)(x)$, $(f - g)(x)$, $(fg)(x)$, and $\left(\dfrac{f}{g}\right)(x)$.

18. Let $f(x) = -4x$ and $g(x) = -3x + 5$. Find $(f + g)(x)$, $(f - g)(x)$, $(fg)(x)$, and $\left(\dfrac{f}{g}\right)(x)$.

In Exercises 19–30, for each pair of functions, f and g, determine the domain of $f + g$.

19. $f(x) = 3x + 7, g(x) = 9x + 10$

20. $f(x) = 7x + 4, g(x) = 5x - 2$

21. $f(x) = 3x + 7, g(x) = \dfrac{2}{x - 5}$

22. $f(x) = 7x + 4, g(x) = \dfrac{2}{x - 6}$

23. $f(x) = \dfrac{1}{x}, g(x) = \dfrac{2}{x - 5}$

24. $f(x) = \dfrac{1}{x}, g(x) = \dfrac{2}{x - 6}$

25. $f(x) = \dfrac{8x}{x - 2}, g(x) = \dfrac{6}{x + 3}$

26. $f(x) = \dfrac{9x}{x - 4}, g(x) = \dfrac{7}{x + 8}$

27. $f(x) = \dfrac{8x}{x - 2}, g(x) = \dfrac{6}{2 - x}$

28. $f(x) = \dfrac{9x}{x - 4}, g(x) = \dfrac{7}{4 - x}$

29. $f(x) = x^2, g(x) = x^3$

30. $f(x) = x^2 + 1, g(x) = x^3 - 1$

In Exercises 31–50, let
$$f(x) = x^2 + 4x \quad \text{and} \quad g(x) = 2 - x.$$
Find each of the following.

31. $(f + g)(x)$ and $(f + g)(3)$

32. $(f + g)(x)$ and $(f + g)(4)$

33. $f(-2) + g(-2)$ 34. $f(-3) + g(-3)$

35. $(f - g)(x)$ and $(f - g)(5)$

36. $(f - g)(x)$ and $(f - g)(6)$

37. $f(-2) - g(-2)$ 38. $f(-3) - g(-3)$

39. $(fg)(x)$ and $(fg)(2)$

40. $(fg)(x)$ and $(fg)(3)$

41. $(fg)(5)$ 42. $(fg)(6)$

43. $\left(\dfrac{f}{g}\right)(x)$ and $\left(\dfrac{f}{g}\right)(1)$

44. $\left(\dfrac{f}{g}\right)(x)$ and $\left(\dfrac{f}{g}\right)(3)$

45. $\left(\dfrac{f}{g}\right)(-1)$ 46. $\left(\dfrac{f}{g}\right)(0)$

47. The domain of $f + g$

48. The domain of $f - g$

49. The domain of $\dfrac{f}{g}$

50. The domain of fg

Practice PLUS

Use the graphs of f and g at the top of the next page to solve Exercises 51–58.

51. Find $(f + g)(-3)$.

52. Find $(g - f)(-2)$.

53. Find $(fg)(2)$.

54. Find $\left(\dfrac{g}{f}\right)(3)$.

55. Find the domain of $f + g$.

56. Find the domain of $\dfrac{f}{g}$.

57. Graph $f + g$.

58. Graph $f - g$.

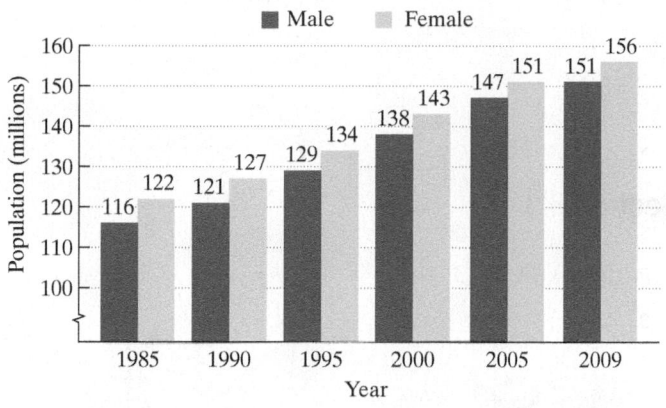

Use the table defining f and g to solve Exercises 59–62.

x	f(x)	g(x)
−2	5	0
−1	3	−2
0	−2	4
1	−6	−3
2	0	1

59. Find $(f + g)(1) - (g - f)(-1)$.

60. Find $(f + g)(-1) - (g - f)(0)$.

61. Find $(fg)(-2) - \left[\left(\dfrac{f}{g}\right)(1)\right]^2$.

62. Find $(fg)(2) - \left[\left(\dfrac{g}{f}\right)(0)\right]^2$.

Application Exercises

The bar graph shows the population of the United States, in millions, for six selected years.

Population of the United States

■ Male   ■ Female

Population (millions) vs *Year*

- 1985: Male 116, Female 122
- 1990: Male 121, Female 127
- 1995: Male 129, Female 134
- 2000: Male 138, Female 143
- 2005: Male 147, Female 151
- 2009: Male 151, Female 156

Source: U.S. Census Bureau

Here are two functions that model the data:

$$M(x) = 1.54x + 114.6$$

Male U.S. population, $M(x)$, in millions, x years after 1985

$$F(x) = 1.48x + 120.6$$

Female U.S. population, $F(x)$, in millions, x years after 1985

Use the functions in the previous column to solve Exercises 63–65.

63. a. Write a function that models the total U.S. population for the years shown in the bar graph.

   b. Use the function from part (a) to find the total U.S. population in 2005.

   c. Does the result in part (b) overestimate or underestimate the actual total U.S. population in 2005 shown by the bar graph? By how much?

64. a. Write a function that models the difference between the female U.S. population and the male U.S. population for the years shown in the bar graph.

   b. Use the function from part (a) to find how many more women than men there were in the U.S. population in 2005.

   c. Does the result in part (b) overestimate or underestimate the actual difference between the female and male population in 2005 shown by the bar graph? By how much?

65. a. Write a function that models the ratio of men to women in the U.S. population for the years shown in the bar graph.

   b. Use the function from part (a) to find the ratio of men to women, correct to three decimal places, in 2000.

   c. Does the result in part (b) overestimate or underestimate the actual ratio of men to women in 2000 shown by the bar graph? By how much?

66. A company that sells radios has yearly fixed costs of $600,000. It costs the company $45 to produce each radio. Each radio will sell for $65. The company's costs and revenue are modeled by the following functions:

$$C(x) = 600,000 + 45x$$ This function models the company's costs.

$$R(x) = 65x$$ This function models the company's revenue.

Find and interpret $(R - C)(20,000)$, $(R - C)(30,000)$, and $(R - C)(40,000)$.

Writing in Mathematics

67. If a function is defined by an equation, explain how to find its domain.

68. If equations for functions f and g are given, explain how to find $f + g$.

69. If the equations of two functions are given, explain how to obtain the quotient function and its domain.

70. If equations for functions f and g are given, describe two ways to find $(f - g)(3)$.

Technology Exercises

In Exercises 71–74, graph each of the three functions in the same $[-10, 10, 1]$ by $[-10, 10, 1]$ viewing rectangle.

71. $y_1 = 2x + 3$
$y_2 = 2 - 2x$
$y_3 = y_1 + y_2$

72. $y_1 = x - 4$
$y_2 = 2x$
$y_3 = y_1 - y_2$

73. $y_1 = x$
$y_2 = x - 4$
$y_3 = y_1 \cdot y_2$

74. $y_1 = x^2 - 2x$
$y_2 = x$
$y_3 = \dfrac{y_1}{y_2}$

75. In Exercise 74, use the $\boxed{\text{TRACE}}$ feature to trace along y_3. What happens at $x = 0$? Explain why this occurs.

Critical Thinking Exercises

Make Sense? *In Exercises 76–79, determine whether each statement "makes sense" or "does not make sense" and explain your reasoning.*

76. There is an endless list of real numbers that cannot be included in the domain of $f(x) = \sqrt{x}$.

77. I used a function to model data from 1980 through 2005. The independent variable in my model represented the number of years after 1980, so the function's domain was $\{0, 1, 2, 3, \dots, 25\}$.

78. If I have equations for functions f and g, and 3 is in both domains, then there are always two ways to determine $(f + g)(3)$.

79. I have two functions. Function f models total world population x years after 2000 and function g models population of the world's more-developed regions x years after 2000. I can use $f - g$ to determine the population of the world's less-developed regions for the years in both function's domains.

In Exercises 80–83, determine whether each statement is true or false. If the statement is false, make the necessary change(s) to produce a true statement.

80. If $(f + g)(a) = 0$, then $f(a)$ and $g(a)$ must be opposites, or additive inverses.

81. If $(f - g)(a) = 0$, then $f(a)$ and $g(a)$ must be equal.

82. If $\left(\dfrac{f}{g}\right)(a) = 0$, then $f(a)$ must be 0.

83. If $(fg)(a) = 0$, then $f(a)$ must be 0.

Review Exercises

84. Solve for b: $R = 3(a + b)$. (Section 2.4, Example 2)

85. Solve: $3(6 - x) = 3 - 2(x - 4)$. (Section 2.3, Example 3)

86. If $f(x) = 6x - 4$, find $f(b + 2)$. (Section 8.1, Example 3)

Preview Exercises

Exercises 87–89 will help you prepare for the material covered in the next section.

87. Let $f(x) = 3x - 4$ and $g(x) = x^2 + 6$.
 a. Find $f(5)$.
 b. Find $g(f(5))$.

88. Simplify: $3\left(\dfrac{x - 2}{3}\right) + 2$.

89. Solve for y: $x = 7y - 5$.

MID-CHAPTER CHECK POINT Section 8.1–Section 8.3

✓ **What You Know:** We learned that a function is a relation in which no two ordered pairs have the same first component and different second components. We represented functions as equations and used function notation. We graphed functions and applied the vertical line test to identify graphs of functions. We determined the domain and range of a function from its graph, using inputs on the x-axis for the domain and outputs on the y-axis for the range. Finally, we developed an algebra of functions to combine functions and determine their domains.

In Exercises 1–6, determine whether each relation is a function. Give the domain and range for each relation.

1. $\{(2, 6), (1, 4), (2, -6)\}$

2. $\{(0, 1), (2, 1), (3, 4)\}$

3.
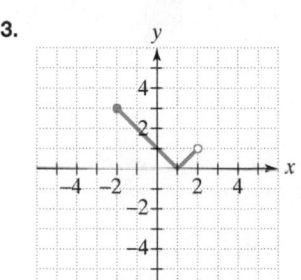

4.
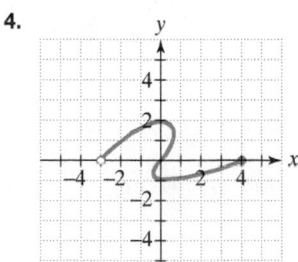

5.

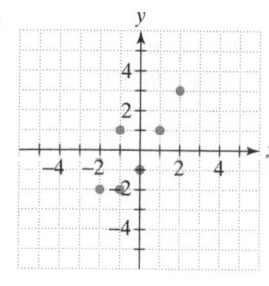

6.

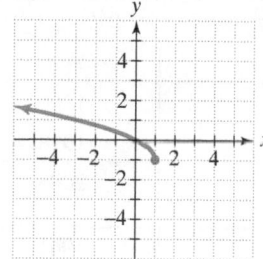

Use the graph of f to solve Exercises 7–12.

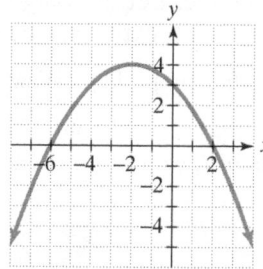

7. Explain why *f* represents the graph of a function.

8. Use the graph to find $f(-4)$.

9. For what value or values of *x* is $f(x) = 4$?

10. For what value or values of *x* is $f(x) = 0$?

11. Find the domain of *f*.

12. Find the range of *f*.

In Exercises 13–14, find the domain of each function.

13. $f(x) = (x + 2)(x - 2)$

14. $g(x) = \dfrac{1}{(x + 2)(x - 2)}$

In Exercises 15–22, let

$$f(x) = x^2 - 3x + 8 \text{ and } g(x) = -2x - 5.$$

Find each of the following.

15. $f(0) + g(-10)$ **16.** $f(-1) - g(3)$

17. $f(a) + g(a + 3)$

18. $(f + g)(x)$ and $(f + g)(-2)$

19. $(f - g)(x)$ and $(f - g)(5)$

20. $(fg)(x)$ and $(fg)(-1)$

21. $\left(\dfrac{f}{g}\right)(x)$ and $\left(\dfrac{f}{g}\right)(-4)$

22. The domain of $\dfrac{f}{g}$

Composite and Inverse Functions

Objectives

1 Form composite functions.

2 Verify inverse functions.

3 Find the inverse of a function.

4 Use the horizontal line test to determine if a function has an inverse function.

5 Use the graph of a one-to-one function to graph its inverse function.

Based on Shakespeare's *Romeo and Juliet*, the film *West Side Story* swept the 1961 Academy Awards with ten Oscars. The top four movies to win the most Oscars are shown in **Table 8.2** on the next page.

Table 8.2	Films Winning the Most Oscars	
Movie	**Year**	**Number of Academy Awards**
Ben-Hur	1960	11
Titanic	1998	11
The Lord of the Rings: The Return of the King	2003	11
West Side Story	1961	10

Source: Russell Ash, *The Top 10 of Everything, 2011*

We can use the information in **Table 8.2** to define a function. Let the domain of the function be the set of four movies shown in the table. Let the range be the number of Academy Awards for each of the respective films. The function can be written as follows:

f: {(*Ben-Hur*, 11), (*Titanic*, 11), (*The Lord of the Rings*, 11), (*West Side Story*, 10)}.

Now let's "undo" f by interchanging the first and second components in each of the ordered pairs. Switching the inputs and outputs of f, we obtain the following relation:

Same first component

Undoing f: {(11, *Ben-Hur*), (11, *Titanic*), (11, *The Lord of the Rings*), (10, *West Side Story*)}.

Different second components

Can you see that this relation is not a function? Three of its ordered pairs have the same first component and different second components. This violates the definition of a function.

If a function f is a set of ordered pairs, (x, y), then the changes produced by f can be "undone" by reversing the components of all the ordered pairs. The resulting relation, (y, x), may or may not be a function. To understand what occurs when we interchange components, we turn to the topics of *composite* and *inverse functions*.

① Form composite functions.

Composite Functions

In the previous section, we saw that functions could be combined using addition, subtraction, multiplication, and division. Now let's consider another way of combining two functions. To help understand this new combination, suppose that your local computer store is having a sale. The models that are on sale cost either $300 less than the regular price or 85% of the regular price. If x represents the computer's regular price, the discounts can be described with the following functions:

$$f(x) = x - 300 \qquad g(x) = 0.85x.$$

The computer is on sale for $300 less than its regular price.

The computer is on sale for 85% of its regular price.

At the store, you bargain with the salesperson. Eventually, she makes an offer you can't refuse. The sale price will be 85% of the regular price followed by a $300 reduction:

$$0.85x - 300.$$

85% of the regular price

followed by a $300 reduction

In terms of functions f and g, this offer can be obtained by taking the output of $g(x) = 0.85x$, namely $0.85x$, and using it as the input of f:

$$f(x) = x - 300$$

Replace x with $0.85x$, the output of $g(x) = 0.85x$.

$$f(0.85x) = 0.85x - 300.$$

Because $0.85x$ is $g(x)$, we can write this last equation as

$$f(g(x)) = 0.85x - 300.$$

We read this equation as "f of g of x is equal to $0.85x - 300$." We call $f(g(x))$ the **composition of the function f with g**, or a **composite function**. This composite function is written $f \circ g$. Thus,

$$(f \circ g)(x) = f(g(x)) = 0.85x - 300.$$

This can be read "f of g of x" or "f composed with g of x."

Like all functions, we can evaluate $f \circ g$ for a specified value of x in the function's domain. For example, here's how to find the value of the composite function describing the offer you cannot refuse at 1400:

$$(f \circ g)(x) = 0.85x - 300$$

Replace x with 1400.

$$(f \circ g)(1400) = 0.85(1400) - 300 = 1190 - 300 = 890.$$

Because $(f \circ g)(1400) = 890$, this means that a computer that regularly sells for $1400 is on sale for $890 subject to both discounts. We can use a partial table of coordinates for each of the discount functions, g and f, to numerically verify this result.

Computer's regular price	85% of the regular price		85% of the regular price	$300 reduction
x	$g(x) = 0.85x$		x	$f(x) = x - 300$
1200	1020		1020	720
1300	1105		1105	805
1400	1190		1190	890

Using these tables, we can find $(f \circ g)(1400)$:

$$(f \circ g)(1400) = f(g(1400)) = f(1190) = 890.$$

The table for g shows that $g(1400) = 1190$. The table for f shows that $f(1190) = 890$.

This verifies that a computer that regularly sells for $1400 is on sale for $890 subject to both discounts.

Before you run out to buy a computer, let's generalize our discussion of the computer's double discount and define the composition of any two functions.

The Composition of Functions

The **composition of the function f with g** is denoted by $f \circ g$ and is defined by the equation

$$(f \circ g)(x) = f(g(x)).$$

The **domain of the composite function $f \circ g$** is the set of all x such that

1. x is in the domain of g and
2. $g(x)$ is in the domain of f.

EXAMPLE 1   Forming Composite Functions

Given $f(x) = 3x - 4$ and $g(x) = x^2 + 6$, find each of the following composite functions:

a. $(f \circ g)(x)$ **b.** $(g \circ f)(x)$.

Solution

a. We begin with $(f \circ g)(x)$, the composition of f with g. Because $(f \circ g)(x)$ means $f(g(x))$, we must replace each occurrence of x in the equation for f with $g(x)$.

$$f(x) = 3x - 4 \qquad \text{This is the given equation for f.}$$

Replace x with $g(x)$.

$$(f \circ g)(x) = f(g(x)) = 3g(x) - 4$$

$$= 3(x^2 + 6) - 4 \qquad \text{Because } g(x) = x^2 + 6, \text{ replace } g(x) \text{ with } x^2 + 6.$$

$$= 3x^2 + 18 - 4 \qquad \text{Use the distributive property.}$$

$$= 3x^2 + 14 \qquad \text{Simplify.}$$

Thus, $(f \circ g)(x) = 3x^2 + 14$.

b. Next, we find $(g \circ f)(x)$, the composition of g with f. Because $(g \circ f)(x)$ means $g(f(x))$, we must replace each occurrence of x in the equation for g with $f(x)$.

$$g(x) = x^2 + 6 \qquad \text{This is the given equation for g.}$$

Replace x with $f(x)$.

$$(g \circ f)(x) = g(f(x)) = (f(x))^2 + 6$$

$$= (3x - 4)^2 + 6 \qquad \text{Because } f(x) = 3x - 4, \text{ replace } f(x) \text{ with } 3x - 4.$$

$$= 9x^2 - 24x + 16 + 6 \qquad \text{Use } (A - B)^2 = A^2 - 2AB + B^2 \text{ to square } 3x - 4.$$

$$= 9x^2 - 24x + 22 \qquad \text{Simplify.}$$

Thus, $(g \circ f)(x) = 9x^2 - 24x + 22$.

Notice that $f \circ g$ is not the same function as $g \circ f$. ∎

✓ **CHECK POINT 1**   Given $f(x) = 5x + 6$ and $g(x) = x^2 - 1$, find each of the following composite functions:

a. $(f \circ g)(x)$ **b.** $(g \circ f)(x)$.

Inverse Functions

Here are two functions that describe situations related to the price of a computer, x:

$$f(x) = x - 300 \qquad g(x) = x + 300.$$

Function f subtracts \$300 from the computer's price and function g adds \$300 to the computer's price. Let's see what $f(g(x))$ does. Put $g(x)$ into f:

$$f(x) = x - 300 \qquad\qquad \textit{This is the given equation for f.}$$

Replace x with $g(x)$.

$$f(g(x)) = g(x) - 300$$
$$= x + 300 - 300 \qquad \textit{Because } g(x) = x + 300,$$
$$\qquad\qquad\qquad\qquad \textit{replace } g(x) \textit{ with } x + 300.$$
$$= x.$$

This is the computer's original price.

By putting $g(x)$ into f and finding $f(g(x))$, we see that the computer's price, x, went through two changes: the first, an increase; the second, a decrease:

$$x + 300 - 300.$$

The final price of the computer, x, is identical to its starting price, x.

In general, if the changes made to x by function g are undone by the changes made by function f, then

$$f(g(x)) = x.$$

Assume, also, that this "undoing" takes place in the other direction:

$$g(f(x)) = x.$$

Under these conditions, we say that each function is the *inverse function* of the other. The fact that g is the inverse of f is expressed by renaming g as f^{-1}, read "f-inverse." For example, the inverse functions

$$f(x) = x - 300 \qquad g(x) = x + 300$$

are usually named as follows:

$$f(x) = x - 300 \qquad f^{-1}(x) = x + 300.$$

We can use partial tables of coordinates for f and f^{-1} to gain numerical insight into the relationship between a function and its inverse function.

Computer's regular price	\$300 reduction		Price with \$300 reduction	\$300 price increase
x	$f(x) = x - 300$		x	$f^{-1}(x) = x + 300$
1200	900		900	1200
1300	1000		1000	1300
1400	1100		1100	1400

Ordered pairs for f:
(1200, 900), (1300, 1000), (1400, 1100)

Ordered pairs for f^{-1}:
(900, 1200), (1000, 1300), (1100, 1400)

The tables illustrate that if a function f is the set of ordered pairs (x, y), then the inverse, f^{-1}, is the set of ordered pairs (y, x). Using these tables, we can see how one function's changes to x are undone by the other function:

$$(f^{-1} \circ f)(1300) = f^{-1}(f(1300)) = f^{-1}(1000) = 1300.$$

The table for f shows that $f(1300) = 1000$.

The table for f^{-1} shows that $f^{-1}(1000) = 1300$.

The final price of the computer, \$1300, is identical to its starting price, \$1300.

With these ideas in mind, we present the formal definition of the inverse of a function:

Definition of the Inverse of a Function

Let f and g be two functions such that

$$f(g(x)) = x \qquad \text{for every } x \text{ in the domain of } g$$

and

$$g(f(x)) = x \qquad \text{for every } x \text{ in the domain of } f.$$

The function g is the **inverse of the function f**, and is denoted by f^{-1} (read "f-inverse"). Thus, $f(f^{-1}(x)) = x$ and $f^{-1}(f(x)) = x$. The domain of f is equal to the range of f^{-1}, and vice versa.

Great Question!

Is the −1 in f^{-1} an exponent?

The notation f^{-1} represents the inverse function of f. The −1 is *not* an exponent. The notation f^{-1} does *not* mean $\dfrac{1}{f}$:

$$f^{-1} \neq \frac{1}{f}.$$

2 Verify inverse functions.

EXAMPLE 2   Verifying Inverse Functions

Show that each function is the inverse of the other:

$$f(x) = 5x \qquad \text{and} \qquad g(x) = \frac{x}{5}.$$

Solution   To show that f and g are inverses of each other, we must show that $f(g(x)) = x$ and $g(f(x)) = x$. We begin with $f(g(x))$.

$$f(x) = 5x \qquad \text{This is the given equation for } f.$$

Replace x with $g(x)$.

$$f(g(x)) = 5g(x) = 5\left(\frac{x}{5}\right) = x \qquad \text{Because } g(x) = \frac{x}{5}, \text{ replace } g(x) \text{ with } \frac{x}{5}. \text{ Then simplify.}$$

Next, we find $g(f(x))$.

$$g(x) = \frac{x}{5} \qquad \text{This is the given equation for } g.$$

Replace x with $f(x)$.

$$g(f(x)) = \frac{f(x)}{5} = \frac{5x}{5} = x \qquad \text{Because } f(x) = 5x, \text{ replace } f(x) \text{ with } 5x. \text{ Then simplify.}$$

Because g is the inverse of f (and vice versa), we can use inverse notation and write

$$f(x) = 5x \qquad \text{and} \qquad f^{-1}(x) = \frac{x}{5}.$$

Notice how f^{-1} undoes the change produced by f: f changes x by multiplying by 5 and f^{-1} undoes this change by dividing by 5. ∎

✓ **CHECK POINT 2**   Show that each function is the inverse of the other:

$$f(x) = 7x \qquad \text{and} \qquad g(x) = \frac{x}{7}.$$

EXAMPLE 3   Verifying Inverse Functions

Show that each function is the inverse of the other:

$$f(x) = 3x + 2 \qquad \text{and} \qquad g(x) = \frac{x - 2}{3}.$$

Solution   To show that f and g are inverses of each other, we must show that $f(g(x)) = x$ and $g(f(x)) = x$. We begin with $f(g(x))$.

$$f(x) = 3x + 2$$

Replace x with $g(x)$.

$$f(g(x)) = 3g(x) + 2 = 3\left(\frac{x-2}{3}\right) + 2 = (x - 2) + 2 = x$$

$$g(x) = \frac{x-2}{3}$$

Next, we find $g(f(x))$.

$$g(x) = \frac{x-2}{3}$$

Replace x with $f(x)$.

$$g(f(x)) = \frac{f(x) - 2}{3} = \frac{(3x+2) - 2}{3} = \frac{3x}{3} = x$$

$$f(x) = 3x + 2$$

Because g is the inverse of f (and vice versa), we can use inverse notation and write

$$f(x) = 3x + 2 \quad \text{and} \quad f^{-1}(x) = \frac{x-2}{3}.$$

Notice how f^{-1} undoes the changes produced by f: f changes x by *multiplying* by 3 and *adding* 2, and f^{-1} undoes this by *subtracting* 2 and *dividing* by 3. This "undoing" process is illustrated in **Figure 8.16**. ∎

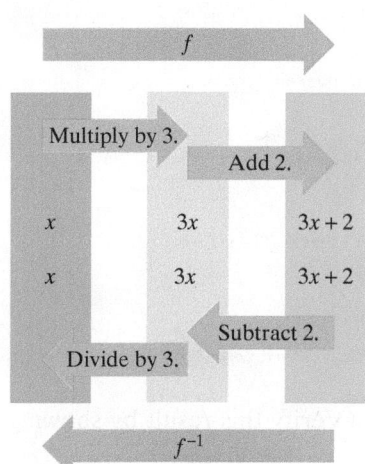

Figure 8.16 f^{-1} undoes the changes produced by f.

✓ **CHECK POINT 3**   Show that each function is the inverse of the other:

$$f(x) = 4x - 7 \quad \text{and} \quad g(x) = \frac{x+7}{4}.$$

---

3 Find the inverse of a function.

Finding the Inverse of a Function

The definition of the inverse of a function tells us that the domain of f is equal to the range of f^{-1}, and vice versa. This means that if the function f is the set of ordered pairs (x, y), then the inverse of f is the set of ordered pairs (y, x). If a function is defined by an equation, we can obtain the equation for f^{-1}, the inverse of f, by interchanging the role of x and y in the equation for the function f.

Finding the Inverse of a Function

The equation for the inverse of a function f can be found as follows:

1. Replace $f(x)$ with y in the equation for $f(x)$.
2. Interchange x and y.
3. Solve for y. If this equation does not define y as a function of x, the function f does not have an inverse function and this procedure ends. If this equation does define y as a function of x, the function f has an inverse function.
4. If f has an inverse function, replace y in step 3 with $f^{-1}(x)$. We can verify our result by showing that $f(f^{-1}(x)) = x$ and $f^{-1}(f(x)) = x$.

The procedure for finding a function's inverse uses a *switch-and-solve* strategy. Switch x and y, then solve for y.

EXAMPLE 4   Finding the Inverse of a Function

Find the inverse of $f(x) = 7x - 5$.

Solution

Step 1. Replace $f(x)$ with y:

$$y = 7x - 5.$$

Step 2. Interchange x and y:

$$x = 7y - 5. \quad \text{This is the inverse function.}$$

Step 3. Solve for y:

$$x + 5 = 7y \qquad \text{Add 5 to both sides.}$$

$$\frac{x + 5}{7} = y. \qquad \text{Divide both sides by 7.}$$

Step 4. Replace y with $f^{-1}(x)$:

$$f^{-1}(x) = \frac{x + 5}{7}. \qquad \text{The equation is written with } f^{-1} \text{ on the left.}$$

Thus, the inverse of $f(x) = 7x - 5$ is $f^{-1}(x) = \dfrac{x + 5}{7}$. (Verify this result by showing that $f(f^{-1}(x)) = x$ and $f^{-1}(f(x)) = x$.)

The inverse function, f^{-1}, undoes the changes produced by f. f changes x by multiplying by 7 and subtracting 5. f^{-1} undoes this by adding 5 and dividing by 7. ∎

☑ **CHECK POINT 4**   Find the inverse of $f(x) = 2x + 7$.

---

4 Use the horizontal line test to determine if a function has an inverse function.

The Horizontal Line Test and One-to-One Functions

Some functions do not have inverses that are functions. For example, consider the function $f(x) = x^2$, or $y = x^2$. We can use a few of the solutions of $y = x^2$ to illustrate numerically that this function does not have an inverse:

Four solutions of $y = x^2$.

$$(-2, 4), \quad (-1, 1), \quad (1, 1), \quad (2, 4),$$

Interchange x and y in each ordered pair.

$$(4, -2), \quad (1, -1), \quad (1, 1), \quad (4, 2).$$

The input 1 is associated with two outputs, −1 and 1.

The input 4 is associated with two outputs, −2 and 2.

A function provides exactly one output for each input. Thus, the ordered pairs in the bottom row do not define a function.

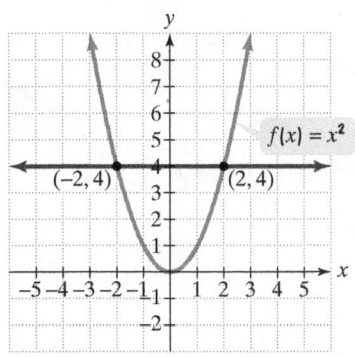

Figure 8.17 The horizontal line intersects the graph twice.

Can we look at the graph of a function and tell if it represents a function with an inverse? Yes. The graph of the quadratic function $f(x) = x^2$ is shown in **Figure 8.17**. Four units above the x-axis, a horizontal line is drawn. This line intersects the graph at two of its points, $(-2, 4)$ and $(2, 4)$. Inverse functions have ordered pairs with the coordinates reversed. We just saw what happened when we interchanged x and y. We obtained $(4, -2)$ and $(4, 2)$, and these ordered pairs do not define a function.

If any horizontal line, such as the one in **Figure 8.17**, intersects a graph at two or more points, the set of these points will not define a function when their coordinates are reversed. This suggests the **horizontal line test** for inverse functions.

The Horizontal Line Test for Inverse Functions

A function f has an inverse that is a function, f^{-1}, if there is no horizontal line that intersects the graph of the function f at more than one point.

EXAMPLE 5 Applying the Horizontal Line Test

Which of the following graphs represent functions that have inverse functions?

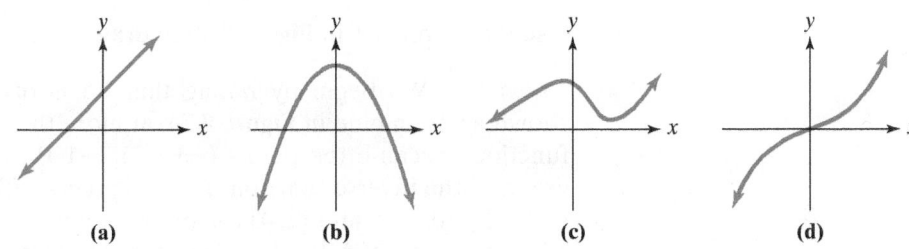

(a) (b) (c) (d)

Solution Notice that horizontal lines can be drawn in graphs **(b)** and **(c)** that intersect the graphs more than once. These graphs do not pass the horizontal line test. These are not the graphs of functions with inverse functions. By contrast, no horizontal line can be drawn in graphs **(a)** and **(d)** that intersects the graphs more than once. These graphs pass the horizontal line test. Thus, the graphs in parts **(a)** and **(d)** represent functions that have inverse functions.

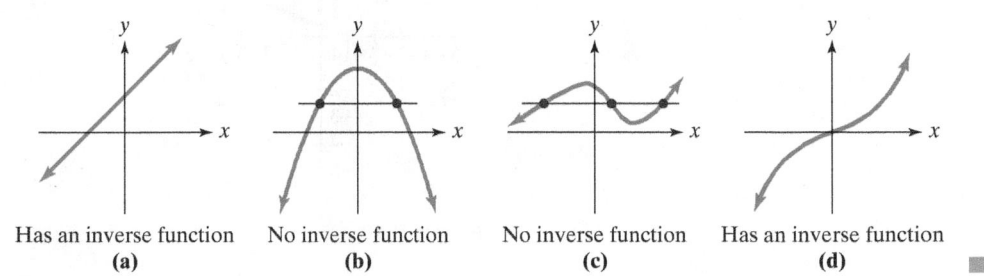

Has an inverse function No inverse function No inverse function Has an inverse function
(a) (b) (c) (d)

✓ **CHECK POINT 5** Which of the following graphs represent functions that have inverse functions?

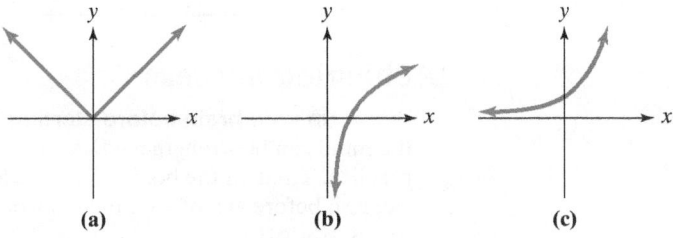

(a) (b) (c)

A function passes the horizontal line test when no two different ordered pairs have the same second component. This means that if $x_1 \neq x_2$, then $f(x_1) \neq f(x_2)$. Such a function is called a **one-to-one function**. Thus, **a one-to-one function is a function in which no two different ordered pairs have the same second component. Only**

one-to-one functions have inverse functions. Any function that passes the horizontal line test is a one-to-one function. Any one-to-one function has a graph that passes the horizontal line test.

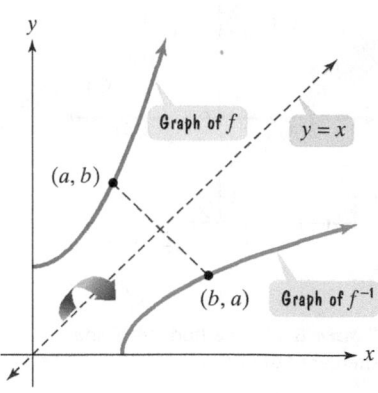

Figure 8.18 The graph of f^{-1} is a reflection of the graph of f about $y = x$.

5 Use the graph of a one-to-one function to graph its inverse function.

Graphs of f and f^{-1}

There is a relationship between the graph of a one-to-one function, f, and its inverse, f^{-1}. Because inverse functions have ordered pairs with the coordinates reversed, if the point (a, b) is on the graph of f, then the point (b, a) is on the graph of f^{-1}. The points (a, b) and (b, a) are symmetric with respect to the line $y = x$. Thus, **the graph of f^{-1} is a reflection of the graph of f about the line $y = x$.** This is illustrated in **Figure 8.18**.

> **EXAMPLE 6**   Graphing the Inverse Function

Use the graph of f in **Figure 8.19** to draw the graph of its inverse function.

Solution We begin by noting that no horizontal line intersects the graph of f, shown again in blue in **Figure 8.20**, at more than one point, so f does have an inverse function. Because the points $(-3, -2)$, $(-1, 0)$, and $(4, 2)$ are on the graph of f, the graph of the inverse function, f^{-1}, has points with these ordered pairs reversed. Thus, $(-2, -3)$, $(0, -1)$, and $(2, 4)$ are on the graph of f^{-1}. We can use these points to graph f^{-1}. The graph of f^{-1} is shown in green in **Figure 8.20**. Note that the green graph of f^{-1} is the reflection of the blue graph of f about the line $y = x$.

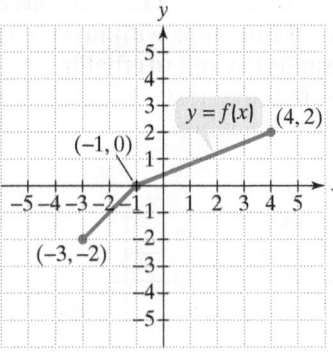

Figure 8.19

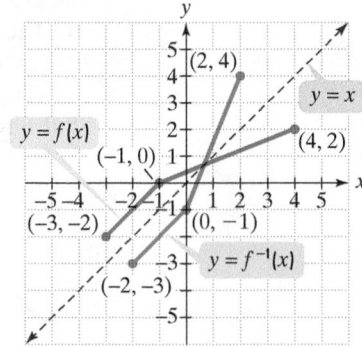

Figure 8.20 The graphs of f and f^{-1}

> ✓ **CHECK POINT 6**   The graph of function f consists of two line segments, one segment from $(-2, -2)$ to $(-1, 0)$ and a second segment from $(-1, 0)$ to $(1, 2)$. Graph f and use the graph to draw the graph of its inverse function.

Achieving Success

Warm up your brain before starting the assigned homework. Researchers say the mind can be strengthened, just like your muscles, with regular training and rigorous practice. Think of the book's Exercise Sets as brain calisthenics. If you're feeling a bit sluggish before any of your mental workouts, try this warmup:

> In the list below say the color the word is printed in, not the word itself. Once you can do this in 15 seconds without an error, the warmup is over and it's time to move on to the assigned exercises.

Blue Yellow Red Green Yellow Green Blue Red Yellow Red

> **CONCEPT AND VOCABULARY CHECK**

Fill in each blank so that the resulting statement is true.

1. The notation $f \circ g$, called the _____ of the function f with g, is defined by $(f \circ g)(x) =$ _____.

2. I find $(f \circ g)(x)$ by replacing each occurrence of x in the equation for _____ with _____.

3. The notation $g \circ f$, called the _____ of the function g with f, is defined by $(g \circ f)(x) =$ _____.

4. I find $(g \circ f)(x)$ by replacing each occurrence of x in the equation for _____ with _____.

5. True or false: $f \circ g$ is the same function as $g \circ f$. _____

6. True or false: $f(g(x)) = f(x) \cdot g(x)$ _____

7. The notation f^{-1} means the _____ of the function f.

8. If the function g is the inverse of the function f, then $f(g(x)) =$ _____ and $g(f(x)) =$ _____.

9. A function f has an inverse that is a function if there is no _____ line that intersects the graph of f at more than one point. Such a function is called a/an _____ function.

10. The graph of f^{-1} is a reflection of the graph of f about the line whose equation is _____.

> **8.4 EXERCISE SET** MyMathLab® Watch the videos in MyMathLab Download the MyDashBoard App

Practice Exercises

In Exercises 1–14, find

a. $(f \circ g)(x)$; b. $(g \circ f)(x)$; c. $(f \circ g)(2)$.

1. $f(x) = 2x$, $g(x) = x + 7$

2. $f(x) = 3x$, $g(x) = x - 5$

3. $f(x) = x + 4$, $g(x) = 2x + 1$

4. $f(x) = 5x + 2$, $g(x) = 3x - 4$

5. $f(x) = 4x - 3$, $g(x) = 5x^2 - 2$

6. $f(x) = 7x + 1$, $g(x) = 2x^2 - 9$

7. $f(x) = x^2 + 2$, $g(x) = x^2 - 2$

8. $f(x) = x^2 + 1$, $g(x) = x^2 - 3$

9. $f(x) = \sqrt{x}$, $g(x) = x - 1$

10. $f(x) = \sqrt{x}$, $g(x) = x + 2$

11. $f(x) = 2x - 3$, $g(x) = \dfrac{x + 3}{2}$

12. $f(x) = 6x - 3$, $g(x) = \dfrac{x + 3}{6}$

13. $f(x) = \dfrac{1}{x}$, $g(x) = \dfrac{1}{x}$

14. $f(x) = \dfrac{1}{x}$, $g(x) = \dfrac{2}{x}$

In Exercises 15–24, find $f(g(x))$ and $g(f(x))$ and determine whether each pair of functions f and g are inverses of each other.

15. $f(x) = 4x$ and $g(x) = \dfrac{x}{4}$

16. $f(x) = 6x$ and $g(x) = \dfrac{x}{6}$

17. $f(x) = 3x + 8$ and $g(x) = \dfrac{x - 8}{3}$

18. $f(x) = 4x + 9$ and $g(x) = \dfrac{x - 9}{4}$

19. $f(x) = 5x - 9$ and $g(x) = \dfrac{x + 5}{9}$

20. $f(x) = 3x - 7$ and $g(x) = \dfrac{x + 3}{7}$

21. $f(x) = \dfrac{3}{x - 4}$ and $g(x) = \dfrac{3}{x} + 4$

22. $f(x) = \dfrac{2}{x - 5}$ and $g(x) = \dfrac{2}{x} + 5$

23. $f(x) = -x$ and $g(x) = -x$

24. $f(x) = -x$ and $g(x) = x$

The functions in Exercises 25–34 are all one-to-one. For each function,

   a. *Find an equation for* $f^{-1}(x)$, *the inverse function.*

   b. *Verify that your equation is correct by showing that*
$f(f^{-1}(x)) = x$ *and* $f^{-1}(f(x)) = x$.

25. $f(x) = x + 3$

26. $f(x) = x + 5$

27. $f(x) = 2x$

28. $f(x) = 4x$

29. $f(x) = 2x + 3$

30. $f(x) = 3x - 1$

31. $f(x) = \dfrac{1}{x}$

32. $f(x) = \dfrac{2}{x}$

33. $f(x) = \dfrac{2x + 1}{x - 3}$

34. $f(x) = \dfrac{2x - 3}{x + 1}$

Which graphs in Exercises 35–40 represent functions that have inverse functions?

35.
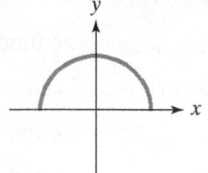

36.
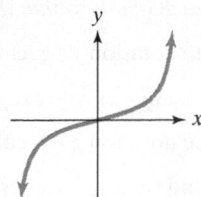

37.
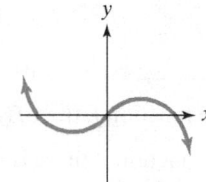

38.
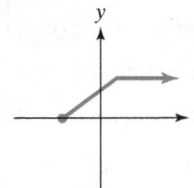

39.
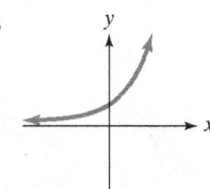

40.
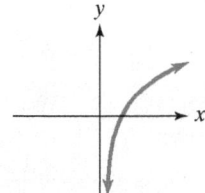

In Exercises 41–44, use the graph of f to draw the graph of its inverse function.

41.
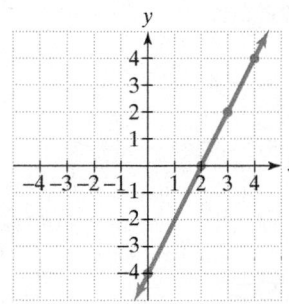

42.
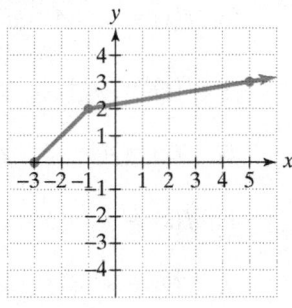

43.
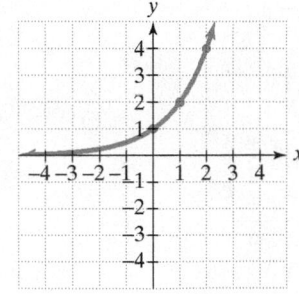

44.
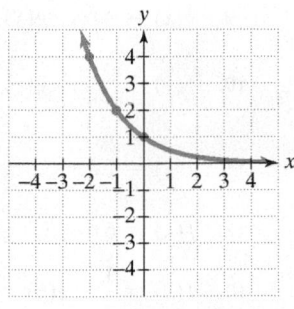

Practice PLUS

In Exercises 45–50, f and g are defined by the following tables. Use the tables to evaluate each composite function.

x	f(x)		x	g(x)
−1	1		−1	0
0	4		1	1
1	5		4	2
2	−1		10	−1

45. $f(g(1))$

46. $f(g(4))$

47. $(g \circ f)(-1)$

48. $(g \circ f)(0)$

49. $f^{-1}(g(10))$

50. $f^{-1}(g(1))$

In Exercises 51–54, use the graphs of f and g to evaluate each composite function.

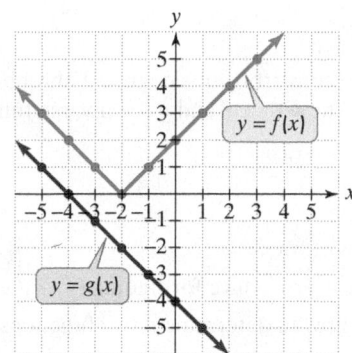

51. $(f \circ g)(-1)$

52. $(f \circ g)(1)$

53. $(g \circ f)(0)$

54. $(g \circ f)(-1)$

In Exercises 55–60, let

$$f(x) = 2x - 5$$
$$g(x) = 4x - 1$$
$$h(x) = x^2 + x + 2.$$

Evaluate the indicated function without finding an equation for the function.

55. $(f \circ g)(0)$

56. $(g \circ f)(0)$

57. $f^{-1}(1)$

58. $g^{-1}(7)$

59. $g(f[h(1)])$

60. $f(g[h(1)])$

Application Exercises

61. The regular price of a computer is x dollars. Let $f(x) = x - 400$ and $g(x) = 0.75x$.

    a. Describe what the functions f and g model in terms of the price of the computer.

    b. Find $(f \circ g)(x)$ and describe what this models in terms of the price of the computer.

    c. Repeat part (b) for $(g \circ f)(x)$.

    d. Which composite function models the greater discount on the computer, $f \circ g$ or $g \circ f$? Explain.

    e. Find f^{-1} and describe what this models in terms of the price of the computer.

62. The regular price of a pair of jeans is x dollars. Let $f(x) = x - 5$ and $g(x) = 0.6x$.

    a. Describe what functions f and g model in terms of the price of the jeans.

    b. Find $(f \circ g)(x)$ and describe what this models in terms of the price of the jeans.

    c. Repeat part (b) for $(g \circ f)(x)$.

    d. Which composite function models the greater discount on the jeans, $f \circ g$ or $g \circ f$? Explain.

    e. Find f^{-1} and describe what this models in terms of the price of the jeans.

Way to Go *Holland was the first country to establish an official bicycle policy. It currently has over 12,000 miles of paths and lanes exclusively for bicycles. The following graph shows the percentage of travel by bike and by car in Holland, as well as in four other selected countries. Use the information in the graph to solve Exercises 63–64.*

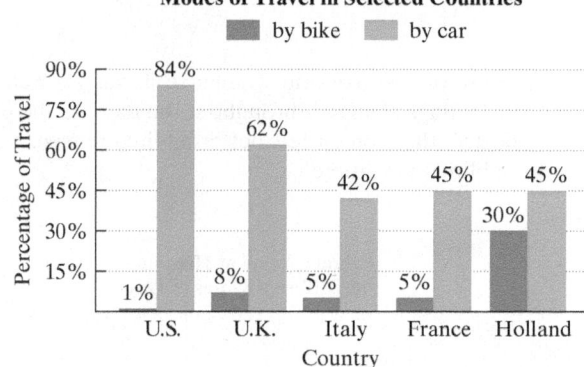

Source: EUROSTAT

63. a. Consider a function, f, whose domain is the set of the five countries shown in the graph. Let the range be the percentage of travel by bike in each of the respective countries. Write function f as a set of ordered pairs.

    b. Write the relation that is the inverse of f as a set of ordered pairs. Is this relation a function? Explain your answer.

64. **a.** Consider a function, f, whose domain is the set of the five countries shown in the graph on the previous page. Let the range be the percentage of travel by car in each of the respective countries. Write function f as a set of ordered pairs.

b. Write the relation that is the inverse of f as a set of ordered pairs. Is this relation a function? Explain your answer.

65. The graph represents the probability that two people in the same room share a birthday as a function of the number of people in the room. Call the function f.

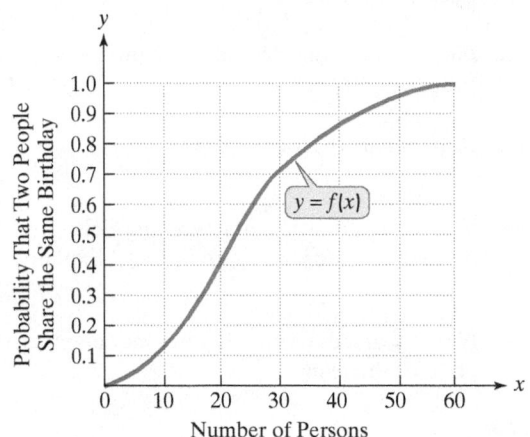

a. Explain why f has an inverse that is a function.

b. Describe in practical terms the meanings of $f^{-1}(0.25)$, $f^{-1}(0.5)$, and $f^{-1}(0.7)$.

66. A study of 900 working women in Texas showed that their feelings changed throughout the day. As the graph indicates, the women felt better as time passed, except for a blip at lunchtime.

Average Level of Happiness at Different Times of Day

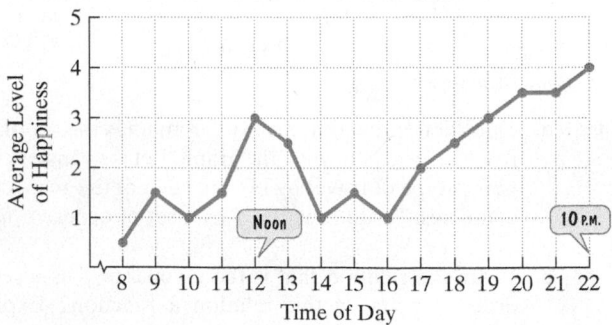

Source: D. Kahneman et al., "A Survey Method for Characterizing Daily Life Experience." *Science,* Vol. 306, No. 5702, Dec. 3, 2004, pp. 1776–1780.

a. Does the graph have an inverse that is a function? Explain your answer.

b. Identify two or more times of day when the average happiness level is 3. Express your answers as ordered pairs.

c. Do the ordered pairs in part (b) indicate that the graph represents a one-to-one function? Explain your answer.

67. The formula

$$y = f(x) = \frac{9}{5}x + 32$$

is used to convert from x degrees Celsius to y degrees Fahrenheit. The formula

$$y = g(x) = \frac{5}{9}(x - 32)$$

is used to convert from x degrees Fahrenheit to y degrees Celsius. Show that f and g are inverse functions.

Writing in Mathematics

68. Describe a procedure for finding $(f \circ g)(x)$.

69. Explain how to determine if two functions are inverses of each other.

70. Describe how to find the inverse of a one-to-one function.

71. What is the horizontal line test and what does it indicate?

72. Describe how to use the graph of a one-to-one function to draw the graph of its inverse function.

73. How can a graphing utility be used to visually determine if two functions are inverses of each other?

Technology Exercises

In Exercises 74–81, use a graphing utility to graph each function. Use the graph to determine whether the function has an inverse that is a function (that is, whether the function is one-to-one).

74. $f(x) = x^2 - 1$

75. $f(x) = \sqrt[3]{2 - x}$

76. $f(x) = \dfrac{x^3}{2}$

77. $f(x) = \dfrac{x^4}{4}$

78. $f(x) = |x - 2|$

79. $f(x) = (x - 1)^3$

80. $f(x) = -\sqrt{16 - x^2}$

81. $f(x) = x^3 + x + 1$

In Exercises 82–84, use a graphing utility to graph f and g in the same viewing rectangle. In addition, graph the line y = x and visually determine if f and g are inverses.

82. $f(x) = 4x + 4$, $g(x) = 0.25x - 1$

83. $f(x) = \dfrac{1}{x} + 2$, $g(x) = \dfrac{1}{x - 2}$

84. $f(x) = \sqrt[3]{x} - 2$, $g(x) = (x + 2)^3$

Critical Thinking Exercises

Make Sense? *In Exercises 85–88, determine whether each statement "makes sense" or "does not make sense" and explain your reasoning.*

85. This diagram illustrates that $f(g(x)) = x^2 + 4$.

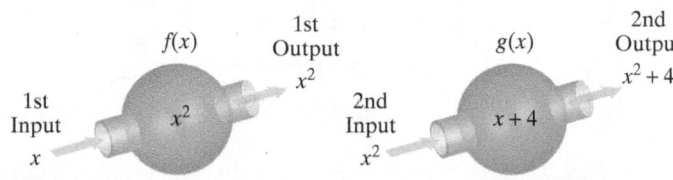

86. I must have made a mistake in finding the composite functions $f \circ g$ and $g \circ f$, because I notice that $f \circ g$ is the same function as $g \circ f$.

87. Regardless of what quadratic function I graph, the graph is shaped like a bowl or an inverted bowl, so this indicates that the quadratic function has an inverse function.

88. I'm working with the linear function $f(x) = 3x + 5$ and I do not need to find f^{-1} in order to determine the value of $(f \circ f^{-1})(17)$.

In Exercises 89–92, determine whether each statement is true or false. If the statement is false, make the necessary change(s) to produce a true statement.

89. The inverse of $\{(1, 4), (2, 7)\}$ is $\{(2, 7), (1, 4)\}$.

90. The function $f(x) = 5$ is one-to-one.

91. If $f(x) = \sqrt{x}$ and $g(x) = 2x - 1$, then $(f \circ g)(5) = g(2)$.

92. If $f(x) = 3x$, then $f^{-1}(x) = \dfrac{1}{3x}$.

93. If $f(x) = 3x$ and $g(x) = x + 5$, find $(f \circ g)^{-1}(x)$ and $(g^{-1} \circ f^{-1})(x)$.

94. Show that
$$f(x) = \frac{3x - 2}{5x - 3}$$
is its own inverse.

95. Consider the two functions defined by $f(x) = m_1x + b_1$ and $g(x) = m_2x + b_2$. Prove that the slope of the composite function of f with g is equal to the product of the slopes of the two functions.

Review Exercises

96. Divide and write the quotient in scientific notation:
$$\frac{4.3 \times 10^5}{8.6 \times 10^{-4}}.$$
(Section 5.7, Example 9)

97. Divide: $\dfrac{x^3 + 7x^2 - 2x + 3}{x - 2}$.
(Section 5.6, Example 2)

98. Solve:
$$\begin{cases} 3x + 2y = 6 \\ 8x + 3y = 1. \end{cases}$$
(Section 4.3, Example 4)

Preview Exercises

Exercises 99–101 will help you prepare for the material covered in the first section of the next chapter.

99. Solve: $2 - 12x = 7(x - 1)$.

100. Solve: $\dfrac{x + 3}{4} = \dfrac{x - 2}{3} + \dfrac{1}{4}$.

101. Solve and express the solution set in interval notation:
$$600x - (500,000 + 400x) > 0.$$

GROUP PROJECT

CHAPTER

8

The bar graphs in Exercises 63 and 64 in Exercise Set 8.4 (see page 631) illustrate that if a relation is a function, reversing the components in each of its ordered pairs may result in a relation that is no longer a function. Group members should find examples of bar graphs, like the ones in Exercises 63 and 64, that illustrate this idea. Consult almanacs, newsapaper, magazines, or the Internet. The group should select the graph with the most intriguing data. For the graph selected, write and solve a problem with two parts similar to Exercise 63 or 64.

Chapter 8 Summary

Definitions and Concepts	Examples

Section 8.1 Introduction to Functions

A relation is any set of ordered pairs. The set of first components of the ordered pairs is the domain and the set of second components is the range. A function is a relation in which each member of the domain corresponds to exactly one member of the range. No two ordered pairs of a function can have the same first component and different second components.

The domain of the relation $\{(1, 2), (3, 4), (3, 7)\}$ is $\{1, 3\}$. The range is $\{2, 4, 7\}$. The relation is not a function: 3, in the domain, corresponds to both 4 and 7 in the range.

If a function is defined by an equation, the notation $f(x)$, read "f of x" or "f at x," describes the value of the function at the number, or input, x.

If $f(x) = 7x - 5$, then
$$f(a + 2) = 7(a + 2) - 5$$
$$= 7a + 14 - 5$$
$$= 7a + 9.$$

Section 8.2 Graphs of Functions

The graph of a function is the graph of its ordered pairs.

The Vertical Line Test for Functions

If any vertical line intersects a graph in more than one point, the graph does not define y as a function of x.

At the left or right of a function's graph, you will often find closed dots, open dots, or arrows. A closed dot shows that the graph ends and the point belongs to the graph. An open dot shows that the graph ends and the point does not belong to the graph. An arrow indicates that the graph extends indefinitely.

The graph of a function can be used to determine the function's domain and its range. To find the domain, look for all the inputs on the x-axis that correspond to points on the graph. To find the range, look for all the outputs on the y-axis that correspond to points on the graph.

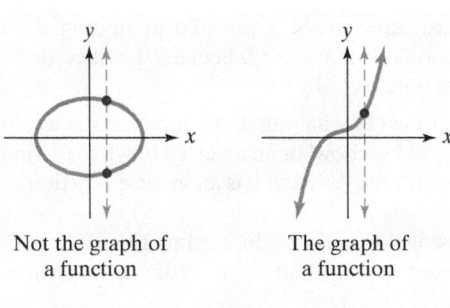

Not the graph of a function   The graph of a function

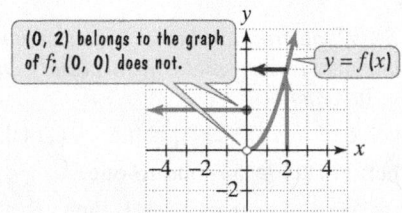

(0, 2) belongs to the graph of f; (0, 0) does not. $y = f(x)$

To find $f(2)$, locate 2 on the x-axis. The graph shows $f(2) = 4$.

Domain of $f = (-\infty, \infty)$

Range of $f = (0, \infty)$

Section 8.3 The Algebra of Functions

A Function's Domain

If a function f does not model data or verbal conditions, its domain is the largest set of real numbers for which the value of $f(x)$ is a real number. Exclude from a function's domain real numbers that cause division by zero and real numbers that result in a square root of a negative number.

$$f(x) = 7x + 13$$

Domain of $f = (-\infty, \infty)$

$$g(x) = \frac{7x}{12 - x}$$

Domain of $g = (-\infty, 12)$ or $(12, \infty)$

Definitions and Concepts	**Examples**

Section 8.3 The Algebra of Functions (continued)

The Algebra of Functions

Let f and g be two functions. The sum $f + g$, the difference $f - g$, the product fg, and the quotient $\dfrac{f}{g}$ are functions whose domains are the set of all real numbers common to the domains of f and g, defined as follows:

1. Sum: $(f + g)(x) = f(x) + g(x)$
2. Difference: $(f - g)(x) = f(x) - g(x)$
3. Product: $(fg)(x) = f(x) \cdot g(x)$
4. Quotient: $\left(\dfrac{f}{g}\right)(x) = \dfrac{f(x)}{g(x)},\ g(x) \neq 0.$

Let $f(x) = x^2 + 2x$ and $g(x) = 4 - x$.

- $(f + g)(x) = (x^2 + 2x) + (4 - x) = x^2 + x + 4$
  $(f + g)(-2) = (-2)^2 + (-2) + 4 = 4 - 2 + 4 = 6$
- $(f - g)(x) = (x^2 + 2x) - (4 - x) = x^2 + 2x - 4 + x$
  $\qquad\qquad\qquad = x^2 + 3x - 4$
  $(f - g)(5) = 5^2 + 3 \cdot 5 - 4 = 25 + 15 - 4 = 36$
- $(fg)(x) = (x^2 + 2x)(4 - x) = 4x^2 - x^3 + 8x - 2x^2$
  $\qquad\qquad = -x^3 + 2x^2 + 8x$
  $(fg)(1) = -1^3 + 2 \cdot 1^2 + 8 \cdot 1 = -1 + 2 + 8 = 9$
- $\left(\dfrac{f}{g}\right)(x) = \dfrac{x^2 + 2x}{4 - x},\ x \neq 4$
  $\left(\dfrac{f}{g}\right)(3) = \dfrac{3^2 + 2 \cdot 3}{4 - 3} = \dfrac{9 + 6}{1} = 15$

Section 8.4 Composite and Inverse Functions

Composite Functions

The composite function $f \circ g$ is defined by
$$(f \circ g)(x) = f(g(x)).$$
The composite function $g \circ f$ is defined by
$$(g \circ f)(x) = g(f(x)).$$

Let $f(x) = x^2 + x$ and $g(x) = 2x + 1$.

- $(f \circ g)(x) = f(g(x)) = (g(x))^2 + g(x)$

  Replace x with $g(x)$.

  $= (2x + 1)^2 + (2x + 1) = 4x^2 + 4x + 1 + 2x + 1$
  $= 4x^2 + 6x + 2$
- $(g \circ f)(x) = g(f(x)) = 2f(x) + 1$

  Replace x with $f(x)$.

  $= 2(x^2 + x) + 1 = 2x^2 + 2x + 1$

Inverse Functions

If $f(g(x)) = x$ and $g(f(x)) = x$, function g is the inverse of function f, denoted f^{-1} and read "f inverse." The procedure for finding a function's inverse uses a switch-and-solve strategy. Switch x and y, then solve for y.

If $f(x) = 2x - 5$, find $f^{-1}(x)$.

$\quad y = 2x - 5 \qquad$ Replace $f(x)$ with y.
$\quad x = 2y - 5 \qquad$ Exchange x and y.
$\quad x + 5 = 2y \qquad$ Solve for y.
$\quad \dfrac{x + 5}{2} = y$

$\quad f^{-1}(x) = \dfrac{x + 5}{2} \qquad$ Replace y with $f^{-1}(x)$.

The Horizontal Line Test for Inverse Functions

A function, f, has an inverse that is a function, f^{-1}, if there is no horizontal line that intersects the graph of f at more than one point. A one-to-one function is one in which no two different ordered pairs have the same second component. Only one-to-one functions have inverse functions. If the point (a, b) is on the graph of f, then the point (b, a) is on the graph of f^{-1}. The graph of f^{-1} is a reflection of the graph of f about the line $y = x$.

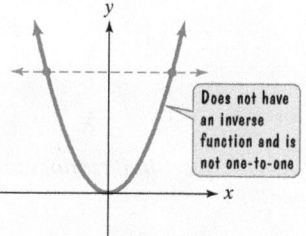

Does not have an inverse function and is not one-to-one

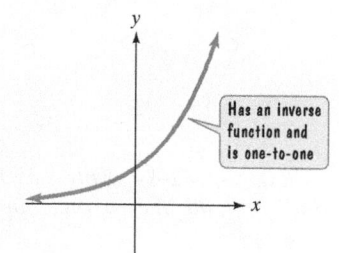

Has an inverse function and is one-to-one

CHAPTER 8 REVIEW EXERCISES

8.1 *In Exercises 1–3, determine whether each relation is a function. Give the domain and range for each relation.*

1. $\{(3, 10), (4, 10), (5, 10)\}$

2. $\{(1, 12), (2, 100), (3, \pi), (4, -6)\}$

3. $\{(13, 14), (15, 16), (13, 17)\}$

In Exercises 4–5, find the indicated function values.

4. $f(x) = 7x - 5$
   a. $f(0)$ b. $f(3)$ c. $f(-10)$
   d. $f(2a)$ e. $f(a + 2)$

5. $g(x) = 3x^2 - 5x + 2$
   a. $g(0)$ b. $g(5)$ c. $g(-4)$
   d. $g(b)$ e. $g(4a)$

8.2 *In Exercises 6–11, use the vertical line test to identify graphs in which y is a function of x.*

6.
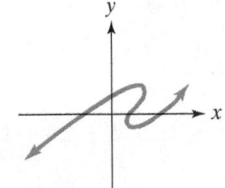

7.
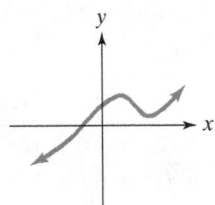

8.
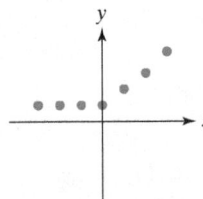

9.
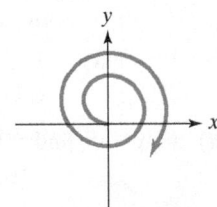

10.
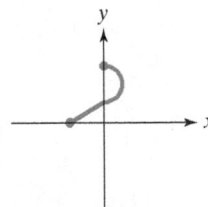

11.
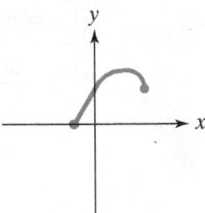

In Exercises 12–14, express each interval in set-builder notation and graph the interval on a number line.

12. $(-2, 3]$

13. $[-1.5, 2]$

14. $(-1, \infty)$

Use the graph of f to solve Exercises 15–19.

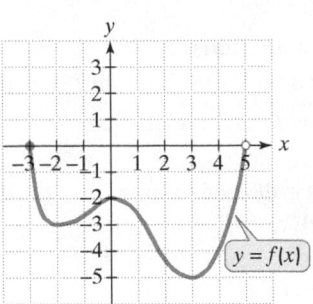

15. Find $f(-2)$.

16. Find $f(0)$.

17. For what value of x is $f(x) = -5$?

18. Find the domain of f.

19. Find the range of f.

20. The graph shows the height, in meters, of an eagle in terms of its time, in seconds, in flight.

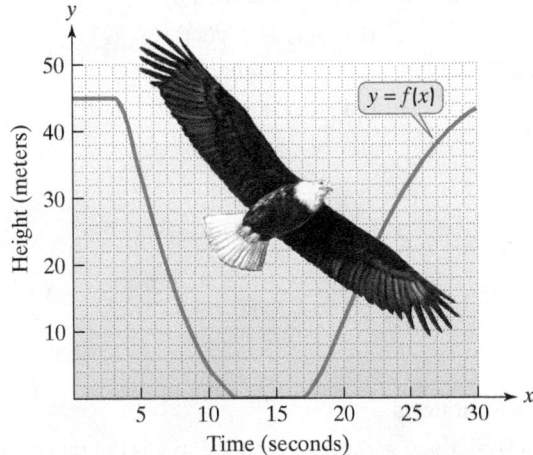

   a. Use the graph to explain why the eagle's height is a function of its time in flight.

   b. Find $f(15)$. Describe what this means in practical terms.

   c. What is a reasonable estimate of the eagle's maximum height?

   d. For what values of x is $f(x) = 20$? Describe what this means in practical terms.

   e. Use the graph of the function to write a description of the eagle's flight.

8.3 *In Exercises 21–23, find the domain of each function.*

21. $f(x) = 7x - 3$

22. $g(x) = \dfrac{1}{x + 8}$

23. $f(x) = x + \dfrac{3x}{x - 5}$

In Exercises 24–25, find **a.** $(f + g)(x)$ *and* **b.** $(f + g)(3)$.

24. $f(x) = 4x - 5, g(x) = 2x + 1$

25. $f(x) = 5x^2 - x + 4, g(x) = x - 3$

In Exercises 26–27, for each pair of functions, f and g, determine the domain of f + g.

26. $f(x) = 3x + 4, g(x) = \dfrac{5}{4 - x}$

27. $f(x) = \dfrac{7x}{x + 6}, g(x) = \dfrac{4}{x + 1}$

In Exercises 28–35, let

$$f(x) = x^2 - 2x \quad \text{and} \quad g(x) = x - 5.$$

Find each of the following.

28. $(f + g)(x)$ and $(f + g)(-2)$

29. $f(3) + g(3)$

30. $(f - g)(x)$ and $(f - g)(1)$

31. $f(4) - g(4)$

32. $(fg)(x)$ and $(fg)(-3)$

33. $\left(\dfrac{f}{g}\right)(x)$ and $\left(\dfrac{f}{g}\right)(4)$

34. The domain of $f - g$

35. The domain of $\dfrac{f}{g}$

8.4 *In Exercises 36–37, find* **a.** $(f \circ g)(x)$; **b.** $(g \circ f)(x)$; **c.** $(f \circ g)(3)$.

36. $f(x) = x^2 + 3, g(x) = 4x - 1$

37. $f(x) = \sqrt{x}, g(x) = x + 1$

In Exercises 38–39, find $f(g(x))$ and $g(f(x))$ and determine whether each pair of functions f and g are inverses of each other.

38. $f(x) = \dfrac{3}{5}x + \dfrac{1}{2}$ and $g(x) = \dfrac{5}{3}x - 2$

39. $f(x) = 2 - 5x$ and $g(x) = \dfrac{2 - x}{5}$

The functions in Exercises 40–41 are all one-to-one. For each function,

  a. *Find an equation of $f^{-1}(x)$, the inverse function.*

  b. *Verify that your equation is correct by showing that $f(f^{-1}(x)) = x$ and $f^{-1}(f(x)) = x$.*

40. $f(x) = 4x - 3$

41. $f(x) = -\dfrac{1}{x}$

Which graphs in Exercises 42–45 represent functions that have inverse functions?

42.

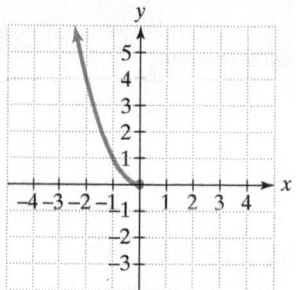

43.

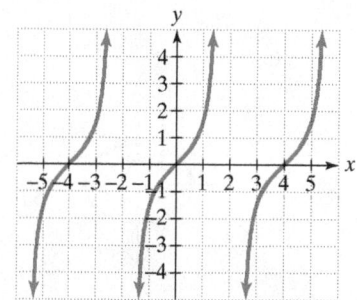

44.

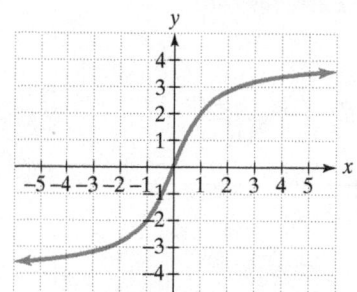

45.

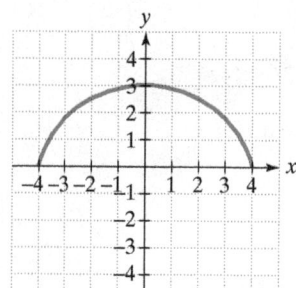

46. Use the graph of f in the figure shown to draw the graph of its inverse function.

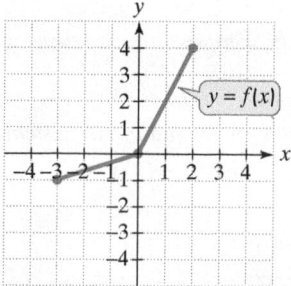

CHAPTER 8 TEST

Step-by-step test solutions are found on the Chapter Test Prep Videos available in MyMathLab® or on YouTube (search "BlitzerCombinedAlg" and click on "Channels").

In Exercises 1–2, determine whether each relation is a function. Give the domain and range for each relation.

1. $\{(1, 2), (3, 4), (5, 6), (6, 6)\}$

2. $\{(2, 1), (4, 3), (6, 5), (6, 6)\}$

3. If $f(x) = 3x - 2$, find $f(a + 4)$.

4. If $f(x) = 4x^2 - 3x + 6$, find $f(-2)$.

In Exercises 5–6, identify the graph or graphs in which y is a function of x.

5.

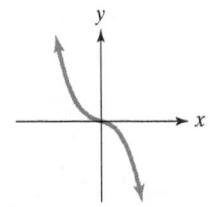

6.

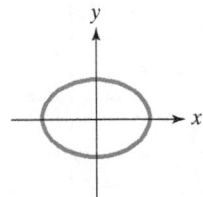

Use the graph of f to solve Exercises 7–10.

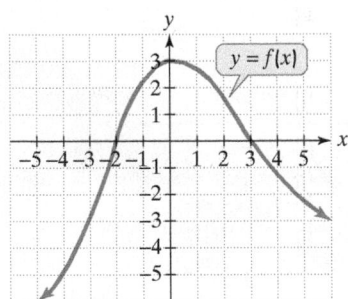

7. Find $f(6)$.

8. List two values of x for which $f(x) = 0$.

9. Find the domain of f.

10. Find the range of f.

11. Find the domain of $f(x) = \dfrac{6}{10 - x}$.

In Exercises 12–16, let

$$f(x) = x^2 + 4x \quad and \quad g(x) = x + 2.$$

Find each of the following.

12. $(f + g)(x)$ and $(f + g)(3)$

13. $(f - g)(x)$ and $(f - g)(-1)$

14. $(fg)(x)$ and $(fg)(-5)$

15. $\left(\dfrac{f}{g}\right)(x)$ and $\left(\dfrac{f}{g}\right)(2)$

16. The domain of $\dfrac{f}{g}$

17. If $f(x) = x^2 + x$ and $g(x) = 3x - 1$, find $(f \circ g)(x)$ and $(g \circ f)(x)$.

18. If $f(x) = 5x - 7$, find $f^{-1}(x)$.

19. A function f models the amount given to charity as a function of income. The graph of f is shown in the figure.

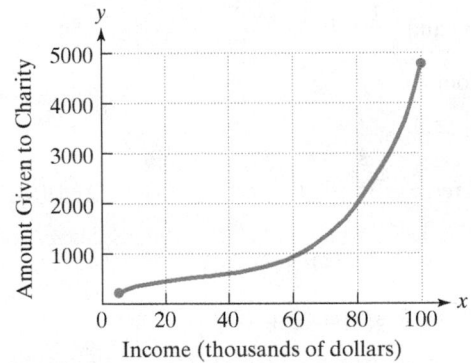

a. Explain why f has an inverse that is a function.

b. Find $f(80)$.

c. Describe in practical terms the meaning of $f^{-1}(2000)$.

CUMULATIVE REVIEW EXERCISES (CHAPTERS 1–8)

In Exercises 1–6, solve each equation or system of equations.

1. $2x + 3x - 5 + 7 = 10x + 3 - 6x - 4$

2. $2x^2 + 5x = 12$

3. $\begin{cases} 8x - 5y = -4 \\ 2x + 15y = -66 \end{cases}$

4. $\dfrac{15}{x} - 4 = \dfrac{6}{x} + 3$

5. $-3x - 7 = 8$

6. If $f(x) = 2x^2 - 5x + 2$ and $g(x) = x^2 - 2x + 3$, find $(f - g)(x)$ and $(f - g)(3)$.

In Exercises 7–11, simplify each expression.

7. $\dfrac{8x^3}{-4x^7}$

8. $-8 - (-3) \cdot 4$

9. $\dfrac{\dfrac{1}{x} - \dfrac{1}{2}}{\dfrac{1}{3} - \dfrac{x}{6}}$

10. $\dfrac{4 - x^2}{3x^2 - 5x - 2}$

11. $-5 - (-8) - (4 - 6)$

In Exercises 12–13, factor completely.

12. $x^2 - 18x + 77$

13. $x^3 - 25x$

In Exercises 14–17, perform the indicated operations. If possible, simplify the answer.

14. $\dfrac{6x^3 - 19x^2 + 16x - 4}{x - 2}$

15. $(2x - 3)(4x^2 + 6x + 9)$

16. $\dfrac{3x}{x^2 + x - 2} - \dfrac{2}{x + 2}$

17. $\dfrac{5x^2 - 6x + 1}{x^2 - 1} \div \dfrac{16x^2 - 9}{4x^2 + 7x + 3}$

18. Solve the system:
$$\begin{cases} x + 3y - z = 5 \\ -x + 2y + 3z = 13 \\ 2x - 5y - z = -8. \end{cases}$$

In Exercises 19–20, graph each equation in a rectangular coordinate system.

19. $2x - y = 4$

20. $y = -\dfrac{2}{3}x$

21. Is $\{(1, 5), (2, 5), (3, 5), (4, 5), (6, 5)\}$ a function? Give the relation's domain and range.

22. Find the slope of the line through $(-1, 5)$ and $(2, -3)$.

23. Write the point-slope form of the equation of the line with slope 5, passing through $(-2, -3)$. Then use the point-slope form of the equation to write the slope-intercept form of the line's equation.

24. Multiply and write the answer in scientific notation:
$$(7 \times 10^{-8})(3 \times 10^2).$$

25. Find the domain of $f(x) = \dfrac{1}{15 - x}$.

Inequalities and Problem Solving

Y ou are in Yosemite National Park in California, surrounded by evergreen forests, alpine meadows, and sheer walls of granite. The beauty of soaring cliffs, plunging waterfalls, gigantic trees, rugged canyons, mountains, and valleys is overwhelming. This is so different from where you live and attend college, a region in which grasslands predominate. What variables affect whether regions are forests, grasslands, or deserts, and what kinds of mathematical models are used to describe the incredibly diverse land that forms the surface of our planet?

The role that temperature and precipitation play in determining whether regions are forests, grasslands, or deserts can be modeled using inequalities in two variables. You will use these models and their graphs in Example 4 of Section 9.4.

9.1

Reviewing Linear Inequalities and Using Inequalities in Business Applications

Objectives

1 Review how to solve linear inequalities.

2 Use linear inequalities to solve problems involving revenue, cost, and profit.

Driving through your neighborhood, you see kids selling lemonade. Would it surprise you to know that this activity can be analyzed using linear inequalities? By doing so, you will view profit and loss in the business world in a new way. In this section, we use linear inequalities to solve problems and model business ventures.

1 Review how to solve linear inequalities.

Great Question!

Where can I find a more detailed presentation on solving linear inequalities?

See Section 2.7, Examples 3 through 8, on pages 186 through 190.

Solving Linear Inequalities in One Variable

We know that a linear equation in x can be expressed as $ax + b = 0$. A **linear inequality in x** can be written in one of the following forms: $ax + b < 0$, $ax + b \leq 0$, $ax + b > 0$, $ax + b \geq 0$. In each form, $a \neq 0$.

If an inequality does not contain fractions, it can be solved using the following procedure. (In Example 3, we will see how to clear fractions.) Notice, again, how similar this procedure is to the procedure for solving a linear equation.

Solving a Linear Inequality

1. Simplify the algebraic expression on each side.
2. Use the addition property of inequality to collect all the variable terms on one side and all the constant terms on the other side.
3. Use the multiplication property of inequality to isolate the variable and solve. Change the sense of the inequality when multiplying or dividing both sides by a negative number.
4. Express the solution set in interval notation and graph the solution set on a number line.

EXAMPLE 1 Solving a Linear Inequality

Solve and graph the solution set on a number line:

$$3x - 5 > -17.$$

Solution

Step 1. Simplify each side. Because each side is already simplified, we can skip this step.

Step 2. Collect variable terms on one side and constant terms on the other side. The variable term, $3x$, is already on the left side of $3x - 5 > -17$. We will collect constant terms on the right side by adding 5 to both sides.

$$3x - 5 > -17 \qquad \text{This is the given inequality.}$$
$$3x - 5 + 5 > -17 + 5 \qquad \text{Add 5 to both sides.}$$
$$3x > -12 \qquad \text{Simplify.}$$

Step 3. Isolate the variable and solve. We isolate the variable, x, by dividing both sides by 3. Because we are dividing by a positive number, we do not reverse the direction of the inequality symbol.

$$\frac{3x}{3} > \frac{-12}{3} \qquad \text{Divide both sides by 3.}$$

$$x > -4 \qquad \text{Simplify.}$$

Step 4. Express the solution set in interval notation and graph the set on a number line. The solution set consists of all real numbers that are greater than -4. The interval notation for this solution set is $(-4, \infty)$. The graph of the solution set is shown as follows:

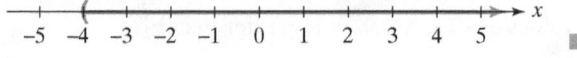

✓ **CHECK POINT 1**   Solve and graph the solution set on a number line:

$$4x - 3 > -23.$$

EXAMPLE 2   Solving a Linear Inequality

Solve and graph the solution set on a number line:

$$-2x - 4 > x + 5.$$

Solution

Step 1. Simplify each side. Because each side is already simplified, we can skip this step.

Step 2. Collect variable terms on one side and constant terms on the other side. We will collect variable terms on the left and constant terms on the right.

$$-2x - 4 > x + 5 \qquad \text{This is the given inequality.}$$

$$-2x - 4 - x > x + 5 - x \qquad \text{Subtract } x \text{ from both sides.}$$

$$-3x - 4 > 5 \qquad \text{Simplify.}$$

$$-3x - 4 + 4 > 5 + 4 \qquad \text{Add 4 to both sides.}$$

$$-3x > 9 \qquad \text{Simplify.}$$

Step 3. Isolate the variable and solve. We isolate the variable, x, by dividing both sides by -3. Because we are dividing by a negative number, we must reverse the direction of the inequality symbol.

$$\frac{-3x}{-3} < \frac{9}{-3} \qquad \begin{array}{l}\text{Divide both sides by } -3 \text{ and}\\ \text{change the sense of the inequality.}\end{array}$$

$$x < -3 \qquad \text{Simplify.}$$

Step 4. Express the solution set in interval notation and graph the set on a number line. The solution set consists of all real numbers that are less than -3. The interval notation for this solution set is $(-\infty, -3)$. The graph of the solution set is shown as follows:

Using Technology

Numeric and Graphic Connections

You can use a graphing utility to check the solution set of a linear inequality. Enter each side of the inequality separately under y_1 and y_2. Then use the table or the graphs. To use the table, first locate the x-value for which the y-values are the same. Then scroll up or down to locate x values for which y_1 is greater than y_2 or for which y_1 is less than y_2. To use the graphs, locate the intersection point and then find the x-values for which the graph of y_1 lies above the graph of y_2 ($y_1 > y_2$) or for which the graph of y_1 lies below the graph of y_2 ($y_1 < y_2$).

Let's verify our work in Example 2 and show that $(-\infty, -3)$ is the solution set of

$$-2x - 4 > x + 5.$$

Enter $y_1 = -2x - 4$ in the $\boxed{y=}$ screen.

Enter $y_2 = x + 5$ in the $\boxed{y=}$ screen.

We are looking for values of x for which y_1 is greater than y_2.

Numeric Check

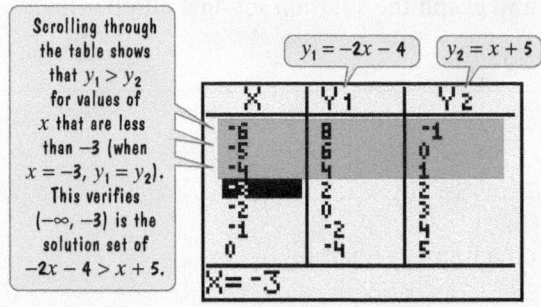

Scrolling through the table shows that $y_1 > y_2$ for values of x that are less than -3 (when $x = -3$, $y_1 = y_2$). This verifies $(-\infty, -3)$ is the solution set of $-2x - 4 > x + 5$.

Graphic Check

Display the graphs for y_1 and y_2. Use the intersection feature. The solution set is the set of x-values for which the graph of y_1 lies above the graph of y_2.

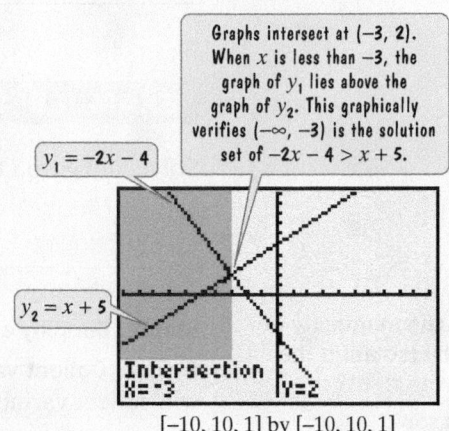

Graphs intersect at $(-3, 2)$. When x is less than -3, the graph of y_1 lies above the graph of y_2. This graphically verifies $(-\infty, -3)$ is the solution set of $-2x - 4 > x + 5$.

$[-10, 10, 1]$ by $[-10, 10, 1]$

✓ **CHECK POINT 2** Solve and graph the solution set: $3x + 1 > 7x - 15$.

If an inequality contains fractions, begin by multiplying both sides by the least common denominator. This will clear the inequality of fractions.

EXAMPLE 3 Solving a Linear Inequality Containing Fractions

Solve and graph the solution set on a number line:

$$\frac{x + 3}{4} \geq \frac{x - 2}{3} + \frac{1}{4}.$$

Solution The denominators are 4, 3, and 4. The least common denominator is 12. We begin by multiplying both sides of the inequality by 12.

$$\frac{x + 3}{4} \geq \frac{x - 2}{3} + \frac{1}{4} \qquad \text{This is the given inequality.}$$

$$12\left(\frac{x + 3}{4}\right) \geq 12\left(\frac{x - 2}{3} + \frac{1}{4}\right) \qquad \text{Multiply both sides by 12. Multiplying by a positive number preserves the sense of the inequality.}$$

$$\frac{12}{1} \cdot \frac{x+3}{4} \geq \frac{12}{1} \cdot \frac{x-2}{3} + \frac{12}{1} \cdot \frac{1}{4}$$ Multiply each term by 12. Use the distributive property on the right side.

$$\frac{\overset{3}{\cancel{12}}}{1} \cdot \frac{x+3}{\underset{1}{\cancel{4}}} \geq \frac{\overset{4}{\cancel{12}}}{1} \cdot \frac{x-2}{\underset{1}{\cancel{3}}} + \frac{\overset{3}{\cancel{12}}}{1} \cdot \frac{1}{\underset{1}{\cancel{4}}}$$ Divide out common factors in each multiplication.

$$3(x+3) > 4(x-2) + 3$$ The fractions are now cleared.

Now that the fractions have been cleared, we follow the four steps that we used in the previous examples.

Step 1. Simplify each side.

$$3(x+3) \geq 4(x-2) + 3$$ This is the inequality with the fractions cleared.

$$3x + 9 \geq 4x - 8 + 3$$ Use the distributive property.

$$3x + 9 \geq 4x - 5$$ Simplify.

Step 2. Collect variable terms on one side and constant terms on the other side. We will collect variable terms on the left and constant terms on the right.

$$3x + 9 - 4x \geq 4x - 5 - 4x$$ Subtract 4x from both sides.

$$-x + 9 \geq -5$$ Simplify.

$$-x + 9 - 9 \geq -5 - 9$$ Subtract 9 from both sides.

$$-x \geq -14$$ Simplify.

Step 3. Isolate the variable and solve. To isolate x, we must eliminate the negative sign in front of the x. Because $-x$ means $-1x$, we can do this by multiplying (or dividing) both sides of the inequality by -1. We are multiplying by a negative number. Thus, we must reverse the direction of the inequality symbol.

$$(-1)(-x) \leq (-1)(-14)$$ Multiply both sides by -1 and change the sense of the inequality.

$$x \leq 14$$ Simplify.

Step 4. Express the solution set in interval notation and graph the set on a number line. The solution set consists of all real numbers that are less than or equal to 14. The interval notation for this solution set is $(-\infty, 14]$. The graph of the solution set is shown as follows:

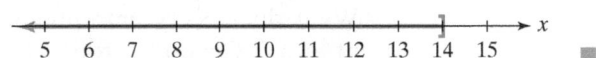

✓ **CHECK POINT 3** Solve and graph the solution set on a number line:
$$\frac{x-4}{2} \geq \frac{x-2}{3} + \frac{5}{6}.$$

---

2 Use linear inequalities to solve problems involving revenue, cost, and profit.

Functions of Business and Linear Inequalities

As a young entrepreneur, did you ever try selling lemonade in your front yard? Suppose that you charged 55 cents for each cup and you sold 45 cups. Your **revenue** is your income from selling these 45 units, or $0.55(45) = $24.75. Your *revenue function* from selling x cups is

$$R(x) = 0.55x.$$

This is the unit price: 55¢ for each cup.   This is the number of units sold.

For any business, the **revenue function**, $R(x)$, is the money generated by selling x units of the product:

$$R(x) = px.$$

Price per unit x units sold

Back to selling lemonade and energizing the neighborhood with white sugar: Is your revenue for the afternoon also your profit? No. We need to consider the cost of the business. You estimate that the lemons, white sugar, and bottled water cost 5 cents per cup. Furthermore, mommy dearest is charging you a $10 rental fee for use of your (her?) front yard. Your *cost function* for selling x cups of lemonade is

$$C(x) = 10 + 0.05x.$$

This is your   This is your variable cost:
$10 fixed cost.   5¢ for each cup produced.

For any business, the **cost function**, $C(x)$, is the cost of producing x units of the product:

$$C(x) = (\text{fixed cost}) + cx.$$

Cost per unit x units produced

The term on the right, cx, represents **variable cost**, because it varies based on the number of units produced. Thus, the cost function is the sum of the fixed cost and the variable cost.

Revenue and Cost Functions

A company produces and sells x units of a product.

Revenue Function

$$R(x) = (\text{price per unit sold})x$$

Cost Function

$$C(x) = \text{fixed cost} + (\text{cost per unit produced})x$$

What does every entrepreneur, from a kid selling lemonade to Donald Trump, want to do? Generate profit, of course. The *profit* made is the money taken in, or the revenue, minus the money spent, or the cost. This relationship between revenue and cost allows us to define the *profit function*, $P(x)$.

The Profit Function

The profit, $P(x)$, generated after producing and selling x units of a product is given by the **profit function**

$$P(x) = R(x) - C(x),$$

where R and C are the revenue and cost functions, respectively.

Figure 9.1 at the top of the next page shows the graphs of the revenue and cost functions for the lemonade business. Similar graphs and models apply no matter how small or large a business venture may be.

$$R(x) = 0.55x \qquad\qquad C(x) = 10 + 0.05x$$

Revenue is 55¢ times the   Cost is $10 plus 5¢ times the
number of cups sold.   number of cups produced.

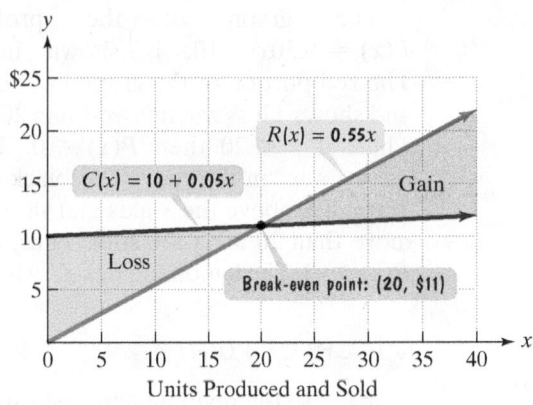

Figure 9.1

The lines intersect at the point (20, 11). This means that when 20 cups are produced and sold, both cost and revenue are $11. In business, this point of intersection is called the **break-even point**. At the break-even point, the money coming in is equal to the money going out. Can you see what happens for x-values less than 20? The red cost graph is above the blue revenue graph: $C(x) > R(x)$. The cost is greater than the revenue and the business is losing money. Thus, if you sell fewer than 20 cups of lemonade, the result is a *loss*. By contrast, look at what happens for x-values greater than 20. The blue revenue graph is above the red cost graph: $R(x) > C(x)$. The revenue is greater than the cost and the business is making money. Thus, if you sell more than 20 cups of lemonade, the result is a *gain*.

EXAMPLE 4   Writing a Profit Function to Determine a Profit

a. Use the revenue and cost functions shown in **Figure 9.1**,

$$R(x) = 0.55x \quad \text{and} \quad C(x) = 10 + 0.05x,$$

   to write the profit function for producing and selling x cups of lemonade.

b. Write and solve a linear inequality to answer this question: More than how many cups of lemonade must be sold for the business to make money?

Solution

a. The profit function is the difference between the revenue function, $R(x) = 0.55x$, and the cost function, $C(x) = 10 + 0.05x$.

$$
\begin{aligned}
P(x) &= R(x) - C(x) &&\text{This is the definition of the profit function.}\\
&= 0.55x - (10 + 0.05x) &&\text{Substitute the given functions.}\\
&= 0.55x - 10 - 0.05x &&\text{Distribute } -1 \text{ to each term in parentheses.}\\
&= 0.50x - 10 &&\text{Simplify.}
\end{aligned}
$$

   The profit function is $P(x) = 0.50x - 10$.

b. A business makes money, or realizes a gain, when $P(x) > 0$. A business loses money, or experiences a loss, when $P(x) < 0$. Because we are interested in making money, we want $P(x) > 0$.

$$
\begin{aligned}
0.50x - 10 &> 0 &&\text{This inequality models a gain.}\\
0.50x &> 10 &&\text{Add 10 to both sides.}\\
\frac{0.50x}{0.50} &> \frac{10}{0.50} &&\text{Divide both sides by 0.50. Division by a}\\
&&&\text{positive number does not reverse the}\\
&&&\text{sense of the inequality.}\\
x &> 20 &&\text{Simplify.}
\end{aligned}
$$

More than 20 cups of lemonade must be sold for the business to make money. ■

Discover for Yourself

A business makes money when revenue exceeds cost:

$$R(x) > C(x).$$

Use $R(x) = 0.55x$ and $C(x) = 10 + 0.05x$ to verify that more than 20 cups of lemonade must be sold for the business to make money.

The graph of the profit function, $P(x) = 0.50x - 10$, is shown in **Figure 9.2**. The red portion of the graph lies below the x-axis and shows a loss when fewer than 20 units are sold. Thus, if $x < 20$ then $P(x) < 0$. The lemonade business is "in the red." The black portion of the graph lies above the x-axis and shows a gain when more than 20 units are sold. Thus, if $x > 20$ then $P(x) > 0$. The lemonade business is "in the black."

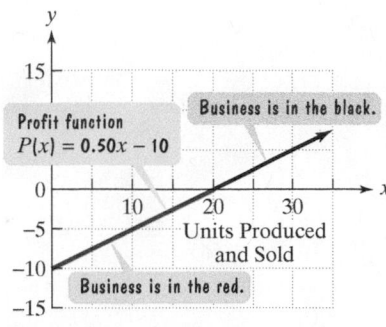

Figure 9.2 In the red and in the black

✓ **CHECK POINT 4**

a. Use the revenue and cost functions

$$R(x) = 200x$$
$$C(x) = 160{,}000 + 75x$$

to write the profit function for producing and selling x units.

b. More than how many units must be produced and sold for the business to make money?

 **EXAMPLE 5** Writing a Profit Function to Determine When a Business Makes Money

Technology is now promising to bring light, fast, and beautiful wheelchairs to millions of disabled people. A company is planning to manufacture these radically different wheelchairs. Fixed cost will be $500,000 and it will cost $400 to produce each wheelchair. Each wheelchair will be sold for $600.

a. Write the cost function, C, of producing x wheelchairs.

b. Write the revenue function, R, from the sale of x wheelchairs.

c. Write the profit function, P, from producing and selling x wheelchairs.

d. More than how many wheelchairs must be produced and sold for the business to make money?

Solution

a. The cost function is the sum of the fixed cost and variable cost.

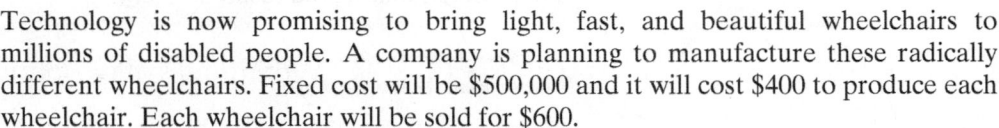

$$C(x) = 500{,}000 + 400x$$

b. The revenue function is the money generated from the sale of x wheelchairs.

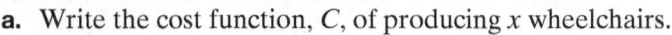

$$R(x) = 600x$$

c. The profit function is the difference between the revenue function and the cost function.

$P(x) = R(x) - C(x)$	This is the definition of the profit function.
$= 600x - (500{,}000 + 400x)$	Substitute the given functions.
$= 600x - 500{,}000 - 400x$	Distribute -1 to each term in parentheses.
$= 200x - 500{,}000$	Simplify: $600x - 400x = 200x$.

The profit function is $P(x) = 200x - 500{,}000$.

d. A business makes money when $P(x) > 0$. We use the profit function $P(x) = 200x - 500,000$ from part (c).

$$200x - 500,000 > 0 \qquad \text{Set the profit function greater than 0.}$$

$$200x > 500,000 \qquad \text{Add 500,000 to both sides.}$$

$$\frac{200x}{200} > \frac{500,000}{200} \qquad \text{Divide both sides by 200.}$$

$$x > 2500 \qquad \text{Simplify.}$$

More than 2500 wheelchairs must be produced and sold for the business to make money. ■

Figure 9.3 shows the graphs of the revenue and cost functions for the wheelchair business in Example 5.

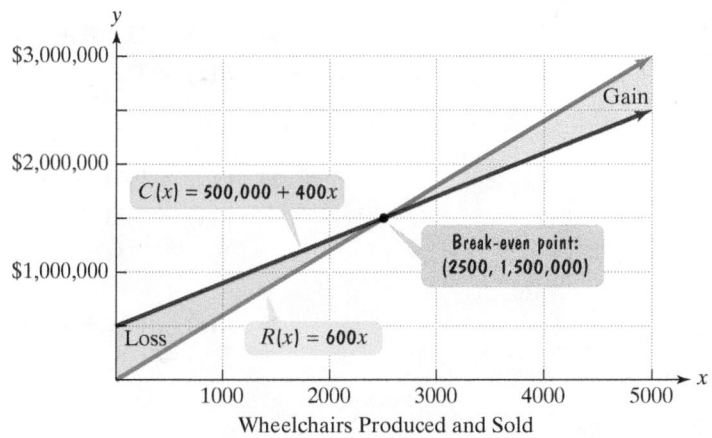

Figure 9.3

The intersection point indicates that the company breaks even by producing and selling 2500 wheelchairs. To the right of the intersection point, the graph of the blue revenue function lies above that of the red cost function. This confirms that the business makes money by producing and selling more than 2500 wheelchairs.

☑ **CHECK POINT 5** A company that manufactures running shoes has a fixed cost of $300,000. Additionally, it costs $30 to produce each pair of shoes. They are sold at $80 per pair.

a. Write the cost function, C, of producing x pairs of running shoes.

b. Write the revenue function, R, from the sale of x pairs of running shoes.

c. Write the profit function, P, from producing and selling x pairs of running shoes.

d. More than how many pairs of running shoes must be produced and sold for the business to make money?

CONCEPT AND VOCABULARY CHECK

Fill in each blank so that the resulting statement is true.

1. The linear inequality $-3x - 4 > 5$ can be solved by first _____ to both sides and then _____ both sides by _____, which changes the _____ of the inequality symbol from _____ to _____.

2. A company's _____ function is the money generated by selling x units of its product. The difference between this function and the company's cost function is called its _____ function.

3. A company has a graph that shows the money it generates by selling x units of its product. It also has a graph that shows its cost of producing x units of its product. The point of intersection of these graphs is called the company's _____.

9.1 EXERCISE SET MyMathLab®

Watch the videos
in MyMathLab

Download the
MyDashBoard App

Practice Exercises

In Exercises 1–26, solve each linear inequality and graph the solution set on a number line.

1. $5x + 11 < 26$
2. $2x + 5 < 17$
3. $3x - 8 \geq 13$
4. $8x - 2 \geq 14$
5. $-9x \geq 36$
6. $-5x \leq 30$
7. $8x - 11 \leq 3x - 13$
8. $18x + 45 \leq 12x - 8$
9. $4(x + 1) + 2 \geq 3x + 6$
10. $8x + 3 > 3(2x + 1) + x + 5$
11. $2x - 11 < -3(x + 2)$
12. $-4(x + 2) > 3x + 20$
13. $1 - (x + 3) \geq 4 - 2x$
14. $5(3 - x) \leq 3x - 1$
15. $\dfrac{x}{4} - \dfrac{1}{2} \leq \dfrac{x}{2} + 1$
16. $\dfrac{3x}{10} + 1 \geq \dfrac{1}{5} - \dfrac{x}{10}$
17. $1 - \dfrac{x}{2} > 4$
18. $7 - \dfrac{4}{5}x < \dfrac{3}{5}$
19. $\dfrac{x - 4}{6} \geq \dfrac{x - 2}{9} + \dfrac{5}{18}$
20. $\dfrac{4x - 3}{6} + 2 \geq \dfrac{2x - 1}{12}$
21. $7(x + 4) - 13 < 12 + 13(3 + x)$
22. $-3[7x - (2x - 3)] > -2(x + 1)$
23. $6 - \dfrac{2}{3}(3x - 12) \leq \dfrac{2}{5}(10x + 50)$
24. $\dfrac{2}{7}(7 - 21x) - 4 > 10 - \dfrac{3}{11}(11x - 11)$
25. $3[3(x + 5) + 8x + 7] + 5[3(x - 6)$
    $- 2(3x - 5)] < 2(4x + 3)$
26. $5[3(2 - 3x) - 2(5 - x)] - 6[5(x - 2)$
    $- 2(4x - 3)] < 3x + 19$
27. Let $f(x) = 3x + 2$ and $g(x) = 5x - 8$. Find all values of x for which $f(x) > g(x)$.
28. Let $f(x) = 2x - 9$ and $g(x) = 5x + 4$. Find all values of x for which $f(x) > g(x)$.
29. Let $f(x) = \dfrac{2}{5}(10x + 15)$ and $g(x) = \dfrac{1}{4}(8 - 12x)$. Find all values of x for which $g(x) \leq f(x)$.
30. Let $f(x) = \dfrac{3}{5}(10x - 15) + 9$ and $g(x) = \dfrac{3}{8}(16 - 8x) - 7$. Find all values of x for which $g(x) \leq f(x)$.
31. Let $f(x) = 1 - (x + 3) + 2x$. Find all values of x for which $f(x)$ is at least 4.
32. Let $f(x) = 2x - 11 + 3(x + 2)$. Find all values of x for which $f(x)$ is at most 0.

In Exercises 33–36, cost and revenue functions for producing and selling x units of a product are given. Cost and revenue are expressed in dollars.

a. Write the profit function from producing and selling x units of the product.

b. More than how many units must be produced and sold for the business to make money?

33. $C(x) = 25,500 + 15x$
    $R(x) = 32x$

34. $C(x) = 15,000 + 12x$
    $R(x) = 32x$

35. $C(x) = 105x + 70,000$
    $R(x) = 245x$

36. $C(x) = 1.2x + 1500$
    $R(x) = 1.7x$

Practice PLUS

In Exercises 37–38, solve each linear inequality and graph the solution set on a number line.

37. $2(x + 3) > 6 - \{4[x - (3x - 4) - x] + 4\}$
38. $3(4x - 6) < 4 - \{5x - [6x - (4x - (3x + 2))]\}$

In Exercises 39–40, write an inequality with x isolated on the left side that is equivalent to the given inequality.

39. $ax + b > c$; Assume $a < 0$.

40. $\dfrac{ax + b}{c} > b$; Assume $a > 0$ and $c < 0$.

In Exercises 41–42, use the graphs of y_1 and y_2 to solve each inequality.

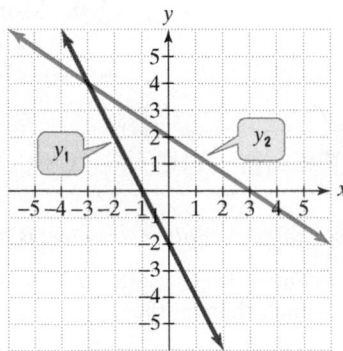

41. $y_1 \geq y_2$
42. $y_1 \leq y_2$

In Exercises 43–44, use the table of values for the linear functions y_1 and y_2 to solve each inequality.

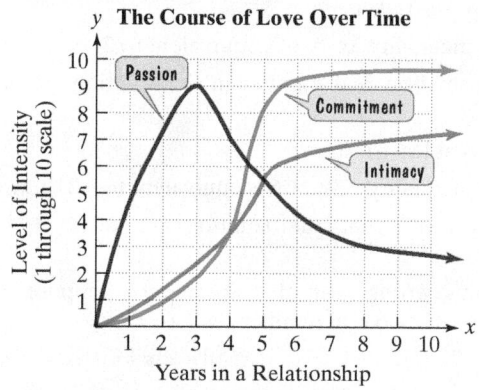

43. $y_1 < y_2$

44. $y_1 > y_2$

Application Exercises

The graphs show that the three components of love, namely passion, intimacy, and commitment, progress differently over time. Passion peaks early in a relationship and then declines. By contrast, intimacy and commitment build gradually. Use the graphs to solve Exercises 45–52.

The Course of Love Over Time

Source: R. J. Sternberg, A Triangular Theory of Love, *Psychological Review,* 93, 119–135.

45. Use interval notation to write an inequality that expresses for which years in a relationship intimacy is greater than commitment.

46. Use interval notation to write an inequality that expresses for which years in a relationship passion is greater than or equal to intimacy.

47. What is the relationship between passion and intimacy on the interval [5, 7)?

48. What is the relationship between intimacy and commitment on the interval [4, 7)?

49. What is the relationship between passion and commitment for $\{x \mid 6 < x < 8\}$?

50. What is the relationship between passion and commitment for $\{x \mid 7 < x < 9\}$?

51. What is the maximum level of intensity for passion? After how many years in a relationship does this occur?

52. After approximately how many years do levels of intensity for commitment exceed the maximum level of intensity for passion?

53. The percentage, P, of U.S. voters who use electronic voting systems, such as optical scans, in national elections can be modeled by the formula

$$P = 3.1x + 25.8,$$

where x is the number of years after 1994. In which years were more than 63% of U.S. voters using electronic systems?

54. The percentage, P, of U.S. voters who use punch cards or lever machines in national elections can be modeled by the formula

$$P = -2.5x + 63.1,$$

where x is the number of years after 1994. In which years were fewer than 38.1% of U.S. voters using punch cards or lever machines?

Exercises 55–58 describe a number of business ventures. For each exercise,

  a. *Write the cost function, C.*

  b. *Write the revenue function, R.*

  c. *Write the profit function, P.*

  d. *More than how many units must be produced and sold for the business to make money?*

55. A company that manufactures small canoes has a fixed cost of $18,000. It costs $20 to produce each canoe. The selling price is $80 per canoe. (In solving this exercise, let x represent the number of canoes produced and sold.)

56. A company that manufactures bicycles has a fixed cost of $100,000. It costs $100 to produce each bicycle. The selling price is $300 per bike. (In solving this exercise, let x represent the number of bicycles produced and sold.)

57. You invest in a new play. The cost includes an overhead of $30,000, plus production costs of $2500 per performance. A sold-out performance brings in $3125. (In solving this exercise, let x represent the number of sold-out performances.)

58. You invested $30,000 and started a business writing greeting cards. Supplies cost 2¢ per card and you are selling each card for 50¢. (In solving this exercise, let x represent the number of cards produced and sold.)

59. A company manufactures and sells blank audiocassette tapes. The weekly fixed cost is $10,000 and it costs $0.40 to produce each tape. The selling price is $2.00 per tape. How many tapes must be produced and sold each week for the company make money?

60. A company manufactures and sells personalized stationery. The weekly fixed cost is $3000 and it costs $3.00 to produce each package of stationery. The selling price is $5.50 per package. How many packages of stationery must be produced and sold each week for the company to make money?

61. You are choosing between two telephone plans. Plan A has a monthly fee of $15 with a charge of $0.08 per minute for all calls. Plan B has a monthly fee of $3 with a charge of $0.12 per minute for all calls. How many minutes of calls in a month make plan A the better deal?

62. A city commission has proposed two tax bills. The first bill requires that a homeowner pay $1800 plus 3% of the assessed home value in taxes. The second bill requires taxes of $200 plus 8% of the assessed home value. What price range of home assessment would make the first bill a better deal?

Writing in Mathematics

63. Describe a revenue function for a business venture.

64. Describe a cost function for a business venture. What are the two kinds of costs that are modeled by this function?

65. What is the profit function for a business venture and how is it determined?

66. If the profit function for a business venture is known, explain how to find how many units must be produced and sold for the business to make money.

67. What is the slope of each model in Exercises 53–54? What does this mean in terms of the percentage of U.S. voters using electronic voting systems and more traditional methods, such as punch cards or lever machines? What explanations can you offer for these changes in vote-counting systems?

Technology Exercises

In Exercises 68–69, solve each inequality using a graphing utility. Graph each side separately. Then determine the values of x for which the graph on the left side lies above the graph on the right side.

68. $-3(x - 6) > 2x - 2$

69. $-2(x + 4) > 6x + 16$

70. Use a graphing utility's $\boxed{\text{TABLE}}$ feature to verify your work in Exercises 68–69.

71. A bank offers two checking account plans. Plan A has a base service charge of $4.00 per month plus 10¢ per check. Plan B charges a base service charge of $2.00 per month plus 15¢ per check.

   a. Write models for the total monthly costs for each plan if x checks are written.

   b. Use a graphing utility to graph the models in the same [0, 50, 1] by [0, 10, 1] viewing rectangle.

   c. Use the graphs (and the intersection feature) to determine for what number of checks per month plan A will be better than plan B.

   d. Verify the result of part (c) algebraically by solving an inequality.

Critical Thinking Exercises

Make Sense? *In Exercises 72–75, determine whether each statement "makes sense" or "does not make sense" and explain your reasoning.*

72. I began the solution of $5 - 3(x + 2) > 10x$ by simplifying the left side, obtaining $2x + 4 > 10x$.

73. It's not a good thing for a business if $R(x) > C(x)$.

74. I determined the profit function for a business by subtracting its revenue function from its cost function.

75. A business is in the red when $P(x) < 0$.

In Exercises 76–79, determine whether each statement is true or false. If the statement is false, make the necessary change(s) to produce a true statement.

76. The inequality $3x > 6$ is equivalent to $2 > x$.

77. The smallest real number in the solution set of $2x > 6$ is 4.

78. If x is at least 7, then $x > 7$.

79. The inequality $-3x > 6$ is equivalent to $-2 > x$.

80. Find a so that the solution set of $ax + 4 \leq -12$ is $[8, \infty)$.

81. What's wrong with this argument? Suppose x and y represent two real numbers, where $x > y$.

$2 > 1$	This is a true statement.
$2(y - x) > 1(y - x)$	Multiply both sides by $y - x$.
$2y - 2x > y - x$	Use the distributive property.
$y - 2x > -x$	Subtract y from both sides.
$y > x$	Add 2x to both sides.

The final inequality, $y > x$, is impossible because we were initially given $x > y$.

Review Exercises

82. If $f(x) = x^2 - 2x + 5$, find $f(-4)$.
(Section 8.1, Example 3)

83. Solve the system:
$$\begin{cases} 2x - y - z = -3 \\ 3x - 2y - 2z = -5 \\ -x + y + 2z = 4. \end{cases}$$
(Section 4.5, Example 2)

84. Factor: $25x^2 - 81$. (Section 6.4, Example 1)

Preview Exercises

Exercises 85–87 will help you prepare for the material covered in the next section.

85. Consider the sets $A = \{1, 2, 3, 4\}$ and $B = \{3, 4, 5, 6, 7\}$.

    a. Write the set consisting of elements common to both set A and set B.

    b. Write the set consisting of elements that are members of set A or of set B or of both sets.

86. a. Solve: $x - 3 < 5$.

    b. Solve: $2x + 4 < 14$.

    c. Give an example of a number that satisfies the inequality in part (a) and the inequality in part (b).

d. Give an example of a number that satisfies the inequality in part (a), but not the inequality in part (b).

87. a. Solve: $2x - 6 \geq -4$.

    b. Solve: $5x + 2 \geq 17$.

    c. Give an example of a number that satisfies the inequality in part (a) and the inequality in part (b).

d. Give an example of a number that satisfies the inequality in part (a), but not the inequality in part (b).

Compound Inequalities

Objectives

1 Find the intersection of two sets.

2 Solve compound inequalities involving *and*.

3 Find the union of two sets.

4 Solve compound inequalities involving *or*.

Among the U.S. presidents, more have had dogs than cats in the White House. Sets of presidents with dogs, cats, dogs and cats, dogs or cats, and neither, with the number of presidents belonging to each of these sets, are shown in **Figure 9.4**.

$17 + 9 + 2 = 28$ presidents had dogs OR cats.

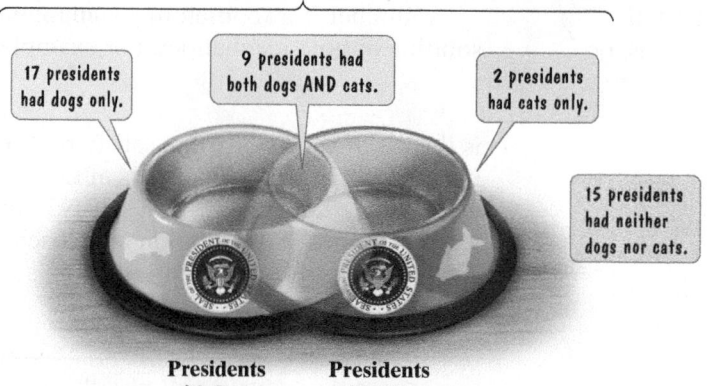

17 presidents had dogs only.

9 presidents had both dogs AND cats.

2 presidents had cats only.

15 presidents had neither dogs nor cats.

Presidents with Dogs　**Presidents with Cats**

Figure 9.4 Presidential pets in the White House

Source: factmonster.com

Figure 9.4 indicates that 9 presidents had dogs *and* cats in the White House. Furthermore, 28 presidents had dogs *or* cats in the White House. In this section, we'll focus on sets joined with the word *and* and the word *or*. The difference is that we'll be using solution sets of inequalities rather than sets of presidents with pets.

A **compound inequality** is formed by joining two inequalities with the word *and* or the word *or*.

Examples of Compound Inequalities

- $x - 3 < 5$ and $2x + 4 < 14$
- $3x - 5 \leq 13$ or $5x + 2 > -3$

Compound inequalities illustrate the importance of the words *and* and *or* in mathematics, as well as in everyday English.

1 Find the intersection of two sets.

Compound Inequalities Involving *And*

If A and B are sets, we can form a new set consisting of all elements that are in both A and B. This set is called the *intersection* of the two sets.

Definition of the Intersection of Sets

The **intersection** of sets A and B, written $A \cap B$, is the set of elements common to both set A **and** set B. This definition can be expressed in set-builder notation as follows:

$$A \cap B = \{x \,|\, x \in A \text{ and } x \in B\}.$$

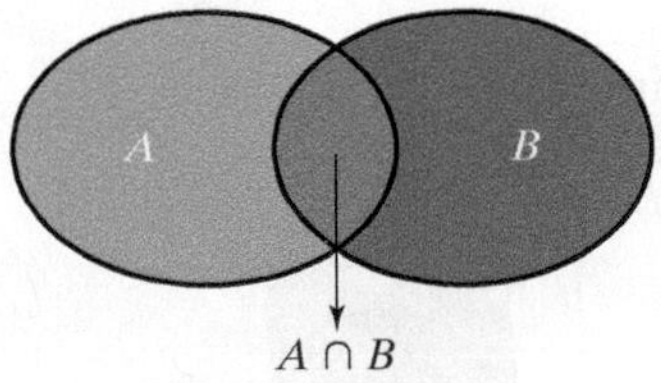

Figure 9.5 shows a useful way of picturing the intersection of sets A and B. The figure indicates that $A \cap B$ contains those elements that belong to both A and B at the same time.

$A \cap B$

Figure 9.5 Picturing the intersection of two sets

EXAMPLE 1 Finding the Intersection of Two Sets

Find the intersection: $\{7, 8, 9, 10, 11\} \cap \{6, 8, 10, 12\}$.

Solution The elements common to $\{7, 8, 9, 10, 11\}$ and $\{6, 8, 10, 12\}$ are 8 and 10. Thus,

$$\{7, 8, 9, 10, 11\} \cap \{6, 8, 10, 12\} = \{8, 10\}. \blacksquare$$

✓ **CHECK POINT 1** Find the intersection: $\{3, 4, 5, 6, 7\} \cap \{3, 7, 8, 9\}$.

2 Solve compound inequalities involving *and*.

A number is a **solution of a compound inequality formed by the word *and*** if it is a solution of both inequalities. For example, the solution set of the compound inequality

$$x \leq 6 \quad \text{and} \quad x \geq 2$$

is the set of values of x that satisfy both $x \leq 6$ and $x \geq 2$. Thus, the solution set is the intersection of the solution sets of the two inequalities.

What are the numbers that satisfy both $x \leq 6$ and $x \geq 2$? These numbers are easier to see if we graph the solution set to each inequality on a number line. These graphs are shown in **Figure 9.6**. The intersection is shown in the third graph.

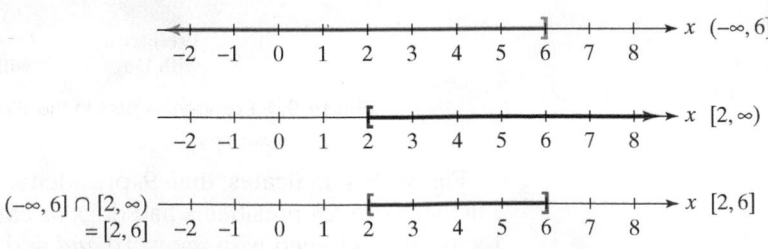

Figure 9.6 Numbers satisfying both $x \leq 6$ and $x \geq 2$

The numbers common to both sets are those that are less than or equal to 6 and greater than or equal to 2. This set in interval notation is [2, 6].

Here is a procedure for finding the solution set of a compound inequality containing the word *and*.

Solving Compound Inequalities Involving *AND*

1. Solve each inequality separately.
2. Graph the solution set to each inequality on a number line and take the intersection of these solution sets. This intersection appears as the portion of the number line that the two graphs have in common.

EXAMPLE 2 Solving a Compound Inequality with *And*

Solve: $x - 3 < 5$ and $2x + 4 < 14$.

Solution

Step 1. Solve each inequality separately.

$$x - 3 < 5 \quad \text{and} \quad 2x + 4 < 14$$
$$x < 8 \qquad\qquad 2x < 10$$
$$\qquad\qquad\qquad x < 5$$

Step 2. Take the intersection of the solution sets of the two inequalities. We graph the solution sets of $x < 8$ and $x < 5$. The intersection is shown in the third graph.

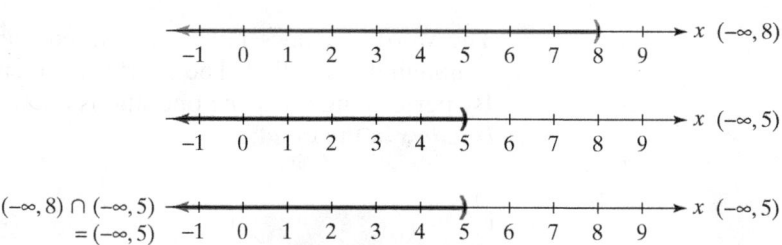

The numbers common to both sets are those that are less than 5. The solution set in interval notation is $(-\infty, 5)$. Take a moment to check that any number in $(-\infty, 5)$ satisfies both of the original inequalities. ∎

✓ **CHECK POINT 2** Solve: $x + 2 < 5$ and $2x - 4 < -2$.

EXAMPLE 3 Solving a Compound Inequality with *And*

Solve: $2x - 7 > 3$ and $5x - 4 < 6$.

Solution

Step 1. Solve each inequality separately.

$$2x - 7 > 3 \quad \text{and} \quad 5x - 4 < 6$$
$$2x > 10 \qquad\qquad 5x < 10$$
$$x > 5 \qquad\qquad\quad x < 2$$

Step 2. Take the intersection of the solution sets of the two inequalities. We graph the solution sets of $x > 5$ and $x < 2$. We use these graphs to find their intersection.

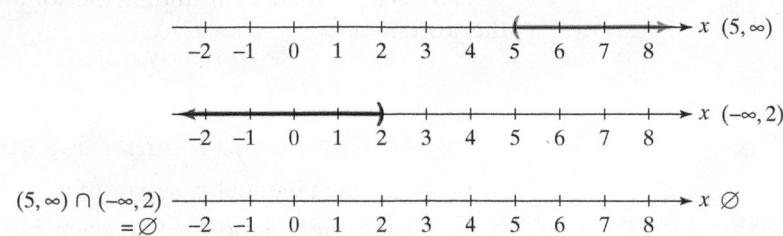

There is no number that is both greater than 5 and at the same time less than 2. Thus, the solution set is the empty set, $\varnothing$. ■

✓ **CHECK POINT 3**   Solve: $4x - 5 > 7$ and $5x - 2 < 3$.

If $a < b$, the compound inequality

$$a < x \text{ and } x < b$$

can be written in the shorter form

$$a < x < b.$$

For example, the compound inequality

$$-3 < 2x + 1 \text{ and } 2x + 1 < 3$$

can be abbreviated

$$-3 < 2x + 1 < 3.$$

The word *and* does not appear when the inequality is written in the shorter form, although it is implied. The shorter form enables us to solve both inequalities at once. By performing the same operations on all three parts of the inequality, our goal is to **isolate x in the middle**.

EXAMPLE 4   Solving a Compound Inequality

Solve and graph the solution set:

$$-3 < 2x + 1 \leq 3.$$

Solution   We would like to isolate x in the middle. We can do this by first subtracting 1 from all three parts of the compound inequality. Then we isolate x from $2x$ by dividing all three parts of the inequality by 2.

$$-3 < 2x + 1 \leq 3 \qquad \text{This is the given inequality.}$$
$$-3 - 1 < 2x + 1 - 1 \leq 3 - 1 \qquad \text{Subtract 1 from all three parts.}$$
$$-4 < 2x \leq 2 \qquad \text{Simplify.}$$
$$\frac{-4}{2} < \frac{2x}{2} \leq \frac{2}{2} \qquad \text{Divide each part by 2.}$$
$$-2 < x \leq 1 \qquad \text{Simplify.}$$

The solution set consists of all real numbers greater than -2 and less than or equal to 1, represented by $(-2, 1]$ in interval notation. The graph is shown as follows:

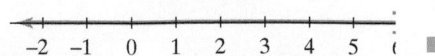

Using Technology

Numeric and Graphic Connections

Let's verify our work in Example 4 and show that $(-2, 1]$ is the solution set of $-3 < 2x + 1 \le 3$.

Numeric Check

To check numerically, enter $y_1 = 2x + 1$.

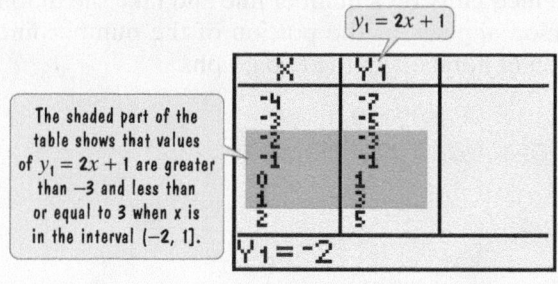

The shaded part of the table shows that values of $y_1 = 2x + 1$ are greater than -3 and less than or equal to 3 when x is in the interval $(-2, 1]$.

Graphic Check

To check graphically, graph each part of

$$-3 < 2x + 1 \le 3.$$

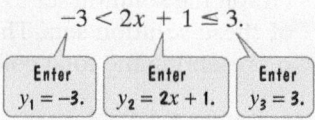

Enter $y_1 = -3.$ Enter $y_2 = 2x + 1.$ Enter $y_3 = 3.$

The figure shows that the graph of $y_2 = 2x + 1$ lies above the graph of $y_1 = -3$ and on or below the graph of $y_3 = 3$ when x is in the interval $(-2, 1]$.

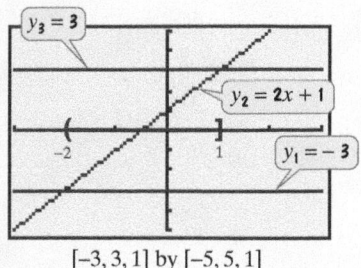

$[-3, 3, 1]$ by $[-5, 5, 1]$

✓ **CHECK POINT 4** Solve and graph the solution set: $1 \le 2x + 3 < 11$.

---

3 Find the union of two sets.

Compound Inequalities Involving *Or*

Another set that we can form from sets A and B consists of elements that are in A or B or in both sets. This set is called the *union* of the two sets.

Definition of the Union of Sets

The **union** of sets A and B, written $A \cup B$, is the set of elements that are members of set A **or** of set B or of both sets. This definition can be expressed in set-builder notation as follows:

$$A \cup B = \{x \,|\, x \in A \text{ or } x \in B\}.$$

Figure 9.7 shows a useful way of picturing the union of sets A and B. The figure indicates that $A \cup B$ is formed by joining the sets together.

We can find the union of set A and set B by listing the elements of set A. Then, we include any elements of set B that have not already been listed. Enclose all elements that are listed with braces. This shows that the union of two sets is also a set.

EXAMPLE 5 Finding the Union of Two Sets

Find the union: $\{7, 8, 9, 10, 11\} \cup \{6, 8, 10, 12\}$.

Solution To find $\{7, 8, 9, 10, 11\} \cup \{6, 8, 10, 12\}$, start by listing all the elements from the first set, namely 7, 8, 9, 10, and 11. Now list all the elements from the second set that are not in the first set, namely 6 and 12. The union is the set consisting of all these elements. Thus,

$$\{7, 8, 9, 10, 11\} \cup \{6, 8, 10, 12\} = \{6, 7, 8, 9, 10, 11, 12\}.$$

Although 8 and 10 appear in both sets,   do not list 8 and 10 twice. ∎

✓ **CHECK POINT 5** Find the union: $\{3, 4, 5, 6, 7\} \cup \{3, 7, 8, 9\}$.

A ∪ B

Figure 9.7 Picturing the union of two sets

Great Question!

How can I use the words *union* and *intersection* to help me distinguish between these two operations?

Union, as in a marriage union, suggests joining things, or uniting them. Intersection, as in the intersection of two crossing streets, brings to mind the area common to both, suggesting things that overlap.

4 Solve compound inequalities involving *or*.

A number is a **solution of a compound inequality formed by the word *or*** if it is a solution of either inequality. Thus, the solution set of a compound inequality formed by the word *or* is the union of the solution sets of the two inequalities.

Solving Compound Inequalities Involving *OR*

1. Solve each inequality separately.
2. Graph the solution set to each inequality on a number line and take the union of these solution sets. This union appears as the portion of the number line representing the total collection of numbers in the two graphs.

EXAMPLE 6 Solving a Compound Inequality with *Or*

Solve: $2x - 3 < 7$ or $35 - 4x \leq 3$.

Solution

Step 1. Solve each inequality separately.

$$2x - 3 < 7 \quad \text{or} \quad 35 - 4x \leq 3$$
$$2x < 10 \qquad \qquad -4x \leq -32$$
$$x < 5 \qquad \qquad \quad x \geq 8$$

Step 2. Take the union of the solution sets of the two inequalities. We graph the solution sets of $x < 5$ and $x \geq 8$. We use these graphs to find their union.

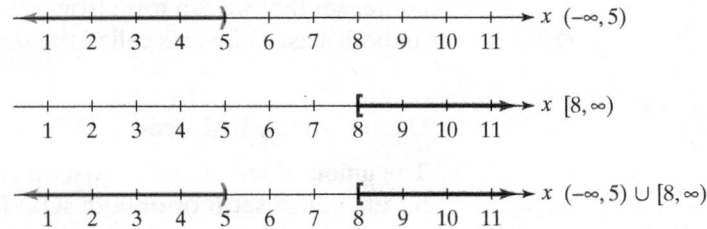

The solution set consists of all numbers that are less than 5 or greater than or equal to 8. The solution set in interval notation is $(-\infty, 5) \cup [8, \infty)$. There is no shortcut way to express this union when interval notation is used. ∎

✓ **CHECK POINT 6** Solve: $3x - 5 \leq -2$ or $10 - 2x < 4$.

EXAMPLE 7 Solving a Compound Inequality with *Or*

Solve: $3x - 5 \leq 13$ or $5x + 2 > -3$.

Solution

Step 1. Solve each inequality separately.

$$3x - 5 \leq 13 \quad \text{or} \quad 5x + 2 > -3$$
$$3x \leq 18 \qquad \qquad 5x > -5$$
$$x \leq 6 \qquad \qquad \quad x > -1$$

Step 2. Take the union of the solution sets of the two inequalities. We graph the solution sets of $x \leq 6$ and $x > -1$. We use these graphs to find their union.

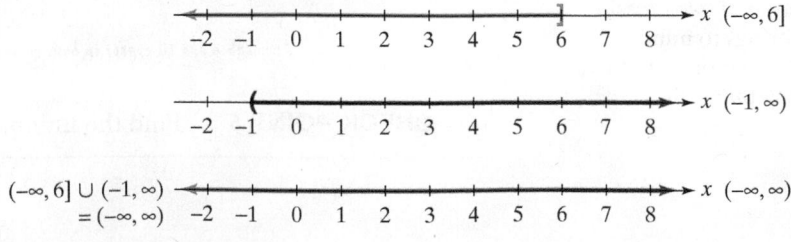

Because all real numbers are either less than or equal to 6 or greater than -1 or both, the union of the two sets fills the entire number line. Thus, the solution set in interval notation is $(-\infty, \infty)$. Any real number that you select will satisfy at least one of the original inequalities. ■

✓ **CHECK POINT 7**   Solve: $2x + 5 \geq 3$ or $2x + 3 < 3$.

CONCEPT AND VOCABULARY CHECK

Fill in each blank so that the resulting statement is true.

1. The set of elements common to both set A and set B is called the _____ of sets A and B, and is symbolized by _____.

2. The set of elements that are members of set A or set B or of both sets is called the _____ of sets A and B, and is symbolized by _____.

3. The set of elements common to both $(-\infty, 9)$ and $(-\infty, 12)$ is _____.

4. The set of elements in $(-\infty, 9)$ or $(-\infty, 12)$ or in both sets is _____.

5. The way to solve $-7 < 3x - 4 \leq 5$ is to isolate x in the _____.

9.2 EXERCISE SET   MyMathLab®

 Watch the videos in MyMathLab

 Download the MyDashBoard App

Practice Exercises

In Exercises 1–6, find the intersection of the sets.

1. $\{1, 2, 3, 4\} \cap \{2, 4, 5\}$
2. $\{1, 3, 7\} \cap \{2, 3, 8\}$
3. $\{1, 3, 5, 7\} \cap \{2, 4, 6, 8, 10\}$
4. $\{0, 1, 3, 5\} \cap \{-5, -3, -1\}$
5. $\{a, b, c, d\} \cap \varnothing$
6. $\{w, y, z\} \cap \varnothing$

In Exercises 7–24, solve each compound inequality. Use graphs to show the solution set to each of the two given inequalities, as well as a third graph that shows the solution set of the compound inequality. Except for the empty set, express the solution set in interval notation.

7. $x > 3$ and $x > 6$
8. $x > 2$ and $x > 4$
9. $x \leq 5$ and $x \leq 1$
10. $x \leq 6$ and $x \leq 2$
11. $x < 2$ and $x \geq -1$
12. $x < 3$ and $x \geq -1$
13. $x > 2$ and $x < -1$
14. $x > 3$ and $x < -1$
15. $5x < -20$ and $3x > -18$
16. $3x \leq 15$ and $2x > -6$
17. $x - 4 \leq 2$ and $3x + 1 > -8$
18. $3x + 2 > -4$ and $2x - 1 < 5$
19. $2x > 5x - 15$ and $7x > 2x + 10$
20. $6 - 5x > 1 - 3x$ and $4x - 3 > x - 9$

21. $4(1 - x) < -6$ and $\dfrac{x - 7}{5} \leq -2$

22. $5(x - 2) > 15$ and $\dfrac{x - 6}{4} \leq -2$

23. $x - 1 \leq 7x - 1$ and $4x - 7 < 3 - x$
24. $2x + 1 > 4x - 3$ and $x - 1 \geq 3x + 5$

In Exercises 25–32, solve each inequality and graph the solution set on a number line. Express the solution set in interval notation.

25. $6 < x + 3 < 8$
26. $7 < x + 5 < 11$
27. $-3 \leq x - 2 < 1$
28. $-6 < x - 4 \leq 1$
29. $-11 < 2x - 1 \leq -5$
30. $3 \leq 4x - 3 < 19$

31. $-3 \leq \dfrac{2x}{3} - 5 < -1$

32. $-6 \leq \dfrac{x}{2} - 4 < -3$

In Exercises 33–38, find the union of the sets.

33. $\{1, 2, 3, 4\} \cup \{2, 4, 5\}$
34. $\{1, 3, 7, 8\} \cup \{2, 3, 8\}$
35. $\{1, 3, 5, 7\} \cup \{2, 4, 6, 8, 10\}$
36. $\{0, 1, 3, 5\} \cup \{2, 4, 6\}$
37. $\{a, e, i, o, u\} \cup \varnothing$
38. $\{e, m, p, t, y\} \cup \varnothing$

In Exercises 39–54, solve each compound inequality. Use graphs to show the solution set to each of the two given inequalities, as well as a third graph that shows the solution set of the compound inequality. Express the solution set in interval notation.

39. $x > 3$ or $x > 6$

40. $x > 2$ or $x > 4$

41. $x \le 5$ or $x \le 1$

42. $x \le 6$ or $x \le 2$

43. $x < 2$ or $x \ge -1$

44. $x < 3$ or $x \ge -1$

45. $x \ge 2$ or $x < -1$

46. $x \ge 3$ or $x < -1$

47. $3x > 12$ or $2x < -6$

48. $3x < 3$ or $2x > 10$

49. $3x + 2 \le 5$ or $5x - 7 \ge 8$

50. $2x - 5 \le -11$ or $5x + 1 \ge 6$

51. $4x + 3 < -1$ or $2x - 3 \ge -11$

52. $2x + 1 < 15$ or $3x - 4 \ge -1$

53. $-2x + 5 > 7$ or $-3x + 10 > 2x$

54. $16 - 3x \ge -8$ or $13 - x > 4x + 3$

55. Let $f(x) = 2x + 3$ and $g(x) = 3x - 1$. Find all values of x for which $f(x) \ge 5$ and $g(x) > 11$.

56. Let $f(x) = 4x + 5$ and $g(x) = 3x - 4$. Find all values of x for which $f(x) \ge 5$ and $g(x) \le 2$.

57. Let $f(x) = 3x - 1$ and $g(x) = 4 - x$. Find all values of x for which $f(x) < -1$ or $g(x) < -2$.

58. Let $f(x) = 2x - 5$ and $g(x) = 3 - x$. Find all values of x for which $f(x) \ge 3$ or $g(x) < 0$.

Practice PLUS

In Exercises 59–60, write an inequality with x isolated in the middle that is equivalent to the given inequality. Assume $a > 0, b > 0$, and $c > 0$.

59. $-c < ax - b < c$

60. $-2 < \dfrac{ax - b}{c} < 2$

In Exercises 61–62, use the graphs of y_1, y_2, and y_3 to solve each compound inequality.

61. $-3 \le 2x - 1 \le 5$

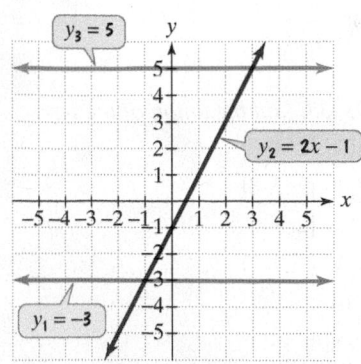

62. $x - 2 < 2x - 1 < x + 2$

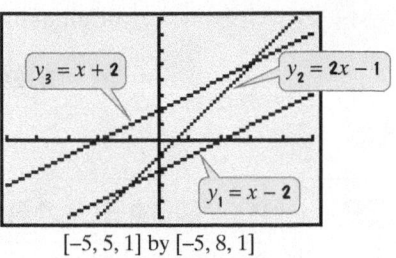

[-5, 5, 1] by [-5, 8, 1]

63. Solve $x - 2 < 2x - 1 < x + 2$, the inequality in Exercise 62, using algebraic methods. (*Hint:* Rewrite the inequality as $2x - 1 > x - 2$ and $2x - 1 < x + 2$.)

64. Use the hint given in Exercise 63 to solve $x \le 3x - 10 \le 2x$.

In Exercises 65–66, use the table to solve each inequality.

65. $-2 \le 5x + 3 < 13$

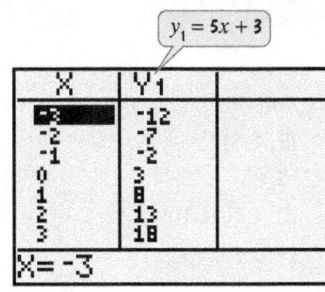

66. $-3 < 2x - 5 \le 3$

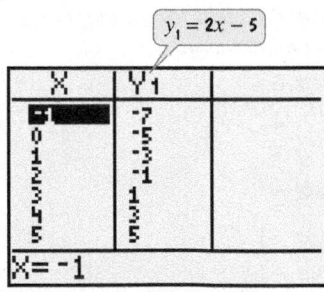

In Exercises 67–68, use the roster method to find the set of negative integers that are solutions of each inequality.

67. $5 - 4x \ge 1$ and $3 - 7x < 31$

68. $-5 < 3x + 4 \le 16$

Application Exercises

In more U.S. marriages, spouses have different faiths. The bar graph shows the percentage of households with an interfaith marriage in 1988 and 2008. Also shown is the percentage of households in which a person of faith is married to someone with no religion.

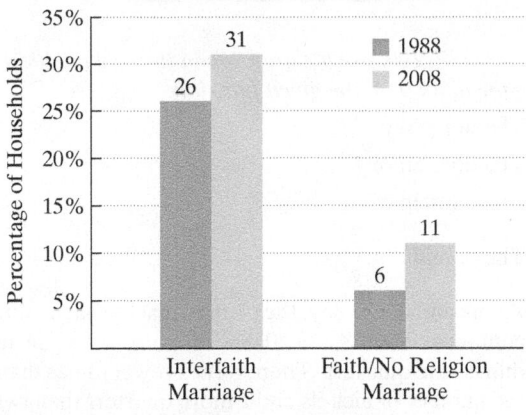

Percentage of U.S. Households in Which Married Couples Do Not Share the Same Faith

Source: General Social Survey, University of Chicago

The formula

$$I = \frac{1}{4}x + 26$$

models the percentage of U.S. households with an interfaith marriage, I, x years after 1988. The formula

$$N = \frac{1}{4}x + 6$$

models the percentage of U.S. households in which a person of faith is married to someone with no religion, N, x years after 1988. Use these models to solve Exercises 69–70.

69. a. In which years will more than 33% of U.S. households have an interfaith marriage?

   b. In which years will more than 14% of U.S. households have a person of faith married to someone with no religion?

   c. Based on your answers to parts (a) and (b), in which years will more than 33% of households have an interfaith marriage and more than 14% have a faith/no religion marriage?

   d. Based on your answers to parts (a) and (b), in which years will more than 33% of households have an interfaith marriage or more than 14% have a faith/no religion marriage?

70. a. In which years will more than 34% of U.S. households have an interfaith marriage?

   b. In which years will more than 15% of U.S. households have a person of faith married to someone with no religion?

c. Based on your answers to parts (a) and (b), in which years will more than 34% of households have an interfaith marriage and more than 15% have a faith/no religion marriage?

   d. Based on your answers to parts (a) and (b), in which years will more than 34% of households have an interfaith marriage or more than 15% have a faith/no religion marriage?

71. A basic cellphone plan costs $20 per month for 60 calling minutes. Additional time costs $0.40 per minute. The formula

$$C = 20 + 0.40(x - 60)$$

gives the monthly cost for this plan, C, for x calling minutes, where x > 60. How many calling minutes are possible for a monthly cost of at least $28 and at most $40?

72. The formula for converting Fahrenheit temperature, F, to Celsius temperature, C, is

$$C = \frac{5}{9}(F - 32).$$

If Celsius temperature ranges from 15° to 35°, inclusive, what is the range for the Fahrenheit temperature? Use interval notation to express this range.

73. On the first four exams, your grades are 70, 75, 87, and 92. There is still one more exam, and you are hoping to earn a B in the course. This will occur if the average of your five exam grades is greater than or equal to 80 and less than 90. What range of grades on the fifth exam will result in earning a B? Use interval notation to express this range.

74. On the first four exams, your grades are 82, 75, 80, and 90. There is still a final exam, and it counts as two grades. You are hoping to earn a B in the course: This will occur if the average of your six exam grades is greater than or equal to 80 and less than 90. What range of grades on the final exam will result in earning a B? Use interval notation to express this range.

75. The toll to a bridge is $3.00. A three-month pass costs $7.50 and reduces the toll to $0.50. A six-month pass costs $30 and permits crossing the bridge for no additional fee. How many crossing per three-month period does it take for the three-month pass to be the best deal?

76. Parts for an automobile repair cost $175. The mechanic charges $34 per hour. If you receive an estimate for at least $226 and at most $294 for fixing the car, what is the time interval that the mechanic will be working on the job?

Writing in Mathematics

77. Describe what is meant by the intersection of two sets. Give an example.

78. Explain how to solve a compound inequality involving *and*.

79. Why is 1 < 2x + 3 < 9 a compound inequality? What are the two inequalities and what is the word that joins them?

80. Explain how to solve $1 < 2x + 3 < 9$.

81. Describe what is meant by the union of two sets. Give an example.

82. Explain how to solve a compound inequality involving *or*.

Technology Exercises

In Exercises 83–86, solve each inequality using a graphing utility. Graph each of the three parts of the inequality separately in the same viewing rectangle. The solution set consists of all values of x for which the graph of the linear function in the middle lies between the graphs of the constant functions on the left and the right.

83. $1 < x + 3 < 9$

84. $-1 < \dfrac{x + 4}{2} < 3$

85. $1 \leq 4x - 7 \leq 3$

86. $2 \leq 4 - x \leq 7$

87. Use a graphing utility's ⟨TABLE⟩ feature to verify your work in Exercises 83–86.

Critical Thinking Exercises

Make Sense? *In Exercises 88–91, determine whether each statement "makes sense" or "does not make sense" and explain your reasoning.*

88. I've noticed that when solving some compound inequalities with *or*, there is no way to express the solution set using a single interval, but this does not happen with *and* compound inequalities.

89. Compound inequalities with *and* have solutions that satisfy both inequalities, whereas compound inequalities with *or* have solutions that satisfy at least one of the inequalities.

90. I'm considering the compound inequality $x < 8$ and $x \geq 8$, so the solution set is $\varnothing$.

91. I'm considering the compound inequality $x < 8$ or $x \geq 8$, so the solution set is $(-\infty, \infty)$.

In Exercises 92–95, determine whether each statement is true or false. If the statement is false, make the necessary change(s) to produce a true statement.

92. $(-\infty, -1] \cap [-4, \infty) = [-4, -1]$

93. $(-\infty, 3) \cup (-\infty, -2) = (-\infty, -2)$

94. The union of two sets can never give the same result as the intersection of those same two sets.

95. The solution set of the compound inequality $x < a$ and $x > a$ is the set of all real numbers excluding a.

96. Solve and express the solution set in interval notation: $-7 \leq 8 - 3x \leq 20$ and $-7 < 6x - 1 < 41$.

The graphs of $f(x) = \sqrt{4 - x}$ and $g(x) = \sqrt{x + 1}$ are shown in a $[-3, 10, 1]$ by $[-2, 5, 1]$ viewing rectangle.

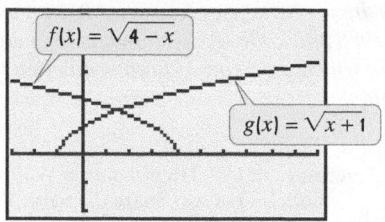

In Exercises 97–100, use the graphs and interval notation to express the domain of the given function.

97. The domain of f

98. The domain of g

99. The domain of $f + g$

100. The domain of $\dfrac{f}{g}$

101. At the end of the day, the change machine at a laundrette contained at least \$3.20 and at most \$5.45 in nickels, dimes, and quarters. There were 3 fewer dimes than twice the number of nickels and 2 more quarters than twice the number of nickels. What was the least possible number and the greatest possible number of nickels?

Review Exercises

102. If $f(x) = x^2 - 3x + 4$ and $g(x) = 2x - 5$, find $(g - f)(x)$ and $(g - f)(-1)$. (Section 8.3, Example 4)

103. Use function notation to write the equation of the line passing through $(4, 2)$ and perpendicular to the line whose equation is $4x - 2y = 8$. (Section 3.5, Example 4)

104. Simplify: $4 - [2(x - 4) - 5]$. (Section 1.8, Example 11)

Preview Exercises

Exercises 105–107 will help you prepare for the material covered in the next section.

105. Find all values of x satisfying $1 - 4x = 3$ or $1 - 4x = -3$.

106. Find all values of x satisfying $3x - 1 = x + 5$ or $3x - 1 = -(x + 5)$.

107. a. Substitute -5 for x and determine whether -5 satisfies $|2x + 3| \geq 5$.

    b. Does 0 satisfy $|2x + 3| \geq 5$?

SECTION

9.3

Objectives

1 Solve absolute value equations.

2 Solve absolute value inequalities of the form $|u| < c$.

3 Solve absolute value inequalities of the form $|u| > c$.

4 Recognize absolute value inequalities with no solution or all real numbers as solutions.

5 Solve problems using absolute value inequalities.

1 Solve absolute value equations.

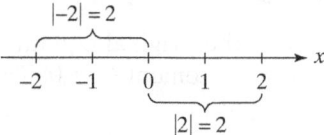

Figure 9.8 If $|x| = 2$, then $x = 2$ or $x = -2$.

Great Question!

Why did you use the variable u when rewriting $|u| = c$ as $u = c$ or $u = -c$? What's wrong with using x and rewriting $|x| = c$ as $x = c$ or $x = -c$?

We could have used x. However, this becomes confusing when the expression inside the absolute value bars is a linear expression given in terms of x. For example, on the right we solved $|2x - 3| = 11$. In this case, think of u as $2x - 3$. This is easier than using x to represent $2x - 3$.

Equations and Inequalities Involving Absolute Value

What activities do you dread? Reading math textbooks with bottle-cap designs on the cover? (Be kind!) No, that's not America's most-dreaded activity. In a random sample of 1000 U.S. adults, 46% of those questioned responded, "Public speaking."

Numerical information, such as the percentage of adults who dread public speaking, is often given with a margin of error. Inequalities involving absolute value are used to describe errors in polling, as well as errors of measurement in manufacturing, engineering, science, and other fields. In this section, you will learn to solve equations and inequalities containing absolute value. With these skills, you will be able to analyze data on the activities U.S. adults say they dread.

Equations Involving Absolute Value

We have seen that the absolute value of a, denoted $|a|$, is the distance from 0 to a on a number line. Now consider **absolute value equations**, such as

$$|x| = 2.$$

This means that we must determine real numbers whose distance from the origin on a number line is 2. **Figure 9.8** shows that there are two numbers such that $|x| = 2$, namely, 2 and -2. We write $x = 2$ or $x = -2$. This observation can be generalized as follows:

Rewriting an Absolute Value Equation Without Absolute Value Bars

If c is a positive real number and u represents any algebraic expression, then $|u| = c$ is equivalent to $u = c$ or $u = -c$.

EXAMPLE 1 Solving an Equation Involving Absolute Value

Solve: $|2x - 3| = 11$.

Solution

$$|2x - 3| = 11$$ This is the given equation.

$$2x - 3 = 11 \quad \text{or} \quad 2x - 3 = -11$$ Rewrite the equation without absolute value bars: $|u| = c$ is equivalent to $u = c$ or $u = -c$. In this case $u = 2x - 3$.

$$2x = 14 \qquad\qquad 2x = -8$$ Add 3 to both sides of each equation.

$$x = 7 \qquad\qquad x = -4$$ Divide both sides of each equation by 2.

Using Technology

Graphic Connections

You can use a graphing utility to verify the solution set of an absolute value equation. Consider, for example,

$$|2x - 3| = 11.$$

Graph $y_1 = |2x - 3|$ and $y_2 = 11$. The graphs are shown in a $[-10, 10, 1]$ by $[-1, 15, 1]$ viewing rectangle. The x-coordinates of the intersection points are -4 and 7, verifying that $\{-4, 7\}$ is the solution set.

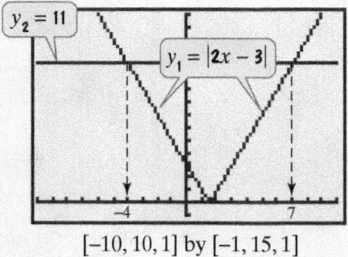

$[-10, 10, 1]$ by $[-1, 15, 1]$

Check 7:

$|2x - 3| = 11$

$|2(7) - 3| \overset{?}{=} 11$

$|14 - 3| \overset{?}{=} 11$

$|11| \overset{?}{=} 11$

$11 = 11$, true

Check -4:

$|2x - 3| = 11$

$|2(-4) - 3| \overset{?}{=} 11$

$|-8 - 3| \overset{?}{=} 11$

$|-11| \overset{?}{=} 11$

$11 = 11$, true

This is the original equation.

Substitute the proposed solutions.

Perform operations inside the absolute value bars.

These true statements indicate that 7 and -4 are solutions.

The solutions are -4 and 7. We can also say that the solution set is $\{-4, 7\}$. ■

✓ **CHECK POINT 1** Solve: $|2x - 1| = 5$.

EXAMPLE 2 Solving an Equation Involving Absolute Value

Solve: $5|1 - 4x| - 15 = 0$.

Solution

$5|1 - 4x| - 15 = 0$ This is the given equation.

> We need to isolate $|1 - 4x|$, the absolute value expression.

$5|1 - 4x| = 15$ Add 15 to both sides.

$|1 - 4x| = 3$ Divide both sides by 5.

$1 - 4x = 3$ or $1 - 4x = -3$ Rewrite $|u| = c$ as $u = c$ or $u = -c$.

$-4x = 2 \qquad\qquad -4x = -4$ Subtract 1 from both sides of each equation.

$x = -\frac{1}{2} \qquad\qquad x = 1$ Divide both sides of each equation by -4.

Take a moment to check $-\frac{1}{2}$ and 1, the proposed solutions, in the original equation, $5|1 - 4x| - 15 = 0$. In each case, you should obtain the true statement $0 = 0$. The solutions are $-\frac{1}{2}$ and 1, and the solution set is $\left\{-\frac{1}{2}, 1\right\}$. ■

✓ **CHECK POINT 2** Solve: $2|1 - 3x| - 28 = 0$.

The absolute value of a number is never negative. Thus, if u is an algebraic expression and c is a negative number, then $|u| = c$ has no solution. For example, the equation $|3x - 6| = -2$ has no solution because $|3x - 6|$ cannot be negative. The solution set is $\varnothing$, the empty set.

The absolute value of 0 is 0. Thus, if u is an algebraic expression and $|u| = 0$, the solution is found by solving $u = 0$. For example, the solution of $|x - 2| = 0$ is obtained by solving $x - 2 = 0$. The solution is 2 and the solution set is $\{2\}$.

Some equations have two absolute value expressions, such as

$$|3x - 1| = |x + 5|.$$

These absolute value expressions are equal when the expressions inside the absolute value bars are equal to or opposites of each other.

Rewriting an Absolute Value Equation with Two Absolute Values Without Absolute Value Bars

If $|u| = |v|$, then $u = v$ or $u = -v$.

EXAMPLE 3   Solving an Absolute Value Equation with Two Absolute Values

Solve: $|3x - 1| = |x + 5|$.

Solution   We rewrite the equation without absolute value bars.

$$|u| = |v| \quad \text{means} \quad u = v \quad \text{or} \quad u = -v$$

$$|3x - 1| = |x + 5| \quad \text{means} \quad 3x - 1 = x + 5 \quad \text{or} \quad 3x - 1 = -(x + 5).$$

We now solve the two equations that do not contain absolute value bars.

$$
\begin{array}{ll}
3x - 1 = x + 5 \quad \text{or} & 3x - 1 = -(x + 5) \\
2x - 1 = 5 & 3x - 1 = -x - 5 \\
2x = 6 & 4x - 1 = -5 \\
x = 3 & 4x = -4 \\
& x = -1
\end{array}
$$

Take a moment to complete the solution process by checking the two proposed solutions in the original equation. The solutions are -1 and 3, and the solution set is $\{-1, 3\}$. ∎

 CHECK POINT 3   Solve: $|2x - 7| = |x + 3|$.

---

2 Solve absolute value inequalities of the form $|u| < c$.

Inequalities Involving Absolute Value

Absolute value can also arise in inequalities. Consider, for example,

$$|x| < 2.$$

This means that the distance from 0 to x on a number line is less than 2, as shown in **Figure 9.9**.

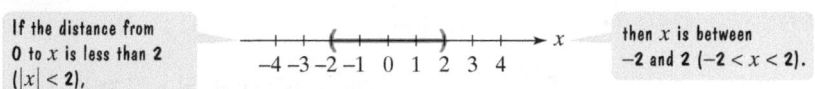

Figure 9.9

Generalizing from the observations in the voice balloons gives us a method for solving inequalities of the form $|u| < c$, where c is a positive number. This method involves rewriting the given inequality without absolute value bars.

> **Solving Absolute Value Inequalities of the Form $|u| < c$**
>
> If c is a positive real number and u represents any algebraic expression, then
>
> $$|u| < c \text{ is equivalent to } -c < u < c.$$
>
> This rule is valid if $<$ is replaced by $\leq$.

EXAMPLE 4   Solving an Absolute Value Inequality of the Form $|u| < c$

Solve and graph the solution set on a number line:

$$|x - 4| < 3.$$

Solution We rewrite the inequality without absolute value bars.

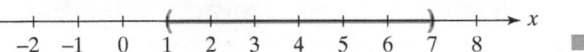

$$|x - 4| < 3 \quad \text{means} \quad -3 < x - 4 < 3.$$

We solve the compound inequality by adding 4 to all three parts.

$$-3 < x - 4 < 3$$
$$-3 + 4 < x - 4 + 4 < 3 + 4$$
$$1 < x < 7$$

The solution set is all real numbers greater than 1 and less than 7, denoted in interval notation by $(1, 7)$. The graph of the solution set is shown as follows:

We can use the rectangular coordinate system to visualize the solution set of

$$|x - 4| < 3.$$

Figure 9.10 shows the graphs of $f(x) = |x - 4|$ and $g(x) = 3$. The solution set of $|x - 4| < 3$ consists of all values of x for which the blue graph of f lies below the red graph of g. These x-values make up the interval $(1, 7)$, which is the solution set.

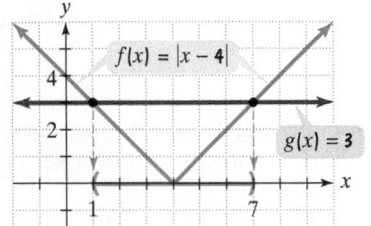

Figure 9.10 The solution set of $|x - 4| < 3$ is $(1, 7)$.

✓ **CHECK POINT 4** Solve and graph the solution set on a number line: $|x - 2| < 5$.

Before rewriting an absolute value inequality without absolute value bars, isolate the absolute value expression on one side of the inequality.

EXAMPLE 5 Solving an Absolute Value Inequality

Solve and graph the solution set on a number line: $-2|3x + 5| + 7 \geq -13$.

Solution

$$-2|3x + 5| + 7 \geq -13 \qquad \text{This is the given inequality.}$$

> We need to isolate $|3x + 5|$, the absolute value expression.

$$-2|3x + 5| + 7 - 7 \geq -13 - 7 \qquad \text{Subtract 7 from both sides.}$$
$$-2|3x + 5| \geq -20 \qquad \text{Simplify.}$$
$$\frac{-2|3x + 5|}{-2} \leq \frac{-20}{-2} \qquad \text{Divide both sides by } -2 \text{ and change the sense of the inequality.}$$
$$|3x + 5| \leq 10 \qquad \text{Simplify.}$$
$$-10 \leq 3x + 5 \leq 10 \qquad \text{Rewrite without absolute value bars:}$$
$$|u| \leq c \text{ means } -c \leq u \leq c.$$

> Now we need to isolate x in the middle.

$$-10 - 5 \leq 3x + 5 - 5 \leq 10 - 5 \qquad \text{Subtract 5 from all three parts.}$$
$$-15 \leq 3x \leq 5 \qquad \text{Simplify.}$$
$$\frac{-15}{3} \leq \frac{3x}{3} \leq \frac{5}{3} \qquad \text{Divide each part by 3.}$$
$$-5 \leq x \leq \frac{5}{3} \qquad \text{Simplify.}$$

The solution set is $\left[-5, \frac{5}{3}\right]$ in interval notation. The graph is shown as follows:

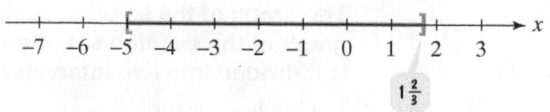

✓ **CHECK POINT 5**   Solve and graph the solution set on a number line: $-3|5x - 2| + 20 \geq -19$.

---

3 Solve absolute value inequalities of the form $|u| > c$.

Now let's consider absolute value inequalities with greater than symbols, such as

$$|x| > 2.$$

This means that the distance from 0 to x on a number line is greater than 2, as shown in **Figure 9.11**.

If the distance from 0 to x is greater than 2 ($|x| > 2$),   then x is less than -2 or greater than 2 ($x < -2$ or $x > 2$).

Figure 9.11

Generalizing from the observations in the voice balloons gives us a method for solving inequalities of the form $|u| > c$, where c is a positive number. This method once again involves rewriting the given inequality without absolute value bars.

Solving Absolute Value Inequalities of the Form $|u| > c$

If c is a positive real number and u represents any algebraic expression, then

$$|u| > c \text{ is equivalent to } u < -c \text{ or } u > c.$$

This rule is valid if $>$ is replaced by $\geq$.

EXAMPLE 6   Solving an Absolute Value Inequality of the Form $|u| \geq c$

Solve and graph the solution set on a number line:

$$|2x + 3| \geq 5.$$

Solution   We rewrite the inequality without absolute value bars.

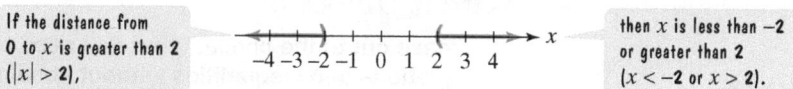

$|2x + 3| \geq 5$ means $2x + 3 \leq -5$ or $2x + 3 \geq 5.$

We solve this compound inequality by solving each of these inequalities separately. Then we take the union of their solution sets.

$2x + 3 \leq -5$ or $2x + 3 \geq 5$ These are the inequalities without absolute value bars.

$2x \leq -8$ $2x \geq 2$ Subtract 3 from both sides.

$x \leq -4$ $x \geq 1$ Divide both sides by 2.

The solution set consists of all numbers that are less than or equal to -4 or greater than or equal to 1. The solution set in interval notation is $(-\infty, -4] \cup [1, \infty)$. The graph of the solution set is shown as follows:

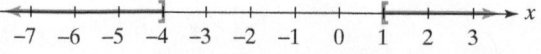

Great Question!

The graph of the solution set in Example 6 consists of two intervals. When is the graph of the solution set of an absolute value inequality a single interval and when is it divided into two intervals?

If u is a linear expression and $c > 0$, the graph of the solution set for $|u| > c$ will be divided into two intervals whose union cannot be represented as a single interval. The graph of the solution set for $|u| < c$ will be a single interval. Avoid the common error of rewriting $|u| > c$ as $-c < u > c$.

✓ **CHECK POINT 6**   Solve and graph the solution set on a number line: $|2x - 5| \geq 3$

Great Question!

Please cut to the chase. What do I need to know when rewriting absolute value equations and inequalities without absolute value bars?

Here's a brief summary. If $c > 0$,

- $|u| = c$ is equivalent to $u = c$ or $u = -c$.
- $|u| < c$ is equivalent to $-c < u < c$.
- $|u| > c$ is equivalent to $u < -c$ or $u > c$.

---

4 Recognize absolute value inequalities with no solution or all real numbers as solutions.

Absolute Value Inequalities with Unusual Solution Sets

We have been working with $|u| < c$ and $|u| > c$, where c is a positive number. Now let's see what happens to these inequalities if c is a negative number. Consider, for example, $|x| < -2$. Because $|x|$ always has a value that is greater than or equal to 0, there is no number whose absolute value is less than -2. The inequality $|x| < -2$ has no solution. The solution set is $\varnothing$.

Now consider the inequality $|x| > -2$. Because $|x|$ is never negative, all numbers have an absolute value greater than -2. All real numbers satisfy the inequality $|x| > -2$. The solution set is $(-\infty, \infty)$.

Absolute Value Inequalities with Unusual Solution Sets

If u is an algebraic expression and c is a negative number,

1. The inequality $|u| < c$ has no solution.
2. The inequality $|u| > c$ is true for all real numbers for which u is defined.

---

5 Solve problems using absolute value inequalities.

Applications

We opened this section with this question:

   What activities do you dread?

In a random sample of 1000 U.S. adults, 46% of those questioned responded, "Public speaking." The problem is that this is a single random sample. Do 46% of adults in the entire U.S. population dread public speaking?

   If you look at the results of a poll like the one in **Figure 9.12**, you will observe that a margin of error is reported.

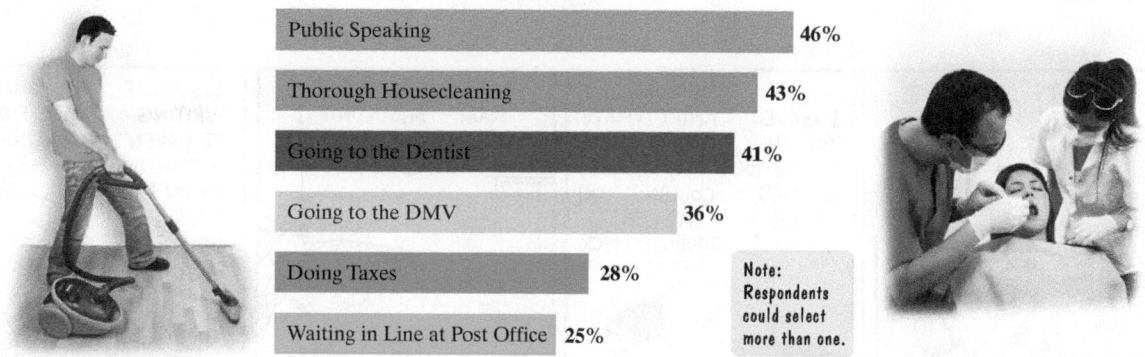

Figure 9.12

Source: TNS survey of 1000 adults, March 2010, Margin of error: ±3.2%

> Note the margin of error.

The margin of error in the dreaded-activities poll is ±3.2%. This means that the actual percentage of U.S. adults who dread public speaking is at most 3.2% greater than or less than 46%. If x represents the percentage of U.S. adults in the population who dread public speaking, then the poll's margin of error can be expressed as an absolute value inequality:

$$|x - 46| \le 3.2.$$

EXAMPLE 7   Analyzing a Poll's Margin of Error

The inequality

$$|x - 28| \le 3.2$$

describes the percentage of U.S. adults in the population who dread doing taxes. Solve the inequality and interpret the solution.

Solution   We rewrite the inequality without absolute value bars.

$$|u| \le c \quad \text{means} \quad -c \le u \le c.$$

$$|x - 28| \le 3.2 \quad \text{means} \quad -3.2 \le x - 28 \le 3.2.$$

We solve the compound inequality by adding 28 to all three parts.

$-3.2 \le x - 28 \le 3.2$	This is $\lvert x - 28\rvert \le 3.2$ without absolute value bars.
$-3.2 + 28 \le x - 28 + 28 \le 3.2 + 28$	Add 28 to all three parts.
$24.8 \le x \le 31.2$	Simplify. The solution is [24.8, 31.2].

The percentage of U.S. adults in the population who dread doing taxes is somewhere between a low of 24.8% and a high of 31.2%. Notice that these percents are 3.2% above and below the given 28% in **Figure 9.12**, and that 3.2% is the poll's margin of error.   ∎

✓ **CHECK POINT 7**   Solve the inequality:

$$|x - 41| \le 3.2.$$

Interpret the solution in terms of the information in **Figure 9.12**.

Achieving Success

FoxTrot

FOXTROT copyright © 2003, 2009 by Bill Amend/Distributed by Universal Uclick

A test-taking tip: Live for partial credit.
Always show your work. If worse comes to worst, write *something* down, anything, even if it's a formula that you think might solve a problem or a possible idea or procedure for solving the problem. As Bill Amend's *FoxTrot* cartoon indicates, partial credit has salvaged more than a few test scores.

CONCEPT AND VOCABULARY CHECK

Fill in each blank so that the resulting statement is true.

1. If $c > 0$, $|u| = c$ is equivalent to $u = $ _____ or $u = $ _____.

2. $|u| = |v|$ is equivalent to $u = $ _____ or $u = $ _____.

3. If $c > 0$, $|u| < c$ is equivalent to _____ $< u <$ _____.

4. If $c > 0$, $|u| > c$ is equivalent to $u < $ _____ or $u > $ _____.

5. $|u| < c$ has no solution if c _____ 0.

6. $|u| > c$ is true for all real numbers if c _____ 0.

Match each absolute value equation or inequality in the left column with an equivalent statement from the right column.

7. $|3x - 1| = 5$ **A.** $-5 \le 3x - 1 \le 5$

8. $|3x - 1| \ge 5$ **B.** $3x - 1 = x$ or $3x - 1 = -x$

9. $|3x - 1| \le 5$ **C.** $3x - 1 = 5$ or $3x - 1 = -5$

10. $|3x - 1| = |x|$ **D.** $\varnothing$

11. $|3x - 1| < -5$ **E.** $3x - 1 \le -5$ or $3x - 1 \ge 5$

12. $|3x - 1| > -5$ **F.** $(-\infty, \infty)$

9.3 EXERCISE SET

MyMathLab®

 Watch the videos in MyMathLab   Download the MyDashBoard App

Practice Exercises

In Exercises 1–38, find the solution set for each equation.

1. $|x| = 8$
2. $|x| = 6$
3. $|x - 2| = 7$
4. $|x + 1| = 5$
5. $|2x - 1| = 7$
6. $|2x - 3| = 11$
7. $\left|\dfrac{4x - 2}{3}\right| = 2$
8. $\left|\dfrac{3x - 1}{5}\right| = 1$
9. $|x| = -8$
10. $|x| = -6$
11. $|x + 3| = 0$
12. $|x + 2| = 0$
13. $2|y + 6| = 10$
14. $3|y + 5| = 12$

15. $3|2x - 1| = 21$

16. $2|3x - 2| = 14$

17. $|6y - 2| + 4 = 32$

18. $|3y - 1| + 10 = 25$

19. $7|5x| + 2 = 16$

20. $7|3x| + 2 = 16$

21. $|x + 1| + 5 = 3$

22. $|x + 1| + 6 = 2$

23. $|4y + 1| + 10 = 4$

24. $|3y - 2| + 8 = 1$

25. $|2x - 1| + 3 = 3$

26. $|3x - 2| + 4 = 4$

27. $|5x - 8| = |3x + 2|$

28. $|4x - 9| = |2x + 1|$

29. $|2x - 4| = |x - 1|$

30. $|6x| = |3x - 9|$

31. $|2x - 5| = |2x + 5|$

32. $|3x - 5| = |3x + 5|$

33. $|x - 3| = |5 - x|$

34. $|x - 3| = |6 - x|$

35. $|2y - 6| = |10 - 2y|$

36. $|4y + 3| = |4y + 5|$

37. $\left|\dfrac{2x}{3} - 2\right| = \left|\dfrac{x}{3} + 3\right|$

38. $\left|\dfrac{x}{2} - 2\right| = \left|x - \dfrac{1}{2}\right|$

In Exercises 39–74, solve and graph the solution set on a number line.

39. $|x| < 3$

40. $|x| < 5$

41. $|x - 2| < 1$

42. $|x - 1| < 5$

43. $|x + 2| \le 1$

44. $|x + 1| \le 5$

45. $|2x - 6| < 8$

46. $|3x + 5| < 17$

47. $|x| > 3$

48. $|x| > 5$

49. $|x + 3| > 1$

50. $|x - 2| > 5$

51. $|x - 4| \ge 2$

52. $|x - 3| \ge 4$

53. $|3x - 8| > 7$

54. $|5x - 2| > 13$

55. $|2(x - 1) + 4| \le 8$

56. $|3(x - 1) + 2| \le 20$

57. $\left|\dfrac{2x + 6}{3}\right| < 2$

58. $\left|\dfrac{3x - 3}{4}\right| < 6$

59. $\left|\dfrac{2x + 2}{4}\right| \ge 2$

60. $\left|\dfrac{3x - 3}{9}\right| \ge 1$

61. $\left|3 - \dfrac{2x}{3}\right| > 5$

62. $\left|3 - \dfrac{3x}{4}\right| > 9$

63. $|x - 2| < -1$

64. $|x - 3| < -2$

65. $|x + 6| > -10$

66. $|x + 4| > -12$

67. $|x + 2| + 9 \le 16$

68. $|x - 2| + 4 \le 5$

69. $2|2x - 3| + 10 > 12$

70. $3|2x - 1| + 2 > 8$

71. $-4|1 - x| < -16$

72. $-2|5 - x| < -6$

73. $3 \le |2x - 1|$

74. $9 \le |4x + 7|$

75. Let $f(x) = |5 - 4x|$. Find all values of x for which $f(x) = 11$.

76. Let $f(x) = |2 - 3x|$. Find all values of x for which $f(x) = 13$.

77. Let $f(x) = |3 - x|$ and $g(x) = |3x + 11|$. Find all values of x for which $f(x) = g(x)$.

78. Let $f(x) = |3x + 1|$ and $g(x) = |6x - 2|$. Find all values of x for which $f(x) = g(x)$.

79. Let $g(x) = |-1 + 3(x + 1)|$. Find all values of x for which $g(x) \le 5$.

80. Let $g(x) = |-3 + 4(x + 1)|$. Find all values of x for which $g(x) \le 3$.

81. Let $h(x) = |2x - 3| + 1$. Find all values of x for which $h(x) > 6$.

82. Let $h(x) = |2x - 4| - 6$. Find all values of x for which $h(x) > 18$.

Practice PLUS

83. When 3 times a number is subtracted from 4, the absolute value of the difference is at least 5. Use interval notation to express the set of all real numbers that satisfy this condition.

84. When 4 times a number is subtracted from 5, the absolute value of the difference is at most 13. Use interval notation to express the set of all real numbers that satisfy this condition.

In Exercises 85–86, solve each inequality. Assume that $a > 0$ and $c > 0$.

85. $|ax + b| < c$

86. $|ax + b| \ge c$

In Exercises 87–88, use the graph of $f(x) = |4 - x|$ to solve each equation or inequality.

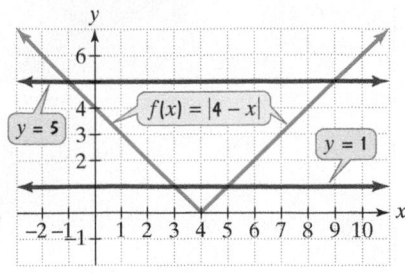

87. $|4 - x| = 1$ **88.** $|4 - x| < 5$

In Exercises 89–90, use the table to solve each inequality.

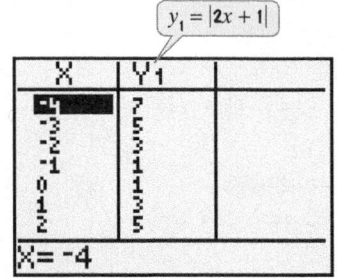

89. $|2x + 1| \leq 3$

90. $|2x + 1| \geq 3$

Application Exercises

How to Blow Your Job Interview *The data in the bar graph are from a random survey of 1910 job interviewers. The graph shows the top interviewer turnoffs and the percentage of surveyed interviewers who were offended by each of these behaviors. In Exercises 91–92, let x represent the actual percentage of interviewers in the entire population of job interviewers.*

Top Interviewer Turnoffs

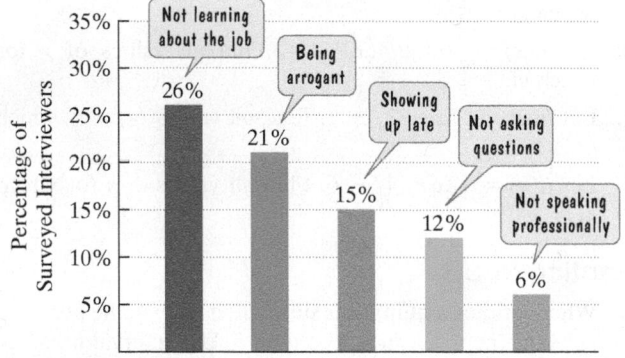

Source: Scott Erker, PhD., and Kelli Buczynski, "Are You Failing the Interview? 2009 Survey of Global Interviewing Practices and Perceptions." Development Dimensions International.

91. Solve the inequality: $|x - 21| \leq 3$. Interpret the solution in terms of the information in the graph. What is the margin of error?

92. Solve the inequality: $|x - 15| \leq 3$. Interpret the solution in terms of the information in the graph. What is the margin of error?

93. The inequality $|T - 57| \leq 7$ describes the range of monthly average temperature, T, in degrees Fahrenheit, for San Francisco, California. Solve the inequality and interpret the solution.

94. The inequality $|T - 50| \leq 22$ describes the range of monthly average temperature, T, in degrees Fahrenheit, for Albany, New York. Solve the inequality and interpret the solution.

The specifications for machine parts are given with tolerance limits that describe a range of measurements for which the part is acceptable. In Exercises 95–96, x represents the length of a machine part, in centimeters. The tolerance limit is 0.01 centimeter.

95. Solve: $|x - 8.6| \leq 0.01$. If the length of the machine part is supposed to be 8.6 centimeters, interpret the solution.

96. Solve: $|x - 9.4| \leq 0.01$. If the length of the machine part is supposed to be 9.4 centimeters, interpret the solution.

97. If a coin is tossed 100 times, we would expect approximately 50 of the outcomes to be heads. It can be demonstrated that a coin is unfair if h, the number of outcomes that result in heads, satisfies $\left| \dfrac{h - 50}{5} \right| \geq 1.645$. Describe the number of outcomes that result in heads that determine an unfair coin that is tossed 100 times.

Writing in Mathematics

98. Explain how to solve an equation containing one absolute value expression.

99. Explain why the procedure that you described in Exercise 98 does not apply to the equation $|x - 5| = -3$. What is the solution set of this equation?

100. Describe how to solve an absolute value equation with two absolute values.

101. Describe how to solve an absolute value inequality of the form $|u| < c$ for $c > 0$.

102. Explain why the procedure that you described in Exercise 101 does not apply to the inequality $|x - 5| < -3$. What is the solution set of this inequality?

103. Describe how to solve an absolute value inequality of the form $|u| > c$ for $c > 0$.

104. Explain why the procedure that you described in Exercise 103 does not apply to the inequality $|x - 5| > -3$. What is the solution set of this inequality?

Technology Exercises

In Exercises 105–107, solve each equation using a graphing utility. Graph each side separately in the same viewing rectangle. The solutions are the x-coordinates of the intersection points.

105. $|x + 1| = 5$

106. $|3(x + 4)| = 12$

107. $|2x - 3| = |9 - 4x|$

In Exercises 108–110, solve each inequality using a graphing utility. Graph each side separately in the same viewing rectangle. The solution set consists of all values of x for which the graph of the left side lies below the graph of the right side.

108. $|2x + 3| < 5$

109. $\left|\dfrac{2x - 1}{3}\right| < \dfrac{5}{3}$

110. $|x + 4| < -1$

In Exercises 111–113, solve each inequality using a graphing utility. Graph each side separately in the same viewing rectangle. The solution set consists of all values of x for which the graph of the left side lies above the graph of the right side.

111. $|2x - 1| > 7$

112. $|0.1x - 0.4| + 0.4 > 0.6$

113. $|x + 4| > -1$

114. Use a graphing utility to verify the solution sets for any five equations or inequalities that you solved by hand in Exercises 1–74.

Critical Thinking Exercises

Make Sense? *In Exercises 115–118, determine whether each statement "makes sense" or "does not make sense" and explain your reasoning.*

115. I solved $|x - 2| = 5$ by rewriting the equation as $x - 2 = 5$ or $x + 2 = 5$.

116. I solved $|x - 2| > 5$ by rewriting the inequality as $-5 < x - 2 > 5$.

117. Because the absolute value of any expression is never less than a negative number, I can immediately conclude that the inequality $|2x - 5| - 9 < -4$ has no solution.

118. I'll win the contest if I can complete the crossword puzzle in 20 minutes plus or minus 5 minutes, so my winning time, x, is modeled by $|x - 20| \le 5$.

In Exercises 119–122, determine whether each statement is true or false. If the statement is false, make the necessary change(s) to produce a true statement.

119. All absolute value equations have two solutions.

120. The equation $|x| = -6$ is equivalent to $x = 6$ or $x = -6$.

121. Values of -5 and 5 satisfy $|x| = 5, |x| \le 5$, and $|x| \ge -5$.

122. The absolute value of any linear expression is greater than 0 for all real numbers except the number for which the expression is equal to 0.

123. Write an absolute value inequality for which the interval shown is the solution.

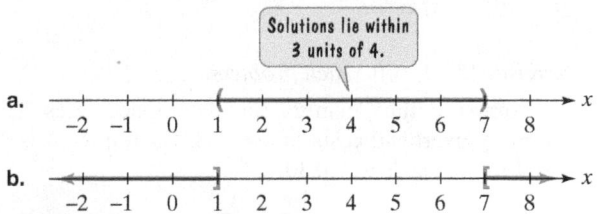

124. The percentage, p, of defective products manufactured by a company is given by $|p - 0.3\%| \le 0.2\%$. If 100,000 products are manufactured and the company offers a \$5 refund for each defective product, describe the company's cost for refunds.

125. Solve: $|2x + 5| = 3x + 4$.

Review and Preview Exercises

Exercises 126–128 will enable you to review graphing linear functions. In addition, they will help you prepare for the material covered in the next section. In each exercise, graph the linear function.

126. $3x - 5y = 15$ (Section 3.2, Example 5)

127. $f(x) = -\dfrac{2}{3}x$ or $y = -\dfrac{2}{3}x$ (Section 3.4, Example 4)

128. $f(x) = -2$ or $y = -2$ (Section 3.2, Example 7)

| MID-CHAPTER CHECK POINT | Section 9.1–Section 9.3 |

✓ **What You Know:** We reviewed how to solve linear inequalities, expressing solution sets in interval notation. We know that it is necessary to change the sense of an inequality when multiplying or dividing both sides by a negative number. We solved compound inequalities with *and* by finding the intersection of solution sets and with *or* by finding the union of solution sets. Finally, we solved equations and inequalities involving absolute value by carefully rewriting the given equation or inequality without absolute value bars. For positive values of c, we wrote $|u| = c$ as $u = c$ or $u = -c$. We wrote $|u| < c$ as $-c < u < c$, and we wrote $|u| > c$ as $u < -c$ or $u > c$.

In Exercises 1–17, solve each inequality or equation.

1. $4 - 3x \ge 12 - x$

2. $5 \le 2x - 1 < 9$

3. $|4x - 7| = 5$

4. $-10 - 3(2x + 1) > 8x + 1$

5. $2x + 7 < -11$ or $-3x - 2 < 13$

6. $|3x - 2| \le 4$

7. $|x + 5| = |5x - 8|$

8. $5 - 2x \ge 9$ and $5x + 3 > -17$

9. $3x - 2 > -8$ or $2x + 1 < 9$

10. $\dfrac{x}{2} + 3 \le \dfrac{x}{3} + \dfrac{5}{2}$

11. $\dfrac{2}{3}(6x - 9) + 4 > 5x + 1$

12. $|5x + 3| > 2$

13. $7 - \left|\dfrac{x}{2} + 2\right| \le 4$

14. $\dfrac{x + 3}{4} < \dfrac{1}{3}$

15. $5x + 1 \ge 4x - 2$ and $2x - 3 > 5$

16. $3 - |2x - 5| = -6$

17. $3 + |2x - 5| = -6$

In Exercises 18–21, solve each problem.

18. A company that manufactures compact discs has fixed monthly overhead costs of $60,000. Each disc costs $0.18 to produce and sells for $0.30.

a. Write the cost function, C, of producing x discs per month.

b. Write the revenue function, R, from the sale of x discs per month.

c. Write the profit function, P, from producing and selling x discs per month.

d. How many discs should be produced and sold each month for the company to have a profit of at least $30,000?

19. A car rental agency rents a certain car for $40 per day with unlimited mileage or $24 per day plus $0.20 per mile. How far can a customer drive this car per day for the $24 option to cost no more than the unlimited mileage option?

20. To receive a B in a course, you must have an average of at least 80% but less than 90% on five exams. Your grades on the first four exams were 95%, 79%, 91%, and 86%. What range of grades on the fifth exam will result in a B for the course?

21. A retiree requires an annual income of at least $9000 from an investment paying 7.5% annual interest. How much should the retiree invest to achieve the desired return?

9.4

Linear Inequalities in Two Variables

Objectives

1 Graph a linear inequality in two variables.

2 Use mathematical models involving linear inequalities.

3 Graph a system of linear inequalities.

This book was written in Point Reyes National Seashore, 40 miles north of San Francisco. The park consists of 75,000 acres with miles of pristine surf-washed beaches, forested ridges, and bays bordered by white cliffs.

Like your author, many people are kept inspired and energized surrounded by nature's unspoiled beauty. In this section, you will see how systems of inequalities model whether a region's natural beauty manifests itself in forests, grasslands, or deserts.

Linear Inequalities in Two Variables and Their Solutions

We have seen that equations in the form $Ax + By = C$ are straight lines when graphed. If we change the symbol $=$ to $>$, $<$, $\ge$, or $\le$, we obtain a **linear inequality in two variables**. Some examples of linear inequalities in two variables are $x + y > 2$, $3x - 5y \le 15$, and $2x - y < 4$.

A **solution of an inequality in two variables**, x and y, is an ordered pair of real numbers with the following property: When the x-coordinate is substituted for x and the y-coordinate is substituted for y in the inequality, we obtain a true statement. For example, (3, 2) is a solution of the inequality $x + y > 1$. When 3 is substituted for x and 2 is substituted for y, we obtain the true statement $3 + 2 > 1$, or $5 > 1$. Because there are infinitely many pairs of numbers that have a sum greater than 1, the inequality $x + y > 1$ has infinitely many solutions. Each ordered-pair solution is said to **satisfy** the inequality. Thus, (3, 2) satisfies the inequality $x + y > 1$.

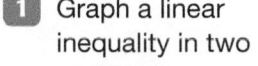

1 Graph a linear inequality in two variables.

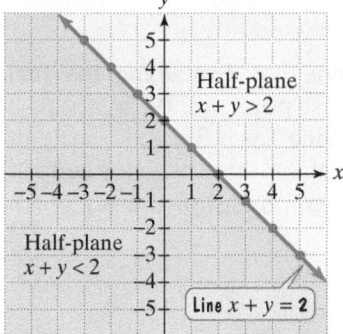

Figure 9.13

The Graph of a Linear Inequality in Two Variables

We know that the graph of an equation in two variables is the set of all points whose coordinates satisfy the equation. Similarly, the **graph of an inequality in two variables** is the set of all points whose coordinates satisfy the inequality.

Let's use **Figure 9.13** to get an idea of what the graph of a linear inequality in two variables looks like. Part of the figure shows the graph of the linear equation $x + y = 2$. The line divides the points in the rectangular coordinate system into three sets. First, there is the set of points along the line, satisfying $x + y = 2$. Next, there is the set of points in the green region above the line. Points in the green region satisfy the linear inequality $x + y > 2$. Finally, there is the set of points in the purple region below the line. Points in the purple region satisfy the linear inequality $x + y < 2$.

A **half-plane** is the set of all the points on one side of a line. In **Figure 9.13**, the green region is a half-plane. The purple region is also a half-plane. A half-plane is the graph of a linear inequality that involves $>$ or $<$. The graph of a linear inequality that involves $\geq$ or $\leq$ is a half-plane and a line. A solid line is used to show that a line is part of a graph. A dashed line is used to show that a line is not part of a graph.

Graphing a Linear Inequality in Two Variables

1. Replace the inequality symbol with an equal sign and graph the corresponding linear equation. Draw a solid line if the original inequality contains a $\leq$ or $\geq$ symbol. Draw a dashed line if the original inequality contains a $<$ or $>$ symbol.

2. Choose a test point from one of the half-planes. (Do not choose a point on the line.) Substitute the coordinates of the test point into the inequality.

3. If a true statement results, shade the half-plane containing this test point. If a false statement results, shade the half-plane not containing this test point.

EXAMPLE 1   Graphing a Linear Inequality in Two Variables

Graph: $2x - 3y \geq 6$.

Solution

Step 1. Replace the inequality symbol by = and graph the linear equation. We need to graph $2x - 3y = 6$. We can use intercepts to graph this line.

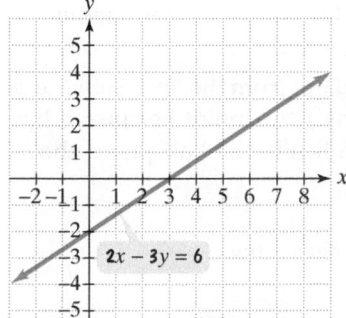

Figure 9.14 Preparing to graph $2x - 3y \geq 6$

We set $y = 0$ to find the x-intercept:	We set $x = 0$ to find the y-intercept:
$2x - 3y = 6$	$2x - 3y = 6$
$2x - 3 \cdot 0 = 6$	$2 \cdot 0 - 3y = 6$
$2x = 6$	$-3y = 6$
$x = 3.$	$y = -2.$

The x-intercept is 3, so the line passes through (3, 0). The y-intercept is -2, so the line passes through $(0, -2)$. Using the intercepts, the line is shown in **Figure 9.14** as a solid line. This is because the inequality $2x - 3y \geq 6$ contains a $\geq$ symbol, in which equality is included.

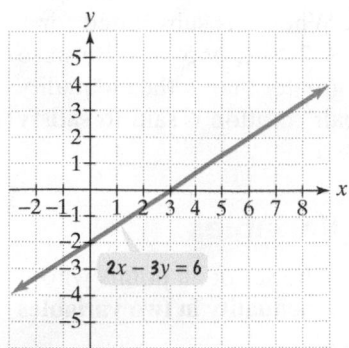

Figure 9.14 The graph of $2x - 3y = 6$ (repeated)

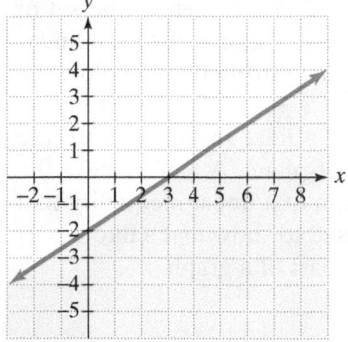

Figure 9.15 The graph of $2x - 3y \geq 6$

Step 2. Choose a test point from one of the half-planes and not from the line. Substitute its coordinates into the inequality. Figure 9.14, repeated in the margin, shows that the line $2x - 3y = 6$ divides the plane into three parts—the line itself and two half-planes. The points in one half-plane satisfy $2x - 3y > 6$. The points in the other half-plane satisfy $2x - 3y < 6$. We need to find which half-plane belongs to the solution of $2x - 3y \geq 6$. To do so, we test a point from either half-plane. The origin, $(0, 0)$, is the easiest point to test.

$$2x - 3y \geq 6 \qquad \text{This is the given inequality.}$$
$$2 \cdot 0 - 3 \cdot 0 \overset{?}{\geq} 6 \qquad \text{Test } (0, 0) \text{ by substituting 0 for } x \text{ and 0 for } y.$$
$$0 - 0 \overset{?}{\geq} 6 \qquad \text{Multiply.}$$
$$0 \geq 6 \qquad \text{This statement is false.}$$

Step 3. If a false statement results, shade the half-plane not containing the test point. Because 0 is not greater than or equal to 6, the test point, $(0, 0)$, is not part of the solution set. Thus, the half-plane below the solid line $2x - 3y = 6$ is part of the solution set. The solution set is the line and the half-plane that does not contain the point $(0, 0)$, indicated by shading this half-plane. The graph is shown using green shading and a blue line in **Figure 9.15.** ∎

✓ **CHECK POINT 1** Graph: $4x - 2y \geq 8$.

When graphing a linear inequality, choose a test point that lies in one of the half-planes and *not on the line dividing the half-planes*. The test point $(0, 0)$ is convenient because it is easy to calculate when 0 is substituted for each variable. However, if $(0, 0)$ lies on the dividing line and not in a half-plane, a different test point must be selected.

EXAMPLE 2 Graphing a Linear Inequality in Two Variables

Graph: $y > -\dfrac{2}{3}x$.

Solution

Step 1. Replace the inequality symbol by = and graph the linear equation. Because we are interested in graphing $y > -\frac{2}{3}x$, we begin by graphing $y = -\frac{2}{3}x$. We can use the slope and the y-intercept to graph this linear function.

$$y = -\frac{2}{3}x + 0$$

Slope $= \dfrac{-2}{3} = \dfrac{\text{rise}}{\text{run}}$ y-intercept $= 0$

The y-intercept is 0, so the line passes through $(0, 0)$. Using the y-intercept and the slope, the line is shown in **Figure 9.16** as a dashed line. This is because the inequality $y > -\frac{2}{3}x$ contains a $>$ symbol, in which equality is not included.

Step 2. Choose a test point from one of the half-planes and not from the line. Substitute its coordinates into the inequality. We cannot use $(0, 0)$ as a test point because it lies on the line and not in a half-plane. Let's use $(1, 1)$, which lies in the half-plane above the line.

$$y > -\frac{2}{3}x \qquad \text{This is the given inequality.}$$
$$1 \overset{?}{>} -\frac{2}{3} \cdot 1 \qquad \text{Test } (1, 1) \text{ by substituting 1 for } x \text{ and 1 for } y.$$
$$1 > -\frac{2}{3} \qquad \text{This statement is true.}$$

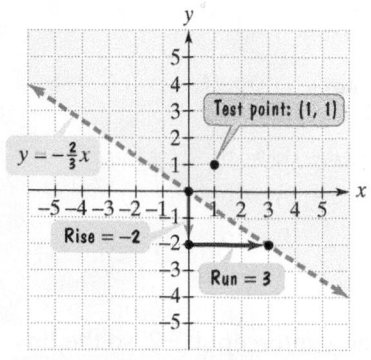

Figure 9.16 The graph of $y > -\frac{2}{3}x$

Step 3. If a true statement results, shade the half-plane containing the test point. Because 1 is greater than $-\frac{2}{3}$, the test point $(1, 1)$ is part of the solution set. All the points on the same side of the line $y = -\frac{2}{3}x$ as the point $(1, 1)$ are members of the solution set. The solution set is the half-plane that contains the point $(1, 1)$, indicated by shading this half-plane. The graph is shown using green shading and a dashed blue line in **Figure 9.16**. ■

Using Technology

Most graphing utilities can graph inequalities in two variables with the $\boxed{\text{SHADE}}$ feature. The procedure varies by model, so consult your manual. For most graphing utilities, you must first solve for y if it is not already isolated. The figure shows the graph of $y > -\frac{2}{3}x$. Most displays do not distinguish between dashed and solid boundary lines.

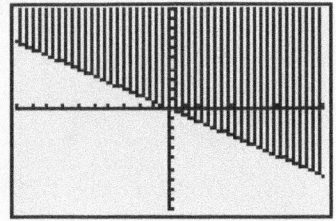

✓ **CHECK POINT 2**   Graph: $y > -\dfrac{3}{4}x.$

Graphing Linear Inequalities without Using Test Points

You can graph inequalities in the form $y > mx + b$ or $y < mx + b$ without using test points. The inequality symbol indicates which half-plane to shade.

- If $y > mx + b$, shade the half-plane above the line $y = mx + b$.
- If $y < mx + b$, shade the half-plane below the line $y = mx + b$.

Observe how this is illustrated in **Figure 9.16**. The graph of $y > -\frac{2}{3}x$ is the half-plane above the line $y = -\frac{2}{3}x$.

It is also not necessary to use test points when graphing inequalities involving half-planes on one side of a vertical or a horizontal line.

Great Question!

When is it important to use test points to graph linear inequalities?

Continue using test points to graph inequalities in the form $Ax + By > C$ or $Ax + By < C$. The graph of $Ax + By > C$ can lie above or below the line given by $Ax + By = C$, depending on the value of B. The same comment applies to the graph of $Ax + By < C$.

For the Vertical Line $x = a$:	For the Horizontal Line $y = b$:
• If $x > a$, shade the half-plane to the right of $x = a$.	• If $y > b$, shade the half-plane above $y = b$.
• If $x < a$, shade the half-plane to the left of $x = a$.	• If $y < b$, shade the half-plane below $y = b$.

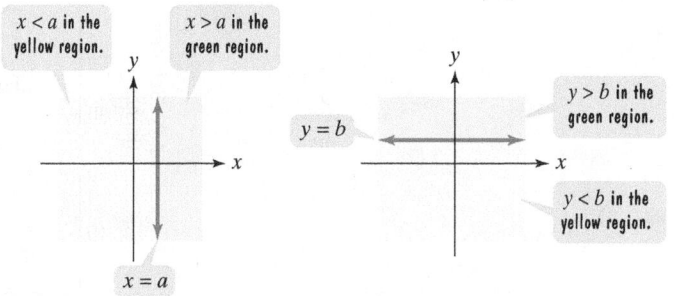

EXAMPLE 3 Graphing Inequalities without Using Test Points

Graph each inequality in a rectangular coordinate system:

a. $y \leq -3$　　　**b.** $x > 2$.

Solution

a. $y \leq -3$

> Graph $y = -3$, a horizontal line with y-intercept -3. The line is solid because equality is included in $y \leq -3$. Because of the less than part of $\leq$, shade the half-plane below the horizontal line.

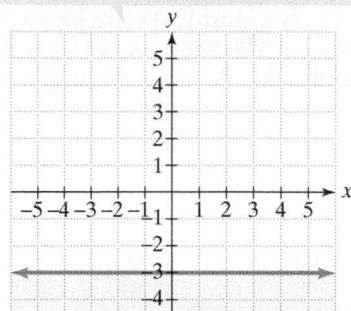

b. $x > 2$

> Graph $x = 2$, a vertical line with x-intercept **2**. The line is dashed because equality is not included in $x > 2$. Because of $>$, the greater than symbol, shade the half-plane to the right of the vertical line.

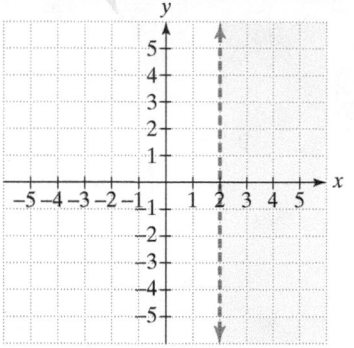

✓ **CHECK POINT 3**　Graph each inequality in a rectangular coordinate system:

a. $y > 1$　　　**b.** $x \leq -2$.

---

2 Use mathematical models involving linear inequalities.

Modeling with Systems of Linear Inequalities

Just as two or more linear equations make up a system of linear equations, two or more linear inequalities make up a **system of linear inequalities**. **A solution of a system of linear inequalities** in two variables is an ordered pair that satisfies each inequality in the system.

EXAMPLE 4 Forests, Grasslands, Deserts, and Systems of Inequalities

Temperature and precipitation affect whether or not trees and forests can grow. At certain levels of precipitation and temperature, only grasslands and deserts will exist. **Figure 9.17** shows three kinds of regions—deserts, grasslands, and forests—that result from various ranges of temperature, T, and precipitation, P.

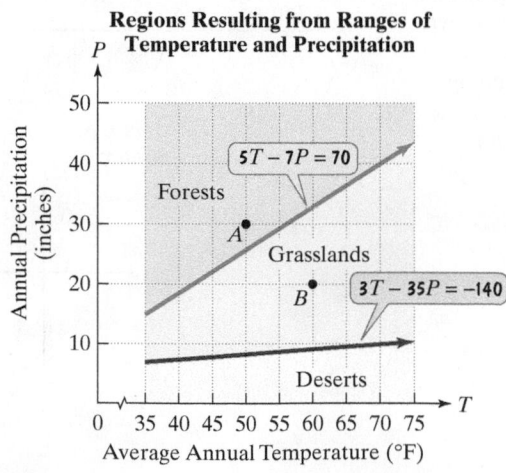

Regions Resulting from Ranges of Temperature and Precipitation

Figure 9.17

Source: Albert Miller et al., *Elements of Meteorology.* © 1983 Merrill Publishing Company

Systems of inequalities can be used to model where forests, grasslands, and deserts occur. Because these regions occur when the average annual temperature, T, is 35°F or greater, each system contains the inequality $T \geq 35$.

Forests occur if	Grasslands occur if	Deserts occur if
$\begin{cases} T \geq 35 \\ 5T - 7P < 70. \end{cases}$	$\begin{cases} T \geq 35 \\ 5T - 7P \geq 70 \\ 3T - 35P \leq -140. \end{cases}$	$\begin{cases} T \geq 35 \\ 3T - 35P > -140. \end{cases}$

Show that point A in **Figure 9.17** is a solution of the system of inequalities that models where forests occur.

Solution   Point A has coordinates $(50, 30)$. This means that if a region has an average annual temperature of 50°F and an average annual precipitation of 30 inches, a forest occurs. We can show that $(50, 30)$ satisfies the system of inequalities for forests by substituting 50 for T and 30 for P in each inequality in the system.

$$T \geq 35 \qquad\qquad\qquad 5T - 7P < 70$$
$$50 \geq 35, \quad \text{true} \qquad\qquad 5 \cdot 50 - 7 \cdot 30 \overset{?}{<} 70$$
$$250 - 210 \overset{?}{<} 70$$
$$40 < 70, \quad \text{true}$$

The coordinates $(50, 30)$ make each inequality true. Thus, $(50, 30)$ satisfies the system for forests. ■

✓ **CHECK POINT 4**   Show that point B in **Figure 9.17** is a solution of the system of inequalities that models where grasslands occur.

---

3 Graph a system of linear inequalities.

Graphing Systems of Linear Inequalities

The **solution set of a system of linear inequalities in two variables** is the set of all ordered pairs that satisfy each inequality in the system. Thus, to graph a system of inequalities in two variables, begin by graphing each individual inequality in the same rectangular coordinate system. Then find the region, if there is one, that is common to every graph in the system. This region of intersection gives a picture of the system's solution set.

EXAMPLE 5   Graphing a System of Linear Inequalities

Graph the solution set of the system:

$$\begin{cases} x - y < 1 \\ 2x + 3y \geq 12. \end{cases}$$

Solution   Replacing each inequality symbol with an equal sign indicates that we need to graph $x - y = 1$ and $2x + 3y = 12$. We can use intercepts to graph these lines.

$$x - y = 1 \qquad\qquad\qquad\qquad 2x + 3y = 12$$

x-intercept: $x - 0 = 1$ Set $y = 0$ in each equation. x-intercept: $2x + 3 \cdot 0 = 12$
$\qquad\qquad\qquad x = 1$ $\qquad\qquad\qquad\qquad\qquad\qquad\qquad 2x = 12$
The line passes through $(1, 0)$. $\qquad\qquad\qquad\qquad\qquad\qquad x = 6$
$\qquad\qquad\qquad\qquad\qquad\qquad\qquad\qquad\qquad$ The line passes through $(6, 0)$.

y-intercept: $0 - y = 1$ Set $x = 0$ in each equation. y-intercept: $2 \cdot 0 + 3y = 12$
$\qquad\qquad\qquad -y = 1$ $\qquad\qquad\qquad\qquad\qquad\qquad\qquad\qquad 3y = 12$
$\qquad\qquad\qquad\quad y = -1$ $\qquad\qquad\qquad\qquad\qquad\qquad\qquad\qquad y = 4$
The line passes through $(0, -1)$ $\qquad\qquad\qquad$ The line passes through $(0, 4)$.

Now we are ready to graph the solution set of the system of linear inequalities.

Graph $x - y < 1$. The blue line, $x - y = 1$, is dashed: Equality is not included in $x - y < 1$. Because (0, 0) makes the inequality true (0 − 0 < 1, or 0 < 1, is true), shade the half-plane containing (0, 0) in yellow.

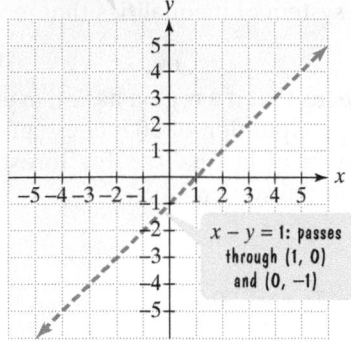

The graph of $x - y < 1$

Add the graph of $2x + 3y \geq 12$. The red line, $2x + 3y = 12$, is solid: Equality is included in $2x + 3y \geq 12$. Because (0, 0) makes the inequality false (2 · 0 + 3 · 0 ≥ 12, or 0 ≥ 12, is false), shade the half-plane not containing (0, 0) using green vertical shading.

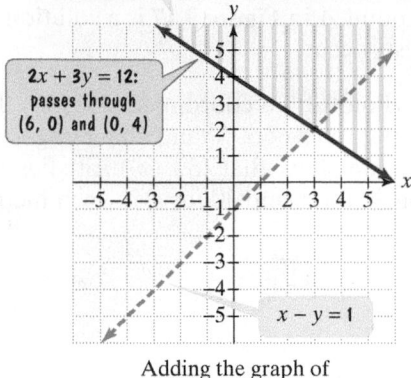

Adding the graph of
$2x + 3y \geq 12$

The solution set of the system is graphed as the intersection (the overlap) of the two half-planes. This is the region in which the yellow shading and the green vertical shading overlap.

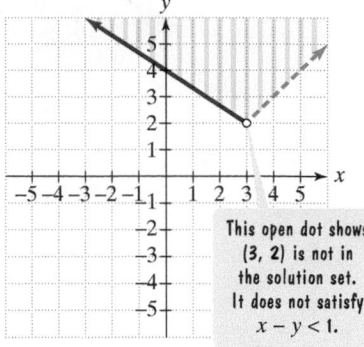

The graph of $x - y < 1$
and $2x + 3y \geq 12$

✓ **CHECK POINT 5**   Graph the solution set of the system:

$$\begin{cases} x - 3y < 6 \\ 2x + 3y \geq -6. \end{cases}$$

A system of inequalities has no solution if there are no points in the rectangular coordinate system that simultaneously satisfy each inequality in the system. For example, the system

$$\begin{cases} 2x + 3y \geq 6 \\ 2x + 3y \leq 0, \end{cases}$$

whose separate graphs are shown in **Figure 9.18**, has no overlapping region. Thus, the system has no solution. The solution set is $\varnothing$, the empty set.

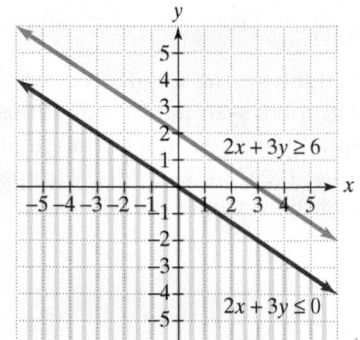

Figure 9.18 A system of inequalities with no solution

EXAMPLE 6   Graphing a System of Inequalities

Graph the solution set of the system:

$$\begin{cases} x - y < 2 \\ -2 \leq x < 4 \\ y < 3. \end{cases}$$

Solution We begin by graphing $x - y < 2$, the first given inequality. The line $x - y = 2$ has an x-intercept of 2 and a y-intercept of -2. The test point (0, 0) makes the inequality $x - y < 2$ true. The graph of $x - y < 2$ is shown in **Figure 9.19**.

Now, let's consider the second given inequality, $-2 \leq x < 4$. Replacing the inequality symbols by =, we obtain $x = -2$ and $x = 4$, graphed as red vertical lines in **Figure 9.20**. The line $x = 4$ is not included. Because x is between -2 and 4, we shade the region between the vertical lines. We must intersect this region with the yellow region in **Figure 9.19**. The resulting region is shown in yellow and green vertical shading in **Figure 9.20**.

Finally, let's consider the third given inequality, $y < 3$. Replacing the inequality symbol by $=$, we obtain $y = 3$, which graphs as a horizontal line. Because of the less than symbol in $y < 3$, the graph consists of the half-plane below the line $y = 3$. We must intersect this half-plane with the region in **Figure 9.20**. The resulting region is shown in yellow and green vertical shading in **Figure 9.21**. This region represents the graph of the solution set of the given system.

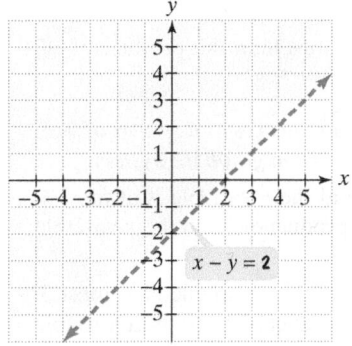

Figure 9.19 The graph of $x - y < 2$

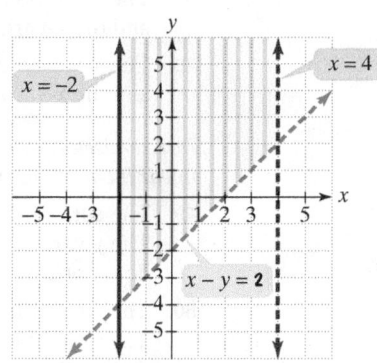

Figure 9.20 The graph of $x - y < 2$ and $-2 \le x < 4$

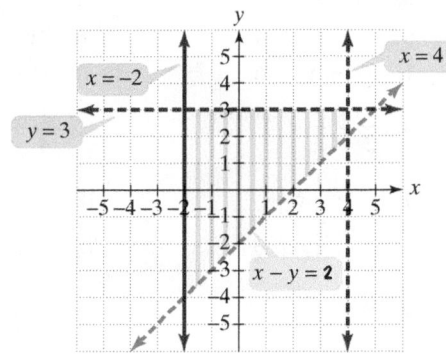

Figure 9.21 The graph of $x - y < 2$ and $-2 \le x < 4$ and $y < 3$

✓ **CHECK POINT 6**   Graph the solution set of the system:

$$\begin{cases} x + y < 2 \\ -2 \le x < 1 \\ \quad\quad y > -3. \end{cases}$$

CONCEPT AND VOCABULARY CHECK

Fill in each blank so that the resulting statement is true.

1. The ordered pair (3, 2) is a/an _____ of the inequality $x + y > 1$ because when 3 is substituted for _____ and 2 is substituted for _____, the true statement _____ is obtained.

2. The set of all points that satisfy a linear inequality in two variables is called the _____ of the inequality.

3. The set of all points on one side of a line is called a/an _____.

4. True or false: The graph of $2x - 3y > 6$ includes the line $2x - 3y = 6$. _____

5. True or false: The graph of the linear equation $2x - 3y = 6$ is used to graph the linear inequality $2x - 3y > 6$. _____

6. True or false: When graphing $4x - 2y \ge 8$, to determine which side of the line to shade, choose a test point on $4x - 2y = 8$. _____

7. The solution set of the system

$$\begin{cases} x - y < 1 \\ 2x + 3y \ge 12 \end{cases}$$

is the set of ordered pairs that satisfy _____ and _____.

8. True or false: The graph of the solution set of the system

$$\begin{cases} x - 3y < 6 \\ 2x + 3y \ge -6 \end{cases}$$

includes the intersection point of $x - 3y = 6$ and $2x + 3y = -6$. _____

9.4 EXERCISE SET

MyMathLab®  Watch the videos in MyMathLab  Download the MyDashBoard App

Practice Exercises

In Exercises 1–22, graph each inequality.

1. $x + y \geq 3$
2. $x + y \geq 2$
3. $x - y < 5$
4. $x - y < 6$
5. $x + 2y > 4$
6. $2x + y > 6$
7. $3x - y \leq 6$
8. $x - 3y \leq 6$
9. $\dfrac{x}{2} + \dfrac{y}{3} < 1$
10. $\dfrac{x}{4} + \dfrac{y}{2} < 1$
11. $y > \dfrac{1}{3}x$
12. $y > \dfrac{1}{4}x$
13. $y \leq 3x + 2$
14. $y \leq 2x - 1$
15. $y < -\dfrac{1}{4}x$
16. $y < -\dfrac{1}{3}x$
17. $x \leq 2$
18. $x \leq -4$
19. $y > -4$
20. $y > -2$
21. $y \geq 0$
22. $x \leq 0$

In Exercises 23–46, graph the solution set of each system of inequalities or indicate that the system has no solution.

23. $\begin{cases} 3x + 6y \leq 6 \\ 2x + y \leq 8 \end{cases}$
24. $\begin{cases} x - y \geq 4 \\ x + y \leq 6 \end{cases}$

25. $\begin{cases} 2x - 5y \leq 10 \\ 3x - 2y > 6 \end{cases}$
26. $\begin{cases} 2x - y \leq 4 \\ 3x + 2y > -6 \end{cases}$

27. $\begin{cases} y > 2x - 3 \\ y < -x + 6 \end{cases}$
28. $\begin{cases} y < -2x + 4 \\ y < x - 4 \end{cases}$

29. $\begin{cases} x + 2y \leq 4 \\ y \geq x - 3 \end{cases}$
30. $\begin{cases} x + y \leq 4 \\ y \geq 2x - 4 \end{cases}$

31. $\begin{cases} x \leq 2 \\ y \geq -1 \end{cases}$
32. $\begin{cases} x \leq 3 \\ y \leq -1 \end{cases}$

33. $-2 \leq x < 5$
34. $-2 < y \leq 5$

35. $\begin{cases} x - y \leq 1 \\ x \geq 2 \end{cases}$
36. $\begin{cases} 4x - 5y \geq -20 \\ x \geq -3 \end{cases}$

37. $\begin{cases} x + y > 4 \\ x + y < -1 \end{cases}$
38. $\begin{cases} x + y > 3 \\ x + y < -2 \end{cases}$

39. $\begin{cases} x + y > 4 \\ x + y > -1 \end{cases}$
40. $\begin{cases} x + y > 3 \\ x + y > -2 \end{cases}$

41. $\begin{cases} x - y \leq 2 \\ x \geq -2 \\ y \leq 3 \end{cases}$
42. $\begin{cases} 3x + y \leq 6 \\ x \geq -2 \\ y \leq 4 \end{cases}$

43. $\begin{cases} x \geq 0 \\ y \geq 0 \\ 2x + 5y \leq 10 \\ 3x + 4y \leq 12 \end{cases}$
44. $\begin{cases} x \geq 0 \\ y \geq 0 \\ 2x + y \leq 4 \\ 2x - 3y \leq 6 \end{cases}$

45. $\begin{cases} 3x + y \leq 6 \\ 2x - y \leq -1 \\ x \geq -2 \\ y \leq 4 \end{cases}$
46. $\begin{cases} 2x + y \leq 6 \\ x + y \geq 2 \\ 1 \leq x \leq 2 \\ y \leq 3 \end{cases}$

Practice PLUS

In Exercises 47–48, write each sentence as a linear inequality in two variables. Then graph the inequality.

47. The y-variable is at least 4 more than the product of -2 and the x-variable.

48. The y-variable is at least 2 more than the product of -3 and the x-variable.

In Exercises 49–50, write the given sentences as a system of linear inequalities in two variables. Then graph the system.

49. The sum of the x-variable and the y-variable is at most 4. The y-variable added to the product of 3 and the x-variable does not exceed 6.

50. The sum of the x-variable and the y-variable is at most 3. The y-variable added to the product of 4 and the x-variable does not exceed 6.

In Exercises 51–52, rewrite each inequality in the system without absolute value bars. Then graph the rewritten system in rectangular coordinates.

51. $\begin{cases} |x| \leq 2 \\ |y| \leq 3 \end{cases}$

52. $\begin{cases} |x| \leq 1 \\ |y| \leq 2 \end{cases}$

*The graphs of solution sets of systems of inequalities involve finding the intersection of the solution sets of two or more inequalities. By contrast, in Exercises 53–54 you will be graphing the **union** of the solution sets of two inequalities.*

53. Graph the union of $y > \frac{3}{2}x - 2$ and $y < 4$.
54. Graph the union of $x - y \geq -1$ and $5x - 2y \leq 10$.

Without graphing, in Exercises 55–58, determine if each system has no solution or infinitely many solutions.

55. $\begin{cases} 3x + y < 9 \\ 3x + y > 9 \end{cases}$

56. $\begin{cases} 6x - y \leq 24 \\ 6x - y > 24 \end{cases}$

57. $\begin{cases} 3x + y \leq 9 \\ 3x + y \geq 9 \end{cases}$

58. $\begin{cases} 6x - y \leq 24 \\ 6x - y \geq 24 \end{cases}$

Application Exercises

Maximum heart rate, H, in beats per minute is a function of age, a, modeled by the formula

$$H = 220 - a,$$

where $10 \leq a \leq 70$. The bar graph at the top of the next column shows the target heart rate ranges for four types of exercise goals in terms of maximum heart rate.

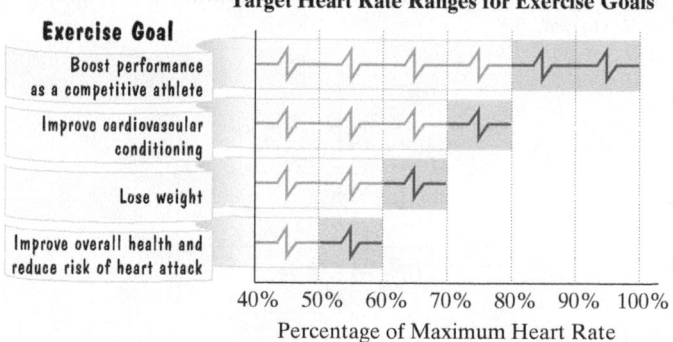

Target Heart Rate Ranges for Exercise Goals

Source: *Vitality* Magazine

In Exercises 59–62, systems of inequalities will be used to model three of the target heart rate ranges shown in the bar graph. We begin with the target heart rate range for cardiovascular conditioning, modeled by the following system of inequalities:

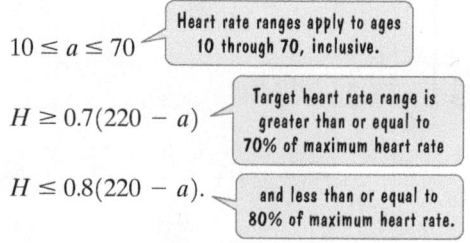

$10 \le a \le 70$ — Heart rate ranges apply to ages 10 through 70, inclusive.

$H \ge 0.7(220 - a)$ — Target heart rate range is greater than or equal to 70% of maximum heart rate

$H \le 0.8(220 - a)$. — and less than or equal to 80% of maximum heart rate.

The graph of this system is shown in the figure. Use the graph to solve Exercises 59–60.

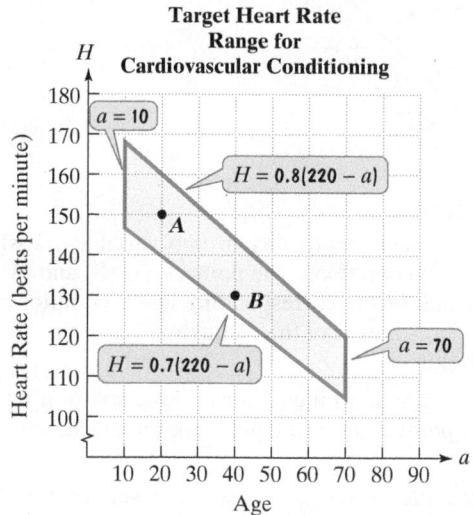

Target Heart Rate Range for Cardiovascular Conditioning

59. a. What are the coordinates of point *A* and what does this mean in terms of age and heart rate?

b. Show that point *A* is a solution of the system of inequalities.

60. a. What are the coordinates of point *B* and what does this mean in terms of age and heart rate?

b. Show that point *B* is a solution of the system of inequalities.

61. Write a system of inequalities that models the target heart rate range for the goal of losing weight.

62. Write a system of inequalities that models the target heart rate range for improving overall health.

63. On your next vacation, you will divide lodging between large resorts and small inns. Let *x* represent the number of nights spent in large resorts. Let *y* represent the number of nights spent in small inns.

a. Write a system of inequalities that models the following conditions:

You want to stay at least 5 nights. At least one night should be spent at a large resort. Large resorts average $200 per night and small inns average $100 per night. Your budget permits no more than $700 for lodging.

b. Graph the solution set of the system of inequalities in part (a).

c. Based on your graph in part (b), how many nights could you spend at a large resort and still stay within your budget?

64. a. An elevator can hold no more than 2000 pounds. If children average 80 pounds and adults average 160 pounds, write a system of inequalities that models when the elevator holding *x* children and *y* adults is overloaded.

b. Graph the solution set of the system of inequalities in part (a).

Writing in Mathematics

65. What is a linear inequality in two variables? Provide an example with your description.

66. How do you determine if an ordered pair is a solution of an inequality in two variables, *x* and *y*?

67. What is a half-plane?

68. What does a solid line mean in the graph of an inequality?

69. What does a dashed line mean in the graph of an inequality?

70. Explain how to graph $x - 2y < 4$.

71. What is a system of linear inequalities?

72. What is a solution of a system of linear inequalities?

73. Explain how to graph the solution set of a system of inequalities.

74. What does it mean if a system of linear inequalities has no solution?

Technology Exercises

Graphing utilities can be used to shade regions in the rectangular coordinate system, thereby graphing an inequality in two variables. Read the section of the user's manual for your graphing utility that describes how to shade a region. Then use your graphing utility to graph the inequalities in Exercises 75–78.

75. $y \leq 4x + 4$

76. $y \geq \dfrac{2}{3}x - 2$

77. $2x + y \leq 6$

78. $3x - 2y \geq 6$

79. Does your graphing utility have any limitations in terms of graphing inequalities? If so, what are they?

80. Use a graphing utility with a $\boxed{\text{SHADE}}$ feature to verify any five of the graphs that you drew by hand in Exercises 1–22.

81. Use a graphing utility with a $\boxed{\text{SHADE}}$ feature to verify any five of the graphs that you drew by hand for the systems in Exercises 23–46.

Critical Thinking Exercises

Make Sense? *In Exercises 82–85, determine whether each statement "makes sense" or "does not make sense" and explain your reasoning.*

82. When graphing a linear inequality, I should always use (0, 0) as a test point because it's easy to perform the calculations when 0 is substituted for each variable.

83. If you want me to graph $x < 3$, you need to tell me whether to use a number line or a rectangular coordinate system.

84. When graphing $3x - 4y < 12$, it's not necessary for me to graph the linear equation $3x - 4y = 12$ because the inequality contains a $<$ symbol, in which equality is not included.

85. Linear inequalities can model situations in which I'm interested in purchasing two items at different costs, I can spend no more than a specified amount on both items, and I want to know how many of each item I can purchase.

In Exercises 86–89, determine whether each statement is true or false. If the statement is false, make the necessary change(s) to produce a true statement.

86. The graph of $3x - 5y < 10$ consists of a dashed line and a shaded half-plane below the line.

87. The graph of $y \geq -x + 1$ consists of a solid line that rises from left to right and a shaded half-plane above the line.

88. The ordered pair $(-2, 40)$ satisfies the following system:
$$\begin{cases} y \geq 9x + 11 \\ 13x + y > 14. \end{cases}$$

89. For the graph of $y < x - 3$, the points $(0, -3)$ and $(8, 5)$ lie on the graph of the corresponding linear equation, but neither point is a solution of the inequality.

90. Write a linear inequality in two variables whose graph is shown.

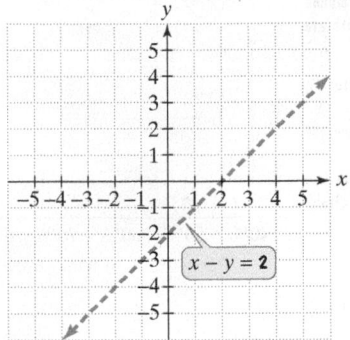

In Exercises 91–92, write a system of inequalities for each graph.

91.

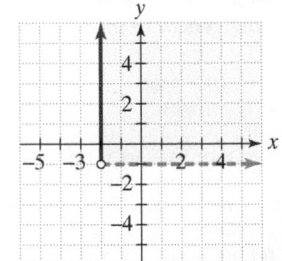

92.

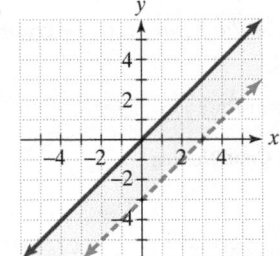

93. Write a linear inequality in two variables satisfying the following conditions: The points $(-3, -8)$ and $(4, 6)$ lie on the graph of the corresponding linear equation and each point is a solution of the inequality. The point $(1, 1)$ is also a solution.

94. Write a system of inequalities whose solution set includes every point in the rectangular coordinate system.

95. Sketch the graph of the solution set for the following system of inequalities:
$$\begin{cases} y \geq nx + b \ (n < 0, b > 0) \\ y \leq mx + b \ (m > 0, b > 0). \end{cases}$$

Review Exercises

96. Solve the system:
$$\begin{cases} 3x - y = 8 \\ x - 5y = -2. \end{cases}$$

(Section 4.3, Example 2)

97. Solve by graphing:

$$\begin{cases} y = 3x - 2 \\ y = -2x + 8. \end{cases}$$

(Section 4.1, Example 3)

98. Factor completely: $2x^6 + 20x^5y + 50x^4y^2$.

(Section 6.5, Example 8)

Preview Exercises

Exercises 99–101 will help you prepare for the material covered in the first section of the next chapter.

99. If $f(x) = \sqrt{3x + 12}$, find $f(-1)$.

100. If $f(x) = \sqrt{3x + 12}$, find $f(8)$.

101. Use the graph of $f(x) = \sqrt{3x + 12}$, to identify the function's domain and its range.

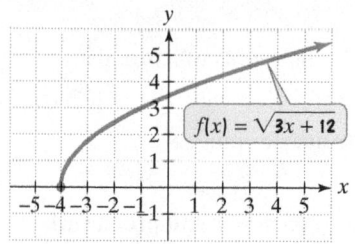

$f(x) = \sqrt{3x + 12}$

GROUP PROJECT

CHAPTER 9

Each group member should research one situation that provides two different pricing options. These can involve areas such as public transportation options (with or without discount passes) or telephone plans or anything of interest. Be sure to bring in all the details for each option. At the group meeting, select the two pricing situations that are most interesting and relevant. Using each situation, write a word problem about selecting the better of the two options. The word problem should be one that can be solved using a linear inequality. The group should turn in the two problems and their solutions.

Chapter 9 Summary

Definitions and Concepts	Examples

Section 9.1 Reviewing Linear Inequalities and Using Inequalities in Business Applications

A linear inequality in one variable can be written in the form $ax + b < 0$, $ax + b \le 0$, $ax + b > 0$, or $ax + b \ge 0$. The set of all numbers that make the inequality a true statement is its solution set, represented using interval notation.

Solving a Linear Inequality

1. Simplify each side.

2. Collect variable terms on one side and constant terms on the other side.

3. Isolate the variable and solve.

If an inequality is multiplied or divided by a negative number, the direction of the inequality symbol must be reversed.

Solve: $2(x + 3) - 5x \le 15$.

$$2x + 6 - 5x \le 15$$

$$-3x + 6 \le 15$$

$$-3x \le 9$$

$$\frac{-3x}{-3} \ge \frac{9}{-3}$$

$$x \ge -3$$

Solution set: $[-3, \infty)$

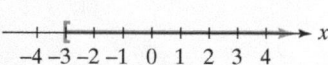

Definitions and Concepts	**Examples**

Section 9.1 Reviewing Linear Inequalities and Using Inequalities in Business Applications (continued)

Functions of Business

A company produces and sells x units of a product.

Revenue Function

$$R(x) = \text{(price per unit sold)}x$$

Cost Function

$$C(x) = \text{fixed cost} + \text{(cost per unit produced)}x$$

Profit Function

$$P(x) = R(x) - C(x)$$

A business makes money, or has a gain, when $P(x) > 0$.
A business loses money, or has a loss, when $P(x) < 0$.

A company that manufactures lamps has a fixed cost of $80,000 and it costs $20 to produce each lamp. Lamps are sold for $70.

a. Write the cost function.

$$C(x) = 80,000 + 20x$$

  Fixed cost   Variable cost: $20 per lamp.

b. Write the revenue function.

$$R(x) = 70x$$

  Revenue per lamp, $70, times number of lamps sold

c. Write the profit function.

$$P(x) = R(x) - C(x)$$
$$= 70x - (80,000 + 20x)$$
$$= 50x - 80,000$$

d. More than how many lamps must be produced and sold for the company to make money?

Solve $\qquad\qquad P(x) > 0.$
$$50x - 80,000 > 0$$
$$50x > 80,000$$
$$x > 1600$$

More than 1600 lamps must be produced and sold for the company to make money.

Section 9.2 Compound Inequalities

Intersection (∩) and Union (∪)

$A \cap B$ is the set of elements common to both set A and set B.
$A \cup B$ is the set of elements that are members of set A or set B or of both sets.

$\{1, 3, 5, 7\} \cap \{5, 7, 9, 11\} = \{5, 7\}$
$\{1, 3, 5, 7\} \cup \{5, 7, 9, 11\} = \{1, 3, 5, 7, 9, 11\}$

A compound inequality is formed by joining two inequalities with the word *and* or *or*.

When the connecting word is *and*, graph each inequality separately and take the intersection of their solution sets.

Solve: $x + 1 > 3$ and $x + 4 \le 8$.
$$x > 2 \quad \text{and} \quad x \le 4$$

Solution set: $(2, 4]$

The compound inequality $a < x < b$ means $a < x$ and $x < b$. Solve by isolating the variable in the middle.

Solve: $-1 < \dfrac{2x + 1}{3} \le 2$.

$$-3 < 2x + 1 \le 6 \qquad \text{Multiply by 3.}$$
$$-4 < 2x \le 5 \qquad \text{Subtract 1.}$$
$$-2 < x \le \frac{5}{2} \qquad \text{Divide by 2.}$$

Solution set: $\left(-2, \dfrac{5}{2}\right]$

Definitions and Concepts	**Examples**

Section 9.2 Compound Inequalities (continued)

When the connecting word in a compound inequality is *or*, graph each inequality separately and take the union of their solution sets.

Solve: $x - 2 > -3$ or $2x \le -6$.

$$x > -1 \quad \text{or} \quad x \le -3$$

Solution set: $(-\infty, -3] \cup (-1, \infty)$

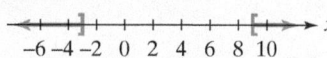

Section 9.3 Equations and Inequalities Involving Absolute Value

Absolute Value Equations

1. If $c > 0$, then $|u| = c$ means $u = c$ or $u = -c$.
2. If $c < 0$, then $|u| = c$ has no solution.
3. If $c = 0$, then $|u| = 0$ means $u = 0$.

Solve: $|2x - 7| = 3$.

$$2x - 7 = 3 \quad \text{or} \quad 2x - 7 = -3$$
$$2x = 10 \qquad\qquad 2x = 4$$
$$x = 5 \qquad\qquad x = 2$$

The solution set is $\{2, 5\}$.

Absolute Value Equations with Two Absolute Value Bars

If $|u| = |v|$, then $u = v$ or $u = -v$.

Solve: $|x - 6| = |2x + 1|$.

$$x - 6 = 2x + 1 \quad \text{or} \quad x - 6 = -(2x + 1)$$
$$-x - 6 = 1 \qquad\qquad x - 6 = -2x - 1$$
$$-x = 7 \qquad\qquad 3x - 6 = -1$$
$$x = -7 \qquad\qquad 3x = 5$$
$$x = \frac{5}{3}$$

The solutions are -7 and $\frac{5}{3}$, and the solution set is $\left\{-7, \frac{5}{3}\right\}$.

Solving Absolute Value Inequalities

If c is a positive number,

1. $|u| < c$ is equivalent to $-c < u < c$.
2. $|u| > c$ is equivalent to $u < -c$ or $u > c$.

In each case, the absolute value inequality is rewritten as an equivalent compound inequality without absolute value bars.

Solve: $|x - 4| < 3$.

$$-3 < x - 4 < 3$$
$$1 < x < 7 \quad \text{Add 4.}$$

The solution set is $(1, 7)$.

Solve: $\left|\dfrac{x}{3} - 1\right| \ge 2$.

$$\frac{x}{3} - 1 \le -2 \quad \text{or} \quad \frac{x}{3} - 1 \ge 2.$$
$$x - 3 \le -6 \quad \text{or} \quad x - 3 \ge 6 \quad \text{Multiply by 3.}$$
$$x \le -3 \quad \text{or} \qquad x \ge 9 \quad \text{Add 3.}$$

The solution set is $(-\infty, -3] \cup [9, \infty)$.

Absolute Value Inequalities with Unusual Solution Sets

If c is a negative number,

1. $|u| < c$ has no solution.
2. $|u| > c$ is true for all real numbers for which u is defined.

- $|x - 4| < -3$ has no solution. The solution set is $\varnothing$.
- $|3x + 6| > -12$ is true for all real numbers. The solution set is $(-\infty, \infty)$.

Definitions and Concepts	**Examples**

Section 9.4 Linear Inequalities in Two Variables

If the equal sign in $Ax + By = C$ is replaced with an inequality symbol, the result is a linear inequality in two variables. Its graph is the set of all points whose coordinates satisfy the inequality. To obtain the graph,

1. Replace the inequality symbol with an equal sign and graph the boundary line. Use a solid line for $\leq$ or $\geq$ and a dashed line for $<$ or $>$.

2. Choose a test point not on the line and substitute its coordinates into the inequality.

3. If a true statement results, shade the half-plane containing the test point. If a false statement results, shade the half-plane not containing the test point.

Graph: $x - 2y \leq 4$.

1. Graph $x - 2y = 4$. Use a solid line because the inequality symbol is $\leq$.

2. Test $(0, 0)$.

$$x - 2y \leq 4$$
$$0 - 2 \cdot 0 \overset{?}{\leq} 4$$
$$0 \leq 4, \quad \text{true}$$

3. The inequality is true. Shade the half-plane containing $(0, 0)$.

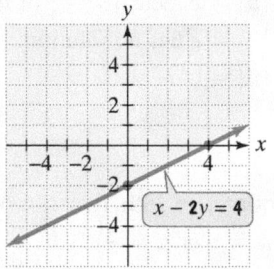

Two or more linear inequalities make up a system of linear inequalities. A solution is an ordered pair satisfying all inequalities in the system. To graph a system of inequalities, graph each inequality in the system. The overlapping region, if there is one, represents the solutions of the system. If there is no overlapping region, the system has no solution.

Graph the solutions of the system:

$$\begin{cases} y \leq -2x \\ x - y \geq 3. \end{cases}$$

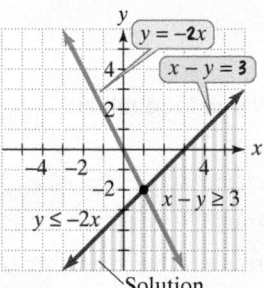

CHAPTER 9 REVIEW EXERCISES

9.1 *In Exercises 1–5, solve each linear inequality and graph the solution set on a number line.*

1. $-6x + 3 \leq 15$

2. $6x - 9 \geq -4x - 3$

3. $\dfrac{x}{3} - \dfrac{3}{4} - 1 > \dfrac{x}{2}$

4. $6x + 5 > -2(x - 3) - 25$

5. $3(2x - 1) - 2(x - 4) \geq 7 + 2(3 + 4x)$

6. The cost and revenue functions for producing and selling x units of a toaster oven are

$$C(x) = 40x + 357{,}000 \quad \text{and} \quad R(x) = 125x.$$

a. Write the profit function, P, from producing and selling x toaster ovens.

b. More than how many toaster ovens must be produced and sold for the business to make money?

Use this information to solve Exercises 7–10: A company is planning to produce and sell a new line of computers. The fixed cost will be $360,000 and it will cost $850 to produce each computer. Each computer will be sold for $1150.

7. Write the cost function, C, of producing x computers.

8. Write the revenue function, R, from the sale of x computers.

9. Write the profit function, P, from producing and selling x computers.

10. More than how many computers must be produced and sold for the business to make money?

11. A person can choose between two charges on a checking account. The first method involves a fixed cost of $11 per month plus 6¢ for each check written. The second method involves a fixed cost of $4 per month plus 20¢ for each check written. How many checks should be written to make the first method a better deal?

12. A salesperson earns $500 per month plus a commission of 20% of sales. Describe the sales needed to receive a total income that exceeds $3200 per month.

9.2 *In Exercises 13–16, let $A = \{a, b, c\}$, $B = \{a, c, d, e\}$, and $C = \{a, d, f, g\}$. Find the indicated set.*

13. $A \cap B$

14. $A \cap C$

15. $A \cup B$

16. $A \cup C$

In Exercises 17–27, solve each compound inequality. Other than $\varnothing$, graph the solution set on a number line.

17. $x \le 3$ and $x < 6$

18. $x \le 3$ or $x < 6$

19. $-2x < -12$ and $x - 3 < 5$

20. $5x + 3 \le 18$ and $2x - 7 \le -5$

21. $2x - 5 > -1$ and $3x < 3$

22. $2x - 5 > -1$ or $3x < 3$

23. $x + 1 \le -3$ or $-4x + 3 < -5$

24. $5x - 2 \le -22$ or $-3x - 2 > 4$

25. $5x + 4 \ge -11$ or $1 - 4x \ge 9$

26. $-3 < x + 2 \le 4$

27. $-1 \le 4x + 2 \le 6$

28. To receive a B in a course, you must have an average of at least 80% but less than 90% on five exams. Your grades on the first four exams were 95%, 79%, 91%, and 86%. What range of grades on the fifth exam will result in a B for the course? Use interval notation to express this range.

9.3 *In Exercises 29–32, find the solution set for each equation.*

29. $|2x + 1| = 7$

30. $|3x + 2| = -5$

31. $2|x - 3| - 7 = 10$

32. $|4x - 3| = |7x + 9|$

In Exercises 33–37, solve each absolute value inequality. Other than $\varnothing$, graph the solution set on a number line.

33. $|2x + 3| \le 15$

34. $\left| \dfrac{2x + 6}{3} \right| > 2$

35. $|2x + 5| - 7 < -6$

36. $-4|x + 2| + 5 \le -7$

37. $|2x - 3| + 4 \le -10$

38. Approximately 90% of the population sleeps h hours daily, where h is modeled by the inequality $|h - 6.5| \le 1$. Write a sentence describing the range for the number of hours that most people sleep. Do *not* use the phrase "absolute value" in your description.

9.4 *In Exercises 39–44, graph each inequality in a rectangular coordinate system.*

39. $3x - 4y > 12$

40. $x - 3y \le 6$

41. $y \le -\dfrac{1}{2}x + 2$

42. $y > \dfrac{3}{5}x$

43. $x \le 2$

44. $y > -3$

In Exercises 45–53, graph the solution set of each system of inequalities or indicate that the system has no solution.

45. $\begin{cases} 2x - y \le 4 \\ x + y \ge 5 \end{cases}$

46. $\begin{cases} y < -x + 4 \\ y > x - 4 \end{cases}$

47. $-3 \le x < 5$

48. $-2 < y \le 6$

49. $\begin{cases} x \ge 3 \\ y \le 0 \end{cases}$

50. $\begin{cases} 2x - y > -4 \\ x \ge 0 \end{cases}$

51. $\begin{cases} x + y \le 6 \\ y \ge 2x - 3 \end{cases}$

52. $\begin{cases} 3x + 2y \ge 4 \\ x - y \le 3 \\ x \ge 0, y \ge 0 \end{cases}$

53. $\begin{cases} 2x - y > 2 \\ 2x - y < -2 \end{cases}$

CHAPTER 9 TEST

CHAPTER
Test Prep
VIDEOS

Step-by-step test solutions are found on the Chapter Test Prep Videos available in MyMathLab® or on YouTube (search "BlitzerCombinedAlg" and click on "Channels").

In Exercises 1–2, solve and graph the solution set on a number line.

1. $3(x + 4) \ge 5x - 12$

2. $\dfrac{x}{6} + \dfrac{1}{8} \le \dfrac{x}{2} - \dfrac{3}{4}$

3. A company is planning to manufacture computer desks. The fixed cost will be $60,000 and it will cost $200 to produce each desk. Each desk will be sold for $450.

   a. Write the cost function, C, of producing x desks.

   b. Write the revenue function, R, from the sale of x desks.

c. Write the profit function, P, from producing and selling x desks.

d. More than how many desks must be produced and sold for the business to make money?

4. Find the intersection: $\{2, 4, 6, 8, 10\} \cap \{4, 6, 12, 14\}$.

5. Find the union: $\{2, 4, 6, 8, 10\} \cup \{4, 6, 12, 14\}$.

In Exercises 6–10, solve each compound inequality. Other than $\varnothing$, graph the solution set on a number line.

6. $2x + 4 < 2$ and $x - 3 > -5$

7. $x + 6 \geq 4$ and $2x + 3 \geq -2$

8. $2x - 3 < 5$ or $3x - 6 \leq 4$

9. $x + 3 \leq -1$ or $-4x + 3 < -5$

10. $-3 \leq \dfrac{2x + 5}{3} < 6$

In Exercises 11–12, find the solution set for each equation.

11. $|5x + 3| = 7$

12. $|6x + 1| = |4x + 15|$

In Exercises 13–14, solve and graph the solution set on a number line.

13. $|2x - 1| < 7$

14. $|2x - 3| \geq 5$

15. The inequality $|b - 98.6| > 8$ describes a person's body temperature, b, in degrees Fahrenheit, when hyperthermia (extremely high body temperature) or hypothermia (extremely low body temperature) occurs. Solve the inequality and interpret the solution.

In Exercises 16–18, graph each inequality in a rectangular coordinate system.

16. $3x - 2y < 6$ 17. $y \geq \dfrac{1}{2}x - 1$ 18. $y \leq -1$

In Exercises 19–21, graph the solution set of each system of inequalities.

19. $\begin{cases} x + y \geq 2 \\ x - y \geq 4 \end{cases}$ 20. $\begin{cases} 3x + y \leq 9 \\ 2x + 3y \geq 6 \\ x \geq 0, y \geq 0 \end{cases}$ 21. $-2 < x \leq 4$

CUMULATIVE REVIEW EXERCISES (CHAPTERS 1–9)

In Exercises 1–2, solve each equation.

1. $5(x + 1) + 2 = x - 3(2x + 1)$

2. $\dfrac{2(x + 6)}{3} = 1 + \dfrac{4x - 7}{3}$

3. Simplify: $\dfrac{-10x^2y^4}{15x^7y^{-3}}$.

4. If $f(x) = x^2 - 3x + 4$, find $f(-3)$ and $f(2a)$.

5. If $f(x) = 3x^2 - 4x + 1$ and $g(x) = x^2 - 5x - 1$, find $(f - g)(x)$ and $(f - g)(2)$.

6. Use function notation to write the equation of the line passing through $(2, 3)$ and perpendicular to the line whose equation is $y = 2x - 3$.

In Exercises 7–10, graph each equation or inequality in a rectangular coordinate system.

7. $f(x) = 2x + 1$ 8. $y > 2x$

9. $2x - y \geq 6$ 10. $f(x) = -1$

11. Solve the system:

$$\begin{cases} 3x - y + z = -15 \\ x + 2y - z = 1. \\ 2x + 3y - 2z = 0 \end{cases}$$

12. If $f(x) = \dfrac{x}{3} - 4$, find $f^{-1}(x)$.

13. If $f(x) = 3x^2 - 1$ and $g(x) = x + 2$, find $f(g(x))$ and $g(f(x))$.

14. A motel with 60 rooms charges $90 per night for rooms with kitchen facilities and $80 per night for rooms without kitchen facilities. When all rooms are occupied, the nightly revenue is $5260. How many rooms of each kind are there?

15. Which of the following are functions?

a.
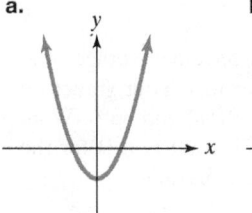

b.
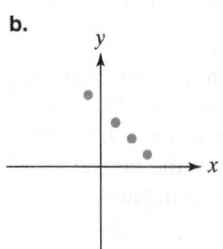

c.
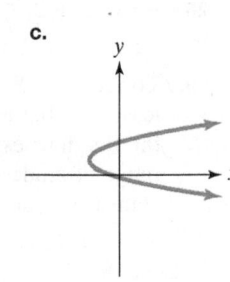

In Exercises 16–20, solve and graph the solution set on a number line.

16. $\dfrac{x}{4} - \dfrac{3}{4} - 1 \leq \dfrac{x}{2}$

17. $2x + 5 \leq 11$ and $-3x > 18$

18. $x - 4 \geq 1$ or $-3x + 1 \geq -5 - x$

19. $|2x + 3| \leq 17$

20. $|3x - 8| > 7$

Radicals, Radical Functions, and Rational Exponents

Can mathematical models be created for events that appear to involve random behavior, such as stock market fluctuations or air turbulence? Chaos theory, a new frontier of mathematics, offers models and computer-generated images that reveal order and underlying patterns where only the erratic and the unpredictable had been observed. Because most behavior is chaotic, the computer has become a canvas that looks more like the real world than anything previously seen. Magnified portions of these computer images yield repetitions of the original structure, as well as new and unexpected patterns. The computer generates these visualizations of chaos by plotting large numbers of points for functions whose domains are nonreal numbers involving the square root of negative one.

We define $\sqrt{-1}$ in Section 10.7 and hint at chaotic possibilities in the Blitzer Bonus on page 759. If you are intrigued by how the operations of nonreal numbers in Section 10.7 reveal that the world is not random [rather, the underlying patterns are far more intricate than we had previously assumed], we suggest reading *Chaos* by James Gleick, published by Penguin Books.

Radical Expressions and Functions

Objectives

1 Evaluate square roots.

2 Evaluate square root functions.

3 Find the domain of square root functions.

4 Use models that are square root functions.

5 Simplify expressions of the form $\sqrt{a^2}$.

6 Evaluate cube root functions.

7 Simplify expressions of the form $\sqrt[3]{a^3}$.

8 Find even and odd roots.

9 Simplify expressions of the form $\sqrt[n]{a^n}$.

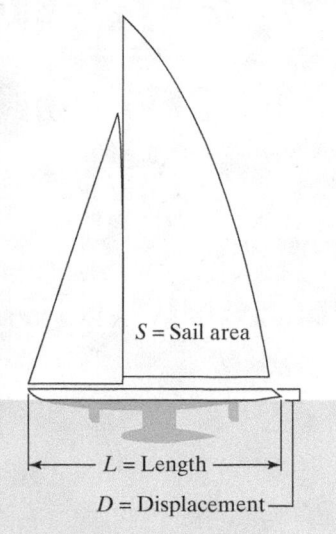

S = Sail area

L = Length

D = Displacement

The America's Cup is the supreme event in ocean sailing. Competition is fierce and the costs are huge. Competitors look to mathematics to provide the critical innovation that can make the difference between winning and losing. The basic dimensions of competitors' yachts must satisfy an inequality containing square roots and cube roots:

$$L + 1.25\sqrt{S} - 9.8\sqrt[3]{D} \le 16.296.$$

In the inequality, L is the yacht's length, in meters, S is its sail area, in square meters, and D is its displacement, in cubic meters.

In this section, we introduce a new category of expressions and functions that contain roots. You will see why square root functions are used to describe phenomena that are continuing to grow but whose growth is leveling off.

1 Evaluate square roots.

Square Roots

From our earlier work with exponents, we are aware that the square of 5 and the square of -5 are both 25:

$$5^2 = 25 \quad \text{and} \quad (-5)^2 = 25.$$

The reverse operation of squaring a number is finding the *square root* of the number. For example,

- One square root of 25 is 5 because $5^2 = 25$.
- Another square root of 25 is -5 because $(-5)^2 = 25$.

In general, **if $b^2 = a$, then b is a square root of a**.

The symbol $\sqrt{}$ is used to denote the *positive* or *principal square root* of a number. For example,

- $\sqrt{25} = 5$ because $5^2 = 25$ and 5 is positive.
- $\sqrt{100} = 10$ because $10^2 = 100$ and 10 is positive.

The symbol $\sqrt{}$ that we use to denote the principal square root is called a **radical sign**. The number under the radical sign is called the **radicand**. Together we refer to the radical sign and its radicand as a **radical expression**.

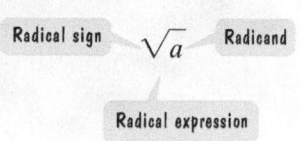

Radical sign $\sqrt{a}$ Radicand

Radical expression

Definition of the Principal Square Root

If a is a nonnegative real number, the nonnegative number b such that $b^2 = a$, denoted by $b = \sqrt{a}$, is the **principal square root** of a.

The symbol $-\sqrt{}$ is used to denote the negative square root of a number. For example,

- $-\sqrt{25} = -5$ because $(-5)^2 = 25$ and -5 is negative.
- $-\sqrt{100} = -10$ because $(-10)^2 = 100$ and -10 is negative.

EXAMPLE 1   Evaluating Square Roots

Evaluate:

a. $\sqrt{81}$ 　　b. $-\sqrt{9}$ 　　c. $\sqrt{\dfrac{4}{49}}$

d. $\sqrt{0.0064}$ 　　e. $\sqrt{36 + 64}$ 　　f. $\sqrt{36} + \sqrt{64}$.

Great Question!

Is $\sqrt{a + b}$ equal to $\sqrt{a} + \sqrt{b}$?

No. In Example 1, parts (e) and (f), observe that $\sqrt{36 + 64}$ is not equal to $\sqrt{36} + \sqrt{64}$. In general,

$$\sqrt{a + b} \neq \sqrt{a} + \sqrt{b}$$

and

$$\sqrt{a - b} \neq \sqrt{a} - \sqrt{b}.$$

Solution

a. $\sqrt{81} = 9$ 　　The principal square root of 81 is 9 because $9^2 = 81$.

b. $-\sqrt{9} = -3$ 　　The negative square root of 9 is -3 because $(-3)^2 = 9$.

c. $\sqrt{\dfrac{4}{49}} = \dfrac{2}{7}$ 　　The principal square root of $\dfrac{4}{49}$ is $\dfrac{2}{7}$ because $\left(\dfrac{2}{7}\right)^2 = \dfrac{4}{49}$.

d. $\sqrt{0.0064} = 0.08$ 　　The principal square root of 0.0064 is 0.08 because $(0.08)^2 = (0.08)(0.08) = 0.0064$.

e. $\sqrt{36 + 64} = \sqrt{100}$ 　　Simplify the radicand.

　　$= 10$ 　　Take the principal square root of 100, which is 10.

f. $\sqrt{36} + \sqrt{64} = 6 + 8$ 　　$\sqrt{36} = 6$ because $6^2 = 36$. $\sqrt{64} = 8$ because $8^2 = 64$.

　　$= 14$ ■

✓ **CHECK POINT 1**   Evaluate:

a. $\sqrt{64}$ 　　b. $-\sqrt{49}$ 　　c. $\sqrt{\dfrac{16}{25}}$

d. $\sqrt{0.0081}$ 　　e. $\sqrt{9 + 16}$ 　　f. $\sqrt{9} + \sqrt{16}$.

Let's see what happens to the radical expression $\sqrt{x}$ if x is a negative number. Is the square root of a negative number a real number? For example, consider $\sqrt{-25}$. Is there a real number whose square is -25? No. Thus, $\sqrt{-25}$ is not a real number. In general, **a square root of a negative number is not a real number**.

2 Evaluate square root functions.

Square Root Functions

Because each nonnegative real number, x, has precisely one principal square root, $\sqrt{x}$, there is a **square root function** defined by

$$f(x) = \sqrt{x}.$$

The domain of this function is $[0, \infty)$. We can graph $f(x) = \sqrt{x}$ by selecting nonnegative real numbers for x. It is easiest to choose perfect squares, numbers that have rational square roots. **Table 10.1** shows five such choices for x and the calculations for the corresponding outputs. We plot these ordered pairs as points in the rectangular coordinate system and connect the points with a smooth curve. The graph of $f(x) = \sqrt{x}$ is shown in **Figure 10.1**.

Table 10.1		
x	$f(x) = \sqrt{x}$	(x, y) or $(x, f(x))$
0	$f(0) = \sqrt{0} = 0$	$(0, 0)$
1	$f(1) = \sqrt{1} = 1$	$(1, 1)$
4	$f(4) = \sqrt{4} = 2$	$(4, 2)$
9	$f(9) = \sqrt{9} = 3$	$(9, 3)$
16	$f(16) = \sqrt{16} = 4$	$(16, 4)$

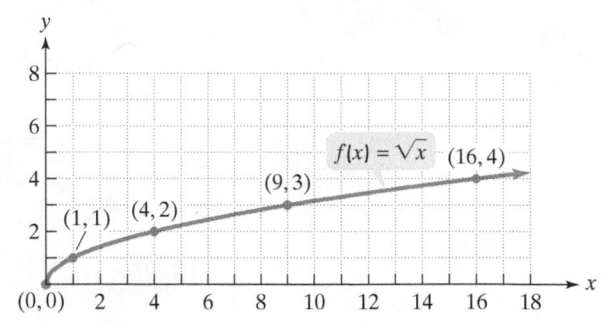

Figure 10.1 The graph of the square root function $f(x) = \sqrt{x}$

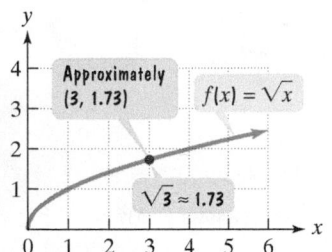

Figure 10.2 Visualizing $\sqrt{3}$ as a point on the graph of $f(x) = \sqrt{x}$

Is it possible to choose values of x for **Table 10.1** that are not squares of integers, or perfect squares? Yes. For example, we can let $x = 3$. Thus, $f(3) = \sqrt{3}$. Because 3 is not a perfect square, $\sqrt{3}$ is an irrational number, one that cannot be expressed as a quotient of integers. We can use a calculator to find a decimal approximation of $\sqrt{3}$.

Many Scientific Calculators

3 $\boxed{\sqrt{}}$

Many Graphing Calculators

$\boxed{\sqrt{}}$ 3 $\boxed{\text{ENTER}}$

Rounding the displayed number to two decimal places, $\sqrt{3} \approx 1.73$. This information is shown visually as a point, approximately $(3, 1.73)$, on the graph of $f(x) = \sqrt{x}$ in **Figure 10.2**.

To evaluate a square root function, we use substitution, just as we have done to evaluate other functions.

EXAMPLE 2 Evaluating Square Root Functions

For each function, find the indicated function value:

a. $f(x) = \sqrt{5x - 6}; f(2)$ **b.** $g(x) = -\sqrt{64 - 8x}; g(-3)$.

Solution

a. $f(2) = \sqrt{5 \cdot 2 - 6}$ Substitute 2 for x in $f(x) = \sqrt{5x - 6}$.
$\quad\quad = \sqrt{4} = 2$ Simplify the radicand and take the square root.

b. $g(-3) = -\sqrt{64 - 8(-3)}$ Substitute -3 for x in $g(x) = -\sqrt{64 - 8x}$.
$\quad\quad = -\sqrt{88} \approx -9.38$ Simplify the radicand:
$\quad\quad\quad\quad\quad\quad\quad\quad\quad$ $64 - 8(-3) = 64 - (-24) = 64 + 24 = 88.$
$\quad\quad\quad\quad\quad\quad\quad\quad\quad$ Then use a calculator to approximate $\sqrt{88}$. ∎

✓ **CHECK POINT 2** For each function, find the indicated function value:

a. $f(x) = \sqrt{12x - 20}; f(3)$
b. $g(x) = -\sqrt{9 - 3x}; g(-5)$.

---

3 Find the domain of square root functions.

We have seen that the domain of a function f is the largest set of real numbers for which the value of $f(x)$ is a real number. Because only nonnegative numbers have real square roots, the domain of a square root function is the set of real numbers for which the radicand is nonnegative.

EXAMPLE 3   Finding the Domain of a Square Root Function

Find the domain of

$$f(x) = \sqrt{3x + 12}.$$

Solution   The domain is the set of real numbers, x, for which the radicand, $3x + 12$, is nonnegative. We set the radicand greater than or equal to 0 and solve the resulting inequality.

$$3x + 12 \geq 0$$
$$3x \geq -12$$
$$x \geq -4$$

The domain of f is $[-4, \infty)$. ■

Figure 10.3 shows the graph of $f(x) = \sqrt{3x + 12}$ in a $[-10, 10, 1]$ by $[-10, 10, 1]$ viewing rectangle. The graph appears only for $x \geq -4$, verifying $[-4, \infty)$ as the domain. Can you see how the graph also illustrates this square root function's range? The graph only appears for nonnegative values of y. Thus, the range is $[0, \infty)$.

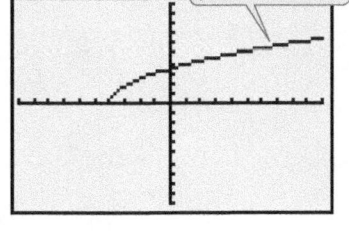

Figure 10.3

✓ **CHECK POINT 3**   Find the domain of
$$f(x) = \sqrt{9x - 27}.$$

4 Use models that are square root functions.

The graph of the square root function $f(x) = \sqrt{x}$ is increasing from left to right. However, the rate of increase is slowing down as the graph moves to the right. This is why square root functions are often used to model growing phenomena with growth that is leveling off.

EXAMPLE 4   Modeling with a Square Root Function

By 2008, the amount of "clutter," including commercials and plugs for other shows, had increased to the point where an "hour-long" drama on cable TV was 45.1 minutes. The graph in **Figure 10.4** shows the average number of nonprogram minutes in an hour of prime-time cable television. Although the minutes of clutter grew from 1996 through 2008, the growth was leveling off. The data can be modeled by the function

$$M(x) = 0.7\sqrt{x} + 12.5,$$

where $M(x)$ is the average number of nonprogram minutes in an hour of prime-time cable x years after 1996. According to the model, in 2002, how many cluttered minutes disrupted cable TV action in an hour? Round to the nearest tenth of a minute. What is the difference between the actual data and the number of minutes that you obtained?

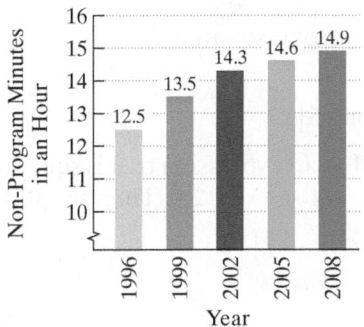

Figure 10.4

Source: The Nielsen Media Company, Monitor-Plus

Solution   Because 2002 is 6 years after 1996, we substitute 6 for x and evaluate the function at 6.

$$M(x) = 0.7\sqrt{x} + 12.5 \qquad \text{Use the given function.}$$
$$M(6) = 0.7\sqrt{6} + 12.5 \qquad \text{Substitute 6 for x.}$$
$$\approx 14.2 \qquad \text{Use a calculator.}$$

The model indicates that there were approximately 14.2 nonprogram minutes in an hour of prime-time cable in 2002. **Figure 10.4** shows 14.3 minutes, so the difference is $14.3 - 14.2$, or 0.1 minute. ■

✓ **CHECK POINT 4**   If the trend from 1996 through 2008 continues, use the square root function in Example 4 to predict how many cluttered minutes, rounded to the nearest tenth, there will be in an hour in 2014.

⑤ Simplify expressions of the form $\sqrt{a^2}$.

Simplifying Expressions of the Form $\sqrt{a^2}$

You may think that $\sqrt{a^2} = a$. However, this is not necessarily true. Consider the following examples:

$$\sqrt{4^2} = \sqrt{16} = 4$$
$$\sqrt{(-4)^2} = \sqrt{16} = 4.$$

The result is not −4, but rather the absolute value of −4, or 4.

Here is a rule for simplifying expressions of the form $\sqrt{a^2}$:

Simplifying $\sqrt{a^2}$

For any real number a,

$$\sqrt{a^2} = |a|.$$

In words, the principal square root of a^2 is the absolute value of a.

Using Technology

Graphic Connections

The graphs of $f(x) = \sqrt{x^2}$ and $g(x) = |x|$ are shown in a $[-10, 10, 1]$ by $[-2, 10, 1]$ viewing rectangle. The graphs are the same. Thus,

$$\sqrt{x^2} = |x|.$$

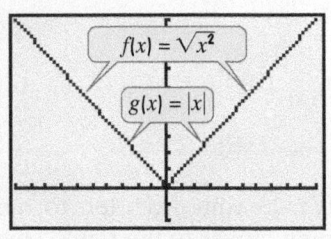

$f(x) = \sqrt{x^2}$

$g(x) = |x|$

EXAMPLE 5 Simplifying Radical Expressions

Simplify each expression:

a. $\sqrt{(-6)^2}$ **b.** $\sqrt{(x+5)^2}$ **c.** $\sqrt{25x^6}$ **d.** $\sqrt{x^2 - 4x + 4}$.

Solution The principal square root of an expression squared is the absolute value of that expression. In parts (a) and (b), we are given squared radicands. In parts (c) and (d), it will first be necessary to express the radicand as an expression that is squared.

a. $\sqrt{(-6)^2} = |-6| = 6$

b. $\sqrt{(x+5)^2} = |x+5|$

c. To simplify $\sqrt{25x^6}$, first write $25x^6$ as an expression that is squared: $25x^6 = (5x^3)^2$. Then simplify.

$$\sqrt{25x^6} = \sqrt{(5x^3)^2} = |5x^3| \quad \text{or} \quad 5|x^3|$$

d. To simplify $\sqrt{x^2 - 4x + 4}$, first write $x^2 - 4x + 4$ as an expression that is squared by factoring the perfect square trinomial: $x^2 - 4x + 4 = (x-2)^2$. Then simplify.

$$\sqrt{x^2 - 4x + 4} = \sqrt{(x-2)^2} = |x-2| \quad\blacksquare$$

✓ **CHECK POINT 5** Simplify each expression:

a. $\sqrt{(-7)^2}$

b. $\sqrt{(x+8)^2}$

c. $\sqrt{49x^{10}}$

d. $\sqrt{x^2 - 6x + 9}$.

In some situations, we are told that no radicands involve negative quantities raised to even powers. When the expression being squared is nonnegative, it is not necessary to use absolute value when simplifying $\sqrt{a^2}$. For example, assuming that no radicands contain negative quantities that are squared,

$$\sqrt{x^6} = \sqrt{(x^3)^2} = x^3$$
$$\sqrt{25x^2 + 10x + 1} = \sqrt{(5x+1)^2} = 5x + 1.$$

6 Evaluate cube root functions.

Cube Roots and Cube Root Functions

Finding the square root of a number reverses the process of squaring a number. Similarly, finding the cube root of a number reverses the process of cubing a number. For example, $2^3 = 8$, and so the cube root of 8 is 2. The notation that we use is $\sqrt[3]{8} = 2$.

Definition of the Cube Root of a Number

The **cube root** of a real number a is written $\sqrt[3]{a}$.

$$\sqrt[3]{a} = b \quad \text{means that} \quad b^3 = a.$$

Great Question!

Should I know the cube roots of certain numbers by heart?

Some cube roots occur so frequently that you might want to memorize them.

$\sqrt[3]{1} = 1$
$\sqrt[3]{8} = 2$
$\sqrt[3]{27} = 3$
$\sqrt[3]{64} = 4$
$\sqrt[3]{125} = 5$
$\sqrt[3]{216} = 6$
$\sqrt[3]{1000} = 10$

For example,

$$\sqrt[3]{64} = 4 \quad \text{because} \quad 4^3 = 64.$$
$$\sqrt[3]{-27} = -3 \quad \text{because} \quad (-3)^3 = -27.$$

In contrast to square roots, the cube root of a negative number is a real number. All real numbers have cube roots. The cube root of a positive number is positive. The cube root of a negative number is negative.

Because every real number, x, has precisely one cube root, $\sqrt[3]{x}$, there is a **cube root function** defined by

$$f(x) = \sqrt[3]{x}.$$

The domain of this function is the set of all real numbers. We can graph $f(x) = \sqrt[3]{x}$ by selecting perfect cubes, numbers that have rational cube roots, for x. **Table 10.2** shows five such choices for x and the calculations for the corresponding outputs. We plot these ordered pairs as points in the rectangular coordinate system and connect the points with a smooth curve. The graph of $f(x) = \sqrt[3]{x}$ is shown in **Figure 10.5**.

Table 10.2

x	$f(x) = \sqrt[3]{x}$	(x, y) or $(x, f(x))$
-8	$f(-8) = \sqrt[3]{-8} = -2$	$(-8, -2)$
-1	$f(-1) = \sqrt[3]{-1} = -1$	$(-1, -1)$
0	$f(0) = \sqrt[3]{0} = 0$	$(0, 0)$
1	$f(1) = \sqrt[3]{1} = 1$	$(1, 1)$
8	$f(8) = \sqrt[3]{8} = 2$	$(8, 2)$

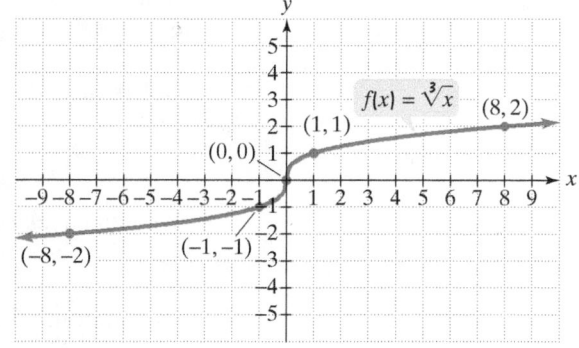

Figure 10.5 The graph of the cube root function $f(x) = \sqrt[3]{x}$

Notice that both the domain and the range of $f(x) = \sqrt[3]{x}$ are the set of all real numbers, $(-\infty, \infty)$.

EXAMPLE 6   Evaluating Cube Root Functions

For each function, find the indicated function value:

a. $f(x) = \sqrt[3]{x - 2}; \quad f(127)$
b. $g(x) = \sqrt[3]{8x - 8}; \quad g(-7).$

Solution

a. $f(x) = \sqrt[3]{x} - 2$ This is the given function.

$f(127) = \sqrt[3]{127} - 2$ Substitute 127 for x.

$= \sqrt[3]{125}$ Simplify the radicand.

$= 5$ $\sqrt[3]{125} = 5$ because $5^3 = 125$.

b. $g(x) = \sqrt[3]{8x} - 8$ This is the given function.

$g(-7) = \sqrt[3]{8(-7)} - 8$ Substitute −7 for x.

$= \sqrt[3]{-64}$ Simplify the radicand: $8(-7) - 8 = -56 - 8 = -64$.

$= -4$ $\sqrt[3]{-64} = -4$ because $(-4)^3 = -64$. ∎

✓ **CHECK POINT 6**   For each function, find the indicated function value:

a. $f(x) = \sqrt[3]{x} - 6$; $f(33)$

b. $g(x) = \sqrt[3]{2x} + 2$; $g(-5)$.

---

7 Simplify expressions of the form $\sqrt[3]{a^3}$.

Because the cube root of a positive number is positive and the cube root of a negative number is negative, absolute value is not needed to simplify expressions of the form $\sqrt[3]{a^3}$.

Simplifying $\sqrt[3]{a^3}$

For any real number a,

$$\sqrt[3]{a^3} = a.$$

In words, the cube root of any expression cubed is that expression.

EXAMPLE 7   Simplifying a Cube Root

Simplify: $\sqrt[3]{-64x^3}$.

Solution Begin by expressing the radicand as an expression that is cubed: $-64x^3 = (-4x)^3$. Then simplify.

$$\sqrt[3]{-64x^3} = \sqrt[3]{(-4x)^3} = -4x$$

We can check our answer by cubing $-4x$:

$$(-4x)^3 = (-4)^3 x^3 = -64x^3.$$

By obtaining the original radicand, we know that our simplification is correct. ∎

✓ **CHECK POINT 7**   Simplify: $\sqrt[3]{-27x^3}$.

---

8 Find even and odd roots.

Even and Odd *n*th Roots

Up to this point, we have focused on square roots and cube roots. Other radical expressions have different roots. For example, the fifth root of a, written $\sqrt[5]{a}$, is the number b for which $b^5 = a$. Thus,

$$\sqrt[5]{32} = 2 \qquad \text{because} \qquad 2^5 = 2 \cdot 2 \cdot 2 \cdot 2 \cdot 2 = 32.$$

The radical expression $\sqrt[n]{a}$ represents the **nth root** of a. The number n is called the **index**. An index of 2 represents a square root and is not written. An index of 3 represents a cube root.

If the index n in $\sqrt[n]{a}$ is an odd number, a root is said to be an **odd root**. A cube root is an odd root. Other odd roots have the same characteristics as cube roots.

- Every real number has exactly one real root when n is odd. An odd root of a positive number is positive and an odd root of a negative number is negative.

$$3^5 = 3 \cdot 3 \cdot 3 \cdot 3 \cdot 3 = 243, \text{ so the fifth root of 243 is 3.}$$

$$(-3)^5 = (-3)(-3)(-3)(-3)(-3) = -243, \text{ so the fifth root of } -243 \text{ is } -3.$$

- The (odd) nth root of a, $\sqrt[n]{a}$, is the number b for which $b^n = a$.

$$\sqrt[5]{243} = 3 \qquad\qquad \sqrt[5]{-243} = -3$$

$3^5 = 243$ $(-3)^5 = -243$

Great Question!

Should I know the higher roots of certain numbers by heart?

Some higher even and odd roots occur so frequently that you might want to memorize them.

Fourth Roots	Fifth Roots
$\sqrt[4]{1} = 1$	$\sqrt[5]{1} = 1$
$\sqrt[4]{16} = 2$	$\sqrt[5]{32} = 2$
$\sqrt[4]{81} = 3$	$\sqrt[5]{243} = 3$
$\sqrt[4]{256} = 4$	
$\sqrt[4]{625} = 5$	

If the index n in $\sqrt[n]{a}$ is an even number, a root is said to be an **even root**. A square root is an even root. Other even roots have the same characteristics as square roots.

- Every positive real number has two real roots when n is even. One root is positive and one is negative.

$$2^4 = 2 \cdot 2 \cdot 2 \cdot 2 = 16 \quad \text{and} \quad (-2)^4 = (-2)(-2)(-2)(-2) = 16,$$

so both 2 and -2 are fourth roots of 16.

- The positive root, called the **principal nth root** and represented by $\sqrt[n]{a}$, is the nonnegative number b for which $b^n = a$. The symbol $-\sqrt[n]{a}$ is used to denote the negative nth root.

$$\sqrt[4]{16} = 2 \qquad\qquad -\sqrt[4]{16} = -2$$

$2^4 = 16$ $(-2)^4 = 16$

- **An even root of a negative number is not a real number.**

$$\sqrt[4]{-16} \text{ is not a real number.}$$

EXAMPLE 8   Finding Even and Odd Roots

Find the indicated root, or state that the expression is not a real number:

a. $\sqrt[4]{81}$ **b.** $-\sqrt[4]{81}$ **c.** $\sqrt[4]{-81}$ **d.** $\sqrt[5]{-32}$.

Solution

a. $\sqrt[4]{81} = 3$ The principal fourth root of 81 is 3 because $3^4 = 3 \cdot 3 \cdot 3 \cdot 3 = 81$.

b. $-\sqrt[4]{81} = -3$ The negative fourth root of 81 is -3 because $(-3)^4 = (-3)(-3)(-3)(-3) = 81$.

c. $\sqrt[4]{-81}$ is not a real number because the index, 4, is even and the radicand, -81, is negative. No real number can be raised to the fourth power to give a negative result such as -81. Real numbers to even powers can only result in nonnegative numbers.

d. $\sqrt[5]{-32} = -2$ because $(-2)^5 = (-2)(-2)(-2)(-2)(-2) = -32$. An odd root of a negative real number is always negative. ■

✓ CHECK POINT 8   Find the indicated root, or state that the expression is not a real number:

a. $\sqrt[4]{16}$ **b.** $-\sqrt[4]{16}$ **c.** $\sqrt[4]{-16}$ **d.** $\sqrt[5]{-1}$.

9 Simplify expressions of the form $\sqrt[n]{a^n}$.

Simplifying Expressions of the Form $\sqrt[n]{a^n}$

We have seen that

$$\sqrt{a^2} = |a| \qquad \text{and} \qquad \sqrt[3]{a^3} = a.$$

Expressions of the form $\sqrt[n]{a^n}$ can be simplified in the same manner. Unless a is known to be nonnegative, absolute value notation is needed when n is even. When the index is odd, absolute value bars are not used.

Simplifying $\sqrt[n]{a^n}$

For any real number a,

1. If n is even, $\sqrt[n]{a^n} = |a|$.
2. If n is odd, $\sqrt[n]{a^n} = a$.

EXAMPLE 9 Simplifying Radical Expressions

Simplify:

a. $\sqrt[4]{(x-3)^4}$ b. $\sqrt[5]{(2x+7)^5}$ c. $\sqrt[6]{(-5)^6}$.

Solution Each expression involves the nth root of a radicand raised to the nth power. Thus, each radical expression can be simplified. Absolute value bars are necessary in parts (a) and (c) because the index, n, is even.

a. $\sqrt[4]{(x-3)^4} = |x-3|$ $\sqrt[n]{a^n} = |a|$ if n is even.
b. $\sqrt[5]{(2x+7)^5} = 2x+7$ $\sqrt[n]{a^n} = a$ if n is odd.
c. $\sqrt[6]{(-5)^6} = |-5| = 5$ $\sqrt[n]{a^n} = |a|$ if n is even. ∎

✓ **CHECK POINT 9** Simplify:

a. $\sqrt[4]{(x+6)^4}$ b. $\sqrt[5]{(3x-2)^5}$ c. $\sqrt[6]{(-8)^6}$.

Achieving Success

Here are some suggestions from other students for improving your success in college. Which of these suggestions do you find most relevant for achieving success in this course? What helpful suggestions would you add to this list?

> This includes preparing for your final exam. We'll give you some strategies to prepare for a final in Section 11.2.

- Recognize that to start studying earlier is better.
  —*Azusa Uchiada, Yokkaichi, Japan*

- Do your work—get it over! There's no magic secret behind it.
  —*James Arnold, Layton, Utah*

- Have smart friends.
  —*Janet, Long Beach, California*

- Have an open mind about all points of view.
  —*Solongo, Ulan Bator, Mongolia*

- Work hard to get better knowledge even if your results are not the highest.
  —*Daniel Tilahun Mezegebu, Addis Ababa, Ethiopia*

- Review your test papers after the tests.
  —*Jim Iantao, Taipei, Taiwan*

- Don't watch TV while you're studying.
  —*Cammy Le, Long Beach, California*

Source: Sean Covey, *The 6 Most Important Decisions You'll Ever Make: A Guide for Teens*, Fireside Press, 2006.

CONCEPT AND VOCABULARY CHECK

Fill in each blank so that the resulting statement is true.

1. The symbol $\sqrt{}$ is used to denote the nonnegative, or _____, square root of a number.

2. $\sqrt{64} = 8$ because _____ $= 64$.

3. The domain of $f(x) = \sqrt{x}$ is _____.

4. The domain of $f(x) = \sqrt{5x - 20}$ can be found by solving the inequality _____.

5. For any real number a, $\sqrt{a^2} = $ _____.

6. $\sqrt[3]{1000} = 10$ because _____ $= 1000$.

7. $\sqrt[3]{-125} = -5$ because _____ $= -125$.

8. For any real number a, $\sqrt[3]{a^3} = $ _____.

9. The domain of $f(x) = \sqrt[3]{x}$ is _____.

10. The radical expression $\sqrt[n]{a}$ represents the _____ root of a. The number n is called the _____.

11. If n is even, $\sqrt[n]{a^n} = $ _____. If n is odd, $\sqrt[n]{a^n} = $ _____.

12. True or false: $-\sqrt{25}$ is a real number. _____

13. True or false: $\sqrt{-25}$ is a real number. _____

14. True or false: $\sqrt[3]{-1}$ is a real number. _____

15. True or false: $\sqrt[4]{-1}$ is a real number. _____

10.1 EXERCISE SET

 MyMathLab®

Watch the videos
in MyMathLab

Download the
MyDashBoard App

Practice Exercises

In Exercises 1–20, evaluate each expression, or state that the expression is not a real number.

1. $\sqrt{36}$
2. $\sqrt{16}$
3. $-\sqrt{36}$
4. $-\sqrt{16}$
5. $\sqrt{-36}$
6. $\sqrt{-16}$
7. $\sqrt{\dfrac{1}{25}}$
8. $\sqrt{\dfrac{1}{49}}$
9. $-\sqrt{\dfrac{9}{16}}$
10. $-\sqrt{\dfrac{4}{25}}$
11. $\sqrt{0.81}$
12. $\sqrt{0.49}$
13. $-\sqrt{0.04}$
14. $-\sqrt{0.64}$
15. $\sqrt{25 - 16}$
16. $\sqrt{144 + 25}$
17. $\sqrt{25} - \sqrt{16}$
18. $\sqrt{144} + \sqrt{25}$
19. $\sqrt{16 - 25}$
20. $\sqrt{25 - 144}$

In Exercises 21–26, find the indicated function values for each function. If necessary, round to two decimal places. If the function value is not a real number and does not exist, so state.

21. $f(x) = \sqrt{x - 2}$; $f(18), f(3), f(2), f(-2)$

22. $f(x) = \sqrt{x - 3}$; $f(28), f(4), f(3), f(-1)$

23. $g(x) = -\sqrt{2x + 3}$; $g(11), g(1), g(-1), g(-2)$

24. $g(x) = -\sqrt{2x + 1}$; $g(4), g(1), g\left(-\dfrac{1}{2}\right), g(-1)$

25. $h(x) = \sqrt{(x - 1)^2}$; $h(5), h(3), h(0), h(-5)$

26. $h(x) = \sqrt{(x - 2)^2}$; $h(5), h(3), h(0), h(-5)$

In Exercises 27–32, find the domain of each square root function. Then use the domain to match the radical function with its graph. [The graphs are labeled (a) through (f) and are shown in [-10, 10, 1] by [-10, 10, 1] viewing rectangles below and on the next page.]

27. $f(x) = \sqrt{x - 3}$
28. $f(x) = \sqrt{x + 2}$
29. $f(x) = \sqrt{3x + 15}$
30. $f(x) = \sqrt{3x - 15}$
31. $f(x) = \sqrt{6 - 2x}$
32. $f(x) = \sqrt{8 - 2x}$

a.

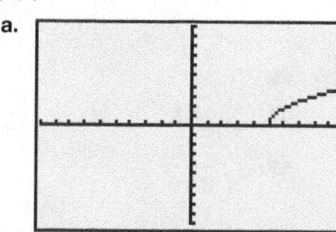

b.

c.

d.

e.

f.

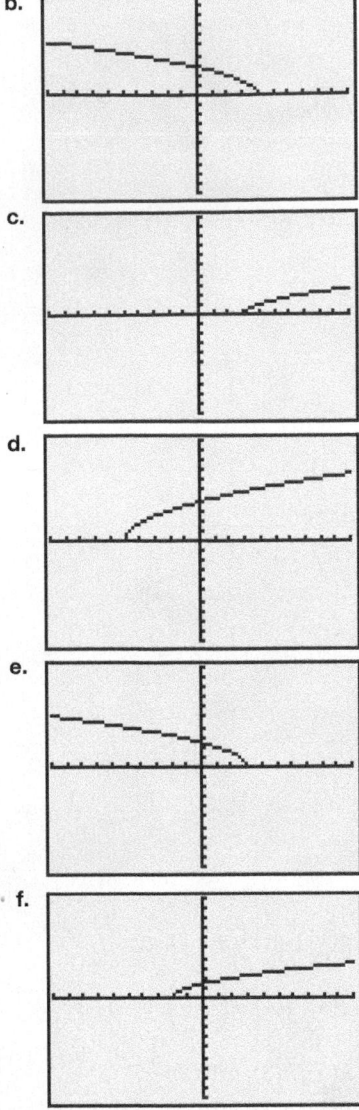

In Exercises 33–46, simplify each expression.

33. $\sqrt{5^2}$ **34.** $\sqrt{7^2}$

35. $\sqrt{(-4)^2}$ **36.** $\sqrt{(-10)^2}$

37. $\sqrt{(x-1)^2}$ **38.** $\sqrt{(x-2)^2}$

39. $\sqrt{36x^4}$ **40.** $\sqrt{81x^4}$

41. $-\sqrt{100x^6}$

42. $-\sqrt{49x^6}$

43. $\sqrt{x^2+12x+36}$

44. $\sqrt{x^2+14x+49}$

45. $-\sqrt{x^2-8x+16}$

46. $-\sqrt{x^2-10x+25}$

In Exercises 47–54, find each cube root.

47. $\sqrt[3]{27}$ **48.** $\sqrt[3]{64}$

49. $\sqrt[3]{-27}$ **50.** $\sqrt[3]{-64}$

51. $\sqrt[3]{\dfrac{1}{125}}$ **52.** $\sqrt[3]{\dfrac{1}{1000}}$

53. $\sqrt[3]{\dfrac{-27}{1000}}$ **54.** $\sqrt[3]{\dfrac{-8}{125}}$

In Exercises 55–58, find the indicated function values for each function.

55. $f(x) = \sqrt[3]{x-1}; f(28), f(9), f(0), f(-63)$

56. $f(x) = \sqrt[3]{x-3}; f(30), f(11), f(2), f(-122)$

57. $g(x) = -\sqrt[3]{8x-8}; g(2), g(1), g(0)$

58. $g(x) = -\sqrt[3]{2x+1}; g(13), g(0), g(-63)$

In Exercises 59–76, find the indicated root, or state that the expression is not a real number.

59. $\sqrt[4]{1}$ **60.** $\sqrt[5]{1}$

61. $\sqrt[4]{16}$ **62.** $\sqrt[4]{81}$

63. $-\sqrt[4]{16}$ **64.** $-\sqrt[4]{81}$

65. $\sqrt[4]{-16}$ **66.** $\sqrt[4]{-81}$

67. $\sqrt[5]{-1}$ **68.** $\sqrt[7]{-1}$

69. $\sqrt[6]{-1}$ **70.** $\sqrt[8]{-1}$

71. $-\sqrt[4]{256}$ **72.** $-\sqrt[4]{10{,}000}$

73. $\sqrt[6]{64}$ **74.** $\sqrt[5]{32}$

75. $-\sqrt[5]{32}$ **76.** $-\sqrt[6]{64}$

In Exercises 77–90, simplify each expression. Include absolute value bars where necessary.

77. $\sqrt[3]{x^3}$ **78.** $\sqrt[5]{x^5}$

79. $\sqrt[4]{y^4}$ **80.** $\sqrt[6]{y^6}$

81. $\sqrt[3]{-8x^3}$ **82.** $\sqrt[3]{-125x^3}$

83. $\sqrt[3]{(-5)^3}$ **84.** $\sqrt[3]{(-6)^3}$

85. $\sqrt[4]{(-5)^4}$ **86.** $\sqrt[6]{(-6)^6}$

87. $\sqrt[4]{(x+3)^4}$ **88.** $\sqrt[4]{(x+5)^4}$

89. $\sqrt[5]{-32(x-1)^5}$

90. $\sqrt[5]{-32(x-2)^5}$

Practice PLUS

In Exercises 91–94, complete each table and graph the given function. Identify the function's domain and range.

91. $f(x) = \sqrt{x} + 3$

x	$f(x) = \sqrt{x}+3$
0	
1	
4	
9	

92. $f(x) = \sqrt{x} - 2$

x	$f(x) = \sqrt{x}-2$
0	
1	
4	
9	

93. $f(x) = \sqrt{x - 3}$

x	$f(x) = \sqrt{x - 3}$
3	
4	
7	
12	

94. $f(x) = \sqrt{4 - x}$

x	$f(x) = \sqrt{4 - x}$
−5	
0	
3	
4	

In Exercises 95–98, find the domain of each function.

95. $f(x) = \dfrac{\sqrt[3]{x}}{\sqrt{30 - 2x}}$

96. $f(x) = \dfrac{\sqrt[3]{x}}{\sqrt{80 - 5x}}$

97. $f(x) = \dfrac{\sqrt{x - 1}}{\sqrt{3 - x}}$ **98.** $f(x) = \dfrac{\sqrt{x - 2}}{\sqrt{7 - x}}$

In Exercises 99–100, evaluate each expression.

99. $\sqrt[3]{\sqrt[4]{16} + \sqrt{625}}$

100. $\sqrt[3]{\sqrt{\sqrt{169} + \sqrt{9}} + \sqrt{\sqrt[3]{1000} + \sqrt[3]{216}}}$

Application Exercises

101. The function $f(x) = 2.9\sqrt{x} + 20.1$ models the median height, $f(x)$, in inches, of boys who are x months of age. The graph of f is shown.

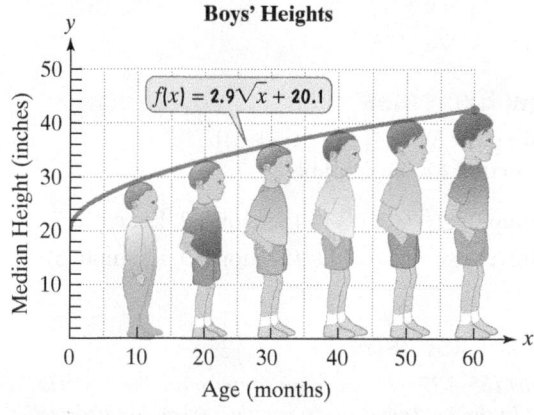

Boys' Heights

$f(x) = 2.9\sqrt{x} + 20.1$

Median Height (inches) / Age (months)

Source: Laura Walter Nathanson, THE PORTABLE PEDIATRICIAN FOR PARENTS, © 1994 Harper Perennial

  a. According to the model, what is the median height of boys who are 48 months, or four years, old? Use a calculator and round to the nearest tenth of an inch. The actual median height for boys at 48 months is 40.8 inches. Does the model overestimate or underestimate the actual height? By how much?

  b. Use the model to find the average rate of change, in inches per month, between birth and 10 months. Round to the nearest tenth.

  c. Use the model to find the average rate of change, in inches per month, between 50 and 60 months. Round to the nearest tenth. How does this compare with your answer in part (b)? How is this difference shown by the graph?

102. The function $f(x) = 3.1\sqrt{x} + 19$ models the median height, $f(x)$, in inches, of girls who are x months of age. The graph of f is shown.

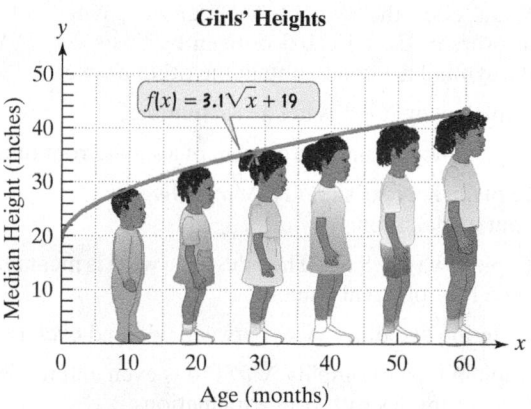

Girls' Heights

$f(x) = 3.1\sqrt{x} + 19$

Median Height (inches) / Age (months)

Source: Laura Walther Nathanson, THE PORTABLE PEDIATRICIAN FOR PARENTS © 1994 Harper Perennial

  a. According to the model, what is the median height of girls who are 48 months, or four years, old? Use a calculator and round to the nearest tenth of an inch. The actual median height for girls at 48 months is 40.2 inches. Does the model overestimate or underestimate the actual height? By how much?

  b. Use the model to find the average rate of change, in inches per month, between birth and 10 months. Round to the nearest tenth.

  c. Use the model to find the average rate of change, in inches per month, between 50 and 60 months. Round to the nearest tenth. How does this compare with your answer in part (b)? How is this difference shown by the graph?

Police use the function $f(x) = \sqrt{20x}$ to estimate the speed of a car, $f(x)$, in miles per hour, based on the length, x, in feet, of its skid marks upon sudden braking on a dry asphalt road. Use the function to solve Exercises 103–104.

103. A motorist is involved in an accident. A police officer measures the car's skid marks to be 245 feet long. Estimate the speed at which the motorist was traveling before braking. If the posted speed limit is 50 miles per hour and the motorist tells the officer he was not speeding, should the officer believe him? Explain.

104. A motorist is involved in an accident. A police officer measures the car's skid marks to be 45 feet long. Use the function described at the bottom of the previous page to estimate the speed at which the motorist was traveling before braking. If the posted speed limit is 35 miles per hour and the motorist tells the officer she was not speeding, should the officer believe her? Explain.

Writing in Mathematics

105. What are the square roots of 36? Explain why each of these numbers is a square root.

106. What does the symbol $\sqrt{}$ denote? Which of your answers in Exercise 105 is given by this symbol? Write the symbol needed to obtain the other answer.

107. Explain why $\sqrt{-1}$ is not a real number.

108. Explain how to find the domain of a square root function.

109. Explain how to simplify $\sqrt{a^2}$. Give an example with your explanation.

110. Explain why $\sqrt[3]{8}$ is 2. Then describe what is meant by the cube root of a real number.

111. Describe two differences between odd and even roots.

112. Explain how to simplify $\sqrt[n]{a^n}$ if n is even and if n is odd. Give examples with your explanations.

113. Explain the meaning of the words *radical*, *radicand*, and *index*. Give an example with your explanation.

114. Describe the trend in a boy's growth from birth through five years, shown in the graph for Exercise 101. Why is a square root function a useful model for the data?

Technology Exercises

115. Use a graphing utility to graph $y_1 = \sqrt{x}$, $y_2 = \sqrt{x+4}$, and $y_3 = \sqrt{x-3}$ in the same $[-5, 10, 1]$ by $[0, 6, 1]$ viewing rectangle. Describe one similarity and one difference that you observe among the graphs. Use the word *shift* in your response.

116. Use a graphing utility to graph $y = \sqrt{x}$, $y = \sqrt{x}+4$, and $y = \sqrt{x}-3$ in the same $[-1, 10, 1]$ by $[-10, 10, 1]$ viewing rectangle. Describe one similarity and one difference that you observe among the graphs.

117. Use a graphing utility to graph $f(x) = \sqrt{x}$, $g(x) = -\sqrt{x}$, $h(x) = \sqrt{-x}$, and $k(x) = -\sqrt{-x}$ in the same $[-10, 10, 1]$ by $[-4, 4, 1]$ viewing rectangle. Use the graphs to describe the domains and the ranges of functions f, g, h, and k.

118. Use a graphing utility to graph $y_1 = \sqrt{x^2}$ and $y_2 = -x$ in the same viewing rectangle.

    a. For what values of x is $\sqrt{x^2} = -x$?

    b. For what values of x is $\sqrt{x^2} \neq -x$?

Critical Thinking Exercises

Make Sense? *In Exercises 119–122, determine whether each statement "makes sense" or "does not make sense" and explain your reasoning.*

119. $\sqrt[4]{(-8)^4}$ cannot be positive 8 because the power and the index cancel each other.

120. If I am given any real number, that number has exactly one odd root and two even roots.

121. I need to restrict the domains of radical functions with even indices, but these restrictions are not necessary when indices are odd.

122. Using my calculator, I determined that $5^5 = 3125$, so 5 must be the fifth root of 3125.

In Exercises 123–126, determine whether each statement is true or false. If the statement is false, make the necessary change(s) to produce a true statement.

123. The domain of $f(x) = \sqrt[3]{x-4}$ is $[4, \infty)$.

124. If n is odd and b is negative, then $\sqrt[n]{b}$ is not a real number.

125. If $x = -2$, then $\sqrt{x^6} = x^3$.

126. The expression $\sqrt[n]{4}$ represents increasingly larger numbers for $n = 2, 3, 4, 5, 6$, and so on.

127. Write a function whose domain is $(-\infty, 5]$.

128. Let $f(x) = \sqrt{x-3}$ and $g(x) = \sqrt{x+1}$. Find the domain of $f + g$ and $\dfrac{f}{g}$.

129. Simplify: $\sqrt{(2x+3)^{10}}$.

In Exercises 130–131, graph each function by hand. Then describe the relationship between the function that you graphed and the graph of $f(x) = \sqrt{x}$.

130. $g(x) = \sqrt{x} + 2$

131. $h(x) = \sqrt{x+3}$

Review Exercises

132. Simplify: $3x - 2[x - 3(x+5)]$. (Section 1.8, Example 11)

133. Simplify: $(-3x^{-4}y^3)^{-2}$. (Section 5.7, Example 6)

134. Solve: $|3x - 4| > 11$. (Section 9.3, Example 6)

Preview Exercises

Exercises 135–137 will help you prepare for the material covered in the next section. In each exercise, use properties of exponents to simplify the expression. Be sure that no negative exponents appear in your simplified expression. (If you have forgotten how to simplify an exponential expression, see the box on page 407.)

135. $(2^3 x^5)(2^4 x^{-6})$

136. $\dfrac{32x^2}{16x^5}$

137. $(x^{-2}y^3)^4$

Objectives

1 Use the definition of $a^{\frac{1}{n}}$.

2 Use the definition of $a^{\frac{m}{n}}$.

3 Use the definition of $a^{-\frac{m}{n}}$.

4 Simplify expressions with rational exponents.

5 Simplify radical expressions using rational exponents.

Rational Exponents

The Galápagos Islands arc a chain of volcanic islands lying 600 miles west of Ecuador. They are famed for over 5000 species of plants and animals, including a rare flightless cormorant, marine iguanas, and giant tortoises weighing more than 600 pounds. Early in 2001, the plants and

Marine iguanas of the Galápagos Islands

wildlife that live in the Galápagos were at risk from a massive oil spill that flooded 150,000 gallons of toxic fuel into one of the world's most fragile ecosystems. The long-term danger of the accident is that fuel sinking to the ocean floor will destroy algae that is vital to the food chain. Any imbalance in the food chain could threaten the rare Galápagos plant and animal species that have evolved for thousands of years in isolation with little human intervention.

At risk on these ecologically vulnerable islands are unique flora and fauna that helped to inspire Charles Darwin's theory of evolution. Darwin made an enormous collection of the island's plant species. The function

$$f(x) = 29x^{\frac{1}{3}}$$

models the number of plant species, $f(x)$, on the various islands of the Galápagos in terms of the area, x, in square miles, of a particular island. But x to the *what* power? How can we interpret the information given by this function? In this section, we turn our attention to rational exponents such as $\frac{1}{3}$ and their relationship to roots of real numbers.

Defining Rational Exponents

We define rational exponents so that their properties are the same as the properties for integer exponents. For example, suppose that $x = 7^{\frac{1}{3}}$. We know that exponents are multiplied when an exponential expression is raised to a power. For this to be true,

$$x^3 = \left(7^{\frac{1}{3}}\right)^3 = 7^{\frac{1}{3} \cdot 3} = 7^1 = 7.$$

We see that $x^3 = 7$. This means that x is the number whose cube is 7. Thus, $x = \sqrt[3]{7}$. Remember that we began with $x = 7^{\frac{1}{3}}$. This means that

$$7^{\frac{1}{3}} = \sqrt[3]{7}.$$

We can generalize $7^{\frac{1}{3}} = \sqrt[3]{7}$ with the following definition:

1 Use the definition of $a^{\frac{1}{n}}$.

The Definition of $a^{\frac{1}{n}}$

If $\sqrt[n]{a}$ represents a real number and $n \geq 2$ is an integer, then

$$a^{\frac{1}{n}} = \sqrt[n]{a}.$$

> The denominator of the rational exponent is the radical's index.

If n is even, a must be nonnegative. If n is odd, a can be any real number.

Using Technology

This graphing utility screen shows that
$\sqrt{64} = 8$ and $64^{\frac{1}{2}} = 8$.

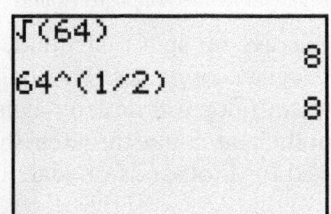

```
√(64)
                    8
64^(1/2)
                    8
```

EXAMPLE 1 Using the Definition of $a^{\frac{1}{n}}$

Use radical notation to rewrite each expression. Simplify, if possible:

a. $64^{\frac{1}{2}}$ **b.** $(-125)^{\frac{1}{3}}$ **c.** $(6x^2y)^{\frac{1}{5}}$.

Solution

a. $64^{\frac{1}{2}} = \sqrt{64} = 8$

> The denominator is the index.

b. $(-125)^{\frac{1}{3}} = \sqrt[3]{-125} = -5$

c. $(6x^2y)^{\frac{1}{5}} = \sqrt[5]{6x^2y}$ ■

✓ **CHECK POINT 1** Use radical notation to rewrite each expression. Simplify, if possible:

a. $25^{\frac{1}{2}}$ **b.** $(-8)^{\frac{1}{3}}$ **c.** $(5xy^2)^{\frac{1}{4}}$.

In our next example, we begin with radical notation and rewrite the expression with rational exponents.

> The radical's index becomes the exponent's denominator.

$$\sqrt[n]{a} = a^{\frac{1}{n}}$$

> The radicand becomes the base.

EXAMPLE 2 Using the Definition of $a^{\frac{1}{n}}$

Rewrite with rational exponents:

a. $\sqrt[5]{13ab}$ **b.** $\sqrt[7]{\dfrac{xy^2}{17}}$.

Solution Parentheses are needed to show that the entire radicand becomes the base.

a. $\sqrt[5]{13ab} = (13ab)^{\frac{1}{5}}$

> The index is the exponent's denominator.

b. $\sqrt[7]{\dfrac{xy^2}{17}} = \left(\dfrac{xy^2}{17}\right)^{\frac{1}{7}}$ ■

| ✓ **CHECK POINT 2** | Rewrite with rational exponents: |

a. $\sqrt[4]{5xy}$ b. $\sqrt[5]{\dfrac{a^3 b}{2}}$.

---

2 Use the definition of $a^{\frac{m}{n}}$.

Can rational exponents have numerators other than 1? The answer is yes. If the numerator is some other integer, we still want to multiply exponents when raising a power to a power. For this reason,

$$a^{\frac{2}{3}} = (a^{\frac{1}{3}})^2 \quad \text{and} \quad a^{\frac{2}{3}} = (a^2)^{\frac{1}{3}}.$$

This means $(\sqrt[3]{a})^2$.   This means $\sqrt[3]{a^2}$.

Thus,

$$a^{\frac{2}{3}} = (\sqrt[3]{a})^2 = \sqrt[3]{a^2}.$$

Do you see that the denominator, 3, of the rational exponent is the same as the index of the radical? The numerator, 2, of the rational exponent serves as an exponent in each of the two radical forms. We generalize these ideas with the following definition:

> **The Definition of $a^{\frac{m}{n}}$**
>
> If $\sqrt[n]{a}$ represents a real number, $\dfrac{m}{n}$ is a positive rational number reduced to lowest terms, and $n \geq 2$ is an integer, then
>
> $$a^{\frac{m}{n}} = (\sqrt[n]{a})^m \quad \text{First take the } n\text{th root of } a.$$
>
> and
>
> $$a^{\frac{m}{n}} = \sqrt[n]{a^m}. \quad \text{First raise } a \text{ to the } m \text{ power.}$$

The first form of the definition shown in the box involves taking the root first. This form is often preferable because smaller numbers are involved. Notice that the rational exponent consists of two parts, indicated by the following voice balloons:

The numerator is the exponent.

$$a^{\frac{m}{n}} = (\sqrt[n]{a})^m$$

The denominator is the radical's index.

EXAMPLE 3 Using the Definition of $a^{\frac{m}{n}}$

Use radical notation to rewrite each expression and simplify:

a. $1000^{\frac{2}{3}}$ b. $16^{\frac{3}{2}}$ c. $-32^{\frac{3}{5}}$.

Solution

a. $(1000)^{\frac{2}{3}} = (\sqrt[3]{1000})^2 = 10^2 = 100$

The denominator of $\frac{2}{3}$ is the root and the numerator is the exponent.

b. $16^{\frac{3}{2}} = (\sqrt{16})^3 = 4^3 = 64$

c. $-32^{\frac{3}{5}} = -(\sqrt[5]{32})^3 = -2^3 = -8$

The base is 32 and the negative sign is not affected by the exponent.

Using Technology

Here are the calculator keystroke sequences for $1000^{\frac{2}{3}}$:

Many Scientific Calculators

1000 $\boxed{y^x}$ $\boxed{(}$ 2 $\boxed{\div}$ 3 $\boxed{)}$ $\boxed{=}$

Many Graphing Calculators

1000 $\boxed{\wedge}$ $\boxed{(}$ 2 $\boxed{\div}$ 3 $\boxed{)}$ ENTER

✓ **CHECK POINT 3** Use radical notation to rewrite each expression and simplify:

a. $8^{\frac{4}{3}}$ **b.** $25^{\frac{3}{2}}$ **c.** $-81^{\frac{3}{4}}$.

In our next example, we begin with radical notation and rewrite the expression with rational exponents. When changing from radical form to exponential form, the index becomes the denominator of the rational exponent.

EXAMPLE 4 Using the Definition of $a^{\frac{m}{n}}$

Rewrite with rational exponents:

a. $\sqrt[3]{7^5}$ **b.** $\left(\sqrt[4]{13xy}\right)^9$.

Solution

a. $\sqrt[3]{7^5} = 7^{\frac{5}{3}}$

> The index is the exponent's denominator.

b. $\left(\sqrt[4]{13xy}\right)^9 = (13xy)^{\frac{9}{4}}$ ∎

✓ **CHECK POINT 4** Rewrite with rational exponents:

a. $\sqrt[3]{6^4}$ **b.** $\left(\sqrt[5]{2xy}\right)^7$.

3 Use the definition of $a^{-\frac{m}{n}}$.

Can a rational exponent be negative? Yes. The way that negative rational exponents are defined is similar to the way that negative integer exponents are defined.

> **The Definition of $a^{-\frac{m}{n}}$**
>
> If $a^{\frac{m}{n}}$ is a nonzero real number, then
>
> $$a^{-\frac{m}{n}} = \frac{1}{a^{\frac{m}{n}}}.$$

EXAMPLE 5 Using the Definition of $a^{-\frac{m}{n}}$

Rewrite each expression with a positive exponent. Simplify, if possible:

a. $36^{-\frac{1}{2}}$ **b.** $125^{-\frac{1}{3}}$ **c.** $16^{-\frac{3}{4}}$ **d.** $(7xy)^{-\frac{4}{7}}$.

Solution

a. $36^{-\frac{1}{2}} = \dfrac{1}{36^{\frac{1}{2}}} = \dfrac{1}{\sqrt{36}} = \dfrac{1}{6}$

b. $125^{-\frac{1}{3}} = \dfrac{1}{125^{\frac{1}{3}}} = \dfrac{1}{\sqrt[3]{125}} = \dfrac{1}{5}$

c. $16^{-\frac{3}{4}} = \dfrac{1}{16^{\frac{3}{4}}} = \dfrac{1}{\left(\sqrt[4]{16}\right)^3} = \dfrac{1}{2^3} = \dfrac{1}{8}$

d. $(7xy)^{-\frac{4}{7}} = \dfrac{1}{(7xy)^{\frac{4}{7}}}$ ∎

Using Technology

Here are the calculator keystroke sequences for $16^{-\frac{3}{4}}$:

Many Scientific Calculators

16 y^x (3 +/− ÷ 4) =

Many Graphing Calculators

16 ^ (((−) 3 ÷ 4) ENTER

☑ **CHECK POINT 5**   Rewrite each expression with a positive exponent. Simplify, if possible:

a. $100^{-\frac{1}{2}}$

b. $8^{-\frac{1}{3}}$

c. $32^{-\frac{3}{5}}$

d. $(3xy)^{-\frac{5}{9}}$.

Properties of Rational Exponents

The same properties apply to rational exponents as to integer exponents. The following is a summary of these properties:

Properties of Rational Exponents

If m and n are rational exponents, and a and b are real numbers for which the following expressions are defined, then

1. $b^m \cdot b^n = b^{m+n}$ When multiplying exponential expressions with the same base, add the exponents. Use this sum as the exponent of the common base.

2. $\dfrac{b^m}{b^n} = b^{m-n}$ When dividing exponential expressions with the same base, subtract the exponents. Use this difference as the exponent of the common base.

3. $(b^m)^n = b^{mn}$ When an exponential expression is raised to a power, multiply the exponents. Place the product of the exponents on the base and remove the parentheses.

4. $(ab)^n = a^n b^n$ When a product is raised to a power, raise each factor to that power and multiply.

5. $\left(\dfrac{a}{b}\right)^n = \dfrac{a^n}{b^n}$ When a quotient is raised to a power, raise the numerator to that power and divide by the denominator to that power.

4 Simplify expressions with rational exponents.

We can use these properties to simplify exponential expressions with rational exponents. As with integer exponents, an expression with rational exponents is **simplified** when:

- No parentheses appear.
- No powers are raised to powers.
- Each base occurs only once.
- No negative or zero exponents appear.

Great Question!

Because $6 \cdot 6 = 36$, should I multiply 6 and 6 when simplifying $6^{\frac{1}{7}} \cdot 6^{\frac{4}{7}}$?

No. To simplify $6^{\frac{1}{7}} \cdot 6^{\frac{4}{7}}$, **do not multiply the numerical bases.**

Incorrect!

$6^{\frac{1}{7}} \cdot 6^{\frac{4}{7}} = 36^{\frac{1}{7} + \frac{4}{7}}$

EXAMPLE 6   Simplifying Expressions with Rational Exponents

Simplify:

a. $6^{\frac{1}{7}} \cdot 6^{\frac{4}{7}}$ **b.** $\dfrac{32x^{\frac{1}{2}}}{16x^{\frac{3}{4}}}$ **c.** $\left(8.3^{\frac{3}{4}}\right)^{\frac{2}{3}}$ **d.** $\left(x^{-\frac{2}{5}} y^{\frac{1}{3}}\right)^{\frac{1}{2}}$.

Solution

a. $6^{\frac{1}{7}} \cdot 6^{\frac{4}{7}} = 6^{\frac{1}{7} + \frac{4}{7}}$ To multiply with the same base, add exponents.

$= 6^{\frac{5}{7}}$ Simplify: $\dfrac{1}{7} + \dfrac{4}{7} = \dfrac{5}{7}$.

b. $\dfrac{32x^{\frac{1}{2}}}{16x^{\frac{3}{4}}} = \dfrac{32}{16}x^{\frac{1}{2}-\frac{3}{4}}$ Divide coefficients. To divide with the same base, subtract exponents.

$= 2x^{\frac{2}{4}-\frac{3}{4}}$ Write exponents in terms of the LCD, 4.

$= 2x^{-\frac{1}{4}}$ Subtract: $\dfrac{2}{4} - \dfrac{3}{4} = -\dfrac{1}{4}$.

$= \dfrac{2}{x^{\frac{1}{4}}}$ Rewrite with a positive exponent: $a^{-\frac{m}{n}} = \dfrac{1}{a^{\frac{m}{n}}}$.

c. $\left(8.3^{\frac{3}{4}}\right)^{\frac{2}{3}} = 8.3^{\left(\frac{3}{4}\right)\left(\frac{2}{3}\right)}$ To raise a power to a power, multiply exponents.

$= 8.3^{\frac{1}{2}}$ Multiply: $\dfrac{3}{4} \cdot \dfrac{2}{3} = \dfrac{6}{12} = \dfrac{1}{2}$.

d. $\left(x^{-\frac{2}{5}}y^{\frac{1}{3}}\right)^{\frac{1}{2}} = \left(x^{-\frac{2}{5}}\right)^{\frac{1}{2}}\left(y^{\frac{1}{3}}\right)^{\frac{1}{2}}$ To raise a product to a power, raise each factor to the power.

$= x^{-\frac{1}{5}}y^{\frac{1}{6}}$ Multiply: $-\dfrac{2}{5} \cdot \dfrac{1}{2} = -\dfrac{1}{5}$ and $\dfrac{1}{3} \cdot \dfrac{1}{2} = \dfrac{1}{6}$.

$= \dfrac{y^{\frac{1}{6}}}{x^{\frac{1}{5}}}$ Rewrite with positive exponents. ∎

✓ **CHECK POINT 6**   Simplify:

a. $7^{\frac{1}{2}} \cdot 7^{\frac{1}{3}}$ **b.** $\dfrac{50x^{\frac{1}{3}}}{10x^{\frac{4}{3}}}$ **c.** $\left(9.1^{\frac{2}{5}}\right)^{\frac{3}{4}}$ **d.** $\left(x^{-\frac{3}{5}}y^{\frac{1}{4}}\right)^{\frac{1}{3}}$.

---

5 Simplify radical expressions using rational exponents.

Using Rational Exponents to Simplify Radical Expressions

Some radical expressions can be simplified using rational exponents. We will use the following procedure:

Simplifying Radical Expressions Using Rational Exponents

1. Rewrite each radical expression as an exponential expression with a rational exponent.
2. Simplify using properties of rational exponents.
3. Rewrite in radical notation if rational exponents still appear.

EXAMPLE 7   Simplifying Radical Expressions Using Rational Exponents

Use rational exponents to simplify. Assume that all variables represent nonnegative numbers.

a. $\sqrt[10]{x^5}$ **b.** $\sqrt[3]{27a^{15}}$ **c.** $\sqrt[4]{x^6y^2}$ **d.** $\sqrt{x} \cdot \sqrt[3]{x}$ **e.** $\sqrt[3]{\sqrt{x}}$

Solution

a. $\sqrt[10]{x^5} = x^{\frac{5}{10}}$ Rewrite as an exponential expression.

$= x^{\frac{1}{2}}$ Simplify the exponent.

$= \sqrt{x}$ Rewrite in radical notation.

b. $\sqrt[3]{27a^{15}} = (27a^{15})^{\frac{1}{3}}$ Rewrite as an exponential expression.

$= 27^{\frac{1}{3}}(a^{15})^{\frac{1}{3}}$ Raise each factor in parentheses to the $\frac{1}{3}$ power.

$= \sqrt[3]{27} \cdot a^{15\left(\frac{1}{3}\right)}$ To raise a power to a power, multiply exponents.

$= 3a^5$ $\sqrt[3]{27} = 3$. Multiply exponents: $15 \cdot \frac{1}{3} = 5$.

c. $\sqrt[4]{x^6 y^2} = (x^6 y^2)^{\frac{1}{4}}$ Rewrite as an exponential expression.

$= (x^6)^{\frac{1}{4}} (y^2)^{\frac{1}{4}}$ Raise each factor in parentheses to the $\frac{1}{4}$ power.

$= x^{\frac{6}{4}} y^{\frac{2}{4}}$ To raise powers to powers, multiply.

$= x^{\frac{3}{2}} y^{\frac{1}{2}}$ Simplify.

$= (x^3 y)^{\frac{1}{2}}$ $a^n b^n = (ab)^n$

$= \sqrt{x^3 y}$ Rewrite in radical notation.

d. $\sqrt{x} \cdot \sqrt[3]{x} = x^{\frac{1}{2}} \cdot x^{\frac{1}{3}}$ Rewrite as exponential expressions.

$= x^{\frac{1}{2} + \frac{1}{3}}$ To multiply with the same base, add exponents.

$= x^{\frac{3}{6} + \frac{2}{6}}$ Write exponents in terms of the LCD, 6.

$= x^{\frac{5}{6}}$ Add: $\frac{3}{6} + \frac{2}{6} = \frac{5}{6}$.

$= \sqrt[6]{x^5}$ Rewrite in radical notation.

e. $\sqrt[3]{\sqrt{x}} = \sqrt[3]{x^{\frac{1}{2}}}$ Write the radicand as an exponential expression.

$= \left(x^{\frac{1}{2}}\right)^{\frac{1}{3}}$ Write the entire expression in exponential form.

$= x^{\frac{1}{6}}$ To raise powers to powers, multiply: $\frac{1}{2} \cdot \frac{1}{3} = \frac{1}{6}$.

$= \sqrt[6]{x}$ Rewrite in radical notation. ∎

✓ **CHECK POINT 7**   Use rational exponents to simplify:

a. $\sqrt[6]{x^3}$ **b.** $\sqrt[3]{8a^{12}}$ **c.** $\sqrt[8]{x^4 y^2}$

d. $\dfrac{\sqrt{x}}{\sqrt[3]{x}}$ **e.** $\sqrt{\sqrt[3]{x}}$.

CONCEPT AND VOCABULARY CHECK

Fill in each blank so that the resulting statement is true.

1. $36^{\frac{1}{2}} = \sqrt{\underline{\hspace{0.5cm}}} = \underline{\hspace{0.5cm}}$

2. $8^{\frac{1}{3}} = \sqrt[3]{\underline{\hspace{0.5cm}}} = \underline{\hspace{0.5cm}}$

3. $a^{\frac{1}{n}} = \underline{\hspace{0.5cm}}$

4. $16^{\frac{3}{4}} = (\sqrt[4]{\underline{\hspace{0.5cm}}})^3 = (\underline{\hspace{0.5cm}})^3 = \underline{\hspace{0.5cm}}$

5. $a^{\frac{m}{n}} = \underline{\hspace{2cm}}$

6. $\sqrt[3]{2^5} = 2^{\underline{\hspace{0.3cm}}}$

7. $16^{-\frac{3}{2}} = \dfrac{1}{\underline{\hspace{0.5cm}}} = \dfrac{1}{(\sqrt{\underline{\hspace{0.3cm}}})^{\underline{\hspace{0.3cm}}}} = \dfrac{1}{(\underline{\hspace{0.3cm}})^{\underline{\hspace{0.3cm}}}} = \dfrac{1}{\underline{\hspace{0.5cm}}}$

10.2 EXERCISE SET

MyMathLab®   Watch the videos in MyMathLab   Download the MyDashBoard App

Practice Exercises

In Exercises 1–20, use radical notation to rewrite each expression. Simplify, if possible.

1. $49^{\frac{1}{2}}$ **2.** $100^{\frac{1}{2}}$

3. $(-27)^{\frac{1}{3}}$ **4.** $(-64)^{\frac{1}{3}}$

5. $-16^{\frac{1}{4}}$ **6.** $-81^{\frac{1}{4}}$

7. $(xy)^{\frac{1}{3}}$ **8.** $(xy)^{\frac{1}{4}}$

9. $(2xy^3)^{\frac{1}{5}}$ **10.** $(3xy^4)^{\frac{1}{5}}$

11. $81^{\frac{3}{2}}$ **12.** $25^{\frac{3}{2}}$

13. $125^{\frac{2}{3}}$ **14.** $1000^{\frac{2}{3}}$

15. $(-32)^{\frac{3}{5}}$ **16.** $(-27)^{\frac{2}{3}}$

17. $27^{\frac{2}{3}} + 16^{\frac{3}{4}}$

18. $4^{\frac{5}{2}} - 8^{\frac{2}{3}}$

19. $(xy)^{\frac{4}{7}}$ **20.** $(xy)^{\frac{4}{9}}$

In Exercises 21–38, rewrite each expression with rational exponents.

21. $\sqrt{7}$ **22.** $\sqrt{13}$

23. $\sqrt[3]{5}$ **24.** $\sqrt[3]{6}$

25. $\sqrt[5]{11x}$ **26.** $\sqrt[5]{13x}$

27. $\sqrt{x^3}$ **28.** $\sqrt{x^5}$

29. $\sqrt[5]{x^3}$ **30.** $\sqrt[7]{x^4}$

31. $\sqrt[5]{x^2 y}$ **32.** $\sqrt[7]{xy^3}$

33. $\left(\sqrt{19xy}\right)^3$ **34.** $\left(\sqrt{11xy}\right)^3$

35. $\left(\sqrt[6]{7xy^2}\right)^5$ **36.** $\left(\sqrt[6]{9x^2 y}\right)^5$

37. $2x\sqrt[3]{y^2}$ **38.** $4x\sqrt[5]{y^2}$

In Exercises 39–54, rewrite each expression with a positive rational exponent. Simplify, if possible.

39. $49^{-\frac{1}{2}}$ **40.** $9^{-\frac{1}{2}}$

41. $27^{-\frac{1}{3}}$

42. $125^{-\frac{1}{3}}$

43. $16^{-\frac{3}{4}}$

44. $81^{-\frac{5}{4}}$

45. $8^{-\frac{2}{3}}$

46. $32^{-\frac{4}{5}}$

47. $\left(\dfrac{8}{27}\right)^{-\frac{1}{3}}$

48. $\left(\dfrac{8}{125}\right)^{-\frac{1}{3}}$

49. $(-64)^{-\frac{2}{3}}$

50. $(-8)^{-\frac{2}{3}}$

51. $(2xy)^{-\frac{7}{10}}$

52. $(4xy)^{-\frac{4}{7}}$

53. $5xz^{-\frac{1}{3}}$

54. $7xz^{-\frac{1}{4}}$

In Exercises 55–78, use properties of rational exponents to simplify each expression. Assume that all variables represent positive numbers.

55. $3^{\frac{3}{4}}\cdot 3^{\frac{1}{4}}$

56. $5^{\frac{2}{3}}\cdot 5^{\frac{1}{3}}$

57. $\dfrac{16^{\frac{3}{4}}}{16^{\frac{1}{4}}}$

58. $\dfrac{100^{\frac{3}{4}}}{100^{\frac{1}{4}}}$

59. $x^{\frac{1}{2}}\cdot x^{\frac{1}{3}}$

60. $x^{\frac{1}{2}}\cdot x^{\frac{2}{3}}$

61. $\dfrac{x^{\frac{4}{5}}}{x^{\frac{1}{5}}}$

62. $\dfrac{x^{\frac{3}{7}}}{x^{\frac{1}{7}}}$

63. $\dfrac{x^{\frac{1}{3}}}{x^{\frac{3}{4}}}$

64. $\dfrac{x^{\frac{1}{4}}}{x^{\frac{3}{5}}}$

65. $\left(5^{\frac{2}{3}}\right)^3$

66. $\left(3^{\frac{4}{5}}\right)^5$

67. $\left(y^{-\frac{2}{3}}\right)^{\frac{1}{4}}$

68. $\left(y^{-\frac{3}{4}}\right)^{\frac{1}{6}}$

69. $\left(2x^{\frac{1}{5}}\right)^5$

70. $\left(2x^{\frac{1}{4}}\right)^4$

71. $(25x^4y^6)^{\frac{1}{2}}$

72. $(125x^9y^6)^{\frac{1}{3}}$

73. $\left(x^{\frac{1}{2}}y^{-\frac{3}{5}}\right)^{\frac{1}{2}}$

74. $\left(x^{\frac{1}{4}}y^{-\frac{2}{5}}\right)^{\frac{1}{3}}$

75. $\dfrac{3^{\frac{1}{2}}\cdot 3^{\frac{3}{4}}}{3^{\frac{1}{4}}}$

76. $\dfrac{5^{\frac{3}{4}}\cdot 5^{\frac{1}{2}}}{5^{\frac{1}{4}}}$

77. $\dfrac{\left(3y^{\frac{1}{4}}\right)^3}{y^{\frac{1}{12}}}$

78. $\dfrac{\left(2y^{\frac{1}{5}}\right)^4}{y^{\frac{3}{10}}}$

In Exercises 79–112, use rational exponents to simplify each expression. If rational exponents appear after simplifying, write the answer in radical notation. Assume that all variables represent positive numbers.

79. $\sqrt[8]{x^2}$

80. $\sqrt[10]{x^2}$

81. $\sqrt[3]{8a^6}$

82. $\sqrt[3]{27a^{12}}$

83. $\sqrt[5]{x^{10}y^{15}}$

84. $\sqrt[5]{x^{15}y^{20}}$

85. $\left(\sqrt[3]{xy}\right)^{18}$

86. $\left(\sqrt[3]{xy}\right)^{21}$

87. $\sqrt[10]{(3y)^2}$

88. $\sqrt[12]{(3y)^2}$

89. $\left(\sqrt[6]{2a}\right)^4$

90. $\left(\sqrt[8]{2a}\right)^6$

91. $\sqrt[9]{x^6y^3}$

92. $\sqrt[4]{x^2y^6}$

93. $\sqrt{2}\cdot\sqrt[3]{2}$

94. $\sqrt{3}\cdot\sqrt[3]{3}$

95. $\sqrt[5]{x^2}\cdot\sqrt{x}$

96. $\sqrt[7]{x^2}\cdot\sqrt{x}$

97. $\sqrt[4]{a^2b}\cdot\sqrt[3]{ab}$

98. $\sqrt[6]{ab^2}\cdot\sqrt[3]{a^2b}$

99. $\dfrac{\sqrt[4]{x}}{\sqrt[5]{x}}$

100. $\dfrac{\sqrt[3]{x}}{\sqrt[4]{x}}$

101. $\dfrac{\sqrt[3]{y^2}}{\sqrt[6]{y}}$

102. $\dfrac{\sqrt[5]{y^2}}{\sqrt[10]{y^3}}$

103. $\sqrt[4]{\sqrt{x}}$

104. $\sqrt[5]{\sqrt{x}}$

105. $\sqrt{\sqrt{x^2y}}$

106. $\sqrt{\sqrt{xy^2}}$

107. $\sqrt[4]{\sqrt[3]{2x}}$

108. $\sqrt[5]{\sqrt[3]{2x}}$

109. $\left(\sqrt[4]{x^3y^5}\right)^{12}$

110. $\left(\sqrt[5]{x^4y^2}\right)^{20}$

111. $\dfrac{\sqrt[4]{a^5b^5}}{\sqrt{ab}}$

112. $\dfrac{\sqrt[4]{a^3b^3}}{\sqrt{ab}}$

Practice PLUS

In Exercises 113–116, use the distributive property or the FOIL method to perform each multiplication.

113. $x^{\frac{1}{3}}\left(x^{\frac{1}{3}}-x^{\frac{2}{3}}\right)$

114. $x^{-\frac{1}{4}}\left(x^{\frac{9}{4}}-x^{\frac{1}{4}}\right)$

115. $\left(x^{\frac{1}{2}}-3\right)\left(x^{\frac{1}{2}}+5\right)$

116. $\left(x^{\frac{1}{3}}-2\right)\left(x^{\frac{1}{3}}+6\right)$

In Exercises 117–120, factor out the greatest common factor from each expression.

117. $6x^{\frac{1}{2}}+2x^{\frac{3}{2}}$

118. $8x^{\frac{1}{4}}+4x^{\frac{5}{4}}$

119. $15x^{\frac{1}{3}}-60x$

120. $7x^{\frac{1}{3}}-70x$

In Exercises 121–124, simplify each expression. Assume that all variables represent positive numbers.

121. $(49x^{-2}y^4)^{-\frac{1}{2}}\left(xy^{\frac{1}{2}}\right)$

122. $(8x^{-6}y^3)^{\frac{1}{3}}\left(x^{\frac{5}{6}}y^{-\frac{1}{3}}\right)^6$

123. $\left(\dfrac{x^{-\frac{5}{4}}y^{\frac{1}{3}}}{x^{-\frac{3}{4}}}\right)^{-6}$

124. $\left(\dfrac{x^{\frac{1}{2}}y^{-\frac{7}{4}}}{y^{-\frac{5}{4}}}\right)^{-4}$

Application Exercises

The Galápagos Islands, lying 600 miles west of Ecuador, are famed for their extraordinary wildlife. The function

$$f(x)=29x^{\frac{1}{3}}$$

models the number of plant species, $f(x)$, on the various islands of the Galápagos chain in terms of the area, x, in square miles, of a particular island. Use the function to solve Exercises 125–126.

125. How many species of plants are on a Galápagos island that has an area of 8 square miles?

126. How many species of plants are on a Galápagos island that has an area of 27 square miles?

The function

$$f(x) = 70x^{\frac{3}{4}}$$

models the number of calories per day, f(x), a person needs to maintain life in terms of that person's weight, x, in kilograms. (1 kilogram is approximately 2.2 pounds.) Use this model and a calculator to solve Exercises 127–128. Round answers to the nearest calorie.

127. How many calories per day does a person who weighs 80 kilograms (approximately 176 pounds) need to maintain life?

128. How many calories per day does a person who weighs 70 kilograms (approximately 154 pounds) need to maintain life?

The way that we perceive the temperature on a cold day depends on both air temperature and wind speed. The windchill is what the air temperature would have to be with no wind to achieve the same chilling effect on the skin. In 2002, the National Weather Service issued new windchill temperatures, shown in the table below. (One reason for this new windchill index is that the wind speed is now calculated at 5 feet, the average height of the human body's face, rather than 33 feet, the height of the standard anemometer, an instrument that calculates wind speed.)

New Windchill Temperature Index

Air Temperature (°F)

	30	25	20	15	10	5	0	−5	−10	−15	−20	−25
5	25	19	13	7	1	−5	−11	−16	−22	−28	−34	−40
10	21	15	9	3	−4	−10	−16	−22	−28	−35	−41	−47
15	19	13	6	0	−7	−13	−19	−26	−32	−39	−45	−51
20	17	11	4	−2	−9	−15	−22	−29	−35	−42	−48	−55
25	16	9	3	−4	−11	−17	−24	−31	−37	−44	−51	−58
30	15	8	1	−5	−12	−19	−26	−33	−39	−46	−53	−60
35	14	7	0	−7	−14	−21	−27	−34	−41	−48	−55	−62
40	13	6	−1	−8	−15	−22	−29	−36	−43	−50	−57	−64
45	12	5	−2	−9	−16	−23	−30	−37	−44	−51	−58	−65
50	12	4	−3	−10	−17	−24	−31	−38	−45	−52	−60	−67
55	11	4	−3	−11	−18	−25	−32	−39	−46	−54	−61	−68
60	10	3	−4	−11	−19	−26	−33	−40	−48	−55	−62	−69

Wind Speed (miles per hour)

▨ Frostbite occurs in 15 minutes or less.

Source: National Weather Service

The windchill temperatures shown in the table can be calculated using

$$C = 35.74 + 0.6215t - 35.74\sqrt[25]{v^4} + 0.4275t\sqrt[25]{v^4},$$

in which C is the windchill, in degrees Fahrenheit, t is the air temperature, in degrees Fahrenheit, and v is the wind speed, in miles per hour. Use the formula to solve Exercises 129–132.

129. a. Rewrite the equation for calculating windchill temperatures using rational exponents.

   b. Use the form of the equation in part (a) and a calculator to find the windchill temperature, to the nearest degree, when the air temperature is 25°F and the wind speed is 30 miles per hour.

130. a. Rewrite the equation for calculating windchill temperatures using rational exponents.

   b. Use the form of the equation in part (a) and a calculator to find the windchill temperature, to the nearest degree, when the air temperature is 35°F and the wind speed is 15 miles per hour.

131. a. Substitute 0 for *t* in the equation with rational exponents from Exercise 129(a) and write a function $C(v)$ that gives the windchill temperature as a function of wind speed for an air temperature of 0°F.

   b. Find and interpret $C(25)$. Use a calculator and round to the nearest degree.

   c. Identify your solution to part (b) on the graph shown.

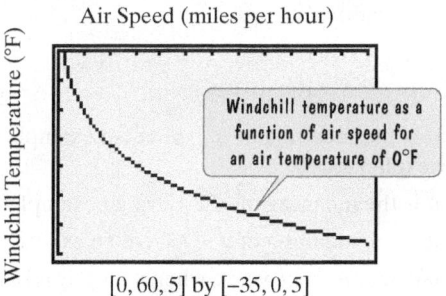

Air Speed (miles per hour)

Windchill temperature as a function of air speed for an air temperature of 0°F

$[0, 60, 5]$ by $[-35, 0, 5]$

132. a. Substitute 30 for *t* in the equation with rational exponents from Exercise 130(a) and write a function $C(v)$ that gives the windchill temperature as a function of wind speed for an air temperature of 30°F. Simplify the function's formula so that it contains exactly two terms.

   b. Find and interpret $C(40)$. Use a calculator and round to the nearest degree.

   c. Identify your solution to part (b) on the graph shown.

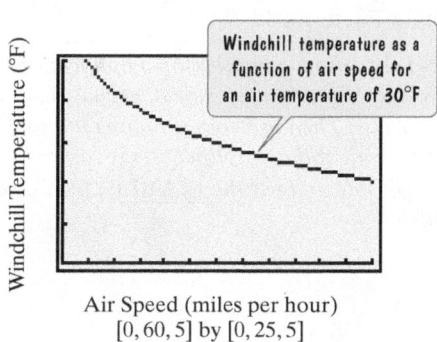

Windchill temperature as a function of air speed for an air temperature of 30°F

Air Speed (miles per hour)
$[0, 60, 5]$ by $[0, 25, 5]$

Your job is to determine whether or not yachts are eligible for the America's Cup, the supreme event in ocean sailing. The basic dimensions of competitors' yachts must satisfy

$$L + 1.25\sqrt{S} - 9.8\sqrt[3]{D} \le 16.296,$$

where L is the yacht's length, in meters, S is its sail area, in square meters, and D is its displacement, in cubic meters. Use this information to solve Exercises 133–134.

133. a. Rewrite the inequality using rational exponents.

    b. Use your calculator to determine if a yacht with length 20.85 meters, sail area 276.4 square meters, and displacement 18.55 cubic meters is eligible for the America's Cup.

134. a. Rewrite the inequality using rational exponents.

    b. Use your calculator to determine if a yacht with length 22.85 meters, sail area 312.5 square meters, and displacement 22.34 cubic meters is eligible for the America's Cup.

Writing in Mathematics

135. What is the meaning of $a^{\frac{1}{n}}$? Give an example to support your explanation.

136. What is the meaning of $a^{\frac{m}{n}}$? Give an example.

137. What is the meaning of $a^{-\frac{m}{n}}$? Give an example.

138. Explain why $a^{\frac{1}{n}}$ is negative when n is odd and a is negative. What happens if n is even and a is negative? Why?

139. In simplifying $36^{\frac{3}{2}}$, is it better to use $a^{\frac{m}{n}} = \sqrt[n]{a^m}$ or $a^{\frac{m}{n}} = (\sqrt[n]{a})^m$? Explain.

140. How can you tell if an expression with rational exponents is simplified?

141. Explain how to simplify $\sqrt[3]{x} \cdot \sqrt{x}$.

142. Explain how to simplify $\sqrt[3]{\sqrt{x}}$.

Technology Exercises

143. Use a scientific or graphing calculator to verify your results in Exercises 15–18.

144. Use a scientific or graphing calculator to verify your results in Exercises 45–50.

Exercises 145–147 show a number of simplifications, not all of which are correct. Enter the left side of each equation as y_1 and the right side as y_2. Then use your graphing utility's $\boxed{\text{TABLE}}$ *feature to determine if the simplification is correct. If it is not, correct the right side and use the* $\boxed{\text{TABLE}}$ *feature to verify your simplification.*

145. $x^{\frac{3}{5}} \cdot x^{-\frac{1}{10}} = x^{\frac{1}{2}}$

146. $\left(x^{-\frac{1}{2}} \cdot x^{\frac{3}{4}}\right)^{-2} = x^{\frac{1}{2}}$

147. $\dfrac{x^{\frac{1}{4}}}{x^{\frac{1}{2}} \cdot x^{-\frac{3}{4}}} = \dfrac{1}{x^{\frac{1}{2}}}$

Critical Thinking Exercises

Make Sense? *In Exercises 148–151, determine whether each statement "makes sense" or "does not make sense" and explain your reasoning.*

148. By adding the exponents, I simplified $7^{\frac{1}{2}} \cdot 7^{\frac{1}{2}}$ and obtained 49.

149. When I use the definition for $a^{\frac{m}{n}}$, I usually prefer to first raise a to the m power because smaller numbers are involved.

150. There's no question that $(-64)^{\frac{1}{3}} = -64^{\frac{1}{3}}$, so I can conclude that $(-64)^{\frac{1}{2}} = -64^{\frac{1}{2}}$.

151. I checked the following simplification and every step is correct:

$$\begin{aligned}
5\left(4 - 5^{\frac{1}{2}}\right) &= 5 \cdot 4 - 5 \cdot 5^{\frac{1}{2}} \\
&= 20 - 25^{\frac{1}{2}} \\
&= 20 - \sqrt{25} \\
&= 20 - 5 = 15.
\end{aligned}$$

In Exercises 152–155, determine whether each statement is true or false. If the statement is false, make the necessary change(s) to produce a true statement.

152. If n is odd, then $(-b)^{\frac{1}{n}} = -b^{\frac{1}{n}}$.

153. $(a + b)^{\frac{1}{n}} = a^{\frac{1}{n}} + b^{\frac{1}{n}}$

154. $8^{-\frac{2}{3}} = -4$

155. $4^{-3.5} = \dfrac{1}{128}$

156. A mathematics professor recently purchased a birthday cake for her son with the inscription

$$\text{Happy}\left(2^{\frac{5}{2}} \cdot 2^{\frac{3}{4}} \div 2^{\frac{1}{4}}\right)\text{th Birthday.}$$

How old is the son?

157. The birthday boy in Exercise 156, excited by the inscription on the cake, tried to wolf down the whole thing. Professor Mom, concerned about the possible metamorphosis of her son into a blimp, exclaimed, "Hold on! It is your birthday, so why not take $\dfrac{8^{-\frac{4}{3}} + 2^{-2}}{16^{-\frac{3}{4}} + 2^{-1}}$ of the cake? I'll eat half of what's left over." How much of the cake did the professor eat?

158. Simplify: $\left[3 + \left(27^{\frac{2}{3}} + 32^{\frac{2}{5}}\right)\right]^{\frac{3}{2}} - 9^{\frac{1}{2}}$.

159. Find the domain of $f(x) = (x - 3)^{\frac{1}{2}}(x + 4)^{-\frac{1}{2}}$.

Review Exercises

160. Write the equation of the linear function whose graph passes through $(5, 1)$ and $(4, 3)$. (Section 3.5, Example 2)

161. Graph $y \le -\dfrac{3}{2}x + 3$. (Section 9.4, Example 2)

162. If $f(x) = 3x^2 - 5x + 4$, find $f(a + h)$.
(Section 8.1, Example 3)

Preview Exercises

Exercises 163–165 will help you prepare for the material covered in the next section.

163. a. Find $\sqrt{16} \cdot \sqrt{4}$.

   b. Find $\sqrt{16 \cdot 4}$.

   c. Based on your answers to parts (a) and (b), what can you conclude?

164. a. Use a calculator to approximate $\sqrt{300}$ to two decimal places.

   b. Use a calculator to approximate $10\sqrt{3}$ to two decimal places.

   c. Based on your answers to parts (a) and (b), what can you conclude?

165. Simplify: **a.** $\sqrt[3]{x^{21}}$ **b.** $\sqrt[6]{y^{24}}$.

SECTION 10.3

Multiplying and Simplifying Radical Expressions

Objectives

1 Use the product rule to multiply radicals.

2 Use factoring and the product rule to simplify radicals.

3 Multiply radicals and then simplify.

Mirror II (1963), George Tooker. Addison Gallery, Phillips Academy, MA. © George Tooker.

There is a mathematical model that describes our improving emotional health as we age. Unfortunately, not everything gets better. The aging process is also accompanied by a number of physical transformations, including changes in vision that require glasses for reading, the onset of wrinkles and sagging skin, and a decrease in heart response. A change in heart response occurs fairly early; after 20, our hearts become less adept at accelerating in response to exercise. In this section's Exercise Set, you will see how a radical function models changes in heart function throughout the aging process, as we turn to multiplying and simplifying radical expressions.

1 Use the product rule to multiply radicals.

The Product Rule for Radicals

A rule for multiplying radicals can be generalized by comparing $\sqrt{25} \cdot \sqrt{4}$ and $\sqrt{25 \cdot 4}$. Notice that

$$\sqrt{25} \cdot \sqrt{4} = 5 \cdot 2 = 10 \quad \text{and} \quad \sqrt{25 \cdot 4} = \sqrt{100} = 10.$$

Because we obtain 10 in both situations, the original radical expressions must be equal. That is,

$$\sqrt{25} \cdot \sqrt{4} = \sqrt{25 \cdot 4}.$$

This result is a special case of the **product rule for radicals** that can be generalized as follows:

> ### The Product Rule for Radicals
> If $\sqrt[n]{a}$ and $\sqrt[n]{b}$ are real numbers, then
> $$\sqrt[n]{a} \cdot \sqrt[n]{b} = \sqrt[n]{ab}.$$
>
> The product of two nth roots is the nth root of the product of the radicands.

Great Question!

Can I use the product rule to simplify radicals with different indices, such as $\sqrt{x} \cdot \sqrt[3]{x}$?

No. The product rule can be used only when the radicals have the same index. If indices differ, rational exponents can be used, as in $\sqrt{x} \cdot \sqrt[3]{x}$, which was Example 7(d) in the previous section.

EXAMPLE 1 Using the Product Rule for Radicals

Multiply:

a. $\sqrt{3} \cdot \sqrt{7}$ **b.** $\sqrt{x+7} \cdot \sqrt{x-7}$ **c.** $\sqrt[3]{7} \cdot \sqrt[3]{9}$ **d.** $\sqrt[8]{10x} \cdot \sqrt[8]{8x^4}$.

Solution In each problem, the indices are the same. Thus, we multiply by multiplying the radicands.

a. $\sqrt{3} \cdot \sqrt{7} = \sqrt{3 \cdot 7} = \sqrt{21}$

b. $\sqrt{x+7} \cdot \sqrt{x-7} = \sqrt{(x+7)(x-7)} = \sqrt{x^2 - 49}$

This is not equal to $\sqrt{x^2} - \sqrt{49}$.

c. $\sqrt[3]{7} \cdot \sqrt[3]{9} = \sqrt[3]{7 \cdot 9} = \sqrt[3]{63}$

d. $\sqrt[8]{10x} \cdot \sqrt[8]{8x^4} = \sqrt[8]{10x \cdot 8x^4} = \sqrt[8]{80x^5}$ ∎

✓ CHECK POINT 1 Multiply:

a. $\sqrt{5} \cdot \sqrt{11}$ **b.** $\sqrt{x+4} \cdot \sqrt{x-4}$

c. $\sqrt[3]{6} \cdot \sqrt[3]{10}$ **d.** $\sqrt[7]{2x} \cdot \sqrt[7]{6x^3}$.

2 Use factoring and the product rule to simplify radicals.

Using Factoring and the Product Rule to Simplify Radicals

In Chapter 6, we saw that a number that is the square of an integer is a **perfect square**. For example, 100 is a perfect square because $100 = 10^2$. A number is a **perfect cube** if it is the cube of an integer. Thus, 125 is a perfect cube because $125 = 5^3$. In general, a number is a **perfect nth power** if it is the nth power of an integer. Thus, p is a perfect nth power if there is an integer q such that $p = q^n$.

A radical of index n is **simplified** when its radicand has no factors other than 1 that are perfect nth powers. For example, $\sqrt{300}$ is not simplified because it can be expressed as $\sqrt{100 \cdot 3}$ and 100 is a perfect square. We can use the product rule in the form

$$\sqrt[n]{ab} = \sqrt[n]{a} \cdot \sqrt[n]{b}$$

to simplify $\sqrt[n]{ab}$ when a or b is a perfect nth power. Consider $\sqrt{300}$. To simplify, we factor 300 so that one of its factors is the greatest perfect square possible.

$$\sqrt{300} = \sqrt{100 \cdot 3} \qquad \text{Factor 300: 100 is the greatest perfect square factor.}$$
$$= \sqrt{100} \cdot \sqrt{3} \qquad \text{Use the product rule: } \sqrt[n]{ab} = \sqrt[n]{a} \cdot \sqrt[n]{b}.$$
$$= 10\sqrt{3} \qquad \text{Write } \sqrt{100} \text{ as 10. We read } 10\sqrt{3} \text{ as "ten times the square root of three."}$$

Using Technology

You can use a calculator to provide numerical support that $\sqrt{300} = 10\sqrt{3}$. First find an approximation for $\sqrt{300}$:

$$300 \ \boxed{\sqrt{\ }} \ \approx 17.32$$

or

$$\boxed{\sqrt{\ }} \ 300 \ \boxed{\text{ENTER}} \ \approx 17.32.$$

Now find an aproximation for $10\sqrt{3}$:

$$10 \ \boxed{\times} \ 3 \ \boxed{\sqrt{\ }} \ \approx 17.32$$

or

$$10 \ \boxed{\sqrt{\ }} \ 3 \ \boxed{\text{ENTER}} \ \approx 17.32.$$

Correct to two decimal places,

$$\sqrt{300} \approx 17.32 \quad \text{and}$$
$$10\sqrt{3} \approx 17.32.$$

This verifies that $\sqrt{300} = 10\sqrt{3}$. Use this technique to support the numerical results for the answers in this section. *Caution:* A simplified radical does not mean a decimal approximation.

Simplifying Radical Expressions by Factoring

A radical expression whose index is n is **simplified** when its radicand has no factors that are perfect nth powers. To simplify, use the following procedure:

1. Write the radicand as the product of two factors, one of which is the greatest perfect nth power.
2. Use the product rule to take the nth root of each factor.
3. Find the nth root of the perfect nth power.

EXAMPLE 2 Simplifying Radicals by Factoring

Simplify by factoring:

a. $\sqrt{75}$ **b.** $\sqrt[3]{54}$ **c.** $\sqrt[5]{64}$ **d.** $\sqrt{500xy^2}$.

Solution

a. $\sqrt{75} = \sqrt{25 \cdot 3}$ 25 is the greatest perfect square that is a factor of 75.

$\quad\quad\quad = \sqrt{25} \cdot \sqrt{3}$ Take the square root of each factor: $\sqrt[n]{ab} = \sqrt[n]{a} \cdot \sqrt[n]{b}$

$\quad\quad\quad = 5\sqrt{3}$ Write $\sqrt{25}$ as 5.

b. $\sqrt[3]{54} = \sqrt[3]{27 \cdot 2}$ 27 is the greatest perfect cube that is a factor of 54: $27 = 3^3$.

$\quad\quad\quad = \sqrt[3]{27} \cdot \sqrt[3]{2}$ Take the cube root of each factor: $\sqrt[n]{ab} = \sqrt[n]{a} \cdot \sqrt[n]{b}$.

$\quad\quad\quad = 3\sqrt[3]{2}$ Write $\sqrt[3]{27}$ as 3.

c. $\sqrt[5]{64} = \sqrt[5]{32 \cdot 2}$ 32 is the greatest perfect fifth power that is a factor of 64: $32 = 2^5$.

$\quad\quad\quad = \sqrt[5]{32} \cdot \sqrt[5]{2}$ Take the fifth root of each factor: $\sqrt[n]{ab} = \sqrt[n]{a} \cdot \sqrt[n]{b}$

$\quad\quad\quad = 2\sqrt[5]{2}$ Write $\sqrt[5]{32}$ as 2.

d. $\sqrt{500xy^2} = \sqrt{100y^2 \cdot 5x}$ $100y^2$ is the greatest perfect square that is a factor of $500xy^2$: $100y^2 = (10y)^2$.

$\quad\quad\quad = \sqrt{100y^2} \cdot \sqrt{5x}$ Factor into two radicals.

$\quad\quad\quad = 10|y| \sqrt{5x}$ Take the square root of $100y^2$. ■

✓ CHECK POINT 2 Simplify by factoring:

a. $\sqrt{80}$ **b.** $\sqrt[3]{40}$

c. $\sqrt[4]{32}$ **d.** $\sqrt{200x^2y}$.

EXAMPLE 3 Simplifying a Radical Function

If

$$f(x) = \sqrt{2x^2 + 4x + 2},$$

express the function, f, in simplified form.

Solution Begin by factoring the radicand. The GCF is 2. Simplification is possible if we obtain a factor that is a perfect square.

$$f(x) = \sqrt{2x^2 + 4x + 2} \qquad \text{This is the given function.}$$
$$= \sqrt{2(x^2 + 2x + 1)} \qquad \text{Factor out the GCF.}$$
$$= \sqrt{2(x + 1)^2} \qquad \text{Factor the perfect square trinomial:}$$
$$\qquad\qquad\qquad\qquad A^2 + 2AB + B^2 = (A + B)^2.$$
$$= \sqrt{2} \cdot \sqrt{(x + 1)^2} \qquad \text{Take the square root of each factor. The factor } (x + 1)^2 \text{ is}$$
$$\qquad\qquad\qquad\qquad \text{a perfect square.}$$
$$= \sqrt{2}|x + 1| \qquad \text{Take the square root of } (x + 1)^2.$$

In simplified form,
$$f(x) = \sqrt{2}|x + 1|. \quad \blacksquare$$

Using Technology

Graphic Connections

The graphs of

$$f(x) = \sqrt{2x^2 + 4x + 2}, \quad g(x) = \sqrt{2}|x + 1|, \quad \text{and} \quad h(x) = \sqrt{2}(x + 1)$$

are shown in **Figure 10.6** in three separate $[-5, 5, 1]$ by $[-5, 5, 1]$ viewing rectangles. The graphs in **Figure 10.6 (a)** and **(b)** are identical. This verifies that our simplification in Example 3 is correct: $\sqrt{2x^2 + 4x + 2} = \sqrt{2}|x + 1|$. Now compare the graphs in **Figure 10.6 (a)** and **(c)**. Can you see that they are not the same? This illustrates the importance of not leaving out absolute value bars:

$$\sqrt{2x^2 + 4x + 2} \neq \sqrt{2}(x + 1).$$

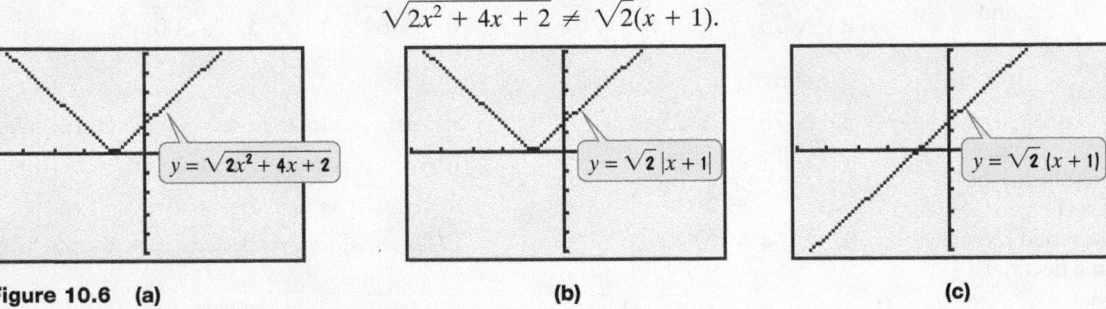

Figure 10.6   **(a)**   **(b)**   **(c)**

✓   **CHECK POINT 3**   If $f(x) = \sqrt{3x^2 - 12x + 12}$, express the function, f, in simplified form.

For the remainder of this chapter, in situations that do not involve functions, we will **assume that no radicands involve negative quantities raised to even powers. Based upon this assumption, absolute value bars are not necessary when taking even roots.**

Simplifying When Variables to Even Powers in a Radicand Are Nonnegative Quantities

For any nonnegative real number a,

$$\sqrt[n]{a^n} = a.$$

In simplifying an nth root, how do we find variable factors in the radicand that are perfect nth powers? The **perfect nth powers have exponents that are divisible by n**. Simplification is possible by observation or by using rational exponents. Here are some examples:

- $\sqrt{x^6} = \sqrt{(x^3)^2} = x^3$ or $\sqrt{x^6} = (x^6)^{\frac{1}{2}} = x^3$

  6 is divisible by the index, 2. Thus, x^6 is a perfect square.

- $\sqrt[3]{y^{21}} = \sqrt[3]{(y^7)^3} = y^7$ or $\sqrt[3]{y^{21}} = (y^{21})^{\frac{1}{3}} = y^7$

  21 is divisible by the index, 3. Thus, y^{21} is a perfect cube.

- $\sqrt[6]{z^{24}} = \sqrt[6]{(z^4)^6} = z^4$ or $\sqrt[6]{z^{24}} = (z^{24})^{\frac{1}{6}} = z^4.$

  24 is divisible by the index, 6. Thus, z^{24} is a perfect 6th power.

EXAMPLE 4 Simplifying a Radical by Factoring

Simplify: $\sqrt{x^5 y^{13} z^7}$.

Solution We write the radicand as the product of the greatest perfect square factor and another factor. Because the index is 2, variables that have exponents that are divisible by 2 are part of the perfect square factor. We use the greatest exponents that are divisible by 2.

$$\sqrt{x^5 y^{13} z^7} = \sqrt{x^4 \cdot x \cdot y^{12} \cdot y \cdot z^6 \cdot z} \qquad \text{Use the greatest even power of each variable.}$$
$$= \sqrt{(x^4 y^{12} z^6)(xyz)} \qquad \text{Group the perfect square factors.}$$
$$= \sqrt{x^4 y^{12} z^6} \cdot \sqrt{xyz} \qquad \text{Factor into two radicals.}$$
$$= x^2 y^6 z^3 \sqrt{xyz} \qquad \sqrt{x^4 y^{12} z^6} = (x^4 y^{12} z^6)^{\frac{1}{2}} = x^2 y^6 z^3 \quad \blacksquare$$

Discover for Yourself

Square the answer in Example 4 and show that it is correct. If it is a square root, you should obtain the given radicand, $x^5 y^{13} z^7$.

✓ **CHECK POINT 4** Simplify: $\sqrt{x^9 y^{11} z^3}$.

EXAMPLE 5 Simplifying a Radical by Factoring

Simplify: $\sqrt[3]{32 x^8 y^{16}}$.

Solution We write the radicand as the product of the greatest perfect cube factor and another factor. Because the index is 3, variables that have exponents that are divisible by 3 are part of the perfect cube factor. We use the greatest exponents that are divisible by 3.

$$\sqrt[3]{32 x^8 y^{16}} = \sqrt[3]{8 \cdot 4 \cdot x^6 \cdot x^2 \cdot y^{15} \cdot y} \qquad \text{Identify perfect cube factors.}$$
$$= \sqrt[3]{(8 x^6 y^{15})(4 x^2 y)} \qquad \text{Group the perfect cube factors.}$$
$$= \sqrt[3]{8 x^6 y^{15}} \cdot \sqrt[3]{4 x^2 y} \qquad \text{Factor into two radicals.}$$
$$= 2 x^2 y^5 \sqrt[3]{4 x^2 y} \qquad \sqrt[3]{8} = 2 \text{ and } \sqrt[3]{x^6 y^{15}} = (x^6 y^{15})^{\frac{1}{3}} = x^2 y^5. \quad \blacksquare$$

✓ **CHECK POINT 5** Simplify: $\sqrt[3]{40 x^{10} y^{14}}$.

EXAMPLE 6 Simplifying a Radical by Factoring

Simplify: $\sqrt[5]{64 x^3 y^7 z^{29}}$.

Solution We write the radicand as the product of the greatest perfect 5th power and another factor. Because the index is 5, variables that have exponents that are divisible by 5 are part of the perfect fifth factor. We use the greatest exponents that are divisible by 5.

$$\sqrt[5]{64 x^3 y^7 z^{29}} = \sqrt[5]{32 \cdot 2 \cdot x^3 \cdot y^5 \cdot y^2 \cdot z^{25} \cdot z^4} \qquad \text{Identify perfect fifth factors.}$$
$$= \sqrt[5]{(32 y^5 z^{25})(2 x^3 y^2 z^4)} \qquad \text{Group the perfect fifth factors.}$$
$$= \sqrt[5]{32 y^5 z^{25}} \cdot \sqrt[5]{2 x^3 y^2 z^4} \qquad \text{Factor into two radicals.}$$
$$= 2 y z^5 \sqrt[5]{2 x^3 y^2 z^4} \qquad \sqrt[5]{32} = 2 \text{ and } \sqrt[5]{y^5 z^{25}} = (y^5 z^{25})^{\frac{1}{5}} = y z^5. \quad \blacksquare$$

✓ **CHECK POINT 6** Simplify: $\sqrt[5]{32 x^{12} y^2 z^8}$.

3 Multiply radicals and then simplify.

Multiplying and Simplifying Radicals

We have seen how to use the product rule when multiplying radicals with the same index. Sometimes after multiplying, we can simplify the resulting radical.

EXAMPLE 7 Multiplying Radicals and Then Simplifying

Multiply and simplify:

a. $\sqrt{15} \cdot \sqrt{3}$ **b.** $7\sqrt[3]{4} \cdot 5\sqrt[3]{6}$ **c.** $\sqrt[4]{8x^3y^2} \cdot \sqrt[4]{8x^5y^3}$.

Great Question!

When should I write an expression under one radical and when should I separate the radicals?

- Use $\sqrt[n]{a} \cdot \sqrt[n]{b} = \sqrt[n]{ab}$, writing under one radical, when *multiplying*.
- Use $\sqrt[n]{ab} = \sqrt[n]{a} \cdot \sqrt[n]{b}$, factoring into two radicals, when *simplifying*.

Solution

a. $\sqrt{15} \cdot \sqrt{3} = \sqrt{15 \cdot 3}$ Use the product rule.

$= \sqrt{45} = \sqrt{9 \cdot 5}$ 9 is the greatest perfect square factor of 45.

$= \sqrt{9} \cdot \sqrt{5} = 3\sqrt{5}$

b. $7\sqrt[3]{4} \cdot 5\sqrt[3]{6} = 35\sqrt[3]{4 \cdot 6}$ Use the product rule.

$= 35\sqrt[3]{24} = 35\sqrt[3]{8 \cdot 3}$ 8 is the greatest perfect cube factor of 24.

$= 35\sqrt[3]{8} \cdot \sqrt[3]{3} = 35 \cdot 2 \cdot \sqrt[3]{3}$

$= 70\sqrt[3]{3}$

c. $\sqrt[4]{8x^3y^2} \cdot \sqrt[4]{8x^5y^3} = \sqrt[4]{8x^3y^2 \cdot 8x^5y^3}$ Use the product rule.

$= \sqrt[4]{64x^8y^5}$ Multiply.

$= \sqrt[4]{16 \cdot 4 \cdot x^8 \cdot y^4 \cdot y}$ Identify perfect fourth factors.

$= \sqrt[4]{(16x^8y^4)(4y)}$ Group the perfect fourth factors.

$= \sqrt[4]{16x^8y^4} \cdot \sqrt[4]{4y}$ Factor into two radicals.

$= 2x^2y\sqrt[4]{4y}$ $\sqrt[4]{16} = 2$ and

$\sqrt[4]{x^8y^4} = (x^8y^4)^{\frac{1}{4}} = x^2y.$ ∎

☑ **CHECK POINT 7** Multiply and simplify:

a. $\sqrt{6} \cdot \sqrt{2}$

b. $10\sqrt[3]{16} \cdot 5\sqrt[3]{2}$

c. $\sqrt[4]{4x^2y} \cdot \sqrt[4]{8x^6y^3}$.

CONCEPT AND VOCABULARY CHECK

Fill in each blank so that the resulting statement is true.

1. If $\sqrt[n]{a}$ and $\sqrt[n]{b}$ are real numbers, then $\sqrt[n]{a} \cdot \sqrt[n]{b} =$ _____.

2. $\sqrt[3]{7} \cdot \sqrt[3]{11} = \sqrt[3]{\_\_}$.

3. $\sqrt[3]{8 \cdot 5} = \sqrt[3]{\_} \cdot \sqrt[3]{\_} =$ _____.

4. $\sqrt{5(x+1)^2} =$ _____.

5. Variable factors in a radicand that are perfect nth powers can be simplified by observation. For example, $\sqrt{x^{10}} =$ _____, $\sqrt[3]{x^{12}} =$ _____, and $\sqrt[6]{x^{30}} =$ _____.

10.3 EXERCISE SET

MyMathLab®

 Watch the videos in MyMathLab

 Download the MyDashBoard App

Practice Exercises

In Exercises 1–20, use the product rule to multiply.

1. $\sqrt{3} \cdot \sqrt{5}$ 2. $\sqrt{7} \cdot \sqrt{5}$

3. $\sqrt[3]{2} \cdot \sqrt[3]{9}$ 4. $\sqrt[3]{5} \cdot \sqrt[3]{4}$

5. $\sqrt[4]{11} \cdot \sqrt[4]{3}$ 6. $\sqrt[5]{9} \cdot \sqrt[5]{3}$

7. $\sqrt{3x} \cdot \sqrt{11y}$ 8. $\sqrt{5x} \cdot \sqrt{11y}$

9. $\sqrt[5]{6x^3} \cdot \sqrt[5]{4x}$ 10. $\sqrt[4]{6x^2} \cdot \sqrt[4]{3x}$

11. $\sqrt{x+3} \cdot \sqrt{x-3}$

12. $\sqrt{x+6} \cdot \sqrt{x-6}$

13. $\sqrt[6]{x-4} \cdot \sqrt[6]{(x-4)^4}$

14. $\sqrt[6]{x-5} \cdot \sqrt[6]{(x-5)^4}$

15. $\sqrt{\dfrac{2x}{3}} \cdot \sqrt{\dfrac{3}{2}}$ 16. $\sqrt{\dfrac{2x}{5}} \cdot \sqrt{\dfrac{5}{2}}$

17. $\sqrt[4]{\dfrac{x}{7}} \cdot \sqrt[4]{\dfrac{3}{y}}$

18. $\sqrt[4]{\dfrac{x}{3}} \cdot \sqrt[4]{\dfrac{7}{y}}$

19. $\sqrt[7]{7x^2y} \cdot \sqrt[7]{11x^3y^2}$

20. $\sqrt[9]{12x^2y^3} \cdot \sqrt[9]{3x^3y^4}$

In Exercises 21–32, simplify by factoring.

21. $\sqrt{50}$

22. $\sqrt{27}$

23. $\sqrt{45}$

24. $\sqrt{28}$

25. $\sqrt{75x}$

26. $\sqrt{40x}$

27. $\sqrt[3]{16}$

28. $\sqrt[3]{54}$

29. $\sqrt[3]{27x^3}$

30. $\sqrt[3]{250x^3}$

31. $\sqrt[3]{-16x^2y^3}$

32. $\sqrt[3]{-32x^2y^3}$

In Exercises 33–38, express the function, f, in simplified form. Assume that x can be any real number.

33. $f(x) = \sqrt{36(x+2)^2}$

34. $f(x) = \sqrt{81(x-2)^2}$

35. $f(x) = \sqrt[3]{32(x+2)^3}$

36. $f(x) = \sqrt[3]{48(x-2)^3}$

37. $f(x) = \sqrt{3x^2 - 6x + 3}$

38. $f(x) = \sqrt{5x^2 - 10x + 5}$

In Exercises 39–60, simplify by factoring. Assume that all variables in a radicand represent positive real numbers and no radicands involve negative quantities raised to even powers.

39. $\sqrt{x^7}$

40. $\sqrt{x^5}$

41. $\sqrt{x^8y^9}$

42. $\sqrt{x^6y^7}$

43. $\sqrt{48x^3}$

44. $\sqrt{40x^3}$

45. $\sqrt[3]{y^8}$

46. $\sqrt[3]{y^{11}}$

47. $\sqrt[3]{x^{14}y^3z}$

48. $\sqrt[3]{x^3y^{17}z^2}$

49. $\sqrt[3]{81x^8y^6}$

50. $\sqrt[3]{32x^9y^{17}}$

51. $\sqrt[3]{(x+y)^5}$

52. $\sqrt[3]{(x+y)^4}$

53. $\sqrt[5]{y^{17}}$

54. $\sqrt[5]{y^{18}}$

55. $\sqrt[5]{64x^6y^{17}}$

56. $\sqrt[5]{64x^7y^{16}}$

57. $\sqrt[4]{80x^{10}}$

58. $\sqrt[4]{96x^{11}}$

59. $\sqrt[4]{(x-3)^{10}}$

60. $\sqrt[4]{(x-2)^{14}}$

In Exercises 61–82, multiply and simplify. Assume that all variables in a radicand represent positive real numbers and no radicands involve negative quantities raised to even powers.

61. $\sqrt{12} \cdot \sqrt{2}$

62. $\sqrt{3} \cdot \sqrt{6}$

63. $\sqrt{5x} \cdot \sqrt{10y}$

64. $\sqrt{8x} \cdot \sqrt{10y}$

65. $\sqrt{12x} \cdot \sqrt{3x}$

66. $\sqrt{20x} \cdot \sqrt{5x}$

67. $\sqrt{50xy} \cdot \sqrt{4xy^2}$

68. $\sqrt{5xy} \cdot \sqrt{10xy^2}$

69. $2\sqrt{5} \cdot 3\sqrt{40}$

70. $3\sqrt{15} \cdot 5\sqrt{6}$

71. $\sqrt[3]{12} \cdot \sqrt[3]{4}$

72. $\sqrt[4]{4} \cdot \sqrt[4]{8}$

73. $\sqrt{5x^3} \cdot \sqrt{8x^2}$

74. $\sqrt{2x^7} \cdot \sqrt{12x^4}$

75. $\sqrt[3]{25x^4y^2} \cdot \sqrt[3]{5xy^{12}}$

76. $\sqrt[3]{6x^7y} \cdot \sqrt[3]{9x^4y^{12}}$

77. $\sqrt[4]{8x^2y^3z^6} \cdot \sqrt[4]{2x^4yz}$

78. $\sqrt[4]{4x^2y^3z^3} \cdot \sqrt[4]{8x^4yz^6}$

79. $\sqrt[5]{8x^4y^6z^2} \cdot \sqrt[5]{8xy^7z^4}$

80. $\sqrt[5]{8x^4y^3z^3} \cdot \sqrt[5]{8xy^9z^8}$

81. $\sqrt[3]{x-y} \cdot \sqrt[3]{(x-y)^7}$

82. $\sqrt[3]{x-6} \cdot \sqrt[3]{(x-6)^7}$

Practice PLUS

In Exercises 83–90, simplify each expression. Assume that all variables in a radicand represent positive real numbers and no radicands involve negative quantities raised to even powers.

83. $-2x^2y \left(\sqrt[3]{54x^3y^7z^2} \right)$

84. $\dfrac{-x^2y^7}{2} \left(\sqrt[3]{-32x^4y^9z^7} \right)$

85. $-3y \left(\sqrt[5]{64x^3y^6} \right)$

86. $-4x^2y^7 \left(\sqrt[5]{-32x^{11}y^{17}} \right)$

87. $\left(-2xy^2\sqrt{3x} \right)\left(xy\sqrt{6x} \right)$

88. $\left(-5x^2y^3z\sqrt{2xyz} \right)\left(-x^4z\sqrt{10xz} \right)$

89. $\left(2x^2y\sqrt[4]{8xy} \right)\left(-3xy^2\sqrt[4]{2x^2y^3} \right)$

90. $\left(5a^2b\sqrt[4]{8a^2b} \right)\left(4ab\sqrt[4]{4a^3b^2} \right)$

Application Exercises

The function

$$d(x) = \sqrt{\dfrac{3x}{2}}$$

models the distance, d(x), in miles, that a person x feet high can see to the horizon. Use this function to solve Exercises 91–92.

91. The pool deck on a cruise ship is 72 feet above the water. How far can passengers on the pool deck see? Write the answer in simplified radical form. Then use the simplified radical form and a calculator to express the answer to the nearest tenth of a mile.

92. The captain of a cruise ship is on the star deck, which is 120 feet above the water. How far can the captain see? Write the answer in simplified radical form. Then use the simplified radical form and a calculator to express the answer to the nearest tenth of a mile.

Paleontologists use the function

$$W(x) = 4\sqrt{2x}$$

to estimate the walking speed of a dinosaur, $W(x)$, in feet per second, where x is the length, in feet, of the dinosaur's leg. The graph of W is shown in the figure. Use this information to solve Exercises 93–94.

Dinosaur Walking Speeds

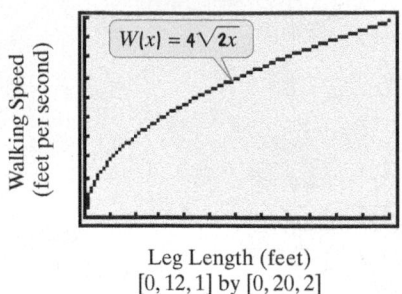

Leg Length (feet)
$[0, 12, 1]$ by $[0, 20, 2]$

93. What is the walking speed of a dinosaur whose leg length is 6 feet? Use the function's equation and express the answer in simplified radical form. Then use the function's graph to estimate the answer to the nearest foot per second.

94. What is the walking speed of a dinosaur whose leg length is 10 feet? Use the function's equation and express the answer in simplified radical form. Then use the function's graph to estimate the answer to the nearest foot per second.

*Your **cardiac index** is your heart's output, in liters of blood per minute, divided by your body's surface area, in square meters. The cardiac index, $C(x)$, can be modeled by*

$$C(x) = \frac{7.644}{\sqrt[4]{x}}, \qquad 10 \le x \le 80,$$

where x is an individual's age, in years. The graph of the function is shown. Use the function to solve Exercises 95–96.

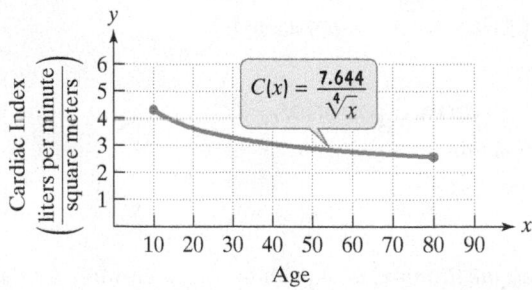

95. **a.** Find the cardiac index of a 32-year-old. Express the denominator in simplified radical form and reduce the fraction.

   b. Use the form of the answer in part (a) and a calculator to express the cardiac index to the nearest hundredth. Identify your solution as a point on the graph.

96. **a.** Find the cardiac index of an 80-year-old. Express the denominator in simplified radical form and reduce the fraction.

   b. Use the form of the answer in part (a) and a calculator to express the cardiac index to the nearest hundredth. Identify your solution as a point on the graph.

Writing in Mathematics

97. What is the product rule for radicals? Give an example to show how it is used.

98. Explain why $\sqrt{50}$ is not simplified. What do we mean when we say a radical expression is simplified?

99. In simplifying an nth root, explain how to find variable factors in the radicand that are perfect nth powers.

100. Without showing all the details, explain how to simplify $\sqrt[3]{16x^{14}}$.

101. As you get older, what would you expect to happen to your heart's output? Explain how this is shown in the graph for Exercises 95–96. Is this trend taking place progressively more rapidly or more slowly over the entire interval? What does this mean about this aspect of aging?

Technology Exercises

102. Use a calculator to provide numerical support for your simplifications in Exercises 21–24 and 27–28. In each case, find a decimal approximation for the given expression. Then find a decimal approximation for your simplified expression. The approximations should be the same.

In Exercises 103–106, determine if each simplification is correct by graphing the function on each side of the equation with your graphing utility. Use the given viewing rectangle. The graphs should be the same. If they are not, correct the right side of the equation and then use your graphing utility to verify the simplification.

103. $\sqrt{x^4} = x^2$; $[0, 5, 1]$ by $[0, 20, 1]$

104. $\sqrt{8x^2} = 4x\sqrt{2}$; $[-5, 5, 1]$ by $[-5, 20, 1]$

105. $\sqrt{3x^2 - 6x + 3} = (x - 1)\sqrt{3}$; $[-5, 5, 1]$ by $[-5, 5, 1]$

106. $\sqrt[3]{2x} \cdot \sqrt[3]{4x^2} = 4x$; $[-10, 10, 1]$ by $[-10, 10, 1]$

Critical Thinking Exercises

Make Sense? *In Exercises 107–110, determine whether each statement "makes sense" or "does not make sense" and explain your reasoning.*

107. Because the product rule for radicals applies when $\sqrt[n]{a}$ and $\sqrt[n]{b}$ are real numbers, I can use it to find $\sqrt[3]{16} \cdot \sqrt[3]{-4}$, but not to find $\sqrt{8} \cdot \sqrt{-2}$.

108. I multiply nth roots by taking the nth root of the product of the radicands.

109. I need to know how to factor a trinomial to simplify $\sqrt{x^2 - 10x + 25}$.

110. I know that I've simplified a radical expression when it contains a single radical.

In Exercises 111–114, determine whether each statement is true or false. If the statement is false, make the necessary change(s) to produce a true statement.

111. $2\sqrt{5} \cdot 6\sqrt{5} = 12\sqrt{5}$

112. $\sqrt[3]{4} \cdot \sqrt[3]{4} = 4$

113. $\sqrt{12} = 2\sqrt{6}$

114. $\sqrt[3]{3^{15}} = 243$

115. If a number is tripled, what happens to its square root?

116. What must be done to a number so that its cube root is tripled?

117. If $f(x) = \sqrt[3]{2x}$ and $(fg)(x) = 2x$, find $g(x)$.

118. Graph $f(x) = \sqrt{(x-1)^2}$ by hand.

Review Exercises

119. Solve: $2x - 1 \le 21$ and $2x + 2 \ge 12$.
(Section 9.2, Example 2)

120. Solve:
$$\begin{cases} 5x + 2y = 2 \\ 4x + 3y = -4. \end{cases}$$
(Section 4.3, Example 4)

121. Factor: $64x^3 - 27$. (Section 6.4, Example 8)

Preview Exercises

Exercises 122–124 will help you prepare for the material covered in the next section.

122. **a.** Simplify: $21x + 10x$.
b. Simplify: $21\sqrt{2} + 10\sqrt{2}$.

123. **a.** Simplify: $4x - 12x$.
b. Simplify: $4\sqrt[3]{2} - 12\sqrt[3]{2}$.

124. Simplify: $\dfrac{\sqrt[4]{7y^5}}{\sqrt[4]{x^{12}}}$.

Adding, Subtracting, and Dividing Radical Expressions

Objectives

1 Add and subtract radical expressions.

2 Use the quotient rule to simplify radical expressions.

3 Use the quotient rule to divide radical expressions.

The future is now: You have the opportunity to explore the cosmos in a starship traveling near the speed of light. The experience will enable you to understand the mysteries of the universe firsthand, transporting you to unimagined levels of knowing and being. The downside: According to Einstein's theory of relativity, close to the speed of light, your aging rate relative to friends on Earth is nearly zero. You will return from your two-year journey to an unknown futuristic world. In this section's Exercise Set, we provide an expression involving radical division that models your return to this unrecognizable world. To make sense of the model, we turn to various operations with radicals, including addition, subtraction, and division.

1 Add and subtract radical expressions.

Adding and Subtracting Radical Expressions

We know that like terms have exactly the same variable factors and can be combined. For example,

$$7x + 6x = (7 + 6)x = 13x.$$

Two or more radical expressions that have the same indices *and* the same radicands are called **like radicals**. Like radicals are combined in exactly the same way that we combine like terms. For example,

$$7\sqrt{11} + 6\sqrt{11} = (7 + 6)\sqrt{11} = 13\sqrt{11}.$$

> 7 square roots of 11 plus 6 square roots of 11 result in 13 square roots of 11.

EXAMPLE 1 Adding and Subtracting Like Radicals

Simplify (add or subtract) by combining like radical terms:

a. $7\sqrt{2} + 8\sqrt{2}$

b. $\sqrt[3]{5} - 4x\sqrt[3]{5} + 8\sqrt[6]{5}$

c. $8\sqrt[6]{5x} - 5\sqrt[6]{5x} + 4\sqrt[3]{5x}$.

Solution

a. $7\sqrt{2} + 8\sqrt{2}$

$= (7 + 8)\sqrt{2}$ Apply the distributive property.

$= 15\sqrt{2}$ Simplify.

b. $\sqrt[3]{5} - 4x\sqrt[3]{5} + 8\sqrt[3]{5}$

$= (1 - 4x + 8)\sqrt[3]{5}$ Apply the distributive property.

$= (9 - 4x)\sqrt[3]{5}$ Simplify.

c. $8\sqrt[6]{5x} - 5\sqrt[6]{5x} + 4\sqrt[3]{5x}$

$= (8 - 5)\sqrt[6]{5x} + 4\sqrt[3]{5x}$ Apply the distributive property to the two terms with like radicals.

$= 3\sqrt[6]{5x} + 4\sqrt[3]{5x}$ The indices, 6 and 3, differ. These are not like radicals and cannot be combined. ∎

✓ CHECK POINT 1 Simplify by combining like radical terms:

a. $8\sqrt{13} + 2\sqrt{13}$

b. $9\sqrt[3]{7} - 6x\sqrt[3]{7} + 12\sqrt[3]{7}$

c. $7\sqrt[4]{3x} - 2\sqrt[3]{3x} + 2\sqrt[3]{3x}$.

In some cases, radical expressions can be combined once they have been simplified. For example, to add $\sqrt{2}$ and $\sqrt{8}$, we can write $\sqrt{8}$ as $\sqrt{4 \cdot 2}$ because 4 is a perfect square factor of 8.

$$\sqrt{2} + \sqrt{8} = \sqrt{2} + \sqrt{4 \cdot 2} = 1\sqrt{2} + 2\sqrt{2} = (1 + 2)\sqrt{2} = 3\sqrt{2}$$

Always begin by simplifying radical terms. This makes it possible to identify and combine any like radicals.

EXAMPLE 2 Combining Radicals That First Require Simplification

Simplify by combining like radical terms, if possible:

a. $7\sqrt{18} + 5\sqrt{8}$ **b.** $4\sqrt{27x} - 8\sqrt{12x}$ **c.** $7\sqrt{3} - 2\sqrt{5}$.

Solution

a. $7\sqrt{18} + 5\sqrt{8}$

$= 7\sqrt{9 \cdot 2} + 5\sqrt{4 \cdot 2}$ Factor the radicands using the greatest perfect square factors.

$= 7\sqrt{9} \cdot \sqrt{2} + 5\sqrt{4} \cdot \sqrt{2}$ Take the square root of each factor.

$= 7 \cdot 3 \cdot \sqrt{2} + 5 \cdot 2 \cdot \sqrt{2}$ $\sqrt{9} = 3$ and $\sqrt{4} = 2$.

$= 21\sqrt{2} + 10\sqrt{2}$ Multiply.

$= (21 + 10)\sqrt{2}$ Apply the distributive property.

$= 31\sqrt{2}$ Simplify.

b. $4\sqrt{27x} - 8\sqrt{12x}$

$= 4\sqrt{9 \cdot 3x} - 8\sqrt{4 \cdot 3x}$ Factor the radicands using the greatest perfect square factors.

$= 4\sqrt{9} \cdot \sqrt{3x} - 8\sqrt{4} \cdot \sqrt{3x}$ Take the square root of each factor.

$= 4 \cdot 3 \cdot \sqrt{3x} - 8 \cdot 2 \cdot \sqrt{3x}$ $\sqrt{9} = 3$ and $\sqrt{4} = 2$.

$= 12\sqrt{3x} - 16\sqrt{3x}$ Multiply.

$= (12 - 16)\sqrt{3x}$ Apply the distributive property.

$= -4\sqrt{3x}$ Simplify.

c. $7\sqrt{3} - 2\sqrt{5}$ cannot be simplified. The radical expressions have different radicands, namely 3 and 5, and are not like terms. ∎

✓ **CHECK POINT 2** Simplify by combining like radical terms, if possible:

a. $3\sqrt{20} + 5\sqrt{45}$
b. $3\sqrt{12x} - 6\sqrt{27x}$
c. $8\sqrt{5} - 6\sqrt{2}$.

EXAMPLE 3 Adding and Subtracting with Higher Indices

Simplify by combining like radical terms, if possible:

a. $2\sqrt[3]{16} - 4\sqrt[3]{54}$ **b.** $5\sqrt[3]{xy^2} + \sqrt[3]{8x^4y^5}$.

Solution

a. $2\sqrt[3]{16} - 4\sqrt[3]{54}$

$= 2\sqrt[3]{8 \cdot 2} - 4\sqrt[3]{27 \cdot 2}$ Factor the radicands using the greatest perfect cube factors.

$= 2\sqrt[3]{8} \cdot \sqrt[3]{2} - 4\sqrt[3]{27} \cdot \sqrt[3]{2}$ Take the cube root of each factor.

$= 2 \cdot 2 \cdot \sqrt[3]{2} - 4 \cdot 3 \cdot \sqrt[3]{2}$ $\sqrt[3]{8} = 2$ and $\sqrt[3]{27} = 3$.

$= 4\sqrt[3]{2} - 12\sqrt[3]{2}$ Multiply.

$= (4 - 12)\sqrt[3]{2}$ Apply the distributive property.

$= -8\sqrt[3]{2}$ Simplify.

b. $5\sqrt[3]{xy^2} + \sqrt[3]{8x^4y^5}$

$= 5\sqrt[3]{xy^2} + \sqrt[3]{(8x^3y^3)xy^2}$ Factor the second radicand using the greatest perfect cube factor.

$= 5\sqrt[3]{xy^2} + \sqrt[3]{8x^3y^3} \cdot \sqrt[3]{xy^2}$ Take the cube root of each factor.

$= 5\sqrt[3]{xy^2} + 2xy\sqrt[3]{xy^2}$ $\sqrt[3]{8} = 2$ and $\sqrt[3]{x^3y^3} = (x^3y^3)^{\frac{1}{3}} = xy$.

$= (5 + 2xy)\sqrt[3]{xy^2}$ Apply the distributive property. ∎

✓ **CHECK POINT 3**   Simplify by combining like radical terms, if possible:

a. $3\sqrt[3]{24} - 5\sqrt[3]{81}$

b. $5\sqrt[3]{x^2y} + \sqrt[3]{27x^5y^4}$.

Dividing Radical Expressions

We have seen that the root of a product is the product of the roots. The root of a quotient can also be expressed as the quotient of roots. Here is an example:

$$\sqrt{\frac{64}{4}} = \sqrt{16} = 4 \quad \text{and} \quad \frac{\sqrt{64}}{\sqrt{4}} = \frac{8}{2} = 4.$$

This expression is the square root of a quotient.   This expression is the quotient of two square roots.

The two procedures produce the same result, 4. This is a special case of the **quotient rule for radicals**.

2 Use the quotient rule to simplify radical expressions.

The Quotient Rule for Radicals

If $\sqrt[n]{a}$ and $\sqrt[n]{b}$ are real numbers and $b \neq 0$, then

$$\sqrt[n]{\frac{a}{b}} = \frac{\sqrt[n]{a}}{\sqrt[n]{b}}.$$

The nth root of a quotient is the quotient of the nth roots of the numerator and denominator.

We know that a radical is simplified when its radicand has no factors other than 1 that are perfect nth powers. The quotient rule can be used to simplify some radicals. Keep in mind that all variables in radicands represent positive real numbers.

EXAMPLE 4   Using the Quotient Rule to Simplify Radicals

Simplify using the quotient rule:

a. $\sqrt[3]{\frac{16}{27}}$ **b.** $\sqrt{\frac{x^2}{25y^6}}$ **c.** $\sqrt[4]{\frac{7y^5}{x^{12}}}$.

Solution   We simplify each expression by taking the roots of the numerator and the denominator. Then we use factoring to simplify the resulting radicals, if possible.

a. $\sqrt[3]{\dfrac{16}{27}} = \dfrac{\sqrt[3]{16}}{\sqrt[3]{27}} = \dfrac{\sqrt[3]{8 \cdot 2}}{3} = \dfrac{\sqrt[3]{8} \cdot \sqrt[3]{2}}{3} = \dfrac{2\sqrt[3]{2}}{3}$

b. $\sqrt{\dfrac{x^2}{25y^6}} = \dfrac{\sqrt{x^2}}{\sqrt{25y^6}} = \dfrac{x}{5(y^6)^{\frac{1}{2}}} = \dfrac{x}{5y^3}$

> Try to do this step mentally.

c. $\sqrt[4]{\dfrac{7y^5}{x^{12}}} = \dfrac{\sqrt[4]{7y^5}}{\sqrt[4]{x^{12}}} = \dfrac{\sqrt[4]{y^4 \cdot 7y}}{\sqrt[4]{x^{12}}} = \dfrac{y\sqrt[4]{7y}}{x^3}$ ∎

☑ **CHECK POINT 4**   Simplify using the quotient rule:

a. $\sqrt[3]{\dfrac{24}{125}}$ **b.** $\sqrt{\dfrac{9x^3}{y^{10}}}$ **c.** $\sqrt[3]{\dfrac{8y^7}{x^{12}}}$.

By reversing the two sides of the quotient rule, we obtain a procedure for dividing radical expressions.

3 Use the quotient rule to divide radical expressions.

Dividing Radical Expressions

If $\sqrt[n]{a}$ and $\sqrt[n]{b}$ are real numbers and $b \neq 0$, then

$$\dfrac{\sqrt[n]{a}}{\sqrt[n]{b}} = \sqrt[n]{\dfrac{a}{b}}.$$

To divide two radical expressions with the same index, divide the radicands and retain the common index.

Great Question!

When should I write a quotient under a single radical and when should I use separate radicals for a quotient's numerator and denominator?

- Use
$$\dfrac{\sqrt[n]{a}}{\sqrt[n]{b}} = \sqrt[n]{\dfrac{a}{b}},$$
writing under one radical (as in Example 5), when *dividing*.

- Use
$$\sqrt[n]{\dfrac{a}{b}} = \dfrac{\sqrt[n]{a}}{\sqrt[n]{b}},$$
with separate radicals for the numerator and the denominator (as in Example 4), when *simplifying* a quotient.

EXAMPLE 5   Dividing Radical Expressions

Divide and, if possible, simplify:

a. $\dfrac{\sqrt{48x^3}}{\sqrt{6x}}$ **b.** $\dfrac{\sqrt{45xy}}{2\sqrt{5}}$ **c.** $\dfrac{\sqrt[3]{16x^5y^2}}{\sqrt[3]{2xy^{-1}}}$.

Solution   In each part of this problem, the indices in the numerator and the denominator are the same. Perform each division by dividing the radicands and retaining the common index.

a. $\dfrac{\sqrt{48x^3}}{\sqrt{6x}} = \sqrt{\dfrac{48x^3}{6x}} = \sqrt{8x^2} = \sqrt{4x^2 \cdot 2} = \sqrt{4x^2} \cdot \sqrt{2} = 2x\sqrt{2}$

b. $\dfrac{\sqrt{45xy}}{2\sqrt{5}} = \dfrac{1}{2} \cdot \sqrt{\dfrac{45xy}{5}} = \dfrac{1}{2} \cdot \sqrt{9xy} = \dfrac{1}{2} \cdot 3\sqrt{xy}$ or $\dfrac{3\sqrt{xy}}{2}$

c. $\dfrac{\sqrt[3]{16x^5y^2}}{\sqrt[3]{2xy^{-1}}} = \sqrt[3]{\dfrac{16x^5y^2}{2xy^{-1}}}$ Divide the radicands and retain the common index.

$= \sqrt[3]{8x^{5-1}y^{2-(-1)}}$ Divide factors in the radicand. Subtract exponents on common bases.

$= \sqrt[3]{8x^4y^3}$ Simplify.

$= \sqrt[3]{(8x^3y^3)x}$ Factor using the greatest perfect cube factor.

$= \sqrt[3]{8x^3y^3} \cdot \sqrt[3]{x}$ Factor into two radicals.

$= 2xy\sqrt[3]{x}$ Simplify. ∎

✓ **CHECK POINT 5**   Divide and, if possible, simplify:

a. $\dfrac{\sqrt{40x^5}}{\sqrt{2x}}$

b. $\dfrac{\sqrt{50xy}}{2\sqrt{2}}$

c. $\dfrac{\sqrt[3]{48x^7y}}{\sqrt[3]{6xy^{-2}}}$.

CONCEPT AND VOCABULARY CHECK

Fill in each blank so that the resulting statement is true.

1. $5\sqrt{3} + 8\sqrt{3} = (\_+\_)\sqrt{3} = \_\_\_$

2. $\sqrt{27} - \sqrt{12} = \sqrt{\_\cdot 3} - \sqrt{\_\cdot 3} = \_\sqrt{3} - \_\sqrt{3} = \_\_$

3. $\sqrt[3]{54} + \sqrt[3]{16} = \sqrt[3]{\_\_\cdot 2} + \sqrt[3]{\_\_\cdot 2} = \_\sqrt[3]{2} + \_\sqrt[3]{2} = \_\_$

4. If $\sqrt[n]{a}$ and $\sqrt[n]{b}$ are real numbers and $b \neq 0$, the quotient rule for radicals states that

$$\sqrt[n]{\dfrac{a}{b}} = \dfrac{\_\_}{\_\_}.$$

5. $\sqrt[3]{\dfrac{8}{27}} = \dfrac{\sqrt[3]{\_\_}}{\sqrt[3]{\_\_}} = \_$

6. If $x > 0$,

$$\dfrac{\sqrt{72x^3}}{\sqrt{2x}} = \sqrt{\dfrac{\_\_}{\_\_}} = \sqrt{\_\_} = \_\_.$$

10.4 EXERCISE SET

MyMathLab®

Watch the videos in MyMathLab

Download the MyDashBoard App

Practice Exercises

In this exercise set, assume that all variables represent positive real numbers.

In Exercises 1–10, add or subtract as indicated.

1. $8\sqrt{5} + 3\sqrt{5}$ **2.** $7\sqrt{3} + 2\sqrt{3}$ $9\sqrt{3}$

3. $9\sqrt[3]{6} - 2\sqrt[3]{6}$ **4.** $9\sqrt[3]{7} - 4\sqrt[3]{7}$

5. $4\sqrt[5]{2} + 3\sqrt[5]{2} - 5\sqrt[5]{2}$

6. $6\sqrt[5]{3} + 2\sqrt[5]{3} - 3\sqrt[5]{3}$ $5\sqrt[5]{3}$

7. $3\sqrt{13} - 2\sqrt{5} - 2\sqrt{13} + 4\sqrt{5}$

8. $8\sqrt{17} - 5\sqrt{19} - 6\sqrt{17} + 4\sqrt{19}$

9. $3\sqrt{5} - \sqrt[3]{x} + 4\sqrt{5} + 3\sqrt[3]{x}$

10. $6\sqrt{7} - \sqrt[3]{x} + 2\sqrt{7} + 5\sqrt[3]{x}$

In Exercises 11–28, add or subtract as indicated. You will need to simplify terms to identify the like radicals.

11. $\sqrt{3} + \sqrt{27}$ **12.** $\sqrt{5} + \sqrt{20}$ $3\sqrt{5}$

13. $7\sqrt{12} + \sqrt{75}$ **14.** $5\sqrt{12} + \sqrt{75}$

15. $3\sqrt{32x} - 2\sqrt{18x}$

16. $5\sqrt{45x} - 2\sqrt{20x}$

17. $5\sqrt[3]{16} + \sqrt[3]{54}$

18. $3\sqrt[3]{24} + \sqrt[3]{81}$

19. $3\sqrt{45x^3} + \sqrt{5x}$

20. $8\sqrt{45x^3} + \sqrt{5x}$

21. $\sqrt[3]{54xy^3} + y\sqrt[3]{128x}$

22. $\sqrt[3]{24xy^3} + y\sqrt[3]{81x}$

23. $\sqrt[3]{54x^4} - \sqrt[3]{16x}$

24. $\sqrt[3]{81x^4} - \sqrt[3]{24x}$

25. $\sqrt{9x - 18} + \sqrt{x - 2}$

26. $\sqrt{4x - 12} + \sqrt{x - 3} -$

27. $2\sqrt[3]{x^4y^2} + 3x\sqrt[3]{xy^2}$

28. $4\sqrt[3]{x^4y^2} + 5x\sqrt[3]{xy^2}$

In Exercises 29–44, simplify using the quotient rule.

29. $\sqrt{\dfrac{11}{4}}$ **30.** $\sqrt{\dfrac{19}{25}}$

31. $\sqrt[3]{\dfrac{19}{27}}$ **32.** $\sqrt[3]{\dfrac{11}{64}}$

33. $\sqrt{\dfrac{x^2}{36y^8}}$ **34.** $\sqrt{\dfrac{x^2}{144y^{12}}}$

35. $\sqrt{\dfrac{8x^3}{25y^6}}$ **36.** $\sqrt{\dfrac{50x^3}{81y^8}}$

37. $\sqrt[3]{\dfrac{x^4}{8y^3}}$ **38.** $\sqrt[3]{\dfrac{x^5}{125y^3}}$

39. $\sqrt[3]{\dfrac{50x^8}{27y^{12}}}$ **40.** $\sqrt[3]{\dfrac{81x^8}{8y^{15}}}$

41. $\sqrt[4]{\dfrac{9y^6}{x^8}}$ **42.** $\sqrt[4]{\dfrac{13y^7}{x^{12}}}$

43. $\sqrt[5]{\dfrac{64x^{13}}{y^{20}}}$ **44.** $\sqrt[5]{\dfrac{64x^{14}}{y^{15}}}$

In Exercises 45–66, divide and, if possible, simplify.

45. $\dfrac{\sqrt{40}}{\sqrt{5}}$

46. $\dfrac{\sqrt{200}}{\sqrt{10}}$

47. $\dfrac{\sqrt[3]{48}}{\sqrt[3]{6}}$

48. $\dfrac{\sqrt[3]{54}}{\sqrt[3]{2}}$

49. $\dfrac{\sqrt{54x^3}}{\sqrt{6x}}$

50. $\dfrac{\sqrt{72x^3}}{\sqrt{2x}}$

51. $\dfrac{\sqrt{x^5y^3}}{\sqrt{xy}}$

52. $\dfrac{\sqrt{x^7y^6}}{\sqrt{x^3y^2}}$

53. $\dfrac{\sqrt{200x^3}}{\sqrt{10x^{-1}}}$

54. $\dfrac{\sqrt{500x^3}}{\sqrt{10x^{-1}}}$

55. $\dfrac{\sqrt{48a^8b^7}}{\sqrt{3a^{-2}b^{-3}}}$

56. $\dfrac{\sqrt{54a^7b^{11}}}{\sqrt{3a^{-4}b^{-2}}}$

57. $\dfrac{\sqrt{72xy}}{2\sqrt{2}}$

58. $\dfrac{\sqrt{50xy}}{2\sqrt{2}}$

59. $\dfrac{\sqrt[3]{24x^3y^5}}{\sqrt[3]{3y^2}}$

60. $\dfrac{\sqrt[3]{250x^5y^3}}{\sqrt[3]{2x^3}}$

61. $\dfrac{\sqrt[4]{32x^{10}y^8}}{\sqrt[4]{2x^2y^{-2}}}$

62. $\dfrac{\sqrt[5]{96x^{12}y^{11}}}{\sqrt[5]{3x^2y^{-2}}}$

63. $\dfrac{\sqrt[3]{x^2 + 5x + 6}}{\sqrt[3]{x + 2}}$

64. $\dfrac{\sqrt[3]{x^2 + 7x + 12}}{\sqrt[3]{x + 3}}$

65. $\dfrac{\sqrt[3]{a^3 + b^3}}{\sqrt[3]{a + b}}$

66. $\dfrac{\sqrt[3]{a^3 - b^3}}{\sqrt[3]{a - b}}$

Practice PLUS

In Exercises 67–76, perform the indicated operations.

67. $\dfrac{\sqrt{32}}{5} + \dfrac{\sqrt{18}}{7}$

68. $\dfrac{\sqrt{27}}{2} + \dfrac{\sqrt{75}}{7}$

69. $3x\sqrt{8xy^2} - 5y\sqrt{32x^3} + \sqrt{18x^3y^2}$

70. $6x\sqrt{3xy^2} - 4x^2\sqrt{27xy} - 5\sqrt{75x^5y}$

71. $5\sqrt{2x^3} + \dfrac{30x^3\sqrt{24x^2}}{3x^2\sqrt{3x}}$

72. $7\sqrt{2x^3} + \dfrac{40x^3\sqrt{150x^2}}{5x^2\sqrt{3x}}$

73. $2x\sqrt{75xy} - \dfrac{\sqrt{81xy^2}}{\sqrt{3x^{-2}y}}$

74. $5\sqrt{8x^2y^3} - \dfrac{9x^2\sqrt{64y}}{3x\sqrt{2y^{-2}}}$

75. $\dfrac{15x^4\sqrt[3]{80x^3y^2}}{5x^3\sqrt[3]{2x^2y}} - \dfrac{75\sqrt[3]{5x^3y}}{25\sqrt[3]{x^{-1}}}$

76. $\dfrac{16x^4\sqrt[3]{48x^3y^2}}{8x^3\sqrt[3]{3x^2y}} - \dfrac{20\sqrt[3]{2x^3y}}{4\sqrt[3]{x^{-1}}}$

In Exercises 77–80, find $\left(\dfrac{f}{g}\right)(x)$ and the domain of $\dfrac{f}{g}$. Express each quotient function in simplified form.

77. $f(x) = \sqrt{48x^5},\ g(x) = \sqrt{3x^2}$

78. $f(x) = \sqrt{x^2 - 25},\ g(x) = \sqrt{x + 5}$

79. $f(x) = \sqrt[3]{32x^6},\ g(x) = \sqrt[3]{2x^2}$

80. $f(x) = \sqrt[3]{2x^6},\ g(x) = \sqrt[3]{16x}$

Application Exercises

*Exercises 81–84 involve the perimeter and area of various geometric figures. Refer to **Table 2.3** on page 169 if you've forgotten any of the formulas.*

In Exercises 81–82, find the perimeter and area of each rectangle. Express answers in simplified radical form.

81.

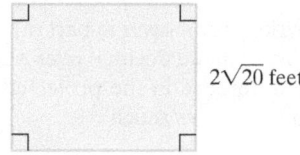

$\sqrt{125}$ feet

$2\sqrt{20}$ feet

82.

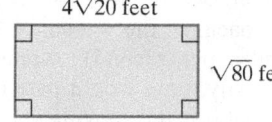

$4\sqrt{20}$ feet

$\sqrt{80}$ feet

83. Find the perimeter of the triangle in simplified radical form.

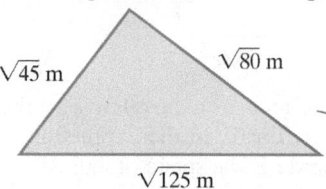

$\sqrt{45}$ m $\sqrt{80}$ m

$\sqrt{125}$ m

84. Find the area of the trapezoid in simplified radical form.

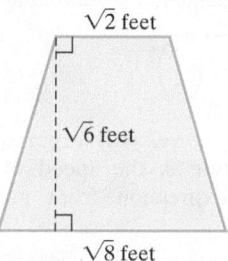

$\sqrt{2}$ feet

$\sqrt{6}$ feet

$\sqrt{8}$ feet

85. America is getting older. The graph shows the projected elderly U.S. population for ages 65–84 and for ages 85 and older.

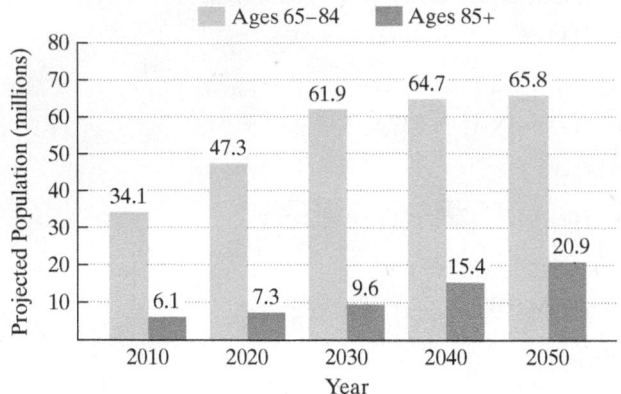

Projected Elderly United States Population

Source: U.S. Census Bureau

The function $f(x) = 5\sqrt{x} + 34.1$ models the projected number of Americans ages 65–84, $f(x)$, in millions, x years after 2010.

a. Use the function to find $f(40) - f(10)$. Express this difference in simplified radical form. What does this simplified radical represent?

b. Use a calculator and write your answer in part (a) to the nearest tenth. Does this rounded decimal overestimate or underestimate the difference in the projected data shown by the bar graph? By how much?

86. What does travel in space have to do with radicals? Imagine that in the future we will be able to travel in starships at velocities approaching the speed of light (approximately 186,000 miles per second). According to Einstein's theory of relativity, time would pass more quickly on Earth than it would in the moving starship. The radical expression

$$R_f \frac{\sqrt{c^2 - v^2}}{\sqrt{c^2}}$$

gives the aging rate of an astronaut relative to the aging rate of a friend, R_f, on Earth. In the expression, v is the astronaut's velocity and c is the speed of light.

a. Use the quotient rule and simplify the expression that shows your aging rate relative to a friend on Earth. Working in a step-by-step manner, express your aging rate as

$$R_f \sqrt{1 - \left(\frac{v}{c}\right)^2}.$$

b. You are moving at velocities approaching the speed of light. Substitute c, the speed of light, for v in the simplified expression from part (a). Simplify

completely. Close to the speed of light, what is your aging rate relative to a friend on Earth? What does this mean?

Writing in Mathematics

87. What are like radicals? Give an example with your explanation.

88. Explain how to add like radicals. Give an example with your explanation.

89. If only like radicals can be combined, why is it possible to add $\sqrt{2}$ and $\sqrt{8}$?

90. Explain how to simplify a radical expression using the quotient rule. Provide an example.

91. Explain how to divide radical expressions with the same index.

92. In Exercise 85, use the data displayed by the bar graph to explain why we used a square root function to model the projected population for the 65–84 age group, but not for the 85+ group.

Technology Exercises

93. Use a calculator to provide numerical support for any four exercises that you worked from Exercises 1–66 that do not contain variables. Begin by finding a decimal approximation for the given expression. Then find a decimal approximation for your answer. The two decimal approximations should be the same.

In Exercises 94–96, determine if each operation is performed correctly by graphing the function on each side of the equation with your graphing utility. Use the given viewing rectangle. The graphs should be the same. If they are not, correct the right side of the equation and then use your graphing utility to verify the correction.

94. $\sqrt{4x} + \sqrt{9x} = 5\sqrt{x}$
[0, 5, 1] by [0, 10, 1]

95. $\sqrt{16x} - \sqrt{9x} = \sqrt{7x}$
[0, 5, 1] by [0, 5, 1]

96. $x\sqrt{8} + x\sqrt{2} = x\sqrt{10}$
[−5, 5, 1] by [−15, 15, 1]

Critical Thinking Exercises

Make Sense? *In Exercises 97–100, determine whether each statement "makes sense" or "does not make sense" and explain your reasoning.*

97. I divide nth roots by taking the nth root of the quotient of the radicands.

98. The unlike radicals $3\sqrt{2}$ and $5\sqrt{3}$ remind me of the unlike terms $3x$ and $5y$ that cannot be combined by addition or subtraction.

99. I simplified the terms of $3\sqrt[3]{81} + 2\sqrt[3]{54}$, and then I was able to identify and add the like radicals.

100. Without using a calculator, it's easier for me to estimate the decimal value of $\sqrt{72} + \sqrt{32} + \sqrt{18}$ by first simplifying.

In Exercises 101–104, determine whether each statement is true or false. If the statement is false, make the necessary change(s) to produce a true statement.

101. $\sqrt{5} + \sqrt{5} = \sqrt{10}$

102. $4\sqrt{3} + 5\sqrt{3} = 9\sqrt{6}$

103. If any two radical expressions are completely simplified, they can then be combined through addition or subtraction.

104. $\dfrac{\sqrt{-8}}{\sqrt{2}} = \sqrt{\dfrac{-8}{2}} = \sqrt{-4} = -2$

105. If an irrational number is decreased by $2\sqrt{18} - \sqrt{50}$, the result is $\sqrt{2}$. What is the number?

106. Simplify: $\dfrac{\sqrt{20}}{3} + \dfrac{\sqrt{45}}{4} - \sqrt{80}$.

107. Simplify: $\dfrac{6\sqrt{49xy}\,\sqrt{ab^2}}{7\sqrt{36x^{-3}y^{-5}}\,\sqrt{a^{-9}b^{-1}}}$.

Review Exercises

108. Solve: $2(3x - 1) - 4 = 2x - (6 - x)$.
(Section 2.3, Example 3)

109. Factor: $x^2 - 8xy + 12y^2$.
(Section 6.2, Example 6)

110. Add: $\dfrac{2}{x^2 + 5x + 6} + \dfrac{3x}{x^2 + 6x + 9}$.
(Section 7.4, Example 7)

Preview Exercises

Exercises 111–113 will help you prepare for the material covered in the next section.

111. a. Multiply: $7(x + 5)$.
  b. Multiply: $\sqrt{7}(x + \sqrt{5})$.

112. a. Multiply: $(x + 5)(6x + 3)$.
  b. Multiply: $(\sqrt{2} + 5)(6\sqrt{2} + 3)$.

113. Multiply and simplify:
$$\frac{10y}{\sqrt[5]{4x^3y}} \cdot \frac{\sqrt[5]{8x^2y^4}}{\sqrt[5]{8x^2y^4}}.$$

MID-CHAPTER CHECK POINT	Section 10.1–Section 10.4

 What You Know: We learned to find roots of numbers. We saw that the domain of a square root function is the set of real numbers for which the radicand is nonnegative. We learned to simplify radical expressions, using $\sqrt[n]{a^n} = |a|$ if n is even and $\sqrt[n]{a^n} = a$ if n is odd. The definition $a^{\frac{m}{n}} = (\sqrt[n]{a})^m = \sqrt[n]{a^m}$ connected rational exponents and radicals. Finally, we performed various operations with radicals, including multiplication, addition, subtraction, and division.

In Exercises 1–23, simplify the given expression or perform the indicated operation(s) and, if possible, simplify. Assume that all variables represent positive real numbers.

1. $\sqrt{100} - \sqrt[3]{-27}$
2. $\sqrt{8x^5y^7}$
3. $3\sqrt[3]{4x^2} + 2\sqrt[3]{4x^2}$
4. $(3\sqrt[3]{4x^2})(2\sqrt[3]{4x^2})$
5. $27^{\frac{2}{3}} + (-32)^{\frac{3}{5}}$
6. $(64x^3y^{\frac{1}{4}})^{\frac{1}{3}}$
7. $5\sqrt{27} - 4\sqrt{48}$
8. $\sqrt{\dfrac{500x^3}{4y^4}}$
9. $\dfrac{x}{\sqrt[4]{x}}$
10. $\sqrt[3]{54x^5}$

11. $\dfrac{\sqrt[3]{160}}{\sqrt[3]{2}}$
12. $\sqrt[5]{\dfrac{x^{10}}{y^{20}}}$
13. $\dfrac{(x^{\frac{2}{3}})^2}{(x^{\frac{1}{4}})^3}$
14. $\sqrt[6]{x^6y^4}$
15. $\sqrt[7]{(x-2)^3} \cdot \sqrt[7]{(x-2)^6}$
16. $\sqrt[4]{32x^{11}y^{17}}$
17. $4\sqrt[3]{16} + 2\sqrt[3]{54}$
18. $\dfrac{\sqrt[7]{x^4y^9}}{\sqrt[7]{x^{-5}y^7}}$
19. $(-125)^{-\frac{2}{3}}$
20. $\sqrt{2} \cdot \sqrt[3]{2}$
21. $\sqrt[3]{\dfrac{32x}{y^4}} \cdot \sqrt[3]{\dfrac{2x^2}{y^2}}$
22. $\sqrt{32xy^2} \cdot \sqrt{2x^3y^5}$
23. $4x\sqrt{6x^4y^3} - 7y\sqrt{24x^6y}$

In Exercises 24–25, find the domain of each function.

24. $f(x) = \sqrt{30 - 5x}$
25. $g(x) = \sqrt[3]{3x - 15}$

10.5

Multiplying with More Than One Term and Rationalizing Denominators

Objectives

1. Multiply radical expressions with more than one term.

2. Use polynomial special products to multiply radicals.

3. Rationalize denominators containing one term.

4. Rationalize denominators containing two terms.

5. Rationalize numerators.

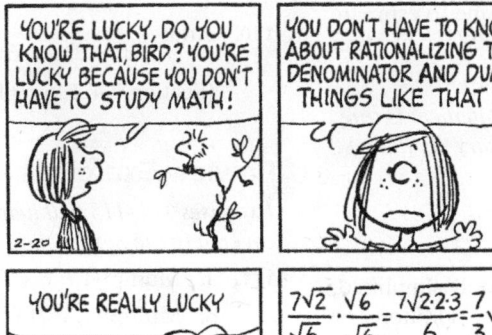

The late Charles Schulz, creator of the "Peanuts" comic strip, transfixed 350 million readers worldwide with the joys and angst of his hapless Charlie Brown and Snoopy, a romantic self-deluded beagle. In 18,250 comic strips that spanned nearly 50 years, mathematics was often featured. Is the discussion of radicals shown above the real thing, or is it just an algebraic scam? By the time you complete this section on multiplying and dividing radicals, you will be able to decide.

1. Multiply radical expressions with more than one term.

Multiplying Radical Expressions with More Than One Term

Radical expressions with more than one term are multiplied in much the same way that polynomials with more than one term are multiplied. Example 1 uses the distributive property and the FOIL method to perform multiplications.

EXAMPLE 1 Multiplying Radicals

Multiply:

a. $\sqrt{7}(x + \sqrt{2})$ **b.** $\sqrt[3]{x}(\sqrt[3]{6} - \sqrt[3]{x^2})$ **c.** $(5\sqrt{2} + 2\sqrt{3})(4\sqrt{2} - 3\sqrt{3})$.

Solution

a.
$\sqrt{7}(x + \sqrt{2})$
$= \sqrt{7} \cdot x + \sqrt{7} \cdot \sqrt{2}$ Use the distributive property.
$= x\sqrt{7} + \sqrt{14}$ Multiply the radicals.

b.
$\sqrt[3]{x}(\sqrt[3]{6} - \sqrt[3]{x^2})$
$= \sqrt[3]{x} \cdot \sqrt[3]{6} - \sqrt[3]{x} \cdot \sqrt[3]{x^2}$ Use the distributive property.
$= \sqrt[3]{6x} - \sqrt[3]{x^3}$ Multiply the radicals: $\sqrt[n]{a} \cdot \sqrt[n]{b} = \sqrt[n]{ab}$.
$= \sqrt[3]{6x} - x$ Simplify: $\sqrt[3]{x^3} = x$.

first $\overbrace{\qquad}^{\text{last}}$

c. $(5\sqrt{2} + 2\sqrt{3})(4\sqrt{2} - 3\sqrt{3})$ *Use the FOIL method.*

inside

outside

$$= \overset{F}{(5\sqrt{2})(4\sqrt{2})} + \overset{O}{(5\sqrt{2})(-3\sqrt{3})} + \overset{I}{(2\sqrt{3})(4\sqrt{2})} + \overset{L}{(2\sqrt{3})(-3\sqrt{3})}$$

$= 20 \cdot 2 - 15\sqrt{6} + 8\sqrt{6} - 6 \cdot 3$ *Multiply. Note that $\sqrt{2} \cdot \sqrt{2} = \sqrt{4} = 2$ and $\sqrt{3} \cdot \sqrt{3} = \sqrt{9} = 3$.*

$= 40 - 15\sqrt{6} + 8\sqrt{6} - 18$ *Complete the multiplications.*

$= (40 - 18) + (-15\sqrt{6} + 8\sqrt{6})$ *Group like terms. Try to do this step mentally.*

$= 22 - 7\sqrt{6}$ *Combine numerical terms and like radicals.* ∎

✓ **CHECK POINT 1**   Multiply:

a. $\sqrt{6}(x + \sqrt{10})$

b. $\sqrt[3]{y}(\sqrt[3]{y^2} - \sqrt[3]{7})$

c. $(6\sqrt{5} + 3\sqrt{2})(2\sqrt{5} - 4\sqrt{2})$.

2 Use polynomial special products to multiply radicals.

Some radicals can be multiplied using the special products for multiplying polynomials.

EXAMPLE 2   Using Special Products to Multiply Radicals

Multiply:

a. $(\sqrt{3} + \sqrt{7})^2$ **b.** $(\sqrt{7} + \sqrt{3})(\sqrt{7} - \sqrt{3})$ **c.** $(\sqrt{a} - \sqrt{b})(\sqrt{a} + \sqrt{b})$.

Solution   Use the special-product formulas shown.

$$(A + B)^2 = A^2 + 2 \cdot A \cdot B + B^2$$

a. $(\sqrt{3} + \sqrt{7})^2 = (\sqrt{3})^2 + 2 \cdot \sqrt{3} \cdot \sqrt{7} + (\sqrt{7})^2$ *Use the special product for $(A + B)^2$.*

$= 3 + 2\sqrt{21} + 7$ *Multiply the radicals.*

$= 10 + 2\sqrt{21}$ *Simplify.*

$$(A + B) \cdot (A - B) = A^2 - B^2$$

b. $(\sqrt{7} + \sqrt{3})(\sqrt{7} - \sqrt{3}) = (\sqrt{7})^2 - (\sqrt{3})^2$ *Use the special product for $(A + B)(A - B)$.*

$= 7 - 3$ *Simplify: $(\sqrt{a})^2 = a$.*

$= 4$

$$(A - B) \cdot (A + B) = A^2 - B^2$$

c. $(\sqrt{a} - \sqrt{b})(\sqrt{a} + \sqrt{b}) = (\sqrt{a})^2 - (\sqrt{b})^2 = a - b$ ∎

Radical expressions that involve the sum and difference of the same two terms are called **conjugates**. For example,

$$\sqrt{7} + \sqrt{3} \quad \text{and} \quad \sqrt{7} - \sqrt{3}$$

are conjugates of each other. Parts (b) and (c) of Example 2 illustrate that the product of two radical expressions need not be a radical expression:

$$(\sqrt{7} + \sqrt{3})(\sqrt{7} - \sqrt{3}) = 4$$

The product of conjugates does not contain a radical.

$$(\sqrt{a} - \sqrt{b})(\sqrt{a} + \sqrt{b}) = a - b.$$

Later in this section, we will use conjugates to simplify quotients.

✓ **CHECK POINT 2** Multiply:

a. $\left(\sqrt{5} + \sqrt{6}\right)^2$ **b.** $\left(\sqrt{6} + \sqrt{5}\right)\left(\sqrt{6} - \sqrt{5}\right)$
c. $\left(\sqrt{a} - \sqrt{7}\right)\left(\sqrt{a} + \sqrt{7}\right)$.

3 Rationalize denominators containing one term.

Rationalizing Denominators Containing One Term

You can use a calculator to compare the approximate values for $\dfrac{1}{\sqrt{3}}$ and $\dfrac{\sqrt{3}}{3}$. The two approximations are the same. This is not a coincidence:

$$\frac{1}{\sqrt{3}} = \frac{1}{\sqrt{3}} \cdot \boxed{\frac{\sqrt{3}}{\sqrt{3}}} = \frac{\sqrt{3}}{\sqrt{9}} = \frac{\sqrt{3}}{3}.$$

Any number divided by itself is 1.
Multiplication by 1 does not change the value of $\frac{1}{\sqrt{3}}$.

This process involves rewriting a radical expression as an equivalent expression in which the denominator no longer contains any radicals. The process is called **rationalizing the denominator**. When the denominator contains a single radical with an nth root, **multiply the numerator and the denominator by a radical of index n that produces a perfect nth power in the denominator's radicand.**

EXAMPLE 3 Rationalizing Denominators

Rationalize each denominator:

a. $\dfrac{\sqrt{5}}{\sqrt{6}}$ **b.** $\sqrt[3]{\dfrac{7}{25}}$.

Solution

a. If we multiply the numerator and the denominator of $\dfrac{\sqrt{5}}{\sqrt{6}}$ by $\sqrt{6}$, the denominator becomes $\sqrt{6} \cdot \sqrt{6} = \sqrt{36} = 6$. The denominator's radicand, 36, is a perfect square. The denominator no longer contains a radical. Therefore, we multiply by 1, choosing $\dfrac{\sqrt{6}}{\sqrt{6}}$ for 1.

$$\frac{\sqrt{5}}{\sqrt{6}} = \frac{\sqrt{5}}{\sqrt{6}} \cdot \frac{\sqrt{6}}{\sqrt{6}}$$

Multiply the numerator and denominator by $\sqrt{6}$ to remove the radical in the denominator.

$$= \frac{\sqrt{30}}{\sqrt{36}}$$

Multiply numerators and multiply denominators. The denominator's radicand, 36, is a perfect square.

$$= \frac{\sqrt{30}}{6}$$

Simplify: $\sqrt{36} = 6$.

b. Using the quotient rule, we can express $\sqrt[3]{\dfrac{7}{25}}$ as $\dfrac{\sqrt[3]{7}}{\sqrt[3]{25}}$. We have cube roots, so we want the denominator's radicand to be a perfect cube. Right now, the denominator's radicand is 25 or 5^2. We know that $\sqrt[3]{5^3} = 5$. If we multiply the numerator and the denominator of $\dfrac{\sqrt[3]{7}}{\sqrt[3]{25}}$ by $\sqrt[3]{5}$, the denominator becomes

$$\sqrt[3]{25} \cdot \sqrt[3]{5} = \sqrt[3]{5^2} \cdot \sqrt[3]{5} = \sqrt[3]{5^3} = 5.$$

The denominator's radicand, 5^3, is a perfect cube. The denominator no longer contains a radical. Therefore, we multiply by 1, choosing $\dfrac{\sqrt[3]{5}}{\sqrt[3]{5}}$ for 1.

$$\sqrt[3]{\frac{7}{25}} = \frac{\sqrt[3]{7}}{\sqrt[3]{25}} \qquad \text{Use the quotient rule and rewrite as the quotient of radicals.}$$

$$= \frac{\sqrt[3]{7}}{\sqrt[3]{5^2}} \qquad \text{Write the denominator's radicand as an exponential expression.}$$

$$= \frac{\sqrt[3]{7}}{\sqrt[3]{5^2}} \cdot \frac{\sqrt[3]{5}}{\sqrt[3]{5}} \qquad \text{Multiply numerator and denominator by } \sqrt[3]{5} \text{ to remove the radical in the denominator.}$$

$$= \frac{\sqrt[3]{35}}{\sqrt[3]{5^3}} \qquad \text{Multiply numerators and denominators. The denominator's radicand, } 5^3, \text{ is a perfect cube.}$$

$$= \frac{\sqrt[3]{35}}{5} \qquad \text{Simplify: } \sqrt[3]{5^3} = 5. \quad \blacksquare$$

Great Question!

What exactly does rationalizing a denominator do to an irrational number in the denominator?

Rationalizing a numerical denominator makes that denominator a rational number.

✓ **CHECK POINT 3**   Rationalize each denominator:

a. $\dfrac{\sqrt{3}}{\sqrt{7}}$ **b.** $\sqrt[3]{\dfrac{2}{9}}.$

Example 3 showed that it is helpful to express the denominator's radicand using exponents. In this way, we can easily find the extra factor or factors needed to produce a perfect nth power. For example, suppose that $\sqrt[5]{8}$ appears in the denominator. We want a perfect fifth power. By expressing $\sqrt[5]{8}$ as $\sqrt[5]{2^3}$, we would multiply the numerator and denominator by $\sqrt[5]{2^2}$ because

$$\sqrt[5]{2^3} \cdot \sqrt[5]{2^2} = \sqrt[5]{2^5} = 2.$$

EXAMPLE 4   Rationalizing Denominators

Rationalize each denominator:

a. $\sqrt{\dfrac{3x}{5y}}$ **b.** $\dfrac{\sqrt[3]{x}}{\sqrt[3]{36y}}$ **c.** $\dfrac{10y}{\sqrt[5]{4x^3y}}.$

Solution   By examining each denominator, you can determine how to multiply by 1. Let's first look at the denominators. For the square root, we must produce exponents of 2 in the radicand. For the cube root, we need exponents of 3, and for the fifth root, we want exponents of 5.

Examine the denominators to determine how to remove each radical.

- $\sqrt{5y}$
- $\sqrt[3]{36y}$ or $\sqrt[3]{6^2y}$
- $\sqrt[5]{4x^3y}$ or $\sqrt[5]{2^2x^3y}$

Multiply by $\sqrt{5y}$:
$\sqrt{5y} \cdot \sqrt{5y} = \sqrt{25y^2} = 5y.$

Multiply by $\sqrt[3]{6y^2}$:
$\sqrt[3]{6^2y} \cdot \sqrt[3]{6y^2} = \sqrt[3]{6^3y^3} = 6y.$

Multiply by $\sqrt[5]{2^3x^2y^4}$:
$\sqrt[5]{2^2x^3y} \cdot \sqrt[5]{2^3x^2y^4} = \sqrt[5]{2^5x^5y^5} = 2xy.$

a. $\sqrt{\dfrac{3x}{5y}} = \dfrac{\sqrt{3x}}{\sqrt{5y}} = \dfrac{\sqrt{3x}}{\sqrt{5y}} \cdot \dfrac{\sqrt{5y}}{\sqrt{5y}} = \dfrac{\sqrt{15xy}}{\sqrt{25y^2}} = \dfrac{\sqrt{15xy}}{5y}$

Multiply by 1. $25y^2$ is a perfect square.

b. $\dfrac{\sqrt[3]{x}}{\sqrt[3]{36y}} = \dfrac{\sqrt[3]{x}}{\sqrt[3]{6^2y}} = \dfrac{\sqrt[3]{x}}{\sqrt[3]{6^2y}} \cdot \dfrac{\sqrt[3]{6y^2}}{\sqrt[3]{6y^2}} = \dfrac{\sqrt[3]{6xy^2}}{\sqrt[3]{6^3y^3}} = \dfrac{\sqrt[3]{6xy^2}}{6y}$

Multiply by 1. 6^3y^3 is a perfect cube.

c. $\dfrac{10y}{\sqrt[5]{4x^3y}} = \dfrac{10y}{\sqrt[5]{2^2x^3y}} = \dfrac{10y}{\sqrt[5]{2^2x^3y}} \cdot \dfrac{\sqrt[5]{2^3x^2y^4}}{\sqrt[5]{2^3x^2y^4}}$

Multiply by 1.

$= \dfrac{10y\sqrt[5]{2^3x^2y^4}}{\sqrt[5]{2^5x^5y^5}} = \dfrac{10y\sqrt[5]{8x^2y^4}}{2xy} = \dfrac{5\sqrt[5]{8x^2y^4}}{x}$

$2^5x^5y^5$ is a perfect 5th power.   Simplify: Divide numerator and denominator by $2y$. ∎

✓ **CHECK POINT 4** Rationalize each denominator:

a. $\sqrt{\dfrac{2x}{7y}}$ **b.** $\dfrac{\sqrt[3]{x}}{\sqrt[3]{9y}}$ **c.** $\dfrac{6x}{\sqrt[5]{8x^2y^4}}.$

4 Rationalize denominators containing two terms.

Rationalizing Denominators Containing Two Terms

How can we rationalize a denominator if the denominator contains two terms with one or more square roots? **Multiply the numerator and the denominator by the conjugate of the denominator.** Here are three examples of such expressions:

- $\dfrac{8}{3\sqrt{2} + 4}$
- $\dfrac{2 + \sqrt{5}}{\sqrt{6} - \sqrt{3}}$
- $\dfrac{h}{\sqrt{x + h} - \sqrt{x}}$

The conjugate of the denominator is $3\sqrt{2} - 4$.   The conjugate of the denominator is $\sqrt{6} + \sqrt{3}$.   The conjugate of the denominator is $\sqrt{x + h} + \sqrt{x}$.

The product of the denominator and its conjugate is found using the formula

$$(A + B)(A - B) = A^2 - B^2.$$

The simplified product will not contain a radical.

EXAMPLE 5 Rationalizing a Denominator Containing Two Terms

Rationalize the denominator: $\dfrac{8}{3\sqrt{2} + 4}.$

Solution   The conjugate of the denominator is $3\sqrt{2} - 4$. If we multiply the numerator and the denominator by $3\sqrt{2} - 4$, the simplified denominator will not contain a radical. Therefore, we multiply by 1, choosing $\dfrac{3\sqrt{2} - 4}{3\sqrt{2} - 4}$ for 1.

$$\frac{8}{3\sqrt{2} + 4} = \frac{8}{3\sqrt{2} + 4} \cdot \frac{3\sqrt{2} - 4}{3\sqrt{2} - 4}$$ 　　Multiply by 1.

$$= \frac{8(3\sqrt{2} - 4)}{(3\sqrt{2})^2 - 4^2}$$ 　　$(A + B)(A - B) = A^2 - B^2$

Leave the numerator in factored form. This helps simplify, if possible.

$$= \frac{8(3\sqrt{2} - 4)}{18 - 16}$$ 　　$(3\sqrt{2})^2 = 9 \cdot 2 = 18$

$$= \frac{8(3\sqrt{2} - 4)}{2}$$ 　　This expression can still be simplified.

$$= \frac{\overset{4}{8}(3\sqrt{2} - 4)}{\underset{1}{2}}$$ 　　Divide the numerator and denominator by 2.

$$= 4(3\sqrt{2} - 4) \quad \text{or} \quad 12\sqrt{2} - 16 \quad \blacksquare$$

✓ **CHECK POINT 5**   Rationalize the denominator: $\dfrac{18}{2\sqrt{3} + 3}$.

EXAMPLE 6   Rationalizing a Denominator Containing Two Terms

Rationalize the denominator: $\dfrac{2 + \sqrt{5}}{\sqrt{6} - \sqrt{3}}$.

Solution   The conjugate of the denominator is $\sqrt{6} + \sqrt{3}$. Multiplication of both the numerator and denominator by $\sqrt{6} + \sqrt{3}$ will rationalize the denominator. This will produce a rational number in the denominator.

$$\frac{2 + \sqrt{5}}{\sqrt{6} - \sqrt{3}} = \frac{2 + \sqrt{5}}{\sqrt{6} - \sqrt{3}} \cdot \frac{\sqrt{6} + \sqrt{3}}{\sqrt{6} + \sqrt{3}}$$ 　　Multiply by 1.

F　O　I　L

$$= \frac{2\sqrt{6} + 2\sqrt{3} + \sqrt{5} \cdot \sqrt{6} + \sqrt{5} \cdot \sqrt{3}}{(\sqrt{6})^2 - (\sqrt{3})^2}$$ 　　Use FOIL in the numerator and $(A - B)(A + B) = A^2 - B^2$ in the denominator.

$$= \frac{2\sqrt{6} + 2\sqrt{3} + \sqrt{30} + \sqrt{15}}{6 - 3}$$

$$= \frac{2\sqrt{6} + 2\sqrt{3} + \sqrt{30} + \sqrt{15}}{3}$$ 　　Further simplification is not possible. $\blacksquare$

✓ **CHECK POINT 6**   Rationalize the denominator: $\dfrac{3 + \sqrt{7}}{\sqrt{5} - \sqrt{2}}$.

⑤ Rationalize numerators.

Rationalizing Numerators

We have seen that square root functions are often used to model growing phenomena with growth that is leveling off. **Figure 10.7** shows a male's height as a function of his age. The pattern of his growth suggests modeling with a square root function.

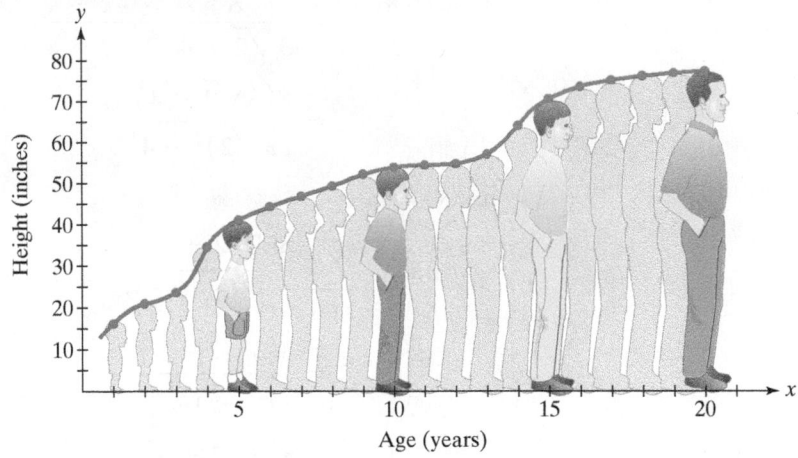

Figure 10.7

If we use $f(x) = \sqrt{x}$ to model height, $f(x)$, at age x, the expression

$$\frac{f(a + h) - f(a)}{h} = \frac{\sqrt{a + h} - \sqrt{a}}{h}$$

describes the man's average growth rate from age a to age $a + h$. Can you see that this expression is not defined if $h = 0$? However, to explore the man's average growth rates for successively shorter periods of time, we need to know what happens to the expression as h takes on values that get closer and closer to 0.

What happens to growth near the instant in time that the man is age a? The question is answered in calculus by **rationalizing the numerator**. The procedure is similar to rationalizing the denominator. **To rationalize a numerator, multiply by 1 to eliminate the radical in the *numerator*.**

EXAMPLE 7   Rationalizing a Numerator

Rationalize the numerator:

$$\frac{\sqrt{a + h} - \sqrt{a}}{h}.$$

Solution   The conjugate of the numerator is $\sqrt{a + h} + \sqrt{a}$. If we multiply the numerator and the denominator by $\sqrt{a + h} + \sqrt{a}$, the simplified numerator will not contain radicals. Therefore, we multiply by 1, choosing $\dfrac{\sqrt{a + h} + \sqrt{a}}{\sqrt{a + h} + \sqrt{a}}$ for 1.

$$\frac{\sqrt{a + h} - \sqrt{a}}{h} = \frac{\sqrt{a + h} - \sqrt{a}}{h} \cdot \frac{\sqrt{a + h} + \sqrt{a}}{\sqrt{a + h} + \sqrt{a}}$$
Multiply by 1.

$$= \frac{(\sqrt{a + h})^2 - (\sqrt{a})^2}{h(\sqrt{a + h} + \sqrt{a})}$$
$(A - B)(A + B) = A^2 - B^2$
Leave the denominator in factored form.

$$= \frac{a + h - a}{h(\sqrt{a + h} + \sqrt{a})}$$
$(\sqrt{a + h})^2 = a + h$
and $(\sqrt{a})^2 = a$.

$$= \frac{h}{h(\sqrt{a + h} + \sqrt{a})}$$
Simplify the numerator.

$$= \frac{1}{\sqrt{a + h} + \sqrt{a}}$$
Simplify by dividing the numerator and the denominator by h. ∎

✓ **CHECK POINT 7**   Rationalize the numerator: $\dfrac{\sqrt{x+3} - \sqrt{x}}{3}$.

CONCEPT AND VOCABULARY CHECK

Fill in each blank so that the resulting statement is true.

1. Consider the following multiplication problem:

$$\left(7\sqrt{5} + 3\sqrt{2}\right)\left(10\sqrt{5} - 6\sqrt{2}\right).$$

   Using the FOIL method, the product of the first terms is _____, the product of the outside terms is _____, the product of the inside terms is _____, and the product of the last terms is _____.

2. $\left(\sqrt{10} + \sqrt{5}\right)\left(\sqrt{10} - \sqrt{5}\right) = (\underline{\ \ })^2 - (\underline{\ \ })^2 = \underline{\ \ } - \underline{\ \ } = \underline{\ \ }$

3. The process of rewriting a radical expression as an equivalent expression in which the denominator no longer contains any radicals is called _____.

4. The number $\dfrac{\sqrt{7}}{\sqrt{5}}$ can be rewritten without a radical in the denominator by multiplying the numerator and denominator by _____.

5. The number $\sqrt[3]{\dfrac{2}{3}}$ can be rewritten without a radical in the denominator by multiplying the numerator and the denominator by _____.

6. The conjugate of $7\sqrt{2} + 5$ is _____.

7. The number $\dfrac{2\sqrt{6} + \sqrt{5}}{3\sqrt{6} - \sqrt{5}}$ can be rewritten without a radical in the denominator by multiplying the numerator and denominator by _____.

10.5 EXERCISE SET

Watch the videos in MyMathLab   Download the MyDashBoard App

Practice Exercises

In this Exercise Set, assume that all variables represent positive real numbers.

In Exercises 1–38, multiply as indicated. If possible, simplify any radical expressions that appear in the product.

1. $\sqrt{2}\left(x + \sqrt{7}\right)$

2. $\sqrt{5}\left(x + \sqrt{3}\right)$

3. $\sqrt{6}\left(7 - \sqrt{6}\right)$

4. $\sqrt{3}\left(5 - \sqrt{3}\right)$

5. $\sqrt{3}\left(4\sqrt{6} - 2\sqrt{3}\right)$

6. $\sqrt{6}\left(4\sqrt{6} - 3\sqrt{2}\right)$

7. $\sqrt[3]{2}\left(\sqrt[3]{6} + 4\sqrt[3]{5}\right)$

8. $\sqrt[3]{3}\left(\sqrt[3]{6} + 7\sqrt[3]{4}\right)$

9. $\sqrt[3]{x}\left(\sqrt[3]{16x^2} - \sqrt[3]{x}\right)$

10. $\sqrt[3]{x}\left(\sqrt[3]{24x^2} - \sqrt[3]{x}\right)$

11. $\left(5 + \sqrt{2}\right)\left(6 + \sqrt{2}\right)$

12. $\left(7 + \sqrt{2}\right)\left(8 + \sqrt{2}\right)$

13. $\left(6 + \sqrt{5}\right)\left(9 - 4\sqrt{5}\right)$

14. $\left(4 + \sqrt{5}\right)\left(10 - 3\sqrt{5}\right)$

15. $\left(6 - 3\sqrt{7}\right)\left(2 - 5\sqrt{7}\right)$

16. $\left(7 - 2\sqrt{7}\right)\left(5 - 3\sqrt{7}\right)$

17. $\left(\sqrt{2} + \sqrt{7}\right)\left(\sqrt{3} + \sqrt{5}\right)$

18. $\left(\sqrt{3} + \sqrt{2}\right)\left(\sqrt{10} + \sqrt{11}\right)$

19. $\left(\sqrt{2} - \sqrt{7}\right)\left(\sqrt{3} - \sqrt{5}\right)$

20. $\left(\sqrt{3} - \sqrt{2}\right)\left(\sqrt{10} - \sqrt{11}\right)$

21. $\left(3\sqrt{2} - 4\sqrt{3}\right)\left(2\sqrt{2} + 5\sqrt{3}\right)$

22. $\left(3\sqrt{5} - 2\sqrt{3}\right)\left(4\sqrt{5} + 5\sqrt{3}\right)$

23. $\left(\sqrt{3} + \sqrt{5}\right)^2$

24. $\left(\sqrt{2} + \sqrt{7}\right)^2$

25. $\left(\sqrt{3x} - \sqrt{y}\right)^2$

26. $\left(\sqrt{2x} - \sqrt{y}\right)^2$

27. $\left(\sqrt{5} + 7\right)\left(\sqrt{5} - 7\right)$

28. $\left(\sqrt{6} + 2\right)\left(\sqrt{6} - 2\right)$

29. $\left(2 - 5\sqrt{3}\right)\left(2 + 5\sqrt{3}\right)$

30. $\left(3 - 5\sqrt{2}\right)\left(3 + 5\sqrt{2}\right)$

31. $\left(3\sqrt{2} + 2\sqrt{3}\right)\left(3\sqrt{2} - 2\sqrt{3}\right)$

32. $\left(4\sqrt{3} + 3\sqrt{2}\right)\left(4\sqrt{3} - 3\sqrt{2}\right)$

33. $\left(3 - \sqrt{x}\right)\left(2 - \sqrt{x}\right)$

34. $\left(4 - \sqrt{x}\right)\left(3 - \sqrt{x}\right)$

35. $\left(\sqrt[3]{x} - 4\right)\left(\sqrt[3]{x} + 5\right)$

36. $\left(\sqrt[3]{x} - 3\right)\left(\sqrt[3]{x} + 7\right)$

37. $\left(x + \sqrt[3]{y^2}\right)\left(2x - \sqrt[3]{y^2}\right)$

38. $\left(x - \sqrt[5]{y^3}\right)\left(2x + \sqrt[5]{y^3}\right)$

In Exercises 39–64, rationalize each denominator.

39. $\dfrac{\sqrt{2}}{\sqrt{5}}$

40. $\dfrac{\sqrt{7}}{\sqrt{3}}$

41. $\sqrt{\dfrac{11}{x}}$

42. $\sqrt{\dfrac{6}{x}}$

43. $\dfrac{9}{\sqrt{3y}}$

44. $\dfrac{12}{\sqrt{3y}}$

45. $\dfrac{1}{\sqrt[3]{2}}$

46. $\dfrac{1}{\sqrt[3]{3}}$

47. $\dfrac{6}{\sqrt[3]{4}}$

48. $\dfrac{10}{\sqrt[3]{5}}$

49. $\sqrt[3]{\dfrac{2}{3}}$

50. $\sqrt[3]{\dfrac{3}{4}}$

51. $\dfrac{4}{\sqrt[3]{x}}$

52. $\dfrac{7}{\sqrt[3]{x}}$

53. $\sqrt[3]{\dfrac{2}{y^2}}$

54. $\sqrt[3]{\dfrac{5}{y^2}}$

55. $\dfrac{7}{\sqrt[3]{2x^2}}$

56. $\dfrac{10}{\sqrt[3]{4x^2}}$

57. $\sqrt[3]{\dfrac{2}{xy^2}}$

58. $\sqrt[3]{\dfrac{3}{xy^2}}$

59. $\dfrac{3}{\sqrt[4]{x}}$

60. $\dfrac{5}{\sqrt[4]{x}}$

61. $\dfrac{6}{\sqrt[5]{8x^3}}$

62. $\dfrac{10}{\sqrt[5]{16x^2}}$

63. $\dfrac{2x^2y}{\sqrt[5]{4x^2y^4}}$

64. $\dfrac{3xy^2}{\sqrt[5]{8xy^3}}$

In Exercises 65–74, simplify each radical expression and then rationalize the denominator.

65. $\dfrac{9}{\sqrt{3x^2y}}$

66. $\dfrac{25}{\sqrt{5x^2y}}$

67. $-\sqrt{\dfrac{75a^5}{b^3}}$

68. $-\sqrt{\dfrac{150a^3}{b^5}}$

69. $\sqrt{\dfrac{7m^2n^3}{14m^3n^2}}$

70. $\sqrt{\dfrac{5m^4n^6}{15m^3n^4}}$

71. $\dfrac{3}{\sqrt[4]{x^5y^3}}$

72. $\dfrac{5}{\sqrt[4]{x^2y^7}}$

73. $\dfrac{12}{\sqrt[3]{-8x^5y^8}}$

74. $\dfrac{15}{\sqrt[3]{-27x^4y^{11}}}$

In Exercises 75–92, rationalize each denominator. Simplify, if possible.

75. $\dfrac{8}{\sqrt{5} + 2}$

76. $\dfrac{15}{\sqrt{6} + 1}$

77. $\dfrac{13}{\sqrt{11} - 3}$

78. $\dfrac{17}{\sqrt{10} - 2}$

79. $\dfrac{6}{\sqrt{5} + \sqrt{3}}$

80. $\dfrac{12}{\sqrt{7} + \sqrt{3}}$

81. $\dfrac{\sqrt{a}}{\sqrt{a} - \sqrt{b}}$

82. $\dfrac{\sqrt{b}}{\sqrt{a} - \sqrt{b}}$

83. $\dfrac{25}{5\sqrt{2} - 3\sqrt{5}}$

84. $\dfrac{35}{5\sqrt{2} - 3\sqrt{5}}$

85. $\dfrac{\sqrt{5} + \sqrt{3}}{\sqrt{5} - \sqrt{3}}$

86. $\dfrac{\sqrt{11} - \sqrt{5}}{\sqrt{11} + \sqrt{5}}$

87. $\dfrac{\sqrt{x} + 1}{\sqrt{x} + 3}$

88. $\dfrac{\sqrt{x} - 2}{\sqrt{x} - 5}$

89. $\dfrac{5\sqrt{3} - 3\sqrt{2}}{3\sqrt{2} - 2\sqrt{3}}$

90. $\dfrac{2\sqrt{6} + \sqrt{5}}{3\sqrt{6} - \sqrt{5}}$

91. $\dfrac{2\sqrt{x} + \sqrt{y}}{\sqrt{y} - 2\sqrt{x}}$

92. $\dfrac{3\sqrt{x} + \sqrt{y}}{\sqrt{y} - 3\sqrt{x}}$

In Exercises 93–104, rationalize each numerator. Simplify, if possible.

93. $\sqrt{\dfrac{3}{2}}$

94. $\sqrt{\dfrac{5}{3}}$

95. $\dfrac{\sqrt[3]{4x}}{\sqrt[3]{y}}$

96. $\dfrac{\sqrt[3]{2x}}{\sqrt[3]{y}}$

97. $\dfrac{\sqrt{x} + 3}{\sqrt{x}}$

98. $\dfrac{\sqrt{x} + 4}{\sqrt{x}}$

99. $\dfrac{\sqrt{a} + \sqrt{b}}{\sqrt{a} - \sqrt{b}}$

100. $\dfrac{\sqrt{a} - \sqrt{b}}{\sqrt{a} + \sqrt{b}}$

101. $\dfrac{\sqrt{x+5} - \sqrt{x}}{5}$

102. $\dfrac{\sqrt{x+7} - \sqrt{x}}{7}$

103. $\dfrac{\sqrt{x} + \sqrt{y}}{x^2 - y^2}$

104. $\dfrac{\sqrt{x} - \sqrt{y}}{x^2 - y^2}$

Practice PLUS

In Exercises 105–112, add or subtract as indicated. Begin by rationalizing denominators for all terms in which denominators contain radicals.

105. $\sqrt{2} + \dfrac{1}{\sqrt{2}}$

106. $\sqrt{5} + \dfrac{1}{\sqrt{5}}$

107. $\sqrt[3]{25} - \dfrac{15}{\sqrt[3]{5}}$

108. $\sqrt[4]{8} - \dfrac{20}{\sqrt[3]{2}}$

109. $\sqrt{6} - \sqrt{\dfrac{1}{6}} + \sqrt{\dfrac{2}{3}}$

110. $\sqrt{15} - \sqrt{\dfrac{5}{3}} + \sqrt{\dfrac{3}{5}}$

111. $\dfrac{2}{\sqrt{2} + \sqrt{3}} + \sqrt{75} - \sqrt{50}$

112. $\dfrac{5}{\sqrt{2} + \sqrt{7}} - 2\sqrt{32} + \sqrt{28}$

113. Let $f(x) = x^2 - 6x - 4$. Find $f(3 - \sqrt{13})$.

114. Let $f(x) = x^2 + 4x - 2$. Find $f(-2 + \sqrt{6})$.

115. Let $f(x) = \sqrt{9 + x}$. Find $f(3\sqrt{5}) \cdot f(-3\sqrt{5})$.

116. Let $f(x) = x^2$. Find $f(\sqrt{a+1} - \sqrt{a-1})$.

Application Exercises

117. The early Greeks believed that the most pleasing of all rectangles were **golden rectangles**, whose ratio of width to height is

$$\frac{w}{h} = \frac{2}{\sqrt{5} - 1}.$$

The Parthenon at Athens fits into a golden rectangle once the triangular pediment is reconstructed.

Rationalize the denominator of the golden ratio. Then use a calculator and find the ratio of width to height, correct to the nearest hundredth, in golden rectangles.

118. In the "Peanuts" cartoon shown in the section opener on page 732, Woodstock appears to be working steps mentally. Fill in the missing steps that show how to go from $\dfrac{7\sqrt{2 \cdot 2 \cdot 3}}{6}$ to $\dfrac{7}{3}\sqrt{3}$.

In Exercises 119–120, write expressions for the perimeter and area of each figure. Then simplify these expressions. Assume that all measures are given in inches.

119.
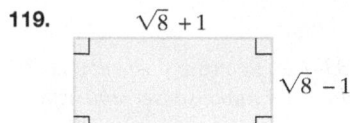

120.
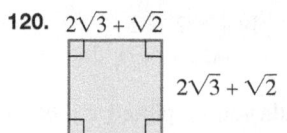

Writing in Mathematics

121. Explain how to perform this multiplication: $\sqrt{2}(\sqrt{7} + \sqrt{10})$.

122. Explain how to perform this multiplication: $(2 + \sqrt{3})(4 + \sqrt{3})$.

123. Explain how to perform this multiplication: $(2 + \sqrt{3})^2$.

124. What are conjugates? Give an example with your explanation.

125. Describe how to multiply conjugates.

126. Describe what it means to rationalize a denominator. Use both $\dfrac{1}{\sqrt{5}}$ and $\dfrac{1}{5 + \sqrt{5}}$ in your explanation.

127. When a radical expression has its denominator rationalized, we change the denominator so that it no longer contains any radicals. Doesn't this change the value of the radical expression? Explain.

128. Square the real number $\dfrac{2}{\sqrt{3}}$. Observe that the radical is eliminated from the denominator. Explain whether this process is equivalent to rationalizing the denominator.

Technology Exercises

In Exercises 129–132, determine if each operation is performed correctly by graphing the function on each side of the equation with your graphing utility. Use the given viewing rectangle. The graphs should be the same. If they are not, correct the right side of the equation and then use your graphing utility to verify the correction.

129. $(\sqrt{x} - 1)(\sqrt{x} - 1) = x + 1$
[0, 5, 1] by [−1, 2, 1]

130. $(\sqrt{x} + 2)(\sqrt{x} - 2) = x^2 - 4$ for $x \geq 0$
[0, 10, 1] by [−10, 10, 1]

131. $(\sqrt{x} + 1)^2 = x + 1$
[0, 8, 1] by [0, 15, 1]

132. $\dfrac{3}{\sqrt{x+3} - \sqrt{x}} = \sqrt{x+3} + \sqrt{x}$
[0, 8, 1] by [0, 6, 1]

139. $\dfrac{4\sqrt{x}}{\sqrt{x} - y} = \dfrac{4x + 4y\sqrt{x}}{x - y^2}$

140. $(\sqrt{x} - 7)^2 = x - 49$

141. Solve:
$$7[(2x - 5) - (x + 1)] = (\sqrt{7} + 2)(\sqrt{7} - 2).$$

142. Simplify: $(\sqrt{2 + \sqrt{3}} + \sqrt{2 - \sqrt{3}})^2$.

143. Rationalize the denominator: $\dfrac{1}{\sqrt{2} + \sqrt{3} + \sqrt{4}}$.

Critical Thinking Exercises

Make Sense? *In Exercises 133–136, determine whether each statement "makes sense" or "does not make sense" and explain your reasoning.*

133. I use the same ideas to multiply $(\sqrt{2} + 5)(\sqrt{2} + 4)$ that I did to find the binomial product $(x + 5)(x + 4)$.

134. I used a special-product formula and simplified as follows: $(\sqrt{2} + \sqrt{5})^2 = 2 + 5 = 7$.

135. In some cases when I multiply a square root expression and its conjugate, the simplified product contains a radical.

136. I use the fact that 1 is the multiplicative identity to both rationalize denominators and rewrite rational expressions with a common denominator.

In Exercises 137–140, determine whether each statement is true or false. If the statement is false, make the necessary change(s) to produce a true statement.

137. $\dfrac{\sqrt{3} + 7}{\sqrt{3} - 2} = -\dfrac{7}{2}$

138. $\dfrac{4}{\sqrt{x + y}} = \dfrac{4\sqrt{x - y}}{x - y}$

Review Exercises

144. Add: $\dfrac{2}{x - 2} + \dfrac{3}{x^2 - 4}$.
(Section 7.4, Example 7)

145. Solve: $3x - 4 \leq 2$ and $4x + 5 \geq 5$.
(Section 9.2, Example 2)

146. Determine whether each relation is a function.
(Section 8.1, Example 2)
   a. $\{(-1, 1), (1, 1), (-2, 4), (2, 4)\}$
   b. $\{(1, -1), (1, 1), (4, -2), (4, 2)\}$

Preview Exercises

Exercises 147–149 will help you prepare for the material covered in the next section.

147. Multiply: $(\sqrt{x + 4} + 1)^2$.

148. Solve: $4x^2 - 16x + 16 = 4(x + 4)$.

149. Solve: $26 - 11x = 16 - 8x + x^2$.

Objectives

1 Solve radical equations.

2 Use models that are radical functions to solve problems.

Radical Equations

One of the most dramatic developments in the U.S. work force has been the increase in the number of women. In 1960, approximately 38% of women belonged to the labor force. By 2000, that number had increased to over 60%. With higher levels of education than ever before, most women now choose careers in the labor force rather than homemaking.
   The function

$$f(x) = 3.5\sqrt{x} + 38$$

models the percentage of U.S. women in the labor force, $f(x)$, x years after 1960. How can we predict the year when,

say, 70% of women will participate in the U.S. work force? Substitute 70 for $f(x)$ in $f(x) = 3.5\sqrt{x} + 38$ and solve for x:

$$70 = 3.5\sqrt{x} + 38.$$

The resulting equation contains a variable in the radicand and is called a *radical equation*. A **radical equation** is an equation in which the variable occurs in a square root, cube root, or any higher root. Some examples of radical equations are

$$\sqrt{2x + 3} = 5, \quad \sqrt{3x + 1} - \sqrt{x + 4} = 1, \quad \text{and} \quad \sqrt[3]{3x - 1} + 4 = 0.$$

Variables occur in radicands.

In this section, you will learn how to solve radical equations. Solving such equations will enable you to solve new kinds of problems using radical functions.

1 Solve radical equations.

Solving Radical Equations

Consider the following radical equation:

$$\sqrt{x} = 9.$$

We solve the equation by squaring both sides:

Squaring both sides eliminates the square root.

$$(\sqrt{x})^2 = 9^2$$
$$x = 81.$$

The proposed solution, 81, can be checked in the original equation, $\sqrt{x} = 9$. Because $\sqrt{81} = 9$, the solution is 81 and the solution set is {81}.

In general, we solve radical equations with square roots by squaring both sides of the equation. We solve radical equations with nth roots by raising both sides of the equation to the nth power. Unfortunately, if n is even, all the solutions of the equation raised to the even power may not be solutions of the original equation. Consider, for example, the equation

$$x = 4.$$

If we square both sides, we obtain

$$x^2 = 16.$$
$$x^2 - 16 = 0 \qquad \text{Subtract 16 from both sides and write the quadratic equation in standard form.}$$
$$(x + 4)(x - 4) = 0 \qquad \text{Factor.}$$
$$x + 4 = 0 \quad \text{or} \quad x - 4 = 0 \qquad \text{Set each factor equal to 0.}$$
$$x = -4 \qquad\qquad x = 4 \qquad \text{Solve the resulting equations.}$$

The equation $x^2 = 16$ has two solutions, -4 and 4. By contrast, only 4 is a solution of the original equation, $x = 4$. For this reason, **when raising both sides of an equation to an even power, always check proposed solutions in the original equation**.

Here is a general method for solving radical equations with nth roots:

Solving Radical Equations Containing nth Roots

1. If necessary, arrange terms so that one radical is isolated on one side of the equation.
2. Raise both sides of the equation to the nth power to eliminate the nth root.
3. Solve the resulting equation. If this equation still contains radicals, repeat steps 1 and 2.
4. Check all proposed solutions in the original equation.

EXAMPLE 1 Solving a Radical Equation

Solve: $\sqrt{2x + 3} = 5$.

Solution

Step 1. Isolate a radical on one side. The radical, $\sqrt{2x + 3}$, is already isolated on the left side of the equation, so we can skip this step.

Step 2. Raise both sides to the *n*th power. Because *n*, the index, is 2, we square both sides.

$$\sqrt{2x + 3} = 5 \qquad \text{This is the given equation.}$$
$$\left(\sqrt{2x + 3}\right)^2 = 5^2 \qquad \text{Square both sides to eliminate the radical.}$$
$$2x + 3 = 25 \qquad \text{Simplify.}$$

Step 3. Solve the resulting equation.

$$2x + 3 = 25 \qquad \text{The resulting equation is a linear equation.}$$
$$2x = 22 \qquad \text{Subtract 3 from both sides.}$$
$$x = 11 \qquad \text{Divide both sides by 2.}$$

Step 4. Check the proposed solution in the original equation. Because both sides were raised to an even power, this check is essential.

Check 11:
$$\sqrt{2x + 3} = 5$$
$$\sqrt{2 \cdot 11 + 3} \overset{?}{=} 5$$
$$\sqrt{25} \overset{?}{=} 5$$
$$5 = 5, \qquad \text{true}$$

The solution is 11 and the solution set is {11}. ■

☑ **CHECK POINT 1** Solve: $\sqrt{3x + 4} = 8$.

EXAMPLE 2 Solving a Radical Equation

Solve: $\sqrt{x - 3} + 6 = 5$.

Solution

Step 1. Isolate a radical on one side. The radical, $\sqrt{x - 3}$, can be isolated by subtracting 6 from both sides. We obtain

$$\sqrt{x - 3} = -1.$$

> A principal square root cannot be negative. This equation has no solution. Let's continue the solution procedure to see what happens.

Step 2. Raise both sides to the *n*th power. Because *n*, the index, is 2, we square both sides.

$$\left(\sqrt{x - 3}\right)^2 = (-1)^2$$
$$x - 3 = 1 \qquad \text{Simplify.}$$

Step 3. Solve the resulting equation.

$$x - 3 = 1 \qquad \text{The resulting equation is a linear equation.}$$
$$x = 4 \qquad \text{Add 3 to both sides.}$$

Step 4. Check the proposed solution in the original equation.

Check 4:

$$\sqrt{x - 3} + 6 = 5$$
$$\sqrt{4 - 3} + 6 \overset{?}{=} 5$$
$$\sqrt{1} + 6 \overset{?}{=} 5$$
$$1 + 6 \overset{?}{=} 5$$
$$7 = 5, \quad \text{false}$$

This false statement indicates that 4 is not a solution. Thus, the equation has no solution. The solution set is $\varnothing$, the empty set. ∎

Example 2 illustrates that extra solutions may be introduced when you raise both sides of a radical equation to an even power. Such solutions, which are not solutions of the given equation, are called **extraneous solutions**. Thus, 4 is an extraneous solution of $\sqrt{x - 3} + 6 = 5$.

✓ **CHECK POINT 2** Solve: $\sqrt{x - 1} + 7 = 2$.

EXAMPLE 3 Solving a Radical Equation

Solve:

$$x + \sqrt{26 - 11x} = 4.$$

Solution

Step 1. Isolate a radical on one side. We isolate the radical, $\sqrt{26 - 11x}$, by subtracting x from both sides.

$$x + \sqrt{26 - 11x} = 4 \qquad \text{This is the given equation.}$$
$$\sqrt{26 - 11x} = 4 - x \qquad \text{Subtract } x \text{ from both sides.}$$

Step 2. Square both sides.

$$\left(\sqrt{26 - 11x}\right)^2 = (4 - x)^2$$
$$26 - 11x = 16 - 8x + x^2 \qquad \text{Simplify. Use the special-product formula}$$
$$\qquad (A - B)^2 = A^2 - 2AB + B^2 \text{ to square}$$
$$\qquad 4 - x.$$

Step 3. Solve the resulting equation. Because of the x^2-term, the resulting equation is a quadratic equation. We need to write this quadratic equation in standard form. We can obtain zero on the left side by subtracting 26 and adding $11x$ on both sides.

$$26 - 26 - 11x + 11x = 16 - 26 - 8x + 11x + x^2$$
$$0 = x^2 + 3x - 10 \qquad \text{Simplify.}$$
$$0 = (x + 5)(x - 2) \qquad \text{Factor.}$$
$$x + 5 = 0 \quad \text{or} \quad x - 2 = 0 \qquad \text{Set each factor equal to zero.}$$
$$x = -5 \qquad\qquad x = 2 \qquad \text{Solve for } x.$$

Step 4. Check the proposed solutions in the original equation.

Check -5:	**Check 2:**
$x + \sqrt{26 - 11x} = 4$	$x + \sqrt{26 - 11x} = 4$
$-5 + \sqrt{26 - 11(-5)} \overset{?}{=} 4$	$2 + \sqrt{26 - 11 \cdot 2} \overset{?}{=} 4$
$-5 + \sqrt{81} \overset{?}{=} 4$	$2 + \sqrt{4} \overset{?}{=} 4$
$-5 + 9 \overset{?}{=} 4$	$2 + 2 \overset{?}{=} 4$
$4 = 4, \quad \text{true}$	$4 = 4, \quad \text{true}$

The solutions are -5 and 2, and the solution set is $\{-5, 2\}$. ∎

Great Question!

Can I square the right side of $\sqrt{26 - 11x} = 4 - x$ by first squaring 4 and then squaring x?

No. Be sure to square *both* sides of an equation. Do *not* square each term.

Correct:
$$\left(\sqrt{26 - 11x}\right)^2 = (4 - x)^2$$

Incorrect!

Using Technology

Graphic Connections

You can use a graphing utility to provide a graphic check that $\{-5, 2\}$ is the solution set of $x + \sqrt{26 - 11x} = 4$.

Use the given equation

$$x + \sqrt{26 - 11x} = 4.$$

Enter $y_1 = x + \sqrt{26 - 11x}$ in the $\boxed{y=}$ screen.

Enter $y_2 = 4$ in the $\boxed{y=}$ screen.

Display graphs for y_1 and y_2. The solutions are the x-coordinates of the intersection points. These x-coordinates are -5 and 2. This verifies $\{-5, 2\}$ as the solution set of $x + \sqrt{26 - 11x} = 4$.

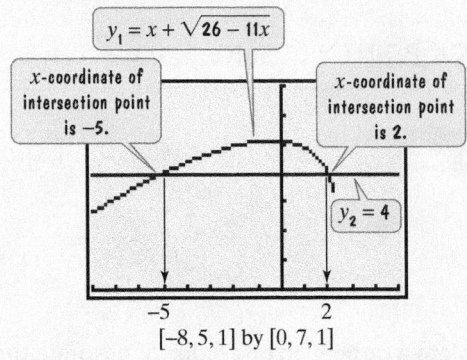

$[-8, 5, 1]$ by $[0, 7, 1]$

Use the equivalent equation

$$x + \sqrt{26 - 11x} - 4 = 0.$$

Enter $y_1 = x + \sqrt{26 - 11x} - 4$ in the $\boxed{y=}$ screen.

Display the graph for y_1. The solutions are the x-intercepts. The x-intercepts are -5 and 2. This verifies $\{-5, 2\}$ as the solution set of $x + \sqrt{26 - 11x} - 4 = 0$.

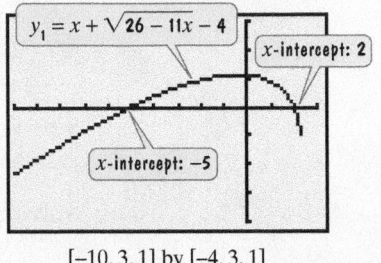

$[-10, 3, 1]$ by $[-4, 3, 1]$

✓ **CHECK POINT 3**   Solve: $\sqrt{6x + 7} - x = 2$.

The solution of a radical equation with two or more square root expressions involves isolating a radical, squaring both sides, and then repeating this process. Let's consider an equation containing two square root expressions.

EXAMPLE 4   Solving an Equation That Has Two Radicals

Solve: $\sqrt{3x + 1} - \sqrt{x + 4} = 1$.

Solution

Step 1. Isolate a radical on one side. We can isolate the radical $\sqrt{3x + 1}$ by adding $\sqrt{x + 4}$ to both sides. We obtain

$$\sqrt{3x + 1} = \sqrt{x + 4} + 1.$$

Step 2. Square both sides.

$$\left(\sqrt{3x + 1}\right)^2 = \left(\sqrt{x + 4} + 1\right)^2$$

Squaring the expression on the right side of the equation can be a bit tricky. We have to use the formula

$$(A + B)^2 = A^2 + 2AB + B^2.$$

Focusing on just the right side, here is how the squaring is done:

$$(A + B)^2 \quad = \quad A^2 \; + \; 2 \; \cdot \; A \; \cdot \; B \; + \; B^2$$

$$\left(\sqrt{x + 4} + 1\right)^2 = \left(\sqrt{x + 4}\right)^2 + 2 \cdot \sqrt{x + 4} \cdot 1 + 1^2 = x + 4 + 2\sqrt{x + 4} + 1.$$

Now let's return to squaring both sides.

$$\left(\sqrt{3x+1}\right)^2 = \left(\sqrt{x+4}+1\right)^2$$

Square both sides of the equation with an isolated radical.

$$3x + 1 = x + 4 + 2\sqrt{x+4} + 1$$

$\left(\sqrt{3x+1}\right)^2 = 3x+1$; Square the right side using the formula for $(A+B)^2$.

$$3x + 1 = x + 5 + 2\sqrt{x+4}$$

Combine numerical terms on the right side: $4 + 1 = 5$.

Can you see that the resulting equation still contains a radical, namely $\sqrt{x+4}$? Thus, we need to repeat the first two steps.

Repeat Step 1. Isolate a radical on one side. We isolate $2\sqrt{x+4}$, the radical term, by subtracting $x + 5$ from both sides. We obtain

$$3x + 1 = x + 5 + 2\sqrt{x+4}$$

This is the equation from our last step.

$$2x - 4 = 2\sqrt{x+4}.$$

Subtract x and subtract 5 from both sides.

Although we can simplify the equation by dividing both sides by 2, this sort of simplification is not always helpful. Thus, we will work with the equation in this form.

Repeat Step 2. Square both sides.

Be careful in squaring both sides. Use $(A - B)^2 = A^2 - 2AB + B^2$ to square the left side. Use $(AB)^2 = A^2B^2$ to square the right side.

$$(2x - 4)^2 = (2\sqrt{x+4})^2$$

Square both sides.

$$4x^2 - 16x + 16 = 4(x+4)$$

Square both 2 and $\sqrt{x+4}$ on the right side.

Step 3. Solve the resulting equation. We solve this quadratic equation by writing it in standard form.

$$4x^2 - 16x + 16 = 4x + 16$$

Use the distributive property.

$$4x^2 - 20x = 0$$

Subtract $4x + 16$ from both sides.

$$4x(x - 5) = 0$$

Factor.

$$4x = 0 \quad \text{or} \quad x - 5 = 0$$

Set each factor equal to zero.

$$x = 0 \qquad\qquad x = 5$$

Solve for x.

Step 4. Check the proposed solutions in the original equation.

Check 0:	**Check 5:**

$$\sqrt{3x+1} - \sqrt{x+4} = 1 \qquad\qquad \sqrt{3x+1} - \sqrt{x+4} = 1$$
$$\sqrt{3 \cdot 0 + 1} - \sqrt{0+4} \overset{?}{=} 1 \qquad \sqrt{3 \cdot 5 + 1} - \sqrt{5+4} \overset{?}{=} 1$$
$$\sqrt{1} - \sqrt{4} \overset{?}{=} 1 \qquad\qquad \sqrt{16} - \sqrt{9} \overset{?}{=} 1$$
$$1 - 2 \overset{?}{=} 1 \qquad\qquad 4 - 3 \overset{?}{=} 1$$
$$-1 = 1, \quad \text{false} \qquad\qquad 1 = 1, \quad \text{true}$$

The check indicates that 0 is not a solution. It is an extraneous solution brought about by squaring each side of the equation. The only solution is 5 and the solution set is {5}. ∎

✓ **CHECK POINT 4**   Solve: $\sqrt{x+5} - \sqrt{x-3} = 2.$

EXAMPLE 5 Solving a Radical Equation

Solve: $(3x - 1)^{\frac{1}{3}} + 4 = 0$.

Solution Although we can rewrite the equation in radical form

$$\sqrt[3]{3x - 1} + 4 = 0,$$

it is not necessary to do so. Because the equation involves a cube root, we isolate the radical term—that is, the term with the rational exponent—and cube both sides.

$$(3x - 1)^{\frac{1}{3}} + 4 = 0 \qquad \text{This is the given equation.}$$

$$(3x - 1)^{\frac{1}{3}} = -4 \qquad \text{Subtract 4 from both sides and isolate the term with the rational exponent.}$$

$$\left[(3x - 1)^{\frac{1}{3}}\right]^3 = (-4)^3 \qquad \text{Cube both sides.}$$

$$3x - 1 = -64 \qquad \text{Multiply exponents on the left side and simplify.}$$

$$3x = -63 \qquad \text{Add 1 to both sides.}$$

$$x = -21 \qquad \text{Divide both sides by 3.}$$

Because both sides were raised to an odd power, it is not essential to check the proposed solution, -21. However, checking is always a good idea. Do so now and verify that -21 is the solution and the solution set is $\{-21\}$. ■

Example 5 illustrates that a radical equation with rational exponents can be solved by

1. isolating the expression with the rational exponent, and

2. raising both sides of the equation to a power that is the reciprocal of the rational exponent.

Keep in mind that it is essential to check proposed solutions when both sides have been raised to even powers. Thus, equations with rational exponents such as $\frac{1}{2}$ and $\frac{1}{4}$ must be checked.

✓ **CHECK POINT 5** Solve: $(2x - 3)^{\frac{1}{3}} + 3 = 0$.

---

2 Use models that are radical functions to solve problems.

Applications of Radical Equations

Radical equations can be solved to answer questions about variables contained in radical functions.

EXAMPLE 6 Women in the Labor Force

The bar graph in **Figure 10.8** shows the percentage of U.S. women in the labor force from 1960 through 2010. The function $f(x) = 3.5\sqrt{x} + 38$ models the percentage of U.S. women in the labor force, $f(x)$, x years after 1960. According to the model, when will 70% of U.S. women participate in the work force?

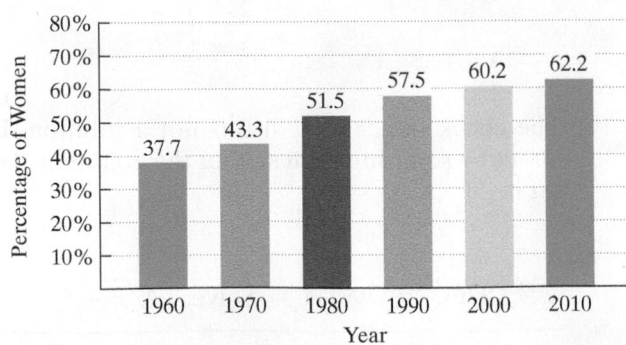

Percentage of United States Women in the Labor Force

Figure 10.8

Source: U.S. Department of Labor

Solution To find when 70% of U.S. women will be in the work force, substitute 70 for $f(x)$ in the given function. Then solve for x, the number of years after 1960.

$$f(x) = 3.5\sqrt{x} + 38 \qquad \text{This is the given function.}$$
$$70 = 3.5\sqrt{x} + 38 \qquad \text{Substitute 70 for } f(x).$$
$$32 = 3.5\sqrt{x} \qquad \text{Subtract 38 from both sides.}$$
$$\frac{32}{3.5} = \sqrt{x} \qquad \text{Divide both sides by 3.5.}$$
$$\left(\frac{32}{3.5}\right)^2 = \left(\sqrt{x}\right)^2 \qquad \text{Square both sides.}$$
$$84 \approx x \qquad \text{Use a calculator.}$$

The model indicates that 70% of U.S. women will be in the labor force approximately 84 years after 1960. Because $1960 + 84 = 2044$, this is projected to occur in 2044. ■

✓ **CHECK POINT 6** Use the function in Example 6 to project when 73% of U.S. women will participate in the work force.

CONCEPT AND VOCABULARY CHECK

Fill in each blank so that the resulting statement is true.

1. An equation in which the variable occurs in a square root, cube root, or any higher root is called a/an _____ equation.

2. Solutions of a squared equation that are not solutions of the original equation are called _____ solutions.

3. Consider the equation
$$\sqrt{2x + 1} = x - 7.$$
Squaring the left side and simplifying results in _____. Squaring the right side and simplifying results in _____.

4. Consider the equation
$$\sqrt{x + 2} = 3 - \sqrt{x - 1}.$$
Squaring the left side and simplifying results in _____. Squaring the right side and simplifying results in _____.

5. Consider the equation
$$(2x + 3)^{\frac{1}{3}} = 2.$$
Cubing the left side and simplifying results in _____. Cubing the right side and simplifying results in _____.

6. True or false: 4 is a solution of $\sqrt{5x + 16} = x + 2.$ _____

7. True or false: -3 is a solution of $\sqrt{5x + 16} = x + 2.$ _____

10.6 EXERCISE SET MyMathLab®

Watch the videos in MyMathLab

Download the MyDashBoard App

Practice Exercises

In Exercises 1–38, solve each radical equation.

1. $\sqrt{3x - 2} = 4$
2. $\sqrt{5x - 1} = 8$
3. $\sqrt{5x - 4} - 9 = 0$
4. $\sqrt{3x - 2} - 5 = 0$
5. $\sqrt{3x + 7} + 10 = 4$
6. $\sqrt{2x + 5} + 11 = 6$
7. $x = \sqrt{7x + 8}$
8. $x = \sqrt{6x + 7}$
9. $\sqrt{5x + 1} = x + 1$
10. $\sqrt{2x + 1} = x - 7$
11. $x = \sqrt{2x - 2} + 1$
12. $x = \sqrt{3x + 7} - 3$
13. $x - 2\sqrt{x - 3} = 3$
14. $3x - \sqrt{3x + 7} = -5$
15. $\sqrt{2x - 5} = \sqrt{x + 4}$
16. $\sqrt{6x + 2} = \sqrt{5x + 3}$
17. $\sqrt[3]{2x + 11} = 3$
18. $\sqrt[3]{6x - 3} = 3$
19. $\sqrt[3]{2x - 6} - 4 = 0$

20. $\sqrt[3]{4x - 3} - 5 = 0$

21. $\sqrt{x - 7} = 7 - \sqrt{x}$

22. $\sqrt{x - 8} = \sqrt{x} - 2$

23. $\sqrt{x + 2} + \sqrt{x - 1} = 3$

24. $\sqrt{x - 4} + \sqrt{x + 4} = 4$

25. $2\sqrt{4x + 1} - 9 = x - 5$

26. $2\sqrt{x - 3} + 4 = x + 1$

27. $(2x + 3)^{\frac{1}{3}} + 4 = 6$

28. $(3x - 6)^{\frac{1}{3}} + 5 = 8$

29. $(3x + 1)^{\frac{1}{4}} + 7 = 9$

30. $(2x + 3)^{\frac{1}{4}} + 7 = 10$

31. $(x + 2)^{\frac{1}{2}} + 8 = 4$

32. $(x - 3)^{\frac{1}{2}} + 8 = 6$

33. $\sqrt{2x - 3} - \sqrt{x - 2} = 1$

34. $\sqrt{x + 2} + \sqrt{3x + 7} = 1$

35. $3x^{\frac{1}{3}} = (x^2 + 17x)^{\frac{1}{3}}$

36. $2(x - 1)^{\frac{1}{3}} = (x^2 + 2x)^{\frac{1}{3}}$

37. $(x + 8)^{\frac{1}{4}} = (2x)^{\frac{1}{4}}$

38. $(x - 2)^{\frac{1}{4}} = (3x - 8)^{\frac{1}{4}}$

Practice PLUS

39. If $f(x) = x + \sqrt{x + 5}$, find all values of x for which $f(x) = 7$.

40. If $f(x) = x - \sqrt{x - 2}$, find all values of x for which $f(x) = 4$.

41. If $f(x) = (5x + 16)^{\frac{1}{3}}$ and $g(x) = (x - 12)^{\frac{1}{3}}$, find all values of x for which $f(x) = g(x)$.

42. If $f(x) = (9x + 2)^{\frac{1}{4}}$ and $g(x) = (5x + 18)^{\frac{1}{4}}$, find all values of x for which $f(x) = g(x)$.

In Exercises 43–46, solve each formula for the specified variable.

43. $r = \sqrt{\dfrac{3V}{\pi h}}$ for V

44. $r = \sqrt{\dfrac{A}{4\pi}}$ for A

45. $t = 2\pi\sqrt{\dfrac{l}{32}}$ for l

46. $v = \sqrt{\dfrac{FR}{m}}$ for m

47. If 5 times a number is decreased by 4, the principal square root of this difference is 2 less than the number. Find the number(s).

48. If a number is decreased by 3, the principal square root of this difference is 5 less than the number. Find the number(s).

In Exercises 49–50, find the x-intercept(s) of the graph of each function without graphing the function.

49. $f(x) = \sqrt{x + 16} - \sqrt{x} - 2$

50. $f(x) = \sqrt{2x - 3} - \sqrt{2x} + 1$

Application Exercises

A basketball player's hang time is the time spent in the air when shooting a basket. The formula

$$t = \frac{\sqrt{d}}{2}$$

models hang time, t, in seconds, in terms of the vertical distance of a player's jump, d, in feet. Use this formula to solve Exercises 51–52.

51. When Michael Wilson of the Harlem Globetrotters slam-dunked a basketball 12 feet, his hang time for the shot was approximately 1.16 seconds. What was the vertical distance of his jump, rounded to the nearest tenth of a foot?

52. If hang time for a shot by a professional basketball player is 0.85 second, what is the vertical distance of the jump, rounded to the nearest tenth of a foot?

Use the graph of the formula for hang time to solve Exercises 53–54.

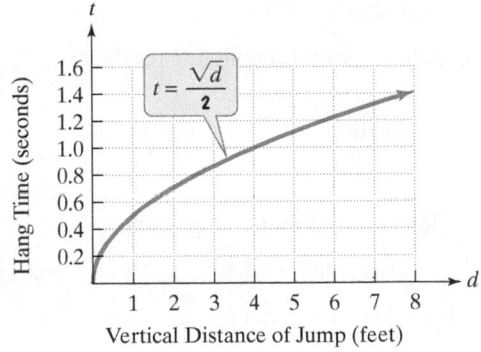

53. How is your answer to Exercise 51 shown on the graph?

54. How is your answer to Exercise 52 shown on the graph?

The graph at the top of the next page shows the less income people have, the more likely they are to report that their health is fair or poor. The function

$$f(x) = -4.4\sqrt{x} + 38$$

models the percentage of Americans reporting fair or poor health, f(x), in terms of annual income, x, in thousands of dollars. Use this function to solve Exercises 55–56.

Americans Reporting Fair or Poor Health, by Annual Income

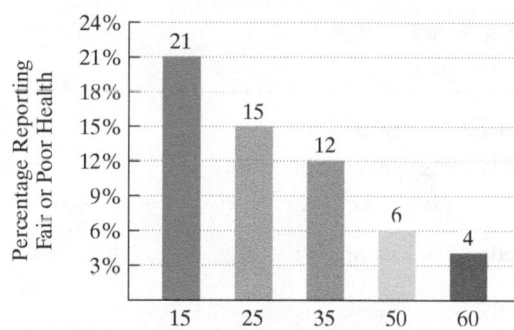

Source: William Kornblum and Joseph Julian, *Social Problems, Twelfth Edition*, Prentice Hall, 2007.

55. a. Find and interpret $f(25)$. Does this underestimate or overestimate the percent displayed by the graph? By how much?

b. According to the model, what annual income corresponds to 14% reporting fair or poor health? Round to the nearest thousand dollars.

56. a. Find and interpret $f(60)$. Round to one decimal place. Does this underestimate or overestimate the percent displayed by the graph? By how much?

b. According to the model, what annual income corresponds to 24% reporting fair or poor health? Round to the nearest thousand dollars.

The function

$$f(x) = 29x^{\frac{1}{3}}$$

models the number of plant species, $f(x)$, on the islands of the Galápagos in terms of the area, x, in square miles, of a particular island. Use the function to solve Exercises 57–58.

57. What is the area of a Galápagos island that has 87 species of plants?

58. What is the area of a Galápagos island that has 58 species of plants?

For each planet in our solar system, its year is the time it takes the planet to revolve once around the sun. The function

$$f(x) = 0.2x^{\frac{3}{2}}$$

models the number of Earth days in a planet's year, $f(x)$, where x is the average distance of the planet from the sun, in millions of kilometers. Use the function to solve Exercises 59–60.

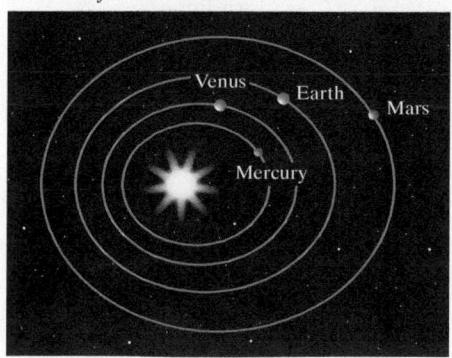

59. We, of course, have 365 Earth days in our year. What is the average distance of Earth from the sun? Use a calculator and round to the nearest million kilometers.

60. There are approximately 88 Earth days in the year of the planet Mercury. What is the average distance of Mercury from the sun? Use a calculator and round to the nearest million kilometers.

Writing in Mathematics

61. What is a radical equation?

62. In solving $\sqrt{2x - 1} + 2 = x$, why is it a good idea to isolate the radical term? What if we don't do this and simply square each side? Describe what happens.

63. What is an extraneous solution to a radical equation?

64. Explain why $\sqrt{x} = -1$ has no solution.

65. Explain how to solve a radical equation with rational exponents.

66. In Example 6 of the section, we used a square root function that modeled an increase in the percentage of U.S. women in the labor force, although the rate of increase in this percentage was leveling off. Describe an event that might occur in the future that could result in an ever-increasing rate in the percentage of women in the labor force. Would a square root function be appropriate for modeling this trend? Explain your answer.

67. The graph for Exercises 55–56 shows that the less income people have, the more likely they are to report fair or poor health. What explanations can you offer for this trend?

Technology Exercises

In Exercises 68–72, use a graphing utility to solve each radical equation. Graph each side of the equation in the given viewing rectangle. The equation's solution set is given by the x-coordinate(s) of the point(s) of intersection. Check by substitution.

68. $\sqrt{2x + 2} = \sqrt{3x - 5}$

    $[-1, 10, 1]$ by $[-1, 5, 1]$

69. $\sqrt{x} + 3 = 5$

    $[-1, 6, 1]$ by $[-1, 6, 1]$

70. $\sqrt{x^2 + 3} = x + 1$

    $[-1, 6, 1]$ by $[-1, 6, 1]$

71. $4\sqrt{x} = x + 3$

    $[-1, 10, 1]$ by $[-1, 14, 1]$

72. $\sqrt{x} + 4 = 2$

    $[-2, 18, 1]$ by $[0, 10, 1]$

Critical Thinking Exercises

Make Sense? *In Exercises 73–76, determine whether each statement "makes sense" or "does not make sense" and explain your reasoning.*

73. When checking a radical equation's proposed solution, I can substitute into the original equation or any equation that is part of the solution process.

74. After squaring both sides of a radical equation, the only solution that I obtained was extraneous, so $\varnothing$ must be the solution set of the original equation.

75. When I raise both sides of an equation to any power, there's always the possibility of extraneous solutions.

76. Now that I know how to solve radical equations, I can use models that are radical functions to determine the value of the independent variable when a function value is known.

In Exercises 77–80, determine whether each statement is true or false. If the statement is false, make the necessary change(s) to produce a true statement.

77. The first step in solving $\sqrt{x+6} = x+2$ is to square both sides, obtaining $x + 6 = x^2 + 4$.

78. The equations $\sqrt{x+4} = -5$ and $x + 4 = 25$ have the same solution set.

79. The equation $-\sqrt{x} = 9$ has no solution.

80. The equation $\sqrt{x^2 + 9x + 3} = -x$ has no solution because a principal square root is always nonnegative.

81. Find the length of the three sides of the right triangle shown in the figure.

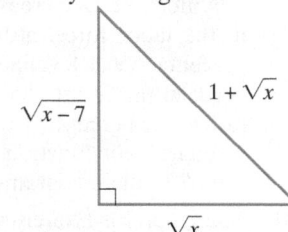

In Exercises 82–84, solve each equation.

82. $\sqrt[3]{x\sqrt{x}} = 9$

83. $\sqrt{\sqrt{x} + \sqrt{x+9}} = 3$

84. $(x-4)^{\frac{2}{3}} = 25$

Review Exercises

85. Divide using synthetic division:
$$(4x^4 - 3x^3 + 2x^2 - x - 1) \div (x+3).$$
(Section 5.6, Example 5)

86. Divide:
$$\frac{3x^2 - 12}{x^2 + 2x - 8} \div \frac{6x + 18}{x+4}.$$
(Section 7.2, Example 6)

87. Factor: $y^2 - 6y + 9 - 25x^2$.
(Section 6.5, Example 8)

Preview Exercises

Exercises 88–90 will help you prepare for the material covered in the next section.

88. Simplify: $(-5 + 7x) - (-11 - 6x)$.

89. Multiply: $(7 - 3x)(-2 - 5x)$.

90. Rationalize the denominator: $\dfrac{7 + 4\sqrt{2}}{2 - 5\sqrt{2}}$.

SECTION

10.7

Complex Numbers

Objectives

1 Express square roots of negative numbers in terms of i.

2 Add and subtract complex numbers.

3 Multiply complex numbers.

4 Divide complex numbers.

5 Simplify powers of i.

Who is this kid warning us about our eyeballs turning black if we attempt to find the square root of -9? Don't believe what you hear on the street. Although square roots of negative numbers are not real numbers, they do play a significant role in algebra. In this section, we move beyond the real numbers and discuss square roots with negative radicands.

1 Express square roots of negative numbers in terms of *i*.

The Imaginary Unit *i*

In Chapter 11, we will study equations whose solutions involve the square roots of negative numbers. Because the square of a real number is never negative, there is no real number x such that $x^2 = -1$. To provide a setting in which such equations have solutions, mathematicians invented an expanded system of numbers, the complex numbers. The *imaginary number i*, defined to be a solution of the equation $x^2 = -1$, is the basis of this new set.

> **The Imaginary Unit *i***
>
> The **imaginary unit *i*** is defined as
>
> $$i = \sqrt{-1}, \quad \text{where} \quad i^2 = -1.$$

Using the imaginary unit *i*, we can express the square root of any negative number as a real multiple of *i*. For example,

$$\sqrt{-25} = \sqrt{25(-1)} = \sqrt{25}\sqrt{-1} = 5i.$$

We can check that $\sqrt{-25} = 5i$ by squaring $5i$ and obtaining -25.

$$(5i)^2 = 5^2 i^2 = 25(-1) = -25$$

> **The Square Root of a Negative Number**
>
> If b is a positive real number, then
>
> $$\sqrt{-b} = \sqrt{b(-1)} = \sqrt{b}\sqrt{-1} = \sqrt{b}i \quad \text{or} \quad i\sqrt{b}.$$

EXAMPLE 1 Expressing Square Roots of Negative Numbers as Multiples of *i*

Write as a multiple of *i*:

a. $\sqrt{-9}$ **b.** $\sqrt{-3}$ **c.** $\sqrt{-80}$.

Solution

a. $\sqrt{-9} = \sqrt{9(-1)} = \sqrt{9}\sqrt{-1} = 3i$

b. $\sqrt{-3} = \sqrt{3(-1)} = \sqrt{3}\sqrt{-1} = \sqrt{3}i$ — Be sure not to write *i* under the radical.

c. $\sqrt{-80} = \sqrt{80(-1)} = \sqrt{80}\sqrt{-1} = \sqrt{16 \cdot 5}\sqrt{-1} = 4\sqrt{5}i$

In order to avoid writing *i* under a radical, let's agree to write *i* before any radical. Consequently, we express the multiple of *i* in part (b) as $i\sqrt{3}$ and the multiple of *i* in part (c) as $4i\sqrt{5}$. ∎

Great Question!

Now that we've introduced square roots of negative numbers, can I still use the rule $\sqrt{ab} = \sqrt{a}\sqrt{b}$?

We allow the use of the product rule $\sqrt{ab} = \sqrt{a}\sqrt{b}$ when a is positive and b is -1. However, you cannot use $\sqrt{ab} = \sqrt{a}\sqrt{b}$ when both a and b are negative.

✓ **CHECK POINT 1** Write as a multiple of *i*:

a. $\sqrt{-64}$ **b.** $\sqrt{-11}$ **c.** $\sqrt{-48}$.

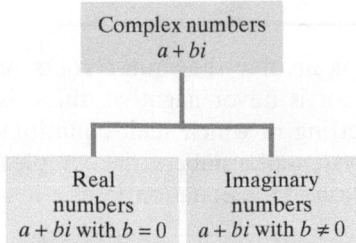

Figure 10.9
The complex number system

A new system of numbers, called *complex numbers*, is based on adding multiples of i, such as $5i$, to the real numbers.

Complex Numbers and Imaginary Numbers

The set of all numbers in the form

$$a + bi,$$

with real numbers a and b, and i, the imaginary unit, is called the set of **complex numbers**. The real number a is called the **real part**, and the real number b is called the **imaginary part** of the complex number $a + bi$. If $b \neq 0$, then the complex number is called an **imaginary number** (**Figure 10.9**).

Here are some examples of complex numbers. Each number can be written in the form $a + bi$.

$-4 + 6i$ $2i = 0 + 2i$ $3 = 3 + 0i$

| a, the real part, is -4. | b, the imaginary part, is 6. | a, the real part, is 0. | b, the imaginary part, is 2. | a, the real part, is 3. | b, the imaginary part, is 0. |

Can you see that b, the imaginary part, is not zero in the first two complex numbers? Because $b \neq 0$, these complex numbers are imaginary numbers. By contrast, the imaginary part of the complex number on the right is zero. This complex number is not an imaginary number. The number 3, or $3 + 0i$, is a real number.

2 Add and subtract complex numbers.

Adding and Subtracting Complex Numbers

The form of a complex number $a + bi$ is like the binomial $a + bx$. Consequently, we can add, subtract, and multiply complex numbers using the same methods we used for binomials, remembering that $i^2 = -1$.

Adding and Subtracting Complex Numbers

1. $(a + bi) + (c + di) = (a + c) + (b + d)i$

   In words, this says that you add complex numbers by adding their real parts, adding their imaginary parts, and expressing the sum as a complex number.

2. $(a + bi) - (c + di) = (a - c) + (b - d)i$

   In words, this says that you subtract complex numbers by subtracting their real parts, subtracting their imaginary parts, and expressing the difference as a complex number.

EXAMPLE 2 Adding and Subtracting Complex Numbers

Perform the indicated operations, writing the result in the form $a + bi$:

a. $(5 - 11i) + (7 + 4i)$ **b.** $(-5 + 7i) - (-11 - 6i)$.

Solution

a. $(5 - 11i) + (7 + 4i)$
    $= 5 - 11i + 7 + 4i$ Remove the parentheses.
    $= 5 + 7 - 11i + 4i$ Group real and imaginary terms.
    $= (5 + 7) + (-11 + 4)i$ Add real parts and add imaginary parts.
    $= 12 - 7i$ Simplify.

Great Question!

Are operations with complex numbers similar to operations with polynomials?

Yes. The following examples, using the same integers as in Example 2, show how operations with complex numbers are just like operations with polynomials.

a. $(5 - 11x) + (7 + 4x)$
$= 12 - 7x$

b. $(-5 + 7x) - (-11 - 6x)$
$= -5 + 7x + 11 + 6x$
$= 6 + 13x$

3 Multiply complex numbers.

b. $(-5 + 7i) - (-11 - 6i)$

$= -5 + 7i + 11 + 6i$ Remove the parentheses. Change signs of real and imaginary parts in the complex number being subtracted.

$= -5 + 11 + 7i + 6i$ Group real and imaginary terms.

$= (-5 + 11) + (7 + 6)i$ Add real parts and add imaginary parts.

$= 6 + 13i$ Simplify. ∎

✓ **CHECK POINT 2**   Add or subtract as indicated:

a. $(5 - 2i) + (3 + 3i)$

b. $(2 + 6i) - (12 - 4i)$.

Multiplying Complex Numbers

Multiplication of complex numbers is performed the same way as multiplication of polynomials, using the distributive property and the FOIL method. After completing the multiplication, we replace any occurrences of i^2 with -1. This idea is illustrated in the next example.

EXAMPLE 3   Multiplying Complex Numbers

Find the products:

a. $4i(3 - 5i)$ **b.** $(7 - 3i)(-2 - 5i)$.

Solution

a. $4i(3 - 5i)$

$= 4i \cdot 3 - 4i \cdot 5i$ Distribute $4i$ throughout the parentheses.

$= 12i - 20i^2$ Multiply.

$= 12i - 20(-1)$ Replace i^2 with -1.

$= 20 + 12i$ Simplify to $12i + 20$ and write in $a + bi$ form.

b. $(7 - 3i)(-2 - 5i)$

     F   O   I   L

$= -14 - 35i + 6i + 15i^2$ Use the FOIL method.

$= -14 - 35i + 6i + 15(-1)$ $i^2 = -1$

$= -14 - 15 - 35i + 6i$ Group real and imaginary terms.

$= -29 - 29i$ Combine real and imaginary terms. ∎

✓ **CHECK POINT 3**   Find the products:

a. $7i(2 - 9i)$ **b.** $(5 + 4i)(6 - 7i)$.

Consider the multiplication problem

$$5i \cdot 2i = 10i^2 = 10(-1) = -10.$$

The problem $5i \cdot 2i$ can also be given in terms of square roots of negative numbers:

$$\sqrt{-25} \cdot \sqrt{-4}.$$

Because the product rule for radicals only applies to real numbers, multiplying radicands is incorrect. **When performing operations with square roots of negative numbers, begin by expressing all square roots in terms of i.** Then perform the indicated operation.

CORRECT:

$$\sqrt{-25} \cdot \sqrt{-4} = \sqrt{25}\sqrt{-1} \cdot \sqrt{4}\sqrt{-1}$$
$$= 5i \cdot 2i$$
$$= 10i^2 = 10(-1) = -10$$

INCORRECT:

$$\cancel{\sqrt{-25} \cdot \sqrt{-4} = \sqrt{(-25)(-4)}}$$
$$\cancel{= \sqrt{100}}$$
$$\cancel{= 10}$$

EXAMPLE 4 Multiplying Square Roots of Negative Numbers

Multiply: $\sqrt{-3} \cdot \sqrt{-5}$.

Solution

$$\sqrt{-3} \cdot \sqrt{-5} = \sqrt{3}\sqrt{-1} \cdot \sqrt{5}\sqrt{-1}$$
$$= i\sqrt{3} \cdot i\sqrt{5} \quad \text{Express square roots in terms of } i.$$
$$= i^2\sqrt{15} \quad \sqrt{3} \cdot \sqrt{5} = \sqrt{15} \text{ and } i \cdot i = i^2.$$
$$= (-1)\sqrt{15} \quad i^2 = -1$$
$$= -\sqrt{15} \quad \blacksquare$$

☑ **CHECK POINT 4** Multiply: $\sqrt{-5} \cdot \sqrt{-7}$.

---

4 Divide complex numbers.

Conjugates and Division

It is possible to multiply imaginary numbers and obtain a real number. Here is an example:

$$\underset{\text{F}}{} \quad \underset{\text{O}}{} \quad \underset{\text{I}}{} \quad \underset{\text{L}}{}$$

$$(4 + 7i)(4 - 7i) = 16 - 28i + 28i - 49i^2$$
$$= 16 - 49i^2 = 16 - 49(-1) = 65.$$

Replace i^2 with -1.

You can also perform $(4 + 7i)(4 - 7i)$ using the formula

$$(A + B)(A - B) = A^2 - B^2.$$

A real number is obtained even faster:

$$(4 + 7i)(4 - 7i) = 4^2 - (7i)^2 = 16 - 49i^2 = 16 - 49(-1) = 65.$$

The **conjugate** of the complex number $a + bi$ is $a - bi$. The **conjugate** of the complex number $a - bi$ is $a + bi$. The multiplication problem that we just performed involved conjugates. The multiplication of conjugates always results in a real number:

$$(a + bi)(a - bi) = a^2 - (bi)^2 = a^2 - b^2i^2 = a^2 - b^2(-1) = a^2 + b^2.$$

The product eliminates i.

Conjugates are used to divide complex numbers. The goal of the division procedure is to obtain a real number in the denominator. This real number becomes the denominator of a and b in the quotient $a + bi$. By multiplying the numerator and the denominator of the division by the conjugate of the denominator, you will obtain this real number in the denominator. Here are two examples of such divisions:

- $\dfrac{7 + 4i}{2 - 5i}$

  The conjugate of the denominator is 2 + 5i.

- $\dfrac{5i - 4}{3i}$ or $\dfrac{5i - 4}{0 + 3i}$.

  The conjugate of the denominator is 0 − 3i, or −3i.

The procedure for dividing complex numbers, illustrated in Examples 5 and 6, should remind you of rationalizing denominators.

EXAMPLE 5   Using Conjugates to Divide Complex Numbers

Divide and simplify to the form $a + bi$:

$$\frac{7 + 4i}{2 - 5i}.$$

Solution   The conjugate of the denominator is $2 + 5i$. Multiplication of both the numerator and the denominator by $2 + 5i$ will eliminate i from the denominator while maintaining the value of the expression.

$\dfrac{7 + 4i}{2 - 5i} = \dfrac{7 + 4i}{2 - 5i} \cdot \dfrac{2 + 5i}{2 + 5i}$ Multiply by 1.

$= \dfrac{\overset{F}{14} + \overset{O}{35i} + \overset{I}{8i} + \overset{L}{20i^2}}{2^2 - (5i)^2}$ Use FOIL in the numerator and $(A - B)(A + B) = A^2 - B^2$ in the denominator.

$= \dfrac{14 + 43i + 20i^2}{4 - 25i^2}$ Simplify.

$= \dfrac{14 + 43i + 20(-1)}{4 - 25(-1)}$ $i^2 = -1$

$= \dfrac{14 + 43i - 20}{4 + 25}$ Perform the multiplications involving −1.

$= \dfrac{-6 + 43i}{29}$ Combine real terms in the numerator and denominator.

$= -\dfrac{6}{29} + \dfrac{43}{29}i$ Express the answer in the form $a + bi$.   ∎

✓ **CHECK POINT 5**   Divide and simplify to the form $a + bi$:

$$\frac{6 + 2i}{4 - 3i}.$$

EXAMPLE 6   Using Conjugates to Divide Complex Numbers

Divide and simplify to the form $a + bi$:

$$\frac{5i - 4}{3i}.$$

Solution The denominator of $\dfrac{5i - 4}{3i}$ is $3i$, or $0 + 3i$. The conjugate of the denominator is $0 - 3i$. Multiplication of both the numerator and the denominator by $-3i$ will eliminate i from the denominator while maintaining the value of the expression.

$$\frac{5i - 4}{3i} = \frac{5i - 4}{3i} \cdot \frac{-3i}{-3i} \qquad \text{Multiply by 1.}$$

$$= \frac{-15i^2 + 12i}{-9i^2} \qquad \text{Multiply. Use the distributive property in the numerator.}$$

$$= \frac{-15(-1) + 12i}{-9(-1)} \qquad i^2 = -1$$

$$= \frac{15 + 12i}{9} \qquad \text{Perform the multiplications involving } -1.$$

$$= \frac{15}{9} + \frac{12}{9}i \qquad \text{Express the division in the form } a + bi.$$

$$= \frac{5}{3} + \frac{4}{3}i \qquad \text{Simplify real and imaginary parts.} \quad \blacksquare$$

✓ **CHECK POINT 6** Divide and simplify to the form $a + bi$:

$$\frac{3 - 2i}{4i}.$$

---

5 Simplify powers of i.

Powers of i

Using the fact that $i^2 = -1$, any integral power of i greater than or equal to 2 can be simplified to either $-i$, i, -1, or 1. Here are some examples:

$$i^3 = i^2 \cdot i = (-1)i = -i$$

$$i^4 = (i^2)^2 = (-1)^2 = 1$$

$$i^5 = i^4 \cdot i = (i^2)^2 \cdot i = (-1)^2 \cdot i = i$$

$$i^6 = (i^2)^3 = (-1)^3 = -1$$

Here is a procedure for simplifying powers of i:

Simplifying Powers of i

1. Express the given power of i in terms of i^2.
2. Replace i^2 with -1 and simplify. Use the fact that -1 to an even power is 1 and -1 to an odd power is -1.

EXAMPLE 7 Simplifying Powers of i

Simplify:

a. i^{12} **b.** i^{39} **c.** i^{50}.

Solution

a. $i^{12} = (i^2)^6 = (-1)^6 = 1$
b. $i^{39} = i^{38}i = (i^2)^{19}i = (-1)^{19}i = (-1)i = -i$
c. $i^{50} = (i^2)^{25} = (-1)^{25} = -1$ $\blacksquare$

✓ **CHECK POINT 7** Simplify:

a. i^{16} **b.** i^{25} **c.** i^{35}.

Blitzer Bonus

The Patterns of Chaos

One of the new frontiers of mathematics suggests that there is an underlying order in things that appear to be random, such as the hiss and crackle of background noises as you tune a radio. Irregularities in the heartbeat, some of them severe enough to cause a heart attack, or irregularities in our sleeping patterns, such as insomnia, are examples of chaotic behavior. Chaos in the mathematical sense does not mean a complete lack of form or arrangement. In mathematics, chaos is used to describe something that appears to be random, but actually contains underlying patterns that are far more intricate than previously assumed. The patterns of chaos appear in images like the one on the right and the one in the chapter opener, called the Mandelbrot set. Magnified portions of this image yield repetitions of the original structure, as well as new and unexpected patterns. The Mandelbrot set transforms the hidden structure of chaotic events into a source of wonder and inspiration.

The Mandelbrot set is made possible by opening up graphing to include complex numbers in the form $a + bi$. Each complex number is plotted like an ordered pair in a coordinate system consisting of a real axis and an imaginary axis. Plot certain complex numbers in this system, add color to the magnified boundary of the graph, and the patterns of chaos begin to appear.

29-Fold M-Set Seahorse. © 2011 Richard F. Voss

CONCEPT AND VOCABULARY CHECK

Fill in each blank so that the resulting statement is true.

1. The imaginary unit i is defined as $i = $ _____, where $i^2 = $ _____.

2. $\sqrt{-16} = \sqrt{16(-1)} = \sqrt{16}\sqrt{-1} = $ _____

3. The set of all numbers in the form $a + bi$ is called the set of _____ numbers. If $b \neq 0$, then the number is also called a/an _____ number. If $b = 0$, then the number is also called a/an _____ number.

4. $-9i + 3i = $ _____

5. $10i - (-4i) = $ _____

6. Consider the following multiplication problem:
   $$(3 + 2i)(6 - 5i).$$
   Using the FOIL method, the product of the first terms is _____, the product of the outside terms is _____, and the product of the inside terms is _____. The product of the last terms in terms of i^2 is _____, which simplifies to _____.

7. The conjugate of $2 - 9i$ is _____.

8. The division
   $$\frac{7 + 4i}{2 - 5i}$$
   is performed by multiplying the numerator and denominator by _____.

9. The division
   $$\frac{3 - 2i}{4i}$$
   is performed by multiplying the numerator and denominator by _____.

10. $i^{16} = (i^2)^8 = (\underline{})^8 = \underline{}$

11. $i^{35} = i^{34}i = (i^2)^{17}i = (\underline{})^{17}i = (\underline{})i = \underline{}$

10.7 EXERCISE SET

MyMathLab®

Watch the videos in MyMathLab

Download the MyDashBoard App

Practice Exercises

In Exercises 1–16, express each number in terms of i and simplify, if possible.

1. $\sqrt{-100}$
2. $\sqrt{-49}$
3. $\sqrt{-23}$
4. $\sqrt{-21}$
5. $\sqrt{-18}$
6. $\sqrt{-125}$
7. $\sqrt{-63}$
8. $\sqrt{-28}$
9. $-\sqrt{-108}$
10. $-\sqrt{-300}$
11. $5 + \sqrt{-36}$
12. $7 + \sqrt{-4}$
13. $15 + \sqrt{-3}$
14. $20 + \sqrt{-5}$
15. $-2 - \sqrt{-18}$
16. $-3 - \sqrt{-27}$

In Exercises 17–32, add or subtract as indicated. Write the result in the form a + bi.

17. $(3 + 2i) + (5 + i)$
18. $(6 + 5i) + (4 + 3i)$
19. $(7 + 2i) + (1 - 4i)$
20. $(-2 + 6i) + (4 - i)$
21. $(10 + 7i) - (5 + 4i)$
22. $(11 + 8i) - (2 + 5i)$
23. $(9 - 4i) - (10 + 3i)$
24. $(8 - 5i) - (6 + 2i)$
25. $(3 + 2i) - (5 - 7i)$
26. $(-7 + 5i) - (9 - 11i)$
27. $(-5 + 4i) - (-13 - 11i)$
28. $(-9 + 2i) - (-17 - 6i)$
29. $8i - (14 - 9i)$
30. $15i - (12 - 11i)$
31. $\left(2 + i\sqrt{3}\right) + \left(7 + 4i\sqrt{3}\right)$
32. $\left(4 + i\sqrt{5}\right) + \left(8 + 6i\sqrt{5}\right)$

In Exercises 33–62, find each product. Write imaginary results in the form a + bi.

33. $2i(5 + 3i)$
34. $5i(4 + 7i)$
35. $3i(7i - 5)$
36. $8i(4i - 3)$
37. $-7i(2 - 5i)$
38. $-6i(3 - 5i)$
39. $(3 + i)(4 + 5i)$
40. $(4 + i)(5 + 6i)$
41. $(7 - 5i)(2 - 3i)$
42. $(8 - 4i)(3 - 2i)$
43. $(6 - 3i)(-2 + 5i)$
44. $(7 - 2i)(-3 + 6i)$
45. $(3 + 5i)(3 - 5i)$
46. $(2 + 7i)(2 - 7i)$
47. $(-5 + 3i)(-5 - 3i)$
48. $(-4 + 2i)(-4 - 2i)$
49. $\left(3 - i\sqrt{2}\right)\left(3 + i\sqrt{2}\right)$
50. $\left(5 - i\sqrt{3}\right)\left(5 + i\sqrt{3}\right)$
51. $(2 + 3i)^2$
52. $(3 + 2i)^2$
53. $(5 - 2i)^2$
54. $(5 - 3i)^2$
55. $\sqrt{-7} \cdot \sqrt{-2}$
56. $\sqrt{-7} \cdot \sqrt{-3}$
57. $\sqrt{-9} \cdot \sqrt{-4}$
58. $\sqrt{-16} \cdot \sqrt{-4}$
59. $\sqrt{-7} \cdot \sqrt{-25}$
60. $\sqrt{-3} \cdot \sqrt{-36}$
61. $\sqrt{-8} \cdot \sqrt{-3}$
62. $\sqrt{-9} \cdot \sqrt{-5}$

In Exercises 63–84, divide and simplify to the form a + bi.

63. $\dfrac{2}{3 + i}$
64. $\dfrac{3}{4 + i}$
65. $\dfrac{2i}{1 + i}$
66. $\dfrac{5i}{2 + i}$
67. $\dfrac{7}{4 - 3i}$
68. $\dfrac{9}{1 - 2i}$
69. $\dfrac{6i}{3 - 2i}$
70. $\dfrac{5i}{2 - 3i}$
71. $\dfrac{1 + i}{1 - i}$
72. $\dfrac{1 - i}{1 + i}$
73. $\dfrac{2 - 3i}{3 + i}$
74. $\dfrac{2 + 3i}{3 - i}$
75. $\dfrac{5 - 2i}{3 + 2i}$
76. $\dfrac{6 - 3i}{4 + 2i}$
77. $\dfrac{4 + 5i}{3 - 7i}$
78. $\dfrac{5 - i}{3 - 2i}$
79. $\dfrac{7}{3i}$
80. $\dfrac{5}{7i}$
81. $\dfrac{8 - 5i}{2i}$
82. $\dfrac{3 + 4i}{5i}$
83. $\dfrac{4 + 7i}{-3i}$
84. $\dfrac{5 + i}{-4i}$

In Exercises 85–100, simplify each expression.

85. i^{10}
86. i^{14}
87. i^{11}
88. i^{15}
89. i^{22}
90. i^{46}
91. i^{200}
92. i^{400}
93. i^{17}
94. i^{21}
95. $(-i)^4$
96. $(-i)^6$
97. $(-i)^9$
98. $(-i)^{13}$
99. $i^{24} + i^2$
100. $i^{28} + i^{30}$

Practice PLUS

In Exercises 101–108, perform the indicated operation(s) and write the result in the form a + bi.

101. $(2 - 3i)(1 - i) - (3 - i)(3 + i)$
102. $(8 + 9i)(2 - i) - (1 - i)(1 + i)$
103. $(2 + i)^2 - (3 - i)^2$
104. $(4 - i)^2 - (1 + 2i)^2$
105. $5\sqrt{-16} + 3\sqrt{-81}$
106. $5\sqrt{-8} + 3\sqrt{-18}$
107. $\dfrac{i^4 + i^{12}}{i^8 - i^7}$
108. $\dfrac{i^8 + i^{40}}{i^4 + i^3}$

109. Let $f(x) = x^2 - 2x + 2$. Find $f(1 + i)$.

110. Let $f(x) = x^2 - 2x + 5$. Find $f(1 - 2i)$.

In Exercises 111–114, simplify each evaluation to the form a + bi.

111. Let $f(x) = x - 3i$ and $g(x) = 4x + 2i$. Find $(fg)(-1)$.

112. Let $f(x) = 12x - i$ and $g(x) = 6x + 3i$. Find $(fg)\left(-\frac{1}{3}\right)$.

113. Let $f(x) = \dfrac{x^2 + 19}{2 - x}$. Find $f(3i)$.

114. Let $f(x) = \dfrac{x^2 + 11}{3 - x}$. Find $f(4i)$.

Application Exercises

Complex numbers are used in electronics to describe the current in an electric circuit. Ohm's law relates the current in a circuit, I, in amperes, the voltage of the circuit, E, in volts, and the resistance of the circuit, R, in ohms, by the formula $E = IR$. Use this formula to solve Exercises 115–116.

115. Find E, the voltage of a circuit, if $I = (4 - 5i)$ amperes and $R = (3 + 7i)$ ohms.

116. Find E, the voltage of a circuit, if $I = (2 - 3i)$ amperes and $R = (3 + 5i)$ ohms.

117. The mathematician Girolamo Cardano is credited with the first use (in 1545) of negative square roots in solving the now-famous problem, "Find two numbers whose sum is 10 and whose product is 40." Show that the complex numbers $5 + i\sqrt{15}$ and $5 - i\sqrt{15}$ satisfy the conditions of the problem. (Cardano did not use the symbolism $i\sqrt{15}$ or even $\sqrt{-15}$. He wrote R.m 15 for $\sqrt{-15}$, meaning "radix minus 15." He regarded the numbers $5 + $ R.m 15 and $5 - $ R.m 15 as "fictitious" or "ghost numbers," and considered the problem "manifestly impossible." But in a mathematically adventurous spirit, he exclaimed, "Nevertheless, we will operate.")

Writing in Mathematics

118. What is the imaginary unit i?

119. Explain how to write $\sqrt{-64}$ as a multiple of i.

120. What is a complex number? Explain when a complex number is a real number and when it is an imaginary number. Provide examples with your explanation.

121. Explain how to add complex numbers. Give an example.

122. Explain how to subtract complex numbers. Give an example.

123. Explain how to find the product of $2i$ and $5 + 3i$.

124. Explain how to find the product of $3 + 2i$ and $5 + 3i$.

125. Explain how to find the product of $3 + 2i$ and $3 - 2i$.

126. Explain how to find the product of $\sqrt{-1}$ and $\sqrt{-4}$. Describe a common error in the multiplication that needs to be avoided.

127. What is the conjugate of $2 + 3i$? What happens when you multiply this complex number by its conjugate?

128. Explain how to divide complex numbers. Provide an example with your explanation.

129. Explain each of the three jokes in the cartoon on page 752.

130. A stand-up comedian uses algebra in some jokes, including one about a telephone recording that announces "You have just reached an imaginary number. Please multiply by i and dial again." Explain the joke.

Explain the error in Exercises 131–132.

131. $\sqrt{-9} + \sqrt{-16} = \sqrt{-25} = i\sqrt{25} = 5i$

132. $\left(\sqrt{-9}\right)^2 = \sqrt{-9} \cdot \sqrt{-9} = \sqrt{81} = 9$

Critical Thinking Exercises

Make Sense? *In Exercises 133–136, determine whether each statement "makes sense" or "does not make sense" and explain your reasoning.*

133. The joke in the cartoon below is based on the math teacher not realizing that the average of complex real numbers is sometimes a complex imaginary number.

ROBOTMAN by Jim Meddick

Monty copyright © 1993 Jim Meddick. Distributed by Universal Uclick. Reprinted with permission. All rights reserved

134. The word *imaginary* in imaginary numbers tells me that these numbers are undefined.

135. By writing the imaginary number $5i$, I can immediately see that 5 is the constant and i is the variable.

136. When I add or subtract complex numbers, I am basically combining like terms.

In Exercises 137–140, determine whether each statement is true or false. If the statement is false, make the necessary change(s) to produce a true statement.

137. Some irrational numbers are not complex numbers.

138. $(3 + 7i)(3 - 7i)$ is an imaginary number.

139. $\dfrac{7 + 3i}{5 + 3i} = \dfrac{7}{5}$

140. In the complex number system, $x^2 + y^2$ (the sum of two squares) can be factored as $(x + yi)(x - yi)$.

In Exercises 141–143, perform the indicated operations and write the result in the form a + bi.

141. $\dfrac{4}{(2 + i)(3 - i)}$

142. $\dfrac{1 + i}{1 + 2i} + \dfrac{1 - i}{1 - 2i}$

143. $\dfrac{8}{1 + \dfrac{2}{i}}$

Review Exercises

144. Simplify:

$$\dfrac{\dfrac{x}{y^2} + \dfrac{1}{y}}{\dfrac{y}{x^2} + \dfrac{1}{x}}.$$

(Section 7.5, Example 3 or Example 6)

145. Solve for x: $\dfrac{1}{x} + \dfrac{1}{y} = \dfrac{1}{z}$.

(Section 7.6, Example 7)

146. If $f(x) = 2x^2 - x$ and $g(x) = x - 6$, find $(g - f)(3)$.
(Section 8.3, Example 4)

Preview Exercises

Exercises 147–149 will help you prepare for the material covered in the first section of the next chapter.

147. Solve by factoring: $2x^2 + 7x - 4 = 0$.

148. Solve by factoring: $x^2 = 9$.

149. Use substitution to determine if $-\sqrt{6}$ is a solution of the quadratic equation $3x^2 = 18$.

GROUP PROJECT

CHAPTER 10

Group members should consult an almanac, newspaper, magazine, or the Internet and return to the group with as much data as possible that show phenomena that are continuing to grow over time, but whose growth is leveling off. Select the five data sets that you find most intriguing. Let x represent the number of years after the first year in each data set. Model the data by hand using

$$f(x) = a\sqrt{x} + b.$$

Use the first and last data points to find values for a and b. The first data point corresponds to $x = 0$. Its second coordinate gives the value of b. To find a, substitute the second data point into $f(x) = a\sqrt{x} + b$, with the value that you obtained for b. Now solve the equation and find a. Substitute a and b into $f(x) = a\sqrt{x} + b$ to obtain a square root function that models each data set. Then use the function to make predictions about what might occur in the future. Are there circumstances that might affect the accuracy of the predictions? List some of these circumstances.

Chapter 10 Summary

Definitions and Concepts	Examples

Section 10.1 Radical Expressions and Functions

If $b^2 = a$, then b is a square root of a. The principal square root of a, designated $\sqrt{a}$, is the nonnegative number satisfying $b^2 = a$. The negative square root of a is written $-\sqrt{a}$. A square root of a negative number is not a real number.

A radical function in x is a function defined by an expression containing a root of x. The domain of a square root function is the set of real numbers for which the radicand is nonnegative.

Let $f(x) = \sqrt{6 - 2x}$.

$f(-15) = \sqrt{6 - 2(-15)} = \sqrt{6 + 30} = \sqrt{36} = 6$

$f(5) = \sqrt{6 - 2 \cdot 5} = \sqrt{6 - 10} = \sqrt{-4}$, not a real number

Domain of f: Set the radicand greater than or equal to zero.

$$6 - 2x \geq 0$$
$$-2x \geq -6$$
$$x \leq 3$$

Domain of $f = (-\infty, 3]$

Definitions and Concepts	**Examples**

Section 10.1 Radical Expressions and Functions (continued)

The cube root of a real number a is written $\sqrt[3]{a}$.

$$\sqrt[3]{a} = b \quad \text{means that} \quad b^3 = a.$$

The nth root of a real number a is written $\sqrt[n]{a}$. The number n is the index. Every real number has one root when n is odd. The odd nth root of a, $\sqrt[n]{a}$, is the number b for which $b^n = a$. Every positive real number has two real roots when n is even. An even root of a negative number is not a real number.

If n is even, then $\sqrt[n]{a^n} = |a|$.

If n is odd, then $\sqrt[n]{a^n} = a$.

- $\sqrt[3]{-8} = -2$ because $(-2)^3 = -8$.
- $\sqrt[4]{-16}$ is not a real number.
- $\sqrt{x^2 - 14x + 49} = \sqrt{(x-7)^2} = |x - 7|$
- $\sqrt[3]{125(x+6)^3} = 5(x+6)$

Section 10.2 Rational Exponents

- $a^{\frac{1}{n}} = \sqrt[n]{a}$

- $a^{\frac{m}{n}} = (\sqrt[n]{a})^m$ or $\sqrt[n]{a^m}$

- $a^{-\frac{m}{n}} = \dfrac{1}{a^{\frac{m}{n}}}$

- $121^{\frac{1}{2}} = \sqrt{121} = 11$
- $64^{\frac{1}{3}} = \sqrt[3]{64} = 4$
- $27^{\frac{5}{3}} = (\sqrt[3]{27})^5 = 3^5 = 3 \cdot 3 \cdot 3 \cdot 3 \cdot 3 = 243$
- $16^{-\frac{3}{4}} = \dfrac{1}{16^{\frac{3}{4}}} = \dfrac{1}{(\sqrt[4]{16})^3} = \dfrac{1}{2^3} = \dfrac{1}{8}$
- $(\sqrt[3]{7xy})^4 = (7xy)^{\frac{4}{3}}$

Properties of integer exponents are true for rational exponents. An expression with rational exponents is simplified when no parentheses appear, no powers are raised to powers, each base occurs once, and no negative or zero exponents appear.

Simplify: $\left(8x^{\frac{1}{3}}y^{-\frac{1}{2}}\right)^{\frac{1}{3}}$.

$$= 8^{\frac{1}{3}}\left(x^{\frac{1}{3}}\right)^{\frac{1}{3}}\left(y^{-\frac{1}{2}}\right)^{\frac{1}{3}}$$

$$= 2x^{\frac{1}{9}}y^{-\frac{1}{6}} = \frac{2x^{\frac{1}{9}}}{y^{\frac{1}{6}}}$$

Some radical expressions can be simplifed using rational exponents. Rewrite the expression using rational exponents, simplify, and rewrite in radical notation if rational exponents still appear.

- $\sqrt[9]{x^3} = x^{\frac{3}{9}} = x^{\frac{1}{3}} = \sqrt[3]{x}$
- $\sqrt[5]{x^2} \cdot \sqrt[4]{x} = x^{\frac{2}{5}} \cdot x^{\frac{1}{4}} = x^{\frac{2}{5}+\frac{1}{4}}$

$$= x^{\frac{8}{20}+\frac{5}{20}} = x^{\frac{13}{20}} = \sqrt[20]{x^{13}}$$

Section 10.3 Multiplying and Simplifying Radical Expressions

The product rule for radicals can be used to multiply radicals

$$\sqrt[n]{a} \cdot \sqrt[n]{b} = \sqrt[n]{ab}.$$

$\sqrt[3]{7x} \cdot \sqrt[3]{10y^2} = \sqrt[3]{7x \cdot 10y^2} = \sqrt[3]{70xy^2}$

The product rule for radicals can be used to simplify radicals:

$$\sqrt[n]{ab} = \sqrt[n]{a} \cdot \sqrt[n]{b}.$$

A radical expression with index n is simplified when its radicand has no factors that are perfect nth powers. To simplify, write the radicand as the product of two factors, one of which is the greatest perfect nth power. Then use the product rule to take the nth root of each factor. If all variables in a radicand are positive, then

$$\sqrt[n]{a^n} = a.$$

Some radicals can be simplified after multiplication is performed.

- Simplify: $\sqrt[3]{54x^7y^{11}}$.

$$= \sqrt[3]{27 \cdot 2 \cdot x^6 \cdot x \cdot y^9 \cdot y^2}$$

$$= \sqrt[3]{(27x^6y^9)(2xy^2)}$$

$$= \sqrt[3]{27x^6y^9} \cdot \sqrt[3]{2xy^2} = 3x^2y^3\sqrt[3]{2xy^2}$$

- Assuming positive variables, multiply and simplify:

$$\sqrt[4]{4x^2y} \cdot \sqrt[4]{4xy^3}.$$

$$= \sqrt[4]{4x^2y \cdot 4xy^3} = \sqrt[4]{16x^3y^4}$$

$$= \sqrt[4]{16y^4} \cdot \sqrt[4]{x^3} = 2y\sqrt[4]{x^3}$$

Definitions and Concepts	**Examples**

Section 10.4 Adding, Subtracting, and Dividing Radical Expressions

Like radicals have the same indices and radicands. Like radicals can be added or subtracted using the distributive property. In some cases, radicals can be combined once they have been simplified.

$4\sqrt{18} - 6\sqrt{50}$
$= 4\sqrt{9 \cdot 2} - 6\sqrt{25 \cdot 2} = 4 \cdot 3\sqrt{2} - 6 \cdot 5\sqrt{2}$
$= 12\sqrt{2} - 30\sqrt{2} = -18\sqrt{2}$

The quotient rule for radicals can be used to simplify radicals:
$$\sqrt[n]{\frac{a}{b}} = \frac{\sqrt[n]{a}}{\sqrt[n]{b}}.$$

$$\sqrt[3]{-\frac{8}{x^{12}}} = \frac{\sqrt[3]{-8}}{\sqrt[3]{x^{12}}} = -\frac{2}{x^4}$$

$$\sqrt[3]{x^{12}} = (x^{12})^{\frac{1}{3}} = x^4$$

The quotient rule for radicals can be used to divide radicals with the same indices:
$$\frac{\sqrt[n]{a}}{\sqrt[n]{b}} = \sqrt[n]{\frac{a}{b}}.$$
Some radicals can be simplified after the division is performed.

Assuming a positive variable, divide and simplify:
$$\frac{\sqrt[4]{64x^5}}{\sqrt[4]{2x^{-2}}} = \sqrt[4]{32x^{5-(-2)}} = \sqrt[4]{32x^7}$$
$$= \sqrt[4]{16 \cdot 2 \cdot x^4 \cdot x^3} = \sqrt[4]{16x^4} \cdot \sqrt[4]{2x^3}$$
$$= 2x\sqrt[4]{2x^3}.$$

Section 10.5 Multiplying with More Than One Term and Rationalizing Denominators

Radical expressions with more than one term are multiplied in much the same way that polynomials with more than one term are multiplied.

- $\sqrt{5}(2\sqrt{6} - \sqrt{3}) = 2\sqrt{30} - \sqrt{15}$

- $(4\sqrt{3} - 2\sqrt{2})(\sqrt{3} + \sqrt{2})$

     F   O   I   L

$= 4\sqrt{3} \cdot \sqrt{3} + 4\sqrt{3} \cdot \sqrt{2} - 2\sqrt{2} \cdot \sqrt{3} - 2\sqrt{2} \cdot \sqrt{2}$
$= 4 \cdot 3 + 4\sqrt{6} - 2\sqrt{6} - 2 \cdot 2$
$= 12 + 4\sqrt{6} - 2\sqrt{6} - 4 = 8 + 2\sqrt{6}$

Radical expressions that involve the sum and difference of the same two terms are called conjugates. Use
$$(A + B)(A - B) = A^2 - B^2$$
to multiply conjugates.

$(8 + 2\sqrt{5})(8 - 2\sqrt{5})$
$= 8^2 - (2\sqrt{5})^2$
$= 64 - 4 \cdot 5$
$= 64 - 20 = 44$

The process of rewriting a radical expression as an equivalent expression without any radicals in the denominator is called rationalizing the denominator. When the denominator contains a single radical with an nth root, multiply the numerator and the denominator by a radical of index n that produces a perfect nth power in the denominator's radicand.

Rationalize the denominator: $\dfrac{7}{\sqrt[3]{2x}}$.

$$= \frac{7}{\sqrt[3]{2x}} \cdot \frac{\sqrt[3]{4x^2}}{\sqrt[3]{4x^2}} = \frac{7\sqrt[3]{4x^2}}{\sqrt[3]{8x^3}} = \frac{7\sqrt[3]{4x^2}}{2x}$$

| **Definitions and Concepts** | **Examples** |

Section 10.5 Multiplying with More Than One Term and Rationalizing Denominators (continued)

If the denominator contains two terms, rationalize the denominator by multiplying the numerator and the denominator by the conjugate of the denominator.

$$\frac{13}{5 - \sqrt{3}} = \frac{13}{5 - \sqrt{3}} \cdot \frac{5 + \sqrt{3}}{5 + \sqrt{3}}$$

$$= \frac{13(5 + \sqrt{3})}{5^2 - (\sqrt{3})^2}$$

$$= \frac{13(5 + \sqrt{3})}{25 - 3} = \frac{13(5 + \sqrt{3})}{22}$$

Section 10.6 Radical Equations

A radical equation is an equation in which the variable occurs in a radicand.

Solving Radical Equations Containing *n*th Roots

1. Isolate one radical on one side of the equation.
2. Raise both sides to the *n*th power.
3. Solve the resulting equation.
4. Check proposed solutions in the original equation. Solutions of an equation to an even power that is radical-free, but not the original equation, are called extraneous solutions.

Solve: $\sqrt{6x + 13} - 2x = 1$.

$\sqrt{6x + 13} = 2x + 1$ Isolate the radical.

$(\sqrt{6x + 13})^2 = (2x + 1)^2$ Square both sides.

$6x + 13 = 4x^2 + 4x + 1$ $(A + B)^2 = A^2 + 2AB + B^2$

$0 = 4x^2 - 2x - 12$ Subtract $6x + 13$ from both sides.

$0 = 2(2x^2 - x - 6)$ Factor out the GCF.

$0 = 2(2x + 3)(x - 2)$ Factor completely.

$2x + 3 = 0$ or $x - 2 = 0$ Set variable factors equal to zero.

$2x = -3$ $x = 2$ Solve for x.

$x = -\frac{3}{2}$

Check both proposed solutions. 2 checks, but $-\frac{3}{2}$ is extraneous. The solution is 2 and the solution set is $\{2\}$.

Section 10.7 Complex Numbers

The imaginary unit i is defined as

$$i = \sqrt{-1}, \quad \text{where} \quad i^2 = -1.$$

The set of numbers in the form $a + bi$ is called the set of complex numbers; a is the real part and b is the imaginary part. If $b = 0$, the complex number is a real number. If $b \neq 0$, the complex number is an imaginary number.

- $\sqrt{-81} = \sqrt{81(-1)} = \sqrt{81}\sqrt{-1} = 9i$
- $\sqrt{-75} = \sqrt{75(-1)} = \sqrt{25 \cdot 3}\sqrt{-1} = 5i\sqrt{3}$

To add or subtract complex numbers, add or subtract their real parts and add or subtract their imaginary parts.

$(2 - 4i) - (7 - 10i)$

$= 2 - 4i - 7 + 10i$

$= (2 - 7) + (-4 + 10)i = -5 + 6i$

Definitions and Concepts	**Examples**

Section 10.7 Complex Numbers (continued)

To multiply complex numbers, multiply as if they were polynomials. After completing the multiplication, replace i^2 with -1. When performing operations with square roots of negative numbers, begin by expressing all square roots in terms of i. Then multiply.

$$\overset{F\quad O\quad I\quad L}{(2 - 3i)(4 + 5i)} = 8 + 10i - 12i - 15i^2$$
$$= 8 + 10i - 12i - 15(-1)$$
$$= 23 - 2i$$
$$\sqrt{-36} \cdot \sqrt{-100} = \sqrt{36(-1)} \cdot \sqrt{100(-1)}$$
$$= 6i \cdot 10i = 60i^2 = 60(-1) = -60$$

The complex numbers $a + bi$ and $a - bi$ are conjugates. Conjugates can be multiplied using the formula
$$(A + B)(A - B) = A^2 - B^2.$$
The multiplication of conjugates results in a real number.

$$(3 + 5i)(3 - 5i) = 3^2 - (5i)^2$$
$$= 9 - 25i^2$$
$$= 9 - 25(-1) = 34$$

To divide complex numbers, multiply the numerator and the denominator by the conjugate of the denominator in order to obtain a real number in the denominator. This real number becomes the denominator of a and b in the quotient $a + bi$.

$$\frac{5 + 2i}{4 - i} = \frac{5 + 2i}{4 - i} \cdot \frac{4 + i}{4 + i} = \frac{20 + 5i + 8i + 2i^2}{16 - i^2}$$
$$= \frac{20 + 13i + 2(-1)}{16 - (-1)}$$
$$= \frac{20 + 13i - 2}{16 + 1}$$
$$= \frac{18 + 13i}{17} = \frac{18}{17} + \frac{13}{17}i$$

To simplify powers of i, rewrite the expression in terms of i^2. Then replace i^2 with -1 and simplify.

Simplify: i^{27}.
$$i^{27} = i^{26} \cdot i = (i^2)^{13}i$$
$$= (-1)^{13}i = (-1)i = -i$$

CHAPTER 10 REVIEW EXERCISES

10.1 *In Exercises 1–5, find the indicated root, or state that the expression is not a real number.*

1. $\sqrt{81}$
2. $-\sqrt{\dfrac{1}{100}}$
3. $\sqrt[3]{-27}$
4. $\sqrt[4]{-16}$
5. $\sqrt[5]{-32}$

In Exercises 6–7, find the indicated function values for each function. If necessary, round to two decimal places. If the function value is not a real number and does not exist, so state.

6. $f(x) = \sqrt{2x - 5}; \quad f(15), f(4), f\left(\dfrac{5}{2}\right), f(1)$

7. $g(x) = \sqrt[3]{4x - 8}; \quad g(4), g(0), g(-14)$

In Exercises 8–9, find the domain of each square root function.

8. $f(x) = \sqrt{x - 2}$
9. $g(x) = \sqrt{100 - 4x}$

In Exercises 10–15, simplify each expression. Assume that each variable can represent any real number, so include absolute value bars where necessary.

10. $\sqrt{25x^2}$
11. $\sqrt{(x + 14)^2}$
12. $\sqrt{x^2 - 8x + 16}$
13. $\sqrt[3]{64x^3}$
14. $\sqrt[4]{16x^4}$
15. $\sqrt[5]{-32(x + 7)^5}$

10.2 *In Exercises 16–18, use radical notation to rewrite each expression. Simplify, if possible.*

16. $(5xy)^{\frac{1}{3}}$
17. $16^{\frac{3}{2}}$
18. $32^{\frac{4}{5}}$

In Exercises 19–20, rewrite each expression with rational exponents.

19. $\sqrt{7x}$
20. $\left(\sqrt[3]{19xy}\right)^5$

In Exercises 21–22, rewrite each expression with a positive rational exponent. Simplify, if possible.

21. $8^{-\frac{2}{3}}$

22. $3x(ab)^{-\frac{4}{5}}$

In Exercises 23–26, use properties of rational exponents to simplify each expression.

23. $x^{\frac{1}{3}} \cdot x^{\frac{1}{4}}$

24. $\dfrac{5^{\frac{1}{2}}}{5^{\frac{1}{3}}}$

25. $(8x^6 y^3)^{\frac{1}{3}}$

26. $\left(x^{-\frac{2}{3}} y^{\frac{1}{4}}\right)^{\frac{1}{2}}$

In Exercises 27–31, use rational exponents to simplify each expression. If rational exponents appear after simplifying, write the answer in radical notation.

27. $\sqrt[3]{x^9 y^{12}}$

28. $\sqrt[9]{x^3 y^9}$

29. $\sqrt{x} \cdot \sqrt[3]{x}$

30. $\dfrac{\sqrt[3]{x^2}}{\sqrt[4]{x^2}}$

31. $\sqrt[5]{\sqrt[3]{x}}$

32. The function $f(x) = 350x^{\frac{2}{3}}$ models the expenditures, $f(x)$, in millions of dollars, for the U.S. National Park Service x years after 1985. According to this model, what were expenditures in 2012?

10.3 *In Exercises 33–35, use the product rule to multiply.*

33. $\sqrt{3x} \cdot \sqrt{7y}$

34. $\sqrt[5]{7x^2} \cdot \sqrt[5]{11x}$

35. $\sqrt[6]{x-5} \cdot \sqrt[6]{(x-5)^4}$

36. If $f(x) = \sqrt{7x^2 - 14x + 7}$, express the function, f, in simplified form. Assume that x can be any real number.

In Exercises 37–39, simplify by factoring. Assume that all variables in a radicand represent positive real numbers.

37. $\sqrt{20x^3}$

38. $\sqrt[3]{54x^8 y^6}$

39. $\sqrt[4]{32x^3 y^{11} z^5}$

In Exercises 40–43, multiply and simplify, if possible. Assume that all variables in a radicand represent positive real numbers.

40. $\sqrt{6x^3} \cdot \sqrt{4x^2}$

41. $\sqrt[3]{4x^2 y} \cdot \sqrt[3]{4xy^4}$

42. $\sqrt[5]{2x^4 y^3 z^4} \cdot \sqrt[5]{8xy^6 z^7}$

43. $\sqrt{x+1} \cdot \sqrt{x-1}$

10.4 *Assume that all variables represent positive real numbers. In Exercises 44–47, add or subtract as indicated.*

44. $6\sqrt[3]{3} + 2\sqrt[3]{3}$

45. $5\sqrt{18} - 3\sqrt{8}$

46. $\sqrt[3]{27x^4} + \sqrt[3]{xy^6}$

47. $2\sqrt[3]{6} - 5\sqrt[3]{48}$

In Exercises 48–50, simplify using the quotient rule.

48. $\sqrt[3]{\dfrac{16}{125}}$

49. $\sqrt{\dfrac{x^3}{100y^4}}$

50. $\sqrt[4]{\dfrac{3y^5}{16x^{20}}}$

In Exercises 51–54, divide and, if possible, simplify.

51. $\dfrac{\sqrt{48}}{\sqrt{2}}$

52. $\dfrac{\sqrt[3]{32}}{\sqrt[3]{2}}$

53. $\dfrac{\sqrt[4]{64x^7}}{\sqrt[4]{2x^2}}$

54. $\dfrac{\sqrt{200x^3 y^2}}{\sqrt{2x^{-2}y}}$

10.5 *Assume that all variables represent positive real numbers.*

In Exercises 55–62, multiply as indicated. If possible, simplify any radical expressions that appear in the product.

55. $\sqrt{3}(2\sqrt{6} + 4\sqrt{15})$

56. $\sqrt[3]{5}(\sqrt[3]{50} - \sqrt[3]{2})$

57. $(\sqrt{7} - 3\sqrt{5})(\sqrt{7} + 6\sqrt{5})$

58. $(\sqrt{x} - \sqrt{11})(\sqrt{y} - \sqrt{11})$

59. $(\sqrt{5} + \sqrt{8})^2$

60. $(2\sqrt{3} - \sqrt{10})^2$

61. $(\sqrt{7} + \sqrt{13})(\sqrt{7} - \sqrt{13})$

62. $(7 - 3\sqrt{5})(7 + 3\sqrt{5})$

In Exercises 63–75, rationalize each denominator. Simplify, if possible.

63. $\dfrac{4}{\sqrt{6}}$

64. $\sqrt{\dfrac{2}{7}}$

65. $\dfrac{12}{\sqrt[3]{9}}$

66. $\sqrt{\dfrac{2x}{5y}}$

67. $\dfrac{14}{\sqrt[3]{2x^2}}$

68. $\sqrt[4]{\dfrac{7}{3x}}$

69. $\dfrac{5}{\sqrt[5]{32x^4 y}}$

70. $\dfrac{6}{\sqrt{3} - 1}$

71. $\dfrac{\sqrt{7}}{\sqrt{5} + \sqrt{3}}$

72. $\dfrac{10}{2\sqrt{5} - 3\sqrt{2}}$

73. $\dfrac{\sqrt{x} + 5}{\sqrt{x} - 3}$

74. $\dfrac{\sqrt{7} + \sqrt{3}}{\sqrt{7} - \sqrt{3}}$

75. $\dfrac{2\sqrt{3} + \sqrt{6}}{2\sqrt{6} + \sqrt{3}}$

In Exercises 76–79, rationalize each numerator. Simplify, if possible.

76. $\sqrt{\dfrac{2}{7}}$

77. $\dfrac{\sqrt[3]{3x}}{\sqrt[3]{y}}$

78. $\dfrac{\sqrt{7}}{\sqrt{5} + \sqrt{3}}$

79. $\dfrac{\sqrt{7} + \sqrt{3}}{\sqrt{7} - \sqrt{3}}$

10.6 *In Exercises 80–84, solve each radical equation.*

80. $\sqrt{2x + 4} = 6$

81. $\sqrt{x - 5} + 9 = 4$

82. $\sqrt{2x - 3} + x = 3$

83. $\sqrt{x - 4} + \sqrt{x + 1} = 5$

84. $(x^2 + 6x)^{\frac{1}{3}} + 2 = 0$

85. According to the Centers for Disease Control and Prevention, the average American adult is 23 pounds overweight and collectively we are 4.6 billion pounds overweight. The graph shows the percentage of obese American adults for selected years from 1980 through 2010.

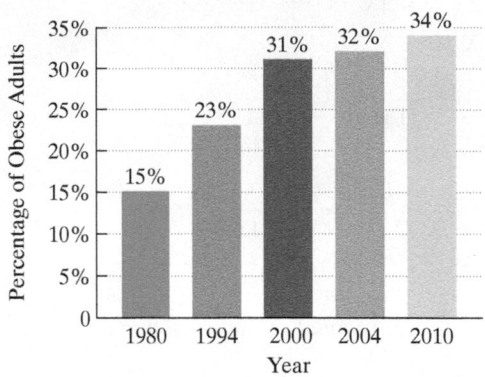

Percentage of Obese Adults in the United States

Source: Centers for Disease Control and Prevention

The function

$$f(x) = 3.5\sqrt{x} + 15$$

models the percentage of obese American adults, $f(x)$, x years after 1980.

a. Find and interpret $f(30)$. Round to the nearest percent. How does this rounded value compare with the percentage of obese adults displayed by the graph?

b. According to the model, in which year will 36% of adults be obese?

86. Out of a group of 50,000 births, the number of people, $f(x)$, surviving to age x is modeled by the function

$$f(x) = 5000\sqrt{100 - x}.$$

To what age will 20,000 people in the group survive?

10.7 *In Exercises 87–89, express each number in terms of i and simplify, if possible.*

87. $\sqrt{-81}$

88. $\sqrt{-63}$

89. $-\sqrt{-8}$

In Exercises 90–99, perform the indicated operation. Write the result in the form a + bi.

90. $(7 + 12i) + (5 - 10i)$

91. $(8 - 3i) - (17 - 7i)$

92. $4i(3i - 2)$

93. $(7 - 5i)(2 + 3i)$

94. $(3 - 4i)^2$

95. $(7 + 8i)(7 - 8i)$

96. $\sqrt{-8} \cdot \sqrt{-3}$

97. $\dfrac{6}{5 + i}$

98. $\dfrac{3 + 4i}{4 - 2i}$

99. $\dfrac{5 + i}{3i}$

In Exercises 100–101, simplify each expression.

100. i^{16}

101. i^{23}

CHAPTER 10 TEST

CHAPTER Test Prep VIDEOS

Step-by-step test solutions are found on the Chapter Test Prep Videos available in MyMathLab® or on YouTube (search "BlitzerCombinedAlg" and click on "Channels").

1. Let $f(x) = \sqrt{8 - 2x}$.

    a. Find $f(-14)$.

    b. Find the domain of f.

2. Evaluate: $27^{-\frac{4}{3}}$.

3. Simplify: $\left(25x^{-\frac{1}{2}}y^{\frac{1}{4}}\right)^{\frac{1}{2}}$.

In Exercises 4–5, use rational exponents to simplify each expression. If rational exponents appear after simplifying, write the answer in radical notation.

4. $\sqrt[8]{x^4}$

5. $\sqrt[4]{x} \cdot \sqrt[5]{x}$

In Exercises 6–9, simplify each expression. Assume that each variable can represent any real number.

6. $\sqrt{75x^2}$

7. $\sqrt{x^2 - 10x + 25}$

8. $\sqrt[3]{16x^4y^8}$

9. $\sqrt[5]{-\dfrac{32}{x^{10}}}$

In Exercises 10–17, perform the indicated operation and, if possible, simplify. Assume that all variables represent positive real numbers.

10. $\sqrt[3]{5x^2} \cdot \sqrt[3]{10y}$

11. $\sqrt[4]{8x^3y} \cdot \sqrt[4]{4xy^2}$

12. $3\sqrt{18} - 4\sqrt{32}$

13. $\sqrt[3]{8x^4} + \sqrt[3]{xy^6}$

14. $\dfrac{\sqrt[3]{16x^8}}{\sqrt[3]{2x^4}}$

15. $\sqrt{3}\left(4\sqrt{6} - \sqrt{5}\right)$

16. $\left(5\sqrt{6} - 2\sqrt{2}\right)\left(\sqrt{6} + \sqrt{2}\right)$

17. $\left(7 - \sqrt{3}\right)^2$

In Exercises 18–20, rationalize each denominator. Simplify, if possible. Assume all variables represent positive real numbers.

18. $\sqrt{\dfrac{5}{x}}$

19. $\dfrac{5}{\sqrt[3]{5x^2}}$

20. $\dfrac{\sqrt{2} - \sqrt{3}}{\sqrt{2} + \sqrt{3}}$

In Exercises 21–23, solve each radical equation.

21. $3 + \sqrt{2x - 3} = x$
22. $\sqrt{x + 9} - \sqrt{x - 7} = 2$
23. $(11x + 6)^{\frac{1}{3}} + 3 = 0$
24. The function
$$f(x) = 2.9\sqrt{x} + 20.1$$
models the average height, $f(x)$, in inches, of boys who are x months of age, $0 \le x \le 60$. Find the age at which the average height of boys is 40.4 inches.
25. Express in terms of i and simplify: $\sqrt{-75}$.

In Exercises 26–29, perform the indicated operation. Write the result in the form $a + bi$.

26. $(5 - 3i) - (6 - 9i)$
27. $(3 - 4i)(2 + 5i)$
28. $\sqrt{-9} \cdot \sqrt{-4}$
29. $\dfrac{3 + i}{1 - 2i}$
30. Simplify: i^{35}.

CUMULATIVE REVIEW EXERCISES (CHAPTERS 1–10)

In Exercises 1–5, solve each equation, inequality, or system.

1. $\begin{cases} 2x - y + z = -5 \\ x - 2y - 3z = 6 \\ x + y - 2z = 1 \end{cases}$
2. $3x^2 - 11x = 4$
3. $2(x + 4) < 5x + 3(x + 2)$
4. $\dfrac{1}{x + 2} + \dfrac{15}{x^2 - 4} = \dfrac{5}{x - 2}$
5. $\sqrt{x + 2} - \sqrt{x + 1} = 1$
6. Graph the solution set of the system:
$$\begin{cases} x + 2y < 2 \\ 2y - x > 4. \end{cases}$$

In Exercises 7–15, perform the indicated operations.

7. $\dfrac{8x^2}{3x^2 - 12} \div \dfrac{40}{x - 2}$
8. $\dfrac{x + \dfrac{1}{y}}{y + \dfrac{1}{x}}$
9. $(2x - 3)(4x^2 - 5x - 2)$
10. $\dfrac{7x}{x^2 - 2x - 15} - \dfrac{2}{x - 5}$

11. $7(8 - 10)^3 - 7 + 3 \div (-3)$
12. $\sqrt{80x} - 5\sqrt{20x} + 2\sqrt{45x}$
13. $\dfrac{\sqrt{3} - 2}{2\sqrt{3} + 5}$
14. $(2x^3 - 3x^2 + 3x - 4) \div (x - 2)$
15. $(2\sqrt{3} + 5\sqrt{2})(\sqrt{3} - 4\sqrt{2})$

In Exercises 16–17, factor completely.

16. $24x^2 + 10x - 4$
17. $16x^4 - 1$
18. The amount of light provided by a light bulb varies inversely as the square of the distance from the bulb. The illumination provided is 120 lumens at a distance of 10 feet. How many lumens are provided at a distance of 15 feet?
19. You invested $6000 in two accounts paying 7% and 9% annual interest. At the end of the year, the total interest from these investments was $510. How much was invested at each rate?
20. Although there are 2332 students enrolled in the college, this is 12% fewer students than there were enrolled last year. How many students were enrolled last year?

Quadratic Equations and Functions

$$W_{e \quad are}$$

surrounded by evidence that the world is profoundly mathematical. After turning a somersault, a diver's path can be modeled by a quadratic function, $f(x) = ax^2 + bx + c$, as can the path of a football tossed from quarterback to receiver or the path of a flipped coin. Even if you throw an object directly upward, although its path is straight and vertical, its changing height over time is described by a quadratic function. And tailgaters beware: whether you're driving a car, a motorcycle, or a truck on dry or wet roads, an array of quadratic functions that model your required stopping distances at various speeds are available to help you become a safer driver.

The quadratic functions surrounding our long history of throwing things appear throughout the chapter, including Example 6 in Section 11.3 and Example 6 in Section 11.5. Tailgaters should pay close attention to the Section 11.5 opener, Exercises 73–74 and 88–89 in Exercise Set 11.5, and Exercises 34–35 in the Chapter Review Exercises.

The Square Root Property and Completing the Square; Distance and Midpoint Formulas

Objectives

1 Solve quadratic equations using the square root property.

2 Complete the square of a binomial.

3 Solve quadratic equations by completing the square.

4 Solve problems using the square root property.

5 Find the distance between two points.

6 Find the midpoint of a line segment.

I'm very well acquainted, too, with matters mathematical,
I understand equations, both simple and quadratical.
About binomial theorem I'm teeming with a lot of news,
With many cheerful facts about the square of the hypotenuse.

—Gilbert and Sullivan,
The Pirates of Penzance

Equations quadratical? Cheerful news about the square of the hypotenuse? You've come to the right place. In this section, you'll enhance your understanding of quadratic equations with two new solution methods, called the *square root method* and *completing the square*. Using these techniques, we introduce (cheerfully, of course) the Pythagorean Theorem and the square of the hypotenuse.

Great Question!

Haven't we already discussed quadratic equations and quadratic functions? Other than not calling them equations and functions quadratical, what am I expected to know about these topics in order to achieve success in this chapter?

Here is a summary of what you should already know about quadratic equations and quadratic functions.

1. A **quadratic equation** in x can be written in the standard form

$$ax^2 + bx + c = 0, \quad a \neq 0.$$

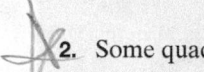2. Some quadratic equations can be solved by factoring.

    Solve: $2x^2 + 7x - 4 = 0.$

$$(2x - 1)(x + 4) = 0$$

$$2x - 1 = 0 \quad \text{or} \quad x + 4 = 0$$

$$2x = 1 \qquad\qquad x = -4$$

$$x = \tfrac{1}{2}$$

The solutions are -4 and $\tfrac{1}{2}$, and the solution set is $\left\{-4, \tfrac{1}{2}\right\}$.

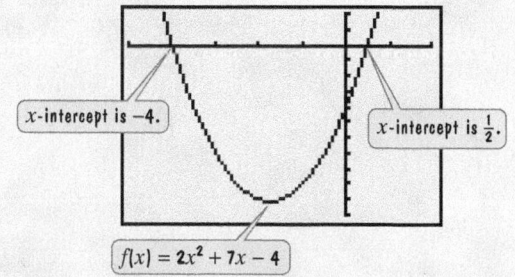

Figure 11.1

3. A polynomial function of the form

$$f(x) = ax^2 + bx + c, \quad a \neq 0$$

is a **quadratic function**. Graphs of quadratic functions are shaped like bowls or inverted bowls, with the same behavior at each end.

4. The real solutions of $ax^2 + bx + c = 0$ correspond to the x-intercepts for the graph of the quadratic function $f(x) = ax^2 + bx + c$. For example, the solutions of the equation $2x^2 + 7x - 4 = 0$ are -4 and $\tfrac{1}{2}$. **Figure 11.1** shows that the solutions appear as x-intercepts on the graph of the quadratic function $f(x) = 2x^2 + 7x - 4$.

Now that we've summarized what we know, let's look at where we go. How do we solve a quadratic equation, $ax^2 + bx + c = 0$, if the trinomial $ax^2 + bx + c$ cannot be factored? Methods other than factoring are needed. In this section and the next, we look at other ways of solving quadratic equations.

① **Solve quadratic equations using the square root property.**

The Square Root Property

Let's begin with a relatively simple quadratic equation:

$$x^2 = 9.$$

The value of x must be a number whose square is 9. There are two numbers whose square is 9:

$$x = \sqrt{9} = 3 \quad \text{or} \quad x = -\sqrt{9} = -3.$$

Thus, the solutions of $x^2 = 9$ are 3 and -3. This is an example of the **square root property**.

Discover for Yourself

Solve $x^2 = 9$, or

$$x^2 - 9 = 0,$$

by factoring. What is the advantage of using the square root property?

The Square Root Property

If u is an algebraic expression and d is a nonzero real number, then $u^2 = d$ is equivalent to $u = \sqrt{d}$ or $u = -\sqrt{d}$:

$$\text{If } u^2 = d, \quad \text{then} \quad u = \sqrt{d} \text{ or } u = -\sqrt{d}.$$

Equivalently,

$$\text{If } u^2 = d, \quad \text{then} \quad u = \pm\sqrt{d}.$$

Notice that $u = \pm\sqrt{d}$ is a shorthand notation to indicate that $u = \sqrt{d}$ or $u = -\sqrt{d}$. Although we usually read $u = \pm\sqrt{d}$ as "u equals plus or minus the square root of d," we actually mean that u is the positive square root of d or the negative square root of d.

EXAMPLE 1 Solving a Quadratic Equation by the Square Root Property

Solve: $3x^2 = 18$.

Solution To apply the square root property, we need a squared expression by itself on one side of the equation.

$$3x^2 = 18$$

We want x^2 by itself.

We can get x^2 by itself if we divide both sides by 3.

$$
\begin{array}{ll}
3x^2 = 18 & \text{This is the original equation.} \\[4pt]
\dfrac{3x^2}{3} = \dfrac{18}{3} & \text{Divide both sides by 3.} \\[4pt]
x^2 = 6 & \text{Simplify.} \\[4pt]
x = \sqrt{6} \quad \text{or} \quad x = -\sqrt{6} & \text{Apply the square root property.}
\end{array}
$$

Now let's check these proposed solutions in the original equation. Because the equation has an x^2-term and no x-term, we can check both values, $\pm\sqrt{6}$, at once.

Check $\sqrt{6}$ **and** $-\sqrt{6}$**:**

$$3x^2 = 18 \qquad \text{This is the original equation.}$$
$$3(\pm\sqrt{6})^2 \stackrel{?}{=} 18 \qquad \text{Substitute the proposed solutions.}$$
$$3 \cdot 6 \stackrel{?}{=} 18 \qquad (\pm\sqrt{6})^2 = 6$$
$$18 = 18, \qquad \text{true}$$

The solutions are $-\sqrt{6}$ and $\sqrt{6}$. The solution set is $\{-\sqrt{6}, \sqrt{6}\}$ or $\{\pm\sqrt{6}\}$. ∎

✓ **CHECK POINT 1**   Solve: $4x^2 = 28$.

In this section, we will express irrational solutions in simplified radical form, rationalizing denominators when possible.

EXAMPLE 2   Solving a Quadratic Equation
by the Square Root Property

Solve: $2x^2 - 7 = 0$.

Solution   To solve by the square root property, we isolate the squared expression on one side of the equation.

$$2x^2 - 7 = 0$$

We want x^2 by itself.

$$2x^2 - 7 = 0 \qquad \text{This is the original equation.}$$
$$2x^2 = 7 \qquad \text{Add 7 to both sides.}$$
$$x^2 = \frac{7}{2} \qquad \text{Divide both sides by 2.}$$
$$x = \sqrt{\frac{7}{2}} \quad \text{or} \quad x = -\sqrt{\frac{7}{2}} \qquad \text{Apply the square root property.}$$

Because the proposed solutions are opposites, we can rationalize both denominators at once:

$$\pm\sqrt{\frac{7}{2}} = \pm\frac{\sqrt{7}}{\sqrt{2}} \cdot \frac{\sqrt{2}}{\sqrt{2}} = \pm\frac{\sqrt{14}}{2}.$$

Substitute these values into the original equation and verify that the solutions are $-\frac{\sqrt{14}}{2}$ and $\frac{\sqrt{14}}{2}$. The solution set is $\left\{-\frac{\sqrt{14}}{2}, \frac{\sqrt{14}}{2}\right\}$ or $\left\{\pm\frac{\sqrt{14}}{2}\right\}$. ∎

✓ **CHECK POINT 2**   Solve: $3x^2 - 11 = 0$.

Some quadratic equations have solutions that are imaginary numbers.

EXAMPLE 3   Solving a Quadratic Equation
by the Square Root Property

Solve: $9x^2 + 25 = 0$.

Solution   We begin by isolating the squared expression on one side of the equation.

$$9x^2 + 25 = 0$$

We need to isolate x^2.

$$9x^2 + 25 = 0$$ This is the original equation.

$$9x^2 = -25$$ Subtract 25 from both sides.

$$x^2 = -\frac{25}{9}$$ Divide both sides by 9.

$$x = \sqrt{-\frac{25}{9}} \quad \text{or} \quad x = -\sqrt{-\frac{25}{9}}$$ Apply the square root property.

$$x = \sqrt{\frac{25}{9}}\sqrt{-1} \qquad x = -\sqrt{\frac{25}{9}}\sqrt{-1}$$

$$x = \frac{5}{3}i \qquad\qquad x = -\frac{5}{3}i \qquad\qquad \sqrt{-1} = i$$

Using Technology

Graphic Connections

The graph of

$$f(x) = 9x^2 + 25$$

has no x-intercepts. This shows that

$$9x^2 + 25 = 0$$

has no real solutions. Example 3 algebraically establishes that the solutions are imaginary numbers.

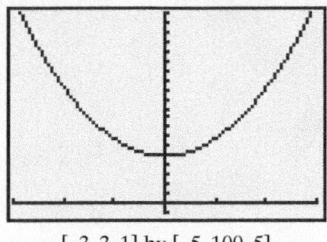

[-3, 3, 1] by [-5, 100, 5]

Because the equation has an x^2-term and no x-term, we can check both proposed solutions, $\pm\frac{5}{3}i$, at once.

Check $\dfrac{5}{3}i$ and $-\dfrac{5}{3}i$:

$$9x^2 + 25 = 0 \quad \text{This is the original equation.}$$

$$9\left(\pm\frac{5}{3}i\right)^2 + 25 \stackrel{?}{=} 0 \quad \text{Substitute the proposed solutions.}$$

$$9\left(\frac{25}{9}i^2\right) + 25 \stackrel{?}{=} 0 \quad \left(\pm\frac{5}{3}i\right)^2 = \left(\pm\frac{5}{3}\right)^2 i^2 = \frac{25}{9}i^2$$

$$25i^2 + 25 \stackrel{?}{=} 0 \quad 9\cdot\frac{25}{9} = 25$$

$$\boxed{i^2 = -1}$$

$$25(-1) + 25 \stackrel{?}{=} 0 \quad \text{Replace } i^2 \text{ with } -1.$$

$$0 = 0, \quad \text{true}$$

The solutions are $-\dfrac{5}{3}i$ and $\dfrac{5}{3}i$. The solution set is $\left\{-\dfrac{5}{3}i, \dfrac{5}{3}i\right\}$ or $\left\{\pm\dfrac{5}{3}i\right\}$. ■

✓ **CHECK POINT 3**   Solve: $4x^2 + 9 = 0$.

Can we solve an equation such as $(x - 1)^2 = 5$ using the square root property? Yes. The equation is in the form $u^2 = d$, where u^2, the squared expression, is by itself on the left side.

$$(x - 1)^2 \qquad = \qquad 5$$

This is u^2 in $u^2 = d$ with $u = x - 1$.   This is d in $u^2 = d$ with $d = 5$.

Discover for Yourself

Try solving

$$(x - 1)^2 = 5$$

by writing the equation in standard form and factoring. What problem do you encounter?

EXAMPLE 4   Solving a Quadratic Equation by the Square Root Property

Solve by the square root property: $(x - 1)^2 = 5$.

Solution

$$(x - 1)^2 = 5 \qquad \text{This is the original equation.}$$

$$x - 1 = \sqrt{5} \quad \text{or} \quad x - 1 = -\sqrt{5} \qquad \text{Apply the square root property.}$$

$$x = 1 + \sqrt{5} \qquad\qquad x = 1 - \sqrt{5} \qquad \text{Add 1 to both sides in each equation.}$$

Check $1 + \sqrt{5}$:	**Check $1 - \sqrt{5}$:**
$(x-1)^2 = 5$	$(x-1)^2 = 5$
$\left(1 + \sqrt{5} - 1\right)^2 \overset{?}{=} 5$	$\left(1 - \sqrt{5} - 1\right)^2 \overset{?}{=} 5$
$\left(\sqrt{5}\right)^2 \overset{?}{=} 5$	$\left(-\sqrt{5}\right)^2 \overset{?}{=} 5$
$5 = 5$, true	$5 = 5$, true

The solutions are $1 \pm \sqrt{5}$, and the solution set is $\left\{1 + \sqrt{5}, 1 - \sqrt{5}\right\}$ or $\left\{1 \pm \sqrt{5}\right\}$. ∎

✓ **CHECK POINT 4** Solve: $(x-3)^2 = 10$.

2 Complete the square of a binomial.

Completing the Square

We return to the question that opened this section: How do we solve a quadratic equation, $ax^2 + bx + c = 0$, if the trinomial $ax^2 + bx + c$ cannot be factored? We can convert the equation into an equivalent equation that can be solved using the square root property. This is accomplished by **completing the square**.

Completing the Square

If $x^2 + bx$ is a binomial, then by adding $\left(\dfrac{b}{2}\right)^2$, which is the square of half the coefficient of x, a perfect square trinomial will result.

The coefficient of x^2 must be 1 to complete the square.

$$x^2 + bx + \left(\frac{b}{2}\right)^2 = \left(x + \frac{b}{2}\right)^2$$

EXAMPLE 5 Completing the Square

What term should be added to each binomial so that it becomes a perfect square trinomial? Write and factor the trinomial.

a. $x^2 + 8x$ **b.** $x^2 - 7x$ **c.** $x^2 + \dfrac{3}{5}x$

Solution To complete the square, we must add a term to each binomial. The term that should be added is the square of half the coefficient of x.

$$x^2 + 8x \qquad x^2 - 7x \qquad x^2 + \frac{3}{5}x$$

Add $\left(\frac{8}{2}\right)^2 = 4^2$. Add 16 to complete the square.

Add $\left(\frac{-7}{2}\right)^2$, or $\frac{49}{4}$, to complete the square.

Add $\left(\frac{1}{2} \cdot \frac{3}{5}\right)^2 = \left(\frac{3}{10}\right)^2$. Add $\frac{9}{100}$ to complete the square.

a. The coefficient of the x-term in $x^2 + 8x$ is 8. Half of 8 is 4, and $4^2 = 16$. Add 16. The result is a perfect square trinomial.

$$x^2 + 8x + 16 = (x+4)^2$$

b. The coefficient of the x-term in $x^2 - 7x$ is -7. Half of -7 is $-\dfrac{7}{2}$, and $\left(-\dfrac{7}{2}\right)^2 = \dfrac{49}{4}$. Add $\dfrac{49}{4}$. The result is a perfect square trinomial.

$$x^2 - 7x + \frac{49}{4} = \left(x - \frac{7}{2}\right)^2$$

c. The coefficient of the x-term in $x^2 + \frac{3}{5}x$ is $\frac{3}{5}$. Half of $\frac{3}{5}$ is $\frac{1}{2} \cdot \frac{3}{5}$, or $\frac{3}{10}$, and $\left(\frac{3}{10}\right)^2 = \frac{9}{100}$. Add $\frac{9}{100}$. The result is a perfect square trinomial.

$$x^2 + \frac{3}{5}x + \frac{9}{100} = \left(x + \frac{3}{10}\right)^2 \quad \blacksquare$$

Great Question!

I'm not accustomed to factoring perfect square trinomials in which fractions are involved. Is there a rule or an observation that can make the factoring easier?

Yes. The constant in the factorization is always half the coefficient of x.

$$x^2 - 7x + \frac{49}{4} = \left(x - \frac{7}{2}\right)^2 \qquad x^2 + \frac{3}{5}x + \frac{9}{100} = \left(x + \frac{3}{10}\right)^2$$

Half the coefficient of x, -7, is $-\frac{7}{2}$. | Half the coefficient of x, $\frac{3}{5}$, is $\frac{3}{10}$.

☑ **CHECK POINT 5** What term should be added to each binomial so that it becomes a perfect square trinomial? Write and factor the trinomial.

a. $x^2 + 10x$

b. $x^2 - 3x$

c. $x^2 + \frac{3}{4}x$

---

3 Solve quadratic equations by completing the square.

Solving Quadratic Equations by Completing the Square

We can solve *any* quadratic equation by completing the square. If the coefficient of the x^2-term is one, we add the square of half the coefficient of x to both sides of the equation. **When you add a constant term to one side of the equation to complete the square, be certain to add the same constant to the other side of the equation.** These ideas are illustrated in Example 6.

Great Question!

When I solve a quadratic equation by completing the square, doesn't this result in a new equation? How do I know that the solutions of this new equation are the same as those of the given equation?

When you complete the square for the binomial expression $x^2 + bx$, you obtain a different polynomial. When you solve a quadratic equation by completing the square, you obtain an equation with the same solution set because the constant needed to complete the square is added to *both sides.*

EXAMPLE 6 Solving a Quadratic Equation by Completing the Square

Solve by completing the square: $x^2 - 6x + 4 = 0$.

Solution We begin by subtracting 4 from both sides. This is done to isolate the binomial $x^2 - 6x$ so that we can complete the square.

$$x^2 - 6x + 4 = 0 \qquad \text{This is the original equation.}$$
$$x^2 - 6x = -4 \qquad \text{Subtract 4 from both sides.}$$

Next, we work with $x^2 - 6x = -4$ and complete the square. Find half the coefficient of the x-term and square it. The coefficient of the x-term is -6. Half of -6 is -3 and $(-3)^2 = 9$. Thus, we add 9 to both sides of the equation.

$$x^2 - 6x + 9 = -4 + 9 \qquad \text{Add 9 to both sides to complete the square.}$$
$$(x - 3)^2 = 5 \qquad \text{Factor and simplify.}$$
$$x - 3 = \sqrt{5} \quad \text{or} \quad x - 3 = -\sqrt{5} \qquad \text{Apply the square root property.}$$
$$x = 3 + \sqrt{5} \qquad x = 3 - \sqrt{5} \qquad \text{Add 3 to both sides in each equation.}$$

The solutions are $3 \pm \sqrt{5}$, and the solution set is $\{3 + \sqrt{5}, 3 - \sqrt{5}\}$ or $\{3 \pm \sqrt{5}\}$. $\blacksquare$

If you solve a quadratic equation by completing the square and the solutions are rational numbers, the equation can also be solved by factoring. By contrast, quadratic equations with irrational solutions cannot be solved by factoring. However, all quadratic equations can be solved by completing the square.

✓ **CHECK POINT 6**   Solve by completing the square: $x^2 + 4x - 1 = 0$.

We have seen that the leading coefficient must be 1 in order to complete the square. If the coefficient of the x^2-term in a quadratic equation is not 1, you must divide each side of the equation by this coefficient before completing the square. For example, to solve $9x^2 - 6x - 4 = 0$ by completing the square, first divide every term by 9:

$$\frac{9x^2}{9} - \frac{6x}{9} - \frac{4}{9} = \frac{0}{9}$$

$$x^2 - \frac{6}{9}x - \frac{4}{9} = 0$$

$$x^2 - \frac{2}{3}x - \frac{4}{9} = 0.$$

Now that the coefficient of the x^2-term is 1, we can solve by completing the square.

EXAMPLE 7   Solving a Quadratic Equation by Completing the Square

Solve by completing the square: $9x^2 - 6x - 4 = 0$.

Solution

$9x^2 - 6x - 4 = 0$	This is the original equation.
$x^2 - \dfrac{2}{3}x - \dfrac{4}{9} = 0$	Divide both sides by 9.
$x^2 - \dfrac{2}{3}x = \dfrac{4}{9}$	Add $\frac{4}{9}$ to both sides to isolate the binomial.
$x^2 - \dfrac{2}{3}x + \dfrac{1}{9} = \dfrac{4}{9} + \dfrac{1}{9}$	Complete the square: Half of $-\frac{2}{3}$ is $-\frac{2}{6}$, or $-\frac{1}{3}$, and $\left(-\frac{1}{3}\right)^2 = \frac{1}{9}$.
$\left(x - \dfrac{1}{3}\right)^2 = \dfrac{5}{9}$	Factor and simplify.
$x - \dfrac{1}{3} = \sqrt{\dfrac{5}{9}}$ or $x - \dfrac{1}{3} = -\sqrt{\dfrac{5}{9}}$	Apply the square root property.
$x - \dfrac{1}{3} = \dfrac{\sqrt{5}}{3}$ $x - \dfrac{1}{3} = -\dfrac{\sqrt{5}}{3}$	$\sqrt{\dfrac{5}{9}} = \dfrac{\sqrt{5}}{\sqrt{9}} = \dfrac{\sqrt{5}}{3}$
$x = \dfrac{1}{3} + \dfrac{\sqrt{5}}{3}$ $x = \dfrac{1}{3} - \dfrac{\sqrt{5}}{3}$	Add $\frac{1}{3}$ to both sides and solve for x.
$x = \dfrac{1 + \sqrt{5}}{3}$ $x = \dfrac{1 - \sqrt{5}}{3}$	Express solutions with a common denominator.

The solutions are $\dfrac{1 \pm \sqrt{5}}{3}$ and the solution set is $\left\{\dfrac{1 \pm \sqrt{5}}{3}\right\}$. ■

✓ **CHECK POINT 7**   Solve by completing the square: $2x^2 + 3x - 4 = 0$.

Using Technology

Graphic Connections

Obtain a decimal approximation for each solution of

$$9x^2 - 6x - 4 = 0,$$

the equation in Example 7.

$$\frac{1 + \sqrt{5}}{3} \approx 1.1$$

$$\frac{1 - \sqrt{5}}{3} \approx -0.4$$

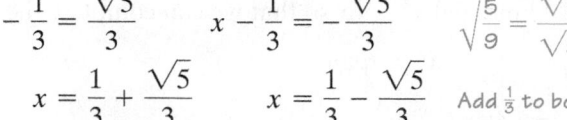

$f(x) = 9x^2 - 6x - 4$

x-intercept ≈ −0.4   x-intercept ≈ 1.1

[−2, 2, 1] by [−10, 10, 1]

The x-intercepts of $f(x) = 9x^2 - 6x - 4$ verify the solutions.

EXAMPLE 8 Solving a Quadratic Equation by Completing the Square

Solve by completing the square: $2x^2 - x + 6 = 0$.

Solution

$$2x^2 - x + 6 = 0$$ This is the original equation.

$$x^2 - \frac{1}{2}x + 3 = 0$$ Divide both sides by 2.

$$x^2 - \frac{1}{2}x = -3$$ Subtract 3 from both sides to isolate the binomial.

$$x^2 - \frac{1}{2}x + \frac{1}{16} = -3 + \frac{1}{16}$$ Complete the square: Half of $-\frac{1}{2}$ is $-\frac{1}{4}$ and $\left(-\frac{1}{4}\right)^2 = \frac{1}{16}$.

$$\left(x - \frac{1}{4}\right)^2 = \frac{-47}{16}$$ Factor and simplify: $-3 + \frac{1}{16} = \frac{-48}{16} + \frac{1}{16} = \frac{-47}{16}$.

$$x - \frac{1}{4} = \pm\sqrt{\frac{-47}{16}}$$ Apply the square root property.

$$x - \frac{1}{4} = \pm i\frac{\sqrt{47}}{4}$$ $\pm\sqrt{\frac{-47}{16}} = \pm\sqrt{\frac{47(-1)}{16}} = \pm i\sqrt{\frac{47}{16}}$

$$= \pm i\frac{\sqrt{47}}{\sqrt{16}} = \pm i\frac{\sqrt{47}}{4}$$

$$x = \frac{1}{4} \pm i\frac{\sqrt{47}}{4}$$ Add $\frac{1}{4}$ to both sides and solve for x.

The solutions are $\frac{1}{4} \pm i\frac{\sqrt{47}}{4}$ and the solution set is $\left\{\frac{1}{4} \pm i\frac{\sqrt{47}}{4}\right\}$. ∎

✓ **CHECK POINT 8** Solve by completing the square: $3x^2 - 9x + 8 = 0$.

---

4 Solve problems using the square root property.

Applications

We all want a wonderful life with fulfilling work, good health, and loving relationships. And let's be honest: Financial security wouldn't hurt! Achieving this goal depends on understanding how money in a savings account grows in remarkable ways as a result of *compound interest*. **Compound interest** is interest computed on your original investment as well as on any accumulated interest. For example, suppose you deposit $1000, the principal, in a savings account at a rate of 5%. **Table 11.1** shows how the investment grows if the interest earned is automatically added on to the principal.

Table 11.1 Compound Interest on $1000

Year	Starting Balance	Interest Earned: $I = Pr$	New Balance
1	$1000	$1000 × 0.05 = $50	$1050
2	$1050	$1050 × 0.05 = $52.50	$1102.50
3	$1102.50	$1102.50 × 0.05 ≈ $55.13	$1157.63

A faster way to determine the amount, A, in an account subject to compound interest is to use the following formula:

A Formula for Compound Interest

Suppose that an amount of money, P, is invested at interest rate r, compounded annually. In t years, the amount, A, or balance, in the account is given by the formula

$$A = P(1 + r)^t.$$

Some compound interest problems can be solved using quadratic equations.

EXAMPLE 9 Solving a Compound Interest Problem

You invested $1000 in an account whose interest is compounded annually. After 2 years, the amount, or balance, in the account is $1210. Find the annual interest rate.

Solution We are given that

P (the amount invested) = $1000

t (the time of the investment) = 2 years

A (the amount, or balance, in the account) = $1210.

We are asked to find the annual interest rate, r. We substitute the three given values into the compound interest formula and solve for r.

$A = P(1 + r)^t$	Use the compound interest formula.
$1210 = 1000(1 + r)^2$	Substitute the given values.
$\dfrac{1210}{1000} = (1 + r)^2$	Divide both sides by 1000.
$\dfrac{121}{100} = (1 + r)^2$	Simplify the fraction.
$1 + r = \sqrt{\dfrac{121}{100}}$ or $1 + r = -\sqrt{\dfrac{121}{100}}$	Apply the square root property.
$1 + r = \dfrac{11}{10}$ $1 + r = -\dfrac{11}{10}$	$\sqrt{\dfrac{121}{100}} = \dfrac{\sqrt{121}}{\sqrt{100}} = \dfrac{11}{10}$
$r = \dfrac{11}{10} - 1$ $r = -\dfrac{11}{10} - 1$	Subtract 1 from both sides and solve for r.
$r = \dfrac{1}{10}$ $r = -\dfrac{21}{10}$	$\dfrac{11}{10} - 1 = \dfrac{11}{10} - \dfrac{10}{10} = \dfrac{1}{10}$ and $-\dfrac{11}{10} - 1 = -\dfrac{11}{10} - \dfrac{10}{10} = -\dfrac{21}{10}.$

Because the interest rate cannot be negative, we reject $-\dfrac{21}{10}$. Thus, the annual interest rate is $\dfrac{1}{10} = 0.10 = 10\%$.

We can check this answer using the formula $A = P(1 + r)^t$. If $1000 is invested for 2 years at 10% interest, compounded annually, the balance in the account is

$$A = \$1000(1 + 0.10)^2 = \$1000(1.10)^2 = \$1210.$$

Because this is precisely the amount given by the problem's conditions, the annual interest rate is, indeed, 10% compounded annually. ■

✓ **CHECK POINT 9** You invested $3000 in an account whose interest is compounded annually. After 2 years, the amount, or balance, in the account is $4320. Find the annual interest rate.

The solution to our next problem relies on knowing the **Pythagorean Theorem**. The theorem relates the lengths of the three sides of a **right triangle**, a triangle with one angle measuring 90°. The side opposite the 90° angle is called the **hypotenuse**. The other sides are called **legs**. The legs form the two sides of the right angle.

The Pythagorean Theorem

The sum of the squares of the lengths of the legs of a right triangle equals the square of the length of the hypotenuse.

If the legs have lengths a and b, and the hypotenuse has length c, then

$$a^2 + b^2 = c^2.$$

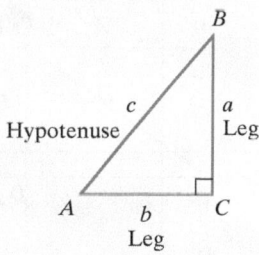

EXAMPLE 10   Using the Pythagorean Theorem and the Square Root Property

a. A wheelchair ramp with a length of 122 inches has a horizontal distance of 120 inches. What is the ramp's vertical distance?

b. Construction laws are very specific when it comes to access ramps for the disabled. Every vertical rise of 1 inch requires a horizontal run of 12 inches. Does this ramp satisfy the requirement?

Solution

a. **Figure 11.2** shows the right triangle that is formed by the ramp, the wall, and the ground. We can find x, the ramp's vertical distance, using the Pythagorean Theorem.

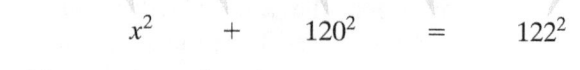

$$x^2 \quad + \quad 120^2 \quad = \quad 122^2$$

We solve this equation using the square root property.

$x^2 + 120^2 = 122^2$ This is the equation resulting from the Pythagorean Theorem.

$x^2 + 14{,}400 = 14{,}884$ Square 120 and 122.

$x^2 = 484$ Isolate x^2 by subtracting 14,400 from both sides.

$x = \sqrt{484}$ or $x = -\sqrt{484}$ Apply the square root property.

$x = 22$ $x = -22$

Because x represents the ramp's vertical distance, we reject the negative value. Thus, the ramp's vertical distance is 22 inches.

b. Every vertical rise of 1 inch requires a horizontal run of 12 inches. Because the ramp has a vertical distance of 22 inches, it requires a horizontal distance of 22(12) inches, or 264 inches. The horizontal distance is only 120 inches, so this ramp does not satisfy construction laws for access ramps for the disabled. ■

✓ **CHECK POINT 10**   A 50-foot supporting wire is to be attached to an antenna. The wire is anchored 20 feet from the base of the antenna. How high up the antenna is the wire attached? Express the answer in simplified radical form. Then find a decimal approximation to the nearest tenth of a foot.

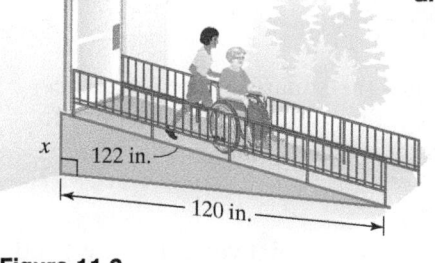

x 122 in.

120 in.

Figure 11.2

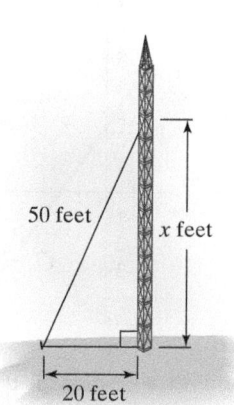

50 feet

x feet

20 feet

5 Find the distance between two points.

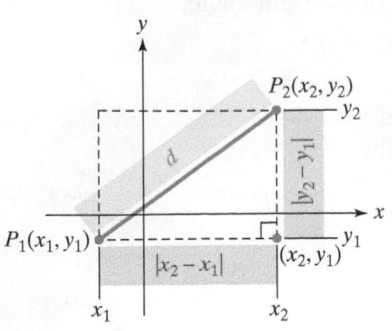

Figure 11.3

The Distance Formula

Using the Pythagorean Theorem, we can find the distance between the two points $P_1(x_1, y_1)$ and $P_2(x_2, y_2)$ in the rectangular coordinate system. The two points are illustrated in **Figure 11.3**.

The distance that we need to find is represented by d and shown in blue. Notice that the distance between two points on the dashed horizontal line is the absolute value of the difference between the x-coordinates of the two points. This distance, $|x_2 - x_1|$, is shown in pink. Similarly, the distance between two points on the dashed vertical line is the absolute value of the difference between the y-coordinates of the two points. This distance, $|y_2 - y_1|$, is also shown in pink.

Because the dashed lines are horizontal and vertical, a right triangle is formed. Thus, we can use the Pythagorean Theorem to find the distance d. Squaring the lengths of the triangle's sides results in positive numbers, so absolute value notation is not necessary.

$$d^2 = (x_2 - x_1)^2 + (y_2 - y_1)^2$$ Apply the Pythagorean Theorem to the right triangle in Figure 11.3.

$$d = \pm\sqrt{(x_2 - x_1)^2 + (y_2 - y_1)^2}$$ Apply the square root property.

$$d = \sqrt{(x_2 - x_1)^2 + (y_2 - y_1)^2}$$ Because distance is nonnegative, write only the principal square root.

This result is called the **distance formula**.

The Distance Formula

The distance, d, between the points (x_1, y_1) and (x_2, y_2) in the rectangular coordinate system is

$$d = \sqrt{(x_2 - x_1)^2 + (y_2 - y_1)^2}.$$

To compute the distance between two points, find the square of the difference between the x-coordinates plus the square of the difference between the y-coordinates. The principal square root of this sum is the distance.

When using the distance formula, it does not matter which point you call (x_1, y_1) and which you call (x_2, y_2).

EXAMPLE 11   Using the Distance Formula

Find the distance between $(-1, 4)$ and $(3, -2)$. Express the answer in simplified radical form and then round to two decimal places.

Solution   We will let $(x_1, y_1) = (-1, 4)$ and $(x_2, y_2) = (3, -2)$.

$$d = \sqrt{(x_2 - x_1)^2 + (y_2 - y_1)^2}$$ Use the distance formula.

$$= \sqrt{[3 - (-1)]^2 + (-2 - 4)^2}$$ Substitute the given values.

$$= \sqrt{4^2 + (-6)^2}$$ Perform operations inside grouping symbols: $3 - (-1) = 3 + 1 = 4$ and $-2 - 4 = -6$.

$$= \sqrt{16 + 36}$$ Square 4 and -6.

> Caution: This does not equal $\sqrt{16} + \sqrt{36}$.

$$= \sqrt{52}$$ Add.

$$= \sqrt{4 \cdot 13} = 2\sqrt{13} \approx 7.21$$ $\sqrt{52} = \sqrt{4 \cdot 13} = \sqrt{4}\sqrt{13} = 2\sqrt{13}$

The distance between the given points is $2\sqrt{13}$ units, or approximately 7.21 units. The situation is illustrated in **Figure 11.4**. ∎

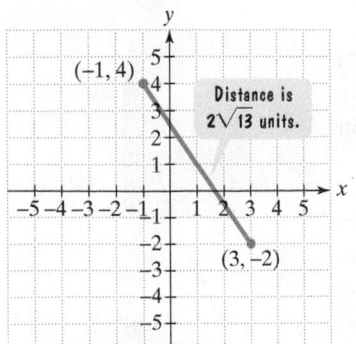

Figure 11.4 Finding the distance between two points

✓ **CHECK POINT 11**   Find the distance between $(-1, -3)$ and $(2, 3)$. Express the answer in simplified radical form and then round to two decimal places.

6 Find the midpoint of a line segment.

The Midpoint Formula

The distance formula can be used to derive a formula for finding the midpoint of a line segment between two points. The formula is given as follows:

The Midpoint Formula

Consider a line segment whose endpoints are (x_1, y_1) and (x_2, y_2). The coordinates of the segment's midpoint are

$$\left(\frac{x_1 + x_2}{2}, \frac{y_1 + y_2}{2} \right).$$

To find the midpoint, take the average of the two x-coordinates and the average of the two y-coordinates.

Great Question!

Do I add or subtract coordinates when applying the midpoint and distance formulas?

The midpoint formula requires finding the *sum* of coordinates. By contrast, the distance formula requires finding the *difference* of coordinates:

Midpoint:
Sum of coordinates

$$\left(\frac{x_1 + x_2}{2}, \frac{y_1 + y_2}{2} \right)$$

Distance:
Difference of coordinates

$$\sqrt{(x_2 - x_1)^2 + (y_2 - y_1)^2}$$

It's easy to confuse the two formulas. Be sure to use addition, not subtraction, when applying the midpoint formula.

EXAMPLE 12   Using the Midpoint Formula

Find the midpoint of the line segment with endpoints $(1, -6)$ and $(-8, -4)$.

Solution   To find the coordinates of the midpoint, we average the coordinates of the endpoints.

$$\text{Midpoint} = \left(\frac{1 + (-8)}{2}, \frac{-6 + (-4)}{2} \right) = \left(\frac{-7}{2}, \frac{-10}{2} \right) = \left(-\frac{7}{2}, -5 \right)$$

Average the
x-coordinates.

Average the
y-coordinates.

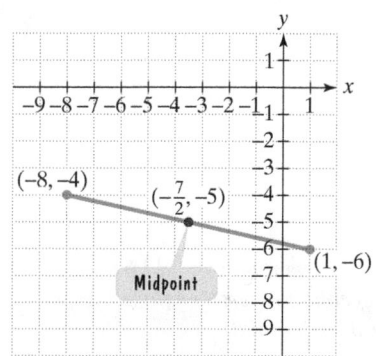

Figure 11.5 Finding a line segment's midpoint

Figure 11.5 illustrates that the point $\left(-\frac{7}{2}, -5 \right)$ is midway between the points $(1, -6)$ and $(-8, -4)$. ■

✓ **CHECK POINT 12**   Find the midpoint of the line segment with endpoints $(1, 2)$ and $(7, -3)$.

CONCEPT AND VOCABULARY CHECK

Fill in each blank so that the resulting statement is true.

1. The square root property states that if $u^2 = d$, then $u = $ _____.

2. If $x^2 = 7$, then $x = $ _____.

3. If $x^2 = \dfrac{11}{2}$, then $x = $ _____. Rationalizing denominators, we obtain $x = $ _____.

4. If $x^2 = -9$, then $x = $ _____.

5. To complete the square on $x^2 + 10x$, add _____.

6. To complete the square on $x^2 - 3x$, add _____.

7. To complete the square on $x^2 - \dfrac{4}{5}x$, add _____.

8. To solve $x^2 + 6x = 7$ by completing the square, add _____ to both sides of the equation.

9. To solve $x^2 - \dfrac{2}{3}x = \dfrac{4}{9}$ by completing the square, add _____ to both sides of the equation.

10. A triangle with one angle measuring 90° is called a/an _____ triangle. The side opposite the 90° angle is called the _____. The other sides are called _____.

11. The Pythagorean Theorem states that in any _____ triangle, the sum of the squares of the lengths of the _____ equals _____.

12. The distance, d, between the points (x_1, y_1) and (x_2, y_2) in the rectangular coordinate system is $d = $ _____.

13. The midpoint of a line segment whose endpoints are (x_1, y_1) and (x_2, y_2) is (_____, _____).

11.1 EXERCISE SET   MyMathLab®

Watch the videos in MyMathLab   Download the MyDashBoard App

Practice Exercises

In Exercises 1–22, solve each equation by the square root property. If possible, simplify radicals or rationalize denominators. Express imaginary solutions in the form $a + bi$.

1. $3x^2 = 75$

2. $5x^2 - 20$

3. $7x^2 = 42$

4. $8x^2 = 40$

5. $16x^2 = 25$

6. $4x^2 = 49$

7. $3x^2 - 2 = 0$

8. $3x^2 - 5 = 0$

9. $25x^2 + 16 = 0$

10. $4x^2 + 49 = 0$

11. $(x + 7)^2 = 9$

12. $(x + 3)^2 = 64$

13. $(x - 3)^2 = 5$

14. $(x - 4)^2 = 3$

15. $2(x + 2)^2 = 16$

16. $3(x + 2)^2 = 36$

17. $(x - 5)^2 = -9$

18. $(x - 5)^2 = -4$

19. $\left(x + \dfrac{3}{4}\right)^2 = \dfrac{11}{16}$

20. $\left(x + \dfrac{2}{5}\right)^2 = \dfrac{7}{25}$

21. $x^2 - 6x + 9 = 36$

22. $x^2 - 6x + 9 = 49$

In Exercises 23–34, determine the constant that should be added to the binomial so that it becomes a perfect square trinomial. Then write and factor the trinomial.

23. $x^2 + 2x$

24. $x^2 + 4x$

25. $x^2 - 14x$

26. $x^2 - 10x$

27. $x^2 + 7x$

28. $x^2 + 9x$

29. $x^2 - \dfrac{1}{2}x$

30. $x^2 - \dfrac{1}{3}x$

31. $x^2 + \dfrac{4}{3}x$

32. $x^2 + \dfrac{4}{5}x$

33. $x^2 - \dfrac{9}{4}x$

34. $x^2 - \dfrac{9}{5}x$

In Exercises 35–58, solve each quadratic equation by completing the square.

35. $x^2 + 4x = 32$

36. $x^2 + 6x = 7$

37. $x^2 + 6x = -2$

38. $x^2 + 2x = 5$

39. $x^2 - 8x + 1 = 0$

40. $x^2 + 8x - 5 = 0$

41. $x^2 + 2x + 2 = 0$

42. $x^2 - 4x + 8 = 0$

43. $x^2 + 3x - 1 = 0$

44. $x^2 - 3x - 5 = 0$

45. $x^2 + \dfrac{4}{7}x + \dfrac{3}{49} = 0$

46. $x^2 + \dfrac{6}{5}x + \dfrac{8}{25} = 0$

47. $x^2 + x - 1 = 0$

48. $x^2 - 7x + 3 = 0$

49. $2x^2 + 3x - 5 = 0$

50. $2x^2 + 5x - 3 = 0$

51. $3x^2 + 6x + 1 = 0$

52. $3x^2 - 6x + 2 = 0$

53. $3x^2 - 8x + 1 = 0$

54. $2x^2 + 3x - 4 = 0$

55. $8x^2 - 4x + 1 = 0$

56. $9x^2 - 6x + 5 = 0$

57. $2x^2 - 5x + 7 = 0$

58. $4x^2 - 2x + 5 = 0$

59. If $g(x) = \left(x - \dfrac{2}{5}\right)^2$, find all values of x for which $g(x) = \dfrac{9}{25}$.

60. If $g(x) = \left(x + \dfrac{1}{3}\right)^2$, find all values of x for which $g(x) = \dfrac{4}{9}$.

61. If $h(x) = 5(x + 2)^2$, find all values of x for which $h(x) = -125$.

62. If $h(x) = 3(x - 4)^2$, find all values of x for which $h(x) = -12$.

In Exercises 63–80, find the distance between each pair of points. If necessary, express answers in simplified radical form and then round to two decimal places.

63. $(2, 3)$ and $(14, 8)$

64. $(5, 1)$ and $(8, 5)$

65. $(4, 1)$ and $(6, 3)$

66. $(2, 3)$ and $(3, 5)$

67. $(0, 0)$ and $(-3, 4)$

68. $(0, 0)$ and $(3, -4)$

69. $(-2, -6)$ and $(3, -4)$

70. $(-4, -1)$ and $(2, -3)$

71. $(0, -3)$ and $(4, 1)$

72. $(0, -2)$ and $(4, 3)$

73. $(3.5, 8.2)$ and $(-0.5, 6.2)$

74. $(2.6, 1.3)$ and $(1.6, -5.7)$

75. $\left(0, -\sqrt{3}\right)$ and $\left(\sqrt{5}, 0\right)$

76. $\left(0, -\sqrt{2}\right)$ and $\left(\sqrt{7}, 0\right)$

77. $\left(3\sqrt{3}, \sqrt{5}\right)$ and $\left(-\sqrt{3}, 4\sqrt{5}\right)$

78. $\left(2\sqrt{3}, \sqrt{6}\right)$ and $\left(-\sqrt{3}, 5\sqrt{6}\right)$

79. $\left(\dfrac{7}{3}, \dfrac{1}{5}\right)$ and $\left(\dfrac{1}{3}, \dfrac{6}{5}\right)$

80. $\left(-\dfrac{1}{4}, -\dfrac{1}{7}\right)$ and $\left(\dfrac{3}{4}, \dfrac{6}{7}\right)$

In Exercises 81–92, find the midpoint of the line segment with the given endpoints.

81. $(6, 8)$ and $(2, 4)$

82. $(10, 4)$ and $(2, 6)$

83. $(-2, -8)$ and $(-6, -2)$

84. $(-4, -7)$ and $(-1, -3)$

85. $(-3, -4)$ and $(6, -8)$

86. $(-2, -1)$ and $(-8, 6)$

87. $\left(-\dfrac{7}{2}, \dfrac{3}{2}\right)$ and $\left(-\dfrac{5}{2}, -\dfrac{11}{2}\right)$

88. $\left(-\dfrac{2}{5}, \dfrac{7}{15}\right)$ and $\left(-\dfrac{2}{5}, -\dfrac{4}{15}\right)$

89. $\left(8, 3\sqrt{5}\right)$ and $\left(-6, 7\sqrt{5}\right)$

90. $\left(7\sqrt{3}, -6\right)$ and $\left(3\sqrt{3}, -2\right)$

91. $\left(\sqrt{18}, -4\right)$ and $\left(\sqrt{2}, 4\right)$

92. $\left(\sqrt{50}, -6\right)$ and $\left(\sqrt{2}, 6\right)$

Practice PLUS

93. Three times the square of the difference between a number and 2 is -12. Find the number(s).

94. Three times the square of the difference between a number and 9 is -27. Find the number(s).

In Exercises 95–98, solve the formula for the specified variable. Because each variable is nonnegative, list only the principal square root. If possible, simplify radicals or rationalize denominators.

95. $h = \dfrac{v^2}{2g}$ for v

96. $s = \dfrac{kwd^2}{l}$ for d

97. $A = P(1 + r)^2$ for r

98. $C = \dfrac{kP_1P_2}{d^2}$ for d

In Exercises 99–102, solve each quadratic equation by completing the square.

99. $\dfrac{x^2}{3} + \dfrac{x}{9} - \dfrac{1}{6} = 0$

100. $\dfrac{x^2}{2} - \dfrac{x}{6} - \dfrac{3}{4} = 0$

101. $x^2 - bx = 2b^2$

102. $x^2 - bx = 6b^2$

103. The ancient Greeks used a geometric method for completing the square in which they literally transformed a figure into a square.

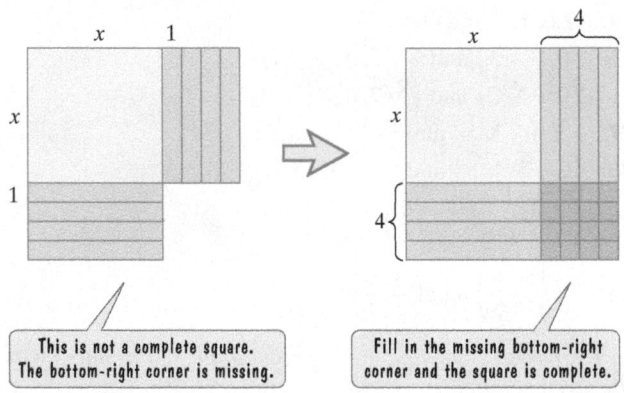

This is not a complete square. The bottom-right corner is missing.

Fill in the missing bottom-right corner and the square is complete.

a. Write a binomial in x that represents the combined area of the small square and the eight rectangular stripes that make up the incomplete square on the left.

b. What is the area of the region in the bottom-right corner that literally completes the square?

c. Write a trinomial in x that represents the combined area of the small square, the eight rectangular stripes, and the bottom-right corner that make up the complete square on the right.

d. Use the length of each side of the complete square on the right to express its area as the square of a binomial.

104. An **isosceles right triangle** has legs that are the same length and acute angles each measuring 45°.

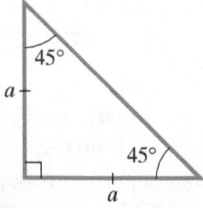

a. Write an expression in terms of a that represents the length of the hypotenuse.

b. Use your result from part (a) to write a sentence that describes the length of the hypotenuse of an isosceles right triangle in terms of the length of a leg.

Application Exercises

In Exercises 105–108, use the compound interest formula

$$A = P(1 + r)^t$$

to find the annual interest rate, r.

105. In 2 years, an investment of $2000 grows to $2880.

106. In 2 years, an investment of $2000 grows to $2420.

107. In 2 years, an investment of $1280 grows to $1445.

108. In 2 years, an investment of $80,000 grows to $101,250.

Charter schools operate outside the constraints of regular public schools. They get public money, but in most cases, their teachers are not unionized. This freedom has allowed a minority of them to shine, building flexible, demanding programs that defy expectations. But only 1 in 6 charter schools significantly outperforms traditional counterparts. And more than a third underperform. The graph shows the number of U.S. students enrolled in charter schools for selected years from 2000 through 2008.

Number of Students in the United States Enrolled in Charter Schools

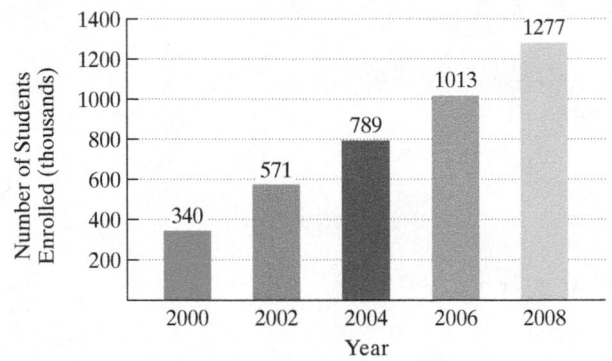

Source: National Center for Education Statistics

The data shown can be modeled by the function

$$f(x) = 15x^2 + 340,$$

where $f(x)$ represents the number of students enrolled in charter schools, in thousands, x years after 2000. Use this information to solve Exercises 109–110.

109. a. According to the model, how many students, in thousands, were enrolled in charter schools in 2008? Does this underestimate or overestimate the number displayed by the graph? By how much?

b. According to the model, in which year will 3280 thousand students be enrolled in charter schools?

110. a. According to the model, how many students, in thousands, were enrolled in charter schools in 2006? Does this underestimate or overestimate the number displayed by the graph? By how much?

b. According to the model, in which year will 3715 thousand students be enrolled in charter schools?

The function $s(t) = 16t^2$ models the distance, $s(t)$, in feet, that an object falls in t seconds. Use this function and the square root property to solve Exercises 111–112. Express answers in simplified radical form. Then use your calculator to find a decimal approximation to the nearest tenth of a second.

111. A sky diver jumps from an airplane and falls for 4800 feet before opening a parachute. For how many seconds was the diver in a free fall?

112. A sky diver jumps from an airplane and falls for 3200 feet before opening a parachute. For how many seconds was the diver in a free fall?

Use the Pythagorean Theorem and the square root property to solve Exercises 113–118. Express answers in simplified radical form. Then find a decimal approximation to the nearest tenth.

113. A rectangular park is 6 miles long and 3 miles wide. How long is a pedestrian route that runs diagonally across the park?

114. A rectangular park is 4 miles long and 2 miles wide. How long is a pedestrian route that runs diagonally across the park?

115. The base of a 30-foot ladder is 10 feet from the building. If the ladder reaches the flat roof, how tall is the building?

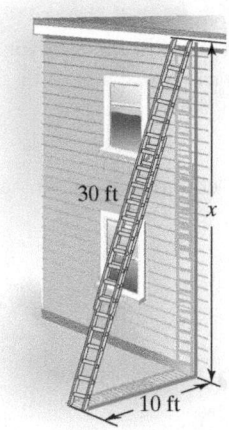

30 ft

x

10 ft

116. The doorway into a room is 4 feet wide and 8 feet high. What is the diameter of the largest circular tabletop that can be taken through this doorway diagonally?

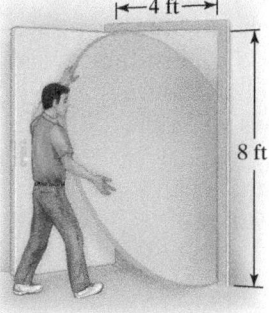

←4 ft→

8 ft

117. A supporting wire is to be attached to the top of a 50-foot antenna. If the wire must be anchored 50 feet from the base of the antenna, what length of wire is required?

118. A supporting wire is to be attached to the top of a 70-foot antenna. If the wire must be anchored 70 feet from the base of the antenna, what length of wire is required?

119. A square flower bed is to be enlarged by adding 2 meters on each side. If the larger square has an area of 196 square meters, what is the length of a side of the original square?

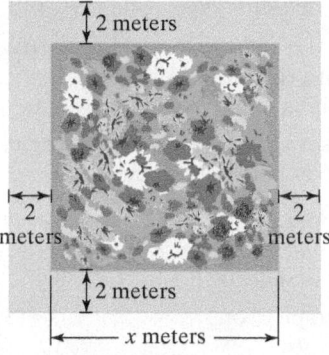

2 meters

2 meters

2 meters

2 meters

— x meters —

120. A square flower bed is to be enlarged by adding 4 feet on each side. If the larger square has an area of 225 square feet, what is the length of a side of the original square?

Writing in Mathematics

121. What is the square root property?

122. Explain how to solve $(x - 1)^2 = 16$ using the square root property.

123. Explain how to complete the square for a binomial. Use $x^2 + 6x$ to illustrate your explanation.

124. Explain how to solve $x^2 + 6x + 8 = 0$ by completing the square.

125. What is compound interest?

126. In your own words, describe the compound interest formula

$$A = P(1 + r)^t.$$

127. In your own words, describe how to find the distance between two points in the rectangular coordinate system.

128. In your own words, describe how to find the midpoint of a line segment if its endpoints are known.

Technology Exercises

129. Use a graphing utility to solve $4 - (x + 1)^2 = 0$. Graph $y = 4 - (x + 1)^2$ in a $[-5, 5, 1]$ by $[-5, 5, 1]$ viewing rectangle. The equation's solutions are the graph's x-intercepts. Check by substitution in the given equation.

130. Use a graphing utility to solve $(x - 1)^2 - 9 = 0$. Graph $y = (x - 1)^2 - 9$ in a $[-5, 5, 1]$ by $[-9, 3, 1]$ viewing rectangle. The equation's solutions are the graph's x-intercepts. Check by substitution in the given equation.

131. Use a graphing utility and x-intercepts to verify any of the real solutions that you obtained for five of the quadratic equations in Exercises 35–54.

Critical Thinking Exercises

Make Sense? *In Exercises 132–135, determine whether each statement "makes sense" or "does not make sense" and explain your reasoning.*

132. When the coefficient of the x-term in a quadratic equation is negative and I'm solving by completing the square, I add a negative constant to each side of the equation.

133. When I complete the square for the binomial $x^2 + bx$, I obtain a different polynomial, but when I solve a quadratic equation by completing the square, I obtain an equation with the same solution set.

134. When I use the square root property to determine the length of a right triangle's side, I don't even bother to list the negative square root.

135. When I solved $4x^2 + 10x = 0$ by completing the square, I added 25 to both sides of the equation.

In Exercises 136–139, determine whether each statement is true or false. If the statement is false, make the necessary change(s) to produce a true statement.

136. The graph of $y = (x - 2)^2 + 3$ cannot have x-intercepts.

137. The equation $(x - 5)^2 = 12$ is equivalent to $x - 5 = 2\sqrt{3}$.

138. In completing the square for $2x^2 - 6x = 5$, we should add 9 to both sides.

139. Although not every quadratic equation can be solved by completing the square, they can all be solved by factoring.

140. Solve for y: $\dfrac{x^2}{a^2} + \dfrac{y^2}{b^2} = 1$.

141. Solve by completing the square:
$$x^2 + x + c = 0.$$

142. Solve by completing the square:
$$x^2 + bx + c = 0.$$

143. Solve: $x^4 - 8x^2 + 15 = 0$.

Review Exercises

144. Simplify: $4x - 2 - 3[4 - 2(3 - x)]$.
(Section 1.8, Example 11)

145. Factor: $1 - 8x^3$. (Section 6.4, Example 8)

146. Divide: $(x^4 - 5x^3 + 2x^2 - 6) \div (x - 3)$.
(Section 5.6, Example 2 or Example 5)

Preview Exercises

Exercises 147–149 will help you prepare for the material covered in the next section.

147. a. Solve by factoring: $8x^2 + 2x - 1 = 0$.

b. The quadratic equation in part (a) is in the standard form $ax^2 + bx + c = 0$. Compute $b^2 - 4ac$. Is $b^2 - 4ac$ a perfect square?

148. a. Solve by factoring: $9x^2 - 6x + 1 = 0$.

b. The quadratic equation in part (a) is in the standard form $ax^2 + bx + c = 0$. Compute $b^2 - 4ac$.

149. a. Clear fractions in the following equation and write in the form $ax^2 + bx + c = 0$:
$$3 + \frac{4}{x} = -\frac{2}{x^2}.$$

b. For the equation you wrote in part (a), compute $b^2 - 4ac$.

The Quadratic Formula

Objectives

1 Solve quadratic equations using the quadratic formula.

2 Use the discriminant to determine the number and type of solutions.

3 Determine the most efficient method to use when solving a quadratic equation.

4 Write quadratic equations from solutions.

5 Use the quadratic formula to solve problems.

Until fairly recently, many doctors believed that your blood pressure was theirs to know and yours to worry about. Today, however, people are encouraged to find out their blood pressure. That pumped-up cuff that squeezes against your upper arm measures blood pressure in millimeters (mm) of mercury (Hg). Blood pressure is given in two numbers: systolic pressure over diastolic pressure, such as 120 over 80. Systolic pressure is the pressure of blood against the artery walls when the heart contracts. Diastolic pressure is the pressure of blood against the artery walls when the heart is at rest.

In this section, we will derive a formula that will enable you to solve quadratic equations more quickly than the method of completing the square. Using this formula, we will work with functions that model changing systolic pressure for men and women with age.

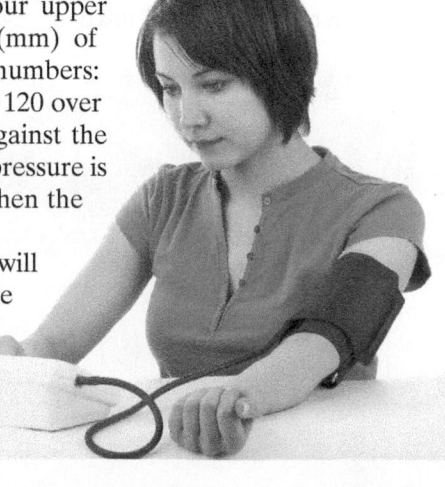

1 Solve quadratic equations using the quadratic formula.

Solving Quadratic Equations Using the Quadratic Formula

We can use the method of completing the square to derive a formula that can be used to solve all quadratic equations. The derivation given below also shows a particular quadratic equation, $3x^2 - 2x - 4 = 0$, to specifically illustrate each of the steps.

Deriving the Quadratic Formula

Standard Form of a Quadratic Equation	Comment	A Specific Example
$ax^2 + bx + c = 0, a > 0$	This is the given equation.	$3x^2 - 2x - 4 = 0$
$x^2 + \dfrac{b}{a}x + \dfrac{c}{a} = 0$	Divide both sides by the coefficient of x^2.	$x^2 - \dfrac{2}{3}x - \dfrac{4}{3} = 0$
$x^2 + \dfrac{b}{a}x = -\dfrac{c}{a}$	Isolate the binomial by adding $-\dfrac{c}{a}$ on both sides.	$x^2 - \dfrac{2}{3}x = \dfrac{4}{3}$
$x^2 + \dfrac{b}{a}x + \left(\dfrac{b}{2a}\right)^2 = -\dfrac{c}{a} + \left(\dfrac{b}{2a}\right)^2$ $\underbrace{\qquad}_{(\text{half})^2}$	Complete the square. Add the square of half the coefficient of x to both sides.	$x^2 - \dfrac{2}{3}x + \left(-\dfrac{1}{3}\right)^2 = \dfrac{4}{3} + \left(-\dfrac{1}{3}\right)^2$ $\underbrace{\qquad}_{(\text{half})^2}$
$x^2 + \dfrac{b}{a}x + \dfrac{b^2}{4a^2} = -\dfrac{c}{a} + \dfrac{b^2}{4a^2}$		$x^2 - \dfrac{2}{3}x + \dfrac{1}{9} = \dfrac{4}{3} + \dfrac{1}{9}$
$\left(x + \dfrac{b}{2a}\right)^2 = -\dfrac{c}{a}\cdot\dfrac{4a}{4a} + \dfrac{b^2}{4a^2}$	Factor on the left side and obtain a common denominator on the right side.	$\left(x - \dfrac{1}{3}\right)^2 = \dfrac{4}{3}\cdot\dfrac{3}{3} + \dfrac{1}{9}$
$\left(x + \dfrac{b}{2a}\right)^2 = \dfrac{-4ac + b^2}{4a^2}$ $\left(x + \dfrac{b}{2a}\right)^2 = \dfrac{b^2 - 4ac}{4a^2}$	Add fractions on the right side.	$\left(x - \dfrac{1}{3}\right)^2 = \dfrac{12 + 1}{9}$ $\left(x - \dfrac{1}{3}\right)^2 = \dfrac{13}{9}$
$x + \dfrac{b}{2a} = \pm\sqrt{\dfrac{b^2 - 4ac}{4a^2}}$	Apply the square root property.	$x - \dfrac{1}{3} = \pm\sqrt{\dfrac{13}{9}}$
$x + \dfrac{b}{2a} = \pm\dfrac{\sqrt{b^2 - 4ac}}{2a}$	Take the square root of the quotient, simplifying the denominator.	$x - \dfrac{1}{3} = \pm\dfrac{\sqrt{13}}{3}$
$x = \dfrac{-b}{2a} \pm \dfrac{\sqrt{b^2 - 4ac}}{2a}$	Solve for x by subtracting $\dfrac{b}{2a}$ from both sides.	$x = \dfrac{1}{3} \pm \dfrac{\sqrt{13}}{3}$
$x = \dfrac{-b \pm \sqrt{b^2 - 4ac}}{2a}$	Combine fractions on the right side.	$x = \dfrac{1 \pm \sqrt{13}}{3}$

The formula shown at the bottom of the left column is called the *quadratic formula*. A similar proof shows that the same formula can be used to solve quadratic equations if a, the coefficient of the x^2-term, is negative.

The Quadratic Formula

The solutions of a quadratic equation in standard form $ax^2 + bx + c = 0$, with $a \neq 0$, are given by the **quadratic formula**:

$$x = \frac{-b \pm \sqrt{b^2 - 4ac}}{2a}.$$

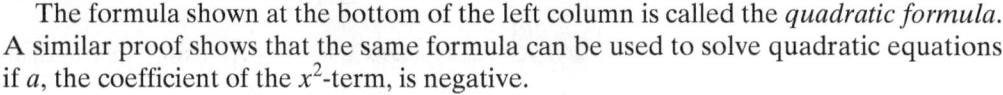
x equals negative b plus or minus the square root of $b^2 - 4ac$, all divided by $2a$.

To use the quadratic formula, write the quadratic equation in standard form if necessary. Then determine the numerical values for a (the coefficient of the x^2-term), b (the coefficient of the x-term), and c (the constant term). Substitute the values of a, b, and c into the quadratic formula and evaluate the expression. The $\pm$ sign indicates that there are two (not necessarily distinct) solutions of the equation.

EXAMPLE 1 Solving a Quadratic Equation Using the Quadratic Formula

Solve using the quadratic formula: $8x^2 + 2x - 1 = 0$.

Solution The given equation is in standard form. Begin by identifying the values for a, b, and c.

$$8x^2 + 2x - 1 = 0$$

$$a = 8 \qquad b = 2 \qquad c = -1$$

Substituting these values into the quadratic formula and simplifying gives the equation's solutions.

$$x = \frac{-b \pm \sqrt{b^2 - 4ac}}{2a}$$

Use the quadratic formula.

$$x = \frac{-2 \pm \sqrt{2^2 - 4(8)(-1)}}{2(8)}$$

Substitute the values for a, b, and c: a = 8, b = 2, and c = −1.

$$= \frac{-2 \pm \sqrt{4 - (-32)}}{16}$$

$2^2 - 4(8)(-1) = 4 - (-32)$

$$= \frac{-2 \pm \sqrt{36}}{16}$$

$4 - (-32) = 4 + 32 = 36$

$$= \frac{-2 \pm 6}{16}$$

$\sqrt{36} = 6$

Using Technology

Graphic Connections

The graph of the quadratic function

$$y = 8x^2 + 2x - 1$$

has x-intercepts at $-\frac{1}{2}$ and $\frac{1}{4}$. This verifies that $\left\{-\frac{1}{2}, \frac{1}{4}\right\}$ is the solution set of the quadratic equation

$$8x^2 + 2x - 1 = 0.$$

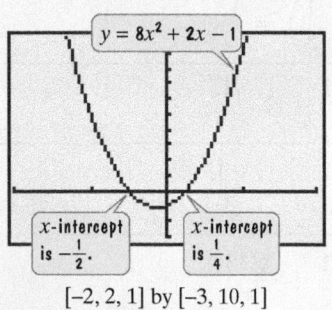

x-intercept is $-\frac{1}{2}$. x-intercept is $\frac{1}{4}$.

$[-2, 2, 1]$ by $[-3, 10, 1]$

Now we will evaluate this expression in two different ways to obtain the two solutions. On the left, we will *add* 6 to −2. On the right, we will *subtract* 6 from −2.

$$x = \frac{-2 + 6}{16} \quad \text{or} \quad x = \frac{-2 - 6}{16}$$

$$= \frac{4}{16} = \frac{1}{4} \qquad\qquad = \frac{-8}{16} = -\frac{1}{2}$$

The solutions are $-\frac{1}{2}$ and $\frac{1}{4}$, and the solution set is $\left\{-\frac{1}{2}, \frac{1}{4}\right\}$. ■

In Example 1, the solutions of $8x^2 + 2x - 1 = 0$ are rational numbers. This means that the equation can also be solved by factoring. The reason that the solutions are rational numbers is that $b^2 - 4ac$, the radicand in the quadratic formula, is 36, which is a perfect square. If a, b, and c are rational numbers, all quadratic equations for which $b^2 - 4ac$ is a perfect square have rational solutions.

✓ **CHECK POINT 1** Solve using the quadratic formula: $2x^2 + 9x - 5 = 0$.

EXAMPLE 2 Solving a Quadratic Equation Using the Quadratic Formula

Solve using the quadratic formula:

$$2x^2 = 4x + 1.$$

Solution The quadratic equation must be in standard form to identify the values for a, b, and c. To move all terms to one side and obtain zero on the right, we subtract $4x + 1$ from both sides. Then we can identify the values for a, b, and c.

$$2x^2 = 4x + 1 \qquad \text{This is the given equation.}$$

$$2x^2 - 4x - 1 = 0 \qquad \text{Subtract } 4x + 1 \text{ from both sides.}$$

$a = 2 \qquad b = -4 \qquad c = -1$

Substituting these values into the quadratic formula and simplifying gives the equation's solutions.

$$x = \frac{-b \pm \sqrt{b^2 - 4ac}}{2a} \qquad \text{Use the quadratic formula.}$$

$$x = \frac{-(-4) \pm \sqrt{(-4)^2 - 4(2)(-1)}}{2(2)} \qquad \begin{array}{l}\text{Substitute the values for } a, b, \text{ and } c:\\ a = 2, b = -4, \text{ and } c = -1.\end{array}$$

$$= \frac{4 \pm \sqrt{16 - (-8)}}{4} \qquad (-4)^2 - 4(2)(-1) = 16 - (-8)$$

$$= \frac{4 \pm \sqrt{24}}{4} \qquad 16 - (-8) = 16 + 8 = 24$$

$$= \frac{4 \pm 2\sqrt{6}}{4} \qquad \sqrt{24} = \sqrt{4 \cdot 6} = \sqrt{4} \cdot \sqrt{6} = 2\sqrt{6}$$

$$= \frac{2(2 \pm \sqrt{6})}{4} \qquad \text{Factor out 2 from the numerator.}$$

$$= \frac{2 \pm \sqrt{6}}{2} \qquad \text{Divide the numerator and denominator by 2.}$$

The solutions are $\dfrac{2 \pm \sqrt{6}}{2}$, and the solution set is $\left\{\dfrac{2 + \sqrt{6}}{2}, \dfrac{2 - \sqrt{6}}{2}\right\}$ or $\left\{\dfrac{2 \pm \sqrt{6}}{2}\right\}$. ∎

Using Technology

You can use a graphing utility to verify that the solutions of $2x^2 - 4x - 1 = 0$ are $\dfrac{2 \pm \sqrt{6}}{2}$. Begin by entering $y_1 = 2x^2 - 4x - 1$ in the $\boxed{Y=}$ screen. Then evaluate this function at each of the proposed solutions.

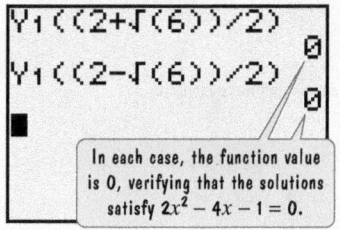

In each case, the function value is 0, verifying that the solutions satisfy $2x^2 - 4x - 1 = 0$.

In Example 2, the solutions of $2x^2 = 4x + 1$ are irrational numbers. This means that the equation cannot be solved by factoring. The reason that the solutions are irrational numbers is that $b^2 - 4ac$, the radicand in the quadratic formula, is 24, which is not a perfect square. Notice, too, that the solutions, $\dfrac{2 + \sqrt{6}}{2}$ and $\dfrac{2 - \sqrt{6}}{2}$, are conjugates.

Great Question!

The simplification of the irrational solutions in Example 2 was kind of tricky. Any suggestions to guide the process?

Many students use the quadratic formula correctly until the last step, where they make an error in simplifying the solutions. Be sure to factor the numerator before dividing the numerator and the denominator by the greatest common factor.

$$\frac{4 \pm 2\sqrt{6}}{4} = \frac{2(2 \pm \sqrt{6})}{4} = \frac{\overset{1}{\cancel{2}}(2 \pm \sqrt{6})}{\underset{2}{\cancel{4}}} = \frac{2 \pm \sqrt{6}}{2}$$

You cannot divide just one term in the numerator and the denominator by their greatest common factor.

Incorrect!

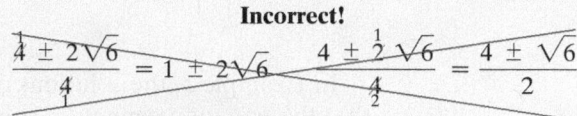

Can all irrational solutions of quadratic equations be simplified? No. The following solutions cannot be simplified:

$$\frac{5 \pm 2\sqrt{7}}{2} \qquad \frac{-4 \pm 3\sqrt{7}}{2}.$$

Other than 1, terms in each numerator have no common factor.

✓ **CHECK POINT 2** Solve using the quadratic formula: $2x^2 = 6x - 1$.

EXAMPLE 3 Solving a Quadratic Equation Using the Quadratic Formula

Solve using the quadratic formula:

$$3x^2 + 2 = -4x.$$

Solution Begin by writing the quadratic equation in standard form.

$3x^2 + 2 = -4x$ This is the given equation.

$3x^2 + 4x + 2 = 0$ Add 4x to both sides.

$a = 3 \quad b = 4 \quad c = 2$

Substituting these values into the quadratic formula and simplifying gives the equation's solutions.

$$x = \frac{-b \pm \sqrt{b^2 - 4ac}}{2a}$$ Use the quadratic formula.

$$x = \frac{-4 \pm \sqrt{4^2 - 4 \cdot 3 \cdot 2}}{2 \cdot 3}$$ Substitute the values for a, b, and c: $a = 3, b = 4$, and $c = 2$.

$$= \frac{-4 \pm \sqrt{16 - 24}}{6}$$ Multiply under the radical.

$$= \frac{-4 \pm \sqrt{-8}}{6}$$ Subtract under the radical.

$$= \frac{-4 \pm 2i\sqrt{2}}{6}$$ $\sqrt{-8} = \sqrt{8(-1)} = \sqrt{8}\sqrt{-1} = i\sqrt{8} = i\sqrt{4 \cdot 2} = 2i\sqrt{2}$

$$= \frac{2(-2 \pm i\sqrt{2})}{6}$$ Factor out 2 from the numerator.

$$= \frac{-2 \pm i\sqrt{2}}{3}$$ Divide the numerator and denominator by 2.

$$= -\frac{2}{3} \pm i\frac{\sqrt{2}}{3}$$ Express in the form $a + bi$, writing i before the square root.

The solutions are $-\dfrac{2}{3} \pm i\dfrac{\sqrt{2}}{3}$, and the solution set is $\left\{ -\dfrac{2}{3} + i\dfrac{\sqrt{2}}{3}, -\dfrac{2}{3} - i\dfrac{\sqrt{2}}{3} \right\}$ or $\left\{ -\dfrac{2}{3} \pm i\dfrac{\sqrt{2}}{3} \right\}$. ■

Using Technology

Graphic Connections

The graph of the quadratic function

$$y = 3x^2 + 4x + 2$$

has no x-intercepts. This verifies that the equation in Example 3

$$3x^2 + 2 = -4x, \quad \text{or}$$
$$3x^2 + 4x + 2 = 0$$

has imaginary solutions.

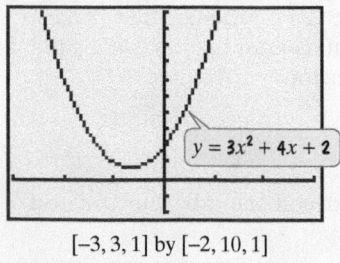

$y = 3x^2 + 4x + 2$

[−3, 3, 1] by [−2, 10, 1]

In Example 3, the solutions of $3x^2 + 2 = -4x$ are imaginary numbers. This means that the equation cannot be solved using factoring. The reason that the solutions are imaginary numbers is that $b^2 - 4ac$, the radicand in the quadratic formula, is -8, which is negative. Notice, too, that the solutions are complex conjugates.

Great Question!

Should I check irrational and imaginary solutions by substitution in the given quadratic equation?

No. Checking irrational and imaginary solutions can be time-consuming. The solutions given by the quadratic formula are always correct, unless you have made a careless error. Checking for computational errors or errors in simplification is sufficient.

2 Use the discriminant to determine the number and type of solutions.

✓ **CHECK POINT 3** Solve using the quadratic formula: $3x^2 + 5 = -6x$.

Some rational equations can be solved using the quadratic formula. For example, consider the equation

$$3 + \frac{4}{x} = -\frac{2}{x^2}.$$

The denominators are x and x^2. The least common denominator is x^2. We clear fractions by multiplying both sides of the equation by x^2. Notice that x cannot equal zero.

$$x^2\left(3 + \frac{4}{x}\right) = x^2\left(-\frac{2}{x^2}\right), \quad x \neq 0$$

$$3x^2 + \frac{4}{x} \cdot x^2 = x^2\left(-\frac{2}{x^2}\right) \qquad \text{Use the distributive property.}$$

$$3x^2 + 4x = -2 \qquad \text{Simplify.}$$

By adding 2 to both sides of $3x^2 + 4x = -2$, we obtain the standard form of the quadratic equation:

$$3x^2 + 4x + 2 = 0.$$

This is the equation that we solved in Example 3. The two imaginary solutions are not part of the restriction that $x \neq 0$.

The Discriminant

The quantity $b^2 - 4ac$, which appears under the radical sign in the quadratic formula, is called the **discriminant**. **Table 11.2** shows how the discriminant of the quadratic equation $ax^2 + bx + c = 0$ determines the number and type of solutions.

Table 11.2 The Discriminant and the Kinds of Solutions to $ax^2 + bx + c = 0$

Discriminant $b^2 - 4ac$	Kinds of Solutions to $ax^2 + bx + c = 0$	Graph of $y = ax^2 + bx + c$
$b^2 - 4ac > 0$	**Two unequal real solutions:** If a, b, and c are rational numbers and the discriminant is a perfect square, the solutions are *rational*. If the discriminant is not a perfect square, the solutions are *irrational* conjugates.	Two x-intercepts
$b^2 - 4ac = 0$	**One solution (a repeated solution) that is a real number:** If a, b, and c are rational numbers, the repeated solution is also a rational number.	One x-intercept
$b^2 - 4ac < 0$	**No real solution; two imaginary solutions:** The solutions are complex conjugates.	No x-intercepts

For each equation, compute the discriminant. Then determine the number and type of solutions:

a. $3x^2 + 4x - 5 = 0$ **b.** $9x^2 - 6x + 1 = 0$ **c.** $3x^2 - 8x + 7 = 0.$

Solution Begin by identifying the values for a, b, and c in each equation. Then compute $b^2 - 4ac$, the discriminant.

a. $3x^2 + 4x - 5 = 0$

$a = 3 \quad b = 4 \quad c = -5$

Substitute and compute the discriminant:

$$b^2 - 4ac = 4^2 - 4 \cdot 3(-5) = 16 - (-60) = 16 + 60 = 76.$$

The discriminant, 76, is a positive number that is not a perfect square. Thus, there are two real irrational solutions. (These solutions are conjugates of each other.)

b. $9x^2 - 6x + 1 = 0$

$a = 9 \quad b = -6 \quad c = 1$

Substitute and compute the discriminant:

$$b^2 - 4ac = (-6)^2 - 4 \cdot 9 \cdot 1 = 36 - 36 = 0.$$

The discriminant, 0, shows that there is only one real solution. This real solution is a rational number.

c. $3x^2 - 8x + 7 = 0$

$a = 3 \quad b = -8 \quad c = 7$

$$b^2 - 4ac = (-8)^2 - 4 \cdot 3 \cdot 7 = 64 - 84 = -20$$

The negative discriminant, -20, shows that there are two imaginary solutions. (These solutions are complex conjugates of each other.) ■

✓ **CHECK POINT 4** For each equation, compute the discriminant. Then determine the number and type of solutions:

a. $x^2 + 6x + 9 = 0$

b. $2x^2 - 7x - 4 = 0$

c. $3x^2 - 2x + 4 = 0.$

Great Question!

Is the square root sign part of the discriminant?

No. The discriminant is $b^2 - 4ac$. It is not $\sqrt{b^2 - 4ac}$, so do not give the discriminant as a radical.

3 Determine the most efficient method to use when solving a quadratic equation.

Determining Which Method to Use

All quadratic equations can be solved by the quadratic formula. However, if an equation is in the form $u^2 = d$, such as $x^2 = 5$ or $(2x + 3)^2 = 8$, it is faster to use the square root property, taking the square root of both sides. If the equation is not in the form $u^2 = d$, write the quadratic equation in standard form ($ax^2 + bx + c = 0$). Try to solve the equation by factoring. If $ax^2 + bx + c$ cannot be factored, then solve the quadratic equation by the quadratic formula.

Because we used the method of completing the square to derive the quadratic formula, we no longer need it for solving quadratic equations. However, we will use completing the square in Chapter 13 to help graph certain kinds of equations.

Table 11.3 summarizes our observations about which technique to use when solving a quadratic equation.

Table 11.3	Determining the Most Efficient Technique to Use When Solving a Quadratic Equation	
Description and Form of the Quadratic Equation	**Most Efficient Solution Method**	**Example**
$ax^2 + bx + c = 0$ and $ax^2 + bx + c$ can be factored easily.	Factor and use the zero-product principle.	$3x^2 + 5x - 2 = 0$ $(3x - 1)(x + 2) = 0$ $3x - 1 = 0$ or $x + 2 = 0$ $x = \dfrac{1}{3}$ $x = -2$
$ax^2 + c = 0$ The quadratic equation has no x-term. $(b = 0)$	Solve for x^2 and apply the square root property.	$4x^2 - 7 = 0$ $4x^2 = 7$ $x^2 = \dfrac{7}{4}$ $x = \pm\dfrac{\sqrt{7}}{2}$
$u^2 = d$; u is a first-degree polynomial.	Use the square root property.	$(x + 4)^2 = 5$ $x + 4 = \pm\sqrt{5}$ $x = -4 \pm \sqrt{5}$
$ax^2 + bx + c = 0$ and $ax^2 + bx + c$ cannot be factored or the factoring is too difficult.	Use the quadratic formula: $x = \dfrac{-b \pm \sqrt{b^2 - 4ac}}{2a}$.	$x^2 - 2x - 6 = 0$ $a = 1$ $b = -2$ $c = -6$ $x = \dfrac{-(-2) \pm \sqrt{(-2)^2 - 4(1)(-6)}}{2(1)}$ $= \dfrac{2 \pm \sqrt{4 - 4(1)(-6)}}{2(1)}$ $= \dfrac{2 \pm \sqrt{28}}{2} = \dfrac{2 \pm \sqrt{4}\sqrt{7}}{2}$ $= \dfrac{2 \pm 2\sqrt{7}}{2} = \dfrac{2(1 \pm \sqrt{7})}{2}$ $= 1 \pm \sqrt{7}$

4 Write quadratic equations from solutions.

Writing Quadratic Equations from Solutions

Using the zero-product principle, the equation $(x - 3)(x + 5) = 0$ has two solutions, 3 and -5. By applying the zero-product principle in reverse, we can find a quadratic equation that has two given numbers as its solutions.

The Zero-Product Principle in Reverse

If $A = 0$ or $B = 0$, then $AB = 0$.

EXAMPLE 5 Writing Equations from Solutions

Write a quadratic equation with the given solution set:

a. $\left\{-\dfrac{5}{3}, \dfrac{1}{2}\right\}$ **b.** $\left\{-2\sqrt{3}, 2\sqrt{3}\right\}$ **c.** $\{-5i, 5i\}$.

Zero (1960–1971), Jasper Johns. © 2011 Jasper Johns/VAGA.

The special properties of zero make it possible to write a quadratic equation from its solutions.

Solution

a. Because the solution set is $\left\{-\dfrac{5}{3}, \dfrac{1}{2}\right\}$, then

$$x = -\frac{5}{3} \quad \text{or} \qquad x = \frac{1}{2}.$$

$x + \dfrac{5}{3} = 0 \quad$ or $\quad x - \dfrac{1}{2} = 0$	Obtain zero on one side of each equation.
$3x + 5 = 0 \quad$ or $\quad 2x - 1 = 0$	Clear fractions, multiplying by 3 and 2, respectively.
$(3x + 5)(2x - 1) = 0$	Use the zero-product principle in reverse: If $A = 0$ or $B = 0$, then $AB = 0$.
$6x^2 - 3x + 10x - 5 = 0$	Use the FOIL method to multiply.
$6x^2 + 7x - 5 = 0$	Combine like terms.

Thus, one equation is $6x^2 + 7x - 5 = 0$. Many other quadratic equations have $\{-\frac{5}{3}, \frac{1}{2}\}$ for their solution sets. These equations can be obtained by multiplying both sides of $6x^2 + 7x - 5 = 0$ by any nonzero real number.

b. Because the solution set is $\left\{-2\sqrt{3}, 2\sqrt{3}\right\}$, then

$$x = -2\sqrt{3} \quad \text{or} \qquad x = 2\sqrt{3}.$$

$x + 2\sqrt{3} = 0 \qquad$ or $\quad x - 2\sqrt{3} = 0$	Obtain zero on one side of each equation.
$\left(x + 2\sqrt{3}\right)\left(x - 2\sqrt{3}\right) = 0$	Use the zero-product principle in reverse: If $A = 0$ or $B = 0$, then $AB = 0$.
$x^2 - \left(2\sqrt{3}\right)^2 = 0$	Multiply conjugates using $(A + B)(A - B) = A^2 - B^2$.
$x^2 - 12 = 0$	$(2\sqrt{3})^2 = 2^2(\sqrt{3})^2 = 4 \cdot 3 = 12$

Thus, one equation is $x^2 - 12 = 0$.

c. Because the solution set is $\{-5i, 5i\}$, then

$$x = -5i \quad \text{or} \qquad x = 5i.$$

$x + 5i = 0 \qquad$ or $\quad x - 5i = 0$	Obtain zero on one side of each equation.
$(x + 5i)(x - 5i) = 0$	Use the zero-product principle in reverse: If $A = 0$ or $B = 0$, then $AB = 0$.
$x^2 - (5i)^2 = 0$	Multiply conjugates using $(A + B)(A - B) = A^2 - B^2$.
$x^2 - 25i^2 = 0$	$(5i)^2 = 5^2 i^2 = 25i^2$
$x^2 - 25(-1) = 0$	$i^2 = -1$
$x^2 + 25 = 0$	This is the required equation. ∎

☑ **CHECK POINT 5** Write a quadratic equation with the given solution set:

a. $\left\{-\frac{3}{5}, \frac{1}{4}\right\}$ **b.** $\left\{-5\sqrt{2}, 5\sqrt{2}\right\}$

c. $\{-7i, 7i\}$.

---

5 Use the quadratic formula to solve problems.

Applications

Quadratic equations can be solved to answer questions about variables contained in quadratic functions.

EXAMPLE 6 Blood Pressure and Age

The graphs in **Figure 11.6** illustrate that a person's normal systolic blood pressure, measured in millimeters of mercury (mm Hg), depends on his or her age. The function

$$P(A) = 0.006A^2 - 0.02A + 120$$

models a man's normal systolic pressure, $P(A)$, at age A.

a. Find the age, to the nearest year, of a man whose normal systolic blood pressure is 125 mm Hg.

b. Use the graphs in **Figure 11.6** to describe the differences between the normal systolic blood pressures of men and women as they age.

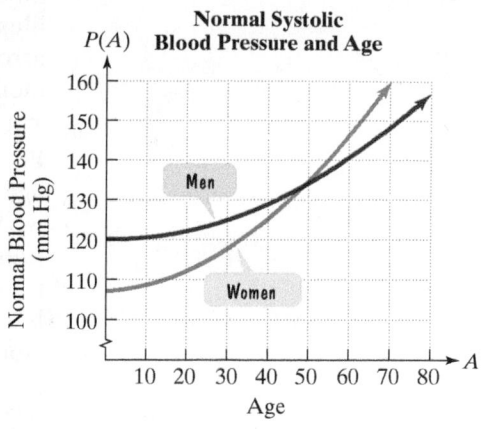

Figure 11.6

Solution

a. We are interested in the age of a man with a normal systolic blood pressure of 125 millimeters of mercury. Thus, we substitute 125 for $P(A)$ in the given function for men. Then we solve for A, the man's age.

$$P(A) = 0.006A^2 - 0.02A + 120 \qquad \text{This is the given function for men.}$$

$$125 = 0.006A^2 - 0.02A + 120 \qquad \text{Substitute 125 for } P(A).$$

$$0 = 0.006A^2 - 0.02A - 5 \qquad \text{Subtract 125 from both sides and write the quadratic equation in standard form.}$$

$$a = 0.006 \quad b = -0.02 \quad c = -5$$

Because the trinomial on the right side of the equation is prime, we solve using the quadratic formula.

Notice that the variable is A, rather than the usual x.

$$A = \frac{-b \pm \sqrt{b^2 - 4ac}}{2a} \qquad \text{Use the quadratic formula.}$$

$$= \frac{-(-0.02) \pm \sqrt{(-0.02)^2 - 4(0.006)(-5)}}{2(0.006)} \qquad \text{Substitute the values for } a, b, \text{ and } c: \\ a = 0.006, \\ b = -0.02, \text{ and} \\ c = -5.$$

$$= \frac{0.02 \pm \sqrt{0.1204}}{0.012} \qquad \text{Use a calculator to simplify the expression under the square root.}$$

$$\approx \frac{0.02 \pm 0.347}{0.012} \qquad \text{Use a calculator:} \\ \sqrt{0.1204} \approx 0.347.$$

$$A \approx \frac{0.02 + 0.347}{0.012} \quad \text{or} \quad A \approx \frac{0.02 - 0.347}{0.012}$$

$$A \approx 31 \qquad\qquad A \approx -27 \qquad \text{Use a calculator and round to the nearest integer.}$$

Reject this solution. Age cannot be negative.

Using Technology

On most calculators, here is how to approximate

$$\frac{0.02 + \sqrt{0.1204}}{0.012}.$$

Many Scientific Calculators

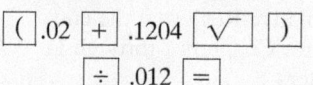

Many Graphing Calculators

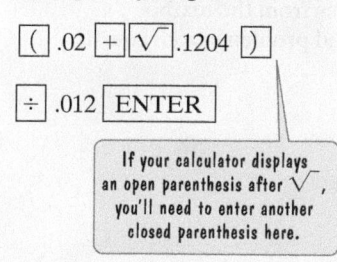

If your calculator displays an open parenthesis after √, you'll need to enter another closed parenthesis here.

The positive solution, $A \approx 31$, indicates that 31 is the approximate age of a man whose normal systolic blood pressure is 125 mm Hg. This is illustrated by the black lines with the arrows on the red graph representing men in **Figure 11.7**.

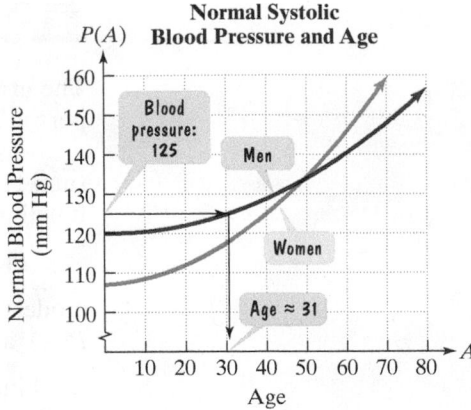

Figure 11.7

b. Take a second look at the graphs in **Figure 11.7**. Before approximately age 50, the blue graph representing women's normal systolic blood pressure lies below the red graph representing men's normal systolic blood pressure. Thus, up to age 50, women's normal systolic blood pressure is lower than men's, although it is increasing at a faster rate. After age 50, women's normal systolic blood pressure is higher than men's. ■

✓ **CHECK POINT 6** The function $P(A) = 0.01A^2 + 0.05A + 107$ models a woman's normal systolic blood pressure, $P(A)$, at age A. Use this function to find the age, to the nearest year, of a woman whose normal systolic blood pressure is 115 mm Hg. Use the blue graph in **Figure 11.7** to verify your solution.

Achieving Success

Many combined introductory and intermediate algebra courses cover only selected sections from this chapter and the book's remaining chapters. Regardless of the content requirements for your course, **it's never too early to start thinking about a final exam.** Here are some strategies to help you prepare for your final:

- Review your back exams. Be sure you understand any error that you made. Seek help with any concepts that are still unclear.

- Ask your professor if there are additional materials to help students review for the final. This includes review sheets and final exams from previous semesters.

- Attend any review sessions conducted by your professor or by the math department.

- Use the strategy first introduced on page 251: Imagine that your professor will permit two 3 by 5 index cards of notes on the final. Organize and create such a two-card summary for the most vital information in the course, including all important formulas. Refer to the chapter summaries in the textbook to prepare your personalized summary.

- For further review, work the relevant exercises in the Cumulative Review at the end of this chapter. The 41 exercises in the Cumulative Review for Chapters 1 through 11 cover most of the objectives in the book's first eleven chapters.

- Write your own final exam with detailed solutions for each item. You can use test questions from back exams in mixed order, worked examples from the textbook's chapter summaries, exercises in the Cumulative Reviews, and problems from course handouts. Use your test as a practice final exam.

CONCEPT AND VOCABULARY CHECK

Fill in each blank so that the resulting statement is true.

1. The solutions of a quadratic equation in standard form $ax^2 + bx + c = 0, a \neq 0$, are given by the quadratic formula

    $x = $ _____.

2. In order to solve $2x^2 + 9x - 5 = 0$ by the quadratic formula, we use $a = $ _____, $b = $ _____, and $c = $ _____.

3. In order to solve $x^2 = 4x + 1$ by the quadratic formula, we use $a = $ _____, $b = $ _____, and $c = $ _____.

4. $x = \dfrac{-(-4) \pm \sqrt{(-4)^2 - 4(1)(2)}}{2(1)}$ simplifies to $x = $ _____.

5. $x = \dfrac{-4 \pm \sqrt{4^2 - 4 \cdot 2 \cdot 5}}{2 \cdot 2}$ simplifies to $x = $ _____.

6. The discriminant of $ax^2 + bx + c = 0$ is defined by _____.

7. If the discriminant of $ax^2 + bx + c = 0$ is negative, the quadratic equation has _____ real solutions.

8. If the discriminant of $ax^2 + bx + c = 0$ is positive, the quadratic equation has _____ real solutions.

9. The most efficient technique for solving $(2x + 7)^2 = 25$ is by using _____.

10. The most efficient technique for solving $x^2 + 5x - 10 = 0$ is by using _____.

11. The most efficient technique for solving $x^2 + 8x + 15 = 0$ is by using _____.

12. True or false: An equation with the solution set $\{2, 5\}$ is $(x + 2)(x + 5) = 0$. _____

11.2 EXERCISE SET

MyMathLab®

Watch the videos in MyMathLab

Download the MyDashBoard App

Practice Exercises

In Exercises 1–18, solve each equation using the quadratic formula. Simplify solutions, if possible.

1. $x^2 + 8x + 12 = 0$
2. $x^2 + 8x + 15 = 0$
3. $2x^2 - 7x = -5$
4. $5x^2 + 8x = -3$
5. $x^2 + 3x - 20 = 0$
6. $x^2 + 5x - 10 = 0$
7. $3x^2 - 7x = 3$
8. $4x^2 + 3x = 2$
9. $6x^2 = 2x + 1$
10. $2x^2 = -4x + 5$
11. $4x^2 - 3x = -6$
12. $9x^2 + x = -2$
13. $x^2 - 4x + 8 = 0$

14. $x^2 + 6x + 13 = 0$
15. $3x^2 = 8x - 7$
16. $3x^2 = 4x - 6$
17. $2x(x - 2) = x + 12$
18. $2x(x + 4) = 3x - 3$

In Exercises 19–30, compute the discriminant. Then determine the number and type of solutions for the given equation.

19. $x^2 + 8x + 3 = 0$
20. $x^2 + 7x + 4 = 0$
21. $x^2 + 6x + 8 = 0$
22. $x^2 + 2x - 3 = 0$
23. $2x^2 + x + 3 = 0$
24. $2x^2 - 4x + 3 = 0$
25. $2x^2 + 6x = 0$
26. $3x^2 - 5x = 0$
27. $5x^2 + 3 = 0$
28. $5x^2 + 4 = 0$
29. $9x^2 = 12x - 4$
30. $4x^2 = 20x - 25$

In Exercises 31–50, solve each equation by the method of your choice. Simplify solutions, if possible.

31. $3x^2 - 4x = 4$

32. $2x^2 - x = 1$

33. $x^2 - 2x = 1$

34. $2x^2 + 3x = 1$

35. $3x^2 = x - 9$

36. $2x^2 = -6x - 7$

37. $(2x - 5)(x + 1) = 2$

38. $(2x + 3)(x + 4) = 1$

39. $(3x - 4)^2 = 16$

40. $(2x + 7)^2 = 25$

41. $\dfrac{x^2}{2} + 2x + \dfrac{2}{3} = 0$

42. $\dfrac{x^2}{3} - x - \dfrac{1}{6} = 0$

43. $(3x - 2)^2 = 10$

44. $(4x - 1)^2 = 15$

45. $\dfrac{1}{x} + \dfrac{1}{x + 2} = \dfrac{1}{3}$

46. $\dfrac{1}{x} + \dfrac{1}{x + 3} = \dfrac{1}{4}$

47. $(2x - 6)(x + 2) = 5(x - 1) - 12$

48. $7x(x - 2) = 3 - 2(x + 4)$

49. $x^2 + 10 = 2(2x - 1)$

50. $x(x + 6) = -12$

In Exercises 51–64, write a quadratic equation in standard form with the given solution set.

51. $\{-3, 5\}$

52. $\{-2, 6\}$

53. $\left\{-\dfrac{2}{3}, \dfrac{1}{4}\right\}$

54. $\left\{-\dfrac{5}{6}, \dfrac{1}{3}\right\}$

55. $\left\{-\sqrt{2}, \sqrt{2}\right\}$

56. $\left\{-\sqrt{3}, \sqrt{3}\right\}$

57. $\left\{-2\sqrt{5}, 2\sqrt{5}\right\}$

58. $\left\{-3\sqrt{5}, 3\sqrt{5}\right\}$

59. $\{-6i, 6i\}$

60. $\{-8i, 8i\}$

61. $\{1 + i, 1 - i\}$

62. $\{2 + i, 2 - i\}$

63. $\left\{1 + \sqrt{2}, 1 - \sqrt{2}\right\}$

64. $\left\{1 + \sqrt{3}, 1 - \sqrt{3}\right\}$

Practice PLUS

Exercises 65–68 describe quadratic equations. Match each description with the graph of the corresponding quadratic function. Each graph is shown in a $[-10, 10, 1]$ by $[-10, 10, 1]$ viewing rectangle.

65. A quadratic equation whose solution set contains imaginary numbers

66. A quadratic equation whose discriminant is 0

67. A quadratic equation whose solution set is $\left\{3 \pm \sqrt{2}\right\}$

68. A quadratic equation whose solution set contains integers

a.

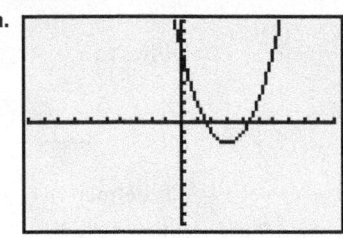

b.

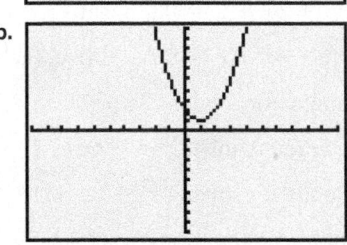

c.

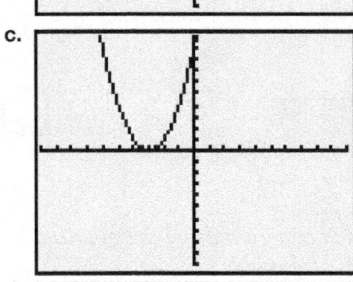

d.

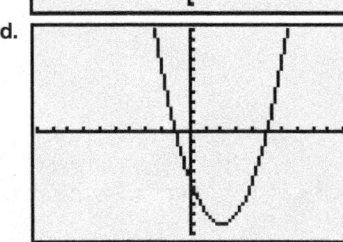

69. When the sum of 6 and twice a positive number is subtracted from the square of the number, 0 results. Find the number.

70. When the sum of 1 and twice a negative number is subtracted from twice the square of the number, 0 results. Find the number.

In Exercises 71–76, solve each equation by the method of your choice.

71. $\dfrac{1}{x^2 - 3x + 2} = \dfrac{1}{x + 2} + \dfrac{5}{x^2 - 4}$

72. $\dfrac{x - 1}{x - 2} + \dfrac{x}{x - 3} = \dfrac{1}{x^2 - 5x + 6}$

73. $\sqrt{2}x^2 + 3x - 2\sqrt{2} = 0$

74. $\sqrt{3}x^2 + 6x + 7\sqrt{3} = 0$

75. $|x^2 + 2x| = 3$

76. $|x^2 + 3x| = 2$

Application Exercises

A driver's age has something to do with his or her chance of getting into a fatal car crash. The bar graph shows the number of fatal vehicle crashes per 100 million miles driven for drivers of various age groups. For example, 25-year-old drivers are involved in 4.1 fatal crashes per 100 million miles driven. Thus, when a group of 25-year-old Americans have driven a total of 100 million miles, approximately 4 have been in accidents in which someone died.

Age of United States Drivers and Fatal Crashes

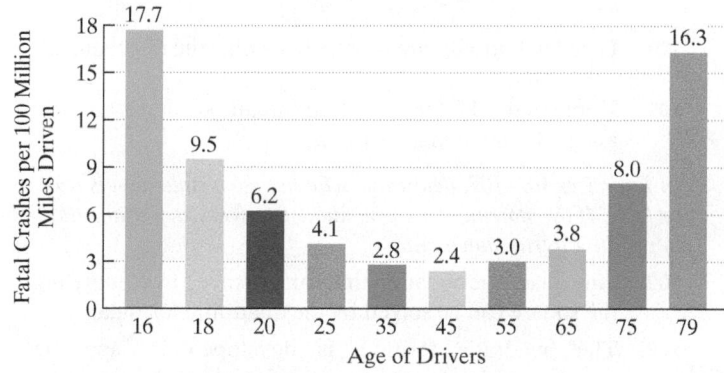

Source: Insurance Institute for Highway Safety

The number of fatal vehicle crashes per 100 million miles, $f(x)$, for drivers of age x can be modeled by the quadratic function

$$f(x) = 0.013x^2 - 1.19x + 28.24.$$

Use the function to solve Exercises 77–78.

77. What age groups are expected to be involved in 3 fatal crashes per 100 million miles driven? How well does the function model the trend in the actual data shown in the bar graph?

78. What age groups are expected to be involved in 10 fatal crashes per 100 million miles driven? How well does the function model the trend in the actual data shown in the bar graph?

Throwing events in track and field include the shot put, the discus throw, the hammer throw, and the javelin throw. The distance that an athlete can achieve depends on the initial velocity of the object thrown and the angle above the horizontal at which the object leaves the hand.

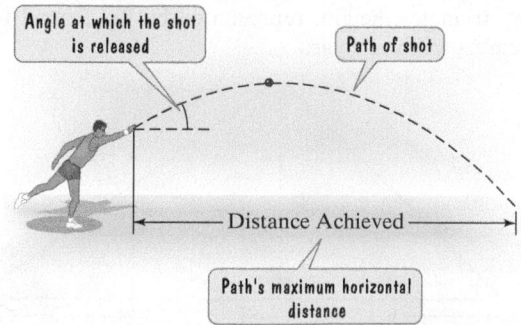

In Exercises 79–80, an athlete whose event is the shot put releases the shot with the same initial velocity, but at different angles.

79. When the shot is released at an angle of 35°, its path can be modeled by the function

$$f(x) = -0.01x^2 + 0.7x + 6.1,$$

in which x is the shot's horizontal distance, in feet, and $f(x)$ is its height, in feet. This function is shown by one of the graphs, (a) or (b), in the figure. Use the function to determine the shot's maximum distance. Use a calculator and round to the nearest tenth of a foot. Which graph, (a) or (b), shows the shot's path?

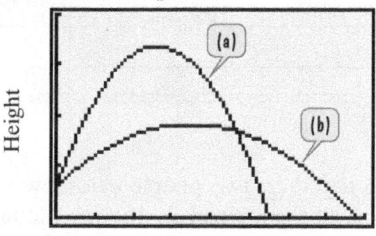

Horizontal Distance
$[0, 80, 10]$ by $[0, 40, 10]$

80. When the shot is released at an angle of 65°, its path can be modeled by the function

$$f(x) = -0.04x^2 + 2.1x + 6.1,$$

in which x is the shot's horizontal distance, in feet, and $f(x)$ is its height, in feet. This function is shown by one of the graphs, (a) or (b), in the figure above. Use the function to determine the shot's maximum distance. Use a calculator and round to the nearest tenth of a foot. Which graph, (a) or (b), shows the shot's path?

81. The length of a rectangle is 4 meters longer than the width. If the area is 8 square meters, find the rectangle's dimensions. Round to the nearest tenth of a meter.

82. The length of a rectangle exceeds twice its width by 3 inches. If the area is 10 square inches, find the rectangle's dimensions. Round to the nearest tenth of an inch.

83. The longer leg of a right triangle exceeds the shorter leg by 1 inch, and the hypotenuse exceeds the longer leg by 7 inches. Find the lengths of the legs. Round to the nearest tenth of a inch.

84. The hypotenuse of a right triangle is 6 feet long. One leg is 2 feet shorter than the other. Find the lengths of the legs. Round to the nearest tenth of a foot.

85. A rain gutter is made from sheets of aluminum that are 20 inches wide. As shown in the figure, the edges are turned up to form right angles. Determine the depth of the gutter that will allow a cross-sectional area of 13 square inches. Show that there are two different solutions to the problem. Round to the nearest tenth of an inch.

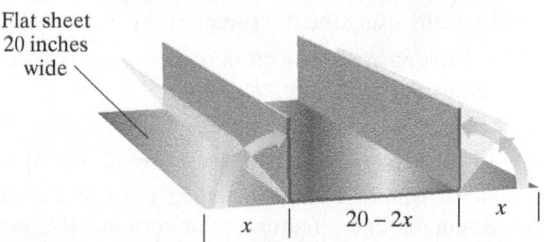

86. A piece of wire is 8 inches long. The wire is cut into two pieces and then each piece is bent into a square. Find the length of each piece if the sum of the areas of these squares is to be 2 square inches.

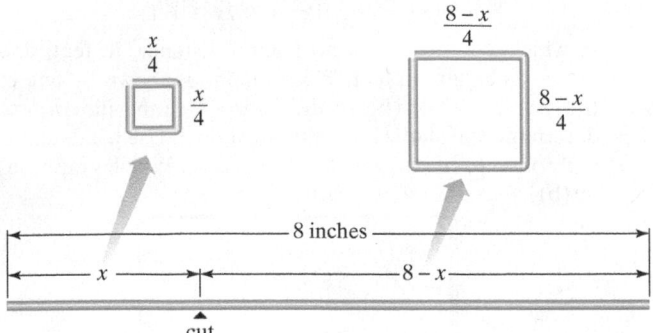

87. Working together, two people can mow a large lawn in 4 hours. One person can do the job alone 1 hour faster than the other person. How long does it take each person working alone to mow the lawn? Round to the nearest tenth of an hour.

88. A pool has an inlet pipe to fill it and an outlet pipe to empty it. It takes 2 hours longer to empty the pool than it does to fill it. The inlet pipe is turned on to fill the pool, but the outlet pipe is accidentally left open. Despite this, the pool fills in 8 hours. How long does it take the outlet pipe to empty the pool? Round to the nearest tenth of an hour.

Writing in Mathematics

89. What is the quadratic formula and why is it useful?

90. Without going into specific details for every step, describe how the quadratic formula is derived.

91. Explain how to solve $x^2 + 6x + 8 = 0$ using the quadratic formula.

92. If a quadratic equation has imaginary solutions, how is this shown on the graph of the corresponding quadratic function?

93. What is the discriminant and what information does it provide about a quadratic equation?

94. If you are given a quadratic equation, how do you determine which method to use to solve it?

95. Explain how to write a quadratic equation from its solution set. Give an example with your explanation.

Technology Exercises

96. Use a graphing utility to graph the quadratic function related to any five of the quadratic equations in Exercises 19–30. How does each graph illustrate what you determined algebraically using the discriminant?

97. Reread Exercise 85. The cross-sectional area of the gutter is given by the quadratic function

$$f(x) = x(20 - 2x).$$

Graph the function in a [0, 10, 1] by [0, 60, 5] viewing rectangle. Then ⃞ TRACE ⃞ along the curve or use the maximum function feature to determine the depth of

the gutter that will maximize its cross-sectional area and allow the greatest amount of water to flow. What is the maximum area? Does the situation described in Exercise 85 take full advantage of the sheets of aluminum?

Critical Thinking Exercises

Make Sense? *In Exercises 98–101, determine whether each statement "makes sense" or "does not make sense" and explain your reasoning.*

98. Because I want to solve $25x^2 - 169 = 0$ fairly quickly, I'll use the quadratic formula.

99. I simplified $\dfrac{3 + 2\sqrt{3}}{2}$ to $3 + \sqrt{3}$ because 2 is a factor of $2\sqrt{3}$.

100. I need to find a square root to determine the discriminant.

101. I obtained -17 for the discriminant, so there are two imaginary irrational solutions.

In Exercises 102–105, determine whether each statement is true or false. If the statement is false, make the necessary change(s) to produce a true statement.

102. Any quadratic equation that can be solved by completing the square can be solved by the quadratic formula.

103. The quadratic formula is developed by applying factoring and the zero-product principle to the quadratic equation $ax^2 + bx + c = 0$.

104. In using the quadratic formula to solve the quadratic equation $5x^2 = 2x - 7$, we have $a = 5, b = 2$, and $c = -7$.

105. The quadratic formula can be used to solve the equation $x^2 - 9 = 0$.

106. Solve for t: $s = -16t^2 + v_0 t$.

107. A rectangular swimming pool is 12 meters long and 8 meters wide. A tile border of uniform width is to be built around the pool using 120 square meters of tile. The tile is from a discontinued stock (so no additional materials are available) and all 120 square meters are to be used. How wide should the border be? Round to the nearest tenth of a meter. If zoning laws require at least a 2-meter-wide border around the pool, can this be done with the available tile?

108. The area of the shaded green region outside the rectangle and inside the triangle is 10 square yards. Find the triangle's height, represented by $2x$. Round to the nearest tenth of a yard.

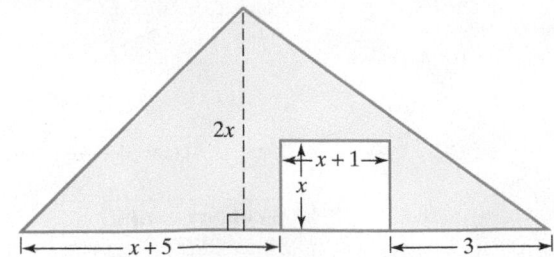

Review Exercises

109. Solve: $|5x + 2| = |4 - 3x|$. (Section 9.3, Example 3)

110. Solve: $\sqrt{2x - 5} - \sqrt{x - 3} = 1$.
(Section 10.6, Example 4)

111. Rationalize the denominator: $\dfrac{5}{\sqrt{3} + x}$. (Section 10.5, Example 5)

Preview Exercises

Exercises 112–114 will help you prepare for the material covered in the next section.

112. Use point plotting to graph $f(x) = x^2$ and $g(x) = x^2 + 2$ in the same rectangular coordinate system.

113. Use point plotting to graph $f(x) = x^2$ and $g(x) = (x + 2)^2$ in the same rectangular coordinate system.

114. Find the x-intercepts for the graph of $f(x) = -2(x - 3)^2 + 8$.

SECTION 11.3

Quadratic Functions and Their Graphs

Objectives

1 Recognize characteristics of parabolas.

2 Graph parabolas in the form $f(x) = a(x - h)^2 + k$.

3 Graph parabolas in the form $f(x) = ax^2 + bx + c$.

4 Determine a quadratic function's minimum or maximum value.

5 Solve problems involving a quadratic function's minimum or maximum value.

We have a long history of throwing things. Before 400 B.C., the Greeks competed in games that included discus throwing. In the seventeenth century, English soldiers organized cannonball-throwing competitions. In 1827, a Yale University student, disappointed over failing an exam, took out his frustrations at the passing of a collection plate in chapel. Seizing the monetary tray, he flung it in the direction of a large open space on campus. Yale students see this act of frustration as the origin of the Frisbee.

In this section, we study quadratic functions and their graphs. By graphing functions that model the paths of the things we throw, you will be able to determine both the maximum height these objects attain and the distance these objects travel.

1 Recognize characteristics of parabolas.

Graphs of Quadratic Functions

The graph of any quadratic function

$$f(x) = ax^2 + bx + c, \quad a \neq 0,$$

is called a **parabola**. Parabolas are shaped like bowls or inverted bowls, as shown in **Figure 11.8** at the top of the next page. If the coefficient of x^2 (the value of a in $ax^2 + bx + c$) is positive, the parabola opens upward. If the coefficient of x^2 is negative, the parabola opens downward. The **vertex** (or turning point) of the parabola is the lowest point on the graph when it opens upward and the highest point on the graph when it opens downward.

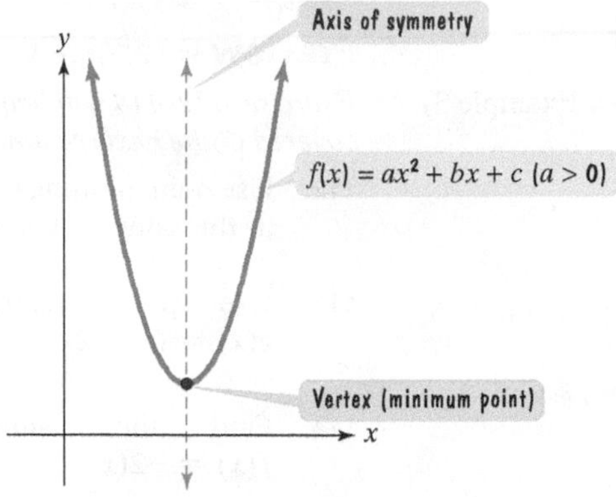

Axis of symmetry

$f(x) = ax^2 + bx + c \ (a > 0)$

Vertex (minimum point)

Vertex (maximum point)

$f(x) = ax^2 + bx + c \ (a < 0)$

Axis of symmetry

$a > 0$: Parabola opens upward.

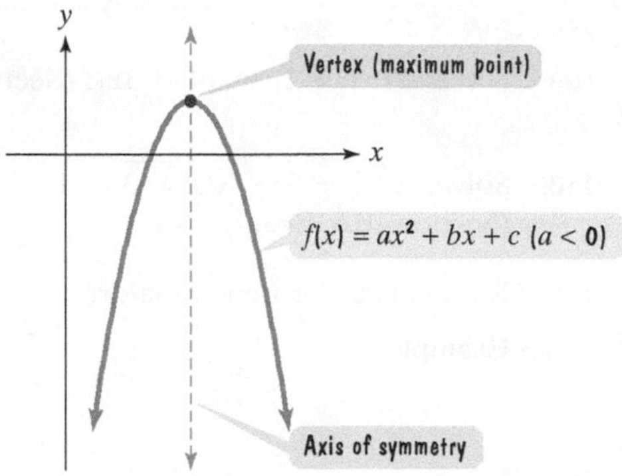

$a < 0$: Parabola opens downward.

Figure 11.8 Characteristics of graphs of quadratic functions

The two halves of a parabola are mirror images of each other. A "mirror line" through the vertex, called the **axis of symmetry**, divides the figure in half. If a parabola is folded along its axis of symmetry, the two halves match exactly.

2 Graph parabolas in the form $f(x) = a(x - h)^2 + k$.

Graphing Quadratic Functions in the Form $f(x) = a(x - h)^2 + k$

One way to obtain the graph of a quadratic function is to use point plotting. Let's begin by graphing the functions $f(x) = x^2$, $g(x) = 2x^2$, and $h(x) = \frac{1}{2}x^2$ in the same rectangular coordinate system. Select integers for x, starting with -3 and ending with 3. A partial table of coordinates for each function is shown below. The three parabolas are shown in **Figure 11.9**.

x	$f(x) = x^2$	(x, y) or $(x, f(x))$	x	$g(x) = 2x^2$	(x, y) or $(x, g(x))$
-3	$f(-3) = (-3)^2 = 9$	$(-3, 9)$	-3	$g(-3) = 2(-3)^2 = 18$	$(-3, 18)$
-2	$f(-2) = (-2)^2 = 4$	$(-2, 4)$	-2	$g(-2) = 2(-2)^2 = 8$	$(-2, 8)$
-1	$f(-1) = (-1)^2 = 1$	$(-1, 1)$	-1	$g(-1) = 2(-1)^2 = 2$	$(-1, 2)$
0	$f(0) = 0^2 = 0$	$(0, 0)$	0	$g(0) = 2 \cdot 0^2 = 0$	$(0, 0)$
1	$f(1) = 1^2 = 1$	$(1, 1)$	1	$g(1) = 2 \cdot 1^2 = 2$	$(1, 2)$
2	$f(2) = 2^2 = 4$	$(2, 4)$	2	$g(2) = 2 \cdot 2^2 = 8$	$(2, 8)$
3	$f(3) = 3^2 = 9$	$(3, 9)$	3	$g(3) = 2 \cdot 3^2 = 18$	$(3, 18)$

x	$h(x) = \dfrac{1}{2}x^2$	(x, y) or $(x, h(x))$
-3	$h(-3) = \dfrac{1}{2}(-3)^2 = \dfrac{9}{2}$	$\left(-3, \dfrac{9}{2}\right)$
-2	$h(-2) = \dfrac{1}{2}(-2)^2 = 2$	$(-2, 2)$
-1	$h(-1) = \dfrac{1}{2}(-1)^2 = \dfrac{1}{2}$	$\left(-1, \dfrac{1}{2}\right)$
0	$h(0) = \dfrac{1}{2} \cdot 0^2 = 0$	$(0, 0)$
1	$h(1) = \dfrac{1}{2} \cdot 1^2 = \dfrac{1}{2}$	$\left(1, \dfrac{1}{2}\right)$
2	$h(2) = \dfrac{1}{2} \cdot 2^2 = 2$	$(2, 2)$
3	$h(3) = \dfrac{1}{2} \cdot 3^2 = \dfrac{9}{2}$	$\left(3, \dfrac{9}{2}\right)$

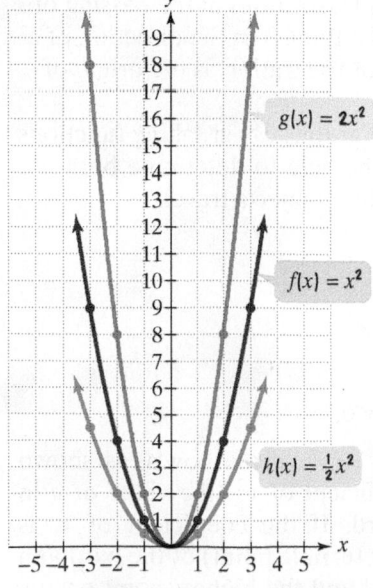

$g(x) = 2x^2$

$f(x) = x^2$

$h(x) = \frac{1}{2}x^2$

Figure 11.9

Can you see that the graphs of f, g, and h in **Figure 11.9** all have the same vertex, $(0, 0)$? They also have the same axis of symmetry, the y-axis, or $x = 0$. This is true for all graphs of the form $f(x) = ax^2$. However, the blue graph of $g(x) = 2x^2$ is a narrower parabola than the red graph of $f(x) = x^2$. By contrast, the green graph of $h(x) = \frac{1}{2}x^2$ is a flatter parabola than the red graph of $f(x) = x^2$.

Is there a more efficient method than point plotting to obtain the graph of a quadratic function? The answer is yes. The method is based on comparing graphs of the form $g(x) = a(x - h)^2 + k$ to those of the form $f(x) = ax^2$.

In **Figure 11.10(a)**, the graph of $f(x) = ax^2$ for $a > 0$ is shown in black. The parabola's vertex is $(0, 0)$ and it opens upward. In **Figure 11.10(b)**, the graph of $f(x) = ax^2$ for $a < 0$ is shown in black. The parabola's vertex is $(0, 0)$ and it opens downward.

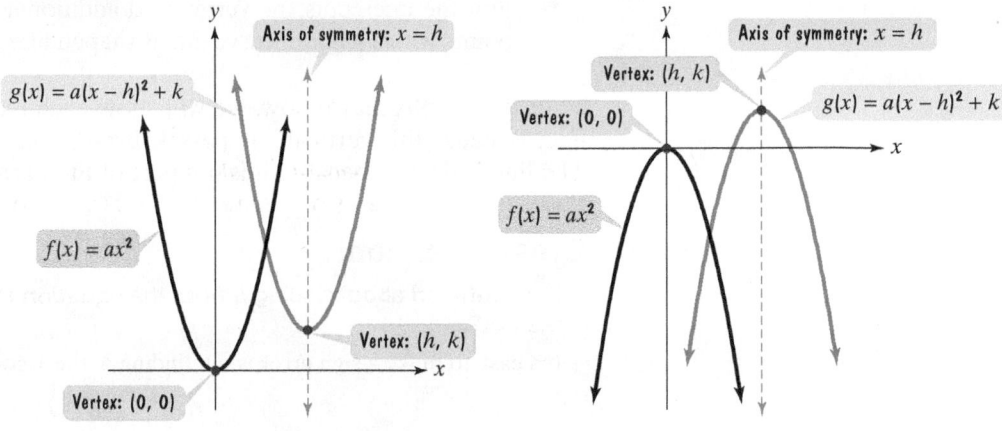

(a) $a > 0$: Parabola opens upward.   **(b)** $a < 0$: Parabola opens downward.

Figure 11.10 Moving, or shifting, the graph of $f(x) = ax^2$

Figures 11.10(a) and **11.10(b)** also show the graph of $g(x) = a(x - h)^2 + k$ in blue. Compare these graphs to those of $f(x) = ax^2$. Observe that h determines a horizontal move, or shift, and k determines a vertical move, or shift, of the graph of $f(x) = ax^2$:

$$g(x) = a(x - h)^2 + k.$$

If $h > 0$, the graph of $f(x) = ax^2$ is shifted h units to the right.

If $k > 0$, the graph of $y = a(x - h)^2$ is shifted k units up.

Consequently, the vertex $(0, 0)$ on the black graph of $f(x) = ax^2$ moves to the point (h, k) on the blue graph of $g(x) = a(x - h)^2 + k$. The axis of symmetry is the vertical line whose equation is $x = h$.

The form of the expression for g is convenient because it immediately identifies the vertex of the parabola as (h, k).

Quadratic Functions in the Form $f(x) = a(x - h)^2 + k$

The graph of

$$f(x) = a(x - h)^2 + k, \quad a \neq 0$$

is a parabola whose vertex is the point (h, k). The parabola is symmetric with respect to the line $x = h$. If $a > 0$, the parabola opens upward; if $a < 0$, the parabola opens downward.

The sign of a in $f(x) = a(x - h)^2 + k$ determines whether the parabola opens upward or downward. Furthermore, if $|a|$ is small, the parabola opens more flatly than if $|a|$ is large. On the next page is a general procedure for graphing parabolas whose equations are in this form.

Graphing Quadratic Functions with Equations in the Form $f(x) = a(x - h)^2 + k$

To graph $f(x) = a(x - h)^2 + k$,

1. Determine whether the parabola opens upward or downward. If $a > 0$, it opens upward. If $a < 0$, it opens downward.
2. Determine the vertex of the parabola. The vertex is (h, k).
3. Find any x-intercepts by solving $f(x) = 0$. The equation's real solutions are the x-intercepts.
4. Find the y-intercept by computing $f(0)$.
5. Plot the intercepts, the vertex, and additional points as necessary. Connect these points with a smooth curve that is shaped like a bowl or an inverted bowl.

In the graphs that follow, we will show each axis of symmetry as a dashed vertical line. Because this vertical line passes through the vertex, (h, k), its equation is $x = h$. The line is dashed because it is not part of the parabola.

Great Question!

I'm confused about finding h from the equation $f(x) = a(x - h)^2 + k$. Can you help me out?

It's easy to make a sign error when finding h, the x-coordinate of the vertex. In

$$f(x) = a(x - h)^2 + k,$$

h **is the number that follows the subtraction sign.**

- $f(x) = -2(x - 3)^2 + 8$

  The number *after* the subtraction is 3: $h = 3$.

- $f(x) = (x + 3)^2 + 1$
  $= (x - (-3))^2 + 1$

  The number *after* the subtraction is −3: $h = -3$.

EXAMPLE 1 Graphing a Quadratic Function in the Form $f(x) = a(x - h)^2 + k$

Graph the quadratic function $f(x) = -2(x - 3)^2 + 8$.

Solution We can graph this function by following the steps in the preceding box. We begin by identifying values for a, h, and k.

$$f(x) = a(x - h)^2 + k$$

$$a = -2 \quad h = 3 \quad k = 8$$

$$f(x) = -2(x - 3)^2 + 8$$

Step 1. Determine how the parabola opens. Note that a, the coefficient of x^2, is -2. Thus, $a < 0$; this negative value tells us that the parabola opens downward.

Step 2. Find the vertex. The vertex of the parabola is at (h, k). Because $h = 3$ and $k = 8$, the parabola has its vertex at $(3, 8)$.

Step 3. Find the x-intercepts by solving f(x) = 0. Replace $f(x)$ with 0 in $f(x) = -2(x - 3)^2 + 8$.

$$0 = -2(x - 3)^2 + 8 \qquad \text{Find x-intercepts, setting } f(x) \text{ equal to 0.}$$
$$2(x - 3)^2 = 8 \qquad \text{Solve for x. Add } 2(x - 3)^2 \text{ to both sides of the equation.}$$

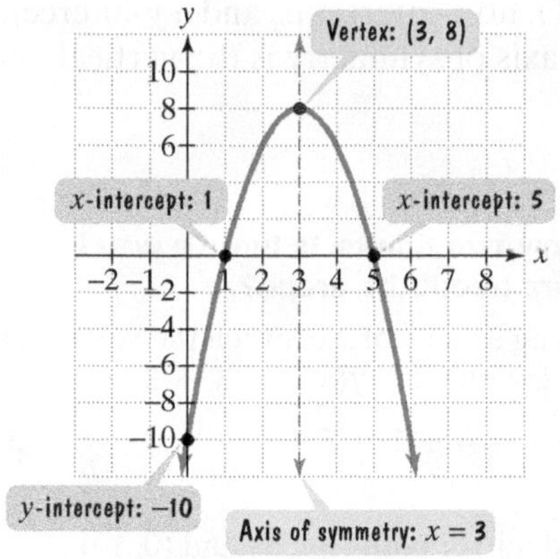

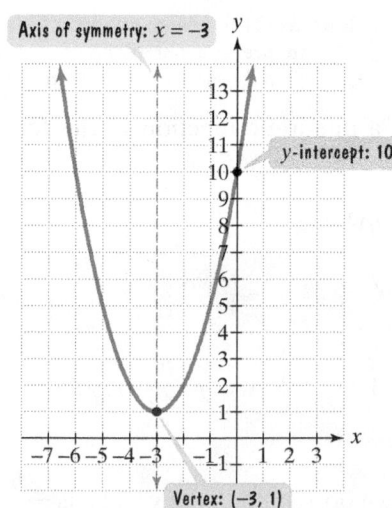

Vertex: (3, 8)

x-intercept: 1 x-intercept: 5

y-intercept: −10

Axis of symmetry: $x = 3$

Figure 11.11 The graph of $f(x) = -2(x - 3)^2 + 8$

$$(x - 3)^2 = 4 \qquad \text{Divide both sides by 2.}$$
$$x - 3 = \sqrt{4} \quad \text{or} \quad x - 3 = -\sqrt{4} \qquad \text{Apply the square root property.}$$
$$x - 3 = 2 \qquad\qquad x - 3 = -2 \qquad \sqrt{4} = 2$$
$$x = 5 \qquad\qquad x = 1 \qquad \text{Add 3 to both sides in each equation.}$$

The x-intercepts are 5 and 1. The parabola passes through $(5, 0)$ and $(1, 0)$.

Step 4. Find the y-intercept by computing f(0). Replace x with 0 in $f(x) = -2(x - 3)^2 + 8$.

$$f(0) = -2(0 - 3)^2 + 8 = -2(-3)^2 + 8 = -2(9) + 8 = -10$$

The y-intercept is -10. The parabola passes through $(0, -10)$.

Step 5. Graph the parabola. With a vertex at $(3, 8)$, x-intercepts at 5 and 1, and a y-intercept at -10, the graph of f is shown in **Figure 11.11**. The axis of symmetry is the vertical line whose equation is $x = 3$. ∎

✓ **CHECK POINT 1**   Graph the quadratic function $f(x) = -(x - 1)^2 + 4$.

EXAMPLE 2   Graphing a Quadratic Function in the Form $f(x) = a(x - h)^2 + k$

Graph the quadratic function $f(x) = (x + 3)^2 + 1$.

Solution   We begin by finding values for $a, h,$ and k.

$$f(x) = a(x - h)^2 + k \qquad \text{Form of quadratic function}$$
$$f(x) = (x + 3)^2 + 1 \qquad \text{Given function}$$
$$f(x) = 1(x - (-3))^2 + 1$$

$$a = 1 \qquad h = -3 \qquad k = 1$$

Step 1. Determine how the parabola opens. Note that a, the coefficient of x^2, is 1. Thus, $a > 0$; this positive value tells us that the parabola opens upward.

Step 2. Find the vertex. The vertex of the parabola is at (h, k). Because $h = -3$ and $k = 1$, the parabola has its vertex at $(-3, 1)$. See **Figure 11.12**.

Step 3. Find the x-intercepts by solving f(x) = 0. Replace $f(x)$ with 0 in $f(x) = (x + 3)^2 + 1$. Because the vertex is at $(-3, 1)$, which lies above the x-axis, and the parabola opens upward, it appears that this parabola has no x-intercepts. We can verify this observation algebraically.

$$0 = (x + 3)^2 + 1 \qquad \text{Find possible x-intercepts, setting } f(x) \text{ equal to 0.}$$

$$-1 = (x + 3)^2 \qquad \text{Solve for x. Subtract 1 from both sides.}$$

$$x + 3 = \sqrt{-1} \quad \text{or} \quad x + 3 = -\sqrt{-1} \qquad \text{Apply the square root property.}$$
$$x + 3 = i \qquad\qquad x + 3 = -i \qquad \sqrt{-1} = i$$
$$x = -3 + i \qquad\qquad x = -3 - i \qquad \text{The solutions are } -3 \pm i.$$

Because this equation has no real solutions, the parabola has no x-intercepts.

Step 4. Find the y-intercept by computing f(0). Replace x with 0 in $f(x) = (x + 3)^2 + 1$.

$$f(0) = (0 + 3)^2 + 1 = 3^2 + 1 = 9 + 1 = 10$$

The y-intercept is 10. The parabola passes through $(0, 10)$.

Axis of symmetry: $x = -3$

y-intercept: 10

Vertex: (−3, 1)

Figure 11.12 The graph of $f(x) = (x + 3)^2 + 1$

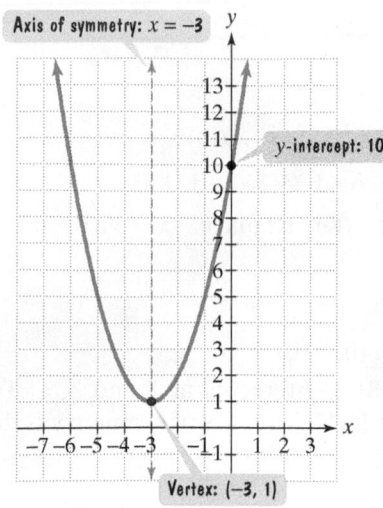

Axis of symmetry: $x = -3$

y-intercept: 10

Vertex: (−3, 1)

Figure 11.12 (repeated)

Step 5. Graph the parabola. With a vertex at $(-3, 1)$, no x-intercepts, and a y-intercept at 10, the graph of f is shown in **Figure 11.12**. The axis of symmetry is the vertical line whose equation is $x = -3$. ∎

Great Question!

You graphed the parabola in Figure 11.12 using only two points. Is there a way I can find some additional points to feel more secure about the graph?

Yes. You can find an additional point or two by choosing a value of x between the vertex and the y-intercept. For example, let $x = -1$. Replace x with -1 in $f(x) = (x + 3)^2 + 1$.

$$f(-1) = (-1 + 3)^2 + 1 = 2^2 + 1 = 4 + 1 = 5$$

The parabola passes through $(-1, 5)$.

More points? You can use the axis of symmetry and mirror both $(-1, 5)$ and $(0, 10)$. Mirroring $(-1, 5)$ gives the point $(-5, 5)$. Mirroring $(0, 10)$ gives the point $(-6, 10)$. Take a moment to identify these points in **Figure 11.12**.

✓ **CHECK POINT 2** Graph the quadratic function $f(x) = (x - 2)^2 + 1$.

---

3 Graph parabolas in the form $f(x) = ax^2 + bx + c$.

Graphing Quadratic Functions in the Form $f(x) = ax^2 + bx + c$

Quadratic functions are frequently expressed in the form $f(x) = ax^2 + bx + c$. How can we identify the vertex of a parabola whose equation is in this form? Completing the square provides the answer to this question.

$$f(x) = ax^2 + bx + c$$

$$= a\left(x^2 + \frac{b}{a}x\right) + c \qquad \text{Factor out } a \text{ from } ax^2 + bx.$$

$$= a\left(x^2 + \frac{b}{a}x + \frac{b^2}{4a^2}\right) + c - a\left(\frac{b^2}{4a^2}\right)$$

Complete the square by adding the square of half the coefficient of x.

By completing the square, we added $a \cdot \dfrac{b^2}{4a^2}$. To avoid changing the function's equation, we must subtract this term.

$$= a\left(x + \frac{b}{2a}\right)^2 + c - \frac{b^2}{4a} \qquad \text{Write the trinomial as the square of a binomial and simplify the constant term.}$$

Now let's compare the form of this equation with a quadratic function in the form $f(x) = a(x - h)^2 + k$.

The form we know how to graph
$$f(x) = a(x - h)^2 + k$$

$$h = -\frac{b}{2a} \qquad k = c - \frac{b^2}{4a}$$

Equation under discussion
$$f(x) = a\left(x - \left(-\frac{b}{2a}\right)\right)^2 + c - \frac{b^2}{4a}$$

The important part of this observation is that h, the x-coordinate of the vertex, is $-\dfrac{b}{2a}$.

The y-coordinate can be found by evaluating the function at $-\dfrac{b}{2a}$.

The Vertex of a Parabola Whose Equation Is $f(x) = ax^2 + bx + c$

Consider the parabola defined by the quadratic function $f(x) = ax^2 + bx + c$. The parabola's vertex is $\left(-\dfrac{b}{2a}, f\left(-\dfrac{b}{2a}\right)\right)$. The x-coordinate is $-\dfrac{b}{2a}$. The y-coordinate is found by substituting the x-coordinate into the parabola's equation and evaluating the function at this value of x.

EXAMPLE 3   Finding a Parabola's Vertex

Find the vertex for the parabola whose equation is $f(x) = 3x^2 + 12x + 8$.

Solution   We know that the x-coordinate of the vertex is $x = -\dfrac{b}{2a}$. Let's identify the numbers a, b, and c in the given equation, which is in the form $f(x) = ax^2 + bx + c$.

$$f(x) = 3x^2 + 12x + 8$$

$$a = 3 \qquad b = 12 \qquad c = 8$$

Substitute the values of a and b into the equation for the x-coordinate:

$$x = -\frac{b}{2a} = -\frac{12}{2 \cdot 3} = -\frac{12}{6} = -2.$$

The x-coordinate of the vertex is -2. We substitute -2 for x into the equation of the function, $f(x) = 3x^2 + 12x + 8$, to find the y-coordinate:

$$f(-2) = 3(-2)^2 + 12(-2) + 8 = 3(4) + 12(-2) + 8 = 12 - 24 + 8 = -4.$$

The vertex is $(-2, -4)$. ∎

✓ **CHECK POINT 3**   Find the vertex for the parabola whose equation is $f(x) = 2x^2 + 8x - 1$.

We can apply our five-step procedure and graph parabolas in the form $f(x) = ax^2 + bx + c$.

Graphing Quadratic Functions with Equations in the Form $f(x) = ax^2 + bx + c$

To graph $f(x) = ax^2 + bx + c$,

1. Determine whether the parabola opens upward or downward. If $a > 0$, it opens upward. If $a < 0$, it opens downward.

2. Determine the vertex of the parabola. The vertex is $\left(-\dfrac{b}{2a}, f\left(-\dfrac{b}{2a}\right)\right)$.

3. Find any x-intercepts by solving $f(x) = 0$. The real solutions of $ax^2 + bx + c = 0$ are the x-intercepts.

4. Find the y-intercept by computing $f(0)$. Because $f(0) = c$ (the constant term in the function's equation), the y-intercept is c and the parabola passes through $(0, c)$.

5. Plot the intercepts, the vertex, and additional points as necessary. Connect these points with a smooth curve.

EXAMPLE 4 Graphing a Quadratic Function in the Form $f(x) = ax^2 + bx + c$

Graph the quadratic function $f(x) = -x^2 - 2x + 1$. Use the graph to identify the function's domain and its range.

Solution

Step 1. Determine how the parabola opens. Note that a, the coefficient of x^2, is -1. Thus, $a < 0$; this negative value tells us that the parabola opens downward.

Step 2. Find the vertex. We know that the x-coordinate of the vertex is $x = -\dfrac{b}{2a}$. We identify a, b, and c in $f(x) = ax^2 + bx + c$.

$$f(x) = -x^2 - 2x + 1$$

$$\boxed{a = -1} \quad \boxed{b = -2} \quad \boxed{c = 1}$$

Substitute the values of a and b into the equation for the x-coordinate:

$$x = -\frac{b}{2a} = -\frac{-2}{2(-1)} = -\left(\frac{-2}{-2}\right) = -1.$$

The x-coordinate of the vertex is -1. We substitute -1 for x into the equation of the function, $f(x) = -x^2 - 2x + 1$, to find the y-coordinate:

$$f(-1) = -(-1)^2 - 2(-1) + 1 = -1 + 2 + 1 = 2.$$

The vertex is $(-1, 2)$.

Step 3. Find the x-intercepts by solving $f(x) = 0$. Replace $f(x)$ with 0 in $f(x) = -x^2 - 2x + 1$. We obtain $0 = -x^2 - 2x + 1$. This equation cannot be solved by factoring. We will use the quadratic formula to solve it.

$$-x^2 - 2x + 1 = 0$$

$$\boxed{a = -1} \quad \boxed{b = -2} \quad \boxed{c = 1}$$

To locate the x-intercepts, we need decimal approximations. Thus, there is no need to simplify the radical form of the solutions.

$$x = \frac{-b \pm \sqrt{b^2 - 4ac}}{2a} = \frac{-(-2) \pm \sqrt{(-2)^2 - 4(-1)(1)}}{2(-1)} = \frac{2 \pm \sqrt{4 - (-4)}}{-2}$$

$$x = \frac{2 + \sqrt{8}}{-2} \approx -2.4 \quad \text{or} \quad x = \frac{2 - \sqrt{8}}{-2} \approx 0.4$$

The x-intercepts are approximately -2.4 and 0.4. The parabola passes through $(-2.4, 0)$ and $(0.4, 0)$.

Step 4. Find the y-intercept by computing $f(0)$. Replace x with 0 in $f(x) = -x^2 - 2x + 1$.

$$f(0) = -0^2 - 2 \cdot 0 + 1 = 1$$

The y-intercept is 1, which is the constant term in the function's equation. The parabola passes through $(0, 1)$.

Step 5. Graph the parabola. With a vertex at $(-1, 2)$, x-intercepts at -2.4 and 0.4, and a y-intercept at 1, the graph of f is shown in **Figure 11.13(a)**. The axis of symmetry is the vertical line whose equation is $x = -1$.

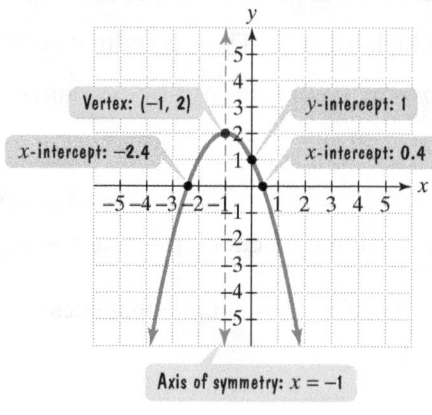

Figure 11.13(a) The graph of $f(x) = -x^2 - 2x + 1$

Figure 11.13(b) Determining the domain and range of $f(x) = -x^2 - 2x + 1$

Now we are ready to determine the domain and range of $f(x) = -x^2 - 2x + 1$. We can use the parabola, shown again in **Figure 11.13(b)**, to do so. To find the domain, look for all the inputs on the x-axis that correspond to points on the graph. As the graph widens and continues to fall at both ends, can you see that these inputs include all real numbers?

$$\text{Domain of } f \text{ is } (-\infty, \infty).$$

To find the range, look for all the outputs on the y-axis that correspond to points on the graph. **Figure 11.13(b)** shows that the parabola's vertex, $(-1, 2)$, is the highest point on the graph. Because the y-coordinate of the vertex is 2, outputs on the y-axis fall at or below 2.

$$\text{Range of } f \text{ is } (-\infty, 2]. \quad \blacksquare$$

✓ **CHECK POINT 4**   Graph the quadratic function $f(x) = -x^2 + 4x + 1$. Use the graph to identify the function's domain and its range.

Great Question!

Are there rules to find domains and ranges of quadratic functions?

Yes. The domain of any quadratic function includes all real numbers. If the vertex is the graph's highest point, the range includes all real numbers at or below the y-coordinate of the vertex. If the vertex is the graph's lowest point, the range includes all real numbers at or above the y-coordinate of the vertex.

Great Question!

I feel overwhelmed by the amount of information required to graph just one quadratic function. Is there a way I can organize the information and gain a better understanding of the graphing procedure?

You're right: The skills needed to graph a quadratic function combine information from many of this book's chapters. Try organizing the items you need to graph quadratic functions in a table, something like this:

Graphing $f(x) = a(x - h)^2 + k$ or $f(x) = ax^2 + bx + c$

1. Opens upward if $a > 0$. Opens downward if $a < 0$.	2. Find the vertex. (h, k) or $\left(-\dfrac{b}{2a}, f\left(-\dfrac{b}{2a}\right)\right)$
3. Find x-intercepts. Solve $f(x) = 0$. Find real solutions.	4. Find the y-intercept. Find $f(0)$.

4 Determine a quadratic function's minimum or maximum value.

Minimum and Maximum Values of Quadratic Functions

Consider the quadratic function $f(x) = ax^2 + bx + c$. If $a > 0$, the parabola opens upward and the vertex is its lowest point. If $a < 0$, the parabola opens downward and the vertex is its highest point. The x-coordinate of the vertex is $-\dfrac{b}{2a}$. Thus, we can find the minimum or maximum value of f by evaluating the quadratic function at $x = -\dfrac{b}{2a}$.

Minimum and Maximum: Quadratic Functions

Consider the quadratic function $f(x) = ax^2 + bx + c$.

1. If $a > 0$, then f has a minimum that occurs at $x = -\dfrac{b}{2a}$. This minimum value is $f\left(-\dfrac{b}{2a}\right)$.

2. If $a < 0$, then f has a maximum that occurs at $x = -\dfrac{b}{2a}$. This maximum value is $f\left(-\dfrac{b}{2a}\right)$.

In each case, the value of x gives the location of the minimum or maximum value. The value of y, or $f\left(-\dfrac{b}{2a}\right)$, gives that minimum or maximum value.

EXAMPLE 5 Obtaining Information about a Quadratic Function from Its Equation

Consider the quadratic function $f(x) = -3x^2 + 6x - 13$.

a. Determine, without graphing, whether the function has a minimum value or a maximum value.

b. Find the minimum or maximum value and determine where it occurs.

c. Identify the function's domain and its range.

Solution We begin by identifying a, b, and c in the function's equation:

$$f(x) = -3x^2 + 6x - 13.$$

$$a = -3 \qquad b = 6 \qquad c = -13$$

a. Because $a < 0$, the function has a maximum value.

b. The maximum value occurs at

$$x = -\frac{b}{2a} = -\frac{6}{2(-3)} = -\frac{6}{-6} = -(-1) = 1.$$

The maximum value occurs at $x = 1$ and the maximum value of $f(x) = -3x^2 + 6x - 13$ is

$$f(1) = -3 \cdot 1^2 + 6 \cdot 1 - 13 = -3 + 6 - 13 = -10.$$

We see that the maximum is -10 at $x = 1$.

c. Like all quadratic functions, the domain is $(-\infty, \infty)$. Because the function's maximum value is -10, the range includes all real numbers at or below -10. The range is $(-\infty, -10]$. ∎

We can use the graph of $f(x) = -3x^2 + 6x - 13$ to visualize the results of Example 5. **Figure 11.14** shows the graph in a $[-6, 6, 1]$ by $[-50, 20, 10]$ viewing rectangle. The maximum function feature verifies that the function's maximum is -10 at $x = 1$. Notice that x gives the location of the maximum and y gives the maximum value. Notice, too, that the maximum value is -10 and not the ordered pair $(1, -10)$.

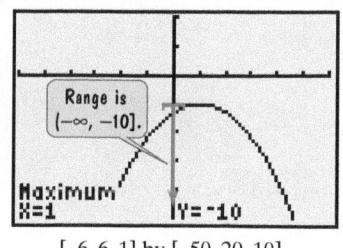

$[-6, 6, 1]$ by $[-50, 20, 10]$

Figure 11.14

✓ **CHECK POINT 5**　Repeat parts (a) through (c) of Example 5 using the quadratic function $f(x) = 4x^2 - 16x + 1000$.

⑤ Solve problems involving a quadratic function's minimum or maximum value.

Applications of Quadratic Functions

Many applied problems involve finding the maximum or minimum value of a quadratic function, as well as where this value occurs.

EXAMPLE 6　Parabolic Paths of a Shot Put

An athlete whose event is the shot put releases the shot with the same initial velocity, but at different angles. **Figure 11.15** shows the parabolic paths for shots released at angles of 35° and 65°.

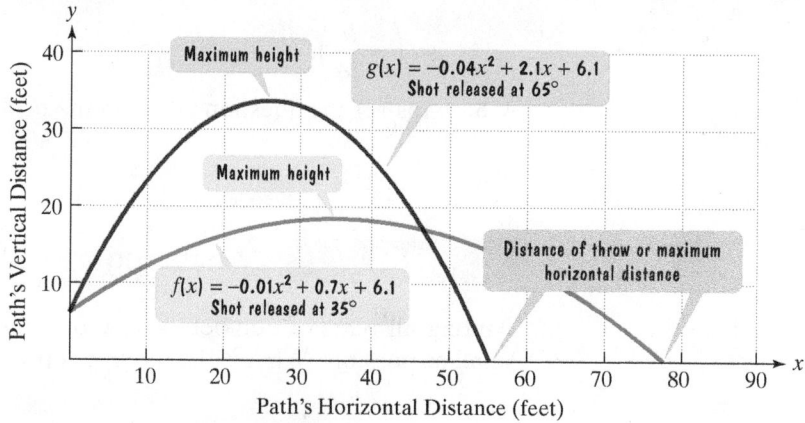

Figure 11.15 Two paths of a shot put

When the shot is released at an angle of 35°, its path can be modeled by the function

$$f(x) = -0.01x^2 + 0.7x + 6.1,$$

in which x is the shot's horizontal distance, in feet, and $f(x)$ is its height, in feet. What is the maximum height of this shot's path?

Solution　The quadratic function is in the form $f(x) = ax^2 + bx + c$, with $a = -0.01$ and $b = 0.7$. Because $a < 0$, the function has a maximum that occurs at $x = -\dfrac{b}{2a}$.

$$x = -\frac{b}{2a} = -\frac{0.7}{2(-0.01)} = -(-35) = 35$$

This means that the shot's maximum height occurs when its horizontal distance is 35 feet. Can you see how this is shown by the blue graph of f in **Figure 11.15**? The maximum height of this path is

$$f(35) = -0.01(35)^2 + 0.7(35) + 6.1 = 18.35$$

or 18.35 feet. ■

✓ **CHECK POINT 6** Use function g, whose equation and graph are shown in **Figure 11.15** on the previous page, to find the maximum height, to the nearest tenth of a foot, when the shot is released at an angle of 65°.

Quadratic functions can also be modeled from verbal conditions. Once we have obtained a quadratic function, we can then use the x-coordinate of the vertex to determine its maximum or minimum value. Here is a step-by-step strategy for solving these kinds of problems:

Strategy for Solving Problems Involving Maximizing or Minimizing Quadratic Functions

1. Read the problem carefully and decide which quantity is to be maximized or minimized.
2. Use the conditions of the problem to express the quantity as a function in one variable.
3. Rewrite the function in the form $f(x) = ax^2 + bx + c$.
4. Calculate $-\dfrac{b}{2a}$. If $a > 0$, f has a minimum at $x = -\dfrac{b}{2a}$. This minimum value is $f\left(-\dfrac{b}{2a}\right)$. If $a < 0$, f has a maximum at $x = -\dfrac{b}{2a}$. This maximum value is $f\left(-\dfrac{b}{2a}\right)$.
5. Answer the question posed in the problem.

EXAMPLE 7 Minimizing a Product

Among all pairs of numbers whose difference is 10, find a pair whose product is as small as possible. What is the minimum product?

Solution

Step 1. Decide what must be maximized or minimized. We must minimize the product of two numbers. Calling the numbers x and y, and calling the product P, we must minimize

$$P = xy.$$

Step 2. Express this quantity as a function in one variable. In the formula $P = xy$, P is expressed in terms of two variables, x and y. However, because the difference of the numbers is 10, we can write

$$x - y = 10.$$

We can solve this equation for y in terms of x (or vice versa), substitute the result into $P = xy$, and obtain P as a function of one variable.

$$-y = -x + 10 \qquad \text{Subtract } x \text{ from both sides of } x - y = 10.$$

$$y = x - 10 \qquad \text{Multiply both sides of the equation by } -1 \text{ and solve for } y.$$

Now we substitute $x - 10$ for y in $P = xy$.

$$P = xy = x(x - 10).$$

Because P is now a function of x, we can write

$$P(x) = x(x - 10).$$

Step 3. Write the function in the form $f(x) = ax^2 + bx + c$. We apply the distributive property to obtain

$$P(x) = x(x - 10) = x^2 - 10x.$$

$a = 1$ $b = -10$

Using Technology

Numeric Connections

The $\boxed{\text{TABLE}}$ feature of a graphing utility can be used to verify our work in Example 7.

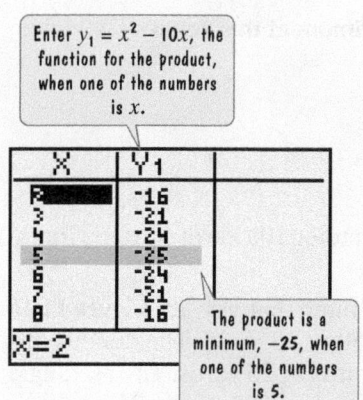

Enter $y_1 = x^2 - 10x$, the function for the product, when one of the numbers is x.

X	Y₁
2	-16
3	-21
4	-24
5	-25
6	-24
7	-21
8	-16

X=2

The product is a minimum, −25, when one of the numbers is 5.

Step 4. Calculate $-\dfrac{b}{2a}$. If $a > 0$, the function has a minimum at this value. The voice balloons show that $a = 1$ and $b = -10$.

$$x = -\frac{b}{2a} = -\frac{-10}{2(1)} = -(-5) = 5$$

This means that the product, P, of two numbers whose difference is 10 is a minimum when one of the numbers, x, is 5.

Step 5. Answer the question posed by the problem. The problem asks for the two numbers and the minimum product. We found that one of the numbers, x, is 5. Now we must find the second number, y.

$$y = x - 10 = 5 - 10 = -5.$$

The number pair whose difference is 10 and whose product is as small as possible is 5, −5. The minimum product is $5(-5)$, or −25. ∎

✓ **CHECK POINT 7**   Among all pairs of numbers whose difference is 8, find a pair whose product is as small as possible. What is the minimum product?

EXAMPLE 8   Maximizing Area

You have 100 yards of fencing to enclose a rectangular region. Find the dimensions of the rectangle that maximize the enclosed area. What is the maximum area?

Solution

Step 1. Decide what must be maximized or minimized. We must maximize area. What we do not know are the rectangle's dimensions, x and y.

Step 2. Express this quantity as a function in one variable. Because we must maximize area, we have $A = xy$. We need to transform this into a function in which A is represented by one variable. Because you have 100 yards of fencing, the perimeter of the rectangle is 100 yards. This means that

$$2x + 2y = 100.$$

We can solve this equation for y in terms of x, substitute the result into $A = xy$, and obtain A as a function in one variable. We begin by solving for y.

$$2y = 100 - 2x \qquad \text{Subtract 2x from both sides.}$$

$$y = \frac{100 - 2x}{2} \qquad \text{Divide both sides by 2.}$$

$$y = 50 - x \qquad \text{Divide each term in the numerator by 2.}$$

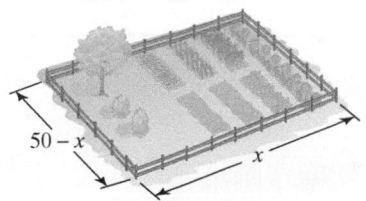

Figure 11.16 What value of *x* will maximize the rectangle's area?

Now we substitute $50 - x$ for y in $A = xy$.

$$A = xy = x(50 - x)$$

The rectangle and its dimensions are illustrated in **Figure 11.16**. Because A is now a function of x, we can write

$$A(x) = x(50 - x).$$

This function models the area, $A(x)$, of any rectangle whose perimeter is 100 yards in terms of one of its dimensions, x.

Step 3. Write the function in the form $f(x) = ax^2 + bx + c$. We apply the distributive property to obtain

$$A(x) = x(50 - x) = 50x - x^2 = -x^2 + 50x.$$

$$\boxed{a = -1} \quad \boxed{b = 50}$$

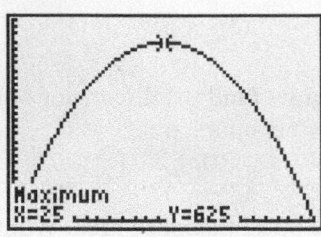

Step 4. Calculate $-\dfrac{b}{2a}$. If $a < 0$, the function has a maximum at this value. The voice balloons show that $a = -1$ and $b = 50$.

$$x = -\frac{b}{2a} = -\frac{50}{2(-1)} = 25$$

This means that the area, $A(x)$, of a rectangle with perimeter 100 yards is a maximum when one of the rectangle's dimensions, x, is 25 yards.

Step 5. Answer the question posed by the problem. We found that $x = 25$. **Figure 11.16** shows that the rectangle's other dimension is $50 - x = 50 - 25 = 25$. The dimensions of the rectangle that maximize the enclosed area are 25 yards by 25 yards. The rectangle that gives the maximum area is actually a square with an area of 25 yards $\cdot$ 25 yards, or 625 square yards. ∎

☑ **CHECK POINT 8** You have 120 feet of fencing to enclose a rectangular region. Find the dimensions of the rectangle that maximize the enclosed area. What is the maximum area?

Achieving Success

A recent government study cited in *Math: A Rich Heritage* (Globe Fearon Educational Publisher) found this simple fact: **The more college mathematics courses you take, the greater your earning potential will be.** Even jobs that do not require a college degree require mathematical thinking that involves attending to precision, making sense of complex problems, and persevering in solving them. No other discipline comes close to math in offering a more extensive set of tools for application and intellectual development. Take as much math as possible as you continue your journey into higher education.

CONCEPT AND VOCABULARY CHECK

Fill in each blank so that the resulting statement is true.

1. The graph of any quadratic function, $f(x) = ax^2 + bx + c, a \neq 0$, is called a/an _____. If $a > 0$, the graph opens _____. If $a < 0$, the graph opens _____.

2. The vertex of a parabola is the _____ point if $a > 0$.

3. The vertex of a parabola is the _____ point if $a < 0$.

4. The vertex of the graph of $f(x) = a(x - h)^2 + k, a \neq 0$, is the point _____.

5. The x-coordinate of the vertex of the graph of $f(x) = ax^2 + bx + c, a \neq 0$, is _____. The y-coordinate of the vertex is found by substituting _____ into the function's equation and evaluating the function at this value of x.

6. If $f(x) = a(x - h)^2 + k$ or $f(x) = ax^2 + bx + c, a \neq 0$, any x-intercepts are found by solving $f(x) =$ _____. The equation's real _____ are the x-intercepts.

7. If $f(x) = a(x - h)^2 + k$ or $f(x) = ax^2 + bx + c, a \neq 0$, the y-intercept is found by computing _____.

11.3 EXERCISE SET

MyMathLab®

 Watch the videos in MyMathLab

 Download the MyDashBoard App

Practice Exercises

In Exercises 1–4, the graph of a quadratic function is given. Write the function's equation, selecting from the following options:

$$f(x) = (x + 1)^2 - 1, g(x) = (x + 1)^2 + 1,$$
$$h(x) = (x - 1)^2 + 1, j(x) = (x - 1)^2 - 1.$$

1.

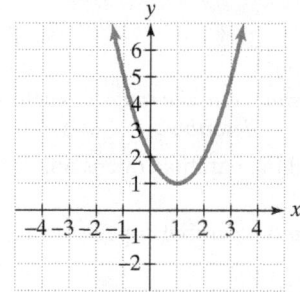

2.

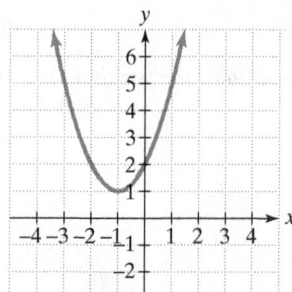

3.

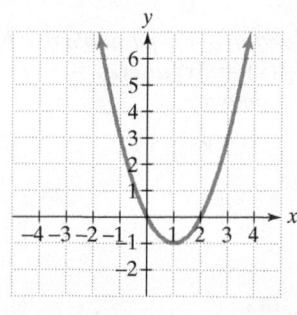

4.

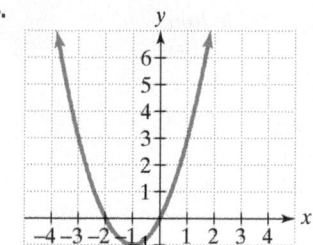

In Exercises 5–8, the graph of a quadratic function is given. Write the function's equation, selecting from the following options:

$$f(x) = x^2 + 2x + 1, g(x) = x^2 - 2x + 1,$$
$$h(x) = x^2 - 1, j(x) = -x^2 - 1.$$

5.

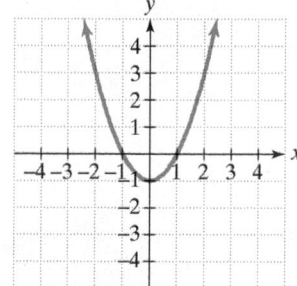

6.

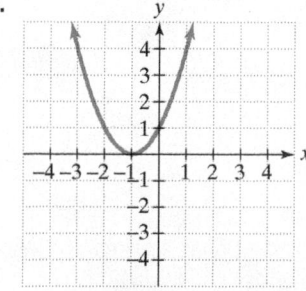

7.

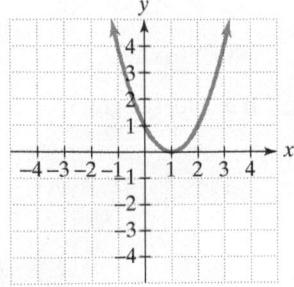

8.

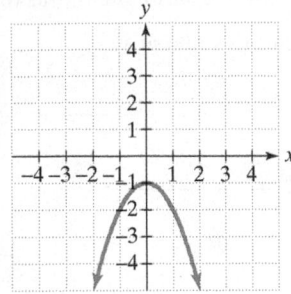

In Exercises 9–16, find the coordinates of the vertex for the parabola defined by the given quadratic function.

9. $f(x) = 2(x - 3)^2 + 1$

10. $f(x) = -3(x - 2)^2 + 12$

11. $f(x) = -2(x + 1)^2 + 5$

12. $f(x) = -2(x + 4)^2 - 8$

13. $f(x) = 2x^2 - 8x + 3$

14. $f(x) = 3x^2 - 12x + 1$

15. $f(x) = -x^2 - 2x + 8$

16. $f(x) = -2x^2 + 8x - 1$

In Exercises 17–38, use the vertex and intercepts to sketch the graph of each quadratic function. Use the graph to identify the function's range.

17. $f(x) = (x - 4)^2 - 1$

18. $f(x) = (x - 1)^2 - 2$

19. $f(x) = (x - 1)^2 + 2$

20. $f(x) = (x - 3)^2 + 2$

21. $y - 1 = (x - 3)^2$

22. $y - 3 = (x - 1)^2$

23. $f(x) = 2(x + 2)^2 - 1$

24. $f(x) = \dfrac{5}{4} - \left(x - \dfrac{1}{2}\right)^2$

25. $f(x) = 4 - (x - 1)^2$

26. $f(x) = 1 - (x - 3)^2$

27. $f(x) = x^2 - 2x - 3$

28. $f(x) = x^2 - 2x - 15$

29. $f(x) = x^2 + 3x - 10$

30. $f(x) = 2x^2 - 7x - 4$

31. $f(x) = 2x - x^2 + 3$

32. $f(x) = 5 - 4x - x^2$

33. $f(x) = x^2 + 6x + 3$

34. $f(x) = x^2 + 4x - 1$

35. $f(x) = 2x^2 + 4x - 3$

36. $f(x) = 3x^2 - 2x - 4$

37. $f(x) = 2x - x^2 - 2$

38. $f(x) = 6 - 4x + x^2$

In Exercises 39–44, an equation of a quadratic function is given.

   a. *Determine, without graphing, whether the function has a minimum value or a maximum value.*

   b. *Find the minimum or maximum value and determine where it occurs.*

   c. *Identify the function's domain and its range.*

39. $f(x) = 3x^2 - 12x - 1$

40. $f(x) = 2x^2 - 8x - 3$

41. $f(x) = -4x^2 + 8x - 3$

42. $f(x) = -2x^2 - 12x + 3$

43. $f(x) = 5x^2 - 5x$

44. $f(x) = 6x^2 - 6x$

Practice PLUS

In Exercises 45–48, give the domain and the range of each quadratic function whose graph is described.

45. The vertex is $(-1, -2)$ and the parabola opens up.

46. The vertex is $(-3, -4)$ and the parabola opens down.

47. Maximum $= -6$ at $x = 10$

48. Minimum $= 18$ at $x = -6$

In Exercises 49–52, write an equation of the parabola that has the same shape as the graph of $f(x) = 2x^2$, but with the given point as the vertex.

49. $(5, 3)$

50. $(7, 4)$

51. $(-10, -5)$

52. $(-8, -6)$

In Exercises 53–56, write an equation of the parabola that has the same shape as the graph of $f(x) = 3x^2$ or $g(x) = -3x^2$, but with the given maximum or minimum.

53. Maximum $= 4$ at $x = -2$

54. Maximum $= -7$ at $x = 5$

55. Minimum $= 0$ at $x = 11$

56. Minimum $= 0$ at $x = 9$

Application Exercises

57. A person standing close to the edge on the top of a 160-foot building throws a baseball vertically upward. The quadratic function

$$s(t) = -16t^2 + 64t + 160$$

models the ball's height above the ground, $s(t)$, in feet, t seconds after it was thrown.

   a. After how many seconds does the ball reach its maximum height? What is the maximum height?

   b. How many seconds does it take until the ball finally hits the ground? Round to the nearest tenth of a second.

   c. Find $s(0)$ and describe what this means.

   d. Use your results from parts (a) through (c) to graph the quadratic function. Begin the graph with $t = 0$ and end with the value of t for which the ball hits the ground.

58. A person standing close to the edge on the top of a 200-foot building throws a baseball vertically upward. The quadratic function

$$s(t) = -16t^2 + 64t + 200$$

models the ball's height above the ground, $s(t)$, in feet, t seconds after it was thrown.

   a. After how many seconds does the ball reach its maximum height? What is the maximum height?

   b. How many seconds does it take until the ball finally hits the ground? Round to the nearest tenth of a second.

   c. Find $s(0)$ and describe what this means.

   d. Use your results from parts (a) through (c) to graph the quadratic function. Begin the graph with $t = 0$ and end with the value of t for which the ball hits the ground.

59. Among all pairs of numbers whose sum is 16, find a pair whose product is as large as possible. What is the maximum product?

60. Among all pairs of numbers whose sum is 20, find a pair whose product is as large as possible. What is the maximum product?

61. Among all pairs of numbers whose difference is 16, find a pair whose product is as small as possible. What is the minimum product?

62. Among all pairs of numbers whose difference is 24, find a pair whose product is as small as possible. What is the minimum product?

63. You have 600 feet of fencing to enclose a rectangular plot that borders on a river. If you do not fence the side along the river, find the length and width of the plot that will maximize the area. What is the largest area that can be enclosed?

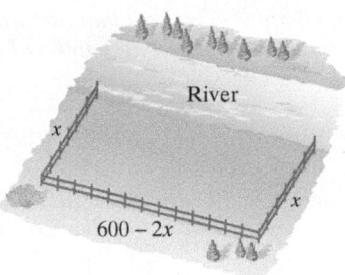

64. You have 200 feet of fencing to enclose a rectangular plot that borders on a river. If you do not fence the side along the river, find the length and width of the plot that will maximize the area. What is the largest area that can be enclosed?

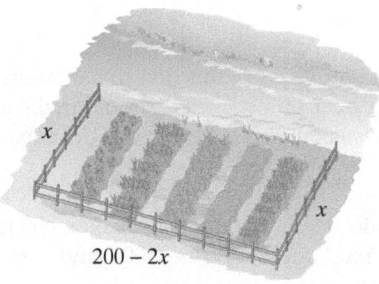

65. You have 50 yards of fencing to enclose a rectangular region. Find the dimensions of the rectangle that maximize the enclosed area. What is the maximum area?

66. You have 80 yards of fencing to enclose a rectangular region. Find the dimensions of the rectangle that maximize the enclosed area. What is the maximum area?

67. A rain gutter is made from sheets of aluminum that are 20 inches wide by turning up the edges to form right angles. Determine the depth of the gutter that will maximize its cross-sectional area and allow the greatest amount of water to flow. What is the maximum cross-sectional area?

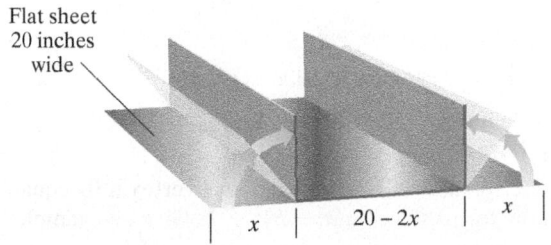

68. A rain gutter is made from sheets of aluminum that are 12 inches wide by turning up the edges to form right angles. Determine the depth of the gutter that will maximize its cross-sectional area and allow the greatest amount of water to flow. What is the maximum cross-sectional area?

In Chapter 9, we saw that the profit, P(x), generated after producing and selling x units of a product is given by the function

$$P(x) = R(x) - C(x),$$

where R and C are the revenue and cost functions, respectively. Use these functions to solve Exercises 69–70.

69. Hunky Beef, a local sandwich store, has a fixed weekly cost of $525.00, and variable costs for making a roast beef sandwich are $0.55.

    a. Let x represent the number of roast beef sandwiches made and sold each week. Write the weekly cost function, C, for Hunky Beef.

    b. The function $R(x) = -0.001x^2 + 3x$ describes the money that Hunky Beef takes in each week from the sale of x roast beef sandwiches. Use this revenue function and the cost function from part (a) to write the store's weekly profit function, P.

    c. Use the store's profit function to determine the number of roast beef sandwiches it should make and sell each week to maximize profit. What is the maximum weekly profit?

70. Virtual Fido is a company that makes electronic virtual pets. The fixed weekly cost is $3000, and variable costs for each pet are $20.

    a. Let x represent the number of virtual pets made and sold each week. Write the weekly cost function, C, for Virtual Fido.

    b. The function $R(x) = -x^2 + 1000x$ describes the money that Virtual Fido takes in each week from the sale of x virtual pets. Use this revenue function and the cost function from part (a) to write the weekly profit function, P.

    c. Use the profit function to determine the number of virtual pets that should be made and sold each week to maximize profit. What is the maximum weekly profit?

Writing in Mathematics

71. What is a parabola? Describe its shape.

72. Explain how to decide whether a parabola opens upward or downward.

73. Describe how to find a parabola's vertex if its equation is in the form $f(x) = a(x - h)^2 + k$. Give an example.

74. Describe how to find a parabola's vertex if its equation is in the form $f(x) = ax^2 + bx + c$. Use $f(x) = x^2 - 6x + 8$ as an example.

75. A parabola that opens upward has its vertex at (1, 2). Describe as much as you can about the parabola based on this information. Include in your discussion the number of x-intercepts (if any) for the parabola.

Technology Exercises

76. Use a graphing utility to verify any five of your hand-drawn graphs in Exercises 17–38.

77. **a.** Use a graphing utility to graph $y = 2x^2 - 82x + 720$ in a standard viewing rectangle. What do you observe?

    b. Find the coordinates of the vertex for the given quadratic function.

    c. The answer to part (b) is (20.5, −120.5). Because the leading coefficient, 2, of the given function is positive, the vertex is a minimum point on the graph. Use this fact to help find a viewing rectangle that will give a relatively complete picture of the parabola. With an axis of symmetry at $x = 20.5$, the setting for x should extend past this, so try Xmin = 0 and Xmax = 30. The setting for y should include (and probably go below) the y-coordinate of the graph's minimum point, so try Ymin = −130. Experiment with Ymax until your utility shows the parabola's major features.

    d. In general, explain how knowing the coordinates of a parabola's vertex can help determine a reasonable viewing rectangle on a graphing utility for obtaining a complete picture of the parabola.

In Exercises 78–81, find the vertex for each parabola. Then determine a reasonable viewing rectangle on your graphing utility and use it to graph the parabola.

78. $y = -0.25x^2 + 40x$

79. $y = -4x^2 + 20x + 160$

80. $y = 5x^2 + 40x + 600$

81. $y = 0.01x^2 + 0.6x + 100$

Critical Thinking Exercises

Make Sense? *In Exercises 82–85, determine whether each statement "makes sense" or "does not make sense" and explain your reasoning.*

82. Parabolas that open up appear to form smiles ($a > 0$), while parabolas that open down frown ($a < 0$).

83. I must have made an error when graphing this parabola because its axis of symmetry is the y-axis.

84. I like to think of a parabola's vertex as the point where it intersects its axis of symmetry.

85. I threw a baseball vertically upward and its path was a parabola.

In Exercises 86–89, determine whether each statement is true or false. If the statement is false, make the necessary change(s) to produce a true statement.

86. No quadratic functions have a range of $(-\infty, \infty)$.

87. The vertex of the parabola described by $f(x) = 2(x - 5)^2 - 1$ is at (5, 1).

88. The graph of $f(x) = -2(x + 4)^2 - 8$ has one y-intercept and two x-intercepts.

89. The maximum value of y for the quadratic function $f(x) = -x^2 + x + 1$ is 1.

In Exercises 90–91, find the axis of symmetry for each parabola whose equation is given. Use the axis of symmetry to find a second point on the parabola whose y-coordinate is the same as the given point.

90. $f(x) = 3(x + 2)^2 - 5$; $(-1, -5)$

91. $f(x) = (x - 3)^2 + 2$; $(6, 11)$

In Exercises 92–93, write the equation of each parabola in $f(x) = a(x - h)^2 + k$ form.

92. Vertex: $(-3, -4)$; The graph passes through the point $(1, 4)$.

93. Vertex: $(-3, -1)$; The graph passes through the point $(-2, -3)$.

94. A rancher has 1000 feet of fencing to construct six corrals, as shown in the figure. Find the dimensions that maximize the enclosed area. What is the maximum area?

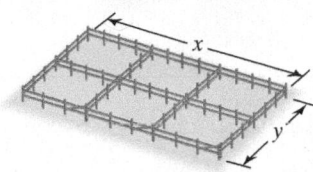

95. The annual yield per lemon tree is fairly constant at 320 pounds when the number of trees per acre is 50 or fewer. For each additional tree over 50, the annual yield per tree for all trees on the acre decreases by 4 pounds due to overcrowding. Find the number of trees that should be planted on an acre to produce the maximum yield. How many pounds is the maximum yield?

Review Exercises

96. Solve: $\dfrac{2}{x + 5} + \dfrac{1}{x - 5} = \dfrac{16}{x^2 - 25}$. (Section 7.6, Example 4)

97. Simplify: $\dfrac{1 + \dfrac{2}{x}}{1 - \dfrac{4}{x^2}}$. (Section 7.5, Example 2 or Example 5)

98. Solve the system:

$$\begin{cases} 2x + 3y = 6 \\ x - 4y = 14. \end{cases}$$

(Section 4.3, Example 2)

Preview Exercises

Exercises 99–101 will help you prepare for the material covered in the next section.

In Exercises 99–100, solve each quadratic equation for u.

99. $u^2 - 8u - 9 = 0$

100. $2u^2 - u - 10 = 0$

101. If $u = x^{\frac{1}{3}}$, rewrite $5x^{\frac{2}{3}} + 11x^{\frac{1}{3}} + 2 = 0$ as a quadratic equation in u. [*Hint:* $x^{\frac{2}{3}} = \left(x^{\frac{1}{3}}\right)^2$.]

MID-CHAPTER CHECK POINT   Section 11.1–Section 11.3

 What You Know: We found the length $\left[d = \sqrt{(x_2 - x_1)^2 + (y_2 - y_1)^2}\right]$ and the midpoint $\left[\left(\dfrac{x_1 + x_2}{2}, \dfrac{y_1 + y_2}{2}\right)\right]$ of the segment with endpoints (x_1, y_1) and (x_2, y_2). We saw that not all quadratic equations can be solved by factoring. We learned three new methods for solving these equations: the square root property, completing the square, and the quadratic formula. We saw that the discriminant of $ax^2 + bx + c = 0$, namely $b^2 - 4ac$, determines the number and type of the equation's solutions. We graphed quadratic functions using vertices, intercepts, and additional points, as necessary. We learned that the vertex of $f(x) = a(x - h)^2 + k$ is (h, k) and the vertex of $f(x) = ax^2 + bx + c$ is $\left(-\dfrac{b}{2a}, f\left(-\dfrac{b}{2a}\right)\right)$. We used the vertex to solve problems that involved minimizing or maximizing quadratic functions.

In Exercises 1–13, solve each equation by the method of your choice. Simplify solutions, if possible.

1. $(3x - 5)^2 = 36$

2. $5x^2 - 2x = 7$

3. $3x^2 - 6x - 2 = 0$

4. $x^2 + 6x = -2$

5. $5x^2 + 1 = 37$

6. $x^2 - 5x + 8 = 0$

7. $2x^2 + 26 = 0$

8. $(2x + 3)(x + 2) = 10$

9. $(x + 3)^2 = 24$

10. $\dfrac{1}{x^2} - \dfrac{4}{x} + 1 = 0$

11. $x(2x - 3) = -4$

12. $\dfrac{x^2}{3} + \dfrac{x}{2} = \dfrac{2}{3}$

13. $\dfrac{2x}{x^2 + 6x + 8} = \dfrac{x}{x + 4} - \dfrac{2}{x + 2}$

14. Solve by completing the square: $x^2 + 10x - 3 = 0$.

In Exercises 15–16, find the length (in simplified radical form and rounded to two decimal places) and the midpoint of the line segment with the given endpoints.

15. $(2, -2)$ and $(-2, 2)$

16. $(-5, 8)$ and $(-10, 14)$

In Exercises 17–20, graph the given quadratic function. Give each function's domain and range.

17. $f(x) = (x - 3)^2 - 4$
18. $g(x) = 5 - (x + 2)^2$
19. $h(x) = -x^2 - 4x + 5$
20. $f(x) = 3x^2 - 6x + 1$

In Exercises 21–22, without solving the equation, determine the number and type of solutions.

21. $2x^2 + 5x + 4 = 0$
22. $10x(x + 4) = 15x - 15$

In Exercises 23–24, write a quadratic equation in standard form with the given solution set.

23. $\left\{-\dfrac{1}{2}, \dfrac{3}{4}\right\}$

24. $\left\{-2\sqrt{3}, 2\sqrt{3}\right\}$

25. A company manufactures and sells bath cabinets. The function
$$P(x) = -x^2 + 150x - 4425$$
models the company's daily profit, $P(x)$, when x cabinets are manufactured and sold per day. How many cabinets should be manufactured and sold per day to maximize the company's profit? What is the maximum daily profit?

26. Among all pairs of numbers whose sum is -18, find a pair whose product is as large as possible. What is the maximum product?

27. The base of a triangle measures 40 inches minus twice the measure of its height. For what measure of the height does the triangle have a maximum area? What is the maximum area?

SECTION

11.4

Objective

1. Solve equations that are quadratic in form.

Equations Quadratic in Form

"My husband asked me if we have any cheese puffs. Like he can't go and lift the couch cushion up himself."
— Roseanne Barr

How important is it for you to have a clean house? The percentage of people who find this to be quite important varies by age. In the Exercise Set, you will work with a function that models this phenomenon. Your work will be based on equations that are not quadratic, but that can be written as quadratic equations using an appropriate substitution. Here are some examples:

Given Equation	Substitution	New Equation
$x^4 - 10x^2 + 9 = 0$ or $(x^2)^2 - 10x^2 + 9 = 0$	$u = x^2$	$u^2 - 10u + 9 = 0$
$5x^{\frac{2}{3}} + 11x^{\frac{1}{3}} + 2 = 0$ or $5\left(x^{\frac{1}{3}}\right)^2 + 11x^{\frac{1}{3}} + 2 = 0$	$u = x^{\frac{1}{3}}$	$5u^2 + 11u + 2 = 0$

An equation that is **quadratic in form** is one that can be expressed as a quadratic equation using an appropriate substitution. Both of the preceding given equations are quadratic in form.

In an equation that is quadratic in form, the variable factor in one term is the square of the variable factor in the other variable term. The third term is a constant. By letting

1. Solve equations that are quadratic in form.

u equal the variable factor that reappears squared, a quadratic equation in *u* will result. Now it's easy. Solve this quadratic equation for *u*. Finally, use your substitution to find the values for the variable in the given equation. Example 1 shows how this is done.

EXAMPLE 1 Solving an Equation Quadratic in Form

Solve: $x^4 - 8x^2 - 9 = 0$.

Solution Can you see that the variable factor in one term is the square of the variable factor in the other variable term?

$$x^4 - 8x^2 - 9 = 0$$

x^4 is the square of x^2: $(x^2)^2 = x^4$.

We will let *u* equal the variable factor that reappears squared. Thus,

let $u = x^2$.

Now we write the given equation as a quadratic equation in *u* and solve for *u*.

$x^4 - 8x^2 - 9 = 0$	This is the given equation.
$(x^2)^2 - 8x^2 - 9 = 0$	The given equation contains x^2 and x^2 squared.
$u^2 - 8u - 9 = 0$	Let $u = x^2$. Replace x^2 with u.
$(u - 9)(u + 1) = 0$	Factor.
$u - 9 = 0$ or $u + 1 = 0$	Apply the zero-product principle.
$u = 9$ $u = -1$	Solve for u.

We're not done! Why not? We were asked to solve for *x* and we have values for *u*. We use the original substitution, $u = x^2$, to solve for *x*. Replace *u* with x^2 in each equation shown, namely $u = 9$ and $u = -1$.

$x^2 = 9$	$x^2 = -1$	
$x = \pm\sqrt{9}$	$x = \pm\sqrt{-1}$	Apply the square root property.
$x = \pm 3$	$x = \pm i$	

Substitute these values into the given equation and verify that the solutions are $-3, 3, -i,$ and i. The solution set is $\{-3, 3, -i, i\}$. The graph in the Using Technology box shows that only the real solutions, -3 and 3, appear as *x*-intercepts. ■

✓ CHECK POINT 1 Solve: $x^4 - 5x^2 + 6 = 0$.

If checking proposed solutions is not overly cumbersome, you should do so either algebraically or with a graphing utility. The Using Technology box shows a check of the two real solutions in Example 1. Are there situations when solving equations quadratic in form where a check is essential? Yes. **If at any point in the solution process both sides of an equation are raised to an even power, a check is required.** Extraneous solutions that are not solutions of the given equation may have been introduced.

EXAMPLE 2 Solving an Equation Quadratic in Form

Solve: $2x - \sqrt{x} - 10 = 0$.

Solution To identify exponents on the terms, let's rewrite $\sqrt{x}$ as $x^{\frac{1}{2}}$. The equation can be expressed as

$$2x^1 - x^{\frac{1}{2}} - 10 = 0.$$

Using Technology

Graphic Connections

The graph of

$$y = x^4 - 8x^2 - 9$$

has *x*-intercepts at -3 and 3. This verifies that the real solutions of

$$x^4 - 8x^2 - 9 = 0$$

are -3 and 3. The imaginary solutions, $-i$ and i, are not shown as intercepts.

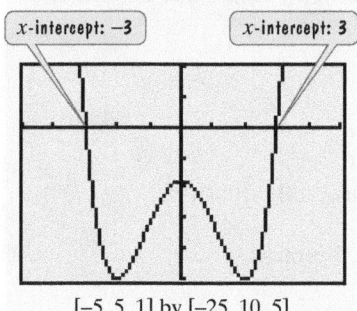

x-intercept: −3   x-intercept: 3

$[-5, 5, 1]$ by $[-25, 10, 5]$

By expressing the equation as $2x^1 - x^{\frac{1}{2}} - 10 = 0$, we can see that the variable factor in one term is the square of the variable factor in the other variable term.

$$2x^1 - x^{\frac{1}{2}} - 10 = 0$$

x^1 is the square
of $x^{\frac{1}{2}}$: $\left(x^{\frac{1}{2}}\right)^2 = x^1$.

We will let u equal the variable factor that reappears squared. Thus,

$$\text{let } u = x^{\frac{1}{2}}.$$

Now we write the given equation as a quadratic equation in u and solve for u.

$2x - \sqrt{x} - 10 = 0$	This is the given equation.
$2x^1 - x^{\frac{1}{2}} - 10 = 0$	This is the given equation in exponential form.
$2\left(x^{\frac{1}{2}}\right)^2 - x^{\frac{1}{2}} - 10 = 0$	The equation contains $x^{\frac{1}{2}}$ and $x^{\frac{1}{2}}$ squared.
$2u^2 - u - 10 = 0$	Let $u = x^{\frac{1}{2}}$. Replace $x^{\frac{1}{2}}$ with u.
$(2u - 5)(u + 2) = 0$	Factor.
$2u - 5 = 0$ or $u + 2 = 0$	Set each factor equal to zero.
$u = \dfrac{5}{2}$ $\quad$ $u = -2$	Solve for u.

Use the original substitution, $u = x^{\frac{1}{2}}$, to solve for x. Replace u with $x^{\frac{1}{2}}$ in each of the preceding equations.

$x^{\frac{1}{2}} = \dfrac{5}{2}$ $\quad$ or $\quad$ $x^{\frac{1}{2}} = -2$	Replace u with $x^{\frac{1}{2}}$.
$\left(x^{\frac{1}{2}}\right)^2 = \left(\dfrac{5}{2}\right)^2$ $\quad$ $\left(x^{\frac{1}{2}}\right)^2 = (-2)^2$	Solve for x by squaring both sides of each equation.

Both sides are raised to even powers. We must check.

$x = \dfrac{25}{4}$ $\quad\quad$ $x = 4$	Square $\dfrac{5}{2}$ and -2.

It is essential to check both proposed solutions in the original equation.

Check $\dfrac{25}{4}$: $\qquad\qquad\qquad\qquad$ **Check 4:**

$$2x - \sqrt{x} - 10 = 0 \qquad\qquad 2x - \sqrt{x} - 10 = 0$$

$$2 \cdot \frac{25}{4} - \sqrt{\frac{25}{4}} - 10 \overset{?}{=} 0 \qquad\qquad 2 \cdot 4 - \sqrt{4} - 10 \overset{?}{=} 0$$

$$\frac{25}{2} - \frac{5}{2} - 10 \overset{?}{=} 0 \qquad\qquad\qquad 8 - 2 - 10 \overset{?}{=} 0$$

$$\frac{20}{2} - 10 \overset{?}{=} 0 \qquad\qquad\qquad\qquad 6 - 10 \overset{?}{=} 0$$

$$0 = 0, \quad \text{true} \qquad\qquad\qquad -4 = 0, \quad \text{false}$$

The check indicates that 4 is not a solution. It is an extraneous solution brought about by squaring each side of one of the equations. The only solution is $\dfrac{25}{4}$ and the solution set is $\left\{\dfrac{25}{4}\right\}$. ∎

✓ **CHECK POINT 2** Solve: $x - 2\sqrt{x} - 8 = 0$.

The equations in Examples 1 and 2 can be solved by methods other than using substitutions.

$$x^4 - 8x^2 - 9 = 0 \qquad\qquad 2x - \sqrt{x} - 10 = 0$$

This equation can be solved directly by factoring:
$(x^2 - 9)(x^2 + 1) = 0.$

This equation can be solved by isolating the radical term:
$2x - 10 = \sqrt{x}.$
Then square both sides.

In the examples that follow, solving the equations by methods other than first introducing a substitution becomes increasingly difficult.

EXAMPLE 3   Solving an Equation Quadratic in Form

Solve: $(x^2 - 5)^2 + 3(x^2 - 5) - 10 = 0.$

Solution   This equation contains $x^2 - 5$ and $x^2 - 5$ squared. We
let $u = x^2 - 5.$

$(x^2 - 5)^2 + 3(x^2 - 5) - 10 = 0$	This is the given equation.
$u^2 + 3u - 10 = 0$	Let $u = x^2 - 5.$
$(u + 5)(u - 2) = 0$	Factor.
$u + 5 = 0 \quad$ or $\quad u - 2 = 0$	Set each factor equal to zero.
$u = -5 \qquad\qquad u = 2$	Solve for u.

Discover for Yourself

Solve Example 3 by first simplifying the given equation's left side. Then factor out x^2 and solve the resulting equation. Do you get the same solutions? Which method, substitution or first simplifying, is faster?

Use the original substitution, $u = x^2 - 5$, to solve for x. Replace u with $x^2 - 5$ in each of the preceding equations.

$x^2 - 5 = -5 \quad$ or $\quad x^2 - 5 = 2$	Replace u with $x^2 - 5.$
$x^2 = 0 \qquad\qquad x^2 = 7$	Solve for x by isolating $x^2.$
$x = 0 \qquad\qquad x = \pm\sqrt{7}$	Apply the square root property.

Although we did not raise both sides of an equation to an even power, checking the three proposed solutions in the original equation is a good idea. Do this now and verify that the solutions are $-\sqrt{7}, 0$, and $\sqrt{7}$, and the solution set is $\{-\sqrt{7}, 0, \sqrt{7}\}$.   ∎

✓ **CHECK POINT 3**   Solve: $(x^2 - 4)^2 + (x^2 - 4) - 6 = 0.$

EXAMPLE 4   Solving an Equation Quadratic in Form

Solve: $10x^{-2} + 7x^{-1} + 1 = 0.$

Solution   The variable factor in one term is the square of the variable factor in the other variable term.

$$10x^{-2} + 7x^{-1} + 1 = 0$$

x^{-2} is the square
of x^{-1}: $(x^{-1})^2 = x^{-2}.$

We will let u equal the variable factor that reappears squared. Thus,

$$\text{let } u = x^{-1}.$$

Now we write the given equation as a quadratic equation in u and solve for u.

$$10x^{-2} + 7x^{-1} + 1 = 0 \qquad \text{This is the given equation.}$$

$$10(x^{-1})^2 + 7x^{-1} + 1 = 0 \qquad \text{The equation contains } x^{-1} \text{ and } x^{-1} \text{ squared.}$$

$$10u^2 + 7u + 1 = 0 \qquad \text{Let } u = x^{-1}.$$

$$(5u + 1)(2u + 1) = 0 \qquad \text{Factor.}$$

$$5u + 1 = 0 \quad \text{or} \quad 2u + 1 = 0 \qquad \text{Set each factor equal to zero.}$$

$$5u = -1 \qquad\qquad 2u = -1 \qquad \text{Solve each equation for } u.$$

$$u = -\frac{1}{5} \qquad\qquad u = -\frac{1}{2}$$

Use the original substitution, $u = x^{-1}$, to solve for x. Replace u with x^{-1} in each of the preceding equations.

$$x^{-1} = -\frac{1}{5} \quad \text{or} \quad x^{-1} = -\frac{1}{2} \qquad \text{Replace } u \text{ with } x^{-1}.$$

$$(x^{-1})^{-1} = \left(-\frac{1}{5}\right)^{-1} \quad (x^{-1})^{-1} = \left(-\frac{1}{2}\right)^{-1} \qquad \text{Solve for x by raising both sides of each}$$
$$\text{equation to the } -1 \text{ power.}$$

$$x = -5 \qquad\qquad x = -2$$

$$\left(-\tfrac{1}{5}\right)^{-1} = \tfrac{1}{-\frac{1}{5}} = -5 \qquad \left(-\tfrac{1}{2}\right)^{-1} = \tfrac{1}{-\frac{1}{2}} = -2$$

We did not raise both sides of an equation to an even power. A check will show that both -5 and -2 are solutions of the original equation. The solution set is $\{-5, -2\}$. ∎

✓ **CHECK POINT 4** Solve: $2x^{-2} + x^{-1} - 1 = 0$.

EXAMPLE 5 Solving an Equation Quadratic in Form

Solve: $5x^{\frac{2}{3}} + 11x^{\frac{1}{3}} + 2 = 0$.

Solution The variable factor in one term is the square of the variable factor in the other variable term.

$$5x^{\frac{2}{3}} + 11x^{\frac{1}{3}} + 2 = 0$$

$$x^{\frac{2}{3}} \text{ is the square}$$
$$\text{of } x^{\frac{1}{3}} : \left(x^{\frac{1}{3}}\right)^2 = x^{\frac{2}{3}}.$$

We will let u equal the variable factor that reappears squared. Thus,

$$\text{let } u = x^{\frac{1}{3}}.$$

Now we write the given equation as a quadratic equation in u and solve for u.

$$5x^{\frac{2}{3}} + 11x^{\frac{1}{3}} + 2 = 0 \qquad \text{This is the given equation.}$$

$$5\left(x^{\frac{1}{3}}\right)^2 + 11\left(x^{\frac{1}{3}}\right) + 2 = 0 \qquad \text{The given equation contains } x^{\frac{1}{3}} \text{ and } x^{\frac{1}{3}} \text{ squared.}$$

$$5u^2 + 11u + 2 = 0 \qquad \text{Let } u = x^{\frac{1}{3}}.$$

$$(5u + 1)(u + 2) = 0 \qquad \text{Factor.}$$

$$5u + 1 = 0 \quad \text{or} \quad u + 2 = 0 \qquad \text{Set each factor equal to 0.}$$

$$u = -\tfrac{1}{5} \qquad\qquad u = -2 \qquad \text{Solve for } u.$$

Use the original substitution, $u = x^{\frac{1}{3}}$, to solve for x. Replace u with $x^{\frac{1}{3}}$ in each of the preceding equations.

$$x^{\frac{1}{3}} = -\frac{1}{5} \quad \text{or} \quad x^{\frac{1}{3}} = -2 \qquad \text{Replace } u \text{ with } x^{\frac{1}{3}} \text{ in } u = -\frac{1}{5} \text{ and } u = -2.$$

$$\left(x^{\frac{1}{3}}\right)^3 = \left(-\frac{1}{5}\right)^3 \quad \left(x^{\frac{1}{3}}\right)^3 = (-2)^3 \qquad \text{Solve for } x \text{ by cubing both sides of each equation.}$$

$$x = -\frac{1}{125} \qquad\qquad x = -8$$

We did not raise both sides of an equation to an even power. A check will show that both -8 and $-\frac{1}{125}$ are solutions of the original equation. The solution set is $\left\{-8, -\frac{1}{125}\right\}$. ∎

✓ **CHECK POINT 5**   Solve: $3x^{\frac{2}{3}} - 11x^{\frac{1}{3}} - 4 = 0$.

CONCEPT AND VOCABULARY CHECK

Fill in each blank so that the resulting statement is true.

1. We solve $x^4 - 13x^2 + 36 = 0$ by letting $u = $ _____.
   We then rewrite the equation in terms of u as _____.

2. We solve $x - 2\sqrt{x} - 8 = 0$ by letting $u = $ _____.
   We then rewrite the equation in terms of u as _____.

3. We solve $(x + 3)^2 + 7(x + 3) - 18 = 0$ by letting $u = $ _____.
   We then rewrite the equation in terms of u as _____.

4. We solve $2x^{-2} - 7x^{-1} + 3 = 0$ by letting $u = $ _____.
   We then rewrite the equation in terms of u as _____.

5. We solve $x^{\frac{2}{3}} + 2x^{\frac{1}{3}} - 3 = 0$ by letting $u = $ _____.
   We then rewrite the equation in terms of u as _____.

11.4 EXERCISE SET

MyMathLab®

Watch the videos
in MyMathLab

Download the
MyDashBoard App

Practice Exercises

In Exercises 1–32, solve each equation by making an appropriate substitution. If at any point in the solution process both sides of an equation are raised to an even power, a check is required.

1. $x^4 - 5x^2 + 4 = 0$
2. $x^4 - 13x^2 + 36 = 0$
3. $x^4 - 11x^2 + 18 = 0$
4. $x^4 - 9x^2 + 20 = 0$
5. $x^4 + 2x^2 = 8$
6. $x^4 + 4x^2 = 5$
7. $x + \sqrt{x} - 2 = 0$
8. $x + \sqrt{x} - 6 = 0$
9. $x - 4x^{\frac{1}{2}} - 21 = 0$
10. $x - 6x^{\frac{1}{2}} + 8 = 0$
11. $x - 13\sqrt{x} + 40 = 0$
12. $2x - 7\sqrt{x} - 30 = 0$

13. $(x - 5)^2 - 4(x - 5) - 21 = 0$
14. $(x + 3)^2 + 7(x + 3) - 18 = 0$
15. $(x^2 - 1)^2 - (x^2 - 1) = 2$
16. $(x^2 - 2)^2 - (x^2 - 2) = 6$
17. $(x^2 + 3x)^2 - 8(x^2 + 3x) - 20 = 0$
18. $(x^2 - 2x)^2 - 11(x^2 - 2x) + 24 = 0$
19. $x^{-2} - x^{-1} - 20 = 0$
20. $x^{-2} - x^{-1} - 6 = 0$
21. $2x^{-2} - 7x^{-1} + 3 = 0$
22. $20x^{-2} + 9x^{-1} + 1 = 0$
23. $x^{-2} - 4x^{-1} = 3$
24. $x^{-2} - 6x^{-1} = -4$
25. $x^{\frac{2}{3}} - x^{\frac{1}{3}} - 6 = 0$
26. $x^{\frac{2}{3}} + 2x^{\frac{1}{3}} - 3 = 0$

27. $x^{\frac{2}{5}} + x^{\frac{1}{5}} - 6 = 0$

28. $x^{\frac{2}{5}} + x^{\frac{1}{5}} - 2 = 0$

29. $2x^{\frac{1}{2}} - x^{\frac{1}{4}} = 1$

30. $2x^{\frac{1}{2}} - 5x^{\frac{1}{4}} = 3$

31. $\left(x - \dfrac{8}{x}\right)^2 + 5\left(x - \dfrac{8}{x}\right) - 14 = 0$

32. $\left(x - \dfrac{10}{x}\right)^2 + 6\left(x - \dfrac{10}{x}\right) - 27 = 0$

In Exercises 33–38, find the x-intercepts of the given function, f. Then use the x-intercepts to match each function with its graph. [The graphs are labeled (a) through (f).]

33. $f(x) = x^4 - 5x^2 + 4$

34. $f(x) = x^4 - 13x^2 + 36$

35. $f(x) = x^{\frac{1}{3}} + 2x^{\frac{1}{6}} - 3$

36. $f(x) = x^{-2} - x^{-1} - 6$

37. $f(x) = (x + 2)^2 - 9(x + 2) + 20$

38. $f(x) = 2(x + 2)^2 + 5(x + 2) - 3$

a.
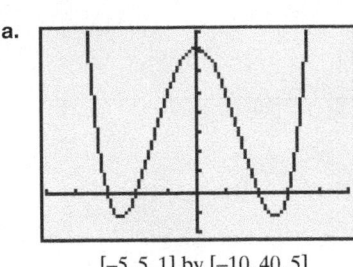
$[-5, 5, 1]$ by $[-10, 40, 5]$

c.
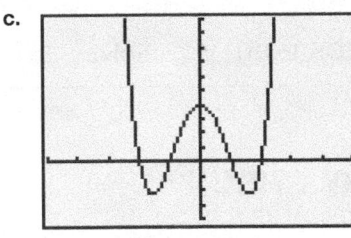
$[-5, 5, 1]$ by $[-4, 10, 1]$

e.
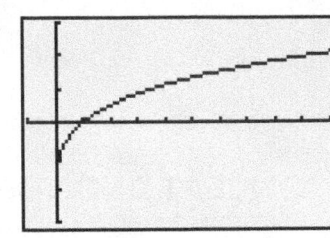
$[-1, 10, 1]$ by $[-3, 3, 1]$

b.
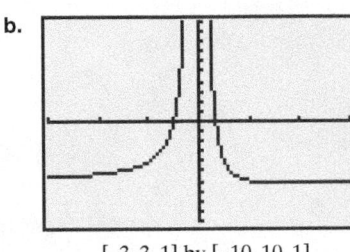
$[-3, 3, 1]$ by $[-10, 10, 1]$

d.
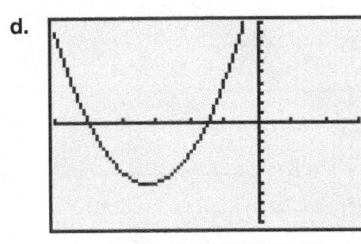
$[-6, 3, 1]$ by $[-10, 10, 1]$

f.
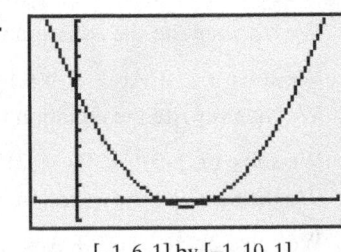
$[-1, 6, 1]$ by $[-1, 10, 1]$

Practice PLUS

39. Let $f(x) = (x^2 + 3x - 2)^2 - 10(x^2 + 3x - 2)$. Find all x such that $f(x) = -16$.

40. Let $f(x) = (x^2 + 2x - 2)^2 - 7(x^2 + 2x - 2)$. Find all x such that $f(x) = -6$.

41. Let $f(x) = 3\left(\dfrac{1}{x} + 1\right)^2 + 5\left(\dfrac{1}{x} + 1\right)$. Find all x such that $f(x) = 2$.

42. Let $f(x) = 2x^{\frac{2}{3}} + 3x^{\frac{1}{3}}$. Find all x such that $f(x) = 2$.

43. Let $f(x) = \dfrac{x}{x - 4}$ and $g(x) = 13\sqrt{\dfrac{x}{x - 4}} - 36$. Find all x such that $f(x) = g(x)$.

44. Let $f(x) = \dfrac{x}{x - 2} + 10$ and $g(x) = -11\sqrt{\dfrac{x}{x - 2}}$. Find all x such that $f(x) = g(x)$.

45. Let $f(x) = 3(x - 4)^{-2}$ and $g(x) = 16(x - 4)^{-1}$. Find all x such that $f(x)$ exceeds $g(x)$ by 12.

46. Let $f(x) = 6\left(\dfrac{2x}{x - 3}\right)^2$ and $g(x) = 5\left(\dfrac{2x}{x - 3}\right)$. Find all x such that $f(x)$ exceeds $g(x)$ by 6.

Application Exercises

How important is it for you to have a clean house? The bar graph indicates that the percentage of people who find this to be quite important varies by age. The percentage, $P(x)$, who find having a clean house very important can be modeled by the function

$$P(x) = 0.04(x + 40)^2 - 3(x + 40) + 104,$$

where x is the number of years a person's age is above or below 40. Thus, x is positive for people over 40 and negative for people under 40. Use the function to solve Exercises 47–48.

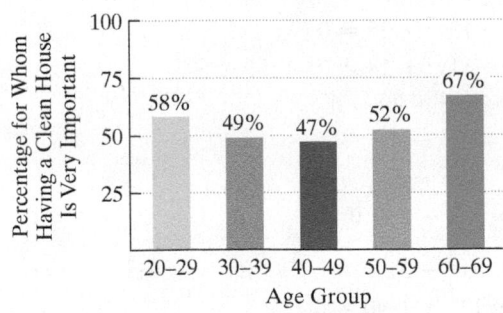
The Importance of Having a Clean House, by Age

Source: Soap and Detergent Association

47. According to the model, at which ages do 60% of us feel that having a clean house is very important? Substitute 60 for $P(x)$ and solve the quadratic-in-form equation. How well does the function model the data shown in the bar graph?

48. According to the model, at which ages do 50% of us feel that having a clean house is very important? Substitute 50 for $P(x)$ and solve the quadratic-in-form equation. How well does the function model the data shown in the bar graph?

Writing in Mathematics

49. Explain how to recognize an equation that is quadratic in form. Provide two original examples with your explanation.

50. Describe two methods for solving this equation:

$$x - 5\sqrt{x} + 4 = 0.$$

Technology Exercises

51. Use a graphing utility to verify the solutions of any five equations in Exercises 1–32 that you solved algebraically. The real solutions should appear as x-intercepts on the graph of the function related to the given equation.

Use a graphing utility to find the real solutions of the equations in Exercises 52–59. Check by direct substitution.

52. $x^6 - 7x^3 - 8 = 0$

53. $3(x-2)^{-2} - 4(x-2)^{-1} + 1 = 0$

54. $x^4 - 10x^2 + 9 = 0$

55. $2x + 6\sqrt{x} = 8$

56. $2(x+1)^2 = 5(x+1) + 3$

57. $(x^2 - 3x)^2 + 2(x^2 - 3x) - 24 = 0$

58. $x^{\frac{1}{2}} + 4x^{\frac{1}{4}} = 5$

59. $x^{\frac{2}{3}} - 3x^{\frac{1}{3}} + 2 = 0$

Critical Thinking Exercises

Make Sense? *In Exercises 60–63, determine whether each statement "makes sense" or "does not make sense" and explain your reasoning.*

60. When I solve an equation that is quadratic in form, it's important to write down the substitution that I am making.

61. Although I've rewritten an equation that is quadratic in form as $au^2 + bu + c = 0$ and solved for u, I'm not finished.

62. Checking is always a good idea, but it's never necessary when solving an equation that is quadratic in form.

63. The equation $5x^{\frac{2}{3}} + 11x^{\frac{1}{3}} + 2 = 0$ is quadratic in form, but when I reverse the variable terms and obtain $11x^{\frac{1}{3}} + 5x^{\frac{2}{3}} + 2 = 0$, the resulting equation is no longer quadratic in form.

In Exercises 64–67, determine whether each statement is true or false. If the statement is false, make the necessary change(s) to produce a true statement.

64. If an equation is quadratic in form, there is only one method that can be used to obtain its solution set.

65. An equation with three terms that is quadratic in form has a variable factor in one term that is the square of the variable factor in another term.

66. Because x^6 is the square of x^3, the equation $x^6 - 5x^3 + 6x = 0$ is quadratic in form.

67. To solve $x - 9\sqrt{x} + 14 = 0$, we let $\sqrt{u} = x$.

In Exercises 68–70, use a substitution to solve each equation.

68. $x^4 - 5x^2 - 2 = 0$

69. $5x^6 + x^3 = 18$

70. $\sqrt{\dfrac{x+4}{x-1}} + \sqrt{\dfrac{x-1}{x+4}} = \dfrac{5}{2}\left(\text{Let } u = \sqrt{\dfrac{x+4}{x-1}}.\right)$

Review Exercises

71. Simplify:

$$\frac{2x^2}{10x^3 - 2x^2}.$$

(Section 7.1, Example 3)

72. Divide: $\dfrac{2+i}{1-i}$. (Section 10.7, Example 5)

73. If $f(x) = \sqrt{x+1}$, find $f(3) - f(24)$.
(Section 8.1, Example 3)

Preview Exercises

Exercises 74–76 will help you prepare for the material covered in the next section.

74. Solve: $2x^2 + x = 15$.

75. Solve: $x^3 + x^2 = 4x + 4$.

76. Simplify: $\dfrac{x+1}{x+3} - 2$.

Polynomial and Rational Inequalities

Objectives

1. Solve polynomial inequalities.

2. Solve rational inequalities.

3. Solve problems modeled by polynomial or rational inequalities.

Copyright © 2011 by Warren Miller/The New Yorker Collection/The Cartoon Bank.

Tailgaters beware: If your car is going 35 miles per hour on dry pavement, your required stopping distance is 160 feet, or the width of a football field. At 65 miles per hour, the distance required is 410 feet, or approximately the length of one and one-tenth football fields. **Figure 11.17** shows stopping distances for cars at various speeds on dry roads and on wet roads.

Stopping Distances for Cars at Selected Speeds

■ Dry Pavement   ■ Wet Pavement

(Bar chart: Stopping Distance (feet) vs. Speed (miles per hour))

- At 35 mph: 160 (dry), 185 (wet)
- At 45 mph: 225 (dry), 275 (wet)
- At 55 mph: 310 (dry), 380 (wet)
- At 65 mph: 410 (dry), 505 (wet)

Figure 11.17

Source: National Highway Traffic Safety Administration

Using Technology

We used the statistical menu of a graphing utility and the quadratic regression program to obtain the quadratic function that models stopping distance on dry pavement. After entering the appropriate data from **Figure 11.17**, namely

(35, 160), (45, 225), (55, 310), (65, 410),

we obtained the results shown in the screen.

```
QuadReg
 y=ax²+bx+c
 a=.0875
 b=-.4
 c=66.5625
```

A car's required stopping distance, $f(x)$, in feet, on dry pavement traveling at x miles per hour can be modeled by the quadratic function

$$f(x) = 0.0875x^2 - 0.4x + 66.6.$$

How can we use this function to determine speeds on dry pavement requiring stopping distances that exceed the length of one and one-half football fields, or 540 feet? We must solve the inequality

$$0.0875x^2 - 0.4x + 66.6 > 540.$$

Required stopping distance   exceeds   540 feet.

We begin by subtracting 540 from both sides. This will give us zero on the right:

$$0.0875x^2 - 0.4x + 66.6 - 540 > 540 - 540$$
$$0.0875x^2 - 0.4x - 473.4 > 0.$$

The form of this inequality is $ax^2 + bx + c > 0$. Such a quadratic inequality is called a *polynomial inequality*.

Definition of a Polynomial Inequality

A polynomial inequality is any inequality that can be put in one of the forms

$$f(x) < 0, \quad f(x) > 0, \quad f(x) \le 0, \quad \text{or} \quad f(x) \ge 0,$$

where f is a polynomial function.

In this section, we establish the basic techniques for solving polynomial inequalities. We will also use these techniques to solve inequalities involving rational functions.

1 Solve polynomial inequalities.

Solving Polynomial Inequalities

Graphs can help us visualize the solutions of polynomial inequalities. For example, the graph of $f(x) = x^2 - 7x + 10$ is shown in **Figure 11.18**. The x-intercepts, 2 and 5, are **boundary points** between where the graph lies above the x-axis, shown in blue, and where the graph lies below the x-axis, shown in red.

Locating the x-intercepts of a polynomial function, f, is an important step in finding the solution set for polynomial inequalities in the form $f(x) < 0$ or $f(x) > 0$.

We use the x-intercepts of f as boundary points that divide the real number line into intervals. On each interval, the graph of f is either above the x-axis [$f(x) > 0$] or below the x-axis [$f(x) < 0$]. For this reason, x-intercepts play a fundamental role in solving polynomial inequalities. The x-intercepts are found by solving the equation $f(x) = 0$.

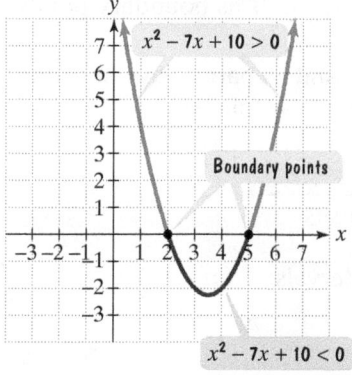

Figure 11.18

Procedure for Solving Polynomial Inequalities

1. Express the inequality in the form

$$f(x) < 0 \quad \text{or} \quad f(x) > 0,$$

   where f is a polynomial function.

2. Solve the equation $f(x) = 0$. The real solutions are the **boundary points**.

3. Locate these boundary points on a number line, thereby dividing the number line into intervals.

4. Choose one representative number, called a **test value**, within each interval and evaluate f at that number.

   a. If the value of f is positive, then $f(x) > 0$ for all numbers, x, in the interval.

   b. If the value of f is negative, then $f(x) < 0$ for all numbers, x, in the interval.

5. Write the solution set, selecting the interval or intervals that satisfy the given inequality.

This procedure is valid if $<$ is replaced by $\le$ or $>$ is replaced by $\ge$. However, if the inequality involves $\le$ or $\ge$, include the boundary points [the solutions of $f(x) = 0$] in the solution set.

EXAMPLE 1   Solving a Polynomial Inequality

Solve and graph the solution set on a real number line: $2x^2 + x > 15$.

Solution

Step 1. Express the inequality in the form $f(x) < 0$ or $f(x) > 0$. We begin by rewriting the inequality so that 0 is on the right side.

Using Technology

Graphic Connections

The solution set for

$$2x^2 + x > 15$$

or, equivalently,

$$2x^2 + x - 15 > 0$$

can be verified with a graphing utility. The graph of $f(x) = 2x^2 + x - 15$ was obtained using a $[-10, 10, 1]$ by $[-16, 6, 1]$ viewing rectangle. The graph lies above the x-axis, representing $>$, for all x in $(-\infty, -3)$ or $(\frac{5}{2}, \infty)$.

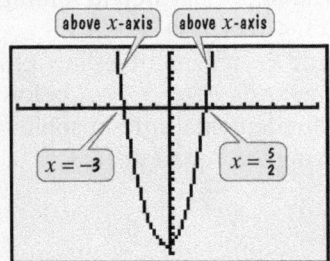

$$2x^2 + x > 15 \qquad \text{This is the given inequality.}$$
$$2x^2 + x - 15 > 15 - 15 \qquad \text{Subtract 15 from both sides.}$$
$$2x^2 + x - 15 > 0 \qquad \text{Simplify.}$$

This inequality is equivalent to the one we wish to solve. It is in the form $f(x) > 0$, where $f(x) = 2x^2 + x - 15$.

Step 2. Solve the equation $f(x) = 0$. We find the x-intercepts of $f(x) = 2x^2 + x - 15$ by solving the equation $2x^2 + x - 15 = 0$.

$$2x^2 + x - 15 = 0 \qquad \begin{array}{l}\text{This polynomial equation is a}\\\text{quadratic equation.}\end{array}$$
$$(2x - 5)(x + 3) = 0 \qquad \text{Factor.}$$
$$2x - 5 = 0 \quad \text{or} \quad x + 3 = 0 \qquad \text{Set each factor equal to 0.}$$
$$x = \tfrac{5}{2} \qquad\qquad x = -3 \qquad \text{Solve for x.}$$

The x-intercepts of f are -3 and $\frac{5}{2}$. We will use these x-intercepts as boundary points on a number line.

Step 3. Locate the boundary points on a number line and separate the line into intervals. The number line with the boundary points is shown as follows:

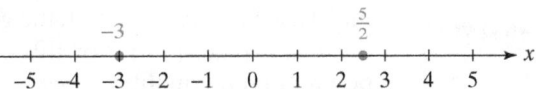

The boundary points divide the number line into three intervals:

$$(-\infty, -3) \quad \left(-3, \tfrac{5}{2}\right) \quad \left(\tfrac{5}{2}, \infty\right).$$

Step 4. Choose one test value within each interval and evaluate f at that number.

Interval	Test Value	Substitute into $f(x) = 2x^2 + x - 15$	Conclusion
$(-\infty, -3)$	-4	$f(-4) = 2(-4)^2 + (-4) - 15$ $= 13$, positive	$f(x) > 0$ for all x in $(-\infty, -3)$.
$\left(-3, \dfrac{5}{2}\right)$	0	$f(0) = 2 \cdot 0^2 + 0 - 15$ $= -15$, negative	$f(x) < 0$ for all x in $\left(-3, \dfrac{5}{2}\right)$.
$\left(\dfrac{5}{2}, \infty\right)$	3	$f(3) = 2 \cdot 3^2 + 3 - 15$ $= 6$, positive	$f(x) > 0$ for all x in $\left(\dfrac{5}{2}, \infty\right)$.

Step 5. Write the solution set, selecting the interval or intervals that satisfy the given inequality. We are interested in solving $f(x) > 0$, where $f(x) = 2x^2 + x - 15$. Based on our work in step 4, we see that $f(x) > 0$ for all x in $(-\infty, -3)$ or $(\frac{5}{2}, \infty)$. Thus, the solution set of the given inequality, $2x^2 + x > 15$, or, equivalently, $2x^2 + x - 15 > 0$, is

$$(-\infty, -3) \cup \left(\tfrac{5}{2}, \infty\right).$$

The graph of the solution set on a number line is shown as follows:

The graph of the solution set on a number line is shown as follows. ■

✓ **CHECK POINT 1** Solve and graph the solution set on a real number line: $x^2 - x > 20$.

EXAMPLE 2   Solving a Polynomial Inequality

Solve and graph the solution set on a real number line: $4x^2 \le 1 - 2x$.

Solution

Step 1. Express the inequality in the form $f(x) \le 0$ or $f(x) \ge 0$. We begin by rewriting the inequality so that 0 is on the right side.

$$4x^2 \le 1 - 2x \qquad \text{This is the given inequality.}$$
$$4x^2 + 2x - 1 \le 1 - 2x + 2x - 1 \qquad \text{Add 2x and subtract 1 on both sides.}$$
$$4x^2 + 2x - 1 \le 0 \qquad \text{Simplify.}$$

This inequality is equivalent to the one we wish to solve. It is in the form $f(x) \le 0$, where $f(x) = 4x^2 + 2x - 1$.

Step 2. Solve the equation $f(x) = 0$. We will find the x-intercepts of $f(x) = 4x^2 + 2x - 1$ by solving the equation $4x^2 + 2x - 1 = 0$. This equation cannot be solved by factoring. We will use the quadratic formula to solve it.

$$4x^2 + 2x - 1 = 0$$

$$\boxed{a = 4} \quad \boxed{b = 2} \quad \boxed{c = -1}$$

$$x = \frac{-b \pm \sqrt{b^2 - 4ac}}{2a} = \frac{-2 \pm \sqrt{2^2 - 4 \cdot 4(-1)}}{2 \cdot 4} = \frac{-2 \pm \sqrt{4 - (-16)}}{8}$$

$$= \frac{-2 \pm \sqrt{20}}{8} = \frac{-2 \pm \sqrt{4}\sqrt{5}}{8} = \frac{-2 \pm 2\sqrt{5}}{8} = \frac{2(-1 \pm \sqrt{5})}{8} = \frac{-1 \pm \sqrt{5}}{4}$$

$$x = \frac{-1 + \sqrt{5}}{4} \approx 0.3 \qquad\qquad x = \frac{-1 - \sqrt{5}}{4} \approx -0.8$$

The x-intercepts of f are $\dfrac{-1 + \sqrt{5}}{4}$ (approximately 0.3) and $\dfrac{-1 - \sqrt{5}}{4}$ (approximately -0.8). We will use these x-intercepts as boundary points on a number line.

Step 3. Locate the boundary points on a number line and separate the line into intervals. The number line with the boundary points is shown as follows:

$$\boxed{\frac{-1 - \sqrt{5}}{4} \approx -0.8} \qquad \boxed{\frac{-1 + \sqrt{5}}{4} \approx 0.3}$$

The boundary points divide the number line into three intervals:

$$\left(-\infty, \frac{-1 - \sqrt{5}}{4}\right) \quad \left(\frac{-1 - \sqrt{5}}{4}, \frac{-1 + \sqrt{5}}{4}\right) \quad \left(\frac{-1 + \sqrt{5}}{4}, \infty\right).$$

Step 4. Choose one test value within each interval and evaluate f at that number.

Interval	Test Value	Substitute into $f(x) = 4x^2 + 2x - 1$	Conclusion
$\left(-\infty, \dfrac{-1 - \sqrt{5}}{4}\right)$	-1	$f(-1) = 4(-1)^2 + 2(-1) - 1$ $= 1$, positive	$f(x) > 0$ for all x in $\left(-\infty, \dfrac{-1 - \sqrt{5}}{4}\right)$.
$\left(\dfrac{-1 - \sqrt{5}}{4}, \dfrac{-1 + \sqrt{5}}{4}\right)$	0	$f(0) = 4 \cdot 0^2 + 2 \cdot 0 - 1$ $= -1$, negative	$f(x) < 0$ for all x in $\left(\dfrac{-1 - \sqrt{5}}{4}, \dfrac{-1 + \sqrt{5}}{4}\right)$.
$\left(\dfrac{-1 + \sqrt{5}}{4}, \infty\right)$	1	$f(1) = 4 \cdot 1^2 + 2 \cdot 1 - 1$ $= 5$, positive	$f(x) > 0$ for all x in $\left(\dfrac{-1 + \sqrt{5}}{4}, \infty\right)$.

Using Technology

Graphic Connections

The solution set for

$$4x^2 \le 1 - 2x$$

or, equivalently,

$$4x^2 + 2x - 1 \le 0$$

can be verified with a graphing utility. The graph of $f(x) = 4x^2 + 2x - 1$ was obtained using a $[-2, 2, 1]$ by $[-10, 10, 1]$ viewing rectangle. The graph lies on or below the x-axis, representing $\le$, for all x in

$$\left[\frac{-1 - \sqrt{5}}{4}, \frac{-1 + \sqrt{5}}{4}\right]$$

$$\approx [-0.8, 0.3].$$

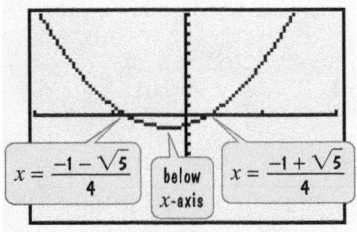

$[-2, 2, 1]$ by $[-10, 10, 1]$

Step 5. Write the solution set, selecting the interval or intervals that satisfy the given inequality. We are interested in solving $f(x) \le 0$, where $f(x) = 4x^2 + 2x - 1$. Based on our work in step 4, we see that $f(x) < 0$ for all x in $\left(\dfrac{-1 - \sqrt{5}}{4}, \dfrac{-1 + \sqrt{5}}{4}\right)$. However, because the inequality involves $\le$ (less than or *equal to*), we must also include the solutions of $4x^2 + 2x - 1 = 0$, namely $\dfrac{-1 - \sqrt{5}}{4}$ and $\dfrac{-1 + \sqrt{5}}{4}$, in the solution set. Thus, the solution set of the given inequality $4x^2 \le 1 - 2x$, or, equivalently, $4x^2 + 2x - 1 \le 0$, is

$$\left[\frac{-1 - \sqrt{5}}{4}, \frac{-1 + \sqrt{5}}{4}\right].$$

The graph of the solution set on a number line is shown as follows:

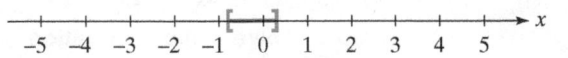

✓ **CHECK POINT 2** Solve and graph the solution set on a real number line: $2x^2 \le -6x - 1$.

EXAMPLE 3 Solving a Polynomial Inequality

Solve and graph the solution set on a real number line: $x^3 + x^2 \le 4x + 4$.

Solution

Step 1. Express the inequality in the form $f(x) \le 0$ or $f(x) \ge 0$. We begin by rewriting the inequality so that 0 is on the right side.

$$x^3 + x^2 \le 4x + 4 \qquad \text{This is the given inequality.}$$
$$x^3 + x^2 - 4x - 4 \le 4x + 4 - 4x - 4 \qquad \text{Subtract } 4x + 4 \text{ from both sides.}$$
$$x^3 + x^2 - 4x - 4 \le 0 \qquad \text{Simplify.}$$

This inequality is equivalent to the one we wish to solve. It is in the form $f(x) \le 0$, where $f(x) = x^3 + x^2 - 4x - 4$.

Step 2. Solve the equation $f(x) = 0$. We find the x-intercepts of $f(x) = x^3 + x^2 - 4x - 4$ by solving the equation $x^3 + x^2 - 4x - 4 = 0$.

$$x^3 + x^2 - 4x - 4 = 0 \qquad \text{This polynomial equation is of degree 3.}$$
$$x^2(x + 1) - 4(x + 1) = 0 \qquad \text{Factor } x^2 \text{ from the first two terms and } -4 \text{ from the last two terms.}$$
$$(x + 1)(x^2 - 4) = 0 \qquad \text{A common factor of } x + 1 \text{ is factored from the expression.}$$
$$(x + 1)(x + 2)(x - 2) = 0 \qquad \text{Factor completely.}$$
$$x + 1 = 0 \quad \text{or} \quad x + 2 = 0 \quad \text{or} \quad x - 2 = 0 \qquad \text{Set each factor equal to 0.}$$
$$x = -1 \qquad\qquad x = -2 \qquad\qquad x = 2 \qquad \text{Solve for } x.$$

The x-intercepts of f are $-2, -1$, and 2. We will use these x-intercepts as boundary points on a number line.

Step 3. Locate the boundary points on a number line and separate the line into intervals. The number line with the boundary points is shown as follows:

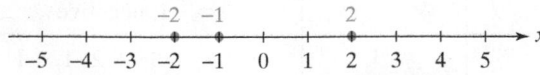

The boundary points divide the number line into four intervals:

$$(-\infty, -2) \quad (-2, -1) \quad (-1, 2) \quad (2, \infty).$$

Step 4. Choose one test value within each interval and evaluate f at that number.

Interval	Test Value	Substitute into $f(x) = x^3 + x^2 - 4x - 4$	Conclusion
$(-\infty, -2)$	-3	$f(-3) = (-3)^3 + (-3)^2 - 4(-3) - 4$ $= -10$, negative	$f(x) < 0$ for all x in $(-\infty, -2)$.
$(-2, -1)$	-1.5	$f(-1.5) = (-1.5)^3 + (-1.5)^2 - 4(-1.5) - 4$ $= 0.875$, positive	$f(x) > 0$ for all x in $(-2, -1)$.
$(-1, 2)$	0	$f(0) = 0^3 + 0^2 - 4 \cdot 0 - 4$ $= -4$, negative	$f(x) < 0$ for all x in $(-1, 2)$.
$(2, \infty)$	3	$f(3) = 3^3 + 3^2 - 4 \cdot 3 - 4$ $= 20$, positive	$f(x) > 0$ for all x in $(2, \infty)$.

Using Technology

Graphic Connections

The solution set for

$$x^3 + x^2 \leq 4x + 4$$

or, equivalently,

$$x^3 + x^2 - 4x - 4 \leq 0$$

can be verified with a graphing utility. The graph of $f(x) = x^3 + x^2 - 4x - 4$ lies on or below the x-axis, representing $\leq$, for all x in $(-\infty, -2]$ or $[-1, 2]$.

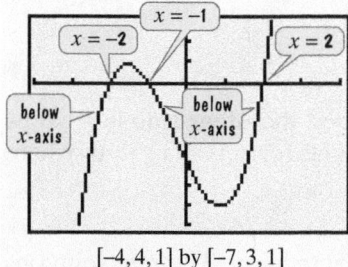

$[-4, 4, 1]$ by $[-7, 3, 1]$

Step 5. Write the solution set, selecting the interval or intervals that satisfy the given inequality. We are interested in solving $f(x) \leq 0$, where $f(x) = x^3 + x^2 - 4x - 4$. Based on our work in step 4, we see that $f(x) < 0$ for all x in $(-\infty, -2)$ or $(-1, 2)$. However, because the inequality involves $\leq$ (less than or *equal to*), we must also include the solutions of $x^3 + x^2 - 4x - 4 = 0$, namely $-2, -1$, and 2, in the solution set. Thus, the solution set of the given inequality, $x^3 + x^2 \leq 4x + 4$, or, equivalently, $x^3 + x^2 - 4x - 4 \leq 0$, is

$$(-\infty, -2] \cup [-1, 2].$$

The graph of the solution set on a number line is shown as follows:

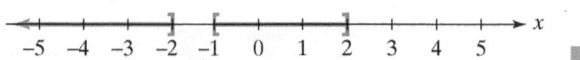

✓ **CHECK POINT 3** Solve and graph the solution set on a real number line: $x^3 + 3x^2 \leq x + 3$.

Solving Rational Inequalities

A **rational inequality** is any inequality that can be put in one of the forms

$$f(x) < 0, \quad f(x) > 0, \quad f(x) \leq 0, \quad \text{or} \quad f(x) \geq 0,$$

where f is a rational function. An example of a rational inequality is

$$\frac{3x + 3}{2x + 4} > 0.$$

This inequality is in the form $f(x) > 0$, where f is the rational function given by

$$f(x) = \frac{3x + 3}{2x + 4}.$$

The graph of f is shown in **Figure 11.19**.

We can find the x-intercept of f by setting the numerator equal to 0:

$$3x + 3 = 0$$
$$3x = -3$$
$$x = -1.$$

f has an x-intercept at -1 and passes through $(-1, 0)$.

2 Solve rational inequalities.

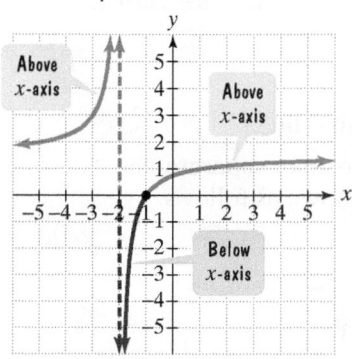

Figure 11.19 The graph of $f(x) = \dfrac{3x + 3}{2x + 4}$

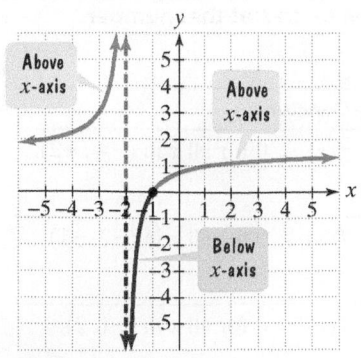

Figure 11.19 The graph of $f(x) = \dfrac{3x + 3}{2x + 4}$ (repeated)

We can determine where f is undefined by setting the denominator of $f(x) = \dfrac{3x + 3}{2x + 4}$ equal to 0:

$$2x + 4 = 0$$
$$2x = -4$$
$$x = -2.$$

> f is undefined at -2. Figure 11.19 shows that the function's vertical asymptote is $x = -2$.

By setting both the numerator and the denominator of f equal to 0, we obtained the solutions -2 and -1. These numbers separate the x-axis into three intervals: $(-\infty, -2), (-2, -1),$ and $(-1, \infty)$. On each interval, the graph of f is either above the x-axis $[f(x) > 0]$ or below the x-axis $[f(x) < 0]$.

Examine the graph in **Figure 11.19** carefully. Can you see that it is above the x-axis for all x in $(-\infty, -2)$ or $(-1, \infty)$, shown in blue? Thus, the solution set of $\dfrac{3x + 3}{2x + 4} > 0$ is $(-\infty, -2) \cup (-1, \infty)$. By contrast, the graph of f lies below the x-axis for all x in $(-2, -1)$, shown in red. Thus, the solution set of $\dfrac{3x + 3}{2x + 4} < 0$ is $(-2, -1)$.

The first step in solving a rational inequality is to bring all terms to one side, obtaining zero on the other side. Then express the rational function on the nonzero side as a single quotient. The second step is to set the numerator and the denominator of f equal to zero. The solutions of these equations serve as boundary points that separate the real number line into intervals. At this point, the procedure is the same as the one we used for solving polynomial inequalities.

EXAMPLE 4 Solving a Rational Inequality

Solve and graph the solution set: $\dfrac{x + 3}{x - 7} < 0.$

Solution

Step 1. Express the inequality so that one side is zero and the other side is a single quotient. The given inequality is already in this form. The form is $f(x) < 0$, where $f(x) = \dfrac{x + 3}{x - 7}$.

Step 2. Set the numerator and the denominator of f equal to zero. The real solutions are the boundary points.

$$x + 3 = 0 \qquad x - 7 = 0$$

> Set the numerator and denominator equal to 0. These are the values that make the quotient zero or undefined.

$$x = -3 \qquad x = 7$$

> Solve for x.

We will use these solutions as boundary points on a number line.

Step 3. Locate the boundary points on a number line and separate the line into intervals. The number line with the boundary points is shown as follows:

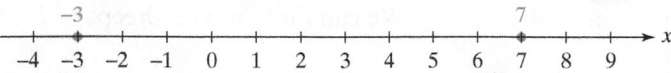

The boundary points divide the number line into three intervals:

$$(-\infty, -3) \quad (-3, 7) \quad (7, \infty).$$

Step 4. Choose one test value within each interval and evaluate f at that number.

Interval	Test Value	Substitute into $f(x) = \dfrac{x + 3}{x - 7}$	Conclusion
$(-\infty, -3)$	-4	$f(-4) = \dfrac{-4 + 3}{-4 - 7}$ $= \dfrac{-1}{-11} = \dfrac{1}{11}$, positive	$f(x) > 0$ for all x in $(-\infty, -3)$.
$(-3, 7)$	0	$f(0) = \dfrac{0 + 3}{0 - 7}$ $= -\dfrac{3}{7}$, negative	$f(x) < 0$ for all x in $(-3, 7)$.
$(7, \infty)$	8	$f(8) = \dfrac{8 + 3}{8 - 7}$ $= 11$, positive	$f(x) > 0$ for all x in $(7, \infty)$.

Step 5. Write the solution set, selecting the interval or intervals that satisfy the given inequality. We are interested in solving $f(x) < 0$, where $f(x) = \dfrac{x + 3}{x - 7}$. Based on our work in step 4, we see that $f(x) < 0$ for all x in $(-3, 7)$.

Because $f(x) < 0$ for all x in $(-3, 7)$, the solution set of the given inequality, $\dfrac{x + 3}{x - 7} < 0$, is $(-3, 7)$.

The graph of the solution set on a number line is shown as follows:

✓ **CHECK POINT 4**   Solve and graph the solution set: $\dfrac{x - 5}{x + 2} < 0$.

Great Question!

Can I begin solving

$$\dfrac{x + 1}{x + 3} \geq 2$$

by multiplying both sides by $x + 3$?

No. We do not know if $x + 3$ is positive or negative. Thus, we do not know whether or not to change the sense of the inequality.

EXAMPLE 5   Solving a Rational Inequality

Solve and graph the solution set: $\dfrac{x + 1}{x + 3} \geq 2$.

Solution

Step 1. Express the inequality so that one side is zero and the other side is a single quotient. We subtract 2 from both sides to obtain zero on the right.

$$\dfrac{x + 1}{x + 3} \geq 2 \quad \text{This is the given inequality.}$$

$$\dfrac{x + 1}{x + 3} - 2 \geq 0 \quad \text{Subtract 2 from both sides, obtaining 0 on the right.}$$

$$\dfrac{x + 1}{x + 3} - \dfrac{2(x + 3)}{x + 3} \geq 0 \quad \text{The least common denominator is } x + 3. \text{ Express 2 in terms of this denominator.}$$

$$\dfrac{x + 1 - 2(x + 3)}{x + 3} \geq 0 \quad \text{Subtract rational expressions.}$$

$$\dfrac{x + 1 - 2x - 6}{x + 3} \geq 0 \quad \text{Apply the distributive property.}$$

$$\dfrac{-x - 5}{x + 3} \geq 0 \quad \text{Simplify.}$$

This inequality is equivalent to the one we wish to solve. It is in the form $f(x) \geq 0$, where $f(x) = \dfrac{-x - 5}{x + 3}$.

Step 2. Set the numerator and the denominator of f equal to zero. We need to solve $f(x) \geq 0$, where $f(x) = \dfrac{-x - 5}{x + 3}$. We use $f(x) = \dfrac{-x - 5}{x + 3}$. The real solutions obtained by setting the numerator and the denominator equal to zero are the boundary points.

$$-x - 5 = 0 \qquad x + 3 = 0$$ Set the numerator and denominator equal to 0. These are the values that make the quotient zero or undefined.

$$x = -5 \qquad x = -3$$ Solve for x.

We will use these solutions as boundary points on a number line.

Step 3. Locate the boundary points on a number line and separate the line into intervals. The number line with the boundary points is shown as follows:

$$
\begin{array}{c}
-5 \qquad -3 \\
\hline
-7 \; -6 \; -5 \; -4 \; -3 \; -2 \; -1 \;\; 0 \;\; 1 \;\; 2 \;\; 3
\end{array} \to x
$$

The boundary points divide the number line into three intervals:

$$(-\infty, -5) \quad (-5, -3) \quad (-3, \infty).$$

Step 4. Choose one test value within each interval and evaluate f at that number.

Interval	Test Value	Substitute into $f(x) = \dfrac{-x - 5}{x + 3}$	Conclusion
$(-\infty, -5)$	-6	$f(-6) = \dfrac{-(-6) - 5}{-6 + 3}$ $= -\frac{1}{3}$, negative	$f(x) < 0$ for all x in $(-\infty, -5)$.
$(-5, -3)$	-4	$f(-4) = \dfrac{-(-4) - 5}{-4 + 3}$ $= 1$, positive	$f(x) > 0$ for all x in $(-5, -3)$.
$(-3, \infty)$	0	$f(0) = \dfrac{-0 - 5}{0 + 3}$ $= -\frac{5}{3}$, negative	$f(x) < 0$ for all x in $(-3, \infty)$.

Step 5. Write the solution set, selecting the interval or intervals that satisfy the given inequality. We are interested in solving $f(x) \geq 0$, where $f(x) = \dfrac{-x - 5}{x + 3}$. Based on our work in step 4, we see that $f(x) > 0$ for all x in $(-5, -3)$. However, because the inequality involves $\geq$ (greater than or *equal to*), we must also include the solution of $f(x) = 0$, namely the value that we obtained when we set the numerator of f equal to zero. Thus, we must include -5 in the solution set. The solution set of the given inequality is $[-5, -3)$.

The graph of the solution set on a number line is shown as follows:

$$
\begin{array}{c}
\hline
-7 \; -6 \; [-5 \; -4 \; -3) \; -2 \; -1 \;\; 0 \;\; 1 \;\; 2 \;\; 3
\end{array} \to x
$$

Great Question!

Which boundary points must I always exclude from the solution set of a rational inequality?

Never include values that cause a rational function's denominator to equal zero. Division by zero is undefined.

Using Technology

Graphic Connections

The solution set for

$$\frac{x + 1}{x + 3} \geq 2$$

or, equivalently,

$$\frac{-x - 5}{x + 3} \geq 0$$

can be verified with a graphing utility. The graph of $f(x) = \dfrac{-x - 5}{x + 3}$ lies on or above the x-axis, representing $\geq$, for all x in $[-5, -3)$.

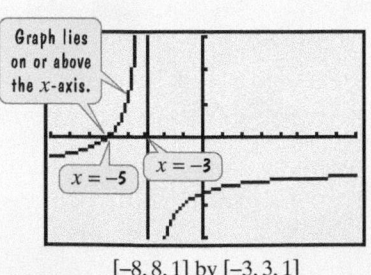

Graph lies on or above the x-axis.

$x = -5$ $x = -3$

$[-8, 8, 1]$ by $[-3, 3, 1]$

✓ **CHECK POINT 5**   Solve and graph the solution set: $\dfrac{2x}{x+1} \geq 1$.

---

3 Solve problems modeled by polynomial or rational inequalities.

Applications

Polynomial inequalities can be solved to answer questions about variables contained in polynomial functions.

EXAMPLE 6   Modeling the Position of a Free-Falling Object

A ball is thrown vertically upward from the top of the Leaning Tower of Pisa (190 feet high) with an initial velocity of 96 feet per second (**Figure 11.20**). The function

$$s(t) = -16t^2 + 96t + 190$$

models the ball's height above the ground, $s(t)$, in feet, t seconds after it was thrown. During which time period will the ball's height exceed that of the tower?

Solution   Using the problem's question and the given model for the ball's height, $s(t) = -16t^2 + 96t + 190$, we obtain a polynomial inequality.

$$-16t^2 + 96t + 190 > 190$$

When will the ball's height   exceed that   of the tower?

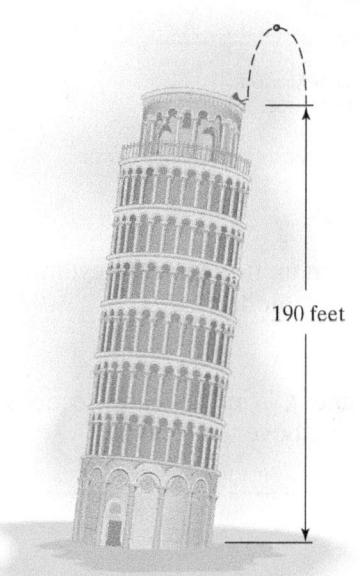

Figure 11.20 Throwing a ball from the top of the Leaning Tower of Pisa

190 feet

$-16t^2 + 96t + 190 > 190$ This is the inequality that models the problem's question.

$-16t^2 + 96t > 0$ Subtract 190 from both sides. This inequality is in the form $f(t) > 0$, where $f(t) = -16t^2 + 96t$.

$-16t^2 + 96t = 0$ Solve the equation $f(t) = 0$.

$-16t(t - 6) = 0$ Factor.

$-16t = 0$ or $t - 6 = 0$ Set each factor equal to 0.

$t = 0$ $t = 6$ Solve for t. The boundary points are 0 and 6.

Locate these values on a number line.

The intervals are $(-\infty, 0)$, $(0, 6)$, and $(6, \infty)$. For our purposes, the mathematical model is useful only from $t = 0$ until the ball hits the ground. (By setting $-16t^2 + 96t + 190$ equal to zero, we find $t \approx 7.57$; the ball hits the ground after approximately 7.57 seconds.) Thus, we use $(0, 6)$ and $(6, 7.57)$ for our intervals.

Interval	Test Value	Substitute into $f(t) = -16t^2 + 96t$	Conclusion
$(0, 6)$	1	$f(1) = -16 \cdot 1^2 + 96 \cdot 1$ $= 80$, positive	$f(t) > 0$ for all t in $(0, 6)$.
$(6, 7.57)$	7	$f(7) = -16 \cdot 7^2 + 96 \cdot 7$ $= -112$, negative	$f(t) < 0$ for all t in $(6, 7.57)$.

We are interested in solving $f(t) > 0$, where $f(t) = -16t^2 + 96t$. We see that $f(t) > 0$ for all t in $(0, 6)$. This means that the ball's height exceeds that of the tower between 0 and 6 seconds. ∎

Using Technology

Graphic Connections

The graphs of

$$y_1 = -16x^2 + 96x + 190$$

and

$$y_2 = 190$$

are shown in a

$$[0, 8, 1] \text{ by } [0, 360, 36]$$

viewing rectangle. The graphs show that the ball's height exceeds that of the tower between 0 and 6 seconds.

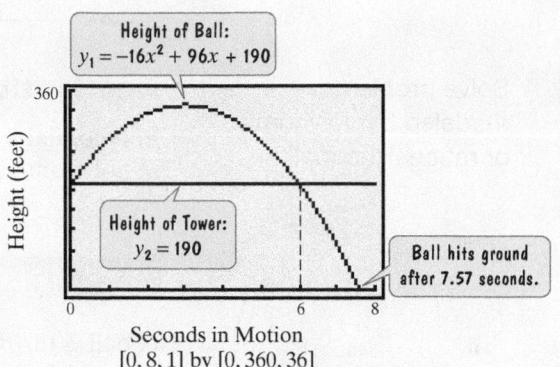

Height of Ball:
$y_1 = -16x^2 + 96x + 190$

Height of Tower:
$y_2 = 190$

Ball hits ground after 7.57 seconds.

Seconds in Motion
$[0, 8, 1]$ by $[0, 360, 36]$

✓ **CHECK POINT 6** An object is propelled straight up from ground level with an initial velocity of 80 feet per second. Its height at time t is modeled by

$$s(t) = -16t^2 + 80t,$$

where the height, $s(t)$, is measured in feet and the time, t, is measured in seconds. In which time interval will the object be more than 64 feet above the ground?

CONCEPT AND VOCABULARY CHECK

Fill in each blank so that the resulting statement is true.

1. We solve the polynomial inequality $x^2 + 8x + 15 > 0$ by first solving the equation _____. The real solutions of this equation, -5 and -3, shown on the number line, are called _____ points.

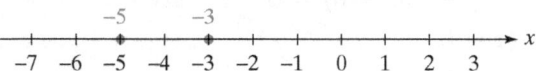

2. The points at -5 and -3 shown above divide the number line into three intervals:

_____, _____, _____.

3. True or false: A test value for the leftmost interval on the number line shown above could be -10. _____

4. True or false: A test value for the rightmost interval on the number line shown above could be 0. _____

5. Consider the rational inequality

$$\frac{x - 1}{x + 2} \geq 0.$$

Setting the numerator and the denominator of $\frac{x - 1}{x + 2}$ equal to zero, we obtain $x = 1$ and $x = -2$. These values are shown as points on the number line. Also shown is information about three test values.

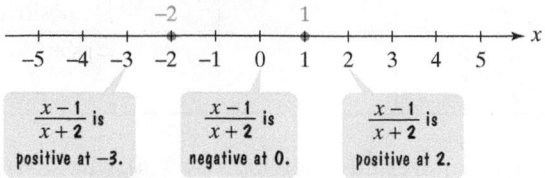

$\frac{x-1}{x+2}$ is positive at -3.

$\frac{x-1}{x+2}$ is negative at 0.

$\frac{x-1}{x+2}$ is positive at 2.

Based on the information shown above, the solution set of $\frac{x - 1}{x + 2} \geq 0$ is _____.

Watch the videos
in MyMathLab   Download the
MyDashBoard App

11.5 EXERCISE SET   MyMathLab®

Practice Exercises

Solve each polynomial inequality in Exercises 1–40 and graph the solution set on a real number line.

1. $(x - 4)(x + 2) > 0$
2. $(x + 3)(x - 5) > 0$
3. $(x - 7)(x + 3) \leq 0$
4. $(x + 1)(x - 7) \leq 0$
5. $x^2 - 5x + 4 > 0$
6. $x^2 - 4x + 3 < 0$
7. $x^2 + 5x + 4 > 0$
8. $x^2 + x - 6 > 0$
9. $x^2 - 6x + 8 \leq 0$
10. $x^2 - 2x - 3 \geq 0$
11. $3x^2 + 10x - 8 \leq 0$
12. $9x^2 + 3x - 2 \geq 0$
13. $2x^2 + x < 15$
14. $6x^2 + x > 1$
15. $4x^2 + 7x < -3$
16. $3x^2 + 16x < -5$
17. $x^2 - 4x \geq 0$
18. $x^2 + 2x < 0$
19. $2x^2 + 3x > 0$
20. $3x^2 - 5x \leq 0$
21. $-x^2 + x \geq 0$
22. $-x^2 + 2x \geq 0$
23. $x^2 \leq 4x - 2$
24. $x^2 \leq 2x + 2$
25. $3x^2 > 4x + 2$
26. $3x^2 > 10x - 5$
27. $2x^2 - 5x \geq 1$
28. $5x^2 + 8x \geq 11$
29. $x^2 - 6x + 9 < 0$
30. $4x^2 - 4x + 1 \geq 0$
31. $(x - 1)(x - 2)(x - 3) \geq 0$
32. $(x + 1)(x + 2)(x + 3) \geq 0$
33. $x^3 + 2x^2 - x - 2 \geq 0$
34. $x^3 + 2x^2 - 4x - 8 \geq 0$
35. $x^3 - 3x^2 - 9x + 27 < 0$
36. $x^3 + 7x^2 - x - 7 < 0$
37. $x^3 + x^2 + 4x + 4 > 0$
38. $x^3 - x^2 + 9x - 9 > 0$
39. $x^3 \geq 9x^2$
40. $x^3 \leq 4x^2$

Solve each rational inequality in Exercises 41–56 and graph the solution set on a real number line.

41. $\dfrac{x - 4}{x + 3} > 0$
42. $\dfrac{x + 5}{x - 2} > 0$
43. $\dfrac{x + 3}{x + 4} < 0$
44. $\dfrac{x + 5}{x + 2} < 0$
45. $\dfrac{-x + 2}{x - 4} \geq 0$
46. $\dfrac{-x - 3}{x + 2} \leq 0$
47. $\dfrac{4 - 2x}{3x + 4} \leq 0$
48. $\dfrac{3x + 5}{6 - 2x} \geq 0$
49. $\dfrac{x}{x - 3} > 0$
50. $\dfrac{x + 4}{x} > 0$
51. $\dfrac{x + 1}{x + 3} < 2$
52. $\dfrac{x}{x - 1} > 2$
53. $\dfrac{x + 4}{2x - 1} \leq 3$
54. $\dfrac{1}{x - 3} < 1$
55. $\dfrac{x - 2}{x + 2} \leq 2$
56. $\dfrac{x}{x + 2} \geq 2$

In Exercises 57–60, use the given functions to find all values of x that satisfy the required inequality.

57. $f(x) = 2x^2, g(x) = 5x - 2; f(x) \geq g(x)$
58. $f(x) = 4x^2, g(x) = 9x - 2; f(x) < g(x)$
59. $f(x) = \dfrac{2x}{x + 1}, g(x) = 1; f(x) < g(x)$
60. $f(x) = \dfrac{x}{2x - 1}, g(x) = 1; f(x) \geq g(x)$

Practice PLUS

Solve each inequality in Exercises 61–66 and graph the solution set on a real number line.

61. $|x^2 + 2x - 36| > 12$
62. $|x^2 + 6x + 1| > 8$
63. $\dfrac{3}{x + 3} > \dfrac{3}{x - 2}$

64. $\dfrac{1}{x+1} > \dfrac{2}{x-1}$

65. $\dfrac{x^2 - x - 2}{x^2 - 4x + 3} > 0$

66. $\dfrac{x^2 - 3x + 2}{x^2 - 2x - 3} > 0$

In Exercises 67–68, use the graph of the polynomial function to solve each inequality.

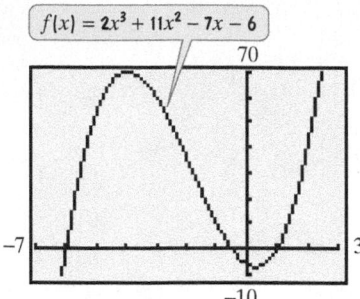

$$f(x) = 2x^3 + 11x^2 - 7x - 6$$

67. $2x^3 + 11x^2 \ge 7x + 6$

68. $2x^3 + 11x^2 < 7x + 6$

In Exercises 69–70, use the graph of the rational function to solve each inequality.

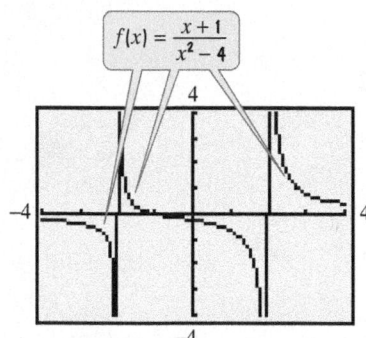

$$f(x) = \dfrac{x+1}{x^2 - 4}$$

69. $\dfrac{1}{4(x+2)} \le -\dfrac{3}{4(x-2)}$

70. $\dfrac{1}{4(x+2)} > -\dfrac{3}{4(x-2)}$

Application Exercises

71. You throw a ball straight up from a rooftop 160 feet high with an initial speed of 48 feet per second. The function
$$s(t) = -16t^2 + 48t + 160$$
models the ball's height above the ground, $s(t)$, in feet, t seconds after it was thrown. During which time period will the ball's height exceed that of the rooftop?

72. Divers in Acapulco, Mexico, dive headfirst from the top of a cliff 87 feet above the Pacific Ocean. The function
$$s(t) = -16t^2 + 8t + 87$$
models a diver's height above the ocean, $s(t)$, in feet, t seconds after leaping. During which time period will the diver's height exceed that of the cliff?

The functions

$$f(x) = 0.0875x^2 - 0.4x + 66.6$$

and

$$g(x) = 0.0875x^2 + 1.9x + 11.6$$

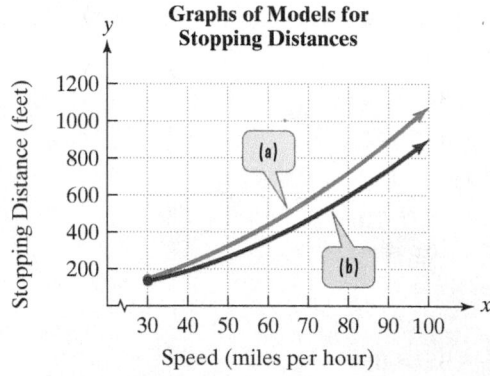

model a car's stopping distance, $f(x)$ or $g(x)$, in feet, traveling at x miles per hour. Function f models stopping distance on dry pavement and function g models stopping distance on wet pavement. The graphs of these functions are shown for speeds of 30 miles per hour and greater. Notice that the figure does not specify which graph is the model for dry roads and which is the model for wet roads. Use this information to solve Exercises 73–74.

73. a. Use the given functions to find the stopping distance on dry pavement and the stopping distance on wet pavement for a car traveling at 35 miles per hour. Round to the nearest foot.

b. Based on your answers to part (a), which rectangular coordinate graph shows stopping distances on dry pavement and which shows stopping distances on wet pavement?

c. How well do your answers to part (a) model the actual stopping distances shown in **Figure 11.17** on page 830?

d. Determine speeds on dry pavement requiring stopping distances that exceed the length of one and one-half football fields, or 540 feet. Round to the nearest mile per hour. How is this shown on the appropriate graph of the models?

74. a. Use the given functions to find the stopping distance on dry pavement and the stopping distance on wet pavement for a car traveling at 55 miles per hour. Round to the nearest foot.

b. Based on your answers to part (a), which rectangular coordinate graph shows stopping distances on dry pavement and which shows stopping distances on wet pavement?

c. How well do your answers to part (a) model the actual stopping distances shown in **Figure 11.17** on page 830?

d. Determine speeds on wet pavement requiring stopping distances that exceed the length of one and one-half football fields, or 540 feet. Round to the nearest mile per hour. How is this shown on the appropriate graph of the models on page 650?

A company manufactures wheelchairs. The average cost function, $\overline{C}$, of producing x wheelchairs per month is given by

$$\overline{C}(x) = \frac{500,000 + 400x}{x}.$$

The graph of the rational function is shown. Use the function to solve Exercises 75–76.

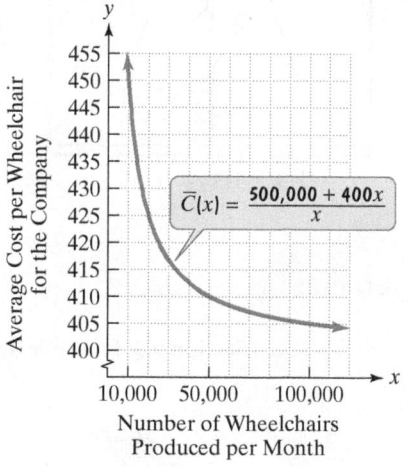

Average Cost per Wheelchair for the Company

Number of Wheelchairs Produced per Month

75. Describe the company's production level so that the average cost of producing each wheelchair does not exceed $425. Use a rational inequality to solve the problem. Then explain how your solution is shown on the graph.

76. Describe the company's production level so that the average cost of producing each wheelchair does not exceed $410. Use a rational inequality to solve the problem. Then explain how your solution is shown on the graph.

77. The perimeter of a rectangle is 50 feet. Describe the possible length of a side if the area of the rectangle is not to exceed 114 square feet.

78. The perimeter of a rectangle is 180 feet. Describe the possible lengths of a side if the area of the rectangle is not to exceed 800 square feet.

Writing in Mathematics

79. What is a polynomial inequality?

80. What is a rational inequality?

81. Describe similarities and differences between the solutions of

$$(x - 2)(x + 5) \geq 0 \quad \text{and} \quad \frac{x - 2}{x + 5} \geq 0.$$

Technology Exercises

Solve each inequality in Exercises 82–87 using a graphing utility.

82. $x^2 + 3x - 10 > 0$

83. $2x^2 + 5x - 3 \leq 0$

84. $\dfrac{x - 4}{x - 1} \leq 0$

85. $\dfrac{x + 2}{x - 3} \leq 2$

86. $\dfrac{1}{x + 1} \leq \dfrac{2}{x + 4}$

87. $x^3 + 2x^2 - 5x - 6 > 0$

The graph shows stopping distances for trucks at various speeds on dry roads and on wet roads. Use this information to solve Exercises 88–89.

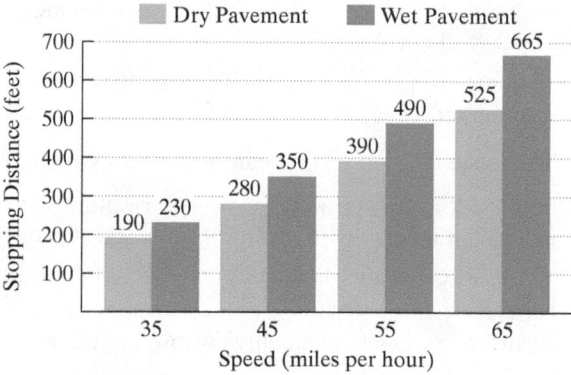

Stopping Distances for Trucks at Selected Speeds

Stopping Distance (feet)

Speed (miles per hour)

Source: National Highway Traffic Safety Administration

88. a. Use the statistical menu of your graphing utility and the quadratic regression program to obtain the quadratic function that models a truck's stopping distance, $f(x)$, in feet, on dry pavement traveling at x miles per hour. Round the x-coefficient and the constant term to one decimal place.

   b. Use the function from part (a) to determine speeds on dry pavement requiring stopping distances that exceed 455 feet. Round to the nearest mile per hour.

89. a. Use the statistical menu of your graphing utility and the quadratic regression program to obtain the quadratic function that models a truck's stopping distance, $f(x)$, in feet, on wet pavement traveling at x miles per hour. Round the constant term to one decimal place.

   b. Use the function from part (a) to determine speeds on wet pavement requiring stopping distances that exceed 446 feet.

Critical Thinking Exercises

Make Sense? In Exercises 90–93, determine whether each statement "makes sense" or "does not make sense" and explain your reasoning.

90. When solving $f(x) > 0$, where f is a polynomial function, I only pay attention to the sign of f at each test value and not the actual function value.

91. I'm solving a polynomial inequality that has a value for which the polynomial function is undefined.

92. Because it takes me longer to come to a stop on a wet road than on a dry road, graph (a) for Exercises 73–74 is the model for stopping distances on wet pavement and graph (b) is the model for stopping distances on dry pavement.

93. I began the solution of the rational inequality $\dfrac{x+1}{x+3} \geq 2$ by setting both $x + 1$ and $x + 3$ equal to zero.

In Exercises 94–97, determine whether each statement is true or false. If the statement is false, make the necessary change(s) to produce a true statement.

94. The solution set of $x^2 > 25$ is $(5, \infty)$.

95. The inequality $\dfrac{x-2}{x+3} < 2$ can be solved by multiplying both sides by $x + 3$, resulting in the equivalent inequality $x - 2 < 2(x + 3)$.

96. $(x+3)(x-1) \geq 0$ and $\dfrac{x+3}{x-1} \geq 0$ have the same solution set.

97. The inequality $\dfrac{x-2}{x+3} < 2$ can be solved by multiplying both sides by $(x+3)^2, x \neq -3$, resulting in the equivalent inequality $(x-2)(x+3) < 2(x+3)^2$.

98. Write a quadratic inequality whose solution set is $[-3, 5]$.

99. Write a rational inequality whose solution set is $(-\infty, -4) \cup [3, \infty)$.

In Exercises 100–103, use inspection to describe each inequality's solution set. Do not solve any of the inequalities.

100. $(x-2)^2 > 0$

101. $(x-2)^2 \leq 0$ **102.** $(x-2)^2 < -1$

103. $\dfrac{1}{(x-2)^2} > 0$

104. The graphing calculator screen shows the graph of $y = 4x^2 - 8x + 7$.

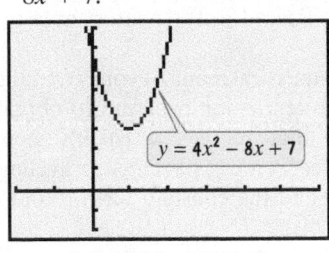

$[-2, 6, 1]$ by $[-2, 8, 1]$

a. Use the graph to describe the solution set for $4x^2 - 8x + 7 > 0$.

b. Use the graph to describe the solution set for $4x^2 - 8x + 7 < 0$.

c. Use an algebraic approach to verify each of your descriptions in parts (a) and (b).

105. The graphing calculator screen shows the graph of $y = \sqrt{27 - 3x^2}$. Write and solve a quadratic inequality that explains why the graph only appears for $-3 \leq x \leq 3$.

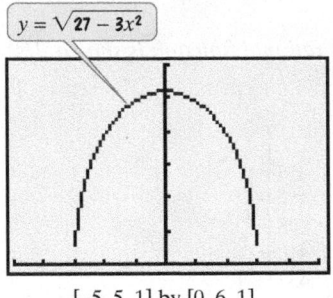

$[-5, 5, 1]$ by $[0, 6, 1]$

Review Exercises

106. Solve: $\left| \dfrac{x-5}{3} \right| < 8$. (Section 9.3, Example 4)

107. Divide:

$$\frac{2x+6}{x^2+8x+16} \div \frac{x^2-9}{x^2+3x-4}.$$

(Section 7.2, Example 6)

108. Factor completely: $x^4 - 16y^4$. (Section 6.5, Example 7)

Preview Exercises

Exercises 109–111 will help you prepare for the material covered in the first section of the next chapter. In each exercise, use point plotting to graph the function. Begin by setting up a table of coordinates, selecting integers from −3 to 3, inclusive, for x.

109. $f(x) = 2^x$

110. $f(x) = 2^{-x}$

111. $f(x) = 2^x + 1$

GROUP PROJECT

CHAPTER 11

Throughout the chapter, we have considered functions that model the position of free-falling objects. Any object that is falling, or vertically projected into the air, has its height above the ground, $s(t)$, in feet, after t seconds in motion, modeled by the quadratic function

$$s(t) = -16t^2 + v_0 t + s_0,$$

where v_0 is the original velocity (initial velocity) of the object, in feet per second, and s_0 is the original height (initial height) of the object, in feet, above the ground. In this exercise, group members will be working with this position function. The exercise is appropriate for groups of three to five people.

a. Drop a ball from a height of 3 feet, 6 feet, and 12 feet. Record the number of seconds it takes for the ball to hit the ground.

b. For each of the three initial positions, use the position function to determine the time required for the ball to hit the ground.

c. What factors might result in differences between the times that you recorded and the times indicated by the function?

d. What appears to be happening to the time required for a free-falling object to hit the ground as its initial height is doubled? Verify this observation algebraically and with a graphing utility.

e. Repeat part (a) using a sheet of paper rather than a ball. What differences do you observe? What factor seems to be ignored in the position function?

f. What is meant by the acceleration due to gravity and how does this number appear in the position function for a free-falling object?

Chapter 11 Summary

| **Definitions and Concepts** | **Examples** |

Section 11.1 The Square Root Property and Completing the Square; Distance and Midpoint Formulas

The Square Root Property

If u is an algebraic expression and d is a real number, then

If $u^2 = d$, then $u = \sqrt{d}$ or $u = -\sqrt{d}$.

Equivalently,

If $u^2 = d$, then $u = \pm\sqrt{d}$.

Solve: $(x - 6)^2 = 50.$

$$x - 6 = \pm\sqrt{50}$$
$$x - 6 = \pm\sqrt{25 \cdot 2}$$
$$x - 6 = \pm 5\sqrt{2}$$
$$x = 6 \pm 5\sqrt{2}$$

The solutions are $6 \pm 5\sqrt{2}$ and the solution set is $\{6 \pm 5\sqrt{2}\}$.

Completing the Square

If $x^2 + bx$ is a binomial, then by adding $\left(\dfrac{b}{2}\right)^2$, the square of half the coefficient of x, a perfect square trinomial will result. That is,

$$x^2 + bx + \left(\frac{b}{2}\right)^2 = \left(x + \frac{b}{2}\right)^2.$$

Complete the square:

$$x^2 + \frac{2}{7}x.$$

Half of $\frac{2}{7}$ is $\frac{1}{2} \cdot \frac{2}{7} = \frac{1}{7}$ and $\left(\frac{1}{7}\right)^2 = \frac{1}{49}$.

$$x^2 + \frac{2}{7}x + \frac{1}{49} = \left(x + \frac{1}{7}\right)^2$$

Solving Quadratic Equations by Completing the Square

1. If the coefficient of x^2 is not 1, divide both sides by this coefficient.

2. Isolate variable terms on one side.

3. Complete the square by adding the square of half the coefficient of x to both sides.

4. Factor the perfect square trinomial.

5. Solve by applying the square root property.

Solve by completing the square:

$$2x^2 + 16x - 6 = 0.$$

$$\frac{2x^2}{2} + \frac{16x}{2} - \frac{6}{2} = \frac{0}{2}. \quad \text{Divide by 2.}$$
$$x^2 + 8x - 3 = 0 \quad \text{Simplify.}$$
$$x^2 + 8x = 3 \quad \text{Add 3.}$$

The coefficient of x is 8. Half of 8 is 4 and $4^2 = 16$. Add 16 to both sides.

$$x^2 + 8x + 16 = 3 + 16$$
$$(x + 4)^2 = 19$$
$$x + 4 = \pm\sqrt{19}$$
$$x = -4 \pm \sqrt{19}$$

| **Definitions and Concepts** | **Examples** |

Section 11.1 The Square Root Property and Completing the Square; Distance and Midpoint Formulas (continued)

The Distance Formula

The distance, d, between the points (x_1, y_1) and (x_2, y_2) is given by

$$d = \sqrt{(x_2 - x_1)^2 + (y_2 - y_1)^2}.$$

Find the distance between $(-3, -5)$ and $(6, -2)$.

$$d = \sqrt{[6 - (-3)]^2 + [-2 - (-5)]^2}$$
$$= \sqrt{9^2 + 3^2} = \sqrt{81 + 9} = \sqrt{90} = 3\sqrt{10} \approx 9.49$$

The Midpoint Formula

The midpoint of the line segment whose endpoints are (x_1, y_1) and (x_2, y_2) is the point with coordinates

$$\left(\frac{x_1 + x_2}{2}, \frac{y_1 + y_2}{2} \right).$$

Find the midpoint of the line segment whose endpoints are $(-3, 6)$ and $(4, 1)$.

$$\text{midpoint} = \left(\frac{-3 + 4}{2}, \frac{6 + 1}{2} \right) = \left(\frac{1}{2}, \frac{7}{2} \right)$$

Section 11.2 The Quadratic Formula

The solutions of a quadratic equation in standard form
$$ax^2 + bx + c = 0, \quad a \neq 0,$$
are given by the quadratic formula
$$x = \frac{-b \pm \sqrt{b^2 - 4ac}}{2a}.$$

Solve using the quadratic formula:

$$2x^2 = 6x - 3.$$

First write the equation in standard form by subtracting $6x$ and adding 3 on both sides.

$$2x^2 - 6x + 3 = 0$$

$a = 2$ $b = -6$ $c = 3$

$$x = \frac{-(-6) \pm \sqrt{(-6)^2 - 4 \cdot 2 \cdot 3}}{2 \cdot 2} = \frac{6 \pm \sqrt{36 - 24}}{4}$$

$$= \frac{6 \pm \sqrt{12}}{4} = \frac{6 \pm \sqrt{4 \cdot 3}}{4} = \frac{6 \pm 2\sqrt{3}}{4}$$

$$= \frac{2(3 \pm \sqrt{3})}{2 \cdot 2} = \frac{3 \pm \sqrt{3}}{2}$$

The Discriminant

The discriminant, $b^2 - 4ac$, of the quadratic equation $ax^2 + bx + c = 0$ determines the number and type of solutions.

$$2x^2 - 7x - 4 = 0$$

$a = 2$ $b = -7$ $c = -4$

$$b^2 - 4ac = (-7)^2 - 4(2)(-4)$$
$$= 49 - (-32) = 49 + 32 = 81$$

Positive perfect square

The equation has two real rational solutions.

Discriminant	Solutions
Positive perfect square, with a, b, and c rational numbers	2 real rational solutions
Positive and not a perfect square	2 real irrational solutions
Zero, with a, b, and c rational numbers	1 real rational solution
Negative	2 imaginary solutions

Definitions and Concepts	**Examples**

Section 11.2 The Quadratic Formula (continued)

Writing Quadratic Equations from Solutions

The zero-product principle in reverse makes it possible to write a quadratic equation from solutions:

If $A = 0$ or $B = 0$, then $AB = 0$.

Write a quadratic equation with the solution set $\{-2\sqrt{3}, 2\sqrt{3}\}$.

$$x = -2\sqrt{3} \qquad\qquad x = 2\sqrt{3}$$
$$x + 2\sqrt{3} = 0 \quad \text{or} \quad x - 2\sqrt{3} = 0$$
$$(x + 2\sqrt{3})(x - 2\sqrt{3}) = 0$$
$$x^2 - (2\sqrt{3})^2 = 0$$
$$x^2 - 12 = 0$$

Section 11.3 Quadratic Functions and Their Graphs

The graph of the quadratic function

$$f(x) = a(x - h)^2 + k, \quad a \neq 0,$$

is called a parabola. The vertex, or turning point, is (h, k). The graph opens upward if a is positive and downward if a is negative. The axis of symmetry is a vertical line passing through the vertex. The graph can be obtained using the vertex, x-intercepts, if any, (set $f(x)$ equal to zero and solve), and the y-intercept (set $x = 0$).

Graph: $f(x) = -(x + 3)^2 + 1$.

$$f(x) = -1(x - (-3))^2 + 1$$

$a = -1 \qquad h = -3 \qquad k = 1$

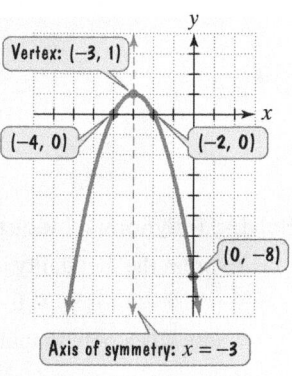

- Vertex $(h, k) = (-3, 1)$
- Opens downward because $a < 0$
- x-intercepts: Set $f(x) = 0$.

$$0 = -(x + 3)^2 + 1$$
$$(x + 3)^2 = 1$$
$$x + 3 = \pm\sqrt{1}$$
$$x + 3 = 1 \quad \text{or} \quad x + 3 = -1$$
$$x = -2 \qquad\qquad x = -4$$

- y-intercept: Set $x = 0$.

$$f(0) = -(0 + 3)^2 + 1 = -9 + 1 = -8$$

A parabola whose equation is in the form

$$f(x) = ax^2 + bx + c, \quad a \neq 0,$$

has its vertex at

$$\left(-\frac{b}{2a}, f\left(-\frac{b}{2a}\right)\right).$$

The parabola is graphed as described in the left column above. The only difference is how we determine the vertex. If $a > 0$, then f has a minimum that occurs at $x = -\frac{b}{2a}$. This minimum value is $f\left(-\frac{b}{2a}\right)$. If $a < 0$, then f has a maximum that occurs at $x = -\frac{b}{2a}$. This maximum value is $f\left(-\frac{b}{2a}\right)$.

Graph:

$$f(x) = x^2 - 6x + 5.$$

$a = 1 \qquad b = -6 \qquad c = 5$

- Vertex:

$$x = -\frac{b}{2a} = \frac{-6}{2 \cdot 1} = 3$$
$$f(3) = 3^2 - 6 \cdot 3 + 5 = -4$$

Vertex is at $(3, -4)$.

- Opens upward because $a > 0$.
- x-intercepts: Set $f(x) = 0$.

$$x^2 - 6x + 5 = 0$$
$$(x - 1)(x - 5) = 0$$
$$x = 1 \quad \text{or} \quad x = 5$$

- y-intercept: Set $x = 0$.

$$f(0) = 0^2 - 6 \cdot 0 + 5 = 5$$

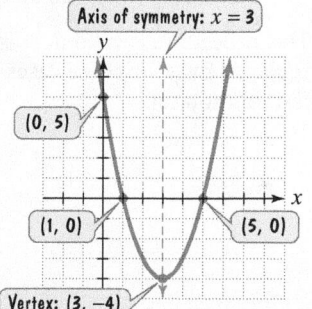

Definitions and Concepts	**Examples**

Section 11.4 Equations Quadratic in Form

An equation that is quadratic in form is one that can be expressed as a quadratic equation using an appropriate substitution. In these equations, the variable factor in one term is the square of the variable factor in the other variable term. Let u = the variable factor that reappears squared. If at any point in the solution process both sides of an equation are raised to an even power, a check is required.

Solve:

$$x^{\frac{2}{3}} - 3x^{\frac{1}{3}} + 2 = 0.$$
$$\left(x^{\frac{1}{3}}\right)^2 - 3x^{\frac{1}{3}} + 2 = 0$$

Let $u = x^{\frac{1}{3}}$.

$$u^2 - 3u + 2 = 0$$
$$(u - 1)(u - 2) = 0$$
$$u - 1 = 0 \quad \text{or} \quad u - 2 = 0$$
$$u = 1 \qquad\qquad u = 2$$
$$x^{\frac{1}{3}} = 1 \qquad\qquad x^{\frac{1}{3}} = 2$$
$$\left(x^{\frac{1}{3}}\right)^3 = 1^3 \qquad \left(x^{\frac{1}{3}}\right)^3 = 2^3$$
$$x = 1 \qquad\qquad x = 8$$

The solutions are 1 and 8, and the solution set is $\{1, 8\}$.

Section 11.5 Polynomial and Rational Inequalities

Solving Polynomial Inequalities

1. Express the inequality in the form
$$f(x) < 0 \quad \text{or} \quad f(x) > 0,$$
where f is a polynomial function.

2. Solve the equation $f(x) = 0$. The real solutions are the boundary points.

3. Locate these boundary points on a number line, thereby dividing the number line into intervals.

4. Choose one representative number, called a test value, within each interval and evaluate f at that number.

   a. If the value of f is positive, then $f(x) > 0$ for all x in the interval.

   b. If the value of f is negative, then $f(x) < 0$ for all x in the interval.

5. Write the solution set, selecting the interval or intervals that satisfy the given inequality.

This procedure is valid if $<$ is replaced by $\leq$ and $>$ is replaced by $\geq$. In these cases, include the boundary points in the solution set.

Solve: $2x^2 + x - 6 > 0$.

The form of the inequality is $f(x) > 0$ with $f(x) = 2x^2 + x - 6$. Solve $f(x) = 0$.
$$2x^2 + x - 6 = 0$$
$$(2x - 3)(x + 2) = 0$$
$$2x - 3 = 0 \quad \text{or} \quad x + 2 = 0$$
$$x = \frac{3}{2} \qquad\qquad x = -2$$

Use $-3, 0$, and 2 as test values.
$$f(-3) = 2(-3)^2 + (-3) - 6 = 9, \text{positive}$$
$$f(x) > 0 \text{ for all } x \text{ in } (-\infty, -2).$$
$$f(0) = 2 \cdot 0^2 + 0 - 6 = -6, \text{negative}$$
$$f(x) < 0 \text{ for all } x \text{ in } \left(-2, \frac{3}{2}\right).$$
$$f(2) = 2 \cdot 2^2 + 2 - 6 = 4, \text{positive}$$
$$f(x) > 0 \text{ for all } x \text{ in } \left(\frac{3}{2}, \infty\right).$$

The solution set is $(-\infty, -2) \cup \left(\frac{3}{2}, \infty\right)$.

Definitions and Concepts

Examples

Section 11.5 Polynomial and Rational Inequalities (continued)

Solving Rational Inequalities

1. Express the inequality in the form
$$f(x) < 0 \quad \text{or} \quad f(x) > 0,$$
where f is a rational function written as a single quotient.

2. Set the numerator and the denominator of f equal to zero. The real solutions are the boundary points.

3. Locate these boundary points on a number line, thereby dividing the number line into intervals.

4. Choose one representative number, called a test value, within each interval and evaluate f at that number.

   a. If the value of f is positive, then $f(x) > 0$ for all x in the interval.

   b. If the value of f is negative, then $f(x) < 0$ for all x in the interval.

5. Write the solution set, selecting the interval or intervals that satisfy the given inequality.

This procedure is valid if $<$ is replaced by $\leq$ and $>$ is replaced by $\geq$. In these cases, include any values that make the numerator of f zero. Always exclude any values that make the denominator zero.

Solve: $\dfrac{x}{x+4} \geq 2$.

$$\frac{x}{x+4} - \frac{2(x+4)}{x+4} \geq 0$$

$$\frac{-x-8}{x+4} \geq 0$$

The form of the inequality is $f(x) \geq 0$ with $f(x) = \dfrac{-x-8}{x+4}$.

Set the numerator and the denominator equal to zero.

$$\begin{array}{ll} -x - 8 = 0 & x + 4 = 0 \\ -8 = x & x = -4 \end{array}$$

Use $-9, -7$, and -3 as test values.

$$f(-9) = \frac{-(-9) - 8}{-9 + 4} = \frac{1}{-5}, \text{ negative}$$

$$f(x) < 0 \text{ for all } x \text{ in } (-\infty, -8).$$

$$f(-7) = \frac{-(-7) - 8}{-7 + 4} = \frac{-1}{-3} = \frac{1}{3}, \text{ positive}$$

$$f(x) > 0 \text{ for all } x \text{ in } (-8, -4).$$

$$f(-3) = \frac{-(-3) - 8}{-3 + 4} = \frac{-5}{1} = -5, \text{ negative}$$

$$f(x) < 0 \text{ for all } x \text{ in } (-4, \infty).$$

Because of $\geq$, include -8, the value that makes the numerator zero, in the solution set.
The solution set is $[-8, -4)$.

CHAPTER 11 REVIEW EXERCISES

11.1 *In Exercises 1–5, solve each equation by the square root property. If possible, simplify radicals or rationalize denominators. Express imaginary solutions in the form $a + bi$.*

1. $2x^2 - 3 = 125$

2. $3x^2 - 150 = 0$

3. $3x^2 - 2 = 0$

4. $(x - 4)^2 = 18$

5. $(x + 7)^2 = -36$

In Exercises 6–7, determine the constant that should be added to the binomial so that it becomes a perfect square trinomial. Then write and factor the trinomial.

6. $x^2 + 20x$

7. $x^2 - 3x$

In Exercises 8–10, solve each quadratic equation by completing the square.

8. $x^2 - 12x + 27 = 0$

9. $x^2 - 7x - 1 = 0$

10. $2x^2 + 3x - 4 = 0$

11. In 2 years, an investment of $2500 grows to $2916. Use the compound interest formula
$$A = P(1 + r)^t$$
to find the annual interest rate, r.

12. The function $W(t) = 3t^2$ models the weight of a human fetus, $W(t)$, in grams, after t weeks, where $0 \leq t \leq 39$. After how many weeks does the fetus weigh 588 grams?

13. A building casts a shadow that is double the length of the building's height. If the distance from the end of the shadow to the top of the building is 300 meters, how high is the building? Express the answer in simplified radical form. Then find a decimal approximation to the nearest tenth of a meter.

In Exercises 14–15, find the distance between each pair of points. If necessary, express answers in simplified radical form and then round to two decimal places.

14. $(-2, -3)$ and $(3, 9)$

15. $(-4, 3)$ and $(-2, 5)$

In Exercises 16–17, find the midpoint of the line segment with the given endpoints.

16. $(2, 6)$ and $(-12, 4)$

17. $(4, -6)$ and $(-15, 2)$

11.2 *In Exercises 18–20, solve each equation using the quadratic formula. Simplify solutions, if possible.*

18. $x^2 = 2x + 4$

19. $x^2 - 2x + 19 = 0$

20. $2x^2 = 3 - 4x$

In Exercises 21–23, without solving the given quadratic equation, determine the number and type of solutions.

21. $x^2 - 4x + 13 = 0$

22. $9x^2 = 2 - 3x$

23. $2x^2 + 4x = 3$

In Exercises 24–30, solve each equation by the method of your choice. Simplify solutions, if possible.

24. $3x^2 - 10x - 8 = 0$

25. $(2x - 3)(x + 2) = x^2 - 2x + 4$

26. $5x^2 - x - 1 = 0$

27. $x^2 - 16 = 0$

28. $(x - 3)^2 - 8 = 0$

29. $3x^2 - x + 2 = 0$

30. $\dfrac{5}{x + 1} + \dfrac{x - 1}{4} = 2$

In Exercises 31–33, write a quadratic equation in standard form with the given solution set.

31. $\left\{-\dfrac{1}{3}, \dfrac{3}{5}\right\}$

32. $\{-9i, 9i\}$

33. $\{-4\sqrt{3}, 4\sqrt{3}\}$

34. The graph shows stopping distances for motorcycles at various speeds on dry roads and on wet roads.

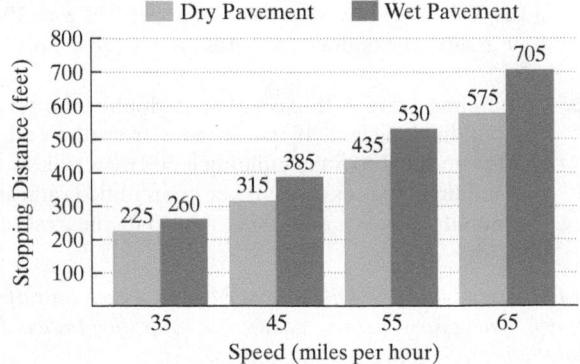

Stopping Distances for Motorcycles at Selected Speeds

Source: National Highway Traffic Safety Administration

The functions

$$f(x) = 0.125x^2 - 0.8x + 99$$

Dry pavement

and

Wet pavement

$$g(x) = 0.125x^2 + 2.3x + 27$$

model a motorcycle's stopping distance, $f(x)$ or $g(x)$, in feet traveling at x miles per hour. Function f models stopping distance on dry pavement and function g models stopping distance on wet pavement.

a. Use function g to find the stopping distance on wet pavement for a motorcycle traveling at 35 miles per hour. Round to the nearest foot. Does your rounded answer overestimate or underestimate the stopping distance shown by the graph? By how many feet?

b. Use function f to determine a motorcycle's speed requiring a stopping distance on dry pavement of 267 feet.

35. The graphs of the functions in Exercise 34 are shown for speeds of 30 miles per hour and greater.

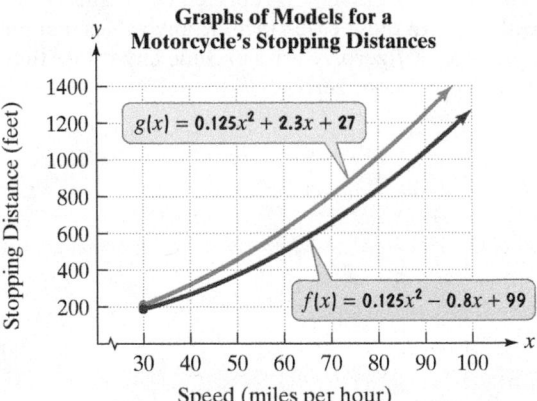

Graphs of Models for a Motorcycle's Stopping Distances

a. How is your answer to Exercise 34(a) shown on the graph of g?

b. How is your answer to Exercise 34(b) shown on the graph of f?

36. A baseball is hit by a batter. The function

$$s(t) = -16t^2 + 140t + 3$$

models the ball's height above the ground, $s(t)$, in feet, t seconds after it is hit. How long will it take for the ball to strike the ground? Round to the nearest tenth of a second.

11.3 *In Exercises 37–40, use the vertex and intercepts to sketch the graph of each quadratic function.*

37. $f(x) = -(x + 1)^2 + 4$ **38.** $f(x) = (x + 4)^2 - 2$

39. $f(x) = -x^2 + 2x + 3$ **40.** $f(x) = 2x^2 - 4x - 6$

41. The function

$$f(x) = -0.02x^2 + x + 1$$

models the yearly growth of a young redwood tree, $f(x)$, in inches, with x inches of rainfall per year. How many inches of rainfall per year result in maximum tree growth? What is the maximum yearly growth?

42. A model rocket is launched upward from a platform 40 feet above the ground. The quadratic function

$$s(t) = -16t^2 + 400t + 40$$

models the rocket's height above the ground, $s(t)$, in feet, t seconds after it was launched. After how many seconds does the rocket reach its maximum height? What is the maximum height?

43. The function

$$f(x) = 104.5x^2 - 1501.5x + 6016$$

models the death rate per year per 100,000 males, $f(x)$, for U.S. men who average x hours of sleep each night. How many hours of sleep, to the nearest tenth of an hour, corresponds to the minimum death rate? What is this minimum death rate, to the nearest whole number?

44. A field bordering a straight stream is to be enclosed. The side bordering the stream is not to be fenced. If 1000 yards of fencing material is to be used, what are the dimensions of the largest rectangular field that can be fenced? What is the maximum area?

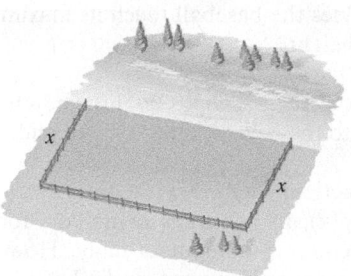

45. Among all pairs of numbers whose difference is 14, find a pair whose product is as small as possible. What is the minimum product?

11.4 *In Exercises 46–51, solve each equation by making an appropriate substitution. When necessary, check proposed solutions.*

46. $x^4 - 6x^2 + 8 = 0$

47. $x + 7\sqrt{x} - 8 = 0$

48. $(x^2 + 2x)^2 - 14(x^2 + 2x) = 15$

49. $x^{-2} + x^{-1} - 56 = 0$

50. $x^{\frac{2}{3}} - x^{\frac{1}{3}} - 12 = 0$

51. $x^{\frac{1}{2}} + 3x^{\frac{1}{4}} - 10 = 0$

11.5 *In Exercises 52–56, solve each inequality and graph the solution set on a real number line.*

52. $2x^2 + 5x - 3 < 0$

53. $2x^2 + 9x + 4 \geq 0$

54. $x^3 + 2x^2 > 3x$

55. $\dfrac{x - 6}{x + 2} > 0$

56. $\dfrac{x + 3}{x - 4} \leq 5$

57. A model rocket is launched from ground level. The function

$$s(t) = -16t^2 + 48t$$

models the rocket's height above the ground, $s(t)$, in feet, t seconds after it was launched. During which time period will the rocket's height exceed 32 feet?

58. The function

$$H(x) = \frac{15}{8}x^2 - 30x + 200$$

models heart rate, $H(x)$, in beats per minute, x minutes after a strenuous workout.

a. What is the heart rate immediately following the workout?

b. According to the model, during which intervals of time after the strenuous workout does the heart rate exceed 110 beats per minute? For which of these intervals has model breakdown occurred? Which interval provides a more realistic answer? How did you determine this?

CHAPTER 11 TEST

CHAPTER
Test Prep
VIDEOS

Step-by-step test solutions are found on the Chapter Test Prep Videos available in MyMathLab® or on YouTube (search "BlitzerCombinedAlg" and click on "Channels").

Express solutions to all equations in simplified form. Rationalize denominators, if possible.

In Exercises 1–2, solve each equation by the square root property.

1. $2x^2 - 5 = 0$

2. $(x - 3)^2 = 20$

In Exercises 3–4, determine the constant that should be added to the binomial so that it becomes a perfect square trinomial. Then write and factor the trinomial.

3. $x^2 - 16x$

4. $x^2 + \dfrac{2}{5}x$

5. Solve by completing the square: $x^2 - 6x + 7 = 0$.

6. Use the measurements determined by the surveyor to find the width of the pond. Express the answer in simplified radical form.

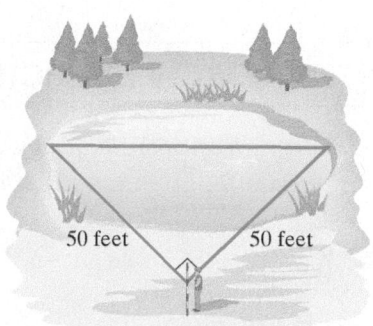

50 feet   50 feet

7. Find the distance between $(-1, 5)$ and $(2, -3)$. Express the answer in radical form and then round to two decimal places.

8. Find the midpoint of the line segment whose endpoints are $(-5, -2)$ and $(12, -6)$.

In Exercises 9–10, without solving the given quadratic equation, determine the number and type of solutions.

9. $3x^2 + 4x - 2 = 0$

10. $x^2 = 4x - 8$

In Exercises 11–14, solve each equation by the method of your choice.

11. $2x^2 + 9x = 5$

12. $x^2 + 8x + 5 = 0$

13. $(x + 2)^2 + 25 = 0$

14. $2x^2 - 6x + 5 = 0$

In Exercises 15–16, write a quadratic equation in standard form with the given solution set.

15. $\{-3, 7\}$

16. $\{-10i, 10i\}$

17. As gas prices surge, more Americans are cycling as a way to save money, stay fit, or both. In 2010, Boston installed 20 miles of bike lanes and New York City added more than 50 miles. The bar graph shows the number of bicycle friendly U.S. communities, as designated by the League of American Bicyclists, for selected years from 2003 through 2011.

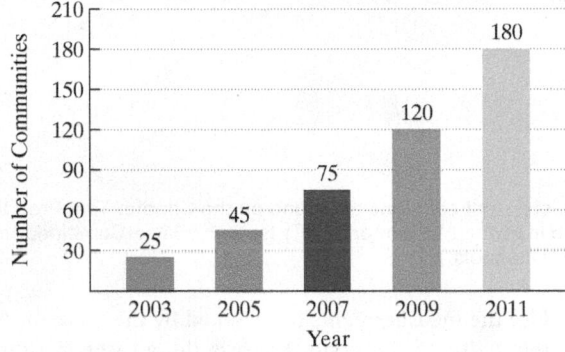

Number of U.S. Communities Designated "Bicycle Friendly" by the League of American Bicyclists

Source: League of American Bicyclists

The function
$$f(x) = 1.7x^2 + 6x + 26$$
models the number of bicycle friendly communities, $f(x)$, x years after 2003.

a. Use the function to find the number of bicycle friendly communities in 2011. Round to the nearest whole number. Does this rounded value underestimate or overestimate the number shown by the graph? By how much?

b. Use the function to determine the year in which 826 U.S. communities will be bicycle friendly.

In Exercises 18–19, use the vertex and intercepts to sketch the graph of each quadratic function.

18. $f(x) = (x + 1)^2 + 4$

19. $f(x) = x^2 - 2x - 3$

A baseball player hits a pop fly into the air. The function
$$s(t) = -16t^2 + 64t + 5$$
models the ball's height above the ground, $s(t)$, in feet, t seconds after it is hit. Use the function to solve Exercises 20–21.

20. When does the baseball reach its maximum height? What is that height?

21. After how many seconds does the baseball hit the ground? Round to the nearest tenth of a second.

22. The function $f(x) = -x^2 + 46x - 360$ models the daily profit, $f(x)$, in hundreds of dollars, for a company that manufactures x computers daily. How many computers should be manufactured each day to maximize profit? What is the maximum daily profit?

In Exercises 23–25, solve each equation by making an appropriate substitution. When necessary, check proposed solutions.

23. $(2x - 5)^2 + 4(2x - 5) + 3 = 0$

24. $x^4 - 13x^2 + 36 = 0$

25. $x^{\frac{2}{3}} - 9x^{\frac{1}{3}} + 8 = 0$

In Exercises 26–27, solve each inequality and graph the solution set on a real number line.

26. $x^2 - x - 12 < 0$

27. $\dfrac{2x + 1}{x - 3} \leq 3$

CUMULATIVE REVIEW EXERCISES (CHAPTERS 1–11)

In Exercises 1–13, solve each equation, inequality, or system.

1. $9(x - 1) = 1 + 3(x - 2)$

2. $\begin{cases} 3x + 4y = -7 \\ x - 2y = -9 \end{cases}$

3. $\begin{cases} x - y + 3z = -9 \\ 2x + 3y - z = 16 \\ 5x + 2y - z = 15 \end{cases}$

4. $7x + 18 \leq 9x - 2$

5. $4x - 3 < 13$ and $-3x - 4 \geq 8$

6. $2x + 4 > 8$ or $x - 7 \geq 3$

7. $|2x - 1| < 5$

8. $\left| \dfrac{2}{3}x - 4 \right| = 2$

9. $\dfrac{4}{x-3} - \dfrac{6}{x+3} = \dfrac{24}{x^2 - 9}$

10. $\sqrt{x+4} - \sqrt{x-3} = 1$

11. $2x^2 = 5 - 4x$

12. $x^{\frac{2}{3}} - 5x^{\frac{1}{3}} + 6 = 0$

13. $2x^2 + x - 6 \le 0$

In Exercises 14–17, graph each function, equation, or inequality in a rectangular coordinate system.

14. $x - 3y = 6$

15. $f(x) = \dfrac{1}{2}x - 1$

16. $3x - 2y > -6$

17. $f(x) = -2(x-3)^2 + 2$

In Exercises 18–28, perform the indicated operations, and simplify, if possible.

18. $4[2x - 6(x - y)]$

19. $(-5x^3y^2)(4x^4y^{-6})$

20. $(8x^2 - 9xy - 11y^2) - (7x^2 - 4xy + 5y^2)$

21. $(3x - 1)(2x + 5)$

22. $(3x^2 - 4y)^2$

23. $\dfrac{3x}{x+5} - \dfrac{2}{x^2 + 7x + 10}$

24. $\dfrac{1 - \dfrac{9}{x^2}}{1 + \dfrac{3}{x}}$

25. $\dfrac{x^2 - 6x + 8}{3x + 9} \div \dfrac{x^2 - 4}{x + 3}$

26. $\sqrt{5xy} \cdot \sqrt{10x^2y}$

27. $4\sqrt{72} - 3\sqrt{50}$

28. $(5 + 3i)(7 - 3i)$

In Exercises 29–31, factor completely.

29. $81x^4 - 1$

30. $24x^3 - 22x^2 + 4x$

31. $x^3 + 27y^3$

In Exercises 32–34, let $f(x) = x^2 + 3x - 15$ and $g(x) = x - 2$. Find each indicated expression.

32. $(f - g)(x)$ and $(f - g)(5)$

33. $\left(\dfrac{f}{g}\right)(x)$ and the domain of $\dfrac{f}{g}$

34. $\dfrac{f(a + h) - f(a)}{h}$

35. Divide:

$$(3x^3 - x^2 + 4x + 8) \div (x + 2).$$

36. Solve for R: $I = \dfrac{R}{R + r}$.

37. Write the slope-intercept form of the equation of the line through $(-2, 5)$ and parallel to the line whose equation is $3x + y = 9$.

38. The price of a computer is reduced by 30% to $434. What was the original price?

39. The area of a rectangle is 52 square yards. The length of the rectangle is 1 yard longer than 3 times its width. Find the rectangle's dimensions.

40. You invested $4000 in two stocks paying 12% and 14% annual interest. At the end of the year, the total interest from these investments was $508. How much was invested at each rate?

41. The current, I, in amperes, flowing in an electrical circuit varies inversely as the resistance, R, in ohms, in the circuit. When the resistance of an electric percolator is 22 ohms, it draws 5 amperes of current. How much current is needed when the resistance is 10 ohms?

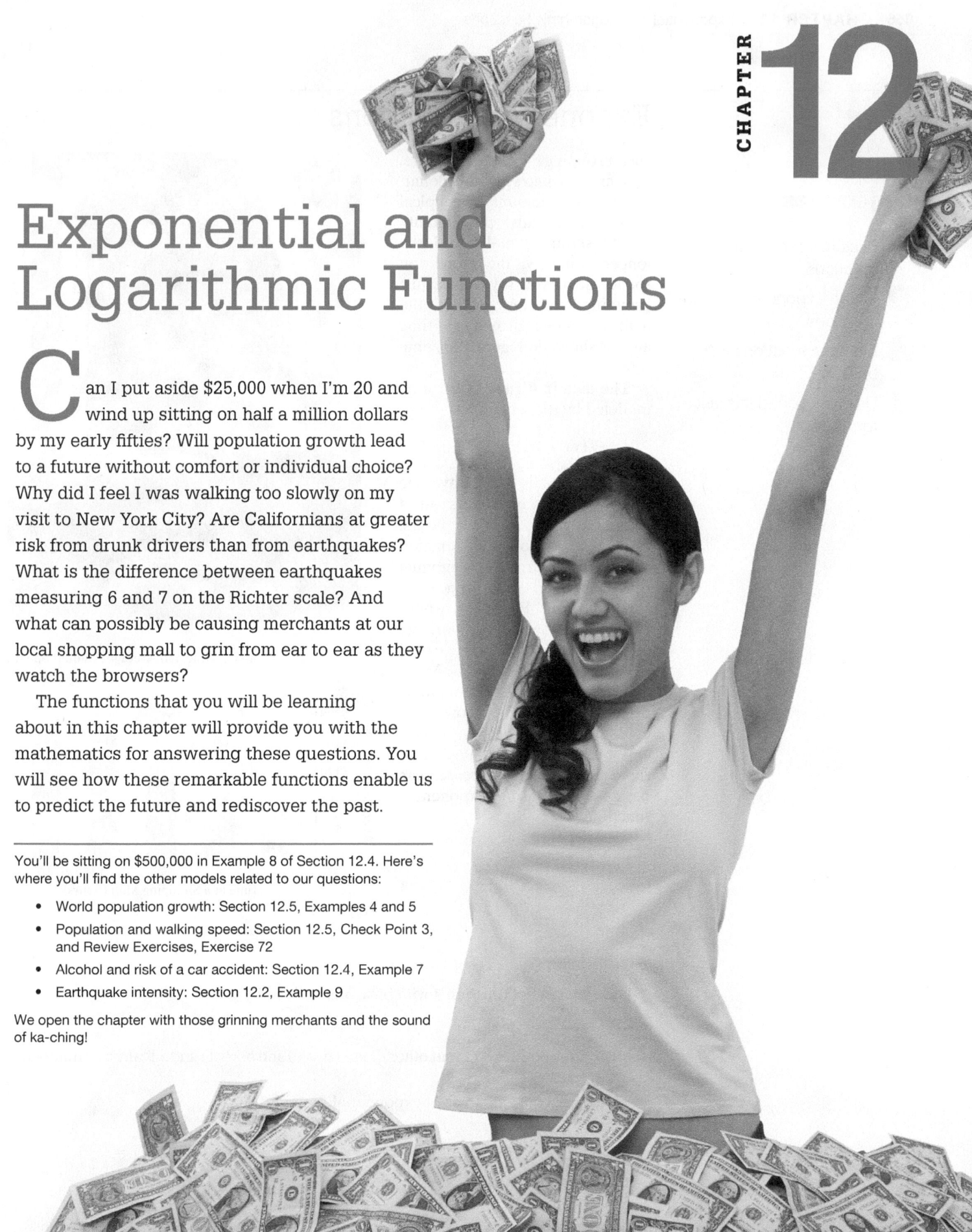

Exponential and Logarithmic Functions

Can I put aside $25,000 when I'm 20 and wind up sitting on half a million dollars by my early fifties? Will population growth lead to a future without comfort or individual choice? Why did I feel I was walking too slowly on my visit to New York City? Are Californians at greater risk from drunk drivers than from earthquakes? What is the difference between earthquakes measuring 6 and 7 on the Richter scale? And what can possibly be causing merchants at our local shopping mall to grin from ear to ear as they watch the browsers?

The functions that you will be learning about in this chapter will provide you with the mathematics for answering these questions. You will see how these remarkable functions enable us to predict the future and rediscover the past.

You'll be sitting on $500,000 in Example 8 of Section 12.4. Here's where you'll find the other models related to our questions:

- World population growth: Section 12.5, Examples 4 and 5
- Population and walking speed: Section 12.5, Check Point 3, and Review Exercises, Exercise 72
- Alcohol and risk of a car accident: Section 12.4, Example 7
- Earthquake intensity: Section 12.2, Example 9

We open the chapter with those grinning merchants and the sound of ka-ching!

SECTION

12.1

Objectives

1 Evaluate exponential functions.

2 Graph exponential functions.

3 Evaluate functions with base e.

4 Use compound interest formulas.

Exponential Functions

Just browsing? Take your time. Researchers know, to the dollar, the average amount the typical consumer spends per minute at the shopping mall. And the longer you stay, the more you spend. So if you say you're just browsing, that's just fine with the mall merchants. Browsing is time and, as shown in **Figure 12.1**, time is money.

The data in **Figure 12.1** can be modeled by the function

$$f(x) = 42.2(1.56)^x,$$

where $f(x)$ is the average amount spent, in dollars, at a shopping mall after x hours. Can you see how this function is different from polynomial functions? The variable x is in the exponent. Functions whose equations contain a variable in the exponent are called **exponential functions**. Many real-life situations, including population growth, growth of epidemics, radioactive decay, and other changes that involve rapid increase or decrease, can be described using exponential functions.

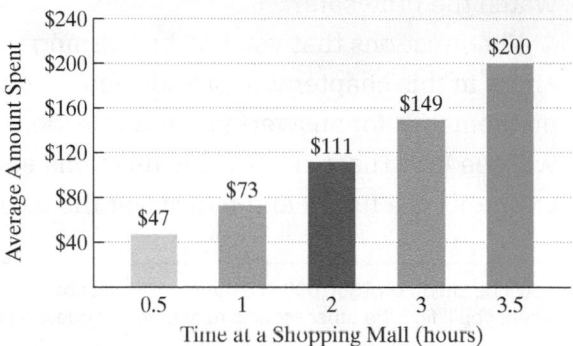

Mall Browsing Time and Average Amount Spent

Figure 12.1

Source: International Council of Shopping Centers Research, 2006

Definition of an Exponential Function

The **exponential function f with base b** is defined by

$$f(x) = b^x \quad \text{or} \quad y = b^x,$$

where b is a positive constant other than 1 ($b > 0$ and $b \neq 1$) and x is any real number.

Here are some examples of exponential functions:

$$f(x) = 2^x \qquad g(x) = 10^x \qquad h(x) = 3^{x+1} \qquad j(x) = \left(\frac{1}{2}\right)^{x-1}.$$

Base is 2.   Base is 10.   Base is 3.   Base is $\frac{1}{2}$.

Each of these functions has a constant base and a variable exponent.

By contrast, the following functions are not exponential functions:

$$F(x) = x^2 \qquad G(x) = 1^x \qquad H(x) = (-1)^x \qquad J(x) = x^x.$$

Variable is the base and not the exponent.

The base of an exponential function must be a positive constant other than 1.

The base of an exponential function must be positive.

Variable is both the base and the exponent.

Why is $G(x) = 1^x$ not classified as an exponential function? The number 1 raised to any power is 1. Thus, the function G can be written as $G(x) = 1$, which is a constant function.

Why is $H(x) = (-1)^x$ not an exponential function? The base of an exponential function must be positive to avoid having to exclude many values of x from the domain that result in nonreal numbers in the range:

$$H(x) = (-1)^x \qquad H\left(\frac{1}{2}\right) = (-1)^{\frac{1}{2}} = \sqrt{-1} = i.$$

Not an exponential function

All values of x resulting in even roots of negative numbers produce nonreal numbers.

1 Evaluate exponential functions.

You will need a calculator to evaluate exponential expressions. Most scientific calculators have a $\boxed{y^x}$ key. Graphing calculators have a $\boxed{\wedge}$ key. To evaluate expressions of the form b^x, enter the base b, press $\boxed{y^x}$ or $\boxed{\wedge}$, enter the exponent x, and finally press $\boxed{=}$ or $\boxed{\text{ENTER}}$.

EXAMPLE 1 Evaluating an Exponential Function

The exponential function $f(x) = 42.2(1.56)^x$ models the average amount spent, $f(x)$, in dollars, at a shopping mall after x hours. What is the average amount spent, to the nearest dollar, after four hours?

Solution Because we are interested in the amount spent after four hours, substitute 4 for x and evaluate the function.

$$f(x) = 42.2(1.56)^x \qquad \text{This is the given function.}$$
$$f(4) = 42.2(1.56)^4 \qquad \text{Substitute 4 for x.}$$

Use a scientific or graphing calculator to evaluate $f(4)$. Press the following keys on your calculator to do this:

Scientific calculator: $42.2 \boxed{\times} 1.56 \boxed{y^x} 4 \boxed{=}$

Graphing calculator: $42.2 \boxed{\times} 1.56 \boxed{\wedge} 4 \boxed{\text{ENTER}}$.

The display should be approximately 249.92566.

$$f(4) = 42.2(1.56)^4 \approx 249.92566 \approx 250$$

Thus, the average amount spent after four hours at a mall is approximately $250. ∎

✓ CHECK POINT 1 Use the exponential function in Example 1 to find the average amount spent, to the nearest dollar, after three hours at a shopping mall. Does this rounded function value underestimate or overestimate the amount shown in **Figure 12.1**? By how much?

2 Graph exponential functions.

Graphing Exponential Functions

We are familiar with expressions involving b^x where x is a rational number. For example,

$$b^{1.7} = b^{\frac{17}{10}} = \sqrt[10]{b^{17}} \quad \text{and} \quad b^{1.73} = b^{\frac{173}{100}} = \sqrt[100]{b^{173}}.$$

However, note that the definition of $f(x) = b^x$ includes all real numbers for the domain x. You may wonder what b^x means when x is an irrational number, such as $b^{\sqrt{3}}$ or b^{π}. Using closer and closer approximations for $\sqrt{3}$ ($\sqrt{3} \approx 1.73205$), we can think of $b^{\sqrt{3}}$ as the value that has the successively closer approximations

$$b^{1.7}, b^{1.73}, b^{1.732}, b^{1.73205}, \dots.$$

In this way, we can graph the exponential function with no holes, or points of discontinuity, at the irrational domain values.

> **EXAMPLE 2**   Graphing an Exponential Function

Graph: $f(x) = 2^x$.

Solution   We begin by setting up a table of coordinates.

x	$f(x) = 2^x$
-3	$f(-3) = 2^{-3} = \dfrac{1}{8}$
-2	$f(-2) = 2^{-2} = \dfrac{1}{4}$
-1	$f(-1) = 2^{-1} = \dfrac{1}{2}$
0	$f(0) = 2^0 = 1$
1	$f(1) = 2^1 = 2$
2	$f(2) = 2^2 = 4$
3	$f(3) = 2^3 = 8$

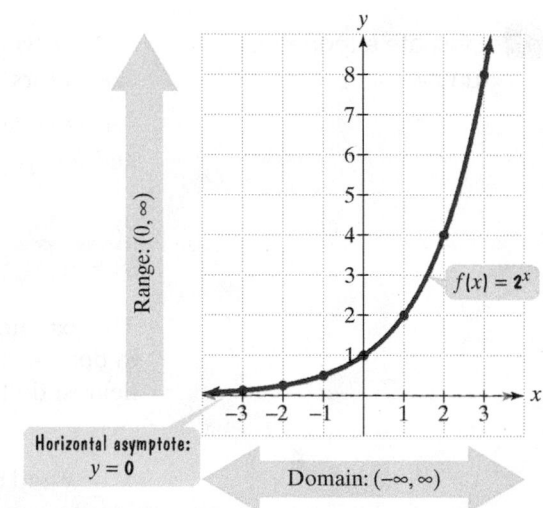

Figure 12.2 The graph of $f(x) = 2^x$

We plot these points, connecting them with a continuous curve. **Figure 12.2** shows the graph of $f(x) = 2^x$. Observe that the graph approaches, but never touches, the negative portion of the x-axis. Thus, the x-axis, or $y = 0$, is a horizontal asymptote. The range is the set of all positive real numbers. Although we used integers for x in our table of coordinates, you can use a calculator to find additional points. For example, $f(0.3) = 2^{0.3} \approx 1.231$ and $f(0.95) = 2^{0.95} \approx 1.932$. The points $(0.3, 1.231)$ and $(0.95, 1.932)$ approximately fit the graph. ∎

✓ **CHECK POINT 2**   Graph: $f(x) = 3^x$.

EXAMPLE 3   Graphing an Exponential Function

Graph: $g(x) = \left(\dfrac{1}{2}\right)^x$.

Solution   We begin by setting up a table of coordinates. We compute the function values by noting that

$$g(x) = \left(\frac{1}{2}\right)^x = (2^{-1})^x = 2^{-x}.$$

x	$g(x) = \left(\dfrac{1}{2}\right)^x$ or 2^{-x}
-3	$g(-3) = 2^{-(-3)} = 2^3 = 8$
-2	$g(-2) = 2^{-(-2)} = 2^2 = 4$
-1	$g(-1) = 2^{-(-1)} = 2^1 = 2$
0	$g(0) = 2^{-0} = 1$
1	$g(1) = 2^{-1} = \dfrac{1}{2^1} = \dfrac{1}{2}$
2	$g(2) = 2^{-2} = \dfrac{1}{2^2} = \dfrac{1}{4}$
3	$g(3) = 2^{-3} = \dfrac{1}{2^3} = \dfrac{1}{8}$

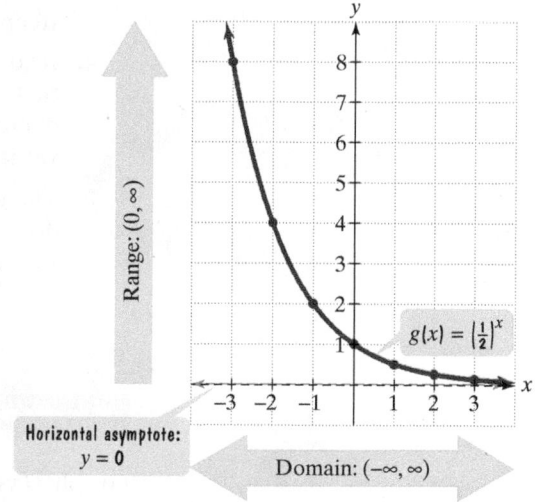

Figure 12.3 The graph of $g(x) = \left(\dfrac{1}{2}\right)^x$

We plot these points, connecting them with a continuous curve. **Figure 9.3** shows the graph of $g(x) = \left(\frac{1}{2}\right)^x$. This time the graph approaches, but never touches, the *positive* portion of the x-axis. Once again, the x-axis, or $y = 0$, is a horizontal asymptote. The range consists of all positive real numbers.   ∎

Do you notice a relationship between the graphs of $f(x) = 2^x$ and $g(x) = \left(\frac{1}{2}\right)^x$ in **Figures 12.2** and **12.3**? The graph of $g(x) = \left(\frac{1}{2}\right)^x$ is a mirror image, or reflection, of the graph of $f(x) = 2^x$ about the y-axis.

✓ **CHECK POINT 3**   Graph: $f(x) = \left(\frac{1}{3}\right)^x$. Note that $f(x) = \left(\frac{1}{3}\right)^x = (3^{-1})^x = 3^{-x}$.

Four exponential functions have been graphed in **Figure 12.4**. Compare the black and green graphs, where $b > 1$, to those in blue and red, where $b < 1$. When $b > 1$, the value of y increases as the value of x increases. When $b < 1$, the value of y decreases as the value of x increases. Notice that all four graphs pass through $(0, 1)$. These graphs illustrate general characteristics of exponential functions, listed in the box on the next page.

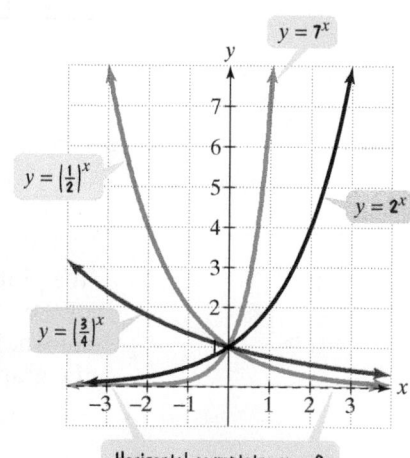

Figure 12.4 Graphs of four exponential functions

Characteristics of Exponential Functions of the Form $f(x) = b^x$

1. The domain of $f(x) = b^x$ consists of all real numbers: $(-\infty, \infty)$. The range of $f(x) = b^x$ consists of all positive real numbers: $(0, \infty)$.

2. The graphs of all exponential functions of the form $f(x) = b^x$ pass through the point $(0, 1)$ because $f(0) = b^0 = 1$ ($b \neq 0$). The y-intercept is 1.

3. If $b > 1$, $f(x) = b^x$ has a graph that goes up to the right and is an increasing function. The greater the value of b, the steeper the increase.

4. If $0 < b < 1$, $f(x) = b^x$ has a graph that goes down to the right and is a decreasing function. The smaller the value of b, the steeper the decrease.

5. The graph of $f(x) = b^x$ approaches, but does not touch, the x-axis. The x-axis, or $y = 0$, is a horizontal asymptote.

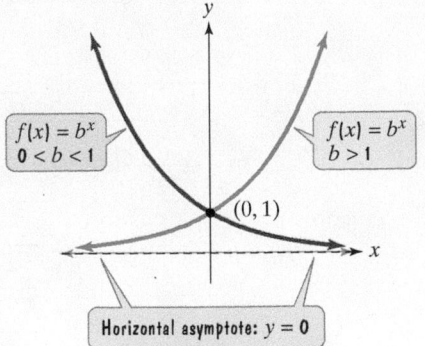

EXAMPLE 4 Graphing Exponential Functions

Graph $f(x) = 3^x$ and $g(x) = 3^{x+1}$ in the same rectangular coordinate system. How is the graph of g related to the graph of f?

Solution We begin by setting up a table showing some of the coordinates for f and g, selecting integers from -2 to 2 for x. Notice that $x + 1$ is the exponent for $g(x) = 3^{x+1}$.

x	$f(x) = 3^x$	$g(x) = 3^{x+1}$
-2	$f(-2) = 3^{-2} = \frac{1}{9}$	$g(-2) = 3^{-2+1} = 3^{-1} = \frac{1}{3}$
-1	$f(-1) = 3^{-1} = \frac{1}{3}$	$g(-1) = 3^{-1+1} = 3^0 = 1$
0	$f(0) = 3^0 = 1$	$g(0) = 3^{0+1} = 3^1 = 3$
1	$f(1) = 3^1 = 3$	$g(1) = 3^{1+1} = 3^2 = 9$
2	$f(2) = 3^2 = 9$	$g(2) = 3^{2+1} = 3^3 = 27$

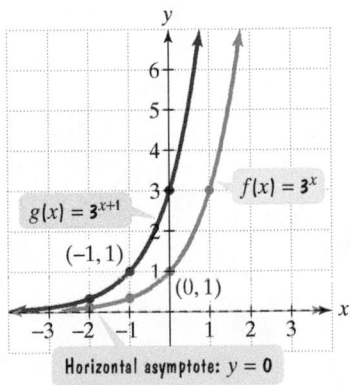

Figure 12.5

We plot the points for each function and connect them with a smooth curve. Because of the scale on the y-axis, the points on each function corresponding to $x = 2$ are not shown. **Figure 12.5** shows the graphs of $f(x) = 3^x$ and $g(x) = 3^{x+1}$. The graph of g is the graph of f shifted one unit to the left. ∎

✓ **CHECK POINT 4** Graph $f(x) = 3^x$ and $g(x) = 3^{x-1}$ in the same rectangular coordinate system. Select integers from -2 to 2 for x. How is the graph of g related to the graph of f?

EXAMPLE 5 Graphing Exponential Functions

Graph $f(x) = 2^x$ and $g(x) = 2^x - 3$ in the same rectangular coordinate system. How is the graph of g related to the graph of f?

Solution We begin by setting up a table showing some of the coordinates for f and g, selecting integers from -2 to 2 for x.

x	$f(x) = 2^x$	$g(x) = 2^x - 3$
-2	$f(-2) = 2^{-2} = \frac{1}{4}$	$g(-2) = 2^{-2} - 3 = \frac{1}{4} - 3 = -2\frac{3}{4}$
-1	$f(-1) = 2^{-1} = \frac{1}{2}$	$g(-1) = 2^{-1} - 3 = \frac{1}{2} - 3 = -2\frac{1}{2}$
0	$f(0) = 2^0 = 1$	$g(0) = 2^0 - 3 = 1 - 3 = -2$
1	$f(1) = 2^1 = 2$	$g(1) = 2^1 - 3 = 2 - 3 = -1$
2	$f(2) = 2^2 = 4$	$g(2) = 2^2 - 3 = 4 - 3 = 1$

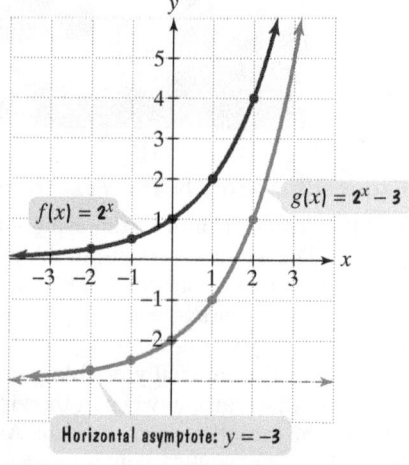

Horizontal asymptote: $y = -3$

Figure 12.6

We plot the points for each function and connect them with a smooth curve. **Figure 12.6** shows the graphs of $f(x) = 2^x$ and $g(x) = 2^x - 3$. The graph of g is the graph of f shifted down three units. As a result, $y = -3$ is the horizontal asymptote for g. ∎

✓ **CHECK POINT 5** Graph $f(x) = 2^x$ and $g(x) = 2^x + 3$ in the same rectangular coordinate system. Select integers from -2 to 2 for x. How is the graph of g related to the graph of f?

---

3 Evaluate functions with base e.

Using Technology

Graphic Connections

As n increases, the graph of $y = \left(1 + \frac{1}{n}\right)^n$ approaches the graph of $y = e$.

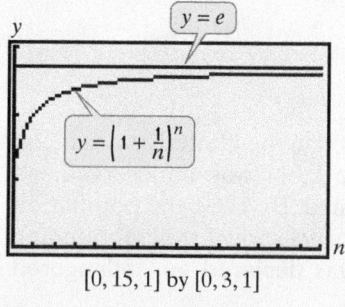

$[0, 15, 1]$ by $[0, 3, 1]$

The Natural Base e

An irrational number, symbolized by the letter e, appears as the base in many applied exponential functions. The number e is defined as the value that $\left(1 + \frac{1}{n}\right)^n$ approaches as n gets larger and larger. **Table 12.1** shows values of $\left(1 + \frac{1}{n}\right)^n$ for increasingly large values of n. As n increases, the approximate value of e to nine decimal places is

$$e \approx 2.718281827.$$

The irrational number e, approximately 2.72, is called the **natural base**. The function $f(x) = e^x$ is called the **natural exponential function**.

Table 12.1

n	$\left(1 + \frac{1}{n}\right)^n$
1	2
2	2.25
5	2.48832
10	2.59374246
100	2.704813829
1000	2.716923932
10,000	2.718145927
100,000	2.718268237
1,000,000	2.718280469
1,000,000,000	2.718281827

As n takes on increasingly large values, the expression $\left(1 + \frac{1}{n}\right)^n$ approaches e.

Blitzer Bonus

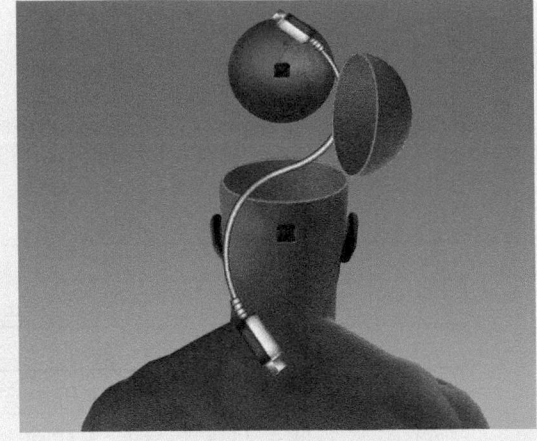

Exponential Growth: The Year Humans Become Immortal

In 2011, *Jeopardy!* aired a three-night match between a personable computer named Watson and the show's two most successful players. The winner: Watson. In the time it took each human contestant to respond to one trivia question, Watson was able to scan the content of one million books. It was also trained to understand the puns and twists of phrases unique to *Jeopardy!* clues.

Watson's remarkable accomplishments can be thought of as a single data point on an exponential curve that models growth in computing power. According to inventor, author, and computer scientist Ray Kurzweil (1948–), computer technology is progressing exponentially, doubling in power each year. What does this mean in terms of the accelerating pace of the graph of $y = 2^x$ that starts slowly and then rockets skyward toward infinity? According to Kurzweil, by 2023, a supercomputer will surpass the brainpower of a human. As progress accelerates exponentially and every hour brings a century's worth of scientific breakthroughs, by 2045, computers will surpass the brainpower equivalent to that of all human brains combined. Here's where it gets exponentially weird: In that year (says Kurzweil), we will be able to scan our consciousness into computers and enter a virtual existence, or swap our bodies for immortal robots. Indefinite life extension will become a reality and people will die only if they choose to.

Use a scientific or graphing calculator with an $\boxed{e^x}$ key to evaluate e to various powers. For example, to find e^2, press the following keys on most calculators:

Scientific calculator: $2\ \boxed{e^x}$

Graphing calculator: $\boxed{e^x}\ 2\ \boxed{\text{ENTER}}$.

The display should be approximately 7.389.

$$e^2 \approx 7.389$$

The number e lies between 2 and 3. Because $2^2 = 4$ and $3^2 = 9$, it makes sense that e^2, approximately 7.389, lies between 4 and 9.

Because $2 < e < 3$, the graph of $y = e^x$ is between the graphs of $y = 2^x$ and $y = 3^x$, shown in **Figure 12.7**.

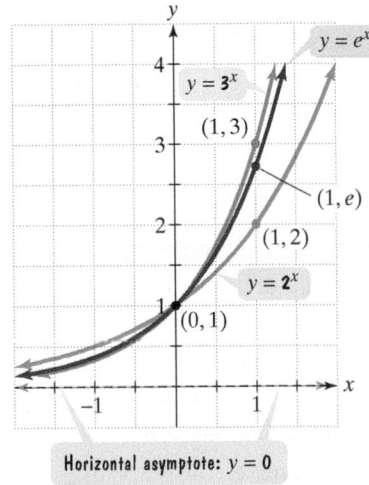

Figure 12.7 Graphs of three exponential functions

EXAMPLE 6   Gray Wolf Population

Insatiable killer. That's the reputation the gray wolf acquired in the United States in the nineteenth and early twentieth centuries. Although the label was undeserved, an estimated two million wolves were shot, trapped, or poisoned. By 1960, the population was reduced to 800 wolves. **Figure 12.8** on the next page shows the rebounding population in two recovery areas after the gray wolf was declared an endangered species and received federal protection.

Gray Wolf Population in Two Recovery Areas for Selected Years

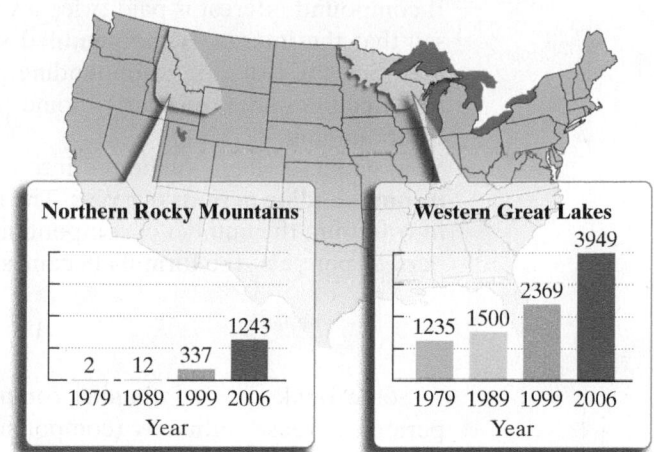

Figure 12.8

Source: U.S. Fish and Wildlife Service

The exponential function

$$f(x) = 1.26e^{0.247x}$$

models the gray wolf population of the Northern Rocky Mountains, $f(x)$, x years after 1978. If the wolf is not removed from the endangered species list and trends shown in **Figure 12.8** continue, project its population in the recovery area in 2010.

Solution Because 2010 is 32 years after 1978, we substitute 32 for x in the given function.

$$f(x) = 1.26e^{0.247x} \quad \text{This is the given function.}$$
$$f(32) = 1.26e^{0.247(32)} \quad \text{Substitute 32 for x.}$$

Perform this computation on your calculator.

Scientific calculator: 1.26 $\boxed{\times}$ $\boxed{(}$ .247 $\boxed{\times}$ 32 $\boxed{)}$ $\boxed{e^x}$ $\boxed{=}$

Graphing calculator: 1.26 $\boxed{\times}$ $\boxed{e^x}$ $\boxed{(}$ .247 $\boxed{\times}$ 32 $\boxed{)}$ $\boxed{\text{ENTER}}$

The display should be approximately 3412.1973. Thus,

> This parenthesis is given on some calculators.

$$f(32) = 1.26e^{0.247(32)} \approx 3412.$$

This indicates that the gray wolf population of the Northern Rocky Mountains in the year 2010 is projected to be 3412. ∎

☑ **CHECK POINT 6** The exponential function $f(x) = 1066e^{0.042x}$ models the gray wolf population of the Western Great Lakes, $f(x)$, x years after 1978. If trends shown in **Figure 12.8** continue, project the gray wolf's population in the recovery area in 2012.

In 2008, using exponential functions and projections like those in Example 6 and Check Point 6, the U.S. Fish and Wildlife Service removed the gray wolf from the endangered species list, a ruling environmentalists vowed to appeal.

4 Use compound interest formulas.

Compound Interest

In Chapter 11, we saw that the amount of money, A, that a principal, P, will be worth after t years at interest rate r, compounded annually, is given by the formula

$$A = P(1 + r)^t.$$

Most savings institutions have plans in which interest is paid more than once a year. If compound interest is paid twice a year, the compounding period is six months. We say that the interest is **compounded semiannually**. When compound interest is paid four times a year, the compounding period is three months and the interest is said to be **compounded quarterly**. Some plans allow for monthly compounding or daily compounding.

In general, when compound interest is paid n times a year, we say that there are **n compounding periods per year**. The formula $A = P(1 + r)^t$ can be adjusted to take into account the number of compounding periods in a year. If there are n compounding periods per year, the formula becomes

$$A = P\left(1 + \frac{r}{n}\right)^{nt}.$$

Some banks use **continuous compounding**, where the number of compounding periods increases infinitely (compounding interest every trillionth of a second, every quadrillionth of a second, etc.). As n, the number of compounding periods in a year, increases without bound, the expression $\left(1 + \frac{1}{n}\right)^n$ approaches e. As a result, the formula for continuous compounding is $A = Pe^{rt}$. Although continuous compounding sounds terrific, it yields only a fraction of a percent more interest over a year than daily compounding.

Formulas for Compound Interest

After t years, the balance, A, in an account with principal P and annual interest rate r (in decimal form) is given by the following formulas:

1. For n compounding periods per year: $A = P\left(1 + \frac{r}{n}\right)^{nt}$

2. For continuous compounding: $A = Pe^{rt}$.

EXAMPLE 7 Choosing Between Investments

You decide to invest $8000 for 6 years and you have a choice between two accounts. The first pays 7% per year, compounded monthly. The second pays 6.85% per year, compounded continuously. Which is the better investment?

Solution The better investment is the one with the greater balance in the account after 6 years. Let's begin with the account with monthly compounding. We use the compound interest model with $P = 8000, r = 7\% = 0.07, n = 12$ (monthly compounding means 12 compounding periods per year), and $t = 6$.

$$A = P\left(1 + \frac{r}{n}\right)^{nt} = 8000\left(1 + \frac{0.07}{12}\right)^{12 \cdot 6} \approx 12{,}160.84$$

The balance in this account after 6 years is $12,160.84.

For the second investment option, we use the model for continuous compounding with $P = 8000, r = 6.85\% = 0.0685$, and $t = 6$.

$$A = Pe^{rt} = 8000e^{0.0685(6)} \approx 12{,}066.60$$

The balance in this account after 6 years is $12,066.60, slightly less than the previous amount. Thus, the better investment is the 7% monthly compounding option. ∎

✓ **CHECK POINT 7** A sum of $10,000 is invested at an annual rate of 8%. Find the balance in the account after 5 years subject to **a.** quarterly compounding and **b.** continuous compounding.

CONCEPT AND VOCABULARY CHECK

Fill in each blank so that the resulting statement is true.

1. The exponential function f with base b is defined by $f(x) = $ _____, $b > 0$ and $b \neq 1$. Using interval notation, the domain of this function is _____ and the range is _____.

2. The graph of the exponential function f with base b approaches, but does not touch, the _____-axis. This axis, whose equation is _____, is a/an _____ asymptote.

3. The value that $\left(1 + \dfrac{1}{n}\right)^n$ approaches as n gets larger and larger is the irrational number _____, called the _____ base. This irrational number is approximately equal to _____.

4. Consider the compound interest formula

$$A = P\left(1 + \frac{r}{n}\right)^{nt}.$$

This formula gives the balance, _____, in an account with principal _____ and annual interest rate _____, in decimal form, subject to compound interest paid _____ times per year.

5. If compound interest is paid twice a year, we say that the interest is compounded _____. If compound interest is paid four times a year, we say that the interest is compounded _____. If the number of compounding periods increases infinitely, we call this _____ compounding.

12.1 EXERCISE SET

  Watch the videos in MyMathLab    Download the MyDashBoard App

Practice Exercises

In Exercises 1–10, approximate each number using a calculator. Round your answer to three decimal places.

1. $2^{3.4}$
2. $3^{2.4}$
3. $3^{\sqrt{5}}$
4. $5^{\sqrt{3}}$
5. $4^{-1.5}$
6. $6^{-1.2}$
7. $e^{2.3}$
8. $e^{3.4}$
9. $e^{-0.95}$
10. $e^{-0.75}$

In Exercises 11–16, set up a table of coordinates for each function. Select integers from -2 to 2, inclusive, for x. Then use the table of coordinates to match the function with its graph. [The graphs are labeled (a) through (f).]

11. $f(x) = 3^x$
12. $f(x) = 3^{x-1}$
13. $f(x) = 3^x - 1$
14. $f(x) = -3^x$
15. $f(x) = 3^{-x}$
16. $f(x) = -3^{-x}$

a.

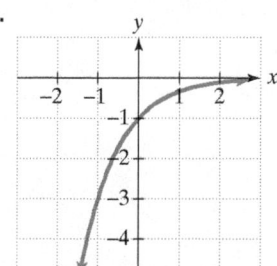

b.

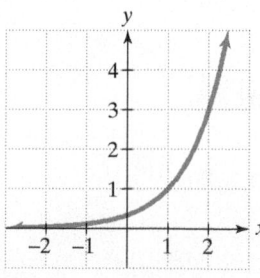

c.

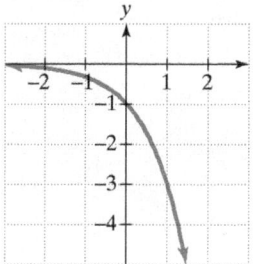

d.

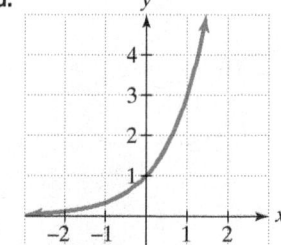

e.

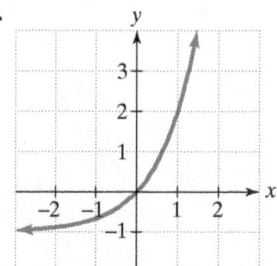

f.

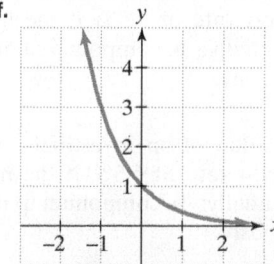

In Exercises 17–24, graph each function by making a table of coordinates. If applicable, use a graphing utility to confirm your hand-drawn graph.

17. $f(x) = 4^x$

18. $f(x) = 5^x$

19. $g(x) = \left(\dfrac{3}{2}\right)^x$

20. $g(x) = \left(\dfrac{4}{3}\right)^x$

21. $h(x) = \left(\dfrac{1}{2}\right)^x$

22. $h(x) = \left(\dfrac{1}{3}\right)^x$

23. $f(x) = (0.6)^x$

24. $f(x) = (0.8)^x$

In Exercises 25–38, graph functions f and g in the same rectangular coordinate system. Select integers from −2 to 2, inclusive, for x. Then describe how the graph of g is related to the graph of f. If applicable, use a graphing utility to confirm your hand-drawn graphs.

25. $f(x) = 2^x$ and $g(x) = 2^{x+1}$

26. $f(x) = 2^x$ and $g(x) = 2^{x+2}$

27. $f(x) = 2^x$ and $g(x) = 2^{x-2}$

28. $f(x) = 2^x$ and $g(x) = 2^{x-1}$

29. $f(x) = 2^x$ and $g(x) = 2^x + 1$

30. $f(x) = 2^x$ and $g(x) = 2^x + 2$

31. $f(x) = 2^x$ and $g(x) = 2^x - 2$

32. $f(x) = 2^x$ and $g(x) = 2^x - 1$

33. $f(x) = 3^x$ and $g(x) = -3^x$

34. $f(x) = 3^x$ and $g(x) = 3^{-x}$

35. $f(x) = 2^x$ and $g(x) = 2^{x+1} - 1$

36. $f(x) = 2^x$ and $g(x) = 2^{x+1} - 2$

37. $f(x) = 3^x$ and $g(x) = \frac{1}{3} \cdot 3^x$

38. $f(x) = 3^x$ and $g(x) = 3 \cdot 3^x$

Use the compound interest formulas, $A = P\left(1 + \dfrac{r}{n}\right)^{nt}$ and $A = Pe^{rt}$, to solve Exercises 39–42. Round answers to the nearest cent.

39. Find the accumulated value of an investment of $10,000 for 5 years at an interest rate of 5.5% if the money is **a.** compounded semiannually; **b.** compounded monthly; **c.** compounded continuously.

40. Find the accumulated value of an investment of $5000 for 10 years at an interest rate of 6.5% if the money is **a.** compounded semiannually; **b.** compounded monthly; **c.** compounded continuously.

41. Suppose that you have $12,000 to invest. Which investment yields the greater return over 3 years: 7% compounded monthly or 6.85% compounded continuously?

42. Suppose that you have $6000 to invest. Which investment yields the greater return over 4 years: 8.25% compounded quarterly or 8.3% compounded semiannually?

Practice PLUS

In Exercises 43–48, use each exponential function's graph to determine the function's domain and range.

43.

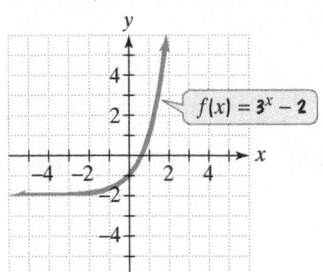

$f(x) = 3^x - 2$

44.

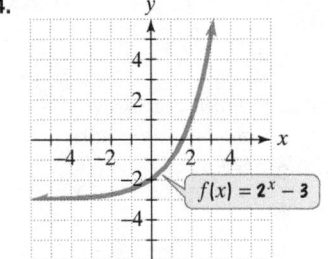

$f(x) = 2^x - 3$

45.

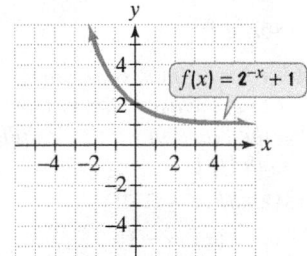

$f(x) = 2^{-x} + 1$

46.

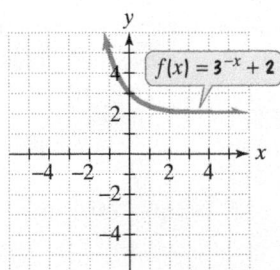

$f(x) = 3^{-x} + 2$

47.

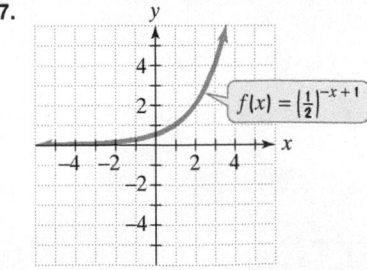

$f(x) = \left|\dfrac{1}{2}\right|^{-x+1}$

48.

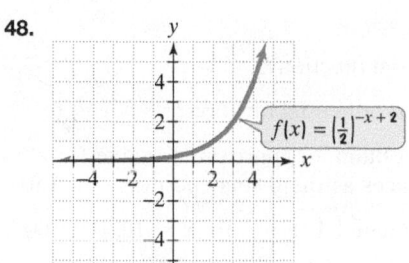

$f(x) = \left[\frac{1}{2}\right]^{-x+2}$

In Exercises 49–50, graph f and g in the same rectangular coordinate system. Then find the point of intersection of the two graphs.

49. $f(x) = 2^x, g(x) = 2^{-x}$

50. $f(x) = 2^{x+1}, g(x) = 2^{-x+1}$

51. Graph $y = 2^x$ and $x = 2^y$ in the same rectangular coordinate system.

52. Graph $y = 3^x$ and $x = 3^y$ in the same rectangular coordinate system.

Application Exercises

Use a calculator with a $\boxed{y^x}$ *key or a* $\boxed{\wedge}$ *key to solve Exercises 53–56.*

53. India is currently one of the world's fastest-growing countries. By 2040, the population of India will be larger than the population of China; by 2050, nearly one-third of the world's population will live in these two countries alone. The exponential function $f(x) = 574(1.026)^x$ models the population of India, $f(x)$, in millions, x years after 1974.

  a. Substitute 0 for x and, without using a calculator, find India's population in 1974.

  b. Substitute 27 for x and use your calculator to find India's population, to the nearest million, in the year 2001 as modeled by this function.

  c. Find India's population, to the nearest million, in the year 2028 as predicted by this function.

  d. Find India's population, to the nearest million, in the year 2055 as predicted by this function.

  e. What appears to be happening to India's population every 27 years?

54. The 1986 explosion at the Chernobyl nuclear power plant in the former Soviet Union sent about 1000 kilograms of radioactive cesium-137 into the atmosphere. The function $f(x) = 1000(0.5)^{\frac{x}{30}}$ describes the amount, $f(x)$, in kilograms, of cesium-137 remaining in Chernobyl x years after 1986. If even 100 kilograms of cesium-137 remain in Chernobyl's atmosphere, the area is considered unsafe for human habitation. Find $f(80)$ and determine if Chernobyl will be safe for human habitation by 2066.

The formula $S = C(1 + r)^t$ models inflation, where $C =$ the value today, $r =$ the annual inflation rate, and $S =$ the inflated value t years from now. Use this formula to solve Exercises 55–56. Round answers to the nearest dollar.

55. If the inflation rate is 6%, how much will a house now worth $465,000 be worth in 10 years?

56. If the inflation rate is 3%, how much will a house now worth $510,000 be worth in 5 years?

Use a calculator with an $\boxed{e^x}$ *key to solve Exercises 57–63.*

As of July 2010, 500 million people worldwide shared versions of their lives on Facebook. The graph shows the number of active Facebook users (users who returned to the site within 30 days) for selected months from 2009 through 2010.

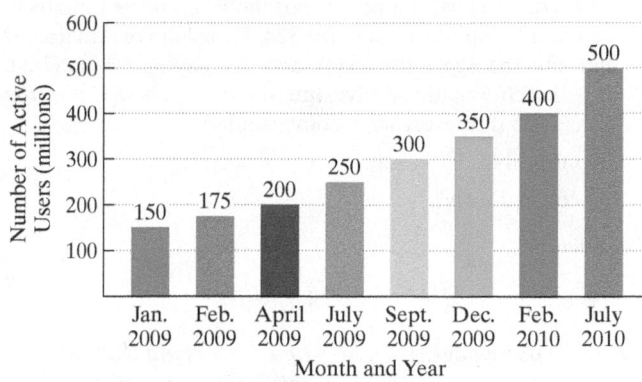

Number of Active Facebook Users

Source: Facebook

The data can be modeled by

$$f(x) = 19x + 127 \quad \text{and} \quad g(x) = 152.6e^{0.0667x},$$

in which $f(x)$ and $g(x)$ represent the number of active Facebook users, in millions, x months after December 2008. Use these functions to solve Exercises 57–58. Round answers to the nearest whole million.

57. a. According to the linear model, how many millions of active Facebook users were there in February 2010, 14 months after December 2008?

  b. According to the exponential model, how many millions of active Facebook users were there in February 2010?

  c. Which function is a better model for the data in February 2010?

58. a. According to the linear model, how many millions of active Facebook users were there in July 2010, 19 months after December 2008?

  b. According to the exponential model, how many millions of active Facebook users were there in July 2010?

  c. Which function is a better model for the data in July 2010?

59. In college, we study large volumes of information— information that, unfortunately, we do not often retain for very long. The function

$$f(x) = 80e^{-0.5x} + 20$$

describes the percentage of information, $f(x)$, that a particular person remembers x weeks after learning the information.

  a. Substitute 0 for x and, without using a calculator, find the percentage of information remembered at the moment it is first learned.

b. Substitute 1 for x and find the percentage of information that is remembered after 1 week.

c. Find the percentage of information that is remembered after 4 weeks.

d. Find the percentage of information that is remembered after one year (52 weeks).

60. In 1626, Peter Minuit persuaded the Wappinger Indians to sell him Manhattan Island for $24. If the Native Americans had put the $24 into a bank account paying 5% interest, how much would the investment have been worth in the year 2005 if interest were compounded

   a. monthly?

   b. continuously?

The function

$$f(x) = \frac{90}{1 + 270e^{-0.122x}}$$

models the percentage, $f(x)$, of people x years old with some coronary heart disease. Use this function to solve Exercises 61–62. Round answers to the nearest tenth of a percent.

61. Evaluate $f(30)$ and describe what this means in practical terms.

62. Evaluate $f(70)$ and describe what this means in practical terms.

63. The function

$$N(t) = \frac{30,000}{1 + 20e^{-1.5t}}$$

describes the number of people, $N(t)$, who become ill with influenza t weeks after its initial outbreak in a town with 30,000 inhabitants. The horizontal asymptote in the graph indicates that there is a limit to the epidemic's growth.

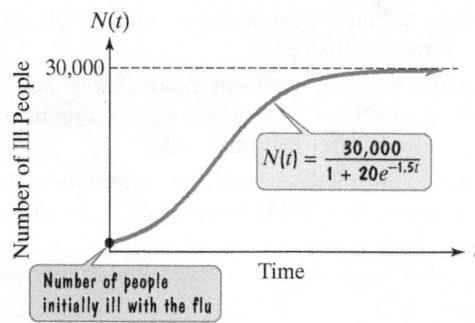

a. How many people became ill with the flu when the epidemic began? (When the epidemic began, $t = 0$.)

b. How many people were ill by the end of the third week?

c. Why can't the spread of an epidemic simply grow indefinitely? What does the horizontal asymptote shown in the graph indicate about the limiting size of the population that becomes ill?

Writing in Mathematics

64. What is an exponential function?

65. What is the natural exponential function?

66. Use a calculator to obtain an approximate value for e to as many decimal places as the display permits. Then use the calculator to evaluate $\left(1 + \dfrac{1}{x}\right)^x$ for $x = 10, 100, 1000, 10,000, 100,000,$ and $1,000,000$. Describe what happens to the expression as x increases.

67. Write an example similar to Example 7 on page 864 in which continuous compounding at a slightly lower yearly interest rate is a better investment than compounding n times per year.

68. Describe how you could use the graph of $f(x) = 2^x$ to obtain a decimal approximation for $\sqrt{2}$.

Technology Exercises

69. You have $10,000 to invest. One bank pays 5% interest compounded quarterly and the other pays 4.5% interest compounded monthly.

   a. Use the formula for compound interest to write a function for the balance in each account at any time t in years.

   b. Use a graphing utility to graph both functions in an appropriate viewing rectangle. Based on the graphs, which bank offers the better return on your money?

70. **a.** Graph $y = e^x$ and $y = 1 + x + \dfrac{x^2}{2}$ in the same viewing rectangle.

   b. Graph $y = e^x$ and $y = 1 + x + \dfrac{x^2}{2} + \dfrac{x^3}{6}$ in the same viewing rectangle.

   c. Graph $y = e^x$ and $y = 1 + x + \dfrac{x^2}{2} + \dfrac{x^3}{6} + \dfrac{x^4}{24}$ in the same viewing rectangle.

   d. Describe what you observe in parts (a)–(c). Try generalizing this observation.

Critical Thinking Exercises

Make Sense? *In Exercises 71–74, determine whether each statement "makes sense" or "does not make sense" and explain your reasoning.*

71. My graph of $f(x) = 3 \cdot 2^x$ shows that the horizontal asymptote for f is $x = 3$.

72. I'm using a photocopier to reduce an image over and over by 50%, so the exponential function $f(x) = \left(\frac{1}{2}\right)^x$ models the new image size, where x is the number of reductions.

73. Taxing thoughts: I'm looking at data that show the number of pages in the publication that explains the U.S. tax code for selected years from 1945 through 2009. A linear function appears to be a better choice than an exponential function for modeling the number of pages in the tax code during this period.

Number of Pages in the Federal Tax Code

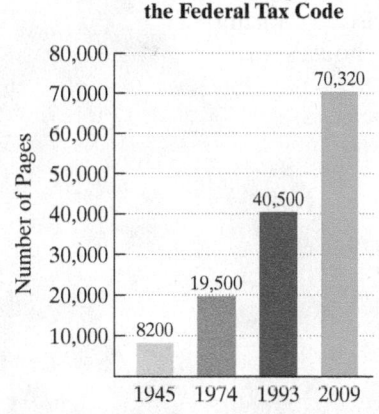

Source: CCH Inc.

74. I use the natural base e when determining how much money I'd have in a bank account that earns compound interest subject to continuous compounding.

In Exercises 75–78, determine whether each statement is true or false. If the statement is false, make the necessary change(s) to produce a true statement.

75. As the number of compounding periods increases on a fixed investment, the amount of money in the account over a fixed interval of time will increase without bound.

76. The functions $f(x) = 3^{-x}$ and $g(x) = -3^x$ have the same graph.

77. If $f(x) = 2^x$, then $f(a + b) = f(a) + f(b)$.

78. The functions $f(x) = \left(\frac{1}{3}\right)^x$ and $g(x) = 3^{-x}$ have the same graph.

79. The graphs labeled (a)–(d) in the figure represent $y = 3^x$, $y = 5^x$, $y = \left(\frac{1}{3}\right)^x$, and $y = \left(\frac{1}{5}\right)^x$, but not necessarily in that order. Which is which? Describe the process that enables you to make this decision.

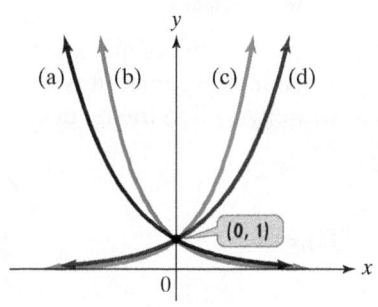

80. The hyperbolic cosine and hyperbolic sine functions are defined by

$$\cosh x = \frac{e^x + e^{-x}}{2} \quad \text{and} \quad \sinh x = \frac{e^x - e^{-x}}{2}.$$

Prove that $(\cosh x)^2 - (\sinh x)^2 = 1$.

Review Exercises

81. Solve for b: $D = \dfrac{ab}{a + b}$.

(Section 7.6, Example 7)

82. Subtract: $\dfrac{2x + 3}{x^2 - 7x + 12} - \dfrac{2}{x - 3}$.

(Section 7.4, Example 7)

83. Solve: $x(x - 3) = 10$.

(Section 6.6, Example 6)

Preview Exercises

Exercises 84–86 will help you prepare for the material covered in the next section.

84. In Section 8.4, we used a *switch-and-solve* strategy for finding a function's inverse. (See the box on page 625.) What problem do you encounter when using this strategy to find the inverse of $f(x) = 2^x$?

85. 25 to what power gives 5? $(25^? = 5)$

86. If $f(x) = 2x - 5$, find $f^{-1}(x)$.

12.2

Objectives

1 Change from logarithmic to exponential form.

2 Change from exponential to logarithmic form.

3 Evaluate logarithms.

4 Use basic logarithmic properties.

5 Graph logarithmic functions.

6 Find the domain of a logarithmic function.

7 Use common logarithms.

8 Use natural logarithms.

Logarithmic Functions

The earthquake that ripped through northern California on October 17, 1989, measured 7.1 on the Richter scale, killed more than 60 people, and injured more than 2400. Shown here is San Francisco's Marina district, where shock waves tossed houses off their foundations and into the street.

A higher measure on the Richter scale is more devastating than it seems because for each increase in one unit on the scale, there is a tenfold increase in the intensity of an earthquake. In this section, our focus is on the inverse of the exponential function, called the logarithmic function. The logarithmic function will help you to understand diverse phenomena, including earthquake intensity, human memory, and the pace of life in large cities.

Great Question!

You mentioned that the inverse of the exponential function is called the logarithmic function. We haven't discussed inverses of functions since Section 8.4. What should I already know about functions and their inverses?

Here's a brief summary:

1. Only one-to-one functions have inverses that are functions. A function, f, has an inverse function, f^{-1}, if there is no horizontal line that intersects the graph of f at more than one point.

2. If a function is one-to-one, its inverse function can be found by interchanging x and y in the function's equation and solving for y.

3. If $f(a) = b$, then $f^{-1}(b) = a$. The domain of f is the range of f^{-1}. The range of f is the domain of f^{-1}.

4. $f(f^{-1}(x)) = x$ and $f^{-1}(f(x)) = x$.

5. The graph of f^{-1} is the reflection of the graph of f about the line $y = x$.

The Definition of Logarithmic Functions

No horizontal line can be drawn that intersects the graph of an exponential function at more than one point. This means that the exponential function is one-to-one and has an inverse. Let's use our switch-and-solve strategy to find the inverse.

> All exponential functions have inverse functions. $f(x) = b^x$

Step 1. Replace $f(x)$ with y: $y = b^x$.

Step 2. Interchange x and y: $x = b^y$.

Step 3. Solve for y: ?

The question mark indicates that we do not have a method for solving $b^y = x$ for y. To isolate the exponent y, a new notation, called *logarithmic notation*, is needed. This notation gives us a way to name the inverse of $f(x) = b^x$. **The inverse function of the exponential function with base b is called the *logarithmic function with base b*.**

Definition of the Logarithmic Function

For $x > 0$ and $b > 0$, $b \neq 1$,

$$y = \log_b x \text{ is equivalent to } b^y = x.$$

The function $f(x) = \log_b x$ is the **logarithmic function with base b**.

The equations

$$y = \log_b x \quad \text{and} \quad b^y = x$$

are different ways of expressing the same thing. The first equation is in **logarithmic form** and the second equivalent equation is in **exponential form**.

Notice that a **logarithm, y, is an exponent. Logarithmic form allows us to isolate this exponent.** You should learn the location of the base and exponent in each form.

Location of Base and Exponent in Exponential and Logarithmic Forms

Logarithmic Form: $y = \log_b x$ Exponential Form: $b^y = x$

Great Question!

Much of what you've discussed so far involves changing from logarithmic form to the more familiar exponential form. Is there a pattern I can use to help me remember how to do this?

Yes. To change from logarithmic form to exponential form, use this pattern:

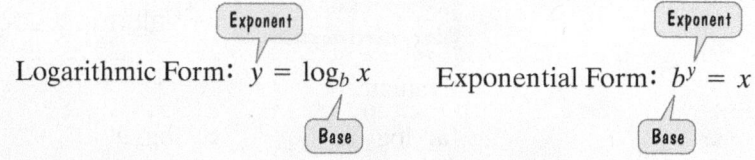

$$y = \log_b x \quad \text{means} \quad b^y = x.$$

1 Change from logarithmic to exponential form.

EXAMPLE 1   Changing from Logarithmic to Exponential Form

Write each equation in its equivalent exponential form:

a. $2 = \log_5 x$ **b.** $3 = \log_b 64$ **c.** $\log_3 7 = y$.

Solution   We use the fact that $y = \log_b x$ means $b^y = x$.

a. $2 = \log_5 x$ means $5^2 = x$.
Logarithms are exponents.

b. $3 = \log_b 64$ means $b^3 = 64$.
Logarithms are exponents.

c. $\log_3 7 = y$ or $y = \log_3 7$ means $3^y = 7$. ∎

✓ **CHECK POINT 1**   Write each equation in its equivalent exponential form:

a. $3 = \log_7 x$ **b.** $2 = \log_b 25$ **c.** $\log_4 26 = y$.

2 Change from exponential to logarithmic form.

EXAMPLE 2   Changing from Exponential to Logarithmic Form

Write each equation in its equivalent logarithmic form:

a. $12^2 = x$ **b.** $b^3 = 8$ **c.** $e^y = 9$.

Solution We use the fact that $b^y = x$ means $y = \log_b x$. In logarithmic form, the exponent is isolated on one side of the equal sign.

a. $12^2 = x$ means $2 = \log_{12} x$.   **b.** $b^3 = 8$ means $3 = \log_b 8$.

Exponents are logarithms.   Exponents are logarithms.

c. $e^y = 9$ means $y = \log_e 9$. ∎

✓ **CHECK POINT 2**   Write each equation in its equivalent logarithmic form:

a. $2^5 = x$ **b.** $b^3 = 27$ **c.** $e^y = 33$.

3 Evaluate logarithms.

Remembering that logarithms are exponents makes it possible to evaluate some logarithms by inspection. The logarithm of x with base b, $\log_b x$, is the exponent to which b must be raised to get x. For example, suppose we want to evaluate $\log_2 32$. We ask, 2 to what power gives 32? Because $2^5 = 32$, we can conclude that $\log_2 32 = 5$.

EXAMPLE 3   Evaluating Logarithms

Evaluate:

a. $\log_2 16$ **b.** $\log_3 9$ **c.** $\log_{25} 5$.

Solution

Logarithmic Expression	Question Needed for Evaluation	Logarithmic Expression Evaluated
a. $\log_2 16$	2 to what power gives 16?	$\log_2 16 = 4$ because $2^4 = 16$.
b. $\log_3 9$	3 to what power gives 9?	$\log_3 9 = 2$ because $3^2 = 9$.
c. $\log_{25} 5$	25 to what power gives 5?	$\log_{25} 5 = \frac{1}{2}$ because $25^{\frac{1}{2}} = \sqrt{25} = 5$.

∎

✓ **CHECK POINT 3**   Evaluate:

a. $\log_{10} 100$ **b.** $\log_3 3$ **c.** $\log_{36} 6$.

4 Use basic logarithmic properties.

Basic Logarithmic Properties

Because logarithms are exponents, they have properties that can be verified using the properties of exponents.

Basic Logarithmic Properties Involving 1

1. $\log_b b = 1$ because 1 is the exponent to which b must be raised to obtain b. ($b^1 = b$)

2. $\log_b 1 = 0$ because 0 is the exponent to which b must be raised to obtain 1. ($b^0 = 1$)

EXAMPLE 4   Using Properties of Logarithms

Evaluate:

a. $\log_7 7$ **b.** $\log_5 1$.

Solution

a. Because $\log_b b = 1$, we conclude $\log_7 7 = 1$.

b. Because $\log_b 1 = 0$, we conclude $\log_5 1 = 0$. ∎

> ✓ **CHECK POINT 4** Evaluate:
>
> **a.** $\log_9 9$　　　**b.** $\log_8 1$.

Now that we are familiar with logarithmic notation, let's resume and finish the switch-and-solve strategy for finding the inverse of $f(x) = b^x$.

Step 1. Replace f(x) with y: $y = b^x$.

Step 2. Interchange x and y: $x = b^y$.

Step 3. Solve for y: $y = \log_b x$.

Step 4. Replace y with $f^{-1}(x)$: $f^{-1}(x) = \log_b x$.

The completed switch-and-solve strategy illustrates that if $f(x) = b^x$, then $f^{-1}(x) = \log_b x$. The inverse of an exponential function is the logarithmic function with the same base.

In Section 8.4, we saw how inverse functions "undo" one another. In particular,

$$f(f^{-1}(x)) = x \quad \text{and} \quad f^{-1}(f(x)) = x.$$

Applying these relationships to exponential and logarithmic functions, we obtain the following **inverse properties of logarithms:**

Inverse Properties of Logarithms

For $b > 0$ and $b \neq 1$,

$\log_b b^x = x$　　The logarithm with base b of b raised to a power equals that power.

$b^{\log_b x} = x$　　b raised to the logarithm with base b of a number equals that number.

EXAMPLE 5 Using Inverse Properties of Logarithms

Evaluate:

a. $\log_4 4^5$　　　**b.** $6^{\log_6 9}$.

Solution

a. Because $\log_b b^x = x$, we conclude $\log_4 4^5 = 5$.

b. Because $b^{\log_b x} = x$, we conclude $6^{\log_6 9} = 9$. ∎

> ✓ **CHECK POINT 5** Evaluate:
>
> **a.** $\log_7 7^8$　　　**b.** $3^{\log_3 17}$.

---

5 Graph logarithmic functions.

Graphs of Logarithmic Functions

How do we graph logarithmic functions? We use the fact that a logarithmic function is the inverse of an exponential function. This means that the logarithmic function reverses the coordinates of the exponential function. It also means that the graph of the logarithmic function is a reflection of the graph of the exponential function about the line $y = x$.

EXAMPLE 6 Graphs of Exponential and Logarithmic Functions

Graph $f(x) = 2^x$ and $g(x) = \log_2 x$ in the same rectangular coordinate system.

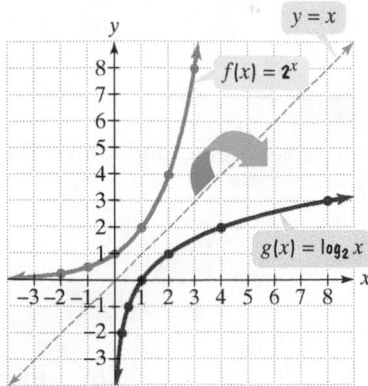

Figure 12.9 The graphs of $f(x) = 2^x$ and its inverse function

Solution We first set up a table of coordinates for $f(x) = 2^x$. Reversing these coordinates gives the coordinates for the inverse function $g(x) = \log_2 x$.

x	-2	-1	0	1	2	3
$f(x) = 2^x$	$\frac{1}{4}$	$\frac{1}{2}$	1	2	4	8

Reverse coordinates.

x	$\frac{1}{4}$	$\frac{1}{2}$	1	2	4	8
$g(x) = \log_2 x$	-2	-1	0	1	2	3

We now plot the ordered pairs from each table, connecting them with smooth curves. **Figure 12.9** shows the graphs of $f(x) = 2^x$ and its inverse function $g(x) = \log_2 x$. The graph of the inverse can also be drawn by reflecting the graph of $f(x) = 2^x$ about the line $y = x$. ∎

Great Question!

You found the coordinates of $g(x) = \log_2 x$ by reversing the coordinates of $f(x) = 2^x$. Do I have to do it that way?

Not necessarily. You can obtain a partial table of coordinates for $g(x) = \log_2 x$ without having to obtain and reverse coordinates for $f(x) = 2^x$. Because $g(x) = \log_2 x$ means $2^{g(x)} = x$, we begin with values for $g(x)$ and compute corresponding values for x:

Use $x = 2^{g(x)}$ to compute x. For example, if $g(x) = -2$, $x = 2^{-2} = \frac{1}{2^2} = \frac{1}{4}$.

Start with values for $g(x)$.

x	$\frac{1}{4}$	$\frac{1}{2}$	1	2	4	8
$g(x) = \log_2 x$	-2	-1	0	1	2	3

✓ **CHECK POINT 6** Graph $f(x) = 3^x$ and $g(x) = \log_3 x$ in the same rectangular coordinate system.

Figure 12.10 illustrates the relationship between the graph of an exponential function, shown in blue, and its inverse, a logarithmic function, shown in red, for bases greater than 1 and for bases between 0 and 1. Also shown and labeled are the exponential function's horizontal asymptote ($y = 0$) and the logarithmic function's vertical asymptote ($x = 0$).

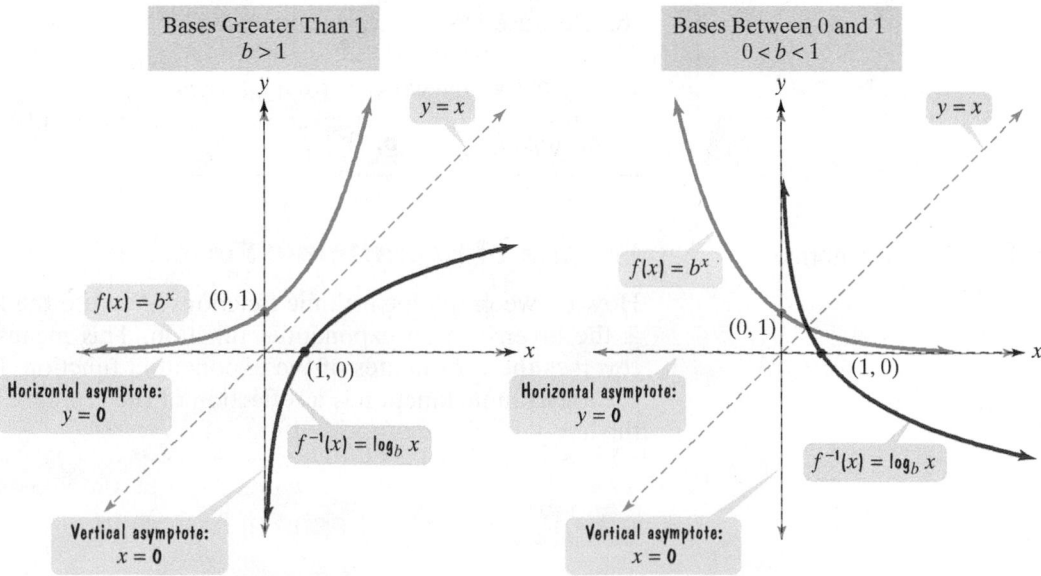

Figure 12.10 Graphs of exponential and logarithmic functions

The red graphs in **Figure 12.10** illustrate the following general characteristics of logarithmic functions:

> ### Characteristics of Logarithmic Functions of the Form $f(x) = \log_b x$
>
> 1. The domain of $f(x) = \log_b x$ consists of all positive real numbers: $(0, \infty)$. The range of $f(x) = \log_b x$ consists of all real numbers: $(-\infty, \infty)$.
> 2. The graphs of all logarithmic functions of the form $f(x) = \log_b x$ pass through the point $(1, 0)$ because $f(1) = \log_b 1 = 0$. The x-intercept is 1. There is no y-intercept.
> 3. If $b > 1$, $f(x) = \log_b x$ has a graph that goes up to the right and is an increasing function.
> 4. If $0 < b < 1$, $f(x) = \log_b x$ has a graph that goes down to the right and is a decreasing function.
> 5. The graph of $f(x) = \log_b x$ approaches, but does not touch, the y-axis. The y-axis, or $x = 0$, is a vertical asymptote.

6 Find the domain of a logarithmic function.

The Domain of a Logarithmic Function

In Section 12.1, we learned that the domain of an exponential function of the form $f(x) = b^x$ includes all real numbers and its range is the set of positive real numbers. Because the logarithmic function reverses the domain and the range of the exponential function, the **domain of a logarithmic function of the form $f(x) = \log_b x$ is the set of all positive real numbers.** Thus, $\log_2 8$ is defined because the value of x in the logarithmic expression, 8, is greater than zero and therefore is included in the domain of the logarithmic function $f(x) = \log_2 x$. However, $\log_2 0$ and $\log_2 (-8)$ are not defined because 0 and -8 are not positive real numbers and therefore are excluded from the domain of the logarithmic function $f(x) = \log_2 x$. In general, **the domain of $f(x) = \log_b g(x)$ consists of all x for which $g(x) > 0$.**

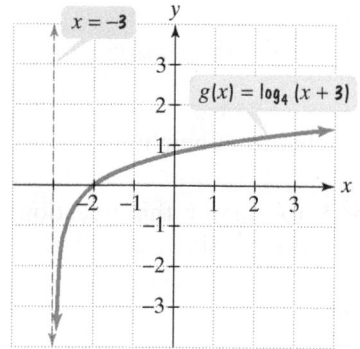

Figure 12.11 The domain of $g(x) = \log_4(x + 3)$ is $(-3, \infty)$.

EXAMPLE 7 Finding the Domain of a Logarithmic Function

Find the domain of $g(x) = \log_4 (x + 3)$.

Solution The domain of g consists of all x for which $x + 3 > 0$. Solving this inequality for x, we obtain $x > -3$. Thus, the domain of g is $(-3, \infty)$. This is illustrated in **Figure 12.11**. The vertical asymptote is $x = -3$ and all points on the graph of g have x-coordinates that are greater than -3. ∎

✓ **CHECK POINT 7** Find the domain of $h(x) = \log_4 (x - 5)$.

7 Use common logarithms.

Common Logarithms

The logarithmic function with base 10 is called the **common logarithmic function.** The function $f(x) = \log_{10} x$ is usually expressed as $f(x) = \log x$. A calculator with a $\boxed{\text{LOG}}$ key can be used to evaluate common logarithms. On the next page are some examples.

Logarithm	Most Scientific Calculator Keystrokes	Most Graphing Calculator Keystrokes	Display (or Approximate Display)
log 1000	1000 LOG	LOG 1000 ENTER	3
$\log \dfrac{5}{2}$	(5 ÷ 2) LOG	LOG (5 ÷ 2) ENTER	0.39794
$\dfrac{\log 5}{\log 2}$	5 LOG ÷ 2 LOG =	LOG 5 ÷ LOG 2 ENTER	2.32193
log(−3)	3 +/− LOG	LOG (−) 3 ENTER	ERROR

Some graphing calculators display an open parenthesis when the LOG key is pressed. In this case, remember to close the set of parentheses after entering the function's domain value: LOG 5) ÷ LOG 2) ENTER .

The error message or NONREAL ANS message given by many calculators for log(−3) is a reminder that the domain of the common logarithmic function, $f(x) = \log x$, is the set of positive real numbers. In general, the domain of $f(x) = \log g(x)$ consists of all x for which $g(x) > 0$.

Many real-life phenomena start with rapid growth and then the growth begins to level off. This type of behavior can be modeled by logarithmic functions.

EXAMPLE 8 Modeling Heights of Children

The percentage of adult height attained by a boy who is x years old can be modeled by

$$f(x) = 29 + 48.8 \log(x + 1),$$

where x represents the boy's age and $f(x)$ represents the percentage of his adult height. Approximately what percentage of his adult height has a boy attained at age eight?

Solution We substitute the boy's age, 8, for x and evaluate the function at 8.

$$f(x) = 29 + 48.8 \log(x + 1) \qquad \text{This is the given function.}$$
$$f(8) = 29 + 48.8 \log(8 + 1) \qquad \text{Substitute 8 for x.}$$
$$= 29 + 48.8 \log 9 \qquad \text{Graphing calculator keystrokes:}$$
$$\qquad\qquad\qquad\qquad\qquad\qquad 29\ +\ 48.8\ \text{LOG}\ 9\ \text{ENTER}$$

$$\approx 76$$

Thus, an 8-year-old boy has attained approximately 76% of his adult height. ■

✓ **CHECK POINT 8** Use the function in Example 8 to answer this question: Approximately what percentage of his adult height has a boy attained at age ten?

The basic properties of logarithms that were listed earlier in this section can be applied to common logarithms.

Properties of Common Logarithms

General Properties		Common Logarithms
1. $\log_b 1 = 0$		1. $\log 1 = 0$
2. $\log_b b = 1$		2. $\log 10 = 1$
3. $\log_b b^x = x$	Inverse properties	3. $\log 10^x = x$
4. $b^{\log_b x} = x$		4. $10^{\log x} = x$

The property $\log 10^x = x$ can be used to evaluate common logarithms involving powers of 10. For example,

$$\log 100 = \log 10^2 = 2, \quad \log 1000 = \log 10^3 = 3, \quad \text{and} \quad \log 10^{7.1} = 7.1.$$

EXAMPLE 9 Earthquake Intensity

The magnitude, R, on the Richter scale of an earthquake of intensity I is given by

$$R = \log \frac{I}{I_0},$$

where I_0 is the intensity of a barely felt zero-level earthquake. The earthquake that destroyed San Francisco in 1906 was $10^{8.3}$ times as intense as a zero-level earthquake. What was its magnitude on the Richter scale?

Solution Because the earthquake was $10^{8.3}$ times as intense as a zero-level earthquake, the intensity, I, is $10^{8.3} I_0$.

$$R = \log \frac{I}{I_0} \qquad \text{This is the formula for magnitude on the Richter scale.}$$

$$R = \log \frac{10^{8.3} I_0}{I_0} \qquad \text{Substitute } 10^{8.3} I_0 \text{ for } I.$$

$$= \log 10^{8.3} \qquad \text{Simplify.}$$

$$= 8.3 \qquad \text{Use the property } \log 10^x = x.$$

San Francisco's 1906 earthquake registered 8.3 on the Richter scale. ∎

✓ **CHECK POINT 9** Use the formula in Example 9 to solve this problem. If an earthquake is 10,000 times as intense as a zero-level quake ($I = 10,000 I_0$), what is its magnitude on the Richter scale?

---

8 Use natural logarithms.

Natural Logarithms

The logarithmic function with base e is called the **natural logarithmic function**. The function $f(x) = \log_e x$ is usually expressed as $f(x) = \ln x$, read "el en of x." A calculator with an $\boxed{\text{LN}}$ key can be used to evaluate natural logarithms. Keystrokes are identical to those shown for common logarithmic evaluations on page 696.

Like the domain of all logarithmic functions, the domain of the natural logarithmic function $f(x) = \ln x$ is the set of all positive real numbers. Thus, the domain of $f(x) = \ln g(x)$ consists of all x for which $g(x) > 0$.

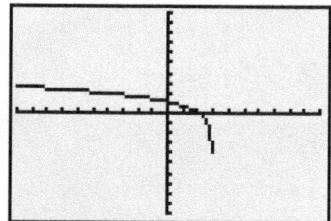

$[-10, 10, 1]$ by $[-10, 10, 1]$

Figure 12.12 The domain of $f(x) = \ln(3 - x)$ is $(-\infty, 3)$.

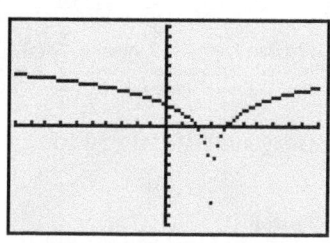

$[-10, 10, 1]$ by $[-10, 10, 1]$

Figure 12.13 The value 3 is excluded from the domain of $g(x) = \ln(x - 3)^2$.

EXAMPLE 10 Finding Domains of Natural Logarithmic Functions

Find the domain of each function:

a. $f(x) = \ln(3 - x)$ **b.** $g(x) = \ln(x - 3)^2$.

Solution

a. The domain of $f(x) = \ln(3 - x)$ consists of all x for which $3 - x > 0$. Solving this inequality for x, we obtain $x < 3$. Thus, the domain of f is $(-\infty, 3)$. This is verified by the graph in **Figure 12.12**.

b. The domain of $g(x) = \ln(x - 3)^2$ consists of all x for which $(x - 3)^2 > 0$. It follows that the domain of g is all real numbers except 3. Thus, the domain of g is $(-\infty, 3) \cup (3, \infty)$. This is shown by the graph in **Figure 12.13**. To make it more obvious that 3 is excluded from the domain, we used a $\boxed{\text{DOT}}$ format. ∎

✓ **CHECK POINT 10** Find the domain of each function:

a. $f(x) = \ln(4 - x)$

b. $g(x) = \ln x^2$.

The basic properties of logarithms that were listed earlier in this section can be applied to natural logarithms.

Properties of Natural Logarithms

General Properties		Natural Logarithms
1. $\log_b 1 = 0$		1. $\ln 1 = 0$
2. $\log_b b = 1$		2. $\ln e = 1$
3. $\log_b b^x = x$	Inverse properties	3. $\ln e^x = x$
4. $b^{\log_b x} = x$		4. $e^{\ln x} = x$

Examine the inverse properties, $\ln e^x = x$ and $e^{\ln x} = x$. Can you see how ln and e "undo" one another? For example,

$$\ln e^2 = 2, \quad \ln e^{7x^2} = 7x^2, \quad e^{\ln 2} = 2, \quad \text{and} \quad e^{\ln 7x^2} = 7x^2.$$

EXAMPLE 11 Dangerous Heat: Temperature in an Enclosed Vehicle

When the outside air temperature is anywhere from 72° to 96° Fahrenheit, the temperature in an enclosed vehicle climbs by 43° in the first hour. The bar graph in **Figure 12.14** shows the temperature increase throughout the hour. The function

$$f(x) = 13.4 \ln x - 11.6$$

models the temperature increase, $f(x)$, in degrees Fahrenheit, after x minutes. Use the function to find the temperature increase, to the nearest degree, after 50 minutes. How well does the function model the actual increase shown in **Figure 12.14**?

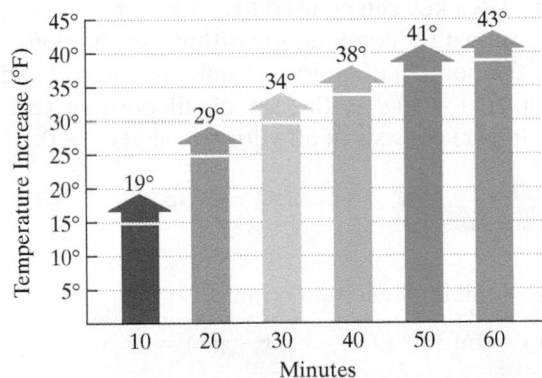

Temperature Increase in an Enclosed Vehicle

Figure 12.14

Source: Lynn I Gibbs and David W. Lawrence, "Heat Exposure in an Enclosed Automobile." *Journal of the Lousiana State Medical Society*, Volume 147(12), 1995

Solution We find the temperature increase after 50 minutes by substituting 50 for x and evaluating the function at 50.

$$f(x) = 13.4 \ln x - 11.6 \qquad \text{This is the given function.}$$

$$f(50) = 13.4 \ln 50 - 11.6 \qquad \text{Substitute 50 for x.}$$

$$\approx 41 \qquad \text{Graphing calculator keystrokes:}$$

13.4 [ln] 50 [−] 11.6 [ENTER]. On some calculators, a parenthesis is needed after 50.

According to the function, the temperature will increase by approximately 41° after 50 minutes. Because the increase shown in **Figure 12.14** is 41°, the function models the actual increase extremely well. ∎

✓ **CHECK POINT 11** Use the function in Example 11 to find the temperature increase, to the nearest degree, after 30 minutes. How well does the function model the actual increase shown in **Figure 12.14**?

Achieving Success

As you get ready to move on to higher-level math courses, this might be a good time to gauge your level of math anxiety. The following questionnaire can be used to get an idea of your math anxiety level.

Are You Anxious About Math?

Rate each of the following statements on a scale of 1 (strongly disagree) to 5 (strongly agree).

1. ————— I cringe when I have to go to math class.
2. ————— I would be uneasy about going to the board in math class and presenting my solution to a homework problem.
3. ————— I am afraid to ask questions in math class.
4. ————— I would be worried if my professor called on me in math class.
5. ————— I understand math now, but I worry that it's going to get really difficult in higher-level math courses.
6. ————— I tend to zone out during math lecture.
7. ————— I fear math tests more than any other kind.
8. ————— I don't know how to study for math tests.
9. ————— Things are clear in math class, but when I get home and read my notes, it's like I was never there.
10. ————— I'm afraid I won't be able to keep up with other students in higher-level math courses.

Scoring Key:

40–50	Sure thing, you have math anxiety.
30–39	You're still somewhat fearful about math.
20–29	You're on the fence.
10–19	Wow! Loose as a goose!

Source: Ellen Freedman, *Do You Have Math Anxiety? A Self-Test* (mathpower.com/anxtest). Reprinted with permission from Ellen Freedman

I hope that you've enjoyed this textbook and your math course, and that you feel confident in your ability to succeed in forthcoming higher-level mathematics.

CONCEPT AND VOCABULARY CHECK

Fill in each blank so that the resulting statement is true.

1. $y = \log_b x$ is equivalent to the exponential form ————————, $x > 0, b > 0, b \neq 1$.

2. The function $f(x) = \log_b x$ is the ————————— function with base —————————.

3. $\log_b b = $ —————————

4. $\log_b 1 = $ —————————

5. $\log_b b^x = $ —————————

6. $b^{\log_b x} = $ —————————

7. Using interval notation, the domain of $f(x) = \log_b x$ is ————————— and the range is —————————.

8. The graph of $f(x) = \log_b x$ approaches, but does not touch, the —————————-axis. This axis, whose equation is —————————, is a/an ————————— asymptote.

9. The domain of $g(x) = \log_2 (5 - x)$ can be found by solving the inequality —————————.

10. The logarithmic function with base 10 is called the ————————— logarithmic function. The function $f(x) = \log_{10} x$ is usually expressed as $f(x) = $ —————————.

11. The logarithmic function with base e is called the ————————— logarithmic function. The function $f(x) = \log_e x$ is usually expressed as $f(x) = $ —————————.

12.2 EXERCISE SET   MyMathLab®

Watch the videos
in MyMathLab

Download the
MyDashBoard App

Practice Exercises

In Exercises 1–8, write each equation in its equivalent exponential form.

1. $4 = \log_2 16$

2. $6 = \log_2 64$

3. $2 = \log_3 x$

4. $2 = \log_9 x$

5. $5 = \log_b 32$

6. $3 = \log_b 27$

7. $\log_6 216 = y$

8. $\log_5 125 = y$

In Exercises 9–20, write each equation in its equivalent logarithmic form.

9. $2^3 = 8$

10. $5^4 = 625$

11. $2^{-4} = \frac{1}{16}$

12. $5^{-3} = \frac{1}{125}$

13. $\sqrt[3]{8} = 2$

14. $\sqrt[3]{64} = 4$

15. $13^2 = x$

16. $15^2 = x$

17. $b^3 = 1000$

18. $b^3 = 343$

19. $7^y = 200$

20. $8^y = 300$

In Exercises 21–42, evaluate each expression without using a calculator.

21. $\log_4 16$

22. $\log_7 49$

23. $\log_2 64$

24. $\log_3 27$

25. $\log_5 \frac{1}{5}$

26. $\log_6 \frac{1}{6}$

27. $\log_2 \frac{1}{8}$

28. $\log_3 \frac{1}{9}$

29. $\log_7 \sqrt{7}$

30. $\log_6 \sqrt{6}$

31. $\log_2 \frac{1}{\sqrt{2}}$

32. $\log_3 \frac{1}{\sqrt{3}}$

33. $\log_{64} 8$

34. $\log_{81} 9$

35. $\log_5 5$

36. $\log_{11} 11$

37. $\log_4 1$

38. $\log_6 1$

39. $\log_5 5^7$

40. $\log_4 4^6$

41. $8^{\log_8 19}$

42. $7^{\log_7 23}$

43. Graph $f(x) = 4^x$ and $g(x) = \log_4 x$ in the same rectangular coordinate system.

44. Graph $f(x) = 5^x$ and $g(x) = \log_5 x$ in the same rectangular coordinate system.

45. Graph $f(x) = \left(\frac{1}{2}\right)^x$ and $g(x) = \log_{\frac{1}{2}} x$ in the same rectangular coordinate system.

46. Graph $f(x) = \left(\frac{1}{4}\right)^x$ and $g(x) = \log_{\frac{1}{4}} x$ in the same rectangular coordinate system.

In Exercises 47–52, find the domain of each logarithmic function.

47. $f(x) = \log_5(x + 4)$

48. $f(x) = \log_5(x + 6)$

49. $f(x) = \log(2 - x)$

50. $f(x) = \log(7 - x)$

51. $f(x) = \ln(x - 2)^2$

52. $f(x) = \ln(x - 7)^2$

In Exercises 53–66, evaluate each expression without using a calculator.

53. $\log 100$

54. $\log 1000$

55. $\log 10^7$

56. $\log 10^8$

57. $10^{\log 33}$

58. $10^{\log 53}$

59. $\ln 1$

60. $\ln e$

61. $\ln e^6$

62. $\ln e^7$

63. $\ln \frac{1}{e^6}$

64. $\ln \frac{1}{e^7}$

65. $e^{\ln 125}$

66. $e^{\ln 300}$

In Exercises 67–72, simplify each expression.

67. $\ln e^{9x}$

68. $\ln e^{13x}$

69. $e^{\ln 5x^2}$

70. $e^{\ln 7x^2}$

71. $10^{\log \sqrt{x}}$

72. $10^{\log \sqrt[3]{x}}$

Practice PLUS

In Exercises 73–76, write each equation in its equivalent exponential form. Then solve for x.

73. $\log_3(x - 1) = 2$

74. $\log_5(x + 4) = 2$

75. $\log_4 x = -3$

76. $\log_{64} x = \frac{2}{3}$

In Exercises 77–80, evaluate each expression without using a calculator.

77. $\log_3(\log_7 7)$

78. $\log_5(\log_2 32)$

79. $\log_2(\log_3 81)$

80. $\log(\ln e)$

In Exercises 81–86, match each function with its graph. The graphs are labeled (a) through (f), and each graph is displayed in a $[-5, 5, 1]$ by $[-5, 5, 1]$ viewing rectangle.

81. $f(x) = \ln(x + 2)$

82. $f(x) = \ln(x - 2)$

83. $f(x) = \ln x + 2$

84. $f(x) = \ln x - 2$

85. $f(x) = \ln(1 - x)$

86. $f(x) = \ln(2 - x)$

a.

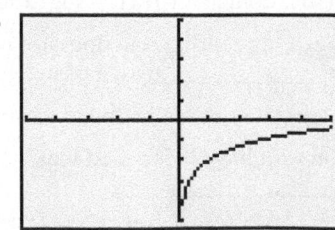

b.

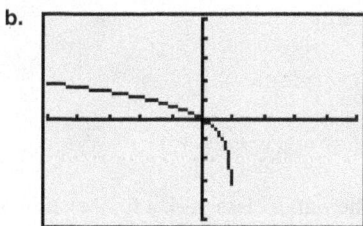

c.

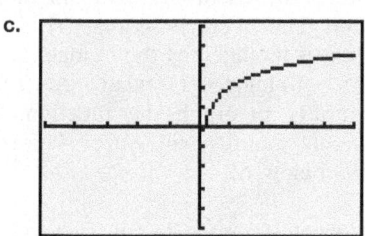

d.

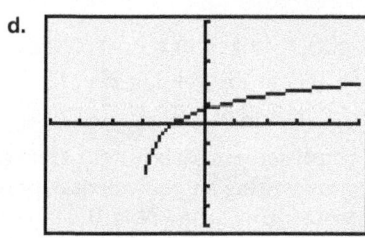

e.

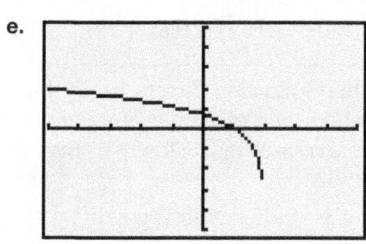

f.

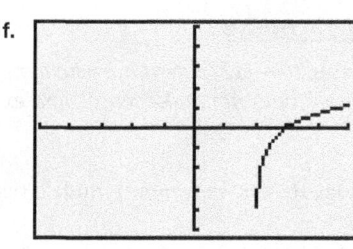

Application Exercises

The percentage of adult height attained by a girl who is x years old can be modeled by

$$f(x) = 62 + 35 \log(x - 4),$$

where x represents the girl's age (from 5 to 15) and f(x) represents the percentage of her adult height. Use the formula to solve Exercises 87–88. Round answers to the nearest tenth of a percent.

87. Approximately what percentage of her adult height has a girl attained at age 13?

88. Approximately what percentage of her adult height has a girl attained at age ten?

The bar graph indicates that the percentage of first-year college students expressing antifeminist views declined after 1970. Use this information to solve Exercises 89–90.

Opposition to Feminism among First-Year United States College Students, 1970–2008

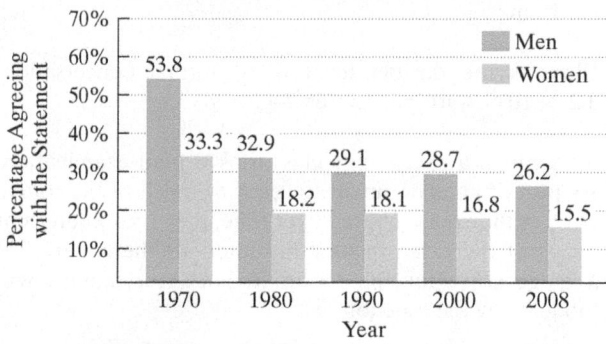

Source: John Macionis, *Sociology, Thirteenth Edition*, Prentice Hall, 2010.

89. The function

$$f(x) = -7.52 \ln x + 53$$

models the percentage of first-year college men, $f(x)$, expressing antifeminist views (by agreeing with the statement) x years after 1969.

  a. Use the function to find the percentage of first-year college men expressing antifeminist views in 2008. Round to one decimal place. Does this function value overestimate or underestimate the percentage displayed by the graph? By how much?

  b. Use the function to project the percentage of first-year college men who will express antifeminist views in 2015. Round to one decimal place.

90. The function

$$f(x) = -4.82 \ln x + 32.5$$

models the percentage of first-year college women, $f(x)$, expressing antifeminist views (by agreeing with the statement) x years after 1969.

  a. Use the function to find the percentage of first-year college women expressing antifeminist views in 2008. Round to one decimal place. Does this function value overestimate or underestimate the percentage displayed by the graph? By how much?

  b. Use the function to project the percentage of first-year college women who will express antifeminist views in 2015. Round to one decimal place.

The loudness level of a sound, D, in decibels, is given by the formula

$$D = 10 \log(10^{12} I),$$

where I is the intensity of the sound, in watts per meter². Decibel levels range from 0, a barely audible sound, to 160, a sound resulting in a ruptured eardrum. Use the formula to solve Exercises 91–92.

91. The sound of a blue whale can be heard 500 miles away, reaching an intensity of 6.3×10^6 watts per meter². Determine the decibel level of this sound. At close range, can the sound of a blue whale rupture the human eardrum?

92. What is the decibel level of a normal conversation, 3.2×10^{-6} watt per meter²?

93. Students in a psychology class took a final examination. As part of an experiment to see how much of the course content they remembered over time, they took equivalent forms of the exam in monthly intervals thereafter. The average score for the group, $f(t)$, after t months was modeled by the function

$$f(t) = 88 - 15 \ln(t + 1), \qquad 0 \le t \le 12.$$

   a. What was the average score on the original exam?

   b. What was the average score, to the nearest tenth, after 2 months? 4 months? 6 months? 8 months? 10 months? one year?

   c. Sketch the graph of f (either by hand or with a graphing utility). Describe what the graph indicates in terms of the material retained by the students.

Writing in Mathematics

94. Describe the relationship between an equation in logarithmic form and an equivalent equation in exponential form.

95. What question can be asked to help evaluate $\log_3 81$?

96. Explain why the logarithm of 1 with base b is 0.

97. Describe the following property using words: $\log_b b^x = x$.

98. Explain how to use the graph of $f(x) = 2^x$ to obtain the graph of $g(x) = \log_2 x$.

99. Explain how to find the domain of a logarithmic function.

100. Logarithmic models are well suited to phenomena in which growth is initially rapid, but then begins to level off. Describe something that is changing over time that can be modeled using a logarithmic function.

101. Suppose that a girl is 4 feet 6 inches at age 10. Explain how to use the function in Exercises 87–88 to determine how tall she can expect to be as an adult.

Technology Exercises

In Exercises 102–105, graph f and g in the same viewing rectangle. Then describe the relationship of the graph of g to the graph of f.

102. $f(x) = \ln x, g(x) = \ln(x + 3)$

103. $f(x) = \ln x, g(x) = \ln x + 3$

104. $f(x) = \log x, g(x) = -\log x$

105. $f(x) = \log x, g(x) = \log(x - 2) + 1$

106. Students in a mathematics class took a final examination. They took equivalent forms of the exam in monthly intervals thereafter. The average score, $f(t)$, for the group after t months is modeled by the human memory function $f(t) = 75 - 10 \log(t + 1)$, where $0 \le t \le 12$. Use a graphing utility to graph the function. Then determine how many months will elapse before the average score falls below 65.

107. In parts (a)–(c), graph f and g in the same viewing rectangle.

   a. $f(x) = \ln(3x), g(x) = \ln 3 + \ln x$

   b. $f(x) = \log(5x^2), g(x) = \log 5 + \log x^2$

   c. $f(x) = \ln(2x^3), g(x) = \ln 2 + \ln x^3$

   d. Describe what you observe in parts (a)–(c). Generalize this observation by writing an equivalent expression for $\log_b(MN)$, where $M > 0$ and $N > 0$.

   e. Complete this statement: The logarithm of a product is equal to _____.

108. Graph each of the following functions in the same viewing rectangle and then place the functions in order from the one that increases most slowly to the one that increases most rapidly.

$$y = x, y = \sqrt{x}, y = e^x, y = \ln x, y = x^x, y = x^2$$

Critical Thinking Exercises

Make Sense? *In Exercises 109–112, determine whether each statement "makes sense" or "does not make sense" and explain your reasoning.*

109. I estimate that $\log_8 16$ lies between 1 and 2 because $8^1 = 8$ and $8^2 = 64$.

110. When graphing a logarithmic function, I like to show the graph of its horizontal asymptote.

111. I can evaluate some common logarithms without having to use a calculator.

112. An earthquake of magnitude 8 on the Richter scale is twice as intense as an earthquake of magnitude 4.

In Exercises 113–116, determine whether each statement is true or false. If the statement is false, make the necessary change(s) to produce a true statement.

113. $\dfrac{\log_2 8}{\log_2 4} = \dfrac{8}{4}$

114. $\log(-100) = -2$

115. The domain of $f(x) = \log_2 x$ is $(-\infty, \infty)$.

116. $\log_b x$ is the exponent to which b must be raised to obtain x.

117. Without using a calculator, find the exact value of
$$\frac{\log_3 81 - \log_\pi 1}{\log_{2\sqrt{2}} 8 - \log 0.001}.$$

118. Without using a calculator, find the exact value of $\log_4[\log_3(\log_2 8)]$.

119. Without using a calculator, determine which is the greater number: $\log_4 60$ or $\log_3 40$.

Review Exercises

120. Solve the system:
$$\begin{cases} 2x = 11 - 5y \\ 3x - 2y = -12. \end{cases}$$
(Section 4.3, Example 4)

121. Factor completely:
$$6x^2 - 8xy + 2y^2.$$
(Section 6.5, Example 8)

122. Solve: $x + 3 \le -4$ or $2 - 7x \le 16$.
(Section 9.2, Example 6)

Preview Exercises

Exercises 123–125 will help you prepare for the material covered in the next section. In each exercise, evaluate the indicated logarithmic expressions without using a calculator.

123. **a.** Evaluate: $\log_2 32$.
 b. Evaluate: $\log_2 8 + \log_2 4$.
 c. What can you conclude about $\log_2 32$, or $\log_2(8 \cdot 4)$?

124. **a.** Evaluate: $\log_2 16$.
 b. Evaluate: $\log_2 32 - \log_2 2$.
 c. What can you conclude about
$$\log_2 16, \text{ or } \log_2\left(\frac{32}{2}\right)?$$

125. **a.** Evaluate: $\log_3 81$.
 b. Evaluate: $2\log_3 9$.
 c. What can you conclude about
$$\log_3 81, \text{ or } \log_3 9^2?$$

12.3

Properties of Logarithms

Objectives

1 Use the product rule.

2 Use the quotient rule.

3 Use the power rule.

4 Expand logarithmic expressions.

5 Condense logarithmic expressions.

6 Use the change-of-base property.

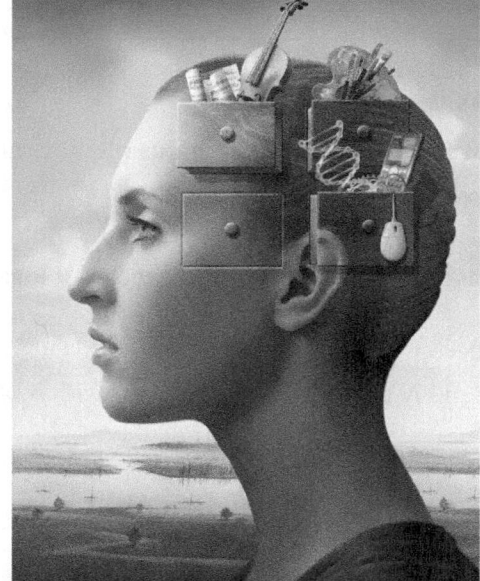

We all learn new things in different ways. In this section, we consider important properties of logarithms. What would be the most effective way for you to learn these properties? Would it be helpful to use your graphing utility and discover one of these properties for yourself? To do so, work Exercise 107 in Exercise Set 12.2 before continuing. Would it be helpful to evaluate certain logarithmic expressions that suggest three of the properties? If this is the case, work Preview Exercises 123–125 in Exercise Set 12.2 before continuing. Would the properties become more meaningful if you could see exactly where they come from? If so, you will find details of the proofs of many of these properties in Appendix D. The remainder of our work in this chapter will be based on the properties of logarithms that you learn in this section.

1 Use the product rule.

The Product Rule

Properties of exponents correspond to properties of logarithms. For example, when we multiply exponential expressions with the same base, we add exponents:
$$b^m \cdot b^n = b^{m+n}.$$

This property of exponents, coupled with an awareness that a logarithm is an exponent, suggests the following property, called the **product rule**:

The Product Rule

Let b, M, and N be positive real numbers with $b \neq 1$.

$$\log_b(MN) = \log_b M + \log_b N$$

The logarithm of a product is the sum of the logarithms.

When we use the product rule to write a single logarithm as the sum of two logarithms, we say that we are **expanding a logarithmic expression**. For example, we can use the product rule to expand $\ln(7x)$:

$$\ln(7x) = \ln 7 + \ln x.$$

| The logarithm of a product | is | the sum of the logarithms. |

EXAMPLE 1 Using the Product Rule

Use the product rule to expand each logarithmic expression:

a. $\log_4(7 \cdot 5)$ **b.** $\log(10x)$.

Solution

a. $\log_4(7 \cdot 5) = \log_4 7 + \log_4 5$ The logarithm of a product is the sum of the logarithms.

b. $\log(10x) = \log 10 + \log x$ The logarithm of a product is the sum of the logarithms. These are common logarithms with base 10 understood.

$\qquad\qquad = 1 + \log x$ Because $\log_b b = 1$, then $\log 10 = 1$. ∎

☑ **CHECK POINT 1** Use the product rule to expand each logarithmic expression:
a. $\log_6(7 \cdot 11)$ **b.** $\log(100x)$.

---

2 Use the quotient rule.

The Quotient Rule

When we divide exponential expressions with the same base, we subtract exponents:

$$\frac{b^m}{b^n} = b^{m-n}.$$

This property suggests the following property of logarithms, called the **quotient rule**:

The Quotient Rule

Let b, M, and N be positive real numbers with $b \neq 1$.

$$\log_b\left(\frac{M}{N}\right) = \log_b M - \log_b N$$

The logarithm of a quotient is the difference of the logarithms.

When we use the quotient rule to write a single logarithm as the difference of two logarithms, we say that we are **expanding a logarithmic expression**. For example, we can use the quotient rule to expand $\log \frac{x}{2}$:

$$\log\left(\frac{x}{2}\right) = \log x - \log 2.$$

| The logarithm of a quotient | is | the difference of the logarithms. |

EXAMPLE 2   Using the Quotient Rule

Use the quotient rule to expand each logarithmic expression:

a. $\log_7\left(\dfrac{19}{x}\right)$ **b.** $\ln\left(\dfrac{e^3}{7}\right)$.

Solution

a. $\log_7\left(\dfrac{19}{x}\right) = \log_7 19 - \log_7 x$ The logarithm of a quotient is the difference of the logarithms.

b. $\ln\left(\dfrac{e^3}{7}\right) = \ln e^3 - \ln 7$ The logarithm of a quotient is the difference of the logarithms. These are natural logarithms with base e understood.

$= 3 - \ln 7$ Because $\ln e^x = x$, then $\ln e^3 = 3$. ∎

☑ **CHECK POINT 2**   Use the quotient rule to expand each logarithmic expression:

a. $\log_8\left(\dfrac{23}{x}\right)$ **b.** $\ln\left(\dfrac{e^5}{11}\right)$.

3 Use the power rule.

The Power Rule

When an exponential expression is raised to a power, we multiply exponents:

$$(b^m)^n = b^{mn}.$$

This property suggests the following property of logarithms, called the **power rule**:

The Power Rule

Let b and M be positive real numbers with $b \neq 1$, and let p be any real number.

$$\log_b M^p = p \log_b M$$

The logarithm of a number with an exponent is the product of the exponent and the logarithm of that number.

When we use the power rule to "pull the exponent to the front," we say that we are **expanding a logarithmic expression**. For example, we can use the power rule to expand $\ln x^2$:

$$\ln x^2 = 2 \ln x.$$

The logarithm of a number with an exponent | is | the product of the exponent and the logarithm of that number.

Figure 12.15 shows the graphs of $y = \ln x^2$ and $y = 2 \ln x$ in $[-5, 5, 1]$ by $[-5, 5, 1]$ viewing rectangles. Are $\ln x^2$ and $2 \ln x$ the same? The graphs illustrate that $y = \ln x^2$ and $y = 2 \ln x$ have different domains. The graphs are only the same if $x > 0$. Thus, we should write

$$\ln x^2 = 2 \ln x \quad \text{for} \quad x > 0.$$

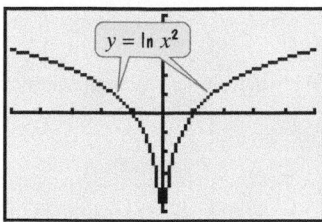

Domain: $(-\infty, 0) \cup (0, \infty)$

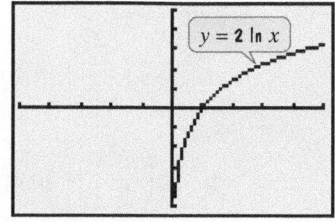

Domain: $(0, \infty)$

Figure 12.15
$\ln x^2$ and $2 \ln x$ have different domains.

When expanding a logarithmic expression, you might want to determine whether the rewriting has changed the domain of the expression. For the rest of this section, assume that all variables and variable expressions represent positive numbers.

4 Expand logarithmic expressions.

EXAMPLE 3 Using the Power Rule

Use the power rule to expand each logarithmic expression:

a. $\log_5 7^4$ **b.** $\ln \sqrt{x}$ **c.** $\log(4x)^5$.

Solution

a. $\log_5 7^4 = 4 \log_5 7$ The logarithm of a number with an exponent is the exponent times the logarithm of that number.

b. $\ln \sqrt{x} = \ln x^{\frac{1}{2}}$ Rewrite the radical using a rational exponent.

     $= \dfrac{1}{2} \ln x$ Use the power rule to bring the exponent to the front.

c. $\log(4x)^5 = 5 \log(4x)$ We immediately apply the power rule because the entire variable expression, 4x, is raised to the 5th power. ∎

 CHECK POINT 3   Use the power rule to expand each logarithmic expression:

    a. $\log_6 8^9$ **b.** $\ln \sqrt[3]{x}$ **c.** $\log(x + 4)^2$.

Expanding Logarithmic Expressions

It is sometimes necessary to use more than one property of logarithms when you expand a logarithmic expression. Properties for expanding logarithmic expressions are as follows:

Properties for Expanding Logarithmic Expressions

For $M > 0$ and $N > 0$:

1. $\log_b(MN) = \log_b M + \log_b N$ Product rule

2. $\log_b\left(\dfrac{M}{N}\right) = \log_b M - \log_b N$ Quotient rule

3. $\log_b M^p = p \log_b M$ Power rule

Great Question!

Are there some common screw-ups that I can avoid when using properties of logarithms?

The graphs show

$$y_1 = \ln(x + 3) \quad \text{and} \quad y_2 = \ln x + \ln 3.$$

The graphs are not the same:

$$\ln(x + 3) \neq \ln x + \ln 3.$$

In general,

$$\log_b(M + N) \neq \log_b M + \log_b N.$$

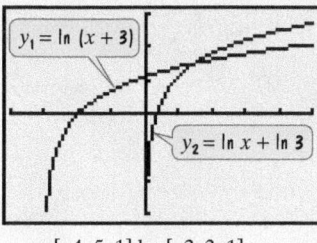

$y_1 = \ln(x + 3)$

$y_2 = \ln x + \ln 3$

$[-4, 5, 1]$ by $[-3, 3, 1]$

Try to avoid the following errors:

Incorrect!

$$\log_b(M + N) = \log_b M + \log_b N$$

$$\log_b(M - N) = \log_b M - \log_b N$$

$$\log_b(M \cdot N) = \log_b M \cdot \log_b N$$

$$\log_b\left(\frac{M}{N}\right) = \frac{\log_b M}{\log_b N}$$

$$\frac{\log_b M}{\log_b N} = \log_b M - \log_b N$$

$$\log_b(MN^p) = p \log_b(MN)$$

EXAMPLE 4 Expanding Logarithmic Expressions

Use logarithmic properties to expand each expression as much as possible:

a. $\log_b\left(x^2\sqrt{y}\right)$

b. $\log_6\left(\dfrac{\sqrt[3]{x}}{36y^4}\right)$.

Solution We will have to use two or more of the properties for expanding logarithms in each part of this example.

a.
$$\log_b\left(x^2\sqrt{y}\right) = \log_b\left(x^2 y^{\frac{1}{2}}\right) \qquad \text{Use exponential notation.}$$

$$= \log_b x^2 + \log_b y^{\frac{1}{2}} \qquad \text{Use the product rule.}$$

$$= 2\log_b x + \frac{1}{2}\log_b y \qquad \text{Use the power rule.}$$

b.
$$\log_6\left(\frac{\sqrt[3]{x}}{36y^4}\right) = \log_6\left(\frac{x^{\frac{1}{3}}}{36y^4}\right) \qquad \text{Use exponential notation.}$$

$$= \log_6 x^{\frac{1}{3}} - \log_6(36y^4) \qquad \text{Use the quotient rule.}$$

$$= \log_6 x^{\frac{1}{3}} - \left(\log_6 36 + \log_6 y^4\right) \qquad \text{Use the product rule on } \log_6(36y^4).$$

$$= \frac{1}{3}\log_6 x - (\log_6 36 + 4\log_6 y) \qquad \text{Use the power rule.}$$

$$= \frac{1}{3}\log_6 x - \log_6 36 - 4\log_6 y \qquad \text{Apply the distributive property.}$$

$$= \frac{1}{3}\log_6 x - 2 - 4\log_6 y \qquad \begin{array}{l}\log_6 36 = 2 \text{ because 2 is the power} \\ \text{to which we must raise 6 to get 36.} \\ (6^2 = 36)\end{array} \quad \blacksquare$$

✓ **CHECK POINT 4** Use logarithmic properties to expand each expression as much as possible:

a. $\log_b\left(x^4\sqrt[3]{y}\right)$

b. $\log_5\left(\dfrac{\sqrt{x}}{25y^3}\right)$.

⑤ Condense logarithmic expressions.

Condensing Logarithmic Expressions

To **condense a logarithmic expression**, we write the sum or difference of two or more logarithmic expressions as a single logarithmic expression. We use the properties of logarithms to do so:

Great Question!

Are the properties listed on the right the same as those in the box on page 886?

Yes. The only difference is that we've reversed the sides in each property from the previous box.

Properties for Condensing Logarithmic Expressions

For $M > 0$ and $N > 0$:

1. $\log_b M + \log_b N = \log_b(MN)$ Product rule

2. $\log_b M - \log_b N = \log_b\left(\dfrac{M}{N}\right)$ Quotient rule

3. $p \log_b M = \log_b M^p$ Power rule

EXAMPLE 5   Condensing Logarithmic Expressions

Write as a single logarithm:

a. $\log_4 2 + \log_4 32$

b. $\log(4x - 3) - \log x.$

Solution

a. $\log_4 2 + \log_4 32 = \log_4(2 \cdot 32)$ Use the product rule.

$\qquad\qquad\qquad\quad = \log_4 64$ We now have a single logarithm. However, we can simplify.

$\qquad\qquad\qquad\quad = 3$ $\log_4 64 = 3$ because $4^3 = 64$.

b. $\log(4x - 3) - \log x = \log\left(\dfrac{4x - 3}{x}\right)$ Use the quotient rule. ■

✓ **CHECK POINT 5**   Write as a single logarithm:

a. $\log 25 + \log 4$ **b.** $\log(7x + 6) - \log x.$

Coefficients of logarithms must be 1 before you can condense them using the product and quotient rules. For example, to condense

$$2 \ln x + \ln(x + 1),$$

the coefficient of the first term must be 1. We use the power rule to rewrite the coefficient as an exponent:

1. Use the power rule to make the number in front an exponent.

$$2 \ln x + \ln(x + 1) = \ln x^2 + \ln(x + 1) = \ln[x^2(x + 1)].$$

2. Use the product rule. The sum of logarithms with coefficients of 1 is the logarithm of the product.

EXAMPLE 6   Condensing Logarithmic Expressions

Write as a single logarithm:

a. $\dfrac{1}{2}\log x + 4 \log(x - 1)$ **b.** $3 \ln(x + 7) - \ln x$

c. $4 \log_b x - 2 \log_b 6 - \frac{1}{2}\log_b y.$

Solution

a. $\frac{1}{2}\log x + 4\log(x - 1)$

$= \log x^{\frac{1}{2}} + \log(x - 1)^4$ Use the power rule so that all coefficients are 1.

$= \log\left[x^{\frac{1}{2}}(x - 1)^4\right]$ Use the product rule. The condensed form can be expressed as $\log\left[\sqrt{x}\,(x - 1)^4\right]$.

b. $3\ln(x + 7) - \ln x$

$= \ln(x + 7)^3 - \ln x$ Use the power rule so that all coefficients are 1.

$= \ln\left[\dfrac{(x + 7)^3}{x}\right]$ Use the quotient rule.

c. $4\log_b x - 2\log_b 6 - \frac{1}{2}\log_b y$

$= \log_b x^4 - \log_b 6^2 - \log_b y^{\frac{1}{2}}$ Use the power rule so that all coefficients are 1.

$= \log_b x^4 - \left(\log_b 36 + \log_b y^{\frac{1}{2}}\right)$ Rewrite as a single subtraction.

$= \log_b x^4 - \log_b\left(36y^{\frac{1}{2}}\right)$ Use the product rule.

$= \log_b\left(\dfrac{x^4}{36y^{\frac{1}{2}}}\right)$ or $\log_b\left(\dfrac{x^4}{36\sqrt{y}}\right)$ Use the quotient rule.   ∎

✓ **CHECK POINT 6**   Write as a single logarithm:

a. $2\ln x + \dfrac{1}{3}\ln(x + 5)$ **b.** $2\log(x - 3) - \log x$

c. $\frac{1}{4}\log_b x - 2\log_b 5 - 10\log_b y$.

---

6 Use the change-of-base property.

The Change-of-Base Property

We have seen that calculators give the values of both common logarithms (base 10) and natural logarithms (base e). To find a logarithm with any other base, we can use the following change-of-base property:

The Change-of-Base Property

For any logarithmic bases a and b, and any positive number M,

$$\log_b M = \frac{\log_a M}{\log_a b}.$$

The logarithm of M with base b is equal to the logarithm of M with any new base divided by the logarithm of b with that new base.

In the change-of-base property, base b is the base of the original logarithm. Base a is a new base that we introduce. Thus, the change-of-base property allows us to change from base b to *any* new base a, as long as the newly introduced base is a positive number not equal to 1.

The change-of-base property is used to write a logarithm in terms of quantities that can be evaluated with a calculator. Because calculators contain keys for common (base 10) and natural (base e) logarithms, we will frequently introduce base 10 or base e.

Change-of-Base Property	Introducing Common Logarithms	Introducing Natural Logarithms
$\log_b M = \dfrac{\log_a M}{\log_a b}$	$\log_b M = \dfrac{\log_{10} M}{\log_{10} b}$	$\log_b M = \dfrac{\log_e M}{\log_e b}$
a is the new introduced base.	10 is the new introduced base.	*e* is the new introduced base.

Using the notations for common logarithms and natural logarithms, we have the following results:

The Change-of-Base Property: Introducing Common and Natural Logarithms

Introducing Common Logarithms

$$\log_b M = \frac{\log M}{\log b}$$

Introducing Natural Logarithms

$$\log_b M = \frac{\ln M}{\ln b}$$

EXAMPLE 7   Changing Base to Common Logarithms

Use common logarithms to evaluate $\log_5 140$.

Solution   Because $\log_b M = \dfrac{\log M}{\log b}$,

$$\log_5 140 = \frac{\log 140}{\log 5}$$

$$\approx 3.07.$$

Use a calculator: 140 [LOG] [÷] 5 [LOG] [=] or [LOG] 140 [÷] [LOG] 5 [ENTER]. On some calculators, parentheses are needed after 140 and 5.

This means that $\log_5 140 \approx 3.07$. ■

Discover for Yourself

Find a reasonable estimate of $\log_5 140$ to the nearest whole number: 5 to what power is 140? Compare your estimate to the value obtained in Example 7.

✓ **CHECK POINT 7**   Use common logarithms to evaluate $\log_7 2506$.

EXAMPLE 8   Changing Base to Natural Logarithms

Use natural logarithms to evaluate $\log_5 140$.

Solution   Because $\log_b M = \dfrac{\ln M}{\ln b}$,

$$\log_5 140 = \frac{\ln 140}{\ln 5}$$

$$\approx 3.07.$$

Use a calculator: 140 [LN] [÷] 5 [LN] [=] or [LN] 140 [÷] [LN] 5 [ENTER]. On some calculators, parentheses are needed after 140 and 5.

We have again shown that $\log_5 140 \approx 3.07$. ■

✓ **CHECK POINT 8**   Use natural logarithms to evaluate $\log_7 2506$.

Using Technology

We can use the change-of-base property to graph logarithmic functions with bases other than 10 or e on a graphing utility. For example, **Figure 12.16** shows the graphs of

$$y = \log_2 x \quad \text{and} \quad y = \log_{20} x$$

in a $[0, 10, 1]$ by $[-3, 3, 1]$ viewing rectangle. Because $\log_2 x = \dfrac{\ln x}{\ln 2}$ and $\log_{20} x = \dfrac{\ln x}{\ln 20}$, the functions are entered as

$$y_1 = \boxed{\text{LN}}\, x \;\boxed{\div}\; \boxed{\text{LN}}\, 2,$$
$$\text{and} \quad y_2 = \boxed{\text{LN}}\, x \;\boxed{\div}\; \boxed{\text{LN}}\, 20.$$

On some calculators, parentheses are needed after x, **2**, and **20**.

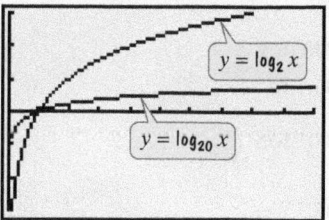

Figure 12.16 Using the change-of-base property to graph logarithmic functions

---

CONCEPT AND VOCABULARY CHECK

Fill in each blank so that the resulting statement is true.

1. The product rule for logarithms states that $\log_b(MN) =$ _____. The logarithm of a product is the _____ of the logarithms.

2. The quotient rule for logarithms states that $\log_b\!\left(\dfrac{M}{N}\right) =$ _____. The logarithm of a quotient is the _____ of the logarithms.

3. The power rule for logarithms states that $\log_b M^p =$ _____. The logarithm of a number with an exponent is the _____ of the exponent and the logarithm of that number.

4. The change-of-base property for logarithms allows us to write logarithms with base b in terms of a new base a. Introducing base a, the property states that

$$\log_b M = \frac{\quad\quad}{\quad\quad}.$$

12.3 EXERCISE SET

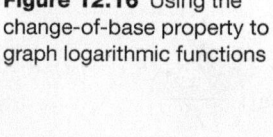

Watch the videos in MyMathLab   Download the MyDashBoard App

Practice Exercises

In all exercises, assume that all variables and variable expressions represent positive numbers.

In Exercises 1–36, use properties of logarithms to expand each logarithmic expression as much as possible. Where possible, evaluate logarithmic expressions without using a calculator.

1. $\log_5(7 \cdot 3)$

2. $\log_8(13 \cdot 7)$

3. $\log_7(7x)$

4. $\log_9(9x)$

5. $\log(1000x)$

6. $\log(10{,}000x)$

7. $\log_7\!\left(\dfrac{7}{x}\right)$

8. $\log_9\!\left(\dfrac{9}{x}\right)$

9. $\log\!\left(\dfrac{x}{100}\right)$

10. $\log\!\left(\dfrac{x}{1000}\right)$

11. $\log_4\!\left(\dfrac{64}{y}\right)$

12. $\log_5\!\left(\dfrac{125}{y}\right)$

13. $\ln\!\left(\dfrac{e^2}{5}\right)$

14. $\ln\!\left(\dfrac{e^4}{8}\right)$

15. $\log_b x^3$

16. $\log_b x^7$

17. $\log N^{-6}$

18. $\log M^{-8}$

19. $\ln \sqrt[5]{x}$

20. $\ln \sqrt[7]{x}$

21. $\log_b(x^2 y)$

22. $\log_b(xy^3)$

23. $\log_4\!\left(\dfrac{\sqrt{x}}{64}\right)$

24. $\log_5\!\left(\dfrac{\sqrt{x}}{25}\right)$

25. $\log_6\!\left(\dfrac{36}{\sqrt{x+1}}\right)$

26. $\log_8\!\left(\dfrac{64}{\sqrt{x+1}}\right)$

27. $\log_b\!\left(\dfrac{x^2 y}{z^2}\right)$

28. $\log_b\left(\dfrac{x^3 y}{z^2}\right)$

29. $\log \sqrt{100x}$

30. $\ln \sqrt{ex}$

31. $\log \sqrt[3]{\dfrac{x}{y}}$

32. $\log \sqrt[5]{\dfrac{x}{y}}$

33. $\log_b\left(\dfrac{\sqrt{xy^3}}{z^3}\right)$

34. $\log_b\left(\dfrac{\sqrt[3]{xy^4}}{z^5}\right)$

35. $\log_5 \sqrt[3]{\dfrac{x^2 y}{25}}$

36. $\log_2 \sqrt[5]{\dfrac{xy^4}{16}}$

In Exercises 37–60, use properties of logarithms to condense each logarithmic expression. Write the expression as a single logarithm whose coefficient is 1. Where possible, evaluate logarithmic expressions.

37. $\log 5 + \log 2$

38. $\log 250 + \log 4$

39. $\ln x + \ln 7$

40. $\ln x + \ln 3$

41. $\log_2 96 - \log_2 3$

42. $\log_3 405 - \log_3 5$

43. $\log(2x + 5) - \log x$

44. $\log(3x + 7) - \log x$

45. $\log x + 3 \log y$

46. $\log x + 7 \log y$

47. $\dfrac{1}{2}\ln x + \ln y$

48. $\dfrac{1}{3}\ln x + \ln y$

49. $2 \log_b x + 3 \log_b y$

50. $5 \log_b x + 6 \log_b y$

51. $5 \ln x - 2 \ln y$

52. $7 \ln x - 3 \ln y$

53. $3 \ln x - \dfrac{1}{3}\ln y$

54. $2 \ln x - \dfrac{1}{2}\ln y$

55. $4 \ln(x + 6) - 3 \ln x$

56. $8 \ln(x + 9) - 4 \ln x$

57. $3 \ln x + 5 \ln y - 6 \ln z$

58. $4 \ln x + 7 \ln y - 3 \ln z$

59. $\dfrac{1}{2}(\log_5 x + \log_5 y) - 2 \log_5(x + 1)$

60. $\dfrac{1}{3}(\log_4 x - \log_4 y) + 2 \log_4(x + 1)$

In Exercises 61–68, use common logarithms or natural logarithms and a calculator to evaluate to four decimal places.

61. $\log_5 13$

62. $\log_6 17$

63. $\log_{14} 87.5$

64. $\log_{16} 57.2$

65. $\log_{0.1} 17$

66. $\log_{0.3} 19$

67. $\log_\pi 63$

68. $\log_\pi 400$

Practice PLUS

In Exercises 69–74, let $\log_b 2 = A$ and $\log_b 3 = C$. Write each expression in terms of A and C.

69. $\log_b \dfrac{3}{2}$

70. $\log_b 6$

71. $\log_b 8$

72. $\log_b 81$

73. $\log_b \sqrt{\dfrac{2}{27}}$

74. $\log_b \sqrt{\dfrac{3}{16}}$

In Exercises 75–88, determine whether each equation is true or false. Where possible, show work to support your conclusion. If the statement is false, make the necessary change(s) to produce a true statement.

75. $\ln e = 0$

76. $\ln 0 = e$

77. $\log_4(2x^3) = 3 \log_4(2x)$

78. $\ln(8x^3) = 3 \ln(2x)$

79. $x \log 10^x = x^2$

80. $\ln(x + 1) = \ln x + \ln 1$

81. $\ln(5x) + \ln 1 = \ln(5x)$

82. $\ln x + \ln(2x) = \ln(3x)$

83. $\log(x + 3) - \log(2x) = \dfrac{\log(x + 3)}{\log(2x)}$

84. $\dfrac{\log(x + 2)}{\log(x - 1)} = \log(x + 2) - \log(x - 1)$

85. $\log_6\left(\dfrac{x - 1}{x^2 + 4}\right) = \log_6(x - 1) - \log_6(x^2 + 4)$

86. $\log_6[4(x + 1)] = \log_6 4 + \log_6(x + 1)$

87. $\log_3 7 = \dfrac{1}{\log_7 3}$

88. $e^x = \dfrac{1}{\ln x}$

In Exercises 89–92,

a. *Evaluate the expression in part (a) without using a calculator.*

b. *Use your result from part (a) to write the expression in part (b) as a single logarithm whose coefficient is 1.*

89. a. $\log_3 9$

    b. $\log_3 x + 4 \log_3 y - 2$

90. a. $\log_2 16$

    b. $\log_2 x + 5 \log_2 y - 4$

91. a. $\log_{25} 5$

    b. $\log_{25} x + \log_{25}(x^2 - 1) - \log_{25}(x + 1) - \dfrac{1}{2}$

92. a. $\log_{36} 6$

    b. $\log_{36} x + \log_{36}(x^2 - 4) - \log_{36}(x + 2) - \dfrac{1}{2}$

Application Exercises

93. The loudness level of a sound can be expressed by comparing the sound's intensity to the intensity of a sound barely audible to the human ear. The formula

$$D = 10(\log I - \log I_0)$$

describes the loudness level of a sound, D, in decibels, where I is the intensity of the sound, in watts per meter^2, and I_0 is the intensity of a sound barely audible to the human ear.

    a. Express the formula so that the expression in parentheses is written as a single logarithm.

    b. Use the form of the formula from part (a) to answer this question. If a sound has an intensity 100 times the intensity of a softer sound, how much larger on the decibel scale is the loudness level of the more intense sound?

94. The formula

$$t = \frac{1}{c}[\ln A - \ln(A - N)]$$

describes the time, t, in weeks, that it takes to achieve mastery of a portion of a task, where A is the maximum learning possible, N is the portion of the learning that is to be achieved, and c is a constant used to measure an individual's learning style.

    a. Express the formula so that the expression in brackets is written as a single logarithm.

    b. The formula is also used to determine how long it will take chimpanzees and apes to master a task. For example, a typical chimpanzee learning sign language can master a maximum of 65 signs. Use the form of the formula from part (a) to answer this question. How many weeks will it take a chimpanzee to master 30 signs if c for that chimp is 0.03?

Writing in Mathematics

95. Describe the product rule for logarithms and give an example.

96. Describe the quotient rule for logarithms and give an example.

97. Describe the power rule for logarithms and give an example.

98. Without showing the details, explain how to condense $\ln x - 2 \ln(x + 1)$.

99. Describe the change-of-base property and give an example.

100. Explain how to use your calculator to find $\log_{14} 283$.

101. You overhear a student talking about a property of logarithms in which division becomes subtraction. Explain what the student means by this.

102. Find $\ln 2$ using a calculator. Then calculate each of the following: $1 - \frac{1}{2}$; $\quad 1 - \frac{1}{2} + \frac{1}{3}$; $\quad 1 - \frac{1}{2} + \frac{1}{3} - \frac{1}{4}$; $1 - \frac{1}{2} + \frac{1}{3} - \frac{1}{4} + \frac{1}{5}$;.... Describe what you observe.

Technology Exercises

103. a. Use a graphing utility (and the change-of-base property) to graph $y = \log_3 x$.

    b. Graph $\quad y = 2 + \log_3 x, \quad y = \log_3(x + 2), \quad$ and $y = -\log_3 x$ in the same viewing rectangle as $y = \log_3 x$. Then describe the change or changes that need to be made to the graph of $y = \log_3 x$ to obtain each of these three graphs.

104. Graph $y = \log x$, $y = \log(10x)$, and $y = \log(0.1x)$ in the same viewing rectangle. Describe the relationship among the three graphs. What logarithmic property accounts for this relationship?

105. Use a graphing utility and the change-of-base property to graph $y = \log_3 x$, $y = \log_{25} x$, and $y = \log_{100} x$ in the same viewing rectangle.

    a. Which graph is on the top in the interval $(0, 1)$? Which is on the bottom?

    b. Which graph is on the top in the interval $(1, \infty)$? Which is on the bottom?

    c. Generalize by writing a statement about which graph is on top, which is on the bottom, and in which intervals, using $y = \log_b x$ where $b > 1$.

Disprove each statement in Exercises 106–110 by

  a. *letting y equal a positive constant of your choice, and*

  b. *using a graphing utility to graph the function on each side of the equal sign. The two functions should have different graphs, showing that the equation is not true in general.*

106. $\log(x + y) = \log x + \log y$

107. $\log \dfrac{x}{y} = \dfrac{\log x}{\log y}$

108. $\ln(x - y) = \ln x - \ln y$

109. $\ln(xy) = (\ln x)(\ln y)$

110. $\dfrac{\ln x}{\ln y} = \ln x - \ln y$

Critical Thinking Exercises

Make Sense? *In Exercises 111–114, determine whether each statement "makes sense" or "does not make sense" and explain your reasoning.*

111. Because I cannot simplify the expression $b^m + b^n$ by adding exponents, there is no property for the logarithm of a sum.

112. Because logarithms are exponents, the product, quotient, and power rules remind me of properties for operations with exponents.

113. I can use any positive number other than 1 in the change-of-base property, but the only practical bases are 10 and e because my calculator gives logarithms for these two bases.

114. I expanded $\log_4 \sqrt{\dfrac{x}{y}}$ by writing the radical using a rational exponent and then applying the quotient rule, obtaining $\dfrac{1}{2}\log_4 x - \log_4 y$.

In Exercises 115–118, determine whether each statement is true or false. If the statement is false, make the necessary change(s) to produce a true statement.

115. $\ln \sqrt{2} = \dfrac{\ln 2}{2}$

116. $\dfrac{\log_7 49}{\log_7 7} = \log_7 49 - \log_7 7$

117. $\log_b(x^3 + y^3) = 3\log_b x + 3\log_b y$

118. $\log_b(xy)^5 = (\log_b x + \log_b y)^5$

119. Use the change-of-base property to prove that
$$\log e = \frac{1}{\ln 10}.$$

120. If $\log 3 = A$ and $\log 7 = B$, find $\log_7 9$ in terms of A and B.

121. Write as a single term that does not contain a logarithm:
$$e^{\ln 8x^5 - \ln 2x^2}.$$

Review Exercises

122. Graph: $5x - 2y > 10$. (Section 9.4, Example 1)

123. Solve: $x - 2(3x - 2) > 2x - 3$. (Section 9.1, Example 2)

124. Divide and simplify: $\dfrac{\sqrt[3]{40x^2y^6}}{\sqrt[3]{5xy}}$. (Section 10.4, Example 5)

Preview Exercises

Exercises 125–127 will help you prepare for the material covered in the next section.

125. Simplify: $16^{\frac{3}{2}}$.

126. Evaluate $3\ln(2x)$ if $x = \dfrac{e^4}{2}$.

127. Solve: $\dfrac{x + 2}{4x + 3} = \dfrac{1}{x}$.

MID-CHAPTER CHECK POINT   Section 12.1–Section 12.3

 What You Know: We evaluated and graphed exponential functions [$f(x) = b^x$, $b > 0$ and $b \neq 1$], including the natural exponential function [$f(x) = e^x$, $e \approx 2.718$]. A function has an inverse that is a function if there is no horizontal line that intersects the function's graph more than once. The exponential function passes this horizontal line test and we called the inverse of the exponential function with base b the logarithmic function with base b. We learned that $y = \log_b x$ is equivalent to $b^y = x$. We evaluated and graphed logarithmic functions, including the common logarithmic function [$f(x) = \log_{10} x$ or $f(x) = \log x$] and the natural logarithmic function [$f(x) = \log_e x$ or $f(x) = \ln x$]. Finally, we used properties of logarithms to expand and condense logarithmic expressions.

In Exercises 1–4, graph the given function. Give each function's domain and range.

1. $f(x) = 2^x - 3$
2. $f(x) = \left(\frac{1}{3}\right)^x$
3. $f(x) = \log_2 x$
4. $f(x) = \log_2 x + 1$

In Exercises 5–8, find the domain of each function.

5. $f(x) = \log_3(x + 6)$
6. $g(x) = \log_3 x + 6$
7. $h(x) = \log_3(x + 6)^2$
8. $f(x) = 3^{x+6}$

In Exercises 9–19, evaluate each expression without using a calculator. If evaluation is not possible, state the reason.

9. $\log_2 8 + \log_5 25$
10. $\log_3 \frac{1}{9}$
11. $\log_{100} 10$
12. $\log \sqrt[3]{10}$
13. $\log_2(\log_3 81)$
14. $\log_3\left(\log_2 \frac{1}{8}\right)$
15. $6^{\log_6 5}$
16. $\ln e^{\sqrt{7}}$
17. $10^{\log 13}$
18. $\log_{100} 0.1$
19. $\log_\pi \pi^{\sqrt{\pi}}$

In Exercises 20–21, expand and evaluate numerical terms.

20. $\log\left(\dfrac{\sqrt{xy}}{1000}\right)$

21. $\ln(e^{19}x^{20})$

In Exercises 22–24, write each expression as a single logarithm.

22. $8\log_7 x - \dfrac{1}{3}\log_7 y$

23. $7\log_5 x + 2\log_5 x$

24. $\dfrac{1}{2}\ln x - 3\ln y - \ln(z-2)$

25. Use the formulas

$$A = P\left(1 + \frac{r}{n}\right)^{nt} \quad \text{and} \quad A = Pe^{rt}$$

to solve this exercise. You plan to invest $8000 for 3 years at an annual rate of 8%. How much more is the return if the interest is compounded continuously than monthly? Round to the nearest dollar.

Exponential and Logarithmic Equations

Objectives

1 Use like bases to solve exponential equations.

2 Use logarithms to solve exponential equations.

3 Use exponential form to solve logarithmic equations.

4 Use the one-to-one property of logarithms to solve logarithmic equations.

5 Solve applied problems involving exponential and logarithmic equations.

At age 20, you inherit $30,000. You'd like to put aside $25,000 and eventually have over half a million dollars for early retirement. Is this possible? In this section, you will see how techniques for solving equations with variable exponents provide an answer to this question.

Exponential Equations

An **exponential equation** is an equation containing a variable in an exponent. Examples of exponential equations include

$$2^{3x-8} = 16, \quad 4^x = 15, \quad \text{and} \quad 40e^{0.6x} = 240.$$

Some exponential equations can be solved by expressing each side of the equation as a power of the same base. All exponential functions are one-to-one—that is, no two different ordered pairs have the same second component. Thus, if b is a positive number other than 1 and $b^M = b^N$, then $M = N$.

1 Use like bases to solve exponential equations.

Solving Exponential Equations by Expressing Each Side as a Power of the Same Base

If $b^M = b^N$, then $M = N$.

> Express each side as a power of the same base.   Set the exponents equal to each other.

1. Rewrite the equation in the form $b^M = b^N$.

2. Set $M = N$.

3. Solve for the variable.

EXAMPLE 1   Solving Exponential Equations

Solve: **a.** $2^{3x-8} = 16$ **b.** $16^x = 64$.

Solution   In each equation, express both sides as a power of the same base. Then set the exponents equal to each other.

a. Because 16 is 2^4, we express each side of $2^{3x-8} = 16$ in terms of base 2.

$$2^{3x-8} = 16 \qquad \text{This is the given equation.}$$
$$2^{3x-8} = 2^4 \qquad \text{Write each side as a power of the same base.}$$
$$3x - 8 = 4 \qquad \text{If } b^M = b^N, b > 0 \text{ and } b \neq 1, \text{ then } M = N.$$
$$3x = 12 \qquad \text{Add 8 to both sides.}$$
$$x = 4 \qquad \text{Divide both sides by 3.}$$

Check 4:

$$2^{3x-8} = 16$$
$$2^{3\cdot4-8} \overset{?}{=} 16$$
$$2^4 \overset{?}{=} 16$$
$$16 = 16, \qquad \text{true}$$

The solution is 4 and the solution set is {4}.

b. Because $16 = 4^2$ and $64 = 4^3$, we express each side of $16^x = 64$ in terms of base 4.

$$16^x = 64 \qquad \text{This is the given equation.}$$
$$(4^2)^x = 4^3 \qquad \text{Write each side as a power of the same base.}$$
$$4^{2x} = 4^3 \qquad \text{When an exponential expression is raised to a power, multiply exponents.}$$
$$2x = 3 \qquad \text{If two powers of the same base are equal, then the exponents are equal.}$$
$$x = \frac{3}{2} \qquad \text{Divide both sides by 2.}$$

Check $\frac{3}{2}$:

$$16^x = 64$$
$$16^{\frac{3}{2}} \overset{?}{=} 64$$
$$(\sqrt{16})^3 \overset{?}{=} 64 \qquad b^{\frac{m}{n}} = (\sqrt[n]{b})^m$$
$$4^3 \overset{?}{=} 64$$
$$64 = 64, \qquad \text{true}$$

The solution is $\frac{3}{2}$ and the solution set is $\left\{\frac{3}{2}\right\}$. ∎

✓ CHECK POINT 1   Solve:
   a. $5^{3x-6} = 125$ **b.** $4^x = 32$.

Using Technology

Graphic Connections

The graphs of

$$y_1 = 2^{3x-8}$$
$$\text{and} \quad y_2 = 16$$

have an intersection point whose x-coordinate is 4. This verifies that {4} is the solution set of $2^{3x-8} = 16$.

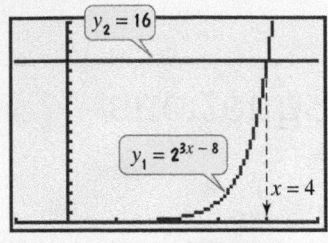

[−1, 5, 1] by [0, 20, 1]

Discover for Yourself

The equation $16^x = 64$ can also be solved by writing each side in terms of base 2. Do this. Which solution method do you prefer?

2   Use logarithms to solve exponential equations.

Most exponential equations cannot be rewritten so that each side has the same base. Here are two examples:

$$4^x = 15 \qquad\qquad 10^x = 120{,}000.$$

We cannot rewrite both sides in terms of base 2 or base 4.

We cannot rewrite both sides in terms of base 10.

Logarithms are extremely useful in solving equations such as $4^x = 15$ and $10^x = 120{,}000$. The solution begins with isolating the exponential expression. Notice that the exponential expression is already isolated in both $4^x = 15$ and $10^x = 120{,}000$. Then we take the logarithm on both sides. Why can we do this? All logarithmic relations are functions. Thus, if M and N are positive real numbers and $M = N$, then $\log_b M = \log_b N$.

The base that is used when taking the logarithm on both sides of an equation can be any base at all. If the exponential equation involves base 10, as in $10^x = 120{,}000$, we'll take the common logarithm on both sides. If the exponential equation involves any other base, as in $4^x = 15$, we'll take the natural logarithm on both sides.

Using Logarithms to Solve Exponential Equations

1. Isolate the exponential expression.
2. Take the common logarithm on both sides of the equation for base 10. Take the natural logarithm on both sides of the equation for bases other than 10.
3. Simplify using one of the following properties:
$$\ln b^x = x \ln b \quad \text{or} \quad \ln e^x = x \quad \text{or} \quad \log 10^x = x.$$
4. Solve for the variable.

EXAMPLE 2 Solving Exponential Equations

Solve: **a.** $4^x = 15$ **b.** $10^x = 120{,}000$.

Solution We will use the natural logarithmic function to solve $4^x = 15$ and the common logarithmic function to solve $10^x = 120{,}000$.

a. Because the exponential expression, 4^x, is already isolated on the left side of $4^x = 15$, we begin by taking the natural logarithm on both sides of the equation.

$$4^x = 15 \qquad \text{This is the given equation.}$$
$$\ln 4^x = \ln 15 \qquad \text{Take the natural logarithm on both sides.}$$
$$x \ln 4 = \ln 15 \qquad \text{Use the power rule and bring the variable exponent to the front: } \ln b^x = x \ln b.$$
$$x = \frac{\ln 15}{\ln 4} \qquad \text{Solve for x by dividing both sides by ln 4.}$$

We now have an exact value for x. We use the exact value for x in the equation's solution set. Thus, the equation's solution is $\dfrac{\ln 15}{\ln 4}$ and the solution set is $\left\{ \dfrac{\ln 15}{\ln 4} \right\}$.
We can obtain a decimal approximation by using a calculator: $x \approx 1.95$. Because $4^2 = 16$, it seems reasonable that the solution to $4^x = 15$ is approximately 1.95.

b. Because the exponential expression, 10^x, is already isolated on the left side of $10^x = 120{,}000$, we begin by taking the common logarithm on both sides of the equation.

$$10^x = 120{,}000 \qquad \text{This is the given equation.}$$
$$\log 10^x = \log 120{,}000 \qquad \text{Take the common logarithm on both sides.}$$
$$x = \log 120{,}000 \qquad \text{Use the inverse property log 10}^x = x \text{ on the left.}$$

The equation's solution is $\log 120{,}000$ and the solution set is $\{\log 120{,}000\}$. We can obtain a decimal approximation by using a calculator: $x \approx 5.08$. Because $10^5 = 100{,}000$, it seems reasonable that the solution to $10^x = 120{,}000$ is approximately 5.08. ∎

Discover for Yourself

Keep in mind that the base used when taking the logarithm on both sides of an equation can be any base at all. Solve $4^x = 15$ by taking the common logarithm on both sides. Solve again, this time taking the logarithm with base 4 on both sides. Use the change-of-base property to show that the solutions are the same as the one obtained in Example 2(a).

☑ **CHECK POINT 2** Solve:

a. $5^x = 134$ **b.** $10^x = 8000$.

Find each solution set and then use a calculator to obtain a decimal approximation to two decimal places for the solution.

EXAMPLE 3   Solving an Exponential Equation

Solve: $40e^{0.6x} - 3 = 237$.

Solution   We begin by adding 3 to both sides and dividing both sides by 40 to isolate the exponential expression, $e^{0.6x}$. Then we take the natural logarithm on both sides of the equation.

$40e^{0.6x} - 3 = 237$	This is the given equation.
$40e^{0.6x} = 240$	Add 3 to both sides.
$e^{0.6x} = 6$	Isolate the exponential factor by dividing both sides by 40.
$\ln e^{0.6x} = \ln 6$	Take the natural logarithm on both sides.
$0.6x = \ln 6$	Use the inverse property $\ln e^x = x$ on the left.
$x = \dfrac{\ln 6}{0.6} \approx 2.99$	Divide both sides by 0.6 and solve for x.

Thus, the solution of the equation is $\dfrac{\ln 6}{0.6} \approx 2.99$. Try checking this approximate solution in the original equation to verify that $\left\{ \dfrac{\ln 6}{0.6} \right\}$ is the solution set. ∎

☑ **CHECK POINT 3**   Solve: $7e^{2x} - 5 = 58$. Find the solution set and then use a calculator to obtain a decimal approximation to two decimal places for the solution.

---

3 Use exponential form to solve logarithmic equations.

Logarithmic Equations

A **logarithmic equation** is an equation containing a variable in a logarithmic expression. Examples of logarithmic equations include

$$\log_4(x + 3) = 2 \quad \text{and} \quad \ln(x + 2) - \ln(4x + 3) = \ln\left(\frac{1}{x}\right).$$

Some logarithmic equations can be expressed in the form $\log_b M = c$. We can solve such equations by rewriting them in exponential form.

Using Exponential Form to Solve Logarithmic Equations

1. Express the equation in the form $\log_b M = c$.

2. Use the definition of a logarithm to rewrite the equation in exponential form:

$$\log_b M = c \quad \text{means} \quad b^c = M.$$

Logarithms are exponents.

3. Solve for the variable.

4. Check proposed solutions in the original equation. Include in the solution set only values for which $M > 0$.

EXAMPLE 4 Solving Logarithmic Equations

Solve: **a.** $\log_4(x + 3) = 2$ **b.** $3\ln(2x) = 12$.

Solution The form $\log_b M = c$ involves a single logarithm whose coefficient is 1 on one side and a constant on the other side. Equation (a) is already in this form. We will need to divide both sides of equation (b) by 3 to obtain this form.

a.

$\log_4(x + 3) = 2$	This is the given equation.
$4^2 = x + 3$	Rewrite in exponential form: $\log_b M = c$ means $b^c = M$.
$16 = x + 3$	Square 4.
$13 = x$	Subtract 3 from both sides.

Check 13:

$\log_4(x + 3) = 2$	This is the given logarithmic equation.
$\log_4(13 + 3) \overset{?}{=} 2$	Substitute 13 for x.
$\log_4 16 \overset{?}{=} 2$	
$2 = 2$, true	$\log_4 16 = 2$ because $4^2 = 16$.

This true statement indicates that the solution is 13 and the solution set is {13}.

b.

$3\ln(2x) = 12$	This is the given equation.
$\ln(2x) = 4$	Divide both sides by 3.
$\log_e(2x) = 4$	Rewrite the natural logarithm showing base e. This step is optional.
$e^4 = 2x$	Rewrite in exponential form: $\log_b M = c$ means $b^c = M$.
$\dfrac{e^4}{2} = x$	Divide both sides by 2.

Check $\dfrac{e^4}{2}$:

$3\ln(2x) = 12$	This is the given logarithmic equation.
$3\ln\left[2\left(\dfrac{e^4}{2}\right)\right] \overset{?}{=} 12$	Substitute $\dfrac{e^4}{2}$ for x.
$3\ln e^4 \overset{?}{=} 12$	Simplify: $\dfrac{2}{1} \cdot \dfrac{e^4}{2} = e^4$.
$3 \cdot 4 \overset{?}{=} 12$	Because $\ln e^x = x$, we conclude $\ln e^4 = 4$.
$12 = 12$, true	

This true statement indicates that the solution is $\dfrac{e^4}{2}$ and the solution set is $\left\{\dfrac{e^4}{2}\right\}$. ∎

✓ **CHECK POINT 4** Solve:

a. $\log_2(x - 4) = 3$ **b.** $4\ln(3x) = 8$.

Logarithmic expressions are defined only for logarithms of positive real numbers. **Always check proposed solutions of a logarithmic equation in the original equation. Exclude from the solution set any proposed solution that produces the logarithm of a negative number or the logarithm of 0.**

To rewrite the logarithmic equation $\log_b M = c$ in the equivalent exponential form $b^c = M$, we need a single logarithm whose coefficient is one. It is sometimes necessary to use properties of logarithms to condense logarithms into a single logarithm. In the next example, we use the product rule for logarithms to obtain a single logarithmic expression on the left side.

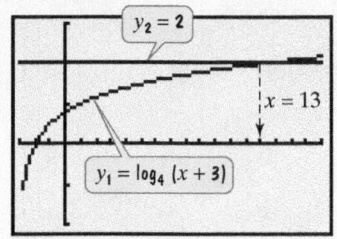

EXAMPLE 5 Solving a Logarithmic Equation

Solve: $\log_2 x + \log_2(x - 7) = 3$.

Solution

$\log_2 x + \log_2(x - 7) = 3$	This is the given equation.
$\log_2[x(x - 7)] = 3$	Use the product rule to obtain a single logarithm: $\log_b M + \log_b N = \log_b(MN)$.
$2^3 = x(x - 7)$	Rewrite in exponential form.
$8 = x^2 - 7x$	Evaluate 2^3 on the left and apply the distributive property on the right.
$0 = x^2 - 7x - 8$	Set the equation equal to 0.
$0 = (x - 8)(x + 1)$	Factor.
$x - 8 = 0$ or $x + 1 = 0$	Set each factor equal to 0.
$x = 8$ $x = -1$	Solve for x.

Check 8:	**Check −1:**
$\log_2 x + \log_2(x - 7) = 3$	$\log_2 x + \log_2(x - 7) = 3$
$\log_2 8 + \log_2(8 - 7) \stackrel{?}{=} 3$	$\log_2(-1) + \log_2(-1 - 7) \stackrel{?}{=} 3$
$\log_2 8 + \log_2 1 \stackrel{?}{=} 3$	The number −1 does not check. It produces
$3 + 0 \stackrel{?}{=} 3$	logarithms of negative numbers. Neither −1 nor
$3 = 3$, true	−8 are in the domain of a logarithmic function.

The solution is 8 and the solution set is {8}. ∎

☑ **CHECK POINT 5** Solve: $\log x + \log(x - 3) = 1$.

---

4 Use the one-to-one property of logarithms to solve logarithmic equations.

Some logarithmic equations can be expressed in the form $\log_b M = \log_b N$. Because all logarithmic functions are one-to-one, we can conclude that $M = N$.

Using the One-to-One Property of Logarithms to Solve Logarithmic Equations

1. Express the equation in the form $\log_b M = \log_b N$. This form involves a single logarithm whose coefficient is 1 on each side of the equation.
2. Use the one-to-one property to rewrite the equation without logarithms: If $\log_b M = \log_b N$, then $M = N$.
3. Solve for the variable.
4. Check proposed solutions in the original equation. Include in the solution set only values for which $M > 0$ and $N > 0$.

EXAMPLE 6   Solving a Logarithmic Equation

Solve: $\ln(x + 2) - \ln(4x + 3) = \ln\left(\dfrac{1}{x}\right)$.

Solution   In order to apply the one-to-one property of logarithms, we need a single logarithm whose coefficient is 1 on each side of the equation. The right side is already in this form. We can obtain a single logarithm on the left side by applying the quotient rule.

$$\ln(x + 2) - \ln(4x + 3) = \ln\left(\frac{1}{x}\right)$$

This is the given equation.

$$\ln\left(\frac{x + 2}{4x + 3}\right) = \ln\left(\frac{1}{x}\right)$$

Use the quotient rule to obtain a single logarithm on the left side:

$$\log_b M - \log_b N = \log_b\left(\frac{M}{N}\right).$$

$$\frac{x + 2}{4x + 3} = \frac{1}{x}$$

Use the one-to-one property:
If $\log_b M = \log_b N$, then $M = N$.

$$x(4x + 3)\left(\frac{x + 2}{4x + 3}\right) = x(4x + 3)\left(\frac{1}{x}\right)$$

Multiply both sides by $x(4x + 3)$, the LCD.

$$x(x + 2) = 4x + 3$$

Simplify.

$$x^2 + 2x = 4x + 3$$

Apply the distributive property.

$$x^2 - 2x - 3 = 0$$

Subtract $4x + 3$ from both sides and set the equation equal to 0.

$$(x - 3)(x + 1) = 0$$

Factor.

$$x - 3 = 0 \quad \text{or} \quad x + 1 = 0$$

Set each factor equal to 0.

$$x = 3 \qquad\qquad x = -1$$

Solve for x.

Substituting 3 for x into the original equation produces the true statement $\ln\left(\frac{1}{3}\right) = \ln\left(\frac{1}{3}\right)$. However, substituting -1 produces logarithms of negative numbers. Thus, -1 is not a solution. The solution is 3 and the solution set is {3}. ■

Using Technology

Numeric Connections

A graphing utility's ┃TABLE┃ feature can be used to verify that {3} is the solution set of

$$\ln(x + 2) - \ln(4x + 3) = \ln\left(\frac{1}{x}\right).$$

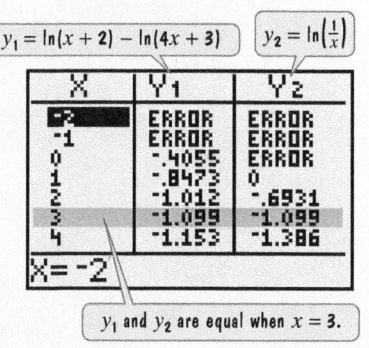

y_1 and y_2 are equal when $x = 3$.

✓ **CHECK POINT 6**   Solve: $\ln(x - 3) = \ln(7x - 23) - \ln(x + 1)$.

---

5 Solve applied problems involving exponential and logarithmic equations.

Applications

Our first applied example provides a mathematical perspective on the old slogan "Alcohol and driving don't mix." In California, where 38% of fatal traffic crashes involve drunk drivers, it is illegal to drive with a blood alcohol concentration of 0.08 or higher. At these levels, drivers may be arrested and charged with driving under the influence.

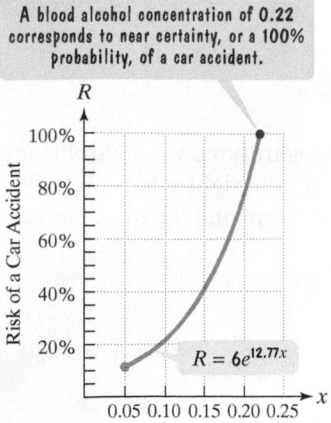

A blood alcohol concentration of 0.22 corresponds to near certainty, or a 100% probability, of a car accident.

Figure 12.17

EXAMPLE 7 Alcohol and Risk of a Car Accident

Medical research indicates that the risk of having a car accident increases exponentially as the concentration of alcohol in the blood increases. The risk is modeled by

$$R = 6e^{12.77x},$$

where x is the blood alcohol concentration and R, given as a percent, is the risk of having a car accident. What blood alcohol concentration corresponds to a 17% risk of a car accident? How is this shown on the graph of R in **Figure 12.17**?

Solution For a risk of 17%, we let $R = 17$ in the equation and solve for x, the blood alcohol concentration.

$$R = 6e^{12.77x}$$ This is the given equation.

$$6e^{12.77x} = 17$$ Substitute 17 for R and (optional) reverse the two sides of the equation.

$$e^{12.77x} = \frac{17}{6}$$ Isolate the exponential factor by dividing both sides by 6.

$$\ln e^{12.77x} = \ln\left(\frac{17}{6}\right)$$ Take the natural logarithm on both sides.

$$12.77x = \ln\left(\frac{17}{6}\right)$$ Use the inverse property $\ln e^x = x$ on the left side.

$$x = \frac{\ln\left(\frac{17}{6}\right)}{12.77} \approx 0.08$$ Divide both sides by 12.77.

For a blood alcohol concentration of 0.08, the risk of a car accident is 17%. This is shown on the graph of R in **Figure 12.17** by the point $(0.08, 17)$ that lies on the blue curve. Take a moment to locate this point on the curve. In many states, it is illegal to drive with a blood alcohol concentration of 0.08. ∎

✓ **CHECK POINT 7** Use the formula in Example 7 to solve this problem. What blood alcohol concentration corresponds to a 7% risk of a car accident? (In many states, drivers under the age of 21 can lose their licenses for driving at this level.)

Suppose that you inherit $30,000 at age 20. Is it possible to invest $25,000 and have over half a million dollars for early retirement? Our next example illustrates the power of compound interest.

EXAMPLE 8 Revisiting the Formula for Compound Interest

The formula

$$A = P\left(1 + \frac{r}{n}\right)^{nt}$$

describes the accumulated value, A, of a sum of money, P, the principal, after t years at annual percentage rate r (in decimal form) compounded n times a year. How long will it take $25,000 to grow to $500,000 at 9% annual interest compounded monthly?

Solution

$$A = P\left(1 + \frac{r}{n}\right)^{nt}$$ This is the given formula.

$$500,000 = 25,000\left(1 + \frac{0.09}{12}\right)^{12t}$$ A(the desired accumulated value) = 500,000, P(the principal) = 25,000, r(the interest rate) = 9% = 0.09, and n = 12 (monthly compounding).

Blitzer Bonus

Playing Doubles: Interest Rates and Doubling Time

One way to calculate what your savings will be worth at some point in the future is to consider doubling time. The following table shows how long it takes for your money to double at different annual interest rates subject to continuous compounding.

Annual Interest Rate	Years to Double
5%	13.9 years
7%	9.9 years
9%	7.7 years
11%	6.3 years

Of course, the first problem is collecting some money to invest. The second problem is finding a reasonably safe investment with a return of 9% or more.

Nicole Bengiveno/The New York Times

Our goal is to solve the equation for t. Let's reverse the two sides of the equation and then simplify within parentheses.

$$25,000\left(1 + \frac{0.09}{12}\right)^{12t} = 500,000$$

Reverse the two sides of the previous equation.

$$25,000(1 + 0.0075)^{12t} = 500,000$$

Divide within parentheses: $\frac{0.09}{12} = 0.0075$.

$$25,000(1.0075)^{12t} = 500,000$$

Add within parentheses.

$$(1.0075)^{12t} = 20$$

Divide both sides by 25,000.

$$\ln(1.0075)^{12t} = \ln 20$$

Take the natural logarithm on both sides.

$$12t \ln(1.0075) = \ln 20$$

Use the power rule to bring the exponent to the front: $\ln b^x = x \ln b$.

$$t = \frac{\ln 20}{12 \ln 1.0075}$$

Solve for t, dividing both sides by $12 \ln 1.0075$.

$$\approx 33.4$$

Use a calculator.

After approximately 33.4 years, the $25,000 will grow to an accumulated value of $500,000. If you set aside the money at age 20, you can begin enjoying a life of leisure at about age 53. ■

✓ **CHECK POINT 8** How long, to the nearest tenth of a year, will it take $1000 to grow to $3600 at 8% annual interest compounded quarterly?

EXAMPLE 9 Modeling Attitudes of College Freshmen

Researchers have surveyed attitudes of college freshmen every year since 1969. **Figure 12.18** shows that since 1980, there has been a decline in first-year college students' opposition to homosexual relationships. This is part of an historical trend among college students toward greater tolerance.

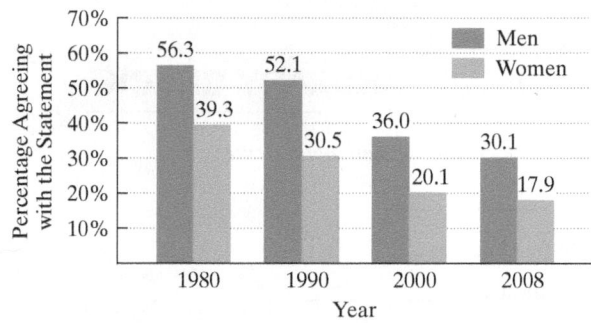

Opposition to Homosexual Relationships among First-Year United States College Students, 1980–2008

Statement: "It is important to have laws prohibiting homosexual relationships."

Figure 12.18

Source: John Macionis, *Sociology, Thirteenth Edition*, Prentice Hall, 2010.

The function

$$f(x) = -6.2 \ln x + 40.5$$

models the percentage of first-year college women, $f(x)$, opposed to homosexual relationships x years after 1979. If trends modeled by the function continue, when will opposition to homosexual relationships among first-year college women diminish to 16%? Round to the nearest year.

Solution To find when 16% of first-year college women will oppose homosexual relationships, we substitute 16 for $f(x)$ and solve for x, the number of years after 1979.

$$f(x) = -6.2 \ln x + 40.5 \qquad \text{This is the given function.}$$
$$16 = -6.2 \ln x + 40.5 \qquad \text{Substitute 16 for } f(x).$$

> Our goal is to isolate ln x and then rewrite the equation in exponential form.

$$16 - 40.5 = -6.2 \ln x + 40.5 - 40.5 \qquad \text{Subtract 40.5 from both sides.}$$

$$-24.5 = -6.2 \ln x \qquad \text{Simplify.}$$

$$\frac{-24.5}{-6.2} = \frac{-6.2 \ln x}{-6.2} \qquad \text{Divide both sides by } -6.2.$$

$$\frac{24.5}{6.2} = \ln x \qquad \text{Simplify.}$$

$$\frac{24.5}{6.2} = \log_e x \qquad \text{Rewrite the natural logarithm showing base } e. \text{ This step is optional.}$$

$$e^{\frac{24.5}{6.2}} = x \qquad \text{Rewrite in exponential form.}$$

$$52 \approx x \qquad \text{Use a calculator.}$$

Approximately 52 years after 1979, in the year 2031, 16% of first-year college women will oppose homosexual relationships. ∎

✓ **CHECK POINT 9** The function $g(x) = -7 \ln x + 59$ models the percentage of first-year college men, $g(x)$, opposed to homosexual relationships x years after 1979. According to this model, when did 40% of male freshmen oppose homosexual relationships? Round to the nearest year.

CONCEPT AND VOCABULARY CHECK

Fill in each blank so that the resulting statement is true.

1. If $b^M = b^N$, then _____.

2. If $2^{4x-1} = 2^7$, then _____ = 7.

3. If $x \ln 9 = \ln 20$, then $x =$ _____.

4. If $e^{0.6x} = 6$, then $0.6x =$ _____.

5. If $\log_5(x + 1) = 3$, then _____ = $x + 1$.

6. If $\log_3 x + \log_3(x + 1) = 2$, then $\log_3$ _____ = 2.

7. If $\ln\left(\dfrac{7x - 23}{x + 1}\right) = \ln(x - 3)$, then _____ = $x - 3$.

8. True or false: $x^4 = 15$ is an exponential equation. _____

9. True or false: $4^x = 15$ is an exponential equation. _____

10. True or false: -3 is a solution of $\log_5 9 = 2 \log_5 x$. _____

12.4 EXERCISE SET

MyMathLab®

Watch the videos
in MyMathLab

Download the
MyDashBoard App

Practice Exercises

Solve each exponential equation in Exercises 1–18 by expressing each side as a power of the same base and then equating exponents.

1. $2^x = 64$ **2.** $3^x = 81$

3. $5^x = 125$ **4.** $5^x = 625$

5. $2^{2x-1} = 32$ **6.** $3^{2x+1} = 27$

7. $4^{2x-1} = 64$ **8.** $5^{3x-1} = 125$

9. $32^x = 8$ **10.** $4^x = 32$

11. $9^x = 27$ **12.** $125^x = 625$

13. $3^{1-x} = \frac{1}{27}$ **14.** $5^{2-x} = \frac{1}{125}$

15. $6^{\frac{x-3}{4}} = \sqrt{6}$ **16.** $7^{\frac{x-2}{6}} = \sqrt{7}$

17. $4^x = \frac{1}{\sqrt{2}}$ **18.** $9^x = \frac{1}{\sqrt[3]{3}}$

Solve each exponential equation in Exercises 19–40 by taking the logarithm on both sides. Express the solution set in terms of logarithms. Then use a calculator to obtain a decimal approximation, correct to two decimal places, for the solution.

19. $e^x = 5.7$

20. $e^x = 0.83$

21. $10^x = 3.91$

22. $10^x = 8.07$

23. $5^x = 17$

24. $19^x = 143$

25. $5e^x = 25$

26. $9e^x = 99$

27. $3e^{5x} = 1977$

28. $4e^{7x} = 10,273$

29. $e^{0.7x} = 13$

30. $e^{0.08x} = 4$

31. $1250e^{0.055x} = 3750$

32. $1250e^{0.065x} = 6250$

33. $30 - (1.4)^x = 0$

34. $135 - (4.7)^x = 0$

35. $e^{1-5x} = 793$

36. $e^{1-8x} = 7957$

37. $7^{x+2} = 410$

38. $5^{x-3} = 137$

39. $2^{x+1} = 5^x$

40. $4^{x+1} = 9^x$

Solve each logarithmic equation in Exercises 41–90. Be sure to reject any value of x that is not in the domain of the original logarithmic expressions. Give the exact answer. Then, where necessary, use a calculator to obtain a decimal approximation, correct to two decimal places, for the solution.

41. $\log_3 x = 4$ **42.** $\log_5 x = 3$

43. $\log_2 x = -4$ **44.** $\log_2 x = -5$

45. $\log_9 x = \frac{1}{2}$ **46.** $\log_{25} x = \frac{1}{2}$

47. $\log x = 2$ **48.** $\log x = 3$

49. $\log_4(x + 5) = 3$ **50.** $\log_5(x - 7) = 2$

51. $\log_3(x - 4) = -3$

52. $\log_7(x + 2) = -2$

53. $\log_4(3x + 2) = 3$

54. $\log_2(4x + 1) = 5$

55. $\ln x = 2$

56. $\ln x = 3$

57. $\ln x = -3$

58. $\ln x = -4$

59. $5 \ln(2x) = 20$

60. $6 \ln(2x) = 30$

61. $6 + 2 \ln x = 5$

62. $7 + 3 \ln x = 6$

63. $\ln \sqrt{x + 3} = 1$

64. $\ln \sqrt{x + 4} = 1$

65. $\log_5 x + \log_5(4x - 1) = 1$

66. $\log_6(x + 5) + \log_6 x = 2$

67. $\log_3(x - 5) + \log_3(x + 3) = 2$

68. $\log_2(x - 1) + \log_2(x + 1) = 3$

69. $\log_2(x + 2) - \log_2(x - 5) = 3$

70. $\log_4(x + 2) - \log_4(x - 1) = 1$

71. $\log(3x - 5) - \log(5x) = 2$

72. $\log(2x - 1) - \log x = 2$

73. $\ln(x + 1) - \ln x = 1$

74. $\ln(x + 2) - \ln x = 2$

75. $\log_3(x + 4) = \log_3 7$

76. $\log_2(x - 5) = \log_2 4$

77. $\log(x + 4) = \log x + \log 4$

78. $\log(5x + 1) = \log(2x + 3) + \log 2$

79. $\log(3x - 3) = \log(x + 1) + \log 4$

80. $\log(2x - 1) = \log(x + 3) + \log 3$

81. $2 \log x = \log 25$

82. $3 \log x = \log 125$

83. $\log(x + 4) - \log 2 = \log(5x + 1)$

84. $\log(x + 7) - \log 3 = \log(7x + 1)$

85. $2 \log x - \log 7 = \log 112$

86. $\log(x - 2) + \log 5 = \log 100$

87. $\log x + \log(x + 3) = \log 10$

88. $\log(x + 3) + \log(x - 2) = \log 14$

89. $\ln(x - 4) + \ln(x + 1) = \ln(x - 8)$

90. $\log_2(x - 1) - \log_2(x + 3) = \log_2\left(\dfrac{1}{x}\right)$

Practice PLUS

In Exercises 91–98, solve each equation.

91. $5^{2x} \cdot 5^{4x} = 125$ **92.** $3^{x+2} \cdot 3^x = 81$

93. $3^{x^2} = 45$

94. $5^{x^2} = 50$

95. $\log_2(x - 6) + \log_2(x - 4) - \log_2 x = 2$

96. $\log_2(x - 3) + \log_2 x - \log_2(x + 2) = 2$

97. $5^{x^2-12} = 25^{2x}$ **98.** $3^{x^2-12} = 9^{2x}$

Application Exercises

99. The formula $A = 36.1e^{0.0126t}$ models the population of California, A, in millions, t years after 2005.

    a. What was the population of California in 2005?

    b. When will the population of California reach 40 million?

100. The formula $A = 22.9e^{0.0183t}$ models the population of Texas, A, in millions, t years after 2005.

    a. What was the population of Texas in 2005?

    b. When will the population of Texas reach 27 million?

The function $f(x) = 20(0.975)^x$ models the percentage of surface sunlight, $f(x)$, that reaches a depth of x feet beneath the surface of the ocean. The figure shows the graph of this function. Use this information to solve Exercises 101–102.

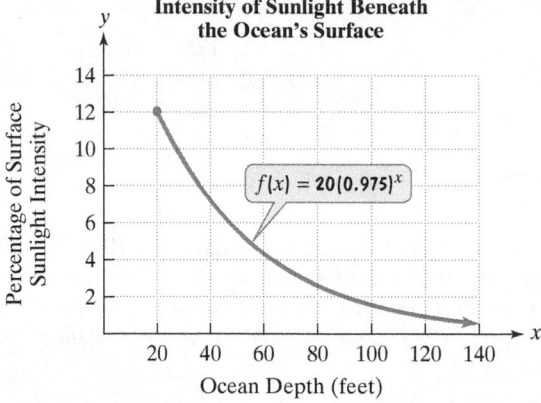

101. Use the function to determine at what depth, to the nearest foot, there is 1% of surface sunlight. How is this shown on the graph of f?

102. Use the function to determine at what depth, to the nearest foot, there is 3% of surface sunlight. How is this shown on the graph of f?

In Exercises 103–106, complete the table for a savings account subject to n compounding periods per year $\left[A = P\left(1 + \dfrac{r}{n}\right)^{nt}\right]$. Round answers to one decimal place.

	Amount Invested	Number of Compounding Periods	Annual Interest Rate	Accumulated Amount	Time t in Years
103.	$12,500	4	5.75%	$20,000	
104.	$7250	12	6.5%	$15,000	
105.	$1000	360		$1400	2
106.	$5000	360		$9000	4

In Exercises 107–110, complete the table for a savings account subject to continuous compounding ($A = Pe^{rt}$). Round answers to one decimal place.

	Amount Invested	Annual Interest Rate	Accumulated Amount	Time t in Years
107.	$8000	8%	Double the amount invested	
108.	$8000		$12,000	2
109.	$2350		Triple the amount invested	7
110.	$17,425	4.25%	$25,000	

By 2019, nearly $1 out of every $5 spent in the U.S. economy is projected to go for health care. The bar graph shows the percentage of the U.S. gross domestic product (GDP) going toward health care from 2007 through 2010, with projections for 2014 and 2019.

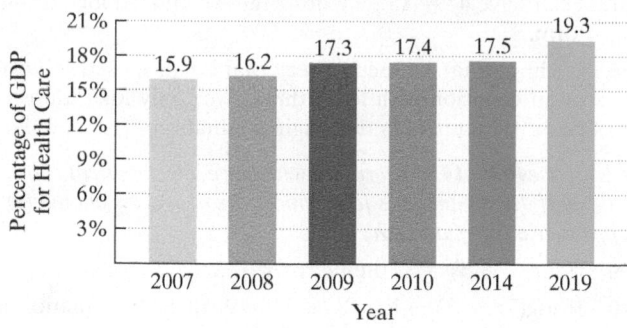

Percentage of the Gross Domestic Product in the United States Going Toward Health Care

Source: Health Affairs (healthaffairs.org)

The data can be modeled by the function $f(x) = 1.2 \ln x + 15.7$, where $f(x)$ is the percentage of the U.S. gross domestic product going toward health care x years after 2006. Use this information to solve Exercises 111–112.

111. a. Use the function to determine the percentage of the U.S. gross domestic product that went toward health care in 2009. Round to the nearest tenth of a percent. Does this underestimate or overestimate the percent displayed by the graph? By how much?

b. According to the model, when will 18.5% of the U.S. gross domestic product go toward health care? Round to the nearest year.

112. a. Use the function to determine the percentage of the U.S. gross domestic product that went toward health care in 2008. Round to the nearest tenth of a percent. Does this underestimate or overestimate the percent displayed by the graph? By how much?

b. According to the model, when will 18.6% of the U.S. gross domestic product go toward health care? Round to the nearest year.

The function $P(x) = 95 - 30 \log_2 x$ models the percentage, $P(x)$, of students who could recall the important features of a classroom lecture as a function of time, where x represents the number of days that have elapsed since the lecture was given. The figure shows the graph of the function. Use this information to solve Exercises 113–114. Round answers to one decimal place.

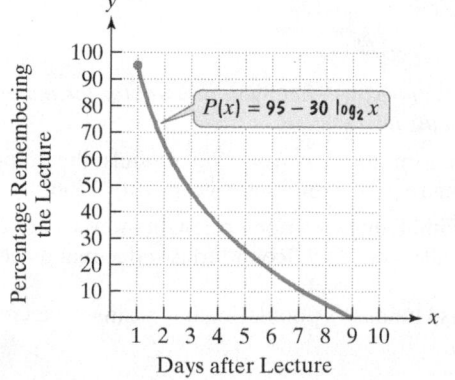

$P(x) = 95 - 30 \log_2 x$

113. After how many days do only half the students recall the important features of the classroom lecture? (Let $P(x) = 50$ and solve for x.) Locate the point on the graph that conveys this information.

114. After how many days have all students forgotten the important features of the classroom lecture? (Let $P(x) = 0$ and solve for x.) Locate the point on the graph that conveys this information.

The pH scale is used to measure the acidity or alkalinity of a solution. The scale ranges from 0 to 14. A neutral solution, such as pure water, has a pH of 7. An acid solution has a pH less than 7 and an alkaline solution has a pH greater than 7. The lower the pH below 7, the more acidic is the solution. Each whole-number decrease in pH represents a tenfold increase in acidity.

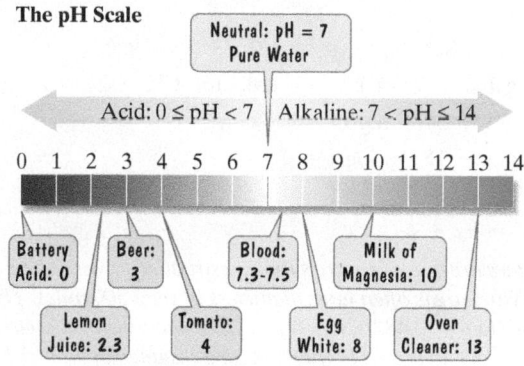

The pH Scale

The pH of a solution is given by
$$pH = -\log x,$$
where x represents the concentration of the hydrogen ions in the solution, in moles per liter. Use the formula to solve Exercises 115–116. Express answers as powers of 10.

115. a. Normal, unpolluted rain has a pH of about 5.6. What is the hydrogen ion concentration?

b. An environmental concern involves the destructive effects of acid rain. The most acidic rainfall ever had a pH of 2.4. What was the hydrogen ion concentration?

c. How many times greater is the hydrogen ion concentration of the acidic rainfall in part (b) than the normal rainfall in part (a)?

116. a. The figure indicates that lemon juice has a pH of 2.3. What is the hydrogen ion concentration?

b. Stomach acid has a pH that ranges from 1 to 3. What is the hydrogen ion concentration of the most acidic stomach?

c. How many times greater is the hydrogen ion concentration of the acidic stomach in part (b) than the lemon juice in part (a)?

Writing in Mathematics

117. What is an exponential equation?

118. Explain how to solve an exponential equation when both sides can be written as a power of the same base.

119. Explain how to solve an exponential equation when both sides cannot be written as a power of the same base. Use $3^x = 140$ in your explanation.

120. What is a logarithmic equation?

121. Explain the differences between solving $\log_3(x - 1) = 4$ and $\log_3(x - 1) = \log_3 4$.

122. In many states, a 17% risk of a car accident with a blood alcohol concentration of 0.08 is the lowest level for charging a motorist with driving under the influence. Do you agree with the 17% risk as a cutoff percentage, or do you feel that the percentage should be lower or higher? Explain your answer. What blood alcohol concentration corresponds to what you believe is an appropriate percentage?

Technology Exercises

In Exercises 123–130, use your graphing utility to graph each side of the equation in the same viewing rectangle. Then use the x-coordinate of the intersection point to find the equation's solution set. Verify this value by direct substitution into the equation.

123. $2^{x+1} = 8$

124. $3^{x+1} = 9$

125. $\log_3(4x - 7) = 2$

126. $\log_3(3x - 2) = 2$

127. $\log(x + 3) + \log x = 1$

128. $\log(x - 15) + \log x = 2$

129. $3^x = 2x + 3$

130. $5^x = 3x + 4$

Hurricanes are one of nature's most destructive forces. These low-pressure areas often have diameters of over 500 miles. The function $f(x) = 0.48 \ln(x + 1) + 27$ models the barometric air pressure, $f(x)$, in inches of mercury, at a distance of x miles from the eye of a hurricane. Use this function to solve Exercises 131–132.

131. Graph the function in a [0, 500, 50] by [27, 30, 1] viewing rectangle. What does the shape of the graph indicate about barometric air pressure as the distance from the eye increases?

132. Use an equation to answer this question: How far from the eye of a hurricane is the barometric air pressure 29 inches of mercury? Use the $\boxed{\text{TRACE}}$ and $\boxed{\text{ZOOM}}$ features or the intersect command of your graphing utility to verify your answer.

133. The function $P(t) = 145e^{-0.092t}$ models a runner's pulse, $P(t)$, in beats per minute, t minutes after a race, where $0 \le t \le 15$. Graph the function using a graphing utility. $\boxed{\text{TRACE}}$ along the graph and determine after how many minutes the runner's pulse will be 70 beats per minute. Round to the nearest tenth of a minute. Verify your observation algebraically.

134. The function $W(t) = 2600(1 - 0.51e^{-0.075t})^3$ models the weight, $W(t)$, in kilograms, of a female African elephant at age t years. (1 kilogram $\approx$ 2.2 pounds) Use a graphing utility to graph the function. Then $\boxed{\text{TRACE}}$ along the curve to estimate the age of an adult female elephant weighing 1800 kilograms.

Critical Thinking Exercises

Make Sense? *In Exercises 135–138, determine whether each statement "makes sense" or "does not make sense" and explain your reasoning.*

135. Because the equations $2^x = 15$ and $2^x = 16$ are similar, I solved them using the same method.

136. Because the equations

$$\log(3x + 1) = 5 \text{ and } \log(3x + 1) = \log 5$$

are similar, I solved them using the same method.

137. I can solve $4^x = 15$ by writing the equation in logarithmic form.

138. It's important for me to check that the proposed solution of an equation with logarithms gives only logarithms of positive numbers in the original equation.

In Exercises 139–142, determine whether each statement is true or false. If the statement is false, make the necessary change(s) to produce a true statement.

139. If $\log(x + 3) = 2$, then $e^2 = x + 3$.

140. If $\log(7x + 3) - \log(2x + 5) = 4$, then the equation in exponential form is $10^4 = (7x + 3) - (2x + 5)$.

141. If $x = \dfrac{1}{k}\ln y$, then $y = e^{kx}$.

142. Examples of exponential equations include $10^x = 5.71$, $e^x = 0.72$, and $x^{10} = 5.71$.

143. If $4000 is deposited into an account paying 3% interest compounded annually and at the same time $2000 is deposited into an account paying 5% interest compounded annually, after how long will the two accounts have the same balance? Round to the nearest year.

Solve each equation in Exercises 144–146. Check each proposed solution by direct substitution or with a graphing utility.

144. $(\ln x)^2 = \ln x^2$.

145. $(\log x)(2 \log x + 1) = 6$

146. $\ln(\ln x) = 0$

Review Exercises

147. Solve: $\sqrt{2x - 1} - \sqrt{x - 1} = 1$.

(Section 10.6, Example 4)

148. Solve: $\dfrac{3}{x + 1} - \dfrac{5}{x} = \dfrac{19}{x^2 + x}$.

(Section 7.6, Example 4)

149. Simplify: $(-2x^3 y^{-2})^{-4}$.

(Section 5.7, Example 6)

Preview Exercises

Exercises 150–152 will help you prepare for the material covered in the next section.

150. The formula $A = 10e^{-0.003t}$ models the population of Hungary, A, in millions, t years after 2006.

a. Find Hungary's population, in millions, for 2006, 2007, 2008, and 2009. Round to two decimal places.

b. Is Hungary's population increasing or decreasing?

151. a. Simplify: $e^{\ln 3}$.

   b. Use your simplification from part (a) to rewrite 3^x in terms of base e.

152. U.S. soldiers fight Russian troops who have invaded New York City. Incoming missiles from Russian submarines and warships ravage the Manhattan skyline. It's just another scenario for the multi-billion-dollar video games *Call of Duty*, which have sold more than 100 million games since the franchise's birth in 2003.

The table shows the annual retail sales for *Call of Duty* video games from 2004 through 2010. Create a scatter plot for the data. Based on the shape of the scatter plot, would a logarithmic function, an exponential function, or a linear function be the best choice for modeling the data?

Annual Retail Sales for *Call of Duty* Games

Year	Retail Sales (millions of dollars)
2004	56
2005	101
2006	196
2007	352
2008	436
2009	778
2010	980

Source: The NPD Group

SECTION

12.5

Exponential Growth and Decay; Modeling Data

Objectives

1 Model exponential growth and decay.

2 Choose an appropriate model for data.

3 Express an exponential model in base e.

The most casual cruise on the Internet shows how people disagree when it comes to making predictions about the effects of the world's growing population. Some argue that there is a recent slowdown in the growth rate, economies remain robust, and famines in North Korea and Ethiopia are aberrations rather than signs of the future. Others say that the 6.9 billion people on Earth is twice as many as can be supported in middle-class comfort, and the world is running out of arable land and fresh water. Debates about entities that are growing exponentially can be approached mathematically: We can create functions that model data and use these functions to make predictions. In this section, we will show you how this is done.

① Model exponential growth and decay.

Exponential Growth and Decay

One of algebra's many applications is to predict the behavior of variables. This can be done with *exponential growth* and *decay models*. With exponential growth or decay, quantities grow or decay at a rate directly proportional to their size. Populations that are growing exponentially grow extremely rapidly as they get larger because there are more adults to have offspring. For example, the **growth rate** for world population is approximately 1.2%, or 0.012. This means that each year world population is 1.2% more than what it was in the previous year. In 2010, world population was 6.9 billion. Thus, we compute the world population in 2011 as follows:

$$6.9 \text{ billion} + 1.2\% \text{ of } 6.9 \text{ billion} = 6.9 + (0.012)(6.9) = 6.9828.$$

This computation indicates that 6.9828 billion people populated the world in 2011. The 0.0828 billion represents an increase of 82.8 million people from 2010 to 2011, the equivalent of the population of Germany. Using 1.2% as the annual growth rate, world population for 2012 is found in a similar manner:

$$6.9828 + 1.2\% \text{ of } 6.9828 = 6.9828 + (0.012)(6.9828) \approx 7.067.$$

This computation indicates that approximately 7.1 billion people populated the world in 2012.

The explosive growth of world population may remind you of the growth of money in an account subject to compound interest. Just as the growth rate for world population is multiplied by the population plus any increase in the population, a compound interest rate is multiplied by your original investment plus any accumulated interest. The balance in an account subject to continuous compounding and world population are special cases of *exponential growth models*.

Exponential Growth and Decay Models

The mathematical model for **exponential growth** or **decay** is given by

$$f(t) = A_0 e^{kt} \quad \text{or} \quad A = A_0 e^{kt}.$$

- **If $k > 0$, the function models the amount, or size, of a *growing* entity.** A_0 is the original amount, or size, of the growing entity at time $t = 0$, A is the amount at time t, and k is a constant representing the growth rate.

- **If $k < 0$, the function models the amount, or size, of a *decaying* entity.** A_0 is the original amount, or size, of the decaying entity at time $t = 0$, A is the amount at time t, and k is a constant representing the decay rate.

Great Question!

Why does the formula for exponential growth look familiar?

You have seen the formula for exponential growth before, but with different letters. It is the formula for compound interest with continous compounding.

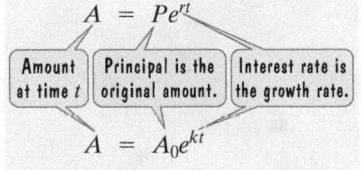

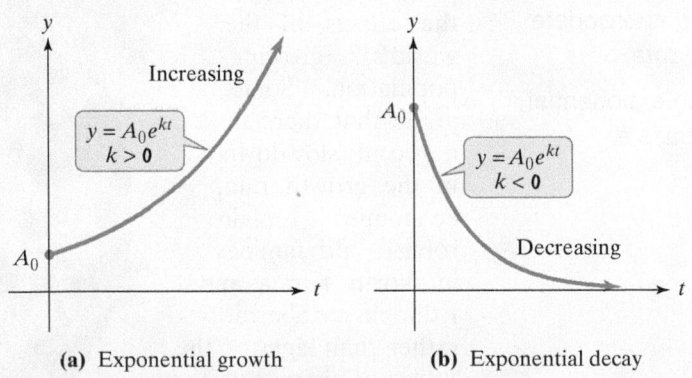

(a) Exponential growth   **(b)** Exponential decay

Sometimes we need to use given data to determine k, the rate of growth or decay. After we compute the value of k, we can use the formula $A = A_0 e^{kt}$ to make predictions. This idea is illustrated in our first two examples.

EXAMPLE 1   Modeling the Growth of the U.S. Population

The graph in **Figure 12.19** on the following page shows the U.S. population, in millions, for five selected years from 1970 through 2009. In 1970, the U.S. population was 203.3 million. By 2009, it had grown to 307.0 million.

a. Find an exponential growth function that models the data for 1970 through 2009.

b. By which year will the U.S. population reach 352 million?

Solution

a. We use the exponential growth model

$$A = A_0 e^{kt}$$

in which t is the number of years after 1970. This means that 1970 corresponds to $t = 0$. At that time the U.S. population was 203.3 million, so we substitute 203.3 for A_0 in the growth model:

$$A = 203.3 e^{kt}.$$

We are given that 307.0 million is the population in 2009. Because 2009 is 39 years after 1970, when $t = 39$ the value of A is 307.0. Substituting these numbers into the growth model will enable us to find k, the growth rate. We know that $k > 0$ because the problem involves growth.

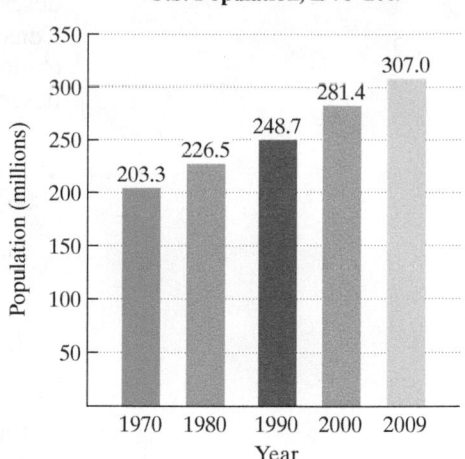

U.S. Population, 1970–2009

Figure 12.19

Source: U.S. Bureau of the Census

$A = 203.3 e^{kt}$	Use the growth model with $A_0 = 203.3$.
$307.0 = 203.3 e^{k \cdot 39}$	When $t = 39$, $A = 307.0$. Substitute these numbers into the model.
$e^{39k} = \dfrac{307.0}{203.3}$	Isolate the exponential factor by dividing both sides by 203.3. We also reversed the sides.
$\ln e^{39k} = \ln\left(\dfrac{307.0}{203.3}\right)$	Take the natural logarithm on both sides.
$39k = \ln\left(\dfrac{307.0}{203.3}\right)$	Simplify the left side using $\ln e^x = x$.
$k = \dfrac{\ln\left(\dfrac{307.0}{203.3}\right)}{39} \approx 0.011$	Divide both sides by 39 and solve for k. Then use a calculator.

The value of k, approximately 0.011, indicates a growth rate of about 1.1%. We substitute 0.011 for k in the growth model, $A = 203.3 e^{kt}$, to obtain an exponential growth function for the U.S. population. It is

$$A = 203.3 e^{0.011t},$$

where t is measured in years after 1970.

b. To find the year in which the U.S. population will reach 352 million, substitute 352 for A in the model from part (a) and solve for t.

$A = 203.3 e^{0.011t}$	This is the model from part (a).
$352 = 203.3 e^{0.011t}$	Substitute 352 for A.
$e^{0.011t} = \dfrac{352}{203.3}$	Divide both sides by 203.3. We also reversed the sides.
$\ln e^{0.011t} = \ln\left(\dfrac{352}{203.3}\right)$	Take the natural logarithm on both sides.
$0.011t = \ln\left(\dfrac{352}{203.3}\right)$	Simplify on the left using $\ln e^x = x$.
$t = \dfrac{\ln\left(\dfrac{352}{203.3}\right)}{0.011} \approx 50$	Divide both sides by 0.011 and solve for t. Then use a calculator.

Because t represents the number of years after 1970, the model indicates that the U.S. population will reach 352 million by 1970 + 50, or in the year 2020. ∎

In Example 1, we used only two data values, the population for 1970 and the population for 2009, to develop a model for U.S. population growth from 1970 through 2009. By not using data for any other years, have we created a model that inaccurately describes both the existing data and future population projections given by the U.S. Census Bureau? Something else to think about: Is an exponential model the best choice for describing U.S. population growth, or might a linear model provide a better description? We return to these issues in Exercises 45–49 in the Exercise Set.

✓ **CHECK POINT 1** In 1990, the population of Africa was 643 million and by 2006 it had grown to 906 million.

a. Use the exponential growth model $A = A_0e^{kt}$, in which t is the number of years after 1990, to find the exponential growth function that models the data.

b. By which year will Africa's population reach 2000 million, or two billion?

Our next example involves exponential decay and its use in determining the age of fossils and artifacts. The method is based on considering the percentage of carbon-14 remaining in the fossil or artifact. Carbon-14 decays exponentially with a *half-life* of approximately 5715 years. The **half-life** of a substance is the time required for half of a given sample to disintegrate. Thus, after 5715 years a given amount of carbon-14 will have decayed to half the original amount. Carbon dating is useful for artifacts or fossils up to 80,000 years old. Older objects do not have enough carbon-14 left to determine age accurately.

Blitzer Bonus

Carbon Dating and Artistic Development

The artistic community was electrified by the discovery in 1995 of spectacular cave paintings in a limestone cavern in France. Carbon dating of the charcoal from the site showed that the images, created by artists of remarkable talent, were 30,000 years old, making them the oldest cave paintings ever found. The artists seemed to have used the cavern's natural contours to heighten a sense of perspective. The quality of the painting suggests that the art of early humans did not mature steadily from primitive to sophisticated in any simple linear fashion.

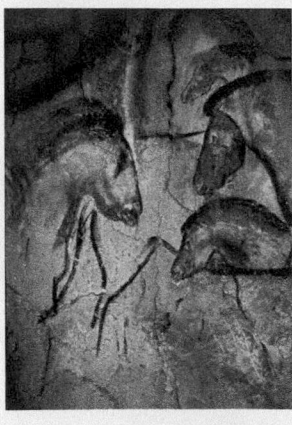

EXAMPLE 2 Carbon-14 Dating: The Dead Sea Scrolls

a. Use the fact that after 5715 years a given amount of carbon-14 will have decayed to half the original amount to find the exponential decay model for carbon-14.

b. In 1947, earthenware jars containing what are known as the Dead Sea Scrolls were found by an Arab Bedouin herdsman. Analysis indicated that the scroll wrappings contained 76% of their original carbon-14. Estimate the age of the Dead Sea Scrolls.

Solution

a. We begin with the exponential decay model $A = A_0e^{kt}$. We know that $k < 0$ because the problem involves the decay of carbon-14. After 5715 years ($t = 5715$), the amount of carbon-14 present, A, is half the original amount, A_0. Thus, we can substitute $\dfrac{A_0}{2}$ for A in the exponential decay model. This will enable us to find k, the decay rate.

$$A = A_0e^{kt} \qquad \text{Begin with the exponential decay model.}$$

$$\frac{A_0}{2} = A_0e^{k \cdot 5715} \qquad \text{After 5715 years } (t = 5715),\ A = \frac{A_0}{2}$$
$$\text{(because the amount present, } A\text{, is half the original amount, } A_0\text{).}$$

$$\frac{1}{2} = e^{5715k} \qquad \text{Divide both sides of the equation by } A_0.$$

$$\ln\left(\frac{1}{2}\right) = \ln e^{5715k} \qquad \text{Take the natural logarithm on both sides.}$$

$$\ln\left(\frac{1}{2}\right) = 5715k \qquad \text{Simplify the right side using } \ln e^x = x.$$

$$k = \frac{\ln\left(\frac{1}{2}\right)}{5715} \approx -0.000121 \qquad \text{Divide both sides by 5715 and solve for } k.$$

Substituting for k in the decay model, $A = A_0e^{kt}$, the model for carbon-14 is

$$A = A_0e^{-0.000121t}.$$

b. In 1947, the Dead Sea Scrolls contained 76% of their original carbon-14. To find their age in 1947, substitute $0.76A_0$ for A in the model from part (a) and solve for t.

$$A = A_0 e^{-0.000121t}$$ This is the decay model for carbon-14.

$$0.76A_0 = A_0 e^{-0.000121t}$$ A, the amount present, is 76% of the original amount, so $A = 0.76A_0$.

$$0.76 = e^{-0.000121t}$$ Divide both sides of the equation by A_0.

$$\ln 0.76 = \ln e^{-0.000121t}$$ Take the natural logarithm on both sides.

$$\ln 0.76 = -0.000121t$$ Simplify the right side using $\ln e^x = x$.

$$t = \frac{\ln 0.76}{-0.000121} \approx 2268$$ Divide both sides by -0.000121 and solve for t.

The Dead Sea Scrolls are approximately 2268 years old plus the number of years between 1947 and the current year. ■

✓ **CHECK POINT 2**   Strontium-90 is a waste product from nuclear reactors. As a consequence of fallout from atmospheric nuclear tests, we all have a measurable amount of strontium-90 in our bones.

a. Use the fact that after 28 years a given amount of strontium-90 will have decayed to half the original amount to find the exponential decay model for strontium-90.

b. Suppose that a nuclear accident occurs and releases 60 grams of strontium-90 into the atmosphere. How long will it take for strontium-90 to decay to a level of 10 grams?

---

2 Choose an appropriate model for data.

Modeling Data

Throughout this chapter, we have been working with models that were given. However, we can create functions that model data by observing patterns in scatter plots. **Figure 12.20** shows scatter plots for data that are exponential or logarithmic.

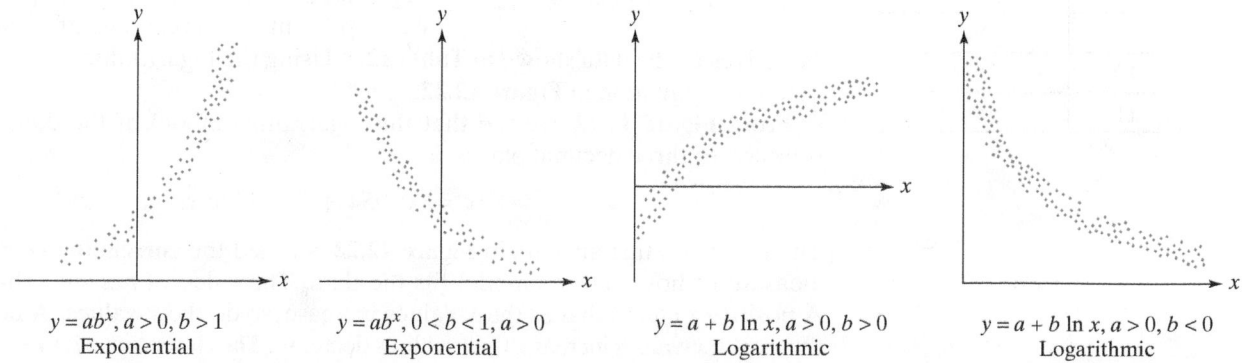

$y = ab^x, a > 0, b > 1$
Exponential

$y = ab^x, 0 < b < 1, a > 0$
Exponential

$y = a + b \ln x, a > 0, b > 0$
Logarithmic

$y = a + b \ln x, a > 0, b < 0$
Logarithmic

Figure 12.20 Scatter plots for exponential or logarithmic models

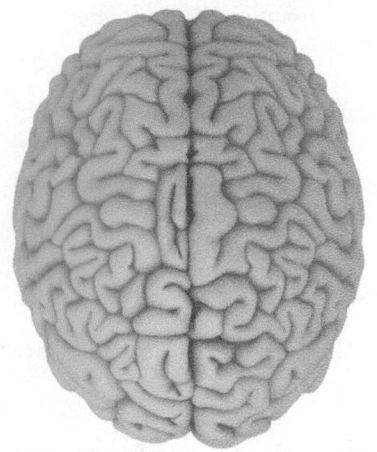

EXAMPLE 3 Choosing a Model for Data

The data in **Table 12.2** indicate that between the ages of 1 and 11, the human brain does not grow linearly, or steadily. A scatter plot for the data is shown in **Figure 12.21**. What type of function would be a good choice for modeling the data?

Table 12.2 Growth of the Human Brain	
Age	**Percentage of Adult Size Brain**
1	30%
2	50%
4	78%
6	88%
8	92%
10	95%
11	99%

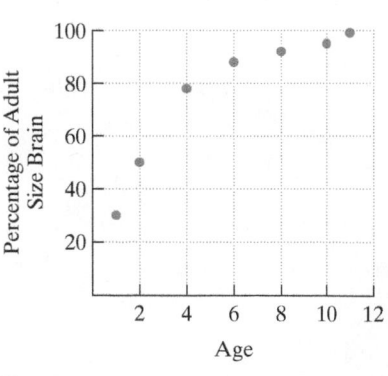

Figure 12.21

Source: Gerrig and Zimbardo, *Psychology and Life*, 18th Edition, Allyn and Bacon, 2008.

Solution Because the data in the scatter plot increase rapidly at first and then begin to level off a bit, the shape suggests that a logarithmic function is a good choice for modeling the data. ■

☑ **CHECK POINT 3** **Table 12.3** shows the populations of various cities, in thousands, and the average walking speed, in feet per second, of a person living in the city. Create a scatter plot for the data. Based on the scatter plot, what type of function would be a good choice for modeling the data?

Table 12.3 Population and Walking Speed	
Population (thousands)	**Walking Speed (feet per second)**
5.5	0.6
14	1.0
71	1.6
138	1.9
342	2.2

Source: Mark H. Bornstein and Helen G. Bornstein, "The Pace of Life." *Nature*, 259, Feb. 19, 1976, pp. 557–559

Table 12.4	
x, Age	**y, Percentage of Adult Size Brain**
1	30
2	50
4	78
6	88
8	92
10	95
11	99

```
LnReg
y=a+blnx
a=31.95404756
b=28.94733911
r²=.9806647799
r=.9902852013
```

Figure 12.22 A logarithmic model for the data in **Table 12.4**

How can we obtain a logarithmic function that models the data for the growth of the human brain? A graphing utility can be used to obtain a logarithmic model of the form $y = a + b \ln x$. **Because the domain of the logarithmic function is the set of positive numbers, zero must not be a value for x.** This is not a problem for the data giving the percentage of an adult size brain because the data begin at age 1. We will assign x to represent age and y to represent the percentage of an adult size brain. This gives us the data shown in **Table 12.4**. Using the logarithmic regression option, we obtain the equation in **Figure 12.22**.

From **Figure 12.22**, we see that the logarithmic model of the data, with numbers rounded to three decimal places, is

$$y = 31.954 + 28.947 \ln x.$$

The number r that appears in **Figure 12.22** is called the **correlation coefficient** and is a measure of how well the model fits the data. The value of r is such that $-1 \le r \le 1$. A positive r means that as the x-values increase, so do the y-values. A negative r means that as the x-values increase, the y-values decrease. **The closer that r is to -1 or 1, the better the model fits the data.** Because r is approximately 0.99, the model fits the data very well.

EXAMPLE 4   Choosing a Model for Data

Figure 12.23(a) shows world population, in billions, for seven selected years from 1950 through 2010. A scatter plot is shown in **Figure 12.23(b)**. Suggest two types of functions that would be good choices for modeling the data.

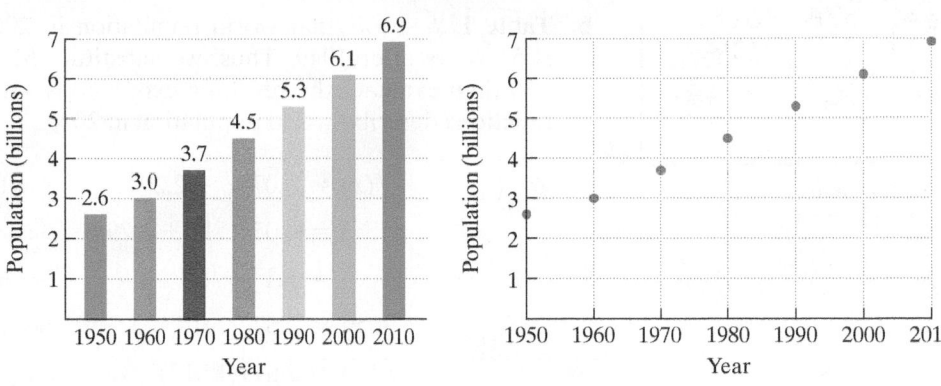

World Population, 1950–2010

Figure 12.23(a)   **Figure 12.23(b)**

Source: U.S. Census Bureau, International Database

Solution   Because the data in the scatter plot appear to increase more and more rapidly, the shape suggests that an exponential model might be a good choice. Furthermore, we can probably draw a line that passes through or near the seven points. Thus, a linear function would also be a good choice for modeling the data.   ■

✓ **CHECK POINT 4**   **Table 12.5** shows the percentage of U.S. men who are married or who have been married, by age. Create a scatter plot for the data. Based on the scatter plot, what type of function would be a good choice for modeling the data?

EXAMPLE 5   Comparing Linear and Exponential Models

The data for world population are shown in **Table 12.6**. Using a graphing utility's linear regression feature and exponential regression feature, we enter the data and obtain the models shown in **Figure 12.24**.

Table 12.5

Percentage of U.S. Men Who Are Married or Who Have Been Married, by Age

Age	Percent
18	2
20	7
25	36
30	61
35	75

Source: National Center for Health Statistics

Although the domain of $y = ab^x$ is the set of all real numbers, some graphing utilities only accept positive values for x. That's why we assigned x to represent the number of years after 1949.

Table 12.6

x, Number of Years after 1949	y, World Population (billions)
1 (1950)	2.6
11 (1960)	3.0
21 (1970)	3.7
31 (1980)	4.5
41 (1990)	5.3
51 (2000)	6.1
61 (2010)	6.9

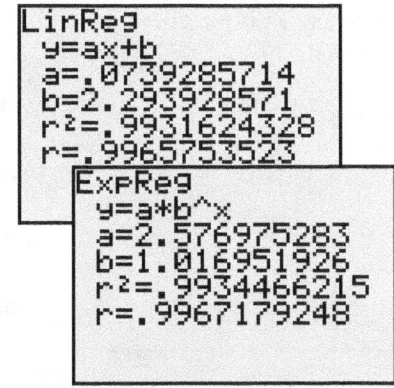

Figure 12.24 A linear model and an exponential model for the data in **Table 12.6**

Because r, the correlation coefficient, is close to 1 in each screen in **Figure 12.24**, the models fit the data very well.

a. Use **Figure 12.24** to express each model in function notation, with numbers rounded to three decimal places.

b. How well do the functions model world population in 2000?

c. By one projection, world population is expected to reach 8 billion in the year 2026. Which function serves as a better model for this prediction?

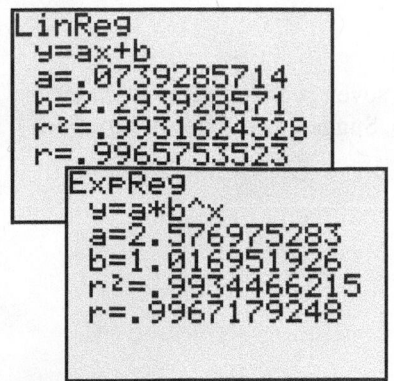

Figure 12.24 (repeated)

Solution

a. Using **Figure 12.24** and rounding to three decimal places, the functions

$$f(x) = 0.074x + 2.294 \quad \text{and} \quad g(x) = 2.577(1.017)^x$$

model world population, in billions, x years after 1949. We named the linear function f and the exponential function g, although any letters can be used.

b. **Table 12.6** shows that world population in 2000 was 6.1 billion. The year 2000 is 51 years after 1949. Thus, we substitute 51 for x in each function's equation and then evaluate the resulting expressions with a calculator to see how well the functions describe world population in 2000.

$f(x) = 0.074x + 2.294$	This is the linear model.
$f(51) = 0.074(51) + 2.294$	Substitute 51 for x.
≈ 6.1	Use a calculator.
$g(x) = 2.577(1.017)^x$	This is the exponential model.
$g(51) = 2.577(1.017)^{51}$	Substitute 51 for x.
≈ 6.1	Use a calculator: 2.577 $\boxed{\times}$ 1.017 $\boxed{y^x}$ (or $\boxed{\wedge}$) 51 $\boxed{=}$.

Because 6.1 billion was the actual world population in 2000, both functions model world population in 2000 extremely well.

c. Let's see which model comes closer to projecting a world population of 8 billion in the year 2026. Because 2026 is 77 years after 1949 ($2026 - 1949 = 77$), we substitute 77 for x in each function's equation.

$f(x) = 0.074x + 2.294$	This is the linear model.
$f(77) = 0.074(77) + 2.294$	Substitute 77 for x.
≈ 8.0	Use a calculator.
$g(x) = 2.577(1.017)^x$	This is the exponential model.
$g(77) = 2.577(1.017)^{77}$	Substitute 77 for x.
≈ 9.4	Use a calculator: 2.577 $\boxed{\times}$ 1.017 $\boxed{y^x}$ (or $\boxed{\wedge}$) 77 $\boxed{=}$.

The linear function $f(x) = 0.074x + 2.294$ serves as a better model for a projected world population of 8 billion by 2026. ■

✓ **CHECK POINT 5** Use the models in Example 5(a) to solve this problem.

a. World population in 1970 was 3.7 billion. Which function serves as a better model for this year?

b. By one projection, world population is expected to reach 9.3 billion by 2050. Which function serves as a better model for this projection?

Great Question!

How can I use a graphing utility to see how well my models describe the data?

Once you have obtained one or more models for the data, you can use a graphing utility's $\boxed{\text{TABLE}}$ feature to numerically see how well each model describes the data. Enter the models as y_1, y_2, and so on. Create a table, scroll through the table, and compare the table values given by the models to the actual data.

When using a graphing utility to model data, begin with a scatter plot, drawn either by hand or with the graphing utility, to obtain a general picture for the shape of the data. It might be difficult to determine which model best fits the data—linear, logarithmic, exponential, quadratic, or something else. If necessary, use your graphing utility to fit several models to the data. The best model is the one that yields the value r, the correlation coefficient, closest to 1 or -1. Finding a proper fit for data can be almost as much art as it is mathematics. In this era of technology, the process of creating models that best fit data is one that involves more decision making than computation.

③ Express an exponential model in base e.

Expressing $y = ab^x$ in Base e

Graphing utilities display exponential models in the form $y = ab^x$. However, our discussion of exponential growth involved base e. Because of the inverse property $b = e^{\ln b}$, we can rewrite any model in the form $y = ab^x$ in terms of base e.

Expressing an Exponential Model in Base e

$$y = ab^x \quad \text{is equivalent to} \quad y = ae^{(\ln b) \cdot x}.$$

EXAMPLE 6 Rewriting the Model for World Population in Base e

We have seen that the function

$$g(x) = 2.577(1.017)^x$$

models world population, $g(x)$, in billions, x years after 1949. Rewrite the model in terms of base e.

Solution   We use the two equivalent equations shown in the voice balloons to rewrite the model in terms of base e.

$$y = ab^x \qquad\qquad y = ae^{(\ln b)\cdot x}$$

$$g(x) = 2.577(1.017)^x \quad \text{is equivalent to} \quad g(x) = 2.577e^{(\ln 1.017)x}.$$

Using $\ln 1.017 \approx 0.017$, the exponential growth model for world population, $g(x)$, in billions, x years after 1949 is

$$g(x) = 2.577e^{0.017x}. \quad \blacksquare$$

In Example 6, we can replace $g(x)$ with A and x with t so that the model has the same letters as those in the exponential growth model $A = A_0 e^{kt}$.

$$A = \boxed{A_o}\ \boxed{e^{kt}} \quad \text{This is the exponential growth model.}$$

$$A = 2.577e^{0.017t} \quad \text{This is the model for world population.}$$

The value of k, 0.017, indicates a growth rate of 1.7%. Although this is an excellent model for the data, we must be careful about making projections about world population using this growth function. Why? World population growth rate is now 1.2%, not 1.7%, so our model will overestimate future populations.

✓ **CHECK POINT 6**   Rewrite $y = 4(7.8)^x$ in terms of base e. Express the answer in terms of a natural logarithm and then round to three decimal places.

CONCEPT AND VOCABULARY CHECK

Fill in each blank so that the resulting statement is true.

1. Consider the model for exponential growth or decay given by

$$A = A_0 e^{kt}.$$

If k _____, the function models the amount, or size, of a growing entity. If k _____, the function models the amount, or size, of a decaying entity.

2. In the model for exponential growth or decay, the amount, or size, at $t = 0$ is represented by _____. The amount, or size, at time t is represented by _____.

For each of the following scatter plots, determine whether an exponential function, a logarithmic function, or a linear function is the best choice for modeling the data.

3.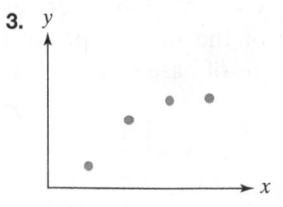

_____

4.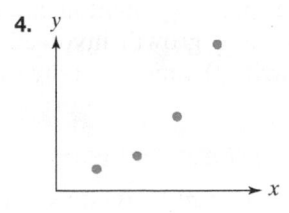

_____

5.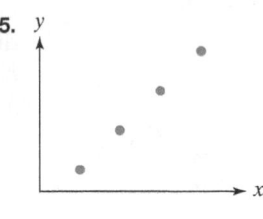

_____

6. $y = 3(5)^x$ can be written in terms of base e as $y = 3e^{(\underline{})\cdot x}$.

12.5 EXERCISE SET MyMathLab®

 Watch the videos in MyMathLab

 Download the MyDashBoard App

Practice Exercises and Application Exercises

The exponential models describe the population of the indicated country, A, in millions, t years after 2006. Use these models to solve Exercise 1–6.

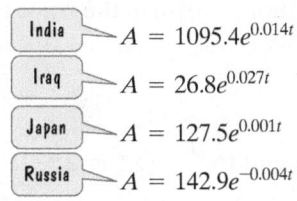

India — $A = 1095.4e^{0.014t}$

Iraq — $A = 26.8e^{0.027t}$

Japan — $A = 127.5e^{0.001t}$

Russia — $A = 142.9e^{-0.004t}$

1. What was the population of Japan in 2006?

2. What was the population of Iraq in 2006?

3. Which country has the greatest growth rate? By what percentage is the population of that country increasing each year?

4. Which country has a decreasing population? By what percentage is the population of that country decreasing each year?

5. When will India's population be 1238 million?

6. When will India's population be 1416 million?

About the size of New Jersey, Israel has seen its population soar to more than 6 million since it was established. The graphs show that by 2050, Palestinians in the West Bank, Gaza Strip, and East Jerusalem will outnumber Israelis. Exercises 7–8 involve the projected growth of these two populations.

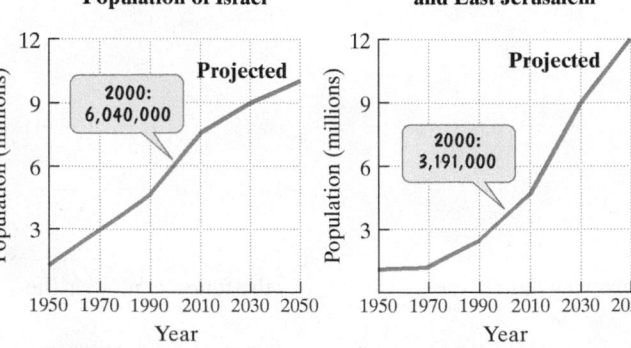

Population of Israel

2000: 6,040,000

Projected

Palestinian Population in West Bank, Gaza, and East Jerusalem

2000: 3,191,000

Projected

Source: Newsweek Magazine

7. a. In 2000, the population of Israel was approximately 6.04 million and by 2050 it is projected to grow to 10 million. Use the exponential growth model $A = A_0e^{kt}$, in which t is the number of years after 2000, to find an exponential growth function that models the data.

b. In which year will Israel's population be 9 million?

8. a. In 2000, the population of the Palestinians in the West Bank, Gaza Strip, and East Jerusalem was approximately 3.2 million and by 2050 it is projected to grow to 12 million. Use the exponential growth model $A = A_0e^{kt}$, in which t is the number of years after 2000, to find an exponential growth function that models the data.

b. In which year will the Palestinian population be 9 million?

In Exercises 9–14, complete the table. Round projected populations to one decimal place and values of k to four decimal places.

	Country	2007 Population (millions)	Projected 2025 Population (millions)	Projected Growth Rate, *k*
9.	Philippines	91.1		0.0147
10.	Pakistan	164.7		0.0157
11.	Colombia	44.4	55.2	
12.	Madagascar	19.4	32.4	
13.	South Africa	44.0	40.0	
14.	Bulgaria	7.3	6.3	

Source: U.S. Census Bureau, International Programs Center

An artifact originally had 16 grams of carbon-14 present. The decay model $A = 16e^{-0.000121t}$ describes the amount of carbon-14 present after t years. Use this model to solve Exercises 15–16.

15. How many grams of carbon-14 will be present in 5715 years?

16. How many grams of carbon-14 will be present in 11,430 years?

17. The half-life of the radioactive element krypton-91 is 10 seconds. If 16 grams of krypton-91 are initially present, how many grams are present after 10 seconds? 20 seconds? 30 seconds? 40 seconds? 50 seconds?

18. The half-life of the radioactive element plutonium-239 is 25,000 years. If 16 grams of plutonium-239 are initially present, how many grams are present after 25,000 years? 50,000 years? 75,000 years? 100,000 years? 125,000 years?

Use the exponential decay model for carbon-14, $A = A_0 e^{-0.000121t}$, *to solve Exercises 19–20.*

19. Prehistoric cave paintings were discovered in a cave in France. The paint contained 15% of the original carbon-14. Estimate the age of the paintings.

20. Skeletons were found at a construction site in San Francisco in 1989. The skeletons contained 88% of the expected amount of carbon-14 found in a living person. In 1989, how old were the skeletons?

21. The August 1978 issue of *National Geographic* described the 1964 find of bones of a newly discovered dinosaur weighing 170 pounds, measuring 9 feet, with a 6-inch claw on one toe of each hind foot. The age of the dinosaur was estimated using potassium-40 dating of rocks surrounding the bones.

  a. Potassium-40 decays exponentially with a half-life of approximately 1.31 billion years. Use the fact that after 1.31 billion years a given amount of potassium-40 will have decayed to half the original amount to show that the decay model for potassium-40 is given by $A = A_0 e^{-0.52912t}$, where t is in billions of years.

  b. Analysis of the rocks surrounding the dinosaur bones indicated that 94.5% of the original amount of potassium-40 was still present. Let $A = 0.945A_0$ in the model in part (a) and estimate the age of the bones of the dinosaur.

22. A bird species in danger of extinction has a population that is decreasing exponentially ($A = A_0 e^{kt}$). Five years ago the population was at 1400 and today only 1000 of the birds are alive. Once the population drops below 100, the situation will be irreversible. When will this happen?

23. Use the exponential growth model, $A = A_0 e^{kt}$, to show that the time it takes a population to double (to grow from A_0 to $2A_0$) is given by $t = \dfrac{\ln 2}{k}$.

24. Use the exponential growth model, $A = A_0 e^{kt}$, to show that the time it takes a population to triple (to grow from A_0 to $3A_0$) is given by $t = \dfrac{\ln 3}{k}$.

Use the formula $t = \dfrac{\ln 2}{k}$ *that gives the time for a population with a growth rate k to double to solve Exercises 25–26. Express each answer to the nearest whole year.*

25. The growth model $A = 4.1e^{0.01t}$ describes New Zealand's population, A, in millions, t years after 2006.

  a. What is New Zealand's growth rate?

  b. How long will it take New Zealand to double its population?

26. The growth model $A = 107.4e^{0.012t}$ describes Mexico's population, A, in millions, t years after 2003.

  a. What is Mexico's growth rate?

  b. How long will it take Mexico to double its population?

Exercises 27–32 present data in the form of tables. For each data set shown by the table,

  a. Create a scatter plot for the data.

  b. Use the scatter plot to determine whether an exponential function, a logarithmic function, or a linear function is the best choice for modeling the data. (If applicable, in Exercise 51 you will use your graphing utility to obtain these functions.)

27. Percent of Miscarriages, by Age

Woman's Age	Percent of Miscarriages
22	9%
27	10%
32	13%
37	20%
42	38%
47	52%

Source: Time Magazine

28. Savings Needed for Health-Care Expenses During Retirement

Age at Death	Savings Needed
80	$219,000
85	$307,000
90	$409,000
95	$524,000
100	$656,000

Source: Employee Benefit Research Institute

29. Intensity and Loudness Level of Various Sounds

Intensity (watts per meter^2)	Loudness Level (decibels)
0.1 (loud thunder)	110
1 (rock concert, 2 yd from speakers)	120
10 (jackhammer)	130
100 (jet takeoff, 40 yd away)	140

30. Temperature Increase in an Enclosed Vehicle

Minutes	Temperature Increase (°F)
10	19°
20	29°
30	34°
40	38°
50	41°
60	43°

31. Dads Raising Kids Alone

Year	Number of Single U.S. Fathers Heading Households with Children Younger Than 18 (millions)
1980	0.6
1990	1.2
2000	1.8
2008	2.2

Source: U.S. Census Bureau

32. Percentage of U.S. Consumers Looking for Trans Fats on Food Labels

Year	Percentage of U.S. Consumers
2004	15%
2005	20%
2006	25%
2007	31%

Source: U.S. Food and Drug Administration (FDA)

In Exercises 33–36, rewrite the equation in terms of base e. Express the answer in terms of a natural logarithm and then round to three decimal places.

33. $y = 100(4.6)^x$

34. $y = 1000(7.3)^x$

35. $y = 2.5(0.7)^x$

36. $y = 4.5(0.6)^x$

Writing in Mathematics

37. Nigeria has a growth rate of 0.025 or 2.5%. Describe what this means.

38. How can you tell if an exponential model describes exponential growth or exponential decay?

39. Suppose that a population that is growing exponentially increases from 800,000 people in 1997 to 1,000,000 people in 2000. Without showing the details, describe how to obtain an exponential growth function that models the data.

40. What is the half-life of a substance?

41. Describe the shape of a scatter plot that suggests modeling the data with an exponential function.

42. You take up weightlifting and record the maximum number of pounds you can lift at the end of each week. You start off with rapid growth in terms of the weight you can lift from week to week, but then the growth begins to level off. Describe how to obtain a function that models the number of pounds you can lift at the end of each week. How can you use this function to predict what might happen if you continue the sport?

43. Would you prefer that your salary be modeled exponentially or logarithmically? Explain your answer.

44. One problem with all exponential growth models is that nothing can grow exponentially forever. (Or can it? See the Blitzer Bonus on page 862.) Describe factors that might limit the size of a population.

Technology Exercises

In Example 1 on page 910, we used two data points and an exponential function to model the population of the United States from 1970 through 2009. The data are shown again in the table. Use all five data points to solve Exercises 45–49.

x, Number of Years after 1969	y, U.S. Population (millions)
1 (1970)	203.3
11 (1980)	226.5
21 (1990)	248.7
31 (2000)	281.4
40 (2009)	307.0

45. a. Use your graphing utility's exponential regression option to obtain a model of the form $y = ab^x$ that fits the data. How well does the correlation coefficient, r, indicate that the model fits the data?

   b. Rewrite the model in terms of base e. By what percentage is the population of the United States increasing each year?

46. Use your graphing utility's logarithmic regression option to obtain a model of the form $y = a + b \ln x$ that fits the data. How well does the correlation coefficient, r, indicate that the model fits the data?

47. Use your graphing utility's linear regression option to obtain a model of the form $y = ax + b$ that fits the data. How well does the correlation coefficient, r, indicate that the model fits the data?

48. Use your graphing utility's power regression option to obtain a model of the form $y = ax^b$ that fits the data. How well does the correlation coefficient, r, indicate that the model fits the data?

49. Use the values of r in Exercises 45–48 to select the two models of best fit. Use each of these models to predict by which year the U.S. population will reach 352 million. How do these answers compare to the year we found in Example 1, namely 2020? If you obtained different years, how do you account for this difference?

50. The figure shows the number of people in the United States age 65 and over, with projected figures for the year 2020 and beyond.

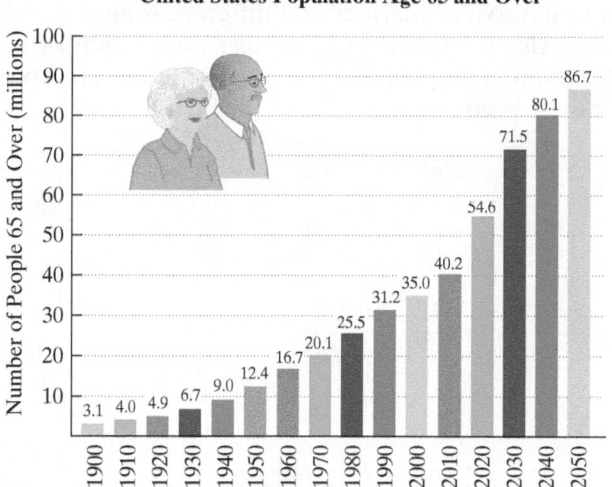

United States Population Age 65 and Over

Source: U.S. Bureau of the Census

a. Let x represent the number of years after 1899 and let y represent the U.S. population age 65 and over, in millions. Use your graphing utility to find the model that best fits the data in the bar graph.

b. Rewrite the model in terms of base e. By what percentage is the 65 and over population increasing each year?

51. In Exercises 27–32, you determined the best choice for the kind of function that modeled the data in the table. For each of the exercises that you worked, use a graphing utility to find the actual function that best fits the data. Then use the model to make a reasonable prediction for a value that exceeds those shown in the table's first column.

Critical Thinking Exercises

Make Sense? *In Exercises 52–55, determine whether each statement "makes sense" or "does not make sense" and explain your reasoning.*

52. I used an exponential model with a positive growth rate to describe the depreciation in my car's value over four years.

53. After 100 years, a population whose growth rate is 3% will have three times as many people as a population whose growth rate is 1%.

54. Because carbon-14 decays exponentially, carbon dating can determine the ages of ancient fossils.

55. When I used an exponential function to model Russia's declining population, the growth rate k was negative.

The exponential growth models describe the population of the indicated country, A, in millions, t years after 2006.

Canada $\ A = 33.1e^{0.009t}$

Uganda $\ A = 28.2e^{0.034t}$

In Exercises 56–59, use this information to determine whether each statement is true or false. If the statement is false, make the necessary change(s) to produce a true statement.

56. In 2006, Canada's population exceeded Uganda's by 4.9 million.

57. By 2009, the models indicate that Canada's population exceeded Uganda's by approximately 2.8 million.

58. The models indicate that in 2013, Uganda's population will exceed Canada's.

59. Uganda's growth rate is approximately 3.8 times that of Canada's.

60. Over a period of time, a hot object cools to the temperature of the surrounding air. This is described mathematically by Newton's Law of Cooling:

$$T = C + (T_0 - C)e^{-kt},$$

where t is the time it takes for an object to cool from temperature T_0 to temperature T, C is the surrounding air temperature, and k is a positive constant that is associated with the cooling object. A cake removed from the oven has a temperature of 210°F and is left to cool in a room that has a temperature of 70°F. After 30 minutes, the temperature of the cake is 140°F. What is the temperature of the cake after 40 minutes?

Review Exercises

61. Divide:

$$\frac{x^2 - 9}{2x^2 + 7x + 3} \div \frac{x^2 - 3x}{2x^2 + 11x + 5}.$$

(Section 7.2, Example 6)

62. Solve: $x^{\frac{2}{3}} + 2x^{\frac{1}{3}} - 3 = 0$.

(Section 11.4, Example 5)

63. Simplify: $6\sqrt{2} - 2\sqrt{50} + 3\sqrt{98}$.

(Section 10.4, Example 2)

Preview Exercises

Exercises 64–66 will help you prepare for the material covered in the first section of the next chapter.

In Exercises 64–65, what term should be add to each binomial so that it becomes a perfect square trinomial? Write and factor the trinomial.

64. $x^2 + 4x$

65. $y^2 - 6y$

66. Use a rectangular coordinate system to graph the circle with center $(1, -1)$ and radius 1.

This activity is intended for three or four people who would like to take up weightlifting. Each person in the group should record the maximum number of pounds that he or she can lift at the end of each week for ten consecutive weeks. Use the logarithmic regression option of a graphing utility to obtain a model showing the amount of weight that group members can lift from week 1 through week 10. Graph each of the models in the same viewing rectangle to observe similarities and differences among weight–growth patterns of each member. Use the functions to predict the amount of weight that group members will be able to lift in the future. If the group continues to work out together, check the accuracy of these predictions.

Chapter 12 Summary

Definitions and Concepts	**Examples**

Section 12.1 Exponential Functions

The exponential function with base b is defined by $f(x) = b^x$, where $b > 0$ and $b \neq 1$. The graph contains the point $(0, 1)$. When $b > 1$, the graph rises from left to right. When $0 < b < 1$, the graph falls from left to right. The x-axis is a horizontal asymptote. The domain is $(-\infty, \infty)$; the range is $(0, \infty)$. The natural exponential function is $f(x) = e^x$, where $e \approx 2.71828$.

Graph $f(x) = 2^x$ and $g(x) = 2^{x-1}$.

x	$f(x) = 2^x$	$g(x) = 2^{x-1}$
-2	$2^{-2} = \frac{1}{4}$	$2^{-3} = \frac{1}{8}$
-1	$2^{-1} = \frac{1}{2}$	$2^{-2} = \frac{1}{4}$
0	$2^0 = 1$	$2^{-1} = \frac{1}{2}$
1	$2^1 = 2$	$2^0 = 1$
2	$2^2 = 4$	$2^1 = 2$

The graph of g is the graph of f shifted one unit to the right.

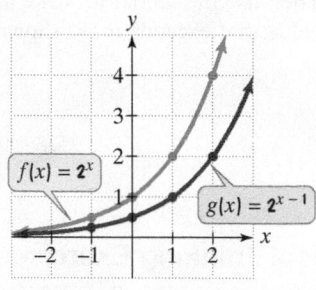

Formulas for Compound Interest

After t years, the balance, A, in an account with principal P and annual interest rate r is given by the following formulas:

1. For n compounding periods per year: $A = P\left(1 + \dfrac{r}{n}\right)^{nt}$

2. For continuous compounding: $A = Pe^{rt}$

Select the better investment for $4000 over 6 years:

- 6% compounded semiannually

$$A = P\left(1 + \frac{r}{n}\right)^{nt}$$

$$= 4000\left(1 + \frac{0.06}{2}\right)^{2\cdot6} \approx \$5703$$

- 5.9% compounded continuously

$$A = Pe^{rt} = 4000e^{0.059(6)} \approx \$5699$$

The first investment is better.

| **Definitions and Concepts** | **Examples** |

Section 12.2 Logarithmic Functions

Definition of the logarithmic function: For $x > 0$ and $b > 0$, $b \neq 1$, $y = \log_b x$ is equivalent to $b^y = x$. The function $f(x) = \log_b x$ is the logarithmic function with base b. This function is the inverse function of the exponential function with base b.

- Write $\log_2 32 = 5$ in exponential form.
$$2^5 = 32 \quad y = \log_b x \text{ means } b^y = x.$$
- Write $\sqrt{49} = 7$, or $49^{\frac{1}{2}} = 7$, in logarithmic form.
$$\frac{1}{2} = \log_{49} 7 \quad b^y = x \text{ means } y = \log_b x.$$

The graph of $f(x) = \log_b x$ can be obtained from $f(x) = b^x$ by reversing coordinates. The graph of $f(x) = \log_b x$ contains the point $(1, 0)$. If $b > 1$, the graph rises from left to right. If $0 < b < 1$, the graph falls from left to right. The y-axis is a vertical asymptote. The domain is $(0, \infty)$; the range is $(-\infty, \infty)$. $f(x) = \log x$ means $f(x) = \log_{10} x$ and is the common logarithmic function. $f(x) = \ln x$ means $f(x) = \log_e x$ and is the natural logarithmic function. The domain of $f(x) = \log_b g(x)$ consists of all x for which $g(x) > 0$.

- Find the domain: $f(x) = \log_6(4 - x)$.
$$4 - x > 0$$
$$4 > x \quad (\text{or } x < 4)$$
The domain is $(-\infty, 4)$.

Basic Logarithmic Properties

Base b ($b > 0, b \neq 1$)	Base 10 (Common Logarithms)	Base e (Natural Logarithms)
$\log_b 1 = 0$	$\log 1 = 0$	$\ln 1 = 0$
$\log_b b = 1$	$\log 10 = 1$	$\ln e = 1$
$\log_b b^x = x$	$\log 10^x = x$	$\ln e^x = x$
$b^{\log_b x} = x$	$10^{\log x} = x$	$e^{\ln x} = x$

- $\log_8 1 = 0$ because $\log_b 1 = 0$.
- $\log_4 4 = 1$ because $\log_b b = 1$.
- $\ln e^{8x} = 8x$ because $\ln e^x = x$.
- $e^{\ln \sqrt[3]{x}} = \sqrt[3]{x}$ because $e^{\ln x} = x$.
- $\log_t t^{25} = 25$ because $\log_b b^x = x$.

Section 12.3 Properties of Logarithms

Properties of Logarithms

For $M > 0$ and $N > 0$:

1. *The Product Rule:* $\log_b(MN) = \log_b M + \log_b N$
2. *The Quotient Rule:* $\log_b\left(\dfrac{M}{N}\right) = \log_b M - \log_b N$
3. *The Power Rule:* $\log_b M^p = p \log_b M$
4. *The Change-of Base Property:*

The General Property	Introducing Common Logarithms	Introducing Natural Logarithms
$\log_b M = \dfrac{\log_a M}{\log_a b}$	$\log_b M = \dfrac{\log M}{\log b}$	$\log_b M = \dfrac{\ln M}{\ln b}$

- Expand: $\log_3(81x^7)$.
$$= \log_3 81 + \log_3 x^7$$
$$= 4 + 7 \log_3 x$$
- Write as a single logarithm: $7 \ln x - 4 \ln y$.
$$= \ln x^7 - \ln y^4 = \ln\left(\frac{x^7}{y^4}\right)$$
- Evaluate: $\log_6 92$.
$$\log_6 92 = \frac{\ln 92}{\ln 6} \approx 2.5237$$

Definitions and Concepts	**Examples**

Section 12.4 Exponential and Logarithmic Equations

An exponential equation is an equation containing a variable in an exponent. Some exponential equations can be solved by expressing both sides as a power of the same base. Then set the exponents equal to each other:

If $b^M = b^N$, then $M = N$.

Solve: $4^{2x-1} = 64$.

$$4^{2x-1} = 4^3$$
$$2x - 1 = 3$$
$$2x = 4$$
$$x = 2$$

The solution is 2 and the solution set is {2}.

If both sides of an exponential equation cannot be expressed as a power of the same base, isolate the exponential expression. Take the natural logarithm on both sides for bases other than 10 and take the common logarithm on both sides for base 10. Simplify using

$\ln b^x = x \ln b$ or $\ln e^x = x$ or $\log 10^x = x$.

Solve: $7^x = 103$.

$$\ln 7^x = \ln 103$$
$$x \ln 7 = \ln 103$$
$$x = \frac{\ln 103}{\ln 7}$$

The solution is $\dfrac{\ln 103}{\ln 7}$ and the solution set is $\left\{ \dfrac{\ln 103}{\ln 7} \right\}$.

A logarithmic equation is an equation containing a variable in a logarithmic expression. Logarithmic equations in the form $\log_b x = c$ can be solved by rewriting in exponential form as $b^c = x$. When checking logarithmic equations, reject proposed solutions that produce the logarithm of a negative number or the logarithm of zero in the original equation.

Solve: $\log_2(3x - 1) = 5$.

$$2^5 = 3x - 1$$
$$32 = 3x - 1 \qquad \boxed{\text{Exponential form}}$$
$$33 = 3x$$
$$11 = x$$

The solution is 11 and the solution set is {11}.

Solve: $3 \ln 2x = 15$.

$$\ln 2x = 5$$
$$\log_e 2x = 5$$
$$e^5 = 2x$$
$$\frac{e^5}{2} = x$$

The solution is $\dfrac{e^5}{2}$ and the solution set is $\left\{ \dfrac{e^5}{2} \right\}$.

Logarithmic equations in the form $\log_b M = \log_b N$, where $M > 0$ and $N > 0$, can be solved using the one-to-one property of logarithms:

If $\log_b M = \log_b N$, then $M = N$.

Solve: $\log(2x - 1) = \log(4x - 3) - \log x$.

$$\log(2x - 1) = \log\left(\frac{4x - 3}{x} \right)$$
$$2x - 1 = \frac{4x - 3}{x}$$
$$x(2x - 1) = 4x - 3$$
$$2x^2 - x = 4x - 3$$
$$2x^2 - 5x + 3 = 0$$
$$(2x - 3)(x - 1) = 0$$
$$2x - 3 = 0 \quad \text{or} \quad x - 1 = 0$$
$$x = \frac{3}{2} \qquad\qquad x = 1$$

Neither number produces the logarithm of 0 or logarithms of negative numbers in the original equation. The solutions are 1 and $\dfrac{3}{2}$, and the solution set is $\left\{ 1, \dfrac{3}{2} \right\}$.

Definitions and Concepts	**Examples**

Section 12.5 Exponential Growth and Decay; Modeling Data

Exponential growth and decay models are given by $A = A_0 e^{kt}$ in which t represents time, A_0 is the amount present at $t = 0$, and A is the amount present at time t. If $k > 0$, the model describes growth and k is the growth rate. If $k < 0$, the model describes decay and k is the decay rate. Scatter plots for exponential and logarithmic models are shown in **Figure 12.20** on page 913. When using a graphing utility to model data, the closer that the correlation coefficient r is to -1 or 1, the better the model fits the data.

The 1970 population of the Tokyo, Japan, urban area was 16.5 million: in 2000, it was 26.4 million. Write an exponential growth function that describes the population, in millions, t years after 1970. Begin with $A = A_0 e^{kt}$.

$A = 16.5 e^{kt}$ In 1970 ($t = 0$), the population was 16.5 million.

$26.4 = 16.5 e^{k \cdot 30}$ When $t = 30$ (in 2000), $A = 26.4$.

$e^{30k} = \dfrac{26.4}{16.5}$ Isolate the exponential factor.

$\ln e^{30k} = \ln\left(\dfrac{26.4}{16.5}\right)$ Take the natural logarithm on both sides.

$30k = \ln\left(\dfrac{26.4}{16.5}\right)$ and $k = \dfrac{\ln\left(\dfrac{26.4}{16.5}\right)}{30} \approx 0.016$

The growth function is $A = 16.5 e^{0.016t}$.

> Growth rate is 0.016 or 1.6%.

Expressing an Exponential Model in Base e

$y = ab^x$ is equivalent to $y = ae^{(\ln b)x}$.

Rewrite in terms of base e: $y = 24(7.2)^x$.

$$y = 24 e^{(\ln 7.2)x} \approx 24 e^{1.974x}$$

CHAPTER 12 REVIEW EXERCISES

12.1 *In Exercises 1–4, set up a table of coordinates for each function. Select integers from −2 to 2, inclusive, for x. Then use the table of coordinates to match the function with its graph. [The graphs are labeled (a) through (d).]*

1. $f(x) = 4^x$

2. $f(x) = 4^{-x}$

3. $f(x) = -4^{-x}$

4. $f(x) = -4^{-x} + 3$

a.

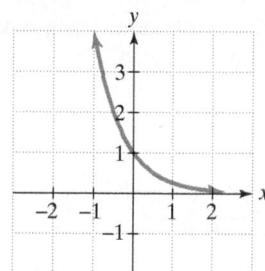

b.

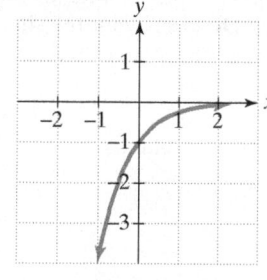

c.

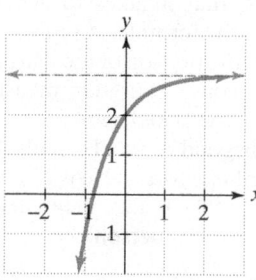

d.

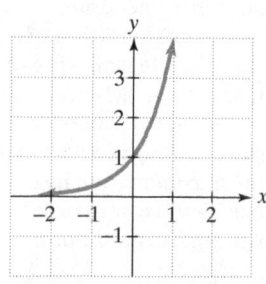

In Exercises 5–8, graph functions f and g in the same rectangular coordinate system. Select integers from −2 to 2, inclusive, for x. Then describe how the graph of g is related to the graph of f. If applicable, use a graphing utility to confirm your hand-drawn graphs.

5. $f(x) = 2^x$ and $g(x) = 2^{x-1}$

6. $f(x) = 2^x$ and $g(x) = \left(\dfrac{1}{2}\right)^x$

7. $f(x) = 3^x$ and $g(x) = 3^x - 1$

8. $f(x) = 3^x$ and $g(x) = -3^x$

Use the compound interest formulas

$$A = P\left(1 + \dfrac{r}{n}\right)^{nt} \quad \text{and} \quad A = Pe^{rt}$$

to solve Exercises 9–10.

9. Suppose that you have \$5000 to invest. Which investment yields the greater return over 5 years: 5.5% compounded semiannually or 5.25% compounded monthly?

10. Suppose that you have $14,000 to invest. Which investment yields the greater return over 10 years: 7% compounded monthly or 6.85% compounded continuously?

11. A cup of coffee is taken out of a microwave oven and placed in a room. The temperature, T, in degrees Fahrenheit, of the coffee after t minutes is modeled by the function $T = 70 + 130e^{-0.04855t}$. The graph of the function is shown in the figure.

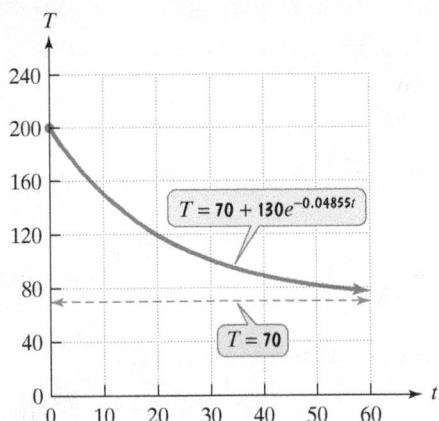

Use the graph to answer each of the following questions.

a. What was the temperature of the coffee when it was first taken out of the microwave?

b. What is a reasonable estimate of the temperature of the coffee after 20 minutes? Use your calculator to verify this estimate.

c. What is the limit of the temperature to which the coffee will cool? What does this tell you about the temperature of the room?

12.2 *In Exercises 12–14, write each equation in its equivalent exponential form.*

12. $\dfrac{1}{2} = \log_{49} 7$ **13.** $3 = \log_4 x$

14. $\log_3 81 = y$

In Exercises 15–17, write each equation in its equivalent logarithmic form.

15. $6^3 = 216$ **16.** $b^4 = 625$

17. $13^y = 874$

In Exercises 18–28, evaluate each expression without using a calculator. If evaluation is not possible, state the reason.

18. $\log_4 64$ **19.** $\log_5 \frac{1}{25}$

20. $\log_3(-9)$

21. $\log_{16} 4$ **22.** $\log_{17} 17$

23. $\log_3 3^8$ **24.** $\ln e^5$

25. $\log_3 \dfrac{1}{\sqrt{3}}$ **26.** $\ln \dfrac{1}{e^2}$

27. $\log \dfrac{1}{1000}$ **28.** $\log_3(\log_8 8)$

29. Graph $f(x) = 2^x$ and $g(x) = \log_2 x$ in the same rectangular coordinate system. Use the graphs to determine each function's domain and range.

30. Graph $f(x) = \left(\frac{1}{3}\right)^x$ and $g(x) = \log_{\frac{1}{3}} x$ in the same rectangular coordinate system. Use the graphs to determine each function's domain and range.

In Exercises 31–33, find the domain of each logarithmic function.

31. $f(x) = \log_8(x + 5)$

32. $f(x) = \log(3 - x)$

33. $f(x) = \ln(x - 1)^2$

In Exercises 34–36, simplify each expression.

34. $\ln e^{6x}$ **35.** $e^{\ln \sqrt{x}}$ **36.** $10^{\log 4x^2}$

37. On the Richter scale, the magnitude, R, of an earthquake of intensity I is given by $R = \log \dfrac{I}{I_0}$, where I_0 is the intensity of a barely felt zero-level earthquake. If the intensity of an earthquake is $1000I_0$, what is its magnitude on the Richter scale?

38. Students in a psychology class took a final examination. As part of an experiment to see how much of the course content they remembered over time, they took equivalent forms of the exam in monthly intervals thereafter. The average score, $f(t)$, for the group after t months is modeled by the function $f(t) = 76 - 18 \log(t + 1)$, where $0 \le t \le 12$.

a. What was the average score when the exam was first given?

b. What was the average score, to the nearest tenth, after 2 months? 4 months? 6 months? 8 months? one year?

c. Use the results from parts (a) and (b) to graph f. Describe what the shape of the graph indicates in terms of the material retained by the students.

39. The formula

$$t = \frac{1}{c}\ln\left(\frac{A}{A - N}\right)$$

describes the time, t, in weeks, that it takes to achieve mastery of a portion of a task. In the formula, A represents maximum learning possible, N is the portion of the learning that is to be achieved, and c is a constant used to measure an individual's learning style. A 50-year-old man decides to start running as a way to maintain good health. He feels that the maximum rate he could ever hope to achieve is 12 miles per hour. How many weeks will it take before the man can run 5 miles per hour if $c = 0.06$ for this person?

12.3 *In Exercises 40-43, use properties of logarithms to expand each logarithmic expression as much as possible. Where possible, evaluate logarithmic expressions without using a calculator. Assume that all variables represent positive numbers.*

40. $\log_6(36x^3)$

41. $\log_4\left(\dfrac{\sqrt{x}}{64}\right)$

42. $\log_2\left(\dfrac{xy^2}{64}\right)$

43. $\ln\sqrt[3]{\dfrac{x}{e}}$

In Exercises 44–47, use properties of logarithms to condense each logarithmic expression. Write the expression as a single logarithm whose coefficient is 1.

44. $\log_b 7 + \log_b 3$

45. $\log 3 - 3\log x$

46. $3\ln x + 4\ln y$

47. $\dfrac{1}{2}\ln x - \ln y$

In Exercises 48–49, use common logarithms or natural logarithms and a calculator to evaluate to four decimal places.

48. $\log_6 72{,}348$

49. $\log_4 0.863$

In Exercises 50–53, determine whether each equation is true or false. Where possible, show work to support your conclusion. If the statement is false, make the necessary change(s) to produce a true statement.

50. $(\ln x)(\ln 1) = 0$

51. $\log(x+9) - \log(x+1) = \dfrac{\log(x+9)}{\log(x+1)}$

52. $(\log_2 x)^4 = 4\log_2 x$

53. $\ln e^x = x\ln e$

12.4 *In Exercises 54–59, solve each exponential equation. Where necessary, express the solution set in terms of natural logarithms and use a calculator to obtain a decimal approximation, correct to two decimal places, for the solution.*

54. $2^{4x-2} = 64$

55. $125^x = 25$

56. $9^x = \dfrac{1}{27}$

57. $8^x = 12{,}143$

58. $9e^{5x} = 1269$

59. $30e^{0.045x} = 90$

In Exercises 60–69, solve each logarithmic equation.

60. $\log_5 x = -3$

61. $\log x = 2$

62. $\log_4(3x-5) = 3$

63. $\ln x = -1$

64. $3 + 4\ln(2x) = 15$

65. $\log_2(x+3) + \log_2(x-3) = 4$

66. $\log_3(x-1) - \log_3(x+2) = 2$

67. $\log_4(3x-5) = \log_4 3$

68. $\ln(x+4) - \ln(x+1) = \ln x$

69. $\log_6(2x+1) = \log_6(x-3) + \log_6(x+5)$

70. The function $P(x) = 14.7e^{-0.21x}$ models the average atmospheric pressure, $P(x)$, in pounds per square inch, at an altitude of x miles above sea level. The atmospheric pressure at the peak of Mt. Everest, the world's highest mountain, is 4.6 pounds per square inch. How many miles above sea level, to the nearest tenth of a mile, is the peak of Mt. Everest?

71. The amount of carbon dioxide in the atmosphere, measured in parts per million, has been increasing as a result of the burning of oil and coal. The buildup of gases and particles traps heat and raises the planet's temperature, a phenomenon called the greenhouse effect. Carbon dioxide accounts for about half of the warming. The function $f(t) = 364(1.005)^t$ projects carbon dioxide concentration, $f(t)$, in parts per million, t years after 2000. Using the projections given by the function, when will the carbon dioxide concentration be double the preindustrial level of 280 parts per million?

72. The function $W(x) = 0.37\ln x + 0.05$ models the average walking speed, $W(x)$, in feet per second, of residents in a city whose population is x thousand. Visitors to New York City frequently feel they are moving too slowly to keep pace with New Yorkers' average walking speed of 3.38 feet per second. What is the population of New York City? Round to the nearest thousand.

73. Use the compound interest formula

$$A = P\left(1 + \frac{r}{n}\right)^{nt}$$

to solve this problem. How long, to the nearest tenth of a year, will it take $12,500 to grow to $20,000 at 6.5% annual interest compounded quarterly?

Use the compound interest formula

$$A = Pe^{rt}$$

to solve Exercises 74–75.

74. How long, to the nearest tenth of a year, will it take $50,000 to triple in value at 7.5% annual interest compounded continuously?

75. What interest rate is required for an investment subject to continuous compounding to triple in 5 years?

12.5

76. According to the U.S. Bureau of the Census, in 1990 there were 22.4 million residents of Hispanic origin living in the United States. By 2008 the number had increased to 46.9 million. The exponential growth function $A = 22.4e^{kt}$ describes the U.S. Hispanic population, A, in millions, t years after 1990.

   a. Find k, correct to three decimal places.

   b. Use the resulting model to project the Hispanic resident population in 2015.

   c. In which year will the Hispanic resident population reach 68 million?

77. Use the exponential decay model, $A = A_0 e^{kt}$, to solve this exercise. The half-life of polonium-210 is 140 days. How long will it take for a sample of this substance to decay to 20% of its original amount?

Exercises 78–79 present data in the form of tables. For each data set shown by the table,

a. *Create a scatter plot for the data.*

b. *Use the scatter plot to determine whether an exponential function or a logarithmic function is a better choice for modeling the data.*

78. Average Savings by Getting More Than One Bid on a Reroofing Job

Number of Bids	Average Amount Saved
2	$1171
3	$1538
4	$1734
5	$1901

Source: www.Checkbook.org

79. Percentage of U.S. Households with HDTV Sets

Year	Percent
2001	1%
2004	10%
2006	26%
2009	47%

Source: Consumer Electronics Association

In Exercises 80–81, rewrite the equation in terms of base e. Express the answer in terms of a natural logarithm and then round to three decimal places.

80. $y = 73(2.6)^x$

81. $y = 6.5(0.43)^x$

82. The figure shows world population projections through the year 2150. The data are from the United Nations Family Planning Program and are based on optimistic or pessimistic expectations for successful control of human population growth. Suppose that you are interested in modeling these data using exponential, logarithmic, linear, and quadratic functions. Which function would you use to model each of the projections? Explain your choices. For the choice corresponding to a quadratic model, would your formula involve one with a positive or negative leading coefficient? Explain.

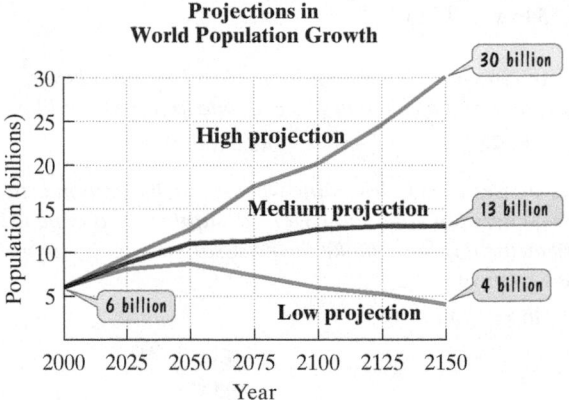

Projections in World Population Growth

CHAPTER 12 TEST

CHAPTER Test Prep VIDEOS

Step-by-step test solutions are found on the Chapter Test Prep Videos available in MyMathLab® or on YouTube (search "BlitzerCombinedAlg" and click on "Channels").

1. Graph $f(x) = 2^x$ and $g(x) = 2^{x+1}$ in the same rectangular coordinate system.

2. Use $A = P\left(1 + \dfrac{r}{n}\right)^{nt}$ and $A = Pe^{rt}$ to solve this problem. Suppose you have $3000 to invest. Which investment yields the greater return over 10 years: 6.5% compounded semiannually or 6% compounded continuously? How much more (to the nearest dollar) is yielded by the better investment?

3. Write in exponential form: $\log_5 125 = 3$.

4. Write in logarithmic form: $\sqrt{36} = 6$.

5. Graph $f(x) = 3^x$ and $g(x) = \log_3 x$ in the same rectangular coordinate system. Use the graphs to determine each function's domain and range.

In Exercises 6–8, simplify each expression.

6. $\ln e^{5x}$ **7.** $\log_b b$ **8.** $\log_6 1$

9. Find the domain: $f(x) = \log_5 (x - 7)$.

10. On the decibel scale, the loudness of a sound, in decibels, is given by $D = 10 \log \dfrac{I}{I_0}$, where I is the intensity of the sound, in watts per meter^2, and I_0 is the intensity of a sound barely audible to the human ear. If the intensity of a sound is $10^{12} I_0$, what is its loudness in decibels? (Such a sound is potentially damaging to the ear.)

In Exercises 11–12, use properties of logarithms to expand each logarithmic expression as much as possible. Where possible, evaluate logarithmic expressions without using a calculator.

11. $\log_4(64x^5)$

12. $\log_3\left(\dfrac{\sqrt[3]{x}}{81}\right)$

In Exercises 13–14, write each expression as a single logarithm.

13. $6 \log x + 2 \log y$

14. $\ln 7 - 3 \ln x$

15. Use a calculator to evaluate $\log_{15} 71$ to four decimal places.

In Exercises 16–23, solve each equation.

16. $3^{x-2} = 81$

17. $5^x = 1.4$

18. $400e^{0.005x} = 1600$

19. $\log_{25} x = \dfrac{1}{2}$

20. $\log_6(4x - 1) = 3$

21. $2 \ln(3x) = 8$

22. $\log x + \log(x + 15) = 2$

23. $\ln(x - 4) - \ln(x + 1) = \ln 6$

24. The function

$$A = 82.4e^{-0.002t}$$

models the population of Germany, A, in millions, t years after 2006.

a. What was the population of Germany in 2006?

b. Is the population of Germany increasing or decreasing? Explain.

c. In which year will the population of Germany be 80.6 million?

Use the formulas

$$A = P\left(1 + \frac{r}{n}\right)^{nt} \quad \text{and} \quad A = Pe^{rt}$$

to solve Exercises 25–26.

25. How long, to the nearest tenth of a year, will it take $4000 to grow to $8000 at 5% annual interest compounded quarterly?

26. What interest rate is required for an investment subject to continuous compounding to double in 10 years?

27. The 1990 population of Europe was 509 million; in 2000, it was 729 million. Write an exponential growth function that describes the population of Europe, in millions, t years after 1990.

28. Use the exponential decay model for carbon-14, $A = A_0 e^{-0.000121t}$, to solve this exercise. Bones of a prehistoric man were discovered and contained 5% of the original amount of carbon-14. How long ago did the man die?

In Exercises 29–32, determine whether the values in each table belong to an exponential function, a logarithmic function, a linear function, or a quadratic function.

29.

x	y
0	3
1	1
2	-1
3	-3
4	-5

30.

x	y
$\frac{1}{3}$	-1
1	0
3	1
9	2
27	3

31.

x	y
0	1
1	5
2	25
3	125
4	625

32.

x	y
0	12
1	3
2	0
3	3
4	12

33. Rewrite $y = 96(0.38)^x$ in terms of base e. Express the answer in terms of a natural logarithm and then round to three decimal places.

CUMULATIVE REVIEW EXERCISES (CHAPTERS 1–12)

In Exercises 1–7, solve each equation, inequality, or system.

1. $8 - (4x - 5) = x - 7$

2. $\begin{cases} 5x + 4y = 22 \\ 3x - 8y = -18 \end{cases}$

3. $\begin{cases} -3x + 2y + 4z = 6 \\ 7x - y + 3z = 23 \\ 2x + 3y + z = 7 \end{cases}$

4. $|x - 1| > 3$

5. $\sqrt{x + 4} - \sqrt{x - 4} = 2$

6. $x - 4 \geq 0$ and $-3x \leq -6$

7. $2x^2 = 3x - 2$

In Exercises 8–12, graph each function, equation, or inequality in a rectangular coordinate system.

8. $3x = 15 + 5y$

9. $2x - 3y > 6$

10. $f(x) = -\dfrac{1}{2}x + 1$

11. $f(x) = x^2 + 6x + 8$

12. $f(x) = (x - 3)^2 - 4$

13. Solve for c: $A = \dfrac{cd}{c + d}$.

In Exercises 14–16, let $f(x) = x^2 + 3x - 15$ and $g(x) = x - 2$. Find each indicated expression.

14. $f(g(x))$

15. $g(f(x))$

16. $g(a + h) - g(a)$

17. If $f(x) = 7x - 3$, find $f^{-1}(x)$.

In Exercises 18–19, find the domain of each function.

18. $f(x) = \dfrac{x - 2}{x^2 - 3x + 2}$

19. $f(x) = \ln(2x - 8)$

20. Write the equation of the linear function whose graph contains the point $(-2, 4)$ and is perpendicular to the line whose equation is $2x + y = 10$.

In Exercises 21–25, perform the indicated operations and simplify, if possible.

21. $\dfrac{-5x^3y^7}{15x^4y^{-2}}$

22. $(4x^2 - 5y)^2$

23. $(5x^3 - 24x^2 + 9) \div (5x + 1)$

24. $\dfrac{\sqrt[3]{32xy^{10}}}{\sqrt[3]{2xy^2}}$

25. $\dfrac{x + 2}{x^2 - 6x + 8} + \dfrac{3x - 8}{x^2 - 5x + 6}$

In Exercises 26–27, factor completely.

26. $x^4 - 4x^3 + 8x - 32$

27. $2x^2 + 12xy + 18y^2$

28. Write as a single logarithm whose coefficient is 1:

$$2 \ln x - \dfrac{1}{2}\ln y.$$

29. The length of a rectangular carpet is 4 feet greater than twice its width. If the area is 48 square feet, find the carpet's length and width.

30. Working alone, you can mow the lawn in 2 hours and your sister can do it in 3 hours. How long will it take you to do the job if you work together?

31. Your motorboat can travel 15 miles per hour in still water. Traveling with the river's current, the boat can cover 20 miles in the same time it takes to go 10 miles against the current. Find the rate of the current.

32. Use the formula for continuous compounding, $A = Pe^{rt}$, to solve this problem. What interest rate is required for an investment of $6000 subject to continuous compounding to grow to $18,000 in 10 years?

Sequences, Series, and the Binomial Theorem

Something incredible has happened. Your college roommate, a gifted athlete, has been given a six-year contract with a professional baseball team. He will be playing against the likes of Alex Rodriguez and Vernon Wells. Management offers him three options. One is a beginning salary of $1,700,000 with annual increases of $70,000 per year starting in the second year. A second option is $1,700,000 the first year with an annual increase of 2% per year beginning in the second year. The third option involves less money the first year—$1,500,000—but there is an annual increase of 9% yearly after that. Which option offers the most money over the six-year contract?

A similar problem appears as Exercise 77 in Exercise Set 14.3, and this problem appears as the Group Project on page 1036.

Sequences and Summation Notation

Objectives

1 Find particular terms of a sequence from the general term.

2 Use factorial notation.

3 Use summation notation.

Sequences

Many creations in nature involve intricate mathematical designs, including a variety of spirals. For example, the arrangement of the individual florets in the head of a sunflower forms spirals. In some species, there are 21 spirals in the clockwise direction and 34 in the counterclockwise direction. The precise numbers depend on the species of sunflower: 21 and 34, or 34 and 55, or 55 and 89, or even 89 and 144.

This observation becomes more interesting when we consider a sequence of numbers investigated by Leonardo of Pisa, also known as Fibonacci, an Italian mathematician of the thirteenth century. The **Fibonacci sequence** of numbers is an infinite sequence that begins as follows:

$$1, 1, 2, 3, 5, 8, 13, 21, 34, 55, 89, 144, 233, \ldots.$$

The first two terms are 1. Every term thereafter is the sum of the two preceding terms. For example, the third term, 2, is the sum of the first and second terms: $1 + 1 = 2$. The fourth term, 3, is the sum of the second and third terms: $1 + 2 = 3$, and so on. Did you know that the number of spirals in a daisy or a sunflower, 21 and 34, are two Fibonacci numbers? The number of spirals in a pinecone, 8 and 13, and a pineapple, 8 and 13, are also Fibonacci numbers.

We can think of the Fibonacci sequence as a function. The terms of the sequence

$$1, 1, 2, 3, 5, 8, 13, 21, 34, 55, 89, 144, 233, \ldots$$

are the range values for a function f whose domain is the set of positive integers.

Domain:	1,	2,	3,	4,	5,	6,	7,	...
	↓	↓	↓	↓	↓	↓	↓	
Range:	1,	1,	2,	3,	5,	8,	13,	...

Thus, $f(1) = 1, f(2) = 1, f(3) = 2, f(4) = 3, f(5) = 5, f(6) = 8, f(7) = 13$, and so on.

The letter a with a subscript is used to represent function values of a sequence, rather than the usual function notation. The subscripts make up the domain of the sequence and they identify the location of a term. Thus, a_1 represents the first term of the sequence, a_2 represents the second term, a_3 the third term, and so on. This notation is shown for the first six terms of the Fibonacci sequence:

$$1, \qquad 1, \qquad 2, \qquad 3, \qquad 5, \qquad 8.$$

$a_1 = 1 \quad a_2 = 1 \quad a_3 = 2 \quad a_4 = 3 \quad a_5 = 5 \quad a_6 = 8$

The notation a_n represents the nth term, or **general term**, of a sequence. The entire sequence is represented by $\{a_n\}$.

Blitzer Bonus

Fibonacci Numbers on the Piano Keyboard

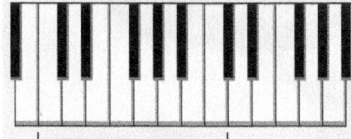

One Octave

Numbers in the Fibonacci sequence can be found in an octave on the piano keyboard. The octave contains 2 black keys in one cluster and 3 black keys in another cluster, for a total of 5 black keys. It also has 8 white keys, for a total of 13 keys. The numbers 2, 3, 5, 8, and 13 are the third through seventh terms of the Fibonacci sequence.

Definition of a Sequence

An **infinite sequence** $\{a_n\}$ is a function whose domain is the set of positive integers. The function values, or **terms**, of the sequence are represented by

$$a_1, a_2, a_3, a_4, \ldots, a_n, \ldots.$$

Sequences whose domains consist only of the first n positive integers are called **finite sequences**.

1 Find particular terms of a sequence from the general term.

EXAMPLE 1 Writing Terms of a Sequence from the General Term

Write the first four terms of the sequence whose nth term, or general term, is given:

a. $a_n = 3n + 4$ **b.** $a_n = \dfrac{(-1)^n}{3^n - 1}.$

Solution

a. We need to find the first four terms of the sequence whose general term is $a_n = 3n + 4$. To do so, we replace n in the formula with 1, 2, 3, and 4.

a_1, 1st term $3 \cdot 1 + 4 = 3 + 4 = 7$ a_2, 2nd term $3 \cdot 2 + 4 = 6 + 4 = 10$

a_3, 3rd term $3 \cdot 3 + 4 = 9 + 4 = 13$ a_4, 4th term $3 \cdot 4 + 4 = 12 + 4 = 16$

The first four terms are $7, 10, 13,$ and 16. The sequence defined by $a_n = 3n + 4$ can be written as

$$7, 10, 13, 16, \ldots, 3n + 4, \ldots.$$

Great Question!

What effect does $(-1)^n$ have on the terms of a sequence?

The factor $(-1)^n$ in the general term of a sequence causes the signs of the terms to alternate between positive and negative, depending on whether n is even or odd.

b. We need to find the first four terms of the sequence whose general term is $a_n = \dfrac{(-1)^n}{3^n - 1}$. To do so, we replace each occurrence of n in the formula with 1, 2, 3, and 4.

a_1, 1st term $\dfrac{(-1)^1}{3^1 - 1} = \dfrac{-1}{3 - 1} = -\dfrac{1}{2}$ a_2, 2nd term $\dfrac{(-1)^2}{3^2 - 1} = \dfrac{1}{9 - 1} = \dfrac{1}{8}$

a_3, 3rd term $\dfrac{(-1)^3}{3^3 - 1} = \dfrac{-1}{27 - 1} = -\dfrac{1}{26}$ a_4, 4th term $\dfrac{(-1)^4}{3^4 - 1} = \dfrac{1}{81 - 1} = \dfrac{1}{80}$

The first four terms are $-\frac{1}{2}, \frac{1}{8}, -\frac{1}{26}$, and $\frac{1}{80}$. The sequence defined by $\dfrac{(-1)^n}{3^n - 1}$ can be written as

$$-\frac{1}{2}, \frac{1}{8}, -\frac{1}{26}, \frac{1}{80}, \ldots, \frac{(-1)^n}{3^n - 1}, \ldots. \quad \blacksquare$$

✓ CHECK POINT 1   Write the first four terms of the sequence whose nth term, or general term, is given:

a. $a_n = 2n + 5$ **b.** $a_n = \dfrac{(-1)^n}{2^n + 1}.$

Although sequences are usually named with the letter a, any lowercase letter can be used. For example, the first four terms of the sequence $\{b_n\} = \left\{ \left(\frac{1}{2} \right)^n \right\}$ are $b_1 = \frac{1}{2}, b_2 = \frac{1}{4}, b_3 = \frac{1}{8},$ and $b_4 = \frac{1}{16}.$

Using Technology

Graphing utilities can write the terms of a sequence and graph them. For example, to find the first six terms of

$$\{a_n\} = \left\{\frac{1}{n}\right\},$$ enter

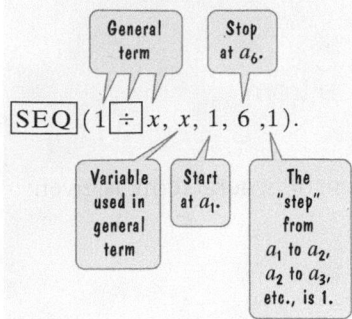

The first few terms of the sequence are shown in the viewing rectangle. By pressing the right arrow key to scroll right, you can see the remaining terms.

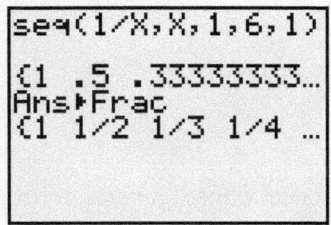

---

2 Use factorial notation.

Using Technology

Calculators have factorial keys. To find 5 factorial, most calculators use one of the following:

Many Scientific Calculators:

$$5 \boxed{x!}$$

Many Graphing Calculators:

$$5 \boxed{!} \boxed{\text{ENTER}}$$

Because $n!$ becomes quite large as n increases, your calculator will display these larger values in scientific notation.

Because a sequence is a function whose domain is the set of positive integers, the **graph of a sequence** is a set of discrete points. For example, consider the sequence whose general term is $a_n = \frac{1}{n}$. How does the graph of this sequence differ from the graph of the rational function $f(x) = \frac{1}{x}$? The graph of $f(x) = \frac{1}{x}$ is shown in **Figure 14.1(a)** for positive values of x. To obtain the graph of the sequence $\{a_n\} = \left\{\frac{1}{n}\right\}$, remove all the points from the graph of f except those whose x-coordinates are positive integers. Thus, we remove all points except $(1, 1), \left(2, \frac{1}{2}\right), \left(3, \frac{1}{3}\right), \left(4, \frac{1}{4}\right)$, and so on. The remaining points are the graph of the sequence $\{a_n\} = \left\{\frac{1}{n}\right\}$, shown in **Figure 14.1(b)**. Notice that the horizontal axis is labeled n and the vertical axis is labeled a_n.

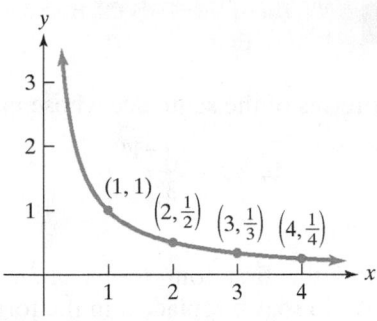 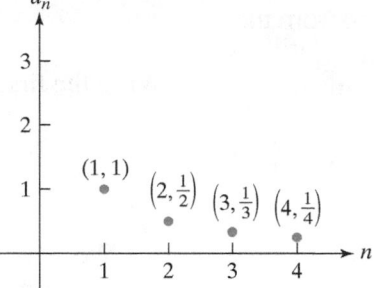

Figure 14.1(a) The graph of $f(x) = \frac{1}{x}, x > 0$

Figure 14.1(b) The graph of $\{a_n\} = \left\{\frac{1}{n}\right\}$

Comparing a continuous graph to the graph of a sequence

Factorial Notation

Products of consecutive positive integers occur quite often in sequences. These products can be expressed in a special notation, called **factorial notation**.

Factorial Notation

If n is a positive integer, the notation $n!$ (read "n factorial") is the product of all positive integers from n down through 1.

$$n! = n(n - 1)(n - 2) \ldots (3)(2)(1)$$

$0!$ (zero factorial), by definition, is 1.

$$0! = 1$$

The values of $n!$ for the first six positive integers are

$$1! = 1$$
$$2! = 2 \cdot 1 = 2$$
$$3! = 3 \cdot 2 \cdot 1 = 6$$
$$4! = 4 \cdot 3 \cdot 2 \cdot 1 = 24$$
$$5! = 5 \cdot 4 \cdot 3 \cdot 2 \cdot 1 = 120$$
$$6! = 6 \cdot 5 \cdot 4 \cdot 3 \cdot 2 \cdot 1 = 720.$$

Factorials affect only the number or variable that they follow unless grouping symbols appear. For example,

$$2 \cdot 3! = 2(3 \cdot 2 \cdot 1) = 2 \cdot 6 = 12$$

whereas

$$(2 \cdot 3)! = 6! = 6 \cdot 5 \cdot 4 \cdot 3 \cdot 2 \cdot 1 = 720.$$

In this sense, factorials are treated similar to exponents in the order of operations.

EXAMPLE 2 Finding Terms of a Sequence Involving Factorials

Write the first four terms of the sequence whose nth term is

$$a_n = \frac{2^n}{(n-1)!}.$$

Solution We need to find the first four terms of the sequence. To do so, we replace each n in the formula with 1, 2, 3, and 4.

a_1, 1st term
$$\frac{2^1}{(1-1)!} = \frac{2}{0!} = \frac{2}{1} = 2$$

a_2, 2nd term
$$\frac{2^2}{(2-1)!} = \frac{4}{1!} = \frac{4}{1} = 4$$

a_3, 3rd term
$$\frac{2^3}{(3-1)!} = \frac{8}{2!} = \frac{8}{2 \cdot 1} = 4$$

a_4, 4th term
$$\frac{2^4}{(4-1)!} = \frac{16}{3!} = \frac{16}{3 \cdot 2 \cdot 1} = \frac{16}{6} = \frac{8}{3}$$

The first four terms are 2, 4, 4, and $\frac{8}{3}$. ∎

✓ **CHECK POINT 2** Write the first four terms of the sequence whose nth term is

$$a_n = \frac{20}{(n+1)!}.$$

③ Use summation notation.

Summation Notation

It is sometimes useful to find the sum of the first n terms of a sequence. For example, consider the cost of raising a child born in the United States in 2006 to a middle-income ($43,200–$72,600 per year) family, shown in **Table 14.1**.

| **Table 14.1** | The Cost of Raising a Child Born in the U.S. in 2006 to a Middle-Income Family | | | | | | | | |

Year	2006	2007	2008	2009	2010	2011	2012	2013	2014
Average Cost	$10,600	$10,930	$11,270	$11,960	$12,330	$12,710	$12,950	$13,350	$13,760
	Child is under 1.	Child is 1.	Child is 2.	Child is 3.	Child is 4.	Child is 5.	Child is 6.	Child is 7.	Child is 8.

Year	2015	2016	2017	2018	2019	2020	2021	2022	2023
Average Cost	$13,970	$14,400	$14,840	$16,360	$16,860	$17,390	$18,430	$19,000	$19,590
	Child is 9.	Child is 10.	Child is 11.	Child is 12.	Child is 13.	Child is 14.	Child is 15.	Child is 16.	Child is 17.

Source: U.S. Department of Agriculture

We can let a_n represent the cost of raising a child in year n, where $n = 1$ corresponds to 2006, $n = 2$ to 2007, $n = 3$ to 2008, and so on. The terms of the finite sequence in **Table 14.1** are given as follows.

10,600, 10,930, 11,270, 11,960, 12,330, 12,710, 12,950, 13,350, 13,760,

a_1 a_2 a_3 a_4 a_5 a_6 a_7 a_8 a_9

13,970, 14,400, 14,840, 16,360, 16,860, 17,390, 18,430, 19,000, 19,590.

a_{10} a_{11} a_{12} a_{13} a_{14} a_{15} a_{16} a_{17} a_{18}

Why might we want to add the terms of this sequence? We do this to find the total cost of raising a child born in 2006 from birth through age 17. Thus,

$$a_1 + a_2 + a_3 + a_4 + a_5 + a_6 + a_7 + a_8 + a_9 + a_{10} + a_{11} + a_{12} + a_{13} + a_{14} + a_{15} + a_{16} + a_{17} + a_{18}$$
$$= 10{,}600 + 10{,}930 + 11{,}270 + 11{,}960 + 12{,}330 + 12{,}710 + 12{,}950 + 13{,}350 + 13{,}760$$
$$+ 13{,}970 + 14{,}400 + 14{,}840 + 16{,}360 + 16{,}860 + 17{,}390 + 18{,}430 + 19{,}000 + 19{,}590$$
$$= 260{,}700.$$

We see that the total cost of raising a child born in 2006 from birth through age 17 is $260,700.

There is a compact notation for expressing the sum of the first n terms of a sequence. For example, rather than write

$$a_1 + a_2 + a_3 + a_4 + a_5 + a_6 + a_7 + a_8 + a_9 + a_{10} + a_{11}$$
$$+ a_{12} + a_{13} + a_{14} + a_{15} + a_{16} + a_{17} + a_{18},$$

we can use *summation notation* to express the sum as

$$\sum_{i=1}^{18} a_i.$$

We read this expression as "the sum as i goes from 1 to 18 of a_i." The letter i is called the *index of summation* and is not related to the use of i to represent $\sqrt{-1}$.

You can think of the symbol Σ (the uppercase Greek letter sigma) as an instruction to add up the terms of a sequence.

Summation Notation

The sum of the first n terms of a sequence is represented by the **summation notation**

$$\sum_{i=1}^{n} a_i = a_1 + a_2 + a_3 + a_4 + \cdots + a_n,$$

where i is the **index of summation**, n is the **upper limit of summation**, and 1 is the **lower limit of summation**.

Any letter can be used for the index of summation. The letters i, j, and k are used commonly. Furthermore, the lower limit of summation can be an integer other than 1.

When we write out a sum that is given in summation notation, we are **expanding the summation**. Example 3 shows how to do this.

EXAMPLE 3 Using Summation Notation

Expand and evaluate the sum:

a. $\sum_{i=1}^{6} (i^2 + 1)$ **b.** $\sum_{k=4}^{7} \left[(-2)^k - 5 \right]$ **c.** $\sum_{i=1}^{5} 3.$

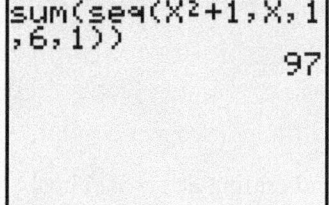
Solution

a. To find $\sum_{i=1}^{6}(i^2 + 1)$, we must replace i in the expression $i^2 + 1$ with all consecutive integers from 1 to 6, inclusive. Then we add.

$$\sum_{i=1}^{6}(i^2 + 1) = (1^2 + 1) + (2^2 + 1) + (3^2 + 1) + (4^2 + 1)$$
$$+ (5^2 + 1) + (6^2 + 1)$$
$$= 2 + 5 + 10 + 17 + 26 + 37$$
$$= 97$$

b. The index of summation in $\sum_{k=4}^{7}\left[(-2)^k - 5\right]$ is k. First we evaluate $(-2)^k - 5$ for all consecutive integers from 4 through 7, inclusive. Then we add.

$$\sum_{k=4}^{7}\left[(-2)^k - 5\right] = \left[(-2)^4 - 5\right] + \left[(-2)^5 - 5\right]$$
$$+ \left[(-2)^6 - 5\right] + \left[(-2)^7 - 5\right]$$
$$= (16 - 5) + (-32 - 5) + (64 - 5) + (-128 - 5)$$
$$= 11 + (-37) + 59 + (-133)$$
$$= -100$$

c. To find $\sum_{i=1}^{5}3$, we observe that every term of the sum is 3. The notation $i = 1$ through 5 indicates that we must add the first five terms of a sequence in which every term is 3.

$$\sum_{i=1}^{5}3 = 3 + 3 + 3 + 3 + 3 = 15 \quad \blacksquare$$

✓ **CHECK POINT 3**   Expand and evaluate the sum:

a. $\sum_{i=1}^{6}2i^2$

b. $\sum_{k=3}^{5}(2^k - 3)$

c. $\sum_{i=1}^{5}4.$

Although the domain of a sequence is the set of positive integers, any integers can be used for the limits of summation. For a given sum, we can vary the upper and lower limits of summation, as well as the letter used for the index of summation. By doing so, we can produce different-looking summation notations for the same sum. For example, the sum of the squares of the first four positive integers, $1^2 + 2^2 + 3^2 + 4^2$, can be expressed in a number of equivalent ways:

$$\sum_{i=1}^{4}i^2 = 1^2 + 2^2 + 3^2 + 4^2 = 30$$

$$\sum_{i=0}^{3}(i + 1)^2 = (0 + 1)^2 + (1 + 1)^2 + (2 + 1)^2 + (3 + 1)^2$$
$$= 1^2 + 2^2 + 3^2 + 4^2 = 30$$

$$\sum_{k=2}^{5}(k - 1)^2 = (2 - 1)^2 + (3 - 1)^2 + (4 - 1)^2 + (5 - 1)^2$$
$$= 1^2 + 2^2 + 3^2 + 4^2 = 30.$$

EXAMPLE 4   Writing Sums in Summation Notation

Express each sum using summation notation:

a. $1^3 + 2^3 + 3^3 + \cdots + 7^3$ **b.** $1 + \dfrac{1}{3} + \dfrac{1}{9} + \dfrac{1}{27} + \cdots + \dfrac{1}{3^{n-1}}.$

Solution   In each case, we will use 1 as the lower limit of summation and i for the index of summation.

a. The sum $1^3 + 2^3 + 3^3 + \cdots + 7^3$ has seven terms, each of the form i^3, starting at $i = 1$ and ending at $i = 7$. Thus,

$$1^3 + 2^3 + 3^3 + \cdots + 7^3 = \sum_{i=1}^{7} i^3.$$

b. The sum

$$1 + \frac{1}{3} + \frac{1}{9} + \frac{1}{27} + \cdots + \frac{1}{3^{n-1}}$$

has n terms, each of the form $\dfrac{1}{3^{i-1}}$, starting at $i = 1$ and ending at $i = n$. Thus,

$$1 + \frac{1}{3} + \frac{1}{9} + \frac{1}{27} + \cdots + \frac{1}{3^{n-1}} = \sum_{i=1}^{n} \frac{1}{3^{i-1}}. \quad \blacksquare$$

✓ **CHECK POINT 4**   Express each sum using summation notation:

a. $1^2 + 2^2 + 3^2 + \cdots + 9^2$ **b.** $1 + \dfrac{1}{2} + \dfrac{1}{4} + \dfrac{1}{8} + \cdots + \dfrac{1}{2^{n-1}}.$

Achieving Success

Don't wait too long after class to review your notes. Reading notes while the classroom experience is fresh in mind will help you to remember what was covered during lecture.

CONCEPT AND VOCABULARY CHECK

Fill in each blank so that the resulting statement is true.

1. $\{a_n\} = a_1, a_2, a_3, a_4, \ldots, a_n, \ldots$ represents an infinite _____, a function whose domain is the set of positive _____. The function values $a_1, a_2, a_3, \ldots$ are called the _____.

2. The nth term of a sequence, represented by a_n, is called the _____ term.

Write the first term of each sequence.

3. $a_n = 5n - 6$ _____   **4.** $a_n = \dfrac{(-1)^n}{4^n - 1}$ _____

5. $5!$, called 5 _____, is the product of all positive integers from _____ down through _____. By definition, $0! =$ _____.

6. $\displaystyle\sum_{i=1}^{n} a_i =$ _____ $+$ _____ $+$ _____ $+ \cdots +$ _____. In this summation notation, i is called the _____ of summation, n is the _____ of summation, and 1 is the _____ of summation.

14.1 EXERCISE SET MyMathLab®

Practice Exercises

In Exercises 1–16, write the first four terms of each sequence whose general term is given.

1. $u_n = 3n + 2$

2. $a_n = 4n - 1$

3. $a_n = 3^n$

4. $a_n = \left(\dfrac{1}{3}\right)^n$

5. $a_n = (-3)^n$

6. $a_n = \left(-\dfrac{1}{3}\right)^n$

7. $a_n = (-1)^n(n + 3)$

8. $a_n = (-1)^{n+1}(n + 4)$

9. $a_n = \dfrac{2n}{n + 4}$

10. $a_n = \dfrac{3n}{n + 5}$

11. $a_n = \dfrac{(-1)^{n+1}}{2^n - 1}$

12. $a_n = \dfrac{(-1)^{n+1}}{2^n + 1}$

13. $a_n = \dfrac{n^2}{n!}$

14. $a_n = \dfrac{(n + 1)!}{n^2}$

15. $a_n = 2(n + 1)!$

16. $a_n = -2(n - 1)!$

In Exercises 17–30, find each indicated sum.

17. $\displaystyle\sum_{i=1}^{6} 5i$

18. $\displaystyle\sum_{i=1}^{6} 7i$

19. $\displaystyle\sum_{i=1}^{4} 2i^2$

20. $\displaystyle\sum_{i=1}^{5} i^3$

21. $\displaystyle\sum_{k=1}^{5} k(k + 4)$

22. $\displaystyle\sum_{k=1}^{4} (k - 3)(k + 2)$

23. $\displaystyle\sum_{i=1}^{4} \left(-\dfrac{1}{2}\right)^i$

24. $\displaystyle\sum_{i=2}^{4} \left(-\dfrac{1}{3}\right)^i$

25. $\displaystyle\sum_{i=5}^{9} 11$

26. $\displaystyle\sum_{i=3}^{7} 12$

27. $\displaystyle\sum_{i=0}^{4} \dfrac{(-1)^i}{i!}$

28. $\displaystyle\sum_{i=0}^{4} \dfrac{(-1)^{i+1}}{(i + 1)!}$

29. $\displaystyle\sum_{i=1}^{5} \dfrac{i!}{(i - 1)!}$

30. $\displaystyle\sum_{i=1}^{5} \dfrac{(i + 2)!}{i!}$

In Exercises 31–42, express each sum using summation notation. Use 1 as the lower limit of summation and i for the index of summation.

31. $1^2 + 2^2 + 3^2 + \cdots + 15^2$

32. $1^4 + 2^4 + 3^4 + \cdots + 12^4$

33. $2 + 2^2 + 2^3 + \cdots + 2^{11}$

34. $5 + 5^2 + 5^3 + \cdots + 5^{12}$

35. $1 + 2 + 3 + \cdots + 30$

36. $1 + 2 + 3 + \cdots + 40$

37. $\dfrac{1}{2} + \dfrac{2}{3} + \dfrac{3}{4} + \cdots + \dfrac{14}{14 + 1}$

38. $\dfrac{1}{3} + \dfrac{2}{4} + \dfrac{3}{5} + \cdots + \dfrac{16}{16 + 2}$

39. $4 + \dfrac{4^2}{2} + \dfrac{4^3}{3} + \cdots + \dfrac{4^n}{n}$

40. $\dfrac{1}{9} + \dfrac{2}{9^2} + \dfrac{3}{9^3} + \cdots + \dfrac{n}{9^n}$

41. $1 + 3 + 5 + \cdots + (2n - 1)$

42. $a + ar + ar^2 + \cdots + ar^{n-1}$

In Exercises 43–48, express each sum using summation notation. Use a lower limit of summation of your choice and k for the index of summation.

43. $5 + 7 + 9 + 11 + \cdots + 31$

44. $6 + 8 + 10 + 12 + \cdots + 32$

45. $a + ar + ar^2 + \cdots + ar^{12}$

46. $a + ar + ar^2 + \cdots + ar^{14}$

47. $a + (a + d) + (a + 2d) + \cdots + (a + nd)$

48. $(a + d) + (a + d^2) + \cdots + (a + d^n)$

Practice PLUS

In Exercises 49–56, use the graphs of $\{a_n\}$ and $\{b_n\}$ to find each indicated sum.

The Graph of $\{a_n\}$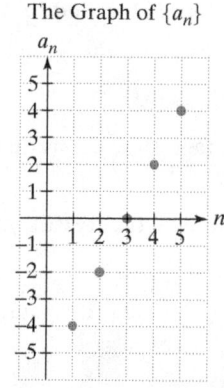

The Graph of $\{b_n\}$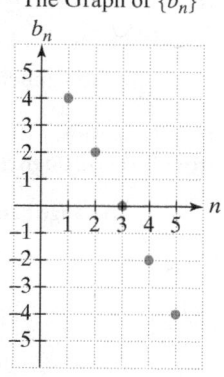

49. $\displaystyle\sum_{i=1}^{5} (a_i^2 + 1)$

50. $\displaystyle\sum_{i=1}^{5} (b_i^2 - 1)$

51. $\displaystyle\sum_{i=1}^{5} (2a_i + b_i)$

52. $\displaystyle\sum_{i=1}^{5} (a_i + 3b_i)$

53. $\displaystyle\sum_{i=4}^{5} \left(\dfrac{a_i}{b_i}\right)^2$

54. $\displaystyle\sum_{i=4}^{5} \left(\dfrac{a_i}{b_i}\right)^3$

55. $\displaystyle\sum_{i=1}^{5} a_i^2 + \sum_{i=1}^{5} b_i^2$

56. $\displaystyle\sum_{i=1}^{5} a_i^2 - \sum_{i=3}^{5} b_i^2$

Application Exercises

57. One in one hundred. That's the ratio of individuals in the U.S. population with autistic disorder, better known as autism. The disorder involves severe deficits in language, social bonding, and imagination, and is often accompanied by mental retardation. The bar graph shows the number of autism cases diagnosed in the United States from 2001 through 2008.

Autism Cases Diagnosed in the United States

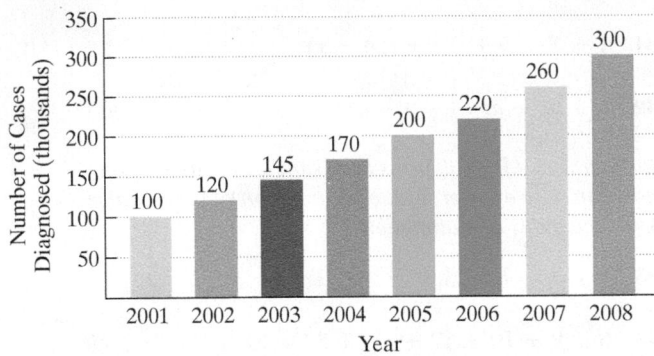

Source: Lilienfeld et al., *Psychology*, Second Edition, Pearson, 2011.

Let a_n represent the number of autism cases diagnosed in the United States, in thousands, n years after 2000.

a. Use the numbers given in the graph to find and interpret $\sum_{i=1}^{8} a_i$.

b. The finite sequence whose general term is
$$a_n = 28n + 63,$$
where $n = 1, 2, 3, \ldots, 8$, models the number of autism cases diagnosed in the United States, in thousands, n years after 2000. Use this model to find $\sum_{i=1}^{8} a_i$. Does this underestimate or overestimate the actual sum in part (a)? By how much?

58. The bar graph shows the average state cigarette tax per pack from 2005 through 2010.

Average State Cigarette Tax per Pack

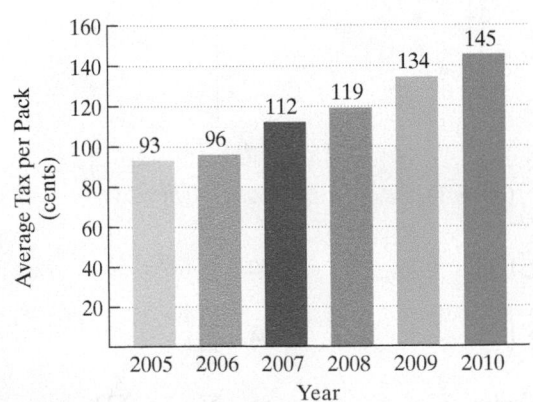

Source: "Trends in Average State Cigarette Tax Rates." Campaign for Tobacco-Free Kids

Let a_n represent the average state cigarette tax per pack, in cents, n years after 2004.

a. Use the numbers given in the graph to find and interpret $\frac{1}{6}\sum_{i=1}^{6} a_i$.

b. The finite sequence whose general term is
$$a_n = 11n + 78,$$
where $n = 1, 2, 3, 4, 5, 6$, models the average state cigarette tax per pack, in cents, n years after 2004. Use this model to find $\frac{1}{6}\sum_{i=1}^{6} a_i$. How does this compare with the actual value you obtained in part (a)?

59. A deposit of $6000 is made in an account that earns 6% interest compounded quarterly. The balance in the account after n quarters is given by the sequence
$$a_n = 6000\left(1 + \frac{0.06}{4}\right)^n, \quad n = 1, 2, 3, \ldots.$$
Find the balance in the account after five years. Round to the nearest cent.

60. A deposit of $10,000 is made in an account that earns 8% interest compounded quarterly. The balance in the account after n quarters is given by the sequence
$$a_n = 10,000\left(1 + \frac{0.08}{4}\right)^n, \quad n = 1, 2, 3, \ldots.$$
Find the balance in the account after six years. Round to the nearest cent.

Writing in Mathematics

61. What is a sequence? Give an example with your description.

62. Explain how to write terms of a sequence if the formula for the general term is given.

63. What does the graph of a sequence look like? How is it obtained?

64. Explain how to find $n!$ if n is a positive integer.

65. What is the meaning of the symbol Σ? Give an example with your description.

66. You buy a new car for $24,000. At the end of n years, the value of your car is given by the sequence
$$a_n = 24,000\left(\frac{3}{4}\right)^n, \quad n = 1, 2, 3, \ldots.$$
Find a_5 and write a sentence explaining what this value represents. Describe the nth term of the sequence in terms of the value of your car at the end of each year.

Technology Exercises

67. Use the $\boxed{\text{SEQ}}$ (sequence) capability of a graphing utility to verify the terms of the sequences you obtained for any five sequences from Exercises 1–16.

68. Use the $\boxed{\text{SUM}}$ $\boxed{\text{SEQ}}$ (sum of the sequence) capability of a graphing utility to verify any five of the sums you obtained in Exercises 17–30.

Many graphing utilities have a sequence-graphing mode that plots the terms of a sequence as points on a rectangular coordinate system. Consult your manual; if your graphing utility has this capability, use it to graph each of the sequences in Exercises 69–72. What appears to be happening to the terms of each sequence as n gets larger?

69. $a_n = \dfrac{n}{n+1}$; n: $[0, 10, 1]$ by a_n: $[0, 1, 0.1]$

70. $a_n = \dfrac{100}{n}$; n: $[0, 1000, 100]$ by a_n: $[0, 1, 0.1]$

71. $a_n = \dfrac{2n^2 + 5n - 7}{n^3}$; n: $[0, 10, 1]$ by a_n: $[0, 2, 0.2]$

72. $a_n = \dfrac{3n^4 + n - 1}{5n^4 + 2n^2 + 1}$; n: $[0, 10, 1]$ by a_n: $[0, 1, 0.1]$

Critical Thinking Exercises

Make Sense?   *In Exercises 73–76, determine whether each statement "makes sense" or "does not make sense" and explain your reasoning.*

73. Now that I've studied sequences, I realize that the joke in this cartoon is based on the fact that you can't have a negative number of sheep.

WHEN MATHEMATICIANS CAN'T SLEEP

74. By writing $a_1, a_2, a_3, a_4, \ldots, a_n, \ldots$, I can see that the range of a sequence is the set of positive integers.

75. It makes a difference whether or not I use parentheses around the expression following the summation symbol, because the value of $\sum\limits_{i=1}^{8}(i + 7)$ is 92, but the value of $\sum\limits_{i=1}^{8} i + 7$ is 43.

76. Without writing out the terms, I can see that $(-1)^{2n}$ in $a_n = \dfrac{(-1)^{2n}}{3n}$ causes the terms to alternate in sign.

In Exercises 77–80, determine whether each statement is true or false. If the statement is false, make the necessary change(s) to produce a true statement.

77. $\sum\limits_{i=1}^{2}(-1)^i 2^i = 0$

78. $\sum\limits_{i=1}^{2} a_i b_i = \sum\limits_{i=1}^{2} a_i \sum\limits_{i=1}^{2} b_i$

79. $\sum\limits_{i=1}^{4} 3i + \sum\limits_{i=1}^{4} 4i = \sum\limits_{i=1}^{4} 7i$

80. $\sum\limits_{i=0}^{6} (-1)^i (i + 1)^2 = \sum\limits_{j=1}^{7} (-1)^j j^2$

In Exercises 81–88, find a general term, a_n, for each sequence. More than one answer may be possible.

81. $1, \dfrac{1}{2}, \dfrac{1}{3}, \dfrac{1}{4}, \ldots$

82. $1, 4, 9, 16, \ldots$

83. $-1, 1, -1, 1, \ldots$

84. $1 \cdot 3, 2 \cdot 4, 3 \cdot 5, 4 \cdot 6, \ldots$

85. $\dfrac{3}{2}, \dfrac{4}{3}, \dfrac{5}{4}, \dfrac{6}{5}, \ldots$

86. $5, 7, 9, 11, \ldots$

87. $\dfrac{4}{1}, \dfrac{9}{2}, \dfrac{16}{3}, \dfrac{25}{4}, \ldots$

88. $4, -8, 16, -32, \ldots$

89. Evaluate without using a calculator: $\dfrac{600!}{599!}$.

In Exercises 90–91, rewrite each expression as a polynomial in standard form.

90. $\dfrac{(n + 4)!}{(n + 2)!}$

91. $\dfrac{n!}{(n - 3)!}$

In Exercises 92–93, expand and write the answer as a single logarithm with a coefficient of 1.

92. $\sum\limits_{i=1}^{4} \log(2i)$

93. $\sum\limits_{i=2}^{4} 2i \log x$

94. If $a_1 = 7$ and $a_n = a_{n-1} + 5$ for $n \geq 2$, write the first four terms of the sequence.

Review Exercises

95. Simplify: $\sqrt[3]{40x^4 y^7}$.
    (Section 10.3, Example 5)

96. Factor: $27x^3 - 8$.
    (Section 6.4, Example 8)

97. Solve: $\dfrac{6}{x} + \dfrac{6}{x + 2} = \dfrac{5}{2}$.
    (Section 7.6, Example 3)

Preview Exercises

Exercises 98–100 will help you prepare for the material covered in the next section.

98. Consider the sequence $8, 3, -2, -7, -12, \ldots$. Find $a_2 - a_1, a_3 - a_2, a_4 - a_3$, and $a_5 - a_4$. What do you observe?

99. Consider the sequence whose nth term is $a_n = 4n - 3$. Find $a_2 - a_1, a_3 - a_2, a_4 - a_3$, and $a_5 - a_4$. What do you observe?

100. Use the formula $a_n = 4 + (n - 1)(-7)$ to find the eighth term of the sequence $4, -3, -10, \ldots$.

Arithmetic Sequences

Objectives

1 Find the common difference for an arithmetic sequence.

2 Write terms of an arithmetic sequence.

3 Use the formula for the general term of an arithmetic sequence.

4 Use the formula for the sum of the first n terms of an arithmetic sequence.

Your grandmother and her financial counselor are looking at options in case an adult residential facility is needed in the future. The good news is that your grandmother's total assets are $500,000. The bad news is that yearly adult residential community costs average $64,130, increasing by $1800 each year. In this section, we will see how sequences can be used to describe your grandmother's situation and help her to identify realistic options.

1 Find the common difference for an arithmetic sequence.

Arithmetic Sequences

The bar graph in **Figure 14.2** shows how much Americans spent on their pets, rounded to the nearest billion dollars, each year from 2001 through 2007.

The graph illustrates that each year spending increased by $2 billion. The sequence of annual spending

$$29, 31, 33, 35, 37, 39, 41, \ldots$$

shows that each term after the first, 29, differs from the preceding term by a constant amount, namely 2. This sequence is an example of an *arithmetic sequence*.

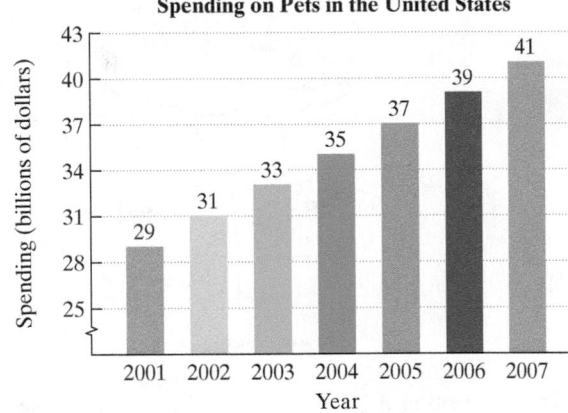

Spending on Pets in the United States

Figure 14.2

Source: American Pet Products Manufacturers Association

Definition of an Arithmetic Sequence

An **arithmetic sequence** is a sequence in which each term after the first differs from the preceding term by a constant amount. The difference between consecutive terms is called the **common difference** of the sequence.

The common difference, d, is found by subtracting any term from the term that directly follows it. In the following examples, the common difference is found by subtracting the first term from the second term, $a_2 - a_1$.

Arithmetic Sequence	**Common Difference**
$142, 146, 150, 154, 158, \ldots$	$d = 146 - 142 = 4$
$-5, -2, 1, 4, 7, \ldots$	$d = -2 - (-5) = -2 + 5 = 3$
$8, 3, -2, -7, -12, \ldots$	$d = 3 - 8 = -5$

Figure 14.3 shows the graphs of the last two arithmetic sequences in our list. The common difference for the increasing sequence in **Figure 14.3(a)** is 3. The common difference for the decreasing sequence in **Figure 14.3(b)** is -5.

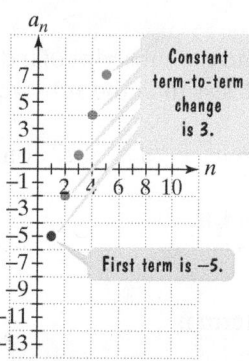

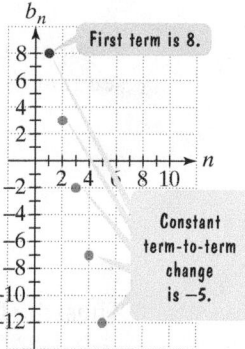

Figure 14.3(a) The graph of $\{a_n\} = -5, -2, 1, 4, 7, \ldots$

Figure 14.3(b) The graph of $\{b_n\} = 8, 3, -2, -7, -12, \ldots$

The graph of each arithmetic sequence in **Figure 14.3** forms a set of discrete points lying on a straight line. This illustrates that **an arithmetic sequence is a linear function whose domain is the set of positive integers**.

If the first term of an arithmetic sequence is a_1, each term after the first is obtained by adding d, the common difference, to the previous term.

2 Write terms of an arithmetic sequence.

EXAMPLE 1   Writing the Terms of an Arithmetic Sequence Using the First Term and the Common Difference

Write the first six terms of the arithmetic sequence with first term 6 and common difference -2.

Solution   To find the second term, we add -2 to the first term, 6, giving 4. For the next term, we add -2 to 4, and so on.

$$
\begin{aligned}
a_1\,(\text{first term}) &= 6 \\
a_2\,(\text{second term}) &= 6 + (-2) = 4 \\
a_3\,(\text{third term}) &= 4 + (-2) = 2 \\
a_4\,(\text{fourth term}) &= 2 + (-2) = 0 \\
a_5\,(\text{fifth term}) &= 0 + (-2) = -2 \\
a_6\,(\text{sixth term}) &= -2 + (-2) = -4
\end{aligned}
$$

The first six terms are

$$6, 4, 2, 0, -2, \text{ and } -4. \quad \blacksquare$$

✓ **CHECK POINT 1**   Write the first six terms of the arithmetic sequence with first term 100 and common difference -30.

③ Use the formula for the general term of an arithmetic sequence.

The General Term of an Arithmetic Sequence

Consider an arithmetic sequence whose first term is a_1 and whose common difference is d. We are looking for a formula for the general term, a_n. Let's begin by writing the first six terms. The first term is a_1. The second term is $a_1 + d$. The third term is $a_1 + d + d$, or $a_1 + 2d$. Thus, we start with a_1 and add d to each successive term. The first six terms are

$$a_1, \qquad a_1 + d, \qquad a_1 + 2d, \qquad a_1 + 3d, \qquad a_1 + 4d, \qquad a_1 + 5d.$$

| a_1, first term | a_2, second term | a_3, third term | a_4, fourth term | a_5, fifth term | a_6, sixth term |

Compare the coefficient of d and the subscript of a denoting the term number. Can you see that the coefficient of d is 1 less than the subscript of a denoting the term number?

a_3: third term $= a_1 + 2d$ $\qquad$ a_4: fourth term $= a_1 + 3d$

One less than 3, or 2, is the coefficient of d. $\qquad$ One less than 4, or 3, is the coefficient of d.

Thus, the formula for the nth term is

$$a_n\text{: }n\text{th term} = a_1 + (n - 1)d.$$

One less than n, or $n - 1$, is the coefficient of d.

General Term of an Arithmetic Sequence

The nth term (the general term) of an arithmetic sequence with first term a_1 and common difference d is

$$a_n = a_1 + (n - 1)d.$$

EXAMPLE 2 Using the Formula for the General Term of an Arithmetic Sequence

Find the eighth term of the arithmetic sequence whose first term is 4 and whose common difference is -7.

Solution To find the eighth term, a_8, we replace n in the formula with 8, a_1 with 4, and d with -7.

$$a_n = a_1 + (n - 1)d$$
$$a_8 = 4 + (8 - 1)(-7) = 4 + 7(-7) = 4 + (-49) = -45$$

The eighth term is -45. We can check this result by writing the first eight terms of the sequence:

$$4, -3, -10, -17, -24, -31, -38, -45. \quad \blacksquare$$

✓ **CHECK POINT 2** Find the ninth term of the arithmetic sequence whose first term is 6 and whose common difference is -5.

EXAMPLE 3	Using an Arithmetic Sequence to Model Teachers' Earnings

According to the National Education Association, teachers in the United States earned an average of \$30,532 in 1990. This amount has increased by approximately \$1472 per year.

a. Write a formula for the nth term of the arithmetic sequence that models teachers' average earnings n years after 1989.

b. How much will U.S. teachers earn, on average, by the year 2020?

Solution

a. We can express teachers' earnings by the following arithmetic sequence:

$$30{,}532, \qquad 32{,}004, \qquad 33{,}476, \qquad 34{,}948, \ldots$$

a_1: earnings in 1990, 1 year after 1989	a_2: earnings in 1991, 2 years after 1989	a_3: earnings in 1992, 3 years after 1989	a_4: earnings in 1993, 4 years after 1989

In this sequence, a_1, the first term, represents the amount teachers earned in 1990. Each subsequent year this amount increases by \$1472, so $d = 1472$. We use the formula for the general term of an arithmetic sequence to write the nth term of the sequence that describes teachers' earnings n years after 1989.

$a_n = a_1 + (n - 1)d$ This is the formula for the general term of an arithmetic sequence.

$a_n = 30{,}532 + (n - 1)1472$ $a_1 = 30{,}532$ and $d = 1472$.

$a_n = 30{,}532 + 1472n - 1472$ Distribute 1472 to each term in parentheses.

$a_n = 1472n + 29{,}060$ Simplify.

Thus, teachers' earnings n years after 1989 can be modeled by $a_n = 1472n + 29{,}060$.

b. Now we need to find teachers' earnings in 2020. The year 2020 is 31 years after 1989: That is, $2020 - 1989 = 31$. Thus, $n = 31$. We substitute 31 for n in $a_n = 1472n + 29{,}060$.

$$a_{31} = 1472 \cdot 31 + 29{,}060 = 74{,}692$$

The 31st term of the sequence is 74,692. Therefore, U.S. teachers are predicted to earn an average of \$74,692 by the year 2020. ∎

☑ **CHECK POINT 3**   Thanks to drive-thrus and curbside delivery, Americans are eating more meals behind the wheel. In 2004, we averaged 32 à la car meals, increasing by approximately 0.7 meal per year. (*Source: Newsweek*)

a. Write a formula for the nth term of the arithmetic sequence that models the average number of car meals n years after 2003.

b. How many car meals will Americans average by the year 2024?

---

4 Use the formula for the sum of the first n terms of an arithmetic sequence.

The Sum of the First n Terms of an Arithmetic Sequence

The sum of the first n terms of an arithmetic sequence, denoted by S_n, and called the **nth partial sum**, can be found without having to add up all the terms. Let

$$S_n = a_1 + a_2 + a_3 + \cdots + a_n$$

be the sum of the first n terms of an arithmetic sequence. Because d is the common difference between terms, S_n can be written forward and backward, as shown on the next page.

$$S_n = a_1 \qquad\quad + (a_1 + d) + (a_1 + 2d) + \cdots + a_n$$
$$\underline{S_n = a_n \qquad\quad + (a_n - d) + (a_n - 2d) + \cdots + a_1}$$
$$2S_n = (a_1 + a_n) + (a_1 + a_n) + (a_1 + a_n) + \cdots + (a_1 + a_n) \qquad \text{Add the two equations.}$$

Because there are n sums of $(a_1 + a_n)$ on the right side, we can express this side as $n(a_1 + a_n)$. Thus, the last equation can be written as follows:

$$2S_n = n(a_1 + a_n).$$

$$S_n = \frac{n}{2}(a_1 + a_n) \qquad \text{Solve for } S_n, \text{ dividing both sides by 2.}$$

We have proved the following result:

The Sum of the First n Terms of an Arithmetic Sequence

The sum, S_n, of the first n terms of an arithmetic sequence is given by

$$S_n = \frac{n}{2}(a_1 + a_n),$$

in which a_1 is the first term and a_n is the nth term.

To find the sum of the terms of an arithmetic sequence using $S_n = \frac{n}{2}(a_1 + a_n)$, we need to know the first term, a_1, the last term, a_n, and the number of terms, n. The following examples illustrate how to use this formula.

EXAMPLE 4 Finding the Sum of n Terms of an Arithmetic Sequence

Find the sum of the first 100 terms of the arithmetic sequence: $1, 3, 5, 7, \ldots$.

Solution   By finding the sum of the first 100 terms of $1, 3, 5, 7, \ldots$, we are finding the sum of the first 100 odd numbers. To find the sum of the first 100 terms, S_{100}, we replace n in the formula with 100.

$$S_n = \frac{n}{2}(a_1 + a_n)$$

$$S_{100} = \frac{100}{2}(a_1 + a_{100})$$

The first term, a_1, is 1.   We must find a_{100}, the 100th term.

We use the formula for the general term of an arithmetic sequence to find a_{100}. The common difference, d, of $1, 3, 5, 7, \ldots$, is 2.

$$a_n = a_1 + (n - 1)d \qquad \text{This is the formula for the nth term of an arithmetic sequence. Use it to find the 100th term.}$$

$$a_{100} = 1 + (100 - 1) \cdot 2 \qquad \text{Substitute 100 for } n, \text{ 2 for } d, \text{ and 1 (the first term) for } a_1.$$

$$= 1 + 99 \cdot 2 \qquad \text{Perform the subtraction in parentheses.}$$

$$= 1 + 198 = 199 \qquad \text{Multiply } (99 \cdot 2 = 198) \text{ and then add.}$$

Now we are ready to find the sum of the 100 terms $1, 3, 5, 7, \ldots, 199$.

$$S_n = \frac{n}{2}(a_1 + a_n)$$ *Use the formula for the sum of the first n terms of an arithmetic sequence. Let n = 100, a₁ = 1, and a₁₀₀ = 199.*

$$S_{100} = \frac{100}{2}(1 + 199) = 50(200) = 10,000$$

The sum of the first 100 odd numbers is 10,000. Equivalently, the 100th partial sum of the sequence $1, 3, 5, 7, \ldots$ is 10,000. ∎

✓ **CHECK POINT 4**   Find the sum of the first 15 terms of the arithmetic sequence: $3, 6, 9, 12, \ldots$.

Using Technology

To find

$$\sum_{i=1}^{25}(5i - 9)$$

on a graphing utility, enter:

SUM SEQ $(5x - 9, x, 1, 25, 1)$.

Then press ENTER.

```
sum(seq(5X-9,X,1
,25,1))
           1400
```

EXAMPLE 5   Using S_n to Evaluate a Summation

Find the following sum: $\displaystyle\sum_{i=1}^{25}(5i - 9)$.

Solution

$$\sum_{i=1}^{25}(5i - 9) = (5 \cdot 1 - 9) + (5 \cdot 2 - 9) + (5 \cdot 3 - 9) + \cdots + (5 \cdot 25 - 9)$$
$$= -4 \quad\quad + 1 \quad\quad + 6 \quad\quad + \cdots + 116$$

By evaluating the first three terms and the last term, we see that $a_1 = -4$, d, the common difference, is $1 - (-4)$ or 5, and a_{25}, the last term, is 116.

$$S_n = \frac{n}{2}(a_1 + a_n)$$ *Use the formula for the sum of the first n terms of an arithmetic sequence. Let n = 25, a₁ = −4, and a₂₅ = 116.*

$$S_{25} = \frac{25}{2}(-4 + 116) = \frac{25}{2}(112) = 1400$$

Thus,

$$\sum_{i=1}^{25}(5i - 9) = 1400. \quad ∎$$

✓ **CHECK POINT 5**   Find the following sum: $\displaystyle\sum_{i=1}^{30}(6i - 11)$.

EXAMPLE 6   Modeling Total Nursing Home Costs over a Six-Year Period

Your grandmother has assets of $500,000. One option that she is considering involves an adult residential community for a six-year period beginning in 2014. The model

$$a_n = 1800n + 64,130$$

describes yearly adult residential community costs n years after 2013. Does your grandmother have enough to pay for the facility?

Solution We must find the sum of an arithmetic sequence whose general term is $a_n = 1800n + 64{,}130$. The first term of the sequence corresponds to the facility's costs in the year 2014. The last term corresponds to costs in the year 2019. Because the model describes costs n years after 2013, $n = 1$ describes the year 2014 and $n = 6$ describes the year 2019.

$$a_n = 1800n + 64{,}130$$ This is the given formula for the general term of the sequence.

$$a_1 = 1800 \cdot 1 + 64{,}130 = 65{,}930$$ Find a_1 by replacing n with 1.

$$a_6 = 1800 \cdot 6 + 64{,}130 = 74{,}930$$ Find a_6 by replacing n with 6.

The first year the facility will cost \$65,930. By year six, the facility will cost \$74,930. Now we must find the sum of the costs for all six years. We focus on the sum of the first six terms of the arithmetic sequence

$$65{,}930, \ 67{,}730, \ \dots, \ 74{,}930.$$

$$a_1 \qquad a_2 \qquad a_6$$

We find this sum using the formula for the sum of the first n terms of an arithmetic sequence. We are adding 6 terms: $n = 6$. The first term is 65,930: $a_1 = 65{,}930$. The last term—that is, the sixth term—is 74,930: $a_6 = 74{,}930$.

$$S_n = \frac{n}{2}(a_1 + a_n)$$

$$S_6 = \frac{6}{2}(65{,}930 + 74{,}930) = 3(140{,}860) = 422{,}580$$

Total adult residential community costs for your grandmother are predicted to be \$422,580. Because your grandmother's assets are \$500,000, she has enough to pay for the facility for the six-year period. ■

✓ **CHECK POINT 6** In Example 6, how much would it cost for the adult residential community for a ten-year period beginning in 2014?

CONCEPT AND VOCABULARY CHECK

Fill in each blank so that the resulting statement is true.

1. A sequence in which each term after the first differs from the preceding term by a constant amount is called a/an _____ sequence. The difference between consecutive terms is called the _____ of the sequence.

2. The nth term of the sequence described in Concept Check 1 is given by the formula $a_n =$ _____, where a_1 is the _____ and d is the _____ of the sequence.

3. The sum, S_n, of the first n terms of the sequence described in Concept Check 1 is given by the formula $S_n =$ _____, where a_1 is the _____ and a_n is the _____.

4. The first term of $\displaystyle\sum_{i=1}^{20}(6i - 4)$ is _____ and the last term is _____.

5. The first three terms of $\displaystyle\sum_{i=1}^{17}(5i + 3)$ are _____, _____, and _____. The common difference is _____.

14.2 EXERCISE SET

MyMathLab®

Practice Exercises

In Exercises 1–6, find the common difference for each arithmetic sequence.

1. $2, 6, 10, 14, \ldots$

2. $3, 8, 13, 18, \ldots$

3. $-7, -2, 3, 8, \ldots$

4. $-10, -4, 2, 8, \ldots$

5. $714, 711, 708, 705, \ldots$

6. $611, 606, 601, 596, \ldots$

In Exercises 7–16, write the first six terms of each arithmetic sequence with the given first term, a_1, and common difference, d.

7. $a_1 = 200, d = 20$

8. $a_1 = 300, d = 50$

9. $a_1 = -7, d = 4$

10. $a_1 = -8, d = 5$

11. $a_1 = 300, d = -90$

12. $a_1 = 200, d = -60$

13. $a_1 = \dfrac{5}{2}, d = -\dfrac{1}{2}$

14. $a_1 = \dfrac{3}{4}, d = -\dfrac{1}{4}$

15. $a_1 = -0.4, d = -1.6$

16. $a_1 = -0.3, d = -1.7$

In Exercises 17–24, use the formula for the general term (the nth term) of an arithmetic sequence to find the indicated term of each sequence with the given first term, a_1, and common difference, d.

17. Find a_6 when $a_1 = 13, d = 4$.

18. Find a_{16} when $a_1 = 9, d = 2$.

19. Find a_{50} when $a_1 = 7, d = 5$.

20. Find a_{60} when $a_1 = 8, d = 6$.

21. Find a_{200} when $a_1 = -40, d = 5$.

22. Find a_{150} when $a_1 = -60, d = 5$.

23. Find a_{60} when $a_1 = 35, d = -3$.

24. Find a_{70} when $a_1 = -32, d = 4$.

In Exercises 25–34, write a formula for the general term (the nth term) of each arithmetic sequence. Then use the formula for a_n to find a_{20}, the 20th term of the sequence.

25. $1, 5, 9, 13, \ldots$

26. $2, 7, 12, 17, \ldots$

27. $7, 3, -1, -5, \ldots$

28. $6, 1, -4, -9, \ldots$

29. $-20, -24, -28, -32, \ldots$

30. $-70, -75, -80, -85, \ldots$

31. $a_1 = -\dfrac{1}{3}, d = \dfrac{1}{3}$

32. $a_1 = -\dfrac{1}{4}, d = \dfrac{1}{4}$

33. $a_1 = 4, d = -0.3$

34. $a_1 = 5, d = -0.2$

35. Find the sum of the first 20 terms of the arithmetic sequence: $4, 10, 16, 22, \ldots$.

36. Find the sum of the first 25 terms of the arithmetic sequence: $7, 19, 31, 43, \ldots$.

37. Find the sum of the first 50 terms of the arithmetic sequence: $-10, -6, -2, 2, \ldots$.

38. Find the sum of the first 50 terms of the arithmetic sequence: $-15, -9, -3, 3, \ldots$.

39. Find $1 + 2 + 3 + 4 + \cdots + 100$, the sum of the first 100 natural numbers.

40. Find $2 + 4 + 6 + 8 + \cdots + 200$, the sum of the first 100 positive even integers.

41. Find the sum of the first 60 positive even integers.

42. Find the sum of the first 80 positive even integers.

43. Find the sum of the even integers between 21 and 45.

44. Find the sum of the odd integers between 30 and 54.

For Exercises 45–50, write out the first three terms and the last term. Then use the formula for the sum of the first n terms of an arithmetic sequence to find the indicated sum.

45. $\displaystyle\sum_{i=1}^{17}(5i + 3)$

46. $\displaystyle\sum_{i=1}^{20}(6i - 4)$

47. $\displaystyle\sum_{i=1}^{30}(-3i + 5)$

48. $\displaystyle\sum_{i=1}^{40}(-2i + 6)$

49. $\displaystyle\sum_{i=1}^{100}4i$

50. $\displaystyle\sum_{i=1}^{50}(-4i)$

Practice PLUS

Use the graphs of the arithmetic sequences $\{a_n\}$ and $\{b_n\}$ to solve Exercises 51–58.

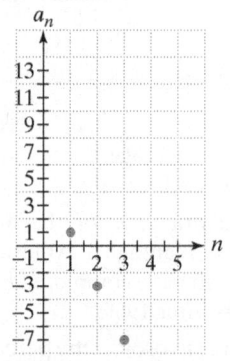

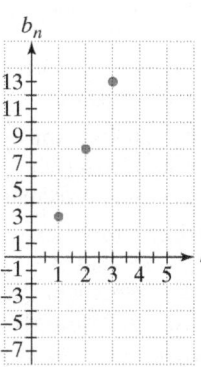

51. Find $a_{14} + b_{12}$.

52. Find $a_{16} + b_{18}$.

53. If $\{a_n\}$ is a finite sequence whose last term is -83, how many terms does $\{a_n\}$ contain?

54. If $\{b_n\}$ is a finite sequence whose last term is 93, how many terms does $\{b_n\}$ contain?

55. Find the difference between the sum of the first 14 terms of $\{b_n\}$ and the sum of the first 14 terms of $\{a_n\}$.

(In Exercises 56–58, continue to refer to the graphs at the bottom of the previous page.)

56. Find the difference between the sum of the first 15 terms of $\{b_n\}$ and the sum of the first 15 terms of $\{a_n\}$.

57. Write a linear function $f(x) = mx + b$, whose domain is the set of positive integers, that represents $\{a_n\}$.

58. Write a linear function $g(x) = mx + b$, whose domain is the set of positive integers, that represents $\{b_n\}$.

Use a system of two equations in two variables, a_1 and d, to solve Exercises 59–60.

59. Write a formula for the general term (the nth term) of the arithmetic sequence whose second term, a_2, is 4 and whose sixth term, a_6, is 16.

60. Write a formula for the general term (the nth term) of the arithmetic sequence whose third term, a_3, is 7 and whose eighth term, a_8, is 17.

Application Exercises

The bar graphs show changes in educational attainment for Americans ages 25 and older from 1970 to 2007. Exercises 61–62 involve developing arithmetic sequences that model the data.

Educational Attainment for Americans Ages 25 and Older

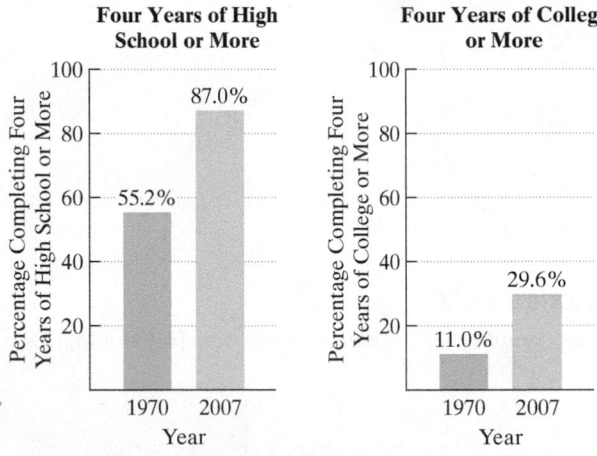

Source: U.S. Census Bureau

61. In 1970, 11.0% of Americans ages 25 and older had completed four years of college or more. On average, this percentage has increased by approximately 0.5 each year.

  a. Write a formula for the nth term of the arithmetic sequence that models the percentage of Americans ages 25 and older who had or will have completed four years of college or more n years after 1969.

  b. Use the model from part (a) to project the percentage of Americans ages 25 and older who will have completed four years of college or more by 2019.

62. In 1970, 55.2% of Americans ages 25 and older had completed four years of high school or more. On average, this percentage has increased by approximately 0.86 each year.

  a. Write a formula for the nth term of the arithmetic sequence that models the percentage of Americans ages 25 and older who had or will have completed four years of high school or more n years after 1969.

  b. Use the model from part (a) to project the percentage of Americans ages 25 and older who will have completed four years of high school or more by 2019.

63. Company A pays $24,000 yearly with raises of $1600 per year. Company B pays $28,000 yearly with raises of $1000 per year. Which company will pay more in year 10? How much more?

64. Company A pays $23,000 yearly with raises of $1200 per year. Company B pays $26,000 yearly with raises of $800 per year. Which company will pay more in year 10? How much more?

Exercises 65–66 are based on the following data showing the average cost of tuition and fees at public and private four-year U.S. colleges.

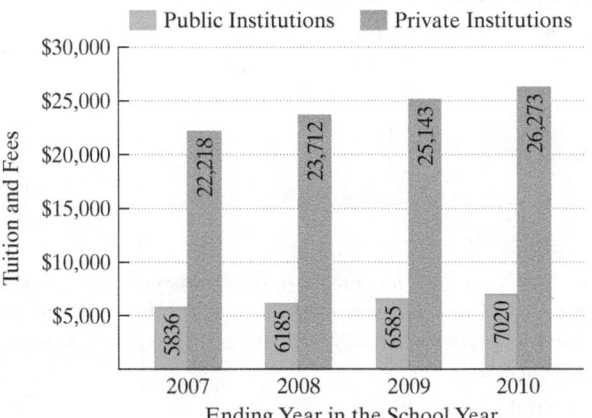

Average Cost of Tuition and Fees at Four-Year United States Colleges

Source: The College Board

65. a. Use the numbers shown in the bar graph to find the total cost of tuition and fees at public colleges for a four-year period from the school year ending in 2007 through the school year ending in 2010.

  b. The model

$$a_n = 395n + 5419$$

describes the cost of tuition and fees at public colleges in academic year n, where $n = 1$ corresponds to the school year ending in 2007, $n = 2$ to the school year ending in 2008, and so on. Use this model and the formula for S_n to find the total cost of tuition and fees at public colleges for a four-year period from the school year ending in 2007 through the school year ending in 2010. How does this compare with the actual sum you obtained in part (a)?

66. a. Use the numbers shown in the bar graph to find the total cost of tuition and fees at private colleges for a four-year period from the school year ending in 2007 through the school year ending in 2010.

  b. The model

$$a_n = 1360n + 20{,}938$$

describes the cost of tuition and fees at private colleges in academic year n, where $n = 1$ corresponds to the school year ending in 2007, $n = 2$ to the school year ending in 2008, and so on. Use this model and the formula for S_n, to find the total cost of tuition and fees at private colleges for a four-year period from the school year ending in 2007 through the school year ending in 2010. Does the model underestimate or overestimate the actual sum that you obtained in part (a)? By how much?

67. Use one of the models in Exercises 65–66 and the formula for S_n to find the total cost of tuition and fees for your undergraduate education. How well does the model describe your anticipated costs?

68. A company offers a starting yearly salary of $33,000 with raises of $2500 per year. Find the total salary over a ten-year period.

69. You are considering two job offers. Company A will start you at $19,000 a year and guarantee a raise of $2600 per year. Company B will start you at a higher salary, $27,000 a year, but will only guarantee a raise of $1200 per year. Find the total salary that each company will pay over a ten-year period. Which company pays the greater total amount?

70. A theater has 30 seats in the first row, 32 seats in the second row, increasing by 2 seats each row for a total of 26 rows. How many seats are there in the theater?

71. A section in a stadium has 20 seats in the first row, 23 seats in the second row, increasing by 3 seats each row for a total of 38 rows. How many seats are in this section of the stadium?

Writing in Mathematics

72. What is an arithmetic sequence? Give an example with your explanation.

73. What is the common difference in an arithmetic sequence?

74. Explain how to find the general term of an arithmetic sequence.

75. Explain how to find the sum of the first n terms of an arithmetic sequence without having to add up all the terms.

Technology Exercises

76. Use the $\boxed{\text{SEQ}}$ (sequence) capability of a graphing utility and the formula you obtained for a_n to verify the value you found for a_{20} in any five exercises from Exercises 25–34.

77. Use the capability of a graphing utility to calculate the sum of a sequence to verify any five of your answers to Exercises 45–50.

Critical Thinking Exercises

Make Sense? *In Exercises 78–81, determine whether each statement "makes sense" or "does not make sense" and explain your reasoning.*

78. Rather than performing the addition, I used the formula $S_n = \frac{n}{2}(a_1 + a_n)$ to find the sum of the first thirty terms of the sequence $2, 4, 8, 16, 32, \ldots$.

79. I was able to find the sum of the first fifty terms of an arithmetic sequence even though I did not identify every term.

80. The sequence for the number of seats per row in our movie theater as the rows move toward the back is arithmetic with $d = 1$ so people don't block the view of those in the row behind them.

81. Beginning at 6:45 A.M., a bus stops on my block every 23 minutes, so I used the formula for the nth term of an arithmetic sequence to describe the stopping time for the nth bus of the day.

In Exercises 82–85, determine whether each statement is true or false. If the statement is false, make the necessary change(s) to produce a true statement.

82. The sum of an arithmetic sequence cannot be negative.

83. The common difference for the arithmetic sequence $1, -1, -3, -5, \ldots$ is 2.

84. An arithmetic sequence is a linear function whose domain is the set of natural numbers.

85. The sequence $\log_2 2, \log_2 4, \log_2 8, \log_2 16, \log_2 32, \ldots$ is arithmetic.

86. Give examples of two different arithmetic sequences whose fourth term, a_4, is 10.

87. In the sequence $21,700, 23,172, 24,644, 26,116, \ldots$, which term is $314,628$?

88. A *degree-day* is a unit used to measure the fuel requirements of buildings. By definition, each degree that the average daily temperature is below 65°F is 1 degree-day. For example, an average daily temperature of 42°F constitutes 23 degree-days. If the average temperature on January 1 was 42°F and fell 2°F for each subsequent day up to and including January 10, how many degree-days are included from January 1 to January 10?

89. Show that the sum of the first n positive odd integers,
$$1 + 3 + 5 + \cdots + (2n - 1),$$
is n^2.

Review Exercises

90. Solve: $\log(x^2 - 25) - \log(x + 5) = 3$.
(Section 12.4, Example 5)

91. Solve: $x^2 + 3x \le 10$.
(Section 11.5, Example 1)

92. Solve for P: $A = \dfrac{Pt}{P + t}$.
(Section 7.6, Example 8)

Preview Exercises

Exercises 93–95 will help you prepare for the material covered in the next section.

93. Consider the sequence $1, -2, 4, -8, 16, \ldots$. Find $\dfrac{a_2}{a_1}, \dfrac{a_3}{a_2}, \dfrac{a_4}{a_3}$, and $\dfrac{a_5}{a_4}$. What do you observe?

94. Consider the sequence whose nth term is $a_n = 3 \cdot 5^n$. Find $\dfrac{a_2}{a_1}, \dfrac{a_3}{a_2}, \dfrac{a_4}{a_3}$, and $\dfrac{a_5}{a_4}$. What do you observe?

95. Use the formula $a_n = a_1 3^{n-1}$ to find the 7th term of the sequence $11, 33, 99, 297, \ldots$.

Geometric Sequences and Series

Objectives

1 Find the common ratio of a geometric sequence.

2 Write terms of a geometric sequence.

3 Use the formula for the general term of a geometric sequence.

4 Use the formula for the sum of the first *n* terms of a geometric sequence.

5 Find the value of an annuity.

6 Use the formula for the sum of an infinite geometric series.

Here we are at the closing moments of a job interview. You're shaking hands with the manager. You managed to answer all the tough questions without losing your poise, and now you've been offered a job. As a matter of fact, your qualifications are so terrific that you've been offered two jobs—one just the day before, with a rival company in the same field! One company offers $30,000 the first year, with increases of 6% per year for four years after that. The other offers $32,000 the

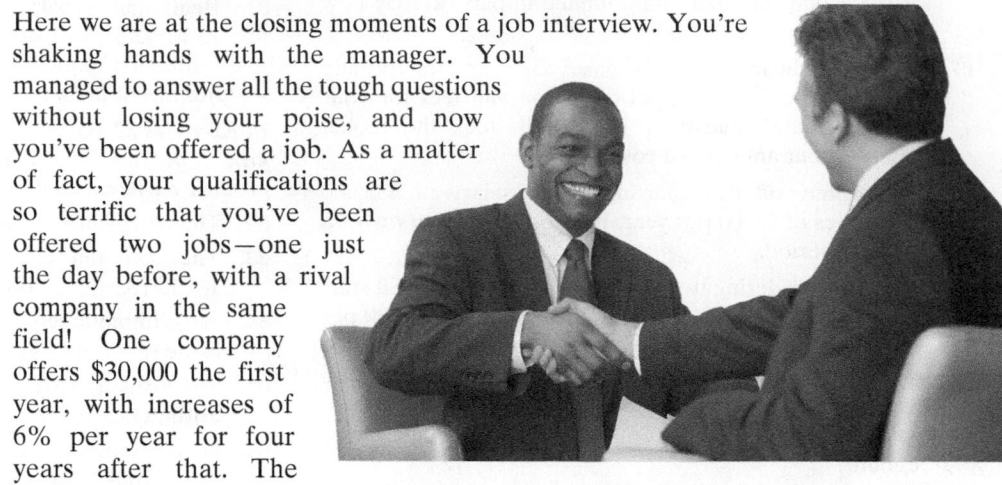

first year, with annual increases of 3% per year after that. Over a five-year period, which is the better offer?

If salary raises amount to a certain percent each year, the yearly salaries over time form a geometric sequence. In this section, we investigate geometric sequences and their properties. After studying the section, you will be in a position to decide which job offer to accept: You will know which company will pay you more over five years.

Geometric Sequences

Figure 14.4 shows a sequence in which the number of squares is increasing. From left to right, the number of squares is 1, 5, 25, 125, and 625. In this sequence, each term after the first, 1, is obtained by multiplying the preceding term by a constant amount, namely 5. This sequence of increasing numbers of squares is an example of a *geometric sequence*.

Figure 14.4 A geometric sequence of squares

Definition of a Geometric Sequence

A **geometric sequence** is a sequence in which each term after the first is obtained by multiplying the preceding term by a fixed nonzero constant. The amount by which we multiply each time is called the **common ratio** of the sequence.

1 Find the common ratio of a geometric sequence.

The common ratio, *r*, is found by dividing any term after the first term by the term that directly precedes it. In the following examples, the common ratio is found by dividing the second term by the first term, $\frac{a_2}{a_1}$.

Geometric Sequence	Common Ratio
$1, 5, 25, 125, 625, \ldots$	$r = \dfrac{5}{1} = 5$
$4, 8, 16, 32, 64, \ldots$	$r = \dfrac{8}{4} = 2$
$6, -12, 24, -48, 96, \ldots$	$r = \dfrac{-12}{6} = -2$
$9, -3, 1, -\dfrac{1}{3}, \dfrac{1}{9}, \ldots$	$r = \dfrac{-3}{9} = -\dfrac{1}{3}$

Great Question!

What happens to the terms of a geometric sequence when the common ratio is negative?

When the common ratio is negative, the signs of the terms alternate.

Figure 14.5 shows a partial graph of the first geometric sequence in our list, $1, 5, 25, 125\ldots$. The graph forms a set of discrete points lying on the exponential function $f(x) = 5^{x-1}$. This illustrates that **a geometric sequence with a positive common ratio other than 1 is an exponential function whose domain is the set of positive integers**.

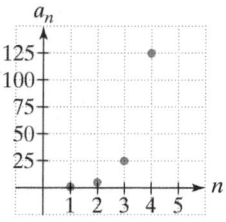

Figure 14.5 The graph of $\{a_n\} = 1, 5, 25, 125, \ldots$

2 Write terms of a geometric sequence.

How do we write out the terms of a geometric sequence when the first term and the common ratio are known? We multiply the first term by the common ratio to get the second term, multiply the second term by the common ratio to get the third term, and so on.

EXAMPLE 1   Writing the Terms of a Geometric Sequence

Write the first six terms of the geometric sequence with first term 6 and common ratio $\frac{1}{3}$.

Solution   The first term is 6. The second term is $6 \cdot \frac{1}{3}$, or 2. The third term is $2 \cdot \frac{1}{3}$, or $\frac{2}{3}$. The fourth term is $\frac{2}{3} \cdot \frac{1}{3}$, or $\frac{2}{9}$, and so on. The first six terms are

$$6, 2, \frac{2}{3}, \frac{2}{9}, \frac{2}{27}, \text{ and } \frac{2}{81}. \quad \blacksquare$$

✓ **CHECK POINT 1**   Write the first six terms of the geometric sequence with first term 12 and common ratio $\frac{1}{2}$.

3 Use the formula for the general term of a geometric sequence.

The General Term of a Geometric Sequence

Consider a geometric sequence whose first term is a_1 and whose common ratio is r. We are looking for a formula for the general term, a_n. Let's begin by writing the first six terms. The first term is a_1. The second term is $a_1 r$. The third term is $a_1 r \cdot r$, or $a_1 r^2$. The fourth term is $a_1 r^2 \cdot r$, or $a_1 r^3$, and so on. Starting with a_1 and multiplying each successive term by r, the first six terms are $a_1, a_1 r, a_1 r^2, a_1 r^3, a_1 r^4$, and $a_1 r^5$.

Compare the exponent on r and the subscript of a denoting the term number:

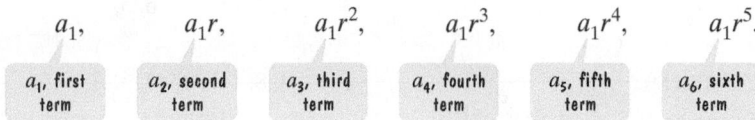

$$a_1, \qquad a_1r, \qquad a_1r^2, \qquad a_1r^3, \qquad a_1r^4, \qquad a_1r^5.$$

a_1, first term $\quad$ a_2, second term $\quad$ a_3, third term $\quad$ a_4, fourth term $\quad$ a_5, fifth term $\quad$ a_6, sixth term

Can you see that the exponent on r is 1 less than the subscript of a denoting the term number?

$$a_3\text{: third term} = a_1r^2 \qquad a_4\text{: fourth term} = a_1r^3$$

One less than 3, or 2, is the exponent on r. $\qquad$ One less than 4, or 3, is the exponent on r.

Thus, the formula for the nth term is

$$a_n = a_1r^{n-1}.$$

One less than n, or $n-1$, is the exponent on r.

General Term of a Geometric Sequence

The nth term (the general term) of a geometric sequence with first term a_1 and common ratio r is

$$a_n = a_1r^{n-1}.$$

EXAMPLE 2 Using the Formula for the General Term of a Geometric Sequence

Find the eighth term of the geometric sequence whose first term is -4 and whose common ratio is -2.

Solution To find the eighth term, a_8, we replace n in the formula with 8, a_1 with -4, and r with -2.

$$a_n = a_1r^{n-1}$$
$$a_8 = -4(-2)^{8-1} = -4(-2)^7 = -4(-128) = 512$$

The eighth term is 512. We can check this result by writing the first eight terms of the sequence:

$$-4, 8, -16, 32, -64, 128, -256, 512. \quad \blacksquare$$

✓ **CHECK POINT 2** Find the seventh term of the geometric sequence whose first term is 5 and whose common ratio is -3.

In Chapter 12, we studied exponential functions of the form $f(x) = b^x$ and used an exponential function to model the growth of the U.S. population from 1970 through 2009 (Example 1 on page 910). In our next example, we revisit the country's population growth over a shorter period of time, 2000 through 2006. Because a geometric sequence is an exponential function whose domain is the set of positive integers, geometric and exponential growth mean the same thing.

EXAMPLE 3   Geometric Population Growth

The table shows the population of the United States in 2000, with estimates given by the Census Bureau for 2001 through 2006.

Year	2000	2001	2002	2003	2004	2005	2006
Population (millions)	281.4	284.5	287.6	290.8	294.0	297.2	300.5

a. Show that the population is increasing geometrically.

b. Write the general term for the geometric sequence modeling the population of the United States, in millions, n years after 1999.

c. Project the U.S. population, in millions, for the year 2009.

Solution

a. First, we use the sequence of population growth, 281.4, 284.5, 287.6, 290.8, and so on, to divide the population for each year by the population in the preceding year.

$$\frac{284.5}{281.4} \approx 1.011, \quad \frac{287.6}{284.5} \approx 1.011, \quad \frac{290.8}{287.6} \approx 1.011$$

Continuing in this manner, we will keep getting approximately 1.011. This means that the population is increasing geometrically with $r \approx 1.011$. The population of the United States in any year shown in the sequence is approximately 1.011 times the population the year before.

b. The sequence of the U.S. population growth is

$$281.4, \ 284.5, \ 287.6, \ 290.8, \ 294.0, \ 297.2, \ 300.5, \ldots .$$

Because the population is increasing geometrically, we can find the general term of this sequence using

$$a_n = a_1 r^{n-1}.$$

In this sequence, $a_1 = 281.4$ and [from part (a)] $r \approx 1.011$. We substitute these values into the formula for the general term. This gives the general term for the geometric sequence modeling the U.S. population, in millions, n years after 1999.

$$a_n = 281.4(1.011)^{n-1}$$

c. We can use the formula for the general term, a_n, in part (b) to project the U.S. population for the year 2009. The year 2009 is 10 years after 1999—that is, $2009 - 1999 = 10$. Thus, $n = 10$. We substitute 10 for n in $a_n = 281.4(1.011)^{n-1}$.

$$a_{10} = 281.4(1.011)^{10-1} = 281.4(1.011)^9 \approx 310.5$$

The model projects that the United States will have a population of approximately 310.5 million in the year 2009. ∎

☑ **CHECK POINT 3**   Write the general term for the geometric sequence

$$3, 6, 12, 24, 48, \ldots .$$

Then use the formula for the general term to find the eighth term.

---

4 Use the formula for the sum of the first n terms of a geometric sequence.

The Sum of the First n Terms of a Geometric Sequence

The sum of the first n terms of a geometric sequence, denoted by S_n, and called the **nth partial sum**, can be found without having to add up all the terms. Recall that the first n terms of a geometric sequence are

$$a_1, a_1r, a_1r^2, \ldots, a_1r^{n-2}, a_1r^{n-1}.$$

We find the sum of $a_1, a_1r, a_1r^2, \ldots, a_1r^{n-2}, a_1r^{n-1}$ as follows:

$$S_n = a_1 + a_1r + a_1r^2 + \cdots + a_1r^{n-2} + a_1r^{n-1}$$
S_n is the sum of the first n terms of the sequence.

$$rS_n = a_1r + a_1r^2 + a_1r^3 + \cdots + a_1r^{n-1} + a_1r^n$$
Multiply both sides of the equation by r.

$$S_n - rS_n = a_1 - a_1r^n$$
Subtract the second equation from the first equation.

$$S_n(1 - r) = a_1(1 - r^n)$$
Factor out S_n on the left and a_1 on the right.

$$S_n = \frac{a_1(1 - r^n)}{1 - r}.$$
Solve for S_n by dividing both sides by $1 - r$ (assuming that $r \neq 1$).

We have proved the following result:

The Sum of the First n Terms of a Geometric Sequence

The sum, S_n, of the first n terms of a geometric sequence is given by

$$S_n = \frac{a_1(1 - r^n)}{1 - r}$$

in which a_1 is the first term and r is the common ratio ($r \neq 1$).

To find the sum of the terms of a geometric sequence, we need to know the first term, a_1, the common ratio, r, and the number of terms, n. The following examples illustrate how to use this formula.

Great Question!

What is the sum of the first n terms of a geometric sequence if the common ratio is 1?

If the common ratio is 1, the geometric sequence is

$$a_1, a_1, a_1, a_1, \ldots.$$

The sum of the first n terms of this sequence is na_1:

$$S_n = \underbrace{a_1 + a_1 + a_1 + \cdots + a_1}_{\text{There are } n \text{ terms.}}$$

$$= na_1.$$

EXAMPLE 4 Finding the Sum of the First n Terms of a Geometric Sequence

Find the sum of the first 18 terms of the geometric sequence: $2, -8, 32, -128, \ldots$.

Solution To find the sum of the first 18 terms, S_{18}, we replace n in the formula with 18.

$$S_n = \frac{a_1(1 - r^n)}{1 - r}$$

$$S_{18} = \frac{a_1(1 - r^{18})}{1 - r}$$

The first term, a_1, is 2.　　We must find r, the common ratio.

We can find the common ratio by dividing the second term of $2, -8, 32, -128, \ldots$ by the first term.

$$r = \frac{a_2}{a_1} = \frac{-8}{2} = -4$$

Now we are ready to find the sum of the first 18 terms of $2, -8, 32, -128, \ldots$.

$$S_n = \frac{a_1(1 - r^n)}{1 - r}$$
Use the formula for the sum of the first n terms of a geometric sequence.

$$S_{18} = \frac{2[1 - (-4)^{18}]}{1 - (-4)}$$
a_1 (the first term) $= 2$, $r = -4$, and $n = 18$ because we want the sum of the first 18 terms.

$$= -27{,}487{,}790{,}694$$
Use a calculator.

The sum of the first 18 terms is $-27{,}487{,}790{,}694$. Equivalently, this number is the 18th partial sum of the sequence $2, -8, 32, -128, \ldots$. ∎

✓ **CHECK POINT 4** Find the sum of the first nine terms of the geometric sequence: $2, -6, 18, -54, \ldots$.

EXAMPLE 5 Using S_n to Evaluate a Summation

Find the following sum: $\displaystyle\sum_{i=1}^{10} 6 \cdot 2^i$.

Solution Let's write out a few terms in the sum.

$$\sum_{i=1}^{10} 6 \cdot 2^i = 6 \cdot 2 + 6 \cdot 2^2 + 6 \cdot 2^3 + \cdots + 6 \cdot 2^{10}$$

Do you see that each term after the first is obtained by multiplying the preceding term by 2? To find the sum of the 10 terms ($n = 10$), we need to know the first term, a_1, and the common ratio, r. The first term is $6 \cdot 2$ or 12: $a_1 = 12$. The common ratio is 2.

$$S_n = \frac{a_1(1 - r^n)}{1 - r} \qquad \text{Use the formula for the sum of the first } n \text{ terms of a geometric sequence.}$$

$$S_{10} = \frac{12(1 - 2^{10})}{1 - 2} \qquad a_1 \text{ (the first term)} = 12, r = 2, \text{ and } n = 10 \text{ because we are adding ten terms.}$$

$$= 12{,}276 \qquad \text{Use a calculator.}$$

Thus,

$$\sum_{i=1}^{10} 6 \cdot 2^i = 12{,}276. \quad \blacksquare$$

✓ **CHECK POINT 5** Find the following sum: $\displaystyle\sum_{i=1}^{8} 2 \cdot 3^i$.

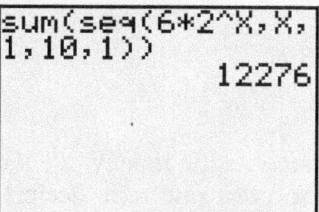
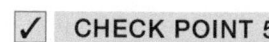

Some of the exercises in the previous Exercise Set involved situations in which salaries increased by a fixed amount each year. A more realistic situation is one in which salaries increase by a certain percent each year. Example 6 shows how such a situation can be described using a geometric sequence.

EXAMPLE 6 Computing a Lifetime Salary

A union contract specifies that each worker will receive a 5% pay increase each year for the next 30 years. One worker is paid $20,000 the first year. What is this person's total lifetime salary over a 30-year period?

Solution The salary for the first year is $20,000. With a 5% raise, the second-year salary is computed as follows:

Salary for year 2 $= 20{,}000 + 20{,}000(0.05) = 20{,}000(1 + 0.05) = 20{,}000(1.05)$.

Each year, the salary is 1.05 times what it was in the previous year. Thus, the salary for year 3 is 1.05 times $20,000(1.05)$, or $20,000(1.05)^2$. The salaries for the first five years are given in the table.

Yearly Salaries

Year 1	Year 2	Year 3	Year 4	Year 5	...
20,000	20,000(1.05)	20,000(1.05)²	20,000(1.05)³	20,000(1.05)⁴	...

The numbers in the bottom row form a geometric sequence with $a_1 = 20{,}000$ and $r = 1.05$. To find the total salary over 30 years, we use the formula for the sum of the first n terms of a geometric sequence, with $n = 30$.

$$S_n = \frac{a_1(1 - r^n)}{1 - r}$$

$$S_{30} = \frac{20{,}000[1 - (1.05)^{30}]}{1 - 1.05}$$

Use $S_n = \dfrac{a_1(1 - r^n)}{1 - r}$ with
$a_1 = 20{,}000$, $r = 1.05$,
and $n = 30$.

Total salary over 30 years

$$\approx 1{,}328{,}777$$

Use a calculator.

The total salary over the 30-year period is approximately \$1,328,777. ∎

✓ **CHECK POINT 6**   A job pays a salary of \$30,000 the first year. During the next 29 years, the salary increases by 6% each year. What is the total lifetime salary over the 30-year period?

5 Find the value of an annuity.

Annuities

The compound interest formula

$$A = P(1 + r)^t$$

gives the future value, A, after t years, when a fixed amount of money, P, the principal, is deposited in an account that pays an annual interest rate r (in decimal form) compounded once a year. However, money is often invested in small amounts at periodic intervals. For example, to save for retirement, you might decide to place \$1000 into an Individual Retirement Account (IRA) at the end of each year until you retire. An **annuity** is a sequence of equal payments made at equal time periods. An IRA is an example of an annuity.

Suppose P dollars is deposited into an account at the end of each year. The account pays an annual interest rate, r, compounded annually. At the end of the first year, the account contains P dollars. At the end of the second year, P dollars is deposited again. At the time of this deposit, the first deposit has received interest earned during the second year. The **value of the annuity** is the sum of all deposits made plus all interest paid. Thus, the value of the annuity after two years is

$$P + P(1 + r).$$

Deposit of P dollars at end of second year

First-year deposit of P dollars with interest earned for a year

The value of the annuity after three years is

$$P \quad + \quad P(1 + r) \quad + \quad P(1 + r)^2.$$

Deposit of P dollars at end of third year

Second-year deposit of P dollars with interest earned for a year

First-year deposit of P dollars with interest earned over two years

The value of the annuity after t years is

$$P + P(1 + r) + P(1 + r)^2 + P(1 + r)^3 + \cdots + P(1 + r)^{t-1}.$$

Deposit of P dollars at end of year t

First-year deposit of P dollars with interest earned over $t - 1$ years

This is the sum of the terms of a geometric sequence with first term P and common ratio $1 + r$. We use the formula

$$S_n = \frac{a_1(1 - r^n)}{1 - r}$$

to find the sum of the terms:

$$S_t = \frac{P[1 - (1 + r)^t]}{1 - (1 + r)} = \frac{P[1 - (1 + r)^t]}{-r} = P\frac{(1 + r)^t - 1}{r}.$$

This formula gives the value of an annuity after t years if interest is compounded once a year. We can adjust the formula to find the value of an annuity if equal payments are made at the end of each of n yearly compounding periods.

Value of an Annuity: Interest Compounded n Times Per Year

If P is the deposit made at the end of each compounding period for an annuity at r percent annual interest compounded n times per year, the value, A, of the annuity after t years is

$$A = \frac{P\left[\left(1 + \dfrac{r}{n}\right)^{nt} - 1\right]}{\dfrac{r}{n}}.$$

EXAMPLE 7   Determining the Value of an Annuity

At age 25, to save for retirement, you decide to deposit $200 at the end of each month into an IRA that pays 7.5% compounded monthly.

a. How much will you have from the IRA when you retire at age 65?

b. Find the interest.

Solution

a. Because you are 25, the amount that you will have from the IRA when you retire at 65 is its value after 40 years.

$$A = \frac{P\left[\left(1 + \dfrac{r}{n}\right)^{nt} - 1\right]}{\dfrac{r}{n}}$$

Use the formula for the value of an annuity.

$$A = \frac{200\left[\left(1 + \dfrac{0.075}{12}\right)^{12 \cdot 40} - 1\right]}{\dfrac{0.075}{12}}$$

The annuity involves month-end deposits of $200: $P = 200$. The interest rate is 7.5%: $r = 0.075$. The interest is compounded monthly: $n = 12$. The number of years is 40: $t = 40$.

$$= \frac{200[(1 + 0.00625)^{480} - 1]}{0.00625}$$

Using parentheses keys, this can be performed in a single step on a graphing calculator.

$$= \frac{200[(1.00625)^{480} - 1]}{0.00625}$$

$$\approx \frac{200(19.8989 - 1)}{0.00625}$$

Use a calculator to find $(1.00625)^{480}$: $1.00625 \; \boxed{y^x} \; 480 \; \boxed{=}$.

$$\approx 604{,}765$$

After 40 years, you will have approximately $604,765 when retiring at age 65.

b. Interest = Value of the IRA − Total deposits

$$\approx \$604{,}765 - \$200 \cdot 12 \cdot 40$$

> **$200 per month × 12 months per year × 40 years**

$$= \$604{,}765 - \$96{,}000 = \$508{,}765$$

The interest is approximately $508,765, more than five times the amount of your contributions to the IRA. ∎

✓ **CHECK POINT 7**   At age 30, to save for retirement, you decide to deposit $100 at the end of each month into an IRA that pays 9.5% compounded monthly.

   a. How much will you have from the IRA when you retire at age 65?

   b. Find the interest.

---

6 Use the formula for the sum of an infinite geometric series.

Geometric Series

An infinite sum of the form

$$a_1 + a_1 r + a_1 r^2 + a_1 r^3 + \cdots + a_1 r^{n-1} + \cdots$$

with first term a_1 and common ratio r is called an **infinite geometric series**. How can we determine which infinite geometric series have sums and which do not? We look at what happens to r^n as n gets larger in the formula for the sum of the first n terms of this series, namely

$$S_n = \frac{a_1(1 - r^n)}{1 - r}.$$

If r is any number between -1 and 1, that is, $-1 < r < 1$, the term r^n approaches 0 as n gets larger. For example, consider what happens to r^n for $r = \frac{1}{2}$:

$$\left(\frac{1}{2}\right)^1 = \frac{1}{2} \quad \left(\frac{1}{2}\right)^2 = \frac{1}{4} \quad \left(\frac{1}{2}\right)^3 = \frac{1}{8} \quad \left(\frac{1}{2}\right)^4 = \frac{1}{16} \quad \left(\frac{1}{2}\right)^5 = \frac{1}{32} \quad \left(\frac{1}{2}\right)^6 = \frac{1}{64}.$$

> **These numbers are approaching 0 as n gets larger.**

Take another look at the formula for the sum of the first n terms of a geometric sequence.

$$S_n = \frac{a_1(1 - r^n)}{1 - r}$$

> **If $-1 < r < 1$, r^n approaches 0 as n gets larger.**

Let us replace r^n with 0 in the formula for S_n. This change gives us a formula for the sum of an infinite geometric series with a common ratio between -1 and 1.

> **The Sum of an Infinite Geometric Series**
>
> If $-1 < r < 1$ (equivalently, $|r| < 1$), then the sum of the infinite geometric series
>
> $$a_1 + a_1 r + a_1 r^2 + a_1 r^3 + \cdots$$
>
> in which a_1 is the first term and r is the common ratio is given by
>
> $$S = \frac{a_1}{1 - r}.$$
>
> If $|r| \geq 1$, the infinite series does not have a sum.

To use the formula for the sum of an infinite geometric series, we need to know the first term and the common ratio. For example, consider

First term, a_1, is $\frac{1}{2}$.

$$\frac{1}{2} + \frac{1}{4} + \frac{1}{8} + \frac{1}{16} + \frac{1}{32} + \cdots.$$

Common ratio, r, is $\frac{a_2}{a_1}$.

$r = \frac{1}{4} \div \frac{1}{2} = \frac{1}{4} \cdot 2 = \frac{1}{2}$

With $r = \frac{1}{2}$, the condition that $|r| < 1$ is met, so the infinite geometric series has a sum given by $S = \frac{a_1}{1 - r}$. The sum of the series is found as follows:

$$\frac{1}{2} + \frac{1}{4} + \frac{1}{8} + \frac{1}{16} + \frac{1}{32} + \cdots = \frac{a_1}{1 - r} = \frac{\frac{1}{2}}{1 - \frac{1}{2}} = \frac{\frac{1}{2}}{\frac{1}{2}} = 1.$$

Thus, the sum of the infinite geometric series is 1. Notice how this is illustrated in **Figure 14.6**. As more terms are included, the sum is approaching the area of one complete circle.

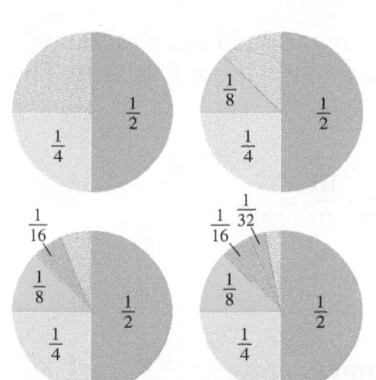

Figure 14.6 The sum $\frac{1}{2} + \frac{1}{4} + \frac{1}{8} + \frac{1}{16} + \frac{1}{32} + \cdots$ is approaching 1.

EXAMPLE 8   Finding the Sum of an Infinite Geometric Series

Find the sum of the infinite geometric series: $\frac{3}{8} - \frac{3}{16} + \frac{3}{32} - \frac{3}{64} + \cdots$.

Solution   Before finding the sum, we must find the common ratio.

$$r = \frac{a_2}{a_1} = \frac{-\frac{3}{16}}{\frac{3}{8}} = -\frac{3}{16} \cdot \frac{8}{3} = -\frac{1}{2}$$

Because $r = -\frac{1}{2}$, the condition that $|r| < 1$ is met. Thus, the infinite geometric series has a sum.

$$S = \frac{a_1}{1 - r}$$

This is the formula for the sum of an infinite geometric series. Let $a_1 = \frac{3}{8}$ and $r = -\frac{1}{2}$.

$$= \frac{\frac{3}{8}}{1 - \left(-\frac{1}{2}\right)} = \frac{\frac{3}{8}}{\frac{3}{2}} = \frac{3}{8} \cdot \frac{2}{3} = \frac{1}{4}$$

Thus, the sum of $\frac{3}{8} - \frac{3}{16} + \frac{3}{32} - \frac{3}{64} + \cdots$ is $\frac{1}{4}$. Put in an informal way, as we continue to add more and more terms, the sum is approximately $\frac{1}{4}$. ∎

✓ **CHECK POINT 8**   Find the sum of the infinite geometric series:
$3 + 2 + \frac{4}{3} + \frac{8}{9} + \cdots$.

We can use the formula for the sum of an infinite geometric series to express a repeating decimal as a fraction in lowest terms.

EXAMPLE 9 Writing a Repeating Decimal as a Fraction

Express $0.\overline{7}$ as a fraction in lowest terms.

Solution

$$0.\overline{7} = 0.7777\ldots = \frac{7}{10} + \frac{7}{100} + \frac{7}{1000} + \frac{7}{10,000} + \cdots$$

Observe that $0.\overline{7}$ is an infinite geometric series with first term $\frac{7}{10}$ and common ratio $\frac{1}{10}$. Because $r = \frac{1}{10}$, the condition that $|r| < 1$ is met. Thus, we can use our formula to find the sum. Therefore,

$$0.\overline{7} = \frac{a_1}{1 - r} = \frac{\frac{7}{10}}{1 - \frac{1}{10}} = \frac{\frac{7}{10}}{\frac{9}{10}} = \frac{7}{10} \cdot \frac{10}{9} = \frac{7}{9}.$$

An equivalent fraction for $0.\overline{7}$ is $\frac{7}{9}$. ∎

✓ **CHECK POINT 9** Express $0.\overline{9}$ as a fraction in lowest terms.

Infinite geometric series have many applications, as illustrated in Example 10.

EXAMPLE 10 Tax Rebates and the Multiplier Effect

A tax rebate that returns a certain amount of money to taxpayers can have a total effect on the economy that is many times this amount. In economics, this phenomenon is called the **multiplier effect**. Suppose, for example, that the government reduces taxes so that each consumer has $2000 more income. The government assumes that each person will spend 70% of this (=$1400). The individuals and businesses receiving this $1400 in turn spend 70% of it (=$980), creating extra income for other people to spend, and so on. Determine the total amount spent on consumer goods from the initial $2000 tax rebate.

Solution The total amount spent is given by the infinite geometric series

$$1400 + 980 + 686 + \cdots.$$

$1400

70% is spent.

$980

70% is spent.

$686

70% of 1400   70% of 980

The first term is 1400: $a_1 = 1400$. The common ratio is 70%, or 0.7: $r = 0.7$. Because $r = 0.7$, the condition that $|r| < 1$ is met. Thus, we can use our formula to find the sum. Therefore,

$$1400 + 980 + 686 + \cdots = \frac{a_1}{1 - r} = \frac{1400}{1 - 0.7} \approx 4667.$$

This means that the total amount spent on consumer goods from the initial $2000 rebate is approximately $4667. ∎

✓ **CHECK POINT 10** Rework Example 10 and determine the total amount spent on consumer goods with a $1000 tax rebate and 80% spending down the line.

CONCEPT AND VOCABULARY CHECK

Fill in each blank so that the resulting statement is true.

1. A sequence in which each term after the first is obtained by multiplying the preceding term by a fixed nonzero constant is called a/an _____ sequence. The amount by which we multiply each time is called the _____ of the sequence.

2. The nth term of the sequence described in Concept Check 1 is given by the formula $a =$ _____, where a_1 is the _____ and r is the _____ of the sequence.

3. The sum, S_n, of the first n terms of the sequence described in Concept Check 1 is given by the formula $S_n =$ _____, where a_1 is the _____ and r is the _____, $r \neq 1$.

4. A sequence of equal payments made at equal time periods is called a/an _____. Its value, A, after t years is given by the formula

$$A = \frac{P\left[(1 + \frac{r}{n})^{nt} - 1\right]}{\frac{r}{n}},$$

where _____ is the deposit made at the end of each compounding period at _____ percent annual interest compounded _____ times per year.

5. An infinite sum of the form

$$a_1 + a_1r + a_1r^2 + a_1r^3 + \cdots$$

is called a/an _____. If $-1 < r < $ ___, its sum, S, is given by the formula $S =$ _____. The series does not have a sum if _____.

6. The first four terms of $\sum_{i=1}^{6} 2^i$ are _____, _____, _____, and _____. The common ratio is _____.

Determine whether each sequence is arithmetic or geometric.

7. $4, 8, 12, 16, 20, \ldots$ _____

8. $4, 8, 16, 32, 64, \ldots$ _____

9. $1, -3, 9, -27, 81, \ldots$ _____

10. $-1, 1, 3, 5, 7, \ldots$ _____

14.3 EXERCISE SET

Watch the videos in MyMathLab   Download the MyDashBoard App

Practice Exercises

In Exercises 1–8, find the common ratio for each geometric sequence.

1. $5, 15, 45, 135, \ldots$

2. $5, 10, 20, 40, \ldots$

3. $-15, 30, -60, 120, \ldots$

4. $-2, 6, -18, 54, \ldots$

5. $3, \dfrac{9}{2}, \dfrac{27}{4}, \dfrac{81}{8}, \ldots$

6. $4, \dfrac{8}{3}, \dfrac{16}{9}, \dfrac{32}{27}, \ldots$

7. $4, -0.4, 0.04, -0.004, \ldots$

8. $7, -0.7, 0.07, -0.007, \ldots$

In Exercises 9–16, write the first five terms of each geometric sequence with the given first term, a_1, and common ratio, r.

9. $a_1 = 2, r = 3$

10. $a_1 = 2, r = 4$

11. $a_1 = 20, r = \dfrac{1}{2}$

12. $a_1 = 24, r = \dfrac{1}{3}$

13. $a_1 = -4, r = -10$

14. $a_1 = -3, r = -10$

15. $a_1 = -\dfrac{1}{4}, r = -2$

16. $a_1 = -\dfrac{1}{16}, r = -4$

In Exercises 17–24, use the formula for the general term (the nth term) of a geometric sequence to find the indicated term of each sequence with the given first term, a_1, and common ratio, r.

17. Find a_8 when $a_1 = 6, r = 2$.

18. Find a_8 when $a_1 = 5, r = 3$.

19. Find a_{12} when $a_1 = 5, r = -2$.

20. Find a_{12} when $a_1 = 4, r = -2$.

21. Find a_6 when $a_1 = 6400, r = -\dfrac{1}{2}$.

22. Find a_6 when $a_1 = 8000, r = -\dfrac{1}{2}$.

23. Find a_8 when $a_1 = 1{,}000{,}000, r = 0.1$.

24. Find a_8 when $a_1 = 40{,}000, r = 0.1$.

In Exercises 25–32, write a formula for the general term (the nth term) of each geometric sequence. Then use the formula for a_n to find a_7, the seventh term of the sequence.

25. $3, 12, 48, 192, \ldots$

26. $3, 15, 75, 375, \ldots$

27. $18, 6, 2, \dfrac{2}{3}, \ldots$

28. $12, 6, 3, \dfrac{3}{2}, \ldots$

29. $1.5, -3, 6, -12, \ldots$

30. $5, -1, \dfrac{1}{5}, -\dfrac{1}{25}, \ldots$

31. $0.0004, -0.004, 0.04, -0.4 \ldots$

32. $0.0007, -0.007, 0.07, -0.7, \ldots$

Use the formula for the sum of the first n terms of a geometric sequence to solve Exercises 33–38.

33. Find the sum of the first 12 terms of the geometric sequence: $2, 6, 18, 54 \ldots$.

34. Find the sum of the first 12 terms of the geometric sequence: $3, 6, 12, 24, \ldots$.

35. Find the sum of the first 11 terms of the geometric sequence: $3, -6, 12, -24, \ldots$.

36. Find the sum of the first 11 terms of the geometric sequence: $4, -12, 36, -108, \ldots$.

37. Find the sum of the first 14 terms of the geometric sequence: $-\frac{3}{2}, 3, -6, 12, \ldots$.

38. Find the sum of the first 14 terms of the geometric sequence: $-\frac{1}{24}, \frac{1}{12}, -\frac{1}{6}, \frac{1}{3}, \ldots$.

In Exercises 39–44, find the indicated sum. Use the formula for the sum of the first n terms of a geometric sequence.

39. $\displaystyle\sum_{i=1}^{8} 3^i$

40. $\displaystyle\sum_{i=1}^{6} 4^i$

41. $\displaystyle\sum_{i=1}^{10} 5 \cdot 2^i$

42. $\displaystyle\sum_{i=1}^{7} 4(-3)^i$

43. $\displaystyle\sum_{i=1}^{6} \left(\dfrac{1}{2}\right)^{i+1}$

44. $\displaystyle\sum_{i=1}^{6} \left(\dfrac{1}{3}\right)^{i+1}$

In Exercises 45–52, find the sum of each infinite geometric series.

45. $1 + \dfrac{1}{3} + \dfrac{1}{9} + \dfrac{1}{27} + \cdots$

46. $1 + \dfrac{1}{4} + \dfrac{1}{16} + \dfrac{1}{64} + \cdots$

47. $3 + \dfrac{3}{4} + \dfrac{3}{4^2} + \dfrac{3}{4^3} + \cdots$

48. $5 + \dfrac{5}{6} + \dfrac{5}{6^2} + \dfrac{5}{6^3} + \cdots$

49. $1 - \dfrac{1}{2} + \dfrac{1}{4} - \dfrac{1}{8} + \cdots$

50. $3 - 1 + \dfrac{1}{3} - \dfrac{1}{9} + \cdots$

51. $\displaystyle\sum_{i=1}^{\infty} 26(-0.3)^{i-1}$

52. $\displaystyle\sum_{i=1}^{\infty} 51(-0.7)^{i-1}$

In Exercises 53–58, express each repeating decimal as a fraction in lowest terms.

53. $0.\overline{5} = \dfrac{5}{10} + \dfrac{5}{100} + \dfrac{5}{1000} + \dfrac{5}{10,000} + \cdots$

54. $0.\overline{1} = \dfrac{1}{10} + \dfrac{1}{100} + \dfrac{1}{1000} + \dfrac{1}{10,000} + \cdots$

55. $0.\overline{47} = \dfrac{47}{100} + \dfrac{47}{10,000} + \dfrac{47}{1,000,000} + \cdots$

56. $0.\overline{83} = \dfrac{83}{100} + \dfrac{83}{10,000} + \dfrac{83}{1,000,000} + \cdots$

57. $0.\overline{257}$

58. $0.\overline{529}$

In Exercises 59–64, the general term of a sequence is given. Determine whether the sequence is arithmetic, geometric, or neither. If the sequence is arithmetic, find the common difference; if it is geometric, find the common ratio.

59. $a_n = n + 5$

60. $a_n = n - 3$

61. $a_n = 2^n$

62. $a_n = \left(\dfrac{1}{2}\right)^n$

63. $a_n = n^2 + 5$

64. $a_n = n^2 - 3$

Practice PLUS

In Exercises 65–70, let

$$\{a_n\} = -5, 10, -20, 40, \ldots,$$
$$\{b_n\} = 10, -5, -20, -35, \ldots,$$
and $\quad \{c_n\} = -2, 1, -\tfrac{1}{2}, \tfrac{1}{4}, \ldots$.

65. Find $a_{10} + b_{10}$.

66. Find $a_{11} + b_{11}$.

67. Find the difference between the sum of the first 10 terms of $\{a_n\}$ and the sum of the first 10 terms of $\{b_n\}$.

68. Find the difference between the sum of the first 11 terms of $\{a_n\}$ and the sum of the first 11 terms of $\{b_n\}$.

69. Find the product of the sum of the first 6 terms of $\{a_n\}$ and the sum of the infinite series containing all the terms of $\{c_n\}$.

70. Find the product of the sum of the first 9 terms of $\{a_n\}$ and the sum of the infinite series containing all the terms of $\{c_n\}$.

In Exercises 71–72, find a_2 and a_3 for each geometric sequence.

71. $8, a_2, a_3, 27$

72. $2, a_2, a_3, -54$

In Exercises 73–74, round all answers to the nearest dollar.

73. Here are two ways of investing $30,000 for 20 years.

• Lump-Sum Deposit	Rate	Time
$30,000	5% compounded annually	20 years

• Periodic Deposits	Rate	Time
$1500 at the end of each year	5% compounded annually	20 years

After 20 years, how much more will you have from the lump-sum investment than from the annuity?

74. Here are two ways of investing $40,000 for 25 years.

• Lump-Sum Deposit	Rate	Time
$40,000	6.5% compounded annually	25 years

• Periodic Deposits	Rate	Time
$1600 at the end of each year	6.5% compounded annually	25 years

After 25 years, how much more will you have from the lump-sum investment than from the annuity?

Application Exercises

Use the formula for the general term (the nth term) of a geometric sequence to solve Exercises 75–78.

In Exercises 75–76 suppose you save $1 the first day of a month, $2 the second day, $4 the third day, and so on. That is, each day you save twice as much as you did the day before.

75. What will you put aside for savings on the fifteenth day of the month?

76. What will you put aside for savings on the thirtieth day of the month?

77. A professional baseball player signs a contract with a beginning salary of $3,000,000 for the first year and an annual increase of 4% per year beginning in the second year. That is, beginning in year 2, the athlete's salary will be 1.04 times what it was in the previous year. What is the athlete's salary for year 7 of the contract? Round to the nearest dollar.

78. You are offered a job that pays $30,000 for the first year with an annual increase of 5% per year beginning in the second year. That is, beginning in year 2, your salary will be 1.05 times what it was in the previous year. What can you expect to earn in your sixth year on the job? Round to the nearest dollar.

In Exercises 79–80, you will develop geometric sequences that model the population growth for California and Texas, the two most-populated U.S. states.

79. The table shows population estimates for California from 2003 through 2006 from the U.S. Census Bureau.

Year	2003	2004	2005	2006
Population in millions	35.48	35.89	36.13	36.46

  a. Divide the population for each year by the population in the preceding year. Round to two decimal places and show that California has a population increase that is approximately geometric.

  b. Write the general term of the geometric sequence modeling California's population, in millions, *n* years after 2002.

  c. Use your model from part (b) to project California's population, in millions, for the year 2010. Round to two decimal places.

80. The table shows population estimates for Texas from 2003 through 2006 from the U.S. Census Bureau.

Year	2003	2004	2005	2006
Population in millions	22.12	22.49	22.86	23.41

  a. Divide the population for each year by the population in the preceding year. Round to two decimal places and show that Texas has a population increase that is approximately geometric.

  b. Write the general term of the geometric sequence modeling Texas's population, in millions, *n* years after 2002.

  c. Use your model from part (b) to project Texas's population, in millions, for the year 2010. Round to two decimal places.

Use the formula for the sum of the first n terms of a geometric sequence to solve Exercises 81–86.

In Exercises 81–82, you save $1 the first day of a month, $2 the second day, $4 the third day, continuing to double your savings each day.

81. What will your total savings be for the first 15 days?

82. What will your total savings be for the first 30 days?

83. A job pays a salary of $24,000 the first year. During the next 19 years, the salary increases by 5% each year. What is the total lifetime salary over the 20-year period? Round to the nearest dollar.

84. You are investigating two employment opportunities. Company A offers $30,000 the first year. During the next four years, the salary is guaranteed to increase by 6% per year. Company B offers $32,000 the first year, with guaranteed annual increases of 3% per year after that. Which company offers the better total salary for a five-year contract? By how much? Round to the nearest dollar.

85. A pendulum swings through an arc of 20 inches. On each successive swing, the length of the arc is 90% of the previous length.

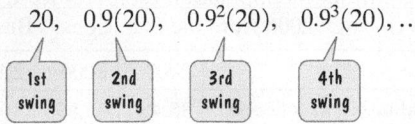

$$20, \quad 0.9(20), \quad 0.9^2(20), \quad 0.9^3(20), \dots$$

After 10 swings, what is the total length of the distance the pendulum has swung? Round to the nearest hundredth of an inch.

86. A pendulum swings through an arc of 16 inches. On each successive swing, the length of the arc is 96% of the previous length.

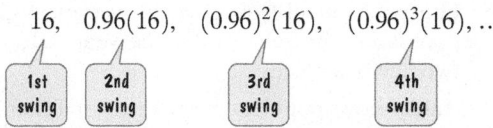

$$16, \quad 0.96(16), \quad (0.96)^2(16), \quad (0.96)^3(16), \dots$$

After 10 swings, what is the total length of the distance the pendulum has swung? Round to the nearest hundredth of an inch.

Use the formula for the value of an annuity to solve Exercises 87–92. Round answers to the nearest dollar.

87. To save money for a sabbatical to earn a master's degree, you deposit $2000 at the end of each year in an annuity that pays 7.5% compounded annually.

  a. How much will you have saved at the end of five years?

  b. Find the interest.

88. To save money for a sabbatical to earn a master's degree, you deposit $2500 at the end of each year in an annuity that pays 6.25% compounded annually.

  a. How much will you have saved at the end of five years?

  b. Find the interest.

89. At age 25, to save for retirement, you decide to deposit $50 at the end of each month in an IRA that pays 5.5% compounded monthly.

  a. How much will you have from the IRA when you retire at age 65?

  b. Find the interest.

90. At age 25, to save for retirement, you decide to deposit $75 at the end of each month in an IRA that pays 6.5% compounded monthly.

  a. How much will you have from the IRA when you retire at age 65?

  b. Find the interest.

91. To offer scholarship funds to children of employees, a company invests $10,000 at the end of every three months in an annuity that pays 10.5% compounded quarterly.

  a. How much will the company have in scholarship funds at the end of ten years?

  b. Find the interest.

92. To offer scholarship funds to children of employees, a company invests $15,000 at the end of every three months in an annuity that pays 9% compounded quarterly.

  a. How much will the company have in scholarship funds at the end of ten years?

  b. Find the interest.

Use the formula for the sum of an infinite geometric series to solve Exercises 93–95.

93. A new factory in a small town has an annual payroll of $6 million. It is expected that 60% of this money will be spent in the town by factory personnel. The people in the town who receive this money are expected to spend 60% of what they receive in the town, and so on. What is the total of all this spending, called the *total economic impact* of the factory, on the town each year?

94. How much additional spending will be generated by a $10 billion tax rebate if 60% of all income is spent?

95. If the shading process shown in the figure is continued indefinitely, what fractional part of the largest square will eventually be shaded?

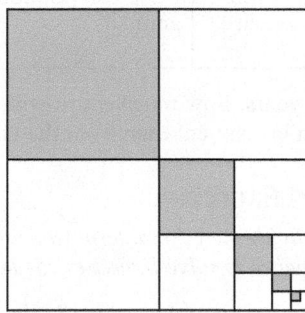

Writing in Mathematics

96. What is a geometric sequence? Give an example with your explanation.

97. What is the common ratio in a geometric sequence?

98. Explain how to find the general term of a geometric sequence.

99. Explain how to find the sum of the first *n* terms of a geometric sequence without having to add up all the terms.

100. What is an annuity?

101. What is the difference between a geometric sequence and an infinite geometric series?

102. How do you determine if an infinite geometric series has a sum? Explain how to find the sum of such an infinite geometric series.

103. Would you rather have $10,000,000 and a brand new BMW or 1¢ today, 2¢ tomorrow, 4¢ on day 3, 8¢ on day 4, 16¢ on day 5, and so on, for 30 days? Explain.

104. For the first 30 days of a flu outbreak, the number of students on your campus who become ill is increasing. Which is worse: the number of students with the flu is increasing arithmetically or is increasing geometrically? Explain your answer.

Technology Exercises

105. Use the $\boxed{\text{SEQ}}$ (sequence) capability of a graphing utility and the formula you obtained for a_n to verify the value you found for a_7 in any three exercises from Exercises 25–32.

106. Use the capability of a graphing utility to calculate the sum of a sequence to verify any three of your answers to Exercises 39–44.

In Exercises 107–108, use a graphing utility to graph the function. Determine the horizontal asymptote for the graph of f and discuss its relationship to the sum of the given series.

107. **Function**   **Series**

$$f(x) = \frac{2\left[1 - \left(\frac{1}{3}\right)^x\right]}{1 - \frac{1}{3}} \qquad 2 + 2\left(\frac{1}{3}\right) + 2\left(\frac{1}{3}\right)^2 + 2\left(\frac{1}{3}\right)^3 + \cdots$$

108. **Function**   **Series**

$$f(x) = \frac{4[1 - (0.6)^x]}{1 - 0.6} \qquad 4 + 4(0.6) + 4(0.6)^2 + 4(0.6)^3 + \cdots$$

Critical Thinking Exercises

Make Sense? *In Exercises 109–112, determine whether each statement "makes sense" or "does not make sense" and explain your reasoning.*

109. There's no end to the number of geometric sequences that I can generate whose first term is 5 if I pick nonzero numbers r and multiply 5 by each value of r repeatedly.

110. I've noticed that the big difference between arithmetic and geometric sequences is that arithmetic sequences are based on addition and geometric sequences are based on multiplication.

111. I modeled California's population growth with a geometric sequence, so my model is an exponential function whose domain is the set of natural numbers.

112. I used a formula to find the sum of the infinite geometric series $3 + 1 + \frac{1}{3} + \frac{1}{9} + \cdots$ and then checked my answer by actually adding all the terms.

In Exercises 113–116, determine whether each statement is true or false. If the statement is false, make the necessary change(s) to produce a true statement.

113. The sequence $2, 6, 24, 120, \ldots$ is an example of a geometric sequence.

114. The sum of the geometric series $\frac{1}{2} + \frac{1}{4} + \frac{1}{8} + \cdots + \frac{1}{512}$ can only be estimated without knowing precisely what terms occur between $\frac{1}{8}$ and $\frac{1}{512}$.

115. $10 - 5 + \frac{5}{2} - \frac{5}{4} + \cdots = \dfrac{10}{1 - \frac{1}{2}}$

116. If the nth term of a geometric sequence is $a_n = 3(0.5)^{n-1}$, the common ratio is $\frac{1}{2}$.

117. In a pest-eradication program, sterilized male flies are released into the general population each day. Ninety percent of those flies will survive a given day. How many flies should be released each day if the long-range goal of the program is to keep 20,000 sterilized flies in the population?

118. You are now 25 years old and would like to retire at age 55 with a retirement fund of $1,000,000. How much should you deposit at the end of each month for the next 30 years in an IRA paying 10% annual interest compounded monthly to achieve your goal? Round to the nearest dollar.

Review Exercises

119. Simplify: $\sqrt{28} - 3\sqrt{7} + \sqrt{63}$. (Section 10.4, Example 2)

120. Solve: $2x^2 = 4 - x$. (Section 11.2, Example 2)

121. Rationalize the denominator: $\dfrac{6}{\sqrt{3} - \sqrt{5}}$. (Section 10.5, Example 5)

Preview Exercises

Exercises 122–124 will help you prepare for the material covered in the next section.

Each exercise involves observing a pattern in the expanded form of the binomial expression $(a + b)^n$.

$(a + b)^1 = a + b$

$(a + b)^2 = a^2 + 2ab + b^2$

$(a + b)^3 = a^3 + 3a^2b + 3ab^2 + b^3$

$(a + b)^4 = a^4 + 4a^3b + 6a^2b^2 + 4ab^3 + b^4$

$(a + b)^5 = a^5 + 5a^4b + 10a^3b^2 + 10a^2b^3 + 5ab^4 + b^5$

122. Describe the pattern for the exponents on a.

123. Describe the pattern for the exponents on b.

124. Describe the pattern for the sum of the exponents on the variables in each term.

MID-CHAPTER CHECK POINT   Section 14.1–Section 14.3

✓ **What You Know:** We learned that a sequence is a function whose domain is the set of positive integers. In an arithmetic sequence, each term after the first differs from the preceding term by a constant, the common difference, d. In a geometric sequence, each term after the first is obtained by multiplying the preceding term by a nonzero constant, the common ratio, r. We found the general term of arithmetic sequences $[a_n = a_1 + (n - 1)d]$ and geometric sequences $[a_n = a_1 r^{n-1}]$ and used these formulas to find particular terms. We determined the sum of the first n terms of arithmetic sequences $\left[S_n = \dfrac{n}{2}(a_1 + a_n) \right]$ and geometric sequences $\left[S_n = \dfrac{a_1(1 - r^n)}{1 - r} \right]$. Finally, we determined the sum of an infinite geometric series, $a_1 + a_1 r + a_1 r^2 + a_1 r^3 + \cdots$, if $-1 < r < 1 \left(S = \dfrac{a_1}{1 - r} \right)$.

In Exercises 1–3, write the first five terms of each sequence. Assume that d represents the common difference of an arithmetic sequence and r represents the common ratio of a geometric sequence.

1. $a_n = (-1)^{n+1} \dfrac{n}{(n - 1)!}$

2. $a_1 = 5, d = -3$

3. $a_1 = 5, r = -3$

In Exercises 4–6, write a formula for the general term (the nth term) of each sequence. Then use the formula to find the indicated term.

4. $2, 6, 10, 14, \ldots ; a_{20}$

5. $3, 6, 12, 24, \ldots ; a_{10}$

6. $\dfrac{3}{2}, 1, \dfrac{1}{2}, 0, \ldots ; a_{30}$

7. Find the sum of the first ten terms of the sequence: $5, 10, 20, 40, \ldots$.

8. Find the sum of the first 50 terms of the sequence: $-2, 0, 2, 4, \ldots$.

9. Find the sum of the first ten terms of the sequence: $-20, 40, -80, 160, \ldots$.

10. Find the sum of the first 100 terms of the sequence: $4, -2, -8, -14, \ldots$.

In Exercises 11–14, find each indicated sum.

11. $\displaystyle\sum_{i=1}^{4} (i + 4)(i - 1)$

12. $\displaystyle\sum_{i=1}^{50} (3i - 2)$

13. $\displaystyle\sum_{i=1}^{6} \left(\dfrac{3}{2} \right)^i$

14. $\displaystyle\sum_{i=1}^{\infty} \left(-\dfrac{2}{5} \right)^{i-1}$

15. Express $0.\overline{45}$ as a fraction in lowest terms.

16. Express the sum using summation notation. Use i for the index of summation.

$$\frac{1}{3} + \frac{2}{4} + \frac{3}{5} + \cdots + \frac{18}{20}$$

17. A skydiver falls 16 feet during the first second of a dive, 48 feet during the second second, 80 feet during the third second, 112 feet during the fourth second, and so on. Find the distance that the skydiver falls during the 15th second and the total distance the skydiver falls in 15 seconds.

18. If the average value of a house increases 10% per year, how much will a house costing $500,000 in year 1 be worth in year 9? Round to the nearest dollar.

SECTION

14.4

The Binomial Theorem

Objectives

1. Evaluate a binomial coefficient.

2. Expand a binomial raised to a power.

3. Find a particular term in a binomial expansion.

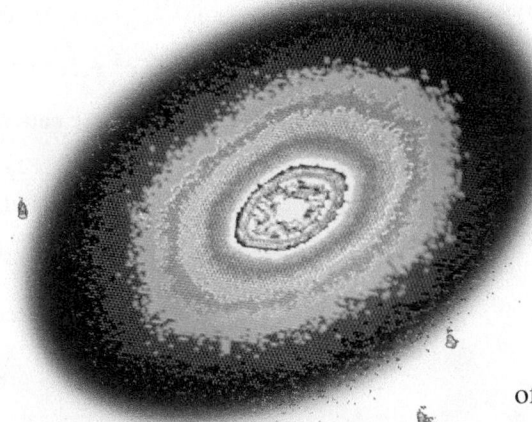

Galaxies are groupings of billions of stars bound together by gravity. Some galaxies, such as the Centaurus galaxy shown here, are elliptical in shape.

Is mathematics discovered or invented? For example, planets revolve in elliptical orbits. Does that mean that the ellipse is out there, waiting for the mind to discover it? Or do people create the definition of an ellipse just as they compose a song? And is it possible for the same mathematics to be discovered/invented by independent researchers separated by time, place, and culture? This is precisely what occurred when mathematicians attempted to find efficient methods for raising binomials to higher and higher powers, such as

$$(x + 2)^3, (x + 2)^4, (x + 2)^5, (x + 2)^6,$$

and so on. In this section, we study higher powers of binomials and a method first discovered/invented by great minds in Eastern and in Western cultures working independently.

1 Evaluate a binomial coefficient.

Binomial Coefficients

Before turning to powers of binomials, we introduce a special notation that uses factorials.

Definition of a Binomial Coefficient $\binom{n}{r}$

For nonnegative integers n and r, with $n \geq r$, the expression $\binom{n}{r}$ (read "n above r") is called a **binomial coefficient** and is defined by

$$\binom{n}{r} = \frac{n!}{r!(n-r)!}.$$

The symbol $_nC_r$ is often used in place of $\binom{n}{r}$ to denote binomial coefficients.

Can you see that the definition of a binomial coefficient involves a fraction with factorials in the numerator and the denominator? When evaluating such an expression, try to reduce the fraction before performing the multiplications. For example, consider $\frac{26!}{21!}$. Rather than writing out 26! as the product of all integers from 26 down to 1, we can express 26! as

$$26! = 26 \cdot 25 \cdot 24 \cdot 23 \cdot 22 \cdot 21!.$$

In this way, we can divide both the numerator and the denominator by the common factor, 21!.

$$\frac{26!}{21!} = \frac{26 \cdot 25 \cdot 24 \cdot 23 \cdot 22 \cdot \cancel{21!}}{\cancel{21!}} = 26 \cdot 25 \cdot 24 \cdot 23 \cdot 22 = 7,893,600$$

EXAMPLE 1   Evaluating Binomial Coefficients

Using Technology

Graphing utilities can compute binomial coefficients. For example, to find $\binom{6}{2}$, many utilities require the sequence

6 | nCr | 2 | ENTER |.

The graphing utility will display 15. Consult your manual and verify the other evaluations in Example 1.

Evaluate:

a. $\binom{6}{2}$ **b.** $\binom{3}{0}$ **c.** $\binom{9}{3}$ **d.** $\binom{4}{4}$.

Solution   In each case, we apply the definition of the binomial coefficient.

a. $\binom{6}{2} = \frac{6!}{2!\,(6-2)!} = \frac{6!}{2!\,4!} = \frac{6 \cdot 5 \cdot \cancel{4!}}{2 \cdot 1 \cdot \cancel{4!}} = 15$

b. $\binom{3}{0} = \frac{3!}{0!(3-0)!} = \frac{\cancel{3!}}{0!\,\cancel{3!}} = \frac{1}{1} = 1$

Remember that 0! = 1.

c. $\dbinom{9}{3} = \dfrac{9!}{3!\,(9-3)!} = \dfrac{9!}{3!\,6!} = \dfrac{9 \cdot 8 \cdot 7 \cdot \cancel{6!}}{3 \cdot 2 \cdot 1 \cdot \cancel{6!}} = 84$

d. $\dbinom{4}{4} = \dfrac{4!}{4!\,(4-4)!} = \dfrac{\cancel{4!}}{\cancel{4!}\,0!} = \dfrac{1}{1} = 1$ ■

✓ **CHECK POINT 1**   Evaluate:

a. $\dbinom{6}{3}$ **b.** $\dbinom{6}{0}$ **c.** $\dbinom{8}{2}$ **d.** $\dbinom{3}{3}$.

---

2 Expand a binomial raised to a power.

The Binomial Theorem

When we write out the *binomial expression* $(a + b)^n$, where n is a positive integer, a number of patterns begin to appear.

$$(a + b)^1 = a + b$$
$$(a + b)^2 = a^2 + 2ab + b^2$$
$$(a + b)^3 = a^3 + 3a^2b + 3ab^2 + b^3$$
$$(a + b)^4 = a^4 + 4a^3b + 6a^2b^2 + 4ab^3 + b^4$$
$$(a + b)^5 = a^5 + 5a^4b + 10a^3b^2 + 10a^2b^3 + 5ab^4 + b^5$$

Each expanded form of the binomial expression is a polynomial. Observe the following patterns:

1. The first term in the expansion of $(a + b)^n$ is a^n. The exponents on a decrease by 1 in each successive term.

2. The exponents on b in the expansion of $(a + b)^n$ increase by 1 in each successive term. In the first term, the exponent on b is 0. (Because $b^0 = 1$, b is not shown in the first term.) The last term is b^n.

3. The sum of the exponents on the variables in any term in the expansion of $(a + b)^n$ is equal to n.

4. The number of terms in the polynomial expansion is one greater than the power of the binomial, n. There are $n + 1$ terms in the expanded form of $(a + b)^n$.

Using these observations, the variable parts of the expansion of $(a + b)^6$ are

$$a^6, \quad a^5b, \quad a^4b^2, \quad a^3b^3, \quad a^2b^4, \quad ab^5, \quad b^6.$$

The first term is a^6, with the exponents on a decreasing by 1 in each successive term. The exponents on b increase from 0 to 6, with the last term being b^6. The sum of the exponents in each term is equal to 6.

We can generalize from these observations to obtain the variable parts of the expansion of $(a + b)^n$. They are

> Exponents on a are decreasing by 1.
> Exponents on b are increasing by 1.

$$a^n, \quad a^{n-1}b, \quad a^{n-2}b^2, \quad a^{n-3}b^3, \dots, \quad ab^{n-1}, \quad b^n.$$

Sum of exponents: $n - 1 + 1 = n$

Sum of exponents: $n - 3 + 3 = n$

Sum of exponents: $1 + n - 1 = n$

If we use binomial coefficients and the pattern for the variable part of each term, a formula called the **Binomial Theorem** can be used to expand any positive integral power of a binomial.

A Formula for Expanding Binomials: The Binomial Theorem

For any positive integer n,

$$(a + b)^n = \binom{n}{0}a^n + \binom{n}{1}a^{n-1}b + \binom{n}{2}a^{n-2}b^2 + \binom{n}{3}a^{n-3}b^3 + \cdots + \binom{n}{n}b^n$$

$$= \sum_{r=0}^{n}\binom{n}{r}a^{n-r}b^r.$$

EXAMPLE 2 Using the Binomial Theorem

Expand: $(x + 2)^4$.

Solution We use the Binomial Theorem

$$(a + b)^n = \binom{n}{0}a^n + \binom{n}{1}a^{n-1}b + \binom{n}{2}a^{n-2}b^2 + \binom{n}{3}a^{n-3}b^3 + \cdots + \binom{n}{n}b^n$$

to expand $(x + 2)^4$. In $(x + 2)^4$, $a = x$, $b = 2$, and $n = 4$. In the expansion, powers of x are in descending order, starting with x^4. Powers of 2 are in ascending order, starting with 2^0. (Because $2^0 = 1$, a 2 is not shown in the first term.) The sum of the exponents on x and 2 in each term is equal to 4, the exponent in the expression $(x + 2)^4$.

$$(x + 2)^4 = \binom{4}{0}x^4 + \binom{4}{1}x^3 \cdot 2 + \binom{4}{2}x^2 \cdot 2^2 + \binom{4}{3}x \cdot 2^3 + \binom{4}{4}2^4$$

These binomial coefficients are evaluated using $\binom{n}{r} = \frac{n!}{r!(n-r)!}$.

$$= \frac{4!}{0!4!}x^4 + \frac{4!}{1!3!}x^3 \cdot 2 + \frac{4!}{2!2!}x^2 \cdot 4 + \frac{4!}{3!1!}x \cdot 8 + \frac{4!}{4!0!} \cdot 16$$

$$\frac{4!}{2!2!} = \frac{4 \cdot 3 \cdot 2!}{2! \cdot 2 \cdot 1} = \frac{12}{2} = 6$$

Take a few minutes to verify the other factorial evaluations.

$$= 1 \cdot x^4 + 4x^3 \cdot 2 + 6x^2 \cdot 4 + 4x \cdot 8 + 1 \cdot 16$$

$$= x^4 + 8x^3 + 24x^2 + 32x + 16 \quad \blacksquare$$

✓ **CHECK POINT 2** Expand: $(x + 1)^4$.

Using Technology

You can use a graphing utility's table feature to find the five binomial coefficients in Example 2.

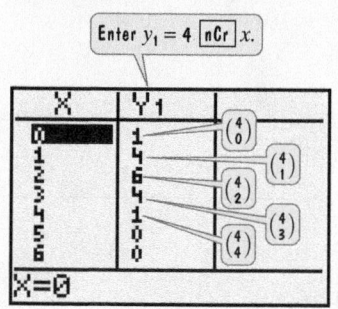

Enter $y_1 = 4$ [nCr] x.

EXAMPLE 3 Using the Binomial Theorem

Expand: $(2x - y)^5$.

Solution Because the Binomial Theorem involves the addition of two terms raised to a power, we rewrite $(2x - y)^5$ as $[2x + (-y)]^5$. We use the Binomial Theorem

$$(a + b)^n = \binom{n}{0}a^n + \binom{n}{1}a^{n-1}b + \binom{n}{2}a^{n-2}b^2 + \binom{n}{3}a^{n-3}b^3 + \cdots + \binom{n}{n}b^n$$

to expand $[2x + (-y)]^5$. In $[2x + (-y)]^5$, $a = 2x$, $b = -y$, and $n = 5$. In the expansion, powers of $2x$ are in descending order, starting with $(2x)^5$. Powers of $-y$ are in ascending order, starting with $(-y)^0$. [Because $(-y)^0 = 1$, a $-y$ is not shown in the first term.] The sum of the exponents on $2x$ and $-y$ in each term is equal to 5, the exponent in the expression $(2x - y)^5$.

$$(2x - y)^5 = [2x + (-y)]^5$$

$$= \binom{5}{0}(2x)^5 + \binom{5}{1}(2x)^4(-y) + \binom{5}{2}(2x)^3(-y)^2 + \binom{5}{3}(2x)^2(-y)^3 + \binom{5}{4}(2x)(-y)^4 + \binom{5}{5}(-y)^5$$

Evaluate binomial coefficients using $\binom{n}{r} = \frac{n!}{r!(n-r)!}$.

$$= \frac{5!}{0!5!}(2x)^5 + \frac{5!}{1!4!}(2x)^4(-y) + \frac{5!}{2!3!}(2x)^3(-y)^2 + \frac{5!}{3!2!}(2x)^2(-y)^3 + \frac{5!}{4!1!}(2x)(-y)^4 + \frac{5!}{5!0!}(-y)^5$$

$$\frac{5!}{2!3!} = \frac{5 \cdot 4 \cdot 3!}{2 \cdot 1 \cdot 3!} = 10$$

Take a few minutes to verify the other factorial evaluations.

$$= 1(2x)^5 + 5(2x)^4(-y) + 10(2x)^3(-y)^2 + 10(2x)^2(-y)^3 + 5(2x)(-y)^4 + 1(-y)^5$$

Raise both factors in these parentheses to the indicated powers.

$$= 1(32x^5) + 5(16x^4)(-y) + 10(8x^3)(-y)^2 + 10(4x^2)(-y)^3 + 5(2x)(-y)^4 + 1(-y)^5$$

Now raise $-y$ to the indicated powers.

$$= 1(32x^5) + 5(16x^4)(-y) + 10(8x^3)y^2 + 10(4x^2)(-y^3) + 5(2x)y^4 + 1(-y^5)$$

Multiplying factors in each of the six terms gives us the desired expansion:

$$(2x - y)^5 = 32x^5 - 80x^4y + 80x^3y^2 - 40x^2y^3 + 10xy^4 - y^5. \quad \blacksquare$$

✓ **CHECK POINT 3**   Expand: $(x - 2y)^5$.

---

3 Find a particular term in a binomial expansion.

Finding a Particular Term in a Binomial Expansion

By observing the terms in the formula for expanding binomials, we can find a formula for finding a particular term without writing the entire expansion.

1st term	2nd term	3rd term

$$\binom{n}{0}a^nb^0 \qquad \binom{n}{1}a^{n-1}b^1 \qquad \binom{n}{2}a^{n-2}b^2$$

The exponent on b is 1 less than the term number.

Based on the observation in the bottom voice balloon, the $(r + 1)$st term of the expansion of $(a + b)^n$ is the term that contains b^r.

Finding a Particular Term in a Binomial Expansion

The $(r + 1)$st term of the expansion of $(a + b)^n$ is

$$\binom{n}{r}a^{n-r}b^r.$$

EXAMPLE 4 Finding a Single Term of a Binomial Expansion

Find the fourth term in the expansion of $(3x + 2y)^7$.

Solution The fourth term in the expansion of $(3x + 2y)^7$ contains $(2y)^3$. To find the fourth term, first note that $4 = 3 + 1$. Equivalently, the fourth term of $(3x + 2y)^7$ is the $(3 + 1)$st term. Thus, $r = 3$, $a = 3x$, $b = 2y$, and $n = 7$. The fourth term is

$$\binom{7}{3}(3x)^{7-3}(2y)^3 = \binom{7}{3}(3x)^4(2y)^3 = \frac{7!}{3!(7-3)!}(3x)^4(2y)^3.$$

Use the formula for the $(r+1)$st term of $(a + b)^n$:
$\binom{n}{r}a^{n-r}b^r$.

We use $\binom{n}{r} = \frac{n!}{r!(n-r)!}$ to evaluate $\binom{7}{3}$.

Now we need to evaluate the factorial expression and raise $3x$ and $2y$ to the indicated powers. We obtain

$$\frac{7!}{3!\,4!}(81x^4)(8y^3) = \frac{7 \cdot 6 \cdot 5 \cdot 4!}{3 \cdot 2 \cdot 1 \cdot 4!}(81x^4)(8y^3) = 35(81x^4)(8y^3) = 22{,}680x^4y^3.$$

The fourth term of $(3x + 2y)^7$ is $22{,}680x^4y^3$. ∎

✓ **CHECK POINT 4** Find the fifth term in the expansion of $(2x + y)^9$.

Blitzer Bonus

The Universality of Mathematics

Pascal's triangle is an array of numbers showing coefficients of the terms in the expansions of $(a + b)^n$. Although credited to French mathematician Blaise Pascal (1623–1662), the triangular array of numbers appeared in a Chinese document printed in 1303. The Binomial Theorem was known in Eastern cultures prior to its discovery in Europe. The same mathematics is often discovered/invented by independent researchers separated by time, place, and culture.

Binomial Expansions

$(a + b)^0 = 1$
$(a + b)^1 = a + b$
$(a + b)^2 = a^2 + 2ab + b^2$
$(a + b)^3 = a^3 + 3a^2b + 3ab^2 + b^3$
$(a + b)^4 = a^4 + 4a^3b + 6a^2b^2 + 4ab^3 + b^4$
$(a + b)^5 = a^5 + 5a^4b + 10a^3b^2 + 10a^2b^3 + 5ab^4 + b^5$

Pascal's Triangle
Coefficients in the Expansions

```
              1
            1   1
          1   2   1
        1   3   3   1
      1   4   6   4   1
    1   5  10  10   5   1
  1   6  15  20  15   6   1
1  7  21  35  35  21   7   1
1 8 28 56 70 56 28 8 1
```

Chinese Document: 1303

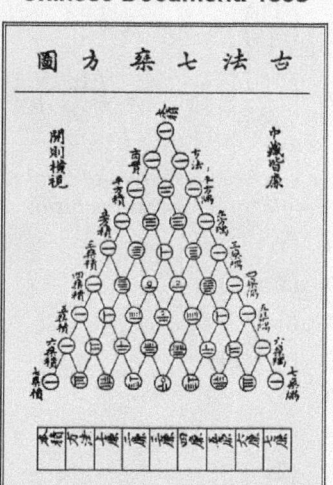

Achieving Success

As you continue taking college math courses, remember that you are expected to study and do homework for at least two hours for each hour of class time. In order to make education a top priority, be consistent in the time and location for studying. Arrange your schedule so that you can work in a quiet location without distractions and at a time when you are alert and best able to study. Do this on a regular basis until the routine of studying and doing homework becomes familiar and comfortable.

CONCEPT AND VOCABULARY CHECK

Fill in each blank so that the resulting statement is true.

1. $\binom{n}{r}$ is called a/an _____ coefficient.

2. $\binom{8}{2} = \dfrac{\_!}{\_!\_!}$

3. $\binom{n}{r} =$ _____

4. $(x + 2)^5 = \binom{5}{0}x^5 +$ _____ $x^4 \cdot 2 +$ _____ $x^3 \cdot 2^2 +$ _____ $x^2 \cdot 2^3 +$ _____ $x \cdot 2^4 +$ _____ $\cdot 2^5$

5. $(a + b)^n = \binom{n}{0}a^n +$ _____ $a^{n-1}b +$ _____ $a^{n-2}b^2 +$ _____ $a^{n-3}b^3 + \cdots +$ _____ b^n

   The sum of the exponents on a and b in each term is _____.

6. The formula in Concept Check 5 is called the _____ Theorem.

7. The $(r + 1)$st term of the expansion of $(a + b)^n$ is $\binom{n}{r}$_____.

14.4 EXERCISE SET

MyMathLab®

Watch the videos in MyMathLab

Download the MyDashBoard App

Practice Exercises

In Exercises 1–8, evaluate the given binomial coefficient.

1. $\binom{8}{3}$
2. $\binom{7}{2}$
3. $\binom{12}{1}$
4. $\binom{11}{1}$
5. $\binom{6}{6}$
6. $\binom{15}{2}$
7. $\binom{100}{2}$
8. $\binom{100}{98}$

In Exercises 9–30, use the Binomial Theorem to expand each binomial and express the result in simplified form.

9. $(x + 2)^3$
10. $(x + 4)^3$
11. $(3x + y)^3$
12. $(x + 3y)^3$
13. $(5x - 1)^3$
14. $(4x - 1)^3$
15. $(2x + 1)^4$
16. $(3x + 1)^4$
17. $(x^2 + 2y)^4$
18. $(x^2 + y)^4$
19. $(y - 3)^4$
20. $(y - 4)^4$
21. $(2x^3 - 1)^4$
22. $(2x^5 - 1)^4$

23. $(c + 2)^5$
24. $(c + 3)^5$
25. $(x - 1)^5$
26. $(x - 2)^5$
27. $(3x - y)^5$
28. $(x - 3y)^5$
29. $(2a + b)^6$
30. $(a + 2b)^6$

In Exercises 31–38, write the first three terms in each binomial expansion, expressing the result in simplified form.

31. $(x + 2)^8$
32. $(x + 3)^8$
33. $(x - 2y)^{10}$
34. $(x - 2y)^9$
35. $(x^2 + 1)^{16}$
36. $(x^2 + 1)^{17}$
37. $(y^3 - 1)^{20}$
38. $(y^3 - 1)^{21}$

In Exercises 39–48, find the indicated term in each expansion.

39. $(2x + y)^6$; third term
40. $(x + 2y)^6$; third term
41. $(x - 1)^9$; fifth term

42. $(x - 1)^{10}$; fifth term
43. $(x^2 + y^3)^8$; sixth term
44. $(x^3 + y^2)^8$; sixth term
45. $\left(x - \frac{1}{2}\right)^9$; fourth term
46. $\left(x + \frac{1}{2}\right)^8$; fourth term
47. $(x^2 + y)^{22}$; the term containing y^{14}
48. $(x + 2y)^{10}$; the term containing y^6

Practice PLUS

In Exercises 49–52, use the Binomial Theorem to expand each expression and write the result in simplified form.

49. $(x^3 + x^{-2})^4$
50. $(x^2 + x^{-3})^4$
51. $\left(x^{\frac{1}{3}} - x^{-\frac{1}{3}}\right)^3$
52. $\left(x^{\frac{2}{3}} - \frac{1}{\sqrt[3]{x}}\right)^3$

Exercises 53–54 involve expressions containing i, where $i = \sqrt{-1}$. Expand each expression and use powers of i to simplify the result.

53. $(-1 + i\sqrt{3})^3$ 54. $(-1 - i\sqrt{3})^3$

In Exercises 55–56, find $\dfrac{f(x + h) - f(x)}{h}$ *and simplify.*

55. $f(x) = x^4 + 7$
56. $f(x) = x^5 + 8$

57. Find the middle term in the expansion of $\left(\dfrac{3}{x} + \dfrac{x}{3}\right)^{10}$.

58. Find the middle term in the expansion of $\left(\dfrac{1}{x} - x^2\right)^{12}$.

Application Exercises

The graph shows that U.S. smokers have a greater probability of suffering from some ailments than the general adult population. Exercises 59–60 are based on some of the probabilities, expressed as decimals, shown to the right of the bars. In each exercise, use a calculator to determine the probability, correct to four decimal places.

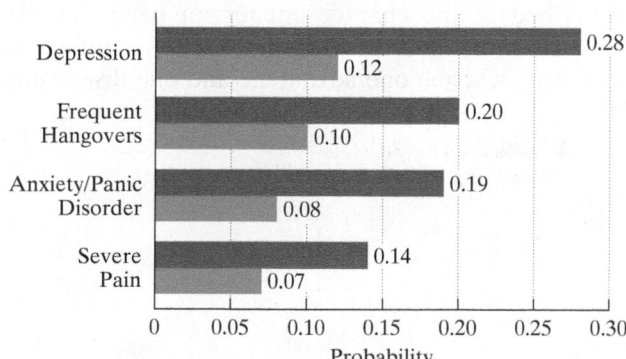

Probability That United States Adults Suffer from Various Ailments

Source: MARS 2005 OTC/DTC Pharmaceutical Study.

If the probability an event will occur is p and the probability it will not occur is q, then each term in the expansion of $(p + q)^n$ represents a probability.

59. The probability that a smoker suffers from depression is 0.28. If five smokers are randomly selected, the probability that three of them will suffer from depression is the third term of the binomial expansion of

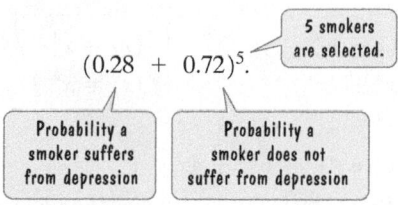

What is this probability?

60. The probability that a person in the general population suffers from depression is 0.12. If five people from the general population are randomly selected, the probability that three of them will suffer from depression is the third term of the binomial expansion of

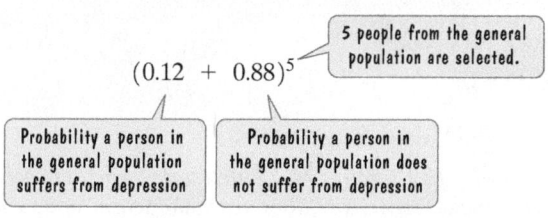

What is this probability?

Writing in Mathematics

61. Explain how to evaluate $\binom{n}{r}$. Provide an example with your explanation.

62. Describe the pattern in the exponents on a in the expansion of $(a + b)^n$.

63. Describe the pattern in the exponents on b in the expansion of $(a + b)^n$.

64. What is true about the sum of the exponents on a and b in any term in the expansion of $(a + b)^n$?

65. How do you determine how many terms there are in a binomial expansion?

66. Explain how to use the Binomial Theorem to expand a binomial. Provide an example with your explanation.

67. Explain how to find a particular term in a binomial expansion without having to write out the entire expansion.

Technology Exercises

68. Use the $\boxed{\text{nCr}}$ key on a graphing utility to verify your answers in Exercises 1–8.

In Exercises 69–70, graph each of the functions in the same viewing rectangle. Describe how the graphs illustrate the Binomial Theorem.

69. $f_1(x) = (x + 2)^3$ $\qquad$ $f_2(x) = x^3$

$f_3(x) = x^3 + 6x^2$ $\qquad$ $f_4(x) = x^3 + 6x^2 + 12x$

$f_5(x) = x^3 + 6x^2 + 12x + 8$

Use a $[-10, 10, 1]$ by $[-30, 30, 10]$ viewing rectangle.

70. $f_1(x) = (x + 1)^4$ $\qquad$ $f_2(x) = x^4$

$f_3(x) = x^4 + 4x^3$ $\qquad$ $f_4(x) = x^4 + 4x^3 + 6x^2$

$f_5(x) = x^4 + 4x^3 + 6x^2 + 4x$

$f_6(x) = x^4 + 4x^3 + 6x^2 + 4x + 1$

Use a $[-5, 5, 1]$ by $[-30, 30, 10]$ viewing rectangle.

In Exercises 71–73, use the Binomial Theorem to find a polynomial expansion for each function. Then use a graphing utility and an approach similar to the one in Exercises 69 and 70 to verify the expansion.

71. $f_1(x) = (x - 1)^3$

72. $f_1(x) = (x - 2)^4$

73. $f_1(x) = (x + 2)^6$

Critical Thinking Exercises

Make Sense? *In Exercises 74–77, determine whether each statement "makes sense" or "does not make sense" and explain your reasoning.*

74. In order to expand $(x^3 - y^4)^5$, I find it helpful to rewrite the expression inside the parentheses as $x^3 + (-y^4)$.

75. Without writing the expansion of $(x - 1)^6$, I can see that the terms have alternating positive and negative signs.

76. I use binomial coefficients to expand $(a + b)^n$, where $\binom{n}{1}$ is the coefficient of the first term, $\binom{n}{2}$ is the coefficient of the second term, and so on.

77. One of the terms in my binomial expansion is $\binom{7}{5}x^2y^4$.

In Exercises 78–81, determine whether each statement is true or false. If the statement is false, make the necessary change(s) to produce a true statement.

78. The binomial expansion for $(a + b)^n$ contains n terms.

79. The Binomial Theorem can be written in condensed form as $(a + b)^n = \sum_{r=0}^{n} \binom{n}{r} a^{n-r} b^r$.

80. The sum of the binomial coefficients in $(a + b)^n$ cannot be 2^n.

81. There are no values of a and b such that

$$(a + b)^4 = a^4 + b^4.$$

82. Use the Binomial Theorem to expand and then simplify the result: $(x^2 + x + 1)^3$. [Hint: Write $x^2 + x + 1$ as $x^2 + (x + 1)$.]

83. Find the term in the expansion of $(x^2 + y^2)^5$ containing x^4 as a factor.

Review Exercises

84. If $f(x) = x^2 + 2x + 3$, find $f(a + 1)$.

(Section 8.1, Example 3)

85. If $f(x) = x^2 + 5x$ and $g(x) = 2x - 3$, find $f(g(x))$ and $g(f(x))$.

(Section 8.4, Example 1)

86. Subtract: $\dfrac{x}{x + 3} - \dfrac{x + 1}{2x^2 - 2x - 24}$.

(Section 7.4, Example 7)

GROUP PROJECT

CHAPTER 14

Group members serve as a financial team analyzing the three options given to the professional baseball player described in the chapter opener on page 991. As a group, determine which option provides the most amount of money over the six-year contract and which provides the least. Describe one advantage and one disadvantage to each option.

Chapter 14 Summary

Definitions and Concepts	**Examples**

Section 14.1 Sequences and Summation Notation

An infinite sequence $\{a_n\}$ is a function whose domain is the set of positive integers. The function values, or terms, are represented by

$$a_1, a_2, a_3, a_4, \ldots, a_n, \ldots.$$

General term: $a_n = \dfrac{(-1)^n}{n^3}$.

$$a_1 = \frac{(-1)^1}{1^3} = -1, \quad a_2 = \frac{(-1)^2}{2^3} = \frac{1}{8},$$

$$a_3 = \frac{(-1)^3}{3^3} = -\frac{1}{27}, \quad a_4 = \frac{(-1)^4}{4^3} = \frac{1}{64}$$

First four terms are $-1, \frac{1}{8}, -\frac{1}{27}$, and $\frac{1}{64}$.

Factorial Notation

$$n! = n(n-1)(n-2)\cdots(3)(2)(1) \quad \text{and} \quad 0! = 1$$

$$6! = 6 \cdot 5 \cdot 4 \cdot 3 \cdot 2 \cdot 1 = 720$$
$$3! = 3 \cdot 2 \cdot 1 = 6$$

Summation Notation

$$\sum_{i=1}^{n} a_i = a_1 + a_2 + a_3 + a_4 + \cdots + a_n$$

In the summation shown here, i is the index of summation, n is the upper limit of summation, and 1 is the lower limit of summation.

$$\sum_{i=3}^{7}(i^2 - 4)$$
$$= (3^2 - 4) + (4^2 - 4) + (5^2 - 4) + (6^2 - 4) + (7^2 - 4)$$
$$= (9 - 4) + (16 - 4) + (25 - 4) + (36 - 4) + (49 - 4)$$
$$= 5 + 12 + 21 + 32 + 45$$
$$= 115$$

Section 14.2 Arithmetic Sequences

In an arithmetic sequence, each term after the first differs from the preceding term by a constant, the common difference. Subtract any term from the term that directly follows it to find the common difference.

General Term of an Arithmetic Sequence

The nth term (the general term) of an arithmetic sequence with first term a_1 and common difference d is

$$a_n = a_1 + (n-1)d.$$

Find the general term and the tenth term:

$$3, 7, 11, 15, \ldots.$$

$$\boxed{a_1 = 3} \quad \boxed{d = 7 - 3 = 4}$$

Using $a_n = a_1 + (n-1)d$,
$$a_n = 3 + (n-1)4 = 3 + 4n - 4 = 4n - 1.$$
The general term is $a_n = 4n - 1$.
The tenth term is $a_{10} = 4 \cdot 10 - 1 = 39$.

The Sum of the First n Terms of an Arithmetic Sequence

The sum, S_n, of the first n terms of an arithmetic sequence is given by

$$S_n = \frac{n}{2}(a_1 + a_n)$$

in which a_1 is the first term and a_n is the nth term.

Find the sum of the first ten terms:

$$2, 5, 8, 11, \ldots.$$

$$\boxed{a_1 = 2} \quad \boxed{d = 5 - 2 = 3}$$

First find a_{10}, the 10th term. Using $a_n = a_1 + (n-1)d$,
$$a_{10} = 2 + (10 - 1) \cdot 3 = 2 + 9 \cdot 3 = 29.$$

Find the sum of the first ten terms using

$$S_n = \frac{n}{2}(a_1 + a_n).$$

$$S_{10} = \frac{10}{2}(a_1 + a_{10}) = 5(2 + 29) = 5(31) = 155$$

Definitions and Concepts	**Examples**

Section 14.3 Geometric Sequences and Series

In a geometric sequence, each term after the first is obtained by multiplying the preceding term by a nonzero constant, the common ratio. Divide any term after the first by the term that directly precedes it to find the common ratio.

Find the general term and the ninth term:

$$12, -6, 3, -\frac{3}{2}, \ldots .$$

$a_1 = 12$ $r = \frac{-6}{12} = -\frac{1}{2}$

General Term of a Geometric Sequence

The nth term (the general term) of a geometric sequence with first term a_1 and common ratio r is

$$a_n = a_1 r^{n-1}.$$

Using $a_n = a_1 r^{n-1}$,

$$a_n = 12\left(-\frac{1}{2}\right)^{n-1} \text{ is the general term.}$$

The ninth term is

$$a_9 = 12\left(-\frac{1}{2}\right)^{9-1} = 12\left(-\frac{1}{2}\right)^{8} = \frac{12}{256} = \frac{3}{64}.$$

The Sum of the First n Terms of a Geometric Sequence

The sum, S_n, of the first n terms of a geometric sequence is given by

$$S_n = \frac{a_1(1 - r^n)}{1 - r}$$

in which a_1 is the first term and r is the common ratio ($r \neq 1$).

Find $\displaystyle\sum_{i=1}^{8} 4 \cdot 3^i$

$$= 4 \cdot 3 + 4 \cdot 3^2 + 4 \cdot 3^3 + \cdots + 4 \cdot 3^8.$$

$a_1 = 12$ $r = \frac{4 \cdot 3^2}{4 \cdot 3} = 3$

Using $S_n = \dfrac{a_1(1 - r^n)}{1 - r}$,

$$S_8 = \frac{12(1 - 3^8)}{1 - 3} = 39{,}360.$$

The Sum of an Infinite Geometric Series

If $-1 < r < 1$ (equivalently, $|r| < 1$), then the sum of the infinite geometric series

$$a_1 + a_1 r + a_1 r^2 + a_1 r^3 + \cdots$$

in which a_1 is the first term and r is the common ratio is given by

$$S = \frac{a_1}{1 - r}.$$

If $|r| \geq 1$, the infinite series does not have a sum.

Find the sum:

$$6 + \frac{6}{3} + \frac{6}{3^2} + \frac{6}{3^3} + \cdots .$$

$a_1 = 6$ $r = \frac{1}{3}$

Using $S = \dfrac{a_1}{1 - r}$, the sum is

$$S = \frac{6}{1 - \dfrac{1}{3}} = \frac{6}{\dfrac{2}{3}} = 6 \cdot \frac{3}{2} = 9.$$

Section 14.4 The Binomial Theorem

Definition of a Binomial Coefficient

$$\binom{n}{r} = \frac{n!}{r!\,(n - r)!}$$

$$\binom{8}{3} = \frac{8!}{3!\,(8 - 3)!} = \frac{8!}{3!\,5!}$$
$$= \frac{8 \cdot 7 \cdot 6 \cdot 5!}{3 \cdot 2 \cdot 1 \cdot 5!} = 56$$

Definitions and Concepts	**Examples**

Section 14.4 The Binomial Theorem (continued)

A Formula for Expanding Binomials: The Binomial Theorem

For any positive integer n,

$$(a + b)^n = \binom{n}{0}a^n + \binom{n}{1}a^{n-1}b$$
$$+ \binom{n}{2}a^{n-2}b^2 + \binom{n}{3}a^{n-3}b^3 + \cdots + \binom{n}{n}b^n.$$

Expand: $(3x - y)^4 = [3x + (-y)]^4$.

$$= \binom{4}{0}(3x)^4 + \binom{4}{1}(3x)^3(-y)$$
$$+ \binom{4}{2}(3x)^2(-y)^2 + \binom{4}{3}(3x)^1(-y)^3 + \binom{4}{4}(-y)^4$$
$$= 1 \cdot 81x^4 + 4 \cdot 27x^3(-y) + 6 \cdot 9x^2y^2 + 4 \cdot 3x(-y^3) + 1 \cdot y^4$$
$$= 81x^4 - 108x^3y + 54x^2y^2 - 12xy^3 + y^4$$

Finding a Particular Term in a Binomial Expansion

The $(r + 1)$st term in the expansion of $(a + b)^n$ is

$$\binom{n}{r}a^{n-r}b^r.$$

The 8th term, or $(7 + 1)$st term $(r = 7)$, of $(x + 2y)^{10}$ is

$$\binom{10}{7}x^{10-7}(2y)^7$$
$$= \binom{10}{7}x^3(2y)^7 = 120x^3 \cdot 128y^7$$
$$= 15,360x^3y^7.$$

CHAPTER 14 REVIEW EXERCISES

14.1 *In Exercises 1–4, write the first four terms of each sequence whose general term is given.*

1. $a_n = 7n - 4$

2. $a_n = (-1)^n \dfrac{n + 2}{n + 1}$

3. $a_n = \dfrac{1}{(n - 1)!}$

4. $a_n = \dfrac{(-1)^{n+1}}{2^n}$

In Exercises 5–6, find each indicated sum.

5. $\displaystyle\sum_{i=1}^{5}(2i^2 - 3)$

6. $\displaystyle\sum_{i=0}^{4}(-1)^{i+1}i!$

In Exercises 7–8, express each sum using summation notation. Use i for the index of summation.

7. $\dfrac{1}{3} + \dfrac{2}{4} + \dfrac{3}{5} + \cdots + \dfrac{15}{17}$

8. $4^3 + 5^3 + 6^3 + \cdots + 13^3$

14.2 *In Exercises 9–11, write the first six terms of each arithmetic sequence.*

9. $a_1 = 7, d = 4$

10. $a_1 = -4, d = -5$

11. $a_1 = \dfrac{3}{2}, d = -\dfrac{1}{2}$

In Exercises 12–14, use the formula for the general term (the nth term) of an arithmetic sequence to find the indicated term of each sequence.

12. Find a_6 when $a_1 = 5, d = 3$.

13. Find a_{12} when $a_1 = -8, d = -2$.

14. Find a_{14} when $a_1 = 14, d = -4$.

In Exercises 15–18, write a formula for the general term (the nth term) of each arithmetic sequence. Then use the formula for a_n to find a_{20}, the 20th term of the sequence.

15. $-7, -3, 1, 5, \ldots$

16. $a_1 = 200, d = -20$

17. $a_1 = -12, d = -\dfrac{1}{2}$

18. $15, 8, 1, -6, \ldots$

19. Find the sum of the first 22 terms of the arithmetic sequence: $5, 12, 19, 26, \ldots$.

20. Find the sum of the first 15 terms of the arithmetic sequence: $-6, -3, 0, 3, \ldots$.

21. Find $3 + 6 + 9 + \cdots + 300$, the sum of the first 100 positive multiples of 3.

In Exercises 22–24, use the formula for the sum of the first n terms of an arithmetic sequence to find the indicated sum.

22. $\displaystyle\sum_{i=1}^{16}(3i + 2)$

23. $\displaystyle\sum_{i=1}^{25}(-2i + 6)$

24. $\displaystyle\sum_{i=1}^{30}(-5i)$

25. The graphic indicates that there are more eyes at school.

Percentage of United States Students Ages 12–18 Seeing Security Cameras at School

Source: U.S. Department of Education

In 2001, 39% of students ages 12–18 reported seeing one or more security cameras at their school. On average, this has increased by approximately 4.75% per year since then.

a. Write a formula for the nth term of the arithmetic sequence that describes the percentage of students ages 12–18 who reported seeing security cameras at school n years after 2000.

b. Use the model to project the percentage of students ages 12–18 who will report seeing security cameras at school by the year 2013.

26. A company offers a starting salary of $31,500 with raises of $2300 per year. Find the total salary over a ten-year period.

27. A theater has 25 seats in the first row and 35 rows in all. Each successive row contains one additional seat. How many seats are in the theater?

14.3 *In Exercises 28–31, write the first five terms of each geometric sequence.*

28. $a_1 = 3, r = 2$

29. $a_1 = \dfrac{1}{2}, r = \dfrac{1}{2}$

30. $a_1 = 16, r = -\dfrac{1}{4}$

31. $a_1 = -5, r = -1$

In Exercises 32–34, use the formula for the general term (the nth term) of a geometric sequence to find the indicated term of each sequence.

32. Find a_7 when $a_1 = 2, r = 3$.

33. Find a_6 when $a_1 = 16, r = \frac{1}{2}$.

34. Find a_5 when $a_1 = -3, r = 2$.

In Exercises 35–37, write a formula for the general term (the nth term) of each geometric sequence. Then use the formula for a_n to find a_8, the eighth term of the sequence.

35. $1, 2, 4, 8, \ldots$

36. $100, 10, 1, \frac{1}{10}, \ldots$

37. $12, -4, \frac{4}{3}, -\frac{4}{9}, \ldots$

38. Find the sum of the first 15 terms of the geometric sequence: $5, -15, 45, -135, \ldots$.

39. Find the sum of the first 7 terms of the geometric sequence: $8, 4, 2, 1, \ldots$.

In Exercises 40–42, use the formula for the sum of the first n terms of a geometric sequence to find the indicated sum.

40. $\displaystyle\sum_{i=1}^{6} 5^i$

41. $\displaystyle\sum_{i=1}^{7} 3(-2)^i$

42. $\displaystyle\sum_{i=1}^{5} 2\left(\dfrac{1}{4}\right)^{i-1}$

In Exercises 43–46, find the sum of each infinite geometric series.

43. $9 + 3 + 1 + \dfrac{1}{3} + \cdots$

44. $2 - 1 + \dfrac{1}{2} - \dfrac{1}{4} + \cdots$

45. $-6 + 4 - \dfrac{8}{3} + \dfrac{16}{9} - \cdots$

46. $\displaystyle\sum_{i=1}^{\infty} 5(0.8)^i$

In Exercises 47–48, express each repeating decimal as a fraction in lowest terms.

47. $0.\overline{6}$

48. $0.\overline{47}$

49. Projections for the U.S. population, ages 85 and older, are shown in the following table.

Year	2000	2010	2020	2030	2040	2050
Projected Population in millions	4.2	5.9	8.3	11.6	16.2	22.7

Actual 2000 population

Source: U.S. Census Bureau

a. Show that the U.S. population, ages 85 and older is projected to increase geometrically.

b. Write the general term of the geometric sequence describing the U.S. population ages 85 and older, in millions, n decades after 1990.

c. Use the model in part (b) to project the U.S. population, ages 85 and older, in 2080.

50. A job pays $32,000 for the first year with an annual increase of 6% per year beginning in the second year. What is the salary in the sixth year? What is the total salary paid over this six-year period? Round answers to the nearest dollar.

In Exercises 51–52, use the formula for the value of an annuity and round to the nearest dollar.

51. You spend $10 per week on lottery tickets, averaging $520 per year. Instead of buying tickets, if you deposited the $520 at the end of each year in an annuity paying 6% compounded annually,

a. How much would you have after 20 years?

b. Find the interest.

52. To save for retirement, you decide to deposit $100 at the end of each month in an IRA that pays 5.5% compounded monthly.

    a. How much will you have from the IRA after 30 years?

    b. Find the interest.

53. A factory in an isolated town has an annual payroll of $4 million. It is estimated that 70% of this money is spent within the town, that people in the town receiving this money will again spend 70% of what they receive in the town, and so on. What is the total of all this spending in the town each year?

14.4 *In Exercises 54–55, evaluate the given binomial coefficient.*

54. $\dbinom{11}{8}$ **55.** $\dbinom{90}{2}$

In Exercises 56–59, use the Binomial Theorem to expand each binomial and express the result in simplified form.

56. $(2x + 1)^3$

57. $(x^2 - 1)^4$

58. $(x + 2y)^5$

59. $(x - 2)^6$

In Exercises 60–61, write the first three terms in each binomial expansion, expressing the result in simplified form.

60. $(x^2 + 3)^8$

61. $(x - 3)^9$

In Exercises 62–63, find the indicated term in each expansion.

62. $(x + 2)^5$; fourth term

63. $(2x - 3)^6$; fifth term

CHAPTER 14 TEST

CHAPTER
Test Prep
VIDEOS

Step-by-step test solutions are found on the Chapter Test Prep Videos available in MyMathLab® or on YouTube (search "BlitzerCombinedAlg" and click on "Channels").

1. Write the first five terms of the sequence whose general term is $a_n = \dfrac{(-1)^{n+1}}{n^2}$.

2. Find the indicated sum: $\displaystyle\sum_{i=1}^{5}(i^2 + 10)$.

3. Express the sum using summation notation. Use i for the index of summation.

$$\frac{2}{3} + \frac{3}{4} + \frac{4}{5} + \cdots + \frac{21}{22}$$

In Exercises 4–5, write a formula for the general term (the nth term) of each sequence. Then use the formula to find the twelfth term of the sequence.

4. $4, 9, 14, 19, \ldots$

5. $16, 4, 1, \frac{1}{4}, \ldots$

In Exercises 6–7, use the formula for the sum of the first n terms of an arithmetic sequence.

6. Find the sum of the first ten terms of the arithmetic sequence: $-7, -14, -21, -28, \ldots$.

7. Find $\displaystyle\sum_{i=1}^{20}(3i - 4)$.

In Exercises 8–9, use the formula for the sum of the first n terms of a geometric sequence.

8. Find the sum of the first ten terms of the geometric sequence: $7, -14, 28, -56, \ldots$.

9. Find $\displaystyle\sum_{i=1}^{15}(-2)^i$.

10. Find the sum of the infinite geometric series:

$$4 + \frac{4}{2} + \frac{4}{2^2} + \frac{4}{2^3} + \cdots.$$

11. Express $0.\overline{73}$ in fractional notation.

12. A job pays $30,000 for the first year with an annual increase of 4% per year beginning in the second year. What is the total salary paid over an eight-year period? Round to the nearest dollar.

13. Evaluate: $\dbinom{9}{2}$.

14. Use the Binomial. Theorem to expand and simplify: $(x^2 - 1)^5$.

15. Use the Binomial Theorem to write the first three terms in the expansion and simplify: $(x + y^2)^8$.

CUMULATIVE REVIEW EXERCISES (CHAPTERS 1–14)

In Exercises 1–10, solve each equation, inequality, or system.

1. $\sqrt{2x + 5} - \sqrt{x + 3} = 2$

2. $(x - 5)^2 = -49$

3. $x^2 + x > 6$

4. $6x - 3(5x + 2) = 4(1 - x)$

5. $\dfrac{2}{x - 3} - \dfrac{3}{x + 3} = \dfrac{12}{x^2 - 9}$

6. $3x + 2 < 4$ and $4 - x > 1$

7. $\begin{cases} 3x - 2y + z = 7 \\ 2x + 3y - z = 13 \\ x - y + 2z = -6 \end{cases}$

8. $\log_9 x + \log_9(x - 8) = 1$

9. $\begin{cases} 2x^2 - 3y^2 = 5 \\ 3x^2 + 4y^2 = 16 \end{cases}$

10. $\begin{cases} 2x^2 - y^2 = -8 \\ x - y = 6 \end{cases}$

In Exercises 11–15, graph each function, equation or inequality in a rectangular coordinate system.

11. $f(x) = (x + 2)^2 - 4$

12. $y < -3x + 5$

13. $f(x) = 3^{x-2}$

14. $\dfrac{x^2}{16} + \dfrac{y^2}{4} = 1$

15. $x^2 - y^2 = 9$

In Exercises 16–19, perform the indicated operations and simplify, if possible.

16. $\dfrac{2x + 1}{x - 5} - \dfrac{4}{x^2 - 3x - 10}$

17. $\dfrac{\dfrac{1}{x - 1} + 1}{\dfrac{1}{x + 1} - 1}$

18. $\dfrac{6}{\sqrt{5} - \sqrt{2}}$

19. $8\sqrt{45} + 2\sqrt{5} - 7\sqrt{20}$

20. Rationalize the denominator: $\dfrac{5}{\sqrt[3]{2x^2y}}$.

21. Factor completely: $5ax + 5ay - 4bx - 4by$.

22. Write as a single logarithm: $5 \log x - \dfrac{1}{2} \log y$.

23. Solve for p: $\dfrac{1}{p} + \dfrac{1}{q} = \dfrac{1}{f}$.

24. Find the distance between $(6, -1)$ and $(-3, -4)$. Round to two decimal places.

25. Find the indicated sum: $\displaystyle\sum_{i=2}^{5} (i^3 - 4)$.

26. Find the sum of the first 30 terms of the arithmetic sequence: $2, 6, 10, 14, \ldots$.

27. Express $0.\overline{3}$ as a fraction in lowest terms.

28. Use the Binomial Theorem to expand and simplify: $(2x - y^3)^4$.

In Exercises 29–31, find the domain of each function.

29. $f(x) = \dfrac{2}{x^2 + 2x - 15}$

30. $f(x) = \sqrt{2x - 6}$

31. $f(x) = \ln(1 - x)$

32. The length of a rectangular garden is 2 feet more than twice its width. If 22 feet of fencing is needed to enclose the garden, what are its dimensions?

33. With a 6% raise, you will earn $19,610 annually. What is your salary before this raise?

34. The function $F(t) = 1 - k \ln(t + 1)$ models the fraction of people, $F(t)$, who remember all the words in a list of nonsense words t hours after memorizing them. After 3 hours, only half the people could remember all the words. Determine the value of k and then predict the fraction of people in the group who will remember all the words after 6 hours. Round to three decimal places and then express the fraction with a denominator of 1000.

Mean, Median, and Mode

Objectives

1 Determine the mean of a set of data items.

2 Determine the median of a set of data items.

3 Determine the mode of a set of data items.

One way to analyze a list of data items is to determine a single value that represents what is "average" or "typical" of the data. Such values are known as **measures of central tendency** because they are located toward the center of the data. Three such measures are discussed in this appendix: the *mean* (or *average*), the *median*, and the *mode*.

The Mean

By far the most commonly used measure of central tendency is the *mean*. The **mean** is obtained by adding all the data items and then dividing the sum by the number of items. The Greek letter sigma, Σ, called **a symbol of summation**, is used to indicate the sum of data items. The notation Σx, read "the sum of x," means to add all the data items in a given data set. We can use this symbol to give a formula for calculating the mean.

1 Determine the mean of a set of data items.

The Mean

The **mean** is the sum of the data items divided by the number of items.

$$\text{Mean} = \frac{\Sigma x}{n},$$

where Σx represents the sum of all the data items and n represents the number of items.

EXAMPLE 1 Calculating the Mean

Table A.1 shows the ten youngest male singers in the United States to have a number 1 single. Find the mean age of these male singers at the time of their number 1 single.

Table A.1	Youngest U.S. Male Singers to Have a Number 1 Single	
Artist/Year	**Title**	**Age**
Stevie Wonder, 1963	"Fingertips"	13
Donny Osmond, 1971	"Go Away Little Girl"	13
Michael Jackson, 1972	"Ben"	14
Laurie London, 1958	"He's Got the Whole World in His Hands"	14
Chris Brown, 2005	"Run It!"	15
Paul Anka, 1957	"Diana"	16
Brian Hyland, 1960	"Itsy Bitsy Teenie Weenie Yellow Polkadot Bikini"	16
Shaun Cassidy, 1977	"Da Doo Ron Ron"	17
Soulja Boy, 2007	"Crank That Soulja Boy"	17
Sean Kingston, 2007	"Beautiful Girls"	17

Source: Russell Ash, *The Top 10 of Everything*

Ages of Youngest U.S. Male Singers (repeated)

Age	Age
13	16
13	16
14	17
14	17
15	17

Solution We find the mean by adding the ages and dividing this sum by 10, the number of data items.

$$\text{Mean} = \frac{\Sigma x}{n} = \frac{13 + 13 + 14 + 14 + 15 + 16 + 16 + 17 + 17 + 17}{10} = \frac{152}{10} = 15.2$$

The mean age of the ten youngest singers to have a number 1 single is 15.2. ■

One and only one mean can be calculated for any set of data items. The mean may or may not be one of the actual data items. In Example 1, the mean was 15.2, although no data item is 15.2.

✓ **CHECK POINT 1** Find the mean for each group of data items:

a. 10, 20, 30, 40, 50   **b.** 3, 10, 10, 10, 117.

2 Determine the median of a set of data items.

The Median

The *median* age in the United States is 35.3. The oldest state by median age is Florida (38.7) and the youngest state is Utah (27.1). To find these values, researchers begin with appropriate random samples. The data items—that is, the ages—are arranged from youngest to oldest. The median age is the data item in the middle of each set of ranked, or ordered, data.

The Median

To find the **median** of a group of data items,

1. Arrange the data items in order, from smallest to largest.
2. If the number of data items is odd, the median is the data item in the middle of the list.
3. If the number of data items is even, the median is the mean of the two middle data items.

EXAMPLE 2   Finding the Median

Find the median for each of the following groups of data:

a. 84, 90, 98, 95, 88   **b.** 68, 74, 7, 13, 15, 25, 28, 59, 34, 47.

Solution

a. Arrange the data items in order, from smallest to largest. The number of data items in the list, five, is odd. Thus, the median is the middle number.

$$84, 88, 90, 95, 98$$

Middle data item

The median is 90. Notice that two data items lie above 90 and two data items lie below 90.

b. Arrange the data items in order, from smallest to largest. The number of data items in the list, ten, is even. Thus, the median is the mean of the two middle data items.

$$7, 13, 15, 25, 28, 34, 47, 59, 68, 74$$

Middle data items are 28 and 34.

$$\text{Median} = \frac{28 + 34}{2} = \frac{62}{2} = 31$$

Great Question!

What exactly does the median do with the data?

The median splits the data items down the middle, like the median strip in a road.

The median is 31. Five data items lie above 31 and five data items lie below 31.

7 13 15 25 28 31 34 47 59 68 74

Five data items lie below 31.

Five data items lie above 31.

Median is 31.

✓ **CHECK POINT 2**   Find the median for each of the following groups of data:

a. 28, 42, 40, 25, 35

b. 72, 61, 85, 93, 79, 87.

Statisticians generally use the median, rather than the mean, when reporting income. Why? Our next example will help to answer this question.

EXAMPLE 3   Comparing the Median and the Mean

Five employees in the assembly section of a television manufacturing company earn salaries of $19,700, $20,400, $21,500, $22,600, and $23,000 annually. The section manager has an annual salary of $95,000.

a. Find the median annual salary for the six people.

b. Find the mean annual salary for the six people.

Solution

a. To compute the median, first arrange the salaries in order:

$19,700, $20,400, $21,500, $22,600, $23,000, $95,000.

Because the list contains an even number of data items, six, the median is the mean of the two middle items.

$$\text{Median} = \frac{\$21,500 + \$22,600}{2} = \frac{\$44,100}{2} = \$22,050$$

The median annual salary is $22,050.

b. We find the mean annual salary by adding the six annual salaries and dividing by 6.

$$\text{Mean} = \frac{\$19,700 + \$20,400 + \$21,500 + \$22,600 + \$23,000 + \$95,000}{6}$$

$$= \frac{\$202,200}{6} = \$33,700$$

The mean annual salary is $33,700.

In Example 3, the median annual salary is $22,050 and the mean annual salary is $33,700. Why such a big difference between these two measures of central tendency? The relatively high annual salary of the section manager, $95,000, pulls the mean salary to a value considerably higher than the median salary. When one or more data items are much greater than the other items, these extreme values can greatly influence the mean. In cases like this, the median is often more representative of the data.

✓ **CHECK POINT 3** **Table A.2** shows 2008 compensation (sum of salary, bonus, perks, stock and option awards) for CEOs of six U.S. companies.

Table A.2 Executive Compensation for Six U.S. Companies

Company	Executive	Total Compensation (millions of dollars)
Apple	Steve Jobs	$0
Coca-Cola	Muhtar Kent	$19.6
IBM	Samuel Palmisano	$21.0
Exxon Mobil	Rex Tillerson	$23.9
Comcast	Brian Roberts	$24.7
Abbott Laboratories	Miles White	$25.1

Source: USA TODAY

a. Find the mean compensation, in millions of dollars, for the six CEOs.

b. Find the median compensation, in millions of dollars, for the six CEOs.

c. Describe why one of the measures of central tendency is greater than the other.

3 Determine the mode of a set of data items.

The Mode

A value that occurs most often in a set of data items is called the *mode*.

The Mode

The **mode** is the data value that occurs most often in a data set. If more than one data value occurs most often, then each of these data values is a mode. If there is no data value that occurs most often, then the data set has no mode.

EXAMPLE 4 Finding the Mode

Find the mode for each of the following groups of data:

a. 7, 2, 4, 7, 8, 10

b. 2, 1, 4, 5, 3

c. 3, 3, 4, 5, 6, 6.

Solution

a. 7, 2, 4, 7, 8, 10

7 occurs most often.

The mode is 7.

b. 2, 1, 4, 5, 3

Each data item occurs the same number of times.

There is no mode.

c. 3, 3, 4, 5, 6, 6

Both 3 and 6 occur most often.

The modes are 3 and 6. ■

✓ **CHECK POINT 4** Find the mode for each of the following groups of data:

a. 3, 8, 5, 8, 9, 10

b. 3, 8, 5, 8, 9, 3

c. 3, 8, 5, 6, 9, 10.

> **EXAMPLE 5** Finding Three Measures of Central Tendency

Suppose your six exam grades in a course are

$$52, \quad 69, \quad 75, \quad 86, \quad 86, \quad \text{and} \quad 92.$$

Compute your final course grade (90–100 = A, 80–89 = B, 70–79 = C, 60–69 – D, below 60 = F) using the

a. mean.　　**b.** median.　　**c.** mode.

Solution

a. The mean is the sum of the data items divided by the number of items, 6.

$$\text{Mean} = \frac{52 + 69 + 75 + 86 + 86 + 92}{6} = \frac{460}{6} \approx 76.67$$

Using the mean, your final course grade is C.

b. The six data items, 52, 69, 75, 86, 86, and 92, are arranged in order. Because the number of data items is even, the median is the mean of the two middle items.

$$\text{Median} = \frac{75 + 86}{2} = \frac{161}{2} = 80.5$$

Using the median, your final course grade is B.

c. The mode is the data value that occurs most frequently. Because 86 occurs most often, the mode is 86. Using the mode, your final course grade is B. ■

✓ **CHECK POINT 5** *Consumer Reports* magazine gave the following data for the number of calories in a meat hot dog for each of eight brands:

$$107, \quad 136, \quad 138, \quad 138, \quad 172, \quad 173, \quad 190, \quad 191.$$

Find the mean, median, and mode for the number of calories in a meat hot dog for the eight brands. If necessary, round answers to the nearest tenth of a calorie.

APPENDIX A EXERCISE SET　MyMathLab®

 Watch the videos in MyMathLab

 Download the MyDashBoard App

Practice Exercises

In Exercises 1–8, find the mean for each group of data items.

1. 7, 4, 3, 2, 8, 5, 1, 3
2. 11, 6, 4, 0, 2, 1, 12, 0, 0
3. 91, 95, 99, 97, 93, 95
4. 100, 100, 90, 30, 70, 100
5. 100, 40, 70, 40, 60
6. 1, 3, 5, 10, 8, 5, 6, 8
7. 1.6, 3.8, 5.0, 2.7, 4.2, 4.2, 3.2, 4.7, 3.6, 2.5, 2.5
8. 1.4, 2.1, 1.6, 3.0, 1.4, 2.2, 1.4, 9.0, 9.0, 1.8

In Exercises 9–16, find the median for each group of data items.

9. 7, 4, 3, 2, 8, 5, 1, 3
10. 11, 6, 4, 0, 2, 1, 12, 0, 0
11. 91, 95, 99, 97, 93, 95

12. 100, 100, 90, 30, 70, 100
13. 100, 40, 70, 40, 60
14. 1, 3, 5, 10, 8, 5, 6, 8
15. 1.6, 3.8, 5.0, 2.7, 4.2, 4.2, 3.2, 4.7, 3.6, 2.5, 2.5
16. 1.4, 2.1, 1.6, 3.0, 1.4, 2.2, 1.4, 9.0, 9.0, 1.8

In Exercises 17–24, find the mode for each group of data items. If there is no mode, so state.

17. 7, 4, 3, 2, 8, 5, 1, 3
18. 11, 6, 4, 0, 2, 1, 12, 0, 0
19. 91, 95, 99, 97, 93, 95
20. 100, 100, 90, 30, 70, 100
21. 100, 40, 70, 40, 60
22. 1, 3, 5, 10, 8, 5, 6, 8
23. 1.6, 3.8, 5.0, 2.7, 4.2, 4.2, 3.2, 4.7, 3.6, 2.5, 2.5
24. 1.4, 2.1, 1.6, 3.0, 1.4, 2.2, 1.4, 9.0, 9.0, 1.8

Application Exercises

Exercises 25–26 present data related to age. For each data set described in boldface, find the mean, median, and mode (or state that there is no mode). If necessary, round answers to the nearest tenth.

25. **Ages of the Six Youngest U.S. Presidents at Inauguration**

    T. Roosevelt (42), Kennedy (43), Clinton (46), Grant (46), Cleveland (47), Obama (47)

26. **Ages of the First Six U.S. Presidents at Inauguration**

    Washington (57), J. Adams (61), Jefferson (57), Madison (57), Monroe (58), J. Q. Adams (57)

27. The annual salaries of four salespeople and the owner of a bookstore are

    $17,500, $19,000, $22,000, $27,500, $98,500.

    Find the mean and the median. Is the mean or the median more representative of the five annual salaries? Briefly explain your answer.

28. In one common system for finding a grade-point average, or GPA,

    $$A = 4, \quad B = 3, \quad C = 2, \quad D = 1, \quad F = 0.$$

    The GPA is calculated by multiplying the number of credit hours for a course and the number assigned to each grade, and then adding these products. Then divide this sum by the total number of credit hours. Because each course grade is weighted according to the number of credits of the course, GPA is called a *weighted mean*. Calculate the GPA, rounded to the nearest tenth, for this transcript:

    Sociology: 3 cr. A;   Biology: 3.5 cr. C;   Music: 1 cr. B;

    Math: 4 cr. B;   English: 3 cr. C.

APPENDIX B

Matrix Solutions to Linear Systems

Objectives

1 Write the augmented matrix for a linear system.

2 Perform matrix row operations.

3 Use matrices to solve linear systems in two variables.

4 Use matrices to solve linear systems in three variables.

5 Use matrices to identify inconsistent and dependent systems.

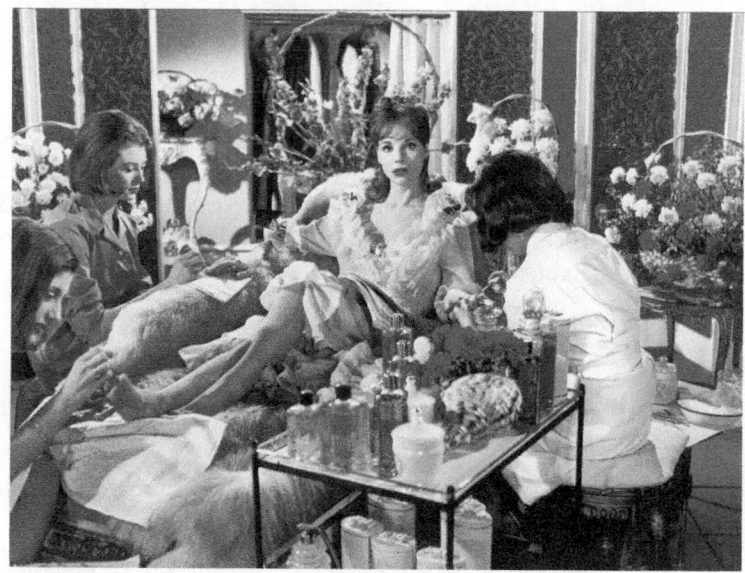

We spend a lot of time sprucing up. The data below show the average number of minutes per day Americans spend on grooming.

Average Number of Minutes per Day Americans Spend on Grooming

	Ages 15–19	Ages 20–24	Ages 45–54	Ages 65+	Married	Single
Men	37	37	34	28	31	34
Women	59	49	46	46	44	50

Source: U.S. Bureau of Labor Statistics, American Time-Use Survey

The 12 numbers inside the brackets are arranged in two rows and six columns. This rectangular array of 12 numbers, arranged in rows and columns and placed in brackets, is an example of a **matrix** (plural: **matrices**). The numbers inside the brackets are called **elements** of the matrix. Matrices are used to display information and to solve systems of linear equations.

1 Write the augmented matrix for a linear system.

Augmented Matrices

A matrix gives us a shortened way of writing a system of equations. The first step in solving a system of linear equations using matrices is to write the *augmented matrix*. An **augmented matrix** has a vertical bar separating the columns of the matrix into two groups. The coefficients of each variable are placed to the left of the vertical line and the constants are placed to the right. If any variable is missing, its coefficient is 0. Here are two examples.

System of Linear Equations   **Augmented Matrix**

$$\begin{aligned} x + 3y &= 5 \\ 2x - y &= -4 \end{aligned} \qquad \left[\begin{array}{cc|c} 1 & 3 & 5 \\ 2 & -1 & -4 \end{array}\right]$$

$$\begin{aligned} 3x + 4y \quad &= 19 \\ 2y + 3z &= 8 \\ 4x \quad - 5z &= 7 \end{aligned} \qquad \left[\begin{array}{ccc|c} 3 & 4 & 0 & 19 \\ 0 & 2 & 3 & 8 \\ 4 & 0 & -5 & 7 \end{array}\right]$$

Our goal in solving a linear system using matrices is to produce a matrix with 1s down the diagonal from upper left to lower right on the left side of the vertical bar, called the **main diagonal**, and 0s below the 1s. In general, the matrix will be one of the following forms.

This is the desired form for systems with two equations.
$$\begin{bmatrix} 1 & a & | & b \\ 0 & 1 & | & c \end{bmatrix}$$

$$\begin{bmatrix} 1 & a & b & | & c \\ 0 & 1 & d & | & e \\ 0 & 0 & 1 & | & f \end{bmatrix}$$
This is the desired form for systems with three equations.

The last row of these matrices gives us the value of one variable. The values of the other variables can then be found by back-substitution.

2 Perform matrix row operations.

Matrix Row Operations

A matrix with 1s down the main diagonal and 0s below the 1s is said to be in **row-echelon form**. How do we produce a matrix in this form? We use **row operations** on the augmented matrix. These row operations are just like what you did when solving a linear system by the addition method. The difference is that we no longer write the variables, usually represented by x, y, and z.

Matrix Row Operations

The following row operations produce matrices that represent systems with the same solution set:

1. Two rows of a matrix may be interchanged. This is the same as interchanging two equations in a linear system.

2. The elements in any row may be multiplied by a nonzero number. This is the same as multiplying both sides of an equation by a nonzero number.

3. The elements in any row may be multiplied by a nonzero number, and these products may be added to the corresponding elements in any other row. This is the same as multiplying an equation by a nonzero number and then adding equations to eliminate a variable.

Two matrices are **row equivalent** if one can be obtained from the other by a sequence of row operations.

Great Question!

Can you clarify what I'm supposed to do to find $kR_i + R_j$? Which row do I work with and which row do I replace?

When performing the row operation

$$kR_i + R_j$$

you use row i to find the products. However, **elements in row i do not change. It is the elements in row j that change:** Add k times the elements in row i to the corresponding elements in row j. Replace elements in row j by these sums.

Each matrix row operation in the preceding box can be expressed symbolically as follows:

1. Interchange the elements in the ith and jth rows: $R_i \leftrightarrow R_j$.
2. Multiply each element in the ith row by k: kR_i.
3. Add k times the elements in row i to the corresponding elements in row j: $kR_i + R_j$.

EXAMPLE 1   Performing Matrix Row Operations

Use the matrix

$$\begin{bmatrix} 3 & 18 & -12 & | & 21 \\ 1 & 2 & -3 & | & 5 \\ -2 & -3 & 4 & | & -6 \end{bmatrix}$$

and perform each indicated row operation:

a. $R_1 \leftrightarrow R_2$ **b.** $\dfrac{1}{3} R_1$ **c.** $2R_2 + R_3$.

Solution

a. The notation $R_1 \leftrightarrow R_2$ means to interchange the elements in row 1 and row 2. This results in the row-equivalent matrix

$$\begin{bmatrix} 1 & 2 & -3 & 5 \\ 3 & 18 & -12 & 21 \\ -2 & -3 & 4 & -6 \end{bmatrix}.$$

This was row 2; now it's row 1.

This was row 1; now it's row 2.

b. The notation $\frac{1}{3}R_1$ means to multiply each element in row 1 by $\frac{1}{3}$. This results in the row-equivalent matrix

$$\begin{bmatrix} \frac{1}{3}(3) & \frac{1}{3}(18) & \frac{1}{3}(-12) & \frac{1}{3}(21) \\ 1 & 2 & -3 & 5 \\ -2 & -3 & 4 & -6 \end{bmatrix} = \begin{bmatrix} 1 & 6 & -4 & 7 \\ 1 & 2 & -3 & 5 \\ -2 & -3 & 4 & -6 \end{bmatrix}.$$

c. The notation $2R_2 + R_3$ means to add 2 times the elements in row 2 to the corresponding elements in row 3. Replace the elements in row 3 by these sums. First, we find 2 times the elements in row 2, namely, 1, 2, −3 and 5:

$$2(1) \text{ or } 2, \qquad 2(2) \text{ or } 4, \qquad 2(-3) \text{ or } -6, \qquad 2(5) \text{ or } 10.$$

Now we add these products to the corresponding elements in row 3. Although we use row 2 to find the products, row 2 does not change. It is the elements in row 3 that change, resulting in the row-equivalent matrix

Replace row 3 by the sum of itself and 2 times row 2.

$$\begin{bmatrix} 3 & 18 & -12 & 21 \\ 1 & 2 & -3 & 5 \\ -2+2=0 & -3+4=1 & 4+(-6)=-2 & -6+10=4 \end{bmatrix} = \begin{bmatrix} 3 & 18 & -12 & 21 \\ 1 & 2 & -3 & 5 \\ 0 & 1 & -2 & 4 \end{bmatrix}.$$

✓ **CHECK POINT 1**   Use the matrix

$$\begin{bmatrix} 4 & 12 & -20 & 8 \\ 1 & 6 & -3 & 7 \\ -3 & -2 & 1 & -9 \end{bmatrix}$$

and perform each indicated row operation:

a. $R_1 \leftrightarrow R_2$

b. $\dfrac{1}{4}R_1$

c. $3R_2 + R_3$.

---

③ Use matrices to solve linear systems in two variables.

Solving Linear Systems in Two Variables Using Matrices

The process that we use to solve linear systems using matrix row operations is often called **Gaussian elimination**, after the German mathematician Carl Friedrich Gauss (1777–1855). On the next page are the steps used in solving linear systems in two variables with matrices.

Solving Linear Systems in Two Variables Using Matrices

1. Write the augmented matrix for the system.

2. Use matrix row operations to simplify the matrix to a row-equivalent matrix in row-echelon form, with 1s down the main diagonal from upper left to lower right, and a 0 below the 1 in the first column.

$$\begin{bmatrix} 1 & * & | & * \\ * & * & | & * \end{bmatrix} \rightarrow \begin{bmatrix} 1 & * & | & * \\ 0 & * & | & * \end{bmatrix} \rightarrow \begin{bmatrix} 1 & * & | & * \\ 0 & 1 & | & * \end{bmatrix}$$

Get 1 in the upper left-hand corner. | Use the 1 in the first column to get 0 below it. | Get 1 in the second row, second column position.

3. Write the system of linear equations corresponding to the matrix from step 2 and use back-substitution to find the system's solution.

EXAMPLE 2 Using Matrices to Solve a Linear System

Use matrices to solve the system:

$$\begin{cases} 4x - 3y = -15 \\ x + 2y = -1. \end{cases}$$

Solution

Step 1. Write the augmented matrix for the system.

Linear System

$$\begin{cases} 4x - 3y = -15 \\ x + 2y = -1 \end{cases}$$

Augmented Matrix

$$\begin{bmatrix} 4 & -3 & | & -15 \\ 1 & 2 & | & -1 \end{bmatrix}$$

Step 2. Use matrix row operations to simplify the matrix to row-echelon form, with 1s down the main diagonal from upper left to lower right, and a 0 below the 1 in the first column. Our first step in achieving this goal is to get 1 in the top position of the first column.

We want 1 in this position.
$$\begin{bmatrix} 4 & -3 & | & -15 \\ 1 & 2 & | & -1 \end{bmatrix}$$

To get 1 in this position, we interchange row 1 and row 2: $R_1 \leftrightarrow R_2$.

$$\begin{bmatrix} 1 & 2 & | & -1 \\ 4 & -3 & | & -15 \end{bmatrix}$$
This was row 2; now it's row 1.

This was row 1; now it's row 2.

Now we want a 0 below the 1 in the first column.

We want 0 in this position.
$$\begin{bmatrix} 1 & 2 & | & -1 \\ 4 & -3 & | & -15 \end{bmatrix}$$

Let's get a 0 where there is now a 4. If we multiply the top row of numbers by -4 and add these products to the second row of numbers, we will get 0 in this position: $-4R_1 + R_2$. *We change only row 2.*

Replace row 2 by $-4R_1 + R_2$.
$$\begin{bmatrix} 1 & 2 & | & -1 \\ -4(1) + 4 & -4(2) + (-3) & | & -4(-1) + (-15) \end{bmatrix} = \begin{bmatrix} 1 & 2 & | & -1 \\ 0 & -11 & | & -11 \end{bmatrix}$$

We move on to the second column. We want 1 in the second row, second column.

We want 1 in this position.
$$\begin{bmatrix} 1 & 2 & | & -1 \\ 0 & -11 & | & -11 \end{bmatrix}$$

To get 1 in the desired position, we multiply -11 by its multiplicative inverse, or reciprocal, $-\frac{1}{11}$. Therefore, we multiply all the numbers in the second row by $-\frac{1}{11}$: $-\frac{1}{11}R_2$.

$$-\frac{1}{11}R_2 \quad \begin{bmatrix} 1 & 2 & -1 \\ -\frac{1}{11}(0) & -\frac{1}{11}(-11) & -\frac{1}{11}(-11) \end{bmatrix} = \begin{bmatrix} 1 & 2 & -1 \\ 0 & 1 & 1 \end{bmatrix}$$

We now have the desired matrix in row-echelon form, with 1s down the main diagonal and a 0 below the 1 in the first column.

Step 3. Write the system of linear equations corresponding to the matrix from step 2 and use back-substitution to find the system's solution. The system represented by the matrix from step 2 is

$$\begin{bmatrix} 1 & 2 & -1 \\ 0 & 1 & 1 \end{bmatrix} \rightarrow \begin{cases} 1x + 2y = -1 \\ 0x + 1y = 1 \end{cases} \text{ or } \begin{cases} x + 2y = -1 & \text{(1)} \\ \qquad y = 1. & \text{(2)} \end{cases}$$

We immediately see from Equation (2) that the value for y is 1. To find x, we back-substitute 1 for y in Equation (1).

$$x + 2y = -1 \quad \text{Equation (1)}$$
$$x + 2 \cdot 1 = -1 \quad \text{Substitute 1 for } y.$$
$$x + 2 = -1 \quad \text{Multiply.}$$
$$x = -3 \quad \text{Subtract 2 from both sides.}$$

With $x = -3$ and $y = 1$, the proposed solution is $(-3, 1)$. Take a moment to show that $(-3, 1)$ satisfies both equations. The solution is $(-3, 1)$ and the solution set is $\{(-3, 1)\}$. ∎

✓ **CHECK POINT 2**   Use matrices to solve the system:

$$\begin{cases} 2x - y = -4 \\ x + 3y = 5. \end{cases}$$

---

4 Use matrices to solve linear systems in three variables.

Solving Linear Systems in Three Variables Using Matrices

Gaussian elimination is also used to solve a system of linear equations in three variables. Most of the work involves using matrix row operations to obtain a matrix with 1s down the main diagonal and 0s below the 1s in the first and second columns.

Solving Linear Systems in Three Variables Using Matrices

1. Write the augmented matrix for the system.

2. Use matrix row operations to simplify the matrix to a row-equivalent matrix in row-echelon form, with 1s down the main diagonal from upper left to lower right, and 0s below the 1s in the first and second columns.

$$\begin{bmatrix} 1 & * & * & * \\ * & * & * & * \\ * & * & * & * \end{bmatrix} \rightarrow \begin{bmatrix} 1 & * & * & * \\ 0 & * & * & * \\ 0 & * & * & * \end{bmatrix} \rightarrow \begin{bmatrix} 1 & * & * & * \\ 0 & 1 & * & * \\ 0 & * & * & * \end{bmatrix} \rightarrow \begin{bmatrix} 1 & * & * & * \\ 0 & 1 & * & * \\ 0 & 0 & * & * \end{bmatrix} \rightarrow \begin{bmatrix} 1 & * & * & * \\ 0 & 1 & * & * \\ 0 & 0 & 1 & * \end{bmatrix}$$

Get 1 in the upper left-hand corner.   Use the 1 in the first column to get 0s below it.   Get 1 in the second row, second column position.   Use the 1 in the second column to get 0 below it.   Get 1 in the third row, third column position.

3. Write the system of linear equations corresponding to the matrix from step 2 and use back-substitution to find the system's solution.

EXAMPLE 3 Using Matrices to Solve a Linear System

Use matrices to solve the system:

$$\begin{cases} 3x + y + 2z = 31 \\ x + y + 2z = 19 \\ x + 3y + 2z = 25. \end{cases}$$

Solution

Step 1. Write the augmented matrix for the system.

Linear System

$$\begin{cases} 3x + y + 2z = 31 \\ x + y + 2z = 19 \\ x + 3y + 2z = 25 \end{cases}$$

Augmented Matrix

$$\begin{bmatrix} 3 & 1 & 2 & | & 31 \\ 1 & 1 & 2 & | & 19 \\ 1 & 3 & 2 & | & 25 \end{bmatrix}$$

Step 2. Use matrix row operations to simplify the matrix to row-echelon form, with 1s down the main diagonal from upper left to lower right, and 0s below the 1s in the first and second columns. Our first step in achieving this goal is to get 1 in the top position of the first column.

We want 1 in this position.

$$\begin{bmatrix} 3 & 1 & 2 & | & 31 \\ 1 & 1 & 2 & | & 19 \\ 1 & 3 & 2 & | & 25 \end{bmatrix}$$

To get 1 in this position, we interchange row 1 and row 2: $R_1 \leftrightarrow R_2$.
We could also interchange row 1 and row 3 to get 1 in the upper left-hand corner.

$$\begin{bmatrix} 1 & 1 & 2 & | & 19 \\ 3 & 1 & 2 & | & 31 \\ 1 & 3 & 2 & | & 25 \end{bmatrix}$$

This was row 2; now it's row 1.

This was row 1; now it's row 2.

Now we want to get 0s below the 1 in the first column.

We want 0 in these positions.

$$\begin{bmatrix} 1 & 1 & 2 & | & 19 \\ 3 & 1 & 2 & | & 31 \\ 1 & 3 & 2 & | & 25 \end{bmatrix}$$

To get a 0 where there is now a 3, multiply the top row of numbers by -3 and add these products to the second row of numbers: $-3R_1 + R_2$. To get a 0 in the bottom of the first column where there is now a 1, multiply the top row of numbers by -1 and add these products to the third row of numbers: $-1R_1 + R_3$. Although we are using row 1 to find the products, the numbers in row 1 do not change.

Replace row 2 by $-3R_1 + R_2$.

Replace row 3 by $-1R_1 + R_3$.

$$\begin{bmatrix} 1 & 1 & 2 & | & 19 \\ -3(1) + 3 & -3(1) + 1 & -3(2) + 2 & | & -3(19) + 31 \\ -1(1) + 1 & -1(1) + 3 & -1(2) + 2 & | & -1(19) + 25 \end{bmatrix} = \begin{bmatrix} 1 & 1 & 2 & | & 19 \\ 0 & -2 & -4 & | & -26 \\ 0 & 2 & 0 & | & 6 \end{bmatrix}$$

We want 1 in this position.

We move on to the second column. To get 1 in the desired position, we multiply -2 by its reciprocal, $-\frac{1}{2}$. Therefore, we multiply all the numbers in the second row by $-\frac{1}{2}$: $-\frac{1}{2}R_2$.

$$-\frac{1}{2}R_2 \quad \begin{bmatrix} 1 & 1 & 2 & | & 19 \\ -\frac{1}{2}(0) & -\frac{1}{2}(-2) & -\frac{1}{2}(-4) & | & -\frac{1}{2}(-26) \\ 0 & 2 & 0 & | & 6 \end{bmatrix} = \begin{bmatrix} 1 & 1 & 2 & | & 19 \\ 0 & 1 & 2 & | & 13 \\ 0 & 2 & 0 & | & 6 \end{bmatrix}.$$

We want 0 in this position.

We are not yet done with the second column. The voice balloon shows that we want to get a 0 where there is now a 2. If we multiply the second row of numbers by -2 and add these products to the third row of numbers, we will get 0 in this position: $-2R_2 + R_3$. Although we are using the numbers in row 2 to find the products, the numbers in row 2 do not change.

Replace row 3 by $-2R_2 + R_3$.

$$\begin{bmatrix} 1 & 1 & 2 & | & 19 \\ 0 & 1 & 2 & | & 13 \\ -2(0)+0 & -2(1)+2 & -2(2)+0 & | & -2(13)+6 \end{bmatrix} = \begin{bmatrix} 1 & 1 & 2 & | & 19 \\ 0 & 1 & 2 & | & 13 \\ 0 & 0 & -4 & | & -20 \end{bmatrix}.$$

We want 1 in this position.

We move on to the third column. To get 1 in the desired position, we multiply -4 by its reciprocal, $-\frac{1}{4}$. Therefore, we multiply all the numbers in the third row by $-\frac{1}{4}$: $-\frac{1}{4}R_3$.

$$-\frac{1}{4}R_3 \quad \begin{bmatrix} 1 & 1 & 2 & | & 19 \\ 0 & 1 & 2 & | & 13 \\ -\frac{1}{4}(0) & -\frac{1}{4}(0) & -\frac{1}{4}(-4) & | & -\frac{1}{4}(-20) \end{bmatrix} = \begin{bmatrix} 1 & 1 & 2 & | & 19 \\ 0 & 1 & 2 & | & 13 \\ 0 & 0 & 1 & | & 5 \end{bmatrix}.$$

We now have the desired matrix in row-echelon form, with 1s down the main diagonal and 0s below the 1s in the first and second columns.

Step 3. Write the system of linear equations corresponding to the matrix from step 2 and use back-substitution to find the system's solution. The system represented by the matrix from step 2 is

$$\begin{bmatrix} 1 & 1 & 2 & | & 19 \\ 0 & 1 & 2 & | & 13 \\ 0 & 0 & 1 & | & 5 \end{bmatrix} \rightarrow \begin{cases} 1x + 1y + 2z = 19 \\ 0x + 1y + 2z = 13 \\ 0x + 0y + 1z = 5 \end{cases} \text{ or } \begin{cases} x + y + 2z = 19 \text{ (1)} \\ y + 2z = 13. \text{ (2)} \\ z = 5 \text{ (3)} \end{cases}$$

We immediately see from Equation (3) that the value for z is 5. To find y, we back-substitute 5 for z in the second equation.

$$y + 2z = 13 \quad \text{Equation (2)}$$
$$y + 2(5) = 13 \quad \text{Substitute 5 for z.}$$
$$y = 3 \quad \text{Solve for y.}$$

Finally, back-substitute 3 for y and 5 for z in the first equation.

$$x + y + 2z = 19 \quad \text{Equation (1)}$$
$$x + 3 + 2(5) = 19 \quad \text{Substitute 3 for y and 5 for z.}$$
$$x + 13 = 19 \quad \text{Multiply and add.}$$
$$x = 6 \quad \text{Subtract 13 from both sides.}$$

The solution of the original system is $(6, 3, 5)$ and the solution set is $\{(6, 3, 5)\}$. Check to see that the solution satisfies all three equations in the given system. ∎

Using Technology

Most graphing utilities can convert an augmented matrix to row-echelon form, with 1s down the main diagonal and 0s below the 1s. However, row-echelon form is not unique. Your graphing utility might give a row-echelon form different from the one you obtained by hand. However, all row-echelon forms for a given system's augmented matrix produce the same solution to the system. Enter the augmented matrix and name it A. Then use the REF (row-echelon form) command on matrix A.

☑ **CHECK POINT 3** Use matrices to solve the system:

$$\begin{cases} 2x + y + 2z = 18 \\ x - y + 2z = 9 \\ x + 2y - z = 6. \end{cases}$$

Modern supercomputers are capable of solving systems with more than 600,000 variables. The augmented matrices for such systems are huge, but the solution using matrices is exactly like what we did in Example 3. Work with the augmented matrix, one column at a time. First, get 1s down the main diagonal from upper left to lower right. Then get 0s below the 1s.

5 Use matrices to identify inconsistent and dependent systems.

Inconsistent Systems and Systems with Dependent Equations

When solving a system using matrices, you might obtain a matrix with a row in which the numbers to the left of the vertical bar are all zeros, but a nonzero number appears on the right. In such a case, the system is inconsistent and has no solution. For example, a system of equations that yields the following matrix is an inconsistent system:

$$\begin{bmatrix} 1 & -2 & | & 3 \\ 0 & 0 & | & -4 \end{bmatrix}.$$

The second row of the matrix represents the equation $0x + 0y = -4$, which is false for all values of x and y.

If you obtain a matrix in which a 0 appears across an entire row, the system contains dependent equations and has infinitely many solutions. This row of zeros represents $0x + 0y = 0$ or $0x + 0y + 0z = 0$. These equations are satisfied by infinitely many ordered pairs or triples.

APPENDIX B EXERCISE SET MyMathLab®   Watch the videos in MyMathLab  Download the MyDashBoard App

Practice Exercises

In Exercises 1–14, perform each matrix row operation and write the new matrix.

1. $\begin{bmatrix} 2 & 2 & | & 5 \\ 1 & -\frac{3}{2} & | & 5 \end{bmatrix} R_1 \leftrightarrow R_2$

2. $\begin{bmatrix} -6 & 9 & | & 4 \\ 1 & -\frac{3}{2} & | & 4 \end{bmatrix} R_1 \leftrightarrow R_2$

3. $\begin{bmatrix} -6 & 8 & | & -12 \\ 3 & 5 & | & -2 \end{bmatrix} -\frac{1}{6} R_1$

4. $\begin{bmatrix} -2 & 3 & | & -10 \\ 4 & 2 & | & 5 \end{bmatrix} -\frac{1}{2} R_1$

5. $\begin{bmatrix} 1 & -3 & | & 5 \\ 2 & 6 & | & 4 \end{bmatrix} -2R_1 + R_2$

6. $\begin{bmatrix} 1 & -3 & | & 1 \\ 2 & 1 & | & -5 \end{bmatrix} -2R_1 + R_2$

7. $\begin{bmatrix} 1 & -\frac{3}{2} & | & \frac{7}{2} \\ 3 & 4 & | & 2 \end{bmatrix} -3R_1 + R_2$

8. $\begin{bmatrix} 1 & -\frac{2}{5} & | & \frac{3}{4} \\ 4 & 2 & | & -1 \end{bmatrix} -4R_1 + R_2$

9. $\begin{bmatrix} 2 & -6 & 4 & | & 10 \\ 1 & 5 & -5 & | & 0 \\ 3 & 0 & 4 & | & 7 \end{bmatrix} \frac{1}{2} R_1$

10. $\begin{bmatrix} 3 & -12 & 6 & | & 9 \\ 1 & -4 & 4 & | & 0 \\ 2 & 0 & 7 & | & 4 \end{bmatrix} \frac{1}{3} R_1$

11. $\begin{bmatrix} 1 & -3 & 2 & | & 0 \\ 3 & 1 & -1 & | & 7 \\ 2 & -2 & 1 & | & 3 \end{bmatrix} -3R_1 + R_2$

12. $\begin{bmatrix} 1 & -1 & 5 & | & -6 \\ 3 & 3 & -1 & | & 10 \\ 1 & 3 & 2 & | & 5 \end{bmatrix} -3R_1 + R_2$

13. $\begin{bmatrix} 1 & 1 & -1 & | & 6 \\ 2 & -1 & 1 & | & -3 \\ 3 & -1 & -1 & | & 4 \end{bmatrix} \begin{matrix} -2R_1 + R_2 \\ -3R_1 + R_3 \end{matrix}$

14. $\begin{bmatrix} 1 & 2 & 1 & | & 2 \\ -2 & -1 & 2 & | & 5 \\ 1 & 3 & -2 & | & -8 \end{bmatrix} \begin{matrix} 2R_1 + R_2 \\ -1R_1 + R_3 \end{matrix}$

In Exercises 15–38, solve each system using matrices. If there is no solution or if there are infinitely many solutions and a system's equations are dependent, so state.

15. $\begin{cases} x + y = 6 \\ x - y = 2 \end{cases}$ **16.** $\begin{cases} x + 2y = 11 \\ x - y = -1 \end{cases}$

17. $\begin{cases} 2x + y = 3 \\ x - 3y = 12 \end{cases}$ **18.** $\begin{cases} 3x - 5y = 7 \\ x - y = 1 \end{cases}$

19. $\begin{cases} 5x + 7y = -25 \\ 11x + 6y = -8 \end{cases}$

20. $\begin{cases} 3x - 5y = 22 \\ 4x - 2y = 20 \end{cases}$

21. $\begin{cases} 4x - 2y = 5 \\ -2x + y = 6 \end{cases}$

22. $\begin{cases} -3x + 4y = 12 \\ 6x - 8y = 16 \end{cases}$

23. $\begin{cases} x - 2y = 1 \\ -2x + 4y = -2 \end{cases}$

24. $\begin{cases} 3x - 6y = 1 \\ 2x - 4y = \dfrac{2}{3} \end{cases}$

25. $\begin{cases} x + y - z = -2 \\ 2x - y + z = 5 \\ -x + 2y + 2z = 1 \end{cases}$

26. $\begin{cases} x - 2y - z = 2 \\ 2x - y + z = 4 \\ -x + y - 2z = -4 \end{cases}$

27. $\begin{cases} x + 3y = 0 \\ x + y + z = 1 \\ 3x - y - z = 11 \end{cases}$

28. $\begin{cases} 3y - z = -1 \\ x + 5y - z = -4 \\ -3x + 6y + 2z = 11 \end{cases}$

29. $\begin{cases} 2x + 2y + 7z = -1 \\ 2x + y + 2z = 2 \\ 4x + 6y + z = 15 \end{cases}$

30. $\begin{cases} 3x + 2y + 3z = 3 \\ 4x - 5y + 7z = 1 \\ 2x + 3y - 2z = 6 \end{cases}$

31. $\begin{cases} x + y + z = 6 \\ x - z = -2 \\ y + 3z = 11 \end{cases}$

32. $\begin{cases} x + y + z = 3 \\ -y + 2z = 1 \\ -x + z = 0 \end{cases}$

33. $\begin{cases} x - y + 3z = 4 \\ 2x - 2y + 6z = 7 \\ 3x - y + 5z = 14 \end{cases}$

34. $\begin{cases} 3x - y + 2z = 4 \\ -6x + 2y - 4z = 1 \\ 5x - 3y + 8z = 0 \end{cases}$

35. $\begin{cases} x - 2y + z = 4 \\ 5x - 10y + 5y = 20 \\ -2x + 4y - 2z = -8 \end{cases}$

36. $\begin{cases} x - 3y + z = 2 \\ 4x - 12y + 4z = 8 \\ -2x + 6y - 2z = -4 \end{cases}$

37. $\begin{cases} x + y = 1 \\ y + 2z = -2 \\ 2x - z = 0 \end{cases}$

38. $\begin{cases} x + 3y = 3 \\ y + 2z = -8 \\ x - z = 7 \end{cases}$

Practice PLUS

In Exercises 39–40, write the system of linear equations represented by the augmented matrix. Use $w, x, y,$ and $z,$ for the variables. Once the system is written, use back-substitution to find its solution set, $\{(w, x, y, z)\}$.

39. $\left[\begin{array}{cccc|c} 1 & -1 & 1 & 1 & 3 \\ 0 & 1 & -2 & -1 & 0 \\ 0 & 0 & 1 & 6 & 17 \\ 0 & 0 & 0 & 1 & 3 \end{array}\right]$

40. $\left[\begin{array}{cccc|c} 1 & 2 & -1 & 0 & 2 \\ 0 & 1 & 1 & -2 & -3 \\ 0 & 0 & 1 & -1 & -2 \\ 0 & 0 & 0 & 1 & 3 \end{array}\right]$

In Exercises 41–42, perform each matrix row operation and write the new matrix.

41. $\left[\begin{array}{cccc|c} 1 & -1 & 1 & 1 & 3 \\ 0 & 1 & -2 & -1 & 0 \\ 2 & 0 & 3 & 4 & 11 \\ 5 & 1 & 2 & 4 & 6 \end{array}\right] \begin{array}{l} \\ \\ -2R_1 + R_3 \\ -5R_1 + R_4 \end{array}$

42. $\left[\begin{array}{cccc|c} 1 & -5 & 2 & -2 & 4 \\ 0 & 1 & -3 & -1 & 0 \\ 3 & 0 & 2 & -1 & 6 \\ -4 & 1 & 4 & 2 & -3 \end{array}\right] \begin{array}{l} \\ \\ -3R_1 + R_3 \\ 4R_1 + R_4 \end{array}$

In Exercises 43–44, solve each system using matrices. You will need to use matrix row operations to obtain matrices like those in Exercises 39 and 40, with 1s down the main diagonal and 0s below the 1s. Express the solution set as $\{(w, x, y, z)\}$.

43. $\begin{cases} w + x + y + z = 4 \\ 2w + x - 2y - z = 0 \\ w - 2x - y - 2z = -2 \\ 3w + 2x + y + 3z = 4 \end{cases}$

44. $\begin{cases} w + x + y + z = 5 \\ w + 2x - y - 2z = -1 \\ w - 3x - 3y - z = -1 \\ 2w - x + 2y - z = -2 \end{cases}$

APPENDIX C

Determinants and Cramer's Rule

Objectives

1. Evaluate a second-order determinant.

2. Solve a system of linear equations in two variables using Cramer's rule.

3. Evaluate a third-order determinant.

4. Solve a system of linear equations in three variables using Cramer's rule.

5. Use determinants to identify inconsistent and dependent systems.

As cyberspace absorbs more and more of our work, play, shopping, and socializing, where will it all end? Which activities will still be offline in 2025?

Our technologically transformed lives can be traced back to the English inventor Charles Babbage (1791–1871). Babbage knew of a method for solving linear systems called *Cramer's rule*, in honor of the Swiss geometer Gabriel Cramer (1704–1752). Cramer's rule was simple, but involved numerous multiplications for large systems. Babbage designed a machine, called the

A portion of Charles Babbage's unrealized Difference Engine

"difference engine," that consisted of toothed wheels on shafts for performing these multiplications. Despite the fact that only one-seventh of the functions ever worked, Babbage's invention demonstrated how complex calculations could be handled mechanically. In 1944, scientists at IBM used the lessons of the difference engine to create the world's first computer.

Those who invented computers hoped to relegate the drudgery of repeated computation to a machine. In this section, we look at a method for solving linear systems that played a critical role in this process. The method uses real numbers, called *determinants*, that are associated with arrays of numbers. As with matrix methods, solutions are obtained by writing down the coefficients and constants of a linear system and performing operations with them.

The Determinant of a 2 × 2 Matrix

A matrix of **order** $m \times n$ has m rows and n columns. If $m = n$, a matrix has the same number of rows as columns and is called a **square matrix**. Associated with every square matrix is a real number, called its **determinant**. The determinant for a 2 × 2 square matrix is defined as follows:

1. Evaluate a second-order determinant.

Great Question!

What does the definition of a determinant mean? What am I supposed to do?

To evaluate a second-order determinant, find the difference of the product of the two diagonals.

$$\begin{vmatrix} a_1 & b_1 \\ a_2 & b_2 \end{vmatrix} = a_1b_2 - a_2b_1$$

Definition of the Determinant of a 2 × 2 Matrix

The determinant of the matrix $\begin{bmatrix} a_1 & b_1 \\ a_2 & b_2 \end{bmatrix}$ is denoted by $\begin{vmatrix} a_1 & b_1 \\ a_2 & b_2 \end{vmatrix}$ and is defined by

$$\begin{vmatrix} a_1 & b_1 \\ a_2 & b_2 \end{vmatrix} = a_1b_2 - a_2b_1.$$

We also say that the **value** of the **second-order determinant** $\begin{vmatrix} a_1 & b_1 \\ a_2 & b_2 \end{vmatrix}$ is $a_1b_2 - a_2b_1$.

Example 1 illustrates that the determinant of a matrix may be positive or negative. The determinant can also have 0 as its value.

EXAMPLE 1   Evaluating the Determinant of a 2 × 2 Matrix

Evaluate the determinant of each of the following matrices:

a. $\begin{bmatrix} 5 & 6 \\ 7 & 3 \end{bmatrix}$ **b.** $\begin{bmatrix} 2 & 4 \\ -3 & -5 \end{bmatrix}$.

Solution   We multiply and subtract as indicated.

a. $\begin{vmatrix} 5 & 6 \\ 7 & 3 \end{vmatrix} = 5 \cdot 3 - 7 \cdot 6 = 15 - 42 = -27$ *The value of the second-order determinant is −27.*

b. $\begin{vmatrix} 2 & 4 \\ -3 & -5 \end{vmatrix} = 2(-5) - (-3)(4) = -10 + 12 = 2$ *The value of the second-order determinant is 2.* ∎

✓ **CHECK POINT 1**   Evaluate the determinant of each of the following matrices:

a. $\begin{bmatrix} 10 & 9 \\ 6 & 5 \end{bmatrix}$ **b.** $\begin{bmatrix} 4 & 3 \\ -5 & -8 \end{bmatrix}$.

Discover for Yourself

Write and then evaluate three determinants, one whose value is positive, one whose value is negative, and one whose value is 0.

2 Solve a system of linear equations in two variables using Cramer's rule.

Solving Systems of Linear Equations in Two Variables Using Determinants

Determinants can be used to solve a linear system in two variables. In general, such a system appears as

$$\begin{cases} a_1 x + b_1 y = c_1 \\ a_2 x + b_2 y = c_2. \end{cases}$$

Let's first solve this system for x using the addition method. We can solve for x by eliminating y from the equations. Multiply the first equation by b_2 and the second equation by $-b_1$. Then add the two equations:

$$\begin{cases} a_1 x + b_1 y = c_1 \quad \xrightarrow{\text{Multiply by } b_2.} \\ a_2 x + b_2 y = c_2 \quad \xrightarrow{\text{Multiply by } -b_1.} \end{cases} \begin{cases} a_1 b_2 x + b_1 b_2 y = c_1 b_2 \\ -a_2 b_1 x - b_1 b_2 y = -c_2 b_1 \end{cases}$$

$$\text{Add:} \quad (a_1 b_2 - a_2 b_1) x = c_1 b_2 - c_2 b_1$$

$$x = \frac{c_1 b_2 - c_2 b_1}{a_1 b_2 - a_2 b_1}$$

Because

$$\begin{vmatrix} c_1 & b_1 \\ c_2 & b_2 \end{vmatrix} = c_1 b_2 - c_2 b_1 \quad \text{and} \quad \begin{vmatrix} a_1 & b_1 \\ a_2 & b_2 \end{vmatrix} = a_1 b_2 - a_2 b_1,$$

we can express our answer for x as the quotient of two determinants:

$$x = \frac{\begin{vmatrix} c_1 & b_1 \\ c_2 & b_2 \end{vmatrix}}{\begin{vmatrix} a_1 & b_1 \\ a_2 & b_2 \end{vmatrix}}.$$

Similarly, we could use the addition method to solve our system for y, again expressing y as the quotient of two determinants. This method of using determinants to solve the linear system, called **Cramer's rule**, is summarized in the box on the next page.

Solving a Linear System in Two Variables Using Determinants

Cramer's Rule

If

$$\begin{cases} a_1x + b_1y = c_1 \\ a_2x + b_2y = c_2 \end{cases}$$

then

$$x = \dfrac{\begin{vmatrix} c_1 & b_1 \\ c_2 & b_2 \end{vmatrix}}{\begin{vmatrix} a_1 & b_1 \\ a_2 & b_2 \end{vmatrix}} \quad \text{and} \quad y = \dfrac{\begin{vmatrix} a_1 & c_1 \\ a_2 & c_2 \end{vmatrix}}{\begin{vmatrix} a_1 & b_1 \\ a_2 & b_2 \end{vmatrix}},$$

where

$$\begin{vmatrix} a_1 & b_1 \\ a_2 & b_2 \end{vmatrix} \neq 0.$$

Here are some helpful tips when solving

$$\begin{cases} a_1x + b_1y = c_1 \\ a_2x + b_2y = c_2 \end{cases}$$

using determinants:

1. Three different determinants are used to find x and y. The determinants in the denominators for x and y are identical. The determinants in the numerators for x and y differ. In abbreviated notation, we write

$$x = \frac{D_x}{D} \quad \text{and} \quad y = \frac{D_y}{D}, \text{where } D \neq 0.$$

2. The elements of D, the determinant in the denominator, are the coefficients of the variables in the system.

$$D = \begin{vmatrix} a_1 & b_1 \\ a_2 & b_2 \end{vmatrix}$$

3. D_x, the determinant in the numerator of x, is obtained by replacing the x-coefficients, in D, a_1 and a_2, with the constants on the right sides of the equations, c_1 and c_2.

$$D = \begin{vmatrix} a_1 & b_1 \\ a_2 & b_2 \end{vmatrix} \quad \text{and} \quad D_x = \begin{vmatrix} c_1 & b_1 \\ c_2 & b_2 \end{vmatrix} \quad \begin{array}{l} \text{Replace the column with } a_1 \text{ and } a_2 \text{ with} \\ \text{the constants } c_1 \text{ and } c_2 \text{ to get } D_x. \end{array}$$

4. D_y, the determinant in the numerator for y, is obtained by replacing the y-coefficients, in D, b_1 and b_2, with the constants on the right sides of the equations, c_1 and c_2.

$$D = \begin{vmatrix} a_1 & b_1 \\ a_2 & b_2 \end{vmatrix} \quad \text{and} \quad D_y = \begin{vmatrix} a_1 & c_1 \\ a_2 & c_2 \end{vmatrix} \quad \begin{array}{l} \text{Replace the column with } b_1 \text{ and } b_2 \text{ with} \\ \text{the constants } c_1 \text{ and } c_2 \text{ to get } D_y. \end{array}$$

EXAMPLE 2 Using Cramer's Rule to Solve a Linear System

Use Cramer's rule to solve the system:

$$\begin{cases} 5x - 4y = 2 \\ 6x - 5y = 1. \end{cases}$$

Solution   Because

$$x = \frac{D_x}{D} \quad \text{and} \quad y = \frac{D_y}{D},$$

we will set up and evaluate the three determinants D, D_x, and D_y.

$$\begin{cases} 5x - 4y = 2 \\ 6x - 5y = 1 \end{cases}$$

The system we are interested in solving (repeated so that you don't have to look back)

1. D, the determinant in both denominators, consists of the x- and y-coefficients.

$$D = \begin{vmatrix} 5 & -4 \\ 6 & -5 \end{vmatrix} = (5)(-5) - (6)(-4) = -25 + 24 = -1$$

Because this determinant is not zero, we continue to use Cramer's rule to solve the system.

2. D_x, the determinant in the numerator for x, is obtained by replacing the x-coefficients in D, 5 and 6, by the constants on the right sides of the equations, 2 and 1.

$$D_x = \begin{vmatrix} 2 & -4 \\ 1 & -5 \end{vmatrix} = (2)(-5) - (1)(-4) = -10 + 4 = -6$$

3. D_y, the determinant in the numerator for y, is obtained by replacing the y-coefficients in D, -4 and -5, by the constants on the right sides of the equations, 2 and 1.

$$D_y = \begin{vmatrix} 5 & 2 \\ 6 & 1 \end{vmatrix} = (5)(1) - (6)(2) = 5 - 12 = -7$$

4. Thus,

$$x = \frac{D_x}{D} = \frac{-6}{-1} = 6 \quad \text{and} \quad y = \frac{D_y}{D} = \frac{-7}{-1} = 7.$$

As always, the ordered pair $(6, 7)$ should be checked by substituting these values into the original equations. The solution is $(6, 7)$ and the solution set is $\{(6, 7)\}$. ∎

✓ **CHECK POINT 2**   Use Cramer's rule to solve the system:

$$\begin{cases} 5x + 4y = 12 \\ 3x - 6y = 24. \end{cases}$$

3 Evaluate a third-order determinant.

The Determinant of a 3 × 3 Matrix

The determinant for a 3×3 matrix is defined in terms of second-order determinants:

Definition of the Determinant of a 3 × 3 Matrix

A third-order determinant is defined by

Subtract.   Add.

$$\begin{vmatrix} a_1 & b_1 & c_1 \\ a_2 & b_2 & c_2 \\ a_3 & b_3 & c_3 \end{vmatrix} = a_1 \begin{vmatrix} b_2 & c_2 \\ b_3 & c_3 \end{vmatrix} - a_2 \begin{vmatrix} b_1 & c_1 \\ b_3 & c_3 \end{vmatrix} + a_3 \begin{vmatrix} b_1 & c_1 \\ b_2 & c_2 \end{vmatrix}.$$

Each a on the right comes from the first column.

$$\begin{vmatrix} a_1 & b_1 & c_1 \\ a_2 & b_2 & c_2 \\ a_3 & b_3 & c_3 \end{vmatrix} =$$

$$a_1 \begin{vmatrix} b_2 & c_2 \\ b_3 & c_3 \end{vmatrix} - a_2 \begin{vmatrix} b_1 & c_1 \\ b_3 & c_3 \end{vmatrix} + a_3 \begin{vmatrix} b_1 & c_1 \\ b_2 & c_2 \end{vmatrix}$$

The definition of a third-order determinant (repeated)

Here are some tips that should be helpful when evaluating the determinant of a 3×3 matrix:

Evaluating the Determinant of a 3 × 3 Matrix

1. Each of the three terms in the definition of a third-order determinant contains two factors—a numerical factor and a second-order determinant.

2. The numerical factor in each term is an element from the first column of the third-order determinant.

3. The minus sign precedes the second term.

4. The second-order determinant that appears in each term is obtained by crossing out the row and the column containing the numerical factor.

$$a_1 \begin{vmatrix} b_2 & c_2 \\ b_3 & c_3 \end{vmatrix} - a_2 \begin{vmatrix} b_1 & c_1 \\ b_3 & c_3 \end{vmatrix} + a_3 \begin{vmatrix} b_1 & c_1 \\ b_2 & c_2 \end{vmatrix}$$

$$\begin{vmatrix} a_1 & b_1 & c_1 \\ a_2 & b_2 & c_2 \\ a_3 & b_3 & c_3 \end{vmatrix} \quad \begin{vmatrix} a_1 & b_1 & c_1 \\ a_2 & b_2 & c_2 \\ a_3 & b_3 & c_3 \end{vmatrix} \quad \begin{vmatrix} a_1 & b_1 & c_1 \\ a_2 & b_2 & c_2 \\ a_3 & b_3 & c_3 \end{vmatrix}$$

The **minor** of an element is the determinant that remains after deleting the row and column of that element. For this reason, we call this method **expansion by minors**.

EXAMPLE 3   Evaluating the Determinant of a 3 × 3 Matrix

Evaluate the determinant of the following matrix:

$$\begin{bmatrix} 4 & 1 & 0 \\ -9 & 3 & 4 \\ -3 & 8 & 1 \end{bmatrix}.$$

Solution We know that each of the three terms in the determinant contains a numerical factor and a second-order determinant. The numerical factors are from the first column of the given matrix. They are shown in red in the following matrix:

$$\begin{bmatrix} 4 & 1 & 0 \\ -9 & 3 & 4 \\ -3 & 8 & 1 \end{bmatrix}.$$

We find the minor for each numerical factor by deleting the row and column of that element.

$$\begin{bmatrix} 4 & 1 & 0 \\ -9 & 3 & 4 \\ -3 & 8 & 1 \end{bmatrix} \quad \begin{bmatrix} 4 & 1 & 0 \\ -9 & 3 & 4 \\ -3 & 8 & 1 \end{bmatrix} \quad \begin{bmatrix} 4 & 1 & 0 \\ -9 & 3 & 4 \\ -3 & 8 & 1 \end{bmatrix}$$

The minor for 4 is $\begin{vmatrix} 3 & 4 \\ 8 & 1 \end{vmatrix}$.

The minor for -9 is $\begin{vmatrix} 1 & 0 \\ 8 & 1 \end{vmatrix}$.

The minor for -3 is $\begin{vmatrix} 1 & 0 \\ 3 & 4 \end{vmatrix}$.

Now we have three numerical factors, 4, -9, and -3, and three second-order determinants. We multiply each numerical factor by its second-order determinant to find the three terms of the third-order determinant:

$$4 \begin{vmatrix} 3 & 4 \\ 8 & 1 \end{vmatrix}, \quad -9 \begin{vmatrix} 1 & 0 \\ 8 & 1 \end{vmatrix}, \quad -3 \begin{vmatrix} 1 & 0 \\ 3 & 4 \end{vmatrix}.$$

Using Technology

A graphing utility can be used to evaluate the determinant of a matrix. Enter the matrix and call it *A*. Then use the determinant command. The screen below verifies our result in Example 3.

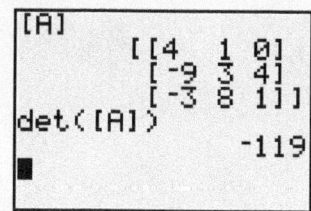

Based on the preceding definition, we subtract the second term from the first term and add the third term.

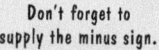

Don't forget to supply the minus sign.

$$\begin{vmatrix} 4 & 1 & 0 \\ -9 & 3 & 4 \\ -3 & 8 & 1 \end{vmatrix} = 4\begin{vmatrix} 3 & 4 \\ 8 & 1 \end{vmatrix} - (-9)\begin{vmatrix} 1 & 0 \\ 8 & 1 \end{vmatrix} - 3\begin{vmatrix} 1 & 0 \\ 3 & 4 \end{vmatrix}$$

Begin by evaluating the three second-order determinants.

$$= 4(3 \cdot 1 - 8 \cdot 4) + 9(1 \cdot 1 - 8 \cdot 0) - 3(1 \cdot 4 - 3 \cdot 0)$$

$$= 4(3 - 32) + 9(1 - 0) - 3(4 - 0)$$
Multiply within parentheses.

$$= 4(-29) + 9(1) - 3(4)$$
Subtract within parentheses.

$$= -116 + 9 - 12$$
Multiply.

$$= -119$$
Add and subtract as indicated.

✓ **CHECK POINT 3**   Evaluate the determinant of the following matrix:

$$\begin{bmatrix} 2 & 1 & 7 \\ -5 & 6 & 0 \\ -4 & 3 & 1 \end{bmatrix}.$$

4 Solve a system of linear equations in three variables using Cramer's rule.

Solving Systems of Linear Equations in Three Variables Using Determinants

Cramer's rule can be applied to solving systems of linear equations in three variables. The determinants in the numerator and denominator of the quotients determining each variable are third-order determinants.

Solving Three Equations in Three Variables Using Determinants

Cramer's Rule

If

$$\begin{cases} a_1x + b_1y + c_1z = d_1 \\ a_2x + b_2y + c_2z = d_2 \\ a_3x + b_3y + c_3z = d_3, \end{cases}$$

then

$$x = \frac{D_x}{D}, \quad y = \frac{D_y}{D}, \quad \text{and} \quad z = \frac{D_z}{D}.$$

These four third-order determinants are given by

$$D = \begin{vmatrix} a_1 & b_1 & c_1 \\ a_2 & b_2 & c_2 \\ a_3 & b_3 & c_3 \end{vmatrix}$$
These are the coefficients of the variables x, y, and z. $D \neq 0$

$$D_x = \begin{vmatrix} d_1 & b_1 & c_1 \\ d_2 & b_2 & c_2 \\ d_3 & b_3 & c_3 \end{vmatrix}$$
Replace x-coefficients in D with the constants on the right of the three equations.

$$D_y = \begin{vmatrix} a_1 & d_1 & c_1 \\ a_2 & d_2 & c_2 \\ a_3 & d_3 & c_3 \end{vmatrix}$$
Replace y-coefficients in D with the constants on the right of the three equations.

$$D_z = \begin{vmatrix} a_1 & b_1 & d_1 \\ a_2 & b_2 & d_2 \\ a_3 & b_3 & d_3 \end{vmatrix}.$$
Replace z-coefficients in D with the constants on the right of the three equations.

EXAMPLE 4 Using Cramer's Rule to Solve a Linear System in Three Variables

Use Cramer's rule to solve:

$$\begin{cases} x + 2y - z = -4 \\ x + 4y - 2z = -6 \\ 2x + 3y + z = 3. \end{cases}$$

Solution Because

$$x = \frac{D_x}{D}, \quad y = \frac{D_y}{D}, \quad \text{and} \quad z = \frac{D_z}{D},$$

we need to set up and evaluate four determinants.

Step 1. Set up the determinants.

1. D, the determinant in all three denominators, consists of the x-, y-, and z-coefficients.

$$D = \begin{vmatrix} 1 & 2 & -1 \\ 1 & 4 & -2 \\ 2 & 3 & 1 \end{vmatrix}$$

2. D_x, the determinant in the numerator for x, is obtained by replacing the x-coefficients in D, 1, 1, and 2, with the constants on the right sides of the equations, -4, -6, and 3.

$$D_x = \begin{vmatrix} -4 & 2 & -1 \\ -6 & 4 & -2 \\ 3 & 3 & 1 \end{vmatrix}$$

3. D_y, the determinant in the numerator for y, is obtained by replacing the y-coefficients in D, 2, 4, and 3, with the constants on the right sides of the equations, -4, -6, and 3.

$$D_y = \begin{vmatrix} 1 & -4 & -1 \\ 1 & -6 & -2 \\ 2 & 3 & 1 \end{vmatrix}$$

4. D_z, the determinant in the numerator for z, is obtained by replacing the z-coefficients in D, -1, -2, and 1, with the constants on the right sides of the equations, -4, -6, and 3.

$$D_z = \begin{vmatrix} 1 & 2 & -4 \\ 1 & 4 & -6 \\ 2 & 3 & 3 \end{vmatrix}$$

Step 2. Evaluate the four determinants.

$$D = \begin{vmatrix} 1 & 2 & -1 \\ 1 & 4 & -2 \\ 2 & 3 & 1 \end{vmatrix} = 1 \begin{vmatrix} 4 & -2 \\ 3 & 1 \end{vmatrix} - 1 \begin{vmatrix} 2 & -1 \\ 3 & 1 \end{vmatrix} + 2 \begin{vmatrix} 2 & -1 \\ 4 & -2 \end{vmatrix}$$

$$= 1(4 + 6) - 1(2 + 3) + 2(-4 + 4)$$
$$= 1(10) - 1(5) + 2(0) = 5$$

Using the same technique to evaluate each determinant, we obtain

$$D_x = -10, \quad D_y = 5, \quad \text{and} \quad D_z = 20.$$

Step 3. Substitute these four values and solve the system.

$$x = \frac{D_x}{D} = \frac{-10}{5} = -2$$

$$y = \frac{D_y}{D} = \frac{5}{5} = 1$$

$$z = \frac{D_z}{D} = \frac{20}{5} = 4$$

The ordered triple $(-2, 1, 4)$ can be checked by substitution into the original three equations. The solution is $(-2, 1, 4)$ and the solution set is $\{(-2, 1, 4)\}$. ∎

✓ **CHECK POINT 4**   Use Cramer's rule to solve the system:

$$\begin{cases} 3x - 2y + z = 16 \\ 2x + 3y - z = -9 \\ x + 4y + 3z = 2. \end{cases}$$

5 Use determinants to identify inconsistent and dependent systems.

Cramer's Rule with Inconsistent and Dependent Systems

If D, the determinant in the denominator, is 0, the variables described by the quotient of determinants are not real numbers. However, when $D = 0$, this indicates that the system is inconsistent or contains dependent equations. This gives rise to the following two situations:

Determinants: Inconsistent and Dependent Systems

1. If $D = 0$ and at least one of the determinants in the numerator is not 0, then the system is inconsistent. The solution set is $\varnothing$.

2. If $D = 0$ and all the determinants in the numerators are 0, then the equations in the system are dependent. The system has infinitely many solutions.

Although we have focused on applying determinants to solve linear systems, they have other applications, some of which we consider in the Exercise Set that follows.

APPENDIX C EXERCISE SET

MyMathLab®

 Watch the videos in MyMathLab

 Download the MyDashBoard App

Practice Exercises

Evaluate each determinant in Exercises 1–10.

1. $\begin{vmatrix} 5 & 7 \\ 2 & 3 \end{vmatrix}$

2. $\begin{vmatrix} 4 & 8 \\ 5 & 6 \end{vmatrix}$

3. $\begin{vmatrix} -4 & 1 \\ 5 & 6 \end{vmatrix}$

4. $\begin{vmatrix} 7 & 9 \\ -2 & -5 \end{vmatrix}$

5. $\begin{vmatrix} -7 & 14 \\ 2 & -4 \end{vmatrix}$

6. $\begin{vmatrix} 1 & 3 \\ -8 & 2 \end{vmatrix}$

7. $\begin{vmatrix} -5 & -1 \\ -2 & -7 \end{vmatrix}$

8. $\begin{vmatrix} \frac{1}{5} & \frac{1}{6} \\ -6 & 5 \end{vmatrix}$

9. $\begin{vmatrix} \frac{1}{2} & \frac{1}{2} \\ \frac{1}{8} & -\frac{3}{4} \end{vmatrix}$

10. $\begin{vmatrix} \frac{2}{3} & \frac{1}{3} \\ -\frac{1}{2} & \frac{3}{4} \end{vmatrix}$

For Exercises 11–26, use Cramer's rule to solve each system or to determine that the system is inconsistent or contains dependent equations.

11. $\begin{cases} x + y = 7 \\ x - y = 3 \end{cases}$

12. $\begin{cases} 2x + y = 3 \\ x - y = 3 \end{cases}$

13. $\begin{cases} 12x + 3y = 15 \\ 2x - 3y = 13 \end{cases}$

14. $\begin{cases} x - 2y = 5 \\ 5x - y = -2 \end{cases}$

15. $\begin{cases} 4x - 5y = 17 \\ 2x + 3y = 3 \end{cases}$

16. $\begin{cases} 3x + 2y = 2 \\ 2x + 2y = 3 \end{cases}$

17. $\begin{cases} x - 3y = 4 \\ 3x - 4y = 12 \end{cases}$

18. $\begin{cases} 2x - 9y = 5 \\ 3x - 3y = 11 \end{cases}$

19. $\begin{cases} 3x - 4y = 4 \\ 2x + 2y = 12 \end{cases}$

20. $\begin{cases} 3x = 7y + 1 \\ 2x = 3y - 1 \end{cases}$

21. $\begin{cases} 2x = 3y + 2 \\ 5x = 51 - 4y \end{cases}$

22. $\begin{cases} y = -4x + 2 \\ 2x = 3y + 8 \end{cases}$

23. $\begin{cases} 3x = 2 - 3y \\ 2y = 3 - 2x \end{cases}$ **24.** $\begin{cases} x + 2y - 3 = 0 \\ 12 = 8y + 4x \end{cases}$

25. $\begin{cases} 4y = 16 - 3x \\ 6x = 32 - 8y \end{cases}$ **26.** $\begin{cases} 2x = 7 + 3y \\ 4x - 6y = 3 \end{cases}$

Evaluate each determinant in Exercises 27–32.

27. $\begin{vmatrix} 3 & 0 & 0 \\ 2 & 1 & -5 \\ 2 & 5 & -1 \end{vmatrix}$ **28.** $\begin{vmatrix} 4 & 0 & 0 \\ 3 & -1 & 4 \\ 2 & -3 & 5 \end{vmatrix}$

29. $\begin{vmatrix} 3 & 1 & 0 \\ -3 & 4 & 0 \\ -1 & 3 & -5 \end{vmatrix}$ **30.** $\begin{vmatrix} 2 & -4 & 2 \\ -1 & 0 & 5 \\ 3 & 0 & 4 \end{vmatrix}$

31. $\begin{vmatrix} 1 & 1 & 1 \\ 2 & 2 & 2 \\ -3 & 4 & -5 \end{vmatrix}$ **32.** $\begin{vmatrix} 1 & 2 & 3 \\ 2 & 2 & -3 \\ 3 & 2 & 1 \end{vmatrix}$

In Exercises 33–40, use Cramer's rule to solve each system.

33. $\begin{cases} x + y + z = 0 \\ 2x - y + z = -1 \\ -x + 3y - z = -8 \end{cases}$ **34.** $\begin{cases} x - y + 2z = 3 \\ 2x + 3y + z = 9 \\ -x - y + 3z = 11 \end{cases}$

35. $\begin{cases} 4x - 5y - 6z = -1 \\ x - 2y - 5z = -12 \\ 2x - y = 7 \end{cases}$ **36.** $\begin{cases} x - 3y + z = -2 \\ x + 2y = 8 \\ 2x - y = 1 \end{cases}$

37. $\begin{cases} x + y + z = 4 \\ x - 2y + z = 7 \\ x + 3y + 2z = 4 \end{cases}$ **38.** $\begin{cases} 2x + 2y + 3z = 10 \\ 4x - y + z = -5 \\ 5x - 2y + 6z = 1 \end{cases}$

39. $\begin{cases} x + 2z = 4 \\ 2y - z = 5 \\ 2x + 3y = 13 \end{cases}$ **40.** $\begin{cases} 3x + 2z = 4 \\ 5x - y = -4 \\ 4y + 3z = 22 \end{cases}$

Practice PLUS

In Exercises 41–42, evaluate each determinant.

41. $\begin{vmatrix} \begin{vmatrix} 3 & 1 \\ -2 & 3 \end{vmatrix} & \begin{vmatrix} 7 & 0 \\ 1 & 5 \end{vmatrix} \\ \begin{vmatrix} 3 & 0 \\ 0 & 7 \end{vmatrix} & \begin{vmatrix} 9 & -6 \\ 3 & 5 \end{vmatrix} \end{vmatrix}$ **42.** $\begin{vmatrix} \begin{vmatrix} 5 & 0 \\ 4 & -3 \end{vmatrix} & \begin{vmatrix} -1 & 0 \\ 0 & -1 \end{vmatrix} \\ \begin{vmatrix} 7 & -5 \\ 4 & 6 \end{vmatrix} & \begin{vmatrix} 4 & 1 \\ -3 & 5 \end{vmatrix} \end{vmatrix}$

In Exercises 43–44, write the system of linear equations for which Cramer's rule yields the given determinants.

43. $D = \begin{vmatrix} 2 & -4 \\ 3 & 5 \end{vmatrix}$, $D_x = \begin{vmatrix} 8 & -4 \\ -10 & 5 \end{vmatrix}$

44. $D = \begin{vmatrix} 2 & -3 \\ 5 & 6 \end{vmatrix}$, $D_x = \begin{vmatrix} 8 & -3 \\ 11 & 6 \end{vmatrix}$

In Exercises 45–48, solve each equation for x.

45. $\begin{vmatrix} -2 & x \\ 4 & 6 \end{vmatrix} = 32$ **46.** $\begin{vmatrix} x + 3 & -6 \\ x - 2 & -4 \end{vmatrix} = 28$

47. $\begin{vmatrix} 1 & x & -2 \\ 3 & 1 & 1 \\ 0 & -2 & 2 \end{vmatrix} = -8$ **48.** $\begin{vmatrix} 2 & x & 1 \\ -3 & 1 & 0 \\ 2 & 1 & 4 \end{vmatrix} = 39$

Application Exercises

Determinants are used to find the area of a triangle whose vertices are given by three points in a rectangular coordinate system. The area of a triangle with vertices (x_1, y_1), (x_2, y_2), and (x_3, y_3) is

$$\text{Area} = \pm \frac{1}{2} \begin{vmatrix} x_1 & y_1 & 1 \\ x_2 & y_2 & 1 \\ x_3 & y_3 & 1 \end{vmatrix},$$

where the $\pm$ symbol indicates that the appropriate sign should be chosen to yield a positive area. Use this information to work Exercises 49–50.

49. Use determinants to find the area of the triangle whose vertices are $(3, -5)$, $(2, 6)$, and $(-3, 5)$.

50. Use determinants to find the area of the triangle whose vertices are $(1, 1)$, $(-2, -3)$, and $(11, -3)$.

Determinants are used to show that three points lie on the same line (are collinear). If $\begin{vmatrix} x_1 & y_1 & 1 \\ x_2 & y_2 & 1 \\ x_3 & y_3 & 1 \end{vmatrix} = 0$,

then the points (x_1, y_1), (x_2, y_2), and (x_3, y_3) are collinear. If the determinant does not equal 0, then the points are not collinear. Use this information to work Exercises 51–52.

51. Are the points $(3, -1)$, $(0, -3)$, and $(12, 5)$ collinear?

52. Are the points $(-4, -6)$, $(1, 0)$, and $(11, 12)$ collinear?

Determinants are used to write an equation of a line passing through two points. An equation of the line passing through the distinct points (x_1, y_1) and (x_2, y_2) is given by

$$\begin{vmatrix} x & y & 1 \\ x_1 & y_1 & 1 \\ x_2 & y_2 & 1 \end{vmatrix} = 0.$$

Use this information to work Exercises 53–54.

53. Use the determinant to write an equation for the line passing through $(3, -5)$ and $(-2, 6)$. Then expand the determinant, expressing the line's equation in slope-intercept form.

54. Use the determinant to write an equation for the line passing through $(-1, 3)$ and $(2, 4)$. Then expand the determinant, expressing the line's equation in slope-intercept form.

Where Did That Come From? Selected Proofs

Section 12.3   Properties of Logarithms

The Product Rule

Let b, M, and N be positive real numbers with $b \neq 1$.

$$\log_b(MN) = \log_b M + \log_b N$$

Proof. We begin by letting $\log_b M = R$ and $\log_b N = S$. Now we write each logarithm in exponential form.

$$\log_b M = R \quad \text{means} \quad b^R = M.$$
$$\log_b N = S \quad \text{means} \quad b^S = N.$$

By substituting and using a property of exponents, we see that

$$MN = b^R b^S = b^{R+S}$$

Now we change $MN = b^{R+S}$ to logarithmic form.

$$MN = b^{R+S} \quad \text{means} \quad \log_b(MN) = R + S.$$

Finally, substituting $\log_b M$ for R and $\log_b N$ for S gives us

$$\log_b(MN) = \log_b M + \log_b N,$$

the property that we wanted to prove.

The quotient and power rules for logarithms are proved using similar procedures.

The Change-of-Base Property

For any logarithmic bases a and b, and any positive number M,

$$\log_b M = \frac{\log_a M}{\log_a b}.$$

Proof. To prove the change-of-base property, we let x equal the logarithm on the left side:

$$\log_b M = x.$$

Now we rewrite this logarithm in exponential form.

$$\log_b M = x \quad \text{means} \quad b^x = M.$$

Because b^x and M are equal, the logarithms with base a for each of these expressions must be equal. This means that

$$\log_a b^x = \log_a M$$
$$x \log_a b = \log_a M \qquad \text{Apply the power rule for logarithms on the left side.}$$
$$x = \frac{\log_a M}{\log_a b}. \qquad \text{Solve for x by dividing both sides by } \log_a b.$$

In our first step we let x equal $\log_b M$. Replacing x on the left side by $\log_b M$ gives us

$$\log_b M = \frac{\log_a M}{\log_a b},$$

which is the change-of-base property.

Section 13.2   The Ellipse

The Standard Form of the Equation of an Ellipse with a Horizontal Major Axis Centered at the Origin

Proof. Refer to **Figure D.1**.

$$d_1 + d_2 = 2a \qquad \text{The sum of the distances from } P \\ \text{to the foci equals a constant, } 2a.$$

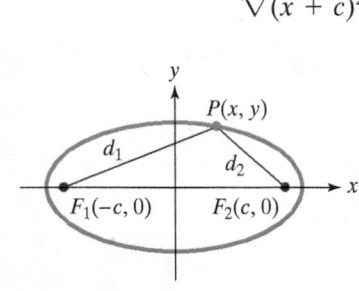

$$\sqrt{(x + c)^2 + y^2} + \sqrt{(x - c)^2 + y^2} = 2a \qquad \text{Use the distance formula.}$$

$$\sqrt{(x + c)^2 + y^2} = 2a - \sqrt{(x - c)^2 + y^2} \qquad \text{Isolate a radical.}$$

$$(x + c)^2 + y^2 = 4a^2 - 4a\sqrt{(x - c)^2 + y^2} \qquad \text{Square both sides.}$$

$$+ (x - c)^2 + y^2$$

$$x^2 + 2cx + c^2 + y^2 = 4a^2 - 4a\sqrt{(x - c)^2 + y^2} \qquad \text{Square } x + c \text{ and } x - c.$$

$$+ x^2 - 2cx + c^2 + y^2$$

$$4cx - 4a^2 = -4a\sqrt{(x - c)^2 + y^2} \qquad \text{Simplify and isolate the radical.}$$

$$cx - a^2 = -a\sqrt{(x - c)^2 + y^2} \qquad \text{Divide both sides by 4.}$$

$$(cx - a^2)^2 = a^2\left[(x - c)^2 + y^2\right] \qquad \text{Square both sides.}$$

$$c^2x^2 - 2a^2cx + a^4 = a^2(x^2 - 2cx + c^2 + y^2) \qquad \text{Square } cx - a^2 \text{ and } x - c.$$

$$c^2x^2 - 2a^2cx + a^4 = a^2x^2 - 2a^2cx + a^2c^2 + a^2y^2 \qquad \text{Use the distributive property.}$$

$$c^2x^2 + a^4 = a^2x^2 + a^2c^2 + a^2y^2 \qquad \text{Add } 2a^2cx \text{ to both sides.}$$

$$c^2x^2 - a^2x^2 - a^2y^2 = a^2c^2 - a^4 \qquad \text{Rearrange the terms.}$$

$$(c^2 - a^2)x^2 - a^2y^2 = a^2(c^2 - a^2) \qquad \text{Factor out } x^2 \text{ and } a^2, \text{ respectively.}$$

$$(a^2 - c^2)x^2 + a^2y^2 = a^2(a^2 - c^2) \qquad \text{Multiply both sides by } -1.$$

Refer to the discussion on page 941 and let $b^2 = a^2 - c^2$ in the preceding equation.

$$b^2x^2 + a^2y^2 = a^2b^2$$

$$\frac{x^2}{a^2} + \frac{y^2}{b^2} = 1 \qquad \text{Divide both sides by } a^2b^2.$$

Section 13.3   The Hyperbola

The Asymptotes of a Hyperbola Centered at the Origin

The hyperbola

$$\frac{x^2}{a^2} - \frac{y^2}{b^2} = 1$$

with a horizontal transverse axis has the two asymptotes

$$y = \frac{b}{a}x \quad \text{and} \quad y = -\frac{b}{a}x.$$

Proof. Begin by solving the hyperbola's equation for y.

$$\frac{x^2}{a^2} - \frac{y^2}{b^2} = 1 \qquad \text{This is the standard form of the equation of a hyperbola.}$$

$$\frac{y^2}{b^2} = \frac{x^2}{a^2} - 1 \qquad \text{We isolate the term involving } y^2 \text{ to solve for } y.$$

$$y^2 = \frac{b^2x^2}{a^2} - b^2 \qquad \text{Multiply both sides by } b^2.$$

$$y^2 = \frac{b^2x^2}{a^2}\left(1 - \frac{a^2}{x^2}\right) \qquad \text{Factor out } \frac{b^2x^2}{a^2} \text{ on the right. Verify that this result is correct by multiplying using the distributive property and obtaining the previous step.}$$

$$y = \pm\sqrt{\frac{b^2x^2}{a^2}\left(1 - \frac{a^2}{x^2}\right)} \qquad \text{Solve for } y \text{ using the square root method: If } u^2 = d \text{ then } u = \pm\sqrt{d}.$$

$$y = \pm\frac{b}{a}x\sqrt{1 - \frac{a^2}{x^2}} \qquad \text{Simplify.}$$

As x increases or decreases without bound, the value of $\frac{a^2}{x^2}$ approaches 0. Consequently, the value of y can be approximated by

$$y = \pm\frac{b}{a}x.$$

This means that the lines whose equations are $y = \frac{b}{a}x$ and $y = -\frac{b}{a}x$ are asymptotes for the graph of the hyperbola.

Answers to Selected Exercises

CHAPTER 1

Section 1.1 Check Point Exercises

1. a. 26 **b.** 32 **2. a.** 37 **b.** 2 **3. a.** $6x$ **b.** $4 + x$ **c.** $3x + 5$ **d.** $12 - 2x$ **e.** $\dfrac{15}{x}$ **4. a.** not a solution **b.** solution

5. a. $\dfrac{x}{6} = 5$ **b.** $7 - 2x = 1$ **6. a.** 65% **b.** 60% **c.** overestimates by 5%

Concept and Vocabulary Check

1. variable **2.** expression **3.** substituting; evaluating **4.** equation; solution **5.** formula; modeling; models

Exercise Set 1.1

1. 12 **3.** 8 **5.** 20 **7.** 7 **9.** 17 **11.** 18 **13.** 5 **15.** 19 **17.** 24 **19.** 13 **21.** 10 **23.** 5 **25.** $x + 4$ **27.** $x - 4$

29. $x + 4$ **31.** $x - 9$ **33.** $9 - x$ **35.** $3x - 5$ **37.** $12x - 1$ **39.** $\dfrac{10}{x} + \dfrac{x}{10}$ **41.** $\dfrac{x}{30} + 6$ **43.** solution **45.** solution

47. not a solution **49.** solution **51.** not a solution **53.** solution **55.** solution **57.** not a solution **59.** $4x = 28$ **61.** $\dfrac{14}{x} = \dfrac{1}{2}$

63. $20 - x = 5$ **65.** $2x + 6 = 16$ **67.** $3x - 5 = 7$ **69.** $4x + 5 = 33$ **71.** $4(x + 5) = 33$ **73.** $5x = 24 - x$ **75.** 8 **77.** 59

79. a. 14 **b.** yes **81. a.** 6 **b.** no **83. a.** \$4.22; underestimates by \$0.01 **b.** underestimates by \$0.63 **85.** overestimates by 0.8%

87. a. 44 **b.** 164 **99.** makes sense **101.** makes sense **103.** true **105.** true **107.** Choices of variables may vary. Example: $h =$ hours

worked, $s =$ salary; $s = 20h$ **109.** $\dfrac{6}{35}$ **110.** $\dfrac{10}{21}$ **111.** $\dfrac{4}{17}$

Section 1.2 Check Point Exercises

1. $\dfrac{21}{8}$ **2.** $1\dfrac{2}{3}$ **3.** $2 \cdot 2 \cdot 3 \cdot 3$ **4. a.** $\dfrac{2}{3}$ **b.** $\dfrac{7}{4}$ **c.** $\dfrac{13}{15}$ **d.** $\dfrac{1}{5}$ **5. a.** $\dfrac{8}{33}$ **b.** $\dfrac{18}{5}$ or $3\dfrac{3}{5}$ **c.** $\dfrac{2}{7}$ **d.** $\dfrac{51}{10}$ or $5\dfrac{1}{10}$ **6. a.** $\dfrac{10}{3}$ or $3\dfrac{1}{3}$

b. $\dfrac{2}{9}$ **c.** $\dfrac{3}{2}$ or $1\dfrac{1}{2}$ **7. a.** $\dfrac{5}{11}$ **b.** $\dfrac{2}{3}$ **c.** $\dfrac{9}{4}$ or $2\dfrac{1}{4}$ **8.** $\dfrac{14}{21}$ **9. a.** $\dfrac{11}{10}$ or $1\dfrac{1}{10}$ **b.** $\dfrac{7}{12}$ **c.** $\dfrac{5}{4}$ or $1\dfrac{1}{4}$ **10.** $\dfrac{53}{60}$

11. a. solution **b.** solution **12. a.** $\dfrac{2}{3}(x - 6)$ **b.** $\dfrac{3}{4}x - 2 = \dfrac{1}{5}x$ **13.** 25°C

Concept and Vocabulary Check

1. numerator; denominator **2.** mixed; improper **3.** 5; 3; 2; 5 **4.** natural **5.** prime **6.** factors; product **7.** $\dfrac{a}{b}$ **8.** $\dfrac{a \cdot c}{b \cdot d}$

9. reciprocals **10.** $\dfrac{d}{c}$ **11.** $\dfrac{a + c}{b}$ **12.** least common denominator

Exercise Set 1.2

1. $\dfrac{19}{8}$ **3.** $\dfrac{38}{5}$ **5.** $\dfrac{135}{16}$ **7.** $4\dfrac{3}{5}$ **9.** $8\dfrac{4}{9}$ **11.** $35\dfrac{11}{20}$ **13.** $2 \cdot 11$ **15.** $2 \cdot 2 \cdot 5$ **17.** prime **19.** $2 \cdot 2 \cdot 3 \cdot 3$ **21.** $2 \cdot 2 \cdot 5 \cdot 7$

23. prime **25.** $3 \cdot 3 \cdot 3 \cdot 3$ **27.** $2 \cdot 2 \cdot 2 \cdot 2 \cdot 3 \cdot 5$ **29.** $\dfrac{5}{8}$ **31.** $\dfrac{5}{6}$ **33.** $\dfrac{7}{10}$ **35.** $\dfrac{2}{5}$ **37.** $\dfrac{22}{25}$ **39.** $\dfrac{60}{43}$ **41.** $\dfrac{2}{15}$ **43.** $\dfrac{21}{88}$

45. $\dfrac{36}{7}$ **47.** $\dfrac{1}{12}$ **49.** $\dfrac{15}{14}$ **51.** 6 **53.** $\dfrac{15}{16}$ **55.** $\dfrac{9}{5}$ **57.** $\dfrac{5}{9}$ **59.** 3 **61.** $\dfrac{7}{10}$ **63.** $\dfrac{1}{2}$ **65.** 6 **67.** $\dfrac{6}{11}$ **69.** $\dfrac{2}{3}$ **71.** $\dfrac{5}{4}$

73. $\dfrac{1}{6}$ **75.** 2 **77.** $\dfrac{7}{10}$ **79.** $\dfrac{9}{10}$ **81.** $\dfrac{19}{24}$ **83.** $\dfrac{7}{18}$ **85.** $\dfrac{7}{12}$ **87.** $\dfrac{41}{80}$ **89.** $1\dfrac{5}{12}$ or $\dfrac{17}{12}$ **91.** solution **93.** solution

95. not a solution **97.** solution **99.** not a solution **101.** solution **103.** $\dfrac{1}{5}x$ **105.** $x - \dfrac{1}{4}x$ **107.** $x - \dfrac{1}{4} = \dfrac{1}{2}x$ **109.** $\dfrac{1}{7}x + \dfrac{1}{8}x = 12$

111. $\dfrac{2}{3}(x + 6)$ **113.** $\dfrac{2}{3}x + 6 = x - 3$ **115.** $\dfrac{3a}{20}$ **117.** $\dfrac{20}{x}$ **119.** $\dfrac{4}{15}$ **121.** not a solution **123.** 20°C **125. a.** 140 beats per minute

b. 160 beats per minute **127. a.** $H = \dfrac{9}{10}(220 - a)$ **b.** 162 beats per minute **129.** 21; underestimates by $\dfrac{1}{2}$ present

141. makes sense   **143.** makes sense   **145.** false   **147.** true
149.

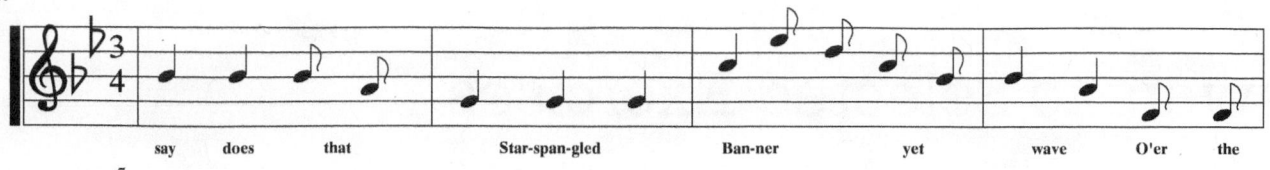

| say | does | that | Star-span-gled | Ban-ner | yet | wave | O'er | the |

150. 5   **151.** $\frac{5}{2}$ **152.** -4

Section 1.3 Check Point Exercises

1. a. -500 **b.** -282 **2.**

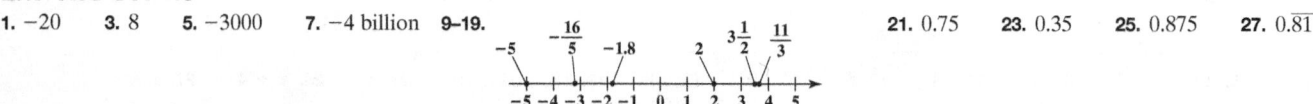

3.

4. a. 0.375 **b.** $0.\overline{45}$ **5. a.** $\sqrt{9}$ **b.** $0, \sqrt{9}$ **c.** $-9, 0, \sqrt{9}$ **d.** $-9, -1.3, 0, 0.\overline{3}, \sqrt{9}$ **e.** $\frac{\pi}{2}, \sqrt{10}$ **f.** $-9, -1.3, 0, 0.\overline{3}, \frac{\pi}{2}, \sqrt{9}, \sqrt{10}$
6. a. $>$ **b.** $<$ **c.** $<$ **d.** $<$ **7. a.** true   **b.** true   **c.** false   **8. a.** 4   **b.** 6^2 **c.** $\sqrt{2}$

Concept and Vocabulary Check

1. natural   **2.** whole   **3.** integers   **4.** rational   **5.** irrational   **6.** rational; irrational   **7.** left   **8.** absolute value; $|a|$

Exercise Set 1.3

1. -20 **3.** 8   **5.** -3000 **7.** -4 billion   **9–19.**   **21.** 0.75 **23.** 0.35 **25.** 0.875 **27.** $0.\overline{81}$

29. -0.5 **31.** $-0.8\overline{3}$ **33. a.** $\sqrt{100}$ **b.** $0, \sqrt{100}$ **c.** $-9, 0, \sqrt{100}$ **d.** $-9, -\frac{4}{5}, 0, 0.25, 9.2, \sqrt{100}$ **e.** $\sqrt{3}$

f. $-9, -\frac{4}{5}, 0, 0.25, \sqrt{3}, 9.2, \sqrt{100}$ **35. a.** $\sqrt{64}$ **b.** $0, \sqrt{64}$ **c.** $-11, 0, \sqrt{64}$ **d.** $-11, -\frac{5}{6}, 0, 0.75, \sqrt{64}$ **e.** $\sqrt{5}, \pi$

f. $-11, -\frac{5}{6}, 0, 0.75, \sqrt{5}, \pi, \sqrt{64}$ **37.** 0   **39.** Answers will vary; $\frac{1}{2}$ is an example.   **41.** Answers will vary; 6 is an example.

43. Answers will vary; π is an example.   **45.** $<$ **47.** $>$ **49.** $>$ **51.** $<$ **53.** $>$ **55.** $<$ **57.** $<$ **59.** $>$

61. $>$ **63.** true   **65.** true   **67.** true   **69.** false   **71.** 6   **73.** 7   **75.** $\frac{5}{6}$ **77.** $\sqrt{11}$ **79.** $>$ **81.** $=$ **83.** $<$ **85.** $=$

87. rational numbers   **89.** integers   **91.** all real numbers   **93.** whole numbers
95. a.   **b.** Rhode Island, Georgia, Louisiana, Florida, Hawaii

109. does not make sense   **111.** makes sense   **113.** false   **115.** false   **117.** false   **119.** $-\frac{1}{2}d$ **121.** $-3.464; -4$ and -3

123. $-0.236; -1$ and 0   **124.** 27; 27; Both expressions have the same value.   **125.** 32; 32; Both expressions have the same value.
126. 28; 28; Both expressions have the same value.

Section 1.4 Check Point Exercises

1. a. 3 terms   **b.** 6   **c.** 11   **d.** $6x$ and $2x$ **2. a.** $14 + x$ **b.** $y7$ **3. a.** $17 + 5x$ **b.** $x5 + 17$ **4. a.** $20 + x$ or $x + 20$ **b.** $30x$
5. $12 + x$ or $x + 12$ **6.** $5x + 15$ **7.** $24y + 42$ **8. a.** $10x$ **b.** $5a$ **9. a.** $18x + 10$ **b.** $14x + 8y$ **10.** $25x + 21$ **11.** $38x + 23y$

Concept and Vocabulary Check

1. like   **2.** $b + a$ **3.** ab **4.** $a + (b + c)$ **5.** $(ab)c$ **6.** $ab + ac$ **7.** simplified

Exercise Set 1.4

1. a. 2   **b.** 3   **c.** 5   **d.** no   **3. a.** 3   **b.** 1   **c.** 2   **d.** yes; x and $5x$ **5. a.** 3   **b.** 4   **c.** 1   **d.** no   **7.** $4 + y$ **9.** $3x + 5$
11. $5y + 4x$ **13.** $5(3 + x)$ **15.** $x9$ **17.** $x + 6y$ **19.** $x7 + 23$ **21.** $(x + 3)5$ **23.** $(7 + 5) + x = 12 + x$ **25.** $(7 \cdot 4)x = 28x$

27. $3x + 15$ **29.** $16x + 24$ **31.** $4 + 2r$ **33.** $5x + 5y$ **35.** $3x - 6$ **37.** $8x - 10$ **39.** $\frac{5}{2}x - 6$ **41.** $8x + 28$ **43.** $6x + 18 + 12y$
45. $15x - 10 + 20y$ **47.** $17x$ **49.** $8a$ **51.** $14 + x$ **53.** $11y + 3$ **55.** $9x + 1$ **57.** $14a + 14$ **59.** $15x + 6$ **61.** $15x + 2$
63. $41a + 24b$ **65.** Distributive property; Commutative property of addition; Associative property of addition; Commutative property of addition
67. $7x + 2x; 9x$ **69.** $12x - 3x; 9x$ **71.** $6(4x); 24x$ **73.** $6(4 + x); 24 + 6x$ **75.** $8 + 5(x - 1); 5x + 3$ **77. a.** $U = 13n + 133$
b. 211 million; overestimates by 1 million   **89.** does not make sense   **91.** makes sense   **93.** false   **95.** false   **97.** 60   **98.** -60 **99.** -5

Mid-Chapter Check Point Exercises

1. 62 **2.** 2 **3.** 43 **4.** $\frac{1}{4}x - 2$ **5.** $\frac{x}{6} + 5 = 19$ **6.** not a solution **7.** not a solution **8. a.** 11,919; overestimates by 131 miles

b. 10,274 **9.** $\frac{1}{6}$ **10.** $\frac{1}{2}$ **11.** $\frac{25}{66}$ **12.** $\frac{2}{3}$ **13.** $\frac{1}{5}$ **14.** $\frac{25}{27}$ **15.** $\frac{37}{18}$ or $2\frac{1}{8}$ **16.** 10°C

17. < **18.** $0.\overline{09}$ **19.** 19.3 **20.** $-11, -\frac{3}{7}, 0, 0.45, \sqrt{25}$ **21.** $(x + 3)5$ **22.** $5(3 + x)$ **23.** $5x + 15$ **24.** $65x + 21$ **25.** $10x + 34y$

Section 1.5 Check Point Exercises

1. $4 + (-7) = -3$ 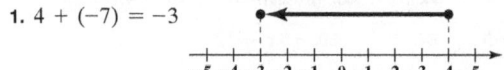 **2. a.** $-1 + (-3) = -4$

b. $-5 + 3 = -2$ 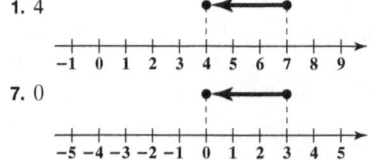 **3. a.** -35 **b.** -1.5 **c.** $-\frac{5}{6}$ **4. a.** -13 **b.** 1.2 **c.** $-\frac{1}{2}$ **5. a.** $-17x$

b. $-7y + 6z$ **c.** $-20x + 15$ **6.** down 3 ft

Concept and Vocabulary Check

1. additive inverses **2.** zero **3.** negative number **4.** positive number **5.** 0 **6.** negative number **7.** positive number **8.** 0

Exercise Set 1.5

1. 4 **3.** -7 **5.** -4

7. 0 **9.** -7 **11.** 0 **13.** -60 **15.** -18 **17.** -1.3 **19.** -1 **21.** -5 **23.** 4 **25.** -3 **27.** -1.5 **29.** -5.7 **31.** $\frac{3}{10}$ **33.** $\frac{1}{8}$ **35.** $-\frac{43}{35}$ **37.** -8 **39.** 62 **41.** 8 **43.** -21

45. 22.1 **47.** $-8x$ **49.** $13y$ **51.** $-23a$ **53.** $-6y + 4z$ **55.** $-8b + 4$ **57.** $-2x + 14y$ **59.** $-3y + 24$ **61.** 12 **63.** -30

65. > **67.** $-6x + (-13x); -19x$ **69.** $\frac{-20}{x} + \frac{3}{x}; \frac{-17}{x}$ **71.** 44°F **73.** 600 ft below sea level **75.** 3°F **77.** the 25-yard line

79. a. $-\$410$ billion **b.** $-\$407$ billion **c.** $-\$817$ billion **89.** makes sense **91.** makes sense **93.** true **95.** false **97.** negative
99. positive **102. a.** $\sqrt{4}$ **b.** $0, \sqrt{4}$ **c.** $-6, 0, \sqrt{4}$ **d.** $-6, 0, 0.\overline{7}, \sqrt{4}$ **e.** $-\pi, \sqrt{3}$ **f.** $-6, \pi, 0, 0.\overline{7}, \sqrt{3}, \sqrt{4}$ **103.** true
104. solution **105.** -3 **106.** -21 **107.** 5

Section 1.6 Check Point Exercises

1. a. -8 **b.** 9 **c.** -5 **2. a.** 9.2 **b.** $-\frac{14}{15}$ **c.** 7π **3.** 15 **4.** $-6, 4a, -7ab$ **5. a.** $4 - 7x$ **b.** $-9x + 4y$ **6.** 19,763 m

Concept and Vocabulary Check

1. (-14) **2.** 14 **3.** 14 **4.** $-8; (-14)$ **5.** $3; (-12); (-23)$ **6.** $(-4y); 6$ **7.** three; addition

Exercise Set 1.6

1. a. -12 **b.** $5 + (-12)$ **3. a.** 7 **b.** $5 + 7$ **5.** 6 **7.** -6 **9.** 23 **11.** 11 **13.** -11 **15.** -38 **17.** 0 **19.** 0 **21.** 26

23. -13 **25.** 13 **27.** $-\frac{2}{7}$ **29.** $\frac{4}{5}$ **31.** -1 **33.** $-\frac{3}{5}$ **35.** $\frac{3}{4}$ **37.** $\frac{1}{4}$ **39.** 7.6 **41.** -2 **43.** 2.6 **45.** 0 **47.** 3π

49. 13π **51.** 19 **53.** -3 **55.** -15 **57.** 0 **59.** -52 **61.** -187 **63.** $\frac{7}{6}$ **65.** -4.49 **67.** $-\frac{3}{8}$ **69.** $-3x, -8y$

71. $12x, -5xy, -4$ **73.** $-6x$ **75.** $4 - 10y$ **77.** $5 - 7a$ **79.** $-4 - 9b$ **81.** $24 + 11x$ **83.** $3y - 8x$ **85.** 9 **87.** $\frac{7}{8}$

89. -2 **91.** $6x - (-5x); 11x$ **93.** $\frac{-5}{x} - \left(\frac{-2}{x}\right); \frac{-3}{x}$ **95.** 19,757 ft **97.** 21°F **99.** 3°F **101.** shrink by 5 years

103. shrink by 11 years **105.** 10 years **107.** no change **109.** 9 years **117.** makes sense **119.** makes sense **121.** false **123.** true
125. negative **127.** positive **131.** not a solution **132.** $15x + 40y$ **133.** Answers will vary; -1 is an example. **134.** -12 **135.** -9 **136.** 12

Section 1.7 Check Point Exercises

1. a. -40 **b.** $-\frac{4}{21}$ **c.** 36 **d.** 5.5 **e.** 0 **2. a.** 24 **b.** -30 **3. a.** $\frac{1}{7}$ **b.** 8 **c.** $-\frac{1}{6}$ **d.** $-\frac{13}{7}$ **4. a.** -4 **b.** 8

5. a. 8 **b.** $-\frac{8}{15}$ **c.** -7.3 **d.** 0 **6. a.** $-20x$ **b.** $10x$ **c.** $-b$ **d.** $-21x + 28$ **e.** $-7y + 6$ **7.** $-y - 26$

8. not a solution **9.** 49.4%; overestimates by 0.4%

Concept and Vocabulary Check

1. positive **2.** negative **3.** negative **4.** positive **5.** 0 **6.** negative **7.** negative **8.** positive **9.** 0
10. undefined **11.** positive

Exercise Set 1.7

1. -45 **3.** 24 **5.** -21 **7.** 19 **9.** 0 **11.** -12 **13.** 9 **15.** $\dfrac{12}{35}$ **17.** $-\dfrac{14}{27}$ **19.** -3.6 **21.** 0.12 **23.** 30 **25.** -72

27. 24 **29.** -27 **31.** 90 **33.** 0 **35.** $\dfrac{1}{4}$ **37.** 5 **39.** $-\dfrac{1}{10}$ **41.** $-\dfrac{5}{2}$ **43. a.** $-32\cdot\left(\dfrac{1}{4}\right)$ **b.** -8 **45. a.** $-60\cdot\left(-\dfrac{1}{5}\right)$

b. 12 **47.** -3 **49.** -7 **51.** 30 **53.** 0 **55.** undefined **57.** -5 **59.** -12 **61.** 6 **63.** 0 **65.** undefined **67.** -4.3

69. $\dfrac{5}{6}$ **71.** $-\dfrac{16}{9}$ **73.** -1 **75.** -15 **77.** $-10x$ **79.** $3y$ **81.** $9x$ **83.** $-4x$ **85.** $-b$ **87.** $3y$ **89.** $-8x+12$

91. $6x-12$ **93.** $-2y+5$ **95.** $y-14$ **97.** solution **99.** solution **101.** not a solution **103.** solution **105.** not a solution

107. solution **109.** $4(-10)+8=-32$ **111.** $(-9)(-3)-(-2)=29$ **113.** $\dfrac{-18}{-15+12}=6$ **115.** $-6-\left(\dfrac{12}{-4}\right)=-3$ **117.** $-30°$C

119. a. underestimates by 1% **b.** 16% **121. a.** 25 hours **b.** 23 hours; it's less than the estimate.
c. $D=-0.4n+39-(0.2n+12)$; $D=-0.6n+27$ **d.** 3 hours; underestimates **131.** does not make sense **133.** makes sense

135. false **137.** false **139.** $5x$ **141.** $\dfrac{x}{12}$ **145.** $1.144x+2.5$ **147.** -9 **148.** -3 **149.** 2 **150.** 36 **151.** -125 **152.** 16

Section 1.8 Check Point Exercises

1. a. 36 **b.** -64 **c.** 1 **d.** -1 **2. a.** $21x^2$ **b.** $8x^3$ **c.** cannot simplify **3.** 15 **4.** 32 **5. a.** 36 **b.** 12 **6.** $\dfrac{3}{4}$ **7.** -40

8. -31 **9.** $\dfrac{5}{7}$ **10.** -5 **11.** $7x^2+15$ **12.** 2662 calories; underestimates by 38 calories **13. a.** \$330 **b.** \$60 **c.** \$33

Concept and Vocabulary Check

1. base; exponent **2.** b to the nth power **3.** multiply **4.** add **5.** divide **6.** subtract **7.** multiply

Exercise Set 1.8

1. 81 **3.** 64 **5.** 16 **7.** -64 **9.** 625 **11.** -625 **13.** -100 **15.** $19x^2$ **17.** $15x^3$ **19.** $9x^4$ **21.** $-x^2$ **23.** x^3
25. cannot be simplified **27.** 0 **29.** 25 **31.** 27 **33.** 12 **35.** 5 **37.** 45 **39.** -24 **41.** 300 **43.** 0 **45.** -32 **47.** 64

49. 30 **51.** $\dfrac{4}{3}$ **53.** 2 **55.** 2 **57.** 3 **59.** 88 **61.** -60 **63.** -36 **65.** 14 **67.** $-\dfrac{3}{4}$ **69.** $-\dfrac{9}{40}$ **71.** $-\dfrac{37}{36}$ or $-1\dfrac{1}{36}$
73. 24 **75.** 28 **77.** 9 **79.** -7 **81.** $15x-27$ **83.** $15-3y$ **85.** $16y-25$ **87.** $-2x^2-9$ **89.** $-10-(-2)^3=-2$
91. $[2(7-10)]^2=36$ **93.** $x-(5x+8)=-4x-8$ **95.** $5(x^3-4)=5x^3-20$ **97.** 1924 calories; underestimates by 76 calories **99. a.** 22%
b. 18.7%; less than the estimate **101. a.** \$2,000,000 (or \$200 tens of thousands) **b.** \$8,000,000 (or \$800 tens of thousands) **c.** Cost inceases.
107. does not make sense **109.** makes sense **111.** false **113.** false **115.** $(2\cdot3+3)\cdot5=45$ **117.** 6 **118.** -24
119. Answers will vary; -3 is an example. **120.** solution **121.** solution **122.** not a solution

Review Exercises

1. 40 **2.** 50 **3.** 3 **4.** 84 **5.** $7x-6$ **6.** $\dfrac{x}{5}-2=18$ **7.** $9-2x=14$ **8.** $3(x+7)$ **9.** not a solution **10.** solution

11. solution **12.** 11, although answers may vary by ±1 **13.** 11 Latin words; very well; It is the same. **14. a.** \$13,569; overestimates by \$194

b. \$22,182 **15.** $\dfrac{23}{7}$ **16.** $\dfrac{64}{11}$ **17.** $1\dfrac{8}{9}$ **18.** $5\dfrac{2}{5}$ **19.** $2\cdot2\cdot3\cdot5$ **20.** $3\cdot3\cdot7$ **21.** prime **22.** $\dfrac{5}{11}$ **23.** $\dfrac{8}{15}$ **24.** $\dfrac{21}{50}$ **25.** $\dfrac{8}{3}$

26. $\dfrac{1}{4}$ **27.** $\dfrac{2}{3}$ **28.** $\dfrac{29}{18}$ **29.** $\dfrac{37}{60}$ **30.** solution **31.** not a solution **32.** $2-\dfrac{1}{2}x=\dfrac{1}{4}x$ **33.** $\dfrac{3}{5}(x+6)$ **34.** 152 beats per minute

35.

36.

37. 0.625 **38.** $0.\overline{27}$ **39. a.** $\sqrt{81}$ **b.** $0,\sqrt{81}$
c. $-17,0,\sqrt{81}$ **d.** $-17,-\dfrac{9}{13},0,0.75,\sqrt{81}$
e. $\sqrt{2},\pi$ **f.** $-17,-\dfrac{9}{13},0,0.75,\sqrt{2},\pi,\sqrt{81}$

40. Answers will vary; -2 is an example. **41.** Answers will vary; $\dfrac{1}{2}$ is an example. **42.** Answers will vary; π is an example. **43.** $<$

44. $>$ **45.** $>$ **46.** $<$ **47.** false **48.** true **49.** 58 **50.** 2.75 **51.** $13y+7$ **52.** $(x+7)9$ **53.** $(6+4)+y=10+y$
54. $(7\cdot10)x=70x$ **55.** $24x-12+30y$ **56.** $7a+2$ **57.** $28x+19$

58. 2

59. -3 **60.** $-\dfrac{11}{20}$ **61.** -7 **62.** $-4x+5y$ or $5y-4x$ **63.** $-10y+40$ or $40-10y$
64. 800 ft below sea level **65.** 23 ft **66.** $9+(-13)$ **67.** 4

68. $-\dfrac{6}{5}$ **69.** -1.5 **70.** -7 **71.** -3 **72.** $-5-8a$ **73.** $27{,}150$ ft   **74.** 84 **75.** $-\dfrac{3}{11}$ **76.** -120 **77.** -9 **78.** undefined

79. 2 **80.** $3x$ **81.** $-x-1$ **82.** not a solution   **83.** solution   **84.** 81%; overestimates by 4% **85.** 36 **86.** -36 **87.** -32

88. $6x^3$ **89.** cannot be simplified   **90.** -16 **91.** -16 **92.** 10 **93.** -2 **94.** 17 **95.** -88 **96.** 14

97. $-\dfrac{20}{3}$ **98.** $-\dfrac{2}{5}$ **99.** 6 **100.** 10 **101.** $28a-20$ **102.** $6y-12$ **103. a.** underestimates by 1% **b.** 35.8%

Chapter Test

1. 4 **2.** -11 **3.** -51 **4.** $\dfrac{1}{5}$ **5.** $-\dfrac{35}{6}$ or $-5\dfrac{5}{6}$ **6.** -5 **7.** 1 **8.** -4 **9.** -32 **10.** 1 **11.** $4x+4$ **12.** $13x-19y$

13. $10-6x$ **14.** $-7, -\dfrac{4}{5}, 0, 0.25, \sqrt{4}, \dfrac{22}{7}$ **15.** $>$ **16.** 12.8 **17.** -15 **18.** 150 **19.** $2(3+x)$ **20.** $(-6\cdot4)x=-24x$

21. $35x-7+14y$ **22.** $17{,}030$ ft   **23.** not a solution   **24.** solution   **25.** $\dfrac{1}{4}x-5=32$ **26.** $5(x+4)-7$

27. underestimates by 1 million   **28.** 144 beats per minute, although answers may vary by ±1 **29.** 144 beats per minute; very well; It is the same.

CHAPTER 2

Section 2.1 Check Point Exercises

1. 17 or $\{17\}$ **2.** 2.29 or $\{2.29\}$ **3.** $\dfrac{1}{4}$ or $\left\{\dfrac{1}{4}\right\}$ **4.** 13 or $\{13\}$ **5.** 12 or $\{12\}$ **6.** 11 or $\{11\}$ **7.** 2100 words

Concept and Vocabulary Check

1. solving   **2.** linear   **3.** equivalent   **4.** $b+c$ **5.** subtract; solution   **6.** adding 7   **7.** subtracting $6x$

Exercise Set 2.1

1. linear   **3.** not linear   **5.** not linear   **7.** linear   **9.** not linear   **11.** 23 or $\{23\}$ **13.** -20 or $\{-20\}$ **15.** -16 or $\{-16\}$

17. -12 or $\{-12\}$ **19.** 4 or $\{4\}$ **21.** -11 or $\{-11\}$ **23.** 2 or $\{2\}$ **25.** $-\dfrac{17}{12}$ or $\left\{-\dfrac{17}{12}\right\}$ **27.** $\dfrac{21}{4}$ or $\left\{\dfrac{21}{4}\right\}$ **29.** $-\dfrac{11}{20}$ or $\left\{-\dfrac{11}{20}\right\}$

31. 4.3 or $\{4.3\}$ **33.** $-\dfrac{21}{4}$ or $\left\{-\dfrac{21}{4}\right\}$ **35.** 18 or $\{18\}$ **37.** $\dfrac{9}{10}$ or $\left\{\dfrac{9}{10}\right\}$ **39.** -310 or $\{-310\}$ **41.** 4.3 or $\{4.3\}$ **43.** 0 or $\{0\}$

45. 11 or $\{11\}$ **47.** 5 or $\{5\}$ **49.** -13 or $\{-13\}$ **51.** 6 or $\{6\}$ **53.** -12 or $\{-12\}$ **55.** $x=\triangle+\square$ **57.** $\triangle-\square=x$

59. $x-12=-2; 10$ **61.** $\dfrac{2}{5}x-8=\dfrac{7}{5}x; -8$ **63.** \$1700   **65.** \$10,384; underestimates by \$307   **67. a.** 52 ± 1 **b.** 51.4; very well

73. does not make sense   **75.** makes sense   **77.** false   **79.** true   **81.** Answers will vary; example: $x-100=-101$.

83. 2.7529 or $\{2.7529\}$ **84.** $\dfrac{9}{x}-4x$ **85.** -12 **86.** $6-9x$ **87.** x **88.** y **89.** yes

Section 2.2 Check Point Exercises

1. 36 or $\{36\}$ **2. a.** 21 or $\{21\}$ **b.** -4 or $\{-4\}$ **c.** -3.1 or $\{-3.1\}$ **3. a.** 24 or $\{24\}$ **b.** -16 or $\{-16\}$ **4. a.** -5 or $\{-5\}$
b. 3 or $\{3\}$ **5.** 6 or $\{6\}$ **6.** -10 or $\{-10\}$ **7.** 6 or $\{6\}$ **8. a.** overestimates by \$7   **b.** 2017

Concept and Vocabulary Check

1. bc **2.** divide   **3.** multiplying; 7   **4.** dividing/multiplying; $-8/-\dfrac{1}{8}$ **5.** multiplying; $\dfrac{5}{3}$ **6.** multiplying/dividing; -1
7. subtracting 2; dividing; 5

Exercise Set 2.2

1. 30 or $\{30\}$ **3.** -33 or $\{-33\}$ **5.** 7 or $\{7\}$ **7.** -9 or $\{-9\}$ **9.** $-\dfrac{7}{2}$ or $\left\{-\dfrac{7}{2}\right\}$ **11.** 6 or $\{6\}$ **13.** $-\dfrac{3}{4}$ or $\left\{-\dfrac{3}{4}\right\}$ **15.** 0 or $\{0\}$

17. 18 or $\{18\}$ **19.** -8 or $\{-8\}$ **21.** -17 or $\{-17\}$ **23.** 47 or $\{47\}$ **25.** 45 or $\{45\}$ **27.** -5 or $\{-5\}$ **29.** 5 or $\{5\}$ **31.** 6 or $\{6\}$

33. -1 or $\{-1\}$ **35.** -2 or $\{-2\}$ **37.** $\dfrac{9}{4}$ or $\left\{\dfrac{9}{4}\right\}$ **39.** -6 or $\{-6\}$ **41.** -3 or $\{-3\}$ **43.** -3 or $\{-3\}$ **45.** 4 or $\{4\}$ **47.** $-\dfrac{3}{2}$ or $\left\{-\dfrac{3}{2}\right\}$

49. 2 or $\{2\}$ **51.** -4 or $\{-4\}$ **53.** -6 or $\{-6\}$ **55.** $x=\square\cdot\triangle$ **57.** $-\triangle=x$ **59.** $6x=10; \dfrac{5}{3}$ **61.** $\dfrac{x}{-9}=5; -45$

63. $4x-8=56; 16$ **65.** $-3x+15=-6; 7$ **67.** 10 sec   **69.** 1502.2 mph   **71. a.** underestimates by \$40   **b.** 2029

77. does not make sense   **79.** does not make sense   **81.** false   **83.** true   **85.** Answers will vary; example: $\dfrac{5}{4}x=-20$.

87. 6.5 or $\{6.5\}$ **88.** 100   **89.** -100 **90.** 3   **91.** $-3x+7$ **92.** yes   **93.** $2x-78$

Section 2.3 Check Point Exercises

1. 6 or {6} **2.** 2 or {2} **3.** 5 or {5} **4.** −2 or {−2} **5.** −15 or {−15} **6.** no solution or ∅ **7.** all real numbers or {$x|x$ is a real number} **8.** 3.7; shown as the point whose corresponding value on the vertical axis is 10 and whose value on the horizontal axis is 3.7

Concept and Vocabulary Check

1. simplifying each side; combining like terms **2.** 30 **3.** 100 **4.** inconsistent **5.** identity **6.** inconsistent **7.** identity

Exercise Set 2.3

1. 3 or {3} **3.** −1 or {−1} **5.** 4 or {4} **7.** 4 or {4} **9.** $\frac{7}{2}$ or $\left\{\frac{7}{2}\right\}$ **11.** −3 or {−3} **13.** 6 or {6} **15.** 8 or {8} **17.** 4 or {4}

19. 1 or {1} **21.** −4 or {−4} **23.** 5 or {5} **25.** 6 or {6} **27.** 1 or {1} **29.** −57 or {−57} **31.** −10 or {−10} **33.** 18 or {18}

35. $\frac{7}{4}$ or $\left\{\frac{7}{4}\right\}$ **37.** 1 or {1} **39.** 24 or {24} **41.** −6 or {−6} **43.** 20 or {20} **45.** −7 or {−7} **47.** 9 or {9} **49.** 30 or {30}

51. 25 or {25} **53.** 20 or {20} **55.** −1 or {−1} **57.** 7700 or {7700} **59.** no solution or ∅ **61.** all real numbers or {$x|x$ is a real number}

63. $\frac{2}{3}$ or $\left\{\frac{2}{3}\right\}$ **65.** all real numbers or {$x|x$ is a real number} **67.** no solution or ∅ **69.** 0 or {0} **71.** no solution or ∅ **73.** 0 or {0}

75. $\frac{4}{3}$ or $\left\{\frac{4}{3}\right\}$ **77.** all real numbers or {$x|x$ is a real number} **79.** $x = \square\$ - \square\triangle$ **81.** 240 **83.** $\frac{x}{5} + \frac{x}{3} = 16$; 30

85. $\frac{3x}{4} - 3 = \frac{x}{2}$; 12 **87.** 85 mph **89.** 142 lb; 13 lb **91.** 409.2 ft **99.** makes sense **101.** does not make sense **103.** false **105.** false

107. 3 or {3} **109.** < **110.** < **111.** −10 **112. a.** $T - D = pm$ **b.** $\frac{T - D}{p} = m$ **113.** 16 or {16} **114.** 0.05 or {0.05}

Section 2.4 Check Point Exercises

1. $l = \frac{A}{w}$ **2.** $l = \frac{P - 2w}{2}$ **3.** $m = \frac{T - D}{p}$ **4.** $x = 15 + 12y$ **5.** 4.5 **6.** 15 **7.** 36% **8.** 35% **9. a.** $1152 **b.** 4% decrease

Concept and Vocabulary Check

1. isolated on one side **2.** $A = lw$ **3.** $P = 2l + 2w$ **4.** $A = PB$ **5.** subtract b; divide by m

Exercise Set 2.4

1. $r = \frac{d}{t}$ **3.** $P = \frac{I}{rt}$ **5.** $r = \frac{C}{2\pi}$ **7.** $m = \frac{E}{c^2}$ **9.** $m = \frac{y - b}{x}$ **11.** $D = T - pm$ **13.** $b = \frac{2A}{h}$ **15.** $n = 5M$ **17.** $c = 4F - 160$

19. $a = 2A - b$ **21.** $r = \frac{S - P}{Pt}$ **23.** $b = \frac{2A}{h} - a$ **25.** $x = \frac{C - By}{A}$ **27.** 6 **29.** 7.2 **31.** 5 **33.** 170 **35.** 20%

37. 12% **39.** 60% **41.** 75% **43.** $x = \frac{y}{a + b}$ **45.** $x = \frac{y - 5}{a - b}$ **47.** $x = \frac{y}{c + d}$ **49.** $x = \frac{y + C}{A - B}$ **51. a.** $z = 3A - x - y$ **b.** 96%

53. a. $t = \frac{d}{r}$ **b.** 2.5 hr **55.** 522 **57.** 500 billion pounds **59. a.** 27% **b.** 46% **c.** 117% **61.** 12.5% **63.** $9

65. a. $1008 **b.** $17,808 **67. a.** $103.20 **b.** $756.80 **69.** 15% **71.** no; 2% loss **75.** makes sense **77.** does not make sense

79. false **81.** false **83.** $C = \frac{100M}{Q}$ **84.** 12 or {12} **85.** 20 or {20} **86.** $0.7x$ **87.** $\frac{13}{x} - 7x$ **88.** $8(x + 14)$ **89.** $9(x - 5)$

Mid-Chapter Check Point Exercises

1. 16 or {16} **2.** −3 or {−3} **3.** $C = \frac{825H}{E}$ **4.** 8.4 **5.** 30 or {30} **6.** $-\frac{8}{9}$ or $\left\{-\frac{8}{9}\right\}$ **7.** $r = \frac{S}{2\pi h}$ **8.** 40

9. 20 or {20} **10.** 5.25 or {5.25} **11.** −1 or {−1} **12.** $x = \frac{By + C}{A}$ **13.** no solution or ∅ **14.** 40 or {40} **15.** 12.5% **16.** $-\frac{6}{5}$ or $\left\{-\frac{6}{5}\right\}$

17. 25% **18.** all real numbers or {$x|x$ is a real number} **19. a.** underestimates by 2% **b.** 24

Section 2.5 Check Point Exercises

1. 12 **2.** English: $38 thousand; computer science; $56 thousand **3.** pages 72 and 73 **4.** 32; 4 mi **5.** 40 ft wide and 120 ft long **6.** $940

Concept and Vocabulary Check

1. $4x - 6$ **2.** $x + 215$ **3.** $x + 1$ **4.** $125 + 0.15x$ **5.** $2x + 2 \cdot 4x$ or $2 \cdot 4x + 2x$ or $10x$ **6.** $x - 0.35x$ or $0.65x$

Exercise Set 2.5

1. $x + 60 = 410; 350$ **3.** $x - 23 = 214; 237$ **5.** $7x = 126; 18$ **7.** $\frac{x}{19} = 5; 95$ **9.** $4 + 2x = 56; 26$ **11.** $5x - 7 = 178; 37$

13. $x + 5 = 2x; 5$ **15.** $2(x + 4) = 36; 14$ **17.** $9x = 3x + 30; 5$ **19.** $\frac{3x}{5} + 4 = 34; 50$ **21.** TV: 9 years; sleeping: 28 years

23. Bachelor's: \$55 thousand; Master's: \$61 thousand   **25.** pages 314 and 315   **27.** Ruth: 59; Maris: 61   **29.** 800 mi
31. 10 years after 2008; 2018   **33.** 50 yd wide and 200 yd long   **35.** 160 ft wide and 360 ft long   **37.** 12 ft long and 4 ft high   **39.** \$400
41. \$46,500   **43.** \$22,500   **45.** 11 hr   **51.** does not make sense   **53.** makes sense   **55.** false   **57.** true   **59.** 5 ft 7 in.

61. The uncle is 60 years old, and the woman is 20 years old.   **63.** -20 or $\{20\}$ **64.** 0 or $\{0\}$ **65.** $w = \frac{3V}{lh}$ **66.** 5   **67.** 91   **68.** 64 or $\{64\}$

Section 2.6 Check Point Exercises

1. 12 ft   **2.** 400π ft^2 $\approx$ 1256 ft^2 or 1257 ft^2; 40π ft $\approx$ 126 ft   **3.** large pizza   **4.** 2 times   **5.** no; About 32 more cubic inches are needed.
6. first: 120°; second: 40°; third: 20°   **7.** 60°

Concept and Vocabulary Check

1. $A = \frac{1}{2}bh$ **2.** $A = \pi r^2$ **3.** $C = 2\pi r$ **4.** radius; diameter   **5.** $V = lwh$ **6.** $V = \pi r^2 h$ **7.** 180°   **8.** complementary
9. supplementary   **10.** $90 - x$; $180 - x$

Exercise Set 2.6

1. 18 m; 18 m^2 **3.** 56 in.^{2} **5.** 91 m^2 **7.** 50 ft   **9.** 8 ft   **11.** 50 cm   **13.** 16π cm^2 $\approx$ 50 cm^2; 8π cm $\approx$ 25 cm
15. 36π yd^2 $\approx$ 113 yd^2; 12π yd $\approx$ 38 yd   **17.** 7 in.; 14 in.   **19.** 36 in.^{3} **21.** 150π cm^3 $\approx$ 471 cm^3 **23.** 972π cm^3 $\approx$ 3052 or 3054 cm^3
25. 48π m^3 $\approx$ 151 m^3 **27.** $h = \frac{V}{\pi r^2}$ **29.** 9 times   **31.** $x = 50; x + 30 = 80; 50°, 50°, 80°$ **33.** $4x = 76; 3x + 4 = 61; 2x + 5 = 43; 76°, 61°, 43°$
35. 40°, 80°, 60°   **37.** 32°   **39.** 2°   **41.** 48°   **43.** 90°   **45.** 75°   **47.** 135°   **49.** 50°   **51.** 72 m^2 **53.** 70.5 cm^2 **55.** 448 in.^{3}
57. \$698.18   **59.** large pizza   **61.** \$2261 or \$2262   **63.** approx 19.7 ft   **65.** 21,000 yd^3 **67.** the can with diameter of 6 in. and height of 5 in.
69. Yes, the water tank is a little over one cubic foot too small.   **79.** does not make sense   **81.** does not make sense   **83.** true   **85.** false
87. 2.25 times   **89.** Volume increases 8 times.   **91.** 35°   **92.** $s = \frac{P - b}{2}$ **93.** 8 or $\{8\}$ **94.** 0   **95.** yes   **96.** yes   **97.** 2 or $\{2\}$

Section 2.7 Check Point Exercises

1. a.    **b.**   **c.**

2. a. $[0, \infty)$;   **b.** $(-\infty, 5)$;

3. $(-\infty, 3)$ or $\{x \mid x < 3\}$ **4.** $[-2, \infty)$ or $\{x \mid x \geq -2\}$

5. a. $(-\infty, 8)$ or $\{x \mid x < 8\}$ **b.** $(-3, \infty)$ or $\{x \mid x > -3\}$

6. $[4, \infty)$ or $\{y \mid y \geq 4\}$ **7.** $[1, \infty)$ or $\{x \mid x \geq 1\}$

8. $[1, \infty)$ or $\{x \mid x \geq 1\}$ **9.** no solution or $\varnothing$ **10.** $(-\infty, \infty)$ or $\{x \mid x \text{ is a real number}\}$

11. at least 83%   **12.** at most 43

Concept and Vocabulary Check

1. $(-\infty, 5)$ **2.** $[2, \infty)$ **3.** $< b + c$ **4.** $< bc$ **5.** $> bc$ **6.** subtracting 4; dividing; -3; direction; $>$; $<$
7. $\varnothing$ or the empty set   **8.** $(-\infty, \infty)$

Exercise Set 2.7

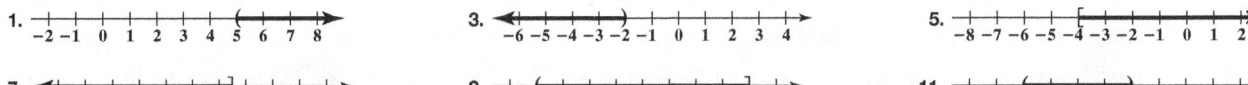

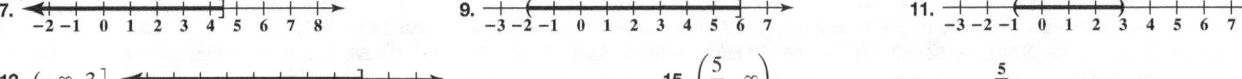

13. $(-\infty, 3]$ **15.** $\left(\frac{5}{2}, \infty\right)$

17. $(-\infty, 0]$ **19.** $(-\infty, 4)$

21. $(7, \infty)$ or $\{x \mid x > 7\}$

23. $(-\infty, 6]$ or $\{x \mid x \le 6\}$

25. $(-\infty, 2)$ or $\{y \mid y < 2\}$

27. $(-\infty, 3]$ or $\{x \mid x \le 3\}$

29. $(-\infty, 16)$ or $\{x \mid x < 16\}$

31. $(4, \infty)$ or $\{x \mid x > 4\}$

33. $\left(\frac{7}{6}, \infty\right)$ or $\left\{x \mid x > \frac{7}{6}\right\}$

35. $\left(-\infty, -\frac{3}{8}\right]$ or $\left\{y \mid y \le -\frac{3}{8}\right\}$

37. $(0, \infty)$ or $\{y \mid y > 0\}$

39. $(-\infty, 8)$ or $\{x \mid x < 8\}$

41. $(-6, \infty)$ or $\{x \mid x > -6\}$

43. $(-\infty, 5)$ or $\{x \mid x < 5\}$

45. $[-7, \infty)$ or $\{x \mid x \ge -7\}$

47. $(-5, \infty)$ or $\{x \mid x > -5\}$

49. $(-\infty, -5]$ or $\{x \mid x \le -5\}$

51. $(-\infty, 3)$ or $\{x \mid x < 3\}$

53. $\left[-\frac{1}{8}, \infty\right)$ or $\left\{y \mid y \ge -\frac{1}{8}\right\}$

55. $(-4, \infty)$ or $\{x \mid x > -4\}$

57. $(5, \infty)$ or $\{x \mid x > 5\}$

59. $(-\infty, 5)$ or $\{x \mid x < 5\}$

61. $[-2, \infty)$ or $\{x \mid x \ge -2\}$

63. $(-3, \infty)$ or $\{x \mid x > -3\}$

65. $[4, \infty)$ or $\{x \mid x \ge 4\}$

67. $\left(\frac{11}{3}, \infty\right)$ or $\left\{x \mid x > \frac{11}{3}\right\}$

69. $(2, \infty)$ or $\{y \mid y > 2\}$

71. $(-\infty, 2)$ or $\{y \mid y < 2\}$

73. $(-\infty, 3)$ or $\{x \mid x < 3\}$

75. $\left(\frac{5}{3}, \infty\right)$ or $\left\{x \mid x > \frac{5}{3}\right\}$

77. $[9, \infty)$ or $\{x \mid x \ge 9\}$

79. $(-\infty, -6)$ or $\{x \mid x < -6\}$

81. no solution or $\varnothing$ **83.** $(-\infty, \infty)$ or $\{x \mid x$ is a real number$\}$ **85.** no solution or $\varnothing$ **87.** $(-\infty, \infty)$ or $\{x \mid x$ is a real number$\}$
89. $(-\infty, 0]$ or $\{x \mid x \le 0\}$ **91.** $x > \dfrac{b - a}{3}$ **93.** $\dfrac{y - b}{m} \ge x$ **95.** x is between -2 and 2; $|x| < 2$
97. x is greater than 2 or less than -2; $|x| > 2$ **99.** weird, cemetery, accommodation **101.** supersede, inoculate **103.** harass
105. 16 years; from 2016 onward **107. a.** at least 96 **b.** if you get less than 66 on the final **109.** up to 1280 mi **111.** up to 29 bags of cement
117. makes sense **119.** makes sense **121.** false **123.** false **125.** more than 720 mi **127.** $(-\infty, 0.4)$ or $\{x \mid x < 0.4\}$ **129.** 20
130. length: 11 in.; width: 6 in. **131.** 4 or $\{4\}$ **132.** yes **133.** no **134.** -3

Review Exercises

1. 32 or $\{32\}$ **2.** -22 or $\{-22\}$ **3.** 12 or $\{12\}$ **4.** -22 or $\{-22\}$ **5.** 5 or $\{5\}$ **6.** 80 or $\{80\}$ **7.** -56 or $\{-56\}$ **8.** 11 or $\{11\}$
9. 4 or $\{4\}$ **10.** -15 or $\{-15\}$ **11.** -12 or $\{-12\}$ **12.** -25 or $\{-25\}$ **13.** 10 or $\{10\}$ **14.** 6 or $\{6\}$ **15.** -5 or $\{-5\}$ **16.** -10 or $\{-10\}$
17. 2 or $\{2\}$ **18.** 1 or $\{1\}$ **19. a.** overestimates by 1% **b.** 2017 **20.** -1 or $\{-1\}$ **21.** 12 or $\{12\}$ **22.** -13 or $\{-13\}$ **23.** -3 or $\{-3\}$
24. -10 or $\{-10\}$ **25.** 2 or $\{2\}$ **26.** 2 or $\{2\}$ **27.** 9 or $\{9\}$ **28.** 4 or $\{4\}$ **29.** no solution or $\varnothing$ **30.** all real numbers or
$\{x \mid x$ is a real number$\}$ **31.** 30 years old **32.** $r = \dfrac{I}{P}$ **33.** $h = \dfrac{3V}{B}$ **34.** $w = \dfrac{P - 2l}{2}$ **35.** $B = 2A - C$ **36.** $m = \dfrac{T - D}{p}$ **37.** 9.6
38. 200 **39.** 48% **40.** 100% **41.** 40% **42.** 12.5% **43.** no; 1% **44. a.** $h = 7r$ **b.** 5 ft 3 in. **45.** 350 gallons **46.** 10
47. Gates: $54 billion; Buffett: $45 billion **48.** pages 46 and 47 **49.** females: 49%; males: 51% **50.** 14 yr; 2015 **51.** 18 checks
52. length: 150 yd; width: 50 yd **53.** $240 **54.** 32.5 ft^2 **55.** 50 cm^2 **56.** 135 yd^2 **57.** 7608 m^2 **58.** 20π m $\approx$ 63 m; 100π m^2 $\approx$ 314 m^2
59. 6 ft **60.** 156 ft^2 **61.** $1890 **62.** medium pizza **63.** 60 cm^3 **64.** 128π yd^3 $\approx$ 402 yd^3 **65.** 288π m^3 $\approx$ 905 m^3 **66.** 4800 m^3
67. 16 fish **68.** $x = 30, 3x = 90, 2x = 60$; $30°, 60°, 90°$ **69.** $85°, 35°, 60°$ **70.** $33°$ **71.** $105°$ **72.** $57.5°$ **73.** $45°$ and $135°$

74.

75.

76. $\left[\frac{3}{2}, \infty\right)$

77. $(-\infty, 0)$

78. $(-\infty, 4)$ or $\{x \mid x < 4\}$ 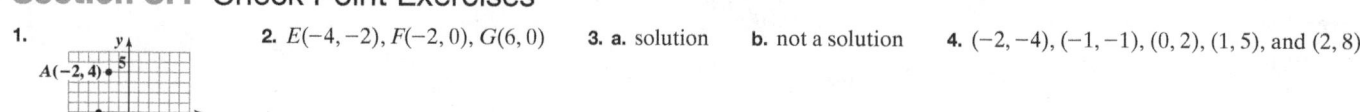   **79.** $(-8, \infty)$ or $\{x \mid x > -8\}$

80. $[-3, \infty)$ or $\{x \mid x \geq -3\}$ **81.** $(6, \infty)$ or $\{x \mid x > 6\}$

82. $[4, \infty)$ or $\{x \mid x \geq 4\}$ **83.** $(-\infty, 2]$ or $\{x \mid x \leq 2\}$

84. $(-\infty, \infty)$ or $\{x \mid x \text{ is a real number}\}$ **85.** no solution or $\varnothing$ **86.** at least 64   **87.** at most 30

Chapter Test

1. $\frac{9}{2}$ or $\left\{\frac{9}{2}\right\}$ **2.** -5 or $\{-5\}$ **3.** $-\frac{4}{3}$ or $\left\{-\frac{4}{3}\right\}$ **4.** 2 or $\{2\}$ **5.** -20 or $\{-20\}$ **6.** $-\frac{5}{3}$ or $\left\{-\frac{5}{3}\right\}$ **7.** -5 or $\{-5\}$ **8.** 60 yr; 2020   **9.** $h = \dfrac{V}{\pi r^2}$

10. $w = \dfrac{P - 2l}{2}$ **11.** 8.4   **12.** 150   **13.** 5%   **14.** 63   **15.** Americans: 13 days; Italians: 42 days

16. 600 min   **17.** length: 150 yd; width: 75 yd   **18.** $35   **19.** 517 m^2 **20.** 525 in.^{2} **21.** 66 ft^2 **22.** 18 in.^{3} **23.** 175π cm$^3 = 550$ cm^3
24. $650   **25.** 14 ft   **26.** 126°, 42°, 12°   **27.** 53°

28. $(-2, \infty)$ **29.** $(-\infty, 3]$

30. $(-\infty, -6)$ or $\{x \mid x < -6\}$ **31.** $(-\infty, -3]$ or $\{x \mid x \leq -3\}$

32. $(7, \infty)$ or $\{x \mid x > 7\}$ **33.** at least 92   **34.** widths greater than 8 in.

Cumulative Review Exercises

1. -4 **2.** -2 **3.** -128 **4.** $-103 - 20x$ **5.** $-4, -\frac{1}{3}, 0, \sqrt{4}, 1063$ **6.** $\dfrac{5}{x} - (x + 2)$ **7.** $<$ **8.** $24x - 6 - 30y$

9. overestimates by 1%   **10.** 2020   **11.** 1 or $\{1\}$ **12.** -15 or $\{-15\}$ **13.** $A = \dfrac{3V}{h}$ **14.** 160   **15.** length: 130 yd; width: 70 yd   **16.** 75,000 gal

17. $\left(-\infty, \dfrac{1}{2}\right]$ **18.** $(-\infty, -3)$ or $\{x \mid x < -3\}$

19. $[-6, \infty)$ or $\{x \mid x \geq -6\}$ **20.** more than $47,500

CHAPTER 3

Section 3.1 Check Point Exercises

1. 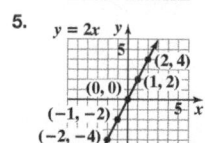   **2.** $E(-4, -2), F(-2, 0), G(6, 0)$ **3. a.** solution   **b.** not a solution   **4.** $(-2, -4), (-1, -1), (0, 2), (1, 5),$ and $(2, 8)$

5. 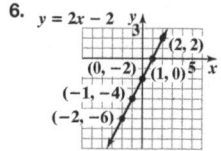   **6.** 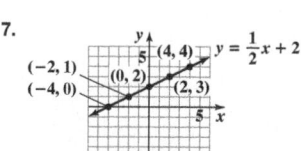   **7.**

8. a. $(0, 1), (5, 8), (10, 15),$ and $(15, 22)$ **b.** 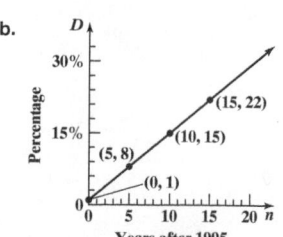   **c.** approximately 29%, although answers may vary by $\pm 1\%$ **d.** 29%

Concept and Vocabulary Check

1. x-axis **2.** y-axis **3.** origin **4.** quadrants; four **5.** x-coordinate; y-coordinate **6.** solution; satisfies **7.** a/one **8.** $mx + b$

Exercise Set 3.1

1. I **3.** II **5.** III **7.** IV

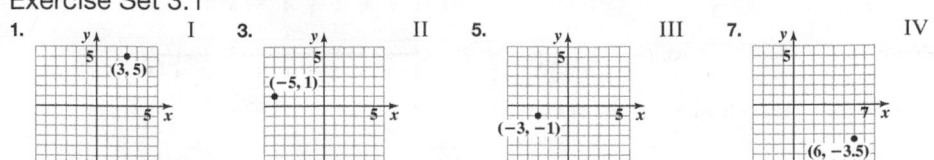

9-23.

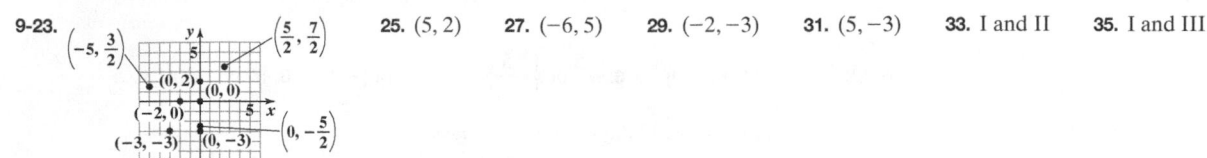

$\left(-5, \frac{3}{2}\right)$ $\left(\frac{5}{2}, \frac{7}{2}\right)$ **25.** $(5, 2)$ **27.** $(-6, 5)$ **29.** $(-2, -3)$ **31.** $(5, -3)$ **33.** I and II **35.** I and III

37. $(2, 3)$ and $(3, 2)$ are not solutions; $(-4, -12)$ is a solution. **39.** $(-5, -20)$ is not a solution; $(0, 0)$ and $(9, -36)$ are solutions.
41. $(2, -2)$ is not a solution; $(0, 6)$ and $(-3, 0)$ are solutions. **43.** $(0, 5)$ is not a solution; $(-5, 6)$ and $(10, -3)$ are solutions.
45. $\left(1, \frac{1}{3}\right)$ is not a solution; $(0, 0)$ and $\left(2, -\frac{2}{3}\right)$ are solutions. **47.** $(3, 4)$ and $(0, -4)$ are not solutions; $(4, 7)$ is a solution.

49.

x	(x, y)
-2	$(-2, -24)$
-1	$(-1, -12)$
0	$(0, 0)$
1	$(1, 12)$
2	$(2, 24)$

51.

x	(x, y)
-2	$(-2, 20)$
-1	$(-1, 10)$
0	$(0, 0)$
1	$(1, -10)$
2	$(2, -20)$

53.

x	(x, y)
-2	$(-2, -21)$
-1	$(-1, -13)$
0	$(0, -5)$
1	$(1, 3)$
2	$(2, 11)$

55.

x	(x, y)
-2	$(-2, 13)$
-1	$(-1, 10)$
0	$(0, 7)$
1	$(1, 4)$
2	$(2, 1)$

57.

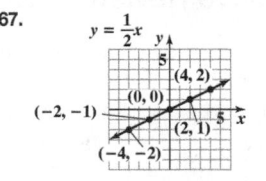

59. **61.** **63.** **65.** **67.**

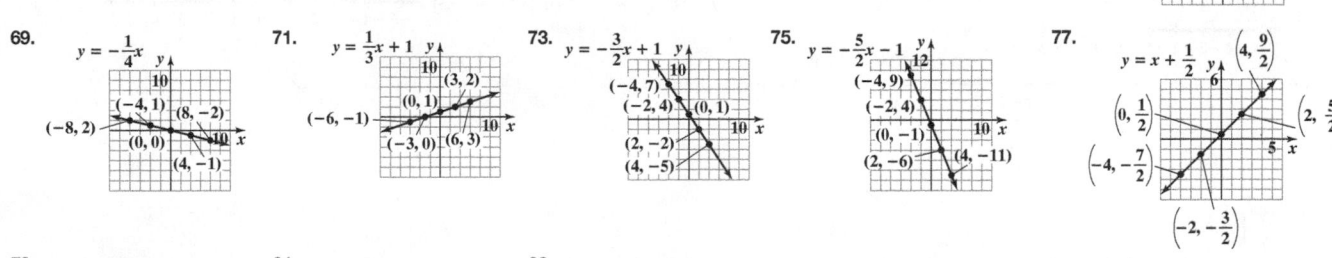

69. **71.** **73.** **75.** **77.**

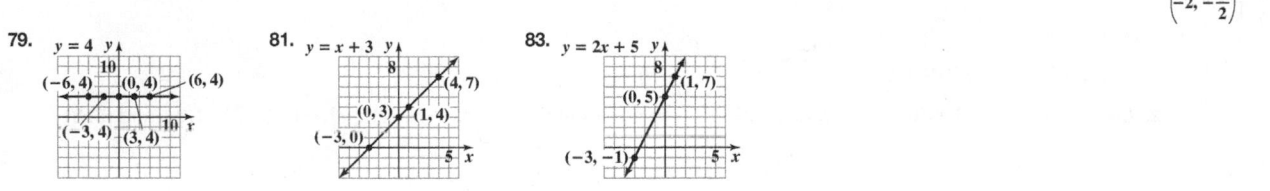

79. **81.** **83.**

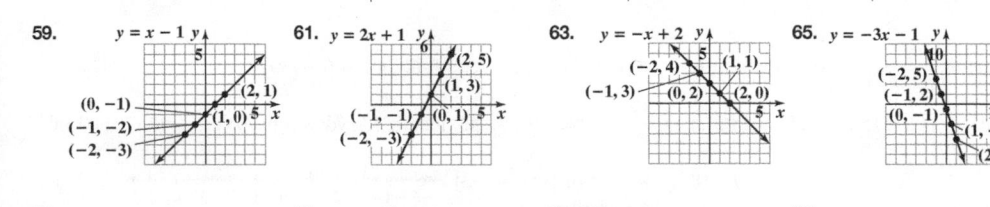

85. a. $8x + 6y = 14.50$ **b.** $1.25 **87.** $(2, 7)$; The football is 7 ft above ground when it is 2 yd from the quarterback. **89.** $(6, 9.25)$

91. 12 ft; 15 yd

93. a. $(0, 31), (5, 43), (10, 55), (15, 67),$ and $(20, 79)$

b.

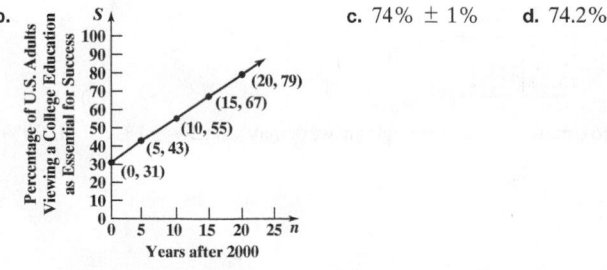

c. $74\% \pm 1\%$ **d.** 74.2%

103. makes sense **105.** does not make sense **107.** false **109.** false

111. a. 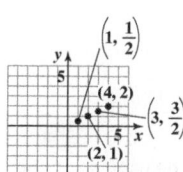 **b.** Change the sign of each x-coordinate. **c.** Change the sign of each y-coordinate.
d. Change the sign of both coordinates.

113. 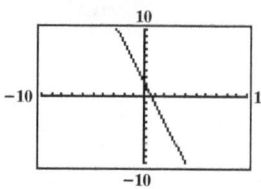 **115.**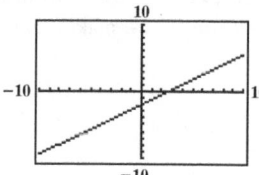

Answers will vary.   Answers will vary.

117. 2 or {2}   **118.** 1   **119.** $h = \dfrac{3V}{A}$ **120.** $(8, 0)$ **121.** $(0, -6)$ **122.** $(0, 0)$

Section 3.2 Check Point Exercises

1. a. x-intercept: -3; y-intercept: 5   **b.** y-intercept: 4; no x-intercept   **c.** x-intercept: 0; y-intercept: 0   **2.** 3   **3.** -4

4. 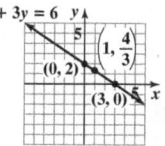   **5.** 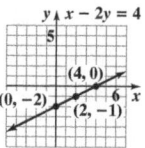   **6.** 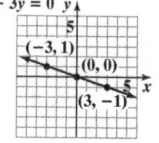   **7.** 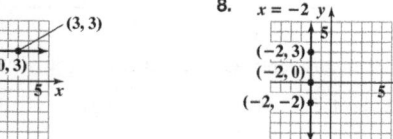   **8.**

Concept and Vocabulary Check

1. x-intercept   **2.** y-intercept   **3.** x-intercept   **4.** y-intercept   **5.** standard   **6.** y; x **7.** x; y **8.** horizontal   **9.** vertical

Exercise Set 3.2

1. a. 3   **b.** 4   **3. a.** -4 **b.** -2 **5. a.** 0   **b.** 0   **7. a.** no x-intercept   **b.** -2 **9.** x-intercept: 10; y-intercept: 4

11. x-intercept: $\dfrac{15}{2}$, or $7\dfrac{1}{2}$; y-intercept: -5 **13.** x-intercept: 8; y-intercept: $-\dfrac{8}{3}$, or $-2\dfrac{2}{3}$ **15.** x-intercept: 0; y-intercept: 0

17. x-intercept: $-\dfrac{11}{2}$, or $-5\dfrac{1}{2}$; y-intercept: $\dfrac{11}{3}$, or $3\dfrac{2}{3}$

19. 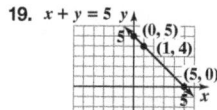   **21.** 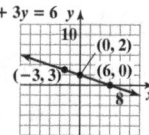   **23.** 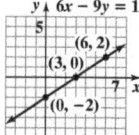   **25.** 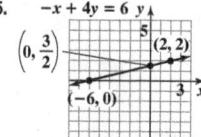   **27.**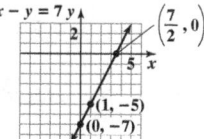

29. 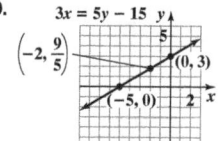   **31.** 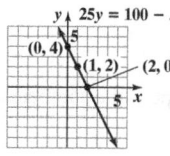   **33.** 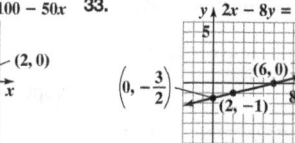   **35.** 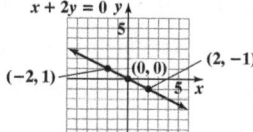   **37.**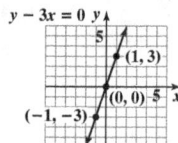

39. 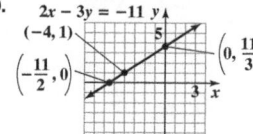   **41.** $y = 3$ **43.** $x = -3$ **45.** $y = 0$ **47.** 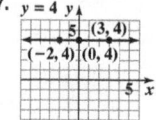   **49.**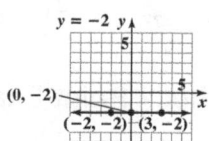

51. 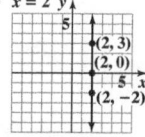   **53.** 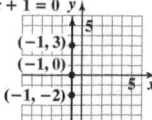   **55.** 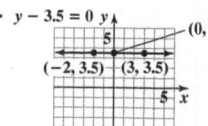   **57.** 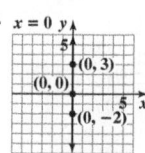   **59.**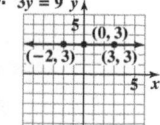

61.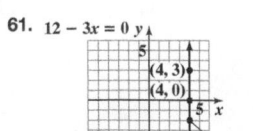
63. Exercise 4   **65.** Exercise 7   **67.** Exercise 1   **69.**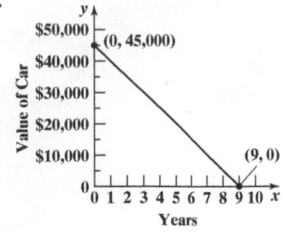

71. from 3 to 12 sec   **73.** 45; The eagle was 45 m above the ground when the observation started.   **75.** 12, 13, 14, 15, 16; The eagle is on the ground at this time.   **77. a.** 9; After 9 years, the car is worth nothing.   **b.** 45,000; The new car is worth $45,000.

c. 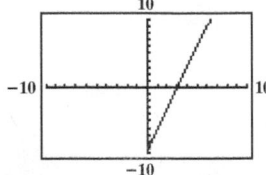   **d.** $20,000; Estimates will vary.   **87.** makes sense   **89.** makes sense   **91.** 2; 5

95. 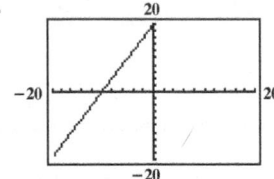   **97.**

x-intercept: 3; *y*-intercept: -9 *x*-intercept: -10; *y*-intercept: 20

98. 13.4   **99.** $4x + 5$ **100.** $[-5, \infty)$ or $\{x|x \geq -5\}$ **101.** 2   **102.** -1 **103.** 0

Section 3.3 Check Point Exercises

1. a. 6   **b.** $-\dfrac{7}{5}$ **2. a.** 0   **b.** undefined   **3.** Both slopes equal 2, so the lines are parallel.   **4.** Product of slopes is $-\dfrac{1}{2}(2) = -1$, so the lines are perpendicular.   **5.** $m \approx 0.32$; The number of men living alone increased by 0.32 million per year. The rate of change is 0.32 million men per year.

Concept and Vocabulary Check

1. $\dfrac{y_2 - y_1}{x_2 - x_1}$ **2.** $y; x$ **3.** positive   **4.** negative   **5.** 0   **6.** undefined   **7.** parallel   **8.** perpendicular

Exercise Set 3.3

1. $\dfrac{3}{4}$; rises   **3.** $\dfrac{1}{4}$; rises   **5.** 0; horizontal   **7.** -5; falls   **9.** undefined; vertical   **11.** $\dfrac{1}{2}$ **13.** $-\dfrac{1}{3}$ **15.** $-\dfrac{1}{2}$ **17.** $-\dfrac{4}{3}$
19. 0   **21.** undefined   **23.** parallel   **25.** not parallel   **27.** perpendicular   **29.** not perpendicular   **31.** parallel
33. neither   **35.** perpendicular

37.   **39.** Slopes of corresponding opposite sides are equal: $-\dfrac{2}{5}$ and $\dfrac{4}{3}$.   **41.** 2
43. 4   **45. a.** 0.16   **b.** 0.16; 0.16; minute of brisk walking   **47.** $\dfrac{1}{3}$
49. 8.3%   **57.** does not make sense   **59.** makes sense   **61.** false   **63.** false
65. m_1, m_3, m_2, m_4 **67.** $m = 2$ **69.** $m = -\dfrac{1}{2}$ **71.** The line's slope is the coefficient of x.

72. 12 in. and 24 in.   **73.** -42 **74.** $(-\infty, 4]$ or $\{x|x \leq 4\}$

75. $(1, 1)$ **76.** $(3, -1)$ **77.** $y = -\dfrac{2}{5}x$

Section 3.4 Check Point Exercises

1. a. $5; -3$ **b.** $\dfrac{2}{3}; 4$ **c.** $-7; 6$ **2.** $y = 3x - 2$ **3.**   **4.**   **5. a.** $y = 0.32x + 8$ **b.** 27.2%

Concept and Vocabulary Check

1. $y = mx + b$; slope; y-intercept **2.** $(0, 3)$; 2; 5 **3.** y

Exercise Set 3.4

1. $3; 2$ **3.** $3; -5$ **5.** $-\dfrac{1}{2}; 5$ **7.** $7; 0$ **9.** $0; 10$ **11.** $-1; 4$ **13.** $y = 5x + 7; 5; 7$ **15.** $y = -x + 6; -1; 6$ **17.** $y = -6x; -6; 0$

19. $y = 2x; 2; 0$ **21.** $y = -\dfrac{2}{7}x; -\dfrac{2}{7}; 0$ **23.** $y = -\dfrac{3}{2}x + \dfrac{3}{2}; -\dfrac{3}{2}; \dfrac{3}{2}$ **25.** $y = \dfrac{3}{4}x - 3; \dfrac{3}{4}; -3$

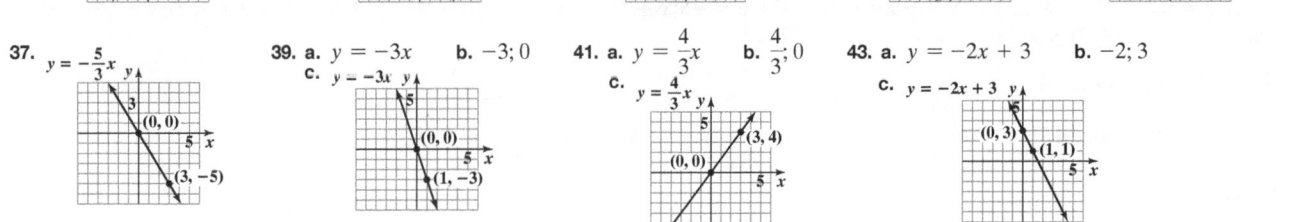

27. $y = 2x + 4$ **29.** $y = -3x + 5$ **31.** $y = \dfrac{1}{2}x + 1$ **33.** $y = \dfrac{2}{3}x - 5$ **35.** $y = -\dfrac{3}{4}x + 2$

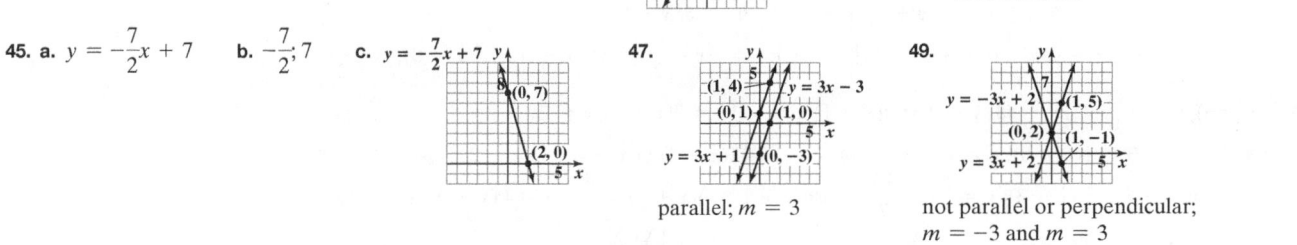

37. $y = -\dfrac{5}{3}x$ **39. a.** $y = -3x$ **b.** $-3; 0$ **41. a.** $y = \dfrac{4}{3}x$ **b.** $\dfrac{4}{3}; 0$ **43. a.** $y = -2x + 3$ **b.** $-2; 3$
c. $y = -3x$ **c.** $y = \dfrac{4}{3}x$ **c.** $y = -2x + 3$

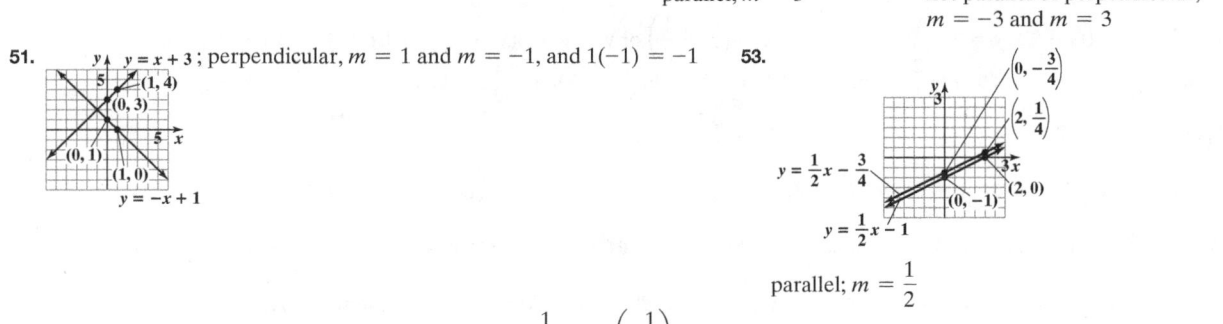

45. a. $y = -\dfrac{7}{2}x + 7$ **b.** $-\dfrac{7}{2}; 7$ **c.** $y = -\dfrac{7}{2}x + 7$ **47.** **49.**

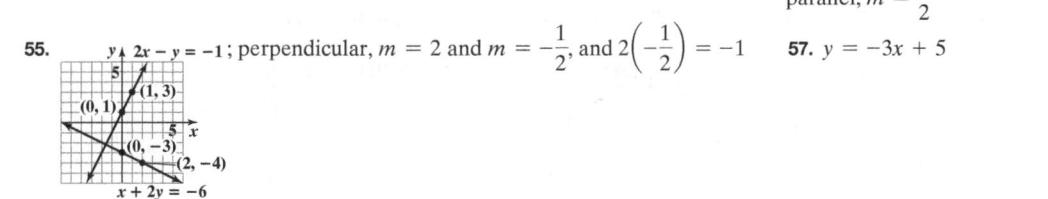

parallel; $m = 3$ not parallel or perpendicular;
                        $m = -3$ and $m = 3$

51. $y = x + 3$; perpendicular, $m = 1$ and $m = -1$, and $1(-1) = -1$ **53.**

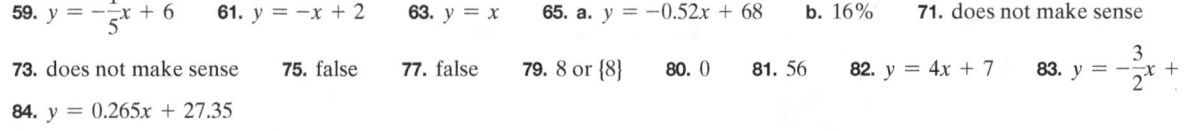

parallel; $m = \dfrac{1}{2}$

55. $2x - y = -1$; perpendicular, $m = 2$ and $m = -\dfrac{1}{2}$, and $2\left(-\dfrac{1}{2}\right) = -1$ **57.** $y = -3x + 5$

59. $y = -\dfrac{1}{5}x + 6$ **61.** $y = -x + 2$ **63.** $y = x$ **65. a.** $y = -0.52x + 68$ **b.** 16% **71.** does not make sense

73. does not make sense **75.** false **77.** false **79.** 8 or $\{8\}$ **80.** 0 **81.** 56 **82.** $y = 4x + 7$ **83.** $y = -\dfrac{3}{2}x + 3$

84. $y = 0.265x + 27.35$

Mid-Chapter Check Point Exercises

1. a. 4 **b.** 2 **c.** $-\dfrac{1}{2}$ **2. a.** -5 **b.** no y-intercept **c.** undefined slope **3. a.** 0 **b.** 0 **c.** $\dfrac{3}{5}$

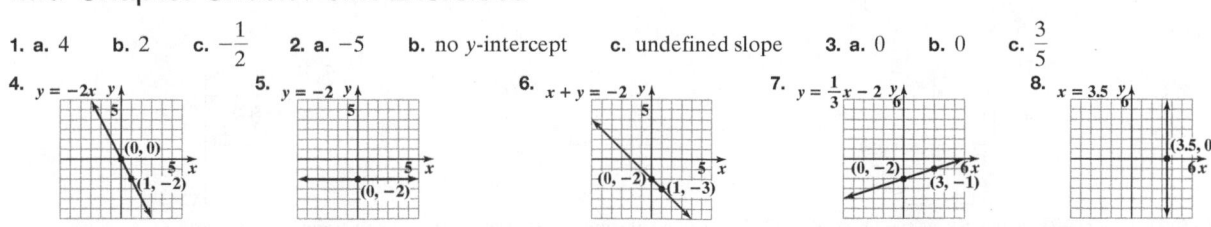

4. $y = -2x$ **5.** $y = -2$ **6.** $x + y = -2$ **7.** $y = \dfrac{1}{3}x - 2$ **8.** $x = 3.5$

9. $4x - 2y = 8$ 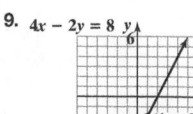 **10.** $y = 3x + 2$ 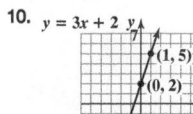 **11.** $3x + y = 0$ 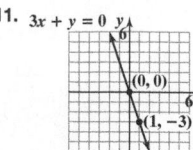 **12.** $y = -x + 4$ 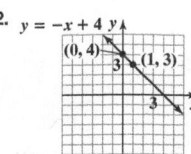 **13.** $y = x - 4$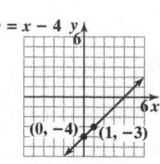

14. $5y = -3x$ 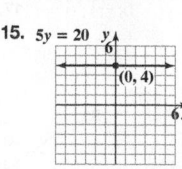 **15.** $5y = 20$ **16.** $\dfrac{5}{2}; -5$ **17.** perpendicular

18. neither **19.** parallel **20. a.** $y = -2.4x + 98$ **b.** 62%

Section 3.5 Check Point Exercises

1. $y + 5 = 6(x - 2); y = 6x - 17$ **2. a.** $y + 1 = -5(x + 2)$ or $y + 6 = -5(x + 1)$ **b.** $y = -5x - 11$ **3.** $y - 5 = 3(x + 2); y = 3x + 11$
4. a. $m = 3$ **b.** $y + 6 = 3(x + 2); y = 3x$ **5.** $y = 0.28x + 27.2; 41.2$

Concept and Vocabulary Check

1. $y - y_1 = m(x - x_1)$ **2.** standard **3.** slope-intercept **4.** point-slope **5.** horizontal **6.** vertical **7.** $-4; -4$ **8.** $\dfrac{1}{2}; -2$

Exercise Set 3.5

1. $y - 5 = 3(x - 2); y = 3x - 1$ **3.** $y - 6 = 5(x + 2); y = 5x + 16$ **5.** $y + 2 = -8(x + 3); y = -8x - 26$
7. $y - 0 = -12(x + 8); y = -12x - 96$ **9.** $y + 2 = -1\left(x + \dfrac{1}{2}\right); y = -x - \dfrac{5}{2}$ **11.** $y - 0 = \dfrac{1}{2}(x - 0); y = \dfrac{1}{2}x$
13. $y + 2 = -\dfrac{2}{3}(x - 6); y = -\dfrac{2}{3}x + 2$ **15.** $y - 2 = 2(x - 1)$ or $y - 10 = 2(x - 5); y = 2x$
17. $y - 0 = 1(x + 3)$ or $y - 3 = 1(x - 0); y = x + 3$ **19.** $y + 1 = 1(x + 3)$ or $y - 4 = 1(x - 2); y = x + 2$
21. $y + 1 = \dfrac{5}{7}(x + 4)$ or $y - 4 = \dfrac{5}{7}(x - 3); y = \dfrac{5}{7}x + \dfrac{13}{7}$ **23.** $y + 1 = 0(x + 3)$ or $y + 1 = 0(x - 4); y = -1$
25. $y - 4 = 1(x - 2)$ or $y - 0 = 1(x + 2); y = x + 2$ **27.** $y - 0 = 8\left(x + \dfrac{1}{2}\right)$ or $y - 4 = 8(x - 0); y = 8x + 4$ **29. a.** 5 **b.** $-\dfrac{1}{5}$
31. a. -7 **b.** $\dfrac{1}{7}$ **33. a.** $\dfrac{1}{2}$ **b.** -2 **35. a.** $-\dfrac{2}{5}$ **b.** $\dfrac{5}{2}$ **37. a.** -4 **b.** $\dfrac{1}{4}$ **39. a.** $-\dfrac{1}{2}$ **b.** 2 **41. a.** $\dfrac{2}{3}$ **b.** $-\dfrac{3}{2}$
43. a. undefined **b.** 0 **45.** $y - 2 = 2(x - 4); y = 2x - 6$ **47.** $y - 4 = -\dfrac{1}{2}(x - 2); y = -\dfrac{1}{2}x + 5$
49. $y + 10 = -4(x + 8); y = -4x - 42$ **51.** $y + 3 = -5(x - 2); y = -5x + 7$ **53.** $y - 2 = \dfrac{2}{3}(x + 2); y = \dfrac{2}{3}x + \dfrac{10}{3}$
55. $y + 7 = -2(x - 4); y = -2x + 1$ **57.** $y = 3x - 2$ **59.** $y = -5x - 20$ **61.** $y = 5$ **63.** $y = -\dfrac{1}{2}x + 1$ **65.** $y = -\dfrac{2}{3}x - 2$
67. $-\dfrac{A}{B}$ **69. a.** $y = 433x + 4138$ **b.** 8468 deaths **73.** does not make sense **75.** makes sense **77.** false **79.** false
81. $E = 2.4M - 20$ **83. b.** 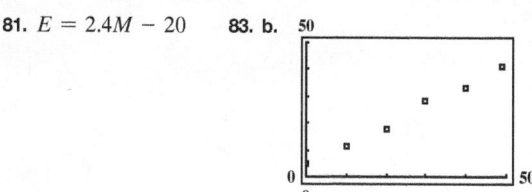 **c.** $a = 0.7417932386; b = 4.245467908; r = 0.9971076103$ **d.**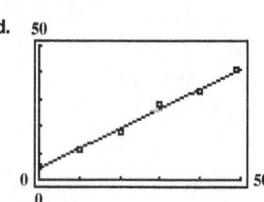

84. at most 12 sheets of paper **85.** $1, \sqrt{4}$ **86.** 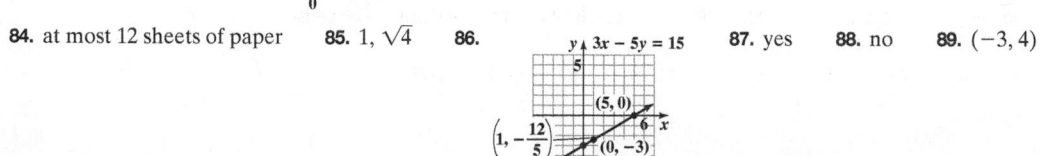 **87.** yes **88.** no **89.** $(-3, 4)$

Review Exercises

1. IV 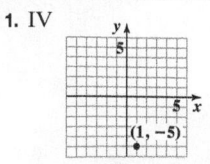 **2.** IV 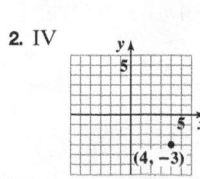 **3.** I 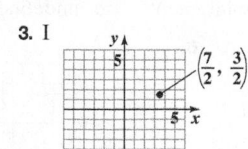 **4.** II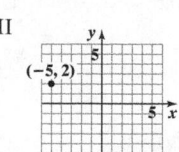

5. $A(5, 6); B(-3, 0); C(-5, 2); D(-4, -2); E(0, -5); F(3, -1)$ **6.** $(-3, 3)$ is not a solution; $(0, 6)$ and $(1, 9)$ are solutions.
7. $(0, 4)$ and $(-1, 15)$ are not solutions; $(4, 0)$ is a solution.

8. a.

x	(x, y)
-2	$(-2, -7)$
-1	$(-1, -5)$
0	$(0, -3)$
1	$(1, -1)$
2	$(2, 1)$

b.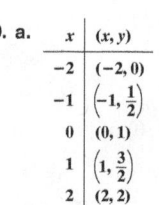

9. a.

x	(x, y)
-2	$(-2, 0)$
-1	$\left(-1, \frac{1}{2}\right)$
0	$(0, 1)$
1	$\left(1, \frac{3}{2}\right)$
2	$(2, 2)$

b.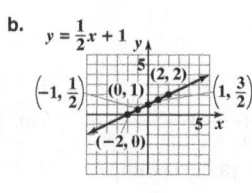

10. a. -2 **b.** -4 **11. a.** no x-intercept   **b.** 2 **12. a.** 0 **b.** 0

13. 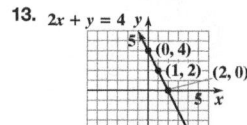   **14.** 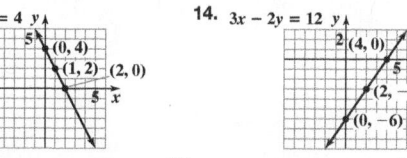   **15.**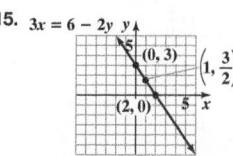

16. 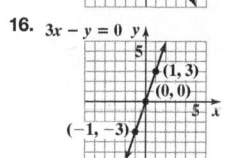   **17.** 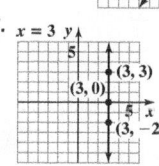   **18.** 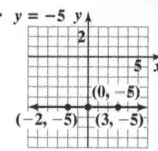   **19.** 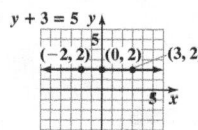   **20.**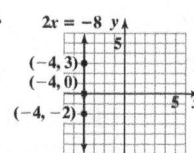

21. a. 5:00 P.M.; $-4°$F.   **b.** 8:00 P.M.; 16°F.   **c.** 4 and 6; At 4:00 P.M. and 6:00 P.M., the temperature was 0°F.   **d.** 12; At noon, the temperature was 12°F.   **e.** The temperature stayed the same, 12°F.   **22.** $-\frac{1}{2}$; falls   **23.** 3; rises   **24.** 0; horizontal   **25.** undefined; vertical

26. $\frac{3}{5}$ **27.** undefined   **28.** $-\frac{1}{3}$ **29.** 0   **30.** neither   **31.** perpendicular   **32.** parallel   **33. a.** -0.48 **b.** $0.48; -0.48\%$; year

34. $5; -7$ **35.** $-4; 6$ **36.** $0; 3$ **37.** $-\frac{2}{3}; 2$

38. 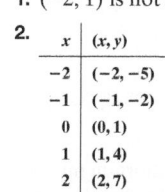   **39.** 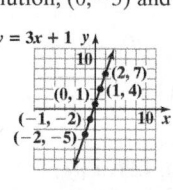   **40.** 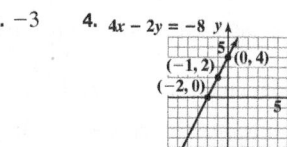   **41.** 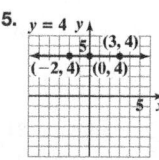   **42.** $y = -\frac{1}{3}x + 2$

43. $y = -\frac{1}{2}x + 4$ 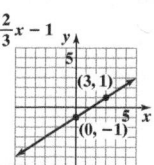   Yes, they are parallel since both have slopes of $-\frac{1}{2}$ and different y-intercepts.

44. a. $y = -1.5x + 30$ **b.** 22.5 million; overestimates by 2.5 million   **45.** $y - 7 = 6(x + 4)$; $y = 6x + 31$
46. $y - 4 = 3(x - 3)$ or $y - 1 = 3(x - 2)$; $y = 3x - 5$ **47.** $y + 7 = -3(x - 4)$; $y = -3x + 5$ **48.** $y - 6 = -3(x + 2)$; $y = -3x$
49. a. $y = 0.08x + 2.1$ **b.** 8.1 billion

Chapter Test

1. $(-2, 1)$ is not a solution; $(0, -5)$ and $(4, 3)$ are solutions.

2.

x	(x, y)
-2	$(-2, -5)$
-1	$(-1, -2)$
0	$(0, 1)$
1	$(1, 4)$
2	$(2, 7)$

3. a. 2   **b.** -3 **4.**   **5.**

6. 3; rises   **7.** undefined; vertical   **8.** $\frac{3}{2}$ **9.** perpendicular   **10.** parallel   **11.** $-1; 10$ **12.** $-2; 6$

13. $y = \frac{2}{3}x - 1$ **14.** $y = -2x + 3$ **15.** $y - 4 = -2(x + 1)$; $y = -2x + 2$
16. $y - 1 = 3(x - 2)$ or $y + 8 = 3(x + 1)$; $y = 3x - 5$
17. $y - 3 = 2(x + 2)$; $y = 2x + 7$
18. $y + 4 = -\frac{1}{2}(x - 6)$; $y = -\frac{1}{2}x - 1$
19. a. -768 **b.** $768; -768$ inmates; year

Cumulative Review Exercises

1. 2   **2.** $4 - 6x$ **3.** $\sqrt{5}$ **4.** 19 or {19}   **5.** $\dfrac{5}{4}$ or $\left\{\dfrac{5}{4}\right\}$ **6.** $x = \dfrac{y - b}{m}$ **7.** 800   **8.** 40 mph

9. $(-\infty, -2]$ or $\{x \mid x \le -2\}$ ⟵|–|–|–|–●|–|–|–|–|–|⟶
−5 −4 −3 −2 −1 0 1 2 3 4 5
10. $(-\infty, 0)$ or $\{x \mid x < 0\}$ ⟵|–|–|–|–|○|–|–|–|–|–⟶
−5 −4 −3 −2 −1 0 1 2 3 4 5

11. 6 hr   **12.** width: 53 m; length: 120 m   **13.** 200 pounds   **14.** 40°, 60°, 80°   **15.** 39   **16.** <

17. $2x - y = 4$ **18.** $x = -5$ **19.** $y = -4x + 3$ **20.** $y = -1$

CHAPTER 4

Section 4.1 Check Point Exercises

1. a. solution   **b.** not a solution   **2.** $(1, 4)$ or $\{(1, 4)\}$ **3.** $(3, 3)$ or $\{(3, 3)\}$ **4.** no solution or $\varnothing$
5. infinitely many solutions; $\{(x, y) \mid x + y = 3\}$ or $\{(x, y) \mid 2x + 2y = 6\}$ **6. a.** $(10, 20)$ or $\{(10, 20)\}$ **b.** If the bridge is used 10 times in a month, the total monthly cost without the discount pass is the same as the monthly cost with the discount pass, namely \$20.

Concept and Vocabulary Check

1. satisfies both equations in the system   **2.** the intersection point   **3.** inconsistent; parallel   **4.** dependent; identical or coincide

Exercise Set 4.1

1. solution   **3.** solution   **5.** not a solution   **7.** solution   **9.** not a solution   **11.** $(4, 2)$ or $\{(4, 2)\}$ **13.** $(-1, 2)$ or $\{(-1, 2)\}$
15. $(3, 0)$ or $\{(3, 0)\}$ **17.** $(1, 0)$ or $\{(1, 0)\}$ **19.** $(-1, 4)$ or $\{(-1, 4)\}$ **21.** $(2, 4)$ or $\{(2, 4)\}$ **23.** $(2, -1)$ or $\{(2, -1)\}$ **25.** no solution or $\varnothing$
27. $(-2, 6)$ or $\{(-2, 6)\}$ **29.** infinitely many solutions; $\{(x, y) \mid x - 2y = 4\}$ or $\{(x, y) \mid 2x - 4y = 8\}$ **31.** $(2, 3)$ or $\{(2, 3)\}$ **33.** no solution or $\varnothing$
35. infinitely many solutions; $\{(x, y) \mid x - y = 0\}$ or $\{(x, y) \mid y = x\}$ **37.** $(2, 4)$ or $\{(2, 4)\}$ **39.** no solution or $\varnothing$ **41.** no solution or $\varnothing$
43. $m = \dfrac{1}{2}$; $b = -3$ and -5; no solution   **45.** $m = -\dfrac{1}{2}$ and 3; $b = 4$; one solution   **47.** $m = 3$; $b = -6$; infinite number of solutions
49. $m = -3$; $b = 0$ and 1; no solution   **51. a.** 40; Both companies charge the same for 40 miles driven.   **b.** Answers will vary: 55.
c. 54; Both companies charge \$54 for 40 miles driven.   **53. a.** $(5, 20)$ or $\{(5, 20)\}$ **b.** Nonmembers and members pay the same amount per month for taking 5 classes, namely \$20.   **63.** does not make sense   **65.** makes sense   **67.** false   **69.** false   **73.** $(1, 4)$ or $\{(1, 4)\}$
75. $(4, 0)$ or $\{(4, 0)\}$ **77.** $(0, -5)$ or $\{(0, -5)\}$ **79.** $(7, 3)$ or $\{(7, 3)\}$ **81.** -12 **82.** 6   **83.** 27   **84.** 3 or {3}   **85.** 4 or {4}
86. no solution or $\varnothing$

Section 4.2 Check Point Exercises

1. $(3, 2)$ or $\{(3, 2)\}$ **2.** $(1, -2)$ or $\{(1, -2)\}$ **3.** no solution or $\varnothing$ **4.** infinitely many solutions; $\{(x, y) \mid y = 3x - 4\}$ or $\{(x, y) \mid 9x - 3y = 12\}$
5. a. $(30, 900)$ or $\{(30, 900)\}$; equilibrium quantity: 30,000; equilibrium price: \$900   **b.** \$900; 30,000; 30,000

Concept and Vocabulary Check

1. $\{(4, 1)\}$ **2.** $\{(-1, 3)\}$ **3.** $\varnothing$ **4.** $\{(x, y) \mid 2x - 6y = 8\}$ or $\{(x, y) \mid x = 3y + 4\}$ **5.** equilibrium

Exercise Set 4.2

1. $(1, 3)$ or $\{(1, 3)\}$ **3.** $(5, 1)$ or $\{(5, 1)\}$ **5.** $(2, 1)$ or $\{(2, 1)\}$ **7.** $(-1, 3)$ or $\{(-1, 3)\}$ **9.** $(4, 5)$ or $\{(4, 5)\}$
11. $\left(-\dfrac{2}{5}, -\dfrac{11}{5}\right)$ or $\left\{\left(-\dfrac{2}{5}, -\dfrac{11}{5}\right)\right\}$ **13.** no solution or $\varnothing$ **15.** infinitely many solutions; $\{(x, y) \mid y = 3x - 5\}$ or $\{(x, y) \mid 21x - 35 = 7y\}$
17. $(0, 0)$ or $\{(0, 0)\}$ **19.** $\left(\dfrac{17}{7}, -\dfrac{8}{7}\right)$ or $\left\{\left(\dfrac{17}{7}, -\dfrac{8}{7}\right)\right\}$ **21.** no solution or $\varnothing$ **23.** $\left(\dfrac{43}{5}, -\dfrac{1}{5}\right)$ or $\left\{\left(\dfrac{43}{5}, -\dfrac{1}{5}\right)\right\}$ **25.** $(200, 700)$ or $\{(200, 700)\}$
27. $(7, 3)$ or $\{(7, 3)\}$ **29.** $(-1, -1)$ or $\{(-1, -1)\}$ **31.** $(5, 4)$ or $\{(5, 4)\}$ **33.** $x + y = 81, y = x + 41$; 20 and 61
35. $x - y = 5, 4x = 6y$; 10 and 15   **37.** $x - y = 1, x + 2y = 7$; 2 and 3   **39.** $(2, 8)$ or $\{(2, 8)\}$ **41. a.** $(4, 4.5)$ or $\{(4, 4.5)\}$; equilibrium number of workers: 4 million; equilibrium hourly wage: \$4.50   **b.** \$4.50; 4; 4   **c.** 2 million   **d.** 5.7 million   **e.** 3.7 million   **49.** does not make sense
51. does not make sense   **53.** true   **55.** false   **57.** $m = \dfrac{5}{2}$

58. $4x + 6y = 12$

59. 12 or {12}   **60.** $-73, 0, \dfrac{3}{1}$
61. 8; The same value of x is obtained in both cases, so $(8, 12)$ is the solution.
62. 12 or {12}   **63.** $-4x + 20y = -12$

Section 4.3 Check Point Exercises

1. $(7, -2)$ or $\{(7, -2)\}$ **2.** $(6, 2)$ or $\{(6, 2)\}$ **3.** $(2, -1)$ or $\{(2, -1)\}$ **4.** $\left(\dfrac{60}{17}, \dfrac{11}{17}\right)$ or $\left\{\left(\dfrac{60}{17}, \dfrac{11}{17}\right)\right\}$
5. no solution or $\varnothing$ **6.** infinitely many solutions; $\{(x, y)|x - 5y = 7\}$ or $\{(x, y)|3x - 15y = 21\}$

Concept and Vocabulary Check

1. -3 **2.** -2 **3.** -2 **4.** 3

Exercise Set 4.3

1. $(4, -7)$ or $\{(4, -7)\}$ **3.** $(3, 0)$ or $\{(3, 0)\}$ **5.** $(-3, 5)$ or $\{(-3, 5)\}$ **7.** $(3, 1)$ or $\{(3, 1)\}$ **9.** $(2, 1)$ or $\{(2, 1)\}$ **11.** $(-2, 2)$ or $\{(-2, 2)\}$
13. $(-7, -1)$ or $\{(-7, -1)\}$ **15.** $(2, 0)$ or $\{(2, 0)\}$ **17.** $(1, -2)$ or $\{(1, -2)\}$ **19.** $(-1, 1)$ or $\{(-1, 1)\}$ **21.** $(3, 1)$ or $\{(3, 1)\}$
23. $(-5, -2)$ or $\{(-5, -2)\}$ **25.** $\left(\dfrac{11}{12}, -\dfrac{7}{6}\right)$ or $\left\{\left(\dfrac{11}{12}, -\dfrac{7}{6}\right)\right\}$ **27.** $\left(\dfrac{23}{16}, \dfrac{3}{8}\right)$ or $\left\{\left(\dfrac{23}{16}, \dfrac{3}{8}\right)\right\}$ **29.** no solution or $\varnothing$
31. infinitely many solutions; $\{(x, y)|x + 3y = 2\}$ or $\{(x, y)|3x + 9y = 6\}$ **33.** no solution or $\varnothing$ **35.** $\left(\dfrac{1}{2}, -\dfrac{1}{2}\right)$ or $\left\{\left(\dfrac{1}{2}, -\dfrac{1}{2}\right)\right\}$
37. infinitely many solutions; $\{(x, y)|x = 5 - 3y\}$ or $\{(x, y)|2x + 6y = 10\}$ **39.** $\left(\dfrac{1}{3}, 1\right)$ or $\left\{\left(\dfrac{1}{3}, 1\right)\right\}$ **41.** $(-10, 21)$ or $\{(-10, 21)\}$
43. $(0, 1)$ or $\{(0, 1)\}$ **45.** $(2, -1)$ or $\{(2, -1)\}$ **47.** $(1, -3)$ or $\{(1, -3)\}$ **49.** $(4, 3)$ or $\{(4, 3)\}$ **51.** no solution or $\varnothing$
53. infinitely many solutions; $\{(x, y)|2(x + 2y) = 6\}$ or $\{(x, y)|3(x + 2y - 3) = 0\}$ **55.** $(3, 2)$ or $\{(3, 2)\}$ **57.** $5x + y = 14, 4x - y = 4$; 2 and 4
59. $4x - 3y = 0, x + y = -7$; -3 and -4 **61.** $(-1, 2)$ or $\{(-1, 2)\}$ **63.** $(-1, 0)$ or $\{(-1, 0)\}$ **65.** $(3, 1)$ or $\{(3, 1)\}$ **67.** dependent
69. week 6; 3.5 symptoms; by the intersection point $(6, 3.5)$ **77.** does not make sense **79.** does not make sense **81.** false **83.** false
85. $A = 2, B = 4$ **86.** 10 **87.** II **88.** 26 or $\{26\}$ **89. a.** $x + y = 28; x - y = 6$ **b.** 17 and 11 **90.** $3x + 2y$ **91. a.** \$30
b. $y = 20 + 0.05x$

Mid-Chapter Check Point Exercises

1. $(2, 0)$ or $\{(2, 0)\}$ **2.** $(1, 1)$ or $\{(1, 1)\}$ **3.** no solution or $\varnothing$ **4.** $(5, 8)$ or $\{(5, 8)\}$ **5.** $(2, -1)$ or $\{(2, -1)\}$ **6.** $(2, 9)$ or $\{(2, 9)\}$
7. $(-2, -3)$ or $\{(-2, -3)\}$ **8.** $(10, -1)$ or $\{(10, -1)\}$ **9.** no solution or $\varnothing$ **10.** $(-12, -1)$ or $\{(-12, -1)\}$ **11.** $(-7, -23)$ or $\{(-7, -23)\}$
12. $\left(\dfrac{16}{17}, -\dfrac{12}{17}\right)$ or $\left\{\left(\dfrac{16}{17}, -\dfrac{12}{17}\right)\right\}$ **13.** infinitely many solutions; $\{(x, y)|y - 2x = 7\}$ or $\{(x, y)|4x = 2y - 14\}$ **14.** $(7, 11)$ or $\{(7, 11)\}$
15. $(6, 10)$ or $\{(6, 10)\}$

Section 4.4 Check Point Exercises

1. men: 65 minutes per day; women: 73 minutes per day **2.** Quarter Pounder: 420 calories; Whopper with cheese: 589 calories
3. length: 100 ft; width: 80 ft **4.** 17.5 yr; \$24,250 **5.** \$3150 at 9%; \$1850 at 11% **6.** 12% solution: 100 oz; 20% solution: 60 oz
7. boat in still water: 35 mph; current: 7 mph

Concept and Vocabulary Check

1. $5x + 6y$ **2.** $10 \cdot 2x + 15 \cdot 2y$ or $20x + 30y$ **3.** $25,600 + 225x$ **4.** $0.04x + 0.05y$ **5.** $0.07x + 0.15y$ **6.** $x + y; x - y$
7. $4(x + y)$

Exercise Set 4.4

1. $x + y = 17, x - y = -3$; 7 and 10 **3.** $3x - y = -1, x + 2y = 23$; 3 and 10 **5.** women: 49 minutes; men: 37 minutes
7. Mr. Goodbar: 264 calories; Mounds bar: 258 calories **9.** 3 Mr. Goodbars and 2 Mounds bars **11.** sweater: \$12; shirt: \$10
13. 44 ft by 20 ft **15.** 90 ft by 70 ft **17. a.** 300 min; \$35 **b.** plan B; Answers will vary. **19.** \$600 of merchandise; \$580
21. adult: 93 tickets; student: 208 tickets **23.** each item in column A: \$3.99; each item in column B: \$1.50
25. 2 servings of macaroni and 4 servings of broccoli **27.** $A: 100°; B: 40°; C: 40°$ **29.** \$2000 at 6%; \$5000 at 8%
31. first fund: \$8000; second fund: \$6000 **33.** \$17,000 at 12% return; \$3000 at 5% loss **35.** California: 100 gal; French: 100 gal
37. 18-karat: 96 g; 12-karat: 204 g **39.** cheaper candy: 30 lb; more expensive candy: 45 lb **41.** 8 nickels and 7 dimes **43.** plane in still air:
130 mph; wind: 30 mph **45.** crew in still water: 6 km/h; current: 2 km/h **47.** in still water: 4.5 mph; current: 1.5 mph
55. does not make sense **57.** does not make sense **59.** 10 birds and 20 lions **61.** There are 5 people downstairs and 7 people upstairs.
63. \$52,500 at 8% and \$17,500 at 12% **65.** $\dfrac{5}{2}$ or $\left\{\dfrac{5}{2}\right\}$ **66.** $6x + 11$ **67.** $y = -2x - 4$
68. yes **69.** $11x + 4y = -3$ **70.** $16a + 4b + c = 1682$

Section 4.5 Check Point Exercises

1. $(-1) - 2(-4) + 3(5) = 22; 2(-1) - 3(-4) - 5 = 5; 3(-1) + (-4) - 5(5) = -32$ **2.** $(1, 4, -3)$ or $\{(1, 4, -3)\}$
3. $(4, 5, 3)$ or $\{(4, 5, 3)\}$ **4.** $y = 3x^2 - 12x + 13$

Concept and Vocabulary Check

1. triple; all **2.** $-2; -4$ **3.** z; add Equations 1 and 3 **4.** quadratic **5.** curve fitting

Exercise Set 4.5

1. not a solution **3.** solution **5.** $(2, 3, 3)$ or $\{(2, 3, 3)\}$ **7.** $(2, -1, 1)$ or $\{(2, -1, 1)\}$ **9.** $(1, 2, 3)$ or $\{(1, 2, 3)\}$ **11.** $(3, 1, 5)$ or $\{(3, 1, 5)\}$
13. $(1, 0, -3)$ or $\{(1, 0, -3)\}$ **15.** $(1, -5, -6)$ or $\{(1, -5, -6)\}$

17. no solution or $\varnothing$ **19.** infinitely many solutions; dependent equations **21.** $\left(\frac{1}{2}, \frac{1}{3}, -1\right)$ or $\left\{\left(\frac{1}{2}, \frac{1}{3}, -1\right)\right\}$ **23.** $y = 2x^2 - x + 3$

25. $y = 2x^2 + x - 5$ **27.** 7, 4, and 5 **29.** $(4, 8, 6)$ or $\{(4, 8, 6)\}$ **31.** $y = -\frac{3}{4}x^2 + 6x - 11$ **33.** $\left\{\left(\frac{8}{a}, -\frac{3}{b}, -\frac{5}{c}\right)\right\}$ **35. a.** $(0, 5), (50, 31), (100, 15)$
b. $\begin{cases} 0a + 0b + c = 5 \\ 2500a + 50b + c = 31 \\ 10{,}000a + 100b + c = 15 \end{cases}$ **37. a.** $y = -16x^2 + 40x + 200$ **b.** 0; After 5 seconds, the ball hits the ground. **39.** housing; $6133;

vehicles/gas: $2269; health care: $5438 **41.** $1200 at 8%; $2000 at 10%; $3500 at 12% **43.** 200 $8 tickets; 150 $10 tickets; 50 $12 tickets
45. 4 oz of food A; 0.5 oz of food B; 1 oz of food C **55.** does not make sense **57.** makes sense **59.** false **61.** true **63.** $60°, 55°, 65°$ **65.** 30 cm

66. 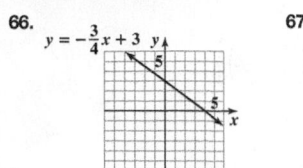 $y = -\frac{3}{4}x + 3$ **67.** 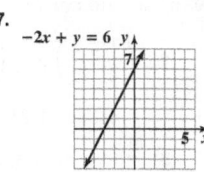 $-2x + y = 6$ **68.** 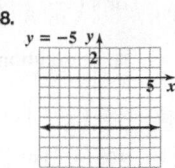 $y = -5$ **69.** $17x^3$ **70.** $-2x^2$ **71.** $-7y^4$

Review Exercises

1. solution **2.** not a solution **3.** no; $(-1, 3)$ does not satisfy $2x + y = -5$. **4.** $(4, -2)$ or $\{(4, -2)\}$ **5.** $(3, -2)$ or $\{(3, -2)\}$
6. $(2, 0)$ or $\{(2, 0)\}$ **7.** $(2, 1)$ or $\{(2, 1)\}$ **8.** $(4, -1)$ or $\{(4, -1)\}$ **9.** no solution or $\varnothing$ **10.** infinitely many solutions;
$\{(x, y)|2x - 4y = 8\}$ or $\{(x, y)|x - 2y = 4\}$ **11.** no solution or $\varnothing$ **12.** $(-2, -6)$ or $\{(-2, -6)\}$ **13.** $(2, 5)$ or $\{(2, 5)\}$
14. no solution or $\varnothing$ **15.** $(2, -1)$ or $\{(2, -1)\}$ **16.** $(5, 4)$ or $\{(5, 4)\}$ **17.** $(-2, -1)$ or $\{(-2, -1)\}$ **18.** $(1, -4)$ or $\{(1, -4)\}$
19. $(20, -21)$ or $\{(20, -21)\}$ **20.** infinitely many solutions; $\{(x, y)|4x + y = 5\}$ or $\{(x, y)|12x + 3y = 15\}$ **21.** no solution or $\varnothing$
22. $(4, 18)$ or $\{(4, 18)\}$ **23.** $\left(-1, -\frac{1}{2}\right)$ or $\left\{\left(-1, -\frac{1}{2}\right)\right\}$ **24. a.** $(20, 1000)$ or $\{(20, 1000)\}$; equilibrium quantity: 20,000; equilibrium price: $1000
b. $1000; 20,000; 20,000 **25.** $(2, 4)$ or $\{(2, 4)\}$ **26.** $(-1, -1)$ or $\{(-1, -1)\}$ **27.** $(2, -1)$ or $\{(2, -1)\}$ **28.** $(3, 2)$ or $\{(3, 2)\}$
29. $(2, 1)$ or $\{(2, 1)\}$ **30.** $(0, 0)$ or $\{(0, 0)\}$ **31.** $\left(\frac{17}{7}, -\frac{15}{7}\right)$ or $\left\{\left(\frac{17}{7}, -\frac{15}{7}\right)\right\}$ **32.** no solution or $\varnothing$
33. infinitely many solutions; $\{(x, y)|3x - 4y = -1\}$ or $\{(x, y)|-6x + 8y = 2\}$ **34.** $(4, -2)$ or $\{(4, -2)\}$ **35.** $(-8, -6)$ or $\{(-8, -6)\}$
36. $(-4, 1)$ or $\{(-4, 1)\}$ **37.** $\left(\frac{5}{2}, 3\right)$ or $\left\{\left(\frac{5}{2}, 3\right)\right\}$ **38.** $(3, 2)$ or $\{(3, 2)\}$ **39.** $\left(\frac{1}{2}, -2\right)$ or $\left\{\left(\frac{1}{2}, -2\right)\right\}$ **40.** no solution or $\varnothing$
41. $(3, 7)$ or $\{(3, 7)\}$ **42.** Klimt: $135 million; Picasso: $104 million **43.** shrimp: 42 mg; scallops: 15 mg **44.** 9 ft by 5 ft **45.** 7 yd by 5 yd
46. room: $80; car: $60 **47.** 200 min; $25 **48.** orchestra: 6 tickets; balcony: 3 tickets **49.** $3000 at 8% and $7000 at 10%
50. 4 gallons of 75% and 6 gallons of 50% **51.** plane in still air: 630 mph; wind: 90 mph **52.** no **53.** $(0, 1, 2)$ or $\{(0, 1, 2)\}$
54. $(2, 1, -1)$ or $\{(2, 1, -1)\}$ **55.** infinitely many solutions; dependent equations **56.** $y = 3x^2 - 4x + 5$
57. war: 124 million; famine: 111 million; tobacco: 71 million

Chapter Test

1. solution **2.** not a solution **3.** $(2, 4)$ or $\{(2, 4)\}$ **4.** $(2, 4)$ or $\{(2, 4)\}$ **5.** $(1, -3)$ or $\{(1, -3)\}$ **6.** $(2, -3)$ or $\{(2, -3)\}$
7. no solution or $\varnothing$ **8.** $(-1, 4)$ or $\{(-1, 4)\}$ **9.** $(-4, 3)$ or $\{(-4, 3)\}$ **10.** infinitely many solutions;
$\{(x, y)|3x - 2y = 2\}$ or $\{(x, y)|-9x + 6y = -6\}$ **11.** Mary: 2.6%; Patricia: 1.1% **12.** 12 yd by 5 yd **13.** 500 min; $40
14. $2000 at 6%; $7000 at 7% **15.** 6% solution: 12 oz; 9% solution: 24 oz **16.** boat in still water: 14 mph; current: 2mph **17** $(1, 3, 2)$ or $\{(1, 3, 2)\}$

Cumulative Review Exercises

1. -36 **2.** $17x - 11$ **3.** -6 or $\{-6\}$ **4.** 20 or $\{20\}$ **5.** $t = \dfrac{A - P}{Pr}$ **6.** $(2, \infty)$ or $\{x|x > 2\}$;

7. 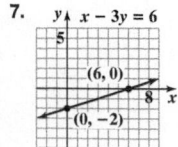 $x - 3y = 6$, $(6, 0)$, $(0, -2)$ **8.** 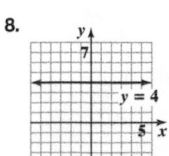 $y = 4$ **9.** 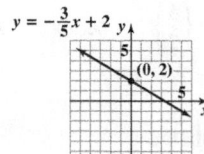 $y = -\frac{3}{5}x + 2$, $(0, 2)$

10. $(0, -2)$ or $\{(0, -2)\}$ **11.** $\left(\frac{3}{2}, -2\right)$ or $\left\{\left(\frac{3}{2}, -2\right)\right\}$ **12.** 1 **13.** $y - 6 = -4(x + 1)$; $y = -4x + 2$ **14.** 10 ft

15. pen: $0.80; pad: $1.20 **16.** $-93, 0, \frac{7}{1}, \sqrt{100}$ **17.** living alone; The percentage of U.S. adults living alone increased from 1960 through 2008.

18. married, living with kids; The percentage of U.S. adults married, living with kids decreased from 1960 through 2008. **19.** 70 years; 2030
20. 2030; In 2030, the same percentage of U.S. adults will be married, living with kids and living alone, namely 19%.

CHAPTER 5

Section 5.1 Check Point Exercises

1. $5x^3 + 4x^2 - 8x - 20$ **2.** $5x^3 + 4x^2 - 8x - 20$ **3.** $7x^2 + 11x + 4$ **4.** $7x^3 + 3x^2 + 12x - 8$ **5.** $3y^3 - 10y^2 - 11y - 8$

6. $y = x^2 - 1$

Concept and Vocabulary Check

1. whole   **2.** standard   **3.** monomial   **4.** binomial   **5.** trinomial   **6.** n **7.** greatest   **8.** like   **9.** opposite

Exercise Set 5.1

1. binomial, 1   **3.** binomial, 3   **5.** monomial, 2   **7.** monomial, 0   **9.** trinomial, 2   **11.** trinomial, 4   **13.** binomial, 3
15. monomial, 23   **17.** $-8x + 13$ **19.** $12x^2 + 15x - 9$ **21.** $10x^2 - 12x$ **23.** $5x^2 - 3x + 13$ **25.** $4y^3 + 10y^2 + y - 2$

27. $3x^3 + 2x^2 - 9x + 7$ **29.** $-2y^3 + 4y^2 + 13y + 13$ **31.** $-3y^6 + 8y^4 + y^2$ **33.** $10x^3 + 1$ **35.** $-\dfrac{2}{5}x^4 + x^3 - \dfrac{1}{8}x^2$

37. $0.01x^5 + x^4 - 0.1x^3 + 0.3x + 0.33$ **39.** $11y^3 - 3y^2$ **41.** $-2x^2 - x + 1$ **43.** $-\dfrac{1}{4}x^4 - \dfrac{7}{15}x^3 - 0.3$ **45.** $-y^3 + 8y^2 - 3y - 14$
47. $-5x^3 - 6x^2 + x - 4$ **49.** $7x^4 - 2x^3 + 4x - 2$ **51.** $8x^2 + 7x - 5$ **53.** $9x^3 - 4.9x^2 + 11.1$ **55.** $-2x - 10$ **57.** $-5x^2 - 9x - 12$
59. $-5x^2 - x$ **61.** $-4x^2 - 4x - 6$ **63.** $-2y - 6$ **65.** $6y^3 + y^2 + 7y - 20$ **67.** $n^3 + 2$ **69.** $y^6 - y^3 - y^2 + y$
71. $26x^4 + 9x^2 + 6x$ **73.** $\dfrac{5}{7}x^3 - \dfrac{9}{20}x$ **75.** $4x + 6$ **77.** $10x^2 - 7$ **79.** $-4y^2 - 7y + 5$ **81.** $9x^3 + 11x^2 - 8$
83. $-y^3 + 8y^2 + y + 14$ **85.** $7x^4 - 2x^3 + 3x^2 - x + 2$ **87.** $0.05x^3 + 0.02x^2 + 1.02x$

89. $y = x^2$

91. $y = x^2 + 1$

93. $y = 4 - x^2$

95. $x^2 + 12x$ **97.** $y^2 - 19y + 16$ **99.** $2x^3 + 3x^2 + 7x - 5$ **101.** $-10y^3 + 2y^2 + y + 3$
103. a. $M - W = -35x^3 + 1373x^2 - 15,995x + 63,210$ **b.** $\$12,348$ **c.** $\$10,923$; overestimates by $\$1425$
105. a. $\$56,995$; underestimates by $\$225$ **b.** $(16, 56,995)$ on the graph for men
c. $\$42,000$, although estimates may vary by $\pm\$1000$ **115.** does not make sense   **117.** does not make sense
119. false   **121.** false   **123.** $\dfrac{2}{3}t^3 - 2t^2 + 4t$ **125.** -10 **126.** 5.6 **127.** -4 or $\{-4\}$ **128.** 7 **129.** $3x^2 + 15x$ **130.** $x^2 + 5x + 6$

Section 5.2 Check Point Exercises

1. a. 2^6 or 64 **b.** x^{10} **c.** y^8 **d.** y^9 **2. a.** 3^{20} **b.** x^{90} **c.** $(-5)^{21}$ **3. a.** $16x^4$ **b.** $-64y^6$ **4. a.** $70x^3$ **b.** $-20x^9$
5. a. $3x^2 + 15x$ **b.** $30x^5 - 12x^3 + 18x^2$ **6. a.** $x^2 + 9x + 20$ **b.** $10x^2 - 29x - 21$ **7.** $5x^3 - 18x^2 + 7x + 6$
8. $6x^5 - 19x^4 + 22x^3 - 8x^2$

Concept and Vocabulary Check

1. b^{m+n}; add   **2.** b^{mn}; multiply   **3.** $a^n b^n$; factor   **4.** distributive; $x^2 + 5x + 7; 2x^2$ **5.** $4x; 7$; like

Exercise Set 5.2

1. x^{18} **3.** y^{12} **5.** x^{11} **7.** 7^{19} **9.** 6^{90} **11.** x^{45} **13.** $(-20)^9$ **15.** $8x^3$ **17.** $25x^2$ **19.** $16x^6$ **21.** $16y^{24}$ **23.** $-32x^{35}$

25. $14x^2$ **27.** $24x^3$ **29.** $-15y^7$ **31.** $\dfrac{1}{8}a^5$ **33.** $-48x^7$ **35.** $4x^2 + 12x$ **37.** $x^2 - 3x$ **39.** $2x^2 - 12x$ **41.** $-12y^2 - 20y$

43. $4x^3 + 8x^2$ **45.** $2y^4 + 6y^3$ **47.** $6y^4 - 8y^3 + 14y^2$ **49.** $6x^4 + 8x^3$ **51.** $-2x^3 - 10x^2 + 6x$ **53.** $12x^4 - 3x^3 + 15x^2$

55. $x^2 + 8x + 15$ **57.** $2x^2 + 9x + 4$ **59.** $x^2 - 2x - 15$ **61.** $x^2 - 2x - 99$ **63.** $2x^2 + 3x - 20$ **65.** $\dfrac{3}{16}x^2 + \dfrac{11}{4}x - 4$

67. $x^3 + 3x^2 + 5x + 3$ **69.** $y^3 - 6y^2 + 13y - 12$ **71.** $2a^3 - 9a^2 + 19a - 15$ **73.** $x^4 + 3x^3 + 5x^2 + 7x + 4$

75. $4x^4 - 4x^3 + 6x^2 - \dfrac{17}{2}x + 3$ **77.** $x^4 + x^3 + x^2 + 3x + 2$ **79.** $x^3 + 3x^2 - 37x + 24$ **81.** $2x^3 - 9x^2 + 27x - 27$

83. $2x^4 + 9x^3 + 6x^2 + 11x + 12$ **85.** $12z^4 - 14z^3 + 19z^2 - 22z + 8$ **87.** $21x^5 - 43x^4 + 38x^3 - 24x^2$
89. $4y^6 - 2y^5 - 6y^4 + 5y^3 - 5y^2 + 8y - 3$ **91.** $x^4 + 6x^3 - 11x^2 - 4x + 3$ **93.** $2x - 2$ **95.** $15x^5 + 42x^3 - 8x^2$

97. $2y^3$ **99.** $16y + 32$ **101.** $2x^2 + 7x - 15 \text{ ft}^2$ **103. a.** $(2x + 1)(x + 2)$ **b.** $2x^2 + 5x + 2$ **c.** $(2x + 1)(x + 2) = 2x^2 + 5x + 2$
113. makes sense   **115.** makes sense   **117.** false   **119.** true   **121.** $8x + 16$ **123.** $-8x^4$ **124.** $(-\infty, -1)$ or $\{x \mid x < -1\}$

125. 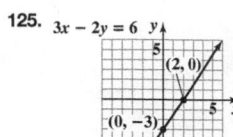   **126.** $-\dfrac{2}{3}$ **127. a.** $x^2 + 7x + 12$ **b.** $x^2 + 25x + 100$ **128. a.** $x^2 - 9$ **b.** $x^2 - 25$
129. a. $x^2 + 6x + 9$ **b.** $x^2 + 10x + 25$

Section 5.3 Check Point Exercises

1. $x^2 + 11x + 30$ **2.** $28x^2 - x - 15$ **3.** $6x^2 - 22x + 20$ **4. a.** $49y^2 - 64$ **b.** $16x^2 - 25$ **c.** $4a^6 - 9$ **5. a.** $x^2 + 20x + 100$
b. $25x^2 + 40x + 16$ **6. a.** $x^2 - 18x + 81$ **b.** $49x^2 - 42x + 9$

Concept and Vocabulary Check

1. $2x^2$; $3x$; $10x$; 15 **2.** $A^2 - B^2$; minus   **3.** $A^2 + 2AB + B^2$; squared; product of the terms; squared
4. $A^2 - 2AB + B^2$; minus; product of the terms; squared   **5.** true   **6.** false

Exercise Set 5.3

1. $x^2 + 10x + 24$ **3.** $y^2 - 4y - 21$ **5.** $2x^2 + 7x - 15$ **7.** $4y^2 - y - 3$ **9.** $10x^2 - 9x - 9$ **11.** $12y^2 - 43y + 35$
13. $-15x^2 - 32x + 7$ **15.** $6y^2 - 28y + 30$ **17.** $15x^4 - 47x^2 + 28$ **19.** $-6x^2 + 17x - 10$ **21.** $x^3 + 5x^2 + 3x + 15$

23. $8x^5 + 40x^3 + 3x^2 + 15$ **25.** $x^2 - 9$ **27.** $9x^2 - 4$ **29.** $9r^2 - 16$ **31.** $9 - r^2$ **33.** $25 - 49x^2$ **35.** $4x^2 - \dfrac{1}{4}$ **37.** $y^4 - 1$

39. $r^6 - 4$ **41.** $1 - y^8$ **43.** $x^{20} - 25$ **45.** $x^2 + 4x + 4$ **47.** $4x^2 + 20x + 25$ **49.** $x^2 - 6x + 9$ **51.** $9y^2 - 24y + 16$

53. $16x^4 - 8x^2 + 1$ **55.** $49 - 28x + 4x^2$ **57.** $4x^2 + 2x + \dfrac{1}{4}$ **59.** $16y^2 - 2y + \dfrac{1}{16}$ **61.** $x^{16} + 6x^8 + 9$ **63.** $x^3 - 1$

65. $x^2 - 2x + 1$ **67.** $9y^2 - 49$ **69.** $12x^4 + 3x^3 + 27x^2$ **71.** $70y^2 + 2y - 12$ **73.** $x^4 + 2x^2 + 1$ **75.** $x^4 + 3x^2 + 2$

77. $x^4 - 16$ **79.** $4 - 12x^5 + 9x^{10}$ **81.** $\dfrac{3}{16}x^4 + 7x^2 - 96$ **83.** $x^2 + 2x + 1$ **85.** $4x^2 - 9$ **87.** $6x + 22$

89. $16x^4 - 72x^2 + 81$ **91.** $16x^4 - 1$ **93.** $x^3 + 6x^2 + 12x + 8$ **95.** $x^2 + 6x + 9 - y^2$ **97.** $(x + 1)(x + 2)$ yd^2 **99.** 56 yd^2; $(6, 56)$
101. $(x^2 + 4x + 4)$ in^2 **109.** makes sense   **111.** makes sense, although answers may vary   **113.** true   **115.** false   **117.** $4x^3 - 36x^2 + 80x$
119. Change $x^2 + 1$ to $x^2 + 2x + 1$.   **121.** Graphs coincide.   **123.** $(2, -1)$ or $\{(2, -1)\}$ **124.** $(1, 1)$ or $\{(1, 1)\}$
125. 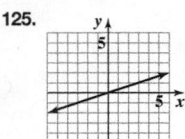   **126.** -72 **127.** $11xy$ **128.** $3x^2 + 11xy + 10y^2$

Section 5.4 Check Point Exercises

1. -9 **2.** polynomial degree: 9;

Term	Coefficient	Degree
$8x^4y^5$	8	9
$-7x^3y^2$	-7	5
$-x^2y$	-1	3
$-5x$	-5	1
11	11	0

3. $2x^2y + 2xy - 4$ **4.** $3x^3 + 2x^2y + 5xy^2 - 10$ **5.** $60x^5y^5$ **6.** $60x^5y^7 - 12x^3y^3 + 18xy^2$
7. a. $21x^2 - 25xy + 6y^2$ **b.** $4x^2 + 16xy + 16y^2$ **8. a.** $36x^2y^4 - 25x^2$ **b.** $x^3 - y^3$

Concept and Vocabulary Check

1. -18 **2.** 6   **3.** a; $n + m$ **4.** 5; 9; 9   **5.** false   **6.** true

Exercise Set 5.4

1. 1 **3.** -47 **5.** -6 **7.** polynomial degree: 9;

Term	Coefficient	Degree
x^3y^2	1	5
$-5x^2y^7$	-5	9
$6y^2$	6	2
-3	-3	0

9. $7x^2y - 4xy$ **11.** $2x^2y + 13xy + 13$ **13.** $-11x^4y^2 - 11x^2y^2 + 2xy$ **15.** $-5x^3 + 8xy - 9y^2$ **17.** $x^4y^2 + 8x^3y + y - 6x$
19. $5x^3 + x^2y - xy^2 - 4y^3$ **21.** $-3x^2y^2 + xy^2 + 5y^2$ **23.** $8a^2b^4 + 3ab^2 + 8ab$ **25.** $-30x + 37y$ **27.** $40x^3y^2$ **29.** $-24x^5y^9$
31. $45x^2y + 18xy^2$ **33.** $50x^3y^2 - 15xy^3$ **35.** $28a^3b^5 + 8a^2b^3$ **37.** $-a^2b + ab^2 - b^3$ **39.** $7x^2 + 38xy + 15y^2$ **41.** $2x^2 + xy - 21y^2$
43. $15x^2y^2 + xy - 2$ **45.** $4x^2 + 12xy + 9y^2$ **47.** $x^2y^2 - 6xy + 9$ **49.** $x^4 + 2x^2y^2 + y^4$ **51.** $x^4 - 4x^2y^2 + 4y^4$ **53.** $9x^2 - y^2$
55. $a^2b^2 - 1$ **57.** $x^2 - y^4$ **59.** $9a^4b^2 - a^2$ **61.** $9x^2y^4 - 16y^2$ **63.** $a^3 - ab^2 + a^2b - b^3$ **65.** $x^3 + 4x^2y + 4xy^2 + y^3$
67. $x^3 - 4x^2y + 4xy^2 - y^3$ **69.** $x^2y^2 - a^2b^2$ **71.** $x^6y + x^4y + x^4 + 2x^2 + 1$ **73.** $x^4y^4 - 6x^2y^2 + 9$ **75.** $x^2 + 2xy + y^2 - 1$
77. $3x^2 + 8xy + 5y^2$ **79.** $2xy + y^2$ **81.** $x^{12}y^{12} - 2x^6y^6 + 1$ **83.** $x^4y^4 - 18x^2y^2 + 81$ **85.** $x^2 - y^2 - 2yz - z^2$
87. no; need 120 more board feet **89.** 192 ft **91.** 0 ft; The ball hits the ground. **93.** 2.5 to 6 sec **95.** $(2, 192)$
97. 2.5 sec; 196 ft **101.** makes sense **103.** does not make sense **105.** false **107.** true **109.** $-x^2 + 18xy + 80y^2$
112. $W = \dfrac{2R - L}{3}$ **113.** 3.8 **114.** -1.6 or $\{-1.6\}$ **115.** 4 **116.** $\dfrac{x^6}{125}$ **117.** $\dfrac{32a^{15}}{b^{20}}$

Mid-Chapter Check Point Exercises

1. $-55x^4y^6$ **2.** $6x^2y^3$ **3.** $12x^2 - x - 35$ **4.** $-x + 12$ **5.** $2x^3 - 11x^2 + 17x - 5$ **6.** $x^2 - x - 4$ **7.** $64x^2 - 48x + 9$ **8.** $70x^9$
9. $x^4 - 4$ **10.** $x^4 + 4x^2 + 4$ **11.** $18a^2 - 11ab - 10b^2$ **12.** $70x^5 - 14x^3 + 21x^2$ **13.** $5a^2b^3 + 2ab - b^2$ **14.** $18y^2 - 50$
15. $2x^3 - x^2 + x$ **16.** $10x^2 - 5xy - 3y^2$ **17.** $-4x^5 + 7x^4 - 10x + 23$ **18.** $x^3 + 27y^3$ **19.** $10x^7 - 5x^4 + 8x^3 - 4$
20. $y^2 - 12yz + 36z^2$ **21.** $-21x^2 + 7$
22.

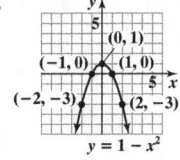

$y = 1 - x^2$

Section 5.5 Check Point Exercises

1. a. 5^8 **b.** x^7 **c.** y^{19} **2. a.** 1 **b.** 1 **c.** -1 **d.** 20 **e.** 1 **3. a.** $\dfrac{x^2}{25}$ **b.** $\dfrac{x^{12}}{8}$ **c.** $\dfrac{16a^{40}}{b^{12}}$
4. a. $-2x^8$ **b.** $\dfrac{1}{5}$ **c.** $3x^5y^3$ **5.** $-5x^7 + 2x^3 - 3x$ **6.** $5x^6 - \dfrac{7}{5}x + 2$ **7.** $3x^6y^4 - xy + 10$

Concept and Vocabulary Check

1. b^{m-n}; subtract **2.** 1 **3.** $\dfrac{a^n}{b^n}$; numerator; denominator **4.** divide; subtract **5.** dividend; divisor; quotient
6. divisor; quotient; dividend **7.** $20x^8 - 10x^4 + 6x^3; 2x^3$

Exercise Set 5.5

1. 3^{15} **3.** x^4 **5.** y^8 **7.** $5^3 \cdot 2^4$ **9.** $x^{75}y^{40}$ **11.** 1 **13.** 1 **15.** -1 **17.** 100 **19.** 1 **21.** 0 **23.** -2 **25.** $\dfrac{x^2}{9}$ **27.** $\dfrac{x^6}{64}$
29. $\dfrac{4x^6}{25}$ **31.** $-\dfrac{64}{27a^9}$ **33.** $-\dfrac{32a^{35}}{b^{20}}$ **35.** $\dfrac{x^8y^{12}}{16z^4}$ **37.** $3x^5$ **39.** $-2x^{20}$ **41.** $-\dfrac{1}{2}y^3$ **43.** $\dfrac{7}{5}y^{12}$ **45.** $6x^5y^4$ **47.** $-\dfrac{1}{2}x^{12}$ **49.** $\dfrac{9}{7}$
51. $-\dfrac{1}{10}x^8y^9z^4$ **53.** $5x^4 + x^3$ **55.** $2x^3 - x^2$ **57.** $y^6 - 9y + 1$ **59.** $-8x^2 + 5x$ **61.** $6x^3 + 2x^2 + 3x$ **63.** $3x^3 - 2x^2 + 10x$
65. $4x - 6$ **67.** $-6z^2 - 2z$ **69.** $4x^2 + 3x - 1$ **71.** $5x^4 - 3x^2 - x$ **73.** $-9x^3 + \dfrac{9}{2}x^2 - 10x + 5$ **75.** $4xy + 2x - 5y$
77. $-4x^5y^3 + 3xy + 2$ **79.** $4x^2 - x + 6$ **81.** $-xy^2$ **83.** $y + 5$ **85.** $3x^{12n} - 6x^{9n} + 2$
87. a. \$5.39 **b.** $\dfrac{3.6x^2 + 158x + 2790}{-0.2x^2 + 21x + 1015}$ **c.** \$5.45; overestimates by \$0.06 **d.** no; The divisor is not a monomial.
97. does not make sense **99.** does not make sense **101.** false **103.** true **105.** $\dfrac{9x^8 - 12x^6}{3x^3}$ **107.** 20.3 **108.** 0.875
109. $y = \dfrac{1}{3}x + 2$

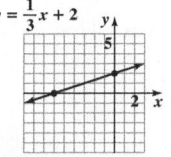

110. quotient: 26; remainder: 0 **111.** quotient: 123; remainder: 6 **112.** quotient: 257; remainder: 1

Section 5.6 Check Point Exercises

1. $x + 5$ **2.** $4x - 3 - \dfrac{3}{2x + 3}$ **3.** $x^2 + x + 1$ **4.** $2x^2 + 7x + 14 + \dfrac{21x - 10}{x^2 - 2x}$ **5.** $x^2 - 2x - 3$

Concept and Vocabulary Check

1. $10x^3 + 4x^2 + 0x + 9$ **2.** $8x^2; 2x; 4x; 10x$ **3.** $5x; 3x - 2; 15x^2 - 10x; 15x^2 - 22x$ **4.** $6x^2 - 10x; 6x^2 + 8x; 18x; -4; 18x - 4$

5. $14; x - 12; 14; x - 12 + \dfrac{14}{x - 5}$ **6.** $4; 1; 5; -74; 1$ **7.** $-5; 4; 0; -8; -2$ **8.** true

Exercise Set 5.6

1. $x + 4$ **3.** $2x + 5$ **5.** $x - 2$ **7.** $2y + 1$ **9.** $x - 5 + \dfrac{14}{x + 2}$ **11.** $y + 3 + \dfrac{4}{y + 2}$ **13.** $x^2 - 5x + 2$ **15.** $6y - 1$

17. $2a + 3$ **19.** $y^2 - y + 2$ **21.** $3x + 5 - \dfrac{5}{2x - 5}$ **23.** $x^2 + 2x + 8 + \dfrac{13}{x - 2}$ **25.** $2y^2 + y + 1 + \dfrac{6}{2y + 3}$

27. $2y^2 - 3y + 2 + \dfrac{1}{3y + 2}$ **29.** $9x^2 + 3x + 1$ **31.** $y^3 - 9y^2 + 27y - 27$ **33.** $2y + 4 + \dfrac{4}{2y - 1}$ **35.** $y^3 + y^2 - y - 1 + \dfrac{4}{y - 1}$

37. $4x^2 + 3x - 8 + \dfrac{18}{x^2 + 3}$ **39.** $5x^2 + x + 3 + \dfrac{x + 7}{3x^2 - 1}$ **41.** $2x + 5$ **43.** $3x - 8 + \dfrac{20}{x + 5}$ **45.** $4x^2 + x + 4 + \dfrac{3}{x - 1}$

47. $6x^4 + 12x^3 + 22x^2 + 48x + 93 + \dfrac{187}{x - 2}$ **49.** $x^3 - 10x^2 + 51x - 260 + \dfrac{1300}{5 + x}$ **51.** $3x^2 + 3x - 3$ **53.** $x^4 + x^3 + 2x^2 + 2x + 2$

55. $x^3 + 4x^2 + 16x + 64$ **57.** $2x^4 - 7x^3 + 15x^2 - 31x + 64 - \dfrac{129}{x + 2}$ **59.** $x^3 - x^2y + xy^2 - y^3 + \dfrac{2y^4}{x + y}$ **61.** $3x^2 + 2x - 1$

63. $4x - 7 + \dfrac{4x + 8}{x^2 + x + 1}$ **65.** $x^3 + x^2 - x - 3 + \dfrac{-x + 5}{x^2 - x + 2}$ **67.** $4x^2 + 5xy - y^2$ **69.** $x^2 + 2x + 3$ **71.** $x^2 + 2x + 3$ units

73. a. $\dfrac{30{,}000x^3 - 30{,}000}{x - 1}$ **b.** $30{,}000x^2 + 30{,}000x + 30{,}000$ **c.** \$94,575 **81.** does not make sense **83.** make sense **85.** true

87. false **89.** -3 **91.** Graphs coincide. **93.** $2x + 3$ should be $2x + 23 + \dfrac{130}{x - 5}$. **95.** $x^2 - 2x + 3$ should be $x^2 + 2x + 3$.

96. $\{(-5, 3)\}$ **97.** 1.2 **98.** -6 or $\{-6\}$ **99. a.** $2; -2$ **b.** $\dfrac{1}{7^2} = 7^{-2}$ **100.** $16x^2$ **101.** x^9

Section 5.7 Check Point Exercises

1. a. $\dfrac{1}{6^2} = \dfrac{1}{36}$ **b.** $\dfrac{1}{5^3} = \dfrac{1}{125}$ **c.** $\dfrac{1}{(-3)^4} = \dfrac{1}{81}$ **d.** $-\dfrac{1}{3^4} = -\dfrac{1}{81}$ **e.** $\dfrac{1}{8^1} = \dfrac{1}{8}$ **2. a.** $\dfrac{7^2}{2^3} = \dfrac{49}{8}$ **b.** $\dfrac{5^2}{4^2} = \dfrac{25}{16}$ **c.** $\dfrac{y^2}{7}$ **d.** $\dfrac{y^8}{x^1} = \dfrac{y^8}{x}$

3. $\dfrac{1}{x^{10}}$ **4. a.** $\dfrac{1}{x^8}$ **b.** $\dfrac{15}{x^6}$ **c.** $-\dfrac{2}{y^6}$ **5.** $\dfrac{36}{x^3}$ **6.** $\dfrac{1}{x^{20}}$ **7. a.** $7{,}400{,}000{,}000$ **b.** 0.000003017 **8. a.** 7.41×10^9 **b.** 9.2×10^{-8}

9. a. 6×10^{10} **b.** 2.1×10^{11} **c.** 6.4×10^{-5} **10.** \$2560

Concept and Vocabulary Check

1. $\dfrac{1}{b^n}$ **2.** false **3.** true **4.** b^n **5.** true **6.** false **7.** a number greater than or equal to 1 and less than 10; integer **8.** true **9.** false

Exercise Set 5.7

1. $\dfrac{1}{8^2} = \dfrac{1}{64}$ **3.** $\dfrac{1}{5^3} = \dfrac{1}{125}$ **5.** $\dfrac{1}{(-6)^2} = \dfrac{1}{36}$ **7.** $-\dfrac{1}{6^2} = -\dfrac{1}{36}$ **9.** $\dfrac{1}{4^1} = \dfrac{1}{4}$ **11.** $\dfrac{1}{2^1} + \dfrac{1}{3^1} - \dfrac{1}{2} + \dfrac{1}{3} - \dfrac{5}{6}$ **13.** $3^2 - 9$ **15.** $(-3)^2 - 9$

17. $\dfrac{8^2}{2^3} = 8$ **19.** $\dfrac{4^2}{1^2} = 16$ **21.** $\dfrac{5^3}{3^3} = \dfrac{125}{27}$ **23.** $\dfrac{x^5}{6}$ **25.** $\dfrac{y^1}{x^8} = \dfrac{y}{x^8}$ **27.** $3 \cdot (-5)^3 = -375$ **29.** $\dfrac{1}{x^5}$ **31.** $\dfrac{8}{x^3}$ **33.** $\dfrac{1}{x^6}$ **35.** $\dfrac{1}{y^{99}}$

37. $\dfrac{3}{z^5}$ **39.** $-\dfrac{4}{x^4}$ **41.** $-\dfrac{1}{3a^3}$ **43.** $\dfrac{7}{5w^8}$ **45.** $\dfrac{1}{x^5}$ **47.** $\dfrac{1}{y^{11}}$ **49.** $\dfrac{16}{x^2}$ **51.** $216y^{17}$ **53.** $\dfrac{1}{x^6}$ **55.** $\dfrac{1}{16x^{12}}$ **57.** $\dfrac{x^2}{9}$ **59.** $-\dfrac{y^3}{8}$

61. $\dfrac{2x^6}{5}$ **63.** x^8 **65.** $16y^6$ **67.** $\dfrac{1}{y^2}$ **69.** $\dfrac{1}{y^{50}}$ **71.** $\dfrac{1}{a^{12}b^{15}}$ **73.** $\dfrac{a^8}{b^{24}}$ **75.** $\dfrac{4}{x^4}$ **77.** $\dfrac{y^9}{x^6}$ **79.** 870 **81.** $923{,}000$ **83.** 3.4

85. 0.79 **87.** 0.0215 **89.** 0.000786 **91.** 3.24×10^4 **93.** 2.2×10^8 **95.** 7.13×10^2 **97.** 6.751×10^3 **99.** 2.7×10^{-3}
101. 2.02×10^{-5} **103.** 5×10^{-3} **105.** 3.14159×10^0 **107.** 6×10^5 **109.** 1.6×10^9 **111.** 3×10^4 **113.** 3×10^6 **115.** 3×10^{-6}

117. 9×10^4 **119.** 2.5×10^6 **121.** 1.25×10^8 **123.** 8.1×10^{-7} **125.** 2.5×10^{-7} **127.** 1 **129.** $\dfrac{y}{16x^8z^6}$ **131.** $\dfrac{1}{x^{12}y^{16}z^{20}}$

133. $\dfrac{x^{18}y^6}{4}$ **135.** 2.5×10^{-3} **137.** 8×10^{-5} **139. a.** 1.35×10^{12} **b.** 3.07×10^8 **c.** \$4.40 × 10^3$; \$4400 **141.** 42,000 years

143. 1.25 sec **153.** makes sense **155.** makes sense **157.** true **159.** false **161.** false **163.** true **169.** $(-\infty, 2)$ or $\{x \mid x < 2\}$
170. 5 **171.** $0, \sqrt{16}$ **172.** $16x^5 - 12x^4 + 4x^3$ **173.** $27x^2y^3 - 9xy^2 + 81xy$ **174.** $x^3 + 3x^2 + 5x + 15$

Review Exercises

1. binomial, 4 **2.** trinomial, 2 **3.** monomial, 1 **4.** $8x^3 + 10x^2 - 20x - 4$ **5.** $13y^3 - 8y^2 + 7y - 5$ **6.** $11y^2 - 4y - 4$
7. $8x^4 - 5x^3 + 6$ **8.** $-14x^4 - 13x^2 + 16x$ **9.** $7y^4 - 5y^3 + 3y^2 - y - 4$ **10.** $3x^2 - 7x + 9$ **11.** $10x^3 - 9x^2 + 2x + 11$

12.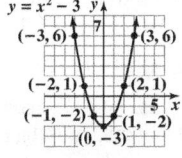

13.

14. x^{23} **15.** y^{14} **16.** x^{100} **17.** $100y^2$ **18.** $-64x^{30}$ **19.** $50x^4$ **20.** $-36y^{11}$ **21.** $30x^{12}$ **22.** $21x^3 + 63x$ **23.** $20x^5 - 55x^4$
24. $-21y^4 + 9y^3 - 18y^2$ **25.** $16y^8 - 20y^7 + 2y^5$ **26.** $x^3 - 2x^2 - 13x + 6$ **27.** $12y^3 + y^2 - 21y + 10$ **28.** $3y^3 - 17y^2 + 41y - 35$
29. $8x^4 + 8x^3 - 18x^2 - 20x - 3$ **30.** $x^2 + 8x + 12$ **31.** $6y^2 - 7y - 5$ **32.** $4x^4 - 14x^2 + 6$ **33.** $25x^2 - 16$ **34.** $49 - 4y^2$
35. $y^4 - 1$ **36.** $x^2 + 6x + 9$ **37.** $9y^2 + 24y + 16$ **38.** $y^2 - 2y + 1$ **39.** $25y^2 - 20y + 4$ **40.** $x^4 + 8x^2 + 16$ **41.** $x^4 - 16$
42. $x^4 - x^2 - 20$ **43.** $x^2 + 7x + 12$ **44.** $x^2 + 50x + 600 \text{ yd}^2$ **45.** 28

46. polynomial degree: 5;

Term	Coefficient	Degree
$4x^2y$	4	3
$9x^3y^2$	9	5
$-17x^4$	-17	4
-12	-12	0

47. $-x^2 - 17xy + 5y^2$ **48.** $2x^3y^2 + x^2y - 6x^2 - 4$ **49.** $-35x^6y^9$ **50.** $15a^3b^5 - 20a^2b^3$ **51.** $3x^2 + 16xy - 35y^2$
52. $36x^2y^2 - 31xy + 3$ **53.** $9x^2 + 30xy + 25y^2$ **54.** $x^2y^2 - 14xy + 49$ **55.** $49x^2 - 16y^2$ **56.** $a^3 - b^3$ **57.** 6^{30} **58.** x^{15}
59. 1 **60.** -1 **61.** 400 **62.** $\dfrac{x^{12}}{8}$ **63.** $\dfrac{81}{16y^{24}}$ **64.** $-5y^6$ **65.** $8x^7y^3$ **66.** $3x^3 - 2x + 6$ **67.** $-6x^5 + 5x^4 + 8x^2$
68. $9x^2y - 3x - 6y$ **69.** $2x + 7$ **70.** $x^2 - 3x + 5$ **71.** $x^2 + 5x + 2 + \dfrac{7}{x - 7}$ **72.** $y^2 + 3y + 9$ **73.** $2x^2 + 3x - 1$
74. $4x^2 - 7x + 5 - \dfrac{4}{x + 1}$ **75.** $3x^3 + 6x^2 + 10x + 10$ **76.** $x^3 - 4x^2 - 16x - 64 + \dfrac{272}{x + 4}$ **77.** $\dfrac{1}{7^2} = \dfrac{1}{49}$ **78.** $\dfrac{1}{(-4)^3} = -\dfrac{1}{64}$
79. $\dfrac{1}{2^1} + \dfrac{1}{4^1} = \dfrac{1}{2} + \dfrac{1}{4} = \dfrac{3}{4}$ **80.** $5^2 = 25$ **81.** $\dfrac{5^3}{2^3} = \dfrac{125}{8}$ **82.** $\dfrac{1}{x^6}$ **83.** $\dfrac{6}{y^2}$ **84.** $\dfrac{30}{x^5}$ **85.** x^8 **86.** $81y^2$ **87.** $\dfrac{1}{y^{19}}$
88. $\dfrac{x^3}{8}$ **89.** $\dfrac{1}{x^6}$ **90.** y^{20} **91.** 23,000 **92.** 0.00176 **93.** 0.9 **94.** 7.39×10^7 **95.** 6.2×10^{-4} **96.** 3.8×10^{-1} **97.** 3.8×10^0
98. 9×10^3 **99.** 5×10^4 **100.** 1.6×10^{-3} **101.** $\$5.36 \times 10^{10}$ **102.** 3.07×10^8 **103.** $\approx \$175$

Chapter Test

1. trinomial, 2 **2.** $13x^3 + x^2 - x - 24$ **3.** $5x^3 + 2x^2 + 2x - 9$ **4.**

5. $-35x^{11}$ **6.** $48x^5 - 30x^3 - 12x^2$ **7.** $3x^3 - 10x^2 - 17x - 6$ **8.** $6y^2 - 13y - 63$ **9.** $49x^2 - 25$ **10.** $x^4 + 6x^2 + 9$
11. $25x^2 - 30x + 9$ **12.** 30 **13.** $2x^2y^3 + 3xy + 12y^2$ **14.** $12a^2 - 13ab - 35b^2$ **15.** $4x^2 + 12xy + 9y^2$ **16.** $-5x^{12}$
17. $3x^3 - 2x^2 + 5x$ **18.** $x^2 - 2x + 3 + \dfrac{1}{2x + 1}$ **19.** $3x^3 - 4x^2 + 7$ **20.** $\dfrac{1}{10^2} = \dfrac{1}{100}$ **21.** $4^3 = 64$ **22.** $-27x^6$ **23.** $\dfrac{4}{x^5}$ **24.** $-\dfrac{21}{x^6}$
25. $16y^4$ **26.** $\dfrac{x^8}{25}$ **27.** $\dfrac{1}{x^{15}}$ **28.** 0.00037 **29.** 7.6×10^6 **30.** 1.23×10^{-2} **31.** 2.1×10^8 **32.** $x^2 + 10x + 16$

Cumulative Review Exercises

1. $\dfrac{35}{9}$ **2.** -128 **3.** 15,050 ft **4.** $-\dfrac{2}{3}$ or $\left\{-\dfrac{2}{3}\right\}$ **5.** $-\dfrac{5}{3}$ or $\left\{-\dfrac{5}{3}\right\}$ **6.** 4 ft by 10 ft

7. $[6, \infty)$ or $\{x \mid x \ge 6\}$; **8.** \$3400 at 12% and \$2600 at 14%

9. 15 liters of 70% and 5 liters of 30% **10.** **11.**

12. $-\dfrac{6}{5}$; falling **13.** $y + 1 = -2(x - 3); y = -2x + 5$ **14.** $(0, 5)$ or $\{(0, 5)\}$ **15.** $(3, -4)$ or $\{(3, -4)\}$ **16.** 500 min; \$40
17. 2.4×10^{-3} **18.** $3x^5 - 6x^3 + 9x + 2$ **19.** $x^2 + 2x + 3$ **20.** $\dfrac{81}{x^2}$

CHAPTER 6

Section 6.1 Check Point Exercises

1. a. $3x^2$ **b.** $4x^2$ **c.** x^2y **2.** $6(x^2 + 3)$ **3.** $5x^2(5 + 7x)$ **4.** $3x^3(5x^2 + 4x - 9)$ **5.** $2xy(4x^2y - 7x + 1)$
6. $-4ab^2(4a^3b^3 - 6a^2b^2 + 5)$ **7. a.** $(x + 1)(x^2 + 7)$ **b.** $(y + 4)(x - 7)$ **8.** $(x + 5)(x^2 + 2)$ **9.** $(y + 3)(x - 5)$

Concept and Vocabulary Check

1. factoring **2.** greatest common factor; smallest/least **3.** false **4.** false

Exercise Set 6.1

1. 4 **3.** $4x$ **5.** $2x^3$ **7.** $3y$ **9.** xy **11.** $4x^4y^3$ **13.** $8(x + 1)$ **15.** $4(y - 1)$ **17.** $5(x + 6)$ **19.** $6(5x - 2)$ **21.** $x(x + 5)$
23. $6(3y^2 + 2)$ **25.** $7x^2(2x + 3)$ **27.** $y(13y - 25)$ **29.** $9y^4(1 + 3y^2)$ **31.** $4x^2(2 - x^2)$ **33.** $4(3y^2 + 4y - 2)$ **35.** $3x^2(3x^2 + 6x + 2)$
37. $50y^2(2y^3 - y + 2)$ **39.** $5x(2 - 4x + x^2)$ **41.** cannot be factored **43.** $3xy(2x^2y + 3)$ **45.** $10xy(3xy^2 - y + 2)$
47. $8x^2y(4xy - 3x - 2)$ **49.** $-6(2x^2 - 3)$ **51.** $-8x^2(x^2 - 4x - 2)$ **53.** $-2ab(2a^2b - 3)$ **55.** $-6x^2y(2xy + 3x - 4)$ **57.** $(x + 5)(x + 3)$
59. $(x + 2)(x - 4)$ **61.** $(y + 6)(x - 7)$ **63.** $(x + y)(3x - 1)$ **65.** $(3x + 1)(4x + 1)$ **67.** $(5x + 4)(7x^2 + 1)$ **69.** $(x + 2)(x + 4)$
71. $(x - 5)(x + 3)$ **73.** $(x^2 + 5)(x - 2)$ **75.** $(x^2 + 2)(x - 1)$ **77.** $(y + 5)(x + 9)$ **79.** $(y - 1)(x + 5)$ **81.** $(x - 2y)(3x + 5y)$
83. $(3x - 2)(x^2 - 2)$ **85.** $(x - a)(x - b)$ **87.** $6x^2yz(4xy^2z^2 + 5y + 3z)$ **89.** $(x^3 - 4)(1 + 3y)$
91. $2x^2(x + 1)(2x^3 - 3x - 4)$ **93.** $(x - 1)(3x^4 + x^2 + 5)$ **95.** $36x^2 - 4\pi x^2; 4x^2(9 - \pi)$ **97. a.** 48 ft **b.** $16x(4 - x)$
c. 48 ft; yes; no; Answers will vary. **99.** $x^3 - 2$ **107.** makes sense **109.** does not make sense
111. false **113.** false **115.** $4(x + 150)$ **119.** $(x - 2)(x - 5)$ should be $(x - 2)(x + 5)$ **121.** $x^2 + 17x + 70$ **122.** $(-3, -2)$ or $\{(-3, -2)\}$
123. $y - 2 = 1(x + 7)$ or $y - 5 = 1(x + 4); y = x + 9$ **124.** 2 and 4 **125.** -3 and -2 **126.** -5 and 7

Section 6.2 Check Point Exercises

1. $(x + 2)(x + 3)$ **2.** $(x - 2)(x - 4)$ **3.** $(x + 5)(x - 2)$ **4.** $(y - 9)(y + 3)$ **5.** cannot factor over the integers; prime
6. $(x - 3y)(x - y)$ **7.** $2x(x - 4)(x + 7)$ **8.** $-2(y - 2)(y + 7)$

Concept and Vocabulary Check

1. $20; -12$ **2.** completely **3.** $+ 10$ **4.** $- 6$ **5.** $+ 5$ **6.** $- 7$ **7.** $- 2y$

Exercise Set 6.2

1. $(x + 6)(x + 1)$ **3.** $(x + 2)(x + 5)$ **5.** $(x + 1)(x + 10)$ **7.** $(x - 4)(x - 3)$ **9.** $(x - 6)(x - 6)$ **11.** $(y - 3)(y - 5)$
13. $(x + 5)(x - 2)$ **15.** $(y + 13)(y - 3)$ **17.** $(x - 5)(x + 3)$ **19.** $(x - 4)(x + 2)$ **21.** prime **23.** $(y - 4)(y - 12)$ **25.** prime
27. $(w - 32)(w + 2)$ **29.** $(y - 5)(y - 13)$ **31.** $(r + 3)(r + 9)$ **33.** prime **35.** $(x + 6y)(x + y)$ **37.** $(x - 3y)(x - 5y)$
39. $(x - 6y)(x + 3y)$ **41.** $(a - 15b)(a - 3b)$ **43.** $3(x + 2)(x + 3)$ **45.** $4(y - 2)(y + 1)$ **47.** $10(x - 10)(x + 6)$
49. $3(x - 2)(x - 9)$ **51.** $2r(r + 2)(r + 1)$ **53.** $4x(x + 6)(x - 3)$ **55.** $2r(r + 8)(r - 4)$ **57.** $y^2(y + 10)(y - 8)$
59. $x^2(x - 5)(x + 2)$ **61.** $2w^2(w - 16)(w + 3)$ **63.** $15x(y - 1)(y + 4)$ **65.** $x^3(x - y)(x + 4y)$ **67.** $-16(t - 5)(t + 1)$
69. $-5(x - 9)(x - 1)$ **71.** $-(x + 8)(x - 5)$ **73.** $-2x(x + 4)(x - 1)$ **75.** $2x^2(y - 15z)(y - z)$ **77.** $(a + b)(x + 5)(x - 4)$
79. $(x + 0.3)(x + 0.2)$ **81.** $\left(x - \dfrac{1}{5}\right)\left(x - \dfrac{1}{5}\right)$ **83. a.** $-16(t - 2)(t + 1)$ **b.** 0; yes; After 2 seconds, you hit the water.

89. does not make sense **91.** does not make sense **93.** false **95.** true **97.** 8, 16 **101.** $x(x + 1)(x + 2)$; the product
of three consecutive integers **103.** correctly factored **105.** $(x + 1)(x - 1)$ should be $(x - 1)(x - 1)$. **107.** 13 or $\{13\}$
108. 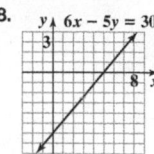 **109.** 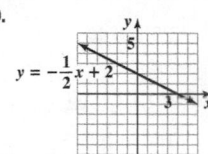 **110.** $2x^2 - x - 6$ **111.** $9x^2 + 15x + 4$ **112.** $(4x - 1)(2x - 5)$

Section 6.3 Check Point Exercises

1. $(5x - 4)(x - 2)$ **2.** $(3x - 1)(2x + 7)$ **3.** $(3x - y)(x - 4y)$ **4.** $(3x + 5)(x - 2)$ **5.** $(2x - 1)(4x - 3)$ **6.** $y^2(5y + 3)(y + 2)$

Concept and Vocabulary Check

1. greatest common factor **2.** $- 3$ **3.** $- 4$ **4.** $2x - 3$ **5.** $3x + 4$ **6.** $x - 2y$

Exercise Set 6.3

1. $(2x + 3)(x + 1)$ **3.** $(3x + 1)(x + 4)$ **5.** $(2x + 3)(x + 4)$ **7.** $(5y - 1)(y - 3)$ **9.** $(3y + 4)(y - 1)$ **11.** $(3x - 2)(x + 5)$
13. $(3x - 1)(x - 7)$ **15.** $(5y - 1)(y - 3)$ **17.** $(3x - 2)(x - 5)$ **19.** $(3w - 4)(2w - 1)$ **21.** $(8x + 1)(x + 4)$
23. $(5x - 2)(x + 7)$ **25.** $(7y - 3)(2y + 3)$ **27.** prime **29.** $(5z - 3)(5z - 3)$ **31.** $(3y + 1)(5y - 2)$ **33.** prime
35. $(5y - 1)(2y + 9)$ **37.** $(4x + 1)(2x - 1)$ **39.** $(3y - 1)(3y - 2)$ **41.** $(5x + 8)(4x - 1)$ **43.** $(2x + y)(x + y)$
45. $(3x + 2y)(x + y)$ **47.** $(2x - 3y)(x - 3y)$ **49.** $(2x - 3y)(3x + 2y)$ **51.** $(3x - 2y)(5x + 7y)$ **53.** $(2a + 5b)(a + b)$
55. $(3a - 2b)(5a + 3b)$ **57.** $(3x - 4y)(4x - 3y)$ **59.** $2(2x + 3)(x + 5)$ **61.** $3(3x + 4)(x - 2)$ **63.** $2(2y - 5)(y + 3)$
65. $3(3y - 4)(y + 5)$ **67.** $x(3x + 1)(x + 1)$ **69.** $x(2x - 5)(x + 1)$ **71.** $3y(3y - 1)(y - 4)$ **73.** $5z(6z + 1)(2z + 1)$

75. $3x^2(5x - 3)(x - 2)$ **77.** $x^3(2x - 3)(5x - 1)$ **79.** $3(2x + 3y)(x - 2y)$ **81.** $2(2x - y)(3x + 4y)$ **83.** $2y(4x - 7)(x + 6)$
85. $2b(2a - 7b)(3a - b)$ **87.** $-4y^4(8x + 3)(x - 1)$ **89.** $10(y + 1)(x + 1)(3x - 2)$ **91. a.** $(2x + 1)(x - 3)$ **b.** $(2y + 3)(y - 2)$
93. $(x - 2)(3x - 2)(x - 1)$ **95. a.** $x^2 + 3x + 2$ **b.** $(x + 2)(x + 1)$ **c.** $x^2 + 3x + 2 = (x + 2)(x + 1)$ **101.** makes sense
103. does not make sense   **105.** true   **107.** false   **109.** $5, 7, -5, -7$ **111.** $(3x^5 + 5)(x^5 - 3)$ **113.** $(25, -5)$ or $\{(25, -5)\}$
114. 8.6×10^{-4} **115.** -1 or $\{-1\}$ **116.** $81x^2 - 100$ **117.** $16x^2 + 40xy + 25y^2$ **118.** $x^3 + 8$

Mid-Chapter Check Point Exercises

1. $x^4(x + 1)$ **2.** $(x + 9)(x - 2)$ **3.** $x^2y(y^2 - y + 1)$ **4.** prime   **5.** $(7x - 1)(x - 3)$ **6.** $(x^2 + 3)(x + 5)$ **7.** $x(2x - 1)(x - 5)$
8. $(x - 4)(y - 7)$ **9.** $(x - 15y)(x - 2y)$ **10.** $(5x + 2)(5x - 7)$ **11.** $2(8x - 3)(x - 4)$ **12.** $(3x + 7y)(x + y)$ **13.** $-2x(3x + 5)(x - 3)$

Section 6.4 Check Point Exercises

1. a. $(x + 9)(x - 9)$ **b.** $(6x + 5)(6x - 5)$ **2. a.** $(5 + 2x^5)(5 - 2x^5)$ **b.** $(10x + 3y)(10x - 3y)$ **3. a.** $2x(3x + 1)(3x - 1)$
b. $18(2 + x)(2 - x)$ or $-18(x + 2)(x - 2)$ **4.** $(9x^2 + 4)(3x + 2)(3x - 2)$ **5. a.** $(x + 7)^2$ **b.** $(x - 3)^2$ **c.** $(4x - 7)^2$ **6.** $(2x + 3y)^2$
7. $(x + 3)(x^2 - 3x + 9)$ **8.** $(1 - y)(1 + y + y^2)$ **9.** $(5x + 2)(25x^2 - 10x + 4)$

Concept and Vocabulary Check

1. $(A + B)(A - B)$ **2.** $(A + B)^2$ **3.** $(A - B)^2$ **4.** $(A + B)(A^2 - AB + B^2)$ **5.** $(A - B)(A^2 + AB + B^2)$
6. $6x; 6x$ **7.** -6 **8.** $4x$ **9.** $+2$ **10.** $-3; +9$ **11.** false   **12.** true   **13.** false   **14.** true   **15.** false

Exercise Set 6.4

1. $(x + 5)(x - 5)$ **3.** $(y + 1)(y - 1)$ **5.** $(2x + 3)(2x - 3)$ **7.** $(5 + x)(5 - x)$ **9.** $(1 + 7x)(1 - 7x)$ **11.** $(3 + 5y)(3 - 5y)$
13. $(x^2 + 3)(x^2 - 3)$ **15.** $(7y^2 + 4)(7y^2 - 4)$ **17.** $(x^5 + 3)(x^5 - 3)$ **19.** $(5x + 4y)(5x - 4y)$ **21.** $(x^2 + y^5)(x^2 - y^5)$
23. $(x^2 + 4)(x + 2)(x - 2)$ **25.** $(4x^2 + 9)(2x + 3)(2x - 3)$ **27.** $2(x + 3)(x - 3)$ **29.** $2x(x + 6)(x - 6)$ **31.** prime   **33.** $3x(x^2 + 9)$
35. $2(3 + y)(3 - y)$ **37.** $3y(y + 4)(y - 4)$ **39.** $2x(3x + 1)(3x - 1)$ **41.** $-3(x + 5)(x - 5)$ **43.** $-5y(y + 2)(y - 2)$
45. $(x + 1)^2$ **47.** $(x - 7)^2$ **49.** $(x - 1)^2$ **51.** $(x + 11)^2$ **53.** $(2x + 1)^2$ **55.** $(5y - 1)^2$ **57.** prime   **59.** $(x + 7y)^2$ **61.** $(x - 6y)^2$
63. prime   **65.** $(4x - 5y)^2$ **67.** $3(2x - 1)^2$ **69.** $x(3x + 1)^2$ **71.** $2(y - 1)^2$ **73.** $2y(y + 7)^2$ **75.** $-6(x - 2)^2$ **77.** $-4y(2y + 1)^2$
79. $(x + 1)(x^2 - x + 1)$ **81.** $(x - 3)(x^2 + 3x + 9)$ **83.** $(2y - 1)(4y^2 + 2y + 1)$ **85.** $(3x + 2)(9x^2 - 6x + 4)$
87. $(xy - 4)(x^2y^2 + 4xy + 16)$ **89.** $y(3y + 2)(9y^2 - 6y + 4)$ **91.** $2(3 - 2y)(9 + 6y + 4y^2)$ **93.** $(4x + 3y)(16x^2 - 12xy + 9y^2)$
95. $(5x - 4y)(25x^2 + 20xy + 16y^2)$ **97.** $\left(5x + \dfrac{2}{7}\right)\left(5x - \dfrac{2}{7}\right)$ **99.** $y\left(y - \dfrac{1}{10}\right)\left(y^2 + \dfrac{y}{10} + \dfrac{1}{100}\right)$ **101.** $x(0.5 + x)(0.5 - x)$
103. $(x + 6)(x - 4)$ **105.** $(x - 3)(x + 1)^2$ **107.** $x^2 - 25 = (x + 5)(x - 5)$ **109.** $x^2 - 16 = (x + 4)(x - 4)$
115. does not make sense   **117.** does not make sense   **119.** false   **121.** false   **123.** $a - b = 0$ and division by 0 is not permitted.
125. $(x^n + 5y^n)(x^n - 5y^n)$ **127.** $[(x + 3) - 1]^2$ or $(x + 2)^2$ **129.** 1   **131.** correctly factored
133. $(x - 1)(x^2 - x + 1)$ should be $(x - 1)(x^2 + x + 1)$.   **134.** $80x^9y^{14}$ **135.** $-4x^2 + 3$
136. $2x + 5$ **137.** $3x(x + 5)(x - 5)$ **138.** $2(x - 5)^2$ **139.** $(x - 2)(x + 1)(x - 1)$

Section 6.5 Check Point Exercises

1. $5x^2(x + 3)(x - 3)$ **2.** $4(x - 6)(x + 2)$ **3.** $4x(x^2 + 4)(x + 2)(x - 2)$ **4.** $(x - 4)(x + 3)(x - 3)$ **5.** $3x(x - 5)^2$
6. $2x^2(x + 3)(x^2 - 3x + 9)$ **7.** $3y(x^2 + 4y^2)(x + 2y)(x - 2y)$ **8.** $3x(2x + 3y)^2$

Concept and Vocabulary Check

1. b   **2.** e   **3.** h   **4.** c   **5.** d   **6.** f   **7.** a   **8.** g

Exercise Set 6.5

1. $-7x(x - 5)$ **3.** $(5x + 7)(5x - 7)$ **5.** $(3x - 1)(9x^2 + 3x + 1)$ **7.** $(5 + x)(x + y)$ **9.** $(7x - 1)(2x - 1)$ **11.** $(x - 1)^2$
13. $(3xy + 2)(9x^2y^2 - 6xy + 4)$ **15.** $(3x + 5)(2x - 3)$ **17.** $5x(x + 2)(x - 2)$ **19.** $7x(x^2 + 1)$ **21.** $5(x - 3)(x + 2)$
23. $2(x^2 + 9)(x + 3)(x - 3)$ **25.** $(x + 2)(x + 3)(x - 3)$ **27.** $3x(x - 4)^2$ **29.** $2x^2(x + 1)(x^2 - x + 1)$ **31.** $2x(3x + 4)$
33. $-2(y - 8)(y + 7)$ **35.** $7y^2(y + 1)^2$ **37.** prime   **39.** $2(4y + 1)(2y - 1)$ **41.** $r(r - 25)$ **43.** $(2w + 5)(2w - 1)$
45. $x(x + 2)(x - 2)$ **47.** prime   **49.** $(9y + 4)(y + 1)$ **51.** $(y + 2)(y + 2)(y - 2)$ **53.** $(4y + 3)^2$ **55.** $-4y(y - 5)(y - 2)$
57. $y(y^2 + 9)(y + 3)(y - 3)$ **59.** $5a^2(2a + 3)(2a - 3)$ **61.** $3x^2(3x^2 + 6x + 2)$ **63.** $(4y - 1)(3y - 2)$ **65.** $(3y + 8)(3y - 8)$ **67.** prime
69. $(2y + 3)(y + 5)(y - 5)$ **71.** $2r(r + 17)(r - 2)$ **73.** $2x^3(2x + 1)(2x - 1)$ **75.** $3(x^2 + 81)$ **77.** $x(x + 2)(x^2 - 2x + 4)$
79. $2y^2(y - 1)(y^2 + y + 1)$ **81.** $2x(3x + 4y)$ **83.** $(y - 7)(x + 3)$ **85.** $(x - 4y)(x + y)$ **87.** $12a^2(6ab^2 + 1 - 2a^2b^2)$
89. $3(a + 6b)(a + 3b)$ **91.** $3x^2y(4x + 1)(4x - 1)$ **93.** $b(3a + 2)(2a - 1)$ **95.** $7xy(x^2 + y^2)(x + y)(x - y)$ **97.** $2xy(5x - 2y)(x - y)$
99. $2b(x + 11)^2$ **101.** $(5a + 7b)(3a - 2b)$ **103.** $-2xy(9x - 2y)(2x - 3y)$ **105.** $(y - x)(a + b)(a - b)$ **107.** $ax(3x + 7)(3x - 2)$
109. $2x^2(x^2 + 3xy + y^2)$ **111.** $y(9x^2 + y^2)(3x + y)(3x - y)$ **113.** $(x + 1)(5x - 6)(2x + 1)$ **115.** $(x^2 + 6)(6x^2 - 1)$
117. $(x - 7 + 2a)(x - 7 - 2a)$ **119.** $(x + 4 + 5a)(x + 4 - 5a)$ **121.** $y(y^2 + 1)(y^4 - y^2 + 1)$ **123.** $16(4 + t)(4 - t)$
125. $\pi b^2 - \pi a^2; \pi(b + a)(b - a)$ **129.** makes sense   **131.** makes sense   **133.** false   **135.** false   **137.** $3x(x^2 + 2)(x + 3)(x - 3)$
139. $(4x^2 - 5)(x + 1)(x - 1)$ **141.** $3(x^n + 3y^n)(x^n - 3y^n)$ **143.** $3x(x + 5)(x - 1)$ should be $3x(x - 5)(x + 1)$.
145. correctly factored   **147.** $(3x + 4)(3x - 4)$ **148.** $5x - 2y = 10$ **149.** $20°, 60°, 100°$ **150.** 0   **151.** 0   **152.** $(x + 4)(x - 3)$

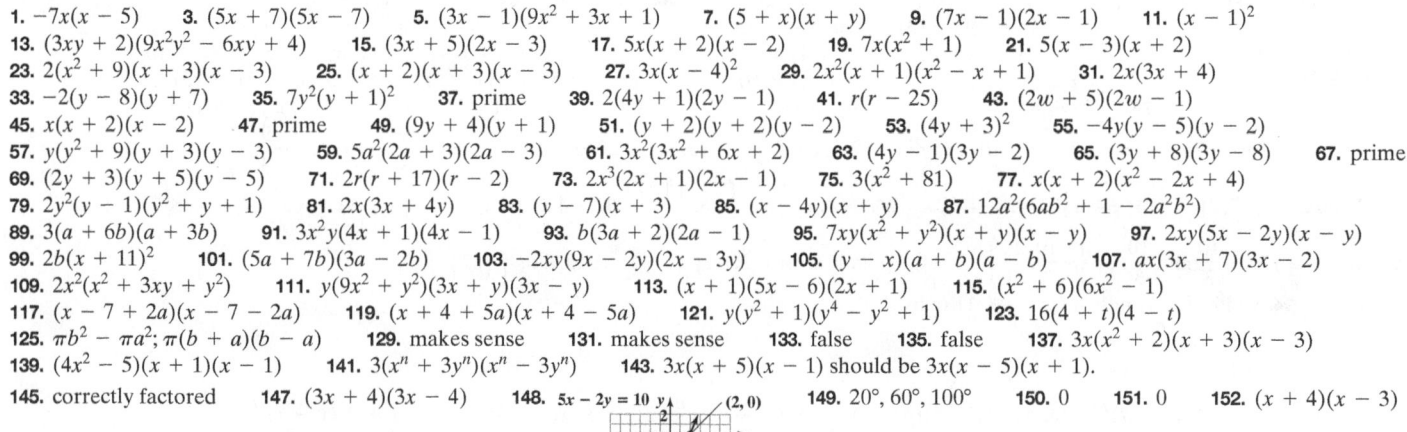

Section 6.6 Check Point Exercises

1. $-\frac{1}{2}$ and 4, or $\left\{-\frac{1}{2}, 4\right\}$ **2.** 1 and 5, or $\{1, 5\}$ **3.** 0 and $\frac{1}{2}$, or $\left\{0, \frac{1}{2}\right\}$ **4.** 5 or $\{5\}$ **5.** $-\frac{5}{4}$ and $\frac{5}{4}$, or $\left\{-\frac{5}{4}, \frac{5}{4}\right\}$ **6.** -2 and 9, or $\{-2, 9\}$
7. 1 sec and 2 sec; $(1, 192)$ and $(2, 192)$ **8.** length: 9ft; width: 6ft

Concept and Vocabulary Check

1. quadratic equation **2.** $A = 0$ or $B = 0$ **3.** x-intercepts **4.** subtracting $5x$ **5.** subtracting $30x$; adding 25

Exercise Set 6.6

1. -7 and 0, or $\{-7, 0\}$ **3.** -4 and 6, or $\{-4, 6\}$ **5.** $-\frac{4}{5}$ and 9, or $\left\{-\frac{4}{5}, 9\right\}$ **7.** $-\frac{9}{2}$ and 4, or $\left\{-\frac{9}{2}, 4\right\}$ **9.** -5 and -3, or $\{-5, -3\}$
11. -3 and 5, or $\{-3, 5\}$ **13.** -3 and 7, or $\{-3, 7\}$ **15.** -8 and -1, or $\{-8, -1\}$ **17.** -4 and 0, or $\{-4, 0\}$ **19.** 0 and 5, or $\{0, 5\}$
21. 0 and 4, or $\{0, 4\}$ **23.** 0 and $\frac{5}{2}$, or $\left\{0, \frac{5}{2}\right\}$ **25.** $-\frac{5}{3}$ and 0, or $\left\{-\frac{5}{3}, 0\right\}$ **27.** -2 or $\{-2\}$ **29.** 6 or $\{6\}$ **31.** $\frac{3}{2}$ or $\left\{\frac{3}{2}\right\}$
33. $-\frac{1}{2}$ and 4, or $\left\{-\frac{1}{2}, 4\right\}$ **35.** -2 and $\frac{9}{5}$, or $\left\{-2, \frac{9}{5}\right\}$ **37.** -7 and 7, or $\{-7, 7\}$ **39.** $-\frac{5}{2}$ and $\frac{5}{2}$, or $\left\{-\frac{5}{2}, \frac{5}{2}\right\}$ **41.** $-\frac{5}{9}$ and $\frac{5}{9}$, or $\left\{-\frac{5}{9}, \frac{5}{9}\right\}$
43. -3 and 7, or $\{-3, 7\}$ **45.** $-\frac{5}{2}$ and $\frac{3}{2}$, or $\left\{-\frac{5}{2}, \frac{3}{2}\right\}$ **47.** -6 and 3, or $\{-6, 3\}$ **49.** -2 and $-\frac{3}{2}$, or $\left\{-2, -\frac{3}{2}\right\}$ **51.** 4 or $\{4\}$
53. $-\frac{5}{2}$ or $\left\{-\frac{5}{2}\right\}$ **55.** $\frac{3}{8}$ or $\left\{\frac{3}{8}\right\}$ **57.** $-3, -2,$ and 4, or $\{-3, -2, 4\}$ **59.** $-6, 0,$ and 6, or $\{-6, 0, 6\}$ **61.** $-2, -1,$ and 0, or $\{-2, -1, 0\}$
63. 2 and 8, or $\{2, 8\}$ **65.** 4 and 5, or $\{4, 5\}$ **67.** 5 sec; Each tick represents one second. **69.** 2 sec; $(2, 276)$ **71.** $\frac{1}{2}$ sec and 4 sec
73. 2002 and 2004 **75.** $(2, 66)$ and $(4, 66)$ **77.** 10 yr **79.** $(10, 7250)$ **81.** 10 teams **83.** length: 15 yd; width: 12 yd
85. base: 5 cm; height: 6 cm **87. a.** $4x^2 + 44x$ **b.** 3 ft **91.** does not make sense **93.** makes sense **95.** false **97.** false
99. $x^2 - 2x - 15 = 0$ **101.** 4 and 5, or $\{4, 5\}$ **103.** c **105.** d **107.** -4 and 1, or $\{-4, 1\}$ **109.** -4 and 3, or $\{-4, 3\}$

113.

$y = -\frac{2}{3}x + 1$

114. $\dfrac{4}{x^6}$ **115.** -2 or $\{-2\}$ **116.** 375 **117.** When x is replaced with 4, the denominator is 0.

118. $\dfrac{(x + 5)(x + 1)}{(x + 5)(x - 5)}; \dfrac{x + 1}{x - 5}$

Review Exercises

1. $15(2x - 3)$ **2.** $-4x(3x^2 + 4x - 100)$ **3.** $5x^2y(6x^2 + 3x + 1)$ **4.** $5(x + 3)$ **5.** $(7x^2 - 1)(x + y)$ **6.** $(x^2 + 2)(x + 3)$
7. $(x + 1)(y + 4)$ **8.** $(x^2 + 5)(x + 1)$ **9.** $(x - 2)(y + 4)$ **10.** $(x - 2)(x - 1)$ **11.** $(x - 5)(x + 4)$ **12.** $(x + 3)(x + 16)$
13. $(x - 4y)(x - 2y)$ **14.** prime **15.** $(x + 17y)(x - y)$ **16.** $3(x + 4)(x - 2)$ **17.** $3x(x - 11)(x - 1)$ **18.** $(x + 5)(3x + 2)$
19. $(y - 3)(5y - 2)$ **20.** $(2x + 5)(2x - 3)$ **21.** prime **22.** $-2(2x + 3)(2x - 1)$ **23.** $x(2x - 9)(x + 8)$ **24.** $4y(3y + 1)(y + 2)$
25. $(2x - y)(x - 3y)$ **26.** $(5x + 4y)(x - 2y)$ **27.** $(2x + 1)(2x - 1)$ **28.** $(9 + 10y)(9 - 10y)$ **29.** $(5a + 7b)(5a - 7b)$
30. $(z^2 + 4)(z + 2)(z - 2)$ **31.** $2(x + 3)(x - 3)$ **32.** prime **33.** $x(3x + 1)(3x - 1)$ **34.** $2x(3y + 2)(3y - 2)$ **35.** $(x + 11)^2$
36. $(x - 8)^2$ **37.** $(3y + 8)^2$ **38.** $(4x - 5)^2$ **39.** prime **40.** $(6x + 5y)^2$ **41.** $(5x - 4y)^2$ **42.** $(x - 3)(x^2 + 3x + 9)$
43. $(4x + 1)(16x^2 - 4x + 1)$ **44.** $2(3x - 2y)(9x^2 + 6xy + 4y^2)$ **45.** $y(3x + 2)(9x^2 - 6x + 4)$ **46.** $(a + 3)(a - 3)$
47. $(a + 2b)(a - 2b)$ **48.** $A^2 + 2A + 1 = (A + 1)^2$ **49.** $x(x - 7)(x - 1)$ **50.** $(5y + 2)(2y + 1)$ **51.** $2(8 + y)(8 - y)$
52. $(3x + 1)^2$ **53.** $-4x^3(5x^4 - 9)$ **54.** $(x - 3)^2(x + 3)$ **55.** prime **56.** $x(2x + 5)(x + 7)$ **57.** $3x(x - 5)^2$
58. $3x^2(x - 2)(x^2 + 2x + 4)$ **59.** $4y^2(y + 3)(y - 3)$ **60.** $5(x + 7)(x - 3)$ **61.** prime **62.** $-2x^3(5x - 2)(x - 4)$
63. $(10y + 7)(10y - 7)$ **64.** $9x^4(x - 2)$ **65.** $(x^2 + 1)(x + 1)(x - 1)$ **66.** $2(y - 2)(y^2 + 2y + 4)$ **67.** $(x + 4)(x^2 - 4x + 16)$
68. $(3x - 2)(2x + 5)$ **69.** $3x^2(x + 2)(x - 2)$ **70.** $(x - 10)(x + 9)$ **71.** $(5x + 2y)(5x + 3y)$ **72.** $x(x + 5)(x^2 - 5x + 25)$
73. $2y(4y + 3)(4y + 1)$ **74.** $-2(y - 4)^2$ **75.** $(x + 5y)(x - 7y)$ **76.** $(x + y)(x + 7)$ **77.** $(3x + 4y)^2$ **78.** $2x^2y(x + 1)(x - 1)$
79. $(10y + 7z)(10y - 7z)$ **80.** prime **81.** $3x^2y^2(x + 2y)(x - 2y)$ **82.** 0 and 12, or $\{0, 12\}$ **83.** $-\frac{9}{4}$ and 7, or $\left\{-\frac{9}{4}, 7\right\}$
84. -7 and 2, or $\{-7, 2\}$ **85.** -4 and 0, or $\{-4, 0\}$ **86.** -8 and $\frac{1}{2}$, or $\left\{-8, \frac{1}{2}\right\}$ **87.** -4 and 8, or $\{-4, 8\}$ **88.** -8 and 7, or $\{-8, 7\}$
89. 7 or $\{7\}$ **90.** $-\frac{10}{3}$ and $\frac{10}{3}$, or $\left\{-\frac{10}{3}, \frac{10}{3}\right\}$ **91.** -5 and -2, or $\{-5, -2\}$ **92.** $\frac{1}{3}$ and 7, or $\left\{\frac{1}{3}, 7\right\}$
93. 2 sec **94.** width: 5 ft; length: 8 ft **95.** 11 m by 11 m

Chapter Test

1. $(x - 3)(x - 6)$ **2.** $(x - 7)^2$ **3.** $5y^2(3y - 1)(y - 2)$ **4.** $(x^2 + 3)(x + 2)$ **5.** $x(x - 9)$ **6.** $x(x + 7)(x - 1)$
7. $2(7x - 3)(x + 5)$ **8.** $(5x + 3)(5x - 3)$ **9.** $(x + 2)(x^2 - 2x + 4)$ **10.** $(x + 3)(x - 7)$ **11.** prime **12.** $3y(2y + 1)(y + 1)$
13. $4(y + 3)(y - 3)$ **14.** $4(2x + 3)^2$ **15.** $2(x^2 + 4)(x + 2)(x - 2)$ **16.** $(6x - 7)^2$ **17.** $(7x - 1)(x - 7)$ **18.** $(x^2 - 5)(x + 2)$
19. $-3y(2y + 3)(2y - 5)$ **20.** $(y - 5)(y^2 + 5y + 25)$ **21.** $5(x - 3y)(x + 2y)$ **22.** -6 and 4, or $\{-6, 4\}$

23. $-\frac{1}{3}$ and 2, or $\left\{-\frac{1}{3}, 2\right\}$ **24.** -2 and 8, or $\{-2, 8\}$ **25.** 0 and $\frac{7}{2}$, or $\left\{0, \frac{7}{2}\right\}$ **26.** $-\frac{9}{4}$ and $\frac{9}{4}$, or $\left\{-\frac{9}{4}, \frac{9}{4}\right\}$

27. -1 and $\frac{6}{5}$, or $\left\{-1, \frac{6}{5}\right\}$ **28.** $x^2 - 4 = (x + 2)(x - 2)$ **29.** 6 sec   **30.** width: 5 ft; length: 11 ft

Cumulative Review Exercises

1. -48 **2.** 0 or $\{0\}$ **3.** 12 or $\{12\}$ **4.** $(-\infty, -2)$ or $\{x \mid x < -2\}$;   ![number line from -5 to 5 with arrow left from -2]   **5.** $38°, 38°, 104°$ **6.** $150

7. 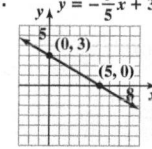 $y = -\frac{3}{5}x + 3$ **8.** $y + 4 = 5(x - 2)$ or $y - 1 = 5(x - 3)$; $y = 5x - 14$ **9.** $\{(2, -2, -1)\}$

10. $(3, -1)$ or $\{(3, -1)\}$ **11.** $\left(-\frac{2}{13}, \frac{23}{13}\right)$ or $\left\{\left(-\frac{2}{13}, \frac{23}{13}\right)\right\}$ **12.** $-\frac{13}{40}$ **13.** $2x^4 - x^3 + 3x + 9$ **14.** $6x^2 + 17xy - 45y^2$

15. $2x^2 + 5x - 3 - \frac{2}{3x - 5}$ **16.** 7.1×10^{-3} **17.** $(3x + 2)(x + 3)$ **18.** $y(y^2 + 4)(y + 2)(y - 2)$ **19.** $(2x + 3)^2$

20. width: 4 ft; length: 6 ft

CHAPTER 7

Section 7.1 Check Point Exercises

1. a. $x = 5$ **b.** $x = -7$ and $x = 4$ **2.** $\frac{x + 4}{3x}$ **3.** $\frac{x^2}{7}$ **4.** $\frac{x - 1}{x + 1}$ **5.** $-\frac{3x + 7}{4}$

Concept and Vocabulary Check

1. polynomials   **2.** 0   **3.** factoring; common factors   **4.** 1   **5.** -1 **6.** false   **7.** false   **8.** true   **9.** false

Exercise Set 7.1

1. $x = 0$ **3.** $x = 8$ **5.** $x = 4$ **7.** $x = -9$ and $x = 2$ **9.** $x = \frac{17}{3}$ and $x = -3$ **11.** $x = -4$ and $x = 3$

13. defined for all real numbers   **15.** $y = -1$ and $y = \frac{3}{4}$ **17.** $y = -5$ and $y = 5$ **19.** defined for all real numbers   **21.** $2x$

23. $\frac{x - 3}{5}$ **25.** $\frac{x - 4}{2x}$ **27.** $\frac{1}{x - 3}$ **29.** $-\frac{5}{x - 3}$ **31.** 3   **33.** $\frac{1}{x - 5}$ **35.** $\frac{2}{3}$ **37.** $\frac{1}{x - 3}$ **39.** $\frac{4}{x - 2}$ **41.** $\frac{y - 1}{y + 9}$

43. $\frac{y - 3}{y - 2}$ **45.** cannot be simplified   **47.** $\frac{x + 6}{x - 6}$ **49.** $x^2 + 1$ **51.** $x^2 + 2x + 4$ **53.** $\frac{x - 4}{x + 4}$ **55.** cannot be simplified

57. cannot be simplified   **59.** -1 **61.** -1 **63.** cannot be simplified   **65.** -2 **67.** $-\frac{2}{x}$ **69.** $-x - 1$ **71.** $-y - 3$ **73.** $-\frac{1}{x}$

75. $\frac{x - y}{2x - y}$ **77.** $\frac{x - 6}{x^2 + 3x + 9}$ **79.** $\frac{3 + y}{3 - y}$ **81.** $\frac{y + 3}{x + 3}$ **83.** $\frac{2}{1 - 2x}$ **85. a.** It costs $86.67 million to inoculate 40% of the population. It

costs $520 million to inoculate 80% of the population. It costs $1170 million to inoculate 90% of the population.   **b.** $x = 100$

c. The cost keeps rising; No amount of money will be enough to inoculate 100% of the population.   **87.** 400 mg

89. a. $300   **b.** $125   **c.** decrease   **91.** 1.5 mg per liter; $(3, 1.5)$ **99.** makes sense   **101.** does not make sense   **103.** false   **105.** false

109. correctly simplified   **111.** $x^2 - 1$ should be $x - 1$.   **113.** $\frac{3}{10}$ **114.** $\frac{1}{6}$ **115.** $(4, 2)$ or $\{(4, 2)\}$ **116.** $\frac{6}{35}$ **117.** $\frac{3}{2}$ **118.** $\frac{2}{3}$

Section 7.2 Check Point Exercises

1. $\frac{9x - 45}{2x + 8}$ **2.** $\frac{3}{8}$ **3.** $\frac{x + 2}{9}$ **4.** $-\frac{5(x + 1)}{7x(2x - 3)}$ **5.** $\frac{(x + 3)(x + 7)}{x - 4}$ **6.** $\frac{x + 3}{x - 5}$ **7.** $\frac{y + 1}{5y(y^2 + 1)}$

Concept and Vocabulary Check

1. numerators; denominators; $\frac{PR}{QS}$ **2.** multiplicative inverse/reciprocal; $\frac{S}{R}$, $\frac{PS}{QR}$ **3.** $\frac{x^2}{15}$ **4.** $\frac{3}{5}$

Exercise Set 7.2

1. $\dfrac{4x-20}{9x+27}$ **3.** $\dfrac{4x}{x+5}$ **5.** $\dfrac{4}{5}$ **7.** $\dfrac{4}{9}$ **9.** 1 **11.** $\dfrac{x+5}{x}$ **13.** $\dfrac{2}{y}$ **15.** $\dfrac{y+2}{y+4}$ **17.** $4(y+3)$ **19.** $\dfrac{x-1}{x+2}$ **21.** $\dfrac{x^2+2x+4}{3x}$

23. $\dfrac{x-2}{x-1}$ **25.** $-\dfrac{2}{x(x+1)}$ **27.** $-\dfrac{y-10}{y-7}$ **29.** $(x-y)(x+y)$ **31.** $\dfrac{4(x+y)}{3(x-y)}$ **33.** $\dfrac{3x}{35}$ **35.** $\dfrac{1}{4}$ **37.** 10 **39.** $\dfrac{7}{9}$ **41.** $\dfrac{3}{4}$

43. $\dfrac{(x-2)^2}{x}$ **45.** $\dfrac{(y-4)(y^2+4)}{y-1}$ **47.** $\dfrac{y}{3}$ **49.** $\dfrac{2(x+3)}{3}$ **51.** $\dfrac{x-5}{2}$ **53.** $\dfrac{y^2+1}{y^2}$ **55.** $\dfrac{y+1}{y-7}$ **57.** 6 **59.** $\dfrac{2}{3}$

61. $\dfrac{(x+y)^2}{32(x-y)^2}$ **63.** $\dfrac{y(x+y)}{(x+1)^2}$ **65.** $\dfrac{y+3}{y-3}$ **67.** $\dfrac{5(x-5)}{4}$ **69.** $\dfrac{(x+y)^2}{x-y}$ **71.** $\dfrac{2}{y^2-by+b^2}$ **73.** $\dfrac{x-2}{x(x-1)}$ sq in. **75.** $\dfrac{3x}{x+2}$ sq in.

81. makes sense **83.** makes sense **85.** true **87.** false **89.** numerator: -3; denominator: $2x-3$ **91.** correct answer

93. $x-3$ should be $x+3$. **95.** $(18, \infty)$ or $\{x|x>18\}$ **96.** $3(x-7)(x+2)$ **97.** -5 and $\dfrac{1}{2}$, or $\left\{-5, \dfrac{1}{2}\right\}$ **98.** $\dfrac{2}{3}$ **99.** x **100.** $\dfrac{x-3}{x+3}$

Section 7.3 Check Point Exercises

1. $x+2$ **2.** $\dfrac{x-5}{x+5}$ **3. a.** $\dfrac{3x+5}{x+7}$ **b.** $3x-4$ **4.** $\dfrac{3y+2}{y-4}$ **5.** $x+3$ **6.** $\dfrac{-4x^2+12x}{x^2-2x-9}$

Concept and Vocabulary Check

1. $\dfrac{P+Q}{R}$; numerators; common denominator **2.** $\dfrac{P-Q}{R}$; numerators; common denominator

3. $\dfrac{-1}{-1}$ **4.** $\dfrac{x+5}{3}$ **5.** $\dfrac{x-5}{3}$ **6.** $\dfrac{x-5+y}{3}$

Exercise Set 7.3

1. $\dfrac{9x}{13}$ **3.** $\dfrac{3x}{5}$ **5.** $\dfrac{x+3}{2}$ **7.** $\dfrac{6}{x}$ **9.** $\dfrac{7}{3x}$ **11.** $\dfrac{9}{x+3}$ **13.** $\dfrac{5x+5}{x-3}$ **15.** 2 **17.** $\dfrac{2y+3}{y-5}$ **19.** $\dfrac{7}{5y}$ **21.** $\dfrac{2}{x+2}$ **23.** $\dfrac{x-2}{x-3}$

25. $\dfrac{3x-4}{5x-4}$ **27.** 1 **29.** 1 **31.** $\dfrac{x}{2x-1}$ **33.** $-\dfrac{1}{y}$ **35.** $\dfrac{y+3}{3y+8}$ **37.** $\dfrac{3y+2}{y-3}$ **39.** $\dfrac{2}{x-3}$ **41.** $\dfrac{3x+7}{x-6}$ **43.** $\dfrac{3x+1}{3x-4}$

45. $x+2$ **47.** 0 **49.** $\dfrac{11}{x-1}$ **51.** $\dfrac{12}{x+3}$ **53.** $\dfrac{y+1}{y-1}$ **55.** $\dfrac{x-2}{x-7}$ **57.** $\dfrac{2x-4}{x^2-25}$ **59.** 1 **61.** $\dfrac{2}{x+y}$ **63.** $\dfrac{x-3}{x-1}$

65. $\dfrac{4b}{4b-3}$ **67.** $\dfrac{y}{y-5}$ **69.** $-\dfrac{1}{c+d}$ **71.** $\dfrac{3y^2+8y-5}{(y+1)(y-4)}$ **73. a.** $\dfrac{100W}{L}$ **b.** round **75.** 10 m **81.** does not make sense

83. makes sense **85.** false **87.** false **89.** $\dfrac{x+1}{x+2}$ **91.** $2x+1$ **93.** -7 **95.** $-4x-1$ **97.** $x-2$ should be $x-3$.

99. $\dfrac{31}{45}$ **100.** $(9x^2+1)(3x+1)(3x-1)$ **101.** $3x^2-7x-5$ **102.** $\dfrac{7}{6}$ **103.** $-\dfrac{17}{24}$ **104.** $\dfrac{y-2}{4y}$

Section 7.4 Check Point Exercises

1. $30x^2$ **2.** $(x+3)(x-3)$ **3.** $7x(x+4)(x+4)$ or $7x(x+4)^2$ **4.** $\dfrac{9+14x}{30x^2}$ **5.** $\dfrac{6x+6}{(x+3)(x-3)}$ **6.** $-\dfrac{5}{x+5}$ **7.** $-\dfrac{5+y}{5y}$

8. $\dfrac{x-15}{(x+5)(x-5)}$

Concept and Vocabulary Check

1. factor denominators **2.** x and $x+3$; $x+3$ and $x+5$; $x(x+3)(x+5)$ **3.** 3 **4.** $x-5$ **5.** -1

Exercise Set 7.4

1. $120x^2$ **3.** $30x^5$ **5.** $(x-3)(x+1)$ **7.** $7y(y+2)$ **9.** $(x+4)(x-4)$ **11.** $y(y+3)(y-3)$ **13.** $(y+1)(y-1)(y-1)$

15. $(x-5)(x+4)(2x-1)$ **17.** $\dfrac{3x+5}{x^2}$ **19.** $\dfrac{37}{18x}$ **21.** $\dfrac{8x+7}{2x^2}$ **23.** $\dfrac{6x+1}{x}$ **25.** $\dfrac{2+9x}{x}$ **27.** $\dfrac{x+1}{2}$ **29.** $\dfrac{7x-20}{x(x-5)}$

31. $\dfrac{5x+1}{(x-1)(x+2)}$ **33.** $\dfrac{11y+15}{4y(y+5)}$ **35.** $-\dfrac{7}{x+7}$ **37.** $\dfrac{3x-55}{(x+5)(x-5)}$ **39.** $\dfrac{x^2+6x}{(x-4)(x+4)}$ **41.** $\dfrac{y+12}{(y+3)(y-3)}$

43. $\dfrac{7x-10}{(x-1)(x-1)}$ **45.** $\dfrac{9y}{4(y-5)}$ **47.** $\dfrac{8y+16}{y(y+4)}$ **49.** $\dfrac{17}{(x-3)(x-4)}$ **51.** $\dfrac{7x-1}{(x+1)(x+1)(x-1)}$ **53.** $\dfrac{x^2-x}{(x+3)(x-2)(x+5)}$

55. $\dfrac{y^2+8y+4}{(y+4)(y+1)(y+1)}$ **57.** $\dfrac{2x^2-4x+34}{(x+3)(x-5)}$ **59.** $\dfrac{5-3y}{2y(y-1)}$ **61.** $-\dfrac{x^2}{(x+3)(x-3)}$ **63.** $\dfrac{2}{y+5}$ **65.** $\dfrac{4x-11}{x-3}$ **67.** $\dfrac{3}{y+1}$

69. $\dfrac{-x^2+7x+3}{(x-3)(x+2)}$ **71.** $\dfrac{x^2+2x-1}{(x+1)(x-1)(x+2)}$ **73.** $\dfrac{-y^2+8y+9}{15y^2}$ **75.** $\dfrac{x^2-2x-6}{3(x+2)(x-2)}$ **77.** $\dfrac{-y-2}{(y+1)(y-1)}$

79. $\dfrac{x + 2xy - y}{xy}$ **81.** $\dfrac{5x + 2y}{(x + y)(x - y)}$ **83.** $\dfrac{-x + 6}{(x + 2)(x - 2)}$ **85.** $\dfrac{6x + 5}{(x - 5)(x + 5)(x - 6)}$ **87.** $\dfrac{9}{(x - 3)(x^2 + 3x + 9)}$

89. $\dfrac{3}{y - 3}$ **91.** $-\dfrac{22y}{(x + 3y)(x + y)(x - 3y)}$ **93.** $\dfrac{D}{5}$ **95.** $\dfrac{D}{24}$ **97.** no   **99.** 5 yr   **101.** $\dfrac{4x^2 + 14x}{(x + 3)(x + 4)}$

109. makes sense   **111.** does not make sense   **113.** false   **115.** true   **117.** $-\dfrac{1}{x(x + h)}$ **119.** $\dfrac{2}{x + 1}$ **120.** $6x^2 - 11x - 35$

121. 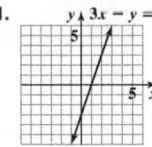   **122.** $y - x - 1$ **123. a.** $\dfrac{11}{15}$ **b.** $\dfrac{1}{15}$ **c.** 11   **124. a.** $\dfrac{y + x}{xy}$ **b.** $\dfrac{1}{y + x}$ **125.** $y + x$

Mid-Chapter Check Point Exercises

1. $x = -2$ and $x = 4$ **2.** $\dfrac{x - 2}{2x + 1}$ **3.** $\dfrac{-3}{y - 2}$ or $\dfrac{3}{2 - y}$ **4.** $\dfrac{2}{w}$ **5.** $\dfrac{4}{x + 4}$ **6.** $\dfrac{4}{(x - 2)^2}$ **7.** $\dfrac{x + 5}{x - 2}$ **8.** $\dfrac{x + 2}{x - 4}$

9. $\dfrac{2x + 1}{(x + 2)(x + 3)(x - 1)}$ **10.** $\dfrac{9 - x}{x - 5}$ **11.** $-\dfrac{2y - 1}{3y}$ or $\dfrac{1 - 2y}{3y}$ **12.** $\dfrac{-y^2}{(y + 1)(y + 2)}$ **13.** $\dfrac{w + 1}{(w - 3)(w + 5)}$

14. $\dfrac{2z^2 - 3z + 15}{(z + 3)(z - 3)(z + 1)}$ **15.** $\dfrac{3z^2 + 5z + 3}{(3z - 1)^2}$ **16.** $\dfrac{3x - 1}{(x + 7)(x - 3)}$ **17.** x **18.** $\dfrac{2x + 1}{(x + 1)(x - 2)(x + 3)}$ **19.** $\dfrac{5x - 5y}{x}$ **20.** $\dfrac{x + 3}{x + 5}$

Section 7.5 Check Point Exercises

1. $\dfrac{11}{5}$ **2.** $\dfrac{2x - 1}{2x + 1}$ **3.** $y - x$ **4.** $\dfrac{11}{5}$ **5.** $\dfrac{2x - 1}{2x + 1}$ **6.** $y - x$

Concept and Vocabulary Check

1. complex; complex   **2.** $\dfrac{18 + 12}{9 - 4} = \dfrac{30}{5} = 6$ **3.** $\dfrac{30 + 2x^2}{24 - x^2}$ **4.** $\dfrac{2x + 3}{5x + x^2}$

Exercise Set 7.5

1. $\dfrac{9}{10}$ **3.** $\dfrac{18}{23}$ **5.** $-\dfrac{4}{5}$ **7.** $\dfrac{3 - 4x}{3 + 4x}$ **9.** $\dfrac{7x - 2}{5x + 1}$ **11.** $\dfrac{2y + 3}{y - 7}$ **13.** $\dfrac{4 - 6y}{4 + 3y}$ **15.** $x - 5$ **17.** $\dfrac{x}{x - 1}$ **19.** $-\dfrac{1}{y}$

21. $\dfrac{xy + 2}{x}$ **23.** $\dfrac{y + x}{x^2 y^2}$ **25.** $\dfrac{x^2 + y}{y(y + 1)}$ **27.** $\dfrac{1}{y}$ **29.** $\dfrac{4 - x}{5x - 3}$ **31.** $\dfrac{2y}{y - 3}$ **33.** $\dfrac{1}{x + 3}$ **35.** $\dfrac{x^2 - 5x + 3}{x^2 - 7x + 2}$ **37.** $\dfrac{2y^2 + 3xy}{y^2 + 2x}$

39. $-\dfrac{6}{5}$ **41.** $\dfrac{1 - x}{(x - 3)(x + 6)}$ **43.** $\dfrac{1}{y(y + 5)}$ **45.** $\dfrac{1}{x - 1}$ **47.** $\dfrac{x + 1}{2x + 1}$ **49.** $\dfrac{2r_1 r_2}{r_1 + r_2}$; $34\dfrac{2}{7}$ mph   **51. a.** $\dfrac{175x^2 - 1585x + 15{,}993}{206x^2 - 349x + 25{,}984}$

b. 67%   **c.** 68%; overestimates by 1%   **57.** makes sense   **59.** does not make sense   **61.** true   **63.** false   **65.** $\dfrac{2y^2}{y + 1}$

67. correct answer   **69.** $x + \dfrac{1}{3}$ should be $3 + x$.   **70.** $2x(x - 5)^2$ **71.** -2 or $\{-2\}$ **72.** $x^3 + y^3$ **73.** 1 or $\{1\}$ **74.** 4 or $\{4\}$

75. $\dfrac{1}{2}$ and 2, or $\left\{\dfrac{1}{2}, 2\right\}$

Section 7.6 Check Point Exercises

1. 4 or $\{4\}$ **2.** 3 or $\{3\}$ **3.** -3 and -2, or $\{-3, -2\}$ **4.** 24 or $\{24\}$ **5.** no solution or $\varnothing$ **6.** 75%   **7.** $x = \dfrac{b - 2a}{a}$ **8.** $x = \dfrac{yz}{y - z}$

Concept and Vocabulary Check

1. LCD   **2.** 0   **3.** $8x$ **4.** $12(x + 3)$ **5.** $1 - x = 3x^2$ **6.** $6 = 5(x + 1) + 3(x - 1)$ **7.** $x \neq -2$
8. $x \neq 3$ and $x \neq 2$ **9.** true   **10.** false

Exercise Set 7.6

1. 12 or $\{12\}$ **3.** 0 or $\{0\}$ **5.** -2 or $\{-2\}$ **7.** 6 or $\{6\}$ **9.** -2 or $\{-2\}$ **11.** 2 or $\{2\}$ **13.** 15 or $\{15\}$ **15.** 4 or $\{4\}$
17. -6 and -1, or $\{-6, -1\}$ **19.** -5 and 5, or $\{-5, 5\}$ **21.** -3 and 3, or $\{-3, 3\}$ **23.** -8 and 1, or $\{-8, 1\}$ **25.** -1 and 16, or $\{-1, 16\}$

27. $\dfrac{3}{2}$ or $\left\{\dfrac{3}{2}\right\}$ **29.** 3 or $\{3\}$ **31.** no solution or $\varnothing$ **33.** $\dfrac{2}{3}$ and 4, or $\left\{\dfrac{2}{3}, 4\right\}$ **35.** no solution or $\varnothing$ **37.** 1 or $\{1\}$ **39.** $\dfrac{1}{5}$ or $\left\{\dfrac{1}{5}\right\}$

41. -3 or $\{-3\}$ **43.** no solution or $\varnothing$ **45.** -6 and $-\dfrac{1}{2}$, or $\left\{-6, -\dfrac{1}{2}\right\}$ **47.** $P_1 = \dfrac{P_2 V_2}{V_1}$ **49.** $f = \dfrac{pq}{q + p}$ **51.** $r = \dfrac{A - P}{P}$ **53.** $m_1 = \dfrac{Fd^2}{Gm_2}$

55. $x = \bar{x} + zs$ **57.** $R = \dfrac{E - Ir}{I}$ **59.** $f_1 = \dfrac{ff_2}{f_2 - f}$ **61.** -3 or $\{-3\}$ **63.** $\dfrac{-6x - 18}{(x - 5)(x + 4)}$ **65.** 1 and 5, or $\{1, 5\}$ **67.** 3 or $\{3\}$

69. 10,000 wheelchairs   **71.** 50%   **73.** 5 yr old   **75.** either 50 or approx 67 cases; $(50, 350)$ or $\left(66\frac{2}{3}, 350\right)$ **83.** does not make sense

85. does not make sense   **87.** false   **89.** false   **91.** -3 and 2, or $\{-3, 2\}$ **93.** $b = 2$ **95.** -5 and 5, or $\{-5, 5\}$ **97.** $(x^3 - 3)(x + 2)$

98. $-\dfrac{12}{x^8}$ **99.** $-20x + 55$ **100.** 2 or $\{2\}$ **101.** $\dfrac{1}{5}, \dfrac{3}{5}, \dfrac{x}{5}$ **102.** $\dfrac{63}{x} = \dfrac{7}{5}$

Section 7.7 Check Point Exercises

1. 2 mph   **2.** $2\frac{2}{3}$ hr or 2 hr 40 min   **3.** \$5880   **4.** 720 deer   **5.** 32 in.   **6.** 32 yd

Concept and Vocabulary Check

1. distance traveled; rate of travel   **2.** 1   **3.** $\dfrac{x}{5}$ **4.** $ad = bc$ **5.** similar

Exercise Set 7.7

1. walking rate: 6 mph; car rate: 9 mph   **3.** downhill rate: 10 mph; uphill rate: 6 mph   **5.** 5 mph
7. walking rate: 3 mph; jogging rate: 6 mph   **9.** 10 mph   **11.** It will take about 8.6 min, which is enough time.
13. It will take about 171.4 hr, which is not enough time.   **15.** 2.4 hr, or 2 hr 24 min   **17.** \$1800   **19.** 20,489 fur seal pups   **21.** 0.97 billion
23. 154.1 in.   **25.** 5 in.   **27.** 6 m   **29.** 16 in.   **31.** 16 ft   **43.** does not make sense   **45.** makes sense   **47.** 30 mph
49. 7 hr   **51.** 2.5 hr   **53.** $(5x + 9)(5x - 9)$ **54.** 6 or $\{6\}$

55.
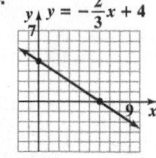
$y = -\dfrac{2}{3}x + 4$
    56. a. 16   **b.** $y = 16x^2$ **c.** 400   **57. a.** 96   **b.** $y = \dfrac{96}{x}$ **c.** 32   **58.** 8

Section 7.8 Check Point Exercises

1. 66 gal   **2.** about 556 ft   **3.** 512 cycles per second   **4.** 24 min   **5.** 96π cubic feet

Concept and Vocabulary Check

1. $y = kx$; constant of variation   **2.** $y = kx^n$ **3.** $y = \dfrac{k}{x}$ **4.** $y = \dfrac{kx}{z}$ **5.** $y = kxz$ **6.** directly; inversely   **7.** jointly; inversely

Exercise Set 7.8

1. 156   **3.** 30   **5.** $\dfrac{5}{6}$ **7.** 240   **9.** 50   **11.** $x = kyz$; $y = \dfrac{x}{kz}$ **13.** $x = \dfrac{kz^3}{y}$; $y = \dfrac{kz^3}{x}$ **15.** $x = \dfrac{kyz}{\sqrt{w}}$; $y = \dfrac{x\sqrt{w}}{kz}$

17. $x = kz(y + w)$; $y = \dfrac{x - kzw}{kz}$ **19.** $x = \dfrac{kz}{y - w}$; $y = \dfrac{xw + kz}{x}$ **21.** 5.4 ft   **23.** 80 in.   **25.** about 607 lb   **27.** 32°

29. a. $L = \dfrac{1890}{R}$ **b.** an approximate model   **c.** 70 yr   **31. a.** 90 beats per minute   **b.** 95 beats per minute   **c.** by the point $(63, 30)$

33. 90 milliroentgens per hour   **35.** This person has a BMI of 24.4 and is not overweight.   **37.** 1800 Btu   **39.** $\dfrac{1}{4}$ of what it was originally
41. a. $C = \dfrac{kP_1P_2}{d^2}$ **b.** $k \approx 0.02$; $C = \dfrac{0.02P_1P_2}{d^2}$ **c.** approximately 39,813 daily phone calls

43. a.
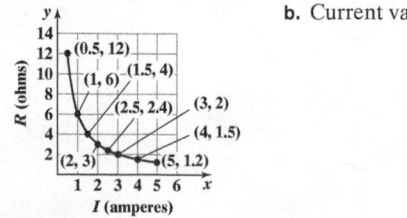
    b. Current varies inversely as resistance.   **c.** $R = \dfrac{6}{I}$

49. z varies directly as the square root of x and inversely as the square of y.   **53.** does not make sense   **55.** makes sense

57. The wind pressure is 4 times more destructive.   **59.** Distance is increased by $\sqrt{50}$, or about 7.07, for the space telescope.   **60.** $\dfrac{16}{3}$ or $\left\{\dfrac{16}{3}\right\}$

61. $9x^2 - 6x + 4 - \dfrac{16}{3x + 2}$ **62.** $6x(x - 5)(x + 4)$ **63.** set 1   **64.** -170 **65.** $5a + 5h + 7$

Review Exercises

1. $x = 4$ **2.** $x = 2$ and $x = -5$ **3.** $x = 1$ and $x = 2$ **4.** defined for all real numbers   **5.** $\dfrac{4x}{3}$ **6.** $x+2$ **7.** x^2 **8.** $\dfrac{x - 3}{x - 6}$

9. $\dfrac{x - 5}{x + 7}$ **10.** $\dfrac{y}{y + 2}$ **11.** cannot be simplified   **12.** $-2(x + 3y)$ **13.** $\dfrac{x - 2}{4}$ **14.** $\dfrac{5}{2}$ **15.** $\dfrac{x + 3}{x + 2}$ **16.** $\dfrac{2y - 1}{5}$ **17.** $\dfrac{-3(y + 1)}{5y(2y - 3)}$

18. $\dfrac{x - 1}{4}$ **19.** $\dfrac{2}{x(x + 1)}$ **20.** $\dfrac{1}{7(y + 3)}$ **21.** $(y - 3)(y - 6)$ **22.** $\dfrac{8}{x - y}$ **23.** 4   **24.** 4   **25.** $3x - 5$ **26.** $3y + 4$

27. $\dfrac{4}{x - 2}$ **28.** $\dfrac{2x + 5}{x - 3}$ **29.** $36x^3$ **30.** $x^2(x - 1)^2$ **31.** $(x + 3)(x + 1)(x + 7)$ **32.** $\dfrac{14x + 15}{6x^2}$ **33.** $\dfrac{7x + 2}{x(x + 1)}$ **34.** $\dfrac{7x + 25}{(x + 3)^2}$

35. $\dfrac{3}{y - 2}$ **36.** $-\dfrac{y}{y - 1}$ **37.** $\dfrac{x^2 + y^2}{xy}$ **38.** $\dfrac{3x^2 - x}{(x + 1)^2(x - 1)}$ **39.** $\dfrac{5x^2 - 3x}{(x + 1)(x - 1)}$ **40.** $\dfrac{4}{(x + 2)(x - 2)(x - 3)}$ **41.** $\dfrac{2x + 13}{x + 3}$

42. $\dfrac{y - 4}{3y}$ **43.** $\dfrac{7}{2}$ **44.** $\dfrac{1}{x - 1}$ **45.** $x + y$ **46.** $\dfrac{3}{x}$ **47.** $\dfrac{3x}{x - 4}$ **48.** 12 or $\{12\}$ **49.** -1 or $\{-1\}$ **50.** -6 and 1, or $\{-6, 1\}$

51. no solution or $\varnothing$ **52.** 5 or $\{5\}$ **53.** -6 and 2, or $\{-6, 2\}$ **54.** 5 or $\{5\}$ **55.** -5 or $\{-5\}$ **56.** 3 yr   **57.** 30%   **58.** $C = R - nP$

59. $T_1 = \dfrac{P_1 V_1 T_2}{P_2 V_2}$ **60.** $P = \dfrac{A}{rT + 1}$ **61.** $R = \dfrac{R_1 R_2}{R_2 + R_1}$ **62.** $n = \dfrac{IR}{E - Ir}$ **63.** 4 mph

64. Slower car's rate is 20 mph; faster car's rate is 30 mph.   **65.** 4 hr   **66.** 324 teachers   **67.** 287 trout   **68.** 5 ft   **69.** $12\dfrac{1}{2}$ ft

70. $4935   **71.** 1600 ft   **72.** 440 vibrations per second   **73.** 112 decibels   **74.** 16 hr   **75.** 800 cubic feet

Chapter Test

1. $x = -9$ and $x = 4$ **2.** $\dfrac{x + 3}{x - 2}$ **3.** $\dfrac{4x}{x + 1}$ **4.** $\dfrac{x - 4}{2}$ **5.** $\dfrac{x}{x + 3}$ **6.** $\dfrac{2x + 6}{x + 1}$ **7.** $\dfrac{5}{(y - 1)(y - 3)}$ **8.** $2y$

9. $\dfrac{2}{y + 3}$ **10.** $\dfrac{x^2 + 2x + 15}{(x + 3)(x - 3)}$ **11.** $\dfrac{8x - 14}{(x + 2)(x - 1)(x - 3)}$ **12.** $\dfrac{-x - 1}{x - 3}$ **13.** $\dfrac{x + 2}{x - 1}$ **14.** $\dfrac{11}{(x - 3)(x - 4)}$ **15.** $\dfrac{4}{y + 4}$

16. $\dfrac{3x - 3y}{x}$ **17.** $\dfrac{5x + 5}{2x + 1}$ **18.** $\dfrac{y - x}{y}$ **19.** 6 or $\{6\}$ **20.** -8 or $\{-8\}$ **21.** 0 and 2, or $\{0, 2\}$ **22.** $a = \dfrac{Rs}{s - R}$ or $a = \dfrac{Rs}{R - s}$ **23.** 2 mph

24. 12 min   **25.** 6000 tule elk   **26.** 3.2 in.   **27.** 52.5 amp

Cumulative Review Exercises

1. 2 or $\{2\}$ **2.** $(-\infty, 2)$ or $\{x|x < 2\}$ **3.** -6 and 3, or $\{-6, 3\}$ **4.** -6 or $\{-6\}$ **5.** $(3, 3)$ or $\{(3, 3)\}$ **6.** $(-2, 2)$ or $\{(-2, 2)\}$

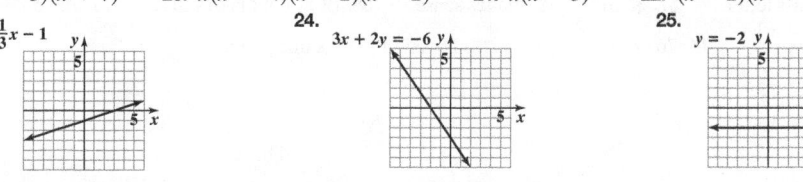

10. -19 **11.** $8x^9$ **12.** $\dfrac{1 - 2x}{4x - 1}$ **13.** $(4x - 1)(x - 3)$ **14.** $(2x - 5)^2$ **15.** $3(x + 5)(x - 5)$ **16.** $-x^2 + 4x + 8$

17. $-2x^4 + 3x^2 - 1$ **18.** $\dfrac{3x^2 + 6x + 16}{(x - 2)(x + 3)}$ **19.** $1225 at 5% and $2775 at 9%   **20.** 17 in. and 51 in.

Mid-Textbook Check Point Exercises

1. 4 or $\{4\}$ **2.** 10 or $\{10\}$ **3.** $(-\infty, 7]$ or $\{x|x \le 7\}$ **4.** $(-3, 4)$ or $\{(-3, 4)\}$ **5.** $(3, 4)$ or $\{(3, 4)\}$ **6.** -8 or $\{-8\}$

7. -3 and -2, or $\{-3, -2\}$ **8.** $\dfrac{4}{x^9}$ **9.** 31   **10.** $10x^2 - 9x + 4$ **11.** $21x^2 - 23x - 20$ **12.** $25x^2 - 20x + 4$

13. $x^3 + 2x^2y + 2xy^2 + y^3$ **14.** $\dfrac{x + 4}{3x^3}$ **15.** $\dfrac{-7x}{(x + 3)(x - 1)(x - 4)}$ **16.** $\dfrac{x}{5}$ **17.** $(2x + 7)(2x - 7)$ **18.** $(x + 3)(x + 1)(x - 1)$

19. $2(x - 3)(x + 7)$ **20.** $x(x^2 + 4)(x + 2)(x - 2)$ **21.** $x(x - 5)^2$ **22.** $(x - 2)(x^2 + 2x + 4)$

23.   **24.**   **25.**

26. -2 **27.** $y - 2 = 2(x - 1)$ or $y - 6 = 2(x - 3)$; $y = 2x$ **28.** 43   **29.** $320   **30.** length: 150 yd; width: 50 yd

31. $12,500 at 7%; $7500 at 9%   **32.** 8 l of 40%; 4 l of 70%   **33.** 16 ft   **34.** 35°; 25°; 120°   **35.** TV: $350; stereo: $370

36. length: 11 m; width: 5 m

CHAPTER 8

Section 8.1 Check Point Exercises

1. domain: $\{0, 10, 20, 30, 38\}$; range: $\{9.1, 6.7, 10.7, 13.2, 19.6\}$ **2. a.** not a function **b.** function **3. a.** 29 **b.** 65 **c.** 46 **d.** $6a + 6h + 9$
4. a. Every element in the domain corresponds to exactly one element in the range. **b.** domain: $\{0, 1, 2, 3, 4\}$; range: $\{3, 0, 1, 2\}$ **c.** 0
d. 2 **e.** $x = 0$ and $x = 4$

Concept and Vocabulary Check

1. relation; domain; range **2.** function **3.** $f; x$ **4.** $r; -2$

Exercise Set 8.1

1. function; domain: $\{1, 3, 5\}$; range: $\{2, 4, 5\}$ **3.** not a function; domain: $\{3, 4\}$; range: $\{4, 5\}$ **5.** function; domain: $\{-3, -2, -1, 0\}$;
range: $\{-3, -2, -1, 0\}$ **7.** not a function; domain: $\{1\}$; range: $\{4, 5, 6\}$ **9. a.** 1 **b.** 6 **c.** -7 **d.** $2a + 1$ **e.** $a + 3$ **11. a.** -2
b. -17 **c.** 0 **d.** $12b - 2$ **e.** $3b + 10$ **13. a.** 5 **b.** 8 **c.** 53 **d.** 32 **e.** $48b^2 + 5$ **15. a.** -1 **b.** 26 **c.** 19
d. $2b^2 + 3b - 1$ **e.** $50a^2 + 15a - 1$ **17. a.** 7 **b.** -7 **c.** 13 **d.** 12 **19. a.** $\dfrac{3}{4}$ **b.** -3 **c.** $\dfrac{11}{8}$ **d.** $\dfrac{13}{9}$ **e.** $\dfrac{2a + 2h - 3}{a + h - 4}$
f. Denominator would be zero. **21. a.** 6 **b.** 12 **c.** 0 **23. a.** 2 **b.** 1 **c.** -1 and 1 **25.** $-2; 10$ **27.** -38 **29.** $-2x^3 - 2x$
31. a. -1 **b.** 7 **c.** 19 **d.** 112 **33. a.** $\{(\text{Iceland}, 9.7), (\text{Finland}, 9.6), (\text{New Zealand}, 9.6), (\text{Denmark}, 9.5)\}$ **b.** Yes; each country
corresponds to exactly one corruption rating. **c.** $\{(9.7, \text{Iceland}), (9.6, \text{Finland}), (9.6, \text{New Zealand}), (9.5, \text{Denmark})\}$ **d.** No; 9.6 in the domain
corresponds to two countries in the range, Finland and New Zealand. **39.** makes sense **41.** makes sense **43.** false **45.** true **47.** true
49. 3 **51.** $f(2) = 6; f(3) = 9; f(4) = 12$; no **52.** 0 **53.** $\dfrac{y^{10}}{9x^4}$ **54.** $\{-15\}$ **55.** -3 and 3 **56.** $(-\infty, \infty)$ **57.** $[1, \infty)$

Section 8.2 Check Point Exercises

1. a. function **b.** function **c.** not a function **2. a.** 400 **b.** 9 **c.** approximately 425
3. a. $\{x | -2 \le x < 5\}$ **b.** $\{x | 1 \le x \le 3.5\}$ **c.** $\{x | x < -1\}$

4. a. domain: $[-2, 1]$; range $[0, 3]$ **b.** domain: $(-2, 1]$; range: $[-1, 2)$ **c.** domain: $[-3, 0)$; range: $\{-3, -2, -1\}$

Concept and Vocabulary Check

1. ordered pairs **2.** more than once; function **3.** $[1, 3)$; domain **4.** $[1, \infty)$; range

Exercise Set 8.2

1. function **3.** not a function **5.** function **7.** not a function **9.** -4 **11.** 4 **13.** 0 **15.** 2 **17.** 2 **19.** -2

21. $\{x | 1 < x \le 6\}$ **23.** $\{x | -5 \le x < 2\}$ **25.** $\{x | -3 \le x \le 1\}$

27. $\{x | x > 2\}$ **29.** $\{x | x \ge -3\}$ **31.** $\{x | x < 3\}$

33. $\{x | x < 5.5\}$ **35.** domain: $[0, 5)$; range: $[-1, 5)$ **37.** domain: $[0, \infty)$; range: $[1, \infty)$ **39.** domain: $[-2, 6]$; range: $[-2, 6]$
41. domain: $(-\infty, \infty)$; range: $(-\infty, 2]$ **43.** domain: $\{-5, -2, 0, 1, 3\}$; range: $\{2\}$ **45. a.** $(-\infty, \infty)$
b. $[-4, \infty)$ **c.** 4 **d.** 2 and 6 **e.** $(1, 0), (7, 0)$ **f.** $(0, 4)$ **g.** $(1, 7)$ **h.** positive

47. a. 57.5; Average global temperature in 1980 was 57.5°F.; $(80, 57.5)$ **b.** overestimates by 0.1°F **49. a.** 368.5; Average carbon dioxide
concentration in 2000 was 368.5 parts per million.; $(50, 368.5)$ on the graph of f **b.** 367.5; Average carbon dioxide concentration in 2000 was 367.5
parts per million.; $(50, 367.5)$ on the graph of g **c.** the quadratic function **51.** 440; For 20-year-old drivers, there are 440 accidents per 50 million
miles driven.; $(20, 440)$ **53.** $x = 45; y = 190$; The minimum number of accidents is 190 per 50 million miles driven and is attributed to 45-year-old
drivers. **55.** 0.78; It costs \$0.78 to mail a 3-ounce first-class letter. **57.** \$0.61 **63.** makes sense **65.** does not make sense **67.** false
69. true **71.** true **73.** -18 **75.** yes **76.** $\dfrac{3}{2}$ or $\left\{\dfrac{3}{2}\right\}$ **77.** 76 yd by 236 yd **78.** Division by 0 is undefined.
79. 19 **80.** $-2x^2 + 42x + 1582$

Section 8.3 Check Point Exercises

1. a. $(-\infty, \infty)$ **b.** $(-\infty, -5)$ or $(-5, \infty)$ **2. a.** $3x^2 + 6x + 6$ **b.** 78 **3. a.** $\dfrac{5}{x} - \dfrac{7}{x - 8}$ **b.** $(-\infty, 0)$ or $(0, 8)$ or $(8, \infty)$ **4. a.** 23
b. $x^2 - 3x - 3; 1$ **c.** $x^3 + x^2 - 6x; -24$ **d.** $\dfrac{x^2 - 2x}{x + 3}; \dfrac{7}{2}$ **5. a.** $(B + D)(x) = -3.2x^2 + 56x + 6406$ **b.** 6545.2 thousand
c. overestimates by 7.2 thousand

Concept and Vocabulary Check

1. zero **2.** negative **3.** $f(x) + g(x)$ **4.** $f(x) - g(x)$ **5.** $f(x) \cdot g(x)$ **6.** $\dfrac{f(x)}{g(x)}; g(x)$ **7.** $(-\infty, \infty)$ **8.** $(2, \infty)$ **9.** $(0, 3); (3, \infty)$

Exercise Set 8.3

1. $(-\infty, \infty)$ **3.** $(-\infty, -4)$ or $(-4, \infty)$ **5.** $(-\infty, 3)$ or $(3, \infty)$ **7.** $(-\infty, 5)$ or $(5, \infty)$ **9.** $(-\infty, -7)$ or $(-7, 9)$ or $(9, \infty)$ **11. a.** $5x - 5$ **b.** 20
13. a. $3x^2 + x - 5$ **b.** 75 **15. a.** $2x^2 - 2$ **b.** 48

17. $(f + g)(x) = 3x - 3; (f - g)(x) = 7x + 3; (fg)(x) = -10x^2 - 15x; \left(\dfrac{f}{g}\right)(x) = \dfrac{5x}{-2x - 3}$ **19.** $(-\infty, \infty)$ **21.** $(-\infty, 5)$ or $(5, \infty)$

23. $(-\infty, 0)$ or $(0, 5)$ or $(5, \infty)$ **25.** $(-\infty, -3)$ or $(-3, 2)$ or $(2, \infty)$ **27.** $(-\infty, 2)$ or $(2, \infty)$ **29.** $(-\infty, \infty)$ **31.** $x^2 + 3x + 2; 20$ **33.** 0

35. $x^2 + 5x - 2; 48$ **37.** -8 **39.** $-x^3 - 2x^2 + 8x; 0$ **41.** -135 **43.** $\dfrac{x^2 + 4x}{2 - x}; 5$ **45.** -1 **47.** $(-\infty, \infty)$ **49.** $(-\infty, 2)$ or $(2, \infty)$

51. 5 **53.** -1 **55.** $[-4, 3]$

57.

59. -4 **61.** -4 **63. a.** $(M + F)(x) = 3.02x + 235.2$ **b.** 295.6 million **c.** underestimates by 2.4 million

65. a. $\left(\dfrac{M}{F}\right)(x) = \dfrac{1.54x + 114.6}{1.48x + 120.6}$ **b.** 0.964 **c.** underestimates by approximately 0.001

71.

73.

75. No y-value is displayed; y_3 is undefined at $x = 0$. **77.** makes sense **79.** makes sense **81.** true **83.** false

84. $b = \dfrac{R - 3a}{3}$ or $b = \dfrac{R}{3} - a$ **85.** 7 or $\{7\}$ **86.** $6b + 8$ **87. a.** 11 **b.** 127 **88.** x **89.** $y = \dfrac{x + 5}{7}$

Mid-Chapter Check Point Exercises

1. not a function; domain: $\{1, 2\}$; range $\{-6, 4, 6\}$ **2.** function; domain: $\{0, 2, 3\}$; range: $\{1, 4\}$ **3.** function; domain: $[-2, 2)$; range: $[0, 3]$
4. not a function; domain: $(-3, 4]$; range: $[-1, 2]$ **5.** not a function; domain: $\{-2, -1, 0, 1, 2\}$; range: $\{-2, -1, 1, 3\}$ **6.** function; domain: $(-\infty, 1]$;
range: $[-1, \infty)$ **7.** No vertical line intersects the graph of f more than once. **8.** 3 **9.** -2 **10.** -6 and 2 **11.** $(-\infty, \infty)$ **12.** $(-\infty, 4]$
13. $(-\infty, \infty)$ **14.** $(-\infty, -2)$ or $(-2, 2)$ or $(2, \infty)$ **15.** 23 **16.** 23 **17.** $a^2 - 5a - 3$ **18.** $x^2 - 5x + 3; 17$ **19.** $x^2 - x + 13; 33$
20. $-2x^3 + x^2 - x - 40; -36$ **21.** $\dfrac{x^2 - 3x + 8}{-2x - 5}; 12$ **22.** $\left(-\infty, -\dfrac{5}{2}\right)$ or $\left(-\dfrac{5}{2}, \infty\right)$

Section 8.4 Check Point Exercises

1. a. $5x^2 + 1$ **b.** $25x^2 + 60x + 35$ **2.** $f(g(x)) = 7\left(\dfrac{x}{7}\right) = x; g(f(x)) = \dfrac{7x}{7} = x$

3. $f(g(x)) = 4\left(\dfrac{x + 7}{4}\right) - 7 = (x + 7) - 7 = x; g(f(x)) = \dfrac{(4x - 7) + 7}{4} = \dfrac{4x}{4} = x$ **4.** $f^{-1}(x) = \dfrac{x - 7}{2}$

5. (b) and (c) **6.**

Concept and Vocabulary Check

1. composition; $f(g(x))$ **2.** $f; g(x)$ **3.** composition; $g(f(x))$ **4.** $g; f(x)$ **5.** false **6.** false
7. inverse **8.** $x; x$ **9.** horizontal; one-to-one **10.** $y = x$

Exercise Set 8.4

1. a. $2x + 14$ **b.** $2x + 7$ **c.** 18 **3. a.** $2x + 5$ **b.** $2x + 9$ **c.** 9 **5. a.** $20x^2 - 11$ **b.** $80x^2 - 120x + 43$ **c.** 69
7. a. $x^4 - 4x^2 + 6$ **b.** $x^4 + 4x^2 + 2$ **c.** 6 **9. a.** $\sqrt{x - 1}$ **b.** $\sqrt{x} - 1$ **c.** 1 **11. a.** x **b.** x **c.** 2 **13. a.** x **b.** x **c.** 2

15. $f(g(x)) = x; g(f(x)) = x$; inverses **17.** $f(g(x)) = x; g(f(x)) = x$; inverses **19.** $f(g(x)) = \dfrac{5x - 56}{9}; g(f(x)) = \dfrac{5x - 4}{9}$; not inverses

21. $f(g(x)) = x; g(f(x)) = x;$ inverses **23.** $f(g(x)) = x; g(f(x)) = x;$ inverses **25. a.** $f^{-1}(x) = x - 3$ **b.** $f(f^{-1}(x)) = (x - 3) + 3 = x$ and $f^{-1}(f(x)) = (x + 3) - 3 = x$ **27. a.** $f^{-1}(x) = \dfrac{x}{2}$ **b.** $f(f^{-1}(x)) = 2\left(\dfrac{x}{2}\right) = x$ and $f^{-1}(f(x)) = \dfrac{2x}{2} = x$ **29. a.** $f^{-1}(x) = \dfrac{x - 3}{2}$ **b.** $f(f^{-1}(x)) = 2\left(\dfrac{x - 3}{2}\right) + 3 = x$ and $f^{-1}(f(x)) = \dfrac{(2x + 3) - 3}{2} = x$

31. a. $f^{-1}(x) = \dfrac{1}{x}$ **b.** $f(f^{-1}(x)) = \dfrac{1}{\frac{1}{x}} = x$ and $f^{-1}(f(x)) = \dfrac{1}{\frac{1}{x}} = x$

33. a. $f^{-1}(x) = \dfrac{3x + 1}{x - 2}$ **b.** $f(f^{-1}(x)) = \dfrac{2\left(\frac{3x + 1}{x - 2}\right) + 1}{\left(\frac{3x + 1}{x - 2}\right) - 3} = x$ and $f^{-1}(f(x)) = \dfrac{3\left(\frac{2x + 1}{x - 3}\right) + 1}{\left(\frac{2x + 1}{x - 3}\right) - 2} = x$ **35.** no inverse **37.** no inverse

39. inverse function **41.** 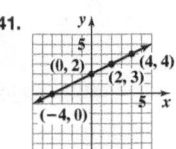 **43.** **45.** 5 **47.** 1 **49.** 2 **51.** 1 **53.** −6 **55.** −7 **57.** 3 **59.** 11

61. a. f represents the price after a \$400 discount, and g represents the price after a 25% discount (75% of the regular price). **b.** $0.75x - 400; f \circ g$ represents an additional \$400 discount on a price that has already been reduced by 25%. **c.** $0.75(x - 400) = 0.75x - 300; g \circ f$ represents an additional 25% discount on a price that has already been reduced \$400. **d.** $f \circ g; 0.75x - 400 < 0.75x - 300$, so $f \circ g$ represents the lower price after the two discounts. **e.** $f^{-1}(x) = x + 400; f^{-1}$ represents the regular price, since the value of x here is the price after a \$400 discount. **63. a.** f: {(U.S., 1%), (U.K., 8%), (Italy, 5%), (France, 5%), (Holland, 30%)} **b.** f^{-1}: {(1%, U.S.), (8%, U.K.), (5%, Italy), (5%, France), (30%, Holland)}; No; The input 5% is associated with two outputs, Italy and France. **65. a.** No horizontal line intersects the graph of f in more than one point. **b.** $f^{-1}(0.25)$, or approximately 15, represents the number of people who would have to be in the room so that the probability of two sharing a birthday would be 0.25; $f^{-1}(0.5)$, or approximately 23, represents the number of people so that the probability would be 0.5; $f^{-1}(0.7)$, or approximately 30, represents the number of people so that the probability would be 0.7.

67. $f(g(x)) = \dfrac{9}{5}\left[\dfrac{5}{9}(x - 32)\right] + 32 = x$ and $g(f(x)) = \dfrac{5}{9}\left[\left(\dfrac{9}{5}x + 32\right) - 32\right] = x$

75. ; inverse function **77.** ; no inverse function

79. ; inverse function **81.** ; inverse function

83. ; inverses **85.** does not make sense **87.** does not make sense **89.** false **91.** true **93.** $(f \circ g)^{-1}(x) = \dfrac{x - 15}{3}; (g^{-1} \circ f^{-1})(x) = \dfrac{x}{3} - 5 = \dfrac{x - 15}{3}$ **95.** $(f \circ g)(x) = m_1(m_2x + b_2) + b_1 = m_1m_2x + m_1b_2 + b_1$; The slope of $f \circ g$ is m_1m_2, which is the product of the slopes of f and g. **96.** 5×10^8 **97.** $x^2 + 9x + 16 + \dfrac{35}{x - 2}$ **98.** $\left\{\left(\dfrac{4}{5}, \dfrac{9}{5}\right)\right\}$ **99.** $\left\{\dfrac{9}{19}\right\}$ **100.** {14} **101.** $(2500, \infty)$

Review Exercises

1. function; domain: {3, 4, 5}; range: {10} **2.** function; domain: {1, 2, 3, 4}; range: {12, 100, π, −6} **3.** not a function; domain: {13, 15}; range: {14, 16, 17} **4. a.** −5 **b.** 16 **c.** −75 **d.** $14a - 5$ **e.** $7a + 9$ **5. a.** 2 **b.** 52 **c.** 70 **d.** $3b^2 - 5b + 2$ **e.** $48a^2 - 20a + 2$ **6.** not a function **7.** function **8.** function **9.** not a function **10.** not a function **11.** function **12.** $\{x \mid -2 < x \le 3\}$ **13.** $\{x \mid -1.5 \le x \le 2\}$ **14.** $\{x \mid x > -1\}$ **15.** −3 **16.** −2 **17.** 3 **18.** $[-3, 5)$ **19.** $[-5, 0]$ **20. a.** For each time, there is only one height. **b.** 0; The eagle was on the ground after 15 seconds. **c.** 45 m **d.** 7 and 22; After 7 seconds and after 22 seconds, the eagle's height is 20 meters. **e.** Answers will vary.

21. $(-\infty, \infty)$ **22.** $(-\infty, -8)$ or $(-8, \infty)$ **23.** $(-\infty, 5)$ or $(5, \infty)$ **24. a.** $6x - 4$ **b.** 14   **25. a.** $5x^2 + 1$ **b.** 46   **26.** $(-\infty, 4)$ or $(4, \infty)$

27. $(-\infty, -6)$ or $(-6, -1)$ or $(-1, \infty)$ **28.** $x^2 - x - 5; 1$ **29.** 1   **30.** $x^2 - 3x + 5; 3$ **31.** 9   **32.** $x^3 - 7x^2 + 10x; -120$ **33.** $\dfrac{x^2 - 2x}{x - 5}; -8$

34. $(-\infty, \infty)$ **35.** $(-\infty, 5)$ or $(5, \infty)$ **36. a.** $16x^2 - 8x + 4$ **b.** $4x^2 + 11$ **c.** 124   **37. a.** $\sqrt{x + 1}$ **b.** $\sqrt{x} + 1$ **c.** 2

38. $f(g(x)) = x - \dfrac{7}{10}; g(f(x)) = x - \dfrac{7}{6};$ not inverses   **39.** $f(g(x)) = x; g(f(x)) = x;$ inverses   **40. a.** $f^{-1}(x) = \dfrac{x + 3}{4}$

b. $f(f^{-1}(x)) = 4\left(\dfrac{x + 3}{4}\right) - 3 = x$ and $f^{-1}(f(x)) = \dfrac{(4x - 3) + 3}{4} = x$ **41. a.** $f^{-1}(x) = -\dfrac{1}{x}$

b. $f(f^{-1}(x)) = -\dfrac{1}{\left(-\dfrac{1}{x}\right)} = x$ and $f^{-1}(f(x)) = -\dfrac{1}{\left(-\dfrac{1}{x}\right)} = x$

42. inverse function   **43.** no inverse function   **44.** inverse function   **45.** no inverse function
46.

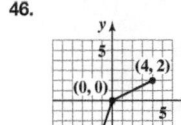

Chapter Test

1. function; domain: $\{1, 3, 5, 6\}$; range: $\{2, 4, 6\}$ **2.** not a function; domain: $\{2, 4, 6\}$; range: $\{1, 3, 5, 6\}$ **3.** $3a + 10$ **4.** 28   **5.** function
6. not a function   **7.** -3 **8.** -2 and 3   **9.** $(-\infty, \infty)$ **10.** $(-\infty, 3]$ **11.** $(-\infty, 10)$ or $(10, \infty)$ **12.** $x^2 + 5x + 2; 26$ **13.** $x^2 + 3x - 2; -4$

14. $x^3 + 6x^2 + 8x; -15$ **15.** $\dfrac{x^2 + 4x}{x + 2}; 3$ **16.** $(-\infty, -2)$ or $(-2, \infty)$ **17.** $(f \circ g)(x) = 9x^2 - 3x; (g \circ f)(x) = 3x^2 + 3x - 1$ **18.** $f^{-1}(x) = \dfrac{x + 7}{5}$

19. a. No horizontal line intersects the graph of f in more than one point.   **b.** 2000   **c.** $f^{-1}(2000)$ represents the income, \$80 thousand, of a family that gives \$2000 to charity.

Cumulative Review Exercises

1. $\{-3\}$ **2.** $\left\{-4, \dfrac{3}{2}\right\}$ **3.** $\{(-3, -4)\}$ **4.** $\left\{\dfrac{9}{7}\right\}$ **5.** $\{-5\}$ **6.** $x^2 - 3x - 1; -1$ **7.** $-\dfrac{2}{x^4}$ **8.** 4   **9.** $\dfrac{3}{x}$ **10.** $-\dfrac{2 + x}{3x + 1}$ **11.** 5

12. $(x - 7)(x - 11)$ **13.** $x(x + 5)(x - 5)$ **14.** $6x^2 - 7x + 2$ **15.** $8x^3 - 27$ **16.** $\dfrac{1}{x - 1}$ **17.** $\dfrac{5x - 1}{4x - 3}$ **18.** $\{(3, 2, 4)\}$

19.

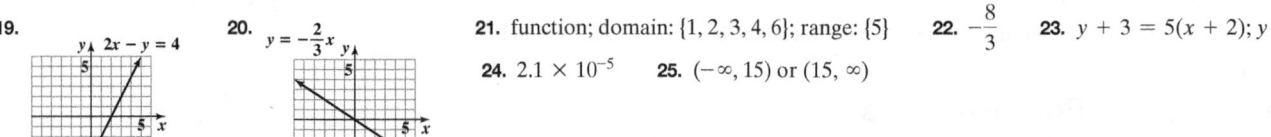

20.   **21.** function; domain: $\{1, 2, 3, 4, 6\}$; range: $\{5\}$ **22.** $-\dfrac{8}{3}$ **23.** $y + 3 = 5(x + 2); y = 5x + 7$

24. 2.1×10^{-5} **25.** $(-\infty, 15)$ or $(15, \infty)$

CHAPTER 9

Section 9.1 Check Point Exercises

1. $(-5, \infty)$ **2.** $(-\infty, 4)$ **3.** $[13, \infty)$

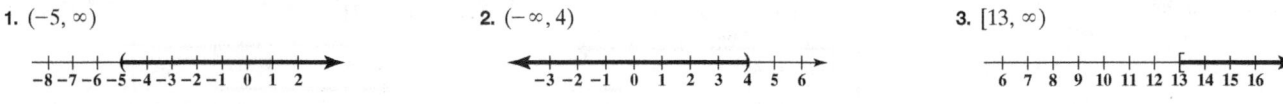

4. a. $P(x) = 125x - 160,000$ **b.** more than 1280 units   **5. a.** $C(x) = 300,000 + 30x$ **b.** $R(x) = 80x$ **c.** $P(x) = 50x - 300,000$
d. more than 6000 pairs

Concept and Vocabulary Check

1. adding 4; dividing; -3; direction; $>$; $<$ **2.** revenue; profit   **3.** break-even point

Exercise Set 9.1

1. $(-\infty, 3)$ **3.** $[7, \infty)$ **5.** $(-\infty, -4]$

7. $\left(-\infty, -\dfrac{2}{5}\right]$ **9.** $[0, \infty)$ **11.** $(-\infty, 1)$

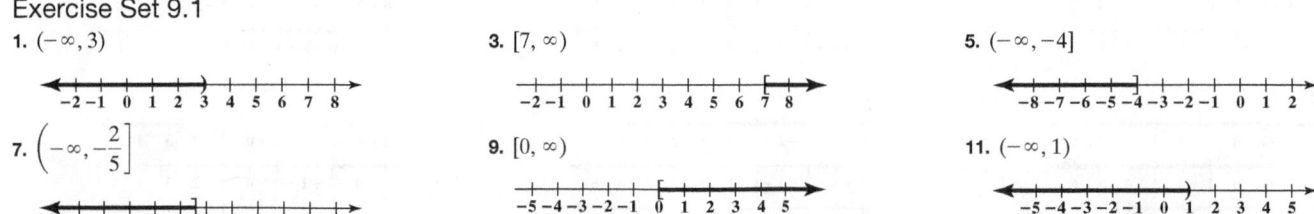

13. $[6, \infty)$

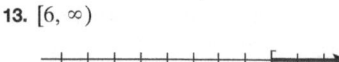

15. $[-6, \infty)$

17. $(-\infty, -6)$

19. $[13, \infty)$

21. $(-6, \infty)$

23. $[-1, \infty)$

25. $(-\infty, -2)$

27. $(-\infty, 5)$

29. $\left[-\dfrac{4}{7}, \infty \right)$

31. $[6, \infty)$ **33. a.** $P(x) = 17x - 25{,}500$ **b.** more than 1500 units **35. a.** $P(x) = 140x - 70{,}000$ **b.** more than 500 units

37. $(-\infty, 2)$ **39.** $x < \dfrac{c - b}{a}$ **41.** $(-\infty, -3]$ **43.** $(-1.4, \infty)$ **45.** $(0, 4)$

47. intimacy $\geq$ passion or passion $\leq$ intimacy **49.** commitment $>$ passion or passion $<$ commitment **51.** 9; after 3 years
53. voting years after 2006 **55. a.** $C(x) = 18{,}000 + 20x$ **b.** $R(x) = 80x$ **c.** $P(x) = 60x - 18{,}000$ **d.** more than 300 canoes
57. a. $C(x) = 30{,}000 - 2500x$ **b.** $R(x) = 3125x$ **c.** $P(x) = 625x - 30{,}000$ **d.** more than 48 sold-out performances
59. more than 6250 tapes **61.** more than 300 minutes

69.

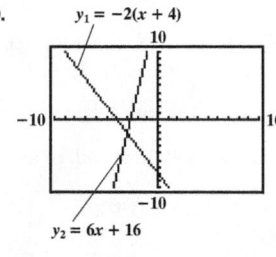

$y_1 = -2(x + 4)$

$y_2 = 6x + 16$

$(-\infty, -3)$

71. a. plan A: $4 + 0.10x$; plan B: $2 + 0.15x$
b.

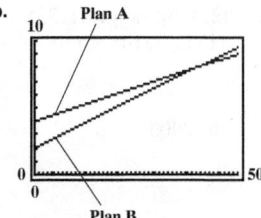

Plan A

Plan B

c–d. more than 40 checks per month

73. does not make sense **75.** makes sense **77.** false **79.** true **81.** Since $x > y$, $y - x < 0$. Thus, when both sides were multiplied by $y - x$, the sense of the inequality should have been changed. **82.** 29 **83.** $\{(-1, -1, 2)\}$ **84.** $(5x + 9)(5x - 9)$

85. a. $\{3, 4\}$ **b.** $\{1, 2, 3, 4, 5, 6, 7\}$ **86. a.** $(-\infty, 8)$ **b.** $(-\infty, 5)$ **c.** any number less than 5 **d.** any number in $[5, 8)$
87. a. $[1, \infty)$ **b.** $[3, \infty)$ **c.** any number greater than or equal to 3 **d.** any number in $[1, 3)$

Section 9.2 Check Point Exercises

1. $\{3, 7\}$ **2.** $(-\infty, 1)$ **3.** $\varnothing$ **4.** $[-1, 4)$; 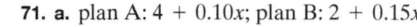 **5.** $\{3, 4, 5, 6, 7, 8, 9\}$ **6.** $(-\infty, 1] \cup (3, \infty)$ **7.** $(-\infty, \infty)$

Concept and Vocabulary Check

1. intersection; $A \cap B$ **2.** union; $A \cup B$ **3.** $(-\infty, 9)$ **4.** $(-\infty, 12)$ **5.** middle

Exercise Set 9.2

1. $\{2, 4\}$ **3.** $\varnothing$ **5.** $\varnothing$ **7.** $(6, \infty)$ **9.** $(-\infty, 1]$

11. $[-1, 2)$

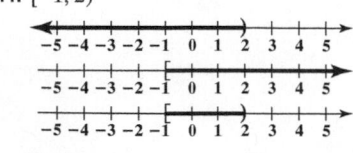

13. $\varnothing$

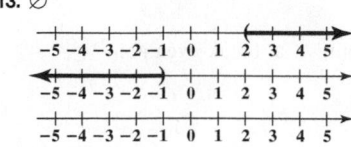

15. $(-6, -4)$

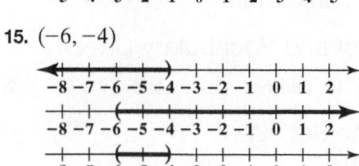

17. $(-3, 6]$

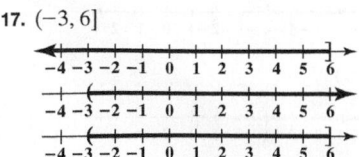

19. $(2, 5)$

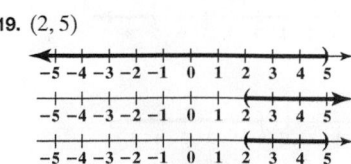

21. $\varnothing$

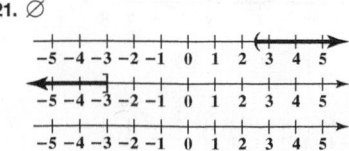

23. $[0, 2)$
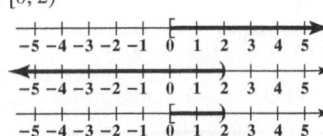

25. $(3, 5)$
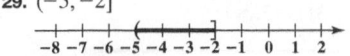

27. $[-1, 3)$
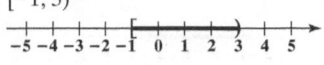

29. $(-5, -2]$
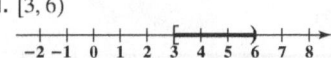

31. $[3, 6)$

33. $\{1, 2, 3, 4, 5\}$ **35.** $\{1, 2, 3, 4, 5, 6, 7, 8, 10\}$ **37.** $\{a, e, i, o, u\}$

39. $(3, \infty)$
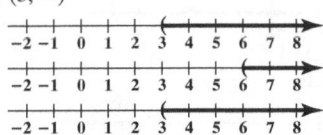

41. $(-\infty, 5]$
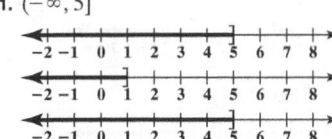

43. $(-\infty, \infty)$
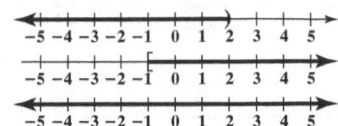

45. $(-\infty, -1) \cup [2, \infty)$
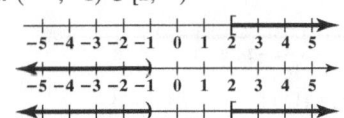

47. $(-\infty, -3) \cup (4, \infty)$
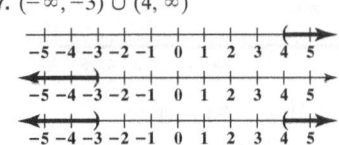

49. $(-\infty, 1] \cup [3, \infty)$
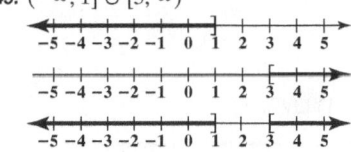

51. $(-\infty, \infty)$
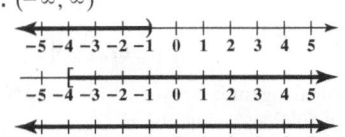

53. $(-\infty, 2)$
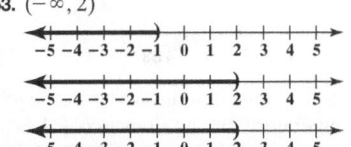

55. $(4, \infty)$ **57.** $(-\infty, 0) \cup (6, \infty)$

59. $\dfrac{b - c}{a} < x < \dfrac{b + c}{a}$ **61.** $[-1, 3]$

63. $(-1, 3)$ **65.** $[-1, 2)$ **67.** $\{-3, -2, -1\}$

69. a. years after 2016   **b.** years after 2020   **c.** years after 2020   **d.** years after 2016   **71.** between 80 and 110 minutes, inclusive
73. $[76, 126)$; If the highest grade is 100, then $[76, 100]$.   **75.** more than 3 and less than 15 crossings per 3-month period

83. $(-2, 6)$;
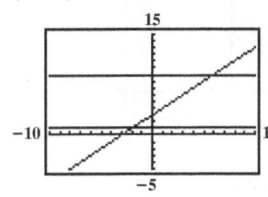

85. $\left[2, \dfrac{5}{2}\right]$;
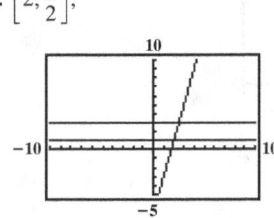

87. Exercise 83:
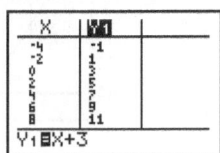

Exercise 85:
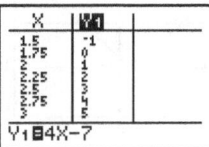

89. makes sense   **91.** makes sense   **93.** false   **95.** false   **97.** $(-\infty, 4]$ **99.** $[-1, 4]$ **101.** least: 4 nickels; greatest: 7 nickels
102. $-x^2 + 5x - 9; -15$ **103.** $f(x) = -\dfrac{1}{2}x + 4$ **104.** $17 - 2x$ **105.** $-\dfrac{1}{2}$ and 1   **106.** -1 and 3
107. a. -5 satisfies the inequality.   **b.** no

Section 9.3 Check Point Exercises

1. -2 and 3, or $\{-2, 3\}$ **2.** $-\dfrac{13}{3}$ and 5, or $\left\{-\dfrac{13}{3}, 5\right\}$ **3.** $\dfrac{4}{3}$ and 10, or $\left\{\dfrac{4}{3}, 10\right\}$

4. $(-3, 7)$

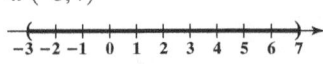

5. $\left[-\dfrac{11}{5}, 3\right]$;
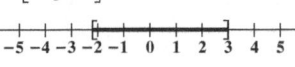

6. $(-\infty, 1] \cup [4, \infty)$
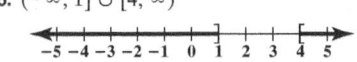

7. $[37.8, 44.2]$; The percentage of U.S. adults in the population who dread going to the dentist is between a low of 37.8% and a high of 44.2%.

Concept and Vocabulary Check

1. $c; -c$ **2.** $v; -v$ **3.** $-c; c$ **4.** $-c; c$ **5.** $<$ **6.** $<$ **7.** C   **8.** E   **9.** A   **10.** B   **11.** D   **12.** F

Exercise Set 9.3

1. $\{-8, 8\}$ **3.** $\{-5, 9\}$ **5.** $\{-3, 4\}$ **7.** $\{-1, 2\}$ **9.** $\varnothing$ **11.** $\{-3\}$ **13.** $\{-11, -1\}$ **15.** $\{-3, 4\}$ **17.** $\left\{-\dfrac{13}{3}, 5\right\}$ **19.** $\left\{-\dfrac{2}{5}, \dfrac{2}{5}\right\}$ **21.** $\varnothing$

23. $\varnothing$ **25.** $\left\{\dfrac{1}{2}\right\}$ **27.** $\left\{\dfrac{3}{4}, 5\right\}$ **29.** $\left\{\dfrac{5}{3}, 3\right\}$ **31.** $\{0\}$ **33.** $\{4\}$ **35.** $\{4\}$ **37.** $\{-1, 15\}$

39. $(-3, 3)$

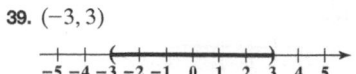

41. $(1, 3)$

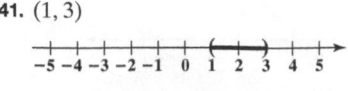

43. $[-3, -1]$

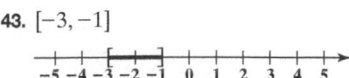

45. $(-1, 7)$

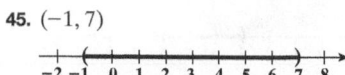

47. $(-\infty, -3) \cup (3, \infty)$

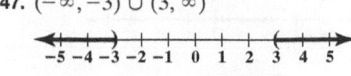

49. $(-\infty, -4) \cup (-2, \infty)$

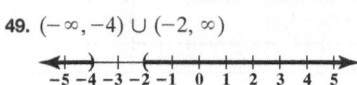

51. $(-\infty, 2] \cup [6, \infty)$

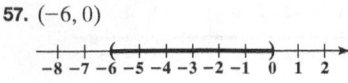

53. $\left(-\infty, \dfrac{1}{3}\right) \cup (5, \infty)$

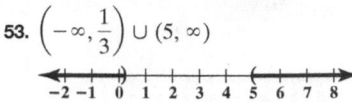

55. $[-5, 3]$

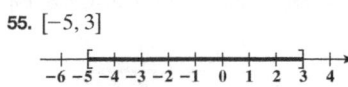

57. $(-6, 0)$

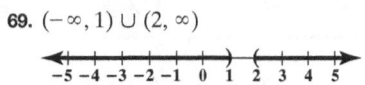

59. $(-\infty, -5] \cup [3, \infty)$

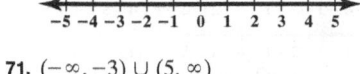

61. $(-\infty, -3) \cup (12, \infty)$

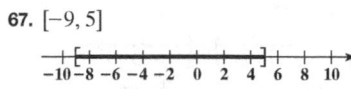

63. $\varnothing$

65. $(-\infty, \infty)$

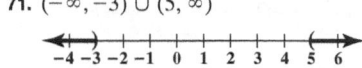

67. $[-9, 5]$

69. $(-\infty, 1) \cup (2, \infty)$

71. $(-\infty, -3) \cup (5, \infty)$

73. $(-\infty, -1] \cup [2, \infty)$

75. $-\dfrac{3}{2}$ and 4 **77.** -7 and -2 **79.** $\left[-\dfrac{7}{3}, 1\right]$ **81.** $(-\infty, -1) \cup (4, \infty)$ **83.** $\left(-\infty, -\dfrac{1}{3}\right] \cup [3, \infty)$ **85.** $\left(\dfrac{-c-b}{a}, \dfrac{c-b}{a}\right)$ **87.** $\{3, 5\}$

89. $[-2, 1]$ **91.** $[18, 24]$; The percentage of interviewers in the population turned off by the job applicant being arrogant is between a low of 18% and a high of 24%.   **93.** $[50, 64]$; The monthly average temperature for San Francisco, CA is between a low of 50°F and a high of 64°F.
95. $[8.59, 8.61]$; A machine part that is supposed to be 8.6 cm is acceptable between a low of 8.59 and a high of 8.61 cm.   **97.** If the number of outcomes that result in heads is 41 or less or 59 or more, then the coin is unfair.
105. $\{-6, 4\}$;

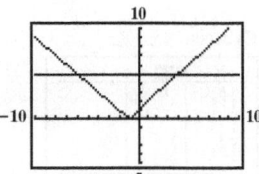

107. $\{2, 3\}$;

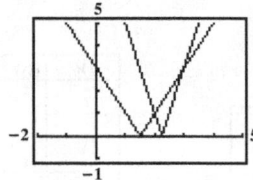

109. $(-2, 3)$;

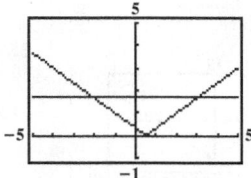

111. $(-\infty, -3) \cup (4, \infty)$;

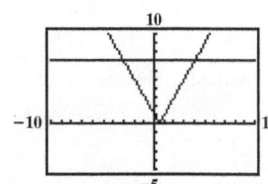

113. $(-\infty, \infty)$;

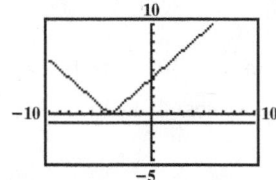

115. does not make sense
117. does not make sense   **119.** false
121. true   **123. a.** $|x - 4| < 3$
b. $|x - 4| \geq 3$ **125.** $\{1\}$

126.

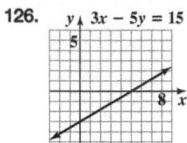

127. $f(x) = -\dfrac{2}{3}x$

128. $f(x) = -2$

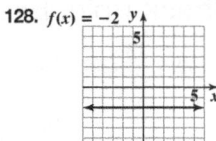

Mid-Chapter Check Point Exercises

1. $(-\infty, -4]$ **2.** $[3, 5)$ **3.** $\left\{\dfrac{1}{2}, 3\right\}$ **4.** $(-\infty, -1)$ **5.** $(-\infty, -9) \cup (-5, \infty)$ **6.** $\left[-\dfrac{2}{3}, 2\right]$ **7.** $\left\{\dfrac{1}{2}, \dfrac{13}{4}\right\}$ **8.** $(-4, -2]$ **9.** $(-\infty, \infty)$

10. $(-\infty, -3]$ **11.** $(-\infty, -3)$ **12.** $(-\infty, -1) \cup \left(-\dfrac{1}{5}, \infty\right)$ **13.** $(-\infty, -10] \cup [2, \infty)$ **14.** $\left(-\infty, -\dfrac{5}{3}\right)$ **15.** $(4, \infty)$ **16.** $\{-2, 7\}$

17. $\varnothing$ **18. a.** $C(x) = 60{,}000 + 0.18x$ **b.** $R(x) = 0.30x$ **c.** $P(x) = 0.12x - 60{,}000$ **d.** at least 750,000 discs each month
19. no more than 80 miles per day   **20.** $[49\%, 99\%]$ **21.** at least \$120,000

Section 9.4 Check Point Exercises

1. $4x - 2y \geq 8$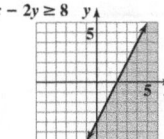

2. $y > -\frac{3}{4}x$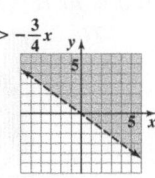

3. a. $y > 1$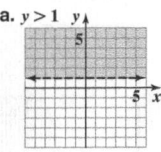

b. $x \leq -2$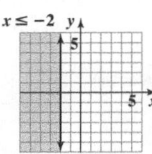

4. $B = (60, 20)$; Using $T = 60$ and $P = 20$, each of the three inequalities for grasslands is true: $60 \geq 35$, true; $5(60) - 7(20) \geq 70$, true; $3(60) - 35(20) \leq -140$, true.

5. $x - 3y < 6$
 $2x + 3y \geq -6$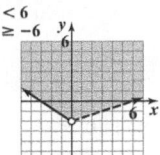

6. $x + y < 2$
 $-2 \leq x < 1$
 $y > -3$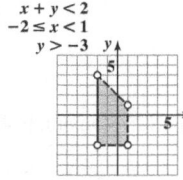

Concept and Vocabulary Check

1. solution; x; y; $5 > 1$ **2.** graph   **3.** half-plane   **4.** false   **5.** true   **6.** false   **7.** $x - y < 1$; $2x + 3y \geq 12$ **8.** false

Exercise Set 9.4

1. $x + y \geq 3$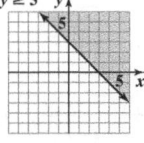

3. $x - y < 5$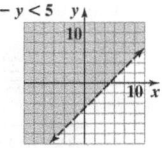

5. $x + 2y > 4$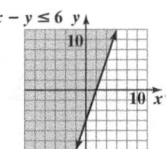

7. $3x - y \leq 6$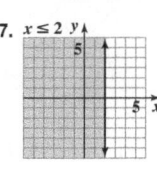

9. $\frac{x}{2} + \frac{y}{3} < 1$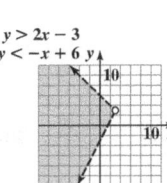

11. $y > \frac{1}{3}x$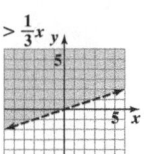

13. $y \leq 3x + 2$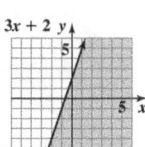

15. $y < -\frac{1}{4}x$

17. $x \leq 2$

19. $y > -4$

21. $y \geq 0$

23. $3x + 6y \leq 6$
 $2x + y \leq 8$

25. $2x - 5y \leq 10$
 $3x - 2y > 6$

27. $y > 2x - 3$
 $y < -x + 6$

29. $x + 2y \leq 4$
 $y \geq x - 3$

31. $x \leq 2$
 $y \geq -1$

33. $-2 \leq x < 5$

35. $x - y \leq 1$
 $x \geq 2$

37. $\varnothing$

39. $x + y > 4$
 $x + y > -1$

41. $x - y \leq 2$
 $x \geq -2$
 $y \leq 3$

43. $x \geq 0$
 $y \geq 0$
 $2x + 5y \leq 10$
 $3x + 4y \leq 12$

45. $3x + y \leq 6$
 $2x - y \leq -1$
 $x \geq -2$
 $y \leq 4$

47. $y \geq -2x + 4$

49. $\begin{cases} x + y \leq 4 \\ 3x + y \leq 6 \end{cases}$
 $x + y \leq 4$
 $3x + y \leq 6$

51. $-2 \leq x \leq 2$; $-3 \leq y \leq 3$

53. $y > \frac{3}{2}x - 2$ or $y < 4$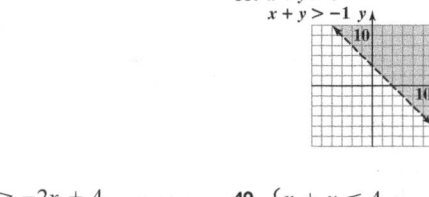

55. no solution   **57.** infinitely many solutions   **59. a.** $A = (20, 150)$; A 20-year-old with a heart rate of 150 beats per minute is within the target range.   **b.** $10 \leq 20 \leq 70$, true; $150 \geq 0.7(220 - 20)$, true; $150 \leq 0.8(220 - 20)$, true   **61.** $10 \leq a \leq 70$; $H \geq 0.6(220 - a)$; $H \leq 0.7(220 - a)$

63. a. 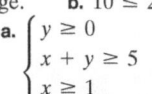   **b.** 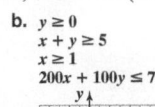 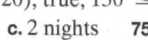   **c.** 2 nights   **75.** 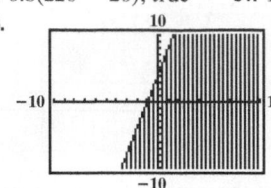   **77.**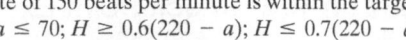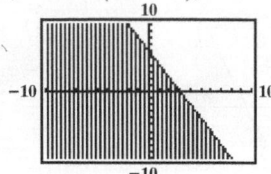

83. makes sense   **85.** makes sense   **87.** false   **89.** true   **91.** $\begin{cases} x \geq -2 \\ y > -1 \end{cases}$ **93.** $y \geq 2x - 2$

95. $\begin{aligned} y &\geq nx + b \\ y &\leq mx + b \end{aligned}$ [graph]   **96.** $\{(3, 1)\}$ **97.** $\{(2, 4)\}$ **98.** $2x^4(x + 5y)^2$ **99.** 3   **100.** 6   **101.** domain: $[-4, \infty)$; range: $[0, \infty)$

Review Exercises

1. $[-2, \infty)$ [number line]

2. $\left[\dfrac{3}{5}, \infty\right)$ [number line]

3. $\left(-\infty, -\dfrac{21}{2}\right)$ [number line]

4. $(-3, \infty)$ [number line]

5. $(-\infty, -2]$ [number line]

6. a. $P(x) = 85x - 357{,}000$ **b.** more than 4200 toaster ovens

7. $C(x) = 360{,}000 + 850x$ **8.** $R(x) = 1150x$ **9.** $P(x) = 300x - 360{,}000$ **10.** more than 1200 computers

11. more than 50 checks   **12.** more than $13,500 in sales   **13.** $\{a, c\}$ **14.** $\{a\}$ **15.** $\{a, b, c, d, e\}$ **16.** $\{a, b, c, d, f, g\}$

17. $(-\infty, 3]$ [number line]

18. $(-\infty, 6)$ [number line]

19. $(6, 8)$ [number line]

20. $(-\infty, 1]$ [number line]

21. $\varnothing$

22. $(-\infty, 1) \cup (2, \infty)$ [number line]

23. $(-\infty, -4] \cup (2, \infty)$ [number line]

24. $(-\infty, -2)$ [number line]

25. $(-\infty, \infty)$ [number line]

26. $(-5, 2]$ [number line]

27. $\left[-\dfrac{3}{4}, 1\right]$ [number line]

28. $[49\%, 99\%)$

29. $\{-4, 3\}$ **30.** $\varnothing$ **31.** $\left\{-\dfrac{11}{2}, \dfrac{23}{2}\right\}$ **32.** $\left\{-4, -\dfrac{6}{11}\right\}$

33. $[-9, 6]$ [number line]

34. $(-\infty, -6) \cup (0, \infty)$ [number line]

35. $(-3, -2)$ [number line]

36. $(-\infty, -5] \cup [1, \infty)$; 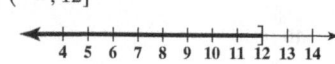   **37.** $\varnothing$ **38.** Approximately 90% of the population sleeps between 5.5 hours and 7.5 hours daily, inclusive.

39. $3x - 4y > 12$ 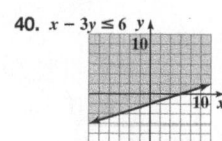   **40.** $x - 3y \le 6$ 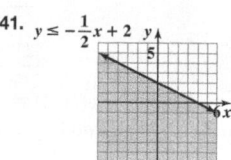   **41.** $y \le -\frac{1}{2}x + 2$ 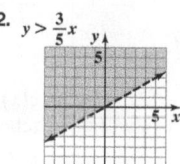   **42.** $y > \frac{3}{5}x$ 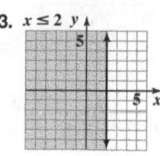   **43.** $x \le 2$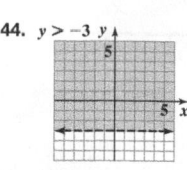

44. $y > -3$ 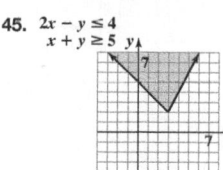   **45.** $2x - y \le 4$ $x + y \ge 5$ 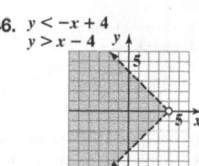   **46.** $y < -x + 4$ $y > x - 4$ 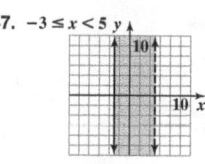   **47.** $-3 \le x < 5$ 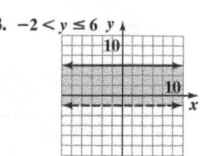   **48.** $-2 < y \le 6$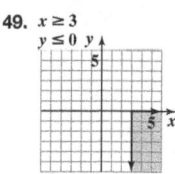

49. $x \ge 3$ $y \le 0$ 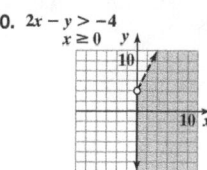   **50.** $2x - y > -4$ $x \ge 0$ 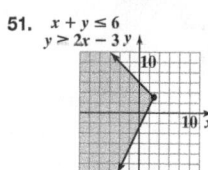   **51.** $x + y \le 6$ $y > 2x - 3$ 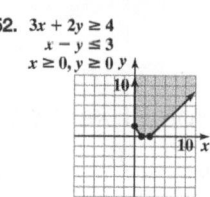   **52.** $3x + 2y \ge 4$ $x - y \le 3$ $x \ge 0, y \ge 0$    **53.** $\varnothing$

Chapter Test

1. $(-\infty, 12]$

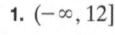

2. $\left[\frac{21}{8}, \infty\right)$

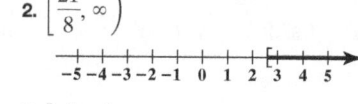

3. a. $C(x) = 60{,}000 + 200x$
  b. $R(x) = 450x$ **c.** $P(x) = 250x - 60{,}000$
  d. more than 240 desks
4. $\{4, 6\}$ **5.** $\{2, 4, 6, 8, 10, 12, 14\}$

6. $(-2, -1)$

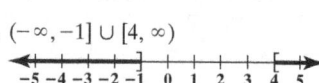

7. $[-2, \infty)$

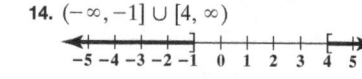

8. $(-\infty, 4)$

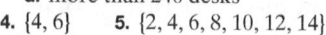

9. $(-\infty, -4] \cup (2, \infty)$

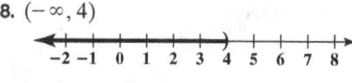

10. $\left[-7, \frac{13}{2}\right)$

11. $\left\{-2, \frac{4}{5}\right\}$ **12.** $\left\{-\frac{8}{5}, 7\right\}$

13. $(-3, 4)$

14. $(-\infty, -1] \cup [4, \infty)$

15. $(-\infty, 90.6) \cup (106.6, \infty)$;
  Hypothermia: Body temperature below 90.6°F; Hyperthermia: Body temperature above 106.6°F

16. $3x - 2y < 6$ 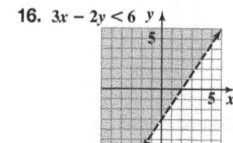   **17.** $y \ge \frac{1}{2}x - 1$ 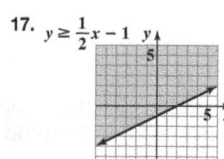   **18.** $y \le -1$ 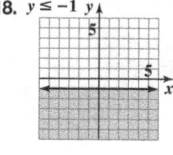   **19.** $x + y \ge 2$ $x - y \ge 4$ 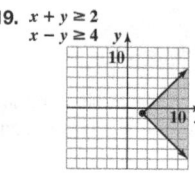   **20.** $3x + y \le 9$ $2x + 3y \ge 6$ $x \ge 0$ $y \ge 0$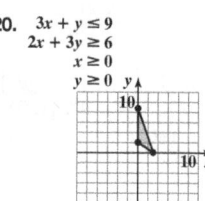

21. $-2 < x \le 4$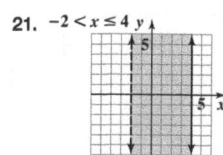

Cumulative Review Exercises

1. $\{-1\}$ **2.** $\{8\}$ **3.** $-\frac{2y^7}{3x^5}$ **4.** $22; 4a^2 - 6a + 4$ **5.** $2x^2 + x + 2; 12$ **6.** $f(x) = -\frac{1}{2}x + 4$

7. $f(x) = 2x + 1$ 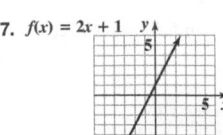   **8.** $y > 2x$ 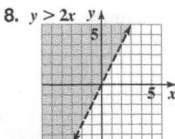   **9.** $2x - y \geq 6$ 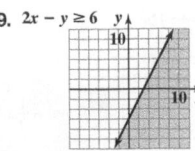   **10.** $f(x) = -1$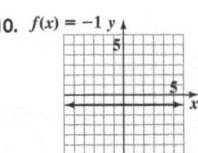

11. $\{(-4, 2, -1)\}$ **12.** $f^{-1}(x) = 3x + 12$ **13.** $f(g(x)) = 3x^2 + 12x + 11; g(f(x)) = 3x^2 + 1$
14. 46 rooms with kitchen facilities and 14 without kitchen facilities   **15.** a. and b. are functions.

16. $[-7, \infty)$

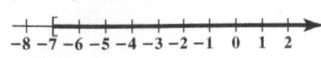

17. $(-\infty, -6)$

18. $(-\infty, 3] \cup [5, \infty)$

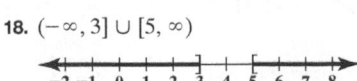

19. $[-10, 7]$

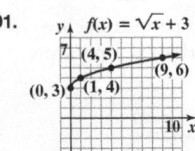

20. $\left(-\infty, \dfrac{1}{3}\right) \cup (5, \infty)$

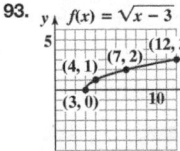

CHAPTER 10

Section 10.1 Check Point Exercises

1. a. 8   **b.** -7 **c.** $\dfrac{4}{5}$ **d.** 0.09   **e.** 5   **f.** 7   **2. a.** 4   **b.** $-\sqrt{24} \approx -4.90$ **3.** $[3, \infty)$
4. approximately 15.5 minutes   **5. a.** 7   **b.** $|x + 8|$ **c.** $|7x^5|$ or $7|x^5|$ **d.** $|x - 3|$ **6. a.** 3   **b.** -2 **7.** $-3x$
8. a. 2   **b.** -2 **c.** not a real number   **d.** -1 **9. a.** $|x + 6|$ **b.** $3x - 2$ **c.** 8

Concept and Vocabulary Check

1. principal   **2.** 8^2 **3.** $[0, \infty)$ **4.** $5x - 20 \geq 0$ **5.** $|a|$ **6.** 10^3 **7.** $(-5)^3$ **8.** a **9.** $(-\infty, \infty)$ **10.** nth; index
11. $|a|; a$ **12.** true   **13.** false   **14.** true   **15.** false

Exercise Set 10.1

1. 6   **3.** -6 **5.** not a real number   **7.** $\dfrac{1}{5}$ **9.** $-\dfrac{3}{4}$ **11.** 0.9   **13.** -0.2 **15.** 3   **17.** 1   **19.** not a real number
21. $4; 1; 0;$ not a real number   **23.** $-5; -\sqrt{5} \approx -2.24; -1;$ not a real number   **25.** $4; 2; 1; 6$ **27.** $[3, \infty); c$
29. $[-5, \infty); d$ **31.** $(-\infty, 3]; e$ **33.** 5   **35.** 4   **37.** $|x - 1|$ **39.** $|6x^2|$ or $6x^2$ **41.** $-|10x^3|$ or $-10|x^3|$
43. $|x + 6|$ **45.** $-|x - 4|$ **47.** 3   **49.** -3 **51.** $\dfrac{1}{5}$ **53.** $-\dfrac{3}{10}$ **55.** $3; 2; -1; -4$ **57.** $-2; 0; 2$
59. 1   **61.** 2   **63.** -2 **65.** not a real number   **67.** -1 **69.** not a real number   **71.** -4 **73.** 2   **75.** -2
77. x **79.** $|y|$ **81.** $-2x$ **83.** -5 **85.** 5   **87.** $|x + 3|$ **89.** $-2(x - 1)$

91. 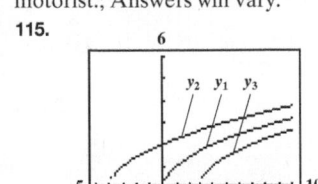   **93.**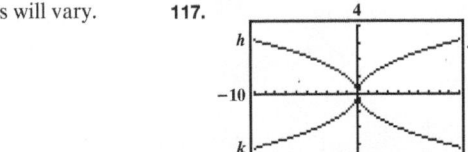

domain: $[0, \infty)$; range: $[3, \infty)$ domain: $[3, \infty)$; range: $[0, \infty)$
95. $(-\infty, 15)$ **97.** $[1, 3)$ **99.** 3   **101. a.** 40.2 in.; underestimates by 0.6 in.   **b.** 0.9 in. per month   **c.** 0.2 in. per month; This is a much
smaller rate of change.; The graph is not as steep between 50 and 60 as it is between 0 and 10.   **103.** 70 mph; The officer should not believe the
motorist.; Answers will vary.

115. ; Answers will vary.   **117.**   domain of f: $[0, \infty)$; range of f: $[0, \infty)$
domain of g: $[0, \infty)$; range of g: $(-\infty, 0]$
domain of h: $(-\infty, 0]$; range of h: $[0, \infty)$
domain of k: $(-\infty, 0]$; range of k: $(-\infty, 0]$

119. does not make sense   **121.** makes sense   **123.** false   **125.** false   **127.** Answers will vary; an example is $f(x) = \sqrt{15 - 3x}$.
129. $|(2x + 3)^5|$ **131.** The graph of h is the graph of f shifted left 3 units.

132. $7x + 30$ **133.** $\dfrac{x^8}{9y^6}$ **134.** $\left(-\infty, -\dfrac{7}{3}\right) \cup (5, \infty)$ **135.** $\dfrac{2^7}{x}$ or $\dfrac{128}{x}$ **136.** $\dfrac{2}{x^3}$ **137.** $\dfrac{y^{12}}{x^8}$

Section 10.2 Check Point Exercises

1. a. $\sqrt{25} = 5$ **b.** $\sqrt[3]{-8} = -2$ **c.** $\sqrt[4]{5xy^2}$ **2. a.** $(5xy)^{1/4}$ **b.** $\left(\dfrac{a^3b}{2}\right)^{1/5}$ **3. a.** $(\sqrt[3]{8})^4 = 16$ **b.** $(\sqrt{25})^3 = 125$ **c.** $-(\sqrt[4]{81})^3 = -27$

4. a. $6^{4/3}$ **b.** $(2xy)^{7/5}$ **5. a.** $\dfrac{1}{100^{1/2}} = \dfrac{1}{10}$ **b.** $\dfrac{1}{8^{1/3}} = \dfrac{1}{2}$ **c.** $\dfrac{1}{32^{3/5}} = \dfrac{1}{8}$ **d.** $\dfrac{1}{(3xy)^{5/9}}$ **6. a.** $7^{5/6}$ **b.** $\dfrac{5}{x}$ **c.** $9.1^{3/10}$ **d.** $\dfrac{y^{1/12}}{x^{1/5}}$

7. a. $\sqrt{x}$ **b.** $2a^4$ **c.** $\sqrt[4]{x^2y}$ **d.** $\sqrt[6]{x}$ **e.** $\sqrt[6]{x}$

Concept and Vocabulary Check

1. $\sqrt{36} = 6$ **2.** $\sqrt[3]{8} = 2$ **3.** $\sqrt[n]{a}$ **4.** $(\sqrt[4]{16})^3 = (2)^3 = 8$ **5.** $(\sqrt[n]{a})^m$ or $\sqrt[n]{a^m}$ **6.** $\dfrac{5}{3}$ **7.** $\dfrac{1}{16^{3/2}} = \dfrac{1}{(\sqrt{16})^3} = \dfrac{1}{(4)^3} = \dfrac{1}{64}$

Exercise Set 10.2

1. $\sqrt{49} = 7$ **3.** $\sqrt[3]{-27} = -3$ **5.** $-\sqrt[4]{16} = -2$ **7.** $\sqrt[3]{xy}$ **9.** $\sqrt[5]{2xy^3}$ **11.** $(\sqrt{81})^3 = 729$ **13.** $(\sqrt[3]{125})^2 = 25$

15. $(\sqrt[5]{-32})^3 = -8$ **17.** $(\sqrt[3]{27})^2 + (\sqrt[4]{16})^3 = 17$ **19.** $\sqrt{(xy)^4}$ **21.** $7^{1/2}$ **23.** $5^{1/3}$ **25.** $(11x)^{1/5}$ **27.** $x^{3/2}$ **29.** $x^{3/5}$

31. $(x^2y)^{1/5}$ **33.** $(19xy)^{3/2}$ **35.** $(7xy^2)^{5/6}$ **37.** $2xy^{2/3}$ **39.** $\dfrac{1}{49^{1/2}} = \dfrac{1}{7}$ **41.** $\dfrac{1}{27^{1/3}} = \dfrac{1}{3}$ **43.** $\dfrac{1}{16^{3/4}} = \dfrac{1}{8}$ **45.** $\dfrac{1}{8^{2/3}} = \dfrac{1}{4}$

47. $\left(\dfrac{27}{8}\right)^{1/3} = \dfrac{3}{2}$ **49.** $\dfrac{1}{(-64)^{2/3}} = \dfrac{1}{16}$ **51.** $\dfrac{1}{(2xy)^{7/10}}$ **53.** $\dfrac{5x}{z^{1/3}}$ **55.** 3   **57.** 4   **59.** $x^{5/6}$ **61.** $x^{3/5}$ **63.** $\dfrac{1}{x^{5/12}}$ **65.** 25

67. $\dfrac{1}{y^{1/6}}$ **69.** $32x$ **71.** $5x^2y^3$ **73.** $\dfrac{x^{1/4}}{y^{3/10}}$ **75.** 3   **77.** $27y^{2/3}$ **79.** $\sqrt[4]{x}$ **81.** $2a^2$ **83.** x^2y^3 **85.** x^6y^6 **87.** $\sqrt[5]{3y}$ **89.** $\sqrt[3]{4a^2}$

91. $\sqrt[3]{x^2y}$ **93.** $\sqrt[6]{2^5}$ or $6\sqrt{32}$ **95.** $\sqrt[10]{x^9}$ **97.** $\sqrt[12]{a^{10}b^7}$ **99.** $\sqrt[20]{x}$ **101.** $\sqrt{y}$ **103.** $\sqrt[8]{x}$ **105.** $\sqrt[4]{x^2y}$ **107.** $\sqrt[12]{2x}$

109. x^9y^{15} **111.** $\sqrt[4]{a^3b^3}$ **113.** $x^{2/3} - x$ **115.** $x + 2x^{1/2} - 15$ **117.** $2x^{1/2}(3 + x)$ **119.** $15x^{1/3}(1 - 4x^{2/3})$ **121.** $\dfrac{x^2}{7y^{3/2}}$ **123.** $\dfrac{x^3}{y^2}$

125. 58 species of plants   **127.** about 1872 calories per day   **129. a.** $C = 35.74 + 0.6215t - 35.74v^{4/25} + 0.4275tv^{4/25}$ **b.** 8°F
131. a. $C(v) = 35.74 - 35.74v^{4/25}$ **b.** $C(25) \approx -24$; When the air temperature is 0°F and the wind speed is 25 miles per hour, the windchill temperature is −24°.   **c.** the point $(25, -24)$ **133. a.** $L + 1.25S^{1/2} - 9.8D^{1/3} \le 16.296$ **b.** eligible
145. simplified correctly 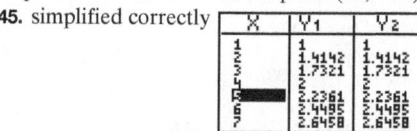   **147.** Right side should be $x^{1/2}$.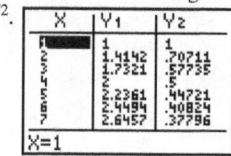

149. does not make sense   **151.** does not make sense   **153.** false   **155.** true   **157.** $\dfrac{1}{4}$ of the cake   **159.** $[3, \infty)$

160. $y = -2x + 11$ or $f(x) = -2x + 11$ **161.** 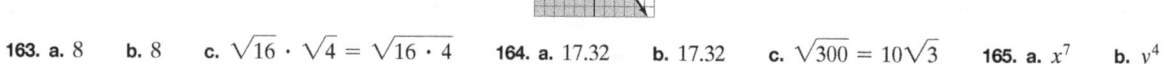   **162.** $3a^2 + 6ah + 3h^2 - 5a - 5h + 4$

$y \le -\dfrac{3}{2}x + 3$

163. a. 8   **b.** 8   **c.** $\sqrt{16} \cdot \sqrt{4} = \sqrt{16 \cdot 4}$ **164. a.** 17.32   **b.** 17.32   **c.** $\sqrt{300} = 10\sqrt{3}$ **165. a.** x^7 **b.** y^4

Section 10.3 Check Point Exercises

1. a. $\sqrt{55}$ **b.** $\sqrt{x^2 - 16}$ **c.** $\sqrt[3]{60}$ **d.** $\sqrt[4]{12x^4}$ **2. a.** $4\sqrt{5}$ **b.** $2\sqrt[3]{5}$ **c.** $2\sqrt[4]{2}$ **d.** $10|x|\sqrt{2y}$
3. $f(x) = \sqrt{3}|x - 2|$ **4.** $x^4y^5z\sqrt{xyz}$ **5.** $2x^3y^4\sqrt[3]{5xy^2}$ **6.** $2x^2z\sqrt[4]{x^2y^2z^3}$ **7. a.** $2\sqrt{3}$ **b.** $100\sqrt[3]{4}$ **c.** $2x^2y\sqrt[4]{2}$

Concept and Vocabulary Check

1. $\sqrt[n]{ab}$ **2.** 77   **3.** $\sqrt[3]{8} \cdot \sqrt[3]{5} = 2\sqrt[3]{5}$ **4.** $\sqrt{5}|x + 1|$ **5.** x^5; x^4; x^5

Exercise Set 10.3

1. $\sqrt{15}$ **3.** $\sqrt[3]{18}$ **5.** $\sqrt[4]{33}$ **7.** $\sqrt{33xy}$ **9.** $\sqrt[5]{24x^4}$ **11.** $\sqrt{x^2 - 9}$ **13.** $\sqrt[6]{(x - 4)^5}$ **15.** $\sqrt{x}$ **17.** $\sqrt[4]{\dfrac{3x}{7y}}$ **19.** $\sqrt{77x^5y^3}$

21. $5\sqrt{2}$ **23.** $3\sqrt{5}$ **25.** $5\sqrt{3x}$ **27.** $2\sqrt[3]{2}$ **29.** $3x$ **31.** $-2y\sqrt[3]{2x^2}$ **33.** $6|x + 2|$ **35.** $2(x + 2)\sqrt[3]{4}$ **37.** $|x - 1|\sqrt{3}$

39. $x^3\sqrt{x}$ **41.** $x^4y^4\sqrt{y}$ **43.** $4x\sqrt{3x}$ **45.** $y^2\sqrt[3]{y^2}$ **47.** $x^4y\sqrt[3]{x^2z}$ **49.** $3x^2y^2\sqrt[3]{3x^2}$ **51.** $(x + y)\sqrt[3]{(x + y)^2}$ **53.** $y^3\sqrt[5]{y^2}$
55. $2xy^3\sqrt[5]{2xy^2}$ **57.** $2x^2\sqrt[4]{5x^2}$ **59.** $(x - 3)^2\sqrt[4]{(x - 3)^2}$ or $(x - 3)^2\sqrt{x - 3}$ **61.** $2\sqrt{6}$ **63.** $5\sqrt{2xy}$ **65.** $6x$ **67.** $10xy\sqrt{2y}$
69. $60\sqrt{2}$ **71.** $2\sqrt[6]{6}$ **73.** $2x^2\sqrt{10x}$ **75.** $5xy^4\sqrt[3]{x^2y^2}$ **77.** $2xyz\sqrt[4]{x^2z^3}$ **79.** $2xy^2z\sqrt[5]{2y^3z}$ **81.** $(x - y)^2\sqrt[3]{(x - y)^2}$
83. $-6x^3y^3\sqrt[3]{2yz^2}$ **85.** $-6y^2\sqrt[5]{2x^3y}$ **87.** $-6x^3y^3\sqrt{2}$ **89.** $-12x^3y^4\sqrt[4]{x^3}$ **91.** $6\sqrt{3}$ miles; 10.4 miles   **93.** $8\sqrt{3}$ ft per sec; 14 ft per sec
95. a. $\dfrac{7.644}{2\sqrt[4]{2}} = \dfrac{3.822}{\sqrt[4]{2}}$ **b.** 3.21 liters of blood per minute per square meter; $(32, 3.21)$

103. Graphs are the same; simplification is correct.

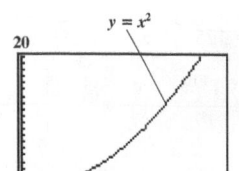

105. Graphs are not the same.; $\sqrt{3x^2 - 6x + 3} = |x - 1|\sqrt{3}$

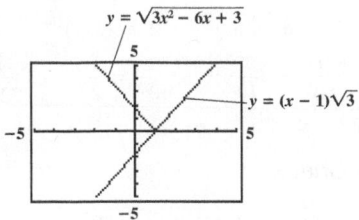

107. makes sense   **109.** makes sense   **111.** false   **113.** false   **115.** Its square root is multiplied by $\sqrt{3}$.   **117.** $g(x) = \sqrt[3]{4x^2}$

119. $[5, 11]$ **120.** $\{(2, -4)\}$ **121.** $(4x - 3)(16x^2 + 12x + 9)$ **122. a.** $31x$ **b.** $31\sqrt{2}$ **123. a.** $-8x$ **b.** $-8\sqrt[3]{2}$ **124.** $\dfrac{y\sqrt[4]{7y}}{x^3}$

Section 10.4 Check Point Exercises

1. a. $10\sqrt{13}$ **b.** $(21 - 6x)\sqrt[3]{7}$ **c.** $5\sqrt[4]{3x} + 2\sqrt[3]{3x}$ **2. a.** $21\sqrt{5}$ **b.** $-12\sqrt{3x}$ **c.** cannot be simplified

3. a. $-9\sqrt[3]{3}$ **b.** $(5 + 3xy)\sqrt[3]{x^2y}$ **4. a.** $\dfrac{2\sqrt[3]{3}}{5}$ **b.** $\dfrac{3x\sqrt{x}}{y^5}$ **c.** $\dfrac{2y^2\sqrt[3]{y}}{x^4}$ **5. a.** $2x^2\sqrt{5}$ **b.** $\dfrac{5\sqrt{xy}}{2}$ **c.** $2x^2y$

Concept and Vocabulary Check

1. $(5 + 8)\sqrt{13} = 13\sqrt{3}$ **2.** $\sqrt{9 \cdot 3} - \sqrt{4 \cdot 3} = 3\sqrt{3} - 2\sqrt{3} = \sqrt{3}$ **3.** $\sqrt[3]{27 \cdot 2} + \sqrt[3]{8 \cdot 2} = 3\sqrt[3]{2} + 2\sqrt[3]{2} = 5\sqrt[3]{2}$

4. $\dfrac{\sqrt[n]{a}}{\sqrt[n]{b}}$ **5.** $\dfrac{\sqrt[3]{8}}{\sqrt[3]{27}} = \dfrac{2}{3}$ **6.** $\sqrt{\dfrac{72x^3}{2x}} = \sqrt{36x^2} = 6x$

Exercise Set 10.4

1. $11\sqrt{5}$ **3.** $7\sqrt[3]{6}$ **5.** $2\sqrt[5]{2}$ **7.** $\sqrt{13} + 2\sqrt{5}$ **9.** $7\sqrt{5} + 2\sqrt[3]{x}$ **11.** $4\sqrt{3}$ **13.** $19\sqrt{3}$ **15.** $6\sqrt{2x}$ **17.** $13\sqrt[3]{2}$

19. $(9x + 1)\sqrt{5x}$ **21.** $7y\sqrt[3]{2x}$ **23.** $(3x - 2)\sqrt[3]{2x}$ **25.** $4\sqrt{x - 2}$ **27.** $5x\sqrt[3]{xy^2}$ **29.** $\dfrac{\sqrt{11}}{2}$ **31.** $\dfrac{\sqrt[3]{19}}{3}$ **33.** $\dfrac{x}{6y^4}$ **35.** $\dfrac{2x\sqrt{2x}}{5y^3}$

37. $\dfrac{x\sqrt[3]{x}}{2y}$ **39.** $\dfrac{x^2\sqrt[3]{50x^2}}{3y^4}$ **41.** $\dfrac{y\sqrt[4]{9y^2}}{x^2}$ **43.** $\dfrac{2x^2\sqrt[5]{2x^3}}{y^4}$ **45.** $2\sqrt{2}$ **47.** 2 **49.** $3x$ **51.** x^2y **53.** $2x^2\sqrt{5}$ **55.** $4a^5b^5$

57. $3\sqrt{xy}$ **59.** $2xy$ **61.** $2x^2y^2\sqrt[4]{y^2}$ or $2x^2y^2\sqrt{y}$ **63.** $\sqrt[3]{x + 3}$ **65.** $\sqrt[3]{a^2 - ab + b^2}$ **67.** $\dfrac{43\sqrt{2}}{35}$ **69.** $-11xy\sqrt{2x}$ **71.** $25x\sqrt{2x}$

73. $7x\sqrt{3xy}$ **75.** $3x\sqrt[3]{5xy}$ **77.** $\left(\dfrac{f}{g}\right)(x) = 4x\sqrt{x}$; domain: $(0, \infty)$ **79.** $\left(\dfrac{f}{g}\right)(x) = 2x\sqrt[3]{2x}$; domain: $(-\infty, 0) \cup (0, \infty)$

81. $P = 18\sqrt{5}$ ft; $A = 100$ sq ft   **83.** $12\sqrt{5}$ m   **85. a.** $5\sqrt{10}$; the projected increase in the number of Americans ages 65–84, in millions, from 2020 to 2050   **b.** 15.8; underestimates by 2.7 million

95. Graphs are not the same.; $\sqrt{16x} - \sqrt{9x} = \sqrt{x}$

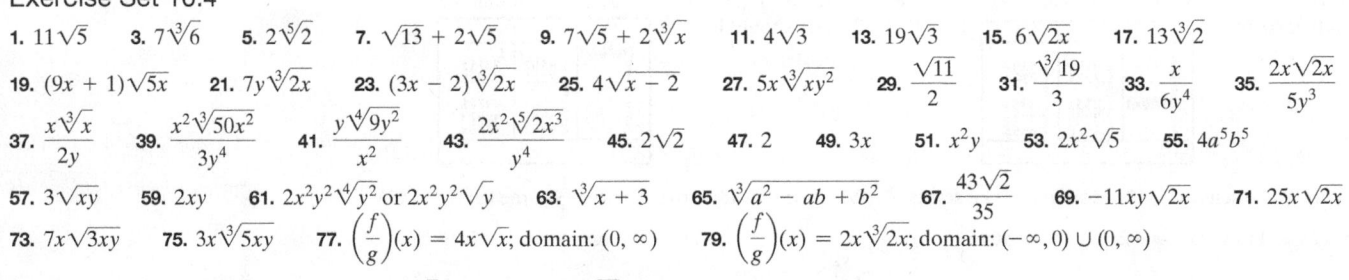

97. makes sense   **99.** does not make sense   **101.** false   **103.** false   **105.** $2\sqrt{2}$ **107.** $x^2y^3a^5b\sqrt{b}$ **108.** $\{0\}$

109. $(x - 2y)(x - 6y)$ **110.** $\dfrac{3x^2 + 8x + 6}{(x + 3)^2(x + 2)}$ **111. a.** $7x + 35$ **b.** $x\sqrt{7} + \sqrt{35}$ **112. a.** $6x^2 + 33x + 15$ **b.** $27 + 33\sqrt{2}$

113. $\dfrac{5\sqrt[5]{8x^2y^4}}{x}$

Mid-Chapter Check Point Exercises

1. 13 **2.** $2x^2y^3\sqrt{2xy}$ **3.** $5\sqrt[3]{4x^2}$ **4.** $12x\sqrt[3]{2x}$ **5.** 1 **6.** $4xy^{1/12}$ **7.** $-\sqrt{3}$ **8.** $\dfrac{5x\sqrt{5x}}{y^2}$ **9.** $\sqrt[4]{x^3}$ **10.** $3x\sqrt[3]{2x^2}$ **11.** $2\sqrt[3]{10}$

12. $\dfrac{x^2}{y^4}$ **13.** $x^{7/12}$ **14.** $x\sqrt[3]{y^2}$ **15.** $(x - 2)\sqrt[7]{(x - 2)^2}$ **16.** $2x^2y^4\sqrt[4]{2x^3y}$ **17.** $14\sqrt[3]{2}$ **18.** $x\sqrt[7]{x^2y^2}$ **19.** $\dfrac{1}{25}$ **20.** $\sqrt[6]{32}$

21. $\dfrac{4x}{y^2}$ **22.** $8x^2y^3\sqrt[3]{y}$ **23.** $-10x^3y\sqrt{6y}$ **24.** $(-\infty, 6]$ **25.** $(-\infty, \infty)$

Section 10.5 Check Point Exercises

1. a. $x\sqrt{6} + 2\sqrt{15}$ **b.** $y - \sqrt[3]{7y}$ **c.** $36 - 18\sqrt{10}$ **2. a.** $11 + 2\sqrt{30}$ **b.** 1 **c.** $a - 7$ **3. a.** $\dfrac{\sqrt{21}}{7}$ **b.** $\dfrac{\sqrt[3]{6}}{3}$

4. a. $\dfrac{\sqrt{14xy}}{7y}$ **b.** $\dfrac{\sqrt[3]{3xy^2}}{3y}$ **c.** $\dfrac{3\sqrt[5]{4x^3y}}{y}$ **5.** $12\sqrt{3} - 18$ **6.** $\dfrac{3\sqrt{5} + 3\sqrt{2} + \sqrt{35} + \sqrt{14}}{3}$ **7.** $\dfrac{1}{\sqrt{x+3} + \sqrt{x}}$

Concept and Vocabulary Check

1. $350; -42\sqrt{10}; 30\sqrt{10}; -36$ **2.** $(\sqrt{10})^2 - (\sqrt{5})^2 = 10 - 5 = 5$ **3.** rationalizing the denominator **4.** $\sqrt{5}$ **5.** $\sqrt[3]{9}$
6. $7\sqrt{2} - 5$ **7.** $3\sqrt{6} + \sqrt{5}$

Exercise Set 10.5

1. $x\sqrt{2} + \sqrt{14}$ **3.** $7\sqrt{6} - 6$ **5.** $12\sqrt{2} - 6$ **7.** $\sqrt[3]{12} + 4\sqrt[3]{10}$ **9.** $2x\sqrt[3]{2} - \sqrt[3]{x^2}$ **11.** $32 + 11\sqrt{2}$ **13.** $34 - 15\sqrt{5}$
15. $117 - 36\sqrt{7}$ **17.** $\sqrt{6} + \sqrt{10} + \sqrt{21} + \sqrt{35}$ **19.** $\sqrt{6} - \sqrt{10} - \sqrt{21} + \sqrt{35}$ **21.** $-48 + 7\sqrt{6}$ **23.** $8 + 2\sqrt{15}$
25. $3x - 2\sqrt{3xy} + y$ **27.** -44 **29.** -71 **31.** 6 **33.** $6 - 5\sqrt{x} + x$ **35.** $\sqrt[3]{x^2} + \sqrt[3]{x} - 20$ **37.** $2x^2 + x\sqrt[3]{y^2} - y\sqrt[3]{y}$ **39.** $\dfrac{\sqrt{10}}{5}$
41. $\dfrac{\sqrt{11x}}{x}$ **43.** $\dfrac{3\sqrt{3y}}{y}$ **45.** $\dfrac{\sqrt[3]{4}}{2}$ **47.** $3\sqrt[3]{2}$ **49.** $\dfrac{\sqrt[3]{18}}{3}$ **51.** $\dfrac{4\sqrt[3]{x^2}}{x}$ **53.** $\dfrac{\sqrt[3]{2y}}{y}$ **55.** $\dfrac{7\sqrt[3]{4x}}{2x}$ **57.** $\dfrac{\sqrt[3]{2x^2y}}{xy}$ **59.** $\dfrac{3\sqrt[4]{x^3}}{x}$
61. $\dfrac{3\sqrt[3]{4x^2}}{x}$ **63.** $x\sqrt[5]{8x^3y}$ **65.** $\dfrac{3\sqrt{3y}}{xy}$ **67.** $-\dfrac{5a^2\sqrt{3ab}}{b^2}$ **69.** $\dfrac{\sqrt{2mn}}{2m}$ **71.** $\dfrac{3\sqrt[4]{x^3y}}{x^2y}$ **73.** $-\dfrac{6\sqrt[3]{xy}}{x^2y^3}$ **75.** $8\sqrt{5} - 16$
77. $\dfrac{13\sqrt{11} + 39}{2}$ **79.** $3\sqrt{5} - 3\sqrt{3}$ **81.** $\dfrac{a + \sqrt{ab}}{a - b}$ **83.** $25\sqrt{2} + 15\sqrt{5}$ **85.** $4 + \sqrt{15}$ **87.** $\dfrac{x - 2\sqrt{x} - 3}{x - 9}$ **89.** $\dfrac{3\sqrt{6} + 4}{2}$
91. $\dfrac{4\sqrt{xy} + 4x + y}{y - 4x}$ **93.** $\dfrac{3}{\sqrt{6}}$ **95.** $\dfrac{2x}{\sqrt[3]{2x^2y}}$ **97.** $\dfrac{x - 9}{x - 3\sqrt{x}}$ **99.** $\dfrac{a - b}{a - 2\sqrt{ab} + b}$ **101.** $\dfrac{1}{\sqrt{x+5} + \sqrt{x}}$ **103.** $\dfrac{1}{(x+y)(\sqrt{x} - \sqrt{y})}$
105. $\dfrac{3\sqrt{2}}{2}$ **107.** $-2\sqrt[3]{25}$ **109.** $\dfrac{7\sqrt{6}}{6}$ **111.** $7\sqrt{3} - 7\sqrt{2}$ **113.** 0 **115.** 6 **117.** $\dfrac{\sqrt{5} + 1}{2}$; 1.62 to 1 **119.** $P = 8\sqrt{2}$ in.; $A = 7$ sq in.
129.

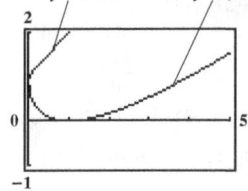

$y = x + 1$ $y = (\sqrt{x} - 1)^2$; Graphs are not the same.; $(\sqrt{x} - 1)(\sqrt{x} - 1) = x - 2\sqrt{x} + 1$

131.

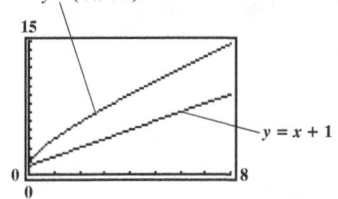

$y = (\sqrt{x} + 1)^2$; Graphs are not the same.; $(\sqrt{x} + 1)^2 = x + 2\sqrt{x} + 1$

$y = x + 1$

133. makes sense **135.** does not make sense **137.** false **139.** true **141.** $\left\{\dfrac{45}{7}\right\}$ **143.** $\dfrac{5\sqrt{2} + 3\sqrt{3} - 4\sqrt{6} + 2}{23}$
144. $\dfrac{2x + 7}{x^2 - 4}$ **145.** $[0, 2]$ **146. a.** function **b.** not a function **147.** $x + 5 + 2\sqrt{x+4}$ **148.** $\{0, 5\}$ **149.** $\{-5, 2\}$

Section 10.6 Check Point Exercises

1. 20 or {20} **2.** no solution or $\varnothing$ **3.** -1 and 3, or $\{-1, 3\}$ **4.** 4 or {4} **5.** -12 or $\{-12\}$ **6.** 2060

Concept and Vocabulary Check

1. radical **2.** extraneous **3.** $2x + 1; x^2 - 14x + 49$ **4.** $x + 2; x + 8 - 6\sqrt{x-1}$ **5.** $2x + 3; 8$ **6.** true **7.** false

Exercise Set 10.6

1. {6} **3.** {17} **5.** no solution or $\varnothing$ **7.** {8} **9.** {0, 3} **11.** {1, 3} **13.** {3, 7} **15.** {9} **17.** {8} **19.** {35} **21.** {16}
23. {2} **25.** {2, 6} **27.** $\left\{\dfrac{5}{2}\right\}$ **29.** {5} **31.** no solution or $\varnothing$ **33.** {2, 6} **35.** {0, 10} **37.** {8} **39.** 4 **41.** -7
43. $V = \dfrac{\pi r^2 h}{3}$ or $V = \dfrac{1}{3}\pi r^2 h$ **45.** $l = \dfrac{8t^2}{\pi^2}$ **47.** 8 **49.** 9 **51.** 5.4 ft **53.** by the point (5.4, 1.16) **55. a.** 16; 16% of Americans earning
$25 thousand annually report fair or poor health.; overestimates by 1% **b.** approximately \$30 thousand **57.** 27 sq mi **59.** 149 million km

69. 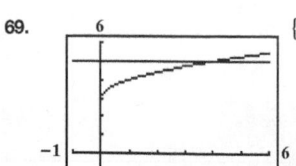 {4}   **71.** 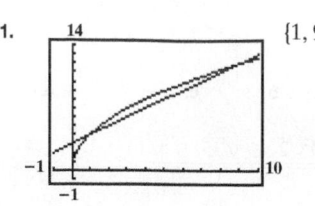 {1, 9}

73. does not make sense   **75.** does not make sense   **77.** false   **79.** true   **81.** $\sqrt{x-7}=3$; $\sqrt{x}=4$; $1+\sqrt{x}=5$ **83.** {16}

85. $4x^3 - 15x^2 + 47x - 142 + \dfrac{425}{x+3}$ **86.** $\dfrac{x+2}{2(x+3)}$ **87.** $(y-3+5x)(y-3-5x)$ **88.** $6+13x$ **89.** $15x^2 - 29x - 14$

90. $\dfrac{54+43\sqrt{2}}{-46}$

Section 10.7 Check Point Exercises

1. a. $8i$ **b.** $i\sqrt{11}$ **c.** $4i\sqrt{3}$ **2. a.** $8+i$ **b.** $-10+10i$ **3. a.** $63+14i$ **b.** $58-11i$ **4.** $-\sqrt{35}$ **5.** $\dfrac{18}{25}+\dfrac{26}{25}i$

6. $-\dfrac{1}{2}-\dfrac{3}{4}i$ **7. a.** 1 **b.** i **c.** $-i$

Concept and Vocabulary Check

1. $\sqrt{-1}$; -1 **2.** $4i$ **3.** complex; imaginary; real   **4.** $-6i$ **5.** $14i$ **6.** 18; $-15i$; $12i$; $-10i^2$; 10 **7.** $2+9i$ **8.** $2+5i$
9. $-4i$ **10.** -1; 1 **11.** -1; -1; $-i$

Exercise Set 10.7

1. $10i$ **3.** $i\sqrt{23}$ **5.** $3i\sqrt{2}$ **7.** $3i\sqrt{7}$ **9.** $-6i\sqrt{3}$ **11.** $5+6i$ **13.** $15+i\sqrt{3}$ **15.** $-2-3i\sqrt{2}$ **17.** $8+3i$ **19.** $8-2i$
21. $5+3i$ **23.** $-1-7i$ **25.** $-2+9i$ **27.** $8+15i$ **29.** $-14+17i$ **31.** $9+5i\sqrt{3}$ **33.** $-6+10i$ **35.** $-21-15i$
37. $-35-14i$ **39.** $7+19i$ **41.** $-1-31i$ **43.** $3+36i$ **45.** 34 **47.** 34 **49.** 11 **51.** $-5+12i$ **53.** $21-20i$ **55.** $-\sqrt{14}$
57. -6 **59.** $-5\sqrt{7}$ **61.** $-2\sqrt{6}$ **63.** $\dfrac{3}{5}-\dfrac{1}{5}i$ **65.** $1+i$ **67.** $\dfrac{28}{25}+\dfrac{21}{25}i$ **69.** $-\dfrac{12}{13}+\dfrac{18}{13}i$ **71.** $0+i$ or i **73.** $\dfrac{3}{10}-\dfrac{11}{10}i$
75. $\dfrac{11}{13}-\dfrac{16}{13}i$ **77.** $-\dfrac{23}{58}+\dfrac{43}{58}i$ **79.** $0-\dfrac{7}{3}i$ or $-\dfrac{7}{3}i$ **81.** $-\dfrac{5}{2}-4i$ **83.** $-\dfrac{7}{3}+\dfrac{4}{3}i$ **85.** -1 **87.** $-i$ **89.** -1 **91.** 1
93. i **95.** 1 **97.** $-i$ **99.** 0 **101.** $-11-5i$ **103.** $-5+10i$ **105.** $0+47i$ or $47i$ **107.** $1-i$ **109.** 0 **111.** $10+10i$
113. $\dfrac{20}{13}+\dfrac{30}{13}i$ **115.** $(47+13i)$ volts   **117.** $(5+i\sqrt{15})+(5-i\sqrt{15})=10$; $(5+i\sqrt{15})(5-i\sqrt{15})=25-15i^2=25+15=40$

131. $\sqrt{-9}+\sqrt{-16}=3i+4i=7i$ **133.** does not make sense   **135.** does not make sense   **137.** false   **139.** false   **141.** $\dfrac{14}{25}-\dfrac{2}{25}i$
143. $\dfrac{8}{5}+\dfrac{16}{5}i$ **144.** $\dfrac{x^2}{y^2}$ **145.** $x=\dfrac{yz}{y-z}$ **146.** -18 **147.** $\left\{-4,\dfrac{1}{2}\right\}$ **148.** $\{-3,3\}$ **149.** $-\sqrt{6}$ is a solution.

Review Exercises

1. 9 **2.** $-\dfrac{1}{10}$ **3.** -3 **4.** not a real number   **5.** -2 **6.** 5; 1.73; 0; not a real number   **7.** 2; -2; -4 **8.** $[2,\infty)$
9. $(-\infty,25]$ **10.** $5|x|$ **11.** $|x+14|$ **12.** $|x-4|$ **13.** $4x$ **14.** $2|x|$ **15.** $-2(x+7)$ **16.** $\sqrt[3]{5xy}$
17. $(\sqrt{16})^3=64$ **18.** $(\sqrt[5]{32})^4=16$ **19.** $(7x)^{1/2}$ **20.** $(19xy)^{5/3}$ **21.** $\dfrac{1}{8^{2/3}}=\dfrac{1}{4}$ **22.** $\dfrac{3x}{a^{4/5}b^{4/5}}=\dfrac{3x}{\sqrt[5]{a^4b^4}}$ **23.** $x^{7/12}$ **24.** $5^{1/6}$
25. $2x^2y$ **26.** $\dfrac{y^{1/8}}{x^{1/3}}$ **27.** x^3y^4 **28.** $y\sqrt[3]{x}$ **29.** $\sqrt[6]{x^5}$ **30.** $\sqrt[6]{x}$ **31.** $\sqrt[15]{x}$ **32.** \$3150 million   **33.** $\sqrt{21xy}$ **34.** $\sqrt[5]{77x^3}$
35. $\sqrt[6]{(x-5)^5}$ **36.** $f(x)=\sqrt{7}|x-1|$ **37.** $2x\sqrt{5x}$ **38.** $3x^2y^2\sqrt[3]{2x^2}$ **39.** $2y^2z\sqrt[4]{2x^3y^3z}$ **40.** $2x^2\sqrt{6x}$ **41.** $2xy\sqrt[3]{2y^2}$
42. $xyz^2\sqrt[5]{16y^4z}$ **43.** $\sqrt{x^2-1}$ **44.** $8\sqrt{3}$ **45.** $9\sqrt{2}$ **46.** $(3x+y^2)\sqrt[3]{x}$ **47.** $-8\sqrt[3]{6}$ **48.** $\dfrac{2}{5}\sqrt{2}$ **49.** $\dfrac{x\sqrt{x}}{10y^2}$ **50.** $\dfrac{y\sqrt[4]{3y}}{2x^5}$
51. $2\sqrt{6}$ **52.** $2\sqrt[3]{2}$ **53.** $2x\sqrt[4]{2x}$ **54.** $10x^2\sqrt{xy}$ **55.** $6\sqrt{2}+12\sqrt{5}$ **56.** $5\sqrt[3]{2}-\sqrt[3]{10}$ **57.** $-83+3\sqrt{35}$
58. $\sqrt{xy}-\sqrt{11x}-\sqrt{11y}+11$ **59.** $13+4\sqrt{10}$ **60.** $22-4\sqrt{30}$ **61.** -6 **62.** 4 **63.** $\dfrac{2\sqrt{6}}{3}$ **64.** $\dfrac{\sqrt{14}}{7}$ **65.** $4\sqrt[3]{3}$
66. $\dfrac{\sqrt{10xy}}{5y}$ **67.** $\dfrac{7\sqrt[3]{4x}}{x}$ **68.** $\dfrac{\sqrt[4]{189x^3}}{3x}$ **69.** $\dfrac{5\sqrt[5]{xy^4}}{2xy}$ **70.** $3\sqrt{3}+3$ **71.** $\dfrac{\sqrt{35}-\sqrt{21}}{2}$ **72.** $10\sqrt{5}+15\sqrt{2}$ **73.** $\dfrac{x+8\sqrt{x}+15}{x-9}$
74. $\dfrac{5+\sqrt{21}}{2}$ **75.** $\dfrac{3\sqrt{2}+2}{7}$ **76.** $\dfrac{2}{\sqrt{14}}$ **77.** $\dfrac{3x}{\sqrt[3]{9x^2y}}$ **78.** $\dfrac{7}{\sqrt{35}+\sqrt{21}}$ **79.** $\dfrac{2}{5-\sqrt{21}}$ **80.** {16}   **81.** no solution or $\varnothing$
82. {2}   **83.** {8}   **84.** {−4, −2}   **85. a.** $f(30)\approx34$; In 2010, approximately 34% of American adults were obese; It's the same.
b. 36 years after 1980, in 2016   **86.** 84 years old   **87.** $9i$ **88.** $3i\sqrt{7}$ **89.** $-2i\sqrt{2}$ **90.** $12+2i$ **91.** $-9+4i$
92. $-12-8i$ **93.** $29+11i$ **94.** $-7-24i$ **95.** $113+0i$ or 113 **96.** $-2\sqrt{6}+0i$ or $-2\sqrt{6}$ **97.** $\dfrac{15}{13}-\dfrac{3}{13}i$
98. $\dfrac{1}{5}+\dfrac{11}{10}i$ **99.** $\dfrac{1}{3}-\dfrac{5}{3}i$ **100.** 1 **101.** $-i$

Chapter Test

1. a. 6   **b.** $(-\infty, 4]$ **2.** $\dfrac{1}{81}$ **3.** $\dfrac{5y^{1/8}}{x^{1/4}}$ **4.** $\sqrt{x}$ **5.** $\sqrt[20]{x^9}$ **6.** $5|x|\sqrt{3}$ **7.** $|x - 5|$ **8.** $2xy^2\sqrt[4]{xy^2}$ **9.** $-\dfrac{2}{x^2}$

10. $\sqrt[3]{50x^2y}$ **11.** $2x\sqrt[4]{2y^3}$ **12.** $-7\sqrt{2}$ **13.** $(2x + y^2)\sqrt[3]{x}$ **14.** $2x\sqrt[3]{x}$ **15.** $12\sqrt{2} - \sqrt{15}$ **16.** $26 + 6\sqrt{3}$ **17.** $52 - 14\sqrt{3}$

18. $\dfrac{\sqrt{5x}}{x}$ **19.** $\dfrac{\sqrt[3]{25x}}{x}$ **20.** $-5 + 2\sqrt{6}$ **21.** $\{6\}$ **22.** $\{16\}$ **23.** $\{-3\}$ **24.** 49 months   **25.** $5i\sqrt{3}$ **26.** $-1 + 6i$

27. $26 + 7i$ **28.** $-6 + 0i$ or -6 **29.** $\dfrac{1}{5} + \dfrac{7}{5}i$ **30.** $-i$

Cumulative Review Exercises

1. $\{(-2, -1, -2)\}$ **2.** $\left\{-\dfrac{1}{3}, 4\right\}$ **3.** $\left(\dfrac{1}{3}, \infty\right)$ **4.** $\left\{\dfrac{3}{4}\right\}$ **5.** $\{-1\}$

6. $x + 2y < 2$
$2y - x > 4$

7. $\dfrac{x^2}{15(x + 2)}$ **8.** $\dfrac{x}{y}$ **9.** $8x^3 - 22x^2 + 11x + 6$ **10.** $\dfrac{5x - 6}{(x - 5)(x + 3)}$ **11.** -64 **12.** 0   **13.** $-\dfrac{16 - 9\sqrt{3}}{13}$

14. $2x^2 + x + 5 + \dfrac{6}{x - 2}$ **15.** $-34 - 3\sqrt{6}$ **16.** $2(3x + 2)(4x - 1)$ **17.** $(4x^2 + 1)(2x + 1)(2x - 1)$

18. about 53 lumens   **19.** \$1500 at 7% and \$4500 at 9%   **20.** 2650 students

CHAPTER 11

Section 11.1   Check Point Exercises

1. $\pm\sqrt{7}$ or $\{\pm\sqrt{7}\}$ **2.** $\pm\dfrac{\sqrt{33}}{3}$ or $\left\{\pm\dfrac{\sqrt{33}}{3}\right\}$ **3.** $\pm\dfrac{3}{2}i$ or $\left\{\pm\dfrac{3}{2}i\right\}$ **4.** $3 \pm \sqrt{10}$ or $\{3 \pm \sqrt{10}\}$ **5. a.** 25; $x^2 + 10x + 25 = (x + 5)^2$

b. $\dfrac{9}{4}$; $x^2 - 3x + \dfrac{9}{4} = \left(x - \dfrac{3}{2}\right)^2$ **c.** $\dfrac{9}{64}$; $x^2 + \dfrac{3}{4}x + \dfrac{9}{64} = \left(x + \dfrac{3}{8}\right)^2$ **6.** $-2 \pm \sqrt{5}$ or $\{-2 \pm \sqrt{5}\}$ **7.** $\dfrac{-3 \pm \sqrt{41}}{4}$ or $\left\{\dfrac{-3 \pm \sqrt{41}}{4}\right\}$

8. $\dfrac{3}{2} \pm i\dfrac{\sqrt{15}}{6}$ or $\left\{\dfrac{3}{2} \pm i\dfrac{\sqrt{15}}{6}\right\}$ **9.** 20%   **10.** $10\sqrt{21}$ ft; 45.8 ft   **11.** $3\sqrt{5} \approx 6.71$ **12.** $\left(4, -\dfrac{1}{2}\right)$

Concept and Vocabulary Check

1. $\pm\sqrt{d}$ **2.** $\pm\sqrt{7}$ **3.** $\pm\sqrt{\dfrac{11}{2}}; \pm\dfrac{\sqrt{22}}{2}$ **4.** $\pm 3i$ **5.** 25   **6.** $\dfrac{9}{4}$ **7.** $\dfrac{4}{25}$ **8.** 9   **9.** $\dfrac{1}{9}$ **10.** right; hypotenuse; legs

11. right; legs; the square of the length of the hypotenuse   **12.** $\sqrt{(x_2 - x_1)^2 + (y_2 - y_1)^2}$ **13.** $\left(\dfrac{x_1 + x_2}{2}, \dfrac{y_1 + y_2}{2}\right)$

Exercise Set 11.1

1. $\{\pm 5\}$ **3.** $\{\pm\sqrt{6}\}$ **5.** $\left\{\pm\dfrac{5}{4}\right\}$ **7.** $\left\{\pm\dfrac{\sqrt{6}}{3}\right\}$ **9.** $\left\{\pm\dfrac{4}{5}i\right\}$ **11.** $\{-10, -4\}$ **13.** $\{3 \pm \sqrt{5}\}$ **15.** $\{-2 \pm 2\sqrt{2}\}$ **17.** $\{5 \pm 3i\}$

19. $\left\{\dfrac{-3 \pm \sqrt{11}}{4}\right\}$ **21.** $\{-3, 9\}$ **23.** 1; $x^2 + 2x + 1 = (x + 1)^2$ **25.** 49; $x^2 - 14x + 49 = (x - 7)^2$ **27.** $\dfrac{49}{4}$; $x^2 + 7x + \dfrac{49}{4} = \left(x + \dfrac{7}{2}\right)^2$

29. $\dfrac{1}{16}$; $x^2 - \dfrac{1}{2}x + \dfrac{1}{16} = \left(x - \dfrac{1}{4}\right)^2$ **31.** $\dfrac{4}{9}$; $x^2 + \dfrac{4}{3}x + \dfrac{4}{9} = \left(x + \dfrac{2}{3}\right)^2$ **33.** $\dfrac{81}{64}$; $x^2 - \dfrac{9}{4}x + \dfrac{81}{64} = \left(x - \dfrac{9}{8}\right)^2$ **35.** $\{-8, 4\}$ **37.** $\{-3 \pm \sqrt{7}\}$

39. $\{4 \pm \sqrt{15}\}$ **41.** $\{-1 \pm i\}$ **43.** $\left\{\dfrac{-3 \pm \sqrt{13}}{2}\right\}$ **45.** $\left\{-\dfrac{3}{7}, -\dfrac{1}{7}\right\}$ **47.** $\left\{\dfrac{-1 \pm \sqrt{5}}{2}\right\}$ **49.** $\left\{-\dfrac{5}{2}, 1\right\}$ **51.** $\left\{\dfrac{-3 \pm \sqrt{6}}{3}\right\}$

53. $\left\{\dfrac{4 \pm \sqrt{13}}{3}\right\}$ **55.** $\left\{\dfrac{1}{4} \pm \dfrac{1}{4}i\right\}$ **57.** $\left\{\dfrac{5}{4} \pm i\dfrac{\sqrt{31}}{4}\right\}$ **59.** $-\dfrac{1}{5}, 1$ **61.** $-2 \pm 5i$ **63.** 13   **65.** $2\sqrt{2} \approx 2.83$ **67.** 5

69. $\sqrt{29} \approx 5.39$ **71.** $4\sqrt{2} \approx 5.66$ **73.** $2\sqrt{5} \approx 4.47$ **75.** $2\sqrt{2} \approx 2.83$ **77.** $\sqrt{93} \approx 9.64$ **79.** $\sqrt{5} \approx 2.24$ **81.** $(4, 6)$

83. $(-4, -5)$ **85.** $\left(\dfrac{3}{2}, -6\right)$ **87.** $(-3, -2)$ **89.** $(1, 5\sqrt{5})$ **91.** $(2\sqrt{2}, 0)$ **93.** $2 \pm 2i$ **95.** $v = \sqrt{2gh}$ **97.** $r = \dfrac{\sqrt{AP}}{P} - 1$

99. $\left\{\dfrac{-1 \pm \sqrt{19}}{6}\right\}$ **101.** $\{-b, 2b\}$ **103. a.** $x^2 + 8x$ **b.** 16   **c.** $x^2 + 8x + 16$ **d.** $(x + 4)^2$ **105.** 20%   **107.** 6.25%

109. a. 1300 thousand; overestimates by 23 thousand   **b.** 2014   **111.** $10\sqrt{3}$ sec; 17.3 sec   **113.** $3\sqrt{5}$ mi; 6.7 mi   **115.** $20\sqrt{2}$ ft; 28.3 ft

117. $50\sqrt{2}$ ft; 70.7 ft   **119.** 10 m

129.

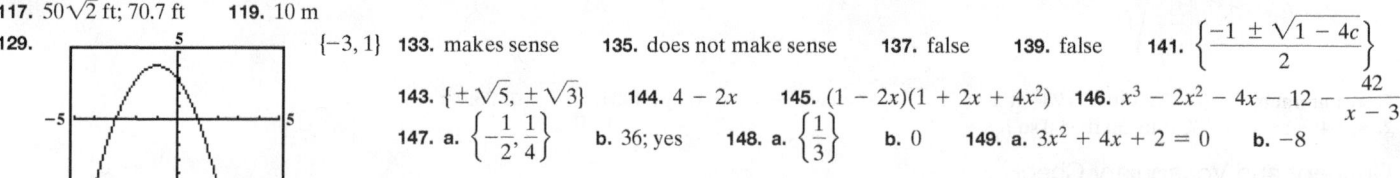

$\{-3, 1\}$ **133.** makes sense   **135.** does not make sense   **137.** false   **139.** false   **141.** $\left\{\dfrac{-1 \pm \sqrt{1 - 4c}}{2}\right\}$

143. $\{\pm\sqrt{5}, \pm\sqrt{3}\}$ **144.** $4 - 2x$ **145.** $(1 - 2x)(1 + 2x + 4x^2)$ **146.** $x^3 - 2x^2 - 4x - 12 - \dfrac{42}{x - 3}$

147. a. $\left\{-\dfrac{1}{2}, \dfrac{1}{4}\right\}$ **b.** 36; yes   **148. a.** $\left\{\dfrac{1}{3}\right\}$ **b.** 0   **149. a.** $3x^2 + 4x + 2 = 0$ **b.** -8

Section 11.2 Check Point Exercises

1. -5 and $\frac{1}{2}$, or $\left\{-5, \frac{1}{2}\right\}$ **2.** $\frac{3 \pm \sqrt{7}}{2}$ or $\left\{\frac{3 \pm \sqrt{7}}{2}\right\}$ **3.** $-1 \pm i\frac{\sqrt{6}}{3}$ or $\left\{-1 \pm i\frac{\sqrt{6}}{3}\right\}$ **4. a.** 0; one real rational solution
b. 81; two real rational solutions **c.** -44; two imaginary solutions that are complex conjugates **5. a.** $20x^2 + 7x - 3 = 0$
b. $x^2 - 50 = 0$ **c.** $x^2 + 49 = 0$ **6.** 26 years old; The point (26, 115) lies approximately on the blue graph.

Concept and Vocabulary Check

1. $\frac{-b \pm \sqrt{b^2 - 4ac}}{2a}$ **2.** 2; 9; -5 **3.** 1; -4; -1 **4.** $2 \pm \sqrt{2}$ **5.** $-1 \pm i\frac{\sqrt{6}}{2}$ **6.** $b^2 - 4ac$ **7.** no **8.** two
9. the square root property **10.** the quadratic formula **11.** factoring and the zero-product principle **12.** false

Exercise Set 11.2

1. $\{-6, -2\}$ **3.** $\left\{1, \frac{5}{2}\right\}$ **5.** $\left\{\frac{-3 \pm \sqrt{89}}{2}\right\}$ **7.** $\left\{\frac{7 \pm \sqrt{85}}{6}\right\}$ **9.** $\left\{\frac{1 \pm \sqrt{7}}{6}\right\}$ **11.** $\left\{\frac{3}{8} \pm i\frac{\sqrt{87}}{8}\right\}$ **13.** $\{2 \pm 2i\}$ **15.** $\left\{\frac{4}{3} \pm i\frac{\sqrt{5}}{3}\right\}$

17. $\left\{-\frac{3}{2}, 4\right\}$ **19.** 52; two real irrational solutions **21.** 4; two real rational solutions **23.** -23; two imaginary solutions

25. 36; two real rational solutions **27.** -60; two imaginary solutions **29.** 0; one (repeated) real rational solution **31.** $\left\{-\frac{2}{3}, 2\right\}$

33. $\{1 \pm \sqrt{2}\}$ **35.** $\left\{\frac{1}{6} \pm i\frac{\sqrt{107}}{6}\right\}$ **37.** $\left\{\frac{3 \pm \sqrt{65}}{4}\right\}$ **39.** $\left\{0, \frac{8}{3}\right\}$ **41.** $\left\{\frac{-6 \pm 2\sqrt{6}}{3}\right\}$ **43.** $\left\{\frac{2 \pm \sqrt{10}}{3}\right\}$ **45.** $\{2 \pm \sqrt{10}\}$

47. $\left\{1, \frac{5}{2}\right\}$ **49.** $\{2 \pm 2i\sqrt{2}\}$ **51.** $x^2 - 2x - 15 = 0$ **53.** $12x^2 + 5x - 2 = 0$ **55.** $x^2 - 2 = 0$ **57.** $x^2 - 20 = 0$

59. $x^2 + 36 = 0$ **61.** $x^2 - 2x + 2 = 0$ **63.** $x^2 - 2x - 1 = 0$ **65.** b **67.** a **69.** $1 + \sqrt{7}$ **71.** $\left\{\frac{-1 \pm \sqrt{21}}{2}\right\}$

73. $\left\{-2\sqrt{2}, \frac{\sqrt{2}}{2}\right\}$ **75.** $\{-3, 1, -1 \pm i\sqrt{2}\}$ **77.** 33-year-olds and 58-year-olds; The function models the actual data well. **79.** 77.8 ft; (b)

81. 5.5 m by 1.5 m **83.** 17.6 in. and 18.6 in. **85.** 9.3 in. and 0.7 in. **87.** 7.5 hr and 8.5 hr

97. 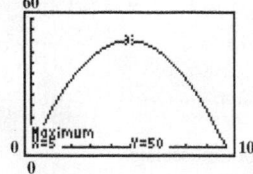 ; depth: 5 in.; maximum area: 50 sq in.

99. does not make sense **101.** does not make sense **103.** false **105.** true **107.** 2.4 m; yes **109.** $\left\{-3, \frac{1}{4}\right\}$ **106.** $\{3, 7\}$

111. $\frac{5\sqrt{3} - 5x}{3 - x^2}$ **112.** 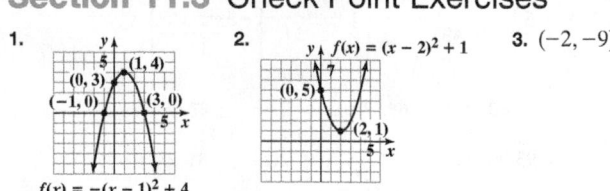 **113.** 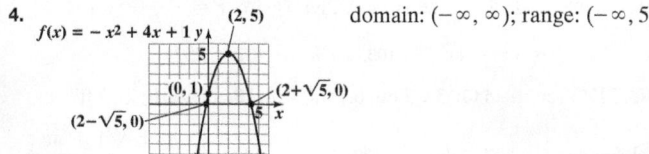 **114.** 1 and 5

Section 11.3 Check Point Exercises

1.

```
      y
   5 (1, 4)
 (0, 3)
(-1, 0)    (3, 0)
         5  x

f(x) = -(x - 1)² + 4
```

2.

```
   y  f(x) = (x - 2)² + 1
      7
(0, 5)
        (2, 1)
       5  x
```

3. $(-2, -9)$

4.

```
         (2, 5)
f(x) = -x² + 4x + 1  y
              5
      (0, 1)      (2+√5, 0)
             5  x
(2-√5, 0)
```

domain: $(-\infty, \infty)$; range: $(-\infty, 5]$

5. a. minimum **b.** Minimum is 984 at $x = 2$. **c.** domain: $(-\infty, \infty)$; range: $[984, \infty)$ **6.** 33.7 ft
7. 4, -4; -16 **8.** 30 ft by 30 ft; 900 sq ft

Concept and Vocabulary Check

1. parabola; upward; downward **2.** lowest/minimum **3.** highest/maximum **4.** (h, k) **5.** $-\frac{b}{2a}$; $-\frac{b}{2a}$ **6.** 0; solutions **7.** $f(0)$

Exercise Set 11.3

1. $h(x) = (x - 1)^2 + 1$ **3.** $j(x) = (x - 1)^2 - 1$ **5.** $h(x) = x^2 - 1$ **7.** $g(x) = x^2 - 2x + 1$ **9.** $(3, 1)$ **11.** $(-1, 5)$ **13.** $(2, -5)$
15. $(-1, 9)$

17.

$y \blacktriangle \; f(x) = (x - 4)^2 - 1$
(3, 0) (5, 0)
(4, −1)
$x = 4$

$[-1, \infty)$

19.

$y \blacktriangle \; f(x) = (x - 1)^2 + 2$
(0, 3) (1, 2)
$x = 1$

$[2, \infty)$

21.

$y \blacktriangle \; y - 1 = (x - 3)^2$
(0, 10)
(3, 1)
$x = 3$

$[1, \infty)$

23.

$y \blacktriangle \; f(x) = 2(x + 2)^2 - 1$
(0, 7)
(−2.7, 0) (−1.3, 0)
(−2, −1) $x = -2$

$[-1, \infty)$

25.

$f(x) = 4 - (x - 1)^2$
(1, 4)
(0, 3) (3, 0)
(−1, 0)
$x = 1$

$(-\infty, 4]$

27.

$y \blacktriangle \; f(x) = x^2 - 2x - 3$
(−1, 0) (3, 0)
(0, −3)
(1, −4)
$x = 1$

$[-4, \infty)$

29.

$y \blacktriangle \; f(x) = x^2 + 3x - 10$
(−5, 0) (2, 0)
(0, −10)
$\left(-\dfrac{3}{2}, -\dfrac{49}{4}\right)$
$x = -\dfrac{3}{2}$

$\left[-\dfrac{49}{4}, \infty\right)$

31.

$f(x) = 2x - x^2 + 3$ (1, 4)
(0, 3)
(−1, 0) (3, 0)
$x = 1$

$(-\infty, 4]$

33. $f(x) = x^2 + 6x + 3$

(0, 3)
$(-3 - \sqrt{6}, 0)$ $(-3 + \sqrt{6}, 0)$
(−3, −6)
$x = -3$

$[-6, \infty)$

35. $f(x) = 2x^2 + 4x - 3$

$\left(\dfrac{-1 - \sqrt{10}}{2}, 0\right)$ $\left(\dfrac{-1 + \sqrt{10}}{2}, 0\right)$
(0, −3)
(−1, −5)
$x = -1$

$[-5, \infty)$

37.

$f(x) = 2x - x^2 - 2$
(0, −2)
(1, −1)
$x = 1$

$(-\infty, -1]$

39. a. minimum   **b.** Minimum is -13 at $x = 2$.   **c.** domain: $(-\infty, \infty)$; range: $[-13, \infty)$ **41. a.** maximum   **b.** Maximum is 1 at $x = 1$.
c. domain: $(-\infty, \infty)$; range: $(-\infty, 1]$ **43. a.** minimum   **b.** Minimum is $-\dfrac{5}{4}$ at $x = \dfrac{1}{2}$.   **c.** domain: $(-\infty, \infty)$; range: $\left[-\dfrac{5}{4}, \infty\right)$
45. domain: $(-\infty, \infty)$; range: $[-2, \infty)$ **47.** domain: $(-\infty, \infty)$; range: $(-\infty, -6]$ **49.** $f(x) = 2(x - 5)^2 + 3$ **51.** $f(x) = 2(x + 10)^2 - 5$
53. $f(x) = -3(x + 2)^2 + 4$ **55.** $f(x) = 3(x - 11)^2$ **57. a.** 2 sec; 224 ft   **b.** 5.7 sec   **c.** 160; 160 feet is the height of the building.
d.

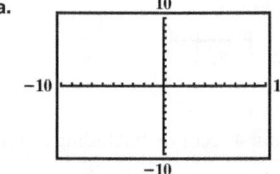

$s(t) \blacktriangle$
250 (2, 224)
$s(t) = -16t^2 + 64t + 160$
(0, 160)
(5.7, 0) t

59. 8 and 8; 64   **61.** $8, -8; -64$ **63.** length: 300 ft; width: 150 ft; maximum area: 45,000 sq ft
65. 12.5 yd by 12.5 yd; 156.25 sq yd   **67.** 5 in.; 50 sq in.   **69. a.** $C(x) = 525 + 0.55x$
b. $P(x) = -0.001x^2 + 2.45x - 525$
c. 1225 sandwiches; $975.63

77. a.

10
−10 ⊢ 10
−10

b. $(20.5, -120.5)$ **c.**

50
0 30
−130

d. Answers will vary.

79. $(2.5, 185)$

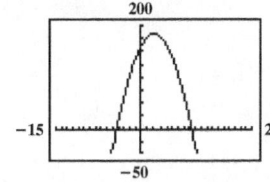

200
−15 ⊢ 20
−50

81. $(-30, 91)$

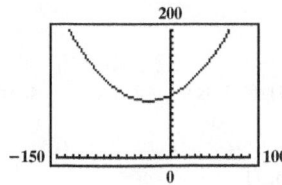

200
−150 ⊢ 100
0

83. does not make sense   **85.** does not make sense   **87.** false   **89.** false   **91.** $x = 3$; $(0, 11)$ **93.** $f(x) = -2(x + 3)^2 - 1$

95. 65 trees; 16,900 pounds   **96.** {7}   **97.** $\dfrac{x}{x - 2}$ **98.** $\{(6, -2)\}$ **99.** $\{-1, 9\}$ **100.** $\left\{-2, \dfrac{5}{2}\right\}$ **101.** $5u^2 + 11u + 2 = 0$

Mid-Chapter Check Point Exercises

1. $\left\{-\frac{1}{3}, \frac{11}{3}\right\}$ **2.** $\left\{-1, \frac{7}{5}\right\}$ **3.** $\left\{\frac{3 \pm \sqrt{15}}{3}\right\}$ **4.** $\{-3 \pm \sqrt{7}\}$ **5.** $\left\{\pm\frac{6\sqrt{5}}{5}\right\}$ **6.** $\left\{\frac{5}{2} \pm i\frac{\sqrt{7}}{2}\right\}$ **7.** $\{\pm i\sqrt{13}\}$ **8.** $\left\{-4, \frac{1}{2}\right\}$

9. $\{-3 \pm 2\sqrt{6}\}$ **10.** $\{2 \pm \sqrt{3}\}$ **11.** $\left\{\frac{3}{4} \pm i\frac{\sqrt{23}}{4}\right\}$ **12.** $\left\{\frac{-3 \pm \sqrt{41}}{4}\right\}$ **13.** $\{4\}$ **14.** $\{-5 \pm 2\sqrt{7}\}$ **15.** $4\sqrt{2} \approx 5.66;\ (0, 0)$

16. $\sqrt{61} \approx 7.81;\ \left(-\frac{15}{2}, 11\right)$

17. $f(x) = (x - 3)^2 - 4$ 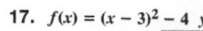   domain: $(-\infty, \infty)$; range: $[-4, \infty)$ **18.** $g(x) = 5 - (x + 2)^2$ 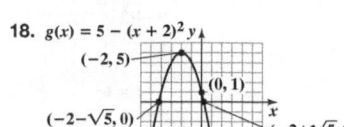   domain: $(-\infty, \infty)$; range: $(-\infty, 5]$

19. $h(x) = -x^2 - 4x + 5$ 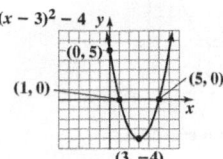   domain: $(-\infty, \infty)$; range: $(-\infty, 9]$ **20.** $f(x) = 3x^2 - 6x + 1$ 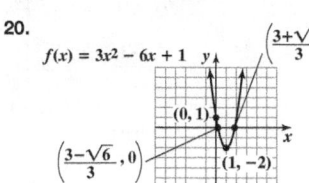   domain: $(-\infty, \infty)$; range: $[-2, \infty)$

21. two imaginary solutions   **22.** two real rational solutions   **23.** $8x^2 - 2x - 3 = 0$ **24.** $x^2 - 12 = 0$ **25.** 75 cabinets per day; \$1200
26. $-9, -9; 81$ **27.** 10 in.; 100 sq in.

Section 11.4 Check Point Exercises

1. $-\sqrt{3}, -\sqrt{2}, \sqrt{2}$, and $\sqrt{3}$ or $\{\pm\sqrt{2}, \pm\sqrt{3}\}$ **2.** 16 or $\{16\}$ **3.** $-\sqrt{6}, -1, 1$, and $\sqrt{6}$, or $\{-\sqrt{6}, -1, 1, \sqrt{6}\}$ **4.** -1 and 2, or $\{-1, 2\}$

5. $-\frac{1}{27}$ and 64, or $\left\{-\frac{1}{27}, 64\right\}$

Concept and Vocabulary Check

1. $x^2; u^2 - 13u + 36 = 0$ **2.** $x^{1/2}$ or $\sqrt{x}; u^2 - 2u - 8 = 0$ **3.** $x + 3; u^2 + 7u - 18 = 0$ **4.** $x^{-1}; 2u^2 - 7u + 3 = 0$
5. $x^{1/3}; u^2 + 2u - 3 = 0$

Exercise Set 11.4

1. $\{-2, 2, -1, 1\}$ **3.** $\{-3, 3, -\sqrt{2}, \sqrt{2}\}$ **5.** $\{-2i, 2i, -\sqrt{2}, \sqrt{2}\}$ **7.** $\{1\}$ **9.** $\{49\}$ **11.** $\{25, 64\}$ **13.** $\{2, 12\}$ **15.** $\{-\sqrt{3}, 0, \sqrt{3}\}$

17. $\{-5, -2, -1, 2\}$ **19.** $\left\{-\frac{1}{4}, \frac{1}{5}\right\}$ **21.** $\left\{\frac{1}{3}, 2\right\}$ **23.** $\left\{\frac{-2 \pm \sqrt{7}}{3}\right\}$ **25.** $\{-8, 27\}$ **27.** $\{-243, 32\}$ **29.** $\{1\}$ **31.** $\{-8, -2, 1, 4\}$

33. $-2, 2, -1$, and $1; c$ **35.** $1; e$ **37.** 2 and $3; f$ **39.** $-5, -4, 1$, and 2   **41.** $-\frac{3}{2}$ and $-\frac{1}{3}$ **43.** $\frac{64}{15}$ and $\frac{81}{20}$ **45.** $\frac{5}{2}$ and $\frac{25}{6}$

47. ages 20 and 55; The function models the data well.   **53.** $\{3, 5\}$ **55.** $\{1\}$ **57.** $\{-1, 4\}$ **59.** $\{1, 8\}$ **61.** makes sense

63. does not make sense   **65.** true   **67.** false   **69.** $\left\{\sqrt[3]{-2}, \frac{\sqrt[3]{225}}{5}\right\}$ **71.** $\frac{1}{5x - 1}$ **72.** $\frac{1}{2} + \frac{3}{2}i$ **73.** -3 **74.** $\left\{-3, \frac{5}{2}\right\}$

75. $\{-2, -1, 2\}$ **76.** $\frac{-x - 5}{x + 3}$

Section 11.5 Check Point Exercises

1. $(-\infty, -4) \cup (5, \infty)$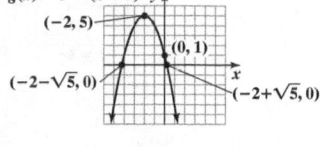

2. $\left[\frac{-3 - \sqrt{7}}{2}, \frac{-3 + \sqrt{7}}{2}\right]$

3. $(-\infty, -3] \cup [-1, 1]$

4. $(-2, 5)$

5. $(-\infty, -1) \cup [1, \infty)$

6. between 1 and 4 seconds, excluding $t = 1$ and $t = 4$

Concept and Vocabulary Check

1. $x^2 + 8x + 15 = 0$; boundary   **2.** $(-\infty, -5), (-5, -3), (-3, \infty)$ **3.** true   **4.** true   **5.** $(-\infty, -2) \cup [1, \infty)$

Exercise Set 11.5

1. $(-\infty, -2) \cup (4, \infty)$ **3.** $[-3, 7]$ **5.** $(-\infty, 1) \cup (4, \infty)$

7. $(-\infty, -4) \cup (-1, \infty)$

9. $[2, 4]$

11. $\left[-4, \dfrac{2}{3}\right]$

13. $\left(-3, \dfrac{5}{2}\right)$

15. $\left(-1, -\dfrac{3}{4}\right)$

17. $(-\infty, 0] \cup [4, \infty)$

19. $\left(-\infty, -\dfrac{3}{2}\right) \cup (0, \infty)$

21. $[0, 1]$

23. $[2 - \sqrt{2}, 2 + \sqrt{2}]$

25. $\left(-\infty, \dfrac{2 - \sqrt{10}}{3}\right) \cup \left(\dfrac{2 + \sqrt{10}}{3}, \infty\right);$

27. $\left(-\infty, \dfrac{5 - \sqrt{33}}{4}\right] \cup \left[\dfrac{5 + \sqrt{33}}{4}, \infty\right);$

29. no solution or $\varnothing$

31. $[1, 2] \cup [3, \infty)$

33. $[-2, -1] \cup [1, \infty)$

35. $(-\infty, -3)$

37. $(-1, \infty)$

39. $\{0\} \cup [9, \infty)$

41. $(-\infty, -3) \cup (4, \infty)$

43. $(-4, -3)$

45. $[2, 4)$

47. $\left(-\infty, -\dfrac{4}{3}\right) \cup [2, \infty)$

49. $(-\infty, 0) \cup (3, \infty)$

51. $(-\infty, -5) \cup (-3, \infty)$

53. $\left(-\infty, \dfrac{1}{2}\right) \cup \left[\dfrac{7}{5}, \infty\right)$

55. $(-\infty, -6] \cup (-2, \infty)$

57. $\left(-\infty, \dfrac{1}{2}\right] \cup [2, \infty)$

59. $(-1, 1)$

61. $(-\infty, -8) \cup (-6, 4) \cup (6, \infty)$

63. $(-3, 2)$

65. $(-\infty, -1) \cup (1, 2) \cup (3, \infty)$

67. $\left[-6, -\dfrac{1}{2}\right] \cup [1, \infty)$ **69.** $(-\infty, -2) \cup [-1, 2)$ **71.** between 0 and 3 seconds, excluding $t = 0$ and $t = 3$ **73. a.** dry: 160 ft; wet: 185 ft
b. dry pavement: graph (b); wet pavement: graph (a)   **c.** extremely well; Function values and data are identical.   **d.** speeds exceeding 76 miles per hour; points on graph (b) to the right of (76, 540)   **75.** The company's production level must be at least 20,000 wheelchairs per month. For values of x greater than or equal to 20,000, the graph lies on or below the line $y = 425$.   **77.** The length of the shorter side cannot exceed 6 ft.
83. $\left[-3, \dfrac{1}{2}\right]$ **85.** $(-\infty, 3) \cup [8, \infty)$ **87.** $(-3, -1) \cup (2, \infty)$ **89. a.** $f(x) = 0.1375x^2 + 0.7x + 37.8$ **b.** speeds exceeding 52 mph
91. does not make sense   **93.** does not make sense   **95.** false   **97.** true   **99.** Answers will vary.; example: $\dfrac{x - 3}{x + 4} \geq 0$ **101.** $\{2\}$
103. $(-\infty, 2) \cup (2, \infty)$ **105.** $27 - 3x^2 \geq 0; [-3, 3]$ **106.** $(-19, 29)$ **107.** $\dfrac{2(x - 1)}{(x + 4)(x - 3)}$ **108.** $(x^2 + 4y^2)(x + 2y)(x - 2y)$

109.

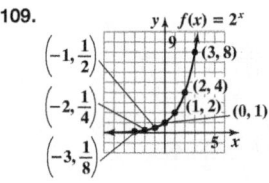

110.

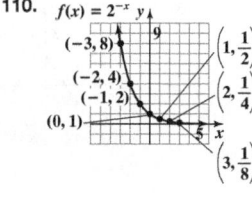

111.

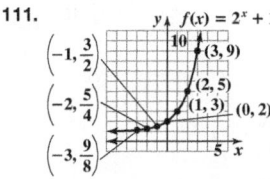

Review Exercises

1. $\{\pm 8\}$ **2.** $\{\pm 5\sqrt{2}\}$ **3.** $\left\{\pm \dfrac{\sqrt{6}}{3}\right\}$ **4.** $\{4 \pm 3\sqrt{2}\}$ **5.** $\{-7 \pm 6i\}$ **6.** $100; x^2 + 20x + 100 = (x + 10)^2$

7. $\dfrac{9}{4}; x^2 - 3x + \dfrac{9}{4} = \left(x - \dfrac{3}{2}\right)^2$ **8.** $\{3, 9\}$ **9.** $\left\{\dfrac{7 \pm \sqrt{53}}{2}\right\}$ **10.** $\left\{\dfrac{-3 \pm \sqrt{41}}{4}\right\}$ **11.** 8%   **12.** 14 weeks   **13.** $60\sqrt{5}$ m; 134.2 m

14. 13   **15.** $2\sqrt{2} \approx 2.83$ **16.** $(-5, 5)$ **17.** $\left(-\dfrac{11}{2}, 2\right)$

18. $\{1 \pm \sqrt{5}\}$ **19.** $\{1 \pm 3i\sqrt{2}\}$ **20.** $\left\{\dfrac{-2 \pm \sqrt{10}}{2}\right\}$ **21.** two imaginary solutions   **22.** two real rational solutions

23. two real irrational solutions   **24.** $\left\{-\dfrac{2}{3}, 4\right\}$ **25.** $\{-5, 2\}$ **26.** $\left\{\dfrac{1 \pm \sqrt{21}}{10}\right\}$ **27.** $\{-4, 4\}$ **28.** $\{3 \pm 2\sqrt{2}\}$ **29.** $\left\{\dfrac{1}{6} \pm i\dfrac{\sqrt{23}}{6}\right\}$

30. $\{4 \pm \sqrt{5}\}$ **31.** $15x^2 - 4x - 3 = 0$ **32.** $x^2 + 81 = 0$ **33.** $x^2 - 48 = 0$ **34. a.** 261 ft; overestimates by 1 ft   **b.** 40 mph
35. a. by the point $(35, 261)$ **b.** by the point $(40, 267)$ **36.** 8.8 sec

37. 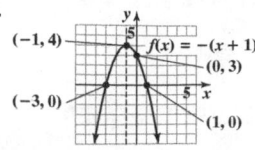   **38.** 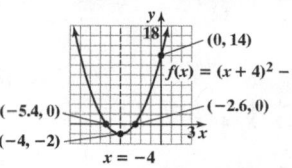   **39.** 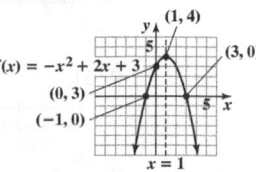   **40.**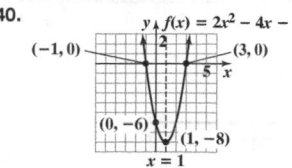

41. 25 in. of rainfall per year; 13.5 in. of growth   **42.** 12.5 sec; 2540 feet   **43.** 7.2 h; 622 per 100,000 males   **44.** 250 yd by 500 yd; 125,000 sq yard
45. -7 and 7; -49 **46.** $\{-\sqrt{2}, \sqrt{2}, -2, 2\}$ **47.** $\{1\}$ **48.** $\{-5, -1, 3\}$ **49.** $\left\{-\dfrac{1}{8}, \dfrac{1}{7}\right\}$ **50.** $\{-27, 64\}$ **51.** $\{16\}$

52. $\left(-3, \dfrac{1}{2}\right)$;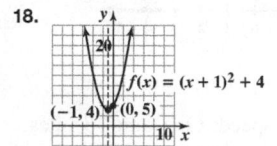

53. $(-\infty, -4] \cup \left[-\dfrac{1}{2}, \infty\right)$;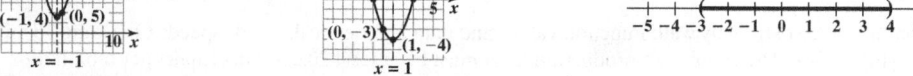

54. $(-3, 0) \cup (1, \infty)$;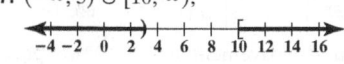

55. $(-\infty, -2) \cup (6, \infty)$;

56. $(-\infty, 4) \cup \left[\dfrac{23}{4}, \infty\right)$;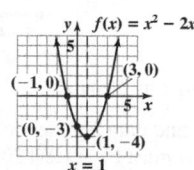

57. between 1 and 2 seconds, excluding $t = 1$ and $t = 2$
58. a. 200 beats per minute
   b. between 0 and 4 minutes and more than 12 minutes after the workout; between 0 and 4 minutes; Answers will vary.

Chapter Test

1. $\left\{\pm\dfrac{\sqrt{10}}{2}\right\}$ **2.** $\{3 \pm 2\sqrt{5}\}$ **3.** $64; x^2 - 16x + 64 = (x - 8)^2$ **4.** $\dfrac{1}{25}; x^2 + \dfrac{2}{5}x + \dfrac{1}{25} = \left(x + \dfrac{1}{5}\right)^2$ **5.** $\{3 \pm \sqrt{2}\}$ **6.** $50\sqrt{2}$ ft

7. $\sqrt{73} \approx 8.54$ **8.** $\left(\dfrac{7}{2}, -4\right)$ **9.** two real irrational solutions   **10.** two imaginary solutions   **11.** $\left\{-5, \dfrac{1}{2}\right\}$ **12.** $\{-4 \pm \sqrt{11}\}$

13. $\{-2 \pm 5i\}$ **14.** $\left\{\dfrac{3}{2} \pm \dfrac{1}{2}i\right\}$ **15.** $x^2 - 4x - 21 = 0$ **16.** $x^2 + 100 = 0$ **17. a.** $182.8 \approx 183$; overestimates by 3   **b.** 2023

18. 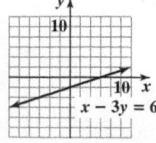   **19.** 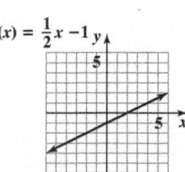   **20.** after 2 sec; 69 ft   **21.** 4.1 sec   **22.** 23 computers; $169 hundreds or $16,900
                               23. $\{1, 2\}$ **24.** $\{-3, 3, -2, 2\}$ **25.** $\{1, 512\}$
                               26. $(-3, 4)$; 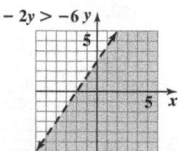   **27.** $(-\infty, 3) \cup [10, \infty)$;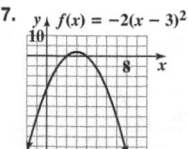

Cumulative Review Exercises

1. $\left\{\begin{matrix}2\\3\end{matrix}\right\}$ **2.** $\{(-5, 2)\}$ **3.** $\{(1, 4, -2)\}$ **4.** $[10, \infty)$ **5.** $(-\infty, -4]$ **6.** $(2, \infty)$ **7.** $(-2, 3)$ **8.** $\{3, 9\}$ **9.** no solution or $\varnothing$ **10.** $\{12\}$

11. $\left\{\dfrac{-2 \pm \sqrt{14}}{2}\right\}$ **12.** $\{8, 27\}$ **13.** $\left[-2, \dfrac{3}{2}\right]$

14.   **15.**   **16.**   **17.**

18. $-16x + 24y$ **19.** $-\dfrac{20x^7}{y^4}$ **20.** $x^2 - 5xy - 16y^2$ **21.** $6x^2 + 13x - 5$ **22.** $9x^4 - 24x^2y + 16y^2$ **23.** $\dfrac{3x^2 + 6x - 2}{(x + 5)(x + 2)}$ **24.** $\dfrac{x - 3}{x}$
25. $\dfrac{x - 4}{3x + 6}$ **26.** $5xy\sqrt{2x}$ **27.** $9\sqrt{2}$ **28.** $44 + 6i$ **29.** $(9x^2 + 1)(3x + 1)(3x - 1)$ **30.** $2x(4x - 1)(3x - 2)$
31. $(x + 3y)(x^2 - 3xy + 9y^2)$ **32.** $x^2 + 2x - 13; 22$ **33.** $x + 5; (-\infty, 2) \cup (2, \infty)$ **34.** $2a + h + 3$
35. $3x^2 - 7x + 18 - \dfrac{28}{x + 2}$ **36.** $R = -\dfrac{Ir}{I - 1}$ or $R = \dfrac{Ir}{1 - I}$ **37.** $y = -3x - 1$ or $f(x) = -3x - 1$ **38.** $620
39. 13 yd by 4 yd   **40.** $2600 at 12% and $1400 at 14%   **41.** 11 amps

CHAPTER 12

Section 12.1 Check Point Exercises

1. approximately $160; overestimates by $11

2.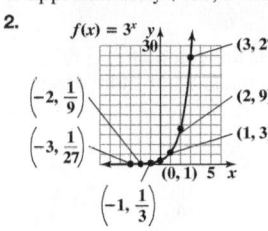

3.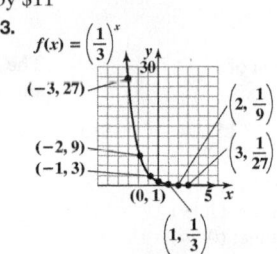

4.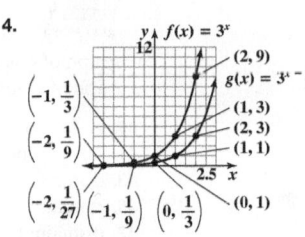

The graph of g is the graph of f shifted 1 unit to the right.

5.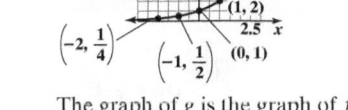

The graph of g is the graph of f shifted up 3 units.

6. approximately 4446   **7. a.** $14,859.47   **b.** $14,918.25

Concept and Vocabulary Check

1. b^x; $(-\infty, \infty)$; $(0, \infty)$ **2.** x; $y = 0$; horizontal   **3.** e; natural; 2.72   **4.** A; P; r; n **5.** semiannually; quarterly; continuous

Exercise Set 12.1

1. 10.556   **3.** 11.665   **5.** 0.125   **7.** 9.974   **9.** 0.387

11. ; d

x	$f(x)$
-2	$\dfrac{1}{9}$
-1	$\dfrac{1}{3}$
0	1
1	3
2	9

13. ; e

x	$f(x)$
-2	$-\dfrac{8}{9}$
-1	$-\dfrac{2}{3}$
0	0
1	2
2	8

15. ; f

x	$f(x)$
-2	9
-1	3
0	1
1	$\dfrac{1}{3}$
2	$\dfrac{1}{9}$

17.

x	$f(x)$
-2	$\dfrac{1}{16}$
-1	$\dfrac{1}{4}$
0	1
1	4
2	16

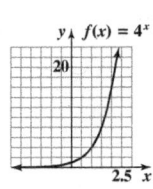

19.

x	$g(x)$
-2	$\dfrac{4}{9}$
-1	$\dfrac{2}{3}$
0	1
1	$\dfrac{3}{2}$
2	$\dfrac{9}{4}$

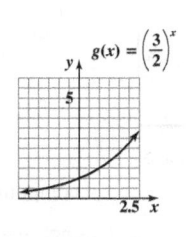

21.

x	$h(x)$
-2	4
-1	2
0	1
1	$\dfrac{1}{2}$
2	$\dfrac{1}{4}$

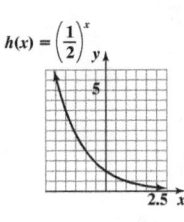

23.

x	$f(x)$
-2	2.78
-1	1.67
0	1
1	0.6
2	0.36

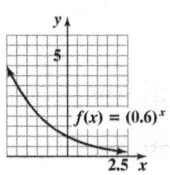

25.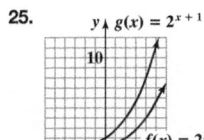

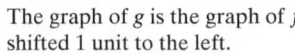

The graph of g is the graph of f shifted 1 unit to the left.

27.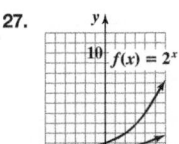

The graph of g is the graph of f shifted 2 units to the right.

29.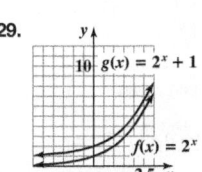

The graph of g is the graph of f shifted up 1 unit.

31.

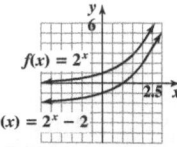

The graph of g is the graph of f shifted down 2 units.

33.

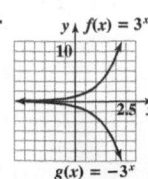

The graph of g is a reflection of the graph of f across the x-axis.

35.

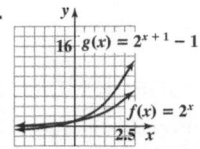

The graph of g is the graph of f shifted 1 unit to the left and 1 unit down.

37.

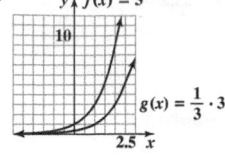

The graph of g is the graph of f stretched vertically by a factor of $\frac{1}{3}$.

39. a. \$13,116.51   **b.** \$13,157.04   **c.** \$13,165.31
41. 7% compounded monthly
43. domain: $(-\infty, \infty)$; range: $(-2, \infty)$
45. domain: $(-\infty, \infty)$; range: $(1, \infty)$
47. domain: $(-\infty, \infty)$; range: $(0, \infty)$

49. $(0, 1)$

51.

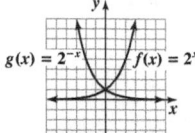

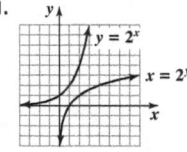

53. a. 574 million   **b.** 1148 million   **c.** 2295 million   **d.** 4590 million
e. It appears to double.   **55.** \$832,744   **57. a.** 393 million
b. approximately 388 million   **c.** the linear model   **59. a.** 100%
b. about 68.5%   **c.** about 30.8%   **d.** about 20%
61. 11.3; About 11.3% of 30-year-olds have some coronary heart disease.

63. a. about 1429 people   **b.** about 24,546 people   **c.** The number of ill people cannot exceed the population.; The asymptote indicates that the number of ill people will not exceed 30,000, the population of the town.
69. a. $f(t) = 10,000\left(1 + \dfrac{0.05}{4}\right)^{4t}$; $f(t) = 10,000\left(1 + \dfrac{0.045}{12}\right)^{12t}$

b.

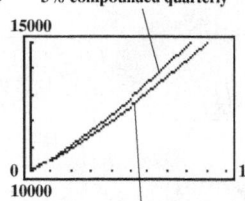

the bank that pays 5% interest compounded quarterly

71. does not make sense   **73.** does not make sense   **75.** false   **77.** false   **79. a.** $y = \left(\dfrac{1}{3}\right)^x$ **b.** $y = \left(\dfrac{1}{5}\right)^x$ **c.** $y = 5^x$ **d.** $y = 3^x$

81. $b = \dfrac{Da}{a - D}$ **82.** $\dfrac{11}{(x - 3)(x - 4)}$ **83.** $\{-2, 5\}$ **84.** There is no method for solving $x = 2^y$ for y.   **85.** $\dfrac{1}{2}$ **86.** $f^{-1}(x) = \dfrac{x + 5}{2}$

Section 12.2 Check Point Exercises

1. a. $7^3 = x$ **b.** $b^2 = 25$ **c.** $4^y = 26$ **2. a.** $5 = \log_2 x$ **b.** $3 = \log_b 27$ **c.** $y = \log_e 33$ **3. a.** 2   **b.** 1   **c.** $\dfrac{1}{2}$ **4. a.** 1   **b.** 0

5. a. 8   **b.** 17   **6.**

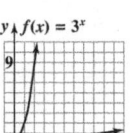

7. $(5, \infty)$ **8.** approximately 80%   **9.** 4
10. a. $(-\infty, 4)$ **b.** $(-\infty, 0) \cup (0, \infty)$ **11.** 34°; extremely well

Concept and Vocabulary Check

1. $b^y = x$ **2.** logarithmic; b **3.** 1   **4.** 0   **5.** x **6.** x **7.** $(0, \infty)$; $(-\infty, \infty)$ **8.** y; $x = 0$; vertical
9. $5 - x > 0$ **10.** common; $\log x$ **11.** natural; $\ln x$

Exercise Set 12.2

1. $2^4 = 16$ **3.** $3^2 = x$ **5.** $b^5 = 32$ **7.** $6^y = 216$ **9.** $\log_2 8 = 3$ **11.** $\log_2 \dfrac{1}{16} = -4$ **13.** $\log_8 2 = \dfrac{1}{3}$ **15.** $\log_{13} x = 2$

17. $\log_b 1000 = 3$ **19.** $\log_7 200 = y$ **21.** 2   **23.** 6   **25.** -1 **27.** -3 **29.** $\dfrac{1}{2}$ **31.** $-\dfrac{1}{2}$ **33.** $\dfrac{1}{2}$ **35.** 1   **37.** 0   **39.** 7

41. 19   **43.** 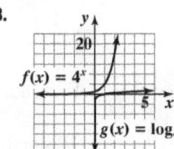   **45.** 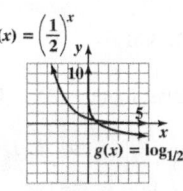   **47.** $(-4, \infty)$ **49.** $(-\infty, 2)$ **51.** $(-\infty, 2) \cup (2, \infty)$ **53.** 2   **55.** 7

57. 33   **59.** 0   **61.** 6   **63.** -6 **65.** 125   **67.** $9x$ **69.** $5x^2$

71. $\sqrt{x}$ **73.** $3^2 = x - 1; \{10\}$ **75.** $4^{-3} = x; \left\{\dfrac{1}{64}\right\}$ **77.** 0   **79.** 2

81. d   **83.** c   **85.** b   **87.** approximately 95.4%

89. a. 25.5%; underestimates by 0.7%   **b.** 24.2%

91. approximately 188 decibels; yes

93. a. 88

b. 71.5; 63.9; 58.8; 55.0; 52.0; 49.5

c.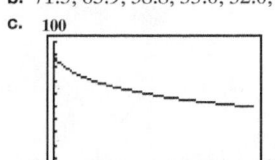

The students remembered less of the material over time.

103.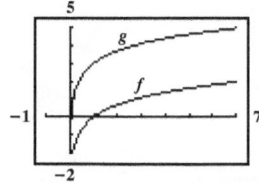

The graph of g is the graph of f shifted up 3 units.

105.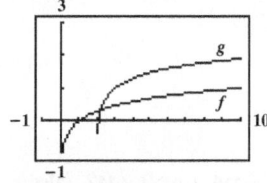

The graph of g is the graph of f shifted 2 units to the right and 1 unit up.

107. a. 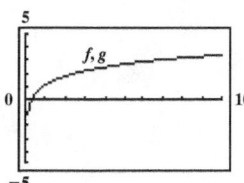   **b.** 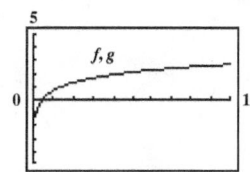   **c.**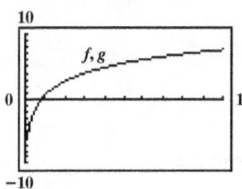

d. In each case, the graphs of f and g are the same.; $\log_b(MN) = \log_b M + \log_b N$

e. the sum of the logarithms of the factors

109. makes sense   **111.** makes sense   **113.** false   **115.** false   **117.** $\dfrac{4}{5}$ **119.** $\log_3 40$ **120.** $\{(-2, 3)\}$ **121.** $2(3x - y)(x - y)$

122. $(-\infty, -7] \cup [-2, \infty)$ **123. a.** 5   **b.** 5   **c.** $\log_2(8 \cdot 4) = \log_2 8 + \log_2 4$ **124. a.** 4   **b.** 4   **c.** $\log_2\left(\dfrac{32}{2}\right) = \log_2 32 - \log_2 2$

125. a. 4   **b.** 4   **c.** $\log_3 9^2 = 2\log_3 9$

Section 12.3 Check Point Exercises

1. a. $\log_6 7 + \log_6 11$ **b.** $2 + \log x$ **2. a.** $\log_8 23 - \log_8 x$ **b.** $5 - \ln 11$ **3. a.** $9 \log_6 8$ **b.** $\dfrac{1}{3} \ln x$ **c.** $2 \log(x + 4)$

4. a. $4 \log_b x + \dfrac{1}{3} \log_b y$ **b.** $\dfrac{1}{2} \log_5 x - 2 - 3 \log_5 y$ **5. a.** $\log 100 = 2$ **b.** $\log\left(\dfrac{7x + 6}{x}\right)$ **6. a.** $\ln(x^2 \sqrt[3]{x + 5})$ **b.** $\log\left[\dfrac{(x - 3)^2}{x}\right]$

c. $\log_b\left(\dfrac{\sqrt[4]{x}}{25y^{10}}\right)$ **7.** $\dfrac{\log 2506}{\log 7} \approx 4.02$ **8.** $\dfrac{\ln 2506}{\ln 7} \approx 4.02$

Concept and Vocabulary Check

1. $\log_b M + \log_b N$; sum   **2.** $\log_b M - \log_b N$; difference   **3.** $p \log_b M$; product   **4.** $\dfrac{\log_a M}{\log_a b}$

Exercise Set 12.3

1. $\log_5 7 + \log_5 3$ **3.** $1 + \log_7 x$ **5.** $3 + \log x$ **7.** $1 - \log_7 x$ **9.** $\log x - 2$ **11.** $3 - \log_4 y$ **13.** $2 - \ln 5$ **15.** $3 \log_b x$

17. $-6 \log N$ **19.** $\dfrac{1}{5} \ln x$ **21.** $2 \log_b x + \log_b y$ **23.** $\dfrac{1}{2} \log_4 x - 3$ **25.** $2 - \dfrac{1}{2} \log_6(x + 1)$ **27.** $2 \log_b x - \log_b y - 2 \log_b z$

29. $1 + \dfrac{1}{2} \log x$ **31.** $\dfrac{1}{3} \log x - \dfrac{1}{3} \log y$ **33.** $\dfrac{1}{2} \log_b x + 3 \log_b y - 3 \log_b z$ **35.** $\dfrac{2}{3} \log_5 x + \dfrac{1}{3} \log_5 y - \dfrac{2}{3}$ **37.** $\log 10 = 1$ **39.** $\ln(7x)$

41. $\log_2 32 = 5$ **43.** $\log\left(\dfrac{2x + 5}{x}\right)$ **45.** $\log(xy^3)$ **47.** $\ln(y\sqrt{x})$ **49.** $\log_b(x^2 y^3)$ **51.** $\ln\left(\dfrac{x^5}{y^2}\right)$ **53.** $\ln\left(\dfrac{x^3}{\sqrt[3]{y}}\right)$ **55.** $\ln\left[\dfrac{(x + 6)^4}{x^3}\right]$

57. $\ln\left(\dfrac{x^3 y^5}{z^6}\right)$ **59.** $\log_5\left[\dfrac{\sqrt{xy}}{(x+1)^2}\right]$ **61.** 1.5937   **63.** 1.6944   **65.** -1.2304 **67.** 3.6193   **69.** $C - A$ **71.** $3A$ **73.** $\dfrac{1}{2}A - \dfrac{3}{2}C$

75. false; $\ln e = 1$ **77.** false; $\log_4(2x)^3 = 3\log_4(2x)$ **79.** true   **81.** true   **83.** false; $\log(x+3) - \log(2x) = \log\left(\dfrac{x+3}{2x}\right)$ **85.** true

87. true   **89. a.** 2   **b.** $\log_3\left(\dfrac{xy^4}{9}\right)$ **91. a.** $\dfrac{1}{2}$ **b.** $\log_{25}\left[\dfrac{x(x^2-1)}{5(x+1)}\right] = \log_{25}\left[\dfrac{x(x-1)}{5}\right]$ **93. a.** $D = 10\log\left(\dfrac{I}{I_0}\right)$ **b.** 20 decibels

103. a. & b. $y = 2 + \log_3 x$ $y = \log_3(x+2)$ $y = \log_3 x$ is shifted up 2 units to obtain $y = 2 + \log_3 x$,
$y = \log_3 x$ is shifted to the left 2 units to obtain $y = \log_3(x+2)$, and
$y = \log_3 x$ is reflected across the x-axis to obtain $y = -\log_3 x$.

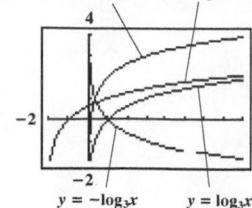

105.

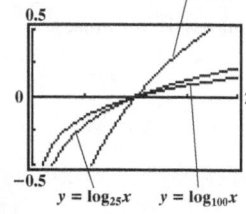

$y = \log_3 x$ **a.** $y = \log_{100} x$ is on the top and $y = \log_3 x$ is on the bottom.
b. $y = \log_3 x$ is on the top and $y = \log_{100} x$ is on the bottom.
c. If $y = \log_b x$ is graphed for two different values of b, the graph of the one with the larger base will be on top in the interval $(0, 1)$ and the one with the smaller base will be on top in the interval $(1, \infty)$.

$y = \log_{25} x$ $y = \log_{100} x$

111. makes sense   **113.** makes sense   **115.** true   **117.** false   **119.** $\log e = \dfrac{\ln e}{\ln 10} = \dfrac{1}{\ln 10}$ **121.** $4x^3$

122.

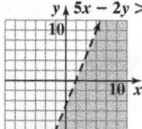

$5x - 2y > 10$ **123.** $(-\infty, 1)$ **124.** $2y\sqrt[3]{xy^2}$ **125.** 64   **126.** 12   **127.** $\{-1, 3\}$

Mid-Chapter Check Point Exercises

1.

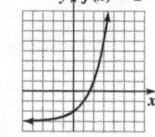

$f(x) = 2^x - 3$ domain: $(-\infty, \infty)$; range: $(-3, \infty)$

2.

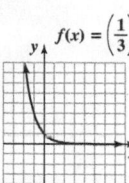

$f(x) = \left(\dfrac{1}{3}\right)^x$ domain: $(-\infty, \infty)$; range: $(0, \infty)$

3.

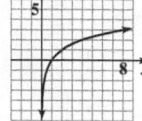

$f(x) = \log_2 x$ domain: $(0, \infty)$; range: $(-\infty, \infty)$

4.

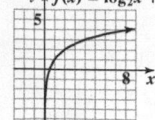

$f(x) = \log_2 x + 1$ domain: $(0, \infty)$; range: $(-\infty, \infty)$

5. $(-6, \infty)$ **6.** $(0, \infty)$ **7.** $(-\infty, -6) \cup (-6, \infty)$ **8.** $(-\infty, \infty)$ **9.** 5   **10.** -2 **11.** $\dfrac{1}{2}$ **12.** $\dfrac{1}{3}$ **13.** 2

14. Evaluation is not possible; $\log_2 \dfrac{1}{8} = -3$ and $\log_3(-3)$ is undefined.   **15.** 5   **16.** $\sqrt{7}$ **17.** 13   **18.** $-\dfrac{1}{2}$ **19.** $\sqrt{\pi}$

20. $\frac{1}{2}\log x + \frac{1}{2}\log y - 3$ **21.** $19 + 20\ln x$ **22.** $\log_7\left(\frac{x^8}{\sqrt[3]{y}}\right)$ **23.** $\log_5 x^9$ **24.** $\ln\left[\frac{\sqrt{x}}{y^3(z-2)}\right]$ **25.** \$8

Section 12.4 Check Point Exercises

1. a. 3 or {3}   **b.** $\frac{5}{2}$ or $\left\{\frac{5}{2}\right\}$ **2. a.** $\frac{\ln 134}{\ln 5} \approx 3.04$ or $\left\{\frac{\ln 134}{\ln 5} \approx 3.04\right\}$ **b.** $\log 8000 \approx 3.90$ or $\{\log 8000 \approx 3.90\}$

3. $\frac{\ln 9}{2} = \ln 3 \approx 1.10$ or $\{\ln 3 \approx 1.10\}$ **4. a.** 12 or {12}   **b.** $\frac{e^2}{3}$ or $\left\{\frac{e^2}{3}\right\}$ **5.** 5 or {5}   **6.** 4 and 5, or {4, 5}

7. blood alcohol concentration of 0.01   **8.** 16.2 years   **9.** 1994

Concept and Vocabulary Check

1. $M = N$ **2.** $4x - 1$ **3.** $\frac{\ln 20}{\ln 9}$ **4.** $\ln 6$ **5.** 5^3 **6.** $x^2 + x$ **7.** $\frac{7x - 23}{x + 1}$

8. false   **9.** true   **10.** false

Exercise Set 12.4

1. {6}   **3.** {3}   **5.** {3}   **7.** {2}   **9.** $\left\{\frac{3}{5}\right\}$ **11.** $\left\{\frac{3}{2}\right\}$ **13.** {4}   **15.** {5}   **17.** $\left\{-\frac{1}{4}\right\}$ **19.** $\{\ln 5.7 \approx 1.74\}$

21. $\{\log 3.91 \approx 0.59\}$ **23.** $\left\{\frac{\ln 17}{\ln 5} \approx 1.76\right\}$ **25.** $\{\ln 5 \approx 1.61\}$ **27.** $\left\{\frac{\ln 659}{5} \approx 1.30\right\}$ **29.** $\left\{\frac{\ln 13}{0.7} \approx 3.66\right\}$ **31.** $\left\{\frac{\ln 3}{0.055} \approx 19.97\right\}$

33. $\left\{\frac{\ln 30}{\ln 1.4} \approx 10.11\right\}$ **35.** $\left\{\frac{1 - \ln 793}{5} \approx -1.14\right\}$ **37.** $\left\{\frac{\ln 410}{\ln 7} - 2 \approx 1.09\right\}$ **39.** $\left\{\frac{\ln 2}{\ln 5 - \ln 2} \approx 0.76\right\}$ **41.** {81}   **43.** $\left\{\frac{1}{16}\right\}$

45. {3}   **47.** {100}   **49.** {59}   **51.** $\left\{\frac{109}{27}\right\}$ **53.** $\left\{\frac{62}{3}\right\}$ **55.** $\{e^2 \approx 7.39\}$ **57.** $\{e^{-3} \approx 0.05\}$ **59.** $\left\{\frac{e^4}{2} \approx 27.30\right\}$ **61.** $\{e^{-1/2} \approx 0.61\}$

63. $\{e^2 - 3 \approx 4.39\}$ **65.** $\left\{\frac{5}{4}\right\}$ **67.** {6}   **69.** {6}   **71.** no solution or $\varnothing$ **73.** $\left\{\frac{1}{e - 1} \approx 0.58\right\}$ **75.** {3}   **77.** $\left\{\frac{4}{3}\right\}$

79. no solution or $\varnothing$ **81.** {5}   **83.** $\left\{\frac{2}{9}\right\}$ **85.** {28}   **87.** {2}   **89.** no solution or $\varnothing$ **91.** $\left\{\frac{1}{2}\right\}$ **93.** $\left\{\pm\sqrt{\frac{\ln 45}{\ln 3}} \approx \pm 1.86\right\}$

95. {12}   **97.** {−2, 6}   **99. a.** 36.1 million   **b.** 2013   **101.** 118 ft; by the point (118, 1)   **103.** 8.2   **105.** 16.8%   **107.** 8.7   **109.** 15.7%   **111. a.** 17.0%; underestimates by 0.3%   **b.** 2016   **113.** about 2.8 days; (2.8, 50)   **115. a.** $10^{-5.6}$ mole per liter   **b.** $10^{-2.4}$ mole per liter   **c.** $10^{3.2}$ times greater

123. {2}
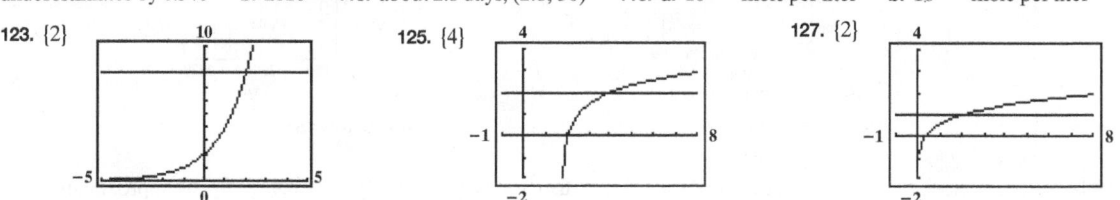

125. {4}

127. {2}

129. {−1.39, 1.69}
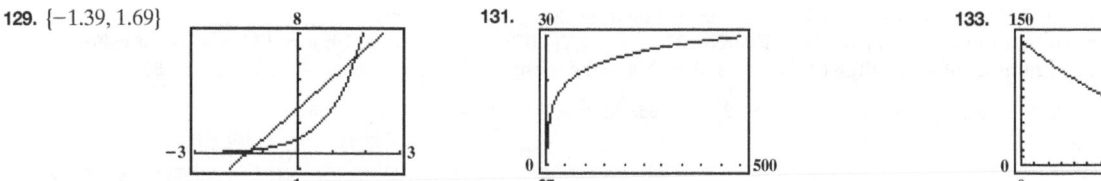

131.
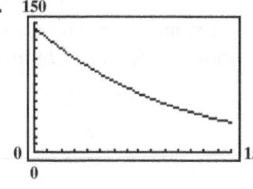

133.

The barometric air pressure increases as the distance from the eye increases.

The runner's pulse will be 70 beats per minute after about 7.9 minutes.

135. does not make sense   **137.** makes sense   **139.** false   **141.** true   **143.** about 36 yr   **145.** $\{10^{-2}, 10^{3/2}\}$

147. {1, 5}   **148.** {−12}   **149.** $\frac{y^8}{16x^{12}}$ **150. a.** 10 million; 9.97 million; 9.94 million; 9.91 million   **b.** decreasing   **151. a.** 3   **b.** $e^{(\ln 3)x}$

152.
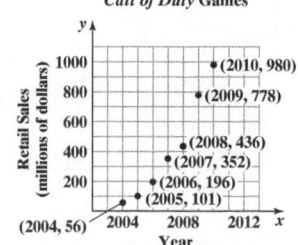

exponential function

Annual Retail Sales of *Call of Duty* Games

Section 12.5 Check Point Exercises

1. a. $A = 643e^{0.021t}$ **b.** 2044   **2. a.** $A = A_0e^{-0.0248t}$ **b.** about 72.2 years

3. 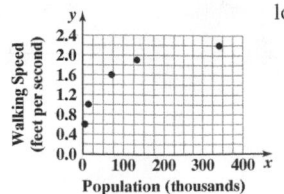   logarithmic function   **4.** 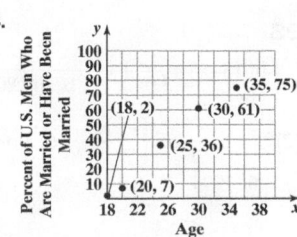   exponential function, although answers may vary

5. a. the exponential function g **b.** the linear function f **6.** $y = 4e^{(\ln 7.8)x}; y = 4e^{2.054x}$

Concept and Vocabulary Check

1. $> 0; < 0$ **2.** $A_0; A$ **3.** logarithmic   **4.** exponential   **5.** linear   **6.** $\ln 5$

Exercise Set 12.5

1. 127.5 million   **3.** Iraq; 2.7%   **5.** 2015   **7. a.** $A = 6.04e^{0.01t}$ **b.** 2040   **9.** 118.7   **11.** 0.0121   **13.** -0.0053
15. approximately 8 grams   **17.** 8 grams after 10 seconds; 4 grams after 20 seconds; 2 grams after 30 seconds; 1 gram after 40 seconds; 0.5 gram

after 50 seconds   **19.** approximately 15,679 years old   **21. a.** $\dfrac{1}{2} = e^{1.31k}$ yields $k = \dfrac{\ln\left(\dfrac{1}{2}\right)}{1.31} \approx -0.52912.$

b. about 0.1069 billion or 106,900,000 years old   **23.** $2A_0 = A_0e^{kt}; 2 = e^{kt}; \ln 2 = \ln e^{kt}; \ln 2 = kt; \dfrac{\ln 2}{k} = t$ **25. a.** 1%   **b.** about 69 years

27.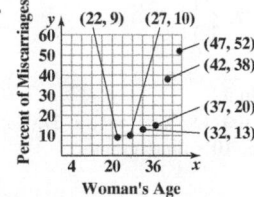

exponential function

29.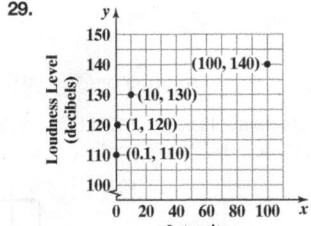

logarithmic function

31.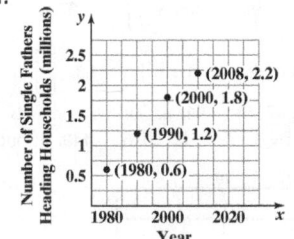

linear function

33. $y = 100e^{(\ln 4.6)x}; y = 100e^{1.526x}$ **35.** $y = 2.5e^{(\ln 0.7)x}; y = 2.5e^{-0.357x}$
45. a. $y = 200.9(1.011)^x; r \approx 0.999;$ Since r is close to 1, the model fits the data well.   **b.** $y = 200.9e^{\ln(1.011)x}; y = 200.9e^{0.0109x};$ by approximately 1%
47. $y = 2.674x + 197.756; r \approx 0.997;$ Since r is close to 1, the model fits the data well.
49. $y = 200.9(1.011)^x; y = 2.674x + 197.756;$ using exponential, by 2020; using linear, by 2027; Answers will vary.
51. Models will vary. Examples are given. Predictions will vary. For Exercise 27: $y = 1.402(1.078)^x;$ For Exercise 29: $y = 120 + 4.343 \ln x;$ For
Exercise 31: $y = 0.058x + 0.558$ ($x =$ number of years after 1979)   **53.** does not make sense   **55.** makes sense   **57.** true   **59.** true
61. $\dfrac{x+5}{x}$ **62.** $\{-27, 1\}$ **63.** $17\sqrt{2}$ **64.** $4; x^2 + 4x + 4 = (x+2)^2$ **65.** $9; y^2 - 6y + 9 = (y-3)^2$ **66.**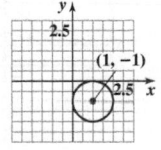

Review Exercises

1.

x	$f(x)$
-2	$\dfrac{1}{16}$
-1	$\dfrac{1}{4}$
0	1
1	4
2	16

; d

2.

x	$f(x)$
-2	16
-1	4
0	1
1	$\dfrac{1}{4}$
2	$\dfrac{1}{16}$

; a

3.

x	$f(x)$
-2	-16
-1	-4
0	-1
1	$-\dfrac{1}{4}$
2	$-\dfrac{1}{16}$

; b

4.

x	$f(x)$
-2	-13
-1	-1
0	2
1	$\dfrac{11}{4}$
2	$\dfrac{47}{16}$

; c

5.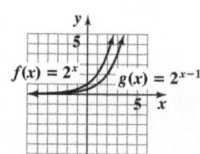

The graph of g is the graph of f shifted to the right 1 unit.

6.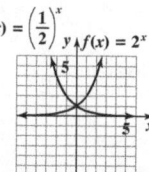

The graph of g is the reflection of f across the y-axis.

7.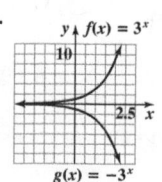

The graph of g is the graph of f shifted down 1 unit.

8.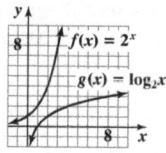

The graph of g is the reflection of f across the x-axis.

9. 5.5% compounded semiannually   **10.** 7% compounded monthly   **11. a.** 200°F   **b.** about 119°F   **c.** 70°F; The temperature of the room is 70°F.   **12.** $49^{1/2} = 7$ **13.** $4^3 = x$ **14.** $3^y = 81$ **15.** $\log_6 216 = 3$ **16.** $\log_b 625 = 4$ **17.** $\log_{13} 874 = y$ **18.** 3

19. -2 **20.** -9 is not in the domain of $y = \log_3 x$.   **21.** $\dfrac{1}{2}$ **22.** 1   **23.** 8   **24.** 5   **25.** $-\dfrac{1}{2}$ **26.** -2 **27.** -3 **28.** 0

29. domain of f: $(-\infty, \infty)$; range of f: $(0, \infty)$; domain of g: $(0, \infty)$; range of g: $(-\infty, \infty)$

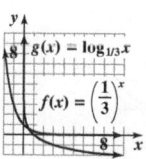

30. domain of f: $(-\infty, \infty)$; range of f: $(0, \infty)$; domain of g: $(0, \infty)$; range of g: $(-\infty, \infty)$

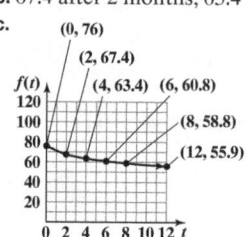

31. $(-5, \infty)$ **32.** $(-\infty, 3)$ **33.** $(-\infty, 1) \cup (1, \infty)$ **34.** $6x$ **35.** $\sqrt{x}$ **36.** $4x^2$ **37.** 3
38. a. 76

b. 67.4 after 2 months; 63.4 after 4 months; 60.8 after 6 months; 58.8 after 8 months; 55.9 after one year

c.

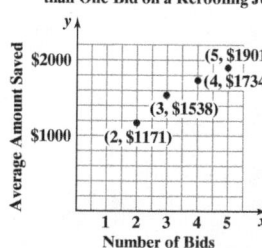

As time increases, the amount of material retained by the students decreases.

39. about 9 weeks   **40.** $2 + 3\log_6 x$ **41.** $\dfrac{1}{2}\log_4 x - 3$ **42.** $\log_2 x + 2\log_2 y - 6$

43. $\dfrac{1}{3}\ln x - \dfrac{1}{3}$ **44.** $\log_b 21$ **45.** $\log\left(\dfrac{3}{x^3}\right)$ **46.** $\ln(x^3 y^4)$ **47.** $\ln\left(\dfrac{\sqrt{x}}{y}\right)$

48. 6.2448   **49.** -0.1063 **50.** true   **51.** false; $\log(x + 9) - \log(x + 1) = \log\left(\dfrac{x + 9}{x + 1}\right)$

52. false; $\log_2 x^4 = 4\log_2 x$ **53.** true   **54.** $\{2\}$ **55.** $\left\{\dfrac{2}{3}\right\}$ **56.** $\left\{-\dfrac{3}{2}\right\}$

57. $\left\{\dfrac{\ln 12{,}143}{\ln 8} \approx 4.52\right\}$ **58.** $\left\{\dfrac{\ln 141}{5} \approx 0.99\right\}$ **59.** $\left\{\dfrac{\ln 3}{0.045} \approx 24.41\right\}$ **60.** $\left\{\dfrac{1}{125}\right\}$

61. $\{100\}$ **62.** $\{23\}$ **63.** $\left\{\dfrac{1}{e}\right\}$ **64.** $\left\{\dfrac{e^3}{2}\right\}$ **65.** $\{5\}$ **66.** no solution or $\varnothing$ **67.** $\left\{\dfrac{8}{3}\right\}$ **68.** $\{2\}$ **69.** $\{4\}$ **70.** 5.5 mi

71. approximately 2086   **72.** approximately 8103 thousand or 8,103,000   **73.** 7.3 years   **74.** 14.6 years   **75.** about 22%
76. a. 0.041   **b.** about 62.4 million   **c.** 2017   **77.** 325 days

78. ; logarithmic function

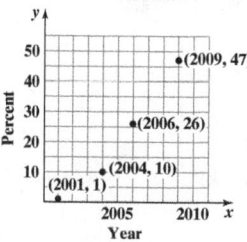

79. ; exponential function

80. $y = 73e^{(\ln 2.6)x}$; $y = 73e^{0.956x}$ **81.** $y = 6.5e^{(\ln 0.43)x}$; $y = 6.5e^{-0.844x}$ **82.** Answers will vary.

Chapter Test

1. 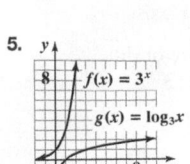 **2.** 6.5% compounded semiannually; $221 **3.** $5^3 = 125$ **4.** $\log_{36} 6 = \dfrac{1}{2}$

5. 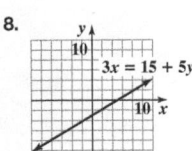 domain of f: $(-\infty, \infty)$; range of f: $(0, \infty)$; domain of g: $(0, \infty)$; range of g: $(-\infty, \infty)$

6. $5x$ **7.** 1 **8.** 0 **9.** $(7, \infty)$ **10.** 120 decibels **11.** $3 + 5\log_4 x$ **12.** $\dfrac{1}{3}\log_3 x - 4$ **13.** $\log(x^6 y^2)$

14. $\ln\left(\dfrac{7}{x^3}\right)$ **15.** 1.5741 **16.** $\{6\}$ **17.** $\left\{\dfrac{\ln 1.4}{\ln 5}\right\}$ **18.** $\left\{\dfrac{\ln 4}{0.005}\right\}$ **19.** $\{5\}$ **20.** $\left\{\dfrac{217}{4}\right\}$ **21.** $\left\{\dfrac{e^4}{3}\right\}$ **22.** $\{5\}$

23. no solution or $\varnothing$ **24. a.** 82.4 million **b.** decreasing; The growth rate, -0.002, is negative. **c.** 2017

25. 13.9 years **26.** about 6.9% **27.** $A = 509e^{0.036t}$ **28.** about 24,758 years ago **29.** linear **30.** logarithmic

31. exponential **32.** quadratic **33.** $y = 96e^{(\ln 0.38)x}$; $y = 96e^{-0.968x}$

Cumulative Review Exercises

1. $\{4\}$ **2.** $\{(2, 3)\}$ **3.** $\{(2, 0, 3)\}$ **4.** $(-\infty, -2) \cup (4, \infty)$ **5.** $\{5\}$ **6.** $[4, \infty)$ **7.** $\left\{\dfrac{3}{4} \pm i\dfrac{\sqrt{7}}{4}\right\}$

8. 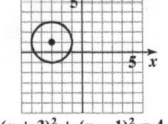 **9.** 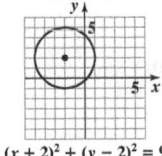 **10.** 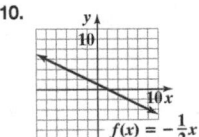 **11.**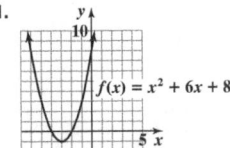

12. 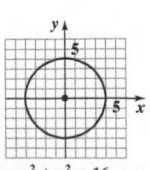 **13.** $c = \dfrac{Ad}{d - A}$ **14.** $x^2 - x - 17$ **15.** $x^2 + 3x - 17$ **16.** h **17.** $f^{-1}(x) = \dfrac{x + 3}{7}$

18. $(-\infty, 1) \cup (1, 2) \cup (2, \infty)$ **19.** $(4, \infty)$ **20.** $y = \dfrac{1}{2}x + 5$ or $f(x) = \dfrac{1}{2}x + 5$ **21.** $-\dfrac{y^9}{3x}$ **22.** $16x^4 - 40x^2y + 25y^2$

23. $x^2 - 5x + 1 + \dfrac{8}{5x + 1}$ **24.** $2y^2\sqrt[3]{2y^2}$ **25.** $\dfrac{4x - 13}{(x - 3)(x - 4)}$ **26.** $(x - 4)(x + 2)(x^2 - 2x + 4)$ **27.** $2(x + 3y)^2$

28. $\ln\left(\dfrac{x^2}{\sqrt{y}}\right)$ **29.** length: 12 ft; width: 4 ft **30.** $\dfrac{6}{5}$ hr or 1 hr and 12 min **31.** 5 mph **32.** approximately 11%

CHAPTER 13

Section 13.1 Check Point Exercises

1. $x^2 + y^2 = 16$
2. $(x - 5)^2 + (y + 6)^2 = 100$

3. center: $(-3, 1)$; radius: 2 units **4.** $(x + 2)^2 + (y - 2)^2 = 9$

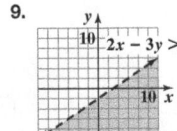

$(x + 3)^2 + (y - 1)^2 = 4$ $(x + 2)^2 + (y - 2)^2 = 9$

Concept and Vocabulary Check

1. circle; center; radius **2.** $(x - h)^2 + (y - k)^2 = r^2$ **3.** general **4.** 4; 16

Exercise Set 13.1

1. $x^2 + y^2 = 49$ **3.** $(x - 3)^2 + (y - 2)^2 = 25$ **5.** $(x + 1)^2 + (y - 4)^2 = 4$ **7.** $(x + 3)^2 + (y + 1)^2 = 3$ **9.** $(x + 4)^2 + y^2 = 100$

11. center: $(0, 0)$; $r = 4$ **13.** center: $(3, 1)$; $r = 6$

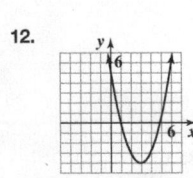   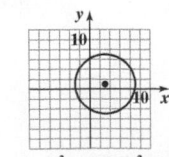

$x^2 + y^2 = 16$ $(x - 3)^2 + (y - 1)^2 = 36$

15. center: $(-3, 2)$; $r = 2$

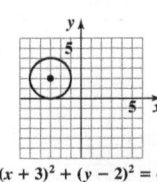

$(x + 3)^2 + (y - 2)^2 = 4$

17. center: $(-2, -2)$; $r = 2$

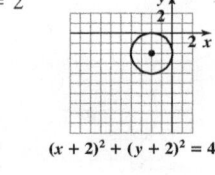

$(x + 2)^2 + (y + 2)^2 = 4$

19. $(x + 3)^2 + (y + 1)^2 = 4$; center: $(-3, -1)$; $r = 2$

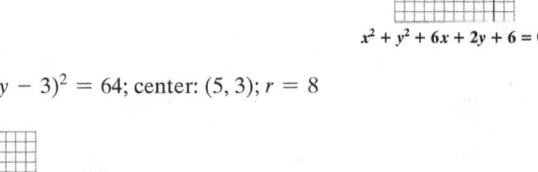

$x^2 + y^2 + 6x + 2y + 6 = 0$

21. $(x - 5)^2 + (y - 3)^2 = 64$; center: $(5, 3)$; $r = 8$

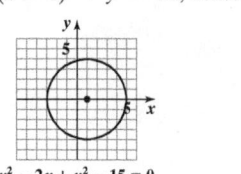

$x^2 + y^2 - 10x - 6y - 30 = 0$

23. $(x + 4)^2 + (y - 1)^2 = 25$; center: $(-4, 1)$; $r = 5$

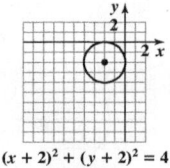

$x^2 + y^2 + 8x - 2y - 8 = 0$

25. $(x - 1)^2 + y^2 = 16$; center: $(1, 0)$; $r = 4$

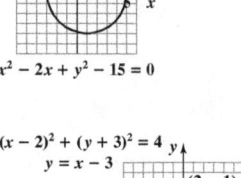

$x^2 - 2x + y^2 - 15 = 0$

27. $x^2 + y^2 = 16$ $\{(0, -4), (4, 0)\}$
$x - y = 4$

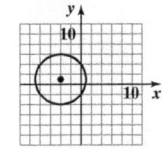

$(4, 0)$
$(0, -4)$

29. $(x - 2)^2 + (y + 3)^2 = 4$
$y = x - 3$

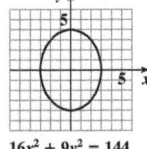

$(2, -1)$
$(0, -3)$

$\{(0, -3), (2, -1)\}$

31. $(x - 2)^2 + (y + 1)^2 = 4$
33. $(x + 3)^2 + (y + 2)^2 = 1$
35. a. $(5, 10)$ **b.** $\sqrt{5}$
c. $(x - 5)^2 + (y - 10)^2 = 5$
37. $(x + 2.4)^2 + (y + 2.7)^2 = 900$

45.

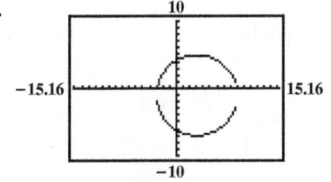

47. does not make sense   **49.** does not make sense   **51.** false   **53.** false   **55.** $11\pi \approx 35$ square units

57. $f(g(x)) = 9x^2 + 24x + 14$; $g(f(x)) = 3x^2 - 2$ **58.** $\{4\}$ **59.** $\left(-\dfrac{5}{2}, \dfrac{15}{2}\right)$ **60.** -3 and 3 **61.** -2 and 2 **62.** $\dfrac{x^2}{16} + \dfrac{y^2}{25} = 1$

Section 13.2 Check Point Exercises

1.

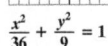

$\dfrac{x^2}{36} + \dfrac{y^2}{9} = 1$

2.

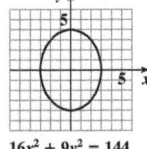

$16x^2 + 9y^2 = 144$

3.

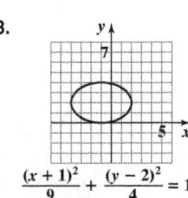

$\dfrac{(x + 1)^2}{9} + \dfrac{(y - 2)^2}{4} = 1$

4. Yes, the height of the archway 6 feet from the center is approximately 9.54 feet.

Concept and Vocabulary Check

1. ellipse; foci; center   **2.** 25; -5; 5; $(-5, 0)$; $(5, 0)$; 9; -3; 3; $(0, -3)$; $(0, 3)$ **3.** 25; -5; 5; $(0, -5)$; $(0, 5)$; 9; -3; 3; $(-3, 0)$; $(3, 0)$
4. $(-1, 4)$ **5.** $(-2, -2)$; $(8, -2)$

Exercise Set 13.2

1.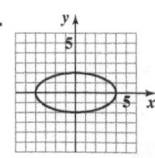

$\frac{x^2}{16} + \frac{y^2}{4} = 1$

3.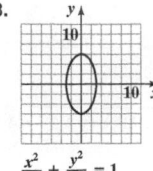

$\frac{x^2}{9} + \frac{y^2}{36} = 1$

5.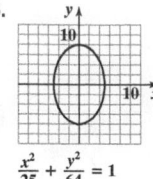

$\frac{x^2}{25} + \frac{y^2}{64} = 1$

7.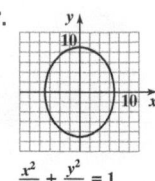

$\frac{x^2}{49} + \frac{y^2}{81} = 1$

9.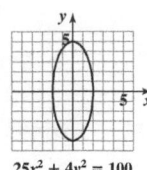

$25x^2 + 4y^2 = 100$

11.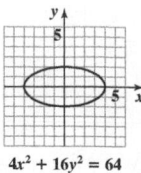

$4x^2 + 16y^2 = 64$

13.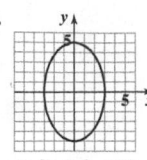

$25x^2 + 9y^2 = 225$

15.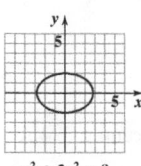

$x^2 + 2y^2 = 8$

17. $\frac{x^2}{4} + \frac{y^2}{1} = 1$ or $\frac{x^2}{4} + y^2 = 1$

19. $\frac{x^2}{1} + \frac{y^2}{4} = 1$ or $x^2 + \frac{y^2}{4} = 1$

21.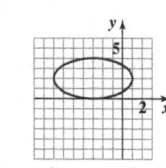

$\frac{(x-2)^2}{9} + \frac{(y-1)^2}{4} = 1$

23.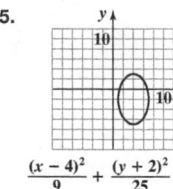

$(x+3)^2 + 4(y-2)^2 = 16$

25.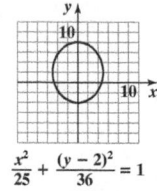

$\frac{(x-4)^2}{9} + \frac{(y+2)^2}{25} = 1$

27.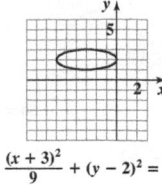

$\frac{x^2}{25} + \frac{(y-2)^2}{36} = 1$

29.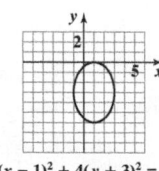

$\frac{(x+3)^2}{9} + (y-2)^2 = 1$

31.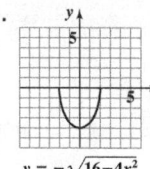

$9(x-1)^2 + 4(y+3)^2 = 36$

33. $\frac{(x+1)^2}{4} + \frac{(y-1)^2}{1} = 1$

35. $\{(0,-1), (0,1)\}$

37. $\{(0,3)\}$

39. $\{(0,-2), (1,0)\}$

41.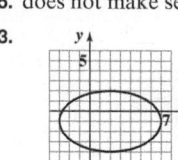

$y = -\sqrt{16 - 4x^2}$

43. Yes, the height of the archway 4 feet from the center is approximately 9.64 feet.

45. a. $\frac{x^2}{48^2} + \frac{y^2}{23^2} = 1$ or $\frac{x^2}{2304} + \frac{y^2}{529} = 1$ **b.** approximately 42 feet

55. does not make sense **57.** makes sense **59.** false **61.** true

63.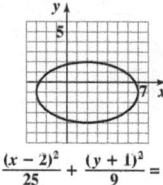

$\frac{(x-2)^2}{25} + \frac{(y+1)^2}{9} = 1$

65. a. 984 mi **b.** 1016 mi **67.** $(x+2)(x+2)(x-2)$ or $(x+2)^2(x-2)$

68. $2xy^2\sqrt[3]{5xy}$ **69.** $\{-1\}$ **70.** $\frac{x^2}{9} - \frac{y^2}{4} = 1$; Terms are separated by subtraction rather than

by addition. **71. a.** -4 and 4 **b.** The equation $y^2 = -9$ has no real solutions. **72. a.** -3 and 3

b. The equation $x^2 = -16$ has no real solutions.

Section 13.3 Check Point Exercises

1. a. $(-5, 0)$ and $(5, 0)$ **b.** $(0, -5)$ and $(0, 5)$

2.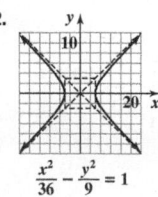

$\frac{x^2}{36} - \frac{y^2}{9} = 1$

3.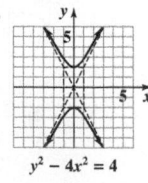

$y^2 - 4x^2 = 4$

Concept and Vocabulary Check

1. hyperbola; vertices; transverse **2.** $(-5, 0)$; $(5, 0)$ **3.** $(0, -5)$; $(0, 5)$ **4.** asymptotes; center **5.** dividing; 36 **6.** ellipse
7. hyperbola **8.** neither **9.** hyperbola **10.** ellipse

Exercise Set 13.3

1. $(-2, 0)$ and $(2, 0)$; b **3.** $(0, -2)$ and $(0, 2)$; a

5.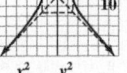
$$\frac{x^2}{9} - \frac{y^2}{25} = 1$$

7.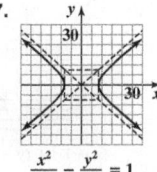
$$\frac{x^2}{100} - \frac{y^2}{64} = 1$$

9.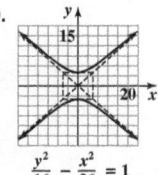
$$\frac{y^2}{16} - \frac{x^2}{36} = 1$$

11.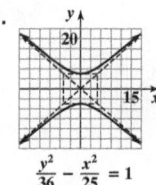
$$\frac{y^2}{36} - \frac{x^2}{25} = 1$$

13.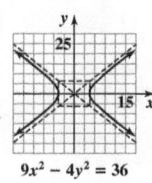
$$9x^2 - 4y^2 = 36$$

15.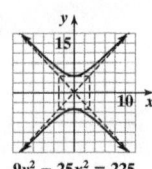
$$9y^2 - 25x^2 = 225$$

17.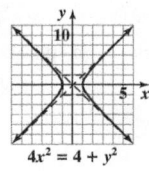
$$4x^2 = 4 + y^2$$

19. $\dfrac{x^2}{9} - \dfrac{y^2}{25} = 1$

21. $\dfrac{y^2}{4} - \dfrac{x^2}{9} = 1$

23.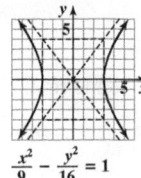
$$\frac{x^2}{9} - \frac{y^2}{16} = 1$$
domain: $(-\infty, -3] \cup [3, \infty)$; range: $(-\infty, \infty)$

25.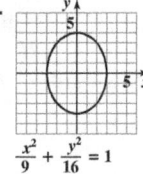
$$\frac{x^2}{9} + \frac{y^2}{16} = 1$$
domain: $[-3, 3]$; range: $[-4, 4]$

27.
$$\frac{y^2}{16} - \frac{x^2}{9} = 1$$
domain: $(-\infty, \infty)$; range: $(-\infty, -4] \cup [4, \infty)$

29. $\{(-2, 0), (2, 0)\}$ **31.** $\{(0, -3), (0, 3)\}$ **33.** 40 yd

41. 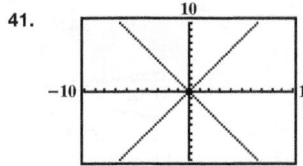   no; two lines; $y = \dfrac{b}{a}x$ and $y = -\dfrac{b}{a}x$ **43.** does not make sense   **45.** makes sense   **47.** false   **49.** true

51.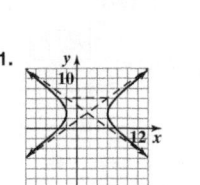
$$\frac{(x-2)^2}{16} - \frac{(y-3)^2}{9} = 1$$

53.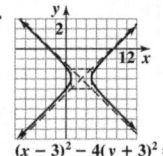
$$(x-3)^2 - 4(y+3)^2 = 4$$

55. $\dfrac{x^2}{36} - \dfrac{y^2}{576} = 1$

57.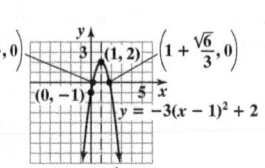
$$y = -x^2 - 4x + 5$$

58. $\left(-\infty, -\dfrac{1}{3}\right] \cup [4, \infty)$ **59.** $\{21\}$ **60.** 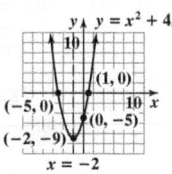   **61.** $\left(1 - \dfrac{\sqrt{6}}{3}, 0\right)$ 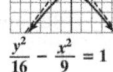 $\left(1 + \dfrac{\sqrt{6}}{3}, 0\right)$
$y = -3(x-1)^2 + 2$ **62.** 0.2 and 1.8

Mid-Chapter Check Point Exercises

1.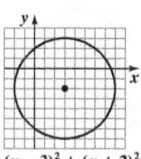
$$x^2 + y^2 = 9$$

2.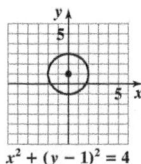
$$(x-3)^2 + (y+2)^2 = 25$$

3.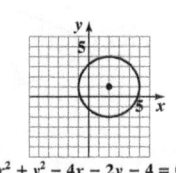
$$x^2 + (y-1)^2 = 4$$

4.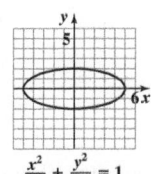
$$x^2 + y^2 - 4x - 2y - 4 = 0$$

5.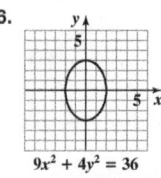
$$\frac{x^2}{25} + \frac{y^2}{4} = 1$$

6.
$$9x^2 + 4y^2 = 36$$

7.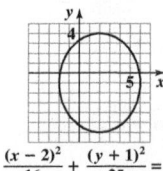
$$\frac{(x-2)^2}{16} + \frac{(y+1)^2}{25} = 1$$

8.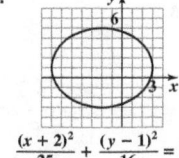
$$\frac{(x+2)^2}{25} + \frac{(y-1)^2}{16} = 1$$

9.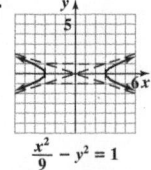
$$\frac{x^2}{9} - y^2 = 1$$

10.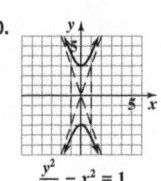
$$\frac{y^2}{9} - x^2 = 1$$

11.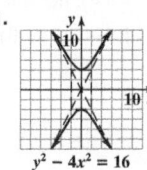
$$y^2 - 4x^2 = 16$$

12.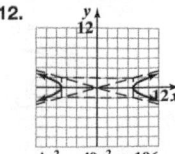
$$4x^2 - 49y^2 = 196$$

13.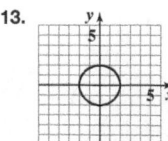
$x^2 + y^2 = 4$

14.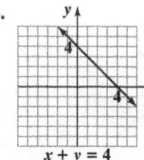
$x + y = 4$

15.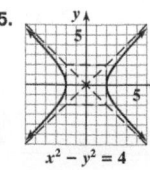
$x^2 - y^2 = 4$

16.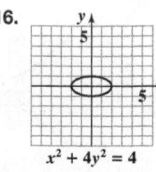
$x^2 + 4y^2 = 4$

17.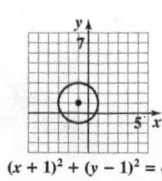
$(x + 1)^2 + (y - 1)^2 = 4$

18.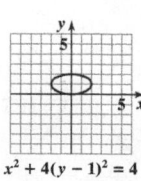
$x^2 + 4(y - 1)^2 = 4$

Section 13.4 Check Point Exercises

1.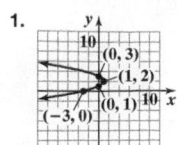
$x = -(y - 2)^2 + 1$

2.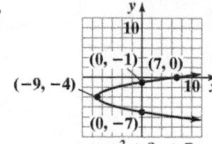
$x = y^2 + 8y + 7$

3. a. hyperbola **b.** ellipse **c.** circle **d.** parabola

Concept and Vocabulary Check

1. parabola; focus **2.** > 0 **3.** < 0 **4.** to the left **5.** to the right **6.** upward **7.** downward **8.** downward **9.** to the right

10. to the left **11.** downward **12.** upward **13.** (h, k) **14.** $-\dfrac{b}{2a}$ **15.** conic **a.** hyperbola **b.** ellipse **c.** circle **d.** parabola

Exercise Set 13.4

1. opens to right; $(-1, 2)$; b **3.** opens to right; $(1, -2)$; f **5.** opens to left; $(1, 2)$; a **7.** $(0, 0)$ **9.** $(3, 2)$ **11.** $(-1, -2)$ **13.** $(0, 6)$
15. $(-3, 3)$ **17.** $(4, -1)$

19.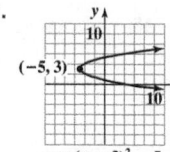
$x = (y - 2)^2 - 4$

21.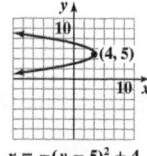
$x = (y - 3)^2 - 5$

23.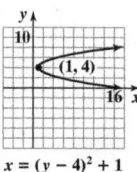
$x = -(y - 5)^2 + 4$

25.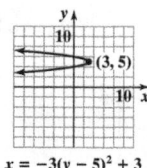
$x = (y - 4)^2 + 1$

27.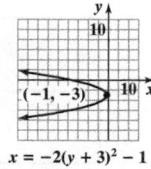
$x = -3(y - 5)^2 + 3$

29.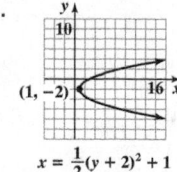
$x = -2(y + 3)^2 - 1$

31.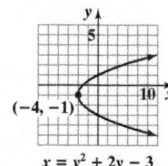
$x = \frac{1}{2}(y + 2)^2 + 1$

33.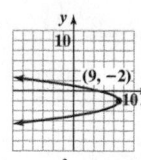
$x = y^2 + 2y - 3$

35.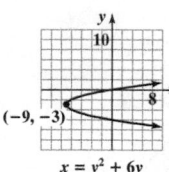
$x = -y^2 - 4y + 5$

37.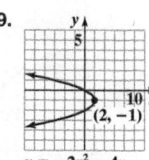
$x = y^2 + 6y$

39.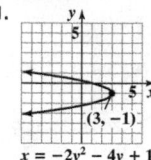
$x = -2y^2 - 4y$

41.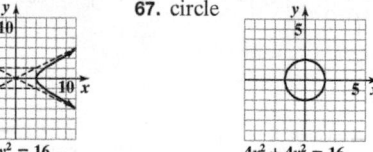
$x = -2y^2 - 4y + 1$

43. a. horizontal **b.** to the right **c.** $(2, 1)$ **45. a.** vertical **b.** upward **c.** $(1, 2)$ **47. a.** vertical **b.** downward **c.** $(-3, 4)$
49. a. horizontal **b.** to the left **c.** $(4, -3)$ **51. a.** vertical **b.** upward **c.** $(2, -5)$ **53. a.** horizontal **b.** to the left **c.** $(5, 2)$
55. parabola **57.** ellipse **59.** hyperbola **61.** circle **63.** hyperbola **65.** hyperbola **67.** circle

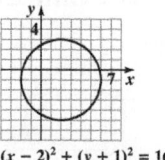
$x^2 - 4y^2 = 16$

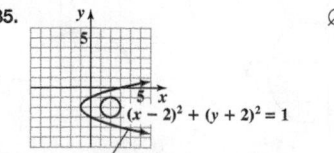
$4x^2 + 4y^2 = 16$

69. ellipse

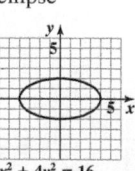
$x^2 + 4y^2 = 16$

71. parabola

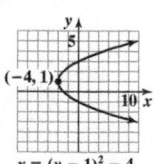
$x = (y - 1)^2 - 4$

73. circle

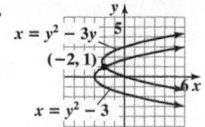
$(x - 2)^2 + (y + 1)^2 = 16$

75. domain: $[-4, \infty)$; range: $(-\infty, \infty)$; not a function **77.** domain: $(-\infty, \infty)$; range: $(-\infty, 1]$; function
79. domain: $(-\infty, 3]$; range: $(-\infty, \infty)$; not a function

81.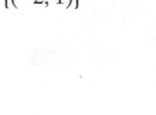
$x = (y - 2)^2 - 4$ $\{(-4, 2), (0, 0)\}$
$y = -\frac{1}{2}x$

83.
$x = y^2 - 3y$ $\{(-2, 1)\}$
$x = y^2 - 3$

85.
$(x - 2)^2 + (y + 2)^2 = 1$
$x = (y + 2)^2 - 1$
$\varnothing$

87. a. $y = 0.0001032x^2$ **b.** 58 ft   **89. a.** $y = \dfrac{1}{18}x^2$ **b.** 4.5 ft   **91. a.** ellipse   **b.** $x^2 + 4y^2 = 4$

101. $y^2 + 2y + (-6x + 13) = 0$; $y = -1 \pm \sqrt{6x - 12}$

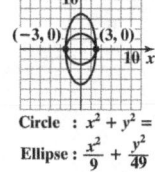

105. makes sense   **107.** makes sense   **109.** true   **111.** false   **113.** Yes; the height of the arch 30 feet from the center is 45.5 feet.
114. 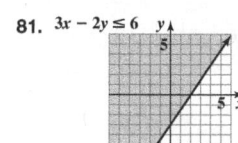   **115.** $f^{-1}(x) = 3x + 15$ **116.** $\{-3, -1\}$ **117.** $\{(2.5, -2)\}$ **118.** $\{(4, -3)\}$ **119.** $\{2, 4\}$

Section 13.5 Check Point Exercises

1. $(0, 1)$ and $(4, 17)$, or $\{(0, 1), (4, 17)\}$ **2.** $(2, -1)$ and $\left(-\dfrac{6}{5}, \dfrac{3}{5}\right)$, or $\left\{(2, -1), \left(-\dfrac{6}{5}, \dfrac{3}{5}\right)\right\}$

3. $(3, 2), (-3, 2), (3, -2),$ and $(-3, -2)$, or $\{(3, 2), (-3, 2), (3, -2), (-3, -2)\}$ **4.** $(0, 5)$ or $\{(0, 5)\}$ **5.** length: 7 feet; width: 3 feet

Concept and Vocabulary Check

1. nonlinear   **2.** $\{(-4, 3), (0, 1)\}$ **3.** 3   **4.** $\{(-2, -\sqrt{3}), (2, -\sqrt{3}), (-2, \sqrt{3}), (-2, -\sqrt{3})\}$ **5.** $-1; y^2 + y = 6$ **6.** $\dfrac{4}{x}; \dfrac{4}{x}; y$

Exercise Set 13.5

1. $\{(2, 0), (-3, 5)\}$ **3.** $\{(2, 0), (1, 1)\}$ **5.** $\{(-3, 11), (4, -10)\}$ **7.** $\{(-3, -4), (4, 3)\}$ **9.** $\left\{(2, 3), \left(-\dfrac{3}{2}, -4\right)\right\}$

11. $\{(3, 0), (-5, -4)\}$ **13.** $\{(3, 1), (-1, -3), (1, 3), (-3, -1)\}$ **15.** $\{(4, -3), (-1, 2)\}$ **17.** $\{(4, -3), (0, 1)\}$

19. $\{(3, 2), (-3, 2), (3, -2), (-3, -2)\}$ **21.** $\{(3, 2), (-3, 2), (3, -2), (-3, -2)\}$ **23.** $\{(2, 1), (-2, 1), (2, -1), (-2, -1)\}$

25. $\{(3, 4), (3, -4)\}$ **27.** $\{(0, 2), (0, -2), (-1, \sqrt{3}), (-1, -\sqrt{3})\}$ **29.** $\{(2, 1), (-2, 1), (2, -1), (-2, -1)\}$

31. $\{(-1, -4), (1, 4), (2\sqrt{2}, \sqrt{2}), (-2\sqrt{2}, -\sqrt{2})\}$ **33.** $\{(4, 1), (2, 2)\}$ **35.** $\{(0, 0), (-1, 1)\}$ **37.** $\{(0, 0), (2, 2), (-2, 2)\}$

39. $\left\{(-4, 1), \left(-\dfrac{5}{2}, \dfrac{1}{4}\right)\right\}$ **41.** $\left\{(-2, 3), \left(\dfrac{12}{5}, -\dfrac{29}{5}\right)\right\}$ **43.** $x + y = 10$; $xy = 24$; 6 and 4

45. $x^2 - y^2 = 3$; $2x^2 + y^2 = 9$; 2 and 1, −2 and 1, −2 and −1, or 2 and −1   **47.** $\{(2, -1), (-2, 1)\}$ **49.** $\{(2, 20), (-2, 4), (-3, 0)\}$

51. $\left\{\left(-1, -\dfrac{1}{2}\right), \left(-1, \dfrac{1}{2}\right), \left(1, -\dfrac{1}{2}\right), \left(1, \dfrac{1}{2}\right)\right\}$

53. Answers will vary.; example:

55. $(0, -4), (2, 0), (-2, 0)$
57. length: 11 feet; width: 7 feet
59. length: 8 inches; width: 6 inches
61. large square: 5 meters by 5 meters; small square: 2 meters by 2 meters

Circle : $x^2 + y^2 = 9$
Ellipse : $\dfrac{x^2}{9} + \dfrac{y^2}{49} = 1$

63. a. between the 1940s and the 1960s   **b.** 1949; 43%; 43%   **c.** 1920; 28%   **d.** 1919; white collar: 27.5%; farmers: 27.4%; fairly well, although answers will vary.   **71.** does not make sense   **73.** makes sense   **75.** true   **77.** false   **79.** $\{(8, 2)\}$

81. $3x - 2y \le 6$ **82.** $m = \dfrac{8}{3}$ **83.** $6x^3 - 16x^2 + 17x - 6$ **84.** $-\dfrac{1}{2}, \dfrac{1}{8}, -\dfrac{1}{27}, \dfrac{1}{80}$ **85.** 120   **86.** 2; 5; 10; 17; 26; 37; Sum is 97.

Review Exercises

1. $x^2 + y^2 = 9$ **2.** $(x + 2)^2 + (y - 4)^2 = 36$

3. center: $(0, 0)$; $r = 1$

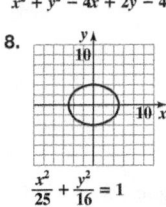

$x^2 + y^2 = 1$

4. center: $(-2, 3)$; $r = 3$

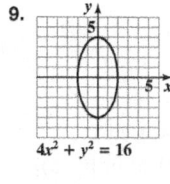

$(x + 2)^2 + (y - 3)^2 = 9$

5. center: $(2, -1)$; $r = 3$

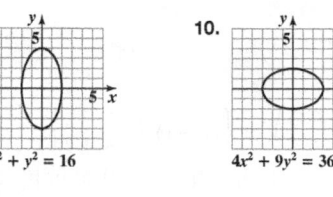

$x^2 + y^2 - 4x + 2y - 4 = 0$

6. center: $(0, 2)$; $r = 2$

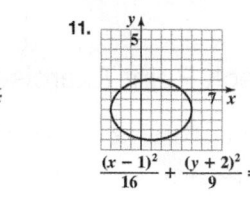

$x^2 + y^2 - 4y = 0$

7.

$\dfrac{x^2}{36} + \dfrac{y^2}{25} = 1$

8.

$\dfrac{x^2}{25} + \dfrac{y^2}{16} = 1$

9.

$4x^2 + y^2 = 16$

10.

$4x^2 + 9y^2 = 36$

11.

$\dfrac{(x - 1)^2}{16} + \dfrac{(y + 2)^2}{9} = 1$

12.

$\dfrac{(x + 1)^2}{9} + \dfrac{(y - 2)^2}{16} = 1$

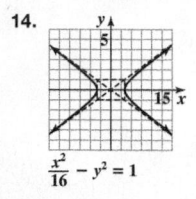

13. Yes, the height of the archway 14 feet from the center is approximately 12.43 feet.

14.

$\dfrac{x^2}{16} - y^2 = 1$

15.

$\dfrac{y^2}{16} - x^2 = 1$

16.

$9x^2 - 16y^2 = 144$

17.

$4y^2 - x^2 = 16$

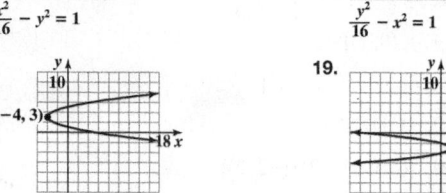

18.

$(-4, 3)$

$x = (y - 3)^2 - 4$

19.

$(2, -3)$

$x = -2(y + 3)^2 + 2$

20.

$(-4, 4)$

$x = y^2 - 8y + 12$

21.

$(10, -2)$

$x = -y^2 - 4y + 6$

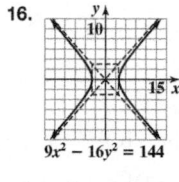 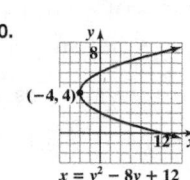 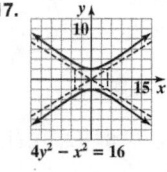 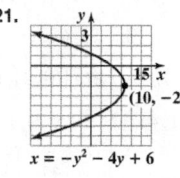

22. parabola   **23.** ellipse   **24.** hyperbola   **25.** circle   **26.** hyperbola   **27.** ellipse   **28.** parabola

29. circle

$5x^2 + 5y^2 = 180$

30. ellipse

$4x^2 + 9y^2 = 36$

31. hyperbola

$4x^2 - 9y^2 = 36$

32. ellipse

$\dfrac{x^2}{25} + \dfrac{y^2}{1} = 1$

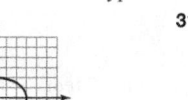 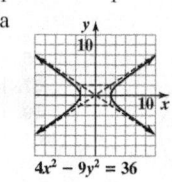 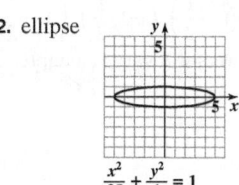

33. parabola

$(-2, 1)$

$x + 3 = -y^2 + 2y$

34. parabola

$(1, 2)$

$y - 3 = x^2 - 2x$

35. ellipse

$\dfrac{(x + 2)^2}{16} + \dfrac{(y - 5)^2}{4} = 1$

36. circle

$(x - 3)^2 + (y + 2)^2 = 4$

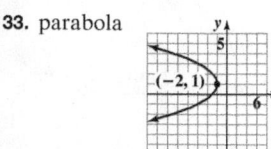 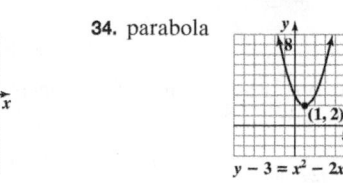 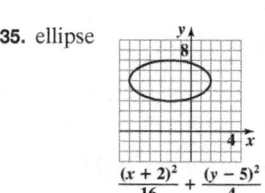 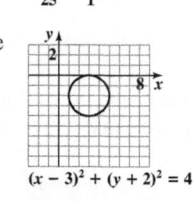

37. circle

$x^2 + y^2 + 6x - 2y + 6 = 0$

38. a. $y = \dfrac{1}{12}x^2$ **b.** $(0, 3)$; 3 inches above the vertex   **39.** $\{(1, 0), (4, 3)\}$

40. $\{(0, 1), (-3, 4)\}$ **41.** $\{(-1, 1), (1, -1)\}$ **42.** $\{(3, \sqrt{6}), (-3, \sqrt{6}), (3, -\sqrt{6}), (-3, -\sqrt{6})\}$

43. $\{(2, 2), (-2, -2)\}$ **44.** $\{(1, 2), (9, 6)\}$ **45.** $\{(-3, -1), (1, 3)\}$ **46.** $\left\{(-1, -1), \left(\dfrac{1}{2}, 2\right)\right\}$

47. $\left\{(0, -1), \left(\dfrac{5}{2}, -\dfrac{7}{2}\right)\right\}$ **48.** $\{(3, 2), (-3, 2), (2, -3), (-2, -3)\}$

49. $\{(3, 1), (-3, 1), (3, -1),$ and $(-3, -1)\}$ **50.** 8 meters by 5 meters   **51.** $(1, 6)$ and $(3, 2)$ **52.** x: 46 ft; y: 28 ft or x: 50 ft; y: 20 ft

Chapter Test

1. $(x - 3)^2 + (y + 2)^2 = 25$ **2.** center: $(5, -3)$; $r = 7$ **3.** center: $(-2, 3)$; $r = 4$ **4.** $(7, -3)$ **5.** $(-2, -5)$

6. hyperbola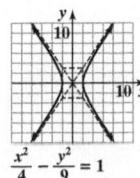
$$\frac{x^2}{4} - \frac{y^2}{9} = 1$$

7. ellipse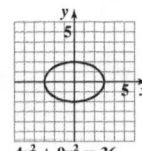
$4x^2 + 9y^2 = 36$

8. parabola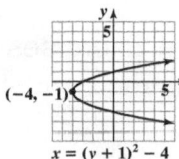
$x = (y + 1)^2 - 4$

9. ellipse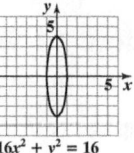
$16x^2 + y^2 = 16$

10. hyperbola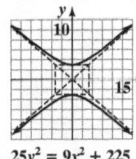
$25y^2 = 9x^2 + 225$

11. parabola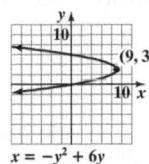
$x = -y^2 + 6y$

12. ellipse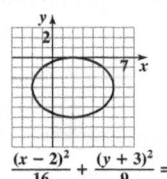
$$\frac{(x - 2)^2}{16} + \frac{(y + 3)^2}{9} = 1$$

13. circle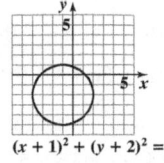
$(x + 1)^2 + (y + 2)^2 = 9$

14. circle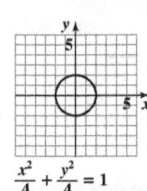
$$\frac{x^2}{4} + \frac{y^2}{4} = 1$$

15. $\{(4, -3), (-3, 4)\}$ **16.** $\{(3, 2), (-3, 2), (3, -2), (-3, -2)\}$
17. 15 feet by 12 feet or 24 feet by 7.5 feet   **18.** 4 feet by 3 feet

Cumulative Review Exercises

1. $(-1, \infty)$ **2.** $\left\{-\frac{1}{2}, 4\right\}$ **3.** $\{2\}$ **4.** $\left(-\frac{5}{3}, -1\right)$ **5.** $\left\{\frac{5}{2}\right\}$ **6.** $\left\{\frac{\ln 8}{0.7} \approx 2.97\right\}$ **7.** $\{(1, 3), (-1, 3), (1, -3), (-1, -3)\}$

8.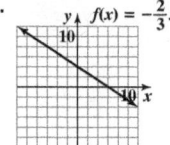
$f(x) = -\frac{2}{3}x + 4$

9. $3x - y > 6$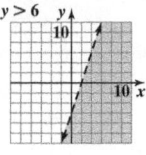

10.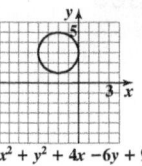
$x^2 + y^2 + 4x - 6y + 9 = 0$

11. $9x^2 - 4y^2 = 36$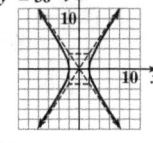

12. 46   **13.** $x^2 - 5x - 1$ **14.** $2xy^2 \sqrt[3]{2y}$ **15.** $11 + 10i$ **16.** $3x(2x - 3)^2$ **17.** $(x - 2)(x + 3)(x - 3)$ **18.** $(-\infty, 2]$
19. $\frac{(1 - \sqrt{x})^2}{1 - x}$ or $\frac{1 - 2\sqrt{x} + x}{1 - x}$ **20.** $\ln(x^{1/3}y^7)$ **21.** $3x^2 + x + 4 + \frac{7}{x - 2}$ **22.** $x^2 - 12 = 0$ **23.** faster car: 50 mph; slower car: 40 mph
24. 175 miles; $67   **25.** apple: 60 calories; banana: 87 calories

CHAPTER 14

Section 14.1 Check Point Exercises

1. a. $7, 9, 11, 13$ **b.** $-\frac{1}{3}, \frac{1}{5}, -\frac{1}{9}, \frac{1}{17}$ **2.** $10, \frac{10}{3}, \frac{5}{6}, \frac{1}{6}$ **3. a.** $2(1)^2 + 2(2)^2 + 2(3)^2 + 2(4)^2 + 2(5)^2 + 2(6)^2 = 182$

b. $(2^3 - 3) + (2^4 - 3) + (2^5 - 3) = 47$ **c.** $4 + 4 + 4 + 4 + 4 = 20$ **4. a.** $\sum_{i=1}^{9} i^2$ **b.** $\sum_{i=1}^{n} \frac{1}{2^{i-1}}$

Concept and Vocabulary Check

1. sequence; integers; terms   **2.** general   **3.** -1 **4.** $-\frac{1}{3}$ **5.** factorial; 5; 1; 1   **6.** $a_1 + a_2 + a_3 + \cdots + a_n$; index; upper limit; lower limit

Exercise Set 14.1

1. $5, 8, 11, 14$ **3.** $3, 9, 27, 81$ **5.** $-3, 9, -27, 81$ **7.** $-4, 5, -6, 7$ **9.** $\frac{2}{5}, \frac{2}{3}, \frac{6}{7}, 1$ **11.** $1, -\frac{1}{3}, \frac{1}{7}, -\frac{1}{15}$ **13.** $1, 2, \frac{3}{2}, \frac{2}{3}$ **15.** $4, 12, 48, 240$

17. 105   **19.** 60   **21.** 115   **23.** $-\frac{5}{16}$ **25.** 55   **27.** $\frac{3}{8}$ **29.** 15   **31.** $\sum_{i=1}^{15} i^2$ **33.** $\sum_{i=1}^{11} 2^i$ **35.** $\sum_{i=1}^{30} i$ **37.** $\sum_{i=1}^{14} \frac{i}{i + 1}$ **39.** $\sum_{i=1}^{n} \frac{4^i}{i}$

41. $\sum_{i=1}^{n} (2i - 1)$ **43.** Answers will vary; examples are: $\sum_{k=1}^{14} (2k + 3)$ or $\sum_{k=2}^{15} (2k + 1)$.   **45.** Answers will vary; an example is: $\sum_{k=0}^{12} ar^k$.

47. Answers will vary; an example is: $\sum_{k=0}^{n} (a + kd)$.   **49.** 45   **51.** 0   **53.** 2   **55.** 80

57. a. 1515; A total of 1515 thousand, or 1,515,000, autism cases were diagnosed in the United States from 2001 through 2008.
b. 1512; underestimates by 3 thousand   **59.** $8081.13   **69.** As n gets larger, the terms get closer to 1.

71. As n gets larger, the terms get closer to 0.   **73.** does not make sense   **75.** makes sense   **77.** false   **79.** true   **81.** $a_n = \frac{1}{n}$

83. $a_n = (-1)^n$ **85.** $a_n = \frac{n + 2}{n + 1}$ **87.** $a_n = \frac{(n + 1)^2}{n}$ **89.** 600   **91.** $n^3 - 3n^2 + 2n$ **93.** $4\log x + 6\log x + 8\log x = \log x^{18}$

95. $2xy^2\sqrt[3]{5xy}$　**96.** $(3x - 2)(9x^2 + 6x + 4)$　**97.** $\left\{-\dfrac{6}{5}, 4\right\}$　**98.** $-5; -5; -5; -5;$ The difference between consecutive terms is always -5.
99. $4; 4; 4; 4;$ The difference between consecutive terms is always 4.　**100.** -45

Section 14.2 Check Point Exercises

1. $100, 70, 40, 10, -20, -50$　**2.** -34　**3. a.** $a_n = 0.7n + 31.3$　**b.** 46　**4.** 360　**5.** 2460　**6.** \$740,300

Concept and Vocabulary Check

1. arithmetic; common difference　**2.** $a_1 + (n - 1)d$; first term; common difference　**3.** $\dfrac{n}{2}(a_1 + a_n)$; first term; nth term

4. $2; 116$　**5.** $8; 13; 18; 5$

Exercise Set 14.2

1. 4　**3.** 5　**5.** -3　**7.** $200, 220, 240, 260, 280, 300$　**9.** $-7, -3, 1, 5, 9, 13$　**11.** $300, 210, 120, 30, -60, -150$　**13.** $\dfrac{5}{2}, 2, \dfrac{3}{2}, 1, \dfrac{1}{2}, 0$

15. $-0.4, -2, -3.6, -5.2, -6.8, -8.4$　**17.** 33　**19.** 252　**21.** 955　**23.** -142　**25.** $a_n = 4n - 3; a_{20} = 77$　**27.** $a_n = 11 - 4n; a_{20} = -69$

29. $a_n = -4n - 16; a_{20} = -96$　**31.** $a_n = \dfrac{1}{3}n - \dfrac{2}{3}; a_{20} = 6$　**33.** $a_n = 4.3 - 0.3n; a_{20} = -1.7$　**35.** 1220　**37.** 4400　**39.** 5050

41. 3660　**43.** 396　**45.** $8 + 13 + 18 + \cdots + 88 = 816$　**47.** $2 + (-1) + (-4) + \cdots + (-85) = -1245$

49. $4 + 8 + 12 + \cdots + 400 = 20{,}200$　**51.** 7　**53.** 22　**55.** 847　**57.** $f(x) = -4x + 5$　**59.** $a_n = 3n - 2$　**61. a.** $a_n = 0.5n + 10.5$

b. 35.5%　**63.** company A; \$1400　**65. a.** \$25,626　**b.** \$25,626; It's the same.　**69.** Company A: \$307,000; Company B: \$324,000; Company B pays the greater total amount.　**71.** 2869 seats　**79.** makes sense　**81.** makes sense　**83.** false　**85.** true　**87.** 200th term

89. $S_n = \dfrac{n}{2}[1 + (2n - 1)] = \dfrac{n}{2}(2n) = n^2$　**90.** $\{1005\}$　**91.** $[-5, 2]$　**92.** $P = \dfrac{At}{t - A}$　**93.** $-2; -2; -2; -2;$ The ratio of a term to the term that directly precedes it is always -2.　**94.** $5; 5; 5; 5;$ The ratio of a term to the term that directly precedes it is always 5.　**95.** 8019

Section 14.3 Check Point Exercises

1. $12, 6, 3, \dfrac{3}{2}, \dfrac{3}{4}, \dfrac{3}{8}$　**2.** 3645　**3.** $a_n = 3(2)^{n-1}; a_8 = 384$　**4.** 9842　**5.** 19,680　**6.** approximately \$2,371,746　**7. a.** \$333,946

b. \$291,946　**8.** 9　**9.** $\dfrac{1}{1}$ or 1　**10.** \$4000

Concept and Vocabulary Check

1. geometric; common ratio　**2.** $a_1 r^{n-1}$; first term; common ratio　**3.** $\dfrac{a_1(1 - r^n)}{1 - r}$; first term; common ratio　**4.** annuity; $P; r; n$

5. infinite geometric series; 1; $\dfrac{a_1}{1 - r}$; $|r| \ge 1$　**6.** $2; 4; 8; 16; 2$　**7.** arithmetic　**8.** geometric　**9.** geometric　**10.** arithmetic

Exercise Set 14.3

1. $r = 3$　**3.** $r = -2$　**5.** $r = \dfrac{3}{2}$　**7.** $r = -0.1$　**9.** $2, 6, 18, 54, 162$　**11.** $20, 10, 5, \dfrac{5}{2}, \dfrac{5}{4}$　**13.** $-4, 40, -400, 4000, -40{,}000$

15. $-\dfrac{1}{4}, \dfrac{1}{2}, -1, 2, -4$　**17.** $a_8 = 768$　**19.** $a_{12} = -10{,}240$　**21.** $a_6 = -200$　**23.** $a_8 = 0.1$　**25.** $a_n = 3(4)^{n-1}; a_7 = 12{,}288$

27. $a_n = 18\left(\dfrac{1}{3}\right)^{n-1}; a_7 = \dfrac{2}{81}$　**29.** $a_n = 1.5(-2)^{n-1}; a_7 = 96$　**31.** $a_n = 0.0004(-10)^{n-1}; a_7 = 400$　**33.** 531,440　**35.** 2049

37. $\dfrac{16{,}383}{2}$ or 8191.5　**39.** 9840　**41.** 10,230　**43.** $\dfrac{63}{128}$　**45.** $\dfrac{3}{2}$　**47.** 4　**49.** $\dfrac{2}{3}$　**51.** 20　**53.** $\dfrac{5}{9}$　**55.** $\dfrac{47}{99}$　**57.** $\dfrac{257}{999}$

59. arithmetic; $d = 1$　**61.** geometric; $r = 2$　**63.** neither　**65.** 2435　**67.** 2280　**69.** -140　**71.** $a_2 = 12, a_3 = 18$　**73.** \$30,000
75. \$16,384　**77.** approximately \$3,795,957　**79. a.** approximately 1.01 for each division　**b.** $a_n = 35.48(1.01)^{n-1}$
c. approximately 38.04 million　**81.** \$32,767　**83.** approximately \$793,583　**85.** approximately 130.26 in.　**87. a.** \$11,617　**b.** \$1617

89. a. \$87,052　**b.** \$63,052　**91. a.** \$693,031　**b.** \$293,031　**93.** \$9 million　**95.** $\dfrac{1}{3}$
107. 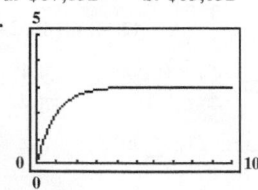　horizontal asymptote: $y = 3$; sum of series: 3

109. makes sense　**111.** makes sense　**113.** false　**115.** false　**117.** 2000 flies　**119.** $2\sqrt{7}$　**120.** $\left\{\dfrac{-1 \pm \sqrt{33}}{4}\right\}$　**121.** $-3(\sqrt{3} + \sqrt{5})$

122. The exponents on a begin with the exponent on $a + b$ and decrease by 1 in each successive term.　**123.** The exponents on b begin with 0, increase by 1 in each successive term, and end with the exponent on $a + b$.　**124.** The sum of the exponents is the exponent on $a + b$.

Mid-Chapter Check Point Exercises

1. $1, -2, \dfrac{3}{2}, -\dfrac{2}{3}, \dfrac{5}{24}$ **2.** $5, 2, -1, -4, -7$ **3.** $5, -15, 45, -135, 405$ **4.** $a_n = 4n - 2; a_{20} = 78$ **5.** $a_n = 3(2)^{n-1}; a_{10} = 1536$

6. $a_n = -\dfrac{1}{2}n + 2; a_{30} = -13$ **7.** 5115   **8.** 2350   **9.** 6820   **10.** $-29,300$ **11.** 44   **12.** 3725   **13.** $\dfrac{1995}{64}$ **14.** $\dfrac{5}{7}$ **15.** $\dfrac{5}{11}$

16. Answers will vary.; an example is $\displaystyle\sum_{i=1}^{18} \dfrac{i}{i+2}$.   **17.** 464 ft; 3600 ft   **18.** \$1,071,794

Section 14.4 Check Point Exercises

1. a. 20   **b.** 1   **c.** 28   **d.** 1   **2.** $x^4 + 4x^3 + 6x^2 + 4x + 1$ **3.** $x^5 - 10x^4y + 40x^3y^2 - 80x^2y^3 + 80xy^4 - 32y^5$ **4.** $4032x^5y^4$

Concept and Vocabulary Check

1. binomial   **2.** $\dfrac{8!}{2!6!}$ **3.** $\dfrac{n!}{r!(n-r)!}$ **4.** $\dbinom{5}{0}x^5 + \dbinom{5}{1}x^4 \cdot 2 + \dbinom{5}{2}x^3 \cdot 2^2 + \dbinom{5}{3}x^2 \cdot 2^3 + \dbinom{5}{4}x \cdot 2^4 + \dbinom{5}{5}2^5$

5. $\dbinom{n}{0}a^n + \dbinom{n}{1}a^{n-1}b + \dbinom{n}{2}a^{n-2}b^2 + \dbinom{n}{3}a^{n-3}b^3 + \cdots + \dbinom{n}{n}b^n$ **6.** Binomial   **7.** $a^{n-r}b^r$

Exercise Set 14.4

1. 56   **3.** 12   **5.** 1   **7.** 4950   **9.** $x^3 + 6x^2 + 12x + 8$ **11.** $27x^3 + 27x^2y + 9xy^2 + y^3$ **13.** $125x^3 - 75x^2 + 15x - 1$
15. $16x^4 + 32x^3 + 24x^2 + 8x + 1$ **17.** $x^8 + 8x^6y + 24x^4y^2 + 32x^2y^3 + 16y^4$ **19.** $y^4 - 12y^3 + 54y^2 - 108y + 81$
21. $16x^{12} - 32x^9 + 24x^6 - 8x^3 + 1$ **23.** $c^5 + 10c^4 + 40c^3 + 80c^2 + 80c + 32$ **25.** $x^5 - 5x^4 + 10x^3 - 10x^2 + 5x - 1$
27. $243x^5 - 405x^4y + 270x^3y^2 - 90x^2y^3 + 15xy^4 - y^5$ **29.** $64a^6 + 192a^5b + 240a^4b^2 + 160a^3b^3 + 60a^2b^4 + 12ab^5 + b^6$
31. $x^8 + 16x^7 + 112x^6$ **33.** $x^{10} - 20x^9y + 180x^8y^2$ **35.** $x^{32} + 16x^{30} + 120x^{28}$ **37.** $y^{60} - 20y^{57} + 190y^{54}$ **39.** $240x^4y^2$ **41.** $126x^5$

43. $56x^6y^{15}$ **45.** $-\dfrac{21}{2}x^6$ **47.** $319{,}770x^{16}y^{14}$ **49.** $x^{12} + 4x^7 + 6x^2 + \dfrac{4}{x^3} + \dfrac{1}{x^8}$ **51.** $x - 3x^{1/3} + \dfrac{3}{x^{1/3}} - \dfrac{1}{x}$ **53.** 8

55. $4x^3 + 6x^2h + 4xh^2 + h^3$ **57.** 252   **59.** 0.1138
69.

$f_2, f_3,$ and f_4 are approaching $f_1 = f_5.$

71. $x^3 - 3x^2 + 3x - 1$
73. $x^6 + 12x^5 + 60x^4 + 160x^3 + 240x^2 + 192x + 64$

75. makes sense   **77.** does not make sense   **79.** true   **81.** false   **83.** $10x^4y^6$ **84.** $a^2 + 4a + 6$

85. $f(g(x)) = 4x^2 - 2x - 6; g(f(x)) = 2x^2 + 10x - 3$ **86.** $\dfrac{2x^2 - 9x - 1}{2(x-4)(x+3)}$

Review Exercises

1. $3, 10, 17, 24$ **2.** $-\dfrac{3}{2}, \dfrac{4}{3}, -\dfrac{5}{4}, \dfrac{6}{5}$ **3.** $1, 1, \dfrac{1}{2}, \dfrac{1}{6}$ **4.** $\dfrac{1}{2}, -\dfrac{1}{4}, \dfrac{1}{8}, -\dfrac{1}{16}$ **5.** 95   **6.** -20 **7.** Answers will vary; an example is $\displaystyle\sum_{i=1}^{15} \dfrac{i}{i+2}$.

8. Answers will vary; examples are $\displaystyle\sum_{i=4}^{13} i^3$ or $\displaystyle\sum_{i=1}^{10} (i+3)^3$.   **9.** $7, 11, 15, 19, 23, 27$ **10.** $-4, -9, -14, -19, -24, -29$ **11.** $\dfrac{3}{2}, 1, \dfrac{1}{2}, 0, -\dfrac{1}{2}, -1$

12. 20   **13.** -30 **14.** -38 **15.** $a_n = 4n - 11; a_{20} = 69$ **16.** $a_n = 220 - 20n; a_{20} = -180$ **17.** $a_n = -\dfrac{23}{2} - \dfrac{1}{2}n; a_{20} = -\dfrac{43}{2}$

18. $a_n = 22 - 7n; a_{20} = -118$ **19.** 1727   **20.** 225   **21.** 15,150   **22.** 440   **23.** -500 **24.** -2325

25. a. $a_n = 4.75n + 34.25$ **b.** 96%   **26.** \$418,500   **27.** 1470 seats   **28.** $3, 6, 12, 24, 48$ **29.** $\dfrac{1}{2}, \dfrac{1}{4}, \dfrac{1}{8}, \dfrac{1}{16}, \dfrac{1}{32}$ **30.** $16, -4, 1, -\dfrac{1}{4}, \dfrac{1}{16}$

31. $-5, 5, -5, 5, -5$ **32.** $a_7 = 1458$ **33.** $a_6 = \dfrac{1}{2}$ **34.** $a_5 = -48$ **35.** $a_n = 1(2)^{n-1}$ or $a_n = 2^{n-1}; a_8 = 128$

36. $a_n = 100\left(\dfrac{1}{10}\right)^{n-1}; a_8 = \dfrac{1}{100{,}000} = 0.00001$ **37.** $a_n = 12\left(-\dfrac{1}{3}\right)^{n-1}; a_8 = -\dfrac{4}{729}$ **38.** 17,936,135   **39.** $\dfrac{127}{8}$ or 15.875   **40.** 19,530

41. -258 **42.** $\dfrac{341}{128}$ **43.** $\dfrac{27}{2}$ **44.** $\dfrac{4}{3}$ **45.** $-\dfrac{18}{5}$ **46.** 20   **47.** $\dfrac{2}{3}$ **48.** $\dfrac{47}{99}$ **49. a.** approximately 1.4 for each division

b. $a_n = 4.2(1.4)^{n-1}$ **c.** 62.0 million   **50.** approximately \$42,823; approximately \$223,210   **51. a.** \$19,129   **b.** \$8729   **52. a.** \$91,361

b. \$55,361   **53.** $\$9\dfrac{1}{3}$ million   **54.** 165   **55.** 4005   **56.** $8x^3 + 12x^2 + 6x + 1$ **57.** $x^8 - 4x^6 + 6x^4 - 4x^2 + 1$
58. $x^5 + 10x^4y + 40x^3y^2 + 80x^2y^3 + 80xy^4 + 32y^5$ **59.** $x^6 - 12x^5 + 60x^4 - 160x^3 + 240x^2 - 192x + 64$ **60.** $x^{16} + 24x^{14} + 252x^{12}$
61. $x^9 - 27x^8 + 324x^7$ **62.** $80x^2$ **63.** $4860x^2$

Chapter Test

1. $1, -\dfrac{1}{4}, \dfrac{1}{9}, -\dfrac{1}{16}, \dfrac{1}{25}$ **2.** 105   **3.** Answers will vary; examples are $\displaystyle\sum_{i=2}^{21} \dfrac{i}{i+1}$ or $\displaystyle\sum_{i=1}^{20} \dfrac{i+1}{i+2}$.   **4.** $a_n = 5n - 1; a_{12} = 59$

5. $a_n = 16\left(\dfrac{1}{4}\right)^{n-1}; a_{12} = \dfrac{1}{262{,}144}$ **6.** -385 **7.** 550   **8.** -2387 **9.** $-21{,}846$ **10.** 8   **11.** $\dfrac{73}{99}$ **12.** approximately \$276,427

13. 36   **14.** $x^{10} - 5x^8 + 10x^6 - 10x^4 + 5x^2 - 1$ **15.** $x^8 + 8x^7y^2 + 28x^6y^4$

Cumulative Review Exercises

1. $\{22\}$ **2.** $\{5 \pm 7i\}$ **3.** $(-\infty, -3) \cup (2, \infty)$ **4.** $\{-2\}$ **5.** no solution or $\varnothing$ **6.** $\left(-\infty, \dfrac{2}{3}\right)$ **7.** $\{(4, 0, -5)\}$ **8.** $\{9\}$

9. $\{(2, 1), (-2, 1), (2, -1), (-2, -1)\}$ **10.** $\{(-14, -20), (2, -4)\}$

11. 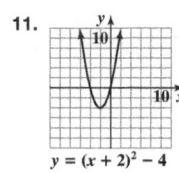 $y = (x + 2)^2 - 4$ **12.** 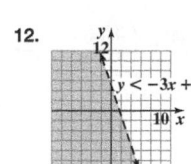 $y < -3x + 5$ **13.** 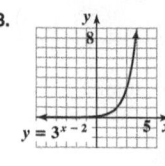 $y = 3^{x-2}$ **14.** 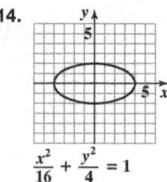 $\dfrac{x^2}{16} + \dfrac{y^2}{4} = 1$ **15.** 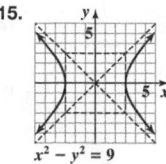 $x^2 - y^2 = 9$

16. $\dfrac{2x^2 + 5x - 2}{(x - 5)(x + 2)}$ **17.** $-\dfrac{x + 1}{x - 1}$ **18.** $2\sqrt{5} + 2\sqrt{2}$

19. $12\sqrt{5}$ **20.** $\dfrac{5\sqrt[3]{4xy^2}}{2xy}$ **21.** $(x + y)(5a - 4b)$ **22.** $\log\left(\dfrac{x^5}{\sqrt{y}}\right)$ **23.** $p = \dfrac{qf}{q - f}$ **24.** $3\sqrt{10} \approx 9.49$ units   **25.** 208   **26.** 1800

27. $\dfrac{1}{3}$ **28.** $16x^4 - 32x^3y^3 + 24x^2y^6 - 8xy^9 + y^{12}$ **29.** $(-\infty, -5) \cup (-5, 3) \cup (3, \infty)$ **30.** $[3, \infty)$ **31.** $(-\infty, 1)$ **32.** 8 ft by 3 ft

33. \$18,500   **34.** $k = \dfrac{1}{2 \ln 4} \approx 0.3607$; about 0.298 or $\dfrac{298}{1000}$

Appendix A Check Point Exercises

1. a. 30   **b.** 30   **2. a.** 35   **b.** 82   **3. a.** \$19.05 million   **b.** \$22.45 million   **c.** One salary, \$0 million, is so much smaller than the other salaries.   **4. a.** 8   **b.** 3 and 8   **c.** no mode   **5.** mean: 155.6; median: 155; mode: 138

Appendix A Exercise Set

1. 4.125   **3.** 95   **5.** 62   **7.** ≈ 3.45 **9.** 3.5   **11.** 95   **13.** 60   **15.** 3.6   **17.** 3   **19.** 95   **21.** 40   **23.** 2.5 and 4.2

25. mean: 45.2; median: 46; mode: 46, 47   **27.** mean: \$36,900; median: \$22,000; The median is more representative.; Answers will vary.

Appendix B Check Point Exercises

1. a. $\begin{bmatrix} 1 & 6 & -3 & | & 7 \\ 4 & 12 & -20 & | & 8 \\ -3 & -2 & 1 & | & -9 \end{bmatrix}$ **b.** $\begin{bmatrix} 1 & 3 & -5 & | & 2 \\ 1 & 6 & -3 & | & 7 \\ -3 & -2 & 1 & | & -9 \end{bmatrix}$ **c.** $\begin{bmatrix} 4 & 12 & -20 & | & 8 \\ 1 & 6 & -3 & | & 7 \\ 0 & 16 & -8 & | & 12 \end{bmatrix}$ **2.** $(-1, 2)$ or $\{(-1, 2)\}$ **3.** $(5, 2, 3)$ or $\{(5, 2, 3)\}$

Appendix B Exercise Set

1. $\begin{bmatrix} 3 & -\dfrac{3}{2} & | & 5 \\ 2 & 2 & | & 5 \end{bmatrix}$ **3.** $\begin{bmatrix} 1 & -\dfrac{4}{3} & | & 2 \\ 3 & 5 & | & -2 \end{bmatrix}$ **5.** $\begin{bmatrix} 1 & -3 & | & 5 \\ 0 & 12 & | & -6 \end{bmatrix}$ **7.** $\begin{bmatrix} 1 & -\dfrac{3}{2} & | & \dfrac{7}{2} \\ 0 & \dfrac{17}{2} & | & -\dfrac{17}{2} \end{bmatrix}$ **9.** $\begin{bmatrix} 1 & -3 & 2 & | & 5 \\ 1 & 5 & -5 & | & 0 \\ 3 & 0 & 4 & | & 7 \end{bmatrix}$ **11.** $\begin{bmatrix} 1 & -3 & 2 & | & 0 \\ 0 & 10 & -7 & | & 7 \\ 2 & -2 & 1 & | & 3 \end{bmatrix}$

13. $\begin{bmatrix} 1 & 1 & -1 & | & 6 \\ 0 & -3 & 3 & | & -15 \\ 0 & -4 & 2 & | & -14 \end{bmatrix}$ **15.** $\{(4, 2)\}$ **17.** $\{(3, -3)\}$ **19.** $\{(2, -5)\}$ **21.** no solution or $\varnothing$ **23.** infinitely many solutions; dependent equations   **25.** $\{(1, -1, 2)\}$ **27.** $\{(3, -1, -1)\}$ **29.** $\{(1, 2, -1)\}$ **31.** $\{(1, 2, 3)\}$ **33.** no solution or $\varnothing$

35. infinitely many solutions; dependent equations   **37.** $\{(-1, 2, -2)\}$ **39.** $\begin{cases} w - x + y + z = 3 \\ x - 2y - z = 0 \\ y + 6z = 17 \\ z = 3 \end{cases}; \{(2, 1, -1, 3)\}$ **41.** $\begin{bmatrix} 1 & -1 & 1 & 1 & | & 3 \\ 0 & 1 & -2 & -1 & | & 0 \\ 0 & 2 & 1 & 2 & | & 5 \\ 0 & 6 & -3 & -1 & | & -9 \end{bmatrix}$

43. $\{(1, 2, 3, -2)\}$

Appendix C Check Point Exercises

1. a. -4 **b.** -17 **2.** $(4, -2)$ or $\{(4, -2)\}$ **3.** 80 **4.** $(2, -3, 4)$ or $\{(2, -3, 4)\}$

Appendix C Exercise Set

1. 1 **3.** -29 **5.** 0 **7.** 33 **9.** $-\dfrac{7}{16}$ **11.** $\{(5, 2)\}$ **13.** $\{(2, -3)\}$ **15.** $\{(3, -1)\}$ **17.** $\{(4, 0)\}$ **19.** $\{(4, 2)\}$ **21.** $\{(7, 4)\}$

23. inconsistent; no solution or $\varnothing$ **25.** dependent equations; infinitely many solutions **27.** 72 **29.** -75 **31.** 0 **33.** $\{(-5, -2, 7)\}$

35. $\{(2, -3, 4)\}$ **37.** $\{(3, -1, 2)\}$ **39.** $\{(2, 3, 1)\}$ **41.** -42 **43.** $\begin{cases} 2x - 4y = 8 \\ 3x + 5y = -10 \end{cases}$ **45.** $\{-11\}$ **47.** $\{4\}$ **49.** 28 sq units **51.** yes

53. $\begin{vmatrix} x & y & 1 \\ 3 & -5 & 1 \\ -2 & 6 & 1 \end{vmatrix} = 0; \; y = -\dfrac{11}{5}x + \dfrac{8}{5}$

Applications Index

Subject Index

Photo Credits

C1

Definitions, Rules, and Formulas

The Real Numbers

Natural Numbers: $\{1, 2, 3, \ldots\}$

Whole Numbers: $\{0, 1, 2, 3, \ldots\}$

Integers: $\{\ldots, -3, -2, -1, 0, 1, 2, 3, \ldots\}$

Rational Numbers: $\{\frac{a}{b} \mid a \text{ and } b \text{ are integers}, b \neq 0\}$

Irrational Numbers: $\{x \mid x \text{ is real and not rational}\}$

Basic Rules of Algebra

Commutative: $a + b = b + a; ab = ba$

Associative: $(a + b) + c = a + (b + c); (ab)c = a(bc)$

Distributive: $a(b + c) = ab + ac; a(b - c) = ab - ac$

Identity: $a + 0 = a; a \cdot 1 = a$

Inverse: $a + (-a) = 0; a \cdot \frac{1}{a} = 1 (a \neq 0)$

Multiplication Properties: $(-1)a = -a;$

$(-1)(-a) = a; a \cdot 0 = 0; (-a)(b) = (a)(-b) = -ab;$

$(-a)(-b) = ab$

Set-Builder Notation, Interval Notation, and Graphs

$(a, b) = \{x \mid a < x < b\}$

$[a, b) = \{x \mid a \leq x < b\}$

$(a, b] = \{x \mid a < x \leq b\}$

$[a, b] = \{x \mid a \leq x \leq b\}$

$(-\infty, b) = \{x \mid x < b\}$

$(-\infty, b] = \{x \mid x \leq b\}$

$(a, \infty) = \{x \mid x > a\}$

$[a, \infty) = \{x \mid x \geq a\}$

$(-\infty, \infty) = \{x \mid x \text{ is a real number}\} = \{x \mid x \in R\}$

Slope Formula

$$\text{slope } (m) = \frac{\text{Change in } y}{\text{Change in } x} = \frac{y_2 - y_1}{x_2 - x_1}, \quad (x_1 \neq x_2)$$

Equations of Lines

1. Slope-intercept form: $y = mx + b$

   m is the line's slope and b is its y-intercept.

2. Standard form: $Ax + By = C$

3. Point-slope form: $y - y_1 = m(x - x_1)$

   m is the line's slope and (x_1, y_1) is a fixed point on the line.

4. Horizontal line parallel to the x-axis: $y = b$

5. Vertical line parallel to the y-axis: $x = a$

Systems of Equations

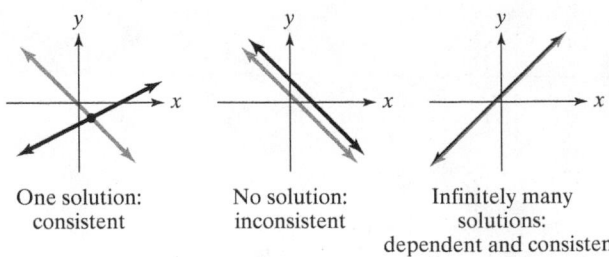

One solution: consistent

No solution: inconsistent

Infinitely many solutions: dependent and consistent

A system of linear equations may be solved: (a) graphically, (b) by the substitution method, (c) by the addition or elimination method, (d) by matrices, or (e) by determinants.

$$\begin{vmatrix} a_1 b_1 \\ a_2 b_2 \end{vmatrix} = a_1 b_2 - a_2 b_1$$

Cramer's Rule:

Given a system of equations of the form

$$\begin{array}{l} a_1 x + b_1 y = c_1 \\ a_2 x + b_2 y = c_2 \end{array}, \quad \text{then } x = \frac{\begin{vmatrix} c_1 b_1 \\ c_2 b_2 \end{vmatrix}}{\begin{vmatrix} a_1 b_1 \\ a_2 b_2 \end{vmatrix}} \text{ and } y = \frac{\begin{vmatrix} a_1 c_1 \\ a_2 c_2 \end{vmatrix}}{\begin{vmatrix} a_1 b_1 \\ a_2 b_2 \end{vmatrix}}.$$

Absolute Value

1. $|x| = \begin{cases} x & \text{if } x \geq 0 \\ -x & \text{if } x < 0 \end{cases}$

2. If $|x| = c$, then $x = c$ or $x = -c$. $(c > 0)$

3. If $|x| < c$, then $-c < x < c$. $(c > 0)$

4. If $|x| > c$, then $x < -c$ or $x > c$. $(c > 0)$

Special Factorizations

1. Difference of two squares:

$$A^2 - B^2 = (A + B)(A - B)$$

2. Perfect square trinomials:

$$A^2 + 2AB + B^2 = (A + B)^2$$
$$A^2 - 2AB + B^2 = (A - B)^2$$

3. Sum of two cubes:

$$A^3 + B^3 = (A + B)(A^2 - AB + B^2)$$

4. Difference of two cubes:

$$A^3 - B^3 = (A - B)(A^2 + AB + B^2)$$

Variation

English Statement	Equation
y varies directly as x.	$y = kx$
y varies directly as x^n.	$y = kx^n$
y varies inversely as x.	$y = \dfrac{k}{x}$
y varies inversely as x^n.	$y = \dfrac{k}{x^n}$
y varies jointly as x and z.	$y = kxz$

Exponents

Definitions of Rational Exponents

1. $a^{\frac{1}{n}} = \sqrt[n]{a}$ **2.** $a^{\frac{m}{n}} = \left(\sqrt[n]{a}\right)^m$ or $\sqrt[n]{a^m}$

3. $a^{-\frac{m}{n}} = \dfrac{1}{a^{\frac{m}{n}}}$

Properties of Rational Exponents

If m and n are rational exponents, and a and b are real numbers for which the following expressions are defined, then

1. $b^m \cdot b^n = b^{m+n}$ **2.** $\dfrac{b^m}{b^n} = b^{m-n}$

3. $\left(b^m\right)^n = b^{mn}$ **4.** $(ab)^n = a^n b^n$

5. $\left(\dfrac{a}{b}\right)^n = \dfrac{a^n}{b^n}$

Radicals

1. If n is even, then $\sqrt[n]{a^n} = |a|$.
2. If n is odd, then $\sqrt[n]{a^n} = a$.
3. The product rule: $\sqrt[n]{a} \cdot \sqrt[n]{b} = \sqrt[n]{ab}$
4. The quotient rule: $\dfrac{\sqrt[n]{a}}{\sqrt[n]{b}} = \sqrt[n]{\dfrac{a}{b}}$

Complex Numbers

1. The imaginary unit i is defined as

$$i = \sqrt{-1}, \quad \text{where} \quad i^2 = -1.$$

The set of numbers in the form $a + bi$ is called the set of complex numbers. If $b = 0$, the complex number is a real number. If $b \neq 0$ the complex number is an imaginary number.

2. The complex numbers $a + bi$ and $a - bi$ are conjugates. Conjugates can be multiplied using the formula

$$(A + B)(A - B) = A^2 - B^2.$$

The multiplication of conjugates results in a real number.

3. To simplify powers of i, rewrite the expression in terms of i^2. Then replace i^2 with -1 and simplify.

Quadratic Equations and Functions

1. The solutions of a quadratic equation in standard form

$$ax^2 + bx + c = 0, \quad a \neq 0,$$

are given by the quadratic formula

$$x = \frac{-b \pm \sqrt{b^2 - 4ac}}{2a}.$$

2. The discriminant, $b^2 - 4ac$, of the quadratic equation $ax^2 + bx + c = 0$ determines the number and type of solutions.

Discriminant	Solutions
Positive perfect square with a, b, and c rational numbers	2 rational solutions
Positive and not a perfect square	2 irrational solutions
Zero, with a, b, and c rational numbers	1 rational solution
Negative	2 imaginary solutions

3. The graph of the quadratic function

$$f(x) = a(x - h)^2 + k, \quad a \neq 0,$$

is called a parabola. The vertex, or turning point, is (h, k). The graph opens upward if a is positive and downward if a negative. The axis of symmetry is a vertical line passing through the vertex. The graph can be obtained using the vertex, x-intercepts, if any, [set $f(x)$ equal to zero], and the y-intercept (set $x = 0$).

4. A parabola whose equation is in the form

$$f(x) = ax^2 + bx + c, \quad a \neq 0,$$

has its vertex at

$$\left(-\frac{b}{2a}, f\left(-\frac{b}{2a}\right)\right).$$

If $a > 0$, then f has a minimum that occurs at $x = -\dfrac{b}{2a}$. If $a < 0$, then f has a maximum that occurs at $x = -\dfrac{b}{2a}$.

Definitions, Rules, and Formulas (continued)

Exponential and Logarithmic Functions

1. Exponential Function: $f(x) = b^x, b > 0, b \neq 1$

Graphs:

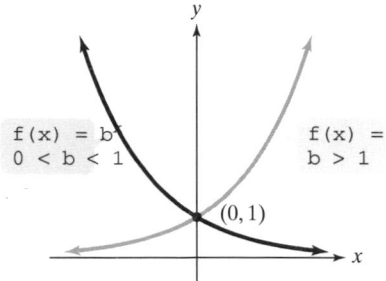

$f(x) = b^x$
$0 < b < 1$

$f(x) = b^x$
$b > 1$

$(0, 1)$

2. Logarithmic Function: $f(x) = \log_b x, b > 0, b \neq 1$
$y = \log_b x$ is equivalent to $x = b^y$.

Graphs:

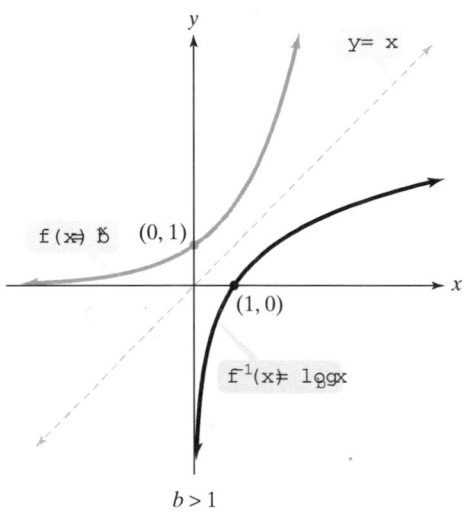

$y = x$

$f(x) = b^x$ $(0, 1)$

$(1, 0)$

$f^{-1}(x) = \log_b x$

$b > 1$

3. Properties of Logarithms

a. $\log_b(MN) = \log_b M + \log_b N$

b. $\log_b\left(\dfrac{M}{N}\right) = \log_b M - \log_b N$

c. $\log_b M^p = p\log_b M$

d. $\log_b M = \dfrac{\log_a M}{\log_a b} = \dfrac{\ln M}{\ln b} = \dfrac{\log M}{\log b}$

e. $\log_b b^x = x; \log 10^x = x; \ln e^x = x$

f. $b^{\log_b x} = x; 10^{\log x} = x; e^{\ln x} = x$

Distance and Midpoint Formulas

1. The distance from (x_1, y_1) to (x_2, y_2) is
$$\sqrt{(x_2 - x_1)^2 + (y_2 - y_1)^2}.$$

2. The midpoint of the line segment with endpoints (x_1, y_1) and (x_2, y_2) is
$$\left(\frac{x_1 + x_2}{2}, \frac{y_1 + y_2}{2}\right).$$

Conic Sections
Circle

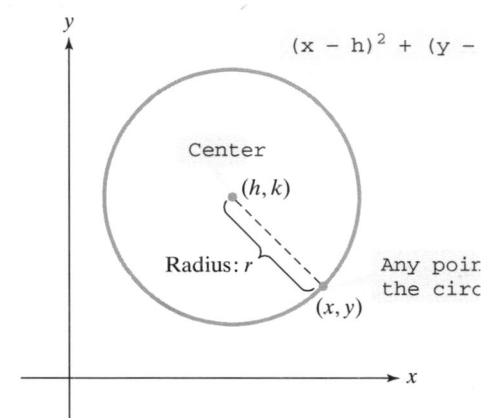

$(x - h)^2 + (y - $

Center
(h, k)

Radius: r

Any poir
the circ

(x, y)

Ellipse

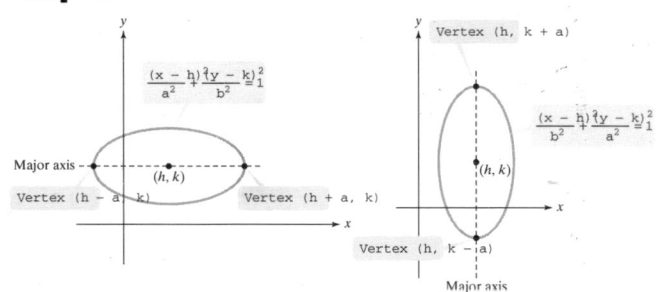

$\dfrac{(x - h)^2}{a^2} + \dfrac{(y - k)^2}{b^2} = 1$

Major axis

(h, k)

Vertex $(h - a, k)$ Vertex $(h + a, k)$

Vertex $(h, k + a)$

$\dfrac{(x - h)^2}{b^2} + \dfrac{(y - k)^2}{a^2} = 1$

(h, k)

Vertex $(h, k - a)$

Major axis

Hyperbola

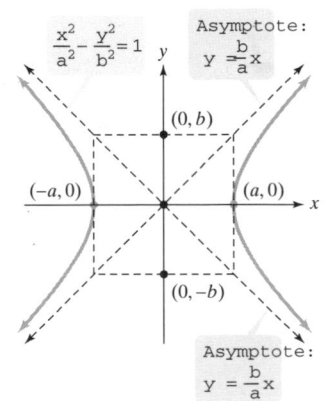

$\dfrac{x^2}{a^2} - \dfrac{y^2}{b^2} = 1$

Asymptote:
$y = \dfrac{b}{a}x$

$(0, b)$

$(-a, 0)$ $(a, 0)$

$(0, -b)$

Asymptote:
$y = \dfrac{b}{a}x$

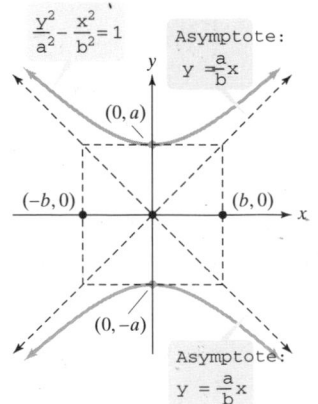

$\dfrac{y^2}{a^2} - \dfrac{x^2}{b^2} = 1$

Asymptote:
$y = \dfrac{a}{b}x$

$(0, a)$

$(-b, 0)$ $(b, 0)$

$(0, -a)$

Asymptote:
$y = \dfrac{a}{b}x$